Structures

CIVIL ENGINEERING PRACTICE

1/Structures

EDITED BY

PAUL N. CHEREMISINOFF

NICHOLAS P. CHEREMISINOFF

SU LING CHENG

IN COLLABORATION WITH

C. W. Bert
D. W. Boyce
Y. T. Chou
A. Coull
R. J. Craig
P. LeP. Darvall
T. G. Davies
E. A. Dickin
P. F. Dux
J. M. Ferritto
R. W. Furlong
K. P. George
P. G. Glockner
L. A. Godoy
C.-T. T. Hsu
R. Hussein
D. H. Jiang
S. Kayal
S. Kitipornchai
S. Kuranishi
K. C. S. Kwok
C. F. Leung
M. S. Mamlouk
S. A. Mirza
M. Morsi
F. W. Muchmore
I. Mungan
G. Onu
M. Papadrakakis
C. A. Prato
T. M. Roberts
H. Scholz
S. Somayaji
T. Stathopoulos
W. Szyszkowski
Y. C. Wong
T. Yabuki

LANCASTER · BASEL

Published in the Western Hemisphere by
Technomic Publishing Company, Inc.
851 New Holland Avenue
Box 3535
Lancaster, Pennsylvania 17604 U.S.A.

Distributed in the Rest of the World by
Technomic Publishing AG

Printed in the United States of America

10 9 8 7 6 5 4 3 2 1

Main entry under title:
Civil Engineering Practice 1—Structures

A Technomic Publishing Company book
Bibliography: p.
Includes index p. 809

Library of Congress Card No. 87-50629
ISBN No. 87762-529-8

TABLE OF CONTENTS

PREFACE

While the designation civil engineering dates back only two centuries, the profession of civil engineering is as old as civilized life. Through ancient times it formed a broader profession, best described as master builder, which included what is now known as architecture and both civil and military engineering. The field of civil engineering was once defined as including all branches of engineering and has come to include established aspects of construction, structures and emerging and newer sub-disciplines (e.g., environmental, water resources, etc.). The civil engineer is engaged in planning, design of works connected with transportation, water and air pollution as well as canals, rivers, piers, harbors, etc. The hydraulic field covers water supply/power, flood control, drainage and irrigation, as well as sewerage and waste disposal.

The civil engineer may also specialize in various stages of projects such as investigation, design, construction, operation, etc. Civil engineers today as well as engineers in all branches have become highly specialized, as well as requiring a multiplicity of skills in methods and procedures. Various civil engineering specialties have led to the requirement of a wide array of knowledge.

Civil engineers today find themselves in a broad range of applications and it was to this end that the concept of putting this series of volumes together was made. The tremendous increase of information and knowledge all over the world has resulted in proliferation of new ideas and concepts as well as a large increase in available information and data in civil engineering. The treatises presented are divided into five volumes for the convenience of reference and the reader:

VOLUME 1 Structures
VOLUME 2 Hydraulics/Mechanics
VOLUME 3 Geotechnical/Ocean Engineering
VOLUME 4 Surveying/Transportation/Energy/Economics & Government/Computers
VOLUME 5 Water Resources/Environmental

A serious effort has been made by each of the contributing specialists to this series to present information that will have enduring value. The intent is to supply the practitioner with an authoritative reference work in the field of civil engineering. References and citations are given to the extensive literature as well as comprehensive, detailed, up-to-date coverage.

To insure the highest degree of reliability in the selected subject matter presented, the collaboration of a large number of specialists was enlisted, and this book presents their efforts. Heartfelt thanks go to these contributors, each of whom has endeavored to present an up-to-date section in their area of expertise and has given willingly of valuable time and knowledge.

PAUL N. CHEREMISINOFF
NICHOLAS P. CHEREMISINOFF
SU LING CHENG

CONTRIBUTORS TO VOLUME 1

C. W. BERT, School of Aerospace, Mechanical and Nuclear Engineering, The University of Oklahoma, Norman, OK

D. W. BOYCE, Department of Civil Engineering, University of Glasgow, Glasgow, Scotland

Y. T. CHOU, Pavement Systems Division, Geotechnical Laboratory, U.S. Army Engineer Waterways Experiment Station, Vicksburg, MS

A. COULL, Department of Civil Engineering, University of Glasgow, Glasgow, Scotland

R. J. CRAIG, Department of Civil and Environmental Engineering, New Jersey Institute of Technology, Newark, NJ

P. LEP. DARVALL, Department of Civil Engineering, Monash University, Clayton, Victoria 3168, Australia

T. G. DAVIES, Department of Civil Engineering, University of Glasgow, Glasgow, Scotland

E. A. DICKIN, Department of Civil Engineering, University of Liverpool, England

P. F. DUX, Department of Civil Engineering, University of Queensland, Australia

J. M. FERRITTO, Naval Civil Engineering Laboratory, Port Hueneme, CA

R. W. FURLONG, Department of Civil Engineering, The University of Texas at Austin, Austin, TX

K. P. GEORGE, Department of Civil Engineering, University of Mississippi, University, MS

P. G. GLOCKNER, Department of Mechanical Engineering, The University of Calgary, Calgary, Alberta, Canada

L. A. GODOY, Structures Department, FCEFyN, National University of Córdoba, Casilla de Correo 916, Córdoba 5000, Argentina

C.-T. T. HSU, Department of Civil and Environmental Engineering, New Jersey Institute of Technology, Newark, NJ

R. HUSSEIN, University of the District of Columbia, Washington, DC

D. H. JIANG, Tongji University, Deptartment of Structural Engineering, Shanghai, China

S. KAYAL, Cement Research Institute of India, M-10, South Extension II, New Delhi - 110 049, India

S. KITIPORNCHAI, Department of Civil Engineering, University of Queensland, Australia

S. KURANISHI, Department of Civil Engineering, Tohoku University, Sendai, Japan

K. C. S. KWOK, School of Civil and Mining Engineering, University of Sydney, Australia

C. F. LEUNG, Department of Civil Engineering, National University of Singapore, Singapore

M. S. MAMLOUK, Department of Civil Engineering, Arizona State University, Tempe, AZ

S. A. MIRZA, Department of Civil Engineering, Lakehead University, Thunder Bay, Ontario, Canada

M. MORSI, Arab Bureau of Design and Tech. Consult., Abbasseya, Cairo, Egypt

F. W. MUCHMORE, Northern Region, USDA Forest Service, Missoula, MT

I. MUNGAN, Mimar Sinan University, Findikli, Istanbul, Turkey

G. ONU, Design Institute for Air, Road and Water Transports. Ministry of Transport and Communications, Bucharest, Romania

M. PAPADRAKAKIS, Institute of Structural Analysis and Aseismic Research, National Technical University, Athens, Greece

C. A. PRATO, Structures Department, FCEFyN, National University of Córdoba, Casilla de Correo 916, Córdoba 5000, Argentina

T. M. ROBERTS, Department of Civil and Structural Engineering, University College, Cardiff, Wales, U.K.

H. SCHOLZ, Department of Civil Engineering, University of the Witwatersrand, Johannesburg, South Africa

S. Somayaji, Department of Civil and Environmental Engineering, California Polytechnic State University, San Luis Obispo, CA

T. Stathopoulos, Centre for Building Studies, Concordia University, Montreal, Quebec, Canada

W. Szyszkowski, Department of Mechanical Engineering, The University of Saskatchewan, Saskatoon, Saskatchewan, Canada

Y. C. Wong, Engineering and Environmental Consultants, Kuala Lumpur, Malaysia

T. Yabuki, Department of Civil Engineering, University of the Ryukyus, Okinawa, Japan

CIVIL ENGINEERING PRACTICE

VOLUME 1
Structures

SECTION 1 Reinforced Concrete Structures

SECTION 2 Structural Analysis

SECTION 3 Stability

SECTION 4 Pavement Design

SECTION 5 Wood Structures

SECTION 6 Composites

VOLUME 2
Hydraulics/Mechanics

SECTION 1 Hydraulics/Open Channel Flow

SECTION 2 Flow in Pipes

SECTION 3 Flow With Bed Load

SECTION 4 Mechanics/Solid Mechanics

SECTION 5 Fluid Mechanics

SECTION 6 Solid-Fluid Interaction

VOLUME 3
Geotechnical/Ocean Engineering

SECTION 1 Soil Mechanics

SECTION 2 Stability

SECTION 3 Bearing Capacity

SECTION 4 Buried Structures

SECTION 5 Waves and Wave Action

SECTION 6 Coastal Structures

VOLUME 4
Surveying/Construction/ Transportation/Energy/ Economics & Goverment/Computers

SECTION 1 Surveying

SECTION 2 Construction

SECTION 3 Transportation

SECTION 4 Energy

SECTION 5 Economics/Government/Data Acquisition

VOLUME 5
Water Resources/Environmental

SECTION 1 Water Supply and Management

SECTION 2 Irrigation

SECTION 3 Environmental

SECTION ONE

Reinforced Concrete Structures

CHAPTER 1

Reinforced Concrete Design

ROBERT JOHN CRAIG*

CONCRETE STRUCTURAL SYSTEMS

The basic components of a building are: (1) floor slabs; (2) beams and girders; (3) columns; (4) walls; and (5) foundation.

Basically, for reinforced concrete floor or roof systems, the system is considered to be one-way or two-way slabs. The one-way slab is where the moments are distributed only in one direction. This is considered when the length of the floor which is supported on all four edges has a long length over short length which is greater than two. If the supported long length over short length is less than two, it is considered to be a two-way slab.

In general, the commonly used reinforced-concrete floor and roof systems can be classified as follows [1]:

1. One-way reinforcing systems.
   - (a) One-way solid slabs.
     - (1) Slab supported on monolithic concrete beams and girders.
     - (2) Slab supported on steel beams (can be composite action or noncomposite action).
     - (3) Slab with light-gage steel decking which serves as form and reinforcement for the concrete.
   - (b) One-way ribbed slab (concrete joist floors).
   - (c) Precast systems, including precast slab or beams or both.
2. Two-way reinforcing systems.
   - (a) Two-way solid slabs.
     - (1) Slab or monolithic concrete beams.
     - (2) Slab on steel beams.
   - (b) Two-way ribbed slab on concrete or steel beam supports.
   - (c) Beamless slabs.
     - (1) Flat slab with drop panels or column capitols or both.
     - (2) Flat plate with no drop panels or column capitols.
     - (3) Ribbed slab.

In this chapter, the design of a one-way slab system will be fully discussed in another section. The different methods of analysis and design for a two-way slab will only be discussed without a numerical example in another section. The different floor and roof slabs are shown in Figure 1.

The beams' and girders' design for one-way or two-way slab systems will be shown in the section on continuity. Figures 1, 2, and 3 show the location of the beams and girders [1,2]. These beams and girders are considered as an indeterminant system and are analyzed by dead load and partial live load situations. The analysis is usually done by moment distribution or matrix analysis. With the slab cast monilithically with the beams and girders, the sections are analyzed and designed as regular rectangular beams, T-beams and doubly-reinforced concrete beams.

The columns will support the floor and roof systems which have just been described. The columns will have bending, shear, and axial loads. The design will reflect this type of behavior. Of importance are the connections of the floor systems and the columns. Much research has been conducted lately to look at this area and its behavior. Both flat slab-column and beam-column connections are of real interest to the designer. These connections must be able to take both the vertical and horizontal loads which are carried by the structure. The columns will be either spiral or tied columns or a combination of both. The load-moment interaction curves are very important in design of columns.

Walls are vertical members which form the enclosures for the building. They may not necessarily be made of concrete and may not be considered as load bearing. Structural walls

*Department of Civil and Environmental Engineering, New Jersey Institute of Technology, Newark, NJ

Building at Northern Kentucky University in Reinforced Concrete (Picture by Robert A. Baumann)

are considered to be: foundation walls, stairwell walls, and shear walls. These types of structural walls will resist horizontal wind and earthquake loadings [3].

The foundation reinforced concrete structural elements take the loads from the columns or walls and place them on the soil or rock foundation. There are six basic types of foundation structures:

1. wall footings
2. independent isolated column footings
3. combined footings
4. cantilever or strap footings
5. pile foundations
6. mat foundations.

These types of footings are shown in Figure 4 [4]. The design and analysis of footings will be shown in another section of this chapter.

MATERIALS—CONCRETE AND STEEL

In the design of reinforced concrete structural members, it is important that the engineer have a basic understanding of the material properties of both concrete and reinforcing. This section will basically be kept short. Other references should be referred to for a more elaborate discussion [5,6].

Basically, there are three requirements of concrete which should be adhered to in developing a concrete mix:

1. The mix must be able to be placed in the form in its plastic state (workable).
2. The mix must have the design strength in its hardened state.
3. The mix must be economical and practical for the job.

At the present time, there are a lot of new types of concrete mixes which are being used in the field such as: (a) fiber

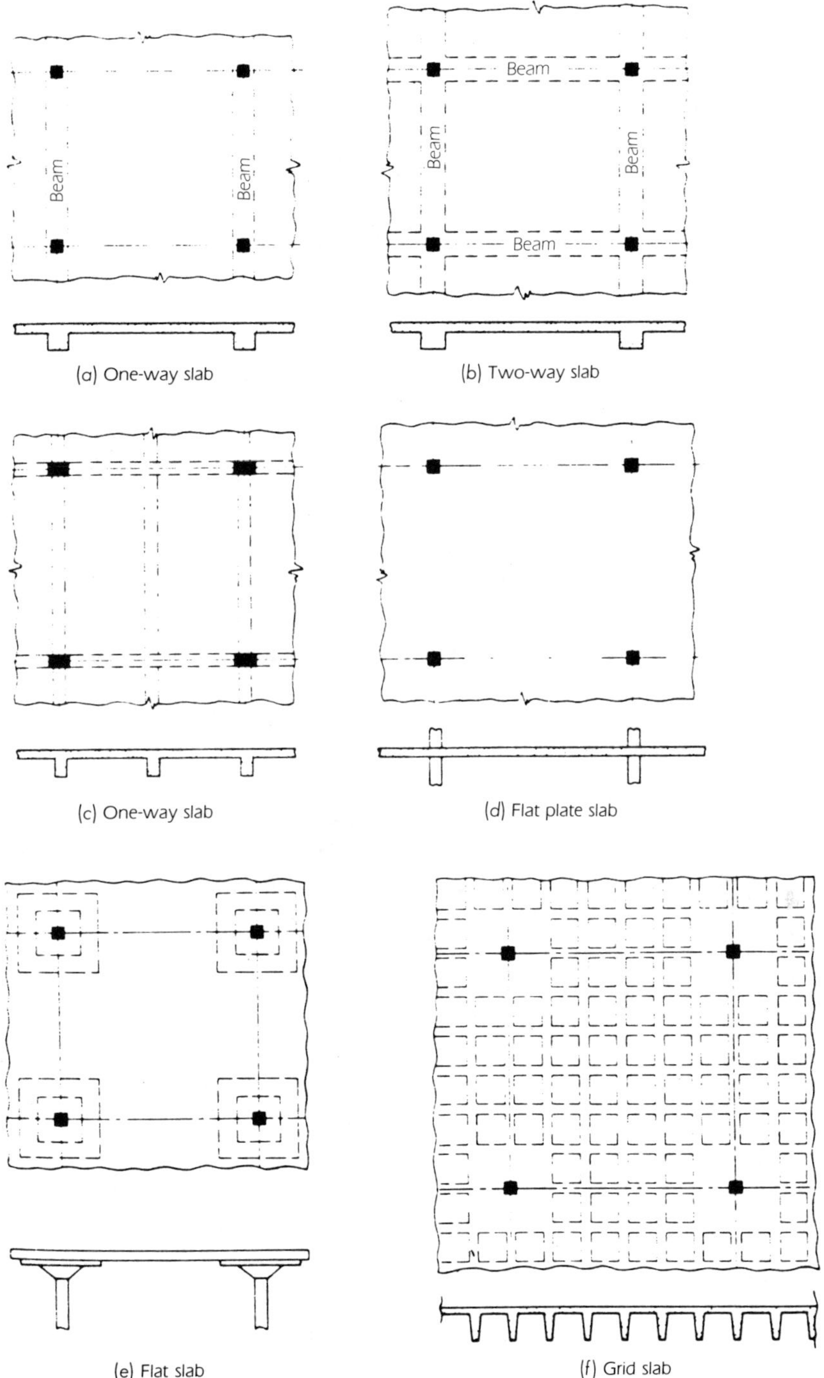

FIGURE 1. Floor slab systems [1]: (a) one-way slab; (b) two-way slab; (c) one-way slab; (d) flat plate slab; (e) flat slab; (f) grid slab.

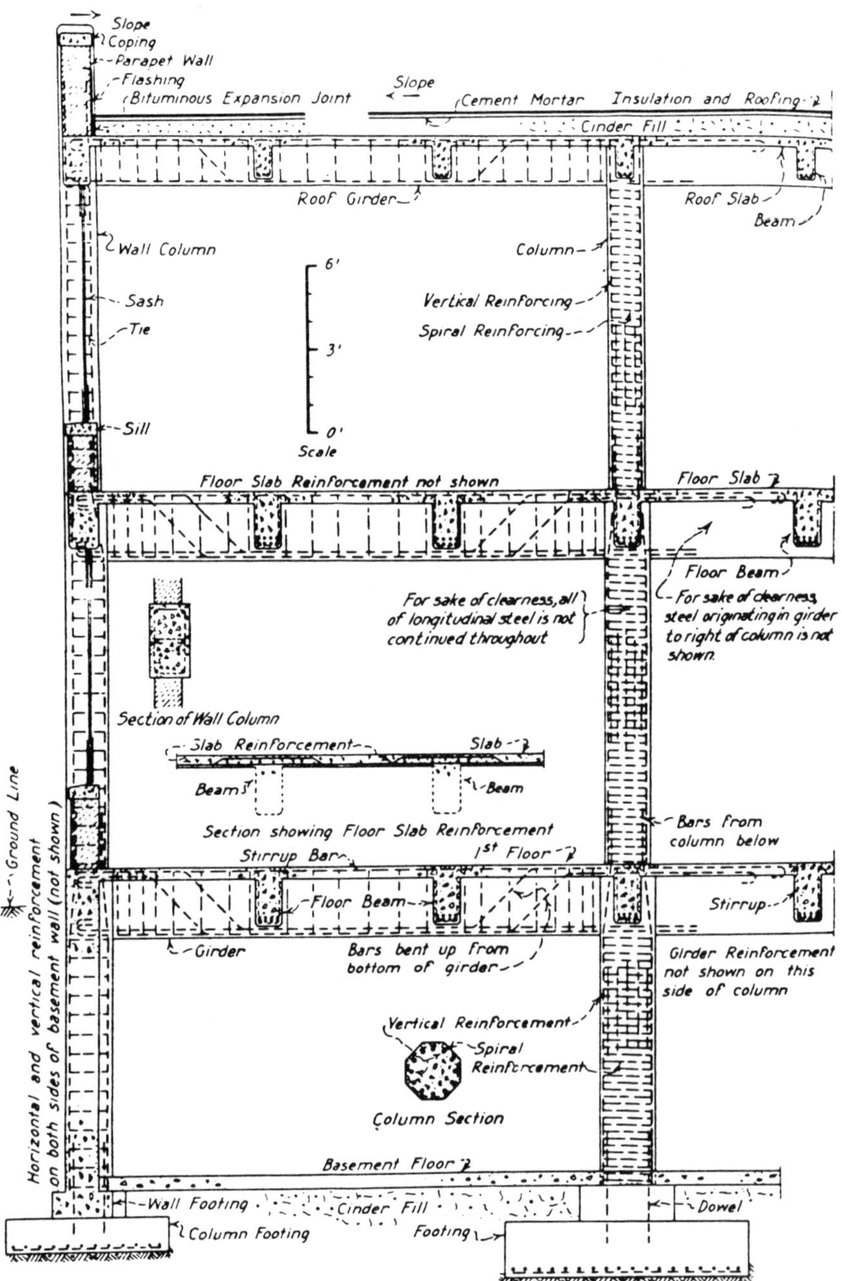

FIGURE 2. Beam and girder construction [2].

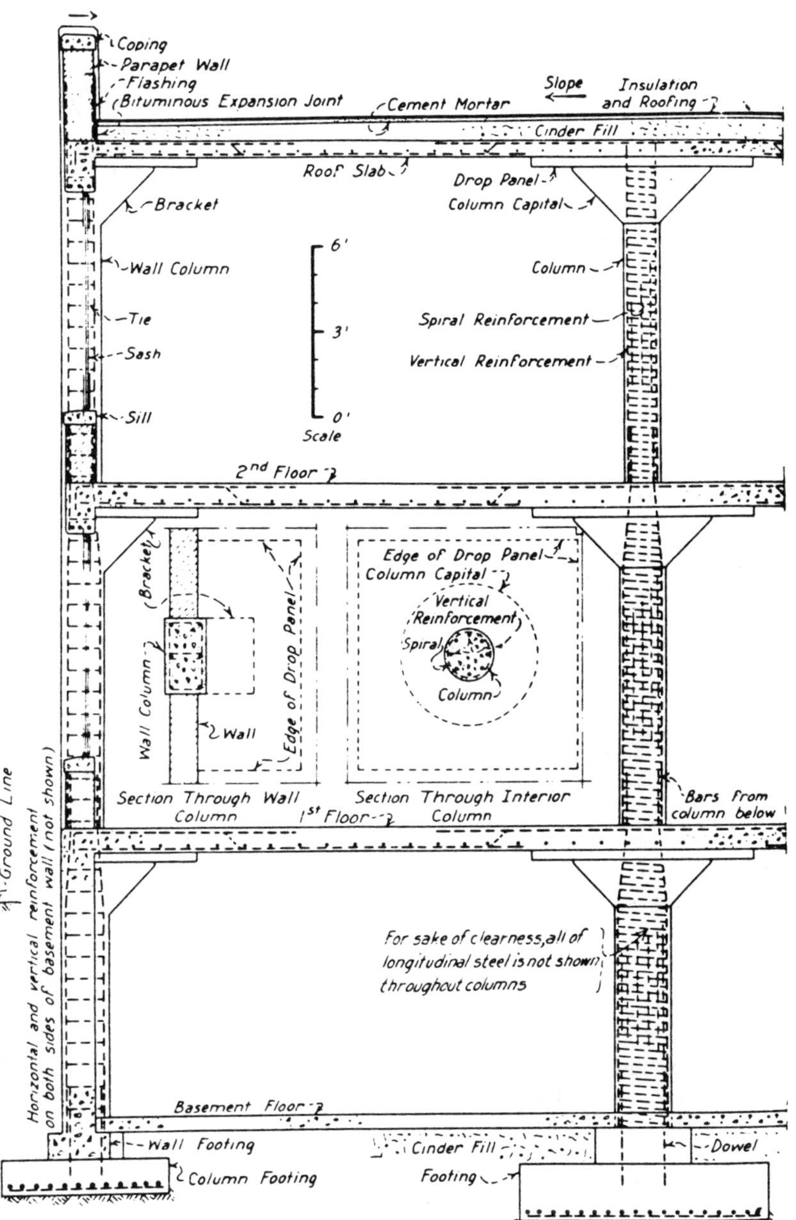

FIGURE 3. Flat-slab construction [2].

reinforced concrete; (b) polymer concrete; (c) sulfur concrete; (d) shotcrete; etc. In this section only normal and lightweight concrete will be considered.

The strength of concrete depends upon the proportion of the ingredients and the temperature and moisture conditions upon placing the concrete mix. The ideal curing and placing conditions for concrete would be 54°F and 100% relative humidity.

When considering concrete mix proportioning and mixing, the different ingredients of concrete must be considered: cement, water, aggregate, and air. Actually, the cement and water act as a cementing agent holding the coarse and fine aggregate together. The aggregate should be well graded to obtain a dense mixture of the aggregate. The aggregate may vary from 60% to 85% depending on the maximum size aggregate being used. The larger the max-

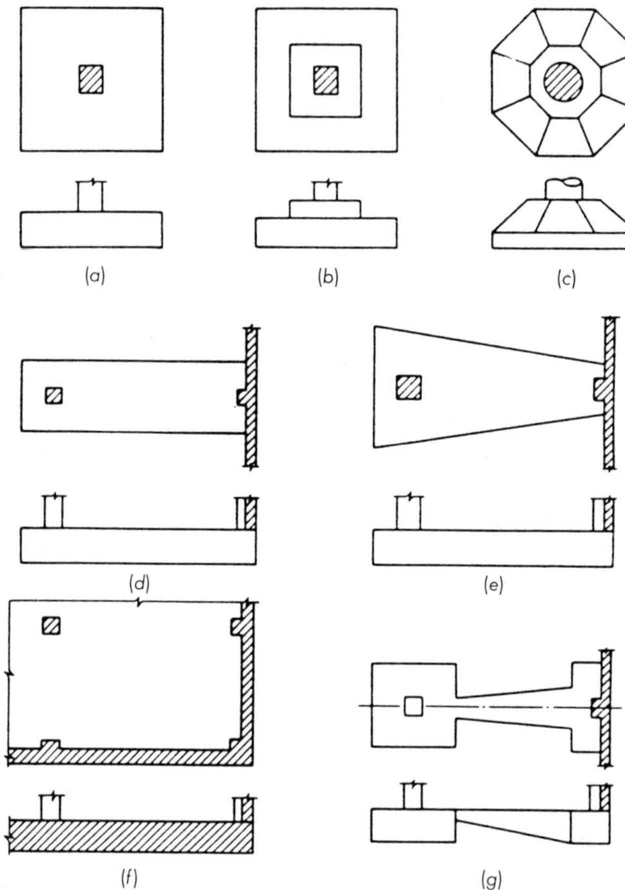

FIGURE 4. Typical column footings [4]: (a) individual square footing; (b) stepped square footing; (c) octagonal sloped footing; (d) rectangular combined footing; (e) trapezoidal combined footing; (f) raft or mat footing; (g) cantilevered footing.

imum aggregate size the more volume of aggregate is necessary. Each ingredient will be discussed separately.

1. Cement—The actual gluing agent is portland cement which is a so-called hydraulic cement. Such cements chemically combine with water (hydration) to form a hardened substance. There are basically five types of portland cement.

Type 1—Ordinary construction were special properties are not required.
Type 2—Ordinary construction exposed to moderate sulfate action, or where moderate heat of hydration is needed.
Type 3—When high early strength is needed.
Type 4—When low heat of hydration is needed.
Type 5—When high sulfate resistance is needed.

There are two other types of portland cement which are used, and they are: (1) portland blast-furnace-slag cement or portland-pozzolan cement, and (2) air-entraining portland cement.

2. Aggregates—Since the aggregate occupies about 75% of the total volume of the concrete mix, its properties are important parameters in strength and durability in the mix. The aggregates are divided into fine aggregate (material passing through a No. 4 sieve), and coarse aggregate (gravel which is larger than a No. 4 sieve). The actual maximum size of the aggregate is governed by the form size and bar sizes and arrangement. Basically, there are three types of aggregates which can be considered for construction:

1. Lightweight	50 to 120 pcf
2. Normal	120 to 160 pcf
3. Heavyweight	200 or more pcf

Normally, the heavier the aggregate, the stronger the ag-

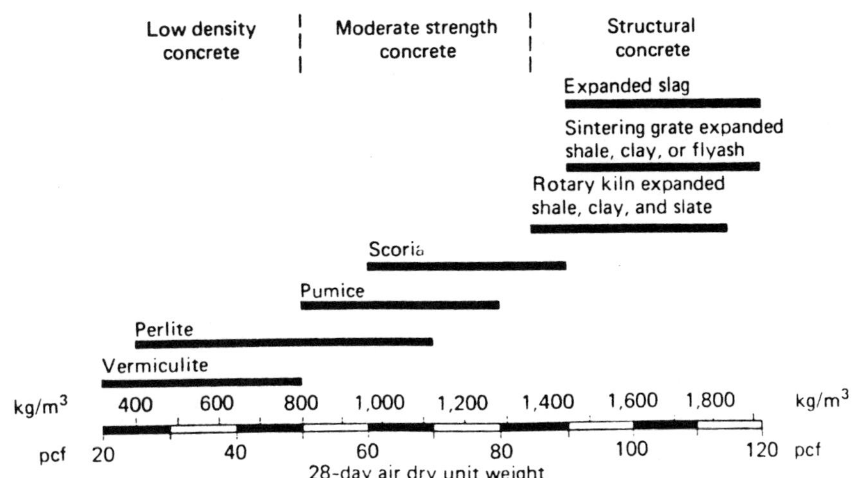

FIGURE 5. Approximate unit weight and use classification of lightweight aggregate concretes [7].

gregate. For normal concrete, the unit weight is considered to be 145 pcf and reinforced concrete is 150 pcf. The lightweight concretes are used for structural and insulating purposes. The expanded clays and shales are used for structural purposes and cellular concretes for insulating purposes and masonry units. If sand is used as the fine aggregate, then it is called sand-lightweight. The approximate unit weight and use classification of lightweight aggregate concretes are shown in Figure 5 [7]. Heavyweight, high-density concrete is used for shielding against gamma and X-radiation in nuclear reactor containers and other structures [8].

3. Water—The amount of water which is introduced into a mix will basically determine the concrete's strength and workability. The strength of a hardened concrete mix is dependent upon the amount of voids. More water in the mix will lead to more voids. The strength of concrete is related to the mixes water/cement ratio (w/c). Figure 6 [1] shows the variation of the strength in compression and flexure for various w/c ratios.

4. Air—Concrete mixes have entrapped air of about 1.5 to 2% (by volume) for non-air-entrained concrete. If air-entraining is used in the mix, then there will be up to 5–6% (by volume) of entrained air bubbles which will be in the mix. Air-entrained bubbles are very small and well dispersed in the cement paste which helps in added durability of regular concrete. The air-entrained concrete will have a reduction in strength of 10–20% strength because of the presence of increased voids.

5. Admixtures—Besides the basic ingredients just described, there are many admixtures which can be added to a mix to help in performance of the wet or hardened concrete. Admixtures help the concrete mixes: (1) workability; (2) resistance to deterioration; (3) retard setting; (4) increase strength; and (5) accelerate strength gain.

Some of the tests used for quality control of concrete in construction are:

1. slump test for consistency
2. unit weight for checking quantities
3. compressive strength for hardened strength.

When looking at concrete properties in a reinforced concrete structure, the structural engineer is interested in the strength of concrete in compression, tension, and flexure. The compressive strength of concrete f'_c is the ultimate compressive strength of a 6″ × 12″ cylinder at 28 days. Typical compressive stress strain curves for different concrete strengths are shown in Figure 7 [5]. Because of the variation in the materials in the concrete mix, the procedures of casting, etc., the compressive strength test values will vary for a given concrete mix. The concrete producers will produce concrete with different variations in strength. The producers can be classified as good, fair, or poor companies, depending upon the coefficient of variation of their test cylinders. Figures 8(a) and 8(b) show a typical plot of number of tests versus strength for various concrete-producing plants. The designer wants the strength of the concrete to be better than what is used in the design. The required compressive strength of the concrete f'_{cr} is dependent upon [1]: design compressive strength of the structure, standard deviation of the compressive strength of the concrete for the plant, and the number of tests used in determining the standard deviation. The required strength f'_{cr} is the largest of the two values.

$$f'_{cr} = K(f'_c + 1.34s) \tag{1}$$

$$f'_{cr} = K(f'_c + 2.33s - 500) \tag{2}$$

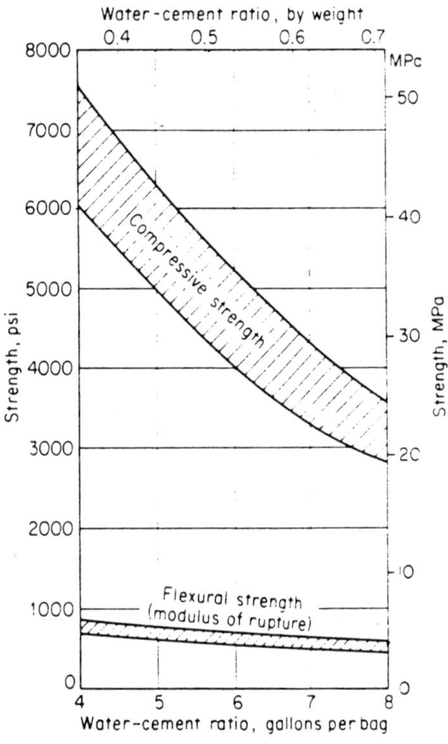

FIGURE 6. Effect of water-cement ratio on 28-day compressive and flexural strength [1].

where s is the standard deviation

$$s = \sqrt{\frac{\sum_{i=1}^{n} (y_i - \bar{y})^2}{n - 1}} \tag{3}$$

where $\bar{y}$ is the mean value of all the n number of tests. The modification factor K is dependent upon the number of tests used to determine the standard deviation. The modification factor K is shown in Chart 1.

When determining the f'_{cr}, the f'_c must be within $f'_c \pm 1000$ psi; use 30 tests or more for best results; and the conditions of evaluation of f'_{cr} must be similar to conditions of casting.

Example 1

A concrete mix producing plant A wants to produce concrete with a compressive strength $f'_c = 3500$ psi. Determine the strength of the concrete strength which must be produced given the following plant A test values from previous mixing.

Test	y_i	$y_i - \bar{y}$	$(y_i - \bar{y})^2$
1	3,940	135	18,225
2	3,840	235	55,225
3	3,960	115	13,225
4	3,910	165	27,225
5	4,280	205	42,025
6	4,220	145	21,025
7	4,510	435	189,225
8	4,320	245	60,025
9	4,260	185	34,225
10	4,360	285	81,225
11	3,710	365	133,225
12	4,120	45	2,025
13	4,030	45	2,025
14	4,070	5	25
15	3,760	315	99,225
16	3,920	155	24,025
	$\Sigma = 65{,}210$		$\Sigma = 802{,}200$

$$\bar{y} = \frac{\Sigma y_i}{n} = \frac{65{,}210}{16} = 4075 \text{ psi}$$

$$s = \sqrt{\frac{\sum_{i=1}^{n} (y_i - \bar{y})^2}{n - 1}} = \sqrt{\frac{802{,}200}{16-1}} = 231 \text{ psi}$$

$$K = 1.16 \text{ from Chart 1}$$

$$f'_{cr} = K(f'_c + 1.34s) = 1.16(3500 + 1.34(231)) = 4420 \text{ psi}$$

$$f'_{cr} = K(f'_c + 2.33s - 500) = 1.16(3500 + 2.33(231) - 500) = 4100 \text{ psi}$$

Thus, $f'_{cr} = 4420$ psi

Thus, plant A must produce 4420 psi concrete for the design $f'_c = 3500$ psi.

Other considerations by the ACI Code 83 are: (1) each test value is the average of at least 2 cylinder strengths; (2) the strength level of an individual class of concrete shall be considered satisfactory if the averages of all sets of three consecutive strength test results equal or exceed the required f'_c; and (3) no individual strength test shall fall below the required f'_c by more than 500 psi.

The tensile strength of a concrete can be determined by three different methods:

1. direct tension
2. split-cylinder
3. modulus of rupture.

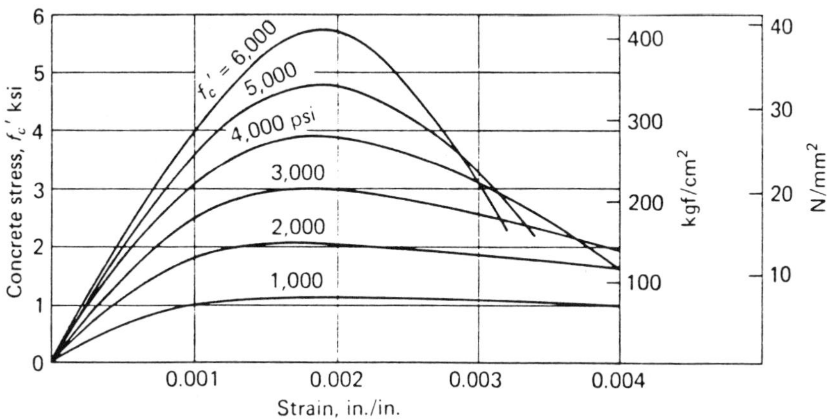

FIGURE 7. Typical stress-strain curves for concrete under short-time loading [5].

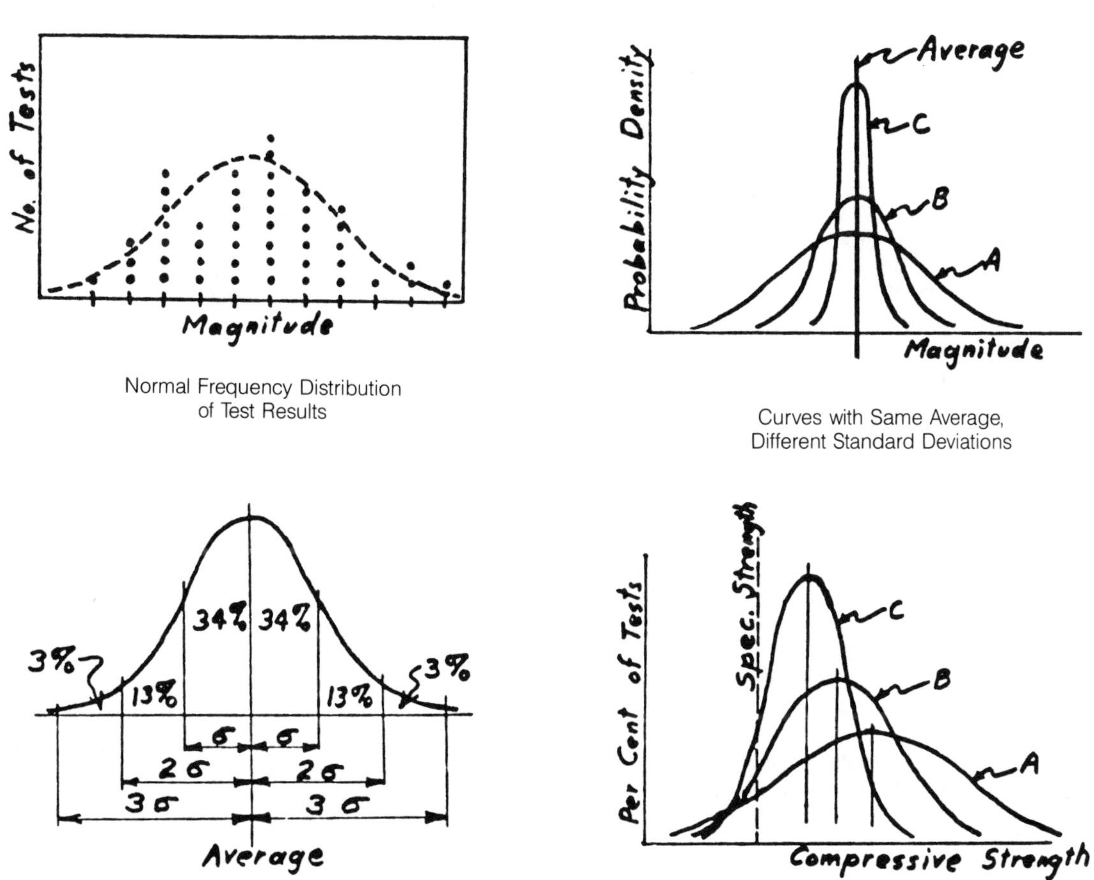

(a) Standard Deviation (b) Effect on Target Strength

FIGURE 8. Compressive strength test results (target strength) [14].

TABLE 1. Required Average Compressive Strength When Data Are Not Available [9].

f'_c	Required f'_{cr}
less than 3000	$f'_c + 1000$
3000 – 5000	$f'_c + 1200$
Over 5000	$f'_c + 1400$

The direct tension can be done by using a large dog-bone specimen which can become very complicated to test. The magnitude of this strength is approximately $.10f'_c$. The split-cylinder test can be performed by setting the 6″ × 12″ cylinder on its side in the testing machine. The cylinder will split in half when the tensile strength is reached. The stress is computed by the formula:

$$f'_{ct} = \frac{2P}{\pi d \ell} \tag{4}$$

The split-cylinder strength is about 10–15% of the concrete f'_c. The flexural strength can also be used to determine the tensile strength. Usually a 6″ × 6″ section beam is loaded using a third point loading with an 18-inch span. The stresses are determined by the flexural formula: f = Mc/I. This is called the modulus of rupture, f_r. The value of f_r is about 15% of the compressive strength f'_c. The ACI Code 83 lets the designer use $f_r = 7.5\sqrt{f'_c}$. Putting these different tests on a comparative scale, the values are:

f'_t	f'_{ct}	f'_r
$2.5-3\sqrt{f'_c}$	$6-7\sqrt{f'_c}$	$9-11\sqrt{f'_c}$

Because of this inherent low strength level of tension stresses, reinforcing bars are put into the concrete members to take up the tensile stresses after the concrete cracks. For lightweight concrete, these values will be smaller. Check ACI Code 83 for permissible values.

The modulus of elasticity of concrete varies with different concretes. The variation with strength of concrete depends on age; properties of aggregate and cement; rate of loading; and type and size of specimen. The usual three different ways of determining the modulus of elasticity are:

1. initial tangent modulus
2. tangent modulus
3. secant modulus.

Figure 9 [5] shows the differences in these methods of the determination of E_c. The ACI Code 83 specifies that the secant modulus may be given as the formula:

$$E_c = 33\, w^{1.5}\sqrt{f'_c} \tag{5}$$

where w is the unit weight of the concrete (lb/ft³). The formula would be $57{,}000\sqrt{f'_c}$ for normal concrete where w = 145 lb/ft³. Looking at typical values for the modulus of elasticity E_c are shown below.

Normal Concrete	Lightweight Concrete
$w = 145$ lb/ft³	$w = 125$ lb/ft³
$f'_c = 33(145)^{1.5}\sqrt{3500}$	$E_c = 33(125)^{1.5}\sqrt{3500}$
$E_c = 3.59 \times 10^6$ psi	$E_c = 2.73 \times 10^6$ psi

When considering the concrete as a material, the designer must look at other properties which are important such as temperature, shrinkage, and creep. The temperature can vary from one section of a structure to another and, if there are restraints in the structure, then temperature induced stresses can be produced. When considering the temperature variations of the steel and concrete internally, there is little problem because the linear coefficient of thermal expansion concrete and steel are about the same (approximately $\alpha_c = \alpha_{steel} = .000005$ in/in/°F).

There are two forms of shrinkage which can occur in a reinforced concrete structure: plastic shrinkage and drying shrinkage. Plastic shrinkage occurs during the first few hours after placing the concrete in the forms. The drying shrinkage occurs after the final set and most of the chemical hydration of the cement has taken place. A concrete specimen can shrink or swell because of moisture movement out of or into the member. The member will want to come into equilibrium with the moisture conditions of its surroundings. Several factors affecting the magnitude of drying shrinkage are: (1) aggregate; (2) water/cement ratio; (3) size of the concrete element; (4) medium ambient conditions; (5) amount of reinforcement; (6) admixtures; (7) type of cement; and (8) carbonation [6]. The shrinkage strain may vary from .0002 to .0007 in/in. The ACI Code 83 makes the designer have a minimum amount of steel to take up temperature and shrinkage stresses. The ACI Code 83 in section 7.12 specifies that for all sections, the $A_s = .0014\, A_g$ where A_g is the gross area of the concrete. Also, for slabs, Grade

Chart 1. Modification Factor for Standard Deviation When Less than 30 Tests Are Available [9].

Number of Tests	K
Less than 15	Use Table Below
15	1.16
20	1.08
25	1.03
30 or more	1.00

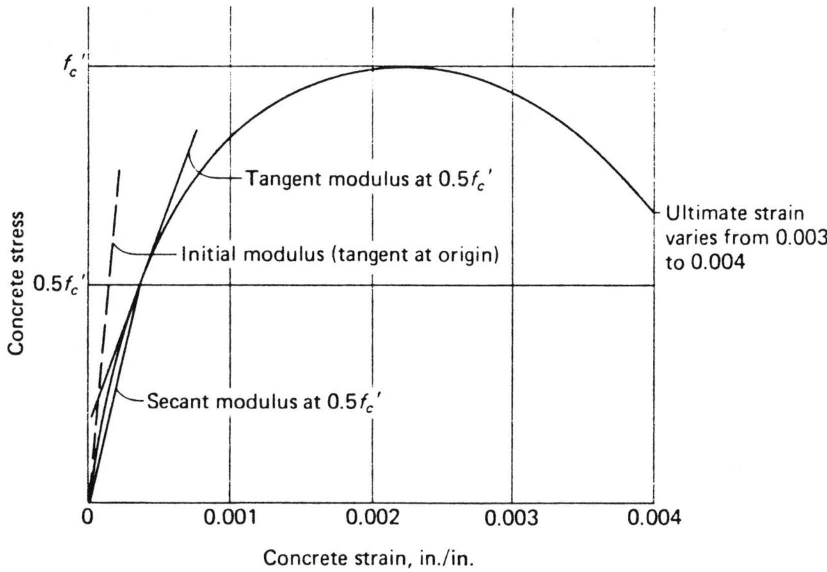

FIGURE 9. Stress-strain curve for concrete [5].

40 to 50 steel, A_s = .0020 A_g; Grade 60 A_s = .0018 A_g; etc., are required.

Creep or "plastic flow," is the increase in strain with time due to a sustained load. The effect of creep on a structural member can be shown in Figure 10 [3]. The effect of creep is similar to shrinkage in that it is dependent upon time. The factors affecting creep are similar to shrinkage. The effect of creep and shrinkage act closely together and are sometimes hard to separate. The effect of creep and shrinkage has to be considered when looking at deflections. This will be considered in that section.

Because concrete is very weak in tension as compared to the compressive strength (f_t is approximately .1 f_c), reinforcing steel is used to take the tensile stresses after cracking occurs in the concrete. The steel which is used is deformed reinforcing bars or standard wire reinforcement. The deformed bars have lugs or protrusions in order to prevent slipping between the bars and concrete. These lugs or protrusions create a bond between the two materials so they work together. The bars in the United States range from ⅜ to 2¼ inches. The bars range from #3, ⅜" in diameter, to #18, 2¼ inches diameter. The bar chart with the numbers and corresponding sizes are given in Chart 2 [9]. There are many different types of deformations depending upon the manufacturer. The bars also have markings which show:

1. the producing mill
2. bar size
3. type of steel
4. Grade.

Figure 11 shows some typical bars with markings [26]. The most common types of steels are Grade 40, 50, and 60. The grade of steel telling the yield of the steel. The most important properties of reinforcing steel are:

1. modulus of elasticity (E_s)
2. yield strength (f_y)
3. ultimate strength (f_u)
4. steel grade designation
5. size of bar or wire.

Table 2 [27] shows the reinforcement grades and strengths for the steel reinforcing. Figure 12 [1] shows some typical stress strain diagrams for steel.

The material in this section is basically what is covered by Wang and Salmon in their materials area [5].

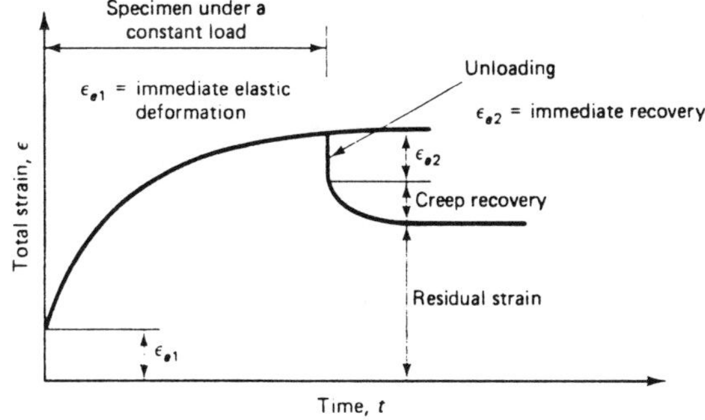

FIGURE 10. Creep recovery versus time [3].

Chart 2. ASTM Standard Reinforcing Bars [9].

Bar size	Nominal diameter, in.	Nominal area, sq. in.	Nominal weight, lb per ft
#3	0.375	0.11	0.376
4	0.500	0.20	0.668
5	0.625	0.31	1.043
6	0.750	0.44	1.502
7	0.875	0.60	2.044
8	1.000	0.79	2.670
9	1.128	1.00	3.400
10	1.270	1.27	4.303
11	1.410	1.56	5.313
14	1.693	2.25	7.650
18	2.257	4.00	13.600

DESIGN METHODS

The American Concrete Institute has a building code for reinforced concrete design—Building Code Requirements for Reinforced Concrete (ACI 318-83). There are two design philosophies: ultimate strength and working stress design methods. The working stress method was used prior to 1963 as the basic design method. After 1963, the ultimate strength method became more popular. The ACI Code 83 is

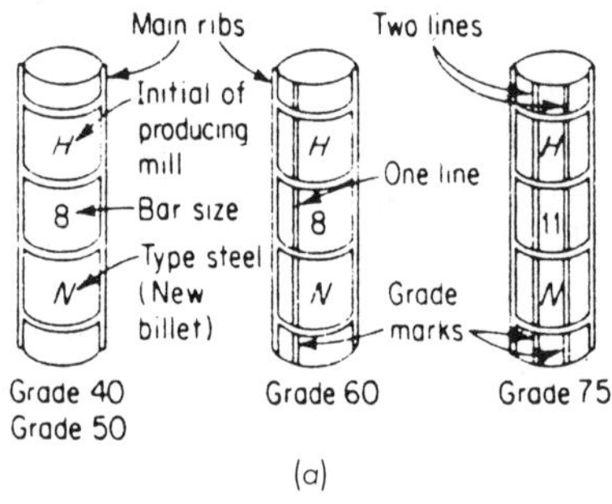

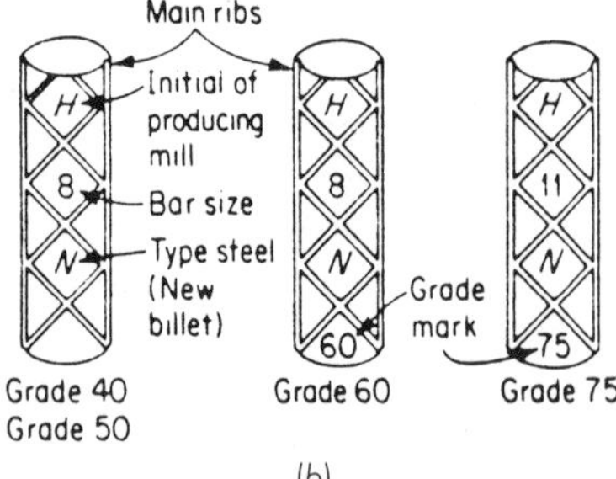

FIGURE 11. Marking system for reinforcing bars [26].

now based on the ultimate strength design with the alternate or service load method explained in an appendix.

The working stress method, also known as service load method, or the alternate method, is described in Appendix B—Alternate Design Method in the ACI Code 83. Stresses computed under the alternate design method do not exceed some predesignated allowable values. The stresses computed under the action of service loads are well within the elastic range. The straight line variation is used in the determination of stresses and strain. The factor of safety is applied to the allowable stresses for the concrete and steel. In working with flexural design the allowable stresses for concrete extreme fiber stress in compression is 0.45 f'_c. The allowable stresses for the reinforcing steel f_s are 20,000 psi for Grade 40 and 50 and 24,000 psi for Grade 60.

The strength design is based on designing the members to fail at a certain determined ultimate load. The service loads are increased by some factor, load factors, to obtain the desired ultimate load capacity. The design considers the nonlinear stress-strain diagram of the concrete. An example of the use of the load factors are as follows:

$$U = \phi N = 1.4D + 1.7L \tag{6}$$

where U is the factored loading, N is the nominal ultimate load, D is the dead load, L is the live load and ϕ is the undercapacity factor. The ϕ factor is to take into account inaccuracies in construction. These inaccuracies are dimensions, position of reinforcement or variations in properties. The undercapacities ϕ are shown in Table 3

The ACI load factors for different load cases will not be described. The different types of loads which can occur are:

1. dead load, D
2. live load, L
3. wind load, W
4. loads due to lateral pressure such as soil, H
5. lateral fluid pressure loads, F
6. earthquake loads, E
7. loads due to settlement, creep, shrinkage, or temperature, T.

The basic factored loading case which must be checked is:

$$U = 1.4D + 1.7L \tag{7}$$

When wind is considered with dead and live loads, the recommended combination is:

$$U = .75(1.4D + 1.7L + 1.7W) \tag{8}$$

also

$$U = .9D + 1.3W \tag{9}$$

If earthquake loads are to be considered, the following com-

TABLE 2. Standardized Reinforcing and Prestressing Steels [27].

Product	ASTM Specification	Grade	Minimum Yield Strength ksi	Minimum Yield Strength MPa	Minimum Tensile Strength ksi	Minimum Tensile Strength MPa
Reinforcing bars	A615	40	40	276	70	483
		60	60	414	90	620
	A616	50	50	345	80	552
		60	60	414	90	620
	A617	40	40	276	70	483
		60	60	414	90	620
	A706	60	60	414	80	552
			(78 max)	(538 max)		
Bar mats	A184,* A704*					
Wire, Smooth	A82		70	483	80	552
Deformed	A496		75	517	85	586
Welded wire fabric,						
Smooth	A185		65	448	75	517
Deformed	A497		70	483	80	552
Prestress bar	A722	Type I	127.5	880	150	1034
		Type II	120	827	150	1034
Prestress wire	A421		188*200	1296*1330	235*250	1620*1725
Prestress strand	A416	250	212.5	1465	250	1725
		270	229.5	1580	270	1860

*Same as reinforcing bars.

binations must be considered:

$$U = .75(1.4D + 1.7L + 1.87E) \quad (10)$$

$$U = .9D + 1.43E \quad (11)$$

or

$$U \geq 1.4D + 1.7L \quad (12)$$

Structures which have to resist lateral pressure due to soil or fluid pressure should be designed for the following factored load combinations:

$$U = 1.4D + 1.7L + 1.7H \quad (13)$$

$$U = .9D + 1.7H \quad (14)$$

$$U = 1.4D + 1.7L \quad (15)$$

$$U = 1.4D + 1.7L + 1.4F \quad (16)$$

$$U = .9D + 1.4F \quad (17)$$

$$U = 1.4D + 1.7L \quad (18)$$

When differential settlement, creep, shrinkage, or temperature may be significant the following factored load combination must be considered:

$$U = .75(1.4D + 1.4T + 1.7L) \quad (19)$$

Any structure or structural element must be designed for the most severe of any of the load combinations just described [3].

Other criteria besides strength which may control the design which must be considered are: excess deflection, cracking, instability, and fatigue.

As explained in PCA notes on ACI Code 83 [11], for good performance of a reinforced concrete structure the structural details are important. The standard practice for reinforcement details is a slow process. The Building Code Committee (ACI 318) works at collecting reports of research and practice with reinforcing materials, suggests new research that is needed, receives reports on new research, and translates the results into specific code provisions for details of reinforcement. The ACI Detailing Manual, SP-66 by ACI Committee 315 [10] provides recommended methods and standards for preparing design drawings, typical details, and drawings for fabrication and placing of reinforcing steel in reinforced concrete structures [11]. Figure 13 shows bar-bending details and Figure 14 shows sizes of hooks [10].

There are also tolerances which are permissible variations from dimensions given on drawings. Accepted tolerances for variation in cross-sectional dimensions of columns and beams and in the thickness of slabs and walls are + ½ in

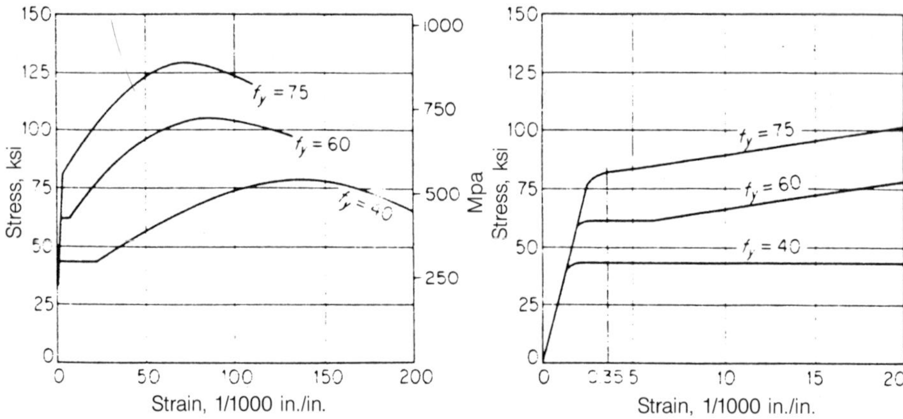

FIGURE 12. Typical stress-strain curves for current reinforcing steels with minimum specified yield points f_y from 40 to 75 ksi [1].

and − ¼ in [12]. For concrete footings accepted variations in plan dimensions are + 2 in and − ½ in [12], whereas the thickness has an accepted tolerance of −5% of specified thickness [5,12]. There are also tolerances on the steel reinforcing bar location.

A reinforced concrete structure can be designed for an earthquake by using the provisions described in Appendix A of the ACI Code 83. The ACI Code 83 has provisions for different areas: high seismic risk (zone 3 and 4), moderate seismic risk (zone 2), and zero or minor (zone 0 and 1). The structural framing systems which are considered are moment frame, structural walls, and frame/wall interaction. There are special provisions for moderate seismicity:

1. limited level of inelastic behavior
2. intermediate level of toughness
3. special reinforcing details (beam/column framing and slab/column framing).

A beam column framing system for stirrups/ties are shown in Figure 15. The special provisions for high seismic risk are in section A.2−A.8 of the ACI Code 83. These provisions help the performance by designing for strength, stiffness, deformation capacity, and details [13].

TABLE 3. Undercapacities ϕ for Design.

Design Area	ϕ Factors
Flexure	.9
Axial tension	.9
Shear and torsion	.85
Compression members, spirally reinforced	.75
Compression members, tied	.70
Bearing on concrete	.70
Bending in plain concrete	.65

BENDING BEHAVIOR

When looking at the flexural bending behavior, it is important to look at the behavior of a beam under a constant moment. Figure 16 shows the behavior of moment and deflection. The important points are moments at:

1. first crack
2. service loads
3. yield
4. ultimate.

The behavior shown in the figure is the desirable behavior in bending. The amount of tensile reinforcement A_s with respect to the compressive strength of the concrete f_c' will determine this behavior.

The ultimate strength of a reinforced concrete member considers a parabolic concrete stress distribution. For ease in computation of the ultimate moment, Whitney rectangular stress block is assumed. It can be shown that the actual parabolic stress block is equivalent to the rectangular Whitney rectangular stress block. It is assumed that the reinforcing bars tensile stress is yielding, f_y. If at the ultimate moment the concrete crushes and the steel is just starting to yield at the same time, this condition is said to be at the balanced condition.

$$\varrho = \frac{A_s}{bd} = \varrho_b \qquad \text{(balanced condition)} \tag{20}$$

If the reinforcing steel amount is less than ϱ_b, then the member is said to be under-reinforced.

$$\varrho < \varrho_b \qquad \text{(under-reinforced)} \tag{21}$$

On the other hand, if the amount of steel is greater than ϱ_b, then the member is said to be over-reinforced.

$$\varrho > \varrho_b \quad \text{(over-reinforced)} \tag{22}$$

The ACI Code 83 requires that the amount of steel in flexural design has to be less than $.75\varrho_b$. Thus, the code requires that the member be under-reinforced. This will produce a member which is ductile. An over-reinforced section would produce a brittle section which fails before M_y is obtained. The value of ϱ_b is given as:

$$\varrho_b = \frac{.85 f'_c \beta_1}{f_y}\left(\frac{87{,}000}{87{,}000 + f_y}\right) \tag{23}$$

where β_1 is the fraction of the compressive zone which defines the depth of the Whitney stress block, $a = \beta_1 c$. The value of $\beta_1 = .85$ $[f'_c - 4000]$ and $\beta_1 = .85 - .05$ $[f'_c - 4000/1000] \geq .65$ for $f'_c \geq 4000$ psi. The ACI Code 83 requires that there be a certain amount of minimum reinforcement $\varrho_{\min} > 200/f_y$. Thus, the Code requires that ϱ be:

$$\frac{200}{f_y} \leq \varrho \leq .75\,\varrho_b \tag{24}$$

which produces a ductile beam and will provide adequate warning before the member fails. The stresses across a section for Whitney's rectangular stress block is shown in Figure 17. The behavior of beams for regular beams with and without compression steel are shown in Figure 18 and Table 4.

The design and analysis for regularly reinforced concrete members shall now be explained. For design the nominal or ultimate moment is:

$$M_n = \frac{M_u}{\phi} = bd^2 f'_c w(1 - .59w) \tag{25}$$

where

$$M_u = 1.4\, M_D + 1.7\, M_L \tag{26}$$

For good design $w \cong .2$ where w is the reinforcement index

$$w = \frac{A_s}{bd}\left(\frac{f_y}{f'_c}\right) = \varrho\,\frac{f_y}{f'_c} \tag{27}$$

Also, for a good design $d \cong 1.5$ to $2.0\ b$ where d is the structural depth and b is the width of the beam. Thus

$$d = \sqrt{\frac{M_n}{bf'_c w(1 - .59w)}} \tag{28}$$

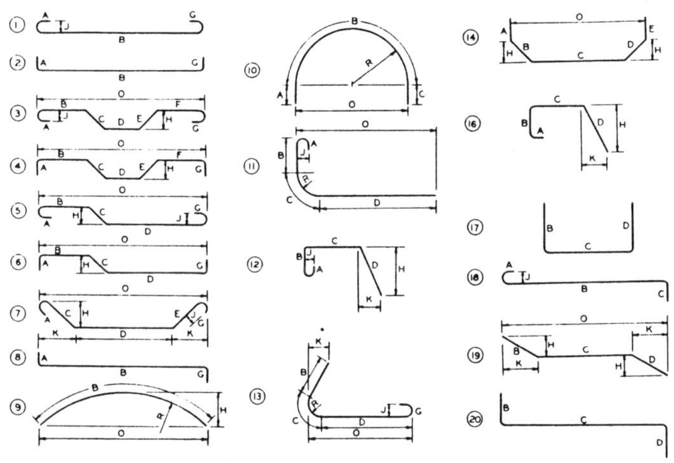

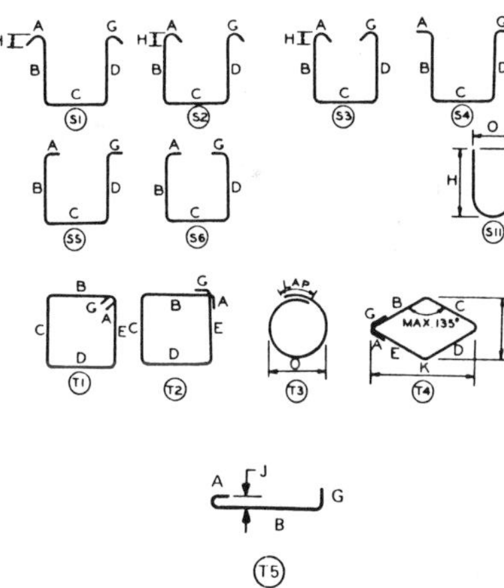

NOTES:
1. All dimensions are out to out of bar except "A" and "G" on standard 180 and 135 deg. hooks.
2. "J" dimension on 180 deg. hooks to be shown only where necessary to restrict hook size, otherwise standard hooks are to be used.
3. Where "J" is not shown "J" will be kept equal to or less than "H" on truss bars. Where "J" can exceed "H," it should be shown.
4. "H" dimension on stirrups to be shown where necessary to fit within concrete.
5. Where bars are to be bent more accurately than standard bending tolerances, bending dimensions which require closer working should have limits indicated.
6. Figures in circles show types.
7. For recommended diameter "D," of bends, hooks, etc., see tables.

Unless otherwise noted diameter D is the same for all bends and hooks on a bar.

Where slope differs from 45° dimensions 'H' and 'K' must be shown.

ENLARGED VIEW SHOWING BAR BENDING DETAILS

FIGURE 13. Typical bar bends [10].

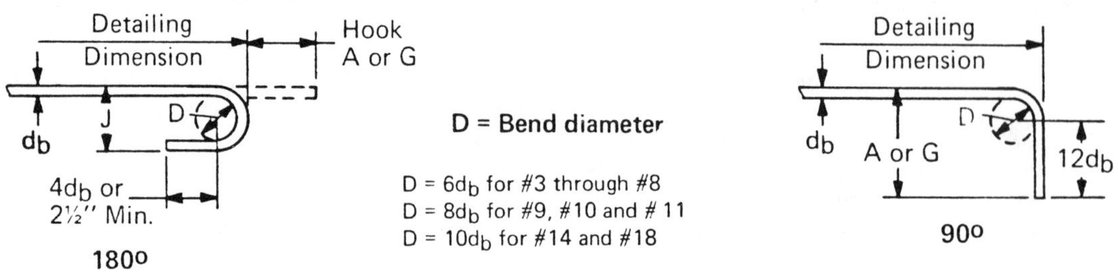

Bar size	Dimensions of standard 180-deg hooks, all grades			Dimensions of standard 90-deg hooks, all grades	
	A or G	J	D	A or G	D
#3	5″	3″	2¼″	6″	2¼″
#4	6	4	3	8	3
#5	7	5	3¾	10	3¾
#6	8	6	4½	1′-0″	4½
#7	10	7	5¼	1-2	5¼
#8	11	8	6	1-4	6
#9	1′-3″	11¼	9	1-7	9
#10	1-5	1′-0¾″	10¼	1-10	10¼
#11	1-7	1-2¼	11¼	2-0	11¼
#14	2-2	1-8½	17	2-7	17
#18	2-11	2-3	22¾	3-5	22¾

NOTE: When available depth is limited, #3 through #11 Grade 40 bars having 180-deg hooks may be bent with $D=5d_b$ and correspondingly smaller A and j dimensions.

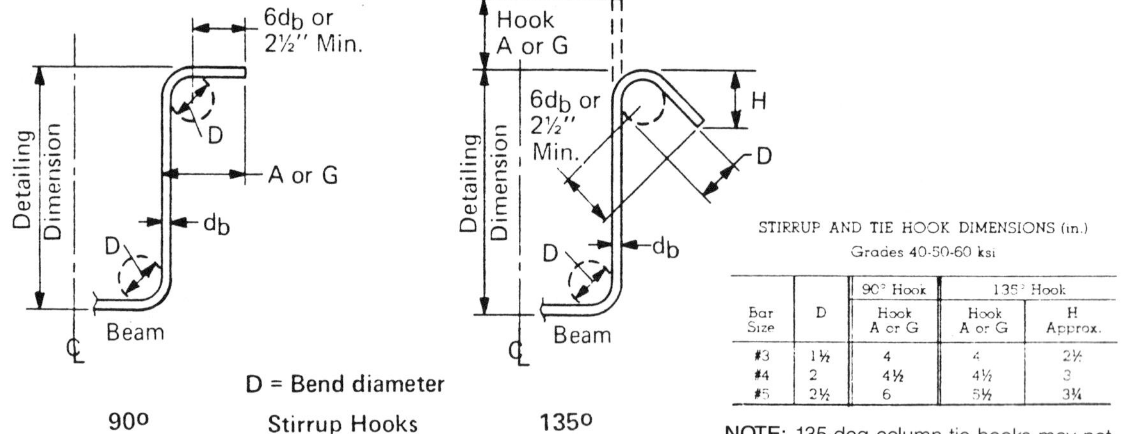

STIRRUP AND TIE HOOK DIMENSIONS (in.)

Grades 40-50-60 ksi

Bar Size	D	90° Hook	135° Hook	
		Hook A or G	Hook A or G	H Approx.
#3	1½	4	4	2½
#4	2	4½	4½	3
#5	2½	6	5½	3¾

NOTE: 135-deg column tie hooks may not be bent to less than diameter of column vertical bar enclosed in hook.

HOOKS AND BENDS OF WELDED WIRE FABRIC

Inside diameter of bends in welded wire fabric, plain or deformed for stirrups and ties shall be at least four wire diameters for wire larger than D6 or W6 and two wire diameters for all other wires. Blends with inside diameter of less than eight wire diameters shall not be less than four wire diameters from nearest welded intersection.

FIGURE 14. Standard hook details [10].

and

$$A_s = bd\,w\left(\frac{f'_c}{f_y}\right) \tag{29}$$

In analysis, if the member is under-reinforced, the moment is

$$M_n = A_s f_y\left(d - \frac{a}{2}\right) \text{ for } \varrho < \varrho_b \tag{30}$$

where

$$a = \frac{A_s f_y}{.85 f'_c b} \tag{31}$$

For a member which is over-reinforced $\varrho > \varrho_b$

$$M_n = A_s f_s\left(d - \frac{a}{2}\right) \tag{32}$$

and

$$f_s A_s = .85 f'_c \beta_1\left(\frac{d\epsilon_u}{\epsilon_u + f_s/E_s}\right) b \tag{33}$$

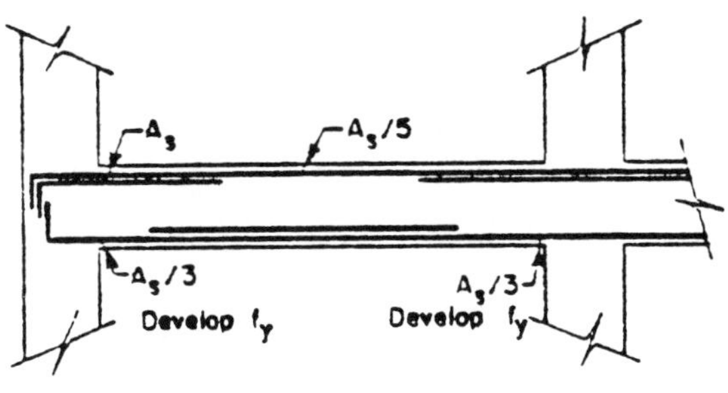

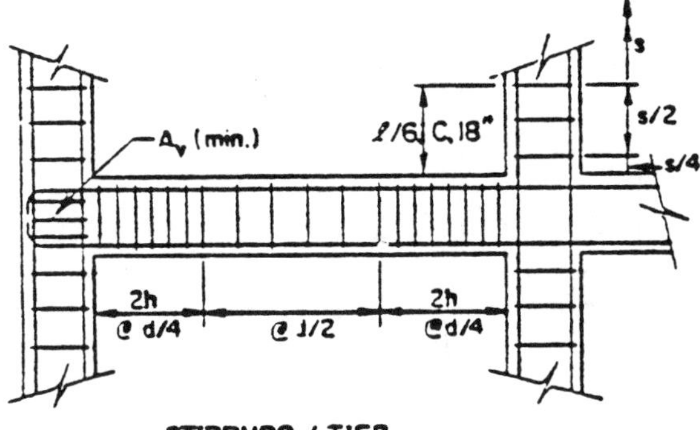

FIGURE 15. Beam/column framing system by ACI Code 83 for zone 2 [13]: (a) longitudinal bars; (b) stirrups/ties.

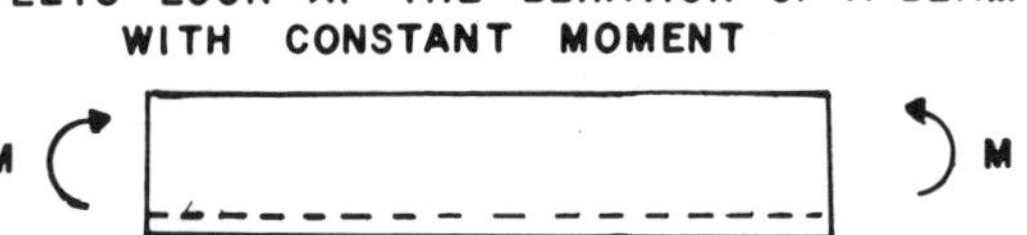

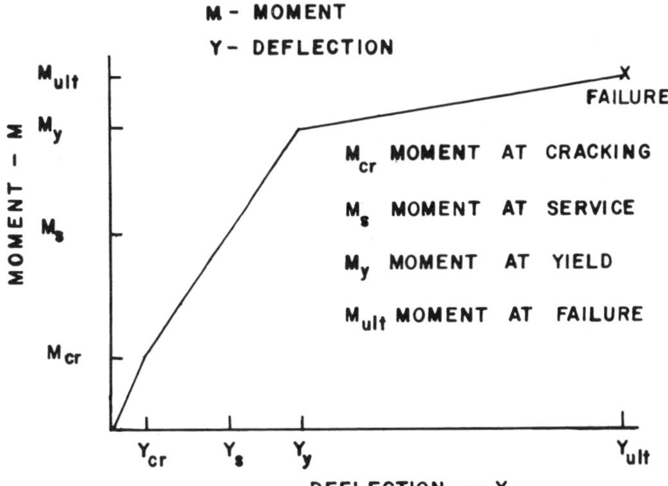

FIGURE 16. Load deflection behavior of beam.

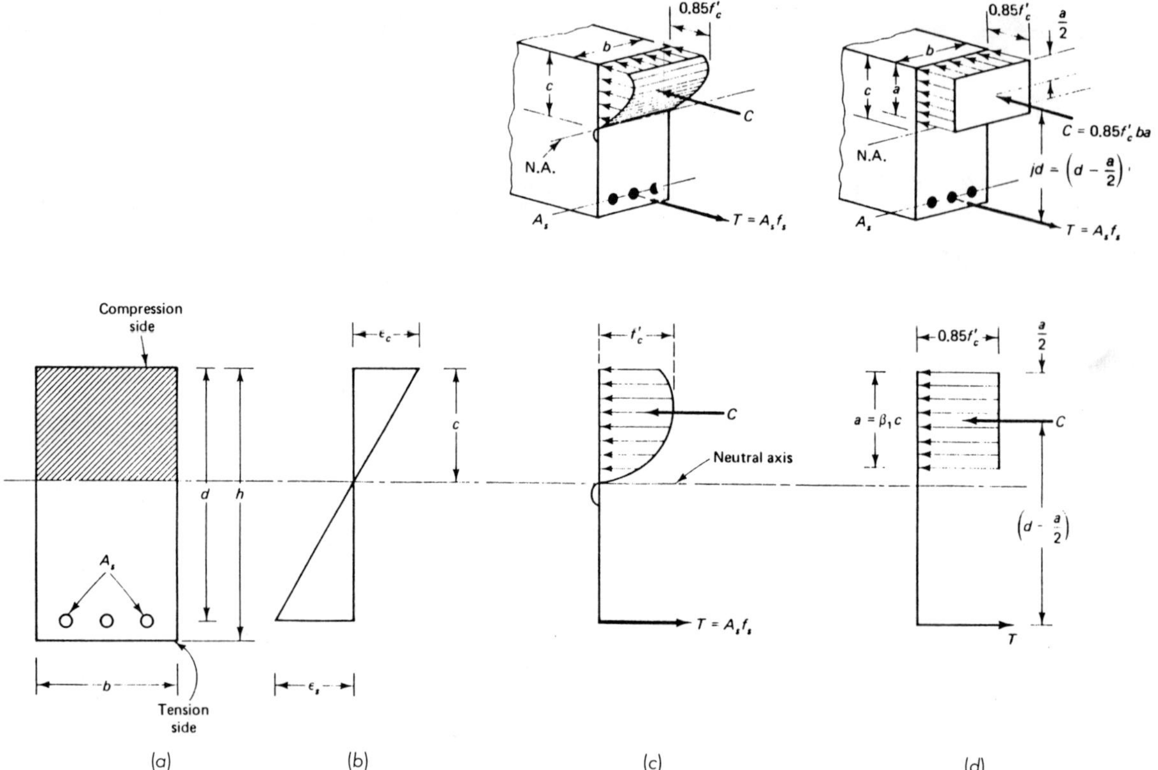

FIGURE 17. Stress and strain distribution across beam depth: (a) beam cross section; (b) strains; (c) actual stress block; (d) assumed equivalent stress block [3].

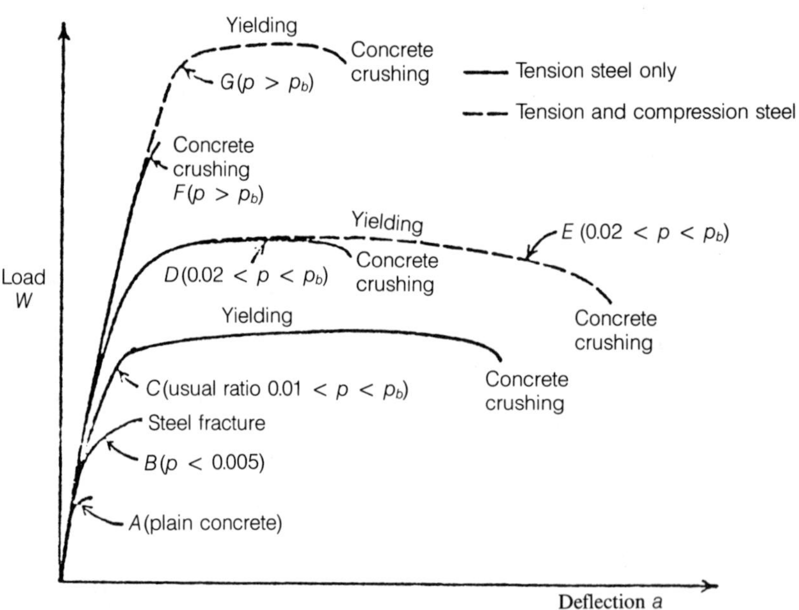

FIGURE 18. Load-deflection curves for beams of a given section with varying reinforcements.

TABLE 4. Characteristics of Members with Different Reinforcement Indexes.

Steel Ratio: Tension $\varrho = A_s/bd$	Steel Ratio: Compression $\varrho' = A_s'/bd$	Reinforcement Index $\omega = (\varrho - \varrho')f_y/f_c'$	Type of Member	Mode of Failure	Ductility Brittleness	Typical Curve
Zero	Zero	Zero	Plain concrete	Concrete in tension	Brittle	A
Very low $p < 0.005$	Zero	Very low	Under-reinforced	Sudden fracture of tension steel	Brittle	B
Low (normal range) $0.01 < p < 0.03$	Zero	Low-normal	Under-reinforced	Crushing after yielding	Very ductile	C
High, $0.02 < p < 0.05$	Zero	High-normal	Under-reinforced	Crushing after yielding	Ductile	D
High (normal range)	Similar to the tension steel	Low-normal	Under-reinforced	Crushing after yielding	Very ductile	E
Very high	Zero	Very high	Over-reinforced	Crushing	Brittle	F
Very high	Similar to the tension steel	Normal	Under-reinforced	Crushing after yielding	Ductile	G

where equation above is used to find f_s. Then a can be found by

$$a = \frac{A_s f_s}{.85 f_c' b} \tag{31}$$

Other items which must be considered in the design process are checking b_{min} so the reinforcement will fit, crack control, deflection and making sure the member is under-reinforced.

Bar spacing is important. The minimum spacing is required for construction and placement ease. The maximum of the three requirements are to be used. These are at least 1 in, 4/3 times the maximum aggregate size, or the diameter of the bar d_b. The cover for cast-in-place and precast concrete members is described in section 7.7 in the ACI Code 83. For beams, the cover requirements are a clear distance of 1½ in. to the reinforcement. Minimum bar charts may be established for checking minimum length. A b_{min} chart is shown in Table 5.

The width of cracks must be looked at to make sure that there is control of corrosion and unsightly appearance. The ACI Code 83 provisions are based on the Gergely-Lutz expression

$$w = C\beta_h f_s \sqrt[3]{d_c A} \tag{34}$$

Looking at Figure 19 [5], the variables are (1) w is the crack width at the tension force; (2) $\beta_h = h_2/h_1$, the ratio of the distances to the working stress neutral axis from the extreme tension fiber and from the controid of the main tension reinforcement; (3) f_s is the service-load stress in the steel (ksi); (4) d_c is thickness of cover; (5) A is A_e/m, effective tension area of concrete surrounding the main tension reinforcing bars; (6) m is the number of bars ($m = A_s/(A_b$ for largest bar); and (7) C is an expeimental constant [5,15]. The ACI Code 83 has simplified equation above (35). The

TABLE 5. Minimum Beam Width (Inches) According to the ACI Code[a] [5].

Size of Bars	Number of Bars in Single Layer of Reinforcement: 2	·3	4	5	6	7	8	Add for Each Added Bar
#4	6.1	7.6	9.1	10.6	12.1	13.6	15.1	1.50
#5	6.3	7.9	9.6	11.2	12.8	14.4	16.1	1.63
#6	6.5	8.3	10.0	11.8	13.5	15.3	17.0	1.75
#7	6.7	8.6	10.5	12.4	14.2	16.1	18.0	1.88
#8	6.9	8.9	10.9	12.9	14.9	16.9	18.9	2.00
#9	7.3	9.5	11.8	14.0	16.3	18.6	20.8	2.26
#10	7.7	10.2	12.8	15.3	17.8	20.4	22.9	2.54
#11	8.0	10.8	13.7	16.5	19.3	22.1	24.9	2.82
#14	8.9	12.3	15.6	19.0	22.4	25.8	29.2	3.39
#18	10.5	15.0	19.5	24.0	28.6	33.1	37.6	4.51

Table shows minimum beam widths when stirrups are used.

For additional bars, add dimension in last column for each added bar.

For bars of different size, determine from table the beam width for smaller size bars and then add last column figure for each larger bar used.

[a]Assumes maximum aggregate size does not exceed three-fourths of the clear space between bars (ACI-3.3.3).

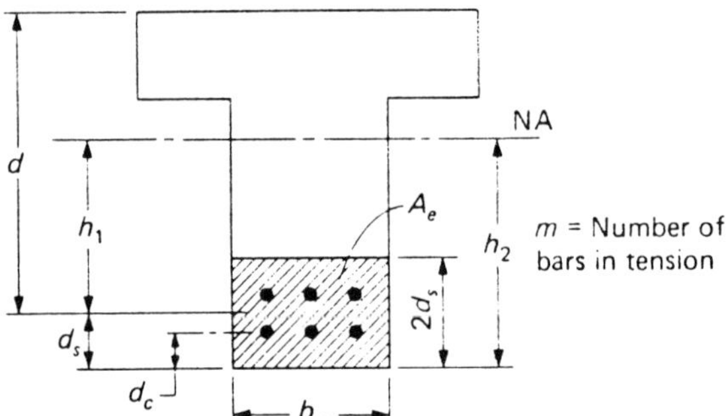

FIGURE 19. Dimensional notation for Gergely-Lutz Equation (34) [5].

equation is

$$Z = f_s \sqrt[3]{d_c A} \tag{35}$$

where Z has to be less than 175 for interior beams and 145 for the exterior beams.

The deflection of a member is important. Deflections may be excessive and cause concern from users or may cause cracking of walls and other members. If deflections are not of importance, Table 9.5a of the code should be applied. If deflection causes concern for walls, etc., Table 9.5b should be used for criteria. The calculation of deflections are described in another section.

The ACI Code 83 also wants to make sure that the reinforcement is between the limits

$$\varrho_{min} \leq \varrho \leq .75\varrho_b \tag{36}$$

or

$$\frac{200}{f_y} \leq \varrho \leq .75\varrho_b \tag{24}$$

This makes sure that the member is under-reinforced and assures that the member will be ductile and have sufficient warning before failure. Also, the member will have some ability for redistribution in the indeterminant structural system.

Example 2

Finding the ultimate moment of a section.

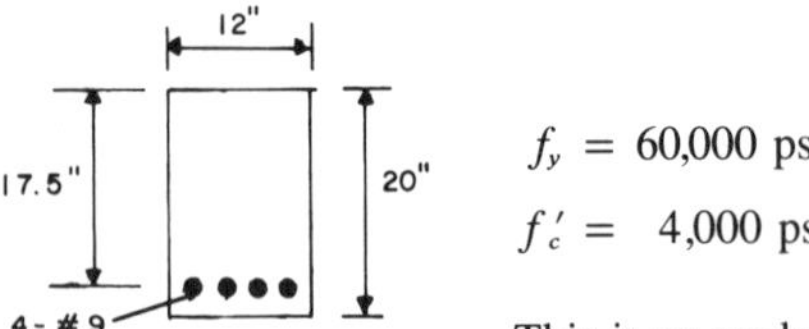

E2.1 Beam section.

$f_y = 60{,}000$ psi

$f'_c = 4{,}000$ psi

This is an analysis problem.

Checking ϱ

$$\varrho = \frac{A_s}{bd} = \frac{4(1.0)}{12(17.5)} = .0190$$

$$\varrho_b = .85\beta_1 \frac{f'_c}{f_y}\left(\frac{87{,}000}{87{,}000 + f_y}\right)$$

$$\varrho_b = .85(.85)\frac{(4000)}{60{,}000}\left(\frac{87{,}000}{87{,}000 + 60{,}000}\right) = .0285$$

$\varrho < \varrho_b$ so member is under-reinforced

$$M_n = A_s f_y \left(d - \frac{a}{2}\right)$$

$$a = \frac{A_s f_y}{.85 f'_c b} = \frac{(4.0)(60{,}000)}{(.85)(4000)(12)} = 5.89 \text{ in}$$

$$M_n = A_s f_y \left(d - \frac{a}{2}\right) = 4.0(60{,}000)\left(17.5 - \frac{5.89}{2}\right)$$

$$M_n = 3{,}490{,}000 \text{ in lb}$$

Example 3

Given the following cross section and moment find the reinforcing steel which is necessary for the design.

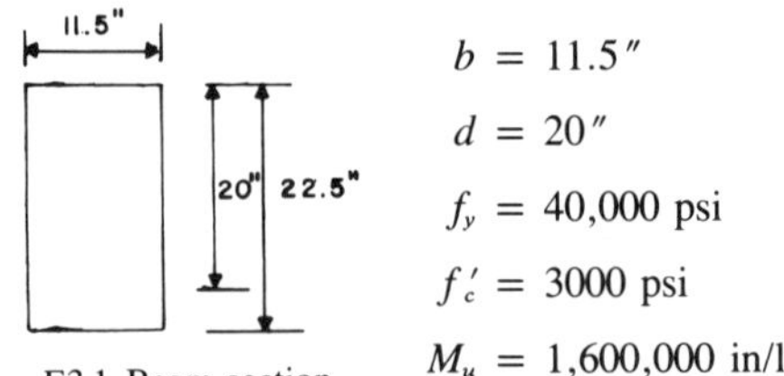

E3.1 Beam section.

$b = 11.5''$

$d = 20''$

$f_y = 40{,}000$ psi

$f'_c = 3000$ psi

$M_u = 1{,}600{,}000$ in/lb

This is a design problem. What is the steel area required? Using a convergent method (there are other approaches), let's assume $a = 4.0$ in.

$$A_s = \frac{M_u}{\phi f_y \left(d - \frac{a}{2}\right)} = \frac{1{,}600{,}000}{(.9)(40{,}000)\left(20 - \frac{4}{2}\right)} = 2.47 \text{ in}^2$$

Check a

$$a = \frac{A_s f_y}{.85 f'_c b} = \frac{2.47(40{,}000)}{.85(3000)(11.5)} = 3.37 \text{ in}$$

Now assume $a = 3.30$ in

$$A_s = \frac{M_u}{\phi f_y\left(d - \frac{a}{2}\right)} = \frac{1{,}600{,}000}{(.9)(40{,}000)\left(20 - \frac{3.30}{2}\right)} = 2.42 \text{ in}^2$$

Check a

$$a = \frac{A_s f_y}{.85 f_c' b} = \frac{2.42(40{,}000)}{.85(3000)(11.5)} = 3.30 \text{ in}$$

Since the assumed a is equal to the actual a, then the design A_s is equal to 2.42 in.^{2}. The ϱ is

$$\varrho = \frac{A_s}{bd} = \frac{2.42}{(11.5)(20)} = .0105$$

Now we must check to make sure $\varrho_{min} \leq \varrho \leq \varrho_{max}$

$$\varrho_{max} = .75\varrho_b = .75\left[.85\beta_1 \frac{f_c'}{f_y}\left(\frac{87{,}000}{87{,}000 + f_y}\right)\right] = .0278$$

$$\varrho_{min} = \frac{200}{f_y} = \frac{200}{40{,}000} = .005$$

Now $\varrho_{min} \leq \varrho \leq \varrho_{max}$ and $.005 \leq .0105 \leq .0278$. Thus the design is adequate. This design procedure is used when an indeterminant beam is designed. The section first is designed for b, d, and A_s at the maximum moment section. Then the remaining sections of "maximum" moments are solved for area of steel A_s using the b and d found earlier for the actual maximum moment and the convergent method.

Example 4

Design of a simply supported beam for flexure.

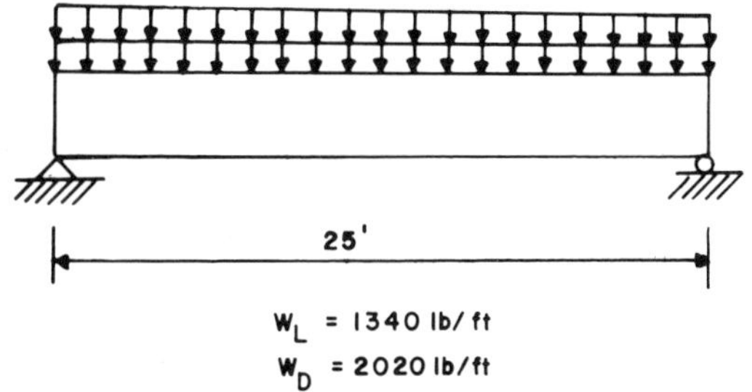

E4.1 Loading diagram for beam.

$f_c' = 4000$ psi

$f_y = 40{,}000$ psi

$w_u = 1.4w_D + 1.7w_L = 1.4(2020) + 1.7(1340)$

$w_u = 5106$ lb/ft

$$M_u = \frac{w_u L^2}{8} = \frac{5106(25^2)}{8} = 399{,}000 \text{ ft lb}$$

Let's assume $w = .3$ (Reinforcement Index)

$$\varrho = \frac{wf_c'}{f_y} = \frac{(.3)(4000)}{40{,}000} = .030$$

$$\varrho_b = \frac{.85_1 f_c'}{f_y}\left(\frac{87{,}000}{87{,}000 + f_y}\right) = .0495$$

$$\varrho_{min} = \frac{200}{f_y} = \frac{200}{40{,}000} = .005$$

Thus, ϱ has to be $\varrho_{min} \leq \varrho \leq .75\ \varrho_b$ hence $.005 \leq .030 \leq .0371$ so reinforcement will be adequate as set up by the ACI Code.

Now

$$M_u = \phi f_c' bd^2 w(1 - .59w)$$

Assume $b = 14$ in

$$d = \sqrt{\frac{M_u}{\phi f_c' b w(1 - .59w)}}$$

$$d = \sqrt{\frac{(399{,}000)(12)}{(.9)(4000)(14)(.3)(1 - .59(.3))}} = 19.61 \text{ in}$$

Thus, $d = 20$ in and $b = 14$ in

$$A_s = \varrho bd = (.03)(20(14)) = 8.40 \text{ in}^2$$

Now 6 – #11 bars will be $1.56(6) = 9.36$ in^2

$$9.36 \text{ in}^2 > 8.4 \text{ in}^2$$

Now check minimum width b_{min}

b_{min} for 6 – #11 is 20.5 in. Not O.K.

Try 2 layers. Thus 3 – #11 – $b_{min} = 12.0$ in < 14.0 in. Thus O.K.

$$\text{Now } h = d + \frac{x}{2} + d_b + \text{Stirrup} + \text{Cover}$$

$$h = 20 + \frac{1.41}{2} + 1.41 + .375 + 1.5 = 23.99 \text{ in}$$

$h = 24$ in

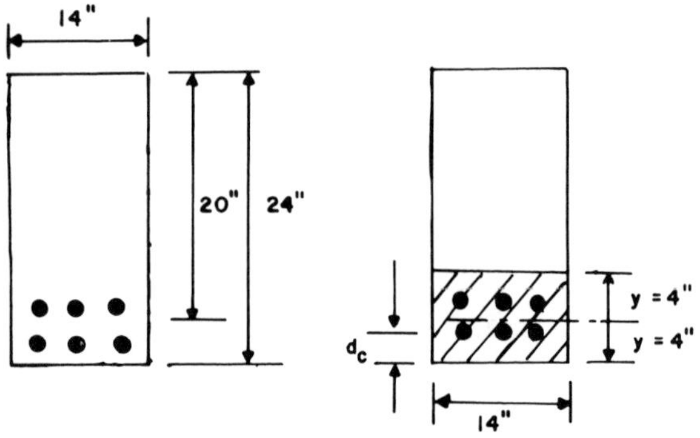

E4.2 Beam section.

Crack Control $Z = f_s \sqrt[3]{d_c A}$

$$f_s = .6f_y \text{ (KSI)}$$

$$d_c = 1.5 + .375 + \frac{1.41}{2} = 2.585 \text{ in}$$

Cover Stirrup $d_b/2$

$$A = \frac{2yb}{\# \text{ Bars}} = \frac{2(4)(14)}{6} = 18.667$$

$$Z = .6(40) \sqrt[3]{(2.585)(18.667)}$$

$$= 87.4 < 145 \;\&\; 175 \quad \text{Thus O.K.}$$

Now let's look at deflection. From Table 9.5a simple supported $h_{min} = \ell/16\ (.4 + f_y/100{,}000)$.

$$h_{min} = \frac{25(12)}{16}\left(.4 + \frac{40{,}000}{100{,}000}\right) = 15 \text{ in} < 24 \text{ in.}$$

Thus O.K.

Example 5

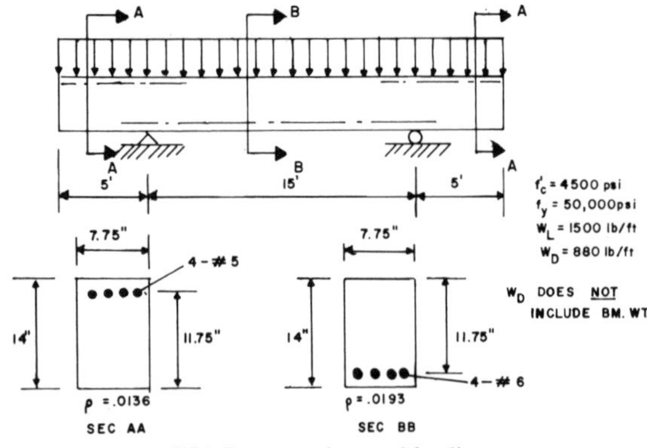

E5.1 Beam sections and loading.

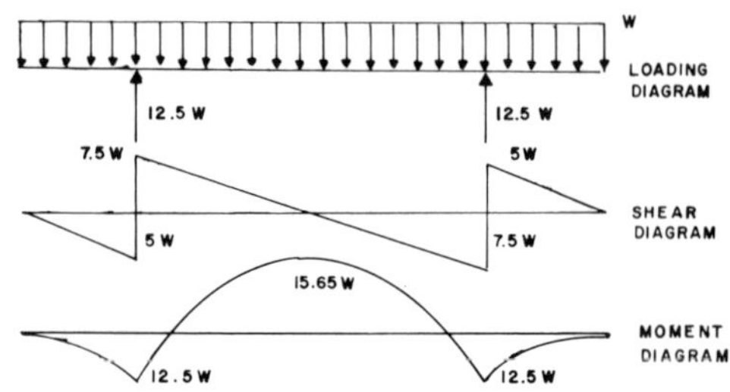

E5.2 Loading, shear, and moment diagram for beam.

For the following beam check for moment only to see if it is adequate.

$$w_u = 1.4\, w_{DL} + 1.7 w_{LL} = 1.4(880) + 1.7(1500)$$

$$+ 1.4\left(\frac{7.75(14)(150)}{144}\right) = 3940 \text{ lb/ft}$$

$$M_u@_{B \text{ or } \mathbb{L}} = 15.625(3940) = 61{,}565 \text{ ft lb}$$

$$M_u@_{A \text{ or } Sup} = 12.5(3940) = 49{,}250 \text{ ft lb}$$

$$\varrho_b = \frac{.85 f'_c \beta_1}{f_y}\left(\frac{87{,}000}{87{,}000 + f_y}\right) = .0401$$

Thus $\varrho < \varrho_b@_{AA}$

$$\varrho < \varrho_b@_{BB}$$

Thus under-reinforced

$$M_u = \phi f_y A_s \left(d - \frac{a}{2}\right)$$

$$a = \frac{A_s f_y}{.85 f'_c b}$$

$$a_{\mathbb{L}} = \frac{4(.44)(50)}{.85(4.5)(7.75)}$$

$$a_{\mathbb{L}} = 2.968 \text{ in}$$

$$M_u@_{\mathbb{L}} = .9(50{,}000)(.44)\left(11.75 - \frac{2.968}{2}\right)$$

$$M_u@_{\mathbb{L}} = 67{,}755 \text{ ft lb}$$

$$a_{Sup} = \frac{(4)(.31)(50)}{.85(4.5)(7.75)}$$

$$a_{Sup} = 2.09$$

$$M_u@_{Support} = .9(50{,}000)(4)(.31)\left(11.75 - \frac{2.09}{2}\right)$$

$$M_u@_{Support} = 49{,}774 \text{ ft lb}$$

$@_{℄}$ 67,755 > 61,565 O.K.

$@_{Sup}$ 49,774 > 49,250 O.K.

Thus O.K. for Moment

DOUBLY REINFORCED CONCRETE BEAMS

A reinforced concrete section may have both tension and compression reinforcement (doubly reinforced). The use of doubly reinforced concrete section can be for deflection control in order to reduce the effects of creep and shrinkage and also when a section is subjected to reverse loading such as found in earthquake design. The balanced percentage of steel is:

$$\bar{\varrho}_b = \varrho_b + \varrho' \frac{f_s'}{f_y} \quad (37)$$

where ϱ_b is the balanced steel ratio of a singly reinforced concrete section; ϱ' is the percentage of compression steel A_s'; and f_s' is the stress of the compression steel if it is not yielding. The code requires that the maximum amount of tensile reinforcement ratio be

$$\varrho \leq 0.75\bar{\varrho}_b + \varrho' \frac{f_s'}{f_y} \quad (38)$$

This is to make sure the section behaves ductily. The value of f_s' is

$$f_s' = 29(10^6)(0.003)\left(1 - \frac{0.85\beta_1 f_c' d'}{(\varrho - \varrho')f_y d}\right) \quad (39)$$

if the compression steel is not yielding. In order to determine whether the compression steel for a section is yielding, the following holds

$$\varrho > \varrho_{min} \text{ steel is yielding } f_s = f_y$$

$$\varrho < \varrho_{min} \text{ steel is not yielding } f_s = f_s'$$

where

$$\varrho_{min} = .85\beta_1 \frac{f_c'}{f_y}\frac{d'}{d}\left(\frac{87{,}000}{87{,}000 - f_y}\right) + \varrho' \quad (40)$$

When analyzing a doubly reinforced concrete section, the steel stresses in the tension and compression steel must be determined. If the tensile and compressive steel stresses are both yielding, the following holds:

$$\varrho > \varrho_{min} \text{ and } \varrho < \bar{\varrho}_b \quad (41)$$

then

$$M_n = A_s' f_y(d - d') + (A_s - A_s')f_y\left(d - \frac{a}{2}\right) \quad (42)$$

where

$$a = \frac{(A_s - A_s')f_y}{.85 f_c' b}$$

If the tension steel stress is yielding and the compressive steel stress is not, then the situation is:

$$\varrho < \varrho_{min} \text{ and } \varrho < \bar{\varrho}_b$$

then

$$M_n = (A_s f_y - A_s' f_s')\left(d - \frac{a}{2}\right) + f_s' A_s (d - d') \quad (43)$$

where f_s' is found from the equation:

$$A_s'\left(\frac{c - d}{c}\right)\epsilon_u E_s + .85 f_c' \beta_1 cb = A_s f_y \quad (44)$$

if c is found and

$$f_s' = \left(\frac{c - d'}{c}\right)\epsilon_u E_s \quad (45)$$

then

$$a = \beta_1 c \text{ or } a = \frac{A_s f_y - A_s' f_s'}{.85 f_c' b}$$

where ϵ_u is the ultimate compressive concrete strain at failure. Basically, the designer is using equilibrium and strain-compatibility.

The design of a doubly reinforced concrete section can be done by a trial and adjustment procedure explained by Nawy [3].

Example 6

Analysis of the following doubly reinforced concrete beam section for nominal moment.

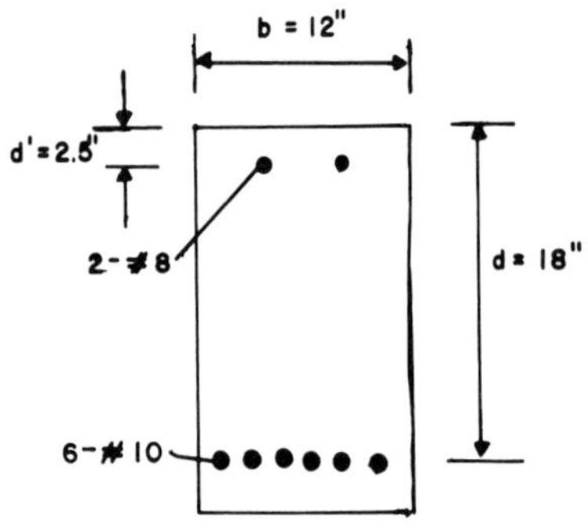

E6.1 Beam section.

$$A_s = 6(1.27) = 7.62 \text{ in}^2$$

$$A_s' = 2(.79) = 1.58 \text{ in}^2$$

$$\varrho = \frac{A_s}{bd} = \frac{7.62}{12(18)} = .03528$$

$$\varrho' = \frac{A_s'}{bd} = \frac{1.58}{12(18)} = .00731$$

$$\varrho_b = \frac{.85f_c'\beta_1}{f_y}\left(\frac{87{,}000}{87{,}000 + f_y}\right)$$

$$= \frac{.85(.85)(4000)}{60{,}000}\left(\frac{87{,}000}{87{,}000 + 60{,}000}\right)$$

$$\varrho_b = .02851$$

$$\bar{\varrho}_b = \varrho_b + \varrho' = .02851 + .00731 = .03582$$

$\bar{\varrho}_b > \varrho$ .03582 > .03528 Thus A_s is yielding

$$\varrho_{min} = .85\beta_1 \frac{f_c'}{f_y}\frac{d'}{d}\left(\frac{87{,}000}{87{,}000 - fy}\right) + \varrho'$$

$$= (.85)(.85)\frac{4000}{60{,}000}\left(\frac{2.5}{18}\right)\left(\frac{87{,}000}{87{,}000 - 60{,}000}\right)$$

$$\varrho_{min} = .02887$$

$\varrho > \varrho_{min}$ Thus A_s' is yielding

Now

$$\varrho > \varrho_{min} \quad \varrho < \bar{\varrho}_b$$

And

$$M_n = A_s' f_y (d - d') + (A_s - A_s') f_y \left(d - \frac{a}{2}\right)$$

$$a = \frac{(A_s - A_s')f_y}{.85f_c'b}$$

$$a = \frac{(7.62 - 1.58)(60{,}000)}{.85(4000)(12)} = 8.882''$$

$$M_n = 1.58(60{,}000)(18 - 2.5)$$

$$+ (7.62 - 1.58)(60{,}000)\left(18 - \frac{8.882}{2}\right)$$

$$M_n = 6{,}379{,}000 \text{ in lb}$$

"T" BEAMS

Flanged beams (T and L) are primarily designed at midspans. Figure 20 shows a span of a continuous beam in a structure. A regular or doubly reinforced concrete section will appear at the face of the supports as shown. At the midsection of the beam, because of the slab, a "T" beam may occur when cracking occurs (depending on the location of the neutral axis). The width of the beam in the slab at the midsection is defined by the code as:

1. Effective overhand $\not> 8\, h_f$.
2. Overhang $\not> \frac{1}{2}\, \ell_n$ clear distance to next beam.
3. b shall not exceed ¼ clear span of beam.

At the midsection, the reinforced concrete section will be one of two cases depending on the location of the neutral axis. *Case I* occurs when the neutral axis c is less than the flange thickness h_f. In this case, the regular rectangular section should be considered. *Case II* occurs when the neutral axis c is more than the flange thickness h_f. "T" beam analysis should then be used. In order to determine the location of the neutral axis c, the following equations are used:

$$a = \frac{A_s f_y}{.85f_c'b} \tag{31}$$

and

$$c = a/\beta_1 \tag{46}$$

In designing or analyzing "T" beams, the balanced condition ϱ_{wb} is:

$$\varrho_{wb} = \varrho_b + \varrho_f \tag{47}$$

where

$$\varrho_f = .85f_c'(b - b_w)\frac{h_f}{f_y b_w d} = \frac{A_{sf}}{b_w d} \tag{48}$$

and ϱ_b is the balanced steel ratio of a singly reinforced concrete section. If the steel percentage $\varrho_w = A_s/b_w d$ is less than ϱ_{wb}, then the section is underreinforced and behaves ductily. The nominal moment M_n is given as:

$$M_n = A_{sf}f_y\left(d - \frac{h_f}{2}\right) + (A_s - A_{sf})f_y\left(d - \frac{a}{2}\right) \quad (49)$$

where

$$a = \frac{(A_s - A_{sf})f_y}{.85f'_c\, b_w} \quad (50)$$

and

$$A_{sf} = \frac{.85f'_c\,(b - b_w)h_f}{f_y} \quad (51)$$

In design a trial and adjustment procedure can be used for the "T" beam as described by Nawy [3]. As in the design of singly and doubly reinforced beams, the maximum ϱ_w has to be less than $.75\varrho_{wb}$. The minimum amount of ϱ_w is $200/f_y$.

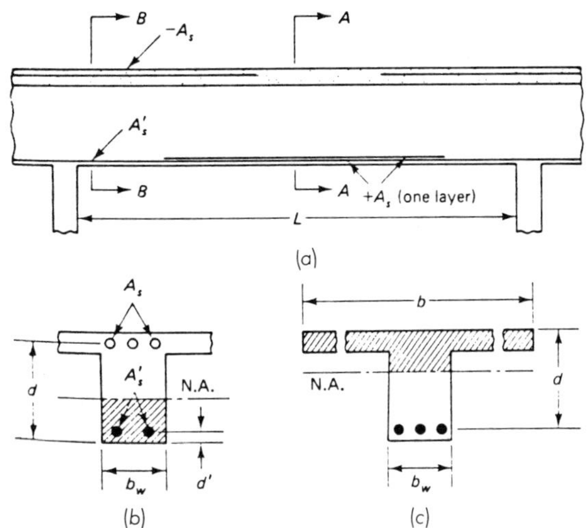

FIGURE 20. Elevation and sections of a monolithic continuous beam: (a) beam elevation; (b) support section B–B (inverted doubly reinforced beam); (c) mid-span section A–A (real T-beam) [3].

Example 7

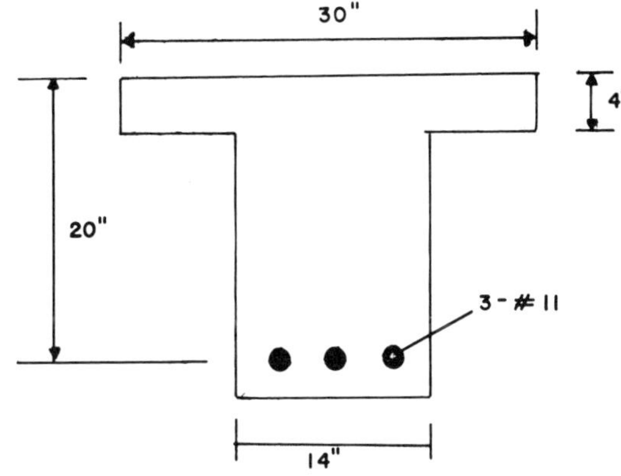

E7.1 "T" beam section.

Analysis of the following "T" beam for its nominal moment.

$f_y = 60{,}000$ psi

$f'_c = 2500$ psi

$A_s - 3 - \#11$

$A_s = 3(1.56) = 4.68\ \text{in}^2$

Check N.A.

$$\varrho_w = \frac{A_s}{b_w d} = \frac{4.68}{(14)(20)} = .0167$$

$$a = \frac{A_s f_y}{.85f'_c b} = \frac{4.68(60{,}000)}{.85(2500)(30)} = 4.405\ \text{in}$$

4.405 > 4.0 "T" Beam Analysis

$$A_{sf} = \frac{.85f'_c(b - b_w)h_f}{f_y} = \frac{.85(2500)(30 - 14)(4)}{60{,}000}$$

$A_{sf} = 2.2667$

$\varrho_w = .0167$

$$\varrho_f = \frac{A_{sf}}{b_w d} = \frac{2.2667}{(14)(20)} = .0081$$

$$\varrho_b = .85\beta_1\frac{f'_c}{f_y}\left(\frac{87{,}000}{87{,}000 + f_y}\right)$$

$$= \frac{.85(.85)(2500)}{60{,}000}\left(\frac{87{,}000}{87{,}000 + 60{,}000}\right)$$

$\varrho_b = .0177$

$\varrho_{wb} = \varrho_b + \varrho_f = .0177 + .0081 = .0258$

$\varrho_w < \varrho_{wb}\ .0167 < .0258$

Under-reinforced

$$M_n = A_{sf}f_y\left(d - \frac{h_f}{2}\right) + (A_s - A_{sf})f_y\left(d - \frac{a}{2}\right)$$

$$M_n = M_{n1} + M_{n2}$$

$$M_{n_1} = A_{sf} f_y \left(d - \frac{h_f}{2} \right)$$

$$M_{n_1} = 2.2667(60{,}000) \left(20 - \frac{4}{2} \right)$$

$$M_{n_1} = 2{,}448{,}000 \text{ in lb}$$

$$M_{n_2} = (A_s - A_{sf}) f_y \left(d - \frac{a}{2} \right)$$

$$a = \frac{(A_s - A_{sf}) f_y}{.85 f'_c b_w} = \frac{(4.68 - 2.2667)60{,}000}{.85(2500)(14)}$$

$$a = 4.867 \text{ in}$$

$$M_{n_2} = (4.68 - 2.2667)(60{,}000) \left(20 - \frac{4.867}{2} \right)$$

$$M_{n_2} = 2{,}542{,}000$$

$$M_n = M_{n_1} + M_{n_2} = 2{,}448{,}000 + 2{,}542{,}000$$

$$M_n = 4{,}990{,}000 \text{ in lb}$$

SHEAR BEHAVIOR AND REINFORCEMENT

No theory of failure proposed yet predicts accurately the failure of reinforced concrete in either tension or compression under all possible states of stress. The following are some considerations:

1. Maximum tensile stress theory has been able to predict tensile failures (shear and bending moment);
2. A limiting value of compression strain has been generally accepted as the criteria for flexural crushing; and
3. Mohr's theory of failure has been shown to be useful in predicting crushing failure when applied to simple states of stress [25].

Stress trajectories can help in explaining shear behavior. When maximum tensile stress applied to the concrete exceeds the tensile strength of the concrete, a crack will appear perpendicular to that stress orientation as shown in Figure 21(a) and (b). The stress trajectories and maximum tensile stress can be developed accurately by means of photoelasticity or mathematical analysis. From the element shown in Figure 21(b), and using elementary elasticity theory, the maximum tensile stress is:

$$f_{t\max} = \tfrac{1}{2} f_t + \sqrt{(\tfrac{1}{2} f_t)^2 + \tau_{xy}^2} \tag{52}$$

Thus, the crack patterns will follow the maximum tensile stress. The crack will appear when the maximum tensile stress reaches the concrete tensile strength. Figure 21(a) and 21(b) also show the cracking of the beam and the compressive stress trajectories. The crack appears along the compressive stress trajectory.

When classifying the different types of shear and flexural failures for beams, the following variables must be considered.

1. type of loading;
2. geometry of section;
3. amount and arrangement of reinforcing; and
4. interaction between steel and concrete.

For a beam subjected to a concentrated load, the major variable affecting the mode of failure is probably the ratio of the distance "a" from the load to the support and the depth of the member "d." The ratio a/d can also be expressed in terms of M/Vd, where the M/Vd expression can be used to describe the different loading conditions. The types of shear failures can be classified according to the a/d ratio, shear span to flexural depth of the beam. The beam failure classifications are:

1. deep beams ($0 < a/d < 1.0$);
2. short beams ($1.0 < a/d < 2.8$);
3. normal beams ($2.8 < a/d < 6.0$); and
4. long beams ($a/d > 6.0$). (See Figure 22.)

Long beams, which are very slender, usually fail in flexure and have an a/d ratio larger than 6.0. Normal beams of rectangular cross section for a/d between 2.8 and 6.0 fail by means of a flexure-shear crack called "diagonal tension failure" (see Figure 22b). Short beams ($1.0 < a/d < 2.8$) generally have two types of failures: shear tension and shear compression. The ultimate shear failure of the beam occurs for short beams as:

1. an anchorage failure of the longitudinal steel ("shear-tension" failure), or
2. a crushing of the concrete in the compression zone ("shear-compression" failure) (see Figure 22c).

For deep beams ($a/d < 1.0$), there is considered strength which can be developed after the inclined crack has occurred. There are several modes of failure which are possible for this tied-arch system which are:

1. an anchorage failure of the tension reinforcement at the support;
2. a crushing failure at the reactions;
3. a flexure failure by either crushing of the concrete near the top of the arch or a yielding of the tension reinforcement;
4. a failure of the arch thrust; and
5. a failure of the arch rib due to a crushing of concrete in the rib (see Figure 22d).

The mechanism of shear transfer in reinforced concrete

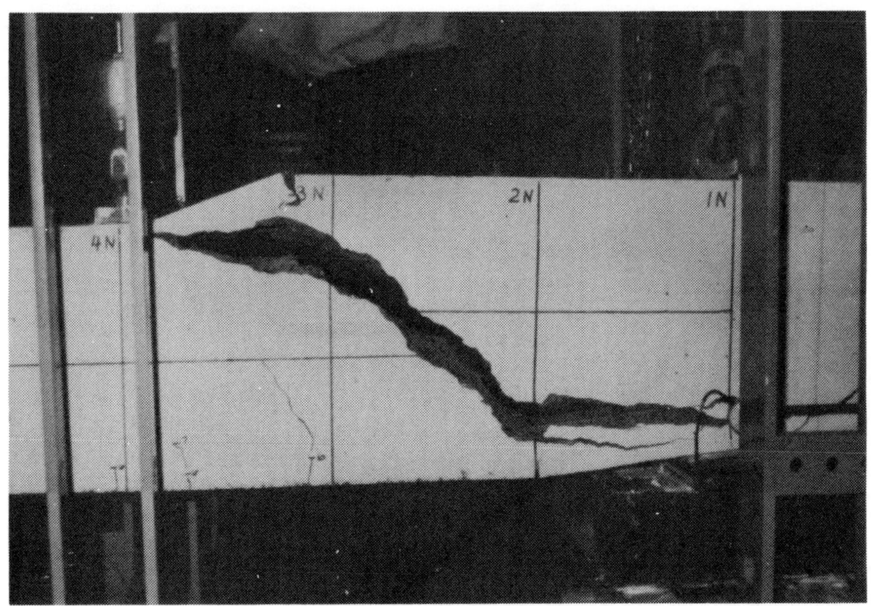

FIGURE 21a. Shear failure pattern.

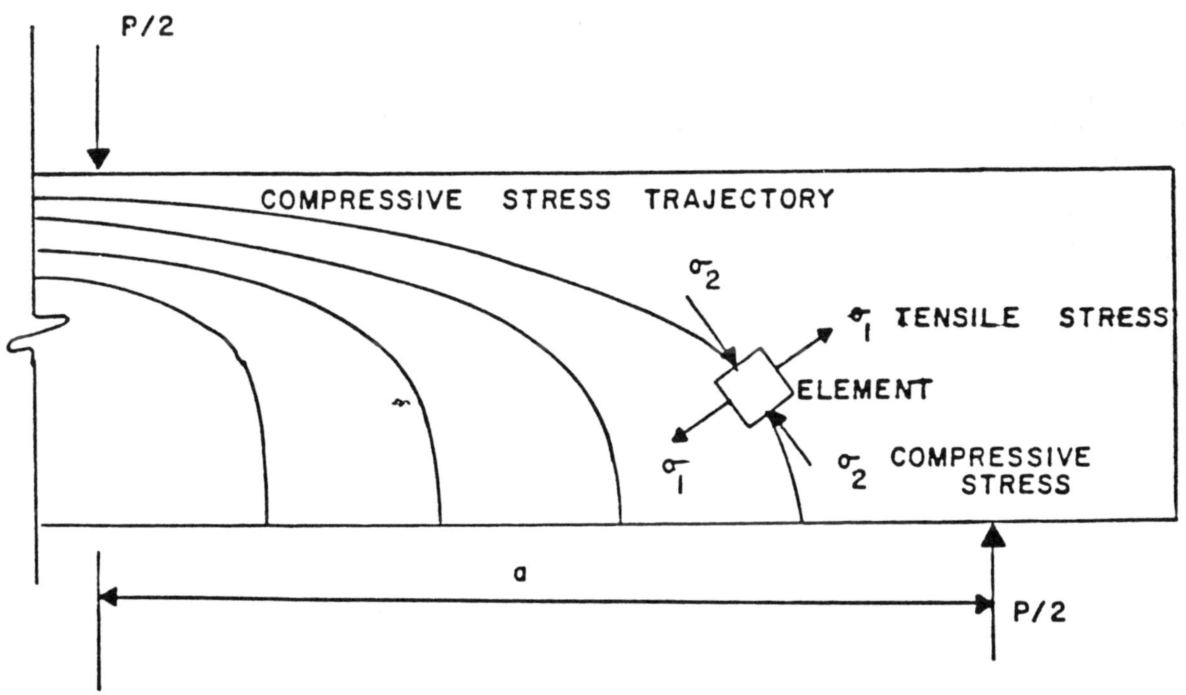

FIGURE 21b. Stress trajectories in homogeneous rectangular beam.

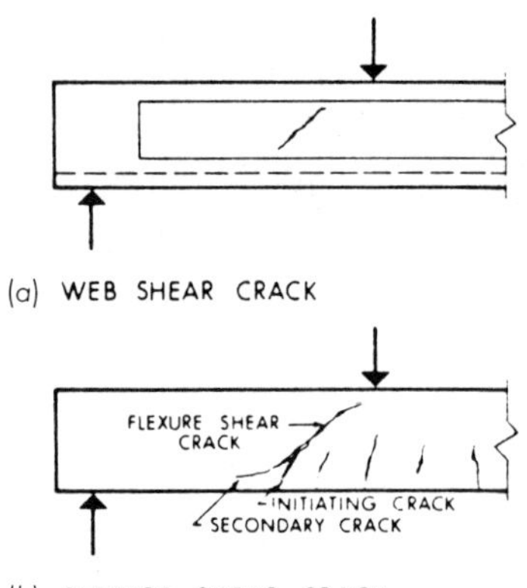

FIGURE 22a. Types of inclined cracks [24]: (a) web shear crack; (b) flexure shear crack.

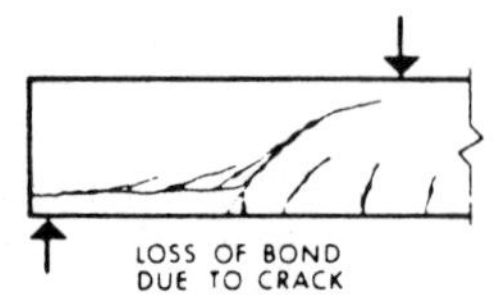

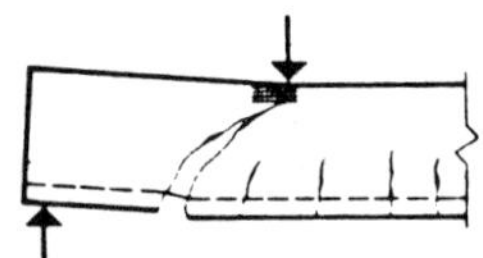

FIGURE 22c. Typical shear failures in short beams [24]: (a) shear-tension failure; (b) shear-compression failure.

with stirrups are the following:

1. shear stress in the uncracked concrete – V_c;
2. interface shear transfer – V_a;
3. dowel action – V_d; and
4. shear reinforcement – V_s.

When designing for shear in a beam, the nominal shear – V_n across a section is:

$$V_n = V_c + V_s \tag{53}$$

$$V_n = V_u/\phi = (1.4V_{DL} + 1.7V_{LL})/\phi \tag{54}$$

The code has two formulas for the shear capacity of the concrete V_c. The simplified method value is:

$$V_c = 2.0\sqrt{f'_c}b_w d \tag{55}$$

The more detailed method value is:

$$V_c = \left(1.9\sqrt{f'_c} + 2500\varrho_w \frac{V_u d}{M_u}\right) b_w d \le 3.5\sqrt{f'_c}b_w d \tag{56}$$

However, stirrups must be provided wherever

$$V_n > V_c/2 \tag{57}$$

except for slabs, footings, joists, and small beams shallower than 10 in., 2½times flange thickness, or $b_w/2$. Allowances must be made for joists and lightweight concrete. The capacity of the stirrups is given as

$$V_s = \frac{A_v f_y d}{s}(\sin\alpha + \cos\alpha) \tag{58}$$

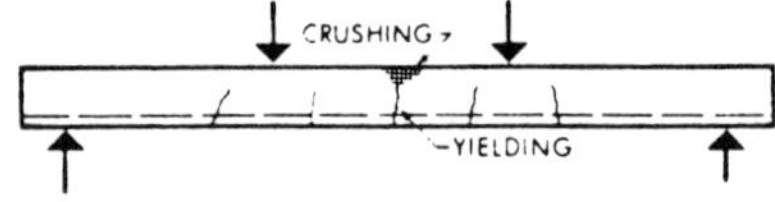

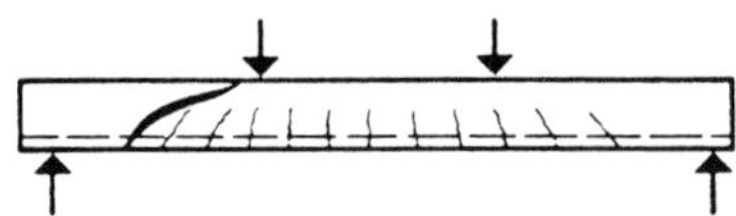

FIGURE 22b. Failures of slender beams [24]: (a) flexural failure; (b) diagonal tension failure.

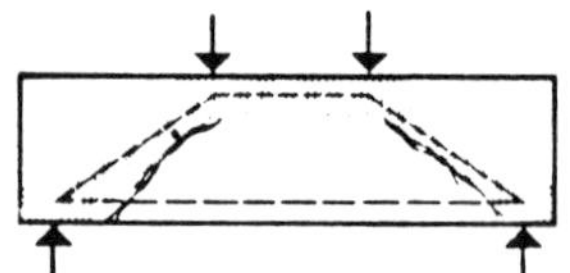

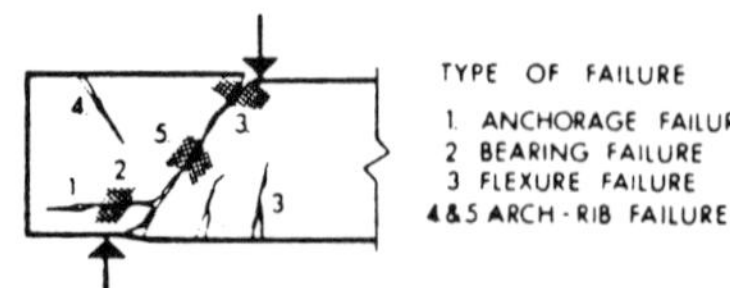

FIGURE 22d. Modes of failure of deep beams [24].

where α is the angle of the stirrup with the longitudinal steel. The stirrups or web reinforcement must not exceed 8 $\sqrt{f'_c}\, b_w d$. The minimum web reinforcing area is min $A_v = 50 b_w s/f_y$. Stirrup spacing must not exceed $d/2 \leq 24$ in. if $V_s \leq 4\sqrt{f'_c}\, b_w d$ and $d/4 \leq 12$ in. if $V_s > 4\sqrt{f'_c}\, b_w d$.

When considering the axial loads, the V_c varies. When axial compression also exists, V_c becomes

$$V_c = 2\left(1 + \frac{N_u}{2000A_g}\right)\sqrt{f'_c}b_w d \tag{59}$$

When significant axial tension exists,

$$V_c = 2\left(1 + \frac{N_u}{500A_g}\right)\sqrt{f'_c}b_w d \tag{60}$$

N_u/A_g is expressed in psi and N_u is negative in tension.

Shear strength of a section is increased if a reaction produces compression in the end region of a member. For this situation, the designer may come out a distance "d" from the face of the support to design for the stirrups. The spacing s must at this point exist back to the support. There are some support conditions where this provision cannot be applied where tension exists. The designer must be aware of these conditions.

Example 8

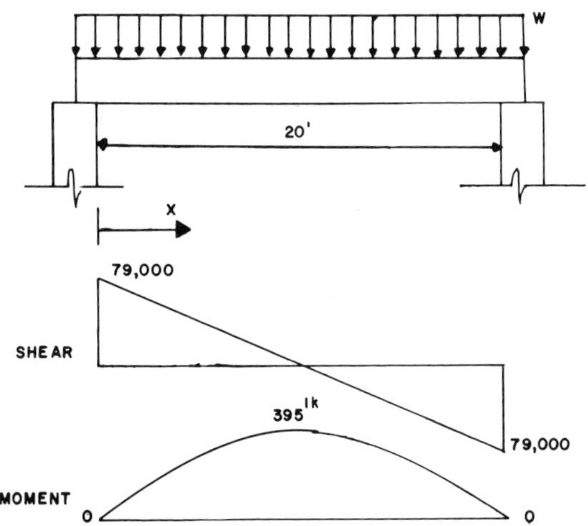

E8.1 Loading, shear, and moment diagrams for beam.

Design stirrups for shear for the beam shown in Figure E8.1. This example is a problem solved by Winter and Nilson [1].

$$f'_c = 3000 \text{ psi}$$

$$f_y = 40{,}000 \text{ psi}$$

$$A_s = 9.86 \text{ in}^2$$

$$d = 22 \text{ in}$$

$$b = 16 \text{ in}$$

$$w_{DL} = 1645 \text{ lb/ft}$$

$$w_{LL} = 3290 \text{ lb/ft}$$

$$w_u = 1.4w_{DL} + 1.7w_{LL}$$

$$w_u = 1.4(1645) + 1.7(3290) = 7900 \text{ lb/ft}$$

$$w_u = 7900 \text{ lb/ft}$$

$$V_u = 79{,}000 - 7{,}900x$$

$$M_u = 79{,}000x - \frac{7900x^2}{2}$$

2 AREA OF BAR

E8.2 Stirrup.

Come out a distance d from face of support.

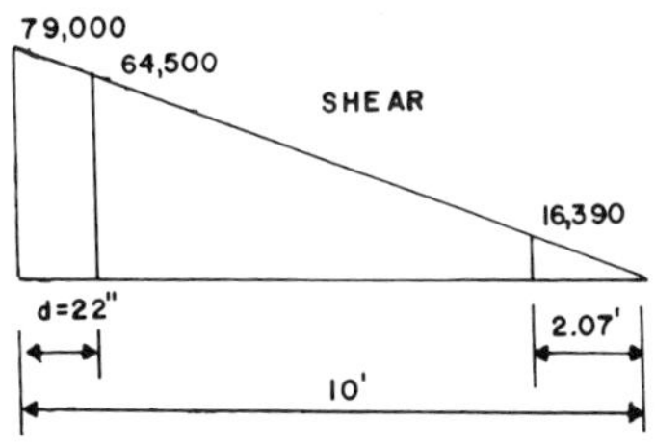

E8.3 Shear diagram.

$$V_c = 2\sqrt{f'_c}\, b_w d = 2\sqrt{3000}\,(22)(16) = 38{,}559 \text{ lb}$$

$$\frac{\phi V_c}{2} = \frac{(.85)(38{,}559)}{2} = 16{,}390 \text{ lb}$$

$$\text{Stirrups stop @ } \frac{16{,}390}{7900} = 2.07 \text{ ft}$$

$$s = \frac{\phi f_y d A_v}{V_u - \phi V_c} = \frac{.85(40{,}000)(22)(.22)}{V_u - \phi V_c} = \frac{164{,}560}{V_u - \phi V_c}$$

Location	V_u	V_c	ϕV_c	$V_u - \phi V_c$	s	s_{ACT}
8	15,800	38,500	32,700	−16,900	—	11
7	23,700	38,500	32,700	−9,000	—	11
6	31,600	38,500	32,700	−1,100	—	11
5	39,500	38,500	32,700	6,800	24	11
4	47,400	38,500	32,700	14,700	11	11
3	55,300	38,500	32,700	22,600	7	7
2	63,200	38,500	32,700	30,500	5.4	5
1.83	64,500	38,500	32,700	31,800	5.2	5

$$V_s = \frac{V_u - \phi V_c}{\phi} = \frac{31{,}800}{.85} = 37{,}400 \text{ lb}$$

$$4\sqrt{f'_c}\,b_w d = 4\sqrt{3000}\,(16)(22) = 77{,}000 \text{ lb} > 37{,}400 \text{ lb}$$

$$\therefore\ s_{MAX} = d/2 = \frac{22}{2} = 11.0$$

$$s_{MAX} = \frac{A_v f_y}{50 b_w} = \frac{(.22)(40{,}000)}{50(16)}$$

$$V_{s_{MAX}} = 8\sqrt{f'_c}\,b_w d = 154{,}000 \text{ lb} \quad \text{O.K.}$$

$$s_{MAX} = 11.0 \text{ in}$$

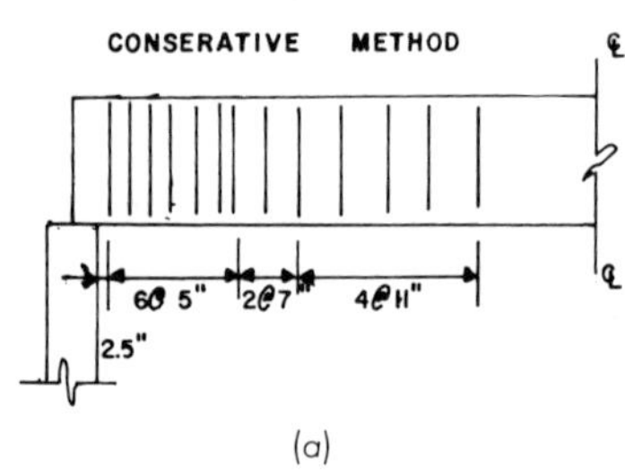

(a)

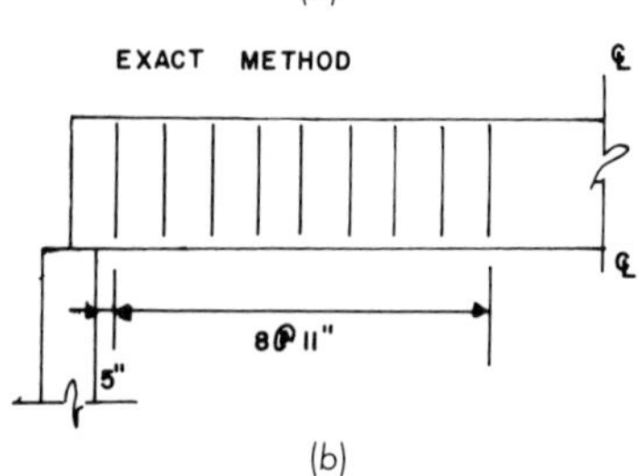

(b)

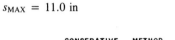

E8.4 Stirrup spacing: (a) conservative method; (b) exact method.

EXACT METHOD

$$V_c = \left(1.9\sqrt{f'_c} + 2500\,\varrho_w \frac{V_u d}{M_u}\right) b_w d$$

$$V_{cMAX} = 3.5\sqrt{f'_c}\,b_w d = 67.5^K$$

$$\varrho_w = \frac{A_s}{b_w d} = .028$$

$$V_u = 79{,}000 - 7900x$$

$$M_u = 79{,}000x - \frac{7900x^2}{2}$$

Location	M_u^{1K}	V_u^{1K}	$\frac{V_u d}{M_u}$	V_c^K	V_u^K	$V_u - \phi V_c$	s	s_{ACT}
8	379	16	.08	37.6	15.8	−16.2	—	11
7	360	23	.12	38.4	23.7	−8.9	—	11
6	332	31	.17	40.8	31.6	−3.2	—	11
5	296	39	.24	42.6	39.5	3.2	51	11
4	252	47	.34	45.0	47.4	9.2	18	11
3	201	55	.50	48.9	55.3	13.7	12	11
2	142	63	.81	56.7	63.2	15.0	11	11
1.83	125	65	.95	60.0	64.5	13.5	12	11

$$\frac{V_c}{2}\ V_u < \frac{V_c}{2}\ @\ 8 \text{ ft}$$

$$s = \frac{\phi f_y d A_v}{V_u - \phi V_c} = \frac{.85(40{,}000)(22)(.22)}{V_u - \phi V_c} = \frac{165^K}{V_u - \phi V_c}$$

$$4\sqrt{f'_c}\,b_w d = 77^K$$

$$V_s = \frac{V_u - \phi V_c}{\phi} = \frac{15}{.85} = 17^K < 77^K \text{ O.K.}$$

$$s_{MAX} = \frac{d}{2} = \frac{22}{2} = 11 \text{ in}$$

$$s_{MAX} = \frac{A_v f_y}{50 b_w} = \frac{.22(40{,}000)}{50(16)} = 11.0 \text{ in}$$

$$V_{sMAX} = 8\sqrt{f'_c}\,b_w d = 154^K > 17^K \text{ O.K.}$$

BOND AND DEVELOPMENT LENGTH

Design of tensile and compressive reinforcement and stirrups is based on assumption that no slip occurs between the steel and the concrete. There are two areas of bond and development length which are bond stress and anchorage. Anchorage bond which is developed at points of maximum moment where the longitudinal bars have to develop the required tensile force of the moment. This can also be applied at points where bars are cut off other than points of inflection within continuous spans. The flexural bond where there are large changes in the tensile force of the longitudinal reinforcing. This can occur in regions of high shear. Examples of this are points of inflection within continuous spans and at simply supported ends of beams. The case of a uniformly loaded simply supported beam illustrates these two cases (see Figure 23).

There are four different points which must be checked for development length for tensile reinforcements:

1. Maximum moment locations.
2. Bar cutoffs other than zero moment regions.
3. Points of inflection.
4. Exterior simple supports.

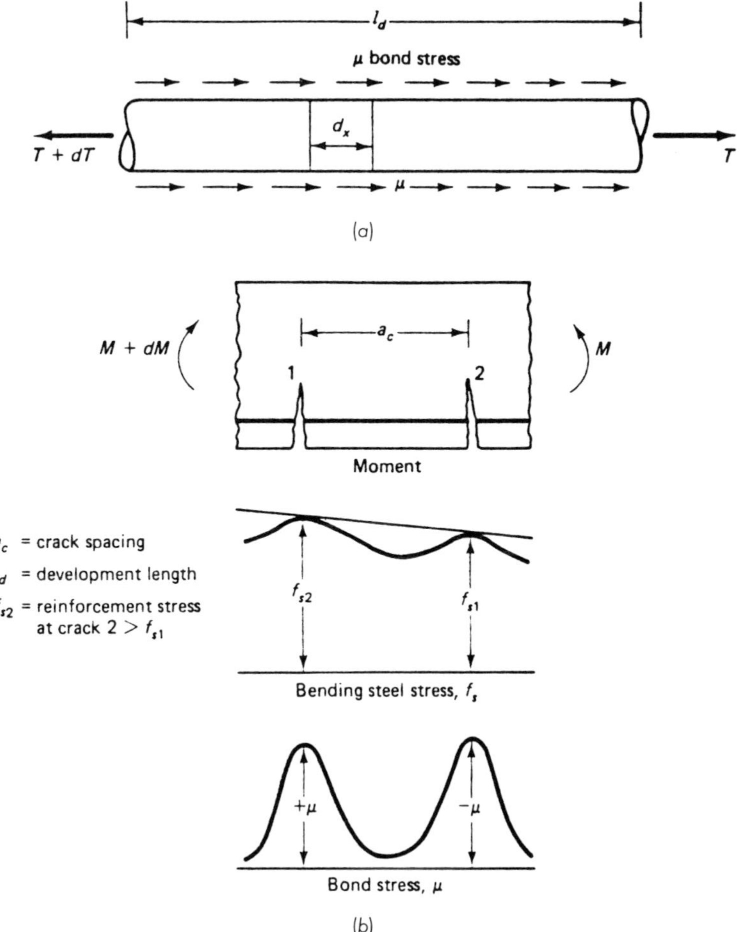

FIGURE 23. Bond stress across a reinforcing bar: (a) pull-out anchorage bond in a bar; (b) flexural bond [3].

The basic mechanics of bond will be looked at briefly. The bond between reinforcing steel and concrete is developed by: (1) bearing of lugs or wedge action; (2) adhesion or shear between steel and concrete at contact; and (3) shear between concrete and concrete at top edge of the lugs. Bond failures result in longitudinal cracks in either horizontal or vertical plane. Research has shown: (1) bond is highly variable due to tensile cracks, cover, spacing between bars, bar diameter, etc.; (2) closely spaced stirrups increase bond strength by preventing spalling due to wedge action; and (3) ultimate bond force is about 30 $\sqrt{f'_c}$.

Essentially, the ACI Code 83 says that bars must have sufficient embedment (ℓ) so that full strength may be developed for each bar. Thus, there is a minimum (ℓ_d) based on bar size. If the geometry is such that $\ell < \ell_d$ then hooks must be used to provide the necessary anchorage.

The requirements for basic tension development length ℓ_{db} of deformed bars and deformed wire which are given in Section 12.2.2:

(A) For bar sizes #3 – #11

$$\ell_{db} = .04 A_b f_y / \sqrt{f'_c} \tag{61a}$$

$$\ell_{db} = .0004 d_b f_y \tag{61b}$$

$$\ell_{db} = 12 \text{ in.} \tag{61c}$$

(B) For #14 bars

$$\ell_{db} = .085\, f_y / \sqrt{f'_c} \tag{61d}$$

(C) For #18 bars

$$\ell_{db} = .11\, f_y / \sqrt{f'_c} \tag{61e}$$

(D) For Deformed Wire

$$\ell_{db} = .03\ d_b f_y / \sqrt{f'_c} \qquad (61f)$$

$$\ell_{db} = 12 \text{ inches}$$

The required development length $\ell = \ell_{db} \times$ applicable modification factors. These factors are:

1. factor for top bars = 1.4 (this is when more than 12 inches of fresh concrete is cast below the bars);
2. factor for lateral spacing = .8 (this is where the spacing of the bars is greater than 6");
3. factor for excess bar area (A_s required/A_s provided);
4. factor when using $f_y > 60{,}000-(2 - 60{,}000/f_y)$;
5. factor for bars enclosed within a spiral not less than ¼ in. in diameter and not more than a 4 in. pitch. Lightweight concrete use must also be considered.

The development length in compression for all bars is:

$$\ell_{db} = .02 d_b f_y / \sqrt{f'_c} \qquad (62a)$$

$$\ell_{db} = .0003 d_b f_y \qquad (62b)$$

$$\ell_{db} = 8 \text{ inches} \qquad (62c)$$

Use the largest of the three values.

The ℓ_{db} may be reduced by a factor of the ratio of the required area to the area provided of steel reinforcing.

If beam dimensions are limited and a bar cannot be extended in a straight line a distance equal to ℓ_d. A hook or bend may be added to reduce ℓ_d straight such that ℓ_h is required.

The ACI Code 83, Section 12.5.2 specifies that the basic developmental length ℓ_{hd} of a Grade 60 hooked bar is:

$$f'_c \ell_{hb} = 1200 d_b / \sqrt{f'_c} \qquad (63)$$

There are factors which must be considered just as for ℓ_d. The ℓ_{hb} must be multiplied by applicable modification factors. These factors are:

1. factor for using reinforcing steel other than Grade 60 or f_y = 60,000 psi-(f_y/60) where f_y is in ksi;
2. factor for concrete cover—For #11 bar and smaller, side cover (normal to plane of hooks) not less than 2½ inches, and for 90 degree hook, cover on bar extension beyond hook not less than 2 inches (.7);
3. factor for ties or stirrups—For #11 bar and smaller, hook enclosed vertically or horizontally within ties or stirrup—ties spaced along the full development length ℓ_{dh} not greater than $3d_b$, where d_b is the diameter of the hooked bar;
4. factor for excess reinforcement—where the factor is A_s (required)/A_s (provided); and
5. factor for lightweight concrete must also be considered (1.3).

Development of reinforcing at exterior simple supports and points of inflection have to also be considered. At the exterior simple support the available embediment length is equal to

$$\frac{1.3M_n}{V_u} + \ell_a \geq \ell_d \qquad (64)$$

where M_n is the nominal flexural strength of bars coming into the support; V_u is the factored load shear at the support, and ℓ_a is the embediment length beyond the centerline of the support (could also be effective length of hook or mechanical anchorage). The 1.3 factor is used when the area is under a compressive stress field caused by the reaction. The points of inflection must also be looked at for the available embediment length which is equal to

$$\frac{M_n}{V_u} + \ell_a \geq \ell_d \qquad (65)$$

The ℓ_a is the actual embediment length which must not exceed the larger of $12d_b$ or d.

Example 9

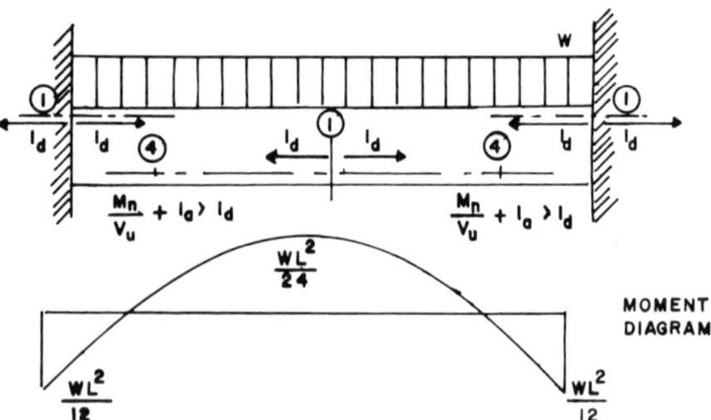

E9.1 Development length check.

Show where you would check development length. Check development length @

1. Maximum moments
2. Bar cut offs other than P.I.
3. Ext. S.S.
4. P.I. ($+A_s$)

Example 10

How much length is required into the wall for the following beam?

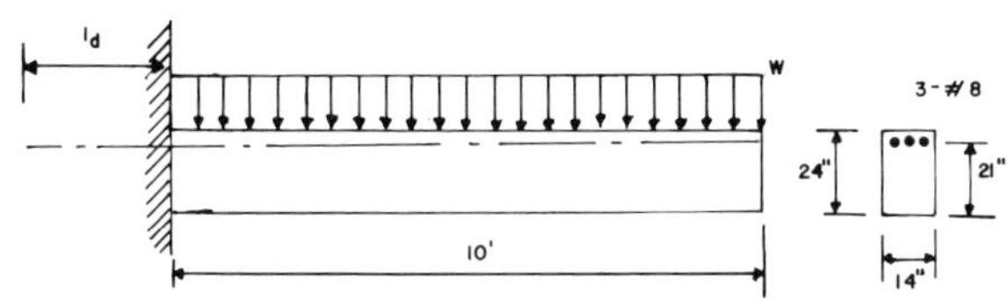

E10.1. Anchorage check.

Given

$$w_{DL} = 500 \text{ lb/ft}$$

$$w_{LL} = 1000 \text{ lb/ft}$$

$$f'_c = 5000 \text{ psi}$$

$$f_y = 40{,}000 \text{ psi}$$

$$\ell_{db} = .04A_b f_y/\sqrt{f'_c} =$$

$$= .04(.79)(40{,}000)/\sqrt{5000} = 17.87''$$

$$\ell_{db} = .0004d_b f_y = .0004(1.0)(40{,}000) = 16''$$

$$\ell_{db} = 12.0 \text{ in} = 12 \text{ in}$$

$$\ell_{db} = 17.87''$$

$$\ell_d = \ell_{db}(1.4)\left(2 - \cancelto{1.0}{\frac{60{,}000}{f_y}}\right)\left(\frac{A_{s\text{REQ.}}}{A_{s\text{ACT.}}}\right)\cancelto{1.0}{(.8)}\ \cancelto{1.0}{(.75)}$$

$$\frac{A_{s\text{REQ.}}}{A_{s\text{ACT.}}} = \frac{A_{s\text{REQ.}}}{2.37 \text{ in}^2}$$

$$M_u = \frac{w_u L^2}{2} = \frac{[(1.4)(500) + 1.7(1000)]10^2(12)}{2}$$

$$= 1{,}440{,}000 \text{ in lb}$$

Finding $A_{s\text{REQ.}}$

Assume $a = 3$ in

$$A_s = \frac{M_u}{\phi f_y\left(d - \frac{a}{2}\right)} = \frac{1{,}440{,}000}{(.9)(40{,}000)\left(21 - \frac{3}{2}\right)}$$

$$= 2.05 \text{ in}^2$$

$$a = \frac{A_s f_y}{.85f'_c b} = \frac{2.05(40{,}000)}{.85(5{,}000)(14)} = 1.37 \text{ in}$$

Assume $a = 1.37$ in

$$A_s = \frac{M_u}{\phi f_y\left(d - \frac{a}{2}\right)} = \frac{1{,}440{,}000}{(.9)(40{,}000)\left(21 - \frac{1.37}{2}\right)}$$

$$= 1.969 \text{ in}^2$$

$$a = \frac{A_s f_y}{.85f'_c b} = \frac{1.969(40{,}000)}{.85(5{,}000)(14)}$$

$$= 1.32 \text{ in (converged close enough)}$$

$$\ell_d = 17.87(1.4)\left(\frac{1.969}{2.37}\right) = 20.76'' \text{ Required}$$

Example 11

Check development length of reinforcing steel @ all points necessary for problem (neglect effect of beam weight).

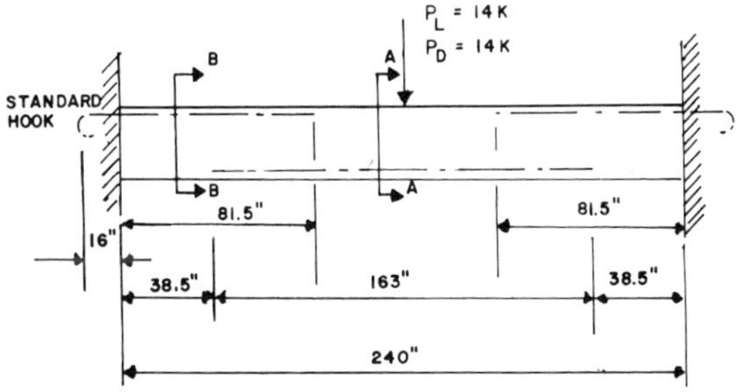

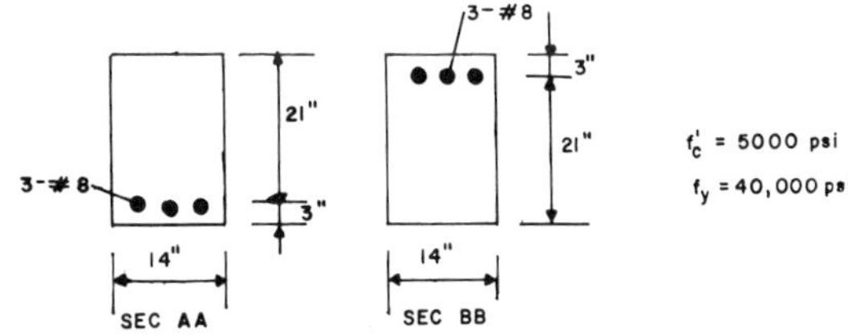

E11.1 Beam and sections.

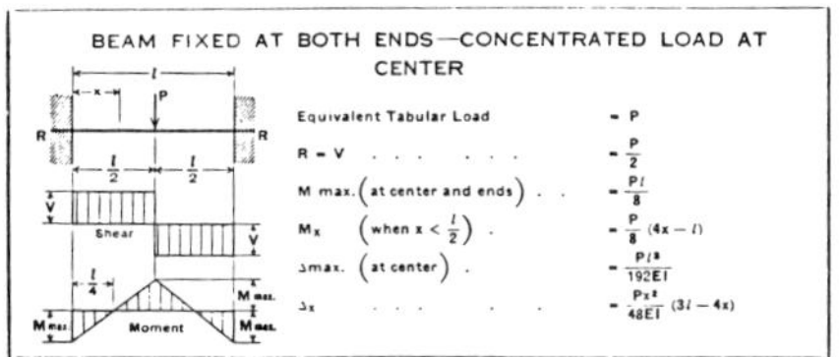

E11.2 Beam loading, shear, and moment diagrams [28].

Checking development length at critical sections:

1. @ Maximum moments—[Points (1)]
2. Bar cutoffs other than P.I.—None
3. @ s.s. Ext. $- \ell_a + \dfrac{M_n}{V_u} \geq \ell_d$—None
4. @ P.I. $(+ A_s) - \ell_a + \dfrac{M_n}{V_u} \geq \ell_d$—[Points (2)]

$$\ell_{db} = .04A_b f_y / f'_c = .04(.79)(40{,}000)\ 5000 = 17.8''$$

$$\ell_{db} = .0004 d_b f_y = .004(1.0)(40{,}000) = 16''$$

$$\ell_{db} = 12''$$

Considering points (1)

$$\ell_d \ @\ \text{℄} = \ell_{db}(\overset{1.0}{\cancel{1.4}})\left(2 - \overset{1.0}{\cancel{\frac{60{,}000}{f_y}}}\right)\left(\frac{A_{s\text{REQ.}}}{A_{s\text{ACT.}}}\right)(\overset{1.0}{\cancel{.8}})(\overset{1.0}{\cancel{.75}})$$

Now

$$\frac{A_{s\text{REQ.}}}{A_{s\text{ACT.}}} = \frac{M_{u\text{REQ.}}}{M_{u\text{ACT.}}}$$

$$M_{u\text{REQ.}} = \frac{[14(1.4) + 14(1.7)](20)(12)}{8} = 1{,}302{,}000 \text{ in lb}$$

$$M_{u\text{ACT.}} = \phi A_s f_y \left(d - \frac{a}{2}\right)$$

$$= (.9)(3)(.79)(40{,}000)\left(21 - \frac{1.59}{2}\right)$$

$$\text{since } a = \frac{A_s f_y}{.85 f'_c b} = \frac{3(.79)(40{,}000)}{(.85)(5000)(14)} = 1.59 \text{ in}$$

$$M_{u\text{ACT.}} = 1{,}724{,}000 \text{ in lb}$$

$$\ell_d = 17.8''\left(\frac{1{,}302{,}000}{1{,}724{,}000}\right) = (17.8'')(.75) = 13.5''$$

$$\ell_d = 13.5 \text{ in}$$

82″ actual O.K.

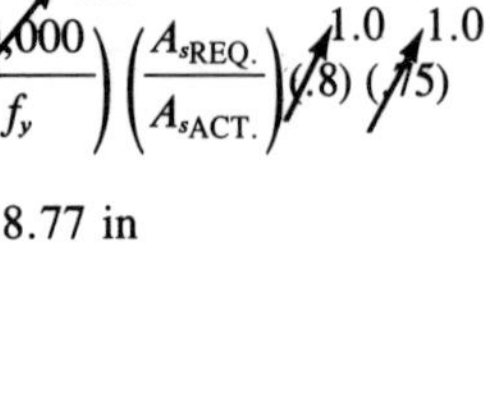

ℓd = 13.5 in.
ℓd ℓd
OK
82" actual 82" actual

E11.3 Development length.

2. ℓ_d @ wall $= \ell_{db}(1.4)\left(2 - \overset{1.0}{\cancel{\dfrac{60{,}000}{f_y}}}\right)\left(\dfrac{A_{s\text{REQ.}}}{A_{s\text{ACT.}}}\right)(\overset{1.0}{\cancel{.8}})(\overset{1.0}{\cancel{.75}})$

$$\ell_d \ @ \text{ wall} = 17.8(1.4)(.75) = 18.77 \text{ in}$$

16″ actual

$$\ell_h = 1200\ d_b / \sqrt{f'_c}\ \text{(Modification factors)}$$

Modifying multipliers $- f_y/60{,}000 = \dfrac{40{,}000}{60{,}000} = .67$

$$\ell_h = \frac{1200(1.0)}{\sqrt{5000}}(.67) = 11.4''$$

$$\ell_h = 11.4'' < 16'' \quad \text{O.K.}$$

Considering points (2)

The following must be true $\ell_a + \dfrac{M_n}{V_u} > \ell_d$

$$\ell_a = 82'' - 60'' = 22''$$

ℓ_a must not exceed $d = 21''$ or $12d_b = 12''$ (greater)

Thus, $\ell_a = 21''$

$$\ell_a + \frac{M_n}{V_u} = 21'' + \frac{1{,}724{,}000}{(.85)(21{,}700)(.9)}$$

$$= 125'' > 17.88'' \quad \text{O.K.}$$

SPLICES

Steel bars are fabricated in lengths of 20, 40, and 60 feet, depending on the bar diameter, transportation facilities, etc. Bars are usually tailored according to the details of the structural members. When bars are short, it is necessary to splice them in the field. Splices may be made by lapping or welding or with mechanical devices that provide positive connection between bars. For bars larger than #11, lap splices should not be used. For noncontact lap splices in flexural members, bars should not be spaced transversely farther apart than one-fifth the required length or 6 inches. A welded splice where the bars are butted and welded is required to carry 125 percent of the specified yield strength. The ACI Code 83 section 12.14, specifies that full positive connections must develop in tension or compression at least 125 percent of the specified yield strength of the bar [16].

Splices of deformed bars and deformed wire in tension shall be explained according to the ACI Code 83. Minimum length of lap slices shall be as required for Class A, B, or C splice, but not less than 12 inches, where: Class A splice is 1.0 ℓ_d; Class B splice is 1.3 ℓ_d; and Class C splice is 1.7 ℓ_d, where ℓ_d is the developmental length specified in an earlier section. Lap splices of deformed bars and deformed wire in tension shall conform to Table 6. In tension tie members, the splices must be staggered and have a lap slice of twice the

TABLE 6. Tension Lap Splices [9].

A_s provided* / A_s required	Maximum percent of A_x spliced within required lap length		
	50	75	100
Equal to or greater than 2	Class A	Class A	Class B
Less than 2	Class B	Class C	Class C

*Ratio of area of reinforcement provided to area of renforcement required by analysis at splice location.

development length ($2\ \ell_d$). The values just stated must be increased by 20 percent for a 3-bar bundle and 33 percent for a 4-bar bundle [16,17].

The splices for deformed bars in compression shall have a minimum lap length of $.02 d_b f_y / \sqrt{f'_c}$, but not less than $.0005 f_y d_b$ when $f_y < 60$ ksi, nor 12 inches. For f'_c less than 3000 psi, length of lap shall be increased by one-third. When bars of different size are lap spliced in compression, splice length shall be the larger of: development length of larger bar, or splice length of smaller bar. Bar sizes #14 and #18 may be lap spliced to #11 and smaller bars. In tied reinforced compression members, where ties throughout the lap splice length have an effective area not less than 0.0015 hs, where h is the overall thickness of the member and s is the tie spacing, lap splice length may be multiplied by .83 but lap length shall not be less than 12 inches. In spirally reinforced compression members, lap splice length of bars within a spiral may be multiplied by 0.75, but lap splice length shall not be less than 12 inches. When longitudinal bars in columns are offset, the slope of the inclined portion of the bar with the axis of the column should not exceed 1 in 6. When column faces are offset 3 inches or more, splices of vertical bars should be made by separate dowels adjacent to the offset face [16]. This section was taken from [16].

DEFLECTIONS

The ACI Code 83 provisions look at the deflections at service loads and not at the nominal or ultimate load of the structural elements. From the ACI Code 83 the Tables 9.5(a) and 9.5(b) show the minimum thickness of nonprestressed beams or one-way slabs unless deflections are computed and maximum permissible computed deflections respectively. Two-way slabs may be considered by designing for a minimum thickness given by Equations (9−11), 9−12), and (9−13) in the ACI Code 83. The designer should be aware of the variation of reinforced concrete structural members. Because of this reason, simple procedures should be used in checking and determining deflections of structural members. An in-depth treatise on the subject of deflection control may be found in [11,17,18].

Two methods are set up by the code for controlling deflections of one-way and two-way flexural systems. Deflections can be controlled by means of minimum thickness which is shown in Table 9.5a, or Equations (9−11), (9−12), and (9−13) or directly by limiting computed deflections, Table 9.5b. In this section the deflection calculations will be limited to one-way flexural systems.

When calculating the deflection of a one-way slab or beam at service loads, the deflection can be considered as given in structural analysis. Considering a simply supported beam which is uniformly loaded the deflection would be

$$y = \frac{5\ w\ \ell^4}{384EI} \tag{66}$$

Now looking at each term: w will be the uniform service loads (do not consider load factors), thus $w = w_{DL} + w_{LL}$; ℓ is the length of the member between supports; E is the modulus of elasticity of concrete (57,000 $\sqrt{f'_c}$ for normal concrete); and I is the moment of inertia of the section. Because of the variation of cracking with the variation of load and the reinforced concrete section, the I can vary between I_g (gross moment of inertia) and I_{cr} (cracked moment of inertia). From research, the moment of inertia will be equal to an equivalent moment of inertia which is I_e.

$$I_e = \left(\frac{M_{cr}}{M_a}\right)^3 I_g + \left(1 - \left(\frac{M_{cr}}{M_a}\right)^3\right) I_{cr} \tag{67}$$

which lies between I_g and I_{cr} for the section under investigation. If $M_{cr}/M_a > 1.0$, then use I_g. The moment cracking $M_{cr} = f_r I_g / y_t$ where I_g is the gross moment inertia of the section; f_r is the modulus of rupture 7.5 $\sqrt{f'_c}$ for normal concrete; y_t is the distance from the neutral axis to the extreme fiber of concrete in tension. M_a is the maximum service load moment acting at the condition under which the deflection is being considered.

The code section 9.5.2.5 considers the long-time deflections due to the combined effects of creep and shrinkage. The equation is

$$y \text{ (creep and shrinkage)} = \lambda \cdot y \text{ (sustained)} \tag{68}$$

where

$$\lambda = \frac{\xi}{1 + 50\varrho'} \tag{69}$$

where ϱ' is the amount of compression steel at midspan for simple and continuous spans, and at support for cantilevers. Time dependent factor ξ for sustained loads may be taken

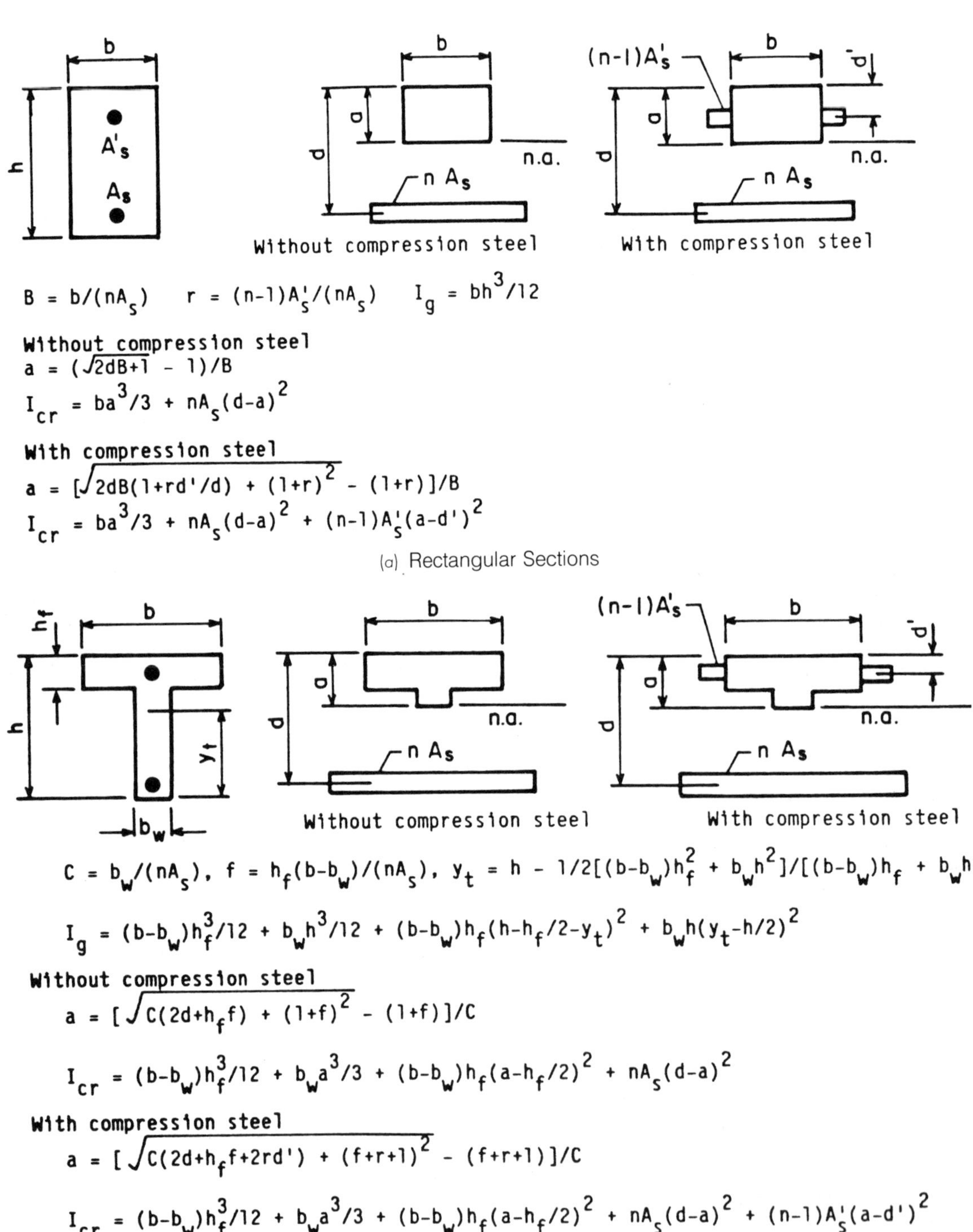

FIGURE 24. Moment of inertia of gross section, I_g, and cracked transformed section, I_{cr} [11].

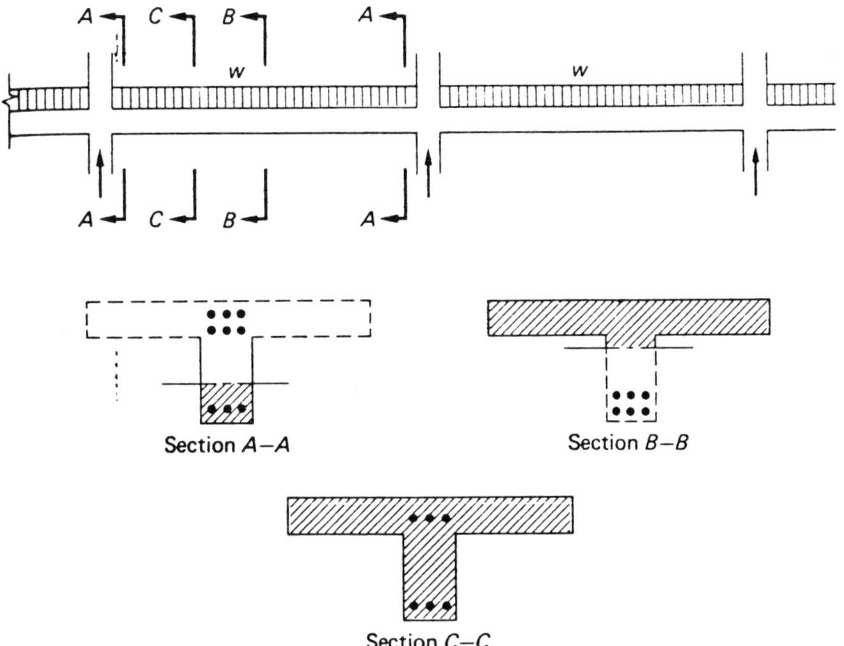

FIGURE 25. Effective moment of inertia for continuous T-shaped sections [5].

equal to

5 years or more	2.0
12 months	1.4
6 months	1.2
3 months	1.0

The values of I_{cr}, the cracked moment of inertia, for the section has to be determined depending on whether the section is a rectangular or "T" beam with or without compression steel. Figure 24 shows the I_g, gross moment of inertia; a, location of neurtral axis from the extreme fiber in compression; and I_{cr}, cracked moment of inertia, for the different sections which can be considered.

In a reinforced concrete structure, the members are continuous and the calculations must reflect this situation. For example, consider a continuous beam or girder in a structural frame. Figure 25 shows the beam or girder with the appropriate sections shown at the ends and center of the member. Notice sections at the end are represented by rectangular members, and at the center the section is a T-beam. Thus, the I_e of each section is different. The I_e becomes

$$\text{Avg. } I_e = 0.70\, I_m + 0.15(I_{e1} + I_{e3}) \tag{70}$$

where I_m refers to the midspan section I_{e2}, and I_{e1} and I_{e3} refers to I_e at the respective ends. Thus, the equivalent moment of inertia at each section are:

$$I_{e1} = \left(\frac{M_{cr1}}{M_{a1}}\right)^3 I_{g1} + \left[1 - \left(\frac{M_{cr1}}{M_{a1}}\right)^3\right] I_{cr1} \tag{71a}$$

$$I_{e2} = \left(\frac{M_{cr2}}{M_{a2}}\right)^3 I_{g2} + \left[1 - \left(\frac{M_{cr2}}{M_{a2}}\right)^3\right] I_{cr2} \tag{71b}$$

$$I_{e3} = \left(\frac{M_{cr3}}{M_{a3}}\right)^3 I_{g3} + \left[1 - \left(\frac{M_{cr3}}{M_{a3}}\right)^3\right] I_{cr3} \tag{71c}$$

The moment diagram for a span of a continuous beam or girder is shown in Figure 26. The structural member is usually subjected to positive moment in the center and negative moments at the ends. From conjugate beam analysis it can be shown that the deflection at center line is

$$y_m = \frac{5\,\ell^2}{48EI}\left[M_s + \frac{1}{10}(M_A + M_B)\right] \tag{72}$$

where

$$M_s = M_o + \frac{1}{2}(M_A + M_B) \tag{73}$$

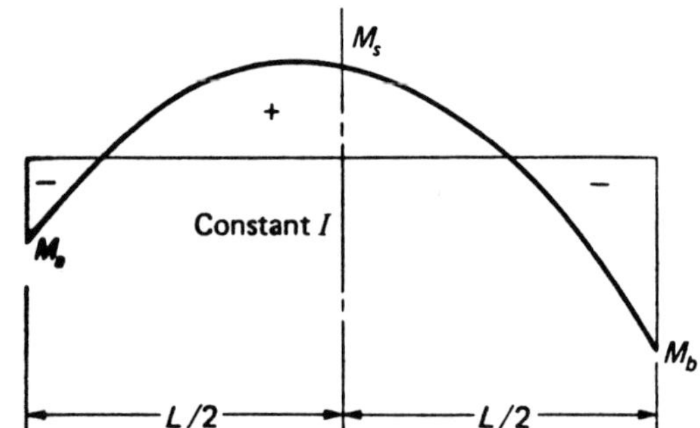

FIGURE 26. Typical bending moment diagram for a uniformly loaded span [5].

and M_o is the moment at center considering the beam to be simply supported.

Example 12

Find Deflection for beam at the center. Problem similar to one presented in Wang and Salmon [5].

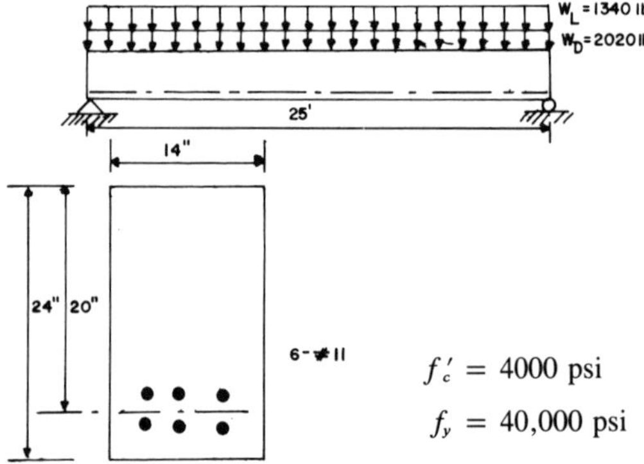

$$f'_c = 4000 \text{ psi}$$

$$f_y = 40{,}000 \text{ psi}$$

E12.1 Beam and section.

$$M_{MAX_{DL}} = \frac{wL^2}{8} = \frac{2020(25)^2}{8} = 158{,}000 \text{ ft lb}$$

$$M_{MAX_{LL}} = \frac{wL^2}{8} = \frac{1340(25)^2}{8} = 105{,}000 \text{ ft lb}$$

$$\text{Now } M_{cr} = \frac{f_r I_g}{y_t}$$

$$f_r = 7.5\sqrt{f'_c} = 7.5\sqrt{4000} = 474 \text{ psi}$$

$$I_g = \frac{bh^3}{12} = \frac{(14)(24)^3}{12} = 16{,}128 \text{ in}^4$$

$$y_t = \frac{h}{2} = \frac{24}{2} = 12 \text{ in}$$

$$M_{cr} = \frac{f_r I_g}{y_t} = \frac{474(16{,}128)}{12}$$

$$M_{cr} = 637{,}000 \text{ in lb} = 53{,}000 \text{ ft lb}$$

$$E_c = 57{,}000\sqrt{f'_c} = 57{,}000\sqrt{4000} = 3.6 \times 10^6$$

$$n = \frac{E_{ST}}{E_c} = \frac{29 \times 10^6}{3.6 \times 10^6} = 8.0$$

$$b\chi\left(\frac{\chi}{2}\right) = n\,A_s(d - \chi)$$

$$14\chi\left(\frac{\chi}{2}\right) = 8(6)(1.56)(20 - \chi)$$

$$\chi = 10.22 \text{ in}$$

E12.2 Section.

Find I_{cr}

$$I_{cr} = \frac{b\chi^3}{3} + n\,A_s(d - \chi)^2$$

$$I_{cr} = \frac{14(10.22)^3}{3} + 8(6)(1.56)(20 - 10.22)^2$$

$$I_{cr} = 12{,}143 \text{ in}^4$$

$$I_e = \left(\frac{M_{cr}}{M_{MAX}}\right)^3 I_g + \left[1 - \left(\frac{M_{cr}}{M_{MAX}}\right)^3\right] I_{cr}$$

$$I_{eDL} = \left(\frac{53}{158}\right)^3 16{,}128 + \left[1 - \left(\frac{53}{158}\right)^3\right] 12{,}143$$

$$I_{eDL} = 12{,}293 \text{ in}^4$$

I_{eLL+DL} $\quad M_{MAX} = 158 + 105 = 263$ ftK

$$I_{eLL+DL} = \left(\frac{53}{263}\right)^3 16{,}128 + \left[1 - \left(\frac{53}{263}\right)^3\right] 12{,}143$$

$$I_{eLL+DL} = 12{,}176 \text{ in}^4$$

$$\Delta = \frac{5wL^4}{384EI}$$

$$\Delta_{DL} = \frac{5wL^4}{384EI_{eDL}} = \frac{5(2020)(25)^4(12^3)}{384(3.6 \times 10^6)(12{,}293)} = .401 \text{ in}$$

$$\Delta_{DL+LL} = \frac{5wL^4}{384EI_{eDL+LL}} = \frac{5(2020 + 1340)(25)^4(12^3)}{384(3.6 \times 10^6)(12{,}176)}$$

$$= .674 \text{ in}$$

Immediate LL deflection $\Delta_{LL} = \Delta_{DL+LL} - \Delta_{DL} = .273$

Long-term deflections $\lambda = \dfrac{\xi}{1 + 50\varrho'} = 2.0$ after 5 years

$\Delta_{\text{LONG TERM}} = \lambda\ \Delta_{i\,\text{SUSTAINED LOADS}}$

$\Delta_{\text{LONG TERM}} = 2.0(.401) = .801$ in

$\Delta_{\text{MAX}} = \Delta_{i\text{LL}} + \Delta_{CR+SH} = .273 \text{ in} + .801 \text{ in} = 1.074$ in

$\Delta_{\text{MAX}} \leq \dfrac{\ell}{480}\ \ \dfrac{25(12)}{480} = .625 < 1.074$ in No Good

CONTINUITY

Reinforced concrete buildings are constructed to form a monolithic system. One of the most common systems used in building construction is the slab-beam-girder floor construction. This typical floor system is shown in Figure 27 [5]. Section A-A shows the actual structural floor slab over the beams. When analyzing the slab, it becomes a continuous beam supported by the beams which are represented by simple supports. The slab beam is considered to have a width of 12 inches. Figure 27 shows the beams monolithic with girders (Bl, B2, and B3). These beams are structurally analyzed similar to the slab system. The beams, B4, B5, and B6, are monolithic with the columns and are analyzed using the frame shown. The columns must be considered. The girders, Gl, G2, and G3, are also monolithic with columns and are analyzed similar to B4, B5, and B6.

For describing the structural analysis procedure, beams Bl, B2, and B3 will be used. Figure 27 shows the structural frame configuration which will be used. The dead load is to be considered to be uniformly distributed along the continuous beam. The live-load needs to be positioned in different pattern loadings to produce the maximum and minimum moments which will occur in the beams. These pattern loadings are determined by looking at the influence lines for the structural beam. The pattern live-loading situations for the beam are shown in Figure 28.

Considering the loading cases just described, the shear

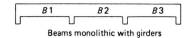

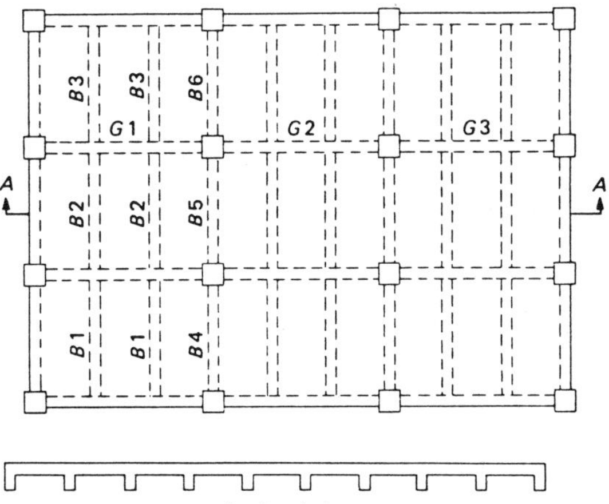

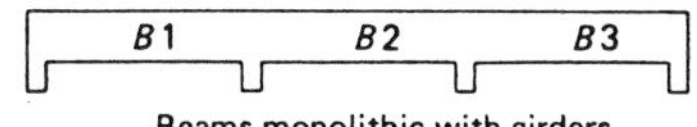

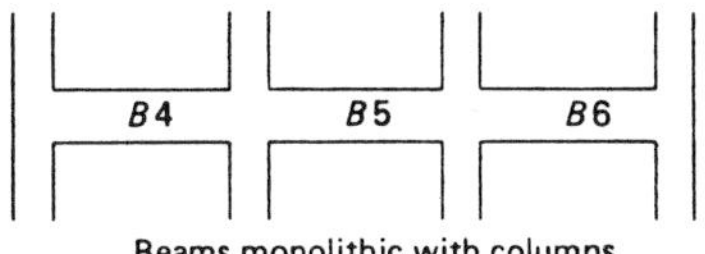

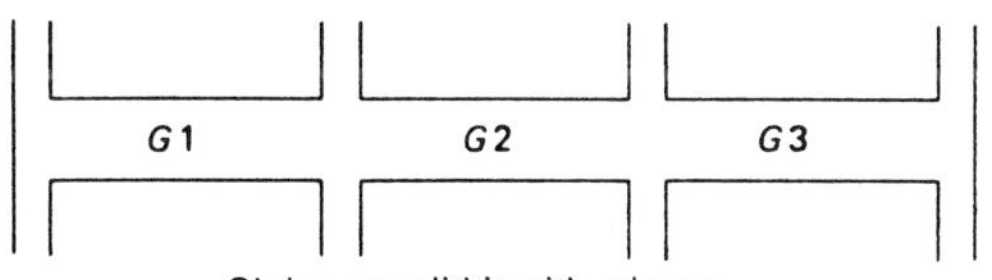

Slab–beam–girder floor.

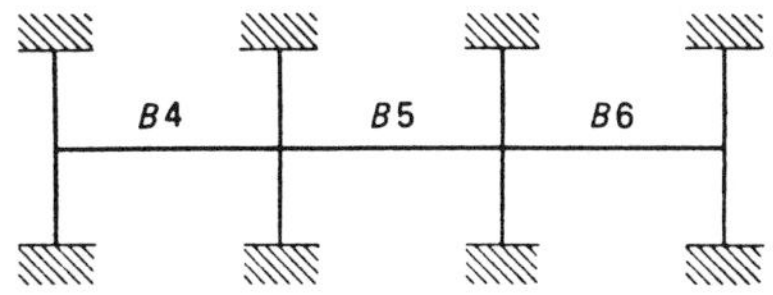

Beams and adjacent columns.

FIGURE 27. Typical one-way slab floor system [5].

and moment diagrams have to be determined for each case. The indeterminant beam can be analyzed by the moment distribution or matrix methods. The ACI Code 83 specifies moment coefficients which can be used to determine the maximum moments and shears. These moment coefficients

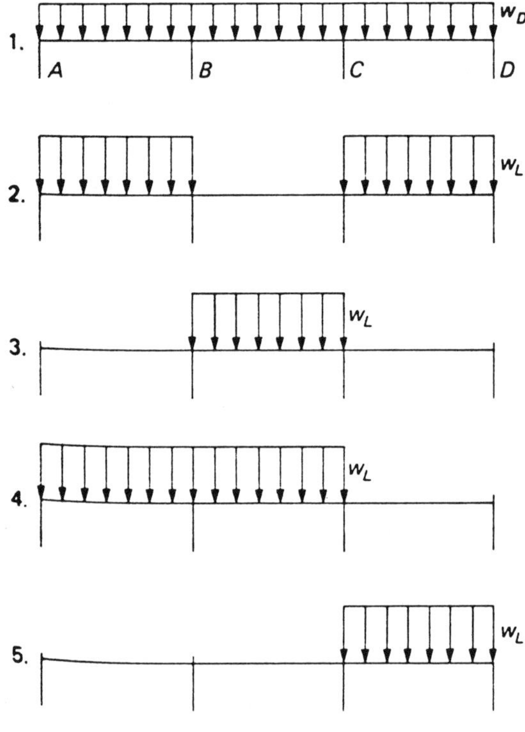

FIGURE 28. Loading conditions [5].

which are shown in Figure 29 can be used only if the live load to dead load is less than 3.0; the span lengths are similar (within 80 percent); and there are no concentrated moments or loads in the span. When determining the moments, the moment redistribution effects can be considered at failure. This is considered in the ACI Code 83 in the limit analysis.

ONE-WAY SLABS

The system shown in Figure 27 has a floor system which is considered a one-way slab. There are two types of basic systems, one-way and two-way, which are described by the way the system takes the moments. Figure 30 shows the two systems. The slab is considered to be one-way if the long to short span is more than two. In the analysis of the one-way system, it will be considered a continuous beam (b = 12 inches) over the support beams. The moments go in the short direction. There are no moments considered in the long direction but temperature and shrinkage steel must be placed in that direction. The moments and shears can be determined by elastic analysis such as moment distribution and matrix analysis or if applicable the moment coefficients can be used (as shown in continuity section).

The thickness of the slab is determined on the basis of deflection, bending, and shear. The slab should not have ex-

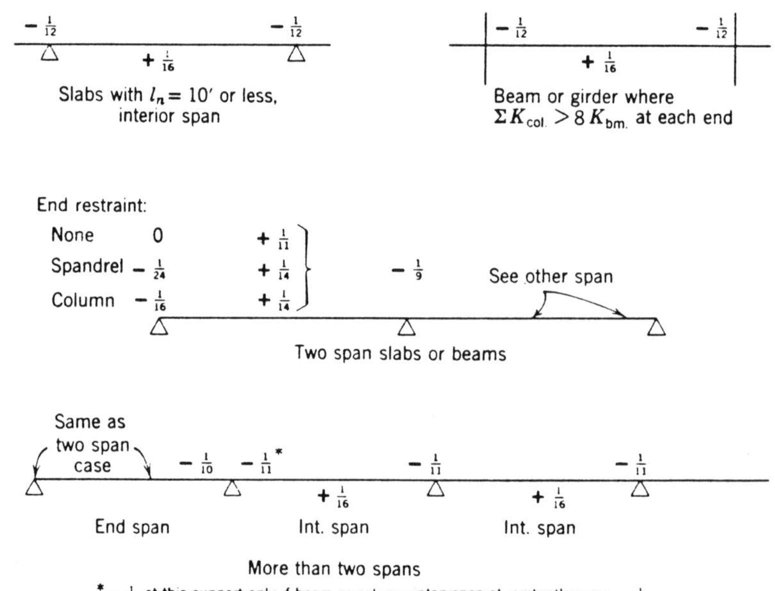

FIGURE 29. ACI Code (8.4.2) moment coefficients for nearly equal spans and live load less than three times the dead load. M = (coef.) wl_n^2 [25].

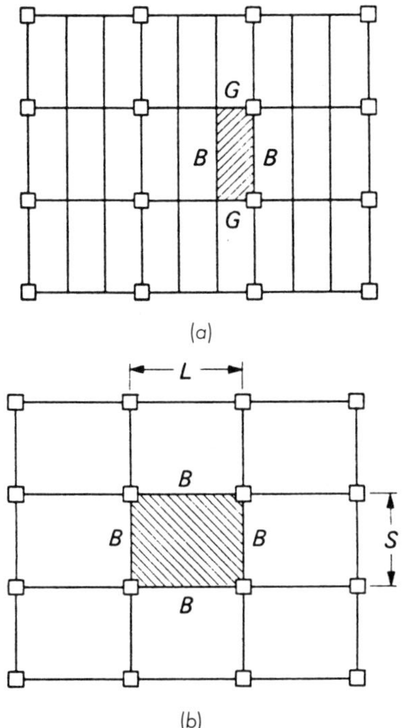

FIGURE 30. One-way versus two-way slabs [5].

cessive deflections. The deflection requirements are described in the ACI Code 83, Table 9.5a and 9.5b. Table 9.5b should be used if the construction is to be damaged by large deflections. The flexural depth, d, should be determined by design procedures for singly reinforced concrete members. The maximum moment should be used in this flexural analysis. The shear also has to be checked even though it does not usually control in the design, $V_u < \phi\ (2\ \sqrt{f_c'}\,bd)$.

The renforcement which are normally about a #4 bar shall be placed according to the flexural or development length requirements. The maximum spacing being: (1) not greater than three times the slab thickness; (2) not more than 18 inches; and (3) not more than the spacing required for temperature and shrinkage. The development length should be checked at all critical points. The bar cutoffs should be at the points of inflection with the added factor of safety for the positive and negative steel requirements. Figure 31 shows a graph which can be used for the cutting of steel bars [1]. Temperature and shrinkage steel shall be put in the long direction.

Example 13

$$f_c' = 3750 \text{ psi}$$

$$f_y = 60{,}000 \text{ psi}$$

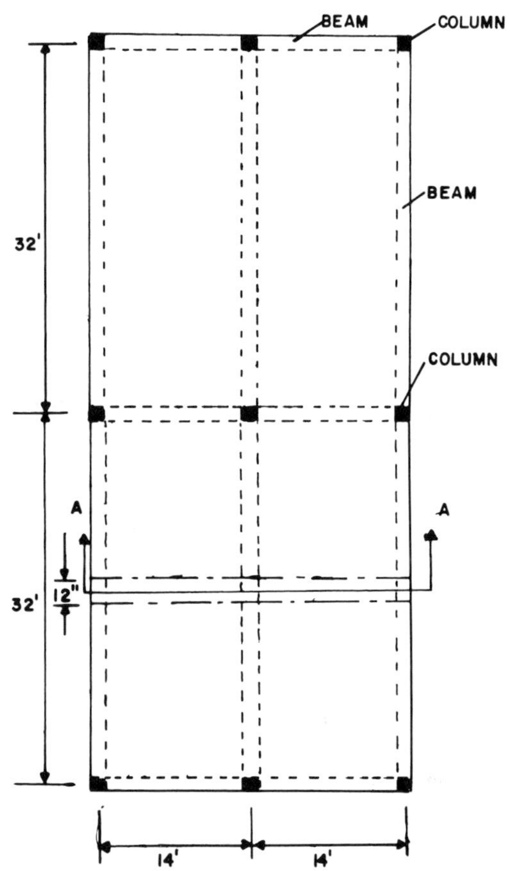

E13.1 Slab dimensions.

$$w_{DL} = 50 \text{ lb/ft}^2$$

$$w_{LL} = 100 \text{ lb/ft}^2$$

One way slab

$$\ell/s = \frac{32}{14} = 2.29 > 2.0$$

Thus one way slab.

Check to see if you can use code coefficients.

1. $\dfrac{L_L}{L_S} = \dfrac{32}{14} = 2.29 > 2.0$ O.K.

2. $\dfrac{w_{LL}}{w_{DL}} = \dfrac{100}{137.5} = .727 < 3.00$ O.K.

3. $\dfrac{L_1}{L_2} = 1.0$ O.K.

4. No concentrated moments or loads. O.K.

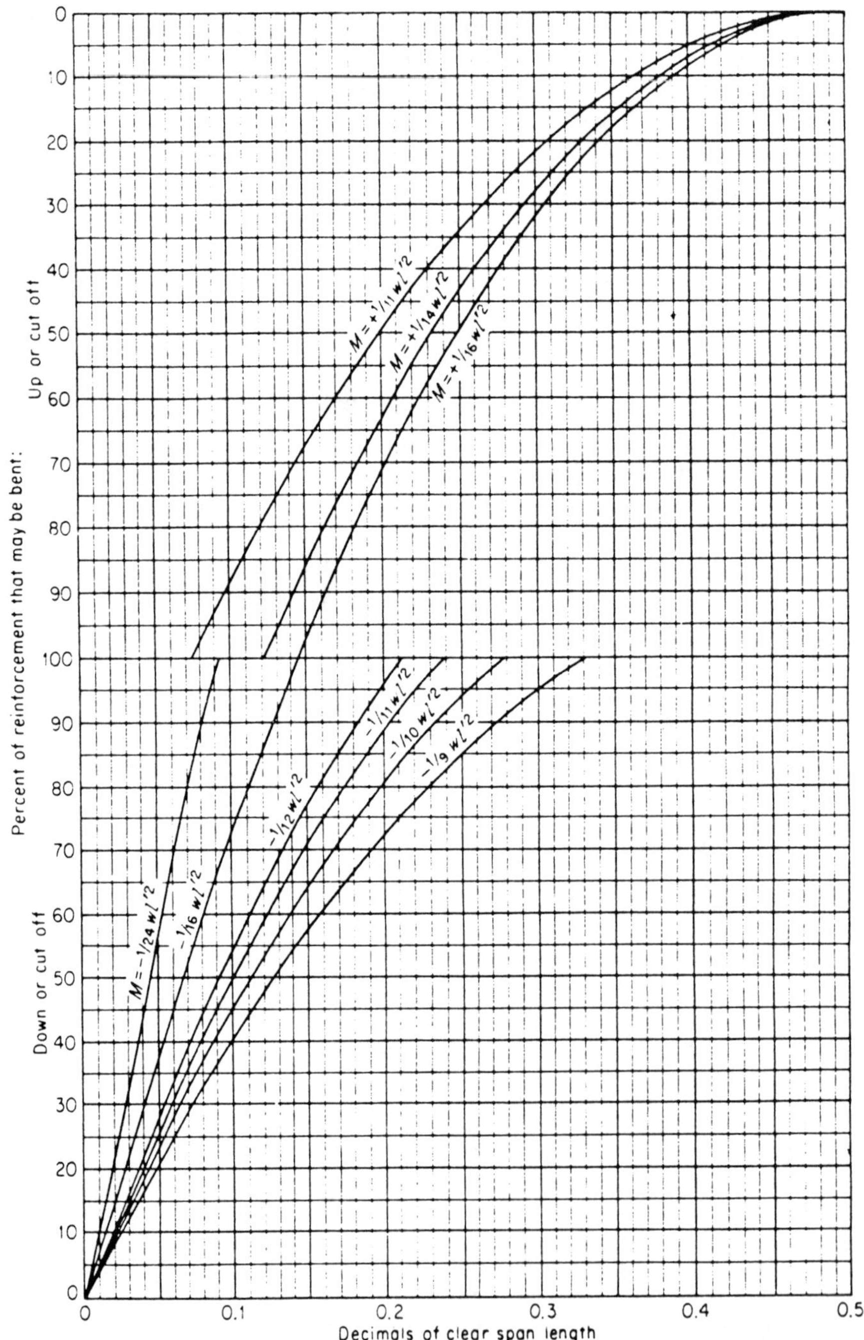

FIGURE 31. Approximate locations of points where bars can be bent up or down or cut off for continuous beams uniformly loaded and built integrally with their supports according to the coefficients of the ACI Code [1].

$$t_{MIN} = \frac{\ell}{24}\left(.4 + \frac{f_y}{100,000}\right) = \frac{\ell}{24} = \frac{14(12)}{24} = 7.0 \text{ in}$$

Assume $t = 7.0$ in

$$w_{DL_{SLAB}} = \frac{7}{12}(150) = 87.5 \text{ lb/ft}^2$$

$$w_{DL} = 50 + 87.5 = 137.50 \text{ lb/ft}^2$$

$$w_{LL} = 100 \text{ lb/ft}^2$$

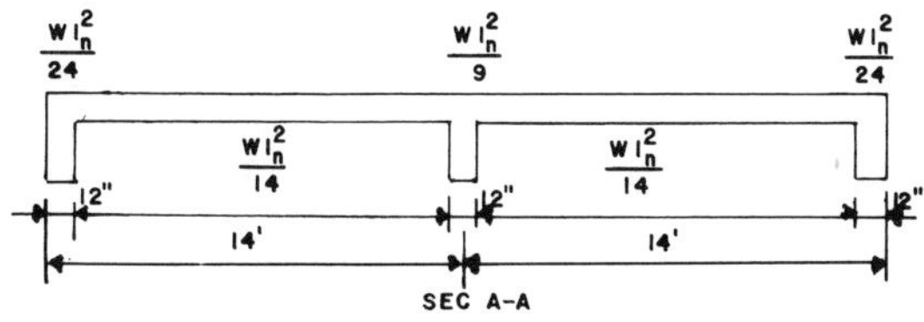

E13.2 Sections and moments.

$$\ell_n = 14' - \frac{12''}{12''} - \frac{6''}{12''} = 12.5'$$

$$w_u = 1.4w_{DL} + 1.7w_{LL} = 1.4(137.5)$$

$$+ 1.7(100) = 362.5$$

$$M_{MAX} = \frac{1}{9}wL_n^2 = \frac{362.5(12.5^2)}{9} = 6293.0 \text{ ft-lb/ft}$$

$$d = \sqrt{\frac{M_u}{\phi f'_c bw(1 - .59w)}}$$

$$= \sqrt{\frac{6293.0(12)}{(.9)(3750)(12)(.2)(1 - .59(.2))}}$$

$$d = 3.25 \text{ in}$$

$$d_{ACT} = t - ¾'' - ½(½) = 7.0 - ¾'' - ¼'' = 6.0 \text{ in}$$

d O.K. $\quad 3.25 < 6.0$ in

$$V_u = \frac{1.15\, wL_n}{2} = \frac{1.15(362.5)(12.5)}{2} = 2605 \text{ lb}$$

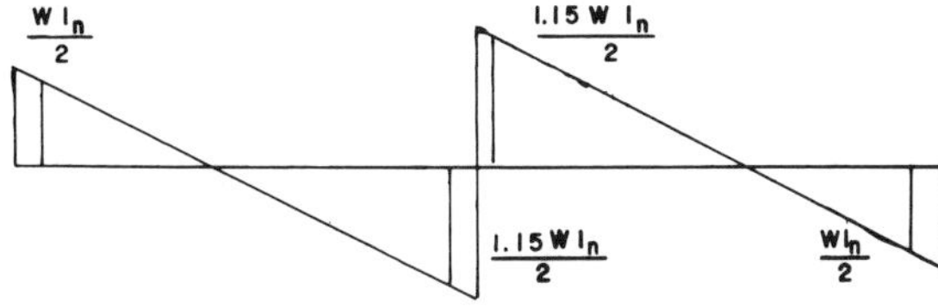

E13.3 Shear Diagram.

$$V_c = 2\sqrt{f'_c}\,bd = 2\sqrt{3750}\,(12)(6) = 2605 \text{ lb}$$

$8820 > 2605$ O.K.

$$M_u = \frac{wL_n^2}{9}$$

Assume $a = .4$

$$A_s = \frac{M_u}{\phi f_y\left(d - \frac{a}{2}\right)} = \frac{6293(12)}{.9(60,000)\left(6 - \frac{.4}{2}\right)} = .24 \text{ in}^2\text{/ft}$$

$$a = \frac{A_s f_y}{.85 f'_c b} = \frac{(.24)(60,000)}{.85(3750)(12)} = .38 \quad \text{O.K.}$$

$$M_u = \frac{wL_n^2}{24} = \frac{362.5(12.5)^2}{24} = 2360 \text{ ft-lb/ft}$$

Assume $a = .15$

$$A_s = \frac{M_u}{\phi f_y\left(d - \frac{a}{2}\right)} = \frac{2360(12)}{.9(60,000)\left(6 - \frac{.15}{2}\right)}$$

$$= .09 \text{ in}^2\text{/ft}$$

$$a = \frac{A_s f_y}{.85 f'_c b} = \frac{.09(60,000)}{.85(3750)(12)} = .14 \quad \text{O.K.}$$

$$M_u = \frac{wL_n^2}{14} = \frac{362.5(12.5)^2}{14} = 4046 \text{ ft-lb/ft}$$

Assume $a = .25$

$$A_s = \frac{M_u}{\phi f_y\left(d - \frac{a}{2}\right)} = \frac{4046(12)}{.9(60,000)\left(6 - \frac{.25}{2}\right)}$$

$$= .15 \text{ in}^2\text{/ft}$$

$$a = \frac{A_s f_y}{.85 f'_c b} = \frac{(.15)(60,000)}{.85(3750)(12)} = .24 \text{ in} \quad \text{O.K.}$$

Location	M_u	M_u	A_s	Spacing s	Act s
1	$\frac{wL_n^2}{24}$	2360	.09 in²	26.67 in	16 in
2	$\frac{wL_n^2}{14}$	4046	.14 in²	17.14 in	16 in

Location	M_u	M_u	A_s	Spacing s	Act s
3	$\frac{wL_n^2}{9}$	6293	.24 in²	10.00 in	10 in
4	$\frac{wL_n^2}{14}$	4046	.14 in²	17.14 in	16 in
5	$\frac{wL_n^2}{24}$	2360	.09 in²	26.67 in	16 in

TEMPERATURE STEEL REQUIRED

$A_s = .0018\,bt = .0018(12)(7) = .15$

$s = 16$ in

$$s_{MAX}\begin{cases} 18'' & \rightarrow 18'' \\ 3t & \rightarrow 21'' \\ \text{Temp.} & \rightarrow 16'' \leftarrow \text{Use } 16'' \text{ as } s_{MAX} \end{cases}$$

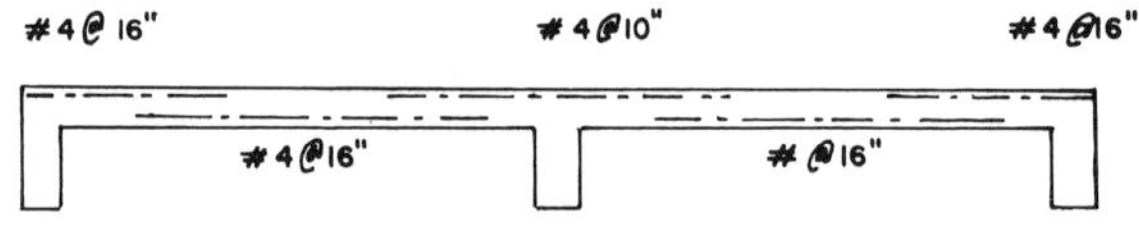

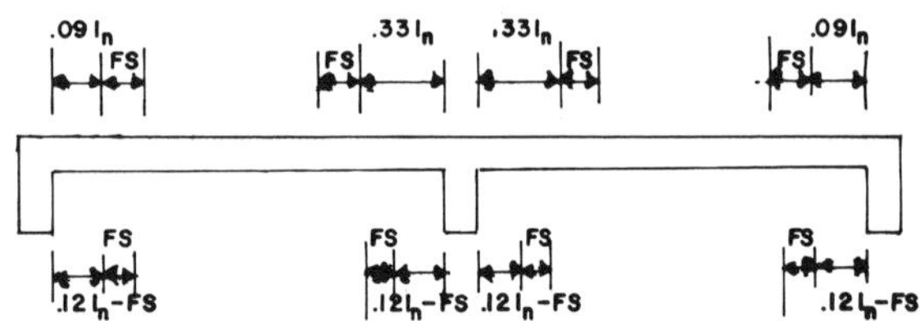

E13.4 Bar spacing and cut-offs.

Neg. steel $-As$

$$\text{F.S.}\ \frac{\ell_n}{16} = \frac{12.5(12)}{16} = 9.38 \text{ in} \leftarrow \text{Use Largest}$$

$d = 6.0$ in

$12d_b = 12(\frac{1}{2}) = 6.0$ in

Pos. steel $+As$

F.S. $d = 6.0$ in

$12d_b = 12(\frac{1}{2}) = 6.0$ in

¼ As goes into support

Check ℓ_d for points also and check deflection if necessary. Put temp. and shrinkage steel in long direction #4 @ 16.

TWO-WAY SLABS

The two-way slab system distributes the moments in two directions. The slab is considered to be two-way if the long to short span is less than two. The two-way floor systems are explained in the concrete structural systems. The analysis for two-way slab systems can be done by the ACI Code 83 approaches, the direct design method; the equivalent frame method; the yield-line analysis; limit-state solutions; and the strip method. In the design, the engineer must consider the moment capacity, the slab-column shear capacity, and serviceability behavior such as deflection and crack control.

The ACI Code 83 approaches are the direct design method and equivalent frame method. The direct design method is an approximate procedure for analyzing the moments in a slab for the gravity loads only. There are six restrictions which must apply in order to use this method. They are:

1. Three continuous spans in each direction minimum.
2. Panels are rectangular $L/s < 2.0$.
3. Successive span lengths in each direction must not differ by more than ⅓ length of longer span.
4. Column offset < 10% of span in direction of the offset.
5. $w_{LL}/w_{DL} < 3.0$.
6. Slabs with beams.

$$.2 < \alpha_1 \ell_2^2/\alpha_2 \ell_1^2 < 5.0$$

$$\alpha = \frac{E_{cb} I_b}{E_{cs} I_s}$$

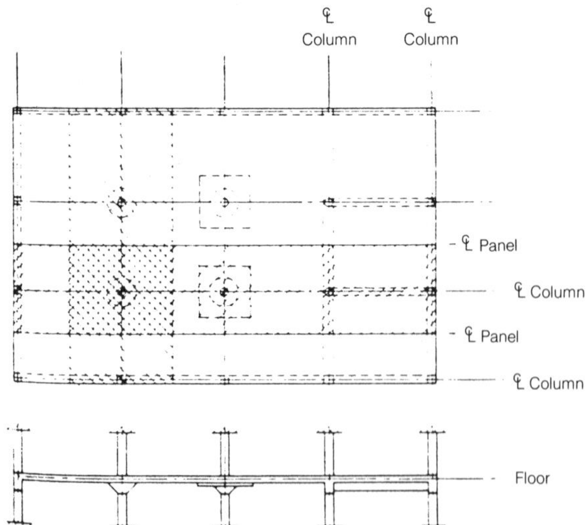

FIGURE 32. Building idealization for equivalent-frame analysis [1].

If these limitations are not satisfied, then the equivalent frame method can be used. While the equivalent frame method defined in section 13.7 of the ACI Code 83 is limited to gravity load analysis, it can be used for lateral load analysis, if modified to account for the loss of stiffness due to cracking in the slab-beams [4]. For analysis the slab system is divided into design strips consisting of a column strip and middle strip. A slab thickness is determined. A total factored static moment is calculated for each span. Then this total factored static moment is distributed to the negative and positive areas of the span. The distributed negative and positive moments are then further distributed to the column and middle strip. There are coefficients and expressions which are set up in the ACI Code 83 for these distributions.

The other method which is defined in the ACI Code 83 is the equivalent frame method. The structure may be considered to be divided into a number of equivalent frames, each consisting of a row of columns and a strip of the supported slab bounded laterally by the centerline of the panel on either side of the centerline of the columns. The equivalent frames shall be taken longitudinally and transversely to the building. This can be seen in Figure 32. Members of the equivalent frame are made up of slab-beams and torsional members in the horizontal direction and the columns in the vertical direction. This is described in Figure 33. The flexural stiffness of each framed member (slab-beams, col-

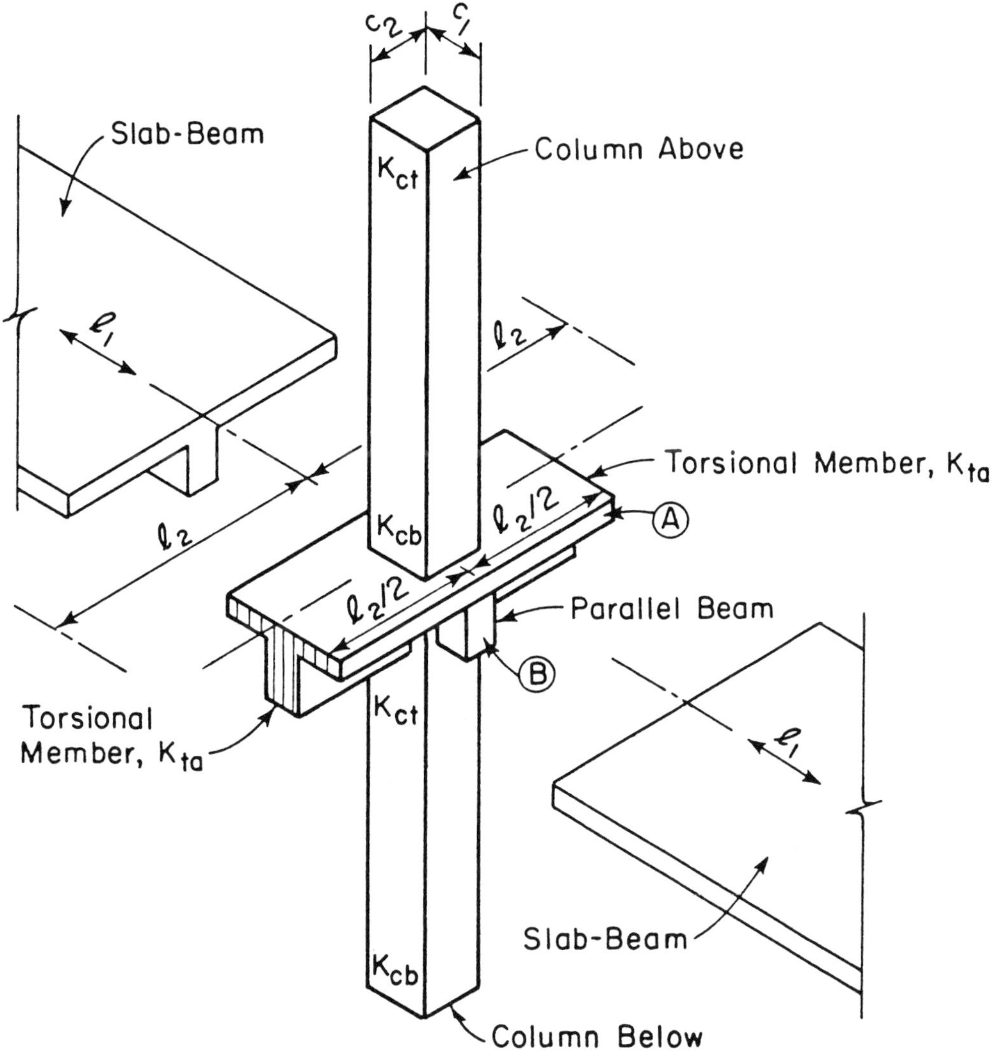

FIGURE 33. Equivalent frame members [11].

umns, and torsional members). These members are considered nonprismatic. The moment distribution method is used in the calculation of dead load and pattern live load moments. Thus, stiffness coefficients, carry-over factors, and fixed-end moment coefficients must be calculated for the different geometric and loading configurations of the nonprismatic members. The moments which are found in the slab-beam are distributed to the column strip and middle strips as in the direct design method.

For both methods, there are minimum thickness requirements given by the ACI Code 83 in section 9.5.3. There are three equations for the thickness given. There are also minimum values and alterations in the formula due to certain types of construction of the two-way slabs. If the slab may cause damage or unsightly aesthetic appearance, an analytical procedure should be used to come up with deflections for both short and long term. These calculated deflections must meet the standards set down in Table 9.5b of the ACI Code 83.

Shear must also be considered in the design of two-way slabs. Shear may be critical in the design, especially in flat-plate floor construction. When checking shear around columns, there are two different types to consider: wide-beam action and two-way action. The wide-beam action will create a diagonal crack. The two-way action will create a punching cone at the joint of the column slab. The concrete shear strength for two-way action is:

$$V_c = (2 + 4/\beta_c)\ \sqrt{f_c'}\ b_o d \quad (74)$$

but not greater than $4\ \sqrt{f_c'}\ b_o d$. β_c is the ratio of the long side to short side of the column and b_o is the perimeter of the critical section defined by section 11.11.1.2 of the ACI Code 83. If the shear strength of the slab is not adequate, the slab thickness may be increased or shear reinforcement supplied. Also, the shear requirements must be checked for the shear introduced by the moment applied at the columns.

The reinforcement for the slabs has to be designed for the calculated factored moments with slab. The reinforcing steel bars are selected for size and spacing. The details of the bar arrangement must be considered to take into account crack control, bar development lengths, and temperature and shrinkage stresses.

COLUMNS

Columns in a building take axial load, shear, and bending moment. The bending and shear may be in one or two directions with respect to the member. These members are analyzed for both gravity load and horizontal loading cases. There are three types of columns: (1) rectangular and cir-

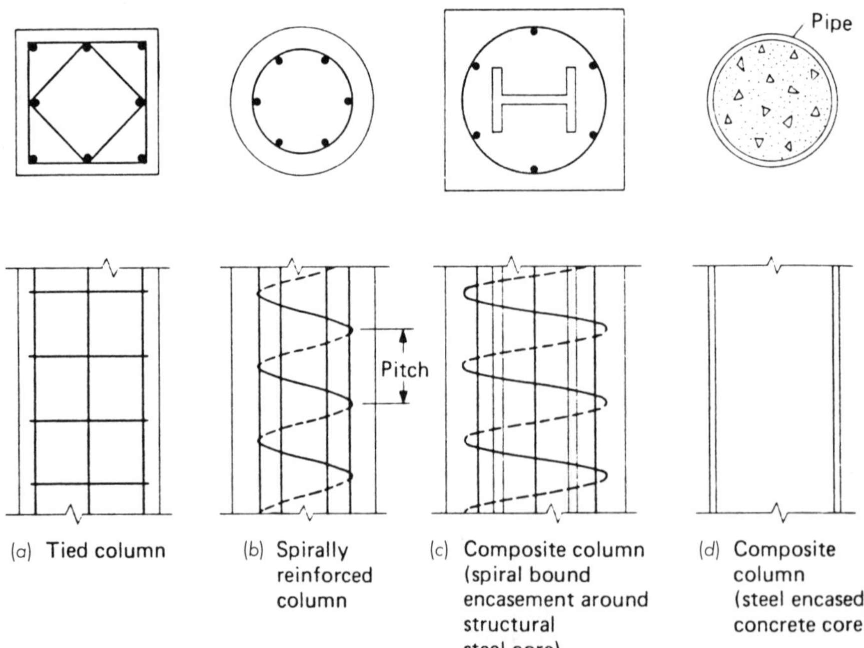

FIGURE 34. Types of columns [5]: (a) tied column; (b) spirally reinforced column; (c) composite column (spiral bound encasement around structural steel core); (d) composite column (steel encased concrete core).

cular columns reinforced with longitudinal bars and lateral ties; (2) circular and square columns reinforced with longitudinal bars; and (3) composite columns with the steel structural shape inside concrete with or without a reinforcing cage. These sections are shown in Figure 34. Basically, there are two types of reinforced concrete columns: tied or spiral. These are based on the lateral confinement reinforcement of the member. The behavior of both of these columns is shown in Figure 35 (load vs. deformation). These curves are for a concentrical load applied to the column only. The spiral column shows a lot more ductility than the tied as shown by the amount of area under the load-deflection curve. The tied column fails unaxially by crushing of the concrete while the longitudinal steel yields. The longitudinal steel will then buckle between the ties. The spiral column will fail by crushing of the concrete while the longitudinal steel yields. The concrete within the spiral is confined by the lateral reinforcement and kept from deteriorating. The concrete cover over the spiral reinforcing will spall and eventually the spiral will break (failure then occurs). The spiral column would be the most ideal type of construction in earthquake areas. Figure 36 shows typical failure of a tied column under a concentrical load.

The design considerations for the ties and spiral are important. The tie reinforcing must be at least ¼ inch in diameter and have a spacing requirement of:

1. less than 16 longitudinal bar diameters,
2. less than 48 tie diameters,
3. less than least dimension of the column.

Also, the spiral reinforcing must be at least ¼ inch in diameter and have a spacing requirement of:

1. $s \leq D_c$ (diameter of the core)
2. $s \leq 3$ inches
3. $s \geq 1\frac{3}{8}$ inch or $1\frac{1}{2}$ (maximum aggregate)

The amount of spiral reinforcing ϱ_s which is A_{sp}/A_c is:

$$\varrho_s = \frac{A_{sp}}{A_c} = \frac{\text{volume of spiral in one loop}}{\text{volume of core for a length}} \tag{75}$$

$$\varrho_s = \frac{a_s\,\pi(D_c - d_b)}{(\pi\,D_c^2/4)s} \tag{76}$$

where a_s is the area of the spiral and d_b is the diameter of the spiral wire. The ACI Code 83 requires that

$$\varrho_s = .45\left(\frac{A_g}{A_c} - 1\right)\frac{f_c'}{f_{sy}} \tag{77}$$

where the f_{sy} must not exceed 60,000 psi.

The slenderness ratio of a column ($K\,\ell_u/r$) determines whether a column behaves as a short column (material failure), a long column (buckling failure), or somewhere in between. The K is the effective length of the column. This value is dependent upon the relative stiffness of the beams and columns and whether the frame is braced or unbraced. The ℓ_u is the clear distance of the column. The r is the radius of gyration of the column. The values of r given by the ACI Code 83 are $.3t$ for a rectangular or square column, and $.25t$ for a circular column, where t is the thickness of the column.

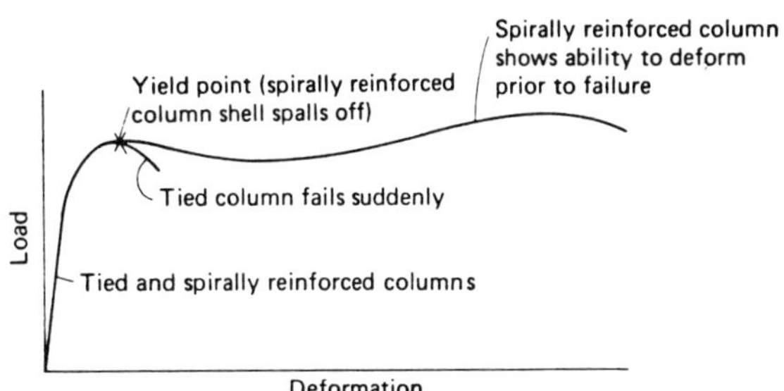

FIGURE 35. Typical load-deformation curves for tied and spirally reinforced columns [5].

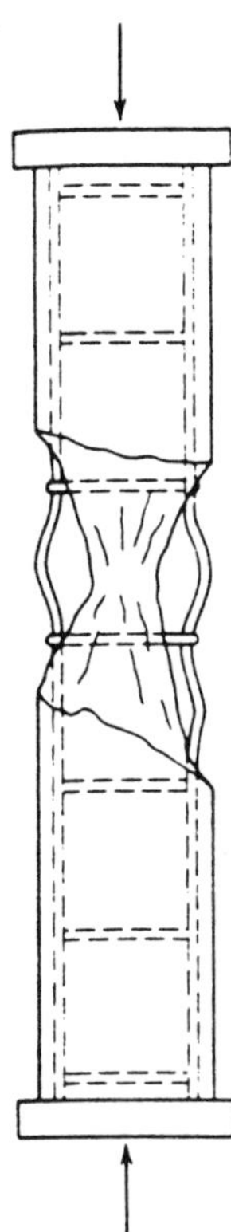

FIGURE 36. Failure of tied column [1].

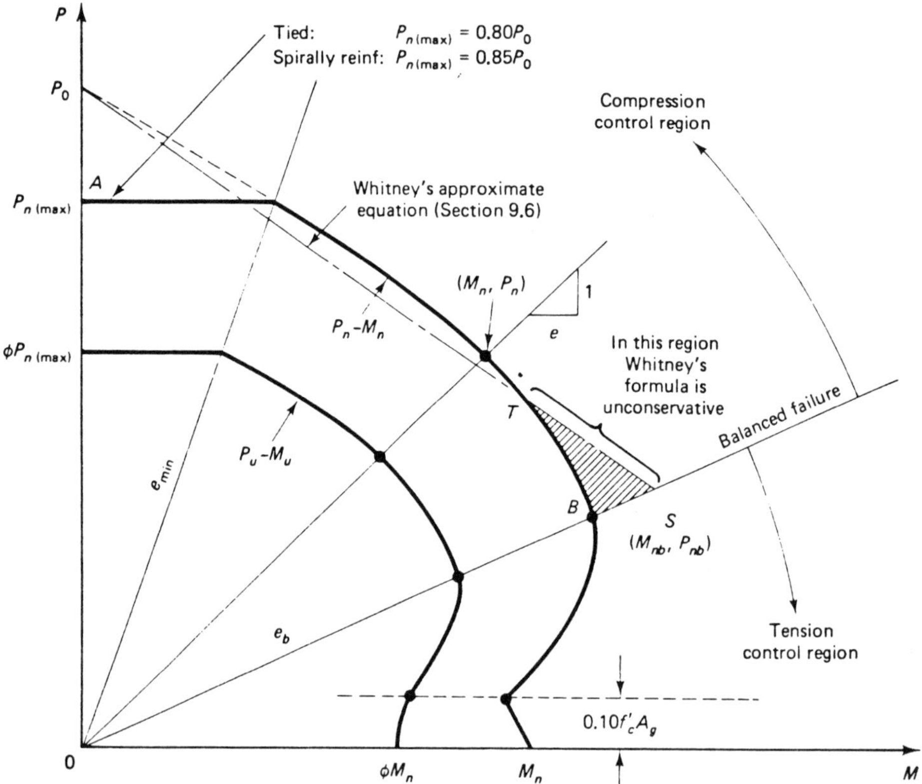

FIGURE 37. Typical load-moment strength (P–M) column interaction diagram [3].

The basic design of a column is performed by using a load-moment interaction curve. Again, as in flexural design, there is a balanced condition where at failure the concrete is crushing and the longitudinal steel just reaches the yield stress. If the longitudinal steel does not reach yielding when the concrete crushes, it is considered to be a compression failure. But if the longitudinal steel is yielding when the concrete crushes, then it is a tension failure. A typical load-moment interaction diagram is shown in Figure 37. The nominal or ultimate load P_n and nominal or ultimate moment, M_n curve, is shown on the diagram. The factored load P_u and factored moment M_u curves are also shown. The strength reduction factors for the different columns are .75 for spiral and .70 for tied. The situation for pure moment has a ϕ factor of .9 and can be determined by flexural behavior as described in an earlier section. The ϕ factor is linear between .9 and .7 or .75 for the column condition where the axial load is $.10f'_cA_g$. Because a column will not have the concentrically loaded case even if the structural analysis predicts it, the ACI Code 83 requires that the maximum axial load P in design be:

$$P_n = .85P_o \text{ (spiral column)} \qquad (78a)$$

$$P_n = .80P_o \text{ (tied column)} \qquad (78b)$$

where P_o is the strength of a short concentrically loaded column

$$P_o = .85f'_c(A_g - A_{st}) + A_{st}f_y \qquad (78c)$$

The area of the longitudinal steel should be between 1 and 8 percent of the gross concrete section (A_g). The interaction curve can be determined by governing equations or a behavioral analysis using force equilibrium and strain compatibility. An interaction curve for a tied column is shown in Figure 38. A typical design and column interaction chart is shown in Figure 39. These charts are used to determine the amount of longitudinal reinforcement required for a column design given a trial section, P_n, and M_n.

The slenderness effects take into account the increase in moment because of the axial load due to deflection in the column due to gravity loads and a story drift due to sidesway. The ACI Code 83 equation for the magnified design moment for unbraced frame considering gravity and lateral load combination

$$M_c = \delta_b M_{2b} + \delta_s M_{2s} \qquad (79a)$$

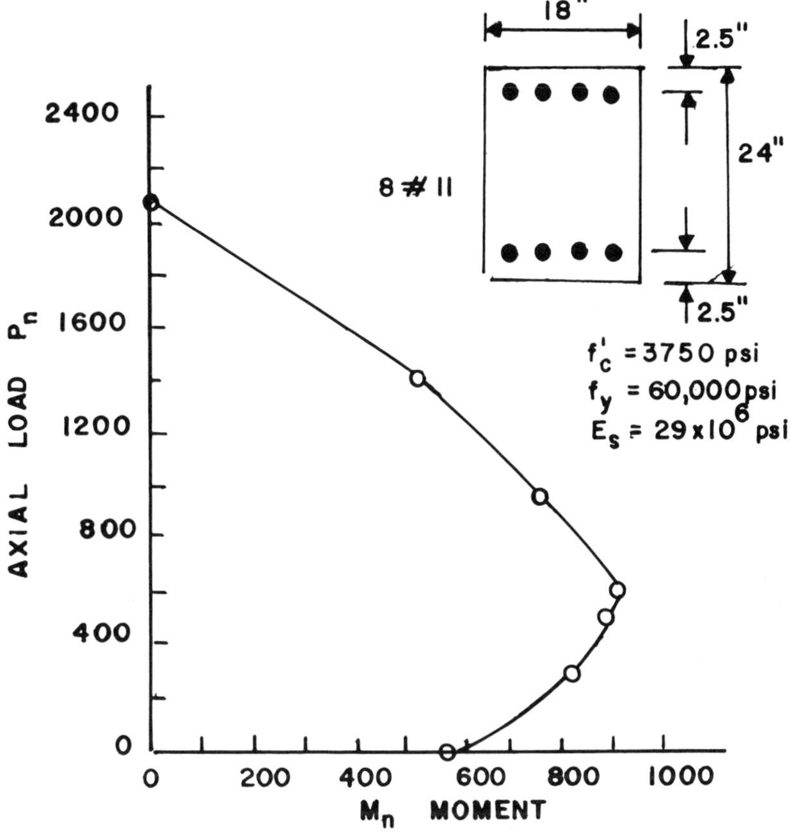

FIGURE 38. Load moment interaction diagram.

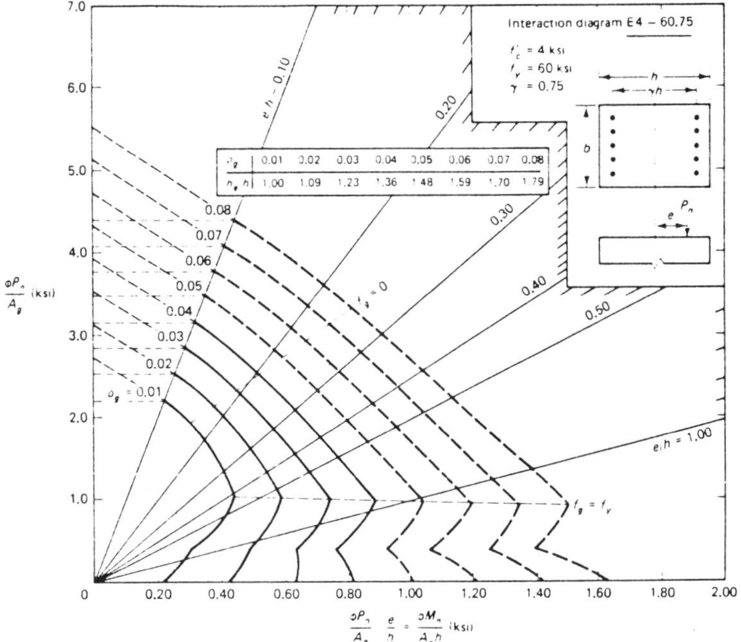

FIGURE 39. Typical nondimensional column interaction charts: chart for small column sizes [3].

The first term is due to the gravity load effect and is the magnified moment due to the members' curvature only. δ_b is the braced frame magnifier

$$\delta_b = \frac{C_m}{1 - P_u/\phi P_{cb}} \tag{79b}$$

where

$$P_{cb} = \frac{\pi^2 EI}{(K_b\, \ell_u)^2} \tag{80}$$

and

$$C_m = .6 + .4(M_{1b}/M_{2b}) \tag{81}$$

The $K_b\, \ell_u$ is the braced effective length. M_{1b} and M_{2b} are the smaller and larger column end moments due to the gravity loads, respectively. M_{2s} is the larger column end moment due to lateral loads. The second term $\delta_s M_{2s}$ is the magnified lateral load moment caused by lateral drift. The sway frame magnifier δ_s is:

$$\delta_s = \frac{1}{1 - \Sigma P_u/\phi \Sigma P_{cs}} \tag{82}$$

where

$$P_{cs} = \frac{\pi^2 EI}{(K_s\, \ell_u)^2} \tag{83}$$

The $K_s\, \ell_u$ is the unbraced effective length. P_{cs} is the critical column load calculated for the unbraced length. ΣP_u and ΣP_{cs} are the summations for all columns within the story being designed.

The gravity load combination magnified design moment for unbraced or braced frames is the same

$$M_c = \delta_b M_{2b} \tag{84}$$

The terms are the same as described for the first term in the gravity and lateral load combination magnified design moment case.

The gravity and lateral load combination magnified design moment for braced frames is

$$M_c = \delta_b M_{2b} + M_{2s} \tag{85}$$

Notice that δ_s is 1.0 [11,9].

The estimate of EI must include effects of cracking and creep under long term loading. From the ACI Code 83, in lieu of a more accurate calculation, EI may be taken as:

$$EI = \frac{(E_c I_g/5) + E_s I_{se}}{1 + \beta_d} \tag{86}$$

or conservatively

$$EI = \frac{E_c I_g/2.5}{1 + \beta_d} \tag{87}$$

where

$$\beta_d = \frac{\text{design dead load moment}}{\text{design total moment}} \tag{88}$$

The slenderness effects as just described are applicable if the slenderness ratio is

$$\frac{K\ell_u}{r} > 34 - 12\left(\frac{M_1}{M_2}\right) \tag{89}$$

for braced frames and

$$\frac{K\ell_u}{r} > 22 \tag{90}$$

for unbraced frames. If the slenderness value is lower than the two above, then the column is considered to be short. Then there will not be any magnification of the design moments. If the slenderness ratio is greather than 100, a second order analysis must be considered.

The effective length $K\ell_u$ is used to modify the basic slenderness effect formulas to account for end restraints other than pinned. It is the length of an auxiliary pinned end column, which has a critical buckling load (Euler load) equal to the column under consideration. Alternately, it is the distance between points of counterflexure of the member under its buckling configuration.

For members in a structural frame, the end restraint lies between the hinged and fixed condition. The effective length coefficient K can be determined from the Jackson and Moreland alignment charts (ACI Code Commentary). See Figure 40.

These charts are based on the following formulas:

$$\frac{\psi_A \psi_B}{4}\left(\frac{\pi}{K}\right)^2 + \frac{\psi_A + \psi_B}{2}\left(1 - \frac{\pi/K}{\tan(\pi/K)}\right) + 2\,\frac{\tan\left(\dfrac{\pi}{2K}\right)}{\pi/K} = 1.0 \tag{91}$$

for braced frames $(.5 < K < 1.0)$.

$$\frac{\psi_A \psi_B \left(\dfrac{\pi}{K}\right)^2 - 36}{6(\psi_A + \psi_B)} = \frac{\pi/K}{\tan(\pi/K)} \tag{92}$$

for unbraced frames $(1.0 < K < \infty)$. ψ_A and ψ_B are degrees of restraint at ends A and B of the column.

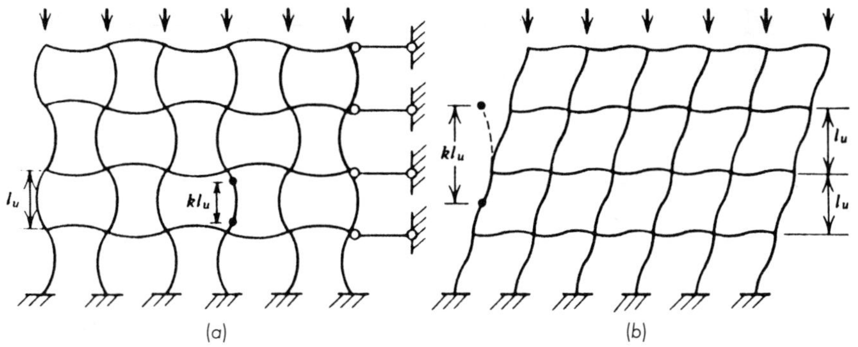

FIGURE 40a. Buckling modes for braced and unbraced frames [25]: (a) braced against sidesway; (b) unbraced against sidesway.

(a) Braced Frames

ψ_A: ∞, 50.0, 10.0, 5.0, 3.0, 2.0, 1.0, 0.9, 0.8, 0.7, 0.6, 0.5, 0.4, 0.3, 0.2, 0.1, 0

k: 1.0, 0.9, 0.8, 0.7, 0.6, 0.5

ψ_B: ∞, 50.0, 10.0, 5.0, 3.0, 2.0, 1.0, 0.9, 0.8, 0.7, 0.6, 0.5, 0.4, 0.3, 0.2, 0.1, 0

(b) Unbraced Frames

ψ_A: ∞, 100.0, 50.0, 30.0, 20.0, 10.0, 9.0, 8.0, 7.0, 6.0, 5.0, 4.0, 3.0, 2.0, 1.0, 0

k: ∞, 20.0, 10.0, 5.0, 4.0, 3.0, 2.0, 1.5, 1.0

ψ_B: ∞, 100.0, 50.0, 30.0, 20.0, 10.0, 9.0, 8.0, 7.0, 6.0, 5.0, 4.0, 3.0, 2.0, 1.0, 0

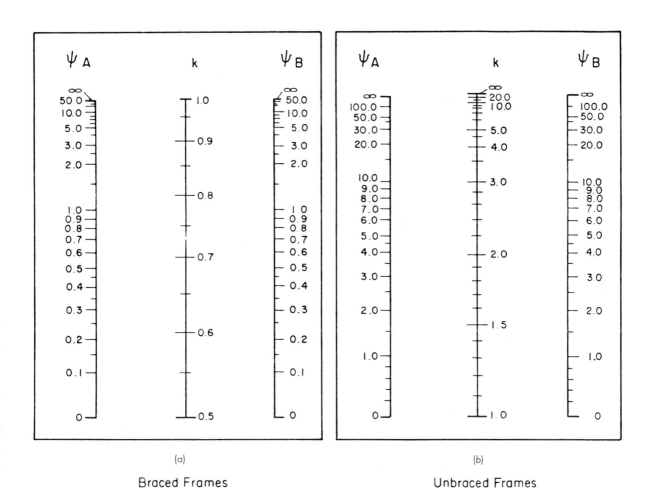

ψ = Ratio of $\sum(EI/\ell_c)$ of compression members to $\sum(EI/\ell)$ of flexural members in a plane at one end of a compression member

k = Effective length factor

FIGURE 40b. Effective length factors [14]: (a) braced frames; (b) unbraced frames.

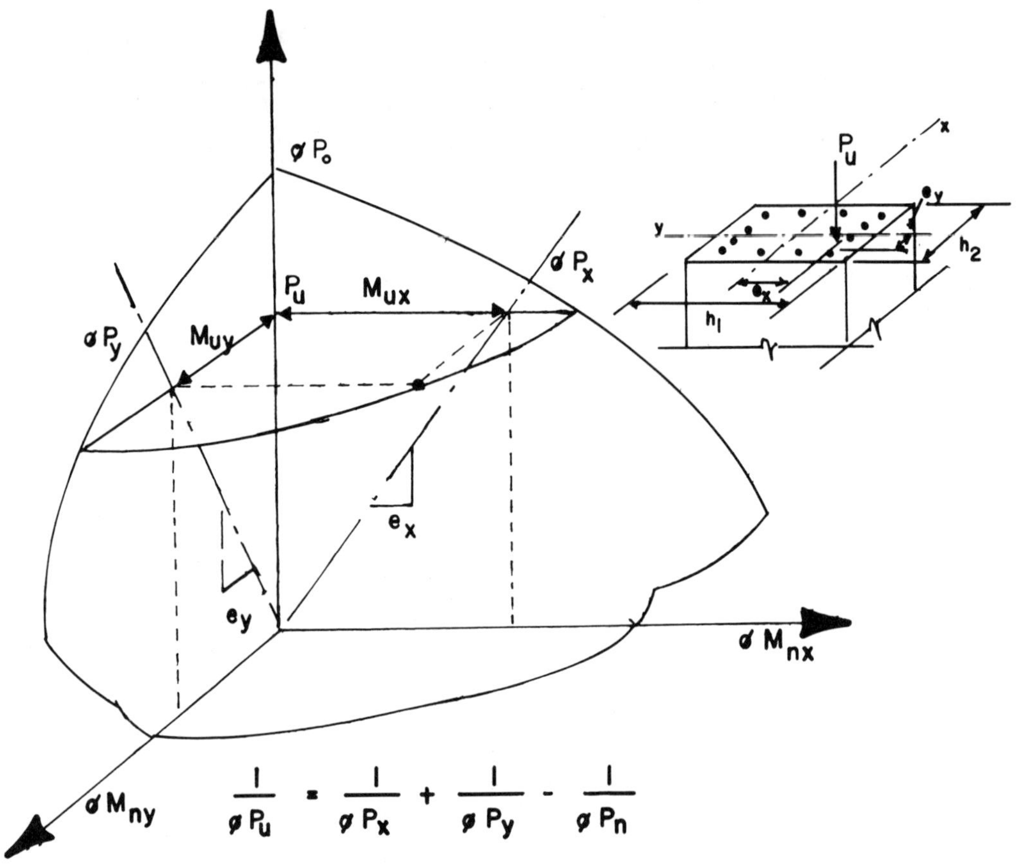

FIGURE 41. Biaxial bending and thrust.

$$\psi = \frac{\Sigma\left(\frac{EI}{\ell}\right) \text{ for all columns}}{\Sigma\left(\frac{EI}{\ell}\right) \text{ for all beams}} \tag{93}$$

In an actual reinforced concrete column in a structure, the member will be subjected to loads in two directions. Thus, in design, the column will be designed for biaxial bending and thrust. There are certain design procedures which must be considered. The biaxial bending and thrust interaction diagram is shown in Figure 41. For a design, the following are given: h_1, h_2, M_{ux}, M_{uy}, P_u, f'_c, and f_y with the interaction curves for P_u and M_{ux} and P_u and M_{uy}. Select steel for a modified "resultant" moment that acts in the plane of the major axis,

$$M_{ur} = 1.1\sqrt{M_{ux}^2 + \left(\frac{h_1}{h_2} M_{uh}\right)^2} \tag{94}$$

With the selected A_{st}, compute ϕP_y, ϕP_x, and ϕP_o for computation of P_u. If the computed value is larger than required P_u, then section and A_{st} are O.K.

Example 14

Design a column which has the following loads from a structural analysis of the building.

$P_D = 150^K$
$P_L = 40^K$
$f'_c = 4000$ psi
$f_y = 60{,}000$ psi
$M_D = 0^K$
$M_L = 250^{\prime K}$
$K = 1$ (sidesway prevented)
$l_u = 8'$

190k
250'k
$l_u = 8'$
250'k
190k

E14.1 Column loading.

Now, try a 24″ × 16″ where the moment is about major axis. The slenderness ratio $K\ell_u/r$ is

$$\frac{K\ell_u}{r} = \frac{(1.0)(8)(12)}{(.3)(24)} = 13.35$$

For braced frames the limit of a short column is

$$\frac{K\ell_u}{r} < 34 - \left(\frac{M_1}{M_2}\right) 12$$

$$34 - (1)(12) = 22 > 13.35$$

Thus, a short column design

$P_u = 1.4P_D + 1.7P_L = 1.4(150) + 1.7(40) = 278^K$
$M_u = 1.4M_D + 1.7M_L = 1.4(0) + 1.7(250) = 425^{1K}$

Consider column dimensions

$h = 24$
$\gamma h = 19$

$$\gamma = \frac{19}{24} = .79$$

(Use $\gamma = .75$ chart)

E14.2 Section.

$$y \text{ axis } \frac{\phi P_n}{A_g} = \frac{P_u}{A_g} = \frac{278}{(24)(16)} = .724$$

$$x \text{ axis } \frac{\phi M_n}{A_g h} = \frac{M_u}{A_g h} = \frac{425(12)}{(24)(16)(24)} = .55$$

Using Figure 39, the ϱ_g which is A_S/A_g is .02. This is between 1 and 8 percent so good.

The $A_S = .02\ A_g = .02(24)(16) = 7.68$ in^2
Try 6−#10 $A_s = 7.62$ in^2 O.K.
Tie spacing design using #3 bars

$$S = \begin{cases} \text{least diam.} = 16'' \leftarrow \text{Use} \\ 16 \text{ long. bar diam.} = 20.3'' \\ 48 \text{ tie bar diam.} = 18'' \end{cases}$$

Thus the section is

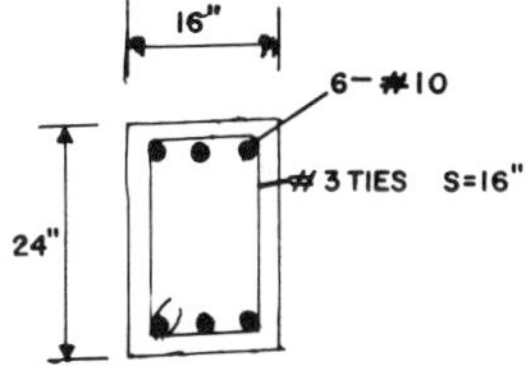

E14.3 Column design.

FOOTINGS

The loads which are put on a structure whether they are live load or dead load which are caused by gravity or horizontal loadings are transmitted to the foundation through the footings. A good foundation analysis is important to be able to predict the behavior of the foundation under the static and dynamic loading situations which the building will transmit to it. For small structures, soil deflections and bearing capacities are sufficient. But for large, important structures, a soil-structure interaction analysis may be required if in an earthquake area.

Basically, there are six types of foundation substructures or footings. These footings must be able to transmit the column loads to the foundation. These footings are: (1) wall footing; (2) independent isolated column footing; (3) combined footing; (4) cantilever or strap footing; (5) pile foundations; and (6) mat or floating foundations [6]. Examples of these footings are shown in Figure 4.

In the design of a footing, they are considered to be rigid and supported by an elastic soil medium. Thus, the soil pressure is assumed to vary linearly. The footing acts similarly to a floor slab. In footing design, the member is proportioned so that it will not fail in bending or shear. In flexural design, the critical section for an isolated column footing is at the face of the column. The design is similar to a rectangular beam whether acting like a cantilever or an indeterminant beam.

In designing for shear, there are two types of failures which must be considered. These are one-way or two-way action (diagonal tension or punching shear). The one-way action or diagonal tension shear failure can occur similar to shear failures in beams. The critical section is a d distance from the face of the column or wall. The two-way or punching shear failure occurs when the column punches through the footing forming a truncated pyramid. The critical section is a distance $d/2$ around the column. The shear capacity of the concrete in the footing is given by

$$V_c = 2\sqrt{f'_c}b_w d \quad \text{one-way action} \tag{55}$$

and

$$V_c = \left(2 + \frac{4}{\beta_c}\right)\sqrt{f'_c}\, b_o d \leq 4\sqrt{f'_c}\, b_o d \tag{74}$$

where β_c is the ratio of the long to short side of the column and b_o is the perimeter of the critical section defined by section 11.11.1.2 of the ACI Code. If the shear capacity is too high, then shear reinforcement may be used. Footings are usually designed with adequate thickness so that shear reinforcement is not necessary. The nominal shear for the section with shear reinforcement is

$$V_n = V_c + V_s \leq 6\sqrt{f'_c}\, b_o d \tag{95}$$

In footing design, the engineer must consider the forces and moments at the base of a column or wall which are transferred to the footing by bearing on the concrete and by reinforcement, dowels, and mechanical connectors [3]. Also, on all reinforcement the development length and anchorage must be checked.

Example 15

Design a rectangular footing for the following problem

$P_D = 240^K$
$P_L = 240^K$
$f'_c = 3000$ psi—footing
$f'_c = 5000$ psi—column
$f_y = 50{,}000$ psi

Column size is 18″ × 18″.
Column longitudinal steel bars are #8.
Permissible soil pressure = 5 ksf.
Because of limitations at the site $B = 7'$.
Assume the thickness of footing is 24 inches $w_D = 300$ psf. Thus, $q = 5000 - 300 = 4700$ psf. Using Service loads for footing size required.

$$A_f = \frac{P_D + P_L}{q} = \frac{240 + 240}{4.7} = 102 \text{ ft}^2$$

$$B \times L = A_f \therefore L = \frac{A_f}{B} = \frac{102}{7} = 14.6 \quad \text{Use } 14'$$

Thus the footing size try will be 7′ × 14′ and

$$q_s = \frac{1.4(240) + 1.7(240)}{7(14)}$$

$$q_s = 7.60 \text{ ksf}$$

Checking for the thickness shear and flexure.

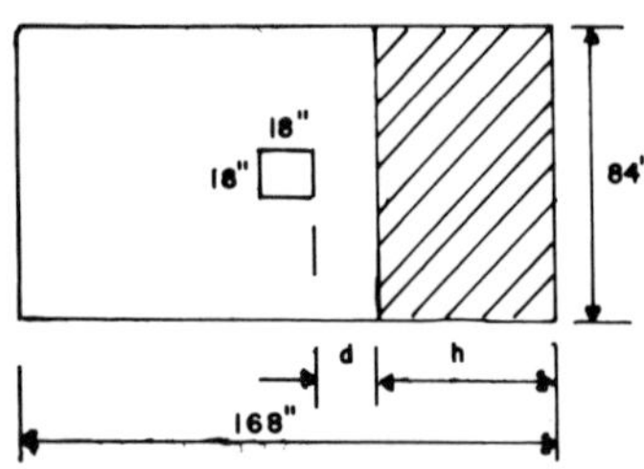

E15.1 Critical section for one way shear.

1. One-way action shear
Assume $d = 24$ in.

$$h = \frac{168}{2} - 9 - 24 = 51 \text{ in}$$

$$V_u = \frac{51(84)(7.60)}{144} = 226 \text{ K}$$

$$V_n = \frac{V_u}{\phi} = \frac{226}{.85} = 266 \text{ K}$$

$$V_c = 2\sqrt{f'_c}\, b_w d$$

$$V_c = 2\sqrt{3000}\,(84)(24) = 221 \text{ K} < 266 \text{ Not O.K.}$$

Assume $d = 30$ in.

$$h = \frac{168}{2} - 9 - 30 = 45 \text{ in}$$

$$V_u = \frac{45(84)(7.60)}{144} = 200 \text{ K}$$

$$V_n = \frac{V_u}{\phi} = \frac{200}{.85} = 235 \text{ K}$$

now

$$V_c = 2\sqrt{3000}\,(84)(30) = 276 \text{ K} > 235 \text{ K} \quad \text{O.K.}$$

assuming one-way action in other side not controlling.
2. Two-way action shear (punching shear)

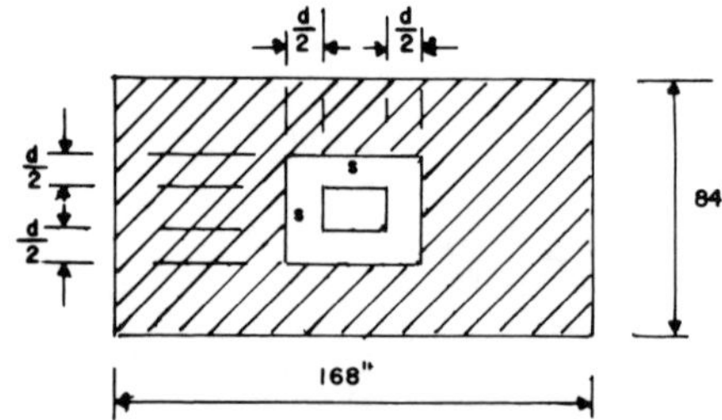

E15.2 Critical section for punching shear.

$$s = 18'' + \frac{30''}{2} + \frac{30''}{2}$$

$$s = 48 \text{ in.}$$

punching cone 48″ × 48″

$$V_u = \frac{[81(168) - 48(48)](7.6)}{144} = 623 \text{ K}$$

$$V_n = \frac{V_u}{\phi} = \frac{623}{.85} = 733 \text{ K}$$

$$V_c = \left(2 + \frac{4}{\beta_c}\right)\sqrt{f'_c}\, b_o d \le 4\sqrt{f'_c}\, b_o d$$

b_o = perimeter of failure zone = 48(4) = 192

$$\beta_c = \frac{18}{18} = 1$$

$$V_c = \left(2 + \frac{4}{1}\right)\sqrt{f'_c}\, b_o d \le 6\sqrt{f'_c}\, b_o d$$

Use maximum $4\sqrt{f'_c}\, b_o d$

$$V_c = 4\sqrt{3000}\,(192)(30) = 1{,}262 \text{ K} > 733 \quad \text{O.K.}$$

3. Bending moment capacity

Checking bending in both directions of footing

$$a = \frac{168''}{2} - 9'' = 75''$$

$$a' = \frac{84''}{2} - 9'' = 33''$$

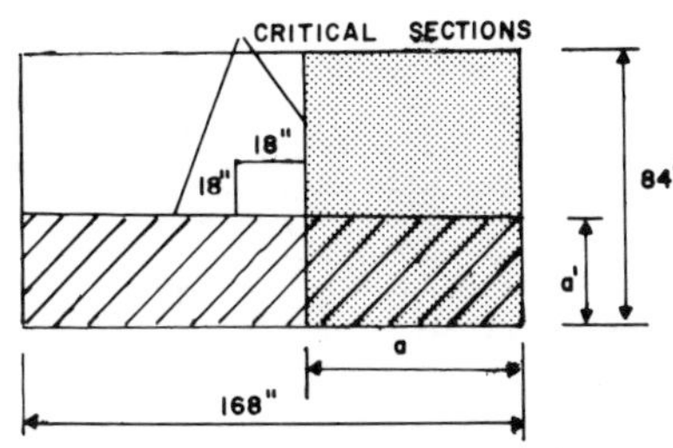

E15.3 Critical sections for flexure.

looking at long direction

$$M_u = 75\left(\frac{75}{2}\right)\frac{(84)(7.60)}{144} = 12{,}468{,}000 \text{ in lb}$$

assume reinforcement ratio $w = .18$ and $\phi = .9$

$$M_u = \phi\, bd^2 f'_c\, w(1 - .59\, w)$$

$$d = \sqrt{\frac{M_u}{\phi\, bf'_c w(1 - .59w)}}$$

$$= \sqrt{\frac{12{,}468{,}000(12)}{.9(84)(3000)(.18)(1 - .59(.18))}} = 18.5''$$

$18.5'' < 30''$ O.K.

Area of steel (using convergent method)

Assume $a = 2.25$ in

$$A_s = \frac{M_u}{\phi f_y\left(d - \frac{a}{2}\right)} = \frac{12{,}468{,}000}{(.9)(50{,}000)\left(30 - \frac{2.25}{2}\right)} = 9.59 \text{ in}^2$$

check a

$$a = \frac{A_s f_y}{.85 f'_c b} = \frac{(9.59)(50{,}000)}{(.85)(3000)(84)} = 2.23 \text{ in} \quad \text{O.K.}$$

$$\varrho_{\text{MIN}} = \frac{200}{f_y} = \frac{200}{50{,}000} = .004$$

$A_s = .004(30)(84) = 10.0 \text{ in}^2$

A_s temperature and shrinkage steel $A_s = .002\, bd$

$A_s = .002(84)(30) = 5.0 \text{ in}^2 < 10.0 \text{ in}^2$

Use $A_s = 10.0 \text{ in}^2$

Try 17 − #7 bars where $A_s = 10.2 \text{ in}^2$

$$\text{Spacing is @ } \frac{84}{17} = 4.94 \text{ in} \quad \text{O.K.}$$

Thus use #7 @ 5 in.

Looking at short direction

$$M_u = \frac{33(168)}{144}\left(\frac{33}{2}\right)(7.6) = 4{,}828{,}000 \text{ in lb}$$

$$d = \sqrt{\frac{M_u}{\phi\, bf'_c w(1 - .59w)}}$$

$$= \sqrt{\frac{4{,}828{,}000}{.9(168)(3000)(.18)(1 - .59(.18))}} = 8.2 \text{ in}$$

Area of steel (using convergent method)

Assume $a = .4$

$$A_s = \frac{M_u}{\phi f_y\left(d - \frac{a}{2}\right)} = \frac{4{,}828{,}000}{.9(50{,}000)\left(30 - \frac{.4}{2}\right)} = 3.60 \text{ in}^2$$

check a

$$a = \frac{A_s f_y}{.85 f'_c b} = \frac{(3.60)(50{,}000)}{(.85)(3000)(168)} = .42 \text{ in} \quad \text{O.K.}$$

$$\varrho_{\text{MIN}} = \frac{200}{f_y} = .004$$

$(A_s = .004(30)(168) = 20.16 \text{ in}^2)$

Could use 4/3 Area which would be $4/3(3.60) = 4.80$ in

Minimum A_s for temperature and shrinkage $= .002\, bd$

$$A_s = .002(30)(168) = 10.08 \text{ in}^2$$

Thus,

$$A_s = \begin{cases} 4.80 \text{ in}^2 & 4/3 \text{ Req.} \\ 10.08 \text{ in}^2 & \text{temp. \& shrink.} \leftarrow \text{Use} \\ 20.16 \text{ in}^2 & \varrho_{\text{MIN}} \end{cases}$$

Reinforcement in short direction — The band width = 7′

$$\beta = \frac{14}{7} = 2$$

and $\dfrac{A_{s1}}{A_s} = \dfrac{2}{\beta + 1}$

$$\therefore A_{s1} = \frac{2(3.60)}{2 + 1} = 2.40 \text{ in}^2$$

Because of temperature and shrinkage A_s, the A_s in the central region is already 5.0 in². Thus the central band width A_s requirement for the short direction is O.K.

Using #7 bars A_b = .6 Use 17 bars (A_s = 10.2 in²) spacing is 168/17 = 9.88 in. Use 9.5 in.

4. Development length check

$$\ell_{db} = \begin{cases} .04 \quad A_b f_y/\sqrt{f'_c} = .04(.6)(50{,}000/\sqrt{3000}) = 22'' \\ .0004 d_b f_y = .0004 \frac{7}{8}(50{,}000) = 17.5'' \\ 12'' = 12'' \end{cases}$$

Use ℓ_{db} for #7 = 22″

$\ell_d = \ell_{db}$ (modifying multipliers) consider the modifying multiplier equal to 1.0

$\ell_d = \ell_{db} = 22''$ consider 3 in cover

$$\ell_{\text{actual in long direction}} = \frac{84}{2} - 9 - 3 = 30 \text{ in}$$

$$\ell_{\text{actual in short direction}} = \frac{168}{2} - 9 - 3 = 72 \text{ in}$$

5. Check force transfer at interface of column and footing.

Bearing strength on column concrete

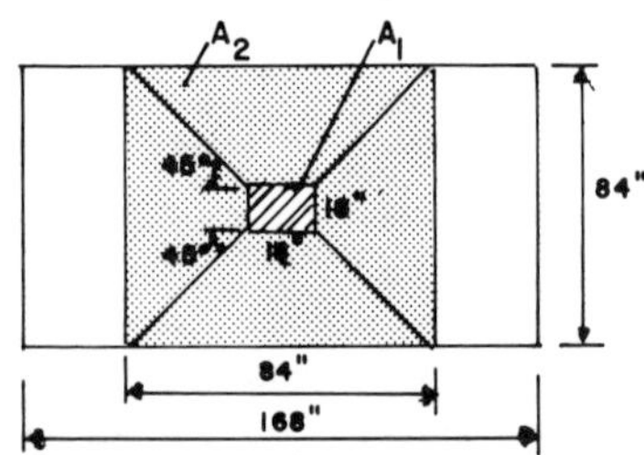

E15.4 Bearing strength on footing concrete.

$$\phi P_{nb} = \phi(.85 f'_c A_1)$$

$$\sqrt{\frac{A_2}{A_1}} = \sqrt{\frac{84(84)}{18(18)}} = 4.672$$

$$\phi P_{nb} = (.70)(.85)(5000)(18)(18) = 964 \text{ K}$$
$$P_u = 1.4(240) + 1.7(240) = 744 \text{ K} < 964 \text{ K} \quad \text{O.K.}$$

Bearing strength on footing concrete.

$$\phi P_{nb} = \sqrt{\frac{A_2}{A_1}}\,[\phi(.85 f'_c A_1)] \text{ where } \sqrt{\frac{A_2}{A_1}} \le 2.0$$

$$\phi P_{nb} = 2.0\,[.70(.85)(3000)(18)(18)] = 1{,}160 \text{ K} > 744 \text{ O.K.}$$

Required dowel bars

A_s (min) = .005 A_g = .005(18)(18) = 1.62 in²
Provide 4 #6 bars as dowels (A_s = 1.76 in²)

Development of dowel reinforcement

column f'_c = 5000 psi	dowel f'_c = 3000 psi
f_y = 50,000 psi	f_y = 50,000 psi
#8 bars	#6 bars

For development into column

$$\ell_{db}\ \#8 \begin{cases} .02 d_b f_y/\sqrt{f'_c} = .02(1)(50{,}000)/\sqrt{5000} = 14'' \\ .0003 d_b f_y = .0003(1)(50{,}000) = 15'' \leftarrow \text{Use} \\ 8'' = 8'' \end{cases}$$

$$\ell_{db}\ \#6 \begin{cases} .02 d_b f_y/\sqrt{f'_c} = .02\left(\frac{3}{4}\right)(50{,}000)/\sqrt{5000} = 10.6'' \\ .0003 d_b f_y = .0003\left(\frac{3}{4}\right)50{,}000 = 11'' \\ 8'' = 8'' \end{cases}$$

ℓ_{db} of #8 controls in column

$\ell_d = \ell_{db}$ considering modifying multipliers = 1.0. Thus, #6 dowel bars must extend 15″ into column. For development into footing

$$\ell_{db}\ \#6 \begin{cases} .02 d_b f_y/\sqrt{f'_c} = .02\left(\frac{3}{4}\right)(50{,}000)/\sqrt{3000} = 13.7'' \\ .0003 d_b f_y = .0003\left(\frac{3}{4}\right)50{,}000 = 11'' \\ 12'' = 12'' \end{cases}$$

$\ell_d = \ell_{db}$ considering modifying multipliers = 1.0. Thus, #6 dowel bars need is 13.7″

$\ell_{actual} = 30'' - \frac{7}{8}'' = 29.1'' > 13.7$ in O.K.

↑ bar diam flexural steel

Note the thickness of the footing should be found

$t = d + d_b + \text{cover} = 30 + \frac{7}{8} + 3 = 33.875''$

Use $t = 34$ in.

Example 16

Design a combined footing, given the following information on the structural analysis, material properties, and construction parameters:

Exterior column: 24″ × 18″ $P_D = 170$ K
$P_L = 130$ K

Interior column: 24″ × 24″ $P_D = 250$ K
$P_L = 200$ K

Center to center of columns is 18′.
Material properties:

footing $f_c' = 3000$ psi column $f_c' = 5000$ psi
$f_y = 60{,}000$ psi $f_y = 60{,}000$ psi

Member is 3′ below grade, 125 psf surcharge, and thickness is 2 feet (300 lb/ft²). Soil allowable pressure is 6000 psf. This problem is similar to that presented by Winter and Nilson [7].

$q_{soil} = 6000 - 300 - (125)(3) - 125 = 5150$ psf

Column load at service stage $= 170 + 130 + 250 + 200 = 750$ K

Area required is $\dfrac{750}{5.15} = 145.5$ ft²

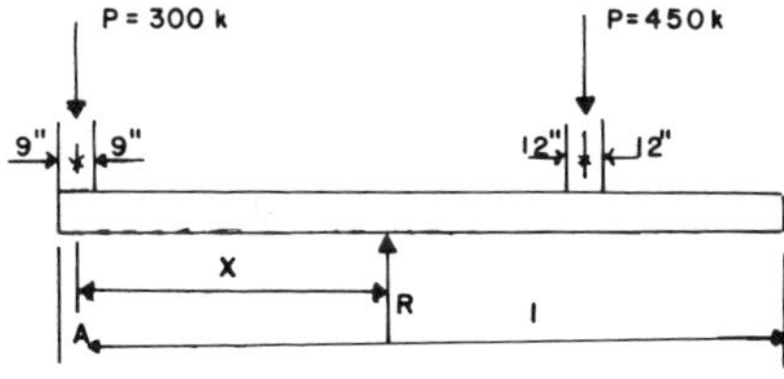

E16.1 Combined footing loads.

$$\curvearrowleft \Sigma M@_A = 0 \curvearrowright$$

$$\frac{R\chi \qquad 450(18)}{}$$

$$\chi = \frac{450(18)}{R}$$

Since $R = 750$
$\chi = 10.8'$
length of member $\ell = 2(10.8 + .75) = 23.1$ ft.

Use 23′3″

$$B = \frac{145.5}{23.25} = 6.3 \text{ ft}$$

Use 6′6″

$$q_s = \frac{P_u}{B(L)} = \frac{1.4(170 + 250) + 1.7(130 + 200)}{(23.25)(6.5)}$$

$$= 7.60 \text{ K/ft}^2$$

The upward pressure per foot is $(7.60)(6.5) = 49.4$ K/ft

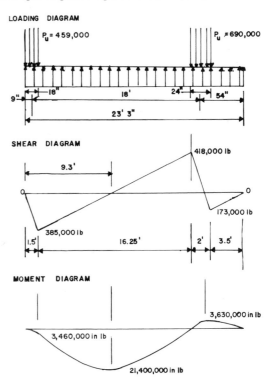

LOADING DIAGRAM
P_u = 459,000
P_u = 690,000
9" 18" 18' 24" 54"
23' 3"

SHEAR DIAGRAM
418,000 lb
9.3'
0
0
173,000 lb
385,000 lb
1.5' 16.25' 2' 3.5'

MOMENT DIAGRAM
3,630,000 in lb
3,460,000 in lb
21,400,000 in lb

E16.2 Combined footing loading, shear, and moment diagrams.

Assume $d = 24$ in

1. Beam shear

Shear is critical at d distance to left from face of internal column.

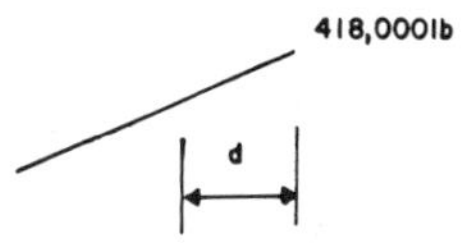

E16.3 One-way shear.

$$V_u = 418{,}000 - \frac{24}{12}(49{,}400) = 319{,}200 \text{ lb}$$

$$V_n = \frac{V_u}{\phi} = \frac{319{,}200}{.85} = 375 \text{ K}$$

$V_c = 2\sqrt{f'_c}\, bd = 2\sqrt{3000}\,(78)(24) = 205 \text{ K} < 375$ Not O.K.
try $d = 37.5$ in,

$$V_u = 418{,}000 - \frac{37.5}{12}(49{,}400) = 264{,}000 \text{ lb}$$

$$V_n = \frac{V_u}{\phi} = \frac{264{,}000}{.85} = 310 \text{ K}$$

$V_c = 2\sqrt{f'_c}\, bd = 2\sqrt{3000}\,(78)(37.5) = 320 \text{ K} < 310$ O.K.

2. Punching shear
Left column (exterior)

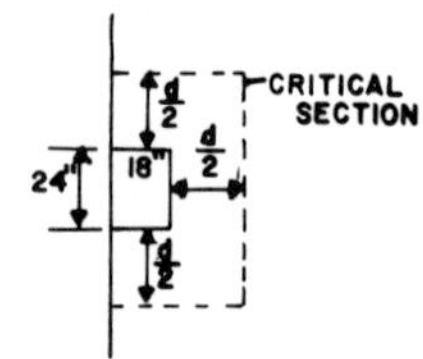

E16.4 Two-way shear left column.

$$b_0 = 24'' + \frac{d}{2} + \frac{d}{2} + 2\left(18 + \frac{d}{2}\right) = 135''$$

$$V_u = 459{,}000 - 7600(3.06)(5.12) = 340{,}000 \text{ lb}$$

$$V_n = \frac{V_u}{\phi} = \frac{340{,}000}{.85} = 400 \text{ K}$$

$$V_c = \left(2 + \frac{4}{\beta_c}\right)\sqrt{f'_c}\, b_o d \le 4\sqrt{f'_c}\, b_o d$$

$$\beta_c = \frac{24}{18} = 1.33$$

$$V_c = \left(2 + \frac{4}{1.33}\right)\sqrt{f'_c}\, b_o d = 5\sqrt{f'_c}\, b_o d$$

Thus, use $4\sqrt{f'_c}\, b_o d$

$V_c = 4\sqrt{3000}\; 135(37.5) = 1{,}110 \text{ K} > 400 \text{ K}$ O.K.

Right column (interior)

$$b_o = \left(24 + \frac{d}{2} + \frac{d}{2}\right)4$$

$$b_o = \left(24 + \frac{37.5}{2} + \frac{37.5}{2}\right)4 = 246 \text{ in}$$

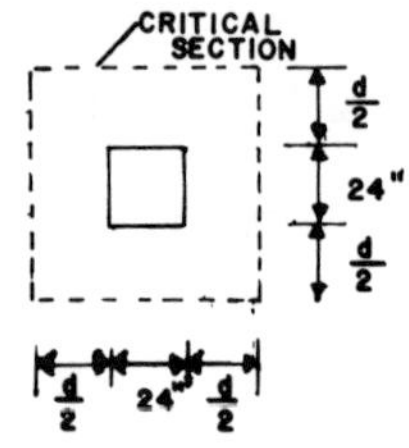

E16.5 Two-way shear right column.

$$V_u = 690{,}000 - \frac{(61.5)(61.5)7600}{144} = 490 \text{ K}$$

$$V_n = \frac{V_u}{\phi} = \frac{490}{.85} = 576 \text{ K}$$

$$V_c = \left(2 + \frac{4}{\beta_c}\right)\sqrt{f'_c}\, b_o d \le 4\sqrt{f'_c}\, b_o d$$

$$\beta_c = \frac{24}{24} = 1 \therefore V_c = \left(2 + \frac{4}{1}\right)\sqrt{f'_c}\, b_o d = 6\sqrt{f'_c}\, b_o d$$

Use

$$V_c = 4\sqrt{f'_c}\, b_o d = 4\sqrt{3000}\,(246)(37.5) = 2{,}020 \text{ K}$$

576 K < 2,020 K O.K.

3. Flexural moment
Design maximum moment at 9.3′ from left

$M_u = 21{,}400{,}000$ in lb
$M_u = \phi b d^2 f'_c w(1 - .59\, w)$

Assume $w = .18$ (reinforcement index)

$$d = \sqrt{\frac{21{,}400{,}000}{(.9)(78)(3000)(.18)[1 - .59(.18)]}}$$

$= 26 \text{ in} < 37.5$ O.K.

Area of steel (using convergent method)
Assume $a = 3.27$ in

$$A_s = \frac{M_u}{\phi f_y\left(d - \frac{a}{2}\right)}$$

$$= \frac{21{,}400{,}000}{(.9)(60{,}000)\left(37.5 - \frac{3.27}{2}\right)} = 11.05 \text{ in}^2$$

$$\varrho_{MIN} = \frac{200}{f_y} = \frac{200}{60{,}000} = .0033$$

$A_{sMIN} = .0033(37.5)(78) = 9.6 \text{ in}^2$

Temperature and shrinkage

$A_s = .0018\, bd$

$A_s = .0018(78)(37.5) = 5.3 \text{ in}^2$ O.K.

Use

$A_s = 11.05 \text{ in}^2$ Thus, 11 − #9 bars ($A_s = 11.0 \text{ in}^2$).

$$\text{spacing } \frac{78}{11} = 7 \text{ in}$$

Check development length

$$\ell_{db} \begin{cases} .04A_b f_y/\sqrt{f'_c} = .04(1.0)(60{,}000)/\sqrt{3000} = 43.8 \text{ in} \\ .0004\, d_b f_y = .0004(1.41)(60{,}000) = 33.6 \text{ in} \\ 12 \text{ in} = 12 \text{ in} \end{cases}$$

$\ell_{db} = 43.8$

$\ell_d = \ell_{db}$ (modifying multipliers)

$\ell_d = 43.8(1.4)(.8) = 49 \text{ in}$

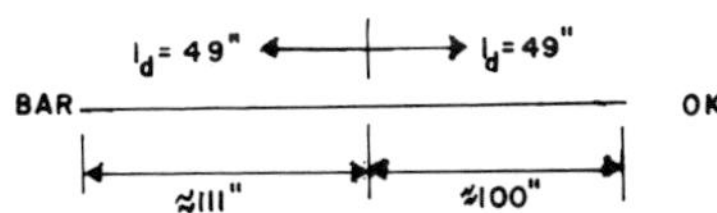

E16.6(a) Development length.

Moment at bottom of footing under the interior column

$$M_u = 3{,}630{,}000 \text{ in lb}$$

Finding reinforcing steel required at section.
Assume $a = 2$ in

$$A_s = \frac{M_u}{\phi f_y \left(d - \frac{a}{2}\right)}$$

$$= \frac{3{,}630{,}000}{(.9)(60{,}000)\left(37.5 - \frac{2}{2}\right)} = 1.8 \text{ in}^2$$

$$\varrho_{MIN} = \frac{200}{f_y} = .0033 \qquad A_s = 9.6 \text{ in}^2$$

$$A_s = \frac{4}{3} A_{sREQ} = \frac{4}{3}(1.8) = 2.4 \text{ in}^2 \text{ (flexure)}$$

Temperature and shrinkage $A_s = .0018bd = 5.3 \text{ in}^2$.

Use $A_s = 5.3 \text{ in}^2$. Thus 12 − #6 spaced 6.5 in. Development length O.K.

Designing steel in short direction

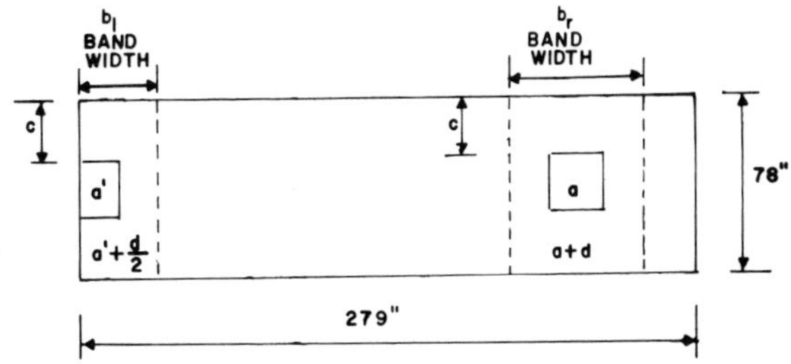

E16.6(b) Designing steel in short direction.

Exterior column

$P_u = 459$ K

$$q'_s = \frac{459}{6.5'} = 70.6 \text{ K/ft}$$

$$c = \frac{6.5 - 2}{2} = 2.25'$$

$$M_u = \frac{1}{2} q'_s c^2$$

$$M_u = \frac{1}{2}(70.6)(2.25)^2 12 = 2{,}144 \text{ in K}$$

Assume $a = 2.0$ in

$$A_s = \frac{M_u}{\phi f_y \left(d - \frac{a}{2}\right)}$$

$$\varrho_{MIN} = \frac{200}{f_y} = .0033$$

$$A_{sMIN} = .0033(d)\left(a' + \frac{d}{2}\right)$$

$$A_{sMIN} = .0033(37.5)\left(18 + \frac{37.5}{2}\right) = 4.55 \text{ in}^2$$

A_{sMIN} also can be $\frac{4}{3} A_{sREQ} = \frac{4}{3}(1.08) = 1.44 \text{ in}^2$

Temperature and shrinkage $A_s = .0018d\, b_\ell$

$$A_s = .0018(37.5)\left(18 + \frac{37.5}{2}\right) = 2.4 \text{ in}^2 \leftarrow \text{Use}$$

Use 8 − #5 bars spaced at $s = 7$ in.

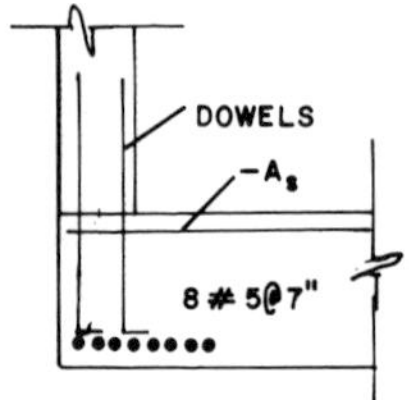

E16.7 Section-left column.

Interior column

$P_u = 690$ K

$$q_s' = \frac{690}{6.5'} = 106.1 \text{ K/ft}$$

$$c = \frac{6.5 - 2}{2} = 2.25'$$

$$M_u = \frac{1}{2} q_s' c^2$$

$$M_u = \frac{1}{2}(106.1)(2.25)^2 12 = 3{,}222 \text{ in K}$$

Assume $a = 2.0$

$$A_s = \frac{M_u}{\phi f_y \left(d - \frac{a}{2}\right)}$$

$$= \frac{3{,}220{,}000}{(.9)(60{,}000)\left(37.5 - \frac{2}{2}\right)} = 1.63 \text{ in}^2$$

$\varrho_{\text{MIN}} = .0033$

$A_{s\text{MIN}} = .0033(d)(a + d)$

$= .0033(37.5)(24 + 37.5) = 7.48 \text{ in}^2$

$A_{s\text{MIN}}$ also can be $\frac{4}{3} A_{s\text{REQ}} = \frac{4}{3}(1.63) = 2.17 \text{ in}^2$

Temperature and shrinkage $A_s = .0018\ db_r$

$A_s = .0018(37.5)(24 + 37.5) = 4.18 \text{ in}^2$ ← Use

Use 14 − #5 bars spaced at $s = 4.4$ in

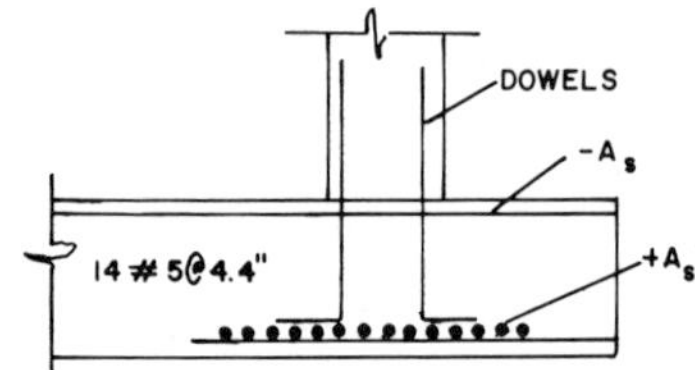

E16.8 Section-right column.

Development length of transverse steel

$$\ell_{db\#5} \begin{cases} .04A_b f_y / \sqrt{f_c'} = .04(.3)(60{,}000)/\sqrt{3000} = 13 \text{ in} \\ .0004 d_b f_y = .0004 \left(\frac{5}{8}\right)(60{,}000) = 15 \text{ in} \leftarrow \text{Use} \\ 12 \text{ in} = 12 \text{ in} \end{cases}$$

$\ell_d = \ell_{db}$ (modifying multipliers) $= \ell_{db}$ (1.0)

$\ell_d = 15$ in

E16.9 Development length.

4. Dowel reinforcing steel needs to be checked and determined as shown in the isolated footing design example.

TORSION

Torsion may arise as a result of primary or secondary actions. The case of primary torsion or equilibrium torsion occurs when the external load has no alternative but to be resisted by torsion. Examples of this type of situation are shown in Figure 42 (eccentric line loading along its span and cantilevers). In statically indeterminate structures, requirements of continuity can also give rise to secondary torsional action. Figure 42 shows typical examples of this situation (edge beams of frames, supporting slabs or secondary beams). In structures, usually flexure, shear, and axial forces are also present [25].

For rectangular plain concrete sections, the ultimate torque can be expressed in terms of the concrete compressive and modulus of rupture strengths [19].

$$T_n = 6(x^2 + 10)y \sqrt[3]{f_c'} \tag{96}$$

and

$$T_n = \frac{x^2 y}{3}(.85 f_r) \tag{97}$$

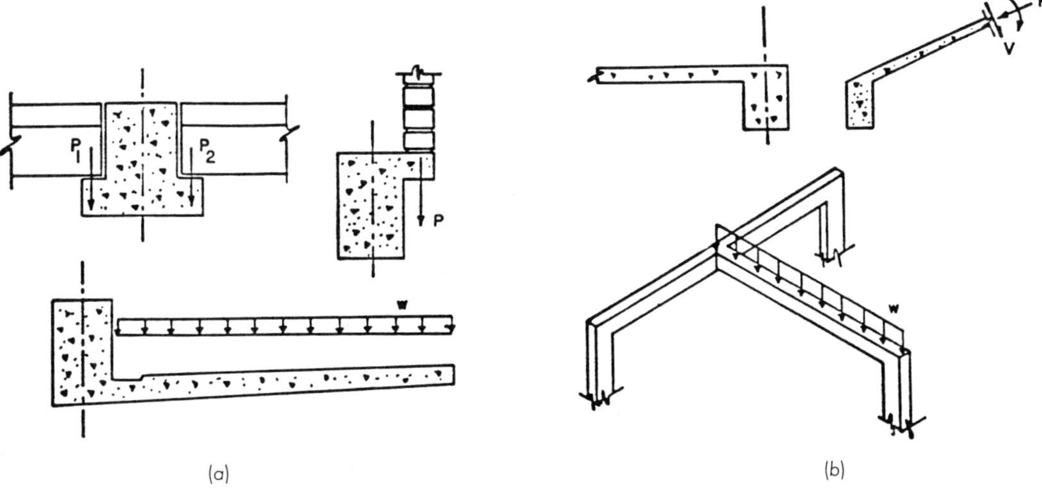

FIGURE 42. Examples of torsion in structures [25]: (a) examples of primary or equilibrium torsion; (b) torsion in statically indeterminate structures.

where x and y are the small and large dimensions of the beam, respectively. If there are T or L sections, the torque strength can be taken as

$$T_n = .8\sqrt{f_c'}\ \Sigma x^2 y \tag{98}$$

according to the ACI Code 83.

If reinforcement (hoop steel) is added then the nominal torque for the section becomes

$$T_n = T_c + T_s \tag{99}$$

where T_c is the part contributed by the concrete and T_s is part contributed by the hoops. The ACI Code 83 is based on the skew-bending theory. The nominal moment becomes

$$T_n = \frac{2.4}{\sqrt{x}} x^2 y \sqrt{f_c'} + \left(.66 + .33 \frac{y_1}{x_1}\right) \frac{x_1 y_1 A_t f_y}{s} \tag{100}$$

where x_1 and y_1 are the shorter and longer center-to-center dimension of the hoop. The behavior of torsion in reinforced concrete beams can also be looked at using the space truss analogy.

In a structure a beam will be subjected to flexure, shear, and axial loads besides torsion. When considering combined bending and torsion, the nominal behavior can be approximated by the interaction curve shown in Figure 43. The mathematical equation for this curve is:

$$\left(\frac{T_n}{T_{no}}\right)^2 + \frac{M_n}{M_{no}} = 1.0 \tag{101}$$

where T_n is the nominal torsional strength in the presence of flexure; T_{no} is the nominal torsional strength acting with torsion alone; M_n is the nominal flexural strength in the presence of torsion; and M_{no} is the nominal flexural strength acting with flexure alone. The strength interaction curve for combined shear and torsion can be represented by a quarter-circle expression

$$\left(\frac{T_n}{T_{no}}\right)^2 + \left(\frac{V_n}{V_{no}}\right)^2 = 1.0 \tag{102}$$

where V_n is the nominal shear strength in the presence of torsion and V_{no} is the nominal shear strength acting with

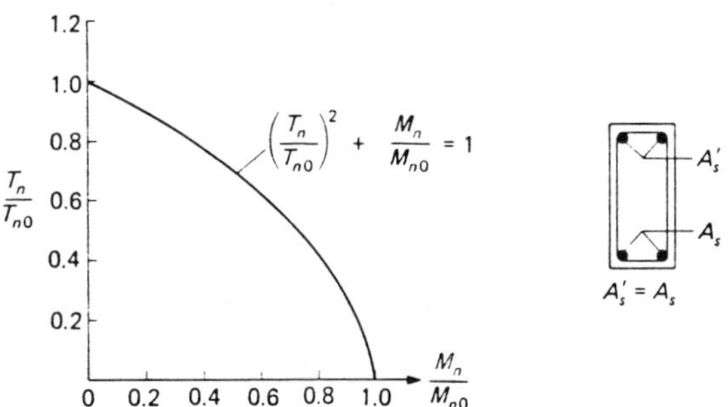

FIGURE 43. Bending-torsion interaction diagram for equal tension and compression longitudinal steel (positive moment yield) [5].

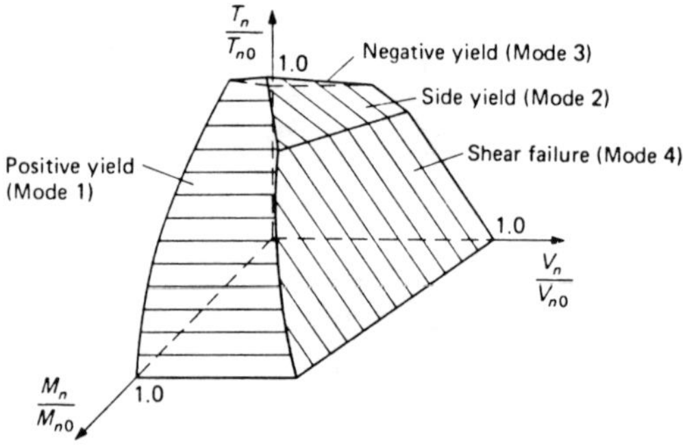

(a) Interaction surface for beam with weaker top steel than bottom steel ($A_s' < A_s$)

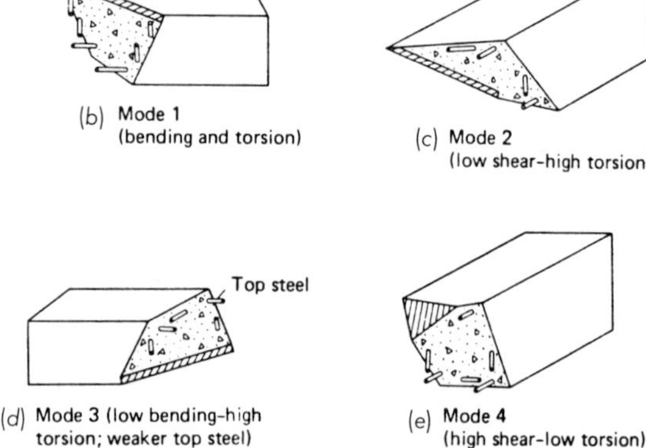

FIGURE 44. Interaction surface and failure modes according to Collins et al. [20] (for members with web reinforcement): (a) interaction surface for beam with weaker top steel than bottom steel ($A_s' < A_s$); (b) mode 1 (bending and torsion); (c) mode 2 (low shear-high torsion); (d) mode 3 (low bending-high torsion; weaker top steel); (e) mode 4 (high shear-low torsion).

shear alone. Strength interaction surface for combined bending, shear, and torsion can be represented by the graph shown in Figure 44 [20,5].

In designing a beam when torsion acts alone, the ACI Code 83 says

$$T_n = T_c + T_s \tag{99}$$

If the torsion on the member is less than

$$T_n \leq .5\sqrt{f_c'}\ \Sigma x^2y \tag{103}$$

then hoops do not have to be used. The torsional strength attributable to the concrete

$$T_c = .8\sqrt{f_c'}\ x^2y \tag{104}$$

for a rectangular section and

$$T_n = .8\sqrt{f_c'}\ \Sigma x^2y \tag{98}$$

for T or L shaped sections. The closed hoop spacing can be found by the following expression

$$s = \frac{A_t\alpha_t x_1 y_1 f_y}{T_n - T_c} \tag{105}$$

where $\alpha_t = .66 + 0.33(y_1/x_1) \leq 1.50$ and A_t is the area of one leg of a closed hoop. The ACI Code requires that the longitudinal steel for torsion is

$$A_\ell = 2\left(\frac{A_t}{s}\right)(x_1 + y_1) \tag{106}$$

The hoop shall not exceed the smaller of $(x_1 + y_1)/4$ or 12 in.

In designing a beam with both shear and torsion, the nominal torsion where T_n can be considered zero is

$$T_n < .5\sqrt{f_c'}\ \Sigma x^2y \tag{103}$$

or

$$T_n = 4\sqrt{f_c'}\ \Sigma\frac{1}{3}x^2y \tag{107}$$

with redistribution. The nominal shear strength of the section for the concrete is

$$V_c = \frac{2\sqrt{f_c'}\ b_w d}{\sqrt{1 + \left(2.5\,C_t\dfrac{T_u}{V_u}\right)^2}} \tag{108}$$

where

$$C_t = \frac{b_w d}{\Sigma x^2 y} \tag{109}$$

The nominal torsional strength of the section for the concrete is

$$T_c = \frac{.8\sqrt{f_c'}\ \Sigma x^2 y}{\sqrt{1 + \left(\dfrac{.4\,V_u}{C_t\,T_u}\right)^2}} \tag{110}$$

When designing the closed hoops, the shear and torsion must both be considered. For the shear, hoop spacing is:

$$\frac{A_v}{s} = \frac{(V_n - V_c)}{f_y d} \tag{111}$$

$$\frac{A_t}{s} = \frac{(T_n - T_c)}{\alpha_t x_1 y_1 f_y} \tag{105}$$

Now the hoop spacing for both is given as

$$\frac{A_{vs}}{s} = \frac{A_v}{s} + \frac{2A_t}{s} \tag{112}$$

The longitudinal steel A_ℓ which is required for torsion will be the greatest value of

$$A_\ell = 2A_t \left(\frac{x_1 + y_1}{s}\right) \tag{106}$$

or

$$A_\ell = \left[\frac{400\, xs}{f_y}\left(\frac{T_u}{T_u + \dfrac{V_u}{3C_t}}\right) - 2A_t\right]\left(\frac{x_1 + y_1}{s}\right) \tag{113}$$

Also

$$2A_t \geq \frac{50 b_w s}{f_y} \tag{114}$$

The minimum requirements for the hoop area are the following

$$\min A_v = \frac{50 b_w s}{f_y} \tag{115}$$

for $T_n < .5\sqrt{f_c'}\Sigma x^2 y$ and $V_n > \frac{1}{2}V_c$

min A_t to satisfy $A_v + 2A_t = \dfrac{50 b_w s}{f_y}$

for $T_n > .5\sqrt{f_c'}\Sigma x^2 y$ and $V_n > \frac{1}{2}V_c$

The minimum requirements for the hoop spacing are the following

$$\min s = \frac{x_1 + y_1}{4}$$

(115)

$$\min s = 12 \text{ in.}$$

Note that $T_s \leq 4T_c$ or the section has to be redesigned [5].

Example 17

Designing the following beam for shear and torsion given the following information.

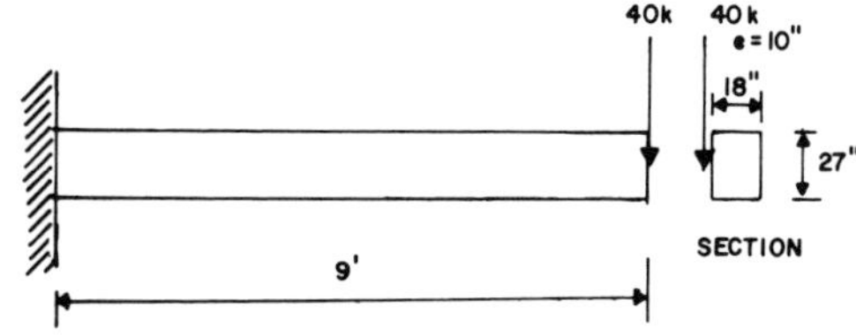

E17.1 Beam and loading.

$f_c' = 3000$ psi
$f_y = 60{,}000$ psi
$B_m w_t = 506$ plf
$d = 27 - 2.75 = 24.25$ in

Critical section will be at a distance d from the face of the column.

$$V_u = 1.7(40{,}000) + 1.4(506)(9 - 2.02) = 72{,}900 \text{ lb}$$

$$V_n = \frac{V_u}{\phi} = \frac{72{,}900}{.85} = 85{,}700 \text{ lb}$$

$$T_u = 1.7(40{,}000)(10) = 680{,}000 \text{ in lb}$$

$$T_n = \frac{T_u}{\phi} = \frac{680{,}000}{.85} = 800{,}000 \text{ in lb}$$

Checking to see if torsion has to be considered

$$.5\sqrt{f_c'}\,\Sigma x^2 y = .5\sqrt{3000}\,(18)(27)$$

$$= 13{,}300 \text{ in lb} < 800{,}000 \text{ in lb}$$

Consider torsion

$$V_c = \frac{2\sqrt{f_c'}\, b_w d}{\sqrt{1 + \left(2.5\, C_t \dfrac{T_u}{V_u}\right)^2}}$$

$$C_t = \frac{b_w d}{\Sigma x^2 y} = \frac{(18)24.25}{(18)^2 27} = .0499$$

$$V_c = \frac{2\sqrt{3000}\,(18)(24.25)}{\sqrt{1 + \left[2.5(.0499)\left(\dfrac{680{,}000}{72{,}900}\right)\right]^2}} = 31.2 \text{ K}$$

$$T_c = \frac{.8\sqrt{f_c'}\,\Sigma x^2 y}{\sqrt{1 + \left(\dfrac{.4\, V_u}{C_t T_u}\right)^2}} \tag{110}$$

$$T_c = \frac{.8\sqrt{3000}\ (18)^2(27)}{\sqrt{1 + \left(\dfrac{.4(72{,}900)}{(.0499)680{,}000}\right)^2}} = 291{,}000 \text{ in lb}$$

$$\frac{A_v}{s} = \frac{(V_n - V_c)}{f_y d} = \frac{(85{,}700 - 31{,}200)}{(60{,}000)24.25} = .0375$$

$$\frac{A_t}{s} = \frac{T_n - T_c}{\alpha_t x_1 y_1 f_y}$$

where $\alpha_t = .66 + .33\ (y_1/x_1) \leq 1.5$

Assume #4 hoops with 1.5 in cover

$x_1 = 18 - 2(1.75) = 14.5$ in
$y_1 = 27 - 2(1.75) = 23.5$ in

Thus,

$$\alpha_t = .66 + .33\left(\frac{23.5}{14.5}\right) = 1.195$$

$$\frac{A_t}{s} = \frac{(800{,}000 - 291{,}000)}{1.195(14.5)(23.5)60{,}000} = .0208$$

Checking $T_s \leq 4T_c$
where $T_s = T_n - T_c$
$(800{,}000 - 291{,}000) \leq 4(291{,}000)$
$509{,}000 \leq 1{,}164{,}000$ O.K.

Spacing

$$\frac{A_v}{s} + \frac{2A_t}{s} = .0375 + (.0208)(2) = .07928$$

$$\frac{A_v}{s} = \frac{2(.2)}{s} = .07928 \text{ Therefore } s = 5.05 \text{ in}$$

Maximum spacing

$$s < \frac{x_1 + y_1}{4} = \frac{14.5 + 23.5}{4} = 9.5 \text{ in}$$

$s < 12$ in

Determining the longitudinal steel

$$A_\ell = 2\frac{A_t}{s}(x_1 + y_1) = 2(.0208)(14.5 + 23.5) = 1.58 \text{ in}^2$$

Also

$$A_\ell = \left[\frac{400\,xs}{f_y}\left(\frac{T_u}{T_u + \dfrac{V_u}{3C_t}}\right) - 2A_t\right]\left(\frac{x_1 + y_1}{s}\right)$$

$$A_\ell = \left[\frac{400(18)(5)}{60{,}000}\left(\frac{680{,}000}{680{,}000 + \dfrac{72{,}900}{3(.0499)}}\right) - 2(.0209)5\right]$$

$$\left(\frac{14.5 + 23.5}{5}\right) = 1.11$$

Use $A_\ell = 1.58$ in^2

$$\frac{1.58}{3} = .53 \text{ in}^2$$

Place .53 in^2 of steel in top, middle, and bottom of beam as shown in Figure E17.2.

Code requires at least #3 bars spacing less than 12 in and one bar each corner. Note flexural steel must be added to top of beam as required by the flexural analysis of the ACI Code 83. Because of beam width three bars are needed across. Look at the section for torsion A_ℓ only the section will be

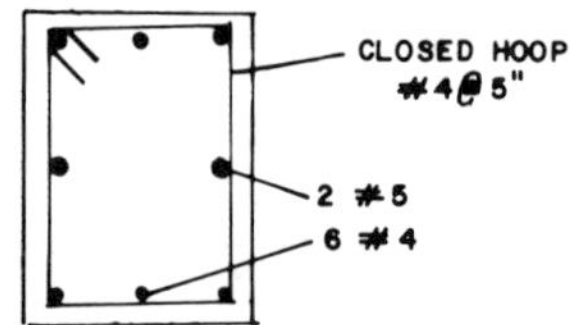

E17.2 Design section.

JOINTS

In a structure, the columns and beams are connected at the joints of the building. This area can be very critical if not designed correctly. The details of design are very important. In this section the beam-column joints in monolithic reinforced concrete will be considered. The material in this area was taken from ACI Committee 352 [21]. The designer should satisfy strength and ductility requirements. Structural joints are classified into two categories: Type 1 and Type 2, based on the loading conditions for the joint and the anticipated deformations of the joint when resisting lateral

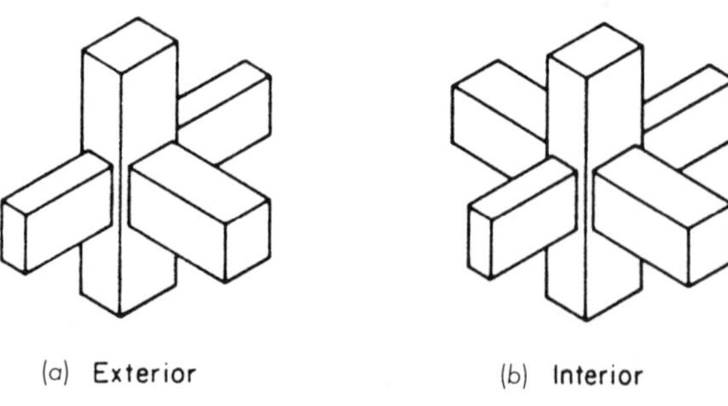

FIGURE 45a. Typical beam-to-column connections [21]: (a) exterior; (b) interior.

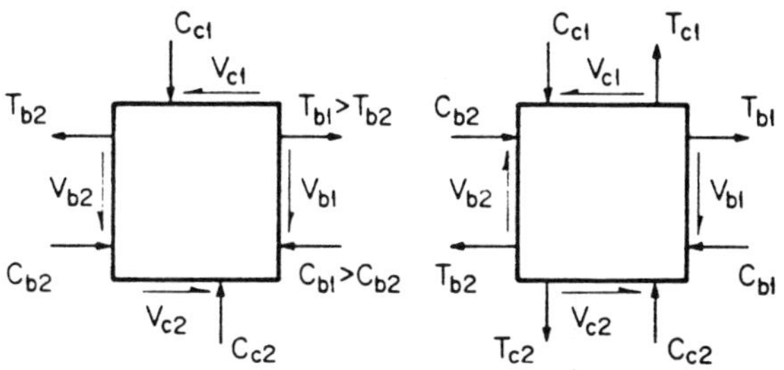

FIGURE 45b. Planar joint forces. T = tension force, C = compression force, V = shear force, subscript b for beam and subscript c for column [21]: (a) due to gravity loads; (b) due to lateral loads.

loads. A Type 1 joint connects members designed to satisfy ACI 318 strength requirements and in which no significant inelastic deformations are anticipated. A Type 2 joint connects members designed to have sustained strength under deformation reversals into the inelastic range [21]. Typical beam-to-column connections are shown in Figure 45 [21]. The forces which are introduced into a typical joint due to gravity and lateral loads are also shown in the figure. The joint should be designed for the critical combination of axial loads, bending moments, torsional moments and shear. Also forces which are generated by creep, shrinkage, temperature, and settlement must be considered. Because of the forces, the reinforcing steel stress at member-joint interface should reflect the increase in stress due to strain hardening caused by excess rotation. The stress is αf_y where

$$\alpha > 1.0 \text{ for Type 1}$$
$$\alpha > 1.25 \text{ for Type 2}$$

The joint should also satisfy the ACI Code 83 requirements for cracking and deflection under service loads [21].

The nominal strength requirements will be briefly described as shown in the committee report from ACI 352 [21]. The longitudinal reinforcement of the columns must pass through the joint adequately in order for the column core to be effectively confined. Transverse reinforcement should be properly designed throughout the joint as required by ACI Code 83 and recommended by committee ACI 352. The primary function of ties in a tied column is to

TABLE 7. Values of γ for Beam-To-Column Joints.

Joint Type	Joint Classification: a Interior	b Exterior	c Corner
1	24	20	15
2	20	15	12

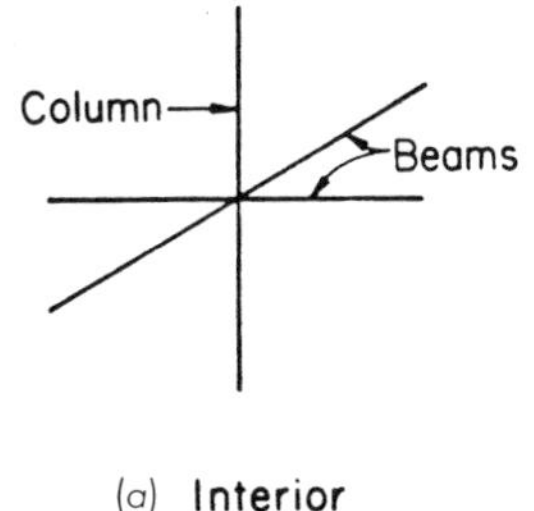

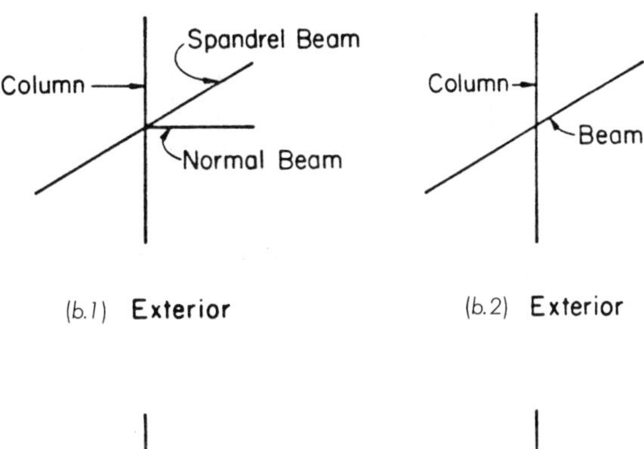

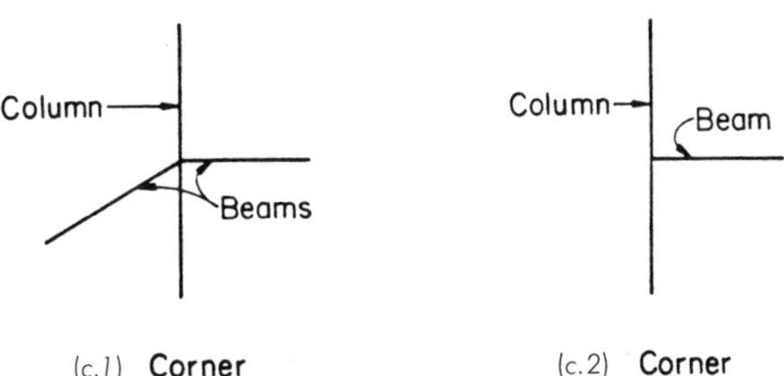

FIGURE 46. Geometric description of joints [21].

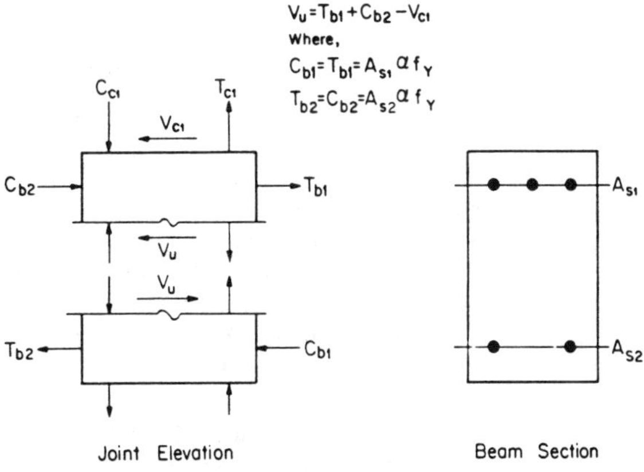

FIGURE 47. Evaluation of horizontal joint shear [21].

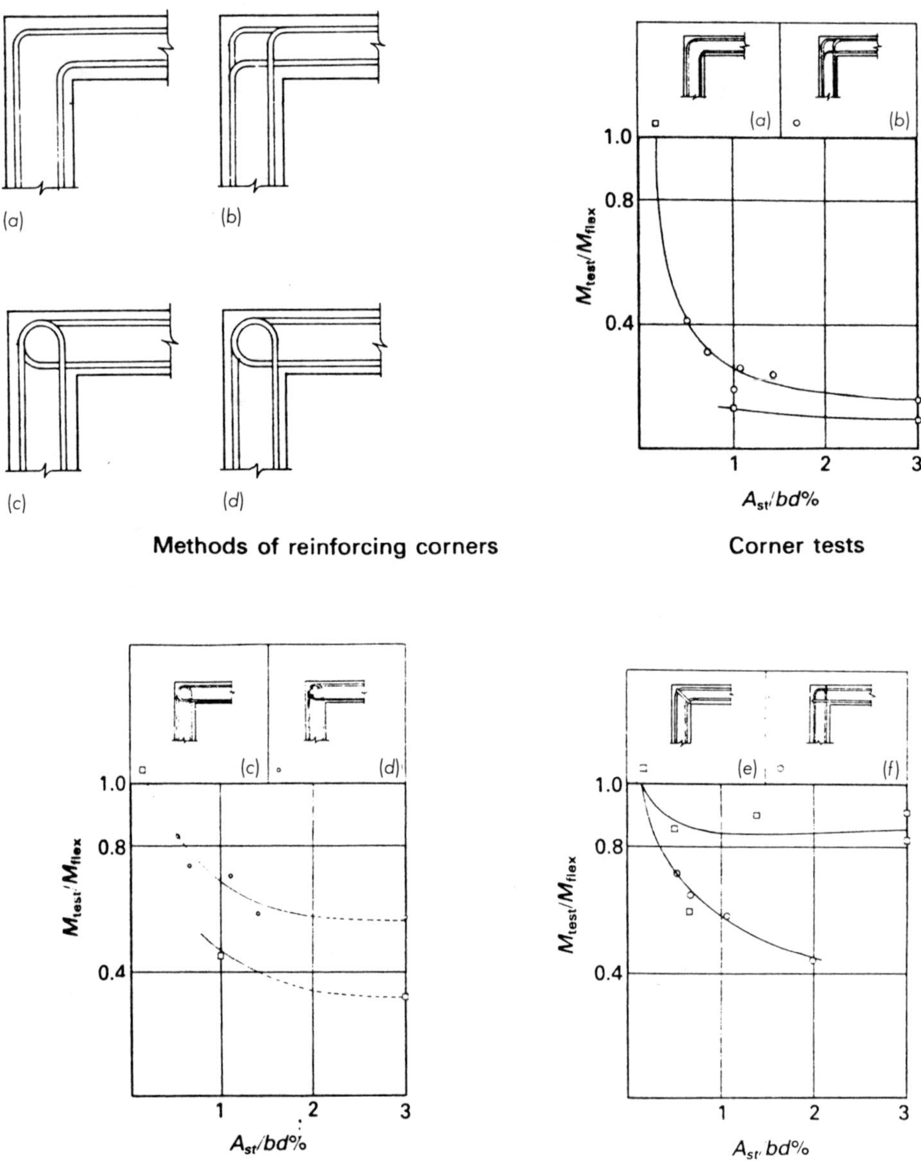

FIGURE 48. Corner joint behavior [22].

prevent the outward buckling of the column longitudinal bars and to provide some confinement to the column core. Transverse reinforcement in the joint can be spiral, single hoop, overlapping hoops, or hoops with crossties. Transverse reinforcement layers which are required in the joint should be extended into the columns to provide confinement of the column core near the joint. Also confinement reinforcement should be provided in the beams adjacent to the column. For joints with beams framing in from two perpendicular directions, the horizontal shear in the joint should be checked independently in each direction. The nominal shear strength of a joint is

$$V_n = \gamma \sqrt{f'_c}\, b_j h \tag{117}$$

where b_j is the effective joint width and h is the thickness of the column in the direction of the load. The values of γ for beam-to-column joints are given in Table 7.

The geometric description of the type of joints is given in Figure 46. The compressive strength is limited to 6000 psi.

The normal procedure for calculating the horizontal design shear in an interior and exterior joint is shown in Figure 47. For high shear forces it is anticipated that the joint is adequately confined as recommended by ACI Code 83 (details for column longitudinal and transverse reinforcement). Vertical shear forces should also be checked in the joint. Flexural design should be carried out according to the ACI Code 83. The design of the ratio of beam to column moments should be done such that flexural hinges are produced in the beams rather than the columns. The details for designing the development length of beam longitudinal reinforcement must follow the ACI Code 83 requirements. There are many provisions and they will not be described. There are reinforcing bars which are terminated and passing through which must be considered in the different joints [21].

Corner joints in a building may be a problem. The detailing of the reinforcement is very important. Opening corners have caused problems in the past. Six common methods of detailing corner reinforcement are shown in Figure 48. The efficiencies of these joints are also shown. For these tests it is evident that proper detailing is critical [22,23].

ACKNOWLEDGEMENTS

The author would like to cite some specific references which were main sources of information for this chapter. These references were: Wang and Salmon [5]; Nawy [3]; Portland Cement Association [11]; and Winter and Nilson [1].

REFERENCES

1. Winter, G. and A. H. Nilson, *Design of Concrete Structures, Ninth Edition*, McGraw-Hill Book Co., p. 647.
2. Huntington, W. C. and R. E. Mickadeit, *Building Construction-Materials and Types of Construction, Fourth Edition*, John Wiley and Sons, p. 768 (1975).
3. Nawy, E. G., "Reinforced Concrete-A Fundamental Approach," Prentice-Hall International Series in Civil Engineering and Engineering Mechanics, Englewood Cliffs, NJ, p. 701.
4. Ree, R. C., "Reinforced-Concrete Design," Section 11, *Structural Engineering Handbook*, E. H. Gaylord, Jr. and C. N. Gaylord, McGraw-Hill, pp. 11-1 – 11-59 (1968).
5. Wang, C. K. and C. G. Salmon, *Reinforced Concrete Design, Third Edition*, Harper & Row, Publishers, p. 918 (1979).
6. Popovics, S., *Concrete-Making-Materials*, Hemisphere Publishing Corp., McGraw-Hill, p. 370 (1979).
7. ACI Committee 213, "A Guide for Structural Lightweight-Aggregate Concrete," *ACI Journal, Proceedings, 64*, 433–469 (August 1967).
8. *Concrete for Radiation Shielding* (Compilation No. 1, 2d ed.), Detroit: American Concrete Institute (1962).
9. ACI-ASCE Committee 318, "Building Code Requirements for Reinforced Concrete," (ACI 318-83), American Concrete Institute, Detroit, MI, p. 111 (1983).
10. *ACI Detailing Manual–1980 (SP-66)*, ACI Committee 315, American Concrete Institute, Detroit (1980).
11. Portland Cement Association, "Notes on ACI-318-83, Building Code Requirements for Reinforced Concrete with Design Applications," PCA, Skokie, IL, Part 1 thru Part 30 (1984).
12. *Recommended Practice for Concrete Formwork* (ACI Std. 347-68), Detroit: American Concrete Institute (1968).
13. Portland Cement Association, "35-mm Slide Set–Building Code Requirements for Reinforced Concrete," Skokie, IL (1983).
14. Cronin, R. C., "The New D.O.T. Concrete Specifications and You," NJ ACI Chapter, NJ D.O.T., and P.C.A., Seminar Notes, Forsgate C.C., NJ (October 6, 1975).
15. Gergely, P. and L. A. Lutz, "Maximum Crack Width in Reinforced Concrete Flexural Members," *Causes, Mechanism, and Control of Cracking in Concrete, SP-20*, Detroit: American Concrete Institute, pp. 87 – 117 (1968).
16. Hassoun, M. N., *Design of Reinforced Concrete Structures*, Prindle, Weber and Schmidt Engineering, Boston, MA, p. 761 (1985).
17. American Concrete Institute, *Deflections of Concrete Structures*, ACI Publication SP-43, p. 637 (1974).
18. Branson, D. E., *Deformation of Concrete Structures*, McGraw-Hill Advanced Book Program, p. 546 (1977).
19. Hsu, T. T. C., "Torsion of Structural Concrete–Plain Concrete Rectangular Sections," *Torsion of Structural Concrete, SP-18*, Detroit, MI: ACI, pp. 203–238 (1968).
20. Collins, M. P., P. F. Walsh, and A. S. Hall, Discussion of "Ultimate Strength of Reinforced Concrete Beams in Combined Torsion and Shear," *ACI Journal, Proceedings, 65*, pp. 786–788 (September 1968).
21. ACI-ASCE Committee 352 Report, "Recommendations for Design of Beam-Column Joints in Monolithic Reinforced Concrete Structures," *ACI Journal, Proceedings, V.82* (3), pp. 266–284 (May–June, 1985).
22. Taylor, H. P. J., "Structural Performance as Influenced by Detailing," Chapter 13, *Handbook of Structural Concrete*, McGraw-Hill, pp. 13-1 – 13-25 (1983).
23. Somerville, G. and H. P. J. Taylor, "The Influence of Reinforcing Detailing on the Strength of Concrete Structures," *Struct. Eng., Vol. 50 (1), pp.* 7– 19 (Jan. 1972).
24. ACI-ASCE Committee 426, "The Shear Strength of Reinforced Concrete Members," *ACI Manual of Concrete Practice 1984,* Part 4–Structural Properties, American Concrete Institute, Detroit, MI, pp. 426-1 to 426-111.
25. Park, R. and T. Paulay, *Reinforced Concrete Structures*, John Wiley and Sons, pp. 1–769 (1975).
26. Concrete Reinforcing Institute, *Manual of Standard Practice, 19th Ed.*, Concrete Reinforcing Institute, Chicago, IL (1968).
27. ACI Committee 439 Report, *Steel Reinforcement Properties and Availability*, Vol. 74, p. 481 (1977).
28. AISC, *Manual of Steel Construction, Eight Edition*, American Institute of Steel Construction, pp. 2–203.

CHAPTER 2

Stress Analysis of Small Precast Concrete Slabs on Grade

YU T. CHOU*

INTRODUCTION

The use of precast concrete structural members is a widely applied, well-established, economical construction technique. Concrete columns, beams, panels, piles, pipes, railroad ties, and other elements for a variety of structures are cast at permanent factories or temporary casting yards, transported to the construction site, and then assembled. Advantages of precasting are as follows: good quality control, economical mass production, rapid construction, reduced congestion at the site, and rapid availability of the structure for use. A study by the U.S. Army Engineer Waterways Experiment Station [1] found that precasting could be applied by U.S. Army Engineers in a theater of operations for construction of structures such as bridges and field fortifications.

Because a concrete pavement consists of a very large number of identical slabs, mass production of precast pavement slabs could be economical. More rapid construction, construction in adverse weather, improved use of materials, and reduced cost are potential benefits of precasting concrete pavements. Basically, small precast concrete slabs are used in rapid emergency repair of road works and in military practice. In the United States the use of precast concrete slabs for roads and airfields has been limited because pavements composed of precast slabs tend to have rough surface and are not suitable for heavy loads traveling at high speeds. However, precast concrete slabs are no doubt the better choice in regions of extreme climate. Precast concrete slabs are used more extensively for roads and for military airfields in countries such as the Soviet Union. Precast concrete slabs have also been used for airfield pavements in China.

Although plain concrete without reinforcing is the most common form of conventional concrete pavement, precast slabs of plain concrete will be relatively thick and difficult to handle because of the low tensile strength of concrete. Consequently, precast concrete slabs are generally reinforced with steel or prestressing. This reinforcing results in a thinner slab, and the reinforcing steel can be positioned in the slab to help carry handling stresses. It is to be noted that reinforcing steel in concrete slabs does not retard the initial crack in the slab, but it does keep the crack closed tightly. The slab continues to perform well in this state, and failure condition for the slab is changed from cracking to spalling along the crack. To increase resistance to spalling of the concrete, steel fibers may be added to increase its flexural and tensile strengths, ductility, toughness, and dynamic strength. The increased dynamic strength of fibers reinforced concrete may also provide additional protection from handling strength.

Lightweight aggregates are sometimes used in precast concrete slabs. The concrete has a compressive strength up to 6,000 psi and a weight of 90-120 pcf. This offers a method of reducing deadweight of slabs for handling but generally at the expense of a reduced modulus of elasticity, lower tensile strength, and lower abrasion resistance.

STATE-OF-THE-ART OF PRECAST CONCRETE PAVEMENTS

Only limited amounts of analytical work can be found in the stress analysis of small concrete slabs on grade in the United States [1-5]. Analytical work is limited mainly because of the lack of good analytical models with the capability of computing stresses in a finite-size slab on grade. Westergaard equations [6,7] were used but were known ap-

*Research Civil Engineer, Pavement Systems Division, Geotechnical Laboratory, U.S. Army Engineer Waterways Experimental Station, Vicksburg, MS

plicable only for a very large slab, and simplifying assumptions were imposed to analyze small slabs. Results of model tests [8] suggest that the slab dimension should be three times the radius of relative stiffness for the slab to ensure compatibility with the assumption of the Westergaard model.

References [9-18] present documents of designs, constructions, and performance of precast concrete pavements in the United States. Considerable work has been documented in Soviet literature [19-30], but the Westergaard solution used by Soviets also requires simplifying assumptions. In the late 1970's a research program studying the potential application of precast concrete slab blocks for airfield pavements was conducted at the U.S. Army Engineer Waterways Experiment Station [31,32]. The finite-element method was used in the stress analysis of small precast concrete slabs on grade [31,33]. Since the finite-element method is applicable to any slab sizes and any loading locations, the results of the analysis are considered more representative than those of Westergaard analysis.

The design, construction, and performance of precast concrete pavements in different countries are briefly presented in the following paragraphs. Details can be found in Reference [31].

Soviet Union

The first concrete airfields in the Soviet Union were constructed in 1931-1932 of precast, unreinforced concrete hexagons 4.1-ft-long sides and 3.9- to 5.5-in. thicknesses. Heavier aircraft introduced after World War II required larger hexagons 4.9 ft long and 5.5 to 8.7 in. thick. The unreinforced hexagons tended to rock and spall; and when modern concrete placing equipment became available after 1950, new pavements were built of rectangular, cast-in-place, reinforced concrete slabs [19,20].

In the Soviet Union where the precast concrete industry is extensively developed, precast prestressed concrete slabs remain an acceptable construction material for air fields. By 1970 precast slabs were considered acceptable for airfields subjected to twin-tandem gear loads of 121 kips and tire pressures of 142 psi and single-wheel gear loads up to 66 kips and tire pressures of 142 psi [19]. Precast slabs were not recommended for pavements subject to the heaviest twin-tandem design gear loads of 154 kips and tire pressures of 142 psi. The use of precast slabs on airfields was reported to be increasing yearly, particularly when nonuniform swelling or settlement was a problem, rapid construction was required, conventional concrete construction was inefficient due to project geometry or size, strengthening of existing pavements was needed, or when construction took place at temperatures below freezing [19,20]. One Soviet source also suggested that precast slabs be an efficient pavement repair material [21].

PRECAST ROAD PAVEMENTS

Precast slabs have been used extensively for roads as well as for airfields in the Soviet Union. By 1962 over 180 miles of temporary roads such as forest roads were constructed with precast slabs, and a 10-year evaluation of a precast road showed favorable results [22]. Further experiences with precast roads in Moscow, under heavy industrial traffic in the Donbass, on the Kiev-Odessa Highway and in other projects were reported in the Soviet technical literature with favorable results [19,23,24,25,26,27, and 28]. It was also stated [19] that precast road construction had not yet proven economical. In contrast to this conclusion, it was reported that an estimated 95,000-120,000 sq yd of precast pavements were placed in Moscow between 1968 and 1974, and that amount was increasing [29].

TYPES OF PRECAST SLABS

Biaxially prestressed slabs are preferred for Soviet precast pavements, particularly for those subjected to heavy aircraft loads because they use material more efficiently than conventional plain and reinforced slabs. Actual size and reinforcing of slabs are dictated by the facilities available at the precasting plant. The designs of the six slabs shown in the top part of Table 1 are approved by the Soviet government for general manufacture [19,20, and 30]. These slabs are 5.5 in. thick, but with minor adjustments at the manufacturing plant, they can vary from 4.7 to 6.3 in. thick. The PAG-IX slab is biaxially prestressed with a longitudinal prestress of 400 psi and a transverse prestress of 300 psi. The remaining slabs are prestressed only in the longitudinal direction. A number of precasting plants have reportedly mastered the manufacture of the PAG-XIV slab [19]. Precast unreinforced slab 3.3 by 3.3 ft, reinforced and unreinforced hexagonal slabs 3.8 by 4.9 ft with 3.8-ft-long sides, and prestressed slabs varying from 5.7 to 9.8 ft wide by 19.7 ft long are used in road construction [19,29].

DESIGN AND CONSTRUCTION

One recommended design procedure for Soviet concrete airfield pavements compares a calculated slab bending moment to an allowable moment for the slab [19]. The design is an iterative procedure to bring the bending moment to within 5 percent of the allowable moment. The bending moment is calculated for a single wheel load multiplied by dynamic and overload factors assuming an elastic plate on a Winkler foundation of independent springs [6,7]. Superposition is used to account for additional wheel loads for multi-wheeled gears. The moments are redistributed to account for uniaxial prestress if needed, and the largest moment is then multiplied by a transfer coefficient to convert it to a bending moment at the edge of the slab. The allowable moment is calculated from the cross-sectional geometry, flexural strength of the concrete, and reinforcing and prestress levels. Load repetition and temperature effects are handled by a variable coefficient, and the flexible strength of the concrete is multiplied by a factor of 0.7 to account for variability in strength [19]. Hexagonal slabs[1] have generally been designed on the assumption of a center load on a circular slab, but an approximate solution for hexagonal slabs with edge loadings has also been presented [29]. The load capacity of precast slabs is varied by changing the foundation strength, since the slab thickness, strength, and reinforcing are already standardized.

Soviet precast slabs have been placed both with and without load transfer devices at the joints between slabs. The PAG-IX slab contains additional reinforcement at the edges and corners of the slab, and there is no load transfer across the joint between slabs. The other PAG slabs have two brackets along the short transverse side of the slab, and brackets between adjacent slabs are welded together. Joints every 59 to 66 ft are left unwelded to allow temperature-induced movements. Other jointing methods reported in use include keyed joints, epoxy-filled joints, and sand and cement grout-filled joints [29]. One Soviet investigator reported that severe rocking of hexagonal slabs occurred under the traffic of a 9,700-lb axle load unless a stabilized base was used [29]. No load transfer devices were used with these slabs.

In the Soviet Union a 1.6- to 2.4-in.-thick layer of sand or sand cement is used as a leveling course for the construction of precast pavement [30]. The slabs are placed with a crane and kept aligned with string lines. The slab is then lifted and the impression is visually checked and corrected for high and low spots. This method may require three to five tries before an acceptable fit is obtained. An alternative method is to vibrate the slabs into place with a large vibrator. Two other less common methods include blowing sand under a suspended slab or mudjacking a slab to obtain the desired elevation. Placement rates are reported to be as high as 950 sq yd per shift per crane. Construction tolerances are a maximum joint width of 0.6 in. and a maximum differential elevation of 0.2 in. between adjacent slabs.

United States

PRESTRESSED CONCRETE MISSILE MAT

In 1956 the Ohio River Division Laboratory, U.S. Army Engineer Division, Ohio River, investigated a precast prestressed sectional mat to prevent erosion and dust from missile firings [9]. This mat was to be capable of being placed by military labor withstanding 90,000-lb thrust of the missile and supporting traffic of missile launchers with wheel loads of 25,000 lb at a 40- to 55-psi tire pressure. The concrete used low-weight sintered shale aggregate and high early-strength portland cement that obtained a 28-day compressive strength of 5,800 psi. The individual beams were 12 in. wide, 18 ft long, 5.5 in. thick, and weighed 550 lb. This was light enough to be handled by a team of eight men. Stirrups were used to reinforce the ribs. Pretensioned prestressed cables along the length of the beam were stressed to 6,000 lb, which provided a prestress in the concrete of 1,200 psi after prestress losses. Both rods and cables were used successfully to posttension the individual beams together transversely and build monolithic mats 18 by 20 ft and 33 by 33 ft.

These mats were subjected to moving wheel loads varying from 5,800 to 24,000 lb. Under traffic some spalling occurred at the edges of the beams and some of the posttensioned rods and cables lost up to 17 percent of their prestress. Strain gages on the surfaces of the beams showed that one wheel load was generally distributed over three beams. Three failures occurred in the thin plank section of the beams under 24,000-lb wheel loads. Although the sectional mat successfully withstood the missile blast tests, further study was not recommended because of the weight of the mat and assembly times. Further work with this concept was recommended for temporary roads or storage areas, but it was not pursued.

PRESTRESSED HIGHWAY, BROOKINGS, SOUTH DAKOTA

In 1968 the South Dakota Department of Highways and the Federal Highway Administration built a 24-ft-wide, 900-ft-long test section of precast prestressed concrete slabs on U.S. Highway 14 near Brookings, South Dakota [10]. The pavement design was based on research sponsored by the South Dakota State Department of Highways and Federal Highway Administration and conducted at the South Dakota State University [2,3,11,12]. The final slab design used in construction was 6 ft wide, 24 ft long, and 4-1/2 in. thick. The 3/8-in.-diam. longitudinal cables were prestressed to provide 400-psi prestress in the slab. A grout key was used on the longitudinal sides to provide load transfer between adjacent slabs. An optional connection joint used on approximately half of the slabs was formed by widening the grout key at the slab longitudinal one-quarter and three-quarter points and welding protruding No. 3 reinforcing bars together. The concrete slabs were overlaid with asphaltic concrete, the depth of which varied from 3-1/2 in. at the road center to 1-1/2 in. at the edge to provide the required surface slope and smoothness.

The slabs were lifted by crane, placed on a 1/2-in.-thick sand bedding layer, and seated by a vibratory roller. Half of the slabs was placed with the long side of the slab parallel to the direction of traffic using the optional connection joints. The remaining slabs were placed with the long side perpendicular to the direction of traffic without connection joints. The South Dakota project was used as a basis or ex-

[1]It should be noted that the methodology used by the Soviets analyzing stress conditions in circular, rectangular, and hexagonal concrete slabs involves simplifying assumptions which could result in erroneous results. The finite-element method presented in the later part of this paper can overcome the difficulties.

ample for recommending precast construction for strengthening airport pavements [13] and urban pavement construction [14].

PRECAST CONCRETE PAVEMENT REPAIR SLABS

The Michigan Highway Department developed a technique of pavement repair using precast concrete slabs [15,16]. Eight standard designs were developed for slabs 12 ft long, varying in 2-ft increments from 6 to 12 ft wide, and either 8 or 9 in. thick [16]. Double layers of No. 3 bars at 1-ft 6-in. spacing provided reinforcing. The weight of the slabs varied from 3.7 to 8.1 tons.

In this technique, the damaged pavement is first cut out and removed. A cement-slurry mortar is placed directly on the subbase to the final elevation of the bottom of the repair slab, and then the slab is lowered into place. Transverse joints over 1/4 in. wide are sealed with bituminous filler strips, and longitudinal joints are grouted to within 2 in. of the surface. Final joint sealing with a hot-poured rubber-asphalt sealant is delayed until the slab has been open to traffic to ensure no tilting or misalignment has occurred. Reported repair times varied from 2-1/2 to 8 hr [15]. Two alternate joint designs using dowels inserted in the existing pavement and then welded to plates cast in the repair slab and epoxy joint sealants were also tested [15].

Pavement repairs on I-29 in South Dakota, on I-95 in Virginia, and on several projects in Great Britain used partial-depth precast slabs [16,17]. In this technique, which is used primarily for localized surface damages, a Klarcrete concrete cutting machine [17] cuts a rectangular hole up to a maximum size of 1 ft 6 in. by 2 ft by 4 in. deep. A thin precast slab is then placed in the hole with an epoxy grout.

The San Diego Unified Port District replaced 116 damaged concrete slabs at San Diego's Lindbergh Field with precast reinforced concrete slabs [18]. The precast slabs were cast to match the existing slabs. Damaged slabs were removed; 6 in. of subgrade was excavated; and lean concrete was placed to the desired elevation of the bottom of the precast repair slab. The precast slab was lowered into place and seated with a 10-ton roller. Patented load transfer devices were used at the slab joints. Precasting the repair slabs allowed the airport to stay in operation while the concrete cured. The airport pavement was strengthened by an 8-in.-thick asphaltic concrete overlay after all repairs to the existing airport pavement were complete.

Europe

Between 1947 and 1958 precast prestressed concrete slabs were used for airfield construction in Europe on several occasions. Orly Airport, Paris, France, was the first airfield application of prestressed pavement, which consisted of 3.3-ft-square and 6.3-in.-thick precast slabs [34,35]. These square slabs were posttensioned together into a unique triangular arrangement with sliding joints. A similar design was adopted for an installation in London in 1949 [36]. During load tests conducted at Orly Airport, a 224-kip load on a 32-in.-diam plate caused a 0.41-in. deflection for interior loading and a 0.45-in. deflection with the load adjacent to the sliding joint. Although the precast Orly pavement was structurally adequate, the surface was notably rough, and construction was costly.

In 1956 a 200- by 200-ft section of airport pavement was constructed in Finningley, England, of 30- by 9-ft by 6-in.-thick precast prestressed slabs posttensioned in place [34]. In 1958 a 75- by 1,148-ft taxiway was constructed at Melsbroek, Brussels, with 4.1- by 39-ft by 3-in.-thick prestressed precast slabs [34,37]. These slabs which were in shape of parallelograms were pretensioned. The joints were caulked with mortar, and then the slabs were posttensioned with transverse cables.

Container terminals require pavements with high load capacity and good durability. Often these terminals are built on fill areas and are subject to large subgrade settlements. Precast concrete slabs provided the required strength for large concentrated loads, durability, and a flexible structure that can be releveled after settlement. For these reasons they have been used for several container terminals in Europe [38]. Steel plates have been used on the edges of these precast slabs in the Netherlands and United Kingdom to prevent spalling. Generally this method performs well, but it is costly. A project in Hamburg used slabs 6.6 to 8.2 ft square and 5.5 in. thick and reinforced with 0.3 to 0.5 percent steel. These slabs were chamfered to prevent spalling and have given good performance for 8 years [38]. The precast slabs in container terminals could have elevation variations of 0.2 in. between adjacent slabs over 10 percent of a new pavement. Generally these slabs have been economical only when large settlements were a problem.

Japan

Six experimental pretensioned precast concrete slabs were constructed and tested in Japan [39]. These slabs were designed for DC-8 aircraft and were 7.5 ft wide, 32.8 ft long, and 7.9 in. thick. The slabs were pretensioned to 412 psi at the casting plant. The slabs were placed on an 0.8-in. leveling course of cement-stabilized sand covered with vinyl sheets.

A unique "horn joint" was developed for this precast application. A 1.5-in.-diam steel bar approximately 27.9 in. long was bent into an arc shape with a radius of 51.2 in. Arc-shaped plastic tubes were cast in the edges of the slabs at 15.8-in. spacings. This allowed the steel bars to be inserted and grouted to provide load transfer across the joint.

Summary

Table 1 summarizes the variety of precast pavement dimensions, types, and usages published in the available

TABLE 1. Summary of Precast Concrete Pavements.

Slab	Dimensions ft by ft by in. Thickness	Length/Width Ratio	Type	Use
		Union of Soviet Socialist Republics		
PAG-III	6.6 × 13.1 × 5.5	2.0	Prestressed	Airfield
PAG-IV	6.6 × 13.1 × 5.5	2.0	Prestressed	Airfield
PAG-IX	10.5 × 19.7 × 5.5	1.9	Prestressed	Airfield
PAG-XIV	6.6 × 19.7 × 5.5	3.0	Prestressed	Airfield
PAG-XV	6.6 × 19.7 × 5.5	3.0	Prestressed	Airfield
PAG-XV-1	6.4 × 19.2 × 5.5	3.0	Prestressed	Airfield
Moscow (Mednikov et al. 1974)	Hexagon 3.8-ft side, 7.1-in. thickness	—	Plain and reinforced	Road
(Glushkow and Rayev-Bogoslovskii 1970)	3.3 ft × 3.3 ft 3.8 ft × 4.9 ft	1.0–1.29	Plain	Road
(Glushkov and Rayev-Bogoslovskii 1970)	5.7 to 9.8 ft × 19.7 ft	2.0–3.5	Prestressed	Road
(Stepuro et al. 1964)	6.6 × 19.7 × 5.5	3.0	Prestressed	Road
(Smulka 1963)	6.6 ft × 19.5 ft	3.0	Prestressed	Road and airfield
(Dubrovin et al. 1962)	9.9 ft × 19.8 ft	2.0	Reinforced	Road
(Birger and Klopovskii 1961)	8.2 × 11.6 × 6.4	1.4	Reinforced	Road
(Mikhovich et al. 1961)	4.9 × 5.7 × 6.7	1.2	Reinforced	Road
(Glushkov and Rayev-Bogoslovskii 1970)	Hexagon 4.9-ft side, 5.5–8.7-in. thick	—	Plain	Airfield (1940's)
(Glushkov and Rayev-Bogoslovskii 1970)	Hexagon 4.1-ft side, 3.9–5.5-in. thick	—	Plain	Airfield (1930's)
		United States		
(Mellinger 1956)	1 × 18 × 5.5	18	Prestressed	Missile
Michigan (Transportation Research Board 1974)	6 to 12 × 12 × 8 or 9	1.0–2.0	Reinforced	Road
Brookings, S. D. (Larson and Hang 1972)	6 × 24 × 4.5	4.0	Prestressed	Road (1968)
		Europe		
Hamburg (Patterson 1976)	6.6 × 6.6 × 5.5 8.2 × 8.2 × 5.5	1.0	Reinforced	Port (1968)
Melsbroek (Vandepitte 1961)	4.1 × 39 × 3	9.5	Prestressed	Airfield (1958)
Finningley (Hanna 1976)	9 × 30 × 6	3.3	Prestressed	Airfield (1956)
London (Stott 1955)	2.9 × 2.9 × 6.5	1.0	Prestressed	Airfield (1949)
Orly (Harris 1956)	3.3 × 3.3 × 6.3	1.0	Prestressed	Airfield (1947)
		Japan		
(Sato, Fukute, Inukai 1981)	7.5 × 32.8 × 7.9	4.4	Prestressed	Airfield

technical literature. Most of the slabs, with several exceptions such as the Michigan or Hamburg slabs, have length to width ratios of 2 or greater and will probably behave more as one-way slabs. Slabs have been relatively thin and have made extensive use of prestressing.

Most of the sources recommended use of precast slabs for special problems rather than as a direct competitor to less costly cast-in-place concrete. Typical reasons suggested for precasting have included aggregate shortage, future pavement settlement or heaving, critical speed of construction, or construction in freezing temperatures. Major problems with precast pavements have been high costs and roughness.

RESPONSES OF LARGER CONCRETE SLABS

Various types of load-transfer devices were used in cast-in-place and precast concrete pavements to transfer the wheel load from one slab to the adjacent slab to reduce stresses in the slabs and to reduce the differential settlements between two adjacent slabs. The latter would reduce the roughness of the pavement and thus improve its riding quality. Analysis using finite element method indicated that the response of the smaller concrete slabs to joint efficiency (the capability of transferring the wheel load from one slab to the adjacent one along the joint) is different from the response of larger slabs.

The finite-element computer programs used in the analysis were originally developed by Huang [40,41] and modified and extended by personnel at the U.S. Army Engineer Waterways Experiment Station [42]. The program has the capability of analyzing stress conditions in a rigid pavement of any size that contains cracks and joints subjected to loads and temperature warping, and the program is able to analyze slabs made up of two layers of materials with different engineering properties. The slabs can have full or partial loss of subgrade support over designated regions. The natural subgrade soil is simulated either by a series of springs, i.e., a Winkler foundation, or an elastic layered

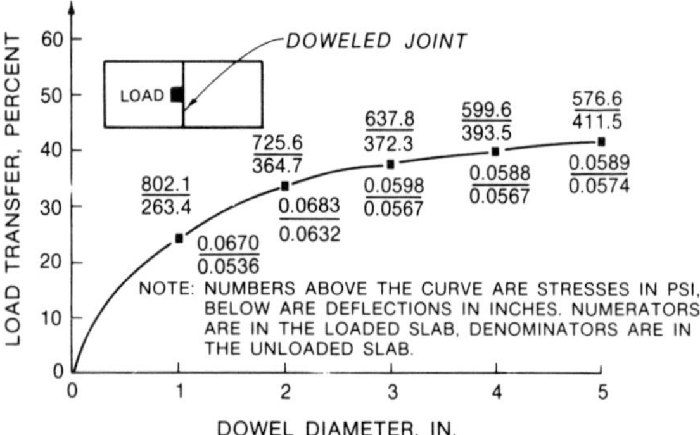

FIGURE 1. Relationship between percent load transfer and dowel diameter, load placed at joint, large slabs.

foundation. Variable slab thickness and modulus of subgrade reaction k are incorporated in the program. Any number of slabs arranged in an arbitrary pattern can be handled in the program. Multiple-wheel loads can also be analyzed, and the number of wheels is not limited. The computer programs are called WESLIQUID and WESLAYER.

The load transferred across a joint in a concrete pavement is accomplished through the transfer of shear forces and moment across the joint. The shear forces are mainly transferred by dowel bars and coarse-aggregate interlock. The amount of moment transfer across a joint or a crack is generally negligible. It is difficult to evaluate the amount of shear force transferred by the coarse-aggregate interlock, but the amount of shear force transferred by the dowel bars can be evaluated using the WESLIQUID program. The slab used in the computation was 25 by 25 ft and 12 in. thick; the modulus of subgrade reaction k was 75 pci. The computation was made for the pavement under the B-47 aircraft loads, which had a dual-wheel-bicycle type landing gear with each dual wheel weighing 94,500 lb. The computation was conducted with the dual wheels placed perpendicular to the joint in which various sizes of dowel bars spaced at 15 in. apart were assumed. The amount of shear forces transferred by the dowel bars across the joint was investigated. In the computation, zero-percent moment transfer across the joint was assumed becaused dowel joints cannot transfer moments.

Computed percentages of load transfer for different dowel-bar diameters are shown in Figure 1. The magnitudes of the maximum tensile stresses and deflections at both sides of the joint for dowel-bar diameters of 1, 2, 3, 4, and 5 in. are presented. The values shown in the numerator and the denominator are those at the loaded and unloaded sides, respectively. The load transfer is defined as the ratio of the stress on the unloaded side of the joint to total stress (the sum of the stresses on both loaded and unloaded sides) as a percentage. At a 50 percent load transfer, the stress at the loaded side is equal to that at the unloaded side. Figure 1 shows that when the load is placed at the joint, the maximum tensile stresses in the loaded slab can be greatly reduced when the efficiency of load transfer across the joint is improved.

Similar computations were made for the dual wheels placed at the center of one of the two slabs. The maximum tensile stresses occurred at the center of the slab and were close to 580 psi for all five dowel bar sizes used in the computations. In other words, when the load is applied at the center of one slab, the maximum stress is not affected by the joint efficiency between adjacent slabs.

RESPONSES OF SMALLER CONCRETE SLABS

To study the response of small concrete slabs to load and to joint efficiency, computations were conducted for three

TABLE 2. Input Data for Analysis of Small Concrete Slabs on Grade.*

System (1)	Slab Size in Feet (2)	Slab Thickness in Inches (3)	Gear Configuration (4)	Joint Condition (5)	Wheel Load in Pounds (6)
Single-slab	3 by 3	3, 6, 12	Single-wheel	NA	30,375
	6 by 6				
	9 by 9				
	12 by 12				
	15 by 15				
	20 by 20				
	25 by 25				
Two-slab	3 by 3	6		Variable shear Transfer, zero Moment transfer	
	6 by 6				
	9 by 9				
	12 by 12				
Nine-slab	6 by 6				
	6 by 6		C141-TT 32.5 by 44 in.		36,000
	9 by 9		C141-TT 32.5 by 44 in.		36,000

*Concrete Modulus E = 4,000,000 psi, Poisson's Ratio $\nu = 0.15$, Tire Contact Area = 400 sq in., Subgrade Modulus k = 100 pci.

different slab systems with varying slab thicknesses and sizes. The three slab systems were single-slab, two-slab, and nine-slab systems; the slab side length (square slab) varied from 3 to 25 ft. Single-wheel loads were used in most analyses, but a C-141 twin-tandem (TT) load was also used in the analysis of the nine-slab system to determine the effects of the interaction of multiple wheels on pavement responses. The input information in the computation is presented in Table 2.

Single-Slab System

Computations were made to find the effects of slab size on the response of smaller slabs to loads. The computed relationships between both maximum stress and the slab size and maximum deflections and slab size are presented in Figure 2. It can be observed that for the considered single-wheel load the smaller the slab, the smaller are the stresses in the slab and the greater the slab deflections. For a 3- by 3-ft concrete slab, a very small area of subgrade support exists, and the slab consequently sinks into the subgrade with very small bending in the slab. Small differences in deflections at the slab's center and at the slab's edge, 0.23653 and 0.23513 in., respectively, are evidence of this behavior; smaller bending ensures smaller stresses in the slab. Explaining it in another manner, for a 3- by 3-ft slab the bending action of the action of the slab is not fully developed, and the wheel load is mainly supported by the subgrade soil.

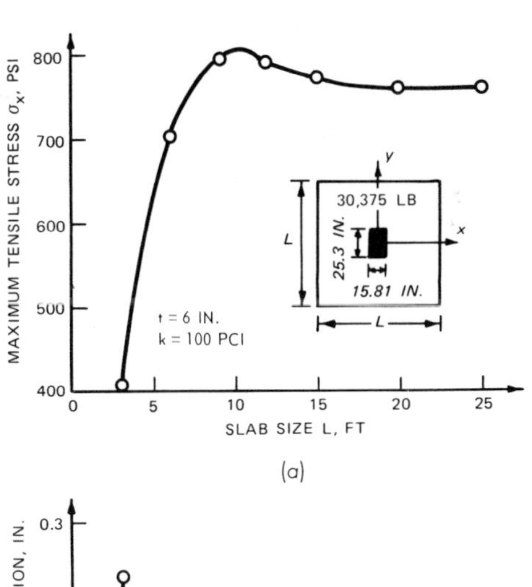

(a)

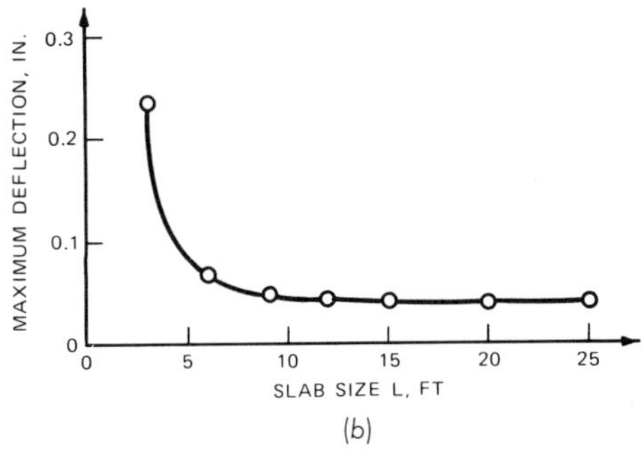

(b)

FIGURE 2. Relationship between maximum tensile stresses and deflections and slab sizes, single-wheel load.

When the slab size is increased, not only the bending action of the slab starts to develop but also the slab is supported by a larger area of subgrade soil. Consequently, the stresses in the slab are increased, and the deflections are decreased. The results presented in Figure 2 show that when the slab size is increased from 3 to about 10 ft, the slab stresses are increased drastically, but the deflections are also decreased drastically. The larger bending in the slab is due to larger differential deflections occurring in the slab. For instance, for a 9- by 9-ft slab, the deflections at the slab's center and at the slab's edge are 0.04946 and 0.01894 in., respectively. The maximum stress is 795.3 psi in this case as compared with the maximum stress of 402.2 psi in the 3-by 3-ft slab previously described (the deflections at the slab's center and at the slab's edge are 0.23653 and 0.23513 in., respectively). When the slab size continues to increase beyond 10 ft, the changes of stresses and deflections become insignificant.

Figure 2 shows that for a given slab rigidity, i.e., thickness $t = 6$ in. and subgrade $k = 100$ pci in this case, there is a particular slab size beyond which the stresses in the slab no longer increase, and deflections no longer decrease with increasing slab size. Below this slab size the stresses decrease and the deflections increase as the slab size is reduced. Computations were also made for other slab thicknesses to check the effect of slab rigidity on the response of smaller slabs. As shown in Figure 3, as the slab rigidity is increased, i.e., the slab thickness is increased, the particular slab size that separates the two distinct behaviors also increases. Figure 3 shows that for slab thicknesses of 3, 6, and 12 in., the particular slab sizes are approximately 6, 10, and 15 ft, respectively. The computations presented in Figure 3 are based on a constant subgrade k value of 100 pci. For a given slab thickness, an increase or decrease of subgrade k value would mean an increase or decrease of the slab relative stiffness. Consequently, an increase or decrease of subgrade k values would increase or decrease the particular slab size.

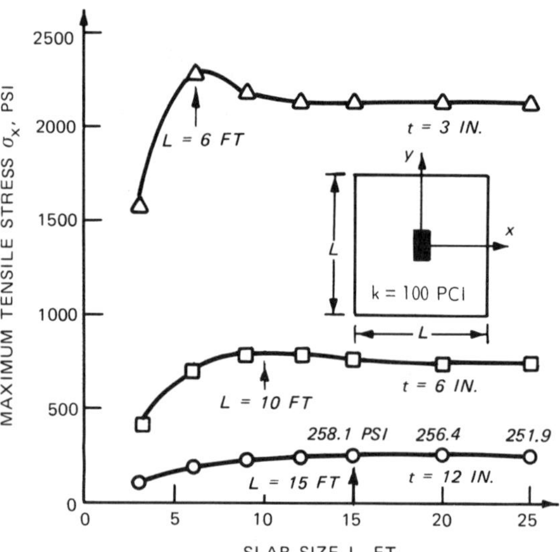

FIGURE 3. Relationships between maximum tensile stresses and deflections and slab size for different slab thicknesses, single-wheel load.

Two-Slab System

The results presented in Figures 2 and 3 are for single slabs of various sizes. Computations were also made for a two-slab system similar to those shown in Figure 1 for a large slab. The same wheel load used in Figure 2 was placed separately at the center of one slab and at the edge of the slab next to the joint. Dowel bars were not used, but various percent shear-transfer values were assumed along the joint. The percent shear transfer is defined as the ratio, as a percentage, of vertical deflections along the joint between the unloaded, or less heavily loaded, slab and the adjacent more heavily loaded slab. One hundred percent shear transfer means the deflections along the joint of the two slabs are equal. Zero moment transfer across the joint is assumed because no moment transfer can take place between two adjacent precast slabs in an airfield pavement. The slabs are also assumed to be square and to have side lengths varying from 3 to 12 ft. The relationships between the maximum stress and the percent shear transfer across the joint are plotted in Figure 4(a). It is to be noted that the definitions of "load transfer" and "shear transfer" are different. The term "load transfer" (or "shear transfer") has been used for years by the U.S. Army Corps of Engineers, and the term "shear transfer" was first used by the Bureau of Public Roads (now the Federal Highway Administration) and is used in the WESLIQUID computer program as a selected option to specify the joint efficiency. The term "shear transfer" may also be called "deflection load transfer" or "deflection transfer."

For large-concrete slabs (25 by 25 ft) used in conventional highway and airfield pavements (Figure 1) when the wheel loads are placed at the center of one slab which is connected to another slab by dowel bars of varying sizes, the maximum stress under the load is not affected by the dowel bar size. However, this is not true for smaller slabs. Figure 4(a) shows that when the load is placed at the center of one slab, the stresses under the load increase when the joint efficiency connecting the two slabs is improved. This effect is more predominant for smaller slabs. This unconventional phenomenon may be explained by the information discussed earlier in Figure 2. Figure 2 shows that when the slab is small, stresses in the slab are increased when slab size is increased. When two small slabs are connected by load-transfer devices, two small slabs are essentially structurally combined into a larger slab. When the slab size becomes greater, the stresses in the slab are consequently increased.

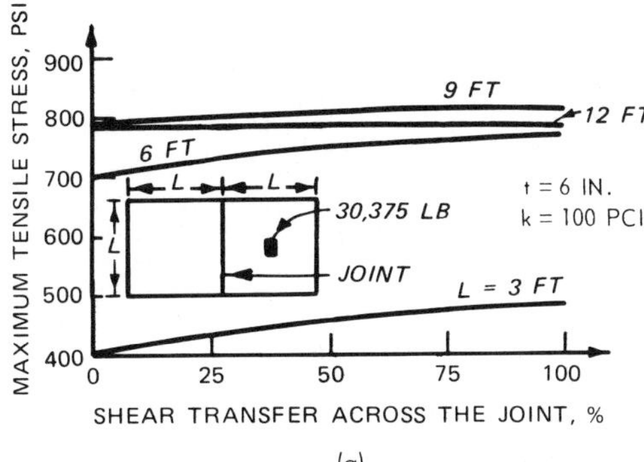

(a)

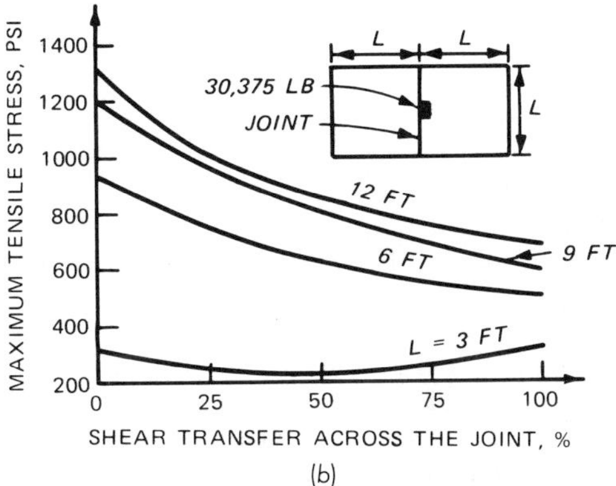

(b)

FIGURE 4. Relationships between maximum tensile stresses (at loaded slab) and percent shear transfer across the joint for various slab sizes.

This is exactly what is shown in Figure 4(a) where the maximum stress increases with the improving joint efficiency of the load-transfer device connecting the two slabs. Figure 4(a) also shows that when the slab size is increased, the stress increases with improving joint efficiency, but the rate of increase decreases with increasing slab size. Figure 4(a) further shows that when the slab size is increased to 12 ft, the maximum stress in the slab is not influenced by the shear transfer at the joint. While the effect of shear transfer across the joint can make two slabs joined by a load-transfer device behave as a larger slab, the 12- by 12-ft slab is already sufficiently large that any further increase in size would not significantly affect the slab's structural behavior.

Figure 4(b) shows the results when the load is placed next to the joint. Similar to the results presented in Figure 1 for a large slab (25 by 25 ft), the stresses in the slab are reduced as the joint efficiency is improved. This is because a part of the load is carried by the other slab through the joint. However, the relationship is not very clear for the smallest slab with an $L = 3$ ft. For such a small slab while the wheel load is placed next to the joint, part of the load is actually near the slab's center. The mixed action of the center load and the edge load may be responsible for the increase in stress when the efficiency of shear transfer is greater than 50 percent.

The results presented in Figures 1 through 4 indicate that the installation of dowel bars and keyed joints would greatly reduce the critical stresses in conventional highway and airfield concrete pavements; however, this is true in small concrete slabs only when the load is placed next to the joint. When the load is placed at the slab's center, the stresses can actually be increased (although not greatly) with increasing joint efficiency. This phenomenon is not critical in the design of small precast concrete slabs because the maximum stress induced by the load placed next to the joint is always greater than the maximum stress induced when the load is placed at the slab's center; however, engineers should be aware of the fact that stresses in small concrete slabs can actually be increased with increasing joint efficiency in some loading cases.

Nine-Slab System

The results presented in Figure 4 are for a two-slab system. The same conclusions are also applicable to precast airfield pavement consisting of multiple slabs. Computations were made for a nine-slab concrete pavement system. Each slab was assumed to be 6 by 6 ft. The slabs are connected by joints with varying shear transfer capability; zero moment transfer was again assumed. Figures 5(a) and 5(b)

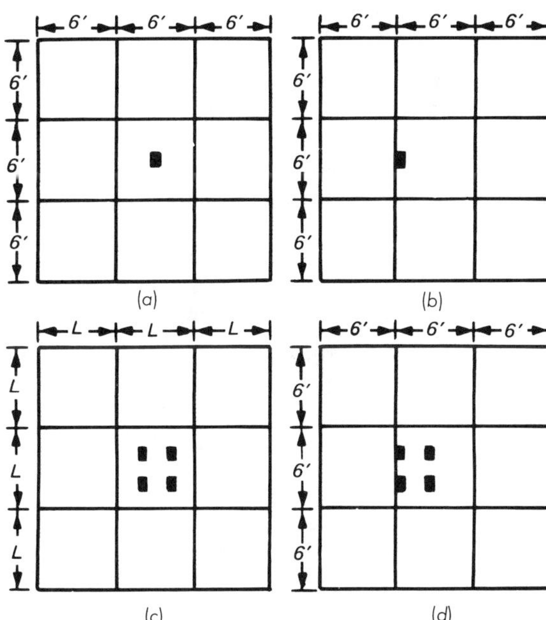

FIGURE 5. Loading position for nine-slab precast airfield pavement system.

TABLE 3. Stresses and Deflections Under Single Wheel Load, 6- by 6-ft Slabs.

	Center Load (Fig. 5(a))		Load Next to Joint (Fig. 5(b))	
Percent Shear Transfer (1)	σ_{max} in psi (2)	W_{max} in in. (3)	σ_{max} in psi (4)	W_{max} in in. (5)
0	699.6	0.06994	940.2	0.21303
25	730.7	0.05973	881.1	0.14000
50	753.4	0.05213	804.2	0.10255
75	768.5	0.04568	731.7	0.08000
100	778.4	0.04244	690.4	0.06506

show the loading positions of the single-wheel load which was placed at the center of the center slab and at the edge of the center slab next to the joint, respectively. The same wheel load used in the two-slab system was used. The maximum stresses and deflections computed at five different values of shear transfer are tabulated in Table 3. Results demonstrated that when the load is placed at the center of the 6- by 6-ft slab, the maximum stress in the slab under the load is increased with increasing efficiency of the joints that connect the slabs together accompanied by a decrease in the deflection. The improvement of the joint efficiency apparently serves the same purpose of increasing the slab size. Since a 6- by 6-ft slab is considered to be a small slab (as shown in Figure 2), the stresses are increased and the deflections are decreased with increasing slab size. When the wheel load is placed at the slab's edge next to the joint (Figure 5(b)), the stress in the loaded slab is greatly reduced when the joint efficiency is increased because the wheel load is partially carried by the adjacent slabs through the joints (Table 3). Similar to the stresses, the deflection of the slab is also greatly reduced when the joint efficiency is increased. This is reasonable because the deflection in a larger slab is smaller (see Figure 2(b)).

TABLE 4. Stresses and Deflections Under Twin-Tandem Load.*

	Slab Size L = 6 ft		Slab Size L = 9 ft	
Percent Shear Transfer (1)	σ_{max} in psi (2)	W_{max} in in. (3)	σ_{max} in psi (4)	W_{max} in in. (5)
0	968.2	0.30495	1,285.3	0.16630
25	965.4	0.23823	1,302.0	0.15646
50	1,038.1	0.18859	1,315.5	0.14904
75	1,070.1	0.16250	1,325.1	0.14331
100	1,086.5	0.13600	1,331.8	0.13878

*Loads placed at center of center slab (Fig. 5(c)).

Results presented in Table 3 are for a single-wheel load. Similar computations were made for a C-141 TT load (32.5 by 44 in.). The center of the TT load was placed at the center of the center slab of the nine-slab pavement system. The loading position is shown in Figure 5(c). Two slab side length, $L = 6$ ft and $L = 9$ ft, were considered, and the results are presented in Table 4. When the TT load is placed at the slab's center similar to the results presented in Table 3 for the single-wheel load, the stresses are increased, and the deflections are decreased with increasing joint efficiency. However, the rate of change is smaller for the 9- by 9-ft slabs than for the 6- by 6-ft slabs. This observation agrees with the finding shown in Figure 4(a) that the rate of stress increases with increasing joint efficiency and decreases with increasing slab size.

Computations were also made for the TT load placed at the edge of the center slab next to the joint (Figure 5(d)). The results are presented in Table 5. As expected, the stresses and deflections are reduced as the joint efficiency is increased. However, the stresses are not reduced very much, as compared with other loading positions. This is because only two wheels of the TT load are placed next to the joint while the other two wheels are closed to the centerline of the slab. While the stresses in the slab induced by the two wheels placed next to the joint are reduced by the improvement of joint efficiency, the stresses induced by the other two wheels close to the centerline of the slab are increased by the improvement of joint efficiency; the effects thus somewhat cancel each other. The results presented in Table 5 indicate that the response of the small-concrete slabs to aircraft with multiple-wheel loads is difficult to predict and has to be analyzed for each loading position.

Conclusions

Based on the computed results of the finite-element method, the following conclusions can be drawn for a precast pavement which consists of small concrete slabs.

When the size of a small concrete slab is increased, the stresses are increased, and the deflections are decreased for any loading position. The change becomes insignificant when the slab size has increased beyond a certain value. This value increases with increasing slab rigidity, tire contact area, and gear spacing for multiple wheels.

In conventional highway and airfield concrete pavements, slab stresses are greatly reduced with increasing joint efficiency when the load is placed near the joint, and stresses are not dependent on the joint efficiency when the load is placed at the slab's center. This is not true, however, for small precast concrete slabs. When the joint efficiency is improved, stresses in the small concrete slab may actually be increased (though not significantly) if the load is placed at the slab's center; the stresses are reduced if the load is placed at the slab's edge next to the joint. This phenomenon is not critical in the design of small precast concrete slabs because the maximum stress induced by the load placed next to the joint is always greater than that when the load is placed at the slab's center, but engineers should be aware of the fact that stresses may actually be increased in small concrete slabs with increasing joint efficiency in some loading cases.

Because the response of the slab to load differs with different loading positions, the response of small concrete slabs to aircraft with multiple-wheel loads is difficult to predict and has to be analyzed for each loading position.

TABLE 5. Stresses and Deflections Under Twin-Tandem Load.*

Percent Shear Transfer (1)	σ_{max} in psi (2)	W_{max} in in. (3)
0	1,165.9	0.46770
25	992.7	0.33200
50	1,005.9	0.25200
75	975.0	0.19989
100	929.1	0.16382

*Loads placed at slab edge next to joint (Fig. 5(d)), slab size L = 6 ft.

STRESS ANALYSIS OF PRECAST CONCRETE SLABS

The finite-element computer programs WESLIQUID and WESLAYER [42] were used in the stress analysis of small precast concrete slabs. As discussed earlier, the Westergaard solution [6,7], which has been widely used in conventional and precast concrete pavements, is applicable only to a very large slab, or the slab dimension should be at least three times the radius of relative stiffness for the slab [8]. The design of precast concrete slabs on grade is controlled by the stresses and deflections in the slab induced by the aircraft or truck load. The horizontal tensile stresses directly influence the required slab thickness, and vertical stresses are proportional to slab deflections. Excessive deflections may not cause the slab to crack but will cause roughness of the pavement on which the vehicles travel. Excessively rough pavement surface will no doubt cause vehicle damages. Localized excessive deflections at the corner and at the edge of the slab can also cause large permanent deformations in the subgrade which in turn cause the concrete slab to break.

Unless otherwise specified, a 10,000-lb wheel load with a 100-sq in. contact area was used. The elastic modulus and Poisson's ratio of the concrete were assumed to be 4,000,000 psi and 0.2, respectively.

Slab Shapes

Precast concrete slabs are square, rectangular, or hexagonal in shape. The general dimensions of the slabs are listed in Table 1. The effect of slab dimensions on the stresses and deflections is illustrated in the following computations.

Five different slabs with the same area (36 sq ft) and the same thickness (6 in.) were studied. The dimensions were 2 by 18 ft, 3 by 12 ft, 4 by 9 ft, 5 by 7.2 ft, and 6 by 6 ft. The wheel load was placed at the center of the slab's edge in the long direction. The maximum tensile stresses and deflections (both occurring at the slab edge) are presented in Table 6 for two different subgrade modulus values (k). Table 6 shows that although all the slabs have the same area, the square slab has the smallest stress and deflection. The stress and deflection in the slab increase with increasing ratio of the length to width of the slab. This effect is less significant for higher subgrade k value or for higher relative stiffness of the slab. The possible explanation for the differences is that

TABLE 6. Stresses and Deflections in Precast Concrete Slabs.*

Slab Dimensions ft	Subgrade Modulus k pci	Relative Stiffness λ in.	Maximum Tensile Stress psi	Maximum Deflection in.
2 by 18	100	29.4	816.0	0.08105
3 by 12	100		734.6	0.07322
4 by 9	100		667.6	0.07119
5 by 7.2	100		599.4	0.07165
6 by 6	100		544.6	0.07301
2 by 18	600	18.8	548.5	0.02015
3 by 12	600		509.7	0.01798
4 by 9	600		502.5	0.01745
5 by 7.2	600		496.1	0.0174.
6 by 6	600		479.9	0.01774

*36 sq ft, E = 4,000,000 psi, ν = 0.2, h = 6 in.

TABLE 7. Stresses and Deflections in Precast Concrete Slabs.*

Slab Dimension ft	Relative Stiffness ℓ in.	Maximum Tensile Stress psi	Maximum Deflection in.
1 by 9	29.4	1366.6	0.1778
2 by 4.5	29.4	533.1	0.2206
3 by 3	29.4	331.5	0.2443

*9 sq ft, E = 4,000,000 psi, ν = 0.2, h = 6 in., k = 100 pci.

TABLE 8. Maximum Stress and Deflection Under Different Loading Positions.*

Loading Position	Maximum Tensile Stress (psi)	Maximum Deflection (in.)
a	691.8	0.05456
b	−291.3	0.12310
c	410.5	0.01967
d	574.3	0.06567

*6 ft by 12 ft by 6 in. concrete slab, k = 100 pci.

the 6- by 6-ft slab has a two-way slab action while the 2- by 18-ft slab possesses only a one-way slab action and is thus less rigid.

The slabs shown in Table 6 have an area of 36 sq ft. Computations were made for smaller slabs to determine whether the same conclusions are valid. Table 7 presents the computed results for the three different slabs which have an area of 9 sq ft. The computations were made only for k = 100 pci. The 3- by 3-ft concrete slab has the smallest tensile stress but the largest deflection. When the length over width ratio increases, the tensile stress in the concrete slab increases, and the deflection decreases. The dimension of the 3- by 3-ft slab is very close to the slab's relative stiffness and therefore the slab action cannot be developed. This produces small tensile stress in the slab but large vertical stress resulting in large deflection. As the slab length increases, slab action starts to develop in the lengthwise direction. The tensile stress in the same direction increases, and the deflection decreases.

Loading Positions

In analyzing stress conditions in highway and airfield concrete pavements, Chou [43] concluded that the most critical condition in a rigid pavement occurs when there is no transverse joint or crack in the pavement, and also when the load is moving along its edge. Under the edge load, the presence of joints and cracks can reduce pavement stresses near the joint and the cracks, but increase the deflections in the same area. In a jointed pavement, the critical stress occurs when the load is halfway between the joints, and the stress can have a magnitude close to that of a pavement with no transverse joint. Therefore, the presence of a joint does not reduce the maximum stress in a pavement. When the load is moving along the center of the pavement, the stresses are smaller and are nearly independent of whether the load is at the center or next to the pavement joint. Under both center and edge loads, maximum deflection occurs when the load is next to the transverse joint; corner deflection induced by the corner load is much greater. Greater pavement deflections induce greater subgrade stresses and, consequently, more severe plastic deformation in the subgrade soil, which may lead to the creation of voids in the subgrade soil along the joint and cause earlier pavement failure.

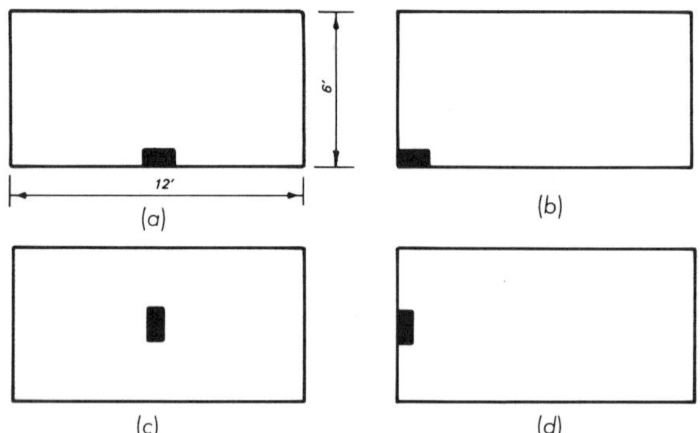

FIGURE 6. Different loading positions for a single-wheel load on a precast concrete slab.

Computations were made for a 6-ft by 12-ft by 6-in. concrete slab with the single-wheel load placed at 4 different positions on the slab as shown in Figure 6. The computed maximum stresses and deflections are presented in Table 8. The maximum stress occurs in loading position a in which the load is placed at the center along the slab edge in the lengthwise direction. The maximum deflection occurs in loading position b in which the load is placed at the slab corner.

Temperature Effects

Temperature gradients cause concrete slab to warp upward (edge up) if it is cooler at the top and downward (edge down) if it is hotter at the top. Because of the concrete's weight, temperature stresses are produced in the warped slab such as tensile stresses at the bottom of the slab in a downward warped pavement. Depending upon the loading position, load stresses in the slab can be additive to thermal stresses.

To illustrate the temperature effect on the concrete slab,

TABLE 9. Effect of Temperature on Stress and Deflections in a Concrete Slab.*

Temperature	Maximum Tensile Stress	Maximum Deflection
No temperature	753.7	0.0484
Positive temperature, 3°F per inch of concrete	572.1	0.0527
Negative temperature, 3°F per inch of concrete	935.3	0.0440

*6.6 ft by 19.7 ft by 5.5-in. slab.

TABLE 10. Effect of Load Transfer on Stresses and Deflections.*

Slabs	Maximum Tensile Stress psi	Maximum Deflection in.
A single slab	753.7	0.0484
Two-slab system	561.9	0.0262
Three-slab system	548.6	0.02578

*1-in. dowel bars spaced 18 in. apart.

computations were made for a 6.6- by 19.7-ft concrete slab (Slab PGA-III in Table 1) for both positive and negative temperature gradients that have a difference of 3°F for each inch of concrete slab, i.e., the difference in temperature between the top and the bottom of the 5.5-in. concrete slab is 16.5°. Table 9 shows the maximum stresses and deflections computed in the slab directly under the load. Under the positive temperature gradient condition, i.e., edges up at the night condition, the stresses are reduced, and the deflections are increased. Under the negative conditions, i.e., edges down at the daytime condition, the stresses are increased, but the deflections are reduced. Clearly thermal stresses and load stresses are additive.

Load Transfer Across the Slabs

The amount of load transfer across the slabs in a precast concrete pavement can affect stress distributions and roughness in the pavement and thus influence the thickness design. The improvement of pavement roughness due to the improvement of load transfer across the slabs in a precast concrete pavement should always be emphasized. Large stresses in the slab can be taken care of by steel reinforcing or prestressing, and large differential settlement between two adjacent slabs can be extremely hazardous to high speed vehicles.

Two computations were made by connecting two and three 6.6- by 19.7-ft slabs (Slab PGA-III in Table 1) together. The connections were placed on the long side of the slabs. One inch dowel bars spaced 18 in. apart were used in the computations. The load was placed at the center next to the joint. In the three-slabs case, the load was placed in the center slab. Table 10 shows the computed maximum tensile stresses and deflections. It is seen that the stress and deflection are greatly reduced in the two-slab system, but the reduction becomes insignificant when the third slab is connected.

Table 10 shows the advantage of providing load transfer between slabs. In conventional pavements this load transfer is provided through aggregate interlock on saw-cut contraction joints, dowel bars across joints, or key joints in the form of tongue and groove connections. Potential methods of providing comparable load transfer for precast slabs are discussed in great length in Reference [31]. Welded joint connections provide load transfer, but a sufficient number of connection joints must be provided on all the slab sides and thus require a significant amount of welding in the field. Insufficient number of connection joints may help maintain alignment of the slabs but will not provide sufficient load transfer, and the welded point may easily be broken under traffic. Dowel bars are provided for load transfer between slabs, and tie bars are used to prevent separation of slabs. The steel bars are welded to steel plate cast into the precast slab [16]. Also, this type of joint will require a large number of field welds under awkward working conditions due to narrow dimensions between slabs. A 12-ft by 15-ft by 8-in.-thick slab with a single-lap welded splice for a joint similar to the one used in South Dakota highway would require 48 field welds of 1-in. bars for load transfer. A more satisfactory double-lap joint would require 192 welds, and the preferred butt splice would be very difficult to fabricate. Welded joint connection poses a number of problems and should probably be avoided if possible.

A grouted shear key is relatively simple to fabricate and construct. The inclined faces of the key ensure mechanical interlock as well as adhesion of the grout to the slab face. Special epoxy and other related polymer grouts can obtain higher adhesion strengths. A key joint can be simple to form, but field assembly would be difficult. Any attempt to slide a slab horizontally will push material into the joint between the slabs. Consequently, one side of the slab must be lifted to allow the male end of the key to be inserted, and then the slab can be lowered onto the base. The Corps of Engineers limits the use of keyed joints to pavements 9 in. or more in thickness and prohibits their use on pavement subject to channelized traffic on medium- and heavy-load airfields.

Subgrade Support Loss

It is known that gaps exist under concrete pavements for a number of reasons such as pumping, plastic deformation

TABLE 11. Stresses and Deflections for Concrete Slabs with Gaps.

Conditions	Maximum Tensile Stress, psi	Maximum Deflection in
No temperature, with gap	583.0	0.03256
No temperature, without gap	561.9	0.02620
Slab curled upward, with gap	398.7	0.03400
Slab curled upward, without gap	380.2	0.03059
Slab curled downward, with gap	767.4	0.03110
Slab curled downward, without gap	743.6	0.02176

of the subgrade, and poor construction practices. The presence of gaps under the pavement is equivalent to the removal of subgrade support for the pavement and consequently can result in greater stresses in the pavement.

A complete subgrade contact condition was assumed in the Westergaard solution. The slab always has full contact with the subgrade soil, and gaps are not allowed between the slab and subgrade no matter how much the slab has deformed due to temperature change or to the applied load. Therefore, the slab is assumed to be supported by a group of springs, and the springs are always connected to the slab. In reality, the pavement can lose subgrade support at some parts due to temperature warping, pumping, and plastic deformation of the subgrade.

When the complete subgrade contact condition is assumed in the Westergaard solution, the principle of elastic behavior holds, i.e., the computed stresses and deflections are linearly proportional to the applied load. However, when the slab and the subgrade are in partial contact, the principle of elastic behavior no longer applies. Also, when the subgrade is in full contact with the slab, the slab weight affects only the deflections but not the stresses in the slab. However, when the slab is in partial contact with the subgrade, slab weight has a significant effect on slab stresses. The effects of the gaps under the slabs are illustrated in the following example computation.

The example computation for load transfer case of two 6.6- by 19.7-ft slabs connected with 1-in. dowels is also given for this purpose. A 1-in. gap in the subgrade is assumed along the joint and both positive and negative temperature gradients are considered. Table 11 presents the maximum stresses and deflections for all the cases.

Table 11 shows that the existence of gaps under the slabs at the joint does not affect the tensile stresses significantly, but it does increase the deflections appreciably, particularly when the slab's edges are curled downward (daytime condition).

Subgrade Types

Because of the difference in the basic assumption of the supporting subgrade soil, the shapes of the theoretical stress and deflection basins of the concrete slab are different under the liquid foundation (represented by WESLIQUID program and Westergaard solution) than under the elastic foundation (represented by the WESLAYER program). In a liquid subgrade, the deflection at a node depends solely on the modulus subgrade reaction, k, at the node and not elsewhere. In an elastic subgrade, however, the deflection at a node depends not only on the elastic modulus, E, of the subgrade, but also on the deflections at other nodes. Consequently, the deflection basin should be steeper in the liquid subgrade than in the elastic subgrade.

Vesic and Saxena [44] presented an extensive theoretical analysis of (infinitely large) slabs on liquid, k, and elastic, E_s, foundations and concluded that no single value of k can yield agreement between the two analyses of all statical influences such as pressures, shearing forces, bending moments, and deflections across the slab. It is therefore not possible to directly compare the responses of a precast concrete slab to loads under a liquid and an elastic subgrade. Analysis was made to illustrate the difference between the two types of subgrade only as a qualitative basis.

Theoretical stress and deflection basins for the 6.6- by 19.7-ft slab (Slab PGA-III in Table 1) are presented in Figure 7. The wheel load is placed at the center and at the edge of the long side of the slab. The stresses and deflections are normalized with respect to the maximum values and are plotted at the center of the wheel load in the direction of the short side of the slab. The stresses are in the direction normal to the plotted direction which are greater than the stresses in the other direction. For slab on elastic foundation, the subgrade elastic modulus, E, was assumed to be 10,000 psi. The corresponding value of modulus of subgrade reaction, k, used for the slab on liquid foundation was determined by matching the maximum stress in the concrete pavement.

Figure 7(a) shows the normalized deflection basins. The basin is much flatter when the subgrade is elastic than when it is liquid. Figure 7(b) shows the shapes of the normalized stress basins. The curvatures of the stress basins for the elastic and liquid subgrades are nearly the same.

Prestressed Concrete Slabs

Because of the requirements of lighter weight of the concrete slabs for transportation and placement in the field and of higher tensile strength of the concrete during handling, precast concrete slabs are generally either steel reinforced or prestressed. A crack forms in a prestressed concrete slab when the tensile stress on the bottom face exceeds the sum

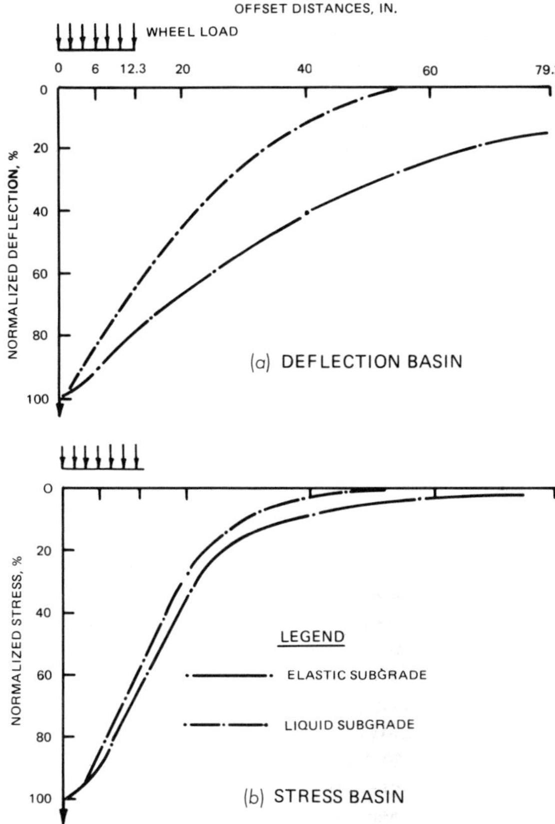

FIGURE 7. Normalized stress and deflection basins for elastic and liquid subgrades.

of the concrete flexural strength and the prestress compressive stresses. This cracking is the limit of elastic behavior of the slab and is the conventional point of slab failure. However, the prestressed slab will redistribute bending moments to increase the negative radial bending moment after this initial crack forms. When the load is removed, the prestress force closes the crack allowing continued service of the slab. Failure of the prestressed slab occurs when tensile cracking develops in the surface of the slab from the negative radial bending moments. A small increase in load or load repetitions then causes a punching shear failure. The Corps uses an empirical design method based on model and accelerated traffic tests to take advantage of this inelastic slab behavior [45]. More detailed information is available on the construction of prestressed concrete structures [46,47].

HANDLING STRESSES

The most severe loading that a precast concrete slab must resist may occur during handling after precasting or during construction rather than under traffic. Every precast concrete slab must be designed to resist handling stresses as well as stresses from traffic loads. In conventional reinforced concrete pavements, reinforcing steel is used to keep cracks in the slab closed, not to carry tension as in structural reinforced concrete. Current Corps of Engineers practice for airfields places the reinforcing 1 in. below the midpoint of the slab [48]. In this position the steel will not contribute to the flexural capacity of the slab while it is being lifted. The designer must proportion the slab and the lifting design to maintain the stresses, modified with appropriate safety factors, below the tensile strength of the concrete, or he must provide properly located reinforcing steel to carry tensile stresses after the concrete cracks.

The slab must be designed for impact loads from handling for which an impact factor of 2 would be prudent. Also, if the lifting cable is at an angle to the slab surface, horizontal as well as vertical loads will be applied to the slab and must be considered. Other factors that must be considered include slabs being handled at the casting yard before the concrete develops full strength, loading from stacking during storage, and detailed design of pickup points. Handling and construction of precast units are discussed in more detail [46,49]. Complex loadings from storage and handling and two-way slabs (length to width ratio less than 2) may be analyzed using general-purpose computer programs such as STRUD developed by the Massachusetts Institute of Technology, NASTRAN developed by the National Aeronautics and Space Administration, or SAP developed by the University of California at Berkeley.

The WESLIQUID computer program [42] has the capability of analyzing stress conditions in a precast concrete slab during lifting. The program can have different subgrade k values at each nodal point. At nodal points where pickup points are located, an extremely large k value, such as 10^{10} pci, is used, and a zero k value is used at all other nodal points. In other words, the entire concrete slab sits on (or is supported by) rigid springs at pickup points. The stresses and deflections in the concrete slabs can thus be computed.

The accuracy of WESLIQUID program for computing stresses in a concrete slab during lifting was verified with available formula. Equations (1) and (2) compute the maximum tensile stresses in a rectangular plate with four sides and two sides (at short sides) supported, respectively [50].

$$\sigma = 0.501\, wb^2/t^2 \quad (1)$$

$$\sigma = 3.048\, wb^2/t^2 \quad (2)$$

where w is the intensity of the uniformly distributed load over the entire slab surface, t is the concrete thickness, and b is the length of the short side of the rectangular slab which is one half of the length of the long side.

WESLIQUID computer program was used to compute the stresses due to dead weight of the concrete in a 6-ft by 12-ft by 6-in. slab supported at four edges and at two edges (at short sides). A k value of 10^{10} pci was used at each nodal point at the edge supported and zero k values were assumed for all the interior nodes. The computed maximum tensile stresses were 36.9 and 221.4 psi for the slab four edges and two edges supported, respectively. The stresses computed using Equations (1) and (2) were 37.6 and 228.6 psi, respectively.

If the lifting cables are at an angle to the slab surface, horizontal forces and moments are produced by the horizontal forces that exist. Horizontal forces cannot be considered in the WESLIQUID program, but their effect on the concrete slab can be estimated. Since horizontal forces apply compression to the concrete slab, the presence of horizontal forces during lifting will reduce the tensile stresses in the slab. However, the compressive forces are generally very small as compared with the maximum tensile stresses and thus can be neglected in computation. Moments applied at nodal points can be considered in the WESLIQUID program. It was found that the effects are also insignificant.

REFERENCES

1. McDonald, J. L. and T. C. Liu, "Precast Concrete Elements for Structures in Selected Theaters of Operation," Technical Report C-78-1, U.S. Army Engineer Waterways Experiment Station, Vicksburg, Miss. (1978).
2. Gorsuch, R. F., "Preliminary Investigation of Precast Prestressed Concrete Pavements," Mississippi State Thesis, South Dakota State University, Brookings, S. Dak. (1962).
3. Jacoby, G. A., "A Study of Expansion and Contraction in a Pavement Consisting of Prestressed Concrete Panels Interconnected in Place," Mississippi State Thesis, South Dakota State University, Brookings, S. Dak. (1967).
4. Ohio River Division Laboratory, "Final Report, Lockbourne No. 2—Experimental Mat," U. S. Army Engineer Division, Ohio River, Cincinnati, Ohio (1950).
5. Ohio River Division Laboratory, "Use of Precast Concrete Block Pavements," Engineer Technical Letter 1110-3-310, Washington, D.C. (1979).
6. Westergaard, H. M., "Stresses in Concrete Pavements Computed by Theoretical Analysis," *Public Roads*, Vol. 7, No. 2, pp. 25–35 (1926).
7. Westergaard, H. M., "New Formulas for Stresses in Concrete Pavements of Airfields," *Transactions, American Society of Civil Engineers*, Vol. 113, pp. 435–444 (1948).
8. Behrmann, R. M., "Small Scale Model Study to Determine Minimum Horizontal Dimensions for Infinite Slab Behavior," Technical Report No. 4-32, Ohio River Division Laboratories, U.S. Army Engineer Division, Ohio River, Cincinnati, Ohio (1964).
9. Mellinger, F. M., "Investigation of Prestressed Concrete Sectional Mats," Technical Report No. 2-4, Ohio River Division Laboratories, U.S. Army Engineer Division, Ohio River, Cincinnati, Ohio (1956).
10. Larson, L. J. and W. Hang, "Construction of a Prefabricated Highway Test Section," Highway Research Record 389, Washington, D.C., pp. 18–24 (1972).
11. Kruse, C. G., "A Laboratory Analysis of a Composite Pavement Consisting of Prestressed and Post-Tensioned Concrete Panels Covered with Asphaltic Concrete," Mississippi State Thesis, South Dakota State University, Brookings, S. Dak. (1966).
12. Hargett, E. R., "Prestressed Concrete Panels for Pavement Construction," *Journal of the Prestressed Concrete Institute*, Vol. 15, No. 1, pp. 43–49 (1970).
13. Hargett, E. R., "Structural Reinforcement for Airport Pavements," *Transportation Engineering Journal*, Vol. 95, No. TE4, American Society of Civil Engineers, pp. 629–637 (1969).
14. Zuk, W., "Prefabricated Highways," *Journal, Transportation Engineering, American Society of Civil Engineers*, Vol. 98, No. TE2 (1972).
15. Jones, S. A. and J. P. Iverson, "Use of Precast Slabs for the Repair of Faulted Joints in Concrete Pavement," Special Report, Federal Highway Administration, Washington, D.C. (1971).
16. Transportation Research Board, "Reconditioning High-Volume Freeways in Urban Areas," Synthesis of Highway Practice No. 25, National Cooperative Highway Research Program, Washington, D.C. (1974).
17. Byrd, L. G., "Precision Concrete Cutting and Repair System for Pavements," *Roadways and Airport Pavements*, SP51, American Concrete Institute, Detroit, Mich. (1975).
18. *Engineering News Record*, "Runway Repair Sets Fast Pace," Vol. 206, No. 10, pp. 34–35 (1981).
19. Glushkov, G. I. and Rayev-Bogoslovskii, B. S., *Construction and Maintenance of Airfields*, Moscow, Foreign Technology Division Translation FTD-MT-24-544-72, Wright-Patterson Air Force Base, Ohio (1970).
20. Rayev-Bogoslovskii, B. S., et al., "Rigid Pavements of Air ports," Moscow, Air Force Systems Command Foreign Technology Division Translation FTD-MT-64-116, Wright-Patterson Air Force Base, Ohio (1961).
21. Mikhno, Y. P., *Restoration of Destroyed Construction*, Moscow, Foreign Technology Division Translation FTD-ID-(RS)T-1214-75, Wright-Patterson Air Force Base, Ohio (1974).
22. Maidel, V. G. and A. A. Timofeev, "Precast Concrete Roads and Footways," *Gorodskoe Khozayaisto Moskvy*, 26(3), pp. 26–29 (1962).
23. Birger, A. and A. Klopovskii, "Prefabricated Roads Made from Vibra-Rolled Concrete Slabs," *Stroitelnye i Arkhitek Moskvy*, 10(3), pp. 13–15 (1961).
24. Dubrovin, E. N., et al., "Precast Reinforced Concrete Slabs in Road Construction," *Gorodskie Khozayaisto Moskvy,* 36(9) (1962).
25. Mikhovich, S. I., L. P. Tarasenko and N. I. Tolmachev, "Prefabricated Concrete Surfaces on Industrial Roads in the Donbass," *Automobil'nyi Dorogoi*, 24(2), pp. 8–9 (1961).

26. Smolka, B. I., "Prefabricated Prestressed Concrete Surfaces," *Automobil'nyi Dorogoi*, 26(11), pp. 9–10 (1963).
27. Timofeev, A. A. and E. F. Leutskii, "Moving and Laying Concrete Slabs with Vacuum Clamps," *Automobil'nyi Dorogoi*, 24(7), pp. 22–24 (1961).
28. Stepuro, N. T., et al., "The Construction of Sectional Concrete Pavements," *Automobil'nyi Dorogoi*, 27(3), pp. 23–25 (1964).
29. Mednikov, I. A., Y. A. Malchanov and L. V. Gorodelskii, "Analysis and Jointing of Polygonal Road Slabs," *Osnovaniya Fundamently i Mekhanika Gruntov*, Vol. 11, No. 5, Translated by Consultants Bureau, New York (1974).
30. Gerberg, A. A. and A. S. Osipon, *Construction of Airports*, Moscow, Air Force Systems Command Foreign Technology Division Translation FTD-TT-64-1273, Wright-Patterson Air Force Base, Ohio (1962).
31. Rollings, R. S. and Y. T. Chou, "Precast Concrete Pavements," Miscellaneous Paper GL-81-10, U.S. Army Engineer Waterways Experiment Station, Vicksburg, Miss. (1981).
32. Rollings, R. S., "Concrete Block Pavements," Technical Report GL-83-3, U.S. Army Engineer Waterways Experiment Station, Vicksburg, Miss. (1983).
33. Chou, Y. T., "Stress Analysis of Small Concrete Slabs on Grade," *Journal of Transportation Engineering*, American Society of Civil Engineering, Vol. 110, No. 5 (1984).
34. Hanna, A. N., et al., "Technological Review of Prestressed Pavements," FHWA-RD-77-8, Federal Highway Administration, Washington, D.C. (1976).
35. Harris, A. J., "Prestressed Concrete Runways: History Practice and Theory, Part I," *Airports and Air Transportation*, Vol. 10, No. 121, London (1956).
36. Stott, J. P., "Prestressed Concrete Roads," *Roads and Road Construction*, Vol. 33, No. 388, London (1955).
37. Vandepitte, D., "Prestressed Concrete Pavements—A Review of European Practice," *Journal, Prestressed Concrete Institution*, Vol. 6, No. 1, Chicago, Ill. (1961).
38. Patterson, W. D. O., "Functional Pavement Design for Container Terminals," *Proceedings, Australian Road Research Board*, Vol. 8, Part 4 (1976).
39. Sato, K., T. Fukute and H. Inukai, "Some New Construction Methods for Prestressed Concrete Airport Pavements," *Proceedings, 2nd International Conference on Concrete Paving Design*, Purdue University, West Lafayette, Ind., pp. 149-159 (1981).
40. Huang, Y. H. and S. T. Wang, "Finite Element Analysis of Concrete Slabs and Its Implications for Rigid Pavement Design," Highway Research Record 466 (1973).
41. Huang, Y. H. and S. T. Wang, "Finite Element Analysis of Concrete Pavements with Partial Subgrade Contact," Transportation Research Record 485, Transportation Research Board (1974).
42. Chou, Y. T., "Structural Analysis Computer Programs for Rigid Multi-component Pavement Structures with Discontinuities—WESLIQUID and WESLAYER," Technical Report GL-81-6, Reports 1, 2, and 3, U. S. Army Engineer Waterways Experiment Station, Vicksburg, Miss. (1981).
43. Chou, Y. T., "Comparative Analysis of Rigid Pavements," *Journal of Transportation Engineering*, American Society of Civil Engineering, Vol. 109, No. 5 (1983).
44. Vesic A. S. and K. Saxena, "Analysis of Structural Behavior of AASHTO Road Test Rigid Pavements," NCHRP Report No. 97, Highway Research Board (1970).
45. Odom, E. C. and P. F. Carlton, "Prestressed Concrete Pavements: Design and Construction Procedures for Civil Airports," Technical Report S-74-10, Vol. II, U.S. Army Engineer Waterways Experiment Station, Vicksburg, Miss. (1974).
46. Gerwick, B. C., Jr., *Construction of Prestressed Concrete Structures*, Wiley-Interscience, New York (1971).
47. Van der Wal, M. L. and H. C. Walker, "Tolerances for Precast Concrete Structures," *Journal, Prestressed Concrete Institute*, Vol. 21, No. 4, Chicago, Ill. (1976).
48. Department of the Army, "Rigid Pavements for Airfields Other Than Army," TM 5-824-3, Washington, D.C. (1979).
49. Waddell, J. J., "Precast Concrete: Handling and Erection," Monograph No. 8, American Concrete Institute, Detroit, Mich. (1974).
50. Roark, R. J., *Formulas for Stress and Strain*, McGraw-Hill Book Company, 4th Edition (1965).

CHAPTER 3

Moment-Curvature Characteristics for Reinforced Concrete

CHENG-TZU THOMAS HSU*

INTRODUCTION

Information on moment-curvature relations of reinforced concrete columns under combined biaxial bending and axial load is relatively scarce. There have been a few papers published to study the experimental moment-curvature relationships under combined bending and axial compression.

The present investigation was aimed at an experimental-analytical study of the behavior of biaxially loaded structural members as the applied load was increased monotonically from zero load until failure; both the ascending and the descending branches of the moment-curvature curve were examined. The experimental results were compared with the computer analysis results to assess accuracy of the proposed analytical model. The moment-curvature (or moment-rotation) characteristics derived can be found useful in the limit analysis studies of two or three dimensional reinforced concrete frames.

The moment-curvature relationship for reinforced concrete columns was investigated by performing the short column tests. The strain distribution across the section was obtained using the data from the strain gages and demec gages. Both sets of strain values were used to establish the curvature values. Both the ascending and the descending branches of the moment-curvature curves were obtained for present study. The experimental results were compared with the analytical results; good agreement was noted through all load stages from zero load up to the maximum moment value.

*Department of Civil and Environmental Engineering, New Jersey Institute of Technology, Newark, NJ

A NEW EXPERIMENTAL TECHNIQUE IN COLUMN TESTS

Introduction

In general, there are two kinds of experimental set-up for column tests. One uses the conventional, universal testing machine, while another type of loading arrangement of columns is similar to that used by Meek [54]. The load deformation curves of reinforced concrete columns obtained from the two above loading arrangements, however, can obtain only from zero load to near the maximum load stage. Beyond that stage, the column specimen gets crushed as soon as the applied load reaches the maximum value. Also, since the column specimen fails rapidly, it is not possible to instrument the specimen to study part of the load-deformation curves beyond the peak. It is therefore not possible to obtain the complete load-deformation characteristics which are so important in the earthquake resistance and limit state design of concrete structures. A new column testing technique, which would allow for the load relaxation of the column specimens by suitably balancing the energy released by the testing machine with the energy absorbed by the specimens and the loading device, was developed by Hsu [33,34,38,40,41] to study the loads and deformation conditions existing in the test specimens near failure loads.

Basic Concept

When a reinforced concrete column is loaded during a compressive test, an increase in the applied load (before the peak is reached) results in loads and mid-span deflection increases. Once the load reaches the maximum load, further mid-span deflection of the column results in an increase in observed mid-span deflections but a decrease in the applied loads. The typical conventional testing machine supplies energy at a constant rate as shown in Figure 2. The increase

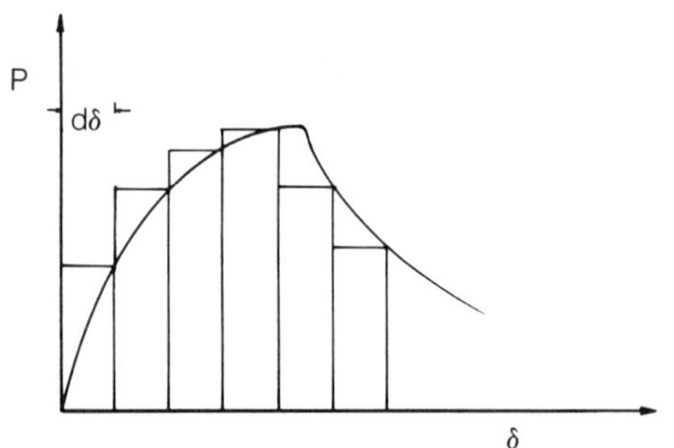

FIGURE 1. Load-deflection curve in column test by the conventional testing machine.

of energy is absorbed by the specimen up to the peak point (Figure 1), but the sudden release of energy upon reaching the peak point causes the sudden collapse of the test columns as shown in Figure 1. The control of machine and the measurement of the mid-span deflection is extremely difficult and would in general vary with the test specimens. This difficulty can be overcome by using a suitable test apparatus such that it would absorb the difference of energies between that supplied by the testing machine and that absorbed by the column specimen.

Use of Hard Rubber Blocks

It was decided to use a loading arrangement with a set of rectangular rubber blocks as shown in Figure 4. Then the energy supplied by the testing machine is stored by the elastic deformation of the elements of the rubber blocks and the column specimen. In Figure 3, the curves A and B indicate the load-deflection characteristics of the column specimen and the rubber blocks respectively. The test arrangements in Figure 4 are based on the following considerations.

$$\int_0^{\delta_c} P_m d\delta_c - \int_0^{\delta_c} P_c d\delta_c < \int_0^{\delta_c} P_r d\delta_c \quad (1)$$

(testing machine) (test specimen) (rubber blocks)

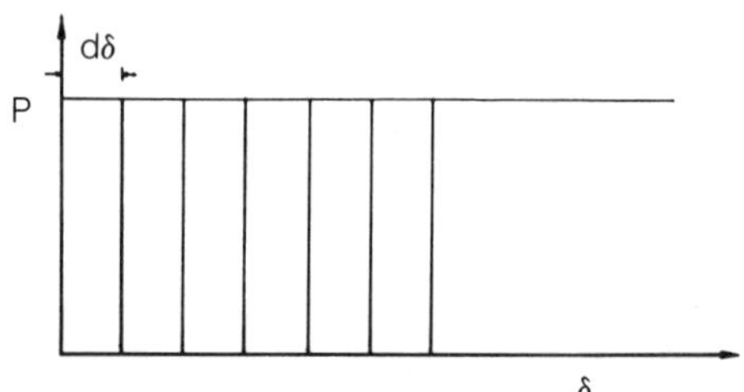

FIGURE 2. Load-deflection curve supplied by the conventional testing machine.

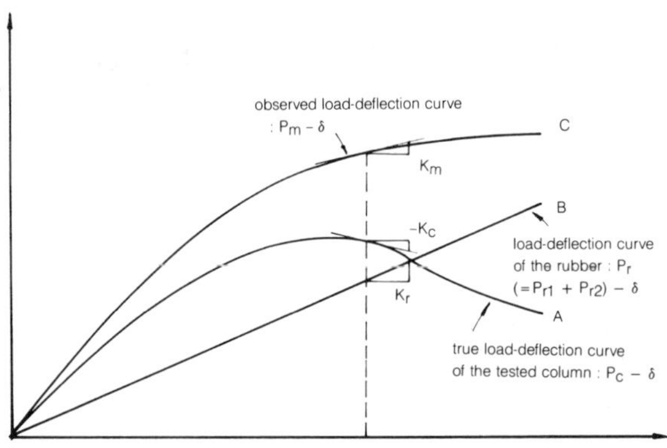

FIGURE 3. Load-deflection curve in column tests.

where $P_r = P_{r1} + P_{r2}$, P_m is the load applied by the testing machine and P_c, P_{r1}, P_{r2} are the reactions due to applied load in column specimen and rubber blocks respectively.

BEHAVIOUR OF REINFORCED CONCRETE COLUMNS UNDER COMBINED UNIAXIAL BENDING AND COMPRESSION

Introduction

A few papers such as Cranston's [16] and Gurfinkel and Robinson's [28] have discussed the load-deformation characteristics of reinforced concrete columns under combined bending moment and axial load. However, no significant effort has been made to correlate the theoretical and experimental results for the moment-curvature and moment-rotation relationships of structural members subjected to combined bending moment and axial load.

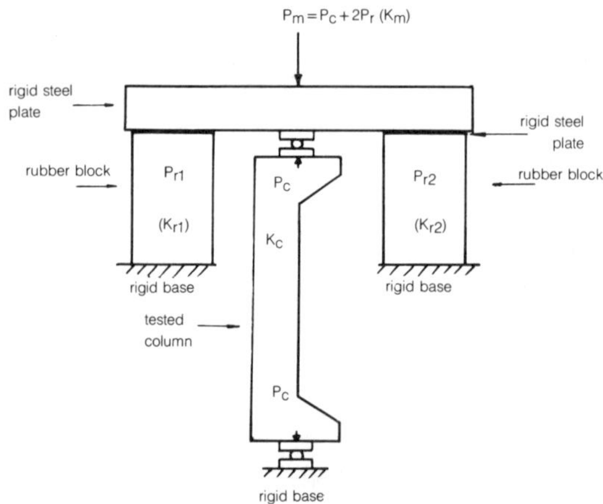

FIGURE 4. Systematic load arrangement of column tests.

Cranston [16] developed a computer program for the analysis of restrained columns to evaluate the moment-curvature relationship under combined bending moment and axial load (using load control). In addition, he has suggested a method for deriving the load-deflection curve by either load or deflection control up to the maximum moment capacity of the given section.

Gurfinkel and Robinson [28] also developed a similar computer approach for the determination of strain distribution and curvature in a reinforced concrete element subjected to the bending moment and axial load (using the load control process).

Both methods evaluate the moment-curvature curves until the maximum moment stage only. Also, the convergence near the maximum moment stage becomes difficult to achieve since the stiffness of the structural member becomes very small.

Hsu and Mirza [33,38] used complete stress-strain curves for the concrete and the reinforcing steel in developing a sub-routine to calculate the complete moment-curvature relationship for a given structural concrete section. This approach was written in such a way that moment-curvature curve can be obtained not only from zero moment to maximum moment but also from maximum to ultimate moment (using the deformation control process).

Hsu [33] also employed an approximate equation for the evaluation of deflection and rotation. The computer program [33] was used to calculate the moment-curvature, load deflection and moment-rotation relationships for pin-ended short column specimens. Test data from Hognestad's work [31] and Hsu's tests [33] have been compared with Hsu's [33,38] computer analysis.

It should be noted that Hsu's [33,38] computer analysis applied to short columns without considering the column lateral buckling. In the experiments by Hognestad [31] and Hsu [33] the columns failed in flexural tension and possessed considerably more curvature or rotation capacity than a member failing in a balanced failure. Moreover, the columns failing in a balanced failure mode, appear to possess a higher degree of curvature or rotation capacity than members failing in compression.

King [47] showed in his work that there was a distinct increase of ultimate strength when the tie spacing was made small, of the order of one third of the least lateral dimension of the column specimen. Later, Hudson [45] and Mirza, Guedelhoefer and Janney [55] concluded in their results that although the tie seemingly had no effect on the ultimate strength of the column, they exerted a strong influence on the mode of failure of the axially loaded members. It should be noted that the concrete encasement apparently provides sufficient restraint against buckling of the reinforcing bars up to the point of compression failure of the concrete.

Furlong and Ferguson [24] showed in their experimental work for the study of the strength of columns as integral parts of frames, suggested that frame action as a restraint to column failure resulted in 5 to 15% more axial load capacity for columns in single curvature than that anticipated for the equivalent isolated columns.

It is well known that the distribution of curvature and the position of neutral axis are governed by the crack pattern. Also the deformations at the steel-concrete interface affect the determination of the distribution of curvature even within the constant moment region.

Fowler [22] measured the moment-curvature relations in his column tests up to the maximum load capacity for the columns governed by compression failure. Obeid [61] tested columns governed by tension failure, yet the latter's measurements of the curvature and rotation were open to question. He also did not obtain the moment-curvature and moment-rotation curves from zero load to the utlimate load. Besides, the evaluation of maximum moment capacity did not take into account the effect of mid-height deflection and the specimens were too short to assess the plastic hinge rotation because of the effects of the brackets. The comparison of Hsu's [33] analytical results with some of the test results obtained by Hognestad [31], Obeid [61] and Hsu [33] can be found in Reference [33].

Moment-Curvature-Axial Load Relationships

ANALYTICAL DERIVATION (DEFORMATION CONTROL)

The following derivation is aimed at analysing the behaviour of a reinforced concrete section subjected to combined bending and axial compression. The computer program was written in FORTRAN IV for tension failures and near balanced failures. The trilinear stress-strain curve for reinforcing steel (including strain-hardening) has been used. The stress-strain curve for unconfined concrete is given in Equation (2) as follows [33]:

$$(1)\ 0 \leq x \leq 1 \tag{2a}$$

$$\eta = 2x - x^2 \text{(Ritter's Parabola)}$$

$$(2)\ 1 \leq x \leq 4 \tag{2b}$$

$$\eta = 1.18 - 0.18x$$

The following three steps comprise the analysis:

Before Cracking

Using the transformed area of steel and assuming that the modulus of elasticity of the reinforcing steel is the same in tension and compression, i.e., $E_s = E_s'$

$$n = \frac{E_s}{E_c} = \frac{E_s'}{E_c} \tag{3}$$

where E_c is the modulus of elasticity in concrete.

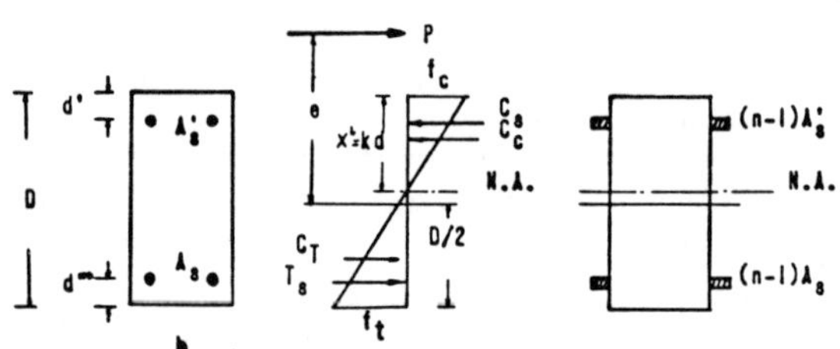

FIGURE 5. Moment-curvature-axial load analysis: before cracking.

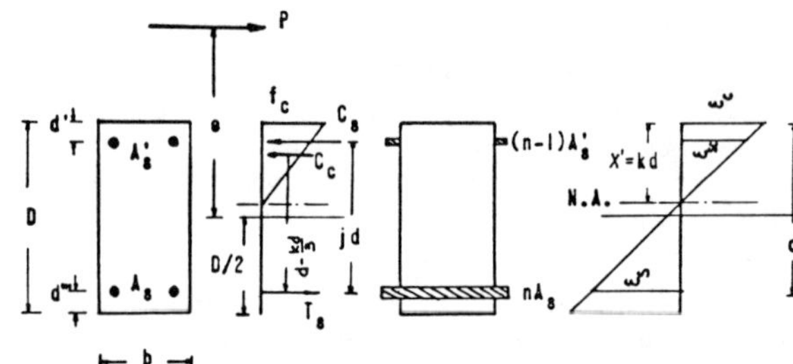

FIGURE 6. Moment-curvature-axial load analysis: immediately after cracking.

The position of the neutral axis is determined by taking moments about P (Figure 5)

$$C_c\left(\frac{x'}{3} + e - \frac{D}{2}\right) + C_s\left(d' + e - \frac{D}{2}\right) = T_s\left(e + \frac{D}{2} - d'''\right) + C_T\left[\frac{2}{3}(D - x') + x' + e - \frac{D}{2}\right] \tag{4}$$

where

$$C_c = \tfrac{1}{2} f_c b x'$$

$$C_s = f_c\left(\frac{x' - d'}{x'}\right)(n - 1)A'_s$$

$$T_s = f_c\left(\frac{d - x'}{x'}\right)(n - 1)A_s$$

$$C_T = \tfrac{1}{2} f_t b(D - x) = \tfrac{1}{2} b(D - x')f_c\left(\frac{D - x'}{x'}\right)$$

then

$$x' = \frac{(12de + 6Dd - 12dd''')(n - 1)A_s + bD^3 + 6D^2be + (12d'^2 + 12d'e - 6Dd')(n - 1)A'_s}{(12d' + 12e - 6D)(n - 1)A'_s + 12Dbe + (12e + 6D - 12d''')(n - 1)A_s} \tag{5}$$

The moment of inertia I for the gross section is given by

$$I = \frac{bD^3}{12} + bD\left(kd - \frac{D}{2}\right)^2 + (n - 1)A'_s(kd - d')^2 + (n - 1)A_s(D - kd)^2 \tag{6}$$

The relationship between the modulus of rupture f_r and the split cylinder strength f_t has been noted to be linear by several investigators. Assuming $f_r = f_t/0.8$, the cracking moment is given by

$$M_{cr} = \frac{f_t I}{0.8(D - kd)} \tag{7}$$

where M_{cr} is the moment about the neutral axis. Therefore,

$$P_c = \frac{M_{cr}}{\left(e - \frac{D}{2} + kd\right)} \tag{8}$$

and the curvature is given by

$$\phi_{cr} = \frac{M_{cr}}{E_c I} \tag{9}$$

Immediately After Cracking

Taking the moment about P again (Figure 6)

$$C_c\left(\frac{x'}{3} + e - \frac{D}{2}\right) + C_s\left(d' + e - \frac{D}{2}\right) = T_s\left(e + \frac{D}{2} - d'''\right) \tag{10}$$

where

$$C_c = \tfrac{1}{2} f_c b x'$$

$$C_s = f_c\left(\frac{x' - d'}{x'}\right)(n - 1)A_s'$$

$$T_s = f_c\left(\frac{d - x'}{x'}\right) nA_s$$

Substitution of the above forces into Equation (10) and the following equation can be solved to determine the position of the neutral axis,

$$\begin{aligned} &2bx'^3 + x'^2(6be - 3bD) + x'\,\{(n - 1)A_s'[12d' + 12e - 6D] + nA_s[12e + 6D - 12d''']\} \\ &+ (n - 1)A_s'(6Dd' - 12d'e - 12d'^2) + nA_s(12dd''' - 6Dd - 12de) = 0 \end{aligned} \tag{11}$$

where $x' = kd$.

Taking moments of internal and external forces about the neutral axis yields the equation

$$A_s f_s\left(\frac{D}{2} - d''' + \frac{D}{2} - kd\right) + \tfrac{1}{2} f_c bkd\left(\frac{2}{3}\,kd\right) + A_s' f_s'(kd - d') = P_c\left(e - \frac{D}{2} + kd\right) \tag{12}$$

Since

$$\frac{\varepsilon_{sc}}{\varepsilon_c} = \frac{kd - d'}{kd} = \frac{(k + j - 1)}{k}$$

therefore

$$\frac{f_s'}{f_c} = \frac{E_s'\varepsilon_{sc}}{E_c\varepsilon_c} = \frac{n(k + j - 1)}{k} \tag{13}$$

Consideration of force equilibrium yields the equation

$$A_s f_s = A_s' f_s' + \tfrac{1}{2} f_c bkd - P_c \tag{14}$$

Substituting Equation (14) into Equations (12) and (13), gives the following two equations:

$$A_s f_s(D - kd - d''') + \frac{1}{3} f_c bk^2d^2 + A_s' \left(\frac{n(k + j - 1)}{k}\right) f_c(kd - d') - P_c \left(e - \frac{D}{2} + kd\right) = 0 \tag{15}$$

and

$$A_s f_s - A_s' \left[\frac{n(k + j - 1)}{k}\right] f_c - ½ f_c bkd + P_c = 0 \tag{16}$$

Solving Equations (15) and (16), one obtains

$$f_c = \frac{P_c \left(e - \frac{D}{2} - d'''\right)}{½ bkd \left(D - d''' - \frac{1}{3} kd\right) - A_s' \left[\frac{n(k + j - 1)}{k}\right] (d' - D + d''')} \tag{17}$$

Substituting Equation (17) into Equation (16) gives f_s and

$$\epsilon_s = \frac{f_s}{E_s}$$

Therefore the curvature immediately after cracking is given by

$$\phi_{ca} = \frac{\epsilon_s}{(1 - k)d} \tag{18}$$

Between Cracking and Ultimate

CASE A

See Figure 7. $\epsilon_c \leq \epsilon_o$

The total force due to concrete in compression is given by

$$C_{c1} = b \int_0^{kd} f dy$$

where f is expressed by Equation (2). Since

$$\frac{\varepsilon}{\varepsilon_0} = \frac{y}{y_1}$$

where

$$y_1 = (1 - k)d \frac{\varepsilon_0}{\varepsilon_s}$$

Therefore

$$C_{c1} = bf_c' \frac{k^2d^2}{y_1} \left(1 - \frac{kd}{3y_1}\right) \tag{19}$$

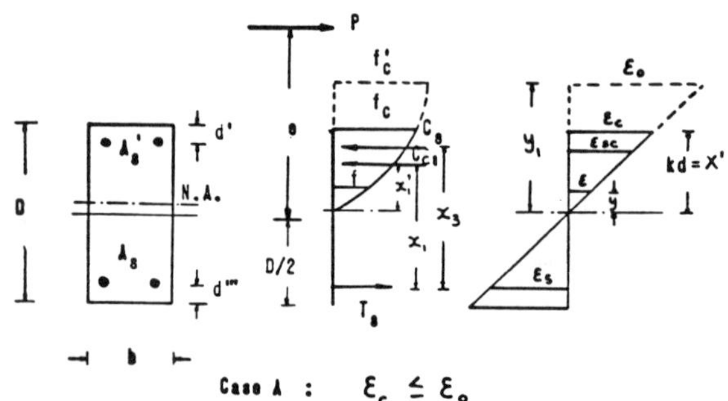

FIGURE 7. Moment-curvature-axial load analysis: between cracking and ultimate.

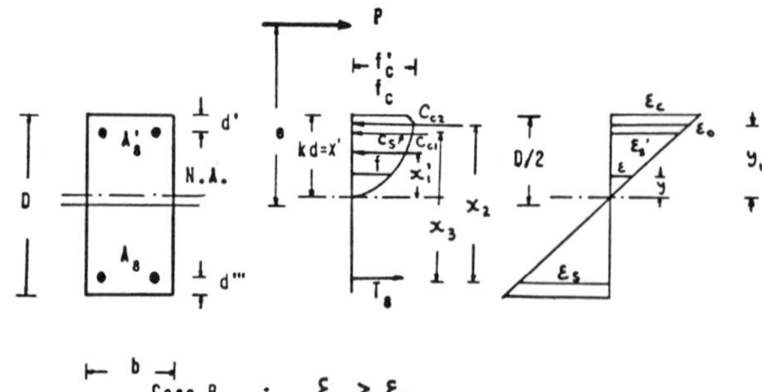

FIGURE 8. Moment-curvature-axial load analysis: between cracking and ultimate.

The distance of the centroid of the concrete compressive block from the neutral axis is given by

$$x_1' = \frac{b\int_0^{kd} fydy}{b\int_0^{kd} fdy} = \frac{bf_c'}{C_{c1}}\frac{k^3d^3}{y_1}\left(\frac{2}{3} - \frac{kd}{4y_1}\right) \quad (20)$$

Also

$$\epsilon_s' = \frac{(kd - d')\epsilon_s}{(1 - k)\,d}\ , \qquad \epsilon_y' = \frac{f_y'}{E_s'}$$

Hence, $f_s' = E_s'\epsilon_s' = E_s\epsilon_s'$ if $\epsilon_s' \le \epsilon_y'$ or $f_s' = f_y'$ if $\epsilon_s' \ge \epsilon_y'$. Let $C_s = A_s'f_s'$ and $f_s = E_s\epsilon_s$ if $\epsilon_s \le \epsilon_y$ or $f_s = f_y$ if $\epsilon_y \le \epsilon_s \le \epsilon_{sh}$ or $f_s = f_y + (\epsilon_s - \epsilon_{sh})E_{ss}$ if $\epsilon_s \ge \epsilon_{sh}$ where ϵ_{sh} and E_{ss} are defined in Figure 9.

Let $T_s = A_sf_s$ then the condition of equilibrium can be which as

$$T = C_{c1} + C_s - P \quad (21)$$

Now, expressing P as function of T_s, C_{c1} and C_s (P is unknown) and taking moment about the neutral axis as shown in Figure 7.

$$P = \frac{T_s(d - kd)}{\left(e - \frac{D}{2} + kd\right)} + C_{c1}\left(\frac{x'_1}{e - \frac{D}{2} + kd}\right) + C_s\left(\frac{x_3 - d + kd}{e - \frac{D}{2} + kd}\right) \quad (22)$$

Substituting Equation (22) into Equation (21) results in the equation

$$C_{c1} + C_s - T_s - T_s\left(\frac{d - kd}{e - \frac{D}{2} + kd}\right) - C_{c1}\left(\frac{x_1'}{e - \frac{D}{2} + kd}\right) - C_s\left(\frac{x_3 - d + kd}{e - \frac{D}{2} + kd}\right) = 0$$

Let $x_1 = x_1' + d - kd$ and $x_3 = d - d'$ and a rearrangement leads to the equation

$$\left(e - \frac{D}{2} + d\right)(C_{c1} + C_s - T_s) - C_{c1}x_1 - C_sx_3 = 0 \quad (23)$$

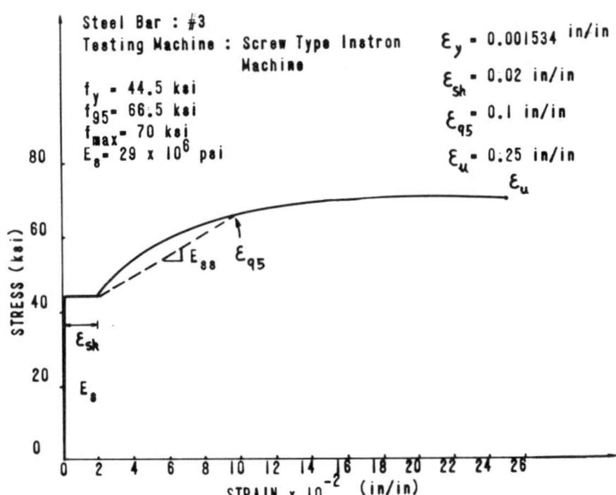

FIGURE 9. Stress-strain curve for reinforcing Bar No. 3.

or

$$\left(e - \frac{D}{2} + kd\right)(C_{c1} + C_s - T_s) - C_{c1}x_1' + C_s(d' - kd) - T_sd(1 - k) = 0 \quad (24)$$

Further simplification results in the following equation to determine k where the coefficients of linear algebraic equation are function of one variable, ϵ_s, i.e.

$$D_1k^4 + D_2k^3 + D_3k^2 + D_4k + D_5 = 0 \quad (25)$$

where D_1, D_2, D_3, D_4, and D_5 can be found in Appendix A. For the case $\epsilon_s' \ge \epsilon_y'$, i.e., $C_s = A_s'f_y'$, the values of coefficients in Equation (25) can be found in Appendix A. Then

$$\epsilon_c = \frac{\epsilon_sk}{(1 - k)}$$

and

$$\phi = \frac{\epsilon_c}{kd} \quad (26)$$

Taking moment about the neutral axis yields the equation:

$$M = C_{c1}x_1' + C_s(x_3 - d + kd) + T_s(d - kd) \quad (27)$$

CASE B

See Figure 8. $\epsilon_c \ge \epsilon_o$

The concrete compressive block is divided into two

parts:

(1) A parabolic shape is assumed for $0 \le \epsilon \le \epsilon_o$, and
(2) A straight line for $\epsilon_o \le \epsilon \le \epsilon_c$

Compressive force corresponding to the first part is given by

$$C_{c1} = b\int_0^{y_1} f dy = \frac{2}{3} by_1 f_c'$$

located at a distance x_1' from the neutral axis, given by

$$x_1' = \frac{5}{12}\frac{by_1^2 f_c'}{C_{c1}} \tag{28}$$

For the case, $\epsilon_o \le \epsilon \le \epsilon_c$, the compressive force is

$$C_{c2} = b\int_{y_1}^{kd} f dy = b\int_{y_1}^{kd} f_c'\left(1.18 - 0.18\frac{\varepsilon}{\varepsilon_0}\right) dy = \tag{29}$$

$$= bf_c'\int_{y_1}^{kd}\left(1.18 - 0.18\frac{y}{y_1}\right) dy$$

where

$$y_1 = (1 - k)\, d\frac{\varepsilon_0}{\varepsilon_s}$$

Hence,

$$C_{c2} = bf_c'\left(1.18kd - \frac{0.09k^2d^2}{y_1} - 1.09y_1\right) \tag{30}$$

and

$$x_2 = \frac{1}{C_{c2}}\left[bf_c'\int_{y_1}^{kd}\left(1.18 - 0.18\frac{y}{y_1}\right) ydy\right] + d(1 - k)$$

$$= \frac{bf_c'}{C_{c2}}\left(-\frac{0.06k^3d^3}{y_1} + 0.59k^2d^2 - 0.53y_1^2\right) + d(1 - k)$$

Equilibrium of forces results in the equation

$$T_s = C_{c1} + C_{c2} + C_s - P \tag{31}$$

which can be solved to determine the value of P as follows:

$$P = T_s\frac{(d - kd)}{\left(e - \frac{D}{2} + kd\right)} + C_{c1}\frac{x_1'}{\left(e - \frac{D}{2} + kd\right)} \tag{32}$$

$$+ C_s\frac{(x_3 - d + kd)}{\left(e - \frac{D}{2} + kd\right)} + C_{c2}\frac{(x_2 - d + kd)}{\left(e - \frac{D}{2} + kd\right)}$$

Substituting Equation (32) into Equation (31), gives

$$\left(e - \frac{D}{2} + d\right)(C_{c1} + C_{c2} + C_s - T_s)$$

$$- [C_{c1}x_1' + C_{c1}d(1 - k) + C_{c2}x_2] - C_s x_3 = 0$$

In the case of $\epsilon_s' \le \epsilon_y'$,

$$C_s = A_s'f_s' = A_s'E_s\,\varepsilon_s' = A_s'E_s\frac{(kd - d')\varepsilon_s}{(1 - k)d}$$

The values of coefficients D_1, D_2, D_3, D_4, and D_5 to be used in Equation (25) can be found in Appendix A. In the case of $\epsilon_s' \ge \epsilon_y'$, $C_s = A_s'f_s' = A_s'f_y'$. The values of coefficients D_1, D_2, D_3, D_4, and D_5 in Equation ,25) are given in Appendix A.

Here

$$\epsilon_c = \frac{\epsilon_s k}{(1 - k)} \tag{33}$$

and

$$\phi = \frac{\epsilon_c}{kd} \tag{34}$$

and

$$M = T_s(d - kd) + C_{c1}x_1' + C_{c2}(x_2 - d + kd) + C_s(kd - d') \tag{35}$$

It should be noted that Equations (2a) and (2b) are two general mathematical equations to describe the complete stress-strain for unconfined concrete. Therefore, the moment-curvature curve derived using the above equations can be used for analysing any structural concrete member in which the influence of lateral reinforcement on plain concrete is small. Any mathematical equations describing the complete stress-strain for confined concrete can be used if the effect of ties or stirrups become significant.

Another method based on the load control process can be used to predict the moment-curvature relationship under combined uniaxial bending and compression. This method is a special case of the moment-curvature studies subjected to combined biaxial bending and compression (see following section), which predicts the moment-curvature characteristics from zero moment up to the maximum capacity. The complete curve can be obtained by using suitable load increments.

EXPERIMENTAL TEST PROGRAMME

The prominent work in this area is due to Hognestad [31], who measured the load-strain curves, strain distribution across the mid-span section and load-deflection curves in columns. Flower [13] also measured the moment-curvature curve for reinforced concrete columns which were governed by concrete compression failure. In all previous work, the investigators were unable to obtain the load-deformation curves beyond the maximum axial load in the column specimens. The primary objectives of this investigation are therefore not only to assess the accuracy of the computer program developed but also to examine the behaviour of these specimens beyond the maximum experimental applied load in columns.

Four short, square, under-reinforced tied columns, HSl, HS2, U.7 and U.8 were tested under combined bending and axial compression. The ratios of the total length of columns to the depth of the overall section depth (i.e. ℓ/t) were 15 for HSl and HS2 and 10 for U.7 and U.8 respectively. Specimen U.8 was the "trial" specimen for the new loading arrangement in column tests using the hard rubber blocks and was proved to be workable. The test specimens were designed according to the ACI Code 318-71 [2] and were controlled by the testing facilities available in the Structural Laboratory at McGill University. The eccentricities used for the three columns tested were calculated to govern the member failure due to flexural tension. The specimen details and the arrangements of strain gages and demec gage points for HSl, HS2, U.7 and U.8 are shown in Figure 10.

Two kinds of loading arrangements for uniaxially loaded column tests were used in the present investigation. Specimens HSl and HS2 were loaded by the Conventional Universal testing machine (Baldwin-T-E Universal Testing Machine) (see Figure 11), while specimens U.7 and U.8 were loaded by means of the special loading set-up (deformation control) as shown in Figure 12. The end conditions were designed to be pinned-ended. The other experiment details such as material properties, fabrication, instrumentation and test procedures etc. can be found in Reference 33.

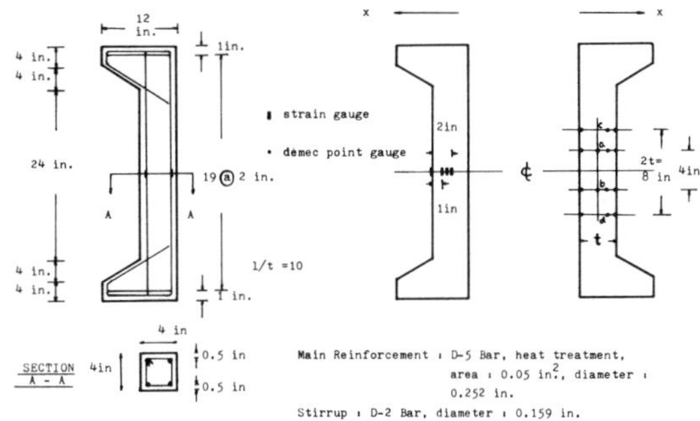

FIGURE 10. Specimen details and arrangements of strain and demec gauge for column U-7.

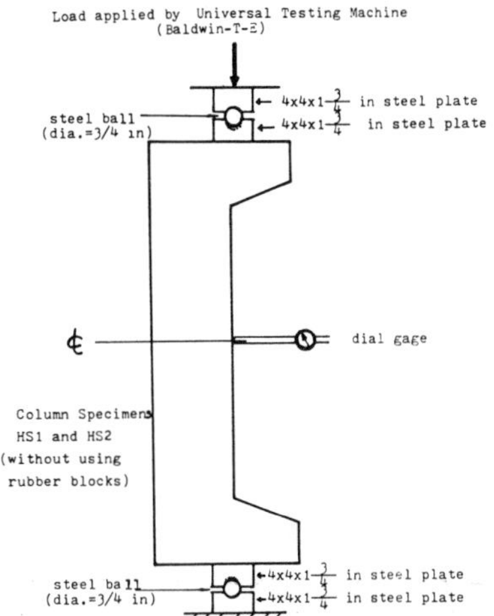

FIGURE 11. Loading and end conditions for pin-ended short columns HS1 and HS2.

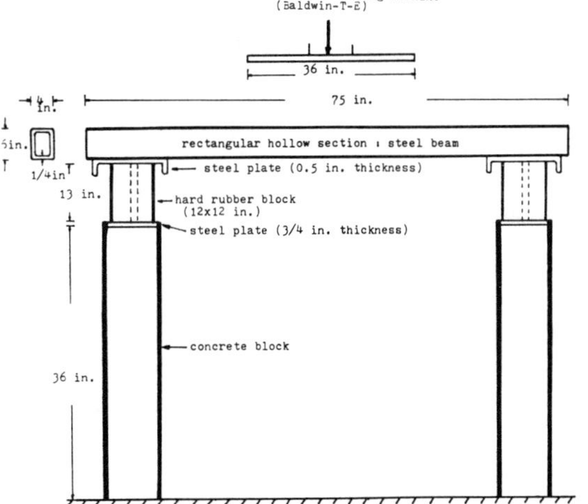

FIGURE 12. Systematic loading arrangement for column tests using hard rubber blocks.

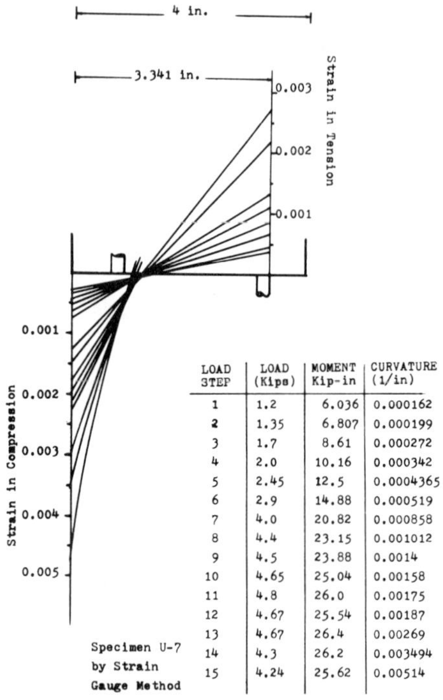

4 in.
3.341 in.
0.003
0.002
0.001
Strain in Tension
0.001
0.002
0.003
0.004
0.005
Strain in Compression

LOAD STEP	LOAD (Kips)	MOMENT Kip-in	CURVATURE (1/in)
1	1.2	6.036	0.000162
2	1.35	6.807	0.000199
3	1.7	8.61	0.000272
4	2.0	10.16	0.000342
5	2.45	12.5	0.0004365
6	2.9	14.88	0.000519
7	4.0	20.82	0.000858
8	4.4	23.15	0.001012
9	4.5	23.88	0.0014
10	4.65	25.04	0.00158
11	4.8	26.0	0.00175
12	4.67	25.54	0.00187
13	4.67	26.4	0.00269
14	4.3	26.2	0.003494
15	4.24	25.62	0.00514

Specimen U-7
by Strain
Gauge Method

FIGURE 13. Strain distribution across mid-height section for specimen U-7.

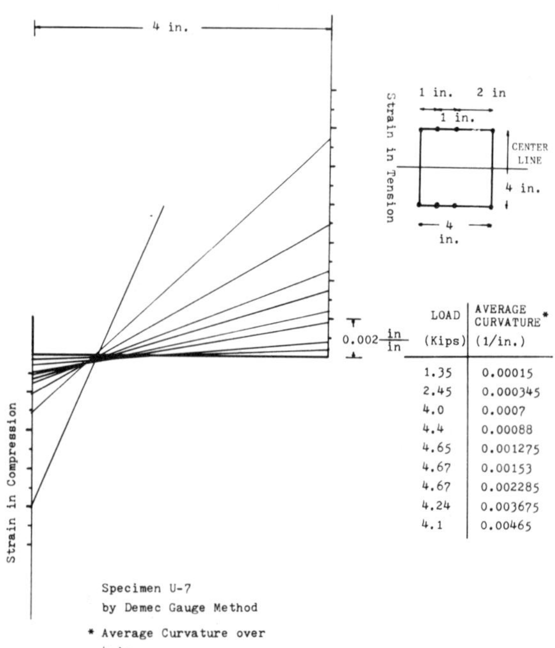

FIGURE 14. Average strain distribution across mid-height section for specimen U-7.

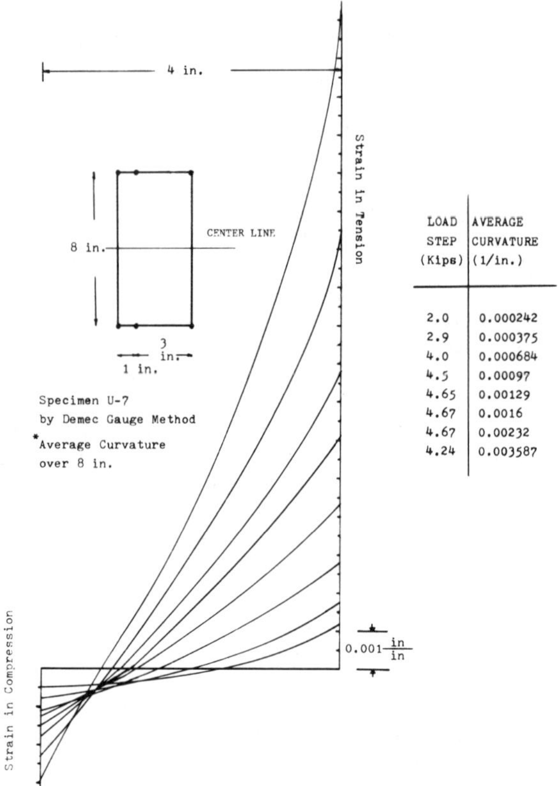

FIGURE 15. Average strain distribution across mid-height section for specimen U-7.

ANALYSIS OF MOMENT-CURVATURE TEST RESULTS

Three short columns HS1, HS2 and U.7 were tested by Hsu [33] to evaluate the strain distribution across the mid-span section in columns (see Figure 13, Figure 14, and Figure 15). The strain results at the measured points were obtained using strain gages and demec gages respectively. The average strain distributions across the mid-span section for U.7 were obtained by 4 in. and 8 in. range demec gage points shown in Figure 14 and 15 respectively.

If the strain distribution across the section is linear or near linear, the curvature is given by

$$\phi = \frac{\epsilon_c + \epsilon_s}{d} \tag{36}$$

where d is the distance between the center of gravity of the tensile reinforcing steel and the extreme compression fibre. If the strain distribution is non-linear, then the curvature is given by

$$\phi = \frac{\epsilon_c}{kd} = \frac{\text{Maximum Concrete Strain}}{\text{Neutral Axis Depth}} \tag{37}$$

where kd is the depth of the compression zone.

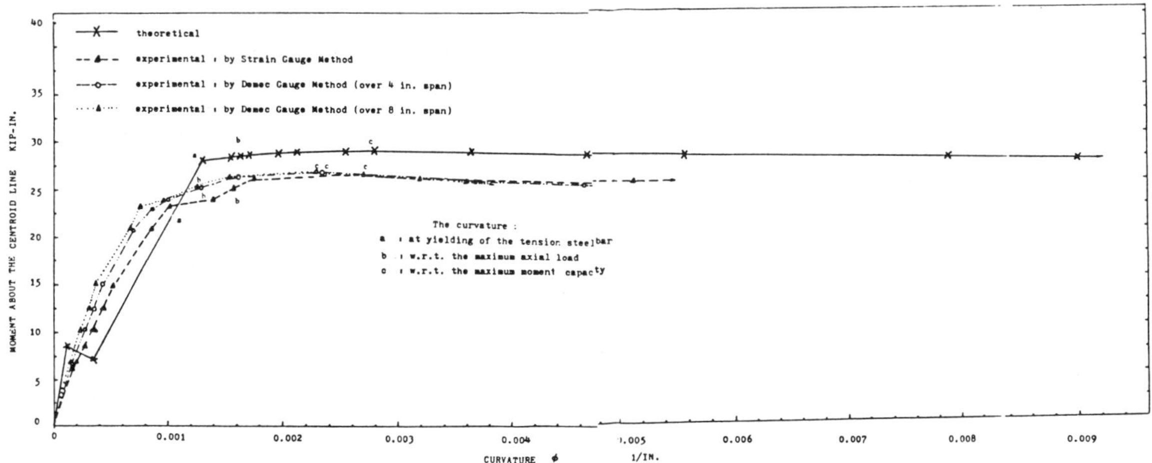

FIGURE 16. Moment-curvature curves at mid-height section for specimen U-7.

Equation (37) has been suggested by Mattock [53] and Corley [15] and several other investigators such as Hsu [33] who computed the deflections for several beams based on the three different definitions of curvature used by Hsu [33]. Hsu [33] compared these values with the experimental results and concluded that the definition of curvature based on Equation (37) was the most realistic.

Figure 16 shows the results of the writer's computer analysis and the three experimental moment-curvature curves for specimen U.7 (the loading arrangement is shown in Figures 4 and 12). The experimental curves shown in Figure 16 have been obtained from zero load and go beyond the maximum experimental axial load and the maximum moment capacity. The termination in the experimental tests was due to the strain gages or demec gages being damaged on account of crushing of the concrete or being overstrained. Curve 1 in Figure 16 was obtained using the strain gages shown in Figure 13. The moment-curvature curves 2 and 3 in Figure 16 were obtained using the strain values obtained using the demec gages (Figures 14 and 15). It is interesting to observe that the curvatures obtained from the "strain gage method" are slightly larger than those obtained from the "demec gage method since the "demec gage method" represents average curvatures over the finite lengths.

Figure 17 shows the writer's analysis and the experimental curves for column HS2. Because of the loading arrangement used (see Figure 11), it is not possible to obtain the complete load-deformation curves. Also, the average curvature obtained from the rotation measurement over 4 in. length at mis-span (level arm method) is in good agreement with the writer's analysis up to near the moment capacity w.r.t. the maximum load capacity. Good agreement was similarly noted with the moment-curvature curve obtained from the "strain-gage method" (see Figure 17). The formula for the calculation of this average curvature from the rotation measurement is given by:

$$\phi = \frac{\theta_{ab}}{d_{ab}} \tag{38}$$

However, the above equation is valid only for linear behaviour of materials under loads.

It has been found that the maximum moment capacity and the moment capacity w.r.t. the maximum load in column specimens were well predicted by the writer's analysis (see Table 1). The theoretical moment was obtained from the above analysis while the experimental one was equal

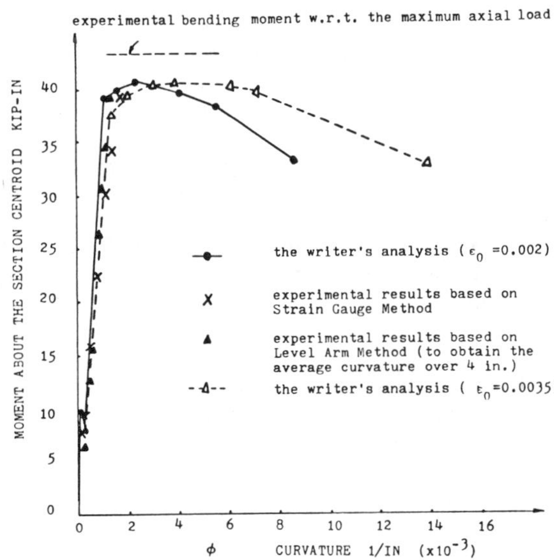

FIGURE 17. Moment-curvature curves for specimen HS2.

TABLE 1. Analysis of Results for Moment Capacities.

	Modes of Failure		Moment Capacity (Kip-in.)			
			test result[5]		writer's analysis	
Test Speci.	writer's analysis	test result	w.r.t. max. load	w.r.t. the max. moment capacity	w.r.t. the max. moment capacity	w.r.t. max. load
A-5a[1]	T[4]	T	621.78	—	602.163	592.4
A-15a[1]	T	T	1136.96	—	1086.49	1067.54
B-5a[1]	T	T	595.612	—	588.6	583.73
B-5b[1]	T	T	589.225	—	596.675	591.66
S1-5[2]	T	T	32.5	32.5	31.6	31.6
S2-5[2]	T	T	46.0	46.0	46.6	46.6
HS1[3]	T	T	36.2	—	33.53	32.285
HS2[3]	T	T	43.45	—	40.98	39.474
U-7[3]	T	T	26.0	26.4	28.795	28.435

[1]Hognestad's test specimens
[2]Obeid's test specimens (simply supported beams under two-point loading)
[3]The writer's test specimens
[4]Denotes tension failure
[5]Based on $P_4x(e + \delta_2)$ for pinned-end short columns where P_4 and δ_2 were obtained from tests

to the experimental load multiplied by (e + central deflection δ_2).

In compression failures, concrete starts crushing and spalling before the farthest bar in tension reaches the yield stress. The tension failures are characterized by large deformations consequent to yielding of tension steel and considerable movement of the neutral axis prior to the crushing of the concrete. This is followed by the buckling of the farthest bar in compression before the crushing of the concrete in compression zone in concrete column. All the column specimens in this investigation were characterized as tension failures by their load-strain curves. Also, the concrete crushing zone plays an important role in forming the discontinuity region or so-called plastic hinge in concrete. The average finite discontinuity length can vary from 3.8 in. to 9 in. based on the above test results. The other experimental and analytical results of load-strain, load-deflection and moment-rotation can be found in Reference 33.

BEHAVIOUR OF REINFORCED CONCRETE COLUMNS OF STANDARD SHAPES UNDER COMBINED BIAXIAL BENDING AND COMPRESSION

Introduction

This section presents a method for the determination of strain and curvature distributions in reinforced and prestressed concrete sections subjected to biaxial bending moments and/or axial load. A closed form solutions which accounts for (1) the inelasticity of concrete, (2) the elastic, plastic and strain-hardening behaviour of steel reinforcement, and (3) the geometrical complexities in the section, etc., is very time consuming and impractical. Hence, a numerical analysis suggested by Hsu [33,39–41] was used in this investigation.

The computer programs developed in this investigation (detailed in Reference 33) have the ability to use any standard reinforced and prestressed concrete section geometry and material properties. These computer programs also give the information for the stress and strain distributions across the section, the ultimate strength and interaction surface of biaxially loaded short columns, and also develop the flexural rigidity coefficients for members of a three-dimensional structural concrete frame for both the elastic and inelastic loading stages and can be incorporated without much difficulty in the existing three-dimensional programs of steel frame analysis in elastic analysis such as Gere and Weaver [25], Weaver [73], Wang [71], Beaufait, Rowan, Hoadley and Hackett [5] and Harrison [29] and Mamet [50] and inelastic analysis such as Briunette and Fenves [11], Morris and Fenves [57,58], Toridis and Khozeimeh [70], Wen and Farhoomand [74], Nigam [60], Porter and Powell [65].

It should be noted that the writer's program can calculate the load-deformation curves from zero to maximum moment capacity using a "load control" process in the case of combined biaxial bending and axial compression. For the case of moment-deformation characteristics under constant axial load, the "deformation control" process was used. Furthermore, the approximate equations for the evaluation of deflection and rotation for pin-ended biaxially loaded short columns have also been derived in this investigation.

Historical Review

BIAXIAL MOMENT-CURVATURE RELATIONSHIPS

Very little information is available on the behaviour of reinforced concrete sections under combined biaxial bending and axial compression. Two investigators have studied the biaxial moment-curvature relationships under constant axial compression. Cranston [16] suggested a numerical method in which biaxial moment curvature relationships under constant axial force can be obtained from zero load to the maximum moment capacity. He realized that difficulties arose in the handling of a strain-softening material when a large part of the cross section has negative stiffness. Warner [72] also proposed another approach for the moment-curvature relationship under constant axial compression. Warner's method cannot be generally used in 3-D structural analysis because he assumed that θ (inclination of the curvature vector to one of the principle axes which is related to the angle position of the neutral axis in section) is given which is impossible to know in these studies. Therefore, Warner's method can be applied to the study of the uniaxial moment-curvature relationship under constant axial load for any geometrical section.

LOAD-DEFLECTION RELATIONSHIPS OF BIAXIALLY LOADED COLUMNS

Drysdale [19] developed a computer approach on the load-deflection relationships for pin-ended long columns. Later Farah and Huggins [20] modified Drysdale's computer approach, solving the simultaneous non-linear equations by Newton-Raphson method suggested by Gurfinkel and Robinson [28]. Farah and Huggins [20] concluded that their approach can determine the column deflected shape for various axial load values. This information can be used to determine the buckling load in the case of a long column or the crushing load in the case of a short column.

ULTIMATE LOAD STUDIES OF BIAXIALLY LOADED SHORT COLUMNS

Gurfinkel [27] extended the methods by Gurfinkel and Robinson [28] to a study of the ultimate load and biaxial moments in column to any materials such as masonry and other materials, weak in tension, under biaxial bending. Some investigators also studied the failure of short columns under combined biaxial bendings and axial compression. Included in the list are Anderson and Lee [4], Chu and Pabarcius [14], Bresler [8], Furlong [23], Czerniak [18], Pannell [62], Meek [54], Aas-Jakobsen [1], Parme, Nieves and Gouwens [64] and Ramamurthy [66], etc. All the methods above, however, did not take into account the effect of mid-span deflection on ultimate load capacity in columns.

Numerical and Computer Method

BASIC ASSUMPTIONS

The following assumptions have been made in this theoretical analysis:

1. The bending moments are applied about the principal axes.
2. Plane sections remain plane.
3. The longitudinal stress at a point is a function only of the longitudinal strain at that point. The effect of creep and shrinkage are ignored.
4. The stress-strain curves for the materials used are known.
5. Strain reversal does not occur.
6. The effect of deformation due to shear and torsion and impact effects are negligible.
7. The section does not buckle before the ultimate load is attained.
8. Perfect bond exists between the concrete and the reinforcing steel.

THEORETICAL ANALYSIS: COMPUTER APPROACH

The general moment-curvature relationship (or moment-rotation or load-deflection relationship) is shown in Figure 18. Typical moment-curvature and load-deflection curves for a uniaxially or a biaxially loaded column are shown in Figure 18 which show that close to the peak of the load deflection curve there can be two equilibrium positions corresponding to the same load (Bleich [7] and Cranston [16]). To avoid difficulties in this region it is convenient to find solutions corresponding in specified deflections. Yet in the initial stage (from zero to the minimum load) it is simpler to find solution corresponding to specified loads. The method developed in this section uses a numerical analysis based on the structural concrete sections subjected to combined biaxial bending and axial load in which bending moments about two principal axes M_x and M_y can be found from zero to the maximum moments (Figure 18).

In the case of the biaxial moment-curvature under constant axial load, deformation control process was used. The details of this method can be found in Reference 33.

Figure 19 shows an arbitrary structural concrete section in column. The cross-section of the structural member is divided into several small elements. Consider an element k with its centroid at point (x_k, y_k) referred to the principal axes (Figure 19). The strain ϵ_k across the element k can be assumed to be uniform and since plane sections

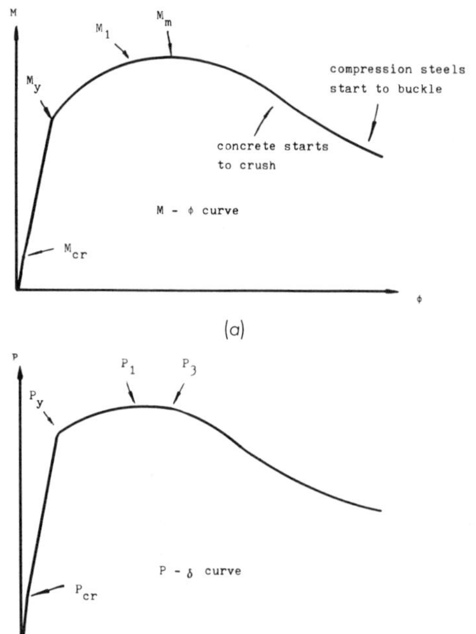

FIGURE 18. Typical relationship between moment-curvature and load-deflection curves for short columns.

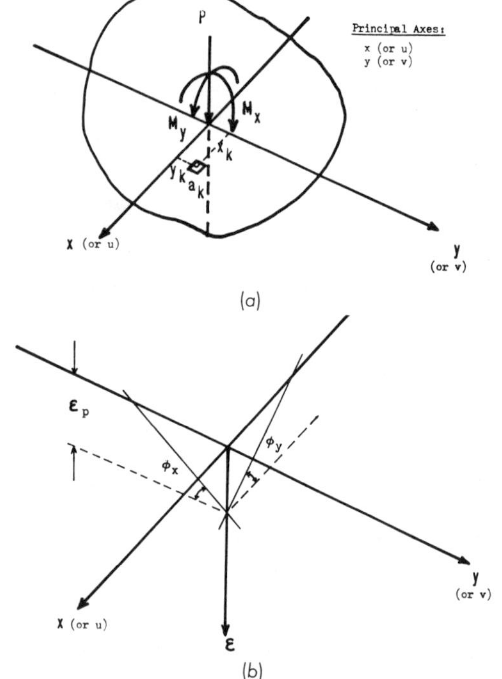

FIGURE 19. Idealization of a cross-section subjected to biaxial bending and axial load.

remain plane during bending,

$$\epsilon_k = \epsilon_p + \phi_x y_k + \phi_y x_k \tag{39}$$

where

ϵ_p = uniform direct strain due to an axial load P

ϕ_x = the curvature produced by the bending moment component M_x and is considered positive when it causes compressive strains in the positive y − direction, and

ϕ_y = the curvature produced by the bending moment component M_y and is considered positive when it causes compression in the positive x − direction.

The writer has modified Cranston and Chatterji's [17] stress-strain curves for the concrete as shown in Figure 20. These curves account for the strain-softening of concrete and the values of ultimate compressive strains for the unconfined and the confined concrete elements have been maintained. The experimental stress-strain curve for steel has been idealized using piece-wise linear approximation to the curve in the strain-hardening region as shown in Figure 21. The value of the tangent modulus of elasticity $(E_t)_k$ for a steel or concrete element to be used in a given iteration cycle can be obtained from Figures 20 and 21.

Once the strain distribution across the cross-section is established, the axial force P and the bending moment components M_x and M_y can be calculated using the following equations:

$$P_{(c)} = \sum_{k=1}^{n} f_k a_k \tag{40a}$$

$$M_{x(c)} = \sum_{k=1}^{n} f_k a_k y_k \tag{40b}$$

$$M_{y(c)} = \sum_{k=1}^{n} f_k a_k x_k \tag{40c}$$

Subscript (c) indicates values of P, M_x and M_y2 calculated in an iteration cycle, and a_k is the area of element k.

For a given section (known geometry and material properties) the stress resultants P, M_x and M_y can be expressed as functions of ϕ_x, ϕ_y and ϵ_p given by the following equations:

$$P = P(\phi_x, \phi_y, \epsilon_p) \tag{41a}$$

$$M_x = M_x(\phi_x, \phi_y, \epsilon_p) \tag{41b}$$

$$M_y = M_y(\phi_x, \phi_y, \epsilon_p) \tag{41c}$$

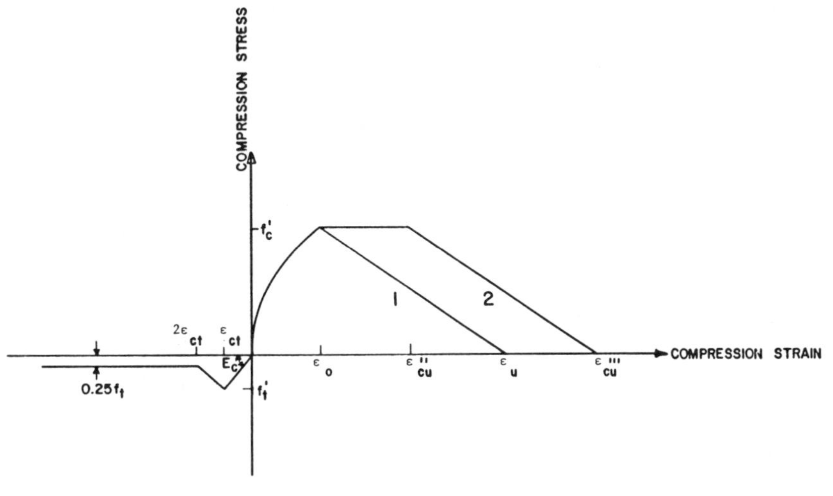

1: UNCONFINED CONCRETE
2: CONFINED CONCRETE

FIGURE 20. Concrete stress-strain curves.

If $P_{(s)}$ is the final value of P for which the equilibrium and the compatibility conditions are satisfied, the convergence of $P_{(c)}$ to $P_{(s)}$ can be accelerated using a modification of the extended Newton-Raphson method. The final values of M_x and M_y can be calculated using the equations in the case of biaxially loaded columns.

$$M_{x(s)} = P_{(s)} e_y \tag{42a}$$

and

$$M_{y(s)} = P_{(s)} e_x \tag{42b}$$

where e_x and e_y are the assumed load eccentricity components.* $P_{(s)}$ $M_{x(s)}$, and $M_{y(s)}$ can be expressed in terms of $P_{(c)}$, $M_{x(s)}$ and $M_{y(c)}$ using Taylor's expansion retaining linear terms, as follows:

$$P_{(s)} = P_{(c)} + \frac{\partial P_{(c)}}{\partial \phi_x} \delta\phi_x + \frac{\partial P_{(c)}}{\partial \phi_y} \delta\phi_y + \frac{\partial P_{(c)}}{\partial \varepsilon_p} \delta\varepsilon_p \tag{43a}$$

$$M_{x(s)} = M_{x(c)} + \frac{\partial M_{x(c)}}{\partial \phi_x} \delta\phi_x + \frac{\partial M_{x(c)}}{\partial \phi_y} \delta\phi_y + \frac{\partial M_{x(c)}}{\partial \varepsilon_p} \delta\varepsilon_p \tag{43b}$$

$$M_{y(s)} = M_{y(c)} + \frac{\partial M_{y(c)}}{\partial \phi_x} \delta\phi_x + \frac{\partial M_{y(c)}}{\partial \phi_y} \delta\phi_y + \frac{\partial M_{y(c)}}{\partial \varepsilon_p} \delta\varepsilon_p \tag{43c}$$

Let

$$u' = P_{(c)} - P_{(s)} \tag{44a}$$

$$v' = M_{x(c)} - M_{x(s)} \tag{44b}$$

$$w' = M_{y(c)} - M_{y(s)} \tag{44c}$$

Then Equation (43) can be written as

$$-u' = \frac{\partial P_{(c)}}{\partial \varepsilon_p} \delta\varepsilon_p + \frac{\partial P_{(c)}}{\partial \phi_x} \delta\phi_x + \frac{\partial P_{(c)}}{\partial \phi_y} \delta\phi_y \tag{45a}$$

$$-v' = \frac{\partial M_{x(c)}}{\partial \varepsilon_p} \delta\varepsilon_p + \frac{\partial M_{x(c)}}{\partial \phi_x} \delta\phi_x + \frac{\partial M_{x(c)}}{\partial \phi_y} \delta\phi_y \tag{45b}$$

$$-w' = \frac{\partial M_{y(c)}}{\partial \varepsilon_p} \delta\varepsilon_p + \frac{\partial M_{y(c)}}{\partial \phi_x} \delta\phi_x + \frac{\partial M_{y(c)}}{\partial \phi_y} \delta\phi_y \tag{45c}$$

An increment in axial load $\delta p_{(c)}$ produces an increment of strain, $\delta\epsilon_p$, at each element in the section. The corresponding stress change at element k is therefore $\delta\epsilon_p(E_t)_k$. The resulting change $\delta P_{(c)}$ in $P_{(c)}$ is given by

$$\delta P_{(c)} = \sum_{k=1}^{n} (E_t)_k a_k \delta\varepsilon_p$$

*If $P_{(s)}$ and $M_{x(s)}$ or $M_{(s)}$ are not linear relations, $P_{(s)}$, $M_{x(s)}$ and $M_{y(s)}$ must be given as individual values in the computer input.

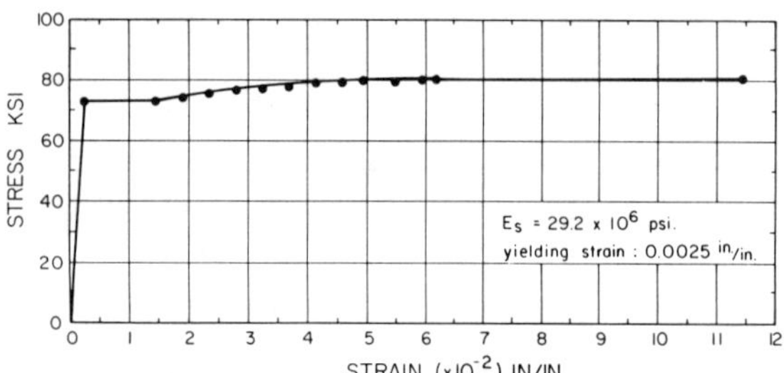

FIGURE 21. Steel stress-strain curve for deformed bar D-5 (1 ksi = 6.895 × 10⁶ N/m²).

Therefore,

$$\frac{\partial P_{(c)}}{\partial \varepsilon_p} = \sum_{k=1}^{n} (E_t)_k a_k \tag{46a}$$

Similarly, the changes $\delta M_{x(c)}$ and $\delta M_{y(c)}$ can be expressed in terms of $\delta\epsilon_p$ and lead to the equations:

$$\frac{\partial M_{x(c)}}{\partial \varepsilon_p} = \sum_{k=1}^{n} (E_t)_k a_k y_k \tag{46b}$$

$$\frac{\partial M_{y(c)}}{\partial \varepsilon_p} = \sum_{k=1}^{n} (E_t)_k a_k x_k \tag{46c}$$

Similar expressions can be derived for $\delta P_{(c)}$, $\delta M_{x(c)}$ and $\delta M_{y(c)}$ in terms of changes $\delta\phi_x$ and $\delta\phi_y$ and yield the following results:

$$\frac{\partial P_{(c)}}{\partial \phi_x} = \sum_{k=1}^{n} (E_t)_k a_k y_k \tag{46d}$$

$$\frac{\partial M_{x(c)}}{\partial \phi_x} = \sum_{k=1}^{n} (E_t)_k a_k y_k^2 \tag{46e}$$

$$\frac{\partial M_{y(c)}}{\partial \phi_x} = \sum_{k=1}^{n} (E_t)_k a_k x_k y_k \tag{46f}$$

$$\frac{\partial P_{(c)}}{\partial \phi_y} = \sum_{k=1}^{n} (E_t)_k a_k x_k \tag{46g}$$

$$\frac{\partial M_{x(c)}}{\partial \phi_y} = \sum_{k=1}^{n} (E_t)_k a_k x_k y_k \tag{46h}$$

$$\frac{\partial M_{y(c)}}{\partial \phi_y} = \sum_{k=1}^{n} (E_t)_k a_k x_k^2 \tag{46i}$$

Equations (45) and (46) can be rearranged in a matrix form as shown in Equation (47) to give the rates of changes of P, M_x, and M_y due to changes in ϵ_p, ϕ_x and ϕ_y,

$$\begin{bmatrix} \sum_{k=1}^{n} (E_t)_k a_k & \sum_{k=1}^{n} (E_t)_k a_k y_k & \sum_{k=1}^{n} (E_t)_k a_k x_k \\ & \sum_{k=1}^{n} (E_t)_k a_k y_k^2 & \sum_{k=1}^{n} (E_t)_k a_k x_k y_k \\ \text{Symmetric} & & \sum_{k=1}^{n} (E_t)_k a_k x_k^2 \end{bmatrix} \begin{Bmatrix} \delta\varepsilon_p \\ \delta\phi_x \\ \delta\phi_y \end{Bmatrix} = - \begin{Bmatrix} u' \\ v' \\ w' \end{Bmatrix} \tag{47a}$$

or

$$[\underset{\sim}{K}] \begin{Bmatrix} \delta\varepsilon_p \\ \delta\phi_x \\ \delta\phi_y \end{Bmatrix} = - \begin{Bmatrix} u' \\ v' \\ w' \end{Bmatrix} \tag{47b}$$

or

$$\begin{Bmatrix} \delta\varepsilon_p \\ \delta\phi_x \\ \delta\phi_y \end{Bmatrix} = -[\underset{\sim}{K}]^{-1} \begin{Bmatrix} u' \\ v' \\ w' \end{Bmatrix} \tag{47c}$$

The values of u', v' and w' can be selected to suit the accuracy required and their substitution in Equation (47c) at the end of mth iteration cycle yields the values of $\delta\epsilon_p$, $\delta\phi_x$ and $\delta\phi_y$ which lead to values of ϵ_p, ϕ_x and ϕ_y for the $(m + 1)^{th}$ iteration cycle as follows:

$$\varepsilon_{p(m+1)} = \varepsilon_{p(m)} + \delta\varepsilon_p \tag{48a}$$

$$\phi_{x(m+1)} = \phi_{x(m)} + \delta\phi_x \tag{48b}$$

$$\phi_{y(m+1)} = \phi_{y(m)} + \delta\phi_y \tag{48c}$$

The iteration at a given load level continues in the computer program developed until convergence is obtained within specified tolerance. Once this is achieved, the computer program takes up the next load level and repeats the entire procedure.

Computer Program shown in Reference 33 was written in "FORTRAN IV" language and prepared by the writer for an IBM Computer 360/75. The flow diagram is shown in Figure 22. This computer program will be discussed in the following sections.

DISCUSSIONS OF ACCURACY AND CONVERGENCE OF COMPUTER ANALYSIS

It must be noted that errors can arise on account of one or more of the following:

(1) The assumption of uniform strain and hence a strain distribution within a small element (into which the structure is divied) makes the accuracy dependent upon the numbers of elements into which the section has been idealized. This cross-section idealization introduces errors in calculating the curvatures corresponding to a given loading. A particular error resulting from the cross-section idealization is that the procedure cannot deal with

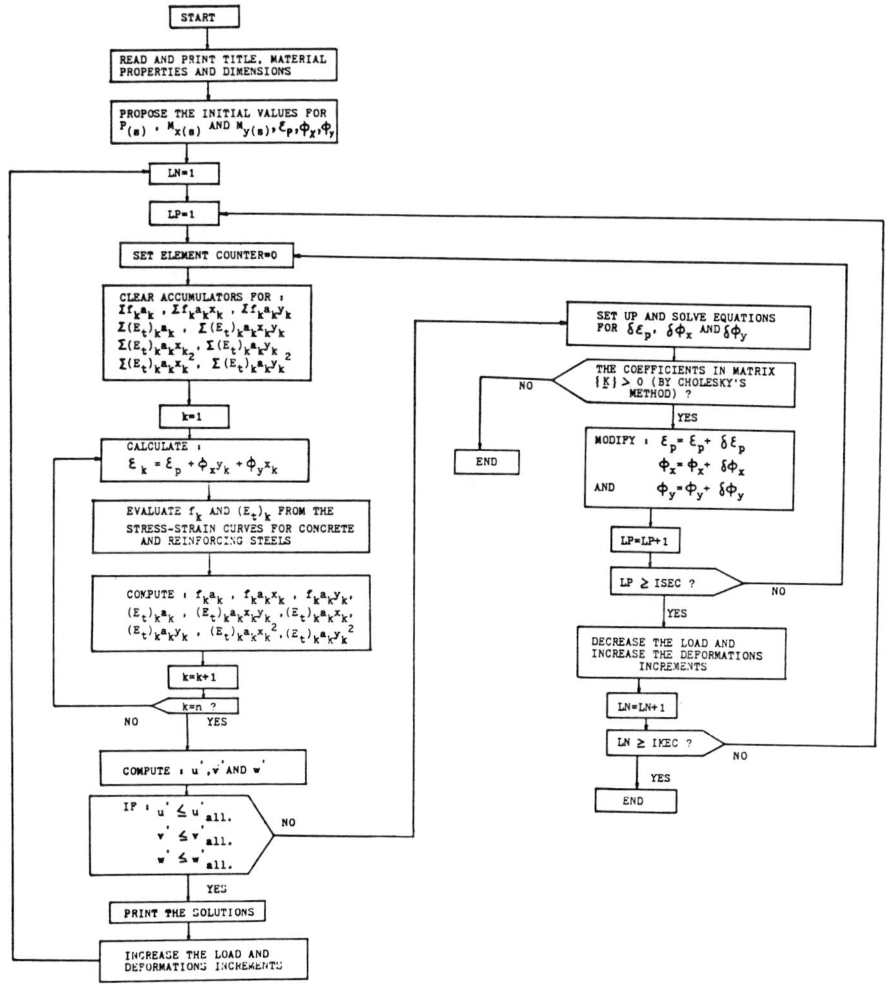

FIGURE 22. Flow diagram: computer program for moment-curvature relationship under combined biaxial bending and axial load.

TABLE 2. Typical Process of Convergence.

Notations	ε_p (in/in)	$M_{x(c)}$ (lbs-in) or $M_{x(s)}$	ϕ_x (l/in)	$M_{y(c)}$ (lbs-in) or $M_{y(s)}$	ϕ_y (l/in)	P_c (lbs.) or P_s
proposed values	0.000034	34500.	0.000021	12590.	0.00001	10000.
Trial 1	-0.00006	269637.625	0.00033	71151.25	0.000092	55553.496
Trial 2	0.000019	-44993.7734	-0.000035	-14772.7461	-0.000012	8901.0898
Trial 3	0.000195	125860.25	0.000092	38017.32	0.00003	58670.9375
Trial 4	0.000019	26955.5859	0.000016	10047.7187	0.000007	4912.60547
Trial 5	0.00004	31939.5898	0.000019	12180.5976	0.000007	11757.07
Trial 6	0.00003	35684.7929	0.000023	11656.0351	0.000009	9545.72656
Expected Value	0.000034	34499.9961	0.000021	12589.9961	0.00001	10000.

plastic hinges [16]. Such hinges can form in columns made up of ideal elastic-plastic material [32]. If full plasticity is present, the tangent moduli of all elements are zero and matrix $[\underset{\sim}{K}]$ becomes singular. Fortunately, the procedure for claculating the curvature at any load level utilizes materials which are either strain-hardening or strain-softening, provided the stiffness of the section remains positive, i.e. provided the coefficients of matrix $[\underset{\sim}{K}]$ remain positive. It may be noted that in the Cholesky's method [68,73] the computation will be terminated if the coefficients of the matrix $[\underset{\sim}{K}]$ become negative.

(2) Due to allowable incompatibilities. There are three allowable incompatibilities in procedures. One is $u'_{all.}$ for P, others are $v'_{all.}$ and $w'_{all.}$ for M_x and M_y respectively. When the inelastic range is entered, the curvatures can become extremely sensitive to small changes in axial load and moments, and to ensure one percent accuracy in the calculation of curvatures, $u'_{all.}$, $v'_{all.}$ and $w'_{all.}$ may have to be 10^{-5} and 10^{-6} times the values of axial load and moments respectively. It should be noted that accuracy in this context refers to the accuracy of determination of the response of the idealized cross-section.

(3) Errors may arise from the assumptions of stress-strain curves both for reinforcing bar and unconfined or confined concrete (see Figure 20).

The convergence of the procedures to calculate the curvatures and the axial strain corresponding for a given axial load and moments is dependent upon the validity of Equations (39) and (47). If the stiffness of the section becomes very small, the procedures will occasionally not converge [16,77]. If loads beyond the ultimate capacity of the section are proposed, the procedures will be unable to reach a solution.

The typical process of convergence for each trial ϵ_p, ϕ_x and ϕ_y corresponding to P, M_x and M_y is shown in Table 2.

Experimental Test Programme

Six short, reinforced tied columns were tested under biaxial bending and axial compression; the parameters studied were the load eccentricity and the various loading arrangements and the end conditions. The test specimens were designed according to the 1971 ACI Building Code [2]. Two loading brackets were provided at each column end to assist with the applications of biaxially eccentric loads. These brackets were heavily reinforced to prevent any premature failure. The primary objective of this investigation was to examine the complete moment-curvature behavior of the column specimens from zero load

TABLE 3. Column Species.

Column Speci.	Size (in x in)	Height/ Breadth	No. and Size of Bars	$p=(A_s + A_s')/A_g$	f_y (Ksi.)	$E_s \times 10^6$ (Psi.)	A_s'' (in²)	f_c' (Psi)	e_y' (in.)	e_x' (in.)	e_y (in.)	e_x (in.)
U-1	4 x 4	10	9-D5	0.0281	73	29.2	0.02	3905	7	5	—	—
U-2	4 x 4	10	9-D5	0.0281	73	29.2	0.02	3806	7	6	—	—
U-3	4 x 4	10	9-D5	0.0281	73	29.2	0.02	3894	7	7	—	—
U-4	4 x 4	10	9-D5	0.0281	73	29.2	0.02	3830	4	4	—	—
U-5	4 x 4	10	9-D5	0.0281	73	29.2	0.02	3715	—	—	5.5	0.5
U-6	4 x 4	10	9-D5	0.0281	73	29.2	0.02	3895	—	—	7	0.5

Note: Conversion Factors—1 in. = 25.4 mm., 1 ksi = 6.895 x 10^6 N/m².

TABLE 4. Mixture of Concrete (Per Batch).

aggregate total weight:	sieve size NO.	percentage of aggregate (%) (total 100%)	weight (lbs.) (total weight 16.2 lbs.)
16.2 lbs.	No. 10	20%	3.24
	No. 16	20%	3.24
	No. 24	25%	4.05
	No. 40	25%	4.05
	No. 70	10%	1.62
cement total weight			4.15 lbs.
fresh water total weight			3.36 lbs.
total weight per batch			23.71 lbs.

to maximum load using load control and from maximum load until failure using deflection control besides assessing the accuracy of the analytical model.

The details of geometry of the specimens are shown in Figure 23 and Table 3. The reinforcement for specimens consisted of D5 size deformed steel bars (diameter = 0.252 in. or 6.4 mm., cross-sectional area = 0.05 in.^{2} or 32.25 mm.^{2}, according to ASTM A496-64) which were heated at 1100°F for a period of 20 minutes (percentage elongation = 11.5 percent). The column ties were fabricated from D2 deformed wire. The typical stress-strain curve for D5 deformed bar is shown in Figure 21. The concrete used for casting the test specimens was prepared from a graded mixture of crushed quartz sand, Portland cement Type III and water. The concrete properties and the maximum concrete compressive stress were determined using 3 × 6 in. (76.2 × 152.4 mm) cylinders. These were reported in Table 4.

Strain gages were installed on the steel bars at mid-height and carefully waterproofed. Type KFC-5-C1-11 strain gages with a gage length of 5mm. were used. Type PL-10-11 strain gages with a gage length of 10mm. were installed on the concrete surface. Typical strain gage locations for steel and concrete for specimens are shown in Figure 24. Some of the specimens were also instrumented with demec gage points (gage length = 4 in. or 101.6 mm.) on both faces to obtain the concrete strain (Figure 25).

The column specimens were tested in the Baldwin-Tate-Energy Testing Machine (capacity = 400 kips. or 1.78 × 10^6 N.). The applied loads were measured using calibrated load cells. The Loading Arrangement I (Figure 26b and 27) was used for specimens U-1, U-2, and U-4 excepting specimens U-5 and U-6 for which the Loading Arrangement II (Figure 26a) was utilized. Specimens U-3, U-4, U-5 and U-6 were pin-ended conditions. Both Loading Arrangements I and II used hard rubber blocks at the ends of a spreader beam as shown in Figure 27 to

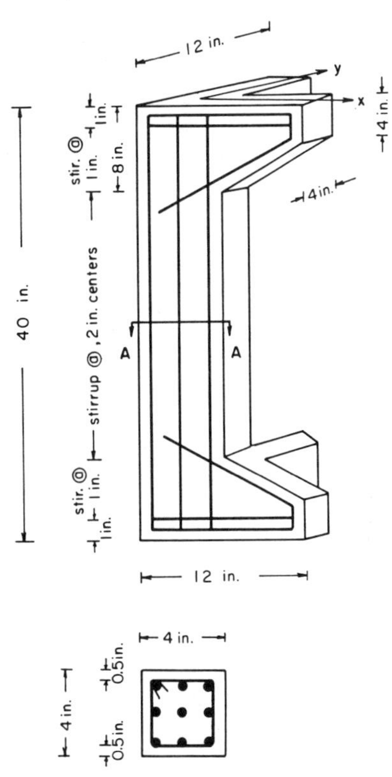

FIGURE 23. Specimen details (1 in. = 25.4 mm).

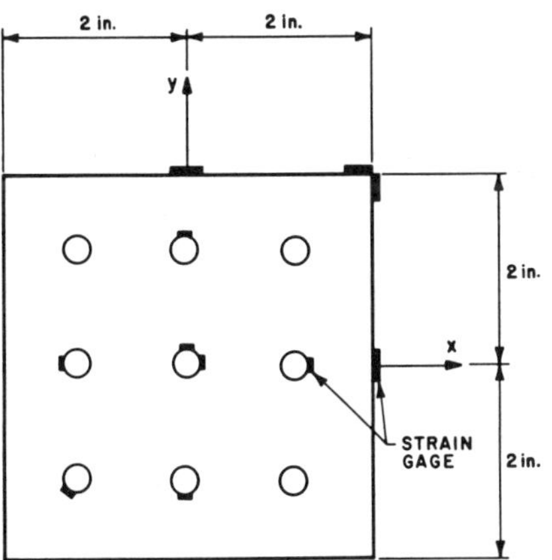

FIGURE 24. Locations of strain gages at mid-height of specimen (1 in. = 25.4 mm).

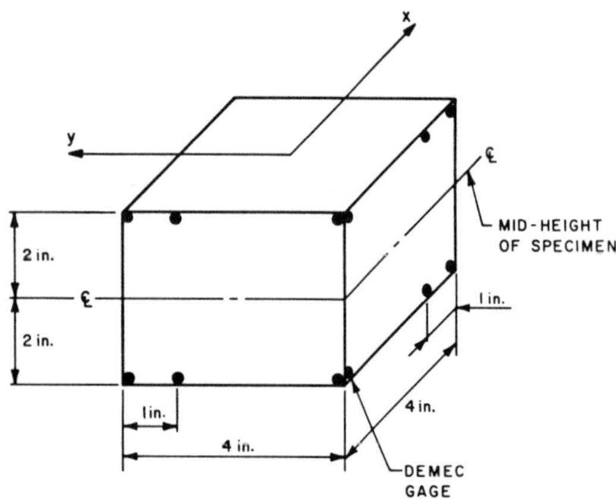

FIGURE 25. Locations of demec gages in specimen (1 in. = 25.4 mm).

FIGURE 27. Experimental set-up for column tests showing Specimen U-3.

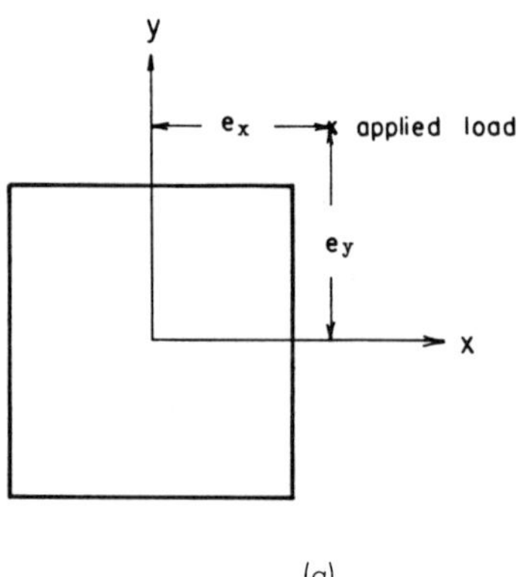

(a)

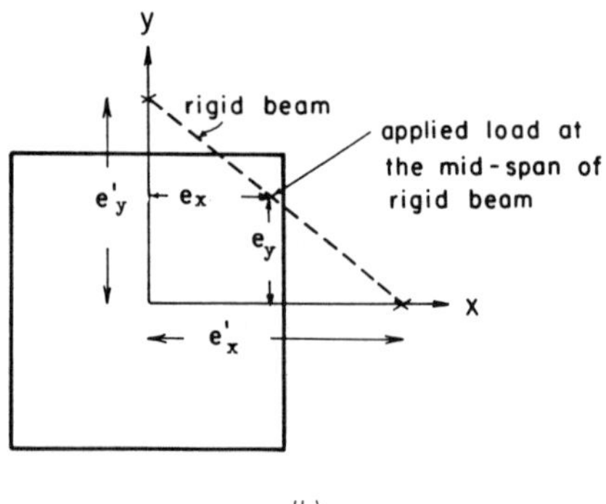

(b)

FIGURE 26. Loading conditions for biaxially loaded short columns.

eliminate the sudden collapse of the specimen at maximum load by eliminating the sudden release of energy upon reaching the maximum load. This enabled the testing of the specimens beyond the maximum load by balancing the energy supplied by the testing machine with the energy absorbed by the column specimen and the loading device including the hard rubber blocks. The Loading Arrangements I and II with hard rubber blocks were successfully used to determine the descending branch of the moment-curvature curve or load-deformation curve beyond the maximum load point by using a deflection control. More details of experimental test programe can be found in References 33 and 41.

Analysis of Test Results

YIELD AND MAXIMUM STRENGTHS

The experimental values of M_x and M_y were computed using the experimental axial load values obtained from the load cell measurements and the load eccentricities were corrected for the mid-height deflection of the column. These experimental values are detailed in Table 5.

The analytical values of bending moment components M_x and M_y were computed for all specimens using the writer's numerical analysis mentioned in the previous section. The cross section of the column specimens was divided into 81 concrete elements and 9 steel elements (Figure 28). Bending moment components were computed for two conditions (a) and (b) as shown in Table 5. Condition (a) relates to the bending moment values in which the remotest tension reinforcing bar reaches the yield point. Condition (b) refers to the maximum moment capacity. An examination of Table 5 shows that good agreement was achieved between the experimental strengths and the computed values.

TABLE 5. Bending Moment Results.

Column Speci.	Analysis: Maximum Moment Capacity (Kip-in.) M_x	M_y	Analysis: Moment at Yield (Kip-in.) M_x	M_y	Failure Mode	Test: Maximum Moment Capacity (Kip-in.) M_x	M_y	Test: Moment at Yield (Kip-in.) M_x	M_y	Failure Mode
U-1	36.63	26.16	30.1	21.5	T*	34.05	25.84	28.05	21.0	T
U-2	33.39	28.62	27.1	23.0	T	32.37	26.85	27.01	23.09	T
U-3	31.14	31.34	24.75	24.75	T	29.29	29.29	23.28	23.28	T
U-4	32.58	32.58	28.1	28.1	T	30.96	30.96	26.0	26.0	T
U-5	48.4	6.05	41.07	5.17	T	49.33	6.22	38.91	4.82	T
U-6	48.61	3.47	39.0	2.85	T	47.74	3.48	38.5	2.83	T

*T denotes "tension failure";
Conversion Factors: 1 kip-in. = 1.13×10^2 N-m.

MOMENT-CURVATURE CURVES

The strain distribution along both axes for all specimens was obtained using the data from both the electrical strain gages and demec gages. Both sets of strain values were first established before evaluating the moment-curvature relationships. Typical strain distributions for specimen U-3 ($e_x' = e_y'$) using the strain gage and the demec gage values are shown in Figures 29 and 30. Typical experimental moment-curvature values using both strain distributions for specimens U-1 through U-6 are shown in Figures 31 through 36 and Table 6. Good agreement was obtained between the theoretical curves and the experimental curves from zero load up to the maximum moment capacity (the ascending branch of the moment-curvature curves). The descending branch of the moment-curvature curves cannot be achieved using the writer's numerical

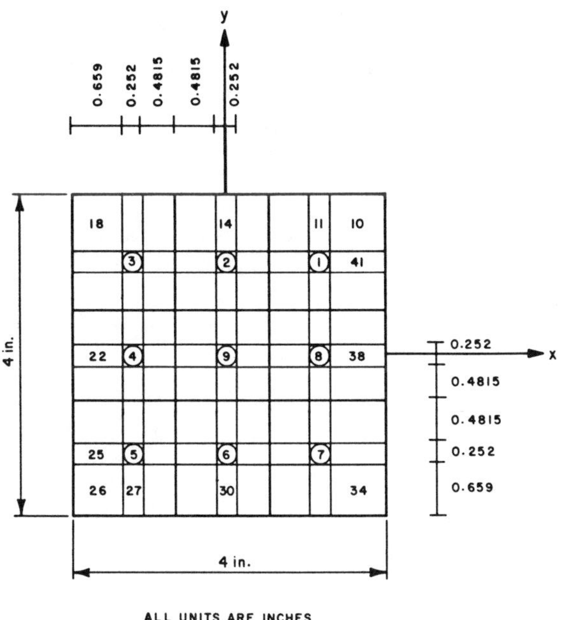

FIGURE 28. Cross section of specimen divided into nine steel elements and eighty-one concrete elements (1 in. = 25.4 mm).

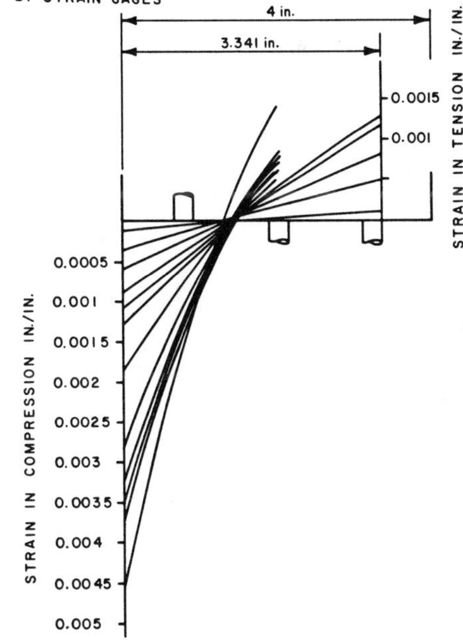

FIGURE 29. Strain distribution across the mid-height section (x- or y-axis) for Specimen U-3 (1 in. = 25.4 mm.).

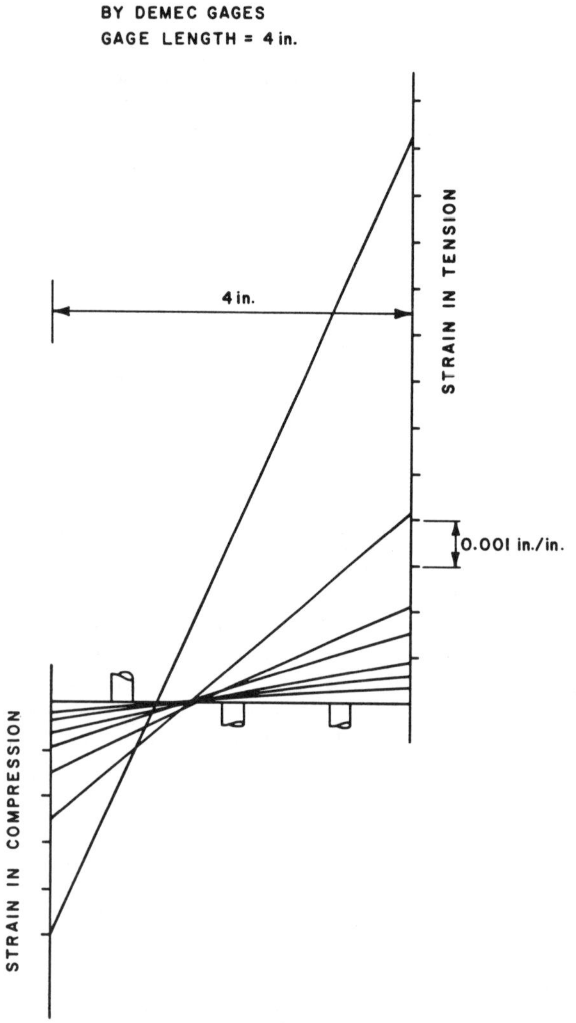

FIGURE 30. Strain distribution across the mid-height section (x- or y-axis) for Specimen U-3 (1 in. = 25.4 mm.).

analysis because of the control of load increments in computations. The terminations in the experimental curvature measurements were due to the strain gages getting damaged or dislodging of demec gages because of the crushing of the concrete or because of severe tension cracks. Figures 31 through 36 show that the bending moment and curvature components were computed for three conditions. Condition (a) relates to the moment and curvature values at yield. Conditions (b) and (c) refer to the maximum axial load in column and the maximum moment capacity in section respectively.

LOAD-CURVATURE AND LOAD-DEFLECTION CURVES

The deflection components along the x- and y-axis were measured using dial gages with a least count of 0.001 in.; the theoretical mid-height deflection components were calculated using a modification of the moment-area theorems. The loads were obtained from the readings in load cells; and the theoretical axial load P_3 was calculated considering the effect of mid-height deflection components. The results of theoretical and experimental load-curvature and load-deflection curves are illustrated in Figures 37 through 43. Good agreement was obtained between the theoretical and experimental curves from zero load until the maximum moment capacity of the column specimen. It must be noted that the experimental curves of Figures 37 through 43 show both ascending and descending branches of the curves. More details of theoretical and experimental studies on this subject can be found in Reference 33 and Reference 41.

BEHAVIOUR OF IRREGULAR REINFORCED CONCRETE COLUMNS UNDER COMBINED BIAXIAL BENDING AND COMPRESSION

Introduction

The behavior of irregular reinforced concrete columns, shearwalls and other structural members has been a con-

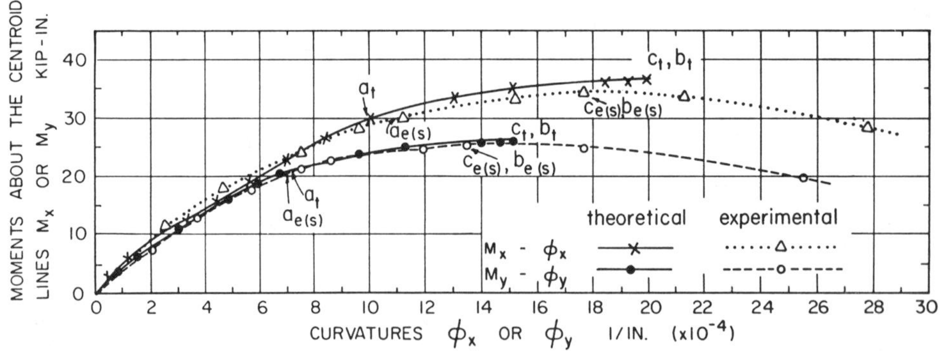

FIGURE 31. Biaxial moment-curvature curves for Specimen U-1 (1 kip-in. = 1.13 × 10² N-m., 1 l/in. = 39.4 l/m.).

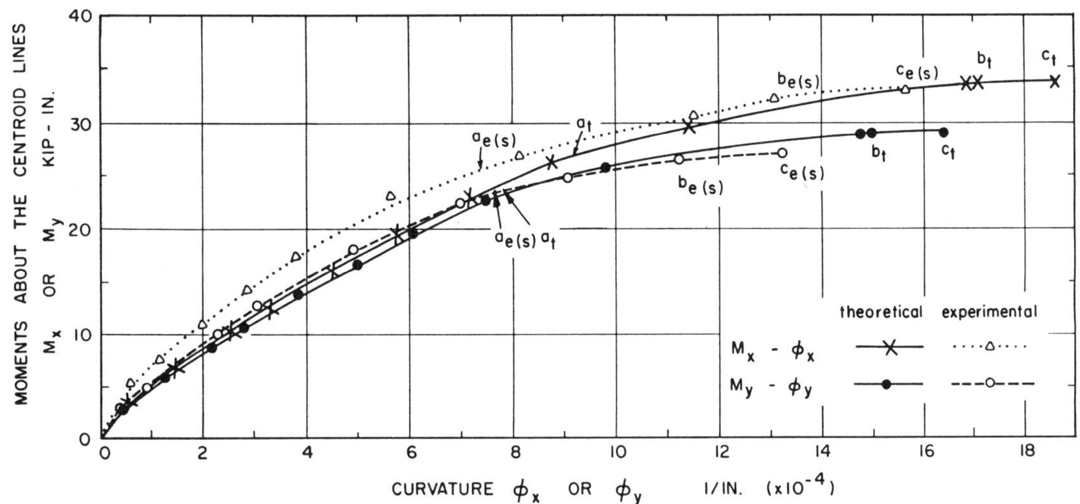

FIGURE 32. Biaxial moment-curvature curves for Specimen U-2.

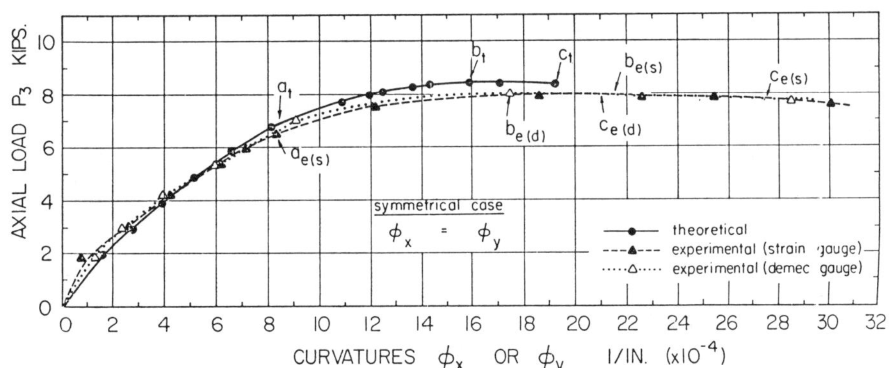

FIGURE 33. Biaxial load-curvature curves for Specimen U-3 (1 kip. = 4448 N., 1 l/in. = 0.394 l/cm.).

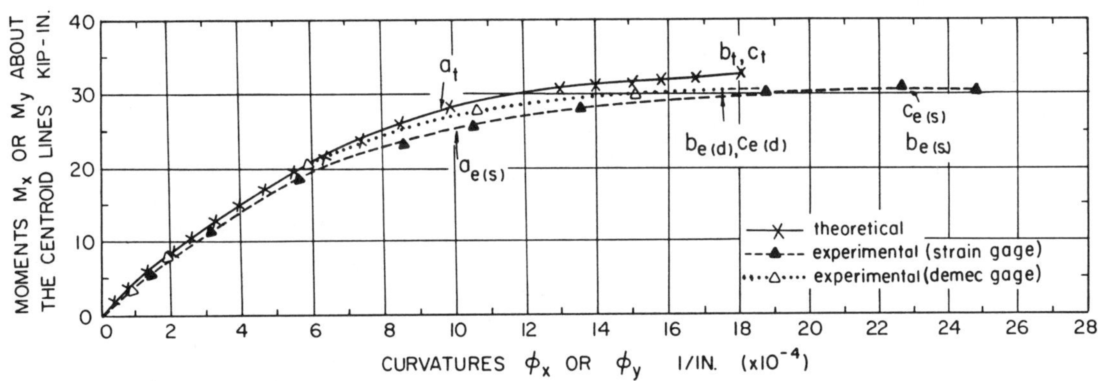

FIGURE 34. Biaxial moment-curvature curves for Specimen U-4 (1 kip-in. = 1.13 × 10² N-m., 1 l/in. = 39.4 l.m.).

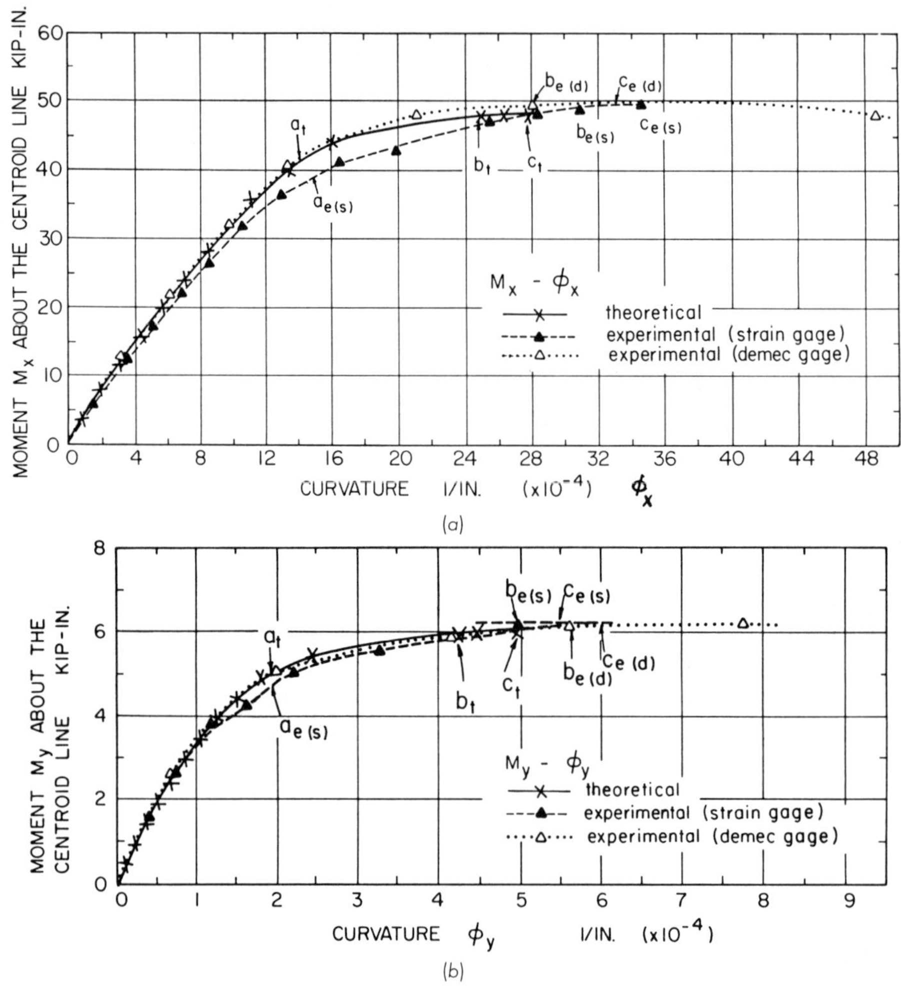

FIGURE 35. Biaxial moment-curvature curves for Specimen U-5 (1 kip-in. = 1.13 × 10² N-m., 1 l/in. = 39.4 l/m.).

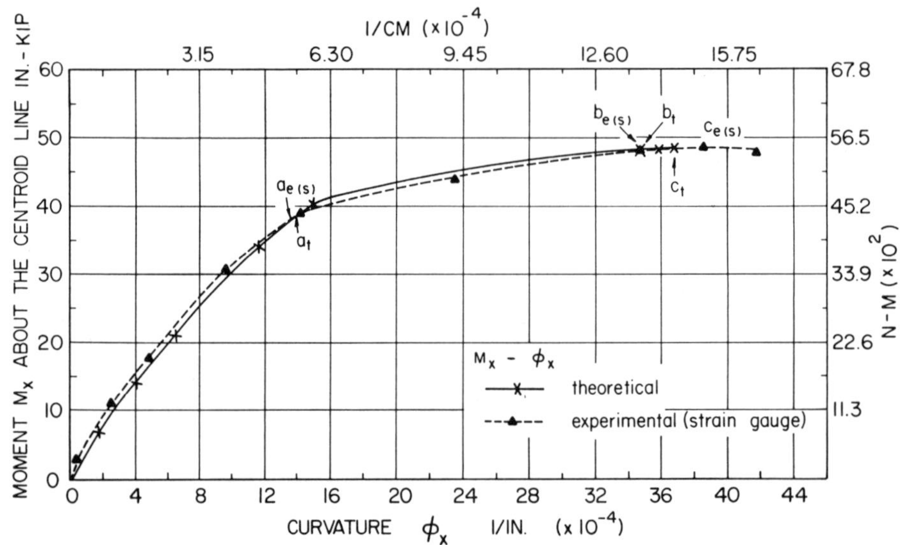

FIGURE 36. Biaxial moment-curvature curves for Specimen U-6 (continued)

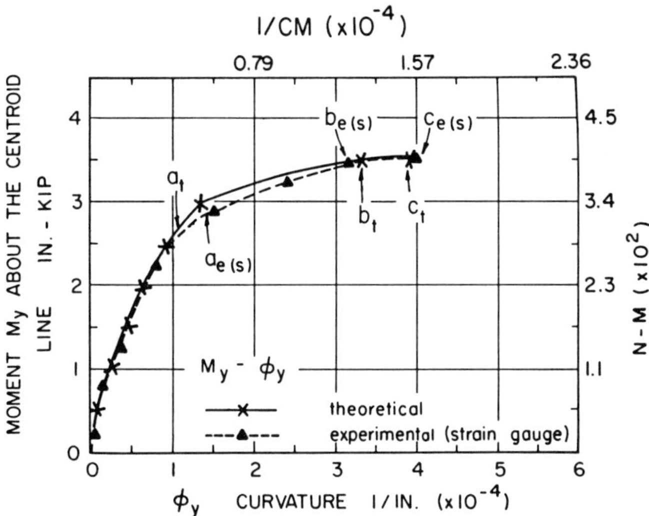

FIGURE 36. Biaxial moment-curvature curves for Specimen U-6.

stant concern for structural engineers to design a safe and economic structure. The shape of the elements in a reinforced concrete structure may be used to optimize its structural strength, to make better use of the available space, to improve the esthetic appearance of the structure, or to facilitate construction. Due to the locations of the structural members, the shapes of the structures and the nature of the applied loads, many members are subjected to combined biaxial bending and axial load.

In the case of reinforced concrete columns and shearwalls, wide flange cross sections have been used to improve the structural strength of the members, L-shaped cross sections are usually located at the corner of the buildings, C-shaped cross sections are commonly used as columns and enclosures of the elevator shafts, S and X-shaped sections have been used for purely architectural reasons, and other irregular sections are used in the precast concrete industry. In concrete foundation construction, the hollow box or round columns are frequently used. Hollow round cross sections are also used in piling and pole construction.

Current methods of analysis are based on the basic governing equations of strain compatibility and summation of forces and moments and the stress-strain relations for both concrete and reinforcing steel. The methods can be classified into three groups; (a) Discrete element method: The cross section of the structural member is divided into several small elements. Square, rectangular

TABLE 6. Curvature Results.

	Analysis				Test					
	Curvature at Maximum Moment (l/in.) x 10^{-3}		Curvature at Yield (l/in.) x 10^{-4}		Curvature at Maximum Moment (l/in.) x 10^{-3}				Curvature at Yield (l/in.) x 10^{-4}	
Column Speci.	ϕ_x	ϕ_y	ϕ_x	ϕ_y	ϕ_x^a	ϕ_x^b	ϕ_y^a	ϕ_y^b	ϕ_x^a	ϕ_y^a
U-1	1.993	1.52	10.0	7.19	1.77	—	1.35	—	8.8	6.8
U-2	1.87	1.647	9.35	7.85	1.58	—	1.33	—	8.4	7.6
U-3	1.92	1.92	8.45	8.45	2.7	2.1*	2.7	2.1*	8.4	8.4
U-4	1.81	1.81	9.65	9.65	2.27	1.8*	2.27	1.8*	10.1	10.1
U-5	2.78	0.5	14.2	2.0	3.47	3.3*	0.55	0.6	14.6	1.95
U-6	3.674	0.394	14.0	1.2	3.85	—	0.4	—	13.8	1.36

*obtained by extrapolating
[a]from strain gage method
[b]from demec gage method
Conversion Factors: 1 l/in. = 39.4 l/m.

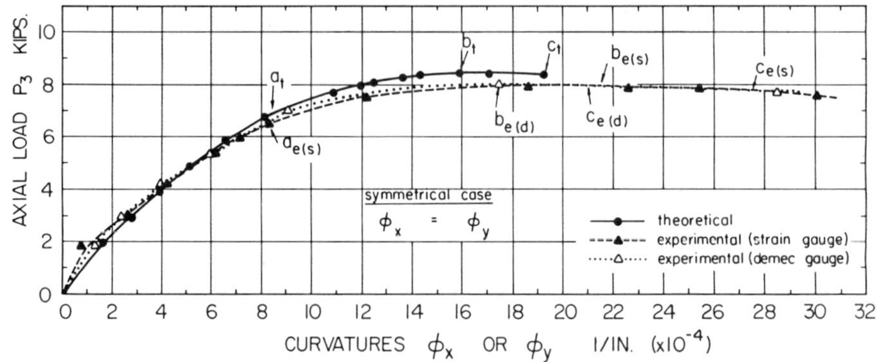

FIGURE 37. Biaxial load-curvature curves for Specimen U-3 (1 kip. = 4448 N., 1 l/in. = 0.394 l/cm.).

or triangular elements were used by Cranston [16], Chen and Ausuta [12], and Gesund [26] among others. As the normal strain distribution is planar, the corresponding stresses are uni-dimensional because they depend only on the distance to the neutral axis. Therefore, an approximation can be obtained by dividing the compression zone in bands parallel to the neutral axis and concentrating the band resultant in its center of gravity. (b) Triangular superposition method: The compression of the structural member is assumed to be triangular, since it is feasible to describe systematically any polygonal section by means of triangular components, and the principle of superposition of individual forces and moments is valid while a planar strain distribution is true, the method can be easily programmed. A general solution to the problem by using the classical Newton-Raphson numerical method has been successfully applied to the design of rectangular footing and reinforced concrete section by Gurfinkel and Robinson [28], Farah and Huggins [20], and Gurfinkel [27]. (c) Line integral method: If the stress block is represented by a polynomial function, it is possible to compute the corresponding stress integrals by converting them into line integrals, or evaluate directly from the vertex coordinate by straight integration or using the Gaussian quadrature technique. The method has been used by Werner [75] and Marin and Martin [52], Marin [51], and Brondum-Nielsen [10].

Hsu [33,35,36,40,41] proposed a computer analysis program by using the rectangular discrete element method and the extended Newton-Raphson numerical technique. The stress-strain curves for concrete and the reinforcing steel were idealized using a piecewise linear approximation. This analytical model developed can account for any section geometry and material properties and also simulate the load-deformation and moment-curvature behaviour of structural member under biaxial bending and axial load (compression and tension). The analysis results have been compared with the rectangular column tests by Hsu [33], Anderson and Lee [4], Bresler [8], and Ramamurthy [66]. Good agreement was obtained between the experimental deformations and strength and those of the analytical results calculated using the above computer pro-

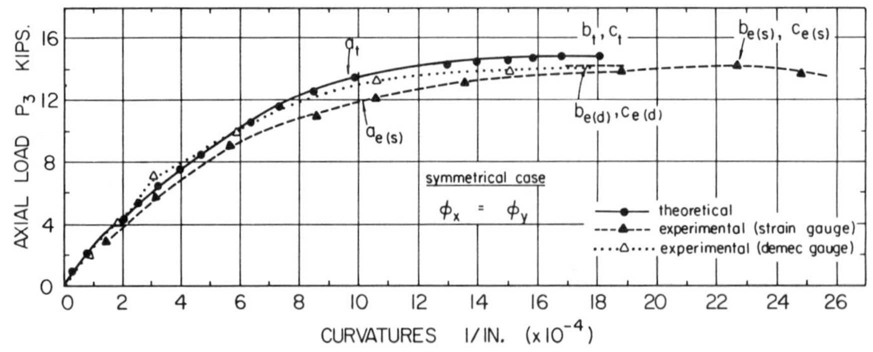

FIGURE 38. Biaxial load-curvature curves for Specimen U-4 (1 kip. = 4448 N., 1 l/in. = 0.394 l/cm.).

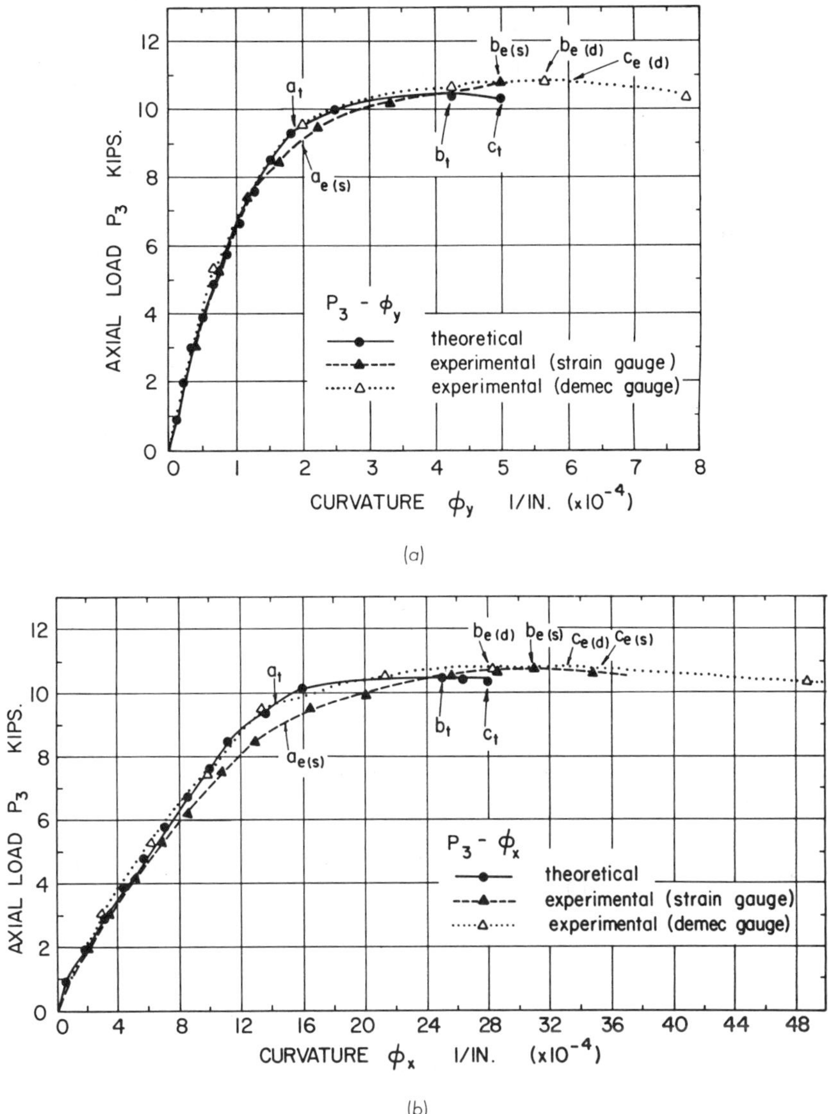

FIGURE 39. Biaxial load-curvature curves for Specimen U-5 (1 kip. = 4448 N., 1 l/in. = 0.394 l/cm.).

gram. Bhattacharyay, et al. [6], and Hsu, et al. [35–37, 42–44] conducted the strength tests of columns having T, L and C shapes subject to biaxial bending and axial compression (monotonic loading). They compared their test results with Hsu's analytical model and concluded that the iteration procedure is a highly convergent one and is fairly good in estimating the ultimate load carrying capacity of the columns.

If the cross section of the columns and shear walls other than the rectangular or the circular are used, the structural engineer must determine the strength of the section, because there are very few design aids available. A few design charts or interaction diagrams of the irregular sections have been published recently. They are: (a) Hollow circular and rectangular sections by Montoya, et al. [56], Anderson and Moustafa [3], and Brettle [9]; (b) X-section by Marin [51]; (c) L-section by Hsu [35], Liu [48] and Marin [51]; (d) C-section by Hsu, et al. [44], Marin [51] and Park and Paulay [63]; and (e) I-section by Jalil, Morisset and Perchat [46].

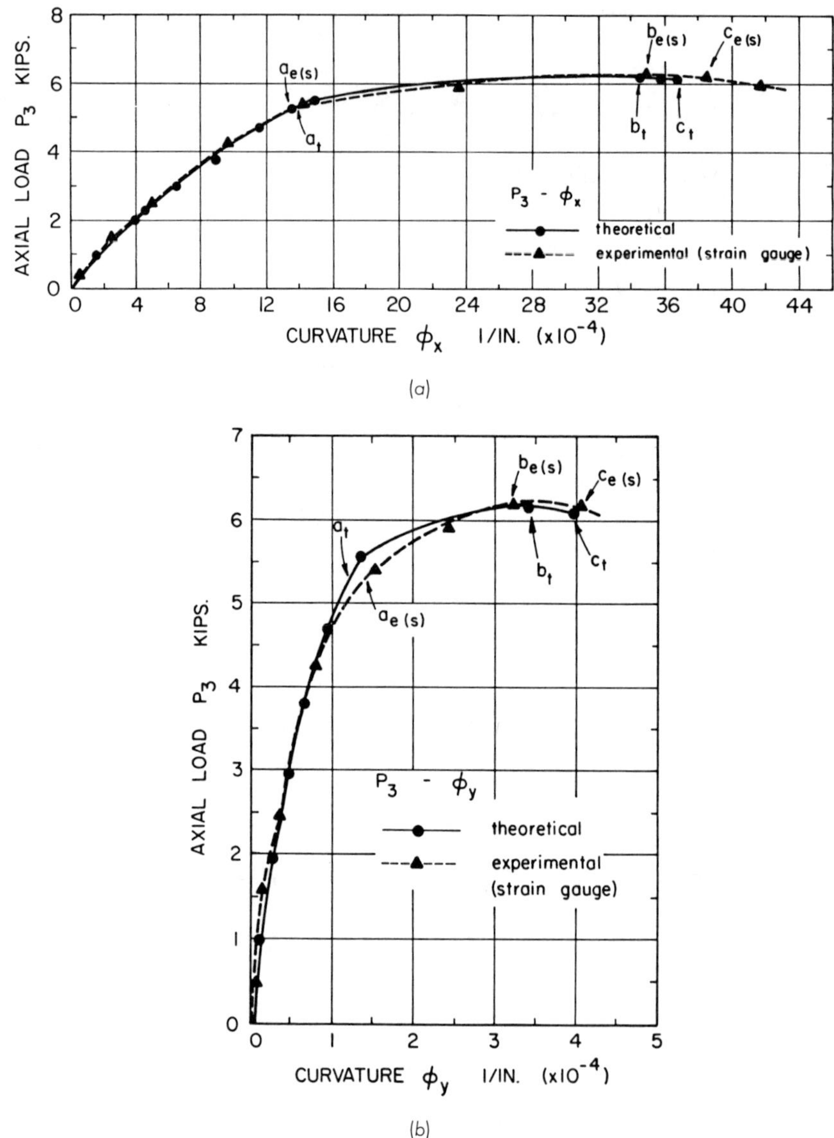

FIGURE 40. Biaxial load-curvature curves for Specimen U-6 (1 kip. = 4448 N., 1 l/in. = 0.394 l/cm.).

The experimental work in column research has been limited almost exclusively to rectangular, circular and octagonal cross sections. Unfortunately, there are few tests of columns with cross sections other than the one previously mentioned. Hsu, et al. [37] and Herrera and Ochoa [30] tested several C-shaped columns under monotonic loading with relative eccentricities of 0.25 and 0.375. Although limited in scope, these tests showed a linear strain distribution across the section and concrete strain up to 0.007. I-shaped sections have been tested under cyclic loading by Fiorato, Oesterle and Corley [21]. They obtained very impressive results. Bhattacharyay, Chattopadhyay, Ray and Som [6], Ramamurthy and Hafeez Kahn [67] and Hsu, et al. [42,43] also tested several L-shaped and T-shaped columns with and without end diaphragms. However, only the ultimate strength of the columns was obtained. In spite of the extensive use of L-shaped cross section in columns and C and T-shaped

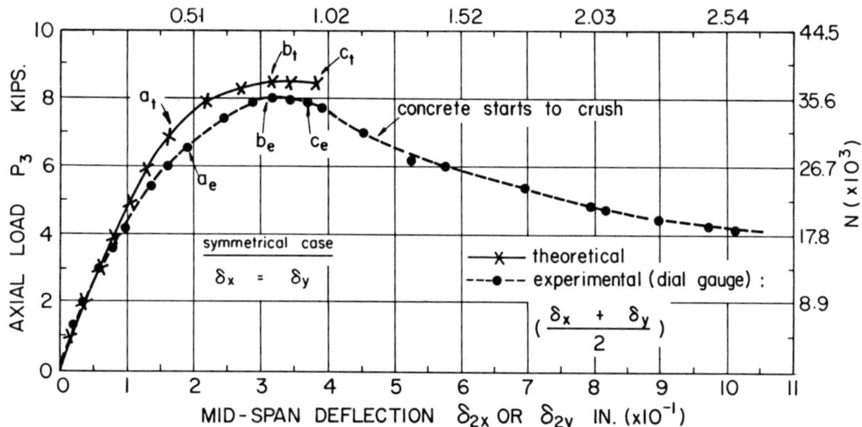

FIGURE 41. Load-deflection curves for Specimen U-3.

cross sections in shearwalls in modern buildings, there is very little experimental data on both the strength and deformation behavior of these sections under monotonic and cyclic loading. It is well known that the deformation behavior of columns such as ductility, load-deformation and moment-curvature-rotation characteristics is of vital importance in limit state design and earthquake engineering studies. The present existing experimental data are not adequate enough to validate the existing analytical models and proposed design methods. In spite of the facts that there is a few research work on irregular reinforced concrete columns, it is expected that the use of columns and shearwalls with cross section other than rectangular or circular will be increased in the years ahead, especially in modern high-rise building constructions.

Analysis and Computer Method for L-Section

Consider an element k with its centroid point u_k, v_k referred to the principal axes. The strain ϵ_k across the element k can be assumed to be uniform and since plane sections remain plane during bending,

$$\epsilon_k = \epsilon_p + \phi_u v_k + \phi_v u_k \tag{39}$$

where

ϵ_p = uniform direct strain due to an axial load P

ϕ_u = the curvature produced by the bending moment component M_u and is considered positive when it causes compressive strains in the positive v-direction, and

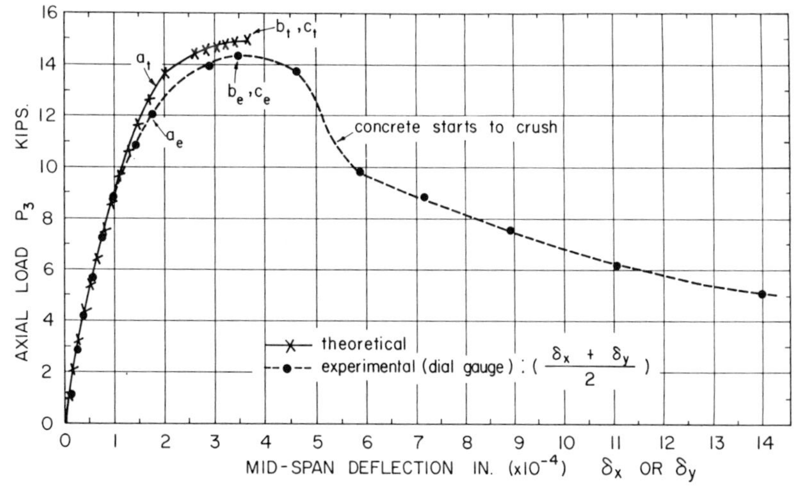

FIGURE 42. Load-deflection curves for Specimen U-4 (1 kip. = 4448 N., 1 l/in. = 2.54 cm.).

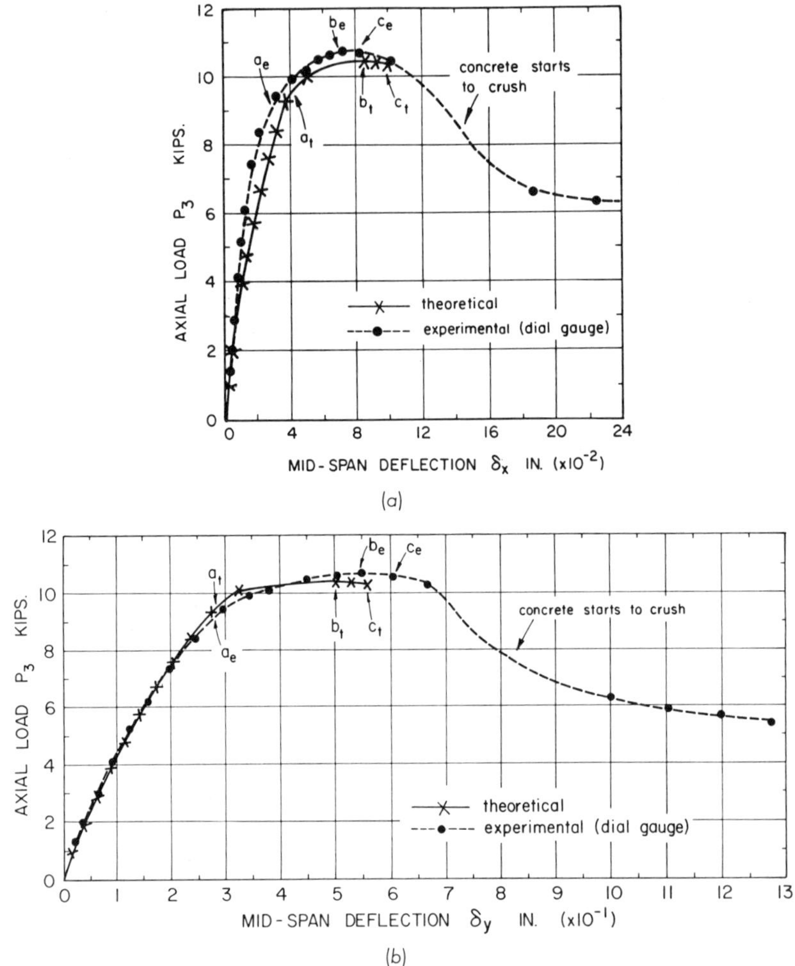

FIGURE 43. Load-deflection curves for Specimen U-5 (1 kip. = 4448 N., 1 l/in. = 2.54 cm.).

ϕ_v = the curvature produced by the bending moment component M_v and is considered positive when it causes compression in the positive u-direction.

Once the strain distribution across the cross section is established, the axial force P and the bending moment components M_u and M_v can be calculated using the following equations:

$$P_{(c)} = \sum_{k=1}^{n} f_k a_k;$$

$$M_{u(c)} = \sum_{k=1}^{n} f_k a_k v_k; \qquad (40)$$

$$M_{v(c)} = \sum_{k=1}^{n} f_k a_k u_k;$$

Subscript (c) indicates values of P, M_u, and M_v, calculated in an iteration cycle, and a_k is the area of the element k.

For a given section (known geometry and material properties) the stress resultants P, M_u, and M_v, can be expressed as functions of ϕ_u, ϕ_v and ϵ_p given by the following equations:

$$P = P\,(\phi_u, \phi_v, \varepsilon_p);$$

$$M_u = M_u\,(\phi_u, \phi_v, \varepsilon_p); \qquad (41)$$

$$M_v = M_v\ (\phi_u,\ \phi_v,\ \varepsilon_p)$$

If $p_{(s)}$ is the final value of P for which the equilibrium and the compatibility conditions are satisfied, the convergence of $P_{(c)}$ to $P_{(s)}$ can be accelerated using a modification of the extended Newton-Raphson method. The final values of M_u and M_v can be calculated using the following equations:

$$M_{u(s)} = P_{(s)}\ e_v \text{ and } M_{v(s)} = P_{(s)}\ e_u \qquad (42)$$

where e_u and e_v are load eccentricity components along u- and v-axis respectively. $P_{(s)}$, $M_{u(s)}$ and $M_{v(s)}$ can be expressed in terms of $P_{(c)}$, $M_{u(c)}$, and $M_{v(c)}$ using Taylor's expansion and retaining linear terms. The process of the iteration at a given load level continues in the computer program developed until convergence is obtained within the specified tolerances. Once this is achieved, the computer program takes up the next load level and repeats the entire procedure. The computer program, its accuracy and the convergence of the procedures was discussed in more detail in References 35 and 48 and earlier in this chapter.

Since the principal axes are taken for analytical purpose, co-ordinates transformation is an important procedure. From the strength of materials, the following steps can be used for transformation of co-ordinates, moments, and curvatures (see Reference 35):

(1) Find moment of inertia I_x, I_y and product moment of inertia I_{xy}.
(2) Use equation $tan2\theta = 2I_{xy}/(I_y - I_x)$ to determine the angle between the centroidal and the principal axes.
(3) Use equation:

$$\begin{bmatrix} u \\ v \end{bmatrix} = [R] \begin{bmatrix} x \\ y \end{bmatrix} \qquad (49)$$

where

$$[R] = \begin{bmatrix} \text{Cos}\theta & \text{-Sin}\theta \\ \text{Sin}\theta & \text{Cos}\theta \end{bmatrix} \qquad (50)$$

Following these steps, the data for the specimens used in this study (see Figure 44) can be determined as follows:

$$I_x = 144.8 \text{ in.}^4$$
$$I_y = 81.5 \text{ in.}^4$$
$$I_{xy} = -43.4 \text{ in.}^4$$
$$\theta = 27°$$

From the above investigation, the load, moment and curvature with respect to the principal axes u and v can be found easily. For practical purpose, these results should be transferred to the centroidal axes x and y.

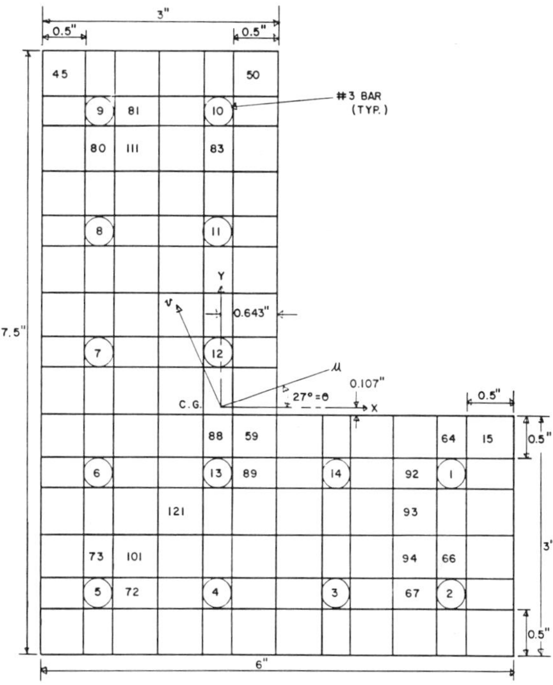

FIGURE 44. L-shaped section divided into elements for computer analysis.

Now consider the centroidal axes x and y as global co-ordinate axes and the principal axes u and v as local co-ordinate axes as shown in Figure 44. The angle of rotation is considered in counter-clockwise direction. The transformation matrix R' can be obtained as follows:

$$[R'] = \begin{bmatrix} \text{Cos}\theta & \text{Sin}\theta \\ \text{-Sin}\theta & \text{Cos}\theta \end{bmatrix} \qquad (51)$$

Moments and curvature about the centroidal axes in terms of the moment and the curvature about the principal axes can be given as follows:

$$\begin{bmatrix} M_x \\ M_y \end{bmatrix} = [R'] \begin{bmatrix} M_u \\ M_v \end{bmatrix} \qquad (52)$$

$$\begin{cases} M_x = M_u\ \text{Cos}\theta + M_v\ \text{Sin}\theta \\ M_y = M_v\ \text{Cos}\theta - M_u\ \text{Sin}\theta \end{cases} \qquad (53)$$

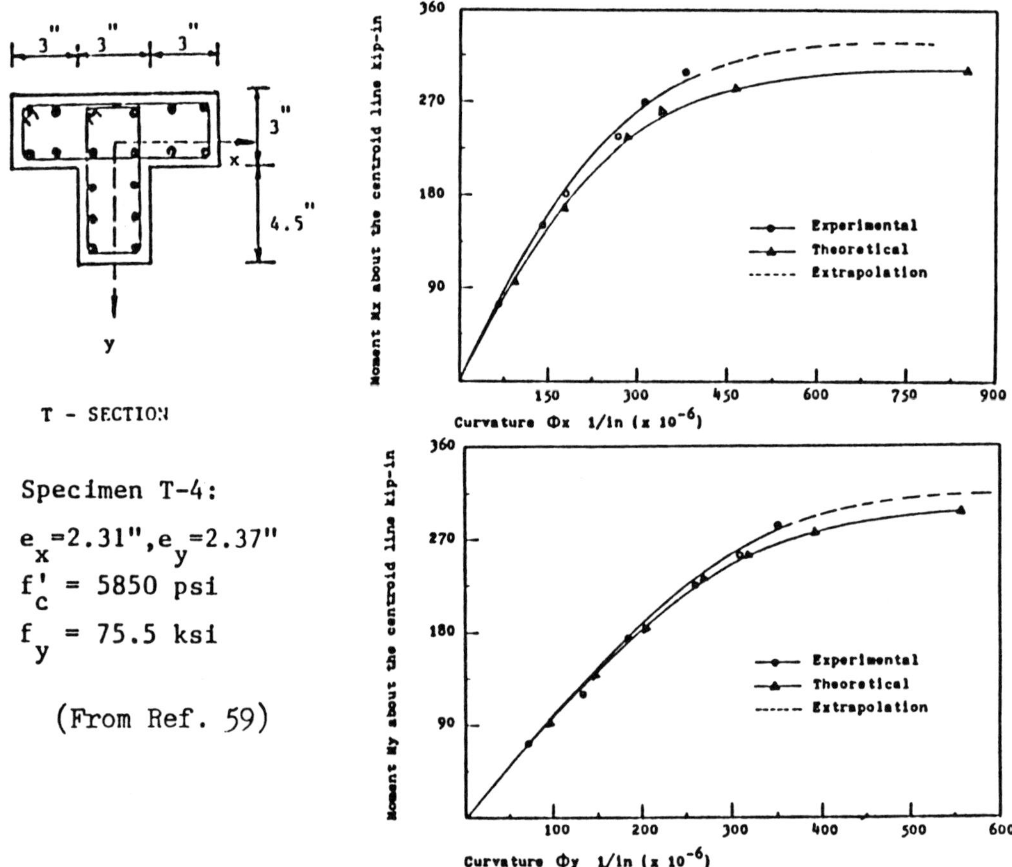

FIGURE 45. Biaxial moment-curvature curves for column with tee section.

and

$$\begin{bmatrix} \phi_x & \phi_{xy} \\ \phi_{xy} & \phi_y \end{bmatrix} = [R']^T \begin{bmatrix} \phi_u & \phi_{uv} \\ \phi_{uv} & \phi_v \end{bmatrix} [R'] \quad (54)$$

Since u and v are the principal axes, the product of curvature, $\phi_{uv} = 0$, then

$$\begin{bmatrix} \phi_x & \phi_{xy} \\ \phi_{xy} & \phi_y \end{bmatrix} = \begin{bmatrix} \text{Cos}\theta & \text{-Sin}\theta \\ \text{Sin}\theta & \text{Cos}\theta \end{bmatrix} \begin{bmatrix} \phi_u & 0 \\ 0 & \phi_v \end{bmatrix} \begin{bmatrix} \text{Cos}\theta & \text{Sin}\theta \\ \text{-Sin}\theta & \text{Cos}\theta \end{bmatrix}$$

$$\begin{bmatrix} (\phi_u \text{Cos}^2\theta + \phi_v \text{Sin}^2\theta) \\ (\phi_u - \phi_v) \text{Sin}\theta \text{Cos}\theta \end{bmatrix} \begin{bmatrix} (\phi_u - \phi_v) \text{Sin}\theta \text{Cos}\theta \\ (\phi_u \text{Sin}^2\theta + \phi_v \text{Cos}^2\theta) \end{bmatrix} \quad (55)$$

Therefore,

$$\begin{cases} \phi_x = \phi_u \text{Cos}^2\theta + \phi_v \text{Sin}^2\theta \\ \phi_y = \phi_u \text{Sin}^2\theta + \phi_v \text{Cos}^2\theta \end{cases} \quad (56)$$

Since, $\theta = 27°$ (see Figure 44),

$$\begin{cases} M_x = 0.891\ M_u + 0.454\ M_v \\ M_y = 0.891\ M_v - 0.454\ M_u \end{cases} \quad (57)$$

$$\begin{cases} \phi_x = 0.794\ \phi_u + 0.2061\ \phi_v \\ \phi_y = 0.794\ \phi_v + 0.2061\ \phi_u \end{cases} \quad (58)$$

The approximate equations proposed by Hsu [33,41] were used to evaluate the central deflections of the column specimens and are as follows:

$$\delta_x = \frac{\phi_y \ell^2}{8}$$

and (59)

$$\delta_y = \frac{\phi_x \ell^2}{8}$$

where δ_x and δ_y represent the mid-height deflections along

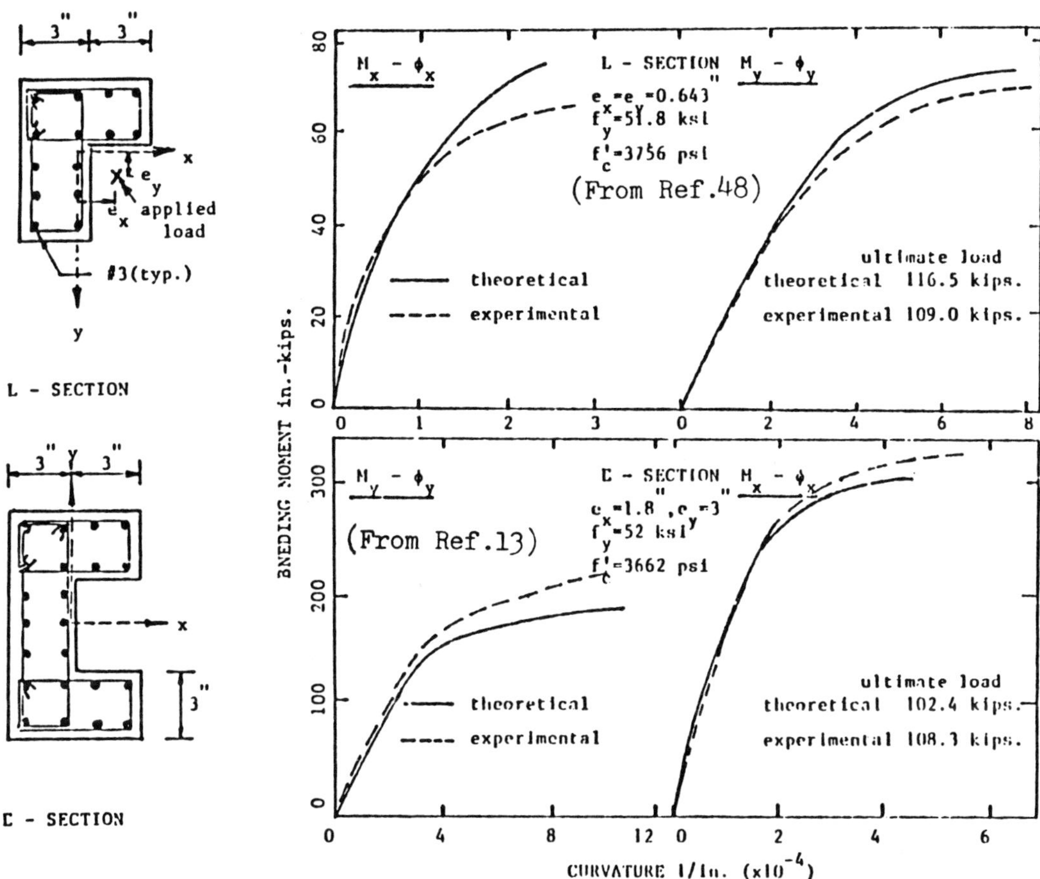

FIGURE 46. Biaxial moment-curvature curves for columns with L- and channel sections.

x and y axis respectively, and ℓ is an overall height of the column.

It should be noted that the analysis method of moment-curvature results for T- and channel sections can be referred to the section on Behavior of Reinforced Concrete Columns of Standard Shapes Under Combined Biaxial Bending and Compression.

Experimental Programme

All together, 18 specimens were tested in the present study. All columns were designed as short tied columns and were each six feet long. Physical characteristics of columns tested are shown in Figures 44, 45 and 46. The brackets were heavily reinforced to prevent local failure. All columns were reinforced longitudinally by #3 bars with different yield strength, f_y, as seen in Figure 47. These longitudinal bars were held together by 1/8 in. ties at spacings of 2 to 4 inches center to center. The ties and longitudinal bars were tied together using 16 gauge binding wire. The reinforcement was assembled into a unit before it was placed in the mold.

Of total 18 specimens, nine specimens each were designed for L-section, T-section and channel section. All the specimens were tested and studied for their complete behaviour under combined axial bending moments and axial compression and were used to examine some of the variables involved, such as steel yield strength, relative eccentricities, and loading variations. The test frames and the experimental set-up were constructed for this experimental programme as shown in Figure 11. The end conditions were pinned-ended, and local control test procedure was used.

The concrete used for casting the test specimens was prepared from graded mixture of crushed quartz and sand. Portland cement Type III and water. The water-cement ratio varied from 0.65–0.8, and the cement-sand ratio varied (3 ~ 3.2). The concrete properties and the stress-strain curve will be determined using 3 × 6 in. cylinders. The reinforcement for the test specimens consists of intermediate grade #3 bars, their stress-strain curves are shown in Figure 47 and the column ties were fabricated using plain bars with a diameter of 0.125 inches.

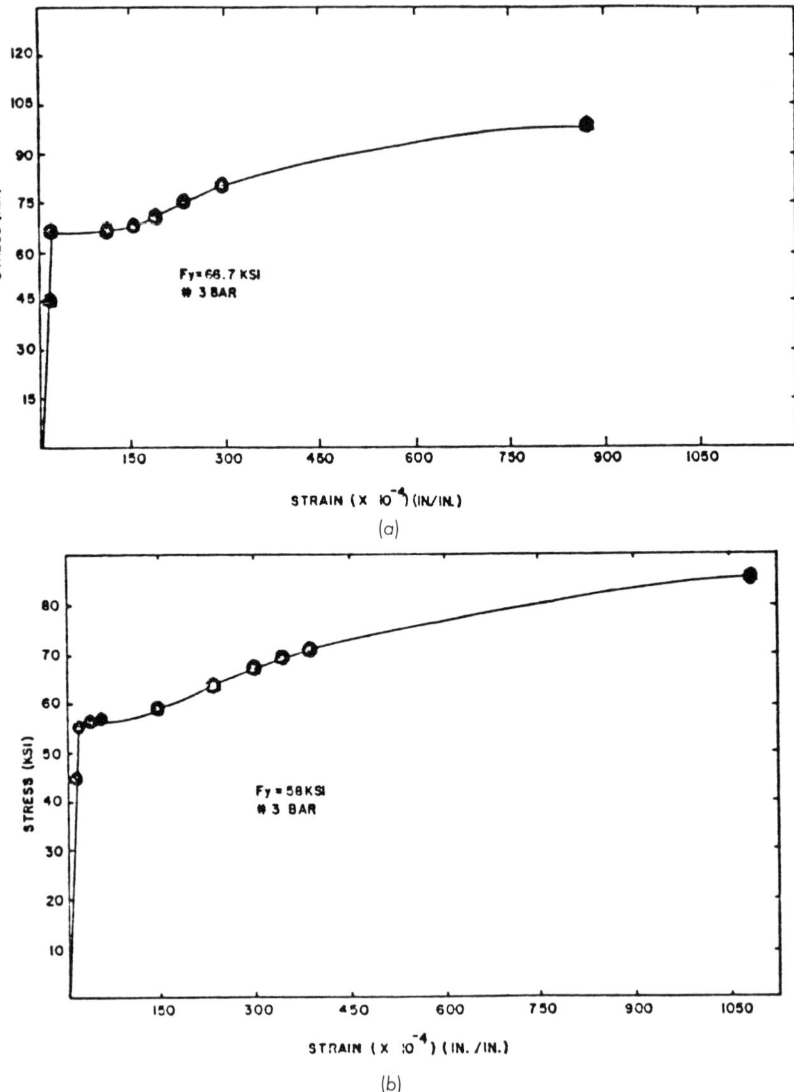

(a)

(b)

FIGURE 47. Experimental stress-strain curve for reinforcement.

Several types of instrumentation were used in the experimental tests. These included bonded strain gages on the outer and inner surfaces of the model, embedded strain gages and demec gages in the concrete, bonded strain gages on the reinforcing steel for strain and curvature measurements, and Ames dial gages for deflection measurements. A detailed instrumentation layout will be worked out for each model with a view to compare experimental data with the computed results at homologous points. The columns were tested in the horizontal position and the specimens were loaded using the Enerpac 100-ton capacity hydraulic cylinder (effective area = 20.63 in.^{2})

Manual Enerpac pump Model PEM 2042 with a maximum pressure of 10,000 psi was used to drive the ram. For the case of Monotonic loading conditions the loads are applied monotonically from zero load until failure of the specimen. For the case of cyclic loading conditions, each specimen was loaded up to 40% and 70% of the ultimate capacity and released the load to 5% of the ultimate capacity and then reloaded until failure of the specimen. The experimental curvature near the critical section was claculated from the measured strains in steels and/or concrete surfaces using both electrical strain gages and mechanical demec gages. The central deflection of the

TABLE 7. Ultimate Moment Capacity of L-Section.

Specimen No. (1)	Analysis Results (kips in.)		Test Results (kips in.)	
	M_{nx} (2)	M_{ny} (3)	M_{nx} (4)	M_{ny} (5)
2a	75.0	75.0	71.6[2]	71.6[2]
3a	25.3	98.4	22.9[2]	89.2[2]
4b	189.0	58.0	190.8[2] / 205.0[1]	58.4[2] / 79.8[1]
5b	192.3	58.8	192.4[2] / 216.0[1]	58.9[2] / 84.8[1]
6b	192.2	44.5	174.3[2] / 188.2[1]	45.0[2] / 61.1[1]

[1]Experimental $M_{nx} = P_n e_y$
$M_{ny} = P_n e_x$
[2]Experimental $M_{nx} = P_n (e_y + \delta_y)$ where δ_y, δ_y obtained from the load-deflection curve.

specimen was measured with dial gages. More details of the experimental set up, and test results can be found in References 13, 49, 59, 69 and 76.

Analysis of Test Results

ULTIMATE MOMENT CAPACITY, INTERACTION DIAGRAM AND LOAD CONTOUR

Table 7 shows a comparative study of analytical and experimental results for ultimate load capacity of the columns. For nominal bending moments, M_{nx}, M_{ny} which were obtained from the following equations:

$$M_{nx} = P_n(e_y + \delta_y)$$
$$M_{ny} = P_n(e_x + \delta_x) \tag{60}$$

A good agreement was noted for the theoretical and experimental values. For M_{nx} and M_{ny} calculated by Equation (60), the experimental values are more than those predicted by the analysis, therefore, the analytical values are on the conservative side.

Figure 48 shows two sets of interaction diagrams for Specimens 46 and 5b for comparative study. The theoretical interaction curves were obtained directly from the analysis, and the maximum compressive capacity of column specimen, P_n, was calculated by:

$$P_n = 0.85 f_c' (A_g - A_{st}) + f_y A_{st} \tag{61}$$

As compared with the experimental values, the theoretical strength interaction diagrams are on the conservative side.

The load contours and three-dimensional failure surfaces have formed the basis of current design procedures for reinforced concrete columns subject to biaxial bending and axial compression in the various national codes. The load contour method involves cutting the failure surface as shown in Figure 49 at a constant value of P_n to give a so-called "Load Contour" interaction curve Figures 50 and 51 show a typical load contours and dimensionless load contours for columns with L-section.

MOMENT-CURVATURE CURVES

The moments and curvatures were obtained about x and y axes where x and y axes are the centroidal axes. Experimental moments were obtained from either Equation (42) or Equation (60). Steel and concrete strains were measured from the strain gages attached to their surfaces. To obtain the strain distributions about x- and y-axes, the demec gage method was used. This method was successfully used previously in References 33, 40 and 41. Two types of demec gage arrangements were used [69] to calculate the strain distribution across the critical section (actually each pair of demec points was installed 3 in. away from the critical section). The strain distributions across the x-z and y-z planes were found at each loading stage and were plotted against the distance between the corresponding pair of demec points. Once the strain distribution across the section was established, the follow-

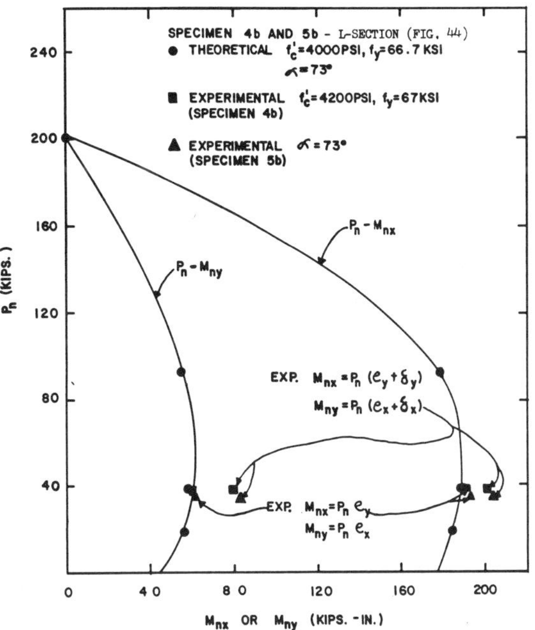

FIGURE 48. Interaction diagrams corresponding to centroidal axes.

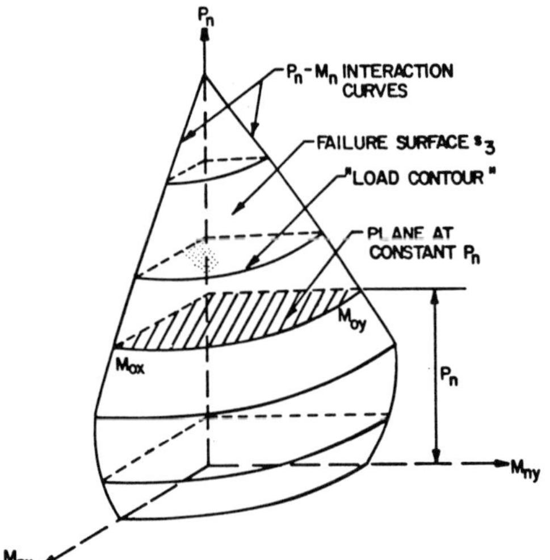

FIGURE 49. Failure surface of combined biaxial bending and axial compression.

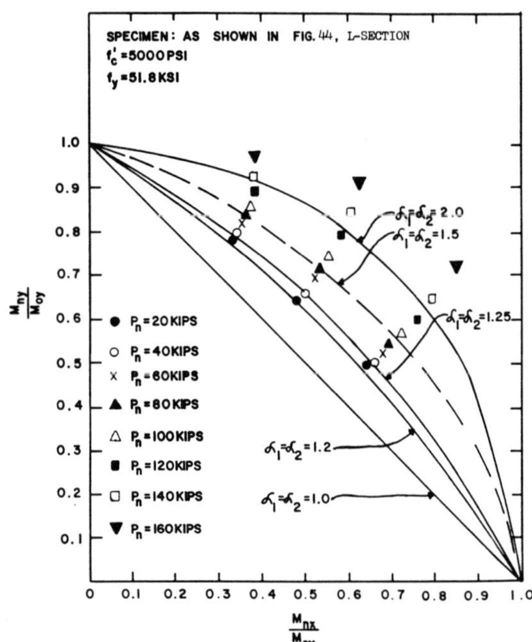

FIGURE 51. Dimensionless load contours.

ing equation was used to calculate the curvature:

$$\phi_x \text{ or } \phi_y = \frac{\epsilon_c}{kd} \tag{37}$$

Equation (37) was used previously in References 33, 40 and 41, respectively, where kd is the distance between the location with the concrete strain ϵ_c and the point of zero strain along the x or y-axis. Typical theoretical and experimental moment-curvature curves are shown in Figures 45, 46, 52 and 53. An excellent agreement can be noted between the theoretical and experimental values from zero load up to the moment at yield, and a satisfactory agreement can also be achieved from the moment at yield up to the maximum moment capacity of the column specimen.

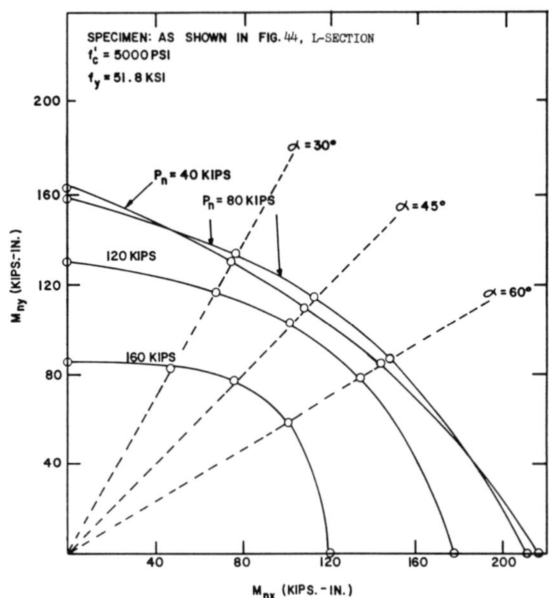

FIGURE 50. Load contours.

SUMMARY AND CONCLUSION

An analytical model is presented to simulate the moment-curvature behavior of reinforced concrete columns under combined uniaxial/biaxial bending and compression. Based on the control of load increments, the algorithm enables determination of moment-curvature-strain relationship with any geometry and material properties up to the maximum moment capacity of the section; however, with a constant axial load or a control of deformation increments, this computer program can be used to compute both the ascending and the descending branches of the moment-curvature curve. The moment-curvature (or moment-rotation) relations play an important part in the study of limit analysis of two or three dimensional reinforced concrete frames.

Good agreement was obtained between the experimental strengths and the analytical values calculated using the

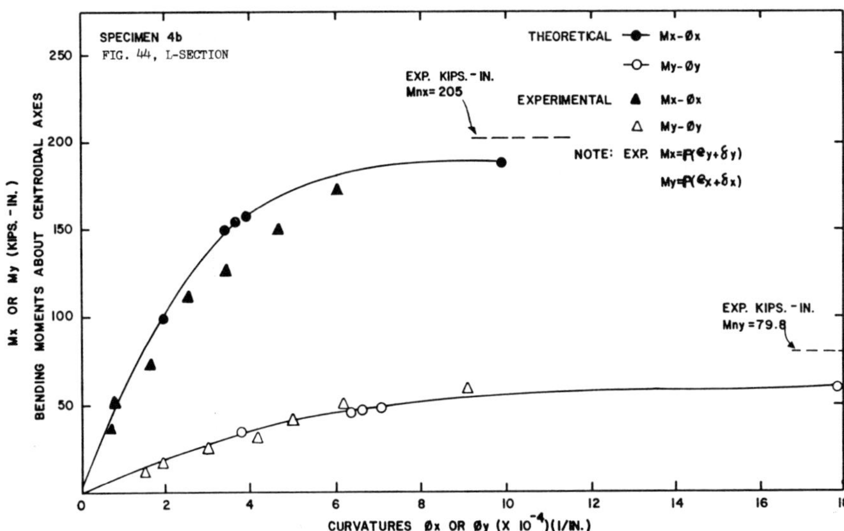

FIGURE 52. Moment-curvature curves about centroidal axes.

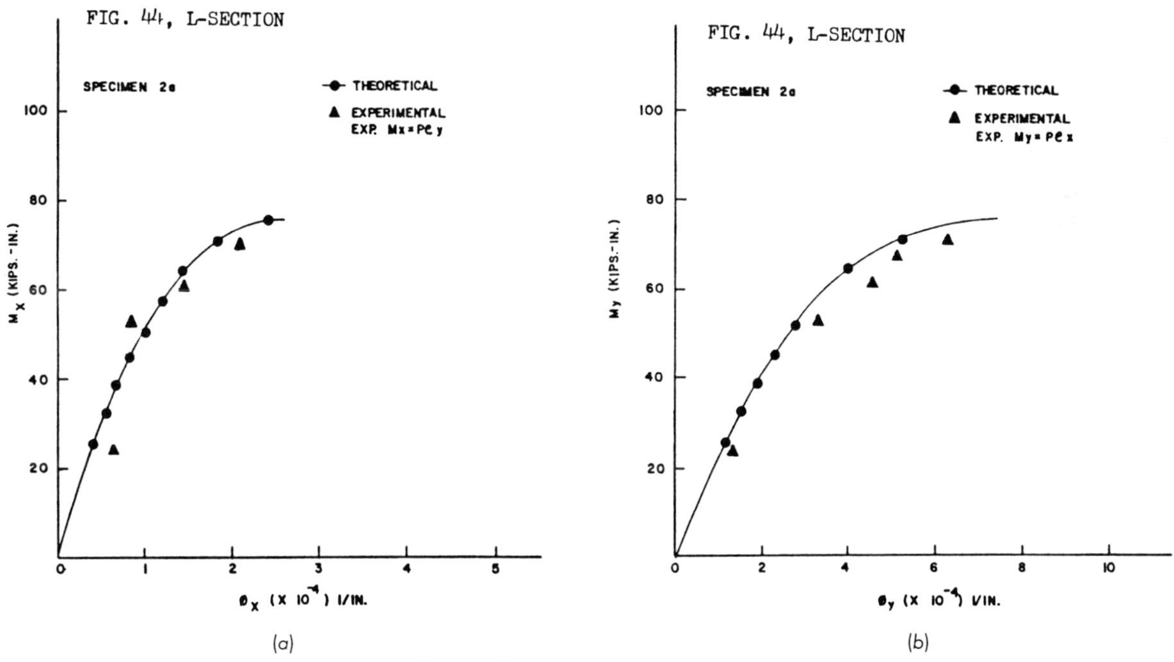

(a)

(b)

FIGURE 53. Moment-curvature diagram.

above computer program. The experimental strain and curvature data obtained from the tests were noted to be in good agreement with the analytical results through all load stages from zero up to the maximum moment value.

The load system incorporating hard rubber blocks was successful in preventing a sudden failure at the maximum load during column tests. Both the ascending and the descending branches of the moment-curvature were obtained experimentally. The curvature results from both the strain gage and the demec gage data showed good agreement compared with the analytical values. The results also showed the values obtained from the demec gage measurements which calculated the average curvature for the section were slightly less than the values obtained by strain gage measurements which calculated the curvature for the section.

ACKNOWLEDGEMENTS

Part of this research was made possible by grants from National Research Council of Canada and the Canada Emergency Measures Organization through their support of the structural concrete frame analysis and behavior studies at McGill University.

Financial support and computer time provided by the New Jersey Institute of Technology are gratefully acknowledged. Some computations and experimental work of the present study were conducted by the writer's graduate students between 1979 and 1984, A Majlesi, A. Yekta, M. Taghechian, M. Saeedi, A. Mesktooli, M. C. Liu, A. M. Shah, S. J. Jou, K. Fitzgerald, D. Chidambarrao, Y. Tarabay, A. Nader, S. Yalamarthy, A. Harb, M. S. Zghondi, N. Patel, C. L. Hsu, and H. Ghadimy; their efforts on irregular column studies are greatly acknowledged.

LIST OF SYMBOLS

a_t = theoretical point at yielding of the farthest tension steel bar
$a_{e(s)}$ = experimental point (by strain gages) at yielding of the farthest tension steel bar
a_k = area of element k
A_s = area of tension reinforcement in section
A_s' = area of compression reinforcement in section
A_s'' = area of lateral reinforcement
A_g = gross area of the section
A_{st} = Total area of main reinforcement
b = beam width
b'' = width of the stirrup in section
b_t = theoretical point with respect to the maximum axial load
$b_{e(d)}$ = experimental point (by demec gages) with respect to the maximum axial load
$b_{e(s)}$ = experimental point (by strain gages) with respect to the maximum axial load
c_t = theoretical point with respect to the maximum moment capacity of the section
$c_{e(d)}$ = experimental point (by demec gages) with respect to the maximum moment capacity of the section
$c_{e(s)}$ = experimental point (by strain gages) with respect to the maximum moment capacity of the section
C_c = compressive force in concrete
C_{c1} = compressive force in concrete ($\epsilon \leq \epsilon_o$)
C_{c2} = compressive force in concrete ($\epsilon > \epsilon_0$)
C_s = force in compression steel reinforcement
d = effective depth of the beam
D = overall depth of the column prismatic section
d' = distance from the extreme compression fibre to centroid of the compression steel reinforcement
D_1, D_2, D_3, D_4, and D_5 = coefficients
d'' = depth of the stirrups in section
d''' = distance from the extreme tension fibre to centroid of the tension steel reinforcement
e = eccentricity for the uniaxially loaded column
e_y = eccentricity along y-axis
e_x = eccentricity along x-axis
e_x' = $2e_x$
e_y' = $2e_y$
$(E_t)_k$ = tangent modulus of elasticity for steel or concrete element k
E_c = initial modulus of concrete
E_s = initial modulus of tension steel
E_s' = initial modulus of compression steel
E_{ss} = modulus of elasticity of tensile steel after strain-hardening
F = compatibility factor
f_c = concrete stress at the extreme compression fibre
f_c' = maximum concrete compressive stress
f_y = yield strength of tension reinforcement
f_y' = yield strength of compression reinforcement
f_y'' = yield strength of lateral reinforcement
f_k = stress across the element k
f_s = strength of tension reinforcement
f_s' = strength of compression reinforcement
f_t = tensile stress in concrete (=500 psi)
k = Element number
kd = Distance from maximum compressive concrete strain to the neutral axis
ℓ = total length of column
ℓ' = Effective length of column
l_x' = the length of column member w.r.t. the ends of the prismatic shaft along x-axis
l_y' = the length of column member w.r.t. the ends of the prismatic shaft long y-axis
M = bending moment
M_{cr} = moment at cracking in which the crack starts to appear in concrete at the extreme tension fibre
M_y = moment at yield in which the farthest tension reinforcement bar reaches the yield point

M_m = M_{max} = maximum moment
M_u = moment at failure
M_x = bending moment about x-axis (left-hand rule)
M_y = bending moment about y-axis (right-hand rule)
M_{nx} = Nominal bending moment about x axis, P_n e_y
M_{ny} = Nominal bending moment about y axis, P_n e_x
M_{ux} = ϕ M_{nx}, ϕ is a strength reduction factor
M_{uy} = ϕ M_{ny}, ϕ is a strength reduction factor
M_{ox} = M_{nx} capacity at axial load P_n when M_{ny} is zero
M_{oy} = M_{ny} capacity at axial load P_n when M_{nx} is zero
n = E_s/E_c
P_{cr} = axial load at cracking in which the crack starts to appear in concrete at the extreme tension fibre
P = axial load
P_n = Nominal axial load
P_3 = the maximum axial load which is obtained from the consideration of the effect of the mid-height deflection δ_2
p_b = reinforcement ratio producing balanced conditions at ultimate strength
p = A_s/bd
p' = A_s'/bd
p'' = the ratio of the total volume of stirrup steel per unit length to the total volume of concrete enclosed by the stirrup per unit length
S = spacing of lateral reinforcement
T_s = force in tension steel
t = overall depth of beam or column sections
u' = allowable compatibility for P
v' = allowable compatibility for M_x
w' = allowable compatibility for M_y
w = weight of concrete, lb. per cu. ft.
x' = distance from the positive of the neutral axis to the extreme concrete compressive fibre
x_3 = distance between C_s and T_s
x = ϵ/ϵ_o = non-dimensional concrete strain factor
x_c = ϵ_c/ϵ_o
x_k = position of the centroid of element k along x-axis
x_1 = distance between the positions of C_{c1} and T_s
x_2 = distance between the positions of C_{c2} and T_s
x_1' = distance between the position of C_{c1} to the position of the neutral axis
y_k = the position of the centroid of element k along y-axis
y see Figure 7
y_1 see Figure 7
η = f/f_c' = non-dimensional concrete stress factor
η_c = f_c/f_c'
ϵ = compressive strain in concrete from the compression test
ϵ_c = compressive strain in concrete at the extreme fibre
ϵ_o = concrete compressive strain w.r.t. f_c'. It is taken to be 0.002 in the present investigation, except as indicated
ϵ_u = ultimate concrete compressive strain
ϵ_{sh} = steel strain at the start of strain-hardening
ϵ_{su} = ultimate steel strain at failure
ϵ_y = yield strain in tension steel
ϵ_y' = yield strain in compression steel
ϵ_s = steel strain in tension steel = f_s/E_s
ϵ_{cu}'' = $\sqrt[3]{p''}/24.5 + 0.002$
ϵ_{cu}''' = $\sqrt[3]{p''}/4.9 + 0.012$
ϵ_s' = steel strain in compressive steel = f_s'/E_s'
ϵ_k = strain across the element k
ϵ_p = uniform direct strain due to an axial load P
ϵ_{ct} = concrete tensile strain
ϕ = curvature as a measure of strain gradient in a section
ϕ_{cr} = curvature just before cracking
$\phi_{cr(a)}$ = curvature immediately after cracking
ϕ_y = curvature w.r.t. M_y
ϕ_m = curvature w.r.t. M_m
ϕ_u = curvature w.r.t. M_u
ϕ_{area} = area of the curvature diagram
ϕ_x = the curvature producted by the bending moment component M_x and is considered positive when it causes compressive strain in the positive y-direction
ϕ_y = the curvature produced by the bending moment component M_y and is considered positive when it causes compression in the positive x-direction
δ_y = the deflection at yielding
δ_{2x} = the deflection at mid-height w.r.t. the knife edge along x-axis
δ_{2y} = the deflection at mid-height w.r.t. the knife edge along y-axis
d_{ab} = the original distance between the position "a" and "b"
δ_2 = the deflection at mid-height w.r.t. the knife edge
θ_{ab} = the relative rotation between the positions "a" and "b"
$\propto$ = $\tan^{-1}(e_y/e_x)$
$\propto_1, \propto_2$ = Exponents in load contour equations
subscript (m) and (m + 1) indicates the "m" and "m + 1" iteration angles
subscript (c) indicates values calculated in an iteration cycle
subscript (s) indicates the expected values in an iteration cycle

REFERENCES

1. Aas-Jakobsen, A., "Biaxial Eccentricities in Ultimate Load Design," *Proceedings, ACI, 61* (1), 293 (1964).
2. ACI Committee 318-71, "Building Code Requirement for Reinforced Concrete (ACI 318-71)," American Concrete Institute, Detroit, U.S.A. (1971).
3. Anderson, A. R. and S. E. Moustafa, "Ultimate Strength of Prestressed Concrete Piles and Columns," *J. of the ACI, 67* (8), 620 (August 1970).

4. Anderson, P. and H. N. Lee, "A Modified Plastic Theory of Reinforced Concrete," Bulletin No. 33, University of Minnesota, V. LIV, No. 19, 44 p. (April 1951).
5. Beaufait, F. W., H. W. Rowan, Jr., P. G. Hoadley, and R. M. Hackett, *Computer Method of Structural Analysis*. Prentice-Hall, Inc., Englewood Cliffs, NJ (1970).
6. Bhattacharyay, S., B. Chattopadhyay, T. C) Ray, and P. Som, "An Investigation on Concrete Columns with Special Reference to L and T Sections With and Without Diaphragm," LABSE, *16*, 333 (1974).
7. Bleich, H. H. *Buckling Strength of Metal Structures*. McGraw-Hill Book Company (1952).
8. Bresler, B., "Design Criteria for Reinforced Columns Under Axial Load and Biaxial Bending," *Journal of the ACI, 32* (1), 481 (1960).
9. Brettle, H. J., "Ultimate Strength Design of Multi-Cell Rectangular Reinforced Concrete Piers," Pub. No. 28-1, LASBE, 41 (1968).
10. Brondum-Nielsen, T., "Ultimate Limit States of Cracked Arbitrary Concrete Sections under Axial Load and Biaxial Bending," *ACI Concrete International, Design and Construction, 4* (11), 41-55 (Nov. 1982).
11. Bruinette, K. E. and S. J. Fenves, "A General Formulation of the Elastic-Plastic Analysis of Space Frameworks," The International Conference on "Space Structures," University of Surrey, England, 92 (1967).
12. Chen, W. F. and R. Atsuta, *Theory of Beam Columns*. McGraw-Hill Book Co., Vol. 2 (1977).
13. Chidambarrao, D., "Behavior of Channel-Shaped Reinforced Concrete Columns Under Combined Biaxial Bending and Compression," M.S. Thesis, New Jersey Institute of Technology, Newark, NJ (Aug. 1983).
14. Chu, K. H. and A. Pabarcius, "Biaxially Loaded Reinforced Concrete columns," *Proceedings, ASCE, 84*(8), 1865 (December 1958).
15. Corley, W. G., "Rotational Capacity of Reinforced Concrete Beams," *Proceedings of the ASCE Structural Division, 92* (4), 121-146 (October 1966).
16. Cranston, W. B., "A Computer Method for the Analysis of Restrained Columns," Cement and Concrete Association, London, Report TRA/402 (April 1967).
17. Cranston, W. B. and A. K. Chatterji, "Computer Analysis of Reinforced Concrete Portal Frames with Fixed Feet," Technical Report TRA 444, Cement and Concrete Association, London (September 1970).
18. Czerniak, E., "Analytical Approach to Biaxial Eccentricity," *Journal of the Structural Division, ASCE, 88* (4), 105 (August 1962).
19. Drysdale, R. G., "The Behaviour of Slender Reinforced Concrete Columns Subjected to Sustained Biaxial Bendings," Ph.D. Thesis, University of Toronto (September 1967).
20. Farah, A. and M. Huggins, "Analysis of Reinforced Concrete Columns Subjected to Load and Biaxial Bending," *Journal of the ACI, 66* (7), 569 (July 1969).
21. Fiorato, A. E., R. G. Oesterle, and W. G. Corley, "Importance of Reinforcement Details in Earthquake-Resistant Structural Walls," *Proc. of Workshop on Earthquake-Resistant R. C. Construction, III,* 1430, University of California (July 1977).
22. Flower, T. J., "Reinforced Concrete Columns Governed by Concrete Compression," Ph.D. Thesis, University of Texas, Austin, Texas (January 1966).
23. Furlong, R. W., "Ultimate Strength of Square Columns Under Biaxially Eccentric Loads," *Proceedings, 32*(2), 1129 (1961).
24. Furlong, R. W. and P. M. Ferguson, "Tests of Frames with Columns in SIngle Curvature," *Symposium on Reinforced Concrete Columns, ACI SP-13*, 55 (1966).
25. Gere, J. M. and W. Weaver, Jr. *Analysis of Framed Structures*. D. Van Nostrand Company, Inc., Princeton, NJ (1965).
26. Gesund, H., "Monograph on Planning and Design of Tall Buildings," *Vol. CB Structural Design of Tall Concrete and Masonry Buildings*, 185 (1978).
27. Gurfinkel, G., "Analysis of Footings Subjected to Biaxial Bending," *Journal of the Structural Division, ASCE, 93* (ST6), 1049 (June 1970).
28. Gurfinkel, G. and A. Robinson, "Determination of Strain Distribution and Curvature in a Reinforced Concrete Section Subjected to Bending Moment and Longitudinal Load," *Journal of the ACI, 64*(7), 398 (July 1967).
29. Harrison, H. B. *Computer Methods in Structural Analysis*. Prentice-Hall, Inc., Englewood Cliffs, NJ (1973).
30. Herrera, G. H. and E. E. Ochoa, "Verificiation Experimental de un Diagramma de Interaccion de una Pantalla de Concreto Armado," Universidad Central de Venezuela, Caracas (March 1976).
31. Hognestad, E., "A Study of Combined Bending and Axial Load in Reinforced Concrete Members," Bulletin No. 399, Engineering Experiment Station, University of Illinois, Urbana, p. 128 (November 1951).
32. Home, M. R., "The Elasticity-Plastic Theory of Compression Members," *Journal of the Mechanisms of and Physics of Solids, 4*, 104 (1956).
33. Hsu, C. T., "Behavior of Structural Concrete Subjected to Biaxial Flexure and Axial Compression," Ph.D. Thesis, McGill University, 479 p. (August 1974).
34. Hsu, C. T. T., "Failure Surface of Structural Concrete Members," *Proc. 8th Conference on Electronic Computation, ASCE*, Houston, Texas, 671-682 (Feb. 21-23, 1983).
35. Hsu, C. T. T., "L-Shaped Reinforced Concrete Column Section," *Proc. of 4th ASCE-EMD Specialty Conference*, Indiana, 557-560 (May 23-25, 1983).
36. Hsu, C. T. T., "Biaxial Moment-Curvature Characteristics for Irregular Shaped Reinforced Concrete Columns," *Proceedings, 5th ASCE-EMD Specialty Conference*, Laramie, Wyoming, 879-882 (Aug. 1-3, 1984).
37. Hsu, C. T. T., D. Chidambarrao, and M. C. Liu, "Inelastic Behavior of Channel-Shaped Reinforced Concrete Columns," *Abstract, 20th Annual Meeting, Society of Eng. Science*, Newark, Delaware, 99-100 (Aug. 22-24, 1983).

38. Hsu, C. T. and M. S. Mirza, "An Investigation of Behavior of Reinforced Concrete Column Under Combined Flexure and Axial Compression," *Structural Concrete Series No. 73-1*, McGill University (March 1973).
39. Hsu, C. T. and M. S. Mirza, "Structural Concrete—Biaxial Bending and Compression," *TN, Journal of the Structural Division, ASCE, 99* (ST2), 285 (Feb. 1973).
40. Hsu, C. T. and M. S. Mirza, "An Experimental-Analytical Study of Complete Load-Deformation Characteristics of Concrete Compression Members Subjected to Biaxial Bending," *LABSE, 16*, 45 (1974).
41. Hsu, C) T. and M. S. Mirza, "Nonlinear Behavior and Analysis of Reinforced Concrete Columns Under Combined Loadings," Study No. 14, Ed. M. S. Cohn, University of Waterloo Press, p. 109 (1980).
42. Hsu, C. T. T. and A. Nader, "Strength and Deformation of T-Shaped RC Short Columns," *Abstract, 21st Annual Meeting, SES*, Blackburg, Virginia, 177 (Oct. 15–17, 1984).
43. Hsu, C) T. T., M. R. Taghechian, and M. Yekta, "Inelastic Behavior of L-Shaped Reinforced Concrete Columns," *Proc. of 9th CANCAM*, Saskatoon, Canada, 287–288 (May 30–June 3, 1983).
44. Hsu, C. T. T. and Hua Wang, "A Numerical Analysis of Arbitrary Structural Concrete Sections Under Combined Loadings," *Proc. 4th International Conf. on Appl. Num. Modeling*, Taiwan, 108–112 (Dec. 1984).
45. Hudson, F. M., "Reinforced Concrete Columns: Effect of Lateral Ties Spacing on Ultimate Strength," *Symposium on Reinforced Columns, ACI SP-13*, 235 (1966).
46. Jalil, W., A. Morisset, and J. Perchat, "Calcul due Beton Arme a l'etal Limit Ultime," *Eyrolles*, Paris (1976).
47. King, J. W. H., "Effect of Lateral Reinforcement in Reinforced Concrete Columns," *Structural Engineer, 24* (7), 355 (London) (July 1946); and *24* (11), 609 (November 1946).
48. Liu, M. C., "Failure Surface for L-Shaped Reinforced Concrete Short Columns," M.S. Thesis, New Jersey Inst. of Tech., Newark, NJ (May 1983).
49. Majlesi, A., "Analysis of L-Shaped Reinforced Concrete Columns Under Combined Biaxial Bending and Axial Compression," M.S. Thesis, New Jersey Institute of Technology, Newark, NJ (May 1983).
50 Mamet, J. C., "Finite Element Analysis of Tall Buildings," Ph.D. Thesis, McGill University (March 1972).
51. Marin, J., "Design Aids for L-Shaped Reinforced Concrete Columns," *J. of ACI, 76* (11), 1197–1216 (Nov. 1979).
52. Marin, J. and I. Martin, "Designing Columns with Non-Rectangular Cross Section," Preprint 3707, ASCE, Atlanta Convention, 21 p. (Oct. 1979).
53. Mattock, A. H., "Rotational Capacity of Hinging Regions in Reinforced Concrete Beams," *Proceedings of the International Symposium on Flexural Mechanics of Reinforced Concrete*, Miami, Florida, November 1964, ASCE–1965–50, ACI SP–12, 143–182 (1965).
54. Meek, J. L., "Ultimate Strength of Columns with Biaxially Eccentric Loads," *Proceedings, ACI, 60* (2), 1053 (1963).
55. Mirza, M. S., O. C. Guedelhoefer, and J. R. Janney, "Correlation of Models and Phototype Structural Elements and Complete Structures," *Structural Concrete Models—A State of the Art Report*, Compiled and Edited by M. S) Mirza, 99 (October 1972).
56. Montoya, J. A., G. A. Meseguer, and M. F. Cabre. *Normigon Armado*. 8th ed., Gustavo Gili, Barcelona (1976).
57. Morris, G. A. and S. J. Fenves, "A General Procedures for the Analysis of Elastic and Plastic Frameworks," S.R.S. No. 305, Department of Civil Engineering, University of Illinois (August 1967).
58. Morris, G. A. and S. J. Fenves, "Elastic-Plastic Analysis of Frameworks," *Journal of the Structural Division, ASCE, 96* (5), 931 (May 1970).
59. Nader, A. R., "Behavior of Tee-Shaped Reinforced Concrete Columns Subjected to Biaxial Bending and Axial Compression," M.S. Thesis, New Jersey Institute of Tech., Newark , NJ (Aug. 1984).
60. Nigam, N. C., "Yielding in Framed Structures Under Dynamic Loads," *Journal of the Engineering Mechanics Division, ASCE, 96* (EM5), 687 (October 1970).
61. Obeid, E. H., "Compression Hinges in Reinforced Concrete Elements," *Structural Concrete Series No. 20*, 66 (September 1969).
62. Pannell, F. N., "Faulure Surfaces for Members in Compression and Biaxial Bending," *Proceedings, ACI, 60* (1), 129 (1963).
63. Park, R. and T. Paulay. *Reinforced Concrete Structures*. John Wiley and Sons (1975).
64. Parme, A) L., J. M. Nienes, and A. Gouwens, "Capacity of Reinforced Rectangular Columns Subjected to Biaxial Bending," *Proceedings, ACI, 63* (2), 911 (1966).
65. Porter, F. L. and G. H. Powell, "Static and Dynamic Analysis of Inelastic Frame Structures," Report No. EERC 71-3, Earthquake Engineering Center, University of California, Berkeley, California (June 1971).
66. Ramamurthy, L. N., "Investigation of the Ultimate Strength of Square and Rectangular Columns Under Biaxially Eccentric Loads," *Symposium on Reinforced Concrete Columns*, ACI, SP-13, 263 (1966).
67. Ramamurthy, L. N. and T. A. Hafeez Kahn, "L-Shaped Column Design for Biaxial Eccentricity," *Journal of Structural Eng., ASCE, 109* (8), 1903–1917 (Aug. 1983).
68. Salvadori, M. C. and M. L. Baron. *Numerical Methods in Engineering*. Prentice-Hall, Inc., Englewood Cliffs, NJ, 2nd Ed. (1961).
69. Shah, A., "Behavior of L-Shaped Reinforced Concrete Columns Under Combined Bending and Compression," M.S. Thesis, New Jersey Institute of Technology, Newark, NJ (May 1984).
70. Toridis, T. G. and K. Khozeimeh, "Computer Analysis of Rigid Frames," *Computers and Structures, 1*, 193 (1971).
71. Wang, C. K. *Matrix Methods of Structural Analysis*. International Textbook Company, PA, 2nd ed. (1970).
72. Warner, R. F., "Biaxial Moment Thrust Curvature Relations," *Journal of the Structural Division, ASCE, 95* (5), 923 (May 1969).

73. Weaver, W., Jr. *Computer Programs for Structural Analysis*. D. Van Nostrand Company, Inc., Princeton, MJ (1967).
74. Wen, R. K. and Farhoomand, F., "Dynamic Analysis of Inelastic Space Frames," *Journal of the Engineering Mechanics Division, ASCE, 96* (EM5), 667 (October 1970).
75. Werner, H., "Schiefe Beigung Polygonal Umrandeter Stahlbeton-Querschmitte," *Benton-Und Stahlbetonbau*, 92 (April 1974).
76. Yalarmarthy, S., "Behavior of Thin-Walled Channel Shaped Reinforced Concrete Columns Under Combined Biaxial Bending and Compression," M.S. Thesis, New Jersey Institute of Tech. (May 1984).
77. Zienkiewicz, O. C. and Y. K. Cheung. *The Finite Element Method in Structural and Continuum Mechanics*. McGraw-Hill Publishing Company Limited, London (1967).

APPENDIX A—COEFFICIENTS FOR EQUATION (25)

Between Cracking and Ultimate (Case A, $\varepsilon_c \leqslant \varepsilon_0$):

$$\text{Let } c = \frac{\varepsilon_s}{\varepsilon_0}$$

$$D_1 = B_1 d - C_1$$

$$D_2 = \left(e - \frac{D}{2}\right) B_1 + B_2 d - C_2 + A'_s \varepsilon_s E_s d + A_s f_s d$$

$$D_3 = \left(e - \frac{D}{2}\right) B_2 + B_3 d - A'_s E_s \varepsilon_s (d + 2d') - 3A_s f_s d$$

$$D_4 = \left(e - \frac{D}{2}\right) B_3 + B_4 d + A'_s E_s \varepsilon_s \left(2d' + \frac{d'^2}{d}\right) + 3A_s f_s d$$

$$D_5 = \left(e - \frac{D}{2}\right) B_4 - A_s f_s d - A'_s E_s \varepsilon_s \frac{d'^2}{d}$$

and

$$C_1 = bf'_c d^2 \left[-\frac{2}{3} c - \frac{1}{4} c^2\right]$$

$$C_2 = bf'_c d^2 \left[\frac{2}{3} c\right]$$

$$B_1 = -b'_c dc - \frac{bf'_c d}{3} c^2$$

$$B_2 = bf'_c d\, c - A'_s E_s \varepsilon_s - A_s f_s$$

$$B_3 = 2A_s f_s + \varepsilon_s A'_s E_s \left(1 + \frac{d'}{d}\right)$$

$$B_4 = -A' \frac{E_s}{d} \varepsilon_s d' - A_s f_s$$

Between Cracking and Ultimate (Case A, $\varepsilon_c \leqslant \varepsilon_0$, $\varepsilon'_s \geqslant \varepsilon'_y$):

$$\text{Let } c = \frac{\varepsilon_s}{\varepsilon_0}$$

$$B_1 = -bf'_c d\, c - \frac{bf'_c d}{3} c^2$$

$$B_2 = bf'_c d\, c + A'_s f'_y - A_s f_s$$

$$B_3 = 2A_s f_s - 2A'_s f'_y$$

$$B_4 = A'_s f'_y - A_s f_s$$

$$C_1 = bf'_c d^2 \left[-\frac{2}{3} c - \frac{1}{4} c^2\right]$$

$$C_2 = bf'_c d^2 \left[\frac{2}{3} c\right]$$

$$D_1 = B_1 d - C_1$$

$$D_2 = \left(e - \frac{D}{2}\right) B_1 + B_2 d - C_2 + A_s f_s d - A'_s f'_y d$$

$$D_3 = \left(e - \frac{D}{2}\right) B_2 + B_3 d - 3A_s f_s d + A'_s f'_y (d' + 2d)$$

$$D_4 = \left(e - \frac{D}{2}\right) B_3 + B_4 d + 3A_s f_s d - A'_s f'_y (d + 2d')$$

$$D_5 = \left(e - \frac{D}{2}\right) B_4 - A_s f_s d + A'_s f'_y d'$$

Between Cracking and Ultimate (Case B, $\varepsilon_c \geqslant \varepsilon_0$, $\varepsilon'_s \leqslant \varepsilon'_y$):

$$\text{Let } c = \frac{\varepsilon_s}{\varepsilon_0}$$

$$D_1 = 0$$

$$D_2 = -U_1 + W_1$$

$$D_3 = G_1 \left(e - \frac{D}{2} + d\right) - W_1 + W_2 - U_2$$

$$D_4 = G_2\left(e - \frac{D}{2} + d\right) + W_3 - U_3$$

$$- A_s'E_s\varepsilon_s\,(d - d') - W_2$$

$$D_5 = G_3\left(e - \frac{D}{2} + d\right) - W_3 - U_4$$

$$+ A_s'E_s\varepsilon_s\,(d - d')\left(\frac{d'}{d}\right)$$

$$U_1 = bf_c'd^2\left[- 0.06\,c - 0.59 + 0.53\left(\frac{1}{c}\right)^2\right]$$

$$+ bf_c'd^2\left[1.18 + 0.09\,c + 1.09\,\frac{1}{c}\right]$$

$$U_2 = bf_c'd^2\left[0.59 - 1.59\left(\frac{1}{c}\right)^2\right]$$

$$+ bf_c'd^2\left[-2.36 - 0.09\,c - 3.27\,\frac{1}{c}\right]$$

$$U_3 = bf_c'd^2\left[1.59\left(\frac{1}{c}\right)^2\right] + bf_c'd^2\left[1.18 + 3.27\,\frac{1}{c}\right]$$

$$U_4 = bf_c'd^2\left[-0.53\left(\frac{1}{c}\right)^2\right] + bf_c'd^2\left[-1.09\,\frac{1}{c}\right]$$

$$W_1 = bf_c'd^2\left[\frac{5}{12}\left(\frac{1}{c}\right)^2 + \frac{2}{3}\,\frac{1}{c}\right]$$

$$W_2 = bf_c'd^2\left[-\frac{5}{6}\left(\frac{1}{c}\right)^2 - \frac{4}{3}\,\frac{1}{c}\right]$$

$$W_3 = bf_c'd^2\,\frac{1}{c}\left[\frac{5}{12}\,\frac{1}{c} + \frac{2}{3}\right]$$

$$G_1 = bdf_c'\left[- 1.18 - 0.09\,c - 1.09\,\frac{1}{c} + \frac{2}{3}\,\frac{1}{c}\right]$$

$$G_2 = bdf_c'\left[2.18\,\frac{1}{c} + 1.18 - \frac{4}{3}\,\frac{1}{c}\right] + A_sf_s + E_sA_s'\varepsilon_s$$

$$G_3 = bdf_c'\left[- 1.09\,\frac{1}{c} + \frac{2}{3}\,\frac{1}{c}\right] - E_sA_s'\varepsilon_s\left(\frac{d'}{d}\right) - A_sf_s$$

Between Cracking and Ultimate (Case B, $\varepsilon_c \geqslant \varepsilon_0$, $\varepsilon_s' \geqslant \varepsilon_y'$):

$$\text{Let } c = \frac{\varepsilon_s}{\varepsilon_0}$$

$$D_1 = 0$$

$$D_2 = W_1 - U_1$$

$$D_3 = \left(e - \frac{D}{2} + d\right)V_1 - W_1 + W_2 - U_2$$

$$D_4 = \left(e - \frac{D}{2} + d\right)V_2 - W_2 + W_3 - D_3 + A_s'f_y'(d - d')$$

$$D_5 = \left(e - \frac{D}{2} + d\right)V_3 - W_3 - U_4 - A_s'f_y'(d - d')$$

$$V_1 = bdf_c'\left[\frac{2}{3}\,\frac{1}{c} - 1.18 - 0.09\,c - 1.09\,\frac{1}{c}\right]$$

$$V_2 = bdf_c'\;\; 2.18\,\frac{1}{c} + 1.18 - \frac{4}{3}\,\frac{1}{c}\Big] + A_sf_s - A_s'f_y'$$

$$V_3 = \frac{2}{3}\,bdf_c'\,\frac{1}{c} + A_s'f_y' - A_sf_s - 1.09\,bdf_c'\,\frac{1}{c}$$

and U_1, U_2, U_3, U_4, W_1, W_2 and W_3 are the same as above case (previous section).

APPENDIX B—CONVERSION FACTORS

1 in. = 25.4 mm
1 lb (mass) = 0.4536 kg
1 lb (force) = 4.4482 N
1 psi = 6.895 kPa
1 kip = 4448.2 N
1 ksi = 6.895 MPa
1 kip-in = 0.113 kN-m

CHAPTER 4

Some Aspects of Softening in Reinforced Concrete Flexural Members

PETER LeP. DARVALL*

ABSTRACT

Moment reduction with increasing curvature, or softening, is a characteristic of reinforced concrete flexural members. The effects of softening on deflections, on static collapse and shakedown loads, and on dynamic response, are examined. Analysis is possible only when it is assumed that softening occurs over a finite hinge length. At critical softening, a structure cannot sustain increased load, however redundant it may still be. Conventional matrix structural analysis is applied to softening structures, using a stiffness matrix developed for flexural elements with softening portions. A steeper softening slope reduces both the number of hinges formed before collapse and the collapse load. Quite small softening parameters may lead to large reductions in static collapse and shakedown loads, especially when residual moments are present. For dynamic loading, the critical softening parameter depends on the severity and the duration of the applied load, and on ductility. Conversely, for a given softening parameter, critical severity and duration of load may be found.

INTRODUCTION

A rigorous load-deflection analysis to collapse for a statically indeterminate reinforced concrete structure, for which failure will occur in a flexural mode, requires that the complete moment-curvature (M–ϕ) relationship be known for each section of the structure.

Following the development of the plastic method of analysis for steel structures, which assumes bilinear elastic-plastic M–ϕ behavior, an enormous number of tests were conducted in many countries to define the influence of various parameters on the idealized elastic-plastic moment-rotation curves for reinforced concrete hinges [1]. Particular attention was paid to the determination of a maximum hinge rotation (point L_2 in Figure 1), dependent upon assigned upper limits on either the steel or concrete strains, since a limited rotation capacity (or "ductility") was considered to be the major difficulty in applying the plastic method of analysis to reinforced concrete structures. Softening, or decrease in bending moment capacity at far advanced curvature (see latter portion of "less ductile" curve in Figure 1), was excluded from consideration in early attempts to apply the plastic method of analysis to concrete structures.

Rosenbleuth and De Cossio [26] pointed out that analyses which did not take into account the post-plastic (or descending) portions of moment-curvature curves could lead to substantial underestimates in the limit loads computed for statically indeterminate structures, that is, the collapse load possibly will not occur until one or more hinges has softened. Using the device of assigning "contamination lengths" over which curvature increased while moment decreased, they demonstrated how the resistance of a structure as a whole could increase while certain parts were softening.

Similarly, Barnard [2] noted that the importance of the falling portions of moment-curvature curves had been neglected because tests had generally been conducted using load-control with flexible testing machines, rather than displacement-control with stiff machines. He proposed the use of a "discontinuity length" to overcome the computational problem of obtaining increased rotations with falling bending moments, as found in tests. He also suggested that the maximum load (or collapse load) was a less arbitrary definition of the failure load for an indeterminate beam than that for a particular rotation, notwithstanding that the falling portion of the M–ϕ curve corresponded to considerable damage such as yielding, spalling, or crushing.

Cranston [7,8] and Nylander and Sahlin [24], among others, recorded flexural softening (moment reduction) at sections of indeterminate concrete structures while total load was increasing. In effect, their indeterminate structures provided for the first formed "hinges" the test conditions

*Department of Civil Engineering, Monash University, Clayton, Victoria 3168, Australia

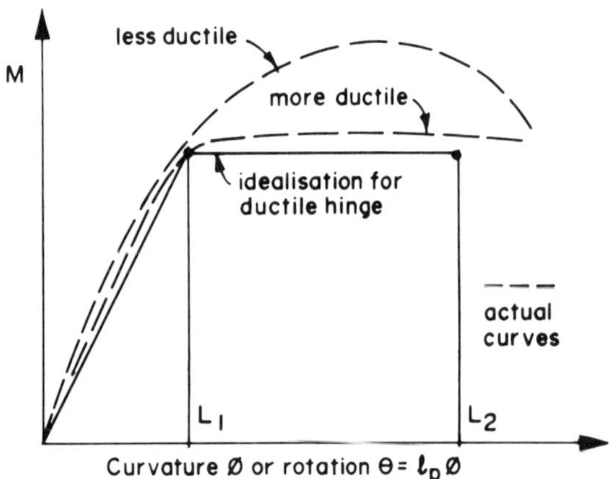

FIGURE 1. Moment-curvature curves for reinforced concrete sections.

specified by Barnard as necessary for the observation of softening behaviour. Softening is most evident for over-reinforced sections such as steel-concrete composite structures [3] and may be a common characteristic for beam-column joints in reinforced concrete frames [30].

Maier et al. [22] examined the stability of a linear-elastic symmetrically loaded fixed-fixed beam with concentrated hinges of rigid, linear-softening characteristics, and demonstrated that there were different stable equilibrium configurations for values of falling slope less (i.e., flatter) than a critical value.

Bazant [4] resolved the continuum mechanics paradox of strain-softening, demonstrating that it could occur only for non-homogeneous materials, such as concrete, where softening occurred within a finite, though small, region. For a symmetrically loaded fixed-fixed beam he demonstrated that the collapse load occurred when the softening gradient for the hinging sections reached a certain critical value, and that this depended on the ratio ℓ_p/L (softening length/span length).

Ghosh [19] obtained good agreement between theoretical and experimental load-deflection curves up to collapse load for several propped cantilevers, basing his analysis on computed M–ϕ curves which had a falling characteristic. Computations involved division of the beams into elements of length, with incremental loading and compatibility checks at each stage. Plastic rotations were assumed to occur over lengths equal to the gauge length on which concrete stress-strain curves were based.

In the examination of the various implications of softening which follow, simple models for the M–ϕ curve for reinforced concrete sections are adopted. Figure 2 shows a typical graph (to scale) of moment versus curvature at the load point of a simply supported beam loaded at midspan [13]. The load was applied by a testing machine in displacement-control mode, and the curvature was taken as an average over a length equal to the effective depth of the

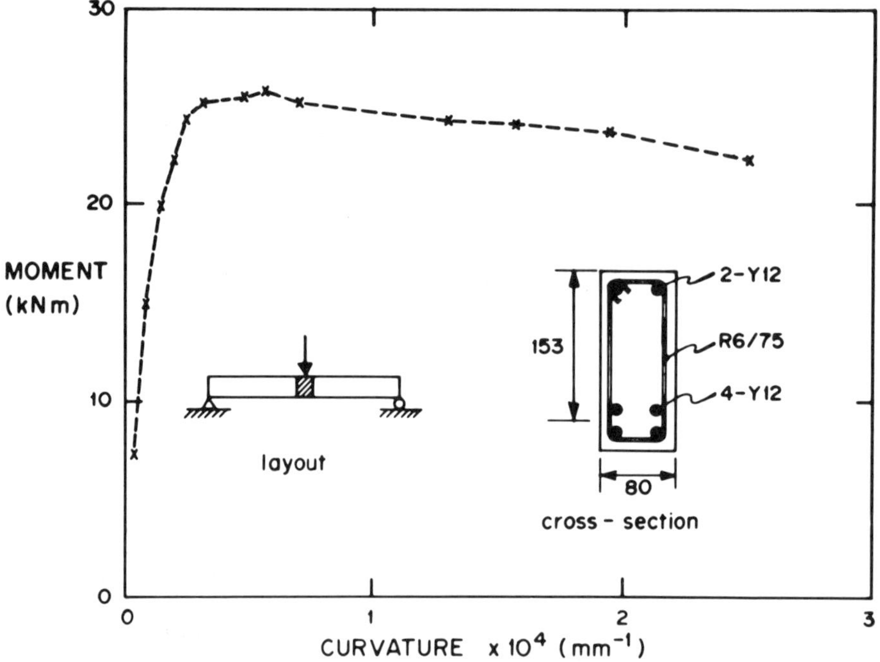

FIGURE 2. Moment-curvature diagram for a midspan hinge.

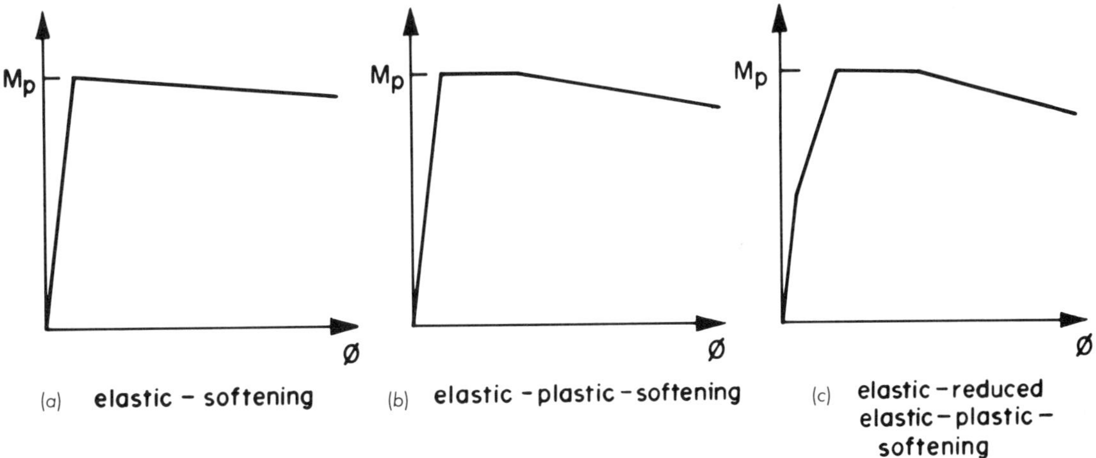

FIGURE 3. Models for moment-curvature curves.

section. The percentage of reinforcement in the cross-section is in accordance with provisions of Codes of Practice for structural concrete.

Figure 3 shows three alternative approximations for a real M–ϕ curve. The model most capable of accurately representing the real M–ϕ curve would be the elastic-reduced elastic-plastic-softening approximation (c), but the elastic-softening approximation (a) and the elastic-plastic-softening model (b) may be used to demonstrate qualitatively the effects of softening without much loss in rigour or accuracy.

It is assumed herein that the elastic slope of the M–ϕ diagram is EI, and the softening slope aEI, where a is a small negative number. To indicate a typical value, the softening parameter a for a straight line approximation to the softening part of the real M–ϕ diagram of Figure 2 is -0.007 (0.7%).

CRITICAL SOFTENING PARAMETERS FOR STATIC LOADS

The important concept of critical softening may be introduced by reference to the propped cantilever of Figure 4(a). It is assumed that any moment-curvature characteristic, Figure 4(b), has a trilinear approximation, as shown. To simplify the analysis it will be assumed that all sections have the same slope EI (this is quite a good approximation for typical reinforcement layouts in beams of constant cross-section size). In this and subsequent examples, it is assumed that flexural action predominates, so that no premature failures, such as from shear, occur.

As the load P is increased it is assumed that the maximum moment, M_p, will be reached at A and rotation independent of moment will then occur over a small length ℓ_p while the moment at C, M_C, increases. If the plateau length $\phi_p - \phi_y$ at A is short, the moment at A will begin to decrease before M_C reaches its value of M_p. The beam at this stage will collapse or sustain constant or increasing load until M_p is reached at C, depending on the downward slope of the M–ϕ diagram at A. Assuming that M_A does enter the falling slope, the load-deflection diagram [P–Δ, Figure 4(c)] will be trilinear to its maximum point, with the moments and curvatures at A and C at the end of each stage shown schematically on one M–ϕ diagram in Figure 4(b). At the critical slope of the softening portion of the M–ϕ diagram for the hinge at A, a_{cr}, the third stage of the P–Δ curve will be horizontal. This leads to the definition of critical softening as that value of the softening parameter for a hinge at which the structure as a whole cannot sustain increased load(s), however redundant the structure may still be.

The analysis to determine the critical value of the parameter a is based on Figure 5. For an incremental displacement in Stage 3, a_{cr} may be determined by satisfying equilibrium and compatibility, or alternatively by the requirement that the second order work Δ^2W is positive for such displacement [4,22]. Using the former approach here, the equilibrium requirement for Stage 3 is expressed by the bending moment increment of Figure 5(d). The corresponding curvature increments are shown in Figure 5(e), and the compatibility condition is that the deflection at B is zero, i.e., that

$$\int_0^L \phi_3(L - x)\, dx = 0 \tag{1}$$

so that

$$\frac{M_A}{LEI}\left[\frac{1}{a_{cr}}\int_0^{\ell_p} (L - x)^2\, dx + \int_{\ell_p}^{L} (L - x)^2\, dx\right] = 0 \tag{2}$$

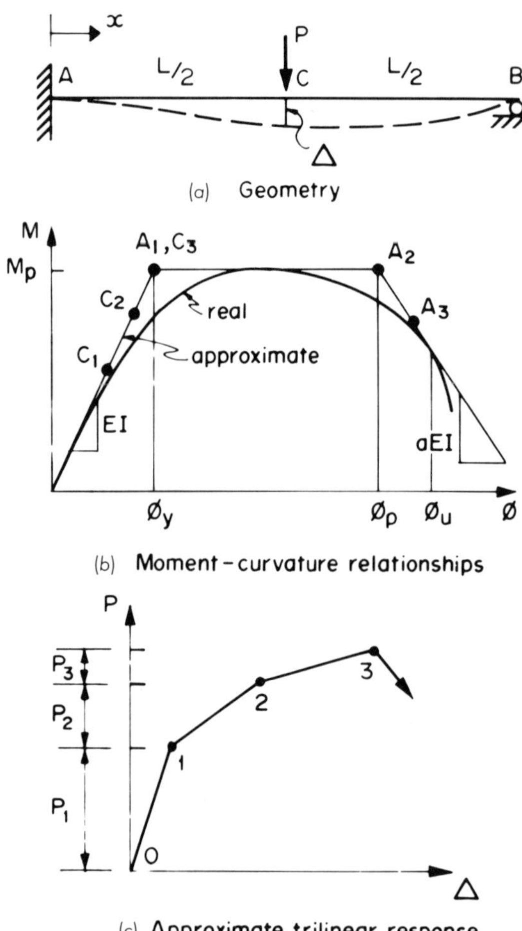

FIGURE 4. Trilinear analysis and response for a propped cantilever.

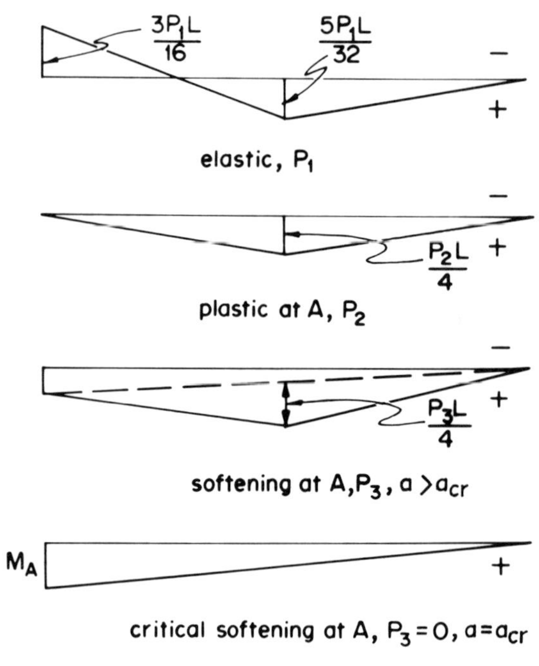

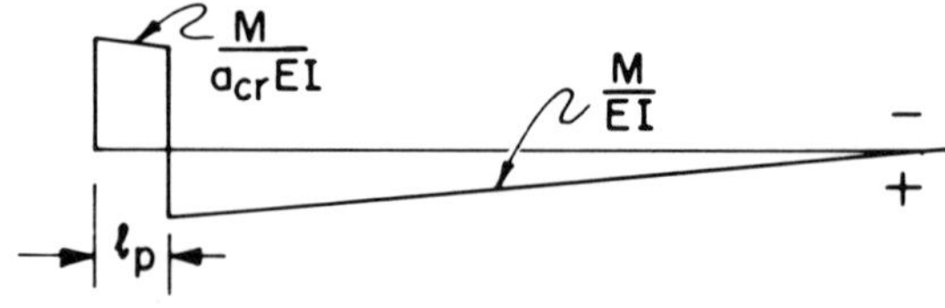

FIGURE 5. Basis of analysis for a_{cr}, propped cantilever.

from which

$$a_{cr} = \frac{-[3m^2 - 3m + 1]}{m^3 - [3m^2 - 3m + 1]} \tag{3}$$

where $m = L/\ell_p$.

When m is large, the solution tends to the value $a_{cr} = 3/m$, which is the approximate solution obtained when it is assumed that all inelastic rotation, θ_A, is concentrated at the point A, so that

$$\theta_A L = \int_0^L \frac{M_3}{EI}(L - x)\,dx \tag{4}$$

and equating this to

$$\phi_A\, \ell_p = \frac{M_A}{aEI}\,\ell_p \tag{5}$$

Exact and approximate values for the critical slope parameter, a_{cr}, for various values of m are compared in Table 1. Typical values of m are likely to be greater than 20 and less than 30, so that the error in calculating a_{cr} by the approximate method will be less than 10%.

It should be noted that for a continuum, where ℓ_p must be assumed to be arbitrarily small, it is not possible to have strain-softening, i.e., a_{cr} tends to zero.

For $a < a_{cr}$ (steeper softening slope) the loss of stored elastic energy in the beam will be greater than the work required in further hinge deformation, as noted by Barnard

TABLE 1. Exact and Approximate Values of a_{cr} for Propped Cantilever.

m	10	20	30
a_{cr} Eq. (3)	−0.372	−0.166	−0.107
a_{cr} = −3/m	−0.300	−0.150	−0.100
% error	19	10	7

[2]. Since the critical falling slope is inversely proportional to $m = L/\ell_p$, stability and ductility are enhanced when the length of the plastic hinge is greater, as noted also by Bazant [4].

A value of m for any particular case may be calculated by using equivalent hinge lengths such as determined by Corley [6]:

$$\ell_p = 0.5d + 0.2\sqrt{d}\left(\frac{z}{d}\right) \qquad (6)$$

where ℓ_p is in inches, d is the effective depth of the section, and z is the distance from the critical section to the adjacent point of contraflexure, or by Sawyer [28],

$$\ell_p = 0.25d + 0.075z \qquad (7)$$

It must be remembered that these values of ℓ_p apply to one side of a maximum moment point. A separate calculation

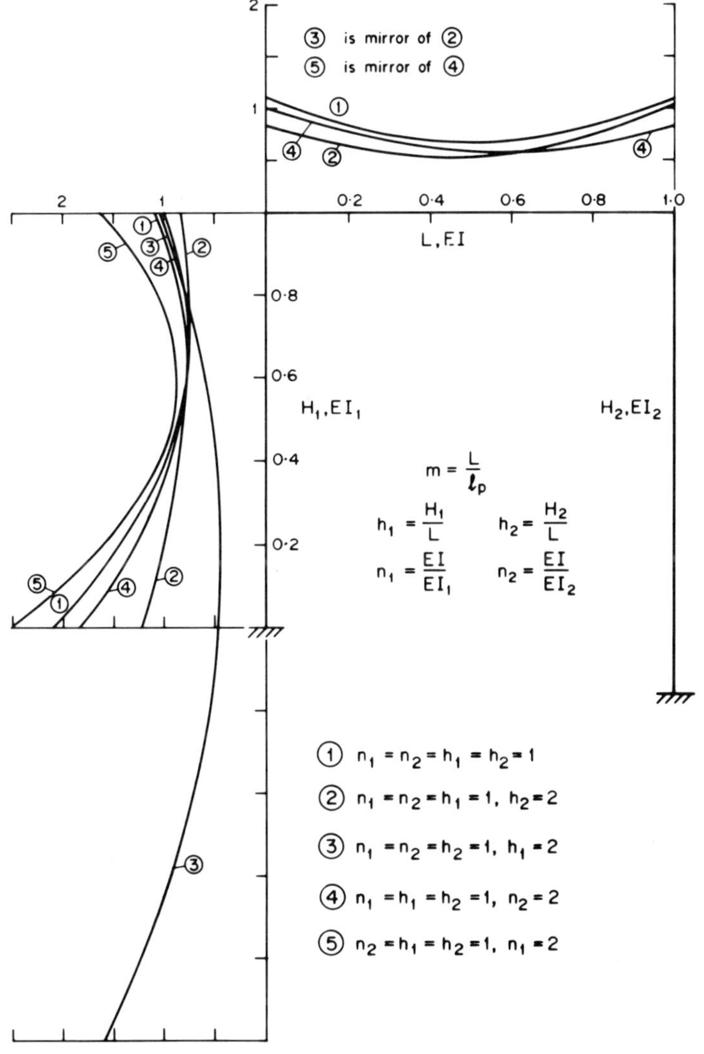

FIGURE 6. K = ma_{cr}, first-formed hinge, fixed-base portal frames.

must be made for the other side of a two-sided hinge, and the total discontinuity length will be the sum of the values for the two sides.

For reinforced concrete elements, values for ℓ_p [6,28] are typically of the order of $2d/3$, where d is the effective depth of a section, and then typical values of the ratio m = member length/total discontinuity length are in the range 15–25.

Critical softening parameters for variable hinge position have been determined similarly for beams of higher indeterminancy and for single bay portal frames [9,10,16,17].

Figure 6 shows graphs of the approximate critical softening parameter of the first-formed hinge versus location in a fixed-base portal frame for different relative dimensions and stiffnesses of the three members.

From the graphs it can be seen that for each member there is a soft point, where the least value of the critical softening parameter occurs. When the degree of indeterminancy is reduced, as for instance by the insertion of a hinge at the base of a column, the critical softening parameter at a given location is reduced. The effect is similar if plasticity or softening are already present elsewhere in the structure [10,16].

LOAD DEFLECTION CURVES FOR ELASTIC-SOFTENING BEAMS

The effect of degree of softening on the load carrying capacity and deflections of a flexural structure may be illustrated by the case of a fixed-fixed beam, which may be thought of as one interior span of a uniformly loaded multi-span beam.

Softening of First-Formed Hinges

In Figure 7 the first-formed hinges of length ℓ_p at the supports are softening, i.e., shedding moment as the bending moment increases at midspan. The moment-curvature behavior is assumed to be simply elastic-softening. All sections other than the hinges are loading or unloading elastically. Using symmetry, with no rotation between midspan and support,

$$\int_0^{L/2} \phi dx = 0 \tag{8}$$

i.e.

$$\frac{1}{EI}\int_0^{L/2-\ell_p}\left(c - \frac{4d}{L^2}x^2\right)dx + \frac{1}{aEI}\int_{L/2-\ell_p}^{L/2}\left(c - \frac{4d}{L^2}x^2\right)dx = 0 \tag{9}$$

leads to

$$a = \frac{-[6m^2(c-d) + 12dm - 8d]}{m^3(3c-d) - [6m^2(c-d) + 12dm - 8d]} \tag{10}$$

in which $m = L/\ell_p$.

Critical softening occurs when $d = 0$, for which

$$a_{cr} = \frac{-2}{m-2} \tag{11}$$

e.g., when $m = 20$, $a = -0.111$.

For sub-critical softening, the load must be increased after softening of the support hinges before the midspan hinge will form and cause collapse. In this case, $d > 0$ and $a > a_{cr}$ (i.e., the softening curve is flatter). For super-critical softening $d < 0$ and $a < a_{cr}$ (the softening curve is steeper). The midspan hinge will form, and collapse will occur even though the load is reduced as deflections increase.

Deflection with Softening Hinges at Supports

The curvature increment shown in Figure 7 may be used to calculate the midspan deflection which occurs during softening of the support hinges.

$$\Delta_2 = \int_0^{L/2} \phi\left(\frac{L}{2} - x\right)dx \tag{12}$$

(Δ_1 is taken to be the initial midspan elastic deflection to onset of plasticity at supports.) Integrating, the midspan deflection (down) is:

$$\Delta_2 = \frac{L^2}{EI}\left\{\left[\frac{c}{8} - \frac{d}{48} - \frac{c-d}{2m^2} - \frac{4d}{3m^3} + \frac{d}{m^4}\right] + \frac{1}{a}\left[\frac{c-d}{2m^2} + \frac{4d}{3m^3} - \frac{d}{m^4}\right]\right\} \tag{13}$$

When $m = \infty$ (uniform beam)

$$\Delta_2 = \frac{L^2}{\mathrm{EI}}\left[\frac{c}{8} - \frac{d}{48}\right] \tag{14}$$

which with the appropriate values of c and d gives the standard midspan deflection increments for a uniformly loaded clamped or simply supported beam.

For a given value of a, Equation (10) provides the c/d ratio to be used in Equation (13). For example, if $m = 20$ and $a = -1/9$, then $d = 0$ (critical softening); and with $c = M_p/2$, so that the midspan hinge is forming, the deflection at constant load during softening before collapse is:

$$\Delta_2 = \frac{M_pL^2}{EI}(0.0563) \tag{15}$$

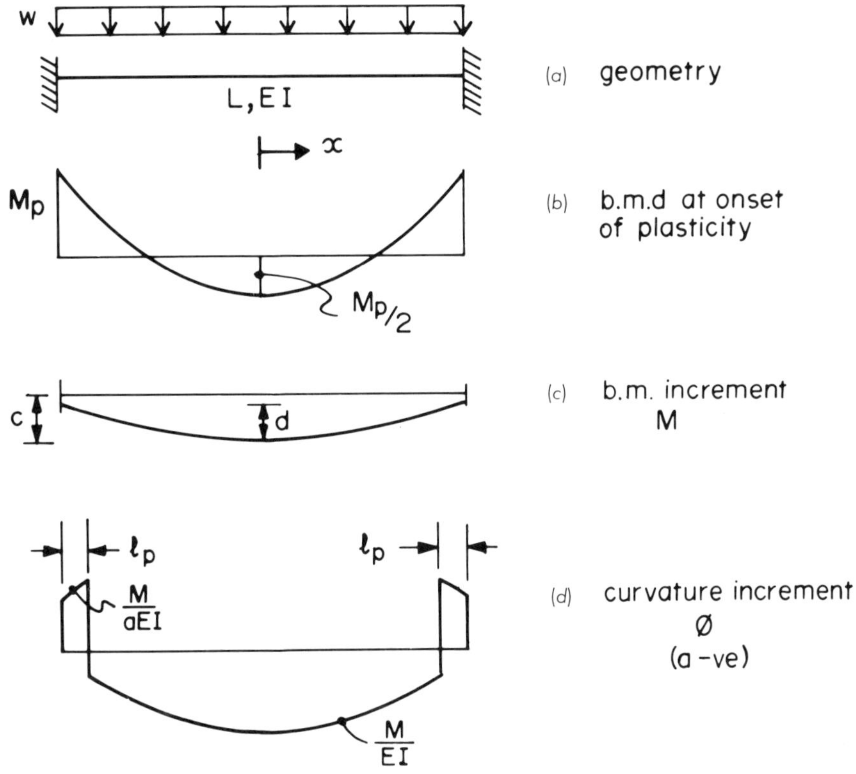

FIGURE 7. Softening of support hinges, fixed-end beam.

Softening of Hinges at Supports and Midspan

Figure 8 provides the basis for analysis when a midspan hinge is softening at the same slope as the support hinges. The midspan hinge has been assigned a length twice that of the support hinge, since the hinge lengths to each side of a maximum moment point are likely to be approximately equal, and a midspan hinge is two-sided, whereas the support hinge is one-sided.

In Figure 8, the bending moment increment has been drawn concave down, corresponding to a load reduction. For simplicity, $c > d$ has been shown, though with both hinges softening $d > c$, corresponding to negative moment shedding at the support, and positive moment shedding at midspan.

Again $\int_0^{L/2} \phi dx = 0$, which leads to

$$a = \frac{-[6m^2(2c - d) + 12dm - 16d]}{m^3(3c - d) - [6m^2(2c - d) + 12dm - 16d]} \quad (16)$$

For $m = 20$ and $a = -1/9$ (critical softening of first formed hinges), $c = 0.46d$.

The expression for the deflection increase at midspan softening is obtained as before:

$$\Delta_3 = \frac{-L^2}{EI}\left\{\left[\frac{c}{8} - \frac{d}{48} - \frac{c}{2m} + \frac{d}{2m^2} - \frac{2d}{3m^3}\right] + \frac{1}{a}\left[\frac{c}{2m} - \frac{d}{2m^2} + \frac{d}{3m^3}\right]\right\} \quad (17)$$

where the minus sign follows from the assumption of a negative moment increment as shown in Figure 8.

Load-Deflection Curves

The elastic deflection in stage 1 before onset of plasticity is $\Delta_1 = M_pL^2/(32EI)$. Assuming elastic-softening behavior, Equations (10) and (13) are used to calculate stage 2 deflections Δ_2 following the elastic stage, when support hinges are softening while the section at midspan is still elastic; and Equations (16) and (17) are used to calculate stage 3 deflections Δ_3 when both support and midspan hinges are softening. The results are shown in Figure 9. Softening in the first formed (support) hinge has not been taken so large as to reduce the hinge moment below zero, i.e., at the extreme right end of curve 4 there is still a negative moment at the supports. Analysis to include plastic stages before softening

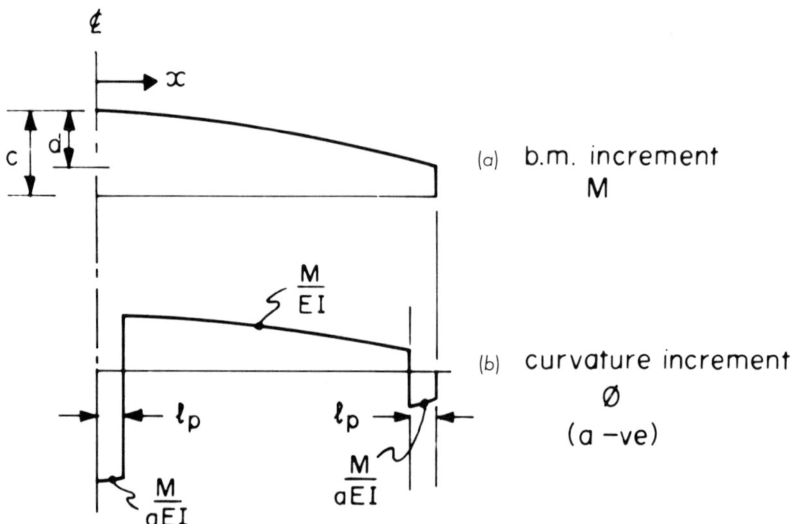

FIGURE 8. Softening of support and midspan hinges, fixed-end beam.

of either support or midspan hinges can be easily included, with the necessary assumption of point hinges.

STIFFNESS MATRIX FOR ELASTIC-SOFTENING BEAMS

The foregoing analyses, while illustrating some implications of softening, have not been in a form suitable for application to beams and frames in general.

The derivation of a stiffness matrix for an elastic beam element with softening portions allows direct application in standard frame analysis programs, extending their range to include softening.

The four standard degrees of freedom for a bending element of stiffness EI are shown in Figure 10, indicating also discontinuity lengths ℓ_{p1} and ℓ_{p2} at the ends of the element, for which the softening parameters are a and b, having the effect on the curvature increment in Figure 10(c) as shown. When a softening parameter > 1, it is the case of a prismatic haunch. For $0 <$ softening parameter < 1, the relevant length has reduced stiffness. For softening parameter < 0, softening is occurring in the relevant length.

The columns of the stiffness matrix are obtained in the usual way: applying each unit displacement in turn and calculating the forces. For example, when unit rotation at the discontinuity end is applied, the determination of stiffness coefficients k_{i2} is based on Figure 10(a), (b) and (c).

Using the two equations,

$$\int_0^L \phi\, dx = 1 \text{ (unit rotation)} \qquad (18)$$

$$\int_0^L \phi\,(L - x)\, dx - 1.L = 0 \text{ (zero displacement)} \qquad (19)$$

we obtain

$$[2bmn^2 - bn^2 + ab(m^2n^2 - 2mn^2 - m^2 + n^2) + am^2]\, k_{22}$$

$$+ [ab(n^2 - m^2 + 2m^2n - m^2n^2) - bn^2 - 2am^2n + am^2]\, k_{42}$$

$$= \frac{2EI}{L} abm^2n^2 \qquad (20)$$

$$[3bm^3n^3 - 3bm(m - 1)^2n^3 - 3bmn^3 + 2bn^3 - 3abm^3n$$

$$+ 3abm(m - 1)^2n^3 - a(b - 1)m^3(n - 1)^2(n + 2)$$

$$+ ab(3m - 2)n^3 + 3am^3n - am^3n^3]\, k_{22}$$

$$+ [2bn^3 - 3bmn^3 - a(b - 1)m^3(n - 1)^2(n + 2)$$

$$+ ab(3m - 2)n^3 - am^3n^3]\, k_{42}$$

$$= \frac{6EI}{L} abm^3n^3 \qquad (21)$$

where

$$m = \frac{L}{\ell_{p1}} \qquad n = \frac{L}{\ell_{p2}}$$

Equations (20) and (21) may be solved for k_{22} and k_{42} for

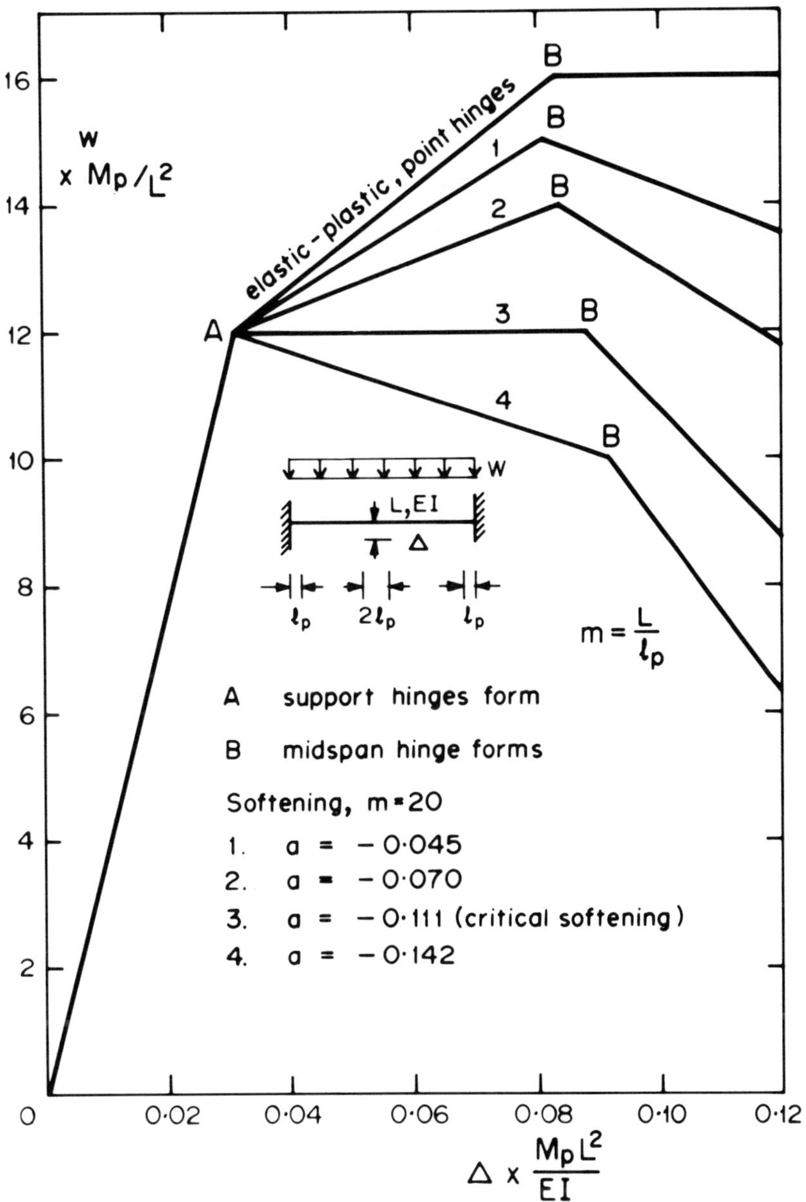

FIGURE 9. Load-deflection curves for beam with softening hinges.

particular input values of a, b, m, and n. Similarly, k_{44} follows from the application of a unit rotation at B. Symmetry and statics provide the other coefficients of the stiffness matrix. When $a = b = 1$, the standard beam stiffness coefficients are obtained:

$$k_{22} = \frac{4EI}{L} \text{ and } k_{42} = \frac{2EI}{L} \tag{22}$$

with $b = 1$, $m = 10$, and varying a, graphs of k_{22}, k_{44}, and k_{42} are shown in Figure 11.

For $a > 1$, k_{22} and k_{42} are stiffness coefficients for a singly haunched beam, tabulated in Reference 25. For $a < 0$, softening is occurring.

The curves are asymptotic vertically to the critical softening value for a softening hinge at one end of a fixed-end beam [18], and horizontally to beam stiffness values for a

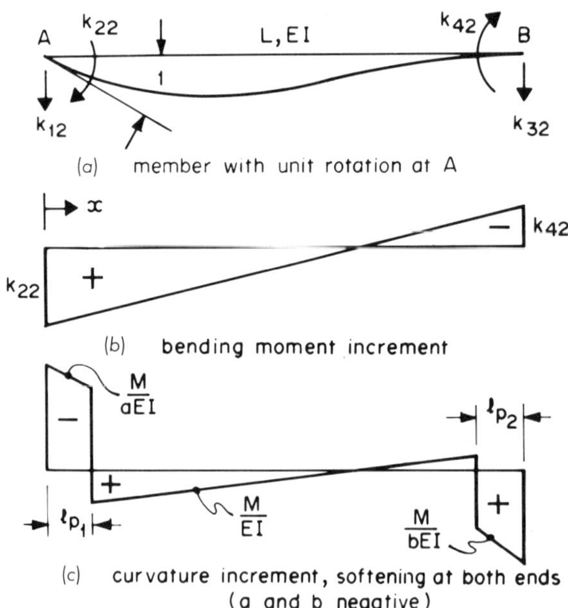

FIGURE 10. Basis for determination of stiffness coefficients k_{i2}.

haunch of infinite stiffness at one end. The curves in the upper left portion of Figure 11 are the solutions for negative *EI* and negative *a*, and thus reflect those for positive *EI* and *a*.

ELASTIC-PLASTIC-SOFTENING ANALYSIS OF PLANE FRAMES

The stiffness coefficients derived above allow use of standard frame analysis programs without any special treatment, except that a node must be provided immediately to one side of any potential softening region exactly as for a haunched element.

Results for examples which follow were obtained using computer program PAWS ("Plastic Analysis with Softening"). PAWS is a modification of ULARC, ("Ultimate Load Analysis of Reinforced Concrete") developed at the University of California at Berkeley [31]. ULARC is an efficient program for elasto-plastic analysis, employing the direct stiffness method. In PAWS, softening (or hardening) are also taken into account. The program can be used for frames of reinforced concrete or steel of arbitrary shape subjected to static point loads and support settlements.

Full details of program PAWS, including input specifications and listing are given elsewhere [23].

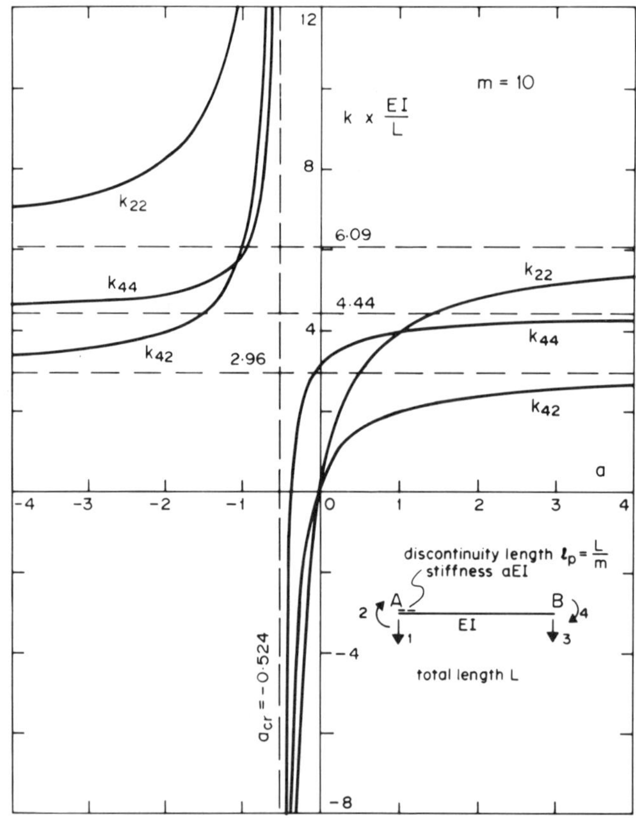

FIGURE 11. Stiffness coefficients versus softening parameter, m = 10.

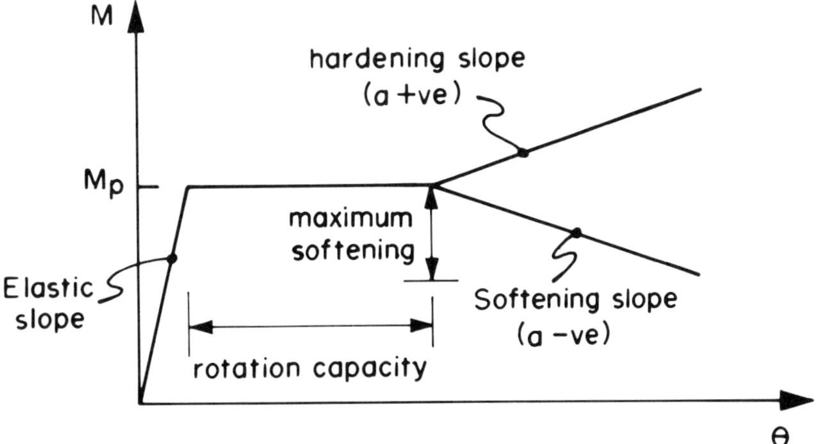

FIGURE 12. Hinge characteristics for program PAWS.

The full range analysis of a frame using the computer program PAWS involves the following preparation, in addition to that required for ULARC.

(1) Identify the locations where perfectly plastic, plastic-softening or plastic-hardening hinges may form. Different types of hinge may occur in the same structure.
(2) For softening or hardening hinges, obtain from M-ϕ diagrams for both positive and negative bending the values of plastic hinge rotation capacity, softening or hardening slope, and the maximum allowable reduction of moments through softening or increase through hardening, as shown in Figure 12.
(3) Estimate L/ℓ_p for each softening or hardening hinge, by using the hinge length equations of Corley [6] or Sawyer [28].

The total response of a structure with plastic-softening hinges to increasing load will be a series of linear stages, each of which is terminated by any of the following:

(1) any hinge location reaching the end of the plastic range and becoming plastic;
(2) any hinge reaching the end of its plastic capacity and beginning to soften (or harden);
(3) any hinge unloading elastically from the plastic or softening range.

At the end of each stage the structure stiffness matrix must be modified for the new hinge condition, and then the load increment for the structure in that state calculated. In the case of softening at one or both ends of a member the stiffness coefficients derived above must be used.

In PAWS, a comparison is made between the load factor to form the next plastic hinge and the least load factor required to bring the plastic rotation of any of the already formed hinges to its rotational capacity. The lesser value gives the load factor for that particular cycle. If this load factor makes any of the displacements at the joints greater than a specified limiting displacement or makes softening more than the specified limit for any hinge, a new load factor is found and the analysis discontinued. The load factor at any cycle must also be less than the maximum load factor specified in the input.

For a structure with plastic hinges which later harden at some slope indefinitely, there is no specific maximum load. Therefore, a displacement limit is imposed to determine the maximum permissible load. With plastic-softening behavior a more general test for failure load is required. As hinges enter plastic and then softening phases, the stiffness of the structure as a whole deteriorates until the structure cannot sustain further loading. A structure in equilibrium is stable only if any compatible infinitesimal geometric change requires positive work from the external loads. A necessary condition for stability is that the stiffness matrix be positive definite. A structure becomes unstable when the determinant of the structure stiffness matrix becomes negative, i.e., when one of the eigen values becomes negative.

In ULARC the structure stiffness matrix is triangularised by the Gausssian elimination process. The signs of the eigen values are given by the signs of the diagonal elements of this upper triangular matrix. As each hinge becomes plastic, or softens, the eigen value signs are checked in PAWS to determine whether the collapse load has been reached.

The output consists of node displacements, member forces, support reactions, plastic hinge rotations, load factor and hinge status at the end of each cycle and the collapse load.

One case of particular interest is that of critical softening of the first hinge, as defined earlier. Program PAWS may be used to find the critical softening parameter for any hinge location in a frame structure by trial and error.

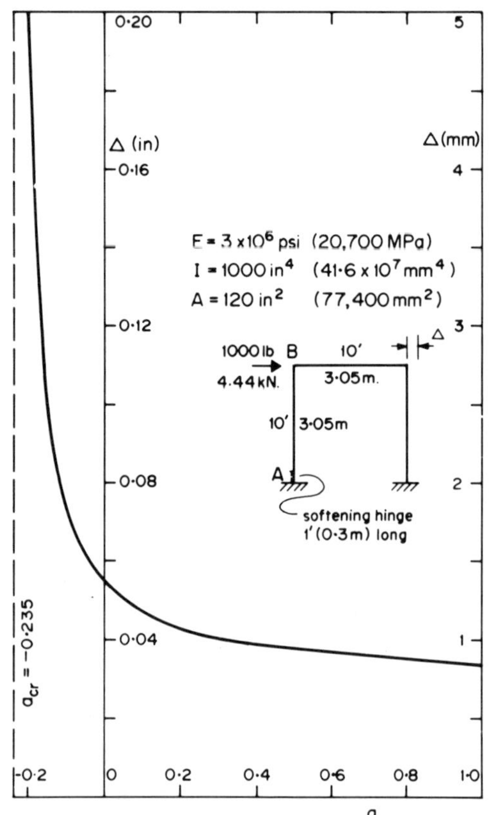

FIGURE 13. Softening at the base of one column of a portal frame.

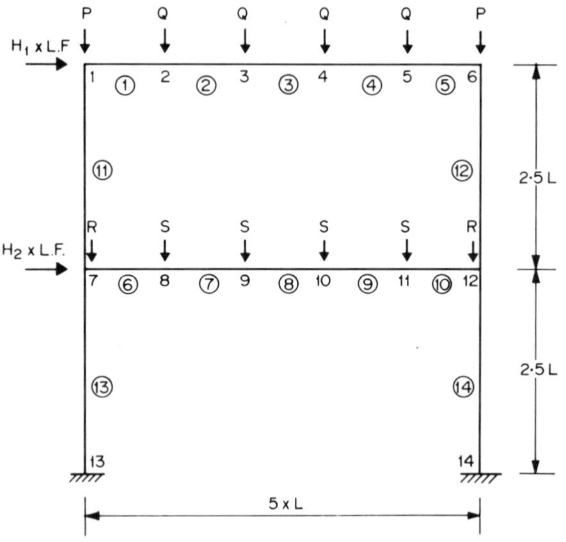

FIGURE 14. Data for elastic-plastic-softening frame.

Example 1. Response of a Frame After One Hinge Softens

Figure 13 shows a portal frame subjected to a sway load increment. As a result of some load or deformation history (e.g., an earthquake) a small length at the base of one column has reduced or softening stiffness. Figure 13 also shows the effect of reduced stiffness or softening on the sway displacement for the given load increment.

The vertical asymptote for total loss of frame stiffness occurs for $a = -0.235$, derived elsewhere [9] as the critical softening parameter for a hinge at this position in the given frame geometry and $m = 10$. At the critical value of a, the overall structure stiffness matrix becomes singular.

Example 2. Full Response of a Two-Storey Frame

Dimensions, loads and labelling for a frame used as an example for ULARC [31] are given in Figure 14. The vertical loads are applied, and then the horizontal loads increased to collapse:

$P = 3.52$ kip (15.64 kN)
$R = 13.36$ kip (59.38 kN)
$H_1 = 17.3$ kip (76.89 kN)
$L = 4$ft (1.22 m)
$Q = 3.24$ kip (14.40 kN)
$S = 8.32$ kip (36.98 kN)
$H_2 = 10$ kip (44.44 kN)

Member stiffness, strength and hinge data are given in Table 2.

As in the original ULARC example, the section strength was made artificially large on one side of joints 3, 4, 9 and 10 to ensure that hinges could form on only one side of these joints.

The results are shown in Figure 15, for various values of softening parameter and rotation capacity.

The critical softening parameter for the first formed hinge in the beam at node 12 was found to be -0.0824 ($m = 15$ for the whole beam). When the stiffening effect of the upper storey is removed, $a_{cr} = -0.0625$ for this hinge, confirming the theoretical value obtained from equations presented previously [9].

It can readily be seen from Figure 15 how a steeper softening slope reduces both the number of hinges formed before collapse and the collapse load. An increase in rotation capacity before softening has the opposite effects. The order of hinge formation may be changed by changing the softening slope [see curves (2) and (3) of Figure 15].

SHAKEDOWN WITH SOFTENING IN BEAMS AND FRAMES

The computation of the response of reinforced concrete frame structures under severe repeated undirectional loads

TABLE 2. Data for Example 2, Figure 14.

Member No.	A $in^2(m^2)$	I $in^4(m^4)$	M_p^+ kip in (kN m)	M_p^- kip in (kN m)	$M = \frac{L}{\ell_p}$
1,5	2000 (1.29)	1440 (5.99×10^{-4})	550 (62.1)	820 (92.6)	6
2–4	"	"	820 (92.6)	550 (62.1)	"
6,10	"	9985 (41.14×10^{-4})	1650 (186.3)	1950 (220.1)	3
7–9	"	"	2440 (275.4)	1650 (186.3)	"
11,12	195 (0.13)	3656 (15.22×10^{-4})	1440 (162.6)	1440 (162.6)	12
13,14	"	"	1680 (189.7)	1680 (189.7)	"

E = 3600 ksi (24800 MPa)

to shakedown or incremental collapse involves consideration of the softening portion of M–ϕ curves unless extended plasticity can be guaranteed at all relevant hinge locations.

The elastic-softening approximation to M–ϕ behavior is again used in what follows to illustrate the effects of softening. A further important assumption is shown in Figure 16. It is assumed that unloading (reduced ϕ) and reloading from the softening section of the M–ϕ curve follow the elastic slope, that reloading peaks at the previous lowest value reached on the softening curve, and softening then continues with further deformation. Although neglecting a narrow envelope between the unloading and reloading curves, the assumptions are closely in accordance with stress-strain curves for plain concrete subjected to compressive load histories [21]. In the examples herein it is assumed that behavior in negative bending mirrors that for positive bending.

Softening at Midspan or Interior Support in a Two-Span Beam

Figure 17 provides the basis for determining changes in the bending moment diagram induced by softening at the position of maximum positive bending moment in a two-span beam. The single point load at midspan is incremented by ΔP. Softening occurs over a hinge length $2\ell_p$, chosen to correspond to ℓ_p on each side of a maximum moment point. Writing compatibility equations expressing no deflection at B or C resulting from the curvature increment, we obtain

$$a = \frac{-\left[3m^2 + \frac{3m}{2}(f-1) - 2(f+1)\right]}{\frac{m^3}{8}(7+f) - \left[3m^2 + \frac{3m}{2}(f-1) - 2(f+1)\right] + \frac{3\theta EIm^3}{cL}} \quad (23)$$

$$a = \frac{-\left[\frac{9}{2}m^2 + \frac{9}{4}m(f-1) - (f+1)\right]}{\frac{3m^3}{16}(11+5f) - \left[\frac{9}{2}m^2 + \frac{9}{4}m(f-1) - (f+1)\right] + \frac{3\theta EIm^3}{cL}} \quad (24)$$

where $m = L/\ell_p, f = (b - c)/c$, and θ is the rotation at A.

Using given values of m, obtained from hinge length, and softening parameter a, θ may be eliminated from Equations (23) and (24) and the resulting equation solved for the ratio b/c.

In this case, $c = b/2$ and $f = 1$ at critical softening. For example, with $m = 20$, $a_{cr} = -0.0391$.

In Figure 18 softening is assumed to occur at a hinge symmetrically placed over the interior support of a two-span beam with equal downward loads at each midspan.

The compatibility equation expressing zero deflection between interior support and end leads to

$$a = \frac{-4[6m^2 - 3m(3-2g) + 4(1-g)]}{m^3(5+6g) - 4[6m^2 - 3m(3-2g) + 4(1-g)]} \quad (25)$$

where

$$g = \frac{c}{b}$$

Shakedown with Softening for a Two-Span Beam

The relationships between bending moment changes and softening parameters derived above may be used to demonstrate the effect of softening on the computation of shakedown loads.

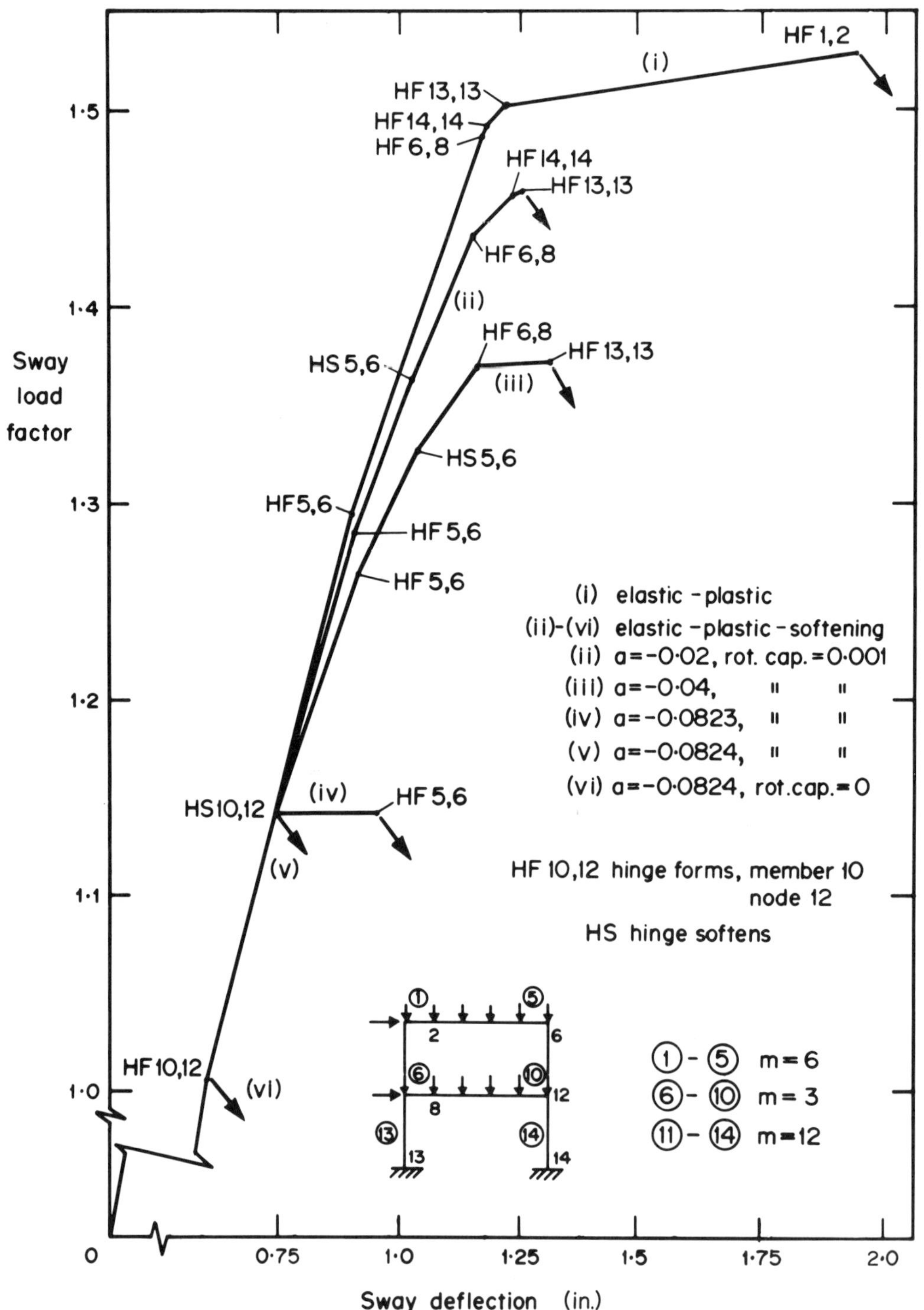

FIGURE 15. Load deflection curves for frame of Example 2.

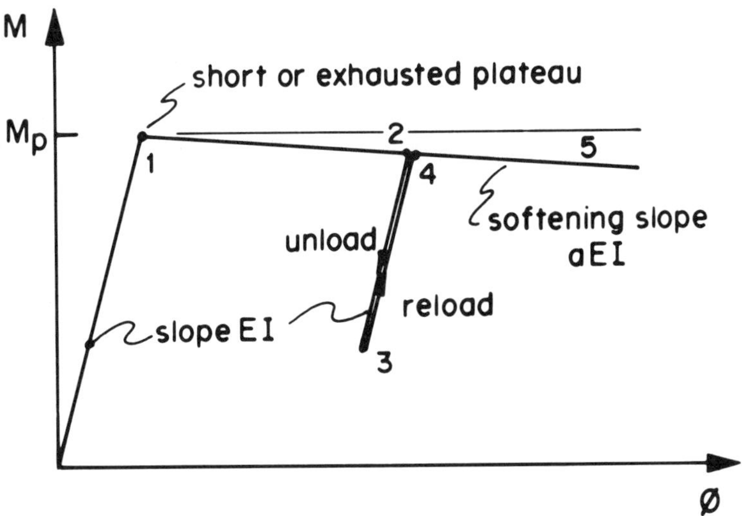

FIGURE 16. Assumptions for shakedown with softening.

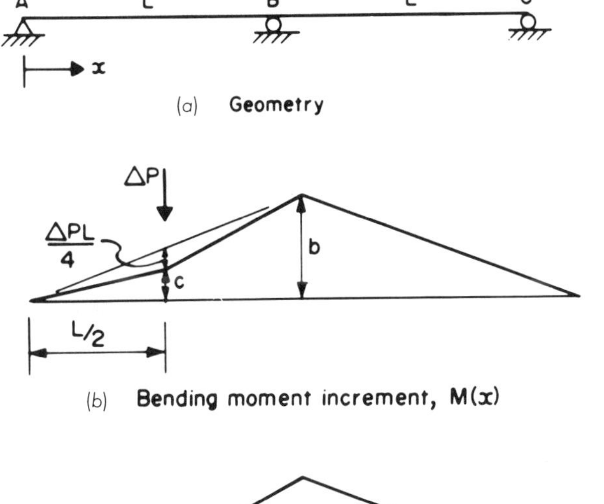

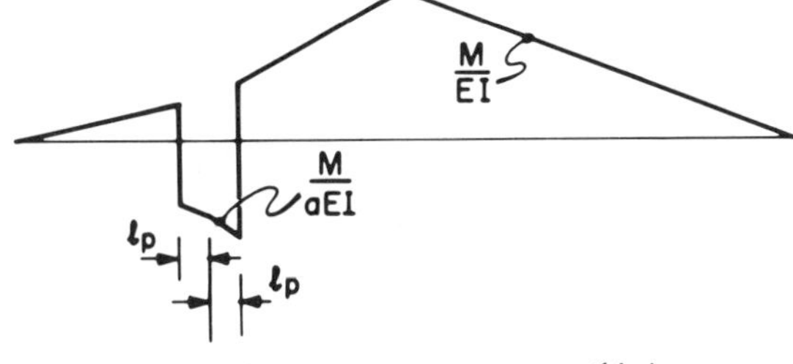

FIGURE 17. Softening at midspan in a two-span beam.

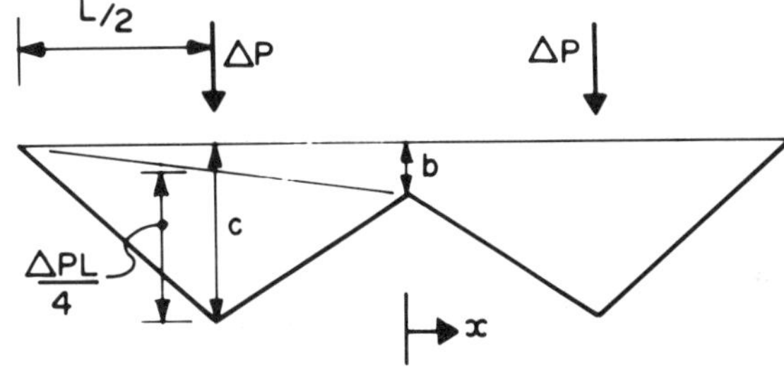

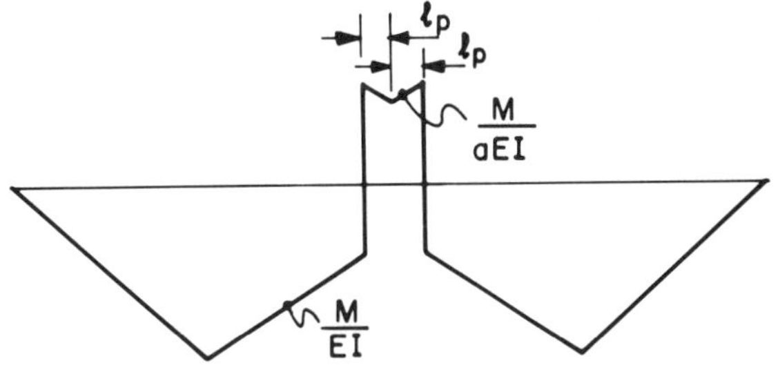

FIGURE 18. Softening at interior support of a two-span beam.

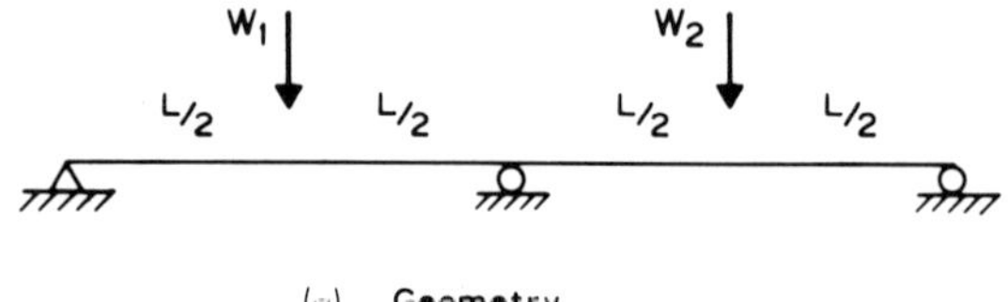

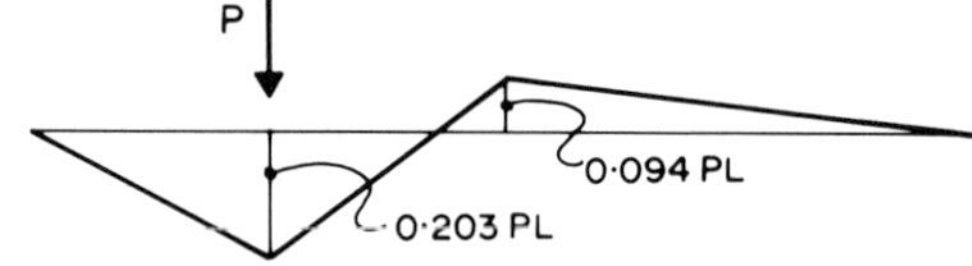

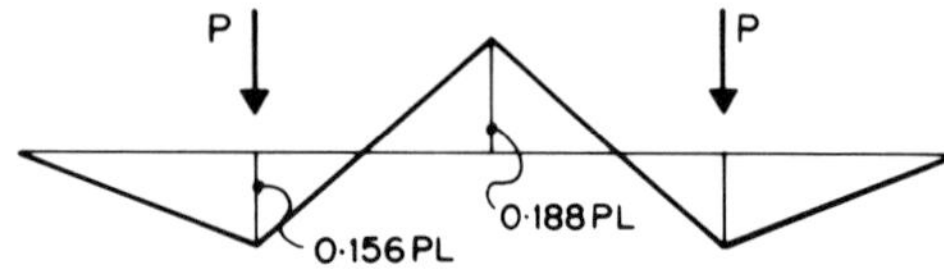

FIGURE 19. Elastic solutions, two-span beam.

Figure 19 provides information for the well known solution when elastic-plastic hinges are located under the loads or at the interior support, and

$$0 \leq W_1 \leq P \qquad\qquad 0 \leq W_2 \leq P$$

Thus, for equal plastic moment M_p in positive or negative bending,

$$\text{first yield load:} \qquad P_y = \frac{M_p}{0.203L} = 4.923\,\frac{M_p}{L}$$

$$\text{static collapse load:} \qquad P_c = 6.000\,\frac{M_p}{L}$$

$$\text{shakedown load:} \qquad P_s = 5.053\,\frac{M_p}{L}$$

and the latter may be easily checked by first applying one load of 5.053 M_P/L with some rotation occurring at the midspan (positive moment) hinge, and then applying both loads together, where some rotation will occur at the negative moment hinge for values of $P > 5.053\ M_P/L$. With one load sustained, incremental collapse will occur for repeated application and removal of the other load when $P > 5.053\ M_P/L$.

The analysis is repeated for the case of softening of the midspan hinge, with $m = 20$ and $a = -0.0282$ (c.f. $a_{cr} = -0.0391$), for which $b/c = 3$ in Figure 17.

Figure 20 shows steps in the analysis:

(a) $P_1 = 4.923\ M_P/L$;
(b) $b = 0.538\ M_P$ and $P_2\ L/4 = b/6$, so $P_2 = 0.359\ M_P/L$ and static collapse load is $5.282\ M_P/L$;
(c) and (d) by trial and error, shakedown load is $(4.923 + 0.048)\ M_P/L = 4.971\ M_P/L$ for the hinge length and softening gradient chosen. A general shakedown theorem to cover softening, similar to that for structures with elastic-plastic hinges, is not evident at this stage.

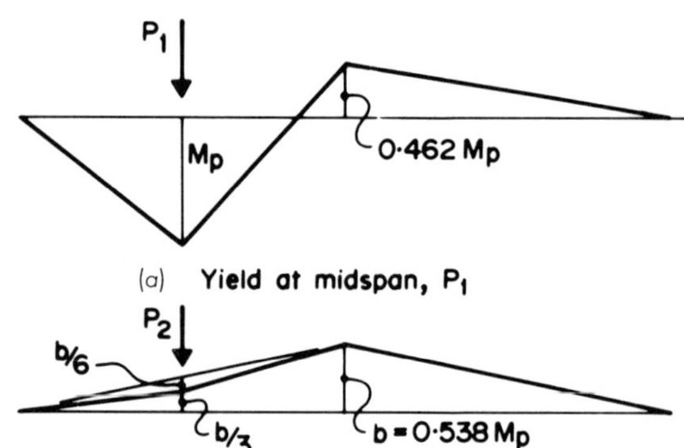

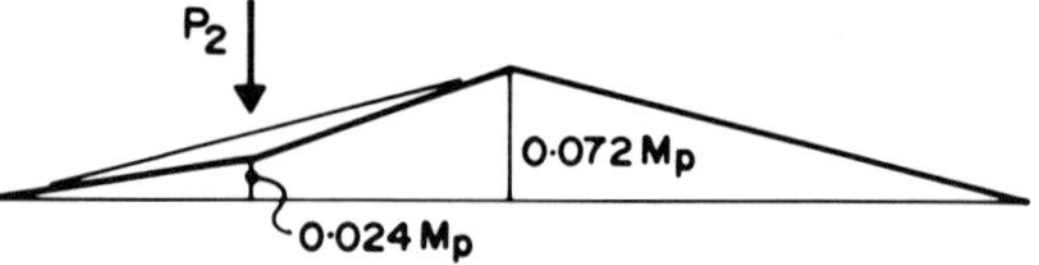

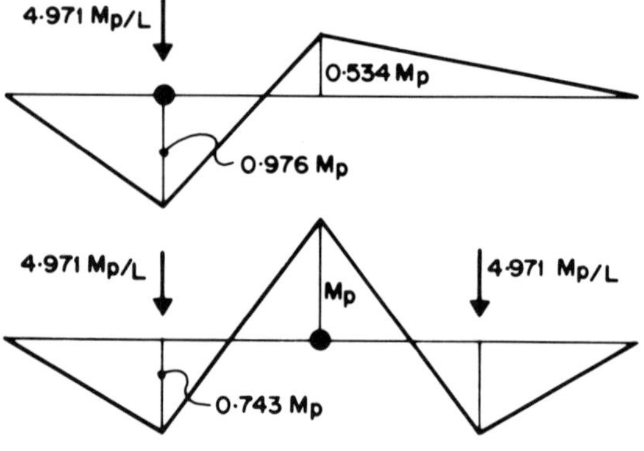

FIGURE 20. Bending moment diagrams for effect of softening on static collapse and shakedown loads.

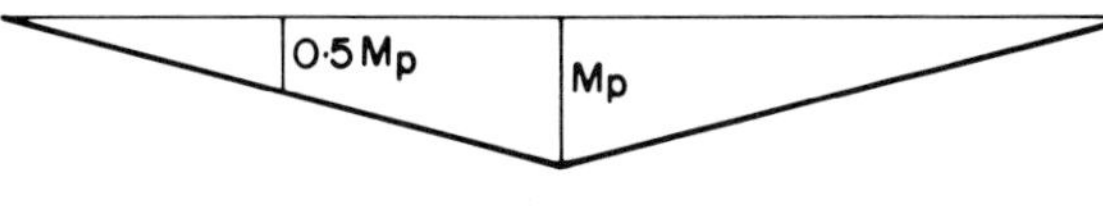

(a) Residual Moments

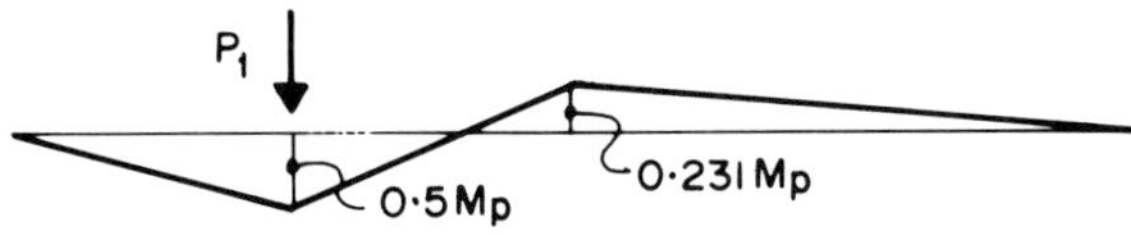

(b) Yield at midspan, P_1

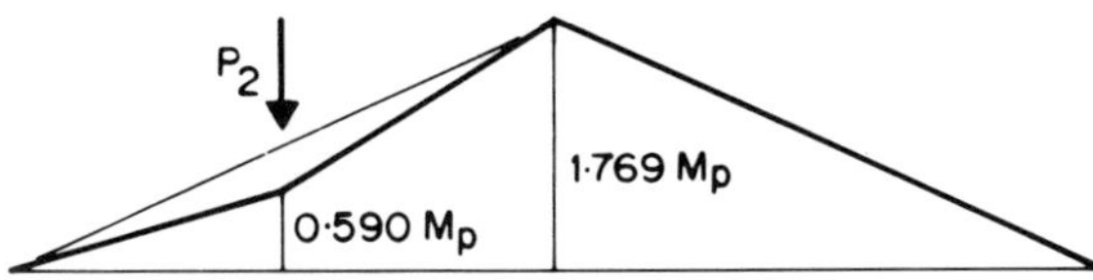

(c) Midspan softening to yield at interior support, P_2

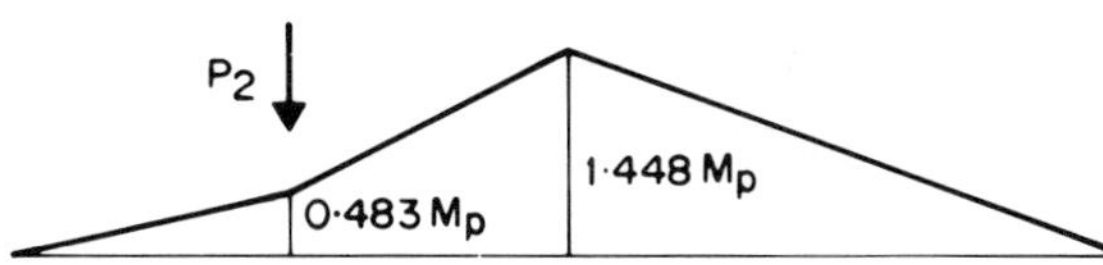

(d) Midspan softening with $P_2 = 0{\cdot}965\ M_p/L$

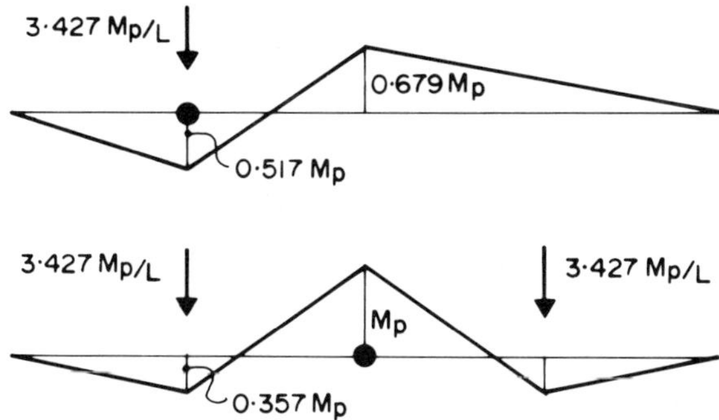

(e) Shakedown loadings, midspan softening

FIGURE 21. Bending moment diagrams for effect of residual moments and softening on static collapse and shakedown loads.

TABLE 3. Summary of Results, Two-Span Beam. Yield Shakedown and Collapse Loads × M_p/L.

	Elastic-Plastic	Elastic-Plastic with Residual	Elastic-Softening	Elastic-Softening with Residual
P_y	4.92	2.46	4.92	2.46
P_s	5.05	5.05	4.97	3.43
P_c	6.00	6.00	5.28	3.64

Shakedown with Softening and Residual Moments for a Two-Span Beam

The effect of softening on static collapse and shakedown loads is more dramatic if unfavourable residual moments such as from differential settlement are present. An extreme case of residual moments resulting from settlement of the interior support is shown in Figure 21(a). The same softening parameter has been assumed.

The steps in the analysis follow Figure 21:

(a) and (b) $P_1 = 2.462\ M_P/L$;
(c) $P_2 = 1.179\ M_P/L$, so static collapse load is $3.641\ M_P/L$ (For elastic-plastic analysis, $(P_2L)/2 = 1.769\ M_p$, so that $P_2 = 3.538\ M_P/L$ and static collapse load is unchanged at $6.000\ M_P/L$);
(d) and (e) by trial and error, shakedown load is 3.427 M_p/L for the hinge length, softening gradient and residual moments chosen. For elastic-plastic analysis the shakedown load is not affected by residual moments.

Static Collapse and Shakedown Loads for a Single-Bay Portal Frame

Data for this example are given in Figure 22(a). $M = L/\ell_p = 40$, and $a = -0.01$ for all possible hinge locations. The exact values of a_{cr}, derived elsewhere [9,10], are: -0.1042 for hinge at column base, -0.0388 for hinge at column top, and -0.0417 for hinge at beam midspan. The rotation capacity of hinges is assumed to be zero, i.e., behaviour is elastic-softening.

Curves of limit load combinations for both proportional loading to static collapse and repeated loads to shakedown are shown in non-dimensional form in Figure 23.

Curve 1 is the locus of load combinations which first produce M_p at some location in the frame and is the same for frames with elastic-plastic and elastic-softening behavior. For an elastic-plastic frame the curve for static collapse loads is horizontal from $(VL)/M_p = 8$, vertical from $(HL)/M_p = 8$, and diagonally as indicated by P_c. Curve 3 is static collapse load combinations for the elastic-softening frame, computed using programs PAWS.

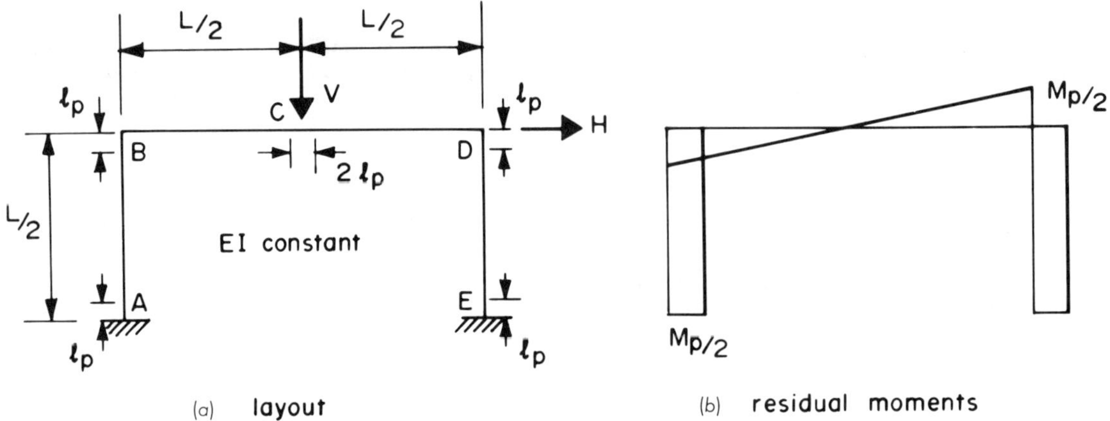

FIGURE 22. Data for elastic-softening frame.

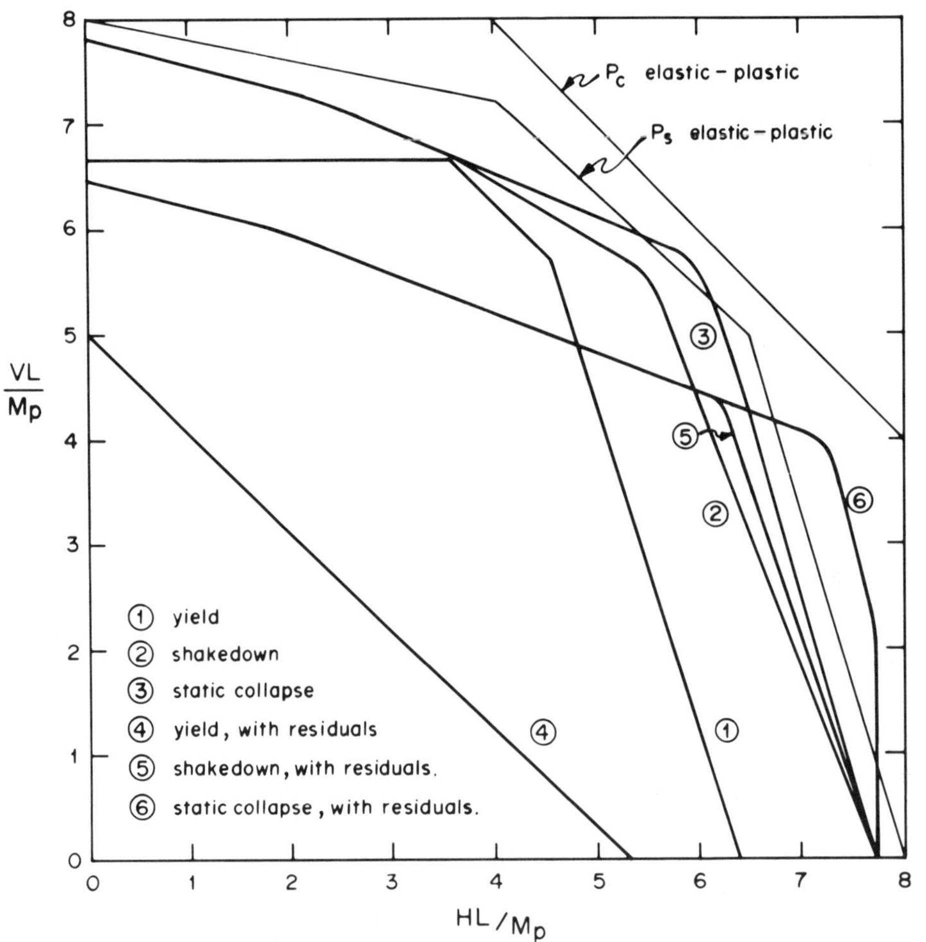

FIGURE 23. Limit loads for single-bay portal frame of Figure 22.

Shakedown loads for the elastic-plastic frame are indicated as P_s in Figure 23. They are obtained as described in textbooks [20].

The shakedown loads for the elastic-softening frame are shown in Curve 2. A typical point on the curve is obtained as follows by trial and error, using $V/H = 1$ as an example, where the shakedown solution is $V \cong 5.5\ M_P/L$.

1. Apply $V = H = 5.55\ M_P/L$. Hinges at D and E are in softening region.
2. Remove V. Hinges at D and E unload and hinge A enters softening region.
3. Reapply V. Hinge A unloads and D and E soften further.
4. Remove V. Hinges D and E unload and A softens further.

After several cycles as above, hinge C reaches plastic moment. Since further cycles of load lead to incremental rotations of four hinges at A, C, D and E, the frame will collapse.

5. Apply $V = H = 5.45\ M_p/L$
6. }
7. } as steps 2,3,4.
8. }

At this load, the softening occurring at hinges A, D, and E becomes less with each cycle and the maximum moments are asymptotic to a stable value. The maximum moment at C is asymptotic to a value less than M_p. In the sense above, the frame has "shaken down."

The curves of Figure 23 show that softening of only 1% leads to significant reductions in static collapse and shakedown loads. These are very little more than the yield load for some load combinations.

Yield, shakedown, and static collapse loads are also shown in Figure 23 for the case when the frame has the pattern of initial or residual moments shown in Figure 22(b), such as caused by differential settlements. With this particular pattern of residual moments, the static collapse load is in general reduced but is increased for certain load combinations, although not to the level of the frame with elastic-plastic hinges. Likewise, whether or not the shakedown load is less than the static collapse load depends on the particular pattern of residual moments and the ratio V/H.

SOME IMPLICATIONS OF SOFTENING ON DYNAMIC RESPONSE

The response of reinforced concrete frames to severe repeated reversible loads or ground motions such as caused by strong earthquakes will depend on the plasticity, softening and hysteresis characteristics of the most severely stressed locations. This is a formidable task, both analytically and in collecting sufficient data on which to predict behaviour over a reasonable range of variables.

What follows is an examination of the effects of softening on the dynamic response of a one degree-of-freedom system. It is hoped thereby to illuminate some implications of softening for more complex structures, and to take a step towards a capacity to include softening in the dynamic analysis of real reinforced concrete frame structures.

Criteria for building design related to the level of loading is specified by the SEAOC Code [29] as follows:

> . . . buildings should be able to resist minor earthquakes without damage, to resist moderate earthquakes without structural damage but with some nonstructural damage, and to resist major earthquakes without collapse but with some structural and nonstructural damage.

In view of this statement, it is perhaps logical to employ elastic design for minor earthquakes, elastic-plastic design for moderate earthquakes and elastic-plastic-softening design for major earthquakes.

The Model

The structure considered is represented in Figure 24. The single degree of freedom is the horizontal displacement, y,

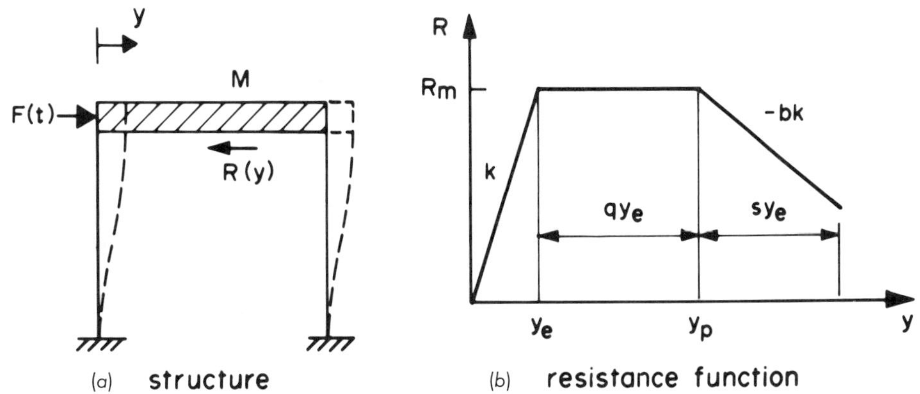

FIGURE 24. Mathematical model of structure.

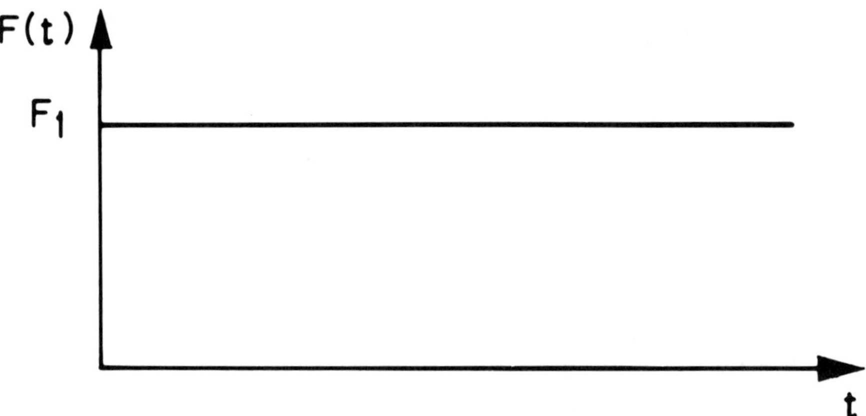

FIGURE 25. Suddenly applied constant load.

of the mass, M. The dynamic load is $F(t)$. The instantaneous resistance function of the structure to displacement is shown as elastic-plastic-softening, with maximum value R_m. The elastic stiffness is k, and the softening stiffness $-bk$ where b is a positive number. The length of the plastic plateau is qy_e, and if there is a limiting range of softening, it may be specified by sy_e.

The resistance model corresponds to the situation where hinges at top and bottom of columns form simultaneously, have the same plastic plateau length, and then soften. More linear stages would be obtained if hinges did not form simultaneously or if plateau lengths varied. The softening slope in the overall structural resistance would correspond to the softening slope and length of the individual hinges. For equal length hinges at top and bottom of the columns, with equal softening slope, the relationship between hinge and structure softening is easily derived [27]. It is also assumed that unloading from the plastic or softening region follows the elastic slope. Likewise, reloading follows the elastic slope.

The response of this structure to a suddenly applied constant load, as defined in Figure 25, is to be computed.

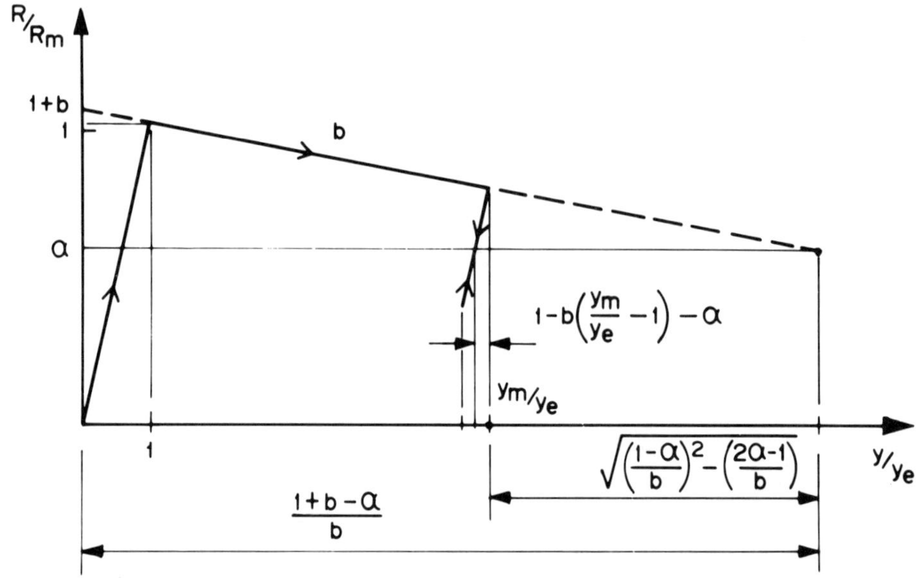

FIGURE 26. Resistance-displacement path for elastic-softening structure with suddenly applied constant load.

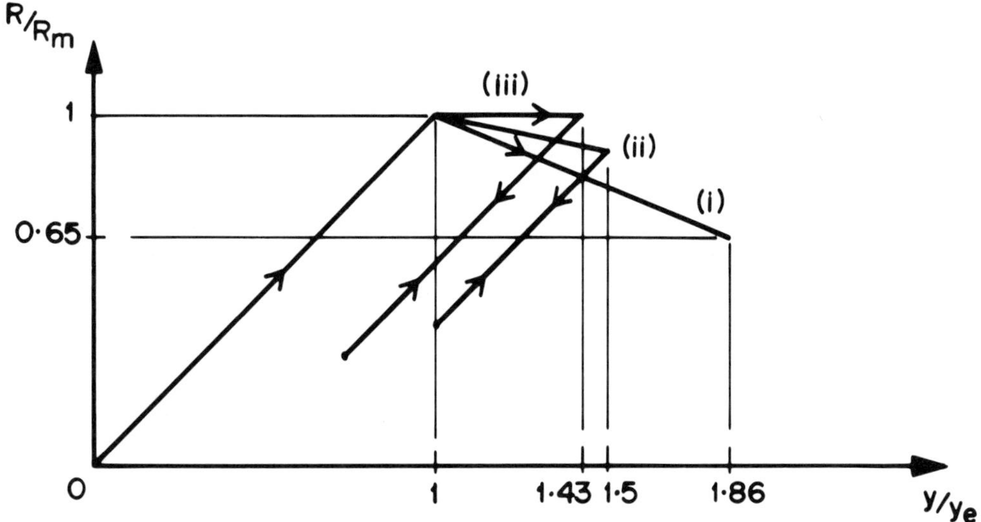

FIGURE 27. Resistance-displacement paths for Example.

We define the load: maximum resistance ratio

$$\alpha = \frac{F_1}{R_m} \tag{26}$$

For simplicity we take q in Figure 24 to be zero, i.e., there is no plastic plateau length.

The Response

In general, the response will involve three separate stages, just as for a system with elastic-plastic response [5]. The first is elastic response up to the elastic limit y_e, reached at time t_e. The second is softening response between the plastic limit and some maximum displacement y_m, reached at time t_m. The third is rebound, or elastic response with displacement decreasing from the maximum.

The mathematical details of the solution may be found elsewhere [27]. For $\alpha < 1/2$, the response is purely elastic:

$$\frac{y_m}{y_e} = \frac{1+b}{b} - \frac{\alpha}{b} - \sqrt{\frac{(1-\alpha)^2}{b} - \frac{(2\alpha - 1)}{b}} \tag{27}$$

For a critical softening parameter,

$$b_{cr} = \frac{(1-\alpha)^2}{2\alpha - 1} \tag{28}$$

the resistance function will drop permanently below the applied load and the structure will accelerate to collapse.

When $b = b_{cr}$,

$$\frac{y_{m,cr}}{y_e} = \frac{1 + b_{cr} - \alpha}{b_{cr}} \tag{29}$$

When $b = 0$, the sub-case of elastic-plastic response applies, and in this case,

$$\frac{y_m}{y_e} = \frac{1}{2(1-\alpha)} \tag{30}$$

The response after the maximum displacement is reached consists of a residual elastic harmonic vibration about a mean position corresponding to the applied force F_1. On the resistance-displacement function this takes place along a line parallel to the elastic curve reaching to y_m on the softening line, and the full amplitude is:

$$2y_e\left(1 + b - \frac{by_m}{y_e} - \alpha\right) \tag{31}$$

The path of the structure through its resistance function is shown in non-dimensionalised form in Figure 26. Equations (27) and (28) provide dimensions for this diagram. When $b = b_{cr}$ the structural response is "soggy" and there is no residual oscillation on reaching $y_{m,cr}$ at $t_{1m} = \infty$.

Example

$F_1 = 130\ kN$ (29.25 kip), $R_m = 200\ kN$ (45 kip), $\alpha = 0.65$, $y_e = 14$ mm (0.55 in), $k = 14.3$ kN mm^{-1}

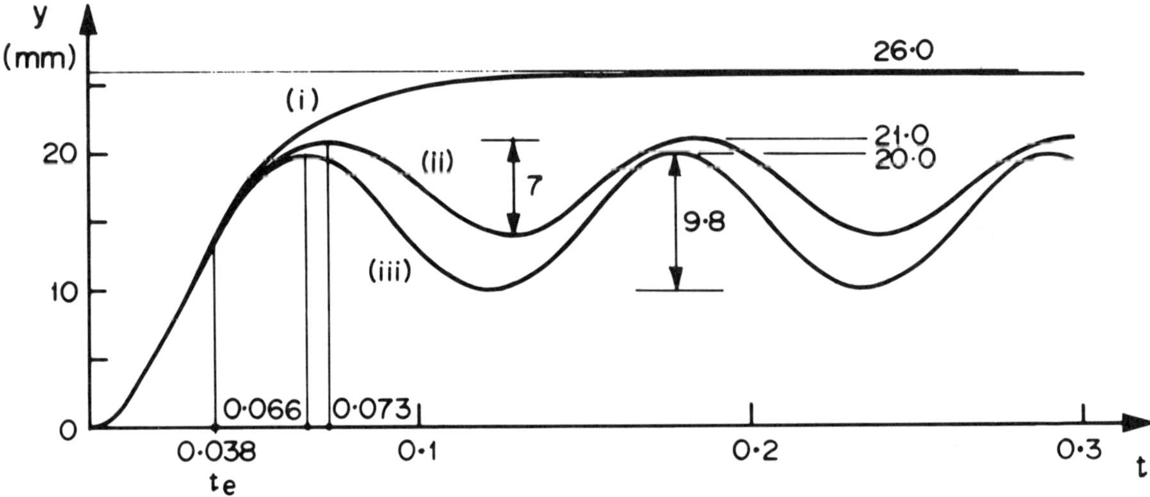

FIGURE 28. Displacement versus time for Example.

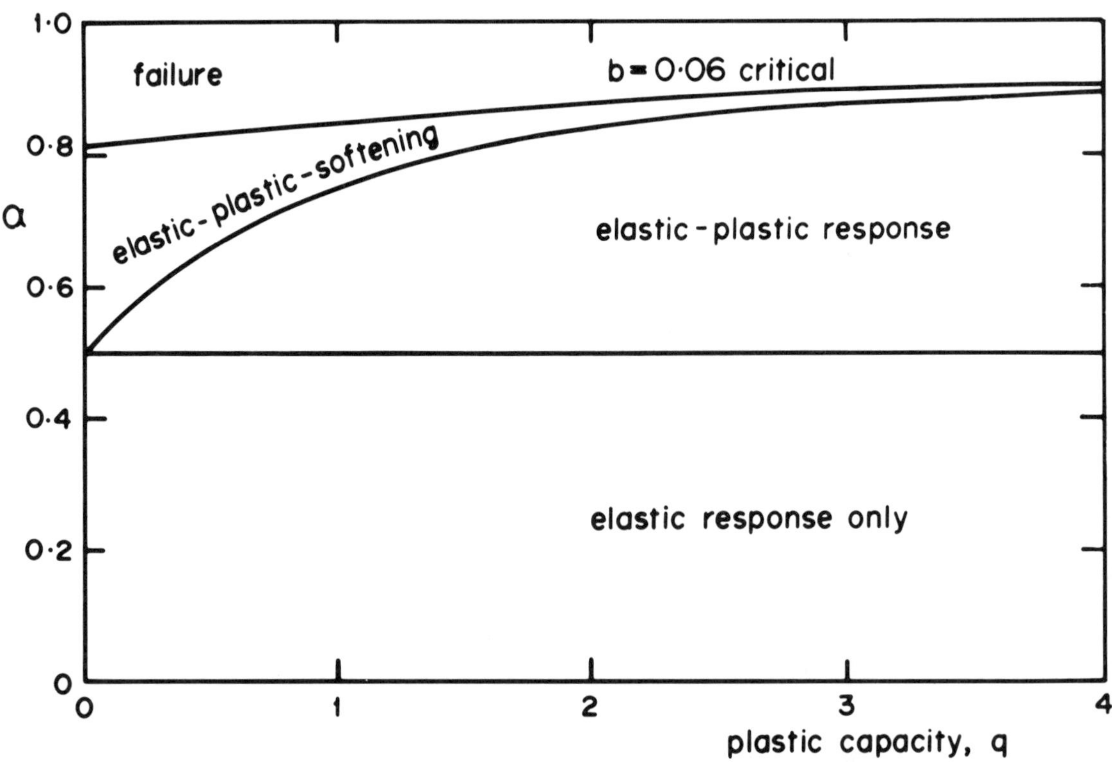

FIGURE 29. Nature of response for various α and q, suddenly applied constant load.

(81.72 kip in^{-1}), M = 0.00453 kN sec^2 mm^{-1} = 4530 kg (9978 lb)

$$\omega = \sqrt{\frac{k}{M}} = 56.2 \text{ rad sec}^{-1},$$

$$T = \frac{2\pi}{\omega} = 0.112\text{sec}, \; y_{st} = \frac{F_1}{k} = 9.1 \text{ mm (0.36 in)}$$

(1) From Equation (28), b_{cr} = 0.408, and from Equation (29), $y_{m.cr}$ = 26.01 mm (1.02 in)
(2) When b = 0.2, from Equation (27), y_m = 21.00 mm (0.83 in) and amplitude of residual vibration is 7 mm (0.28 in) from Equation (31)
(3) When b = 0, from Equation (30), y_m = 20.00 mm (0.79 in) and amplitude of residual vibration is 9.8 mm (0.39 in) from Equation (31)

Figure 27 shows the path of the structure through its resistance function for cases (1)–(3). Details of the working may be found elsewhere [27].

The variation of displacement with time for cases (1)–(3) is shown in Figure 28.

For an elastic-plastic-softening resistance function ($q \neq 0$), the critical softening parameter [27] is

$$b_{cr} = \frac{(1 - \alpha)^2}{1 - 2(1 - \alpha)(1 + q)} \tag{32}$$

or, collapse will not occur as long as

$$\alpha \leq [1+b(1+q)] - \sqrt{[1+b(1+q)]^2 - [1+b(1+2q)]} \tag{33}$$

Figure 29 shows the nature of the response for various values of q and α, with b also taken as 0.06 to indicate the critical α for various values of q.

Critical softening parameters have also been identified for a rectangular load pulse on an elastic-softening structure, and for an impulsive load on an elastic-plastic-softening structure [27].

CONCLUSIONS

1. When it is assumed that hinge softening will occur over finite discontinuity lengths, or hinge lengths, formulae for critical softening parameters may be derived for indeterminate beam or frame structures.
2. Critical softening parameters are inversely proportional to hinge length, so that means whereby hinge lengths are increased will lead to greater reserves of overall strength.
3. "Soft points," where the least value of the critical softening parameter occurs, may in general be found for each member of an indeterminate structure.
4. The critical softening parameter at a given location is reduced if plasticity exists elsewhere, and is reduced still further if softening exists elsewhere.
5. The use of finite discontinuity lengths allows the conventional application of matrix structural analysis to softening situations, via stiffness coefficients for beam elements with softening regions.
6. A computer program, such as PAWS, incorporating a sub-routine for the computation of stiffness when softening is present, provides an efficient and economical method for a complete analysis of concrete frames, for proportional, non-proportional and repeated loads, at all loads up to collapse.
7. A steeper softening slope reduces both the number of hinges formed before collapse, and the collapse load. An increase in rotation capacity before softening (i.e., ductility) has the opposite effect.
8. A computer program such as PAWS may be used to assess rapidly the ductility required at any particular section to maintain any desired proportion of the full elastic-plastic collapse load.
9. Critical softening parameters for first or later formed hinges may be determined by using program PAWS.
10. Quite small softening parameters may lead to large reductions in static collapse and shakedown loads in indeterminate beams and frames.
11. When hinges soften, the presence of residual moments changes the values of static collapse and shakedown loads, whereas for elastic-plastic hinges these loads are unchanged by residual moments.
12. When both unfavourable residual moments and softening are present, dramatic reductions in static collapse and shakedown loads may be expected.
13. Computation of the full response of concrete frame structures to severe dynamic loads requires consideration of softening in addition to plasticity and hysteresis. A number of features of the response of a softening structure with one degree of freedom have been identified, and may be prominent also for multi-degree-of-freedom structures.
14. A critical softening parameter or slope at which collapse will occur may be identified for each type of dynamic loading (constant, rectangular or impulse) and depends on the severity of the applied load as represented by the ratio of maximum applied force to maximum resistance (or by energy of impulse to maximum elastic strain energy), on the plastic plateau length (ductility), on any limit to the softening region, and on duration of load in the case of a rectangular load function.
15. Conversely, for a given softening slope, a critical

severity of load (represented by α), critical plateau length, or critical duration of load may be identified.

16. A steeper softening slope increases the maximum displacement for a given dynamic load, increases the time to maximum displacement, and decreases the amplitude of residual elastic vibration. For critical softening the time to maximum displacement becomes infinite, and there is no residual vibration.
17. The criterion for structural response to enter the softening range may be expressed in terms of severity of load and length of plastic plateau (ductility).

NOTATION

A = member area
a,b = ratio of falling slope to elastic slope on M–ϕ diagram
b = softening parameter of structural resistance
a_{cr}, b_{cr} = critical value of softening parameter
b,c = bending moment increments
d = effective depth of cross-section
EI = elastic flexural stiffness
$F(t)$ = an applied force
F_1 = maximum applied force
f,g = bending moment ratios
H = a horizontal load
H_1, H_2 = horizontal loads, or column heights
k = elastic stiffness of a structure
k_{ij} = element of stiffness matrix
L = a length
ℓ_p, ℓ_{p1}, ℓ_{p2} = discontinuity length, softening length, hinge length
m = L/ℓ_p, L/ℓ_{p1}
M = bending moment or mass
M_y = yield moment
M_p = plastic moment
M_u = ultimate moment
n = L/ℓ_{p2}
n_1 = flexural stiffness ratio EI/EI_1
n_2 = flexural stiffness ratio EI/EI_2
P,Q,R,S = concentrated loads
P_c = static collapse load
P_s = shakedown load
P_y = first yield load
q = parameter defining plateau length of structural resistance
$R(y)$ = resistance function
R_m = maximum resistance
s = parameter defining maximum softening range
T = period of vibration
t = time
t_e = time at end of elastic range
t_m = time at maximum displacement
V = a vertical load
W_1, W_2 = concentrated loads
x = co-ordinate along a member
y = displacement of structure
y_d = displacement at time t_d
y_e = maximum displacement in elastic range
y_m = maximum displacement
$y_{m,cr}$ = maximum displacement corresponding to b_{cr}
y_p = displacement at end of plastic range
y_{st} = elastic displacement for static load of F_1
z = distance from load point to point of contraflexure
α = severity of load F_1/R_m
α_{cr} = critical severity of load
Δ = a displacement
ΔP = a load increment
$\Delta^2 W$ = second order work increment
ϕ = curvature
ϕ_y = curvature at beginning of plateau on trilinear M–ϕ diagram
ϕ_p = curvature at end of plateau on trilinear M–ϕ diagram
ϕ_u = ultimate curvature
θ = a rotation
ω = undamped natural circular frequency

REFERENCES

1. Baker, A. L. L. and A. M. N., Amarakone, (1965) "Inelastic Hyperstatic Frames Analysis," *Proc. Int. Symp. on the Flexural Mechanics of Reinforced Concrete,* Miami, ACI SP-12, pp. 85–142 (1964).
2. Barnard, P. R., "The Collapse of Reinforced Concrete Beams," *Proceedings International Symposium on the Flexural Mechanics of Reinforced Concrete,* Miami, 1964, ACI SP-12, pp. 501–520 (1965).
3. Barnard, P. R. and R. P. Johnson, "Plastic Behaviour of Contínous Composite Beams," *Proceedings I.C.E.*, Vol. 32, London, pp. 180–187 (October 1965).
4. Bazant, Z. P., "Instability, Ductility and Size Effect in Strain-Softening Concrete," *Journal of the Engineering Mechanics Division,* ASCE, Vol. 102, EM2, pp. 331–344 (April 1976).
5. Biggs, J. M., "*Introduction to Structural Dynamics,*" McGraw-Hill (1964).
6. Corley, W. G., "Rotational Capacity of Reinforced Concrete Beams," *Journal of the Structural Division,* ASCE, Vol. 92, ST5, pp. 121–146 (October 1966).
7. Cranston, W. B., "Tests on Reinforced Concrete Frames; 1: Pinned Portal Frames," *Technical Report TRA/392, Cement and Concrete Association,* London (August 1965).
8. Cranston, W. B. and J. A. Cracknell, "Tests on Reinforced Concrete Frames; 2: Portal Frames with Fixed Feet," *Technical Report TRA/420, Cement and Concrete Association,* London (September 1969).

9. Darvall, P. LeP., "Critical Softening of Hinges in Portal Frames," *Journal of Structural Engineering,* ASCE Vol. 110, No. 1, pp. 157-162 (January 1984).
10. Darvall, P. LeP., "Critical Softening of Hinges in Indeterminate Beams and Portal Frames," *Civ. Eng. Trans., I.E. Aust.,* Vol. CE 25, No. 3, pp. 199-210 (1983).
11. Darvall, P. LeP., "Load Deflection Curves for Elastic-Softening Beams," *Journal of Structural Engineering,* ASCE, Vol. 110, No. 10, pp. 2536-2541 (October 1984).
12. Darvall, P. LeP., "Stiffness Matrix for Elastic-Softening Beams," *Journal of Structural Engineering,* ASCE, Vol. 111, No. 2, pp. 469-473 (February 1985).
13. Darvall, P. LeP., "Shakedown with Softening in Reinforced Concrete Beams," *Materials and Structures, Research and Testing,* Vol. 17, No. 102, RILEM, Paris, pp. 421-426 (Nov./Dec. 1984).
14. Darvall, P. LeP., "Shakedown with Softening in Reinforced Concrete Beams and Frames," *Proceedings 9th Australasian Conference on the Mechanics of Structures and Materials,* Sydney, pp. 173-178 (August 1984).
15. Darvall, P. LeP. and P. A. Mendis, "Elastic-Plastic-Softening Analysis of Plane Frames," *Journal of Structural Engineering,* ASCE, Vol. 111, No. 4, pp. 871-888 (April 1985).
16. Darvall, P. LeP., "Critical Softening Parameters and Full-Range Analysis for Indeterminate Beams", *Research Report No. 2/1982. Dept. of Civil Engineering, Monash University,* Clayton, Victoria, Australia.
17. Darvall, P. LeP., "Critical Softening of Hinges in Portal Frames," *Research Report No. 5/1982. Dept. of Civil Engineering, Monash University,* Clayton, Victoria, Australia.
18. Darvall, P. LeP., "Some Aspects of Softening in Flexural Members," *Research Report No. 3/1983, Dept. of Civil Engineering, Monash University,* Clayton, Victoria, Australia.
19. Ghosh, S. K., "Analysis of Concrete Structures Allowing for Strain-Softening," *Indian Concrete Journal,* V51, No. 4, pp. 108-116 (April 1977).
20. Heyman, J., "*Plastic Design of Frames: 2 Applications,*" Cambridge Univ. Press, p. 292 (1971).
21. Karsan, I. and J. O. Jirsa, "Behavior of Concrete Under Compressive Loadings," *Journal of the Structural Division,* ASCE, Vol 95, ST12, pp. 2543-2563 (December 1969).
22. Maier, G., A. Z. Rossi, and J-C. Dotreppe, "Equilibrium Branching Due to Flexural Softening," Technical Note, *Journal of the Engineering Mechanics Division,* ASCE, Vol. 99, EM4, pp. 897-902 (August 1973).
23. Mendis, P. A. and P. LeP. Darvall, "Elastic-Plastic-Softening Analysis of Plane Frames," *Research Report 5/1984, Dept. of Civil Engineering, Monash University,* Clayton, Victoria, Australia.
24. Nylander, H. and S. Sahlin, "Investigation of Continuous Concrete Beams at Far Advanced Compressive Strains in Concrete," *Betong,* 40 (3), Translation No. 66, Cement and Concrete Association, London (1955).
25. Portland Cement Association, *Handbook of Frame Constants,* Chicago, Illinois (1958).
26. Rosenbleuth, E. and R. Diaz de Cossio, "Instability Considerations in Limit Design of Concrete Frames," *Proceedings International Symposium on Flexural Mechanics of Reinforced Concrete,* Miami, 1964, ACI SP-12, pp. 439-463 (1965).
27. Sanjayan, G. and P. LeP. Darvall, "Dynamic Response of a Single Degree-of-Freedom Elastic-Plastic-Softening Structure," *Research Report No. 8/1984, Dept. of Civil Engineering, Monash University,* Clayton, Victoria, Australia.
28. Sawyer, H. A., "Design of Concrete Frames for Two Failure Stages," *Proceedings International Symposium on the Flexural Mechanics of Reinforced Concrete,* Miami, 1964, ACI SP-12, pp. 405-431 (1965).
29. SEAOC, *Recommended Lateral Force Requirements and Commentary,* Seismology Committee, Structural Engineers Association of California, San Francisco, p. 146 (1973).
30. Taylor, H. P. J., "The Behaviour of In-Situ Beam-Column Joints," *Technical Report 42.492, Cement and Concrete Association,* London (May 1974).
31. "ULARC – Simple Elasto-Plastic Analysis of Plane Frames," Programmed by A. Sudhakar, G. H. Powell, G. Orr, and R. Wheaton, University of California, Berkeley, California, Distributed by NISEE/Computer Applications (1972).

CHAPTER 5

Finite Element Method for Uniaxially Loaded RC Columns

SULATA KAYAL*

INTRODUCTION

Two methods have been commonly used for the numerical solution of the governing differential equation of a column, subjected to an axial load and end moments. Both these approaches require the moment-axial thurst-curvature relationships of the cross-section.

Of these methods, the first one, originally due to Von Karman, is based on the procedure used for the computation of column deflection curves with a given axial load and end rotation.

The second method is based on the numerical approach given by Newmark. Cranston [3] and Pfrang and Siess [6] have developed methods based on the features of this approach for the analysis of restrained reinforced concrete columns.

In these iterative methods for column analysis using integration approach the bulk of computer-time is spent in the computation of nodal curvatures.

Pfrang used tabulated curvature values. His method of arriving at a proposed deflected shape as well as the method of modification of the proposal requires a large number of curvature values and hence a large storage. Cranston used a subroutine to generate curvatures when required but this method of modifying the proposed deflected shape in two stages (first stage being for slope and the second being for deflection) as well as the procedure for the modification of slopes itself requires large number of curvature calculations per node per cycle.

Basu and Suryanarayana [2] has developed a method including the features of Newmark's approach and avoiding the tedious computational effort, otherwise involved in such analysis, to the extent possible. Besides this all these methods except the method given by Cranston are restricted to the analysis of braced columns only. These facts make it a difficult proposition to extend these methods for the analysis of plane frames or frame-wall systems, where the analysis of many such members are required.

The method described in this paper was developed to overcome this difficulty. The present method is developed with such a view that its extended version to tall buildings requires only reasonable computer time and storage. Another versatility of this method is that it can handle columns under all possible combinations of loading and end conditions (pin ended, braced and unbraced).

The present method is based on finite element technique.

METHOD

The method is presented in the following sequence. First, the system analysed, the material properties adopted and the assumptions involved are explained. A brief description of the method of analysis based on finite element technique is given. Then the selection of the beam-column element and the generation of its different elastic properties are presented. The two parts of the algorithm, namely, the analysis for specified load and the analysis for specified edge strain are then explained.

This is followed by the description of the important subroutines. Finally, the steps for obtaining the complete load deflection characteristics are described.

SYSTEM ANALYSED

The columns, shown in Figures 1(a) and 1(b) are of length h, and are restrained by beams providing rotational restraints of K_A and K_B at the ends A and B respectively. The

*Cement Research Institute of India, M-10, South Extension II, New Delhi - 110 049, India

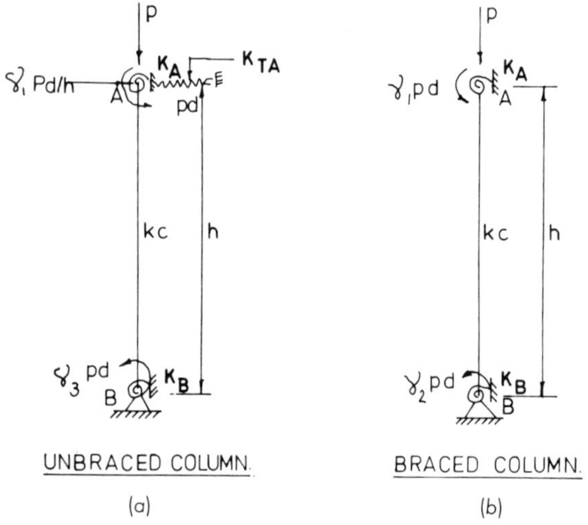

FIGURE 1. System Analysed.

unbraced column, shown in Figure 1(a), is provided also with a translational restraints K_{TA} at the end A. The unbraced column is subjected to an axial load P and moments $\gamma_2 Pd$ and $\gamma_3 Pd$ acting in the anticlockwise direction at the ends A and B respectively. Here, d is the depth of the column-cross section in the plane of bending. The column is also subjected to a lateral load γ_1 Pd/h at the top. These loads are to be proportionally increased up to the point of material failure. The braced column, shown in Figure 1(b), is subjected to an axial load P and moments γ_1 Pd and γ_2 Pd acting in the anticlock-wise direction at the ends A and B respectively.

The applied moments and slopes in anticlock-wise direction are assumed to be positive, as shown in Figure 2.

MATERIAL PROPERTIES

The stress-strain curve of concrete is taken in the form of a fourth degree polynomial, proposed by Basu [1]. The equation of the curve is given by

$$f_c/f_{cp} = \sum_{i=1}^{4} c_i(\epsilon/\epsilon_o)^i \tag{1}$$

where, f_{cp} is the peak stress. The constants c_1, c_2, c_3 and c_4 used for the analyses presented in this text are 2.41, −1.865, 0.5, and −0.045 respectively [1]. The strain at peak stress (ϵ_o) and the crushing strain (ϵ_u) are taken as 0.0025 and 0.0035 respectively (Figure 3(a)).

For steel reinforcements, and elasto-plastic stress-strain curve Figure 3(b) is used.

ASSUMPTIONS

The analysis is based on the following assumptions:

1. Plane sections remain plane during bending.
2. The tensile strength of concrete is negligible.
3. Strain hardening in steel reinforcements is not significant for strains expected before failure.
4. The nonlinear stress-strain curves are reversible, i.e. elastic.
5. The longitudinal stress at any point in the column is dependent only upon the longitudinal strain at that point.
6. Adequate lateral reinforcement is provided to prevent local failure.
7. The lateral deflections of the column are small compared to its length.
8. The deformation due to shear force is negligible compared to that due to bending.
9. Failure is by excessive bending in the plane of flexure, lateral buckling being prevented by bracing etc.
10. The loading is proportional.

FINITE ELEMENT METHOD

The method of nonlinear analysis for an individual member developed here is based on the finite element method. A brief description of this method for linear systems is given in reference [7].

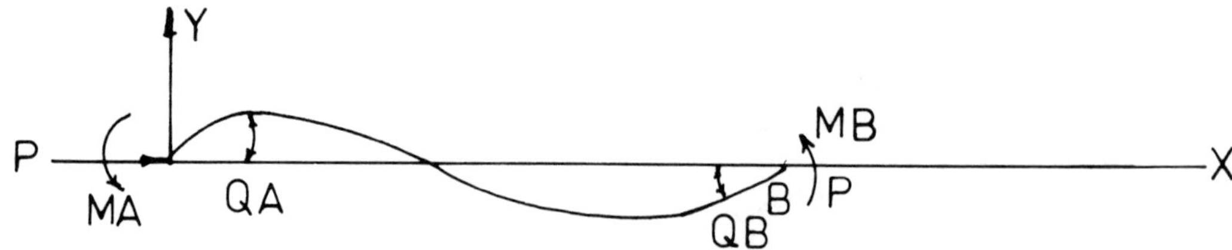

FIGURE 2. Positive moments and rotations.

NONLINEAR SYSTEMS

The force-displacement relationship of a structure becomes nonlinear due to (a) nonlinearity in the stress-strain relationship of the constitutive materials (material nonlinearity) and (b) interaction between high axial loads and large displacements (geometric nonlinearity).

SOLUTION OF NONLINEAR EQUILIBRIUM EQUATIONS

The solution of nonlinear problems by the finite element method is usually attempted by one of the following three basic techniques: (a) incremental procedures based on piecewise linear material property assumption, (b) iterative procedures (e.g. Newton-Raphson method), and (c) mixed procedures utilizing the combination of the above two.

In the present study, an iterative method has been adopted which is a variation of the Newton-Raphson method.

The Newton-Raphson method for solving nonlinear equations with a single unknown δ of the form

$$\psi(\delta) = 0 \tag{2}$$

is as follows.

If a trial value of $\delta = \delta_i$ is found which is sufficiently close to the correct one but for which $\psi(\delta_i) \neq 0$, an improved value for δ is given by

$$\delta_{i+1} = \delta_i + \Delta\delta_i \tag{3a}$$

where

$$\Delta\delta_i = -\phi(\delta_i)/\left(\frac{d\phi}{d\delta}\right)_i \tag{3b}$$

In the case of a structure with only one unknown displacement the equation (2) has the form

$$\phi(\delta) = F - S(\delta) = 0$$

where $S(\delta)$ is the internal force in the elastic system corresponding to displacement δ and is a non-linear function of S, and F is the given external load. Defining $dS/d\delta$ as the tangent stiffness K_T of the elastic system equation (3b) becomes:

$$\Delta\delta_i = \frac{F-S(\delta_i)}{K_T(\delta_i)} = \frac{R(\delta_i)}{K_T(\delta_i)}$$

where $R(\delta_i)$ is the residual. The iterative method is illustrated in Figure 5. The iterations are to be continued till either the residual or the corrections $\Delta\delta$ becomes negligible.

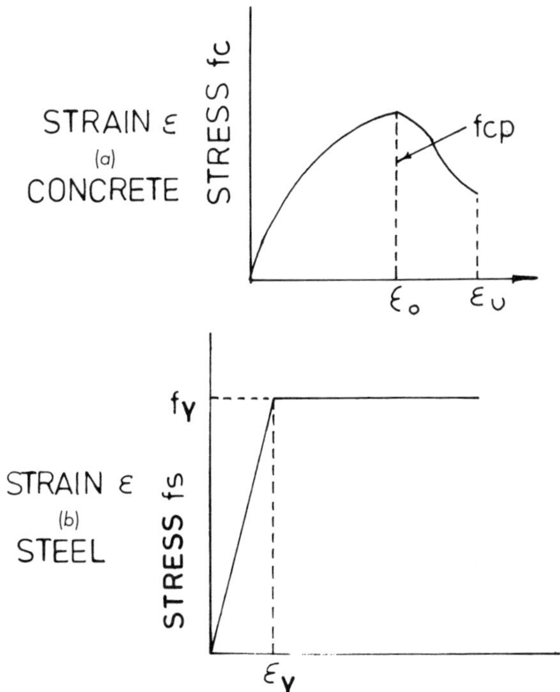

FIGURE 3. Stress strain curves.

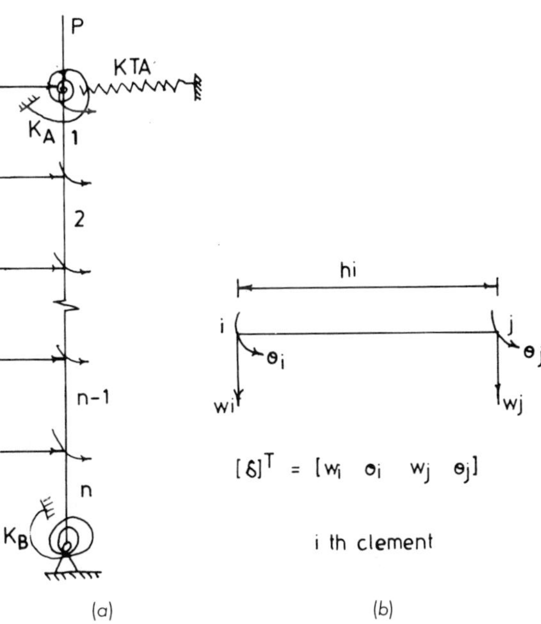

FIGURE 4. Division of a typical column (unbraced) length into n-number of elements.

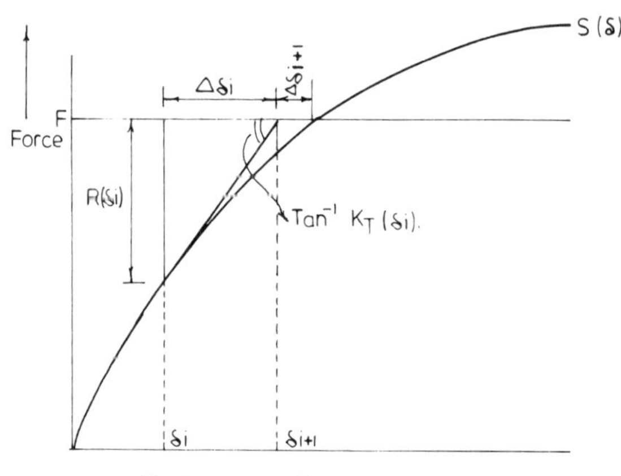

FIGURE 5. Newton-Raphson method.

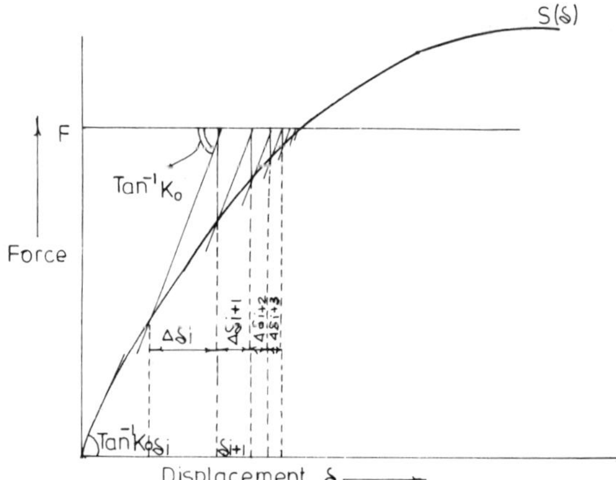

FIGURE 6. Modified Newton-Raphson method.

The above procedure can be extended to solve the nonlinear equilibrium equations of multi-degree-of-freedom systems.

The governing equations in terms of displacement for an m degree-of-freedom system can be written as

$$\begin{aligned} F_1 - S_1(\delta_1, \delta_2, \delta_3, \ldots, \delta_m) &= 0 \\ F_2 - S_2(\delta_1, \delta_2, \delta_3, \ldots, \delta_m) &= 0 \\ &\vdots \\ F_m - S_m(\delta_1, \delta_2, \delta_3, \ldots, \delta_m) &= 0. \end{aligned} \tag{4}$$

The recurrence relation for the solution of equation (4) is given by

$$\{\delta\}_{i+1} = \{\delta\}_i + \{\Delta\}_i \tag{5a}$$

with

$$\{\Delta\delta\}_i = [K_T]^{-1} \{R\}_i \tag{5b}$$

where

$\{\delta\}_i$ is the displacement vector after the ith iteration. $\{\Delta\delta\}_i$ is the correction vector. $\{R\}_i$ is the vector of residuals and $[K_T]_i$ is the tangent stiffness matrix of the elastic system (also called a Jacobian) at $\{\delta\} = \{\delta\}_i$.

Here, the residual vector or the vector of unbulanced nodal loads is found from

$$\{R\}_i = \{F\} - [\{S(\delta)\}_i - P(K_G)\{S_i\}] \tag{5c}$$

The components of $\{S(\delta)\}_i$ are found by considering each element in turn and evaluating its nodal forces in equilibrium with internal stresses corresponding to the strains caused by the imposed nodal displacements $\{\delta\}_i$.

The tangent stiffness matrix is given by

$$[K_T]_i = [K_T^1]_i - P[K_G] \tag{5d}$$

where

$$[K_T^1]_i = \begin{bmatrix} \dfrac{\partial S_1}{\partial \delta_1} & \dfrac{\partial S_1}{\partial \delta_2} & \cdots\cdots & \dfrac{\partial S_1}{\partial \delta_m} \\ \\ \dfrac{\partial S_m}{\partial \delta_1} & \dfrac{\partial S_m}{\partial \delta_2} & \cdots\cdots & \dfrac{\partial S_m}{\partial \delta_m} \end{bmatrix}_i \tag{5e}$$

the derivatives being evaluated at $\{\delta\} = \{\delta\}_i$. As can be seen the numerical evaluation of $[K_T^1]$ involves generation of m additional sets of nodal force vector $\{\delta\}$ corresponding to increments $\Delta\delta_1$, $\Delta\delta_2$ . . . . $\Delta\delta_m$ in turn on the current displacements.

$P[K_G]$ is the geometrical stiffness matrix, when P is the axial load acting on individual member.

The matrix K_G is a property of the element dependent on the geometry of the deformations. Each column of this matrix gives the nodal forces for the element corresponding to the unit value of a particular nodal displacement in the presence of a unit axial load.

The stiffness matrix of an element (Figure 4(b) including geometrical nonlinearity can be generated by making use of the principle of stationary potential energy. Here, the potential energy of an element will consist of the bending strain energy, the potential energy of the axial load and the potential energy of the nodal loads. The derivation of the expressions of stiffness matrices of an element are given in reference [5].

The stiffness matrix (K) for the column (Figure 4(a) at

load level P corresponding to large deflection and linear elastic assumption can be obtained by assembling the individual element stiffness matrix and including the elastic restraints at the appropriate location.

This iterative procedure based on Newton-Raphson technique has the computational disadvantage that it requires evaluation of the tangent stiffness matrix (K_T) and its inversion in each iteration.

In the present study a constant stiffness method, called modified Newton-Raphson technique (Figure 6) has been adopted together with the delta-square extrapolation on individual displacement components as described later. For small load levels, the initial stiffness matrix (K) corresponding to linear elastic assumption and large deflection at load level P has been used for liquidation of residuals. After a number of load levels this matrix has been replaced by $[K_M]$ given by

$$[K_M] = [K_S] - P[K_G] \tag{6}$$

where $[K_S]$ is the "initial" (linear elastic) stiffness matrix like $[K_o]$ but based on the secant flexural rigidity of the cross sections, being equal to the current value of the ratio of moment and curvature at the cross sections, and P is the current axial load level (to be taken as constant for a number of subsequent load levels to avoid reinversion|factorization of $[K_M]$). For the sake of convenience, $[K_M]$ has also been referred to as initial stiffness matrix in the method of analysis described in the following.

SELECTION OF BEAM-COLUMN ELEMENT AND DERIVATION OF ITS PROPERTIES

In the present work instead of dividing the column into a number of discrete parts as shown in Figure 7, a single beam-column element with five degrees of freedom (viz. two end rotations, two end deflections and one mid-length deflection) has been adopted (Figure 7). Two other elements having six degrees of freedom each are shown in Figures 8 and 9. These higher-degree elements can give better results but the advantage of the five-degree beam-column element is that it can describe the nonlinear variation of curvature

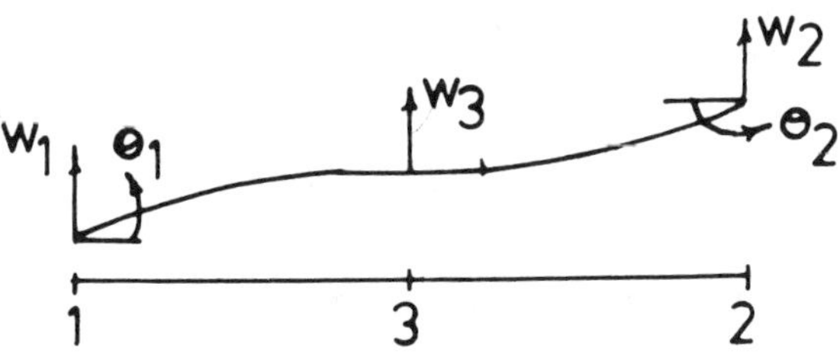

FIGURE 7. Five degree element.

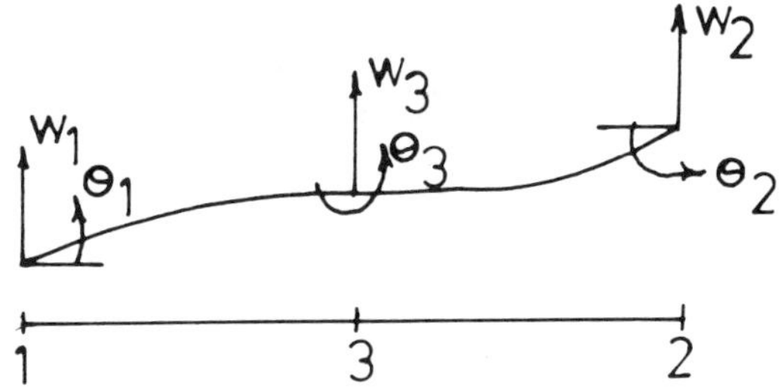

FIGURE 8. Six degree element.

along the length of the member with the least number of displacement co-ordinates.

Two four degree elements can also be used as an alternative. The use of such elements involves an additional unknown rotation at the mid-length and constrain the curvature to vary linearly along the length with a peak at the mid-length, whereas the actual variation will be nonlinear and may have the peak at some other point.

The adoption of the five-degree element has been found justified from the point of view of numerical accuracy and computational economy as discussed later.

The derivation of the elastic properties of the five-degree beam-column element has been presented in reference [5].

The initial stiffness matrix (K_O) corresponding to small deflection linear elastic assumptions and the geometric matrix (K_G) are reproduced in the following.

$$[K_o] = \frac{2EI}{5h^3}\begin{bmatrix} 158 & & & \text{symmetric} & \\ 47h & 18h^2 & & & \\ 98 & 17h & 158 & & \\ -17h & -3h^2 & -47h & 18h^2 & \\ -256 & -64h & -256 & 64h & 512 \end{bmatrix} \tag{7a}$$

$$[K_G] = \frac{1}{210h}\begin{bmatrix} 508 & & & & \\ 29h & 16h^2 & & & \\ 4 & -13h & 508 & & \\ 13h & 5h^2 & -29h & 16h^2 & \\ -512 & -16h & -512 & 16h & 1024 \end{bmatrix}$$

For the present analysis $[K_S]$, the initial stiffness matrix

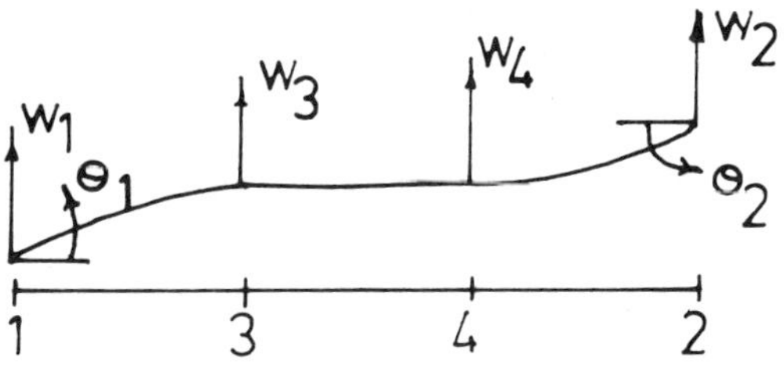

FIGURE 9. Six degree element.

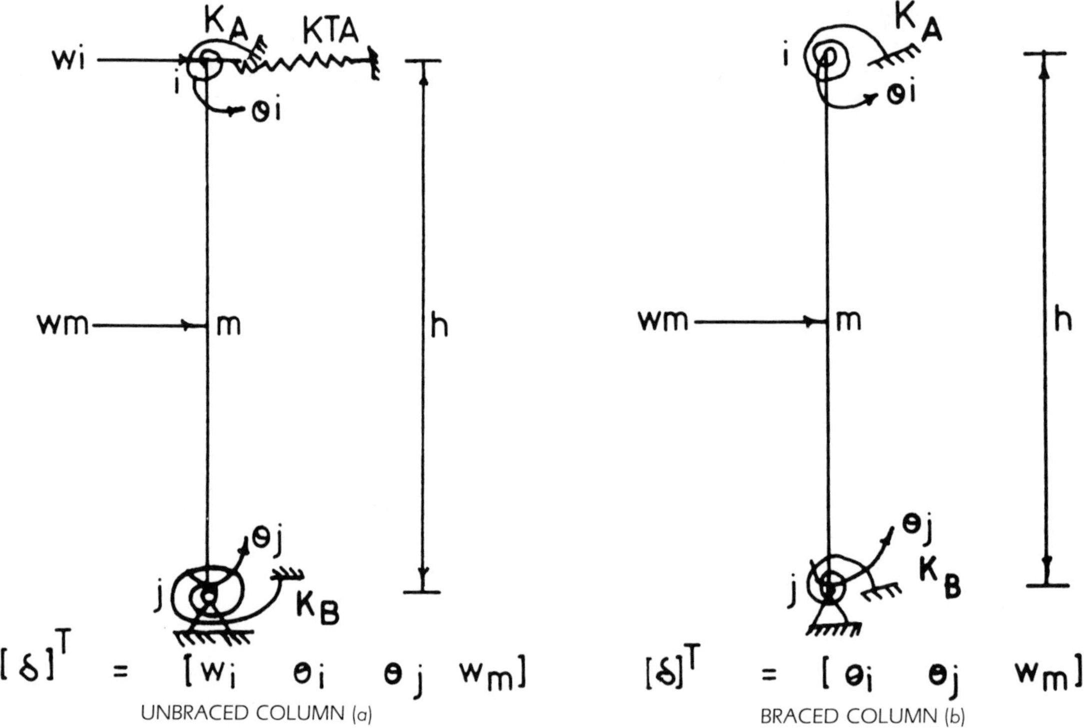

FIGURE 10. Number of degrees-of-freedom involved in the analysis of columns.

based on the secant flexural rigidity, is required to be calculated. When calculating $[K_S]$ the element has variable (EI) that cannot be expressed analytically as a function of x. Thus numerical integration is the only means by which $[K_S]$ can be evaluated. By employing Gaussian quadrature formula the $[K_S]$ matrix for the element is given by

$$\underset{(n \times n)}{[K_S]} = \frac{h}{2} \times \underset{(n \times m)}{[B]} \times \underset{(m \times m)}{[EIS]} \times \underset{(m \times m)}{[GW]} \times \underset{= (m \times m)}{[B]^T} \tag{8a}$$

where $[B]$ is a rectangular matrix each row of which gives, for a sampling point, the curvatures due to unit value of each of the n degrees of feedom in turn, $[GW]$ is a diagonal matrix of weightages at the sampling points, $[EIS]$ is a diagonal matrix of the secant flexural rigidities at the sampling points, m is the total number of sampling points, and h is the length of the element.

Each element of this matrix $[K_S]$ is given by

$$K_{S_{ij}} = \frac{h}{2} \sum_{p=1}^{m} EIS(p) \times GW(p) \times N_i''(p) \times N_j''(p) \tag{8b}$$

where p is a sampling point, m is the total number of sampling points, $GW(p)$ is the Gaussian weightage, and $EIS(p)$ is the secant flexural rigidity of the cross-section at the sampling point p.

The new stiffness matrix $[K_M]$ including the effect of axial load is given by

$$[K_M] = [K_S] - P[K_G] \tag{8c}$$

where $P[K_G]$ is the geometrical stiffness matrix as described before and given by equation (7b) for five degree beam-column element and P is the axial load acting in the column.

Depending on the actual degrees-of-freedom involved in an individual column its elastic properties can be derived from the corresponding elastic properties of a five-degree element [5].

The degrees-of-freedom for an unbraced column and a braced column are shown in Figures 10(a) and 10(b) respectively. The various matrices used in the analysis of these columns are summarised in Appendix 1.

ANALYSIS FOR SPECIFIED LOAD

Steps in the analysis for braced and unbraced columns are essentially the same. The sole point of dissimilarity lies in the number of degrees-of-freedom involved. Since the different properties required subsequently for the analysis are already given separately for braced and unbraced columns, the different steps in the method of analysis can be described together for both these types of columns.

The analysis for a given load level consists of the determination of three sets of displacements and slopes and the extrapolation of these for obtaining a better set.

1. A set of values are proposed for the unknown displacements.

2. The curvature values at the sampling points are evaluated by post-multiplying the matrix of second derivatives of the shape functions (Equation 23 and 27) at the sampling points with the vector of nodal displacements.

3. The moment at a sampling point corresponding to the known curvature and the axial load is evaluated by a subroutine A. described later.

4. Ignoring the additional moments caused by the interaction of the axial load and the deflections for the time being the nodal force vector $\{S(\delta)\}$ in equilibrium with the moment distribution $M(x)$ can be obtained by applying unit displacement in turn corresponding to the nodal forces, and equating the virtual external work done by the nodal loads to the virtual internal work done by the moments $M(x)$ because of the curvatures produced by the unit displacements. This gives

$$\{S(\delta)\} = \int_0^h [H(x)]^T M(x)dx \qquad (9)$$

where $[H(x)]^T$ is given by equations (23) and (27) for unbraced and braced columns respectively.

Using the Gaussian quadrature formula the above integral can be written as

$$\{S(\delta)\} = \frac{h}{2} \underset{(n \times m)}{[B]^T} \underset{(m \times m)}{[GW]} \underset{(m \times 1)}{\{M\}} \qquad (10)$$

Here n is the number of degrees of freedom (3 for a braced and 4 for an unbraced column), m is the number of sampling points, $[B]$ and $[GW]$ are as defined before and $\{M\}$ is the vector of moments $M(x)$ at the sampling points.

5. The residual force (or moment) vector is now found from equation (5c), i.e.

$$\{R\} = \{F\} - (\{S(\delta)\} - P[K_G]\{\delta\}) \qquad (11)$$

where $\{F\}$ is the vector of net external nodal forces (or moments) on the columns obtained by subtracting from the applied forces (or moments) the forces (or moments) carried by the end restraints, and $P[K_G]$ is the geometrical stiffness matrix as given by equations (25) and (29) for an unbraced and a braced column respectively.

6. If the residuals are within the specified tolerances, the assumed displacements are taken to have converged and the analysis proceeds to the next load level. Otherwise, a correction vector to these assumed displacements is obtained by pre-multiplying the residual vector with the inverse of the initial stiffness matrix. This correction vector is then multiplied by a factor (C_L) before it is added to the assumed displacements, to get the improved displacements for starting the next cycle. The C_L factor is used to reflect the change in the stiffness of the column from its initial value.

7. After every two such iterations, (using the same value of C_L), Aitken's delta-square extrapolation process is applied on each component of the displacement vector to accelerate the convergence.

Consider three sets of displacements $\delta_j^{(1)}$, $\delta_j^{(2)}$ and $\delta_j^{(3)}$ where: $\delta_j^{(1)}$ is the proposed value of the displacement node j; $\delta_j^{(2)}$ is the value of the displacement at node j after the first cycle of iteration; and $\delta_j^{(3)}$ is the value of the displacement at the node j after the second cycle of iteration.

Let the difference between the consecutive sets be defined by

$$d_j^{(1)} = \delta_j^{(2)} - \delta_j^{(1)}$$

and

$$d_j^{(2)} = \delta_j^{(3)} - \delta_j^{(2)}$$

The extrapolated displacement is then given by

$$\delta_j^{(1)} = \delta_j^{(1)} + \alpha_j \, d_j^{(1)}$$

where

$$\alpha_j = \frac{1}{1 - (d_j^{(2)}/d_j^{(1)})} \qquad (12)$$

α_j is taken to be zero if $d_j^{(1)}$ is small compared to $\delta_j^{(1)}$.

To avoid the possibility of occurence of very large values of α_j from Equation (12), an upper limit is to be fixed for α_j. For the present investigation, a limit of 4 has been adopted throughout, the sign being given by the equation (12).

ANALYSIS FOR A SPECIFIED NODAL EDGE STRAIN

In this section, maximum strain in the column is specified. The load in equilibrium with this strain and the curvature corresponding to the proposed nodal deflections is found by the subroutine A2. The rest of the steps are similar to those given above. Note that the calculated load will now change with the deflections in each iterative cycle. The multiplying factor used here (C_S) may be different from (C_L).

SUBROUTINES

A1 – Determination of moment at a cross section for a given axial load and curvature. To evaluate the moment at a cross section from the known curvature and the given axial load P, the following procedure has been adopted:

1. A value for the edge strain (maximum compressive) is assumed to fully define the strain distribution in the cross section for the known curvature (as shown in Figures 11b or 11e).
2. The value of axial load ($\bar{P}$) and moment for this strain distribution is evaluated by the subroutine A2.
3. If the calculated axial load P agrees with the given axial load P, the assumed edge strain is correct, and the corresponding moment value is accepted. Otherwise, a correction to the assumed edge strain is obtained by the Newton-Raphson method:

$$\Delta \in_1 = (P - \bar{P})/(d\bar{P}/d\in_1) \tag{13}$$

where

$$dP/d\in_1 = \frac{P^1 - \bar{P}}{\Delta\in} \tag{14}$$

P^1 being obtained for the given curvature, ϕ but with an incremented edge strain ($\epsilon_1 + \Delta\epsilon$).

This correction ($\Delta\epsilon_1$) is then added to the assumed edge strain and the analysis proceeds through steps (1) and (3) till the error in the calculated axial load is negligible.

A2 – Determination of axial load and moment at a cross-

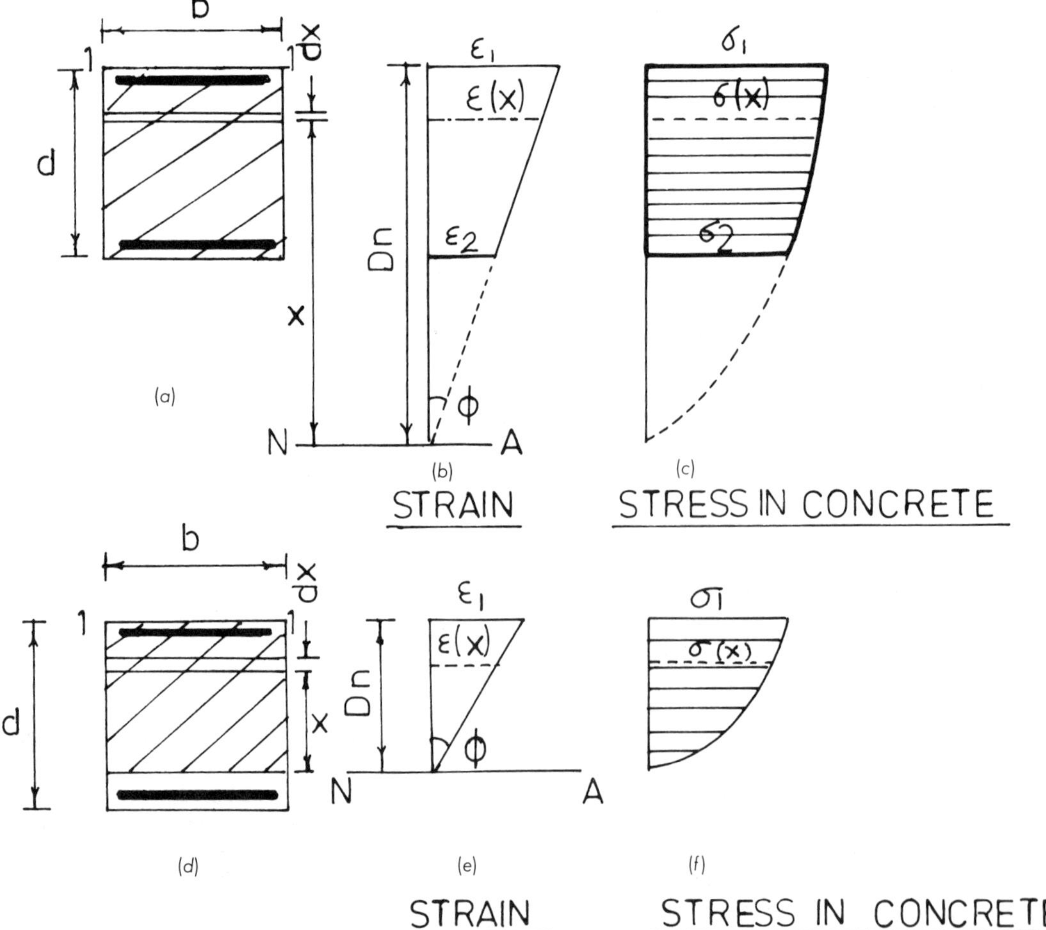

FIGURE 11. Strain and stress distribution on a cross-section under axial load and uniaxial bending.

section for a given edge strain (maximum compressive) and curvature.

Figure 11 shows a reinforced concrete section under uniaxial loading with two possible positions of neutral axis, assuming that the maximum compressive strain occurs at edge 1 − 1. Let the area under compression (shown hatched) be designated as *C*.

The load on the section is given by

$$P = P_c + P_s - P_{sc} \tag{15}$$

where, P_c is the contribution of concrete and is given by

$$P_c = b \int_0^{D_n} \sigma[x) dx \tag{16a}$$

with $a = O$ for $D_n \leq d$ and $a = D_n - d$ for $D_n \geq d$, P_s is the contribution of the steel reinforcement being equal to

$$P_s = \sum_{i=1}^{m} A_{si} f_{si} \tag{16b}$$

and P_{sc} is the deduction for the concrete areas in compression replaced by steel and is given by

$$P_{sc} = \sum_{i=1}^{m} A_{si} f_{sci} \tag{16c}$$

Denoting by M_c, M_s and M_{sc} the contributions to moment about about the neutral axis of the concrete and steel areas and the concrete area replaced by steel respectively, the moment M about the centre of the column is given by

$$M = M_c + M_s - M_{sc} - P\,(D_n - d/2) \tag{17}$$

where

$$M_c = b \int_0^{D_n} \sigma[x)x\, dx \tag{18a}$$

with $a \neq 0$ for $D_n \leq d$ and $a = D_n - d$ for $D_n \geq d$;

$$M_s = \sum_{i=1}^{m} A_{si} f_{si} (D_n - Y_i) \tag{18b}$$

$$M_{sc} = \sum_{i=1}^{m} A_{si} f_{sci} (D_n - Y_i) \tag{18c}$$

in equations (15) to (18c)

A_{si} = Areas of steel in the *i*th layer
m = number of steel layer in a cross-section
Y_i = Co-ordinates of the *i*th layer of steel
f_{si} = the stress in the *i*th layer of steel
f_{sci} = the concrete stress in the area replaced by the steel and is taken as zero if the steel layer is outside the compression zone
D_n = the depth of the neutral axis
d = the depth of the cross section

The choice of the analytical form (Equation 1) for the stress strain relation for concrete allows the closed form expressions to be obtained for the contribution of concrete to the axial load and moment acting on the cross section. These expressions are given in Appendix 2.

STEPS IN THE COMPLETE ANALYSIS

1. The steps in the computer program are given below: For the first load level, the starting displacement vector is obtained by post multiplying the inverse of the relevant stiffness matrix K_0 (as given in equations (24) and (28) with the specified load vector.

2. The "analysis for specified load" is carried out until the required degree of accuracy is obtained, provided the permissible maximum number of cycles per load level is not exceeded. If the permissible number of cycles is exceeded, the analysis is prematurely terminated.

3. The load is given the chosen increment. The proposed displacement vector for the second load level is obtained by the linear extrapolation of the converged displacement vectors at the first load level. Step 2 is then carried out for the second load level.

4. The cumulative total of the number of cycles required up to the previous load level is compared with the corresponding specified limit and the analysis is terminated if this limit is exceeded. Otherwise the number of load levels already treated is checked against the corresponding specified number and the computation switches over to "strain increment algorithm" or "strain controlled process" in step 6 if the specified number has been reached.

5. The load is incremented as in step 3 and the proposed displacement vector for the new load level is obtained by linear extrapolation of the converged displacement vectors at the preceding two load levels. The analysis and checks are carried out as in steps 2 and 4.

6. The maximum compressive edge strain in the column and its location are determined. The difference between the crushing strain and the maximum edge strain is divided by the number of strain intervals to obtain the value of the strain increment.

7. The maximum compressive strain is given this increment and the "analysis for specified edge strain" is carried

out till the required degree of accuracy is obtained, provided that the permissible number of cycles per strain level is not exceeded.

8. The analysis is terminated if the total number of cycles required so far is found to exceed the corresponding limit. Otherwise the maximum compressive strain in the column and also its location are determined. If the strain is found to have reached the crushing value, after the analysis of step 7, the programme comes to proper conclusion due to material failure. If the critical strain after adding the increment is in excess of the crushing strain, the analysis is carried out after making the critical strain equal to the crushing strain and the program comes to its proper conclusion.

9. The checks are carried out as listed in step 7.

NUMERICAL RESULTS

The method has been checked by analysing five braced columns (restrained or pin-ended) and four unbraced columns previously analysed by Cranston [4]. The details of the columns are given in Table 1.

These columns are analysed using a stress-strain curve of concrete given by

$$f_c/f_{c_p} = 2(\epsilon/\epsilon_o) - .6732(\epsilon/\epsilon_o)^2 - .6536(\epsilon/\epsilon_o)^3 + .3268(\epsilon/\epsilon_o)^4 \tag{19}$$

This polynomial curve and the curve used by Cranston practically coincide up to $\epsilon = \epsilon_o = .002$. For $.002 < \epsilon < .0035$, the stresses given by the polynomial expression are lower, the maximum deviation being 6.5%. The polynomial expression is inapplicable for strains beyond 0.0035.

In the present analysis about sixty percent of the estimated failure load has been covered by the load increment algorithm, both for the braced and the unbraced columns. In the case of braced columns the complete load-deflection curve is traced in three equal load increments and six equal strain increments. For unbraced columns, four equal load increments have been used, with the number of strain increments varying from four to ten.

In the iterative process for displacement evaluation, convergence has been assumed to occur when each element of the residual load vector becomes less than the corresponding element in the tolerance vector. For the numerical examples presented in this chapter the tolerance vectors used for braced and unbraced columns are given below

Unbraced column (Figure 10a)

$$\{T\} = \begin{Bmatrix} 0.001\ H \\ 0.001\ Hh \\ 0.001\ Hh \\ 0.001\ H \end{Bmatrix} \tag{20}$$

Braced column (Figure 10b)

$$\{T\} = \begin{Bmatrix} 0.001\ M_1 \\ 0.001\ M_2 \\ 0.001\ M_1/h \end{Bmatrix} \tag{21}$$

where $\{T\}$ is the tolerance vector, H is the applied lateral load on the unbraced column, M_1 and M_2 are the applied

TABLE 1. Details of the Cases Subjected to Uniaxial Bending.

Column Designation	Area of cross-section sq. in (cm²)	Percentage of reinforcement	Length/Depth ratio	Rotational at ends A	Restraint B B	Applied moments/lateral load at end A	B	Remark
C_1	100 (645)	1	15	10 K_c	10 K_c	1.5 Pd	1.5 Pd	
C_2	Do	1	40	10 K_c	24 K_c	1.5 Pd	—	Ref. Fig. 11
C_3	Do	1	40	10 K_c	10 K_c	1.5 Pd	1.5 Pd	
C_4	Do	1	15	2 K_c	8 K_c	.5 Pd	—	
C_5	Do	6	40	—	—	.1 Pd	—	Pin-ended
C_6	Do	1	10	15 K_c	15 K_c	.5 Pd/h	—	Unbraced (Fig. 12)
C_7	Do	1	40	15 K_c	15 K_c	.5 Pd/h	—	
C_8	Do	1	25	3 K_c	3 K_c	.1 Pd/h	—	
C_9	Do	1	25	3 K_c	3 K_c	.5 Pd/h	—	

end moments on the braced column and h is the height of the column.

In the iterative procedure for calculating the moment for a given axial load and curvature, the solution is taken to have converged when the error in the calculated axial load becomes less than 0.1 of the specified load on the section. A limit of 15 cycles has also been specified.

For the convergence of the displacements a limit of 15 cycles per load level and 35 cycles per strain level has been specified.

All these nine columns have been solved with Gaussian quadrature formulae of different order and with different C_L and C_S factors.

The symmetrically restrained unbraced columns have been analysed in two ways. In the first case the analysis has been carried out on the original length of the column; and in the second case, on the half length of the column taken as pinned at one end and restrained at the other end. The magnitude of this restraint is the same as that in the original column. It has been observed that the computer time taken by the analysis using the second procedure is less than that taken by the analysis using the first procedure for the same number of Gauss sampling point.

The failure loads of the columns obtained by the present method and by the method used by Cranston are listed below in Table 2.

TABLE 2. Comparison of Failure Loads of the Columns.

Columns Designation	Cranston's method KIPS (KN)	No. of sampling points used	Present method KIPS (KN)
C_1	250.0 (1111.0)	7	254.0 (1129.0)
C_2	173.0 (769.74)	4	172.46 (767.44)
C_3	142.36 (633.5)	6	140.0 (623.0)
C_4	213.4 (949.6)	4	214.56 (954.8)
C_5	185.0 (823.25)	4	180.0 (801.0)
C_6	179.0 (796.4)	5*	185.06 (823.5)
C_7	46.28 (206.0)	4*	46.11 (205.2)
C_8	71.54 (318.35)	5	72.61 (323.0)
C_9	127.5 (567.35)	4	127.72 (568.4)

*Over half length

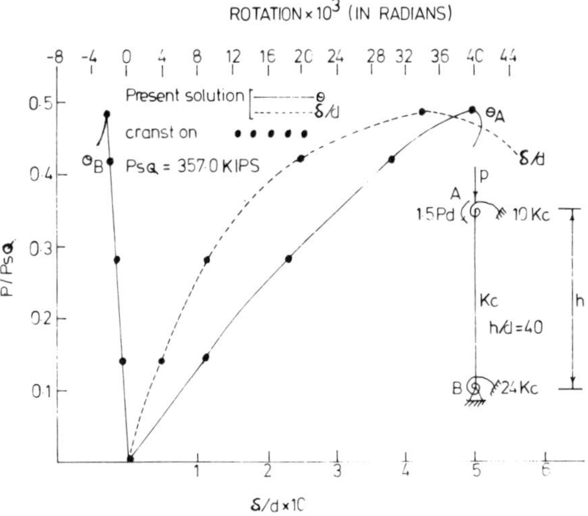

FIGURE 12. Load vs. central deflection/rotation for non-symmetric braced column.

The convergence characteristics of these columns using $C_L = 1$ are given in Table 3.

For the braced columns, the load-mid length deflection and load-end rotations are compared with the results given by Cranston in Figure 12. For unbraced columns, load midlength deflection curves are compared in Figure 13.

The order of Gaussian quadrature method used for solving these columns is mentioned in Table 3. The agreement between the present solution and Cranston's results is satisfactory.

It may be noted for none of the uniaxially loaded columns analysed in this chapter was it necessary to use a modified initial stiffness matrix based on secant flexural rigidity.

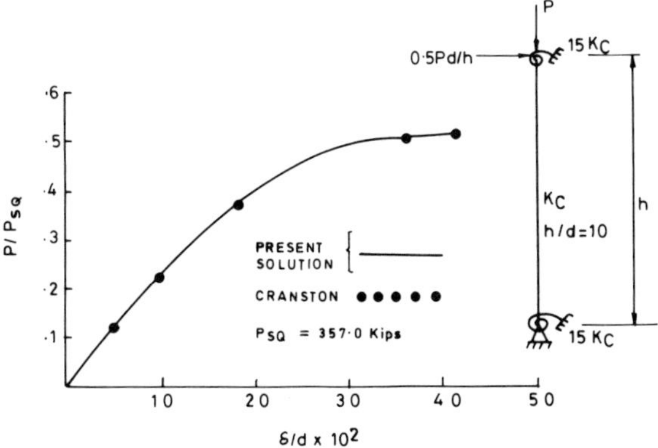

FIGURE 13. Load central deflection characteristic of unbraced column.

TABLE 3. Convergence Characteristics of the Columns Under Uniaxial Bending.

Column designation	No. of Gauss point used	Load-increment algorithm: No. of increments	Load-increment algorithm: Av. No. of cycles/ increments	Strain-increment algorithm: C_S factor used	Strain-increment algorithm: No. of increments	Strain-increment algorithm: Av. No. of cycles/ increments	Av. No. of cycles for moment calculation for a given axial load and curvature	Failure load KIP (KN)
C_1	7	3	5	0.5	3	5	2	254.0(1129.0)
	6	3	5	0.5	3	5	2	257.62(1146.41)
	5	3	5	0.5	3	5	2	254.87(1134.0)
	4	3	6	0.5	3	7	2	261.82(1165.1)
C_2	4	3	7	0.5	6	9	2	172.46(767.45)
C_3	6	3	8	0.5	6	13	2	139.98(622.91)
	5	3	10	0.10	6	18	2	138.05(614.32)
	4	3	9	0.5	6	11	2	133.47(594.0)
C_4	4	3	6	0.5	3	10	2	214.56(954.8)
C_5	4	3	6	0.5	3	10	2	180.0(801.0)
C_6	5*	4	6	0.8	4	18	2	185.06(823.53)
	4	4	7	0.8	4	23	2	185.2(824.14)
C_7	4*	1	7	0.8	10	32	2	46.11(205.2)
C_8	5	4	6	0.8	4	28	3	72.61(323.11)
	4	4	6	0.8	4	29	2	75.21(334.68)
C_9	4	4	4	0.8		19	2	127.72(568.4)

*Over half length

SUMMARY

The present method for the solution of the nonlinear equations is based on the modified Newton-Raphson technique. This being a linearly convergent process, it has been possible to accelerate convergence using Aitken's delta square extrapolation technique applied to individual displacement components.

To evaluate the integral (Equation (9)) Gaussian quadrature formula has been applied. Different order integration formulas have been tried to assess the computation time requirements and the accuracy of the solution and the four-point Gaussian quadrature is found to be the most suitable, except in cases with high rotational restraints or with high slenderness ratios as columns designated as C_1, C_5, C_6 and C_8.

Besides the extrapolation technique, some factors (C_L and C_S) have been used on the displacement increments for accelerating the convergence. In order to select the suitable C_L and C_S factors, four braced columns of different characteristics (C_1, C_2, C_3 and C_4) were analysed with five values for these factors. These values are 0.2, 0.4, 0.6, 0.8 and 1. In these analyses both four-point and five-point Gaussian quadrature formula have been used. The total number of cycles required for generating the load deflection curve up to certain specified load level is monitored. The investigation suggests that a value of C_L and C_S around 1.0 is likely to be found suitable in most cases.

CONCLUSIONS

A method based on the finite element technique has been presented for the analysis of braced and unbraced columns. This method has been verified against available solutions and is found to be satisfactory. It has been found that a single beam-column element with five degrees of freedom (i.e. two end rotations, two end deflections and one deflection at mid-length position) is quite satisfactory for nonlinear analysis of both braced and unbraced columns. The optimum values of the acceleration parameters used in the present modified Newton-Raphson method have been indicated. The suitability of the four-point Guassian quadrature formula has also been established.

REFERENCES

1. Basu, A. K. "Computation of Failure Loads of Composite Columns," *Proceedings, Institution of Civil Engineers,* London, Vol. 36 pp. 557–578 (Mar. 1967).
2. Basu, A. K., and P. Suryanarayana, "A New Method for Inelastic Analysis of Restrained Reinforced Concrete Columns", *The Bridge and Structural Engineer,* New Delhi, Vol. 7, No. 2 (June 1977).
3. Cranston, W. B., "A Computer Method for the Analysis of Restrained Columns", TRA 402, Cement and Concrete Association, 52, Grosvenor Gardens.
4. Cranston, W. B., "Analysis and Design of Reinforced Concrete Columns" *Research Report 20,* Cement and Concrete Association, 52, Grosvenor Gardens, London, SWI (1972).
5. Kayal, S., "Nonlinear Analysis of Plane Frame-Wall Systems," thesis presented to the Indian Institute of Technology, at Delhi, India, in partial fulfilment of the requirements for the degree of Doctor of Philosophy (1978).
6. Pfrang, E. O., and C. P. Siess, "Predicting Structural Behavior Analytically," *Journal of Structural Division, ASCE, Vol. 90,* No. ST5, pp. 99–111 (October, 1964).
7. Zienkiewicz, O. C., *The Finite Element Method in Engineering Science,* McGraw-Hill, London, Second Edition, p. 147.

APPENDIX 1. ELASTIC PROPERTIES FOR UNBRACED AND BRACED COLUMN

Unbraced column (Figure 10a)

Shape functions and second derivatives of shape functions:

$$[N]^T = \begin{Bmatrix} N_1 \\ N_2 \\ N_3 \\ N_4 \end{Bmatrix} = \begin{bmatrix} 1 - \frac{11x^2}{h^2} + \frac{18x^3}{h^3} - \frac{8x^4}{h^4} \\ x - \frac{4x^2}{h} + \frac{5x^3}{h^2} - \frac{2x^4}{h^3} \\ \frac{x^2}{h} - \frac{3x^3}{h^2} + \frac{2x^4}{h^3} \\ \frac{16x^2}{h^2} - \frac{32x^3}{h^3} + \frac{16x^4}{h^4} \end{bmatrix} \quad (22) \qquad [H]^T = \begin{Bmatrix} H_1 \\ H_2 \\ H_3 \\ H_4 \end{Bmatrix} = \begin{bmatrix} -\frac{11}{h^2} + \frac{108x}{h^3} - \frac{96x^2}{h^4} \\ -\frac{8}{h} + \frac{30x}{h^2} - \frac{24x^2}{h^3} \\ \frac{2}{h} - \frac{18x}{h^2} + \frac{24x^2}{h^3} \\ \frac{32}{h^2} - \frac{192x}{h^3} + \frac{192x^2}{h^4} \end{bmatrix} \quad (23)$$

Stiffness matrices:

$$[K_o] = \frac{2EI}{5h^3} \begin{bmatrix} \left(158 + \frac{5h^3}{2EI} \times K_{TA}\right) & & & \text{symmetric} \\ 47h & \left(18h^2 + \frac{5h^3}{2EI} K_A\right) & & \\ -17h & -3h^2 & \left(18h^2 + \frac{5h^3}{2EI} K_B\right) & \\ -256 & -64h & 64h & 512 \end{bmatrix} \quad (24)$$

$$P[K_G] = \frac{P}{210h} \begin{bmatrix} 508 & & & \text{symmetric} \\ 29h & 16h^2 & & \\ 13h & +5h^2 & 16h^2 & \\ -512 & -16h & 16h & 1024 \end{bmatrix} \quad (25)$$

Braced Column (Figure 10b)

Shape functions and second derivatives of shape functions:

$$[N]^T = \begin{Bmatrix} N_1 \\ N_2 \\ N_3 \end{Bmatrix} = \begin{bmatrix} x - \dfrac{4x^2}{h} + \dfrac{5x^3}{h^2} - \dfrac{2x^4}{h^3} \\ \dfrac{x^2}{h} - \dfrac{3x^3}{h^2} + \dfrac{2x^4}{h^3} \\ \dfrac{16x^2}{h^2} - \dfrac{32x^3}{h^3} + \dfrac{16x^4}{h^4} \end{bmatrix} \quad (26) \qquad [H]^T = \begin{Bmatrix} H_1 \\ H_2 \\ H_3 \end{Bmatrix} \begin{bmatrix} -\dfrac{8}{h} + \dfrac{30x}{h^2} - \dfrac{24x^2}{h^3} \\ \dfrac{2}{h} - \dfrac{18x}{h^2} + \dfrac{24x^2}{h^3} \\ \dfrac{32}{h^2} - \dfrac{192x}{h^3} + \dfrac{192x^2}{h^4} \end{bmatrix} \quad (27)$$

Stiffness Matrices

$$[K_o] = \frac{2EI}{5h^3} \begin{bmatrix} \left(18h^2 + \dfrac{5h^3}{2EI} K_A\right) & & \text{symmetric} \\ -3h^2 & \left(18h^2 + \dfrac{5h^3}{2EI} K_B\right) & \\ -64h & 64h & 512 \end{bmatrix} \quad (28)$$

$$P[K_G] = \frac{P}{210h} \begin{bmatrix} 16h^2 & & \text{symmetric} \\ 5h^2 & 16h^2 & \\ -16h & 16h & 1024 \end{bmatrix} \quad (29)$$

APPENDIX 2. EXPRESSIONS FOR CONCRETE-CONTRIBUTION TO AXIAL LOAD AND MOMENT

Referring to Figure 11, when $D_n \leq d$

$$\frac{P_c}{bf_{cp}} = D_n \left[\sum_{i=1}^{4} \frac{C_i}{(i+1)} \frac{\epsilon_1^i}{\epsilon_o^i} \right] \quad (30) \qquad \frac{M_c}{bf_{cp}} = D_n^2 \left[\sum_{i=1}^{4} \frac{C_i}{(i+2)} \frac{\epsilon_1^i}{\epsilon_o^i} \right] \quad (31)$$

and when $D_n \geq d$

$$\frac{P_c}{bf_{cp}} = D_n \left[\sum_{i=1}^{4} \frac{C_i}{(i+1)} \frac{\epsilon_1^i}{\epsilon_0^i} \right] - (D_n - d) \left[\sum_{i=1}^{4} \frac{C_i}{(i+1)} \frac{\epsilon_2^i}{\epsilon_0^i} \right] \quad (32)$$

$$\frac{M_c}{bf_{cp}} = D_n^2 \left[\sum_{i=1}^{4} \frac{C_i}{(i+2)} \frac{\epsilon_1^i}{\epsilon_0^i} \right] - (D_n - d)^2 \left[\sum_{i=1}^{4} \frac{C_i}{(i+2)} \frac{\epsilon_2^i}{\epsilon_0^i} \right] \quad (33)$$

in which C_1, C_2, C_3, and C_4 are the coefficients as given in Equation (1); f_{cp} (Equation (1)).

This procedure is much more economical than the alternative approach of dividing the cross section into a number of strips and then employing numerical integration.

CHAPTER 6

Limit Analysis of Reinforced Concrete Beams and Frames

CHENG-TZU THOMAS HSU*

INTRODUCTION

Inelastic behaviour of structural concrete has played an important part in recent limit design recommendations both in this country and abroad. It is understood that the plastic design of structural steel and limit design of structural concrete differ in several important respects. The two most important differences were suggested by Yu and Hognestad [133] as follows:

1. Rotation Capacity: In structual steel, the various design methods concentrate on the formation of a sufficient number of plastic hinges to transform the whole or part of the structure into a mechanism, thus precipitating collapse. It is noted that little attention is paid to what extent any hinge deforms before all other hinges are formed. However, in the limit design of structural concrete, rotation capacity of sections must be studied in great detail. Also to meet serviceability criteria, excessive flexural cracking must be avoided and therefore it is desirable to limit hinge rotations for structural concrete even when considerable rotation capacity is present after extensive cracking.
2. Distribution of Moment Resistance: In structural steel, the positive and negative moment resistance are equal and constant along the entire length of a member unless cover plates are used over some part of the span. However, the positive and negative moment resistance of structural concrete members can easily be made different by varying the amount and location of reinforcement, and the moment capacity can be varied along the length of a prismatic member.

Present design practice for statically indeterminate reinforced concrete structures as specified by the provisions of the American Concrete Institute (1971, 1983) [45,37] and the National Building Code of Canada (1965) [142] is based on the inconsistent procedure of analyzing the structure elastically and proportioning the section inelastically [117,149]. As a consequence, the ultimate strength of the structure as a whole and hence, the factor of safety against failure remains undefined. In countries prone to earthquakes, it is important that structures behave in a ductile manner when subjected to the action of severe seismic disturbances. Strength and ductility of reinforced concrete frames are dependent on the moment-curvature and therefore the moment-rotation characteristics for the component members.

Energy absorption capacity of a structure is a major criterion in the design of earthquake resistant or nuclear-defence structures. The energy absorbed by a structure can be evaluated from the load-deformation characteristics of its component members. It is well known that the moment-rotation curve for a reinforced concrete section consists of a straight line up to the cracking load followed by another approximately straight line. The curve then flattens off indicating a gradual loss in flexural stiffness or rigidity due to propagation and widening of cracks and exhibits considerable ductility for most of the practically used under-reinforced sections until the ultimate moment M_u is reached. The increase in moment capacity from M_y to M_m can be attributed to the strain-hardening of steel and the rising of the neutral axis after the steel has yielded. The moment capacity decreases after the peak moment M_m because of the secondary crushing of concrete and the resulting decrease in lever arm caused by readjusting of the neutral axis. This is accompanied by a further increase in rotations or curvatures up to collapse. It is obvious that the area of regions between M_y and M_u constitutes a major part of the total energy absorbed by the structure and is therefore important in any limit state study (see Figure 1 or Figure 2).

*Department of Civil and Environmental Engineering,
New Jersey Institute of Technology, Newark, NJ

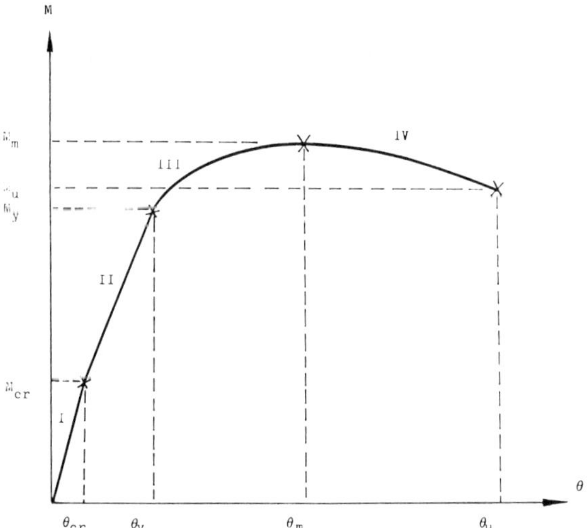

FIGURE 1. Typical moment-deformation curve.

It is now well established that the inelastic behaviour of reinforced concrete sections leads to a redistribution of moments which can bring about increased load bearing capacity of indeterminate structures. Utilization of this characteristics could be of important economic advantage (Yu and Hognestad [133]).

The ACI Commitee 428 [36] has presented empirical limit equations based on the methods suggested by Sawyer [136], Baker [6] and Cohen [135]. However, the following questions still remain unanswered (Mirza and Hsu [117]):

(1) The "Limit Design Method" would normally be used for continuous beams and rigid frames. The application of equations based on beam tests under one point and two point loadings can well be applied to continuous beams at critical sections, e.g. supports, concentrated loads, etc. However, the application of the method to the rigid frames whose members will in general be subjected to varying combinations of bending moments, shearing forces, axial forces and torsion needs further study. In a portal frame with stiffer beams, the so-called plastic hinge (regions over which the member has become inelastic) are more likely to occur in columns which may have significant axial forces besides the bending moments.

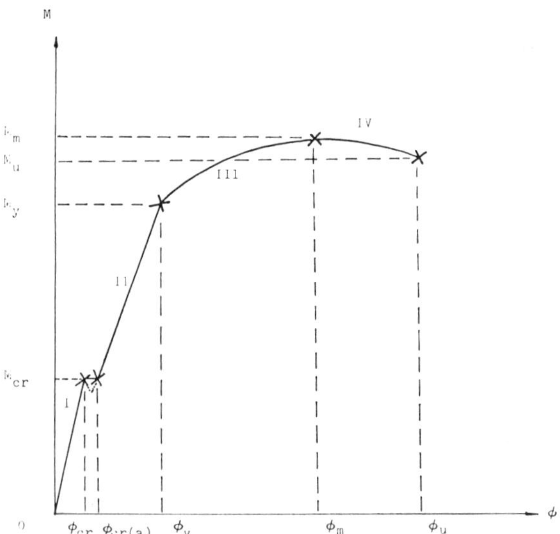

FIGURE 2. Typical moment-curvature curve.

(2) Experimental work at McGill University (Sader [112], Adenot [113]) has suggested that the so-called plastic hinges in a reinforced concrete frame (whose members are under-reinforced ($p \leq 0.75\ p_b$) show much larger rotations than their counterparts tested under a central point loading on a simple span. There is a need for more test data on moment-curvature characteristics for the different types of frame members.

(3) According to Mattock [19], equations derived from the well known principles of force equilibrium and strain compatibility can be used to arrive at a "close estimate" of the moments as well as the "safe limiting estimate" of the curvatures and rotations occurring at a section where the bending moment is a maximum. It appears quite reasonable to calculate the ultimate strength of the critical sections reasonably accurately, however what is more important is the overall ductility of the structure. There will be many sections which will exhibit rotations far exceeding those specified by the proposed clauses [36]. Before placing any limitations on maximum curvatures and rotations, one must be clear about the physical state of the structure when subjected to large deformations required by the "Limit Design" method and the probability of occurrence of effects causing these deformations. Would the structure be used again after it has reached this state? If so, what sort of deflections and rotations will be acceptable to the user of the structure?

(4) Roy and Sozen [58] suggested that the rotation capacity of a reinforced concrete member is a function of the response of concrete in the immediate vicinity of the connection, and therefore information pertinent to rotation capacity should come not from the overall deflection of the test specimen but from the shortening measured in the zone of failure. Roy and Sozen also observed that although the load carrying capacity of the concrete in axially loaded specimens was not significantly improved by the use of rectilinear ties, there was considerable increase of its ductility. The writer is of the opinion that the moment-curvature characteristic of the so-called plastic hinges should be obtained from the stress-strain characteristics of the concrete and the steel used.

The degree of moment redistribution achieved depends primarily upon the rotation capacity of hinging regions which is inherently limited for reinforced concrete members. However, Mattock [44], Ernst [137] and Yu and Hognestad [133] have shown that properly designed members possess adequate rotation capacity so that a successful redistribution of moments can occur. The validity of "adequate rotation capacity" and hence of a successful redistribution of moments is based on experimental evidence of the behaviour of simple specimens such as simply supported or continuous beams and isolated beam-column joints. Several limit design methods have been put forward based on experimental evidence avialable from these and similar tests such as Yu and Hognestad [133], Mattock [141], Ernst [137], etc.

A survey of the literature reveals a scarcity of experimental evidence regarding the load-deformation behaviour of reinforced concrete columns and frames. It also reveals a scarcity of analytical (computer technique) and experimental evidence regarding the load-deformation behaviour of three-dimensional concrete frame structures. This can be attributed to the lack of comprehensive analytical and experimental information on the load-deformation behaviour of structural concrete under uniaxial or biaxial bending and/or compression.

The following specific research areas resulted from the deliberations of West Virginia University Conference (1970) [140] to improve the knowledge of the behaviour of structural concrete framing systems:

1. More specific information on the behaviour of joints with emphasis on joint stiffness, shear transfer through joints and the influence of column thrust and transverse framing members on joint performance;
2. More strength evaluations of existing structures which would include gathering data on structural geometry, material variations, age of frame when loaded and other parameters needed in developing a probabilistic approach to the strength of concrete frames;
3. Determination of suitable E and I values to be used in analysis;
4. Response of building frames, especially joints, to earthquake and other dynamic loads;
5. Development of serviceability criteria based upon human response;
6. In the ensuing discussion it was generally agreed that limit design theories in its present form was a questionable design method.

HISTORICAL REVIEW OF EXISTING LIMIT DESIGN METHODS OF REINFORCED CONCRETE STRUCTURES

An historical review of limit design of structural concrete has been made by Yu and Hognestad [133], Mattock [44], Cohn [139], etc. The limit design theories have been proposed by Ernst [137], Sawyer [116], Baker [6], and Cohn [138,150], etc., who tried to modify the theory of plastic design in steel structures for special application to reinforced concrete structures.

The plastic theory in steel structures [e.g. 89,114] is based on the assumption that a statically indeterminate structure fails by formation of plastic hinges which make the structure kinematically unstable by transforming it into a mechanism. It is assumed that the critical sections (potential hinges) possess infinite rotation capacity and thus a complete redistribution of moments is possible. All plastic deformation is assumed to be concentrated at the hinged section of the member. However these assumptions are not valid for a reinforced concrete structure which has limited ductility. Recent design procedures account for the limited rotation capacity of a reinforced concrete section. However, as compared to plastic theory which is applied widely to metal structure, the present codes do not recognize any of the limit design methods completely. The reluctance on the part of building code authorities to recognize the proposed limit design methods is related to their mathematical complexity and lack of generality as compared to the plastic theory of steel structures.

A brief review of the existing limit design theories in reinforced concrete structures is in the following:

Baker's Limit State Design Method [6]

Baker's method is one of the several proposed limit design methods which recognize the limited rotation capacity of reinforced concrete sections. Briefly, the method is as follows:

It is postulated that a structure N times statically indeterminate will develop N hinges before final collapse. The actual moment-curvature diagram of the reinforced concrete sections is approximated by a bi-linear curve with a flat top. This means that once yielding starts at any critical section, the moment stays constant up to the rupture of the section. The ultimate strength of the structure is assumed to be reached when a very small additional increment of load would cause the formation of one or more additional hinges causing collapse of the structure.

Once N hinges have developed, the structure becomes statically determinate with known moments acting at the hinges. At this instant it is possible to calculate the plastic rotation of the hinges using the standard influence coefficient equations of the form:

$$\begin{aligned} &----- \\ &d_{01} + \bar{x}_1 d_{11} + \bar{x}_2 d_{12} + ----- + \bar{x}_n d_{1n} = -\theta_1 \\ &----- \\ &d_{02} + \bar{x}_1 d_{21} + \bar{x}_2 d_{22} + ----- + \bar{x}_n d_{2n} = -\theta_2 \qquad (1) \\ &----- \\ &d_{0n} + \bar{x}_1 d_{n1} + \bar{x}_2 d_{n2} + ----- + \bar{x}_n d_{nn} = -\theta_n \\ &----- \end{aligned}$$

where

$\theta_1, \theta_2 \ldots \theta_n$ are the plastic hinge rotations in radians

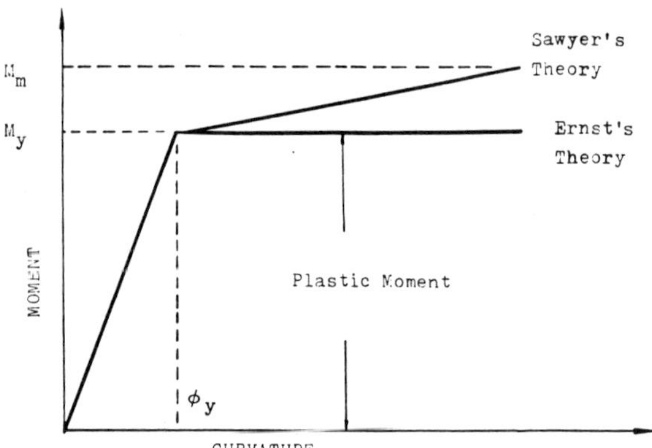

FIGURE 3. Idealized moment-curvature curve.

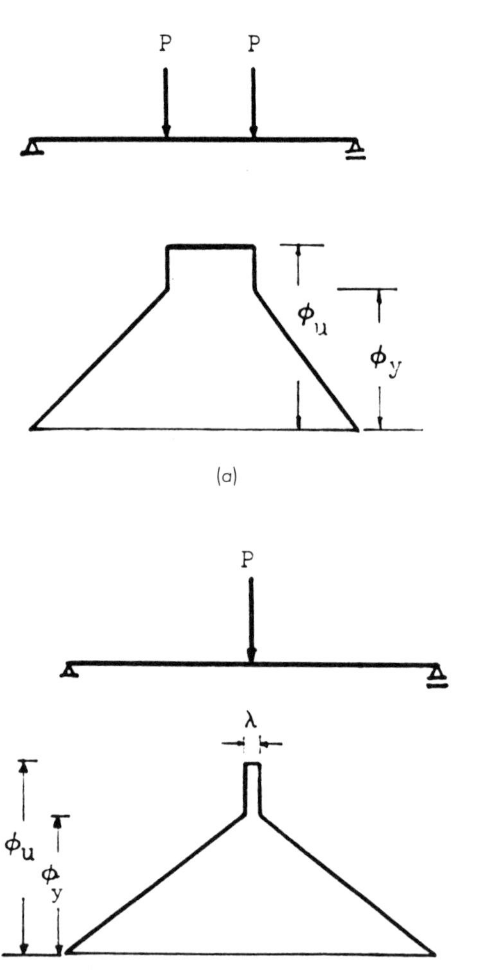

FIGURE 4. Idealized curvature diagrams: (a) beam with finite yield length; (b) beam with localized yield.

$d_{12}, d_{13} \ldots d_{1n}$ are the influence coefficients for hinge rotations

$\bar{x}_1$, etc. are rotations at hinges 1, 2 etc. due to external load.

The plastic hinge rotations thus determined are then compared with the available rotation capacity of the hinges. If the calculated plastic rotations are less than the rotation capacity, the design is acceptable.

Cohn's Optimum Limit Design Method [138]

Cohn proposed three fundamental conditions which are specific to the limit design of reinforced concrete structures, namely: limit equilibrium, rotation compatibility and serviceability.

The first condition postulates the existence of one or more collapse mechanisms. The second condition implies that all plastic hinges necessary for a structural collapse may actually occur, without premature local fracture of the concrete.

The third condition requires a reasonable factor of safety against yielding on which the magnitude of the crack openings and deflections at working loads are finally depending.

The object of Cohn's design method is to provide a simple and rational technique to check the rotation compatibility of plastic hinges in reinforced concrete indeterminate structures proportioned using the optimum criteria suggested by him.

From the practical viewpoint, it follows that for (1) given material properties, (2) loading conditions, and (3) amount of accepted redistribution, the rotation compatibility condition requires limitation of the steel percentages at critical sections.

Ernst's Limit Design Method [137]

Ernst restated the moment-area theorems to include the behaviour of structures in the inelastic range, using a unit rotation diagram instead of the conventional *M/EI* diagram applicable in the elastic case (see Figure 3).

At a section developing its ultimate strength, the unit rotation is $\phi_u = (\epsilon_u + \epsilon_s)/d$, where "$\epsilon_u$" and "$\epsilon_s$" are the maximum concrete and steel strains at ultimate load, and "d" is the effective depth of the section.

The unit rotation diagram of a member with a definite yield length would be as shown in Figure 4a, and for a member in which yield is localized would be shown in Figure 4b. In the second case the yield is considered to have spread over a length "λ;" the total concentrated rotation at a yielding section is therefore $\theta_{p1} = \lambda(\phi_u - \phi_y)$, where ϕ_y is the unit rotation at commencement of yield of steel.

In the analysis of continuous reinforced concrete beams, Ernst has applied this adaptation of the moment-area theorems to check the concentrated rotations necessary at all hinging sections if full redistribution of moments is to occur. In this calculation, the concentrated plastic rotations are considered to act as concentrated loads on the conjugate beam. The concentrated rotations so calculated are compared with experimentally determined values to check whether they are admissible.

Sawyer's Limit Design Method [116]

The theory of Sawyer differs from that of Ernst in that the idealized moment curvature diagram is not used (Figure 3). In its place another idealized diagram is used (see Figure 3), in which the moment in the plastic region increases linearly from the moment at commencement of yield (M_y) up to the maximum moment of the section (M_m). This idealized diagram is used to calculate the total plastic rotation in sections of beams and frames adjacent to points of maximum moment.

Continuous beams and single-bay portal frames were analyzed by Sawyer by a trial and error adaptation of the elastic center method of analysis. In this analysis the total plastic rotation in any particular region of maximum moment is considered to be concentrated at the point of maximum moment. All parts of the beam or frame away from the points of maximum moment are assumed to have constant stiffness. The trial and error approach is continued until the components of moment of the elastic center due to the effect of both elastic and plastic deformation is zero.

Cohn [134] has worked out examples using the three limit design methods given by Sawyer [136], Baker [6] and Cohn [135] to show that differences between three methods are relatively slight and that savings in flexural steel of 10 to 20 percent are possible.

Sheikh, Mirza and McCutcheon [111] analysed the reinforced concrete frames previously tested at McGill University by using the concepts in the plastic analysis of steel structures (Beedle [114], Neal [89]). The collapse load was calculated using virtual work. If there is a heavy axial load on the column, the ultimate moment capacity of the column section is modified. They concluded that the ultimate load capacity of the frames can be predicted with reasonable accuracy by the plastic theory and Baker's limit design method [6]. The elastic theory was seen to be grossly conservative. Ultimate load values predicted by the plastic theory were reasonably accurate, with a maximum difference of 20%. Baker's limit design procedure underestimated the strength of the frame, particularly where axial loads existed.

Cranston [103] and Cranston and Cracknell [91] calculated the ultimate load for one-story portal frame based on the formation of a mechanism at ultimate load. They concluded that the experimental collapse load is considerably below the value of calculated values based on the method of mechanism. This is because the frame was reinforced with a high percentage of tension steel. It should be noted that Sheikh, Mirza and McCutcheon [111] also used the method of mechanism method and compared with Sader's [112] experimental results from under-reinforced one-story plane frame. Also Cranston [103] calculated the ultimate strength of the frames by the method given by Baker [6] which was considerably below the experimental collapse load (similar to that of Sheikh, Mirza and McCutcheon [111]).

Sheikh, Mirza and McCutcheon [111] and Cranston's [103] conclusions were not altogether unexpected, since the method of mechanism was the basis of the infinitely large curvatures and therefore larger rotation capacities at the plastic hinges. The Baker's method was based on the limited rotation capacities at the plastic hinges and embodies a number of conservative assumptions, which tend to give results on the safe side.

CONCRETE STRESS-STRAIN RELATIONSHIPS

Concrete is a complex multi-phase material consisting of aggregates and cement paste. It mainly resists compressive forces; investigators such as Hsu and Slate [51], Newman [52] and Vile [53] have shown that bond between cement paste and aggregate is a weak link in the concrete and bond micro-cracks therefore influence the strength of the concrete. Even before loading, concrete is not completely continuous because it contains a high proportion of small fissures or micro-cracks. In general, concrete is considered as the macroscopically heterogeneous and anisotropic material. However, practically from the viewpoint of engineering, the concrete may be assumed to have statistical homogeneity and isotropy if the representative volume element is sufficiently large [54]. It has also been found that concrete exhibits inelastic or creep behaviour even at low stress levels and under the influence of simultaneous shrinkage.

The structure of concrete and its behaviour under load are extensively complex. Recent developments have provided information concerning the internal structure of the concrete besides the details of failure mechanism and failure criterion of concrete. Recent investigations have shown that the instantaneous deformation of a concrete under load is influenced considerably by the testing condition and composition of the concrete. From a practical point of view, it is the uniaxial stress-strain relationship of the concrete which has received more attention.

There are different kinds of concrete test specimens such as prism, cylinder, cube, besides direct measurements from beam or column tests. Furthermore, there are a lot of different testing machines and methods which have been developed. Recently, concrete specimens reinforced with stirrups or other materials such as fiberglass, have been studied by several investigators in many countries. Critical reviews of this area have been presented by Hognestad [1], Popovics [55], Sargin [2], Kent and Park [17], Khan [20] and Hsu [56].

In standard compression tests of concrete cylinders using a universal testing machine, the cylinder is crushed almost instantaneously, because the energy produced by the machine during the test is much greater than can be absorbed by the concrete cylinder. Since the time required to crush the concrete cylinder is infinitesimal, the common instrumentation in the above experiment fails to record the data for the descending branch of the stress-strain curve. An experimental device which can read the relaxation of the compressive strength of concrete by balancing the two energies was developed at the University of Tokyo and elsewhere. The design of the test machine and experimental set-up were worked out by You [40]. The ascending branch of the stress-strain curve is determined by applying suitable load increments to the test cylinder. However, after a certain maximum value of stress (called the maximum stress point), a further increase in strain is accompanied by a decrease in stress because of the strain softening of the concrete. The test cylinder gets crushed in compression if the energy supplied to the concrete is greater than the energy it can absorb. Considering the stress variation in the specimen it is possible to design a suitable control for the machine to balance the supplied and absorbed energies. This control is complicated because the energy absorption capacity varies from one specimen to another. To overcome this difficulty the frame utilized to load the test cylinder is used to absorb the unbalanced energy and the test specimen will not be crushed as rapidly. You [40] used this modified testing machine and obtained an experimental stress-strain curve shown in Figure 5. Hsu [28,29] developed a mathematical equation expressing the above experimentally obtained stress-strain curve of concrete to study the complete moment-curvature and load deflection characteristics for reinforced concrete members.

The descending branch of the stress-strain curve constitutes the large inelastic deformations involved in the ultimate strength design and the limit design studies of reinforced concrete. It is also useful in understanding the dynamic behaviour of structures. Later studies by Barnard [57,120], Roy and Sozen [58] and others [151] show that the descending branch approaches the strain axis asymptotically. These studies appear to agree with the test results produced using the modified testing machine.

Investigations regarding the influence of confinement upon strength and deformation of concrete have been due to Richard et al. [59], Chan [62], Rüsch and Stöckl [60], Szulczynski and Sozen [63], Roy and Sozen [58,61], Bertero and Felippa [64], Soliman and Yu [65], Martin [66], Base and Read [67] Nawy, Denesi and Grosko [35], etc.

Studies of Richart et al., deal with the increase in the strength of circular concrete columns due to the provision of spiral steel binders. Martin studied the stress-strain characteristics of the concrete using spirally prestressed concrete cylinders. Other studies deal with the effect of confinement due to circular and/or rectangular ties. The properties determined in compression tests have been used in the analyses of reinforced concrete beams having concrete confined in the compression zone by web stirrups by Chan [62], Rüsch and Stöckl [60], Baker and Amarakone [68] and Corley [38].

Some of the test results of Bertero and Felippa [64] show the influence of stirrups and compression steel upon the rotation capacity of reinforced concrete sections. Investigations by Base and Read [67] demonstrate the large rotations which can be realized in reinforced and prestressed concrete beam sections on account of the increased ductility due to the confinement of the concrete.

The results of the above investigations indicate that concrete can undergo considerable deformations when it is con-

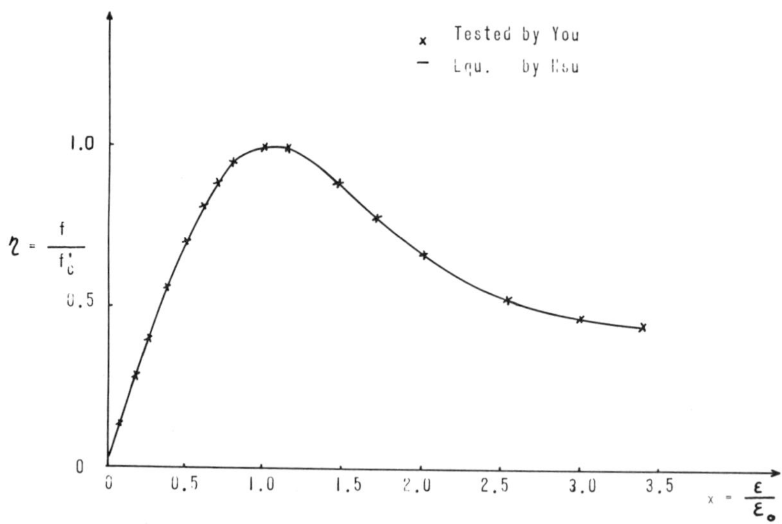

FIGURE 5. Complete stress-strain relationship in concrete.

fined in steel spirals. A proper analysis of the flexure characteristics of such reinforced concrete sections reinforced with spiral steel (which provides confinement to the concrete) basically requires information on the stress-strain characteristics of the concrete. This information has been presented by Roy and Sozen [61], Szulczynski and Sozen [63], Roy and Sozen [58], Bertero and Felippa [64], Ahmad and Shah [151], Sargin and Handa [39], Sargin [2] and Iyengar, Desayi and Reddy [69]. They concluded that confinement increases both the compressive strength and the deformation capacity of the concrete. The increase in strain capacity is considerably greater than the increase in strength. Also the circular spiral is the most effective, while the open stirrups are the least effective.

Concrete Stress-Strain Relationships in Compression

Various empirical stress-strain relationships for concrete have been derived by investigators as follows.

UNCONFINED CONCRETE

1. Hsu [28,29] (Figure 5):

$$\eta = \sin\left(\frac{\pi x}{2}\right) + 0.2x\,(x - 1)(e^{1-x} - 1) \qquad (2a)$$

for $0 \leq x \leq 1$

and

$$\eta = 0.226 + 2.157x - 1.907x^2 + 0.596x^3 - 0.063x^4 \qquad (2b)$$

for $1 \leq x \leq 4$

where

$$\eta = \frac{f}{f'_c} \text{ and } x = \frac{\epsilon}{\epsilon_o}$$

2. Desayi and Krisnan [50]:

$$f = \frac{E_c \epsilon}{\left[1 + \left(\frac{\epsilon}{\epsilon_o}\right)^2\right]} \qquad (3)$$

where

$$E_c = \frac{2f_o}{\epsilon_o},\ f_o = k_3 f'_c$$

$$\epsilon_u = 0.003,\ K_3 = 7/8$$

3. Ban (see Reference by Okayama):

$$f = E_c\epsilon + B\epsilon^2 + C\epsilon^3 \qquad (4)$$

where E_c, B and C are constants which depend on the curing days of the concrete.

4. Okayama [49]:

$$\eta = \sin\frac{\pi}{2}\,(-0.27\,|x - 1| + 0.73x + 0.27) \qquad (5)$$

5. Umemura [46]:

$$\eta = 6.75\,(e^{-0.812x} - e^{-1.218x}) \qquad (6)$$

6. Hognestad, Hanson and McHenry [5] and Kriz and Lee [25]:

$$f^2 + A\epsilon^2 + Bf\epsilon + Cf + D\epsilon = 0 \qquad (7)$$

where A, B, C and D are constants, which correspond to the experiments by Hognestad et al. [5].

7. Hognestad [1]:

$$f = f_o\left[2\frac{\epsilon}{\epsilon_o} - \left(\frac{\epsilon}{\epsilon_o}\right)^2\right] \qquad (8a)$$

for $\epsilon \leq \epsilon_o$

$$f = f_o[\epsilon_u - 0.85\,\epsilon_o - 0.15\epsilon]/(\epsilon_u - \epsilon_o) \qquad (8b)$$

for $\epsilon > \epsilon_o$

where

$$\epsilon_u = 0.0038 \text{ in/in}$$

$$\epsilon_o = \frac{2f_o}{E_c}$$

$$E_c = 18000 + 0.46f'_c$$

$$f_o = 0.85f'_c = K_3 f'_c$$

where K_3 is a stress block parameter.

8. Lee [21]:

$$f = f_o\left[2\frac{\epsilon}{\epsilon_o} - \left(\frac{\epsilon}{\epsilon_o}\right)^2\right] \qquad (9)$$

for $0 < \epsilon < 2\epsilon_o$

where

$$f_o = K_3 f_c'$$

K_3 is a stress block parameter, and ϵ_o is a compressive concrete strain corresponding to maximum compressive stress.
Note that if $K_3 = 1$,

$$\eta = 2x - x^2 \text{ (Ritter's parabola)}$$

9. Sahlin [22] and Smith and Young [4]:

$$f = f_o \left(\frac{\epsilon}{\epsilon_o}\right) e^{(1 - \epsilon/\epsilon_o)} \quad (10a)$$

On the basis of experimental results, Smith and Young proposed,

$$\epsilon_o = 0.0017 \sim 0.002$$

$$f_o = f_c' \quad (10b)$$

$$\text{i.e. } \eta = x\, e^{1-x}$$

10. Young [27]:

$$\text{(a) } f = (E_c\epsilon_o - 2f_c')\left(\frac{\epsilon}{\epsilon_o}\right)^3 - (2E_c\epsilon_o - 3f_c')\left(\frac{\epsilon}{\epsilon_o}\right)^2 + E_c\epsilon \quad (11a)$$

$$\text{(b) } f = f_c' \sin\left(\frac{\pi\,\epsilon}{2\epsilon_o}\right) \quad (11b)$$

where

$$E_c = 1000\, f_c'$$

$$\epsilon_o = 0.002$$

11. Rüsch [26] and Rüsch, Grasser and Rao [70] (CEB Recommendation) (Levi [71]):

$$f = f_o\left[2\frac{\epsilon}{\epsilon_o} - \left(\frac{\epsilon}{\epsilon_o}\right)^2\right] \quad (12a)$$

$$\text{for } \epsilon < \epsilon_o$$

$$f = f_o, \text{ for } \epsilon_o \le \epsilon < \epsilon_u$$

where

$$\epsilon_o = 0.002,\ f_o = f_c' \text{ and } \epsilon_u = 0.0035$$

or

$$\eta = 2x - x^2 \text{ (Ritter's Parabola)} \quad (12b)$$

$$\text{for } x < 1$$

and

$$\eta = 1 \quad (12c)$$

$$\text{for } 1 \le x < 1.75$$

12. Liebenberg [72]:

$$f = E_c\epsilon - a\,\epsilon^{n+1} \quad (13)$$

where

$$E_c = 67000\sqrt{f_c'}$$

$$\epsilon_o = 0.002$$

$$f_o = 0.9f_c' \text{ or } \frac{2}{3} f_{cu}$$

where f_{cu} is standard cube strength of concrete,

$$a = \frac{1}{(n+1)\epsilon_o^n}$$

and

$$n = \frac{f_o}{(E_c\epsilon_o - f_o)}$$

$$\epsilon_u = 0.001\left[0.4 - \frac{f_c'}{(6.5 \times 10^4)}\right]$$

13. Saenz [73]:

$$f = E_c\left\{\epsilon \Big/ \left[1 + \left(\frac{E_c}{E_o} - 2\right)\frac{\epsilon}{\epsilon_o} + \left(\frac{\epsilon}{\epsilon_o}\right)^2\right]\right\} \quad (14a)$$

for the ascending branch.

where

$$E_o = \frac{f_o}{\epsilon_o}$$

$$E_c = \sqrt{f_o} \times 10^5/(1 + 0.006\sqrt{f_o})$$

$$\epsilon_o = 0.001\,[f_o^{1/4}(3.15 - f_o^{1/4}) \times 10^{-3}]$$

and

$$f = \epsilon/[a + a_1\epsilon + a_2\epsilon^2 + a_3\epsilon^3] \quad (14b)$$

for the desending branch, where a, a_1, a_2 and a_3 are determined by experiments.

14. Sturman, Shaw and Winter [74]:

$$f = A\epsilon + B\epsilon^n \quad (15)$$

where $A = 4.0$, $B = -517.0$, $n = 1.9$ for concentric loading, where $A = 4.13$, $B = -80.4$, $n = 1.6$ for eccentric loading.

CONFINED CONCRETE

1. Sargin and Handa [39] and Sargin [2]:

$$\eta = K_3\left[\frac{A\,x + (D - 1)x^2}{1 + (A - 2)x + Dx^2}\right] \quad (16)$$

where $K_3 = 1$ for concrete cylinder tests.

These parameters, A, D and K_3 depend on the following factors: concrete strength, confinement due to lateral reinforcement, creep, strain gradient, cyclic loading, size of specimen and type of loading, etc.

2. Szulczynski and Sozen [63] and Roy and Sozen [61]:

Peak point is given by

$$f = f_o \text{ and } \epsilon = 0.002 \quad (17a)$$

Termination point is given by

$$f = 0.5f_o \text{ and } \epsilon = 0.75p''' \frac{h_e}{s} \quad (17b)$$

where $f_o = h_3 f_c'$; p''' is volumetric ratio; s is spacing of the stirrups; h_e is the distance inside-to-inside of ties (see Figure 6).

3. Baker and Amarakone [68]:

$$f_o = f_c'\left(0.8 + \frac{1}{K}\right) \quad (18a)$$

$$\epsilon_u = 0.0015\,\{1 + 1.5\,p''' + [(0.7 - 0.1\,p''')/K] - 0.1f_c'\} \quad (18b)$$

where K is an experimental parameter and p''' is volumetric ratio.

4. Kent and Park [17]: Kent and Park used the tests results from Soliman and Yu [65], Roy and Sozen [61] and Bertero and Felippa [64] to derive the stress-strain curves shown in Figure 7.

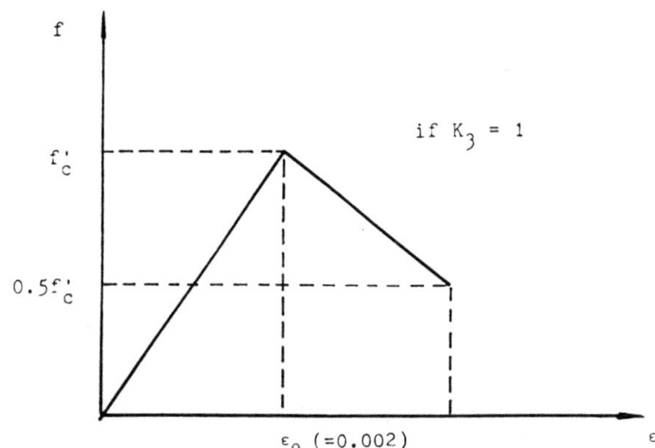

FIGURE 6. Concrete stress-strain curve by Szulczynski and Sozen and Roy and Sozen.

For confined concrete:

$$\text{a) If } f \le f_c',\ f = f_o\left[2\,\frac{\epsilon}{\epsilon_o} - \left(\frac{\epsilon}{\epsilon_o}\right)^2\right] \text{(Ritter's Parabola)} \quad (19a)$$

$$\text{b) If } f > f_c',\ f = f_c'[1 - Z(\epsilon - \epsilon_o)] \quad (19b)$$

where

$$\epsilon_o = 0.002$$

$$Z = \frac{0.5}{\epsilon_{50h} + \epsilon_{50u} - \epsilon_o}$$

$$\epsilon_{50u} = \frac{3 + 0.002f_c'}{f_c' - 1000}$$

$$\epsilon_{50h} = \frac{3}{4}p'''\sqrt{\frac{b''}{s}}$$

$$p''' = \frac{2(b'' + d'')As''}{b''d''s}$$

where s, b'', d'' and A_s'' are shown in Figure 7.

5. Chan [62] (Figure 8):

a) If $f \le f_o$, $f_o = E\epsilon_o$ (straight line) and

$$\epsilon_o = 0.002 \quad (20a)$$

E is the module of elasticity in concrete

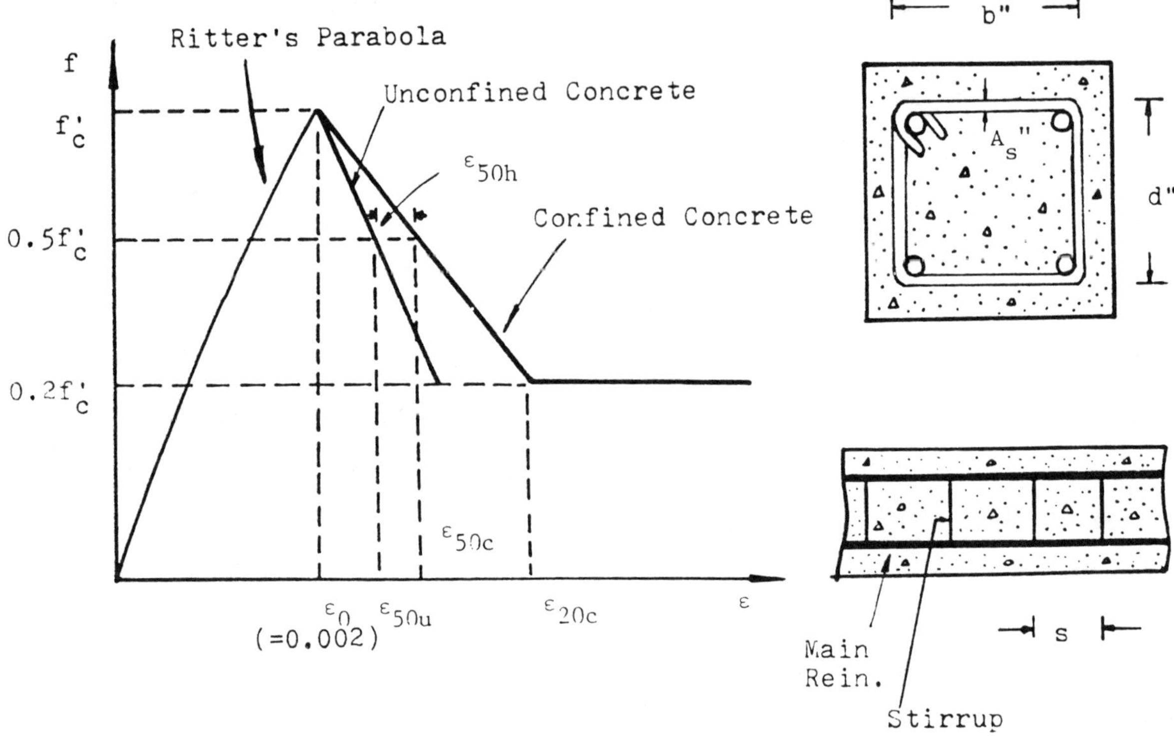

FIGURE 7. Concrete stress-strain curve by Kent and Park.

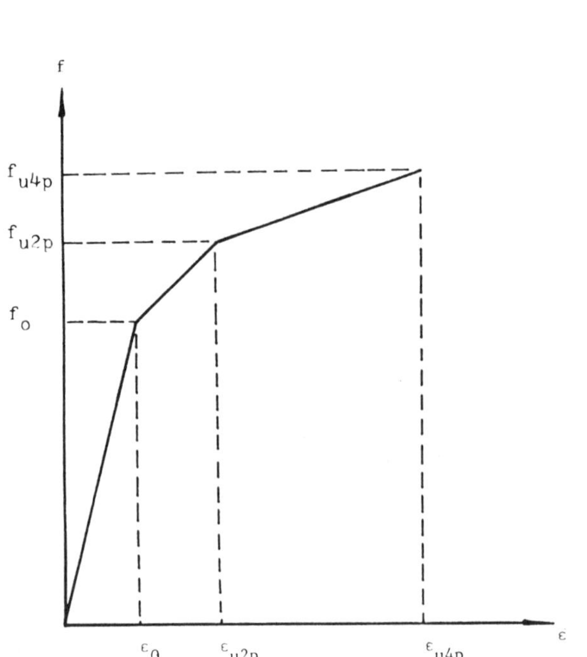

FIGURE 8. Concrete stress-strain curve by Chan.

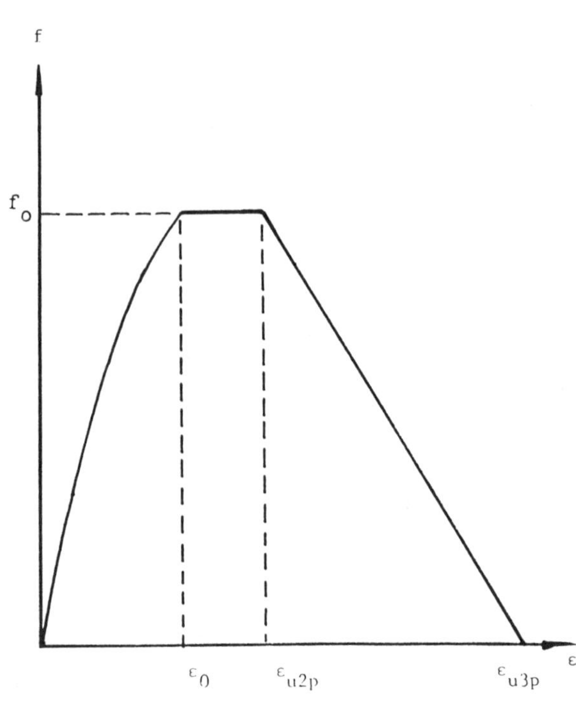

FIGURE 9. Concrete stress-strain curve by Soliman and Yu.

b) If $f > f_o$,

$$f_{u2p} = f_o + f_o \frac{\sqrt{p'''}}{0.238} \tag{20b}$$

$$\epsilon_{u2p} = \frac{\sqrt[3]{p'''}}{24.5} + 0.002 \tag{20c}$$

$$f_{u4p} = f_o + f_o \frac{\sqrt{p''' + 0.05}}{0.238} \tag{20d}$$

$$\epsilon_{u4p} = 0.002 + \frac{\sqrt[3]{p''' + 0.05}}{24.5} \tag{20e}$$

and

$$p''' = \frac{2(b'' + d'')A_s}{b''d''s}$$

6. Soliman and Yu [65] (see Figure 9):

a) If $f \leq f_o$, (Ritter's Parabola)

$$\eta = 2x - x^2 \tag{21}$$

$$\epsilon_o = 0.002 \text{ and}$$

f_o can be related to f_c' for unconfined concrete.

b) If $f > f_o$

$$f = f_o \tag{22a}$$

$$\epsilon_{u2p} = \frac{\sqrt[3]{p'''}}{24.5} + 0.002 \tag{22b}$$

$$\epsilon_{u3p} = \frac{\sqrt[3]{p'''}}{4.9} + 0.012 \tag{22c}$$

$$p''' = \frac{2(b'' + d'')A_s}{b''d''s}$$

7. Cranston and Chatterji [33,129]: Cranston and Chatterji referred to the stress-strain curves analysis by Chan [62] and Szulczynski and Sozen [63] and obtained three different stress-strain relations for unconfined, confined by stirrups and concrete at hinges and lying over 0.5 in. inside stirrups, respectively as shown in Figure 10. The equations in terms

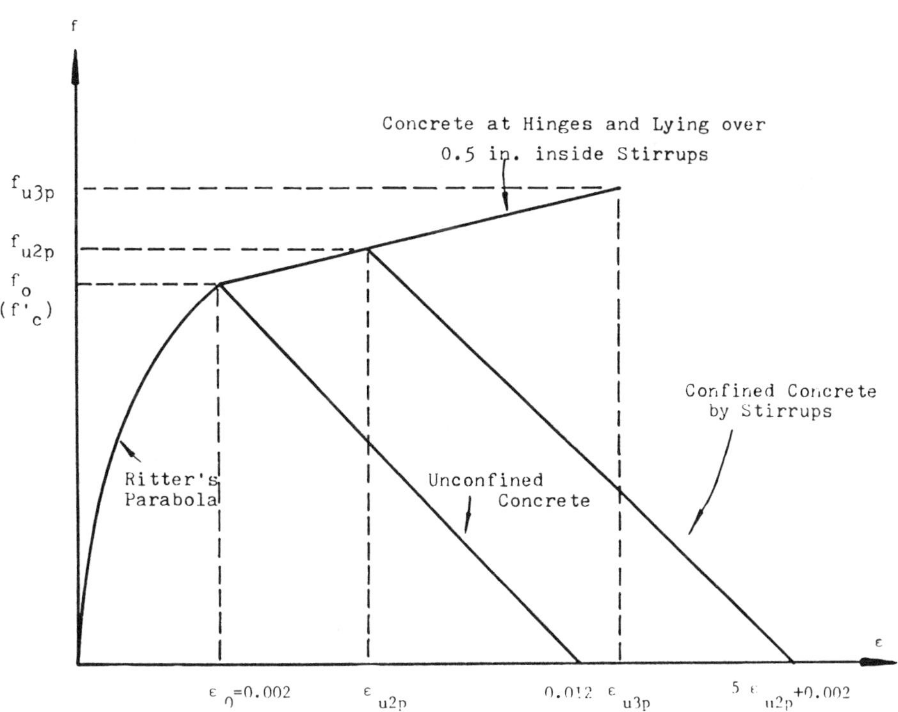

FIGURE 10. Concrete stress-strain curve by Cranston and Chatterji.

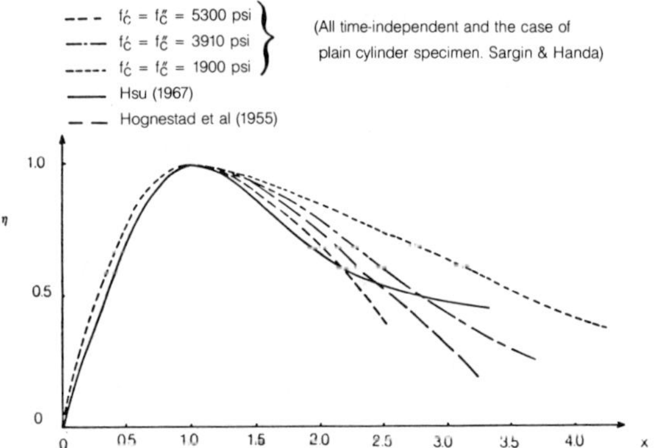

FIGURE 11a. Comparison of stress-strain curves in concrete.

of the notations used in Figure 10 are:

Ritter's Parabola:

$$f = f_o \left[2\frac{\epsilon}{\epsilon_o} - \left(\frac{\epsilon}{\epsilon_o}\right)^2 \right] \quad (23a)$$

$$\text{or } \eta = 2x - x^2 \quad (23b)$$

$$f_{u2p} = f_o + f_o \frac{\sqrt{p'''}}{0.238} \quad (23c)$$

$$\epsilon_{u2p} = \frac{\sqrt[3]{p'''}}{24.5} + 0.002 \quad (23d)$$

$$f_{u3p} = f_o + f_o \frac{\sqrt{p''' + 0.05}}{0.238} \quad (23e)$$

$$\epsilon_{u3p} = 0.002 + \frac{\sqrt[3]{p''' + 0.05}}{24.5} \quad (23f)$$

It must be noted that f_o is f'_c for an unconfined concrete test. The comparison of some of the uniaxial compressive stress-strain curves for unconfined concrete is shown in Figure 11.

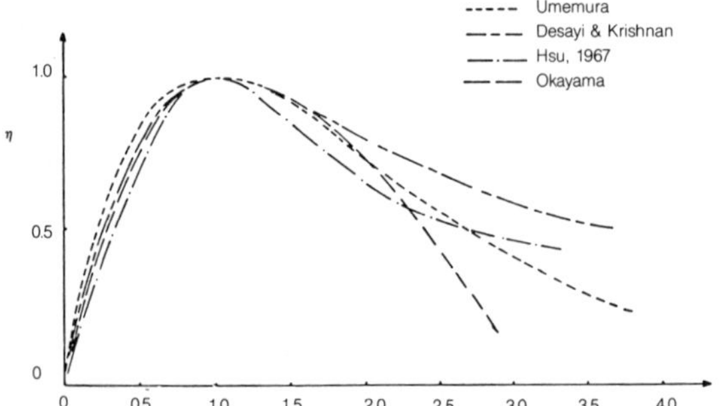

FIGURE 11b. Comparison of stress-strain curves in concrete.

Concrete Stress-Strain Relationships in Tension

1. Hughes and Chapman [75]:

$$f_t = E_c\epsilon_t \qquad \text{for } \epsilon_t \leq \epsilon_{tr} \quad (24a)$$

$$f_t = 0 \qquad \text{for } \epsilon_t > \epsilon_{tr} \quad (24b)$$

where f_t is flexural tensile stress corresponding to a strain ϵ_t, ϵ_{tr} is cracking strain $= f_{tr}/E_c$, and f_{tr} is modules of rupture.

2) Lalonde and Janes [76]:

$$f_t = E_c\epsilon_t \qquad \text{for } \epsilon_t \leq \epsilon_{tr} \quad (25a)$$

where

$$\epsilon_{tr} = \frac{f_{tr}}{E_c}$$

and

$$f_{tr} = a(f'_c)^b \qquad (a = 2 \sim 2.3,\ b = 2/3)$$

$$f_t = 0 \qquad \text{for } \epsilon_t > \epsilon_{tr} \quad (25b)$$

3. Guerrin [77]:

$$f_t = E_c\epsilon_t \qquad \text{for } \epsilon_t \leq \epsilon_{tr} \quad (26a)$$

$$f_t = 0 \qquad \text{for } \epsilon_t > \epsilon_{tr} \quad (26b)$$

where

$$\epsilon_{tr} = \frac{f_{tr}}{E_c},\ f_{tr} = 2.3(f'_c)^{2/3}$$

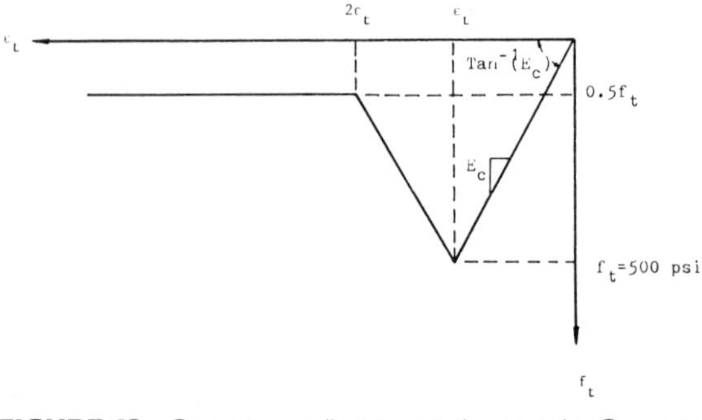

FIGURE 12. Concrete tensile stress-strain curve by Cranston.

4. Cranston and Chatterji [33] and Ferry-Borges and Arga E Lima [78] (see Figure 12):

$$f_t = 500 \text{ psi} \tag{27a}$$

$$E_c = \frac{500}{\epsilon_t}, \; \epsilon_t = \frac{500}{1000 f_c'} = \frac{1}{2 f_c'} \tag{27b}$$

$$E_c = (1000 f_c') \tag{27c}$$

GENERALIZED MOMENT-CURVATURE CHARACTERISTICS FOR STRUCTURAL CONCRETE SECTIONS SUBJECTED TO UNIAXIAL BENDING WITH OR WITHOUT CONSTANT AXIAL LOAD

Introduction

This section presents the development of the generalized moment-curvature relationships for structural concrete sections under flexure and with or without the effect of a constant axial load. Three approaches have been suggested along with their limitations. These approaches are either based on the assumption that the failure at a section occurs when the force equilibrium equation and strain compatibility equations are violated or when the maximum concrete strain exceeds a certain predefined value.

Review of Previous Work

A review of the theories for the analysis of structural concrete sections, that appeared before 1950 and 1967, has been given elsewhere by Hognestad [1] and Sargin [2]. These investigations can be divided into the following groups.

ULTIMATE STRENGTH THEORIES–GROUP A

These include investigations by Whitney [3], Hognestad [1], Smith and Young [4], Hognestad, Hanson and McHenry [5], Baker [6], Whitney and Cohen [7], Janney, Hognestad and McHenry [8], Viest, Elstner and Hognestad [9], Mattock and Kriz [10], Mattock, Kriz and Hognestad [11], etc.

The prime objective of these investigations is a satisfactory agreement between calculated and measured ultimate strength values. The ultimate strength theories are simple, and sufficiently accurate for most practical purposes as suggested by Mattock, Kriz and Hognestad [11] and have been used for some time in the various codes of practice. The ultimate strength theories cannot be considered to be sound because (1) failure is defined in terms of a limiting strain or stress in the concrete and the reinforcing steel. Also the stress distribution in the compression zone of a section is defined in terms of the stress block parameters k_1, k_2, k_3 and ϵ_u, where these parameters are determined experimentally and may not be applicable to conditions other than the test conditions under which its parameters are determined; (2) ultimate strength theories deal with only one loading stage, i.e. failure stage, and cannot explain the behaviour at any other stage of loading; (3) the definition of failure is open to question; (4) the properties of a section, e.g. its shape, the amount of steel reinforcement, etc., and the effect of factors such as the rate of loading, lateral reinforcement, cannot explicitly be taken into account in such a theory; (5) a good strength prediction for under-reinforced sections is not sufficient proof of the soundness of a theory, because it fails to estimate the ductility of an under-reinforced section.

METHODS BASED ON A MAXIMUM CONCRETE STRAIN–GROUP B

The methods in this group consider a section to fail when the strain in the extreme compression fibre reaches a certain value, ϵ_u. The numerical approach in this group is based on the equations of equilibrium and strain compatibility at a section and the given stress-strain curves for the concrete and the reinforcing steel, selects extreme concrete compressive strain ϵ_c or tensile strain ϵ_s as a trial value, then follows to determine the position of neutral axis, bending moment and curvature. The following investigators constitute the group: Pfrang, Siess and Sozen [12], Fowler [13], Green [14], Warner [15], Eiklid, Gerstle and Tulin [16], Sargin [2], Kent and Park [17], Cohn and Ghosh [18], Ghosh [130], Mattock [19] and Khan [20]. Other investigators involved in developing the basic equations for generalized moment-curvature relationship at a section are as follows: Lee [21], Sahlin [22], Kriz [23], Ali [24], Kriz and Lee [25], Rüsch [26], Young [27], Mattock, Kriz and Hognestad [11], Smith and Matthew [47], Todeschini, Bianchini and Kesler [48], Hsu [28], Hsu and Mirza [29], etc.

The main difference between ultimate strength theories and this group is that the latter uses generalized stress-strain curves and functional failure criteria. By using the general theory, the strength and deformation properties of a section can be obtained for a complete spectrum of loading from zero to the maximum and beyond, if this has a physical meaning. Also, the effect of loading rate, lateral reinforcement etc. can be taken into account by the adoption of a general stress-strain relationship for concrete. Failure or ultimate is defined mostly as the loading condition at which a section reaches its maximum capacity, i.e. maximum moment or maximum load. Hsu [28], Hsu and Mirza [29], Kent and Park [17], Cohn and Ghosh [18], and Ghosh [130] concluded that it is possible to predict the descending branch of the moment-curvature curves beyond the maximum moment. But the termination of computation of the moment-curvature curve is dependent upon the value of ϵ_u, therefore, no physical significance should be given to the terminal point of the curve.

According to Barnard [30], concrete is a strain-softening material and a certain length of a beam can continue to rotate when moment is falling off, and the rupture will not occur unless the energy balance in the beam ceases to be satisfied. He showed that in indeterminate reinforced concrete beams there is a distinct maximum load which such a beam can withstand and that it is possible for some hinging regions of a beam to be falling off in moment while the total load in the beam is increasing. Also, the moment redistribution occurs through fall off in moment at some sections as well as through inelastic action.

The work of Rosenblueth and Diaz de Cossio [31], Barnard [30] and other investiagors implies that there is a descending branch of the moment-curvature or the load-deflection curve, however the terminal point of these curves is hard to establish experimetnally or analytically. Most of the investigators have avoided a rational derivation of this point and have instead used simplifying assumptions (Hsu and Mirza [34]).

METHODS BASED ON MAXIMUM CONCRETE AND STEEL STRAINS–GROUP C

A modification of the extended Newton-Raphson method or method of successive approximation has been used by investigators in this group for determination of strain and curvature distributions at a reinforced concrete section. These computer programs have the ability to use any standard reinforced concrete section geometry and material properties. Cranston [32] observed that the difficulty in this method may arise when a strain-softening material is being considered, and results are being sought when a large part of the section has a negative stiffness. In this case, the solution cannot generally be attained. Later, Cranston et al. [33] extended their early method to include the descending branch of the uniaxial moment-curvature curves. Hsu and Mirza [131,132,145] modified and extended Cranston's numerical approach [32,33] and the stress-strain curves to include the descending branch of the biaxial moment-curvature curve in this study.

The typical definition of failure was suggested by Cranston [32] who considered that if the maximum strains in the concrete or the steel reinforcement exceed certain predefined maximum values, the section is considered to have failed.

METHODS BASED ON CONTROL OF DEFORMATION–GROUP D

This group is based on the control of applied deformation and is able to obtain the descending branch of the moment-curvature curve. The approach is similar to the method in Group B. However, the ultimate curvature ϕ_u is not controlled by ϵ_u, instead it is based on the following approach by Hsu and Mirza [29,34]: using the strain and stress distributions obtained, the equilibrium is checked across the section. Generally there will be good agreement on the ascending branch and some part of the descending branch of the curve, however there will be some violation of the equilibrium equations on the latter parts of the descending branch. The computer program continues to evaluate the moment-curvature characteristics if this difference is within suitable prescribed limits. A larger discrepancy automatically stops the computer analysis and the section is assumed to have attained the ultimate loading stage.

Basic Assumptions and Principles

The following well known principles of mechanics in structural concrete are employed.

CONTINUUM MECHANICS

As in most cases of the mechanics of a deformable body, a conceptual model of macroscopic structure both in reinforcing bar and concrete has been introduced rather than the complex microscopic structure. Therefore, in continuum mechanics [121], the concepts on continuous media in engineering aspects are that: stress is defined as the average force intensity on a unit area in or on the medium; strain is the change per unit length in a linear dimension; it is also an average value over the gauge length, when it is defined experimentally (e.g. strain gauge and demec gauge measurements).

A STEADY AND STATIC EQUILIBRIUM

In a plane section such as beam members acted upon by uni-axial stresses, there are two equations of equilibrium: moment equilibrium about any axis in the plane, and force equilibrium in the direction of stresses.

EQUATION OF COMPATIBILITY

The compatibility of strains provides the continuity of deformations in a section. In engineering aspects such as beam members, the Bernoulli-Euler Theory that plane sections remain planar after deformations is usually adopted to ensure strain compatibility.

If there are shear stresses together with bending moment in the structure member, then a natural consequence of these shear stresses is shearing strain, which causes cross sections, initially planar, to become warped [82]. A more elaborate investigation of the problem shows the warping of cross sections also does not substantially affect the strain in longitudinal fibers if a distributed load acts on the beam and the shearing force varies continuously along the beam. In the case of concentrated loads the stress distribution near the loads is more complicated, but this deviation from the straight line law is of a local type.

It is true that deviations from Bernoulli-Euler Theory can be noted in certain cases, which often depend on a certain slip between the concrete and the reinforcing bars, but for normal cases, i.e. for reinforcement with good bonding, these deviations are small and have no practical significance. For certain types of concrete, for example reinforced light weight concrete, where the bond between the concrete and the reinforcement is often weak, Bernoulli-Euler Theory does not hold true. The same applies to ordinary concrete when the reinforcement consists of heavy smooth bars with poor bond. The deviation is particularly marked if the bond is elimiated in one way or another, for example by oiling the reinforcing bars. In the laboratory, bond can be eliminated by wrapping the reinforcing bars with oiled paper. This procedure makes it possible to study the pure case where a bond does not exist. Beams with the reinforcement oiled, or wrapped with oiled paper, show properties which deviate in a marked way from beams with bonded reinforcement.

Beams with defectively bonded reinforcement, which are sometimes met with in practice, occupy an intermediate position between those with wrapped reinforcement and those with fully bonded reinforcement. Since the concrete and the reinforcement do not deform together if the bond is not sufficient, Bernoulli-Euler Theory cannot be applied to such structures. Up till now there have not been any satisfactory experimental and theoretical studies concerning this interesting deviation and its consequences in practice. The random experiments which have been carried out show, however, that beams with poorly bonded reinforcement differ considerably from normal structures where the concrete and the reinforcement co-operate.

In practice, the problem of insufficient bond is encountered in prestressed concrete where the reinforcement sometimes, due to incomplete grouting, is not sufficiently bonded to the concrete. Reinforcement which has been cold worked by passing it through a draw plate, has often a greasy surface due to the lubricate used. If the thin layer of grease is not removed, the bond will only be a fraction of the normal.

In such cases, Bernoulli-Euler Theory cannot be used, as deviations from it can be very great. In deep beams the deviation can also be so great that the theorem cannot be used. Several theoretical investigations which have been carried out show that large deivations from Bernoulli-Euler Theory are to be expected. Experimental investigations have also been carried out. A similar case is that of T-beam with wide flanges. It is well known that the concrete regulations in various countries include a certain limitation of the "effective flange width." For certain cases of loading, for example concentrated loading on a simply supported T-beam, it is found that the calculated effective flange is surprisingly small (Granholm [122]).

CONTINUITY AND EQUILIBRIUM AT THE BOUNDARIES

According to the Saint-Venant's Principle, the strains that are produced in a body by the application of a system of forces statically equivalent to zero force and zero couple, are of negligible magnitude at distances (from the region where the load is applied) which are large compared with the linear dimensions of the part. In other words, the effects of a given loading system, irrespective of its type, are negligible at sections of the structure which are at some distance from the applied load. Therefore, there are no boundary condition problems in an analysis of reinforced concrete sections which are perpendicular to the direction of uniaxial stress and far from the supports (Sargin [2]).

Moment-Curvature Characteristics – The State of the Art

Limit design of indeterminate reinforced concrete structures is based on the rotation capacity of "plastic hinges" which in turn is dependent on the moment-rotation and hence the moment-curvature characteristics of the individual elements (see Figure 2). Nawy et al. [35] have shown that an increase in ductility and therefore the rotation capacity of the hinges can improve moment redistribution. ACI-ASCE Committee 428 [36] has suggested that the curvature corresponding to the ultimate moment M_u be determined using the sections of the ACI Code [37] except that the ultimate compressive strain ϵ_u shall be suitably modified as follows to account for parameters affecting it (Corley [38]):

$$\epsilon_u = 0.003 + 0.02\frac{b}{Z} + \left(\frac{P''' fy''}{20}\right)^2 \quad (28)$$

A study of behaviour of reinforced or prestressed concrete elements requires satisfaction of two criteria at all loading stages – equilibrium of forces and moments, and compatibility of deformations. Sufficient information is available on the stress-strain characteristics of reinforcing bars. Most of the ultimate flexural strength design equations have been derived using an "equivalent stress-block" for concrete. These equations give satisfactory "ultimate strength" values however they cannot be used for studying the deformation behaviour in ranges III and IV of Figure 2. The idealized stress distributions account for the magnitude and position of the resultant compressive force without any consideration for the strain distribution. The experimental load-deflection or moment-rotation curves of under-reinforced and over-reinforced beams have been known to reflect the stress-strain characteristics of steel and concrete respectively. Recent investigations into complete stress-strain characteristics of concrete have studied the influence of several parameters, e.g., strength, lateral reinforcement, creep, strain gradient, size of specimen and type of loading.

You [40] studied the complete stress-strain characteristics of plain concrete using the testing machine developed at the University of Tokyo and developed the complete stress-strain curve shown in Figure 2. Hsu [28] derived the following equations to this curve:

$$\text{for } 0 \leq x \leq 1$$

(where $x = \epsilon/\epsilon_o$, $\epsilon_0 = 0.002$ is taken in the present investigation),

$$\eta = \frac{f}{f'_c} = \sin\left(\frac{\pi x}{2}\right) + 0.2\,x(x-1)(e^{1-x}-1) \quad (2a)$$

and

for $1 \leq x \leq 4$

$$\eta = \frac{f}{f'_c} = 0.226 + 2.157x - 1.907x^2 + 0.596x^3 - 0.063x^4 \tag{2b}$$

It must be noted that different stress-strain curves for unconfined and confined concrete can be used for an investigation of moment-curvature characteristics, e.g. Cranston [33] and Kent and Park [17]. Also, if the beam is statically determinate and the load is applied statically, collapse will ensue as soon as the bending moment at a section reaches M_m. However, in an indeterminate structure, considerable moment redistribution would occur due to rotation of the plastic hinges at load levels near M_m. This is normally accompanied by a drop in the bending strength of the section (beyond M_m) and the structure has the ability to achieve a new equilibrium configuration until such time as all plastic hinges necessary to transform the structures into a mechanism have developed.

The flexural capacity of concrete structure is a three-dimensional phenomenon and depends not only on the cross-sectional properties at a section along the span but also on the development length in directions parallel to the beam axis. Bond, shear and moment resistance cannot be regarded as independent responses to given loads.

There are some discontinuities in the true behaviour of the moment-curvature curve such as flexural cracks, tension yielding, crushing and spalling of the concrete. For example, the flexural cracks in the tension zone cause the stiffness of the structure to decrease, and a corresponding readjustment in the location of the neutral axis. Also a local bond slip occurs between the reinforcing steel and the surrounding concrete. A study of bond stress-strain relation in moment-curvature curve study has been done by Priestly, Park and Lu [41] and Hsu and Mirza [42]. Most investigators (including the present computer analysis) assumed a perfect bond between the reinforcing steel and concrete, or suggested using a coefficient to relate concrete and steel strain, e.g. Baker [6]. In the case of axial load-moment curvature relationship, the flexural crack causes an axial stress redistribution which influences the development of subsequent tension cracking and motion of the neutral axis which controls the distribution of strain and hence the curvature (Broms [43]). Some investigators such as Kent and Park [17] and Cranston et al. [33] reduced the geometric dimension of the critical section after extreme concrete compression strain reaches $\epsilon_c = 0.004$ (they assumed that the concrete cover spalls off at $\epsilon_c = 0.004$). Furthermore, the curvature distribution along the structural member is not linear even in the constant moment region (Priestly et al. [41]).

The different energy absorption characteristics of under-reinforced and over-reinforced concrete sections have already been recognized (Mattock [44], Barnard [30], etc.). The under-reinforced concrete section behaves in a near plastic manner while the over-reinforced concrete section exhibits a sharp falling branch in the moment-curvature curve.

Derivation of Generalized Moment-Curvature Relationships for Singly and Doubly Reinforced Rectangular Section Under Pure Flexure—Method 1

INTRODUCTION

Hsu [28] and Hsu and Mirza [29] derived the moment-curvature equations to study the general behaviour of moment-curvature curves for under-reinforced and over-reinforced concrete section. This study also investigated into the effect of the concrete stress-strain curve. The results were compared with the current ACI Code (1983, 1971) [37,45] and the Japanese Code.

The generalized formula developed can be applied to any complete stress-strain curve for concrete, and is based on the control of the compressive strain in the extreme concrete fibre (Group B in the previous section).

DERIVATION OF EQUATIONS AND DISCUSSIONS OF RESULTS

Generalized moment-curvature relationship is derived for a rectangular beam under pure flexure with both tension and compression steel reinforcement (Figure 13). It is assumed that (1) plane section remains plane, (2) concrete does not resist any tension, and (3) perfect bond exists between the steel and the concrete.

Any complete stress-strain curve (including the descending branch) for concrete can be expressed by the general equation:

$$\eta = f(x) \tag{29}$$

where

$x = \dfrac{\epsilon}{\epsilon_0}$ = normalized compressive strain

ϵ = concrete compressive strain at a given stress f

ϵ_0 = concrete compressive strain corresponding to the maximum compressive strength f'_c

$\eta = \dfrac{f}{f'_c}$

Area under the generalized stress-strain curve (Figure 14) for any strain level ϵ_c is given by

$$A_{\eta x} = \int_0^{x_c} f(x)dx \tag{30}$$

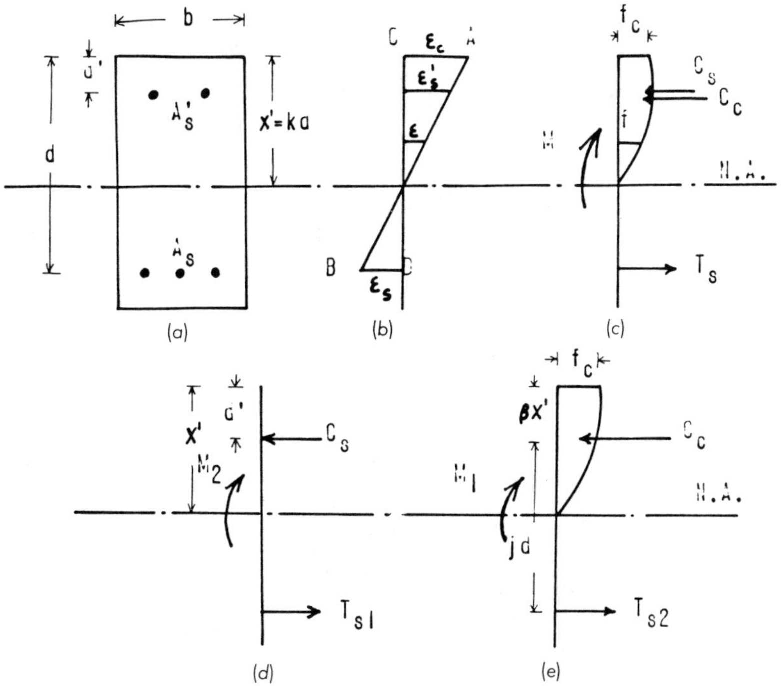

FIGURE 13. Analysis of beam cross-section with both tension and compression reinforcement under pure flexure.

This area represents the energy stored per unit volume in the compression block. The first moment of this area about the η-axis is

$$B_{\eta x} = A_{\eta x}\bar{x} = \int_0^{x_c} xf(x)dx \qquad (31)$$

Therefore

$$\bar{x} = \frac{\int_0^{x_c} xf(x)dx}{\int_0^{x_c} f(x)dx} \qquad (32)$$

One obtains from Figure 13,

$$\frac{kd - \beta x'}{kd} = \frac{\bar{x}}{x_c} = \frac{B_{\eta x}}{x_c A_{\eta x}}$$

which gives

$$\beta = 1 - \frac{B_{\eta x}}{x_c A_{\eta x}} \qquad (33)$$

Define

$$Z' = \frac{A_{\eta x}}{x_c} \qquad (34a)$$

$$Z'' = \frac{Z'}{\eta_c} \qquad (34b)$$

where x_c and η_c refer to generalized stress and strain components, respectively, at the centroid. It may be noted that x_c denotes the area of the rectangle $a'd'f'c'$ in Figure 14, while $\eta_c x_c$ represents the rectangular area $b'e'f'c'$. From similar triangles in Figure 13b,

$$k = \frac{x'}{d} = \frac{\epsilon_c}{\epsilon_c + \epsilon_s}$$

and the curvature ϕ is the given by (Figure 15)

$$\phi = \frac{1}{r} = \frac{\epsilon_c + \epsilon_s}{d} = \frac{\epsilon_c}{x'} \qquad (35)$$

Let

$$\lambda = \phi d = \frac{d}{r} = \epsilon_s + \epsilon_c$$

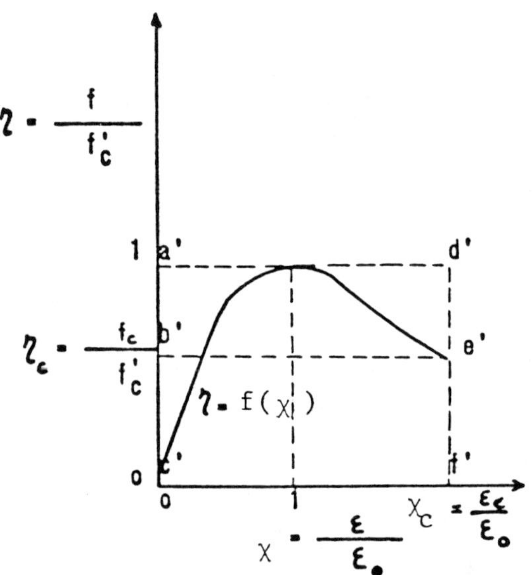

FIGURE 14. Dimensionless stress-strain curve for concrete.

then

$$\lambda = \frac{\epsilon_c}{k} \tag{36}$$

The moment capacity M of the section in Figure 13 is given by

$$M = M_1 + M_2$$

where M_1 is the moment of resistance of the singly reinforced section with tension steel $A_{s2} = A_s - A_{s1}$ and M_2 is the contribution of the steel couple due to forces $A_{s1} f_s$ in the remaining tension steel (A_{s1}) and force $A_s' f_s'$ in the compression steel (A_s') at top. Using equilibrium of forces, it can be shown that

$$k = \frac{x'}{d} = \frac{f_s p - f_s' p'}{Z' f_c'}$$

where

$$p = \frac{A_s}{bd}$$

$$p' = \frac{A_s'}{bd}$$

Therefore,

$$\lambda = \frac{\epsilon_c Z' f_c'}{f_s p - f_s' p'} = \frac{\epsilon_0 f_c' A_{\eta x}}{f_s p - f_s' p'} \tag{37a}$$

For singly reinforced beams, $p' = 0$ and, therefore,

$$\lambda = \frac{\epsilon_0 f_c' A_{\eta x}}{f_s p} \tag{37b}$$

It can be shown from a consideration of moments (Figures 13d and 13e) that

$$M_1 = (pf_s - p'f_s')\, jbd^2$$

and

$$M_2 = p'bdf_s'(d - d')$$

Let

$$M = M_1 + M_2 = R\, bd^2$$

then

$$R = p'f_s'\left(1 - \frac{d'}{d}\right) + (pf_s - p'f_s')\, j \tag{38}$$

It can be shown that

$$j = 1 - \left(\frac{f_s p - f_s' p'}{f_c'}\right) \frac{\beta x'}{Z' kd} \tag{39}$$

From Equations (33) and (34a), one obtains

$$\frac{Bx'}{Z' kd} = \frac{x_c A_{\eta x} - B_{\eta x}}{A_{\eta x}^2} \tag{40}$$

Substituting from Equation (40) in Equation (39) and from

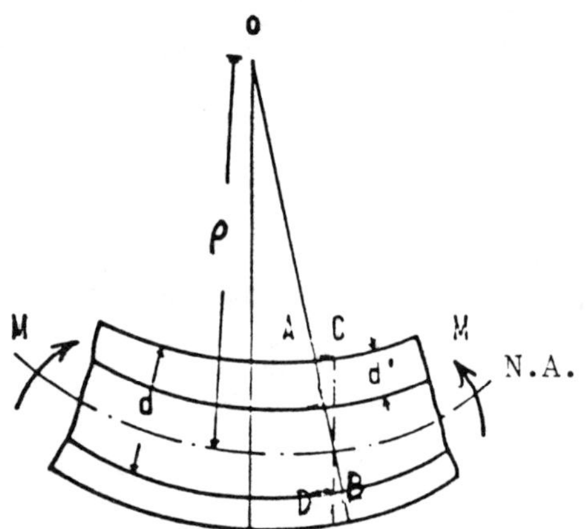

FIGURE 15. A beam segment under pure flexure.

Equation (39) in Equation (38), one gets

$$R = \left\{ pf_s \left[1 - \left(\frac{f_s p - f_s' p'}{f_c'} \right) \left(\frac{x_c A_{\eta x} - B_{\eta x}}{A_{\eta x}^2} \right) \right] + p' f_s' \left[-\frac{d}{d'} + \left(\frac{f_s p - f_s' p'}{f_c'} \right) \left(\frac{x_c A_{\eta x} - B_{\eta x}}{A_{\eta x}^2} \right) \right] \right\} \tag{41a}$$

and for singly reinforced beams, $p' = 0$, therefore,

$$R = pf_s \left[1 - \left(\frac{f_s p}{f_c'} \right) \left(\frac{A_{\eta x} x_c - B_{\eta x}}{A_{\eta x}^2} \right) \right] \tag{41b}$$

Equation (41a) can be rewritten as

$$R = pf_s - p' f_s' \left(\frac{d}{d'} \right) - \left(\frac{f_s p - f_s' p'}{\sqrt{f_c'}} \right)^2 \left(\frac{x_c A_{\eta x} - B_{\eta x}}{A_{\eta x}^2} \right) \tag{41c}$$

It is obvious that for a given section R will be a maximum if $(x_c A_{\eta x} - B_{\eta x})/A_{\eta x}^2$ is a minimum.

Hsu and Mirza [29] used the generalized expression for the complete stress-strain curves for plain concrete (Equation (2)) and developed a plot of $(x_c A_{\eta x} - B_{\eta x})/A_{\eta x}^2$ against x_c (Figure 16). It is seen that $(x_c A_{\eta x} - B_{\eta x})/A_{\eta x}^2$ has a minimum value of 0.5025 at $x_c = 1.35$ (Figure 16), which leads to the following equation for a singly reinforced beam:

$$R = \left(1 - 0.5025 \frac{pf_y}{f_c'} \right) pf_y \tag{42}$$

where f_s is taken as f_y in the ultimate strength design as compared with the corresponding ACI equation from the ACI Code [37,45] and the corresponding equation from the Japanese Code (Umemura [46]):

$$R = \left(1 - 0.59 \frac{pf_y}{f_c'} \right) pf_y \tag{43}$$

$$R = \left(1 - 0.527 \frac{pf_y}{f_c'} \right) pf_y \tag{44}$$

For doubly reinforced sections the maximum value of R is given by

$$R = pf_y \left[1 - 0.5025 \frac{f_y (p - p')}{f_c'} \right] + p' f_y' \left[-\frac{d}{d'} + 0.5025 \frac{f_y (p - p')}{f_c'} \right] \tag{45}$$

as compared with the corresponding expression from the ACI Code [37,45] which gives

$$R = pf_y \left[1 - 0.59 \frac{f_y (p - p')}{f_c'} \right] + p' f_y' \left[-\frac{d}{d'} + 0.59 \frac{f_y (p - p')}{f_c'} \right] \tag{46}$$

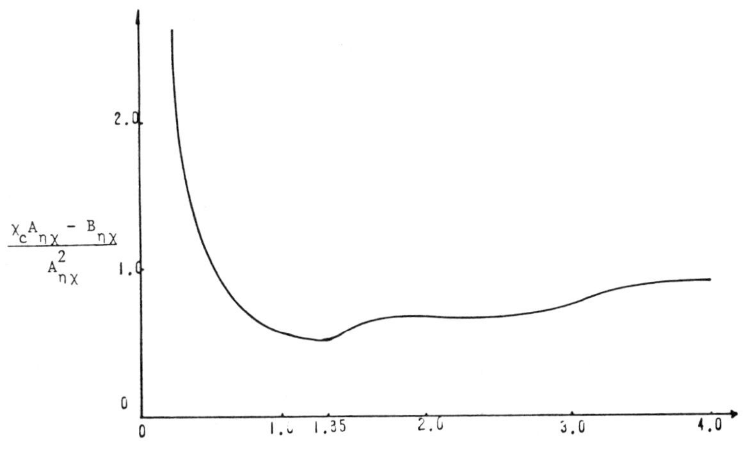

FIGURE 16. Plot of $[(x_c A_{\eta x} - B_{\eta x})/A_{\eta x}^2]$ vs. x_c.

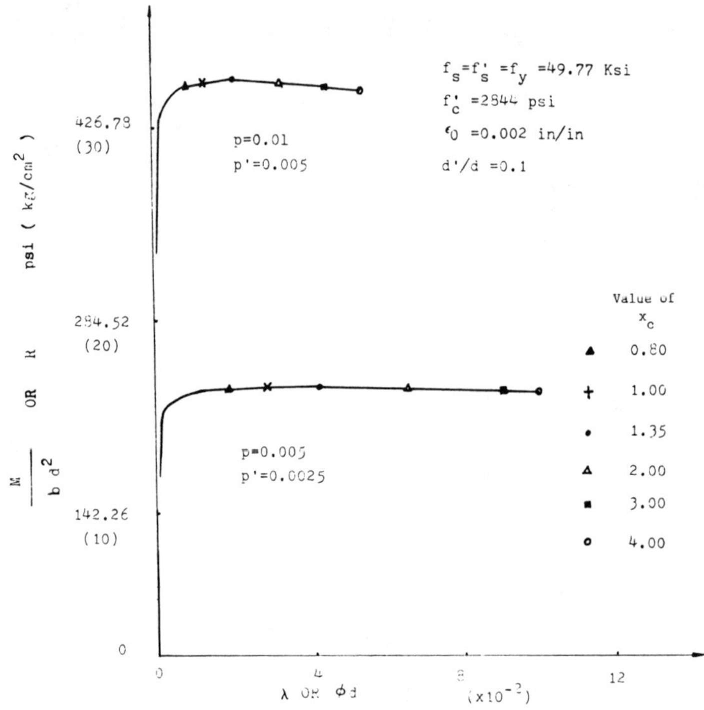

FIGURE 17a. Theoretical moment-curvature diagrams for doubly reinforced section.

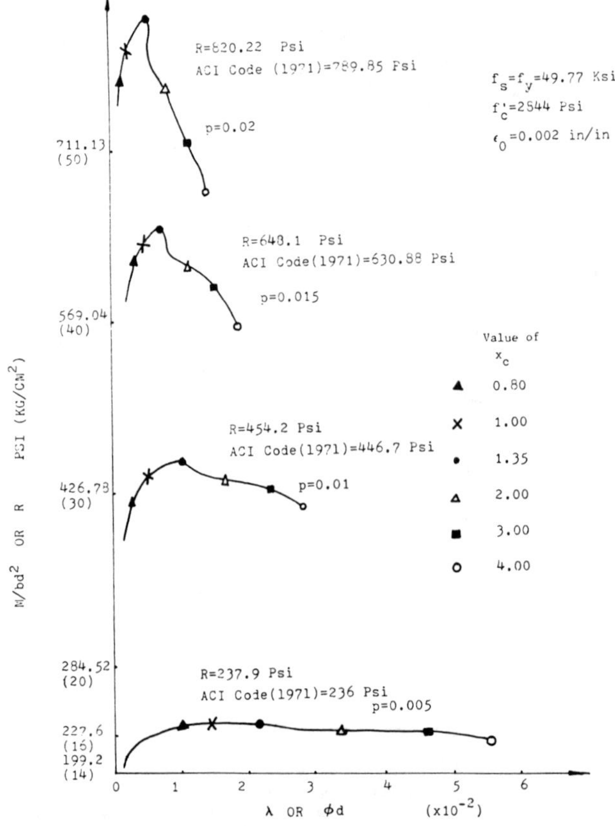

FIGURE 17b. Theoretical moment-curvature curves for singly-reinforced section.

Typical examples of the moment-curvature expression for this method can be seen in Figure 17 in which the stress-strain curve for reinforcing steel was assumed to be rigid-plastic for simplicity. Equations (37) and (41) define a point on the moment-curvature diagram of reinforced concrete section by assigning different values to x_c and f_s. The computation will be terminated automatically if ϵ_c reaches ϵ_u where ϵ_u is the maximum extreme compressive concrete strain.

Computer Analysis of Generalized Moment-Curvature Relationships for Singly and Doubly Reinforced Rectangular Sections Under Pure Flexure and/or Constant Axial Load—Method 2

INTRODUCTION

Hsu and Mirza [29] modified the numerical analysis presented by Mattock [19] and Khan [20] and extended the computer program (see Reference [152]) to include the descending branch of the moment-curvature curve which had been ignored by Mattock [19] and Khan [20]. Also, the effect of the constant axial load on the section is included. The principal features of the computer analysis are as follows:

1. The complete stress-strain curve for plain concrete, i.e., Equation (2) was used.
2. The steel stress-strain curve was used as shown in either Figure 18a or Figure 18b.
3. The modulus of elasticity of concrete was taken to be $33w^{1.5}\sqrt{f_c'}$.
4. The following step-by-step procedures were used: (1) Select any steel strain, ϵ_s, in tension. (2) Locate the neutral axis by solving the force equilibrium and strain energy equations. (3) Obtain the concrete compressive strain ϵ_c at the extreme fiber assuming that plane section remains plane. (4) Evaluate the bending moment and the curvature values corresponding to the selected value of ϵ_s. (5) Using the strain and stress distributions obtained, check equilibrium across the section. Generally there will be good agreement on the ascending branch and some part of the descending branch of the curve. However, there will be some violation of the equilibrium equations on the latter part of the descending branch. The computer program continues to evaluate the moment-curvature characteristics until this difference is within suitable prescribed limits. A larger discrepancy will automatically stop the computer program and section is assumed to have attained the ultimate loading stage.

It should be noted that Equation (2) expresses the complete stress-strain curve for unconfined concrete. Any complete stress-strain equations of the concrete can easily be used in the computer program instead of Equation (2) if the effect of confinement of stirrups on concrete becomes significant.

(a)

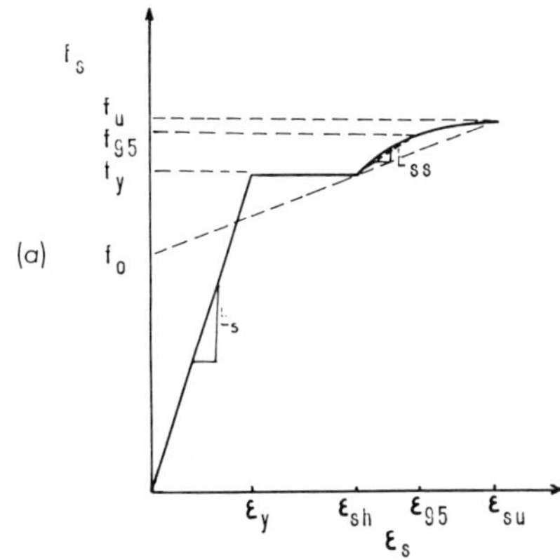

(b)

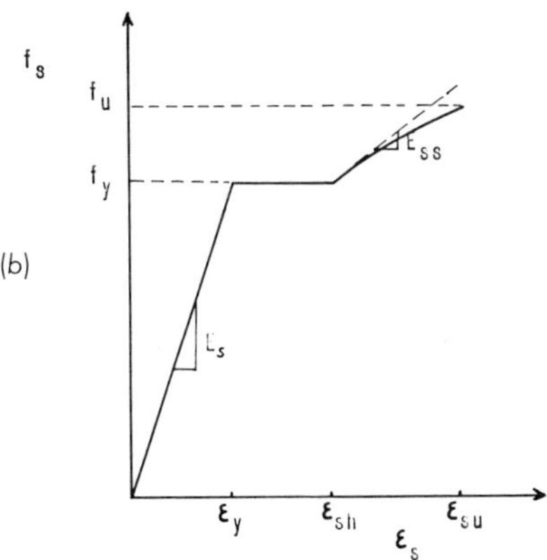

FIGURE 18. Steel stress-strain curves: (a) Experimental curve; (b) Idealized curve.

ANALYSIS

The analysis is involved in the three stages as follows:

Before Cracking (see Figure 19)

$$C_c = \tfrac{1}{2} f_c b x' = \tfrac{1}{2} b x' f_t \left(\frac{x'}{D - x'}\right) \tag{47}$$

$$C_s = f_c \left(\frac{x' - d'}{x'}\right) (n - 1)A_s'$$

$$= \left(\frac{x' - d'}{x'}\right) (n - 1)A_s' f_t \left(\frac{x'}{D - x'}\right) \tag{48}$$

$$T_s = f_c \left(\frac{d - x'}{x'}\right) (n - 1)A_s$$

$$= \left(\frac{d - x'}{x'}\right) (n - 1)A_s f_t \left(\frac{x'}{D - x'}\right) \tag{49}$$

$$C_T = \tfrac{1}{2} b(D - x')f_t \tag{50}$$

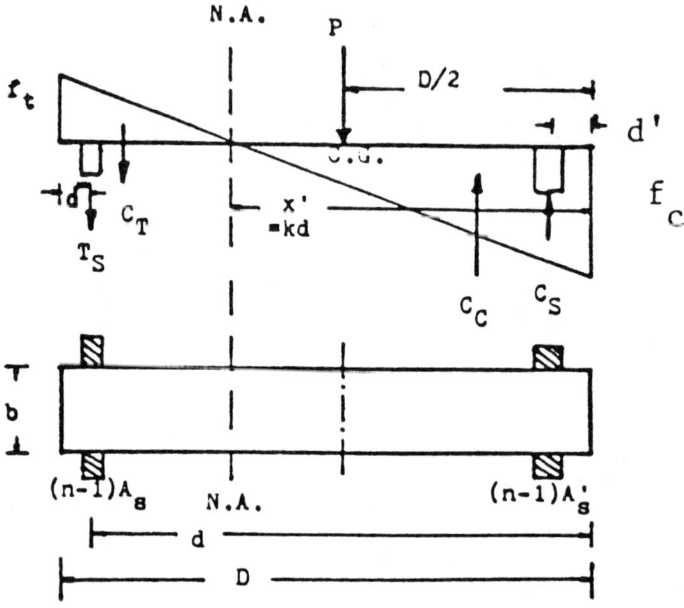

FIGURE 19. Section analysis: before cracking.

Then the condition of equilibrium can be written as

$$C_c + C_s = T_s + C_T + P \tag{51}$$

where P is a constant axial load.

Substituting from Equations (47), (48), (49) and (50) into Equation (51) gives:

$$x' = kd = \frac{(n-1)\, f_t(A_s'd' + A_sd) + \tfrac{1}{2} f_tbD^2 + PD}{(n-1)f_t(A_s' + A_s) + bf_tD + P} \tag{52}$$

The resisting moment about the neutral axis is then

$$M = C_c\frac{2}{3}\,x' + C_s(x' - d') + C_T\,\frac{2}{3}\,(D - x') + T_s(d - x') - P\left(x' - \frac{D}{2}\right) \tag{53}$$

where $P = C_c + C_s - T_s - C_T$.

Substituting from Equations (47), (48), (49) and (50) into Equation (53) gives:

$$M = \frac{1}{3}\,bf_t\left[\frac{(x')^3}{D - x'}\right] + (n-1)A_s'f_t\left[\frac{(x' - d')^2}{D - x'}\right] \tag{54a}$$

$$+ \frac{1}{3}\,bf_t(D - x')^2 + (n-1)A_sf_t\left[\frac{(d - x')^2}{D - x'}\right] - P\left(x' - \frac{D}{2}\right)$$

and

$$\phi_{cr} = \frac{\epsilon_s}{(1 - k)d} \tag{54b}$$

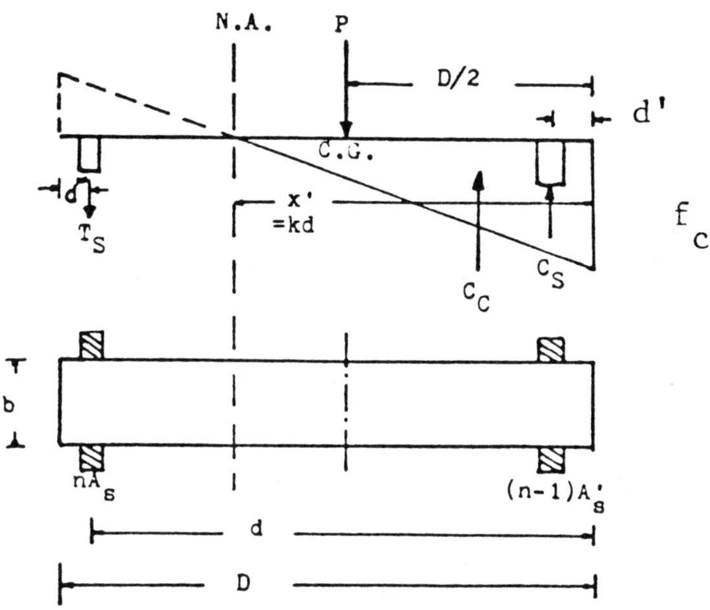

FIGURE 20. Section analysis: immediately after cracking.

Immediately After Cracking

As shown in Figure 20, force equilibrium equation can be written as:

$$C_c + C_s = T_s + P$$

where

$$C_c = \tfrac{1}{2} bx' f_t \left(\frac{x'}{D - x'}\right) = \tfrac{1}{2} f_c bx'$$

$$C_s = (n - 1)A_s' f_t \left(\frac{x' - d'}{D - x'}\right) = f_c \left(\frac{x' - d'}{x'}\right) (n - 1)A_s'$$

and

$$T_s = \left(\frac{d - x'}{D - x'}\right) nA_s f_t = f_c \left(\frac{d - x'}{x'}\right) nA_s$$

then substituting from the above equations into the force equilibrium equation which gives

$$B_1k^7 + B_2k^6 + B_3k^5 + B_4k^4 + B_5k^3 + B_6k^2 + B_7k + B_8 = 0 \tag{55}$$

where

$$k = \frac{x'}{d}$$

and

$$B_1 = B_2 = B_3 = B_4 = B_5 = 0$$

$$B_6 = \tfrac{1}{2} bf_t d^2$$

$$B_7 = [(n - 1)A_s' f_t + nA_s f_t + P]\, d$$

$$B_8 = -\,[(n - 1)A_s' f_t d' + nA_s f_t d + PD]$$

The resisting moment about the neutral axis can then be obtained as:

$$M = \frac{bf_t}{3}\left[\frac{(x')^3}{D - x'}\right] + (n - 1)A_s' f_t \left[\frac{(x' - d')^2}{D - x'}\right] + nA_s f_t \left[\frac{(d - x')^2}{D - x'}\right[- P\left(x' - \frac{D}{2}\right) \tag{56}$$

Since

$$\frac{\epsilon_s'}{\epsilon_c} = \frac{(k + j - 1)}{k}$$

$$\frac{\epsilon_s}{\epsilon_c} = \frac{(d - kd)}{kd} = \frac{1 - k}{k}$$

$$\frac{f_s'}{f_c} = \frac{n(k + j - 1)}{k}$$

$$\frac{f_s}{f_c} = \frac{E_s\epsilon_s}{E_c\epsilon_c} = n\left(\frac{1 - k}{k}\right)$$

Equation (56) gives

$$M = f_c\left[\frac{b(x')^2}{3}\right] + A_s'(x' - d')f_c\left[\frac{n(k + j - 1)}{k}\right] + A_s(d - x')f_c\left[\frac{n(1 - k)}{k}\right] - P\left(x' - \frac{D}{2}\right) \tag{57}$$

or

$$f_c = \frac{M + P\left(x' - \frac{D}{2}\right)}{\frac{b(x')^2}{3} + A_s'(x' - d')\left[\frac{n(k + j - 1)}{k}\right] + A_s(d - x')\, n\left[\frac{(1 - k)}{k}\right]}$$

Since $A_s f_s = A_s' f_s' + \frac{1}{2} f_c bkd - P$
Therefore,

$$f_s = \left[\frac{(p' f_s' + \frac{1}{2} k f_c)}{p}\right] - \frac{P}{A_s}$$

or

$$f_s = \left[\frac{n(1 - k)}{k}\right] f_c$$

Since

$$\epsilon_s = \frac{f_s}{E_s}$$

Therefore

$$\phi_{cr(a)} = \frac{\epsilon_s}{(1 - k)d} \tag{58}$$

Between Cracking and Ultimate

CASE A: $(\epsilon_c \le \epsilon_0)$

The compressive force C_{c1} in the concrete (Figure 21c) is given by:

$$C_{c1} = b\int_0^{kd} f\,dy$$

Substituting for f from Equation (2),

$$C_{c1} = bf_c'\left\{0.6366\, y_1\left[\frac{\left(1.5708\,\frac{kd}{y_1}\right)^2}{2} - \frac{\left(1.5708\,\frac{kd}{y_1}\right)^4}{24} + \frac{\left(1.5708\,\frac{kd}{y_1}\right)^6}{720}\right] -\right.$$

$$- 0.2 \left(\frac{k^2 d^2}{y_1} + kd + y_1 \right) \left[2 - \frac{kd}{y_1} + \frac{\left(1 - \frac{kd}{y_1}\right)^2}{2} + \frac{\left(1 - \frac{kd}{y_1}\right)^3}{6} + \frac{\left(1 - \frac{kd}{y_1}\right)^4}{24} \right.$$

$$\left. + \frac{\left(1 - \frac{kd}{y_1}\right)^5}{120} \right] + 0.5437\, y_1 - 0.0667 \frac{(kd)^3}{y_1^2} + 0.1 \frac{(kd)^2}{y_1} \Bigg\}$$

where

$$\frac{\epsilon}{\epsilon_o} = \frac{y}{y_1}$$

(see Figure 21b), and

$$y_1 = (1 - k)\, d \frac{\epsilon_0}{\epsilon_s}$$

The distance to the centroid of the concrete compression block from the neutral axis is given by:

$$x_1 = \frac{bf_c'}{C_{c1}} \Bigg\{ 0.4053\, y_1^2 \left[- \frac{\left(1.5708 \frac{kd}{y_1}\right)^3}{6} + \frac{\left(1.5708 \frac{kd}{y_1}\right)^5}{120} - \frac{\left(1.5708 \frac{kd}{y_1}\right)^7}{5040} \right.$$

$$\left. + \frac{\left(1.5708 \frac{kd}{y_1}\right)^3}{2} - \frac{\left(1.5708 \frac{kd}{y_1}\right)^5}{24} + \frac{\left(1.5708 \frac{kd}{y_1}\right)^7}{720} \right] + 0.3437\, y_1^2$$

$$\left(- \frac{k^3 d^3}{y_1^3} - \frac{2k^2 d^2}{y_1^2} - \frac{4kd}{y_1} - 4 \right) \left[1 - \frac{kd}{y_1} + \frac{\left(\frac{kd}{y_1}\right)^2}{2} - \frac{\left(\frac{kd}{y_1}\right)^3}{6} - \frac{\left(\frac{kd}{y_1}\right)^5}{120} + \frac{\left(\frac{kd}{y_1}\right)^4}{24} \right]$$

$$+ 1.375\, y_1^2 - 0.05 \frac{k^4 d^4}{y_1^2} + 0.0667 \frac{k^3 d^3}{y_1} \Bigg\}$$

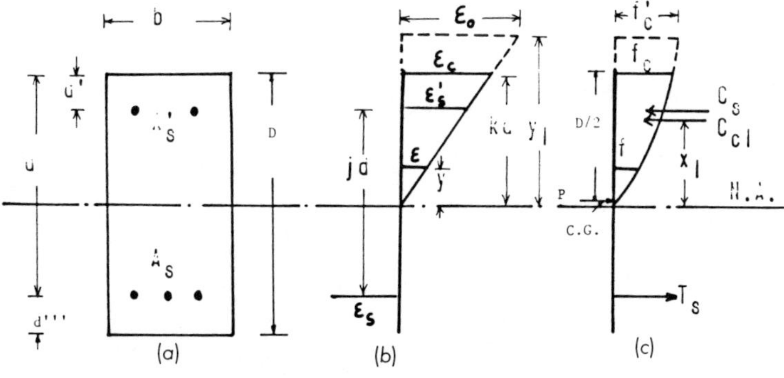

FIGURE 21. Section analysis: Case A.

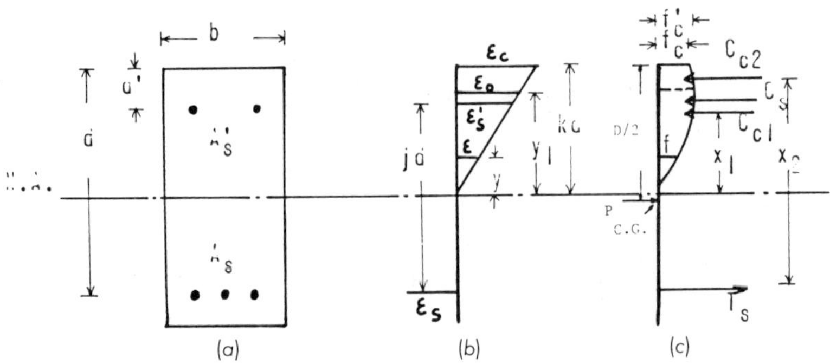

FIGURE 22. Section analysis: Case B.

Also,

$$\epsilon_s' = \frac{(kd - d')\epsilon_s}{(1 - k)d}$$

and

$$\epsilon_y = \frac{f_y}{E_s}, \epsilon_y' = \frac{f_y'}{E_s'} = \frac{f_y'}{E_s}$$

(assume $E_s = E_s'$)

so $f_s' = E_s'\epsilon_s' = E_s\epsilon_s'$ if $\epsilon_s' \leq \epsilon_y'$

or $f_s' = f_y'$ if $\epsilon_s' \geq \epsilon_y'$

Let $C_s = A_s'f_s'$

$f_s = E_s\epsilon_s$ if $\epsilon_s \leq \epsilon_y$

or $f_s = f_y$ if $\epsilon_y \leq \epsilon_s \leq \epsilon_{sh}$

or $f_s = f_y + (\epsilon_s - \epsilon_{sh}) E_{ss}$ if $\epsilon_s \geq \epsilon_{sh}$

where ϵ_{sh} and E_{ss} are defined in Figure 18.

Let $T_s = A_sf_s$

then the condition of equilibrium can be written as

$$P + T_s = C_{c1} + C_s$$

which gives

$$B_1k^7 + B_2k^6 + B_3k^5 + B_4k^4 + B_5k^3 + B_6k^2 + B_7k + B_8 = 0 \quad (55)$$

where B_1, B_2, B_3, B_4, B_5, B_6, B_7, and B_8 can be found in the Appendix.

Equation (55) is a polynomial in k and can be solved for k. From the compatibility equation,

$$\frac{F\epsilon_c}{k} = \frac{\epsilon_s}{1 - k} \text{ or } \epsilon_c = \frac{\epsilon_s k}{F(1 - k)} \quad (59)$$

Here the compatibility factor F is assumed to be one (i.e. there is a perfect bond between steel and concrete). The curvature ϕ can then be obtained from the equation:

$$\phi = \frac{\epsilon_c}{kd} \quad (60)$$

and the moment M about the neutral axis is given by:

$$M = C_{c1}\,[x_1 + d(1 - k)] + C_sjd - P\left(d - \frac{D}{2}\right) \quad (61)$$

CASE B: ($\epsilon_c \geq \epsilon_0$)

The concrete compressive block is divided into two parts as shown in Figure 22c. The calculations for the concrete compressive forces C_{c1} and C_{c2} are as follows:

$$C_{c1} = b\int_0^{y_1} f dy = 0.6168\, bf_c'y_1 \quad (62a)$$

and

$$C_{c2} = b\int_{y_1}^{kd} f dy$$

$$= bf_c'\left\{0.226\,kd + 1.0785\frac{k^2d^2}{y_1} - 0.6357\frac{k^3d^3}{y_1^2} + 0.149\frac{k^4d^4}{y_1^3} - 0.0126\frac{k^5d^5}{y_1^4} - 0.8052y_1\right\} \quad (62b)$$

From Figure 22c the distance between the compressive force resultant and the tension steel is given by:

$$x_2 = d(1 - k) + \left\{\frac{bf_c'}{C_{c2}}\left[0.113k^2d^2 + 0.719\frac{k^3d^3}{y_1} - \right.\right.$$

$$- 0.47675 \frac{k^4 d^4}{y_1^2} + 0.1192 \frac{k^5 d^5}{y_1^3} - 0.0105 \frac{k^6 d^6}{y_1^4} - 0.46395 y_1^2 \Bigg]\Bigg\}$$

The equilibrium of forces gives:

$$P + T_s = C_{c1} + C_{c2} + C_s \quad (63)$$

from which k can be evaluated.

In case the compression reinforcement has not yielded, i.e., $\epsilon_s' < \epsilon_y'$, Equation (63) would have the same mathematical form as Equation (55) except that the coefficients B_1, B_2, B_3, B_4, B_5, B_6, B_7 and B_8 are different and are detailed in the Appendix.

If the compression steel has yielded, i.e., $\epsilon_s' \geq \epsilon_y'$ then $C_s = A_s' f_y'$ and the coefficients B_1, B_2, B_3, B_4, B_5, B_6, B_7 and B_8 in Equation (55) are as shown in the Appendix. The values of ϵ_c and ϕ are determined from Equations (59) and (60) and the resisting moment M about $N.A.$ is given by:

$$M = C_{c1}[x_1 + d(1 - k)] + C_{c2}x_2 + C_s jd - P\left(d - \frac{D}{2}\right) \quad (64)$$

The typical example of the effect of the constant axial load on the moment-curvature curves is shown in Figure 23. As indicated in Figure 23, the larger the axial load, the smaller the ductility presents at the structural concrete section. Figure 24 shows the flow diagram for the calculation of the moment-curvature relationships using Method 2.

ANALYSIS OF TEST RESULTS

Mattock [19] reported the results of tests on simply-supported reinforced concrete beams under two-point loading. He analyzed the data and proposed a method to calculate the rotation capacity of a hinging region in a reinforced concrete beam. Hsu [152] analyzed the results of six beams for which the moment-curvature diagrams were reported in detail. Details of these beams are reproduced in Table 1.

Table 2 shows Mattock's [19] experimental results for the moment capacities at yield and "ultimate" load (called "maximum" load here) along with calculated values using the following methods: (1) Mattock's [19] method and (2) Hsu's method [29].

It should be noted that the experimental moment-curvature curves obtained by Mattock were based on the measurement from the linear differential transformer gages mounted on the tension and compression sides of concrete over half of effective depth ($d/2$) on either side of the mid-span. Therefore, the curvature obtained by Mattock is an average curvature over a length equal to the effective depth. Mattock calculated the values of curvature from experimental data using the equation

$$\frac{\epsilon_c + \epsilon_s}{d} = \frac{\epsilon_c}{kd}$$

The results reported by Mattock [19] correspond to the stage when the maximum load was attained during the test under an increasing load, or to the stage at which the beam failed after some time at a constant load. The value of maximum concrete compressive strain used was

$$\epsilon_u = 0.003 + \frac{0.5}{Z} \quad (65)$$

where Z is a shear span of the beam. The second term on the right-hand side was dropped for the two-point loading case, giving

$$\epsilon_u = 0.003 \quad (66)$$

It should be noted that the present computer program developed can calculate not only the maximum moment M_m resisted by the beam section but also the value of the moment of resistance at failure M_u. The means of the ratios of the experimental yield moment capacity of the beams subjected to two-point load to the values calculated using Mattock's method and Hsu's computer programs are 0.998 and 1.004, respectively (see Table 2). The standard deviation for these cases is 0.029 and 0.028, respectively.

The ratios of the experimental ultimate bending strength to values calculated using Mattock's method, and the maximum moment M_m and the ultimate moment M_u capacities from Hsu's computer program are detailed in Table 2. The means of ratios for the two-point loading case are 1.10 and 1.038, while the standard deviations are 0.104 and 0.086.

The curvatures at three stages corresponding to (1) the yield moment, (2) the maximum, and (3) the ultimate moment are detailed in Table 3 for the cases of the two-point loading.

The experimental values of curvatures reported by Mattock correspond to a stage when the maximum load was attained by the beam or the instant at which the beam failed at constant load after some time.

The ratios of the experimental curvatures (measurement at mid-span) and the values calculated using Mattock's method and Hsu's computer program are detailed in Table 3. The means of these ratios for the two-point loading case work out to be 1.14 and 1.029, respectively, while the standard deviations are 0.080 and 0.049, respectively.

The details of results for curvature at maximum load and at failure have been shown in Table 3. The ductility ratio which is defined as the ratio of the curvature at failure to the curvature at yield, has been detailed in Table 4.

The comparisons among Mattock's experimental and cal-

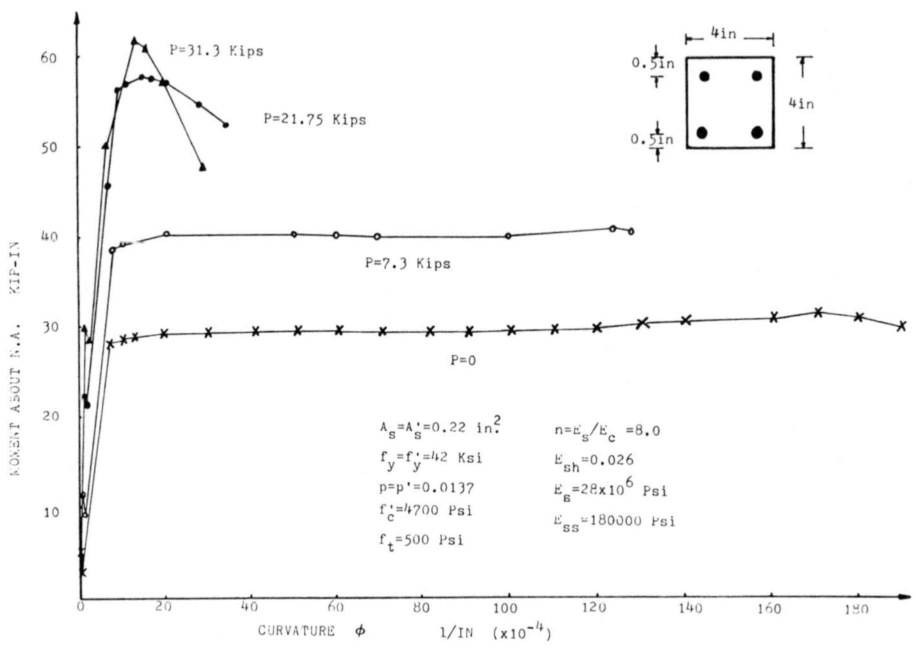

FIGURE 23. Effect of constant axial load on moment-curvature curves.

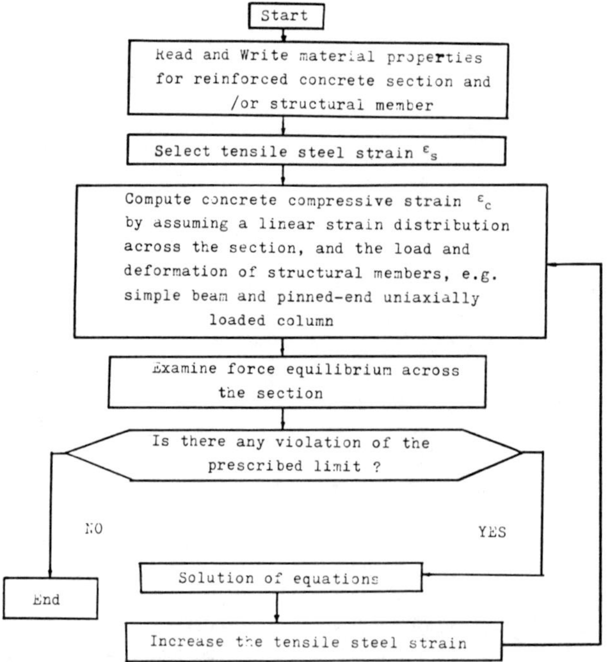

FIGURE 24. Flow diagram: generalized moment-curvature relationship (Method 2: deformation method).

TABLE 1. Properties of Mattock's Test Specimens

Loading Case	Speci. No.	b (in)	t (in)	d (in)	d′ (in)	Z^+ (in)	f'_c (psi)	f^*_t (psi)	A_s (in)	A'_s (in^2)	$E_s \times 10^6$ (psi)	f_y (ksi)	ϵ_{sh}	E_{ss} (ksi)	f'_y (ksi)	$n = E_s/E_c$
Two-Point Loading Case	C2A	6	11	10	0.625	36	4140	500	0.88	0.22	28.2	47.3	0.01	910	49.5	7.58
	C2B	6	11	10	0.625	36	4060	500	0.88	0.	28.2	46.8	0.01	910	0.	7.66
	C5B	6	11	10	0.625	36	4110	500	1.76	0.	28.2	47.0	0.01	910	0.	7.65
	C5A	6	11	10	0.625	36	4190	500	1.76	0.22	28.2	46.6	0.01	910	51.1	7.54
	D2A	6	21	20	0.625	72	3695	500	1.76	0.22	28.2	44.6	0.01	920	51.1	8.05
	D4A	6	21.7	20	0.625	72	4010	500	3.52	0.22	28.2	44.6	0.01	920	51.2	7.69
Single Point Loading Case	C3	6	11	10	0.625	110	3710	500	0.88	0.22	28.2	47.8	0.01	910	49.5	8.15
	C5	6	11	10	0.625	55	3390	500	1.76	0.22	28.2	47.6	0.01	910	48.5	8.40
	C6	6	11	10	0.625	110	3970	500	1.76	0.22	28.2	46.3	0.01	910	48.5	7.78
Remark	* f_t = assumed to be 500 psi. + 'Z' denotes shear span in simply supported reinforced concrete beams															

TABLE 2. Moments at Yield and at Maximum and at Ultimate (Two-Point Loading Case).

	Tested by Mattock		Calculated by Mattock		Hsu's Analysis*			Comparisons				
								At Maximum			At Yield	
Speci. No.	M_y (1)	M_m (2)	M_y (3)	M_u (4)	M_y (5)	M_m (6)	M_u (7)	$\frac{M_m(2)}{M_u(4)}$	$\frac{M_m(2)}{M_m(6)}$	$\frac{M_m(2)}{M_u(7)}$	$\frac{M_y(1)}{M_y(3)}$	$\frac{M_y(1)}{M_y(5)}$
C2A	368	474	369	404	371.6	454.8	453.8	1.17	1.042	1.045	0.997	0.991
C2B	380	476	361	371	360.6	398.3	394.9	1.28	1.195	1.205	1.053	1.054
C5B	677	709	696	663	675.6	680.3	598.3	1.07	1.04	1.185	0.98	1.002
C5A	685	720	699	693	694.3	708.3	650.3	1.04	1.016	1.107	0.98	0.986
D2A	1353	1505	1386	1472	1388.3	1608.	1603.3	1.02	0.936	0.939	0.976	0.975
D4A	2666	2666	2666	2610	2624.2	2665.	2416.3	1.02	1.00	1.103	1.006	1.016
Mean Value								1.10	1.038	1.097	0.998	1.004
Standard Deviation								0.104	0.086	0.097	0.029	0.028
Remark	Unit of moment: Kip-in. * Hsu's analysis: based on Method 2											

TABLE 3. Curvatures at Yield and at Maximum and at Ultimate (Two-Point Loading Case).

	Tested by Mattock ($\times 10^{-4}$)		Calculated by Mattock ($\times 10^{-4}$)		Hsu's Analysis ($\times 10^{-4}$)			Comparisons				
								At Maximum and Ultimate			At Yield	
Speci. No.	ϕ_y (1)	ϕ_m (2)	ϕ_y (3)	ϕ_u (4)	ϕ_y (5)	ϕ_m (6)	ϕ_u (7)	$\frac{\phi_m(2)}{\phi_u(4)}$	$\frac{\phi_m(2)}{\phi_m(6)}$	$\frac{\phi_m(2)}{\phi_u(7)}$	$\frac{\phi_y(1)}{\phi_y(3)}$	$\frac{\phi_y(1)}{\phi_y(5)}$
C2A	2.9	26.3	2.6	16.2	2.8	30.62	31.17	1.62	0.86	0.844	1.12	1.06
C2B	2.9	26.8	2.6	12.8	2.9	25.3	26.4	2.09	1.06	1.015	1.12	1.00
C5B	3.7	11.7	3.1	6.4	3.7	5.2	14.9	1.83	2.25	0.785	1.19	1.00
C5A	3.4	10.5	3.1	7.6	3.4	6.2	17.63	1.38	1.69	0.596	1.10	1.00
D2A	1.4	8.7	1.3	6.9	1.4	13.5	13.8	1.26	0.644	0.6304	1.04	1.00
D4A	1.9	4.0	1.5	3.6	1.7	2.84	8.1	1.13	1.55	0.494	1.27	1.118
Mean Value								1.552	1.342	0.727	1.14	1.029
Standard Deviation								0.364	0.598	0.19	0.08	0.049
Remark	Unit of Curvature:(1/in). The Curvature obtained by Mattock (1965) in tests can be referred to his paper.											

TABLE 4. Ductility Ratio (Two-Point Loading Case).

	$\phi_m(2)/\phi_y(1)$	$\phi_u(4)/\phi_y(3)$	$\phi_u(7)/\phi_y(5)$	$\phi_m(6)/\phi_y(5)$
Speci. No.	Tested by Mattock	Calculated by Mattock	Hsu's analysis	
C2A	9.1	6.2	11.13	10.93
C2B	9.2	4.92	9.103	8.72
C5B	3.2	2.065	4.03	1.41
C5A	3.1	2.45	5.185	1.82
D2A	6.2	5.307	9.86	9.64
D4A	2.1	2.4	4.765	1.67
Mean Value	5.483	3.895	7.345	5.69

culated results and the results computed by the presented analysis, Khan [20] and Kent and Park [17] are shown in Figures 25 and 26. The effect of strain-hardening (see Figure 18) is also shown in Figures 25 and 26.

Numerical method of Generalized Moment-Curvature Relationships for Arbitrary Reinforced Concrete Sections Under Pure Flexure and/or with Constant Axial Load—Method 3

INTRODUCTION

The numerical and computer method presented in this section is similarr to the one developed by Cranston [32]. However, Cranston's approach can determine the moment-curvature curves from zero load up to the maximum moment capacity only. The present method can calculate not

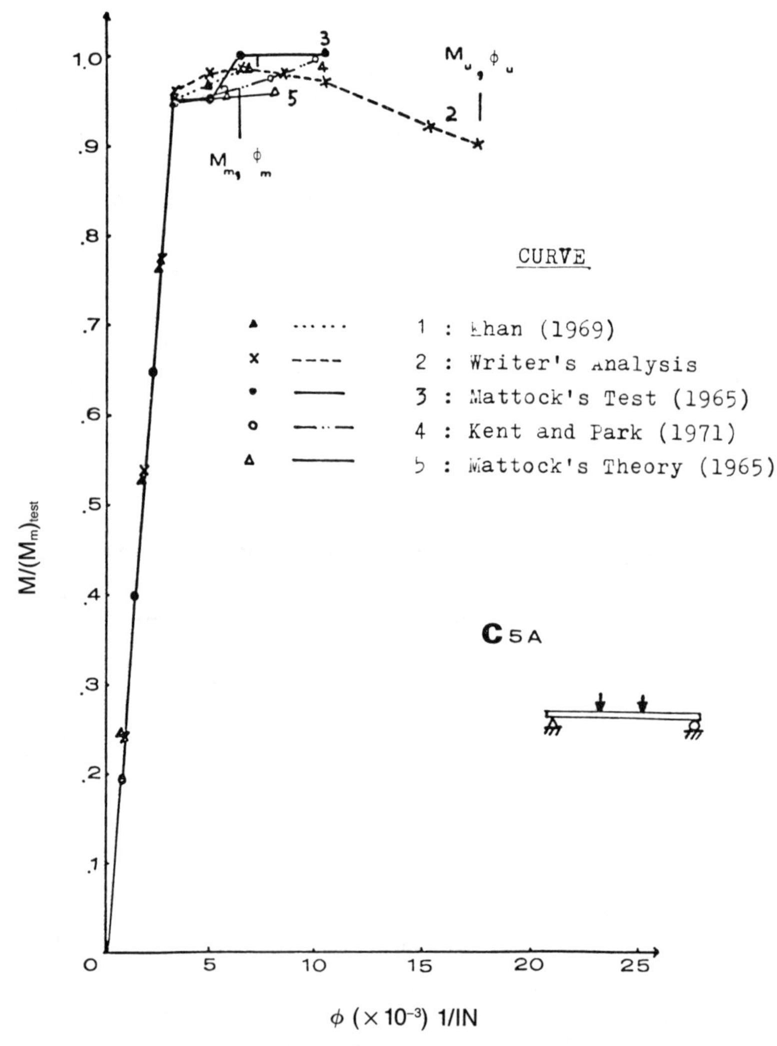

FIGURE 25. M – ϕ curves.

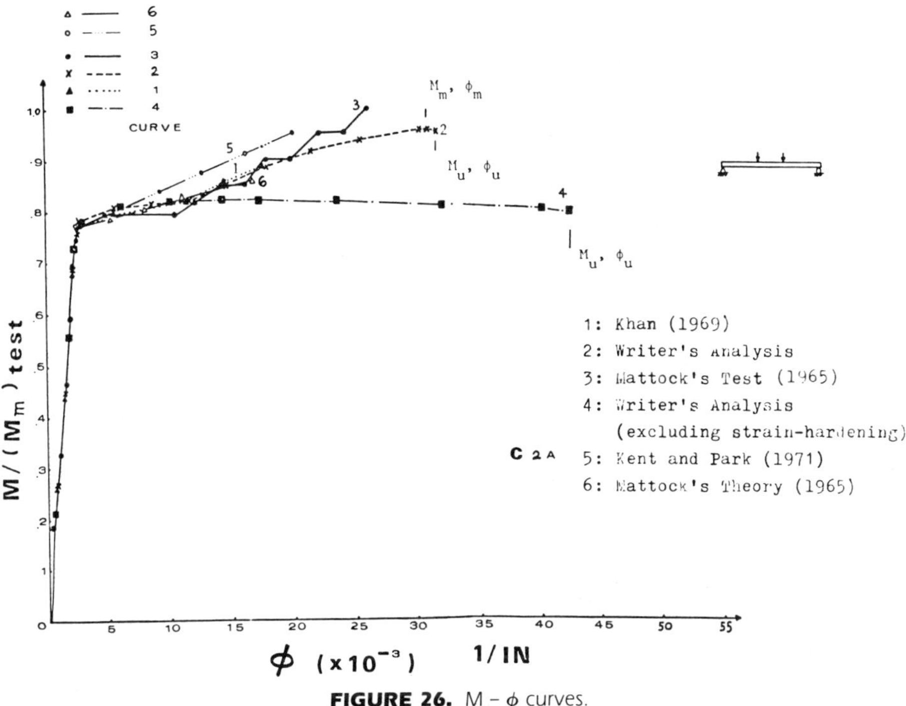

FIGURE 26. M – ϕ curves.

only the ascending branch of the curve but also the descending branch of the curve. The method developed here has an ability to use any arbitrary cross-sections and material properties for concrete and reinforcing steel. However, the numerical method is sometime handicapped by the convergence problem when the concrete strain reaches the descending branch of the concrete stress-strain diagram. The convergence problem can be solved by suitably modifying the proposed initial strain and curvature values.

THEORETICAL DEVELOPMENT

The following assumptions are made in this numerical and computer method: (1) the bending moments are applied about the principal axes of cross sections; (2) the longitudinal stress at a point is a function only of the longitudinal strain at that point. The effect of creep and shrinkage is ignored; (3) the stress-strain curves for the materials are known; (4) strain reversal does not occur; (5) the effect of deformation due to shear and torsion and impact effects are negligible; (6) plane section remains plane before and after bending; (7) the section does not buckle before the ultimate load is attained; (8) perfect bond exists between concrete and reinforcing steel.

A modification of Cranston et al. [33] stress-strain curves for concrete is shown in Figure 27. These curves account for confined and unconfined concrete elements, including the strain softening. The stress-strain curve for the reinforcing steel including the strain hardening effect is shown in Figure 28. This curve is idealized using a piecewise linear approximation. The cross section of the structural concrete member is divided into several elemental areas. Consider an element k with its centroid at point (x_k, y_k) referred to the principal axes of cross section. The strain ϵ_k across the element k can be assumed as follows (see [131, 132 and 145]):

$$\epsilon_k = \epsilon_p + \phi_x y_k + \phi_y x_k \tag{67}$$

and

$$P_{(c)} = \sum_{k=1}^{n} f_k a_k \tag{68a}$$

$$M_{x(c)} = \sum_{k=1}^{n} f_k a_k y_k \tag{68b}$$

$$M_{y(c)} = \sum_{k=1}^{n} f_k a_k x_k \tag{68c}$$

$P_{(c)}$ must now be compared with $P_{(s)}$ where $P_{(s)}$ is given constant axial load. This is done by calculating u' where u' is given by

$$u' = P_{(c)} - P_{(s)} \tag{69}$$

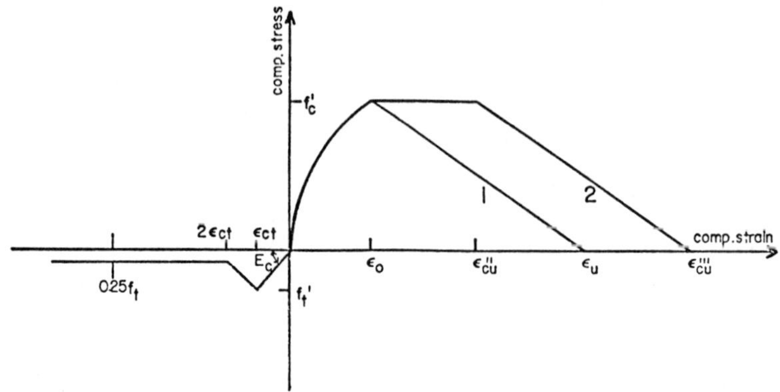

FIGURE 27. Concrete stress-strain curves.

where $P_{(s)}$ is a constant axial load. If $|u'|$ is sufficiently small, the values of ϵ_p and $M_{x(c)}$, $M_{y(c)}$ are taken to be correct and a solution has been found. Where $|u'|$ is not within the prescribed allowable limits, ϵ_p must be modified as follows:

$$\frac{dP_{(c)}}{d\,\epsilon_p} = \sum_{k=1}^{n} (E_t)_k a_k \tag{70}$$

where

$$P = P(\epsilon_p) \tag{71}$$

Using Taylor's expansion and retaining linear terms, one obtains:

$$P_{(s)} = P_{(c)} + \frac{dP_{(c)}}{d\epsilon_p}\,\delta\epsilon_p$$

Therefore,

$$-u' = \frac{dP_{(c)}}{d\epsilon_p}\,\delta\epsilon_p \tag{72}$$

The values of $\delta\epsilon_p$ can then be found from Equations (70) and (72), and the modified value for ϵ_p is given by:

$$\epsilon_{p\,(m+1)} = \epsilon_{p\,(m)} + \delta\epsilon_p \tag{73}$$

A special remark should be made for Equations (72) as follows: Based on the "deformation control" process, the termination of the computation as indicated in Group B and Group C in the section on Generalized Moment-Curvature Characteristics does not have any physical significance. The only definition, as suggested by Cranston [32], is that if a strain outside the range of a particular stress-strain curve, either in unconfined and confined concrete or reinforcing bars, the specimens are defined to have failed. Figure 29

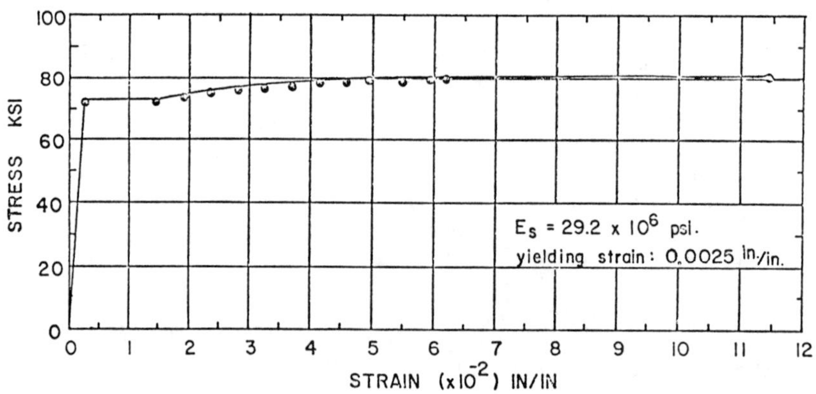

FIGURE 28. Typical steel stress-strain curve.

shows the histories of maximum concrete compressive strain at element 5 at the corner of the column section based on this approach. Figure 29 also shows a descending branch of the M-ϕ curve which is ignored by Cranston [32].

ANALYSIS OF TEST RESULTS

Mattock [19] reported the test results on simply-supported reinforced concrete beams under single-point loading (beam details are shown in Table 1). Hsu [152] used Method 2 and Method 3 to compare with the test results by Mattock. As mentioned before, the effect of stirrup in beam member on concrete stress-strain characteristics was studied in Method 3 and this effect was ignored in Method 2.

Figures 30, 31 and 32 show the different theoretical analyses and Mattock's experimental results for beams C_3, C_5 and C_6. It can be seen that Hsu's Method 3 is very close to Mattock's experimental results.

In general, the deformation behaviour of simply-supported beams under single-point loading is different from that of the two-point loading case. The critical section for the former case is exactly under the loading point where the steel plate confines the concrete on the compression side. Also the confinement of the concrete by stirrup near the critical section affects the deformation behaviour of beams.

INELASTIC ROTATION AND CURVATURE

Inelastic Rotation and Curvature in a Flexural Beam

INELASTIC ROTATION OVER A FINITE LENGTH

In dealing with under-reinforced concrete sections, it is extremely difficult to predict the value of the curvature at or very near the critical section after yielding of the steel reinforcement. The increment in moment between yield and maximum strength is small, usually less than 10%, and in view of the large increments in strain over this range it presents a practical difficulty to satisfy compatibility criterion in methods of analysis or design which require a knowledge of the curvature distribution at all sections. For the purposes of calculation it has been shown [123,124] that it is sufficient and convenient to consider that the inelastic rotation occurring in the immediate vicinity of any critical section may be concentrated at that section. This is equivalent to assuming that the critical sections are capable of being discontinuous, coupled with the assumption that the increment in moment resistance after the steel has yielded is negligible. This demonstrates the concept of plastic hinge in reinforced concrete members. The properties of the plastic hinge are not actually the properties of the individual critical section. Its behaviour is representative of either the member as a whole, or the behaviour of a length of member in any structural system. Burnett [125] stated that the actual moment-curvature for the critical section could not be used to evaluate the curvature distribution over the length of the member and hence to determine the total rotation; nor could this moment-curvature relation be used to determine the curvature for any identical cross-section in any different structural system. This, of course, increases the difficulties of applying any limit analysis method and puts forward the use of moment-rotation characteristics as the most important parameter involved. This limitation on moment-curvature characteristics can be seen very easily from the present investigation (previous section).

In an actual test of a beam under two-point loading, the "concrete crushing zones," which indicate the location of the discontinuity, invariably occur over only a part of the constant moment region. Secondly, if the above description were followed strictly for a beam under single-point loading, it would require that a finite curvature be present along an infinitely short length of beam and hence no rotation within the discontinuity. However, in actual tests, the concrete crushing zones occur over a finite length of the beam either underneath or to one side of the load.

The first of these occurrences can be explained by the fact

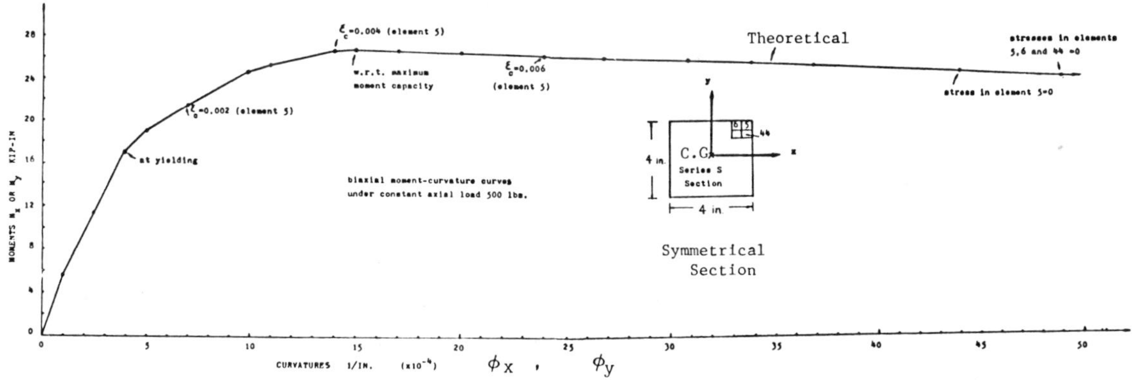

FIGURE 29. Moment-curvature curve under biaxial bending with a constant load.

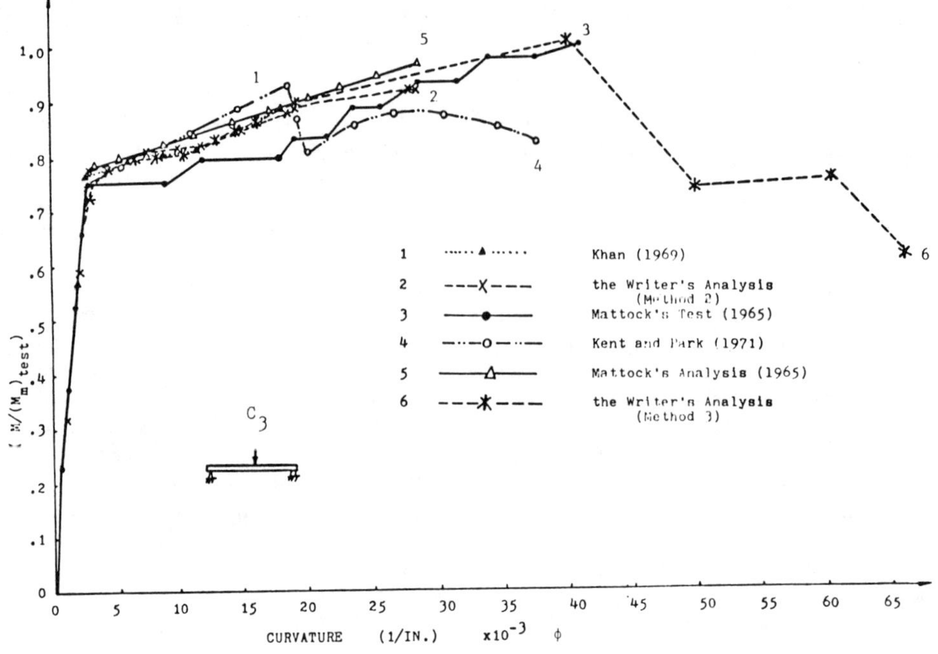

FIGURE 30. M – ϕ curves.

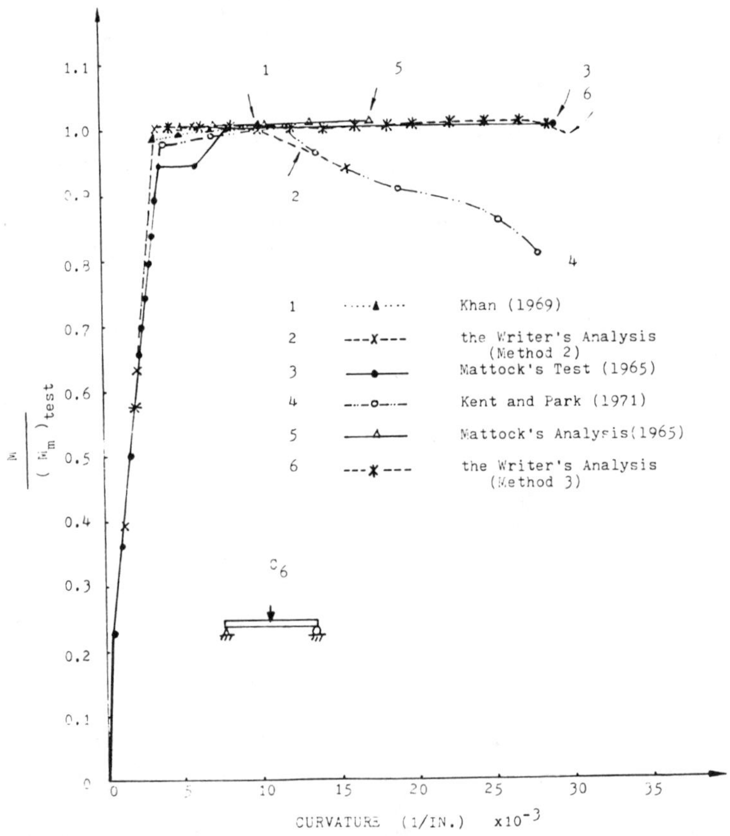

FIGURE 31. M – ϕ curves.

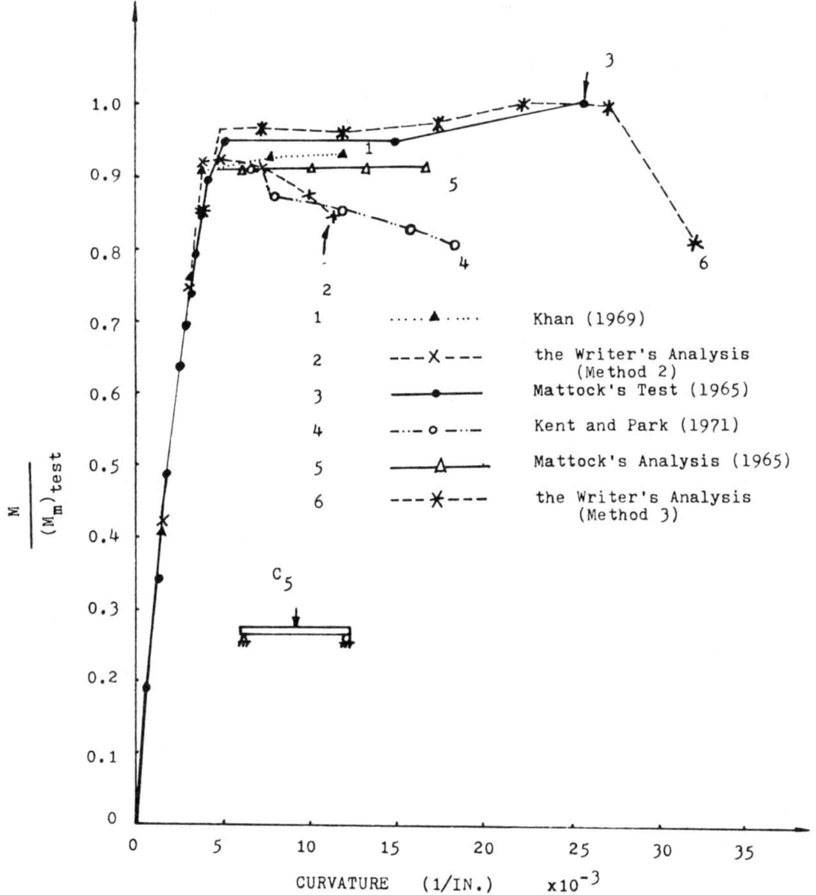

FIGURE 32. M – ϕ curves.

that, while the moment is constant throughout the region between the loads, the curvature varies to some extent depending on the location of cracks and other weaknesses [119]. Therefore only the weakest section in the length between the loads will undergo large rotations, the remainder of the length unloads like the rest of the beam outside the loads. It seems reasonable to explain the finite discontinuity length, under point loading as a property of the concrete failure mechanism—that crushing of the concrete must occur over a finite length.

Very little research has gone into studying the lengths of crushing zones, but intuition combined with whatever results are available can shed some light on this subject. Obviously, the external conditions have some effect, in particular the slope of the moment diagram and the restraint offered by such things as reinforcement and platens. From available test results, it appears that the greater the neutral axis depth, the longer the discontinuity. Barnard [120] and Mattock [19] measured the rotational capacity in reinforced concrete beams and developed an empirical equation which assumed a finite length of the hinging rotation occurring over a length $d/2$ to one side of the section of maximum moment. This assumed hinging length "d" is the same as other investigators such as Baker [126] and Macchi [127]. In the Report [128] of the Institution of Civil Engineering Committee, it is stated that the hinging length may vary from $0.4d$ (where d is effective length in section) to $2.4d$ which is dependent upon the loading condition on the structural member, sectional and member properties, etc.

ROTATIONAL BEHAVIOUR IN BEAMS

In general, two entirely different types of plastic hinges can be developed when a reinforced concrete beam is tested to failure. These have been described by Bachmann [83] as "flexural crack hinge" and "shear crack hinge." The first type of hinge develops in a beam zone in which the bending is predominant. The second hinge develops from the diagonal flexural-shear cracks generated by the relatively large shearing stresses in addition to the bending moment. In this investigation, the primary interest was to study the behaviour of the flexural crack hinge in which the bending moment is predominant.

It is generally assumed that the idealized curvature as shown in Figure 33d can be distributed along the member for a continuous beam. For a given moment-curvature relationship (Figure 33a) the determination of the rotation along a length of beam for any distribution of moment is very straight forward up to the stage where one section of the length reaches its maximum moment of resistance. If each element of the portion of a continuous beam under two-point loading (Figure 33b) can be assumed to follow the moment-curvature curve as shown in Figure 33a, and if the constant moment region is at M_m, then the distribution of curvature can be derived from the moment diagram and the moment-curvature relationship. As shown in Figure 33d, the curvature in the region between the loads is ϕ_m, and the elastic and inelastic rotations or angle changes along the length of the beam are given by the areas under the curvature diagram.

Now if the beam continues to deform, the curvature of the constant moment region will increase and pass ϕ_m, and, because this region is now on the falling branch of the moment-curvature curve, the bending moment must decrease. Since moment varies directly as shear force, the moment diagram must remain linear. As a result, a discontinuity must appear in the curvature diagram and regions outside and between the loads will begin to follow separate paths on the moment-curvature relationship [30].

In the constant momemt region, the moment decreases from M_m to M_1, and the curvature increases correspondingly from ϕ_m to ϕ_1 along the falling branch. Each section in the portions of the beam outside the loads exhibits decrease in both moment and curvature along its own individual unloading path, which is assumed to be approximately parallel to the initial elastic slope M_y/ϕ_y. With curvatures increasing between the loads and decreasing elsewhere, the total rotation (given by the integral of the curvature along the length of the beam) becomes more and more concentrated within the constant moment region. This behaviour is illustrated in Figure 33d.

Because the regions outside the loads are actually undergoing a decrease in rotation, a further requirement is imposed on the beam in order for this behaviour to occur; the decrease in strain energy of these regions must not exceed the increase in strain energy within the discontinuity. If this condition is not true, then there will be a sudden release of energy into the region between the loads sufficient to destroy the beam. This is the one condition when the sudden failure of a concrete beam in flexure can occur independently of the loading system. However, this does not automatically occur when a beam enters the descending portion of its moment-curvature curve.

Physically, when the beam is bent beyond maximum moment, increased cracking, often called "crushing," will appear in the constant moment region while outside the loads the beam will tend to straighten out. If the energy balance condition is not satisfied, then these outside portions will appear to unload into the region between the loads, resulting in a sudden disintegration of the concrete and a drastic, if not total, decrease in load with virtually no change in deflec-

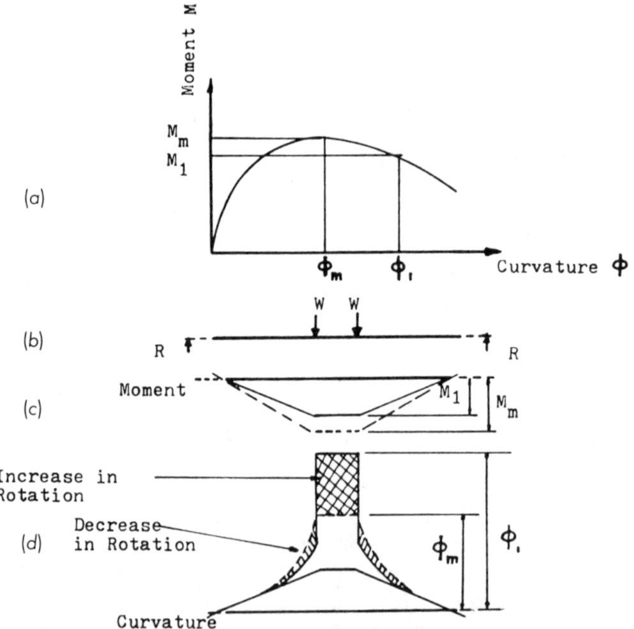

FIGURE 33. Rotations in a beam when moment falls off.

tion. This assumes that the points of contraflexure remain stationary. In a continuous beam, the moments in adjacent regions (e.g. over the supports in Figure 33) are also changing in some manner unless an end span is being considered. These regions may be one of three types: strain-hardening, plastic or strain-softening, depending on whether the moment at the section is increasing, constant or decreasing, respectively. The effect of either a strain-hardening or a plastic region adjacent to a strain-softening region such as that described in Figure 33 is to move the point of contraflexure inwards, hence shortening the region over which the moment is decreasing but making this decrease more pronounced than in the case where the point of contraflexure remains stationary. If the adjacent region is strain-softening, the direction of movement of this point depends on the relative magnitudes of the two moment decreases.

From the above discussion, since the behaviour of a beam in an actual building is not affected by the flexibility of any loading appratus, its ability to rotate is restricted only by the possible occurrence of a release of energy from the unloading regions into the discontinuity. The likelihood of such an energy release is governed by the inter-relationship between the bending moment distribution, the length of the discontinuity, the crack pattern, the overall dimensions of the beam, the stress-strain curves both for concrete and reinforcing bar and then the shape of the moment-curvature relationship and the behaviour of sections adjacent to the section where the bending moment is a maximum. An important factor affecting the ability of a concrete beam to rotate appears to be the slope of the falling branches of the moment-curvature and the stress-strain curve for concrete. When the slope of the falling branch of the moment-curvature curve is small, as in under-reinforced concrete beams, then not only are large curvatures possible within the discontinuity, and hence large rotations, but also the moment decrease is slight and hence the unloading of the regions outside the discontinuity is small [30]. Conversely, when the slope is steep, the likelihood of sudden failure due to energy release will be greater because the curvatures within the discontinuity will be less than in the former case and the moment decrease will be larger.

Definition and Evaluation of Curvature in a Beam

The conventional definition of curvature in reinforced concrete flexural member has been suggested by Mattock [19] and later by Corley [38] as ϕ (see Figure 34):

$$\phi = \frac{\epsilon_c}{kd} = \frac{\text{maximum concrete strain}}{\text{neutral axis depth}} \tag{60}$$

Later Ghosh [130] computed the deflections for several beams based on the three different definitions of curvature, and compared these values with the experimental results. He concluded that the definition of curvature based on

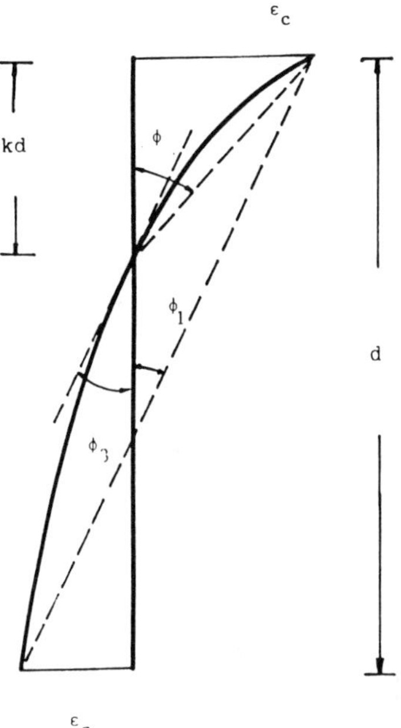

BY STRAIN GAUGE METHOD

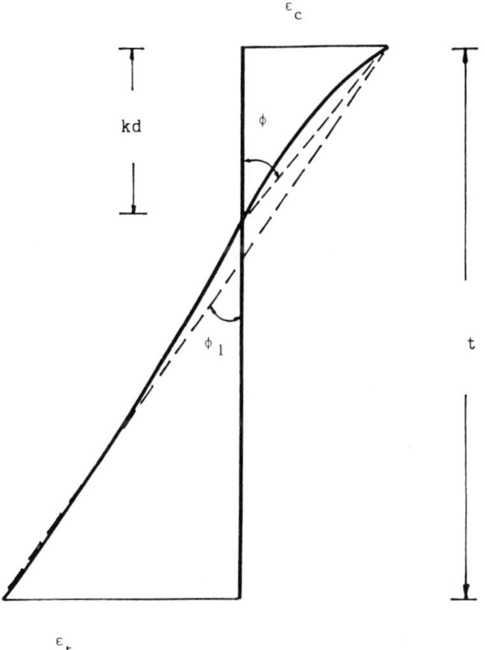

BY DEMEC GAUGE METHOD

FIGURE 34. Non-linear strain distributions in reinforced concrete section.

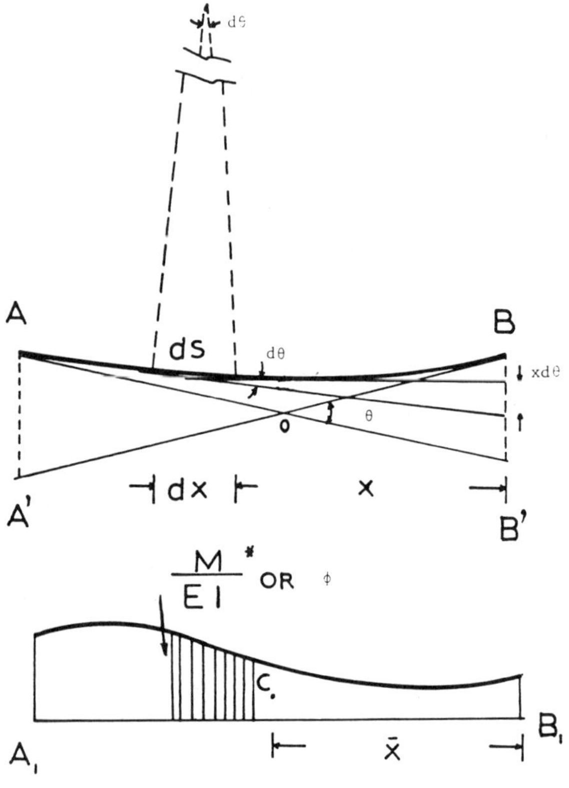

FIGURE 35. Revised moment-area method.

Equation (60) was the most realistic. Hsu used a similar definition of curvature (Equation (60)) to compare the computed deflections with the experimental values for columns subjected to combined uniaxial or biaxial bending moment and axial load [152] and reached the same conclusion as Mattock [19], Corley [38] and Ghosh [130].

Figure 34 shows the typical examples of non-linear strain distributions in reinforced concrete beam or column section in which Figure 34a was obtained by strain gauge method while Figure 34b used the demec gauge method.

AN ANALYTICAL INVESTIGATION OF POST-YIELDING DEFLECTION IN REINFORCED CONCRETE BEAMS

Introduction

For the past decades, researchers such as Ernst [79], Ernst et al. [80], and Gaston et al. [81] have shown that the deflection and the rotation at a point in a reinforced concrete beam can be evaluated using the suitable modifications of the well-known moment-area theorems. It is necessary to know the moment-curvature characteristics for the beam section. The curvature ϕ (unit rotation) is defined by the ratio M/EI where M is the bending moment at a beam section and EI is the flexural rigidity of the section.

In this section, the equations for deflection and rotation for the flexural behaviour of the simply supported beams under single-point and two-point loadings were developed by Hsu [85,152]. This method can be extended to analyze the continuous beams. The continuous beam is considered as a series of the simply-supported beams spanning between the inflection points. Analysis results have been compared with available experimental data.

A Study of Post-Yielding Deflection in Simply Supported Beams Under Two-Point Loading

INTRODUCTION

The computer program developed [84,85,152] consists of one subject to compute the moment-curvature characteristics of the section along with another subject to evaluate the load-deflection and the moment-rotation relationships. This program was used for calculating rotation and deflection at selected stations of simply supported beams tested under two-point loading by Gaston et al. [81]. The analysis techniques used by Gaston et al. [81] and Ernst [79] were approximate and lacked the sophistication of numerical methods now available. The analysis developed by Hsu can predict complete load-deformation characteristics from zero load until failure, and therefore the behaviour of simply supported singly and doubly reinforced beams can be studied through all loading stages and then to failure (with a large increase in deformations and a drop in the applied loads). Failure discussed here is a special type in which the failure is controlled by the deformation rather than the load.

The computer program is detailed in Reference [84]. The flow chart is similar to that shown in Figure 24. The only difference is that the constant axial load on the section is assumed to be zero.

THEROETICAL DEVELOPMENT

The following two familiar moment-area theorems have been used by many investigators to evaluate rotations and deflections in linear elastic systems [82].

(1) *Theorem I:* The angle θ between the tangents to the deflected elastic curve at two points A and B (Figure 35) is equal to the total area of the M/EI-diagram between these two points.

$$\theta = \int_A^B \frac{M}{EI}\,dx \tag{61}$$

(2) *Theorem II:* The deflection of a point B on the deflected elastic curve from the tangent to the elastic curve

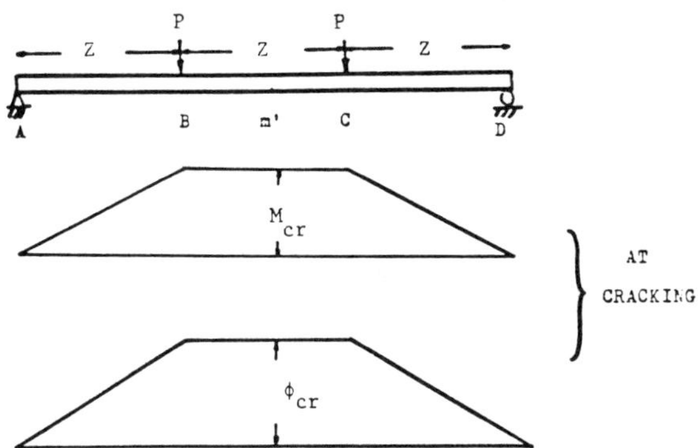

FIGURE 36. Simply supported beam under two-point loading (at cracking).

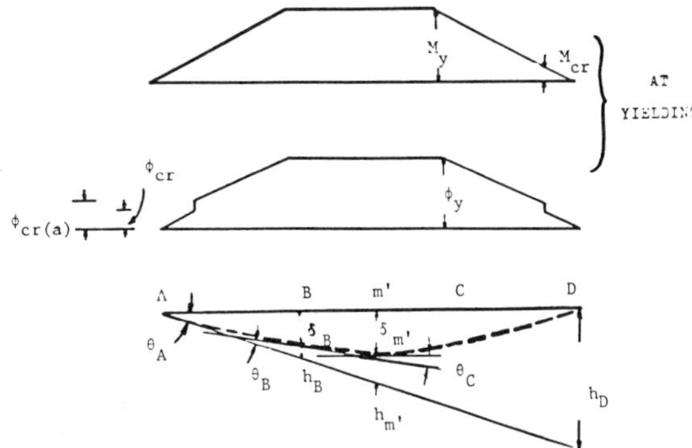

FIGURE 37. Simply supported beam under two-point loading (at yielding).

at another point A (Figure 35) is equal to the statical moment of the area of the M/EI-diagram between A and B about the point B.

$$\delta = \int_A^B x d\theta = \int_A^B \frac{Mx}{EI} dx \tag{62}$$

Equations (61) and (62) are valid for linear elastic behaviour only. However, these equations can be expressed in terms of curvature to be valid in both linear and non-linear ranges of behaviour as follows:

$$\theta = \int_A^B \phi dx \tag{63}$$

$$\delta = \int_A^B \phi x dx \tag{64}$$

The calculation of rotation and deflection is performed in three stages: at cracking, at yielding, and between yielding and ultimate [84,85]:

At Cracking

The first crack appears in tension zone at the point of the maximum moment in a member (M_{cr} = bending moment at cracking when $(f_t)_{max}$ = maximum tensile stress in concrete. Assumed to be 500 psi). Before and at cracking the relationship between moment and curvature is assumed to be linear. Also the curvature distribution along the member is taken to be linear from zero to ϕ_{cr} as shown in Figure 36. The following quantities are evaluated from the data shown in Figure 36.

$$\theta_B = \frac{\phi_{cr} Z}{2} \tag{65}$$

$$\theta_A = \phi_{cr} Z \tag{66}$$

$$\delta_B = \frac{5}{6} \phi_{cr} Z^2 \tag{67}$$

$$\delta_{m'} = \frac{23}{24} \phi_{cr} Z^2 \tag{68}$$

At Yielding

Tension reinforcement yields in tension zone at the point of the maximum moment. The curvature distribution along the member is shown in Figure 37. Total area of the curvature diagram at yielding is equal to

$$\phi_{AREA} = \phi_y Z + \phi_{cr} Z\left(\frac{M_{cr}}{M_y}\right) + (\phi_{cr(a)} + \phi_y)\left(1 - \frac{M_{cr}}{M_y}\right) Z$$

And the intercept h_D (Figure 37) is given by

$$h_D = \phi_{AREA} \frac{3Z}{2}$$

Therefore,

$$\theta_A = \frac{h_D}{3Z} = \frac{\phi_{AREA}}{2} \tag{69}$$

Since

$$\theta_c = \phi_y \frac{Z}{2}$$

Therefore

$$\theta_B = \theta_A - \theta_c = \frac{\phi_{AREA}}{2} - \frac{\phi_y Z}{2} \tag{70}$$

Also

$$\theta_A\left(Z + \frac{Z}{2}\right) - \frac{3Z}{4}\,\phi_{AREA} = \delta_{m'} + h_{m'}$$

and

$$h_{m'} = \frac{Z^2}{8}\,\phi_y + \tfrac{1}{2}\,\phi_{cr}\left(\frac{M_{cr}}{M_y}\right)\left[\frac{3}{2} - \frac{2}{3}\left(\frac{M_{cr}}{M_y}\right)\right] Z^2 + \phi_{cr(a)}\left(1 - \frac{M_{cr}}{M_y}\right)$$
$$\left(1 - \frac{M_{cr}}{2M_y}\right) Z^2 + \tfrac{1}{2}\,(\phi_y - \phi_{cr(a)})\left(1 - \frac{M_{cr}}{M_y}\right)\left(\frac{5}{6} - \frac{1}{3}\,\frac{M_{cr}}{M_y}\right) Z^2$$

Therefore

$$\delta_{m'} = \frac{3Z}{4}\,\phi_{AREA} - h_{m'}$$

$$\theta_A Z = \frac{\phi_{AREA}}{2}\,Z = \delta_B + h_B$$

and (71)

$$h_B = \tfrac{1}{2}\,\phi_{cr}\left(\frac{M_{cr}}{M_y}\right)\left(1 - \frac{2}{3}\,\frac{M_{cr}}{M_y}\right) Z^2 + \tfrac{1}{2}\,\phi_{cr(a)}\left(1 - \frac{M_{cr}}{M_y}\right)^2 Z^2$$
$$+ \frac{1}{6}\,(\phi_y - \phi_{cr(a)})\left(1 - \frac{M_{cr}}{M_y}\right)^2 Z^2$$

Therefore, the deflection δ_B at B is given by

$$\delta_B = \frac{\phi_{AREA}}{2}\,Z - h_B \tag{72}$$

Between Yielding and Ultimate

The equation for the evaluation of rotation and deflection has been derived in the following from the general Equations (63) and (64), which are valid for both linear and nonlinear behaviour of the beams.

In the constant moment region (Figure 38), $dM = 0$ and therefore $d\phi = 0$. Here, $\phi_m{}'$ is the curvature at constant moment region. Along AB (if P is constant), the shearing force at a point distant x from A is given by

$$P = \frac{dM}{dx}$$

$$\text{or } dM = P\,dx$$

At B (or C), $x = Z$; $P = M/Z$.

The rotations θ_A, θ_B, θ_c (Figure 38) are given by

$$\theta_c = \frac{\phi_{m'} Z}{2} \tag{73}$$

$$\theta_B = \int_B^A \phi\,dx = \frac{Z}{M}\int_B^A \phi\,dM \tag{74}$$

$$\theta_A = \frac{Z}{M}\int_B^A \phi\,dM + \frac{\phi_{m'} Z}{2} \tag{75}$$

Since

$$\delta_1{}' = \frac{\phi_{m'} Z^2}{8}$$

and

$$\theta_c = \frac{\delta_1{}'}{k'}$$

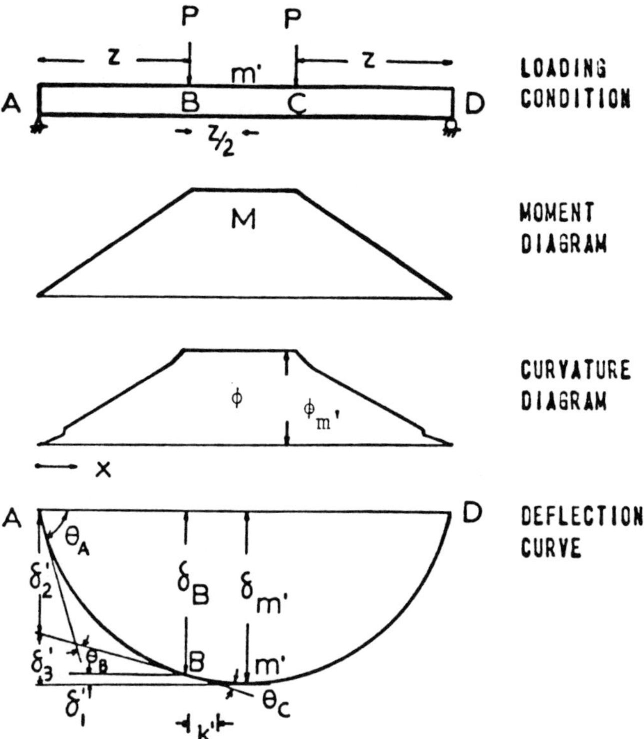

FIGURE 38a. Simply supported beam under two-point loading (between yielding and the maximum moment).

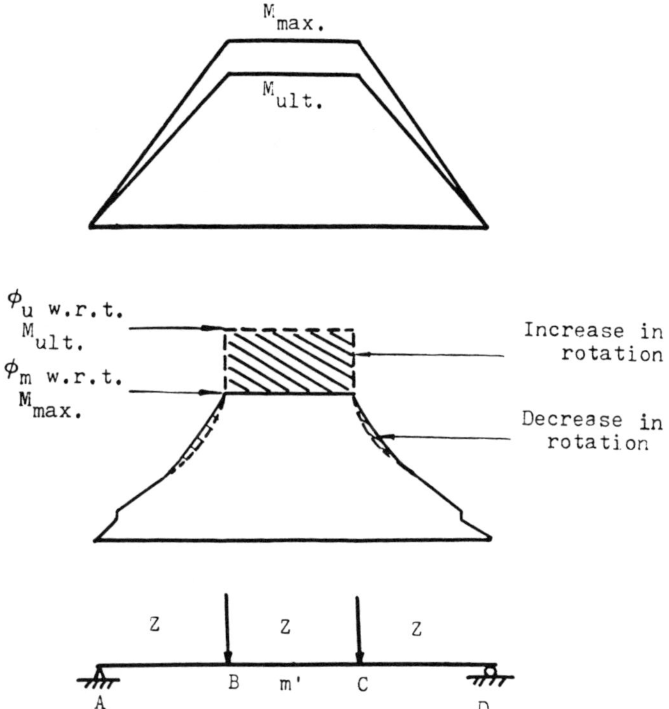

FIGURE 38b. Simply supported beam under two-point loading (between the maximum and ultimate moments).

TABLE 5. Properties of Beams (For Beams with Tension Reinforcement Only).

Test Specimens⁺	b in.	t in.	d in.	d′ in.	Z in.	f'_c psi	f_t psi	A_s in.^{2}	A'_s in.^{2}	$E_s \times 10^6$ psi	f_y ksi	ϵ_{sh}	E_{ss} ksi	f'_y ksi	$n = \frac{E_s}{E_c}$	Modes of Failure
T1La	6	12	10.79	—	36	2150	500	0.2214	0	28.2	54.3	0.0198	855	0	10.67	T*
T1Lb	6	12	10.72	—	36	2520	500	0.4088	0	28.2	46.0	0.0182	875	0	9.79	T
T2La	6	12	10.65	—	36	2120	500	0.6198	0	30.0	40.4	0.0175	1060	0	11.36	T
T2Lb	6	12	10.65	—	36	2440	500	0.6198	0	30.0	55.4	0.0148	998	0	10.59	T
T4La	6	12	10.51	—	36	2380	500	1.1980	0	32.2	44.1	0.0137	1095	0	11.49	T
T4Lb	6	12	10.44	—	36	2810	500	1.5785	0	29.2	43.3	0.0186	800	0	9.59	I
T5L	6	12	10.37	—	36	2500	500	2.0035	0	28.8	40.2	0.0160	822	0	10.03	T
T1Ma	6	12	10.72	—	36	4600	500	0.4088	0	28.2	46.2	0.0175	920	0	7.23	T
T1Mb	6	12	10.58	—	36	4750	500	0.8720	0	29.0	42.9	0.0270	600	0	7.33	T
T3Ma	6	12	10.37	—	36	4800	500	2.0035	0	28.8	41.0	0.0202	708	0	7.23	T
T3Mb	6	12	10.37	—	36	4110	500	2.0035	0	28.8	41.7	0.0171	767	0	7.83	T
T1Ha	6	12	10.58	—	36	5880	500	0.8720	0	30.0	44.2	0.0182	682	0	6.82	T
T1Hb	6	12	10.58	—	36	5180	500	0.8720	0	30.0	52.2	0.0190	690	0	7.26	T
T2H	6	12	10.44	—	36	5400	500	1.5785	0	29.2	45.6	0.0144	500	0	6.92	T
T3H	6	12	9.52	—	36	5920	500	2.3990	0	32.2	43.2	0.0134	930.5	0	7.29	T
T4H	6	12	9.38	—	36	5260	500	3.1573	0	29.2	42.0	0.0154	784	0	7.02	T
T5H	6	12	9.23	—	36	5900	500	3.9984	0	28.8	40.6	0.0172	788	0	6.52	T

Remark

(1) $(f_t)_{max}$ is taken as 500 psi

(2) $E_c = W^{1.5} \times 33\,(f'_c)^{1/2}$ where W = 145 lbs/cu ft

(3) Loading Condition: two-point loadings

Z P Z P Z

(4) Steel Stres-Strain Curve: see text

(5) Typical Reinforced Concrete Section: see text

⁺Gaston, Siess and Newmark's Specimens (1952)

*Denotes Tension Failure

therefore

$$k' = \frac{Z}{4}$$

Also

$$\theta_c = \frac{\delta_3'}{Z + k'}$$

and

$$\delta_3' = \left(Z + \frac{Z}{4}\right)\theta_c = \frac{5\phi_{m'}Z^2}{8}$$

therefore

$$\delta_{m'} = \delta_2' + \delta_3' = \frac{Z^2}{M^2}\int_B^A \phi\, M\, dM + \frac{5}{8}\phi_{m'}Z^2 \qquad (76)$$

and

$$\delta_B = \delta_2' + \delta_3'\left(-\delta_1' = \frac{Z^2}{M^2}\int_B^A\right)\phi\, MdM + \frac{\phi_{m'}Z^2}{2} \qquad (77)$$

It should be noted that the above case applies only to the beam member for which the moment-curvature relationship is the same for the entire span. Equations (74), (75), (76) and (77) can be applied with relative ease up to the maximum bending moment. These equations are also applicable in later stages when the moment is decreasing while the accompanying curvature is increasing (Figure 38b).

ANALYSIS OF UNIVERSITY OF ILLINOIS TEST RESULTS

Gaston, Siess and Newmark [81] tested several under-reinforced and over-reinforced simply supported beams under two-point loading. They used single and double rows of steel reinforcement in their tests. A significant feature of their tests was the experimental determination of the load-deflection curves for all beams from zero load until failure. The experimental set-up and the arrangement for measuring deflections can be found in Reference [81].

Tables 5 and 6 detail the section properties and the modes of failure of beams tested by Gaston et al. [81]. Figures 39, 40 and 41 present the computed moment-curvature relationships for some typical beams. The experimental load-deflection curves obtained by Gaston et al. have been compared with the results obtained by Hsu [152] in Figures 42 through 44. It can be noted that the computed results are in good agreement with the experimental data for most of the specimens.

The first point in Hsu's load-deflection analysis (marked

TABLE 6. Properties of Beams (For Beams with Both Tension and Compresion Reinforcements).

Test Specimens+	b in.	t in.	d in.	d′ in.	Z in.	f'_c psi	f_t psi	A_s in.²	A'_s in.²	$E_s \times 10^6$ psi	f_y ksi	ϵ_{sh}	E_{ss} ksi	f'_y ksi	$n = \frac{E_s}{E_c}$	Modes of Failure
C2W	6	12	10.58	1.125	36	3940	500	0.8760	0.3936	30.0	45.4	0.0154	698	44.5	8.32	T
C2XM	6	12	10.58	1.125	36	4070	500	0.8760	0.3936	30.0	53.3	0.0142	925	47.0	8.19	T
C3W	6	12	10.37	1.125	36	4310	500	2.0035	1.2010	28.8	41.8	0.0192	773	46.7	7.63	T
C3XM	6	12	10.37	1.125	36	3890	500	2.0035	1.2010	28.8	41.8	0.0173	815	42.5	8.04	T
C3YNa	6	12	10.51	1.125	36	3330	500	1.1980	0.6180	32.2	45.2	0.0181	730	56.1	9.69	T
C3YNb	6	12	10.37	1.125	36	4860	500	2.0035	1.2010	28.8	42.1	0.0185	657	47.4	8.69	T
C4XNa	6	12	10.51	1.125	36	2450	500	1.1980	0.6180	32.2	45.5	0.0172	853	41.4	11.32	T
C4XNb	6	12	10.51	1.125	36	2430	500	1.1980	0.8765	32.2	46.4	0.0181	640	44.1	11.39	T
C4ZN	6	12	10.37	1.125	36	2570	500	2.0035	1.2010	28.8	41.3	0.0201	683	46.4	8.41	T
C5YN	6	12	9.38	1.125	36	4480	500	3.1560	1.5760	29.2	44.0	0.01695	750	43.4	8.49	T
C7W	6	12	9.38	1.125	36	3480	500	3.1560	1.5760	29.2	41.6	0.01495	851.5	43.6	8.59	T
Remark	+ Gaston, Siess and Newmark's Specimens (1952) The Notations are shown in Table 5.															

with an (x) in Figure 42, etc.) corresponds to the appearance of the first flexural crack. The second point indicates the yielding of tension reinforcement. The last point shows that the ultimate stage is reached when the equilibrium equation in section cannot be satisfied any further. Beyond the last point, most of the test results obtained by Gaston et al. show further increase in deflections accompanied by a drop in the applied loads.

Tables 7 and 8 show the yield and maximum moments and mid-span deflection at yield, respectively, for analysis and test results. The mid-span deflection at the maximum load capacity has also been shown in Figure 42 through Figure 44. The comparisons of the load-strain curves obtained by Gaston et al. and Hsu are indicated in Figure 45 through Figure 47. It can be noted that the computed results are in good agreement with the experimental data.

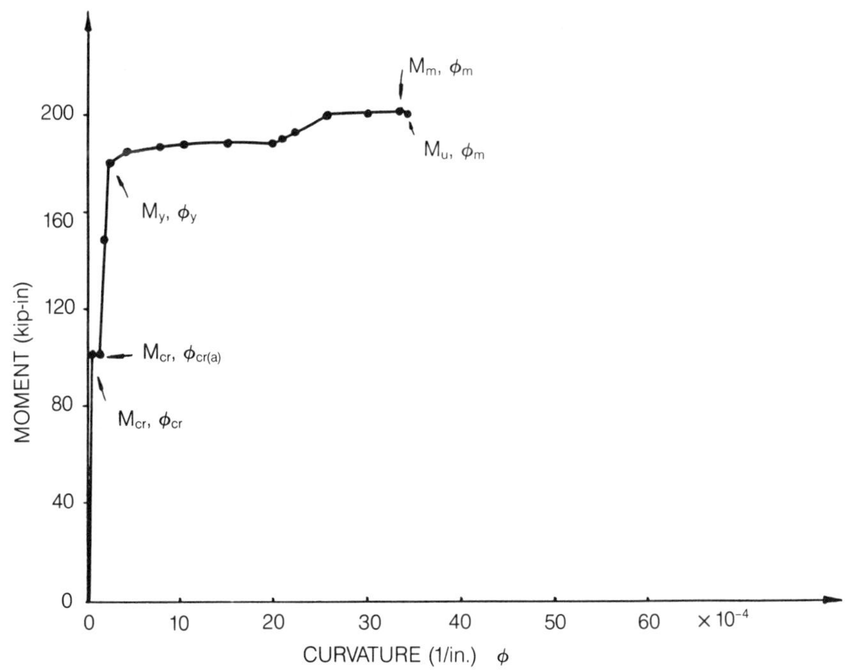

FIGURE 39. Moment-curvature curve for specimen T1Lb.

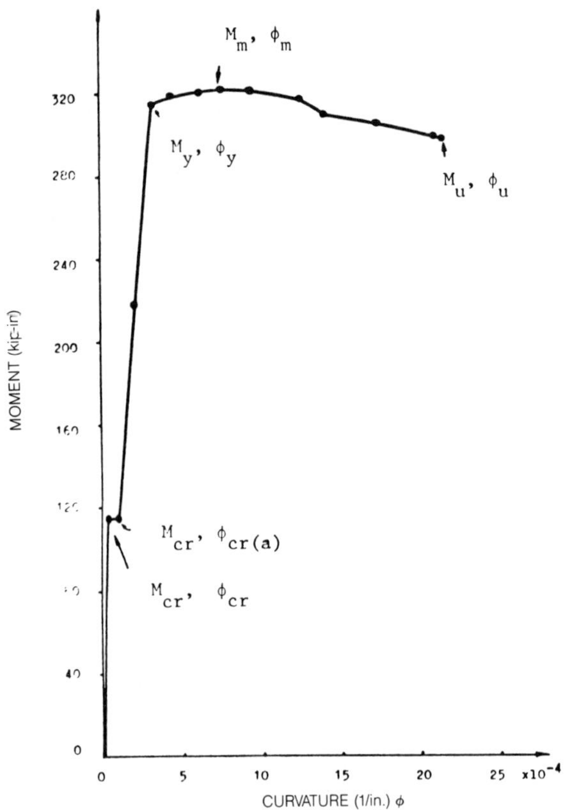

FIGURE 40. Moment-curvature curve for specimen T2Lb.

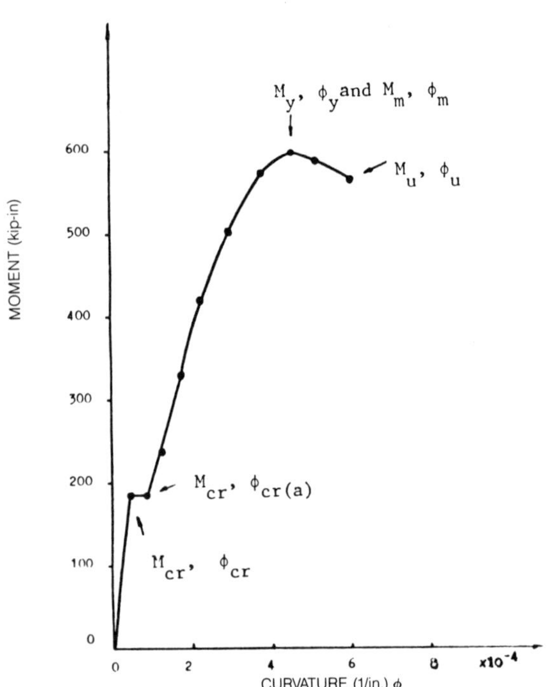

FIGURE 41. Moment-curvature curve for speciment T5L.

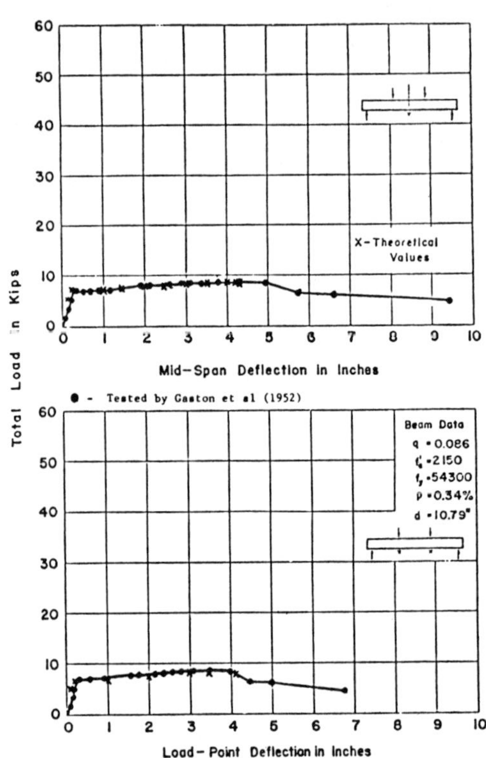

FIGURE 42. Load-deflection curves for beam No. T1La.

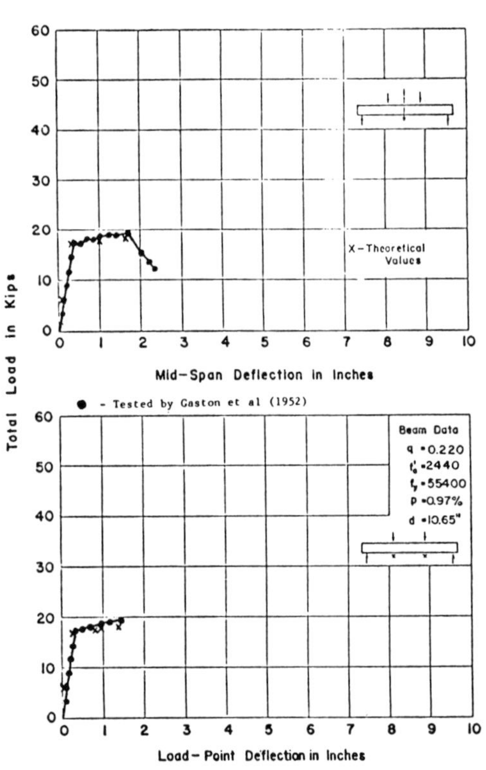

FIGURE 43. Load-deflection curves for beam No. T2Lb.

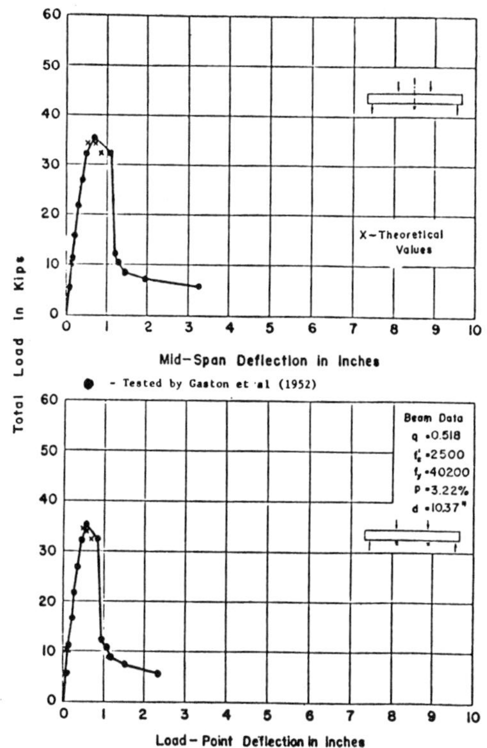

FIGURE 44. Load-deflection curves for beam No. T5L.

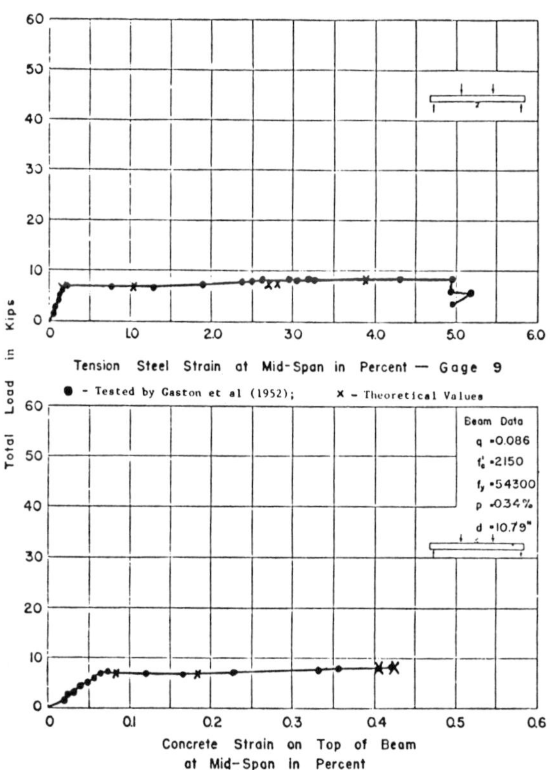

FIGURE 45. Load-strain curves for beam No. T1La.

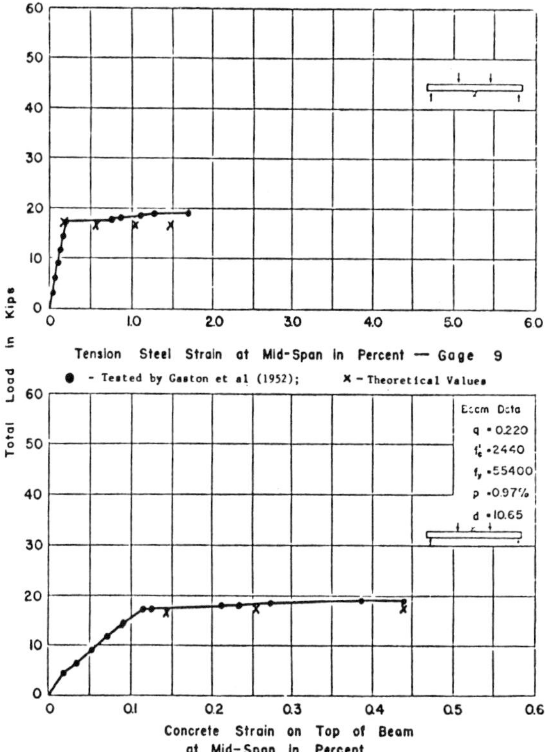

FIGURE 46. Load-strain curves for beam No. T2Lb.

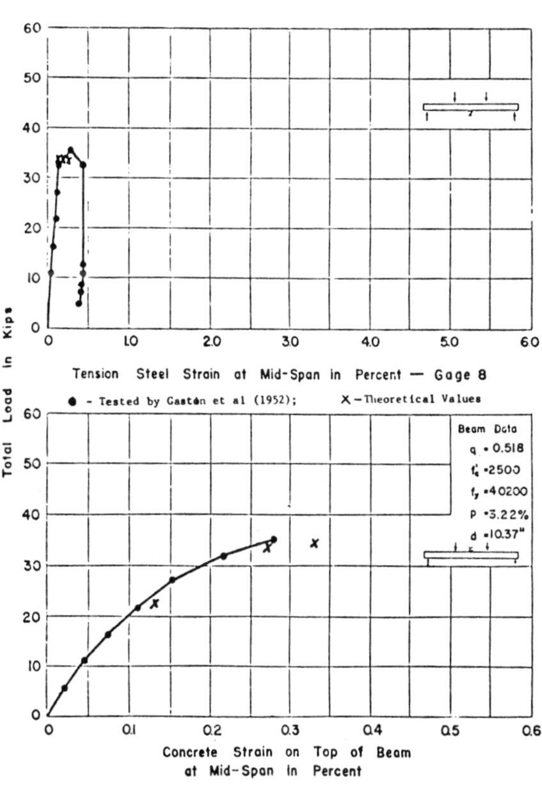

FIGURE 47. Load-strain curves for beam No. T5L.

TABLE 7. Moment Resistance at Yielding of Steel Reinforcement and at Maximum Load.

Test Speci.*	M_{max} (1)	M_{max} (2)	M_{max} (3)	(1)/(2) (4)	(1)/(3) (5)	M_y (6)	M_y (7)	M_y (8)	(6)/(7) (9)	(6)/(8) (10)
T1La	13.1	14.4	12.573	0.91	1.042	10.2	9.9	9.91	1.03	1.029
T1Lb	19.4	18.8	17.74	1.03	1.0935	15.4	14.8	15.07	1.04	1.0218
T2La	23.5	23.8	20.02	0.99	1.1738	20.6	19.6	19.17	1.05	1.07
T2Lb	29.1	28.4	26.9	1.02	1.0818	26.1	26.8	26.2	0.97	0.996
T4La	37.8	38.2	37.66	0.99	1.0037	37.5	39.2	37.3	0.96	1.00
T4Lb	47.0	47.9	47.27	0.98	0.9942	47.0	49.7	37.3	0.95	1.26
T5L	53.1	51.8	49.83	1.03	1.0656	48.4	56.8	49.83	0.85	0.9713
T1Ma	18.7	21.2	22.66	0.88	0.825	15.5	15.1	15.6	1.03	0.993
T1Mb	32.0	31.0	30.9	1.03	1.0356	30.2	29.2	29.36	1.03	1.029
T3Ma	62.1	61.5	60.61	1.01	1.0246	60.5	59.2	59.12	1.02	1.023
T3Mb	61.1	60.7	59.71	1.01	1.0233	58.1	59.9	58.93	0.97	0.985
T1Ha	34.3	34.8	36.99	0.99	0.9272	29.8	30.3	30.576	0.98	0.974
T1Hb	39.9	37.2	38.13	1.07	1.046	34.2	35.7	35.69	0.96	0.96
T2H	53.1	55.9	55.5	0.95	0.957	51.3	53.4	53.69	0.96	0.9555
T3H	67.0	69.7	68.86	0.96	0.973	65.1	68.1	67.43	0.96	0.965
T4H	77.2	80.4	79.17	0.96	0.9751	—	83.7	79.17	—	—
T5H	85.5	100.2	91.56	0.85	0.9341	81.7	99.5	91.56	0.82	0.8923
Average Value				0.98	1.01				0.974	1.008
Standard Deviation				0.057	0.078				0.064	0.078
C2W	40.9	38.1	39.04	1.07	1.048	32.4	30.9	31.23	1.05	1.037
C2XM	44.1	43.1	43.972	1.02	1.0029	38.0	36.3	36.594	1.05	1.038
C3W	69.3	68.6	72.92	1.01	0.9504	62.4	61.3	63.24	1.02	0.986
C3XM	69.5	66.5	69.46	1.05	1.0007	66.3	61.5	63.09	1.08	1.05
C3YNa	49.8	46.3	49.47	1.08	1.0067	41.0	40.9	41.54	1.00	0.987
C3YNb	73.4	69.4	75.46	1.06	0.973	62.4	62.1	63.90	1.00	0.9765
C4XNa	43.2	45.0	42.80	0.96	1.009	40.1	41.3	41.43	0.97	0.968
C4XNb	48.2	50.1	47.77	0.96	1.009	42.8	42.3	42.72	1.01	1.002
C4ZN	71.3	66.1	68.49	1.08	1.041	59.3	60.7	62.24	0.98	0.953
C5YN	90.1	91.0	93.69	0.99	0.96	87.8	89.7	92.27	0.98	0.952
C7W	84.5	86.0	87.90	0.98	0.9613	84.0	84.3	86.997	1.00	0.966
Average Value				1.024	0.9965				1.013	0.9923
Standard Deviation				0.047	0.032				0.034	0.035

Remark: *Gaston, Siess and Newmark's Specimens (1952)
(6) & (1) Tested by Gaston et al.
(7) & (2) Analysed by Gaston et al.
(8) & (3) Analysed by the writer.
Unit: M_{max} (Maximum bending moment) kip.-ft.
M_y (Bending moment at yielding) kip.-ft.

TABLE 8. Mid-Span Deflections at Yielding of Steel Reinforcement.

Test Speci.*	δ_y (1) in.	δ_y (2) in.	δ_y (3) in.	δ_y (4) in.	(1)/(2) (5)	(1)/(3) (6)	(1)/(4) (7)
T1La	0.30	0.27	0.378	0.266	1.11	0.9433	1.128
T1Lb	0.30	0.25	0.2975	0.28	1.20	1.00	1.07
T2La	0.27	0.25	0.294	0.281	1.08	0.92	0.961
T2Lb	0.36	0.34	0.389	0.381	1.06	0.93	0.945
T4La	0.38	0.32	0.3899	0.381	1.19	0.975	1.00
T4Lb	0.46	0.34	0.39	0.385	1.35	0.179	1.19
T5L	0.49	0.35	0.5626	0.544	1.40	0.871	0.90
T1Ma	0.25	0.24	0.266	0.2551	1.04	0.94	0.98
T1Mb	0.34	0.26	0.284	0.2814	1.31	1.197	1.208
T3Ma	0.50	0.32	0.3613	0.359	1.56	1.38	1.392
T3Mb	0.38	0.34	0.394	0.3912	1.12	0.9645	0.971
T1Ha	0.37	0.27	0.2714	0.27	1.37	1.36	1.37
T1Hb	0.37	0.32	0.33	0.329	1.15	1.12	1.12
T2H	0.37	0.33	0.35	0.3471	1.12	1.057	1.065
T3H	0.48	0.39	0.397	0.3953	1.23	1.21	1.214
T4H	—	0.44	0.531	0.527	—	—	—
T5H	0.55	0.47	0.5972	0.59	1.17	0.92	0.93
Average Value					1.22	1.06	1.09
Standard Deviation					0.145	0.161	—
C2W	0.36	0.27	0.29	0.2875	1.31	1.24	1.25
C2XM	0.45	0.32	0.34	0.3365	1.39	1.32	1.337
C3W	0.46	0.30	0.3215	0.3208	1.53	1.43	1.433
C3XM	0.40	0.31	0.33	0.328	1.31	1.21	1.219
C3YNa	0.38	0.30	0.3016	0.30	1.27	1.26	1.267
C3YNb	0.42	0.30	0.316	0.31	1.39	1.32	1.355
C4XNa	0.37	0.31	0.3276	0.322	1.19	1.129	1.149
C4XNb	0.38	0.31	0.315	0.3125	1.22	1.206	1.216
C4ZN	0.40	0.30	0.3303	0.33	1.31	1.21	1.212
C5YN	0.50	0.41	0.4316	0.43	1.22	1.158	1.163
C7W	0.46	0.40	0.433	0.4314	1.15	1.06	1.066
Average Value					1.299	1.232	1.2424
Standard Deviation					0.108	0.102	—

Remark: *Gaston, Siess and Newmark's Specimens (1952)
(1) Tested by Gaston et al.
(2) Analysed by Gaston et al.
(3) Analysed by the writer; in which the curvature distribution was assumed to be linear along member z (see Figure 37).
(4) Analysed by the writer; in which the curvature distribution was assumed to be bi-linear as shown in Figure 37.

UNDER-REINFORCED CONCRETE CONTINUOUS BEAMS UNDER CONCENTRATED LOADS

Introduction

Redistribution of bending moment in continuous reinforced concrete beams is widely recognized as the most useful tool in the hands of the designer and investigators of reinforced concrete structures. Also, the redistribution of the bending moment can result in a reduction of the maximum bending moments in the span or at the supports. This, of course, leads to smaller sections through the beam. Used with discretion, arbitrary redistribution of bending moments in continuous reinforced concrete beams can therefore result in sounder and more economic structures.

Results of experimental investigations on reinforced concrete continuous beams have been reported by Macchi [127], Mattock [141], George and Mirza [144] and Hsu [148].

Mattock [19] and Hawkins, Sozen and Siess [143] proposed in their papers that the behaviour of the continuous beams was similar to that of simply supported beams, i.e., continuous beam could be considered as a series of simply supported beams spanning between the points of inflection.

It should be noted that in the above discussion it was assumed that the points of inflection remain stationary for simplicity of analysis. Barnard [30] also discussed the behaviour of continuous beams and examined the redistribution of bending moments. In order to investigate this movement of the points of inflection in continuous beams, two under-reinforced continuous beams were tested at McGill University [144], as shown in Figure 48. The distance between the loading point and the point of inflection (called K) (see Figure 48) in under-reinforced continuous beams tests was noted to be 2.65 ft. (an average value) while K is 2.39 ft. using the elastic theory (the three-moment theorem). The variation of K's value as shown in Figure 49 is so small that one can assume it to be stationary. The moment-curvature relationship using the method detailed previously and the equations for deflection and rotation for the simply supported beams under single-point loading are developed in this section.

Once the moment-curvature, load-deflection and moment-rotation relationships had been obtained for each of the simply supported beams, Hsu developed an approximate method [148,152] of meeting the slope and deflection continuity conditions at the points of inflection for continuous beams.

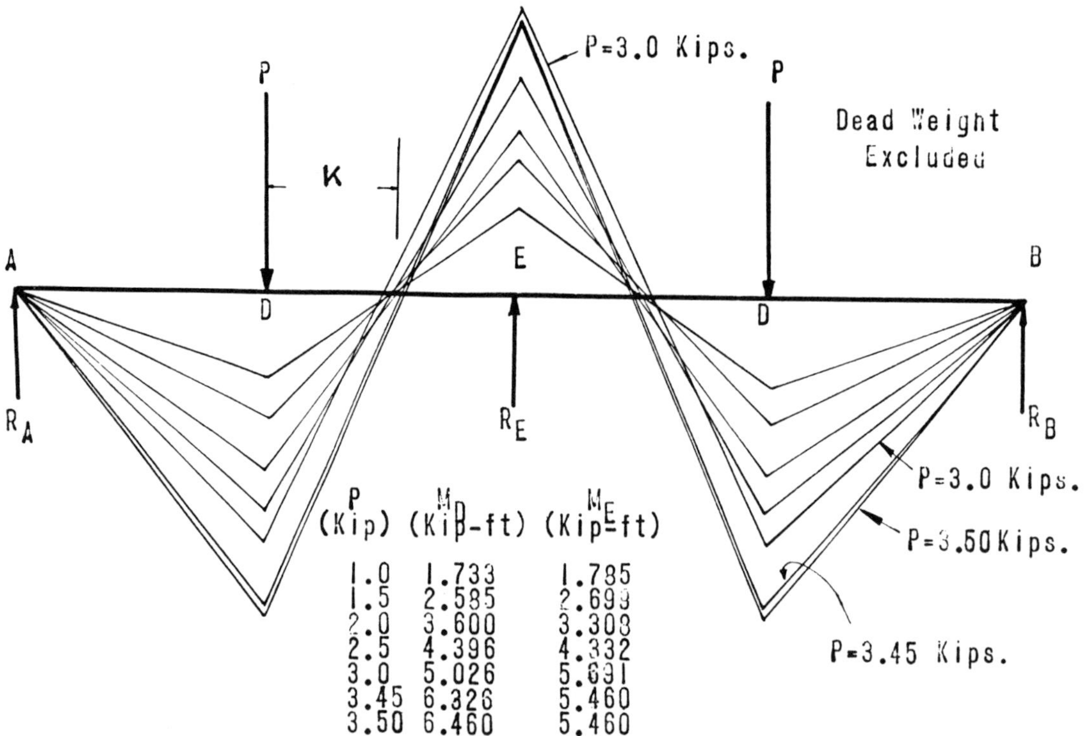

FIGURE 48. Moment distribution along the continuous beam span.

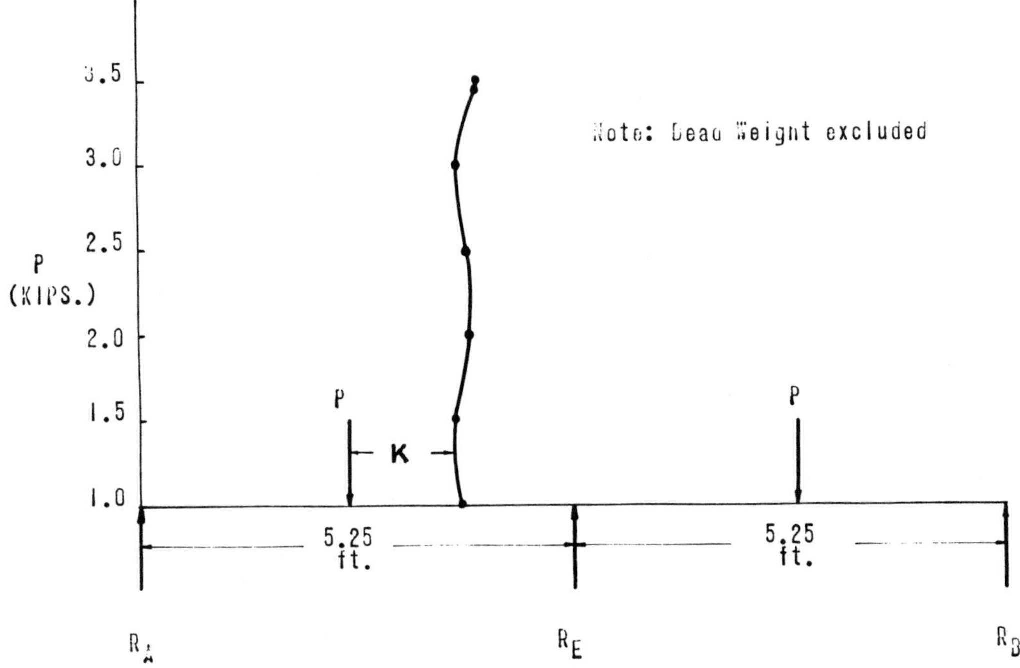

FIGURE 49. The variation of K's value.

Analytical Investigation

Based on the theorem of three-moment method, the value of K can be obtained:

(1) For Macchi's Continuous Beam (Figure 50)

$$K = \frac{3}{20} A \tag{78}$$

(2) For McGill Continuous Beam (Figure 51)

$$K = \frac{5}{11} A \tag{79}$$

The rotation and deflection for a simply supported beam under single-point loading shown in Figure 52 can be analyzed as follows:

JUST BEFORE CRACKING

$$\delta_b = \left(\frac{A + 2K}{3}\right) \phi \left(\frac{A + K}{2}\right)$$

$$\theta_1 = \frac{\phi}{6}(A + 2K) \text{ and } \delta_c' = \frac{\phi}{6} A^2$$

Hence

$$\delta_c = A\theta_1 - \delta_c' = \frac{A}{3} \phi K \tag{80}$$

since

$$\theta_3 = \frac{\phi_1(A - g)}{2}, \ \theta^2 = \frac{\phi A}{2}$$

Hence

$$\theta_c = \theta_2 - \theta_3 = \frac{\phi A}{2} - \frac{\phi_1(A - g)}{2}$$

From Figure 52a, a linear behaviour just before cracking yields

$$\frac{\phi_1}{\phi} = \frac{A - g}{A}$$

Therefore

$$\theta_c = \frac{\phi}{2}\left[A - \frac{(A - g)^2}{A}\right] \tag{81}$$

where θ_c is the local rotation over the gauge length g.

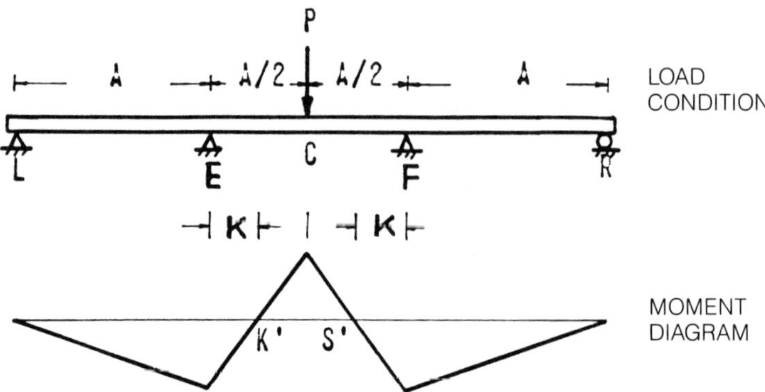

FIGURE 50. Macchi's continuous beam.

AT YIELDING

Figure 52b shows the bending moment and the curvature diagrams at yielding. A linear behaviour is used to approximate the curvature distribution at yielding. This yields the following equation:

$$\theta_c = \frac{\phi}{2}\left[A - \frac{(A-g)^2}{A} \right] \tag{82}$$

$$\delta_c = \frac{A}{3}\,\phi\,K \tag{83}$$

This simplification results in a small increase (approximately 1.5%) in the rotation and deflection at yielding.

BETWEEN YIELDING AND MAXIMUM MOMENTS

From Figure 53 and Figure 54,

$$\theta_3 = \int_C^B \phi d\bar{x} = \int_C^B \phi \frac{dM}{p_B}$$

Since $P_B\bar{x} = M$, $P_B\, d\bar{x} = dM$ if P_B is constant at $\bar{x} = A - g$, $M = M_1$. Therefore

$$\theta_3 = \frac{A-g}{M_1}\int_C^B \phi\, dM = \frac{A}{M}\int_C^B \phi\, dM$$

since

$$p_B = \frac{M_1}{A-g} = \frac{M}{A}$$

Similarly,

$$\theta_2 = \frac{A}{M}\int_D^B \phi\, dM$$

Therefore,

$$\theta_c = \theta_2 - \theta_3 = \frac{A}{\text{M}}\left[\int_D^B \phi\, dM - \int_C^B \phi dM \right] \tag{84}$$

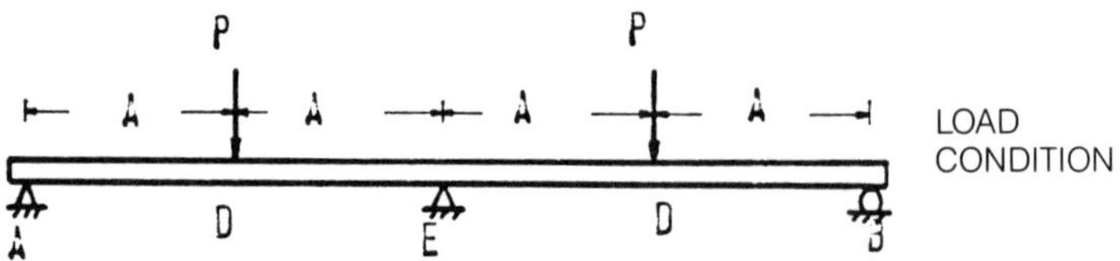

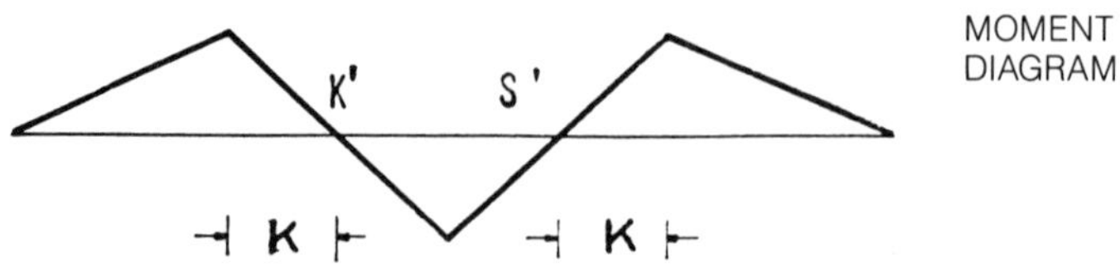

FIGURE 51. McGill University continuous beam.

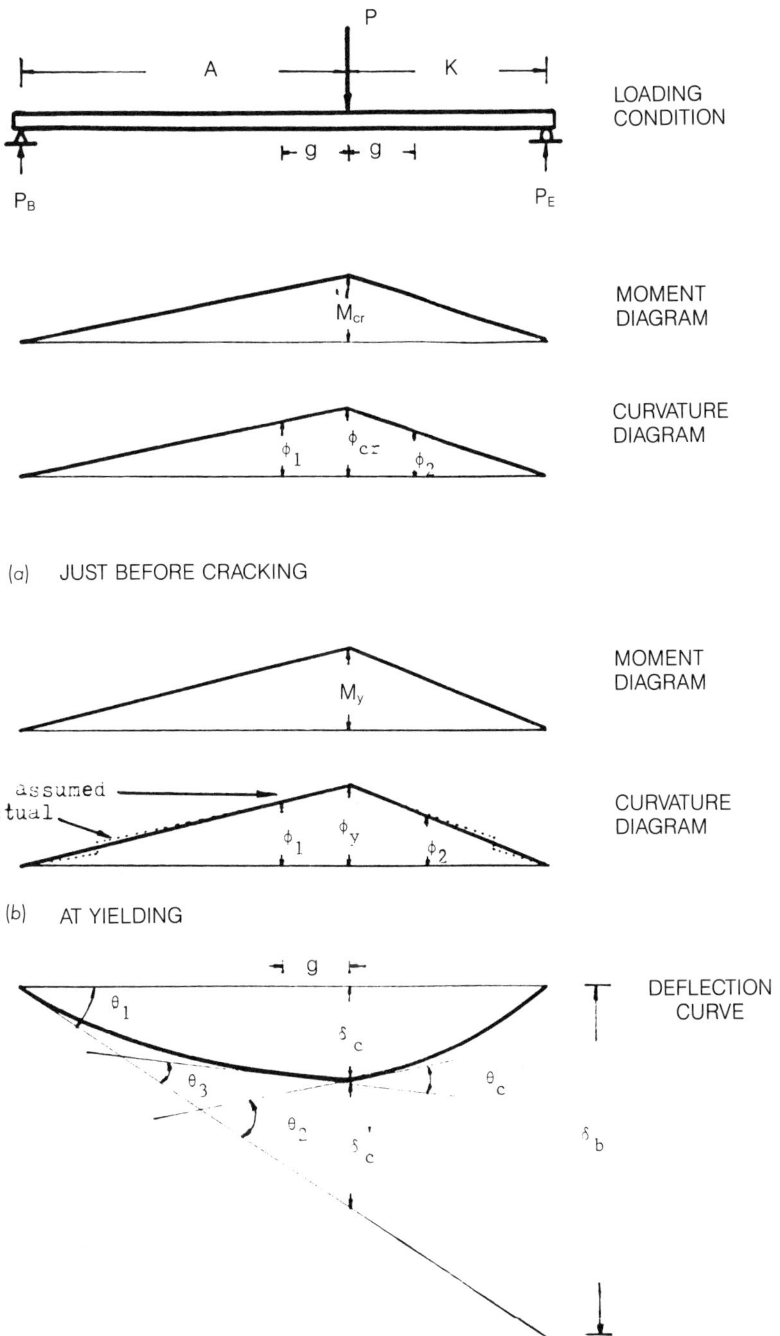

FIGURE 52. Rotation and deflection analysis.

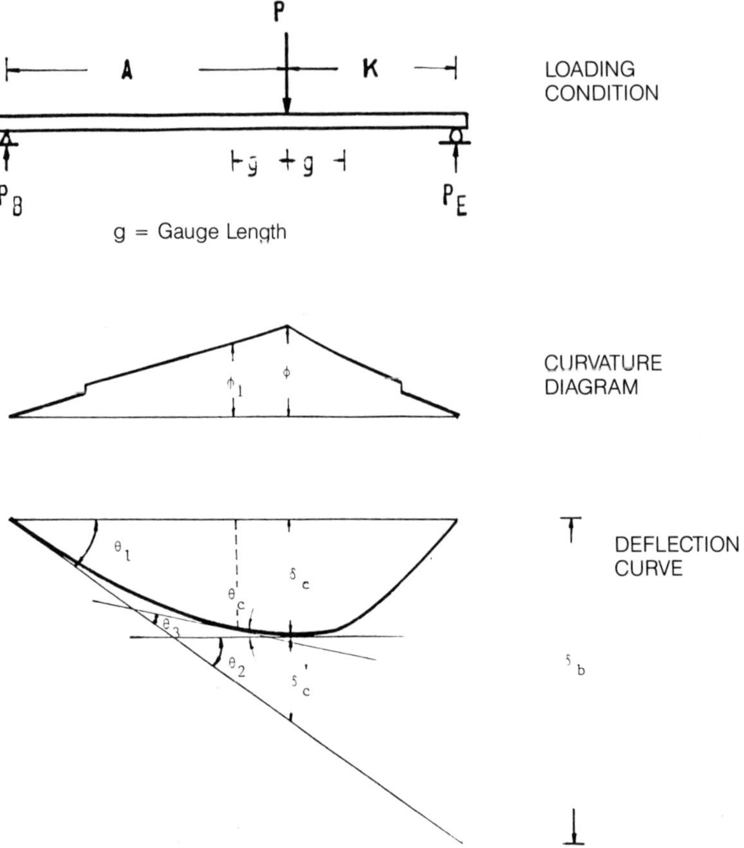

FIGURE 53. Rotation and deflection analysis: between yielding and maximum.

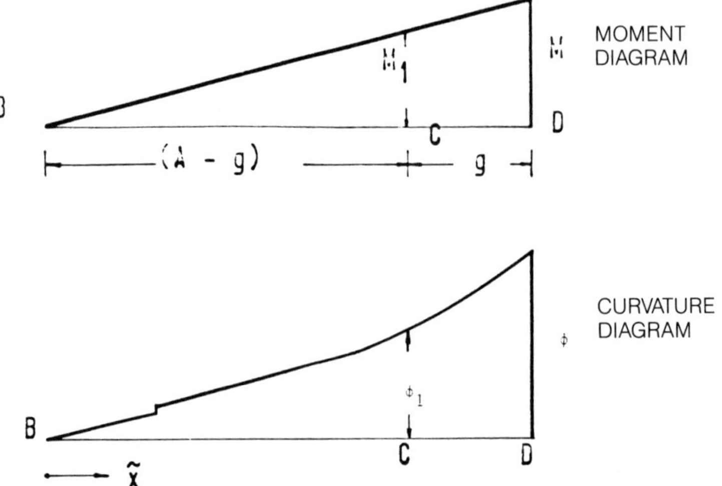

FIGURE 54. Rotation and deflection analysis: between yielding and maximum.

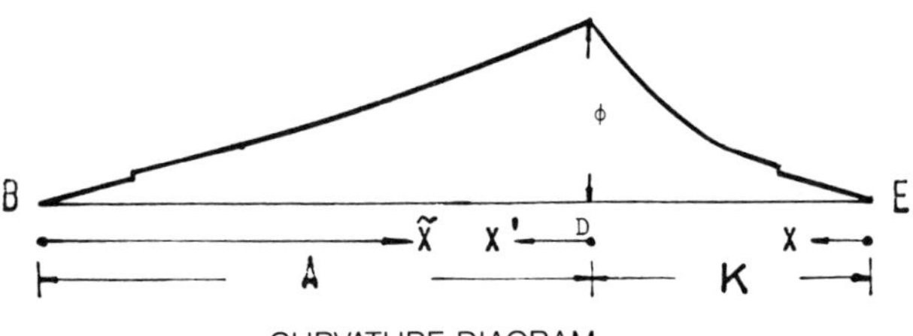

FIGURE 55. Rotation and deflection analysis: between yielding and maximum.

From Figure 53, the following equation is obtained for δ_b:

$$\delta_b = \int_B^E \phi x dx = \int_D^E \phi x dx + \int_B^D \phi x dx = \frac{K^2}{M^2}\int_D^E \phi M dM + \int_B^D \phi(K + x')\, dx'$$

Since

$$\bar{x} = -(A - x') \text{ (see Figure 55)}$$
$$d\bar{x} = dx'$$

Therefore,

$$\delta_b = \frac{K^2}{M^2}\int_D^E \phi\, M dM + \int_B^D \phi\, \bar{x} d\bar{x} + (K + A)\int_B^D \phi d\bar{x}$$

$$= \frac{K^2}{M^2}\int_D^E \phi\, M dM - \frac{A^2}{M^2}\int_D^B \phi M dM - \frac{(K + A)A}{M}\int_D^B \phi dM \tag{85}$$

Also,

$$\theta_1 = \frac{\delta_b}{A + K}$$

and

$$\delta_c + \delta_c' = A\theta_1 \text{ (see Figure 53)}$$

Hence

$$\delta_c' = \int_B^D \phi x' dx' = \int_B^D \phi\, (-A + x' + A)\, dx'$$

$$= \int_B^D \phi \bar{x} d\bar{x} + A\int_B^D \phi d\bar{x} \tag{86}$$

$$= -\frac{A^2}{M^2}\int_D^B \phi M dM - \frac{A^2}{M}\int_D^B \phi dM$$

Therefore,

$$\delta_c = A\theta_1 - \delta_c' = \left(\frac{A}{A + K}\right)\left(\frac{K^2}{M^2}\right)\int_D^E \phi M dM$$
$$+ \frac{A^2 K}{M^2(A + K)}\int_D^B \phi M dM \tag{87}$$

Assuming the moment-curvature characteristics for all the sections in beam are the same, then,

$$\delta_c = \frac{A(K)}{M^2}\int_O^M \phi M dM \tag{88}$$

and

$$\theta_c = \frac{A}{M}\left[\int_O^M \phi dM - \int_O^{M_1} \phi dM\right] \tag{89}$$

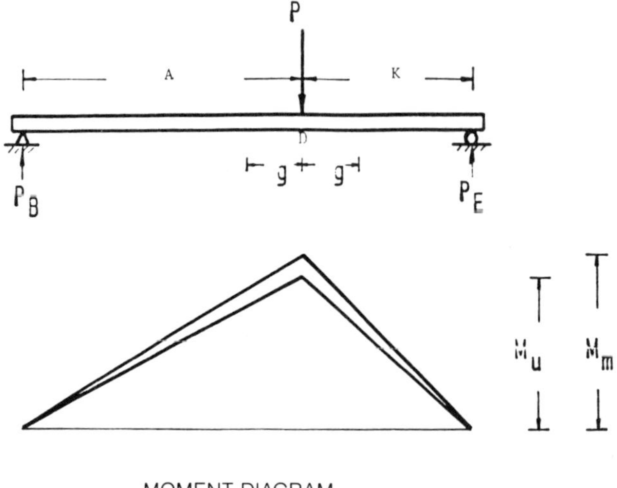

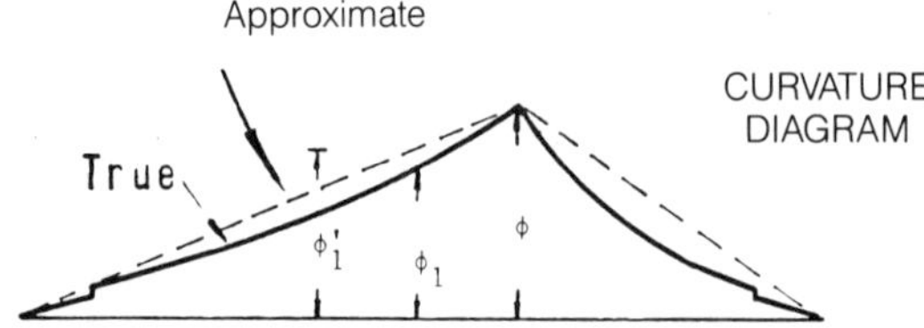

FIGURE 56. Rotation and deflection analysis: between maximum and ultimate.

BETWEEN MAXIMUM AND ULTIMATE MOMENTS

After the point of maximum moment, the bending moment decreases while the curvature increases. The rotation θ_c and the deflection δ_c can be evaluated using the following approximation (Figure 56):

$$\theta_c = \frac{\phi A}{2} - \frac{\phi_1' (A - g)}{2} = \frac{\phi}{2}\left[A - \frac{(A - g)^2}{A}\right] \tag{90}$$

and

$$\delta_c = \frac{A}{3}\phi K \tag{91}$$

It may be noted that Equations (90) and (91) are similar to the formulas shown in Equations (80), (81), (82), and (83).

Analysis of Test Results

INSTITUTE OF CONSTRUCTION SCIENCE TEST RESULTS

Macchi [127] tested a continuous beam as shown in Figure 57. Since Hsu [152] considered the continuous beam as the combination of three simply supported beams (shown in Figure 58a), the approximate method to assess the continuity of slope and deflection at K' and S' (Figure 58a) can be given in the following:

(1) Try a suitable θ_R (see Figure 58a).
(2) Let $\theta_R = \theta_R'$, i.e., $\phi'(2A + K)/6 = \theta_R$. Find ϕ'.
(3) Using the value of ϕ', find δ_1 and $\delta_2 = (A + K)\delta_1)/A$ (see Figure 58b).
(4) Then $\Delta\theta = \theta_2 - \theta_1 \approx (\delta_2 - \delta_1)/K$. Find $\Delta\theta$
(5) Let $\theta_{R(C)} = \theta_R + \Delta\theta = \theta_R' + \Delta\theta$. Find $\theta_{R(C)}$ where $\theta_{R(C)}$ is the corrected value of θ_R or θ_R'.

The analytical results and Macchi's experimental load-deflection curves are shown in Figure 59. It shows that the analysis is satisfactory as compared with the tests.

McGILL UNIVERSITY TEST RESULTS

George et al. [144] tested two continuous beams as shown in Figure 60. The stress-strain curves for reinforcing bars are shown in Figure 61. The location of the point of contraflexure remains relatively fixed as the applied loads are increased (Figure 49). The beam is divided into three simply supported beams as shown in Figure 62, and the following approximate method is used to ensure the compatibility of deformation at K' and S':

(1) Obtain ϕ and P in beam AK (see Figure 62).
(2) Then let $\theta_K = \theta_K'$, i.e., $\phi(K + 2A)/6 = \phi' K_1/2$ and calculate ϕ'.
(3) Using ϕ', calculate δ_{c2}.
(4) Calculate $\delta_c = \delta_{c1} + \delta_{c2} [A/(A + K)]$
(5) Plot P and δ_c as shown in Figure 63.

The analytical results and the experimental load-deflection curves tested by George et al. [144] are shown in Figure 63. The approximate analysis gives a maximum load of $P = 4.063$ kips, while the experimental values in Series 1 and 2 are noted to be 3.876 kips and 3.724 kips, respectively. The bending moment at D is calculated to be 80 kip-in. as compared with the experimental value of 82 kip-in. from Series 2.

COMPUTER ANALYSIS OF INELASTIC REINFORCED CONCRETE PLANE FRAMES

Introduction

Several computer programs based on the matrix displacement method have been developed over the past decades to analyze the two-dimensional structures [86,87,118,119]. Most of these analyses are aimed at linear elastic structural systems which are assumed to behave linear elastically at the working load level, for example for steel structures. The behaviour of concrete structural systems under loadings is generally non-linear. These non-linearities can generally be classified as (1) material non-linearities (non-linear constitutive relationships of materials used) and (2) geometrical non-linearities (resulting from the large deflections occurring at higher load levels). Recent work on elastic-plastic structural analysis is based on either the initial stress or the initial strain approach, although the modified stiffness method has also been used [88–90]. Analysis of reinforced concrete systems is more complicated because of the material non-linearities arising out of inelasticity and shrinkage of concrete and the non-linear behaviour at the steel-concrete interface along with the changing topology introduced by cracking of concrete.

This section presents a brief historical review of existing computer programs for analysis of steel and reinforced concrete frames. Hsu also developed a computer program for the elastic-plastic analysis of frames [95]. Essential features of the formulation and the computer program are as follows [95,147].

The computer program is capable of complete analysis of

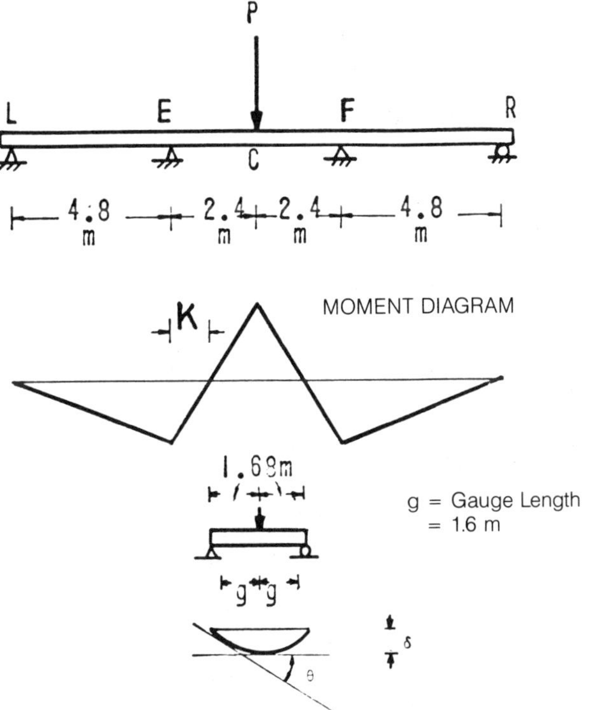

FIGURE 57. Macchi's continuous beam details.

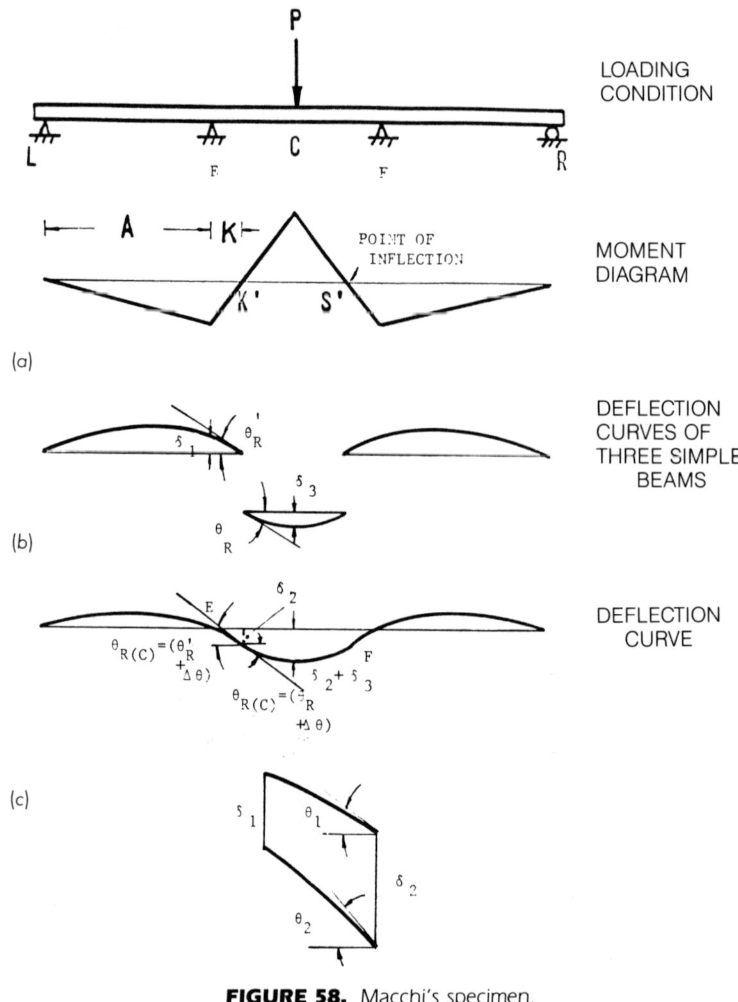

FIGURE 58. Macchi's specimen.

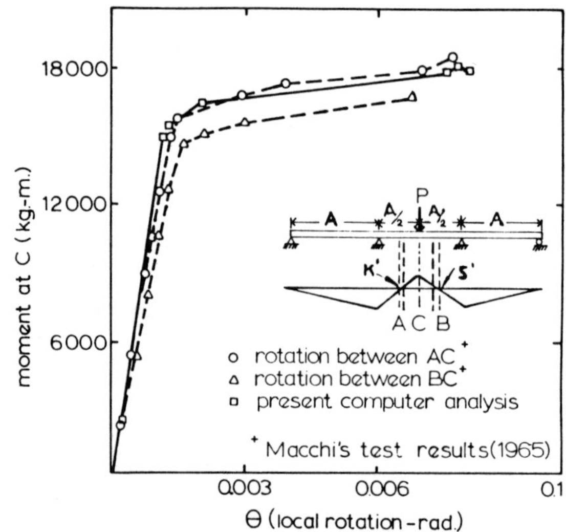

FIGURE 59. Moment-rotation curve for continuous reinforced concrete beam.

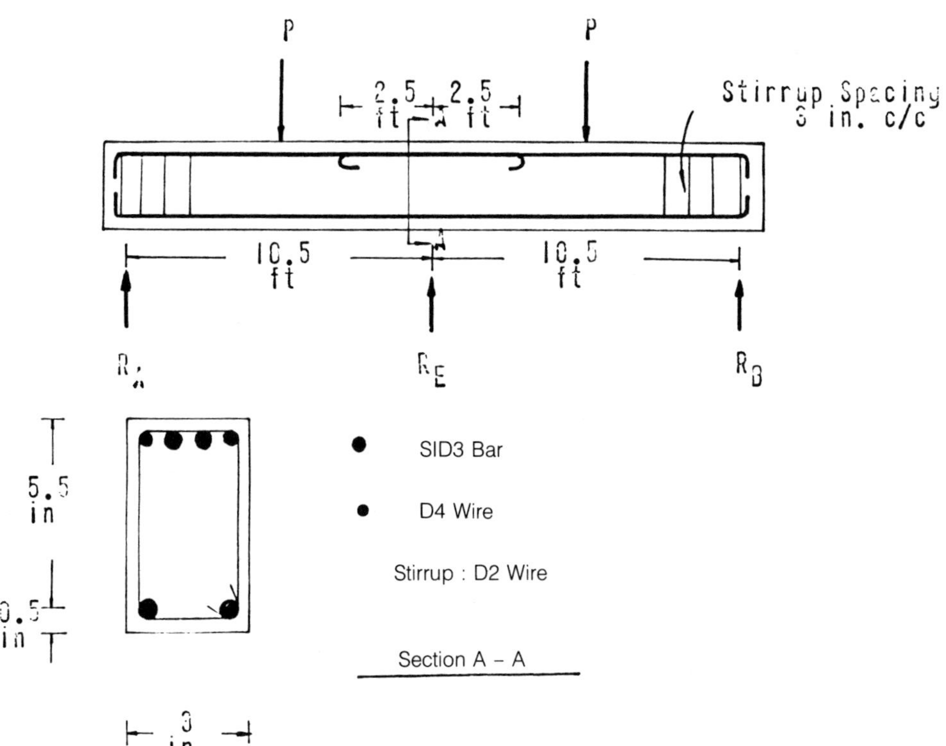

FIGURE 60. McGill University continuous beam details.

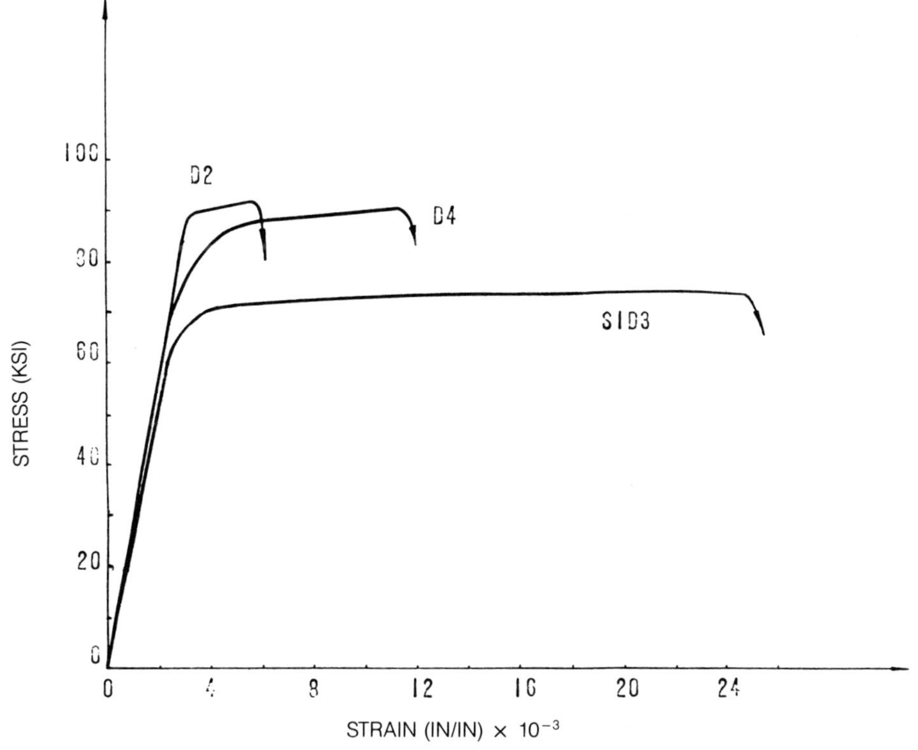

FIGURE 61. Stress-strain curves for reinforcing bars.

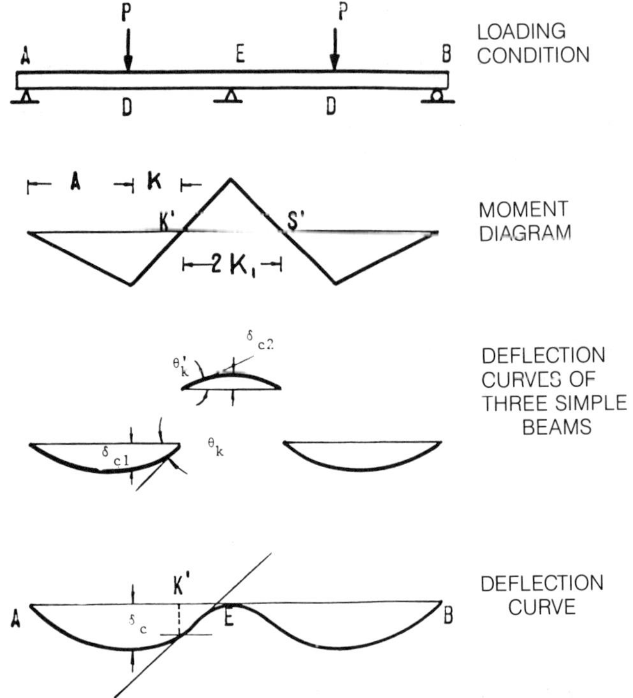

FIGURE 62. McGill University specimen.

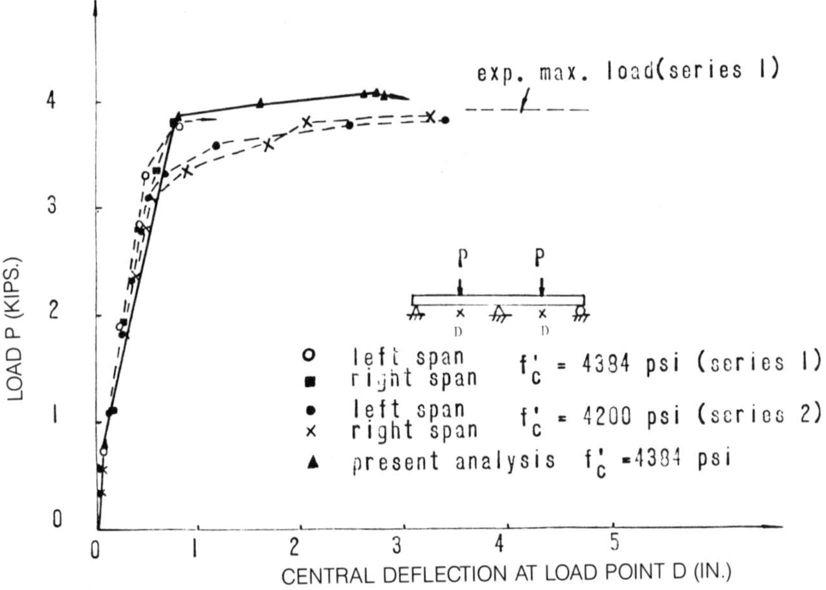

FIGURE 63. McGill University load-deflection tests.

steel and concrete frames from zero load until failure under any system of loads. The computer output gives the complete load-deflection behaviour, the location and the sequence of the formation of hinges until a collapse mechanism is attained. The program can cater for any given geometry (structural layout, member length, section geometry, and properties) and end conditions (fixed and/or pinned connections). The analysis of reinforced concrete frames uses a subroutine (presented in a previous section) to calculate the moment-curvature characteristics under a constant load applied at the section centroid. A reinforced concrete frame tested by Cranston and Cracknell [91] is analyzed using the computer program developed and the results are compared with the experimental and the analytical results obtained by Cranston et al. [33,91].

Survey of Literature

There are two basic matrix computer frame analysis methods: (1) The Flexibility Method and (2) The Displacement Method. The first method treats forces as unknowns and was generally used until the advent of computers. The displacement method is simpler, more systematic and convenient for use on a computer. A brief review of the numerical methods available for the analysis of steel and reinforced concrete structures are as follows.

STEEL FRAME

The first work in the field was due to Livesley [92,93] who developed a method for elastic analysis of plane frames, which accounted for the axial strains and the secondary bending moments (P–δ effects) produced by the axial forces. He extended his approach to analyze the behaviour of the member assuming it to be elastic-perfectly plastic. This method considers the joint displacements as unknowns and examines the bending moments in the frame as the applied loads are increased. Livesley used an iterative approach to calculate the bending moments produced by the axial forces by suitably modifying the member stiffnesses at each loading stage. A fictitious plastic hinge is inserted at a point where the computed bending moment exceeds the plastic moment capacity and the bending moments on either side of the hinge are made equal to the plastic moment capacity. The frame is then reanalyzed at the same load level, and the loading continued until a sufficient number of hinges is formed to transform the structure into a mechanism. At this stage, the computer program stops automatically and the resulting data can be examined to assess the frame behaviour.

Wang [88,94] extended the matrix displacment to establish a limit analysis method for rigid frames. As the bending capacity of the section exceeds the plastic moment capacity of the section, a hinge is introduced at the appropriate location. The member stiffness matrix is then modified to account for the plastic hinge. The procedure is repeated until the bending moment exceeds the plastic moment capacity at another section and a second hinge is then introduced. This procedure is repeated until a collapse mechanism is formed. Wang lumps all applied loads into concentrated loads at suitably selected nodes in the frame. All applied loads are increased proportionally and load factors at each stage when an additional hinge is installed are automatically calculated and printed out. The computer program developed by Wang is efficient and calculates the joint rotations, plastic hinge rotations and the joint deflections at all loading stages including the onset of collapse.

Both Livesley and Wang ignored the effects of axial thrust and assumed that the effects of strain-hardening and of the spread of the plastic zones along the members can be neglected. This led them to the assumption of ideal elastic-perfectly plastic behaviour of members and the "localization of the plastic hinge at a point along the member length."

Wilson [90] used the non-dimensional moment-curvature relationships to analyze the inelastic plane frames. By varying certain parameters, the moment-curvature characteristics can be varied between linear and ideal elastic-plastic behaviour. Non-linear expressions are derived for uniform members giving end slopes in terms of applied end moments, and a differentiation of these expressions yields the member stiffness which is dependent on the load level. The member stiffnesses are modified in stages as the applied loads are increased from zero to failure load in suitable increments.

Gray [97] extended Livesley's analysis to account for strain-hardening of steel. He introduced hinges when the plastic moment capacity was exceeded and used modified curvature characteristics which were functions of the actual hinge rotation and the bending moment diagram for the member.

Jennings and Majid [98] and Yamada et al. [99] used the discontinuities of rotations at the plastic hinges to modify the overall load displacement equations. They accounted for the effects of strain-hardening and computed the inverse of the stiffness matrix only once, which makes the computer program more efficient.

Armen, Pifco and Levine [100] used a finite element technique for the plastic analysis of structures subjected to out-of-plane bending alone, or in combination with in-plane membrane stresses. They accounted for material non-linearities by using the initial strain concept, and this analysis was combined with an incremental technique to account for the effects of geometric non-linear behaviour. This analysis was applied to beam and arch structures and both material and geometric non-linearities were considered.

REINFORCED CONCRETE FRAME

Many rigorous and computationally complex methods for predicting the behaviour of reinforced concrete members have been developed. Most of these methods have been devised to include some of the following aspects [101,146]:

(1) Nonlinear concrete stress-strain relationship
(2) Varying cross section due to non-uniform cracking and inelastic behaviour
(3) Effect of axial load
(4) Effects of creep and shrinkage
(5) Post yielding behaviour of sections
(6) Residual effects of overloading.

Most of the effort in this area has been concentrated on developing suitable computer programs to provide information on strength, ductility, and elastic and plastic behaviour including deflections at specific load levels. A few of these methods are briefly described below.

Borges and Oliveira [102] developed an approach specifically for reinforced concrete plane frames. Individual member response to applied end moments is computed using numerical integration. Member stiffness values are then computed and used to analyze the frame for the applied loads. The resulting moments are compared with the values assumed initially and if the values do not agree within prescribed limits, the procedure is repeated.

Cranston [103], Cranston and Chatterji [33] and Becker [104] developed computer programs to analyze the portal frames tested by Cranston and Cracknell [91]. These analyses treat forces as unknowns; and these forces are being the moments at released hinges inserted in the frame [33,103], or supports in the frame [104]. As far as possible, the hinges are located at points where inelastic behaviour is expected to develop with an increase in applied loads. The frame is divided into segments and the curvature is assumed to vary linearly within each segment. Moment-curvature relations are computed at points within each segment, based on the assumption that the curvature is dependent only on the bending moment. The effects of axial and shear deformations are neglected.

Analysis starts under the action of dead load only and proceeds automatically as the load is applied in stages. As the collapse or maximum load is approached, the analysis switches to deflection control and solutions are obtained as the deflection is increased in stages. Thus, the complete load deflection response of the frame is obtained. Lazero and Richards [115] used an incremental approach with one-dimensional prismatic elements in their analytic model. At each load level, the structure is elastic with stresses on certain elements having reached the specified ultimate capacity. Thus the model is used to illustrate successive concrete cracking, hinge formation, moment redistribution and the eventual collapse.

The methods developed by Drysdale [105], Drysdale and Mirza [146] and Tan [106] to analyze reinforced concrete frame consist of dividing the structure into a number of segments and a further division of the cross sections at the ends of each segment into small elements. An initial set of axial load and bending moments is assigned to each cross section using an elastic analysis. Then for the load and moment on each cross section the strain conditions necessary to provide equilibrium are found using non-linear stress-strain relationships for the concrete and steel. From these strains, effective EI and EA values are obtained and substituted into a stiffness matrix for the segmented structure. Use of suitable techniques to modify the stiffness matrix increases the rate of convergence of this iterative sequence.

The use of cross section elements facilitates accounting for non-linear creep and shrinkage strains and accumulated residual strains produced by unloading of concrete and steel elements which are stressed beyond the elastic range. Where creep, shrinkage, or residual strains are present, the total strains, which are required to produce internal stresses for equilibrium, are composed of a stress dependent component and an independent component resulting from the stress history. The effective stiffness of the segment lengths is computed from the total strains calculated for equilibrium conditions and the iterative procedure described above is employed. When a plastic hinge is formed in the structure, the description of the structure is changed to contain a real hinge with an applied external moment. The magnitude of the applied external moment is related to the amount of rotation required at the hinge for compatible displacements in the structure. The rotation is distributed over the hinge length which is defined as being bounded by the yielding moment at each end.

The methods used by Drysdale [105] and Drysdale and Mirza [146] require a great deal of computer time and memory space and are therefore intended as a research tool. The study of the behaviour of portions of representative structures is accomplished by dividing into elements only the member length near the region in question and allowing the remainder of the structure to remain elastic. Thus localized effects of non-linear behaviour can be simulated.

Similar programs incorporating various components on non-linear behaviour have been developed simultaneously at many other centers including McGill University and the Universities of Texas, Colorado, California at Berkeley, the Building Research Station in England and others. Among the more recent studies, Mufti et al. [107] have studied the behaviour of reinforced concrete elements using a finite element approach, and Franklin [108] has also completed a non-linear analysis of reinforced concrete frames and panels.

Although there are some differences in the assumptions and the mathematical formulations, comparisons with test data have shown that these methods of analysis are able to predict the behaviour at working loads with reasonable accuracy.

The problem of evaluating or devising analytical procedures for practice has received considerable attention. Beeby [109] prepared an extensive report evaluating existing methods of computing deflection and developed a series of empirical formulae for moment-curvature relationships from which to calculate deflections. Lehman [110] used the results of 25 single storey, two bay, 1/8 scale test frames to evaluate the validity of strength reduction factors and moment magnifying procedures for column design. Shaikh, Mirza and McCutcheon [111] studied the limit analysis of reinforced concrete frames using the results of studies by Sader [112] and Adenot [113] for comparison purposes. The analysis used the provision of the 1971 ACI Code (ACI 318-71) [45] and exhibited very good correlation with test results (within 5 percent) for loads up to formation of the first hinge.

The two analytical methods developed in this section [95,96] are devised to take into account a number of the factors contributing to non-linear behaviour of reinforced concrete structures. These methods are used to predict the load-deflection characteristics and are compared with the experimental results. The relatively good agreement obtained indicates that the researchers have developed the ability to accurately predict the behaviour of frames. The differences which are observed can be largely explained by the fact that rotations in joint are ignored. This factor could be partially accomodated by arbitrarily increasing the lengths of members to extend into the joints. However, the joint behaviour is still not well understood in quantitative terms. Additional information on the stress-strain properties of the materials used, particularly with reference to the effects of load history, must be obtained. In this respect even the creep occurring during the time required for testing can create discrepancies with predicted values which use concrete stress-strain relationships obtained from relatively short duration tests (Drysdale and Mirza [146]).

Elastic-Perfectly Plastic Computer Analysis of Reinforced Concrete Plane Frames—Method 1

Plastic design of steel frames and limit design of reinforced concrete frames have been well covered in standard textbooks and the literature. In any inelastic analysis, it is necessary to consider the rotation capacity of the hinging regions before establishing the redistribution of bending moments in the frame. Also, most of the important structures in earthquake regions must have adequate ductility. The post-elastic behaviour of the concrete frames for both of these aspects is dependent on the shape of the moment-curvature curves for the member sections.

MOMENT-CURVATURE RELATIONSHIPS AND ROTATION CAPACITY OF THE PLASTIC HINGES

Neal [89] and Beedle [114] have shown that the hinges formed in a steel frame have a very large rotation capacity. However, the rotation capacity at the hinging regions in a structural concrete member is dependent on several parameters, e.g., material properties, member geometry, applied loads, etc. Recent limit design research in structural concrete has led to the definition of the hinge length; the rotation capacity of the hinge can then be calculated using material constitutive relationships and the section geometry. Hsu [147,152] has developed computer programs to determine the moment curvature characteristics of a reinforced concrete section subjected to a constant axial load applied at the section centroid. The inelastic rotation capacity of the hinge can be calculated using Mattock's equation [19]:

$$\theta_{p1} = \left[\phi_u - \phi_y \frac{M_u}{M_y} \right] \frac{d}{2} \tag{92a}$$

where

θ_{p1} = the inelastic rotation of the hinge
d = the effective depth
ϕ_y = the curvature at yield
M_y = the bending moment at yield
ϕ_u = the curvature at ultimate load
M_u = the ultimate bending moment

For an under-reinforced section, the moment-curvature curve can be approximated by an elastic-perfectly plastic relationship without any serious error. The plastic rotation capacity is then given by

$$\theta_{p1} = \left[\phi_u - \phi_y \right] \frac{d}{2} \tag{92b}$$

The following points must be considered in developing a computer program for an elastic-plastic analysis of steel or reinforced concrete frames.

(1) The inelastic rotation capacity of the hinging region in a reinforced concrete section is dependent on the section geometry and material properties and can be limited in some cases; however, the rotation capacity of a structural steel section can be assumed to be reasonably large.

(2) The appearance of cracks in a structural concrete member gives rise to a varying flexural rigidity (EI) along its length. Also, the descending branch of the moment-curvature curve at a section or the moment-rotation curve at a hinge provides the section with added ductility although its bending strength decreases.

(3) For symmetrical structural steel sections such as the I-section and the rectangular section, the neutral axis and the plastic centroid are coincident and this situation remains unaltered with an increase or a decrease in the applied loads. In a reinforced concrete section, the location of the neutral axis changes with the applied loads, even if the section is symmetrically reinforced. Also, the location of the neutral axis varies along the length of the member due to the presence of cracks.

(4) There are no jumps or local discontinuities in the curvature diagram for a structural steel member; however, discontinuities occur in the curvature distribution along the

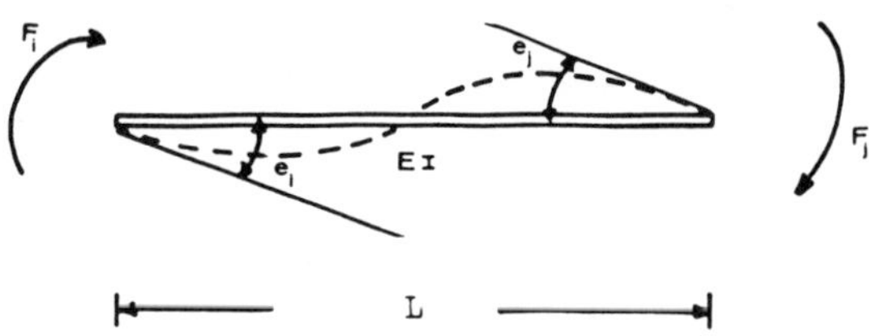

FIGURE 64. Elastic curve of member ij.

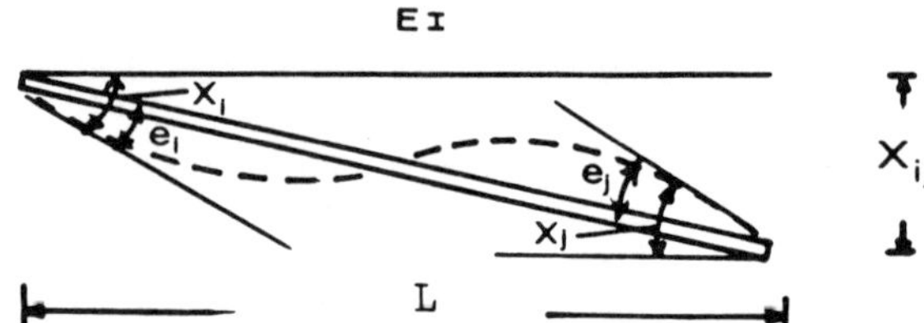

FIGURE 65. External joint rotations and displacement.

length of a reinforced concrete member on account of cracking and crushing of the concrete and the accompanying stress redistribution.

(5) The stress-strain curve for structural steel is generally identical in tension and compression and therefore a structural steel member can be expected to have the same strength if the applied bending moments are reversed. However, the strength of a reinforced section in positive, and negative bending is dependent on the quantity and the arrangement of tension and compression steels and the extent of cracking at the different sections. The bending strength of a reinforced concrete section can therefore be significantly different under reversal of applied loads. Similarly, the ductility of a reinforced concrete section is dependent on the reinforcement details.

(6) The provision of stirrups in a reinforced concrete beam does not only add to the shear strength of the beam but also increases the ductility of the concrete. This increased compressive strain capacity significantly improves the rotation capacity of the hinging region although it does not add to the strength of the section.

MATHEMATICAL FORMULATION

The stiffness matrix of a member $\underset{\sim}{S}$ relates the joint moments $\underline{F}$ and the corresponding rotations $\underline{e}$ as follows:

$$\underline{F} = \underset{\sim}{S}\underline{e} \tag{93a}$$

Equation (93a) can be written explicitly for a member ij (Figure 64) as

$$\underline{F}_i = \underset{\sim}{S}_{ii}\,\underline{e}_i + \underset{\sim}{S}_{ij}\,\underline{e}_j \tag{93b}$$

$$\underline{F}_j = \underset{\sim}{S}_{ji}\,\underline{e}_i + \underset{\sim}{S}_{jj}\,\underline{e}_j \tag{93c}$$

It must be noted that Equation (93) accounts only for flexural deformations in the member ij, while the axial and shear deformations are neglected. This follows an earlier formulation by Wang [88,94]. A more complete formulation for an elastic-plastic material has been developed by Livesley [92,93] and Jennings and Majid [98].

If a simple or a plastic hinge is introduced at the jth end of the member ij as shown in Figure 64, the stiffness matrix $\underset{\sim}{S}$ gets modified as [88,94]:

$$\underset{\sim}{S} = \begin{bmatrix} S'_{ii} & 0 \\ 0 & 0 \end{bmatrix} \tag{94a}$$

Similarly, if a simple or a plastic hinge is introduced at the ith end of the member ij as shown in Figure 64, then $\underset{\sim}{S}$ gets modified as

$$\underset{\sim}{S} = \begin{bmatrix} 0 & 0 \\ 0 & S'_{jj} \end{bmatrix} \tag{94b}$$

For a typical member ij in bending (Figure 65),

$$\underline{F} = \underset{\sim}{S}\,\underline{e}$$

or

$$\underline{e} = \underset{\sim}{S}^{-1}\,\underline{F} = \underset{\sim}{D}\underline{F} \tag{95}$$

where $\underset{\sim}{D}$ is the flexibility matrix for the member ij.

The external rotations X_i, X_j, and the deflection X_{ij} of end j relative to an end i (Figure 65) are related to the internal rotations e_i and e_j by the

$$\underline{e} = \underset{\sim}{B}\underline{X} \tag{96a}$$

where $\underset{\sim}{B}$ = the deformation matrix, and

$$\underline{X} = \begin{Bmatrix} X_i \\ X_j \\ X_{ij} \end{Bmatrix} \tag{96b}$$

Equation (96) can be rewritten as

$$\underline{e} = \underset{\sim}{A}^T\,\underline{X} \tag{97a}$$

where $\underset{\sim}{A}^T = \underset{\sim}{B}$ and $\underset{\sim}{A}$ is the statics matrix which relates the externally applied forces $\underline{P}$ to internal end moments $\underline{F}$ as follows:

$$\underline{P} = \underset{\sim}{A}\,\underline{F} \tag{97b}$$

If there are no hinges either at i or at j, then continuity requires that the internal end rotations $\underline{e}$, as caused by the end moments $\underline{F}$, be equal to those caused by the external joint rotations or displacements. Therefore it follows from Equations (95) and (97) that

$$\underset{\sim}{D}\,\underline{F} = \underset{\sim}{A}^T\,\underline{X} \tag{98}$$

whence

$$\underline{F} = \underset{\sim}{D}^{-1}\,\underset{\sim}{A}^T\,\underline{X} = \underset{\sim}{S}\,\underset{\sim}{A}^T\,\underline{X} \tag{99}$$

which is a familiar equation from the displacement method of rigid frame analysis (see References [88] and [94] by Wang).

If there is a hinge at any member end, then the hinge rotation $\underline{H}$ is given by the angle from the direction of the member as required by the external joint rotations or displacements in Figure 65 to that caused by the end moments in Figure 64. Therefore,

$$\underline{H} = \underset{\sim}{D}\,\underline{F} - \underset{\sim}{A}^T\,\underline{X} \tag{100}$$

Wang's program [88] was developed as a general purpose computer approach for limit analysis of steel plane frames but it lacks the sophistication to handle the concrete structures. Thus the computer program shown in Reference [95] differs from Wang's program in some of its details, but the major difference is in the philosophy of handling the collapse stage. Wang [88] used the conventional mechanism approach to handle the collapse stage in plane frame. Hsu's [152] approach was based on the deformation (i.e., rotation capacity of the plastic hinges) and mechanism approaches to control the failure stage. Also, the program and flow chart shown in References [95] and [152] is less computer time consuming than other published programs and can be applied to the multi-storey plane frames and continuous beams in cases where the moment-curvature characteristics of member sections are reasonably elastic-perfectly plastic.

ANALYSIS OF CEMENT AND CONCRETE ASSOCIATION TEST RESULTS

The problem analyzed using the above computer program is frame FP4 (see Table 9) tested by Cranston and Cracknell [91]. The frame details are shown in References [95] and [152]. It must be noted that the value of flexural rigidity *(EI)* used in the main computer program [95] is the slope of the moment-curvature curve (under a constant axial force at the section centroid) obtained using the computer program in a previous section, Method 2.

The computer analysis results are compared with the experimental and analytical data from References [33] and [91]. Moment-curvature curves for two typical sections are shown in Figure 66. The idealization of the frame for the computer analysis is shown in Figure 67 and Figure 68.

The location and the sequence of formation of hinges and the final collapse mechanism obtained from computer analyses are shown in Figure 69. The failure mechanism obtained experimentally is shown in Figure 69 and is in good agreement with the results of above computer analysis.

The experimental and the computed load-deflection curves are shown in Figure 75. The computed ultimate strengths of the frames is lower than the experimental value because the moment-curvature relationship used in analysis is assumed to be a conservative elastic-perfectly plastic idealization of the moment-curvature results. However, both the horizontal and vertical deflections at the maximum load agree well with the test results showing that the curvature formulation used is satisfactory. Moreover, there is an excellent agreement between the experimental and the computed $\Delta_v - \Delta_s$ curves.

It must be noted that the "collapse mechanism" method has been successfully used in the ultimate load analysis of frame structures. Dependent on the relative values of loads W_1 and W_2, it is possible to obtain four different collapse mechanisms as shown in Figure 70 [91]. The relationship between W_1 and W_2 for these four mechanisms is shown in Figure 70 which also shows the experimental results and the above computer analysis results. A comparison of Cranston's computer analysis results with results obtained from the "collapse mechanism" approach is shown in Figure 70.

Non-Linear Computer Analysis of Reinforced Concrete Plane Frames—Method 2

COMPUTER METHOD

Perhaps the more sophisticated computer analysis to handle simple concrete frame structures has been developed by Cranston [103] who used the flexibility approach as the basis. Cranston's program is general in its theory and predicts the behaviour of a simple framed structure as it is loaded to failure. It has been used to predict the behaviour of simple frames, when moment-curvature relationships for the various cross-sections were available. The most important feature of Cranston's work is that his computer program can handle not only the complete moment-curvature curve (including the descending branch of the curve) but also the complete load-deflection curve by load and deflection control processes (including the descending branch of the curve). However, if (1) there is a lengthy segment of the moment-curvature curve approaching a zero slope or (2) when there is a sharp transition in the slope of linear segment of the curve, the solution may be impeded [104].

Becker [104] modified Cranston's [103] computer program for a single bay one storey plane frame in which the difference is in the philosopy of how to handle the regions that the "plastic hinges" are formed. Becker used the conditions of deformational compatibility at the supports as suggested by Sawyer [131]. It must be noted that Baker [126] and Cranston [103] handled the solution by using released hinges.

TABLE 9. Cranston's Specimen FP4 (Loading II).

Notations	b (in)	t (in)	d (in)	d′ (in)	Z (s.s.) (in)	fc′ (psi)	f_t (psi)	As (in²)	As′ (in²)	Es (psi)	f_y (psi)	E_{sh}	E_{ss} (psi)	f'_y (psi)	$n = \frac{Es}{Ec}$	P (lbs)	EI kip-ft²	M_y kip-ft	Inelastic hinge rotation
Computer Program Notation / Section	B	OD	D	DC	Z	FCU	FT	AS	ASC	E	FY	ESH	ESS	FY1	N	P1	EI	PM	HRC
A1	4	6	5.25	0.75	20	3699	500	0.3925	0.3925	30425000	63700	0.011	652000	63700	8.7	0			
A2	4	6	5.25	0.75	20	3699	500	0.3925	0.3925	30425000	63700	0.011	652000	63700	8.7	5300	1180.6	10.5	0.01834
A3	4	6	5.25	0.75	20	3699	500	0.3925	0.3925	30425000	63700	0.011	652000	63700	8.7	7000	1160.4	10.94	0.0158
B1	4	6	5.25	0.75	20	3699	500	0.5888	0.3925	30425000	63700	0.011	652000	63700	8.7	0			
B2	4	6	5.25	0.75	20	3699	500	0.5888	0.3925	30425000	63700	0.011	652000	63700	8.7	5300	1497.9	14.85	0.00843
B3	4	6	5.25	0.75	20	3699	500	0.5888	0.3925	30425000	63700	0.011	652000	63700	8.7	7000	1473.1	15.05	0.007542
B4	4	6	5.25	0.75	20	3699	500	0.5888	0.3925	30425000	63700	0.011	652000	63700	8.7	4500	1507.4	14.83	0.00869

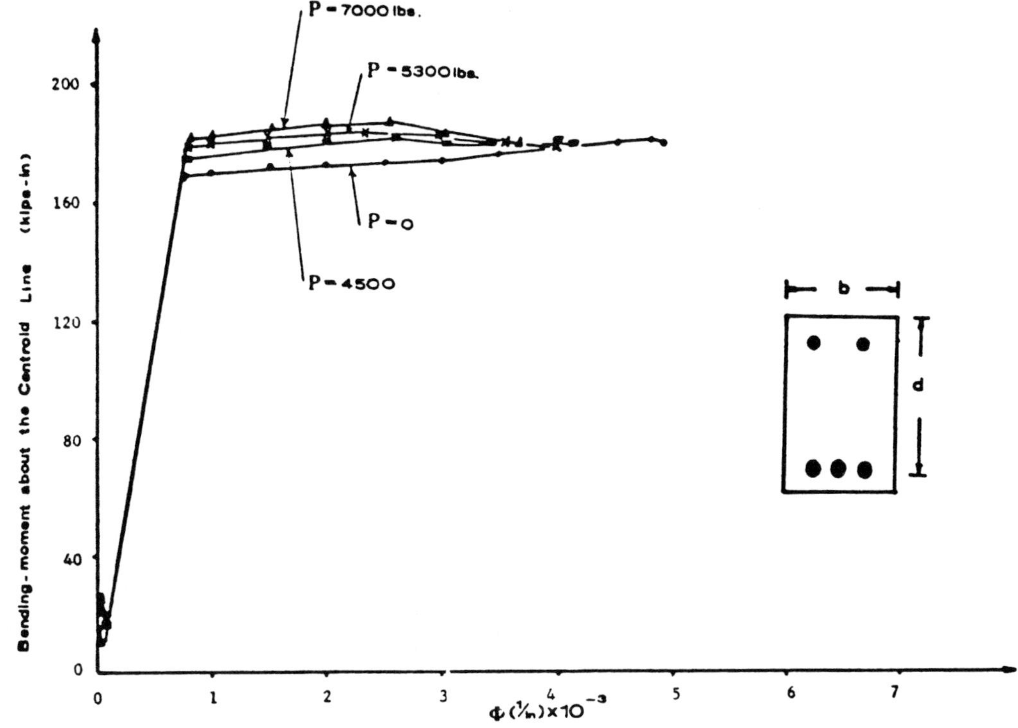

FIGURE 66a. Moment-curvature curves for Section B series

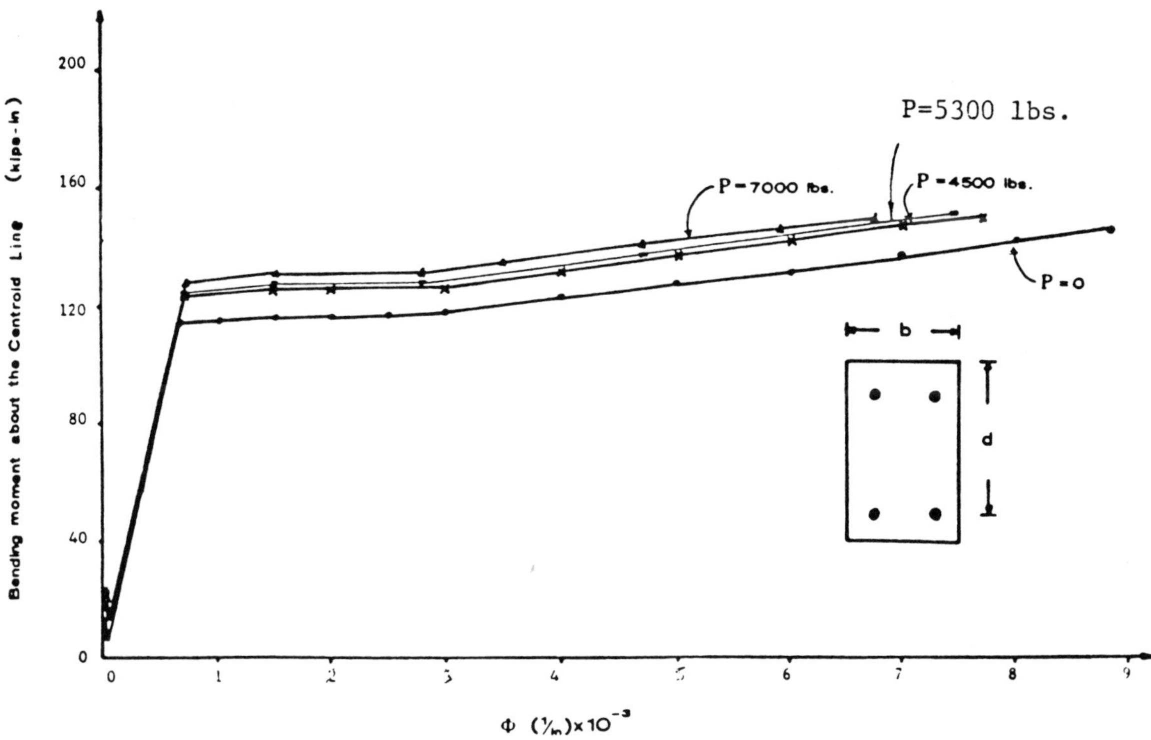

FIGURE 66b. Moment-curvature curves for Section A series.

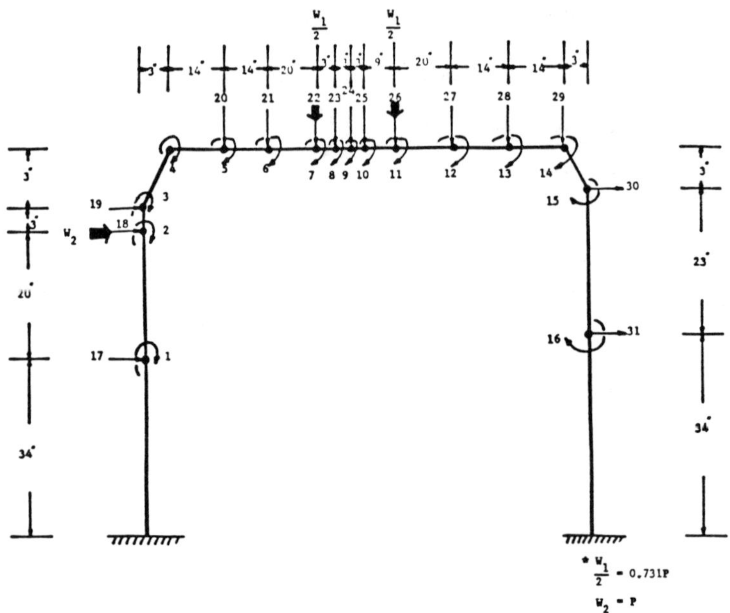

FIGURE 67. The P-X diagram for frame FP4.

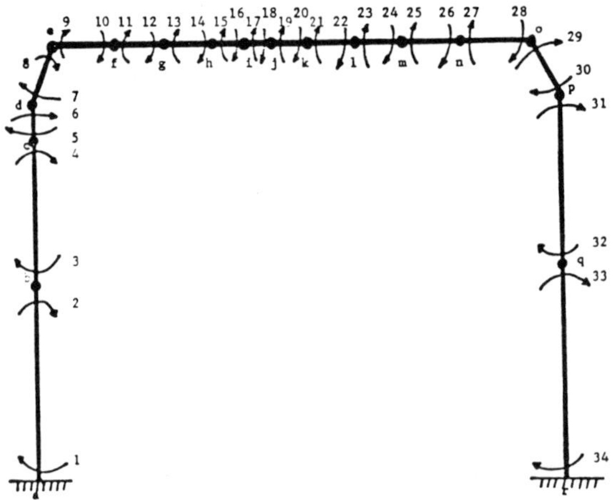

FIGURE 68. The F-e diagram for frame FP4.

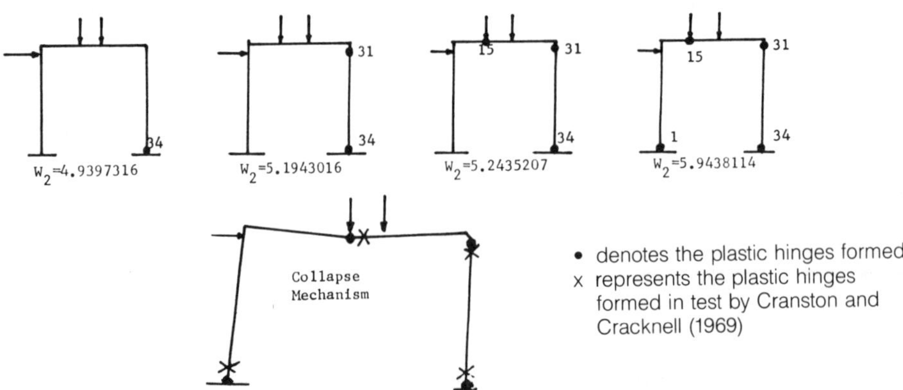

FIGURE 69. Process of mechanism formed by Hsu's analysis.

Cranston's and Becker's programs are reasonably complete and the analysis results agree reasonably with the test data [33,103,104].

The computer program shown in Reference [152] is a modification of Becker's computer program [104] and has been written in the FORTRAN language. Also, the subroutine SHAPE has been modified to the basic development of this program which can be found in References [103] and [104].

ANALYSIS OF CORNELL UNIVERSITY COMPUTER RESULTS

Becker [104] presented several computer results for his single bay one storey rectangular frames. The typical simple frame with fixed end conditions is shown in Figure 71. The moment-curvature curve shown in Figure 72 is assumed to apply to all sections in frame. Hsu [152] divided the moment-curvature curve into five linear segments as shown in Figure 72.

Figure 73a compares Hsu's computer results with Becker's results in which the influence of the hinge length has been accounted for in determining the load-deflection curves. The computer load-lateral deflection curves are compared with Becker's results for moment-curvature curves (Figure 72) idealized using bilinear, trilinear and other approximations.

ANALYSIS OF CEMENT AND CONCRETE ASSOCIATION TEST RESULTS

Cranston and Cracknell [91] and Cranston and Chatterji [33] tested and analyzed single bay reinforced concrete frames with the column ends fixed as shown in Figure 74. The idealization of the frame for the computer analysis is shown in Figure 74, and the load-deflection curves obtained using Hsu's computer program are compared with the experimental data [33,91] as seen in Figure 75. Note that a value of 6 in. was used for D2 in Method 2. Figure 75 indicates that Hsu's Method 2 agrees better with the test results than Hsu's Method 1 (because of the elastic-perfectly plastic assumption in moment-curvature curves). However, the hinging length of the plastic hinges D2 taken as 6 in. seems to be conservative and results in the discrepancies in the sideways deflection as shown in Figure 75. Another improvement in Hsu's computer analysis is the inclusion of the effect of the axial load on the moment-curvature characteristics as shown in Figure 66. The results of computer analysis by Cranston et al. [91,33], shown in Figure 75, overestimate both the strength and the deflection at any load level. This is due to their analytical moment-curvature curves for sections and the assumption of the hinging length for plastic hinges.

It can be concluded that Hsu's Method 1 can predict deflection reasonably accurately while Method 2 gives better strength prediction.

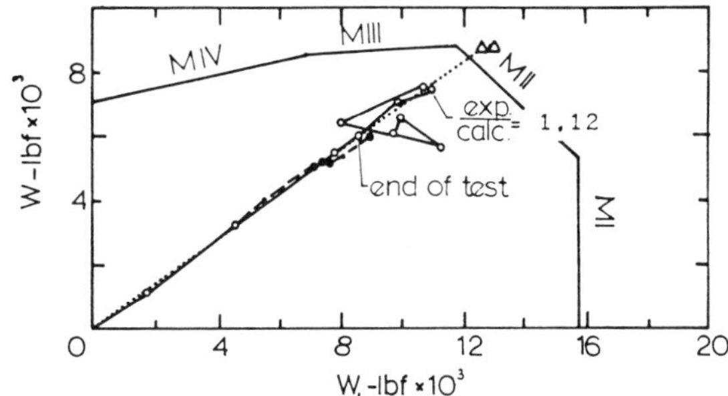

...△... Cranston & Chatterji's computer analysis
—○— experimental test by Cranston & Cracknell for specimen no. FP4
-+- the writer's computer analysis (Method 1)
—— MI MII MIII MIV–mechanisms calculated by Cranston & Cracknell (1969)

FIGURE 70. Maximum loads for reinforced concrete frame FP4.

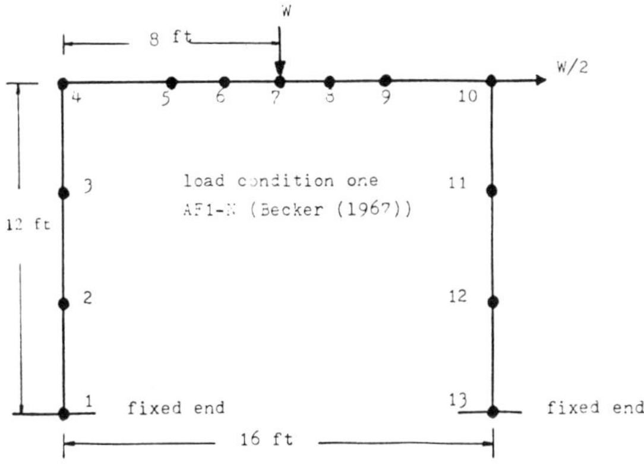

FIGURE 71. Load condition for a plane frame AF1-N.

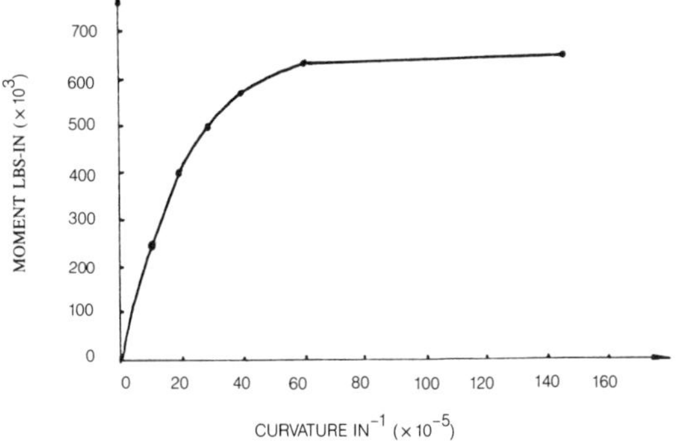

FIGURE 72. Moment-curvature relationship for all sections in plane frame AF1-N.

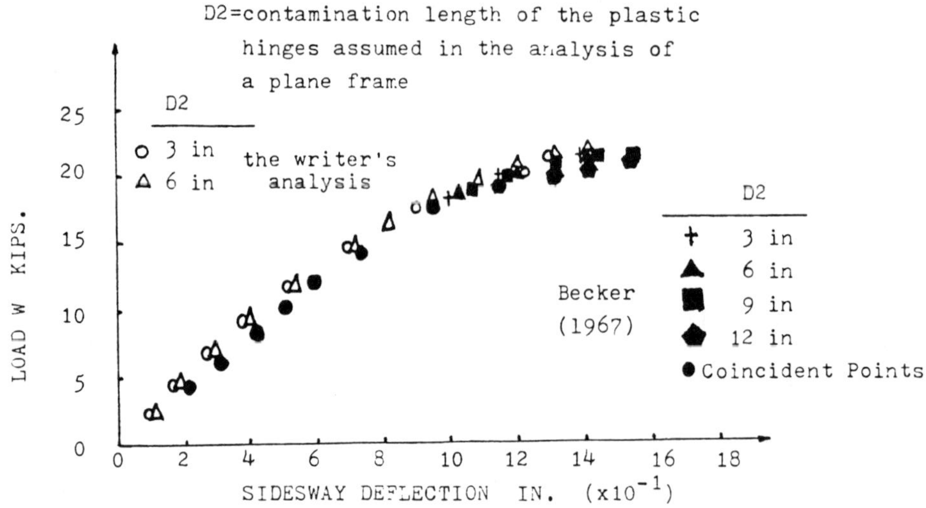

FIGURE 73a. Load-deflection curves for a plane frame AFI-N.

CONCLUSIONS

The results of this investigation can be summarized and conclusions drawn as follows.

Three approaches were developed in the present investigation to study the generalized moment-curvature characteristics for a given reinforced concrete section subjected to uniaxial bending and/or with constant axial load. These are based on a deformation analysis which is detailed earlier. The analytical results were compared with the current ACI Building Code, the National Building Code of Canada, the Japanese Building Code and the Portland Cement Association test results. The results show that the moment-curvature curve for a given reinforced concrete section includes not only an ascending but also a descending branch of the curve. These constitute a basis of the limit analysis of two-dimensional concrete structures. The results also discuss the ductility of a given section—a subject which has not received sufficient attention in the current design practice.

A numerical analysis was also developed in the present investigation to evaluate the moment-curvature relationship subjected to biaxial bending and/or axial load (shown in another chapter in this book). This computer program has the ability to use any standard reinforced concrete section geoemetry and material properties. For a given section under combined biaxial bending and axial compression, the

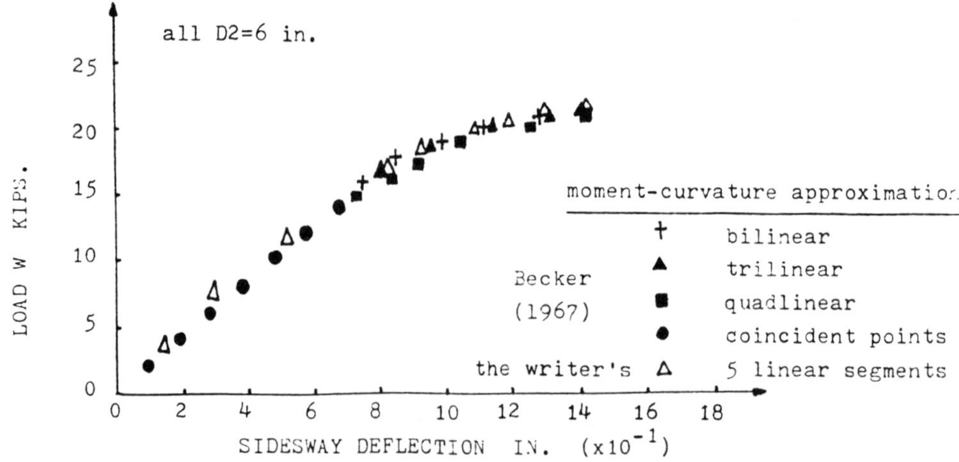

FIGURE 73b. Load-deflection curves for a plane frame AF1-N.

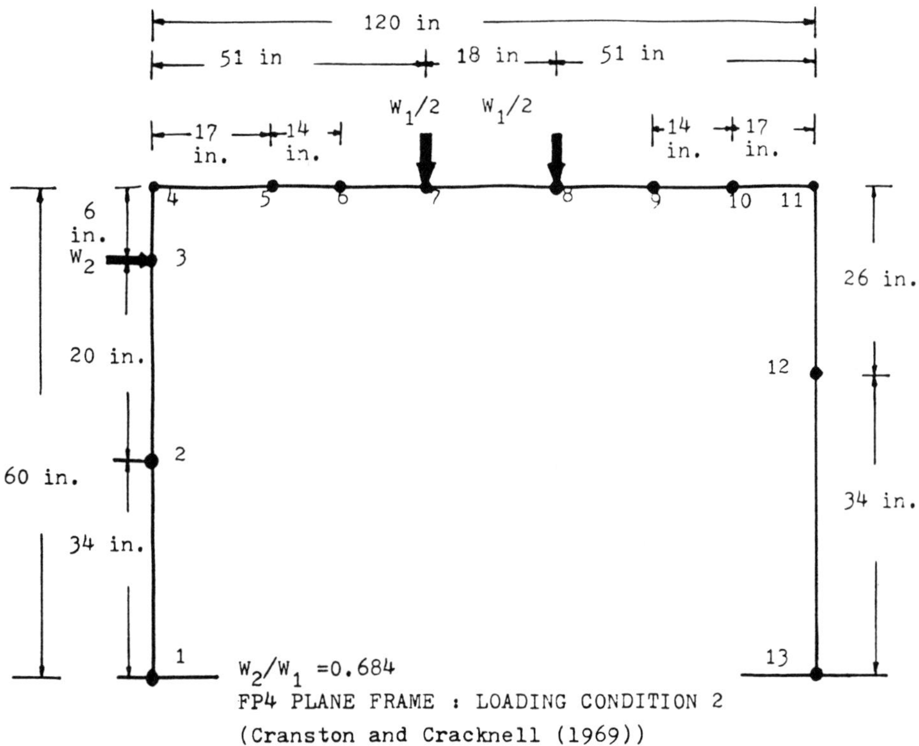

FIGURE 74. Reinforced concrete plane frame FP4 with fixed ends.

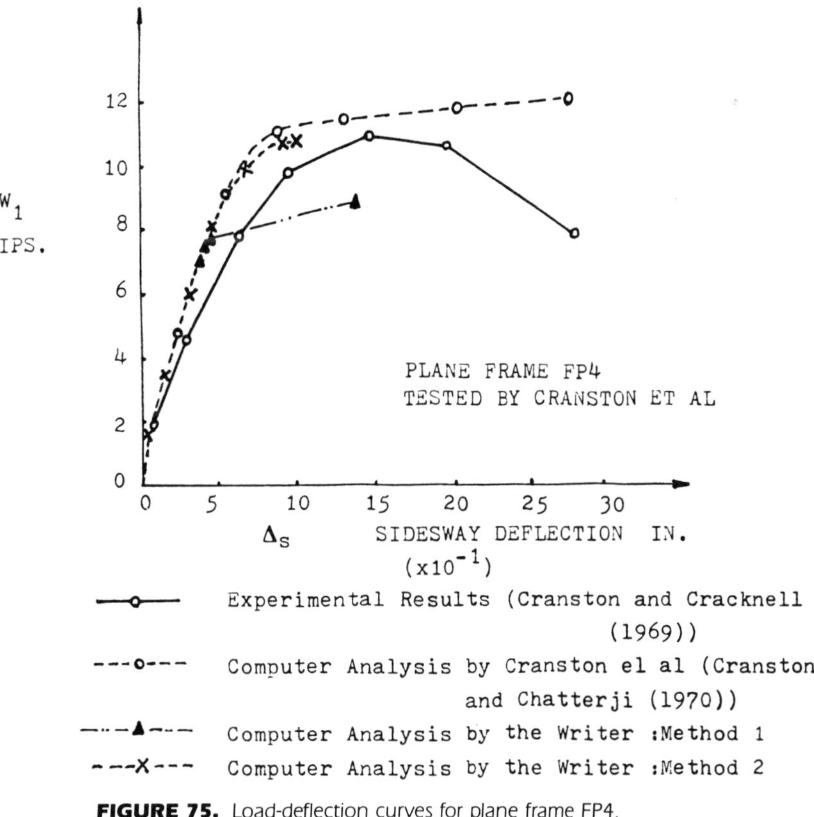

FIGURE 75. Load-deflection curves for plane frame FP4.

algorithm is based on control of loads and therefore the moment-curvature results can be obtained until the maximum moment capacity of the section is reached. For a given section under biaxial bending with a constant axial load, the analysis is based on a deformation control and therefore the biaxial moment-curvature curve gives the ascending and descending branches of the curve. The biaxial moment-curvature analytical results have been compared with the experimental results for biaxially loaded column tests and a satisfactory agreement was obtained with the calculated values [152,153].

An approximate method based on the modified moment-area theorem was developed to evaluate the rotations and deflections in statically determinate concrete structures. The analytical results were compared with the experimental results from specimens tested at the University of Illinois and McGill University and a satisfactory agreement was noted. The analysis of reinforced concrete continuous beams, based on an assumed idealization of a series of the simply supported beams spanning between the point of zero moments, was found useful in predicting the behaviour of under-reinforced concrete structures.

The reinforced concrete plane frames were analyzed using the moment-curvature curves developed and assumed plastic hinge regions. Computer results were compared with the test data from the Cement and Concrete Association and elsewhere. The results show that the ultimate strength and the ductility of the structure as a whole and the factor of safety against failure remain undefined in the present design practice for an indeterminate concrete structure.

ACKNOWLEDGEMENTS

The author wishes to express his appreciation and gratitude to the following persons and organizations.

Professor J. O. McCutcheon, Dr. P. J. Harris and Dr. M. S. Mirza of McGill University and Dr. A. A. Mufti of Nova Scotia Tech. Univ.: their suggestions and encouragements in the early years of this study are acknowledged.

Messrs. M. Adaszkiewicz and Y. Faucher, who assisted the author in most of the experimental tests. The staff of the Structural Laboratory at McGill University, and in particular Mr. J. Bachelor, who operated the testing machine in most of the author's experimental tests. Messrs. C. W. Tan and G. Pitcher, McGill undergraduates, who assisted the author in making the specimens during the summer of 1973.

Mr. D. Price of Uniroyal Ltd. of Montreal provided hard rubber blocks; Dr. J. Houde of Ecole Polytechnique loaned load cells to the author.

The financial assistance of the National Research Council of Canada, the Canada Emergency Measures Organization and the Department of Education, Government of Quebec to continuing research programs at McGill University is gratefully acknowledged. Thanks are due for the fellowship awarded to the author by Miron Company Ltd. of Montreal during the years 1971–72 and 1972–73.

LIST OF SYMBOLS

$\underset{\sim}{A}$ = static matrix
$A_{\eta x}$ = $\int_o^{x_c} f(x)dx$ = area of normalized stress-strain curve for concrete
a_k = area of element k
A_s = area of tension reinforcement in section
A_s' = area of compression reinforcement in section
A_s'' = area of lateral reinforcement
A_g = gross area of the section
$\underset{\sim}{B}$ = deformation matrix
$B_{\eta x}$ = $\int_o^{x_c} f(x)xdx$ = first moment of the normalized stress-strain curve area about the stress axis
b = beam width
B_1, B_2, B_3, B_4, B_5, B_6, B_7, and B_8 = coefficients
C_c = compressive force in concrete
C_{c1} = compressive force in concrete ($\epsilon \leq \epsilon_0$)
C_{c2} = compressive force in concrete ($\epsilon > \epsilon_0$)
C_s = force in compression steel reinforcement
d = effective depth of the beam
D = overall depth of the column prismatic section
d' = distance from the extreme compression fibre to centroid of the compression steel reinforcement
$\underset{\sim}{D}$ = flexibility matrix
$D2$ = contamination length of the plastic hinges assumed in the analysis of a plane frame
d''' = distance from the extreme tension fibre to centroid of the tension steel reinforcement
$\underline{e}$ = rotation matrix
E_0 = f_0/ϵ_0
$(E_t)_k$ = tangent modulus of elasticity for steel or concrete element k
E = the modulus of elasticity of the member ij
E_c = modulus of elasticity for concrete
E_s = initial modulus of elasticity for tension steel
E_s' = initial modulus of elasticity for compression steel
E_{ss} = modulus of elasticity of tensile steel after strain-hardening
F = compatibility factor
$\underline{F}$ = joint moment matrix
f_t = tensile stress in concrete
f_0 = intercept on the stress axis
f_u = ultimate tensile strength of steel
f_{95} = $0.95f_u$
f = concrete stress
f_c = concrete stress at the extreme compression fibre
f_c' = maximum concrete compressive stress
f_y = yield strength of tension reinforcement
f_y' = yield strength of compression reinforcement
f_y'' = yield strength of lateral reinforcement
f_k = stress across the element k

f_s = strength of tension reinforcement
f_s' = strength of compression reinforcement
h_B, $h_{m'}$ and h_D = see Figure 37
$\underline{H}$ = hinge rotation matrix
I = the moment of inertia of the member ij
$j = (d - \beta c)/d$
k_1, k_2 and k_3 = stress block parameters
$k = x'/d$
$\underline{K}$ = a symmetric matrix
$\tilde{k'}$ see Figure 38a
L = the length of the member ij
M = bending moment
M_{cr} = moment at cracking in which the crack starts to appear in concrete at the extreme tension fibre
M_y = moment at yield in which the farthest tension reinforcing bar reaches the yield point
$M_m = M_{max}$ = maximum moment
M_u = moment at failure
M_x = bending moment about x-axis (left-hand rule)
M_y = bending moment about y-axis (right-hand rule)
$n = E_s/E_c$
P_{cr} = axial load at cracking in which the crack starts to appear in concrete at the extreme tension fibre
P = axial load
$\underline{P}$ = externally applied forces matrix
p_b = reinforcement ratio producing balanced conditions at ultimate strength
$p = A_s/bd$
$p' = A_s'/bd$
p''' is the ratio of the total volume of stirrup steel per unit length to the total volume of concrete enclosed by the stirrup per unit length
$= [2(b'' + d'')A_s'']/b''d''S$
$R = M/bd^2$
$\underline{S}$ = stiffness matrix of a member
$\underline{\tilde{S}}_{ij}$ = stiffness matrix of a member, ij
S = spacing of lateral reinforcement
T_s = force in tension steel
t = overall depth of the beam section
u' = allowable compatibility for P
W_1 = vertical applied load
W_2 = horizontal applied load
w = weight of concrete, lb. per cu. ft.
x' = distance from the position of the central axis to the extreme concrete compressive fibre
x_3 = distance between C_s and T_s
$x = \epsilon/\epsilon_0$ = non-dimensional concrete strain factor
$x_c = \epsilon_c/\epsilon_0$
x_k = the position of the centroid of element k along x-axis
$\bar{x} = B_{\eta x}/A_{\eta x}$
x_1 = distance between the positions of C_{c1} and T_s
x_2 = distance between the positions of C_{c2} and T_s
x_1' = distance between the position of C_{c1} to the position of the neutral axis
X_i = external rotation
X_j = external rotation

$$\underline{X} = \begin{Bmatrix} X_i \\ X_j \\ X_{ij} \end{Bmatrix}$$

X_{ij} = the deflection of end j relative to end i
y_k = the position of the centroid of element k along y-axis
y see Figure 21
y_1 see Figure 21
Z = shear span of the beam
$Z'' = Z'/\eta_c$
$Z' = A_{\eta x}/x_c$
Δ_s = sideways deflection
Δ_v = vertical deflection
$\eta = f/f_c'$ = non-dimensional concrete stress factor
$\eta_c = f_c/f_c'$
ϵ = compressive strain in concrete from the compression test
ϵ_c = compressive strain in concrete at the extreme fibre
ϵ_0 = concrete compressive strain w.r.t. f_c'. It is taken to be 0.002 in the present investigation, or as indicated
ϵ_u = ultimate concrete compressive strain
ϵ_{sh} = steel strain at the start of strain-hardening
ϵ_{su} = ultimate steel strain at failure
ϵ_y = yield strain in tension steel
ϵ_y' = yield strain in compression steel
ϵ_{95} = steel strain w.r.t. f_{95}
ϵ_s = steel strain in tension steel $= f_s/E_s$
ϵ_{u2p} = concrete compressive strain shown in Figure 10
ϵ_{u3p} = concrete compressive strain shown in Figure 10
ϵ_s' = steel strain in compressive steel $= f_s'/E_s'$
ϵ_k = strain across the element k
ϵ_p = uniform direct strain due to an axial load P
ϵ_{cu}'' = concrete compressive strain shown in Figure 27
ϵ_{cu}''' = concrete compressive strain shown in Figure 27
ϵ_{ct} = concrete tensile strain shown in Figure 27
$\lambda = \phi d$
$\beta = 1 - (B_{\eta x}/x_c A_{\eta x})$
r = radius of curvature of the deflected axis of the bar
ϕ = curvature as a measure of strain gradient in a section
ϕ_{cr} = curvature just before cracking
$\phi_{cr(a)}$ = curvature immediately after cracking
ϕ_y = curvature w.r.t. M_y
ϕ_m = curvature w.r.t. M_m
ϕ_u = curvature w.r.t. M_u
ϕ_{area} = area of the curvature diagram
ϕ_x = the curvature produced by the bending moment component M_x and is considered positive when it causes compressive strain in the positive y-direction
ϕ_y = the curvature produced by the bending moment component M_y and is considered positive when it causes compression in the positive x-direction
δ_1', δ_2' and δ_3' see Figure 38a

δ_B = the deflection at B
$\delta_{m'}$ = the deflection at m'
δ_y = the deflection at yielding
θ = relative rotation = the angle between the tangents to the deflected curve at two points
θ_A = slope at A (see Figure 37)
θ_B = the relative rotation between A and B (see Figure 37)
θ_C = the relative rotation between B and m' (see Figure 38a)
θ_{p1} = the inelastic rotation of the plastic hinge
subscripts (m) and ($m + 1$) indicates the "m" and "$m + 1$" iteration cycles
subscript (c) indicates values calculated in an iteration cycle
subscript (s) indicates the expected values in an iteration cycle
bar"–" indicates row or column matrix
superscript (T) indicates the transpose of a matrix
Tilde "∼" indicates a general matrix

REFERENCES

1. Hognestad, E., "A Study of Combined Bending and Axial Load in Reinforced Concrete Members," Bulletin No. 399, Engineering Experiment Station, University of Illinois, Urbana, (November 1951).
2. Sargin, M., "Stress-Strain Relationships for Concrete and the Analysis of Structural Concrete Sections," Study No. 4, Solid Mechanics Division, University of Waterloo, Waterloo, Ontario, Canda (1971).
3. Whitney, C. S., "Plastic Theory of Reinforced Concrete Design," *Transactions, ASCE, 107,* 251, Proceedings (Dec. 1940).
4. Smith, G. M. and L. E. Young, "Ultimate Theory in Flexure by Exponential Function," *ACI Journal, Proceedings, 52*(3), 349–360 (November 1955).
5. Hogenstad, E., N. W. Hanson, and D. McHenry, "Concrete Stress Distribution in Ultimate Strength Design," *ACI Journal, Proceedings, 52*(4), 455–479 (December 1955).
6. Baker, A. L. L., "The Ultimate-Load Theory Applied to the Design of Reinforced and Prestressed Concrete Frames," Concrete Publications Limited, London (1956).
7. Whitney, C. S. and E. Cohen, "Guide for Ultimate Strength Design of Reinforced Concrete," *ACI Journal, Proceedings, 28*(5) (November 1956).
8. Janney, J. R., E. Hognestad, and D. McHenry, "Ultimate Flexural Strength of Prestressed and Conventionally Reinforced Concrete Beams," *ACI Journal, Proceedings, 52*(6), 601–620 (February 1956).
9. Viest, I. M., R. C. Elstner, and E. Hognestad, "Sustained Load Strength of Eccentrically Loaded Short Reinforced Concrete Columns," *ACI Journal, Proceedings, 52*(7), 727–755 (March 1956).
10. Mattock, A. H. and L. B. Kriz, "Ultimate Strength of Non-Rectangular Structural Concrete Members," *ACI Journal, Proceedings, 57*(7), 737–766 (January 1961).
11. Mattock, A. H., L. B. Kriz, and E. Hognestad, "Rectangular Concrete Stress Distribution in Ultimate Strength Design," *ACI Journal, Proceedings, 57*(8) 875–928 (February 1961).
12. Pfrang, E. O., C. P. Siess, and M. A. Sozen, "Load-Moment-Curvature Characteristics of Reinforced Concrete Cross-Sections," *ACI Journal, Proceedings, 61*(7), 763–777 (1964).
13. Fowler, T. J., "Reinforced Concrete Columns Governed by Concrete Compression," Ph.D. Thesis, University of Texas, Austin, Texas (January 1966).
14. Green, R., "Behaviour of Unrestrained Reinforced Concrete Columns under Sustained Load," Ph.D. Thesis, University of Texas, Austin, Texas (January 1966).
15. Warner, R. F., "Biaxial Moment Thrust Curvature Relations," *Journal of the Structural Division, ASCE, ST. 5,* 923 (May 1969).
16. Eiklid, G. E., K. H. Gerstle, and L. G. Tulin, "Strain-hardening Effects in Reinforced Concrete," *Magazine of Concrete Research, 21*(69), 211 (December 1969).
17. Kent, D. C. and K. Park, "Flexural Members with Confined Concrete," *Journal of the Structural Division, ASCE, ST. 7,* 1969 (July 1971).
18. Cohn, M. Z. and S. K. Ghosh, "The Flexural Ductility of Reinforced Concrete Sections," Report No. 100, Solid Mechanics Division, University of Waterloo, Waterloo, Ontario, Canada (November 1971).
19. Mattock, A. H., "Rotational Capacity of Hinging Regions in Reinforced Concrete Beams," *Proceedings of the International Symposium on Flexural Mechanics of Reinforced Concrete,* Miami, Florida, November 1964, ASCE -1965-50, ACI SP-12, pp. 143–182 (1965).
20. Khan, A. Q., "An Investigation of the Behaviour of a Three-Dimensional Reinforced Concrete Connection," M. Eng. Thesis, McGill University (July 1969).
21. Lee, L. H. N., "Inelastic Behaviour of Reinforced Concrete Members," *Transactions, ASCE, 120,* 181–202 (1955).
22. Sahlin, S., "Effect of Far-Advanced Compressive Strains of Concrete in Reinforced Concrete Beams Subjected to Bending Moments," *Betong., 40*(3), Library Translation No. 65, Cement and Concrete Association, London (1955).
23. Kriz, L. B., "Ultimate Strength Criteria for Reinforced Concrete," *Journal of the Engineering Mechanics Division, ASCE, 85*(EM3), Proc. Paper 2095, 95–110 (July 1959).
24. Ali, I., "Ultimate Flexural Strength of Reinforced Concrete—A New Approach," *Indian Concrete Journal, 33*(3), Bombay, 83–104 (March 1959).
25. Kriz, L. B. and S. L. Lee, "Ultimate Strength of Over-reinforced Beams," *Journal of the Engineering Mechanics Division, ASCE, 86*(EM3), Proc. Paper 2502, 95–105 (June 1960).

26. Rüsch, H., "Researches Toward a General Flexural Theory for Structural Concrete," *ACI Journal, Procceedings, 57*(1), 1-28 (July 1960).
27. Young, L. E., "Simplifying Ultimate Flexural Theory by Minimizing the Moment of the Stress Block," *ACI Journal, Proceedings, 57*(5), 549-556 (November 1960).
28. Hsu, C. T., "Study of Plastic Conditions in Simply Supported Reinforced Concrete Beams," M.Sc. Thesis, College of Chinese Culture, Taiwan, China (In Chinese) (May 1967).
29. Hsu, C. T. and Mirza, M. S., "Generalized Moment-Curvature Relationships for Singly and Doubly Reinforced Beams," Structural Concrete Series No. 24, McGill University, 48 (November 1969).
30. Barnard, P. R., "The Collapse of Reinforced Concrete Beams," *Proceedings of the International Symposium—Flexural Mechanics of Reinforced Concrete,* ACI SP-12, 501-520 (1965).
31. Rosenblueth, E. and R. Díaz de Cossío, "Instability Considerations in Limit Design of Concrete Frames," *Proceedings of the International Symposium, Flexural Mechanics of Reinforced Concrete,* Miami, Florida, November 1964, ACI SP-12 p. 439 (1965).
32. Cranston, W. B., "Determining the Relation Between Moment, Axial Load and Curvature for Structural Members," Technical Report TRA 395, Cement and Concrete Association, London (June 1966).
33. Cranston, W. B. and Chatterji, A. K., "Computer Analysis of Reinforced Concrete Portal Frames with Fixed Feet," Technical Repot TRA 444, Cement and Concrete Asociation, London (September 1970).
34. Hsu, C. T. and M. S. Mirza, "Discussion of the Paper—Flexural Member with Confined Concrete by Kent and Park," *Journal of the Structural Division, ASCE, 98*(ST.4), 937 (April 1972).
35. Nawy, E. G., R. F. Denesi, and J. J. Grosko, "Rectangular Spiral Effects on Plastic Hinge Rotation Capacity in Reinforced Concrete Beams," *ACI Journal, 65*(12) (December 1968).
36. ACI-ASCE Committee 428—"Progress Report on Code Clauses for 'Limit Design'," *ACI Journal, 65*(8) (September 1968).
37. ACI Committee 318, "Building Code Requirements for Reinforced Concrete—ACI 318-83," American Concrete Institute, Detroit (1983).
38. Corley, W. G., "Rotation Capacity of Reinforced Concrete Beams," *Proceedings of the ASCE Structural Division, 92*(ST.4) (October 1966).
39. Sargin, M. and V. K. Handa, "A General Formulation for the Stress-Strain Properties of Concrete," Solid Mechanics Division, University of Waterloo Report No. 3, Waterloo, Ontario, Canada (May 1969).
40. You, H. T., "Study of Stress-Strain Relationship in Concrete and the Plastic Design in Reinforced Concrete," M.Sc. Thesis, University of Tokyo, Japan (In Japanese) (1966).
41. Priestley, M. J. N., R. Park, and F. P. S. Lu, "Moment-Curvature Relationships for Prestressed Concrete in Constant-Moment Zone," *Magazine of Concrete Research, 23*(75-76), 69 (June-September 1971).
42. Hsu, C. T. and Mirza, M. S., "Discussion of the Paper—Moment-Curvature Relationships for Prestressed Concrete in Constant-Moment Zone," *Magazine of Concrete Research, 24*(79), 101 (June 1972).
43. Broms, B., "Cracking in Reinforced Concrete Members," *ACI Journal, Proceedings, 61*(12), 1635 (December 1964).
44. Mattock, A. H., "Limit Design for Structural Concrete," Portland Cement Association, Research and Development Lab., Vol. 1, No. 2, Bulletin D38 (May 1959).
45. ACI Committee 318-71, "Building Code Requirement for Reinforced Concrete (ACI 318-71)," ACI, Detroit, U.S.A. (1971).
46. Umemura, H., "Plasticity of Reinforced Concrete Beams—Illustration of e-function," (Japanese book).
47. Smith, R. G. and G. D. Matthew, "Behaviour of Reinforced Concrete Beams in Flexure," *The Engineer, 212*(5527), London, 1067-1071 (December 19, 1961).
48. Todeschini, C. E., A. C. Bianchini, and C. E. Kesler, "Behaviour of Concrete Columns Reinforced with High Strength Steels," *ACI Journal, Proceedings, 61*(6), 701-716 (June 1964).
49. Okayama, A., "Stress-strain in Light-weight Concrete Using Artificial Gravels," Chapter 6 (Japanese book).
50. Desayi, P. and S. Krishnan, "Equation for the Stress-strain Curve of Concrete," *ACI Journal, Proceedings, 61*(3), 345 (March 1964).
51. Hsu, T. C. T. and F. O. Slate, "Tensile Bond Strength Between Aggregate and Cement Paste or Mortar," *ACI Journal, Proceedings, 61*(4), 465 (April 1963).
52. Newman, K., "Criteria for the Behaviour of Plain Concrete," *Proceedings of an International Conference, the Structure of Concrete and its Behaviour under Load,* London, September 1965, Cement and Concrete Association, 255 (1968).
53. Vile, G. W. D., "The Strength of Concrete under Short-Term Static Biaxial Stress," *Proceedings of an International Conference, the Structure of Concrete and its Behaviour under Load,* London, September 1965, Cement and Concrete Association, London, 275 (1968).
54. Freudenthal, A. M., *The Inelastic Behaviour of Engineering Materials and Structures,* John Wiley and Sons, Inc. (1950).
55. Popovics, S., "A Review of Stress-Strain Relationships for Concrete," *ACI Journal, Proceedings, 67*(3), 249 (March 1970).
56. Hsu, C. T., "Discussion of the Paper by Sandor Popovics—A Review of Stress-Strain Relationships for Concrete," *ACI Journal Proceedings, 67*(9) (September 1970).
57. Barnard, P. R., "Researchers into the Complete Stress-Strain Curve of Concrete," *Magazine of Concrete Research, 16*(49) (December 1964).
58. Roy, H. E. H. and M. A. Sozen, "Ductility of Concrete," ASCE-ACI Symposium on Flexural Mechanics of Reinforced Concrete, Miami, Florida, November 1964, ACI SP-12 (1965).

59. Richart, F. E., A. Brandtzaeg, and R. L. Brown, "A Study of the Failure of Concrete under Combined Compressive Stresses," University of Illinois Engineering Experiment Station, Bulletin No. 185 (1928).
60. Rüsch, H. and S. Stöckl, "Der Einfluss von Bügeln und Druckstäben auf das Verhalten der Biegedruckzone von Stahlbetonbalken," (The Effect of Stirrups and Compression Reinforcement on the Flexural Compressive Zone of Reinforced Concrete Beams), Berlin, *Deutscher Ausschuss für Stahlbeton,* 148, 75 (1963).
61. Roy, H. E. H. and M. A. Sozen, "A Model to Simulate the Response of Concrete to Multi-axial Loading," Structural Research Series, No. 268, Civil Engineering Studies, University of Illinois, Urbana (June 1963).
62. Chan, W. W. L., "The Ultimate Strength and Deformation of Plastic Hinges in Reinforced Concrete Frameworks," *Magazine of Concrete Research,* 7(21), London, 121–132 (November 1955).
63. Szulczynski, T. and M. A. Sozen, "Load-deformation Characteristics of Reinforced Concrete Prisms with Rectilinear Transverse Reinforcement," Structural Research Series No. 224, Civil Engineering Studies, University of Illinois, Urbana, (September 1961).
64. Bertero, V. V. and C. Felippa, "Discussion of Ductility of Concrete," *Flexural Mechanics of Reinforced Concrete: Proceedings of the International Symposium,* Miami, Florida, November 10–12, 1964. New York. American Society of Civil Engineers, ASCE 1965-50 and ACI SP-12, 227–234 (1965).
65. Soliman, M. T. M. and C. W. Yu, "The Flexural Stress-strain Relationship of Concrete Confinement by Rectangular Transverse Reinforcement," *Magazine of Concrete Research,* 19(61), 223–238 (December 1967).
66. Martin, C. W., "Spirally Prestressed Concrete Cylinders," *ACI Journal. Proceedings,* 65(10), 837–845 (October 1968).
67. Base, G. D. and J. B. Read, "Effectiveness of Helical Binding in the Compression Zone of Concrete Beams," *ACI Journal, Proceedings,* 62(7), 763 (July 1965).
68. Baker, A. L. L. and A. M. N. Amarakone, "Inelastic Hyperstatic Frames Analysis," *Proceedings of the International Symposium on the Flexural Mechanics of Reinforced Concrete,* Miami, Florida, November 10–12, 1964, ASCE 1965-50, ACI SP-12, 85–136 (1965).
69. Sundara Raja Iyengar, K. T., P. Desiyi, and K. Nagi Reddy, "Stress-Strain Characteristics of Concrete Confined in Steel Binder," *Magazine of Concrete Research,* 22(72) (September 1970).
70. Rüsch, H., E. Gasser, and P. S. Rao, "Principes de Calcul du Béton Armé sous des Etats de Contraintes Monoaxiaux," Bulletin d'Information No. 36, Comité Européen du Béton, Luxembourg, 1–112 (June 1962).
71. Levi, F., "The Work of the European Concrete Committee," *ACI Journal, Proceedings,* 57(9), 1041–1070 (March 1961).
72. Liebenberg, A. C., "A Stress-strain Function for Concrete Subjected to Short-term Loading," *Magazine of Concrete Research,* 14(41), London, 85–99 (July 1962).
73. Saenz, L. P., "Discussion of a Paper by P. Desayi and S. Krishnan, 'Equation for the Stress-strain Curve of Concrete'," *ACI Journal, Proceedings,* 61(9), 1229–1235 (September 1964).
74. Sturman, G. M., S. P. Shah, and G. Winter, "Effect of Flexural Strain Gradients on Microcracking and Stress-strain Behaviour of Concrete," *ACI Journal, Proceedings,* 62(7), 805–822 (July 1965).
75. Hughes, B. P., and G. P. Chapman, "The Complete Stress-strain Curve for Concrete in Tension," RILEM Bulletin, New Series, No. 30, 95–97 (March 1966).
76. Lalonde, W. S. and M. F. Janes, *Concrete Engineering Handbook,* First Ed., McGraw-Hill Book Co., New York (1961).
77. Guerrin, A., *Traité de Béton Armé,* (4 vols.), Ed. Dunod, Paris (1959).
78. Ferry-Borges, J. and J. Arga E Lima, "Formation of Cracks in Beams with Low Percentage of Reinforcement," RILEM Symposium on Bond and Crack Formation in Reinforced Concrete, Stockholm, Vol. 2 (1957).
79. Ernst, G. C., "A Brief for Limit Design," *Transactions of ASCE, 121,* 605 (1956).
80. Ernst, G. C. and A. R. Riverland, "Ultimate Load and Deflection for Limit Design of Continuous Structural Concrete," *ACI Journal, 31* (October 1959).
81. Gaston, J. R., C. P. Siess, and N. M. Newmark, "An Investigation of the Load-Deformation Characteristics of Reinforced Concrete Beams up to the Point of Failure," Civil Engineering Studies, Structural Research Series No. 40, University of Illinois, Urbana, IL (December 1952).
82. Timosenko, S. and G. H. MacCullough, *Elements of Strength of Materials,* Third Edition, D. Van Nostrand Company, Inc., Princeton, NJ (May 1959).
83. Bachmann, H., "Influence of Shear and Bond on Rotational Capacity of Reinforced Concrete Beams," International Association for Bridge and Structural Engineering, 30-II, 11–28 (1970).
84. Hsu, C. T. and M. S. Mirza, "An Analytical Investigation of the Load-deflection Characteristics of Reinforced Concrete Beams," Structural Concrete Series, No. 71-9 (October 1971).
85. Hsu, C. T. and M. S. Mirza, "A Study of Post-Yielding Deflection in Simply Supported Reinforced Concrete Beams," *Symposium on 'Deflection of Concrete Structures',* ACI Publication SP-43, 333–356 (1974).
86. Zienciewicz, O. *The Finite Element Method in Engineering Science.* McGraw-Hill Book Company Limited, London (1971).
87. Meek, J. L. *Matrix Structural Analysis.* McGraw-Hill Book Company (1971).
88. Wang, C. K. *Matrix Methods of Structural Analysis.* International Textbook Company, PA, 2nd Ed. (1970).

89. Neal, B. G. *The Plastic Methods of Structural Analysis.* Chapman and Hall Ltd. (1965).
90. Wilson, E. L., "Matrix Analysis of Non-Linear Structures," *Proceedings of the American Society of Civil Engineers Conference on Electronic Computation,* Pittsburgh, 415–428 (September 1960).
91. Cranston, W. B. and J. A. Cracknell, "Tests on Reinforced Concrete Frames 2: Portal Frames with Fixed Feet," Cement and Concrete Association, London, TRA/420 (September 1969).
92. Livesley, R. K., "Automatic Design of Structural Frames," *Quarterly Journal of Mechanics and Applied Mathematics,* 9(3), 257–278 (September 1956).
93. Livesley, R. K. *Matrix Methods of Structural Analysis.* Pergamon and MacMillan (1964).
94. Wang, C. K., "General Computer Program for Limit Analyis," *Journal of the Structural Division, ASCE, 89*(ST.6), 101–118 (December 1963).
95. Hsu, C. T., M. S. Mirza, and A. A. Mufti, "An Elastic-plastic Computer Analysis of Steel and Reinforced Concrete Plane Frames," Structural Concrete Series No. 72-5 (February 1972).
96. Hsu, C. T. and M. S. Mirza, "Limit Analysis of Structural Concrete Frames," paper No. 652, *Proceedings of the 4th Canadian Congress of Applied Mechanics,* Ecole Polytechnique, Quebec, Canada, 11 (May 1973).
97. Gray, P., "The Behaviour of Multi-storey Plane Frame," Ph.D. Thesis, University of Cambridge, England (1965).
98. Jennings, A. and K. Majid, "An Elastic-plastic Analysis by Computer for Framed Structures Loaded up to Collapse," *The Structural Engineer, 43*(12), 407–412 (Dec. 1965).
99. Yamada, Y., T. Kawai, N. Yoshimura, and T. Sakurai, "Analysis of the Elastic-Plastic Problems by the Matrix Displacement Method," *Proceedings of the Second Conference on Matrix Methods in Structural Mechanics,* Air Force Flight Dynamics Laboratory, Wright-Patterson Air Force Base, Ohio, October 1968, 1271–1300 (December 1969).
100. Armen, H., A. Pifko, and H. S. Levine, "A Finite Element Method for the Plastic Bending Analysis of Structures," AFFDL-TR-68-150, 1301–1339 (1969).
101. Drysdale, R. G., M. S. Mirza, and J. O. McCutcheon, "Recent Research in Deflection of Concrete Structures," *Proceedings of the Symposium in Deflections of Structural Concrete,* ACI Canadian Capital Chapter, Montreal, 161–182 (October 1971).
102. Ferry-Borges, J. and E. R. Arantes E Oliveira, "Non-linear Analysis of Reinforced Concrete Structures," *I.A.B.S.E., 23,* 51–69 (1963).
103. Cranston, W. B., "A Computer Method for Inelastic Analysis of Plane Frames," TRA 386, Cement and Concrete Association, London (March 1965).
104. Becker, J. M., "Inelastic Analysis of Reinforced Concrete Frames," M.Sc. Thesis, Cornell Univerity (February 1967).
105. Drysdale, R. G., "Prediction of the Behaviour of Concrete Frames," I.A.B.S.E. Symposium, Madrid, Design of Concrete Structures for Creep, Shrinkage and Temperature Changes (1970).
106. Tan, K. B., "Short-term and Sustained Load Analysis of Concrete Frames," M.Eng. Thesis, McMaster University (1972).
107. Mufti, A. A., M. S. Mirza, J. O. McCutcheon, and J. O. Houde, "A Study of the Behaviour of Reinforced Concrete Elements Using Finite Elements," Structural Concrete Research Series, No. 70-5, McGill University, Montreal (September 1970).
108. Franklin, H. A., "Non-linear Analysis of Reinforced Concrete Frames and Panels," Ph.D. Thesis, University of California, Berkeley (March 1970).
109. Beeby, A. W., "Short-term Deformations of Reinforced Concrete Members," Technical Report TRA 408, Cement and Concrete Association (March 1968).
110. Lehman, D. J., "Short-Term Behaviour of Long Columns in Reinforced Concrete Frames Subjected to Sidesway," M.Eng. Thesis, University of Toronto (October 1968).
111. Shaikh, M. F., M. S. Mirza, and J. O. McCutcheon, "Limit Analysis of Reinforced Concrete Frames," *Transactions of the Engineering Institute of Canada, 14*(A-6) (July 1971).
112. Sader, W. H., "Ultimate Strength of Single Bay One Storey Reinforced Concrete Frames Subjected to Horizontal and Vertical Loading," M.Eng. Thesis, McGill University (August 1967).
113. Adenot, A., "Behaviour of a Double Bay One Storey Reinforced Concrete Frame Subjected to Horizontal and Vertical Loading," Structural Concrete Series No. 70-4, McGill Univerity (September 1970).
114. Beedle, L. S. *Plastic Design of Steel Frames.* John Wiley and Sons, New York (1958).
115. Lazaro, A. L., III and R. Richardo, Jr., "Full-range Analysis of Concrete Frames," *Journal of the Structural Division, 99*(ST.8), 9934 (August 1973).
116. Sawyer, H. A., "Elastic-plastic Design of Single-span Beams and Frames," *Proceedings of ASCE,* paper No. 851 New York (December 1955).
117. Mirza, M. S. and C. T. Hsu, "Discussion of the paper: Progress Report on Code Clauses for Limit Design, by ACI-ASCE Committee 428," *ACI Journal, Proceedings, 66*(3), 221–223 (March 1969).
118. Franchi, A., D. E. Grierson, and M. Z. Cohn, "User-Oriented Nonlinear Analysis Computer System," Nonlinear Design of Concrete Structures, Study No. 14, Solid Mechanics Div., Univ. of Waterloo, Canada, Ed. M. Z. Cohn, 139–172 (1980).
119. Gerlein, M. A. and F. W. Beaufait, "Nonlinear Analysis of Reinforced Concrete Frames," Nonlinear Design of Concrete Structures, Study No. 14, Solid Mechanics Div., Univ. of Waterloo, Canada, Ed. M.Z. Cohn, 173–195 (1980).
120. Barnard, P. R., "On the Collapse of Composite Beam,"

Ph.D. Dissertation, University of Cambridge, England (1963).

121. Truesdell, C. and W. Noll, "The Non-linear Field Theory of Mechanics," *Handbück des Physik Bd. III/3*, Springer-Verlag, Berlin (1965).

122. Granholm, H. *A General Flexural Theory of Reinforced Concrete*, Almquist and Wiksell, Stockholm (1965).

123. Guyon, Y., "Note on the Ultimate-load Design of Statically Indeterminate Structures," Comité Européen du Béton, Bulletin d'Information No. 21, 1 (January 1960).

124. Macchi, G., "Proposed Method of Analysis Based on the Theory of Imposed Rotation," C.E.B. Bulletin d'Information No. 21, 15 (January 1960).

125. Burnett, E. F. P., "The Ultimate Load Design of Statically Indeterminate Concrete Structures," M.Sc. Thesis, University of London (1963).

126. Baker, A. L. L., "Ultimate Load Theory for Concrete Frame Analysis," Transaction ASCE, paper No. 3386, Vol. 127 (1962).

127. Macchi, G., "Elastic Distribution of Moments on Continuous Beams," *Proceedings of the International Symposium 'Flexural Mechanics of Reinforced Concrete,'* Miami, Florida, 1964, ACI SP-12, 237 (1965).

128. I.C.E. Institution Research Committee, "Ultimate Load Design of Concrete Structures," *Proceedings, Institution of Civil Engineering* (February 1962).

129. Cranston, W. B., "A Computer Method for the Analysis of Restrained Columns," TRA 402, Cement and Concrete Association, London (April 1967).

130. Ghosh, S. K., "Some Aspects of the Nonlinear Analysis of Reinforced Concrete Structures," Ph.D. Thesis, University of Waterloo, Waterloo, Ontario, Canada (September 1972).

131. Hsu, C. T. and M. S. Mirza, "A Numerical Analysis of Reinforced and Prestressed Concrete Sections under Combined Flexure and Axial Compression," Structural Concrete Series, No. 72-1, McGill University (February 1972).

132. Hsu, C. T. and M. S. Mirza, "Structural Concrete—Biaxial Bending and Compression," *Journal of the Structural Division, ASCE, 99*(ST.2), 285 (February 1973).

133. Yu, C. W. and E. Hognestad, "Review of Limit Design for Structural Concrete," *Journal of the Structural Division, ASCE, 84*(ST.8), Proc. Paper 1978 (December 1958).

134. Cohn, M. Z., "Application of Limit Design to Reinforced Concrete Structures," No. 1 Solid Mechanics Division, University of Waterloo, Ontario (April 1969).

135. Cohn, M. Z., "Limit Design of Reinforced Concrete Frames," *Proceedings, ASCE, 94*(ST.10), 2467 (October 1968).

136. Sawyer, H. A., Jr., "Design of Concrete Frames for Two Failure Stages," International Symposium on Flexural Mechanics of Reinforced Concrete, ASCE-ACI SP-12, Miami, Florida, November 1964, 405 (1965).

137. Ernst, G. C., "Plastic Hinging at the Intersection of Beams and Columns," *ACI Journal, 28*(12) (June 1957).

138. Cohn, M. Z., "Optimum Limit Design of Reinforced Concrete Continuous Beams," Institution of Civil Engineers, London, 675 (April 1965).

139. Cohn, M. Z., "Limit Design for Reinforced Concrete Structures," ACI Bibliography No. 8 (1970).

140. West Virginia University, "A conference on the behaviour of concrete structural system," Dept. of Civil Engineering, West Virginia University, Mont Chateau Lodge, West Virginia (September 1971).

141. Mattock, A. H., "Redistribution of Design Bending Moments in Reinforced Concrete Continuous Beams," Development Department Bulletin, D30, Portland Cement Association, Skokio, Illinois (1959).

142. Associate Committee on the National Building Code, "National Building Code of Canada," National Research Council, Ottawa (1965).

143. Hawkins, N. M., M. A. Sozen, and C. P. Siess, "Behaviour of Continuous Prestressed Concrete Beams," *Flexural Mechanics of Reinforced Concrete, Proceedings of the International Symposium*, Miami, Florida, 1964, ACI SP-12, 259 (1965).

144. George, A. and M. S. Mirza, "Tests on Reinforced Concrete Continuous Beams," unpublished report, McGill University (1971).

145. Hsu, C. T. and M. S. Mirza, "An Experimental-analytical Study of Complete Load-Deformation Characteristics of Concrete Compression Members Subjected to Biaxial Bending," Work Commission III, Preliminary Report, International Association for Bridge and Structural Engineering, Symposium on 'Design and Safety of Reinforced Concrete Compression Member', Quebec City (August 1974).

146. Drysdale, R. G. and M. S. Mirza, "A Computer Analysis of Deflections of Statically Indeterminate Concrete Structures," ACI SP-43, American Concrete Institute, Symposium on 'Deflection of Concrete Structures', 179 (1974).

147. Hsu, C. T. T., M. S. Mirza, and C. S. S. Sea, "Nonlinear Analysis of Reinforced Concrete Frames," *International Journal of Computer and Structures, 13,* 223–227 (June 1981).

148. Hsu, C. T. T., "A Simple Nonlinear Analysis of Continuous Reinforced Concrete Beams," *Journal of Engineering and Applied Science, 2*(4), 267–276 (1983).

149. MacGregor, J. G., "Challenges and Changes in the Design of Concrete Structures," *Concrete International, Design and Construction, 6*(2), American Concrete Institute, 49–52 (Feb. 1984).

150. Cohn, M. Z., "Nonlinear Design of Concrete Structures: Problems and Prospects," Nonlinear Design of Concrete Structures, Study No. 14, Solid Mechanics Div., Univ. of Waterloo, Canada, Ed. M. Z. Cohn, 3–42 (1980).

151. Ahmad, S. H. and S. P. Shah, "Complete Stress-Strain Curve of Concrete and Nonlinear Design," Nonlinear Design of Concrete Structures, Study No. 14, Solid Mechanics Div., Univ. of Waterloo, Canada, Ed. M. Z. Cohn, 61–81 (1980).

152. Hsu, C. T., "Behaviour of Structural Concrete Subjected to Biaxial Flexure and Axial Compression," Ph.D. Thesis, McGill Univ., 479 p. (August 1974).
153. Hsu, C. T. T. and M. S. Mirza, "Non-Linear Behavior and Analysis of Reinforced Concrete Columns under Combined Loadings," Nonlinear Design of Concrete Structures, Study No. 14, Solid Mechanics Div., Univ. of Waterloo, Canada, Ed., M. Z. Cohn, 109–135 (1980).

APPENDIX–COEFFICIENTS FOR EQUATION (55)

Equation (55) gives:

$$B_1K^7 + B_2K^6 + B_3K^5 + B_4K^4 + B_5K^3 + B_6K^2 + B_7K + B_8 = 0$$

Between Cracking and Ultimate (Case A, $\epsilon_c \le \epsilon_0$):

$$\text{Let } c = \frac{\epsilon_s}{\epsilon_0}$$

$$B_1 = bdf_c'\Big\{0.00453029\, c^6 + 0.0048422\, c^5 + 0.072487\, c^4 + 0.365358\, c^3 + 0.29636915\, c^2 - 0.6135486\, c - 0.000044\,\frac{1}{c}\Big\}$$

$$B_2 = bdf_c'\Big\{-0.0048422\, c^5 - 0.144974\, c^4 - 1.096074\, c^3 -1.185466\, c^2 + 3.067743\, c + 0.000308\,\frac{1}{c}\Big\} - A_s'E_s\epsilon_s - A_sf_s - P$$

$$B_3 = bdf_c'\Big\{0.072487\, c^4 + 1.096074\, c^3 + 1.7782149\, c^2 - 6.135486\, c - 0.000924\,\frac{1}{c}\Big\} + A_s'E_s\epsilon_s\left(5 + \frac{d'}{d}\right) + 6A_sf_s + 6P$$

$$B_4 = bdf_c'\Big\{-\,0.365358\, c^3 - 1.1854766\, c^2 + 6.13546\, c + 0.00154\,\frac{1}{c}\Big\} - 5A_s'E_s\epsilon_s\left(2 + \frac{d'}{d}\right) - 15A_sf_s - 15P$$

$$B_5 = bdf_c'\Big\{0.29636915\, c^2 - 3.067743\, c - 0.00154\,\frac{1}{c}\Big\} + 10A_s'E_s\epsilon_s\left(1 + \frac{d'}{d}\right) + 20A_sf_s + 20P$$

$$B_6 = bdf_c'\Big\{0.6135486\, c + 0.000924\,\frac{1}{c}\Big\} - 5A_s'E_s\epsilon_s\left(1 + 2\,\frac{d'}{d}\right) - 15A_sf_s - 15P$$

$$B_7 = A_s'E_s\epsilon_s\left(1 + 5\frac{d'}{d}\right) + 6A_sf_s - 0.000308\frac{1}{c} + 6P$$

$$B_8 = -A_s'E_s\epsilon_s\frac{d'}{d} - A_sf_s + 0.000044\,(dbf_c')\left(\frac{1}{c}\right) - P$$

Between Cracking and Ultimate (Case B, $\epsilon_c \geq \epsilon_0$, $\epsilon_s' \leq \epsilon_y'$):

$$\text{Let } c = \frac{\epsilon_s}{\epsilon_0}$$

$$B_1 = 0$$

$$B_2 = 0$$

$$B_3 = bdf_c'\left\{-0.6167787\frac{1}{c} + 0.226 - 1.0785\,c - 0.6356667\,c^2 - 0.149\,c^3 - 0.0126\,c^4 + 0.8052333\frac{1}{c}\right\}$$

$$B_4 = bdf_c'\left\{3.0838935\frac{1}{c} - 0.904 + 3.2355\,c + 1.2713334\,c^2 + 0.149\,c^3 - 4.0261665\frac{1}{c}\right\} - A_sf_s - A_s'E_s\epsilon_s - P$$

$$B_5 = bdf_c'\left\{-6.167787\frac{1}{c} + 1.356 - 3.2355\,c - 0.6356667\,c^2 + 8.052333\frac{1}{c}\right\} + 4A_sf_s + A_s'E_s\epsilon_s\left(3 + \frac{d'}{d}\right) + 4P$$

$$B_6 = bdf_c'\left\{6.167787\frac{1}{c} - 0.904 + 1.0785\,c - 8.05233\frac{1}{c}\right\} - 6A_sf_s - 3A_s'E_s\epsilon_s\left(1 + \frac{d'}{d}\right) - 6P$$

$$B_7 = bdf_c'\left\{-3.0838935\frac{1}{c} + 0.226 + 4.0261665\frac{1}{c}\right\} + 4A_sf_s + A_s'E_s\epsilon_s\left(1 + 3\frac{d'}{d}\right) + 4P$$

$$B_8 = bdf_c'\left\{0.6167787\frac{1}{c} - 0.8052333\frac{1}{c}\right\} - A_sf_s - A_s'E_s\epsilon_s\frac{d'}{d} - P$$

Between Cracking and Ultimate (Case B, $\epsilon_c \geq \epsilon_0$, $\epsilon_s' \geq \epsilon_y'$):

$$\text{Let } c = \frac{\epsilon_s}{\epsilon_0}$$

$$B_1 = 0$$

$$B_2 = 0$$

$$B_3 = bdf_c' \left\{ -0.6167787 \frac{1}{c} + 0.226 - 1.0785\,c - 0.6356667\,c^2 \right.$$

$$\left. -0.149\,c^3 - 0.0126\,c^4 + 0.8052333 \frac{1}{c} \right\}$$

$$B_4 = bdf_c' \left\{ 3.03838935 \frac{1}{c} - 0.904 + 3.2355\,c + 1.2713334\,c^2 \right.$$

$$\left. + 0.149\,c^3 - 4.0261665 \frac{1}{c} \right\} - A_s f_s + A_s' f_y' - P$$

$$B_5 = bdf_c' \left\{ -6.167787 \frac{1}{c} + 1.356 - 3.2355\,c - 0.6356667\,c^2 \right.$$

$$\left. + 8.052333 \frac{1}{c} \right\} + 4A_s f_s - 4A_s' f_y' + 4P$$

$$B_6 = bdf_c' \left\{ 6.167787 \frac{1}{c} - 0.904 + 1.0785\,c - 8.05233 \frac{1}{c} \right\}$$

$$- 6A_s f_s + 6A_s' f_y' - 6P$$

$$B_7 = bdf_c' \left\{ -3.0838935 \frac{1}{c} + 0.226 + 4.0261665 \frac{1}{c} \right\}$$

$$+ 4A_s f_s - 4A_s' f_y' + 4P$$

$$B_8 = bdf_c' \left\{ 0.6167787 \frac{1}{c} - 0.8052333 \frac{1}{c} \right\} - A_s f_s + A_s' f_y' - P$$

CHAPTER 7

Building with Fibrous Concrete

R. John Craig*

SUMMARY

The use of fibers in fiber reinforced cement and concrete is gaining increasing importance in the building materials area. Hydraulic cements used in the building area are weak in tension and against impact. Thus with the addition of suitable fibers, it is obvious that these difficulties can be overcome. Applications of steel, polypropylene, glass and asbestos fibers will be described with the main emphasis being on steel fibers.

Examples will be described emphasising the areas of construction where reinforced fibrous concrete is used today. The problems which have been encountered in the construction and use of these products as a building material will be discussed. The benefits of using this material will be brought out along with some economic aspects.

The possible future uses of this type of building material along with some possible research needs will be presented with special emphasis on the use of steel fibers in reinforced fibrous concrete construction in seismic areas. Research which has already been carried out will be shown which emphasizes these future needs.

INTRODUCTION

Fibers have been used to reinforce brittle materials since ancient times such as straws in sunbaked bricks and horse hairs in reinforced plaster. At various intervals since the turn of the century short pieces of fiber have been included within concrete in an attempt to endow it with greater tensile strength and ductility. It was not, however, until 1963 when Romualdi and Batson published the results of an investigation carried out on steel fiber reinforced concretes that any substantial interest was shown either by research organizations or by the construction industry. The claims made by Romualdi and Batson and subsequently by the Battelle Development Corporation [2], which filed a patent for the material later known as Wirand, were far reaching [3]. Other research has been conducted on the various types of fibers in United States, United Kingdom and Russia in late 1950's and early 1960's.

Fiber reinforced concrete is concrete made of hydraulic cements containing fine or fine and coarse aggregate and discontinuous discrete fibers, and reinforced fibrous concrete is regular reinforced concrete which has used fiber reinforced concrete instead of plain concrete. The types of fibers which have been produced in various shapes and sizes are steel, plastic, glass and natural materials. Several properties of various types of fibers are listed in Table 1. Figure 1 shows various shapes and sizes of fibers [4].

MATERIAL PROPERTIES

Material properties of fiber reinforced concrete are influenced to a large amount by the type, volume percentage, aspect ratio, nature of deformation and orientation of the fibers. Some general trends which show the benefits of using the fibers will be mentioned. The fibrous reinforced concretes show substantial increases in flexural strength, moderate increases in direct tension and torsion and only slight increases in compressive strength over plain concrete. The dynamic strengths for concrete reinforced with various types of fibers and subjected to explosive charges, dropped weights and dynamic tensile and compression loads were five to ten times greater than plain concrete. The energy requirements to strip or pull out the fibers (greater toughness) provides the impact strength and the resistance to spalling and fragmentation [4]. In order for test results on measuring

*Department of Civil and Environmental Engineering, New Jersey Institute of Technology, Newark, NJ

TABLE 1. Typical Properties of Fibers [4].

Type of fiber	Tensile strength, ksi	Young's modulus, 10^3ksi	Ultimate elongation, percent	Specific gravity
Acrylic	30-60	0.3	25-45	1.1
Asbestos	80-140	12-20	~0.6	3.2
Cotton	60-100	0.7	3-10	1.5
Glass	150-550	10	1.5-3.5	2.5
Nylon (high tenacity)	110-120	0.6	16-20	1.1
Polyester (high tenacity)	105-125	1.2	11-13	1.4
Polyethylene	~100	0.02-0.06	~10	0.95
Polypropylene	80-110	0.5	~25	0.90
Rayon (high tenacity)	60-90	1.0	10-25	1.5
Rock Wool (Scandinavian)	70-110	10-17	~0.6	2.7
Steel	40-600	29	0.5-35	7.8

1 ksi = 70.31 kgf/cm²

properties of fiber reinforced concrete to be effective, a standard procedure must be used. "Measurement of Properties of Fiber Reinforced Concrete" [5] is recommended.

APPLICATIONS

Present applications of steel fiber reinforced concrete have been in the areas of refractories, pavements, overlays, patching, hydraulic structures, thin shells, armor for jetties, rock slope stabilization, mine tunnel linings and precast products.

Using fibers in paving applications has shown that the thickness of the slab can be half that of plain concrete for the same wheel load coverage. The properties of fibrous concrete which are important are increased flexural strength, improved post-cracking ductility, increased resistance to impact and repeated loading, and improved spalling resistance. Testing was performed by U.S. Army Construction Engineering Research Laboratory at U.S. Army Waterways Experimental Station, Vicksburg, Miss. on runway slabs of fiber reinforced concrete (2 percent of fibers by volume) [6]. Some paving projects were the John F. Kennedy, McCarran International and other airports. Industrial flooring and precast slabs are also an application along these lines. An example of this type of construction is shown in Figure 2.

Fiber reinforced concrete is being used for both repair and in new construction of dams and other hydraulic structures. On Lower Monumental Dam and on Little Goose Dam on the Snake River, Washington, spillway deflectors were constructed of steel fibrous concrete to resist cavitation and erosion damage.

Steel fibrous shotcrete has been used for mine and tunnel linings, in dam construction, for repair of deteriorated surfaces and arches, for rock slope stabilization, for fire protection coatings and in thin shell dome construction. Steel fibers on the order of 1 inch long by .016 inch in diameter (25.4 mm × .41 mm) are used up to 2 percent by volume in shotcreting. The inclusion of steel fibers in shotcrete improves especially the toughness, impact resistance, shear strength, flexural strength, durability factor and the fatigue endurance limit. Also this material provides a large measure of ductility. That is, it requires large deformations for failure to occur and the material continues to carry a significant load after cracking.

In 1973, R. A. Kaden of the U.S. Corps of Engineers supervised the first practical application of steel fibrous shotcrete in a tunnel adit at Ririe Dam, Idaho. Other applications are: 1) Corps of Engineers—Snake River Rock Slope Stabilization; Joint Nordic Program (Nordforsk)—Oil Refinery—Brofjorden, Sweden; U.S. Bureau of Mines—coal mine applications; Bolidens Gruv AB—mines and ore shaft, Sweden; British Rail—arch and tunnel relining, England; Swedish State Power Board—Ringhals Nuclear Power Sta-

FIGURE 1. Shapes of steel fibers: (a) round, (b) rectangular, (c) indented, (d) crimped, (e) hooked ends, (f) melt extract process and (g) enlarged ends [9].

FIGURE 2. Rear view of slipform paver laying steel fiber reinforced concrete as an overlay [9].

tion; roadway tunnels—Japan; etc. [7]. Some examples of this type of construction can be seen in Figure 3.

The introduction of refractories reinforced with metal fibers has increased process efficiency and decreased turn around frequency. Improved refractory life has been experienced in FCCU transfer lines and cyclones, furnace and combustor linings, and slide valves, and further applications are being investigated. The fibers which are typically 1 inch to 1½ inch in length (25.4 mm to 38.1 mm), .013 inch to .030 inch (.33 mm to .76 mm) in diameter and comprise approximately 1.0 percent by volume of the lining are usually a type 300 or 400 series stainless steel. The fiber-reinforced material is applied by gunning, casting or hand packing. Analysis of lining performance implies that crack distribution and internal anchorage are the most important benefits achieved with fiber additions. Increased lining toughness and erosion resistance are attributed to these two beneficial mechanisms. Fibers themselves do not change the erosion resistance of castable material, but allow elimination of hex-mesh support and reduce spalling which can be factors in initiating erosion. Exxon's first commercial application of metal fiber-reinforced refractory took place in 1973 during repair of a section of the Dartmouth, Nova Scotia FCCU feed riser. Their first major use of the material was in 1975 when 14 regenerator secondary cyclones at the Sarnia unit were replaced. These and other examples have been successful and more uses are being found [8].

The use of reinforced fibrous concrete has been applied to hemispherical domes using an inflated membrane process. A layer of fiber shotcrete is applied to the inside of a layer of foam first applied to the interior of a membrane [10]. Batson, Naus and Williamson developed a construction technique for inflation forming steel fiber concrete shell structures suitable for shelters and fortification [11]. Pictures of both types of shell are shown in Figure 4.

There are two main types of glass fibers which have been used in practical applications, borosilicate glass, also known as E-glass, in conjunction with low-alkali and high alumina cements and alkalie resistant glass, often known as Cem-FIL in conjunction with ordinary Portland Cement. Because of unknown factors associated with long-term performance of both of these fibers, they are likely used for non-structural or semi-structural situations [9]. Wall panels constructed of glass fiber reinforced cement (GFRC) use the "spray-up" process. These thin panels are lightweight and easy to erect. An example of this type construction is shown in Figure 5. The exhibiton pavilion at the Stuttgart Federal Gardening Fair is an excellent example of a glass fiber reinforced concrete spray-up thin shell. Glass-reinforced cement also performs well in form-work applications [4].

Besides the steel and glass fibers there is the polypropylene fiber. Some of the applications of this type of fiber are pile shells, cladding panels, flotation units for marines, gunited polypropylene mortor, decorative cladding, timber substitute, manholes, and sheet products.

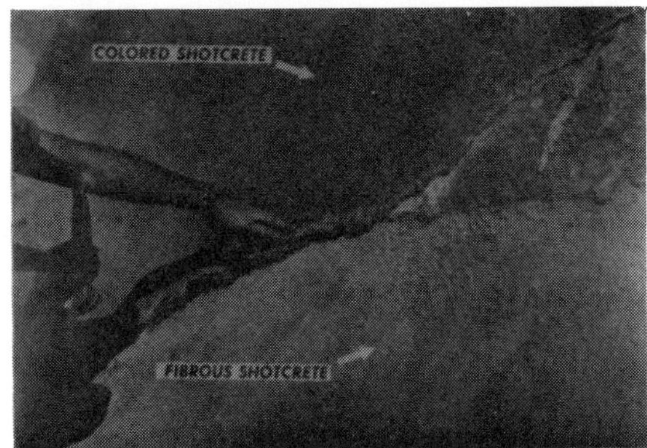

(a)

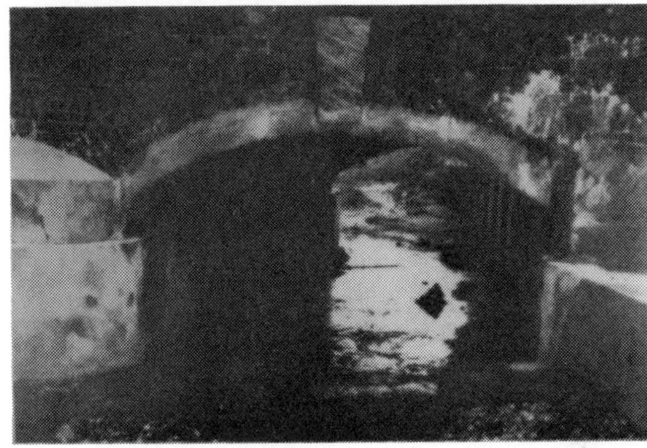

(b)

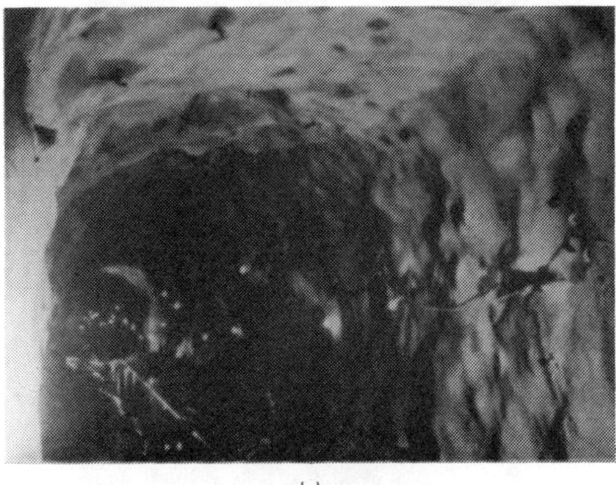

(c)

FIGURE 3. Shotcrete examples with fibers [7]: (a) fibrous shotcrete rock surface stabilization; (b) brick railway bridge reinforced with about 6 in. of steel fibrous shotcrete; (c) emergency coal water tunnel lining at ringhals nuclear power station, Sweden.

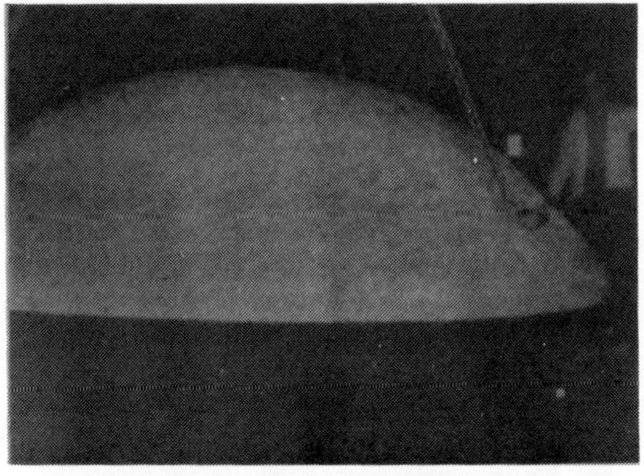

(a)

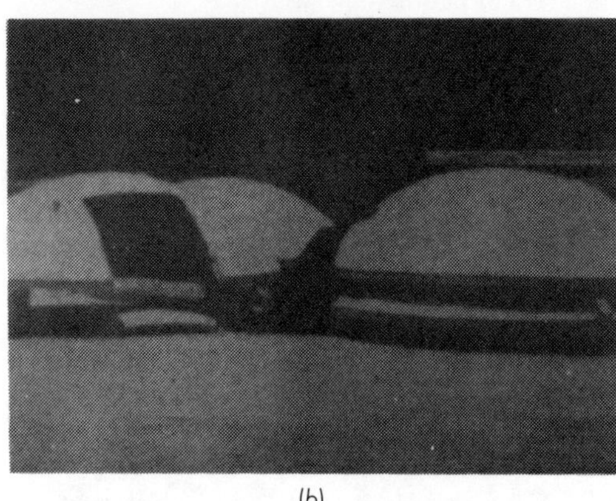

(b)

FIGURE 4. Steel fiber reinforced concrete domes [11]: (a) lifting dome; (b) storing domes.

FIGURE 5. Glass reinforced cement cladding panels [9].

SEISMIC APPLICATIONS

Two exploratory tests on joints by C. H. Henager [12] demonstrated the feasibility of using steel fiber reinforced concrete in ductile concrete joints. The full-size tests on a modified joint using steel fiber reinforced concrete in place of the joint region hoops has good ductility, indicated that the former was stronger and more damage-tolerant than a conventional ductile concrete joint; was some-what stiffer; and required fewer stirrups than in the reinforced concrete beams. The joint in the testing set-up which C. H. Henager tested is shown in Figure 6. The earthquake representation was the same one as used by PCA Structural Laboratory. The results as shown in moment-rotation diagrams for the two specimens are shown in Figures 7a and 7b. In his initial evaluation, Henager concluded that this type of fiber reinforced concrete joint demonstrates the potential for a superior, less costly joint. He also recognized that additional evaluations are needed to investigate optimum cost savings by substitution of steel fibers for hoops in building joints. Various types of joint sizes also need evaluation together with development of design methods [2].

Repeated loading tests at the Sumitomo Metal Industries Ltd. research laboratories of cantilever beams, using fiber reinforced concrete with plastic deformations (deflection ductility factor of 4) showed increased resistance to deterioration by preventing slipping along the plane of maximum shear stress. It could be expected that this superior strength against repeated plastic deformation could be used in preventing the failure of reinforced concrete structures subjected to severe seismic loads [13]. Results of these tests are shown in Figure 8.

The research carried out at New Jersey Institute of Technology by a group under R. John Craig found that fiber reinforced concrete under monotomic loading in beams are by far more ductile and more confining than regular concrete [14]. These are two primary objectives of good seismic construction. These concrete members are also stiffer and more energy absorbent than regular concrete. The crimped end fibers increase the shear strength of concrete by more than 100 percent [15]. The results of some of these tests are reported elsewhere [16].

From Henager's tests and those of other researchers, evidence indicates that the replacement of regular concrete with fiber reinforced concrete can be beneficial to structures in seismic areas. The cumulate results to date indicate that use of reinforced fibrous concrete will increase: 1) the ductility of columns and beams; 2) shear capacities; 3) resistance to cracking; 4) bond; 5) damage tolerance; and 6) stiffness of normal reinforced concrete. With these benefits, the use of fiber reinforced concrete will provide a structure which is more resistant to the impact of earthquake tremors than a conventional reinforced concrete building. There would be a saving in cost of repair of buildings after an earthquake and, more importantly, a better chance of saving lives.

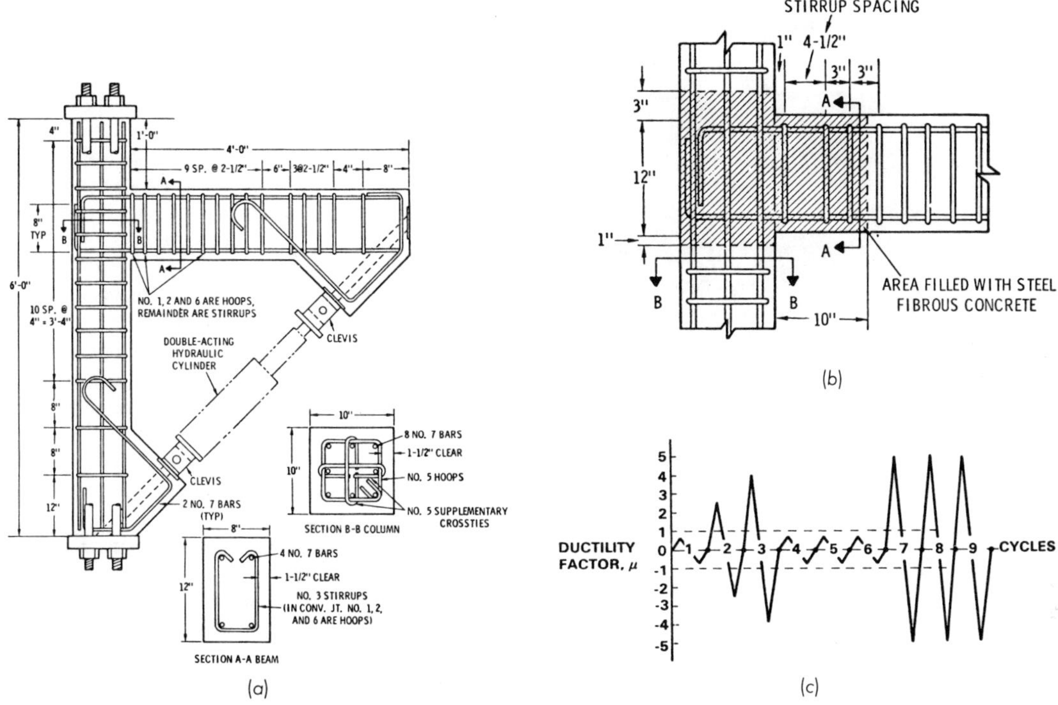

FIGURE 6. Henager's steel fibrous concrete joint for seismic resistant structures [12]: (a) specimen for seismic joint study; (b) modified seismic joint; (c) earthquake representation.

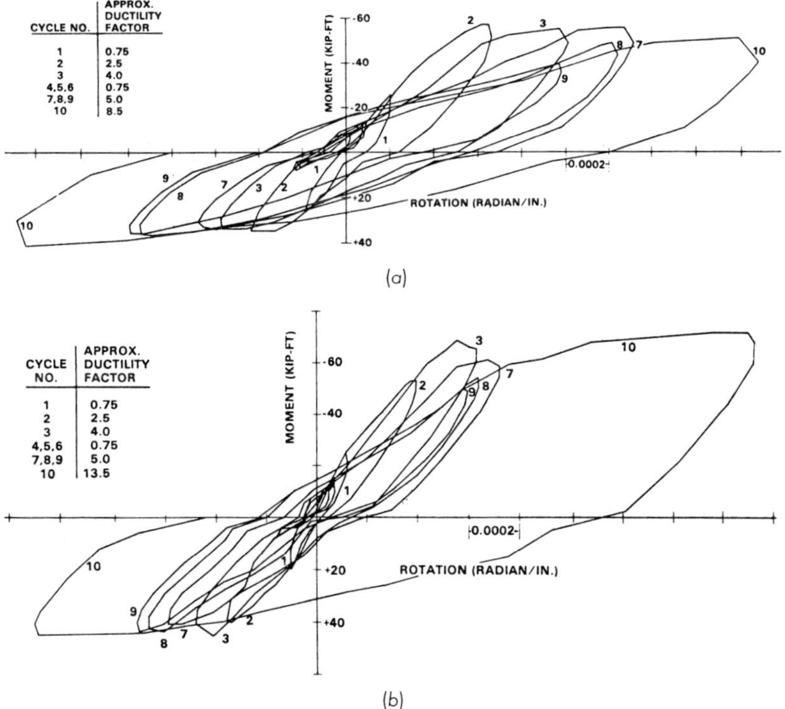

FIGURE 7. Moment-rotation diagrams for steel fibrous concrete joint study [12]: (a) moment-rotation diagram—conventional (ACI 318 71) reinforced concrete ductile joint (Henager); (b) moment-rotation diagrams for steel fibrous concrete joint study.

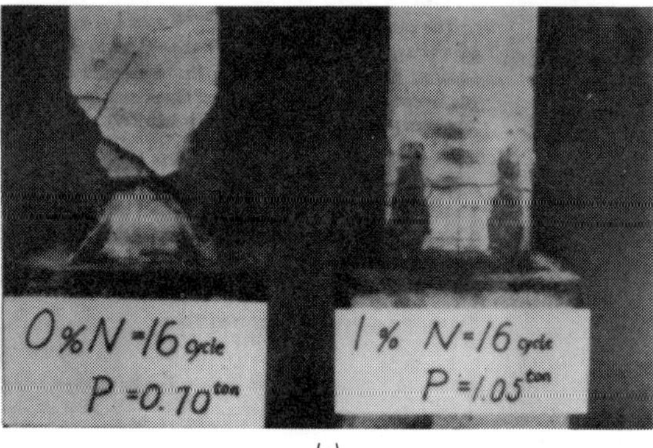

(a)

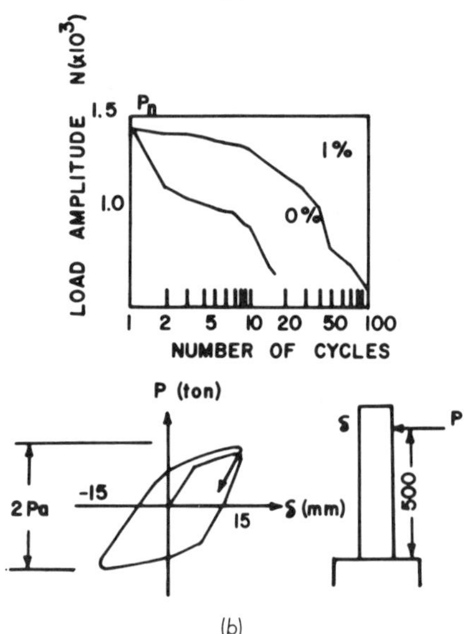

(b)

FIGURE 8. Cyclic loading of reinforced fiber concrete beams [13]: (a) crack pattern in fixed end at sixteenth cycles; (b) load amplitude vs. repeated cycles.

The research which has been carried out thus far in reinforced fiber concrete structures, however, is not sufficient to adequately determine its behavior. Nor does the data provide design criteria for its use in seismic reinforced fiber concrete structural members in seismic areas. Some research is being carried out at the New Jersey Institute of Technology to further the development of reinforced fiber concrete in seismic construction.

CONSTRUCTION

There have been some problems using fiber reinforced concrete in construction by some contractors. Most of these problems could have been eliminated if the proper technology of this material would have been applied. The use of fibers requires more deliberate planning and workmanship than the established concrete construction procedures now used. Present practices now used in production and handling may or may not be appropriate for fiber reinforced concrete.

Segregation or balling and decrease in workability are two major problems encountered. Most fiber balling occurs during the fiber addition process. This can be eliminated by proper sequence and a controlled rate of fiber addition for the fibers. The use of collated or bundled fibers has been very successful. The use of admixtures such as superplasticizers are useful in producing high quality fiber reinforced concrete which has a high degree of workability using a low water/cement ratio. Fly ash can be used to reduce the cement content and improve the workability. Decreasing the corase aggregate content increases the workability also. The volume and type of fibers selected determine the maximum aggregate size and volume of paste. With these factors known, the techniques of good concrete proportioning can be applied to obtain workable and economical mixes. Fiber reinforced concrete mixes have higher cement factors than conventional concrete, and an upper limit of 2 percent by volume should be set on the fiber content.

When placing fiber reinforced concrete the following should be noted: 1) more vibration is recommended and 2) use forks and rakes instead of shovels and hoes (depends on slump) [4].

One problem which may be of concern is the corrosion of fibers. From results of testing [17], it has been shown that after long periods there is no or little adverse effect on the strength properties of steel fibrous mortar. Fiber corrosion was confined to fibers actually exposed on a surface. Internal fibers showed no corrosion.

RESEARCH

The use of reinforced fiber concrete, reinforcing steel with fiber reinforced concrete has a potential of being a superior construction material than conventional reinforced concrete both economically and mechanically if proper engineering is used to take into effect the advantages of the fibers.

Workability and placement studies need to be made on fiber reinforced concrete placement in locations of reinforcing bars. A study should look into the efficient use of fibers by placing them only where they are needed, since placing fibrous reinforced concrete in an entire structure would be uneconomical. More research is needed to develop a formulation of a general methodology that will predict the response of structures with fiber reinforced concrete. Recommendations are needed for changes in the present design code provisions which would facilitate the use of fiber reinforced concrete. The use of fibers in reinforced concrete structural members would help to eliminate the existing problems of confinement and ductility. The present state of the art for design and detailing requires members which are very difficult and costly to construct in the field. The specific use of reinforced fibers concrete in structural elements such as ductile joints and shear walls for seismic areas has good potential.

CONCLUSIONS

The use of fibers in plain concrete and reinforced concrete is just starting. There are many advantages of using this type of material over the conventional construction methods presently being used. If the proper engineering knowledge is applied to fiber reinforced concrete construction, the problems which now exist can be eliminated. Many present-day examples show that the fiber reinforced concrete does perform well when used. However, more research is needed to be able to fully develop this method of construction into areas which have potential.

ACKNOWLEDGEMENT

The author expresses his sincere gratitude to the members of ACI committee 544 Fibrous Concrete. Many of the conclusions and recommendations contained in the chapter are a direct result of their work in fibrous concrete.

REFERENCES

1. Romualdi, J. P. and G. B. Batson, "Mechanics of Crack Arrest in Concrete," *Proceedings of the American Society of Civil Engineers*, 89(EM3), 147–168 (June 1963).
2. Battelle Development Corporation, "Concrete and Steel Material," British Patent No. 1068163 (December 1963).
3. British Building Research Establishment (BRE). Concrete. BRE Building Research Series, Practical Studies from the Building Research Establishment, The Construction Press, Vol. 1, pp. 154–170 (1978).
4. "State-of-the-Art Report on Fiber Reinforced Concrete," ACI 544 Committee Report, ACI Journal (November 1973).
5. ACI Committee 544, "Measurement of Properties of Fibre Reinforced Concrete," *ACI Journal, Proceedings*, 75(7), 283–289 (July 1978).
6. "Fibrous Concrete, Construction Material for the Seventies," Conference Proceedings M-28, U.S. Army Construction Engineering Research Laboratory, Champaign, Ill.
7. Henager, C. H., "Steel Fibrous Shotcrete: A Summary of the State-of-the-Art," *Concrete International, ACI*, 3(1), 50–58 (January 1981).
8. Peterson, J. R. and F. H. Vaughan, "Metal Fiber-Reinforced Refractory for Petroleum Refinery Applications," Corrosion-80, Paper No. 51, Chicago, Ill. (March 3–7, 1980).
9. Hannant, D. J. *Fibre Cements and Fibre Concretes.* John Wiley and Sons, New York, pp. 219 (1978).
10. Henager, C. H., "New Developments in Steel Fibrous Shotcrete," *Concrete Construction*, 25(3), 189–194 (March 1980).
11. Batson, G. B., D. J. Naus, and G. R. Williamson, "Inflation of Steel Fibre Reinforced Concrete Domes," Rilem Symposium 1975, *Fibre Reinforced Cement and Concrete*. The Construction Press LTD, pp. 375–382 (1975).
12. Henager, C. H., "Steel Fibrous Ductile Concrete Joint for Seismic Resistance Structures," *ACI SP53-14, Reinforced Concrete Structures in Seismic Zones*, pp. 371–386 (1977).
13. Nishioka, K., N. Kakimi, S. Yamakawa, and K. Shirakawa, "Effective Applications of Steel Fibre Reinforced Concrete," RILEM Symposium 1975, *Fibre Reinforced Cement and Concrete, Vol. 1*, The Construction Press Ltd., pp. 425–433 (1975).
14. Decker, J. E., "The Inelastic Behavior of Reinforced Fibrous Concrete Beams," M.S. Thesis, Department of Civil and Environmental Engineering, New Jersey Institute of Technology (Spring 1979).
15. Love, J., "Shear Behavior of Reinforced Fibrous Concrete Beams," M.S. Thesis, Department of Civil and Environmental Engineering, New Jersey Institute of Technology (Spring 1981).
16. Craig, R. J., "Using Fibers in Reinforced Concrete Construction-Moment, Shear and Torsion," presented at the 1981 International Symposium on Building Materials, Miami Beach, Fla. (March 1981).
17. Lankard, D. R. and A. J. Walter, "Laboratory and Field Investigations of the Ductility of Wirand Concrete Exposed to Various Service Environments," to Battelle Development Corporation, 26 p. (June 28, 1978).

CHAPTER 8

Structural Behavior and Design of Concrete Inverted T-Beams

S. A. MIRZA* AND R. W. FURLONG**

ABSTRACT

Structural behavior of inverted T-beams that support precast stringers on the flanges of the inverted T differs from that of conventional beams. This is because the loads on the T-section are applied near the bottom rather than at the sides or at the top of the web. This article describes structural behavior of inverted T-beams and provides design guidance and reinforcement details applicable to such beams. The scope of the paper is limited to reinforced concrete and posttentioned prestressed concrete.

INTRODUCTION

Prefabricated stringers with cast-in-place slab are frequently used to achieve economical and speedy construction schemes. Beams constructed in the form of an inverted T possess on each side of the web a bracket or flange overhang that provides a convenient shelf or supporting surface for the precast stringers. Hence, cast-in-place posttensioned prestressed concrete and reinforced concrete inverted T-beams are frequently used in bridges as bentcap girders as indicated in Figure 1. Inverted T-beams can be simply supported, cantilevered over simple supports, or they could be constructed monolithically with piers. They reduce overall floor depth by avoiding deep cross members beneath prefabricated stringers, resulting in lower abutments and shorter approaches.

Stringer bearings on the top face of the flange of an inverted T-beam impose vertical tensile forces (hanger tension) near the bottom of the web as indicated in Figure 2. Such forces are not ordinarily encountered in conventional beams, where vertical forces are applied at the top of the web. Furthermore, the longitudinal and lateral bending of the flange of an inverted T-beam produce a very complex stress distribution in the flange. Hence, the design of reinforcement for the web and for the flange of an inverted T-beam imposes special problems. However, no guidance for handling design problems specifically associated with inverted T-sections is available in the current American standards [1,9].

This paper summarizes design recommendations that are based on observations and analyses of cast-in-place normal weight prestressed and reinforced concrete test specimens reported in studies of inverted T-beams [4,5,6,7,8]. These studies indicate that six modes of failure plus at least one service load condition should be considered as a part of the design of inverted T-girders. The six modes of failure involve the possibility of failure due to: (a) flexure of the overall inverted T-beam, (b) flexural shear acting on the overall inverted T-beam, (c) torsional shear on the overall cross-section, (d) hanger tension on web stirrups, (e) flange punching shear at stringer bearings, and (f) bracket type shear friction in flange at face of the web. The service load condition involves the possible wide cracking at the interface of the web and the flange due to premature yielding of stirrups acting as hangers nearest the concentrated loads. Typical forces and stress types acting on inverted T-beams are illustrated in Figure 2.

This paper is directed toward the bridge engineering audience with well-founded experience in structural design. Hence, only the highlights from the reported research that are associated with design aspects of the inverted T-beam bentcaps are presented here. Further information and details are fully available elsewhere [3,4,5,7,8].

*Department of Civil Engineering, Lakehead University, Thunder Bay, Ontario, Canada

**Department of Civil Engineering, The University of Texas at Austin, Austin, TX

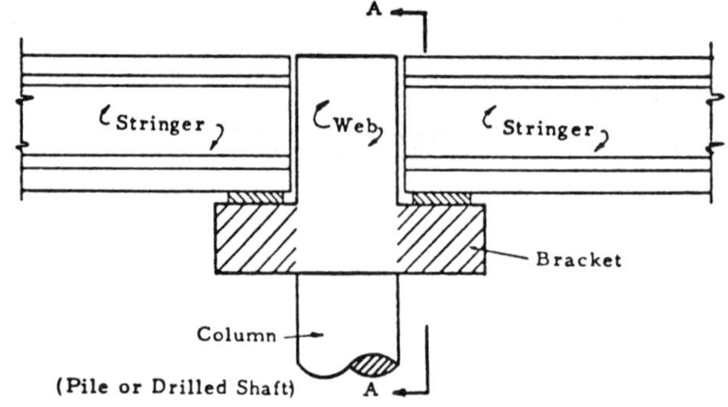

(a) ELEVATION

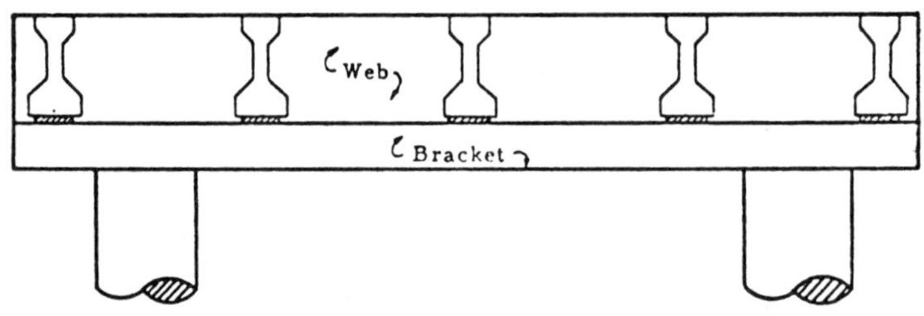

(b) SECTION A-A

FIGURE 1. Highway pier cap.

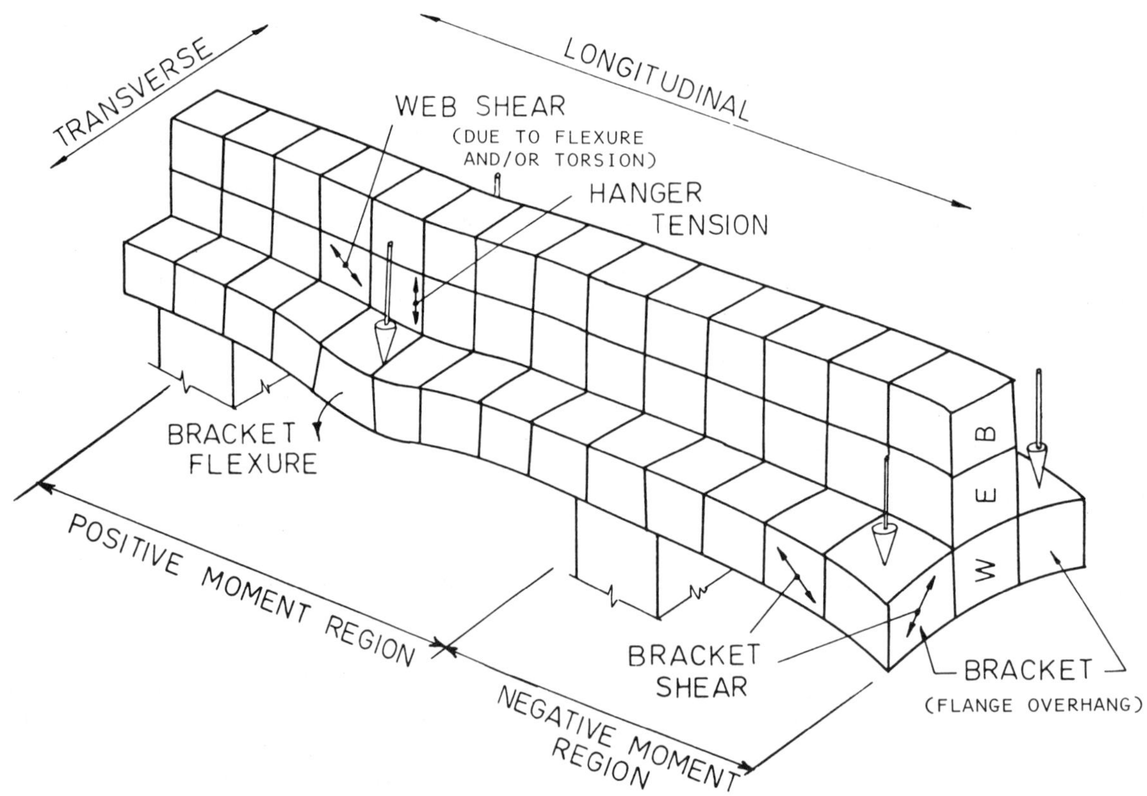

FIGURE 2. Structural actions on inverted T-beams.

DESCRIPTION OF PHYSICAL TESTS

Laboratory test specimens that formed the basis of design recommendations presented in this paper represented model bentcap girders at a scale of approximately 1/3 of the size of the prototype members used by the Texas State Department of Highways and Public Transportation. Strengths of normal weight concrete, reinforcing steel, and prestressing strand used were typical of those currently employed in the industry. The prestressing strands were straight and each strand was posttensioned and grouted.

The principal variables for the specimens involved reinforcement details associated with the web and the flange. The observations included service load twist reversals to simulate the passage of traffic and subsequent loads that caused failure of a region of the test specimens. Hence, the typical load sequences involved three phases of observations: (a) loads simulating service dead load stringer reactions were applied and maintained, (b) service live load forces were applied and removed to produce several cycles of simulated passage of traffic, and (c) loads were increased until failure occurred in a region of the test specimen. Other details are reported elsewhere [4,5,7,8] and will not be repeated here.

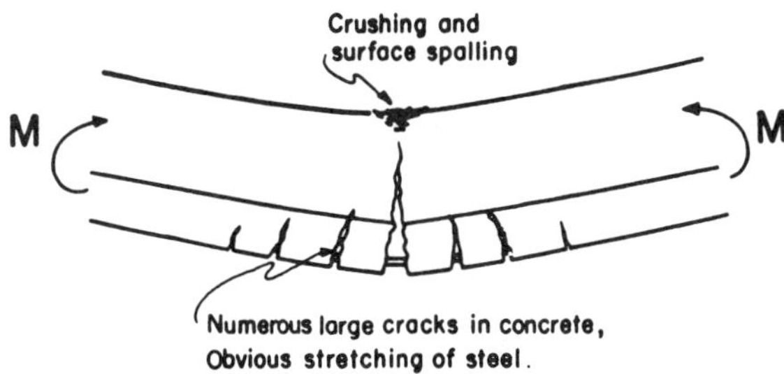

(a) FLEXURAL FAILURE - POSITIVE MOMENT

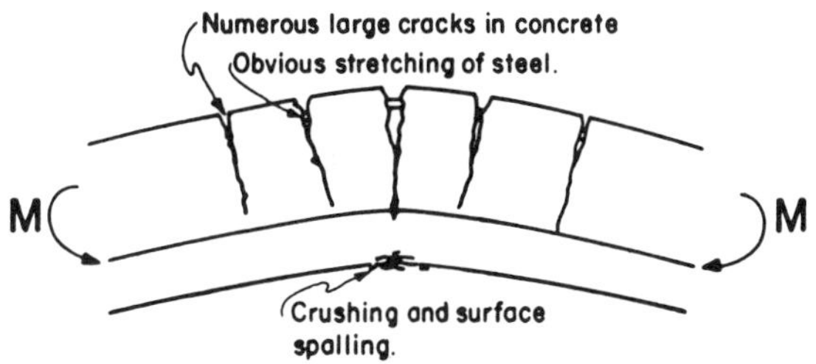

(b) FLEXURAL FAILURE - NEGATIVE MOMENT.

FIGURE 3. Flexural failure modes.

OBSERVED MODES OF FAILURE

The strength of concrete inverted T-beams is limited to the strength of the weakest of several possible components or combinations of components that participate in the retention of the applied loads. The components will be defined in terms of failure modes identified from the physical tests [7,8].

Based on these tests, the strength of an inverted T-beam can be interpreted in terms of resistance to flexure, flexural shear, or torsion. Any of the three or combination of three will be reached only if there exists adequate local strength for hanger action, flange punching shear, and flange bracket response. Sketches of typical failure modes are provided in Figures 3 through 8 with descriptions given in the following paragraphs:

1. A cross-section is considered to have failed in flexure when its resistance to flexural deformation begins to decrease. Under the application of large flexural deformations, reinforcing bars and prestressing strands tend to yield without a reduction in tensile force. Simultaneously, the resistance of concrete and compressive reinforcement can remain relatively constant, but the amount of compressive "yielding" is limited by the amount of stress that can be redistributed by concrete before concrete cracks and spalls off at the surface of maximum compressive strain as shown in Figure 3.

2. Flexural shear failure for specimens with an effective shear span as short as that employed for the reported tests [7,8] involves the yielding of all stittups that cross a large crack which extends diagonally along the side faces of the web until the shear strength or compressive strength of uncracked concrete is exhausted (Figure 4). The maximum shear span near supports for any test region was only 18 percent greater than the effective depth of the member. Additional strength due to short shear span is at least partially offset by application of the stringer loads near the bottom rather than at the sides or at the top of the web.

3. Torsional distress in concrete beams appears in diagonal cracks that extend in a spiral pattern from one face of the member to an adjacent face. Failure begins in the form of diagonal tension cracks across which both longitudinal and stirrup reinforcement must transmit the tensile forces

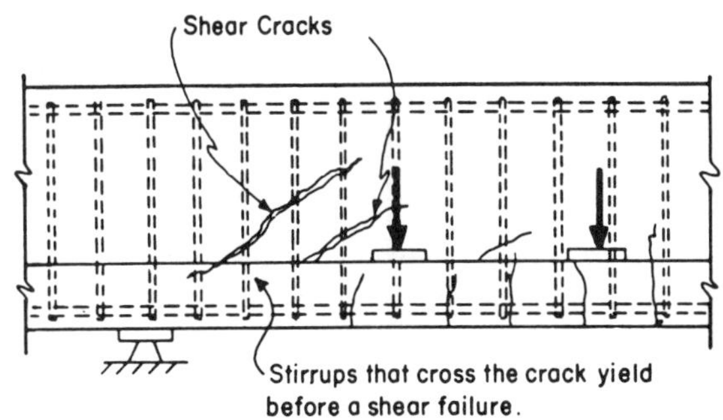

FIGURE 4. Flexural shear failure mode.

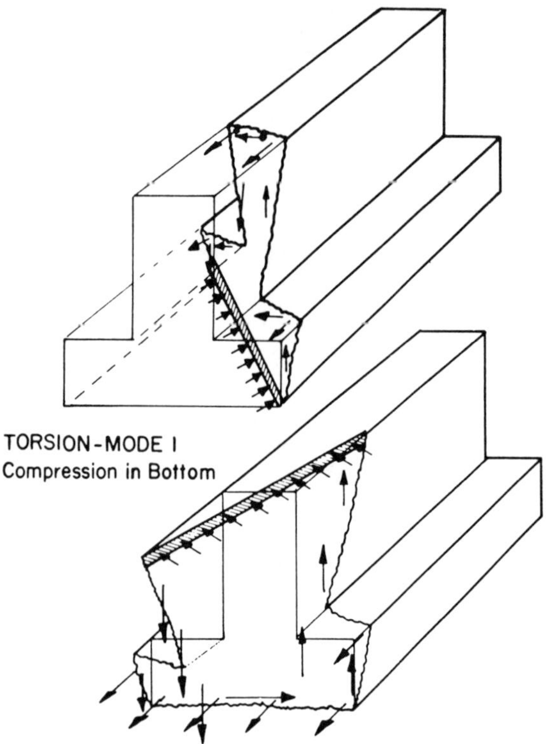

FIGURE 5. Torsion failure modes.

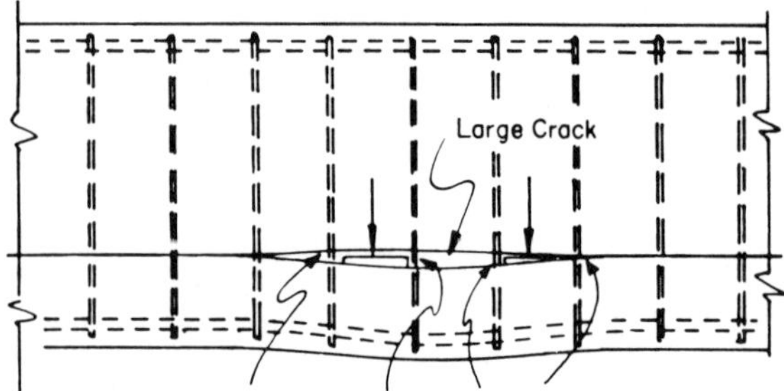

FIGURE 6. Hanger failure mode.

that were lost where the concrete cracked. The diagonal cracks extend in length and in width as torsional forces increase until reinforcement across the crack yields. If the anchorage strength of reinforcing bars is not lost due to edge cracking, torsional deformation continues until there is a compression failure along the face nearest the center of torsional rotation (Figure 5).

4. Hanger failure of stirrups is revealed by the vertical separation that occurs between the flange and the web at the top of the flange. The separation begins as a local phenomenon near the stringer bearing pad, but as the stirrups closest to the bearing pad yield, the flange deflects and causes more hangers to share the concentrated stringer load. The failure occurs after all available stirrups acting as hangers have yielded provided that the flange is strong enough to distribute longitudinally the stringer loads (Figure 6).

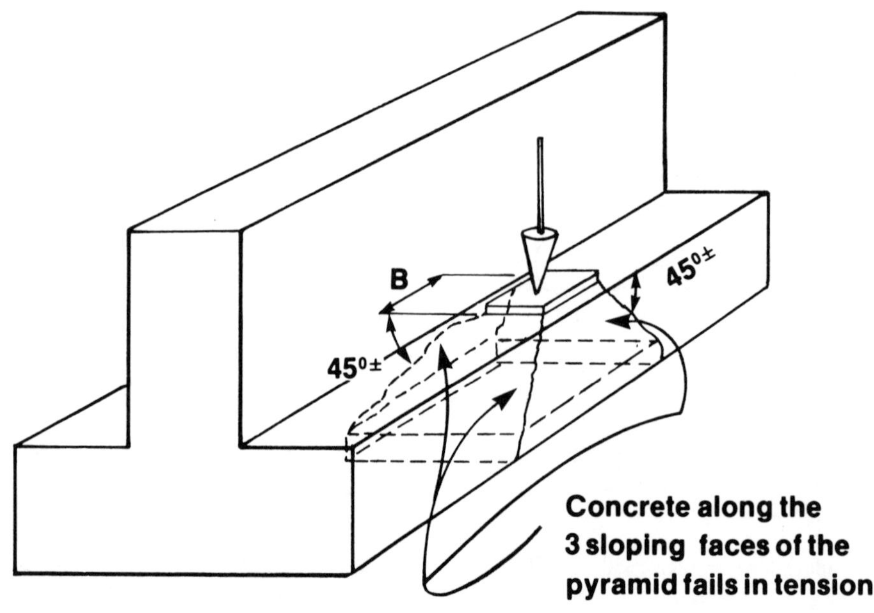

FIGURE 7. Flange punching shear failure mode.

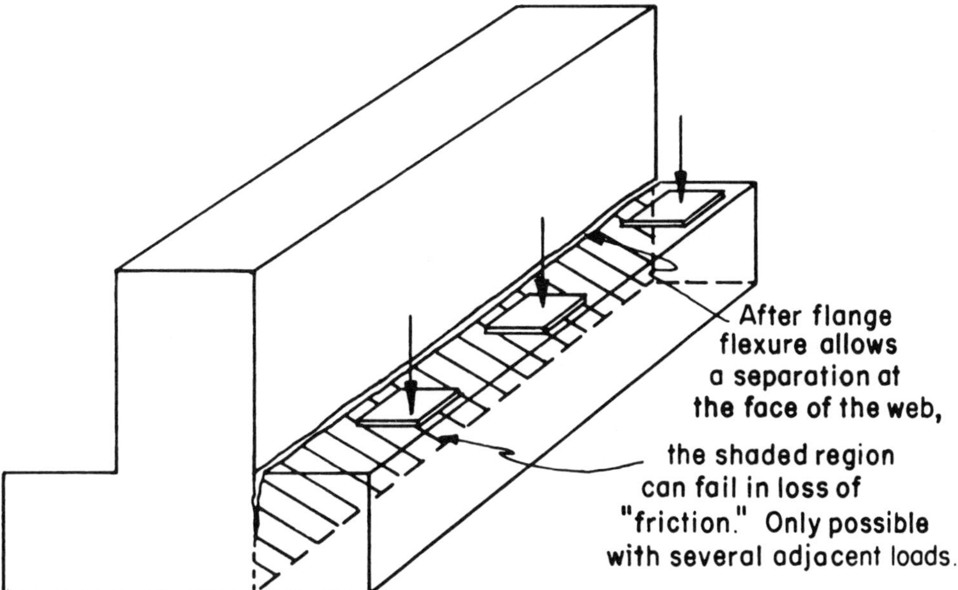

After flange
flexure allows
a separation at
the face of the web,
the shaded region
can fail in loss of
"friction." Only possible
with several adjacent loads.

FIGURE 8. Flange bracket failure mode.

5. Flange punching failure can take place if stringer forces are large enough to "punch out' the truncated pyramid of concrete beneath a bearing pad. The failure is evidenced by the appearance of diagonal tension cracks emanating from the edges of bearing pad as indicated in Figure 7.

6. The bracket failure mode is used here to identify a local loss of resistance to load because the flange acting as a bracket tends to deform outward and away from the web while also deforming downward from loss of shear strength along the face of the web. As the transverse bars acting in flexure across the top of the flange yield, the flange resistance in "shear friction" along the face of the web is unable to sustain the stringer forces and may allow sliding of the flange downward along the face of the web (Figure 8).

DESIGN RECOMMENDATIONS

The design recommendations presented here are based on the observed modes of failure briefly described in the previous section. The detailed behavior and analysis of inverted T-beams has been documented in earlier studies [3,4,5,7,8].

The overall strength of an inverted T-beam must be adequate to support ultimate flexure, flexural shear, and torsional shear forces and any possible combination of such forces. The local strength of inverted T-beam components must be adequate to support forces that are applied as concentrated loads on flange overhangs. Locally, the flange must be deep enough to avoid punching shear weakness, the transverse flange reinforcement must be strong enough to maintain shear friction resistance at the face of the web, and the web stirrups must be sufficient to act as hangers to transmit flange loads into the web.

Service load conditions of deflection and crack control may be more significant than strength requirements for some components of design. Decisions regarding the overall depth of web and the distribution of tensile reinforcement both for flexure and for stirrups acting as hangers may involve service load conditions of behavior.

The height of the web above the top face of the flange of an inverted T-beam is determined by the required depth of the stringer to be supported on the flange. A minimum depth of the flange itself can be derived from punching shear requirements, but additional depth may be appropriate to provide enough flexural stiffness for the overall member. The width of the web can be selected for adequate strength in shear alone and in combined shear and torsion, or it may be determined by placement requirements of flexural reinforcement. The length of the flange overhang is controlled by the size of the bearing pads used to support stringers and should not exceed the flange thickness. In addition to accomodating the bearing pad width, the flange overhang has to provide for the edge distance, the sweep tolerance of the T-beam, the length tolerance of the stringers, and the erection and placing tolerances. The design of an inverted T-beam can be divided into three parts: (a) the design of flange, (b) the design of web stirrups acting as hangers to deliver flange forces into the web, and (c) the overall design of beam itself. Hence, the design recommendations are presented in three parts as well. Note that these recommendations apply to cast-in-place posttensioned concrete and reinforced concrete inverted T-beams employed in bridge construction.

Design of Flange

The strength of the flange should be adequate to sustain the punching shear action of the stringer loads applied on the flange. In addition, the flange should be able to resist the shear friction forces at the face of the web caused by the bracket action of the flange.

PUNCHING SHEAR IN FLANGE

The flange should be deep enough to prevent punching shear failure. This can be achieved by satisfying the following equation for effective flange depth (d_f) from the top face of the flange to the top of bottom layer of transverse reinforcement in the flange:

$$4\phi \sqrt{f'_c}\ (B_p + 2d_f)\ d_f \geq P_u$$

or

$$d_f \geq \frac{B_p}{4}\left[\sqrt{\left(1 + \frac{2P_u}{\phi B_p^2 \sqrt{f'_c}}\right)} - 1\right] \qquad (1)$$

For Equation (1), d_f is defined in Figure 9; $B_p = B + 2B_w$; B = length of bearing pad along the edge

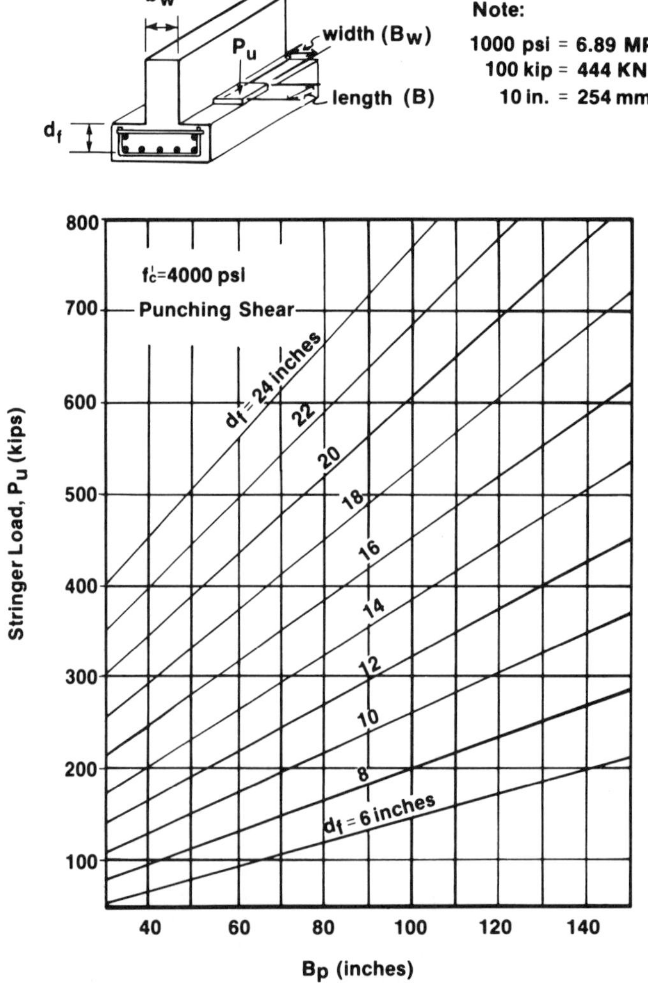

FIGURE 9. Punching shear capacity of flange of an inverted T-beam; f'_c = 4000 psi (27.6 MPa).

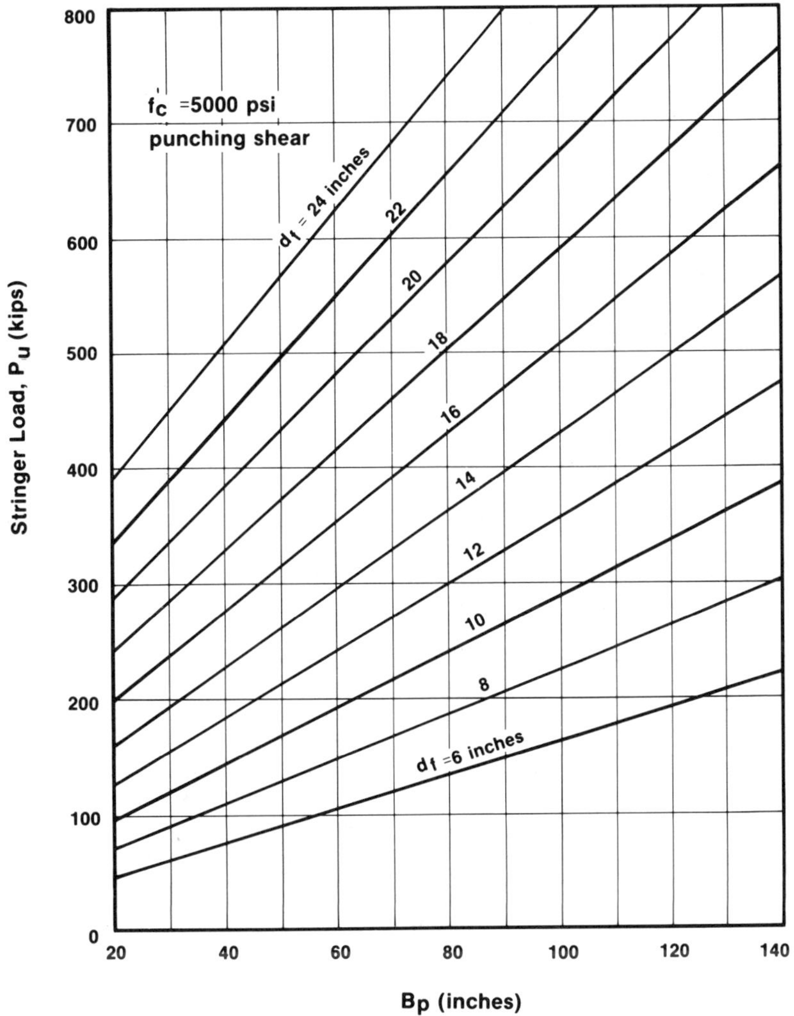

FIGURE 10. Punching shear capacity of flange of an inverted T-beam; f'_c = 5000 psi (34.5 MPa).

of flange; B_w = width of bearing pad perpendicular to the beam axis; P_u = ultimate concentrated load acting on a bearing pad; and ϕ = capacity reduction factor for shear and equals 0.85. This cumbersome equation can be applied readily as a graph with B_p versus P_u for various values of d_f. Such graphs for specified concrete strength f'_c taken as 4000 and 5000 psi (27.6 and 34.5 MPa) are shown in Figures 9 and 10, respectively. Similar graphs can be prepared for other concrete strengths. It may be pointed out that the most common strengths of concrete used for bentcap girders are 4000 and 5000 psi for reinforced and posttensioned prestressed concrete construction, respectively.

Equation (1) is based on a tensile strength of concrete equal to $4\sqrt{f'_c}$ acting on the surface of a truncated pyramid under a bearing pad and is supported by test results [8]. Stirrups that intersect a face of the truncated pyramid can help support the concentrated load if the anchorage of the stirrups can be developed above and below the face of the truncated pyramid. However, no such help from the stirrups was included in Equation (1), because this would require cumbersome checks on design and detailing of stirrups.

The surfaces of truncated pyramids resisting ultimate punching shear under adjacent stringers should not overlap. This can be achieved by providing enough longitudinal and transverse distance between stringers. Hence, the web width b_w should be such that the center-to-center transverse distance between the two stringer reactions acting on opposite sides of the web [$(2a + b_w)$ in Figure 11(a)] at least equals $(2d_f + B_w)$, where B_w is the width of the bearing pad perpendicular to the beam axis. Furthermore, stringers

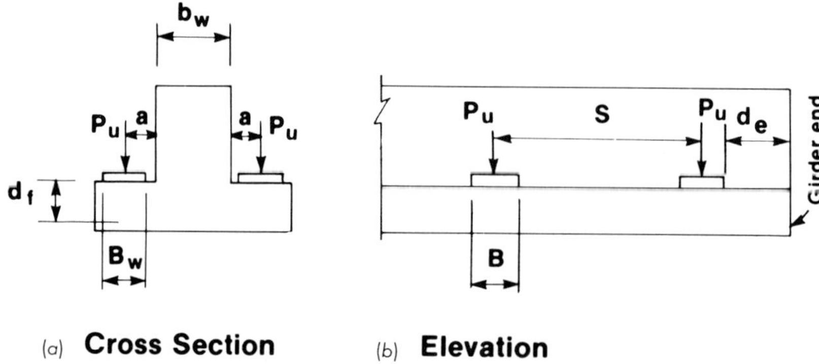

(a) **Cross Section** (b) **Elevation**

FIGURE 11. Stringer spacings required for punching shear.

along the beam axis on each side of the web should be placed at a center-to-center spacing [shown as S in Figure 11(b)] that exceeds $(2d_f + B)$, where B = length of bearing pad along the edge of the flange.

For an end stringer, the distance from the edge of the bearing pad to the longitudinal end of the inverted T-beam [shown as d_e in Figure 11(b)] should be a minimum of $(d_f + B_w)$. This ensures the development of flange punching strength at end stringers at least as great as that developed at interior stringers. If the reaction from the end stringer is less than that from an interior stringer, an end distance smaller than $(d_f + B_w)$ may be provided and can be calculated from the following expression:

$$d_e \geq \left[\frac{P_u}{4\phi\sqrt{f'_c}d_f} - (B + B_w + d_f) \right] \tag{2}$$

in which $\phi = 0.85$. The distance d_e should always be greater than zero in order to accomodate the bearing pad placed near the end of the inverted T-beam.

BRACKET-TYPE SHEAR FRICTION IN FLANGE

The effective depth of flange (d_f) from centroid of top layer of flange transverse reinforcement to the bottom of flange shown in Figure 12(a) and required to fulfill shear friction requirements should satisfy the following equation:

$$0.2\,\phi f'_c d_f(B + 4a) \geq P_u$$

or

$$d_f \geq \frac{6P_u}{f'_c(B + 4a)} \tag{3}$$

in which P_u = ultimate concentrated load acting on a bearing pad; B = length of the bearing pad along the edge of the flange; a = distance from the face of the web to the center of a bearing pad; $0.2 f'_c$ = shear strength of concrete resisting shear friction [1,9]; and $\phi = 0.85$. Use $f'_c = 4000$ psi for $f'_c > 4000$ psi (27.6 MPa) in computation of d_f from Equation (3). This upper limit on f'_c is specified to limit the shear strength of concrete, because Equation (4) used later for computation of A_{vf} will become unconservative for higher values of f'_c [2]. Since the shear friction seldom controls the flange depth of inverted T-beams, this limit on f'_c will not affect most of the practical cases.

In Equation (3), the effective flange length resisting shear friction has been taken as $(B + 4a)$. In most cases, the stringer spacing along the beam axis will be large enough to permit the full effective flange length. However, if the longitudinal spacing of stringers is less than $(B + 4a)$, the stringer spacing should be used in lieu of $(B + 4a)$ in computation of d_f from Equation (3). For a stringer placed near the longitudinal end of an inverted T-beam, the stringer spacing for use in Equation (3) should be taken as twice the distance from the center of the bearing pad to the end of the inverted T-beam or as the longitudinal distance between two adjacent stringers, whichever is smaller.

The transverse reinforcement should be placed perpendicular to the web near the top of the flange to resist flexural tension and to ensure enough pressure for sustaining shear friction force as indicated in Figure 12(a). The transverse reinforcement required to satisfy shear friction should be placed within a distance $2a$ each side from the edge of a bearing pad as shown in Figure 12(b). The area of cross-section of such reinforcement (A_{vf}) should satisfy the following expression [1,9]:

$$\phi\mu A_{vf} f_y \geq P_u$$

or

$$A_{vf} \geq \frac{P_u}{1.2 f_y} \tag{4}$$

in which μ = coefficient of sliding friction taken as 1.4 for normal weight concrete cast monolithically [1,9]; f_y = specified yield strength of reinforcement; and $\phi = 0.85$. This reinforcement should be placed in two or more layers in the top half of the flange thickness; the area of reinforcement in the top layer should equal $2A_{vf}/3$ as indicated in Figure 12(b).

The flange transverse reinforcement placed in the top layer also resists flexural tension, which is caused by the cantilever action of the flange, and should satisfy the following equation:

$$0.8\ \phi d_f A_{sf} f_y \geq P_u a$$

or

$$A_{sf} \geq \frac{1.4 P_u a}{f_y d_f} \tag{5}$$

in which A_{sf} = cross-section area of transverse reinforcement required to resist flexural tension; $0.8\ d_f$ = effective distance between the centroid of compression and the centroid of tension (jd_f) for calculation of flexural reinforcement in the flange; and $\phi = 0.9$. The value of jd_f in a deep cantilever is expected to be smaller than that of an ordinary depth, shallower beam. The suggested value of $0.8\ d_f$ for bracket design is based on a finite element analysis [3]. All transverse reinforcement within a longitudinal distance $2.5a$ each side of the bearing pad can be taken as A_{sf} as shown in Figure 12(c). Thus, the reinforcement placed in the top layer should at least equal A_{sf} or $2A_{vf}/3$, whichever is greater, as indicated in Figure 12(a). If $a \leq 0.4\ d_f$, A_{sf} will not control the design. A_{vf} and distribution of shear friction reinforcement should always be checked through Equation (4).

The use of the term $(B + 5a)$ as the effective length of bracket for the distribution of flexural steel in the top of a bracket is based on an elastic continuum analysis and is supported by test results [3]. The same holds for the use of the

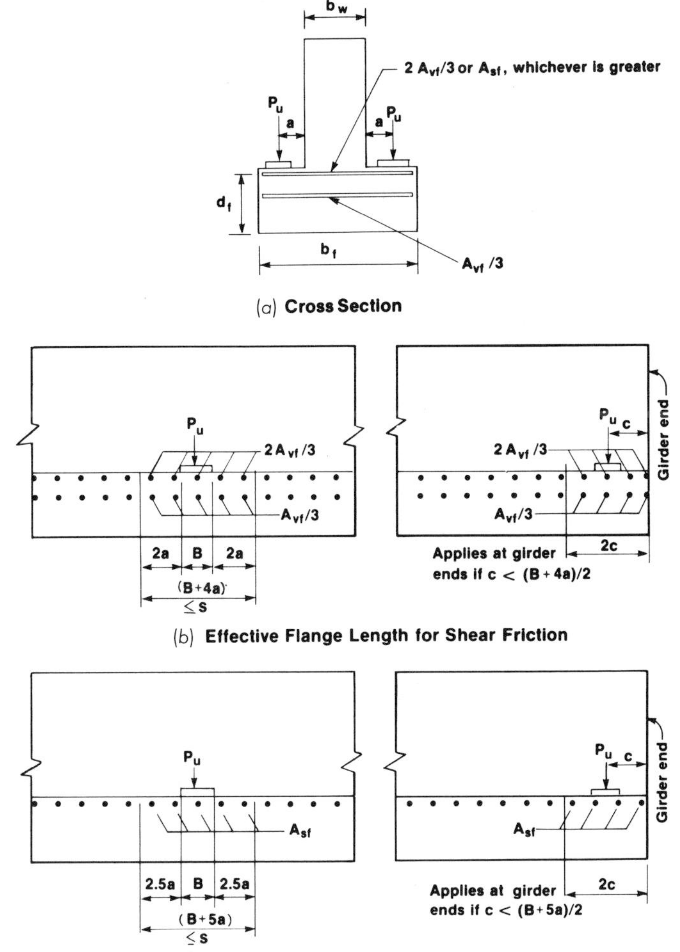

FIGURE 12. Effective flange length for design of bracket reinforcement.

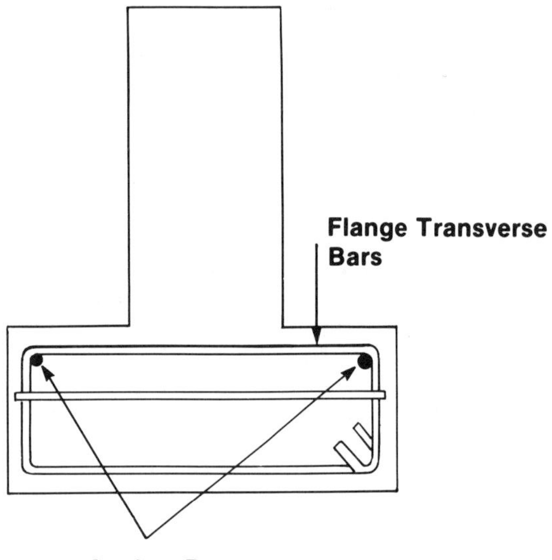

FIGURE 13. Anchorage of flange transverse bars without the use of welding.

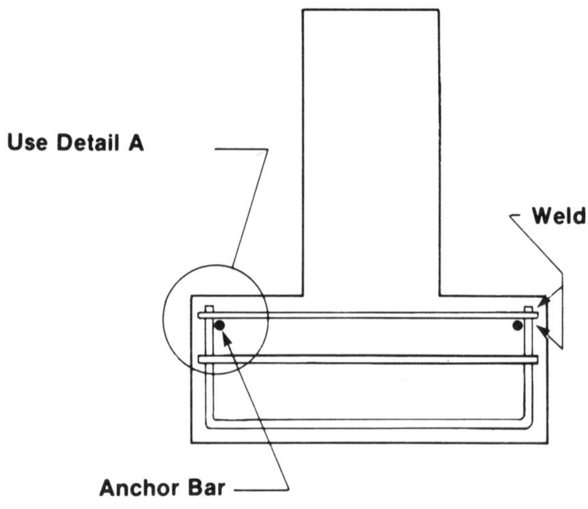

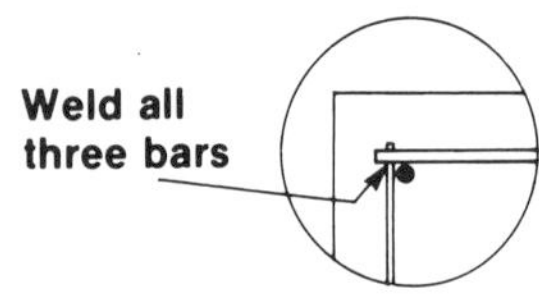

Detail A

FIGURE 14. Anchorage of flange transverse bars with the use of welding.

term $(B + 4a)$ as the effective length of bracket suggested for shear friction. The suggested effective flange length $[(B + 4a)$ for A_{vf} and $(B + 5a)$ for $A_{sf}]$ should not overlap for adjacent stringers. If the distance c between the center of a concentrated load and the longitudinal end of the girder is less than one-half the effective flange length illustrated in Figure 12(b) or 12(c), the effective flange length should be taken as $2c$ or as the spacing of stringers, whichever is smaller.

Near the cantilever end of a bracket where the effective bracket length is controlled by the end distance [Figure 12(c)], jd_f for calculating flexural reinforcement in the flange should be reduced to $0.7d_f$. For such cases, Equation (5) should be modified by substituting $(0.8/0.7) \times 1.4 = 1.6$ in place of the factor 1.4.

The longitudinal forces due to sudden braking of a vehicle are transmitted from its wheels to the deck of the bridge. The magnitude of this longitudinal force depends on the weight, the velocity, and the braking time of the vehicle. The AASHTO specifications [9] call for a longitudinal force of five percent of the live load in all lanes carrying traffic headed in the same direction. This longitudinal force will add little stress to the deck and the stringers but may be important for the design of stringer bearings and the supporting brackets of the inverted T-beams. Consequently, a 5–10 percent longitudinal component of live load stringer reactions could be included for the design of flexural steel in the top of the bracket. However, the rigidity of the bridge deck tends to spread such loads among all stringers, and the approximations used in estimating the effective flange length for flexure hardly justify the superposition of tension due to longitudinal component of live load stringer reactions and that due to flange flexure unless stringers are spaced more closely than $5a$ plus the length of the bearing pad.

The ASSHTO specifications [9] also require that an additional longitudinal force due to friction at expansion bearings should be included in design. The magnitude of this frictional force depends on the type of bearing used. Unless special provisions are made in design of stringer bearings to avoid the frictional force, this force should be computed and additional reinforcement should be added to the flexural steel near the top of the bracket to resist this force. In place of an exact computation of the longitudinal forces due to friction, the ACI Code [1] provisions for bracket design may be used. These clauses require the use of a longitudinal force that at least equals 20 percent of the stringer reaction due to dead plus live load. These clauses also regard the longitudinal force acting on a bracket as a live load even when this force results from creep, shrinkage, or temperature change.

The anchorage of the flange transverse bars may impose a problem, because the flange overhang is usually too short to accommodate the development length of reinforcing bars of usual sizes employed in bridge construction. The detail shown in Figure 13 is recommended wherever the reinforcement size would permit the development of yield strength in the flange transverse bars. However, it may be necessary in some cases to weld the ends of transverse bars to an anchor bar at the exterior face of the flange and perpendicular to the transverse bars. A welding detail used by the Texas State Department of Highways and Public Transportation is illustrated in Figure 14. The development of flange transverse bars can also be achieved by furnishing a continuous steel angel (or plate) along the top corners of the flange and connecting these bars to the angle (or plate). The welding should conform to the American Welding Society D1.4 Code for reinforcing steel [10].

Design of Web Stirrups Acting as Hangers

Stirrups in the web of an inverted T-beam act as hangers to deliver the concentrated loads applied on the flange into the body of the web. The maximum hanger stresses at ultimate load exist when both sides of the web are subjected to maximum live loads simultaneously. The longitudinal distance over which hanger forces can be distributed, defined as effective hanger distance in Figure 15, is limited either by the shear capacity of concrete in the flange each side of a bearing pad or by the longitudinal center-to-center spacing of stringers. Hence, to achieve safe delivery of flange loads into the web, the following strength relationships (Equation (6) and Equation (7)) should be satisfied:

$$\phi A_v f_y [(B + 2d_f)/s] \geq [2P_u - 2(2\phi\sqrt{f'_c}b_f d_f)]$$

or

$$\frac{A_v}{s} \geq \left[\frac{\frac{2P_u}{\phi} - 4\sqrt{f'_c}b_f d_f}{f_y(B + 2d_f)}\right] \tag{6}$$

and

$$\phi A_v f_y\ (S/s) \geq 2P_u$$

or

$$\frac{A_v}{s} \geq \frac{2P_u}{\phi f_y S} \tag{7}$$

In Equations (6) and (7), b_f = overall flange width (Figure 15); d_f = flange depth from top of flange to center of bottom longitudinal reinforcement as indicated in Figure 15(a); S = stringer spacing along the beam axis; P_u = ultimate concentrated load acting on a bearing pad; A_v/s = cross-section area of both legs of a web stirrup divided by the spacing of stirrups; and $\phi = 0.85$. For a stringer load placed near the end of an inverted T-beam, the distance S in Equation (7) should be taken as twice the distance from the center of the bearing pad to the longitudinal end of the inverted T-beam or as the longitudinal distance between two adjacent stringers, whichever is smaller.

Equation (6) controls the design of hangers in cases where the stringer spacing is large enough to permit a failure mode in which flange strength becomes effective in resisting hanger forces as shown in Figure 15(a). For these cases, the hanger reinforcement should be provided for twice the stringer load minus the shear strength of the flange each side of the stringer as indicated by Equation (6). If the stringers are too closely spaced, the hanger failure will take place by separation of the flange from the web over the entire loaded length of the beam. Hence the hangers should be designed for the full stringer loads as indicated by Equation (7). However, both of these equations should be satisfied to insure safe transfer of hanger forces.

The premature yielding of hangers nearest the concentrated loads applied to the flange and the size of cracks that form at the junction of the web and the flange when these hangers reach high stress levels should be controlled at service load conditions. This can be achieved by satisfying the following equation:

$$A_v f_s [(B + 3a)/s] \geq 2P_s$$

or

$$\frac{A_v}{s} \geq \frac{3P_s}{f_y(B + 3a)} \tag{8}$$

in which P_s = concentrated service load acting on a bearing pad; and f_s = hanger stress at service loads limited to a maximum value of $2f_y/3$.

Reduced cracking and hanger steel stresses lower than $2f_y/3$ may be desired at service loads for reasons such as aggressive environment, fatigue control, etc. This can be achieved by simply increasing the numerator $3P_s$ in Equation (8). For example, if the limit service stress were $f_y/2$ instead of $2f_y/3$, the numerator would become $(0.66/0.5) \times 3P_s = 4P_s$.

In Equation (8), the effective distance over which hanger forces can be distributed at service loads has been taken as $(B + 3a)$. If $(B + 3a)$ exceeds the longitudinal stringer spacing, stringer spacing S should be used in lieu of $(B + 3a)$ in computation of A_v/s from Equation (8). For end stringers, S should be taken the same as that defined for Equation (7). Equation (8) is based on test results and is documented in Ref. 7.

The largest value of A_v/s from Equations (6), (7), and (8) should be used. Only vertical stirrups anchored to develop by bond their tension yield forces above and below the top surface of the flange should be considered to carry hanger forces. Since most flanges have a depth inadequate for the development of stirrup bar yield forces, hangers should be

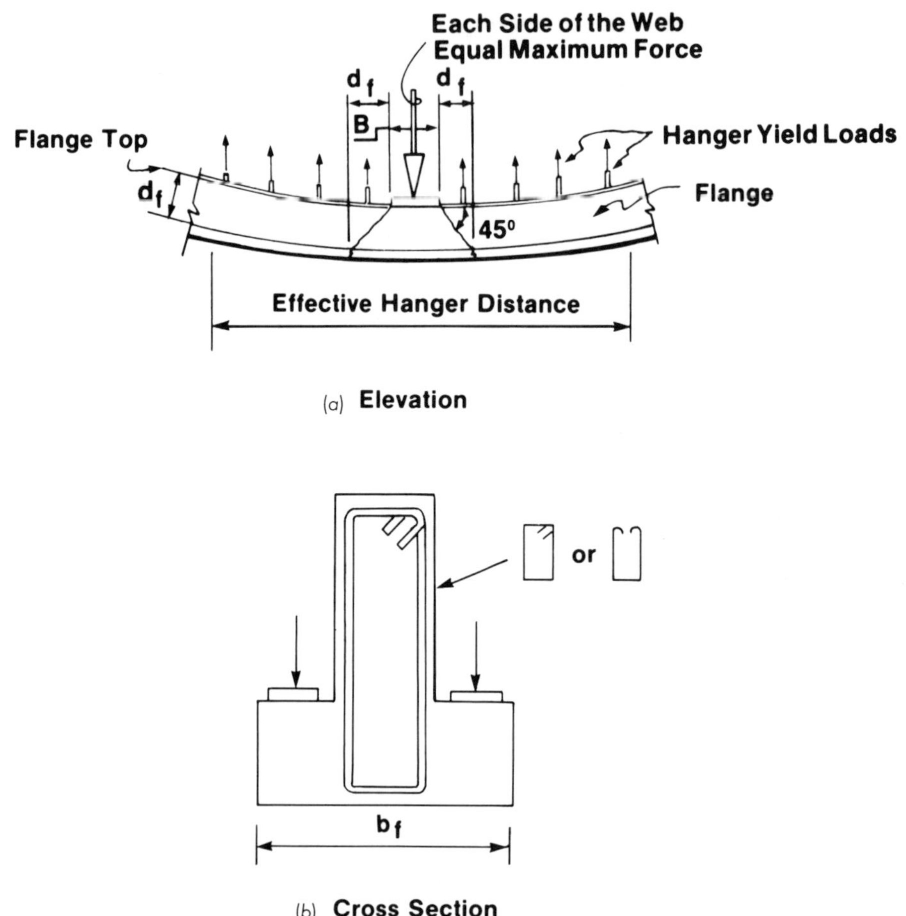

FIGURE 15. Hanger forces in response to flange loads shown on the beam elevation and details of hanger reinforcement shown in the beam cross-section.

closed across the bottom of the inverted T-beam as indicated in Figure 15(b).

It is not necessary to superimpose loads on stirrups acting as hangers and loads on stirrups acting as shear (and torsion) reinforcement. This will be discussed further in a subsequent section.

Overall Design of Inverted T-Beam

The overall strength of an inverted T-beam should be adequate to support ultimate flexure, flexural shear, and torsional shear forces and any possible combination of such forces.

DESIGN FOR MAXIMUM FLEXURAL MOMENT AND SHEAR

The most likely design condition for bridge bentcap girders involves flexural moments and flexural shear forces that are largest when torsion is absent, because traffic loads stringers fully both sides of the girder web. Consequently, a logical procedure for overall design begins with the proportioning of the cross-section and reinforcement solely on the basis of maximum flexural moment and maximum flexural shear force.

Requirements of flexural reinforcement for overall design of the beam are not altered by the location of the T-beam flange. The ultimate strength and serviceability design requirements of the American Concrete Institute Building Code for Reinforced Concrete [1] and the American Association of State Highway and Transportation Officials Specifications [9] are quite safe and adequate for flexure. This applies to both posttensioned prestressed concrete and reinforced concrete inverted T-beams.

The ACI Code [1] permits the designer to consider the maximum end shear as that occuring at a distance d from the face of the support for nonprestressed members and at a distance $h/2$ from the face of the support for prestressed members, where d and h are, respectively, the effective depth and overall thickness of the member. Of course, this is allowed only if no stringer load is placed between the face of the support and the critical section at a distance d or $h/2$ from the face of support. While this is reasonable for conventional beams, this is not appropriate for inverted T-beams unless the terms d_f and $h_f/2$ related to the flange are substituted in lieu of d and $h/2$, respectively. To simplify the shear calculations, however, the critical section for the maximum end shear in inverted T-beams may be taken at the face of support. This simplification will not cause a significant loss of accuracy in shear design of bridge bentcap girders.

With a flange overhang to thickness ratio equal to or less than 1.0, the flange of an inverted T-beam is expected to be stiff enough to fully participate in retention of the shear force. This seems particularly valid for a cross-section subjected to negative moment, creating flexural compression in the flange. The tests on inverted T-beams support this hypothesis [8]. Hence, the flexural shear strength in absence of torsion can be based on modified versions of the ACI Code equations [8]:

$$V_o = \phi\left[2\sqrt{f_c'}A_e + A_v f_y\left(\frac{d}{s}\right)\right] \text{ for reinforced concrete} \tag{9}$$

and

$$V_o = \phi\left[n\sqrt{f_c'}A_e + A_v f_y\left(\frac{d}{s}\right)\right] \text{ for prestressed concrete} \tag{10}$$

In both equations, $A_v f_y d/s \leq 8\sqrt{f_c'}\ b_w d$; $n = 5$ for $M_u/V_u d = 1$, decreasing linearly to 2 as $M_u/V_u d$ increases from 1 to 5; A_e = area of all concrete between compression face and centroid of flexural tension (or prestressing) reinforcement as indicated in Figure 16; $\phi = 0.85$; and d is defined in Figure 16. The suggested approximation for n applies only to the beam with $f_c' \leq 6400$ psi (44 MPa) and having an effective prestressing force greater than or equal to 0.4 times the tensile strength of flexural reinforcement. Stirrups should be designed to resist all applied ultimate shear force above that resisted by the concrete section illustrated in Figure 16. In addition, all other shear clauses of the Code [1,9] should be satisfied.

The shear area A_e used in Equations (9) and (10) does not meet the ACI or AASHTO requirements. Hence, engineers may use $b_w d$ in lieu of A_e when constrained to employ the current code rules.

The tests on inverted T-beams of usual proportions employed in bridge construction have shown that the hanger failure cracks occur at the junction of the web and the flange, whereas the flexural shear failure cracks occur in the web above the flange [7]. Consequently, the yielding of stirrups acting as hanger reinforcement and as shear reinforcement takes place at different locations in the stirrups. The stirrups designed as hangers can then be used as part of the web reinforcement resisting flexural shear. This can be further justified by the following argument. The failure of an inverted T-beam due to flexural shear is an overall failure, whereas the hanger failure is a local failure of stirrups acting as hangers nearest an applied concentrated load to transmit flange forces into the web. Hence, the hanger forces delivered to the web are carried to the beam supports by the web as flexural shear and it is not necessary to superimpose loads on stirrups acting as hangers and loads on stirrups acting as shear reinforcement. However, the web reinforcement in an inverted T-beam should be proportioned

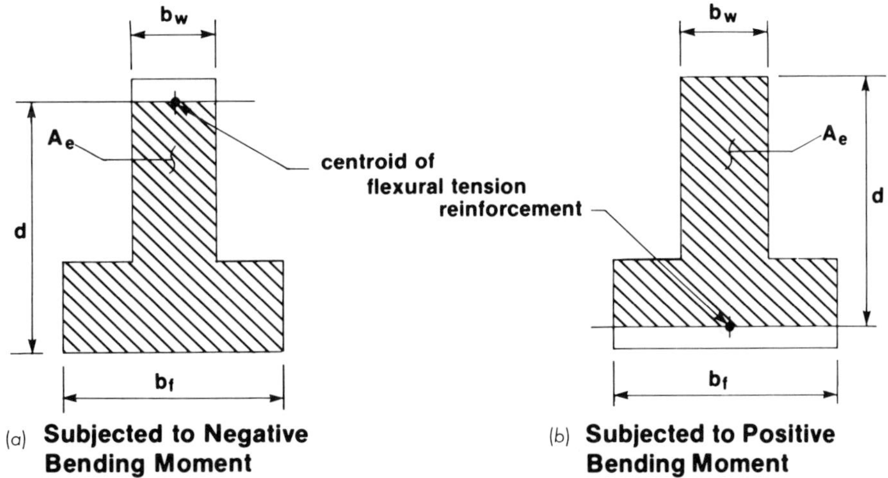

FIGURE 16. Effective area resisting shear force A_e (cross-hatched).

on the basis of hanger requirement or shear strength requirement, whichever is greater.

DESIGN FOR COMBINED EFFECT OF FLEXURAL SHEAR AND TORSION

Loading conditions involving torsion on the inverted T-section might create more severe requirements for proportioning reinforcement. Torsion of inverted T bentcap girders occurs with every passage of design vehicles across the bentcap. As traffic approaches the bentcap, stringer reactions cause twisting or torsion of bentcap toward the approaching load. The direction of twist reverses after the passage of traffic imposes loads on stringers that react on the opposite flange overhang of the inverted T-beam.

When torsion is a maximum, traffic loads stringers on only one side of the web and flexural shear is less than the maximum value. Hence, stirrups also serve as vertical reinforcement of the web that is subjected to combined flexural and torsional shear. The local hanger forces need not be superimposed on the web shear forces for designing stirrups as explained earlier, but the cross-section should be designed for more critical of the two forces.

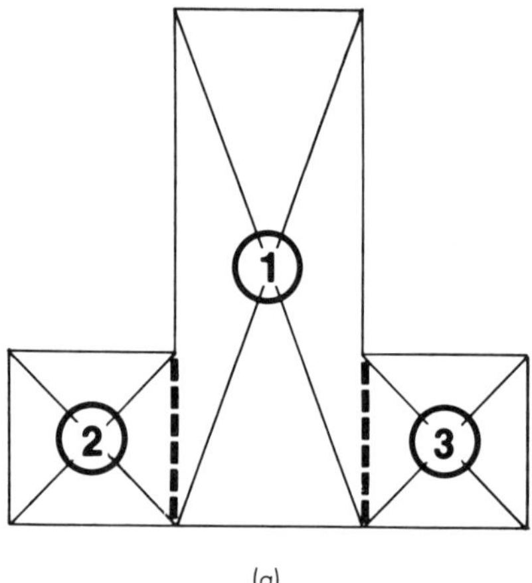

(a)

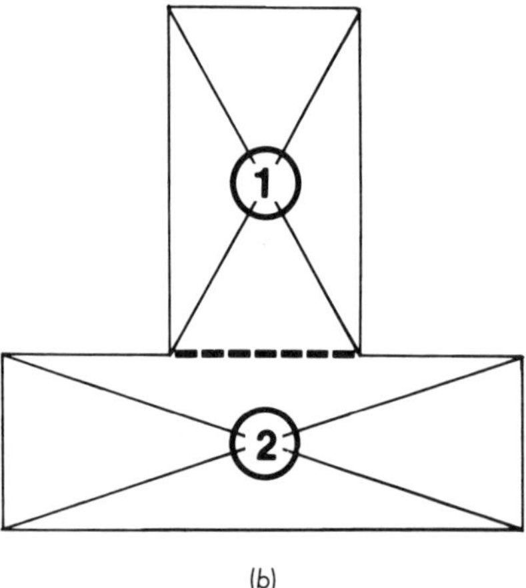

(b)

FIGURE 17. Component rectangles of inverted T-section for torsion analysis [use larger of the two values of $\Sigma x^2 y$ from (a) and (b)].

For the cross-section to be adequate when the combined torsional and flexural shear acts, the following interaction expression should be satisfied:

$$\left(\frac{V_u}{V_o}\right)^2 + \left(\frac{T_u}{T_o}\right)^2 \leq 1.0 \tag{11}$$

in which V_u and T_u are applied ultimate shear force and applied ultimate torque, respectively. Flexural shear capacity V_o can be taken from Equation (9) or 10 and the ultimate pure torsion strength T_o for both reinforced and prestressed concrete members can be calculated from an adaptation [8] of the pure torsion strength equation recommended by the ACI Code [1]:

$$T_o = \phi\left[4\sqrt{f'_c}\,\frac{\Sigma x^2 y}{3} + A_t f_y\left(\frac{\alpha_t x_1 y_1}{s}\right)\right] \leq 18\phi\sqrt{f'_c}\,\frac{\Sigma x^2 y}{3} \tag{12}$$

In Equation (12), A_t = area of one leg of a web stirrup; s = spacing of web stirrups; $\alpha_t = [0.66 + 0.33(y_1/x_1)] \leq 1.5$; x_1 and y_1 = shorter and longer center-to-center dimension, respectively, of closed web stirrups; and $\phi = 0.85$. The contribution to torsion strength of the cross-section of each component rectangle should be computed separately using smaller dimension x and larger dimension y for the rectangle under consideration as indicated in Figure 17. Only closed rectangular web stirrups should be considered effective in resisting torsion. This assumption of neglecting the contribution of flange transverse reinforcement to torsional strength may lead to a slightly conservative design for combined effect of flexural shear and torsion, but will greatly simplify the calculations. Although the ACI Code [1] permits a value of $2.4\sqrt{f'_c}$ for torsional shear strength of concrete, a value of $4\sqrt{f'_c}$ is used in Equation (12). This value seems to be justified for inverted T-beams and is documented in Ref. [8].

Alternately, the combined effect of flexural shear and torsional shear on the cross-section can be satisfied by using the following equations:

$$6\sqrt{f'_c}\,\Sigma x^2 y \geq \frac{T_u}{\beta} \tag{13}$$

and

$$\frac{A_v}{s} \geq \frac{2}{\alpha_t x_1 y_1 f_y}\left[\frac{T_u}{\beta} - 1.33\sqrt{f'_c}\,\Sigma x^2 y\right] \tag{14}$$

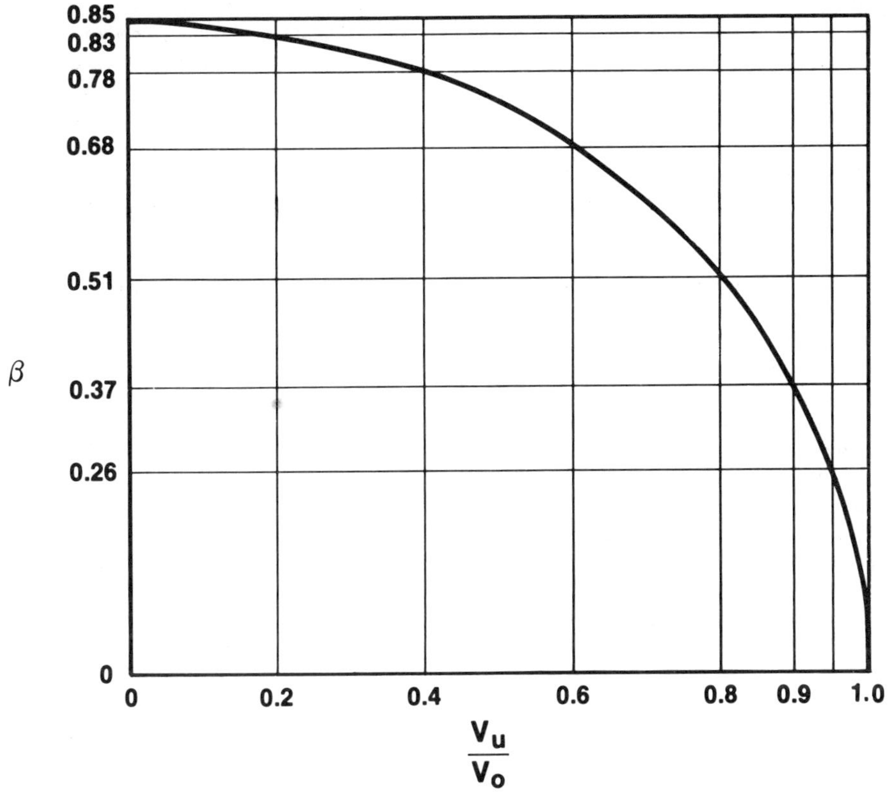

FIGURE 18. Value of β for different ratios of V_u/V_o (linear interpolation of β may be conservatively used between plotted values of V_u/V_o).

in which $\beta = \phi\sqrt{[1 - (V_u/V_o)^2]}$ and can be easily determined from the plot shown in Figure 18. Equations (13) and (14) were derived by solving Equation (11) for T_o and equating it to Equation (12), and can be used in lieu of Equations (11) and (12). Equation (13) ensures yielding of stirrups before crushing of concrete takes place. Hence, the cross-section should be revised if Equation (13) is not satisfied. Equation (14) determines A_v/s, the area of cross-section of both legs of a closed rectangular web stirrup divided by stirrup spacing, required for combined effect of shear and torsion. Note that only one set of equations for combined effect of flexural shear and torsion needs to be checked, i.e. either check Equations (13) and (14) or check Equations (11) and (12).

The torsion shear strength of $4\sqrt{f_c'}$ and the limiting strength of $18\sqrt{f_c'}$ used for concrete in Equation (12) [and for Equations (13) and (14)] do not conform to the ACI requirements. For engineers required to use the code rules, the ACI values may be substituted in these equations.

If the area of transverse reinforcement in the web is controlled by the requirements of maximum flexural shear, there is apparently no need to check for longitudinal reinforcement required for torsion. The supplemental longitudinal reinforcement should be provided to help flexural reinforcement resist torsion for cases in which stirrup design is controlled by the combined effect of torsion and flexural shear. If the area of web stirrups is increased to satisfy Equation (14) or Equation (11), the supplemental longitudinal steel (A_ℓ) with a volume at least equal to the volume of extra web transverse reinforcement (A_t') should be provided. When the yield strength of A_ℓ and A_t' is the same, $sA_\ell = 2(x_1 + y_1)\, A_t' = A_v'\,(x_1 + y_1)$; and

$$A_\ell = \frac{A_v'}{s}\,(x_1 + y_1) \qquad (15)$$

in which $A_v'/s = A_v/s$ required for combined effect of flexural shear and torsion minus A_v/s required for maximum flexural shear acting alone. The area of longitudinal reinforcement A_ℓ should be distributed among the corners of the web plus the four corners of the flange, and it should be added to the flexural reinforcement both for prestressed and non-prestressed members. Note that all other relevant shear and torsion clauses of the Code [1,9] should be satisfied.

DESIGN PROCEDURE

The design procedure for inverted T-beam bentcap girders based on the criteria proposed in this paper can be summarized as follows:

1. Compute the flange thickness for punching shear requirements using Equation (1) (or Figures 9 and 10) and select a value that at least equals the flange overhang. Establish the web width, so that the center-to-center transverse distance between the two stringer reactions acting on the opposite sides of the web ($2a + b_w$) at least equals twice the flange depth plus the width of the bearing pad ($2d_f + B_w$). The web width must be able to accomodate the longitudinal reinforcement required for negative bending moment. Check the requirements for minimum stringer spacings and end distances controlled by punching shear.

The height of the web above the top face of the flange is determined by the depth of the stringers to be supported on the flange. The overall depth of the beam can then be computed and must provide the required flexural stiffness. The nominal span-to-depth ratios for inverted T-beam bentcap girders appear to be from 2 to 4 for cantilevered spans, and 4 to 8 for spans supported at each end [4].

2. Using Equation 3 compute the flange thickness required to resist shear friction and revise the furnished value if needed. Equations (4) and (5), respectively, calculate the flange transverse reinforcement required for shear friction (A_{vf}) and that required for flexural tension (A_{sf}) in the flange. Place A_{sf} or $2A_{vf}/3$, whichever is greater, near the top of the flange and $A_{vf}/3$ in one or more layers below the top layer within the top half of the flange thickness.

Include the effect of longitudinal forces at stringer bearings due to friction in the design of top transverse reinforcement in the flange unless provisions are made to avoid the frictional forces at stringer bearings. The flange transverse reinforcement in the top layer is also subjected to longitudinal component of stringer reactions due to live and impact loads. Anchor the flange transverse bars as per details shown in Figure 13 or 14.

3. Provide stirrups in the web for the most critical effect from hanger tension, maximum flexural shear, and maximum torque plus corresponding flexural shear:

(a) Determine the area of stirrup reinforcement required to resist hanger forces as the largest value of A_v/s obtained from Equations (6), (7), and (8). Note that the maximum stresses due to hanger action act when both sides of the web are subjected to maximum live loads simultaneously.

(b) The maximum stresses due to flexural shear acting alone occur when traffic loads the stringers fully both sides of the web. Determine the required area of stirrups in terms of A_v/s from Equations (9) or (10). Note that the critical section for the maximum end shear in an inverted T-beam may be taken at the face of the support.

(c) For maximum torsion to act on an inverted T-beam, traffic loads the stringers on only one side of the web and the corresponding flexural shear will be less than its maximum value. Again, the critical section near the end of the member may be taken at the face of the support. Satisfy Equation (13) to ensure yielding of stirrups prior to crushing of concrete under ultimate loads. Note that a larger cross-section will be needed to resist combined shear and torsion if Equation (13) is not satisfied. Calculate the required area of stirrups (A_v/s) using Equation (14). The term β used in Equations (13) and (14) may be determined from Figure 18.

The superposition of stirrup reinforcement (A_v/s) required for (a), (b), and (c) is not needed. However, design stirrups for the maximum of the three effects at all critical sections. Provide either closed rectangular stirrups or stirrups that are closed at least across the bottom of the beam as indicated in Figure 15(b). Use only closed rectangular stirrups required to resist forces due to torsion.

4. The most critical section for flexure occurs at the face of the support in cantilever beams (near midspan in simple beams) and the maximum bending moment on an inverted T-beam acts when full live loads are applied both sides of the web. Determine the longitudinal reinforcement required to resist flexural tension at all critical sections. If the web reinforcement (A_v/s) computed for maximum flexural shear acting alone is greater than that calculated for combined effect of torsion and flexural shear, no check on longitudinal reinforcement required for torsion is necessary. Provide supplemental longitudinal reinforcement as per Equation (15) to help flexural reinforcement resist torsion in cases where stirrup design is controlled by the combined effect of torsion and flexural shear. Distribute the supplemental longitudinal reinforcement along the perimeter of the cross-section, particularly at the corners of the web and the flange.

SUMMARY

Reinforced concrete and posttensioned prestressed concrete inverted T-beams are frequently used for bridges. The structural behavior of inverted T-beams differs from that of conventional top-loaded beams, because the loads are introduced into the bottom rather than into the sides or the top of the web. The application of loads near the bottom of the web in inverted T-beams imposes special problems, which are not addressed to in the current American structural codes.

This paper provides recommendations for proportioning cross-section dimensions and reinforcement of cast-in-place normal weight concrete inverted T-beams employed in bridge structures. These beams should be designed to have adequate strength against possible failure due to flexure, flexural shear, torsion and any possible combination of these forces. Reinforcement details for the flanges of inverted T-beams should accommodate flexure, shear friction, and punching shear on the short cantilevered shelf. The transverse reinforcement in the webs of inverted T-beams should resist hanger tension forces caused by loads applied to the lower part of the web. The step-by-step procedure based on the proposed criteria summarizes in the previous section the design of inverted T-beam bentcap girders. A design example given in the Appendix elaborates the application of the proposed criteria.

The paper is intended for use by experienced structural engineers engaged in bridge design. It only provides the highlights from the reported research on inverted T-girders. The background material and other related information is available from the references provided in the paper.

ACKNOWLEDGEMENTS

A major part of the material presented herein was taken from Ref. [6]. The authors wish to thank the Prestressed Concrete Institute for permission to include that material in this paper. Figures 3 through 8 were reproduced from Ref. [7] with permission from the American Concrete Institute.

NOTATION

The following symbols are used in this paper:

A_e = area of cross-section of all concrete between compression face and centroid of flexural tension (or prestressing) reinforcement;
A_ℓ = total area of cross-section of supplemental longitudinal reinforcement required to resist torsion;
A_{sf} = area of cross-section of transverse reinforcement required to resist flexural tension in a flange overhang;
A_t = area of cross-section of one leg of a web stirrup;
A_v = area of cross-section of both legs of a web stirrup;
A_{vf} = area of cross-section of transverse reinforcement required to resist shear friction in a flange overhang;
$A_v'/s = A_v/s$ required for combined effect of flexural shear and torsion minus A_v/s required for maximum flexural shear acting alone;
a = distance from the face of web to the center of a bearing pad taken perpendicular to the beam axis;
B = length of bearing pad along the edge of flange;
$B_p = B + 2B_W$;
B_w = width of bearing pad perpendicular to the beam axis;
b_f = overall width of flange of an inverted T-beam;
b_w = width of web;
d = effective depth of an inverted T-beam between compression face and centroid of flexural tension (or prestressing) steel;
d_e = distance from the edge of bearing pad to the longitudinal end of an inverted T-beam;
d_f = effective depth of flange as defined in Equations (1), (3), and (6);
f_c' = specified strength of concrete;
f_s = service load stress limited to a maximum value of $2f_y/3$ in stirrups acting as hangers;
f_y = specified yield strength of reinforcement;
h = overall thickness of an inverted T-beam;
h_f = overall thickness of flange of an inverted T-beam;
jd_f = effective distance between centroid of compression and centroid of tension for calculation of flexural reinforcement in flange overhang;
M_u = applied ultimate bending moment acting on overall cross-section of an inverted T-beam;
n = a coefficient defined after Equation (10);
P_s = concentrated service load acting on a bearing pad;
P_u = concentrated ultimate load acting on a bearing pad;
S = stringer spacing along the beam axis;
s = spacing of web stirrups;
T_o = ultimate pure torsion strength of overall cross-section of an inverted T-beam;
T_u = applied ultimate torque acting on overall cross-section of an inverted T-beam;
V_o = ultimate flexural shear strength (in absence of torsion) of overall cross-section of an inverted T-beam;
V_u = applied ultimate flexural shear force acting on overall cross-section of an inverted T-beam;
x = smaller dimension of a rectangle;
x_1 = shorter center-to-center dimension of a closed web stirrup;
y = larger dimension of a rectangle;
y_1 = longer center-to-center dimension of a closed web stirrup;
$\alpha_t = [0.66 + 0.33\,(y_1/x_1)] \leq 1.5$;
$\beta = \phi\sqrt{[1 - (V_u/V_o)^2]}$;
μ = coefficient of sliding friction taken as 1.4 for normal weight concrete cast monolithically;
ϕ = strength reduction factor taken as 0.9 for flexure and 0.85 for shear and torsion.

REFERENCES

1. "Building Code Requirements for Reinforced Concrete," ACI 318–83, American Concrete Institute, Detroit, MI (1983).
2. "Commentary on Building Code Requirements for Reinforced Concrete", ACI 318–83, American Concrete Institute, Detroit, Michigan (1983).
3. Furlong, R. W., P. M. Ferguson, and J. S. Ma, "Shear and Anchorage Study of Reinforcement in Inverted T-Beam Bent Cap Girders," Report 113-4, Center for Highway Research, The University of Texas at Austin, Austin, Texas (July 1971).
4. Furlong, R. W. and S. A. Mirza, "Strength and Serviceability of Inverted T-Beam Bent Caps Subject to Combined Flexure, Shear and Torsion," Report 153-1F, Center for Highway Research, The University of Texas at Austin, Austin, Texas (August 1974).
5. Mirza, S. A., "Concrete Inverted T-Beams in Combined Tor-

sion, Shear, and Flexure," Dissertation presented to The University of Texas at Austin, Austin, Texas, in partial fulfillment of the requirements for the degree of Doctor of Philosophy (May 1974).

6. Mirza, S. A. and R. W. Furlong, "Design of Reinforced and Prestressed Concrete Inverted T Beams for Bridge Structures," *Prestressed Concrete Institute Journal*, 30(4), 23–48 (July/August 1985).
7. Mirza, S. A. and R. W. Furlong, "Serviceability Behavior and Failure Mechanisms of Concrete Inverted T-Beam Bridge Bentcaps," *American Concrete Institute Journal*, 80(4), 294–304 (July/August 1983).
8. Mirza, S. A. and R. W. Furlong, "Strength Criteria for Concrete Inverted T-Girders," *Journal of Structural Engineering, American Society of Civil Engineers*, 109(8), 1836–1853 (August 1983).
9. "Standard Specifications for Highway Bridges", American Association of State Highway and Transportation Officials, 12th edition, Washington, D.C. (1977).
10. "Structural Welding Code—Reinforcing Steel", AWS D1.4-79, American Welding Society, New York, N.Y. (1979).

APPENDIX—DESIGN EXAMPLE

Consider a reinforced concrete inverted T-beam shown in Figure 19(a). The beam acts as a bentcap girder in a bridge superstructure to support 4 interior and 2 exterior precast prestressed concrete stringers placed on each side of the web. Each interior stringer exerts a total force of 130,000 lbs and 221,000 lbs (580 kN and 985 kN) at service load and ultimate load conditions, respectively. The corresponding reactions from each of the exterior stringers are 90,000 lbs and 143,000 lbs (400 kN and 636 kN). The ultimate live load plus impact reactions included in the foregoing total loads are 130,000 lbs (580 kN) for each of the interior stringers and 65,000 lbs (290 kN) for each of the exterior stringers. All stringers are placed on 20 in. × 15 in. (508 mm × 381 mm) bearing pads spaced at 84 in. (2134 mm) on centers with an end distance of 21 in. (533 mm) as indicated in Figure 19(a).

Assume that the frictional forces produced at the stringer bearings are negligible. Specified concrete strength (f_c') and specified yield strength of reinforcement (f_y) are 4000

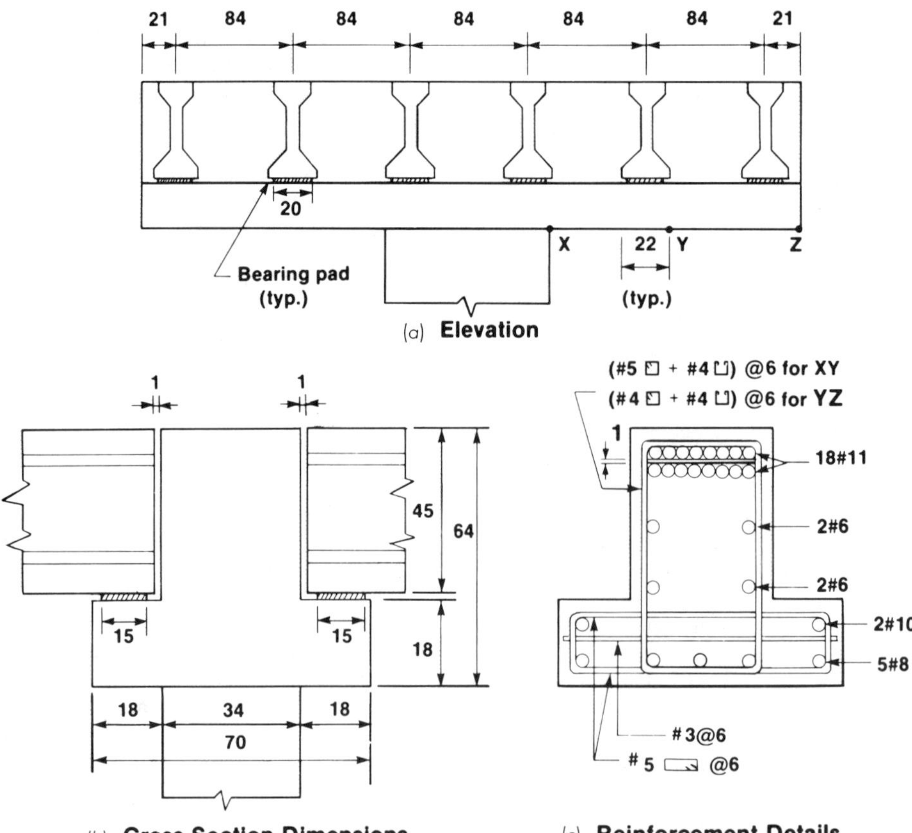

FIGURE 19. Details of an inverted T-beam designed in the example (all dimensions are in inches; 1 in. = 25.4 mm).

lbs/in.^{2} and 60,000 lbs/in.^{2} (27.6 MPa and 414 MPa), respectively. A minimum clear cover of 2 in. (50 mm) is used for all reinforcement and the maximum aggregate size is 1½ in. (38 mm).

Flange Design for Punching Shear

Since B_p equals (20 + 2 × 15) = 50 in., P_u = 221,000 lbs, and f'_c = 4000 lbs/in.^{2}, the required value of d_f is 13.4 in. from Equation (1) or Figure 9. This gives a flange thickness (= 13.4 + 0.625 + 2.0) = 16.1 in. Use a flange thickness of 18 in. with d_f furnished = 15.4 in. These values are acceptable, because the length of the flange overhang (= width of bearing pad + 1 + 2 in. = 18 in.) is not greater than the flange thickness.

The required minimum transverse distance between stringers on the opposite sides of the web is equal to $2d_f + B_w$ (= 2 × 15.4 + 15) = 45.8 in. and a (= 1/2 of the bearing pad width plus 2 in.) = 9.5 in. Hence, the required web width equals (45.8 − 2 × 9.5 =) 26.8 in. Use b_w = 34 in. This width will be required to accommodate the longitudinal reinforcement near the top of the web. The required minimum longitudinal spacing of stringers is ($2d_f + B$ = 2 × 15.4 + 20 =) 50.8 in., which is smaller than the spacing furnished.

At exterior stringers, P_u is equal to 143,000 lbs. Hence, the minimum end distance required for punching shear and calculated from Equation (2) is zero, which is less than the furnished value (= 21 − 10) = 11 in.

Flange Design for Bracket Shear

B + 4a equals (20 + 4 × 9.5 =) 58 in., which is less than S = 84 in. for interior stringers. Hence, use an effective flange length of 58 in. in computation of d_f from Equation (3). From Equation (3), the required d_f is calculated as 5.7 in., which is less than the actual d_f (= 18 − 2 − 0.625/2) = 15.6 in. For end stringers, the effective flange length equals (2 × 21 =) 42 in. and the required d_f = 5.1 in., which is again less than the actual d_f.

From Equation (4), the required A_{vf} equals 3.1 in.^{2} and 2.0 in.^{2} for interior and exterior stringers, respectively. The effective flange length associated with Equation (4) is 58 in. for interior stringers and 42 in. for exterior stringers. This makes the required A_{vf} = 0.65 and 0.57 in.^{2}/ft. for the interior and exterior stringers, respectively. Note that two-thirds of this reinforcement ($2A_{vf}/3$ = 0.43 or 0.38 in.^{2}/ft.) is required for the top layer.

From Equation (5), the required A_{sf} is calculated to be 3.2 in.^{2} and 2.3 in.^{2} for interior and exterior stringers, respectively. Considering the effective flange lengths associated with Equation (5) (67.5 in. for interior and 42 in. for exterior stringers), A_{sf} is equal to 0.57 and 0.65 in.^{2}/ft. for the interior and exterior stringers.

The flange transverse reinforcement placed in the top layer equals $2A_{vf}/3$ or A_{sf}, whichever is greater. Use #5 bars at 6 in. on centers in the top layer over the entire length of the beam. This provides a steel area of 0.62 in.^{2}/ft., which satisfies all the requirements for reinforcement in the top layer. In order to develop the yield strength of flange transverse reinforcement, these bars will be furnished in the shape of closed rectangular stirrups as shown in Figures 13 and 19(c). These bars will also provide support for longitudinal reinforcement in the flange.

The flange transverse steel required in other layers below the top layer equals ($A_{vf}/3$ =) 0.22 and 0.19 in.^{2}/ft. for interior and exterior stringers, respectively. Provide #3 bars at 6 in. on centers in a layer that is placed 4 in. below the top layer over the entire length of the beam. This satisfies the requirements on vertical spacing and area of reinforcement for these bars.

The longitudinal components of the live load stringer reactions were not included for the design of top reinforcement in the flange. Since the stringer spacing exceeds $B + 5a$, there is sufficient reserve strength in the flange to resist such forces. Similarly, the effect of longitudinal forces due to friction at stringer bearings were not considered in the design since these forces are given as negligible in the problem statement. However, it must be realized that in many practical cases, frictional forces can be quite significant and additional flange reinforcement will be required to counter these forces.

Web Design for Hanger Action

Since b_f equals 70 in., d_f (= 18 − 2 − 0.625 − 0.5) = 14.9 in., P_u = 221,000 lbs with S = 84 in. for interior stringers, and B = 20 in., calculated A_v/s for strength requirements is 0.086 in. and 0.103 in. from Equation (6) and (7), respectively. P_s equals 130,000 lbs and $B + 3a$ (= 20 + 3 × 9.5) = 48.5 in., which is less than S = 84 in. Hence, A_v/s required for serviceability considerations and calculated from Equation (8) is 0.134 in. The largest value of A_v/s obtained from Equations (6), (7), and (8) is 0.134 in. and will be used in design. Note that a higher value of A_v/s could be needed if fatigue requirements or aggressive environment is taken into consideration.

At exterior stringers, P_u equals 143,000 lbs, P_s = 90,000 lbs, and S (= 2 × 21) = 42 in. The largest value of A_v/s obtained from Equations (6), (7), and (8) is again 0.134 in. Hence, A_v/s = 0.134 in. is required over the length XZ of the beam for the design of hangers. The final selection of stirrups acting as hangers will be delayed until the design for shear and that for shear plus torsion has been completed.

Web Design for Maximum Flexural Shear

The critical sections occur at X and Y as indicated in Figure 19(a). Since d equals (64 − 2 − 0.625 − 1.41 − 0.5 =) 59.5 in., A_e = 2670 in.^{2}, and the applied ultimate shear force (V_u) at the face of the support or section X [= 2 × (221,000 + 143,000) + (57,000 for the self-weight of the beam)] = 785,000 lbs, the required value of $A_v f_y d/s$ is calculated to be 585,000 lbs from Equation (9). Because this value is less than $8\sqrt{f'_c}b_w d$ (= 1,020,000 lbs), the stirrups can be provided to resist the required force and the cross-section need not be revised. The required A_v/s for shear force at X is then calculated as 0.164 in., which will control the design of stirrups required for shear force in distance XY.

The applied ultimate shear force at section Y is [2 × 143,000 + (30,000 for the self-weight) =] 316,000 lbs and the calculated A_v/s from Equation (9) is 0.01 in. This is lower than the minimum value required. Hence, A_v/s equals $50\ b_w/f_y$ (= 50 × 34/60,000) = 0.03 in., which will control the design of stirrups for shear force in distance YZ. The final selection of stirrups will be delayed until the check for shear plus torsion has been done.

Web Design for Combined Flexural Shear and Torsion

The most critical section for torsion design occurs at X, where Σx^2y = 85,650 in.^{3}, x_1 = 29.4 in., y_1 = 59.4 in., and α_t [= 0.66 + 0.33 × (59.4/29.4)] = 1.33. The applied ultimate torque (T_u) at Section X [= (130,000 + 65,000) × (9.5 + 34/2)] = 5,168,000 lb-in., the applied ultimate shear force (V_u) at Section X [= 785,000 − (130,000 + 65,000)] = 590,000 lbs., and the flexural shear capacity of the cross-section (V_o) at X will at least equal the applied maximum ultimate shear force acting alone which is 785,000 lbs. From Figure 18, β is calculated as 0.56 which satisfies Equation (13) and the cross-section need not be revised. A_v/s required over distance XY for the effect of flexural shear plus torsion is then computed from Equation (14) as 0.03 in.

Since the applied ultimate torque at critical section Y is very small, only minimum reinforcement is required. Hence, A_v/s equals $50\ b_w/f_y$ = 0.03 in., which will control the design of stirrups for the combined effect of flexural shear and torsion in distance YZ.

Selection of Web Stirrups

The following shows the summary of A_v/s required for the web under different actions:

Action on Web	Distance XY	Distance YZ
(a) Hanger tension	0.134 in.	0.134 in. (controls)
(b) Maximum flexural shear	0.164 in. (controls)	0.03 in.
(c) Torsion plus flexural shear	0.03 in.	0.03 in.

The stirrups required for hanger action or flexural shear alone will be closed at least across the bottom of the beam. Closed rectangular stirrups are required for resisting torsional forces. Use (1# 5 closed + 1# 4) stirrups at 6 in. on centers in distance XY and (1# 4 closed + 1# 4) stirrups at 6 in. on centers in distance YZ as shown in Figure 19(c). These stirrups will satisfy requirements for all actions summarized above.

Longitudinal Reinforcement

The most critical section for flexure is at the face of the support and the maximum bending moment occurs when full live loads act on both sides of the web. Applied ultimate bending moment (M_u) at X equals (2 × 221,000 × 6.1 + 2 × 143,000 × 13.1 + 3800 × 14.8^2/2 =) 6,859,000 lb-ft. and d = 59.5 in. This gives the required area of steel = 27.2 in.^{2} Use 18 #11 bars placed in two layers near the top of the beam as indicated in Figure 19(c). Other longitudinal bars shown in Figure 19(c) are required to provide stiffness to the steel cage for handling purposes and to resist the longitudinal forces that occur when torsion acts. Further check on longitudinal reinforcement required for torsion is not necessary since the design of web stirrups is not controlled by torsion plus flexural shear. The longitudinal bars in the top corners of the flange also act as anchor bars for flange transverse reinforcement.

The intent of this example was merely to elaborate the design requirements specifically associated with inverted T-beams and recommended in the body of this paper. In addition to these requirements, all Code clauses [1,9], especially those for spacing and development of reinforcement, should be satisfied. Another area of major consideration is the connection of the pier and the inverted T-beam.

SI Conversions

1 ft. = 12 in. = 305 mm;
1 in.^{2} = 645 mm^2;
1 lb = 4.45 N;
1000 lbs/in.^{2} = 6.9 MPa;
1000 lb-ft. = 12,000 lb-in. = 1356 N.m.

CHAPTER 9

Ultimate Strength of Reinforced Concrete Members for Combined Biaxial Bending and Tension

Cheng-Tzu Thomas Hsu*

INTRODUCTION

The three-dimensional failure surfaces and strength interaction diagrams have formed the basis of current design procedures in the various National Codes. However, a survey of the literature reveals a scarcity of comprehensive analytical and experimental data on the failure surfaces of structural members subject to biaxial bending and axial load. Most of the investigations have emphasized the failure surfaces of the reinforced concrete members under combined biaxial bending and axial compression [4,7,8,19,20].

For the design and analysis of structural concrete section strength subject to combined axial tension and bending moment, several investigators [11,18] have used the same principles as for eccentrically loaded columns (under combined bending and axial compression) by merely extending their interaction diagrams into the range of tensile axial loads. Such curves would be helpful; however, the interaction curves for tensile values would not be the simple graphical extensions of curves for compression values. Harris [11] proposed a simple theory on which the strength method (ultimate strength design) for beams was based. Formulas were developed giving required steel areas for five cases of combined tension and bending. The resulting formulas involved only the already available theory used for flexural members under ACI Building Code 1971 (ACI 318-71). Moreadith [18] presented the strength design method of reinforced concrete cross sections subject to combined axial tension and moment in terms of fundamental section behavior. He developed the basic equations governing section strength (i.e., strain compatibility, summation of forces and moments) that must be satisfied for a section to be capable of generating combined axial tension and moment capacity in response to applied loads. He concluded that the methods used by Harris [11] were satisfactory for applied axial tension force but should not be used in designing cross sections of a structure for response to applied loads. For a structural member in response to applied axial tensile force and bending moment, Moreadith proposed a simple strength design method for practical use.

A survey of the above literature reveals a scarcity of comprehensive analytical and experimental data on the load-deformation (or moment-curvature) behavior of structural concrete member subject to bending and axial tension. Besides, there have been no papers and literature (except the writer [14–17]) as yet published to study the analytical and experimental load-deformation characteristics and the definite design recommendations for members under combined biaxial bending and axial tension.

To study the ultimate strength behavior of reinforced concrete members subject to biaxial bending moments combined with axial tension, a computer analysis based on a modification of the Newton-Raphson method was employed. The resulting strength interaction diagrams and failure surfaces can account for both axial compression and axial tension combined with biaxial bending moments. Also, the analysis can be used to determine the strength, strain and curvature distributions in structural concrete elements. The computer program developed can also be used for any section geometry as well as material properties. Based on the above computer analysis and failure surfaces, a set of design formulas for reinforced concrete members subject to combined biaxial bending and axial load are derived.

NUMERICAL AND COMPUTER ANALYSIS

Hsu, et al. [12–17] modified the extended Newton-Raphson method [9,10] for the determination of strength,

*Department of Civil and Environmental Engineering, New Jersey Institute of Technology, Newark, NJ

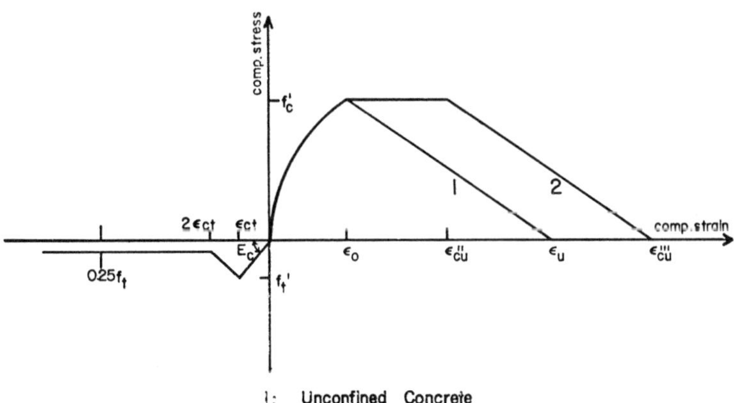

FIGURE 1. Concrete stress-strain curves.

strain and curvature distributions in reinforced concrete members subject to biaxial bending moment and axial load. For analysis purpose, the cross section of the reinforced concrete member is divided into several small elemental areas. A modified form of the Cranston-Chatterji stress-strain curves (see Figure 1) for concrete in tension and in compression was used (Cranston, et al. [5]). These curves can be accounted for the strain-softening of concrete as well as the ultimate compressive strain in confined or unconfined concrete elements. The stress-strain curve for steel reinforcement has been idealized using piece-wise linear approximation to the curve in the strain-hardening region (see Figure 2). Besides, the present computer analysis and program are based on the following assumptions: (1) the bending moments are applied about the principal axes, (2) plane sections remain plane before and after bending, (3) the longitudinal stress at an element is dependent only on the longitudinal strain at that point in which the effect of creep and shrinkage are neglected, (4) strain reversal does not occur, (5) buckling does not occur before the ultimate load is attained.

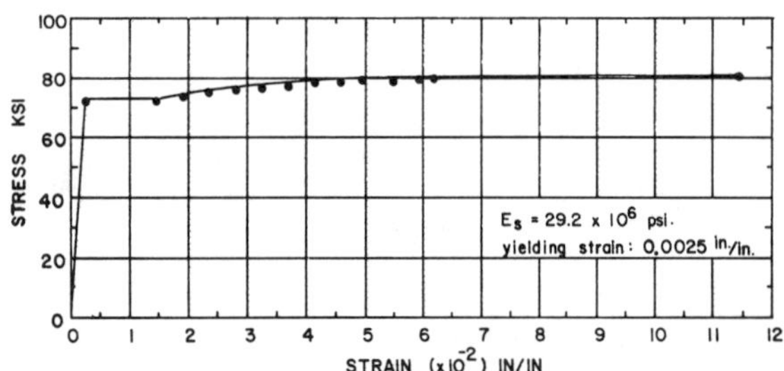

FIGURE 2. Typical steel stress-strain curve.

The cross section of the structural member is divided into several small elements. Consider an element k with its centroid point x_k, y_k referred to the principal axes. The strain ϵ_k across the element k can be assumed to be uniform and since plane sections remain plane during bending,

$$\varepsilon_k = \varepsilon_p + \phi_x\, y_k + \phi_y\, x_k \tag{1}$$

where ϵ_p = uniform direct strain due to an axial load P; ϕ_x = the curvature produced by the bending moment component Mx and is considered positive when it causes compressive strains in the positive y-direction; and ϕ_y = the curvature produced by the bending moment component M_y and is considered positive when it causes compression in the positive x-direction.

Once the strain distribution across the cross section is established, the axial force P and the bending moment components M_x and M_y can be calculated using the following equations:

$$P_{(c)} = \sum_{k=1}^{n} f_k\, a_k \tag{2a}$$

$$M_{x(c)} = \sum_{k=1}^{n} f_k\, a_k\, y_k \tag{2b}$$

$$M_{y(c)} = \sum_{k=1}^{n} f_k\, a_k\, x_k \tag{2c}$$

Subscript (c) indicates values of P, M_x, and M_y calculated in an iteration cycle, and a_k is the area of element k.

For a given section (known geometry and material properties) the stress resultants P, M_x, and M_y can be expressed as functions of ϕ_x, ϕ_y, and ϵ_p given by the following equations:

$$P = P(\phi_x, \phi_y, \varepsilon_p) \tag{3a}$$

$$M_x = M_x(\phi_x, \phi_y, \varepsilon_p) \tag{3b}$$

$$M_y = M_y(\phi_x, \phi_y, \varepsilon_p) \tag{3c}$$

If $P_{(s)}$ is the final value of P for which the equilibrium and the compatibility conditions are satisfied, the convergence of $P_{(c)}$ to $P_{(s)}$ can be accelerated using a modification of the extended Newton-Raphson method. The final values of M_x and M_y can be calculated using the equations:

$$M_{x(s)} = P_{(s)}\, e_y \tag{4a}$$

and

$$M_{y(s)} = P_{(s)}\, e_x \tag{4b}$$

where e_x and e_y are the assumed load eccentricity components. $P_{(s)}$, $M_{x(s)}$, and $M_{y(s)}$ can be expressed in terms of $P_{(c)}$, $M_{x(c)}$, and $M_{y(c)}$ using Taylor's expansion and retaining linear terms, as follows:

$$P_{(s)} = P_{(c)} + \frac{\partial P_{(c)}}{\partial \phi_x}\delta\phi_x + \frac{\partial P_{(c)}}{\partial \phi_y}\delta\phi_y + \frac{\partial P_{(c)}}{\partial \varepsilon_p}\delta\varepsilon_p \tag{5a}$$

$$M_{x(s)} = M_{x(c)} + \frac{\partial M_{x(c)}}{\partial \phi_x}\delta\phi_x + \frac{\partial M_{x(c)}}{\partial \phi_y}\delta\phi_y + \frac{\partial M_{x(c)}}{\partial \varepsilon_p}\delta\varepsilon_p \tag{5b}$$

$$M_{y(s)} = M_{y(c)} + \frac{\partial M_{y(c)}}{\partial \phi_x}\delta\phi_x + \frac{\partial M_{y(c)}}{\partial \phi_y}\delta\phi_y + \frac{\partial M_{y(c)}}{\partial \varepsilon_p}\delta\varepsilon_p \tag{5c}$$

and

$$u = P_{(c)} - P_{(s)} \tag{6a}$$

$$v = M_{x(c)} - M_{x(s)} \tag{6b}$$

$$w = M_{y(c)} - M_{y(s)} \tag{6c}$$

An increment in calculated axial load $\delta P_{(c)}$ produces an increment of strain, $\delta\epsilon_p$, at each element in the section. The corresponding stress change at element k is therefore $\delta\epsilon_p(E_t)k$. The resulting change $\delta P_{(c)}$ in $P_{(c)}$ is given by

$$\delta P_{(c)} = \sum_{k=1}^{n} (E_t)_k\, a_k\, \delta\varepsilon_p$$

Therefore,

$$\frac{\partial P_{(c)}}{\partial \varepsilon_p} = \sum_{k=1}^{n} (E_t)_k\, a_k \tag{7a}$$

Similarly, the changes $\delta M_{x(c)}$ and $\delta M_{y(c)}$ can be expressed in terms of $\delta\, \epsilon_p$ and lead to the equations:

$$\frac{\partial M_{x(c)}}{\partial \varepsilon_p} = \sum_{k=1}^{n} (E_t)_k\, a_k\, y_k \tag{7b}$$

$$\frac{\partial M_{y(c)}}{\partial \varepsilon_p} = \sum_{k=1}^{n} (E_t)_k\, a_k\, x_k \tag{7c}$$

Similar expressions can be derived for $\delta P_{(c)}$, $\delta M_{x(c)}$, and $\delta M_{y(c)}$ in terms of changes $\delta\phi_x$ and $\delta\phi_y$ and yield the following results:

$$\frac{\partial P_{(c)}}{\partial \phi_x} = \sum_{k=1}^{n} (E_t)_k\, a_k\, y_k \tag{7d}$$

$$\frac{\partial M_{x(c)}}{\partial \phi_x} = \sum_{k=1}^{n} (E_t)_k\, a_k\, y_k^2 \tag{7e}$$

$$\frac{\partial M_{y(c)}}{\partial \phi_x} = \sum_{k=1}^{n} (E_t)_k\, a_k\, x_k\, y_k \tag{7f}$$

$$\frac{\partial P_{(c)}}{\partial \phi_y} = \sum_{k=1}^{n} (E_t)_k\, a_k\, x_k \tag{7g}$$

$$\frac{\partial M_{x(c)}}{\partial \phi_y} = \sum_{k=1}^{n} (E_t)_k\, a_k\, x_k\, y_k \tag{7h}$$

$$\frac{\partial M_{y(c)}}{\partial \phi_y} = \sum_{k=1}^{n} (E_t)_k\, a_k\, x_k^2 \tag{7i}$$

Equation (7) can be rearranged in a matrix form (Equation (8)) and given the rates of change of P, M_x, M_y due to changes in ϵ_p, ϕ_x, and ϕ_y

$$\begin{bmatrix} \sum_{k=1}^{n} (E_t) a_k & \sum_{k=1}^{n} (E_t)_k a_k y_k & \sum_{k=1}^{n} (E_t)_k a_k x_k \\ & \sum_{k=1}^{n} (E_t)_k a_k y_k^2 & \sum_{k=1}^{n} (E_t)_k a_k x_k y_k \\ \text{symmetric} & & \sum_{k=1}^{n} (E_t)_k a_k x_k^2 \end{bmatrix}$$

$$\begin{Bmatrix} \delta\varepsilon_p \\ \delta\phi_x \\ \delta\phi_y \end{Bmatrix} = - \begin{Bmatrix} u \\ v \\ w \end{Bmatrix} \quad (8a)$$

where $(E_t)_k$ is the value of the tangent modulus of elasticity for a steel or concrete element.

$$\text{or} \quad [K] \begin{Bmatrix} \delta\varepsilon_p \\ \delta\phi_x \\ \delta\phi_y \end{Bmatrix} = - \begin{Bmatrix} u \\ v \\ w \end{Bmatrix} \quad (8b)$$

$$\begin{Bmatrix} \delta\varepsilon_p \\ \delta\phi_x \\ \delta\phi_y \end{Bmatrix} = - [K]^{-1} \begin{Bmatrix} u \\ v \\ w \end{Bmatrix} \quad (8c)$$

The values of u, v, and w can be selected to suit the accuracy required, and their substitution in Equation (8c) at the end of mth iteration cycle yields the values of $\delta\epsilon_p$, $\delta\phi_x$, and $\delta\phi_y$ which lead to values of ϵ_p, ϕ_x, and ϕ_y for the $(m + 1)$th iteration cycle as follows:

$$\varepsilon_{p(m+1)} = \varepsilon_{p(m)} + \delta\varepsilon_p \quad (9a)$$

$$\phi_{x(m+1)} = \phi_{x(m)} + \delta\phi_x \quad (9b)$$

$$\phi_{y(m+1)} = \phi_{y(m)} + \delta\varepsilon_y \quad (9c)$$

The iteration at a given load level continues in the computer program developed until convergence is obtained within specified tolerances. Once this is achieved, the computer program takes up the next load level and repeats the entire procedure. The computer program (written in FORTRAN IV for an IBM Computer), the flow chart, its accuracy and the convergence of the procedure have been discussed in more detail in Reference 12.

FAILURE SURFACES OF REINFORCED CONCRETE MEMBERS

Present ACI Building Code (1983) [1], ACI Design Handbook (1978) [2], and literature (Bresler [4], Pannell [19], Ford et al. [7], Furlong [8], and Hsu et al. [13]) on design of reinforced concrete member abound with methods and design aids for use in designing the member with combined bending (uniaxial or biaxial) and axial compression. But the subject of reinforced concrete member strength for the design condition of combined bending and axial tension is neither trivial nor undeserving of attention. Combined bending and axial tension is a common problem in the design of reinforced concrete structures for industrial buildings and nuclear power plant facilities. Even in the design of more standard civil engineering structures such as coliseum frames, high rise buildings subject to earthquake or wind loads, or in walls of rectangular or polygonal silos, bunkers, or tanks, there can be significant bending-tension force combination to be considered in the design task [11,18].

The research reported herein is to study the ultimate strength behavior of reinforced concrete members subject to biaxial bending moments combined with axial tension. A

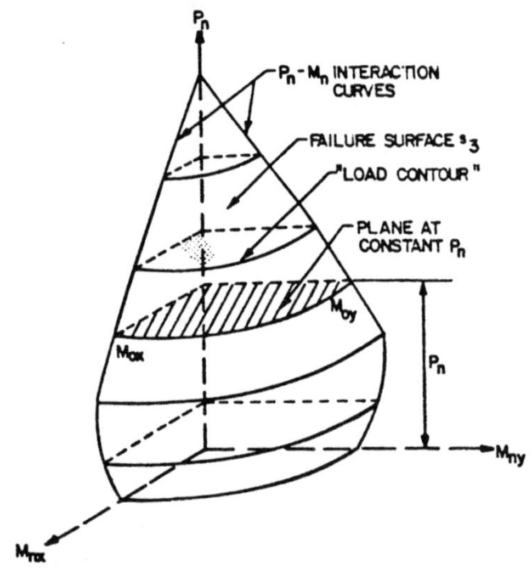

FIGURE 3a. Failure surface of combined biaxial bending and axial compression.

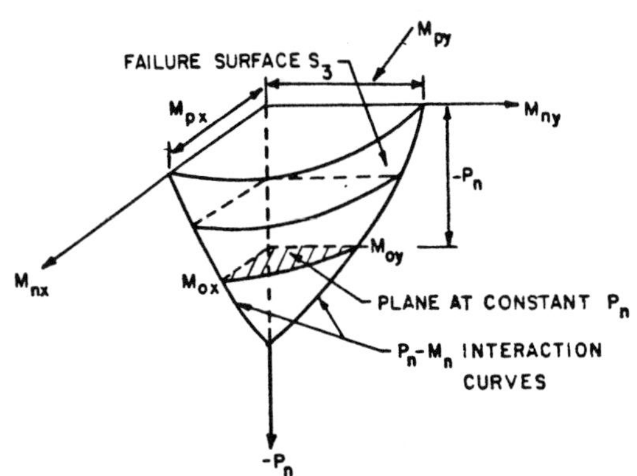

FIGURE 3b. Failure surface of combined biaxial bending and axial tension.

computer program based on a modification of the Newton-Raphson method was developed by Hsu for the analysis of members under combined biaxial bending and axial compression (Hsu et al. [12]). This method has been extended to study the three-dimensional strength interaction diagrams and failure surfaces for members under combined biaxial bending and axial tension [14]. The resulting strength interaction diagrams and failure surfaces can account for both axial compression and axial tension combined with uniaxial or biaxial bending moments. Also, the analysis can be used to determine the strength, strain and curvature distributions in structural concrete elements. The computer program developed can also be used for any section geometry as well as material properties.

Based on the above numerical analysis, the three-dimensional strength interaction diagrams and failure surfaces for structural concrete members under combined biaxial bending and axial load can be obtained. Figure 3 shows the typical failure surfaces for reinforced concrete members with rectangular, square or circular cross sections. To obtain the formulas for design purposes, three rectangular and two cross sections of the reinforced concrete members have been analysed by using the above computer program. Three values of α ($\alpha = \tan^{-1}(e_x/e_y)$ where e_x = eccentricity along x-axis and e_y = eccentricity along y-axis) were used to plot three interaction diagrams for bending moments, M_{nx}, M_{ny} and axial tension, $-P_n$. Figure 4 shows two sections with 8- #4 steel bars (f_y = 40 ksi, f_c' = 4500 psi and f_t' = 500 psi where f_y = steel yield strength, f_c' = ultimate compressive strength in concrete and f_t' = ultimate tensile strength in concrete). The sections are divided into 85 and 89 elemental areas, in which eight for steel elements, thirty-two for unconfined concrete elements, and the rest of the elements are the confined concrete elements. Figures 5 and 6 show the typical interaction curves between M_{nx} or M_{ny} and axial tension, $-P_n$ with α = 0°, 15°, 45°, and 90°, respectively. These curves are used to establish the relation-

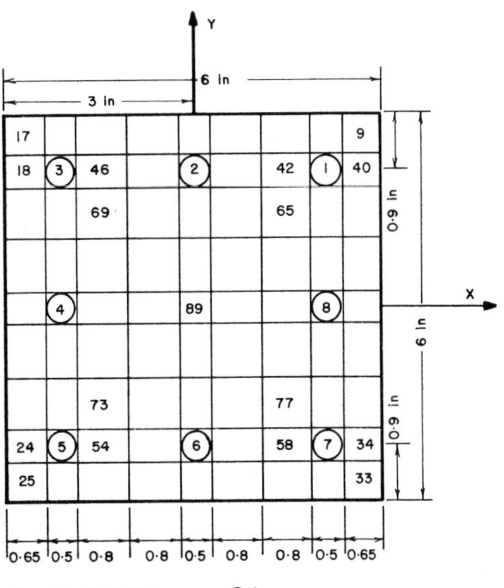

FIGURE 4a. Square cross section.

ships between M_{nx} and $-P_n$, and M_{ny}, and $-P_n$ for the design formulas. To utilize the load contour and extend the failure surface to the area of axial tension combined with biaxial bending, two load contours diagrams were developed for both square and rectangular sections with various values of $-P_n$. Figure 7 shows the various load contours for a square section S1 while Figure 8 presents the various load contours for a rectangular section R1. To establish the mathematical equations for M_{nx} and M_{ny}, a non-dimensional

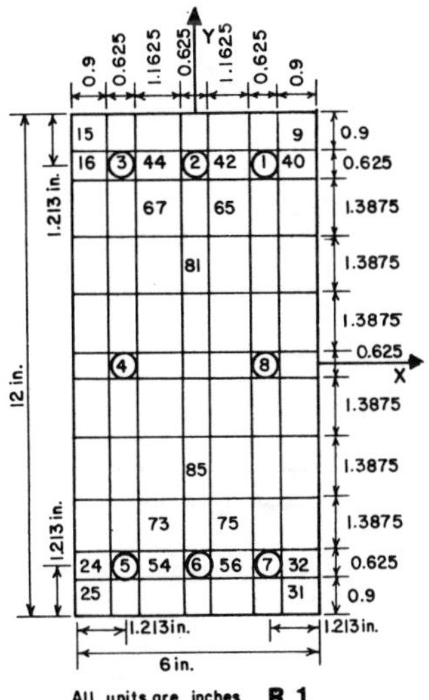

FIGURE 4b. Rectangular cross section.

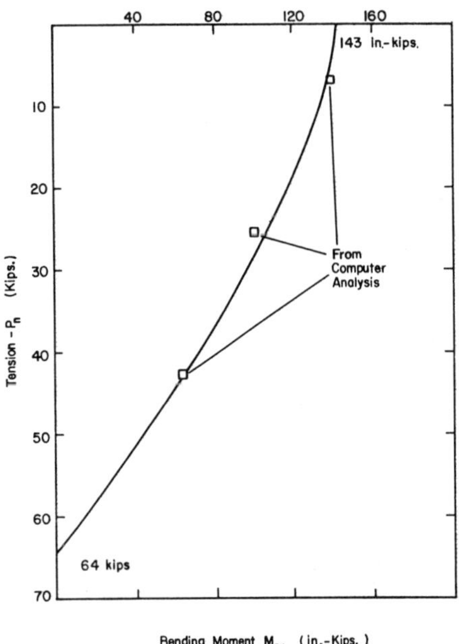

FIGURE 5a. Interaction diagram for a square section ($\alpha = 90°$).

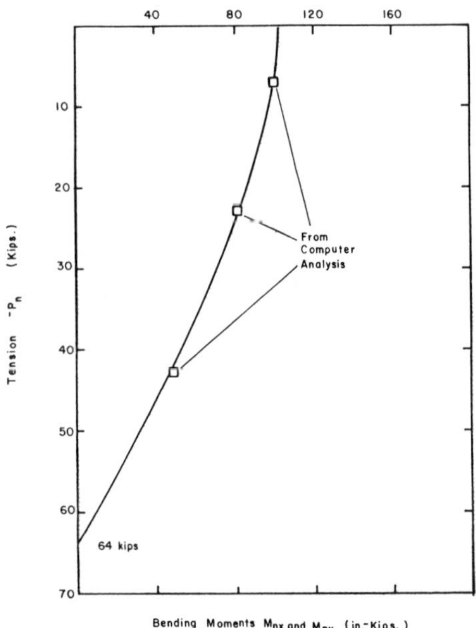

FIGURE 5c. Interaction diagram for a square section ($\alpha = 45°$).

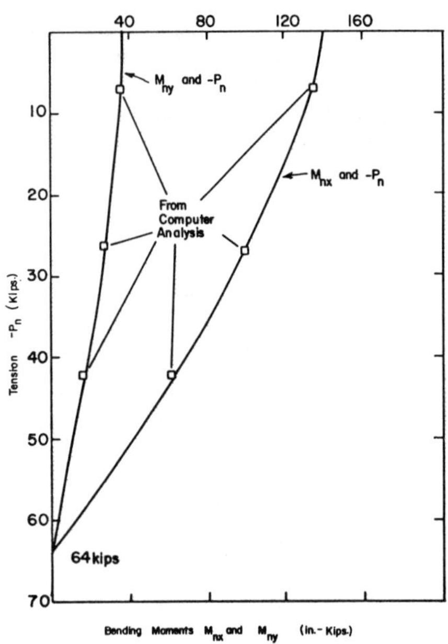

FIGURE 5b. Interaction diagram for a square section ($\alpha = 15°$).

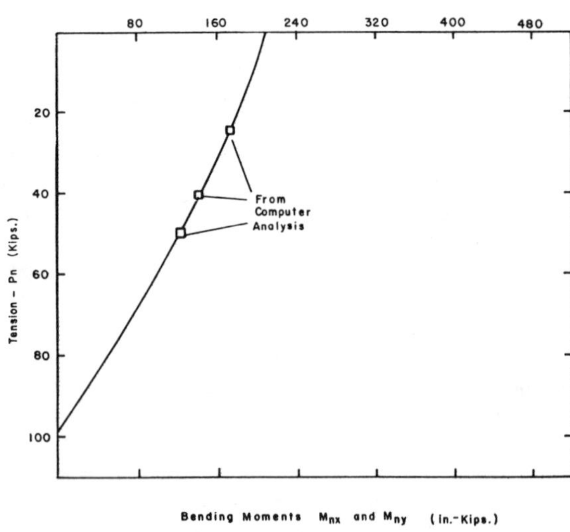

FIGURE 6a. Interaction diagram for a rectangular section ($\alpha = 45°$).

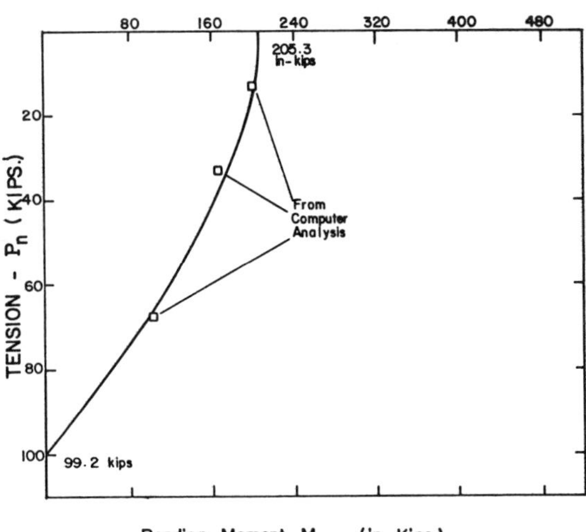

FIGURE 6b. Interaction diagram for a rectangular section ($\alpha = 0°$).

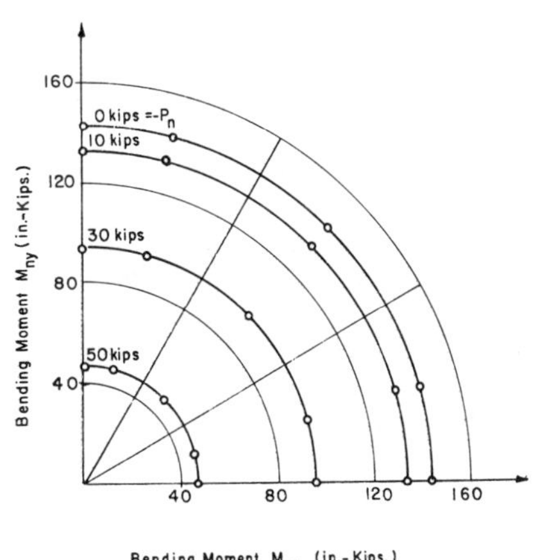

FIGURE 7. Load contours for a square section.

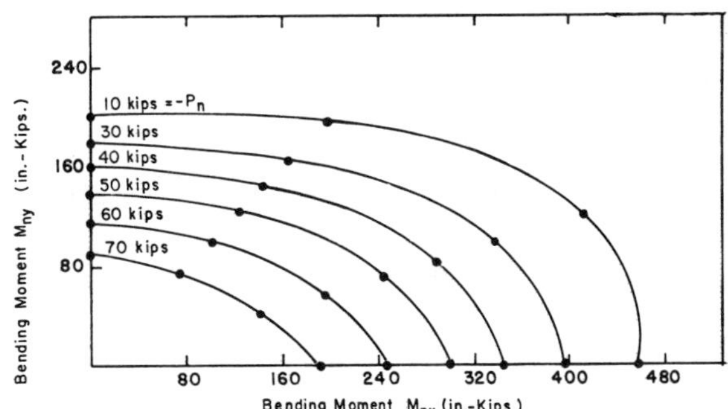

FIGURE 8. Load contours for a rectangular section.

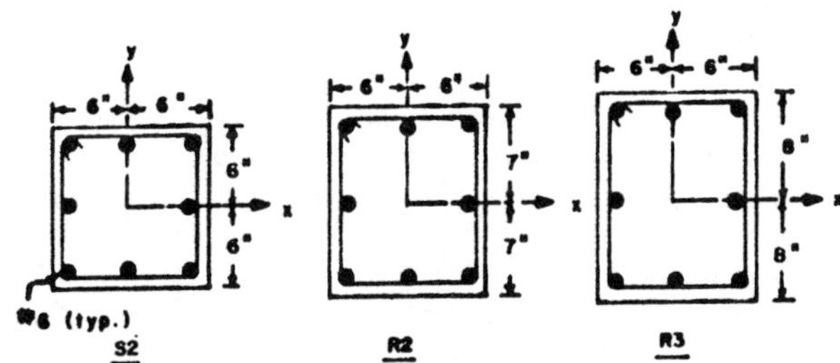

FIGURE 9. Cross sections of reinforced concrete members.

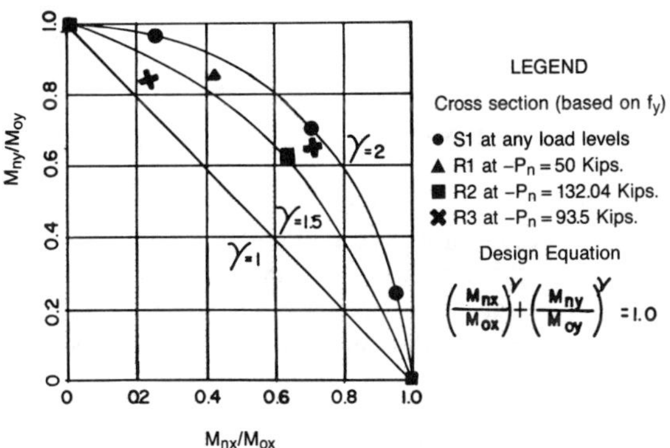

FIGURE 10. Dimensionless load cotours.

TABLE 1. Comparison of Analytical and Experimental Results.

Eccentricity e_y (in)	f'_c (psi)	f_y (ksi)	Analytical Results Based on Ref. 3 & 6		Experimental Results Based on Ref. 3		Experimental Results Based on Ref. 6		Present Results	
			$-P_n$ (kips)	M_{nx} (in-kip)	$-P_n$ (kips)	M_{nx} (in-kip)	$-P_n$ (kips)	M_{nx} (in-kip)	$-P_n$ (kips)	M_{nx} (in-kip)
2	4482	40	17.98	35.96	19	39	—	—	23.07	46.14
5	3981	40	11.63	58.15	13	65	—	—	14.96	74.82
6	4055	60	8.83	53.00	—	—	8.25	49.50	10.07	60.45
29	2817	60	2.52	73.00	—	—	1.90	55.10	2.83	82.11

load contour diagram is needed. Figure 10 shows the results of dimensionless load contours for the reinforced concrete sections as illustrated in Figure 4 and 9, respectively. As seen in Figure 10, the calculated values of γ vary from 1.2 to 1.8 for rectangular sections and are approximately equal to 2.0 for square sections. For practical purposes, it seems satisfactory to assume γ as 1.75 for square sections and as 1.5 for rectangular sections, respectively.

The above numerical and computer method was used to compare with other existing analysis methods and experimental work. All the previous analytical and experimental work on this subject, however, is limited almost exclusively to the square and rectangular sections under combined uniaxial bending and axial tension [see References 3, 6, 21, and 22]. Although the comparisons (see Table 1) are limited in scope, these comparisons show a good agreement in strength interaction curves between the above method and the analytical and experimental work by others.

DESIGN EQUATIONS

Based on the above numerical analysis and failure surfaces, two design formulas for reinforced concrete members subject to combined biaxial bending and axial tension are proposed for square and rectangular sections, respectively. The design equation (for square section) is given by:

$$\left(\frac{M_{nx}}{M_{ox}}\right)^{1.75} + \left(\frac{M_{ny}}{M_{oy}}\right)^{1.75} \leqslant 1 \qquad (10)$$

For a rectangular section, the equation may be expressed in the form:

$$\left(\frac{M_{nx}}{M_{ox}}\right)^{1.5} + \left(\frac{M_{ny}}{M_{oy}}\right)^{1.5} \leqslant 1 \qquad (11)$$

where

$M_{nx} = -P_n e_y$
$M_{ny} = -P_n e_x$
$M_{ox} = M_{nx}$ capacity at axial load $-P_n$ when M_{ny} (or e_x) is zero
$M_{oy} = M_{ny}$ capacity at axial load $-P_n$ when M_{nx} (or e_y) is zero
e_x = eccentricity along x-axis
e_y = eccentricity along y-axis
$-P_n$ = axial tension
$-P_o$ = pure tensile capacity

Equations (10) and (11) represent two non-dimensional design formulas which are an extension of the load contour methods developed by Bresler [4] and Parme, et al. [20] for members under combined biaxial bending and axial compression. To use these two equations, a set of interaction diagrams for reinforced concrete members under combined bending moment M_{nx} or M_{ny} and axial tension $-P_n$ or axial compression P_n must be available. As a result, the design procedure is a cumbersome process. In order to leviate the availability of the interaction diagrams, two design equations for an improvement of the load contour method without using any set of interaction diagrams is quite desirable for the designers. For the case of combined biaxial bending and axial tension, the design equation for square section is proposed as follows:

$$\left(\frac{-P_n}{-P_o}\right) + \left(\frac{M_{nx}}{M_{px}}\right)^{1.75} + \left(\frac{M_{ny}}{M_{py}}\right)^{1.75} \leqslant 1 \qquad (12)$$

For a rectangular section, the design equation may be expressed in the following form:

$$\left(\frac{-P_n}{-P_o}\right) + \left(\frac{M_{nx}}{M_{px}}\right)^{1.5} + \left(\frac{M_{ny}}{M_{py}}\right)^{1.5} \leqslant 1 \qquad (13)$$

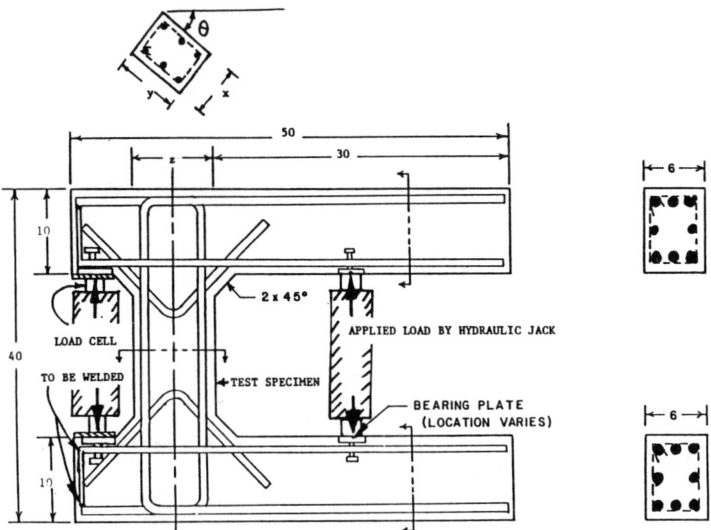

FIGURE 11. Diagrammatic plan of test specimen.

where $M_{py} = M_{ny}$ capacity when M_{nx} (or e_y) and $-P_n$ are zero; $M_{px} = M_{nx}$ capacity when M_{ny} (or e_x) and $-P_n$ are zero. Thus, Equations (10) and (11) provide a design solution for combined biaxial bending and axial tension or axial compression once the interaction diagram is available. However, Equations (12) and (13) could be used for the case of combined biaxial bending and axial tension in lieu of the interaction diagram.

POSSIBLE EXPERIMENTAL SET-UP FOR MEMBERS UNDER COMBINED BIAXIAL BENDING AND AXIAL TENSION

Figure 11 presents a possible diagrammatic plan of test specimen under combined biaxial bending and axial tension. The specimen can be tested in the horizontal positions. The applied loads may be done with hydraulic rams (10-ton or 20-ton capacity) and a hand operated pump. A test gage will be connected to the system for determining the pressure in the rams. Each gage will be calibrated against a MTS system so that gage Psi can be converted to load in pounds. Gage Psi readings during a test will be later converted to load in pounds for the purpose of comparing test results to the theoretical analysis developed.

The configuration of test specimens as illustrated in Figure 11 can be applied with both small and large eccentricities. Also, the tested portion of the specimens being considered is designed in such a way that the tested specimens will be subjected to axial tension combined with biaxial bending moments. The values of x, y, z, and θ as shown are varied and they are dependent upon the sizes of specimens and amounts of M_x and M_y to be applied.

Several types of instrumentation can be used in the experimental inner surfaces of the model, embedded strain gages and demec gages in the concrete, bonded strain gages on the reinforcing steel, crack detection gages, LDVT, x-y recorder and dial gages for deflection measurements. A detailed instrumentation layout will be worked out for each model with a view to compare experimental data with the computed results at homologous points. The loads are to be applied and increased monotonically from zero load until failure of the specimen.

The above experimental set-up proposed by the writer has been used to study the strength and deformation behavior of reinforced concrete members under combined loadings. The experimental work is being conducted at the New Jersey Institute of Technology at present and the preliminary report shows the above experimental set-up is capable of simulating the above loading combinations.

NUMERICAL EXAMPLE

Determine the area of reinforcing steel required of a square section (6 in. × 6 in.) subjected to tension $P_u = 22$ kips and biaxial bending $M_{ux} = M_{uy} = 65$ in-kip. Use $f'_c = 4500$ psi and $f_y = 40{,}000$ psi. The distance between the centroid of the steel and the edge of the section is 0.9 in.

Solution

Step 1: The eccentricity with respect to x- and y-axis is:

$$e_x = e_y = 65/22 = 2.96 \text{ in.}$$

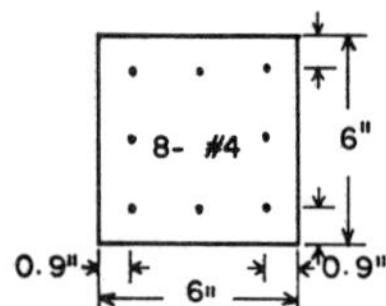

FIGURE 12. Square cross section.

Step 2: Each layer of the reinforcement for a tensile load P_u (with uniform arrangement of reinforcement) is:

$$A_{st1} = \frac{P_u}{\phi f_y} \cdot \frac{1}{4} = \frac{22}{0.9 \cdot 40} \cdot \frac{1}{4} = 0.153 \text{ in.}^2$$

Step 3: The reinforcement for a bending moment M_{ux} by using the ACI 318-83 [1] Building Code Requirement is:

$$M_{ux} = \phi\, \varrho f_y\, bd^2 (1 - 0.59\, \varrho f_y/f'_c)$$

$$65 = 0.9 \cdot \varrho \cdot 40 \cdot 6 \cdot 5.1^2 (1 - 0.59\, \varrho \cdot 40/4.5)$$

$$65 = 5618.2\varrho - 29464.1\, \varrho^2$$

$$\varrho = 0.01237 > \varrho_{min} \quad \text{(O.K.)}$$

$$A_{s1} = 6 \cdot 5.1 \cdot 0.01237 = 0.38 \text{ in.}^2$$

The required steel area is:

$$A_s = A_{s1} + A_{st1} = 0.533 \text{ in.}^2$$

Since M_{uy} is equal to M_{ux}, a symmetrical reinforcement can be used. (8- #4 bars can be used for this section. See Figure 12.)

Step 4: Check the section capacity by using Equation (10) or (12). Since a 6 × 6 in. section is the same as the one shown in Figure 4a, the interaction diagrams (Figure 5a) can be used, as shown in Figure 5a.

$$M_{px} = M_{py} = 143.0 \text{ in-kip. and } P_o = 64 \text{ kips. (tension)}$$

$$P_n = P_u/\phi = 22/0.9 = 24.5 \text{ kips. (tension)}$$

$$M_{nx} = M_{ny} = \frac{M_{ux}}{\phi} = \frac{M_{uy}}{\phi} = 65/0.9 = 72.3 \text{ in-kip.}$$

Substitution of the above values into Equation (12) which yields

$$\left(\frac{-24.5}{-64}\right) + \left(\frac{72.3}{143.0}\right)^{1.75} + \left(\frac{72.3}{143.0}\right)^{1.75} = 0.99 < 1 \text{ (O.K.)}$$

by using Equation (10) one obtains

$$\left(\frac{72.3}{110}\right)^{1.75} + \left(\frac{72.3}{110}\right)^{1.75} = 0.96 < 1 \text{ (O.K.)}$$

where $M_{ox} = M_{oy} = 110$ in-kip (see Figure 5a) at $-P_n = 24.5$ kips. Also, since the loading condition is within the interaction curve at $M_{nx} = M_{ny} = 72.3$ in.-kip, and $-P_n = 24.5$ kips, the results obtained from the above design equations are satisfactory (see Figure 7).

Step 5: Check the steel ratio, ϱ_g where

$$\varrho_g = \frac{A_{st}}{A_g}$$

A_{st} = total steel area in the section

A_g = gross area of concrete section

$$\varrho_g = \frac{8 \times 0.2}{6 \times 6} = 0.044$$

$$0.01 < \varrho_g < 0.08 \text{ (O.K.)}$$

Thus, the section with 8- #4 bars is satisfactory. It should be noted that #4 bars were used for this example instead of other bar sizes, due to the availability of the interaction diagrams. Also the ties must be designed as required by the ACI 318-83 [1].

SUMMARY AND CONCLUSION

The research reported herein is to study the ultimate strength behavior of reinforced concrete members subject to biaxial bending and axial tension. A computer program is developed to study the three-dimensional strength interaction diagrams and failure surfaces for members under these combined actions. Based on the above computer analysis and resulting failure surfaces, two design formulas for reinforced concrete members under combined biaxial bending and axial tension are proposed for square and rectangular sections, respectively. These two equations represent two non-dimensional design formulas which are a further improvement of the load contour methods and can be found useful for practical design uses.

The results of this investigation can be concluded as follows: Equations (10) through (13) are formulated by computer analysis. These equations may be found useful for the design of rectangular and square reinforced concrete members under combined biaxial bending and axial tension. However, further experimental studies are desirable. The shape of the failure surface for a square section is different from that of a rectangular section. It is recommended that further studies of deviation of failure surfaces be conducted, especially for those of irregular cross sections. The above computer program may be used for further development for failure surfaces to validate the test results and may be used to predict the percentage of deviation with different percentage of reinforcements or cross sections. For practical design purpose, it is recommended that a symmetrically reinforced section be chosen for column members; therefore, the center of gravity of the concrete section coincides with the plastic centroid of the steel reinforcement. As a result, deformation due to axial tension will be uniformly distributed. Also, the stability and serviceability requirements, especially the width of concrete cracking, have to be checked with the ACI Building Code Requirements [1].

ACKNOWLEDGEMENTS

Grateful thanks is due to the New Jersey Institute of Technology for supporting computer time. The numerical examples and interaction diagrams in this report were partly solved by a group of the writer's graduate students using the writer's computer program for their M.S. Reports between 1980–1984, M. L. Ling, M. S. Zghondi, A. T. Harb and M. Zargarelahi, and their efforts are greatly appreciated.

LIST OF SYMBOLS

a_k	cross-sectional area of element k
A_g	gross area of the cross section
A_s	$A_{s1} + A_{st1}$
A_{s1}	tensile steel area required due to pure flexure
A_{st}	total steel area in the cross section
A_{st1}	$A_{st}/4$
b	width of the section
d	effective depth of the section
e_x	eccentricity along x-axis
e_y	eccentricity along y-axis
E_t	tangential modulus
f_k	stress across the element k
f_y	yield strength of tension reinforcement
f'_c	maximum concrete compressive stress
f'_t	maximum concrete tensile stress
K	a matrix shown in Equation (8)
$m, m + 1$	indicate iteration cycles
M_x, M_y	bending moments about x- and y-axis, respectively
M_{nx}, M_{ny}	nomial bending moments about x- and y-axis, respectively
M_{ox}	M_{nx} capacity at axial load $-P_n$ when M_{ny} (or e_x) is zero
M_{oy}	M_{ny} capacity at axial load $-P_n$ when M_{nx} (or e_y) is zero
M_{px}	M_{nx} capacity when M_{ny} (or e_x) and $-P_n$ are zero
M_{py}	M_{ny} capacity when M_{nx} (*oe* e_y) and $-P_n$ are zero
M_{ux}, M_{uy}	ultimate bending moments about x- and y-axis, respectively.
P	axial load
P_u	ultimate axial tension
$-P_n$	nomial axial tension
$-P_o$	pure tensile capacity in section
x, y, z	coordinate system
x_k, y_{kx}	x-, y-axis for an element k with its centroidal points referred to the principal axes
u	$P_{(c)} - P_{(s)}$
v	$M_{x(c)} - M_{x(s)}$
w	$M_{y(c)} - M_{y(s)}$
ϵ_k	strain across the element k
ϵ_p	uniform direct strain due to an axial load P
ϕ	strength reduction factor
ϕ_x	the curvature produced by the bending moment component M_x and is considered positive when it causes compressive strain in the positive y-direction
ϕ_y	the curvature produced by the bending moment component M_y and is considered positive when it causes compressive strain in the positive x-direction
ϱ	tensile steel ratio due to pure flexure
ϱ_g	A_{st}/A_g
θ	angle as shown in Figure 11
α	$\tan^{-1} (e_x/e_y)$
γ	an exponent in the dimensionless load contour equation
subscript "c"	indicates values of P, M_x, and M_y calculated in an iteration cycle
subscript "s"	final values of P, M_x, and M_y for which the equilibrium and the compatability conditions are satisfied

REFERENCES

1. ACI Committee 318, Building Code Requirements for Reinforced Concrete (ACI 381-83), Detroit, Michigan (1983).
2. *ACI Design Handbook, Vol. 2, Columns,* Publication SP-17A (78), Detroit, Michigan (1978).
3. Ansari, F., "Design of Reinforced Concrete Members Subjected to Tension with Small Eccentricities," M.S. Thesis, The University of Colorado (1978).

4. Bresler, B., "Design Criteria for Reinforced Columns under Axial Load and Biaxial Bending," *ACI Journal, Proceedings,* 57, 481 (Nov. 1960).
5. Cranston, W. B. and A. K. Chatterji, "Computer Analysis of Reinforced Concrete Portal Frames with Fixed Feet," Technical Report TRA 444, Cement and Concrete Association, London (Sept. 1970).
6. Carnell, D. A., "Design of Reinforced Concrete Members for Combined Bending and Tension," M.S. Thesis, The University of Colorado (1978).
7. Ford, J. S., D. C. Chang, and J. E. Breen, "Behavior of Concrete Columns under Controlled Lateral Deformation," Part I, Final Report, RCRC Project #31, 48 (Aug 1978).
8. Furlong, R. W., "Concrete Columns under Biaxially Eccentric Thrust," *ACI Journal*, 1093 (Oct. 1979).
9. Gurfinkel, G., "Analysis of Footings Subjected to Biaxial Bending," *Journal of the Structural Division, ASCE, Vol. 93* (No. ST6), 1049 (June 1970).
10. Gurfinkel, G. and A. Robinson, "Determination of Strain Distribution and Curvature in a Reinforced Concrete Section Subjected to Bending Moment and Longitudinal Load," *Proceedings, ACI Journal Vol. 64* (No. 7), 398 (July 1967).
11. Harris, E. C., "Design of Members Subjected to Combined Bending and Tension," *ACI Journal,* Vol. 72 (No. 9), 491 (1975).
12. Hsu, C. T. and M. S. Mirza, "Structural Concrete–Biaxial Bending and Compression," TN, *Journal of the Structural Division, ASCE,* 285 (Feb. 1973).
13. Hsu, C. T. T. and M. S. Mirza, "Non-linear Behavior and Analysis of Reinforced Concrete Columns under Combined Loadings," *Study No. 14, SMD, Nonlinear Design of Concrete Structures*, Univ. of Waterloo Press, Ed. M. Z. Cohn, p. 109 (1980).
14. Hsu, C. T. T., "Design and Analysis of Reinforced Concrete Members under Combined Biaxial Bending and Axial Tension," *Proceedings, Annual Conference, Canadian Society for Civil Engineering,* Vol. 1, p. 573, Edmonton, Alberta, Canada (May 27–28 1982).
15. Hsu, C. T. T., "Failure Surface of Structural Concrete Members," *Proceedings of the 8th Conference on Electronic Computation, ASCE,* Houston, Texas, pp., 671–682 (Feb. 21–23, 1983).
16. Hsu, C. T. T., C. S. S. Sea, and H. P. Lee, "A Further Study of Interaction Diagrams and Failure Surfaces of Structural Concrete Members under Combined Biaxial Flexure and Axial Tension," *Proceedings of the 5th ASCE-EMD Specialty Conference,* Laramie, Wyoming, pp. 879–882 (Aug. 1–3, 1984).
17. Hsu, C. T. T. and H. Wang, "A Numerical Analysis of Arbitrary Structural Concrete Sections under Combined Loadings," *Proceedings of the 4th International Conference on Applied Numerical Modeling,* Tainan, Taiwan, R.O.C., p. 108 (Dec. 28–31, 1984).
18. Moreadith, F. L., "Design of Reinforced Concrete for Combined Bending and Tension," *ACI Journal*, V. 75 (No. 6), p. 251 (June 1978).
19. Pannell, F. N., "Failure Surfaces for Members in Compression and Biaxial Bending," *ACI Journal, Proceedings,* 60, p. 129 (Jan. 1963).
20. Parme, A. L., J. M. Nieves, and A. Gouwens, "'Capacity of Reinforced Rectangular Columns Subjected to Biaxial Bending," *ACI Journal*, V. 63, pt. 2, p. 911 (1966).
21. Ramirez, H. and J. O. Jirsa, "Effect of Axial Load on Shear Behavior of Short RC Columns Under Cyclic Lateral Deformations," PMFSEL Report No. 80-1, Dept. of Civil Engineering, The University of Texas at Austin, 162 pages (June 1980).
22. Wang, C. K. and C. G. Salmon, *Reinforced Concrete Design,* 3rd Ed., Harper & Row (1979).

APPENDIX

Conversion Factors–SI Equivalent

1 in.	=	25.4 mm
1 lb (mass)	=	0.4536 kg
1 lb (force)	=	4.4482 N
1 psi	=	6.895 kPa
1 kip	=	4448.2 N
1 ksi	=	6.895 MPa
1 in-kip	=	0.113 kN-m

SECTION TWO

Structural Analysis

CHAPTER 10

Analysis/Design of Concrete Tall-Structures on Micro-Computers

R HUSSEIN* AND M. MORSI**

INTRODUCTION

The human dream of tall structures dates back to the early civilization era when the Egyptians built 481 foot high pyramids in the year 2900 B.C. and the Indians constructed 240 foot high Kutb Minar in the year 1199 A.D. Another magnificient feature of engineering enterprise is the construction of 985 foot high Eiffel Tower in 1889.

The main concept of inventing tall structurescan be attributed to Williams Le Baron Jenny when he first built the 10-story home insurance building in Chicago in 1885. One hundred years have passed since that time and today with the aid of high technology and advanced computers, there are structures as high as 110 stories.

In typical residential and office buildings, the terms used for structural systems are: frame-shear walls, tube-like walls, perforated walls, and tube-in-tube systems. Most of the systems have been built using reinforced concrete.

APPLICATIONS OF COMPUTERS IN THE AREA OF STRUCTURAL ENGINEERING

The analysis and design of tall structures used to be carried out manually. This procedure is cumbersome and time consuming and as a result, the methods adopted for analysis were approximate.

The advent of mainframe computers in the sixties and the matrix/finite element technique permitted the analysis of tall structures to reach the desired degree of accuracy. The development of mini-computers facilitated the adoption of this technology in many engineering firms by either developing inhouse programs or using commercial systems.

At present, with the widespread use of computers, many kinds of structural problems can be solved. The effect of computer applations in structural engineering practice and research has gone beyond generating tools for design processes. The introduction of graphics, for example, has given structural engineers a deeper understanding of structural behaviors. Also, the increased ability for data processing allowed engineers to consider structures as physical systems, with the environment and heirarchial constraints among components and their responses defined much more explicitly than in the past. This factor has resulted in database representations where all analyses, design and detailing information are consistently organized in a single body.

Unfortunately, all these features can be implemented only on mainframes or mini-computers. This has resulted in depriving small and medium sized firms from the advantages of using up-to-date technology. Also, the available softwares are expensive and not easy to use. As a result, a new avenue of research has opened to promoted the applications of micro-computers in solving large-scale problems. Advanced research on micro-processor and other related hardware produced cheaper and efficient micro-computers which are within the reach of any individual interested in the computer-assisted applications.

ANALYSIS OF STRUCTURES ON MICRO-COMPUTERS

The basic concept of static, wind, and earthquake loads and a brief theoretical background of the method of analysis using the stiffness approach are presented in the following:

*University of the District of Columbia, Washington, DC
**Arab Bureau of Design and Technical Consulting, Abbasseya, Cairo, Egypt

Loads

Loads on a tall structure can be classified into three different categories:

1. Vertical static loads due to self-weight of the structure and the weigth of living beings within it.
2. Lateral dynamic loads caused by wind and earthquakes.
3. Loads caused by changes in the temperature, moisture content of the materials, and by the settlement of foundations.

VERTICAL LOADS

The common types of vertical loads are dead loads which always act on the structure and live loads which are not stationary with time.

The computation of dead loads is based on trial and error procedures because it involves the assumption of sizes of members.

The live loads are specified in building codes and are of two kinds: a uniformly distributed load (due to moving people, furniture, etc.) and a concentrated load (due to heavy equipment and accessories).

LATERAL LOADS

There are two main types of lateral loads: (1) wind loading and (2) earthquake loading. A known approach to lateral load analysis starts with an assessment of the magnitude and character of loads, followed by determination of the response of the structure.

Wind Loading

The determination of wind loads and their effects on a structure form a dynamic problem. However, it has been a usual practice to treat the wind as an equivalent statically applied pressure. Some of the considerations required for a design wind pressure are:

a. The anticipated lifetime of the structure
b. The duration and value of gusts
c. Variation of wind speed with height
d. Angle of incidence of the wind
e. The ground effect
f. Architectural features
g. Internal pressures

An equivalent static design wind loading can be obtained from the following expression:

$$p = \tfrac{1}{2} C_a \, C_c \, C_p \varrho \quad V_a^2 \, (H/h)^{2/\alpha} \tag{1}$$

in which

C_a = coefficient dependent on the shape of the structure
C_c = coefficient dependent upon nearby topographic features
C_p = gust coefficient which is dependent upon the magnitude of gust velocities and the size of the structure
ϱ = air density
V_a = basic design wind velocity at height h
H = height above ground at which p is evaluated
h = height at which base velocity was determined
α = an exponent for velocity increase with height determined by the surface roughness in the vicinity of the site

Earthquake Loading

The earthquake loads result from the ground motion. The base motion of a structure is characterized by displacements, velocities, and accelerations which are erratic in direction, magnitude, duration, and sequence.

A way to calculate equivalent horizontal loads due to earthquake is the quasi static approach or the static design method. It has been recommended by ANSI 58.1-1982. It specifies a percentage of the total dead load of the structure as a horizontal force known as base shear and is applied at the base. The method includes an estimate of the fundamental period of vibration based on simple formulas that involve only a general description of the building type and overall dimensions. Ground motion and other parameters included are incorporated in a set of coefficients which convert the dead load into an eqvuialent lateral force.

The equivalent for base shear is:

$$V = ZIKCSW \tag{2}$$

in which

V = total lateral shear at the base
Z = numerical coefficient depending on earthquake zone; (1/8) for Zone 0; (3/16) for Zone 1; (3/8) for Zone 2; (3/4) for Zone 3; 1.0 for Zone 4
I = occupancy importance factor
K = numerical coefficient depending on framing type
C = numerical coefficient depending on natural frequency
S = numerical coefficient for soil profile
W = total dead load

Method of Analysis—The Direct Stiffness Method

The direct stiffness method is a common approach in the development of computer programs. The matrix equation is general in nature but the details of analysis vary with the type of structure. Framed structures can be classified into the six categories: (1) continuous beams, (2) plane trusses, (3) plane frames, (4) grids, (5) space trusses, and (6) space frames.

A plane frame consists of members lying in a single plane. Figure 1 illustrates a typical plane frame of single story in x-y plane.

A typical member i is considered to be rigidly connected to other members at joints j and k. Support restraints may be fixed, pinned, or rollers. The loads may consist of any combination of vertical, horizontal, or moments.

In the stiffness method, equations of joint equilibrium are expressed in terms of independent displacements, known as degrees of freedom. The number of degrees of freedom represents the number of unknowns to be found in the analysis.

In the following mathematical presentation, the joint translations and forces acting on a member are considered to be positive along the positive directions of the coordinate axes, while the positive directions of the joint rotations and moments are determined in accordance with the right hand screw rule.

The member end forces and deformations refer to an individual set of coordinate axes called "member axes" as shown in Figure 2.

In order to develop the equilibrium equation of a system, all the joint deformations and forces on the structure must be expressed with respect to a single fixed coordinate system called "global axes" (designated by X, Y, Z in Figure 2).

The orientation of the member axes with respect to the global axes is described by the following direction cosine matrix:

$$[k] = \begin{bmatrix} \ell_x & m_x \\ \ell_y & m_y \end{bmatrix} \quad (3)$$

in which

$$\ell_x = \frac{x_j - x_i}{L}; \qquad m_x = \frac{Y_j - Y_i}{L} \quad (4)$$

L = the length of the member
i,j = ends of a member

The forces vector (q) and displacements vector (d) in the member axes are related to the corresponding values in global areas as follows:

$$\{q\}_{xyz} = [K]\,\{q\}_{XYZ}$$
$$\{d\}_{xyz} = [K]\,\{d\}_{XYZ} \quad (5)$$

The matrix $[K]$ in Equation (5) is known as the transformation matrix and is given by:

$$[K] = \begin{bmatrix} [k] & \cdot & \cdot & \cdot \\ \cdot & [1] & \cdot & \cdot \\ \cdot & \cdot & [k] & \cdot \\ \cdot & \cdot & \cdot & [1] \end{bmatrix} \quad (6)$$

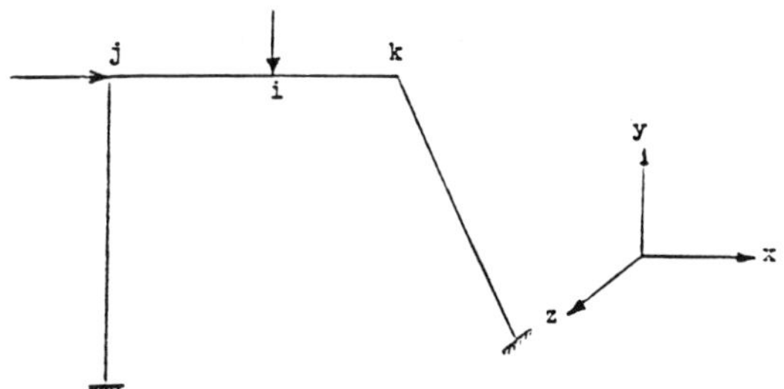

FIGURE 1. Typical plane frame of a single story structure.

The member stiffness matrix in global axes can be evaluated by using:

$$[s]_{xyz} = [K]^T \quad [s]_{xyz} \quad [K] \quad (7)$$

where $[s]_{xyz}$ = stiffness matrix referred to the member axes

$$\begin{bmatrix} S & \cdot & \cdot & -S & \cdot & \cdot \\ \cdot & D & C_i & \cdot & -D & C_j \\ \cdot & C_i & A_i & \cdot & -C_i & B \\ -S & \cdot & \cdot & S & \cdot & \cdot \\ \cdot & -D & -C_i & \cdot & D & -C_j \\ \cdot & C_j & B & \cdot & -C_j & A_j \end{bmatrix} \quad (8)$$

in which

$D = (C_i + C_j)/L$
$C_i = (A_i + B)/L$
$C_j = (A_j + B)/L$
$A = 4EI/L$
$B = 2EI/L$

and the transformation matrix of a frame member in XY-

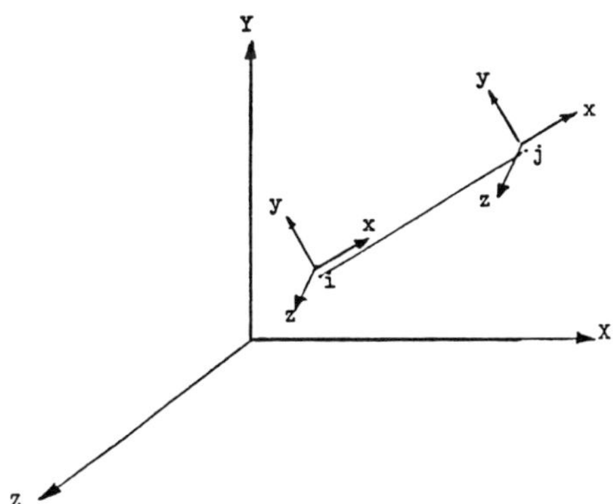

FIGURE 2. Member and global coordinate axes.

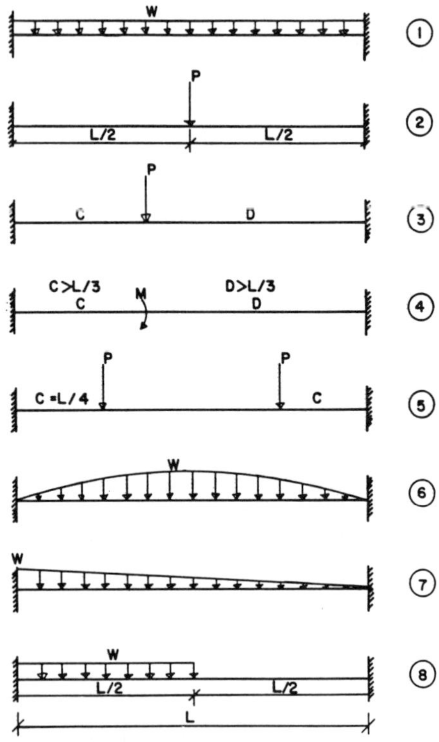

FIGURE 3. Eight loading cases considered.

plane is given by:

$$[k] = \begin{bmatrix} \ell & m & \cdot & \cdot & \cdot & \cdot \\ -m & \ell & \cdot & \cdot & \cdot & \cdot \\ \cdot & \cdot & 1 & \cdot & \cdot & \cdot \\ \cdot & \cdot & \cdot & \ell & m & \cdot \\ \cdot & \cdot & \cdot & -m & \ell & \cdot \\ \cdot & \cdot & \cdot & \cdot & \cdot & 1 \end{bmatrix} \tag{9}$$

The global stiffness matrix of the entire system is assembled from the element stiffness matrices. This is achieved by means of code number approach. This approach requires lesser data preparation, is easy to program, and provides a substantial savings in time and memory space.

In evaluating the optimum band width, the joints of a frame are numbered consecutively along the longest dimension, covering all the joints across the width at each stage.

The fixed end forces of a frame member can be calculated within a system for the eight general types of loading cases given in Figure 3. All the joint loads can also be generated using the member code system.

Finally, there are many methods for solving a large number of equations. An approach which is suited to develop an efficient computer program is called the "method of decomposition." Since the stiffness matrices of linearly elastic structures are always symmetric, a method of decomposi-

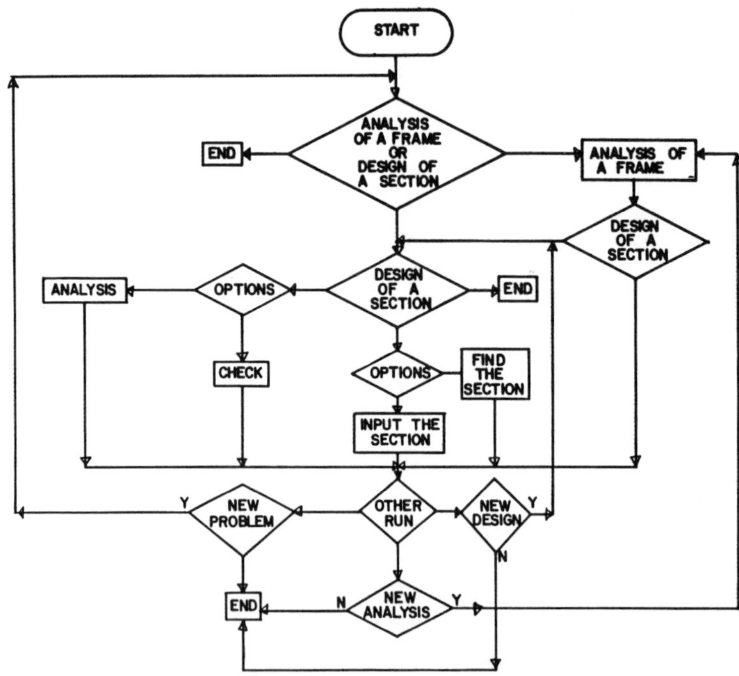

FIGURE 4. Flowchart for the analysis and design phases.

tion, known as Choleskey's square root method, is widely used and takes dvantage of both the symmetrical and the band form of a stiffness matrix. This method is found to be useful for problems of medium size, where all the equations can be retained in the core. In addition, the decomposition may be divided into a forward and a later process so that if many solutions are required, only the later part need be repeated. It is also useful where iteration on residuals is necessary for extra precision in the solution.

Computer Program Development

A program for analyzing a planar frame can now be developed based on the theory presented previously.

A basic flowchart is shown in Figure 4 which is self-explanatory. The program should calculate the half band width, fixed end moments for eight different and widely occurring loading cases.

There should be no restrictions on boundary conditions or the geometry of structures. Any frame with several loading cases should be analyzed in the same run. The READ/DATA statements can be used to store the large amount of data that is required in the input of structural geometry. One of their advantages is that the DATA can be saved and checked and corrected again at any stage. In addition to this, the DATA can be used many times during a particular run of the program.

The first block of DATA should include input data, such as member number, the X and Y projections, the area, the moment of inertia, and the code numbers which indicate the degrees of freedom of joints. The band width is then calculated. The next step is to enter (input) the joint loads and direct loads on the members. The program will calculate fixed end moments for the eight loading cases shown in Figure 3.

The program should also calculate the horizontal distribution of equivalent lateral loads due to wind and earthquake effects.

DESIGN OF CONCRETE STRUCTURES ON MICRO-COMPUTERS

The two main aspects of reinforced concrete design to be presented next are:

1. analysis and design of a compression member
2. analysis and design of a member under flexure, shear and diagonal tension, shear and torsion.

Analysis and Design of a Compression Member

For a given reinforced concrete section, if the depth to neutral axis at the ultimate strength condition is known, the corresponding axial load and eccentricity or moment can be computed by following the ACI 318-83 strength design method. By adopting an iterative technique, the the solution to a wide number of problems can be achieved.

To elaborate on developing a computer program for reinforced concrete design, consider now the uniaxial bending case. For any given cross section there can be an infinite number of strength combinations at which the nominal axial load, P_n, and the nominal capacity, M_n, act together. These combinations of strength can be represented on a curve known as the strength interaction diagram.

The capacity represented by the interactive curve is nominal for any section. The capacities must be reduced by an under capacity factor equal to 0.7 for design purposes as required by the ACI code.

The analysis and design of a section can be carried out as follows:

1. Analysis verifying a section: the interaction diagram for a particular cross section should be developed first, then check whether the load and moment lies inside it.
2. Design: here, the cross section is first assumed and then checked using step (1). If the section is not adequate, it can be revised until the capacity requirements are met with. If this revised section is adequate, either one can be accepted or the adequacy of another section may be considered.

To implement the previous discussion on microcomputer, the first step is to determine the interaction diagram. The parameters required are:

1. Section geometry
2. Compressive strength of concrete
3. Yield strength of reinforcement steel
4. Amount of reinforcement and bar layout

Since a moment can be interpreted as an axial load time and eccentricity, a load and moment combination can be treated as an eccentric axial load. In this manner, a depth to the neutral axis can be assumed and the resultant axial load and bending moment are then calculated. A sub-routine to the main design program should be developed to calculate the resultant load and eccentricity for the assumed depth to the neutral axis. This depth can be adjusted by successive approximations until the load capacity at a desired eccentricity is obtained.

Design of a Member for Flexure; Shear and Diagonal Tension; Shear and Torsion

In a similar way as explained before and in accordance with the ACI code, the design of section can be divided into three parts. These are:

1. flexure for singly-reinforced and doubly-reinforced sections
2. shear and diagonal tension
3. shear and torsion including the design of web steel.

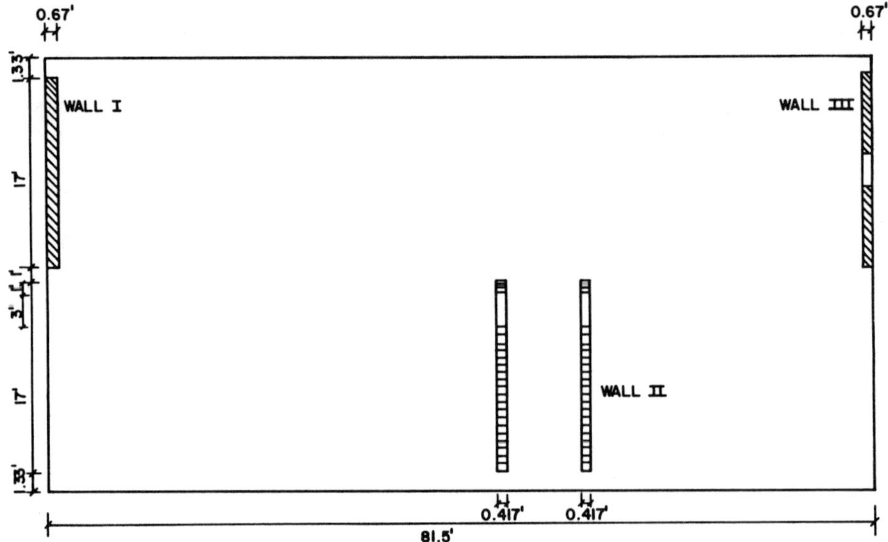

FIGURE 5. The geometry of the three shear walls analyzed.

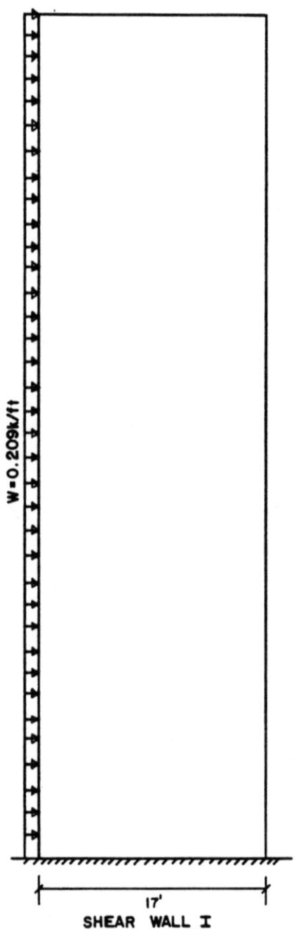

FIGURE 6a. Solid shear wall.

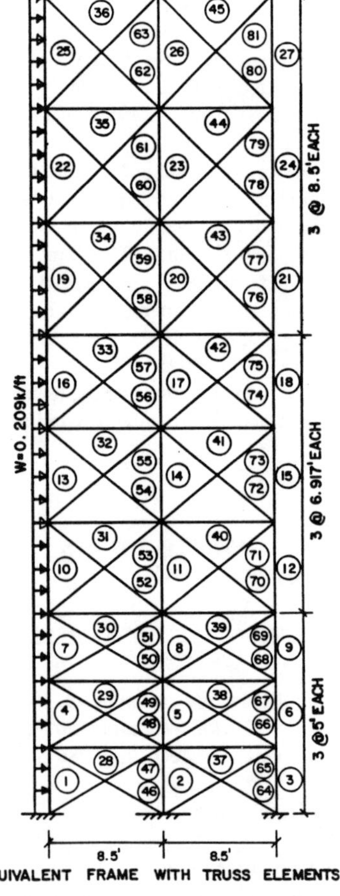

FIGURE 6b. Equivalent frame.

MEMB NO	X-ORD.	Y-ORD.	AREA	MOM.OF INERTIA	CODE NUMBERS
1	0	5	.824	.997	0 0 0 1 2 3
2	0	5	1.648	1.994	0 0 0 4 5 6
3	0	5	.824	.997	0 0 0 7 8 9
4	0	5	.824	.997	1 2 3 10 11 12
5	0	5	1.648	1.994	4 5 6 13 14 15
6	0	5	.824	.997	7 8 9 16 17 18
7	0	5	.824	.997	10 11 12 19 20 21
8	0	5	1.648	1.994	13 14 15 22 23 24
9	0	5	.824	.997	16 17 18 25 26 27
10	0	6.917	.784	1.909	19 20 21 28 29 30
11	0	6.917	1.568	3.818	22 23 24 31 32 33
12	0	6.917	.784	1.909	25 26 27 34 35 36
13	0	6.917	.784	1.909	28 29 30 37 38 39
14	0	6.917	1.568	3.818	31 32 33 40 41 42
15	0	6.917	.784	1.909	34 35 36 43 44 45
16	0	6.917	.784	1.909	37 38 39 46 47 48
17	0	6.917	1.568	3.818	40 41 42 49 50 51
18	0	6.917	.784	1.909	43 44 45 52 53 54
19	0	8.5	.739	2.88	46 47 48 55 56 57
20	0	8.5	1.478	5.76	49 50 51 58 59 60
21	0	8.5	.739	2.88	52 53 54 61 62 63
22	0	8.5	.739	2.88	55 56 57 64 65 66
23	0	8.5	1.478	5.76	58 59 60 67 68 69
24	0	8.5	.739	2.88	61 62 63 70 71 72
25	0	8.5	.739	2.88	64 65 66 73 74 75
26	0	8.5	1.478	5.76	67 68 69 76 77 78
27	0	8.5	.739	2.88	70 71 72 79 80 81
28	8.5	0	.58	3.388	1 2 3 4 5 6
29	8.5	0	.58	3.388	10 11 12 13 14 15
30	8.5	0	.837	4.042	19 20 21 22 23 24
31	8.5	0	1.094	4.876	28 29 30 31 32 33
32	8.5	0	1.094	4.876	37 38 39 40 41 42
33	8.5	0	1.286	5.228	46 47 48 49 50 51
34	8.5	0	1.478	5.76	55 56 57 58 59 60
35	8.5	0	1.478	5.76	64 65 66 67 68 69
36	8.5	0	.739	2.88	73 74 75 76 77 78
37	8.5	0	.58	3.388	4 5 6 7 8 9
38	8.5	0	.58	3.388	13 14 15 16 17 18
39	8.5	0	.837	4.042	22 23 24 25 26 27
40	8.5	0	1.094	4.876	31 32 33 34 35 36
41	8.5	0	1.094	4.876	40 41 42 43 44 45
42	8.5	0	1.286	5.228	49 50 51 52 53 54
43	8.5	0	1.478	5.76	58 59 60 61 62 63
44	8.5	0	1.478	5.76	67 68 69 70 71 72
45	8.5	0	.739	2.88	76 77 78 79 80 81
46	-8.5	5	.345	0	0 0 1 2 3
47	8.5	5	.345	0	0 0 4 5 6
48	-8.5	5	.345	0	4 5 6 10 11 12
49	8.5	5	.345	0	1 2 3 13 14 15
50	-8.5	5	.345	0	13 14 15 19 20 21
51	8.5	5	.345	0	10 11 12 22 23 24
52	-8.5	6.917	.34	0	22 23 24 28 29 30
53	8.5	6.917	.34	0	19 20 21 31 32 33
54	-8.5	6.917	.34	0	31 32 33 37 38 39
55	8.5	6.917	.34	0	28 29 30 40 41 42
56	-8.5	6.917	.34	0	40 41 42 46 47 48
57	8.5	6.917	.34	0	37 38 39 49 50 51
58	-8.5	8.5	.369	0	49 50 51 55 56 57
59	8.5	8.5	.369	0	46 47 48 58 59 60
60	-8.5	8.5	.369	0	58 59 60 64 65 66
61	8.5	8.5	.369	0	55 56 57 67 68 69
62	-8.5	8.5	.369	0	67 68 69 73 74 75
63	8.5	8.5	.369	0	64 65 66 76 77 78
64	-8.5	5	.345	0	0 0 0 4 5 6
65	8.5	5	.345	0	0 0 0 7 8 9
66	-8.5	5	.345	0	7 8 9 13 14 15
67	8.5	5	.345	0	4 5 6 16 17 18
68	-8.5	5	.345	0	16 17 18 22 23 24
69	8.5	5	.345	0	13 14 15 25 26 27
70	-8.5	6.917	.345	0	25 26 27 31 32 33
71	8.5	6.917	.345	0	22 23 24 34 35 36
72	-8.5	6.917	.345	0	34 35 36 40 41 42
73	8.5	6.917	.345	0	31 32 33 43 44 45
74	-8.5	6.917	.345	0	43 44 45 49 50 51
75	8.5	6.917	.345	0	40 41 42 52 53 54
76	-8.5	8.5	.369	0	52 53 54 58 59 60
77	8.5	8.5	.369	0	49 50 51 61 62 63
78	-8.5	8.5	.369	0	61 62 63 67 68 69
79	8.5	8.5	.369	0	58 59 60 70 71 72
80	-8.5	8.5	.369	0	70 71 72 76 77 78
81	8.5	8.5	.369	0	67 68 69 79 80 81

FIGURE 6c. Member properties and code numbers for the frame.

DISPLACEMENTS AND ROTATIONS		DISPLACEMENTS AND ROTATIONS	
1	-1E-04	41	-1E-04
2	-2E-04	42	0
3	0	43	-1E-03
4	-1E-04	44	4E-04
5	0	45	0
6	0	46	-1.4E-03
7	-1E-04	47	-5E-04
8	1E-04	48	0
9	0	49	-1.4E-03
10	-2E-04	50	-1E-04
11	-2E-04	51	0
12	0	52	-1.4E-03
13	-2E-04	53	4E-04
14	-1E-04	54	0
15	0	55	-1.9E-03
16	-2E-04	56	-5E-04
17	1E-04	57	0
18	0	58	-1.9E-03
19	-4E-04	59	-1E-04
20	-3E-04	60	0
21	0	61	-1.9E-03
22	-4E-04	62	4E-04
23	-1E-04	63	0
24	0	64	-2.4E-03
25	-4E-04	65	-5E-04
26	2E-04	66	0
27	0	67	-2.4E-03
28	-7E-04	68	-1E-04
29	-4E-04	69	0
30	0	70	-2.4E-03
31	-7E-04	71	4E-04
32	-1E-04	72	0
33	0	73	-2.8E-03
34	-7E-04	74	-5E-03
35	3E-04	75	0
36	0	76	-2.8E-03
37	-1E-03	77	-1E-04
38	-5E-04	78	0
39	0	79	-2.8E-03
40	-1E-03	80	4E-04
		81	0

FIGURE 6d. Displacements of the frame of Figure 6b.

MEM	X	Y	MOM.LEFT	MOM.RIGHT
1	2.869	20.076	-10.994	2.849
2	3.537	-.076	-15.721	3.537
3	2.487	-19.917	-9.889	2.487
4	1.784	17.047	-6.595	1.784
5	3.826	.153	-14.685	3.826
6	1.749	-16.991	-6.611	1.749
7	1.598	14.06	-5.938	1.598
8	3.403	.191	-12.501	3.403
9	1.599	-13.998	-6.007	1.599
10	1.455	10.396	-7.788	1.455
11	2.927	.135	-15.47	2.927
12	1.439	-10.277	-7.698	1.439
13	1.225	7.357	-6.206	1.225
14	2.524	.129	-12.664	2.524
15	1.217	-7.235	-6.178	1.217
16	.997	4.865	-4.702	.997
17	2.027	.136	-9.563	2.027
18	.973	-4.751	-4.657	.973
19	.74	2.636	-4.165	.74
20	1.469	.101	-8.15	1.469
21	.719	-2.509	-4.033	.719
22	.428	1.129	-2.129	.428
23	.896	.094	-4.436	.896
24	.44	-1.002	-2.182	.44
25	.052	.387	-.513	.052
26	.309	.049	-1.325	.309
27	.147	-.197	-.651	.147
28	-1.111	2.161	9.947	-1.111
29	-1.307	1.963	8.264	-1.307
30	-1.601	2.279	9.445	-1.601
31	-1.755	2.026	8.488	-1.755
32	-1.494	1.668	4.97	-1.494
33	-1.456	1.439	5.937	-1.456
34	-1.4	1.017	4.258	-1.4
35	-1.186	.492	2.023	-1.186
36	-.67	.22	1.188	-.67
37	.71	2.045	8.228	.71
38	.715	1.958	8.502	.715
39	.72	2.323	10.058	.72
40	.622	2.016	8.7	.622
41	.339	1.652	7.148	.339
42	.11	1.461	6.313	.11
43	-.112	1.017	4.387	-.112

FIGURE 6e. Member forces in global coordinate systems for the frame of Figure 6b. (continued)

MEM	X	Y	MOM.LEFT	MOM.RIGHT
44	-.277	.536	2.346	-.277
45	-.2	.144	.626	-.2
46	.161	-.095	0	.161
47	1.597	.939	0	1.597
48	.644	-.38	0	.644
49	1.312	.772	0	1.312
50	.699	-.412	0	.699
51	1.092	.642	0	1.092
52	.352	-.287	0	.352
53	1.196	.973	0	1.196
54	.368	-.3	0	.368
55	.891	.725	0	.891
56	.378	-.309	0	.378
57	.643	.523	0	.643
58	.277	-.278	0	.277
59	.48	.48	0	.48
60	.232	-.233	0	.232
61	.212	.212	0	.212
62	.166	-.167	0	.166
63	.016	.016	0	.016
64	1.604	-.944	0	1.604
65	.023	.014	0	.023
66	1.472	-.867	0	1.472
67	.446	.262	0	.446
68	1.312	-.773	0	1.312
69	.485	.285	0	.485
70	1.368	-1.112	0	1.365
71	.207	.169	0	.207
72	1.052	-.857	0	1.052
73	.219	.178	0	.219
74	.802	-.653	0	.802
75	.229	.186	0	.229
76	.593	-.594	0	.593
77	.161	.161	0	.161
78	.328	-.329	0	.328
79	.125	.125	0	.125
80	.142	-.143	0	.142
81	.052	.052	0	.052

FIGURE 6e. Member forces in global coordinate systems for the frame of Figure 6b.

MEM	MOM.LEFT	MOM.RIGHT	AXIAL FORCE	SHEAR FORCE
45	.626	-.602	-.2	-.145
46	0	0	-.188	-1E-03
47	0	0	1.853	0
48	0	0	-.748	-1E-03
49	0	0	1.522	0
50	0	0	-.812	-1E-03
51	0	0	1.267	0
52	0	0	-.454	0
53	0	0	1.542	-1E-03
54	0	0	-.475	-1E-03
55	0	0	1.149	-1E-03
56	0	0	-.489	-1E-03
57	0	0	.829	-1E-03
58	0	0	-.392	0
59	0	0	.679	0
60	0	0	-.33	0
61	0	0	.3	0
62	0	0	-.236	0
63	0	0	.023	0
64	0	0	-1.862	0
65	0	0	.027	-1E-03
66	0	0	-1.709	0
67	0	0	.518	0
68	0	0	1.523	0
69	0	0	.563	-1E-03
70	0	0	-1.762	0
71	0	0	.268	0
72	0	0	-1.358	-1E-03
73	0	0	.282	0
74	0	0	-1.035	0
75	0	0	.295	-1E-03
76	0	0	-.84	0
77	0	0	.227	0
78	0	0	-.465	0
79	0	0	.177	0
80	0	0	-.202	0
81	0	0	.074	0

FIGURE 6f. Member forces in the member coordinate system for the frame of Figure 6b.

MEM	MOM.LEFT	MOM.RIGHT	AXIAL FORCE	SHEAR FORCE
1	-10.558	3.789	20.076	2.869
2	-15.721	1.967	-.076	3.537
3	-9.889	2.547	-19.917	2.487
4	-6.159	2.762	17.047	1.784
5	-14.685	4.445	.153	3.826
6	-6.611	2.135	-16.991	1.749
7	-5.502	2.492	14.06	1.598
8	-12.501	4.515	.191	3.403
9	-6.007	1.99	-13.998	1.599
10	-6.954	3.116	10.396	1.455
11	-15.47	4.776	.135	2.927
12	-7.698	2.261	-10.277	1.439
13	-5.373	3.102	7.357	1.225
14	12.664	4.8	.129	2.524
15	-6.178	2.241	-7.235	1.217
16	-3.869	3.031	4.865	.997
17	-9.563	4.464	.136	2.027
18	-4.657	2.077	-4.731	.973
19	-2.907	3.388	2.634	.74
20	-8.15	4.343	.101	1.469
21	-4.033	2.08	-2.509	.719
22	-.871	2.768	1.129	.428
23	-4.436	3.188	.094	.896
24	-2.182	1.562	-1.002	.44
25	.745	1.188	.387	.052
26	-1.325	1.31	.049	.309
27	-.651	.601	-.197	.147
28	9.947	-8.424	-1.111	-2.162
29	8.264	-8.444	-1.307	-1.966
30	9.445	-9.927	-1.601	-2.28
31	8.488	-8.741	-1.755	-2.027
32	6.97	-7.214	-1.494	-1.669
33	5.937	-6.3	-1.455	-1.44
34	4.258	-4.393	-1.4	-1.018
35	2.023	-2.168	-1.186	-.493
36	1.188	-.683	-.67	-.221
37	8.228	-9.158	.71	-2.046
38	8.502	-8.143	.715	-1.959
39	10.058	-9.688	.72	-2.324
40	8.7	-8.439	.622	-2.017
41	7.148	-6.898	.339	-1.653
42	6.313	-6.111	.11	-1.462
43	4.387	-4.263	-.112	-1.018
44	2.346	-2.214	-.277	-.537

FIGURE 6f. Member forces in the member coordinate system for the frame of Figure 6b. (continued)

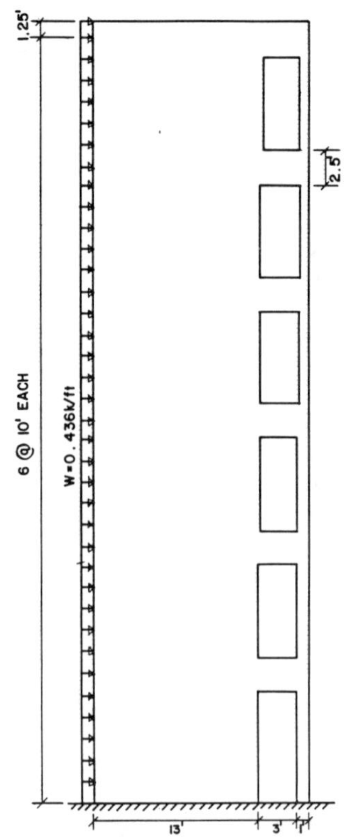

FIGURE 7a. Shear wall (II).

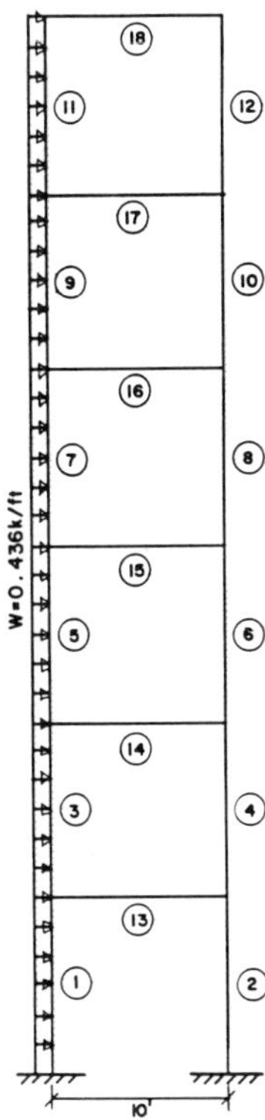

FIGURE 7b. Equivalent frame.

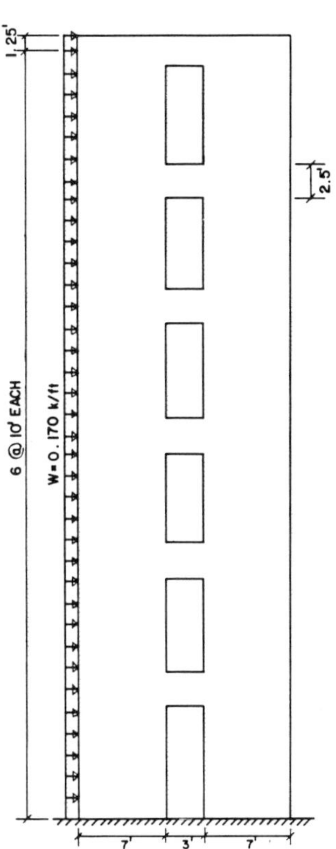

FIGURE 8a. Shear wall III.

MEMB NO	X-ORD.	Y-ORD.	AREA	MOM. OF INERTIA	CODE NUMBERS
1	0	10	5.421	76.346	0 0 0 1 2 3
2	0	10	.417	.0348	0 0 0 4 5 6
3	0	10	5.421	76.346	1 2 3 7 8 9
4	0	10	.417	.0348	4 5 6 10 11 12
5	0	10	5.421	76.346	7 8 9 13 14 15
6	0	10	.417	.0348	10 11 12 16 17 18
7	0	10	5.421	76.346	13 14 15 19 20 21
8	0	10	.417	.0348	16 17 18 22 23 24
9	0	10	5.421	76.346	19 20 21 25 26 27
10	0	10	.417	.0348	22 23 24 28 29 30
11	0	10	5.421	76.346	25 26 27 31 32 33
12	0	10	.417	.0348	28 29 30 34 35 36
13	10	0	1.043	.543	1 2 3 4 5 6
14	10	0	1.043	.543	7 8 9 10 11 12
15	10	0	1.043	.543	13 14 15 16 17 18
16	10	0	1.043	.543	19 20 21 22 23 24
17	10	0	1.043	.543	25 26 27 28 29 30
18	10	0	1.043	.543	31 32 33 34 35 36

FIGURE 9. Member properties and code numbers for the frame of Figure 7b.

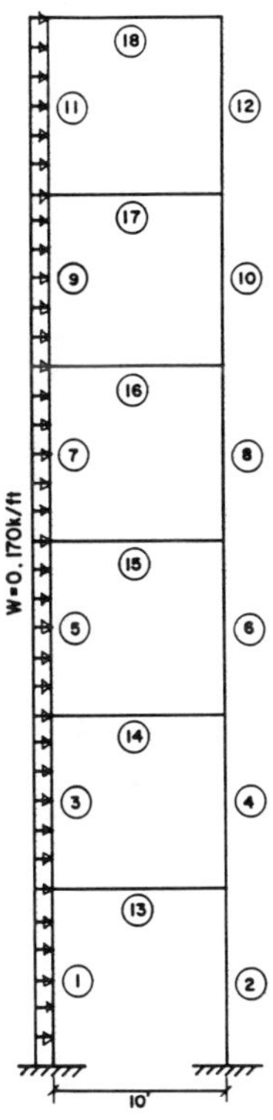

FIGURE 8b. Equivalent frame.

MEMB NO	X-ORD.	Y-ORD.	AREA	MOM. OF INERTIA	CODE NUMBERS
1	0	10	4.69	19.15	0 0 0 1 2 3
2	0	10	4.69	19.15	0 0 0 4 5 6
3	0	10	4.69	19.15	1 2 3 7 8 9
4	0	10	4.69	19.15	4 5 6 10 11 12
5	0	10	4.69	19.15	7 8 9 13 14 15
6	0	10	4.69	19.15	10 11 12 16 17 18
7	0	10	4.69	19.15	13 14 15 19 20 21
8	0	10	4.69	19.15	16 17 18 22 23 24
9	0	10	4.69	19.15	19 20 21 25 26 27
10	0	10	4.69	19.15	22 23 24 28 29 30
11	0	10	4.69	19.15	25 26 27 31 32 33
12	0	10	4.69	19.15	28 29 30 34 35 36
13	10	0	1.675	.872	1 2 3 4 5 6
14	10	0	1.675	.872	7 8 9 10 11 12
15	10	0	1.675	.872	13 14 15 16 17 18
16	10	0	1.675	.872	19 20 21 22 23 24
17	10	0	1.675	.872	25 26 27 28 29 30
18	10	0	1.675	.872	31 32 33 34 35 36

FIGURE 10. Member properties and code numbers for the frame of Figure 8b.

DISPLACEMENTS AND ROTATIONS	
1	-2E-04
2	-1E-04
3	0
4	-2E-04
5	0
6	0
7	-5E-04
8	-1E-04
9	0
10	-5E-04
11	1E-04
12	0
13	-9E-04
14	-1E-04
15	0
16	-9E-04
17	1E-04
10	0
19	-1.4E-03
20	-1E-04
21	0
22	-1.4E-03
23	2E-04
24	0
25	-1.9E-03
26	-1E-04
27	0
28	-1.9E-03
29	2E-04
30	0
31	-2.4E-03
32	-1E-04
33	0
34	-2.4E-03
35	2E-04
36	0

FIGURE 11a. Displacements of the frame of Figure 7(b).

MEM	X	Y	MOM.LEFT	MOM.RIGHT
1	23.839	10.042	-687.26	23.839
2	.14	-10.043	-.748	.14
3	19.29	8.897	-457.947	19.29
4	.329	-8.898	-1.709	.329
5	14.847	7.154	-278.767	14.847
6	.412	-7.155	-2.122	.412
7	10.471	5.202	-145.614	10.471
8	.428	-5.203	-2.192	.428
9	6.131	3.274	-56.009	6.131
10	.408	-3.275	-2.081	.408
11	1.781	1.457	-8.878	1.781
12	.398	-1.458	-1.98	.398
13	.189	1.144	9.079	.189
14	.082	1.743	13.721	.082
15	0.15	1.951	15.319	.015
16	-.02	1.928	15.108	-.02
17	-.011	1.816	14.179	-.011
18	-.399	1.457	12.572	-.399

FIGURE 11b. Member forces in the global coordinate for the frame Figure 7b.

MEM	MOM.LEFT	MOM.RIGHT	AXIAL FORCE	SHEAR FORCE
1	-683.627	-445.234	10.042	23.839
2	-.748	.659	-10.043	.14
3	-454.313	-261.412	8.897	19.29
4	-1.709	1.59	-8.898	.329
5	-275.134	-126.662	7.154	14.847
6	-2.122	2.006	-7.155	.412
7	-141.981	-37.267	5.202	10.471
8	-2.192	2.094	-5.203	.428
9	-52.376	8.935	3.274	6.131
10	-2.081	2.009	-3.275	.408
11	-5.244	12.572	1.457	1.781
12	-1.98	2.004	-1.458	.398
13	9.079	-2.368	.189	-1.145
14	13.721	-3.712	.082	-1.744
15	15.319	-4.199	.015	-1.952
16	15.108	-4.175	-.02	-1.929
17	14.179	-3.99	-.011	-1.817
18	12.572	-2.005	-.399	-1.458

FIGURE 11c. Member forces in the member coordinate of the frame of Figure 7b.

DISPLACEMENTS AND ROTATIONS	
1	-2E-04
2	-1E-04
3	0
4	-2E-04
5	0
6	0
7	-5E-04
8	-1E-04
9	0
10	-5E-04
11	0
12	0
13	-9E-04
14	-1E-04
15	0
16	-9E-04
17	0
18	0
19	-1.3E-03
20	-1E-04
21	0
22	-1.3E-03
23	0
24	0
25	-1.7E-03
26	-1E 04
27	0
28	-1.7E-03
29	0
30	0
31	-2E-03
32	-1E-04
33	0
34	-2E-03
35	0
36	0

FIGURE 12a. Displacements of the frame of Figure 8b.

MEM	X	Y	MOM.LEFT	MOM.RIGHT
1	4.87	14.97	-80.206	4.87
2	4.479	-14.971	-77.511	4.479
3	3.778	12.606	-43.349	3.778
4	3.871	-12.607	-44.505	3.871
5	2.96	9.449	-21.351	2.96
6	2.989	-9.45	-21.567	2.989
7	2.133	6.355	-7.217	2.133
8	2.116	-6.356	-7.146	2.116
9	1.293	3.732	1.003	1.293
10	1.256	-3.733	.908	1.256
11	.329	1.671	3.677	.329
12	.52	-1.678	3.178	.52
13	-.608	2.363	11.852	-.608
14	-.883	3.156	15.784	-.883
15	-.873	3.094	15.471	-.873
16	-.861	2.622	13.111	-.861
17	-.737	2.055	10.263	-.737
18	-.521	1.677	8.392	-.521

FIGURE 12b. Member forces in the global coordinate system for the frame of Figure 8b.

MEM	MOM.LEFT	MOM.RIGHT	AXIAL FORCE	SHEAR FORCE
1	-78.789	-30.081	14.97	4.87
2	-77.511	-32.72	-14.971	4.479
3	-41.933	-4.15	12.606	3.778
4	-44.505	-5.787	-12.607	3.871
5	-19.935	9.671	9.449	2.96
6	-21.567	8.328	-9.45	2.989
7	-5.801	15.531	6.355	2.133
8	-7.146	14.023	-6.356	2.116
9	2.419	15.357	3.732	1.293
10	.908	13.47	-3.733	1.256
11	5.093	8.392	1.677	.329
12	3.178	8.38	-1.678	.52
13	11.852	-11.786	-.608	-2.364
14	15.784	-15.78	-.883	-3.157
15	15.471	-15.474	-.873	-3.095
16	13.111	-13.115	-.861	-2.623
17	10.263	-10.293	-.737	-2.056
18	8.392	-8.381	-.521	-1.678

FIGURE 12c. Member forces in the local coordinate system for the frame of Figure 8b.

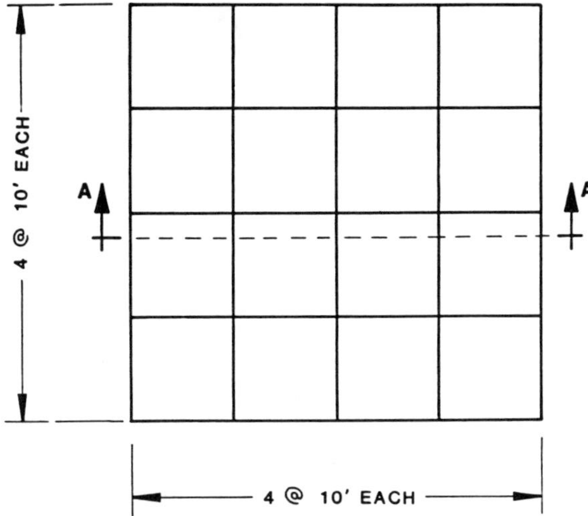

PLAN

FIGURE 13a. Plan view of a four bay, six story frame analyzed for earthquake effect.

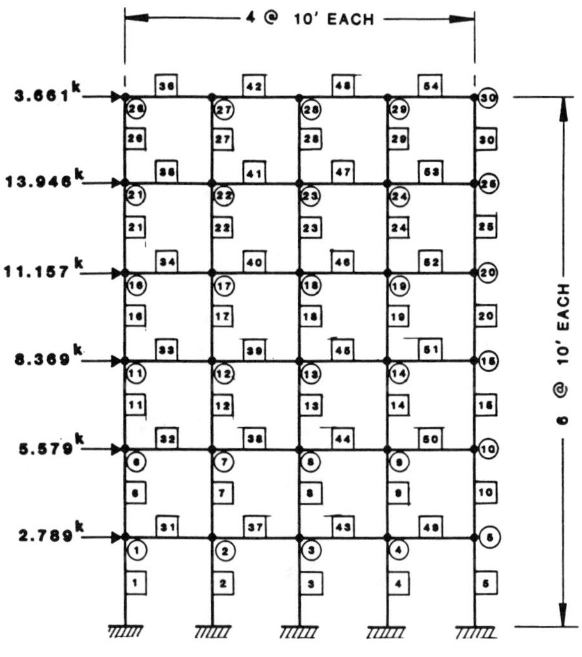

SECTION A–A

EQUIVALENT HORIZONTAL LOADS DUE TO EARTHQUAKE

☐ ELEMENT NUMBER
◯ NODE NUMBER

FIGURE 13b. A four bay, six story frame under earthquake load.

MEMB NO	X-ORD.	Y-ORD.	AREA	MOM.OF INERTIA	CODE NUMBERS
1	0	10	.151	.038	0 0 0 1 2 3
2	0	10	.151	.038	0 0 0 4 5 6
3	0	10	.151	.038	0 0 0 7 8 9
4	0	10	.151	.038	0 0 0 10 11 12
5	0	10	.151	.038	0 0 0 13 14 15
6	0	10	.151	.038	1 2 3 16 17 18
7	0	10	.151	.038	4 5 6 19 20 21
8	0	10	.151	.038	7 8 9 22 23 24
9	0	10	.151	.038	10 11 12 25 26 27
10	0	10	.151	.038	13 14 15 28 29 30
11	0	10	.151	.038	16 17 18 31 32 33
12	0	10	.151	.038	19 20 21 34 35 36
13	0	10	.151	.038	22 23 24 37 38 39
14	0	10	.151	.038	25 26 27 40 41 42
15	0	10	.151	.038	28 29 30 43 44 45
16	0	10	.151	.038	31 32 33 46 47 48
17	0	10	.151	.038	34 35 36 49 50 51
18	0	10	.151	.038	37 38 39 52 53 54
19	0	10	.151	.038	40 41 42 55 56 57
20	0	10	.151	.038	43 44 45 58 59 60
21	0	10	.151	.038	46 47 48 61 62 63
22	0	10	.151	.038	49 50 51 64 65 66
23	0	10	.151	.038	52 53 54 67 68 69
24	0	10	.151	.038	55 56 57 70 71 72
25	0	10	.151	.038	58 59 60 73 74 75
26	0	10	.151	.038	61 62 63 76 77 78
27	0	10	.151	.038	64 65 66 79 80 81
28	0	10	.151	.038	67 68 69 82 83 84
29	0	10	.151	.038	70 71 72 85 86 87
30	0	10	.151	.038	73 74 75 88 89 90
31	15	0	.127	.064	1 2 3 4 5 6
32	15	0	.127	.064	16 17 18 19 20 21
33	15	0	.127	.064	31 32 33 34 35 36
34	15	0	.127	.064	46 47 48 49 50 51
35	15	0	.127	.064	61 62 63 64 65 66
36	15	0	.127	.064	76 77 78 79 80 81
37	15	0	.127	.064	4 5 6 7 8 9
38	15	0	.127	.064	19 20 21 22 23 24
39	15	0	.127	.064	34 35 36 37 38 39
40	15	0	.127	.064	49 50 51 52 53 54
41	15	0	.127	.064	64 65 66 67 68 69
42	15	0	.127	.064	79 80 81 82 83 84
43	15	0	.127	.064	7 8 9 10 11 12
44	15	0	.127	.064	22 23 24 25 26 27
45	15	0	.127	.064	37 38 39 40 41 42
46	15	0	.127	.064	52 53 54 55 56 57
47	15	0	.127	.064	67 68 69 70 71 72
48	15	0	.127	.064	82 83 84 85 86 87
49	15	0	.127	.064	10 11 12 13 14 15
50	15	0	.127	.064	25 26 27 28 29 30
51	15	0	.127	.064	40 41 42 43 44 45
52	15	0	.127	.064	55 56 57 58 59 60
53	15	0	.127	.064	70 71 72 73 74 75
54	15	0	.127	.064	85 86 87 88 89 90

FIGURE 14. Member properties and code numbers for the frame of Figure 13b.

```
FLOOR DEAD LOAD (FDL)                 = 100
NO. OF FLOORS                         = 5
FLOOR LOAD (FL)                       = 1900000
ROOF DEAD LOAD (RDL)                  = 20
ROOF LOAD (RL)                        = 72000
STORY HEIGHT (SH)                     = 10
WALL HEIGHT (WL)                      = 10
NO. OF SIDES (NS)                     = 4
WALL LOAD (WL)                        = 144000
TOTAL DEAD LOAD (W)                   = 2016000
BASE SHEAR (V)                        = 182 KIPS

EQUIVALENT HORIZONTAL LOAD DUE TO EARTHQUAKE EFFECT

ZONE COEFFICIENT                      = .375
OCCUPANCY IMPORTANCE FACTOR (I)       = 1.5
HORIZONTAL FOPCE FACTOR (K)           = 1

CALCULATION OF NATURAL FREQUENCY FACTOR

HEIGHT ABOVE THE BASE,FT (HN)         = 60
CALCULATE FUNDAMENTAL PERIOD

      1)  CASE ... 1
      2)  CASE ... 2
      3)  CASE ... 3

?2

LONGEST DIMENSION OF A SHEAR WALL OR BRACED FRAME IN A
DIRECTION PARALLEL TO AP

NATURAL FREQUENCY FACTOR (C)          = .107123
SOIL PROFILE COEFFICIENT (S)          = 1.5
LENGTH (L)                            = 60
BREADTH (B)                           = 60
```

FIGURE 15. Input data and equivalent static lateral loads due to earthquake of the frame in Figure 13b. (continued)

LEVEL	FLOOR /ROOF WEIGHT	WALL WEIGHT	WX	HX	WX.HX	FX
-	KIP	KIP	KIP	FT.	FT.KIP	KIP
1	360	24	384	10	3840	2.789
2	360	24	384	20	7680	5.378
3	360	24	384	30	11520	8.367
4	360	24	384	40	13360	11.157
5	360	24	384	50	19200	13.946
6	72	12	94	60	5040	3.660

FIGURE 15. Input data and equivalent static lateral loads due to earthquake of the frame in Figure 13b.

DISPLACEMENTS AND ROTATIONS		DISPLACEMENTS AND ROTATIONS	
1	-7.3E-03	46	-.0319
2	-4E-04	47	-1E-03
3	6E-04	48	3E-04
4	-7.2E-03	49	-.0316
5	0	50	0
6	4E-04	51	2E-04
7	-7.2E-03	52	-.0314
8	0	53	-1E-04
9	4E-04	54	2E-04
10	-7.2E-03	55	-.0313
11	-1E-04	56	-1E-04
12	4E-04	57	2E-04
13	-7.1E-03	58	-.0313
14	3E-04	59	9E-04
15	6E-04	60	3E-04
16	-.0167	61	-.0361
17	-7E-04	62	-1E-03
18	6E-04	63	1E-04
19	-.0166	64	-.0358
20	0	65	-1E-04
21	4E-04	66	1E-04
22	-.0165	67	-.0356
23	0	68	-1E-04
24	4E-04	69	1E-04
25	-.0165	70	-.0355
26	-1E-04	71	0
27	4E-04	72	1E-04
28	-.0164	73	-.0354
29	6E-04	74	9E-04
30	6E-04	75	2E-04
31	-.0252	76	-.0374
32	9E-04	77	-1E-03
33	5E-04	78	0
34	-.025	79	-.0373
35	0	80	-1E-04
36	3E-04	81	0
37	-.0249	82	-.0372
38	-1E-04	83	-1E-04
39	3E-04	84	0
40	-.0246	85	-.0371
41	-1E-04	86	0
42	3E-04	87	0
43	-.0247	88	-.0371
44	8E-04	89	9E-04
45	5E-04	90	1E-04

FIGURE 16. Displacements for the frame of Figure 13b.

MEM	X	Y	MOM.LEFT	MOM.RIGHT
1	7.869	25.574	-49.922	7.869
2	10.166	-1.5	-57.479	10.166
3	9.905	-.014	-56.494	9.905
4	10.055	1.435	-56.866	10.055
5	7.704	-25.498	-48.883	7.704
6	6.183	18.248	-30.58	6.183
7	10.358	-.309	-52.011	10.358
8	10.02	.013	-50.052	10.02
9	10.262	.336	-51.531	10.262
10	6.087	-18.29	-30.136	6.087
11	5.295	11.256	-24.971	5.295
12	8.931	.467	-43.635	8.931
13	8.837	.038	-43.069	8.837
14	8.852	-.363	-43.243	8.852
15	5.216	-11.4	-24.585	5.216
16	4.045	5.562	-17.966	4.045
17	6.946	.913	-33.19	6.946
18	6.968	.062	-33.19	6.968
19	6.848	-.732	-32.71	6.848
20	3.954	-5.808	-17.477	3.954
21	2.51	1.648	-9.303	2.51
22	4.247	1.029	-19.295	4.247
23	4.408	.08	-20.053	4.408

FIGURE 17. Member forces in the global coordinate system for the frame of Figure 13b. (continued)

MEM	X	Y	MOM.LEFT	MOM.RIGHT
24	4.146	-.747	-18.907	4.146
25	2.293	-2.012	-8.668	2.293
26	-.215	.077	3.039	-.215
27	.869	.585	-2.97	.869
28	1.512	.043	-5.945	1.512
29	1.217	-.394	-4.722	1.217
30	.276	-.312	.346	.276
31	-1.103	7.325	59.353	-1.103
32	-4.891	6.991	56.225	-4.891
33	-7.12	5.693	45.947	-7.12
34	-9.622	3.913	31.792	-9.622
35	-11.222	1.571	12.761	-11.222
36	-3.876	.077	.893	-3.876
37	-1.296	6.135	45.659	-1.296
38	-3.464	6.215	46.557	-3.464
39	-5.136	5.247	39.407	-5.136
40	-6.923	3.798	28.659	-6.923
41	-7.843	2.015	15.341	-7.843
42	-3.007	.662	5.459	-3.007
43	-1.41	6.109	46.24	-1.41
44	-2.281	6.19	46.546	-2.281
45	-3.267	5.223	39.187	-3.267
46	-4.364	3.78	28.229	-4.364
47	-4.947	2.052	15.085	-4.947
48	-1.494	.705	4.703	-1.494
49	-1.617	7.207	49.819	-1.617
50	-.872	6.889	48.025	-.872
51	-1.263	5.592	38.826	-1.263
52	-1.661	3.796	26.209	-1.661
53	-2.018	1.699	11.577	-2.018
54	-.277	.311	1.571	-.277

FIGURE 17. Member forces in the global coordinate system for the frame of Figure 13b.

MEM	MOM.LEFT	MOM.RIGHT	AXIAL FORCE	SHEAR FORCE
1	-49.922	28.774	25.574	7.869
2	-57.479	44.185	-1.5	10.166
3	-56.494	42.563	-.014	9.905
4	-56.866	43.686	1.435	10.055
5	-48.883	28.157	-25.498	7.704
6	-30.58	31.255	18.248	6.183
7	-52.011	51.576	-.309	10.358
8	-50.052	50.151	.013	10.02
9	-51.531	51.09	.336	10.262
10	-30.136	30.74	-18.29	6.087
11	-24.971	27.981	11.256	5.295
12	-43.635	45.679	.467	8.931
13	-43.069	45.304	.038	8.837
14	-43.243	45.284	-.363	8.852
15	-24.585	27.58	-11.4	5.216
16	-17.966	22.489	5.562	4.045
17	-33.19	36.28	.913	6.946
18	-33.19	36.495	.062	6.968
19	-32.71	35.778	-.732	6.848
20	-17.477	22.065	-5.808	3.954
21	-9.303	15.8	1.648	2.51
22	-19.295	23.184	1.029	4.247
23	-20.053	24.036	.08	4.408
24	-18.907	22.559	-.747	4.146
25	-8.668	14.264	-2.012	2.293
26	3.039	.893	.077	-.215
27	-2.97	5.725	.585	.869
28	-5.945	9.178	.043	1.512
29	-4.722	7.455	-.394	1.217
30	.346	3.106	-.312	.276
31	59.353	-50.536	-1.103	-7.326
32	56.225	-48.654	-4.891	-6.992
33	45.947	-39.462	-7.12	-5.694
34	31.792	-26.916	-9.622	-3.914
35	12.761	-10.813	-11.222	-1.572
36	.893	-.266	-3.876	-.078
37	45.659	-46.375	-1.296	-6.136
38	46.557	-46.673	-3.464	-6.216
39	39.407	-39.308	-5.136	-5.248
40	28.659	-28.319	-6.923	-3.799
41	15.341	-14.896	-7.843	-2.016
42	5.459	-4.476	-3.007	-.663
43	46.24	-45.398	-1.41	-6.11
44	46.546	-46.31	-2.281	-6.191
45	39.187	-39.168	-3.267	-5.224
46	28.229	-28.476	-4.364	-3.781
47	15.085	-15.704	-4.947	-2.053
48	4.703	-5.884	-1.494	-.706
49	49.819	-58.294	-1.617	-7.208
50	48.023	-55.325	-.872	-6.89
51	38.826	-45.057	-1.263	-5.593
52	26.209	-30.733	-1.661	-3.797
53	11.577	-13.919	-2.018	-1.7
54	1.571	-3.107	-.277	-.312

FIGURE 18. Member forces in the member coordinate axes system for the frame of Figure 13b.

```
INPUT AXIAL LOAD AND MOMENT

AXIAL LOAD (PN)           = 708
MOMENT (MX)               = 16471

INPUT MATERIAL PROPERTIES

CONCRETE STRENGTH, KSI  = 4
STEEL YIELD STRESS, KSI = 60

INPUT THE SECTION GEOMETRY

SECTION WIDTH             = 30
SECTION DEPTH             = 30

INPUT THE REINFORCEMENT LAYOUT

CONCRETE COVER TO CENTERLINE OF BAR = 2.5

SIZE OF BAR                         = 18

NUMBER OF BARS                      = 12

NUMBER OF SIDE BARS                 = 10

OUTPUT RESULTS

B                                   = 30 IN.
H                                   = 30 IN.
D                                   = 2.5 IN.
F'C                                 = 4000 PSI
FY                                  = 60000 PSI
12-#18's, 10 SIDE BARS

ANALYSIS OF A COMPRESSION MEMBER

APPLIED LOAD              = 708 KIPS
PHI                       = .7
CAPACITY                  = 855 KIPS

CAPACITY ADEQUATE
```

FIGURE 19. Analysis of a compression section.

```
INPUT AXIAL LOAD AND MOMENT

AXIAL LOAD (PN)           = 10
MOMENT (MX)               = 5343

INPUT MATERIAL PROPERTIES

CONCRETE STRENGTH, KSI  = 4
STEEL YIELD STRESS, KSI = 60

INPUT THE SECTION GEOMETRY

MINIMUM DEPTH             = 150
MAXIMUM DEPTH             = 160

INPUT THE REINFORCEMENT LAYOUT

MINIMUM STEEL RATIO       = .01
MAXIMUM STEEL RATIO       = .04
FRACTION OF STEEL ON SIDES OF SECTION          = 1.0
CONCRETE COVER TO CENTERLINE OF BAR            = 2.5

OUTPUT RESULTS

B                         = 4 IN.
H                         = 150 IN.
D'                        = 2.5 IN.
F'C                       = 4000 PSI
FY                        = 60000 PSI

DESIGN OF COMPRESSION MEMBER

APPLIED LOAD                        = 10
MOMENT                              = 5342
RHO                                 = .04
TOTAL AREA OF STEEL                 = 24
AREA OF STEEL ON EACH FACE          = 0
```

FIGURE 21. Find a compression section.

```
INPUT AXIAL LOAD AND MOMENT

AXIAL LOAD (PN)           = 193
MOMENT (MX)               = 21861

INPUT MATERIAL PROPERTIES

CONCRETE STRENGTH, KSI  = 4
STEEL YIELD STRESS, KSI = 60

INPUT THE SECTION GEOMETRY

SECTION WIDTH             = 30
SECTION DEPTH             = 30

INPUT THE REINFORCEMENT LAYOUT

MINIMUM STEEL RATIO                 = .01
MAXIMUM STEEL RATIO                 = .04
FRACTION OF STEEL ON SIDES OF SECTION      = 1.0
CONCRETE COVER TO CENTERLINE OF BAR        = 2.5
CAPACITY NOT SUFFICIENT AT MAX. REINFORCEMENT

OPTIONS:
     A)  CHANGE SECTION
     B)  CHANGE REINFORCEMENT LIMIT
     C)  RETURN TO DESIGN MENU 1

YOUR CHOICE? ......... B

MINIMUM STEEL RATIO                 = .01
MAXIMUM STEEL RATIO                 = .06
FRACTION OF STEEL ON SIDES OF SECTION      = 1.0
CONCRETE COVER TO CENTERLINE OF BAR        = 2.5

OUTPUT RESULTS

B                         = 30 IN.
H                         = 40 IN.
D'                        = 2.5 IN.
F'C                       = 4000 PSI
FY                        = 60000 PSI

DESIGN OF A COMPRESSION MEMBER

APPLIED LOAD                        = 193
MOMENT                              = 21861
RHO                                 = .031
TOTAL AREA OF STEEL                 = 37.59
AREA OF STEEL ON EACH FACE          = 0
```

FIGURE 20. Design of an input compression section.

```
INPUT

F'C                       = 4000
FY                        = 60000
B                         = 36
D                         = 48
D'                        = 2.5
AS                        = 8.0
AS'                       = 4.0

FINAL RESULTS

MU (LB-IN)                = 10156235
BETA1                     = .85
A                         = 1.961
.75*ROW BAL               = .0214
ROW                       = 4.630 E-03
ROW MINIMUM               = 3.333 E-03
```

FIGURE 22. Design of a section under flexure.

```
BEAMS UNDER SHEAR AND DIAGONAL TENSION

F'C                       = 4000
FY                        = 6000
B                         = 5
D                         = 30
VU                        = 1952
MU                        = 183828
AS                        = .75
AT                        = .4

SHEAR REINFORCEMENT IS NOT NECESSARY
```

FIGURE 23. Design of a section under shear and diagonal tension.

For flexure, the design moment (M_u) should be calculated within the maximum and minimum permissible reinforcement ratios. For diagonal tension, the spacing of stirrups should be considered. For combined diagonal tension and torsion, both the spacing of stirrups and the area of longitudinal reinforcement should be calculated.

EXAMPLE 1

The geometry and the position of concrete shear walls in a building are shown in Figure 5. The following material and structural properties are considered:

Number of stories = 6
Modulus of elasticity of walls I & III = 3275914 Ksf
Modulus of elasticity of wall II = 114657 Ksf
Wind presure = 0.02 Ksf

An equivalent wind load on each wall was calculated by using a flexibility approach and it was found that:

$W_I = 0.209$ k/ft
$W_{II} = 0.436$ k/ft
$W_{III} = 0.170$ k/ft

Once the equivalent, uniformly distributed lateral load on each wall is calculated, the structural analysis is carried out using the theory presented here.

The solid shear wall I is analyzed using the framework method. The wall (Figure 6a) is divided into two vertical and nine horizontal segments as shown in Figure 6b. Figure 6c presents the elements, code numbers, and properties of the system. The output results: displacements and rotations, member forces in the global, and member coordinates are tabulated in Figures 6d to 6f.

The walls with openings are analyzed using the equivalent frame method. The geometry and the equivalent frames of the two systems (shear wall II and shear wall III) are shown in Figures 7a, 7b, 8a and 8b. The input data are presented in Figures 9 and 10, whereas the output results of displacements and rotations, member forces in global, and member coordinates are tabulated in Figures 11 and 12.

EXAMPLE 2

For the analysis of frames under earthquake in zone 2, a four bay, six story frame (Figures 13a and 13b) is considered.

The input data and a part of the output which is consisting of the equivalent lateral load at each floor level are shown in Figures 14 and 15, respectively. The output displacements, code numbers, and member forces in global and member coordinate systems are shown in Figures 16, 17 and 18.

EXAMPLE 3

The concrete sections considered for analysis/design are selected from examples 1 and 2. The types of problems solved are:

1. analysis of a compression section (Figure 19)
2. design of a given compression section (Figure 20)
3. find a compression section (Figure 21)
4. design of a section under flexure (Figure 22)
5. design of a section for shear and diagonal tension (Figure 23)

CHAPTER 11

Tuned Mass Dampers for Tall Buildings and Structures

K. C. S. KWOK*

INTRODUCTION

The response of a building to wind excitation depends on the intensity of the wind, building size, shape, mass, stiffness, and the ability of the structural system to dissipate energy. Modern tall buildings and structures have reached such heights and sizes that occasionally the natural structural damping is not high enough to suppress wind-induced motions. In such cases, damping has to be added to reduce excessive swaying motions which could cause human discomfort, cracked partitions and broken glasses.

Damping can be increased by the addition of a damper (or dampers). Some dampers are designed as real damping devices so that part of the mechanical energy of building motions is converted into heat through the use of either a viscoelastic material or a fluid. Another common device which is increasingly being used is tuned mass damper (TMD), or vibrations absorber, system. These devices have recently been incorporated in the design and construction of a number of major building projects to reduce wind-induced motions. TMDs have also been considered for reduction of seismic response of structures [9].

VISCOELASTIC DAMPERS

Damping devices can be used in trusses, girders, or beams where there is relative motion. Although conventional hydraulic dampers are applicable, they seldom are practical in buildings. A more practical approach is the use of viscoelastic materials which resist forces in shear and dissipate energy as heat. A typical viscoelastic damper [4,6,14] consists of two viscoelastic layers bonded between three parallel rigid surfaces, as shown in Figure 1. The position of the damper, with respect to the application of load, is such that the viscoelastic material undergoes virtually pure shear deformation. The force versus displacement characteristic of such a damper is in the form of a hysteresis loop as shown in Figure 2. The enclosed area of the loop is a measure of the energy dissipated and, therefore, a measure of the physical performance of the damper, which is dependent on factors such as stiffness, geometry, operating temperature, and the heat transfer to the connecting structures. The World Trade Centre in New York City was one of the first major buildings to utilise viscoelastic damper system of the type shown in Figure 1. Approximately 10,000 dampers were installed in each 110 storey tower, with about 100 dampers at the ends of floor trusses at each floor from the 7th to 107th.

TUNED MASS DAMPERS (TMD)

The effectiveness of the use of tuned mass dampers or vibration absorbers for vibration control in mechanical engineering systems, including machinery, automotive and aircraft engines, and ships, has been well documented [3,5]. In recent years, tuned mass damper systems have increasingly been used to reduce wind-induced vibration of tall buildings and structures. Basically, a TMD is a device consisting of a mass attached to a building or structure in such a way that it oscillates at the same frequency of the structure but with a phase shift. The mass is attached to the building via a spring-dashpot system and energy is dissipated by the dashpot as relative motion develops between the mass and the structures. Experimental results from aeroelastic model studies in wind tunnel [8,17,20] and full scale measurements [11] have clearly indicated the effectiveness of TMD system

*School of Civil and Mining Engineering, University of Sydney, Australia

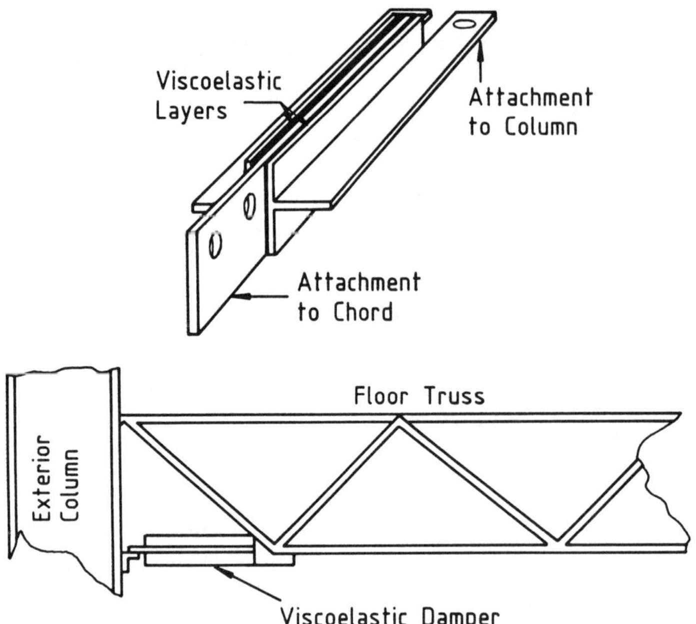

FIGURE 1. Viscoelastic damper used in the World Trade Centre.

in reducing the wind-induced dynamic responses of tall buildings and structures.

Theory of Tuned Mass Damper

Consider a single degree of freedom system fitted with a tuned mass damper, as shown in Figure 3, the equations of motion of the resultant two degree of freedom system subjected to an external excitation $F(t)$ can be written in the form:

$$M\ddot{x}_1 + Kx_1 + k(x_1 - x_2) + c(\dot{x}_1 - \dot{x}_2) = F(t) \quad (1)$$

$$m\ddot{x}_2 + k(x_2 - x_1) + c(\dot{x}_2 - \dot{x}_1) = 0 \quad (2)$$

in which

- M = mass of main system
- m = mass of tuned mass damper
- K = stiffness of main system
- k = stiffness of tuned mass damper
- c = damping capacity of tuned mass damper
- $x_1 = x_1(t)$ = response of main system
- $x_2 = x_2(t)$ = response of tuned mass damper
- $\dot{}$ = 1st time derivative
- $\ddot{}$ = 2nd time derivative

These equations of motion can be analysed in the time domain using an analog computer [15] or more practically in the frequency domain [3,8,15,20].

The form of external excitation varies quite considerably; for example, wind loading on buildings and structures is essentially a random process which can be identified in the frequency domain by the power spectral density function (or spectrum) of varying frequency bandwidth. In parametric study of tuned mass damper system, it is convenient to consider sinusoidal excitation and white noise excitation which are regarded as the two extremes of narrow-band and wide-band excitation respectively. A more accurate assessment of the effectiveness of a TMD system can be obtained if the

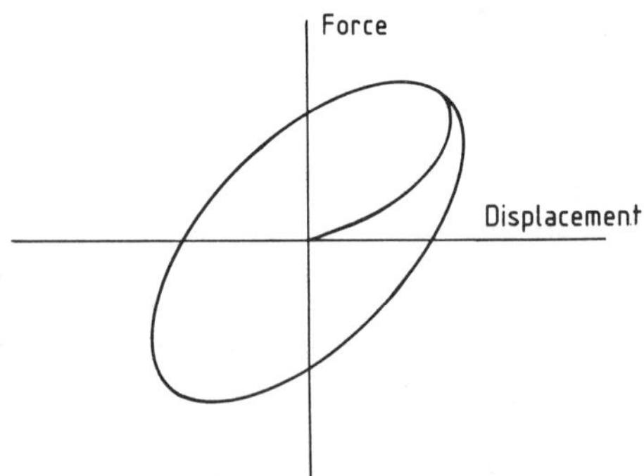

FIGURE 2. A typical hysteresis loop of a viscoelastic damper.

precise excitation spectrum is known. Alternatively, band-limited random noise with a bandwidth which corresponds closely to the real excitation may be used in the analysis.

The response of the system shown in Figure 3 to a sinusoidal excitation is described by the following equations:

$$\left|\frac{x_1}{x_{st}}\right|^2 = \frac{(2\zeta g)^2 + (g^2 - f^2)^2}{(2\zeta g)^2(g^2 - 1 + \mu g^2)^2 + [\mu f^2 g^2 - (g^2 - 1)(g^2 - f^2)]^2} \tag{3}$$

$$\left|\frac{x_1 - x_2}{x_{st}}\right|^2 = \frac{g^4}{(2\zeta g)^2(g^2 - 1 + \mu g^2)^2 + [\mu f^2 g^2 - (g^2 - 1)(g^2 - f^2)]^2} \tag{4}$$

$$\left|\frac{x_1 - x_2}{x_1}\right|^2 = \frac{g^4}{(2\zeta g)^2 + (g^2 - f^2)^2} \tag{5}$$

in which

ω = circular frequency of sinusoidal excitation $F(t) = F_o \sin \omega t$
ω_1 = natural frequency of main system = $\sqrt{K/M}$
ω_2 = natural frequency of tuned mass damper = $\sqrt{k/m}$
g = excitation frequency ratio = ω/ω_1
f = frequency tuning ratio (natural frequencies) = ω_2/ω_1
μ = mass ratio = m/M
ζ = damping of tuned mass damper as fraction of critical damping = $c/2m\omega_1$
x_{st} = static deflection of main system = F_o/K

Under white noise excitation which has a spectral density function of the form:

$$S_F(\omega) = \text{constant} = S_o$$

the variance or mean square response of the main mass can be expressed as [3,8,20]:

$$\overline{x_1^2} = \frac{\omega_1 S_o}{K^2} \int_0^{\infty} |H_{x_1}(g)|^2 dg \tag{6}$$

in which

$$|H_{x_1}(g)|^2 = \left|\frac{x_1}{x_{st}}\right|^2$$

The root-mean-square relative movement of the tuned mass damper, normalised by the root-mean-square move-

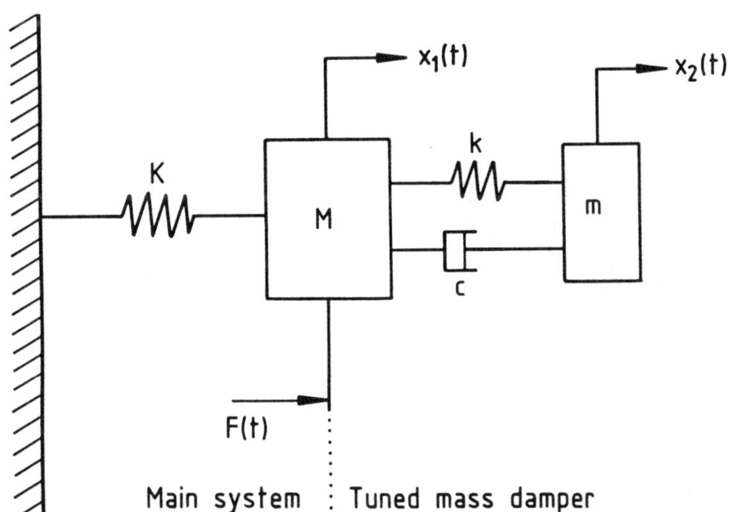

FIGURE 3. A single degree of freedom system fitted with a tuned mass damper.

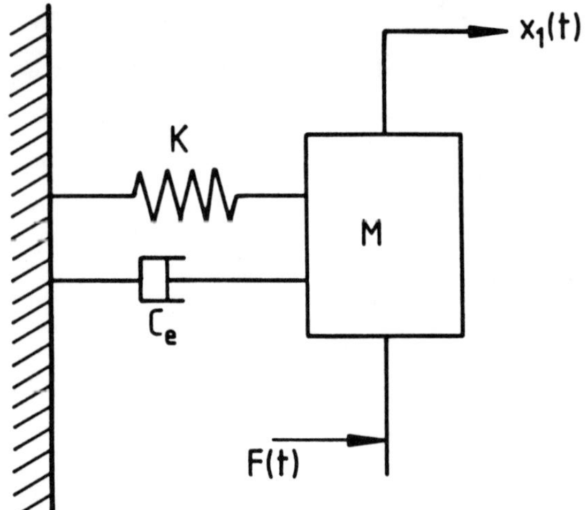

FIGURE 4. Equivalent one degree of freedom system for a main system fitted with a tuned mass damper.

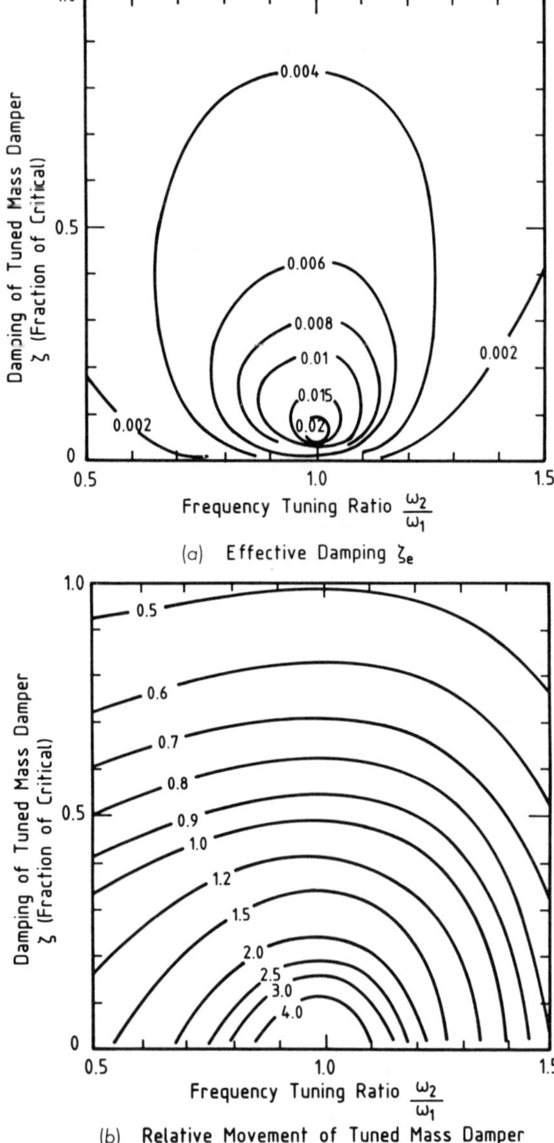

FIGURE 5. Effective damping and relative movement of tuned mass damper for a mass ratio of 0.01.

ment of the main system, becomes:

$$\frac{\sqrt{\overline{(x_1 - x_2)^2}}}{\sqrt{\overline{x_1^2}}} = \left[\frac{\int_0^\infty |H_{x_1-x_2}(g)|^2 dg}{\int_0^\infty |H_{x_1}(g)|^2 dg}\right]^{1/2} \tag{7}$$

in which

$$|H_{x_1-x_2}(g)|^2 = \left|\frac{x_1 - x_2}{x_{st}}\right|^2$$

The effectiveness of a tuned mass damper can be evaluated in terms of the effective damping. By replacing the two degree of freedom system (with the tuned mass damper) with an equivalent one degree of freedom system shown in Figure 4, the effective damping of the equivalent system is defined as the viscous damping required to sustain the same magnitude of response under the same excitation. The variance response of such a lightly-damped one degree of freedom system under white noise excitation is approximately as follows [3,8,20]:

$$\overline{x_1^2} = \frac{\omega_1 S_o}{K} \frac{\pi}{4\zeta_e} \tag{8}$$

in which

ζ_e = effective damping added by the tuned mass damper = $C_e/2M\omega_1$

By equating Equations (6) and (8), the effective damping added by the tuned mass damper becomes:

$$\zeta_e = \frac{\pi}{4\int_o^\infty |H_{x_1}(g)|^2 dg} \tag{9}$$

The effective damping and the relative movement of the tuned mass damper can be determined readily by numer-

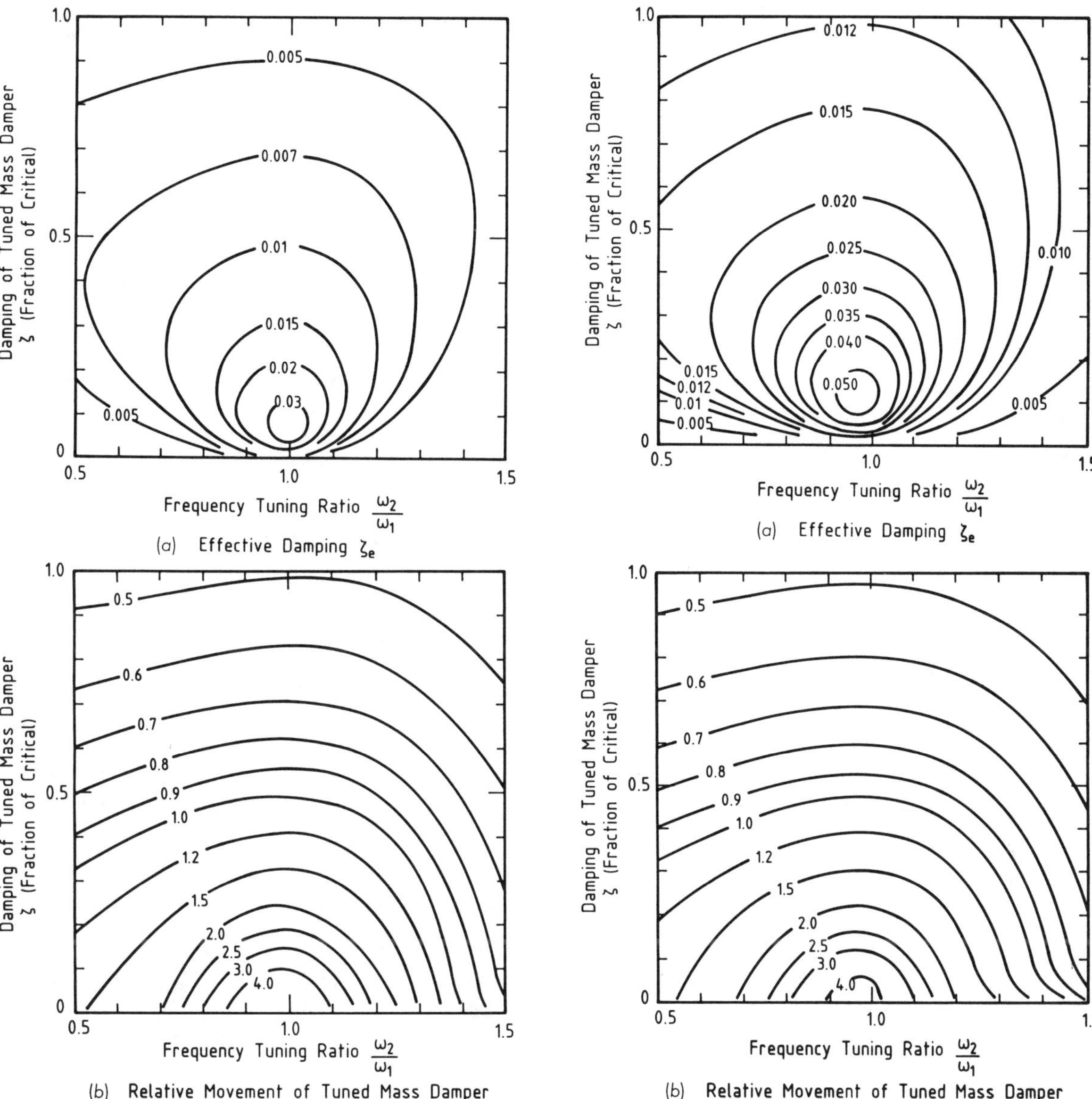

FIGURE 6. Effective damping and relative movement of tuned mass damper for a mass ratio of 0.02.

FIGURE 7. Effective damping and relative movement of tuned mass damper for a mass ratio of 0.05.

ically integrating Equations (7) and (9). The values of effective damping are shown in Figures 5, 6, and 7 as a function of the frequency tuning ratio and the damping capacity of the tuned mass damper for mass ratios of 0.01, 0.02, and 0.05, respectively. The corresponding normalised relative movements of the tuned mass damper are also shown in Figures 5, 6, and 7.

The above approach of estimating the effectiveness of a tuned mass damper on the response of a one degree of freedom system can be extended to continuous structures such as tall buildings [8]. For a tall building shown in Figure 8 of height H and with a uniform mass distribution $m(z)$ (mass per unit height), the mode-generalised mass of the building in the ith mode is

$$M_i = \int_o^H m(z)\psi_i^2(z)dz \tag{10}$$

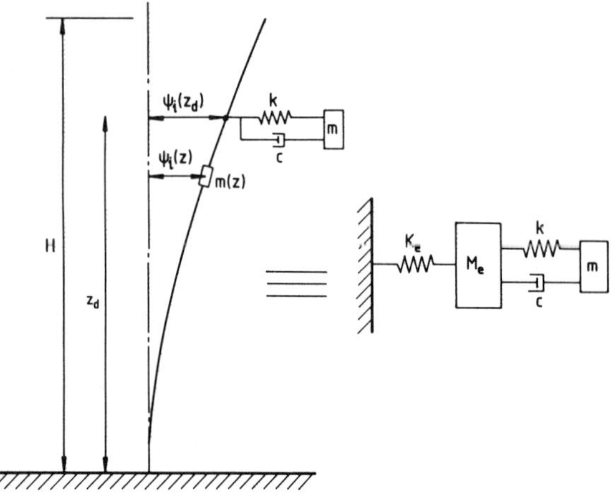

FIGURE 8. Equivalent two degree of freedom system for *i*th mode of a tall building fitted with a tuned mass damper at height z_d [8].

in which

$\psi_i\,(z)$ = mode shape of the *i*th mode

With a tuned mass damper located at height z_d, the equivalent mass M_e and the equivalent stiffness K_e fort the *i*th mode become:

$$M_{e_i} = \frac{M_i}{\psi_i^2(z_d)} \tag{11}$$

$$K_{e_i} = \omega_i^2 M_{e_i} \tag{12}$$

The corresponding mass ratio of the tuned mass damper for the *i*th mode becomes:

$$\mu_i = \frac{m}{M_{e_i}} = \frac{m\psi_i^2(z_d)}{M_i} \tag{13}$$

The analysis presented so far is based on the assumption that the damping capacity of the main mass is zero. For a lightly-damped building fitted with a tuned mass damper, the total effective damping ζ_t of the building can be taken as the linear sum of the structural damping ζ_s and the effective damping ζ_e provided by the tuned mass damper which may be obtained from design charts such as those shown in Figures 5, 6, and 7. That is,

$$\zeta_t = \zeta_s + \zeta_e \tag{14}$$

Tuned Mass Damper with Active Control

The effectiveness of a tuned mass damper can be enhanced by the addition of an active control capacity [1,2]. The equations of motion of a single degree of freedom system fitted with an active TMD are similar to Equations (1) and (2) but with an additional control force $u(t)$ which is a function of the building motion variables; for example, $u(t)$ may be a linear function of the building displacement. The optimal $u(t)$ can be determined by modern control theory. Numerical simulations [2] using realistic parameters showed that significant reduction in building displacement and acceleration can be achieved if a TMD is operated in an active mode. Furthermore, reduction in the relative displacement of the damper mass to the building, and reduction in the mass ratio can also be achieved while keeping building motion within acceptable limits.

Practical Design Considerations

In the engineering design of a tuned mass damper system, several practical design considerations immediately impose constraints on the amount of dynamic response reduction that can be economically achieved. First and foremost is the amount of additional mass that can be practically placed at the top of a building, usually no more than 2% of the mode-generalised mass (about 0.6% of the building mass). Once the TMD is specified, the effectiveness of the damper is determined and the general range of response reduction is set. For example, at a mass ratio of 2%, an additional damping capacity of more than 3% of critical (see Figure 6) could be achieved. But this would require fine tuning and comparatively large motions of the damper mass. It is not difficult to design a TMD system which is tuned accurately to a prescribed frequency. However, the natural frequencies of the building are sensitive to design changes and are often difficult to predict accurately. It would be ideal if the frequency of the TMD system is adjustable in the field. Otherwise, it may be advisable, particularly if fine tuning is essential, to delay detail specification of the TMD system until reliable information becomes available.

The travel of the tuned mass damper mass relative to the building is another important design parameter. This can be reduced by the selection of a higher damper damping, but with a corresponding decrease in the effective damping.

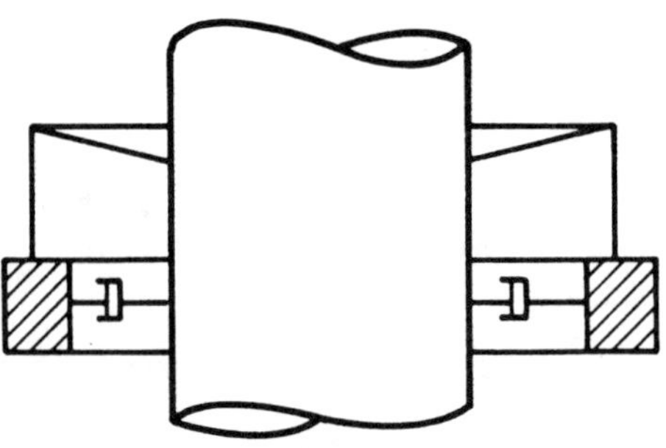

FIGURE 9. Pendulum tuned mass damper.

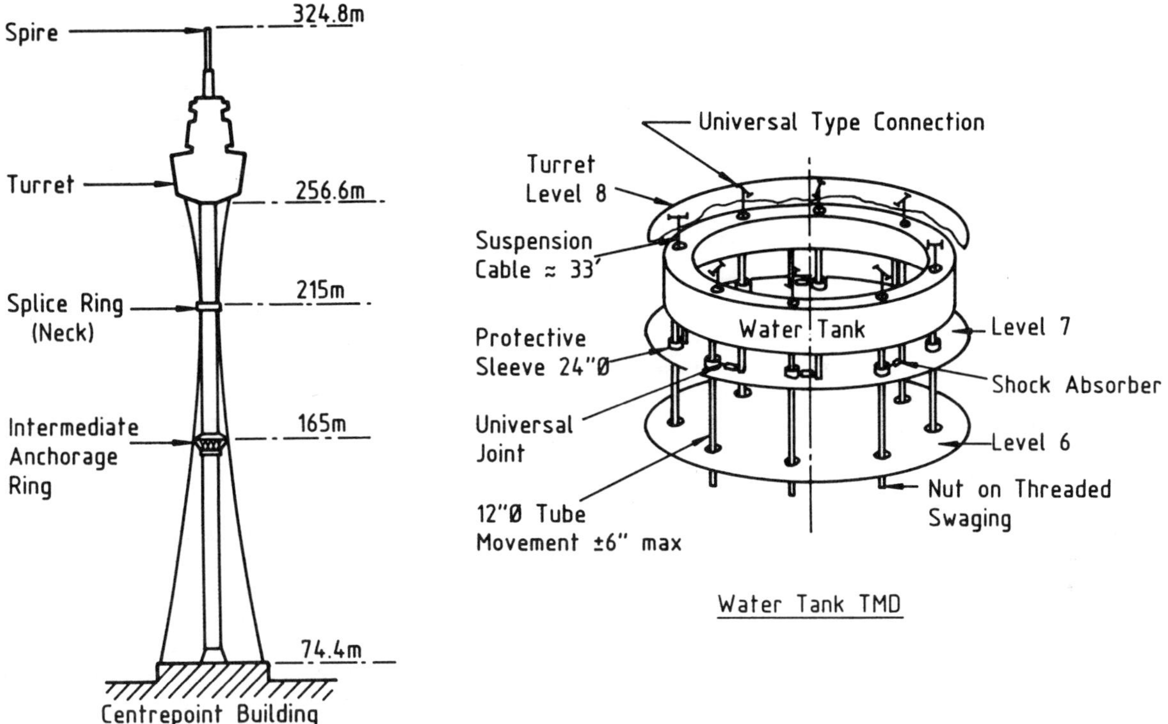

FIGURE 10. Some features and principal dimensions of Sydney Tower.

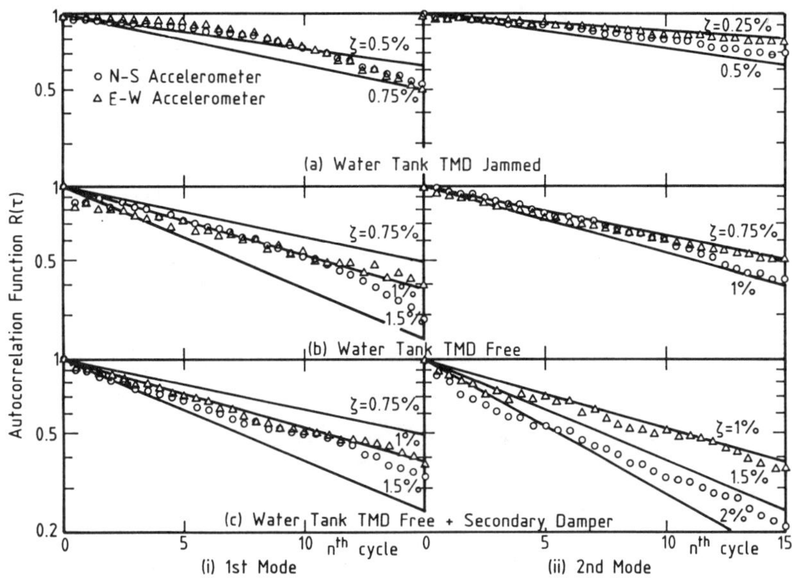

FIGURE 11. Envelopes of damping traces for different damper configuration in Sydney Tower.

TABLE 1. Damping ratios (in % of critical) of first mode and second mode vibrations.

	Water tank TMD jammed	Water tank TMD free	Water tank TMD free + secondary damper
First mode	0.7%	1.0%	1.2%
Second mode	0.4%	0.7%	1.5%

Nevertheless, fairly large movements of the TMD mass relative to the building must be accommodated, or some safety devices can be installed to limit the excessive travel of the mass.

Another major engineering problem is to provide a very low friction bearing surface for the damper mass, so that the damper mass can respond to the building movement at low levels of excitation. This can be overcome by suspending the damper mass, say with cables, provided that the floor height has a sufficient clearance to accommodate the length of suspension.

EXAMPLES OF TUNED MASS DAMPER SYSTEM IN TALL BUILDINGS AND STRUCTURES

The 335 ft (102m) steel antenna mast on top of the 1,815 ft (553m) CN Tower in Toronto has two doughnut-shaped pendulum dampers similar to the type shown in Figure 9 to reduce the second and fourth modes of vibration [13]. The circular steel rings are 8ft (2.45m) and 10ft (3.05m) in diameter, 14in. (0.36m) wide and 12in. (0.31m) deep, and together hold 20 ton (22 tonnes) of lead. Each ring is supported via universal joints by three steel beams attached to the side of the antenna mast, which allows pivoted motions in all directions. Shock absorbers are anchored on the side of the mast and attached to the centre of each universal joint to dissipate energy.

The 1,000ft (305m) tall Sydney Tower in Australia is one of the first buildings with the installation of a large scale tuned mass damper. The doughnut-shaped water tank near the top of the Turret, as shown in Figure 10, which normally serves as the Tower's water and fire protection supply was incorporated into the design of a TMD to reduce wind-induced motions [18,20]. The tank is 7ft (2.1m) deep and 7ft (2.1m) from inner to outer radius, weighs 163 tons (180 tonnes) and hangs 33ft (10m) from the top radial members of the Turret. Energy associated with relative movements between the Tower and water tank is dissipated by 8 shock absorbers installed tangentially to the tank and anchored to the floor of the Turret. Prior to the completion of the Tower, full scale measurements were carried out to determine the natural frequencies and damping level of the Tower [10]. Subsequently, a secondary tuned mass damper of similar design, in the form of a steel ring and with a suspended mass of 36 tons (40 tonnes), was installed on the Intermediate Anchorage Ring to further increase the damping level, particularly in the second mode of vibration. Typical envelopes of damping traces (in the form of autocorrelation function) for different damper configurations are presented in Figure 11 and the damping ratios are given in Table 1 [12].

The water tank TMD was originally designed to suppress first mode vibrations and was tuned to 98% of the computed first mode natural frequency. Because of the substantial differences in the actual and computed values of natural frequencies, which resulted from substantial changes to the mass distribution, there are only moderate increases in damping values in both the first and second modes. On the

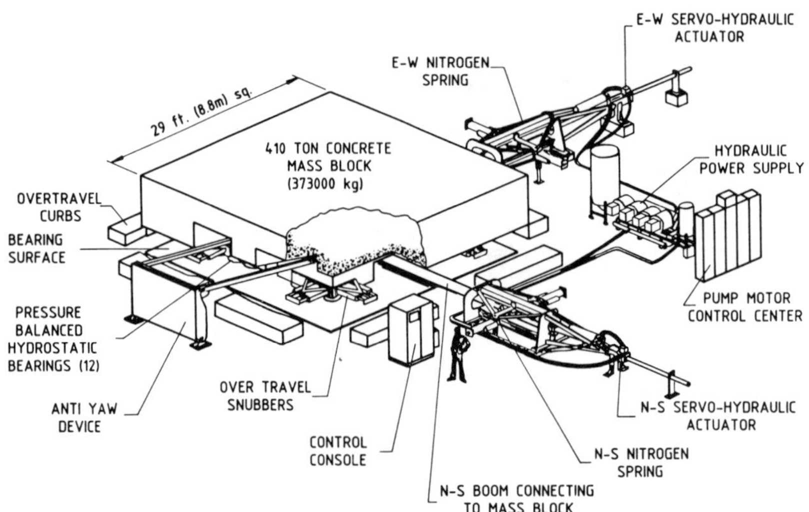

FIGURE 12. Tuned mass damper system in Citicorp Center [21].

other hand, the secondary damper, which is tuned to the actual second mode natural frequency, produces a significant increase in damping level in the second mode, and a marginal increase in the first mode. Results of full scale measurement of acceleration responses [11] showed that there were noticeable reductions in the wind-induced acceleration responses after the installation of the secondary damper.

The 914ft (278m) tall Citicorp Centre in New York City employed a tuned mass damper system, as shown in Figure 12, designed jointly by Le Messurier Associates/SCI and MTS Systems Corporation [8,16,19,21]. The TMD system is located on the 63rd floor, with a 410 ton (452 tonnes) reinforced concrete mass measuring approximately 30 ft (9.1m) square and 8.5ft (2.6m) high. The mass rides on 12 low friction hydrostatic bearings and has a travel range of $\pm$45in. ($\pm$1.14m) in both the north–south and east–west directions. The damper stiffness is provided by nitrogen-charged pneumatic springs and the spring rate and hence the tuning frequency can be varied by adjusting the pre-charge pressure. The damper damping is provided by two hydraulic actuators. The TMD facility was used as an excitor to determine the natural frequency (to which the TMD was subsequently tuned) and damping of the building. The operational parameters of the TMD are a mass ratio of 2%, damper damping of 14% of critical, and tuning ratio of 1. The effective damping, which can be obtained from Figure 6, is about 3% of critical, and a displacement ratio of about 3.5. This gives a total damping (structural damping plus added damping from TMD) of about 4% of critical, which should reduce the dynamic response by about 50%.

The 60 storey Hancock Tower in Boston uses a tuned mass damper system similar to that in the Citicorp Centre [7,19]. Two TMD's were installed 220ft (67m) apart at either ends of the 58th floor. Each TMD consists of a 300 ton (331 tonnes) lead-filled box about 17ft (5.2m) square and 3ft (0.91m) deep which rides on a 30ft (9.1m) long steel plate on which a thin layer of oil is forced through holes in the plate. The boxes are restrained by stiff springs anchored to interior columns of the building, and are connected to shock absorbers. The dampers were designed to move only in the east–west direction and can be induced to work together to counteract swaying motions or in opposition to resist torsional motions.

REFERENCES

1. Abdel-Rohman, M. and H. H. Leipholz, "Active Control of Tall Buildings," *Journal of Structural Engineering*, ASCE, Vol. 109, No. 3, pp. 628–645 (March, 1983).
2. Chang, J. C. H. and T. T. Soong, "Structural Control Using Active Tuned Mass Dampers," *Journal of the Engineering Mechanics Division*, ASCE, Vol. 106, No. EM6, pp. 1091–1098 (December, 1980).
3. Crandall, S. H. and W. D. Marks, *Random Vibration in Mechanical Systems*, Academic Press, New York, N.Y. (1963).
4. "Dampers Blunt the Wind's Force on Tall Buildings," *Architectural Record*, pp. 155–158 (Sept., 1971).
5. Den Hartog, J. P., *Mechanical Vibrations*, McGraw-Hill Book Co., Inc., New York, N.Y. (1956).
6. Feld, L. S., "Superstructure for 1350ft World Trade Centre," *Civil Engineering*, ASCE, pp. 66–70 (June, 1971).
7. "Hancock Tower Now to Get Dampers," *Engineering News Record*, p. 11 (Oct., 1975).
8. Isyumov, N., J. D. Holmes, D. Surry, and A. G. Davenport, "A Study of Wind Effects for the First National City Corporation Project—New York, U.S.A.," Boundary Layer Wind Tunnel Laboratory Special Study Report—BLWT-SS1-75, Univ. of Western Ontario, Canada (April, 1975).
9. Kaynia, A. M., I. Veneziano, and J. M. Biggs, "Seismic Effectiveness of Tuned Mass Dampers," *Journal of the Structural Division*, ASCE, Vol. 107, No. ST8, pp. 1465–1484 (August, 1981).
10. Kwok, K. C. S., "Achieving and Measurement of Damping in Tall Buildings and Structures," Chapter 14, *Course Notes on the Structural and Environmental Effects of Wind on Buildings and Structures*, W. H. Melbourne (ed.), Dept. of Mech. Engineering, Monash Univ. (May, 1981).
11. Kwok, K. C. S., "Full-Scale Measurements of Wind-Induced Response of Sydney Tower," *Journal of Wind Engineering and Industrial Aerodynamics*, 14, pp. 307–318 (1983).
12. Kwok, K. C. S., "Damping Increase in Building with Tuned Mass Damper," *Journal of Engineering Mechanics, ASCE*, Vol. 110, No. 11, pp. 1645–1649 (November, 1984).
13. "Lead Hula-Hoops Stabilise Antenna," *Engineering News Record*, p. 10 (July, 1976).
14. Mahmoodi, P., "Structural Dampers," *Journal of the Structural Division*, ASCE, Vol. 95, No. ST8, pp. 1661–1672 (August, 1969).
15. McNamara, R. J., "Tuned Mass Dampers for Buildings," *Journal of the Structural Division*, ASCE, Vol. 103, No. ST9, pp. 1785–1798 (September, 1977).
16. Peterson, N. R., "Design of Large Scale Tuned Mass Dampers," ASCE Convention and Exposition, Boston, Mass., preprint 3578 (April, 1979).
17. Tanaka, H. and C. Y. Mak, "Effect of Tuned Mass Dampers on Wind-Induced Response of Tall Buildings," *Journal of Wind Engineering and Industrial Aerodynamics*, 14, pp. 357–368 (1983).
18. "Tower Cables Handle Wind, Water Tank Damps It," *Engineering News Record*, p. 23 (December, 1971).
19. "Tuned Mass Dampers Steady Sway of Skyscrapers in Wind," *Engineering News Record*, pp. 28–29 (August, 1977).
20. Vickery, B. J. and A. G. Davenport, "An Investigation of the Behaviour in Wind of the Proposed Centrepoint Tower, in Sydney, Australia," Engineering Science Report No. BLWT-1-70, Univ. of Western Ontario (February, 1970).
21. Weisner, K. B., "Tuned Mass Dampers to Reduce Building Wind Motion," ASCE Convention and Exposition, Boston, Mass., preprint 3510 (April, 1979).

CHAPTER 12

Analysis of Asymmetric Sway Subassemblies

H. Scholz*

INTRODUCTION

It has become an accepted fact that attention must be given to additional P-Delta effects in the elastic–plastic analysis of frameworks subjected to torsion (Figure 1). Several techniques have been developed to deal with this problem. Methods of important practical significance applied to subassemblages were presented by Hibbard and Adams [7] and Rutenberg [15], as well as by Scholz [20].

The subassemblage approach by Hibbard and Adams incorporates a technique developed for plane frames [12] and involves an iterative solution for each load increment and an adjustment of the torsional in-plane displacements of the rigid floor diaphragms. Rutenberg has extended the sway-subassemblage technique for plane frames [12] to three-dimensional problems by applying his direct fictitious column approach coupled to a successive center-of-rigidity adjustment procedure. This effectively eliminates P-Delta iterations in regard to individual load increments, but a stiffness evaluation is required at each level of loading, and the location of the center of rigidity needs constant updating.

Thus, the aforementioned subassemblage techniques constitute a step-by-step solution in one way or another. A subassemblage analysis using a completely different principle was proposed by Scholz.

The fundamental concept is based on an interaction approach incorporating, as the boundary parameters, a framework unaffected by P-Delta effects and a related similar structure failing completely elastically. The actual frame is analyzed with reference to these two extremes by employing a suitable interaction technique. The procedure is an extension of the multicurve interaction method previously applied to plane frames [17,18]. In the following sections the above subassemblage methods are briefly explained and are then applied to the same story in a 20-story sway frame.

SUBASSEMBLAGE TECHNIQUE BY HIBBARD AND ADAMS

This technique employs the subassemblage method previously used to analyze multistory planar frames [12,13]. In the planar approach the structure is divided into one-story slices, as shown by the dashed lines in Figure 1, by passing imaginary planes through midheight of the columns above and below a particular floor level. Points of inflection are assumed to occur at the ends of each column stub. The model is further simplified by assuming that the moments developed in the column stubs above and below a particular joint are equal.

The response of the single-story model under a lateral load applied at the girder level can be obtained by forcing each column into a swayed position and then adding the reactions produced. As each column sways, it is restrained by the adjacent girders. Charts are available to predict the lateral load-sway displacement relationship for a column having restraining girders of a prescribed stiffness [4,12].

By adjusting the stiffnesses of the restraining girders to account for the deterioration in the structure due to plastic hinging, the complete load-deflection relationship for a restrained column is obtained. The lateral loads developed in all columns in the story are then added to determine the load carrying capacity of the story at a particular sway displacement [5].

If two or more dissimilar bents act in parallel, the load carrying capacity of the individual bents may be obtained as described previously. The total capacity of the structure may be determined by adding the responses of the two individual bents, provided that the bents are coupled to deform as a

*Department of Civil Engineering, University of the Witwatersrand, Johannesburg, South Africa

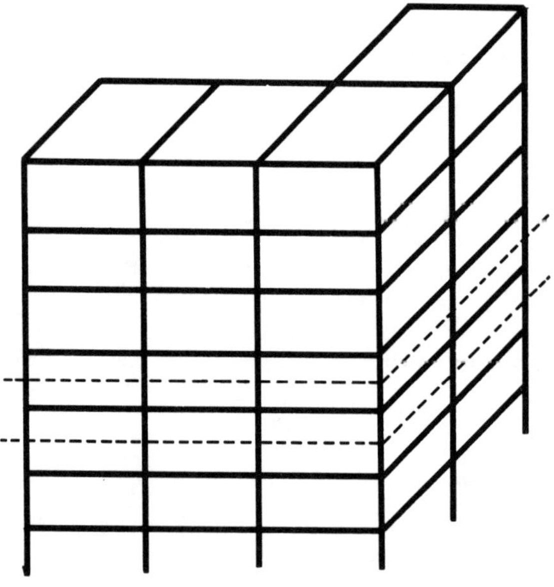

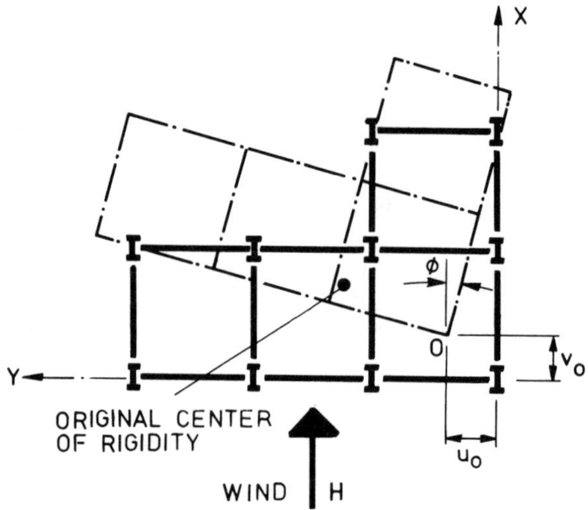

FIGURE 1. Framework subjected to torsion.

unit [22]. However, in a structure which is able to deform torsionally, the deformations of the individual bents are related only through the stiffness of the floor diaphragm. If the diaphragm can be considered rigid, then, the rigid body motion of the floor may be imposed on the individual bents in the structure and the complete response may be obtained.

Assumptions in Asymmetric Analysis

The plan view of the type of structure to be considered is shown in Figure 1. The structure is assumed to consist of two orthogonal series of bents as shown. It is assumed that the response of an individual bent to an imposed sway deformation may be determined by the use of the sway subassemblage technique [13]. Thus, all the assumptions relevant to the sway subassemblage method are implied in this technique. Additionally it is assumed that the floor diaphragms are infinitely rigid in their own planes, thus the deformations of the individual bents must be compatible with the rigid body motions of the floor diaphragms.

In the example to be presented, the torsional resistance (both warping and St. Venant) developed by the individual members will be neglected. Common structural members are of "open" cross section, and the resistance to torsional moments is relatively small. In addition, the effects due to coupling among the various modes of deformation, and to biaxial bending, are ignored.

The uniformly distributed floor loads are assumed to be transferred to the girders of one series of bents, while the girders in the bents in the perpendicular direction, simply stiffen the frame against lateral load.

Analysis

To analyze a given structure torsionally, each bent with its share of the vertical loads is subjected to a second order analysis by the planar subassemblage technique previously mentioned. The relationship between the sway rotation, Δ/h, in the plane of the frames, and the lateral load, H, developed by each bent is obtained. Typical curves are shown in Figure 2 for frames contained in the example structure shown in Figures 3 and 4.

If the structure typified in Figure 1 is forced to translate in the x direction and the floor diaphragms are assumed rigid, the response of the complete structure can be obtained by summing the lateral load capacities of the frames spanning in the x direction for particular values of Δ/h.

However, the structure shown in Figure 1 will actually twist and translate as the lateral loads are applied to the long face of the building. At any given level of lateral load, the shear developed by the frames spanning in the x direction must balance the lateral loads. The shear developed by the frames spanning in the y direction and the net torque developed in the story must vanish.

In each step of the analysis of the structure (Figure 1), point O is assumed to translate through distances u_o (in the y direction) and v_o (in the x direction) and to rotate through the angle ϕ (positive clockwise) relative to the floor below.

Thus, the relative sway of frames spanning in the y direction is

$$\Delta_y = u_o - \phi x \tag{1}$$

in which x denotes the perpendicular distance from the frame to point O. Similarly, the relative sway displacement for frames spanning in the x direction is

$$\Delta_x = v_o + \phi y \tag{2}$$

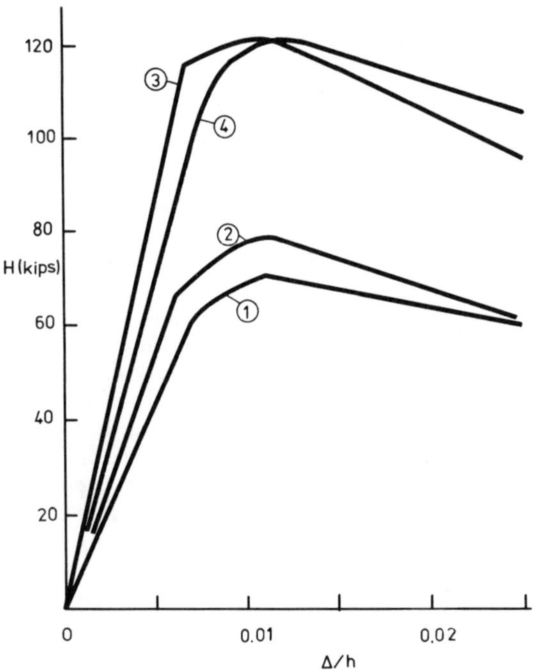

FIGURE 2. Bent load-displacement relationships (1 kip = 4.45 kN).

in which y = the perpendicular distance from the frame to point O.

The lateral force developed by each frame as it is forced into the swayed position can be determined from the bent load-deformation curves. The assumed value of u_o is adjusted so that the net force in the y direction is zero, and the assumed value of ϕ is adjusted so that the net torque about point O vanishes.

Finally, for the selected value of v_o, the lateral forces developed by the bents in the x direction are summed in order to determine the applied load. Proceeding with new values of v_o, the lateral force-sway displacement relationship of the story considering the effects of torsion is obtained. This is an iterative process at each stage of loading and well suited to a computer analysis.

Due to differential displacements of the frames, some will be forced into the unloading range (Figure 2) before others reach their maximum lateral load resistance. This method accounts for the shift in the center of torsional resistance as various portions of the structure yield inelastically.

Example

As an example of the application of the proposed technique, the adequacy of the third floor level of the 20-story steel frame shown in Figures 3 and 4 will be checked. The frame is subjected to a 20-psf (approx. 960 N/m²) lateral

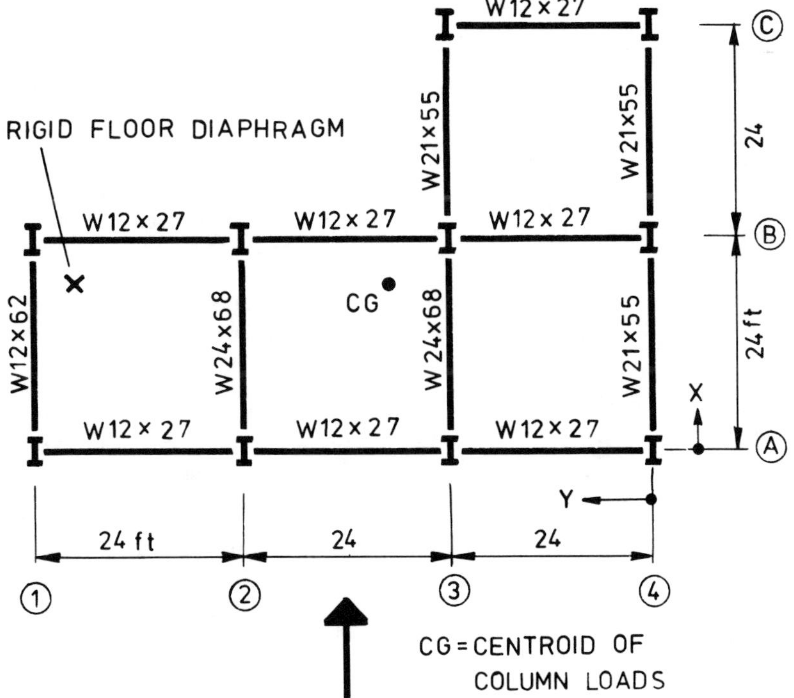

FIGURE 3. Plan of 20-story structure (1 ft = 0.305 m).

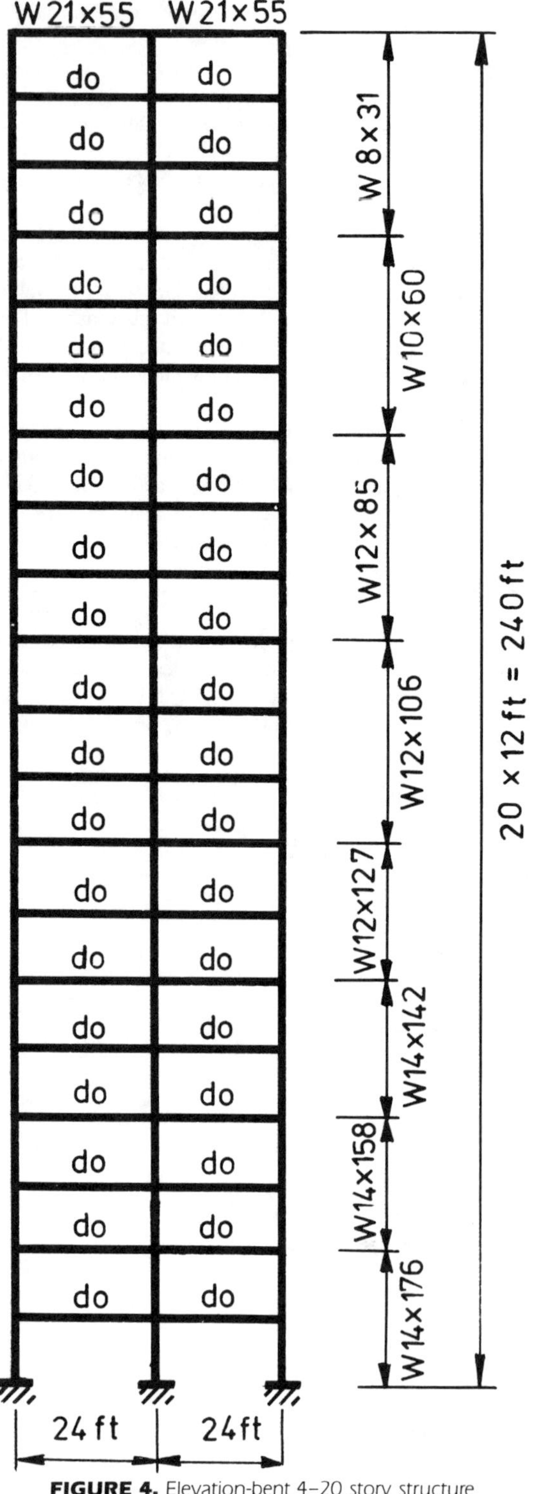

FIGURE 4. Elevation-bent 4–20 story structure (1 ft = 0.305 m).

load in the x direction, and the uniformly distributed floor loads [100 psf (approx. 480 N/m^2) at working load level] are assumed to be transferred to the girders of frames 1 to 4. This results in a load of 1.56 kips/ft (approx. 22.7 $\times$ 10^3 N/m) (at LF = 1.30) on the exterior girders spanning in the x direction and 3.12 kips/ft (approx. 45.5 $\times$ 10^3 N/m) on the interior girders. The W 12 $\times$ 27 girders spanning in the y direction simply stiffen the frame for lateral loads. The frame is similar to that described in Reference 23, however, some girder sizes have been increased for better lateral stiffness.

The equivalent one-story frames used in the subassemblage analysis are shown in Figure 5. The column loads shown are computed on the basis of tributary areas, with no live load reduction factor considered. Each equivalent frame is then analyzed by the subassemblage method [5], and the relationship between the sway rotation, Δ/h, in the plane of the frame and the lateral loads, H, is obtained. These curves for bents 1 to 4 are shown in Figure 2.

If the twisting motion of the structure is ignored, the lateral shear resistance to movement in the x direction can be obtained by summing the lateral force of each of bents 1 to 4 at selected values of Δ/h. The maximum lateral shear resistance developed at this level is 392 kips (approx. 17.5 $\times$ 10^5 N), as shown by the dashed curve in Figure 6. The total working load story shear applied at this level is 302 kips (approx. 13.4 $\times$ 10^5 N), or 302 $\times$ 1.3 = 393 kips (approx. 17.5 $\times$ 10^{5}N) at the factored load level. Thus this story would be considered adequate if the influence of torsion is neglected.

However, this story will twist along with the translation; thus, assuming rigid floor diaphragms, some bents will be displaced further than others. The structure is now analyzed by the previously described technique.

The floor diaphragm is given a displacement, v_o, in the x direction and values for u_o and ϕ are assumed. The y displacement, u_o, is adjusted until the net y force is zero. Then, the net torque about point O is found. If it is not zero, a new value of ϕ is assumed and u_o is adjusted again. When the net y force and the net torque are zero, the lateral shear resistance of the entire story to the displacement, v_o, may be obtained. New values of v_o are assumed and the previous process is repeated until several lateral load-displacement values are obtained. This relationship is shown by the solid curve in Figure 6.

The ultimate value of the resistance to lateral load, H, developed by the story is 328 kips (approx. 14.6 $\times$ 10^5 N). This is below the (factored) applied load of 393 kips (approx. 17.5 $\times$ 10^5 N), thus the structure would be deemed inadequate. The resistance developed considering torsion is, therefore, significantly reduced from the value obtained by neglecting the twisting action of the structure.

The broken curve in Figure 6 shows the results of the analysis proposed by Wynhoven and Adams [25] on the full frame in Figures 3 and 4. In this case, the plastic moment

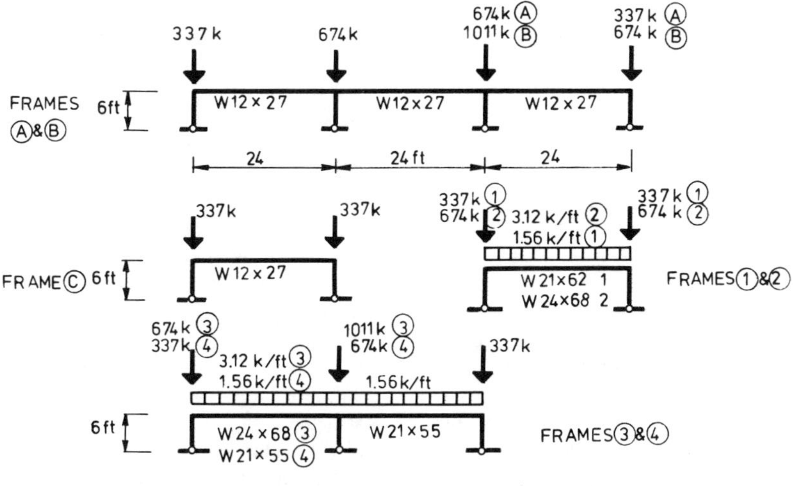

FIGURE 5. Equivalent frames (1 ft = 0.305 m; 1 kip = 4.45 kN).

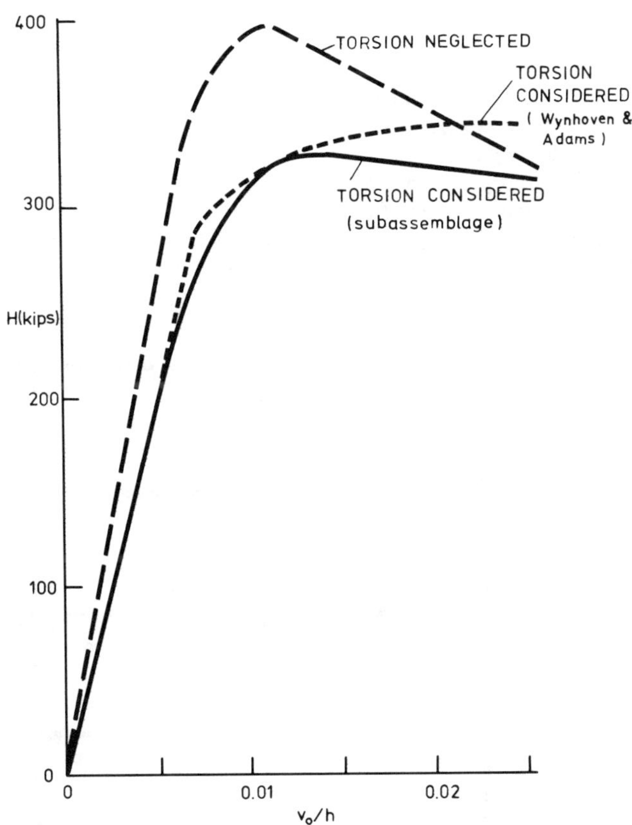

FIGURE 6. Load-displacement relationships (1 kip = 4.45 kN).

capacities of the girders were adjusted to account for the effect of uniformly distributed girder loads on the hinging pattern. Both methods yield results in close agreement.

SUBASSEMBLAGE TECHNIQUE BY RUTENBERG

General Concept

This technique shows that three-dimensional structures, in many instances, can be analyzed for the P-Delta effect by means of first order procedures without iteration. The proposed approach is a natural extension of the direct method proposed recently for planar structures [16]. The introduction of a fictitious column having negative stiffness enables a direct second order analysis to be performed by means of standard first order computer programs. The sway-subassemblage technique for multistory planar frames [4] is extended to 3-D problems by applying center-of-rigidity procedures to the elastic-plastic analysis of the bents of the story.

Shear and Torsion Model with Negative Stiffness Properties

The gravity loads acting on the columns of a framed building structure are usually relatively low compared with their Euler buckling loads. Also, the effects of additional axial forces in the columns due to overturning moments are usually small, and to a certain extent tend to cancel each other. This situation permits applying simplifying approximations to the geometric-stiffness matrix of the structure, which enable the linearization of the problem. Indeed, approximate procedures have been developed which account either directly [2,8,14] or iteratively [10] for P-Delta effects in planar structures. A short description of these methods is given in Reference 10.

For these structures, the constant geometric-stiffness matrix of a column line is tridiagonal. With the usual assumption that the floor diaphragms are rigid in their own plane, the geometric-stiffness of the complete structure is obtained by adding together the sway effects of all the columns. The resulting geometric-stiffness matrix is again tridiagonal. This tridiagonal matrix can be considered as the stiffness matrix of a fictitious shear beam having negative stiffness properties and acting in parallel with the actual structure, as shown in Figure 7. This analogy permits a direct evaluation of second order effects by means of first order plane frame computer programs.

As can be seen from Figure 8, the story shears, H_i, are proportional to the axial forces, P_i, in the columns. Denoting the shear rigidity of the fictitious column in story i by GA_i and summing over all the columns,

$$\frac{1}{h_i}\sum_j P_{ij} = -\frac{P_i}{h_i} = \frac{GA_i}{h_i} \tag{3}$$

so that

$$GA_i = GA_{ix} = GA_{iy} = -P_i \tag{4}$$

It is seen that the negative shear rigidity of the fictitious column is numerically equal to the total axial load, P_i, in the

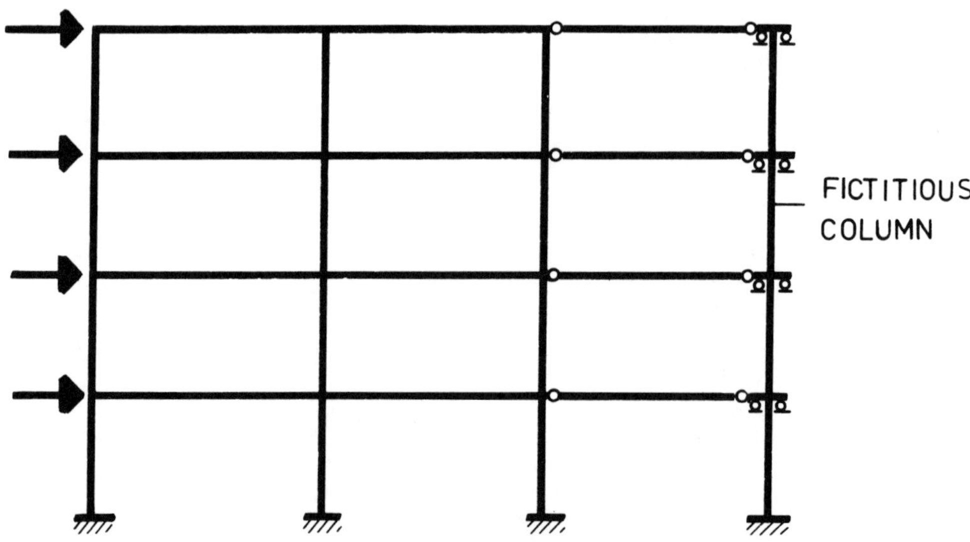

FIGURE 7. Fictitious column model.

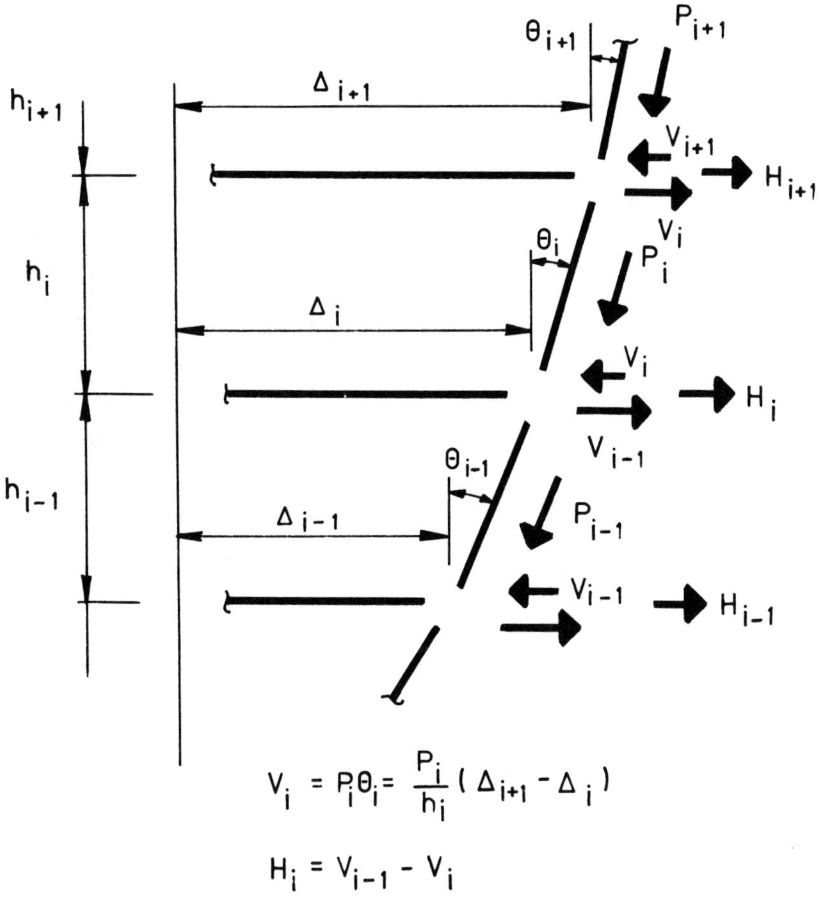

$$V_i = P_i\theta_i = \frac{P_i}{h_i}(\Delta_{i+1} - \Delta_i)$$

$$H_i = V_{i-1} - V_i$$

FIGURE 8. Sway forces due to lateral loads.

story. When the computer program is not capable of modeling shear displacements, a rotation fixed translation free flexural column may be used:

$$-\frac{1}{h_i}\sum_j P_{ij} = -\frac{P_i}{h_i} = \frac{12EI_i}{h_i^3} \tag{5}$$

so that

$$EI_i = -\frac{P_i h_i^2}{12} \tag{6}$$

The formulae are based on the assumption that the deflected shape of the column axis within a story is a straight line. Since, in fact, these columns are bent due to the presence of end moments, additional sway forces are generated due to the eccentricity of the axial force about the deflected centroidal axis of the column. These effects are not large and can be accounted for approximately by a suitable amplification factor, γ, ($1.0 < \gamma < 1.22$, e.g., [14]). With these factors, Equations (4) and (6) read, respectively [16]

$$GA_{ix} = -\gamma_x P_i;\ GA_{iy} = -\gamma_y P_i \tag{7}$$

$$EI_{ix} = -\gamma_x \frac{P_i h_i^2}{12};\ EI_{iy} = \gamma_y \frac{P_i h_i^2}{12} \tag{8}$$

in which γ_x and γ_y represent the amplification factors in the x and y directions, respectively.

In the three-dimensional case, the building undergoes lateral translation, as well as rotation about a vertical axis, and the linearized geometric-stiffness matrix of the structure as a whole is no longer tridiagonal. Yet, when the lateral displacement vector and the corresponding matrices are partitioned into translations in the x and y directions and rotation about the reference axis, each of the partitioned geometric-stiffness submatrices retains its tridiagonal form.

Thus, it is still possible to model the geometric-stiffness properties of an asymmetric structure as those of a column with negative stiffness properties. In this case, however, in addition to the previously computed shear properties of the fictitious column, its planar coordinates, as well as its torsional properties, must be determined for every story.

The fictitious column is to be located at the centroid of the column loads in the story, i.e., point *CG* in Figure 9. The negative torsional rigidity of this column about *CG* can be determined from the following expression:

$$\frac{1}{h_i}\sum_j P_{ij}\,(d_{yij}^2 + d_{xij}^2) = \frac{1}{h_i}\,P_i r_{GI}^2 = -\,\frac{GJ_i}{h_i} \tag{9}$$

in which d_x and d_y denote the *x* and *y* distances of the columns from *CG*, and

$$r_{Gi}^2 = \frac{\sum_j P_{ij}(d_{yij}^2 + d_{xij}^2)}{\sum_j P_{ij}} \tag{10}$$

so that

$$GJ_i = -\,P_i\,r_{Gi}^2 \tag{11}$$

It is seen that the torsional rigidity of the fictitious columns can be modeled as the product of the gravity load in the story and the square of its radius of inertia about *CG*. The amplification factor for torsion γ_ϕ can be obtained from an equation similar to Equation (9) in which the components are suitably factored.

It will be observed that the location of CG and the magnitude of *GJ* depend on the distribution of the axial forces in the story, which is not the case for *GA*. Neglecting the effect of the lateral and second order forces on this distribution is thus another source of error. However, it has been shown by Nair [11] that this approximation still provides an excellent estimate of the displacement caused by second order effects.

When the computer program available to the engineer is not capable of modeling the torsional properties of beam elements, the equivalent flexural element defined in Equation (12) can be used to represent these properties of the fictitious column. However, in this situation, two such columns per story are required to model the negative torsional rigidity, each having one half of the negative story stiffness, namely:

$$\frac{1}{2}\,EI_i = \frac{1}{2}\,EI_{ix} = \frac{1}{2}\,EI_{iy} = -\,\frac{P_i h_i^2}{24} \tag{12}$$

Each of the two columns should be located a distance r_G on opposite sides of *CG*, as shown in Figure 9. In regular buildings, the fictitious column (or the two equivalent columns) in every story will tend to be on one (or two) vertical axis (axes).

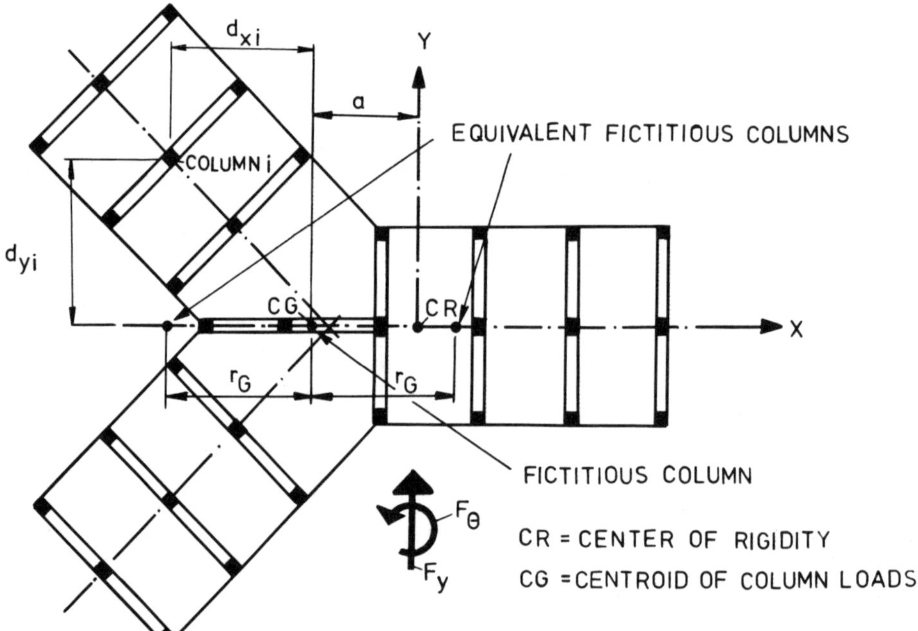

FIGURE 9. Typical floor plan and location of fictitious column.

However, when the distribution of axial forces along the column height varies from story to story, this will not be the case. Therefore, the fictitious column should be modeled so as to ensure continuity from story to story. For example, the columns could be connected at floor levels by means of rigid horizontal links.

Once an additional column with given negative stiffness properties is incorporated into the model of the structure, a first-order space frame computer program automatically will generate all the terms of the second order stiffness matrix. Difficulties will arise when stability checks on the member stiffness matrix are built into the program [9]. However, it appears that most programs for in-house computers have no such checks. The proposed procedure thus obviates the need to resort to an iterative analysis in order to compute the P-Delta sway forces.

Modified Story Stiffness

Many building structures behave approximately like shear beams. This fact permits the distribution of the lateral forces among the frames on a story-by-story basis. In such cases, the first order procedures proposed by Wilbur [24] or Cheong-Siat-Moy [1] may be followed. Once the lateral stiffness values K_{Li} are computed for all the frames in the story, the center of rigidity, *CR*, can be determined from simple statics, and the story shear and torque are distributed among the frames using the standard expressions:

$$H_i = H \frac{K_{Li}}{\Sigma K_{Lx,y}} \pm He \frac{K_{Li}e_i}{\Sigma(K_{Li}e_i^2)} \qquad (13)$$

in which H_i = shear on individual frame; H = shear on story; $\Sigma K_{Lx,y}$ = sum of lateral frame stiffnesses K_{Li} in the x- or y-direction depending on the orientation of H; and e and e_i denote the eccentricity of the shear, H, and frame, i, respectively, in regard to the center of rigidity of the story. Note that in the foregoing formulation, the frames are assumed to form a rectangular grid in plan.

It is now evident that incorporating second order effects in this procedure merely requires adding to the analytical model of the story a column with negative stiffness properties, as outlined earlier, and locating it at the center of the gravity loads in the story (*CG* in Figure 9). The modified stiffnesses marked by an asterisk can easily be evaluated and are given by

$$\Sigma K_{Li}^{*} = \Sigma K_{Li} - \frac{P}{h} \qquad (14)$$

$$\Sigma(K_{Li}e_i^2)^{*} = \Sigma(K_{Li}e_i^2) - \frac{P}{h}(r_G^2 + a^2) \qquad (15)$$

in which a = distance between *CG* and the new *CR*. Note that e and e_i also have to be modified due to the shift in *CR*. The present approach thus obviates the need to resort to the more lengthy formulation given by Rosenblueth [14]. In the following section, this procedure will be generalized to incorporate second order effects in the elastic-plastic frames.

Stability Effects in Elastic-Plastic Frames

The technique described in the preceding section can be applied to a story-by-story analysis of elastic-plastic structures, as follows. First, the lateral force-displacement relationship for the story subassemblage of every frame is computed using the hinge-by-hinge procedure (historical analysis). Since the P-Delta effects are already incorporated in the first order formulation, this is a straightforward task which can be performed easily either by hand following, for example, the simple procedure proposed by Disque [6], or with an in-house microcomputer, or even a programmable calculator. From the force-displacement relationships, the initial lateral and torsional stiffnesses of the story are derived, and the center of rigidity is computed. Since the force-displacement relationships are assumed to be piecewise linear, the story shear has to be applied in increments, each corresponding to a load increment within which the force-displacement curves for all the frames are linear, i.e., when a kink on the curve is reached, the stiffness of the story and the location of the rigidity center are updated.

As will be illustrated in the following example, the subassemblages chosen for the elastic-plastic analysis comprise one floor girder and two rows of columns, half a story in height, pinned at midheight above and below the girder, rather than one row of columns extending the full story height together with the girders above and below.

Several simplifying assumptions are implicit in the sway-subassemblage technique, as well as in the modified procedure outlined in the preceding paragraphs, which may limit its applicability. First, the assumption that gradual penetration of yielding through the cross section, as well as along the member, may be ignored. This does not appear to be a serious limitation in many practical instances [13]. Secondly, in interconnected perpendicular frames, it is tacitly assumed that plastic hinges are formed in girders rather than in columns, so that the effect of the biaxial yield condition can be disregarded [26]. As is well known, this "strong-column-weak-beam" design is advocated for earthquake resistant structures. Thirdly, it is assumed that changes in the axial column forces are relatively small or, alternatively, that they can be predicted with sufficient accuracy so that the elastic-plastic lateral force-deflection relationship is known in advance. Thus, the procedure is particularly suitable for the standard nonproportional loading analysis of multistory frames in which the vertical forces are kept constant (at their factored load value [10]), while the lateral loads are being increased, and where changes in axial col-

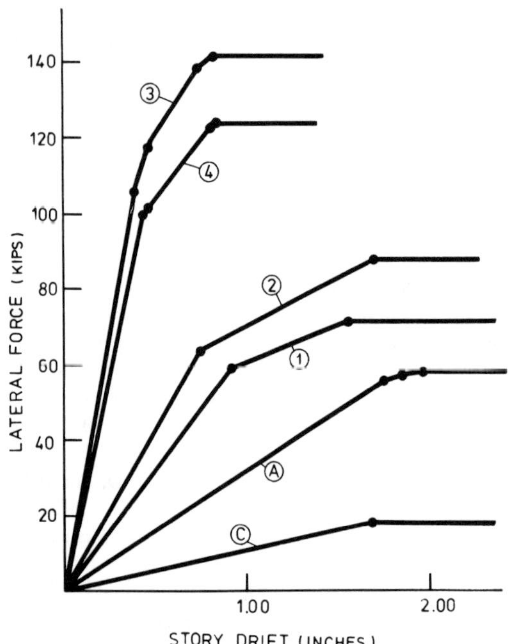

FIGURE 10. Load displacement relationships for third story bents (1 in = 25.4 mm; 1 kip = 4.45 kN).

umn forces resulting from the increasing overturning moment are usually quite small. However, an approximation to the less common case of proportional gravity and lateral loading can be made by using suitably adjusted force-displacement relationships. Another assumption, which is tacitly made in the analysis, is that the effect of twist on yielding of columns and girders may be ignored [26]. Despite these limitations, it is believed that the subassemblage approach is likely to yield useful results.

Example

The steel frame of the 20-story structure shown in Figures 3 and 4 was chosen for this example. This building is the one analyzed by Hibbard and Adams for second order effects using subassemblages. The modulus of elasticity of steel is taken as 29,000 ksi (200,000 MPa). The geometric-stiffness properties of the structure are computed as follows. Assuming for simplicity $\gamma = 1.0$,

$$GA_i = -(21 - i) \times 4 \times 24.0 \times 24.0 \times 0.13 = -(21 - i)\ 299.52 \text{ kips} \tag{16}$$

in which i = story number.

In order to compute GJ_i, it is necessary first to evaluate the axial forces acting on all the columns. Since these are not known exactly, they may be estimated on the basis of tributary areas. This is usually a conservative assumption in the sense that, in fact, the radius of gyration, r_G, is somewhat smaller in view of a higher concentration of axial loads on inner columns due to continuity. On this basis, it was found that the center of the axial loading is located at CG in Figure 3. Using Equations (10) and (11)

$$GJ_i = -(21 - i) \sum_j P_{ij}(d_{jx}^2 + d_{jy}^2) = -(21 - i)\ 34.16 \times 10^6 \text{kip-sq in} \tag{17}$$

Once the properties of the fictitious column have been obtained, it is possible to apply the proposed procedures in order to evaluate the response of the structure in the elastic and elastic-plastic ranges. The analysis in this example is confined to the response of the third story. The single story equivalent frames used in the subassemblage analysis are shown in Figure 5. Since the subassemblage is symmetric about the floor level, the lateral force-displacement relationships were evaluated based on half-story subassemblages. The effect of axial forces on the plastic moment capacity of wide flange steel columns was considered using the following expressions:

$$M_{pc} = 1.18 \left(1 - \frac{P}{P_y}\right) M_p \le M_p \tag{18}$$

for strong axis bending, and

$$M_{pc} = 1.19 \left[1 - \left(\frac{P}{P_y}\right)^2\right] M_p \le M_p \tag{19}$$

for weak axis bending, in which P_y denotes the axial yield force; and M_p and M_{pc} denote, respectively, the plastic moment and the reduced plastic moment due to the effect of axial compression. Note that the preceding equations describe the capacity at the fully plastified cross section, i.e., it is assumed that slenderness effects within a story are small. When this is not the case, appropriate interaction equations should be used [3,10,13].

The force-displacement relationships for all the bents are shown in Figure 10. Note that the response of Bent 3 was found to be slightly affected by the direction of the lateral force, and the curve shown is for wind acting in the direction of the arrow in Figure 3.

Every column in the building belongs to two perpendicular frames, and thus they are all subjected to biaxial bending. Although plastic moment capacities are affected by biaxial interaction, this effect was not found to be significant in this case. This is mainly due to the fact that, on the whole, plastic hinges form much earlier in the girders than in the columns, and to the relatively low level of stresses in

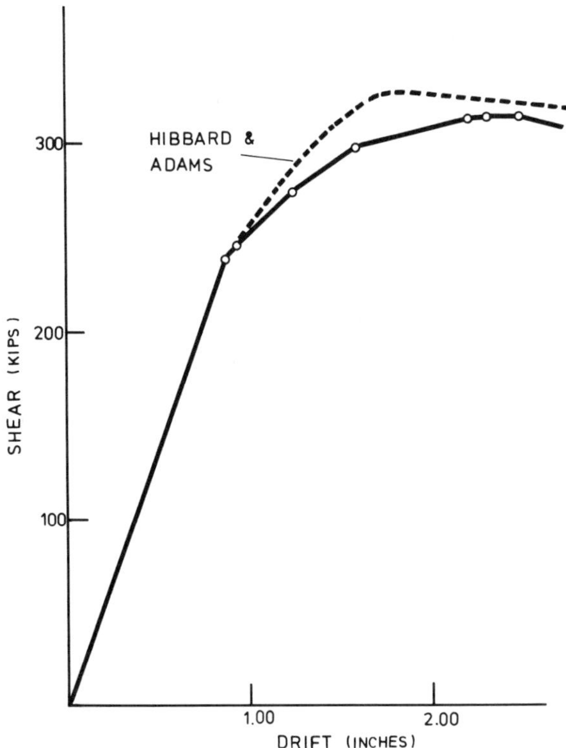

FIGURE 11. Story shear versus drift of bent 1: third story (1 in = 25.4 mm; 1 kip = 4.45 kN).

the *y* direction. Similar findings were reported by Wynhoven and Adams [26].

Applying the procedure outlined previously, the lateral and torsional rigidities of the story, including the P-Delta effects, were computed using the hinge-by-hinge procedure. The resulting shear displacement curve for Bent 1, which is the most critical, is shown in Figure 11. The ultimate shear carried by the structure at this level is 313 kips (1,392kN), compared with 328 kips (1,459 kN) given by Hibbard and Adams.

INTERACTION ANALYSIS BY SCHOLZ

Interaction Method Applied to Asymetric Structures

A typical story of a multistory asymmetric structure is shown in Figure 12 (3rd story of frame of Figures 3 and 4). Applied loadings consist of gravity load as well as lateral load. The resultant of the horizontal load is acting eccentrically with respect to the center of rigidity of the story. Vertical and horizontal loads are accumulated values accounting for stories above the story under investigation. The story model shown is separated from the remainder of the structure at points of inflection in the column members, assumed to occur at midheight of the story.

In applying the principles of the interaction method derived for plane structures [19] to this case, the then formulated three fundamental statements found to be valid for

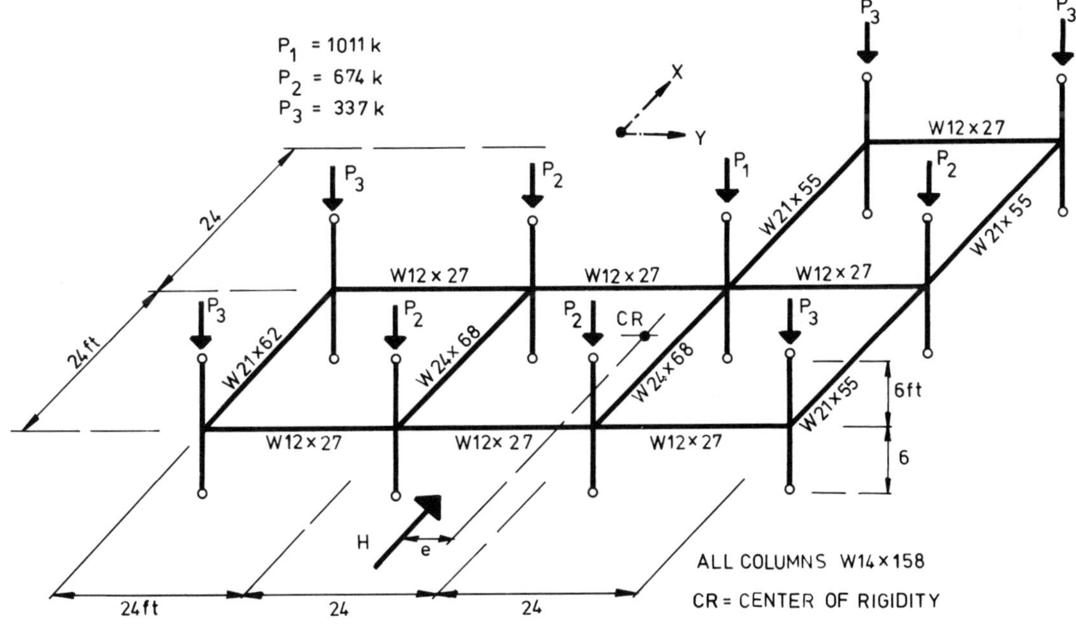

FIGURE 12. Story subassemblage (1 kip = 4.45 kN; 1 ft = 0.305 m).

plane frames are rephrased as follows:

1. The maximum possible failure load of a story is given by the load factor applicable to the weakest rigid-plastic collapse mode of the story. In nondimensional format this condition coincides with the ratio $\gamma_F/\gamma_P = 1$ in Figure 13.
2. The smallest possible elastic failure load factor of the story is derived from the load factor corresponding to the weakest elastic buckling mode of the story. The load factor associated with that mode, γ_C, is used to identify a so-called "limiting story" for which elastic failure and first-yield coincide. This "limiting story" shows the smallest ratio of inelastic failure load to rigid-plastic collapse load of frames located on a specific failure curve such as that shown in Figure 13. The intersection of the curve shown with the right vertical axis corresponds to the "limiting story." It has been verified [19] that for each given story a unique "limiting story" can be found and that the parameters of this configuration can be used to identify that family of stories to which the story under investigation belongs.
3. The ratio of the lowest elastic buckling load to the lowest rigid-plastic collapse load of the actual story, γ_C/γ_P, will identify the failure load factor of the given story on the relevant failure curve. The ratio γ_C/γ_P appears as a straight line through the origin of Figure 13. The factor α will be explained at a later stage.

The "limiting story," identified by the procedure briefly outlined under point 2 above, signifies the transition from inelastic failure to elastic failure. The entire curve of Figure

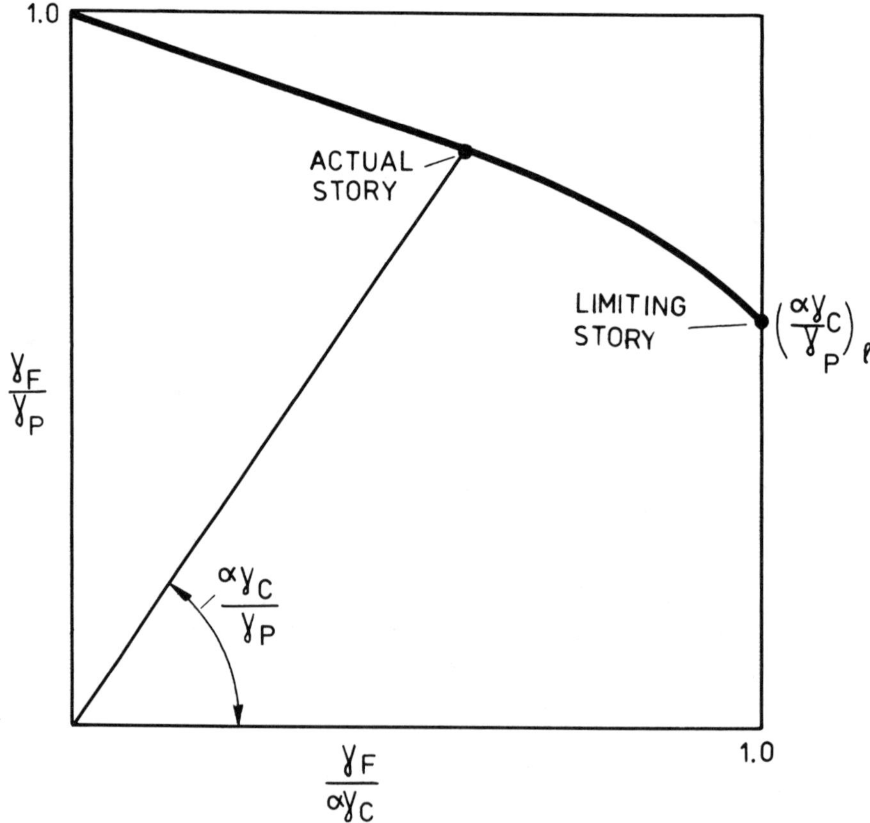

FIGURE 13. Interaction principle.

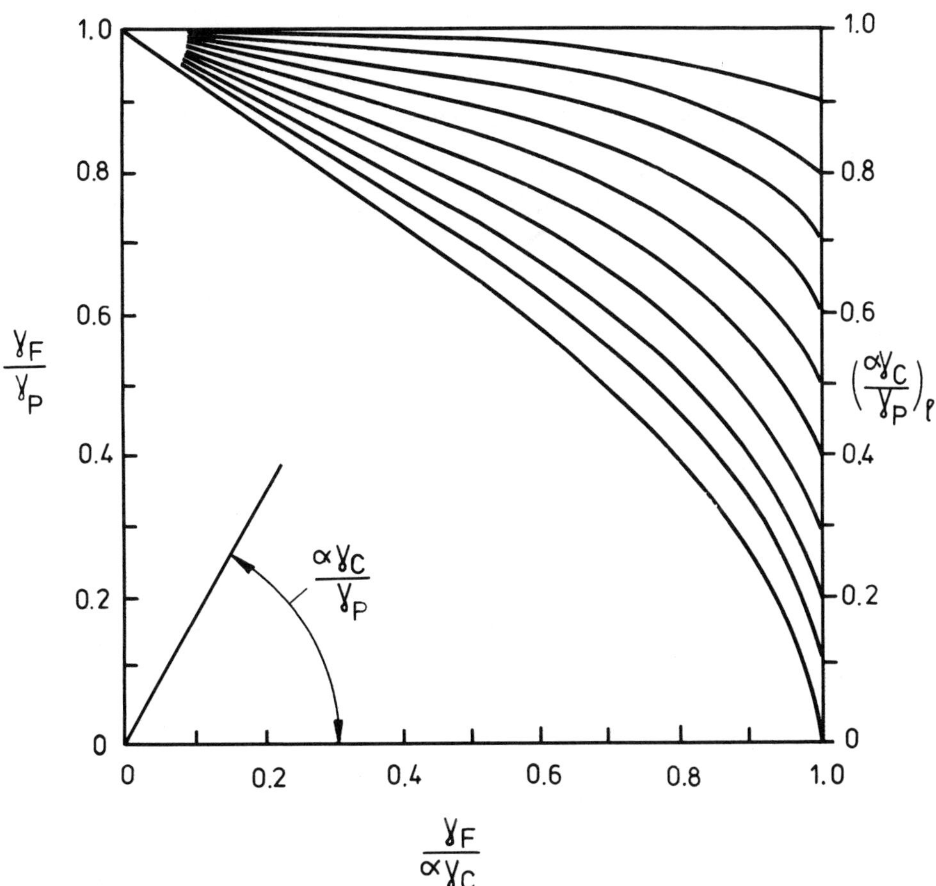

FIGURE 14. Basic interaction curves.

13 between $\gamma_F/\gamma_P = 1$ and $\gamma_F/(\alpha\gamma_C) = 1$ covers an infinite number of structures, which all belong to the same family characterized by their common "limiting story" with the same parameter, $(\alpha\gamma_C/\gamma_P)_\ell$. The presence of torsion is recognized by seeking the lowest possible elastic and rigid-plastic failure modes contained in the story and by incorporating the torsional load effects when computing the parameters of the "limiting story."

Interaction Curves and Required Parameters

In Figure 14 the complete multicurve interaction diagram from Reference 19 has been reproduced. Full details can be found in this reference. However, a brief summary and explanation of the significant parameters are given here. To analyze a story in accordance with these curves, the factor α and the two ratios, $\alpha\gamma_C/\gamma_P$ and $(\alpha\gamma_C/\gamma_P)_\ell$, need to be known. The factor α is found from Equation (20) and assists in determining the "limiting story":

$$\alpha = \frac{0.4}{1 - 0.6\left(\dfrac{700 - \lambda_{HO.4}}{700}\right)^3} \tag{20}$$

The term $\lambda_{HO.4}$ is representative of the nonsymmetrical

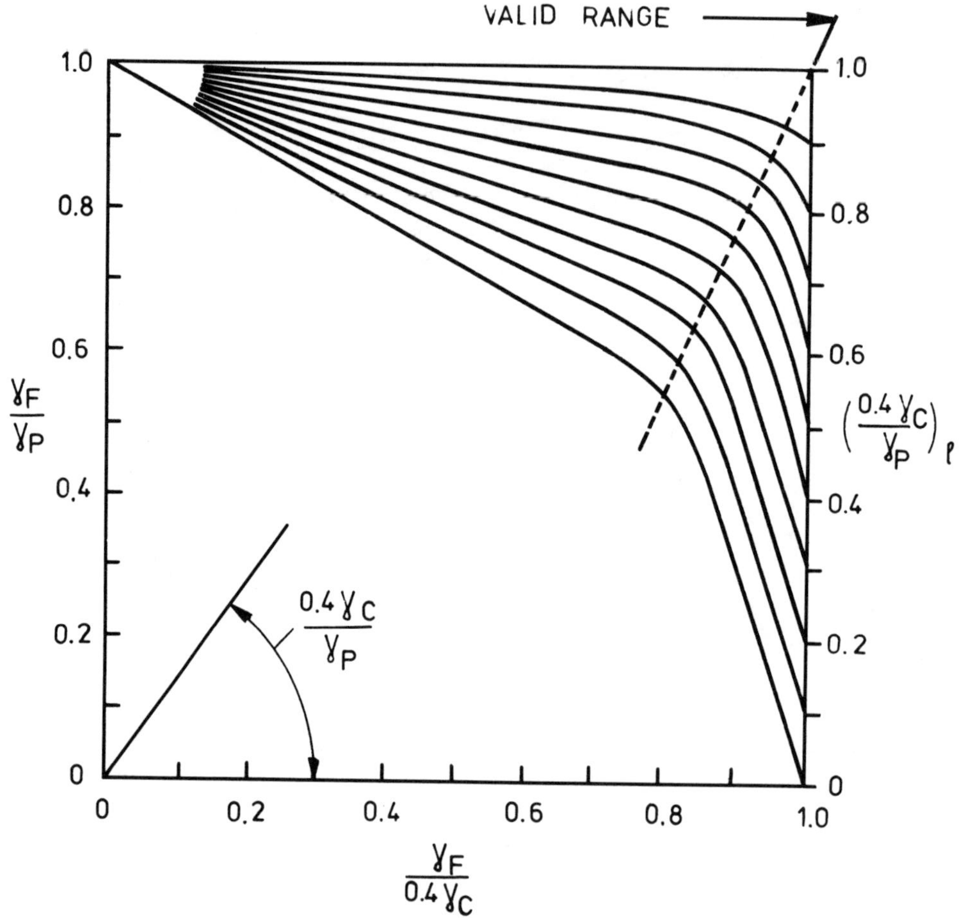

FIGURE 15. Revised interaction curves.

conditions regarding geometry and loading and takes the form of a slenderness ratio. It is evaluated from Equation (21) using elastic second-order moments and axial forces applying horizontal loading related to 40% of the vertical elastic buckling load, i.e., 0.4 γ_C.

$$\lambda_{HO.4} \text{ or } \lambda_\ell = \frac{y}{r}\frac{M}{I}\frac{L}{f_p}\left[\frac{1}{2} + \frac{1}{2}\sqrt{1 + \frac{4f_p PI}{\left(\frac{y}{r}M\right)^2}}\right] \tag{21}$$

in which λ_ℓ = "limiting slenderness ratio" of "limiting frame"; y = distance from centroid of section to extreme fiber; r = radius of gyration; M = elastic second-order bending moments; I = second moment of area; L = member length; f_P = stress at onset of yield; and P = axial member force.

In computing $\lambda_{HO.4}$, the magnitude of the horizontal load is kept in proportion to the vertical load, 0.4 γ_C, as in the case of the applied factored load. Furthermore, in Equation (21) f_P is set equal to 36 ksi (250 MPa) and y/r = 1. For the entire story, that value $\lambda_{HO.4}$ is sought for which the ratio $\lambda/\lambda_{HO.4}$ becomes the smallest, in which λ refers to the slenderness ratio of the member in the actual story.

For the general case Equation (21) is evaluated twice: first for $\alpha = 0.4$, to obtain $\lambda_{HO\cdot4}$, in order to calculate the final value for α, which will give λ_ℓ. For the latter computation, the appropriate stress at the onset of yield and the prevailing ratio y/r need to be substituted. For the story as a whole, the lowest ratio λ/λ_ℓ is significant. The "limiting slenderness ratio," λ_ℓ, defines a "limiting story" that just reaches first-yield at a critical section as all loading is applied related to its elastic buckling load.

Once the "limiting story" is known, the ratio $(\alpha\gamma_C/\gamma_P)_\ell$ of this configuration must be established. This can be done either from first principles as for the actual story, or by using the simplified relationship given by

$$\left(\frac{\alpha\gamma_C}{\gamma_P}\right)_\ell = \frac{\alpha\gamma_C}{\gamma_P}\frac{\lambda}{\lambda_\ell}R \tag{22}$$

If first yielding occurs in a beam section, the simplified expression of Equation (23) applies:

$$\left(\frac{\alpha\gamma_C}{\gamma_P}\right)_\ell = \frac{\alpha\gamma_C}{\gamma_P}\frac{Zf_P}{M}R \tag{23}$$

By comparing Equations (22) and (23), the ratio $\lambda/\lambda_\ell = Zf_P/M$, with Z = elastic section modulus. The subscript ℓ indicates parameters relating to the "limiting story."

The factor R in Equations (22) and (23) recognizes that the reduction due to axial forces to the fully-plastic moment capacity of column sections will be different for the actual story and the "limiting story." The value R can easily be estimated; however, $R = 1$ will often give satisfactory results.

It has been found that in many practical cases of combined lateral and vertical loading, the need to calculate the factor α can be waived by setting $\alpha = 0.4$ and computing the "limiting slenderness ratio," λ_ℓ, from Equation (21) on this basis. The slightly adjusted curves of Figure 15 have been suggested for this approach [17]. However, acceptable results are obtained only within the marked range, for which $0.4\ \gamma_C/\gamma_P \geq 1$.

Assumptions and Required Analyses

It is generally accepted that several simplifying assumptions are associated with sway-subassemblage techniques. In addition, in the presented approach the relatively small torsional and warping resistances of individual members are disregarded, and the horizontal floor diaphragm is considered to be infinitely rigid.

For the rigid-plastic analysis, story failure modes in direction of the x-axis and y-axis of Figure 12 are examined as all applied loading is proportionately increased. Changes in the axial column forces as the lateral loads increase are usually small and are neglected. The reduction to the fully-plastic moment capacity of the column sections are considered, using the well-known expressions applicable to I-sections [Equations (18) and (19)].

A typical story subassemblage that might have to be analyzed on a rigid-plastic basis is shown in Figure 12. The second-order elastic analyses required to find α and λ_ℓ are performed on the same subassemblage. In the latter case, simplified approaches to the solutions are acceptable. For the individual frame, it is usually sufficient to take the results of a first-order analysis for vertical and horizontal loading and magnify the bending moments due to the lateral loads by the term $1/(1 - \gamma_i/\gamma_C)$. The ratio γ_i/γ_C combines the vertical load portion acting on a particular individual frame contained in the story and the elastic buckling load of the same frame. For the elastic analyses, lateral load and torsion are distributed among all frames of the story on the basis of their elastic lateral and rotational stiffness contributions, respectively. This approach likens the building structure to the behaviour of a shear beam.

The elastic lateral story stiffness of an internal subassemblage is conveniently computed from the expression suggested by Cheong-Siat-Moy [1]:

$$K_{Li} = \frac{12E}{h^2}\sum\frac{K_C}{1 + \dfrac{UK_C}{\Sigma K_B}} \tag{24}$$

in which K_{Li} = lateral stiffness of frame i; h = story height; K_C = column stiffness; ΣK_B = sum of beam stiffness at joint; and U = sum of column load above and below joint, divided by column load below joint. For each plane frame contained in the story, the foregoing expression needs to be summarized over all columns in that frame. For bottom stories, slightly different expressions are valid to account for the degree of fixity at the base.

The position of the center of rigidity can be determined by simple statics once the lateral stiffnesses of all frames in the story are known. Subsequently, the shear on individual frames, H_i, can be computed as in Equation (13).

Strictly, second-order elastic stiffness should be used, but it is tacitly assumed that a distribution of shear and torque on the basis of first-order elastic stiffnesses is adequate for the purpose of this approximate analysis procedure.

Example

The application of the multicurve interaction procedure is demonstrated by evaluating the third story (Figure 12) of the 20-story framework shown in Figure 4. Usually the factored design loadings are kept at the same proportion when the parameters are calculated to find the failure load factor from the curves of Figures 14 and 15.

However, the purpose of this example is to demonstrate that similar results are obtained both here and for the subassemblage procedure by Hibbard and Adams, which gives a floor load of 130 psf (6.23 kN/m²), all carried by frames in the x-direction and a corresponding story shear of $H = 328$ kips(1,460 kN) as the failure loads of the third story. These values are thus assigned the load factor $\gamma_F = 1$ for the example.

For the calculation of the rigid-plastic collapse load factor, two significant collapse modes were identified: (1) A sway mode in the x-direction involving all frames orientated that way; and (2) column yielding for the y-direction, since, by nature of the floor spans, no superimposed loads are directly applied to the beams spanning that way.

It must be recalled that, despite the added torsional effect, the total sum of lateral load in the direction of H remains unchanged. Moreover, the sum of the lateral load actions at right angles to H (y-direction) must vanish to zero from simple statics. The relevant plastic collapse mode for the example is schematically shown in Figure 16 and results in a load factor of $\gamma_P = 1.23$, taking $\gamma_F = 1$ as the basis.

The load factor associated with elastic story buckling is determined considering vertical load only. A mode involving buckling in the y-direction is governing, and a load factor $\gamma_C = 1.72$ is obtained for buckling of the entire story in this direction. The term γ_C simply represents the ratio of the sum of the elastic buckling loads of all frames in the y-direction to the sum of the vertical story loads at $\gamma_F = 1$, i.e., the column loads given in Figure 12. To allow for the deflected shape of the columns an average amplification factor $\gamma = 1.11$ has been included.

The next step in the analysis is the calculation of the ratio $0.4\gamma_C/\gamma_P$. For the example, a value of 0.56 is obtained. It is evident that this value falls outside the range of acceptability marked in Figure 15. This is a rather unusual occurrence since most practical frames are covered by the curves of Figure 15. However, for the example, the low ratio $0.4\gamma_C/\gamma_P$ signifies that the given story is highly sensitive in regard to torsional effects due to the low lateral stiffness in the y-direction.

It is thus required to use Figure 14 to find the failure load of the story, and it becomes necessary to calculate the factor α from Equation (20). For this purpose, the frames in the story are subjected to the horizontal loading given in Figure 17. The lateral loads on individual frames were computed on the basis of Equation (13) using $\alpha = 0.4$ for this cycle of calculations. The position of the center of rigidity was obtained from simple statics incorporating the relevant lateral frame stiffnesses. The section within all frames that yields first is sought under the application of the loading of Figure 17. Subsequently, $\lambda_{HO.4}$ must be calculated from Equation (21). This task appears laborious, but it is often possible to identify the critical frame and section by inspection. Alternatively, a methodical check of all possible sections by computer gives rapid results. Usually, such a critical section occurs in a beam rather than in a column member.

In this example, the beam section of frame 1 marked in Figure 18 was found to be critical. The bending moments given in Figure 18 are the result of a first-order elastic analysis of this frame subjected to the lateral loading indicated in the same figure. Magnifying the moments of Figure 18 by $1/(1 - \gamma_i/\gamma_C)$ yields the critical value of max $M = 318$ kft(432 kNm). Evaluating Equation (21) leads to $\lambda_{HO.4} = 23$. Substituting $\lambda_{HO.4} = 23$ into Equation (20) results in $\alpha = 0.87$.

The value $\alpha = 0.87$ is now used in a repeat cycle of the previous calculations for $\lambda_{HO.4}$, except that now, vertical as well as horizontal loads are applied to the structure. Again, first yield was observed for frame 1, for which the applied load and the final bending moments are shown in Figure 19.

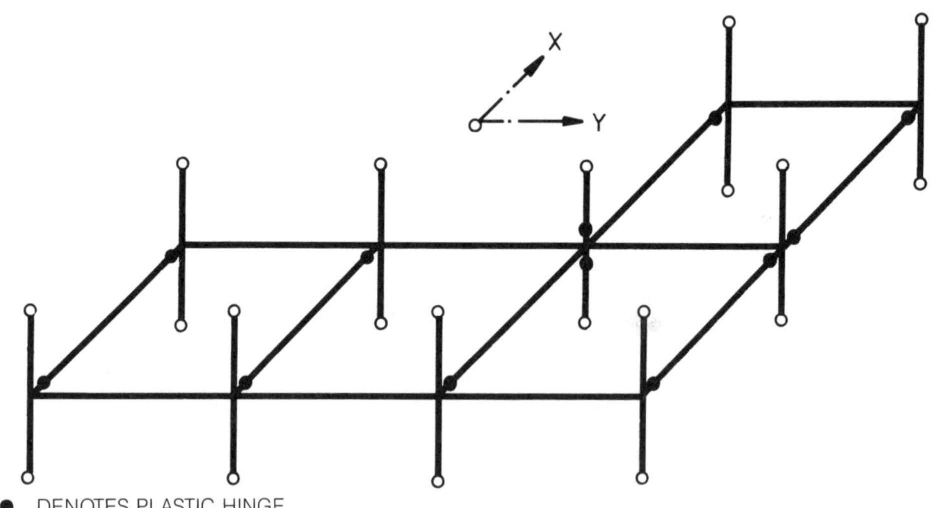

FIGURE 16. Plastic collapse of story subassemblage.

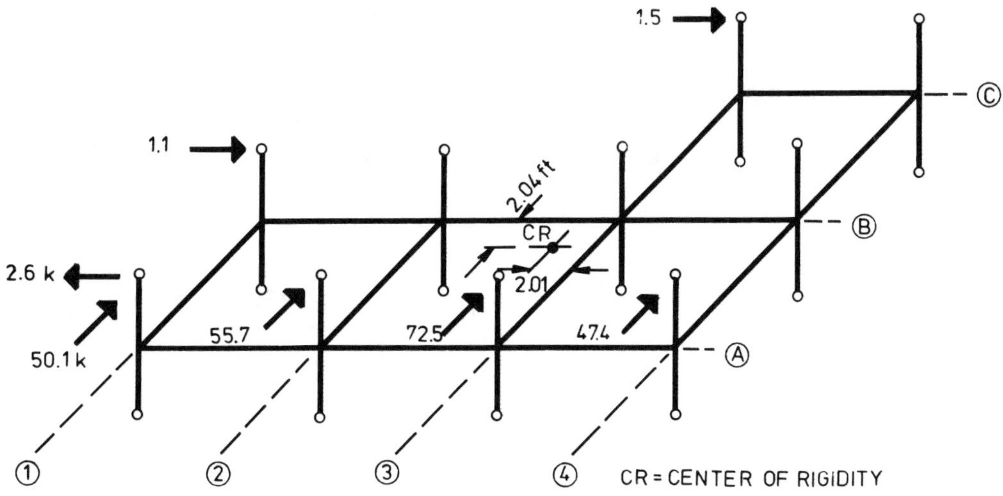

FIGURE 17. Lateral load on story for λ_{HO4} (1 kip = 4.45 kN; 1 ft = 0.305 m).

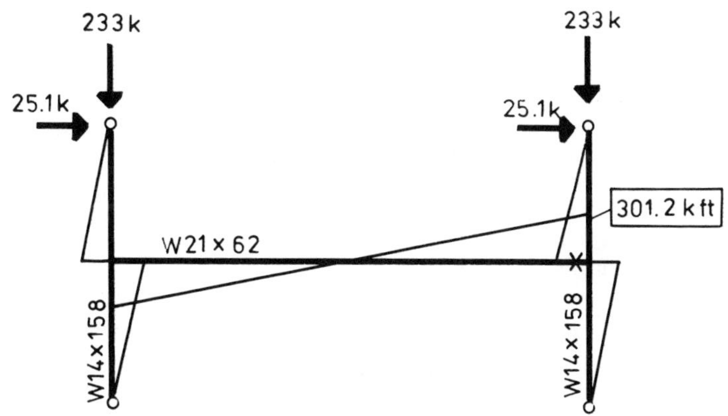

FIGURE 18. Bending moments on frame 1 for λ_{HO4} (1 kip = 4.45 kN; 1 ft = 0.305 m).

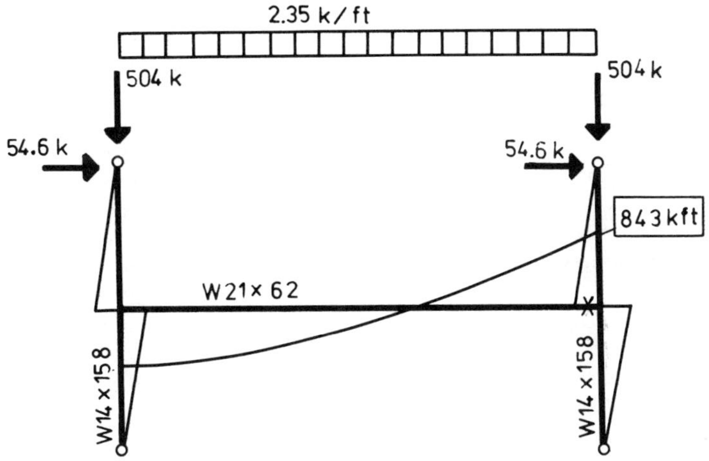

FIGURE 19. Bending moments on frame 1 for λ_ℓ (1 kip = 4.45 kN; 1 ft = 0.305 m).

Using the maximum bending moment value of 843 kft (1,144 kNm), the ratio $\lambda/\lambda_\ell = Zf_P/M = 0.45$ is obtained.

The data established so far can now be used to compute the parameters required to find the failure load factor from Figure 14. The significant terms are summarized here:

$$\frac{\alpha\gamma_C}{\gamma_P} = \frac{0.87 \times 1.72}{1.23} = 1.22 \qquad (25)$$

Equation (23):

$$\left(\frac{\alpha\gamma_C}{\gamma_P}\right)_\ell = 1.22 \times 0.45 \times 0.92 = 0.51 \qquad (26)$$

(including $R = 0.92$)

Figure 14:

$$\frac{\gamma_F}{\gamma_P} = 0.81 \rightarrow \gamma_F = 0.81 \times 1.23 = 1.0 \qquad (27)$$

The value $\gamma_F = 1.0$ indicates that the failure load is identical with the result obtained by Hibbard and Adams. The corresponding story shear, $H = 0.81 \times 1.23 \times 328 = 327$ kips (1,455 kN). The factor $R = 0.92$ is relevant, since the rigid-plastic failure load of the actual story shown in Figure 16 varies from that of the "limiting story" in that the plastic moment capacities in the columns of Figure 16 are differently affected by the presence of axial forces.

Boundary Values

It is of interest to discuss some boundary values obtained for the example when using the multicurve interaction approach for three-dimensional structures in conjunction with torsional load. The range of possible cases can be described by varying the elastic buckling load factor in the *y*-direction of Figure 12. For $\gamma_{Cy} \rightarrow \infty$, the procedure reduces to a problem without torsion effects, i.e., when the failure load is that of a plane frame analysis in direction of *x*. For $\gamma_{Cy} \rightarrow 0$, the result obtained by the multicurve interaction method tends to suggest zero failure load as eccentric lateral load is applied, i.e., no torque can effectively be resisted since immediate failure of the structure would occur by "elastic buckling" in the unrestrained *y*-direction.

Story-by-Story Analysis of Multi-Story Structure

Subassemblage methods can be used on a story-by-story basis for multi-story structures, thereby greatly reducing the number of variables compared with an investigation of the full frame. The load factor of the weakest story is then taken as the load factor applicable to the entire framework.

A framework similar to the structure shown in Figures 3 and 4 was analysed in Reference 21. It is usually sufficient to consider only certain critical stories. One must bear in mind that often for practical reasons members are changed only every second or third story so that the lowest stories in such cases are the most critical ones.

Experiments

Very little experimental work is available on torsionally loaded frames. A single-story model framework subjected to torsion was recently tested by Scholz [21]. The experimental failure load exceeded the value predicted by the interaction technique by less then three percent.

CONCLUSION

A number of approximate elastic-plastic subassemblage analyses for sway frames subjected to torsion have been described. In contrast to full-frame methods it is argued that subassemblage procedures are useful for checking critical stories of a larger structure or even for a story-by-story analysis as a substitute for a full-frame investigation. It must be remembered that the investigation of the complete frame is a much larger and more expensive problem to analyze.

NOTATION

a = distance between CG and CR
CG = centroid of column loads
CR = center of rigidity
d = distance of columns from CG
E = modulus of elasticity
e = eccentricity between H and CR
e_i = eccentricity of frame i
f_p = stres at onset of yield
GA = shear rigidity
GJ = torsional rigidity
H = lateral load
h = story height
I = second moment of area
i = story number
K_L = lateral frame stiffness
L = member length
M = bending moment
M_p = fully-plastic moment
M_{PC} = plastic moment in presence of axial force P
P = axial compression
P_y = yield load
R = reduction factor
r = radius of gyration of section
r_G = distance from equivalent fictitious columns to CG
u_o, v_o = co-ordinates of translation
V = shear force
x,y = co-ordinates of distance

y = distance from centroid of section to extreme fiber
Z = elastic section modulus
α = factor defined by Equation (20)
γ = amplification factor to account for column bending
γ_C = elastic buckling load factor
γ_F = failure load factor
γ_P = plastic collapse load factor
Δ = sway
ϕ = story twist
θ = sway rotation
$\lambda_{HO.4}$ = parameter defined by Equation (21)
λ_ℓ = limiting slenderness ratio in limiting story defined by Equation (21)

REFERENCES

1. Cheong-Siat-Moy, F., "Control of Deflections in Unbraced Steel Frames," *Proceedings of the Institution of Civil Engineers,* London, England, Vol 57, Part 2, pp. 619–634 (Dec 1974).
2. Cheong-Siat-Moy, F., "Multistory Frame Design Using Story Stiffness Concept," *Journal of the Structural Division,* ASCE, Vol 102, No ST6, pp. 1197–1212 (June, 1976).
3. Cheong-Siat-Moy, F., "New Interaction Equation for Steel Beam-Columns," *Journal of the Structural Division,* ASCE, Vol 106, No ST5, pp. 1047–1061 (May, 1980).
4. Daniels, J. H. and L. W. Lu, "Design Charts for the Subassemblage Method of Designing Unbraced Multi-Story Frames," Fritz Laboratory Report No 273.54, Lehigh University, Bethlehem, Pa. (Nov 1966).
5. Driscoll, G. C., Jr., J. O. Armacost, and W. C. Hansell, "Plastic Design of Multi-Story Frames by Computer," *Journal of the Structural Division,* ASCE, Vol 96, No ST1, Proc. Paper 6995, pp. 17–33 (Jan 1970).
6. Disque, D. O., "Applied Plastic Design of Unbraced Multistory Frames," *Engineering Journal,* American Institute of Steel Construction, pp. 124–127 (Oct 1971).
7. Hibbard, J. R. and P. F. Adams, "Subassemblage Technique for Asymmetric Structures," Journal of the Structural Division, ASCE, Vol 99, No ST11, pp. 2259–2268 (Nov 1973).
8. LeMessurier, W. J., "Practical Method of Second Order Analysis," Parts I and II, Engineering Journal, American Institute of Steel Construction, No 4, pp. 89–96 (1976) and No 2, pp. 49–67 (1977).
9. Logcher, R. D., et al., *ICES STRUDL II, Engineering Users Manual,* Dept of Civil Engineering, Massachusetts Institute of Technology, Cambridge, Mass. (June 1969).
10. "Monograph on Planning and Design of Tall Buildings," Council on Tall Buildings, Committee 16, Stability, Vol SB ASCE, New York, NY, Chap. SB-4 (1979).
11. Nair, R. S., "Overall Elastic Stability of Multistory Buildings," Journal of the Structural Division, ASCE, Vol 101, No ST12, pp. 2487–2503 (Dec 1975).
12. "Plastic Design of Multistory Frames," Fritz Engineering Laboratory Report No 273.20, Lehigh University, Bethlehem, Pa. (1965).
13. *Plastic Design in Steel—A Guide and Commentary,* Manual No 41, 2nd Ed., ASCE, New York, NY (1971).
14. Rosenblueth, E., "Slenderness Effects in Buildings," *Journal of the Structural Division,* ASCE, Vol 91, No ST1, pp. 229–252 (Feb 1965).
15. Rutenberg A., "Simplified P-Delta Analysis for Asymmetric Structures," *Journal of the Structural Division,* ASCE, Vol 108, No ST9, pp. 1995–2013 (Sept 1982).
16. Rutenberg, A., "A Direct P-Delta Analysis using Standard Plane Frame Computer Programs," *Computers & Structures,* Vol 14, No 1-2, pp. 97–102 (1981).
17. Scholz, H., "A New Multi-Curve Interaction Method for the Plastic Analysis and Design of Sway Frames," *Proceedings of the Third International Colloquium on Stability of Metal Structures,* Toronto, Canada, pp. 491–495 (May 1983).
18. Scholz, H., "Overall Stability of Multi-Story Sway Frames Using a Story-by-Story Approach," *Proceedings of the Third International Colloquium on Stability of Metal Structures,* Paris, France, pp. 491–495 (Nov 1983).
19. Scholz, H., "Evolution of Approximate Analysis Technique for Unbraced Steel Frames," *The Civil Engineer in South Africa,* SAICE, Vol 26, No 12, pp. 587–594 (Dec 1984).
20. Scholz, H., "Interaction Analysis of Asymmetric Sway Subassemblages," *Journal of Structural Engineering,* ASCE, Vol 110, No 10, pp. 2412–2413 (Oct 1984).
21. Scholz, H., "Interaction Analysis of Asymmetric Sway Frames," Final Report, 12th IABSE Congress, Vancouver, p. 1080 (Sept 1984).
22. Springfield, J. and P. F. Adams, "Some Aspects of Column Design in Tall Steel Buildings," *Journal of the Structural Division,* ASCE, Vol 98, ST5, Proc. Paper 8911, pp. 1069–1083 (May 1972).
23. Weaver, W. and M. F. Nelsen, "Three-Dimensional Analysis of Tier Buildings," *Journal of the Structural Division,* ASCE, Vol 92, No ST6, Proc. Paper 5019, pp. 385–404 (Dec 1966).
24. Wilbur, J. B., *Distribution of Wind Loads to the Bents of a Building,* Boston Society of Civil Engineers, Vol 22, pp. 253–259 (1935).
25. Wynhoven, J. H., and P. F. Adams, "Analysis of Three-Dimensional Structures," *Journal of the Structural Division,* ASCE, Vol 98, No ST1, Proc. Paper 8643, pp. 233–248 (Jan 1972).

CHAPTER 13

Shear Effect in Beam Stiffness

G. ONU*

INTRODUCTION

The effect of the shearing force on the deflections of the beam in bending is expressed by reciprocal sliding of adjacent cross-sections along each other. Thus, the total rotation of the beam cross-sections depends not only on the slope of the deflection curve but also on the transverse deformation due to the shearing force. Additionally, as a result of the parabolic variation of the shearing strain along the depth of the beam, the cross-sections become warped. Due to these reasons a cross-section originally plane and normal to the beam axis remains neither plane during bending nor normal to the curved axis of the beam, as the basic hypothesis of the simple bending theory stipulates.

The effect of transverse shear deformation is immaterial for slender beams, but if the length/depth ratio is relatively small, this effect becomes important so that the results obtained in the simple bending theory are unsatisfactory. Also, when this ratio is small, the mutual sliding of neighbouring cross-sections due to the shearing force, significantly decreases the buckling resistance of the beam. Moreover, in the case of beams loaded with inertia forces, at high vibration modes, the shear effect is important even if the length/depth ratio is large. On the other hand, if one takes into account the fact that the deflection of a deep beam is affected by the transverse normal stresses and also by the manner in which the boundary conditions are defined, it should be necessary to resort to the elasticity theory to gather a rigorous solution [13,22,6].

However, it is possible to determine an acceptable solution in many practical problems by giving up one of the constraints imposed by Bernoulli's hypothesis [24,25]. In spite of that, indeed, the beam problem continues to remain within the limits of a one-dimensional problem. The modified hypothesis is: a straight line originally normal to the neutral axis of the beam will be straight but not normal after deformation. The angle γ in Figure 1, indicating the switched position of the cross-section with respect to the perpendicular to the neutral axis, emphasizes the shear effect when the modified hypothesis is taken into consideration. However, inclusion of both transverse shear deformation and rotatory inertia on the bases of the preceding modified hypothesis yields the "Timoshenko beam."

A literature review indicates that various formulations leading to beam finite elements having four or more degrees of freedom, have been derivated on the bases of the modified hypothesis of plane but not-normal cross-sections [1,3,5,8,11,12,14–18,20,21,23,27,28]. Among them, the four degrees of freedom elements are characterized by a constant shear deformation along the whole beam element. The main objective of this section is to show how a beam element like these last ones may be formulated by means of assumed displacement fields. The manner to take into account the effect of shear deformation without resorting to the finite element technique is shown too. However, the stiffness matrix gathered by using the slope-deflection method serves here as a benchmark solution for the other formulations.

PRELIMINARIES

The flexural or bending deflection W_f for the simple span beam under end moments shown in Figure 2a is governed by the differential equation [20,6]

$$EI\,\frac{d^2W_f}{dx^2} = V1\,x - M1 \tag{1a}$$

*Design Institute for Air, Road and Water Transports, Ministry of Transport and Communications, Bucharest, Romania

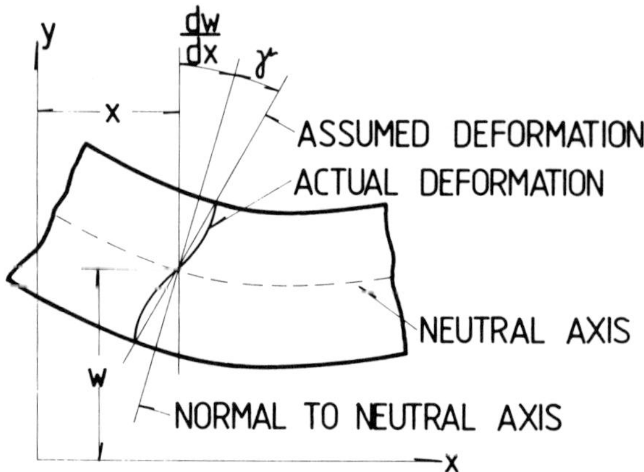

FIGURE 1. Deformation of beam cross-section.

or

$$EI\frac{d^2W_f}{dx^2} = \frac{x-1}{1}M1 + \frac{x}{1}M2 \tag{1b}$$

From integration of the above equation it follows that

$$EI\,W_f = \frac{(M1 + M2)\,x^3}{61} - \frac{M1\,x^2}{2} + C1\,x + C2 \tag{2}$$

in which $C1$ and $C2$ are constants of integration. Using the end conditions $W_f = 0$ at $x = 0$ and $x = 1$, then Equation (2) leads to the well known equation for the deflection curve of simple span beam under end moments when the shear effect is neglected:

$$EI\,W_f = \frac{1}{61}[31x(1 - x)M1 - (1^2 - x^2)x(M1 + M2)] \tag{3}$$

The total lateral deflection W of the beam of uniform cross-section subjected to bending moments and shearing forces is given [25] by

$$W = W_f + W_s \tag{4}$$

where W_f is the flexural deflection considered in the simple bending theory, due to bending strains and W_s is the additional deflection due to shearing strains alone, calculated on the approximation that the shear stress is uniformly distributed over the cross-section. As is shown by Donnell [6], the total lateral deflection of the beam is governed by the differential equation

$$W = \left(1 - \frac{EI}{nGA}\frac{d^2}{dx^2}\right)W_f + C1\,x + C2 \tag{5}$$

in which EI denotes the flexural rigidity of the beam;

A is the cross-sectional area;

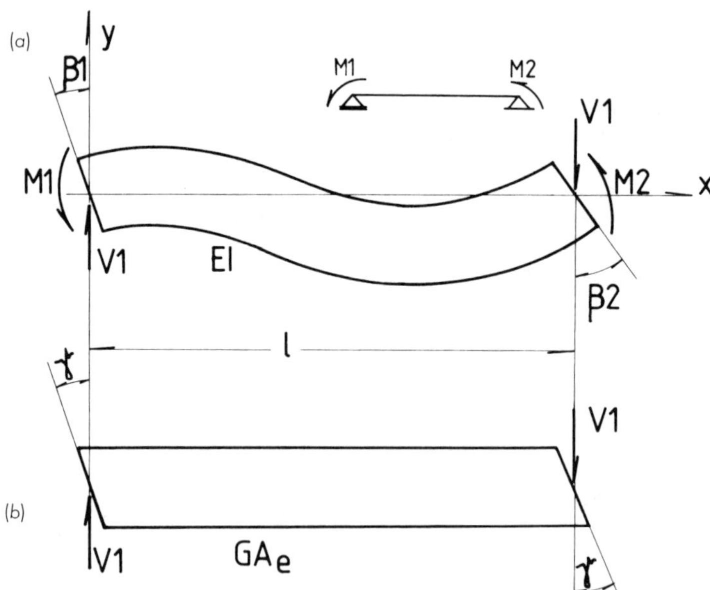

FIGURE 2. Deformations of simply supported beam under end moments: (a) the flexure effect; (b) the shear effect.

n is the shear coefficient or shape factor (see Appendix I);
G is modulus of elasticity in shear; and
$C1\ x$, $C2$ represent a rigid body rotation and translation, respectively.

In order to include the shear deformation in the equation for elastic line of simple span beam, the deflection W_f from Equation (3) is substituted for into Equation (5). After performing the required derivation Equation (5) becomes

$$W = \frac{1}{61EI}[31x(1 - x)M1 - (1^2 - x^2)x(M1 + M2) - \frac{6EI}{nGA}(xM1 - 1M1 + xM2)] + C1\ x + C2 \qquad (6)$$

Using the same boundary conditions as before, that is, $W = 0$ at $x = 0$ and $x = 1$, it results that the rigid body motion $C1\ x + C2$ is as follows:

$$C1\ x + C2 = \frac{1}{1nGA}(xM1 - 1M1 + xM2) \qquad (7)$$

Putting this value in Equation (6) then one obtains

$$W = W_f \qquad (8)$$

so that it can be concluded that in the case shown in Figure 2a the only effect of shearing force is a constant rotation of cross-sections along the whole beam as indicated in Figure 2b [13]. The angle of this shear rotation is provided by

$$\gamma = \frac{V1}{nGA} = \frac{M1 + M2}{1nGA} \qquad (9)$$

The above results suggest that the stiffness coefficients of the beam element, with shear effect included, can be derived directly from the slope-deflection equation if the shear deformation is included in the end rotations of the beam.

If β denotes the slope of the deflection curve when the shearing force is neglected and γ is an extra rotation due to the transverse shear effect, then the total rotation of cross-sections is [5].

$$\phi = \beta + \gamma \qquad (10)$$

The cross-sections rotations of the beam ends (see Figure 3a) are given by

$$\phi1 = \beta1 + \gamma$$
$$\phi2 = \beta2 + \gamma \qquad (11)$$

in which γ is defined by Equation (9) and the two end slopes, $\beta1$ and $\beta2$, can be obtained from the derivative of the equation of simply supported beam under end moments, i.e., Equation (3), at $x = 0$ and $x = 1$:

$$\beta1 = \frac{1}{6EI}(2M1 - M2)$$
$$\beta2 = \frac{1}{6EI}(-M1 + 2M2) \qquad (12)$$

Substituting for γ from Equation (9) and $\beta1$, $\beta2$ from Equations (12) into Equations (11) provides

$$\phi1 = \frac{1}{6EI}[2M1 - M2 + \frac{6EI}{1^2nGA}(M1 + M2)]$$
$$\phi2 = \frac{1}{6EI}[-M1 + 2M2 + \frac{6EI}{1^2nGA}(M1 + M2)] \qquad (13)$$

If the shear-deformation parameter t is defined by

$$t = \frac{12EI}{1^2GA_e} \qquad (14)$$

in which $A_e = nA$, A_e being the beam cross-sectional area effective in shear, then Equations (13) become

$$\phi1 = \frac{1}{12EI}[(4 + t)M1 - (2 - t)M2]$$
$$\phi2 = \frac{1}{12EI}[-(2 - t)M1 + (4 + t)M2] \qquad (15)$$

The solution of Equations (15) for $M1$ and $M2$ yields the end moments due to end cross-sections rotations, namely

$$M1 = \frac{EI}{(1 + t)1}[(4 + t)\phi1 + (2 - t)\phi2]$$
$$M2 = \frac{EI}{(1 + t)1}[(2 - t)\phi1 + (4 + t)\phi2] \qquad (16)$$

In order to determine the stiffness coefficients associated with the rotation $\phi1$, one must take $\phi1 = 1$ and $\phi2 = 0$ in Equations (16), so that

$$M1 = k^*_{11} = \frac{4 + t}{1 + t}\frac{EI}{1}$$
$$M2 = k^*_{12} = \frac{2 - t}{1 + t}\frac{EI}{1} \qquad (17)$$

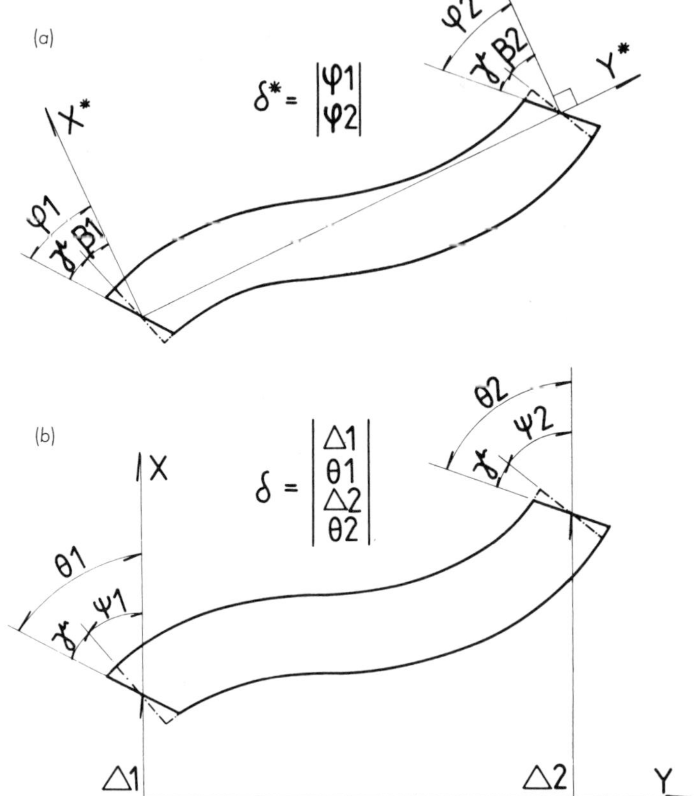

FIGURE 3. Characteristic displacements of simply supported beam: (a) with respect to the local coordinate system attached to the beam element; (b) with respect to the Global Coordinate System.

Similarly, for the right-hand end of beam, if $\phi 2 = 1$ and $\phi 1 = 0$ in Equations (16), then it is readily verified that

$$
\begin{aligned}
k^*_{22} &= k^*_{11} \\
k^*_{21} &= k^*_{12}
\end{aligned}
\tag{18}
$$

The two by two stiffness matrix in terms of the end cross-sections rotations, i.e., $\phi 1$ and $\phi 2$ (Figure 3a) is given by

$$
K^* = \frac{EI}{(1+t)1}\begin{bmatrix} 4+t & 2-t \\ 2-t & 4+t \end{bmatrix} \tag{19}
$$

The four by four stiffness matrix in terms of the nodal displacements, i.e., $\Delta 1$, $\theta 1$, $\Delta 2$, $\theta 2$ (Figure 3b), can be obtained from the stiffness matrix K^* by applying the following coordinate transformation:

$$
K = \lambda^T K^* \lambda \tag{20}
$$

The transformation matrix λ relates the nodal displacements (Figure 3b) to the end cross-sections rotations (Figure 3a) by the following matrix equation [2]:

$$
\delta^* = \lambda\, \delta \tag{21a}
$$

The above transformation in fully extended form may be written as

$$
\begin{bmatrix} \phi 1 \\ \phi 2 \end{bmatrix} = \begin{bmatrix} \frac{1}{1} & 1 & -\frac{1}{1} & 0 \\ \frac{1}{1} & 0 & -\frac{1}{1} & 1 \end{bmatrix} \begin{bmatrix} \Delta 1 \\ \theta 1 \\ \Delta 2 \\ \theta 2 \end{bmatrix} \tag{21b}
$$

where 1 is the beam element length.

The stiffness matrix, obtained from Equation (20), for the four degrees of freedom beam element, including the shear effect is as follows:

$$K = \frac{EI}{1^3(1+t)}\begin{bmatrix} 12 & 61 & -12 & 61 \\ 61 & (4+t)1^2 & -61 & (2-t)1^2 \\ -12 & -61 & 12 & -61 \\ 61 & (2-t)1^2 & -61 & (4+t)1^2 \end{bmatrix} \tag{22}$$

The location and positive directions of the nodal forces and their corresponding displacements related by the stiffness matrix given in Equation (22) are shown in Figure 4. The nodal displacements $U1$, $U2$, $U3$, and $U4$ in Figure 4 clearly correspond to $\Delta1$, $\theta1$, $\Delta2$, and $\theta2$, respectively, in Figure 3b.

However, if the total end slopes, namely $\psi1$ and $\psi2$ in Figure 3b, are taken for nodal rotations instead of $\theta1$ and $\theta2$, then a stiffness matrix different from the one in Equation (22) may be obtained. This stiffness matrix can be gathered from Equation (19) by a matrix transformation as it will be shown in Appendix 2.

PREVIOUS FORMULATIONS

The total strain energy in the Euler-Bernoulli theory for the beam element shown in Figure 4 can be written [8] as follows:

$$\pi = \frac{EI}{2}\left[\int_o^\ell \left(\frac{d^2W_f}{dx^2}\right)^2 dx + \frac{EI}{GA_e}\int_o^\ell \left(\frac{d^3W_f}{dx^3}\right)^2 dx\right] \tag{23}$$

In the above expression, W_f, the only descriptor of bending displacements, is represented by a cubic polynomial expansion. The first integral represents the contribution of flexural deformation and the second one the shear contribution. By minimization of the strain energy in Equation (23) a four by four stiffness matrix is gathered. This stiffness matrix is different from the one given in Equation (22) because the cross-section has been constrained to remain plane and perpendicular to the neutral axis of the beam. In fact, a beam element stiffer than the actual one is characterized by this matrix. For example, the stiffness coefficient k_{11} which relates $S1$ to $U1$ (see Figure 4) is obtained as $12EI(1 + t)/1^3$ instead of $12EI[1 - t/(1 + t)]/1^3$ as it is in Equation (22).

Another approximate solutions may be gathered either by introduction of separate assumptions for lateral displacement and cross-section rotation [26], or by adding the separate effects produced by flexure only and shear only analysis. Additional details regarding the above approximate solutions may be found in Reference [8].

However, there are several known ways of deriving the stiffness matrix in Equation (22) which consider the shear

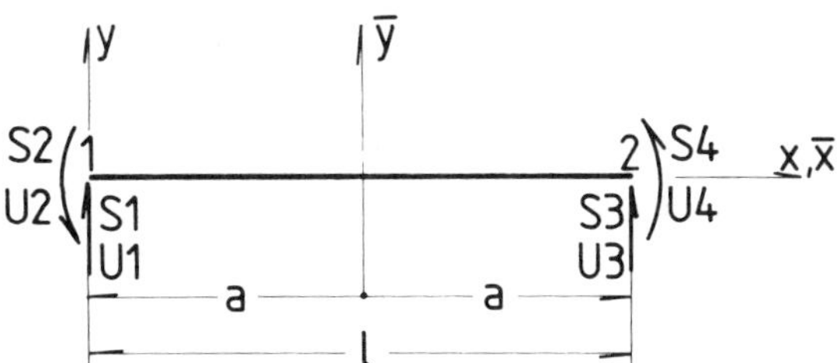

FIGURE 4. Nodal forces and displacements of a beam element.

contribution on the basis of the preceding modified hypothesis of plane but net-normal cross-sections. In this article, formulations leading to this stiffness matrix will be described in some details. Additionally, brief explanations of other formulations based on the modified hypothesis will be presented.

Flexibility-Stiffness Formulation

It is likely that the first approach to the formulation of the stiffness matrix given in Equation (22) is to form a flexibility matrix with shear contribution included [27]. Then, the stiffness matrix is obtained from the inverse of this flexibility matrix and an auxiliary matrix T proceeded from the element static equilibrium relationships. Thus, if the element shown in Figure 4 is fixed only at point 2, then the flexural deflection W_{1f} and the shear deflection W_{1s} at point 1 due to shearing force alone are given by

$$W_{1f} = \frac{S1\ 1^3}{3EI} \tag{24a}$$

$$W_{1s} = \gamma 1 = \frac{S1\ 1}{GA_e} = W_{1f}\,\frac{t}{4} \tag{24b}$$

in which γ and t are defined in Equations (9) and (14), respectively. Since the total flexibility matrix of the element is [8]:

$$d = \frac{1}{12EI}\begin{bmatrix} (4+t)1^2 & -61 \\ -61 & 12 \end{bmatrix} \tag{25}$$

then, the stiffness matrix in Equation (22) can be represented as

$$K = \begin{bmatrix} d^{-1} & d^{-1}T^T \\ Td^{-1} & Td^{-1}T^T \end{bmatrix} \tag{26}$$

The matrix T in the above equation derives from the static equilibrium of the element shown in Figure 4 as follows:

$$\begin{bmatrix} S3 \\ S4 \end{bmatrix} = \begin{bmatrix} -1 & 0 \\ 1 & -1 \end{bmatrix} \begin{bmatrix} S1 \\ S2 \end{bmatrix} = [T] \begin{bmatrix} S1 \\ S2 \end{bmatrix} \quad (27)$$

The interested reader may find further details of this procedure in Reference [8].

Formulation Based on the Differential Equations for Beam Deflections

The total lateral deflection W of a built-in beam subjected to shearing force and bending moments can be provided by Equation (4). The two constituent parts of W, namely, W_f the flexure deflection and W_s the deflection due to shearing force alone, are obtained from integration of the following differential Equations [20]:

$$\frac{dW_s}{dx} = -\frac{S1}{GA_e} \quad (28a)$$

$$EI \frac{d^2W_f}{dx^2} = S1\, x - S2 \quad (28b)$$

where $S1$ and $S2$ are as can be seen in Figure 4. Substituting for W_s and W_f from the above equations into Equation (4) and rearranging, gives

$$EI\, W = \frac{S1\, x^3}{6} - \frac{S2\, x^2}{2} + (C1 - \frac{S1\, EI}{GA_e})x + C2 \quad (29)$$

in which $C1$ and $C2$ are the constants of integration. Using proper end conditions in Equation (29), the stiffness coefficients in Equation (22) are determined. It must be underlined that the boundary condition for the built-in end in the engineering beam theory is taken as $dW_f/dx = 0$. Przemieniecki [20] provides detailed derivations of the stiffness coefficients.

The first formulations based on the differential equations for beam deflections were by McCalley [14,15] (uniform beam element) and by Archer [1] (tapered beam element).

Assumed Stress Hybrid Approach

For a beam element having a bending moment with linear variation and a constant shearing force, the stress field may be defined by the following matrix relationship:

$$\begin{bmatrix} m \\ f \end{bmatrix} = \begin{bmatrix} 1 & x \\ 0 & 1 \end{bmatrix} \begin{bmatrix} \tilde{C}1 \\ \tilde{C}2 \end{bmatrix} \quad (30)$$

where m and f denote the bending moment and shearing force, respectively, along the beam element, and $\tilde{C}1$, $\tilde{C}2$ are generalized parameters. The element nodal forces (see Figure 4) can be expressed in terms of $\tilde{C}1$ and $\tilde{C}2$ by suitable evaluation of Equation (30) at the two ends of element:

$$\begin{bmatrix} S1 \\ S2 \\ S3 \\ S4 \end{bmatrix} = \begin{bmatrix} 0 & -1 \\ 1 & 0 \\ 0 & 1 \\ -1 & -1 \end{bmatrix} \begin{bmatrix} \tilde{C}1 \\ \tilde{C}2 \end{bmatrix} \quad (31a)$$

or, in shorthand form

$$S = L^T\tilde{C} \quad (31b)$$

The complementary strain energy for the beam element shown in Figure 4 can be written as

$$\pi^* = \frac{1}{2EI}\int_o^\ell m^2dx + \frac{1}{2GA_e}\int_o^\ell f^2dx \quad (32a)$$

where the first integral represents the bending contribution and the second one the shear contribution. The last equation in shorthand form is

$$\pi^* = \frac{1}{2}\tilde{C}^T\, d\, \tilde{C} \quad (32b)$$

in which

$$d = \int_o^\ell \begin{bmatrix} 1 & 0 \\ x & 1 \end{bmatrix} \begin{bmatrix} \frac{1}{EI} & 0 \\ 0 & \frac{1}{GA_e} \end{bmatrix} \begin{bmatrix} 1 & x \\ 0 & 1 \end{bmatrix} dx$$

$$= \frac{1}{12EI}\begin{bmatrix} 12 & 6l \\ 6l & (4 + t)l^2 \end{bmatrix} \quad (33)$$

and t is defined by Equation (14). It is readily verified that the stiffness matrix in Equation (22) can be obtained from

$$K = L^Td^{-1}L \quad (34)$$

where L is given by Equation (31b). For further details see Reference [8].

The first formulation based on this approach was by Severn [21]. He has derived both the stiffness matrix in Equation (22) and the one in Appendix 2.

Formulation Based on the Differential Equations of an Infinitesimal Element

This formulation has been presented by Davis, Henshell and Warburton in Reference [5]. An infinitesimal element of beam is in static equilibrium under forces and moments shown in Figure 5. The exact equilibrium equations are as follows:

$$\frac{dM}{dx} = F$$
$$\frac{dF}{dx} = 0 \tag{35}$$

The shear angle γ is given by

$$\gamma = \frac{F}{GA_e} \tag{36}$$

The stress-strain relation can be written as

$$\frac{M}{EI} = \frac{d^2W}{dx^2} + \frac{d\gamma}{dx} \tag{37}$$

In the last two equations GA_e and EI are the shear rigidity and flexural rigidity, respectively. After integrations and suitable substitutions, the above four equations become

$$F = C1$$
$$\gamma = \frac{C1}{GA_e}$$
$$M = C1\,x + C2 \tag{38}$$
$$EI\,W = (C1\,\frac{x^3}{6} + C2\,\frac{x^2}{2} + C3\,x + C4)$$

These equations and Equation (10) may be used to represent the nodal forces and displacements of the beam finite element in Figure 4 in terms of constants $C1$ to $C4$.

The nodal forces can be written as follows:

$$S = T1\,C \tag{39}$$

where

$$T1 = \begin{bmatrix} 1 & 0 & 0 & 0 \\ 0 & -1 & 0 & 0 \\ -1 & 0 & 0 & 0 \\ 1 & 1 & 0 & 0 \end{bmatrix} \tag{40}$$

and $S = [S1, S2, S3, S4]^T$ is as can be seen in Figure 4. The nodal displacements are given by the following matrix equation

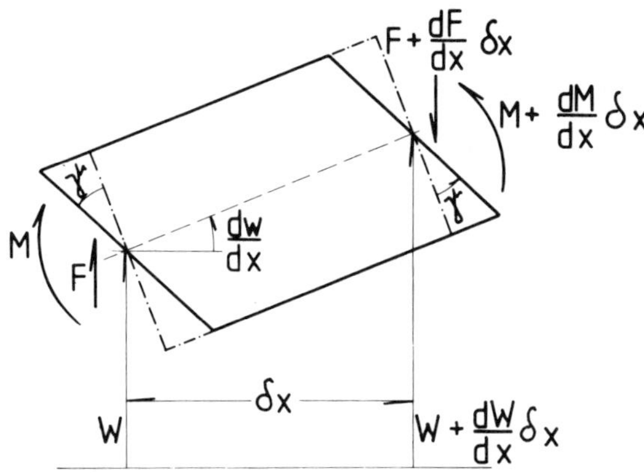

FIGURE 5. Infinitesimal element of beam under end forces and moments.

$$U = \frac{1}{EI}\,T2\,C \tag{41}$$

where

$$T2 = \frac{1}{12}\begin{bmatrix} 0 & 0 & 0 & 12 \\ t\,1^2 & 0 & 12 & 0 \\ 21^3 & 61^2 & 121 & 12 \\ (6+t)1^2 & 121 & 12 & 0 \end{bmatrix} \tag{42}$$

t is defined in Equation (14) and $U = [U1, U2, U3, U4]^T$ is as can be seen in Figure 4. Substituting for C from Equation (41) in Equation (39) then provides

$$S = EI\,T1\,T2^{-1}U = K\,U \tag{43}$$

where K is the stiffness matrix defined in Equation (22).

Other Formulations

The beam finite element with two nodal points derived by Kapur [12] is based on a transverse displacement function made up of two components, that is, W_f (flexural deflection) and W_s (shear deflection) as given in Equation (4). A cubic polynomial expansion for each of the two components is assumed. Four degrees of freedom (d.o.f.) at each of the nodal points, i.e., W_f, W_s, dW_f/dx and dW_s/dx, characterize the behaviour of the element. The expression for the total strain energy is [15] as follows:

$$\pi = \frac{EI}{2}\int_o^\ell \left(\frac{d^2W_f}{dx^2}\right)^2 dx + \frac{GA_e}{2}\int_o^\ell \left(\frac{dW_s}{dx}\right)^2 dx \tag{44}$$

Therefore, the resulting eight by eight stiffness matrix is uncoupled.

Carnegie, Thomas and Dokumaci [3] propose an element with four nodal points equally spaced, having two d.o.f. per node namely W (transverse displacement) and θ (cross-section rotation). The displacement functions for both W and θ are described by cubic expansions. The eight by eight stiffness matrix is obtained by using the following form of total strain energy:

$$\pi = \frac{EI}{2}\int_o^\ell \left(\frac{d\theta}{dx}\right)^2 dx + \frac{GA_e}{2}\int_o^\ell \left(\frac{dW}{dx} - \theta\right)^2 dx \tag{45}$$

As is shown by Zienkiewicz [28] the second integral in Equation (45) may be regarded as a constraint introduced in the total potential energy functional for simple beam when the basic descriptor of flexural displacement is θ. This constraint, namely, $dW/dx - \theta = 0$, is inserted in the functional by the penalty functions method, so that the strain energy ends to be defined by two independent variables θ and W.

In their formulation Nickel and Secor [16] use a functional derived from the variational principles introduced by Gurtin [10]. One of the functional terms represents the total strain energy due to flexural and shear deformation. The basic descriptors for strain energy are W (transverse displacement) and θ (cross-section rotation) as in Equation (45). Using this functional, the authors derive two finite beam elements, the first of them called TIM7 having seven d.o.f. and the second one TIM4 having four. The displacement functions of TIM7 element use a cubic polynomial expansion for W and a quadratic one for θ. The nodal displacements at the end nodes are W, dW/dx and θ. The seventh d.o.f. is the total cross-section rotation at the middle of the beam element. The resulting seven by seven stiffness matrix is given explicitly. TIM4 element is formulated on the bases of a constraint relationship consistent with Bernoulli's hypothesis. The constraint requires the shearing force F were obtainable from the bending moment derivative, so that

$$F = -EI\frac{d^2\theta}{dx^2} = GA_e\left(\frac{dw}{dx} - \theta\right) \tag{46}$$

where W (flexural deformation) and θ (cross-section rotation) are described by cubic and quadratic expansions, respectively. The generalized coordinates in θ function are related to nodal displacements in W, i.e., transverse displacement and slope, through the above constraint equation. A four by four stiffness matrix is derived from the strain energy given in Equation (23) by taking θ instead of dW/dx. This stiffness matrix is identical to the one derived in Appendix 2.

The six d.o.f. beam element proposed by Thomas, J. M. Wilson and R. R. Wilson [23] has as nodal variables W (transverse displacement), θ (cross-section rotation) and γ (shear angle) at each of the two nodes of the element. The displacement functions are as follows:

$$\begin{bmatrix} W \\ \theta \\ \gamma \end{bmatrix} = \begin{bmatrix} 1 & x & x^2 & x^3 & 0 & 0 \\ 0 & 1 & 2x & 3x^2 & 1 & x \\ 0 & 0 & 0 & 0 & 1 & x \end{bmatrix} \begin{bmatrix} \tilde{C}1 \\ \tilde{C}2 \\ \tilde{C}3 \\ \tilde{C}4 \\ \tilde{C}5 \\ \tilde{C}6 \end{bmatrix} \tag{47}$$

The corresponding stiffness matrix for the beam element is derived by using the following form of the total strain energy:

$$\pi = \frac{1}{2}\int_o^\ell EI\left(\frac{d\theta}{dx}\right)^2 dx + \frac{1}{2}\int_o^\ell GA_e\gamma^2\, dx \tag{48}$$

where I and A_e are taken for an element which is tapered such that the sectional properties vary linearly with distance along the beam, i.e.,

$$I = I1 + (I2 - I1)\frac{x}{1}$$
$$A_e = A1 + (A2 - A1)\frac{x}{1} \tag{49}$$

Using a constraint equation in conjunction with a static condensation procedure the six by six stiffness matrix given in Reference [23] may be converted into the four by four stiffness matrix given in Equation (22). All these will be shown in Appendix 3.

The shear effect may be included in the beam stiffness matrix by means of the isoparametric technique. Hinton and Owen [11] give detailed explanations of how the quadratic isoparametric element can be formulated. Each of the three nodal points of the beam element described in [11] has two d.o.f., namely, W (transverse displacement) and θ (cross-section rotation). An interpolation function is assigned to each node so that W, θ and x-coordinate are defined in a common manner. For example, the current cross-section rotation $\theta(\xi)$ is given by

$$\theta(\xi) = \Sigma N_i\,\theta_i \qquad (i = 1, 2, 3) \tag{50}$$

where N_i and θ_i are the interpolation function and cross-section rotation, respectively, at the same node i. ξ is the natural coordinate. The six by six stiffness matrix is calculated by numerical integration from the strain energy given in Equation (45).

A NEW FORMULATION FOR THE BEAM ELEMENT

The purpose of this article is to describe the procedure to obtain the stiffness matrix given in Equation (22) by means of assumed displacement fields [17,18]. Additionally, explanations of how the geometric stiffness matrix, load vector and mass matrix can be derived from the same displacement fields are included.

Interpolation Functions

The behaviour of the beam element, as seen in Figure 4, is described by four d.o.f., i.e., $U1$, $U2$, $U3$, and $U4$. There are two characteristic modes of flexure deformation (Figure 6) for this element having uniform properties along its length and loaded only at nodes. The polynomials used for the analytical representation of the two deformed configurations must be consistent with the force representation established in Figure 4. This implies the possibility of obtaining a bending moment with linear variation and a constant shear along the beam for the antisymmetrical component of the load, as well as a constant bending moment and a zero shear for the symmetrical component. Polynomials

$$W1 = -\frac{\bar{x}}{4} + \frac{\bar{x}^3}{4a^2}$$
$$W2 = -\frac{a}{4} + \frac{\bar{x}^2}{4a} \tag{51}$$

fulfill these requirements.

The rigid body motions of the element, a translation normal to the beam axis and a rotation, may be adequately represented by

$$W3 = \frac{1}{2}$$
$$W4 = -\frac{\bar{x}}{2} \tag{52}$$

If Bernoulli's hypothesis is entirely accepted, the two characteristic modes of deformations (Equation 51) and the rigid body motions (Equation 52) are sufficient for the construction of the shape functions of the transverse displacement field, i.e.:

$$N1 = W3 + \frac{W1 + W4}{a} = \frac{1}{2} - \frac{3\bar{x}}{4a} + \frac{\bar{x}^3}{4a^3}$$
$$N2 = W1 + W2 = \frac{a}{4} - \frac{\bar{x}}{4} - \frac{\bar{x}^2}{4a} + \frac{\bar{x}^3}{4a^2}$$
$$N3 = W3 - \frac{W1 + W4}{a} = \frac{1}{2} + \frac{3\bar{x}}{4a} - \frac{\bar{x}^3}{4a^3}$$
$$N4 = W1 - W2 = -\frac{a}{4} - \frac{\bar{x}}{4} + \frac{\bar{x}^2}{4a} + \frac{\bar{x}^3}{4a^2} \tag{53}$$

Renouncing the requirement according to which the elastic axis slope must be equal to the cross-sectional rota-

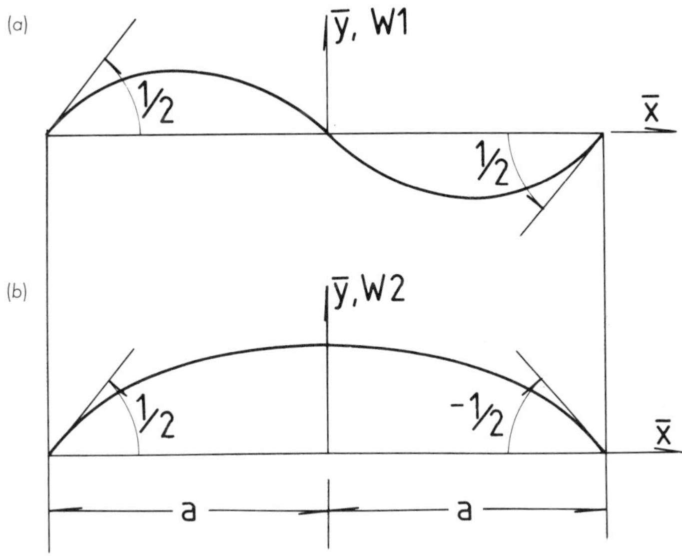

FIGURE 6. Deformated configurations of beam element under nodal rotations: (a) antisymmetrical flexure mode; (b) symmetrical flexure mode.

tion, the deformation mode characterizing the shear must be considered. Since the shearing force is constant along the beam and the cross-sections remain planes (by virtue of the modified hypothesis of plane but not-normal cross-sections), the displaced shape of beam element due to shear loading is the one shown in Figure 2b. As the shear producing this displaced shape is the consequence of an antisymmetrical flexure, it results in the characteristic deformation mode for shearing always accompanying the antisymmetrical flexure mode, and it is resonable to admit that these two deformation modes are defined by the same parameter, $U5$. This is illustrated in Figure 7. In contrast with the parameters $U1$ to $U4$ that point out independent nodal displacements, the parameter $U5$ emphasizes a reciprocal rotation of the end cross-sections due to both the antisymmetrical flexure and shearing force. Under these conditions, the deformed configuration shown in Figure 7a represents the deflection increase obtained when the shear deformation is allowed, $U5$ gets a character of internal d.o.f. and

$$N5 = -\frac{W1}{a} = \frac{\bar{x}}{4a} - \frac{\bar{x}^3}{4a^3} \qquad (54)$$

becomes the fifth shape function of the transverse displacement field.

The two independent variables of the problem, that is, the lateral displacement W and the rotation γ due to the shearing force, can be expressed in matrix form as

$$\begin{bmatrix} W \\ \gamma \end{bmatrix} = \begin{bmatrix} N_w \\ N_\gamma \end{bmatrix} [Z] \qquad (55)$$

in which

$$N_w = (N1, N2, N3, N4, N5] \qquad (56a)$$

$$N_\gamma = [0\ ,0\ ,0\ ,0\ ,N6] \qquad (56b)$$

$$Z^T = [U1, U2, U3, U4, U5] \qquad (57)$$

As it has been assumed, the angular deformation γ is constant along the depth of cross-section ($\gamma = S1/(GA_e)$). Because the shear force $S1$ (see Figure 4) is also constant along the beam, it follows that $N6$ is a constant too and its value may be carried out from a boundary condition, e.g., for $\bar{x} = a$, $dW/d\bar{x} + \gamma - U4 = 0$, where $U4$ is the total rotation of cross-section at the right-hand end of beam. Therefore

$$U4 = \gamma + \frac{dW}{d\bar{x}} = N6\ U5 + U4 - \frac{U5}{2a} \qquad (58)$$

Thus,

$$N6 = \frac{1}{2a} \qquad (59)$$

A reduced form of the displacement field given in equation (55) will be derived in Appendix 4.

Stiffness Matrix

Disregarding Bernoulli's hypothesis, it is permissible to append a separate term representing the strain energy due to transverse shear to the element energy. Such an addition of the energy of shear to the energy due to normal stresses is justified because the shearing stresses do not change the amount of energy due to tension or compression [25]. In this way, the strain energy of the beam element is defined through the agency of the two independent variables W and γ. Taking into consideration the fact that the shear deformation is constant along the beam element, the strain energy expression for the case studied here becomes:

$$\pi = \frac{1}{2}\int_{-a}^{a}\left(\frac{d^2W}{d\bar{x}^2}\right)^2 EI\,d\bar{x} + \frac{1}{2}\int_{-a}^{a} GA_e\,\gamma^2\,d\bar{x} + \frac{P}{2}\int_{-a}^{a}\left(\frac{dW}{d\bar{x}}\right)^2 d\bar{x} \qquad (60)$$

The first integral leads to the conventional stiffness matrix of the beam element. The second integral is the contribution from the shear component of the strain. The third integral gives the geometric stiffness matrix and accounts for the work of axial tensile force P acting through bending displacement. Substituting for W and γ from Equation (55) into Equation (60) provides

$$\pi = \frac{Z^T}{2}\left(\int_{-a}^{a} EI\left[\frac{d^2N_w}{d\bar{x}^2}\right]^T\left[\frac{d^2N_w}{d\bar{x}^2}\right] d\bar{x} + \int_{-a}^{a} GA_e\,[N_\gamma]^T[N_\gamma]\,d\bar{x} + P\int_{-a}^{a}\left[\frac{dN_w}{d\bar{x}}\right]^T\left[\frac{dN_w}{d\bar{x}}\right] d\bar{x}\right) Z \qquad (61)$$

which, after performing the integration, becomes

$$\pi = \frac{1}{2} Z^T \bar{K}_t Z \tag{62}$$

The five by five stiffness matrix $\bar{K}_t$ relates the nodal forces and their corresponding displacements, S and Z, respectively, by the following matrix relationship:

$$S = \bar{K}_t Z \tag{63}$$

The first four entries of the vector S are shown in Figure 4 and the last one represents the force acting on the direction of the degree of freedom $U5$ in Figure 7.

For convenience of presentation, the above stiffness matrix can be separated into two parts, so that

$$\bar{K}_t = \bar{K} + \bar{K}_G \tag{64}$$

The stiffness matrix $\bar{K}$ groups together the stiffness coefficients that hang on the elastic properties of the element. $\bar{K}$ is the elastic stiffness matrix. However, $\bar{K}_G$ is independent of elastic properties and is instead a function of the element length and its axial force P. This matrix, termed geometric stiffness matrix, accounts for the effect of compressive or tensive axial force in reducing or increasing, respectively, the beam resistance to lateral loading. The two stiffness matrices, i.e., $\bar{K}$ and $\bar{K}_G$, will be examined in turn.

Elastic Stiffness Matrix

The extended form of elastic stiffness matrix in Equation (64) is as follows:

$$\bar{K} = \frac{EI}{8a^3}\begin{bmatrix} 12 & 12a & -12 & 12a & -12 \\ 12a & 16a^2 & -12a & 8a^2 & -12a \\ -12 & -12a & 12 & -12a & 12 \\ 12a & 8a^2 & -12a & 16a^2 & -12a \\ -12 & -12a & 12 & -12a & 12 + \dfrac{4GA_e a^2}{EI} \end{bmatrix} \tag{65}$$

The above can be partitioned into

$$\bar{K} = \begin{bmatrix} K11 & K12 \\ K12^T & K22 \end{bmatrix} \tag{66}$$

where $K11$ is in fact the conventional stiffness matrix for the four-d.o.f. beam element (assuming that no shear occurs).

Eliminating the internal d.o.f. $U5$, by static condensation

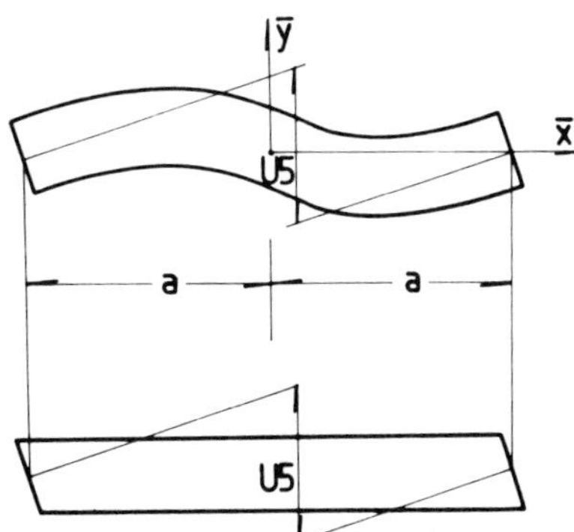

FIGURE 7. Additional degree of freedom.

with zero force, a four by four reduced matrix, including the effect due to the shear deformation, is defined by

$$K = R^T \bar{K} R \tag{67}$$

The stiffness matrix K gathered in the above is identical to the one in Equation (22).

The matrix R in Equation (67) may be given in the following matrix notation:

$$R^T = [I4, \ -K12\ K22^{-1}] \tag{68}$$

in which $I4$ is the four by four unity matrix and $K12$, $K22$ are submatrices in Equation (66). The matrix R in expanded form is

$$R = \frac{1}{1+t}\begin{bmatrix} 1+t & 0 & 0 & 0 \\ 0 & 1+t & 0 & 0 \\ 0 & 0 & 1+t & 0 \\ 0 & 0 & 0 & 1+t \\ t & at & -t & at \end{bmatrix} \tag{69}$$

where t denotes the shear-deformation parameter employed before (Equation 14).

Geometric Stiffness Matrix

The five by five geometric stiffness matrix $\bar{K}_G$ in Equation (64) is presented by

$$\bar{K}_G = \frac{P}{30a}\begin{bmatrix} 18 & 3a & -18 & 3a & -3 \\ 3a & 8a^2 & -3a & -2a^2 & -3a \\ -18 & -3a & 18 & -3a & 3 \\ 3a & -2a^2 & -3a & 8a^2 & -3a \\ -3 & -3a & 3 & -3a & 3 \end{bmatrix} \tag{70}$$

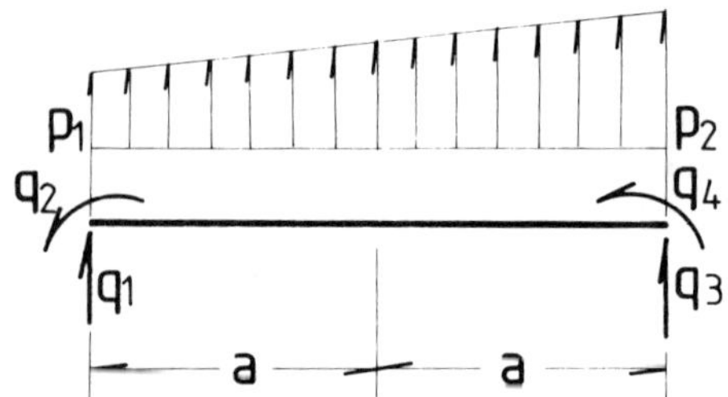

FIGURE 8. Beam element under distributed loading and equivalent concentrated forces.

$\bar{K}_G$ matrix may be partitioned into

$$\bar{K}_G = \begin{bmatrix} \check{K}11 & \check{K}12 \\ \check{K}12^T & \check{K}22 \end{bmatrix} \tag{71}$$

where $\check{K}11$ represents the geometric stiffness matrix for the four-d.o.f. beam element without shear effect [7].

Eliminating the unwanted d.o.f., i.e., $U5$, by static condensation, a four by four reduced matrix, including the effect due to shear, is defined by

$$K_G = R^T \bar{K}_G R \tag{72}$$

Performing the required operations, the matrix K_G may be represented as

$$K_G = \frac{P}{60(1+t)^2 1} \begin{bmatrix} G1 & G2 & -G1 & G2 \\ G2 & G3 & -G2 & -G4 \\ -G1 & -G2 & G1 & -G2 \\ G2 & -G4 & -G2 & G3 \end{bmatrix} \tag{73}$$

in which the entries are as follows:

$$\begin{aligned} G1 &= 72 + 120t + 60t^2 \\ G2 &= 61 \\ G3 &= (8 + 10t + 5t^2)\, 1^2 \\ G4 &= (2 + 10t + 5t^2)\, 1^2 \end{aligned} \tag{74}$$

In the above relationships, $1 = 2a$ and t is given in Equation (14).

The geometric stiffness matrix in Equation (73) and the finite displacement matrix in Reference [1] are identical.

Consistent Load Vector

The stiffness equations for the five-d.o.f. beam element under a distributed load (excluding other types of special forces) can be written [8] as

$$\bar{S} = \bar{K} Z - \bar{Q} \tag{75}$$

in which $\bar{K}$ and Z have been defined previously in Equation (65) and Equation (57), respectively, and $\bar{Q}$ is an equivalent load vector. This vector can be determined [20] from the following relationship

$$\bar{Q} = \int_{-a}^{a} [N_w]\, p(\bar{x})\, d\bar{x} \tag{76}$$

in which $p(\bar{x})$ stands for a distributed pressure acting on the beam element. The load vector $\bar{Q}$ is consistent with the basic stiffness matrix $\bar{K}$ if the same set of shape functions, i.e., $[N_w]$, are used both in $\bar{K}$ and in $\bar{Q}$.

Equation (76) may be used to transform into equivalent concentrated forces any distributed loading $p(\bar{x})$ acting on the beam element. As an example one considers a beam element subjected to a distributed loading as shown in Figure 8. Using Equation (76), the consistent load vector can be obtained from

$$\bar{Q} = \int_{-a}^{a} [N_w] \left(\frac{p2 + p1}{2} + \frac{p2 - p1}{2} \frac{\bar{x}}{a} \right) d\bar{x} \tag{77}$$

where $[N_w]$ is defined in Equation (56a). If necessary integrations are carried out in Equation (77) then $\bar{Q}$ becomes:

$$\bar{Q} = \frac{a}{30} \begin{bmatrix} 15 & -6 \\ 5a & -a \\ 15 & 6 \\ -5a & -a \\ 0 & 1 \end{bmatrix} \begin{bmatrix} p2 + p1 \\ p2 - p1 \end{bmatrix} \tag{78}$$

Eliminating the unwanted d.o.f., i.e., $U5$, a reduced consistent load vector, including the effect of shear deformation, is defined by the following matrix relationship

$$Q = R^T \bar{Q} \tag{79a}$$

where the transformation matrix R is given in Equation (69). By matrix multiplication Equation (79a) yields the components of reduced consistent load vector Q as

$$\begin{bmatrix} q1 \\ q2 \\ q3 \\ q4 \end{bmatrix} = \frac{1}{120(1+t)} \begin{bmatrix} (42 + 40t) & (18 + 20t) \\ (6 + 5t)1 & (4 + 5t)1 \\ (18 + 20t) & (42 + 40t) \\ -(4 + 5t)1 & -(6 + 5t)1 \end{bmatrix} \begin{bmatrix} p1 \\ p2 \end{bmatrix} \tag{79b}$$

In the above, $1 = 2a$ and t is the shear-deformation parameter in Equation (14).

The result obtained in Equation (79b) is identical to the one gathered by Archer [1].

Consistent Mass Matrix

The kinetic energy of a prismatic beam element, vibrating at circular frequency ω, including the effects of the shear deformation and rotatory inertia, can be written [5] as

$$\Omega = \frac{\omega^2}{2}\varrho A \int_{-a}^{a}\left[W^2 + r^2\left(\frac{dW}{d\bar{x}} + \gamma\right)^2\right]d\bar{x} \qquad (80)$$

in which r is the radius of gyration of the beam cross-section and ϱ represents the material density.

Substituting for W and γ from Equation (55) into Equation (80) then provides

$$\Omega = Z\frac{\omega^2}{2}\varrho A\left(\int_{-a}^{a}[N_w]^T[N_w]\,d\bar{x} + r^2 \int_{-a}^{a}\left[\frac{dN_w}{d\bar{x}} + N_\gamma\right]^T\left[\frac{dN_w}{d\bar{x}} + N_\gamma\right]d\bar{x}\right)Z \qquad (81)$$

Integrating with respect to $\bar{x}$ then gives

$$\Omega = \frac{\omega^2}{2}Z^T\,\bar{M}\,Z \qquad (82)$$

in which $\bar{M}$, a five by five mass matrix, can be written in partitioned form as

$$\bar{M} = \varrho A\left(\begin{bmatrix} M11 & M12 \\ M12^T & M22 \end{bmatrix} + r^2\begin{bmatrix} \tilde{M}11 & \tilde{M}12 \\ \tilde{M}12 & \tilde{M}22 \end{bmatrix}\right) \qquad (83)$$

The first matrix in Equation (83) represents the translational mass inertia, while the second one represents the rotatory inertia. The four by four mass matrix

$$M_c = \varrho A(M11 + r^2\tilde{M}11) \qquad (84)$$

in which $M11$ and $\tilde{M}11$ are submatrices in Equation (83), is the conventional mass matrix for the four-d.o.f. beam element without shear effect.

The fully expanded form of the mass matrix $\bar{M}$ in Equation (83) is presented by

$$\bar{M} = \frac{\varrho Aa}{210}\begin{bmatrix} 156 & 44a & 54 & -26a & -9 \\ 44a & 16a^2 & 26a & -12a^2 & -2a \\ 54 & 26a & 156 & -44a & 9 \\ -26a & -12a^2 & -44a & 16a^2 & -2a \\ -9 & -2a & 9 & -2a & 2a \end{bmatrix} + \frac{\varrho Ar^2}{30a}\begin{bmatrix} 18 & 3a & -18 & 3a & -18 \\ 3a & 8a^2 & -3a & -2a^2 & -3a \\ -18 & -3a & 18 & -3a & 18 \\ 3a & -2a^2 & -3a & 8a^2 & -3a \\ -18 & -3a & 18 & -3a & 18 \end{bmatrix} \qquad (85)$$

Eliminating the unwanted d.o.f. $U5$, a four by four consistent mass matrix, including the effects of shear deformation and rotary inertia, is defined, in compact form, by the following matrix product:

$$M = R^T\,\bar{M}R \qquad (86)$$

In the above, R is the same matrix as the one used to obtain the four by four stiffness matrix (Equation (69)) and $\bar{M}$ is presented in Equation (85). If necessary matrix multiplications are carried out in Equation (86), the consistent mass matrix is obtained as follows:

$$M = \frac{\varrho A1}{(1+t)^2}\left(\begin{bmatrix} \bar{N}1 & \bar{N}2 & \bar{N}3 & -\bar{N}4 \\ \bar{N}2 & \bar{N}5 & \bar{N}4 & -\bar{N}6 \\ \bar{N}3 & \bar{N}4 & \bar{N}1 & -\bar{N}2 \\ -\bar{N}4 & -\bar{N}6 & -\bar{N}2 & \bar{N}5 \end{bmatrix} + \left(\frac{r}{1}\right)^2\begin{bmatrix} \bar{N}7 & \bar{N}8 & -\bar{N}7 & \bar{N}8 \\ \bar{N}8 & \bar{N}9 & -\bar{N}8 & \bar{N}10 \\ -\bar{N}7 & -\bar{N}8 & \bar{N}7 & -\bar{N}8 \\ \bar{N}8 & \bar{N}10 & -\bar{N}8 & \bar{N}9 \end{bmatrix}\right) \qquad (87)$$

The entries in Equation (87) are given by

$$\bar{N}1 = \frac{13}{35} + \frac{7}{10}t + \frac{1}{3}t^2$$

$$\bar{N}2 = \left(\frac{11}{210} + \frac{11}{210}t + \frac{1}{24}t^2\right)1$$

$$\bar{N}3 = \frac{9}{70} + \frac{3}{10}t + \frac{1}{6}t^2$$

$$\bar{N}4 = \left(\frac{13}{420} + \frac{3}{40}t + \frac{1}{24}t^2\right)1$$

$$\bar{N}5 = \left(\frac{1}{105} + \frac{1}{60} t + \frac{1}{120} t^2 \right) l^2$$

$$\bar{N}6 = \left(\frac{1}{140} + \frac{1}{60} t + \frac{1}{120} t^2 \right) l^2$$

$$\bar{N}7 = \frac{6}{5}$$

$$\bar{N}8 = \left(\frac{1}{10} - \frac{1}{2} t \right) l$$

$$\bar{N}9 = \left(\frac{2}{15} + \frac{1}{6} t + \frac{1}{3} t^2 \right) l^2$$

$$\bar{N}10 = \left(- \frac{1}{30} - \frac{1}{6} t + \frac{1}{6} t^2 \right) l^2 \qquad (88)$$

The consistent mass matrix in Equation (87) is the same as the mass matrix in References [1,15,20].

CONCLUSIONS

A procedure for the formulation of the stiffness matrix for the four-d.o.f. beam element, in terms of lateral displacements and total rotations of end cross-sections, including the shear effect by coupling the transverse displacement field and the shear angular displacement field, has been presented. The mass matrix, the geometric stiffness matrix and the load vector, formulated consistent with the basic stiffness matrix, by using the same procedure, have been carried out. As it has been proved in Reference [19], this procedure may be extended to other finite elements in so far as the characteristic deformation modes for shearing and proper flexure modes can be emphasized.

Other formulations leading to the stiffness matrices expressed both in terms of lateral displacements and total rotations of end cross-sections and in terms of lateral displacements and slopes of a beam element ends, have been included in this section too.

NOTATION

The following symbols are used in this chapter:

A = cross-sectional area
A_e, $A1$, $A2$ = effective shear areas
a = half length of beam element
C = vector of integration constants
$C1$, $C2$, $C3$, $C4$ = components of vector C
$\tilde{C}$ = vector of generalized parameters
$\tilde{C}1$, $\tilde{C}2$, . . . $\tilde{C}6$ = components of vector $\tilde{C}$
d = flexibility matrix
E = modulus of elasticity
F, f = shearing forces
G = shear modulus
I, $I1$, $I2$ = cross-sectional moments of inertia
K, $\underline{K}$ = four by four stiffness matrices
$\bar{K}$ = five by five elastic stiffness matrix
K^*, $\tilde{K}^*$ = two by two stiffness matrices
K_G = four by four geometric stiffness matrix
$\bar{K}_G$ = five by five geometric stiffness matrix
$\bar{K}_t$ = five by five stiffness matrix
$K11$, $K12$, $K22$ = submatrices of stiffness matrix $\bar{K}$
$\check{K}11$, $\check{K}12$, $\check{K}22$ = submatrices of stiffness matrix $\bar{K}_G$
k^*_{11}, k^*_{12}, k^*_{22} = stiffness coefficients
L = transformation matrix
l = length of beam element
M, m = bending moments
$M1$, $M2$ = end moments
M, M_c = four by four mass matrices
$\bar{M}$ = five by five mass matrix
$M11$, $M12$, $M22$ = submatrices of translational mass matrix
$\tilde{M}11$, $\tilde{M}12$, $\tilde{M}22$ = submatrices of rotatory mass matrix
$N1$, $N2$, . . .$N5$ = components of vectors N_w and N_γ
N_w, N_γ = vectors of interpolation functions
n = shear coefficient
$p1$, $p2$, $p(\bar{x})$ = distributed loading
P = axial force
Q, $\bar{Q}$ = equivalent load vectors
$q1$, $q2$, $q3$, $q4$ = components of vector Q
R = transformation matrix
r = radius of gyration
S = vector of nodal forces
$S1$, $S2$, $S3$, $S4$ = components of vector S
T, $T1$, $T2$, $T3$ = connected matrices
t = shear-deformation parameter
U = vector of nodal displacements
$U1$, $U2$, $U3$, $U4$, $U5$ = components of vector U
$V1$, $V2$ = beam reactions
W, W_f, W_s = lateral displacements
x, y, $\bar{x}$, $\bar{y}$ = cartesian coordinates
$W1$, $W2$, $W3$, $W4$ = polynomials

$Z, \hat{Z}, Z_r$ = vectors of displacements
β = slope of beam deflection curve
$\beta 1, \beta 2$ = end slopes (Figure 3a)
Γ = constraint matrix
γ = shear deformation
$\Delta 1, \Delta 2$ = lateral displacements of beam ends
δ, δ^* = vectors of nodal displacements
θ = cross-section rotation
$\theta 1, \theta 2$ = end cross-sections rotations (Figure 3b)
λ = transformation matrix
ν = Poisson's ratio
π = strain energy
π^* = complementary strain energy
π_t = total potential energy
ϱ = mass density
$\phi 1, \phi 2$ = end cross-sections rotations (Figure 3a)
$\Psi 1, \Psi 2$ = end slopes (Figure 3b)
Ω = kinetic energy and
ω = circular frequency.

REFERENCES

1. Archer, J. S. "Consistent Matrix Formulation for Structural Analysis Using Finite-Element Techniques," *American Institute of Aeronautics and Astronautics Journal*, 3(10):1910–1918 (1965).
2. Avram, C., C. Bob, R. Friedrich and V. Stoian. *Reinforced Concrete Structures—Finite Element Method and Theory of Equivalences*, Editura Academiei R. S. R., Bucharest (1984).
3. Carnegie, W., J. Thomas and E. Dokumaci. "An Improved Method of Matrix Displacement Analysis in Vibration Problems," *The Aeronautical Quarterly*, 20:321–332 (1969).
4. Cowper, G. R. "The Shear Coefficient in Timoshenko's Beam Theory," *Journal of Applied Mechanics*, 33:335–340 (1966).
5. Davis, R., R. Henshell and G. B. Warburton. "A Timoshenko Beam Element," *Journal of Sound and Vibration*, 22(4): 475–487 (1972).
6. Donnell, L. H. *Beams, Plates and Shells*, McGraw-Hill, New York (1976).
7. Gallagher, R. H. and J. Padlog. "Discrete Element Approach to Structural Instability Analysis," *American Institute of Aeronautics and Astronautics Journal*, 1(6):1437–1438 (1963).
8. Gallagher, R. H. *Finite Element Analysis—Fundamentals*, Prentice-Hall, Inc., Englewood Cliffs, N.J. (1975).
9. Goodman, L. E. and J. G. Sutherland. Discussion of paper "Natural Frequencies of Continuous Beams of Uniform Span Length" by R. S. Ayre and L. S. Jacobsen, appeared in *Journal of Applied Mechanics*, 18:217–218 (1951).
10. Gurtin, M. E. "Variational Principles in Linear Elastodynamics," *Arch. Rat. Mech. Anal.*, 16:34–50 (1964).
11. Hinton, E. and D. R. J. Owen. *Finite Element Programming*, Academic Press, London (1977).
12. Kapur, K. K. "Vibrations of a Timoshenko Beam, Using Finite-Element Approach," *The Journal of the Acoustical Society of America*, 40(5):1058–1063 (1966).
13. Mazilu, P. *Statics of Structures*, Vol. 2, Ed. Technica, Bucharest (1959).
14. McCalley, R. B. "Mass Lumping for Beams," *Report DIG/SA 63-68* General Electric Knolls Atomic Power Laboratory, Schenectady, New York (1963).
15. McCalley, R. B. "Rotary Inertia Correction for Mass Matrices," *Report DIG/SA 63-73*, General Electric Knolls Atomic Power Laboratory, Schenectady, New York (1963).
16. Nickel, R. E. and Secor, G. A. "Convergence of Consistently Derived Timoshenko Beam Finite Elements," *International Journal for Numerical methods in Engineering*, 5:243–253 (1972).
17. Onu, G. "Shear Effect in Beam Stiffness Matrix," *Journal of Structural Engineering*, 109(9):2216–2221 (1983).
18. Onu, G. "Inclusion of Shear Effect in the Matrices of the Stright Beam Element," *Studii si cercetari de mecanica aplicata*, 41(4):459–469 (1982).
19. Onu, G. "Inclusion of Shear Effect in the ACM Element," *Computers & Structures*, 18(3):459–464 (1984).
20. Przemieniecki, J. S. *Theory of Matrix Structural Analysis*, McGraw-Hill Book Co., New York (1968).
21. Severn, R. T. "Inclusion of Shear Deflection in the Stiffness Matrix for a Beam Element," *Journal of Strain Analysis*, 5(4):239–241 (1970).
22. Teodorescu, P. P., *Theory of Elasticity*, Vol. 1, Editura Academiei R.S.R., Bucharest (1961).
23. Thomas, D. L., J. M. Wilson and R. R. Wilson. "Timoshenko Beam Finite Element," *Journal of Sound and Vibration*, 31(3):315–330 (1973).
24. Timoshenko, S. S. *Collected Papers*, McGraw-Hill Pub. Comp. Ltd., London (1953).
25. Timoshenko, S. S. *Strength of Materials*, Part I, Second Edition, D. Van Nostrand Co., New York (1951).
26. Tong, P. "New Displacement Hybrid Finite Element Models for Solid Continua," *International Journal of Numerical Methods in Engineering*, 2(1):73–83 (1970).
27. Williams, D. *An Introduction to the Theory of Aircraft Structures*, E. Arnold Pub., London (1960).
28. Zienkiewicz, O. C. *The Finite Element Method*, McGraw-Hill Book Co., London (1977).

APPENDIX 1. SHEAR COEFFICIENT

As a result of the nonlinear variation of the shearing strain along the depth of the beam the cross-sections become warped. In these conditions the shear coefficient accommodates the hypothesis of plane but non-normal cross-sections with the fact that actually the shear stress and strain are parabolically distributed over the cross-section. At first, the shear coefficient has been regarded as the ratio of the average shearing stress on a beam cross-section to the shearing stress at the centroid [25]. Some authors [4,9] have noted that when this ratio is used as shear coefficient the accuracy of Timoshenko's dynamic equilibrium Equations (24) is less satisfactory. A few formulas have been proposed for improving this coefficient. Three of them are shown here for the case of a homogeneous beam of rectangular cross-section.

Goodman and Sutherland [9] derived an expression in which n must satisfy

$$\frac{16(1-\nu)}{n^3} - \frac{8(3-2\mu)}{n^2} + \frac{8}{n} - 1 = 0 \tag{89}$$

where $\mu = (1 - 2\nu)/(2 - 2\nu)$ and ν is Poisson's ratio. If the above formula is used then the frequency given by the Timoshenko's equation is correct in the limit of zero wave length [5].

Cowper [4] proved that the formula

$$n = \frac{10 + 10\nu}{12 + 11\nu} \tag{90}$$

must be used if one desires to obtain the Timoshenko's beam equations from integration of three-dimensional elasticity equations.

The Donnell's relationship for the shear coefficient is [6] as follows:

$$n = \frac{20 + 20\nu}{24 + 15\nu} \tag{91}$$

If one takes $\nu = 0$, then the last two formulas lead to the same shear coefficient, i.e., $n = 5/6$. This value is used in many practical problems.

APPENDIX 2. STIFFNESS MATRIX IN TERMS OF END SLOPES

By referring to Figure 3a when $\phi 1 \neq 0$ and $\phi 2 = 0$, one notes that the shear deformation (see also Equation 9) and the two end slopes can be written as

$$\gamma = \frac{M1 + M2}{lnGA} = \frac{(k^*_{11} + k^*_{21})\phi 1}{lnGA} = \frac{6EI\phi 1}{(1+t)l^2 nGA} = \frac{t\phi 1}{2(1+t)} \tag{92}$$

$$\beta 1 = \phi 1 - \gamma = \phi 1 - \frac{t\phi 1}{2(1+t)} = \frac{2+t}{2(1+t)}\phi 1 \tag{93a}$$

$$\beta 2 = -\gamma = -\frac{t\phi 1}{2(1+t)} \tag{93b}$$

where k^*_{11} and k^*_{21} are entries in Equation (19).

Likewise, for $\phi 2 \neq 0$ and $\phi 1 = 0$, the end slopes are as follows

$$\beta 1 = -\frac{t\phi 2}{2(1+t)} \tag{94a}$$

$$\beta 2 = \frac{(2+t)\phi 2}{2(1+t)} \tag{94b}$$

The last four equations may be arranged in the following matrix form:

$$\begin{bmatrix} \beta 1 \\ \beta 2 \end{bmatrix} = \frac{1}{2(1+t)} \begin{bmatrix} 2+t & -t \\ -t & 2+t \end{bmatrix} \begin{bmatrix} \phi 1 \\ \phi 2 \end{bmatrix} \tag{95}$$

The inverse of the above connected matrix can be found readily as

$$T3 = \begin{bmatrix} \frac{2+t}{2(1+t)} & \frac{-t}{2(1+t)} \\ \frac{-t}{2(1+t)} & \frac{2+t}{2(1+t)} \end{bmatrix}^{-1} = \frac{1}{2}\begin{bmatrix} 2+t & t \\ t & 2+t \end{bmatrix} \tag{96}$$

The stiffness matrix in Equation (19), given in terms of $\phi 1$ and $\phi 2$ (total end rotations), may be transformed by applying a coordinate transformation based on $T3$ matrix. Hence, one obtains a two by two stiffness matrix in terms of $\beta 1$ and $\beta 2$ (end slopes) claculated by the following matrix relationship:

$$\tilde{K}^* = T3^T K^* T3 = \frac{EI}{l}\begin{bmatrix} 4+3t & 2+3t \\ 2+3t & 4+3t \end{bmatrix} \tag{97}$$

where K^* is defined in Equation (19).

The four by four stiffness matrix in terms of the nodal displacements, i.e., $\Delta 1$, $\psi 1$, $\Delta 2$ and $\psi 2$ (see Figure 3b) can be obtained from $\tilde{K}^*$ stiffness matrix by the equation

$$\tilde{K} = \lambda^T \tilde{K}^* \lambda \tag{98}$$

where the transformation matrix λ has been employed before in Equation (21b). The stiffness matrix obtained from the above transformation, for the four-d.o.f. beam element, including the effect of shear deformation, can be written in fully extended from as follows:

$$\tilde{K} = \frac{(1+t)EI}{1^3}\begin{bmatrix} 12 & 61 & -12 & 61 \\ 61 & \dfrac{(4+3t)^2}{1+t} & -61 & \dfrac{(2+3t)1^2}{1+t} \\ -12 & -61 & 12 & -61 \\ 61 & \dfrac{(2+3t)1^2}{1+t} & -61 & \dfrac{(4+3t)1^2}{1+t} \end{bmatrix} \quad (99)$$

This stiffness matrix refers to the ordered vector [$U1$, $U2$, $U3$, $U4$] in Figure 4 which clearly corresponds to end displacements Δ, $\psi 1$, $\Delta 2$ and $\psi 2$ shown in Figure 3b. Because of the nodal rotations $\psi 1$ and $\psi 2$ standing for total end slopes, this stiffness matrix cannot represent clamped boundary conditions correctly. On the other hand, the above stiffness matrix and the one given in Equation (22) should give identical results for cases where no boundary conditions are imposed on the rotations [23].

APPENDIX 3. REDUCED STIFFNESS MATRIX OF THOMAS ET AL.'S ELEMENT

Referred to the ordered vector [$W1$, $\theta 1$, $\gamma 1$, $W2$, $\theta 2$, $\gamma 2$], the six by six stiffness matrix derivated in Reference [23], written out here for a uniform beam element, is as follows:

$$\hat{K} = \frac{EI}{1^3}\begin{bmatrix} 12 & 61 & -61 & -12 & 61 & -61 \\ 61 & 41^2 & -31^2 & -61 & 21^2 & -31^2 \\ -61 & -31^2 & 31^2 + \dfrac{41^2}{t} & 61 & -31^2 & 31^2 + \dfrac{21^2}{t} \\ -12 & -61 & 61 & 12 & -61 & 61 \\ 61 & 21^2 & -31^2 & -61 & 41^2 & -31^2 \\ -61 & -31^2 & 31^2 + \dfrac{21^2}{t} & 61 & -31^2 & 31^2 + \dfrac{41^2}{t} \end{bmatrix} \quad (100)$$

The beam element characterized by this stiffness matrix is different from the one characterized by Equation (22) due to the fact that the shear deformation is assumed to vary linearly. In order to eliminate this variation a constraint is introduced, namely, the two end rotations due to shear are obliged to be dependent on a common parameter. In using matrix symbolism, this constraint can be written down as follows:

$$\hat{Z} = \Gamma Z \quad (101)$$

where $\hat{Z}$ stands for the nodal displacements vector for the six-d.o.f. beam element [23], Z is the nodal displacements vector in Equation (57) and the constraint matrix Γ is given by

$$\Gamma = \begin{bmatrix} 1 & 0 & 0 & 0 & 0 \\ 0 & 1 & 0 & 0 & 0 \\ 0 & 0 & 0 & 0 & \dfrac{1}{1} \\ 0 & 0 & 1 & 0 & 0 \\ 0 & 0 & 0 & 1 & 0 \\ 0 & 0 & 0 & 0 & \dfrac{1}{1} \end{bmatrix} \quad (102)$$

It is easily verified that the following matrix relationship

$$\overline{K} = \Gamma^T K \Gamma \quad (103)$$

leads to the stiffness matrix in Equation (65), so that by taking into account Equation (67), the stiffness matrix in Equation (22) can be gathered from

$$\hat{K} = R^T \Gamma^T \hat{K} \Gamma R \quad (104)$$

in which R is defined in Equation (69).

APPENDIX 4. REDUCED FORM OF THE DISPLACEMENT FIELD

The total potential energy Π_r of the five-d.o.f. beam element, disregarding the axial force effect, can be written as follows:

$$\Pi_r = \Pi - Z^T\overline{Q} = \frac{1}{2} Z^T \overline{K} Z - Z^T \overline{Q}$$

$$= \frac{1}{2}[Z_r^T, U5]\begin{bmatrix} K11 & K12 \\ K12^T & K22 \end{bmatrix}\begin{bmatrix} Z_r \\ U5 \end{bmatrix} - [Z_r^T, U5]\begin{bmatrix} Q_r \\ 0 \end{bmatrix} \quad (105)$$

in which Z, $\overline{Q}$, $\overline{K}$ and its submatrices have been defined previously in Equations (57), (76), (65) and (66), respectively. Z_r is a reduced nodal displacement vector, namely,

$$Z_r = [U1,\ U2,\ U3,\ U4] \qquad (106)$$

and Q_r is a load vector corresponding with the displacement vector Z_r.

The condition of minimum potential energy requires that

$$\frac{d\Pi_t}{dZ} = 0 \qquad (107)$$

leading to

$$\begin{bmatrix} K11 & K12 \\ K12^T & K22 \end{bmatrix} \begin{bmatrix} Z_r \\ U5 \end{bmatrix} - \begin{bmatrix} Q_r \\ 0 \end{bmatrix} = \begin{bmatrix} 0 \\ 0 \end{bmatrix} \qquad (108)$$

The unwanted d.o.f. $U5$ may be expressed in terms of Z_r by solving the second row of the above equations, so that

$$\begin{aligned} U5 &= -\ K22^{-1}\ K12^T\ Z_r \\ &= \frac{t}{1+t}(U1 + aU2 - U3 + aU4) \end{aligned} \qquad (109)$$

in which t denotes the shear-deformation parameter (Equation 14), a is half of the element length and $U1$ to $U4$ are nodal displacements as shown in Figure 4. Substituting for $U5$ from the above equation into the first row in Equations (55) leads to the following matrix relationship:

$$W_r = N_w\ R\ Z_r \qquad (110)$$

where W_r stands for the reduced form of the displacement field. The vectors N_w and Z_r are given in Equation 56a and Equation 106, respectively, and the matrix R is presented in Equation (69). Substituting for $\bar{x} = x - a$ in Equation (110), then, if the necessary matrix multiplications are carried out, the matrix product $N_w R$ becomes:

$$N_w R = \frac{1}{1+t}\begin{bmatrix} (1 - 3\dfrac{x^2}{1^2} + 2\dfrac{x^3}{1^3} + 1 - \dfrac{x}{1}t) \\ (\dfrac{x}{1} - 2\dfrac{x^2}{1^2} + \dfrac{x^3}{1^3} + \dfrac{1}{2}(\dfrac{x}{1} - \dfrac{x^2}{1^2})t)1 \\ (3\dfrac{x^2}{1^2} - 2\dfrac{x^3}{1^3} + \dfrac{x}{1}t) \\ (-\dfrac{x^2}{1^2} + \dfrac{x^3}{1^3} - \dfrac{1}{2}(\dfrac{x}{1} - \dfrac{x^2}{1^2})t)1 \end{bmatrix} \qquad (111)$$

The entries in the above vector are identical to the coordinate displacement functions used by Archer [1] in his formulation.

The reduced form of the displacement field may be used directly for the derivation of geometric stiffness matrix, mass matrix or load vector. As an example, the mass matrix given in Equation (87) may be carried out by using the following expression (20):

$$M = \varrho \int_o^{\ell} (N_w R)^T\ (N_w R)\ dx \qquad (112)$$

in which, ϱ, N_w and R have been defined previously.

CHAPTER 14

Design of Floor Slabs Coupling Shear Walls

A. COULL,* D. W. BOYCE,* AND Y. C. WONG**

1. INTRODUCTION

A common form of construction for multi-storey residential buildings consists of assemblies of shear walls and floor slabs, in which the coupling of the cross-walls by the floor slabs results in an efficient and economical structural system for resisting lateral forces. Figure 1(a) shows a typical idealised floor plan of a slab block in which self-contained apartment units are arranged on opposite sides of a central corridor along the length of the building. This arrangement naturally results in parallel assemblies of division walls running perpendicular to the face of the building, with intersecting longitudinal walls along the corridor and facade enclosing the living spaces. In addition to serving functional requirements, the cross-walls are employed as load-bearing walls, since their disposition favours an efficient distribution of both gravitational and lateral loads to the structural elements. Any longitudinal corridor and facade walls, if designed to be load-bearing, act effectively as flanges for the primary cross-walls. In addition to the partition walls, shear walls may also be used to enclose lift shafts and stair wells to form open-section box structures which act as strong points in the building. Thus, in practice, shear walls of various shapes may be coupled together in cross-wall structures (Figure 2).

The structural walls resist the lateral loads on the structure, due to wind or seismic actions, by cantilever bending action, which results in rotations of the wall cross-sections. The free bending of a pair of shear walls is resisted by the connecting floor slab, which is forced to rotate and bend out of plane where it is attached rigidly to the walls (Figure 1(b)). As a result of the large depth of the wall, the rotations impose considerable differential shearing actions on the connecting slab, which develops transverse reactions to resist the wall deformations [Figure 1(c)], thereby inducing tensile and compressive axial forces into the walls. Due to the large lever arm involved, relatively small axial forces can produce relatively large moments which can reduce greatly the wind moments in the walls and the resulting tensile stresses at the windward edges. The lateral stiffness of the structure may also be greatly increased.

A similar situation arises if relative vertical deformation of the walls occurs, due to unequal vertical loading on the walls or to differential foundation settlement, and the action on the slab is similar to that produced by parallel wall rotation caused by bending [Figure 1(d),(e)].

The structural analysis and design of a slab-coupled shear wall system can be performed using the techniques developed for beam-coupled shear wall systems, provided that either the equivalent width of the slab, assumed to act effectively as a wide coupling beam, or the associated bending stiffness, can be determined. In a coupled-wall system, the coupling medium resists the independent cantilever bending action of the walls by developing shearing actions in the medium. In a slab, the coupling stresses are not uniform across the width, tending to be concentrated in the regions near the inner corridor edges of the cross-walls, and, in order to be able to design the slab safely, it is necessary to assess the magnitude and distribution of the bending stresses developed through the coupling action. In detailing the joints, it is also essential to determine accurately the interactive shear forces developed at the slab–wall junction.

This chapter provides information for a design engineer on the effective coupling stiffness of a floor slab connecting a pair of structural walls of various cross-sectional shapes, and on the peak bending and shear stresses in the slab. Curves are given to illustrate the variation of these quantities with a range of the more important structural parameters for both plane and flanged walls.

*Department of Civil Engineering, University of Glasgow, Glasgow, Scotland

**Engineering and Environmental Consultants, Kuala Lumpur, Malaysia

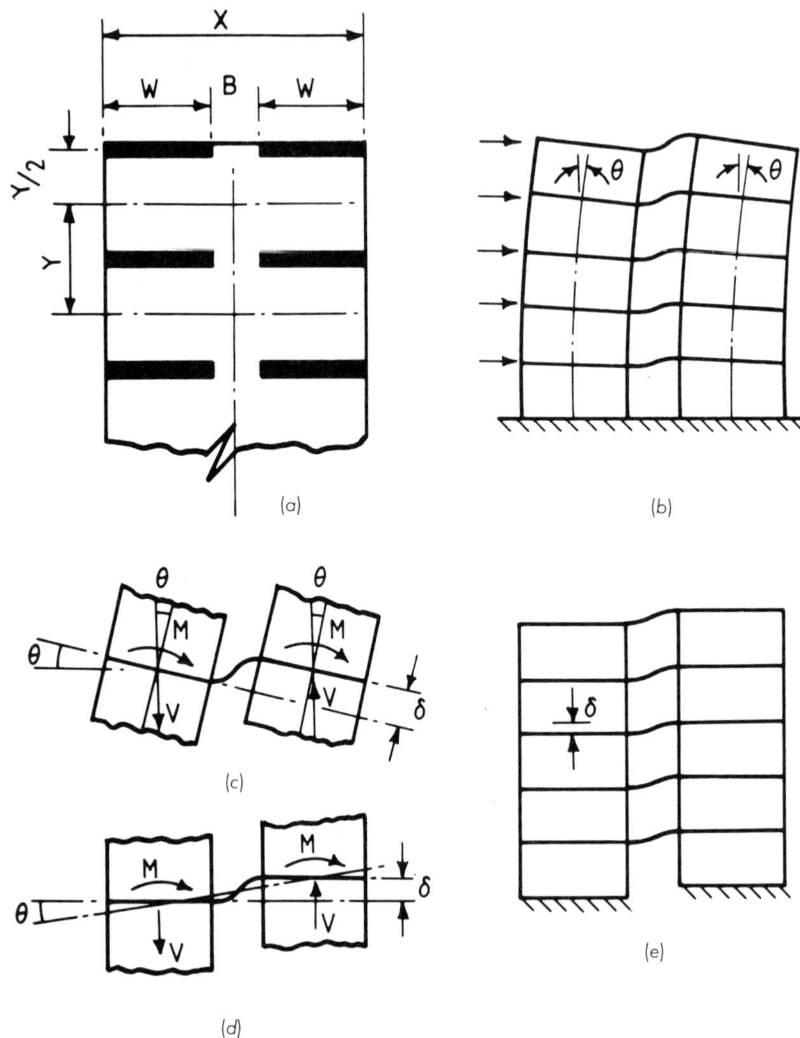

FIGURE 1. Structural action of coupled shear wall-slab structure.

In Section 2, a comprehensive set of generalised design curves and simple empirical formulae are presented to enable the effective bending stiffness of a floor slab coupling a pair of walls in a cross-wall structure to be estimated. In Section 3, contour diagrams and design curves showing the variation of critical bending moment factors are presented to enable a rapid evaluation of slab stresses induced by coupling actions. Based on the calculated interactive shear forces between wall and slab, a design method is suggested for checking against punching shear failure in the slab. Finally, in Section 4, a brief consideration is given to some other related topics which may be of interest and importance in the design of slab and wall structures. The chapter concludes with a brief summary of the information presented, with a caveat regarding its use in a design office.

The Notation used in the various formulae is defined at the end of the chapter.

2. EFFECTIVE STIFFNESS OF FLOOR SLAB

This section presents a comprehensive set of design curves and associated simple empirical formulae to enable the effective bending stiffness of a floor slab coupling a pair of cross-walls to be determined quickly and accurately. The curves and formulae apply to a range of common wall cross-sections in practical structures.

This has been achieved by considering the results of a large number of finite element analyses, which cover a wide range of structural parameters. By eliminating the parameters which have negligible influence on the effective stiffness, sets of generalised curves have been produced. Furthermore, a simple curve-fitting formula has been derived which defines closely the generalised curves for plane walls, and this has been extended to cover the wider range of common wall cross-sectional shapes considered.

The finite element method of analysis employed by the authors is described in Reference [20], and the accuracy of the technique is checked initially by comparison with published data in the literature.

The development of simplified design equations is treated fully for the case of a slab connecting a pair of plane walls, and this enables the corresponding development for flanged walls to be described briefly.

The resistance of the floor slab against the displacements imposed by the shear walls is a measure of its coupling stiffness, which can be defined in terms of the displacements at its ends and the forces producing them. Thus, referring to Figures 1(c) and 1(d), the stiffness of the slab may be defined either as a rotational stiffness M/θ or as a translational stiffness V/δ, since the two are related. Due to the non-uniform bending across the width of slab, the force- displacement relationship must be evaluated from a two-dimensional plate-bending analysis. For convenience, the rotational and translational slab stiffnesses are expressed in the form of non-dimensional stiffness factors K_θ and K_δ given by

$$K_\theta = \frac{M}{\theta}\frac{1}{D} \qquad K_\delta = \frac{V}{\delta}\frac{B^2}{D} \tag{1}$$

where B is the clear corridor opening between walls, and D is the flexural rigidity of the slab ($Et^3/12\,(1 - \nu^2)$ for isotropic slabs).

For the purpose of the overall analysis, it is assumed that a strip of slab acts effectively as a beam in coupling a pair of walls. The effective stiffness of the slab may then be defined in terms of the geometric and material characteristics of this equivalent beam. The effective width can be established by equating the rotational and translational stiffnesses of the slab to those of the equivalent beam. These may be expressed as

$$\frac{M}{\theta} = \frac{6EI}{B^3}(B + W)^2 \tag{2}$$

and

$$\frac{V}{\delta} = \frac{12EI}{B^3} \tag{3}$$

where W is the length of the wall, and I ($= Y_e t^3/12$) is the second moment of area of the beam, of effective width Y_e and thickness t.

The effective width may then be expressed in terms of the rotational and translational stiffness factors, in non-dimensional form, as

$$\frac{Y_e}{Y} = \frac{K_\theta}{6(1 - \nu^2)}\left(\frac{B}{Y}\right)\left(\frac{B}{B + W}\right)^2 \tag{4}$$

or

$$\frac{Y_e}{Y} = \frac{K_\delta}{12(1 - \nu^2)}\left(\frac{B}{Y}\right) \tag{5}$$

where Y is the bay width or longitudinal wall spacing [Figure 1(a)], and ν is Poisson's ratio for the slab material.

2.1 Plane Wall Configurations

Consider the action of two plane shear walls coupled by a uniform floor slab [Figure 2(a)]. It has been found both experimentally and theoretically that the main coupling action between the slab and walls occurs near the inner corridor edges of the walls, and the influence of any slab overhang at the outer edges may be neglected [19]. The influences of the other important parameters have been investigated by a finite element analysis.

2.1.1 INFLUENCE OF WALL LENGTH (W)

In examining the effect of wall length on the effective stiffness of the floor slab, two of the other variables concerned, namely the wall opening B and the slab width Y, were kept constant while the wall length W was varied. As a result of varying W, the floor length X is varied accordingly. As the rotational stiffness K_θ of the equivalent beam is affected directly by the change in the lever arm distance between wall centres when the wall length is varied, the results for the stiffness factor K_θ do not give a clear picture of the actual influence of wall length on the effective coupling slab width. Thus reference is made only to the results obtained for the effective width ratio Y_e/Y.

The variations of Y_e/Y with the ratio of wall length to wall opening, W/B, are shown in Figure 3 for the two ratios B/Y considered. The trend of results is seen to be very similar for both $B/Y = 0.5$ and $B/Y = 1.5$, the effective width ratio increasing with the increase in wall length ratio. The variation of Y_e/Y is relatively rapid for W/B less than 0.3, but becomes insignificant when W/B is larger than 0.5. In practical cross-wall structures, W/B is unlikely to be less than 0.5, and so the effect of variations in wall length may be disregarded in the evaluation of effective slab width, as long as the influence of the ratio B/Y is considered. Similarly, the effect of dissimilar wall lengths in a pair of coupled walls may also be disregarded if the ratio of the shorter wall length to the wall opening is greater than 0.5.

2.1.2 INFLUENCE OF SLAB WIDTH (Y)

To illustrate the effect of slab width more clearly, the effective width Y_e and the slab width Y are normalised with

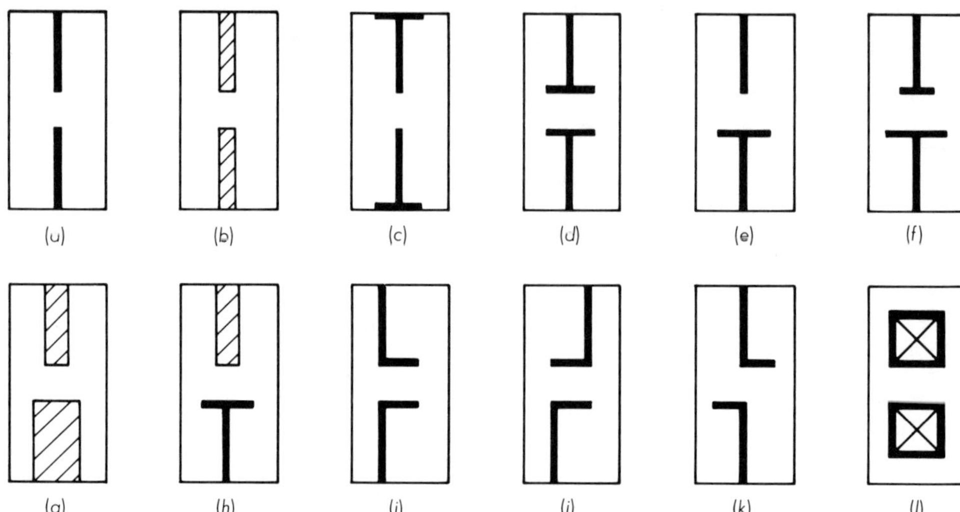

FIGURE 2. Coupled shear wall configurations analysed.

respect to the standard panel length X. Figure 4(a) shows graphically the variation of Y_e/X as a function of Y/X for various wall opening ratios B/X. As expected, the effective width is seen to increase with the slab width, since a wider slab will provide a greater restraint than a narrower slab against bending induced by the coupled walls. The influence of slab width is strongly felt when the width ratio Y/X is smaller than the wall opening width B/X (i.e., when $Y/B < 1$), but when Y/X is larger than B/X the influence of slab width diminishes rapidly. Increasing the slab width beyond a value of about three times the wall opening width appears to have practically no effect on the effective slab width for a particular wall opening width.

2.1.3 INFLUENCE OF WALL OPENING WIDTH (B)

The data from the preceding section are re-plotted in Figure 4(b) to show the variation of effective width Y_e/X as a function of the wall opening width ratio B/X for various

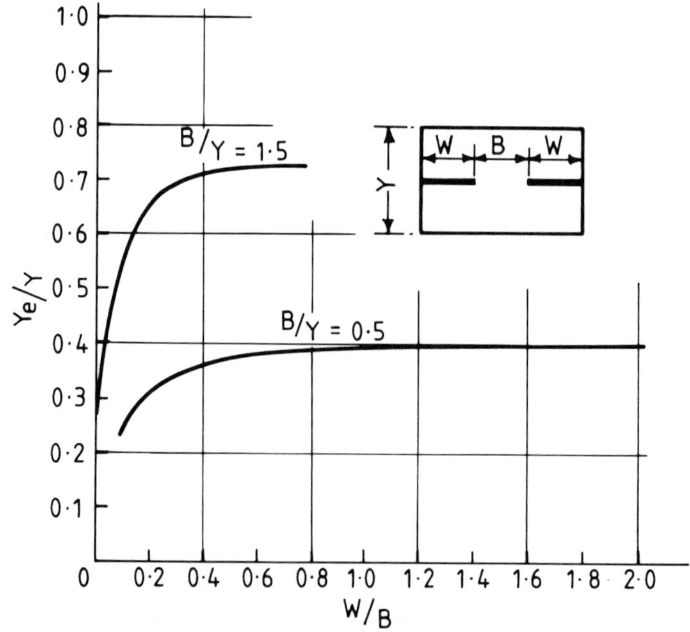

FIGURE 3. Variation of effective slab width with wall length.

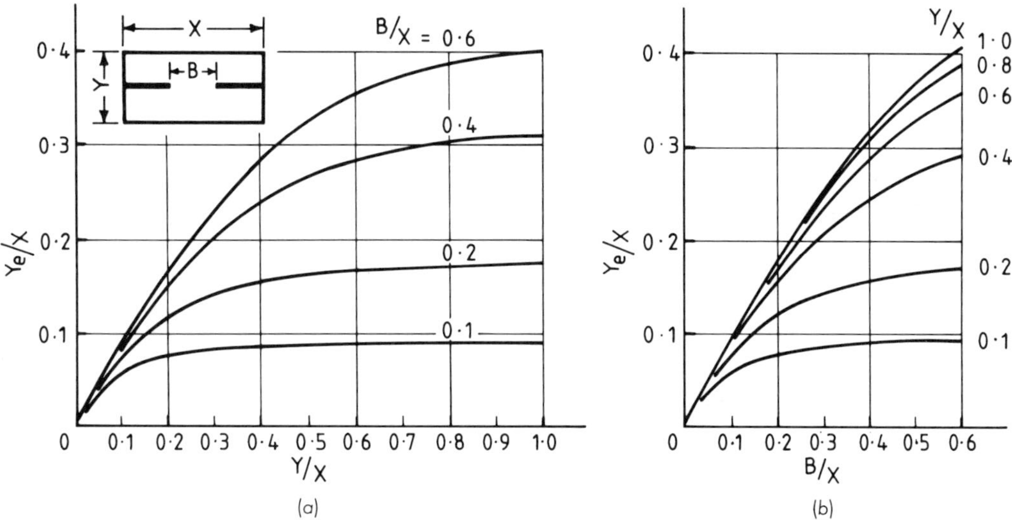

FIGURE 4. Variation of effective slab width with (a) bay width and (b) wall opening width.

slab width ratios Y/X. A comparison of Figure 4(b) with Figure 4(a) shows that the two sets of curves are virtually the same for the range of variables considered. The influence of B/X on Y_e/X for a particular value of Y/X is seen to be practically identical to the influence of Y/X on Y_e/X for the same value of B/X. (The reciprocal influence of B and Y is also clearly indicated if the values of the ratio Y_e/X are related in tabular form to the ratios B/X and Y/X along the horizontal and vertical axes, when the resulting array is virtually symmetrical about the leading diagonal [19]). This interesting reciprocal relationship between B and Y may be exploited to obtain a simple generalised design curve and an empirical effective width equation.

2.1.4 DESIGN CURVES FOR EFFECTIVE SLAB WIDTH (Y_e)

Curves to enable the rapid calculation of effective widths to be achieved are presented in Figure 5, which shows the variation of effective width Y_e/Y as a function of wall opening ratio B/X for various slab aspect ratios Y/X. The numerical results from which the curves have been drawn are given in Reference [19]. The results presented have been obtained by assuming zero wall thickness in the analysis, and for relatively thin walls, the effective slab widths obtained from Figure 5 should prove sufficiently accurate for most practical situations. However, for relatively thick walls, the effective slab widths can still be evaluated satisfactorily from the same set of curves, with the modifications described in Sections 2.1.7 and 2.1.8.

The curves show the influence of slab width, wall opening width and wall length. The curves for various slab aspect ratios Y/X have been spaced at fairly close intervals to facilitate accurate interpolation when the actual value of Y/X for the particular problem falls between any two curves. However, interpolation between the curves is not strictly necessary, since, as shown earlier, the influence of variations in wall length can generally be disregarded when the wall length ratio W/B is larger than 0.5, or, alternatively, when B/X is less than 0.5. The value of the slab length X for the particular problem can thus be arbitrarily adjusted so that

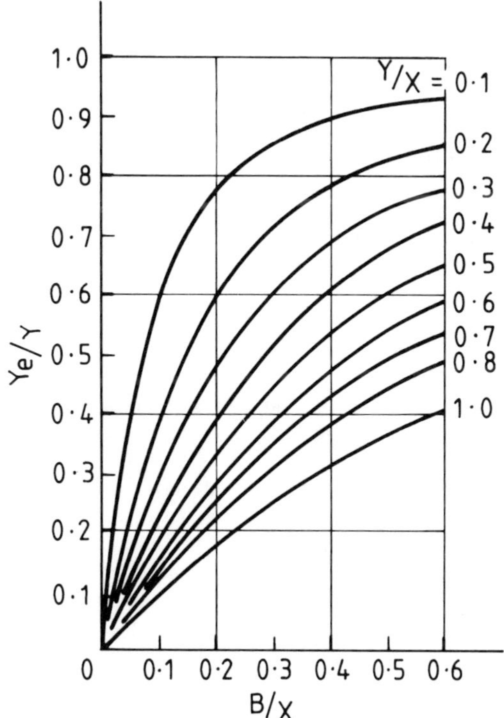

FIGURE 5. Design curves for effective width (planar walls).

the new value of Y/X corresponds to the value for the nearest curve, and the correct effective width Y_e/Y is given by this curve at the new value of B/X.

The effective width values have been evaluated throughout assuming a value of Poisson's ratio of 0.15 for concrete. The slab stiffness factor is not sensitive to small differences in values of Poisson's ratio, but the effective width Y_e/Y is influenced by the value of ν, as can be seen from Equations (4) and (5). Thus, if desired, the value of Y_e/Y obtained from the design curves may be corrected approximately for the actual value of ν by multiplying by a factor $(1 - 0.15^2)/(1 - \nu^2)$.

2.1.5 GENERALISED DESIGN CURVE FOR EFFECTIVE SLAB WIDTH

Although the design curves shown in Figure 5 allow a rapid determination of the effective slab width, it is possible to simplify further the presentation of design information to allow an even more rapid evaluation of the coupling width, although the procedure will be found more appropriate for flanged walls where additional parameters are involved. Since the effective width ratio Y_e/Y is essentially a function of the wall-opening to slab-width ratio B/Y only, the series of design curves presented in Figure 5 can be approximated by a single generalised curve of Y_e/Y versus B/Y, as shown in Figure 6(a). This generalised curve is sufficiently accurate for most practical purposes. With slab and wall proportions with $(B/X + Y/X) \le 1$, which covers most practical cases, the values of Y_e/Y obtained from the generalised curve and from the more accurate series of design curves are practically identical. For such cases the generalised curve should prove more convenient to use than the previous series of curves. In exceptional cases where $(B/X + Y/X) > 1$ and B/X is also larger than 0.4, the effective widths Y_e/Y evaluated by the generalised curve are overestimated, generally by less than 10%. For such cases, a more accurate estimate of Y_e/Y may be obtained from the more accurate series of design curves of Figure 5.

2.1.6 EMPIRICAL DESIGN EQUATION FOR EFFECTIVE SLAB WIDTH

The generalised design curve in Figure 6(a) can be separated into two distinct sections defined by the limits $0 \le B/Y \le 1$ and $1 \le B/Y < \infty$. Values of Y_e/Y from the second section of the generalised curve, which may be termed the reciprocal section, may be obtained from the first section of the curve, termed the normal section, by making use of the reciprocal relationship between B and Y. It was shown earlier that the values of B and Y may be interchanged without sensibly affecting the value of Y_e for the slab. Hence, when the ratio B/Y exceeds unity, the reciprocal of this ratio can be used to evaluate from the normal section the effective width ratio Y_e/B for the reciprocal section of the generalised curve. If the normal section of the curve is then represented by the equation $Y_e/Y = f(B/Y)$, where f denotes a function, the reciprocal section will be represented correspondingly by $Y_e/B = f(Y/B)$.

The normal section of the curve can be represented reasonably accurately by the simple empirical relationship

$$\frac{Y_e}{Y} = \frac{B}{Y}\left(1 - 0.4\frac{B}{Y}\right) \qquad \text{for } 0 \le \frac{B}{Y} \le 1 \tag{6}$$

The reciprocal section is then represented, on interchanging B and Y, by the equation

$$\frac{Y_e}{B} = \frac{Y}{B}\left(1 - 0.4\frac{Y}{B}\right) \qquad \text{for } 0 \le \frac{Y}{B} \le 1 \tag{7}$$

On multiplying Equation (7) by B/Y, the reciprocal equation becomes

$$\frac{Y_e}{Y} = 1 - 0.4\left(\frac{B}{Y}\right)^{-1} \qquad \text{for } 1 \le \frac{B}{Y} < \infty \tag{8}$$

Equations (6) and (8) now represent the generalised design curve over the complete range $0 \le B/Y < \infty$. The value of Y_e/Y given by the reciprocal equation approaches unity when B/Y becomes infinitely large. This is consistent with the fact that a long narrow strip of slab will behave as a beam of flexural rigidity D.

To illustrate the accuracy of the empirical representation, the curve given by Equations (6) and (8) is compared with the generalised design curve in Figure 6(b). It can be seen that the empirical design curve gives an almost perfect fit to the generalised curve. By comparing numerical values of Y_e/Y obtained by the empirical equations with the corresponding accurate values obtained by the finite element technique over a wide range of parameters [19], discrepancies were found to be generally less than 4% for a range of B/Y from 0.1 to 6.0.

2.1.7 EFFECT OF FINITE WALL THICKNESS ON EFFECTIVE WIDTH

A set of curves was presented in Reference [20] to illustrate the influence of wall thickness ratio h/Y on effective width ratio Y_e/Y for a range of wall opening ratios B/X and slab aspect ratios Y/X. The results are also applicable to the case of coupled box cores without openings on the inner edges, since the displacements imposed on the slab by a thick solid wall and by a box core of the same external dimensions are identical.

The results indicated that the thickness of the wall can have a considerable stiffening effect on the slab, the in-

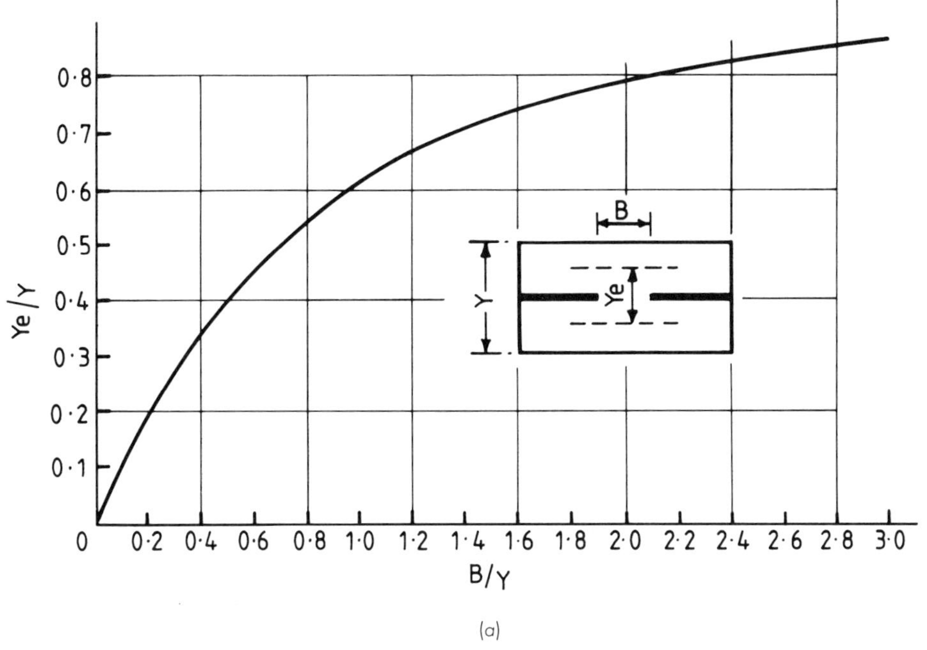

(a)

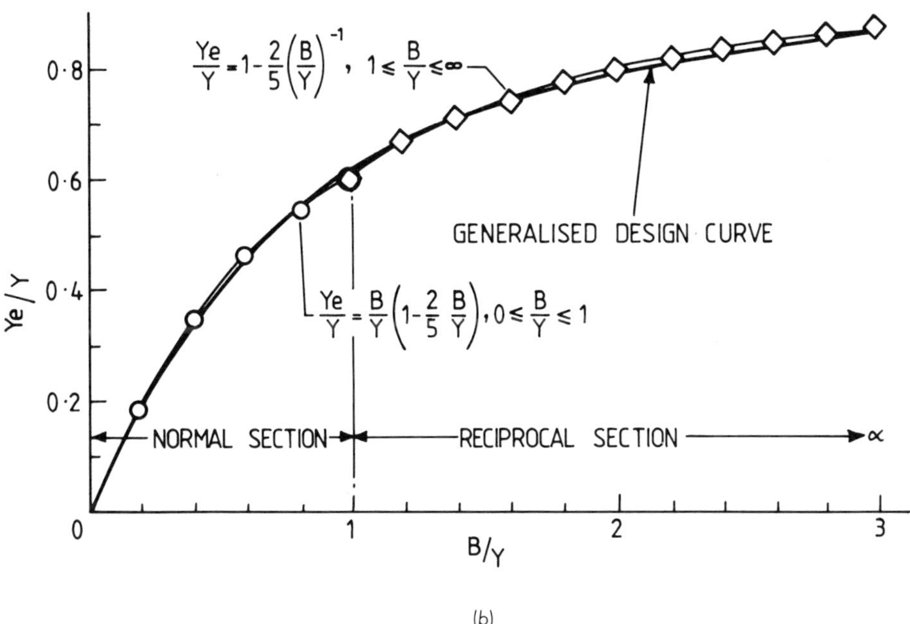

(b)

FIGURE 6. Generalised design curve (a) and empirical curve (b) for effective slab width for plane wall configuration.

TABLE 1. Comparison Between Empirical and Finite Element Results for Thick Wall Configuration.

B/X	h/Y	Y'/Y	B/Y'	Y_e/Y Empirical	Y_e/Y F. Element	Percentage Difference
0.1	0.000	1.000	0.250	0.225	0.217	+3.45
	0.125	0.875	0.286	0.346	0.340	+1.73
	0.250	0.750	0.333	0.466	0.465	+0.24
	0.375	0.625	0.400	0.585	0.589	−0.70
	0.500	0.500	0.500	0.700	0.710	−1.39
	0.750	0.250	1.000	0.900	0.927	−2.92
0.4	0.000	1.000	1.000	0.600	0.610	−1.66
	0.125	0.875	1.143	0.694	0.703	−1.32
	0.250	0.750	1.333	0.775	0.782	−0.96
	0.375	0.625	1.600	0.844	0.852	−1.00
	0.500	0.500	2.000	0.900	0.911	−1.20
	0.750	0.250	4.000	0.975	0.995	−2.06

fluence being relatively more significant with smaller wall opening ratios, and it becomes important to modify the previous empirical equations to take account of the finite wall thickness.

2.1.8 EMPIRICAL DESIGN EQUATIONS TO INCLUDE FINITE WALL THICKNESS

The empirical Equations (6) and (8), which represent the generalised design curves for walls of zero thickness, may be modified to include the influence of finite wall thickness, due to the high in-plane rigidity of the wall itself, by adding an additional slab width equal to the wall thickness h to the effective width for walls of zero thickness. The coupling width Y_e of the actual slab panel may then be considered as being made up of an effective wall strip of width h plus the effective width Y'_e of a reduced slab panel of width Y', equal to $(Y - h)$, yielding

$$\frac{Y_e}{Y} = \frac{h}{Y} + \frac{Y'}{Y}\left(\frac{Y'_e}{Y'}\right) \tag{9}$$

The effective width Y'_e/Y' of the reduced slab panel may be represented by the empirical equations (6) and (8) with Y'_e and Y' in place of Y_e and Y, respectively.

Equation (9) may then be written as

$$\frac{Y_e}{Y} = \frac{h}{Y} + \frac{Y'}{Y} \cdot \frac{B}{Y'}\left(1- 0.4\frac{B}{Y'}\right) \tag{10}$$

$$\text{for } 0 \leq \frac{B}{Y'} \leq 1$$

and

$$\frac{Y_e}{Y} = \frac{h}{Y} + \frac{Y'}{Y}\left(1 - 0.4\left(\frac{B}{Y'}\right)^{-1}\right) \tag{11}$$

$$\text{for } 1 \leq \frac{B}{Y'} < \infty$$

To demonstrate the accuracy of the empirical equations, values of Y_e/Y obtained from them and directly from the finite element method [19] are compared for a slab of aspect ratio $Y/X = 0.4$, with various wall thickness ratios h/Y, and two wall opening ratios B/X of 0.1 and 0.4, in Table 1, which also shows the relevant quantities Y'/Y and B/Y' required in the empirical equations. Although the results cover a wide range of ratios B/X and h/Y, the values obtained by the simple empirical equations lie within 3% of the accurate finite element values for walls of finite thickness.

2.1.9 END BAY RESULTS

End bays occur at the two gable ends of the building block where the gable walls are coupled by the floor slab on one side of the wall only. With the asymmetric coupling of the slab, gable walls will generally undergo some out-of-plane bending which will affect the coupling stiffness of the slab to some extent which will be dependent on the relative wall thickness. Since the gable edge of the slab is less restrained against transverse rotation than a continuous edge, the coupling stiffness of the end bay should be less than half the stiffness of an internal bay slab.

From calculations performed on a series of typical end bay configurations, of varying proportions, with gable walls either so stiff that they do not bend out-of-plane, or with a finite stiffness corresponding to a wall of 12 in. (305 mm) thickness, it was found that the effective width for an end bay generally varies between 44% and 47% of the value for an interior bay. Hence, as a convenient practical design rule, it is suggested that the effective width and stiffness for an end bay should be taken to be 45% of the corresponding interior value.

2.2 Flanged Wall Configurations

Flanged shear walls of T or L cross-section occur frequently in cross-wall structures as a result of making the corridor or facade longitudinal walls of similar construction to the cross walls to satisfy the need for additional load-bearing area or longitudinal stiffness, or simply for the purpose of convenience in building construction. Depending on planning requirements, various arrangements of flanged walls may be encountered in cross-wall structures, and empirical design equations are presented for the various possible configurations.

2.2.1 COUPLED FLANGED WALLS WITH EXTERNAL FLANGES

In this arrangement, the flanges are situated at the exterior or facade ends of the cross-walls [Figure 2(c)], whilst the slab actions induced by the coupled walls are confined mainly to the corridor area and its immediate vicinity. Finite element studies have shown that the presence of external wall flanges increases the effective width of the slab by less than 4% for the most extreme case considered, and for practical purposes the influence of external wall flanges may be safely disregarded.

2.2.2 COUPLED T-SHAPED FLANGED WALLS WITH INTERNAL FLANGES

Consider the pair of identical T-shaped walls shown in Figure 2(d) which are symmetrical with respect to the panel centre-lines. It has been demonstrated experimentally [11] and theoretically [19] that practically no bending of the floor slab occurs in the regions behind the flanges. As with the thick plane wall configuration, however, the width of the wall flange has a major influence on the effective width of the slab, and empirical design formulae have been devised in a similar manner to those developed for plane walls. Since these were presented in detail, only the final results are given here.

A finite element analysis enabled a series of curves to be produced [19], showing the variation of the effective width ratio Y_e/Y as a function of the wall opening ratio B/X for various flange width ratios Z/Y, and also as a function of Z/Y for various ratios B/X. A comparison of these curves with the ones produced earlier for thick walls showed that the influence of the flange width and the influence of the wall thickness were practically identical, the former being only marginally less for the same wall thickness and flange width. By disregarding the negligibly small influence of the wall length on the effective width, the series of curves were generalised into the single set of Figure 7(a) which shows the variation of effective width ratio Y_e/Y as a function of B/Y for various flange width ratios, Z/Y. The generalised curves are accurate compared to those which account for the influence of wall length when the wall and slab proportions are in the practical range $(B/X + (Y - Z)/X) \leq 1$. In unusual cases where $(B/X + (Y - Z)/X)$ is greater than unity and B/X is also greater than 0.4, the curves over-estimate the effective width, but generally by less than 10%.

Since the influences of the flange width and finite wall thickness are practically identical, the simple method suggested earlier to account for the effect of wall thickness may also be used to include the flange width, using the basic design curves presented for plane walls of zero thickness. The effective slab width Y_e may thus be considered to be composed of an effective wall strip equal in width to the flange width Z, plus the effective width Y_e' of a reduced slab panel with plane walls of zero thickness.

The empirical equations (10) and (11) developed for coupled thick planar walls are thus also applicable to coupled flanged walls with a change of the width variable h to the new flange variable Z. The empirical equations for a flanged wall configuration may then be written as

$$\frac{Y_e}{Y} = \frac{Z}{Y} + \frac{Y'}{Y} \cdot \frac{B}{Y'}\left(1 - 0.4\frac{B}{Y'}\right) \qquad \text{for } 0 \leq \frac{B}{Y'} \leq 1 \tag{12}$$

and

$$\frac{Y_e}{Y} = \frac{Z}{Y} + \frac{Y'}{Y}\left(1 - 0.4\left(\frac{B}{Y'}\right)^{-1}\right) \qquad \text{for } 1 \leq \frac{B}{Y'} < \infty \tag{13}$$

To illustrate the accuracy of these empirical equations, the results previously compared with "thick wall" finite element results in Table 1 are compared in Table 2 with the finite element results obtained for flanged wall configurations. It is found that the empirical results agree with the accurate finite element values to within 2½%.

2.2.3 COUPLED PLANE WALLS AND T-SHAPED FLANGED WALLS

For plane walls coupled to T-shaped flanged walls [Figure 2(e)], the rotational stiffnesses evaluated at the inner edges will be unequal because of the dissimilarity between the walls, and, in order to replace the slab by a uniform equiva-

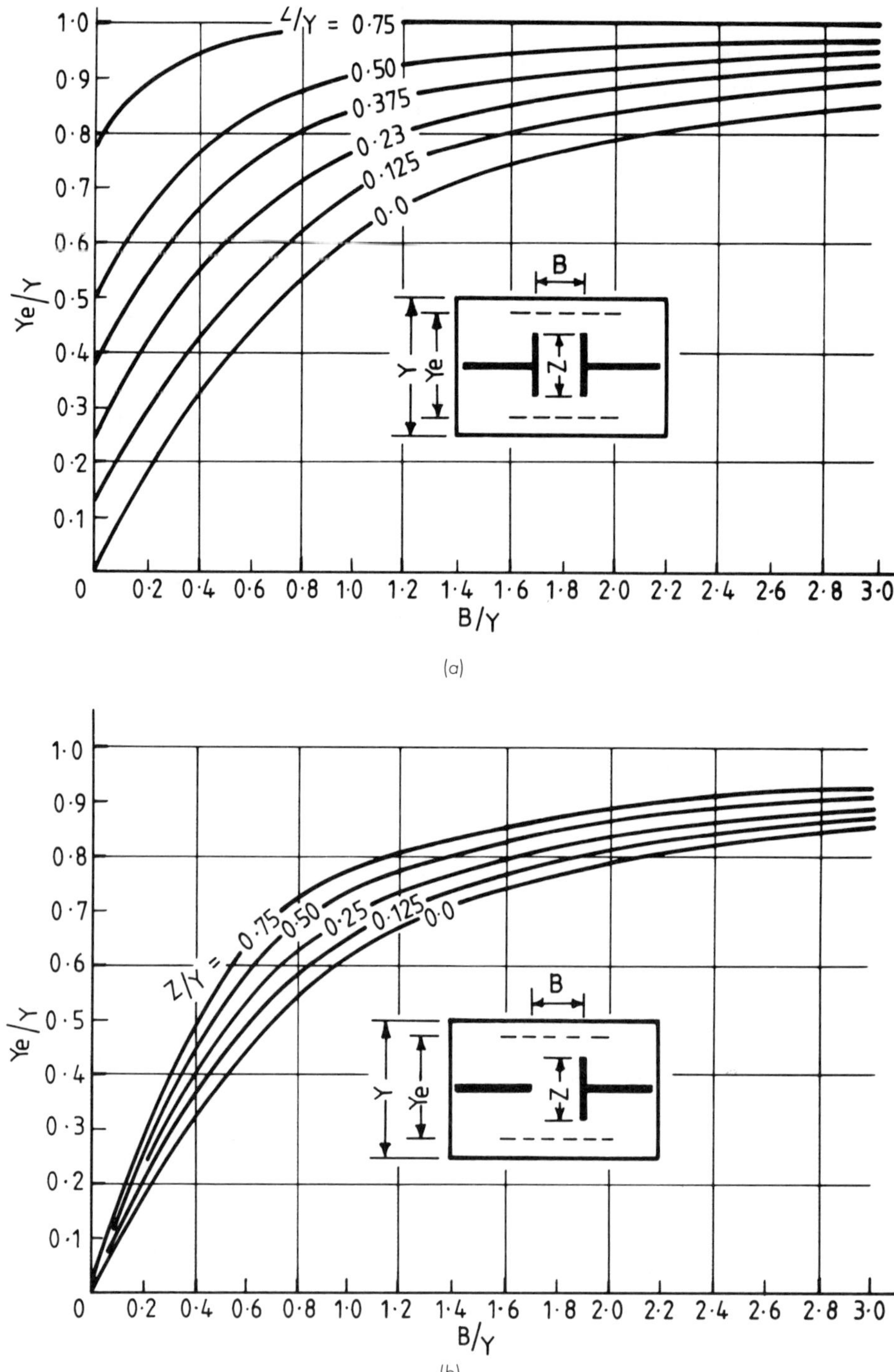

FIGURE 7. Generalised design curves for effective slab width for (a) flanged wall and (b) planar-flanged wall configuration.

TABLE 2. Comparison Between Empircal and Finite Element Results for Flanged Wall Configuration.

B/X	Z/Y	Y_e/Y Empirical	Y_e/Y F. Element	Percentage Difference
0.1	0.125	0.346	0.339	+2.24
	0.250	0.467	0.460	+1.39
	0.375	0.585	0.580	+0.79
	0.500	0.700	0.698	+0.21
	0.750	0.900	0.918	−1.92
0.4	0.125	0.694	0.695	−0.17
	0.250	0.775	0.767	+1.10
	0.375	0.844	0.832	+1.36
	0.500	0.900	0.892	+0.92
	0.750	0.975	0.991	−1.57

lent beam, the average rotational stiffness is taken to be the effective value.

Based on a study of a wide range of parametric ratios, a single set of generalised design curves has again been produced, Figure 7(b) showing the variation of Y_e/Y as a function of B/Y for various ratios Z/Y.

The curves approach a zero value for Y_e/Y when B/Y approaches zero, and are asymptotic to the line $Y_e/Y = 1$. On comparing these curves with the corresponding set in Figure 7(a) for the configuration with two flanged walls, it is seen that the omission of one flange from a wall results in a disproportionately large reduction in the effective width of the slab when the ratio B/Y is small.

The generalised curves are accurate, as compared to precise design curves which account for the influence of wall length, when the slab and wall proportions are such that $(B/X + Y/X) \leq 1$. Where the sum of $(B/X + Y/X)$ exceeds 1.0, and B/X is also greater than 0.4, the effective widths given by the generalised curves are generally higher than the values obtained from the more accurate sets of curves, by less than 10%. In most practical cases, the wall opening ratio is unlikely to exceed 0.4 and the generalised set of design curves should be sufficiently accurate for design situations.

2.2.4 COUPLED T-SHAPED WALLS WITH UNEQUAL FLANGE WIDTHS

When a slab connects two walls with internal flanges of unequal width [Figure 2(f)], the effective width may be determined empirically from the results presented for the planar-flanged wall configurations using a technique similar to that for coupled thick walls.

Based on a number of parametric studies [19], the effective width Y_e of the slab coupling a pair of flanged walls with flange widths Z_1 and Z_2 may be assumed to be made up of an effective wall strip equal in width to the smaller flange

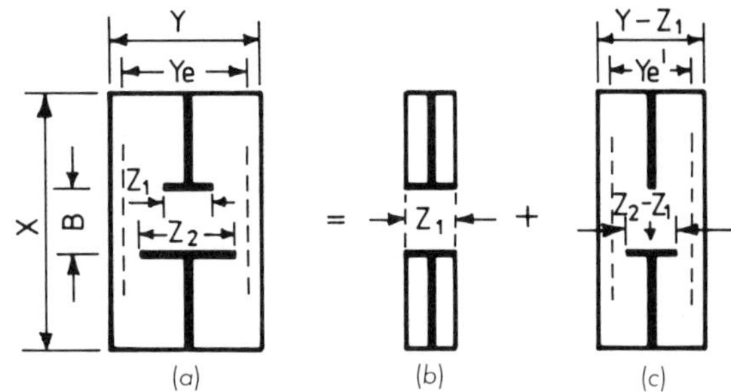

FIGURE 8. Replacement of actual slab panel (a) by effective wall strip (b) and reduced slab panel (c) for coupled flanged walls with unequal flanges.

width Z_1, plus the effective width Y_e' evaluated for a slab panel with a reduced width $Y' = (Y - Z_1)$ coupling a plane wall to a flanged wall of reduced flange width $Z' = (Z_2 - Z_1)$ as shown in Figure 8. The effective width Y_e' for the reduced slab panel can be directly obtained from the generalised design curves of Figure 7(b).

To illustrate the general accuracy of the empirical method, Table 3 compares values of Y_e/Y for various flange combinations with the accurate results obtained directly from a finite element calculation. It is seen that the proposed empirical method gives effective width ratios Y_e/Y which are within 1% of the accurate finite element results. Since the influence of finite wall thickness in plane walls and the influence of flange width are essentially similar, the proposed empirical approach is also applicable to configurations consisting of coupled plane walls of unequal thickness [Figure 2(g)] or a plane wall of finite thickness coupled to a flanged wall [Figure 2(h)].

2.2.5 COUPLED L-SHAPED FLANGED WALLS

A pair of L-shaped flanged walls may be coupled in three

TABLE 3. Comparison Between Empirical and Finite Element Results for Coupled Flanged Walls with Unequal Flanges.
$Y/X = 0.6$ $B/X = 0.4$

		Values of Y_e/Y		
Z_1/Y	Z_2/Y	Empirical Method	Finite Element	Percentage Difference
0.083	0.167	0.578	0.577	+0.17
0.083	0.250	0.606	0.605	+0.17
0.083	0.333	0.633	0.632	+0.16
0.167	0.250	0.642	0.641	+0.16
0.167	0.333	0.675	0.670	+0.75
0.250	0.333	0.700	0.702	−0.28

different ways, as shown in Figures 2(i), (j) and (k). If the cross walls, or web walls, are coupled in-line with opposing flanges [Figure 2(i)], or if the webs are coupled off-line with opposing flanges [Figure 2(j)], the effective widths of the slabs in these two configurations may be obtained directly from the results presented for coupled T-shaped flanged walls or from the results presented for coupled thick planar walls or box cores since both sets of results are essentially similar. However, in the third wall configuration shown in Figure 2(k), the webs are coupled in-line but the flanges are off-set or skew-coupled, and the same curves cannot be used. It has been shown both theoretically [20] and experimentally, [18], that, although the flanges are not directly cross-coupled, they still have a considerable influence on the coupling stiffness for the slab. A comparison of the results obtained for the L-shaped wall configuration [20] with the results for the planar-flanged wall configuration [Figure 7(b)] revealed an interesting and practically useful feature. It was found that with the same total flange width in the coupled walls, the effective slab widths in the coupled L-shaped wall configuration and in the planar-flanged wall configurations are practically identical for any set of ratios B/X, Y/X and Z/Y. The generalised design curves presented for the planar-flanged wall configuration may therefore be used to evaluate also the effective slab widths for L-shaped wall systems if the flange width ratios Z/Y indicated on the curves are halved to correspond to the flange width of the L-shaped walls.

2.2.6 *INFLUENCE OF ORTHOTROPIC SLAB PROPERTIES*

In the elastic analysis of slab behavior, it has been normal practice to regard the floor slab as an isotropic plate. In a practical concrete floor slab, however, the different steel ratios in two directions result in different stiffnesses in the two orthogonal directions. In other forms of floor construction, such as voided slabs, ribbed slabs or pseudo slabs (consisting of precast beams and in situ concrete infill), the orthotropic properties of the floor arise principally from differences in structural geometry of the floor cross-sections. The results which have been obtained assuming isotropic slab properties are not strictly correct when applied to an orthotropic slab, and the degree of approximation involved with actual floor slabs will depend on the degree of orthotropy present.

The flexural behavior of an orthotropic plate is governed by four elastic constants D_x, D_y, D_1, and D_{xy}, representing, respectively, the flexural rigidities in two orthogonal directions, the cross-coupling rigidity and the torsional rigidity [17]. These constants are frequently difficult to determine accurately for practical structural forms, and their evaluation has been discussed by, for example, Timoshenko and Woinowsky-Krieger [17] and by Cusens and Pama [10]. For reinforced concrete, the cross-coupling and torsional rigidities D_1 and D_{xy} may be evaluated in terms of D_x and D_y as follows [17]:

$$D_1 = \nu\sqrt{D_x D_y} \text{ and } D_{xy} = \frac{1-\nu}{2}\sqrt{D_x D_y}$$

These expressions are strictly valid only for solid slabs (which is a common form of floor construction in cross wall buildings) but may be used also as approximate values for other forms of quasi-slab construction for which D_1 and D_{xy} are difficult to evaluate accurately.

An orthotropic plate with sides of lengths a and b in the x- and y-directions, and with these properties, may be treated as an equivalent isotropic plate [17] having a rigidity D and sides $a_o = a \cdot \sqrt[4]{D/D_x}$ and $b_o = b \cdot \sqrt[4]{D/D_y}$. Thus, if the corridor slab aspect ratio B/Y of the orthotropic slab is transformed into an effective isotropic ratio $\varrho\ (B/Y)$, where $\varrho = \sqrt[4]{D_y/D_x}$, the design curves and empirical equations can still be used to evaluate the effective width for the orthotropic slab. The empirical equations (6) and (8) and (12) and (13), may be extended to include orthotropic effects by being rewritten in the following forms:

(a) Plane Wall Configuration (Zero Wall Thickness)

$$Y_e/Y = \varrho\frac{B}{Y}\left(1 - 0.4\varrho\frac{B}{Y}\right) \quad \text{for } 0 \le \varrho\frac{B}{Y} \le 1 \tag{14}$$

and

$$Y_e/Y = 1 - 0.4\left(\varrho\frac{B}{Y}\right)^{-1} \quad \text{for } 1 \le \varrho\frac{B}{Y} < \infty$$

(b) Flanged Wall Configuration and Thick Wall Configuration

$$Y_e/Y = \frac{Z}{Y} + \frac{Y'}{Y}\cdot\varrho\frac{B}{Y'}\left(1 - 0.4\varrho\frac{B}{Y'}\right) \quad \text{for } 0 \le \varrho\frac{B}{Y'} \le 1$$

and

$$Y_e/Y = \frac{Z}{Y} + \frac{Y'}{Y}\left(1 - 0.4\left(\varrho\frac{B}{Y'}\right)^{-1}\right) \quad \text{for } 1 \le \varrho\frac{B}{Y'} \le \infty \tag{15}$$

To examine the accuracy of these extended empirical equations, values of orthotropic effective width were evaluated for a wide range of stiffness ratios D_y/D_x (from ⅛ to 8), of ratios B/X (from 0.1 to 0.4) and B/Y (from 0.25 to 1.0), for both plane and flanged walls, and compared with the values obtained directly from a finite element analysis [19]. It was found that the discrepancies were everywhere less than 3½% and 1% for the plane and flanged walls, respectively.

2.3 Numerical Examples

To illustrate the application of the proposed method of evaluating the effective slab width, two examples are given for flanged wall configurations, for both equal and unequal flanges.

2.3.1 EXAMPLE 1–EQUAL FLANGES

It is required to find the effective width of a slab of aspect ratio $Y/X = 0.6$ coupling a pair of flanged walls with flange width ratio $Z/Y = 0.75$ and wall-opening ratio $B/X = 0.2$.

The reduced slab width $Y'/Y = 1 - Z/Y = 0.25$

The reduced span/width ratio $B/Y' = 0.2/0.6/0.25 = 1.33$

From Figure 7(a), the effective width $Y_e'/Y = 0.69$

The effective width $Y_e'/Y = 0.69 \times 0.25 = 0.173$

The effective wall strip $Z/Y = 0.750$

$\therefore$ The total effective width $Y_e/Y = 0.923$

It may be noted that this figure is within 1.8% of the accurate value of 0.939 obtained directly from a finite element analysis.

The absolute values of the effective width and stiffness may then be determined for any given slab dimensions.

2.3.2 EXAMPLE 2–UNEQUAL WALLS

It is required to evaluate the effective slab width for a structure with the following geometrical ratios:

$Y/X = 0.6$, $B/X = 0.4$, $Z_1/Y = 0.167$, $Z_2/Y = 0.333$

The reduced slab width $Y'/Y = 1 - Z_1/Y = 0.833$

The reduced flange width $Z'/Y = 0.333 - 0.167 = 0.167$

The reduced flange width ratio $Z'/Y = 0.167/0.833 = 0.20$

The reduced span width ratio $B/Y' = 0.4 \times 1/0.6 \times 1/0.833 = 0.8$

From Figure 7(b), the effective width $Y_e'/Y' = 0.61$

The effective width $Y_e'/Y' = 0.61 \times 0.833 = 0.508$

The effective wall strip $Z_1/Y = 0.167$

$\therefore$ The total effective width $Y_e/Y = 0.675$

This figure is within 0.8% of the value of 0.670 obtained directly from a finite element analysis.

3. STRESSES IN SLABS COUPLING SHEAR WALLS

This section describes an elastic analysis of the induced bending moments and shear forces in slab coupling pairs of shear walls. Contour diagrams and design curves showing the variation of critical bending moment factors are presented to enable a rapid evaluation of slab stresses induced by coupling actions. Once the maximum moment per unit width is known, the design of the slab cross-section follows from the standard procedure based on the particular Design Code employed. Based on the calculated interactive shear forces between wall and slab, a design method is suggested for checking against punching shear failure in the slab.

The finite element method of analysis [4] was again used to produce parametric studies from which design curves were devised for both plane and flanged walls.

3.1 Slabs Coupling Plane Walls

3.1.1 STRESS-RESULTANTS IN SLAB

General Distribution Pattern

Based on a finite element analysis, Figures 9(a) to (e) show in perspective the general distribution of bending and twisting moments and shearing forces in a typical quadrant of a slab coupling a pair of plane walls of zero thickness undergoing an arbitrary unit rotation as shown in Figure 9(f). The coupling action results in very large moments and shears being induced in the slab around the coupled end of the wall, but away from this critical region, the stress-resultants diminish rapidly in both the longitudinal and transverse directions. The significant coupling actions are confined to the corridor area, and the portion of the slab some distance away from the corridor remains practically unstressed and unaffected by the coupling action. This explains why an increase in wall length beyond a certain range does not affect significantly the coupling stiffness of the slab (cf Section 2.1.1.)

Problem of Singular Stress-Resultants

The presence of severe stress concentrations calls for careful interpretation of calculated critical stress-resultant values. If the calculated values are to be used directly for the design of the slab section, it is essential to ensure that the peak stress-resultants are theoretically finite and have been evaluated with sufficient accuracy in the numerical solution. It is therefore necessary to examine the convergence of calculated critical stress-resultant values in the slab around the inner corridor edge of the wall. For a typical geometrical configuration, Figure 10 shows quantitatively the variations of bending moment M_x along a critical transverse slab section passing through the inner edge of the wall, evaluated using various mesh-divisions of the slab. It can be seen that whereas values of M_x calculated at other points converge to

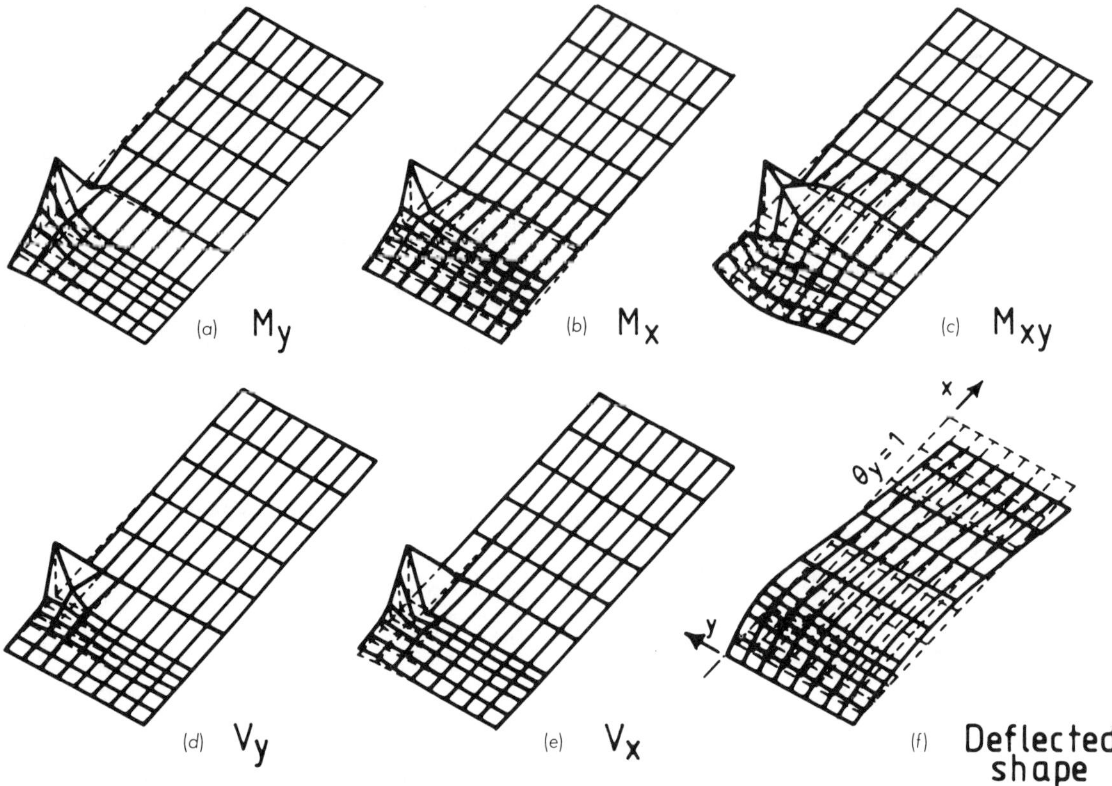

FIGURE 9. Perspective view of distributions of bending and twisting moments and shear forces induced in typical slab quadrant due to unit rotation of wall.

specific finite values, the bending moment calculated at the tip of the wall diverges with mesh refinement, and it may be concluded that a theoretical stress singularity exists in the slab at the end of the wall. Corresponding stress singularities arise in other similar situations, such as plates partially supported along an edge [12] or with an internal line support [15].

The results shown were obtained by assuming that the wall was represented by a line element. When the finite thickness is included, the results of a similar convergence study [19] indicate that large and apparently divergent stress-resultants are now obtained in the slab at the corners of the wall. Again, similar stress singularities arise in closely related problems such as the bending of a slab supported by a rigid column [17] or the bending of a slab with a rectangular opening [13].

The singular stress-resultants are predicted in the slab at the coupled end corners of the wall as a consequence of the use of thin-plate theory and boundary conditions associated with an infinitely rigid wall-support. Since the stress resultants in the slab at the singular points are theoretically unbounded, any finite stress-resultant values furnished at these points by a numerical solution based on thin-plate theory must be considered incorrect or meaningless.

Interpretation of Results at Singular Points

The apparent difficulty of obtaining meaningful results at the singular points in the coupling slab is not of great practical consequence, for, while the stress concentrations are a physical reality, the existence of infinite stresses at a point is obviously a mathematical fiction. The infinitely large stresses predicted will in reality be limited by local elastic deformations, material yielding or stress redistribution in the critically stressed areas in the wall and slab. It is therefore more practical to consider the net distributed forces in a small slab strip, rather than the extremely large stresses at a point, as a basis for the design of the slab section. With this approach, it is possible to interpret more meaningfully the finite element results for stress resultants in the critical region. Although the peak longitudinal bending moment M_x has no physical meaning if it is considered by itself at the point concerned, when combined with calculated bending moment values at other points along a transverse section it gives a correct statical balance between the integrated internal bending moment and the externally applied moment at the section. The integrated bending moment evaluated in the critical zone using calculated moment values is therefore correct, although the peak value by itself is not. The calculated peak bending moment M_x at the critical node should therefore only be used in conjunction with the calculated value at the adjacent node to give an estimate of the net bending moment distribution in the critical strip bounded by the nodes.

The peak critical transverse bending moment M_y is also evaluated in the analysis as a finite value which provides equilibrium between the internal transverse bending moment and the external moment, at the critical longitudinal slab section, and should therefore be interpreted in the same manner as that suggested for the primary bending moment M_x.

Stress-Resultant Factors

The moments and shear forces in the slab, calculated for the arbitrary unit wall displacements assumed in the analysis, may be expressed in the form of non-dimensional stress-resultant factors to facilitate the calculation of the stress-resultants due to any other wall displacements. These factors define the coupling stress-resultants in a slab of unit corridor width B and unit flexural rigidity D, induced either by a unit relative wall rotation $\varnothing$ at the coupled wall end or by a unit relative axial wall displacement δ, where $\delta = B\,\varnothing$. (cf Figure 1).

If M_i and V_i represent, respectively, the calculated bending moment and shear force in the slab, the corresponding stress-resultant factors $\bar{M}_i$ and $\bar{V}_i$ may be defined as

$$\bar{M}_i = \frac{M_i}{D}\frac{B}{\varnothing} \text{ and } \bar{V}_i = \frac{V_i}{D}\frac{B^2}{\varnothing} \tag{16}$$

The stress-resultant factors may be defined alternatively in terms of the relative axial wall displacement δ by replacing $\varnothing$ by δ/B. These factors may be expressed in a form involving the effective coupling width of the slab, Y_e, and the wall

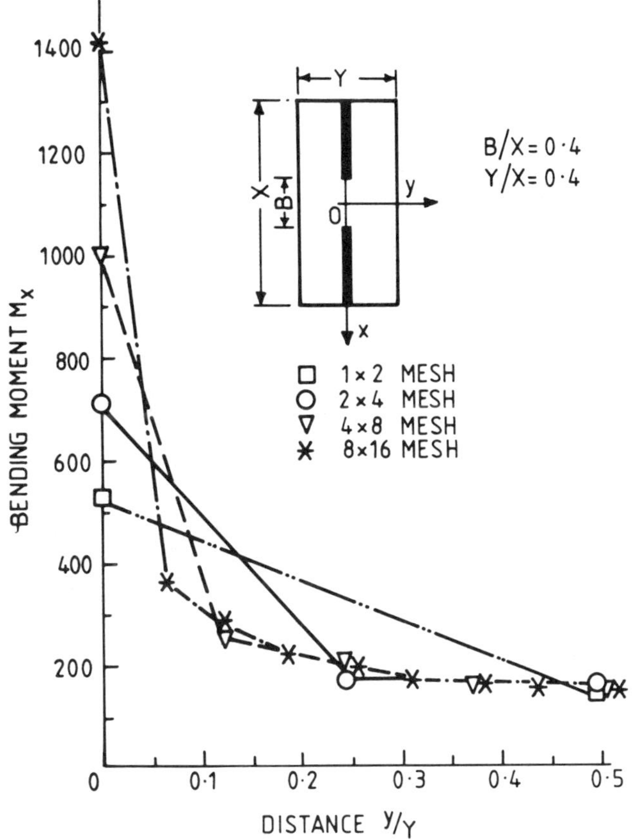

FIGURE 10. Convergence results for longitudinal bending moments at critical transverse slab section.

reaction or lintel beam shear V, [3],

$$\varnothing = \frac{VB^2}{12(1 - \nu^2) D\, Y_e} \qquad (17)$$

where D is the flexural rigidity for the slab, $[Et^3/12(1 - \nu^2)]$, and ν is Poisson's ratio.

Hence,

$$\bar{M}_i = \frac{12(1 - \nu^2) Y_e}{VB} M_i$$

and

$$\bar{V}_i = \frac{12(1 - \nu^2) Y_e}{V} V_i \qquad (18)$$

Since $VB/2\ Y_e$ and V/Y_e represent, respectively, the maximum average bending moment M_a and average shear forces V_a in the equivalent connecting beam, the expressions [18] may be rewritten as

$$\bar{M}_1 = 6(1 - \nu^2) \frac{M_i}{M_a}\ ;\ \bar{V}_i = 12(1 - \nu^2) \frac{V_i}{V_a} \qquad (19)$$

Once the stress-resultant factors $\bar{M}_i$ and $\bar{V}_i$ are determined for the slab, the true stress-resultants M_x, V_x, etc. produced by the system of forces from a coupled shear wall analysis may be determined. In an analysis based on the continuous medium approach, for example, the shear forces V in the lintel beams may be evaluated directly from established equations or design curves [2]. The critical slab bending moments and shear forces then follow directly from the preceding equations.

3.1.2 STRESS-RESULTANT CONTOUR DESIGN DIAGRAMS

It was shown in Section 2.1.1 that, provided the wall-length to wall-opening ratio W/B is greater than 0.5, the effective slab width is practically unaffected by variations in wall length. Considering the form of typical distribution of stress-resultants in the slab shown in Figure 9, it is clear that the stress-resultant factors which are of significance are also unaffected by variations in wall lengths. Hence, when evaluating stress-resultant factors for practical connecting slabs, it is necessary to consider only the influence of the wall-opening to slab-width ratio B/Y.

Figures 11(a) to (e) show the contours of bending moment factors $\bar{M}_x$ and $\bar{M}_y$ for slabs with practical ratios Y/B ranging from 1 to 4 and with a wall thickness ratio h/B of 0.1. The contours in each diagram are shown for a typical quadrant of the slab, and have been reproduced from the computer plots of the finite element results obtained with a fine mesh in the region of stress concentration. The contours of stress-resultant factors allow a rapid and accurate evaluation of bending moments M_x and M_y at any point on the slab, induced by any coupled wall-action, and may be used for the design and detailing of the slab reinforcement.

In all contour diagrams, the stress-resultant value at the critical wall corner has been indicated. Since the stress at this point is theoretically infinite, the value indicated in each diagram should be interpreted in the manner described earlier. When measuring distances to the respective contours, it should be noted that from the definition of stress-resultant factors, the wall-opening width B is represented as a unit distance in the contour diagrams.

Generalised Design Curve for Critical Bending Moments

In the early stages of a design study of the structural system, the stresses at critical slab sections are of immediate interest to the designer, who is required to determine realistic member sizes for the structure. A knowledge of the approximate variation of longitudinal bending moment M_x at a critical transverse slab section is usually sufficient to allow the adequacy of the chosen slab thickness to be established.

Figure 12 shows, for slabs of various normalised widths Y/B, the variation of bending-moment factor M_x along the critical transverse slab section through the inner edge of the plane wall. It can be seen from the curves for various ratios of Y/B that the variation of M_x within a distance given approximately by $y/B = 0.25$ from the wall is practically unaffected by variations in the slab width. Outside this critical region, the variation of M_x is influenced by the width, but the bending-moment factors are so small compared to values in the critical region that the influence of slab width becomes of little practical significance. For practical purposes, therefore, the bending-moment curves for various ratios of Y/B may be approximated by the generalised curve shown by the broken line in Figure 12. This generalised curve allows the variation of M_x to be obtained conveniently for any slab by truncating the tail of the curve at the limit of the slab width. The generalised curve for M_x is presented in more detail in Figure 13. Values of ordinates at various non-dimensional distances $\xi\ (= y/B)$ have been inserted along the curve for convenience in the evaluation of M_x. The value of M_x at $\xi = 0$, which is given as 20.0, should only be used in conjunction with the adjacent values shown to give an estimate of the bending moment in the most critically stressed slab-strip as explained earlier. The integrated bending moments at various slab strips, $M_x . \Delta\xi$, have been evaluated and tabulated below the curve of M_x to facilitate the calculation of steel ratios, if required, for the slab.

It may be noted that the generalised curve of M_x may also be used to evaluate approximately the effective width of any slab for checking purposes. Since the integration of the bending stress-resultant M_x at the transverse section must equal the value of the external moment, the area under the curve of M_x may be evaluated to obtain the moment-rotation relationship leading to the calculation of effective slab

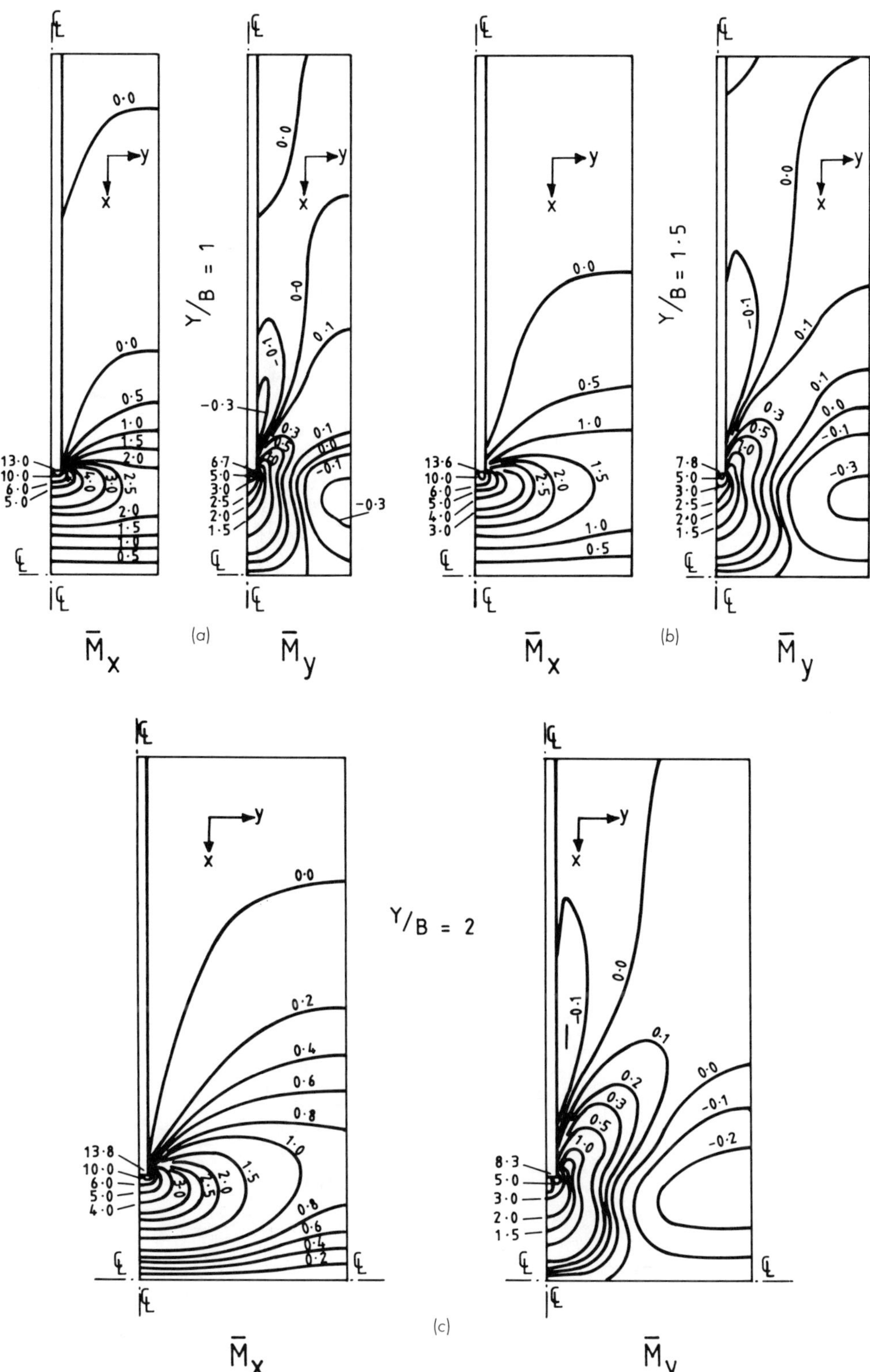

FIGURE 11. Bending moment contours for slab with panel ratios Y/B = 1(a), 1.5(b), 2(c), 3(d), 4(e). (continued)

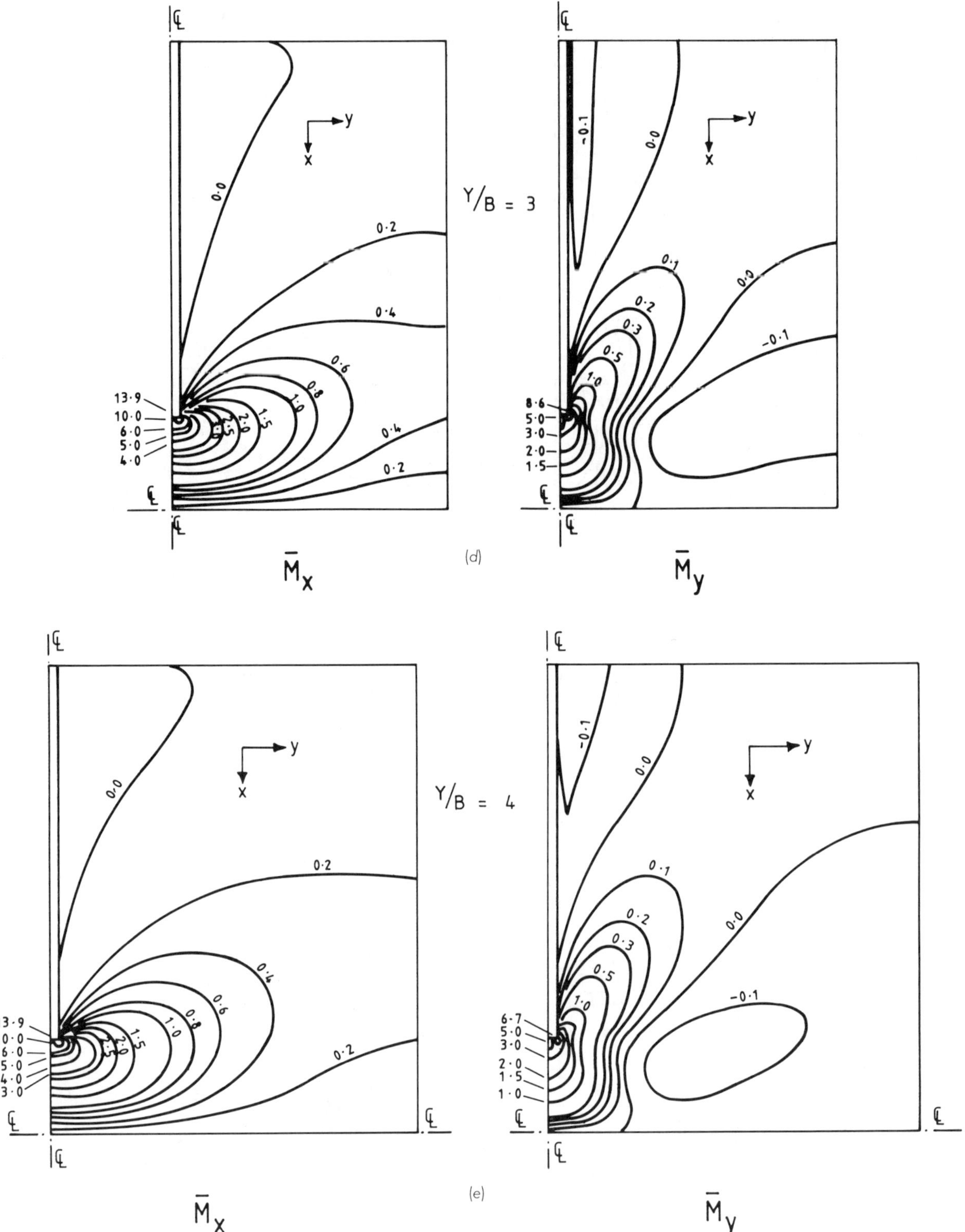

FIGURE 11. Bending moment contours for slab with panel ratios Y/B = 1(a), 1.5(b), 2(c), 3(d), 4(e).

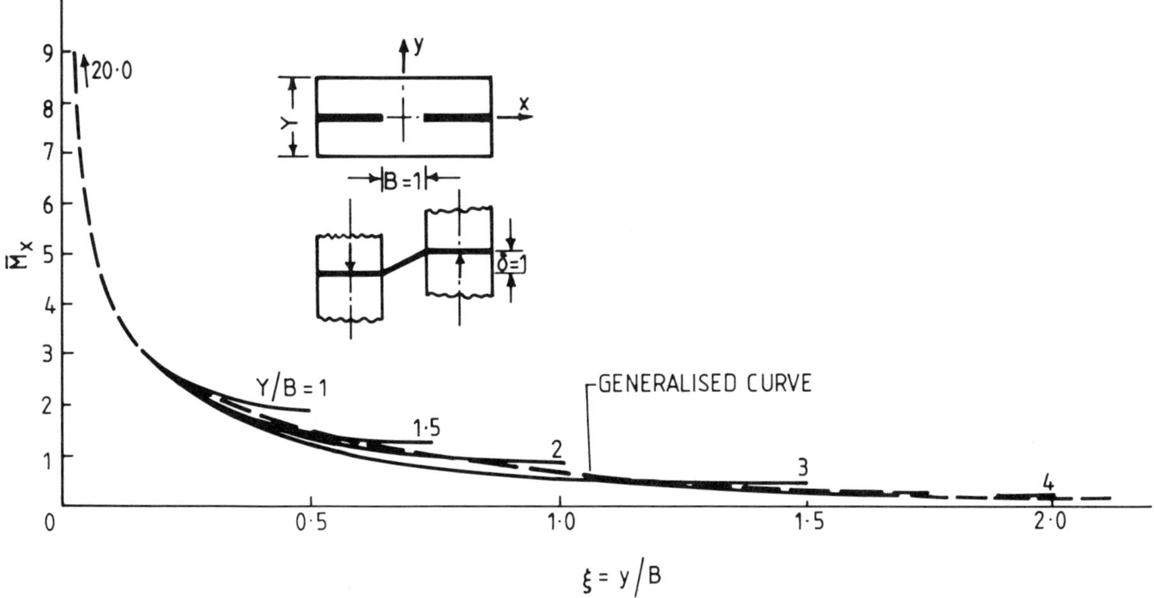

FIGURE 12. Variation of longitudinal bending moments at critical transverse slab section for various slab widths.

width. If the double-area under the truncated curve of M_x for the slab of width Y is K_a, then it can be shown using equation [19] that

$$\frac{Y_e}{Y} = \frac{1}{6(1 - \nu^2)}\left(\frac{B}{Y}\right) K_a \qquad (20)$$

To facilitate the calculation of effective width in this way, the curve for K_a has been included in Figure 13 with the generalised curve for M_x.

An illustration of the use of the curve for K_a for the calculation of effective widths was given in Reference [4].

3.1.3 SHEAR TRANSFER BETWEEN WALL AND SLAB

The general forms of distribution of shearing forces V_x and V_y induced by coupling action throughout the slab have been exemplified in Figures 9(d) and (e). The shearing force distribution in the vicinity of the wall boundary is now examined in more detail for the typical slab geometry considered.

Figures 14(b) and (c) show, respectively, the variations of the shearing forces (stress-resultants) V_y and V_x along two perpendicular sections AB and BC passing through the toe of the wall [Figure 14(a)], the numerical values having been evaluated at the nodal points in the finite element analysis. The shearing forces V_y along the wall boundary represent the continuously distributed shear transferred from the wall to the slab.

The shearing forces V_x and V_y at the inner corridor edge of the wall are also theoretically infinite, but have been evaluated as arbitrary finite values which vary according to the fineness of the mesh used. It is not possible to use the same averaging procedure adopted with the moments M_x and M_y, since a convergence study [19] has shown that a shear discontinuity always exists between the singular and the adjacent node and that the calculated positive and negative shearing stress-resultants at the respective nodes are divergent (apparently to maintain equilibrium) with refinement of element mesh around the singular point. The apparent finite width over which the critical positive shearing forces are distributed is also arbitrary because in the limit of mesh refinement this will be reduced to a point. With these limitations on the calculated shear distribution, it is not possible to evaluate quantitatively the actual form of shear transfer between wall and slab, though it may be concluded from the form of distribution indicated that the shear transfer must be effected essentially as a very large reaction over a very short length at the inner edge of the wall, together with much smaller opposite reactions distributed over the rest of the wall. This form of shear transfer is consistent with observed punching shear failures in coupling slabs [14].

Critical Peripheral Slab Section for Shear

For design purposes, it is not absolutely necessary to know the exact values of shearing forces at points in the slab very close to the wall since at these points the shear forces

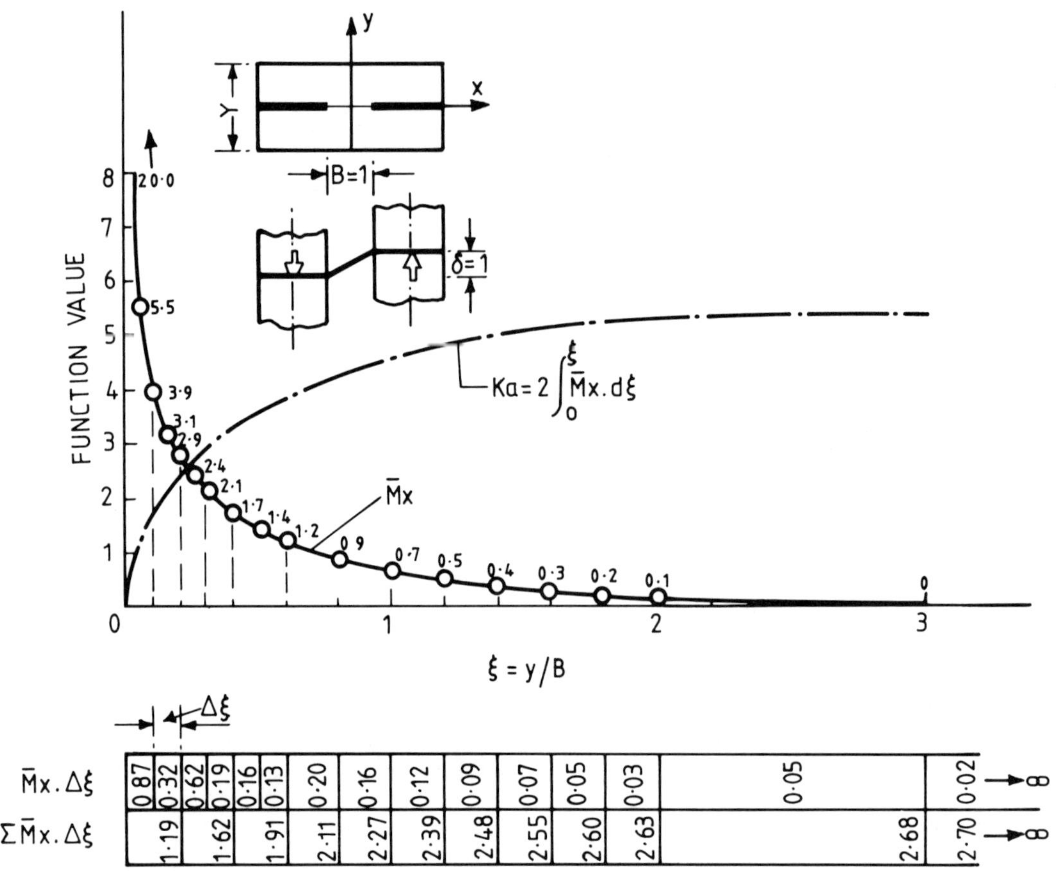

FIGURE 13. Generalised design curve for longitudinal bending moments at critical transverse slab section.

are carried primarily by a strut and tie action in the concrete and steel, with the shear resistance of the slab acting only as a secondary load-carrying medium [Figure 14(d)]. The critical slab section which has to be checked against shear failure is located at some distance U from the face of the wall where frame action in carrying shear is no longer possible. At such sections in the slab, the shear forces predicted by thin-plate theory are finite, and values calculated by the finite element method using sufficiently fine mesh divisions are quite satisfactory for design purposes.

Recommendations for the location of the critical section for shear near column supports in flat slab structures are given in various design Codes [1,16], and these recommendations are assumed applicable for wall-slab structures since both types of structure involve the same form of punching shear failure in the slab. The critical distance U for ultimate strength design is usually given in terms of the effective depth d or the overall depth t of the slab depending on the Code in use, and the shear at failure is assumed uniformly distributed over a peripheral section located at this critical distance from the face of the support.

Shear Distribution at a Typical Peripheral Section

Figure 15 shows the distribution of normalised shearing force factors V_x and V_y along a peripheral section located at a distance of $U/B = 0.1$ from the face of the wall in a slab of normalised width $Y/B = 2$. This distance represents a practical location for the critical shear section required by most codes when applied to the design of cross wall-slab structures.

An equilibrium check at this section showed that the internally and externally applied moments and shears were equal to within acceptable practical levels of accuracy of 2.3% and 0.1%, respectively. The check also indicated that the applied wall moment was resisted primarily by vertical

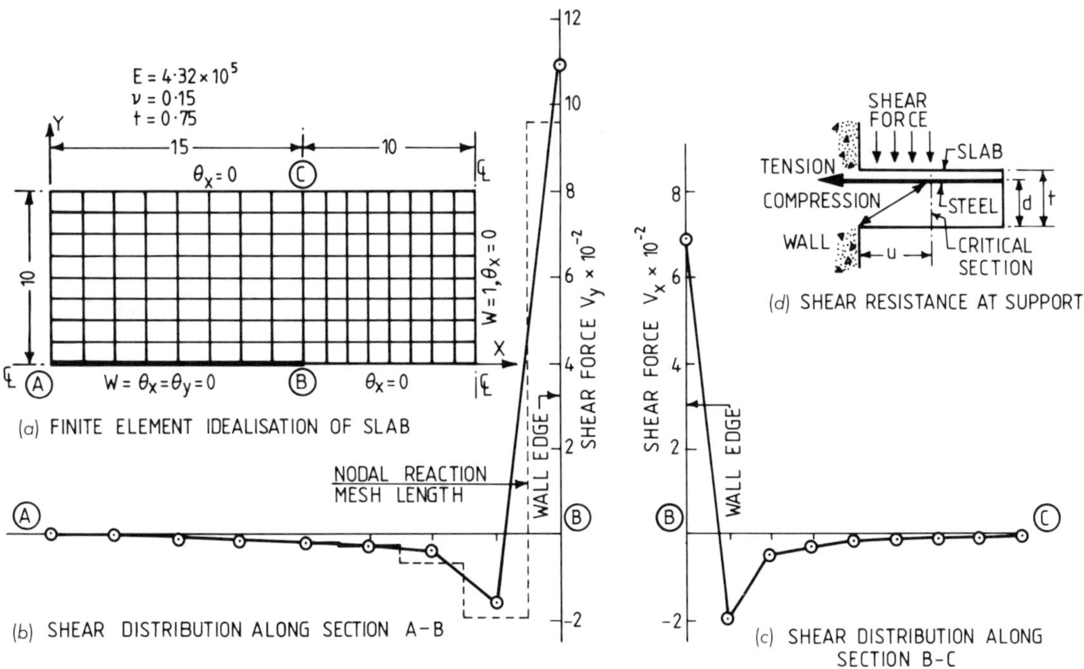

FIGURE 14. Shear force distribution at wall support.

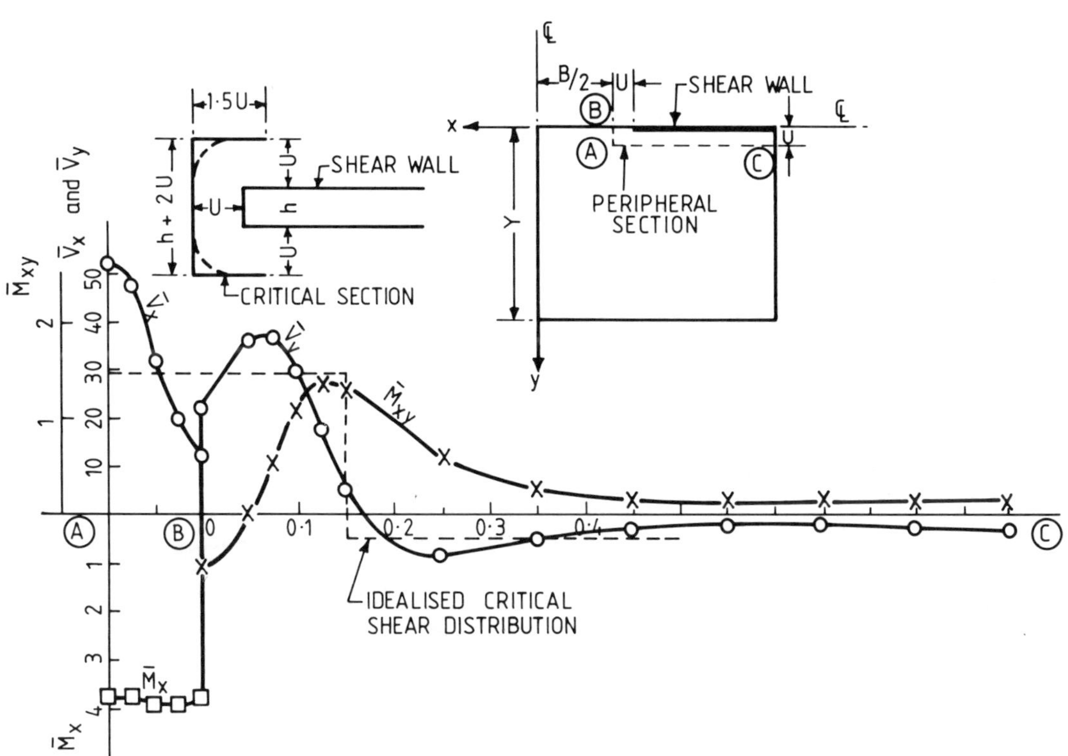

FIGURE 15. Shear force distribution at critical peripheral section.

shearing actions (more than 80%) and very little by bending and twisting actions.

Dimensions of Critical Shear Section

It is obvious from the shear force distributions that the design of the slab against possible shear failure will be governed critically by the substantially higher positive shear distributed over a relatively small section at the front of the peripheral section. For the purpose of design, it is convenient to assume that the critical shear at failure is uniformly distributed over a critical section. With the shear distribution shown in Figure 15, an obvious choice is to consider the whole U-shaped section acted on by the positive shearing forces around the inner edge of the wall as the critical section. Depending on the design Code adopted in practice, the corners of the U-shaped critical section may be taken as square or rounded (Figure 15) to be consistent with similar Code provisions for flat-slab design. The latter case provides a more realistic approximation of uniform shear distribution, since an approximately radial dispersion of the critical shear is indicated by the shape of the actual shear distribution curve. The open ends of the U-shaped section

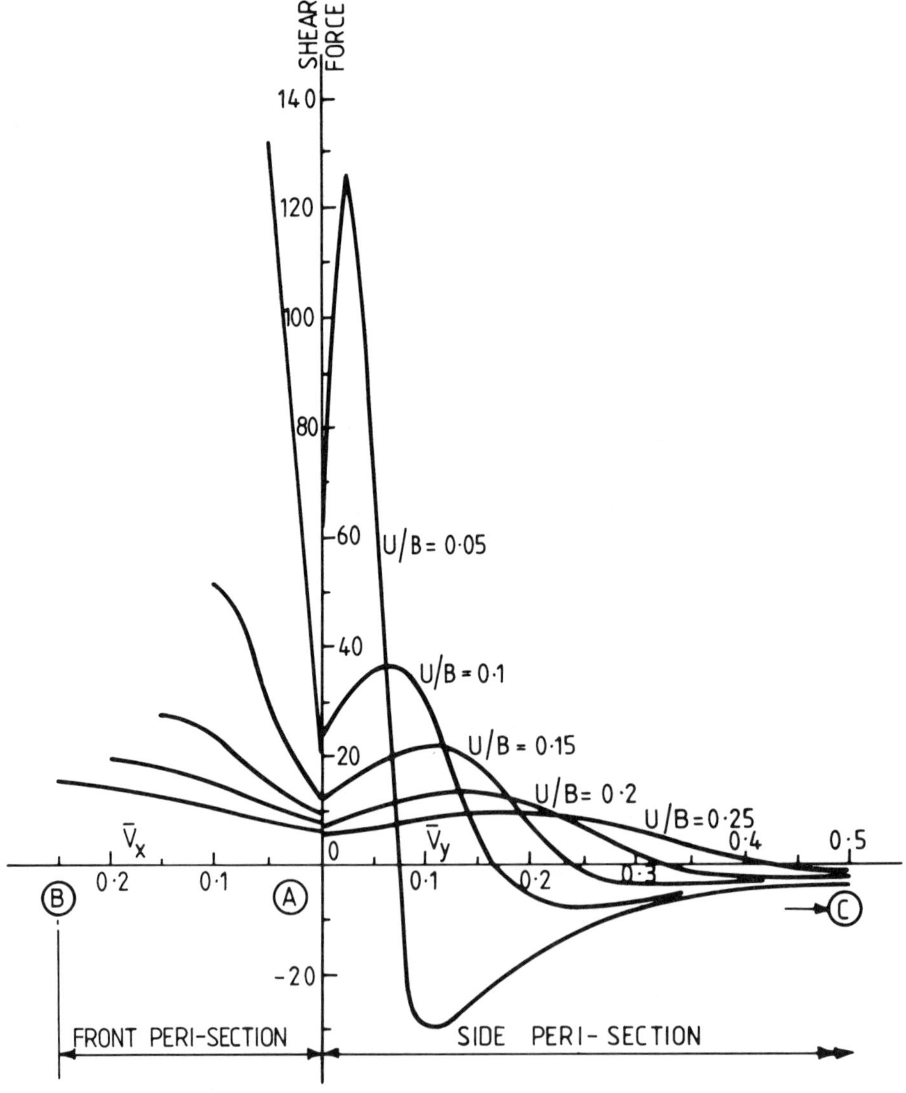

FIGURE 16. Shear force distributions at various peripheral sections.

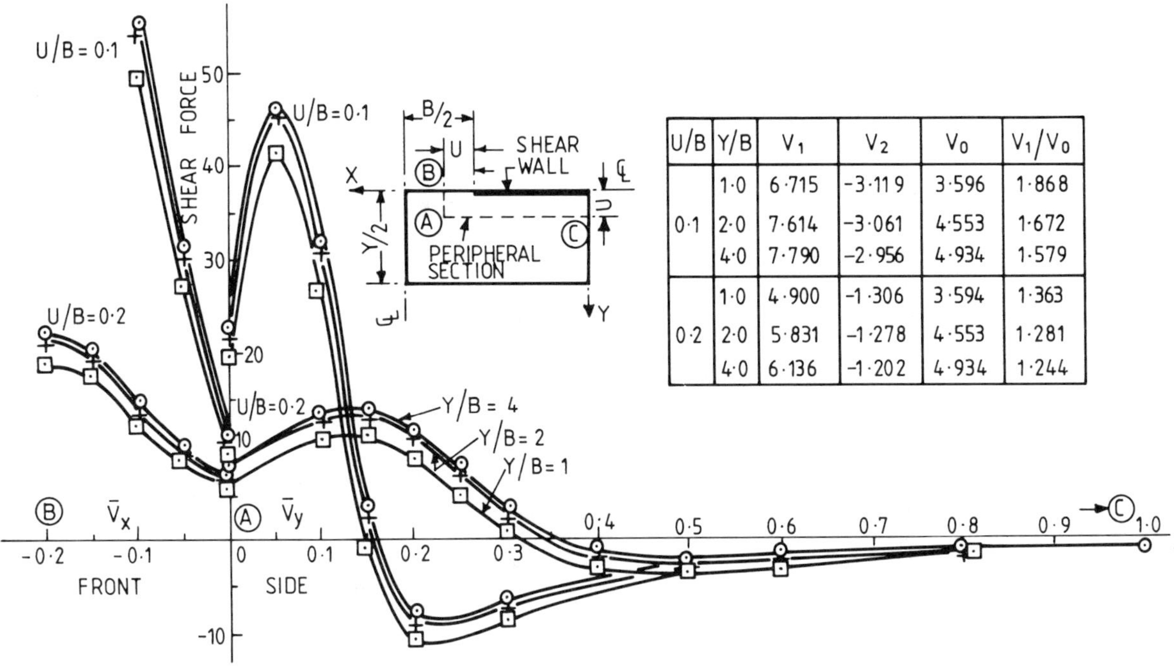

U/B	Y/B	V_1	V_2	V_o	V_1/V_o
0·1	1·0	6·715	-3·119	3·596	1·868
	2·0	7·614	-3·061	4·553	1·672
	4·0	7·790	-2·956	4·934	1·579
0·2	1·0	4·900	-1·306	3·594	1·363
	2·0	5·831	-1·278	4·553	1·281
	4·0	6·136	-1·202	4·934	1·244

FIGURE 17. Peripheral shear force distributions for slabs of various widths.

can be extended to the point where the calculated shearing force changes sign, at a distance of approximately 0.5 U measured back from the front (inner) edge of the wall, the length of each leg of the U-shaped section being taken as 1.5 U. As shown by the shear distributions for various peripheral sections (Figure 16) and for various slab widths Y/B (Figure 17), this convenient side-length is consistent for all possible locations of the critical section. The overall length of the critical section can therefore be generally taken as $h + 5U$ for the square-cornered critical section and $h + (1 + \pi)U$ for the round-cornered section, h being the wall thickness.

Shear Modification Factors

The positive shear force developed in the critical shear section as shown by the various shear distribution curves in Figures 16 and 17 is substantially greater than the applied shear force and varies in value with the location of the peripheral section and with the slab width Y/B. The ratio of the critical positive shear force V_1 to the applied shear force V_o, which may be considered as a "shear modification factor," K_q, has been calculated for various locations of the peripheral section in slabs of normalised widths $Y/B = 1,2$ and 4 for walls of zero thickness. The curves showing the variation of the shear modification factor K_q with the peripheral distance U/B are presented in Figure 18, and may conveniently be used as design curves for evaluating the critical shear at any location of the critical section.

Influence of Finite Wall Thickness on Shear Force Distribution.

The shearing force distributions on which the derivation of the critical shear section and the shear modification factors have been based do not account for the influence of finite wall thickness. To illustrate the influence of wall thickness, the shear distributions for walls of zero thickness and a thickness ratio $h/B = 0.1$ are compared at a peripheral section at $U/B = 0.1$ in Figure 19. The distributions of $\bar{V}_y$ along the sides of the peripheral section for the two cases are almost identical, while the distributions of $\bar{V}_x$ along the front end of the peripheral section show some significant differences. The inclusion of the wall thickness and a wider peripheral section produces lower shearing forces in front of the wall, although the total integrated positive shear distributed over the front ends of the two peripheral sections are not significantly different. The shear modification factor, which is of more relevance for practical shear strength design, is approximately 6% higher than the value for the wall of zero thickness. Since neglecting the effects of wall thickness in the calculation of shear modification factors will result in a slightly conservative estimate of the critical design shear force in the slab, the design curve presented in

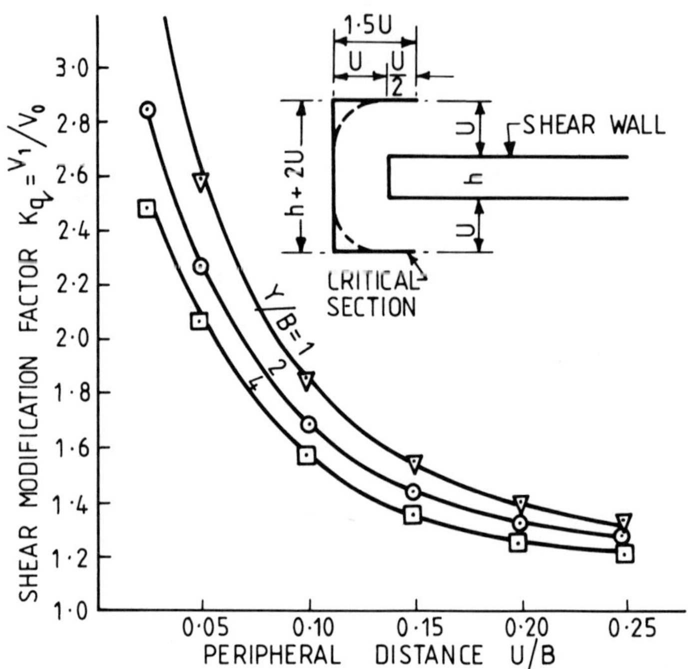

FIGURE 18. Design curve for shear modification factor.

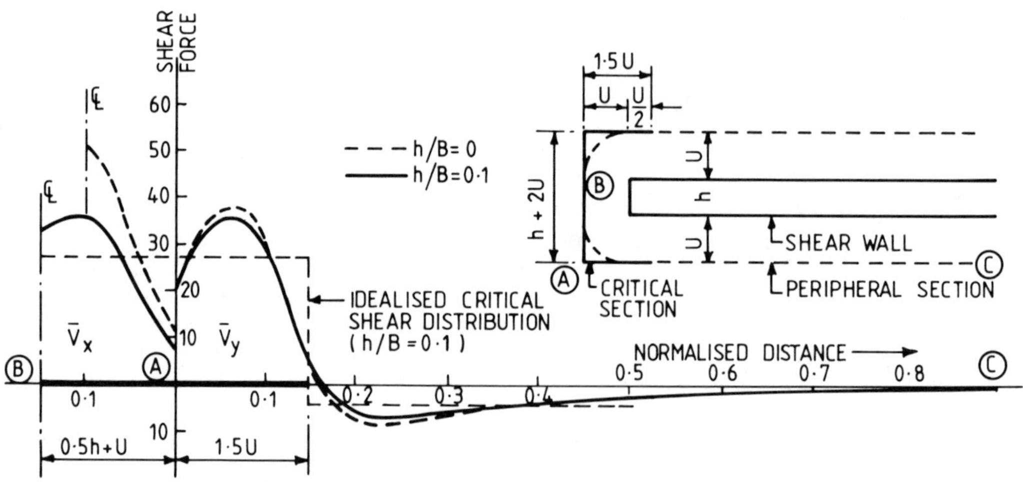

FIGURE 19. Effect of finite wall thickness on peripheral shear force distribution.

Figure 19 should be satisfactory for design purposes when walls are not exceptionally thick (say $h < U$).

Example of Critical Section and Stress Modification Factor

To illustrate the application of the critical section and the shear modification factor, typical punching shear calculations consistent with the American design Code ACI 318-83 [1] and the British design Code CP110 [16] are given in the following example:

The floor slab coupling a pair of plane walls in a 20-storeyed, 170.6 ft (52 m) high cross-wall structure, which has been analysed separately for wind effects, is considered. The typical dimensions of the slab are

$$Y = 20 \text{ ft } (6.1 \text{ m}), X = 50 \text{ ft } (15.25 \text{ m}), B = 5 \text{ ft } (1.525 \text{ m})$$

The wall thickness h and the slab thickness t are both 7.88 in (200 mm). The effective depth of the slab d is 6.69 in (170 mm). The maximum lintel shear at an ultimate wind load of 35.1 lb/ft² (1.68 kN/m²) has been calculated as $V_o = 15.0 \times 10^3$lbf (66.8 kN).

DESIGN TO ACI 318-83

The critical section is located at $U = 0.5d$ and the corners of the critical section are square. The shear area is then $(h + 2.5d)d$. Working in U.S. Customary units, the critical distance $U/B = 0.5 \times 6.69/60 = 0.06$. The shear modification factor from Figure 18 is $K_q = 1.92$. The critical shear stress is, then,

$$v = \frac{1.92 \times 15.0 \times 10^3}{(7.88 + 2.5 \times 6.69)6.69} = 175.0 \text{ lb/in}^2$$

This should not exceed the permissible ultimate shear stress of $4\sqrt{f_c'}$; otherwise, special provision has to be made to increase the shear capacity of the slab.

Assuming a Grade 25 concrete (25 N/mm²) and taking the cylinder crushing strength f_c' as 0.78 times the cube strength of concrete,

$$f_c' = 0.78 \times 25 \times 145 = 2827 \text{ lb/in}^2$$

The permissible ultimate shear stress is, then,

$$v_c = 4\sqrt{f_c'} = 4\sqrt{2827} = 212.7 \text{ lb/in}^2$$

which is 1.22 times the design shear stress.

DESIGN TO CP.110

The critical section is located at $U = 1.5\,t$ from the face of the wall, and corners for the critical section are rounded. The shear area of the critical section is, then,

$$(h + (l + \pi)\, 1.5t)d.$$

Critical distance $U/B = 1.5 \times 7.88/60 = 0.2$

Slab width ratio $Y/B = 20/5 = 4$

Shear modification factor K_q from Figure 18 is

$$K_q = 1.25$$

Critical shear stress is, then,

$$v = \frac{K_q \times V_o}{[h + (1 + \pi)\, 1.5t]\, d}$$

$$= \frac{1.25 \times 15.0 \times 10^3}{(7.88 + 4.14 \times 1.5 \times 7.88)\, 6.69}$$

$$= 49.3 \text{ lb/in}^2 \ (0.34 \text{ N/mm}^2)$$

This value should not exceed the permissible shear stress determined in accordance with C1.3.4.5 and Tables 5 and 14 (CP110). Assuming a Grade 25 concrete and 0.25% steel area, the permissible shear stress is given by

$$\xi_s v_c = 1.10 \times 51.5 = 56.6 \text{ lb/in}^2 \ (0.39 \text{ N/mm}^2)$$

which is 1.15 times the design shear stress.

It is seen from the above calculations that although the shape and location of the critical shear section assumed in accordance with ACI 318-83 are distinctly different from those assumed in CP 110, the ratios of design ultimate shear stress to permissible ultimate shear stress from the two different calculations are not substantially different.

3.2 Stresses in Slabs Connecting Flanged Walls

3.2.1 STRESS-RESULTANTS IN SLAB

It was stated earlier that in a laterally loaded slab-coupled wall structure, virtually all coupling actions between walls take place in the corridor area, and a large portion of the slab some distance back from the inner edges is subjected to negligible bending actions. In particular, for walls with corridor flanges, the slab regions behind the flanges are practically stress-free.

For T-shaped walls with external facade flanges (Figure 20a) the flanges play little part in the coupling actions and may be ignored. The results presented in Section 3.1 for plane walls may then be used for the design of the slab and the wall-slab junction.

For T-shaped walls with interior flanges (Figure 20b), the significant coupling actions arise mainly in the corridor area, with high concentrations of bending moments and shear forces occurring around the tips of the flanges. The bending moments M_x and M_y, and the longitudinal shear force V_x, have their greatest values in the strip of corridor slab directly connecting the wall flanges. Because of the

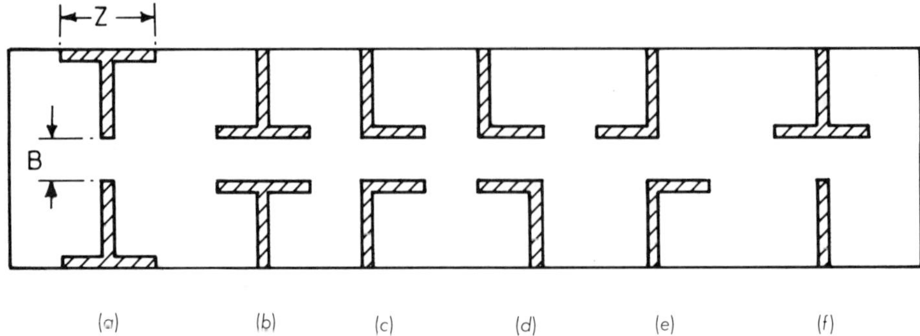

FIGURE 20. Coupled flanged shear wall configurations.

strong coupling action, which tends to produce cylindrical bending of the slab between the flanges, the longitudinal stress-resultants M_x and V_x are much larger than the transverse moments M_y and shears V_y, and the twisting moment M_{xy}, and dominate the slab design.

As in the case of thick plane walls, it may be deduced that a theoretical stress-resultant singularity occurs at the tip of the wall flange. If the distribution of longitudinal bending moments along the critical transverse slab section at the inner edge of the flange is examined in a convergence study, it is found that whereas values of M_x calculated at other points converge to finite values, the value at the flange tip diverges with mesh refinement, indicating the presence of a singularity at that point. Consequently, it becomes necessary to interpret the local results in the manner suggested in Section 3.1 and consider the net distributed moments in a small but finite slab strip at that position, rather than the extremely large values at the tip node point, as a basis for the design of the slab section.

As before, it is convenient to express all stress-resultant values as ratios of the average bending moment M_a or shear force V_a in the equivalent connecting beam. The stress-resultant factors $\bar{M}_i$ and $\bar{V}_i$ are then defined as,

$$\bar{M}_i = 6(1 - \nu^2)\frac{M_i}{M_a} \qquad (i = x \text{ or } y) \qquad (21)$$

$$\bar{V}_i = 12(1 - \nu^2)\frac{V_i}{V_a} \qquad (i = x \text{ or } y) \qquad (22)$$

where M_i and V_i are the bending moment and shear force stress-resultants, and ν is Poisson's ratio. The values of M_a and V_a may be derived from an overall analysis of the structure using the effective coupling stiffness of the slab established earlier.

Generalised Design Curve for Critical Bending Moment Factors

Curves showing the variations of the longitudinal bending moment factor M_x along the critical transverse slab section at the inner edge of the flange (x = −0.5B) for slabs with various wall-opening to slab-width ratios B/Y and flange width ratios Z/Y are presented in Figure 21 (a) to (f). The six sets of curves have been drawn with distance normalised with respect to the half panel width [$\eta = y/(0.5Y)$]. On comparing the sets, it is seen that the curves for different flange width ratios are very similar in form. If the curves are re-drawn with the tip of the flange as the origin for the η-axis, all the curves in a set practically coincide over the greater part of the curve. For the first five sets of curves with B/Y less than 1.0, M_x generally remains constant at a value of 6.0 along the flange until close to the flange tip where it increases rapidly, reaching a peak at the tip. Beyond the end of the flange, M_x decreases very rapidly to a relatively low value and thereafter the decrease becomes very gradual. For the last set of curves with $B/Y = 1$, values of M_x along the flange are generally higher than 6.0 when the flange width ratio Z/Y is small. The peak values of M_x for each set of curves are also higher with smaller ratios of Z/Y. Comparing values of M_x at corresponding points in the slabs with different corridor or wall-opening widths B, it can be seen that at points outside the flange area, M_x is higher for larger ratios of B/Y.

The six sets of curves for M_x may be used directly as design curves to evaluate the critical bending moments which govern the design of the floor slab. However, because of the large number of individual curves involved and the need to interpolate between them, these curves become less convenient when critical moments have to be evaluated rapidly, but only approximately, in the preliminary stages of design. Since the curves of M_x for various ratios of B/Y and Z/Y are similar in form, particularly when they are plotted with the flange tip as origin and the distances are normalised with respect to B instead of $0.5Y$, it is possible to approximate the whole series by a single generalised curve which is sufficiently accurate for practical purposes for slabs with normal ratios of B/Y and Z/Y, and is therefore more convenient to use. The generalised design curve which shows the variation of $\bar{M}_x$ with the normalised distances ξ ($\xi = y/B$) from the flange tip is presented in Figure 22. To obtain the

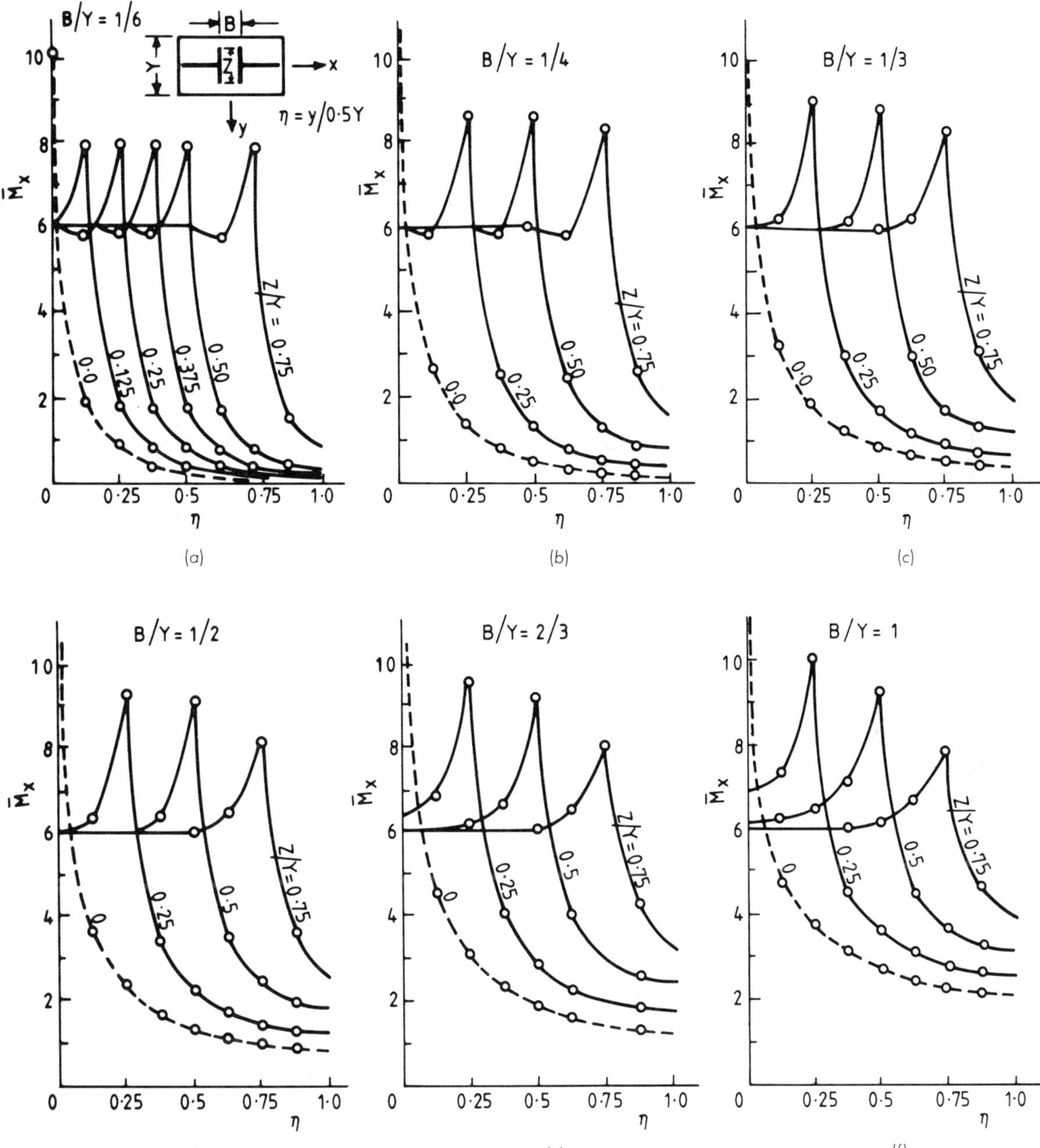

FIGURE 21. Variation of longitudinal bending moments at critical transverse slab section for T-shaped flanged wall configuration.

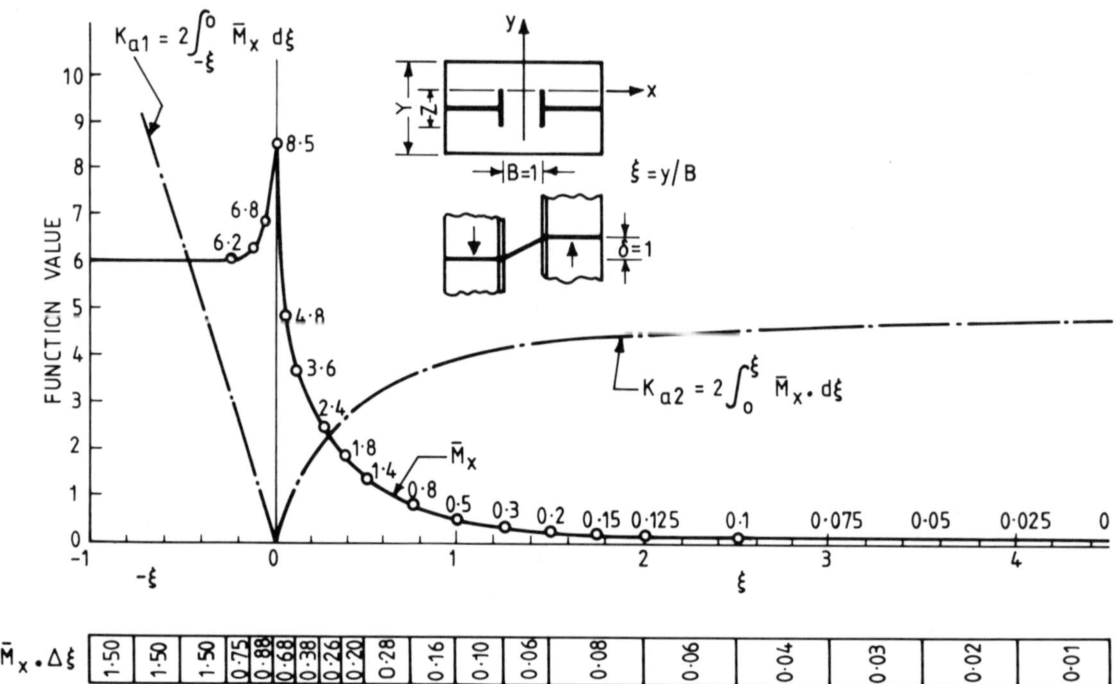

FIGURE 22. Generalised design curve for bending moments at critical transverse slab section for T-shaped walls.

distribution of $\bar{M}_x$ for a slab of width Y with flange width Z and wall opening B the generalised curve is simply truncated at the points $\xi_1 = -0.5Z/B$ and $\xi_2 = 0.5\,(Y - Z)/B$ these values corresponding, respectively, to the centreline and edge of the slab.

The peak value of M_x at the origin of the generalised design curve has been taken as 8.5, this being the approximate average for the range of slabs considered. Although this value may appear to differ considerably from individually calculated peaks in some cases, it must be remembered that any such calculated peak on its own is quite arbitrary in absolute value because of the stress singularity discussed earlier. The average peak value of 8.5, when used in conjunction with the adjacent values indicated on the generalised curve, will give an acceptable estimate of the total moment in the region of the stress concentration.

As in the case of plane walls, the generalised curve for M_x may be used to evaluate the effective slab width Y_e for the slab. Since the integration of the bending stress-resultant M_x at the transverse section must equal the externally applied moment, the area under the curve of M_x may be evaluated to obtain the moment-rotation relationship and hence Y_e. If the double-area under the truncated curve of M_x for the slab of width Y is K_a, then it may be shown that

$$\frac{Y_e}{Y} = \frac{1}{6(1 - \nu^2)}\left(\frac{B}{Y}\right) K_a \qquad (23)$$

For this purpose, the segmental axes $M_x\Delta\xi$ have been tabulated beneath each segment of the curve of M_x, and these may be easily summed over the appropriate length of the curve for the particular slab. The area integrals, $K_{a1} = 2\int_{\xi_1}^{0} \bar{M}_x\, d\xi$ and $K_{a2} = 2\int_{0}^{\xi_2} \bar{M}_x\, d\xi$, evaluated numerically from the summation of segmental areas have been plotted as supplementary design curves along with the generalised curve in Figure 22.

A numerical example illustrating the use of the curve for K_a for the evaluation of the slab effective width was given in Reference [5].

Critical Shear Distribution

Concentrations of shearing forces are developed in the slab around the interior edges of the wall flanges, and, although these may not under normal circumstances lead to punching shear failure in the slab, it is desirable to assess the stress levels which may arise in regions of stress concentration under working or ultimate load conditions. Using the same concept of a critical section and shear modification factors proposed for slabs coupling plane walls, a convenient design procedure for punching shear calculations is again possible.

Figure 23 shows the normalised shear force distribution at various peripheral slab sections around a flanged wall with a normalised flange width of $Z/B = 1$ and slab width of $Y/B = 2$. Referring, for illustration, to the shear distribution curve for the case of $U/B = 0.05$, it is seen that the highest positive shearing forces $\bar{V}_x$ and $\bar{V}_y$ for the particular peripheral section are confined within a small region at the corner of the section. Over a considerable length of this section in front of the flange, the positive shear force $\bar{V}_x$ remains relatively low, at an approximate value of 12.0 (this value being incidentally the value of the effective shear force obtained by distributing the applied shear uniformly over the effective slab width). Around the corner and along the side of the peripheral section, the critical positive shear force $\bar{V}_y$ changes to a negative shear force at a point approximately 1.25 U from the corner.

For design purposes, it is convenient to assume that the positive shear in each half of the slab is uniformly distributed in two distinct zones instead of being distributed according to the theoretical shear distribution curve. The critical design shear force may be assumed to be distributed uniformly over a critical section near the end of the flange, and the shear force in the remaining front portion of the peripheral section assumed to have a value equal to the effective shear force, i.e., $12(1 - \nu^2)$. The critical shear section can be taken as L-shaped, with maximum dimensions of 2U × 1.5U (Figure 23), the recommended dimensions being consistent for other peripheral sections. The corner of the L-shaped critical section may be taken as square or rounded to be consistent with similar recommendations for flat slab or flat plate structures in existing design Codes.

The critical section suggested for shear design has been derived by considering the shear distribution at peripheral sections in a slab with a particular slab width ratio Y/B and flange width ratio Z/B. To illustrate the influence of slab width and flange width on the shear distribution at a peripheral section, the peripheral shear distribution curves defined by $U/B = 0.1$ for slabs with different half flange-width ratios $\xi_1 = 0.5Z/B$ and flange-opening ratios $\xi_2 = 0.5(Y - Z)/B$ are compared in Figure 24. By comparing the shear distribution for the case of $\xi_1 = 0.25$ and $\xi_2 = 0.5$ with the shear distribution for the case of $\xi_1 = 0.5$ and $\xi_2 = 0.5$, it is seen that changing the flange width ratio ξ_1, merely alters the length of the shear distribution curve for $\bar{V}_x$ over the uniform effective shear force region, the shear distribution curve for the critical shear force region remaining practically unaffected by the change in the flange width. By comparing the case of $\xi_1 = 0.5$ and $\xi_2 = 0.5$ with the case of $\xi_1 = 0.5$ and $\xi_2 = 1.0$, on the other hand, it is seen that increasing the slab width increases substantially the positive shear forces in the critical section. The forms of shear distribution and the shapes of the critical shear distribution curve in all three cases are, however, very similar, indicating that the proposed critical design shear section is quite general and may be applied to slabs of any width with walls of any flange width Z which is not smaller than twice the critical peripheral distance U.

Shear Modification Factors

The total positive shear at the peripheral section is in excess of the applied shear and only a portion of the positive shear is distributed over the critical section. In order to facilitate the calculation of critical shears for design, shear modification factors may be calculated for various locations

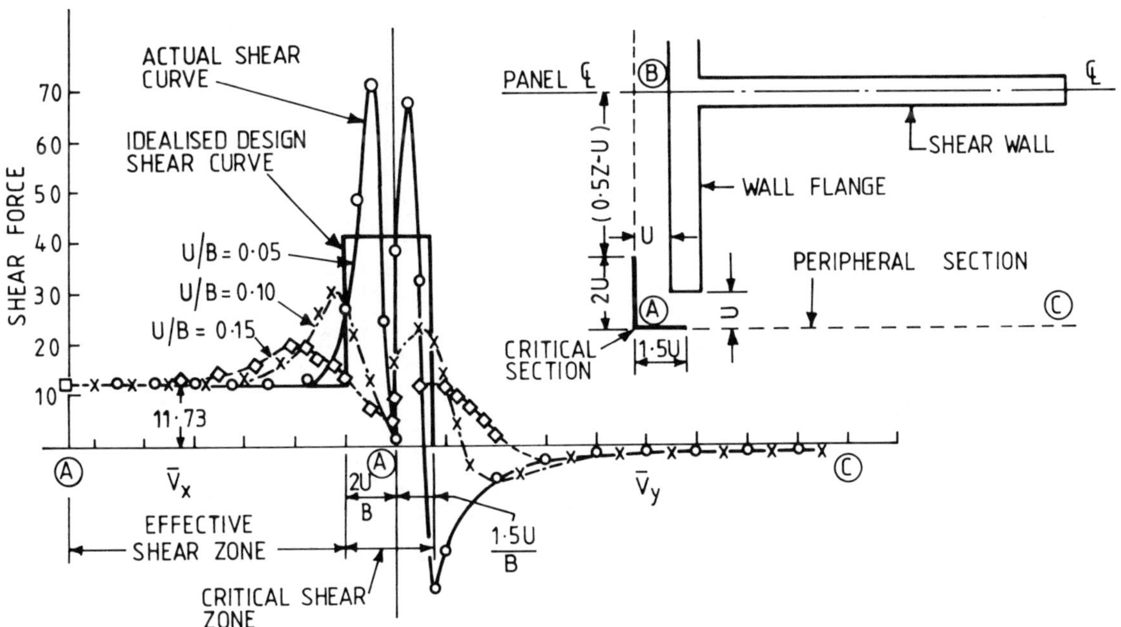

FIGURE 23. Shear force distributions at peripheral sections for T-shaped walls.

of the critical section using the peripheral shear distribution curves. It has been shown that along the middle front portion of the peripheral section, the positive shear forces are essentially uniformly distributed, with a value approximately equal to the effective shear force. Since this portion of the slab is fully effective in coupling the flanges, a change in the flange width produces a change only in the positive shear in this "effective shear zone", equal in value to the change in the applied shear force required to induce unit overall wall displacements. The positive shear in the "critical shear zone" remains practically unchanged as long as the flange is sufficiently wide ($Z > 2U$). Thus, if the critical shear is expressed in terms of the portion of the applied shear with which it is associated, the resulting shear modification factor will be influenced less by the flange width.

Denoting the shear in the "effective shear zone" as V_3, the shear modification factor K_q can be expressed as

$$K_q = (V_1 - V_3)/(V_o - V_3)$$

where V_1 and V_o represent, respectively, the total positive shear in the peripheral section and the total external applied shear.

In the calculation of shear modification factors, V_1 is obtained from the numerical integration of the positive areas under the actual curves of V_x and V_y for a peripheral section, and V_3 is obtained simply from the idealised effective shear distribution.

The shear modification factors evaluated for various peripheral sections in slabs with flange opening width ratios $\xi_2 = 0.25$, 0.5 and 1 are shown in Figure 25. The shear modification factor is seen to be influenced more by the flange opening width ξ_2 for smaller peripheral distances U/B. Although results are not shown, the influence of flange width ratio ξ_1 has been found to be negligible in the evaluation of shear modification factors.

Example Showing Application of Critical Section and Shear Modification Factor

To demonstrate the application of the critical shear section and the shear modification factors, typical punching shear calculations consistent with the requirements of the American and British Codes, ACI 318-83 and CP 110, are shown for a slab coupling a pair of flanged walls in a 30-storey, 255 ft (78m) high cross-wall structure which has already been analysed for wind load effects.

The assumed dimensions for the slab are

B = 5 ft (1.525 m), Y = 20 ft (6.1 m), X = 50 ft (15.25 m) and Z = 10 ft (3.05 m).

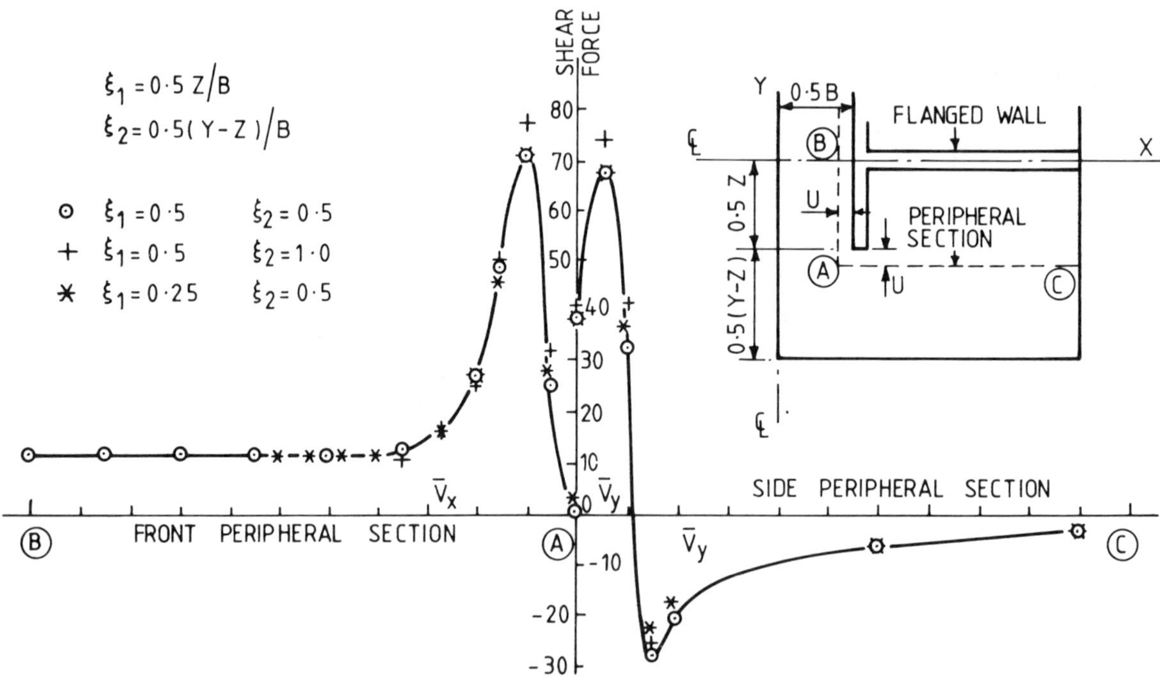

FIGURE 24. Influence of flange width and slab width on peripheral shear distribution for T-shaped walls.

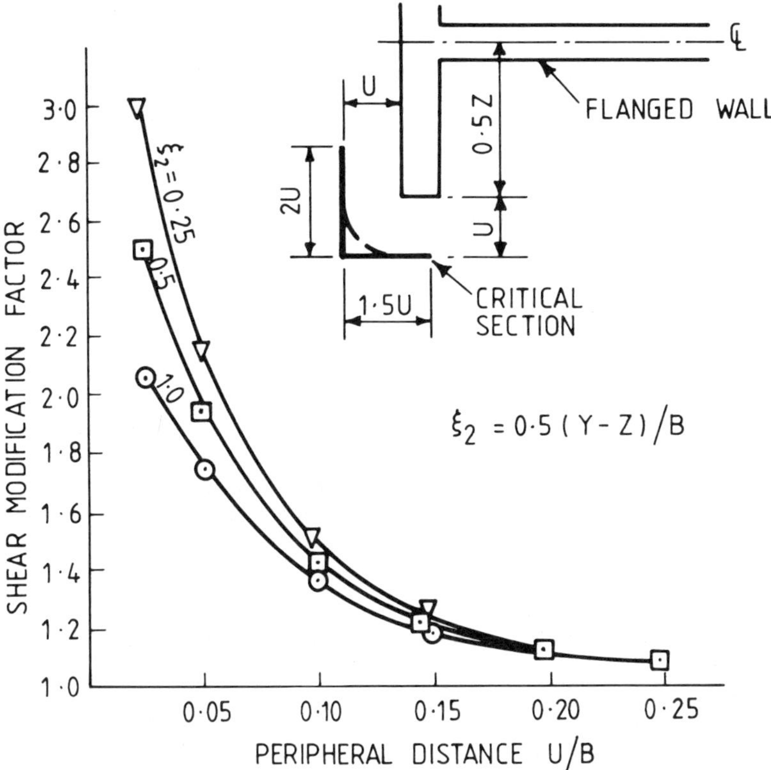

FIGURE 25. Design curve for shear modification factor for T-shaped walls.

The walls are 8 in (200 mm) thick throughout. The effective slab width Y_e is 14 ft (4.2 m), (cf. Section 2). The overall and effective depth of slab are t = 8 in (200 mm) and d = 6.75 in (170 mm), respectively.

The calculated maximum lintel shear under an ultimate wind load of 35 lb/ft² (1.68 kN/m²) is V_o = 30.73 kip (136.6 kN).

DESIGN TO ACI 318-83

The critical section for shear is located at

$U = 0.5d = 0.5 \times 6.75 = 3.375$ in. (85.7 mm)

$U/B = 3.38/60 = 0.06$

The shear in the effective shear zone is

$V_3 = (10 - 2 \times 0.281) \times 30.73/14 = 20.72$ kip (92.16 kN)

The flange opening ratio $\xi_2 = 1.0$

From Figure 25, the shear modification factor $K_q = 1.70$. The critical design shear is, therefore,

$V_c = 1.7 \times (30.73 - 20.72) = 17.02$ kip (75.70 kN)

The length of the critical section is 3.5U = 11.81 in. (0.3 m). The critical shear stress is, then,

$$v = \frac{17.02 \times 10^3}{2 \times 11.81 \times 6.75} = 106.8 \text{ lb/in}^2 \ (0.74 \text{ N/mm}^2)$$

which is about half the value of the permissible ultimate shear stress of $4\sqrt{f_c'}$ for a grade 25 concrete.

DESIGN TO CP110

The critical shear section is located at U = 1.5t = 1 ft (0.3 m). The shear in the effective shear zone is

$V_3 = (Z - 2U)V_o/Y_e = (10 - 2 \times 1) \times 30.73/14$
$= 17.56$ kip (78.1 kN)

The normalised critical distance $U/B = 1/5 = 0.2$.

The flange opening ratio $\xi_2 = 1.0$.

From Figure 25, the shear modification factor $K_q = 1.1$. The critical design shear is, therefore,

$V_c = 1.1 \times (30.73 - 17.56) = 14.48$ kip (64.4 kN)

The length of the critical section on one half of the slab is

$$p = \frac{U}{2}(3 + \pi) = \frac{1}{2}(3 + \pi) = 3.07 \text{ ft } (0.92 \text{ m})$$

The critical punching shear stress is, then,

$$v = \frac{14.48 \times 10^3}{2 \times 36.84 \times 6.75} - 29.11 \text{ lb/in}^2 \ (0.21 \text{ N/mm}^2)$$

which again is only about half the value of the permissible ultimate shear stress for a Grade 25 concrete and 0.25% steel area.

It would appear from the above example calculations that under normal circumstances punching shear failure in slabs coupling wide-flanged walls is unlikely.

3.3. Stresses in Slabs Connecting Plane and Flanged Walls

For a slab coupling a plane and a tee-shaped wall (Figure 20f), large bending moments and shear forces are concentrated around the inner coupled end of the plane wall, as would be expected. The stress-resultants on the flanged-wall side of the slab are comparatively low, and the longitudinal bending moment M_x decreases gradually along the flange wall away from the panel centre-line as coupling of the slab becomes less effective. It is also found that the zero-value contour for M_x, or the line of inflection of the slab, does not deviate significantly from the mid-span position over the portion of the slab where coupling is most effective. Thus the assumption of a uniform equivalent coupling beam, with a mid-span point of contraflexure, assumed for overall analysis, does not appear to be unreasonable although the actual width of slab effectively stressed varies considerably across the wall opening.

GENERALISED DESIGN CURVES FOR CRITICAL BENDING MOMENT FACTORS

The curves showing the variation of longitudinal bending moment factors M_x along two critical transverse sections passing through the coupled ends of the plane wall and flanged wall are shown in Figures 26 (a) to (f), for slabs with various ratios B/Y. The curves for $x/B = 0.5$ and -0.5 refer, respectively, to the sections at the inner edges of the plane wall and flanged wall, and illustrate the higher moments which are developed at the former section. The distribution of M_x along the critical section on the side of the plane wall is seen to be effected very little by the variation of flange width on the opposite wall. For walls with small flange width ratios, a distinct peak is observed in the bending moment distribution curve at the tip of the flange. Since it is likely that the bending moments at this point and also at the tip of the plane wall are theoretically infinite, it is logical to interpret the peak bending moment values in the same manner as before.

Although the curves may be used directly for the design of the critical slab sections, they are not very convenient in use because of the large number involved. Since the bending moment distribution in the most effective coupling zone of the slab is practically unaffected by variations in slab width, the six sets of curves for various ratios of B/Y may be generalised by a single set of curves which is applicable for all ratios B/Y, provided that these ratios are not greater than 1.0, which is almost always the case for practical structures. The generalised set of curves showing the variation of critical bending moment M_x with normalised distance ζ ($= y/B$) for various flange width ratios Z/B are presented in Figure 27. To obtain the approximate distribution of M_x for any slab, the tail of the generalised curve for the appropriate flange width ratio is again truncated at the point corresponding to the actual limit of the slab width.

Checking against punching shear failure may again be carried out approximately using the results presented earlier for plane walls and tee-shaped flanged walls.

3.4 Stresses in Slabs Connecting L-Shaped Flanged Walls

In the configuration shown in Figure 20(c), the cross-walls or wall webs are coupled in-line with opposing flanges, while in the configuration shown in Figure 20(d), the webs are coupled off-line with opposing flanges. Since the slab behaviour is controlled almost entirely by the flange dimensions, the actions in the slabs in these two configurations will be essentially the same as those for slabs connecting T-shaped flanged walls, and thus the critical bending moments and shear stresses may be evaluated approximately from the curves presented earlier for T-shaped flanged walls.

In the third configuration shown in Figure 20(e), the webs are coupled in-line but the flanges are off-set or skew coupled, and the stress distributions are quite different.

GENERALISED DESIGN CURVES FOR CRITICAL BENDING MOMENT FACTORS

In the third configuration, large coupling actions are concentrated around the junction between the flange and web walls on each side of the corridor opening [5]. At the ends of the wall flanges, the slab actions are lower, in contrast to the actions in coupled T-shaped walls, reflecting the lack of direct flange wall coupling across the corridor opening.

To facilitate the calculation of the critical bending moments required for the design of the slab section, curves showing the variation of bending moment factors M_x along the critical transverse slab section immediately in front of the flange wall are presented in Figures 28(a) to (f), for various slab-width and flange-width ratios. It can be seen

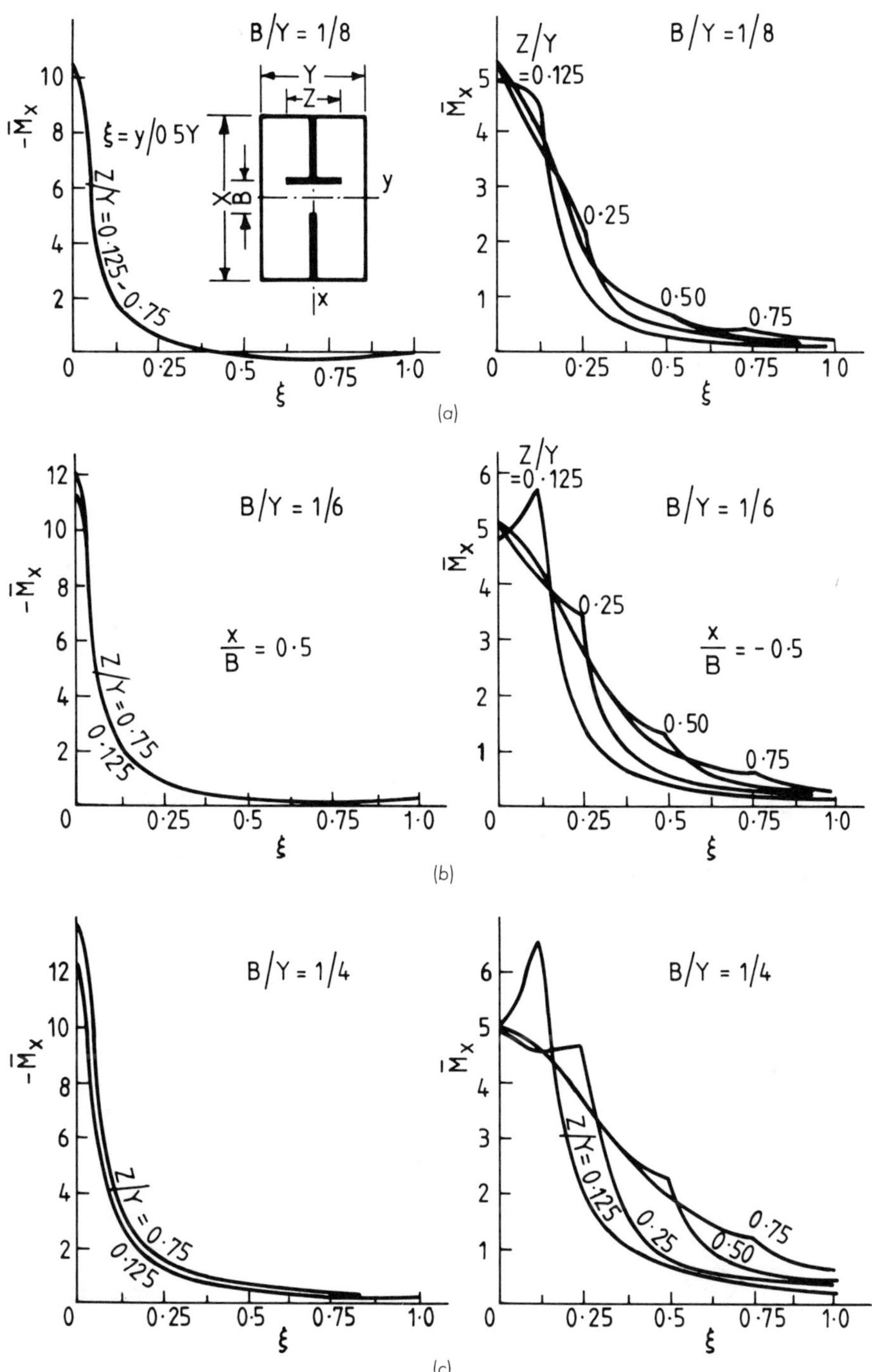

FIGURE 26. Variation of longitudinal bending moments at critical transverse slab sections for planar-flanged wall configuration. (continued)

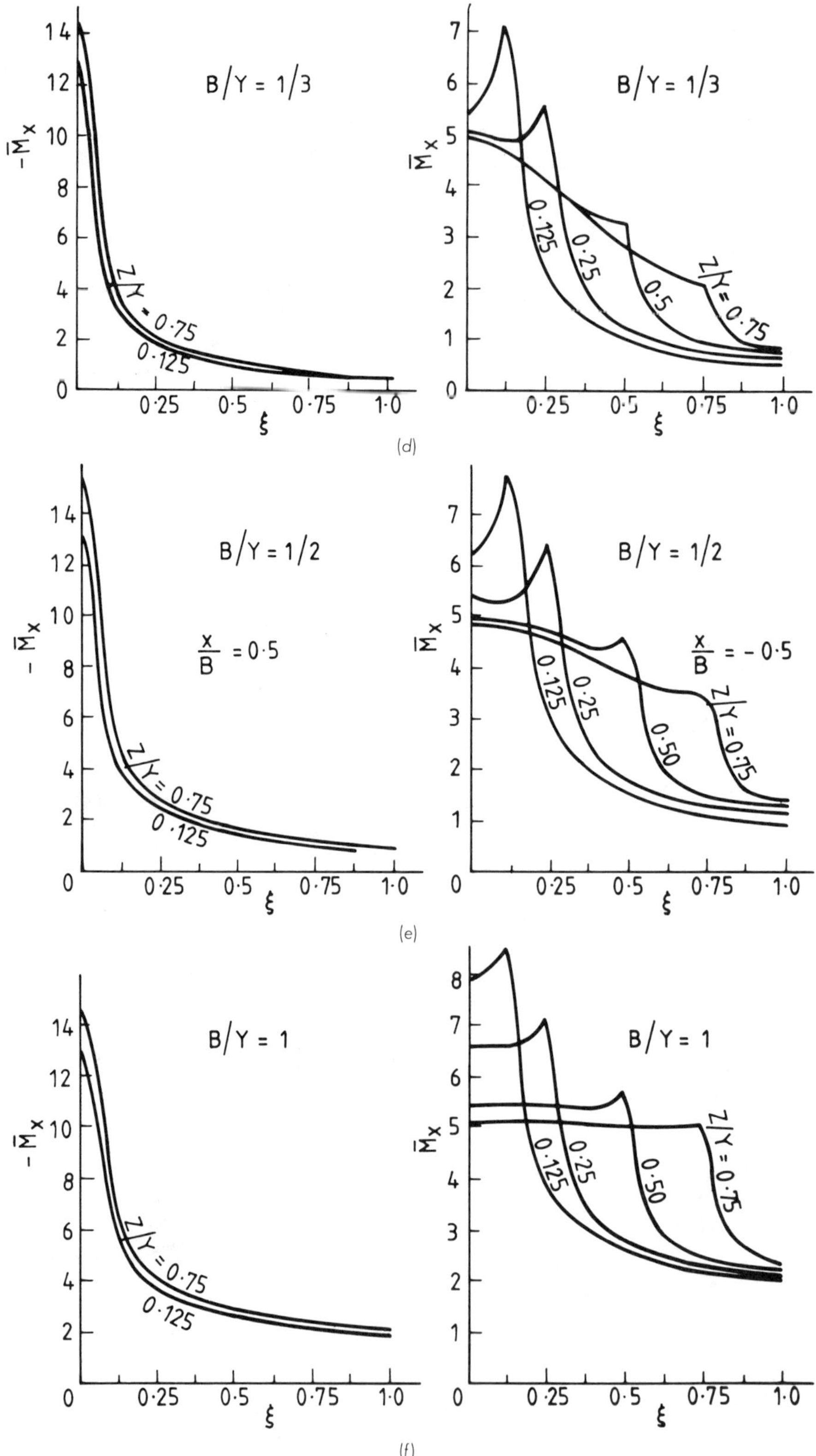

FIGURE 26. Variation of longitudinal bending moments at critical transverse slab sections for planar-flanged wall confituration.

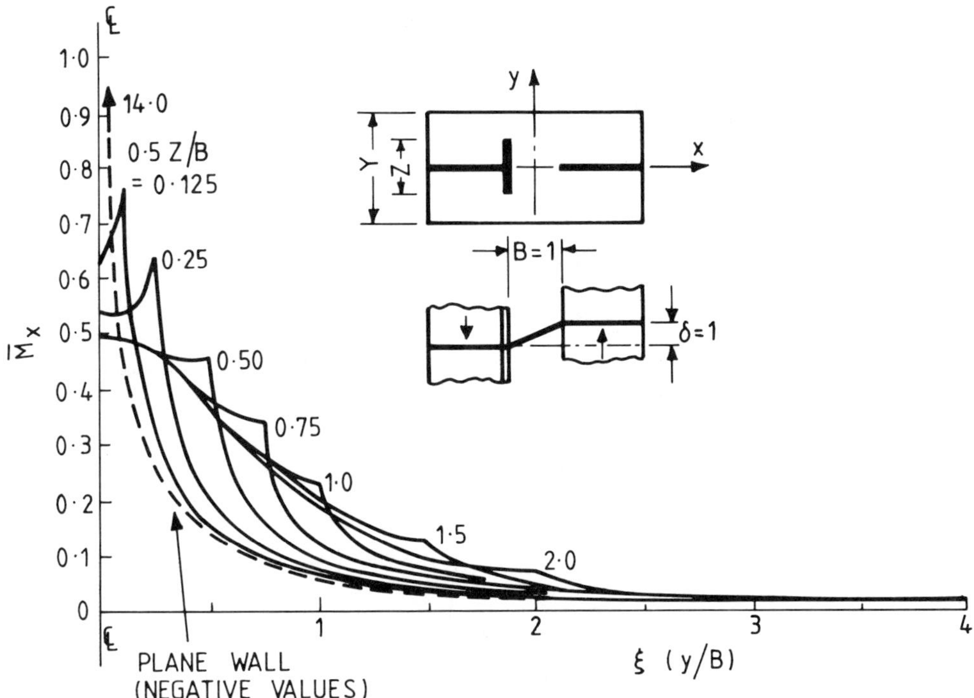

FIGURE 27. Generalised design curve for longitudinal bending moments at critical transverse slab section for planar-flanged walls.

from the curves that the bending moment distribution in the half of the critical section defined by positive values of ξ is practically unaffected by the variation of flange width when the ratio B/Y is less than 1.0. Two distinct peaks may be seen in each curve, occurring at the ends of the flange wall. The bending moments at these points are likely to be theoretically singular judging from the shape of the curve, and from the results presented earlier for other flanged wall configurations. By comparing the various sets of curves it is seen that with a constant slab width Y the bending moment factors M_x are higher with larger wall opening widths B.

With a small degree of approximation, the six sets of curves for M_x presented in Figure 28 may be generalised to the more convenient set of curves shown in Figure 29. The generalised curves show the variation of M_x with the normalised distance ξ ($= y/B$) for various flange width ratios Z/B and may be used to obtain the distribution of M_x in a slab of any width by truncating the tail of the appropriate generalised curve at the points corresponding to the actual limits of the slab width.

The effective width of the slab may again be estimated approximately from Equation (23) by integrating numerically the area under the curve to give K_a.

Checking against punching shear failure may be carried out approximately using the earlier results for coupled T-shaped walls.

4. RELATED DESIGN SITUATIONS

The previous two sections have presented design information for both the effective stiffness of, and the bending and shear stresses in, a slab connecting a pair of solid walls of different cross-sectional shapes. The configurations treated are those which are encountered most frequently in practice.

However, several other similar related situations are of practical interest, and, since design information exists, a brief description is given of them for completeness in this section.

4.1 Influence of Doorway Openings in Walls

The results given in Section 2 may be affected if door openings are provided in the cross-walls for internal circulation within the apartments. Such openings are usually small in width compared to the dimensions of the wall, and should not significantly affect the overall lateral stiffness of the wall. When a pair of perforated walls is coupled by floor slabs, each wall may be assumed to behave as a composite unit in the lateral load analysis of the structure. If cross sections of the perforated wall at the wall/slab junctions are assumed to remain plane when the slab interacts with the wall, then the effective coupling stiffness will be similar to

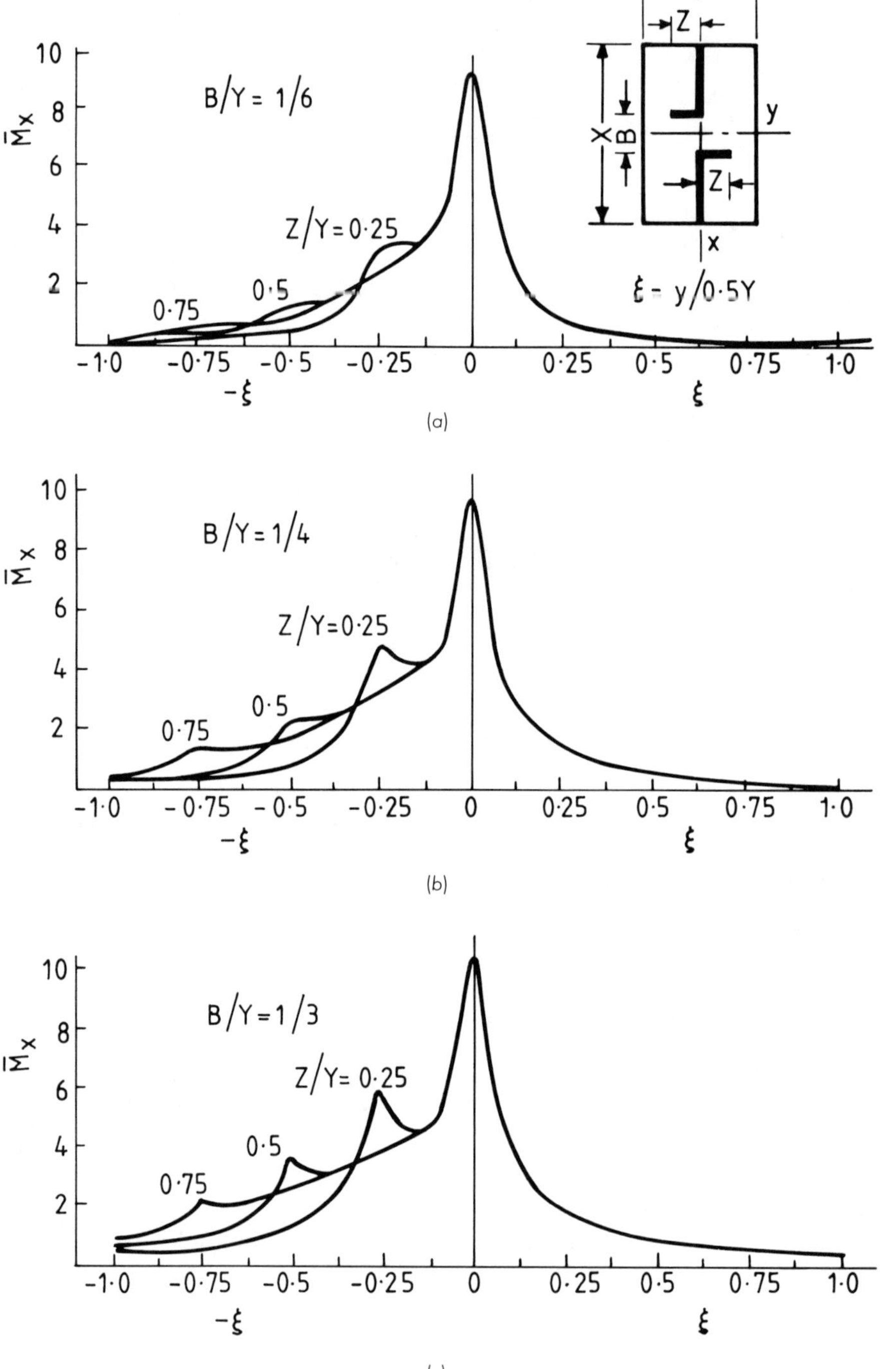

FIGURE 28. Variation of longitudinal bending moments at critical transverse slab section for L-shaped walls. (continued)

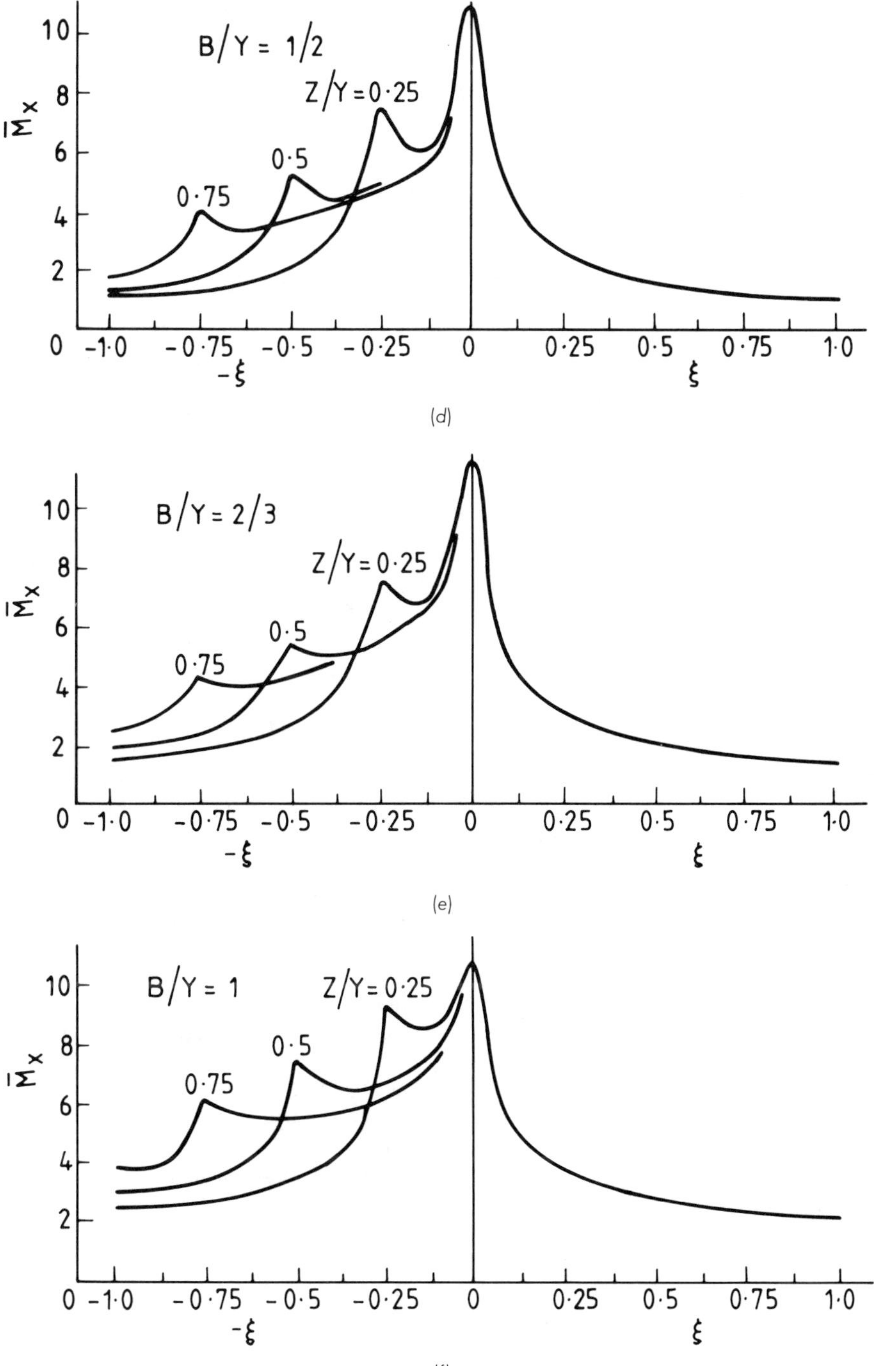

FIGURE 28. Variation of longitudinal bending moments at critical transverse slab section for L-shaped walls.

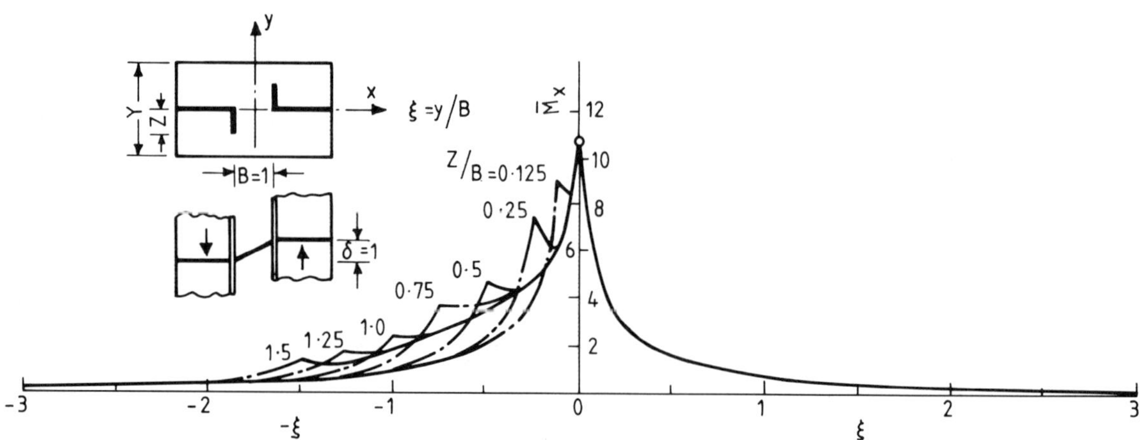

FIGURE 29. Generalised design curve for longitudinal bending moments at critical transverse slab section for L-shaped walls.

that for unperforated walls. However, when local elastic wall deformations are considered, the effective slab width will be influenced to some extent by the presence of an internal doorway opening.

A finite element study [8] was made of the interaction between floor slabs and laterally loaded perforated plane walls, and the influence of the size and position of the doorway opening was examined for a typical wall-slab configuration. It was found that if the opening did not extend over the full floor height, and a lintel of nominal depth 2 ft (0.61 m) was provided, the effective width was reduced by only 6% in the most extreme case when the opening was very near the inner edge of the wall, in the region of high stress concentrations. In most practical cases, the influence of door openings is unlikely to be significant in the calculation of effective slab width.

4.2 Composite Action Between Slabs and Lintel Beams

Although cross-walls are frequently coupled solely by concrete floor slabs, cases arise in which the shear walls are connected by lintel beams monolithic with the floor slabs. A common practice in the analysis of such coupled shear walls is to disregard the contribution of the slab and assume that the walls are coupled only by a prismatic lintel beam. However, in gravity load design it is standard practice to include a portion of the slab as a flange for the beam, so that a greater moment of resistance is obtained by the composite action. While under ultimate load conditions it may be sound practice to ignore the contribution of the slab on account of cracked flange sections at points of negative bending moment, there is no reason why under working load conditions, in which the structural behaviour is sensibly linearly elastic, the beneficial stiffening effect of the slab should not be included in the analysis of the coupled shear walls.

In Reference [22], using the finite element method, a study was made of the composite action of a lintel and slab coupling a pair of plane shear walls, and the relative influences of a range of structural parameters on the stiffness and effective flange width of the composite coupling beam were evaluated. It was found that for the normal range of structural dimensions encountered in practice, the slab width, lintel width and lintel depth have little influence on the effective flange width, but the composite stiffness ratio and the effective flange width increase significantly with the wall opening width. Curves were presented to allow the rapid evaluation of composite stiffness ratio, or the effective flange width, in a practical situation. As in the case of lintel beams coupling plane shear walls, the influence of elastic lintel-wall junction deformation may be allowed for in the calculation of the composite coupling stiffness by increasing the clear span of the lintel by a length equal to the lintel depth, [22]. The influence of the floor slab is significant, however, only when the lintel beam is relatively flexible, when considerable reductions in the wind stresses and deflections may be achieved by including the composite action.

4.3 Effect of Local Elastic Wall Deformations on Floor Slab–Shear Wall Interaction

When investigating the effective coupling width of slab, and the peak bending and shear stresses, it was assumed that the actions in the slender walls obeyed ordinary engineers' beam theory, so that plane sections remained plane after bending occurred due to lateral loading. The slab segments connected to the walls were then assumed to undergo rigid body displacements or rotations. When subjected to bending due to lateral loads, the interactive forces between wall and slab are concentrated over a small area near the inner corridor edges of the walls, while the rear edges are relatively stress free. Consequently, axial deformations of the walls will occur locally in these heavily loaded regions, which may be of significance in reducing the effective stiffness and the high peak stresses. The earlier results, based on the concept of a rigid wall, will thus tend to overestimate the effective width and maximum bending stresses.

A finite element study [7] was made of the effects of local elastic wall deformations in plane cross wall-slab structures, and numerical results were presented to illustrate the relative influence of various structural parameters on the flexibility of the wall/slab junction, and on the effective coupling of plane walls by the slab. It was shown that the slab stiffness ratio varies principally with the single parameter $t/(B^2h)^{1/3}$, and design curves were presented for the rapid evaluation of the stiffness ratio, or the effective span extension, in a practical situation. Either of these quantities could be used to correct the "rigid wall" slab stiffness so as to include the local elastic wall deformation effects.

When the local elastic wall deformation is disregarded, although the effective slab width may be over-estimated, the resulting errors in the calculation of maximum wall deflection and stresses and maximum lintel shear are generally relatively much smaller [7].

The critical bending moment factors in the slab are reduced by the effects of local elastic wall deformation, the amount depending primarily on the slab/wall thickness ratio, so that the values obtained from the 'rigid' wall analysis will tend to be over-safe.

4.4 Stiffening of Structural Cores by Floor Slabs

The lateral resistance against wind or seismic forces of modern high-rise buildings may be provided wholly or partly by one or more structural cores which house elevator shafts, stairwells and service systems. These cores consist essentially of open-section shear walls connected by floor slabs, and possibly lintel beams, at each floor level to form a thin-walled perforated box structure. Due to asymmetric structural layout or eccentric loading, the core may be subjected to torsion as well as bending under the action of the lateral loads.

In contrast to a closed box structure, an open section core has a low torsional stiffness and undergoes large warping deformations under torsion. The presence of floor slabs and lintel beams, which act effectively as cross bracing, stiffens the core against torsional deformations and reduces the warping stresses below the levels arising in the unbraced core.

Two common forms of central core, a single U-shaped core and a twin-channel core, are shown in Figure 30. In the torsional analysis of such core structures, simplifications are possible if the connecting floor slabs are again replaced by an equivalent system of connecting beams at the core openings.

As a result of the high in-plane stiffness of the floor slabs, the core will suffer rigid-body displacements in plan when subjected to torsional loading. Consequently, if it is assumed that the component panels of the core act as slender beam elements which are connected to adjacent panels along the corners, the actions in the two wall panels on opposite sides of the core openings will correspond to those in the walls of a coupled wall system. Thus, provided there is no lobby slab within the profile of the core, the slab will behave in a fashion analogous to that of the end bay of a cross-wall structure (Section 2.1.9). The effective stiffness may then be taken approximately as 45% of the value of the corresponding interior value for the dimensions of the walls and slab concerned.

Based on the finite element method in conjunction with Vlasov's theory of thin-walled beams, studies have been made [9,21] of the torsional stiffening by the surrounding floor slabs of thin-walled open-section cores of the planforms shown in Figure 30. This has led to more accurate estimates of the effective width for cores with and without interior lobby slabs. The studies examined the relative influences of the most important geometrical parameters and slab support conditions, and design curves were presented to allow the equivalent width and effective warping stiffness of the slab to be evaluated rapidly. The absolute effective width for the equivalent beam was shown to be influenced strongly by the support conditions of the slab edges and by the core opening ratio, and less by the core aspect ratio and slab width ratio. The influence of the axial and flexural stiffness of the external columns were also considered in the study.

The results indicated that effective coupling of the core walls can be achieved even with a relatively flexible slab system, and considerable economies may be gained in the design of the core by taking into account the contribution of the coupling slab.

4.5 Coupling Action of Slabs in Hull-Core or Tube-in-Tube Structures

The hull-core or tube-in-tube structure has found considerable favour in the construction of tall commercial buildings since it provides large column-free areas suitable for different types of operations. This efficient structural system confines all services within a central core which also acts as the interior support to the floor slabs. The latter are supported along the exterior by a series of columns which are normally closely spaced and connected by deep spandrel beams forming an exterior framed tube. A typical planform is shown in Figure 31, exemplifying the uninterrupted rent-

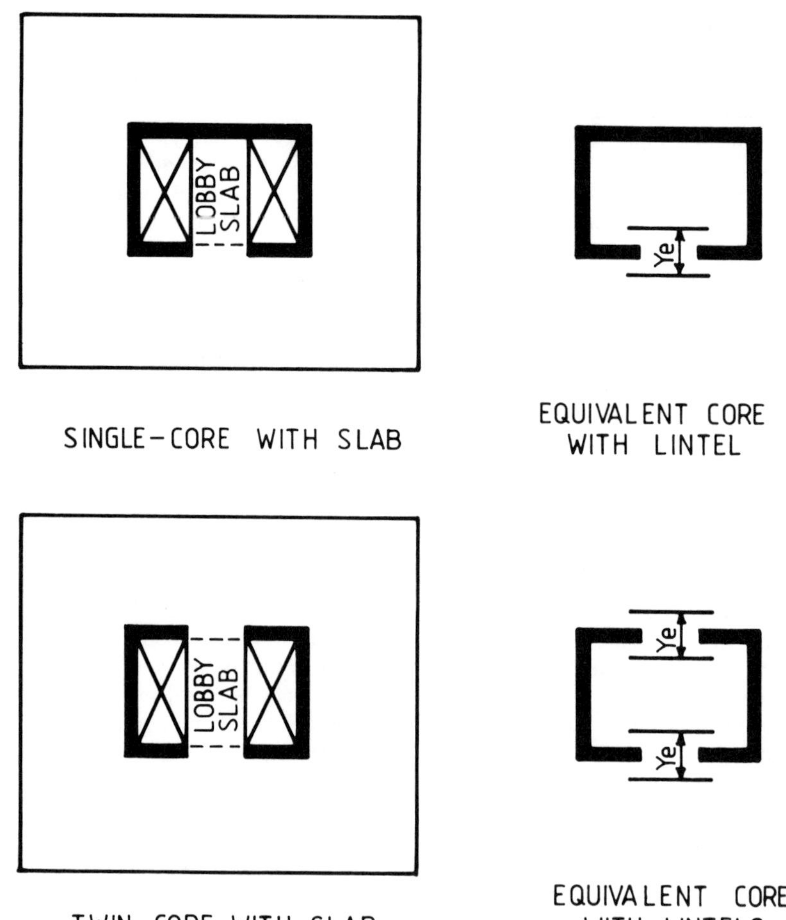

FIGURE 30. Cross-sections of cores with surrounding floor slabs and equivalent lintel beams.

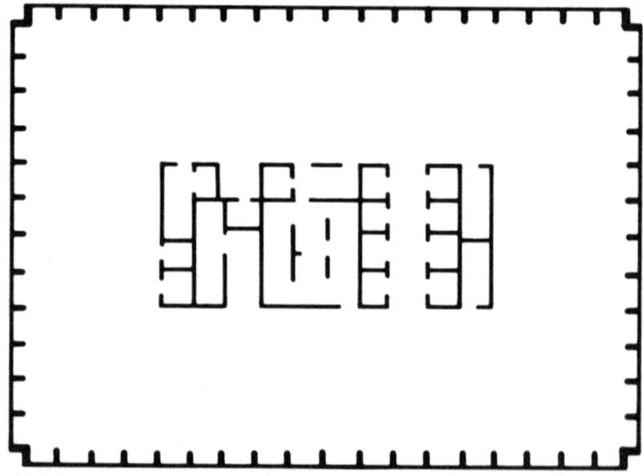

FIGURE 31. Typical floor plan of tube-in-tube structure.

able space which exists between the internal core and external facade.

The success of the system lies in the utilisation of both the inner core and outer hull to resist lateral loads by tubular bending action. Due to the flexibility of the frame members, the behaviour of the outer framed tube is strongly influenced by frame racking and shear lag effects, while the lateral behaviour of the central core is essentially that of a vertical cantilever. Because of the differences in structural action between the two components, a complex interaction is effected through the floor slabs when the structure is subjected to lateral forces, and the amount of composite action which ensues depends on the stiffness of the floor system and the rigidity of its connection with the vertical elements. If the floors are effectively pin-connected to the core and hull, then under lateral loading the floor system transmits only horizontal forces between the two. A more efficient composite system can be achieved by having moment connections between core and hull so that axial forces are induced in the exterior columns by the cantilever bending of the core. These axial column forces not only reduce the shear lag effect but also increase considerably the moment of resistance of the structure due to the large lever arm between the normal frame panels.

Using techniques and assumptions similar to those employed in the investigations on wall-slab interaction, a study [6] was made of the coupling action of slabs in hull-core systems and the relative influence of a range of parameters evaluated.

It was found that the column-to-slab flexural stiffness ratio has a significant influence on the rotational stiffness and effective width of the slab. With a constant column–slab flexural stiffness ratio, the influence of column spacing and spandrel beam stiffness are negligible, while practical variations in column axial stiffnesses have negligible effects on the rotational stiffness of the slab. The effective width of slab increases with the relative core width and core depth. Curves were presented to facilitate the evaluation of rotational stiffness and effective width of slab for practical situations.

5. CONCLUSIONS

A comprehensive set of design curves and associated simple empirical formulae and procedures have been presented to enable the effective width, and hence the coupling stiffness, to be evaluated rapidly for a slab connecting a pair of laterally loaded shear walls. The curves apply to all commonly encountered plane and flanged wall configurations in practical cross-wall structures. The results apply strictly to a typical interior bay of the building, but a simple rule-of-thumb approximation has been suggested for the case of an end bay. Although the curves have been developed for isotropic slabs, an approximate technique has been put forward to allow the results to be applied to other slab forms, such as voided or ribbed slabs, which may be treated as orthotropic plates. The accuracy of the curves and the empirical formulae has been checked by comparing the results with values obtained directly from finite element analyses for the range of structural parameters likely to be encountered in practice.

Based on a finite element analysis, the distributions of bending moments and shear forces induced in a slab coupling a pair of laterally loaded shear walls have been examined. For a unit wall rotation, moment contour diagrams for a range of wall slab plan ratios have been given for plane walls, and a single comprehensive design curve, showing the variation of critical bending moment factors, was devised to allow a rapid evaluation of the slab stresses induced. Generalised design curves have also been presented for the evaluation of the critical bending moments for a slab coupling a pair of laterally loaded T-shaped flanged walls, a pair of plane and flanged walls, and a pair of L-shaped walls. An investigation of the distributions of interactive shear forces between slab and wall, in association with the usual design assumptions regarding shear failure of similar flat plate structures, has enabled a design technique to be suggested for checking against punching shear failure in the slab.

The information contained in this chapter is based on a linear elastic analysis of the interaction between a floor slab and a pair of laterally loaded walls. Although its limitations with respect to reinforced concrete structures are well known, a linear elastic analysis still forms the basis of design of tall building structures. The walls are relatively stiff compared to the slab, and the out-of-plane deformations of the latter are much more significant than any in-plane actions in the walls. Large moments and stress concentrations exist in the flexible slab near the inner corridor edges and local cracking of the concrete or yielding of the steel may occur, with an attendant redistribution of moment and loss of stiffness. This is likely to be more significant in the case of plane walls, where the stress concentrations are greater than for flanged walls. The latter are concentrated near the tips of the flanges only, and a fairly uniform stress field occurs along the remainder of the flange lengths. It may be expected that the curves presented do show the relative effects of the different parameters involved. In design offices, an arbitrary factorisation of the effective width or stiffness, by as much as 50%, is carried out to attempt to take account of local cracking at the wall–slab junction. It must be expected that the design values quoted here represent the maximum moments and bending stiffness attainable, and this factor should be borne in mind when using them.

ACKNOWLEDGEMENT

The material contained in this chapter is based largely on the contents of three research papers [3,4,5]. Thanks are due to the American Society of Civil Engineers and to the Institution of Civil Engineers, London, for permitting the reproduction of material from these papers.

NOTATION

The following symbols are used in this paper.

B = Clear opening between walls or corridor width
D = Flexural rigidity of slab
E = Elastic modulus

h = Wall thickness
i = Subscript denoting x or y
K = Rotational coupling stiffness of slab
K_a = Factor denoting double area under curve for longitudinal bending moment factor
K_q = Shear modification factor
M_a = Average longitudinal bending moment per unit width
$M_x M_y M_{xy}$ = Bending and twisting moments per unit length of slab
$\bar{M}_i$ = Non-dimensional bending moment factor for slab
$O(x,y)$ = Coordinate system
t = Slab thickness
U = Critical peripheral distance for punching shear
v = Shear stress
v_c = Ultimate punching shear stress
V_a = Average shear force per unit width of slab
$V_x V_y$ = Shear forces per unit length of slab in x- and y-direction
$\bar{V}_i$ = Non-dimensional shear force factor for slab
w = Lateral deflection of slab
$w_x w_y$, etc. = $\partial w/\partial x$, $\partial w/\partial y$, etc.
W = Length of cross-wall
X = Total length of slab panel or depth of cross-wall building
Y = Width of slab panel or bay width
Y_e = Effective width of slab
Z = Width of wall flange
δ = Relative displacement at ends of coupling beam
$\varnothing$ = Relative rotation of ends of coupling beam
ξ,η = Non-dimensional coordinates
ξ_s = Depth of slab factor for shear
ν = Poisson's ratio

REFERENCES

1. "Building Code Requirements of Reinforced Concrete." American Concrete Institute, ACI 318-83, Detroit (1983).
2. Coull, A. and J. R. Choudhury, "Stresses and Deflections in Coupled Shear Walls," *Jnl. ACI, Proc.V.64,* No.2, pp. 65-72 (Feb. 1967).
3. Coull, A. and Y. C. Wong, "Bending Stiffness of Floor Slabs in Cross-Wall Structures," *Proc. ICE, London, England, Vol.71,* Part 2, pp. 17–35 (March 1981).
4. Coull, A. and Y. C. Wong, "Design of Floor Slabs Coupling Shear Walls," *Jnl. Struct. Div., ASCE, Vol.109,* No.ST 1, Proc. Paper 17620, pp. 109–125 (Jan. 1983).
5. Coull, A. and Y. C. Wong, "Stresses in Slabs Coupling Flanged Shear Walls," *Jnl. of Structural Engineering, ASCE, Vol.110,* No.1, pp. 105–119 (Jan. 1984).
6. Coull, A. and Y. C. Wong, "Coupling Action of Slabs in Hull-Core Structures," *Jnl. of Structural Engineering, ASCE, Vol.110,* No.3, pp. 575–588 (March 1984).
7. Coull, A. and Y. C. Wong, "Effect of Local Elastic Wall Deformations on the Interaction Between Floor Slabs and Shear Walls," *Proc. ICE, Vol.77,* Part 2, pp. 195–210 (June 1984).
8. Coull, A. and Y. C. Wong, "Influence of Door Openings on Effective Slab Width," *Jnl. of Structural Engineering, ASCE, Vol.110,* No.10, pp. 2531–2535 (Oct. 1984).
9. Coull, A. and Y. C. Wong, "Stiffening of Structural Cores by Floor Slabs," *Jnl. Struct. Eng., ASCE,* Vol. 112, No. 5, pp. 977–994 (May 1986).
10. Cusens, A. R. and Pama, R. P. *Bridge Deck Analysis,* Wiley, London, (1975).
11. El-Hag, A. A. "Coupling of Shear Walls by Floor Slabs in Tall Buildings," Thesis presented to the University of Strathclyde, Glasgow, Scotland, in partial fulfillment of the requirements for the degree of Master of Science (in 1973).
12. Kiattikomol, K, L. M. Keer, and J. Dundurs, "Application of Dual Series to Rectangular Plates," *Proc. ASCE, Vol.100,* No.EM2, pp. 433–444 (April 1974).
13. Salvadori, M. G. and K. C. Reginni, "Simply Supported Corner Plates," *Proc. ASCE, Vol.68,* No.ST11, pp. 141–154 (Nov. 1960).
14. Schwaighofer, J. and M. P. Collins, "Experimental Study of the Behaviour of Reinforced Concrete Coupling Slabs," *Jnl. ACI, Proc.V.74,* No.3, pp. 123–127 (March 1977).
15. Stahl, B. and L. M. Keer, "Vibration and Buckling of a Rectangular Plate with an Internal Support," *Int. Jnl of Solids and Structures, Vol.8,* pp. 69–91 (1972).
16. "The Structural Use of Concrete," *Code of Practice CP 110: Part 1,* British Standards Institute, London (1972).
17. Timoshenko, S. P. and S. Woinowsky-Krieger, *Theory of Plates and Shells,* 2nd Ed., McGraw Hill, New York (1959).
18. Tso, W. K. and A. A. Mahmoud, "Effective Width of Coupling Slabs in Shear Wall Buildings," *Proc. ASCE, Vol.103,* No. ST3, pp. 573–586 (March 1977).
19. Wong, Y. C. "Interaction Between Floor Slabs and Shear Walls in Tall Buildings," Thesis presented to the University of Strathclyde, Glasgow, Scotland, in partial fulfillment of the requirements for the degree of Doctor of Philosophy (in 1978).
20. Wong, Y. C. and A. Coull, "Structural Behaviour of Floor Slabs in Shear Wall Buildings," *Proc. of Int. Conf. on Concrete Slabs.* University of Dundee (April 1979).
21. Wong, Y. C. and A. Coull, "Torsional Stiffening of Structural Cores by Surrounding Floor Slabs," *Proc. of Int. Conf. on Tall Buildings, Hong Kong and Guangzhou,* pp. 229–235 (Dec. 1984).
22. Wong, Y. C. and A. Coull, "Composite Action Between Slabs and Lintel Beams," *Jnl. Structural Engineering, Vol. 110,* No.3, pp. 575–588 (March 1983).

BIBLIOGRAPHY

1. Black, D., V. A. Pulmano, and A. P. Kabaila, "Stiffness of Flat Plates in Cross-Wall Structures," UNICIV Report No. R-133, University of New South Wales, Kennington, Australia (1974).

2. Chang, Y. C., "Slabs in Shear Wall Buildings," M.Sc. Thesis, Dept. of Civil Engineering, University of Toronto, Canada (Sept. 1969).
3. Coull, A. and A. A. El Hag, "Effective Coupling of Shear Walls by Floor Slabs," *ACI Jnl. Proc.*, Vol. 72, No. 8, pp. 429–431 (Aug. 1975).
4. El Buluk, M. A., "Interaction Between Shear Walls and Floor Slabs," M.Sc. Thesis, University of Strathclyde, Glasgow, Scotland (Dec. 1975).
5. Michael, D., Discussion of Reference 7, *ACI Jnl.*, *Proc.* Vol. 66, No. 12, p. 1021 (Dec. 1969).
6. Petersson, H., "Bending Stiffness of Slabs in Shear Wall Structures," Draft Report, Division of Structural Design, Chalmers University of Technology, Gothenberg, Sweden (1972).
7. Qadeer, A. and B. Stafford Smith, "The bending Stiffness of Slabs Connecting Shear Walls," *ACI Jnl. Proc.*, Vol. 66 No. 6, pp. 464–472 (June 1969).
8. Qadeer, and B. Stafford Smith, "Actions in Slabs Connecting Shear Walls," *Proc. of Symposium on Tall Buildings*, Vanderbilt Univerity, Nashville, TN, U.S.A., pp. 315–338 (Nov. 1974).
9. Wong, Y. C. and A. Coull, "Interaction Between Floor Slabs and Shear Walls in Tall Buildings," *Proc. Symposium on Reinforced Concrete Structures Subjected to Wind and Earthquake Forces.* ACI Special Publication 1980.
10. Zienkiwicz, O. C. and Y. K. Cheung, "The Finite Element Method for Analysis of Elastic Isotropic and Orthotropic Slabs," *Proc. ICE, London, England*, Vol. 28, pp. 471–488 (Aug. 1964).

CHAPTER 15

Lateral Load Effect on Steel Arch Bridge Design

TETSUYA YABUKI* AND SHIGERU KURANISHI**

INTRODUCTION

Steel arch bridge structures which are referenced herein are composed of twin, parallel, parabolic main ribs connected by a lateral bracing system as shown in Figure 1. The load combination that is considered consists of vertical loading and practical lateral loading to which bridge structures are actually subjected. The effect of deck systems is also briefly described.

For the practical lateral loads such as wind loads the arch is a curved member in the plane normal to the direction of vertical loading (out-of-plane action). The out-of-plane deformation of arches occurs in a coupled action of its bending and twisting by reason of this curvature. The exact analysis of the coupled effect is quite complicated. Several wind analyses (wind stress and wind deflection) are introduced herein. The laterals in a bracing system carry the transverse wind shear. A general concept of the conventional design method for the wind shear is described herein. It is also described that the transverse beams in the bracing system are extremely important members for the sway or distorsion of the transverse cross-sectional form of the arch structure as a whole. There will also be wind lateral loads (hereafter these are expressed as "practical lateral load") acting directly on the roadway deck. Since the roadway deck and the arch are not free to deflect laterally, there will be lateral force transfers between them through the columns or crown plate. This behavior is briefly examined from a viewpoint of its practical designing.

Arch bridge structures have a spatially curved slender form and are subjected to highly compressive stress. Hence, if a sufficient capacity of the out-of-plane strength or rigidity or both of the structures is not ensured in designing them, their load carrying capacity will be significantly affected by the lateral loads and the designer must be faced with a complex design rule under which the effect of the interaction of the in-plan and out-of-plane behavior of the arch should be considered. Otherwise an ultimate strength analysis of the arch bridge as a spatial structure will be required in each case. Namely, in a sense, the lateral loads must be evaluated as a principal load and the lateral bracing members must be designed also as the main load carrying members of the structure. However, the conventional design is founded on the conception in which the strength and proportioning of bridge structures should be determined prinicpally by the vertical loads. From the practical designing viewpoint, arch bridge structures which are practically subjected to vertical and lateral loads should be ensured such required out-of-plane rigidities as to be able to make rational estimation of the ultimate strength of the bridge as in-plane structures.

Following such design philosophy, several typical numerical results for the load carrying capacity of the arch bridge structures analyzed by a nonlinear three-dimensional finite element method are examined and the influence of several important parameters on the ultimate strength of the arch bridge structures is discussed. On the basis of the numerical results, it has been proposed that steel arch bridges easily can be provided with a practical lateral bracing system (between arch ribs) which will give such sufficient out-of-plane stiffness that the arch strength will be determined primarily by the in-plane behavior. For such a case lateral loads can be taken into account for practical purposes as a set of additional in-plane vertical loads. Thus, the entire design can be made on the basis of a quasi-planar model subjected to the principal vertical loads with a small modification of these loads resulting from wind loads. The required out-of-plane stiffness to ensure that the arch bridge acts basically as an in-plane structure is defined. Ways of in-

*Department of Civil Engineering, University of the Ryukyus, Okinawa, Japan

**Department of Civil Engineering, Tohoku University, Sendai, Japan

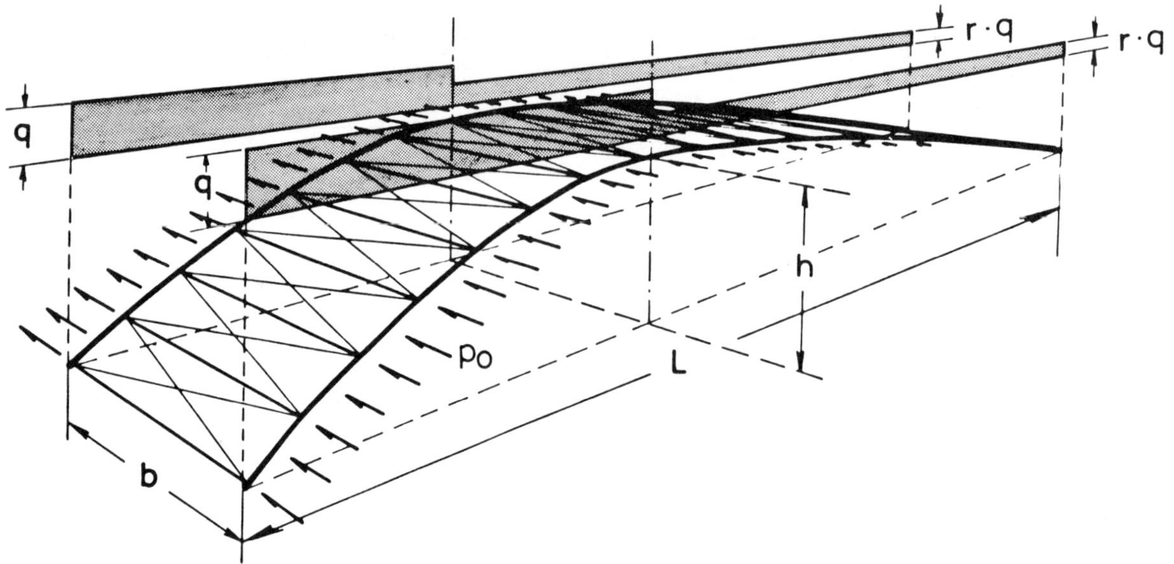

FIGURE 1. General view and applied loads of the reference arch structure.

corporating the practical lateral loading effects in the ultimate strength designing of steel arch bridge structures that may be useful as a guide to the design of practical structures are presented. Some of the available solutions for their lateral-torsional buckling strength are also introduced.

FUNDAMENTALS ON LATERAL BEHAVIOR

Arch Rib Action

The practical lateral loads are carried out to the abutments through relatively lateral bracing systems placed between arch ribs. The arch ribs also play a role in the chords of a wind truss. If the arch ribs are not severely curved and connected by single laterals (light truss system) at the mid-depth of the ribs, their torsion is of minor importance and can be neglected. Analytical expression of this concept is described in the following paragraph. In this case the lateral load design of the ribs follows standard procedure. Namely, the forces acting on the arch rib are applied at the connections of the laterals to the ribs and they act tangentially to the arch curve (so-called axial thrust by wind loads). The tangential forces N_t acting on the arch ribs (axial thrust) are equal to the transverse wind shear, in the panel of lateral bracing system multiplied by the ratio of the panel length to the distance between ribs. Namely:

$$N_t = \frac{Sc}{b} \tag{1}$$

in which S = wind shear force; c = a lateral bracing panel length; b = distance between two arch ribs.

This method is illustrated in Figure 2. These axial thrusts act toward the windward rib and opposite the leeward rib. The stresses by these loads are to resolve the axial thrusts into vertical and horizontal components.

Generally, since the arch ribs are curved members in the plane normal to the direction of in-plane loading, in the strict sense, the effect of curvature on stresses acting on the arch ribs should be considered in the lateral load design. The out-of-plane deformation occurs in a coupled action of the bending and twisting. The exact analysis of the torsional effects is quite complicated because of not only the coupled action but also the interaction of simple torsion of arch ribs with warping torsion as the overall transverse cross-sectional form of the arch bridge structure. These analytical expressions are also described in the following paragraph.

Lateral Web Action

The laterals in the bracing carry the transverse wind shear. An important role of the laterals is to act as lateral web members for the wind shear. In ordinary conventional designing, the laterals are proportioned by the following standard procedure. Namely, the structure is flattened out into a horizontal plane through the supports and analyzed as a plane truss under wind load over the full span as shown in Figure 2. The diagonals are frequently crossed and only the diagonals acting in tension are assumed to be effective. As a result, these members become slender. On the other hand, since the members are ordinary subjected to cyclically varying load (aerodynamic effect of wind), designers should take care of their fatigue problems.

Transverse-Sectional Form

If the two arch ribs are connected by a single lateral bracing system at the mid-depth of the ribs, the simple torsion is of minor importance and can be neglected, practically. The torsional effects are resisted in this case by equal and opposite bending of the arch ribs in the vertical plane, i.e., in-plane bending. Namely, the overall transverse cross-sectional form of the arch is twisted by the difference between the vertical displacements of two arch ribs as shown in Figure 3.

With lateral systems at both the upper and lower flange on chord levels, the simple torsion becomes predominant because the overall structure has a closed torsional section as shown in Figure 4. In these cases, the arch may act as a curved beam with thin-walled closed cross section for the lateral loadings practically.

The transverse beams are frequently crossed between arch ribs. Even if arch ribs have themselves high torsional rigidity, an arch structure composed of single lateral bracing and transverse beams connected with pins may be not able to maintain this torsional rigidity against the out-of-plane deformations as shown in Figure 3. While the transverse beams are rigidly connected to the arch ribs, the torsional rigidity of arch structure is reduced by the deformation of transverse beams. This reduction factor k_T is expressed mathematically as follows:

$$GJ_T = 2GI_T K_T; \quad K_T = \left[1 + \frac{GI_T}{6EI_b} \cdot \frac{bc}{R^2}\left(\frac{\pi}{\alpha}\right)^2\right]^{-1} \tag{2}$$

in which GJ_T, GI_T = torsional rigidity of the arch structure and that of an arch rib; R = the radius of curvature of a circular arch having the same span and rise as the arch under study; EI_b = flexural rigidity of a transverse beam. The results are shown in Figure 5. From this figure if the flexural rigidity of transverse beams is zero, the torsional rigidity of the arch structure becomes also zero. This results from eval-

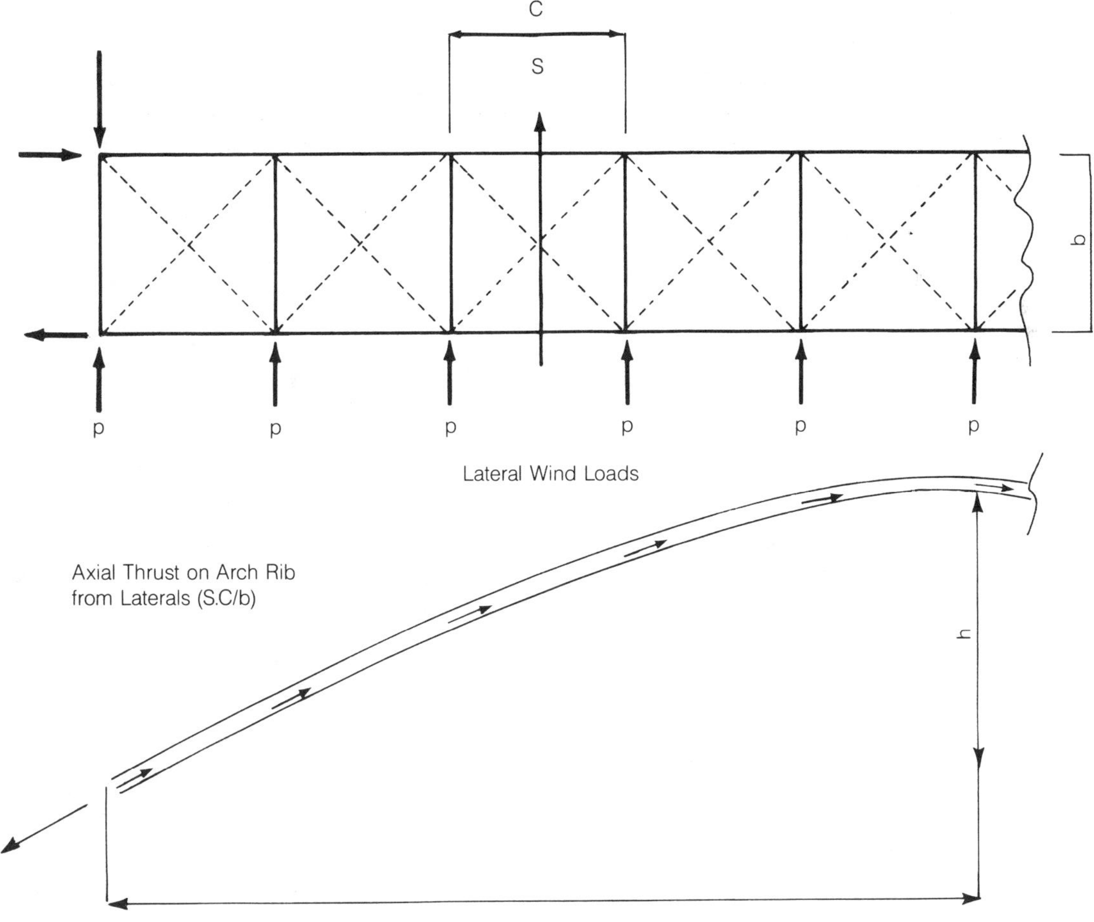

FIGURE 2. Lateral wind analysis by rib bending method.

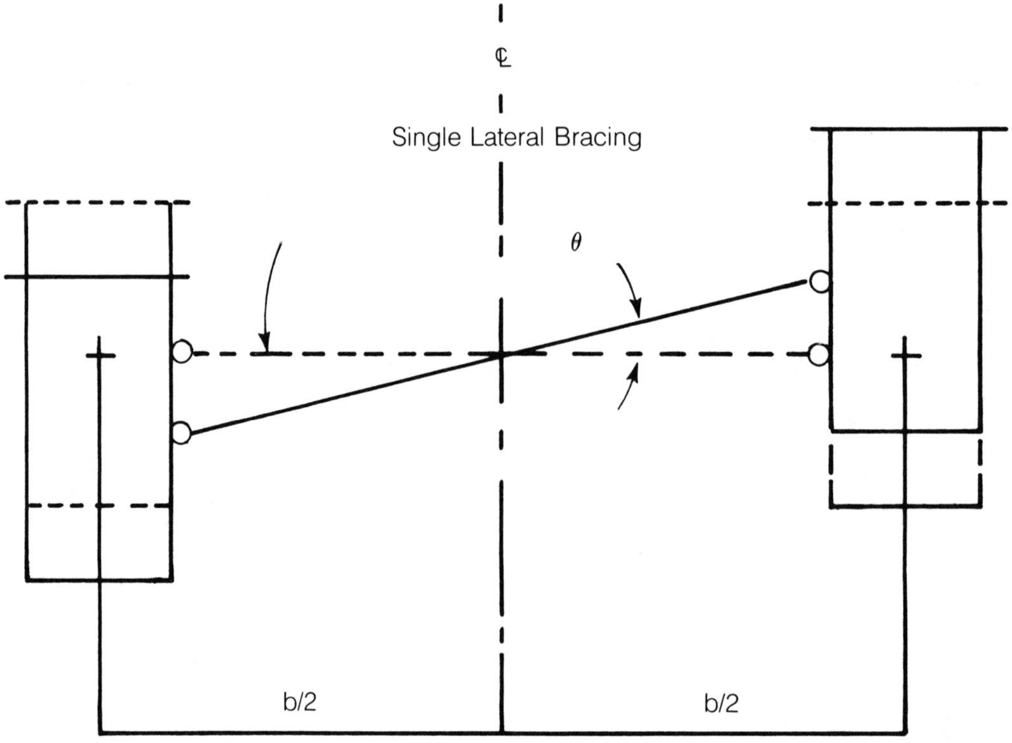

FIGURE 3. Transverse sectional deformation of overall arch structure with single lateral bracing system.

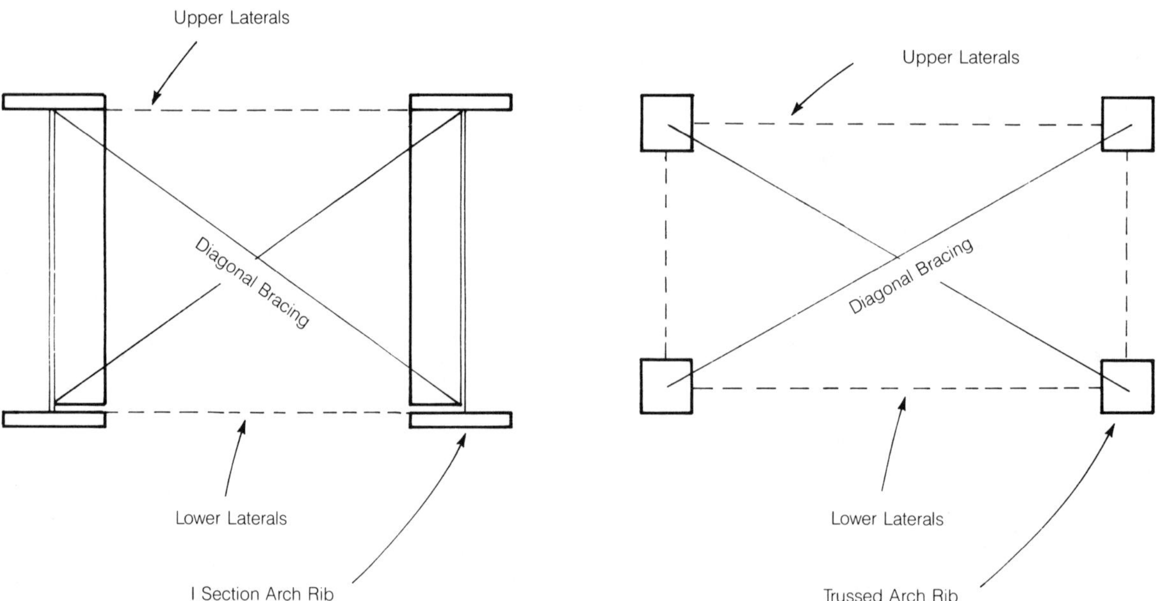

FIGURE 4. Transverse sectional form of overall arch structure with double lateral bracing system.

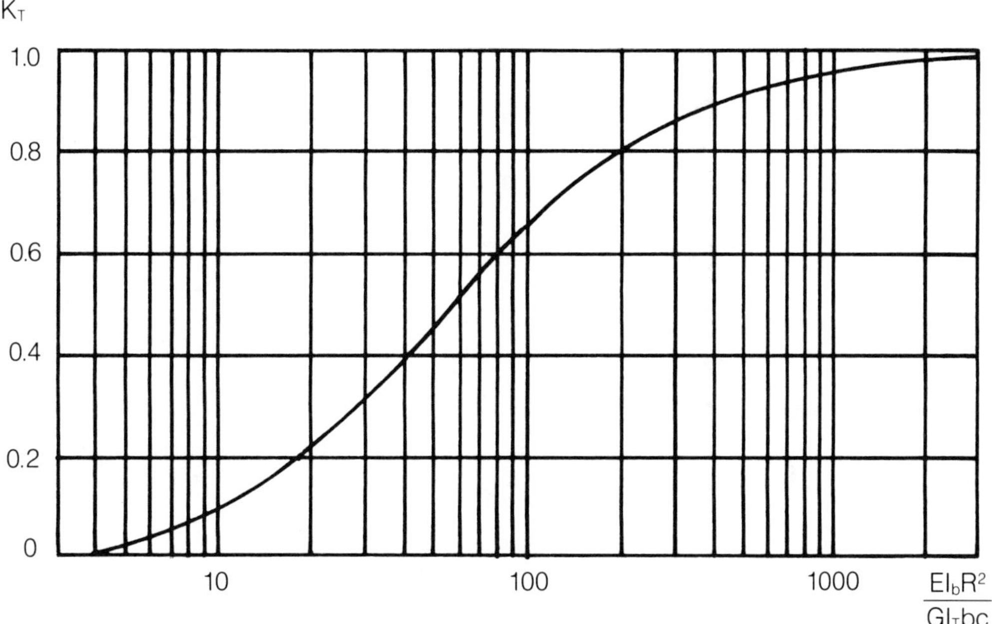

FIGURE 5. Effect of transverse beam flexure on torsional rigidity of overall arch structure.

uating the reduction factor in Equation (1), where the afore-mentioned coupled action of the out-of-plane bending and torsion is neglected. On the other hand, Figure 6 shows typical computed results on the relationship between torsional deformation and stiffness parameters of the arch structure under uniformly distributed lateral load in which the influence of the coupled action is considered – the analytical method is introduced in the following paragraph [15]. The following conclusions can be described chiefly from these computed results [15].

1. Even if an arch structure has ribs with high flexural and torsional rigidity but transverse beams with low flexural rigidity, the arch structure can be expected to have out-of-plane resistance for the torsional deformation, according to the out-of-plane flexural rigidity of arch ribs.

2. The out-of-plane behavior of the arch structure approaches that of an equivalent circular beam as the out-of-plane flexural rigidity of arch ribs decreases, and as the flexural rigidity of transverse beams increases.

Eventually, it can be said that designers must pay attention to the restraint effect of transverse beams on the torsion of arch structure. Although, in the strict sense, the torsional rigidity of arch structure is affected by the coupled action of transverse beam flexure and arch rib flexure in the out-of-plane, Equation (2) may evaluate the reduction of torsional rigidity of the arch structure caused by the deformation of overall transverse-sectional form of the arch structure. In addition, it should be considered that the transverse beams in this case will always be in secondary compression because of their roles as transverse strut members in the bracing system.

Deck System Effect

The lateral actions in the afore-mentioned are for the two arch ribs braced together and subjected to the lateral forces by wind blowing directly against the arch. Generally, it is recommended that the arch in a deck-type bridge is rigidly connected with the stiffened girder in the deck system at the crown. By the connection, the roadway deck and the arch are not free to deflect laterally and the out-of-plane rigidity of the overall structure is increased considerably. The roadway deck will probably have a lateral system of its own and the roadway slab may also act as a lateral system. If there is no diagonal bracing between the columns within the arch span, the main lateral force transfer will be through the short columns at, or the crown plate at, or columns adjacent to, the center of span. The idealized analysis for the lateral force transfer is described in the following paragraph. Use of such diagonal bracing is unusual and is not recommended because of the ineffective mechanical function of the bracing compared with considerably high value of the out-of-plane flexural stiffness of the deck system and also because of the rotation of the arch as the overall transverse-sectional form. Designers should pay attention to the fact that the short stiff columns near the crown of the arch or the crown plate may result in considerable interaction between the deck and the arch rib.

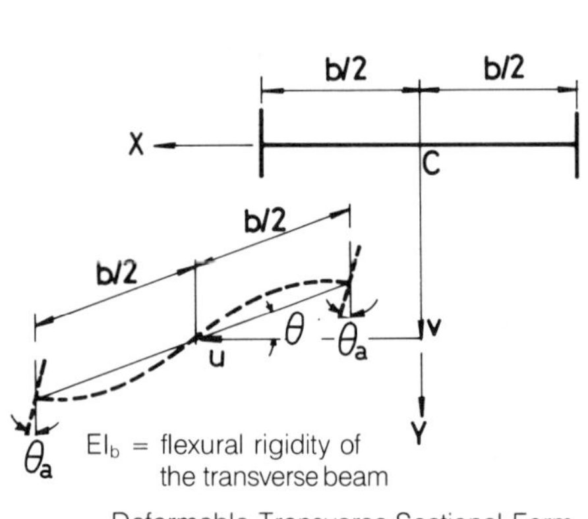

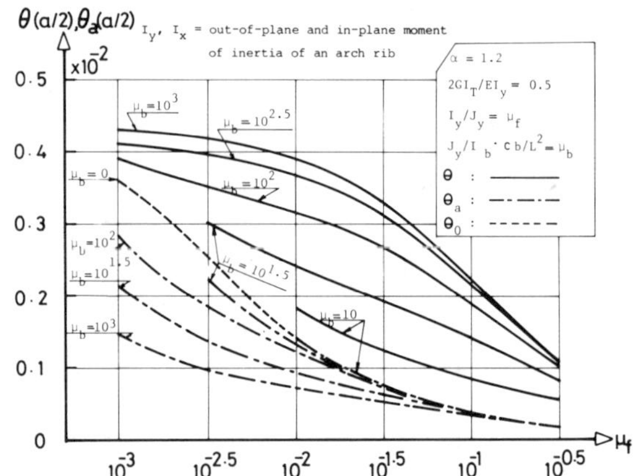

θ_O = torsional angle of arch structure with undeformable transverse-sectional form;
θ = torsional angle of arch structure with deformable transverse-sectional form;
θ_a = torsional angle of arch rib

FIGURE 6. Effects of arch rib-flexural parameter in out-of-plane and transverse beam-flexural parameter on torsional deformation of arch.

LINEAR ANALYSES ON LATERAL BEHAVIOR

Undeformable Transverse-Sectional Form

An equivalent circular beam for the arch structure, as shown in Figure 7, is analyzed under the assumptions that the transverse beams are infinitesimally stiff and rigidly connected to the arch ribs. Furthermore, the influence of out-of-plane flexural stiffness of an arch rib is neglected. Thus it is considered that the arch structure as a whole behaves as a curved beam (equivalent circular beam) under distributed loading, in which the beam has the same span and rise as the arch under study as shown in Figure 7. The equation of the elastic curve of the equivalent circular beam can be set up as follows:

$$2GI_T\left(\theta_o'' - \frac{u''}{R}\right) - \frac{b^2EI_x}{2R^2}\left(\theta_o^{IV} - \frac{u^{IV}}{R}\right) + RM_{c.y} = 0$$
$$EJ_y\left(\theta_o + \frac{u''}{R}\right) + RM_{c.y} = 0 \qquad (3)$$

in which J_y = out-of-plane moment of inertia of the arch structure; $M_{c.y}$ = external out-of-plane bending moment of the arch structure; u = lateral displacement; θ_o = torsional angle; the primes represent differentiation with respect to the angular coordinate ψ along the beam axis. By applying the Laplace Transformation on Equations (3), the solution for the deformation is obtained as follows:

$$u = \frac{pR^3}{bEA_a}\Big[-C_1\cos\psi + \frac{C_2}{m}\cosh m\psi + C_3\sin\psi + \frac{C_4}{m}\sinh m\psi + B_1\{\cos(\alpha - \psi) - \varrho\sin(\alpha - \psi)\}$$
$$+ B_2(\cos\psi + \varrho\sin\psi) + B_3\psi + \frac{1-n}{\sin\alpha}(\cosh m\psi - \cos\psi) - \psi - C_5\Big] \qquad (4a)$$

$$\theta_o = \frac{pR^3}{bEA_a}\Big[\frac{C_2(1+m^2)}{m}\cosh m\psi + \frac{C_4(1+m^2)}{m}\sinh m\psi + 2B_1\cos(\alpha - \psi)$$
$$+ 2B_2\cos\psi + B_3\psi + \frac{1-n}{\sin\alpha}\{\cos(\alpha-\psi) - \cos\psi\} - \psi - C_5\Big]$$

in which

$$M_1 = 1 + \frac{4GI_T}{EA_a b^2}$$

$$M_2 = \frac{J_y}{A_a R^2}$$

$$m = \sqrt{(M_1 - 1)/M_2}$$

$$A_1 = (1 - \cos\alpha)\left(1 - \frac{\alpha}{\sin\alpha} + \frac{2}{1 + m^2}\right) - \frac{2(\cosh m\alpha - 1)\sin\alpha}{m(1 + m^2)\sinh m\alpha}$$

$$A_2 = \left[2(1 - \cos\alpha) + \frac{2(\cosh m\alpha - 1)\sin\alpha}{m^3 \sinh m\alpha} - \frac{\alpha(1 + m^2)\sin\alpha}{m^2}\right]\frac{M_1}{M_1 + M_2}$$

$$A_3 = \frac{\cosh m\alpha - 1}{m(1 + m^2)\sinh m\alpha} - \frac{1 - \cos\alpha}{(1 + m^2)\sin\alpha}$$

$$A_4 = \frac{M_2}{M_1 + M_2}$$

$$n = \frac{A_1 + A_2 + 2A_4(1 + m^2)(0.5\alpha\sin\alpha - 1 + \cos\alpha)}{A_1 + 2A_4(1 + m^2)(A_3\sin\alpha - 1 + \cos\alpha)}$$

$$B_1 = \frac{(n - 1)(M_1 + M_2)}{2(M_1 + M_2 - 1)\sin\alpha} \tag{4b}$$

$$B_2 = -B_1$$

$$B_3 = \frac{M_1}{M_1 - 1}$$

$$B_4 = n$$

$$C_1 = \frac{B_1\{2\sin\alpha\cos\alpha + \alpha(1 + m^2)\} + B_2(m^2\alpha\cos\alpha + \alpha\cos\alpha + 2\sin\alpha) + B_3 m^2(1 - \cos\alpha) + B_4(1 - \cos\alpha)}{(1 + m^2)\sin\alpha}$$

$$C_2 = \frac{2B_1\sin\alpha\cosh m\alpha + 2B_2\sin\alpha - B_3(1 - \cosh m\alpha) + B_4(1 - \cosh m\alpha)}{(1 + m^2)\sinh m\alpha}$$

$$C_3 = \frac{2B_1\sin\alpha - B_3 m^2 - B_4}{1 + m^2}$$

$$C_4 = \frac{-2B_1\sin\alpha - B_3 + B_4}{1 + m^2}$$

$$C_5 = \frac{(B_3 - 1)\alpha}{2}$$

Typical computed results for the torsional angle are shown in Figure 6 by a dotted line.

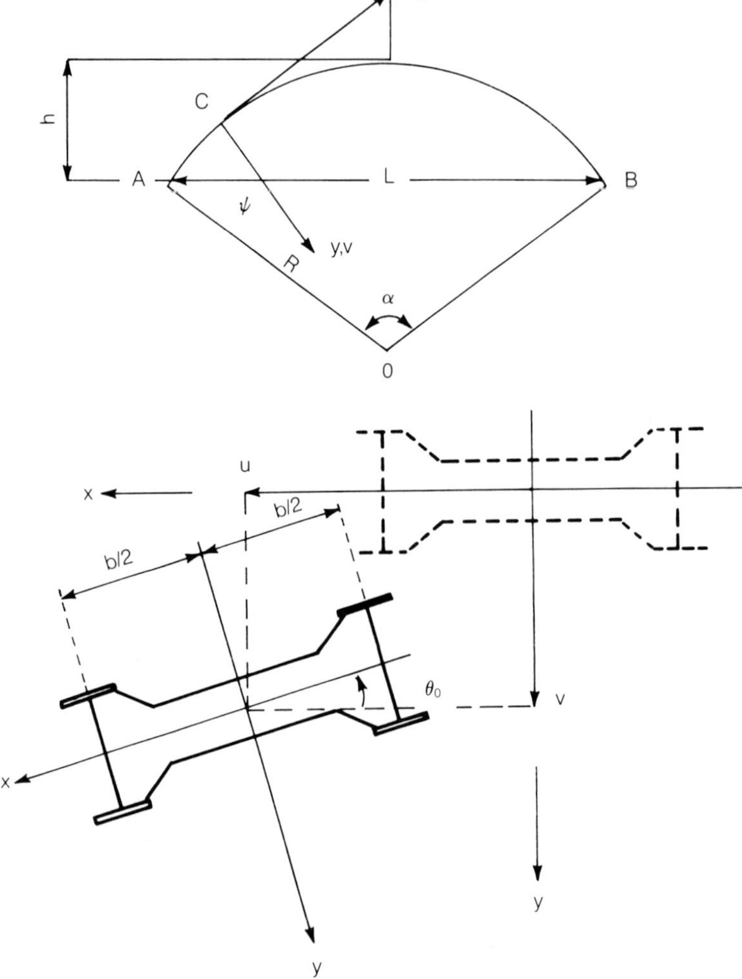

FIGURE 7. Coordinate system and set of displacements of equivalent circular beam.

Deformable Transverse-Sectional Form

The transverse beams are rigidly connected with arch ribs and are deformable. The coupled action of the out-of-plane bending and torsion of arch ribs is considered. Thus the equation of the elastic curve of the arch can be set up as follows:

$$
\begin{aligned}
&2GI_T\left(\theta_a'' - \frac{u''}{R}\right) - \frac{b^2EI_b}{2R^2}\left(\theta^{IV} - \frac{u^{IV}}{R}\right) + RM_{c,y} = 0 \\
&2EI_y\left(\theta_a + \frac{u''}{R}\right) + EJ_y\left(\theta + \frac{u''}{R}\right) + RM_{c,y} = 0 \\
&2GI_T\left(\theta_a'' - \frac{u''}{R}\right) - 2EI_y\left(\theta_a + \frac{u''}{R}\right) + \frac{12ER^2EI_b}{bc}(\theta - \theta_a) = 0
\end{aligned}
\tag{5}
$$

in which an arch is idealized as a curved beam which has the same span and rise as the arch. By applying the Laplace Transformation on Equations (5), the solution could be obtained [15]. Typical computed results for θ and θ_a are also plotted in Figure 6. However, for practical designing viewpoint it would be sufficient to know only the effect of deformation of the transverse beams on the torsional rigidity of the arch structure. For this purpose, the out-of-plane rigidity of individual rib is neglected and out-of-plane bending moment is simply idealized as $M_{c,y} = M_o \sin\pi\psi/\alpha$ in Equations (5). The answer obtained in result is Equation (2).

Lateral Shear

The transverse shear by lateral wind forces is carried by the lateral bracing system (relatively light truss system) in the arch structure. The design of the truss system follows standard procedures. Namely, the structure is flattened out into a horizontal plane through the supports and analyzed as a plane truss (equivalent rolled out length L_S) under wind load over the full span as shown in Figure 2. Plane truss analyses are quite common and hence omitted here.

Interaction Between Arch Rib and Stiffened Girder

By the connection with the stiffened girder and arch rib at the crown, there will be lateral force transfers between them chiefly through the short columns or crown plate. A rough method of analyzing this lateral transfer could be derived by assuming that the arch rib and the stiffened girder, i.e., the deck lateral system and the arch lateral system have the same deflection at the center of the arch span but are free to deflect independently at the other columns within the span. The following will illustrate a method of solving for the value of the lateral force required at the center of the span, between the arch rib and stiffened girder, to equalize the deflection. In the method it is assumed that the torsional deformation is practically negligible insofar as any effect on lateral deflection. The arch is calculated to have a lateral deflection δ_a at the crown for the wind directly against it. It is considered that the lateral deflection δ_a is composed of three components; δ_{ah} from arch rib axial stress by the flexure in the horizontal plane EJ_y, δ_{av} from that in the vertical plane, and δ_{as} from shear stress in the laterals. Since the value of M_{cc} for an arch of average rise to span ratio is about six tenths of the center moment in a fixed end beam of span equal to the rolled out length of the arch axis L_S, the lateral deflection component δ_{ah} will be given by an approximate formula eventually as follows (Figure 8):

$$\delta_{ah} = \frac{1.75pL_S^4}{384EJ_y} \tag{6}$$

The lateral deflection component δ_{av} is approximately calculated as follows:

$$\delta_{av} = \sum \frac{\bar{M}M}{EI_x} \tag{7}$$

in which $\bar{M}$ = in-plane bending moment in 3-hinged system produced by axial thrust under applied unit lateral load at crown as shown in Figure 9b; M = the actual in-plane bending moment produced by the axial thrust N_t under applied actual wind loads as shown in Figure 9a. N_t is given by Equation (1). The lateral deflection component δ_{as} from

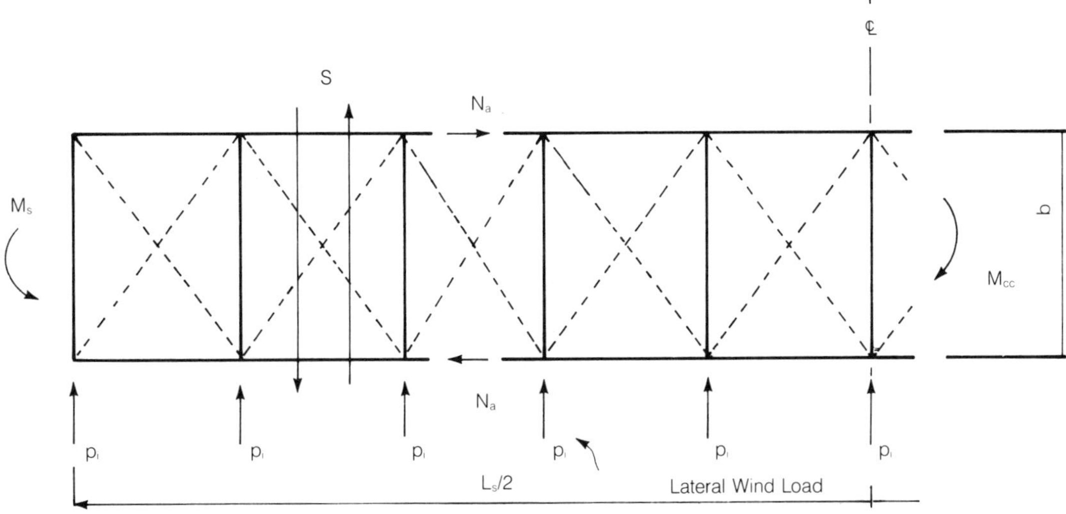

FIGURE 8. Out-of-plane bending moment by arch rib axial thrusts produced by wind loads.

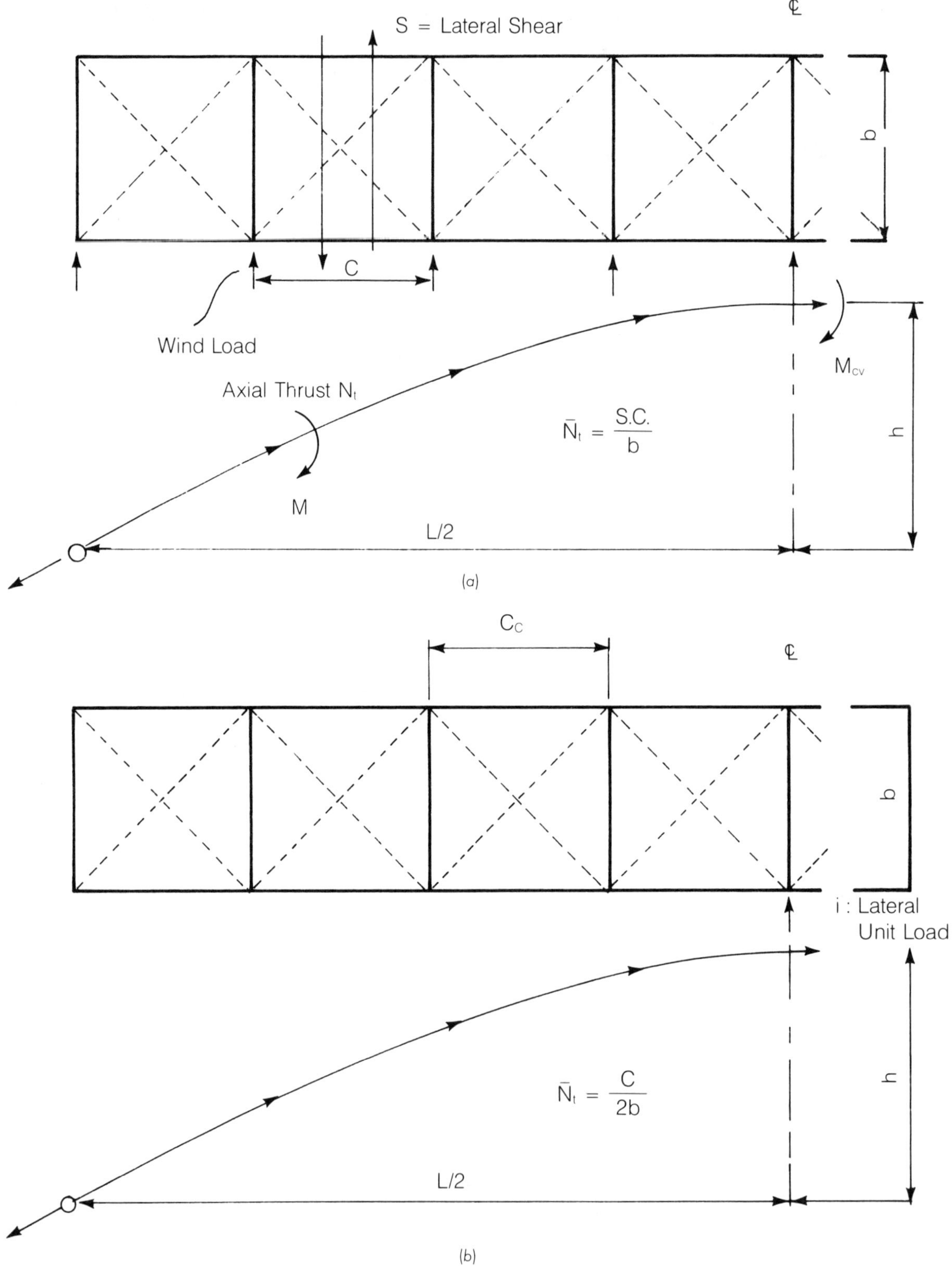

FIGURE 9. In-plane bending moment by arch rib-axial thrusts produced by lateral loads: (a) by lateral wind loads; (b) by a lateral unit load at the crown.

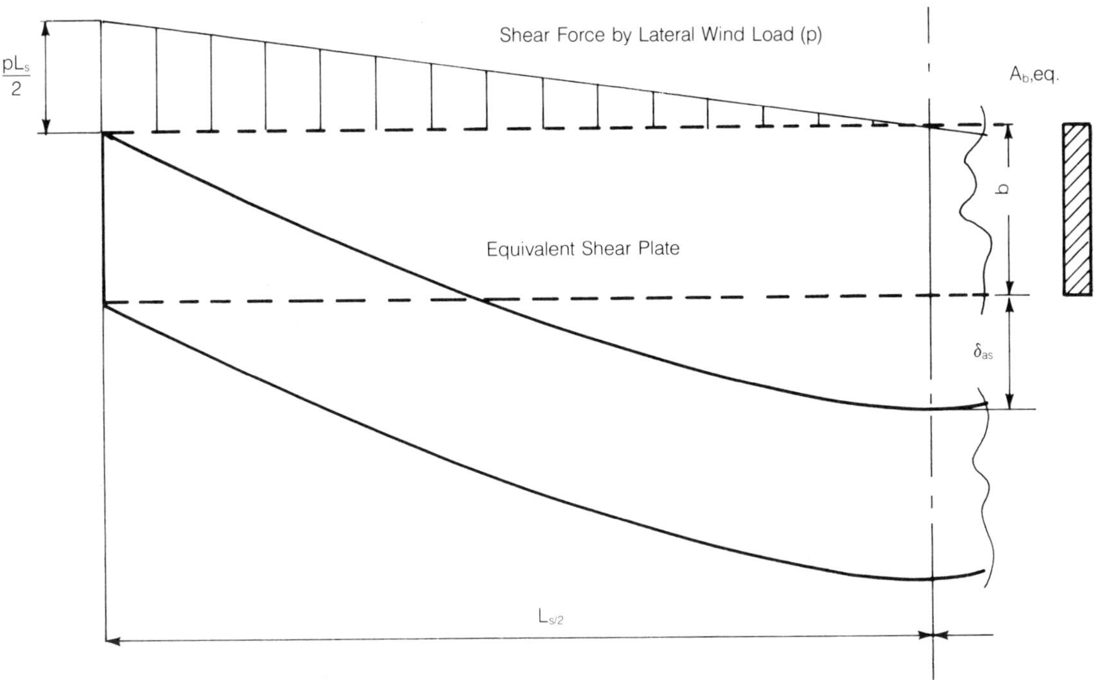

FIGURE 10. Lateral shearing deformation model of overall arch system.

shear stress in the laterals can be roughly given by the fictitious shear plate theory [3] for lateral bracings. The shear plate which has equivalent shear rigidity to that of the lateral bracings is shown in Figure 10. Thus:

$$\delta_{as} = \frac{pL_S^2}{8GA_{b.eq}} \tag{8a}$$

in which $A_{b.eq}$ = cross-sectional area of the equivalent shear plate; for example, a double Warren truss has

$$A_{b.eq} = \frac{E}{G} \cdot \frac{2A_b b^2 c}{d^3} \tag{8b}$$

and K-truss has

$$A_{b.eq} = \frac{E}{G} \cdot \frac{A_b b^2 c}{2d^3} \tag{8c}$$

in which A_b = cross-sectional area of a lateral bracing. To get the lateral deflection δ_a, the lateral deflection δ_{as} from shear and δ_{ah} from horizontal bending flexure must be added to the lateral deflection δ_{av} from vertical bending flexure. The value of δ_{as} for an arch in practical steel bridge structures is about one fifteenth of the center deflection in a built-in beam of span equal to the rolled out length of the arch axis under lateral wind loads.

The stiffened girder (including deck lateral system) is simply calculated to have a lateral deflection δ_g at the center of span for the wind directly against it. Therefore, at the connection between the arch crown and the center of the stiffened girder span, the following deformation compatibility is composed.

$$\delta_g + X_L\overline{\delta_g} = \delta_a - X_L\overline{\delta_a} \tag{9}$$

in which X_L = unknown lateral force transfer from arch to deck at center of span; $\overline{\delta_g}$ = lateral deformation of the girder by unit lateral load at the center of span; $\overline{\delta_a}$ = lateral deformation of the arch by unit lateral load at the crown. From Equation (9), the unknown force, i.e., the lateral force transfer is:

$$X_L = \frac{\delta_a - \delta_g}{\overline{\delta_a} + \overline{\delta_g}} \tag{10}$$

Spatial Structure

In the afore-mentioned analyses, the arch bridge structures which have sufficient out-of-plane rigidities or strengths, or both, are referenced. Therefore, the behavior of arch structure could be evaluated as that of a planar struc-

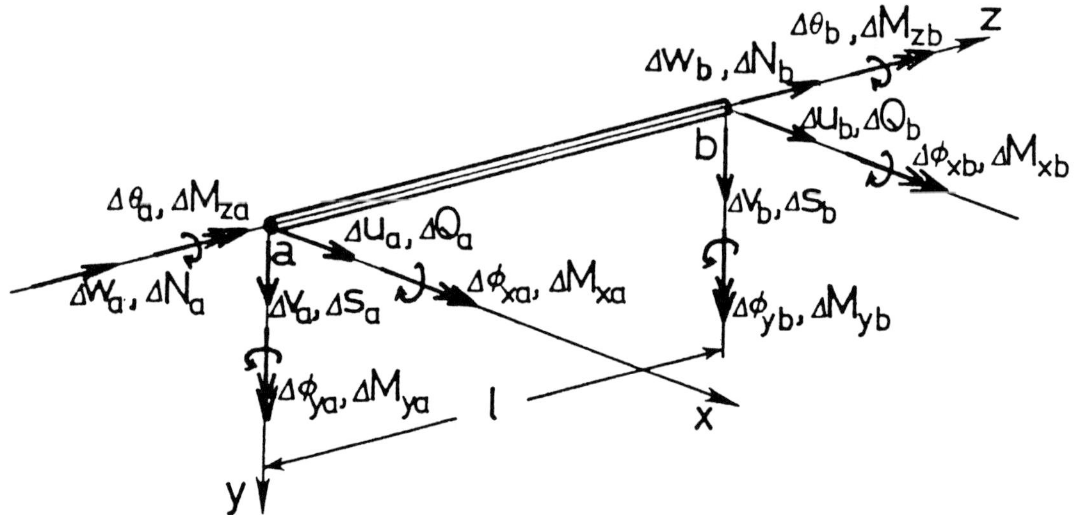

FIGURE 11. Coordinate system, displacements and external member forces applied to the ends of a member element.

ture instead of that of a spatial one. However, in a case where insufficient out-of-plane rigidity or strength (or both) of the structure is provided, the structure will be affected by the coupled action of in-plane and out-of-plane deformation. In this case the structure should be analyzed as a spatial one. The spatial structure analysis can be performed using the finite element method (FEM) conducted by general matrix forms. The strain energy U in a member element of a structure as shown in Figure 11 is given as follows:

$$U = \frac{1}{2}\int_0^l \int_A \sigma \epsilon dAdz + \frac{1}{2}\int_0^l \int_A \tau \sigma dAdz \qquad (11)$$

in which l = a member element length; A = cross-sectional area of a member element; z = longitudinal axis of a member element. By ignoring high order terms in strains, the normal and shear strain ϵ, γ will be expressed mathematically in the matrix form:

$$\epsilon = (B_1)\{\alpha_i\} - y(B_2)\{\alpha_i\} + x(B_3)\{\alpha_i\} \qquad (12a)$$

$$\gamma = \bar{r}(B_4)\{\alpha_i\} \qquad (12b)$$

in which;

$$\begin{aligned}(B_1) &= (0,1,0,0,0,0,0,0,0,0,0,0,)\\ (B_2) &= (0,0,0,0,2,6z,0,0,0,0,0,0)\\ (B_3) &= (0,0,0,0,0,0,0,0,-2,-6z,0,0)\\ (B_4) &= (0,0,0,0,0,0,0,0,0,0,0,1)\end{aligned} \qquad (12c)$$

$\{\alpha\}$ is the coefficient column element which is defined in expressing the displacement components as a matrix form and is indicated in Equations (13).

According to a conventional technique used in the finite element method, the displacement components on a member-element are assumed by the functions of the longitudinal coordinate z. In this analysis, the longitudinal displacement w and rotation of cross section about the z-axis ϕ_z are assumed by the linear functions, respectively, and the deflection u, v in local coordinate (x,y) are approximated by the cubic polynomials. Expressed summarily, it is

$$\{d_i\} = [C]\{\alpha_i\} \qquad (13a)$$

in which $\{d\}$ is the vector of nodal displacements and is given as follows:

$$\{d_i\} = (w_a, v_a, u_a, \phi_{za}, \phi_{ya}, \phi_{xa}, w_b, v_b, u_b, \phi_{zb}, \phi_{yb}, \phi_{xb})^T \qquad (13b)$$

$$\{\alpha_1\} = (\alpha_1, \alpha_2, \alpha_3, \alpha_4, \alpha_5, \alpha_6, \alpha_7, \alpha_8, \alpha_9, \alpha_{10}, \alpha_{11}, \alpha_{12})^T \qquad (13c)$$

Moreover, [c] is the coordinate matrix with respect to z and its reverse matrix $[c^{-1}]$ is indicated hereafter, concretely. Relationship between stress and strain is undoubtedly:

$$\sigma = E\epsilon \qquad \tau = G\gamma \qquad (14)$$

By substituting Equations (12,13,14) into Equation (11) and by applying the stationary condition theorem of the potential energy, the following equilibrium equation is obtained:

$$[k]\,\{d_i\} = \{f_i\} \tag{15}$$

in which $[k]$ denotes the stiffness matrix of a member element and is given as follows:

$$[k] = \int_0^l [c^{-1}]^T \{(B_1)^T (B_1)EA + (B_2)^T(B_2)EI_x + (B_3)^T (B_3)EI_y + (B_4)^T (B_4)GI_T\} [c^{-1}]\, dz \tag{16}$$

and the nodal force vector $\{f\}$ is

$$[f_i] = (N_a, S_a, Q_a, M_{za}, M_{ya}, M_{xa}, N_b, S_b, Q_b, M_{zb}, M_{yb}, M_{xb})^T \tag{17}$$

$[c^{-1}]$ means the reverse matrix of $[c]$ in Equation (13a) and nonzero elements of it are shown as follows:

$$\begin{array}{llllll}
C^{-1}_{1,1} = 1 & C^{-1}_{2,1} = -1/l & C^{-1}_{2,7} = 1/l & C^{-1}_{3,2} = 1 & C^{-1}_{4,6} = 1 & C^{-1}_{5,2} = -3/l^2 \\
C^{-1}_{5,6} = -2l & C^{-1}_{5,8} = 3/l^2 & C^{-1}_{5,12} = -1/l & C^{-1}_{6,2} = 2/l^3 & C^{-1}_{6,6} = 1/l^2 & \\
C^{-1}_{6,8} = -2l^3 & C^{-1}_{6,12} = 1/l^2 & C^{-1}_{7,3} = 1 & C^{-1}_{8,5} = -1 & C^{-1}_{9,3} = -3/l^2 & \\
C^{-1}_{9,5} = 2l & C^{-1}_{9,9} = 3/l^2 & C^{-1}_{9,11} = 1/l & C^{-1}_{10,3} = 2/l^3 & C^{-1}_{10,5} = -1/l^2 & \\
C^{-1}_{10,91} = -2l^3 & C^{-1}_{10,11} = -1/l^2 & C^{-1}_{11,4} = 1 & C^{-1}_{12,4} = -1/l & C^{-1}_{12,10} = 1/l &
\end{array} \tag{18}$$

Using a conventional technique of the stiffness matrix method, Equation (15) can be assembled to a whole structural system based on the global coordinate system, where transformation of Equation (15) to the global coordinate system will be performed by applying the transformation matrix derived on the three-dimensional problem [11].

ULTIMATE STRENGTH BEHAVIOR

Method of Analysis

The spatial behavior and load carrying capacity of steel arch bridge structures under the combined vertical and lateral loading as shown in Figure 12 are discussed in what follows. The spatial behavior and load carrying capacity are determined by the nonlinear, finite element method using the tangent stiffness approach. In the numerical calculation the effect of finite deformation, yielding of the material under combined normal and shear stress, spread of yielding in the cross section and along the length of the arch ribs, unloading caused by strain reversal, and residual stresses due to welding are all taken into account. The stability limit load analysis is carried out by an incremental load method combined with an improved Newton-Raphson iteration procedure. The arch structures which are adopted herein are generally composed of single lateral bracing system and cross beams connected rigidly to the arch ribs. Thin-walled, rectangular, box sections usually adopted on steel arch bridges are considered. The main assumptions made in the analysis are as follows:

1. Material behavior is idealized elastic-perfectly plastic and strain hardening is neglected.
2. Only St. Venant's torsion is considered for the arch ribs, i.e., warping torsion of the ribs is disregarded.
3. Yielding is governed by the normal stress and the shearing stress due to St. Venant's torsion and is defined by the Von Mises criterion.
4. The stress-strain relationship in the plastic range is estimated by Prandtl-Reuss theory.
5. Shear flows due to St. Venant's torsion are uniformly distributed over the cross section even when the cross section is partially yielded.
6. The strains are small but finite deformations are acceptable.
7. Shearing strains due to bending are negligible.
8. The cross section retains in original shape after deformation.
9. The structure is assumed composed of straight finite elements.

Assuming the linearity from the reference stage in which the equilibrium condition has been satisfied to the subsequent stage in which the condition after incremental deformation is taken into consideration, the incremental strain energy ΔU stored from the reference stage to the subsequent

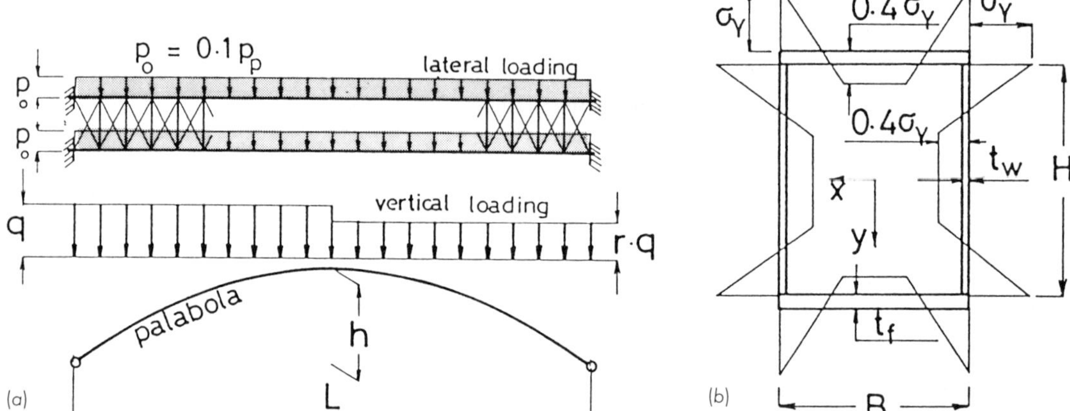

FIGURE 12. Arch geometry, loading and reference cross section: (a) configuration of an arch rib and loading; (b) profiles of cross section and assumed distribution pattern of residual stress.

stage in a member element of a structure is given as follows:

$$\Delta U = \int_0^l \int_A \sigma_0 \Delta\epsilon dA dz + \int_0^l \int_A \tau_0 \Delta\gamma dA dz + \frac{1}{2}\int_0^l \int_A \Delta\sigma\Delta\epsilon dA dz + \frac{1}{2}\int_0^l \int_A \Delta\tau\Delta\gamma dA dz \tag{19}$$

By ignoring higher order terms in strains, the normal and shear strain increments $\Delta\epsilon$, $\Delta\gamma$ will be expressed mathematically in the matrix form:

$$\begin{aligned}\Delta\epsilon = {} & (B_1)\{\alpha\} - y[(B_2) - \psi_{y0}(B_6) - \varrho_{z0}(B_3) - \{\alpha\}^T(B_6)^T(B_3)]\{\alpha\} \\ & + x[(B_3) + \psi_{x0}(B_6) + \varrho_{z0}(B_2) + \{\alpha\}^T(B_6)^T(B_2)]\{\alpha\} + \frac{1}{2}\{\alpha\}^T(B_4)^T(B_4)\{\alpha\} \\ & + \frac{1}{2}\{\alpha\}^T(B_5)^T(B_5)\{\alpha\} + (y^2 + x^2)\left[\frac{1}{2}\{\alpha\}^T(B_7)^T(B_7)\{\alpha\} + \psi_{z0}(B_7)\{\alpha\}\right]\end{aligned} \tag{20a}$$

$$\Delta\gamma = \bar{r}(B_7)\{\alpha\} \tag{20b}$$

in which

$$\begin{aligned}(B_1) &= (0,1,0,\phi_{z0},2z\phi_{z0},3z^2\phi_{z0},0,-\phi_{y0},-2z\phi_{y0},-3z^2\phi_{y0},0,0) \\ (B_2) &= (0,0,0,0,2,6z,0,0,0,0,0,0) \\ (B_3) &= (0,0,0,0,0,0,0,0,-2,-6z,0,0) \\ (B_4) &= (0,0,0,1,2z,3z^2,0,0,0,0,0,0) \\ (B_5) &= (0,0,0,0,0,0,0,-1,-2z,-3z^2,0,0) \\ (B_6) &= (0,0,0,0,0,0,0,0,0,0,1,z) \\ (B_7) &= (0,0,0,0,0,0,0,0,0,0,0,1)\end{aligned} \tag{20c}$$

superscript T = transportation of matrix; x,y = local coordinate system located on a section of a member element and originated at the shear center of a cross section; $\bar{r}$ = distance from the shear center to a tangent drawn at any point on the middle line of the thin wall at the cross section; ϕ_{yo}, ϕ_{zo} = rotation of cross section about y-axis and z-axis, respectively, at the reference stage; ψ_{zo} = torsional rate at the reference stage; ψ_{xo}, ψ_{yo} = curvature with respect to x- and y-axis, respectively, at the reference stage. Moreover, $\{\alpha\}$ is the coefficient column element which is defined in expressing the displacement components as matrix form as shown by Equation (13c) and is indicated in Equation (21a).

According to a conventional technique used in the finite element method the displacement components on a member element are assumed by the functions of the longitudinal coordinate and they are defined in the section "Spatial Structure" above. Expressed summarily, the incremental form is

$$\{\Delta d\} = [c]\,\{\alpha\} \tag{21a}$$

in which $\{\Delta d\}$ is the vector of incremental nodal displacements and given as follows (see Figure 11):

$$\{\Delta d\} = (\Delta w_a, \Delta v_a, \Delta u_a, \Delta\phi_{za}, \Delta\phi_{ya}, \Delta\phi_{xa}, \Delta w_b, \Delta v_b, \Delta u_b, \Delta\phi_{zb}, \Delta\phi_{yb}, \Delta\phi_{xb})^T \tag{21b}$$

$[c]$ and $\{\alpha\}$ are also referred to in the afore-mentioned paragraph.

For the incremental stress-strain relationship in plastic range the Prandtl-Reuss relation assuming linear relationship [20] is employed and if the strain reverse occurs in the plastic range the elastic unloading path is taken into calculation for the equivalent stress. The yield criterion of von Mises is applied to judging yield of the small segments of area in a cross section. The incremental stress-strain relationship in the plastic range will be expressed as follows:

$$\Delta\epsilon = \frac{2}{3}\,\sigma_0\Delta H + \frac{\Delta\sigma}{E} \tag{22a}$$

$$\Delta\gamma = 2\tau_0\Delta H + \frac{\Delta\tau}{G} \tag{22b}$$

in which ΔH means the inclination of the tangent to the equivalent stress-strain curve and is given as follows:

$$\Delta H = \frac{\sigma_0 E\Delta\epsilon + 3\tau_0 G\Delta\gamma}{\frac{2}{3}\,E\sigma_0^2 + 6G\tau_0^2} \tag{22c}$$

The strain reverse under the combined condition of normal and shear stresses can be evaluated by $\Delta H < 0$, using Equation (22c).

Substituting Equations (20) into Equations (22) and summarizing them yields the stress increments expressed as the matrix form: in elastic range

$$\Delta q_e = \frac{s_6}{s_1}\,(B_7)\,\{\alpha\} \tag{23a}$$

$$\Delta\sigma_e = E\Delta\epsilon \tag{23b}$$

and in plastic range

$$\Delta q_p = \frac{1}{s_1 + s_2}\left[-s_3\,(B_1)\,\{\alpha\} + s_4\,(B_2)\,\{\alpha\} - s_5\,(B_3)\,\{\alpha\} - \frac{s_2 s_6}{s_1}\,(B_7)\,\{\alpha\}\right] \tag{24a}$$

$$\Delta\sigma_p = \left(\frac{-3\tau_0}{\sigma_0}\right)\frac{\Delta q_e + \Delta q_p}{t} \tag{24b}$$

where Δq_e and Δq_p mean the shear flow increments in the elastic range and plastic range, respectively, given as $\Delta q_e + \Delta q_p = \Delta\tau t$; moreover, t = thickness of a steel plate which constitutes a cross section of a member element;

$$s_1 = \frac{1}{G}\oint \frac{\overline{ds}}{t}$$

$$s_2 = \frac{1}{E}\int_p \left(\frac{3\tau_0}{\sigma_0}\right)^2 \frac{\overline{ds}}{t}$$

$$s_3 = \int_p \left(\frac{3\tau_0}{\sigma_0}\right)\overline{ds}$$

$$s_4 = \int_p \left(\frac{3\tau_0}{\sigma_0}\right)y\overline{ds} \qquad (24c)$$

$$s_5 = \int_p \left(\frac{3\tau_0}{\sigma_0}\right)x\overline{ds}$$

$$s_6 = \oint r\overline{ds}$$

where

$\int_p \overline{ds}$ = linear integral with respect to a curvilinear coordinate $\bar{s}$ along the middle line of the thin wall in plastic ranges.

In the incremental strain theory, it should be noted that vector direction of the equivalent stress increment is decided according to the stress condition at the reference stage defined as so-called initial stress stage. When the strain path reaches to the critical condition of the reference stage just prior to yield, the strain state in this reference stage is remaining in the elastic condition, but after strain increment corresponding to the incremental process from this stage, the subsequent stage determined by the incremental strain state from the reference stage will be in the plastic condition as shown in Figure 13. In such a case if the incremental strain theory is directly applied to the stress-strain relationship, the stress increment may not be properly evaluated as indicated by the dotted line in Figure 14. In this analysis, thereby, the above-mentioned problem is practically fixed by linearising the incremental strain path from the reference stage to the subsequent one, namely by dividing the strain increment in the interpolation stage on the yielding line as shown by the solid line in Figure 14 [17]. Referring to Figure 13 the normal strain component $\Delta\epsilon_f$ and shear strain $\Delta\gamma_f$ from the reference stage to the interpolation stage will be given by distributing the strain increments $\Delta\epsilon$, $\Delta\gamma$ as follows:

$$\Delta\epsilon_f = \omega\Delta\epsilon$$
$$\Delta\gamma_f = \omega\Delta\gamma \qquad (25)$$

in which ω is a divisionarized coefficient to be calculated by making use of the following equation.

$$\omega = \frac{-(\sigma_0 E\Delta\epsilon + 3\tau_0 G\Delta\gamma) + \sqrt{(\sigma_0 E\Delta\epsilon + 3\tau_0 G\Delta\gamma)^2 - \{(E\Delta\epsilon)^2 + 3(G\Delta\gamma)^2\}(\sigma_0^2 + 3\tau_0^2 - \sigma_y^2)}}{(E\Delta\epsilon)^2 + 3(G\Delta\gamma)^2} \qquad (26)$$

where σ_o = normal stress component at the reference stage; τ_o = shear stress component at the reference stage; σ_y = yield stress level; $\Delta\epsilon, \Delta\gamma$ = normal and shear strain increments, respectively, from reference stage to subsequent one. The strain state corresponding to $\Delta\epsilon_f$ and $\Delta\gamma_f$ is remaining in an elastic state so that the stress-strain relationship can be evaluated by the Hooke's law. Thus, the normal strain component $\Delta\epsilon_s$ and the shear one $\Delta\gamma_s$ from the interpolation stage to the subsequent stage will be given as follows:

$$\delta\epsilon_s = (1 - \omega)\Delta\epsilon$$
$$\Delta\gamma_s = (1 - \omega)\Delta\gamma \qquad (27)$$

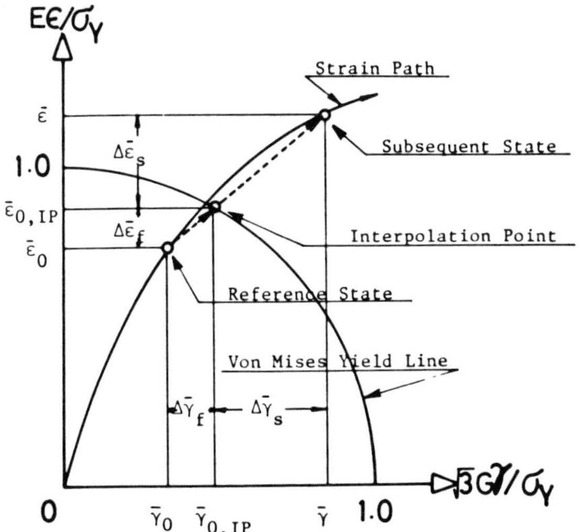

FIGURE 13. Evaluation of incremental strain path in the neighborhood of the yielding.

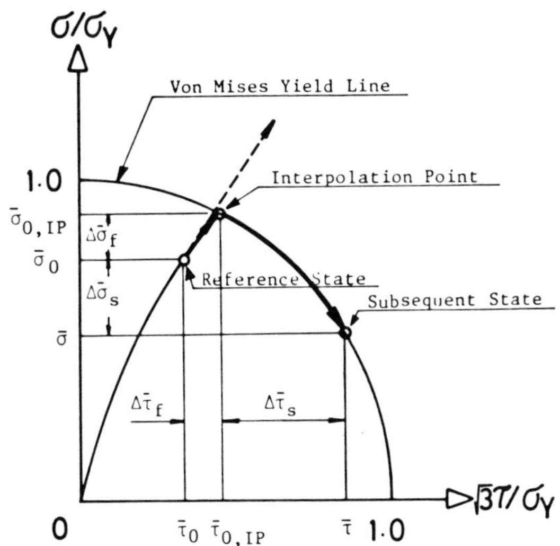

FIGURE 14. Evaluation of incremental stress path in the neighborhood of the yielding.

By applying the Prandtl-Reuss stress-strain relationship as expressed in Equations (24) to Equation (27), the stress increments in the strain defined by Equation (27) can be evaluated practically.

By substituting Equations (20), (23), and (24) to (19) and by applying the stationary condition theorem of the potential energy, the following incremental equilibrium equation is obtained:

$$[k]\{\Delta d\} = \{\delta\}\{f\} + \{f_o\} - \{r_o\} \tag{28}$$

in which $[k]$ denotes the incremental tangent stiffness matrix of a member element evaluated as the reference stage and is given as follows:

$$\begin{aligned}
[k] = \int_o^l & [C^{-1}]^T\{(B_1)^T(B_1)EA_e - (B_1)^T(B_2)ES_{x.e} + (B_1)^T(B_3)ES_{y.e} \\
& - (B_2)^T(B_1)ES_{x.e} + (B_2)^T(B_2)EI_{z.e} - (B_2)^T(B_3)EI_{yx.e} \\
& + (B_3)^T(B_1)ES_{y.e} - (B_3)^T(B_2)EI_{yx.e} + (B_3)^T(B_3)EI_{y.e} \\
& + (B_7)^T(B_7)GI_{z.e} + (B_6)^T(B_2)M_{yo} + (B_6)^T(B_3)M_{xo} + (B_2)^T(B_6)M_{yo} + (B_3)(B_6)^TM_{xo} \\
& + (B_4)^T(B_4)N_0 + (B_5)^T(B_5)N_0 + (B_7)^T(B_7)\Omega_0 + R_1(B_1)^T(B_1) \\
& + R_2(B_1)^T(B_2) + R_3(B_1)^T(B_3) + R_4(B_1)^T(B_7) + R_2(B_2)^T(B_1) \\
& + R_5(B_2)^T(B_2) + R_6(B_2)^T(B_3) + R_7(B_2)^T(B_7) + R_3(B_3)^T(B_1) \\
& + R_6(B_3)^T(B_2) + R_8(B_3)^T(B_3) + R_9(B_3)^T(B_7) + R_4(B_7)^T(B_1) \\
& + R_7(B_7)^T(B_2) + R_9(B_7)^T(B_3) + R_{10}(B_7)^T(B_7)\}[C^{-1}]\,dz
\end{aligned} \tag{29}$$

and the incremental nodal force vector $\{\Delta f\}$ is (see Figure 11):

$$\{\Delta f\} = (\Delta N_a, \Delta S_a, \Delta Q_a, \Delta M_{za}, \Delta M_{ya}, \Delta M_{xa}, \Delta N_b, \Delta S_b, \Delta Q_b, \Delta M_{zb}, \Delta M_{yb}, \Delta M_{xb})^T \tag{30}$$

$\{r_o\}$ means the internal nodal force components corresponding to the external ones $\{f_o\}$ at the reference stage and is given as follows:

$$\{r_0\} = \int_0^l [C^{-1}]^T \{N_0(B_1)^T + M_{x0}(B_2)^T + M_{y0}(B_3)^T + M_{z0}(B_7)^T - \psi_{y0}M_{x0}(B_6)^T$$

$$+ \psi_{x0}M_{y0}(B_6)^T - \phi_{x0}M_{x0}(B_3)^T + \phi_{x0}M_{y0}(B_4)^T + \Omega_0\psi_{x0}(B_7)^T\}dz \quad (31)$$

In these equations

$$N_0 = \int_A \sigma_0 dA$$

$$M_{x0} = -\int_A \sigma_0 y dA$$

$$M_{y0} = \int_A \sigma_0 x dA$$

$$\Omega_0 = \int_A \sigma_0(y^2 + x^2)dA$$

$$A_e = \int_{A,e} dA$$

$$S_{y.e} = \int_{A,e} x dA \quad (32)$$

$$S_{x.e} = \int_{A,e} y dA$$

$$I_{y.e} = \int_{A,e} x^2 dA$$

$$I_{x.e} = \int_{A,e} y^2 dA$$

$$I_{yx.e} = \int_{A,e} yx dA$$

$$I_z = \left(\oint \bar{r}d\bar{s}\right)^2 \Big/ \oint d\bar{s}/t$$

where the suffixes A and e denote the integral over the cross-sectional area and over the elastic ranges, respectively. Furthermore,

$$R_1 = \frac{s_3^2}{s_1 + s_2}$$

$$R_2 = -\frac{s_3 s_4}{s_1 + s_2}$$

$$R_3 = \frac{s_3 s_5}{s_1 + s_2}$$

$$R_4 = -\frac{s_3 s_6}{s_1 + s_2}$$

$$R_5 = \frac{s_4^2}{s_1 + s_2}$$

$$R_6 = -\frac{s_4 s_5}{s_1 + s_2} \quad (33)$$

$$R_7 = \frac{s_4 s_6}{s_1 + s_2}$$

$$R_8 = \frac{s_5^2}{s_1 + s_2}$$

$$R_9 = -\frac{s_5 s_6}{s_1 + s_2}$$

$$R_{10} = -\frac{s_2 s_6^2}{s_1(s_1 + s_2)}$$

In Equation (28), $\{f_o\} - \{r_o\}$ means so-called unbalanced residual nodal forces at the reference stage produced by the linearity assumption and the Newton-Raphson iteration process [21] is repeated until $\{f_o\} - \{r_o\}$ becomes sufficiently small, by substituting $\{\Delta f\} = 0$ into Equation (28).

Generally speaking, the major structural properties of an arch rib that affect the ultimate strength are its rise/span ratio, yield stress level, cross-sectional properties, and distribution pattern and intensity of residual stresses. Among them, the cross-sectional properties are standardized here because of the smaller influence on the ultimate strength as determined by the investigation results on this problem [5,16]. The distribution and intensity of residual stresses are also fixed as shown in Figure 12 based on the results of [5] and [16]. The dimensions of a thin-walled box cross section adopted as reference are as follows: $H/t_w = 190$, $B/H = 0.6$, $A_f/A_w = 0.9$, $r_x/H = 0.4$, and $k/H = 0.32$; in which H,B = depth and width of the reference cross section, t_w = thickness of a web plate, A_f, A_w = cross-sectional area of a flange plate and a web plate respectively, r_x = radius of gyration, and k = core radius. For the range of the structural parameters adopted in the analysis, the proposed formulas are valid in the following range, $\lambda_{IN} = 100 - 300$, $h/L = 0.1 - 0.3$, $\sigma_Y = 240 - 480$ N/mm², $E = 210$ KN/mm², and $G = 81$ KN/mm²; where λ_{IN} = in-plane slenderness ratio of an arch rib which is given by the ratio of the curvilinear length of arch axis to the radius of gyration of the cross section; and h/L = rise to span ratio. The numerical values selected are those generally found for steel arch bridges. The out-of-plane slenderness ratio λ_{OUT} of the arch structure as an overall transverse sectional form is taken into account within the range from 10 to 40 and expressed mathematically as

$$\lambda_{OUT} = L_S \left[\left(\frac{b}{2} \right)^2 + \frac{I_y}{A_a} \right]^{-1/2} \tag{34}$$

The vertical loading used in the calculation (Figure 12) consists of a uniform nodal load, q, over one half of the span and a reduced uniform nodal load, rq, over the other half of the span. The ratio of loads r, could be varied from 0 to 1, the latter value corresponding to the symmetric case of a uniformly loading over the entire span. Uniformly lateral loads, p, are also placed on the nodes of the ribs, as shown in Figure 12. In what follows, the vertical load is nondimensionalized by dividing it by q_p in which q_p = the intensity of full plastic load that produces the squash axial force at the springing

$$q_p = A_a \sigma_y \left[\left\{ \frac{n-2}{2} \right\}^2 + \left\{ \sum_{i=1}^{n} \frac{5Ll_1\, l_2}{8h} (l_1^2 + 3l_1 l_2 + l_2^2 \right\}^2 \right]^{-1/2} \tag{35}$$

in which n = number of the nodes of an arch rib, $l_1 = (i - 1)/n$, $l_2 = 1 - l_1$, i = order of the nodes of an arch rib, A_a = cross-sectional area of an arch rib. Further, the lateral load intensity, p, is also expressed nondimensionally by dividing it by p_p in which p_p = the load that gives full plastic axial thrust at the springings of the arch ribs as calculated by first order elastic analysis of the arch structure flattened out into a horizontal plane through the supports and given by:

$$p_p = \frac{6A_a \sigma_Y b}{L_S(n-2)} \tag{36}$$

Effect of Loading Path

The effect of the loading path on the load carrying capacity of an arch bridge structure subjected to both the vertical and lateral loads shown in Figure 12 is examined [16,17]. In the numerical analysis adopted in what follows, the laterals are taken into account as additional shear springs placed between the two arch ribs in each lateral bracing panel as shown in Figure 15. So, the shear plate which has equivalent shear rigidity to that of the lateral bracing is shown in Figure 10. This shear spring constant k_s can be defined as

$$k_S = \frac{GA_{b.eq}}{b} \tag{37}$$

Typical examples of $A_{b.eq}$ are given by Equations (8). The out-of-plane shear stiffness of lateral bracing as an overall structure is expressed here by a nondimensional parameter which is called shear stiffness parameter of laterals and defined by

$$\mu_S = \frac{\text{central deflection of an equivalent fixed beam by bending}}{\text{central deflection of an equivalent fixed beam by shearing}} = \frac{GA_{b.eq}}{24EA_a} \left(\frac{L_S}{b} \right)^2 \tag{38}$$

For the transverse beams in the lateral bracing system, the following function is taken into consideration; the transverse beams are idealized as torsional springs added to the arch ribs as shown in Figure 16. Using the parameter which expresses the stiffness for the sway or distorsion or both of the transverse sectional form of the arch as an overall structure furnished by the

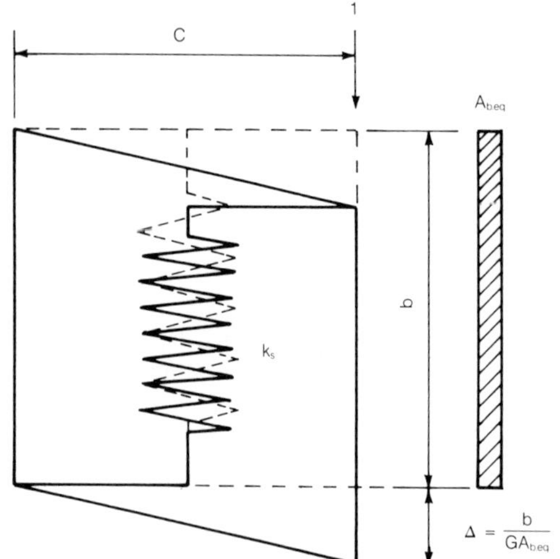

FIGURE 15. Shear spring model of laterals ($k_s s$: shear spring factor).

transverse beams, the torsional spring factor k_T can be evaluated as follows:

$$k_T = \frac{6EI_b}{b} = \frac{6GI_T c}{L^2}\,\mu_T \tag{39}$$

and

$$\mu_T = \frac{EI_b L^2}{GI_T bc} \tag{40}$$

The actual loading path followed by the vertical and lateral loads of bridges is so intricate that it may be impossible to simulate them exactly in numerical analysis. Thus, in order to understand the basic behavior, the effect of the loading path is investigated by adopting the three fundamental paths shown in Figure 17. Loading path I is the path in which the certain lateral load is applied first and then the vertical load is increased proportionally to the collapse load. Loading path II is an inverse path from path I. In path III the vertical

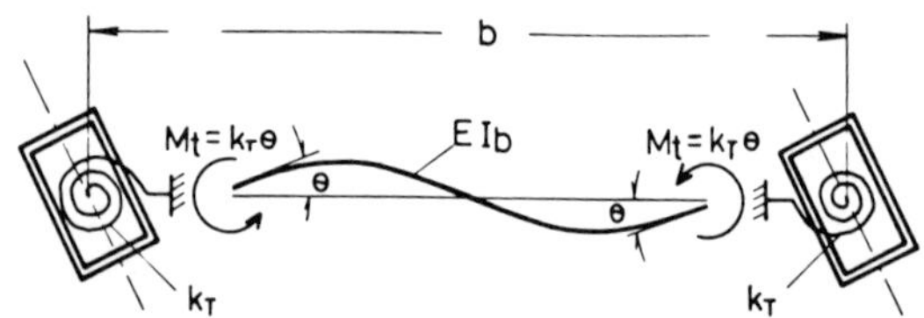

FIGURE 16. Torsional spring model provided by transverse beams (k_T: torsional spring factor).

and lateral loads are applied simultaneously, keeping a proportional relationship between them. Figure 17 shows the load carrying capacity of an arch bridge structure under vertical and lateral loads following the three fundamental loading paths. For the analysis, an arch bridge structure having the standard sectional properties stated earlier and structural proportions shown in Figure 17 is adopted. Judging from the results in this figure, it can be said that the loading path I gives the lowest ultimate strength and the difference between loading paths I and III is within 2% if they are compared to the vertical load carrying capacity for structures subjected to the practical lateral load $p = 0.1p_p$ [4,6,7,16]. From this figure it appears that the load carrying capacity of steel arch bridges is almost independent of the loading path. Thus, in what follows, only loading path I under the certain lateral loading $p = 0.1p_p$ is used for all parametric studies.

Effect of Out-of-Plane Shear and Sway Rigidities

The function of the diagonal members in the lateral bracing system of the arch structures is principally to resist the out-of-plane shearing deformation and to afford the arch ribs sufficient strength for the lateral load. The functions of the transverse beams are to resist the sway or distorsion or both of the transverse sectional form as the overall structure and to develop the inherent torsional rigidity of the arch ribs. The out-of-plane rigidities should be firstly provided so as to keep the above-mentioned functions up to the ultimate state and should contribute to the spatial ultimate strength. From this point of view the parametric analyses have been performed and the functional properties have been discussed [6,7,16,17,18].

Some selected results of the parametric survey performed are shown in Figures 18 and 19, which show the sensitivity of the parameter (μ_s, μ_T) to the ultimate load capacity. The axial thrust versus bending moment relationships at the quarter point of an arch rib of a bridge structure with conventional in-plane and out-of-plane slenderness ratios and subjected to the practical lateral loading are shown in Figure 20. The curves are shown for two different values of μ_s, out-of-plane shearing rigidity of the structure due to the diagonal bracing members. In this figure, the in-plane and out-of-plane bending moments M_{IN}, M_{OUT}, are plotted on the abscissa, while the ordinates represent the axial thrust, N. These quantities are given in nondimensional form by dividing them by the in-plane yield moment, $M_{IN.Y}$, the out-of-plane yield moment, $M_{OUT.Y}$, and the squash load, N_Y, of the arch rib, respectively. In the case of a steel arch bridge with standard out-of-plane shearing rigidity, $\mu_s = 1/15$ [16,18], it can be seen from Figure 20 that the out-of-plane bending moments remain practically small from the initial lateral loading state to the ultimate state, while the in-plane bending moments become large showing nonlinear behavior as the axial thrust increases. On the other hand, in the case of an arch bridge with an extremely small value for the out-

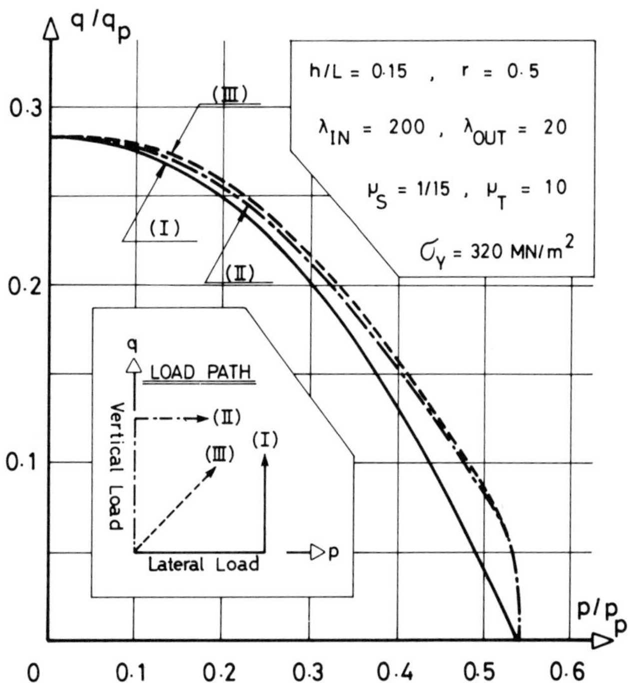

FIGURE 17. Effect of loading path between vertical and lateral load on the ultimate strength.

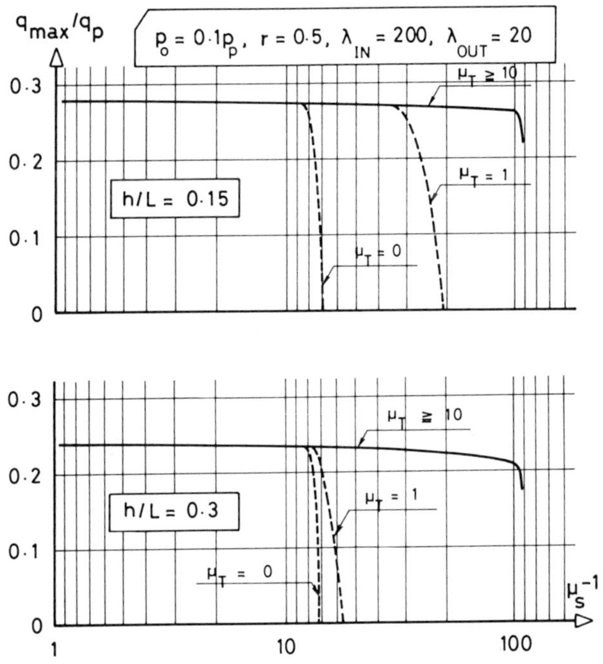

FIGURE 18. Effect of lateral bracing rigidity and rise-to-span ratio on maximum load intensity.

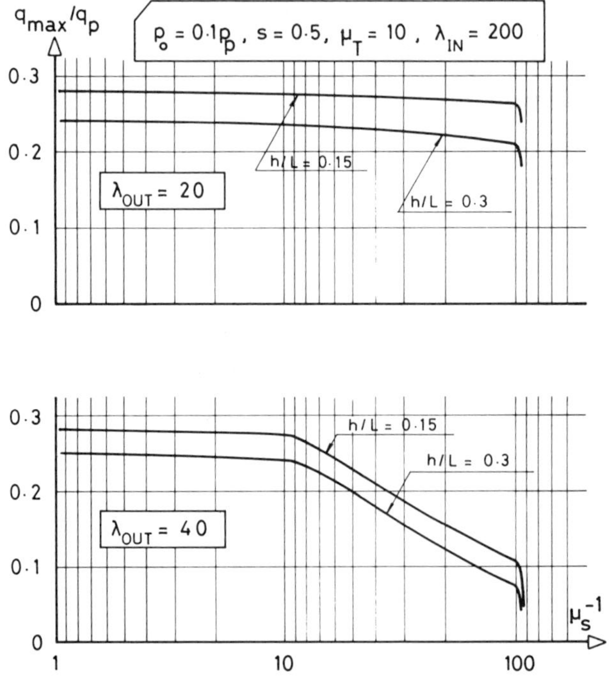

FIGURE 19. Effect of lateral bracing rigidity and transverse beam rigidity on maximum load intensity.

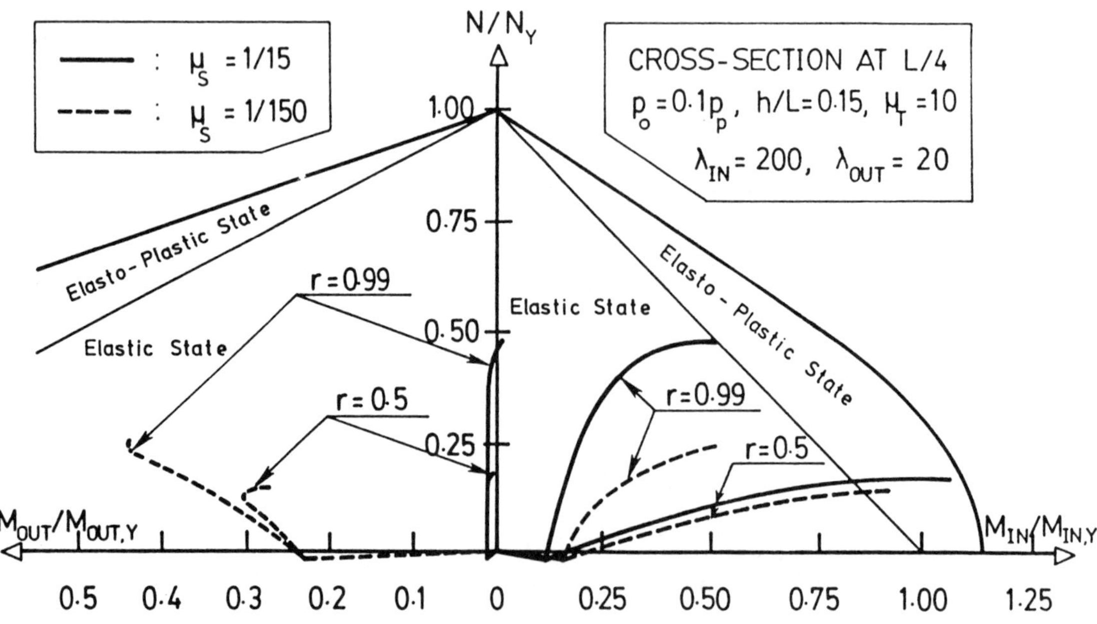

FIGURE 20. Variation of axial thrust, in-plane and out-of-plane bending moments at quarter point of arch rib.

of-plane shearing rigidity (e.g., with $\mu_s = 1/150$ which is only 10% of the standard value for the rigidity analyzed previously), the out-of-plane bending moments also become large, compared to those with standard rigidity. Furthermore, just prior to the ultimate state, the out-of-plane bending moments show extremely intricate nonlinear behavior. From the parametric studies conducted, in which the vertical load distribution ratio, r, was varied, it was observed that the loading case $r = 0.99$ gives the maximum torsional moments for the arch ribs, within the range investigated ($r = 0$ to 0.99, $h/L = 0.1$ to 0.3, $\mu_s = 1/150$ to 1/1.5). Table 1 shows the nondimensional torsional moments ($M_T/M_{T,Y}$) at the ultimate state of the arch ribs under vertical loading with $r = 0.99$. Here

$$M_{T,Y} = 2\sigma_Y t_w \bar{F}/\sqrt{3} \tag{41}$$

in which $\bar{F}$ = the area bounded by the middle line of the thin-walled box section of the arch rib. From Table 1, for the case of $\mu_s = 1/150$ (10% of the standard value), the torsional moment at the ultimate state becomes a maximum corresponding to $r = 0.99$ and $h/L = 0.3$, and has a value of about $0.49M_{T,Y}$ at the springing and $0.03M_{T,Y}$ at the quarter point. On the contrary, with $\mu_s = 1/15$ (standard value for the rigidity), the value of the torsional moment is about $0.1M_{T,Y}$ at the springing and less than $0.01M_{T,Y}$ at the quarter point.

The effect of transverse beams on the rigidity of the arch as a spatial structure is examined [6,16] by considering an arch bridge structure with conventional properties ($h/L = 0.15$, $\lambda_{IN} = 200$). However, a low value of 1/30 for μ_s is chosen, for which the effect of the transverse beams appears distinctly. The change in stiffness of the arch rib, for an incremental value of the vertical load from zero to $0.025q_p$ while the lateral load is maintained at the initial value of $0.1p_p$ is studied. The incremental value of the vertical deflection at the quarter point of the ribs (windward and leeward) is calculated numerically for two values (10, 0) for the transverse rigidity of the bracing system, μ_T. The difference between the increments of vertical deflections of the two ribs was less than 1% in the case of the arch bridge having transverse beams with standard rigidity ($\mu_T = 10$). Thus, in this case, it seems that the stiffnesses of both the ribs are of a similar order of magnitude. The vertical deflection increment of the windward side rib in the case of transverse beams with negligible transverse rigidity ($\mu_T = 0$) is about the same as that for $\mu_T = 10$. On the other hand, the difference between them on the leeward side rib is about 16%. From these results, it can be seen that, in the case of arch bridge structures without transverse beams, not only their out-of-plane rigidities but also the in-plane rigidities decrease under the effect of the lateral loads, so that their load carrying capacity decreases compared to that of the arch structure having transverse beams with conventional rigidities.

TABLE 1. Nondimensional Torsional Moment at the Ultimate State ($M_T/M_{T,Y}$). ($\mu_T = 10$, $p_0 = 0.1\ p_p$, $r = 0.99$, $\lambda_{IN} = 200$)

μ_s	Section \ h/L	0.1	0.2	0.3
1/1.5	Q	0.001	0.005	0.012
	S	0.008	0.029	0.056
1/15	Q	0.003	0.006	0.009
	S	0.037	0.061	0.116
1/150	Q	0.004	0.011	0.034
	S	0.405	0.460	0.489

From the numerically calculated results within the range of the parameters adopted the following behaviors with respect to the bracing effects of the lateral bracing members of the arch bridge structures could be revealed as main tendency:

1. In case of the arch bridge structure having conventional proportions except diagonal members of lateral bracings, the deformations of the arch increase spatially as the applied vertical load increases, when the stiffness parameter μ_s is smaller than the extent of 1/15. On the other hand, when μ_s is larger than the extent of 1/15, the effect of the lateral loading on the ultimate strength is not significant.
2. When the stiffness parameter μ_s is larger than the extent of 1/15, the out-of-plane bending moment and torsional moment which are produced in the arch ribs at the ultimate state are approximately equivalent to the moments given by the first order elastic analysis. The member forces which are produced in the lateral bracing system at the ultimate state are also equivalent to them given by the first order analysis.
3. In the arch having the lateral bracing whose stiffness parameter μ_s is smaller than the extent of 1/15 the arch is advisable to ensure the transverse cross-sectional stiffness as an overall structure for the lateral load, using the transverse beams with the rigidity specified by $\mu_T \geq 10$.

Lateral Load Effect

Typical vertical load versus deformation curves of arch bridge structures under vertical loading and practical lateral loading are shown in Figure 21, for various values of λ_{OUT}, for the out-of-plane slenderness ratio of the arch structure as a whole. The structural properties used in the calculation, shown in the inset of Figure 21, are representative of steel arch bridge structures actually constructed and using lateral bracing members with standard rigidities. This figure shows the variation of the vertical displacement, $v(L/4)$, at a quarter point of the arch rib and the lateral displacement, $u(L/2)$, at the crown of the arch rib, until the ultimate state of the arch structure is attained. From Figure 21 it can be

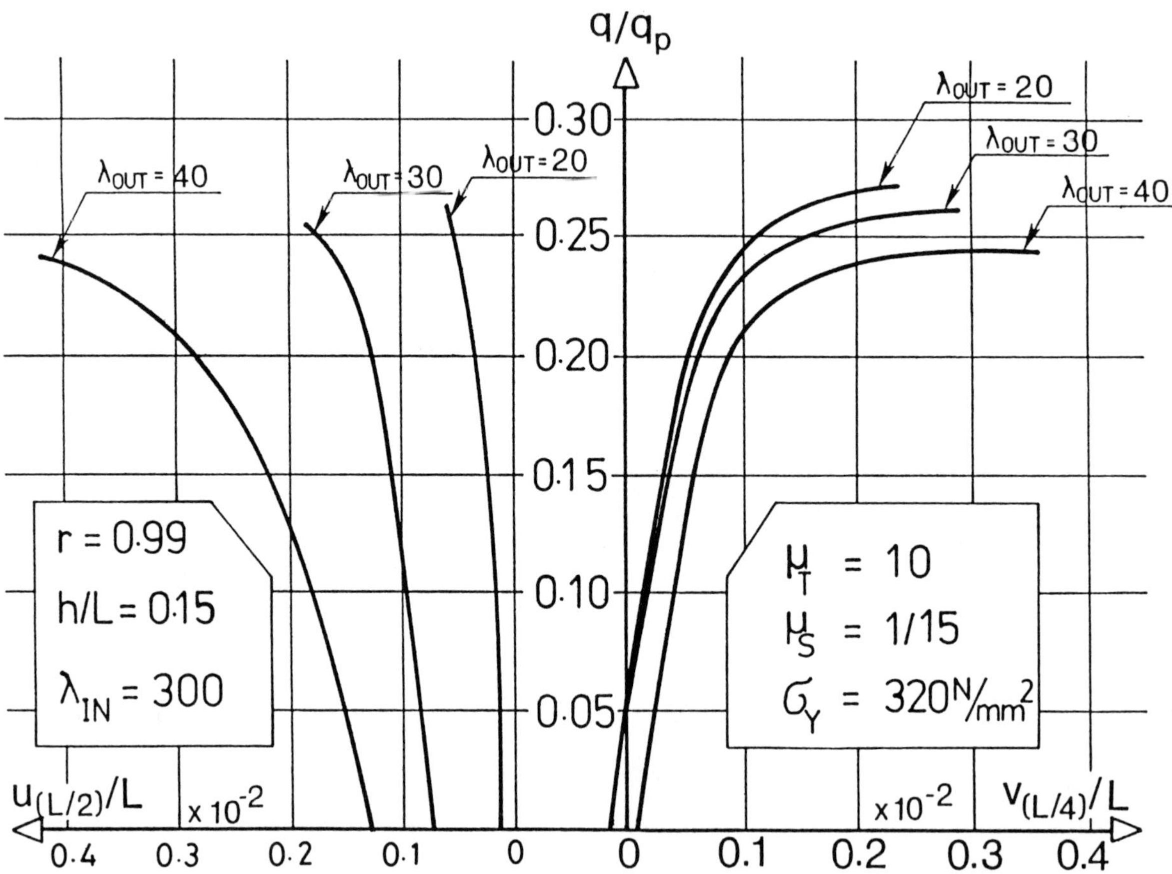

FIGURE 21. Load-deflection curves at quarter point of arch rib in vertical deflection and at crown in lateral direction.

seen that, in the case of a steel arch bridge with conventional out-of-plane slenderness ratio ($\lambda_{OUT} = 20$) and with standard rigidity for the lateral bracing members ($\mu_s = 1/15$ and $\mu_T = 10$), the out-of-plane deformations of the arch ribs remain practically unchanged from the initial lateral loading state to the ultimate state, even for an extremely slender arch ($\lambda_{IN} = 300$). On the other hand, the in-plane deflections become large, showing nonlinear response as the vertical load increases. In what follows, this type of collapse condition will be called "in-plane collapse." It means that the arch, in this case, essentially collapses in the plane that contains the arch axis. Conversely, in the case of an extremely narrow arch bridge—e.g., for $\lambda_{OUT} = 40$, or double the conventional value mentioned earlier, both the out-of-plane and in-plane displacements show a tendency to become fairly large as the vertical load increases and the collapse occurs under complex, spatial, nonlinear behavior.

In a general way, even for arch bridge structures under combined vertical and practical lateral loads, the critical vertical loading pattern that results in collapse is unsymmetrical. The lateral loading effect on the ultimate load carrying capacity under the unsymmetrical vertical loading is estimated herein by using a strength reduction factor ϕ to show the degree of reduction of the ultimate strength induced by the practical lateral load. It is expressed as follows:

$$\Phi = \frac{\text{spatial ultimate strength under vertical and lateral loads}}{\text{planar ultimate strength for same arch under vertical load}} \tag{42}$$

Here, the ultimate strength under the combined loads is evaluated by keeping constant the lateral loads at $p = 0.1p_p$, while increasing the vertical loads to collapse. The arch structures are laterally braced by diagonal members and transverse beams having standard rigidities ($\mu_s = 1/15$, $\mu_T = 10$), and are subjected to practical lateral loading. Typical results for the influence of lateral loading on the load carrying capacity of an arch structure are shown in Figure 22. From Figure 22, it is seen that the reduction in the load carrying capacity due to the presence of the lateral loading becomes remarkable as the rise-to-span ratio becomes large and as the in-plane slenderness ratio of the arch rib becomes large. Furthermore, from similar numerical studies conducted, it was observed that this reduction increases as the out-of-plane slenderness ratio of the arch bridge structure as a whole, γ_{OUT}, becomes large.

From the viewpoint of conventional designing, it is desirable that the structure behaves as a planar structure until collapse. Consequently, from the results analyzed as the spatial structures, the cases of in-plane collapse are selected out and their behaviors are investigated. The results are summarized as shown in Figure 23. In this figure the ratio of the initial lateral deflection W_o at the crown of the arch under the practical lateral load to the span length L is plotted on the abscissa and is considered as an index factor to evaluate the out-of-plane stiffness of the arch structure as a whole. W_o is the lateral displacement calculated by the 1st order spatial structural analysis (explained in the afore-mentioned paragraph) and the typical results are listed in Table 2. By applying the regression analysis to the results in Table 2, the relationship between the nondimentional, initial, lateral displacement W_o/L and the lateral load intensity p_o/p_p could be expressed as a functional formula

$$\frac{W_0}{L} = K_{OUT}^{-1} \frac{p_0}{p_p} \tag{43}$$

in which K_T is the nondimensional out-of-plane stiffness of the arch structure as a whole and can be approximated by

$$K_{OUT} = \frac{1}{\lambda_{OUT}} \cdot \frac{2 \times 10^6}{(\lambda_{IN} - 100)(44.2h/L + 7.84) + 9.05(300 - \lambda_{IN})} \tag{44}$$

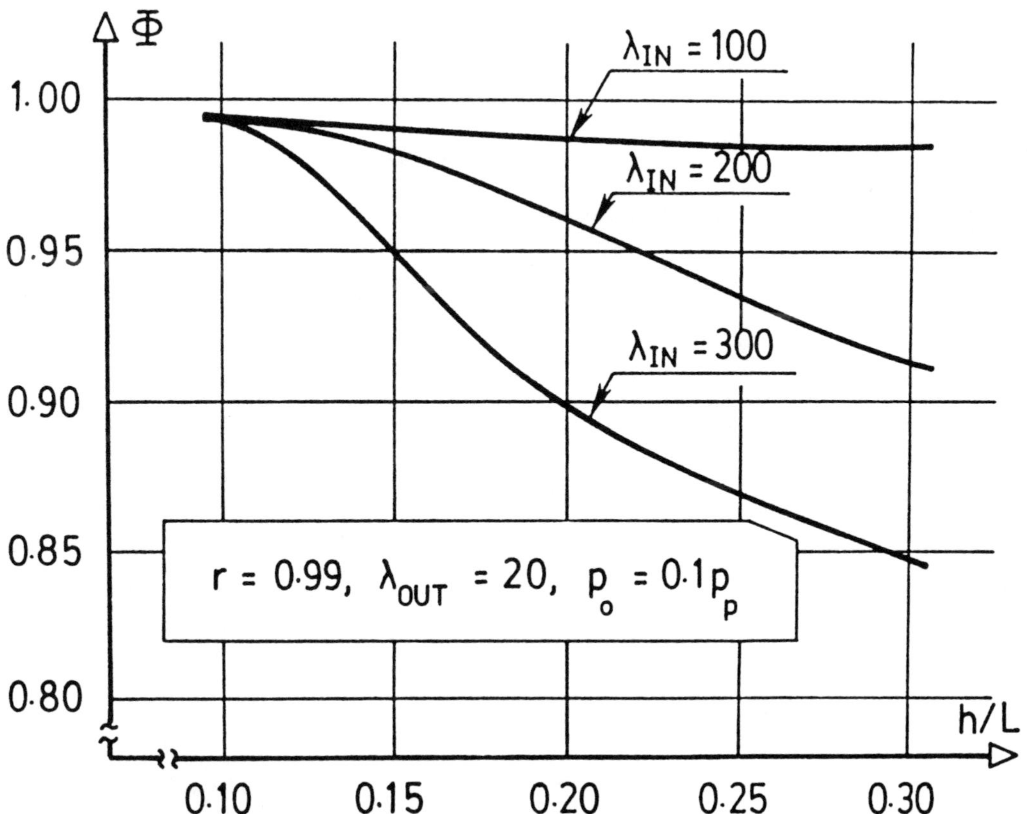

FIGURE 22. Effect of slenderness and rise-to-span ratios of arch on strength reduction factor ϕ.

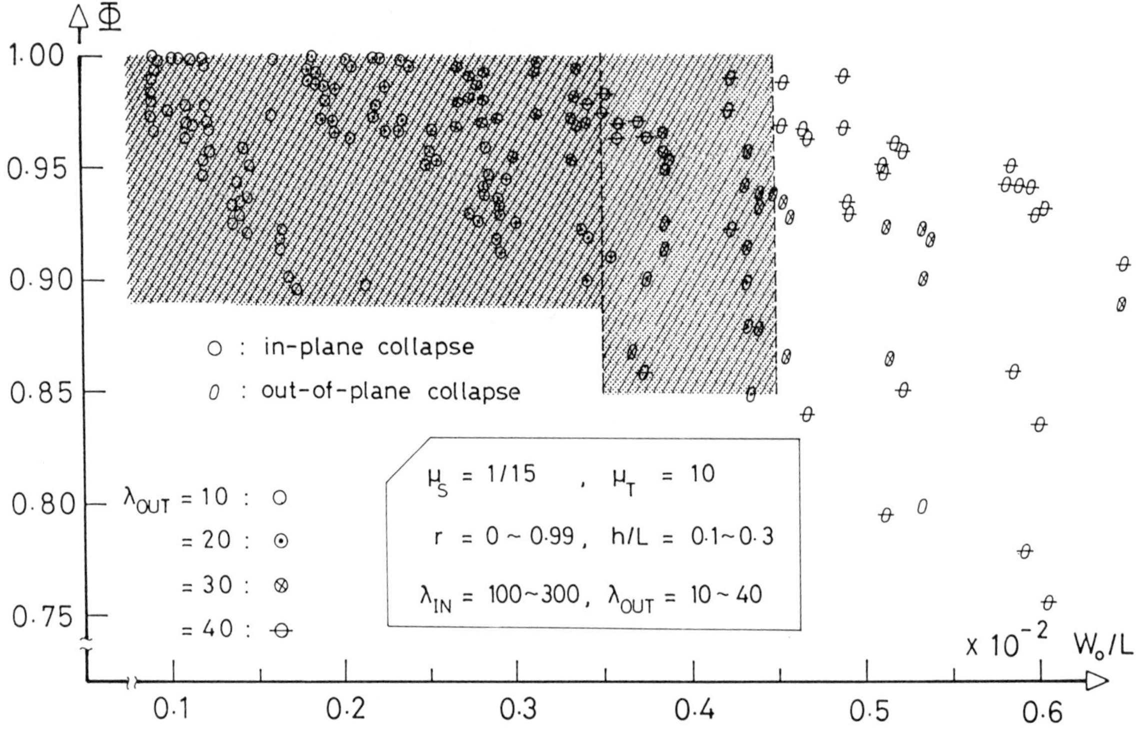

FIGURE 23. Influence of collapse configurations and out-of-plane stiffness on strength reduction factor ϕ.

Quasi-Planar Model

From the results obtained [4,6,7,15,16,17,18], the following statements can be made: If the arch structure has sufficient out-of-plane rigidities, as is the case with well-designed steel arch bridges, the effect of the out-of-plane deformations and stresses on the load carrying capacity of the structure due to the practical lateral loads may not be substantial. The lateral loads may play a role only in the reduction of the rigidity of the arch ribs due to the increased normal stress and early yielding. With this view, a practical analytical method for the load carrying capacity which can estimate

TABLE 2. Nondimensional, Initial, Lateral Deflection (W_o/L) Calculated by the 1st Order, Spatial, Structural Analysis.

Rise to span ratio (1)	In-Plane slenderness ratio (2)	Out-of-Plane Slenderness Ratio			
		10 (3)	20 (4)	30 (5)	40 (6)
0.1	100	0.092×10^{-2}	0.183×10^{-2}	0.270×10^{-2}	0.360×10^{-2}
	200	0.101	0.206	0.314	0.424
	300	0.115	0.234	0.370	0.490
0.15	100	0.092	0.186	0.275	0.360
	200	0.109	0.226	0.350	0.467
	300	0.138	0.284	0.434	0.570
0.2	100	0.092	0.190	0.283	0.358
	200	0.119	0.249	0.380	0.512
	300	0.165	0.339	0.514	0.699
0.3	100	0.091	0.195	0.339	0.371
	200	0.137	0.288	0.430	0.580
	300	0.214	0.435	0.649	0.844

the effect of the lateral loads and rigidities by a planar analysis of the structure has been proposed in [4,16 and 17]. The outline of this method follows:

Since the arch bridge structures considered are assumed to have a rigid lateral bracing system, it was considered that the structure as a whole behaves as a curved beam under distributed lateral wind loading in the section "Linear Analyses on Lateral Behavior". The resulting in-plane cross-sectional forces in the ribs can be evaluated from the following relation, if the case of undeformable sectional form—an equivalent circular beam-model—is taken into consideration

$$N^0 = \pm \frac{EI_x b}{2R}\left(\theta + \frac{u''}{R}\right)$$
$$S^0 = \pm \frac{EI_x b}{2R^3}\left(\theta''' - \frac{u'''}{R}\right) \quad (45)$$
$$M^0 = \pm \frac{EI_x b}{2R^2}\left(\theta'' - \frac{u''}{R}\right)$$

in which + and − in the sign ± stand for the cross-sectional force in windward and leeward side rib, respectively; the primes represent differentiation with respect to the angular coordinate along the arch axis; N^o, S^o, M^o = the axial thrust, in-plane shear, and in-plane bending moment, respectively. From the in-plane forces (N^o, S^o, M^o) acting on the elements in i-1 and i, and from the equilibrium of forces acting on joint j connecting these two elements (Figure 24), the additional in-plane nodal loads (F_z^o, F_y^o, M_{IN}^o) that are equivalent to them can be calculated as follows:

$$\begin{bmatrix} F_z^0 \\ F_y^0 \\ M_{IN}^0 \end{bmatrix}_j = \begin{bmatrix} N_b^0 \\ S_b^0 \\ M_b^0 \end{bmatrix}_{i-1} - \begin{bmatrix} N_a^0 \\ S_a^0 \\ M_b^0 \end{bmatrix}_i \quad (46)$$

In the analysis, N^o and S^o were assumed to be uniform over the length of a finite element, while M_{IN}^o is assumed to be a linear polynomial of z. Therefore, Equation (46) can be rewritten as

$$\begin{bmatrix} F_z^0 \\ F_y^0 \\ M_{IN}^0 \end{bmatrix}_j = \begin{bmatrix} N_a^0 \\ S_a^0 \\ M_a^0 + lS_a^0 \end{bmatrix}_{i-1} - \begin{bmatrix} N_a^0 \\ S_a^0 \\ M_b^0 \end{bmatrix}_i \quad (47)$$

in which l = the length of the $(i - 1)$th finite element. To sum up, the deformations u, θ, etc., are determined by the first-order elastic analysis of the curved beam as aforementioned. Then, using Equation (45), the forces N^o, S^o, M_{IN}^o at the end, a, of each element shown in Figure 24 are calculated. The additional in-plane nodal forces are next obtained from Equation (47). By applying these additional in-plane loads (equivalent to the lateral loads) to the separated planar (i.e., quasi-planar) arch ribs (see Figure 25), the influence of the lateral loads can be approximated. Thus, applying the equivalent additional loads, the load carrying capacity of the arch structures can be determined by planar structural analysis. The numerical analysis method for planar structures [4,5,8,16,18,19] is based on theory similar to that of the spatial analyses described earlier.

Figure 26 shows typical results from the axial thrust and in-plane bending moment for the equivalent curved beam-model under lateral wind loads Equation (45) and for the spatial structure-model under the equivalent additional in-plane loads Equation (47). As is obvious from Figure 26, close agreement between them is obtained. The results are given in Tables 3-5. In these tables, ϕ_s indicates the strength reduction factor as obtained by a spatial inelastic analysis for the load carrying capacity, and ϕ_p indicates the strength reduction factor calculated by the quasi-planar analysis for the load carrying capacity. In these examples, the two arch ribs are braced together laterally by diagonal members and transverse beams with standard rigidities (μ_s = 1/15, μ_T = 10), and the ribs are spaced apart at a conventional distance resulting in λ_{OUT} = 20. From these tables, it appears that the reduction of the load carrying capacity of steel arch bridge structures due to practical lateral loads is less

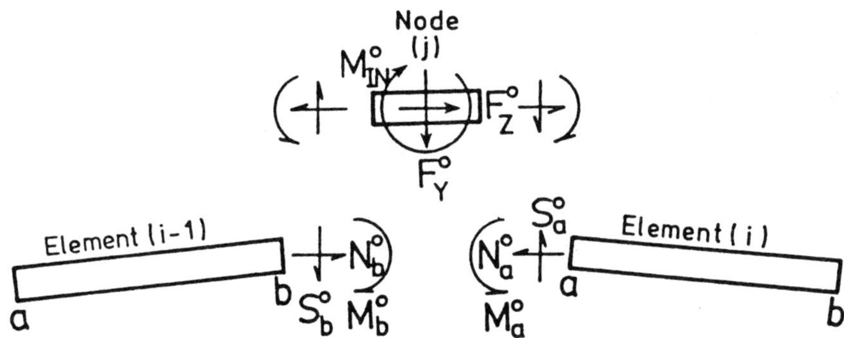

FIGURE 24. In-plane forces acting on node, j, under lateral loads.

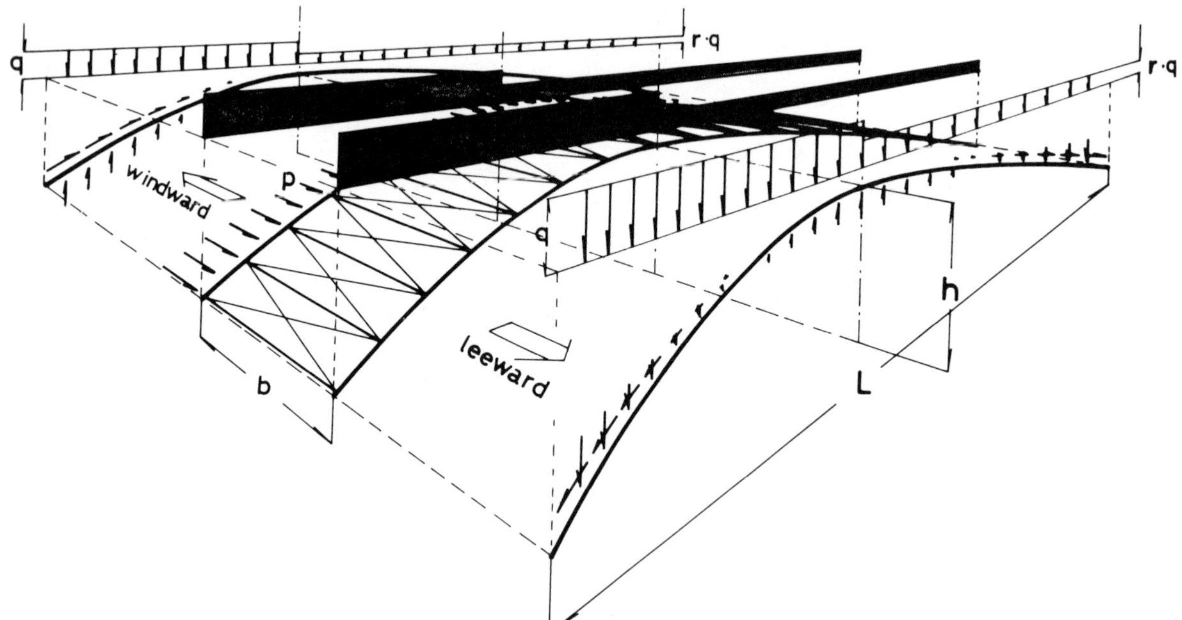

FIGURE 25. Separated planar arch rib model.

than 6%. All the examples showed in-plane collapse even when analyzed as spatial structures. These tables show that, compared to the spatial analysis method, the quasi-planar analysis method gives more conservative values for the load carrying capacity (15% on the maximum and 7% on the average). The reason for this conservatism could be explained by the difference between the load carrying capacity and deformations of the windward and leeward side arch ribs. As shown in Figure 27, the in-plane rigidity of the separated windward side arch rib increases owing to the so-called prestress effect, while the in-plane rigidity of the leeward side arch rib decreases owing to the initial stress effect produced by the lateral loads. In the real structure, however, the difference between the vertical displacements of the windward side rib and the leeward side rib will decrease because of the restraining effect of the transverse beams on the torsion of the arch. This effect is not taken into account in the quasi-planar analysis. The torsional rigidity developed by

TABLE 3. Variation of Strength Reduction Factor as a Function of r.
(h/L = 0.15, λ_{IN} = 200, λ_{OUT} = 20, p_o = 0.1 p_p)

r	ϕ_S	ϕ_P	$(\phi_S - \phi_P)/\phi_S$ %
0.99	0.986	0.924	6.3
0.5	0.978	0.902	7.8
0.0	0.967	0.896	7.3

TABLE 4. Variation of Strength Reduction Factor as a Function of h/L.
(λ_{IN} = 200, λ_{OUT} = 20, p_o = 0.1 p_p)

h/L	r	ϕ_S	ϕ_P	$(\phi_S - \phi_P)/\phi_S$ %
0.10	0.50	0.997	0.919	7.8
	0.99	1.000	0.936	6.4
0.15	0.50	0.978	0.902	7.8
	0.99	0.986	0.924	6.3
0.20	0.50	0.952	0.878	7.8
	0.99	0.980	0.911	7.0

TABLE 5. Variation of Strength Reduction Factor as a Function of λ_{IN}.
(h/L = 0.15, λ_{OUT} = 20, p_o = 0.1 p_p)

λ_{IN}	r	ϕ_S	ϕ_P	$(\phi_S - \phi_P)/\phi_S$ %
100	0.50	0.991	0.968	2.3
	0.99	0.993	0.891	10.3
200	0.50	0.978	0.902	7.8
	0.99	0.986	0.924	6.3
300	0.50	0.959	0.810	15.5
	0.99	0.947	0.840	11.3

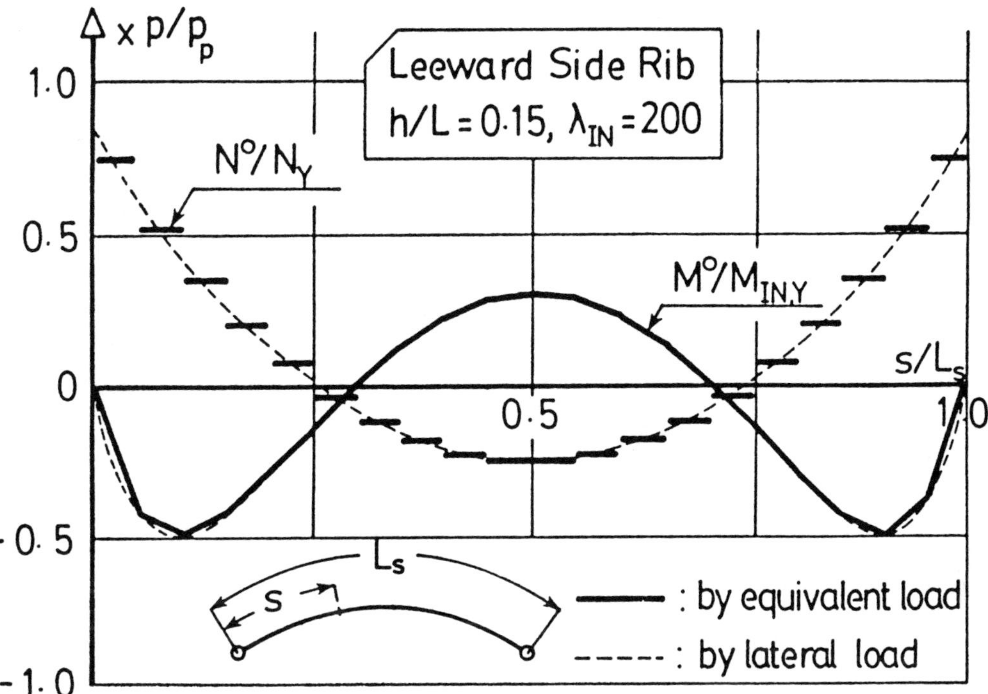

FIGURE 26. In-plane cross sectional forces caused by lateral loads.

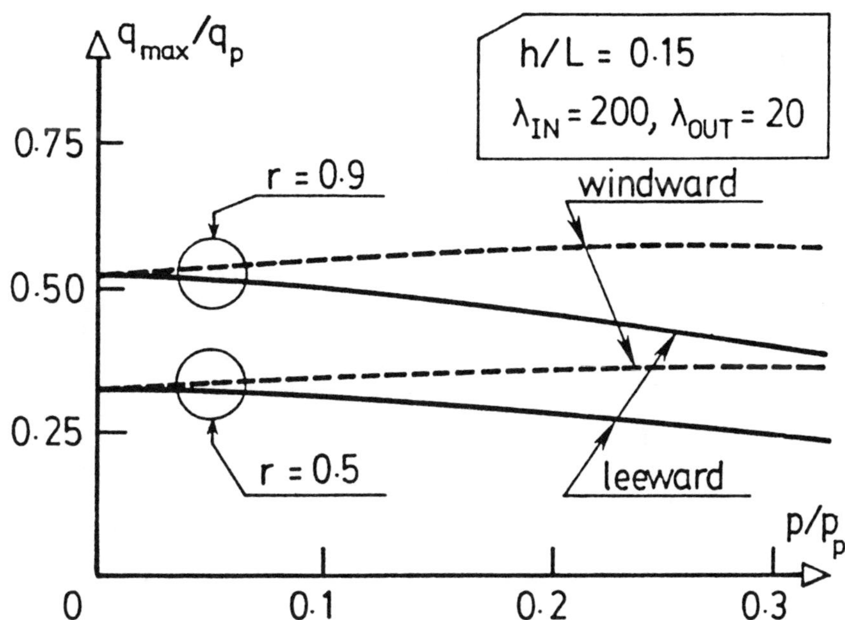

FIGURE 27. Relationship between maximum load intensity and lateral load intensity for separated windward and leeward ribs.

TABLE 6. Comparison of Analytical Results of ϕ for r = 0.75 with the Predicted ϕ Using Equation (52).

h/L (1)	λ_{IN} (2)	q_{max}/q_p (3)	ϕ Computed (4)	ϕ Predicted (5)	(4)/(5) as a percentage (6)
0.1	100	0.700	0.942	0.959	+1.8
	200	0.370	0.994	0.965	−3.0
	300	0.196	1.000	0.972	−2.9
0.2	100	0.668	0.936	0.951	+1.6
	200	0.321	0.944	0.925	−2.1
	300	0.169	0.928	0.900	−3.1
0.3	100	0.676	0.940	0.944	+0.4
	200	0.312	0.929	0.894	−3.9
	300	0.162	0.891	0.844	−5.6

Note: $r = 0.75$; $\lambda_{OUT} = 40$; $\sigma_Y = 320$ N/mm^2;
q_{max} = ultimate load carrying capacity; q_p = Equation (35)

the flexural rigidity of the transverse beams gives the arch bridge structure higher in-plane rigidity than that without transverse beams. This fact has already been confirmed earlier in "Effect of Out-of-Plane Shear and Sway Rigidities" by the analytical results using the spatial structural analysis method. This advantage becomes noticeable as the relative vertical displacement of the arch ribs increases, i.e., as the in-plane slenderness ratio of the ribs becomes large or the unsymmetry of the vertical loads becomes dominant, or both, as is evident from Table 6.

By using the quasi-planar analysis method, some qualitative results have been presented in [16] and [17]. It was observed there that the effect of the lateral loads on the load-carrying capacity of the arch bridge structures becomes critical as the rise to span ratio or the in-plane slenderness ratio, or both, of the arch rib increase. Those fundamentally qualitative tendencies are confirmed by the results obtained by the rigorous spatial analysis method. Other observations are summarized as: If the arch structure has sufficient out-of-plane rigidities as is the case with well-designed steel arch bridges, the lateral rigidities and loads can be taken into account practically as a set of in-plane additional loads that are mechanically equivalent to them, as given by Equation (47), acting on a quasi-planar model. By comparing the results from the quasi-planar model with those obtained by the exact, spatial analysis method, it is observed that the load carrying capacity of arch bridge structures can be simply and accurately evaluated by the quasi-planar model, if the structure has conventional out-of-plane rigidities.

ULTIMATE STRENGTH DESIGNING

Required Out-of-Plane Rigidities

Generally speaking, although an arch is rather a stable structure for loads in the vertical plane, it is not so stiff for lateral loading in the out-of-plane because of the spatial properties. Therefore, in some cases, it may be necessary in designing arch bridge structures to consider their behavior as spatially slender members subjected to high compressive thrust, bi-axial bending moments and torque. In such a case, as insufficient out-of-plane rigidity or strength or both of the structures is provided, the load carrying capacity for the principal loads will be significantly affected by the subsidiary loads such as wind load. Consequently, designers might be forced to proportion them by complex design rules in which the intricate effect of the interaction of the in-plane and out-of-plane behaviors of the arch structures are taken into account. On the other hand, conventional design method for bridge structures is based on the concept that the proportioning of the structures should fundamentally be determined by the principal loads and then modified by the subsidiary loads and it should be as simplified as possible. Following this design philosophy, such required out-of-plane rigidities have been presented [6,16,18] as the ultimate strength of the arch structures can be evaluated as the strength of a planar structure instead of a spatial one.

The required out-of-plane rigidities should be discussed from two phases. One is how to determine the required rigidities of the lateral bracing members. The other is a question as to how to determine the out-of-plane stiffness as an overall transverse-sectional form of the structure system. Even in the ultimate strength design, it will be desirable that the lateral bracing members are designed so as to keep their functions up to the ultimate state. From the point of view, based on results obtained by the numerical parametric analyses, the practical design formulas of these required rigidities have been specified as follows: for the out-of-plane shear rigidity

$$\frac{GA_{b,eq}}{24EA_a}\left(\frac{L_S}{b}\right)^2 \geq \frac{1}{15} \tag{48}$$

for the sway or distorsional rigidity or both

$$\frac{EI_b}{GI_T} \cdot \frac{L_s^2}{bc} \geq 10 \tag{49}$$

The cases characterized as in-plane collapse are scattered mostly within the shaded area in Figure 23. From these results, it can be observed that the out-of-plane stiffness should be such that the relation of $W_o/L \leq 0.0045$ is satisfied for the collapse to be regarded as in-plane collapse. Therefore, by considering this relation, Equation (43), and Equation (44), the required out-of-plane stiffness in order to provide the basis for the conventional design method, which checks the strength of arch bridge structures by analyzing it as planar structures, has been recommended in the form of a simple formula:

$$\frac{L_S}{b} \leq \frac{4.5 \times 10^4}{(\lambda_{IN} - 100)(44.2h/L + 7.84) + 9.05(300 - \lambda_{IN})} \tag{50}$$

Here, it is considered that λ_{OUT} may be taken to be $2L_s/b$ since I_y/A_a is generally negligible in Equation (34). The range of the structural parameters for the practical steel arch bridge are: $h/L = 0.1-0.3$; $L_s/L = 1.02-1.20$; $\lambda_{IN} = 100-300$. Substituting these values into Equation (50), $L/b = 10-25$. Eventually, the required out-of-plane stiffness will be given conservatively by a simple formula as follows:

$$\frac{L}{b} \leq 10 \tag{51}$$

As seen from the strength reduction factors shown in Figure 23 for the afore-mentioned cases, it is advisable to assume that the practical lateral load reduces the planar ultimate strength by 10%.

Lateral Load Effect Evaluation

The arch bridge structures having sufficient out-of-plane shear and sway, or distorsional rigidities, or both – recommended by Equations (48), (49), and (51) – and having structural dimensions within the range examined herein may not fail locally in the out-of-planes up to the overall collapse of the structure [6,7,16,17,18], provided that a check on the compressive member strength should be initially performed for the lateral bracing members of an equivalent rolled out planar truss introduced in the section "Linear Analyses on Lateral Behavior" under lateral wind loads. Consequently, the lateral loading effects on the ultimate strength of such arch bridge structures may be evaluated by reducing their planar ultimate strengths by using the strength reduction factor even if the overall collapses occure spatially. Figure 28 shows a typical relationship between the strength reduc-

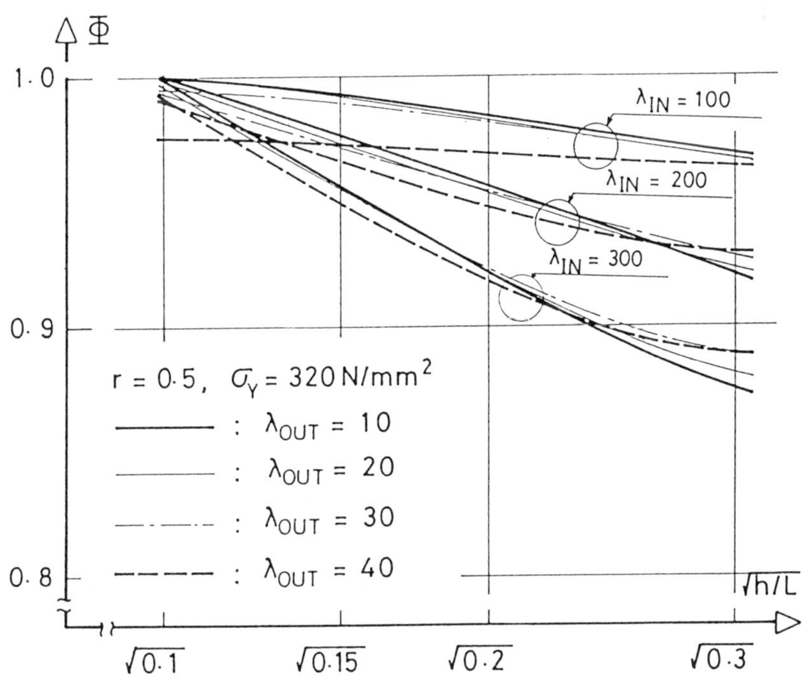

FIGURE 28. Relationship between strength reduction factor and rise/span ratio.

tion factor (for $r = 0.5$) and rise/span ratio varying the in-plane and out-of-plane slenderness ratios. The strength reduction factor ϕ has the tendency to decrease as the in-plane slenderness ratio becomes higher, and is hardly affected by the out-of-plane slenderness ratio. Further, the numerical results show that ϕ is almost independent of the yield stress level of the material. Thus, it can be said that, in the practical range, ϕ has linear functional relationship with the square root of the rise/span ratio and with the in-plane slenderness ratio. Expressed mathematically, it is [7,16]:

$$\phi = A - B\sqrt{h/L} \qquad (52a)$$

Based on the linear regression analysis for the calculated values of ϕ,

$$A = 0.837 \times 10^{-3}\lambda_{IN} + 0.895$$
$$B = 0.244 \times 10^{-2}\lambda_{IN} - 0.181 \qquad (52b)$$

Table 6 shows comparison of some analytical results of the reduction factor for $r = 0.75$ with predicted values using Equation (52). The two show fairly good agreement from a practical viewpoint. Therefore, Equation (52) may be applicable to usual, practical, steel arch bridges.

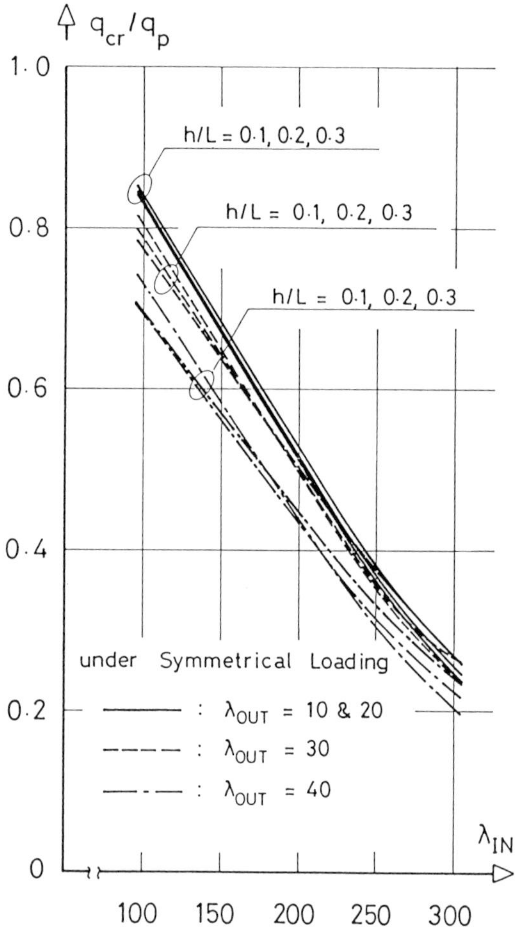

FIGURE 29. Relationship between ultimate vertical loads (uniformly distributed over full span) and in-plane slenderness ratio.

Lateral-Torsional Buckling Evaluation

The spatial ultimate strength of the arch structures subjected to uniformly distributed vertical loads over the full span and the practical lateral loads may be regarded as their lateral-torsional buckling strengths, q_{cr}, practically. Figure 29 shows the relationship between the vertical loads which give the afore-mentioned ultimate strength and the in-plane slenderness ratio for different out-of-plane slenderness ratios and rise/span ratios, for $\sigma_Y = 320$ N/mm². From the figure it can be seen that the ultimate vertical loads have a fairly good linear relationship with the in-plane slenderness ratio and are less influenced by the rise/span ratios within the range examined herein. Figure 30 shows a typical relationship between the strength reduction factor and the yield stress level of the material. As is obvious from the figure, ϕ has a linear functional relationship with the square root of the yield stress level. Consequently, q_{cr} may be expressed in the form of a simple formula as follows [7,16]:

$$\frac{q_{cr}}{q_p} = (A - B\lambda_{IN})(a - b\sqrt{\sigma_Y/E}) \qquad (53a)$$

Based on the regression analysis for the values q_{cr} calculated by the spatial ultimate strength analysis, the coefficients A, B, a, and b are determined as follows:

$$A = 1.08 - 2.65 \times 10^{-4}\lambda_{OUT}(\lambda_{OUT} - 25)$$
$$B = 2.9 \times 10^{-3} - 8 \times 10^{-7}\lambda_{OUT}(\lambda_{OUT} - 25)$$
$$a = 2.43 \qquad (53b)$$
$$b = 36.6$$

The difference between the spatial ultimate strength predicted by Equation (53) and the results calculated by the ultimate strength analysis is less than 3% even in the extreme case ($\lambda_{OUT} = 40$, $\lambda_{IN} = 300$, and $h/L = 0.3$) examined herein.

The effect of braced length on the lateral instability for through-type twin arch was clarified by extensive paramet-

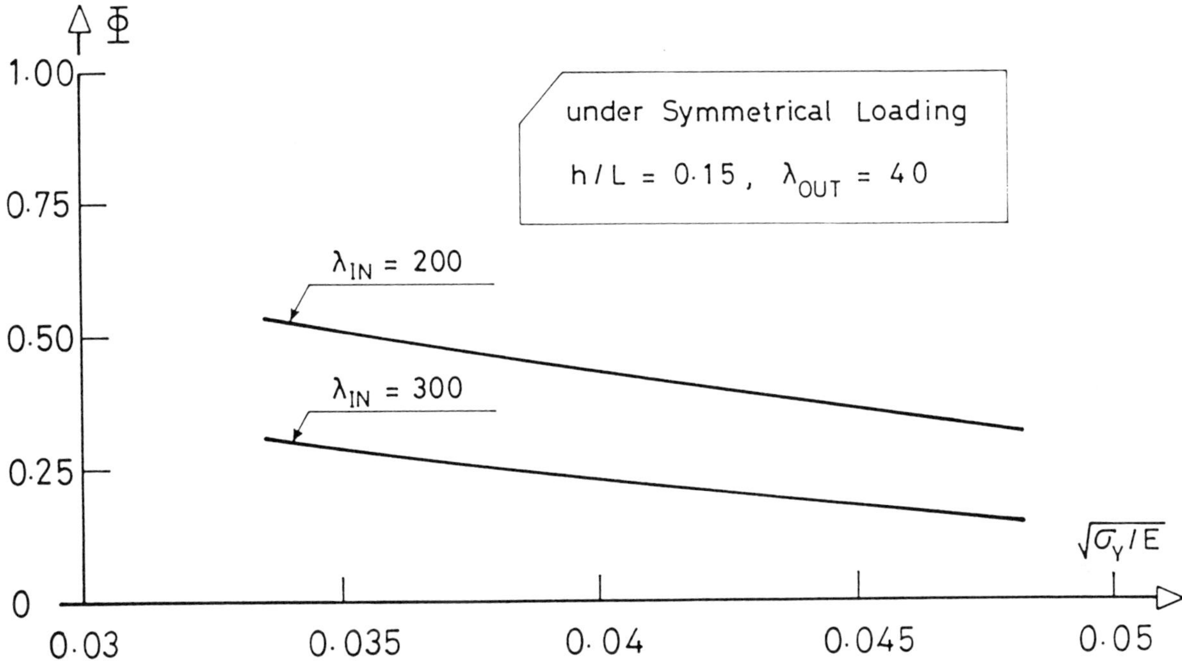

FIGURE 30. Relationship between strength reduction factor and yield stress level (under uniformly distributed vertical loading over full span).

ric studies [9,19]. Reference [10] has specified the effect by incorporating the coefficient K_β into the out-of-plane slenderness ratio of the arch rib. Namely, using an analogy between an arch and a column, an equivalent slenderness function $\bar{\lambda}_a$ for the determination of the ultimate strength of through-type steel arches is defined as follows [1,10]:

$$\bar{\lambda}_a = \frac{1}{\pi} \sqrt{\sigma_y/E} \cdot \frac{L_S}{r_y} \cdot K_\beta$$

$$K_\beta = 1 - \beta + \frac{2 r_y \beta}{b} \tag{54b}$$

in which r_y = out-of-plane radius of gyration of an arch rib section about y-axis; β = a ratio of the length of braced portion to the total length of the arch rib.

Information on the linear, lateral-torsional, bifurcation buckling on arches is easily available by referring to [1,2,9,10,14] and so on, and thus is omitted here.

Application of Design Criteria

Practical design formulation based on the ultimate strength were introduced for steel arch bridge structures subjected to principal vertical loads and practical lateral loads. These design formulas were derived within the range of conventional structural dimensions of steel arch bridges. Application of the derived design criteria are explained summary as follows:

1. The required out-of-plane rigidities of diagonal members and transverse beams which assemble the lateral bracing of steel arch bridge structures have been proposed in Equations (48) and (49) [6,16,18]. Namely, for the design recommendations proposed herein to be valid, it is required that the out-of-plane shear and sway rigidities proposed by Equations (48) and (49) should be provided for the arch structures.

2. It may be permitted to design a steel arch bridge structure using the conventional design method in which the proportioning of bridge structure is fundamentally determined by the principal loads and then modified by the subsidiary loads such as wind load, if the out-of-plane stiffness recommended by Equation (50) or (51) is provided for the arch [7,16,17,18].

3. Based on the ultimate strength design introduced herein, for steel arch bridge structures under the unsymmetrical vertical loads and the practical lateral loads, the effect of the practical lateral loads can be approximated by using the strength reduction factor defined by Equation (52) and by reducing their in-plane ultimate strengths specified in [8,16, or 19] [7,16]. If the out-of-plane stiffness specified by Equation (51) is provided for the arch, it is advisable to esti-

mate the effect of the practical lateral load by reducing the in-plane ultimate strength by 10% [4,7,16,17].

4. The spatial ultimate strength of arch structures subjected to uniformly distributed vertical loads over the full span and practical lateral loads may be regarded as their lateral-torsional buckling strength, practically. The design criterion for the spatial ultimate strength of these cases can be formulated by the in-plane and out-of-plane slenderness ratios and the yield stress level of the material as shown in Equation (53) [7,16]. The effect of the length of braced portion (through-type arch bridge) on the lateral instability can be approximately defined using Equation (54) [9,10].

The lateral restraint effects to be available by rigid framing at the portal bracing portion (case of through-type arch) or rigidly connecting with the stiffened girder and arch rib at the crown (case of deck-type arch) should also be examined, in the strict sense. Since these restraints, however, produce a stiffening effect on the spatial ultimate strength of the arch structures, it can be said that the design formulas introduced herein are available as conservative evaluation for the lateral load effect on steel arch bridge design.

REFERENCES

1. Galambos, T. V. (ed), *Guide to Stability Design Criteria for Metal Structures,* Structural Stability Research Council, 4th ed., John Wiley and Sons, New York, N.Y. (1987).
2. Johnston, B. J. (ed), *Guide to Stability Design Criteria for Metal Structures,* Structural Stability Research Council, 3rd ed., John Wiley and Sons, New York, N.Y. (1976).
3. Kollbrunner, C. F., and K. Basler, *Torsion in Structures,* Springer-Verlag, Berlin Heidelberg/New York (July, 1969).
4. Kuranishi, S., and T. Yabuki, "Ultimate In-plane Strength of 2-Hinged Steel Arches Subjected to Lateral Loads", *Transaction of the Japan Society of Civil Engineers,* Vol.10, pp. 350–353 (1978).
5. Kuranishi, S., and T. Yabuki, "Some Numerical Estimation of Ultimate In-Plane Strength of Two-Hinged Steel Arches," *Proceedings of the Japan Society of Civil Engineers,* No. 287, pp. 155–158 (1979).
6. Kuranishi, S., and T. Yabuki, "Effect of Lateral Bracing Rigidities on the Ultimate Strength of Steel Arch Bridges," *Proceedings of the Japan Society of Civil Engineers,* No. 305, pp. 47–58 (1981) (in Japanese).
7. Kuranishi, S., and T. Yabuki, "Lateral Load Effect on Steel Arch Bridge Design," *Journal of Structural Engineering, ASCE, Vol. 110,* No. 9, pp. 2263–2274 (Sept., 1984).
8. Kuranishi, S., and T. Yabuki, "Ultimate Strength Design Criteria for Two-Hinged Steel Arch Structures," *Proceedings of the Japan Society of Civil Engineers, Structural Eng./Earthquake Eng., Vol. 1,* No. 2, pp. 229s–237s (Oct., 1984).
9. Ojalvo, M., and M. Newman, "Buckling of Naturally Curved and Twisted Beams," *Journal of the Engineering Mechanics Division, ASCE, Vol. 94,* No. EM5, pp. 1067–1087 (1968).
10. Ojalvo, M., E. Demuts, and F. Tokarz, "Out-of-Plane Buckling of Curved Elements," *Journal of the Structural Division, ASCE, Vol. 95,* No. ST10, pp. 2305–2316 (Oct., 1969).
11. Oran, C., "Tangent Stiffness in Space Frames," *Journal of the Structural Division, ASCE, Vol. 99,* No. ST6, pp. 987–1001 (June)
12. Sakimoto, T., and S. Komatsu, "Ultimate Strength of Arches with Bracing Systems," *Journal of the Structural Division, ASCE, Vol. 108,* No. ST5, pp. 1064–1076 (May, 1982).
13. Sakimoto, T., and S. Komatsu, "Ultimate Strength Formula for Steel Arches," *Journal of the Structural Division, ASCE, Vol. 109,* No. 3, pp. 613–627 (March, 1983).
14. Shukla, S. N., and M. Ojalvo, "Lateral Buckling of Parabolic Arches with Tilting Loads," *Journal of the Structural Division, ASCE, Vol. 97,* No. ST6, pp. 1763–1773 (1971).
15. Yabuki, T., and S. Kuranishi, "Out-of-Plane Behavior of Circular Arches Under Side Loadings," *Proceedings of the Japan Society of Civil Engineers,* No. 214, pp. 71–82 (June, 1973).
16. Yabuki, T., "Study on Ultimate Strength Designing for Steel Arch Bridge Structures," thesis presented to Tohoku University, at Sendai, Japan, 1981, in partial fulfillment of the requirement for the degree of Doctor of Engineering (in Japanese).
17. Yabuki, T., S. Vinnakota, and S. Kuranishi, "Lateral Load Effect on Load Carrying Capacity of Steel Arch Bridge Structures," *Journal of Structural Engineering, ASCE, Vol. 105,* ST10, pp. 2434–2449 (Oct., 1983).
18. Yabuki, T., and S. Vinnakota, "Stability of Steel Arch Bridges: A State-of-the-Art Report," *Solid Mechanics Archives, Vol. 9,* Noordhoff International Publishers, Leyden, Netherlands, pp. 115–158 (1984).
19. Yabuki, T., and S. Kuranishi, "Ultimate Strength Design of Steel Arch Bridge Structures," *IABSE Proceedings,* P-84/85, IABSE PERIODICA, pp. 57–64 (Feb., 1985).
20. Yamada, Y., N. Yoshimura, and T. Sakurai, "Plastic Stress-Strain Matrix and its Application for the Solution of Elastic Plastic Problems by Finite Element Method," *International Journal for Mechanical Science,* No. 10, pp. 343–354 (1968).

CHAPTER 16

Flexural Strength of Square Spread Footing

DA HUA JIANG*

INTRODUCTION

Spread footing is an important structural element which distributes a column load to the subgrade soil in order not to yield significant settlement. The area or the dimension of the spread footing in plan is determined from a consideration of the allowable bearing capacity of the soil mass.

Under the column load on top of the footing and the upward subgrade reaction from the bottom, the spread footing acts essentially as a two-way cantilever slab. Therefore it may fail by bending or shear if not properly designed.

Being made of reinforced concrete, the spread footing is usually reinforced at the bottom by two horizontal layers of steel bar mutually perpendicular to each other in the form of a net (Figure 1). The thickness of the footing slab is usually controlled by the punching shear strength of the concrete while the amount of steel is mainly determined from the flexural strength of the footing slab.

ELASTIC ANALYSIS

If the underlying soil is looked upon as an elastic half-space, then the subgrade reaction, or the contact pressure, will be either as shown in Figure 2a or as in Figure 2b, depending upon the material properties of the soil bed and the stiffness of the footing. However, in general engineering practice, the bearing pressure is assumed to be uniformly distributed (Figure 1).

As a two-way thick slab, the bending moments along both the longitudinal and transverse directions of the footing slab can be calculated by using the theory of elasticity.

For the footing shown in Figure 3 a three dimensional finite element analysis is made in which isoparametric eight-node hexahedron elements are used. The bowl-like deflected shape of the footing gives a good picture of two-way bending.

The flexural stress (Figure 4), as well as the bending moment distribution along a section near the face of the column obtained from integration of the stresses over the cross section (Figure 5), shows that the central portion of the footing slab is more heavily stressed than the outer area. For a spread footing with a dimension (slab width versus thickness) as shown in Figures 3 and 4, the flexural stress distribution is nearly linear from the 3-D analysis.

The maximum tensile stress is located at the bottom face of the footing slab and at the center line of the column. If it exceeds the tensile strength of the concrete, cracks will appear.

Since the slab is relatively thick, and the steel percentage low, the spread footing often remains uncracked under service load.

PLASTIC ANALYSIS

Under factored load, the spread footing should be designed, or rather, reinforced to resist the ultimate bending moment developed in the slab.

Inasmuch as the elastic theory can give internal forces in a reinforced concrete structure when the load is small, the ultimate strength of the structure can best be estimated by using the plastic theory.

Upper Bound Solutions

For a two-way slab, the plastic theory, or Johansen's yield line theory, has been widely applied in evaluating the load

*Tongji University, Department of Structural Engineering, Shanghai, China

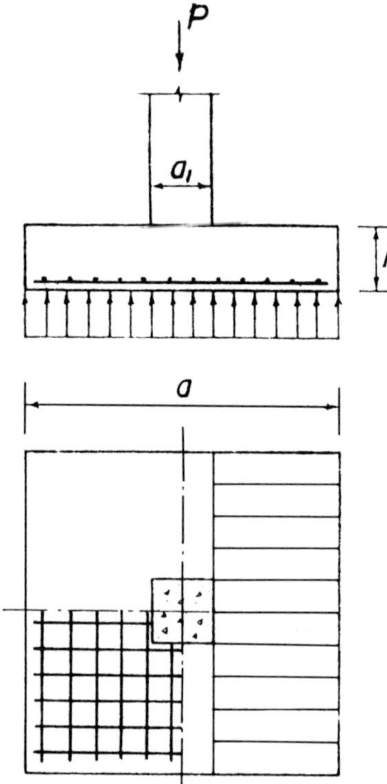

FIGURE 1. Spread footing under column.

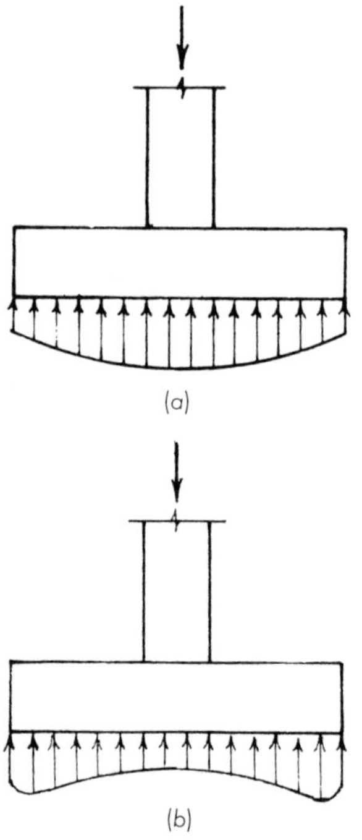

FIGURE 2. Different forms of contact pressure.

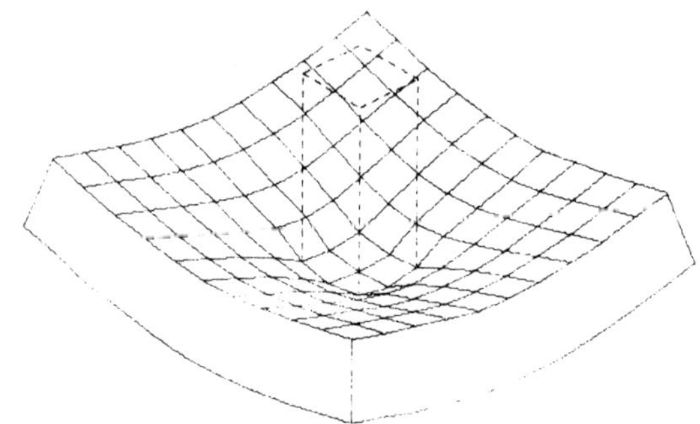

FIGURE 3. Deflected shape of spread footing.

carrying capacity under various boundary conditions. It can also be used to predict the ultimate load of a spread footing. The solution is an upper bound when a system of yield lines is assigned satisfying geometrically permissible velocity field. The material is taken as rigid-plastic; that is, the slab is divided into rigid pieces joined by plastic hinge lines or yield lines.

The yield line patterns considered are shown in Figure 6. It is assumed that the column is monolithically cast with the footing slab and that no yield line will pass through the center of the column section. In other words, only the so-called rigid collapse solutions are considered.

The expressions for the collapse loads P are as follows:

Case 1

$$\frac{P}{m} = \frac{8}{\left(1 - \frac{a_1}{a}\right)^2} \tag{1}$$

Case 2

$$\frac{P}{m} = \frac{24}{\left(1 - \frac{a_1}{a}\right)^2\left(2 + \frac{a_1}{a}\right)} \tag{2}$$

Case 3

$$\frac{P}{m} = \frac{24}{\left(1 - \frac{a_1}{a}\right)\left(3 - 3\,\frac{a_1}{a} - 2\,\frac{a_1^2}{a^2}\right)} \tag{3}$$

Case 4

$$\frac{P}{m} = \frac{8}{1 - 2\,\frac{a_1}{a} + \frac{4}{3}\,\frac{a_1^3}{a^3}} \tag{4}$$

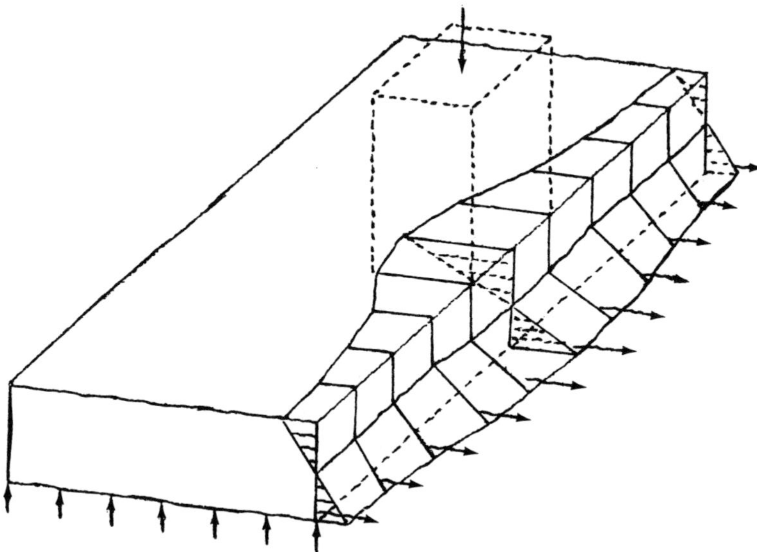

FIGURE 4. Flexural stress distribution from 3D finite element elastic analysis.

Case 5

$$\frac{P}{m} = \frac{8}{\frac{1}{3} + \frac{\sqrt{2}}{3} - \sqrt{2}\,\frac{a_1}{a} + \frac{2}{3}\frac{a_1^3}{a^3}} \tag{5}$$

Case 6

$$\frac{P}{m} = 48 \tag{6}$$

in which m is the flexural strength of the footing slab per unit width, i.e., $m = M/a$; M is the total moment at face of column; a is the width of the square slab, a_1 is the width of the column section.

The above expressions can be derived by establishing virtual work equations for corresponding yield line cases. As an example, for Case 1, the virtual work done by the column load W_P, by the uniform upward subgrade reaction W_p, and by the plastic hinge moment along all yield lines W_m are, respectively (Figure 7):

$$W_P = P\Delta$$

$$W_p = 4W_{pb} + W_{pc}$$

$$= -4\,\frac{P}{a^2}\cdot a_1\left(\frac{a - a_1}{2}\right)\frac{\Delta}{2} - \frac{P}{a^2}\cdot a_1^2\cdot\Delta$$

$$= -P\Delta\,\frac{a_1}{a}$$

$$W_m = 8W_{mj} + 4W_{mk}$$

$$= 8m\cdot\frac{a - a_1}{2}\cdot\frac{\Delta}{\frac{a - a_1}{2}}$$

$$+ 4m\cdot a_1\cdot\frac{\Delta}{\frac{a - a_1}{2}} = \frac{8m\Delta}{1 - \frac{a_1}{a}}$$

where W_{pb} and W_{pc} represent the work done by the contact pressure within the areas b and c, respectively. The work

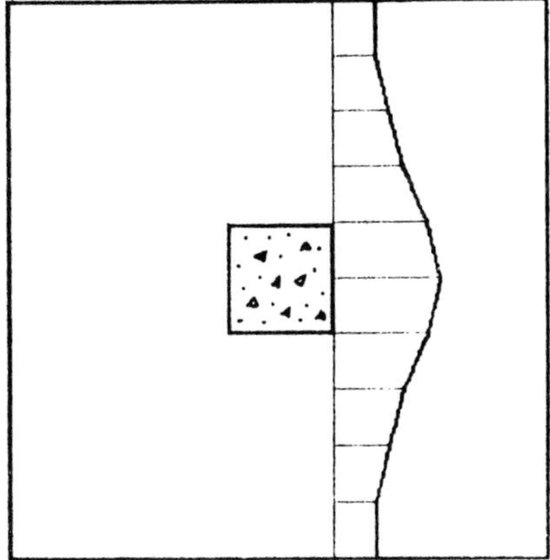

FIGURE 5. Bending moment distribution from 3D finite element elastic analysis.

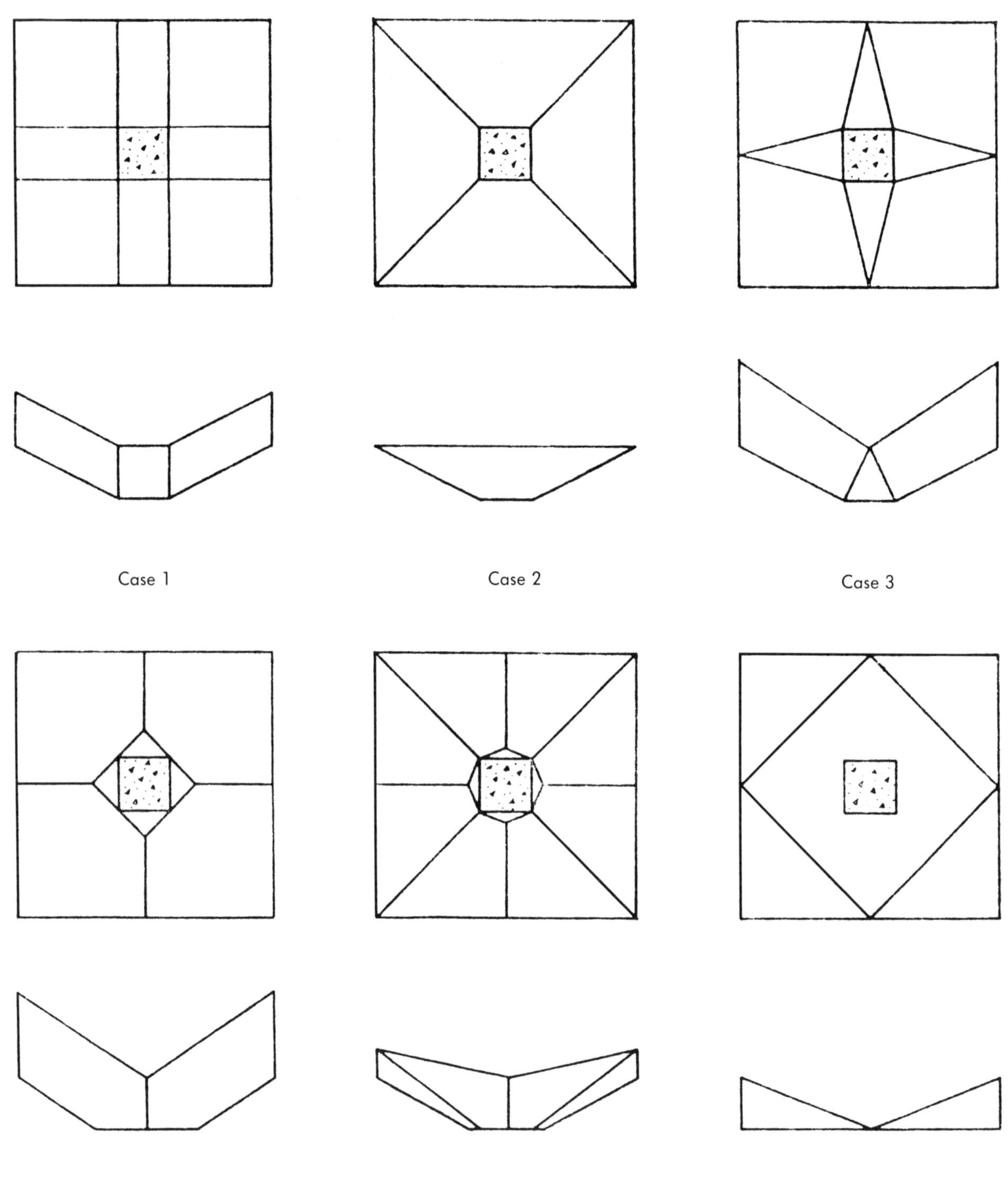

FIGURE 6. Yield line patterns for two-way cantilever slab.

done within the area d is zero. W_{mj} and W_{mk} represent the work done by the yield lines j and k, respectively.

The ultimate load of the footing slab is then obtained from

$$W_P + W_p = W_m$$

or

$$P\Delta - P\Delta \frac{a_1}{a} = \frac{8m\Delta}{1 - \frac{a_1}{a}}$$

which turns out to be Equation (1).

As another example, for Case 2, we have (Figure 8)

$$W_P = P\Delta\left(1 - \frac{a_1}{a}\right)$$

$$W_p = W_{pa} + W_{pb}$$

$$= -\frac{1}{3}\frac{P}{a^2}\cdot a^2\Delta + \frac{1}{3}\frac{P}{a^2}\cdot a_1^2 \cdot \frac{a_1}{a}\Delta$$

$$= -\frac{P\Delta}{3}\left(1 - \frac{a_1^3}{a^3}\right)$$

$$W_m = 4W_{mj} + 4W_{mk}$$

$$= 4m \cdot a_1 \frac{\left(1 - \frac{a_1}{a}\right)\Delta}{\frac{a - a_1}{2}}$$

$$+ 4m \cdot \sqrt{2}\left(\frac{a - a_1}{2}\right) \cdot \frac{2\sqrt{2}\,\Delta}{a} = 8m\Delta$$

$$\therefore \quad P\Delta\left(1 - \frac{a_1}{a}\right) - \frac{P\Delta}{3}\left(1 - \frac{a_1^3}{a^3}\right) = 8m\Delta$$

which is exactly Equation (2).

The correlation between the collapse load and the ratio a_1/a for different yield line patterns is shown graphically in Figure 9, which indicates that Case 1 gives the lowest collapse load.

Other collapse modes, such as that shown in Figure 10, have also been investigated, but none of them gives a smaller load than Case 1.

According to Building Code Requirements for Reinforced Concrete (ACI 318-83), Section 15.4, external moment on any section of a footing shall be determined by passing a vertical plane through the footing, and computing the moment of forces acting over the entire area of footing on

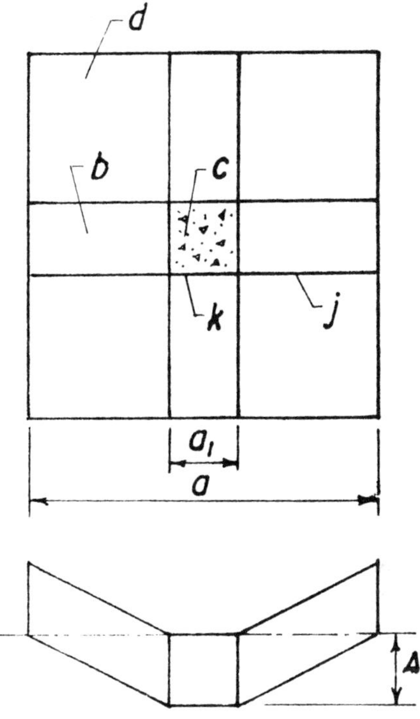

FIGURE 7. Case 1 derivation.

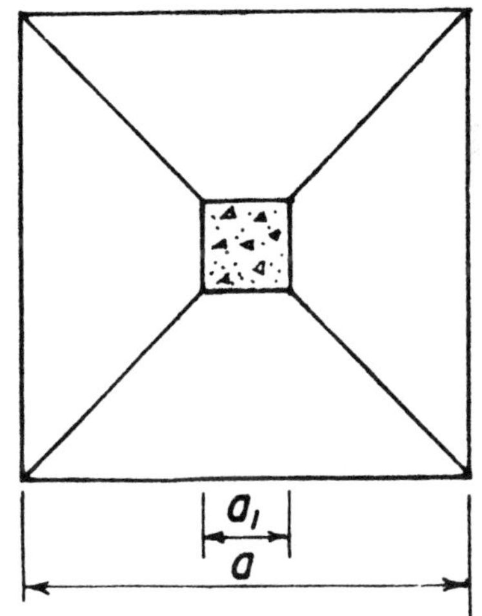

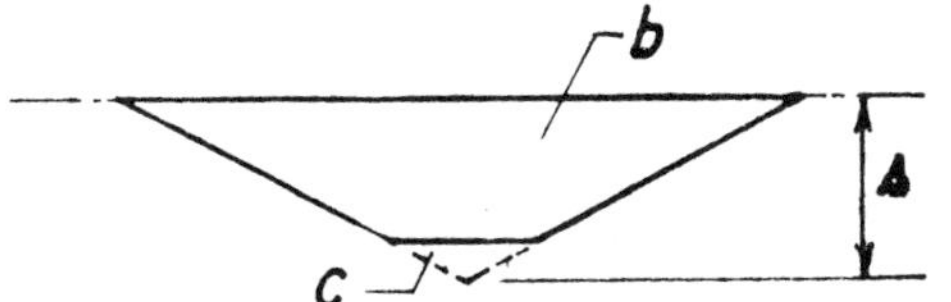

FIGURE 8. Case 2 derivation.

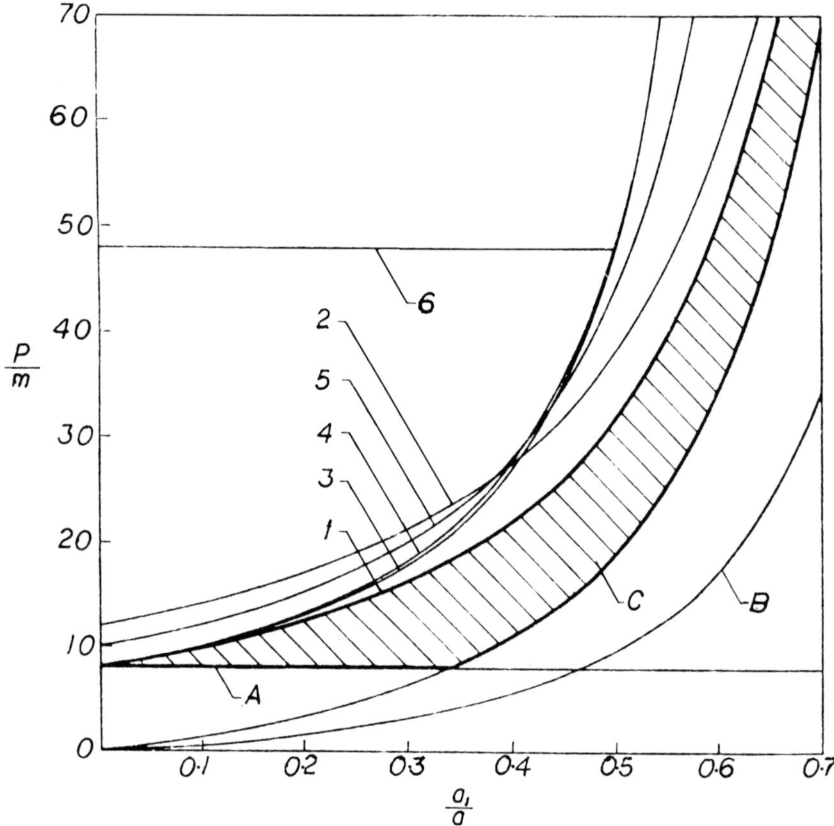

FIGURE 9. Correlation between collapse load and ratio a_1/a for different yield line patterns.

one side of that vertical plane. The Code also prescribes that the critical section for maximum moment is located at the face of column.

The result of Case 1 conforms exactly with the ACI provisions. Hence, the upper bound approach serves as a theoretical basis for the Code requirements.

The distribution of bending moment, as well as the flexural stress, along the critical section is nonuniform from the elastic analysis (Figures 4 and 5); however, the yield moment is taken as uniformly distributed when following the plastic approach. For a square spread footing, it is a general practice that the reinforcement be distributed uniformly across the entire width of footing, as stated in the ACI Code, Section 15.4.3.

Lower Bound Solutions

A lower bound solution consists in finding a statically admissible moment field that corresponds to moments within or on the yield surface, i.e., a safe moment field.

In the following, three solutions for the spread footing are developed.

Case A. The load-distribution element introduced by Hillerborg is quite similar to the spread footing. The difference between them lies in the fact that the column load acting onto the load distribution element is concentrated at a point while the column above the spread footing has a finite width. The bending moment field in the load-distribution element, sketched in Figure 11a, can be accepted as a lower bound solution for the spread footing. The collapse load is

$$\frac{P}{m} = 8 \qquad (7)$$

Case B. The slab is subdivided into a system of cantilever beams of trapezoidal shape in plan (Figure 11b). Under the upward earth pressure, the bending moments in the cantilevers are all radial, the tangential moments being equal to zero. The most critical cantilever is the one running in the diagonal direction. The fixed end is at the face of the col-

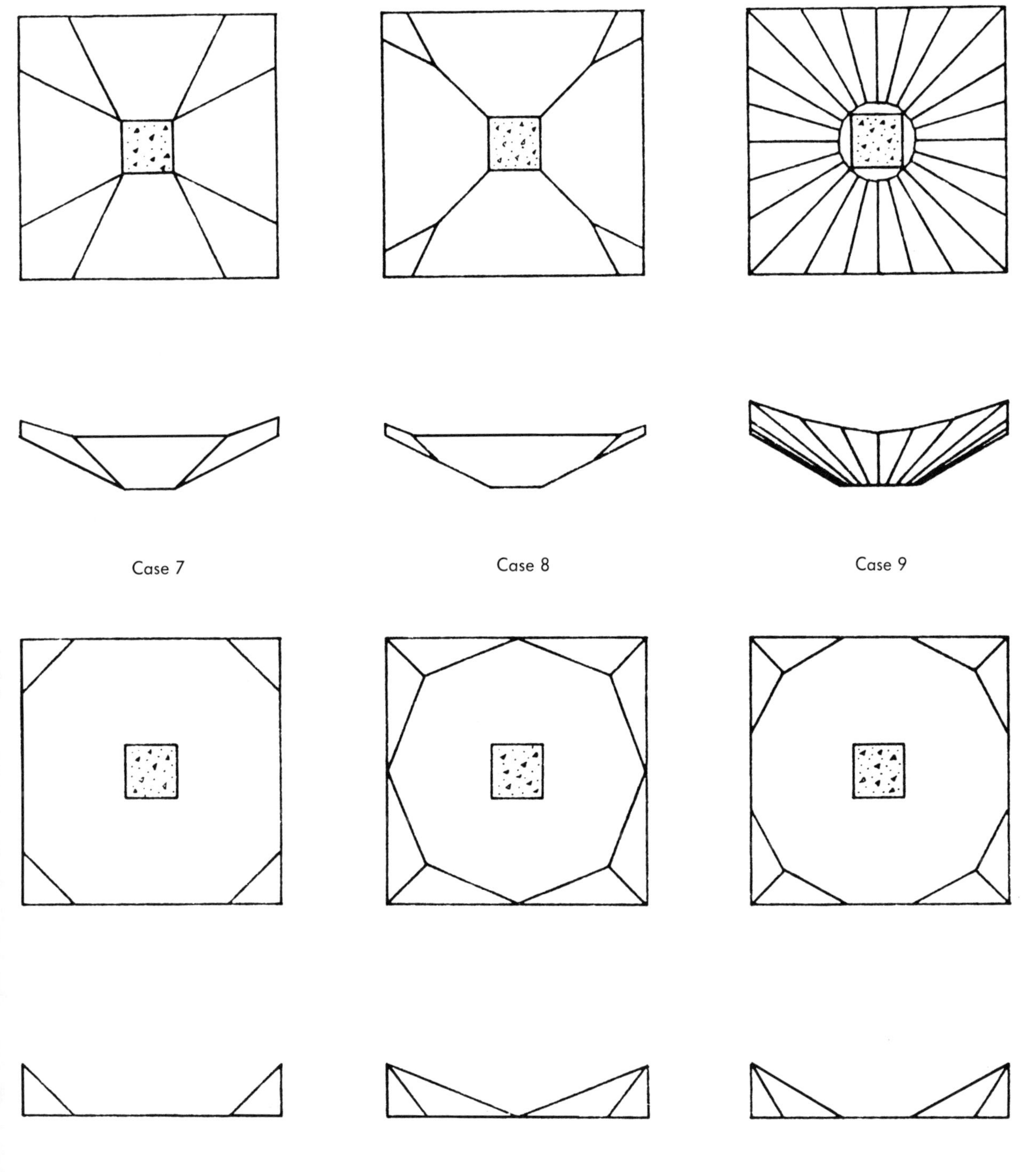

FIGURE 10. Other collapse modes.

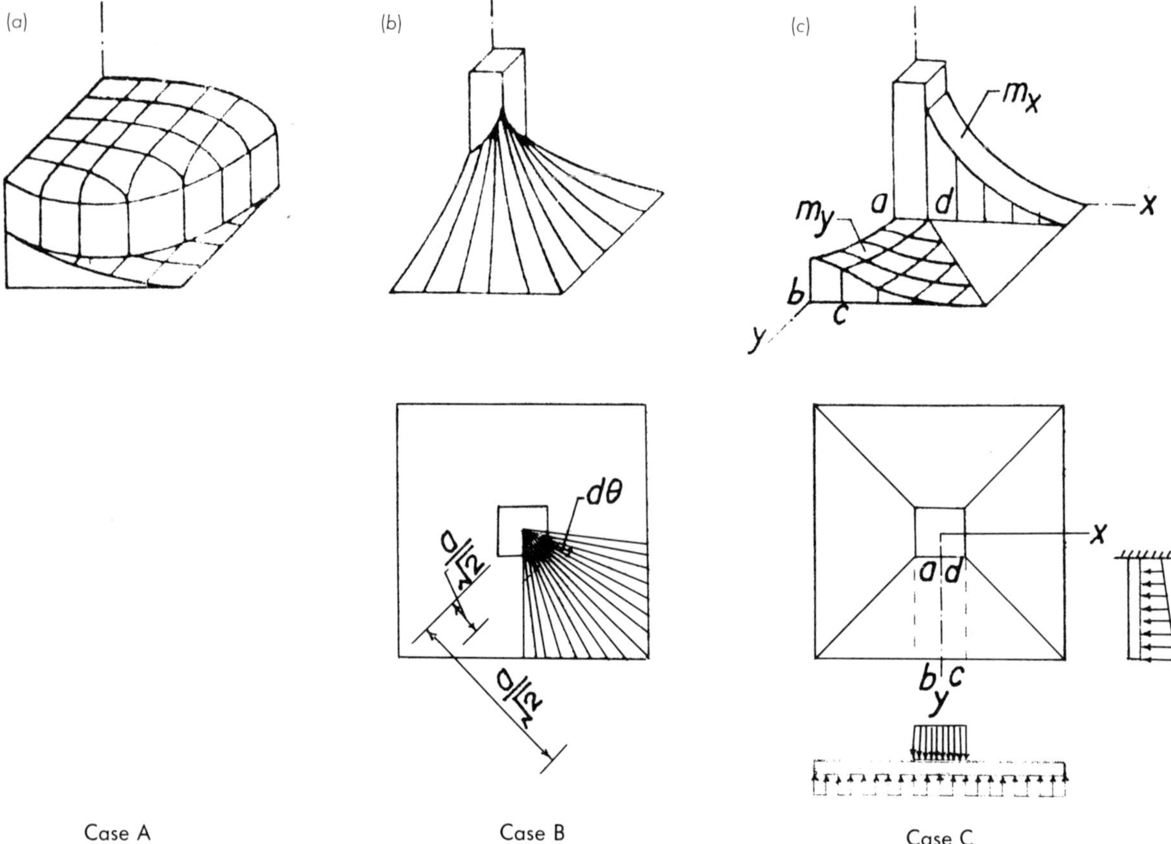

FIGURE 11. Lower bound moment fields.

umn where the ultimate moment per unit width is

$$m = \frac{P}{a^2}\left[\frac{1}{2} \cdot \frac{a}{\sqrt{2}} \cdot \frac{a}{\sqrt{2}}\, d\theta \cdot \left(\frac{2}{3} \cdot \frac{a}{\sqrt{2}} - \frac{a_1}{\sqrt{2}}\right) + \frac{1}{2} \cdot \frac{a_1}{\sqrt{2}} \cdot \frac{a_1}{\sqrt{2}}\, d\theta \cdot \frac{1}{3}\frac{a_1}{\sqrt{2}}\right] \cdot \frac{1}{\frac{a_1}{\sqrt{2}}\, d\theta}$$

From the above expression we obtain the collapse load as

$$\frac{P}{m} = \frac{6\frac{a_1}{a}}{1 - \frac{3}{2}\frac{a_1}{a} + \frac{1}{2}\frac{a_1^3}{a^3}} \tag{8}$$

Case C. The slab is cut along the diagonals (Figure 11c). The load on the trapezoidal area is first transmitted to the rectangular area *abcd* in the *x*-direction. The maximum moment is at point *b* on the center line and equal to

$$m_y = \frac{1}{2} \cdot \frac{P}{a^2}\left(\frac{a}{2}\right)^2 - \frac{1}{2}\frac{P}{a^2}\left(\frac{a}{a_1}\right)\left(\frac{a_1}{2}\right)^2 - \frac{P}{8}\left(1 - \frac{a_1}{a}\right)$$

The load is then transmitted to the column in the *y*-direction, producing a maximum moment at point *a* at the column face with a value of

$$m_x = \frac{P}{24}\left(1 - \frac{a_1}{a}\right)^2\left(2 + \frac{a_1}{a}\right)\frac{a}{a_1}$$

The moment diagram in Figure 11 shows that m_y is smaller than m_x, so that the collapse load of the footing is controlled by the latter moment and equals

$$\frac{P}{m} = \frac{24}{\left(1 - \frac{a_1}{a}\right)^2\left(2\frac{a}{a_1} + 1\right)} \tag{9}$$

The loads calculated from Equations (7), (8), and (9) for the three lower bounds considered have already been shown in Figure 9. It is expected that the exact solution will lie somewhere in the hatched area.

NONUNIFORM SUBGRADE REACTION

If the subgrade reaction is not uniformly distributed, the collapse load of the footing will be changed.

For the yield line pattern of the lowest upper bound solution, Case 1, two different forms of subgrade upward pressure are investigated. If the footing is upon a sand base, the subgrade pressure will be close to that shown in Figure 12a. The collapse load is as follows:

$$\frac{P'}{m} = \frac{32\left[1 + \frac{a_1}{a} + \frac{a_1^2}{a^2} + \frac{p_1}{p_2}\left(2 - \frac{a_1}{a} - \frac{a_1^2}{a^2}\right)\right]}{\left(1 - \frac{a_1}{a}\right)^2\left[3 + \frac{a_1}{a} + \frac{p_1}{p_2}\left(9 - \frac{a_1}{a}\right)\right]} \tag{10}$$

The value of P' is larger than the P from Equation (1).

On the contrary, for the soil pressure shown in Figure

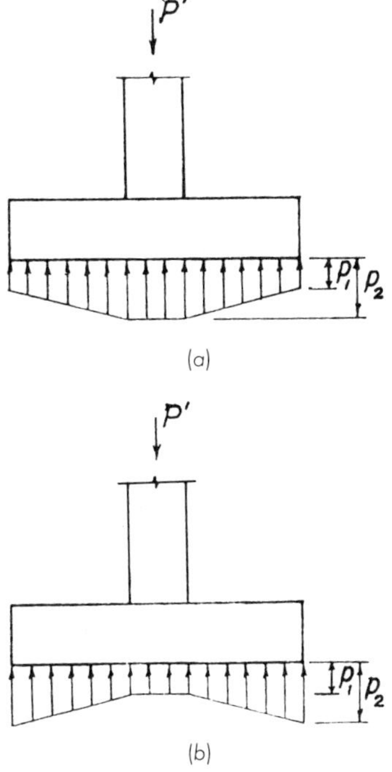

FIGURE 12. Nonuniform soil pressure.

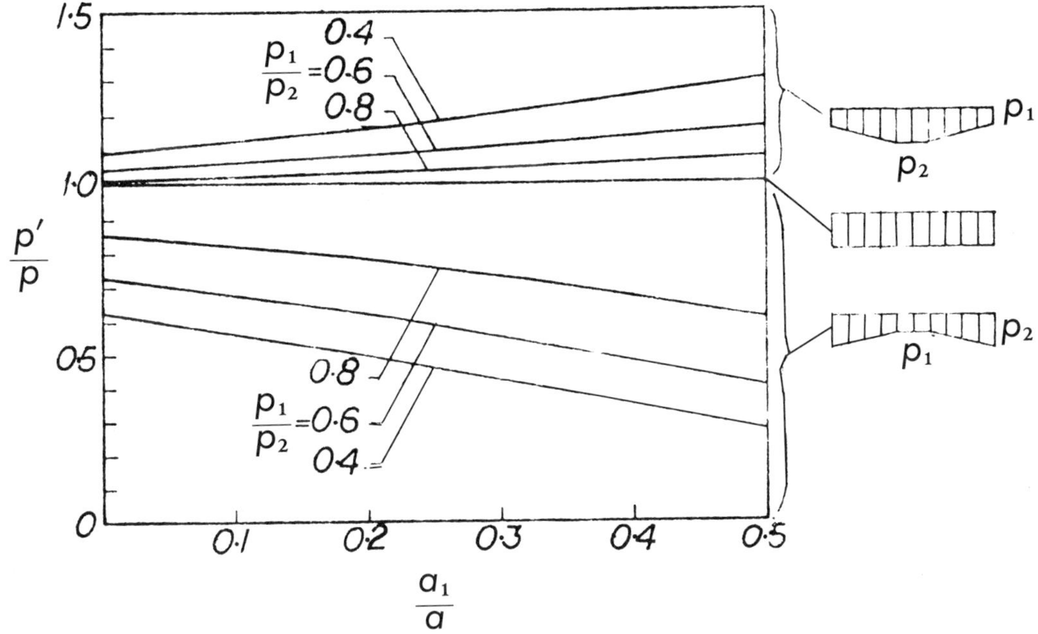

FIGURE 13. Increase and decrease in collapse load, compared with the case of uniform soil pressure, depending upon ratios p_1/p_2 and a_1/a.

12b, the collapse load is smaller:

$$\frac{P'}{m} = \frac{32\left[2 - \frac{a_1}{a} - \frac{a_1^2}{a^2} + \frac{p_1}{p_2}\left(1 + \frac{a_1}{a} + \frac{a_1}{a}\right)^2\right]}{\left(1 - \frac{a_1}{a}\right)\left[18 - \frac{a_1}{a} + 10\frac{a_1^2}{a^2} - \frac{p_1}{p_2}\left(6 + 11\frac{a_1}{a} + 10\frac{a_1^2}{a^2}\right)\right]} \quad (11)$$

The increase and decrease in collapse load, as compared with the case of uniform soil pressure, depend upon the ratios p_1/p_2 and a_1/a (Figure 13).

REFERENCES

1. ACI Committee 318, "Building Code Requirements for Reinforced Concrete (ACI 318-83)," American Concrete Institute, Detroit, 112 pp. (1983).
2. Ferguson, P. M., *Reinforced Concrete Fundamentals,* John Wiley and Sons, Inc., New York, NY (1979).
3. Gesund, H., "Flexural Limit Analysis of Concentrically Loaded Column Footings," *Journal of the American Concrete Institute,* Proc. V. 80, No. 3, pp. 223–228 (May–June 1983).
4. Hillerborg, A., *Strip Method of Design,* A Viewpoint Publication, London, England (1975).
5. Jiang, D. H., "Flexural Strength of Square Spread Footing," *Journal of the Structural Engineering,* ASCE, Vol. 109, No. 8, pp. 1812–1819 (August 1983).
6. Meyerhof, G. G. and K. S. S. Rao, "Collapse Load of Reinforced Concrete Footings," *Journal of the Structural Division,* ASCE, Vol. 100, No. ST5, Proc. Paper 10531, pp. 1001–1018 (May 1974).
7. Richart, F. E., "Reinforced Concrete Wall and Column Footings," *Journal of the American Concrete Institute,* Proceedings, Vol. 45, Part 1, pp. 97–127 (October 1948); Part 2, pp. 237–260 (November 1948).
8. Wang, C. K. and C. G. Salmon, *Reinforced Concrete Design,* Harper International Edition (1979).

CHAPTER 17

Economics of Seismic Design

J. M. FERRITTO*

INTRODUCTION

The seismic design of a new facility can be a complex task. It represents a decision-making process to invest a portion of wealth and resources in providing current seismic lateral resistance capability to resist the potential damage from ground shaking of a future earthquake. There exists an economic balance between current initial seismic strengthening costs, the probability of future earthquakes occurring, and the cost of damage and repair to the structure over its useful life.

In general it is not possible to design and construct a structure with a full 100% certainty that a failure will not occur under some extreme loading condition resulting in loss of life and structure damage. Resources are limited and the anticipated benefits of various risk reducing plans must be weighed against cost of implementation. In earthquake engineering it is impossible to specify with exact certainty the strongest earthquake ground shaking that might possibly occur at a specified location. Even if it were possible, the return time for this event would greatly outweigh the expected life of a typical structure. As long as uncertainties exist a decision needs to be made on which building capacity is established. Building codes establish minimum acceptable levels and are intended to prevent widespread loss of life. Should we do more? Should a prudent building owner invest more of his assets in a building than the minimum code requirements? Both the time of occurrence and magnitude of the event are unknown. Future losses are uncertain and a decision must be made in an environment of uncertainty. A probabilistic procedure can give us a picture of alternative choices and outcomes. It is clear that there are conditions where consideration of the outcome in terms of economics is not appropriate such as in the design of structures which interact with the surroundings such as nuclear power plants, LNG facilities and other critical facilities. In these cases, we are saying that it is unacceptable to expose a large number of people to a risk of death. Society has determined exposure of large segments of the population to risks to be unacceptable. There are other categories of facilities that are required in the recovery process, such as hospitals, and these are valued by society in its codes so as to produce stronger structures intended to minimize damage. The damage evaluation methods presented in this chapter are applicable to all structures. Economic analyses techniques may be applied to all structures where it is possible to characterize items in terms of costs of strengthening and expected damage/losses. This can cover most buildings. The user can develop his own attribute list to add subjective utility values.

This chapter will briefly try to give the authors' experiences in economic analysis of structures to seismic loads. To perform an economic analysis we shall suggest developing several different designs for a structure, each at a higher level of lateral resistance. For each design level we shall analyze the expected damage over the whole distribution of expected shaking, summing up the expected damage over the life of the structure. We shall then economically compare the expected future damage with the present costs of construction and our value for money.

SEISMIC RISK ASSESSMENT

The first step in developing an understanding of the seismic risk at site risk is the assessment or characterization of

*Naval Civil Engineering Laboratory, Port Hueneme, CA

its seismic activity. An engineer may choose many approaches to characterize site motion including acceleration, velocity, and spectrum. The peak ground acceleration is often the most used value to quantify a site and perhaps the most abused. The peak ground acceleration by itself does not give an indication of the duration of shaking and its effect on repeated elastic-inelastic cyclic activity. Often single spikes of high acceleration do not impart significant energy into a structure and these are of little importance. Response spectra are often used to show design levels of acceleration, velocity and displacement. These represent envelope type values and cannot be used to convey duration of shaking and the effects of cyclic activity. However, they improve the picture for design purposes. From the point of an analysis, it is important that we be able to analyze the response of a series of structures of various seismic design levels. To clarify, we must first produce a series of different seismic strength designs that we will examine for damage by exposing the structure to a distribution of earthquake excitations expected over its life. Thus, we must first select various design levels. This part is easy and arbitrary. Let us assume various acceleration levels (say 0.1, 0.15, etc.) and a response spectra for each load level. The response spectra will be discussed below.

In order to be able to evaluate damage, a distribution of expected ground motion is required in terms of the probability of some engineering terms upon which structural response can be related. For simplicity's sake, let us assume the damage evaluation will utilize a nonlinear finite element calculation of the response of a mathematical model of the structure represented by beams and shear walls. Input to this nonlinear response calculations will be a time history of ground acceleration. Thus, to characterize the site it is necessary to predict a probability distribution of peak ground acceleration and associated causative magnitude. This will shape the peak levels, durations and the frequency of excitation of the ground motion.

To summarize the above for the purposes of analysis, it is desired to have a series of response spectra upon which to establish the structural design of alternative levels of seismic resistance. Then it is desired to have a probability distribution of ground acceleration and associated causative magnitudes from which we will construct a series of time histories at different load levels to use in the damage analysis of each structural design. This will be explained below.

RESPONSE SPECTRA

To design a structure it is necessary to specify a loading condition and some measure of response such as stress level, drift level, or ductility level. For the purposes of this exercise and for easy comparison let us specify that we wish to design a structure which remains essentially at yield under the prescribed load level. We size the structure by preliminary design techniques and iterate to a solution. At this point we check the design by model analysis using the response spectra. Note the that response spectra may be site specific to include the effects of soil type and source characteristics or may be a standard spectral shape.

Newmark (1970) has proposed a set of standard spectral values which may be scaled to reflect various ground acceleration levels. This forms a standard elastic spectral shape which may be adjusted for damping values associated with the type of structure.

The elastic spectrum may be adjusted to approximate inelastic behavior of a structure. The displacement region and the velocity region are divided by the structure ductility μ to obtain yield displacement D' and velocity V' (Figure 1). The acceleration region (right side) is relocated by choosing it at a level which corresponds to the same energy absorption for elastoplastic behavior as for elastic for the same period of vibration.

The extreme right-hand portion of the spectrum, where the response is governed by the maximum ground acceleration, remains at the same acceleration level as for the elastic case and, therefore, at a corresponding increased total displacement level. The frequencies at the corners are kept at the same values as in the elastic spectrum. The acceleration transition region of the response spectrum is now drawn also as a straight line transition from the newly located amplified acceleration line to the ground acceleration line, using the same frequency points of intersection as in the elastic response spectrum.

In all cases, the "inelastic maximum acceleration" spectrum and the "inelastic maximum displacement" spectrum differ by the factor μ at the same frequencies. The design spectrum so obtained is shown in Figure 1. Both the maximum displacement and maximum acceleration bounds are shown for comparison with the elastic response spectrum.

The solid line $DVAA_o$ shows the elastic response spectrum. The heavy circles at the intersections of the various branches show the frequencies that remain constant in the construction of the inelastic design spectrum.

The line $D'V'A'A_o$ shows the inelastic acceleration, and the line DVA "A_o" shows the inelastic displacement. These two differ by a constant factor μ for the construction shown, but A and A' differ by the factor $\sqrt{2\mu - 1}$, since this is the factor that corresponds to constant energy.

A study by Newmark and Riddel (1979) investigated the response of elastoplastic systems to numerous earthquake records. Based on this effort the preceding work was found to be unconservative for damping larger than 5% and for ductilities greater than 3.

A data base of existing ground motion records has been assembled and analyzed by the California Institute of Technology. The same nominal ground motion level can be

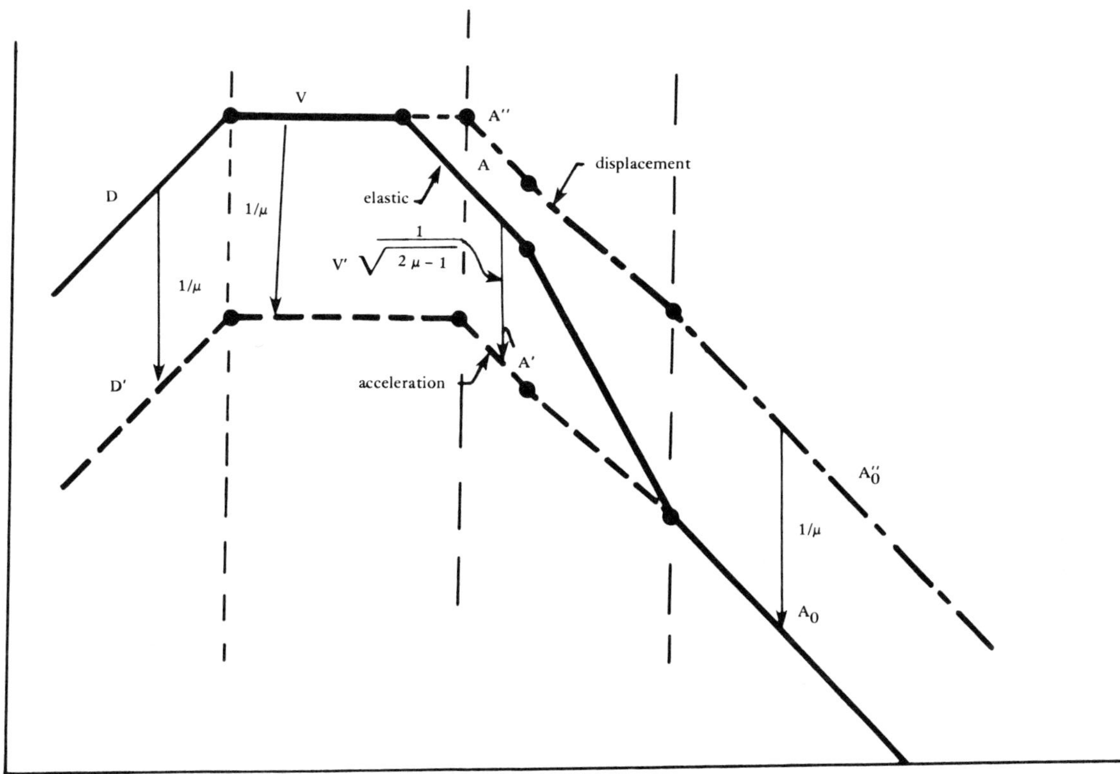

FIGURE 1. Design spectra for inelastic analysis.

produced by different magnitude earthquakes at different distances. As an example of this, a nominal ground motion of 0.15g was caused by the following earthquakes:

Earthquake	Magnitude	Separation Distance (miles)
A	5	14
B	6	18
C	7	30

The closest matching ten spectral records for each earthquake (intermediate soil class) were used by the author to generate average and maximum spectra (Figure 2). Each spectrum represented the characteristics of the magnitude and separation distance. Note the difference magnitude makes in the frequency content in the range of 1.0 to 10.0 seconds period.

The California Institute of Technology response spectra and accelerograms have been categorized by soil site conditions: alluvium, intermediate, or rock. It is of interest to evaluate this effect. Ten spectra of each soil type for a magnitude 6.6 event at 20 miles having a 0.15g nominal ground motion were combined to produce average and maximum spectra (Figure 3). Note the increased frequency content in the alluvium site.

SITE SEISMICITY STUDIES

In order to capture the seismic activity and assess the risk to the site it is necessary to build a mathematical model of the process. The generation of earthquakes as related to the geotectonic mechanisms of faults and the timing, intensity and location of future earthquakes, cannot be predicted with certainty in a deterministic way. The generation of earthquakes in space and time falls under the general category of stochastic processes. Both the Markov and Poisson representations are used to simulate the occurrence of earthquakes generated in time. These representations are stochastic point processes. The Markov model differs from the Poisson model in that in the former the occurrences of new events depend on past events, whereas in the latter model these occurrences are independent of past events.

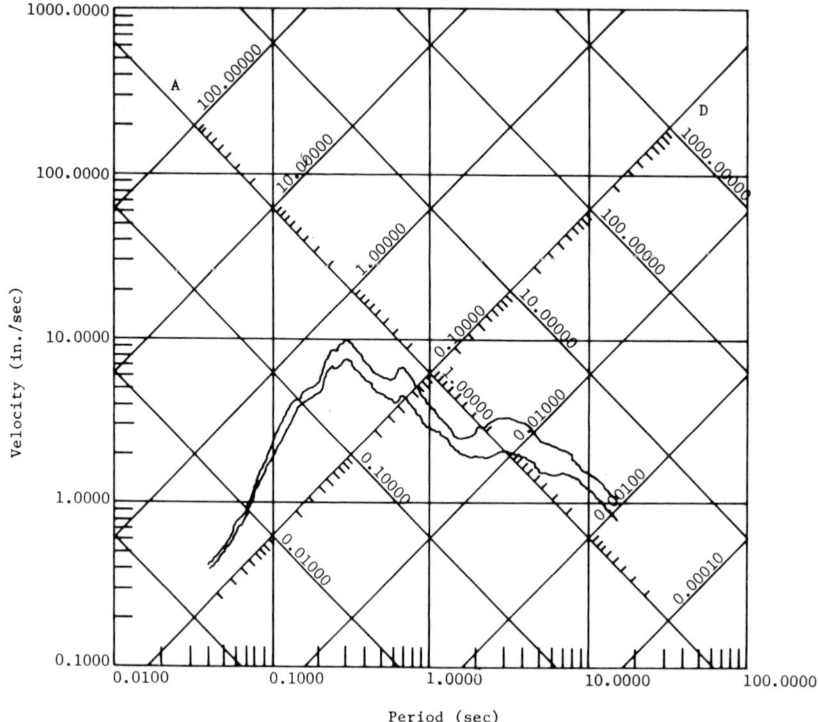

FIGURE 2a. Magnitude 5 earthquake at 14 miles.

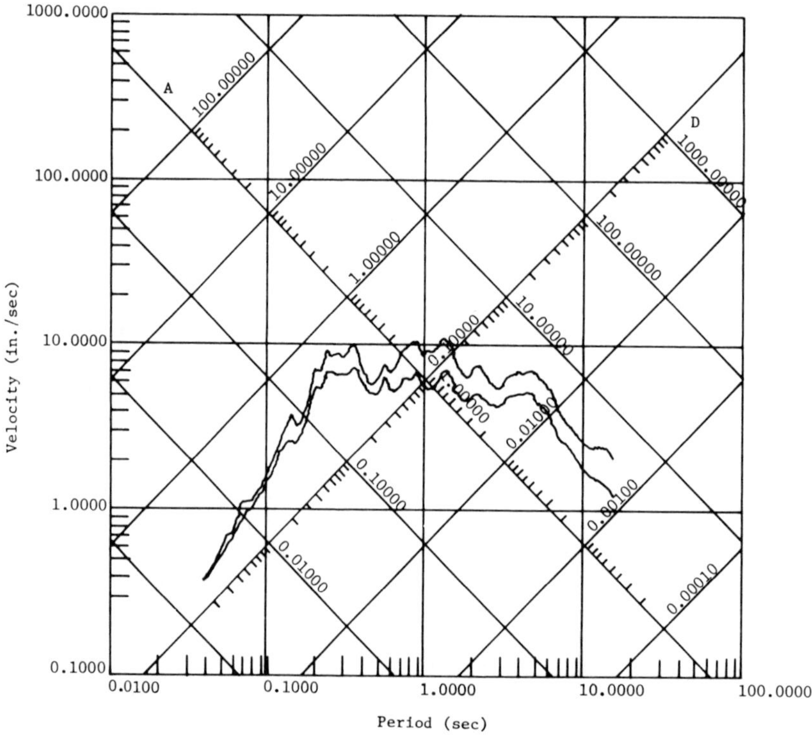

FIGURE 2b. Magnitude 6 earthquake at 18 miles.

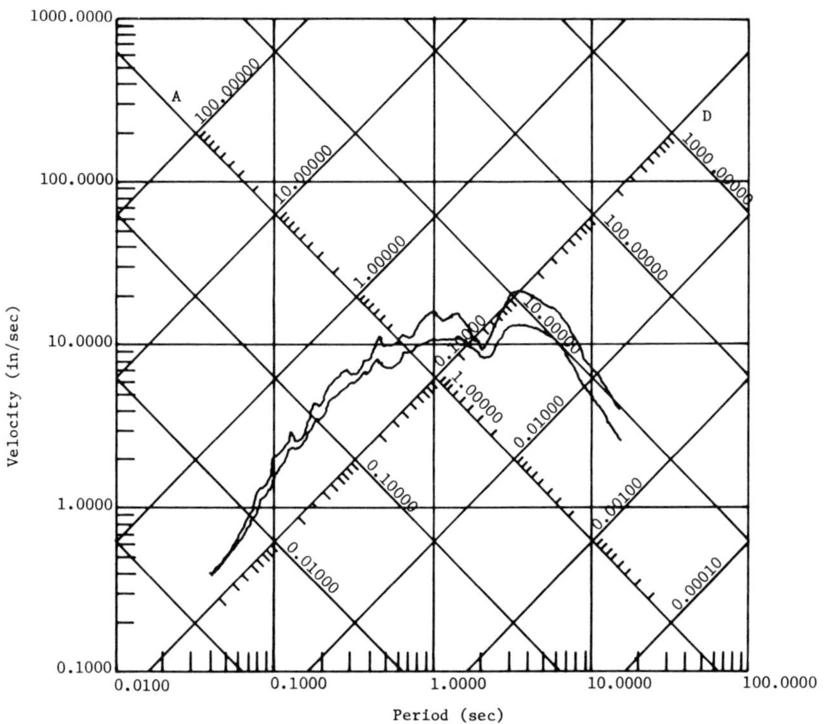

FIGURE 2c. Magnitude 7 earthquake at 30 miles.

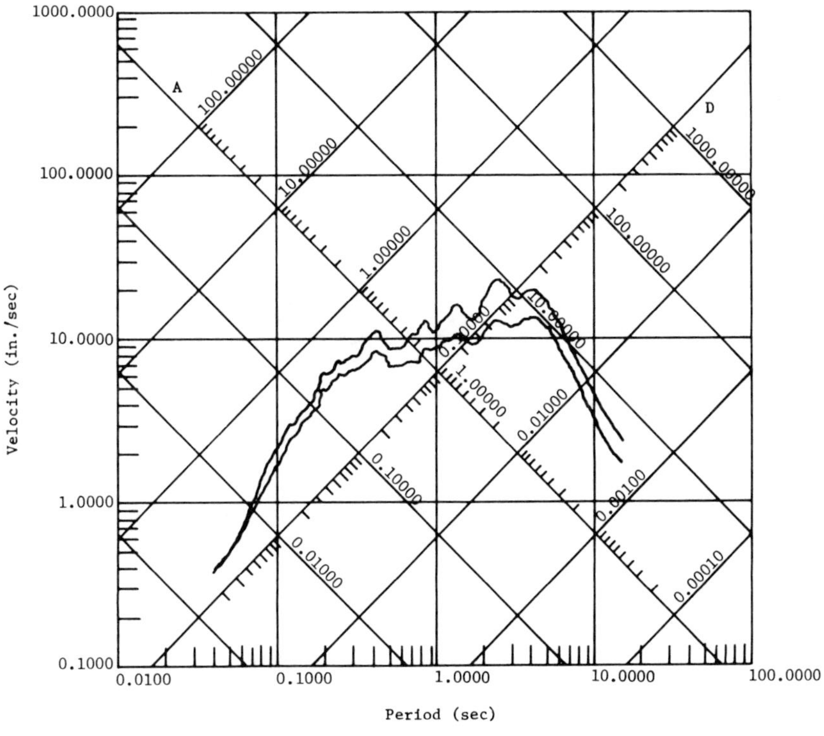

FIGURE 3a. Magnitude 6.6 earthquake, 20 miles, 0.15 g acceleration, alluvium site.

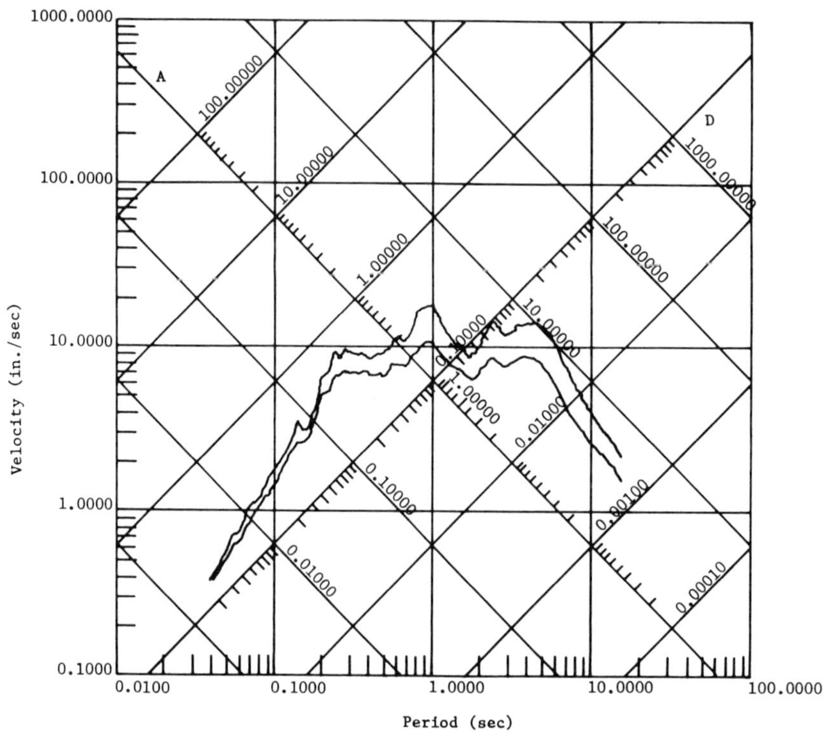

FIGURE 3b. Magnitude 6.6 earthquake, 20 miles, 0.15 g acceleration, intermediate site.

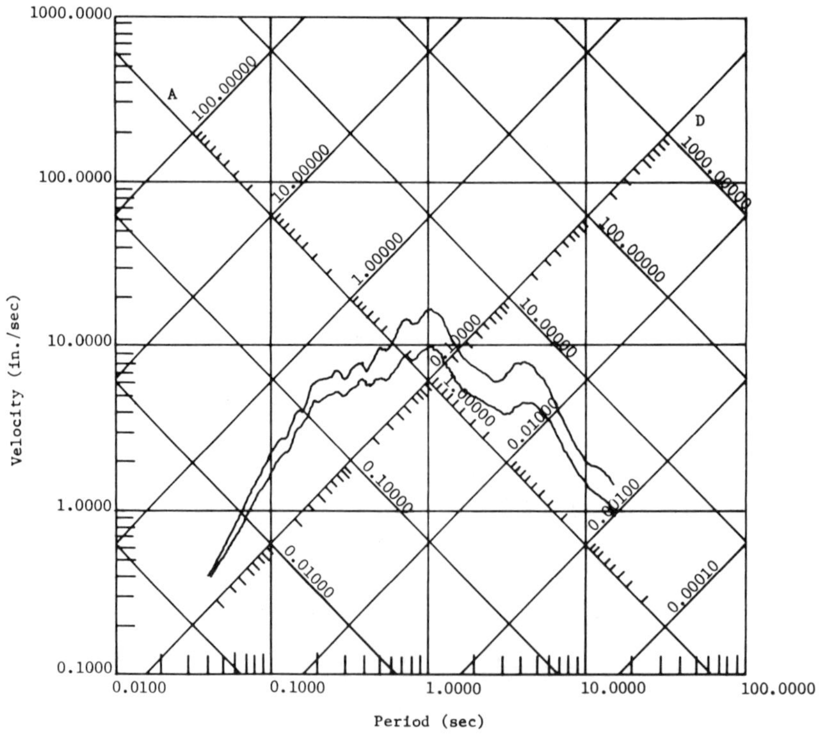

FIGURE 3c. Magnitude 6.6 earthquake, 20 miles, 0.15 g acceleration, rock site.

Results obtained using these two models differ somewhat. The Markov model is better adjusted to data experiences but has some drawbacks due to difficulty in setting the initial conditions and requires more numerical treatment. The Poisson model, on the other hand, does not always agree with experimental data for small magnitude earthquakes, because it ignores the tendency of earthquakes to cluster in space and time. Improvements in the Poisson model have been acheived using the Weibul distribution. Der Kiureghian and Ang (1975), Patwardhan et al. (1980), Sykes et al. (1980), Woodward-Clyde (1978) and Knopoff et al. (1977) are but a few references on this topic.

As a minimum, the process must consider the randomness of the earthquake process by constructing a model to reflect the major areas of uncertainity. Generally, a data base of epicenters is used as the starting point. Events and activity may be associated with known faulting and a recurrence relationship be established computing the recurrence of earthquake magnitude in time. Usually, if faults can be defined, a mathematical model is defined in which faulting regions are assigned and the recurrence relation for each region determined. Earthquakes are then computed for each region as determined by its recurrence relation, distances from the site to earthquake epicenters computed, and a resulting distribution of site acceleration determined. This distribution relates probability of occurrence with site ground acceleration. The parameters required to compute a seismicity study can vary from simple to very complex. Geologic data from paleoseismicity studies can be used to confirm or augment recurrence data computed from epicentral data bases which are generally time limited to several decades since the installation of seismic monitoring stations. Evaluations of the complex time dependency of groups of earthquakes or recent large events is very complex and is an area where advanced models involving a Markov renewal process are required to offset limitations in simpler Poisson models. However, general usage of such models is often limited from lack of data required to quantify the model.

Figure 4 is a typical recurrence relationship for a region based on a study of the regional seismicity. Figure 5 is the recurrence relationship based on a study of a subset of epicenters whose activity is associated with the fault. Also shown is a plot of known geologic data. In this case the geologic data tends to support the epicentral data. Figure 6 gives a plot of probability of not exceeding acceleration at a

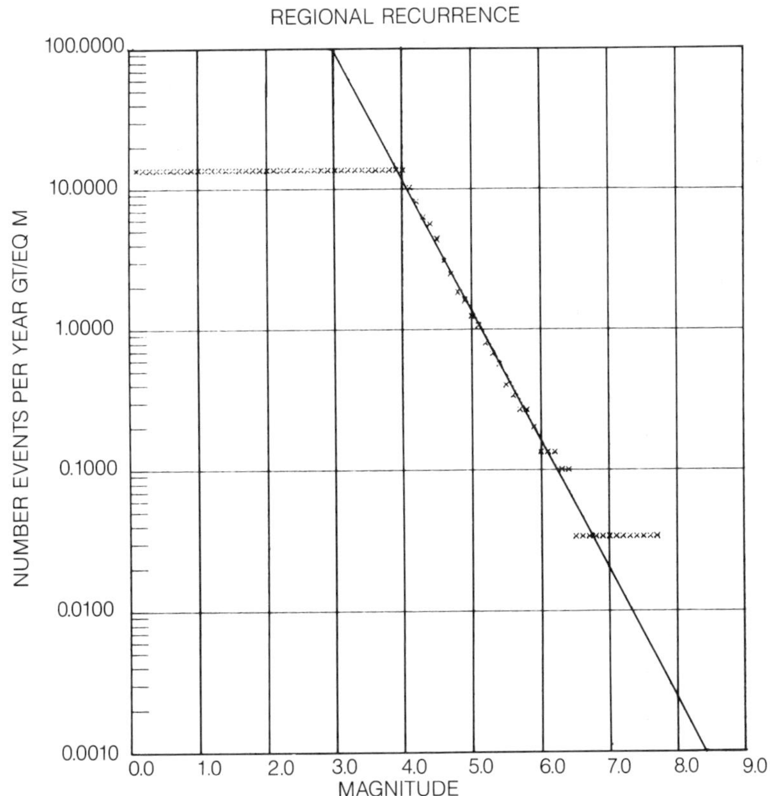

FIGURE 4. Earthquake recurrence relationship.

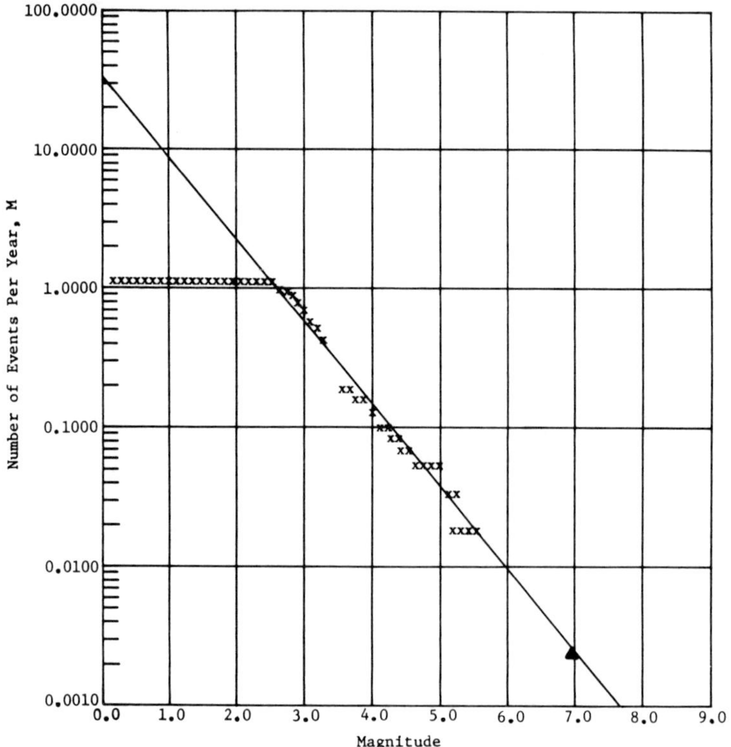

FIGURE 5. Recurrence San Gabriel fault.

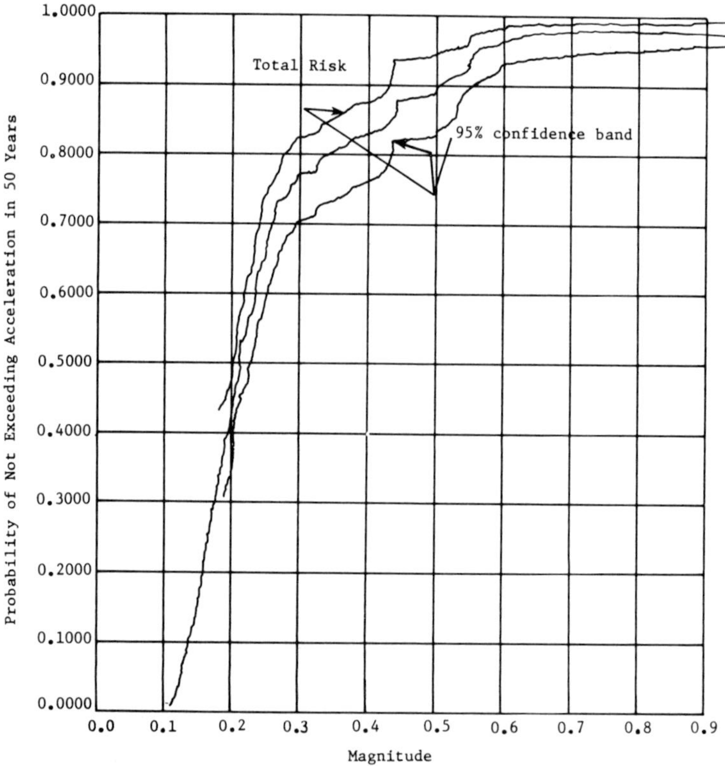

FIGURE 6. Total probability of not exceeding acceleration at the site, all faults contributing.

site for a 50-year exposure. Also shown are confidence bounds for the estimate.

SPECTRA RESPONSE AND ERRORS

For the purpose of an economic analysis it was recommended that a ductility = 1.0 condition be used as the basis for comparison. This simplifies the design process and eliminates the need to produce an inelastic spectra. Specifying a ductility of 1.0 is the same as specifying a higher acceleration and some ductility greater than1.0. Use of a ductility of 1.0 allows the structural engineer to make use of all elastic computer codes without need for a nonlinear analysis in the design phase.

It is of interest to take a brief diversion and look at errors which crop up in elastic and inelastic spectral analysis. A typical frame structure (Figure 7) was selected for analysis. The structure was 40 by 80 feet in plan, having three floors and a roof. An eigenvalue analysis of the east-west end frame yielded the following results:

Mode	Period (sec)
1	1.06
2	0.373
3	0.226
4	0.178

An actual earthquake record was scaled to 0.29 and used as the basis for the response calculations. The spectrum was smoothed using a five-part line segment on a tripartite plot based on visual averaging.

The frame data were prepared for the computer program ETABS. Input consisted of the same approximate spectral representation used previously. The displacement results are as follows:

Level	Displacement (in)
4	2.498
3	2.187
2	1.6175
1	0.9494

The exact spectra were input by defining 50 data points. The results were:

Level	Displacement (in)
4	3.824
3	3.355
2	2.473
1	1.464

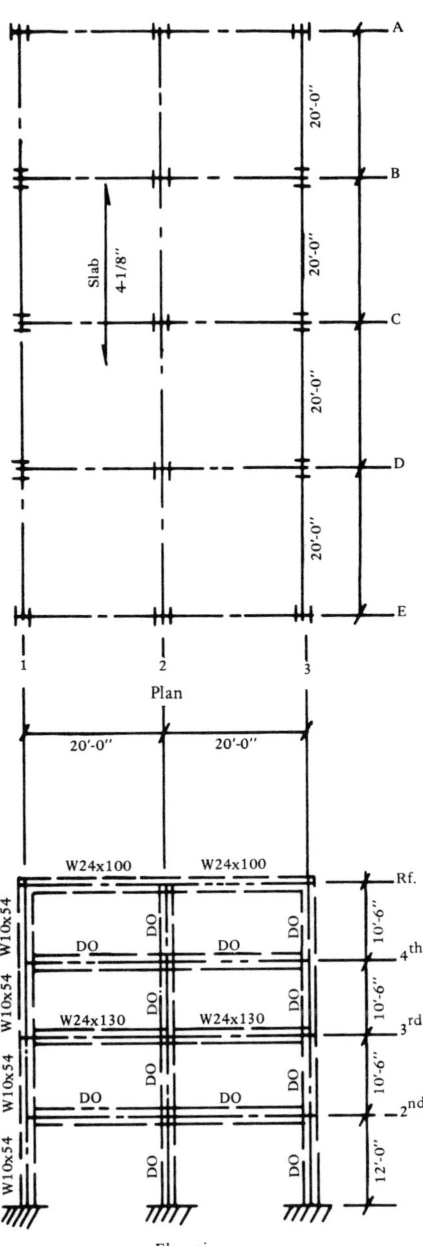

FIGURE 7. Building frame and plan.

These displacements from the exact spectrum are significantly greater than those from previous approximate spectrum. The average line used previously is much lower than the actual spectrum in the vicinity of the first mode period of 1.06 seconds. This difference is very important since the majority of the response is controlled based on first mode behavior. Note that line segments on tripartite plots imply logarithmic interpolation; however, the program ETABS uses arithmetic interpolation. It is essential to input sufficient spectral definition in the regions of structural periods to insure proper accurate computation. The actual time history of the same earthquake scaled to the same ground motion (0.2g) was used as input in a repeated analysis. The results are:

Building Level	Displacement (in) Time History Solution	Spectra Solution
4	3.863	3.824
3	3.324	3.355
2	2.516	2.473
1	1.486	1.463

Both the time history solution and the response spectrum solution computed by square root of the sum of the squares agree closely in this case. Agreement is better than would normally be expected.

Associated with each California Institute of Technology earthquake accelerogram is a set of initial conditions. The initial conditions occur because the record does not begin at zero level, but rather at some instrument trip level. Initial conditions are evaluated as part of the base line correction process. Any program using an accelerogram as input to prescribe loading must incorporate the initial conditions or an error will result. Programs such as SAP which do not allow for initial conditions cannot be used with accelerogram records without creating an error. The lack of initial conditions results in an underestimation of response by about 15% for a typical accelerogram.

The entire structure was next used for comparison in a three-dimensional computation using ETABS. The results for the same forcing function are:

Building Level	Displacement (in) Time History Solution	Response Spectra
4	2.635	2.725
3	2.335	2.343
2	1.755	1.691
1	0.953	0.894

The three-dimensional solution included secondary frames omitted in the two-dimensional representation; this resulted in increased stiffness in comparison with the two-dimensional simplified approximation.

An inelastic two-dimensional structural analysis was performed using the code DRAINTABS. The same time history forcing function used in the elastic analysis was used here. The results are:

Building	Displacement (in) Elastic	Inelastic
4	3.863	4.473
3	3.324	4.151
2	2.516	3.519
1	1.486	2.816

The inelastic structural analysis shows the base columns yielding; this resulted in increased deflections in comparison with the results of the elastic analysis.

The inelastic structural analysis is a more accurate tool; however, the cost of its use is about six times that of the elastic analysis. It is interesting to note the wide range of story displacements that can be computed based on the refinement of the model and the computational procedure used.

To make the results complete, the three response spectra (Figure 2) showing effects of magnitude were applied to the two-dimensional frame. Results of the structural analysis for the same nominal ground motion (0.15g) are:

Building Level	Displacement (in) A Earthquake	B Earthquake	C Earthquake
4	0.968	2.378	3.629
3	0.837	2.081	3.184
2	0.633	1.540	2.346
1	0.397	0.921	1.387

The results show the significance of the variations in spectral shape as a function of magnitude and separation distance. This clearly demonstrates the need for carefully matching magnitude and separation distance for the response spectrum. Similar results would occur with a time history analysis.

The response spectra in Figure 3 showing the effects of soil type were applied to the two-dimensional frame. The results are:

Building Level	Displacement (in) Alluvium Spectrum	Intermediate Spectrum	Rock Spectrum
4	3.29	4.27	5.15
3	2.88	3.75	4.53
2	2.13	2.76	3.33
1	1.25	1.63	1.96

The results again show the need for carefully selecting the

site conditions for the response spectrum for use in the structural analysis.

Use of an inelastic spectra to predict inelastic defamation can produce errors and is not precise. Figure 8 gives the range of response which might be indicated in a typical case.

An MIT study by Biggs et al. (1975) of a four-story building evaluated interstory displacements. Table 1 gives responses from elastic analysis using 39 real earthquakes, 15 artificial earthquakes, and the average response spectrum. Although the mean values agree for this elastic analysis, the coefficient of variation is large. Table 2 gives interstory ductility ratios for the inelastic analysis. Again the variation is large. This illustrates the difficulty in designing for a specified level of yielding. Current procedures for equivalent static forces and elastic analyses are lacking exactness. It is important to point out that ductility ratio or yielding varies throughout the building.

The use of spectra for multi-degree-of-freedom systems when hysteretic behavior in the structure occurs can significantly underestimate damage. The hysteretic behavior results in strength degradation not accounted for in elastic-plastic models. The use of inelastic design response spectra determined from linear elastic spectra has serious limitations in that the duration of the ground shaking and the number and characteristics of the acceleration pulses are omitted. Repeated large acceleration pulses can lead to accumulation of large strains. Furthermore, the type of excitation which induces dynamic response in a linear elastic system is very different from the type of excitation which is critical to an elasto-plastic system (Bertero, 1975). Resonance phenomenon is of major significance in an elastic system; however, small inelastic deformations in a yielded system are equivalent to large values of damping. The natural period of a structure changes with deformation. Use of linear elastic response spectra in elastic design is controlled by the resonance phenomenon induced by single acceleration pulses with the same periodicity as the structure. Considerably larger deformations can be produced by just one long pulse with an effective acceleration exceeding

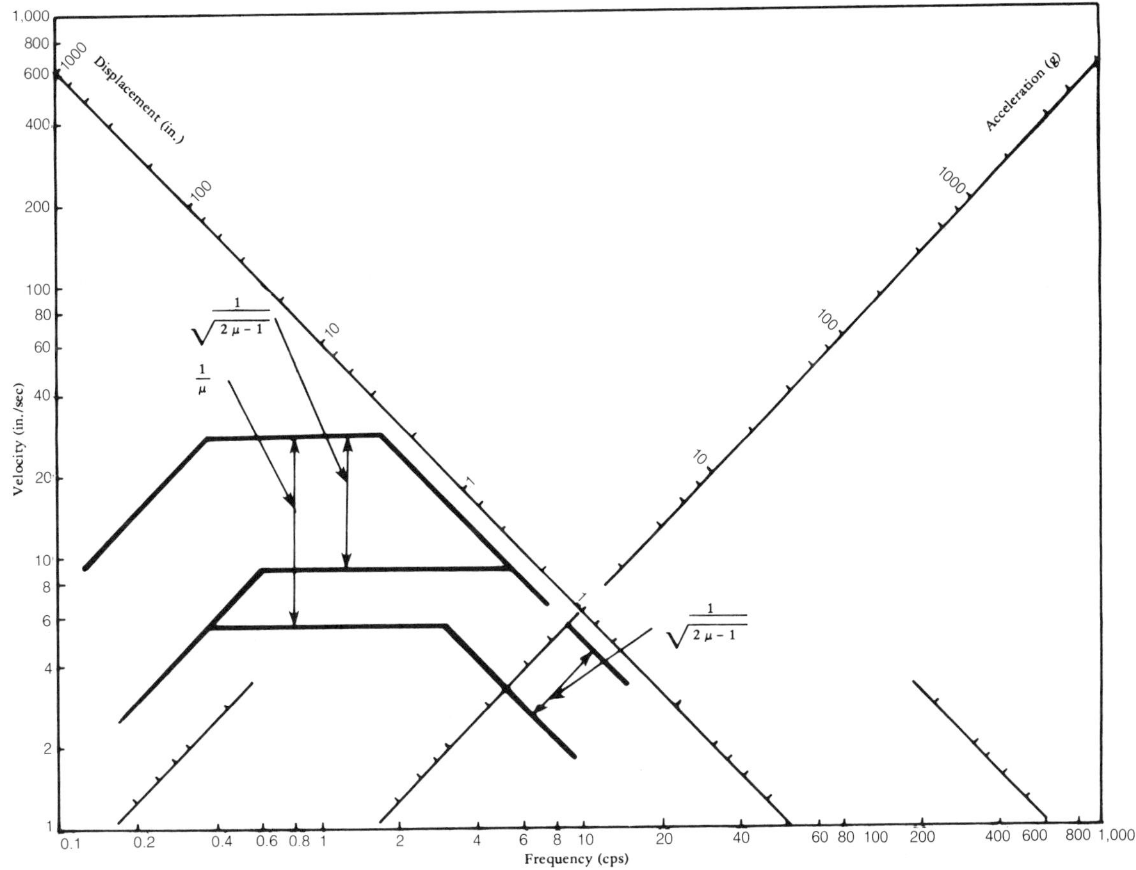

FIGURE 8. Typical range of expected response.

TABLE 1. Peak Interstory Elastic Displacement of Four-Story Building (from Biggs, Hansen, and Holley, 1975) [Fundamental period = 1.13 sec].

Story	Time-History Analysis of 39 Earthquakes[a]	SRSS Modal Analysis-Mean (or Mean + σ) Response Spectrum	Time-History Analysis of 15 Artificial Motions[b]
	Mean		
1	0.122	0.126	0.133
2	0.107	0.104	0.115
3	0.092	0.088	0.093
4	0.063	0.059	0.064
	Mean + σ		
1	0.194	0.193	—
2	0.169	0.166	—
3	0.137	0.131	—
4	0.089	0.083	—
	Coefficient of Variation		
1	0.58	—	0.25
2	0.57	—	0.29
3	0.48	—	0.29
4	0.40	—	0.24

[a]Normalized to 0.3g peak ground accelerations.
[b]All generated from mean response spectrum.

TABLE 2. Peak Interstory Ductility Ratios of Four-Story Building (from Biggs, Hansen, and Holley, 1975) [Fundamental period = 1.13 sec].

Story	Analysis of 39 Earthquakes[a]	Analysis of 15 Artificial Motions[b]
	Mean	
1	5.7	4.4
2	2.6	3.2
3	4.0	5.0
4	9.7	13.8
	Coefficient of Variation	
1	1.23	0.42
2	0.48	0.31
3	0.48	0.28
4	0.49	0.39
	Maximum-Minimum	
1	38.6 – 0.8	10.2 – 2.6
2	5.9 – 0.8	5.6 – 1.9
3	8.9 – 1.0	7.6 – 2.9
4	27.8 – 2.2	21.1 – 7.0

[a]Normalized to 0.3g peak ground accelerations.
[b]All generated from mean response spectrum.

the yield strength of the structure. For inelastic response, the largest incremental velocity rather than the largest peak acceleration is of importance (Bertero, 1975). Thus, the type of ground motion which is critical depends on the type behavior of the structure.

The frame in Figure 7 was reanalyzed using inelastic procedures. This structural analysis showed the base columns yielding; this resulted in increased deflections in comparison with the results of the elastic analysis. The analysis was repeated for other load levels, and the elastic limit of the structure was found to be 0.15g with a story drift of 0.01. The results of the series of analyses are shown in Figure 9. This structure, with a first mode period of 1.06, is in the velocity range of the spectrum. The response lies between μ and $\sqrt{2\mu - 1}$ lines as would be expected.

The series of analyses was repeated to study the response of the structure in the acceleration region. The stiffness of the columns was increased to reduce the period; the yield capacity was not changed. Figure 10 shows the response. The elastic limit was reduced substantially and, although the story drift was reduced, the ductility demand increased. The response exceeds the $\sqrt{2\mu - 1}$ over a portion of the load range.

In the stiff structure, the ductility demand was substantially increased. Story drift was reduced by the stiffening; thus, nonstructural damage based on story drift should be

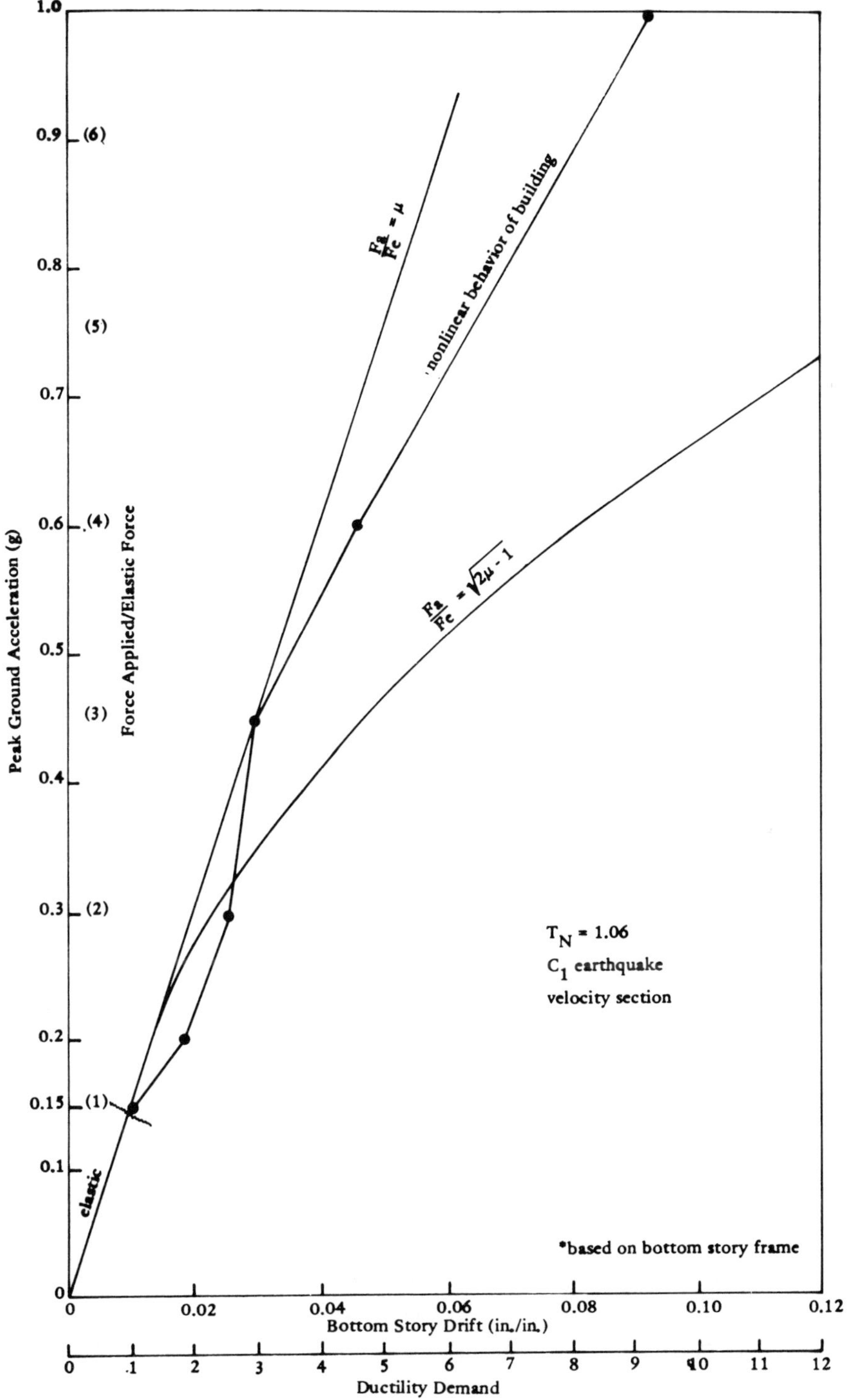

FIGURE 9. Response in velocity region.

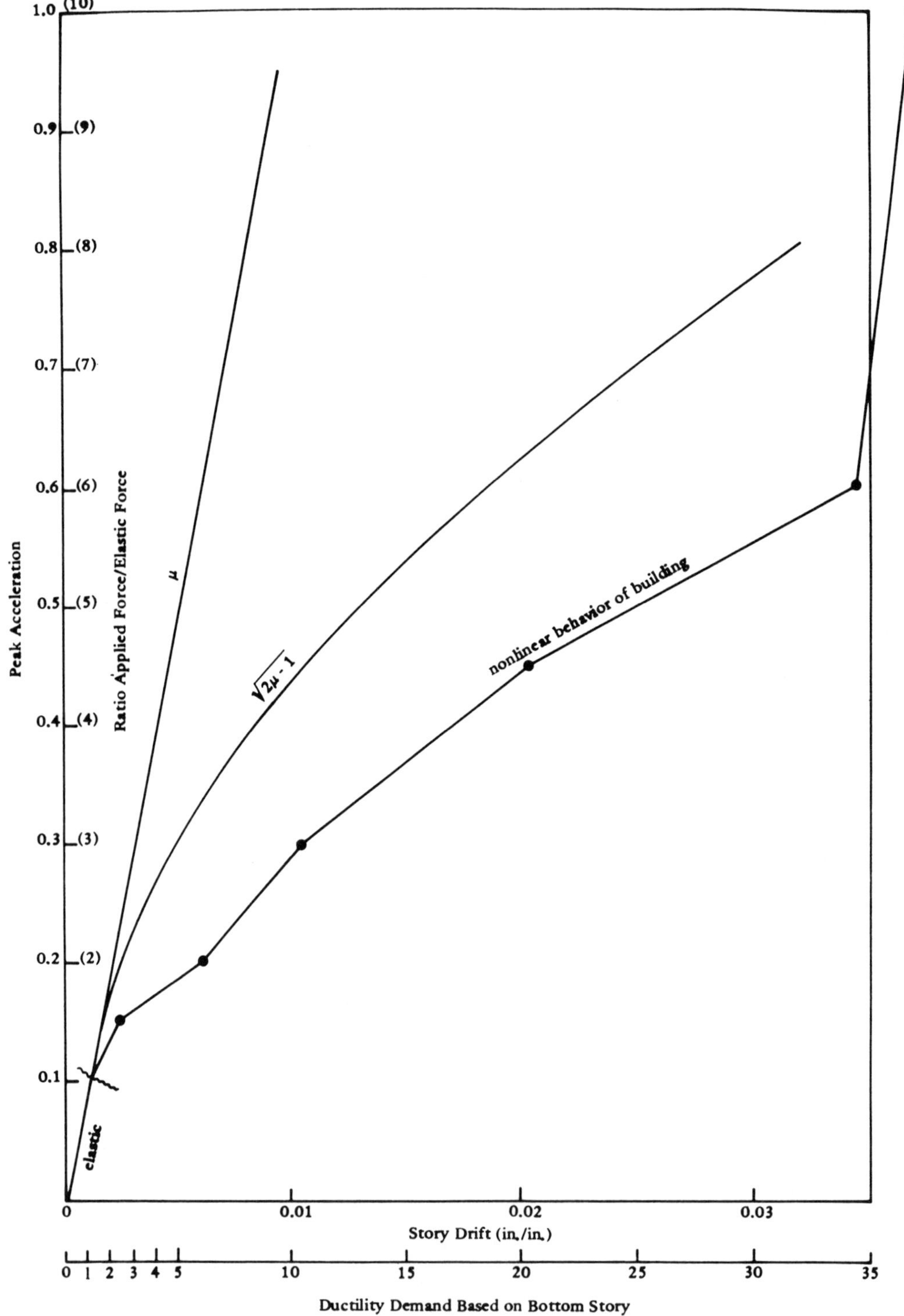

FIGURE 10. Response in acceleration region.

reduced. However, structural damage should be a larger factor.

It is interesting to note the behavior of the structure as it goes through elastic response into inelastic. At low levels of excitation the structure responds elastically, with the maximum displacement (of the first floor) occurring late in the excitation. The structure responds to resonances. As the excitation increases inelastic behavior is noted first in the bottom columns, and then progresses upward as excitation levels increase. At moderate levels, the structure becomes inelastic early in the excitation and leans slightly; maximum displacement occurs earlier in the excitation. At higher levels, the lean increases and is not recovered. The lean offsets later displacement (purely by random loading), so the maximum displacement occurs still earlier in the loading.

In the inelastic range, the response of the structure is based largely on the order and size of the loading pulses. These determine lean direction, which may or may not be reversed while under load and may or may not be recovered at the termination of loading. As stated earlier, the nonlinear response depends on the largest incremental velocity. The results indicate that wide variations in response are possible. Prediction of inelastic behavior based on elastic response spectra must be viewed as a crude approximate technique and may not be applicable in the acceleration range. The use of μ and $\sqrt{2\mu - 1}$ may lead to an unconservative estimation of response.

COST OF SEISMIC STRENGTHENING

To develop an understanding of seismic lateral resistance costs, a typical low rise building was selected for study. The structure chosen was representative of a class of structures used for office work, light industrial work or living quarters. It was a three-story building, for which detailed cost data and drawings were available, which had been recently constructed at an eastern nonseismic area; thus, the nonseismic baseline cost condition was established. The building was a frame structure, 185 by 185 feet (53m by 53m) in plan, which utilized a structural frame system.

The structure selected was redesigned, and it was assumed that it was newly constructed and located in a seismically active area. Seismic design concepts were typical of West Coast standard engineering design practice. Elements of the various framing systems were proportioned using the strength design provision contained in the AISC Specifications (1980) and the ACI Building Code Requirements (1983). Design forces in the individual members were obtained by applying suitable load factors to the loads obtained from the dynamic response analysis. Load factors used in the design analysis are as follows: dead load plus live load, 1.7; dead load plus live load plus static lateral load, 1.3; and dead load plus live load plus seismic response spectrum, 1.0. Ultimate strength of the masonry was estimated as being twice the allowable value which is consistent with the work reported by the National Bureau of Standards (1973). The structure was designed for six levels of peak ground acceleration—0.1g to 0.35g with 0.05g increments. Elastic design spectra utilizing Newmark standard spectral shapes were implemented. Five concepts of seismic lateral resistance were individually utilized: (1) steel moment frame; (2) steel braced frame; (3) steel frame and concrete shear wall; (4) concrete moment frame; and (5) concrete frame and shear wall. For the purposes of this analysis, to facilitate design comparisons, the performance level of the structure under the specified spectra was set at a ductility equal to 1.0 design so that members would be at yield. This performance level was set for several reasons. First, a ductility of 1.0 produces the same response as a ductility value greater than 1.0 at higher acceleration. Second, the use of ductility equalling the 1.0 criteria allows the structural design engineer to use all elastic computer codes without conducting a nonlinear analysis; further, nonlinear spectral techniques did not need to be implemented.

A detailed weight analysis of the structure was performed to determine the mass distribution and the distribution of vertical gravity loads. The two peripheral frames were assumed to each take half of the total mass acting in the directum parallel to the frames. Draft criteria for the frames were established using criteria in ATC-3 (1978). The Effective Peak Velocity (EPV) was defined as the maximum spectral velocity from the response spectra divided by 2.5. The velocity related acceleration coeficient (A_v) was then calculated using the ATC-3 relation for A_v in terms of EPV. The seismic coeficients were then calculated and used to determine the base shear. This value was increased by a detection coefficient to obtain pseudo static-forces to evaluate drift. The maximum relative story drift was limited to 0.015. The designed frames were reviewed and changes were made to reflect framing consideration, connections, and column splicing. In the moment frames, a grade beam was incorporated at the foundation line to transmit the column base moments into the foundation. Also, it was decided to splice the columns just above the second floor level on all of the frames. For the braced frames, it was decided to use simple connections for the girders in the unbraced bays. From the preliminary analysis, it was determined that it would be necessary to brace only two of the five bays.

Based on the preliminary analysis of the braced frames, a preliminary design for the shear wall frames was determined. It was decided to place shear walls in only two bays to be compatible with the braced frame design and the functional requirements for the structure. Preliminary analysis indicated that reinforced concrete walls would have to be used for the four highest base accelerations whereas reinforced masonry walls could be used for the lower two base accelerations.

The required designs for the five concepts of lateral resis-

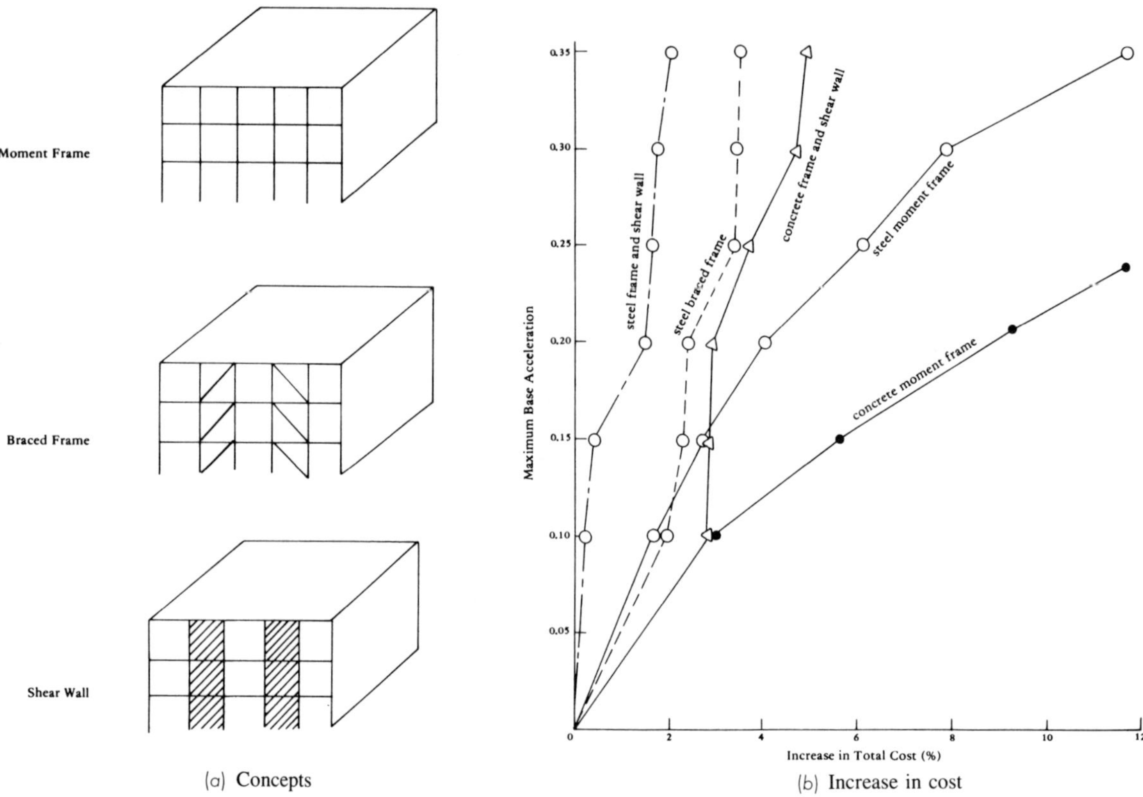

(a) Concepts (b) Increase in cost

FIGURE 11. Seismic strengthening concepts and increase in cost for each type of strengthening.

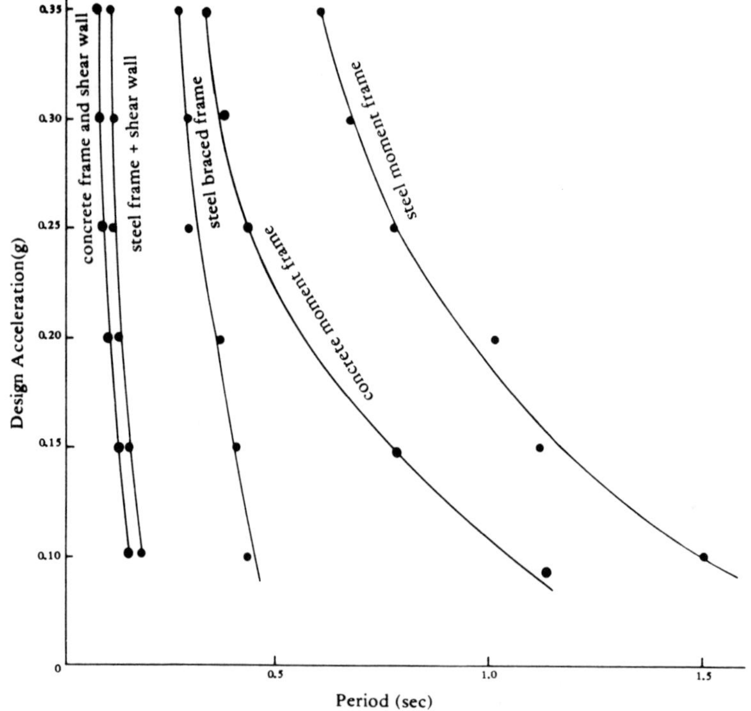

FIGURE 12. First mode period.

tance and six load levels were performed by a firm having significant experience in seismic design. As part of that effort, the detailed cost estimates were provided for seismic lateral resistance systems by treating the structure as new construction, using the available cost data on the existing structure, and making adjustments for West Coast construction practice. Results of the designs are presented in Ferritto (1982).

The detailed structural costs estimated were based on the results of the six design cases for the five concepts of strengthening. The cost of the existing exterior frame construction was deducted from the total building cost, and the cost of each new seismic framing system was then added to obtain a new total building cost. Concrete or masonry seismic shear wall configurations, when utilized, were assumed to replace the existing 6-inch (15cm) concrete block. The cost of foundation redesign was included, and all costs were adjusted to 1981 Los Angeles levels. Figure 11a shows the seismic strengthening concepts, and Figure 11b the increase in cost for seismic strengthening. Details of the cost estimating are given in Ferritto (1982). Figure 12 gives the first-mode periods of the structure for the strengthening concepts as a function of design level. The moment-frames period shows greatest variation with design acceleration.

DEMAND/CAPACITY RATIO

One of the basic elements in a damage function for a building is the expression of the damage or ductility level as a function of the demand/capacity ratio. The demand/capacity ratio is an expression applied to design acceleration. The basic question is whether a given demand/capacity ratio predicts the same ductility for various levels of design

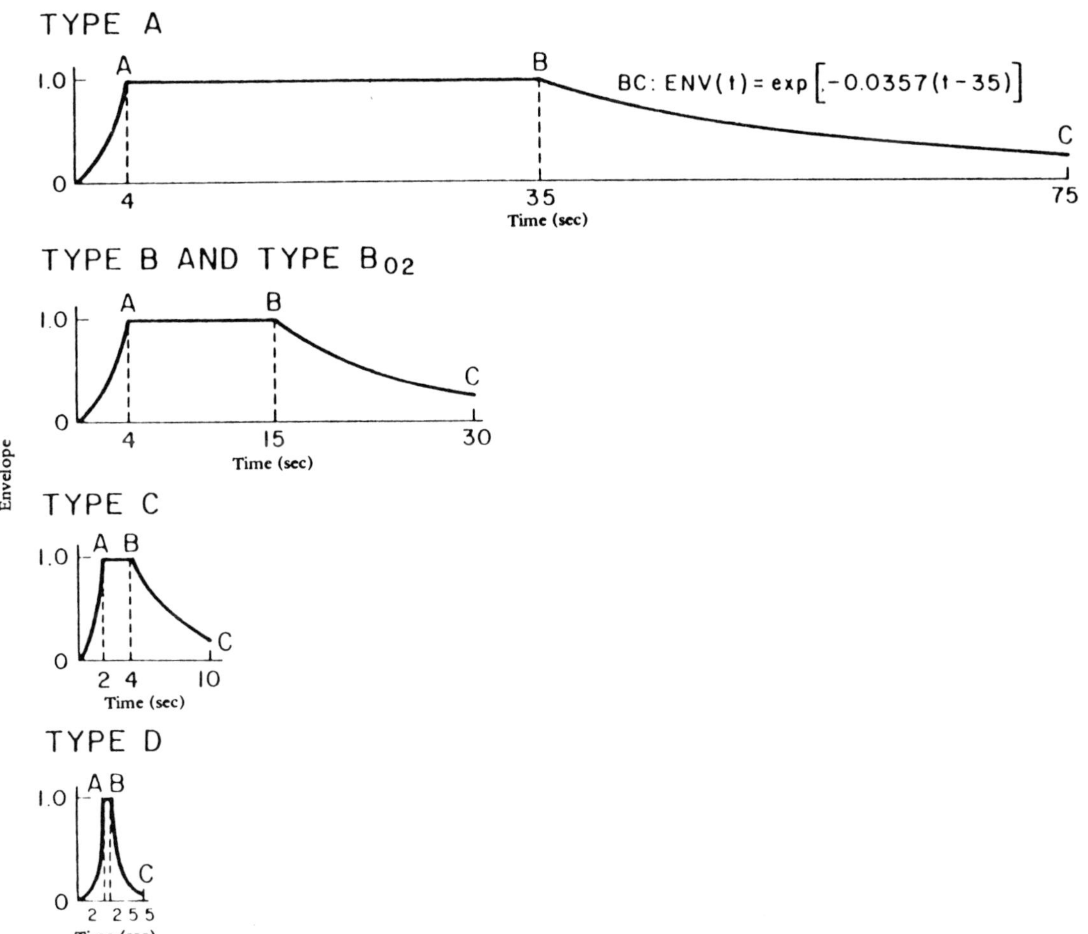

FIGURE 13. Time intensity functions.

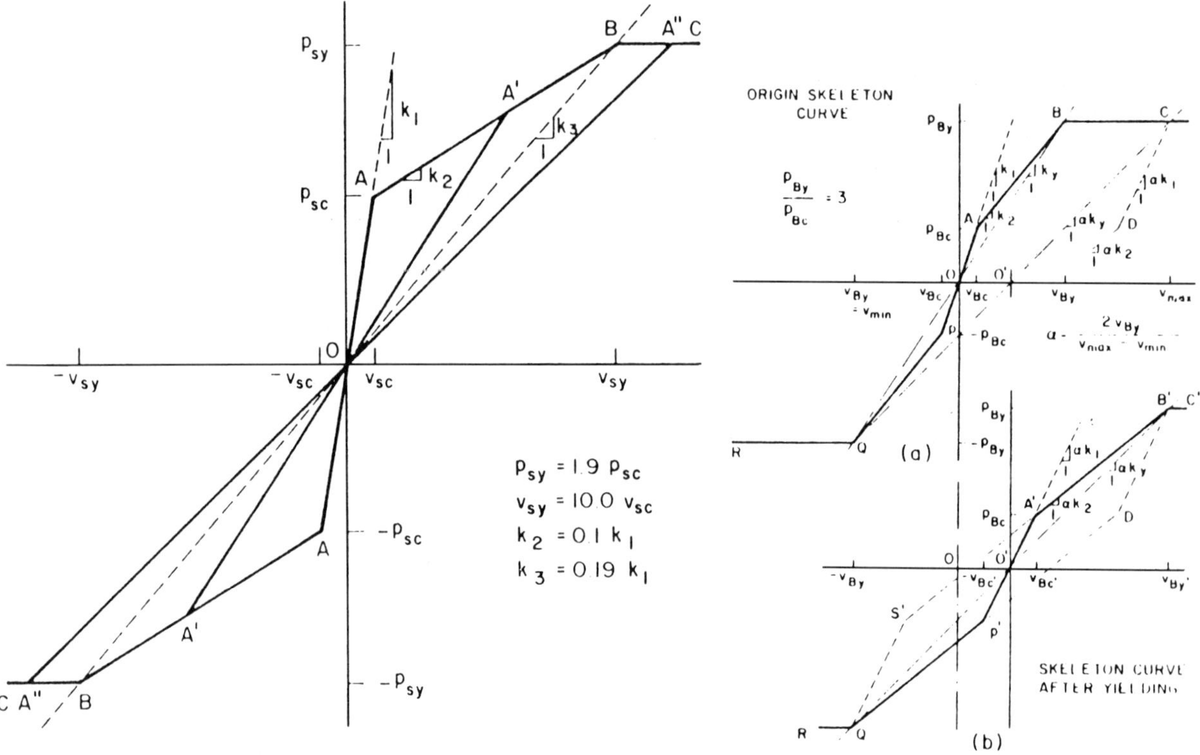

(a) Origin-oriented hysteretic model.
(b) Tri-linear hysteretic model.

FIGURE 14. Structural models.

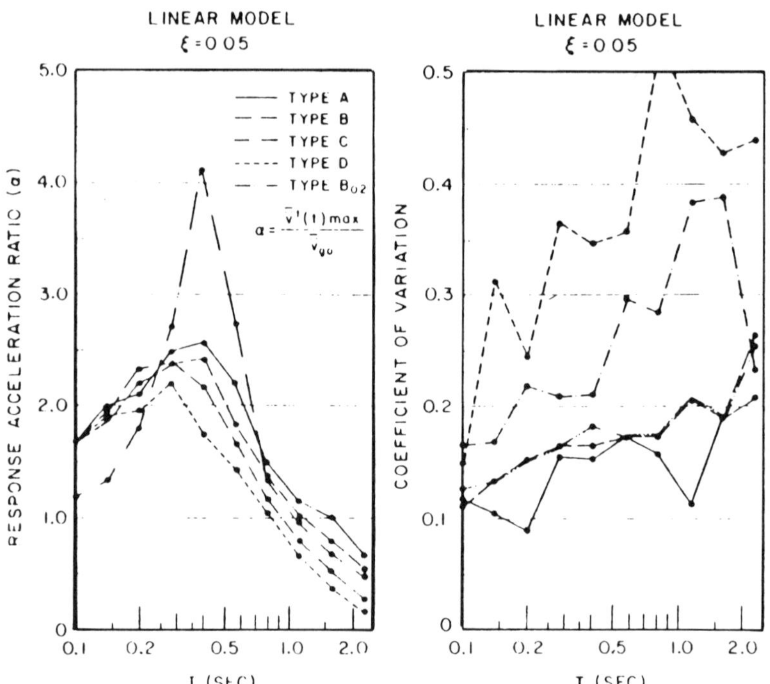

FIGURE 15. Response acceleration ratios for linear model.

capacity. Murakami and Penzien (1975) investigated nonlinear response from excitation of single-degree-of-freedom hysteretic models.

Figure 13 shows the envelope of earthquake records they used; Figure 14 shows the models. Figures 15 through 18 show the linear and nonlinear response for the properties indicated. By use of the data in Figure 16, the demand/capacity ratio can be plotted for mean ductilities of 1.0 and 5.0 (Figure 19). Figure 20, based on Figure 19, shows the ratio of demand to capacity for a ductility of 5.0 to that for a ductility of 1.0. The ratio is seen to vary with the natural period of the structure. The data indicate that as long as the natural period of a structure remains unchanged, a given demand/capacity ratio will produce the same ductility.

This may be restated as follows: given two elastoplastic single-degree-of-freedom systems with the same mass, stiffness, natural period, and excitation accelerogram shape, the same ductility will be produced in both systems if the ratio of peak acceleration to yield resistance acceleration is maintained.

An alternative formulation has the resistance stiffness and mass of one elastoplastic single-degree-of-freedom system, and a multiple of the other. The natural periods of both would be the same and, given the same shape excitation with amplitudes, a multiple of the same ductility would result.

It is known from Culver et al. (1975) that damping and period change with ductility as inelastic behavior progresses (Figures 21 and 22). However, as noted in Figure 20, these changes are small and would not influence the ratio of design applied to acceleration for a constant ductility level.

Changes in a structure's design when strengthening a structure primarily influence the stiffness and, to a lesser

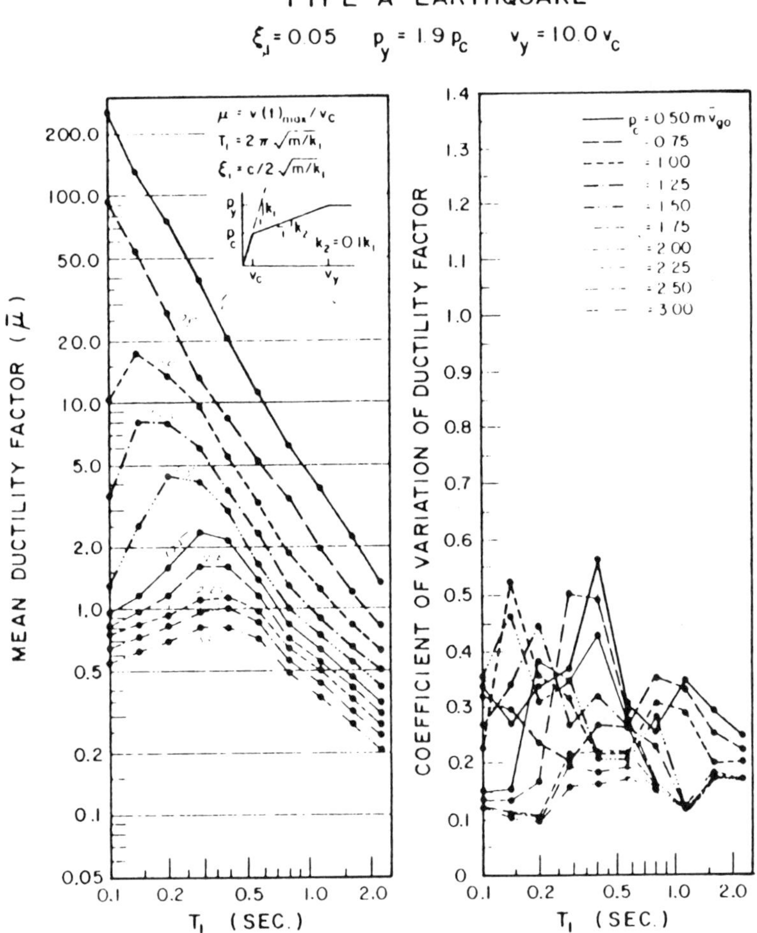

FIGURE 16. Mean ductility factors and corresponding coefficients of variation for origin-oriented model having different strength levels for type A earthquake.

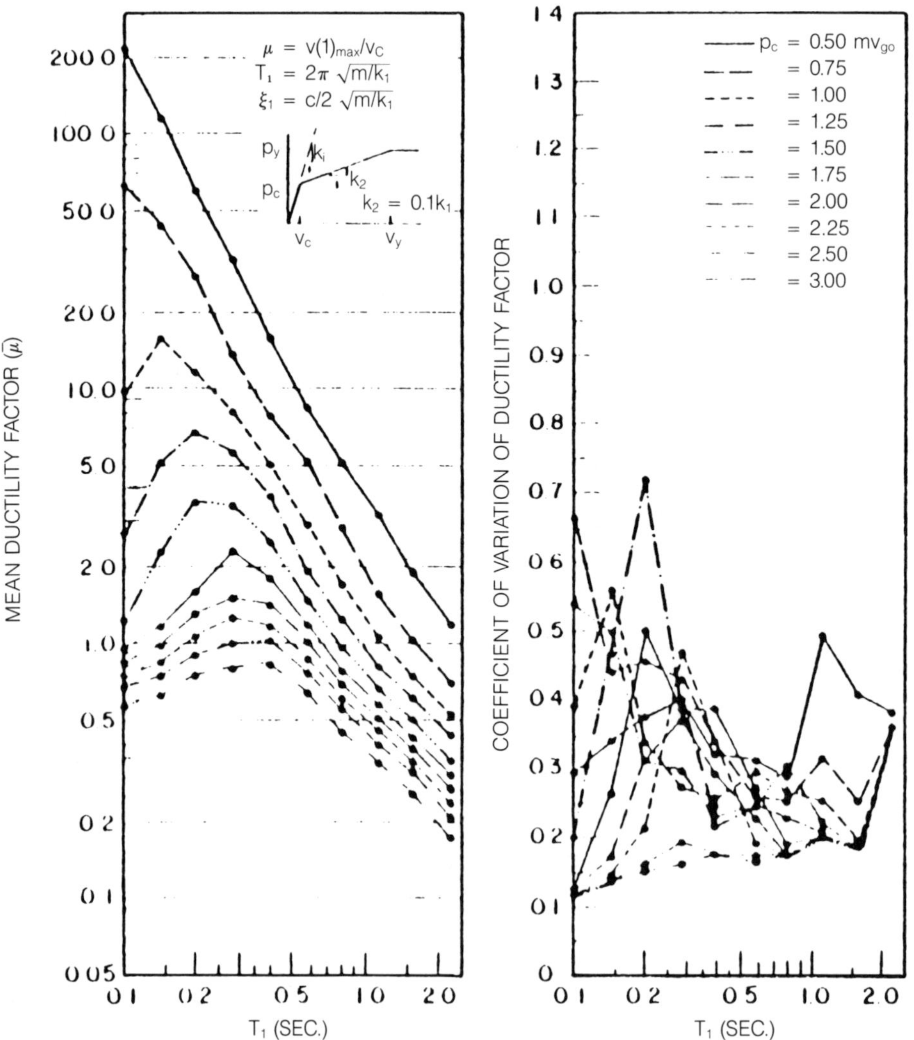

FIGURE 17. Mean ductility factors and corresponding coefficients of variation for origin-oriented model having different strength levels for type B earthquake.

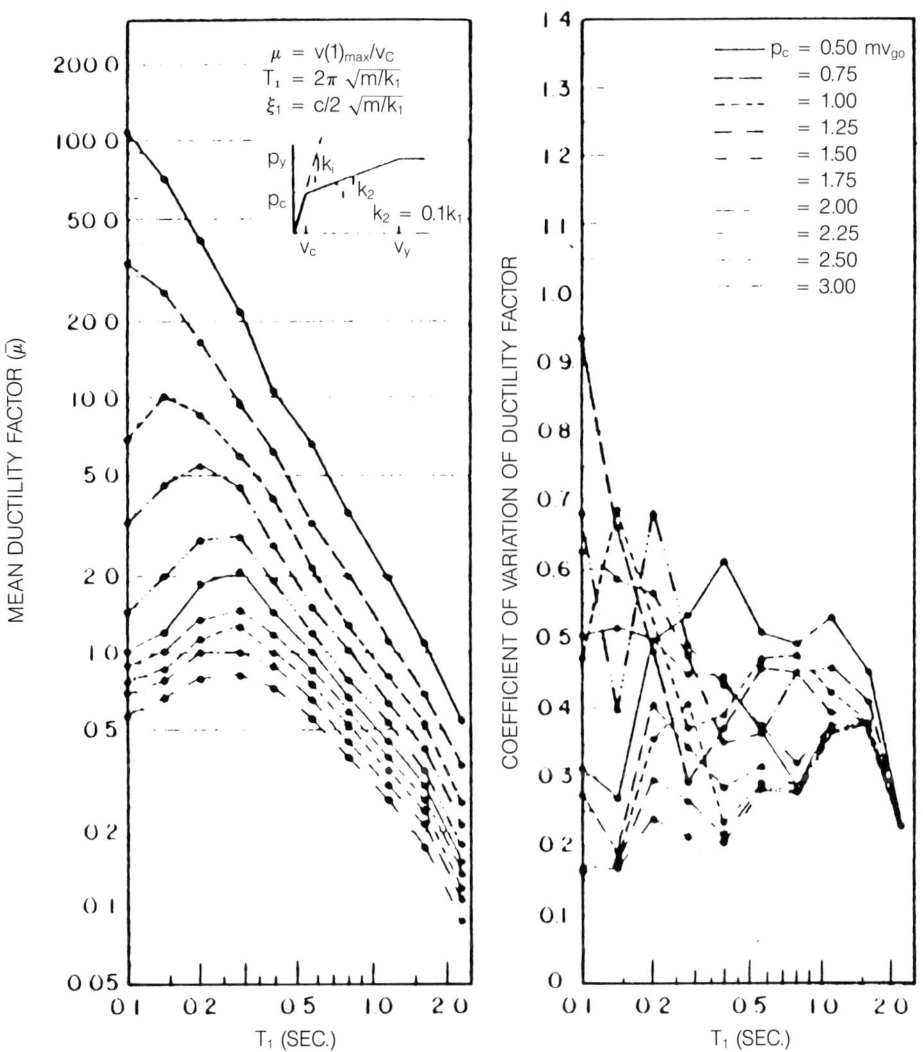

FIGURE 18. Mean ductility factors and corresponding coefficients of variations for origin-oriented model having different strength levels for type C earthquake.

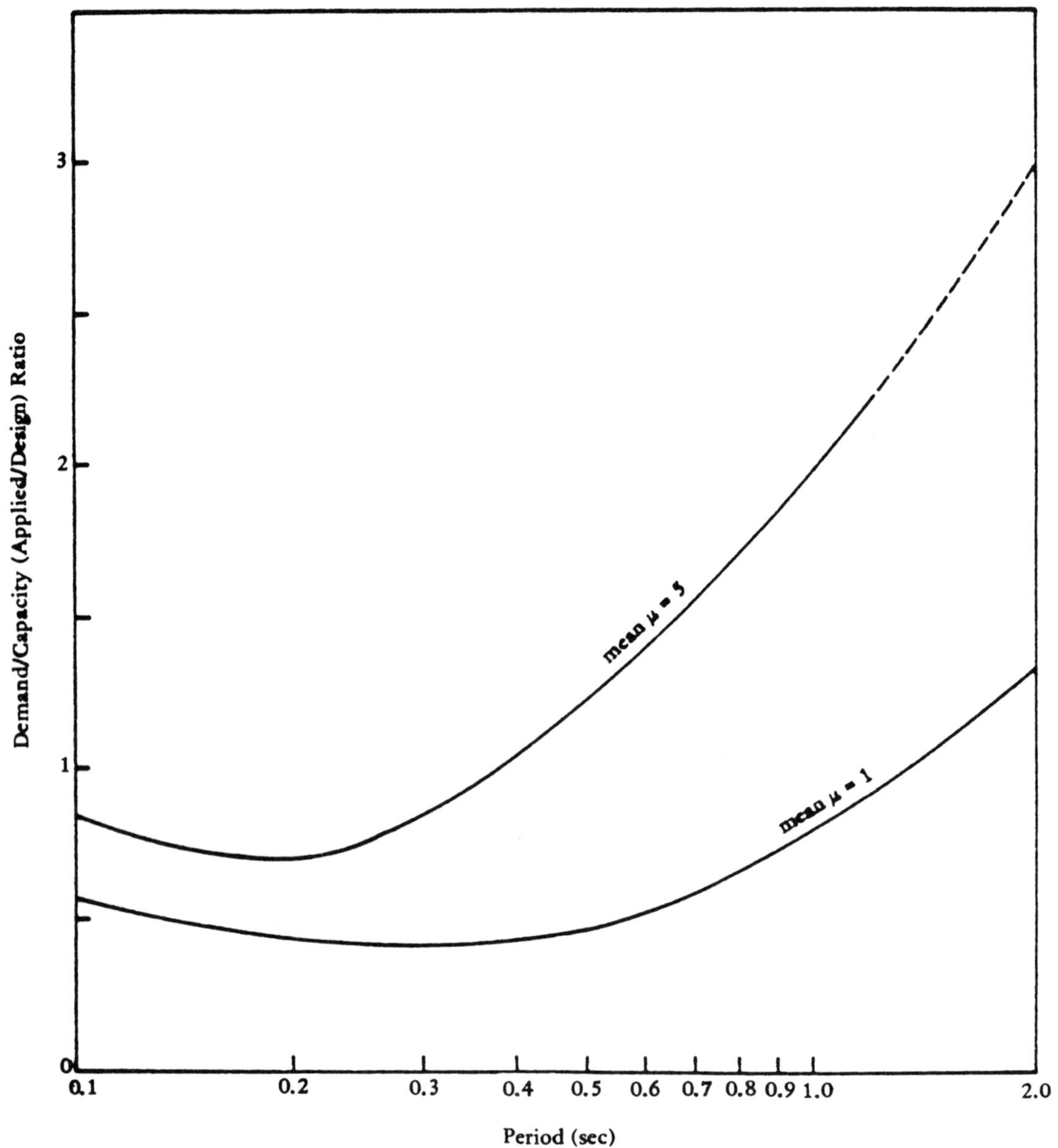

FIGURE 19. Mean ductility.

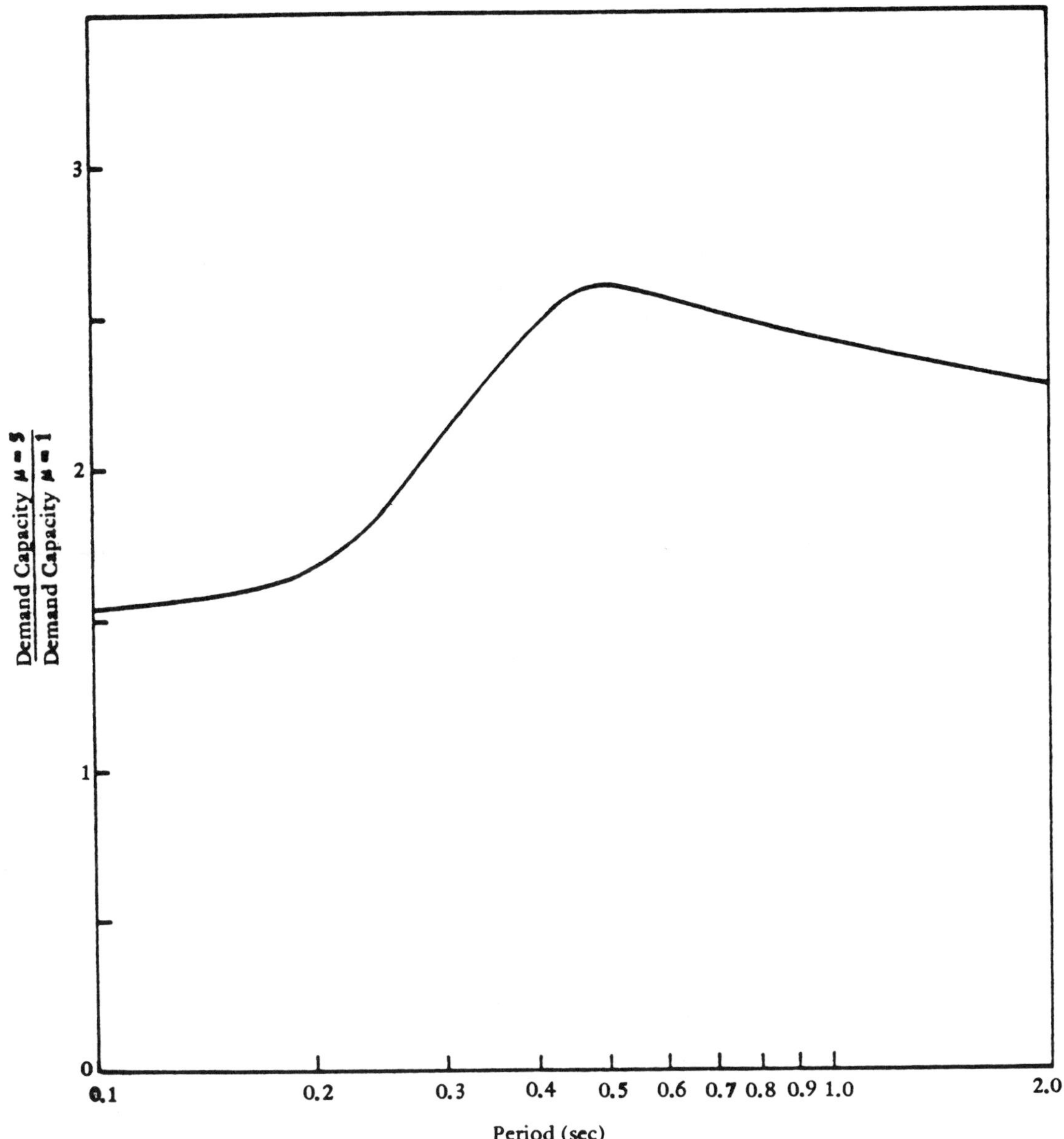

FIGURE 20. Ratio demand/capacity $\mu = 5$ to demand/capacity $\mu = 1$.

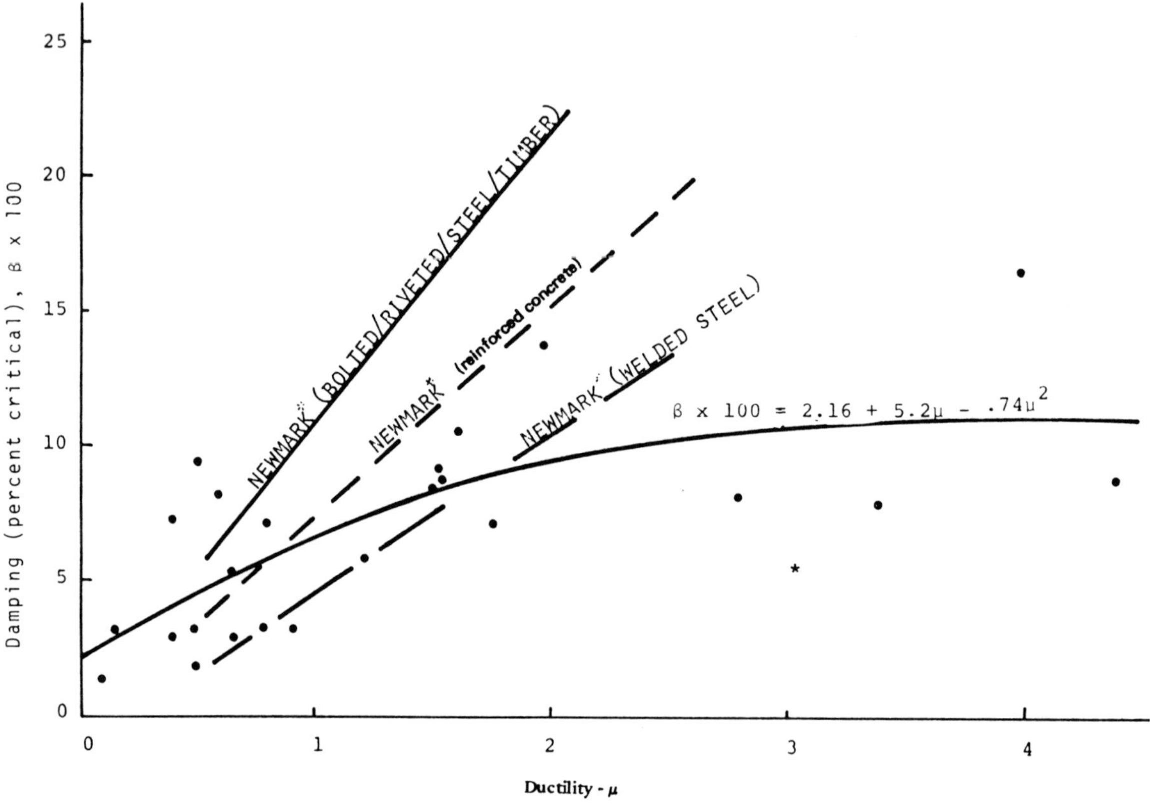

FIGURE 21. Variation of damping as a function of ductility.

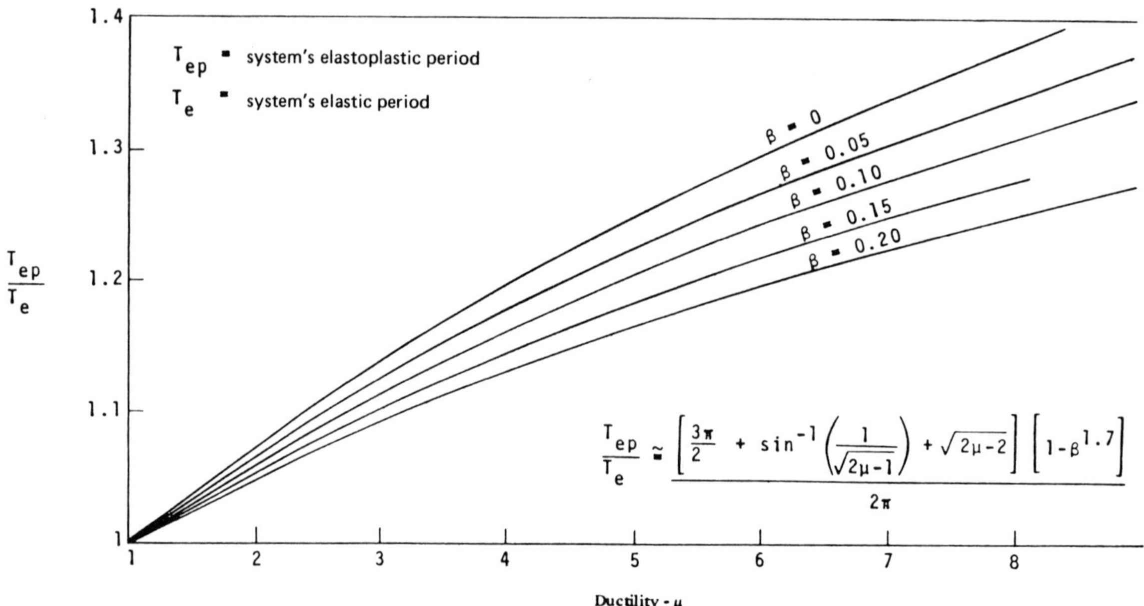

FIGURE 22. Ratio of elastoplastic to elastic periods.

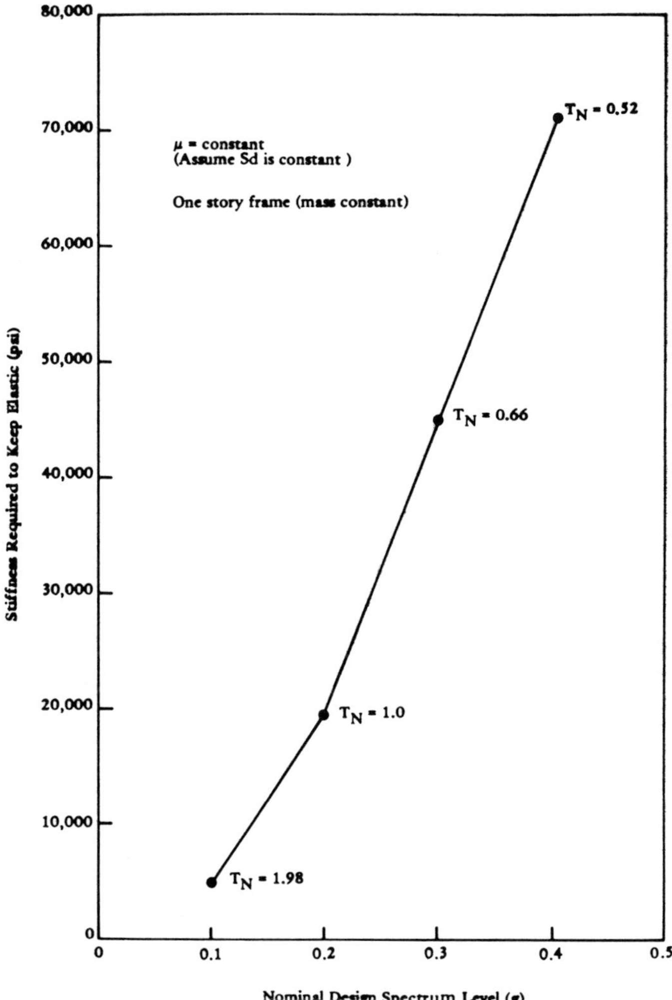

FIGURE 23. Stiffness required to remain elastic.

extent, the mass. Strengthening may be accomplished by substitution of a stronger, or higher yield stress, material of the same stiffness or by an increase in the size of the member, which, of course, would increase the stiffness. These changes in design usually affect the natural period. Figure 23 shows the stiffness required for elastic ($\mu = 1$) design for nominal design spectrum levels. The data over a range show stiffness required may be approximately taken to be linearly proportional to applied spectral loading.

In actual construction, strengthening will take place by increasing column and shear wall cross sections resulting in some minor increase in mass and a large increase in stiffness. Generally, the optimal design range can be narrowed to approximately the 0.1g to 0.2g range; stiffness might be expected to double and the period to halve over the full range. The ductility can be expected to vary by a factor of 2 over this wide range. Thus, a structure designed for 0.2g with 0.2g applied might have applied. Damage estimation relies on estimation of design-to-collapse ratios.

Both the design level and collapse level resistance required depend on the natural period. The ratio of design level (acceleration for $\mu = 1$) to collapse level (acceleration for $\mu = 5$) over a period range of 0.5 to 2.0 seconds varies from 2.25 to 2.60 (Figure 20). Figure 24 shows ductility resulting from an application of load acceleration 2.0 times the acceleration for $\mu = 1.0$. Again, between the range of 0.5 and 2.0 seconds, the resulting ductility is fairly constant. This variation is within the accuracy of the data in general and should not have a major effect on the results of the general damage expression formulated. An alternative approach would be to formulate a damage matrix rather than a function.

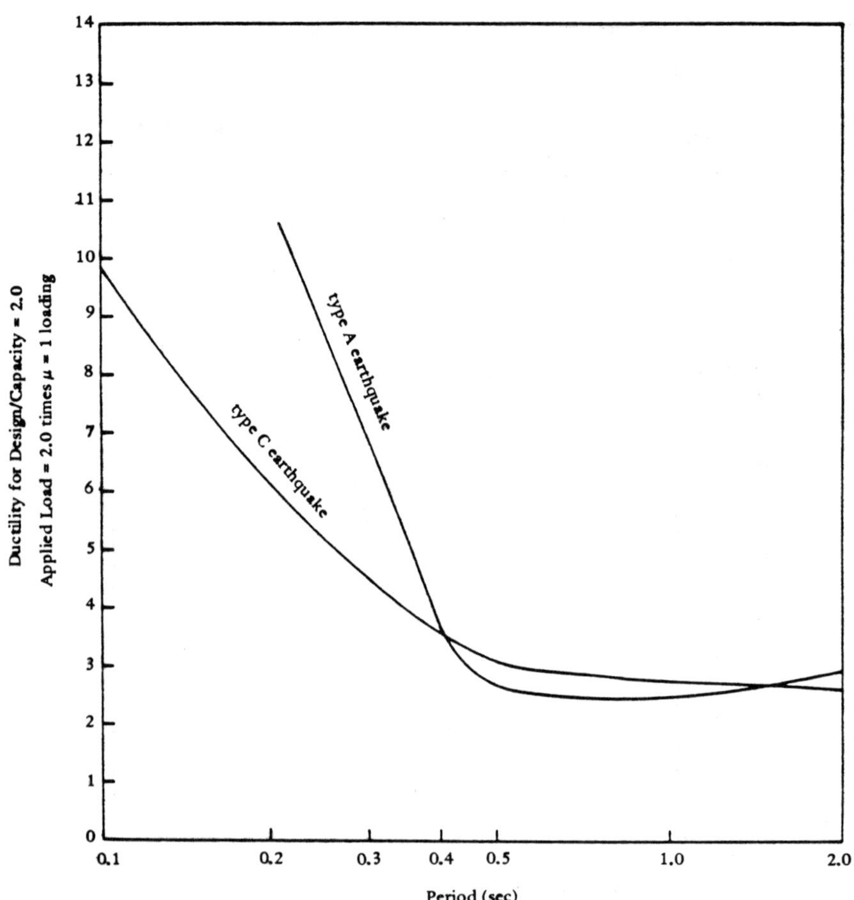

FIGURE 24. Ductility variation with period for load 2.0 times $\mu = 1$ load.

DAMAGE FUNCTION AND DAMAGE MATRIX

The damage function is an expedient approximate procedure for estimating damage. For existing construction, it is possible to estimate no-damage and collapse-damage levels by scaling response from a single modal analysis. For new construction, it is possible to predict damage by establishing the type of construction and the ratio of elastic to collapse response. These are relatively simple terms requiring minimal analysis. As shown earlier, modal analysis may be an inexact solution to nonlinear response to a random earthquake. Further, changes in natural period affect the damage function.

A further refinement is possible by creating a damage matrix in which the damage condition is given as a function of various design acceleration levels and various nominal[1] applied acceleration levels.

[1]The term nominal is used to indicate that the acceleration level is based on a spectrum and will contain higher values as a function of frequency.

Damage to structural frame members, shear walls and other elements associated with displacement are influenced by interstory drift. Elements tied to the floors, such as equipment or contents, are influenced by interstory drift. Elements tied to the floors, such as equipment or contents, are influenced by floor acceleration. Ferritto (1980) is a detailed study of work previously done in damage evaluation, as well as an extensive literature review. The use of a damage matrix allows computation of both drift and acceleration damage to individual building systems such as structural, architectural, mechanical or electrical.

DAMAGE EVALUATIONS

The detailed cost estimate was formulated to identify key elements of the structure to which dollar values could be assigned. Repair factors for damage were estimated, with key elements divided into driftor acceleration-sensitive components, and values of draft and acceleration were then related to damage for each element.

TABLE 3. Damage Ratios—Drift

Element	Cost ($)	Repair Miltiplier	Damage Ratios for Following Interstory Drift in/in								
			0.001	0.005	0.010	0.020	0.030	0.040	0.070	0.100	0.140
1a. Rigid Frames	117,500[a]	2.0	0	0.10	0.02	0.05	0.10	0.25	0.35	0.50	1.00
b. Braced Frames	[a]	2.0	0	0.03	0.14	0.22	0.40	0.85	1.0	1.0	1.0
c. Shear Walls	[a]	2.0	0	0.05	0.30	0.30	0.60	0.85	1.0	1.0	1.0
2a. Nonseismic Structural Frame	625,500	1.5	0	0.005	0.01	0.02	0.10	0.30	1.0	1.0	1.0
3. Masonry	417,600	2.0	0	0.10	0.20	0.50	1.0	1.0	1.0	1.0	1.0
4. Windows and Frames	120,600	1.5	0	0.30	0.80	1.0	1.0	1.0	1.0	1.0	1.0
5. Partitions, Architectural Elements	276,200	1.25	0	0.10	0.30	1.0	1.0	1.0	1.0	1.0	1.0
6. Floor	301,200	1.5	0	0.01	0.04	0.12	0.20	0.35	0.80	1.0	1.0
7. Foundation	412,100	1.5	0	0.01	0.04	0.10	0.25	0.30	0.50	1.0	1.0
8. Building Equipment and Plumbing	731,600	1.25	0	0.02	0.07	0.15	0.35	0.45	0.80	1.0	1.0
9. Contents	500,000	1.0	0	0.02	0.07	0.15	0.35	0.45	0.80	1.0	1.0

[a]Varies with design.

Tables 3 and 4 give damage ratios for each key element for the steel framing concepts as a function of drift and acceleration experienced by the structure. A similar table was developed for the concrete-framing concepts. For example, if a masonry element were to experience 0.01 drift, damage could equal 20%. Note that a value is assigned to contents, and that the utilization of repair multipliers, to be applied to original costs for the repair of damage from seismic shaking, can result in costs exceeding the total cost of the structure. This is reasonable, since demolition and removal costs would be required for major repairs.

To evaluate the damage expected to occur to the structure, each of the six design loading levels for each of the five lateral resistance design concepts was analyzed for a series of applied seismic load levels. Non-linear finite element techniques were employed. The program DRAINTABS was

TABLE 4. Damage Ratios—Acceleration

Element	Cost ($)	Repair Miltiplier	Damage Ratios for Following Floor Acceleration				
			0.08g	0.18g	0.50g	1.2g	1.4g
1. Floor and Roof	301,200	1.5	0.10	0.02	0.10	0.50	1.0
2. Ceilings and Lights	288,500	1.25	0.01	0.10	0.60	0.95	1.0
3. Building Equipment and Plumbing	731,600	1.25	0.01	0.10	0.45	0.60	1.0
4. Elevators	57,000	1.5	0.01	0.10	0.50	0.70	1.0
5. Foundations (Slab on Grade, Sitework)	412,100	1.5	0.01	0.02	0.10	0.50	1.0
6. Contents	500,000	1.05	0.05	0.20	0.60	0.90	1.0

TABLE 5. Damage Ratio—Moment Frame

Applied Load (g)	Damage Ratios for Following Design Acceleration						
	0.00g	0.10g	0.15g	0.20g	0.25g	0.30g	0.35g
0.00	0.00	0.00	0.00	0.00	0.00	0.00	0.00
0.10	0.10	0.08	0.08	0.09	0.11	0.11	0.11
0.20	0.17	0.11	0.13	0.15	0.17	0.19	0.21
0.30	0.24	0.16	0.20	0.23	0.22	0.22	0.24
0.40	0.37	0.24	0.26	0.27	0.27	0.26	0.29
0.50	0.56	0.37	0.29	0.32	0.29	0.29	0.31
0.60	0.82	0.55	0.35	0.36	0.33	0.33	0.32
0.70	1.03	0.68	0.40	0.39	0.35	0.34	0.35
0.80	1.13	0.75	0.43	0.43	0.38	0.38	0.38
0.90	1.25	0.83	0.46	0.45	0.42	0.40	0.39
1.00	1.50	0.89	0.50	0.48	0.46	0.42	0.41

TABLE 6. Damage Ratio—Braced Frame

Applied Load (g)	Damage Ratios for Following Design Acceleration						
	0.00g	0.10g	0.15g	0.20g	0.25g	0.30g	0.35g
0.00	0.00	0.00	0.00	0.00	0.00	0.00	0.00
0.10	0.10	0.08	0.08	0.07	0.07	0.07	0.06
0.20	0.17	0.14	0.14	0.14	0.12	0.12	0.11
0.30	0.24	0.23	0.21	0.20	0.17	0.16	0.16
0.40	0.37	0.29	0.28	0.27	0.24	0.23	0.23
0.50	0.56	0.32	0.31	0.30	0.27	0.26	0.26
0.60	0.82	0.34	0.34	0.32	0.29	0.29	0.28
0.70	1.03	0.37	0.37	0.35	0.31	0.34	0.35
0.80	1.13	0.42	0.41	0.41	0.36	0.37	0.39
0.90	1.25	0.45	0.43	0.43	0.47	0.45	0.52
1.00	1.50	0.49	0.48	0.47	0.56	0.61	0.71

TABLE 7. Damage Ratio—Shear Wall

Applied Load (g)	Damage Ratios for Following Design Acceleration						
	0.00g	0.10g	0.15g	0.20g	0.25g	0.30g	0.35g
0.00	0.00	0.00	0.00	0.00	0.00	0.00	0.00
0.10	0.10	0.04	0.04	0.04	0.04	0.02	0.02
0.20	0.17	0.08	0.07	0.07	0.07	0.07	0.07
0.30	0.24	0.11	0.11	0.10	0.09	0.07	0.10
0.40	0.37	0.15	0.14	0.13	0.12	0.11	0.11
0.50	0.56	0.20	0.18	0.16	0.14	0.11	0.11
0.60	0.82	0.24	0.23	0.18	0.17	0.14	0.14
0.70	1.03	0.26	0.26	0.20	0.17	0.17	0.17
0.80	1.13	0.29	0.28	0.24	0.21	0.12	0.20
0.90	1.25	1.20	0.31	0.25	0.23	0.20	0.19
1.00	1.50	1.50	0.32	0.28	0.25	0.23	0.22

TABLE 8. Damage Ratios Moment Frame

Applied Load (g)	Damage Ratio at the Following Design Acceleration (g)—						
	0.00	0.10	0.15	0.20	0.25	0.30	0.35
0.00	0.00	0.00	0.00	0.00	0.00	0.00	0.00
0.10	0.05	0.01	0.01	0.03	0.04	0.05	0.06
0.20	0.06	0.05	0.04	0.04	0.06	0.07	0.08
0.30	0.18	0.11	0.11	0.10	0.10	0.10	0.09
0.40	0.22	0.14	0.14	0.12	0.12	0.12	0.12
0.50	0.25	0.18	0.18	0.16	0.15	0.14	0.14
0.60	0.35	0.24	0.12	0.19	0.18	0.17	0.17
0.70	0.46	0.28	0.26	0.25	0.21	0.21	0.20
0.80	0.60	0.33	0.30	0.26	0.24	0.23	0.23
0.90	0.82	0.39	0.36	0.30	0.27	0.27	0.26
1.00	1.05	0.44	0.42	0.34	0.30	0.29	0.28

TABLE 9. Damage Ratios From Shear Wall

Applied Load (g)	Damage Ratio at the Following Design Acceleration (g)—						
	0.00	0.10	0.15	0.20	0.25	0.30	0.35
0.00	0.00	0.00	0.00	0.00	0.00	0.00	0.00
0.10	0.05	0.01	0.04	0.03	0.03	0.03	0.03
0.20	0.06	0.04	0.06	0.05	0.10	0.10	0.09
0.30	0.18	0.07	0.08	0.10	0.10	0.09	0.09
0.40	0.22	0.11	0.13	0.14	0.15	0.14	0.13
0.50	0.25	0.15	0.15	0.17	0.16	0.15	0.15
0.60	0.35	0.21	0.20	0.18	0.20	0.19	0.19
0.70	0.46	0.26	0.25	0.22	0.20	0.23	0.23
0.80	0.60	0.31	0.29	0.27	0.24	0.23	0.28
0.90	0.82	0.35	0.34	0.31	0.28	0.27	0.24
1.00	1.05	0.44	0.38	0.35	0.32	0.30	0.29

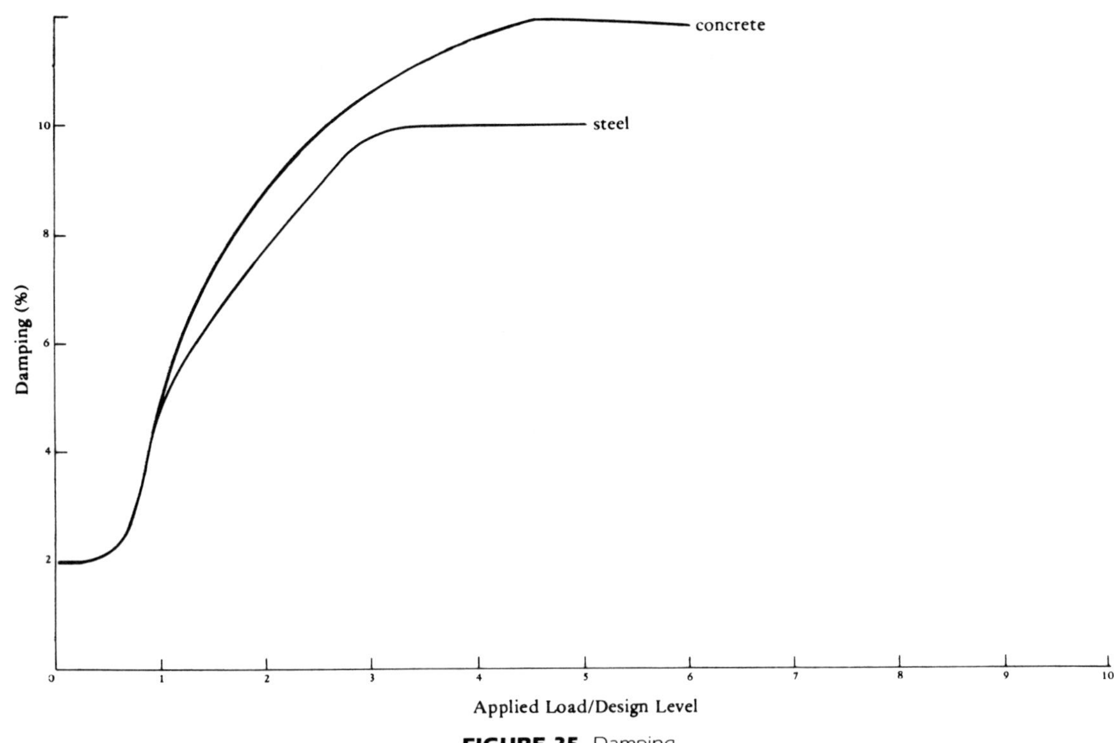

FIGURE 25. Damping.

used to conduct the analysis. Figure 25 gives the damping used for the analysis, based on engineering practice; damping increased with the ratio of applied load to design level. Drift and floor acceleration time-history responses were computed in the analysis. Effective response levels representing sustained loading were selected at 65% of peak values and used in the damage prediction. The rate of 65% has been used in past studies to equate effective peak ground acceleration to sustained levels of loading.

Tables 3 and 4 used in conjunction with the drift and acceleration values computed from the nonlinear analysis, resulted in Tables 5–9 which present damage matrices giving damage to the structure, noted as a function of design level, and applied loading summing up all the elements shown in Tables 3 and 4. Included in the damage matrix is damage to the structure and contents using the noted repair factors. Thus, Tables 5–9 show structural, nonstructural, mechanical and electrical damages, as well as damage to contents, and the cost of restoring the building to operation.

Steel Moment Frame

The response of the moment frame structure is in the constant velocity region of the spectra for all six design ranges. Note that as the structure is stiffened, displacement decreases while acceleration increases. Damage is contingent upon both displacement and acceleration. Note also that for a given applied load level, each of the six design cases is at a different damping level, with the weakest structure being most heavily damped. In the low applied loading level, the strong structures are lightly damped, responding elastically with floor accelerations higher than those of weaker designs at the same load level. The weaker structures are more heavily damped, responding inelastically with lower floor accelerations than stiffer and stronger structures at the same load level. In this range, stiffer structures receive greater damage, a condition which exists to about 0.5g for the range of structures studied. Over 0.5g the stiffer structures experience lower levels of damage, as might be expected. The use of a single time-history event with its unique frequency content results in minor response variations. Any single time history has unique frequency gaps and high points. Since the period of the structure changes with strengthening (stiffening), secondary interactions occur between the frequency high points and structure periods so that responses at a particular design level may be slightly reduced or amplified over the response of an ideal time history without gaps and high points. Further, the six design cases are not exact multiples, but depend on the human selection of available structural shapes. These factors induce very minor dispersion in the results. A clear conclusion, however, is that stiffening in the low applied acceleration region does not

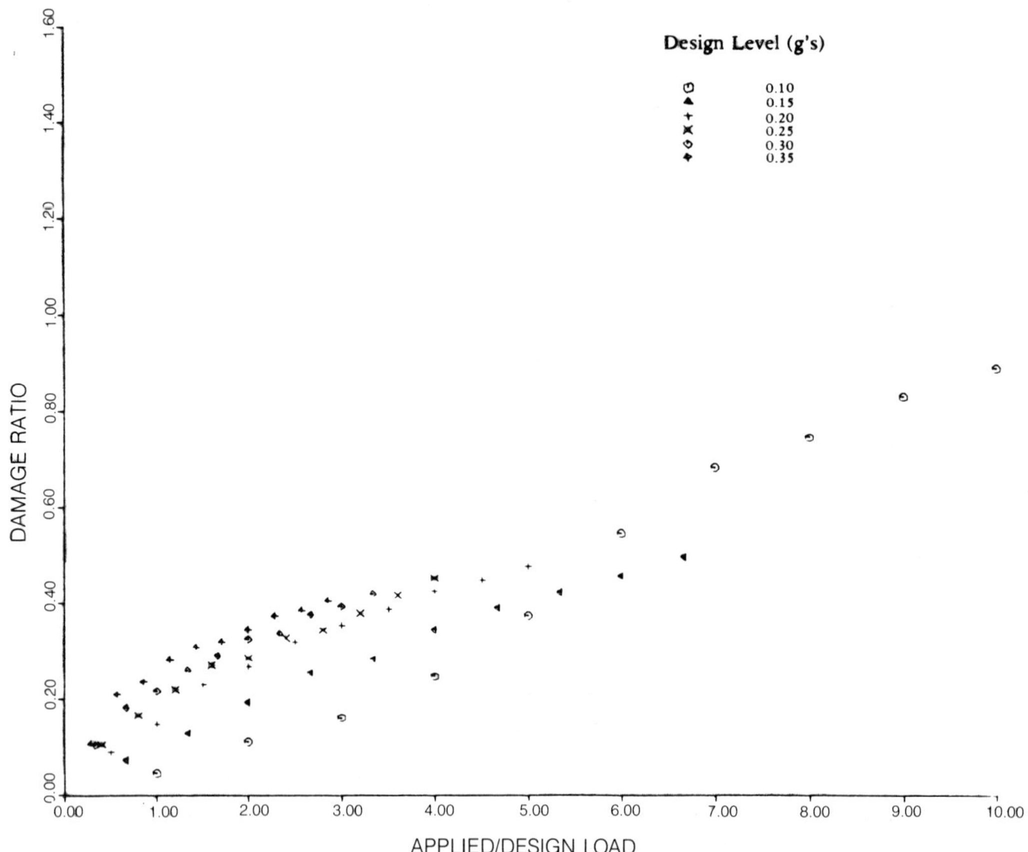

FIGURE 26. Damage ratio as a function of design and load, moment frame.

reduce total damage. Figure 26 shows a plot of damage ratio as a function of applied load to design level. The data illustrate the effects of variation, in period of the structure, on the response. The damage ratio is a complex function of period, damping, range of nonlinear behavior, and the mix of total damage caused by drift and acceleration.

Steel Braced Frame

The response of the braced-frame structure is in the constant acceleration region of the spectra for all six design ranges. The structure in its basic configuration with bracing is much stiffer than the corresponding moment frame, pushing the response from the constant velocity region to the acceleration region. The resulting floor accelerations, produced by the applied loading, are higher than those of the moment frame, while story drifts are reduced. In the medium- and low-level applied loading range, damage decreases with stiffening; however, at high load levels the acceleration dominates, resulting in higher damage with stiffening. Again, note that damping varies with the ratio of applied to design load level. Note also that three of the designs utilized two-bay bracing, and three of the designs utilized three-bay bracing.

Steel Frame Shear Wall

The shear wall/frame structure was the stiffest of the three steel concepts studied. Damage was generally least with this structure; however, collapse did occur for the 0.1g design at the 0.9g applied load. The brittle nature and sudden shear failure are shown by the 0.29 damage ratio at 0.8g loading, and the collapse at 0.9g loading (Table 7). In general, because of the low period of the structure, floor acceleration resulting from amplification of base motion was least, and in high applied acceleration load levels, attenuation occurred.

Concrete Moment Frame

The response of this structure is similar to the steel moment frame. In the low applied load range, stiffer structures receive greater damage; this condition exists to about 0.25g for the range of structures studied. Over 0.25g, the stiffer structures exhibit lower damage, as might be expected.

Concrete Shear Wall

The shear wall/frame structure was stiffer than the concrete moment frame. Less damage was incurred, generally, with this structure. In general, because of the low period of the structure, floor acceleration resulted from the amplification of base motion. In high applied acceleration load levels, attenuation occurred.

ECONOMIC ANALYSIS

There are numerous references which give the fundamentals of engineering economics such as the Naval Facilities Engineering Command Handbook (1975). This reference specifies procedures for the economic analysis of facilities. This analysis must take into consideration the fact that earthquake strengthening is expressed as a current cost increase to protect against future dollar losses. The real world is complicated by cost increases through inflation; to repair or replace a damaged building will cost more in the future than today. Work in the previous sections expresses the costs of seismic design levels and damage as a percentage of (new) building value in order to maintain a common reference. That premise recognizes increased value of the building and increased costs of repair as damage occurs. In an economic sense, this may be expressed as letting the discount rate (the value of return on investment) equal the inflation rate. The selecting of an interest rate establishes the value of money in time. For a corporation it reflects the competing forces at work such as the procurement of additional machines to increase productivity and profits. A dollar spent on seismic resistance is a dollar removed from other alternative productivity-enhancing investments or insurance; or perhaps simply profits. The selection of an interest rate is an expression of the worth or value of the present possession of money rather than future money. Let us for this example select a value of 10% as the return we seek on our investment.

When the present worth of the annual expected damage is considered, using a discount rate of 10%, the present worth estimate of damage would be effectively reduced by a factor of about five. In other words, the earthquake could occur at any point during the life of the structure; the best estimate considers an equivalent series of annual expected losses. The assumed life of the structure is 50 years. The present worth of this accumulated loss series can be computed, and its value is about one-fifth the total expected loss.

Note that the discount rate specified for use is actually a differential rate of 10% over the rate of inflation. One could use the differential rate without taking inflation into account, or one could consider the rate of inflation when projecting increased repair costs, and then discount that cost using a rate of 10% plus inflation. The results for modest inflation rates are approximately the same. The differential cost approach was used in this study.

Included in the economic analysis is a value for injury and loss of life. There may be considerable difficulty in assigning value to a life; however, an economic analysis must represent all elements of the problem in common terms in order to ensure completeness (see, e.g., Ferritto 1981). The value assigned to the loss of life is $300,000 in the analysis. As mentioned in later sections, results were not sensitive to the value selected. For moderate damage levels, neither loss of life nor injury was found to be high, based on a review of previous earthquakes with similar construction.

COMBINING SITE SEISMIC PROBABILITY WITH ECONOMICS

An automated procedure can be used to conduct a site seismic activity study as discussed above. As stated above, the outcome from the seismicity study is a measure of the site activity and could be expressed as say probability of acceleration at a site. For this study five sites—Bremerton, WA; Memphis, TN; San Diego, Port Hueneme, and Long Beach, CA—were examined in light of the cost and damage data presented earlier, and in light of the probability of site acceleration distributions.

Figure 27 shows a relationship of acceleration of any probability to the 80% probability level of not being exceeded in 50-year exposure. This figure was constructed from an average of 5 sites and is only approximate. Figure 28 shows the relationship of probability to return time for any exposure period. Thus, one can use the 80%-probability-of-not-being-exceeded acceleration level in 50-year exposure as an approximate means to characterize a site and define its distribution.

Figure 29 gives typical results of the increased cost of seismic design levels on new construction for the steel moment frame structure. The moment frame is the most expensive lateral resistance system. Costs for the braced frame and shear wall systems do not vary as significantly with design level.

Figure 30 indicates the least-cost design acceleration in terms of the 225-year return-time acceleration (80%-probability-of-not-being-exceeded in 50 years) based on the probability distribution data from the five sites.

Seismic strengthening costs are assumed to be dependent on the type of lateral resistance system utilized; damage is correlated both to drift and acceleration. Stiffening alone limits drift damage but increases acceleration damage. Damage to a structure is complex, and is influenced by damping level, degree of inelastic behavior, acceleration and drift levels, and the spectral region of response. The

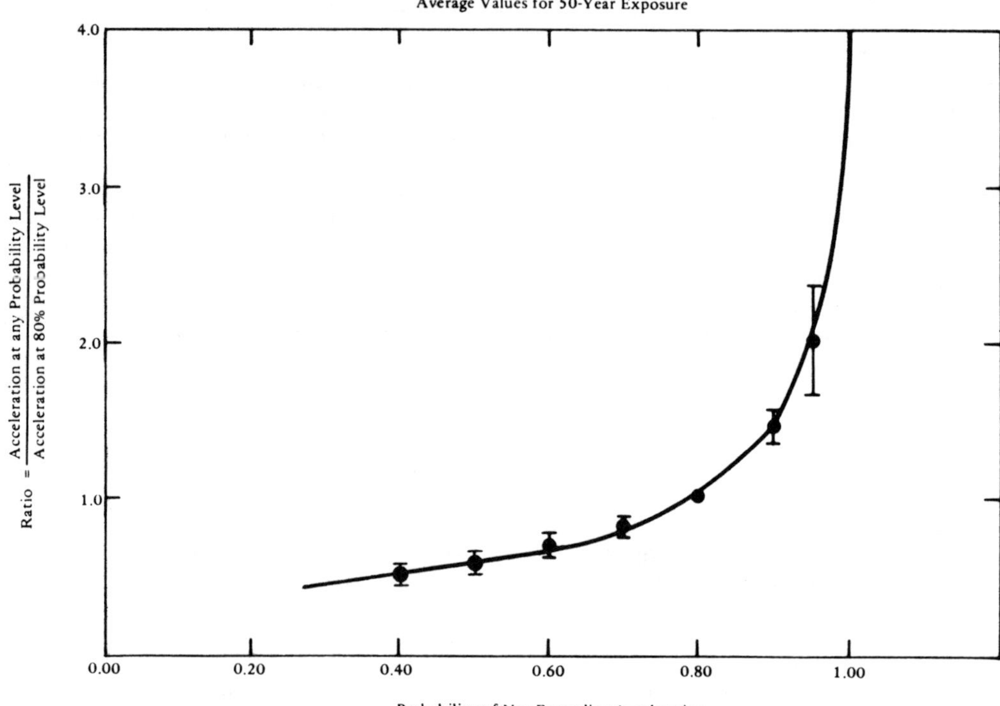

FIGURE 27. Relationship of acceleration at any probability to 80% probability.

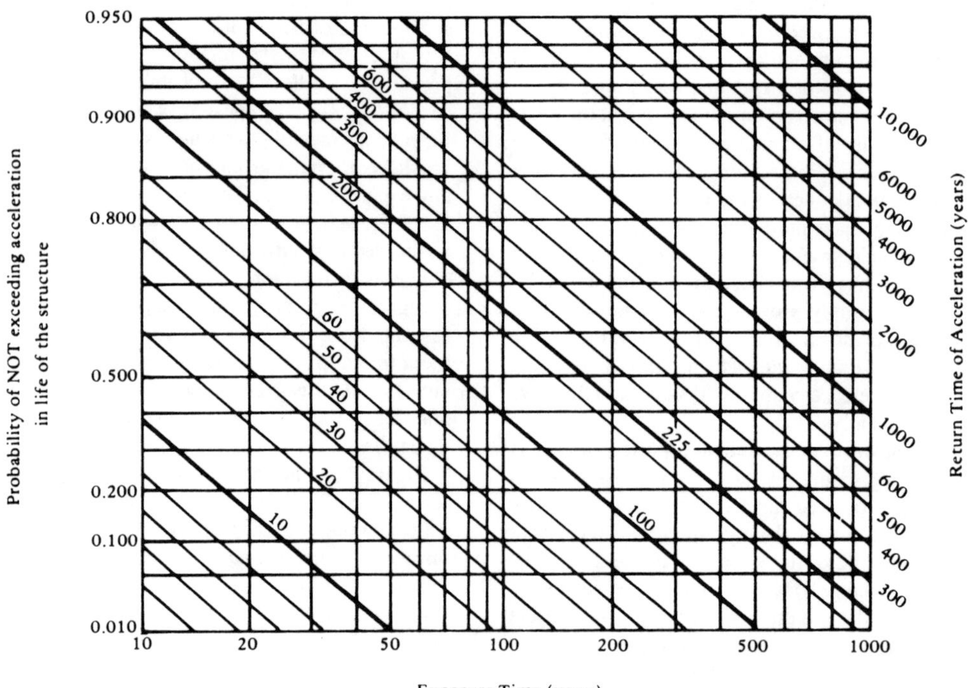

FIGURE 28. Relationship of probability to return time.

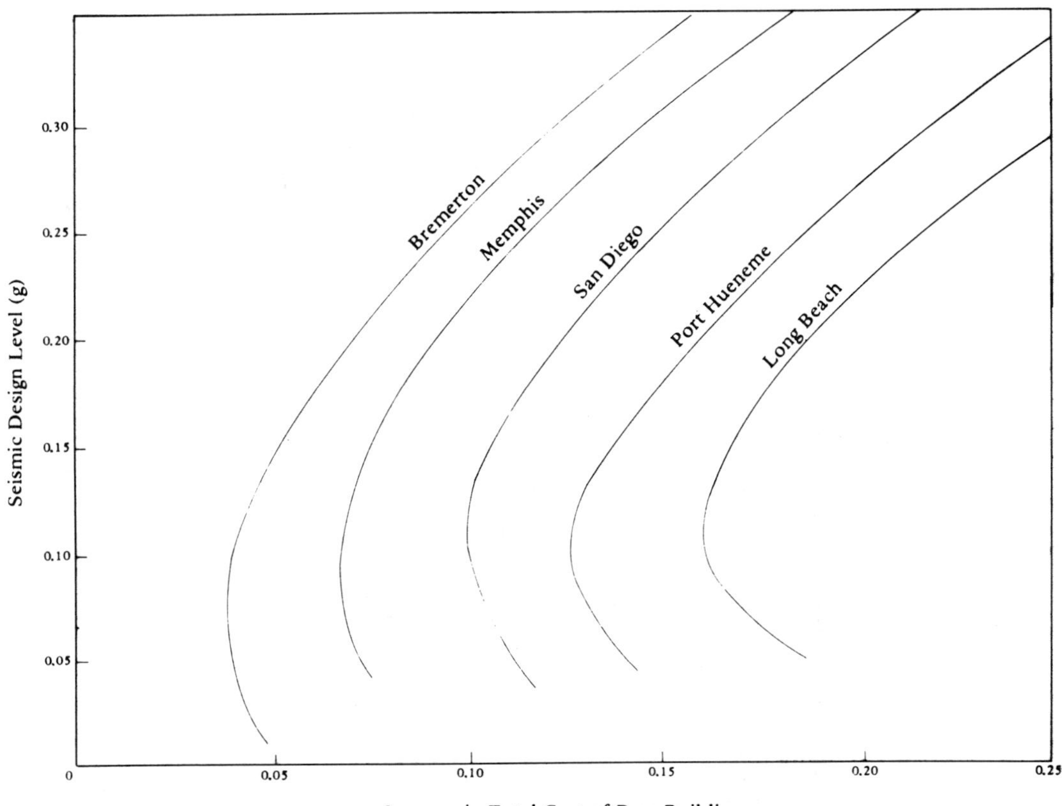

FIGURE 29. Increase in cost in various locations for moment frame structure.

most cost-effective design acceleration is a function of construction type and site seismic exposure.

Since acceleration produces a significant amount of damage, special care should be taken to design ceilings and lights that will withstand it. Acceleration causes equipment to overturn—a factor accounting for most mechanical and electrical losses. Since stiffening produces increased acceleration, consideration should be given to the development and utilization of isolation techniques. A value of design acceleration with a 60- to 100-year return time appears reasonable. A 100-year return-time acceleration would represent a design level of about 70% of the 225-year return-time level, and a 60-year return-time acceleration would be about 50% of the 225-year return-time level. Figure 31 shows a histogram of the probability distribution of acceleration, eration, normalized on the basis that the 225-year return-time acceleration is 0.25g. These data are based on a composite of a number of sites studied. Also shown in the figure is the damage ratio for a steel-moment-frame structure. An examination of the computed results of the probabilistic expected damage analysis over the life of the structure (probability times damage in increments) shows that most damage comes from the exposure to low-level acceleration. Structures which respond elastically in this range, having been designed for high acceleration, exhibit high floor accelerations causing much of the damage. As noted, strengthening produces little or no reduction in damage at low acceleration levels, which are most probable since floor acceleration increases with the resulting stiffening of the structure.

DISCUSSION

Before an economic analysis is employed consideration must be given to the function of the structure. Some structures may be grouped into these categories:

Critical Structures

Those which contain substances which if released would cause a catastrophe, such as nuclear power plants, LNG facilities, etc.

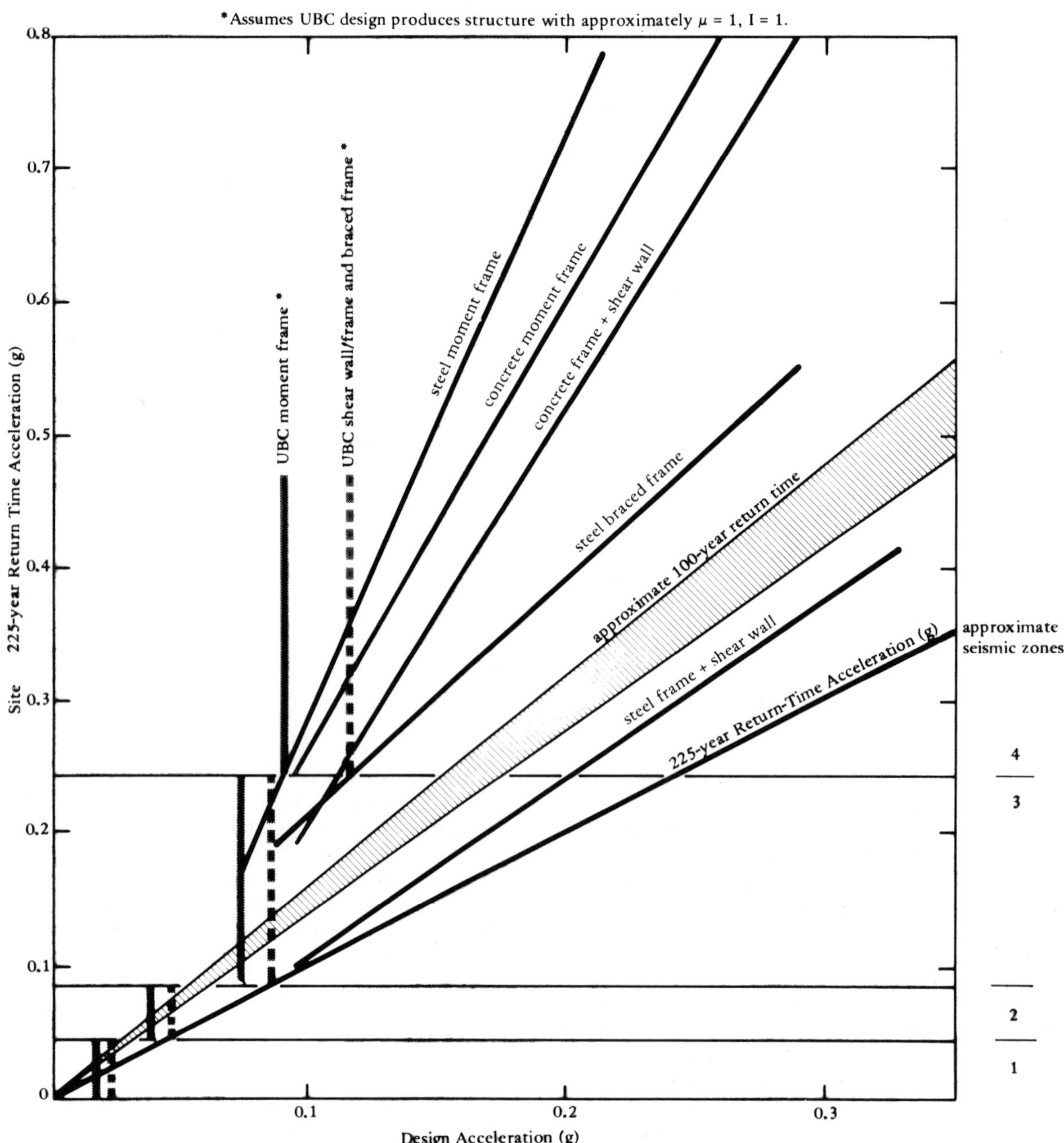

FIGURE 30. Least cost design acceleration for important structures (ductility = 1).

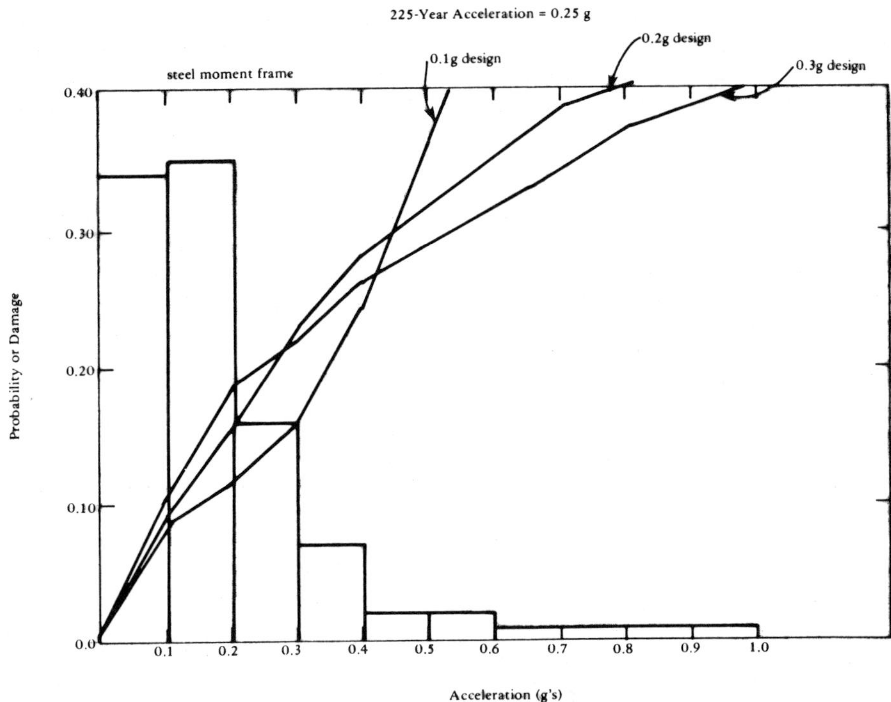

FIGURE 31. Histogram of probability distribution of acceleration and distribution of damage ratio with acceleration.

Essential

Structures in this category must remain functional after exposure to the design earthquake. The following structures qualify as essential: (a) power plants and electrical distribution systems; (b) water tanks; (c) medical facilities; and (d) fire stations.

For the above classes of structure use of economics could not easily be applied because the subjective value of life safety could not be easily expressed in terms of economics. To insure survival of these special structures a more deterministic worst case design is usually applied. Survival of critical and essential structures should be expressed in terms of operational requirements. These operational requirements then can be converted into upper limit bounds on allowable structural response of members. It is essential to fully study the non-structural elements and the mechanical and electrical systems to insure adequate restraint.

EXISTING CONSTRUCTION

Simple structural performance will not guarantee operability. The seismic upgrade of existing construction differs from new seismic design in that the presence of an existing structure forces the upgrade to be incremental in nature, applying units of additional strengthening to improve an existing structure rather than a more continuous set of alternatives in new construction design such as column size or shear wall thickness. The cost of seismic strengthening of existing construction must be greater than those of new construction; and probably since the strengthening takes place as an add-on repair or alteration, its effectiveness will be less than that of new construction. Thus, as a general conclusion, the level to upgrade existing construction cannot be greater than that of new construction.

There are several important considerations in determining upgrade levels for existing construction:

- Level of initial seismic design
- Occupancy level and building importance
- Remaining useful life

It is clear that, for short-lived facilities, those of low occupancy or low importance, and those where the expected damage is low, seismic upgrading should not be undertaken until higher priority structures are upgraded. The decision to upgrade existing facilities is complicated by the variations in types of construction, initial seismic design level, condition of the structure, and site seismic hazard.

Existing construction can be analyzed in the same manner as new construction. The first option is the do-nothing option: this produces a zero expenditure for strengthening and a future damage loss. Several options for upgrading to different seismic levels of resistance can be evaluated. Each carries a certain initial cost and each will reduce the future damage.

The following example may clarify the procedure. The present value of a replacement cost of a hypothetical structure is estimated to be $11,000,000. Based on a site seismicity study, the probability of site acceleration is determined. Results of a structural analysis indicate that 0.05g nominal spectral acceleration would not cause any damage, and a value of 0.20g would cause 100% damage. For simplicity assume a linear damage function between 0 and 100% damage.

Consider the building as existing construction with a 25-year remaining life with a present worth of $11,000,000. Over the remaining life of the building, a loss of $6,960,000 is expected to occur, including loss of life and injuries. This is based on the site acceleration probability in 25 years and the damage function developed. Considering that this might occur at any time during the 25-year exposure period, the present worth of the expected loss is $2,530,000.

Present building:

Do nothing	
Base cost	$11,000,000.
Present worth expected damage	2,530,000.
	$13,530,000.

Upgrading the building to 0.06g:

Upgrade 1: 0.06g, 25-year exposure	
Base cost	$11,000,000
Seismic cost	550,000
Present worth expected damage	2,462,900
	$14,012,900

Upgrading the building to 0.11g:

Upgrade 2: 0.11g, 25-year exposure	
Base cost	$11,000,000
Seismic cost	1,235,000
Present worth expected damage	549,900
	$12,785,200

Upgrading the building to 0.15g:

Upgrade 3: 0.15g, 25-year exposure	
Base cost	$11,000,000
Seismic cost	1,659,900
Present worth expected damage	294,460
	$12,954,360

Results for this hypothetical example indicate that Upgrade 2 produces the least total cost. The total expenditure difference in upgrading cost from Upgrade 2 to Upgrade 3 is about 30%. Damage is reduced $255,440, but cost is increased $424,600. Obviously, the benefits minus cost is negative.

The results of the analysis show that it is justifiable to upgrade the structure to a level of 0.11g resistance and that an expenditure of about $1,200,000 would be about optimal. This procedure is especially useful in allocating limited resources and avoids expenditures with limited payback.

REFERENCES

American Concrete Institute, ACI 318-77: Building code requirements for reinforced concrete, Detroit, Mich.

American Institute of Steel Construction, Inc., *Manual of Steel Construction, Eighth Edition,* Chicago, Ill. (1980).

Applied Technology Council, *NBS Special Publication 510,* "Tentative Provisions for the Development of Seismic Regulations for Buildings," (ATC 3-06) (June 1978).

Bertero, V. V., "Identification of Research Needs for Improving Aseismic Design of Building Structures," University of California, Earthquake Engineering Research Center, *EERC 7527,* Berkeley, Calif. (Sept. 1975).

Biggs, J. M., R. J. Hansen, and M. J. Holley, "On Methods of Structural Analysis and Design for Earthquakes," paper presented at Newmark Symposium on Structural and Geotechnical Mechanics, University of Illinois, Urbana, Ill. (Oct 2–3, 1975).

Culver, L., et al., "Natural Hazards Evaluation of Existing Buildings," National Bureau of Standards, *NBS BSS 61.* Washington, D.C. (Jan. 1975).

Department of Commerce, National Bureau of Standards. "Procedures and Criteria for Earthquake Resistant Design (Part II)," in *Building Practices for Disaster Mitigation, BSS 46.* Washington, D.C. (Feb 1973).

Der Kiureghian, A., and H. S. Ang, "A Line Source Model for Seismic Risk Analysis," University of Illinois at Urbana-Champaign, Urbana, *Structural Research Series 419,* p. 134 (1975).

Ferritto, J. M., "An Economic Analysis of Seismic Upgrading for Existing Buildings," *Civil Engineering Laboratory Technical Memorandum 51-80-17.* Port Hueneme, Calif. (Jul. 1980).

Ferritto, J. M., "Procedure for Conducting Site Seismicity Studies at Naval Shore Facilities and an Economic Analysis of Risk," *Civil Engineering Laboratory Technical Report R-885,* Port Hueneme, Calif. (Feb. 1981).

Ferritto, J. M., *Civil Engineering Laboratory Technical Note N-1640,* "An Economic Analysis of Earthquake Design Levels," Port Hueneme, Calif. (Jul, 1982).

Knopoff, L., and Y. Kagan, "Analysis of the Theory of Extremes as Applied to Earthquake Problems," *Journal of Geophysical Research, Vol. 82,* no. 36 (1977).

Murakami, M., and J. Penzien, "Nonlinear Response Spectra for Probabilistic Seismic Design and Damage Assessment of Reinforced Concrete Structures," University of California, Earthquake Engineering Research Center, EERC 7538. Berkeley, Calif. (Nov. 1975).

Naval Facilities Engineering Command, NAVFAC P422, *Command Handbook.* Alexandria, Va. (revised Oct. 1975).

Newmark, N. M., Chapter 16, "Current Trends in the Seismic Analysis and Design of High-Rise Structures in Earthquake Engineering," R. L. Wiegel, editor. Englewood Cliffs, N.J., Prentice-Hall (1970).

Newmark, N. M., and R. Riddell, "A Statistical Study of Inelastic Response Spectra," in *Proceedings of the Second U.S. National Conference on Earthquake Engineering,* Stanford University, Stanford, Calif. (Aug. 1979).

Patwardhan, A. S., R. B. Kulkarni, and D. Tocher, "A Semi-Markov Model for Characterizing Recurrence of Great Earthquakes," *Seismological Society of America Bulletin, Vol. 70,* no. 1, pp 323–347 (1980).

Sykes, L. R., and R. C. Quittmeyer, "Repeat Times of Great Earthquakes Along Simple Plate Boundaries," *Earthquake Prediction—An International Review,* D. W. Simpson and P. G. Richards, eds., American Geophysical Union, Maurice Ewing, Series 4, pp 217–247 (1980).

Woodward-Clyde Consultants, "Offshore Alaska Seismic Exposure Study," Prepared for Alaska Subarctic Operators Committee (ASOC) (6 volumes) (March 1978).

SECTION THREE

Stability

CHAPTER 18

Buckling of Thin-Walled Beams Under Water Loading

LUIS A. GODOY* AND CARLOS A. PRATO*

INTRODUCTION

Buckling phenomena are usually active constraints in design of thin-walled beams for all types of loading conditions and devices. Local and global instability modes are to be accounted for separately, and the possibilities that their interaction become critical must be anticipated to achieve a safe design. In this pursuit, theoretical, experimental, and numerical work is required to fully understand the mechanisms involved. While these general remarks are valid for loading states originated in self weight and accumulation of solid type non-structural weight, it is even more pertinent when the loading device or mass deforms as a consequence of structural deformations. This effect may lead to additional deformations setting stage for lower buckling loads.

Such is the case of thin-walled open channel sections subjected to water load with a free fluid surface. Prefabricated prestressed concrete beams with thin-walled channel sections used for structural roof closures may be subjected to accidental water accumulation due to congestion of the drainage system. Under this assumption, analytical and experimental results [1–3] have shown that buckling loads can be a limiting factor in the load carrying capacity of the beam due to the peculiarities arising from the water loading. Typical commercially available sections are often designed to have an ultimate strength sufficiently high to account for design dead and live load, as well as buckling modes. It must be noticed, however, that buckling loads under water loading can be significantly lower than those associated with solid type live loads and this may turn out to be lower than that associated with a strength failure.

*Structures Department, FCEFyN, National University of Córdoba, Casilla de Correo 916, Córdoba 5000, Argentina

In the following sections, a general analysis procedure for this class of water induced static instability is presented together with some numerical examples and experimental results. This may orient the designer evaluating actual buckling loads when part of the live load arises from water accumulation on the beam cross section.

FINITE ELEMENT FORMULATION

A finite element formulation, suitable for the numerical evaluation of buckling loads in thin-walled beams, including the effects associated to gravity and water loading, will be outlined in this section. Details of particular elements which may be used are omitted because the reader interested in performing this kind of analysis may find it easier to modify existing computer programmes.

Bifurcation Analysis

The determination of buckling loads of elastic structures leads to an eigenvalue equation of the form [4]

$$\underset{\sim}{K}\,\underset{\sim}{U} = \lambda\,\underset{\sim}{K}_\sigma\,\underset{\sim}{U} \tag{1}$$

where $\underset{\sim}{K}$ is the stiffness matrix of the structure, obtained from a finite element model; $\underset{\sim}{U}$ is the vector of unknown incremental small displacements; $\underset{\sim}{K}_\sigma$ is the load-geometry matrix, dependent on the stresses in the fundamental state and on the location of the point of application of the external load; and λ is the buckling load factor.

The lowest eigenvalue λ satisfying Equation (1) is associated to the lowest buckling load of the structure, and the displacement pattern $\underset{\sim}{U}$ related to λ is the buckling mode.

The load geometry matrix, $\underset{\sim}{K}_\sigma$, should be carefully con-

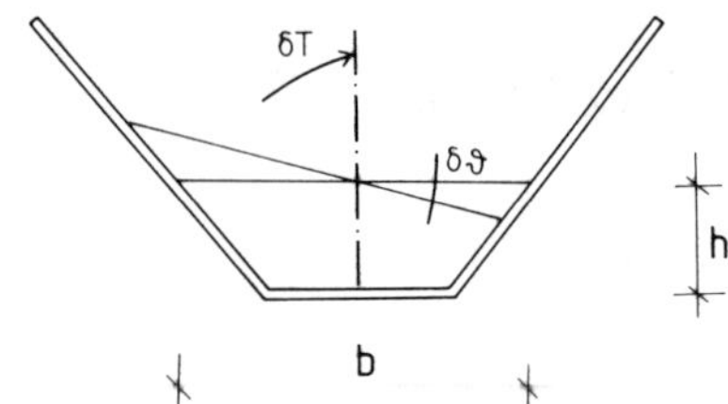

FIGURE 1. Torque due to movement of liquid.

structed in this case since two different types of buckling may occur in a thin-walled beam of open cross section: local buckling of the plates and lateral-torsional buckling of the cross section. Terms which should be included in $\underset{\sim}{K}_\sigma$ for the present problem have been discussed by the authors in Reference [1].

Water Loading

In the classical linear eigenvalue problem defined by Equation (1), the loads are usually considered as unaffected by the deformation of the structure in the buckling mode. Although this is a good approximation for local buckling modes of the plates (even for water load), torsional buckling modes of the structure are stimulated by the movement of the liquid which accompanies a rotation of the cross section. Thus, an additional torque is induced by the liquid loading and should be included in the formulation.

In the prebuckling state, the resultant force of the weight of the liquid acts through the shear center if the cross section has one axis of symmetry. As the structure deforms in a lateral-torsional mode, the resultant force acting in the deformed configuration no longer passes through the shear center due to two independent and cumulative effects:

1. The external gravitational force acting on the beam can be assumed acting in the vertical direction and passing through the center of gravity of the masses (both structural and additional external masses). When these masses do not change their relative position or shape, as occurs when the loading is of "solid" type, the torsional deformations will generate an eccentricity of the resultant weight with respect to the shear center. Terms arising from this effect in the equilibrium equation are already included in Equation (1) through appropriate definition of the load-geometry matrix.
2. The change in shape of the water mass will introduce an additional eccentricity in the deformed configuration. The deformed mass will exhibit a shift in the horizontal position of its center of gravity with respect to that of its undeformed or "solid" c.g. that will cause an additional torque δT.

If the thin-walled beam undergoes a rotation $\delta\theta$ about its longitudinal axis x, as indicated in Figure 1, the additional torque per unit length of beam can be approximated by [1]

$$\delta T = \gamma \frac{b^3}{12} d\theta \tag{2}$$

in which γ is the specific weight of the liquid and b is the free water width parameter. In calculating δT through Equation (2), deformations of the cross section in the buckling mode have been rejected. This does not imply, however, that elastic response to δT is restricted to the global beam theory assumptions, as it has been indicated that Equation (1) is formulated with the more general finite element method.

Notice that for small rotations $\delta\theta$, the torque δT does not depend on the inclination of the lateral walls of the beam.

A linear relation between δT and $\delta\theta$ has been established by Equation (2). The factor $\gamma b^3/12$ could thus be interpreted as an effective decrease in the torsional stiffness of the complete cross section.

Condensation of Eigenvalue Problem for Water Loading

Consideration in detail of the effects associated to the movement of the water in the finite element bifurcation analysis of Equation (1) is a difficult task, and a simplified way based on Equation (2) has been proposed in Reference [1]. The effects due to water loading are incorporated in a global sense, for which the eigenvalue problem (1) is considered.

Let $\underset{\sim}{U}_1$ be the lateral-torsional mode for solid weight determined by solution of Equation (1), and λ_1 the associated load factor. Equation (1) may be condensed to the form

$$(k - k_f) - \Lambda_1 k_\sigma = 0 \tag{3}$$

in which k and k_σ are the condensed forms of the stiffness $\underset{\sim}{K}$ and load-geometry $\underset{\sim}{K}_\sigma$ matrices with respect to $\underset{\sim}{U}_1$, and may be calculated as

$$k = \underset{\sim}{U}_1^T \, K \, \underset{\sim}{U}_1 \tag{4}$$

$$k_\sigma = \underset{\sim}{U}_1^T \, K_\sigma \, \underset{\sim}{U}_1 \tag{5}$$

The influence of the water load as a decrease in torsional

stiffness has been incorporated in k_f, given by

$$k_f = \frac{\gamma}{12} \int_o^\ell b^3 \, \bar{\theta}_1 \, dx \qquad (6)$$

in which $\bar{\theta}_1$ is the average rotational component of the critical buckling mode, and is a function of coordinate x, and ℓ is the span of the beam between supports.

For the condensed one degree of freedom problem (3), the new buckling load Λ_1 is computed as

$$\Lambda_1 = \frac{k - k_f}{k_\sigma} \lambda_1 \qquad (7)$$

It must be noticed that the free water width, b, is not known a priori, since it depends on the level of water which produces buckling. A procedure may be used in which Λ_1 is calculated for several values of b, until the critical load coincides with the required level of water to produce it. This will be illustrated graphically in Figure 2.

A warning must be made about consistency between the values of k_σ and k_f. The membrane stresses are calculated from a fundamental path in which dead loading is considered. Inclusion of water loading in k_f means that the fundamental state should be calculated with part of dead weight and part of water load. Thus, for consistency of the formulation, an iterative solution should be used until the fundamental state and the buckling state are computed with the same level of water. However, the change in buckling load due to such iterative technique is not significant since the longitudinal prebuckling stresses in thin-walled beams (which have the largest influence on $\underset{\sim}{K}_\sigma$) are not sensitive to the specific distribution of load in the cross section. Thus, a good approximation is obtained if $\underset{\sim}{K}_\sigma$ is calculated from the fundamental stresses due to dead load alone.

Example of Bifurcation Load Calculation for Water Loading

The use of Equations (1–7) will be illustrated in this section for the angle section beam shown in Figure 2 and taken from Reference [1]. The finite strip method has been used in this case for the evaluation of the stiffness ($\underset{\sim}{K}$) and load-geometry ($\underset{\sim}{K}_\sigma$) matrices in Equation (1). Solution of the eigenvalue problem considering a fundamental state of stresses produced by unit dead weight yields a value of $\lambda = 7.27$, and a critical dead weight $p_c = 7.27$ kN/m is obtained. The buckling mode corresponds to a rotation of the cross section with very small out of plane deformations, and the angle of rotation follows a half sine wave between supports.

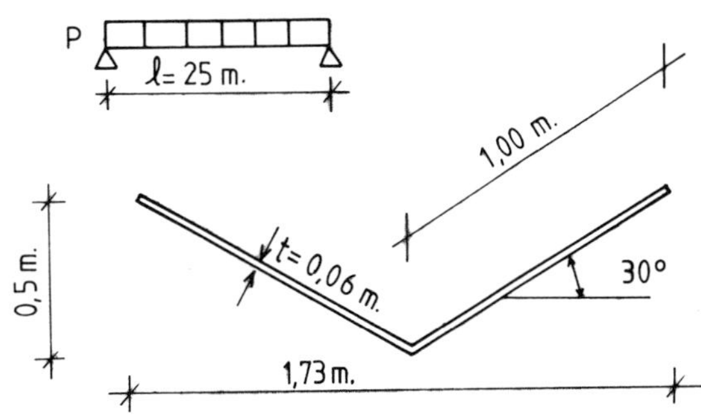

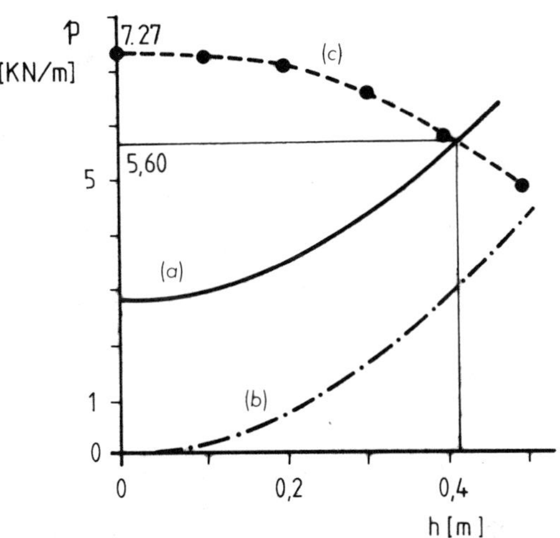

FIGURE 2. Critical load for V-section: (a) Dead weight and water load, Equation (9); (b) Water load only; (c) Critical load E = 21000 MPa, ν = 0.15 (From Reference 1).

Let h be the water level at a certain loading condition, and α the angle of the plates, as shown in Figure 2. The water width b can be evaluated as

$$b = 2h \tan \alpha \qquad (8)$$

and the total load per unit length applied to the beam is given by

$$p = p_d + \gamma h^2 \tan \alpha \qquad (9)$$

where p_d is the self weight of the beam. Equation (9) has been plotted on Figure 2 and represents the curve of applied load.

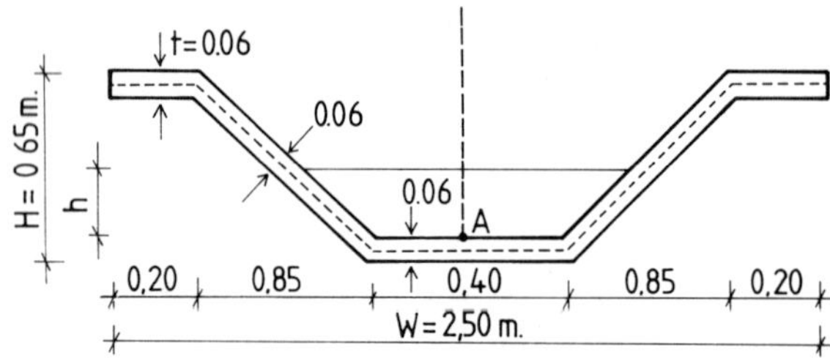

	BUCKLING LOAD
FINITE ELEMENT	7,70
EXPERIMENTAL, MODEL I	7,44
EXPERIMENTAL, MODEL II	7,18

FIGURE 3. Dimensions and results of the models tested $\ell = 25$ m (From Reference 2).

Next, the influence of water loading on buckling loads is determined. For $h = 0$, the applied loading is $p_d = 2.88$ kN/m, and the buckling load corresponding to dead weight is 7.27 kN/m, so that buckling cannot occur. As h increases, the applied load given by (9) is also increased, while the bifurcation load (Equation 7) decreases. Both curves intersect at a value of $h = 0.42$ m, for which buckling is predicted at

$$p = p_c = 5.6 \text{ kN/m}$$

The influence of water loading in this particular case is to reduce the buckling load due to dead weight by 23%.

For the angle section beam studied, if dead weight of the beam is neglected, and only the load due to the liquid is considered, the curve of applied load is reduced and buckling cannot occur.

The reduction in buckling load when movements of water are taken into account depends on a number of factors, such as the geometry of the cross section, the length between supports, and the self weight of the beam.

Other examples of buckling studies including water loading effects may be seen in References [1,3]. Of particular interest to the designer is the parametric study of different section profiles, ranging from an angle section to a U section beam, in which it is seen that buckling loads for U beams (both dead weight and water loaded) are higher than for V beams, but maximum reduction in buckling loads due to water influence is produced in U section beams.

EXPERIMENTAL WORK

Experimental determination of buckling loads for water loaded thin-walled beams has been reported in Reference [2] for precast concrete beams with overall width $W = 2.5$ m and length between supports $\ell = 24.55$ m.

Two models with nominally the same material characteristics and dimensions were tested under increasing water loading, and failure of the structure occurred due to torsional buckling of the cross section, with one half wave in the longitudinal direction. Linear relations between load and vertical deflections were measured during the tests in the prebuckled state, and virtually no cracking was detected until the structure buckled, so that basically elastic buckling was observed. Once the beam buckled, the rotation of the cross section continued without any increase in the load, and the structure collapsed. Thus, buckling and actual collapse of the structure occurred at the same load level. The dimensions and results for the beams tested are summarised in Figure 3.

The experimental buckling loads were compared with bifurcation loads obtained from a finite strip model of Equations (1–7) [1]. Good agreement between both loads was obtained, with numerical results being 8–17% higher than experimental measures. For the tested beams, the numerical models showed that the buckling load with water loading was 40% lower than the buckling load for dead weight [2].

REFERENCES

1. Godoy, L. A. and C. A. Prato, "Buckling of Thin-Walled Beams Under Water Loading," *Journal of the Structural Division*, American Society of Engineers, Vol. 110, No. 11, 2667–2676 (1984).
2. Godoy, L. A., C. A. Larsson, and C. A. Prato, "Experimental Buckling Loads of Water Loaded Beams," in press.
3. Godoy, L. A. and C. A. Prato, "Torsional Buckling of Thin-Walled Open Section Beams: Bifurcation Results," submitted for publication.
4. Zienkiewicz, O. C., *The Finite Element Method*, 3rd ed., McGraw-Hill, London (1977).

CHAPTER 19

Techniques for Estimating Buckling Loads

CHARLES W. BERT*

INTRODUCTION

Buckling is a geometric instability phenomenon to which slender structures, such as columns, rings, plates, and shells, are susceptible. Generally such an instability phenomenon occurs due to compressive strains in the structure, although in some cases the applied external *loading* may be tensile, cf. [1–5].

There are numerous kinds of buckling, such as classical (bifurcation) buckling, finite-disturbance buckling (post-buckling), and snap-through (or durchschlag) buckling. The present work is limited to the first type, which is the one most commonly encountered in engineering structures. For treatment of the latter two types, the reader is referred to Timoshenko and Gere's treatise [6].

Another way of characterizing buckling is static versus dynamic. Here the treatment is limited to static buckling. For background on dynamic stability of structures, the reader is referred to [7]. Still another classification is by the nature of the loading system: conservative or nonconservative. Only the most commonly encountered case, conservative loading, is covered here. For background on the case of nonconservative loading, which requires the use of a dynamic criterion of instability, the reader is referred to [6], p. 152, and [8].

Buckling is also a function of the material behavior of the structural element. Thus, one can have elastic or inelastic buckling. Inelastic behavior includes elastic-plastic, fully plastic, viscoelastic, etc. Here attention is limited to the most important category in the design of structures, namely elastic. For an introduction to inelastic buckling, reference is made to Chajes [9], p. 35.

In formulating classical buckling problems, there is an important fundamental difference from the formulation of problems in the linear theory of elasticity. In buckling theory, the equilibrium equations must be written for the deformed configuration (geometry) of the structural element. This leads to linear terms involving the product of the applied load and the deflection, which are essential to existence of the eigenvalue problem.

Most buckling problems can be *formulated* in differential equation form. For one-dimensional structural elements, such as a slender column or a thin ring, the governing equation is an ordinary differential equation. For two-dimensional elements, such as thin plates or shells, the governing equations are partial differential equations. In certain simple cases of loading and geometry, the governing differential equations can be solved exactly in either closed form or in series form, cf. [6,9–11]. In more complicated cases, it is extremely laborious (as for stepped columns, for example) or even impossible (as for many classes of varying-cross-section columns) to obtain exact solutions. Thus, it becomes necessary to resort to approximate techniques.

Approximate techniques for the analysis of elastic buckling problems may be categorized as simple or numerical. The simple methods include various versions of the energy and Galerkin techniques, while the numerical ones include finite differences [9], p. 106 and [12–13], segmentation (transfer matrix) [13], and finite elements [9], p. 122. Here emphasis is placed on simple methods.

In short, this chapter is concerned with simple approximate methods for predicting buckling loads of elastic columns and plates. Emphasis is placed on applications to various classes of structures, rather than upon general theoretical derivations. However, ample references are given to the underlying theoretical developments.

*School of Aerospace, Mechanical and Nuclear Engineering, The University of Oklahoma, Norman, OK

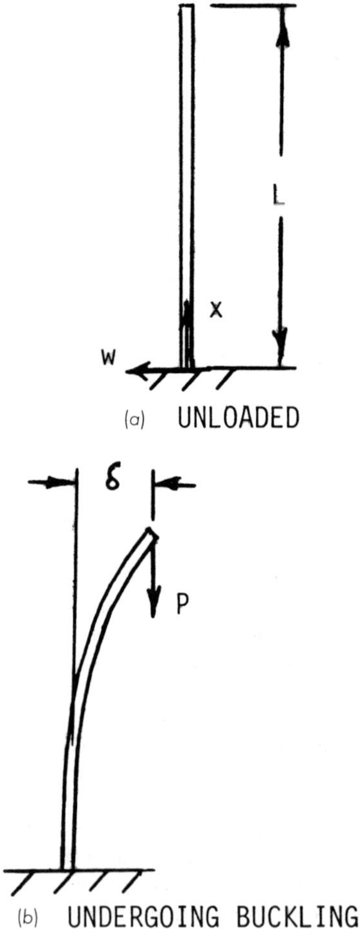

FIGURE 1. A cantilevered prismatic column.

APPLICATIONS TO PRISMATIC COLUMNS

To introduce the various approximate methods, the simplest kind of column is considered: a slender prismatic (uniform-cross-section) column, clamped at one end and free at the other (see Figure 1). Of course, the exact, closed-form solution for this problem is known. This permits one to evaluate the accuracy of the approximate methods.

The column has a flexural rigidity EI and is loaded by a conservative axial load P applied at zero eccentricity. The dead weight of the column itself is neglected. Thus, the governing differential equations of equilibrium for moments and transverse force components are

$$M' - V = 0 \tag{1}$$

$$V' - (Pw')' = 0 \tag{2}$$

Here M is the bending moment, V is the transverse shear force, w is the transverse deflection, and a prime denotes differentiation with respect to x. Eliminating V between Equations (1) and (2), one obtains the following single equation of equilibrium

$$M'' - (Pw')' = 0 \tag{3}$$

For an elastic beam, the constitutive relation is the Bernoulli-Euler equation

$$M = EI\varkappa = -EIw'' \tag{4}$$

where w'' is the linear approximation to the exact, nonlinear expression for the curvature $\varkappa$:

$$\varkappa = -w''/[1 + (w')^2]^{3/2} \tag{5}$$

Finally, combining Equations (3) and (4) yields the equilibrium equation expressed in terms of deflection only:

$$(EIw'')'' + (Pw')' = 0 \tag{6}$$

Since the column is prismatic, EI is independent of x. Also P is independent of x and Equation (6) can be simplified to

$$w^{iv} + \lambda^2 w'' = 0; \; \lambda^2 \equiv P/EI \tag{7}$$

which has the general solution

$$w = C_1 \sin \lambda x + C_2 \cos \lambda x + C_3 x + C_4 \tag{8}$$

For the coordinate system and support shown in Figure 1, the boundary conditions can be expressed as follows. Clamped at station $x = 0$:

$$w(0) = 0; \; w'(0) = 0 \tag{9a}$$

Free at station $x = L$:

$$M(L) = 0 \rightarrow w''(L) = 0; \; V(L) = 0 \rightarrow w'''(L) = 0 \tag{9b}$$

Application of the boundary conditions (9a,b) to the general solution (8) yields the following closed-form expressions for the deflection curve and dimensionless load associated with the critical (lowest) mode:

$$w = \delta[1 - \cos(\pi x/2L)] \tag{10}$$

$$\overline{P} \equiv P_{cr}L^2/EI = \pi^2/4 \approx 2.4674 \tag{11}$$

The latter result will be used as the yardstick of comparison for the approximate results obtained in the ensuing subsections.

Potential Energy Methods

The Rayleigh principle is based upon variational calculus [14–15]. In the case of elastic columns, it can be expressed as

$$d(U_s + V)/dC = 0 \tag{12}$$

where $U_s \equiv$ strain energy $= (1/2) \int_0^L EI(w'')^2 dx$ (13)

$V \equiv$ potential energy of the external load
$= -(P/2) \int_0^L EI(w')^2 dx$ (14)

$C \equiv$ coefficient of the assumed deflection function

$$w = Cf(x) \tag{15}$$

The classical one-term power-law deflection expression is

$$w = Cx^2 \tag{16}$$

Substitution of Equation (16) into Equations (13) and (14) and thence applying Equation (12) yields $\bar{P} = 3$, which is 21.6% higher than the exact value.

The Ritz form of the energy method, known today as the Rayleigh-Ritz method, was introduced by Walther Ritz in 1909 [16]. In terms of the present notation it can be expressed as

$$\partial(U_s + V)/\partial C_i = 0; \quad i = 1, 2, \ldots, n \tag{17}$$

where the C_i are the coefficients in the deflection expression

$$w = C_1 f_1(x) + \ldots + C_n f_n(x) \tag{18}$$

and the functions $f_i(x)$ must each *individually* satisfy the *geometric* boundary conditions, i.e. either $w =$ const. or $w' = 0$. It is noted that it is not necessary that the f_i satisfy the natural or force-type boundary conditions. However, if they do satisfy them, convergence to the exact value is enhanced.

Application of Equations (17) leads to a set of "n" simultaneous homogeneous, linear algebraic equations in the C_i. This is the classical linear eigenvalue problem and to guarantee a nontrivial solution, the determinant of the coefficients of the C_i must vanish. This leads to an nth-degree algebraic equation in $\bar{P}$, the minimum value of which is the dimensionless critical load. For example, the following expression, used by Chajes [9], pp. 97–99, is taken:

$$w = C_1 x^2 + C_2 x^3 \tag{19}$$

Then application of Equations (17) yields a quadratic equation in $\bar{P}$, which has a minimum root $\bar{P} = 2.486$, approximately 0.75% higher than the exact value.

Recently Schmidt [17] introduced a modified form of the Rayleigh method. In the Rayleigh-Schmidt technique, instead of using an a-priori integer power, the power (k) is kept arbitrary through application of Equation (12). Then $\bar{P}$ is a function of k and the "optimum" value of k is the one resulting in the smallest value of $\bar{P}$, since the Rayleigh principle leads to an upper bound. For example, let

$$w = Cx^k \tag{20}$$

Substitution of this expression into Equation (12) leads to the following expression

$$\bar{P} = \frac{2k - 1}{2k - 3}(k - 1)^2 \tag{21}$$

The minimum value is $\bar{P} = 2.7725$, which corresponds to $k = 1.81$ and is 12.4% higher than the exact value. This is a considerable improvement over the 21.6% error of the traditional Rayleigh solution. It is noted that the optimum value of k obtained here is only slightly higher than $k = 1.73$ found by Schmidt [18] to minimize the fundamental natural frequency of a cantilever beam with a mode shape given by Equation (20).

The Rayleigh-Ritz-Schmidt technique was applied by the present investigator [19] using

$$w = C_1 x^2 + C_2 x^k \tag{22}$$

This resulted in the following quadratic equation

$$a\bar{P}^2 + b\bar{P} + c = 0 \tag{23}$$

where

$$a \equiv \frac{1}{3(2k - 1)} - \frac{1}{(k + 1)^2}$$

$$b \equiv \frac{2}{k + 1} - \frac{1}{2k - 1} - \frac{(k - 1)^2}{3(2k - 3)}$$

$$c \equiv (k - 2)^2/(2k - 3)$$

The minimum value is $\bar{P} = 2.4676$, which corresponds to $k = 3.75$ and is only 0.008% higher than the exact value.

Complementary Energy Methods

The so-called Timoshenko quotient is derivable from the complementary energy principle [6,9,15]. It can be ex-

TABLE 1. Summary of Numerical Results for Dimensionless Buckling Load of a Prismatic Cantilever Column.

Analysis Method	Deflection Shape[a]	$\bar{P}$	% Error
Exact closed-form	$\delta[1 - \cos(\pi x/2L)]$	$\pi^2/4$	0
Potential energy:			
Rayleigh	Cx^2	3.0000	21.6
Rayleigh-Schmidt	$Cx^{1.81}$	2.7725	12.4
Rayleigh	$(\delta/2)(3 - \bar{x})\bar{x}^2$	2.5000	1.32
Rayleigh-Ritz	$C_1x^2 + C_2x^3$	2.486	0.75
Rayleigh-Ritz-Schmidt	$C_1x^2 + C_2x^{3.75}$	2.4676	0.008
Complementary energy:			
Timoshenko	Cx^2	2.5000	1.32
Timoshenko-Schmidt	$Cx^{1.725}$	2.4748	0.30
Timoshenko	$(\delta/2)(3 - \bar{x})\bar{x}^2$	2.4706	0.13

[a] $\bar{x} \equiv x/L$.

pressed by

$$d(U_c + V)/dC = 0 \tag{24}$$

This relation is the analog of Equation (12) with the strain energy U_s replaced by the complementary strain energy U_c given by

$$U_c = \int_0^L \frac{M^2}{2EI}\,dx = \int_0^L \frac{P^2(\delta - w)^2}{2EI}\,dx \tag{25}$$

For a linear elastic material, the complementary strain energy is theoretically equal to the strain energy. This equality, of course, does not hold in the case of approximate solutions.

Use of a deflection function of the form given by Equation (16) in conjunction with Equations (14), (24), and (25) yields $\bar{P} = 5/2 = 2.5$, which is 1.32% higher than the exact value. It is noted that this is a considerable improvement over the 21.6% error associated with the use of the same deflection function in conjunction with the PEM (potential energy method). Timoshenko and Gere [6] attributed this improvement to the lack of differentiation involved with the CEM (complementary energy method). Although the CEM results in lower values of $\bar{P}$ than the PEM, certain conditions, studied in [20–21], must be satisfied in order to assure that the value obtained is an upper bound.

The Schmidt technique was first used in conjunction with the CEM by Schmidt himself in 1981 [22]. Using the deflection curve expressed by Equation (20) in conjunction with Equations (14), (24), and (25), one obtains the following algebriac expression for $\bar{P}$ as a function of the power k:

$$\bar{P} = \left(\frac{k^2}{2k - 1}\right) \Big/ \left(\frac{1}{2k + 1} - \frac{2}{k + 1} + 1\right) \tag{26}$$

Minimization of $\bar{P}$ leads to a value of 2.4748 (only 0.3% higher than the exact value) at $k = 1.725$. Schmidt also treated the pinned-pinned prismatic column [22].

In conjunction with the complementary energy method, Timoshenko and Gere [6], p. 88, investigated the following deflection curve for the present problem:

$$w = (x^2\delta/2L^3)(3L - x) \tag{27}$$

The motivation for this expression is that it is the deflection curve for a cantilever beam with a *transverse* load at its tip. They obtained $\bar{P} = 42/17 \approx 2.4706$, which is only 0.13% higher than the exact value. It is interesting to note that use of Equation (27) in conjunction with the PEM, Equation (12), leads to $\bar{P} = 2.5$ (1.32% above exact), which was the identical value obtained by the CEM for a much simpler deflection curve, Equation (16).

Table 1 summarizes the results of all of the investigations in Section 1. It can be concluded that:

1. The conventional CEM is more accurate than the conventional PEM, for the same deflection curve.
2. The Schmidt version of the PEM is also more accurate than the conventional PEM, although the former involves slightly more computational effort.
3. The Schmidt version of the CEM is more accurate than the conventional CEM, although the former involves more computational effort.

The Schmidt versions of certain methods have not been implemented here because of their impracticality from an algebraic or computational viewpoint:

1. A two-term Rayleigh-Ritz-Schmidt version of the CEM
2. A "Schmidt-type" version of the Galerkin method.

For additional information on new developments in the CEM approach to buckling analysis, the reader is referred to [23–25].

Lower Bounds

There are several methods of obtaining lower bounds to the buckling load, based solely upon a knowledge of the Rayleigh (PEM) and Timoshenko (CEM) upper bounds. The following expressions were introduced by Schreyer and Shih [26] and by Ku [27], respectively.

$$\bar{P}_{SS} = \bar{P}_T - [\bar{P}_T(\bar{P}_R - \bar{P}_T)]^{1/2} \tag{28}$$

$$\bar{P}_K = \bar{P}_R - (\bar{P}_R(\bar{P}_R - \bar{P}_T))^{1/2} \tag{29}$$

TABLE 2. Numerical Results for Some Lower Bounds Associated with Table 1.

Deflection Curve[a]	$\bar{P}_R$	$\bar{P}_T$	$\bar{P}_{SS}$	% Error	$\bar{P}_K$	% Error
Cx^2	3.0000	2.5000	1.3820	−44.0	1.7753	−28.0
$(\delta/2)(3 - \bar{x})\bar{x}^2$	2.5000	2.4706	2.2011	−10.8	2.2289	− 9.7

[a] $\bar{x} \equiv x/L$.

Numerical results for lower-bound dimensionless buckling loads based on some deflection curves from Table 1 are listed in Table 2. It is clear that the Ku lower bound is more accurate. Additional information on estimating lower bounds for buckling loads is found in [28–31].

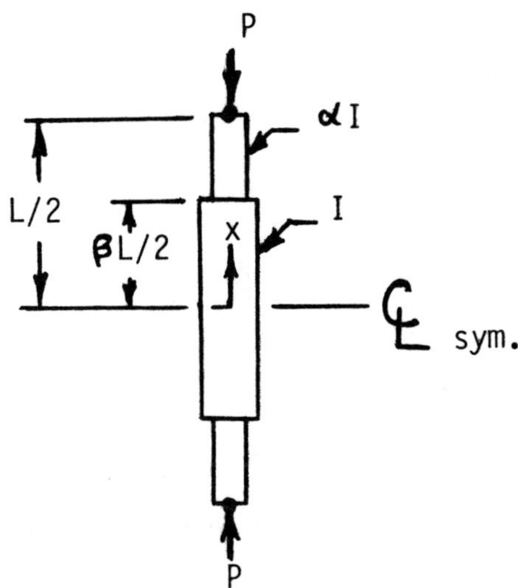

FIGURE 2. Symmetrically stepped column.

APPLICATIONS TO NONUNIFORM COLUMNS

The considerable power of certain approximate methods becomes important in connection with nonuniform columns, for which very few exact solutions exist. These include:

1. Stepped columns, first analyzed by Dinnik [32] in 1932 and widely used in machinery structures. See Figure 2.
2. Smoothly tapered columns, the early history of which was reviewed by Timoshenko and Gere [6], pp. 125–130. See Figure 3.

Stepped Columns

The governing differential equations for an n-section stepped column are of the same form as Equation (6), namely

$$(E_iI_iw'')'' + (P_iw')' = 0; \quad i = 1, \ldots, n \tag{30}$$

In addition to the boundary conditions at each end, section-to-section deflection and slope continuity must be satisfied at each of the $(n - 1)$ junctions:

$$w_{j+1} = w_j\,; \quad w'_{j+1} = w'_j\,; \quad j = 1, \ldots, n - 1 \tag{31}$$

For the case of a symmetrically stepped, three-section column (see Figure 2), refer to Dinnik [32]; see also [6], p. 113.

For the potential energy method, the strain-energy integral can be written, taking advantage of the symmetry (Figure 2), as

$$U_s = (E_1I_1/2)\int_0^{\beta L/2} (w'')^2dx + (E_2I_2/2)\int_{\beta L/2}^{L/2} (w'')^2dx \tag{32}$$

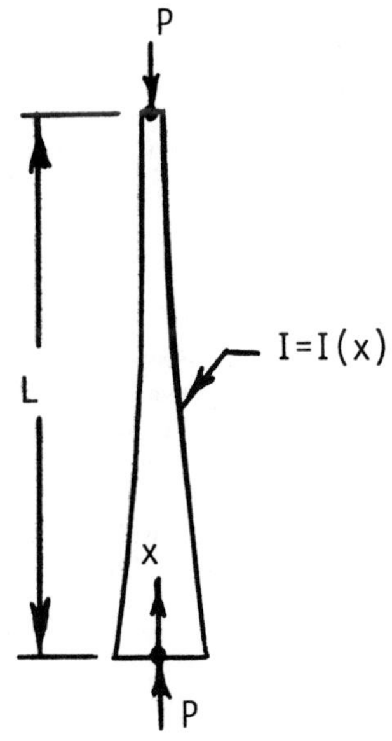

FIGURE 3. Smoothly tapered column.

TABLE 3. Dimensionless Critical Loads for Symmetrically Stepped, Three-Section, Pinned-Pinned Columns.

Case	Method	Ref.	Deflection Eq.	$\bar{P}$	% Error
A	Exact	[32]	—	4.22	0
	Timoshenko	[6]	(33)	4.51	6.87
	FEM[a]	[33]	—		
	10 elements			4.28	1.42
	20 elements			4.23	0.24
	Rayleigh-Schmidt	[19]	(35)	5.85	38.6
	Timoshenko-Schmidt	[19]	(35)	4.26	0.95
B	Exact	[32]	—	6.00[b]	0
	Rayleigh	[9]	(33)	8.53	42.2
	Timoshenko	[6]	(33)	8.05	34.2
	Rayleigh-Ritz	[9]	(34)	7.25	20.8
	Rayleigh-Schmidt	[19]	(35)	7.44	24.0
	Timoshenko-Schmidt	[19]	(35)	6.10	1.67

[a]FEM denotes the finite-element method.
[b]Interpolated from the work of Dinnik.

In the usual case in which the load is uniform along the length, the potential energy due to the applied load is the same as for prismatic columns, Equation (14). using the following trigonometric buckled shape which is the exact shape for a prismatic column, Chajes [9], p. 100, treated a pinned-pinned column by the Rayleigh method:

$$w = C \sin (\pi x/L) \tag{33}$$

Substituting Equation (33) into Equations (14) and (32) and thence in Equation (12), he obtained $\bar{P} = 8.53$ for a column having $\alpha \equiv E_2I_2/E_1I_1 = 0.25$ and a value of 0.5 for β. This was 42.2% higher than Dinnik's exact value.

For the same problem, Chajes [9], p. 101, used the following two-term expression with the Rayleigh-Ritz method, Equation (17):

$$w = C_1 \sin (\pi x/L) + C_2 \sin (3\pi x/L) \tag{34}$$

This led to a set of two homogeneous algebraic equations and thence through the eigenvalue procedure to a quadratic equation in $\bar{P}$. The result was $\bar{P} = 7.25$ which exceeded the exact value by 20.8%.

The present investigator [19] used the Rayleigh-Schmidt PEM in conjunction with the following deflection mode shape:

$$w = C[1 - (2x/L)^k] \tag{35}$$

The expression obtained for $\bar{P}$ as a function of given geometric parameters α and β and the power k was

$$\bar{P} = 4(k-1)^2 \frac{2k-1}{2k-3} [\alpha + (1-\alpha)\beta^{2k-3}] \tag{36}$$

The complementary strain energy integral for the present problem is

$$U = (P^2/2)\left[\frac{1}{E_1I_1} \int_0^{\beta L/2} w^2 dx + \frac{1}{E_2I_2} \int_{\beta L/2}^{L/2} w^2 dx \right] \tag{37}$$

Using the mode shape of Equation (35) in Equations (14), (37), and (24) for the Timoshenko-Schmidt CEM, one obtains [19]

$$\bar{P} = \frac{4k^2}{2k-1} (1/F) \tag{38}$$

where

$$F \equiv \alpha^{-1} + (1-\alpha^{-1})\beta - [2(k+1)^{-1}][(1-\alpha^{-1})\beta^{k+1} + \alpha^{-1}] + [(2k+1)^{-1}][\alpha^{-1} + (1-\alpha^{-1})\beta^{2k+1}] \tag{39}$$

Numerical results for the following two geometric cases are listed in Table 3:

Case A: $\alpha = 0.20$, $\beta = 0.40$
Case B: $\alpha = 0.25$, $\beta = 0.50$

Inspection of Table 3 shows that the conventional (Timoshenko) CEM is more accurate than the conventional (Rayleigh) PEM. Further the one-term Timoshenko-Schmidt CEM gives accuracy comparable to a ten-element FEM analysis.

Returning now to lower-bound methods, the work of Pnueli [34], applicable only to nonuniform columns, should be mentioned. He introduced the concept of an "equivalent prismatic column" having a flexural rigidity $\overline{EI}$ given by

$$\overline{EI} = L/ \int_0^L (EI)^{-1} dx \tag{40}$$

For the present symmetrically stepped column, Equation (40) yields

$$(\overline{EI})^{-1} = (2/L)[\int_0^{\beta L} (E_1I_1)^{-1} dx + \int_{\beta L/2}^{L/2} (E_2I_2)^{-1} dx]$$

or

$$\overline{EI} = E_1 I_1 / \left(\beta + \frac{1 - \beta}{\alpha} \right) \tag{41}$$

Thus, for pinned-pinned ends,

$$\overline{P}_P = \pi^2 / \left(\beta + \frac{1 - \beta}{\alpha} \right) \tag{42}$$

Numerical results for the Schreyer-Shih, Ku, and Pnueli lower bounds ($\overline{P}_{SS}$, $\overline{P}_K$, and $\overline{P}_P$) are listed in Table 4. It is seen that for Case A, the Pnueli bound is most accurate, while for Case B, $\overline{P}_K$ is most accurate.

In [19], the case of an unsymmetrically stepped, two-section pinned-pinned column was treated by the Schmidt version of PEM and CEM using the following deflection curve

$$w = C(\bar{x}^k - \bar{x}) \quad ; \quad \bar{x} \equiv x/L \tag{43}$$

The resulting expressions for P as a function of power k and two geometric parameters were given. When minimized with respect to k, they resulted in respective $\overline{P}$ values 70.1% and 3.09% above the exact value, obtained by recalculating from Thomson's exact analysis [35].

Smoothly Tapered Columns

In [19], a cantilever column with a smoothly varying cross section having a moment of inertia given as follows was treated.

$$I = I_o(1 + n\epsilon\bar{x}^{n-1})^{-1} \tag{44}$$

Use of deflection Equation (43) in conjunction with the Timoshenko-Schmidt CEM gave

$$\overline{P} = \left(\frac{2k^2}{2k - 1} \right) \left[\left(1 - \frac{2}{k + 1} + \frac{1}{2k + 1} \right) + n\epsilon \left(\frac{1}{n} - \frac{2}{k + n} + \frac{1}{2k + n} \right) \right]^{-1} \tag{45}$$

For $n = 2$, Equation (45) yields a minimum value $\overline{P} = 2.335$ at $k = 1.76$, which is only 4.19% above Pnueli's *lower bound* value.

The moment-of-inertia distribution given by

$$I = I_o(1 + \bar{x})^n \tag{46}$$

was also considered in [19]. Using Equation (43) for the

TABLE 4. Numerical Results for Some Lower Bounds Associated with the Schmidt-Type Results of Table 3.

		Case A		Case B	
Method	Eq.	$\overline{P}$	% Error	$\overline{P}$	% Error
Schreyer-Shih ($\overline{P}_{SS}$)	(28)	1.66	−60.7	3.24	−46.0
Ku ($\overline{P}_K$)	(29)	2.80	−33.6	4.28	−28.7
Pnueli ($\overline{P}_P$)	(42)	2.90	−31.3	3.94	−34.3

deflection curve in conjunction with the Rayleigh-Schmidt PEM yielded, for $n = 1$:

$$\overline{P} = (k^2/2)(2k - 1)(4k - 5)/(2k - 3) \tag{47}$$

The minimum value is $\overline{P} = 15.310$, only 7.06% higher than Swenson's result [36] obtained by network analyzer, at $k = 1.755$.

For $n = 2$:

$$\overline{P} = k^2 \left[1 + (2k - 1) \left(\frac{1}{2k - 3} + \frac{1}{k - 1} \right) \right] \tag{48}$$

The minimum is $\overline{P} = 27.455$ at $k = 1.878$. This is 32.0% higher than the exact value reported by Bleich [37]. By numerical computation based on the Collatz enclosure theorem, Appl and Zorowski [38] obtained 20.8014 and 20.7860 for the respective upper and lower bounds of $\overline{P}$.

Unfortunately, for the cases of $n = 1$ and $n = 2$, it was not possible to obtain closed-form integrals in the Timoshenko-Schmidt CEM, except for $n = 2$ and $k = 2$, for which $\overline{P} = 21.412$, only 2.98% higher than Bleich's exact value.

APPLICATIONS TO BUCKLING OF THIN PLATES

The analysis of buckling of thin plates is considerably more complicated then for columns in that the former is a two-dimensional problem rather than one-dimensional. Nevertheless the approximate methods are quite useful in predicting plate buckling loads.

Rectangular Plates

In the case of plates, the general expressions for the Rayleigh and Rayleigh-Ritz methods, Equations (12) and (17), respectively still hold. However, the expressions for the potential-energy components, U_s and V, are more complicated.

For a thin, uniform-thickness, rectangular plate of isotropic material (see Figure 4), the strain energy and potential energy of the external load may be expressed as

$$U_s = (D/2)\int_0^a\int_0^b [w_{,xx}^2 + 2\nu w_{,xx}w_{,yy} + w_{,yy}^2 + 2(1-\nu)w_{,xy}^2]dx\,dy \tag{49}$$

$$V = -(1/2)\int_0^a\int_0^b (N_x^0 w_{,x}^2 + 2N_{xy}^0 w_{,x}w_{,y} + N_y^0 w_{,y}^2)dx\,dy \tag{50}$$

Here w = plate deflection, x and y are rectangular Cartesian coordinates on the plate mid-plane, a and b are the plate dimensions in the x and y directions, $\nu \equiv$ Poisson's ratio, N_x and N_y are the prebuckling in-plane normal membrane stress resultants (force/unit width) applied along the x and y axes, $N_{xy}^0 \equiv$ prebuckling in-plane shear stress resultant, $(\ )_{,xy} \equiv \partial^2(\ )/\partial x\partial y$, etc., and D is the plate flexural rigidity given by

$$D = Eh^3/[12(1-\nu^2)] \tag{51}$$

As an example of the use of the above equations, a square plate, clamped on all four edges and subjected to uniaxial compressive loading in the x direction, is considered. The edges are all completely unrestrained in the plane of the plate. The boundary conditions can be expressed mathematically as

$$w = w_{,n} = 0 \tag{52}$$

on all boundaries. Here n denotes the in-plane direction normal to the boundary on the edge considered, i.e. Equation (52) can be expanded to become

$$\begin{aligned} w(0,y) &= w(a,y) = w(x,0) = w(x,b) = 0 \\ w_{,x}(0,y) &= w_{,x}(a,y) = w_{,y}(x,0) = w_{,y}(x,b) = 0 \end{aligned} \tag{53}$$

Since all of the boundary conditions are geometric, one must choose a form for the deflection function which satisfies all of them. Here the following one-term expression is used:

$$w = W[1 - \cos(2\pi x/a)][1 - \cos(2\pi y/b)] \tag{54}$$

where W is a constant. Before substitution of Equation (54) into Equations (49) and (50), the following intermediate steps are performed

$$\begin{aligned} w_{,x} &= (2\pi/a)W\sin(2\pi x/a)[1-\cos(2\pi y/b)] \\ w_{,y} &= (2\pi/b)W[1-\cos(2\pi x/a)]\sin(2\pi y/b) \\ w_{,xx} &= (2\pi/a)^2 W\cos(2\pi x/a)[1-\cos(2\pi y/b)] \\ w_{,yy} &= (2\pi/b)^2 W[1-\cos(2\pi x/a)]\cos(2\pi y/b) \\ w_{,xy} &= (2\pi/a)(2\pi/b)W\sin(2\pi x/a)\sin(2\pi y/b) \end{aligned} \tag{55}$$

Substitution of Equation (55) into Equations (49) and (50) require that the following definite integrals be evaluated:

$$\int_0^a \cos^2(2\pi x/a)dx = \int_0^a \sin^2(2\pi x/a)dx = a/2$$

$$\int_0^b \cos^2(2\pi y/b)dy = \int_0^b \sin^2(2\pi y/b)dy = b/2$$

$$\int_0^a [1-\cos(2\pi x/a)]^2 dx = 3a/2$$

$$\int_0^b [1-\cos(2\pi y/b)]^2 dy = 3b/2$$

$$\int_0^a \cos(2\pi x/a)[1-\cos(2\pi x/a)]dx = -a/2$$

$$\int_0^b \cos(2\pi y/b)[1-\cos(2\pi y/b)]dx = -b/2$$

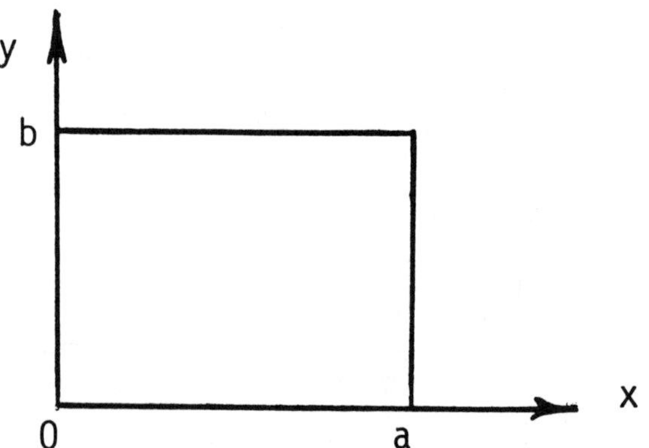

FIGURE 4. Rectangular plate coordinate system and geometry.

Thus, Equations (49) and (50) become

$$U_s = 8\pi^4 DW^2\left[\frac{(a/2)(3b/2)}{a^4} + \frac{2\nu(-a/2)(-b/2)}{a^2b^2} + \frac{(3a/2)(b/2)}{b^4} + \frac{2(1-\nu)(a/2)(b/2)}{a^2b^2}\right]$$

$$V = -2\pi^2(N_x^0/a^2)W^2(a/2)(3b/2)$$

or

$$U_s = 6\pi^4 DW^2 ab[a^{-4} + b^{-4} + (2/3)a^{-2}b^{-2}]$$
$$V = -(3\pi^2/2)N_x^0 W^2(b/a) \tag{56}$$

Substitution of Equations (56) into Equation (12) yields

$$\bar{N} \equiv N_x^0 b^2/\pi^2 D = 4[\alpha^{-2} + \alpha^2 + (2/3)] \tag{57}$$

where $\alpha \equiv a/b$, the plate aspect ratio. It is interesting to note that $\bar{N}$ is independent of Poisson's ratio ν.

For the case of a square plate ($\alpha = 1$), $\bar{N} = 10.67$, while the exact value given in [6] is 10.07; thus, the potential energy method gives a value 5.96% higher than the exact one. It should be pointed out that the buckle mode shape used here, Equation (54), is valid only for aspect ratios in the vicinity of unity or less. The reason for this that Equation (54) yields only the first mode in each direction (x and y), yet in actuality for $\alpha > \sim 1.2$, there are higher modes (one or more nodal lines) in the x direction.

It should be mentioned that expression (49) for U_s can be rewritten as

$$U_s = (D/2)\int_0^a\int_0^b [(w_{,xx} + w_{,yy})^2 - 2(1-\nu)(w_{,xx}w_{,yy} - w_{,xy}^2)]dx\,dy \tag{58}$$

Furthermore, it can be shown that for all rectangular plates either clamped or guided on all edges, the term $(w_{,xx}w_{,yy} - w_{,xy}^2)$ vanishes upon integration. For such cases, Equation (58) simplifies considerably to

$$U_s = (D/2)\int_0^a\int_0^b (w_{,xx} + w_{,yy})^2 dx\,dy \tag{59}$$

The complementary energy method was applied to rectangular plates, simply supported ones in particular, by Oran [39]. The Schmidt version of Rayleigh method was applied by Ercoli and Laura [40] to vibration of stepped rectangular plate free vibration. The same approach could easily be adapted to buckling.

Circular Plates

Usually the lowest buckling mode of axisymmetrically compressed solid circular plates is the axisymmetric one, i.e. the deflection is independent of angular position. Further, in such a case, U_s and V can be expressed as

$$U_s = \pi D\int_0^1 [(w_{,\varrho\varrho} + \varrho^{-1} w_{,\varrho})^2 - 2(1-\nu)\varrho^{-1}w_{,\varrho}w_{,\varrho\varrho}]\,\varrho\, d\varrho \tag{60}$$

$$V = -\int_0^1 N_r^0 w_{,\varrho}^2\, \varrho\, d\varrho \tag{61}$$

Here N_r^0 is the radial compressive prebuckling stress resultant and ϱ is the dimensionless radial position defined by

$$\varrho \equiv r/R \tag{62}$$

where $r \equiv$ radial position and $R \equiv$ edge radius (see Figure 5).

As the first example, the Schmidt version of the Rayleigh potential energy method will be used in conjunction with the following buckle mode shape:

$$w = C(1 - \varrho^k)^2 \tag{63}$$

which satisfies the boundary conditions

$$w(1) = w_{,\rho}(1) = 0 \tag{64}$$

and the regularity conditions at the center

$$w_{,\varrho}(0) = w_{,\varrho\varrho}(0) = \text{finite} \tag{65}$$

provided that $k > 1$.

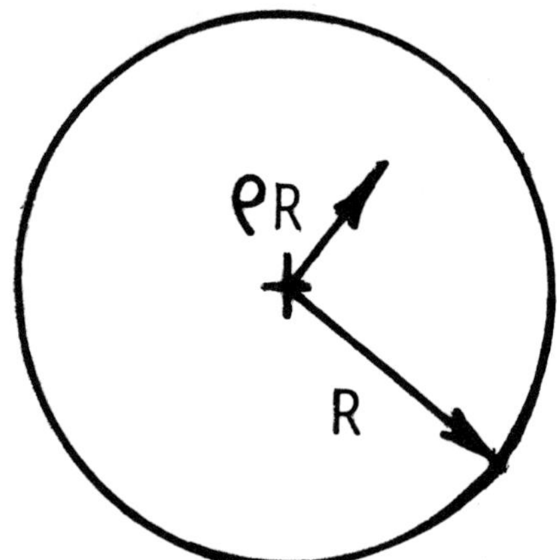

FIGURE 5. Circular plate coordinate system and geometry.

Use of Equation (63) in Equations (60), (61), and (12) leads to the following expression for the dimensionless buckling load:

$$\bar{N}_r \equiv N_r^0 R^2/D = 6k^3\left(\frac{1}{k-1} - \frac{8}{3k-2} + \frac{4}{2k-1}\right) \tag{66}$$

Use of $k = 2$ (conventional Rayleigh technique) yields $\bar{N}_r = 16$, which is 8.96% higher than the exact value (14.684) obtained by Bryan in 1891 by use of Bessel functions, according to [6], p. 390. However, application of the Schmidt approach shows that the minimum value occurs at $k = 1.722$ and is $\bar{N}_r = 15.161$, only 3.25% higher than the exact value. It is interesting to note that for this same mode shape, Schmidt [18] showed that the fundamental frequency of flexural vibration is minimized by $k = 1.85$.

It is noted that only the first and second derivatives of w with respect to ϱ appear in Equations (60) and (61). Thus, it is not necessary to specify w itself, only $w_{,\varrho}$. A convenient form for $w_{,\varrho}$, used by Schmidt [17,22], is

$$w_{,\varrho} = C(\varrho - \varrho^k) \tag{67}$$

The resulting expression for $\bar{N}_r$ is

$$\bar{N}_r = (2/k)(k+1)(k+3) \tag{68}$$

For $k = 2$ (conventional Rayleigh method), $\bar{N}_r = 15$. However, the minimum value occurs when $k^2 = 3$ or $k \approx 1.732$ for which $\bar{N}_r = 14.928$, which is 1.66% higher than the exact value. Thus, using a more accurate buckle mode, Equation (67) rather than Equation (63), cuts the error in approximately one half.

For the case of buckling of a radially compressed clamped plate resting on a Winkler elastic foundation, use of Equation (63) for the mode shape yields

$$\bar{N}_r^! = \bar{N}_r(k) + (3/2)(KR^4/D)f(k) \tag{69}$$

where $\bar{N}_r(k)$ = buckling of clamped plate without an elastic foundation, as given by Equation (66) above, $K \equiv$ Winkler foundation modulus, and $f(k)$ is given by

$$f(k) \equiv (1/k)\left[1 - \frac{8}{k+2} + \frac{6}{k+1} - \frac{8}{3k+2} + \frac{1}{2k+1}\right] \tag{70}$$

For $KR^4/D = 100$, $k = 2$ gives $\bar{N}_r^! = 31$, which is 11.5% higher than the value (27.8) as read from a curve obtained by Kline and Hancock [41] from an exact Bessel-function solution. However, use of the Schmidt idea gives a minimum value of 29.2 (at $k = 1.63$), which is only 5.07% higher than the exact one.

For a simply supported plate, Schmidt [17] used

$$w_{,\varrho} = C\left[\frac{k+\nu}{1+\nu}\varrho - \varrho^k\right] \tag{71}$$

This expression satisfies the regularity conditions, Equations (65), at the center and the geometric boundary condition at the edge

$$w(1) = 0 \tag{72}$$

provided that $\nu > -1$ and $k > 0$. Of course, for all isotropic materials, the theoretical minimum value for ν is -1 and actual isotropic materials usually satisfy $\nu > 0$. This leads to an expression for $\bar{N}_r$ as a function of ν and k. For $\nu = 0.3$, the minimum occurs at $k = 2.5$ for which $\bar{N}_r = 4.20$, which agrees precisely with the exact value, ([6], p. 391) to three significant figures. This same value of $\bar{N}_r$, for $\nu = 0.3$, was also obtained by Oran [39] and Ku [42], using the complementary energy method. Using converged upper and lower bounds, per Equation (29), Ku [42] obtained a curve of $\bar{N}_r$ versus ν, which can be approximated very accurately by

$$\bar{N}_r = 3.4 + 2.67\nu \tag{73}$$

ACKNOWLEDGEMENTS

The author acknowledges helpful suggestions by Professor Isaac Elishakoff of the Technion-Israel Institute of Technology.

REFERENCES

1. Haringx, J. A., *De Ingenieur*, *63*:39 (1951).
2. Yao, J. C., *AIAA J.*, *1*:2316 (1963).
3. Pagano, N. J., J. C. Halpin and J. M. Whitney, *J. Comp. Matls.*, *2*:154 (1968).
4. Zielsdorff, G. J. and R. L. Carlson. *Eng. Frac. Mech.*, *4*:939 (1972).
5. Durban, D., *AIAA J.*, *15*:360 (1977).
6. Timoshenko, S. P. and J. M. Gere, *Theory of Elastic Instability*, 2nd ed., McGraw-Hill, New York (1961).
7. Herrmann, G., ed., *Dynamic Stability of Structures*, Pergamon, Oxford (1967).
8. Ziegler, H., *Advances in Appl. Mech.*, *4*:357 (1956).
9. Chajes, A., *Principles of Structural Stability Theory*, Prentice-Hall, Englewood Cliffs, New Jersey (1974).
10. Brush, D. O. and B. O. Almroth, *Buckling of Bars, Plates, and Shells*, McGraw-Hill, New York (1975).
11. Simitses, G. J., *An Introduction to the Elastic Stability of Structures*, Prentice-Hall, Englewood Cliffs, New Jersey (1976).
12. Seide, P., *J. Eng. Mech. Div.*, *Proc. ASCE*, *101*:549 (1975).
13. Bert, C. W. and C. L. Ko, *Int. J. Eng. Sci.*, *23*: 641 (1985).
14. Temple, G. and W. G. Bickley, *Rayleigh's Principle and Its Applications to Engineering*, Dover, New York (1956).
15. Libove, C., Chap. 44 in *Handbook of Engineering Mechanics* (W. Flügge, ed.), McGraw-Hill, New York (1962).
16. Ritz, W., *Zeits. reine u. angew. Math.*, *135*: 1 (1909).
17. Schmidt, R., *J. Appl Mech.*, *49*: 639 (1982).
18. Schmidt, R., *J. Eng. Mech.*, *109*: 654 (1983).
19. Bert, C. W., *J. Eng. Mech.*, *110*: 1655 (1984).
20. Popelar, C. H., *J. Eng. Mech. Div.*, *Proc. ASCE*, *100*: 623 (1974).
21. Masur, E. F. and C. H. Popelar, *Int. J. Solids Structures*, *12*: 203 (1976).
22. Schmidt, R., *Indust. Math.*, *31*: 37 (1981).
23. Oran, C., *J. Eng. Mech. Div.*, *Proc. ASCE*, *93*: 57 (1967).
24. Angelilillo, M. and L. Dodard, *Archiwum Mechaniki Stosowanej*, *33*: 879 (1981).
25. Murakawa, H., K. W. Reed, S. N. Atluri and R. Rubenstein, *Computers and Structures*, *13*: 11 (1981).
26. Schreyer, H. L. and P. Y. Shih, *J. Eng. Mech. Div.*, *Proc. ASCE*, *99*: 1011 (1973).
27. Ku, A. B., *J. Sound Vib.*, *53*: 183 (1977).
28. Holden, J. T., *Arch. Rational Mech. and Analysis*, *17*: 171 (1983).
29. Hanna, S. Y. and C. D. Michalopoulos, *J. Appl. Mech.*, *46*: 696 (1979).
30. Del Piero, G., *J. Elasticity*, *10*: 135 (1980).
31. Schmidt, R., *Indust. Math.*, *33*: 163 (1983).
32. Dinnik, A. N., *Trans. ASME*, *54*, Appl. Mech. Sec.: 165 (1932).
33. Yettram, A. L. and E. S. Awadalla, *Int. J. Mech. Sci.*, *9*: 315 (1967).
34. Pnueli, D., *J. Appl. Mech.*, *39*: 1139 (1972).
35. Thomson, W. T., *J. Appl. Mech.*, *17*: 132 (1950).
36. Swenson, G. W., Jr., *J. Aeronaut. Sci.*, *19*: 273 (1952).
37. Bleich, F., *Buckling Strength of Metal Structures*, McGraw-Hill, New York (1952).
38. Appl, F. C. and C. F. Zorowski, *J. Appl. Mech.*, *26*: (1959).
39. Oran, C., *J. Eng. Mech. Div.*, *Proc. ASCE*, *94*: 621 (1968).
40. Ercoli, L. and P. A. A. Laura, *Proc.*, *1985 Pressure Vessels and Piping Conference*, PVP Vol. 98-6, ASME, NY, p. 195.
41. Kline, L. V. and J. O. Hancock, *J. Eng. for Industry, Trans. ASME*, *87B*: 232 (1965).
42. Ku, A. B., *Int. J. Mech. Sci.*, *20*: 593 (1978).

CHAPTER 20

Buckling and Bracing of Elastic Beams and Cantilevers

SRITAWAT KITIPORNCHAI* AND PETER F. DUX*

INTRODUCTION

Beams of open cross-section such as I-beams are usually loaded so as to bend about the major axis. Design of such beams usually requires consideration of bending strength and deflections. Bending strength may be limited by the plastic moment capacity, M_p, for heavily braced beams, by local buckling loads for thin flanged fabricated beams or by the flexural-torsional or lateral buckling load for beams not so heavily braced. Beam failure by flexural-torsional buckling is shown in Figure 1.

Classification of Buckling Failures

A typical classification of buckling failure types is presented in Figure 2 which shows possible load-deflection curves for a beam with a mid-point load. Beam slenderness decreases from A to D. Curve A is for a slender beam which buckles while fully elastic. Curve B is for a beam of intermediate slenderness which is partially yielded at failure. Geometrical and material imperfections influence the yielding pattern and hence the capacity. Curve C applies to a beam with a fully plastic zone but without the lateral support necessary for plastic hinge action. Most practical beams would exhibit type B or C failure under increasing loads. Beams of adequate rotational capacity under fully plastic conditions are described as compact beams. Curve D is intended to illustrate this behaviour. Compact beams may exhibit lateral buckling deformations before eventual collapse following local buckling.

*Department of Civil Engineering, University of Queensland, Australia

Influence of Imperfections

The phenomenon of beam buckling has been studied theoretically and experimentally. Most theoretical analyses assume an initially straight beam and seek the load at which the beam enters a state of neutral equilibrium. These are referred to as bifurcation-of-equilibrium analyses. At the bifurcation load the beam is theoretically capable of maintaining a buckling mode shape characterised by lateral displacement and twist. Experiments have shown that real beams with geometrical imperfections exhibit buckling type displacements from the onset of loading. The experimental buckling load is essentially the maximum attainable load at which buckling type displacements become large. Figure 3 shows the capacity versus slenderness behaviour of a beam which indicates that geometrical and material imperfections reduce capacities from the ideal, particularly as slenderness reduces.

Description of Following Sections

The following sections address the problem of elastic buckling. This mode of failure should be considered for slender beams which often occur during construction before final bracing is in place. As well, many codes of practice and design proposals in the literature employ an empirical reduction from ideal elastic capacities to obtain buckling loads for inelastic beams [1−2]. The sections summarise many available solutions obtained from bifurcation analyses. The solutions are applicable to uniform beams of doubly symmetric cross-section such as I-beams. Information on the buckling of monosymmetric beams and non-uniform beams can be obtained from References [3] to [7].

The next section considers simply supported beams and cantilevers under moment gradient loading, point loading

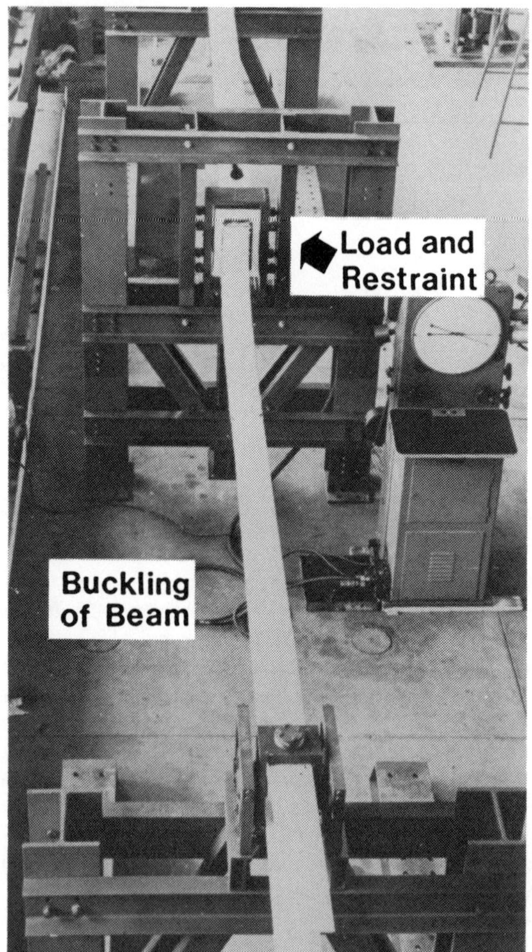

FIGURE 1. Lateral buckling of I-Beam.

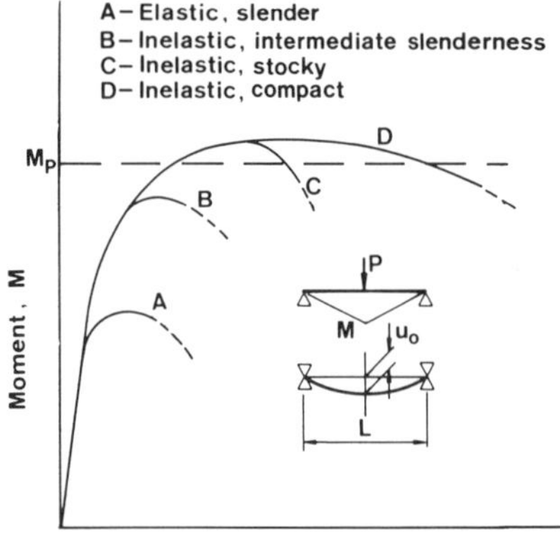

FIGURE 2. Load-deformation behaviour of typical I-Beam.

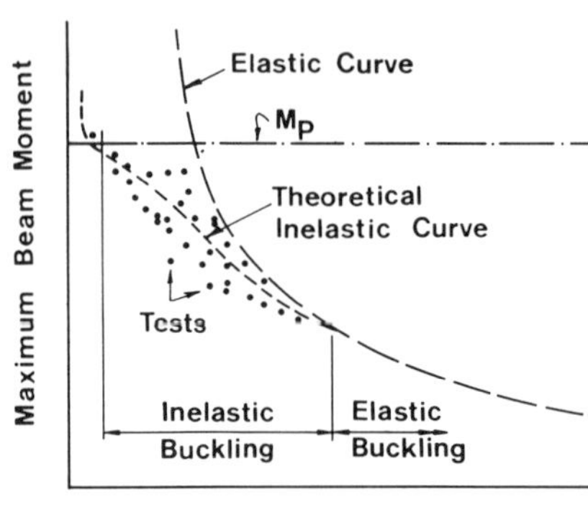

FIGURE 3. Beam capacity versus slenderness.

and uniformly distributed loading. Solutions are also presented for single span beams with a variety of end restraint conditions. The influence of the level of application of transverse loading is also examined.

The section following this examines elastic beams and cantilevers with internal bracing. The influence of brace type and location is investigated along with that of the level of application of load above and below the shear centre. Moment gradient loading, point loading and uniformly distributed loading are considered. The section provides data on optimum internal brace location.

The final section presents a general method for determining the elastic buckling loads of laterally continuous structures. These range from beams with multiple supports and braces to grid structures. An analysis procedure and worked examples are presented.

BUCKLING OF END RESTRAINED BEAMS

Introduction

This section gives lateral buckling capacities for single segment beams under a variety of major axis loading and end restraint conditions. The data applies to elastic beams which are initially straight and are of doubly symmetric cross-section. Additional information can be found from the references cited.

Simply Supported Beams with End Moments

The basic loading configuration is shown in Figure 4. Also shown are characteristic buckling displacements: lateral displacement, u, and twist, ϕ. The simple supports prevent twist and lateral movement at ends A and B but warping and minor axis rotation are free to occur. The moment ratio, β, lies in the range $-1 \leq \beta \leq +1$ with $\beta = -1$ for uniform bending. The elastic buckling solu-

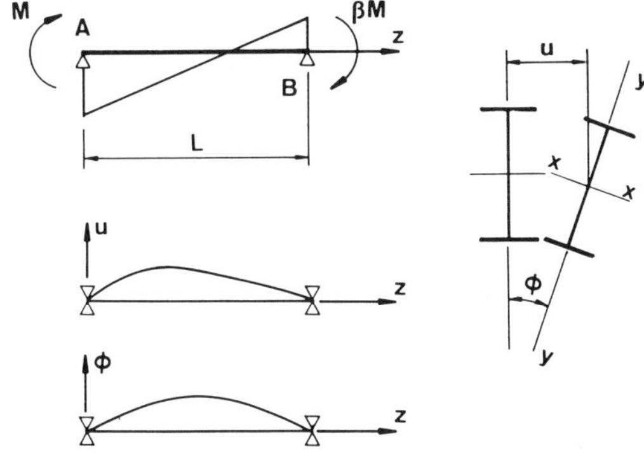

FIGURE 4. Simply supported beam with end moments.

tions limit the value of the larger end moment, M, which occurs at end A.

UNIFORM MOMENT LOADING ($\beta = -1$)

The elastic buckling moment for a beam in uniform bending is

$$M_o = \frac{\pi}{L}\sqrt{EI_yGJ}\left[\sqrt{1 + K^2}\right] \tag{1}$$

in which EI_y = minor axis bending rigidity; GJ = St. Venant torsional rigidity; EI_ω = warping rigidity; L = beam length and K = beam parameter,

$$K = \sqrt{\frac{\pi^2 EI_\omega}{GJL^2}} \tag{2}$$

The beam parameter, K, is a measure of the beam's ability to resist non-uniform torsion via internal warping restraint along its length. For practical purposes it ranges from $K = 0$ (narrow rectangular beam) to $K = 3.0$ (stocky I-beam). Figure 5 indicates the variation of K with cross-sectional properties and length for I-beams.

MOMENT GRADIENT LOADING ($-1 \leq \beta \leq +1$)

No closed form expressions are available for $\beta > -1$. Approximate capacities can be expressed in terms of Equation (1) and a moment modification factor, m, found by numerical solution of the governing differential equations (e.g. Reference [8]).

$$M = M_E = mM_o \tag{3}$$

in which

$$m = 1.75 + 1.05\,\beta + 0.3\,\beta^2 \leq 2.56 \tag{4}$$

and M is the larger end moment. Figure 6 lists moment

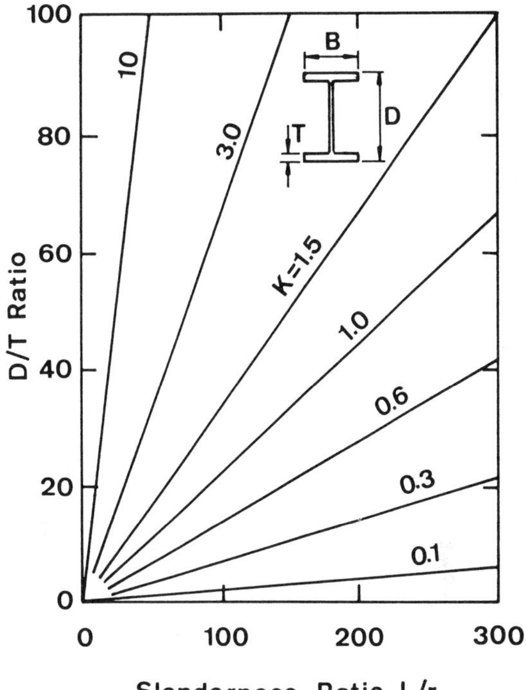

FIGURE 5. Variation of beam parameter, K.

Loading	Modification Factor, m★	Notes
M, βM	$1.75 + 1.05\beta + 0.3\beta^2$	≤ 2.56 $-1 \leq \beta \leq +1$
aL, P	$1.9 - 2.2a(1-a)$	$0 \leq a \leq .5$
aL, P, P, aL	$1 - 0.4a(1 - 5.5a)$	$0 \leq a \leq .5$
$q\frac{PL}{8}$, P, $q\frac{PL}{8}$	$1.35 + 0.36q$	$0 \leq q \leq 1$
$q\frac{wL^2}{12}$, w, $q\frac{wL^2}{12}$	$1.13 + 0.12q$ $4.8q - 2.38$	$0 \leq q \leq .75$ $.75 \leq q \leq 1$
M	0.25	K = 0
P	1.28	K = 0
w	2.05	K = 0

★Shear Centre Loading

FIGURE 6. Moment modification factors for beams and cantilevers.

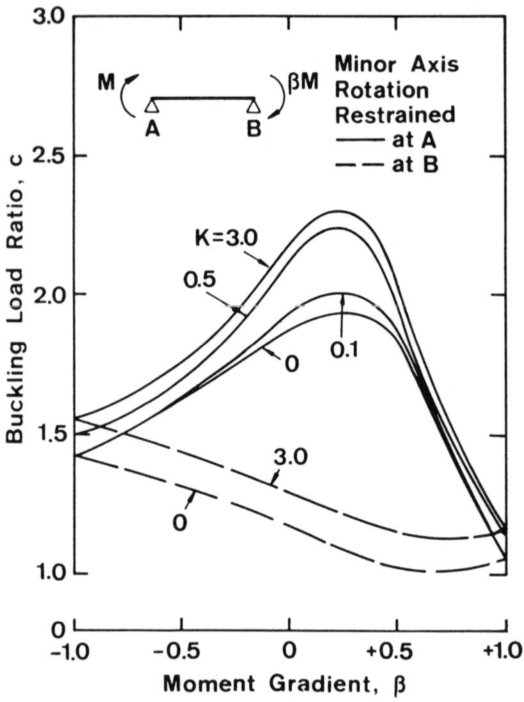

FIGURE 7. Buckling load ratios for moment gradient beams with one end restrained against minor axis rotation.

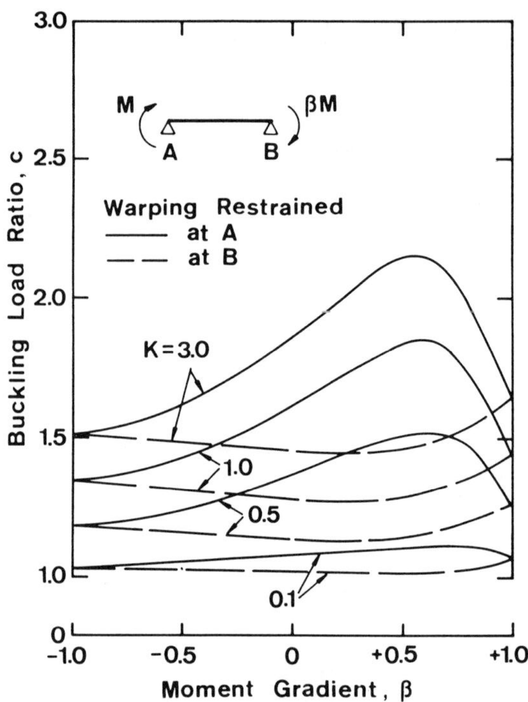

FIGURE 9. Buckling load ratios for moment gradient beams with one end restrained against warping.

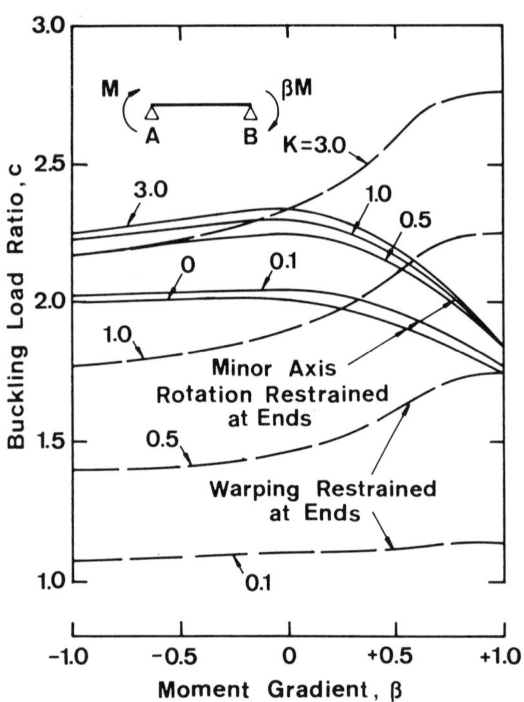

FIGURE 8. Buckling load ratios for moment gradient beams with both ends restrained against minor axis rotation or warping.

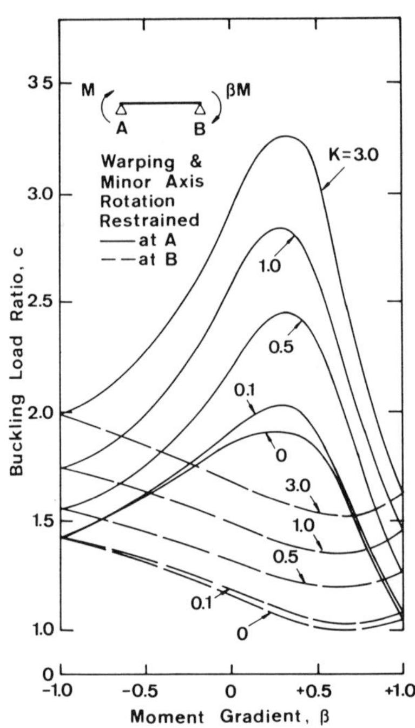

FIGURE 10. Buckling load ratios for moment gradient beams with one end restrained against warping and minor axis rotation.

modification factors for a range of beams and cantilevers. A more detailed presentation of m is offered in the following sections.

Beams with End Moments and Additional End Restraints

This section considers moment gradient beams with end restraints in addition to simple supports. End warping restraints and minor axis rotational restraints are examined separately and together at one or both ends. The restraint stiffness is assumed to be large enough to fully restrain the action in question.

Moment capacities can be found from

$$M = c.M_E \tag{5}$$

in which c = buckling load ratio from Figures 7 to 11.

MINOR AXIS ROTATIONAL RESTRAINT

Figure 7 indicates buckling load ratios due to a rotational restraint at either end. The restraint is most effective when placed at end A (moment = M) and when $-0.25 \leq \beta \leq +0.5$. A restraint at end B (moment = βM) becomes less effective as β increases. When $\beta \approx 0.75$ the buckling mode shape, u, has a zero derivative, du/dz, at end B and a restraint at B has no influence. Buckling load ratios in excess of unity in this region of β reflect inaccuracies in Equation (4) rather than any beneficial effect of the restraint.

Figure 8 shows that minor axis rotational restraints at both ends increase the buckling load by a factor ≈ 2 over the full β and K ranges. Figures 7 and 8 indicate a moderate variation of c with K.

WARPING RESTRAINT

Buckling load ratios for beams with a warping restraint at either A or B are given in Figure 9. The restraint is best placed at end A and is most effective for higher moment gradients. A warping restraint at end B produces a lesser and more uniform buckling load ratio. The modal warping displacement distribution, $d\phi/dz$, and its derivative, $d^2\phi/dz^2$, are less sensitive to changes in β than are u and du/dz. Of particular influence is the beam parameter K. Small K implies a beam which resists torsion mainly by developing St. Venant shearing stresses. The restraint of endwarping has thus only a marginal effect on torsional stiffness and hence on buckling capacity.

Warping restraints at both ends result in the buckling load ratios in Figure 8. Again K is of importance. References [9] and [10] provide warping restraint details and design methods.

COMBINED WARPING AND ROTATIONAL RESTRAINT

Figures 10 and 11 show the effect of combined restraints at either or both ends. For a single combined restraint, the restraint location, moment gradient and beam parameter are of influence. Combined restraints at both ends produce buckling load ratios which depend mainly on K.

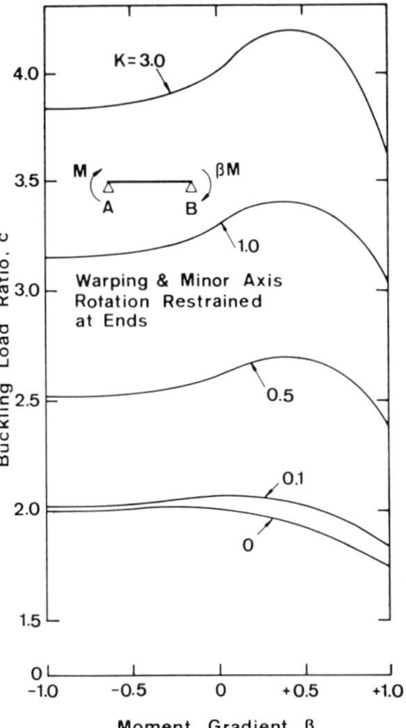

FIGURE 11. Buckling load ratios for moment gradient beams with both ends restrained against warping and minor axis rotation.

Simply Supported Beams with Transverse Loading

CENTRAL POINT LOAD

Moment modification factors, m, for loading at three levels; top flange (TF), shear centre (SC) and bottom flange (BF), are given in Figure 12. These factors are to be used with Equation (3) where M_o is found from Equation (1). The buckling moment thus obtained limits the maximum moment which in this instance, occurs at mid-span. The loads are free to sway laterally hence load application level is of importance as is the beam parameter. In comparison to a load at the shear centre, top flange loading is more severe as it accentuates twisting whereas a bottom flange load opposes it. The range of m is therefore large.

UNIFORMLY DISTRIBUTED LOAD

Moment modification factors for top flange, shear centre and bottom flange loading are given in Figure 13. Equation (3) provides a limit to the mid-span moment. Again, top flange loading is seen to be most severe and a substantial range of m is evident. This range increases with increasing

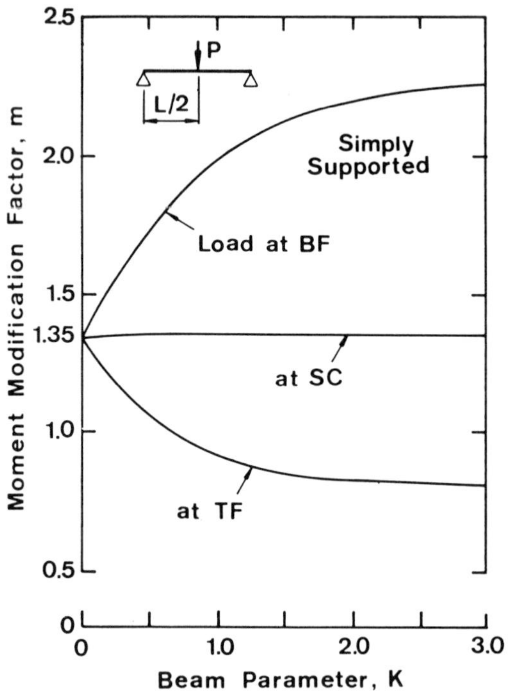

FIGURE 12. Moment modification factors for simply supported beams with central point load.

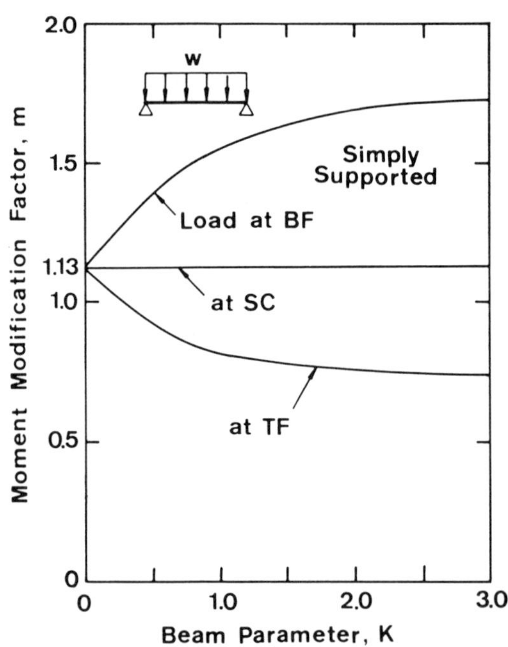

FIGURE 13. Moment modification factors for simply supported beams with uniformly distributed load.

K but is reasonably constant for $K > 2.0$. References [11] and [12] provide additional data.

Beams with Transverse Loading and Additional End Restraints

Beams with transverse loading and end restraints in addition to simple supports are considered here. Warping restraints and minor axis rotational restraints are examined separately or together at both ends. Reference [11] provides additional data.

MINOR AXIS ROTATIONAL RESTRAINT

Figures 14 and 15 give buckling load ratios for beams with both ends restrained against minor axis rotation and carrying the transverse loadings considered earlier. The ratios are sensitive to load level and beam parameter. They are to be used with Equation (5).

WARPING RESTRAINT

End warping restraints in addition to simple supports produce the buckling load ratios in Figures 16 and 17. As K increases, the additional torsional stiffness afforded to the beam by the restraint of end warping has a most pronounced influence when loading is at top flange level. The buckling load ratios are highest for loading at this level. This occurs for both transverse load types. The combined factor, $c \times m$, is highest for bottom flange loading.

COMBINED WARPING AND MINOR AXIS ROTATIONAL RESTRAINT

An effect similar to that described in the previous paragraph is seen in Figures 16 and 18 for combined warping and rotational restraints at both ends. The buckling load ratio is sensitive to K for $K < 1.0$ and approaches a more uniform value as K tends to 3.0.

Propped Cantilevers with Transverse Loading

Moment modification factors are presented in Figure 19 for propped cantilevers carrying a central point load or a uniformly distributed load. Warping and minor axis rotation are prevented at the fixed end along with major axis rotation. Conditions at the prop are assumed to prevent twist and lateral movement. Equation (3) limits the moment at the fixed end for both loadings. Three loading levels are considered and top flange loading is seen to be the most severe. The total distributed load is larger than the point load in all corresponding instances. Further information can be found in Reference [12].

Fixed-Ended Beams with Transverse Loading

A central point load and uniformly distributed loading are considered at top and bottom flange levels and at the shear

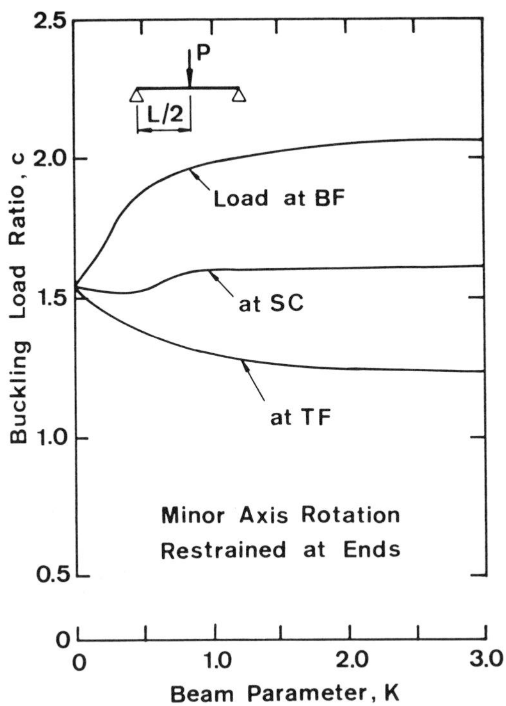

FIGURE 14. Buckling load ratios for minor axis rotation end restrained beams with central point load.

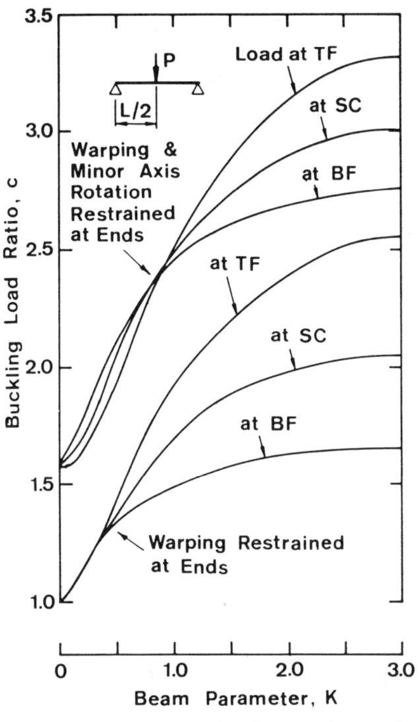

FIGURE 16. Buckling load ratios for warping and minor axis rotation end restrained beams with central point load.

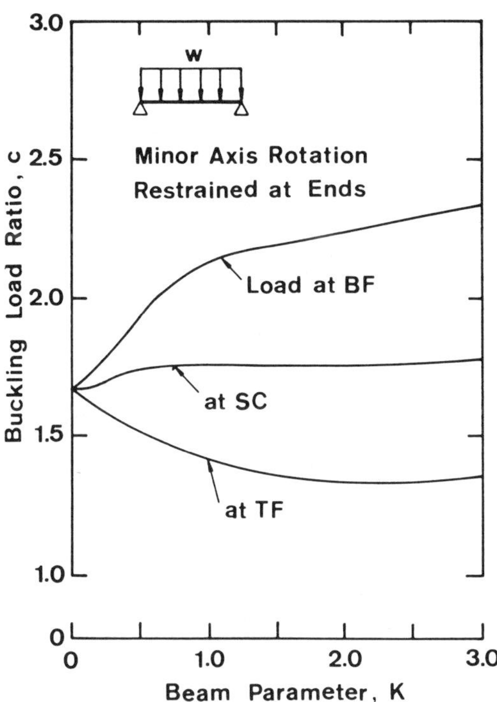

FIGURE 15. Buckling load ratios for minor axis rotation end restrained beams with uniformly distributed load.

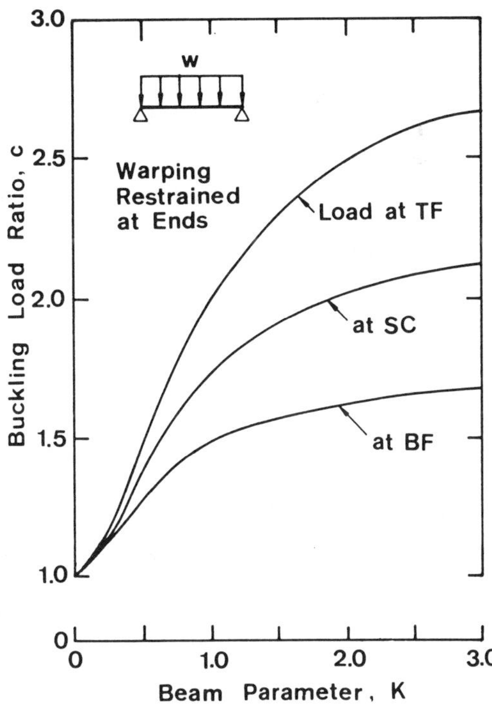

FIGURE 17. Buckling load ratios for warping end restrained beams with uniformly distributed load.

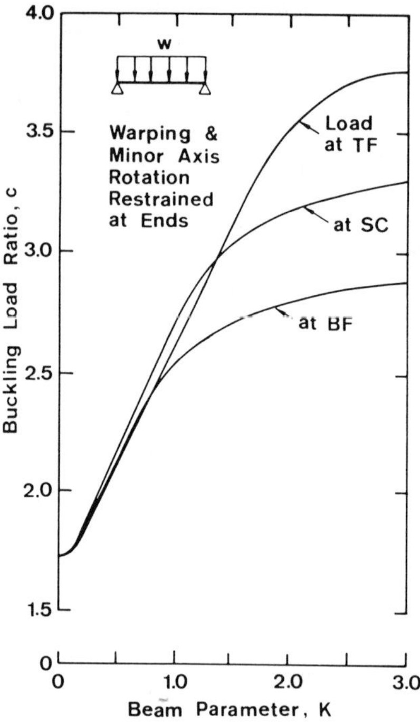

FIGURE 18. Buckling load ratios for warping and minor axis rotation end restrained beams with uniformly distributed load.

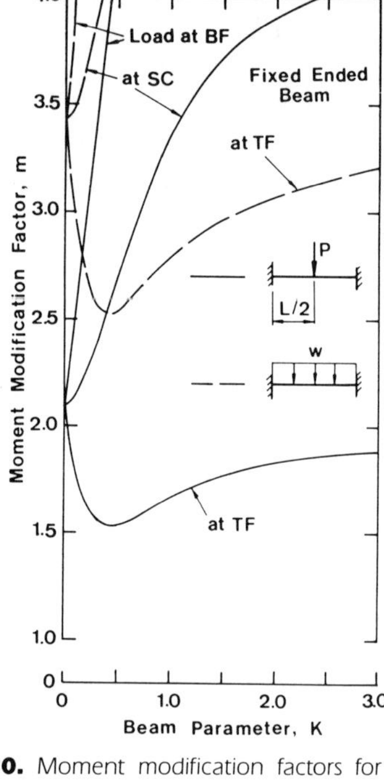

FIGURE 20. Moment modification factors for fixed ended beams.

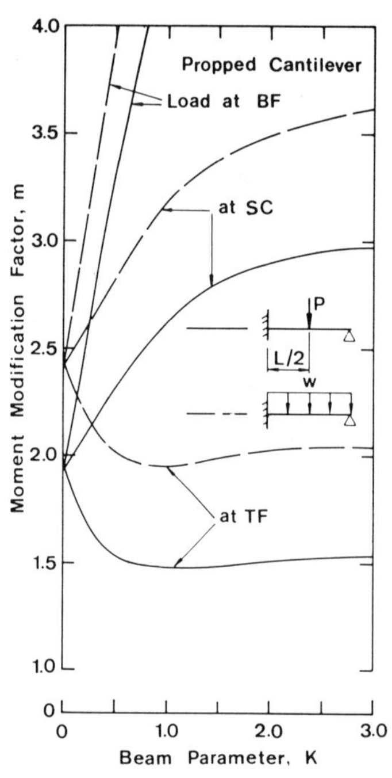

FIGURE 19. Moment modification factors for propped cantilevers.

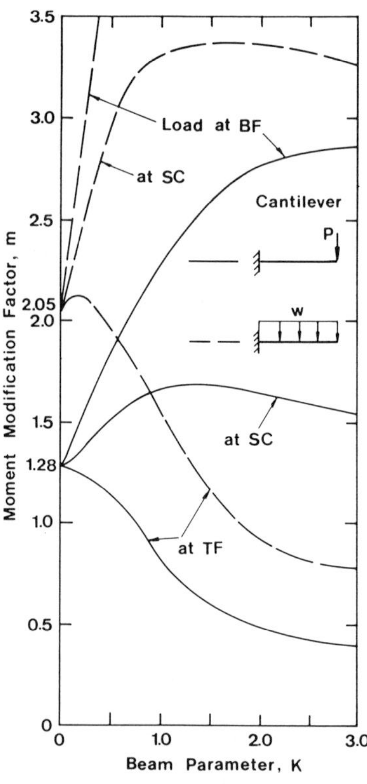

FIGURE 21. Moment modification factors for cantilevers.

centre. All movement is prevented at the fixed ends. Moment modification factors in Figure 20 along with Equation (3) limit the moments at these ends. As expected top flange loading is most severe and point loading is less favourable for stability than is distributed loading. Modification factors for shear centre and bottom flange loading increase rapidly with K. Reference [13] provides capacities for beams under variable major axis rotational end restraints.

Cantilevers with Transverse Loading

Figure 21 provides moment modification factors for cantilevers with tip loading or uniformly distributed loading at three levels of application. The cantilevers are free to move at all locations other than at the fixed end. Buckling capacities from Equation (3) limit the moment at this end. The modification factors vary with K and reduce dramatically for top flange loading. Note that the modification factors for transversely loaded cantilevers in Figure 6 apply only when $K = 0$. Additional data are presented in References [14] and [15].

Concluding Remarks

This section has dealt with the stability of simply supported beams, cantilevers and beams with a variety of additional end restraints and under a number of different loadings. Several parameters are of influence, notably the beam parameter, moment gradient and level of application of transverse loading. The benefits of judiciously chosen end restraints are obvious from the many figures. The next section discusses the stability of beams and cantilevers with braces along their length. Buckling load ratios from that section are to be used with modification factors from this section when assessing buckling loads.

BRACING OF BEAMS

Types of Bracing

Bracing may be provided by either a continuous medium such as a diaphragm or by one or more discrete braces along the beam length. Usually, it is not economical to provide anything less than "rigid bracing" for a member. To achieve a condition of rigid bracing, the brace must possess certain minimum stiffness. A typical relationship between the ratio of the elastic buckling load of a braced beam and the stiffness of the brace, k_s, is shown in Figure 22. For members under uniform moment with a central brace, there is a limiting value of the brace stiffness. In general the curve rises slowly with increasing k_s, and there is no distinct discontinuity in the curve. However, the limiting stiffness can be defined as one which is close to that corresponding to rigid bracing. In practice, it is not difficult to achieve adequate bracing stiffness and code rules for assessing this requirement are available [1,2,16]. In this chapter, adequate stiffness is assumed for the types of braces considered.

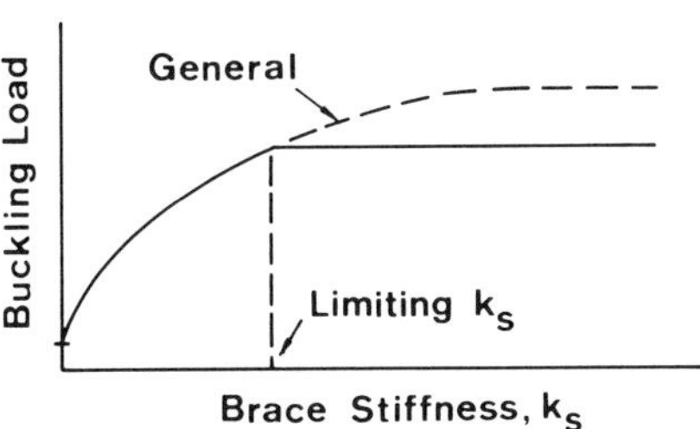

FIGURE 22. Typical relationship between buckling load and brace stiffness.

While beams may be braced in many different ways, most arrangements can be represented by an idealised system comprising an elastic lateral brace acting at a distance $\bar{b}$ above the shear centre of the beam cross-section and an elastic rotational brace (see Figure 23). Full bracing effectively prevents lateral deflection and twist at the cross-section, while partial bracing allows some limited twisting or lateral deflection to occur. The term lateral bracing is used to describe bracing which prevents lateral deflection at the point where the brace is placed while rotational bracing prevents twisting only.

The increase in the buckling capacity in a beam due to the particular brace is defined by the buckling load ratio c,

$$c = \frac{\text{buckling capacity of the braced beam}}{\text{buckling capacity of a similar unbraced beam}} \tag{6}$$

Unbraced beam capacities can be found in the previous sec-

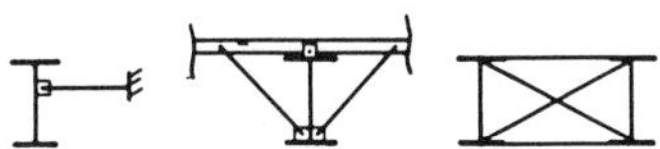

(a) Typical Braces

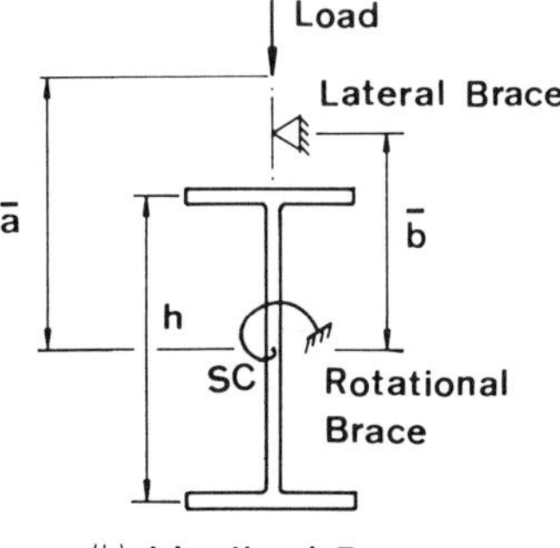

(b) Idealised Braces

FIGURE 23. Cross-section at brace.

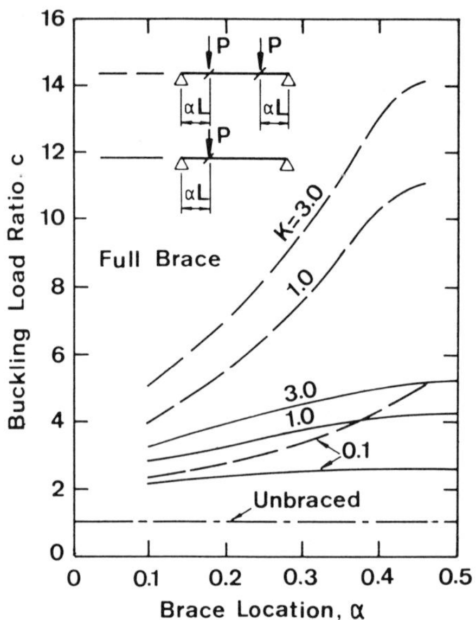

FIGURE 24. Buckling load ratios for simply supported beams with one or two point loads and full braces.

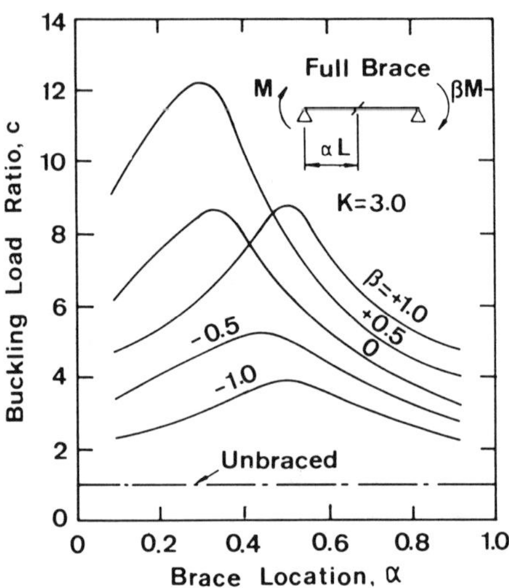

FIGURE 26. Buckling load ratios for simply supported beams with end moments and a full brace, K = 3.0.

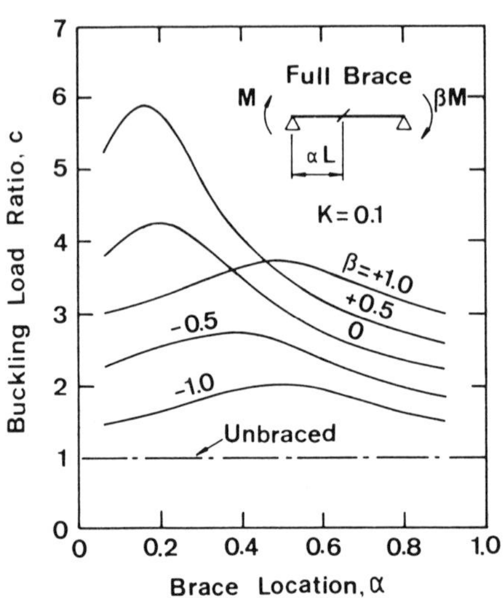

FIGURE 25. Buckling load ratios for simply supported beams with end moments and a full brace, K = 0.1.

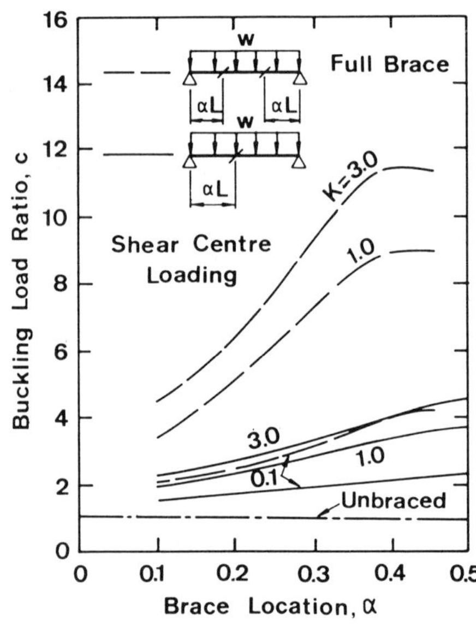

FIGURE 27. Buckling load ratios for simply supported beams with uniformly distributed load and one or two full braces.

tion. Results presented in the following sections have been obtained using the method of finite integrals [17] to solve the governing differential equations for bending and torsion [18]. Braces were included by imposing appropriate boundary conditions involving lateral deflection, u, and cross-sectional rotation, ϕ.

Simply Supported Beams

BRACED BEAMS WITH POINT LOADS

The buckling of simply supported beams with one or two point loads acting at the shear centre is examined in Figure 24. It is common in practice for load points to be braced and hence the height of application of the load above the shear centre, $\bar{a}$, does not affect beam buckling capacity. Buckling load ratios, c, are shown in Figure 24 for three values of beam parameter, K, and for one or two point loads at location, α.

BRACED BEAMS WITH END MOMENTS ($-1 \leq \beta \leq +1$)

Simply supported beams with end moments are considered in this section. The effect of a full brace at various locations along a beam is shown respectively, in Figures 25 and 26 for $K = 0.1$ and 3.0. The curves indicate an optimum brace location for each in-plane moment distribution.

BRACED BEAMS WITH UNIFORMLY DISTRIBUTED LOAD

Simply supported beams with uniformly distributed loads and with one or two intermediate braces are considered in Figures 27 and 28. This type of loading and bracing is common in roof structures where the distributed load may arise from wind or live loading. The load may act at the top flange, shear centre, bottom flange or at any other level. It is usual to provide fly-bracing to the bottom flange of a roof beam under uplift in order to increase its lateral buckling capacity.

Solutions for simply supported beams with distributed loads acting at the shear centre are shown in Figure 27. These results are similar to those obtained for beams with point loads (see Figure 24) and demonstrate the considerable influence of the beam parameter K on the value of c. The effect of the level of load application is shown in Figure 28 for $K = 1.0$. The increase in the buckling load ratio is more pronounced for a beam with top flange loading. Unbraced loads applied above the shear centre contribute to the destabilising influence of in-plane moments, hence appropriately placed braces are particularly effective.

Figures 27 and 28 show an increase in the buckling load ratio as the brace is moved towards the beam centre. A single brace is most effective when placed at mid-span. Two braces in close proximity combine to restrain minor axis rotation and this is evident as the two approach mid-span. Contrary to the common practice of bracing the third points of a beam with uniformly distributed load, it is better to locate braces at the 2/5 points ($\alpha \approx 0.4$). This leads to increases in the elastic buckling loads of 10 to 15 percent.

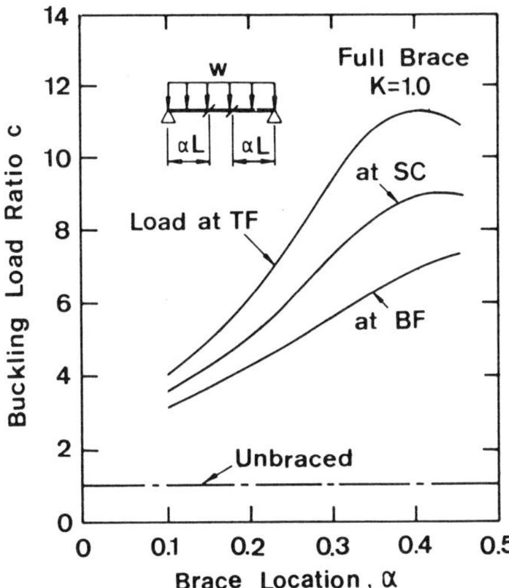

FIGURE 28. Influence of load height on simply supported beams with uniformly distributed load and two full braces.

EFFECT OF LATERAL OR ROTATIONAL BRACING

Simply supported beams with lateral bracing or rotational bracing alone are considered. Results for beams with uniformly distributed load are compared in Figure 29. Beams with a lateral brace attached to the top flange, the shear centre, or the bottom flange are compared to beams with a rota-

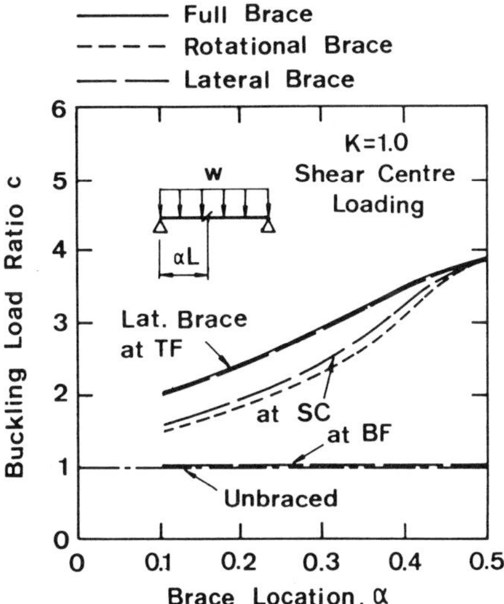

FIGURE 29. Comparison of brace types of simply supported beams with uniformly distributed load, K = 1.0.

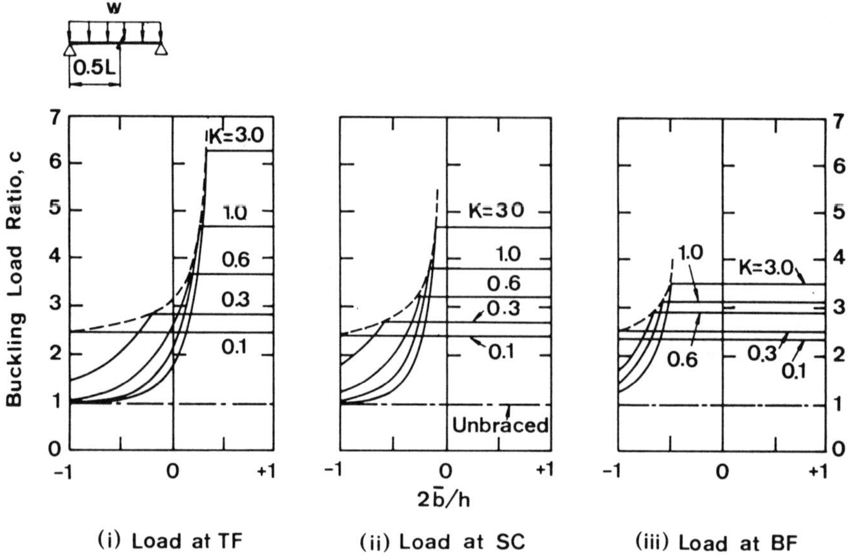

FIGURE 30. Influence of lateral brace and load height for simply supported beams.

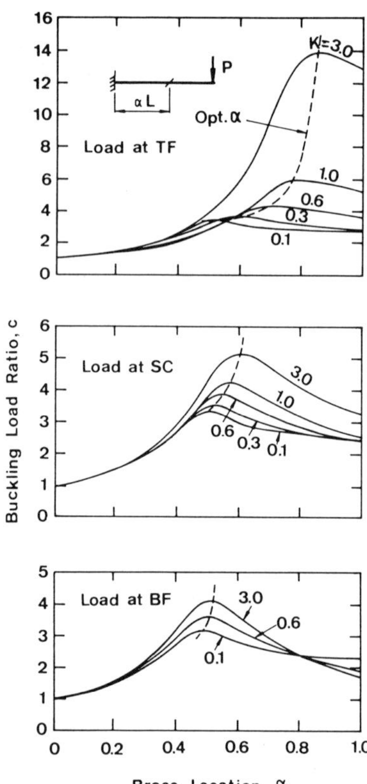

FIGURE 31. Buckling load ratios for cantilevers with a tip load and a full brace.

tional brace at the shear centre, and to beams with a full brace. Lateral bracing is most effective when acting at the top or compression flange level. When attached to the bottom or tension flange, the brace in this instance is complately ineffective. A rotational brace is best located near mid-span.

In Figure 30, the effect of the position of the lateral brace above and below the shear centre is considered with load acting at the top flange, the shear centre or the bottom flange. Solutions have been obtained for a number of beam parameters and with the brace at mid-span ($\alpha = 0.5$). Results indicate that the influence of the brace depends on the brace level ($2\bar{b}/h$), the load height ($2\bar{a}/h$) and also the beam parameter K. Generally it is best to attach the lateral brace to the compression side. For small values of K, the position of attachment is unimportant, and the brace is quite effective even if placed a the tension flange level. However, for larger values of K, the region suitable for brace attachment diminishes. This reduction becomes more severe if the load acts on the compression flange.

Cantilevers

BRACED CANTILEVERS

The influence of the position of a full brace is examined. The buckling load ratio, c, for values of the beam parameters $K = 0.1$ to 3.0 are shown in Figures 31 and 32 for cantilevers with a tip load and uniformly distributed load respectively. The loads are applied at top flange, shear centre or bottom flange. It can be seen that the increases in the buckling load are greatest for large values of the beam

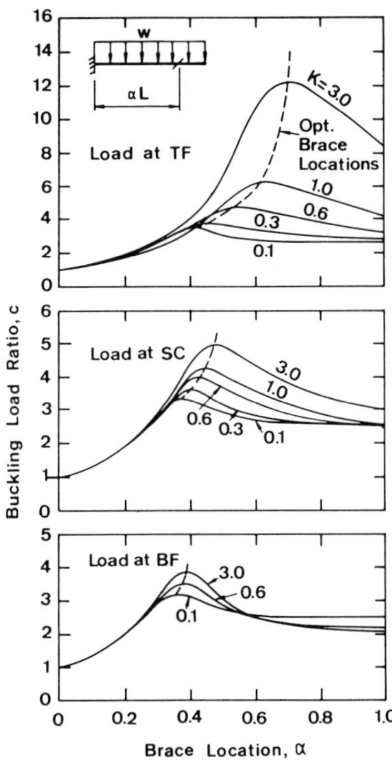

FIGURE 32. Buckling load ratios for cantilevers with uniformly distributed load and a full brace.

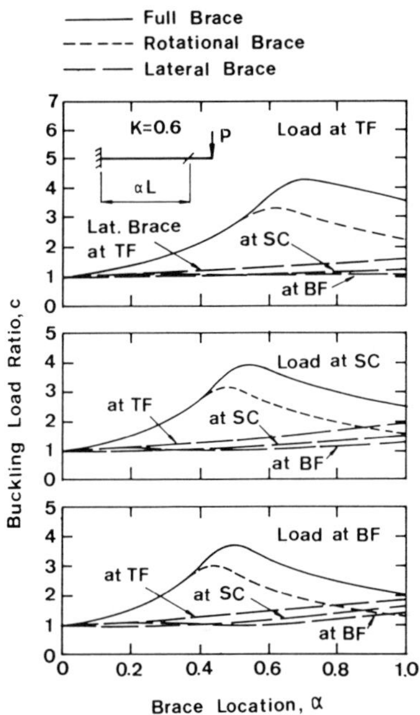

FIGURE 33. Comparison of brace types for cantilevers with a tip load, K = 0.6.

parameter and more so for top flange loading. The maximum value of c that may be achieved ranges from 3 for small values of K to 14 for large values of K.

The results show that for small values of K, the optimum brace location is near mid-span for a tip load and near 0.4 of the length from the fixed end for uniformly distributed load. For higher values of K, the optimum brace locations approach the cantilever tip as the height of load application moves toward the top flange. For a tip load, the optimum location varies between $\alpha = 0.5$ and 0.8 and for a uniformly distributed load, between $\alpha = 0.4$ and 0.7.

EFFECTS OF LATERAL AND/OR ROTATIONAL BRACING

The effectiveness of lateral bracing at various levels of attachment is compared with that of rotational bracing and of full bracing for top flange, shear centre and bottom flange loading, for values of $K = 0.6$ and 3.0. The results are shown respectively, in Figures 33 and 34 for tip load and in Figures 35 and 36 for uniformly distributed load.

The various braces have different influences on the buckling load depending on the level of load height ($2\bar{a}/h$). In all cases full bracing is by far the best. If full bracing cannot be achieved, rotational bracing is the next best as can be seen from Figures 33 to 36. Optimum brace locations are clearly evident.

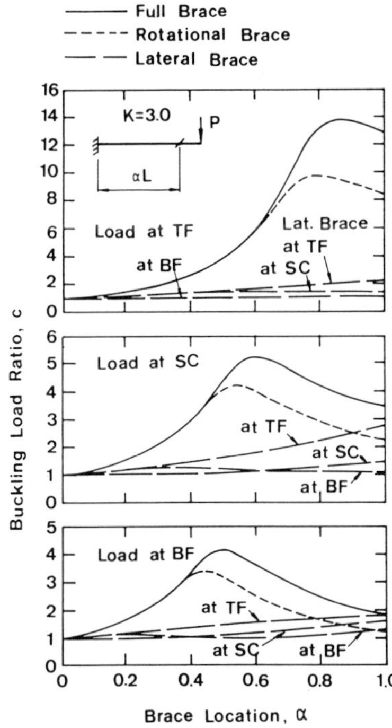

FIGURE 34. Comparison of brace types for cantilevers with a tip load, K = 3.0.

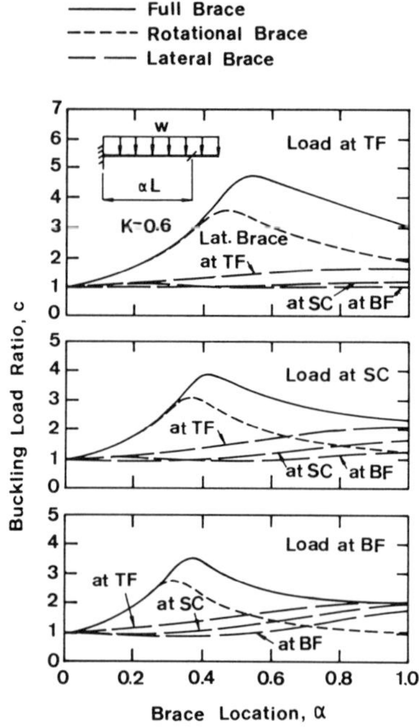

FIGURE 35. Comparison of brace types for cantilevers with uniformly distributed load, K = 0.6.

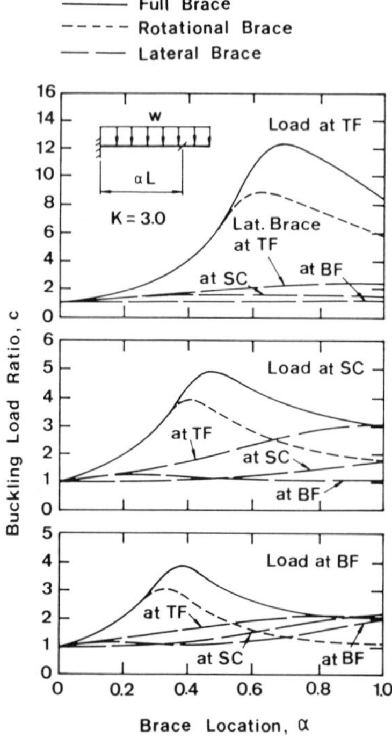

FIGURE 36. Comparison of brace types for cantilevers with uniformly distributed load, K = 3.0.

For lateral bracing only, the buckling load ratios increase slowly as the brace moves towards the tip, irrespective of the level of the brace. Varying the value of K has only little effect on the maximum value of c for top flange and bottom flange loadings. However, for shear centre loading and with a top flange brace, the effect of increasing K is a marked improvement in the value of c.

Lateral braces should be placed as close as possible to the cantilever tip. The effectiveness increases as the level of application of load moves towards the bottom flange. In all cases if lateral bracing alone is used, it should be placed near the top flange and as close as possible to the cantilever tip. Braces placed less than 0.4L from the fixed end are practically useless.

Concluding Remarks

The beam parameter K, has a significant influence on the elastic buckling strength of simply supported beams and cantilevers with intermediate braces. The effectiveness of any type of brace on a beam depends on both K and the location of the brace.

For simply supported beams with a single lateral brace only, the brace is generally effective when acting above the shear centre. When the brace acts below the shear centre the effectiveness depends on the beam parameter K. In general the buckling resistance is signiffcantly increased when loads act below the shear centre.

For cantilevers, the optimum location of a full brace for most cases varies between 0.4L to 0.7L from the fixed end support. For cantilevers with single lateral brace, the brace is best placed near the top (tension) flange level. However, this arrangement is not as effective as a rotational brace or a full brace.

BUCKLING OF LATERALLY CONTINUOUS BEAMS

Introduction

This section presents a simple, approximate method for determining the elastic buckling strength of laterally continuous beams. The term "laterally continuous" implies indeterminacy in the lateral (minor axis) direction imposed by braces both at and away from the beam ends. Typical examples of laterally continuous structures amenable to solution by this method are shown in Figure 37. The basic requirements are:

(1) the structures are loaded by point loads at braced cross-sections;
(2) the braces prevent lateral deflection and twist (full brace); and
(3) elements of the structure are straight and major axis curvature effects are negligible.

The approximate analysis follows the principles of interaction buckling as set out in References [19] and [20] and for grids in Reference [21]. A brief review of these principles is given in the following section.

Interaction Buckling

In an interaction buckling analysis a laterally continuous beam is modelled as an assemblage of segments connected at braced points. If the beam is disturbed from its initially straight position it adopts patterns of minor axis displacement and twist consistent with the brace constraints of $u = \phi = 0$. Compatibility at braced points also requires minor axis bending and warping interaction between segments at their ends. Reference [20] shows that minor axis bending and warping stiffnesses at segment ends are major axis load dependent and generally deteriorate parabolically with increasing load. While these stiffnesses are positive the segment taken separately is stable. If end stiffnesses become negative the segment requires end restraint. At loads less than the buckling load, some segments are able to offer minor axis bending and warping end restraint to more severely loaded segments and the beam as a whole is able to resist disturbance. At the buckling load the reserve of stiffness is zero. The beam enters a state of neutral equilibrium in which it is unable to resist a vanishingly small disturbance.

Model for Analysis

CRITICAL SEGMENT AND SUBASSEMBLAGE

Buckling analysis of the isolated segments (e.g. with Equation (3) can indicate whether a segment is likely to provide restraint or to require it as the beam approaches instability. Reference 19 introduced the term "critical segment" to identify that segment giving the lowest beam load factor for failure from an analysis excluding segment interaction. In the approximate method it is assumed that the critical segment undergoes an earlier deterioration of stiffness and places a higher restraint demand than does any other segment. It therefore dominates in limiting beam capacity. An estimate of the restraint available to the critical segment can be made by examining the immediately adjacent segments. A subassemblage comprising the critical segment and the two adjacent segments is thus identified. It is assumed that the behaviour of the subassemblage adequately reflects that of the beam. A typical beam and subassemblage is shown in Figure 38.

BOUNDARY CONDITIONS

Boundary conditions are required at the far ends of the restraining segments. Support conditions such as a simple support or fixed end are retained. If the far end continues on to another segment it is assumed that a restraint demand equal to that from the critical segment occurs at this end. Alternative conditions are discussed in Reference [20].

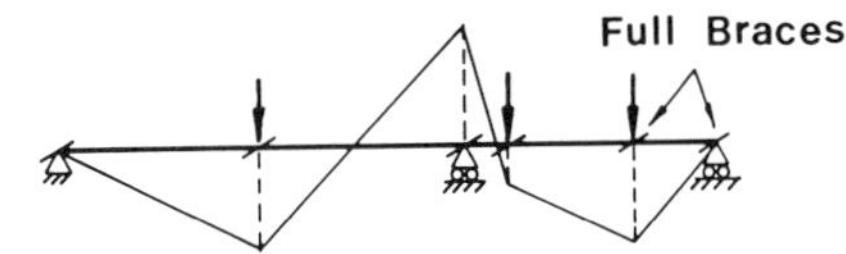

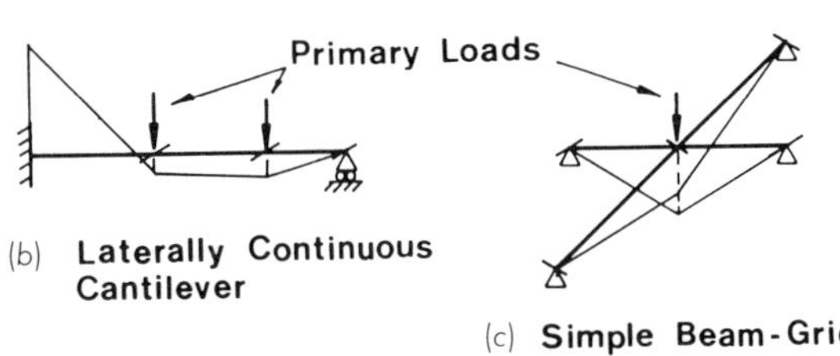

FIGURE 37. Laterally continuous structures.

RESTRAINT STIFFNESS

It is assumed that both the warping and minor axis bending stiffness at the end of a restraining segment vary in the manner given for the bending stiffness in Equation (7) (see Reference [20]).

$$\text{Minor axis bending stiffness at segment end} = n\left(\frac{EI_y}{L}\right)_R\left(1 - \left(\frac{M}{M_E}\right)^2\right)_R \tag{7}$$

where the subscript R refers to the restraining segment and $n = 3$ if the far end is simply supported, $n = 4$ if fixed and $n = 2$ if continuous to another segment. The moment M_E is the elastic buckling moment of the restraining segment found from Equation (3) and M is the larger end moment in the segment. The variation in stiffness is parabolic and results in zero stiffness when $M/M_E = 1.0$. The restraint

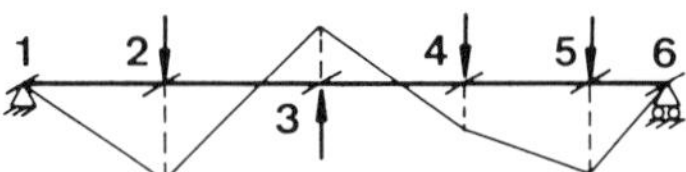

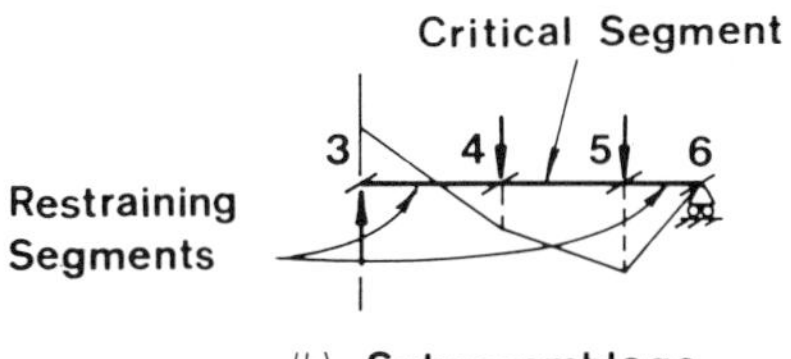

FIGURE 38. Typical beam and subassemblage.

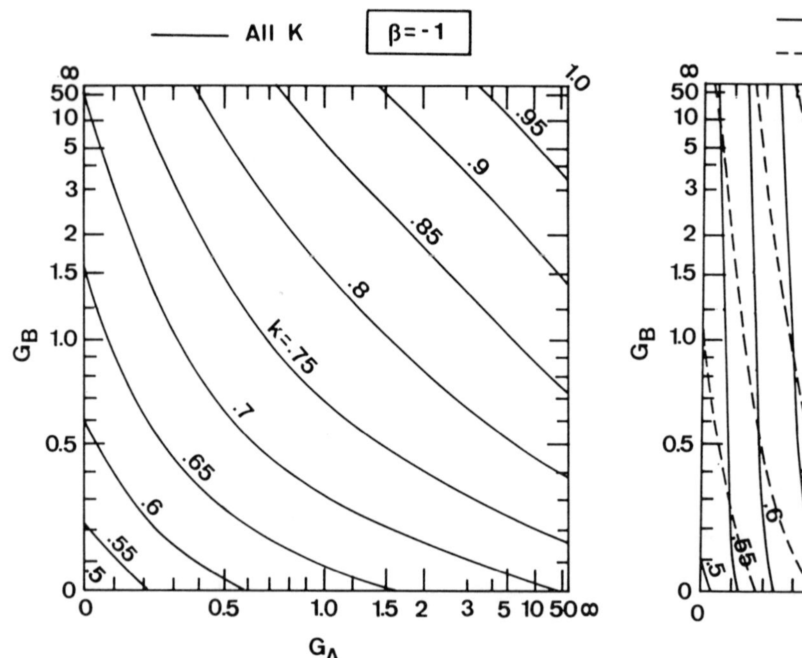

FIGURE 39. Effective length factor chart for $\beta = -1$.

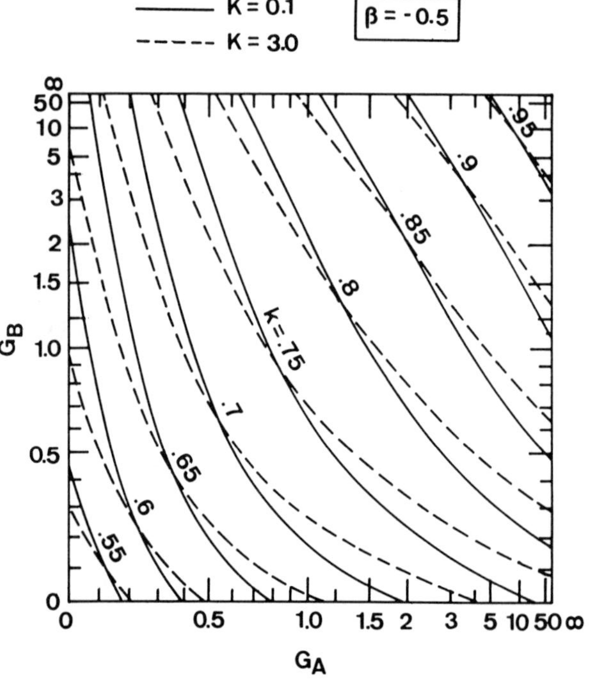

FIGURE 40. Effective length factor chart for $\beta = -.5$.

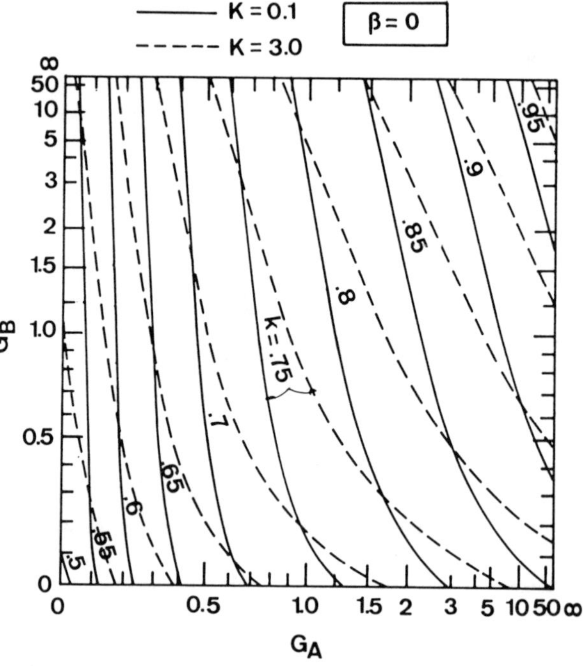

FIGURE 41. Effective length factor chart for $\beta = 0$.

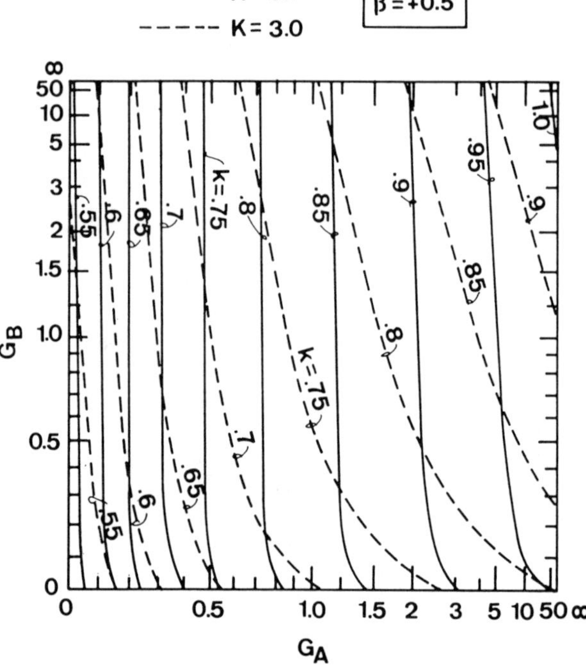

FIGURE 42. Effective length factor chart for $\beta = +.5$.

stiffness is described in a restraint parameter, G, where

$$G_{A,B} = \frac{2\left(\frac{EI_y}{L}\right)_C}{n\left(\frac{EI_y}{L}\right)_R\left(1 - \left(\frac{M}{M_E}\right)^2\right)_R} \tag{8}$$

Subscript C refers to the critical segment and subscripts A and B refer to ends A and B of the critical segment (see Figure 4 for A and B definition). The restraint parameter at an end describes equally the warping and minor axis bending stiffnesses available at that end.

Effective Length Factors

The load factor associated with the buckling of the restrained critical segment is assumed to approximate the load factor for beam failure.

The buckling moment of the critical segment can be expressed as:

$$M_F = m\frac{\pi}{kL}\sqrt{EI_yGJ}\sqrt{1 + \left(\frac{K}{k}\right)^2} \tag{9}$$

where k = effective length factor. When k = 1.0, Equation (9) reduces to Equation (3). Figures 39 to 43 present effective length charts for restrained critical segments for the full range of β and for K varying from 0.1 to 3.0 (see References [20] and [22] for details and additional charts).

Analysis Procedure and Worked Examples

SUMMARY OF STEPS

(1) Determine the major axis bending moment distribution.
(2) Determine K and β for each segment.
(3) For each segment calculate M_E from Equation (3) and the corresponding beam load factor, λ, to produce M_E. The segment with the lowest load factor, λ_C, is the critical segment. The two (at most) adjacent segments have higher laod factors, λ_R.
(4) Assume a trial value of λ_F, the load factor at subassemblage buckling, and calculate G_A and G_B from Equation (8). Note that

$$\frac{\lambda_F}{\lambda_R} = \left(\frac{M}{M_E}\right)_R \tag{10}$$

and this substitution can be made in Equation (8). The trial value λ_F should lie between λ_C and λ_R (min).
(5) Find the critical segment effective length factor, k, using the appropriate chart from Figures 39 to 43, extrapolating linearly if necessary.

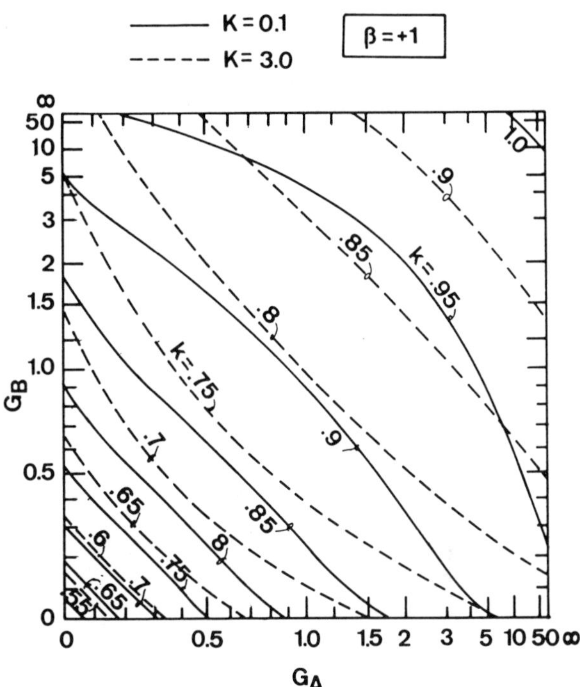

FIGURE 43. Effective length factor chart for $\beta = +1$.

(6) Calculate the revised critical segment buckling moment, M_F, from Equation (9) and obtain a new load factor, λ_F (new). Note that

$$\frac{\lambda_F(\text{new})}{\lambda_C} = \left(\frac{M_F}{M_E}\right)_C \tag{11}$$

(7) Compare the new load factor, λ_F (new), with the assumed factor, λ_F (Step 4), and repeat Steps 4 to 6 if necessary until good agreement is obtained.

The process of cycling ensures consistency between assumed values at Step 4 and calculated values at Step 6. Usually only two or three cycles are required if a reasonable initial guess for λ_F is made at Step 4. Converging upper and lower bounds are found by choosing an initial value of λ_F equal to λ_C and subsequent of λ_F equal to those calculated at Step 6.

WORKED EXAMPLE 1

The analysis procedure is applied to the beam in Figure 44a. Much of the data is summarised in Table 1.

STEP 1 Calculate bending moments (see Figure 44a).
STEP 2 Find β and K for each segment (see Table 1).
STEP 3 Calculate M_E and load factors, λ, to produce buckling in simply supported segments (see Table 1).

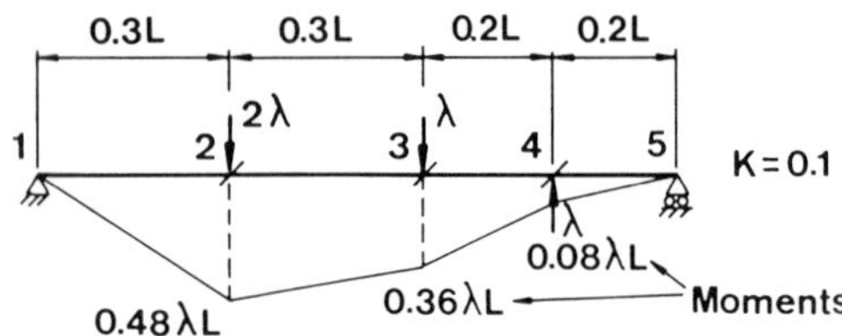

(a) **Laterally Continuous Beam**

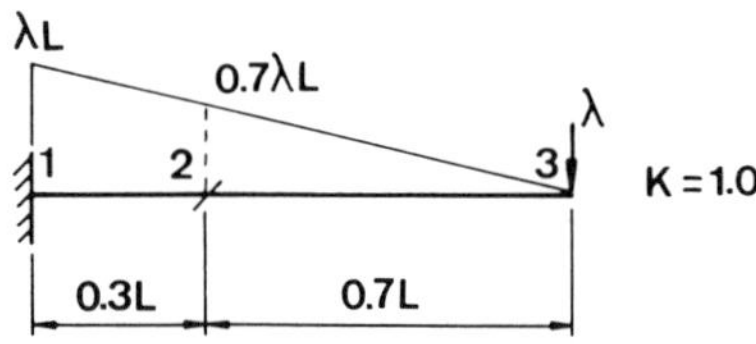

(b) **Braced Cantilever**

FIGURE 44. Worked examples.

Segment 2-3 is the critical segment. The subassemblage comprises segments 1-2, 2-3, 3-4.

STEP 4 Assume a value of λ_F and calculate G_A and G_B. End 2 of the critical segment has the higher end moment and is taken as end A. Usually a close guess can be made for λ_F from the information gathered at Step 3, but for the purposes of this example, an initial value of 25.96 $\sqrt{EI_yGJ}/L$ will be used. Therefore

$$G_A = \frac{2}{3} \times \frac{.3L}{.3L} \times \frac{1}{\left(1 - \left(\frac{25.96}{40.21}\right)^2\right)} = 1.14$$

and

$$G_B = \frac{2}{2} \times \frac{.2L}{.3L} \times \frac{1}{1 - \left(\frac{25.96}{70.72}\right)^2} = .77$$

STEP 5 Find the critical segment effective length factor by interpolation between Figures 39 and 40. Effective length factor, $k \approx 0.76$.

STEP 6 Calculate the revised critical segment buckling moment, M_F, and corresponding load factor, λ_F.

$$M_F = \frac{1.13\pi}{.3L \times .76}\sqrt{EI_yGJ}\sqrt{1 + \left(\frac{.33}{.76}\right)^2}$$

$$= 17.0\sqrt{EI_yGJ}/L$$

and

$$\lambda_F = 35.36\sqrt{EI_yGJ}/L^2$$

Table 2 shows the convergence of λ_F which is taken as the mean value of the estimations at cycles 3 and 4, i.e.

$$\lambda_F = 33.4\sqrt{EI_yGJ}/L$$

A finite integral analysis gives

$$\lambda_F = 34.91\sqrt{EI_yGJ}/L^2$$

TABLE 1. Analysis Data.

Segments	1–2	2–3	3–4	4–5
β =	0.0	−.75	−.22	0.0
K =	.33	.33	.5	.5
*M_E =	19.3	12.46	26.9	30.7
**λ =	40.21	25.96	70.72	384.12

*all multiplied by $\sqrt{EI_yGJ}/J$

**all multiplied by $\sqrt{EI_yGL}/L^2$.

WORKED EXAMPLE 2

The analysis procedure can be refined to obtain better estimates of M_E for restraining segments at Step 3. This modification is advisable when the restraining segment at end A of the critical segment has a well restrained or fixed far end. Analysis of the cantilever in Figure 44b illustrates this. Table 3 summarises analysis data.

STEP 1 Calculate bending moments (see Figure 44b).

STEP 2 Find β and K for each segment (see Table 3).

STEP 3 Calculate M_E and load factors λ for buckling of simply supported segments (see Table 3). Segment 2-3 is critical and is restrained at end A.

STEP 4 (revised). Calculate M_F and load factor λ for buckling of segment 1-2 when simply supported at end 2 and fully restrained at end 1.

$$G_1 = G_A = 0.0$$

$$G_2 = G_B = \infty \quad \text{and } k = 0.68.$$

Hence, the contents of Table 3 are revised to Table 4.

STEP 5 Estimate $\lambda_F \approx 38$. Therefore

$$G_A = \frac{2}{4} \times \frac{.3L}{.7L}\frac{1}{\left(1 - \left(\frac{38}{89.4}\right)^2\right)} = 2.61$$

$$G_B = \infty$$

TABLE 2. Cycles of Analysis.

Cycle	λ_F	G_A	G_B	k	λ_F(new)
1	25.96	1.14	.77	.76	35.36
2	35.36	2.94	.89	.82	32.40
3	32.40	1.90	.84	.79	33.80
4	33.80	2.27	.86	.81	33.0

STEP 6 Find the critical segment effective length factor from Figure 41 ($\beta = 0$).

$$k = .675$$

STEP 7 Calculate the revised critical segment buckling moment and corresponding load factor

$$M_F = 27.24 \sqrt{EI_yGJ}/L$$

and

$$\lambda_F = 38.8 \sqrt{EI_yGJ}/L^2 \qquad \text{cf. } 38\sqrt{EI_yGJ}/L^2$$

i.e. λ_F is approximately 38.45. The standard procedure leads to a value of 32.5 whereas a finite integral analysis gives

$$\lambda_F = 37.6 \sqrt{EI_yGJ}/L^2$$

Concluding Remarks

The approximate method produces accurate results for a wide range of laterally continuous structures as shown by comparison with rigorous solutions (finite element and finite integral) presented in References 20 and 21. Reference 19 outlines a simpler but less accurate method. Reference 23 develops an analysis procedure which accounts for additional warping and minor axis bending interaction which sometimes arises in beams with concentrated moment loading.

TABLE 3. Analysis Data.

Segment	=	1–2	2–3
β	=	−0.7	0.0
K	=	3.33	1.43
*M_E	=	42.34	13.69
**λ	=	42.34	19.56

*all multiplied by $\sqrt{EI_yGJ}/L$.
**all multiplied by $\sqrt{EI_yGJ}/L^2$.

TABLE 4. Revised Data.

Segment	=	1–2	2–3
*M_E	=	89.4	13.69
**λ	=	89.4	19.56

*all multiplied by $\sqrt{EI_yGJ}/L$.
**all multiplied by $\sqrt{EI_yGJ}/L^2$.

NOTATION

- $\bar{a}$ = height of point of application of load above shear centre
- BF = bottom flange
- $\bar{b}$ = height of lateral brace above shear centre
- C = subscript referring to critical segment
- c = ratio of buckling load of braced beam and similar unbraced beam
- E = Young's modulus of elasticity
- F = subscript referring to subassemblage at buckling;
- G = shear modulus of elasticity
- G_A, G_B = restraint parameters
- h = distance between flange centroids
- I_y = minor axis second moment of area
- I_ω = warping section constant
- J = torsion section constant
- K = beam parameter
- k = effective length factor
- k_s = brace stiffness
- L = length of beam
- M, M_E = elastic buckling moment
- M_o = elastic buckling moment of beam under uniform moment ($\beta = -1$)
- M_F = elastic buckling moment of substructure
- M_p = plastic moment
- m = moment modification factor
- n = stiffness coefficient, commonly 2, 3, 4
- R = subscript referring to restraining segment
- r_y = minor axis radius of gyration
- SC = shear centre
- TF = top flange
- u = lateral deflection of shear centre
- x,y = major and minor principal axes
- z = centroidal axis
- α = location of brace
- β = moment gradient or ratio of major axis end moments
- λ = load factor
- ϕ = angle of twist

REFERENCES

1. Standards Association of Australia, "AS 1250-1981 SAA Steel Structures Code," SAA, Sydney (1981).
2. "Specification for the Design, Fabrication and Erection of Structural Steel for Buildings," AISC, New York (1976).
3. Kitipornchai, S. and N. S. Trahair, "Buckling Properties of Momosymmetric I-Beams," *Journal of the Structural Division*, ASCE, Vol. 106, No. ST5, pp. 941 – 957 (May 1980).
4. Kitipornchai, S., C. M. Wang, and N. S. Trahair, "Buckling of Monosymmetric I-Beams under Moment Gradient," *Journal of Structural Engineering*, ASCE, Vol. 112, No. 4, pp. 781-799 (April 1986).
5. Wang, C. M. and S. Kitipornchai, "Buckling Capacities of Monosymmetric I-Beams," Research Report No. CE62, Department of Civil Engineering, University of Queensland (October 1985).
6. Trahair, N. S. and S. Kitipornchai, "Elastic Lateral Buckling of Stepped I-Beams," *Journal of the Structural Division*, ASCE, Vol. 97, No. ST10, pp. 2535 – 2548 (Oct. 1971).
7. Kitipornchai, S. and N. S. Trahair, "Elastic Stability of Tapered I-Beams," *Journal of the Structural Division*, ASCE, Vol. 98, No. ST3, pp. 713 – 728 (March 1973).
8. Trahair, N. S., *The Behaviour and Design of Steel Structures*, Champan and Hall, London, England (1977).
9. Vacharajittiphan, P. and N. S. Trahair, "Warping and Distortion at I-Section Joints," *Journal of the Structural Division*, ASCE, Vol. 100, No. ST8, pp. 547 – 564 (March 1974).
10. Dux, P. F. and D. B. Peters, "Buckling of I-Beams with Discrete Warping Restraints," *Proceedings*, 9th Australasian Conference on the Mechanics of Structures and Materials, Sydney, pp. 13 – 17 (1984).
11. Trahair, N. S., "Stability of I-Beams with Elastic End Restraints," *Journal of the Institution of Engineers, Australia*, Vol. 37, No. 6, pp. 157 – 168 (June 1965).
12. Trahair, N. S., "Elastic Stability of Propped Cantilevers," *The Civil Engineering Transactions*, The Institution of Engineers, Australia, Vol. CE10, No. 1, pp. 94 – 100 (April 1968).
13. Trahair, N. S., "Elastic Stability of I-Beam Elements in Rigid-Jointed Structures," *Journal of the Institution of Engineers, Australia*, Vol. 38, No. 7-8, pp. 171 – 180 (July-August 1966).
14. Anderson, J. M. and N. S. Trahair, "Stability of Monosymmetric Beams and Cantilevers," *Journal of the Structural Division*, ASCE, Vol. 98, No. ST1, pp. 269 – 286 (January 1972).
15. Nethercot, D. A., "Effective Length of Cantilevers as Governed by Lateral Buckling," *The Structural Engineer*, Vol. 51, No. 5, pp. 161 – 168 (May 1973).
16. Mutton, B. R. and N. S. Trahair, "Stiffness Requirements for Lateral Bracing," *Journal of the Structural Division*, ASCE, Vol. 99, No. ST10, pp. 2167 – 2182 (October 1973).
17. Brown, P. T. and N. S. Trahair, "Finite Integral Solution of Differential Equations," *The Civil Engineering Transactions*, The Institution of Engineers, Australia, Vol. CE10, No. 2, pp. 193 – 196 (1968).
18. Kitipornchai, S. and Richter, N. J., "Elastic Lateral Buckling of I-Beams with Discrete Intermediate Restraints," *Civil Engineering Transactions*, The Institution of Engineers, Australia, Vol. CE20, No. 2, pp. 105 – 111 (1978).
19. Nethercot, D. A. and Trahair, N. S., "Lateral Buckling Approximations for Elastic Beams," *The Structural Engineer*, Vol. 54, No. 6, pp. 197 – 204 (June 1976).
20. Dux, P. F. and S. Kitipornchai, "Elastic Buckling of Laterally Continuous I-Beams," *Journal of the Structural Division*, ASCE, Vol. 108, No. ST9, pp. 2099 – 2116 (September 1982).
21. Kitipornchai, S. and P. F. Dux, discussion of "Lateral Buckling of Intersecting Connected Beams" by S. P. Morchi and E. G. Lovell, *Journal of the Engineering Mechanics Division*, ASCE, Vol. 105, No. EM3, pp. 490 – 492 (June 1979).
22. Dux, P. F. and S. Kitipornchai, "Buckling Approximations for Laterally Continuous Elastic I-Beams," Research Report No. CE11, Dept. of Civil Engineering, University of Queensland (April 1980).
23. Cuk, P. E. and N. S. Trahair, "Buckling of Beams with Concentrated Moments," *Journal of Structural Engineering*, ASCE, Vol. 109, No. 6, pp. 1387 – 1401 (June 1983).

CHAPTER 21

Instability of Monosymmetric I-Beams and Cantilevers

T. M. Roberts*

ABSTRACT

The elastic flexural-torsional instability of monosymmetric I-beams and cantilevers is discussed and solutions are presented for a variety of loading conditions, which incorporate the influence of monosymmetry, height of load application relative to the shear centre and the influence of pre-buckling displacements. Later sections deal with the calculation of section properties and current design practice.

INTRODUCTION

Thin walled bars of open cross-section such as angles, channels, T and I-sections are used extensively in practice since they provide an economical use of material for the provision of stiffness and strength. However, the structural behaviour of such members is complex since they are susceptible to instability in a variety of modes, depending on the loading and boundary conditions, and therefore to geometrically non-linear behaviour arising from initial imperfections, residual stresses and material nonlinearity.

In particular, laterally unrestrained thin walled beams of open cross-section, subjected to bending about their major axis are susceptible to flexural-torsional or lateral instability. This mode of inbstability is characterised by lateral displacement and rotation of the cross-section as illustrated in Figure 1. Therefore, current design practice is to base the maximum permissible stress on the elastic critical stress for flexural-torsional instability, modified to allow for plasticity effects in short span beams and with an appropriate factor of safety to allow for initial imperfections and residual stresses.

*Department of Civil and Structural Engineering, University College, Cardiff, Wales, United Kingdom

The elastic flexural-torsional instability of doubly symmetric I-beams is well documented and many closed form solutions for relatively simple loading and boundary conditions appear in standard texts [6,24]. The more general theory of thin walled bars of open cross-section has also been established [6,8,9,24,26].

In comparison with doubly symmetric I-beams, the flexural torsional instability of monosymmetric I-beams i.e. I-beams possessing only one axis of symmetry as shown in Figure 2, has received little attention. Figure 2(b) is typical of steel sections used for crane gantry girders while Figure 2(c) is typical extruded aluminum sections used as glazing bars. The earliest studies of monosymmetric I-beams were made by Vlasov [26] and Goodier [9] who obtained a closed form solution for a simply supported beam subjected to uniform moment. Since that time a number of solutions, some approximate and some theoretically suspect, have been proposed [10,11,14,15,27]. These solutions are discussed by Anderson and Trahair [1] who presented numerical solutions based on the governing differential equations, for monosymmetric I-beams and cantilevers subjected to uniformly distributed and concentrated loads. Anderson and Trahair's numerical solutions were also verified experimentally and are now generally accepted as being correct. The instability of monosymmetric I-beams is complicated by the fact that the centroid and shear centre of the cross-section do not coincide. This leads to an imbalance in the Wagner effect whereby the torsional stiffness of the section is influenced by axial stresses [1].

In a recent series of articles, Roberts et al. [16–21] used energy methods to study the instability and geometrically nonlinear behaviour of thin walled bars of open cross-section. The basis of the analysis is a set of nonlinear expressions for the strains occurring in thin walled bars of open cross-section where subjected to axial, flexural and torsional displacements, derived in accordance with beam theory and the Vlasov assumption that shear deformation

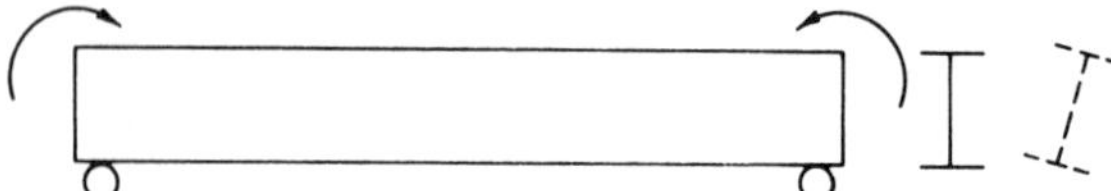

FIGURE 1. Flexural-torsional or lateral instability of a thin walled beam.

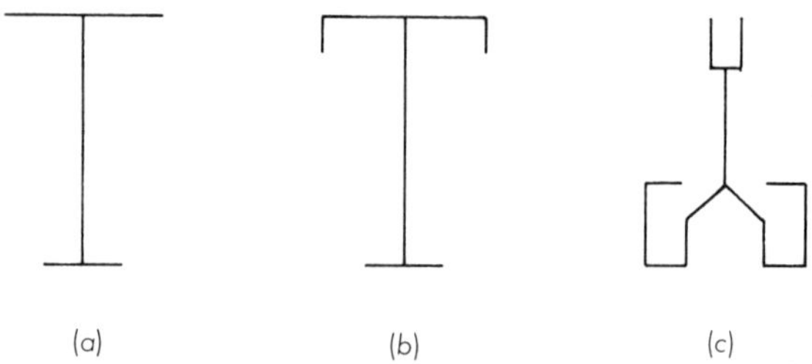

FIGURE 2. Monosymmetric I-sections: (a) simple geometry; (b) with lipped flange; (c) complex aluminum extrusion.

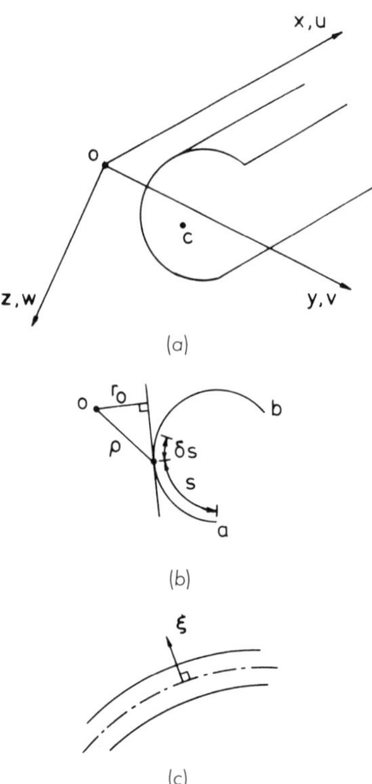

FIGURE 3. Thin walled bar of open cross-section.

due to non-uniform bending and torsion is negligible [16]. It has been shown that the nonlinear expressions for the strains have general application [18–21] and when incorporated in a general instability analysis based on the vanishing of the second variation of the total potential energy, the influence of pre-buckling displacements is automatically incorporated [17,25]. Closed form solutions have also been obtained for the flexural-torsional instability of monosymmetric I-beams subjected to various loading conditions and which are valid for a wide range of section properties [20,21]. Where the closed form solutions are inaccurate due to the complexity of the buckled shape, and for cantilevers, recourse has to be made to numerical solutions.

In this chapter, the energy solutions developed by Roberts, et al. for predicting the elastic critical loads of monosymmetric I-beams and cantilevers are discussed. A major difficulty in applying these and other similar solutions in design is the determination of the section properties, in particular those associated with warping of the cross-section and monosymmetry. A section is devoted therefore to this aspect of the problem and a final section is devoted to current design philosophy.

NONLINEAR STRAINS

Nonlinear expressions for the strains occurring in thin walled bars of open cross-section when subjected to axial, flexural and torsional displacements have been derived by Roberts [16] and only the final expressions are reproduced herein. A typical thin walled bar of variable thickness t is shown in Figure 3(a). The x, y, and z axes pass through the shear centre o and are parallel to the principal axes which pass through the centroid c. The shear centre is the point through which the resultant shear force acts when the bar is subjected to nonuniform bending. The centre of rotation is defined as the point about which the section rotates when the bar is subjected to a pure torque and coincides with the shear centre for linear strains. u denotes the displacement of the centroid in the x direction, v and w denote displacements of the shear centre in the y and z directions and Θ denotes the clockwise rotation of the section about the shear centre.

The displacements u, v, and w and rotation Θ produce nonlinear axial strains ϵ and St. Venant shear strains γ which vary over the cross-section and which are given by

$$\begin{aligned} \epsilon = & - y(v_{xx}\cos\Theta + w_{xx}\sin\Theta) - z(w_{xx}\cos\Theta - v_{xx}\sin\Theta) \\ & + \alpha_w(\Theta_{xx} - w_{xxx} v_x + v_{xxx} w_x) \qquad (1) \\ & + u_x + 0.5(v_x^2 + x_x^2 + \varrho^2\Theta_x^2) \end{aligned}$$

$$\gamma = 2\xi\,(\Theta_x - w_{xx} v_x + v_{xx} w_x) \qquad (2)$$

In Equations (1) and (2) suffices x, xx, and xxx denote differentiation with respect to x, y and z are distances measured from the centroid in the positive y and z directions and ϱ is the distance from the shear centre to any point on the cross-section. α_w is a parameter which relates warping of the cross-section and, for the particular section shown in Figure 3(b), is defined by the equations [24]:

$$\alpha_w = \bar{\alpha}_s - \alpha_s \tag{3}$$

$$\alpha_s = \int_a^s r_o ds$$

$$\bar{\alpha}_s = \frac{1}{\bar{m}} \int_a^b t\, \alpha_s\, ds \tag{4}$$

$$\bar{m} = \int_a^b t\, ds$$

In Equations (4), $r_o\, ds$ is considered positive if the direction of integration is clockwise about the shear centre and vice versa. ξ is the distance measured from and perpendicular to the mean profile of the cross-section as shown in Figure 3(c).

For assumed elastic material, axial stress σ and St. Venant shear stresses τ are related to ϵ and γ by the equations

$$\sigma = E\, \epsilon \qquad \text{and} \qquad \tau = G\, \gamma \tag{5}$$

in which E and G denote Young's modulus and shear modulus, respectively.

ENERGY EQUATIONS

Equilibrium states are defined by the condition that the first variation of the total potential energy V, denoted by δV, is zero [3]. For a thin walled bar subjected to external forces P_i, the corresponding displacements being q_i, and having internal stresses σ and τ and corresponding strains ϵ and γ, this can be expressed by the equation

$$\delta V = -\, P_i\, \delta q_i + \int [\sigma \cdot \delta\epsilon + \tau \cdot \delta\gamma] dvol = 0 \tag{6}$$

in which the integration is over the volume of the structure.

For stable equilibrium, $\delta V = 0$ corresponds to a minimum value and the second variation of V for stationary values of the external forces, denoted by $\delta^2 V$, is positive definite, that is, positive for all admissible variations in displacements and corresponding strains. From Equation (6), $\delta^2 V$ is given by

$$\delta^2 V = -\, P_i\, \delta^2 q_i + \int \Big[\sigma \cdot \delta^2\epsilon + \frac{1}{2}\, \delta\sigma \cdot \delta\epsilon + \tau \cdot \delta^2\gamma + \frac{1}{2}\, \delta\tau \cdot \delta\gamma \Big] dvol \tag{7}$$

NB. If V can be expressed as a function of variables q_i, δ and δ^2 are defined as

$$\delta = \frac{\partial}{\partial q_i} \cdot \delta q_i\ ;\ \delta^2 = \frac{1}{2}\, \frac{\partial^2}{\partial q_i \partial q_j} \cdot \delta q_i\, \delta q_j \tag{8}$$

Critical conditions occur when $\delta^2 V$ changes from positive definite to zero, indicating a transition from stable equilibrium to instability.

For the monosymmetric I-beam shown in Figure 4(a), which prior to buckling is subjected to positive moment M, i.e., moment producing positive curvature about its principal y-axis, the pre-buckling stresses σ and τ are

$$\sigma = -\, \frac{Mz}{I_y}\ ;\ \tau = 0 \tag{9}$$

in which I_y is the second moment of area about the principal y-axis.

Variations in the axial and shear stresses due to flexural torsional buckling, $\delta\sigma$ and $\delta\tau$, are related to variations in the axial and shear strains, $\delta\epsilon$ and $\delta\gamma$, by the equations

$$\delta\sigma = E\, \delta\epsilon \qquad \text{and} \qquad \delta\tau = G\, \delta\gamma \tag{10}$$

The first and second variations of the axial and shear strains defined by Equations (1) and (2) are

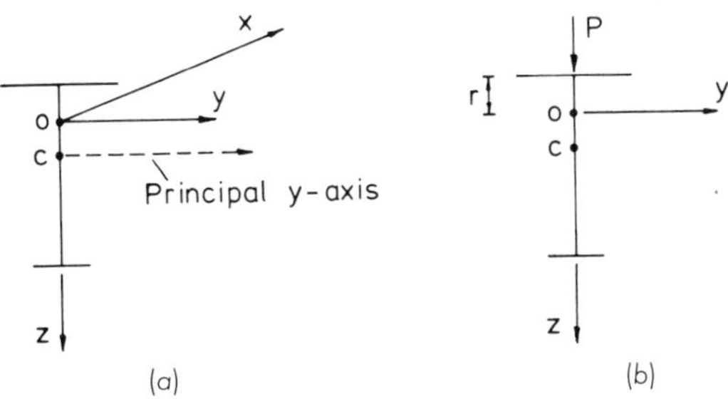

FIGURE 4. Monosymmetric I-beam: (a) axes and principal y-axis; (b) load applied above shear centre.

$$
\begin{aligned}
\delta\varepsilon = & - y(- v_{xx} \cdot \sin\Theta \cdot \delta\Theta + \delta v_{xx} \cdot \cos\Theta \\
& + w_{xx} \cdot \cos\Theta \cdot \delta\Theta + \delta w_{xx} \cdot \sin\Theta) \\
& - z(- w_{xx} \cdot \sin\Theta \cdot \delta\theta + \delta w_{xx} \cdot \cos\Theta \\
& - v_{xx} \cdot \cos\Theta \cdot \delta\theta - \delta v_{xx} \cdot \sin\Theta) \\
& + \alpha_w(\delta\Theta_{xx} - w_{xxx} \cdot \delta v_x - \delta w_{xxx} \cdot v_x \\
& + v_{xxx} \cdot \delta w_x + \delta v_{xxx} \cdot w_x) \\
& + \delta u_x + v_x \cdot \delta v_x + w_x \cdot \delta w_x + \varrho^2 \cdot \Theta_x \cdot \delta\Theta_x
\end{aligned} \tag{11}
$$

$$
\begin{aligned}
\delta^2\varepsilon = & - \frac{y}{2}(- v_{xx}\cos\Theta \cdot \delta\Theta \cdot \delta\Theta - 2\delta v_{xx}\sin\Theta \cdot \delta\Theta \\
& - w_{xx}\sin\Theta \cdot \delta\Theta \cdot \delta\Theta + 2\delta w_{xx} \cdot \cos\Theta \cdot \delta\Theta) \\
& - \frac{z}{2}(- w_{xx} \cdot \cos\Theta \cdot \delta\Theta \cdot \delta\Theta - 2\delta w_{xx} \cdot \sin\Theta \cdot \delta\Theta \\
& + v_{xx} \cdot \sin\Theta \cdot \delta\Theta \cdot \delta\Theta - 2\delta v_{xx} \cdot \cos\Theta \cdot \delta\Theta) \\
& + \frac{\alpha_w}{2}(- 2\, \delta w_{xxx} \cdot \delta v_x + 2\delta v_{xxx} \cdot \delta w_x) \\
& + 0.5(\delta v_x \cdot \delta v_x + \delta w_x \cdot \delta w_x + \varrho^2\, \delta\Theta_x \cdot \delta\Theta_x)
\end{aligned} \tag{12}
$$

$$
\begin{aligned}
\delta\gamma = & \, 2\xi(\delta\Theta_x - w_{xx} \cdot \delta v_x - \delta w_{xx} \cdot v_x \\
& + v_{xx} \cdot \delta w_x + \delta v_{xx} \cdot w_x)
\end{aligned} \tag{13}
$$

$$
\delta^2\gamma = 2\xi(- \delta w_{xx} \cdot \delta v_x + \delta v_{xx} \cdot \delta w_x) \tag{14}
$$

Prior to buckling, the displacements u, v, and rotation Θ are zero. Hence substituting Equations (9) to (14) into Equation (7), and integrating over the area of the cross-section A gives

$$
\begin{aligned}
\delta^2 V = & - P_i\, \delta^2 q_i \\
& - \int M \left\{ \frac{w_{xx}}{2}\, \delta\Theta \cdot \delta\Theta + \delta v_{xx} \cdot \delta\Theta \right. \\
& \left. + \frac{C_y}{2I_y}\, \delta\Theta_x \cdot \delta\Theta_x \right\} dx \\
& + \frac{GJ}{2} \int (\delta\Theta_x - w_{xx} \cdot \delta v_x + w_x \cdot \delta v_{xx})^2 dx \\
& + \frac{EI_z}{2} \int (\delta v_{xx} + w_{xx}\delta\Theta)^2 dx \\
& + \frac{EC_w}{2} \int (\delta\Theta_{xx} - w_{xxx} \cdot \delta v_x + w_x \cdot \delta v_{xxx})^2 dx \\
& + \frac{EA}{2} \int (\delta u_x + w_x \delta w_x)^2 dx \\
& + \frac{EI_y}{2} \int \delta w_{xx} \cdot \delta w_{xx} dx
\end{aligned} \tag{15}
$$

In arriving at Equation (15) the following integrations have been used

$$
\int y\, dA = \int z\, dA = \int yz\, dA = 0 \tag{16}
$$

$$
\int \alpha_w dA = \int y\alpha_w\, dA = \int z\, \alpha_w\, dA = 0 \tag{17}
$$

$$
\begin{aligned}
& \int dA = A \\
& \int y^2 dA = I_z \\
& \int z^2 dA = I_y
\end{aligned} \tag{18}
$$

$$
\begin{aligned}
& \int \alpha_w^2 dA = C_w \\
& \int 4\, \xi^2 dA = J \\
& \int z\varrho^2 dA = C_y
\end{aligned} \tag{19}
$$

Equations (16) are the first and product second moments of area about the centroid and Equations (17) are a consequence of the reciprocal theorem. The cross-section area is A and I_z and I_y are the second moments of area about the principal axes passing through the centroid. C_w and J are the warping constant and St. Venant torsion constant, respectively, and C_y is the monosymmetry constant.

Equation (15) can be simplified as follows. The last two integrals are positive definite and, since they contain displacement variations δu and δw which do not appear in any of the other integrals, can only increase the calculated critical loads. These last two integrals can therefore be omitted. Third and higher order terms associated with GJ and EC_w

can also be omitted since their influence on the calculated critical loads is negligible (17). Critical conditions occur, therefore, when

$$\delta^2 V = -P_i\,\delta^2 q_i - \int M\left\{\frac{w_{xx}}{2}\cdot\delta\Theta\cdot\delta\Theta + \delta v_{xx}\cdot\delta\Theta + \frac{C_y}{2I_y}\,\delta\Theta_x\cdot\delta\Theta_x\right\}dx \qquad (20)$$

$$+ \frac{GJ}{2}\int \delta\Theta_x\cdot\delta\Theta_x\,dx + \frac{EC_w}{2}\int \delta\Theta_{xx}\cdot\delta\Theta_{xx}\cdot dx + \frac{EI_z}{2}\int(\delta v_{xx} + w_{xx}\delta\Theta)^2 dx = 0$$

The displacements, q_i, of the external forces can often be expressed as linear functions of the displacement variables, and since the second variation of any linear function vanishes, the term $-P_i\delta^2 q_i$ vanishes also. An important exception to this statement concerns a force P applied at a distance r above the shear centre as shown in Figure 4(b). The potential of such a force, VP, is

$$VP = -P\{w + r(1 - \cos\Theta)\} \qquad (21)$$

in which w and Θ are evaluated at the point of application of the force. Hence, if Θ is initially zero,

$$\delta^2 VP = -\frac{Pr}{2}\,\delta\Theta\cdot\delta\Theta \qquad (22)$$

Similarly, for a distributed load of p per unit length,

$$\delta^2 Vp = -\int\frac{pr}{2}\,\delta\Theta\cdot\delta\Theta\cdot dx \qquad (23)$$

SIMPLY SUPPORTED MONOSYMMETRIC I-BEAMS

For simply supported beams subjected to uniform moment, uniformly distributed and central concentrated loads, closed form solutions of Equation (20) can be obtained which are valid for a wide range of section properties.

Prior to buckling, the internal moment, M, about the major principal y-axis, is related to the curvature of the beam by the equation

$$M = EI_y\,w_{xx} \qquad (24)$$

As the beam buckles, the y and z axes rotate through a small angle $\delta\Theta$ as shown in Figure 5. If $\bar{v}$ denotes displacement in the $\bar{y}$ direction, equilibrium in the buckled state requires that

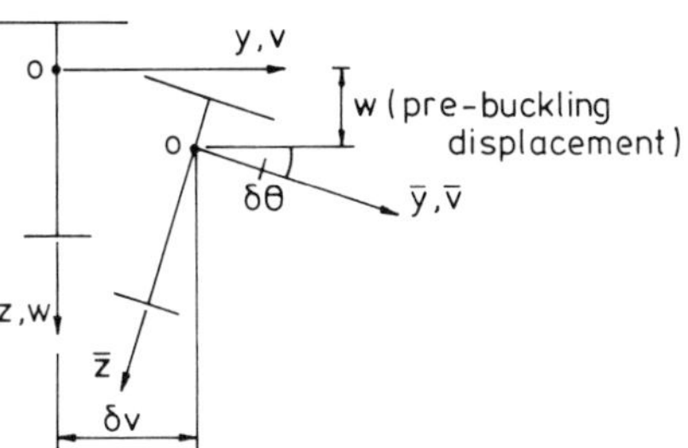

FIGURE 5. Pre-buckling and buckling displacements.

$$M\,\delta\Theta = EI_z\cdot\delta\bar{v}_{xx} \qquad (25)$$

Also for small angular rotations $\delta\Theta$,

$$\delta\bar{v} = \delta v + w\,\delta\Theta \qquad (26)$$

Hence, asuming $\delta\Theta$ to be a constant defining the orientation of the $\bar{y}$, $\bar{z}$ axes,

$$\delta\bar{v}_{xx} = \delta v_{xx} + w_{xx}\cdot\delta\Theta \qquad (27)$$

Substituting Equations (24), (25), and (27) into Equation (20) gives

$$\delta^2 V = -P_i\,\delta^2 q_i - \int\frac{M^2}{2EI_z}\left\{1 - \frac{I_z}{I_y}\right\}\delta\Theta\,\delta\Theta\,dx$$

$$-\int\frac{M\,C_y}{2\,I_y}\cdot\delta\Theta_x\cdot\delta\Theta_x\,dx \qquad (28)$$

$$\frac{GJ}{2}\int\delta\Theta_x\cdot\delta\Theta_x\,dx + \frac{EC_w}{2}\int\delta\Theta_{xx}\cdot\delta\Theta_{xx}dx = 0$$

For simply supported beams of length a, solutions of Equation (28) can be obtained by assuming $\delta\Theta$ to be represented by a single half sine wave [3,5,20,21]

$$\delta\Theta = \delta\Theta_m \sin\frac{\pi x}{a} \qquad (29)$$

in which Θ_m defines the maximum value of Θ at mid span.

Simply Supported Beam—Uniform Moment

For a simply supported beam subjected to a uniform moment as shown in Figure 6(a), a solution of Equation (28)

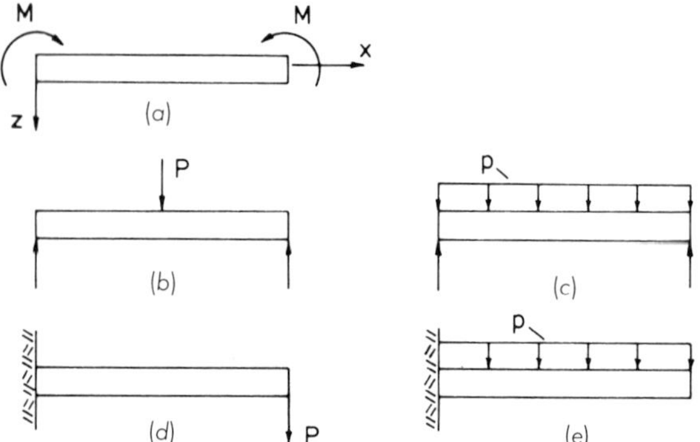

FIGURE 6. Support and loading conditions.

can be obtained in the form

$$M_{cr} = \frac{\lambda_1 (EI_z\, GJ)^{0.5}}{a} \tag{30}$$

in which M_{cr} denotes the critical moment and

$$\lambda_1 = \frac{1}{\alpha}\left[\frac{\pi^2}{2}\beta + \left\{\left(\frac{\pi^2\beta}{2}\right)^2 + \pi^2\alpha(1 + \varkappa^2)\right\}^{0.5}\right] \tag{31}$$

$$\alpha = \left\{1 - \frac{I_z}{I_y}\right\} \tag{32}$$

$$\beta = \frac{C_y}{a\, I_y}\left\{\frac{E\, I_z}{GJ}\right\}^{0.5} ; \varkappa = \left\{\frac{E\, C_w}{GJ} \cdot \frac{\pi^2}{a^2}\right\}^{0.5} \tag{33}$$

α represents the influence of the pre-buckling displacements. If I_z/I_y is small so that $\alpha \cong 1$, the solution agrees exactly with the solution based on the governing differential equations first obtained by Goodier [9] and by Anderson and Trahair [1].

Simply Supported Beam—Central Point Load

For a simply supported beam subjected to a central point load P applied at a distance r above the shear centre, as shown in Figure 6(b), a solution of Equation (28) can be obtained in the form

$$P_{cr} = \frac{\lambda_2 (E\, I_z\, GJ)^{0.5}}{a^2} \tag{34}$$

in which P_{cr} denotes the critical load and

$$\lambda_2 = \frac{1}{\alpha}\,[(10.9\beta - 29.8\mu) + \{(10.9\beta - 29.8\mu)^2 + 294\alpha(1 + \varkappa^2)\}^{0.5}] \tag{35}$$

$$\mu = \frac{r}{\alpha}\left\{\frac{EI_z}{GJ}\right\}^{0.5} \tag{36}$$

Simply Supported Beam—Uniformly Distributed Load

For a simply supported beam subjected to a uniformly distributed load of p per unit length applied at a distance r above the shear centre as shown in Figure 6(c), a solution of Equation (28) can be obtained in the form

$$p_{cr} = \frac{\lambda_3 (E\, I_z\, GJ)^{0.5}}{a^3} \tag{37}$$

in which p_{cr} denotes the critical uniformly distributed load and

$$\lambda_3 = \frac{1}{\alpha}[(23.6\beta - 41.1\mu) + \{(23.6\beta - 41.1\mu)^2 + 810\alpha(1 + \varkappa^2)\}^{0.5}] \tag{38}$$

Numerical Solutions

If the influence of the pre-buckling displacements is neglected, i.e., α assumed equal to unity, the closed form solutions presented herein are in close agreement with numerical solutions of the governing differential equations obtained by Anderson and Trahair [1], for $\varkappa \geq 1$. However, for $\varkappa = 0$ and negative β and μ the agreement is not so good since the buckled shape is complex and variable and cannot be represented by a single half sine wave for $\delta\Theta$.

Numerical solutions of Equation (20), neglecting the influence of pre-buckling displacements, have been obtained by Roberts and Azizian [3,20] using the finite element method, and by Roberts and Burt [5,21] using a trigonometric series representation of δv and $\delta\Theta$. The numerical solutions confirmed the validity of the closed form solutions for $\varkappa \geq 1$ and the validity of the solution for uniform moment for all values of $\varkappa$ and β. The numerical solutions for concentrated and uniformly distributed loads ($\varkappa = 0, 0.5, 1.0$) are summarised in Tables 1 and 2. Tables 3 and 4 compare the numerical solutions obtained by Roberts et al. with the closed form solutions and with the numerical solutions obtained by Anderson and Trahair [1]. As can be seen, both the closed form solutions and the numerical solutions obtained by Anderson and Trahair tend to overestimate the critical loads for $\varkappa = 0$ and negative β and μ.

CANTILEVER MONOSYMMETRIC I-BEAMS

Due to the complexity and variability of the buckled shape, simple closed form solutions cannot be obtained for

TABLE 1. λ_2 Values for Simply Supported Beams with Central Concentrated Loads.

κ	μ	β = 1.0	0.5	0	−0.5	−1.0
	1.0	6.1	4.9	3.9	3.1	2.5
	0.5	12.9	9.6	6.9	4.9	3.5
0	0	28.3	23.0	16.9	8.6	4.4
	−0.5	48.2	41.9	31.7	8.9	4.4
	−1.0	67.7	58.9	38.1	9.0	4.5
	1.0	7.9	6.5	5.4	4.6	3.98
	0.5	15.4	11.8	9.1	7.2	5.8
0.5	0	31.7	25.7	19.1	13.8	10.1
	−0.5	53.2	45.9	38.3	30.1	20.6
	−1.0	74.8	66.6	57.0	44.2	27.6
	1.0	11.8	9.9	8.5	7.3	6.5
	0.5	20.5	16.5	13.4	11.0	9.3
1	0	37.0	30.3	24.2	19.1	15.2
	−0.5	58.8	51.1	43.2	35.3	27.8
	−1.0	81.7	73.8	65.6	57.2	48.4

TABLE 2. λ_2 Values for Simply Supported Beams with Uniformly Distributed Loads.

κ	μ	β = 1.0	0.5	0	−0.5	−1.0
	1.0	14.3	11.3	8.9	7.0	5.6
	0.5	26.9	20.2	14.6	10.2	7.2
0	0	52.5	40.8	28.3	15.4	8.1
	−0.5	88.5	73.7	54.1	16.2	8.1
	−1.0	128.9	111.9	85.5	16.4	8.2
	1.0	18.0	13.9	10.9	8.8	7.3
	0.5	32.7	23.9	17.3	12.9	9.9
0.5	0	59.4	45.1	31.8	21.4	14.8
	−0.5	94.6	77.1	57.8	37.5	23.0
	−1.0	133.6	114.2	90.7	56.9	31.4
	1.0	25.9	20.4	16.4	13.5	11.4
	0.5	42.3	32.3	24.6	19.2	15.4
1.0	0	68.5	53.5	40.2	29.8	22.5
	−0.5	101.9	83.7	65.5	48.0	35.1
	−1.0	139.2	119.3	97.6	74.9	54.2

TABLE 3. λ_2 Values for Simply Supported Beams with Central Concentrated Loads—Comparison with Existing Numerical and Closed Form Solutions.

κ	μ	β = 0.6		0		−0.6	
	0.6	8.7	8.6* 9.2**	6.0	5.9* 6.9**	4.1	4.1* 5.4**
0	0	24.1	24.1 24.9	16.9	16.9 17.1	7.1	8.6 11.8
	−0.6	46.9	46.5 54.4	33.2	32.6 42.7	7.2	14.3 31.9
	0.6	15.4	15.4 15.4	12.1	12.1 12.2	9.7	9.7 10.0
1.0	0	31.6	31.6 31.7	24.2	24.2 24.3	18.2	18.2 18.6
	−0.6	57.1	57.1 58.9	47.6	47.6 48.1	37.8	37.8 38.1

*Reference [1]
**Closed form solution—Equation (35)

TABLE 4. λ_2 Values for Simply Supported Beams with Uniformly Distributed Loads—Comparison with Existing Numerical and Closed Form Solutions.

		β					
x	μ	0.6		0		−0.6	
	0.6	18.8	18.8*	13.0	13.1*	8.8	8.8*
			19.8**		13.0**		9.3**
0	0	43.1	43.0	28.3	28.3	13.2	13.9
			45.9		28.4		17.6
	−0.6	84.0	83.9	60.2	60.2	13.4	18.6
			86.9		62.3		40.8
	0.6	30.9	30.9	22.6	22.6	17.0	17.0
			31.1		22.5		17.1
1	0	56.4	56.4	40.2	40.2	28.1	28.1
			56.8		40.2		28.5
	−0.6	94.3	94.3	71.5	71.5	50.1	50.1
			94.7		71.8		52.1

*Reference [1]
**Closed form solution—Equation (38)

TABLE 5. λ_2 Values for Cantilevers with Concentrated End Loads.

		β				
x	μ	1.0	0.5	0	−0.5	−1.0
	1.0	0.63	0.78	1.00	1.30	1.74
	0.5	0.84	1.20	1.82	2.98	5.17
0	0	1.13	1.96	4.12	8.75	14.7
	−0.5	1.06	2.02	5.14	10.79	17.36
	−1.0	1.06	2.05	5.40	11.35	18.2
	1.0	1.04	1.22	1.47	1.83	2.37
	0.5	1.56	2.00	2.74	4.12	6.87
1	0	2.86	4.46	7.65	12.59	18.5
	−0.5	4.83	7.20	10.87	15.89	21.77
	−1.0	5.43	7.82	11.61	16.86	23.02
	1.0	1.74	2.02	2.40	2.94	3.76
	0.5	2.63	3.30	4.38	6.26	9.69
2	0	4.94	7.29	11.29	16.69	22.57
	−0.5	10.03	13.54	17.7	22.5	27.86
	−1.0	12.34	15.59	19.66	24.52	30.02
	1.0	2.83	3.27	3.86	4.68	5.88
	0.5	4.23	5.24	6.76	9.22	13.21
3	0	7.67	10.65	15.15	20.87	27.02
	−0.5	14.97	19.29	23.94	28.89	34.15
	−1.0	19.37	23.1	27.34	32.06	37.23

TABLE 6. λ_2 Values for Cantilevers with Uniformly Distributed Loads.

		β				
x	μ	1.0	0.5	0	−0.5	−1.0
	1.0	1.42	1.84	2.48	3.49	5.17
	0.5	1.84	2.75	4.64	8.93	19.3
0	0	2.21	4.16	13.39	37.6	67.9
	−0.5	2.33	4.64	21.29	55.8	90.3
	−1.0	2.36	4.74	25.8	67.2	105.1
	1.0	3.52	4.15	5.04	6.38	8.48
	0.5	5.25	6.80	9.51	14.8	26.8
1	0	9.74	15.85	29.93	53.69	81.8
	−0.5	19.6	31.8	51.3	77.5	107.0
	−1.0	25.5	39.6	62.0	91.2	123.1
	1.0	6.59	7.66	9.11	11.19	14.34
	0.5	9.99	12.59	16.8	24.35	38.91
2	0	19.3	29.1	47.2	72.7	100.99
	−0.5	43.1	60.4	81.2	105.4	132.1
	−1.0	57.6	75.1	96.9	122.5	150.6
	1.0	11.0	12.8	15.0	18.3	22.96
	0.5	16.7	20.7	26.9	37.0	54.2
3	0	31.1	44.0	64.4	90.9	119.9
	−0.5	65.1	85.6	107.9	132.0	157.7
	−1.0	88.7	107.7	129.3	153.4	179.5

cantilever monosymmetric I-beams. Numerical solutions of Equation (20), neglecting the influence of the prebuckling displacements, have therefore been obtained by Roberts et al. [3,20,21,25].

Cantilever–Concentrated End Load

For a cantilever subjected to a concentrated end load applied at a distance r above the shear centre as shown in Figure 6(d), it is assumed that the solution can be expressed in the form

$$P_{cr} = \lambda_4 \frac{(E\,I_z\,GJ)^{0.5}}{a^2} \tag{39}$$

Values of λ_4 are given in Table 5 and show close agreement with numerical solutions of the governing differential equations presented by Anderson and Trahair [1].

Cantilevers–Uniformly Distributed Load

For a cantilever subjected to a uniformly distributed load of p per unit length applied at a distance r above the shear centre as shown in Figure 6(e) it is assumed that the solution can be expressed in the form

$$p_{cr} = \lambda_5 \frac{(E\,I_z\,GJ)^{0.5}}{a^3} \tag{40}$$

Values of λ_5 are given in Table 6 and show close agreement with numerical solutions of the governing differential equations presented by Anderson and Trahair [1].

INFLUENCE OF PRE-BUCKLING DISPLACEMENTS

The numerical results presented in Tables 1 to 5 neglect the influence of pre-buckling displacements, i.e., α defined by Equation (32) is assumed equal to unity. However, for many monosymmetric I-beams, I_z/I_y is not small compared with unity and the pre-buckling displacements may result in a significant increase in the elastic critical load.

Inspection of the closed form solutions indicates that the influence of pre-buckling displacements depends upon the relative magnitudes of $\varkappa$, β, and μ. Consider, for example, Equation (35) which defines λ_2 for a simply supported beam subjected to a central concentrated load. This equation can be expressed as

$$\lambda_2 = (10.9\bar{\beta} - 29.8\bar{\mu}) + \{(10.9\bar{\beta} - 29.8\bar{\mu})^2 + 294(1 + \bar{\varkappa}^2)\}^{0.5} \tag{41}$$

in which

$$\bar{\beta} = \frac{\beta}{\alpha}\,;\ \bar{\mu} = \frac{\mu}{\alpha}\,;\ \bar{\varkappa}^2 = \frac{1 + \varkappa^2 - \alpha}{\alpha} \tag{42}$$

It can be deduced, therefore, that the influence of pre-buckling displacements can be incorporated by choosing numerical λ values corresponding to $\bar{\varkappa}$, $\bar{\beta}$ and $\bar{\mu}$ defined by Equations (42).

SECTION PROPERTIES

A major difficulty associated with the calculation of elastic critical loads for monosymmetric beams of complex open cross-section, such as shown in Figure 2(c), is the determination of the section properties, in particular those associated with warping of the cross-section and monosymmetry. General procedures have been developed for evaluating such properties [7,22,24,28] which in general require automatic computation.

For the simpler sections, such as shown in Figure 2(a) and (b), exact closed form solutions can be obtained. Such solutions may require lengthy calculations which in the past have been considered prohibitive in routine design. Simpler approximate solutions have therefore been developed [12] which are adequate for the majority of sections likely to be used in practice. Herein, however, only the exact solutions are presented since they are valid for all beams of similar geometry, and the benefit of using the simpler approximate solutions is considered marginal.

A typical monosymmetric I-beam with a lipped top flange is shown in Figure 7. The dimensions indicated are the dimensions of the mean profile of the cross-section and

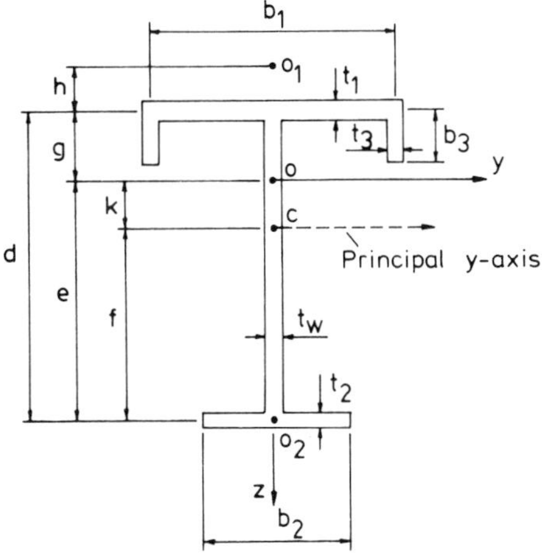

FIGURE 7. Monosymmetric I-beam with lipped flange.

thicknesses are assumed small compared with other cross-section dimensions.

The various section properties are defined by Equations (3), (4), (18), and (19). Definition of C_w is in accordance with general procedures for determining warping constants, but for the section considered, C_w can be deduced by an alternative simpler procedure. The St. Venant torsion constant J can be expressed alternatively as

$$J = \int \frac{t^3}{3}\, ds \tag{43}$$

in which the integration is around the contour of the cross-section. The monosymmetry constant C_y can also be expressed in the alternative forms. Noting that distances y and z are measured from the centroid c while ϱ is measured from the shear centre o, ϱ^2 can be expressed as

$$\varrho^2 = (z + k)^2 + y^2 \tag{44}$$

in which k is the z-coordinate of the centroid relative to the y-z axes passing through the shear centre. Hence,

$$C_y = \int z\,\varrho^2\, dA = \int (z^3 + zy^2)dA + 2\,k\,I_y \tag{45}$$

Returning now to Figure 7, the area of the cross-section A is given by

$$A = b_1t_1 + b_2t_2 + 2b_3t_3 + dt_w \tag{46}$$

The distance f of the centroid c from the bottom flange is given by

$$f = \left\{ b_1t_1d + 2b_3t_3\left(d - \frac{b_3}{2}\right) + \frac{d^2t_w}{2} \right\} \frac{1}{A} \tag{47}$$

Points o_1 and o_2 are the shear centres of the top and bottom flanges respectively. The distance h of o_1 from the top of the beam is given by

$$h = \frac{3\,b_3^2\,t_3}{6\,b_3t_3 + b_1t_1} \tag{48}$$

The second moments of area of the top and bottom flanges about the z axis, denoted by I_{z1} and I_{z2}, respectively, are

$$I_{z1} = \frac{t_1b_1^3}{12} + \frac{b_3t_3b_1^2}{2} \tag{49}$$

$$I_{z2} = \frac{t_2b_2^3}{12} \tag{50}$$

The distance e of the shear centre of the entire section from the bottom flange is given by

$$e = \frac{(d + h)\,I_{z1}}{I_{z1} + I_{z2}} \tag{51}$$

and

$$g = d - e \quad \text{and} \quad k = e - f \tag{52}$$

The second moments of area of the entire section about the principal y and z axes passing through the centroid are

$$I_y = b_1t_1(d - f)^2 + b_2t_2f^2 + \frac{t_wd^3}{12} + dt_w\left(f - \frac{d}{2}\right)^2 + \frac{t_3b_3^3}{6} + 2\,b_3t_3\left(d - f - \frac{b_3}{2}\right)^2 \tag{53}$$

$$I_z = I_{z1} + I_{z2} \tag{54}$$

The St. Venant torsion constant J is given by

$$J = \frac{b_1t_1^3 + b_2t_2^3 + 2b_3t_3^3 + dt_w^3}{3} \tag{55}$$

The warping constant C_w is given by

$$C_w = I_{z1}(g + h)^2 + I_{z2}\,e^2 + \frac{t_3b_3^3b_1^2}{12}\left\{\frac{3b_3t_3 + 2b_1t_1}{6b_3t_3 + b_1t_1}\right\} \tag{56}$$

The first two terms relate to lateral displacement of the shear centres of the top and bottom flanges, respectively, while the third term relates to rotation of the top flange about its own shear centre o_1. Finally, the monosymmetry constant C_y is given by

$$C_y = (f - d)^3b_1t_1 + f^3b_2t_2 + \frac{t_3}{4}\{(f - d + b_3)^4 - (f - d)^4\} + \frac{t_w}{4}\{f^4 - (f - d)^4\} + (f - d)\frac{t_1b_1^3}{12} + \frac{f\,t_2b_2^3}{12} + \frac{t_3b_1^2}{4}\{(f - d + b_3)^2 - (f - d)^2\} + 2\,(e - f)\,I_y \tag{57}$$

DESIGN

Current design practice for laterally unrestrained beams is based on the elastic critical stress for flexural-torsional buckling, modified to allow for plasticity effects in short span beams [6,12,13]. For specific details, reference should be made to particular codes of practice [2,4,23] as only a brief outline is presented herein.

The starting point is the elastic critical moment M_{cro} for a doubly symmetric simply supported beam subjected to uniform moment, which can be expressed as

$$M_{cro} = \frac{\pi}{a}\sqrt{EI_z\, GJ\left\{1 + \frac{EC_w\,\pi^2}{GJ\,a^2}\right\}} \qquad (58)$$

Equation (58) can be deduced from Equations (30) and (31) with $C_y = 0$ and α assumed equal to unity, i.e., neglecting the influence of pre-buckling displacements.

For other forms of loading resulting in non-uniform moment along the span of the beam, the critical value of the maximum moment, M_{cr}, can be expressed as

$$M_{cr} = m\, M_{cro} \qquad (59)$$

For many types of loading, the factor m is approximately independent of the proportions of the beam and a number of typical examples for simply supported beams subjected to moment gradient or loaded through their shear centres by uniformly distributed or concentrated loads are given in Figure 8.

When uniformly distributed or concentrated loads are applied above or below the shear centre and are free to move laterally as the beam buckles, the critical maximum moment may be significantly different from that predicted by Equation (59). Loading above the shear centre decreases the critical moment while loading below the shear centre increases the critical moment. Such effects can be allowed for by modifying the m factors approximately to take account of the position of application of the load [13] but such factors are seldom incorporated in codes of practice.

Modification of the basic equation for M_{cro} is requird to take account of monosymmetry and to allow for different conditions of end restraint.

Having established the appropriate value of M_{cr} to be used in design, the critical compressive and tensile stresses, σ_{crc} and σ_{crt} can be calculated from the equations

$$\sigma_{crc} = \frac{M_{cr}\, z_c}{I_y}\ ;\ \sigma_{crt} = \frac{M_{cr}\, z_t}{I_y} \qquad (60)$$

in which z_c and z_t are the distances of the extreme compression and tension fibres from the centroid of the beam. The maximum allowable compressive stress σ_{ac} under working

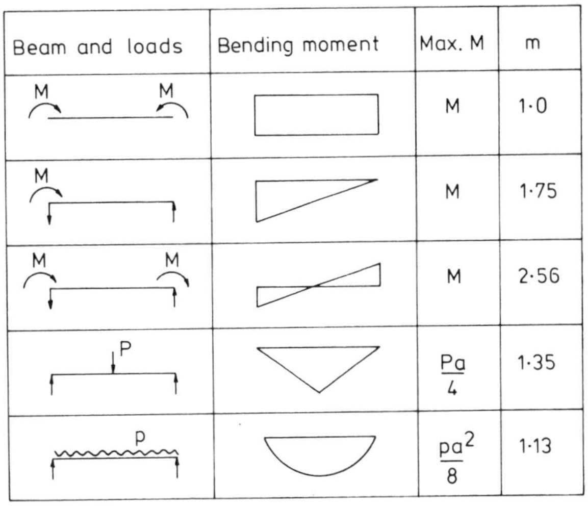

Beam and loads	Bending moment	Max. M	m
M ... M (equal end moments)		M	1·0
M (one end moment)		M	1·75
M ... M (opposite end moments)		M	2·56
P (central point load)		$\frac{Pa}{4}$	1·35
p (uniformly distributed load)		$\frac{pa^2}{8}$	1·13

FIGURE 8. Equivalent uniform moment factors for simply supported beams with shear centre loading.

loads is then related to the critical compressive stress σ_{crc} or yield stress σ_o, either in tabular form or via semi-empirical equations, as indicated in Figure 9, in which η is a slenderness parameter for the beam. Factors of safety vary from approximately 1.6 stocky beams of low slenderness to 2 or more for beams of intermediate or high slenderness. The higher factors of safety for beams of intermediate and high slenderness take account of the influence of initial imperfections and residual stresses. The maximum allowable tensile stress is related to the yield stress via a factor of safety of approximately 1.6.

The basic equations for elastic critical moments, such as Equation (58), are generally considered too complex for routine design due to the lengthy expressions required to calculate the section properties. The basic equations are replaced, therefore, by simplified approximations which are adequate for the majority of sections used in practice.

British Standard

The British Standard BS 449:1969 [4] approximates the elastic critical compressive stress for a simply supported monosymmetric I-beam subjected to uniform moment as

$$\sigma_{crc} = \left(\frac{1675}{a/r_z}\right)^2\sqrt{1 + \frac{1}{20}\left(\frac{a\,T}{r_z D}\right)^2} + k_2 \quad (\mathrm{N/mm^2}) \qquad (61)$$

in which a is the span, r_z is the radius of gyration of the beam about its minor axis, D is the overall depth of the beam and T is the thickness of the compression flange,

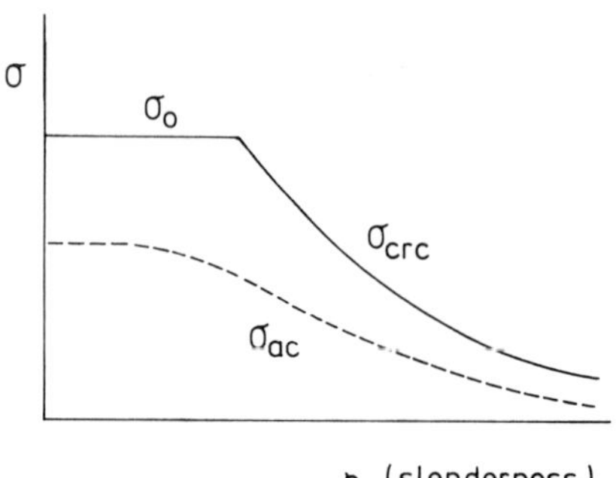

FIGURE 9. Allowable stresses related to yield stress and critical compressive stresses.

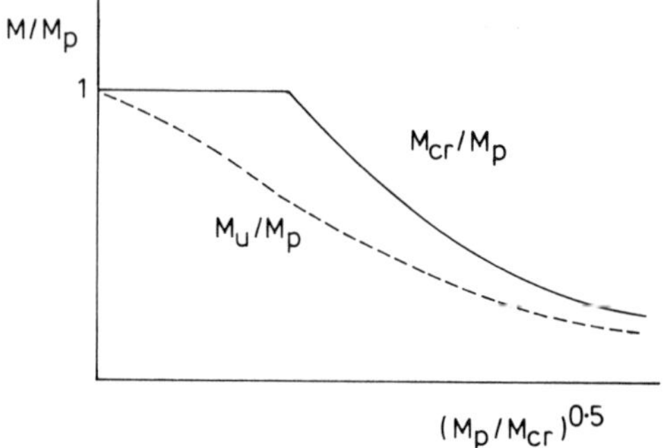

FIGURE 10. Lower bound failure envelope for laterally unrestrained beams.

which is taken as the thickness of the web for T-beams with the stalk in compression. K_2 is a parameter which allows for monosymmetry and which can be expressed in terms of the ratio I_{zc}/I_z, in which I_{zc} is the second moment of area of the compression flange about the minor axis of the beam, as [12]

$$K_2 = 0.5\left(2\,\frac{I_{zc}}{I_z} - 1\right) \text{ for } \frac{I_{zc}}{I_z} > 0.5$$

$$K_x = 2\,\frac{I_{zc}}{I_z} - 1 \text{ for } \frac{I_{zc}}{I_z} \leqslant 0.5 \qquad (62)$$

K_2 is equal to zero for a doubly symmetric beam for which $I_{zc}/I_z = 0.5$. Equation (61) is used when I_{zc} is greater than the corresponding second moment of area of the tension flange I_{zt}. When I_{zt} is greater than I_{zc}, the value of σ_{crc} given by Equation (61) is increased by a factor z_c/z_t.

The influence of different forms of loading is included only in the case of cantilevers, for which effective length factors are provided. For simply supported beams subjected to top flange loading which is free to move laterally as the beam buckles, an effective length of 1.2a is specified. Other effective length factors are provided to allow for different forms of end restraint.

Maximum allowable compressive stresses under working loads are given in terms of the critical compressive stress σ_{crc} and yield stress σ_o in tabular form, the values differing slightly for rolled sections and plate girders.

Australian Standard

The Australian Standard AS 1250:1965 [2] approximates the elastic critical compressive stress for a simply supported monosymmetric I-beam subjected to uniform bending, as

$$\sigma_{crc} = \left(\frac{1628}{a/r_z}\right)^2 \sqrt{1 + \frac{1}{20}\left(\frac{aT}{r_z D}\right)^2} + K_2 \quad (\text{N/mm}^2) \quad (63)$$

This equation is similar to Equation (61) with the factor 1675 replaced by 1628. The thickness T, however, is taken as the thickness of the flange which has the greatest second moment of area about the minor axis of the beam. When I_{zt} exceeds I_{zc}, σ_{crc} is increased by a factor z_c/z_t. Allowances are made for different forms of loading, top flange loading and end restraint which are similar to those of BS 449:1969.

Alternatively, the critical compressive stress can be determined from a rigorous analysis of flexural-torsional buckling, or from published information, in which the type of loading, height of load application, etc., are taken into account. This alternative approach is likely to lead to more economical designs.

Maximum allowable stresses are expressed in terms of critical stresses and the yield stress via a number of semi-empirical equations.

U.S. Specification

In the AISC 1969 Specification [23], no procedures are used for monosymmetric I-beams which are specifically different from those for doubly symmetric I-beams. The elastic critical compressive stress for a simply supported beam subjected to uniform moment is approximated as

$$\sigma_{crc} = \sqrt{\left[\frac{1{,}975{,}000}{(a/r_c)^2}\right]^2 + \left[\frac{130{,}400\,BT}{aD}\right]^2} \quad (\text{N/mm}^2) \qquad (64)$$

in which r_c is the radius of gyration of the compression flange plus one sixth of the web about the minor axis of the beam. B and T are the total width and thickness, respectively, of the compression flange. The value of σ_{crc} is again increased by a factor z_c/z_t when I_{zt} is greater than I_{zc}.

The specification incorporates moment factors to allow for moment gradients along the span, but no allowance is made for the effects of end restraint or height of load application.

Maximum permissible stresses arc related to critical stresses and the yield stress via a number of semi-empirical equations.

Recent Design Criteria

In recent years, considerable effort has been directed towards limit state design with particular emphasis on predicting the ultimate strength of structures and structural components. In particular, it has been proposed [13] that the ultimate moment of resistance M_u of laterally unrestrained beams be expressed in terms of the full plastic moment M_p and the elastic critical moment for flexural-torsional buckling M_{cr}.

A typical graphical representation of the problem is shown in Figure 10 in which $(M_p/M_{cr})^{0.5}$ is adopted as a representative slenderness parameter since it incorporates all the material and section properties of the beam and also the loading and support conditions via appropriate values of M_{cr}. In general, due to initial imperfections and residual stresses, M_u is less than both M_p and M_{cr} and therefore limiting values of M_u/M_p are represented by semi-empirical curves. One of the simplest forms is a Perry type curve which can be expressed as

$$(M_{cr} - M_u)(M_p - M_u) = \bar{\eta}\, M_{cr} M_u \qquad (65)$$

The position of the curve is governed by $\bar{\eta}$ which can be expressed as a function of M_p/M_{cr} so that the design curve represents a lower bound for available test data.

CONCLUDING REMARKS

Solutions are now available for predicting the elastic critical loads for flexural-torsional buckling of monosymmetric I-beams and cantilevers which include such factors as the type of loading, the height of load application relative to the shear centre and the influence of prebuckling displacements. It is anticipated that the use of such solutions will be permitted, or even recommended, by future codes of practice and lead to more economical designs.

NOTATION

A = cross-section area
a = span of beam or length of cantilever
B = overall width of compression flange
$b_1\ b_2\ b_3$ = dimensions shown in Figure 7
C_w = warping constant
C_y = monosymmetry constant
c = centroid
D = overall depth of beam or cantilever
d = depth of web—see Figure 7
E = Young's modulus
e,f = dimensions shown in Figure 7
G = shear modulus
g,h = dimensions shown in Figure 7
I_y,I_z = second moments of area about principal y and z axes
I_{zc},I_{zt} = second moments of area of compression and tension flanges about z axis
I_{z1},I_{z2} = second momtns of area of top and bottom flanges about z axis
J = St. Venant torsion constant
K_2 = approximate monosymmetry parameter
k = dimension shown in Figure 7
M = moment
M_{cr} = elastic critical moment
M_{cro} = elastic critical moment for simply supported doubly symmetric beam under uniform moment
M_p = plastic moment
M_u = ultimate or failure moment
m = moment modification factor
$\bar{m}$ = defined by Equation (4)
o,o_1,o_2 = shear centres of entire cross-section and top and bottom flanges
P = load or force
P_{cr} = elastic critical load
P_i = external forces
p = uniformly distributed load
p_{cr} = elastic critical uniformly distributed load
q_i = displacements of external forces
r = height of load application above shear centre
r_o = as defined in Figure 3(b)
r_c,r_z = radius of gyration of compression flange and entire cross-section about z axis
s = distance measured around cross-section
T = thickness of compression flange or thickness of flange with greatest second moment of area about z-axis
t = thickness
t_w = web thickness
$t_1\ t_2\ t_3$ = thickness shown in Figure 7
u = displacement of centroid in x-direction
V = total potential energy

VP = potential energy of load P
Vp = potential energy of uniformly distributed load p
v = displacement of shear centre in y direction
$\bar{v}$ = displacement of shear centre in $\bar{y}$ direction
w = displacement of shear centre in z direction
x,y,z = coordinate axes passing through shear centre
y,z = distances from centroid in y and z directions
$\bar{y},\bar{z}$ = direction of y,z axes after rotation of cross-section
z_c,z_t = distances from centroid to extreme compression and tension fibres
α = $1 - I_z/I_y$
$\alpha_s,\bar{\alpha}_s,\alpha_w$ = defined by Equations (3) and (4)
β = defined by Equation (33)
$\bar{\beta}$ = β/α
γ = St. Venant shear strain
δ,δ^2 = first and second variations
ϵ = axial strain
$\eta,\bar{\eta}$ = slenderness parameters
Θ = rotation of cross-section about shear centre
Θ_m = maximum value of Θ at mid span
$\varkappa$ = defined by Equation (33)
$\bar{\varkappa}$ = defined by Equation (42)
λ_1 to λ_5 = buckling coefficients
μ = defined by Equation (36)
$\bar{\mu}$ = μ/α
ξ = distance measured from and normal to mean profile of cross-section
ϱ = distance from shear centre to point on cross-section
σ = axial stress
σ_{ac} = allowable compressive stress
$\sigma_{crc},\sigma_{crt}$ = critical compressive and tensile stress
σ_o = yield stress
τ = St. Venant shear stress

REFERENCES

1. Anderson, J. M. and N. S. Trahair, "Stability of Monosymmetric Beams and Cantilevers," *Journal of the Structural Division, ASCE, Vol. 98* (No. ST1), Proc. Paper 8648, 269–286 (Jan. 1972).
2. AS 1250–1975 SAA Steel Structures Code, Standards Association of Australia, Sydney, Australia (1975).
3. Azizian, Z. G., "Instability and Nonlinear Analysis of Thin Walled Structures," Ph.D. Thesis, Department of Civil and Structural Engineering, University College Cardiff, Wales, U.K. (1983).
4. BS 449:1969, "Specification for the Use of Structural Steel in Buildings," British Standards Institution, London, England (1969).
5. Burt, C., "The Flexural-Torsional Instability of Monosymmetric I-Beams," M.Sc. Dissertation, Department of Civil and Structural Engineering, University College Cardiff, Wales, U.K. (1984).
6. Chen, W. F. and T. Atsua, "Theory of Beam-Columns," *Volume 2, Space Behaviour and Design,* McGraw-Hill Co., Inc., New York (1977).
7. Galambos, T. V., *Structural Members and Frames,* Prentice-Hall, Englewood Cliff,s NJ (1968).
8. Goodier, J. N., "The Buckling of Compressed Bars by Torsion and Flexure," Bulletin 27, Cornell University Engineering Experimental Station (Dec. 1941).
9. Goodier, J. N., "Flexural-Torsional Buckling of Bars of Open Section Under Bending, Eccentric Thrust or Torsional Loads," Bulletin 28, Cornell University Engineering Experimental Station (Jan. 1942).
10. Hill, H. N., "The Lateral Instability of Unsymmetrical I-beams," *Journal of Aeronautical Sciences, Vol. 9,* 175 (March 1942).
11. Kerensky, O. A., A. R. Flint, and W. C. Brown, "The Basis for Design of Beams and Plate Girders in the Revised British Standard 153," *Proceedings of Institution of Civil Engineers, Part III, Vol. 5,* 396 (Aug. 1956).
12. Kitipornchai, S. and N. S. Trahair, "Buckling Properties of Monosymmetric I-beams," *Journal of the Structural Division, Proc. ASCE, Vol. 106* (No. ST5) (May 1980).
13. Narayanan, R. (Editor), *Beams and Beam Columns–Stability and Strength,* Applied Science Publishers Ltd., London (1983).
14. O'Connor, C., "The Buckling of a Monosymmetric Beam Loaded in the Plane of Symmetry," *Australian Journal of Applied Science, Vol. 15,* 191 (1964).
15. Pettersson, O., "Combined Bending and Torsion of I-beams of Monosymmetrical Cross-Section," Bulletin No. 10, Division of Building Statics and Structural Engineering, Royal Institute of Technology, Stockholm, Sweden (1952).
16. Roberts, T. M., "Second Order Strains and Instability of Thin Walled Bars of Open Cross-Section," *Int. J. Mech. Sci., Vol. 23,* 297–306 (1981).
17. Roberts, T. M. and Z. G. Aziaian, "Influence of Pre-Buckling Displacements on the Elastic Critical Loads of Thin Walled Bars of Open Cross-Section," *Int. J. Mech. Sci., Vol. 25* (No. 2), 93–104 (1983).
18. Roberts, T. M. and Azizian, Z. G., "Instability of Thin Walled Bars," *Journal of Engineering Mechanics, Proc. ASCE, Vol. 109* (No. 3), 781–794 (June 1983).
19. Roberts, T. M. and Z. G. Azizian, "Nonlinear Analysis of Thin Walled Bars of Open Cross-Section," *Int. J. Mech. Sci., Vol. 25*(No. 8), 565–577 (1983).
20. Roberts, T. M. and Z. G. Azizian, "Instability of Monosymmetric I-beams," *Journal of Structural Engineering, Proc. ASCE, Vol. 110* (No. 6), 1415–1419 (June 1984).
21. Roberts, T. M. and C. Burt, "Instability of Monosymmetric I-beams and Cantilevers," *Int. J. Mech. Sci., Vol. 27,* No. 5, 313–324 (1985).
22. Roberts, T. M., "Section Properties of Thin Walled Bars of

Open Cross-section," *The Structural Engineer, Vol. 63B,* No. 3, 63–67 (1985).

23. *Specification for the Design, Fabrication and Erection of Structural Steel for Buildings,* American Institute of Steel Construction, New York, NY (1969).
24. Timoshenko, S. P. and J. M. Gere, *Theory of Elastic Stability, 2nd ed.*, McGraw-Hill Book Co., Inc., New York (1961).
25. Vacharajittiphan, P., S. T. Woolcock, and N. S. Trahair, "Effect of In-Plane Deformation on Lateral Buckling," *J. Struct. Mech.*, *3*(1), 29–60 (1974).
26. Vlasov, V. Z., *Thin Walled Elastic Beams, 2nd ed.*, Israel Program for Scientific Translations, Jerusalem (1961) (Published in Russia, 1940).
27. Winter, G., "Lateral Instability of Unsymmetrical I-beams and Trusses in Bending," *Proceedings of the American Society of Civil Engineers, Vol. 67* (No. 10), 1851–1864 (Dec. 1941).
28. Yu, W. W., *Cold-Formed Steel Structures,* McGraw-Hill Book Co., New York (1973).

CHAPTER 22

Inelastic Post-Buckling Analysis of Space Frames

MANOLIS PAPADRAKAKIS*

INTRODUCTION

During the last twenty years and especially during the last decade, space frame structures have influenced the architectural scene all over the world. Architects turned their attention to space frame structures because these systems gave them greater freedom of design and in many instances led to lower costs through prefabrication and standardization of component parts. In their search for new forms, architects and engineers have discovered that space frame structures not only offer them many structural advantages, but also produce a striking simplicity of form as well as a very pleasant appearance. In short, space frame structures have become a part of Modern Architecture and Progressive Engineering.

The reason for the present interest in space frame structures is due to several factors; the most important are: (1) introduction and widespread use of high-speed electronic computers; (2) development of highly efficient standardized connections; and (3) a remarkable amount of recent scientific research into the elastic and non-elastic behavior of space frame structures and the determination of modes of their failure under excessive loading. The introduction of digital computers completely changed the approach used in the analysis of complex structures. Modern space frame structures consist, as a rule, of so many members that analysis by "traditional" techniques becomes extremely time consuming and in most practical cases virtually impossible. One of the problems involved in the analysis of complex space frame structures is the sheer amount of data preparation for the computer analysis. The work of Nooshin [116] at the Space Structures Research Centre of the University of Surrey in developing the original concept of formex algebra is considered as a breakthrough in the algebraic representation and processing of highly sophisticated structural configurations.

It has long been recognized that in practice the response of space frame structures can be significantly affected by both geometric and inelastic material properties. Accordingly, the necessity of providing effective tools for the analysis of simultaneous plastic collapse and buckling behavior of space frame structures is self-evident. This creates a challenging engineering problem, a solution of which can only be expected through the use of sophisticated numerical methods of nonlinear structural mechanics.

Definition of a Space Frame

A structure may be defined as "space frame" when it is realized only on the basis of its spatial effect. Figure 1 shows two structural examples each forming a dome. The dome (a) is a combination of arches and bents every one of which can constitute a stable system by itself. Structure (b), however, cannot be understood as a combination of stable plane elements, but its stability is assured by its spatial nature as a whole. This may be shown clearly when their central cross sections are compared. This difference between a complex of plane frames and a space frame can also be understood from the viewpoint of force-flow. In (a) there is a hierarchical contribution on the bearing capacity of the members. On the other hand in (b) all members contribute to support the load according to the three-dimensional topology of the whole structure. Thus, by this second description, a space frame is defined as a structure without (any appreciable) ranks among its constituent members. The second definition is more intuitive than the first one and classifies certain types of structures (for example, double layer grids are among them) as space frames, which do not necessarily satisfy the conditions of the first definition.

*Institute of Structural Analysis and Aseismic Research, National Technical University, Athens, Greece

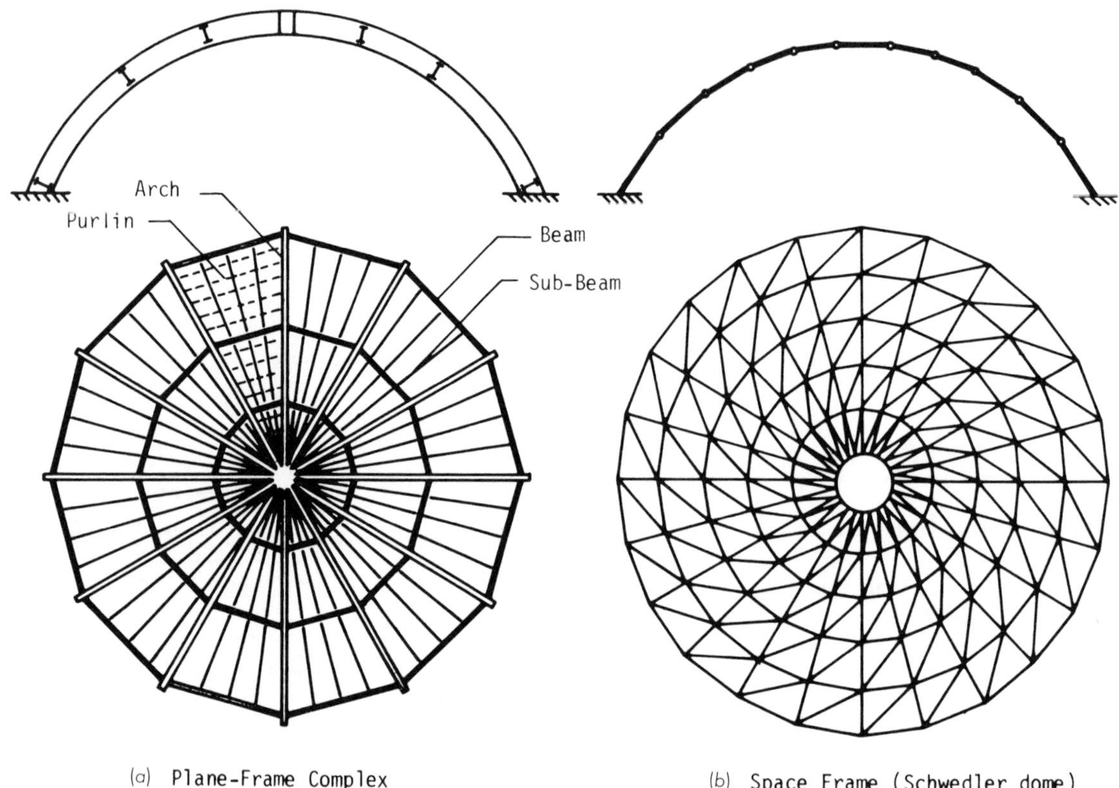

FIGURE 1. Plane-frame complex and space frame.

An alternative definition which describes the same family of structures is referred to in the ASCE report [170] as "lattice structure." A latticed structure is defined by a structural system in a form of a network of elements with its load-carrying capacity mechanism being three-dimensional in nature. Another term which is frequently used is "reticulated structure." For an extensive philology on this subject the reader is referred to [79].

Characteristic Features of Space Frames

The special nature of a space frame by which it is a spatial constitution of structural members without ranks among them, means separation from the object of the classical structural design that aims at giving a minimum necessary section to every structural member. Hence, only some members in the whole structure function with their full loading capacity while all the rest are over-dimensioned. The argument which was stated in the earlier stage of development, that the above-mentioned waste of material would be compensated by the mass-production of the structural members was not fully justified. For this reason, to compete economically with conventional structures, the current design trend of space frames admits some ranking in the cross-sections of their element members to keep the amount of required material as low as possible. On the other hand there are some advantages resulting from the fact that the members have no ranks among them, such as a high stiffness for a loading of unexpected type, since adjacent members cooperate on an equal basis in resisting the deformation of that area. The reserved strength in the members brought about by overdimensioning may also give capacity to a space frame to absorb overloadings.

One of the most important advantages, however, that makes this family of structures so popular is the ease of construction. Space frames consisting of homogeneous light members have in most cases the advantage over plane frame complexes in which members of higher ranks easily exceed the extent of human control both in size and weight. Moreover, the connecting systems of space frames do not need skilled labor to the same extent as other types of structures. There are, however, certain aspects, such as the inaccurate fabrication and erection of the members, or loose connections, which for other types of structures produce less acute problems and have to be given particular attention during the stages of fabrication and erection of a space frame. Problems of accuracy seem to have been alleviated in many space frame systems by the development of machine tools, while problems of falsework have been solved by adopting the principle of lifting the whole or a portion of the structure that has been assembled on the ground.

Form–Arrangement–Connections

The word "form" means the whole of the structural surface regarded as a continuum. The following three approaches are used in the conception of a surface: (1) Geometrical surface classified by the Gaussian curvature; (2) mathematical surface when it is expressed by a mathematical function; (3) equilibrium surface derived from experiment where either the inverted surface of catenary net or shallow soap film is an effective experimental model. A number of variations and combinations of forms based on the above categories can be obtained with various techniques. In many cases the curvature has a great influence on arrangement and connection of the structural elements in space frames, because it dominates the flow of forces in the structure as a whole.

Member arrangement for constituting a three-dimensional surface may generally be operated by two approaches: The "dividing method" and the "assembling method". In the first the whole shape of the surface must be provided at the first stage of the design, whereas in the second the starting point will be the units of the pattern from which assemblage is to be made. Basically, however, the member arrangement may be classified into two systems: (1) grid systems; and (2) unit systems. The former is the network system, component elements of which are single members or plane trussed beams, and the latter is the system in which the surface is covered by the spatial units composed of a truss or a panel.

The connection system remains of tremendous importance in the development of space frames and that is why an enormous amount of energy has been devoted to the research and development of this issue. The joints must not only be strong but also simple both from structural and mechanical viewpoints, and yet they must be easy to produce without sophisticated processing. The development of several patented connectors based on the principles of simplicity and precision has been influential in the increased use of space frame structures. Many systems have been proposed with intent to satisfy these objectives. Some of the most popular systems are, among others, the CDC, Unistrat, Triodetic, Mero, etc. For more details on this section the interested reader is referred to [79,104,117,170].

THE COLLAPSE ANALYSIS PROBLEM

Methods of Analysis

Methods which have been used for the investigation of the structural behavior of space frames may be divided into four broad categories: (1) discrete field analysis; (2) equivalent continuum analysis; (3) matrix analysis; and (4) experimental analysis.

DISCRETE FIELD ANALYSIS

The term discrete field analysis denotes the body of concepts used to obtain field or functional solutions for structures most accurately represented as a lattice or pattern of elements. For the plate-like or shell-like lattice structure, which is composed of the same structural units, the position of any of the units can be described in terms of two natural numbers in two coordinates which are called discrete field variables. The approach is closely analogous to that employed in classical plate or shell theory for continuous surfaces although the mathematical model now consists of difference equations as opposed to differential equations which are associated with the continua. The procedures for a discrete field analysis are distinguished into micro and macro approaches by Dean [50,51]. The method to solve difference equations directly is called micro approach, because the analysis is obtained by analysing the basic unit element and by relating its behavior to that of adjoining elements. On the other hand, macro approach uses the form of summation equations which are constructed as the condition of compatibility on the joint displacements of one directional large structural system and those of the other directional system.

EQUIVALENT CONTINUUM ANALYSIS

The structural state of a regular space frame structure can be approximately represented as functions of continuous variables instead of discrete variables even when a discrete field analysis is not tractable. Two approaches have been used to determine the appropriate relations. The first involves the conversion of the governing difference equations of the system into approximating differential equations by means of Taylor expansions [73,140]. The second method is to relate force or deformation characteristics, or both, of a small segment of the structure to those of a small element of a continuum [54,72,74,75,149,190]. From these relationships equivalent properties such as modulus of elasticity, Poisson's ratio, and plate or shell thickness can be found. Engineering merits of this approach are: the results from existing continuum theories can be applied; it gives a better engineering feeling of the response of the structure; and large systems with fine mesh are analysed without any problem regarding computer storage and cost. There are, however, certain limitations in the applicability of this approach which result from the requirements that the network should be very fine, with regularized shapes, with continuous loads and supports and boundary conditions of a certain type. A detailed study of the merits and limitations of the equivalent shell analysis when applied to the investigation of the elastic buckling response of reticulated shell structures has been carried out by Forman and Hutchinson [55].

However, the most severe limitation of the equivalent continuum analysis and the discrete field analysis remains their inability to simulate and predict the material nonlinearity in conjunction with the geometrical nonlinearity as they are encountered in the inelastic post-buckling behavior of space frames. For this reason only the two subsequent approaches have been examined in this work for the investigation of the inelastic post-buckling behavior of space frames.

MATRIX ANALYSIS

Methods under this classification express the fundamental equations in matrix form and proceed to the solution by utilizing suitable numerical methods via computer implementation. Early difficulties existing in the application of this approach, when large structures with complex topology are encountered, have been almost completely alleviated through the astounding expansion of computer capabilities and the immense development of numerical methods in recent years. The presentation of these methods will be the subject of subsequent sections.

EXPERIMENTAL ANALYSIS

Theoretical results are always vulnerable to the uncertain accuracy associated with various numerical and approximate analysis techniques, particularly in the investigation of the ultimate load behavior of the structures where complicated nonlinear phenomena dominate structural response. Experimental results could provide a useful means of verification of theoretical analyses and could sometimes lead to the discovery of new phenomena which existing theoretical approaches or numerical procedures are unable to foresee. This approach, however, is extremely time-consuming and expensive and for this reason descriptions in the literature of experimental work have almost invariably been related to specific design projects and, as such, it is difficult to make general statements concerning methodology.

Collapse Analysis Approaches

In space frame design, in order to obtain a realistic evaluation of the safety of the structure, it is very important to accurately predict ultimate strength and post-ultimate strength response until final collapse of the structure under static and dynamic loads is observed. Large deflections, plasticity and post-buckling behavior of members should be taken into consideration. We can classify the various approaches currently used in the collapse analysis of space frames into three main categories: (1) plastic mechanism approach; (2) elastic stability approach; and (3) numerical approach.

PLASTIC MECHANISM APPROACH

Essentially this involves finding a plastic mechanism within a structure such that the structure is in equilibrium and no member carries a load greater than its squash load. This approach is best known when used as the yield line method for slabs or for the plastic design of portal frames. Modern methods of structural analysis can be used by utilizing mathematical programming and optimization techniques [38,39,65,184]. Plastic mechanism analysis, however, can only accommodate stability effects in a very approximate manner. The buckling of a strut can be approximated rather crudely by using a lower stress in compression, but this technique has considerable limitations when applied to space frames in general, as we shall see in the following section. Also, the solution of the mathematical programming problem is not trivial and can be very expensive. The plastic mechanism approach should really be restricted to structures such as single layer structures where failure is dominated by member bending.

ELASTIC STABILITY APPROACH

Stability analysis concerns the calculation of certain critical states at which the stiffness of a structure, with respect to a small disturbing force, becomes zero or negative. Typical elastic stability load-displacement curves are shown in Figure 2 which characterize the corresponding problems; (a) the more conventional bifurcation buckling which involves switching from one equilibrium path to another; (b) snap-through buckling; and (c) submission buckling. The "initial" or "linearized" stability problem which is valid

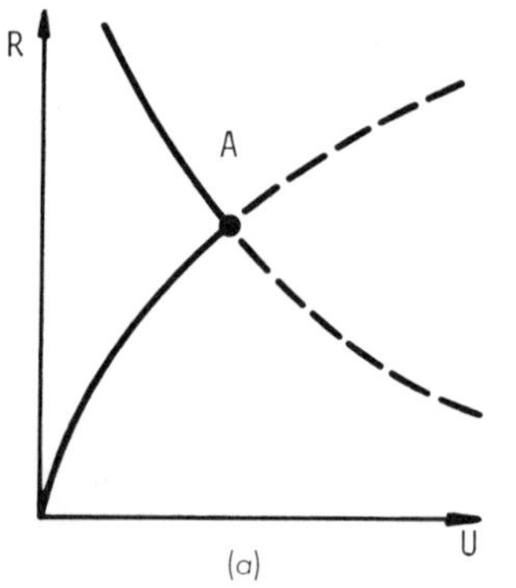

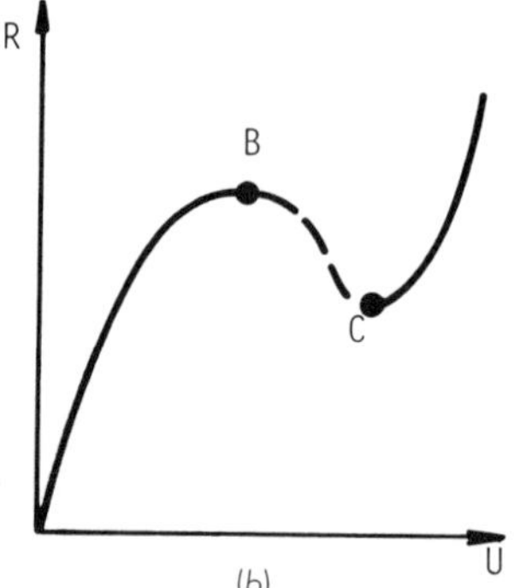

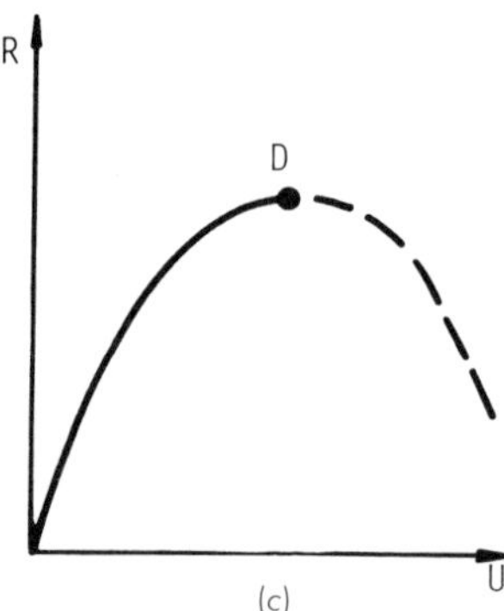

FIGURE 2. Typical load-displacement curves.

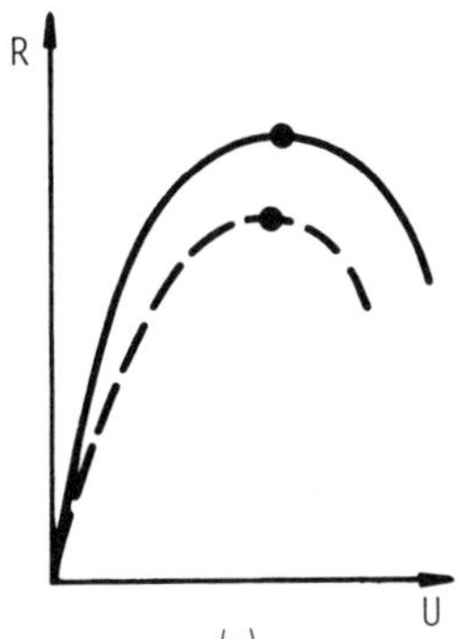

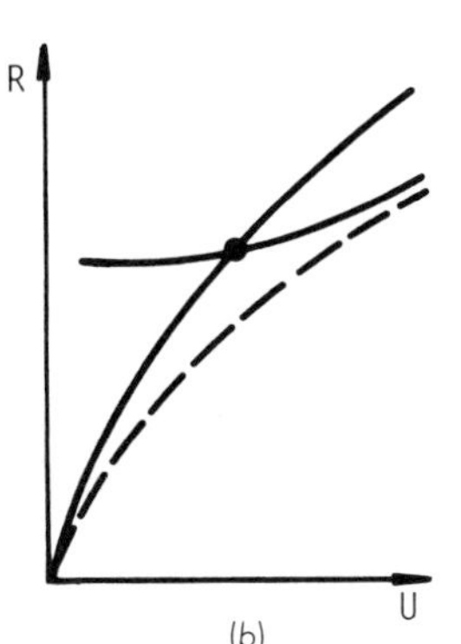

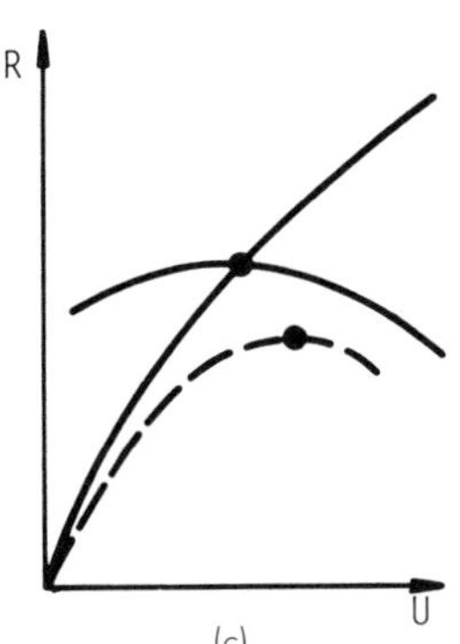

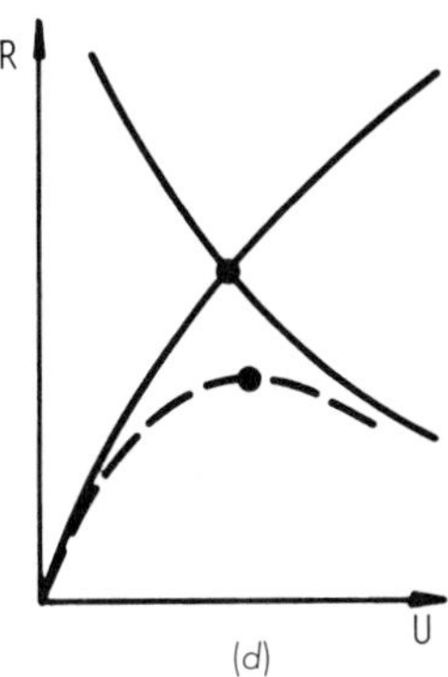

FIGURE 3. Classification of critical points.

under small displacement theory is much more simplified towards the "nonlinear" stability problem with the assumption of large displacement theory. Engineers are generally most familiar with the initial stability problem which can be posed as a linear eigenvalue problem. The lowest few critical loads and their corresponding modes can be computed at a reasonable cost. An initial stability analysis will not necessarily produce the correct critical loads since in general the equilibrium and kinematic equations must refer to the deformed geometry of the structure immediately prior to buckling.

In practice some structures fail at loads well below their theoretical critical load while other structures carry loads which are well above the theoretical load. This leads to interest in post-buckling behavior, imperfection sensitivity, and the magnitude of initial imperfections. A solution to this nonlinear stability problem can be categorized according to Figure 3 as either a "limit" type critical load (*a*) or "bifurcation" type (*b,c,d*). Curves (*b*), (*c*) and (*d*) correspond to stable symmetric, unstable symmetric and unsymmetric bifurcation, respectively. Broken lines represent load-displacement curves with initial imperfections. Theoretical studies and laboratory tests on the imperfection sensitivity problem have been extensively investigated [12,29,43,71,97, 100,145,164,172], but field research on actual structures has not been sufficiently carried out.

Local instability of space frames is associated with member buckling and snap-through of a single joint. The investigations of joint instability are either based on an isolated portion of the structure containing the joint or on a discrete model of the complete structure. The effect of the surrounding structure on the isolated joint model is idealized, ranging from zero to complete constraints. Intermediate restraints are expressed in the form of influence coefficients [99,190]. Local buckling often leads to general buckling as a result of the skeletal dome being unable to resist the shear loads. These loads are transmitted by the buckled unit to the neighboring members causing the area of buckling to increase. Thus, general buckling and eventual failure could be caused by a series of local bucklings occurring at different locations affecting more than one mode. In conclusion, we may say that the stability of such structures is affected by several factors such as geometry, boundary conditions, type of loading, and material properties. Plastic yielding can only be considered in a stability analysis very approximately by reducing the modulus of elasticity. In all cases critical loads need to be interpreted with care.

NUMERICAL APPROACH

This approach can fully take into consideration all kinds of nonlinearities that are present in the structure as a result of the change of geometry or of the material properties of the members of the structure. Therefore, it is the only approach which can predict the inelastic post-buckling response to any desirable degree of accuracy and to any loading or displacement stage. For most problems, what is required is the limit or bifurcation load without the need to trace the post-buckling behavior. However, in some cases, for example in studying the effects of a concentrated load on a restricted part of the structure, it is important to obtain information of the nature of the load shedding after the occurrence of a local instability, in order to assess the response of the whole structure. In a numerical nonlinear analysis when load-displacement paths are traced it is usual to use a combination of incremental and iterative techniques. An extended investigation of all current trends of numerical methods applied to the nonlinear analysis of space frames is presented in a following section.

Dynamic Loads

Usually the structural design of space frames is based on static loads. Space frame structures which are in most cases realized as wide span and high rise structures, are relatively light in weight and low in rigidity compared with other structures. In these cases, the natural periods of the structures become long and the dynamic phenomena could affect the structural design. The important loads to which attention must be paid in design are winds, earthquakes or equipment vibrations. As an average, the magnitude of the vertical component of the ground motion is known to be about, or less than half the horizontal one for earthquake loads. Wind and earthquake loads, however, do not appear to have been a significant factor of collapse failures of space frame structures, while equipment vibrations may have contributed to serviceability failures in roof structures. This conclusion is reached by a recently published ASCE report [171]. It is mainly for this reason that published literature on ultimate dynamic response of space frames is almost non-existent. Methods of dynamic analysis currently used are time history, normal mode, and response spectrum analyses. Response spectrum and normal mode analyses give a good estimation of maximum response when the structure behaves elastically with nearly proportional damping. But where large inelastic deformations are to occur or the dynamic model has a non-proportional damping, it is desirable to adopt time history analyses.

Another phenomenon which may occur during the collapse analysis of a space frame is the dynamic stability problem. This phenomenon usually refers to the snap-through buckling situation where any further increase of load from the critical point B will cause a dynamic jump from position B to position C (Figure 2b). The problem can be modelled as a transient analysis problem using small increment in time. Recent studies in this subject are presented in [37,64,91,121].

BEHAVIOR OF MEMBERS

Member modelling of space frames is primarily affected by the design of the connections and secondarily by the form and arrangement of the structure. The displacement of the nodes cause bending and torsion of the members in addition to axial effects. In many connection designs the bending and torsional stresses could be treated as secondary stresses which do not affect the behavior of the structure substantially. Then, the connections can be approximated by pin joints. When the connections are designed so that they can transfer bending and twisting moments in addition to axial forces, then their effects should be considered. Engineers are accustomed to analysing rigid-jointed triangulated structures as pin-jointed as the rigid joints have little effect on the distribution of forces in these structures. The buckling loads of space frames usually become greater than that of the corresponding pin-jointed members by the effect of joints. But if the rotational rigidities of the joints with finite dimensions are too small, then the buckling loads become smaller than that of the pin-jointed case [69,128].

When inelastic post-buckling behavior is considered, then irrespective of the end conditions of the member, the flexural characteristics should be considered. Several types of models have been developed for treating materially non-linear beam-type problems. In the "section" type of model, the inelastic behavior is defined for the cross-section as a whole, and not for individual points. The bending and twisting moments and the resultant axial force at a cross-section interact with each other to produce yielding of the cross-section. The stiffness of a cross-section is determined by the multi-dimensional action-deformation relationships which relate action quantities such as moments and axial force to the deformation quantities such as curvature and axial strain. In the "section" type, the yielding may be treated in either a distributed or a lumped manner. In the former manner, the yielding is distributed over the length of the element by considering several cross-sections and obtaining the element stiffness using numerical integration. In the latter manner, the plasticity effects occur only at generalized plastic hinges with the beam element between the hinges remaining linearly elastic. The stiffness of the hinges requires the definition of multi-dimensional action-deformation relationships relating moments and axial forces to hinge rotations and axial extensions.

In the "fibre" type of model, the member cross-section is divided into a number of areas (or layers). Each area is assumed to be uniaxially stressed and possesses a stiffness determined by the stress–strain characteristics of the material. The element stiffness is obtained by numerical integration throughout the volume of the beam. The "fibre" type of models provide a better simulation of spread of plasticity in an element, whereas the "section" type of models are computationally more efficient.

Superelements have also been considered in the analysis of space frames [82,167]. These structures, such as double-layer grids, cylindrical roofs and truss-like beams, are assumed to be composed of a repeated basic unit constituted from a number of simple elements. This collective element is then treated as a single element. This idealization was shown to be computationally more efficient than the conventional one.

Pin-Jointed member

In many patented space truss systems, the method of connection of members at the joints approaches the pin-ended connections so that there is little moment interaction between adjacent members. The form of the real load versus axial deformation relationship for the compressive member will therefore be known *ab initio,* whether by experiment or detailed analysis. The investigation of such relationship in struts with different and conditions has been mainly motivated from the role they play as bracing members in building steel frames and offshore steel platforms for severe earthquake motion or extreme environmental loadings. Extensive studies regarding the cyclic inelastic behavior of columns have been reported, mainly in the USA and Japan, and

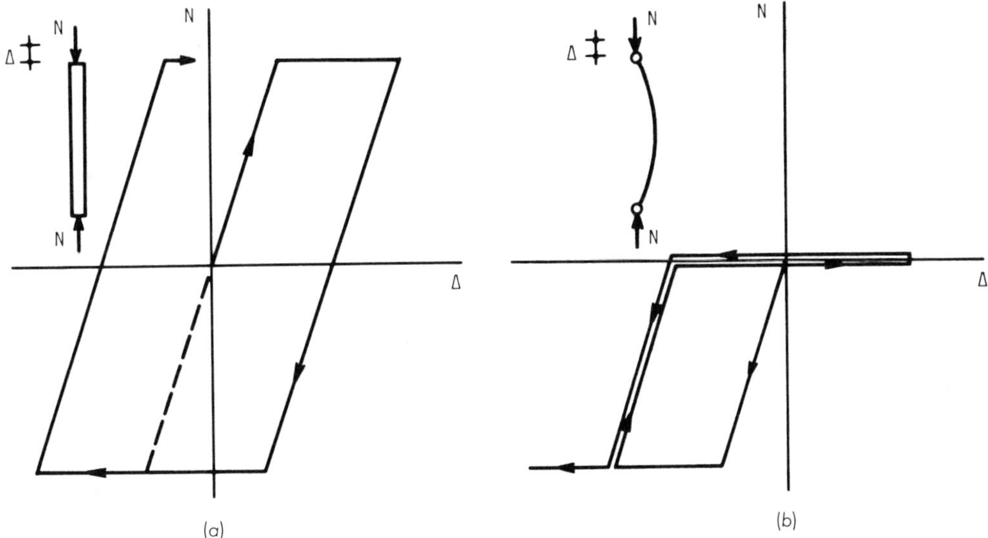

FIGURE 4. N-Δ relationship for stocky and slender struts.

have been directed toward both experimental and theoretical investigations.

GENERAL BEHAVIOR

Figures 4a and b illustrate the axial force N versus axial deflection Δ relationship of two simply supported members with a very small and very large slenderness ratio under reversed loading. Figure 5 indicates a more complex hysteretic behavior which is characteristic in struts with intermediate slenderness ratios, often used in practical designs. Range OA represents the uniaxial compression. At point A the strut buckles elastically until point B when a

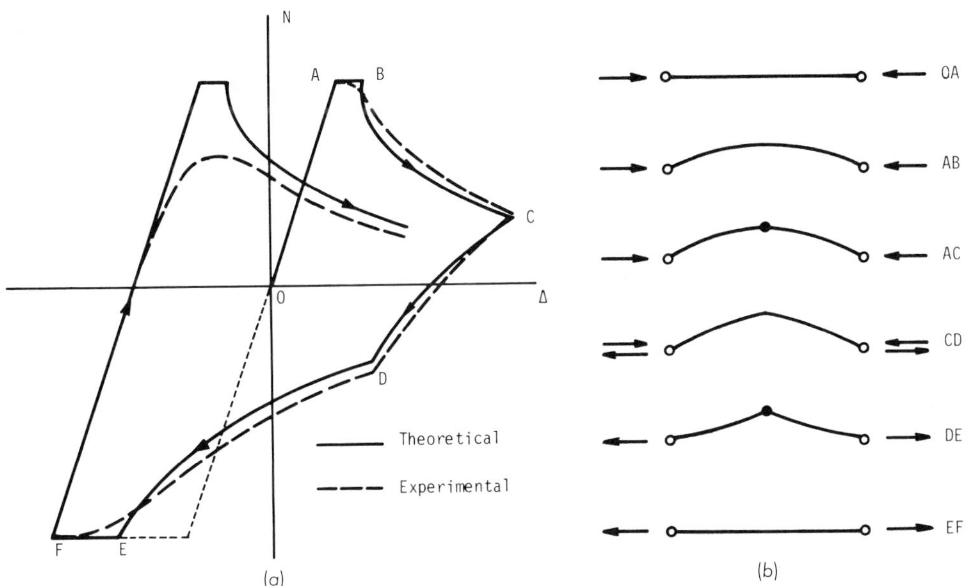

FIGURE 5. N-Δ relationship for a strut with intermediate slenderness ratio; (a) hysteresis loops; (b) shapes on various stages.

plastic hinge takes place at mid-span. Change in geometry and plastic axial strain causes a reduction of the axial load and the strut further deforms along the BC branch. Secondary moment equal to the lateral deflection multiplied by the axial force is developed and since this moment cannot exceed the modified plastic moment due to the axial load, the axial force must decrease as the lateral deflection increases. Once unloading begins at point C, the deflection of the strut is released elastically. The stiffness in this range, however, is less than the initial stiffness because of the lateral deformation. During loading in tension, the stiffness is recovered gradually as the lateral deflection decreases. A plastic hinge occurs at point D, where the stiffness is suddenly reduced because of the plastic rotation and elongation of the plastic hinge. The strut is yielded in tension at point E while it continuous to deform at constant stress. The elongation decreases elastically with the initial stiffness once unloading occurs at point F.

THEORETICAL STUDIES

Theoretical studies followed three main approaches: (1) the plastic hinge concept where all plastic deformation is represented by a plastic hinge formation [76,110,111,128,136, 165,174,179]; (2) the direct analysis by means of the finite element method [9,35,52,57,93,102,168,180]; and (3) intermediate approaches based on properly modelling the mechanical and material behavior of struts, without referring to either the plastic hinge concept or to the finite element formulation [33,156,173].

The plastic hinge concept, which was originally developed by Nonaka [110], has led, under certain assumptions, to an elegant closed form solution for any history of axial loading in the range of small deformations. The bar is idealized as a one-dimensional continuum with uniform material and cross-section along its length. The load is composed of a pair of equal and opposite forces acting at the ends along the original direction and is applied in a quasi-static manner. Lateral deflections occurs about a principal axis without twist, local instabilities do not occur and shear deformation is neglected. The material of the bar is sufficiently ductile and has an elastic-perfectly plastic property under the combined action of the axial force and the bending moment. It is also assumed that the compressed straight bar buckles under Euler's elastic buckling load $N_E = \pi^2 EI/L^2$ or the squash load $N_o = A\sigma_y$, where EI = the flexural rigidity, A = the cross-sectional area and σ_y is the yield stress in tension and compression. This approach can handle any form of cross-section without destroying its analytical nature, by approximating the M-N interaction diagram of the cross-section by piecewise linear interpolation [111,125].

On the basis of the above assumptions the axial displacement can be written as the sum of four components. Thus,

$$\Delta = \Delta^e + \Delta^g + \Delta^p + \Delta^t \tag{1}$$

in which Δ^e is due to uniform elastic axial deformation, Δ^g to change in geometry associated with lateral deflection, Δ^p to plastic axial deformation at the yield hinge, and Δ^t to plastic elongation in a straight configuration distributed along the bar axis. The hysteretic behavior is determined through the cyclical variation of the axial load and the evaluation of the corresponding component of the axial displacement during the three distinctive stages; (a) OA-AB; (b) BC-DE; and (d) CD of Figure 5. The following analysis is based on an ideal I-section and on a compressive axial force. When the axial force is tensile and thus negative, the trigonometric functions have to be replaced by the corresponding hyperbolic functions.

The response of the bar in the initial elastic phase involves only two components of Δ which are attributed to pure axial deformation and to change of geometry due to bending. The first term exists in all loading phases and is related to the axial force by the elastic linear low

$$\Delta^e = \frac{NL}{AE} \tag{2}$$

The second component Δ^g exists when bending moments act on the bar and is evaluated at point B of Figure 5 by considering the basic moment–curvature relation of the small deflection theory for the deflected bar (Figure 6):

$$\Delta^g = -\frac{u^2}{L}\frac{(kL/2)^2}{\left(\sin\frac{kL}{2}\right)^2}\left(\frac{\sin kL}{kL} + 1\right) \tag{3}$$

where $k^2 = N/EI$; $u = M/N$, while an additional equation is obtained from the yield condition. After the formation of the plastic hinge the third component Δ^p is activated and is given by

$$\Delta^p = -2\theta M_o/N_o \tag{4}$$

where

$$\theta = \frac{k}{\tan\frac{kL}{2}}u \tag{5}$$

and M_o = is the limit moment in pure bending; and θ is the plastic hinge rotation. When the bar enters the elastic recovery stage then θ and Δ^p remain constant and Δ^g is now given by

$$\Delta^g = -\frac{\theta^2}{\left(\cos\frac{kL}{2}\right)^2}\frac{L}{4}\left(\frac{\sin kL}{kL} + 1\right) \tag{6}$$

Details of this theory can be found in [110]. Large deformation studies have been also performed by Nonaka [112]. It was confirmed that for practical cases the influence of the additional terms when large deformation theory is considered could be neglected.

The main deficiency of the hinge method, however, remains its inability to anticipate the drop of the critical load after the first cycle, which has been observed during experimental studies (Figure 5). This phenomenon is attributed to the strain hardening and the Bauschinger effects which prevent the bar from regaining its initial straightness after the tensile strength has been reached. The kink which remains in the vicinity of the plastic hinge acts as an initial imperfection during subsequent loading and thus this drop of the critical load appears. Shibata [156] used a modified Shanley's model so that these effects could be included. He introduced a part having a finite length and sustaining a uniform stress in the vicinity of the mid-span of the bar (Figure 7). Then the bar was analysed by representing the flexural deformation of this part as an equivalent hinge deformation and assuming the remaining part of the bar to be rigid with respect to flexural deformations.

For stocky bars, errors produced by the above methodologies may become significant unless plastic deflection is assumed to spread over a finite region. This phenomenon is fairly easily modelled with the use of the finite element analysis. Fujimoto, et al. [57] discretized the bar into twenty segments along the span, the cross-section was divided into twenty pieces as well, and a bilinear stress–strain relationship was adopted. The results are very much in conformity with experimental tests. The finite element method is versatile because of its capacity to include various conditions, but the formulation and computation effort become voluminous. Another approach which leads to a computer code for the prediction of the cyclic inelastic behavior of columns has been developed by Toma and Chen [175].

The inelastic response of imperfect columns is of particular interest because imperfect column behavior could significantly influence the load-displacement path of the entire structure. Theoretical and experimental results in columns with imperfections have been reported in [68,125,128,136, 161,173,175,179].

EXPERIMENTAL STUDIES

A great deal of experimental work has been reported in the literature. The objectives of the tests were to compare them with theoretical results, to investigate the influence of certain complicated phenomena such as local buckling, cracking of the bar, various kinds of imperfections, and finally to develop simplified analytical models based on experimental tests for practical use. Various types of loading programs have been adopted in the experiments. Some were tested under monotonically increasing load until struts experienced excessive failure [41,138,152,154,167]. The majority of tests, however, were performed with cyclic loading.

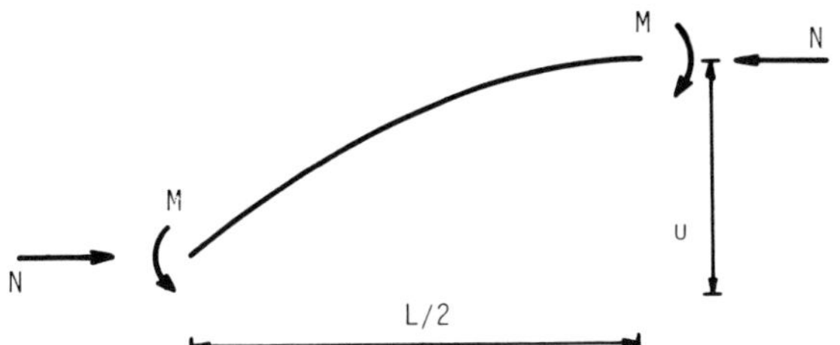

FIGURE 6. Deflected half-strut and notation.

Wakabayashi, et al. [157,181] tested struts having square and H-shaped cross-sections. They first observed the decrease of the critical load after the second cycle (Figure 5). They also observed that end restraints can increase the areas of hysteresis loops, providing a safer evaluation when the strut is represented as simply supported. Jain, et al. [85,86] carried out series of tests of square tube and single angle struts with gusset plates installed at the ends. They concluded that local buckling can occur even if a tube strut satisfies the AISC requirement on the diameter-to-thickness ratio. Other tests by Popov, et al. [24,133] revealed that the decrease in resistance due to severe cyclic load reversals is less appreciable for a strut having a smaller diameter-to-thickness ratio and that round tube struts are the best in performance compared with other types of cross-sections. Gugerli and Goel [67] tested full scale tube and H-shaped struts. They also observed local buckling despite the fact that the AISC requirement is satisfied and that cracking is retarded for a strut with great slenderness ratio. Other tests were reported in [76,80,136,155,178].

ANALYTICAL MODELS

Theoretical results are obtained through sophisticated numerical calculations and, unless gross oversimplifications in the geometry and the stress distribution of the deformed column are made, closed form solutions are not available. What an analyst needs is a formula to provide the axial and/or the flexural stiffness of the member at each displacement level. As this is an almost impossible task when theoretical expressions are followed, researchers have estab-

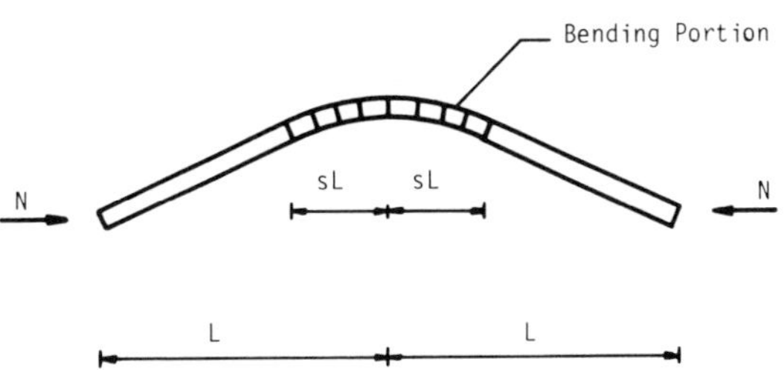

FIGURE 7. A strut, with a finite plastic portion; Shibata's model [156].

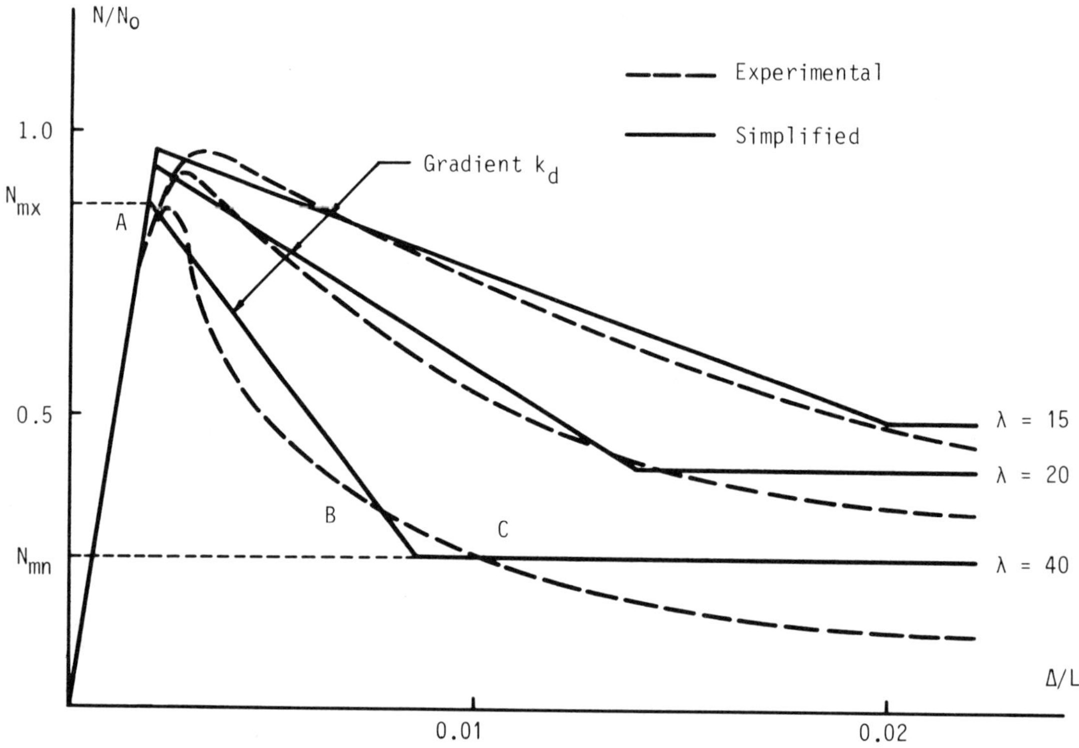

FIGURE 8. N-Δ experimental and simplified curves; Suzuki et al. [166].

lished analytical simplified expressions, based on theoretical and experimental results, which take into consideration all the necessary characteristic parameters. Past studies used the gross idealization of Figure 4a. This model cannot reflect possible change in the behavior according to the slenderness ratio of the struts. A more realistic model that takes into account the buckling characteristics is used in [152] (Figure 9b). Suzuki, et al. [166] noticed that the reduction of the carrying capacity of the member from point A to point B of Figure 8, depends on the slenderness ratio of stub columns, and the deformation at point C depends mainly on the width-thickness ratio of members. Thus, replacing the experimental curve with an analytical curve (broken line), the following can be obtained:

$$N_{mx} = \left[1 - 0.4\left(\frac{\lambda_e}{\Lambda}\right)^2\right] \Big/ \left[1 + \frac{4}{9}\left(\frac{\lambda_e}{\Lambda}\right)^2\right] \quad (7)$$

$$5\lambda_e^2 = N_{mn}^{-2} \cos^2\left(\frac{\pi N_{mn}}{2}\right) \quad (8)$$

$$K_d = -0.15(\lambda_e - 0.5) \quad (9)$$

where λ = slenderness ratio; $\lambda_e = \sqrt{e_y} \cdot \lambda$; e_y = yield strain; and $\Lambda = \pi^2/0.6$. Figure 8 shows the analytical and experimental correlation as well as analytical diagrams for different slenderness ratios. Collins [40] and Schmidt and Gregg [150] used the model (a) shown in Figure 9 in which the load-displacement behavior is modelled by a series of linear approximations.

More sophisticated models were proposed by researchers working on the cyclic inelastic behavior of braces. Higginbotham and Hanson [76] proposed a complete second degree equation with respect to the axial force and axial deflection after buckling. The five constants in the equations are so determined that the curve can approximately simulate the curve derived from more accurate analysis. Curves in other stages also are determined in the same manner. Singh and Goel [158] obtained a piecewise linear model that can properly duplicate the experimental curves. Similar models are proposed in [84,87,103,144]. Wakabayashi, et al.'s [181] model follows analytically the five stages that characterize the cyclic behavior: two linear elastic portions, one plastic portion, and two inelastic mechanism portions. All portions are piecewise linear except the two inelastic portions which are curved. These models are capable of taking into account the degradation in load carrying capacity due to load reversals and expressing the restoring force in a closed form with respect to axial deformations.

Models based on experiments may accurately represent

the post-buckling behavior of members. However, experiments are needed for each member geometry and material used in a structure. Such experiments are expensive and time-consuming. It is also pointed out that models based on assumed load shortening curves are approximate and sometimes risky to use.

End-Restrained Member

The nonlinear inelastic analysis of a rigid jointed space frame is a relatively complicated and expensive procedure. The most accurate way of treating this problem is to consider all sorts of nonlinearities that are inherent in this phenomenon. The effects of change of geometry may take into account; (1) change of member lateral stiffness due to the effect of axial force; (2) change of member length due to bowing; (3) finite deflection of joints; and (4) finite rotation of joints and rigid body rotations. The buckled shape of a rigidly jointed strut cannot be even approximated by the single cubic polynomial used in a linear analysis. A single member must be modelled by using at least two cubic elements jointed in line or alternatively a special element must be devised using at least one internal node. To model the effects of material yield, the member, as already mentioned, must first be divided along its length into subelements. Within each subelement the spread of plasticity through the cross-section is assumed to be constant. There are two distinct methods in use for handling the partially plastic cross-section: subdividing of the cross-section and assuming constant stress within each subdivision, and the use of an approximate yield criterion and flow rule expressed directly as a function of the stress resultants, that is, the axial force, the bending moments etc. The latter method tends to suppress the effects of the spread of plasticity inside the

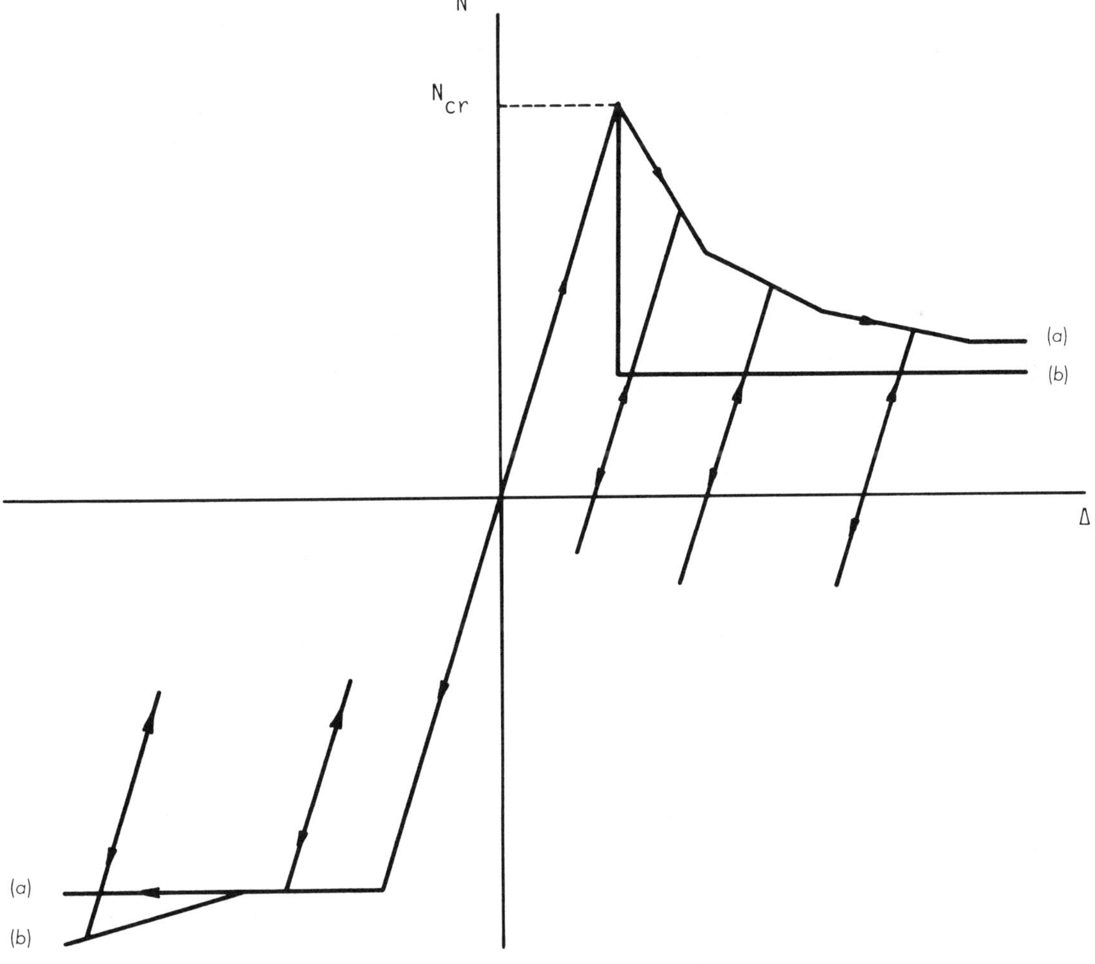

FIGURE 9. N-Δ simplified curves.

member (from initial yield until collapse) and possibly local unloading from playing their role in affecting the results of the analysis. It cannot, also, accommodate torsional buckling, and a different stress resultant yield criterion is required for different cross-section geometries. The use of this method can, however, lead to useful savings in computer time, storage requirements and data preparation.

The inelastic large displacement behavior of a beam in space is not an easy task. Any rigorous analysis which attempts to cover this behavior is complicated. The governing differential equations are often intractable and recourse must be made to numerical methods to obtain solutions. The finite element method is the most capable and popular approach for handling this problem. There are, however, cases in which the specific form, arrangement, connections, and loading of the structure allows the introduction of additional assumptions by which a 3-D member can be approximated as functioning like a 2-D member or like a nonsway member. If these degradations of the role of the members can be substantiated then simplified methods of solutions, very popular with structural engineers, are available. The "approximate approaches" are generally found quite accurate in many cases and can thus be used efficiently to generate a large volume of data from which practical design methods can be developed.

NONSWAY MEMBERS

When the relative transverse displacements of the two ends of a member may be assumed negligible then it is called a nonsway member. Closed form solutions for the cyclic inelastic response based on the plastic hinge concept have been reported by Wakabayashi, et al. [179]. They considered elastic end restraints and initial load eccentricity. Prathuangsit, et al. [136] and Jain, et al. [85,86] have studied the same problem but the ends were restrained with an elastic-perfectly plastic low, simulating in this way the effect of the gusset plate connections. Papadrakakis and Loukakis [128] have studied a more complicated model which takes into account initial eccentricity of the load, initial out-of-straightness, elastoplastic connection of the member to the joint, partially restrained joint and derived closed form analytic expressions for any type of cross-section. Only in the case of the load eccentricity a simple iterative numerical procedure is needed to determine the new position of the load application point after the formation of plastic hinges at the supports. Figure 10 shows the cyclic response of a characteristic case.

The study of the nonsway beam–columns has been the subject of extensive research by Chen and Atsuta [31]. In a recent paper [33] a methodology is proposed which is based on a concept similar to the finite element method. The actual beam is physically replaced by an assembly of finite segments. As a result, the beam-column can be formulated and solved approximately in terms of the behavior of these segments. This approach may be considered as a physical interpretation of the finite difference method as applied numerically to solve differential equations. In other works Toma and Chen have presented closed form solutions of the fixed ended beam-column based on the hinge-by-hinge procedure [173] and on the assumed deflection method [175]. In the latter they considered a two-straight line approximation of the deflected shape and compared the results with other more exact shape approximations. Initial imperfections are taken into account but no plastic axial deformation as at the plastic hinges are considered.

A finite element model is developed by Madi and Smith [102] for tube sections in which a first order geometrically nonlinear analysis is used. The member is divided along its length into a number of straight small elements. The section of the member is divided into regions, and a control point in every region is chosen as representative of the state of stress and strain in every part of that region. Each subelement is also divided into two equal end elements. The current sectional properties of these end elements are to be based on the elastic regions of the cross-section since the plastic regions, under perfect plasticity, do not offer any resistance to further plastic deformations. Finally, experimental studies on end-restrained nonsway columns have been reported by a number of investigators [10,32,76,84,85,86,103,134,155,177, 179].

2-D MEMBERS

Recently, the use of incremental variational principles together with the finite element method made it possible to develop finite element models and computational algorithms which can handle both types of nonlinearities in any desired degree of accuracy. Details of this general formulation are presented in the following subsection. Nevertheless, the development of new and efficient two-dimensional finite element models is still a live problem. Certain simplified assumptions can drastically reduce the mathematical formulation and the computer time, while not affecting substantially the results obtained. In these lines a number of papers have appeared in the last two years.

Kassimali [93] extends an Eulerian large displacement formulation [118] to include material nonlinearities as well. Local member force deformations are based on the beam–column approach and changes in member chord lengths due to axial strains and flexural bowing are taken into account. The material is assumed to be ideally elastic–plastic, and yielding is considered to be concentrated at member ends. Reversals of plastic hinge rotations are not anticipated. Another approximate formulation, which is also based on the plastic hinge concept at critical sections, is presented by Creus [44] et al. The method, however, uses a generalized yield criterion that includes the effects of normal and shear forces.

A conventional six d.o.f. beam for fast and efficient solution is proposed by Yang and Saigal [193]. The material nonlinearity is treated here by including its effect in the

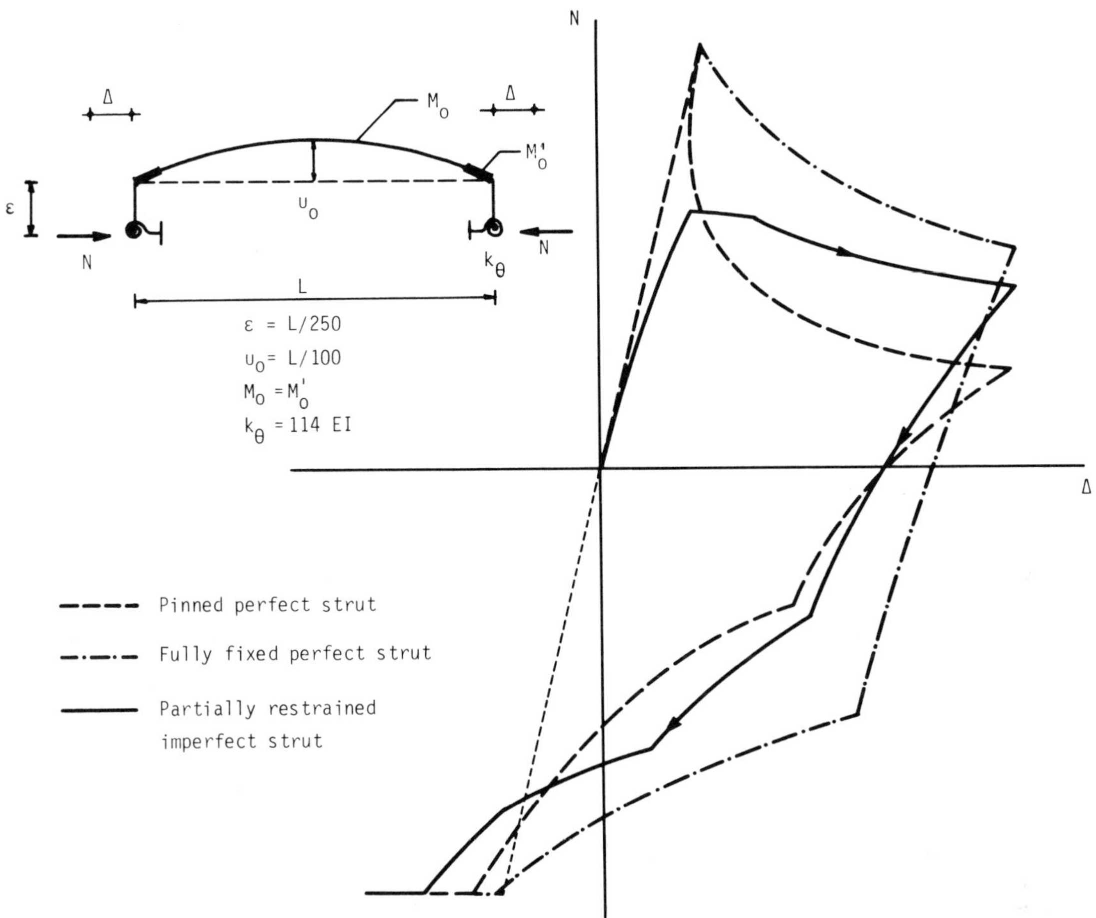

FIGURE 10. N-Δ curves for struts with different end conditions and imperfections; Papadrakakis–Loukakis [128].

governing equations. The stiffness matrix of each element is formed by using a two-dimensional grid Gauss points and considering the material properties at each point corresponding to the uniaxial strain at that point. The geometric nonlinear formulation is based on a previous work by Yang [192] in which a very simple yet highly efficient way of treating the elastic problem, using finite elements, is developed. The final stiffness matrix is composed of two parts, each one containing separately the effects of geometric and material nonlinearities. El-Zanaty and Murray [52] presented an element with five d.o.f. per node in which in addition to the two displacements and rotation, the two first derivatives of the orthogonal displacements at the nodes are included. The effects of residual stresses, strain hardening, gradual penetration of yield through the cross-section and the spread of inelastic zones along the member length are considered. To obtain a more accurate description of the spread of plasticity along the length, each element is subdivided into three elements with unequal lengths to account for the irregular distribution of the sampling points of the Gaussian numerical integration. The formulation is based on a simple geometric approximation in the evaluation of the slope at any point along member reference axis, which permits the virtual work equations to be derived in a manner consistent with the full nonlinear strain-displacement equations without introducing further kinematic approximations. A similar approach was presented by Cichon [35] in which a beam-element is adopted where both local and global d.o.f. are considered for each element.

Kayal [94] used a single beam-column element with five d.o.f. (two end rotations, two end deflections and one midlength deflection). This formulation uses any analytical expression for the stress-strain curve and the application of a strain increment or a displacement increment can handle the dropping brach. The inclusion of the fifth d.o.f. makes unnecessary the subdivision of the member into smaller elements. A hybrid-type large displacement model with three d.o.f. per mode is developed by Bäcklund [11]. The element stiffness matrix is obtained by inversion of a flexibility matrix which is computed from an assumed distribution of internal forces along the element axis. This approach has the advantage that the approximate stress fields better imitate varying stiffness effects along the beam than the traditional approach with assumed displacement fields. Partial yielding is considered across the height and along the element axis of the beam.

3-D ELEMENT

As already mentioned, the problem of analysis of a beam element in space with the inclusion of the geometric and material nonlinearities is extremely complicated and the only tool which can deal adequately with such problems is the finite element method. Under this approach the constitutive equations for elastic or elastoplastic materials are considered. The general three-dimensional nonlinear beam formulation is not a simple extension of a two-dimensional formulation, because in 3-D analysis large rotations have to be accounted for that are not vector quantities. Thus, the development of general nonlinear beam elements is probably one of the most challenging and unresolved research tasks of finite element structural analysis and receives intensive attention from researchers all over the world.

In describing the motion of a body, there are four methods of formulations. Truesdell [176] called them the material, the referential, the spatial and the relative formulations. All four methods are equivalent for a smooth motion of the body: (1) Material description: independent variables are the particle, or the body point X and the time t. This description is rarely used in FEM. (2) Referential description: independent variables are the position $\bar{X}$ of the particle X in an arbitrary reference configuration and the time t. When the reference configuration is chosen at the particular time $t = 0$ then this description is called Lagrangian in the literature although it was first introduced by Euler. (3) Spacial description: independent variables are the current position $\bar{x}$ of a particle X, and the time t. This description was introduced by Bernoulli and D'Alembert but it is usually called Eulerial formulation. This Spatial description is suited to the study of fluids. (4) Relative description: independent variables are the position $\bar{x}$ in a current configuration and a variable time τ. The variable time τ is the time when the particle X occupied a position $\bar{\xi}$, where $\bar{\xi} = \bar{\xi}\ (\bar{x},\tau)$. It is important to realize that the relative description is referential or Lagrangian in nature, in the sense that the reference position is now denoted by $\bar{x}$ at time t rather than $\bar{X}$ at time $t = 0$. A special case of the independent description is the updated Lagrangial formulation (UL) where we describe the future with respect to the present. In UL the independent variables are $\bar{x}$ and τ, where $\bar{x}$ is the position occupied by the material point X at time t. In finite element applications to continuum mechanics problems, only the last three descriptions are used. For a complete and concise survey of formulation methods of geometrical and material nonlinearity problems the interested reader is referred to the work by Gadala, et al. [58,59].

In what follows we will try to describe briefly two types of 3-D beam elements which belong to two different families of element stiffness matrix formulations. The first is the two node Hermitial element [13,14] which is consistently treated on the basis of the equations of continuum mechanics, and the second is the natural element [7,9], which is based on the natural mode technique.

THE HERMITIAN 3-D ELEMENT

This element is introduced by Bathe and Bolourchi [14] through a consistent development of a total Lagrangian (TL) and UL formulation on the basis of the equation of continuum mechanics. These formulations must be implemented using appropriate displacement interpolation functions. Considering the choice of these functions it is recognized that for a beam of constant cross-section in small displacement analysis the Hermitian functions should be employed to interpolate the transverse bending displacements, and linear interpolation must be used to interpolate the torsional and longitudinal displacements. Therefore, the same functions are employed for a beam element that can undergo large rotations (with small strains), but referred to the beam convected coordinate axes. In this way the usual beam kinematic assumptions used are referred to the current beam geometry. The beam element is assumed to be straight.

Consider the motion of a body in a fixed Cartesian coordinate system, as shown in Figure 11, where the solutions at $0, \Delta t, 2\Delta t, \ldots t$, time steps, are known and it is required to solve for the unknown static and kinematic variables in the configuration at time $t + \Delta t$. In the TL formulation all static and kinematic variables are referred to the initial configuration 0. Considering the equilibrium of the body at $t + \Delta t$ the principle of virtual displacement gives

$$\int_{^{o}V} {}^{t+\Delta t}_{o}S_{ij}\, \delta\, {}^{t+\Delta t}_{o}\epsilon_{ij}\, {}^{o}dV = {}^{t+\Delta t}Q \tag{10}$$

where ${}^{t+\Delta t}Q$ is the total external virtual work expression, $\delta\, {}^{t+\Delta t}_{o}\epsilon_{ij}$ is a (virtual) variation in the Cartesian components of the Green-Lagrange strain tensor, ${}^{t+\Delta t}_{o}S$ are the Cartesian components of the second Piola-Kirchhoff stress tensor. Left superscripts refer to the configuration of the body in which the quantity occurs and left subscripts indicate with respect to which configuration the quantity is measured. If both these scripts are the same then the subscript is omitted. The right subscripts refer to the configuration axes. Since the stresses ${}^{t+\Delta t}_{o}S_{ij}$ and strains ${}^{t+\Delta t}_{o}\epsilon_{ij}$ are unknown, for solution the following incremental decompositions are used:

$${}^{t+\Delta t}_{o}S_{ij} = {}^{t}_{o}S_{ij} + {}_{o}S_{ij} \tag{11}$$

$${}^{t+\Delta t}_{o}\epsilon_{ij} = {}^{t}_{o}\epsilon_{ij} + {}_{o}\epsilon_{ij} \tag{12}$$

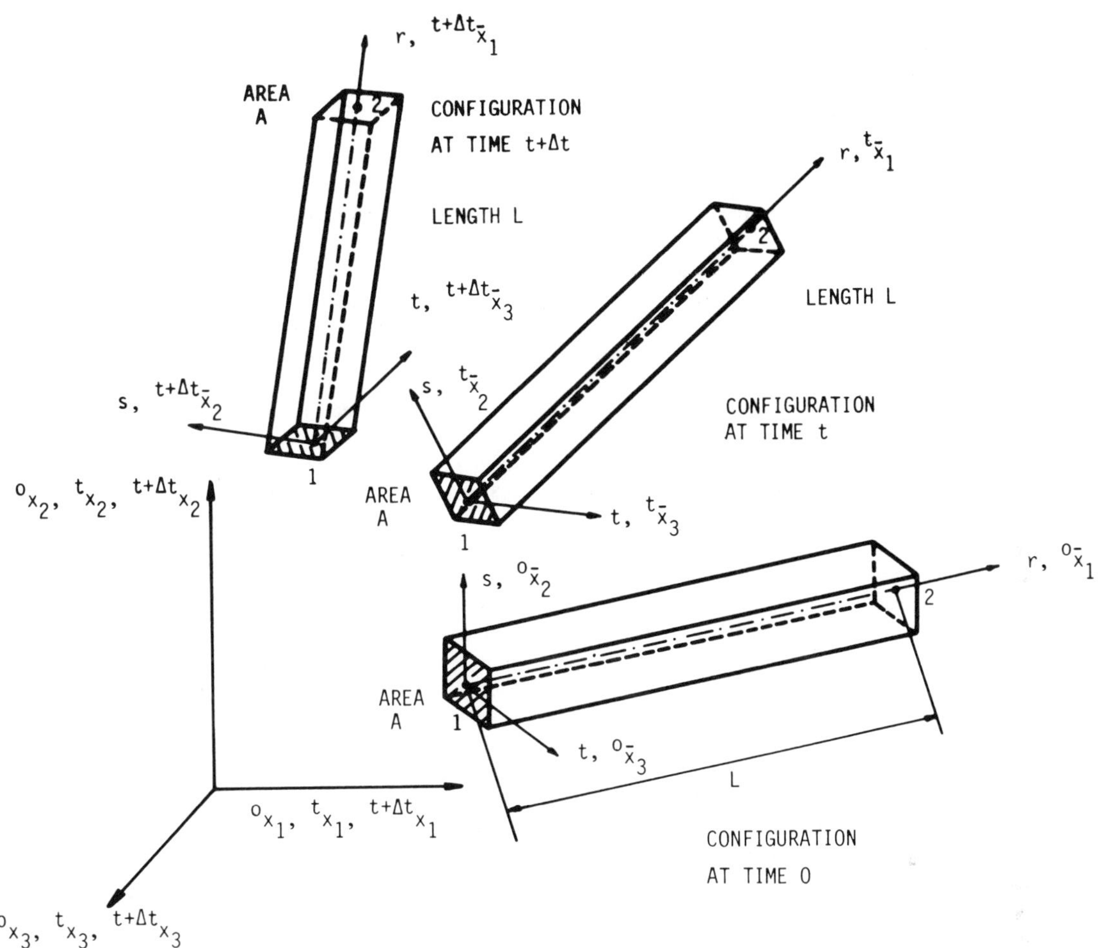

FIGURE 11. Hermitian beam element: motion of the three-dimensional element and its local co-ordinate axes shown in global co-ordinate system.

where ${}^t_oS_{ij}$, ${}^t_o\epsilon_{ij}$ are the known stresses and strains in the configuration at time t. After separating the strain increment components into linear and nonlinear parts

$$ {}_o\epsilon_{ij} = {}_oe_{ij} + {}_o\eta_{ij} \tag{13}$$

using the tensor components ${}_oC_{ijrs}$ for the linearised incremental constitutive relations

$$ {}_oS_{ij} = {}_oC_{ijrs}\, {}_oe_{rs} \tag{14}$$

and using the linearized approximation $\delta_o{}^{t+\Delta t}\epsilon_{ij} = \delta_o e_{ij}$, Equation (10) is transformed to

$$ \int_{^oV} {}_oC_{ijrs}\, {}_oe_{rs}\delta_o e_{ij}{}^o dV + \int_{^oV} {}^t_oS_{ij}\delta_o\eta_{ij}{}^o dV = {}^{t+\Delta t}Q - \int_{^oV} {}^t_oS_{ij}\delta_o\epsilon_{ij}{}^o dV \tag{15}$$

In the UL formulation the same incremental stress and strains as in TL formulation are employed, but all variables are referred to the configuration at time t. The corresponding equation for the UL formulation takes the form

$$ \int_{^tV} {}_tC_{ijrs}\, {}_te_{rs}\delta_t e_{ij}{}^t dV + \int_{^tV} {}^t_t\tau_{ij}\delta_t\eta_{ij}{}^t dV = {}^{t+\Delta t}Q - \int_{^tV} {}^t_t\tau_{ij}\delta_t e_{ij}{}^t dV \tag{16}$$

where the ${}^{t}\tau_{ij}$ are the Cartesian components of the Caushy stress tensor and ${}_{t}e_{ij}$ and ${}_{t}\eta_{ij}$ are the Cartesian components of the linear and nonlinear strain increments, respectively. It was confirmed [14] that both formulations give identical results for geometrical nonlinear problems and that the UL formulation is computationally more effective. The subsequent analysis is confined to the UL formulation while similar expressions are valid for the TL formulation.

The beam element is assumed to behave like the Timoshenko beam, can undergo large deflections and rotations, but small strains are assumed. Thus, the cross-sectional area and the length of the beam do not change during deformation. In Equations (15) and (16) the incremental equilibrium equations of a body in motion are given corresponding to the global coordinate system ${}^{\tau}x_i, \tau = 0$ or t. Considering a typical beam element it is more effective to first evaluate the finite element matrices corresponding to the local principal axes ${}^{\tau}\bar{x}_i$ of the element (Figure 11), and then to transform them to the global axes prior to the element assemblage process. The finite element matrices corresponding to the axes ${}^{\tau}\bar{x}$ are simply obtained by measuring all static and kinematic quantities in this coordinate system. Thus, using Equation (16), the equilibrium equation for a single beam element is obtained

$$({}^{t}_{t}K_L + {}^{t}_{t}K_{NL})u = {}^{t+\Delta t}R - {}^{t}_{t}F \tag{17}$$

where ${}^{t}_{t}K_L$ is the linear strain incremental stiffness matrix; ${}^{t}_{t}K_{NL}$ is the nonlinear or geometric or initial stress incremental stiffness matrix; ${}^{t+\Delta t}R$ is the vector of externally applied element nodal forces; ${}^{t}_{t}F$ is the vector of nodal point forces equivalent to the element stresses at time t; and u is the vector of incremental nodal displacements.

The element matrices in Equation (17) are evaluated using the displacement interpolation functions of the beam element. Table 1 summarises these calculations with the following notation: ${}^{t}_{t}\bar{B}_L$ = linear strain displacement transformation matrices; ${}^{t}_{t}\bar{B}_{NL}$ = nonlinear strain-displacement transformation matrices; ${}_{t}\bar{C}$ = incremental stress-strain material property matrices; ${}^{t}\bar{\tau}$, ${}^{t}\hat{\tau}$ = matrix and vector of Caushy stresses. The kinematic assumptions used in defining the interpolation functions hold for small strains, small rigid body incremental rotations in each solution step, but any size of translational displacements. The solution of the respective Equation (17) for the whole structure will yield nodal point displacement increments, from which the corresponding strain increments ${}_{t}\bar{e}_{ij}$ are calculated. With the increment of strains known, the corresponding stress increments can be calculated which determine the total stresses from which subsequently the vector of internal nodal point forces is evaluated.

TABLE 1. Finite Element Matrices

Integral	Matrix Evaluation
$\int_{{}^{t}V} {}_{t}\bar{C}_{ijrs}\, {}_{t}\bar{e}_{rs}\, \delta_{t}\bar{e}_{ij}\, {}^{t}dV$	${}^{t}_{t}\bar{K}_L\, \bar{u} = \left(\int_{{}^{t}V} {}^{t}_{t}\bar{B}_L^T\, {}_{t}\bar{C}\, {}^{t}_{t}\bar{B}_L\, {}^{t}dV \right) \bar{u}$
$\int_{{}^{t}V} \bar{\tau}_{ij}\, \delta_{t}\bar{\eta}_{ij}\, {}^{t}dV$	${}^{t}_{t}\bar{K}_{NL}\, \bar{u} = \left(\int_{{}^{t}V} {}^{t}_{t}\bar{B}_{NL}^T\, {}^{t}\bar{\tau}\, {}^{t}_{t}\bar{B}_{NL}\, {}^{t}dV \right) \bar{u}$
$\int_{{}^{t}V} {}^{t}\bar{\tau}_{ij}\, \delta_{t}\bar{e}_{ij}\, {}^{t}dV$	${}^{t}_{t}\bar{F} = \int_{{}^{t}V} {}^{t}_{t}\bar{B}_L^T\, {}^{t}\hat{\tau}\, {}^{t}dV$

Material nonlinearities are taken into account by considering appropriate constitutive descriptions [13,17]. A fundamental observation comparing elastic and inelastic analysis is that in the former the total stress can be evaluated from the total strain alone, whereas in the latter the total stress at time t also depends on the stress and strain history. In the TL formulation with the assumptions of small strains the material model implementation used for materially-nonlinear-only analysis can directly be employed by simply substituting in the material characterization the second Piola-Kirchhoff stresses and Green-Lagrange strains for the small displacement engineering stress and strain measures. The use of the TL formulation in inelastic analysis is therefore a direct extension of the elastic case. When, however, the UL formulation is employed then the incremental virtual work expression of Equation (16) and the Jaumann stress rate equation are now used to calculate the Caushy stresses.

In what follows a brief description of the construction of the stress-strain matrix in elasto-plasticity is attempted. Using the usual approach of flow theory to describe the elastic-plastic material behavior, three properties in addition to the elastic stress-strain relations characterize the material behavior: (1) a yield condition which specifies the state of multiaxial stress corresponding to the start of the plastic flow; (2) a flow rule which relates the plastic strain increments to the current stresses and the stress increments subsequent to yielding; and (3) a hardening rule which specifies how the yield condition is modified during plastic flow. The yield condition can be written at time t as

$${}^{t}F({}^{t}\sigma_{ij}, {}^{t}\varkappa) = 0 \tag{18}$$

where ${}^{t}\varkappa$ is a state variable which depends on the plastic strains ${}^{t}e^{P}_{ij}$. If in addition the analysis is restricted to an associated flow rule, the function ${}^{t}F$ is used to calculate the plastic strain increments

$$de^{P}_{ij} = {}^{t}\lambda \frac{\partial {}^{t}F}{\partial {}^{t}\sigma_{ij}} \tag{19}$$

where ${}^{t}\lambda$ is a scalar still to be determined. The relation in Equation (19) is referred to as the normality rule. The stress increments are evaluated using

$$d\sigma_{ij} = C^{E}_{ijrs}(de_{rs} - de^{P}_{rs}) \quad (20)$$

which becomes

$$d\sigma_{ij} = C^{EP}_{ijrs}\, de_{rs} \quad (21)$$

where C^E, C^{EP} represent the elastic and the instantaneous elastic–plastic stress–strain matrices, respectively. The latter matrix depends on the yield function ${}^{t}F$ used and on the material characteristics to be modeled. The final stresses at time $t + dt$ are given by

$$^{t+dt}\sigma = {}^{t}\sigma_{ij} + de_{ij} \quad (22)$$

Equations (18), (19), (20), (22) in conjunction with Equations (15) or (16) serve to specify the problem of the combined geometrical and material nonlinearities for the TL and UL formulations, respectively. Equations (20) and (22) need, however, to be expressed in terms of the Jaumann stress rates for UL formulation [13].

When material nonlinearities are considered, a higher integration order than that used for linear or geometric nonlinear response is frequently required. Since the material nonlinearities are only measured at the integration points of the elements, the use of a relatively low integration order may mean that the spread of the materially nonlinear conditions through the element is not represented accurately. This leads to the conclusion that the Newton-Cotes integration formulas may be effective, because in contrast to the Gaussian quadrature, the integration points for stiffness and stress evaluations are on the boundaries of the elements.

Based on the above formulation Bathe [13] has also developed a general isoparametric solid beam element. This element has certain advantages over the Hermitian element, particularly when material nonlinearities are concerned, but needs for the geometrical nonlinear problem only almost five times greater formulation time than that of the Hermitian element [2,13,19]. Bathe and Wiener [19] used a four node isoparametric beam element to model an I-beam. The web and flanges are represented by individual elements and their response is tied together using constrained equations. Finally, all these models, and others not mentioned here, are incorporated into the multi-purpose computer program ADINA [2].

THE NATURAL 3-D ELEMENT

This element is based on the natural mode technique developed by Argyris and his co-workers at the University of Stuttgart. On the simplest level of explanation the terminology "natural method" refers to the use of separate rigid body displacements and natural deformations for the description of the current state of a finite element. The natural modes of deformation imply a contemplation of the history of the element via a convected system of axes. Individual deformations of elements are considered through the subelement concept, which constitutes the backbone on the natural mode technique. The natural modes are then defined for the simplest possible finite element types such as straight beam in space to cope for the analysis of a curved beam in space. Closely associated with this aspect of the natural modes is a definition of stresses and strains which is in harmony with the element or subelement in question. This formulation handles the noncommutative nature of rotations by the introduction of a new consistent set of rotations called semitangential rotations since they correspond to the conservative semitangential torques introduced by Ziegler [194]. In contrast to rotations about axes fixed in space, semitangential rotations are commutative, even when large. Thus, they entail unique strain-displacement relations and unique potential energy functions which lead directly to symmetric geometric stiffness matrices. The natural formulation permits the use of the natural elastic stiffness matrix of the linear theory for the large displacement–small strain problems as long as the appropriate nonlinear relations between natural deformations and modal Cartesian freedoms are observed. In the end, the complete derivation of the geometric stiffness matrix is in principle reduced to a simple transformation of the nodal freedoms by taking into account the specific properties of the semitangential rotations.

The natural mode technique is extended into three element levels: (1) The urelements are particularly simple basic elements in which the definition of the natural modes is presumed to be given. (2) The subelements are differential elements derived by a limiting process from the urelements. (3) The general elements are constructed from the subelements and belong to the same family as the respective urelements. The urelement of Figure 12 is a beam in space with its local Cartesian coordinate systems. The displacement modes are described by the displacement and rotations of the nodal points at the two ends in conjunction with the well known linear Lagrangian and third-order Hermitian interpolation functions. The nodal freedoms referred to the local Cartesian system of reference are represented here by ϱ. As long as an isolated beam element is considered, the displacement state can also be determined by an arbitrary set of twelve generalized displacements. A set of twelve generalized displacements appropriate for a beam element comprises six components ϱ_{01} to ϱ_{06} rigid body motions and ϱ_{N1} to ϱ_{N6} independent natural deformation modes of the element. A relationship between the rigid body and natural deformation modes and the Cartesian components of the generalized displacement vector can be written

$$\varrho = A_e \tilde{\varrho} = [A_o \; A_N] \begin{bmatrix} \varrho_o \\ \varrho_N \end{bmatrix} \quad (23)$$

where A_e is a 12×12 transformation matrix.

A generalized force vector (rigid forces and natural gener-

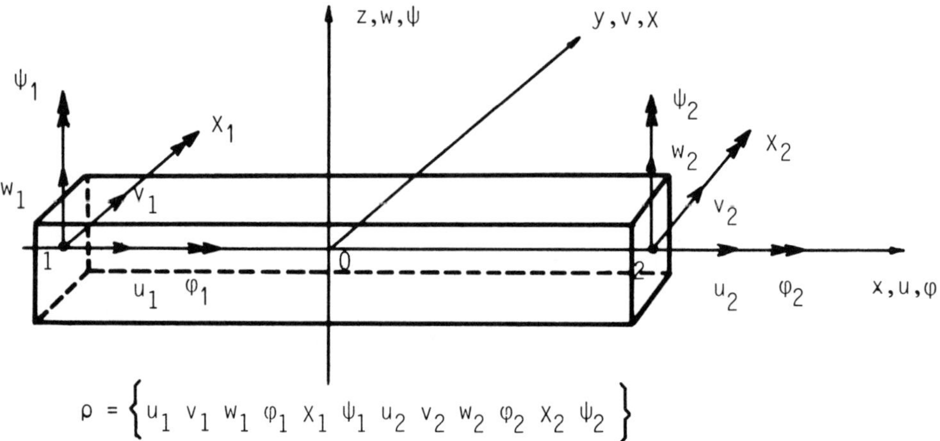

FIGURE 12. Natural beam element: local cartesian system and cartesian nodal freedoms.

alized forces) acting upon the elements are defined in a manner similar to the definition of generalized displacement vectors.

$$p = \alpha_e^T \tilde{p} = [\alpha_o^T \alpha_N^T] \begin{bmatrix} p_o \\ p_N \end{bmatrix} \tag{24}$$

where $\alpha_e = A_e^{-1}$. Application of the principle of virtual work to the assembly of the subelements and using the beam interpolation scheme leads to the natural stiffness matrix k_N as

$$p_N = k_N \, \varrho_N \tag{25}$$

which in the elastic case has a diagonal form and is denoted by k_{NE}. In the nonlinear theory, the nonlinear natural displacements $\bar{\varrho}_N$ are expressed with respect to Cartesian displacements ϱ or ϱ_o, ϱ_N as follows:

$$\bar{\varrho}_{Ni} = \varrho_{Ni} + \frac{1}{2} \varrho^T \Lambda_i \varrho$$

$$\begin{aligned} \bar{\varrho}_{Ni} = \varrho_{Ni} &+ \frac{1}{2} \varrho_o^T \Lambda_{ioo} \varrho_o + \frac{1}{2} \varrho_o^T \Lambda_{ioN} \varrho_N \\ &+ \frac{1}{2} \varrho_N^T \Lambda_{iNo} \varrho_o + \frac{1}{2} \varrho_N^T \Lambda_{iNN} \varrho_N \end{aligned} \tag{26}$$

Matrices $\Lambda_i, \Lambda_{ioo}, \Lambda_{ioN}, \Lambda_{iNo}, \Lambda_{iNN}$, are constant and can be derived explicitly. The stiffness matrix of the geometrical nonlinear case can be deduced by forming the second derivatives of the strain energy with respect to the element displacements ϱ or ϱ_o, ϱ_N. Thus the tangent stiffness matrix k_T is obtained as the sum of the usual elastic stiffness matrix k_E of the linear theory and the so-called geometric stiffness matrix k_G

$$k_T = \alpha_N^T k_{NE} \alpha_N + \sum_i P_{Ni} \Lambda_i = k_E + k_G \tag{27}$$

The tangent stiffness matrix $\tilde{k}_T$ referred to $\tilde{\varrho}$ is deduced similarly

$$\tilde{k}_T = \tilde{k}_E + \tilde{k}_G = \begin{bmatrix} 0 & 0 \\ 0 & k_{NE} \end{bmatrix} + \begin{bmatrix} k_{GOO} & k_{GON} \\ k_{GNO} & k_{GNN} \end{bmatrix} \tag{28}$$

where the components of $\tilde{k}_G$ are functions of $\varrho_N, \Lambda_{ioo}, \Lambda_{ioN}, \Lambda_{iNo}$ and Λ_{iNN}, and is related to k_T by

$$k_T = \alpha_e^T \tilde{k}_T \alpha_e \tag{29}$$

The tangent stiffness of a subelement $\tilde{k}_{LT}$ is derived by considering the limiting cases of Equations (26) and (28).

$$\tilde{k}_{LT} = \tilde{k}_{LE} + \tilde{k}_{LG} = \begin{bmatrix} 0 & 0 \\ 0 & k_{LNE} \end{bmatrix} + \begin{bmatrix} k_{LGOO} & k_{LGON} \\ k_{LGNO} & 0 \end{bmatrix} \tag{30}$$

Due to the fact that in large displacement-small strain analysis k_{LGNN} of the subelement does not appear, the derivation of the complete geometrical stiffness of the subelement becomes extremely simple. The stiffness matrices of the finite element may be obtained as in the linear theory from the integration of the strain energy of the local subelements using the interpolation scheme of the linear theory. The difficulty which arises with the question of the symmetry of the tangential stiffness matrix with rotational degrees of freedom is circumvented by consideration of the semi-tangential rotations, and the local geometrical stiffness of the beam in space can be obtained explicitly. A complete exposition of the above mentioned theory can be found in References 5, 6, and 7. A similar approach which results in almost the same geometrical stiffness matrix is presented by Argyris, et al. [8]. The formulation is based on the consideration of strain and stresses instead of natural displacement and forces. The natural mode technique is hence not an essential requirement for this formulation.

Argyris, et al. [9] extended the geometrical nonlinear analysis previously described to include material nonlinearities by considering the plastic effects to be concentrated at critical cross-sections at the ends of the members. This implies that external loads act only upon the joints. The elastic constitutive law of Equation (25) will now be used in its incremental form

$$\Delta P_N = k_{NE} \Delta \varrho_N \tag{31}$$

In order to obtain the corresponding constitutive relation for the generalized natural quantities, valid in the elastoplastic range of the material, further assumptions are made. Under the assumption of small deformations the increment of the natural deformation vector is split to

$$\Delta \varrho_N = \Delta \varrho_{NE} + \Delta \varrho_{NP} \tag{32}$$

where $\Delta \varrho_{NE}$ and $\Delta \varrho_{NP}$ denote the increments of the elastic and plastic natural deformation vectors. An elasto-ideally plastic material model is also assumed. Further, the existence is assumed of a yield function at the cross-section of each end of the beam which has the form

$$f_o^{(\alpha)} = f_o^{(\alpha)}(P_N, P_L) \tag{33}$$

where P_L is the vector of the limit generalized forces. According to the assumptions, these cross-sections ($\alpha = 1$ or 2) are the only ones likely to go plastic during the deformation process. For the beam element considered four different forms of natural stiffness matrices are distinguished corresponding to the existence of plastic hinges as shown in Figure 13. Explicit expressions for $k_{NEP}^{(1)}$, $k_{NEP}^{(2)}$ and $k_{NEP}^{(1,2)}$ are derived by treating the yield function as a potential and forming its gradient. Thus, to obtain the final tangential stiffness matrix for geometric nonlinearities too, the constitutive matrices $k_{NEP}^{(\alpha)}$ must replace the natural stiffness matrices k_{NE}. In [9] several plastic interaction surfaces are explicitly quoted which have proved to be effective in the numerical analysis of problems considered.

Figure 14 shows comparative results for a beam clamped at both ends using two elements based on the aforementioned analysis and 10 and 50 "layered" beam elements. In this specific case the analysis with the described element terminated at point A due to the singularity of the global stiffness matrix, while this is not the case for the layered element since with this element the structure theoretically never becomes a mechanism. In comparing the computation time it was found that this element needs only a tenth of the CPU time required for the computation with 50 layered elements. Further tests performed in [9] confirmed the effectiveness of this element. For a complete exposition of the theory on this family of elements the reader is referred to [5,6,7,8,9].

An updated Lagrangian formulation has been used by Tang, et al. [168] and Kani, et al. [92] to obtain expressions for the inelastic stiffness matrices in large displacements and rotational three dimensional problems. Geometrically nonlinear elastic response of space frames have been reported among others by Wen and Rahimzadeh [187], Wood and Zienkiewicz [189], Remseth [139], and Boswell [25] where Lagrangian formulations have been used, while Oran [119] used an Eulerian formulation and the conventional "beam-column" theory.

SOLUTION PROCEDURES

As we have seen, the modelling and the mathematical formulation of the problem is an important task for analysing a structure, but of equal importance is the use of efficient procedures for the solution of the resulting system of equations.

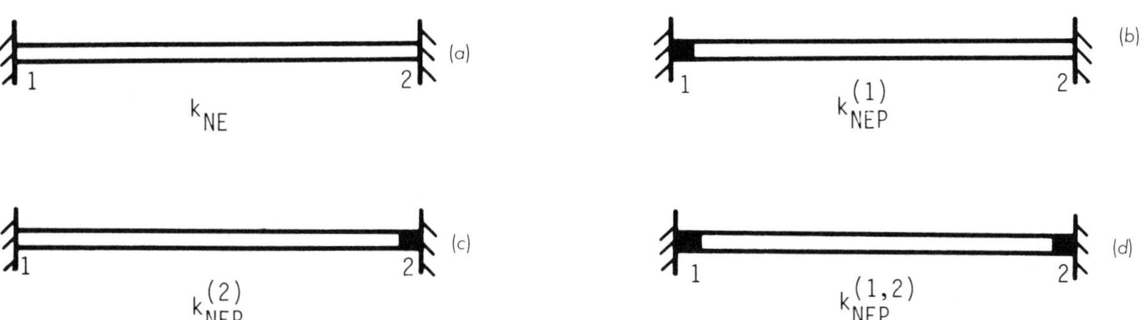

FIGURE 13. Four cases for natural beam element stiffness (elastic and elasto-plastic).

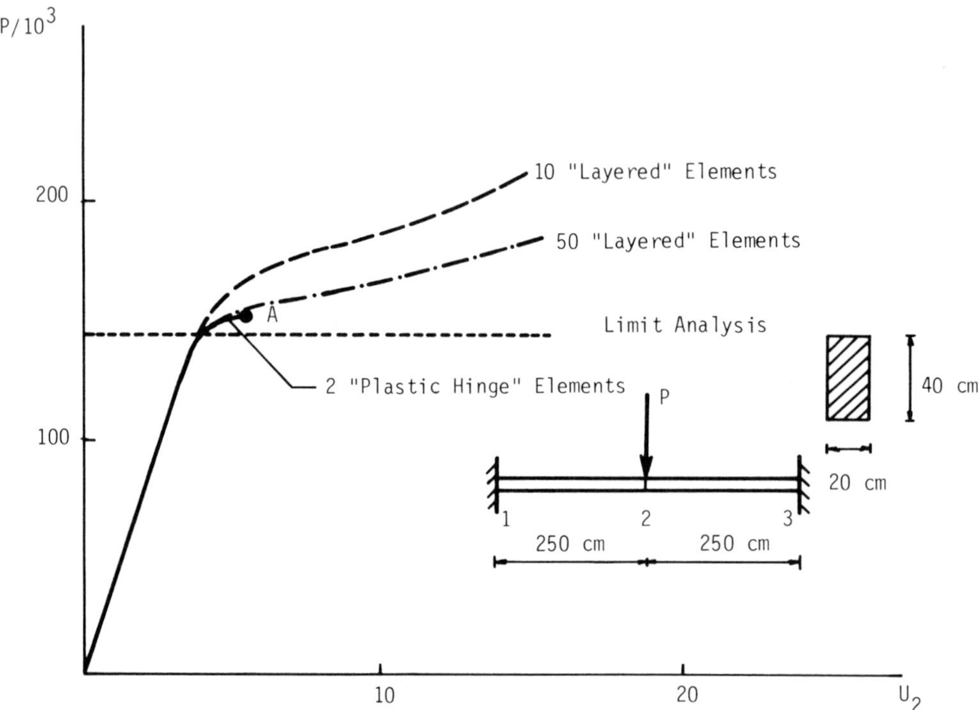

FIGURE 14. Results obtained using the "plastic-hinge" element and "layered" elements; Argyris et al. [9].

Depending on the chosen mode of analysis, the discretization of the physical model results in a linear equation system, an eigenvalue problem, a nonlinear equation system, or, in the case of a transient response analysis, an initial value problem. For any of these problems different solution procedures are available in the computerized structural analysis. The ideal choice among solution procedures of the same purpose is not only dependent on their case and their efficiency, but in some cases hardware restrictions such as available core size may enter into consideration. This means that the cost of an analysis and, in fact, its practical feasibility depend to a considerable degree on the algorithms available for the solution of the equations.

Linear Solution Techniques

The solution of the equations that arise in linear analysis problems

$$KU = R \tag{34}$$

where K = the stiffness matrix of the whole structure, U = the displacement vector, and R = the load vector, is of great importance not only in the case of linear analysis but in the case of nonlinear and dynamic analyses where the solution of linear systems of equations is frequently required at intermediate stages. Essentially, there are two different families of methods for the solution of Equations (34): direct solution techniques and iterative solution methods. In a direct solution Equations (34) are solved using a number of steps and operations that are predetermined in an exact manner, whereas an unspecified number of iterations are required when an iterative solution method is employed.

The most effective direct solution techniques are basically applications of Gauss elimination which can take advantage of the specific properties of the finite element stiffness matrix such as symmetry, positive definiteness, and bandness. Efficient computer implementation of the Gauss elimination method has led to improved techniques for handling the storage requirements of the stiffness matrix. Static condensation, substructuring, and the frontal solution method are among them. Compacted storage schemes of the stiffness matrix have also been used very efficiently [18,89]. In such schemes all matrices are stored as one-dimensional arrays and only the elements below the "skyline" of the matrices are processed. It should be noted that zero elements within the skyline in general do not remain zero during the solution procedure and are therefore stored, whereas all elements outside the skyline are not considered.

Although elimination methods have been used extensively for the computer solution of large sparse sets of linear simultaneous equations, they have the disadvantage that they require the formation and storage of the global stiffness matrix. In contrast, iterative methods do not require any global array and operations are performed on the "element" level of the structure. The most commonly used iterative methods are probably the successive over-relaxation and the conjugate gradient method. The latter, moreover, has been found to possess a remarkable property in that the correct solution is theorectically obtained, apart from round off errors, in no more than n iterations, where n is the order of the matrix. A renewed interest in iterative methods has been observed in recent years because of the widespread application of microcomputers and the special features of the new generation of parallel computers in which iterative methods render their advantages of performing the operations on the "element" level. Thus, a number of techniques have been proposed aiming at accelerating the convergence properties of iterative methods. Preconditioning and block elimination have considerably improved the speed of convergence at the expense of using extra computer storage [78,90,126,185].

Nonlinear Solution Techniques

Significant advancements have been made in the last decade in the development and application of numerical solution procedures to the large deflection, inelastic response of complex structures. This has been brought about primarily for two reasons: a better understanding of the physical properties involved and the over-increasing advances in computer technology providing faster execution times, greater accuracy, larger storage capacity at a smaller cost and greater availability. Of course, this rapid development has been somewhat accelerated by the need to understand what happens to structures when subjected to intense loading conditions. For the solution of nonlinear algebraic equation systems a large number of algorithms are available. The classification of these methods is not an easy task since all may be considered iterative or can be used in an incremental way, and sometimes there is a combination of iterative and incremental. Classification attempts have been reported in [3,22,23,163] among others.

The basic equation to be solved in nonlinear analysis may be written as

$$f = R - F = 0 \tag{35}$$

where f is the force unbalance, R stores the vector of externally applied nodal loads, and F is the vector of nodal point forces that are equivalent to the element stresses. For an exact solution, the force unbalance is zero. However, many of the solution procedures only approximate to the condition $f = 0$. There are several solution procedures which are based on this condition. The method of successive substitutions is characterized by the recursion relation

$$KU^{(i+1)} = R + F^{*(i)} \qquad i = 0,1,2, \ldots \tag{36}$$

where F^* is a pseudo force due to nonlinear effects. Another method is the Newton Raphson (NR) or modified Newton Raphson (mNR) method. The first can be written as

$$K_T^{(i)} \Delta U^{(i+1)} = f^{(i)} \tag{37}$$

where $K_T{}^{(i)}$ is the tangent stiffness matrix in iteration i. The displacement increment $\Delta U^{(i+1)}$ is used to determine an improved displacement vector $U^{(i+1)}$ where

$$U^{(i+1)} = U^{(i)} + \Delta U^{(i+1)} \tag{38}$$

For geometrical nonlinear analysis the NR method has proven itself as one of the best methods of solution available. The method has been successfully applied in a "mixed" formulation where incremental load steps and iterations are combined. The most significant drawback associated with NR is the large amount of computational effort required to compute and "invert" at each iteration the coefficient matrix in Equation (37). This process is computationally expensive and, for solution effectiveness, requires the use of relatively large load increments. However, in material nonlinear analysis or dynamic nonlinear response the load steps that can be employed are restricted in size by other stability and accuracy considerations. For these reasons, most analysts now use the mNR method wherein the coefficient matrix is held constant for several iterations and is updated only when the rate of convergence begins to deteriorate. In many applications this significantly reduces the average amount of computational effort per load step without adversely affecting convergence characteristics. The mNR method is particularly appealing when, as is the case in most analyses, it is required that the complete load deflection path of a complicated structure be determined. The mNR seeks to minimize the number of times the nonlinear stiffness matrix must be calculated.

It is most important for an analyst to understand the manner in which a numerical solution technique works. For some analysts, this understanding is often achieved by noting the form of the recursion relations. For others, a "feel" for the technique may be most easily gained by noting graphically the response of a one d.o.f. problem. Figure 15 demonstrates the convergence of the methods for one d.o.f. for an incremental loading ${}^{t+\Delta t}R - {}^{t}R$. In the analysis of inelastic behavior with path-dependent materials, the method of successive substitutions exhibits a very slow rate of convergence and sometimes tends to converge to an incorrect answer. This is attributed to the fact that during iterations the intermediate configurations may be quite far off from the true equilibrium path. This effect is not depicted in the one dimensional plot but is illustrated in Figure 16.

As an alternative to forms of Newton iteration, a class of methods known as matrix update methods or quasi-Newton methods have been developed for iteration on a nonlinear system of equations. These methods involve updating the inverse of the coefficient matrix to provide a secant approxi-

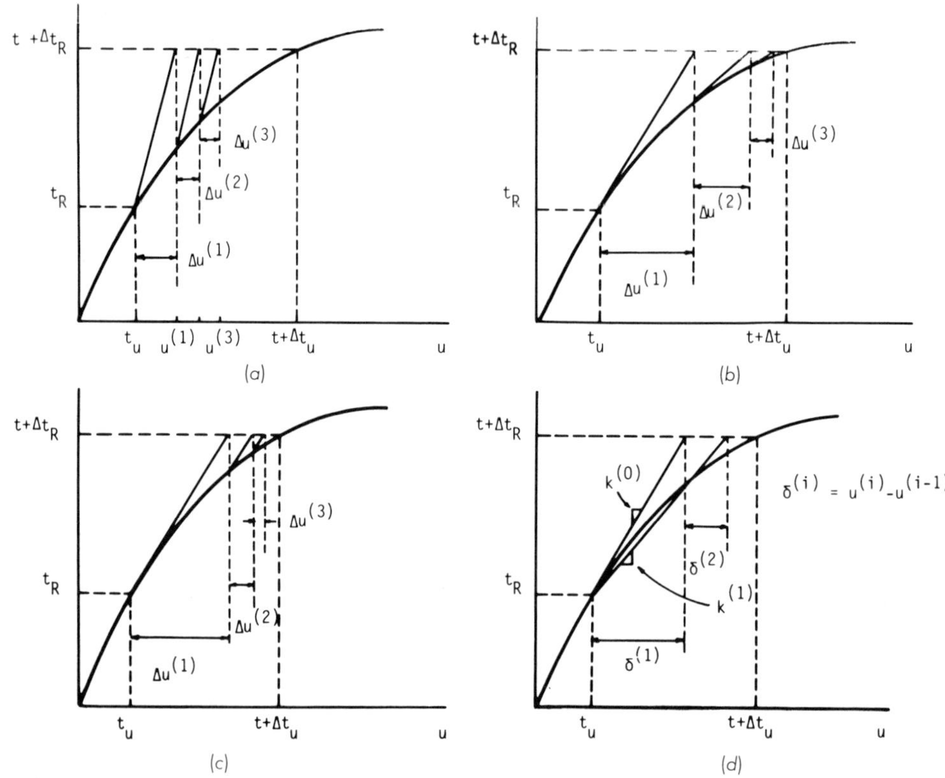

FIGURE 15. Solution techniques for one-dimensional problem; (a) successive approximation; (b) Newton-Raphson; (c) modified Newton-Raphson; (d) BFGS.

mation to the matrix from iteration i to $i + 1$ [15,63,105]. These quasi-Newton methods provide a compromise between the full re-formulation of the stiffness matrix performed in NR method and the use of a stiffness matrix from a previous configuration as is done in mNR method. Among the quasi-Newton methods available the BFGS (Broyden-Fletcher-Goldfarb-Shanno) method appears to be the most effective. In the BFGS method the following procedure is employed in iteration $i + 1$ to evaluate $U^{(i+1)}$ and $K^{(i+1)}$ (Figure 15d): Evaluate a displacement increment

$$\Delta\bar{U} = (K^{-1})^{(i)} f^{(i)} \tag{39}$$

Perform a line search in the direction $\Delta\bar{U}$ to satisfy equilibrium in this direction. The new displacement vector is given by

$$U^{(i+1)} = U^{(i)} + \beta\Delta\bar{U} \tag{40}$$

where β is a scalar multiplier dependent on the linear search. Finally evaluate a correction to the coefficient matrix. In the BFGS method the updated matrix can be expressed in product form

$$K^{(i+1)} = (A^{(i+1)})^T (K^{-1})^{(i)} A^{(i+1)} \tag{41}$$

where the matrix $A^{(i+1)}$ is an nxn matrix of a simple form.

Another family of methods which satisfy the condition $f = 0$ are based on iterative methods for the solution of linear systems. They are called vector iteration methods [123] and are based on the three-term recursion formulas for solving the linear problem introduced by Rutishauser [53]. The general form of the recursion expressions is

$$U^{(1+1)} = -\frac{f^{(i)}}{q_i} + \left(\frac{e_{i-1}}{q_i} + 1\right) U^{(i)} - \frac{e_{i-1}}{q_i} U^{(i-1)} \tag{42}$$

where q_i and e_{i-1} are iteration coefficients and they are characteristic of the method. It can be shown that according to the expressions which define the coefficients q_i and e_{i-1} the recursion formula of Equation (42) can be transformed to the conjugate gradient algorithm, the 2nd order Richardson process, the Chebyshev semi-iterative method or the dynamic relaxation method. The basic advantage of these methods over elimination methods is their vectorial form. Only four vectors with dimension n need be stored since $f^{(i)}$ is the unbalance force vector in iteration i. Other characteristic features of this category of methods is that they do not linearize the response within a load or time step. This family of methods has proved to be extremely effective in large geometrically nonlinear problems [122].

A nonlinear method which chooses the load increment as a path parameter between two displacement states ${}^{t}U$ and ${}^{t+\Delta t}U$ is the static perturbation method. This method seeks to determine the displacement response at the load step ${}^{t+\Delta t}R$ from that of ${}^{t}R$ through a Taylor expansion [61,70,143]. Self-correcting solution procedures have been developed to alleviate the drifting tendency from the exact equilibrium path of the pure incremental methods [59].

Since the need for nonlinear analysis of structures has significantly increased during the recent years and will continue to do so, and since ever more complex problems are being tackled, much emphasis is currently being placed on improving the convergence properties of the methods and developing more general and automatic solution schemes. In recent years intensive research has been directed toward this trend and numerous techniques are reported which improve the basic nonlinear methods previously described. The "alpha" constant approach presented by Nayak and Zienkiewicz [108] introduced a diagonal accelerating matrix for the mNR method. A modification to Aitken's acceleration is proposed by Jennings [88], while a more advanced scheme is proposed by Grisfield [45,46] based on the variable metric advancement toward the solution, and is called secant-Newton method. The conjugate gradient (CG) method has been combined with Newton methods to form the conjugate-Newton [81] and modified conjugate-Newton [62]. Preconditioned CG methods have been also used with great success [28,90,126].

The coupling between Newton and quasi-Newton methods is found effective for highly nonlinear problems [16,63]. The use of line searches to "damp" or scale-down the solution vector is proposed in [48,63]. Noor and Peters [114] introduced a reduced basis technique which reduces the size of the problem and enhances the convergence properties of the method. An element-by-element approximate factorization technique which obviates the need for a global coefficient matrix in conjunction with the CG and BFGS methods has been developed by Hughes, et al. [78]. Block iterations are used by Wellford and Vahdani [185] which combine direct and iterative solution schemes.

Tracing the Nonlinear Response near Limit Points

In the previous section a brief exposition of various solution procedures was attempted in which two major assumptions were tacitly made. First, we assumed that the analyst prescribes the various load levels for which the equilibrium configurations are to be calculated. This can be difficult without an approximate knowledge of the load carrying capacity of the structure. Secondly, we assumed that only the response up to collapse of the structure is sought. For most practical problems, it is quite unnecessary to trace such a convoluted load-deflection path with snap-through and snap-back buckling phenomena as shown in Figure 17 which possess some of the most difficult problems in nonlinear structural analysis. For some problems, all that may appear to be required is the load level at the first limit point. However, without analysis techniques that allow the limit points to be passed, even this information may be unavailable or unreliable. "Collapse loads" are often associated with a failure to achieve convergence with the iterative solution procedure. However, it may only be the iterative solution procedure that has collapsed (possibly as a concequence of round-off errors). For other problems, the analysis may be performed on an individual component of a completed structure. In such a situation, it may be important to obtain information on the nature of the load shedding, following the limit points, in order to assess the performance of the complete structure.

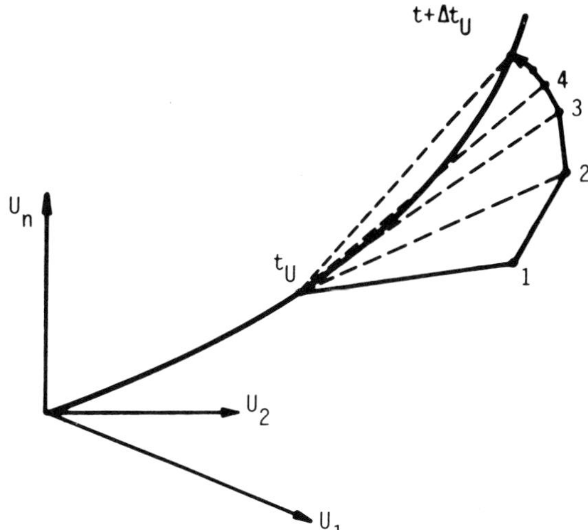

FIGURE 16. Path-effect during equilibrium equations.

When analysing relative simple structures it is tempting to try to avoid the full complexities of a "snap analysis" by applying a simple form of displacement control. Such a "displacement control" may simulate a physical testing procedure. For instance, referring to Figure 17, if the displacement U were to be prescribed, the limit point B could be passed and the load-shedding curve BC could be traced. However, a similar procedure would fail at, or just before, the limit point G. This failure might not matter if the analyst

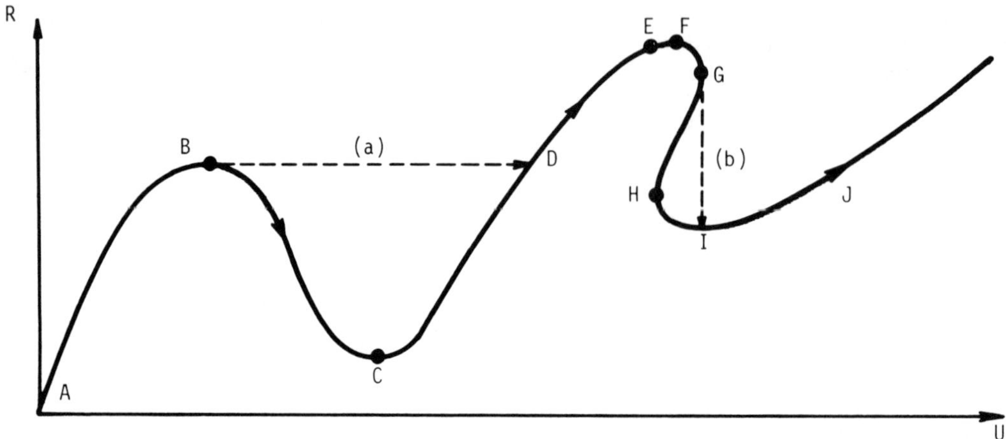

FIGURE 17. "Snap-buckling"; (a) dynamic snap under load control; (b) dynamic snap under displacement control.

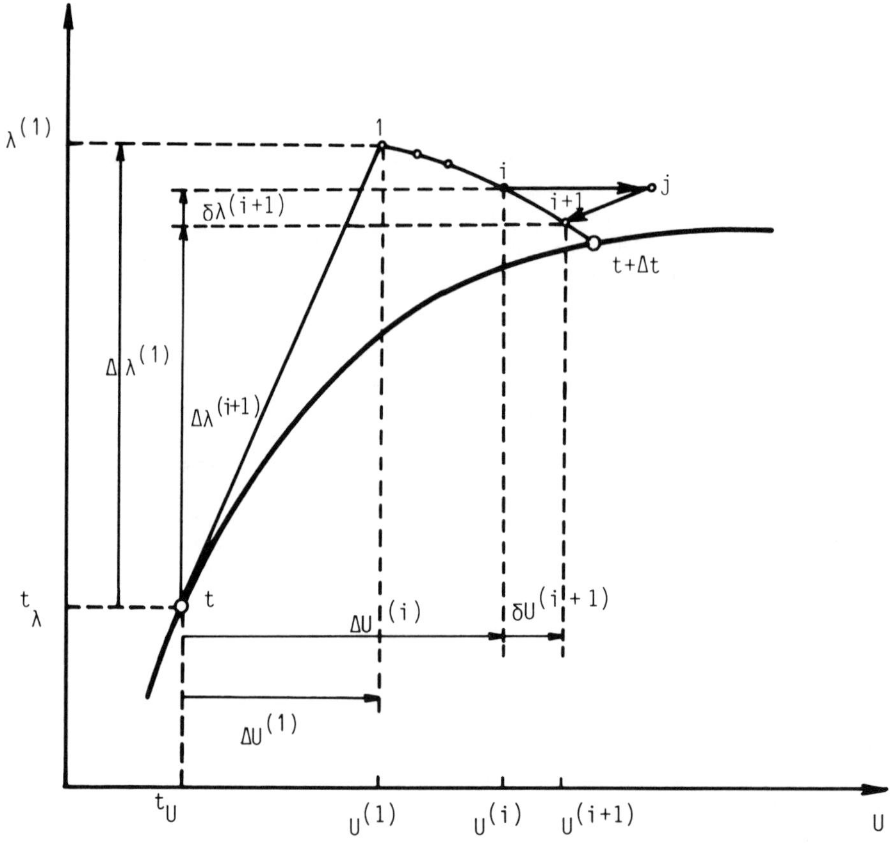

FIGURE 18. Tracing the equilibrium path-notation.

could conclude, "following the (local) maximum at F, there is a very sharp drop-off in load". However, the dramatic nonlinear behavior associated with the limit point G may induce a failure in the incremental/iterative solution procedure at point E. In such a situation the analyst is left with no information on the nature of the failure and may not even be sure that he has a structural (rather than numerical) collapse. Consequently, nonlinear finite element computer programs should be provided with solution procedures that will handle such "snapping phenomena" automatically without resorting to many interventions.

It is well known that the Newton iteration methods are not very efficient and often fail in the neighborhood of limit points. The stiffness matrix approaches singularity, resulting in an increasing number of iterations and a decreasing number of load steps, and finally the solution diverges. One of the earliest techniques to circumvent limit points is the artificial-spring method developed by Wright and Gaylord [191]. The technique is based on the observation that a snap-through problem may be transformed into one with a positive definite characteristic if linear artificial springs are added to the system. The method cannot be recommended for structures with local buckling or when a tendency to bifurcation is present. Bergan [21] introduced the "current stiffness parameter" to guide the algorithm near the limit point. At a prescribed value of the stiffness parameter the iteration procedure is discontinued and a pure incrementation is used. The iteration procedure is resumed when the stiffness parameter again reaches its prescribed value. The technique requires very small load increments to avoid drifting away from the equilibrium path.

The method most often used to avoid singularity is the interchange of dependent and independent variables. In the following a single displacement component selected as a controlling parameter is prescribed and the corresponding load level is taken as unknown. The procedure was introduced first by Argyris [4] but in the meantime has been modified by several authors. As we have seen in the case of an incremental-iterative solution procedure, the nonlinear problem is stepwise linearized and the linearization error is corrected by additional equilibrium equations. For proportional loading the loads may be expressed by one load factor ${}^{t}\lambda$:

$$ {}^{t}R = {}^{t}\lambda R \tag{43} $$

within one increment from configuration t to $t + \Delta t$. The positions i and $i + 1$, before and after an arbitrary iteration cycle are distinguished as in Figure 18. The total increments between positions t and i are denoted by $\Delta U^{(i)}$, $\Delta R^{(i)}$ and $\Delta\lambda^{(i)}$ whereas the changes in increments from i to $i + 1$ are denoted by $\delta U^{(i+1)}$, $\delta R^{(i+1)}$ and $\delta\lambda^{(i+1)}$, respectively.

$$ R^{(i+1)} = {}^{t}R + \Delta R^{(i)} + \delta R^{(i+1)} \tag{44} $$

$$ \lambda^{(i+1)} = {}^{t}\lambda + \Delta\lambda^{(i)} + \delta\lambda^{(i+1)} \tag{45} $$

$$ U^{(i+1)} = {}^{t}U + \Delta U^{(i)} + \delta U^{(i+1)} \tag{46} $$

Supposedly configuration i has already been determined and the incremental equilibrium equations may be expressed by the linearized stiffness expressions

$$ K^{(i)}\delta U^{(i+1)} = R^{(i+1)} - F^{(i)} \tag{47} $$

If $f^{(i)} = {}^{t}R + \Delta R^{(i)} - F^{(i)}$ is considered as the force unbalance, then

$$ K^{(i)}\delta U^{(i+1)} = \delta\lambda^{(i)}R + f^{(i)} \tag{48} $$

The tangent stiffness matrix $K^{(i)}$ at position i may include all possible nonlinear effects. It may be kept unchanged through several iteration cycles following the mNR method. In view of the fact that iteration takes place in the displacement and load space, the load level may change from one iteration to the other. In this case an intermediate position j for the same load level $\lambda^{(j)} = \lambda^{(i)}$ is introduced before the final stage $i + 1$ is reached (Figure 18).

To describe the displacement incremental procedure let us now assume that Equation (48) is reordered so that the prescribed component $\delta U_2^{(i+1)} = \hat{U}_2$ is the last one in the displacement vector $\Delta U^{(i+1)}$. Then, Equation (47) may be composed in two parts

$$ \begin{bmatrix} K_{11} & K_{12} \\ K_{21} & K_{22} \end{bmatrix}^{(i)} \begin{bmatrix} \delta U_1 \\ \delta U_2 \end{bmatrix}^{(i+1)} = \delta\lambda^{(i+1)} \begin{bmatrix} R_1 \\ R_2 \end{bmatrix} + \begin{bmatrix} f_1 \\ f_2 \end{bmatrix}^{(i)} \tag{49} $$

interchanging the variables

$$ \begin{bmatrix} K_{11} & -R_1 \\ K_{21} & -R_2 \end{bmatrix} \begin{bmatrix} \delta U_1 \\ \delta\lambda \end{bmatrix}^{(i+1)} = \begin{bmatrix} f_1 \\ f_2 \end{bmatrix}^{(i)} - \begin{bmatrix} K_{12} \\ K_{22} \end{bmatrix}^{(i)} \hat{U}_2 \tag{50} $$

It is obvious that the loss of the symmetrical and banded structure of the stiffness matrix is a severe handicap of this procedure. Later it was recognized that the solution of Equation (50) could be formed in two parts. The first line of Equation (50) is linear in the unknown increment at the load parameter $\delta\lambda^{(i+1)}$. Therefore, its solution may be decomposed into two parts which could be obtained simultaneously using two different "load" vectors.

$$ K_{11}^{(i)}[\delta U_1^{(i+1)}]^{I} = R_1 \tag{51} $$

$$ K_{11}[\delta U_1^{(i+1)}]^{II} = f_1^{(i)} - k_{12}\hat{U}_2 \tag{52} $$

and

$$\delta U_1^{(i+1)} = \delta\lambda^{(i+1)} [\delta U_1^{(i+1)}]^I + [\delta U_1^{(i+1)}]^{II} \quad (53)$$

The displacement increment $\delta U_1^{(i+1)}$ of Equation (53) is now introduced into the second part of Equation (50). This allows the determination of the load parameter $\delta\lambda^{(i+1)}$. Thus, instead of solving an unsymmetrical equation, Equations (51) and (52) are solved for the same symmetric matrix $K_{11}^{(i)}$ for two right-hand sides.

The previously described modified displacement control method was presented first by Pian and Tong [132] without mentioning the out-of-balance terms. Other researchers have also reported different variations of this technique [101,109,195]. A valuable simplification was utilized by Batoz and Dhatt [20]. The displacement control method is usually employed only in the neighborhood of the critical point although it may be applied throughout the entire load range. Obviously the method fails whenever the structure snaps back from one load level to a lower one. Crisfield [47] and Sabir and Lock [148] used a combination of displacement and load control technique that can handle snap-back phenomena. However, some knowledge of the failure mode is required for a proper choice of the controlling displacement.

Therefore, an obvious modification is to relate the procedure to a measure including all displacements rather than the one single component. Based on this idea Riks [141] and Wempner [186] independently introduced a constraint equation to limit the load step $\delta\lambda^{(1)}$

$$[\delta U^{(1)}]^T \delta U^{(1)} + [\delta\lambda^{(1)}]^2 = dS^2 \quad (54)$$

That is, the generalized "arc-length" of the tangent at t is fixed to a prescribed value dS. Then the iteration path follows a "plane" normal to the tangent (Figure 19(a)); so the scalar product of the tangent $\vec{t}^{(1)}$ and the vector $\delta\vec{U}^{(i+1)}$ containing the unknown load and displacement increments must vanish:

$$[\delta U^{(1)}]^T \delta U^{(i+1)} + \delta\lambda^{(1)} \delta\lambda^{(i+1)} = 0 \quad (55)$$

then from Equations (51), (52), and (55) an expression for $\delta\lambda^{(i+1)}$ is obtained. A modification to the Riks-Wempner method was reported by Crisfield [47] (Figure 20(a)). Instead of iterating in the "plane" normal to the tangent at the used a "spherical" path, and Equation (54) is replaced by:

$$[\Delta U^{(i+1)}]^T \Delta U^{(i+1)} = dS^2 \quad (56)$$

Alternatively the "normal plane" may be updated in every iteration cycle [137].

Crisfield has successfully applied the modified Riks' approach to problems involving both horizontal tangent (limit) points, which could otherwise be passed using displacement control and vertical tangent points, which otherwise could be passed using load control. Such switching is no longer required and here is no need to suppress the equilibrium iterations in the vicinity of the limit points. When applied with the standard mNR method, only one extra vector need be stored. The added computation is negligible. Not only can the limit points be passed, but also the convergence characteristics are significantly improved so that the technique can be beneficially applied to problems for which limit points are not anticipated. The same approach was adapted in [47] to an accelerated mNR method and to the secant-Newton method [46]. In [49] Crisfield describes a method for introducing line searches into the arc-length solution procedures. Such line searches may be used at each iteration to calculate an optimum scalar step-length which

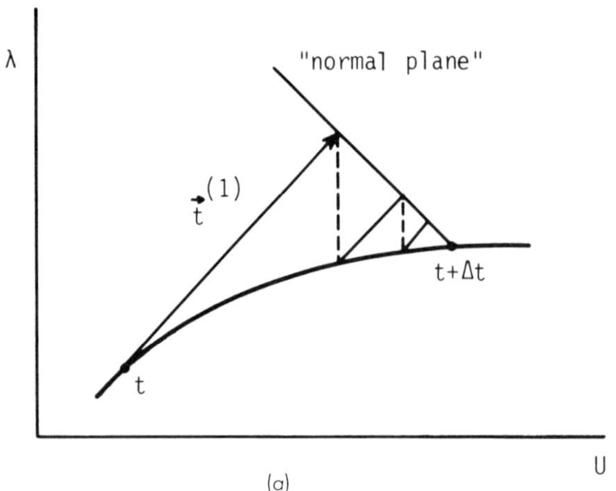

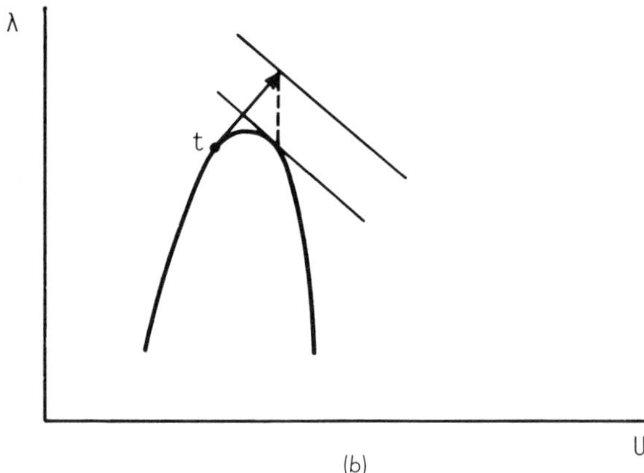

FIGURE 19. Success (a) and failure (b) of the Riks-Wempner method.

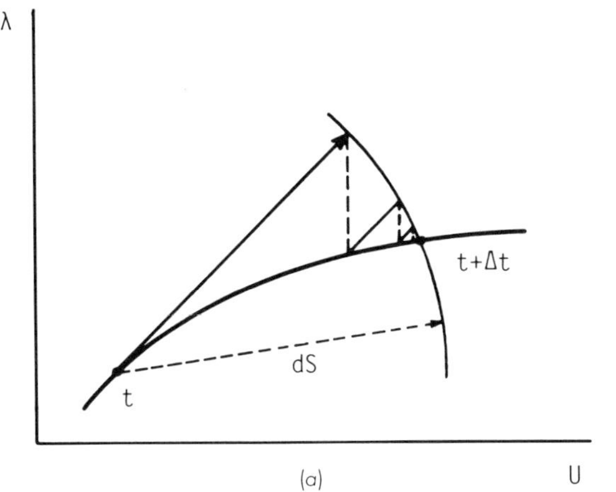

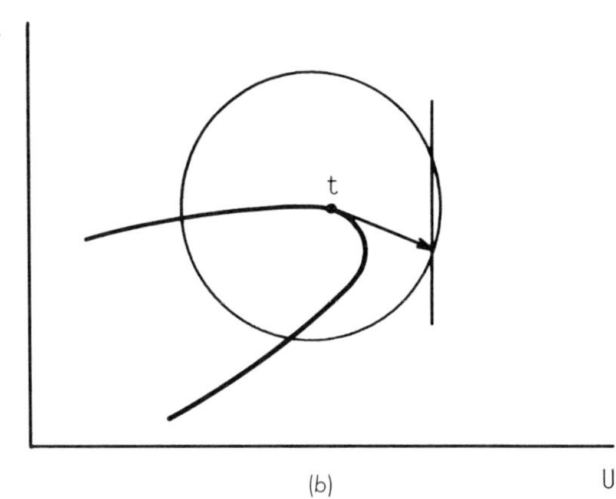

FIGURE 20. Success (a) and failure (b) of the Crisfield method.

scales the normal iterative vector. This technique leads to a substantial improvement in the convergence characteristics particular on "difficult iterations."

Park [130] used the generalized constraint equation

$$[\Delta U^{(i+1)}]^T \Delta U^{(i+1)} + b[\Delta\lambda^{(i+1)}]^2 R^T R = dS^2 \quad (57)$$

where b is a scalar parameter which is related to Bergan's [22] current stiffness parameter. In this way as the stiffness of the structure increases, so the contribution of the load term in Equation (57) increases and the constraint tends towards standard load control. A multi-phase self-adaptive predictor-corrector type algorithm for highly nonlinear structural responses was developed by Padovan and Tovichakchaikul [120]. The hierarchy of this strategy involves three main phases. The first features the use of a warpable hyperelliptic constrained surface. The second corrector phase uses an energy constrained to scale successive iterates. The third involves the use of quality of convergence checks which allow of various self-adaptive modifications of the algorithmic structure when necessary. An alternative to the Riks-Wempner-Crisfield iterative correction scheme is presented by Fried [56]. He considered an orthogonal trajectory approach to the equilibrium curve without the need for an explicit constraint, and without the involvement of an equilibrium point $({}^tU, {}^t\lambda)$. This technique avoids failure when the Riks-Wempner-Crisfield approaches fail to describe the solution path (Figures 19(b) and 20(b)). Finally, other techniques for tracing nonlinear equilibrium paths may be found in [16,22,114,135,142,182].

Truss Analysis Techniques

The aforementioned numerical solution procedures are general and can handle any set of governing nonlinear equations which describe the response of any structure and naturally any space frame. For pin-jointed truss problems, however, several simplified but effective numerical procedures have been reported which can handle the limit state analysis and predict the behavior of the structure in the post-buckling range. Schmidt and Gregg [150] have reported a dual load method that allows the inelastic nonlinear member behavior to be followed. The method assumes that the inelastic strut response is known and the piecewise linearization of the behavior is acceptable. The dual load method has some relationship to the initial stress method which was first used by Wolf [188] for the same problem. Depending on the axial deformation of each member, an appropriate stiffness and residual force is known from the assumed piecewise linear relationships. A first-order linear analysis for two separate loading conditions is then carried out on the structure whose members possess these newly determined stiffnesses. The two loading conditions is the unit external load and the residual forces for each member. A load factor is then evaluated by searching the response of all members. Then, the overall deflections are obtained from the simple addition of the deflections obtained by those of the unit load multiplied by the load factor, and of the residual forces. Thus, the overall truss load-deflection behavior is found essentially by a secant type of approach, but by following each segment of the piecewise-linear deflection behavior of the members, the dependence on prior history of loading can be followed. One of the basic differences of the dual load method from that of the initial stress method is that for the latter method the accuracy of the solution will be dependent on the size of the external load increments, whereas for the dual load method the accuracy will be dependent on the degree of piecewise linearization of the element behavior.

A similar approach called the "fictitious force technique" is presented by Klimke and Posch [96] to the solution of the large deflection combined problem of ultimate load and fire resistance analysis. Rashed, et al. [138] used an analytical model of truss tubular members to form the incremental element stiffness matrix. The usual incremental load analysis is combined with the NR equilibrium iterations. When buckling and/or plastification of a member or more, or fracture of a bolt are detected within a loading step, the load increment is scaled down to that necessary to cause such failure. This prevents the internal force vectors from shooting out of the limiting strength interaction surfaces. Papadrakakis [124] used the bending moment-curvature relationship to model the cyclic response of the members together with the dynamic relaxation method for the solution of the geometrically and materially nonlinear equations. The unloading characteristics of the struts in the inelastic range are determined and the irreversibility of plastic strains are taken into consideration. The dynamic relaxation method which is a "vector iteration" method does not require the evaluation of any stiffness matrix but only direct use of the relatively simple force-displacement relations of the individual discrete elements is needed.

Smith [159] developed a nonlinear stepwise linearization analysis which does not require the repeated updating of the structural stiffness matrix. The method can handle the chordal snap-through phenomenon and member displacement sense reversal. As in the case of [150] the method was developed and applied to the double layer flat grid space truss in which system geometric nonlinearities are not significant. However, the method could be adapted to include such nonlinearities with little difficulty. The method is based on the "residual force" system which together with the external load acting on the original structure will produce a set of internal forces and displacements that correspond to the nonlinear response. The determination of the fictitious load system generally requires an iterative procedure, but can be determined directly if a proper modelling by stepwise linearization of the member response is employed. Other specified solutions have been reported in [36,106,131].

For a complete survey of up-to-date computer programs for solution of nonlinear structural and solid mechanics problems the reader is referred to the work of Noor [113].

NUMERICAL AND EXPERIMENTAL TESTS

In this chapter a brief review of numerical and experimental tests found in the literature is attempted. Examples which will be reported are meant to illustrate the merits of the mathematical formulation and the numerical solution procedure used to obtain the nonlinear response and also to study the structural behavior in increasing load or deformation. Comparative results are reported whenever they exist. Scanning the literature one can find only few publications dealing with the dynamic behavior of space frame structures

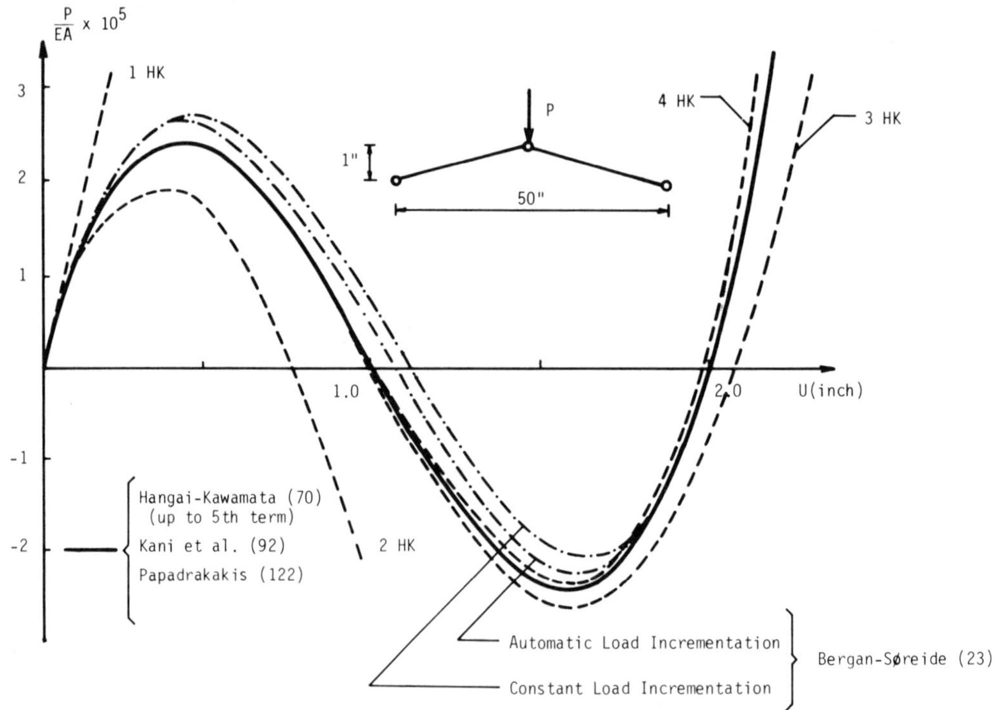

FIGURE 21. Example 1: geometry and elastic load-deflection curves.

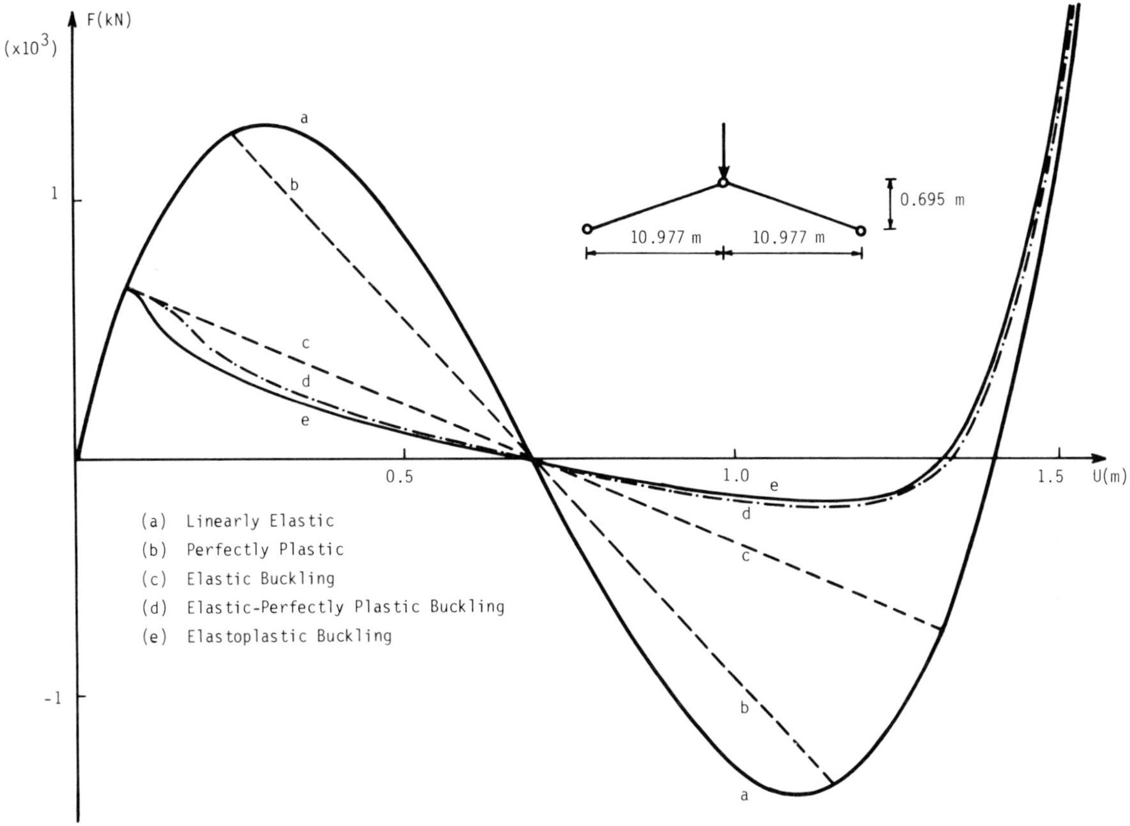

FIGURE 22. Example 1: geometry and inelastic load-deflection curves; Papadrakakis [124].

as compared to the voluminous data that exist for static loads. The main reason is that the dynamic characteristics of space frames have not been identified as a major factor in structural failures in the history of these structures [171].

Example 1.—The simple plane arch structure shown in Figure 21 was used by many investigators to test the mathematical formulation and the numerical solution procedure. In Figure 21 only geometric nonlinearities are considered. Hangai and Kawamata [70] used the static perturbation technique with various degrees of approximations in the nonlinear force-displacement relation. Curves 1 HK up to 5 HK are obtained by taking into account up to the fifth order term in the strain-displacement relations respectively. Bergan and Soreide [23] obtained curves 1 BS and 2 BS with an automatic load incrementation and a constant load incrementation respectively. Kani, et al. [92] employed 2 and 3 node isoparametric beam elements with an updated Lagrangian formulation coupled with Bergan's current stiffness parameter to circumvent the limit point. Papadrakakis [122] used two "vector iteration" methods, namely the conjugate gradient method and the dynamic relaxation method, for the solution of equations. A displacement incrementation technique is used to trace the equilibrium path. Figure 22 shows the response of the structure when material nonlinearities are considered in addition to the geometric ones. The results depicted are obtained in [124] by considering linearly elastic, perfectly plastic behavior (curve b), elastic buckling (curve c), elastic-perfectly plastic buckling (curve d), and elastoplastic buckling (curve e). Kani, et al. [92] also reported results identical with curve b using the previously mentioned formulation.

Example 2.—(24-member hexagonal star-shaped shallow dome). For this space truss it becomes clear (Figure 23) that Hangai-Kawamata's static perturbation technique, with retention of up to the third order terms of displacements, is inadequate to predict the post-buckling behavior of such structures. Jagannathan, et al. [83] used a total Lagrangian formulation with a displacement incrementation of the NR method, and Rothert, et al. [147] a modification to the Riks-Wempner method. They too failed to predict the accurate behavior of the structure because of the large member rotations that occur. It is clear from Figure 23 that their formulation failed to satisfy the geometric constraints of the problem at $U = 2$ cm. The results obtained by Papa-

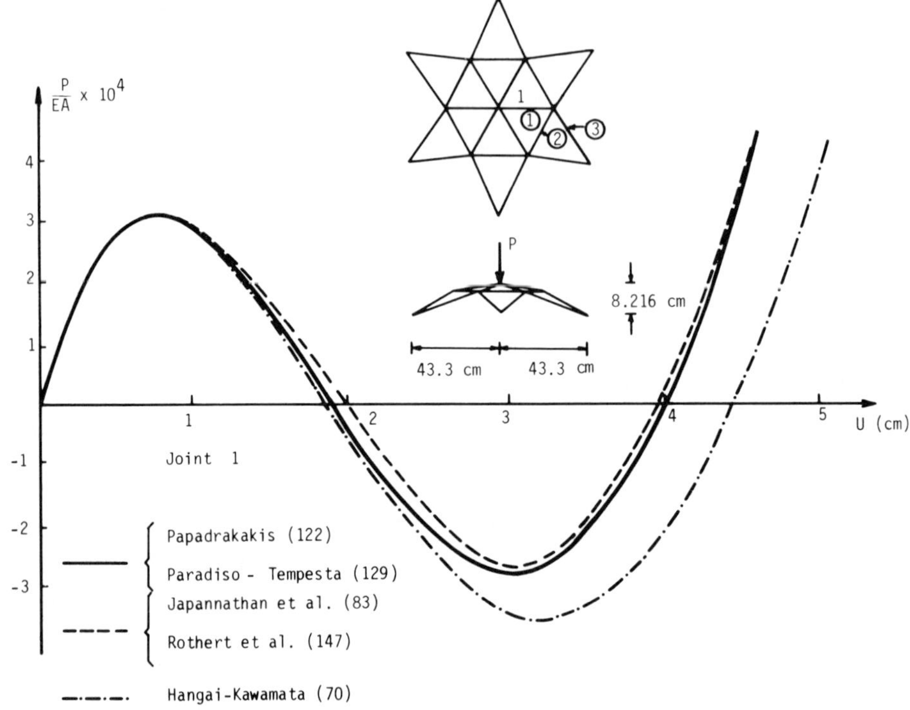

FIGURE 23. Example 2: geometry and elastic load-deflection curves.

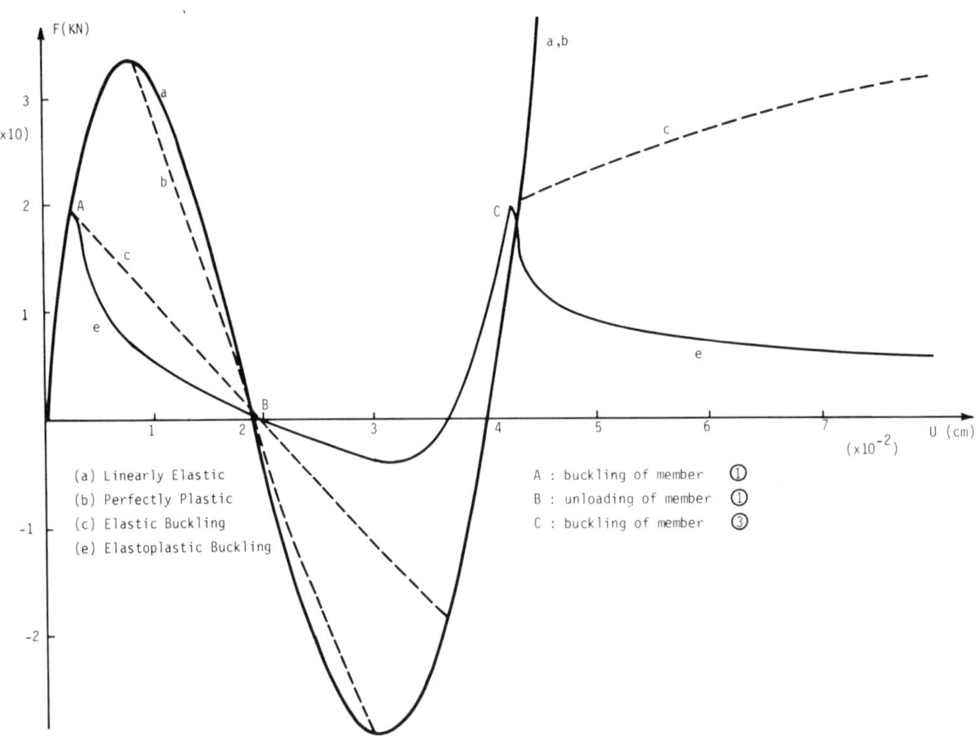

FIGURE 24. Example 2: inelastic load-deflection curves; Papadrakakis [124].

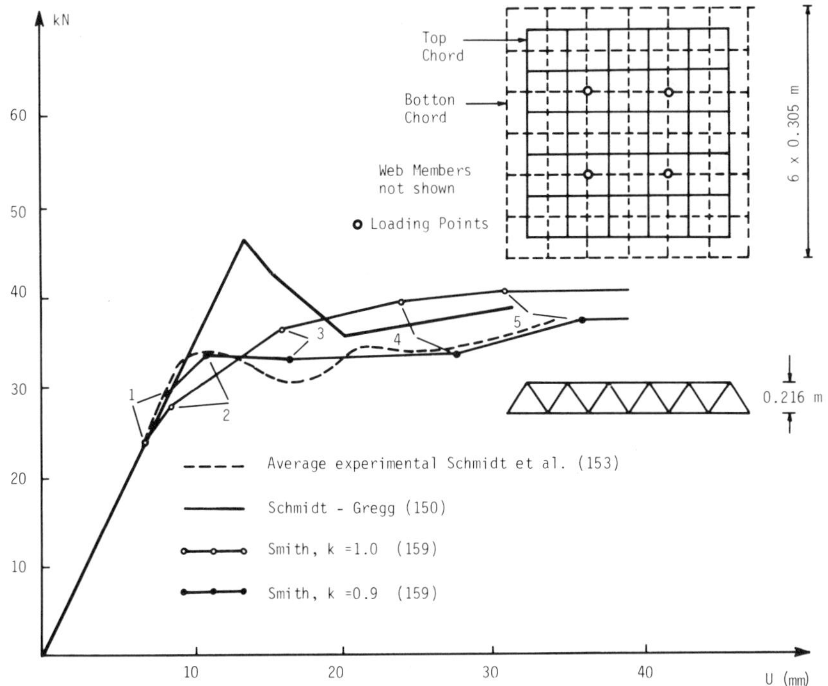

FIGURE 25. Example 3: geometry and inelastic load-deflection curves.

drakakis [122] appeared to be in complete agreement with those obtained by Paradiso and Tempesta [129], where a secant-tangent approach is used by successive linear approximation. Abadan and Holzer [1] and Holzer, et al. [77] utilized a nonlinear programming routine coupled with a perturbation analysis after the critical point. Watson and Holzer [183] used a TL formulation and Crisfield's method, Meek and Tan [107] used a beam-column large rotation formation and Crisfield's method and finally Cichon [36] employed a TL formulation and a numerical procedure based on Waszszyszyn [182].

Figure 24 shows the response of the structure when material nonlinearities are considered [124]. Both types of nonlinearities are also considered in [95] but the results are not compatible with those in Figure 24 due to different structural characteristics.

Example 3.—(6 × 6 bay double layer offset grid space truss). The truss shown in Figure 25 is simply supported to each perimeter joint and loaded symmetrically at the four points at the corners of the central 2 ft square of the bottom chord. The truss was fabricated from aluminium tube, coined at the ends to fit patented "Triodetic" hubs. Schmidt, et al. [69,150,151,152,153] in a number of tests reported the load deflection responses, external load-member responses, and compression and tension test responses on the members. The authors point out the reduction in the moment of inertia of the members at the joint face and because of the rigidity of the joints they used the joint face to joint face distance for the effective length of the members. Schmidt's truss tests show that the compression members did reach force values greater than the pin-ended buckling load but did not achieve the values obtained in the individual member tests before system failure. Figure 25 shows a comparison of theoretical and experimental results for the truss [150,152,153]. The effects of the joint slip, initial force systems caused by lack of fit or welding, combined with the inherent scatter of peak load capacities of the brittle-type compression elements, are considered to furnish the reductions of the experimentally derived ultimate loads when compared with the theoretical predictions. The theoretical analysis was made by the dual load method [150].

In a study performed by Smith [159] for the same truss, the effective length was taken between the end joint faces multiplied by an effective length factor, taken to be 1.0 and 0.9 in two different computer runs. The smaller effective length factor, 0.9, qualitatively accounts for the small but non-zero moment of inertia and plastic moment of resistance at the joint face and gives an effective length similar to that obtained by a classical elastic stability analysis. In the structural analysis, the joints were assumed to be rigid and thus the stiffness of the members was modified accordingly. In Figure 25 analytical results are also plotted and are compared with those obtained by Schmidt, et al. As can be seen, good correlation has been obtained, both for the initial

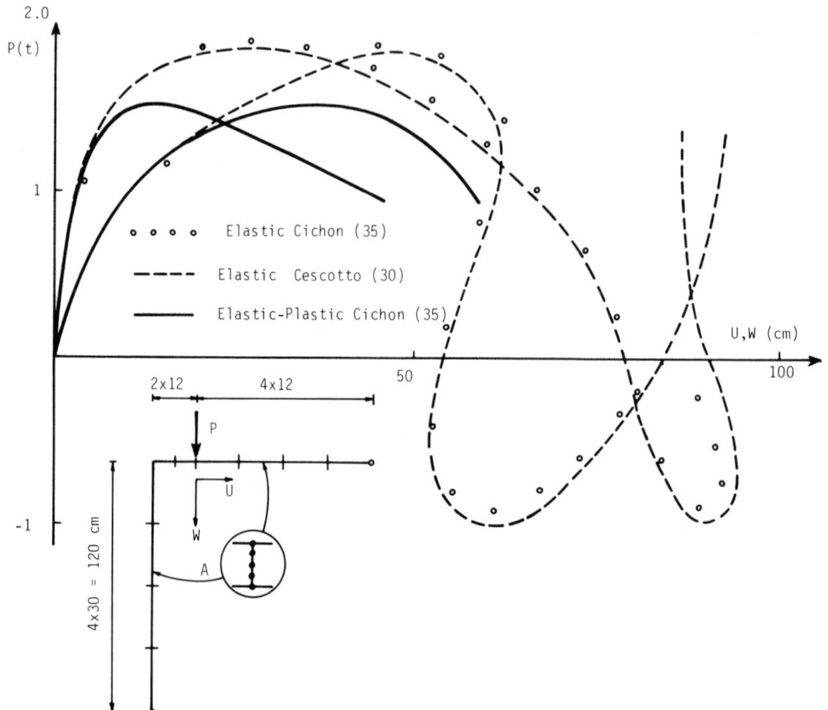

FIGURE 26. Example 4: geometry, load-deflection curves and points of numerical integration.

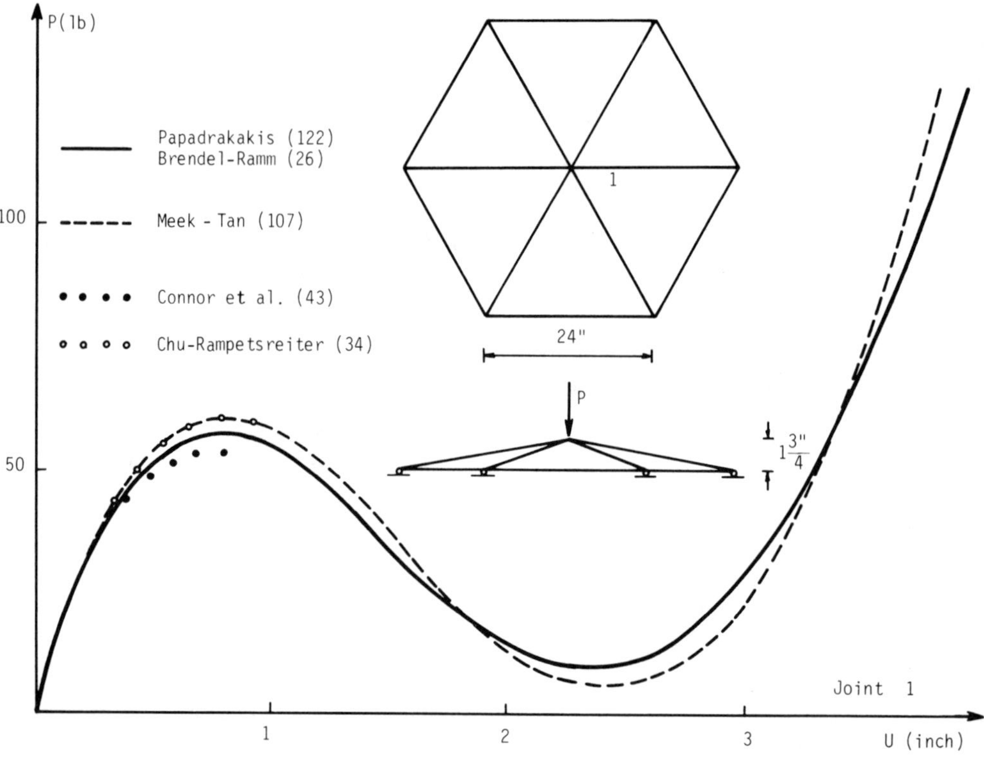

FIGURE 27. Example 5: geometry and elastic load-deflection curves.

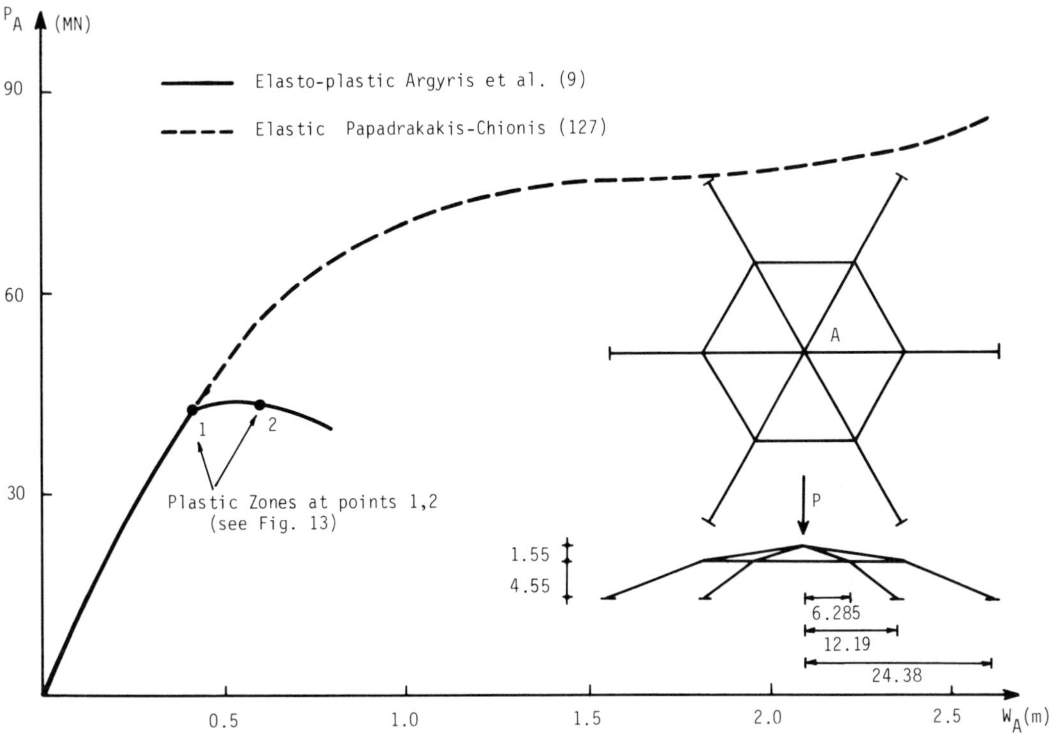

FIGURE 28. Example 6: geometry, elastic and inelastic load-deflection curves.

linear response and for post-linear response. The stiffness of the system is accurately modeled through the full response and the capacity of the system is accurately determined. It can be seen that the pin-ended assumption underestimates the member capacity and the onset of nonlinear behavior of the system. In the analysis which used an effective length factor of 0.9, better correlation was obtained. Initial nonlinearity was predicted close to that obtained. Following early nonlinearities, an unstable, negative stiffness regime was detected, section 2-3 of the curve initiated by "cordal snap-through" of the member, then hardening was predicted in section 4-5 of the curve. The good correlation of the theoretical results of Smith with the experimental results of Schmidt, et al., indicates that the program developed by the former investigator can accurately predict space truss response.

Results obtained by Collins [40] on a similar double layer grid which was constructed in the laboratory with extreme care showed good agreement between theoretical and experimental results without taking into account the effects which were considered by Smith. Other tests on space truss roofs were reported in [27,138,149,160,161,167].

Example 4. – (Lee's Portal frame). The fourth example deals with a snap-through of a frame shown in Figure 26. A finite element solution has been presented by Cescotto [30] where the Henky-type beam element with twelve d.o.f. was used. Cichon used a 2-D beam element previously described and obtained both elastic and elastic-plastic results. An elastic-plastic material curve with kinematic strain hardening has been assumed. Further tests on elastic-plastic two dimensional structures have been reported in [93,168,193].

Example 5. – (12-member hexagonal frame). This space frame (Figure 27) has been experimentally studied by Griggs [66]. The pre-buckling behavior has been solved in [34,42], while in [26,107,122] the solution was traced into the elastic post-buckling range. Brendel and Ram [26] used finite element large rotation Lagrangian formulation with four elements per member. Connor, et al. [42] used a small rotation derivation of the equilibrium equations and the NR method, while Chu and Rampetsreiter [34] employed a combined secant tangent stiffness approach with the mNR method and neglected the effect of length shortening due to bowing. Papadrakakis [122] used the beam column approach with the effect of axial force in the flexural stiffness, of flexural shortening of members, and of large deflections with moderate rotations taken into account. A displacement incrementation of the vector iteration methods was employed to trace the equilibrium path. Meek and Tan [107] used a beam column geometric formulation with arbitrary large rotations which are accommodated by making use of an updated joint orientated matrix [119]. The NR method and Crisfield constant arc-length technique are employed.

Example 6. –(18-member space framed dome). As a final example the space frame shown in Figure 28 was considered under the prescribed downward displacement at the top joint. The material and geometric data were taken from [9]. Papadrakakis and Chionis [127] used the same methodology discussed for example 5, while Argyris, et al. used the 3-D natural element discussed previously, taking into consideration both geometric and material nonlinearities.

Other studies on the elastic static and dynamic stability of space frames can be found in [37,77,115,121].

REFERENCES

1. Abadan, A. O. and S. M. Holzer, "Degree of Stability of Geodesic Domes with Independent Loading Parameters," *Computers and Structures,* Vol. 9, pp. 43–51 (1978).
2. "ADINA-A Finite Element Program for Automatic Dynamic Incremental Nonlinear Analysis-System Theory and Modelling Guide," Report AE 83-4, ADINA Engineering (Sept. 1983).
3. Almroth, B. O. and F. A. Brogan, "Automated Choice of Procedures in Computerized Structural Analysis," Vol. 7, pp. 335–342 (1977).
4. Argyris, J. H., "Continua and Discontinua, Matrix Methods in Structural Mechanics," *Proceedings of the Conference on Matrix Methods,* Wright-Patterson Air Force Base, Ohio (1965).
5. Argyris, J. H., P. C. Dunne, and D. W. Scharpf, "On Large Displacement-Small Strain Analysis of Structures with Rotational Degrees of Freedom," *Computer Methods in Applied Mechanics and Engineering,* Vol. 14, pp. 401–451 (1978).
6. Argyris, J. H., et al., "On Large Displacement–Small Strain Analysis of Structures with Rotational Degrees of Freedom," *Computer Methods in Applied Mechanics and Engineering,* Vol. 15, pp. 99–135 (1978).
7. Argyris, J. H., et al., "Finite Element Method–The Natural Approach," *Computer Methods in Applied Mechanics and Engineering,* Vol. 17/18, pp. 1–106 (1979).
8. Argyris, J. H., et al., "On the Geometrical Stiffness of a Beam in Space-A Consistence V.W. Approach," *Computer Methods in Applied Mechanics and Engineering,* Vol. 20, pp. 105–131 (1979).
9. Argyris, J. H., et al., "Finite Element Analysis to Two-and Three-Dimensional Elasto-Plastic Frames–The Natural Approach," *Computer Methods in Applied Mechanics and Engineering,* Vol. 35, pp. 221–248 (1982).
10. Astaneh-Asl, A. and S. C. Goel, "Cyclic In-Plane Buckling of Double Angle Bracing," *Journal Structural Engineering,* ASCE, Vol. 110, No. 9, pp. 2036–2055 (1984).
11. Bäcklund, J., "Large Deflection Analysis of Elasto-Plastic Beams and Frames," *International Journal of Mechanical Science,* Vol. 18, pp. 269–277 (1976).
12. Ball, R. E. and J. A. Burt, "Dynamic Buckling of Shallow Spherical Shells," *Journal of Applied Mechanics,* pp. 411–416 (June, 1973).
13. Bathe, K.-J., *Finite Element Procedures in Engineering Analysis,* Prentice-Hall, Inc., New Jersey, (1982).
14. Bathe, K.-J. and S. Bolourchi, "Large Displacement Analysis of Three-Dimensional Beam Structures," *International Journal for Numerical Methods in Engineering,* Vol. 14 pp. 961–986 (1979).
15. Bathe, K.-J. and A. P. Cimento, "Some Practical Procedures for the Solution of Nonlinear Finite Element Equations," *Computer Methods in Applied Mechanics and Engineering,* Vol. 22, pp. 59–85 (1980).
16. Bathe, K.-J. and E. Dvorkin, "On the Automatic Solution of Nonlinear Finite Element Equations," Computers and Structures, Vol. 17, No. 5–6, pp. 871–879 (1983).
17. Bathe, K.-J. and H. Ozdemir, "Elastic–Plastic Large Deformation Static and Dynamic Analysis," *Computers and Structures,* Vol. 6, pp. 81–92 (1976).
18. Bathe, K.-J., E. Ramm, and E. L. Wilson, "Finite Element Formulations for Large Deformation Dynamic Analysis," *International Journal for Numerical Methods in Engineering,* Vol. 9, pp. 353–386 (1975).
19. Bathe, K.-J. and P. M. Wiener, "On Elastic-Plastic Analysis of I-Beams in Bending and Torsion," *Computers and Structures, Vol. 17,* No. 5–6, pp. 711–718 (1983).
20. Batoz, J. L. and G. Dhatt, "Incremental Displacement Algorithms for Nonlinear Problems," *International Journal for Numerical Methods in Engineering,* Vol. 14, pp. 1262–1267 (1979).
21. Bergan, P.G., "Solution Algorithms for Nonlinear Structural Problems," *Proceedings of the Conference on Engineering Applications of the Finite Element Method,* Hovik, Norway, published by A. S. Computas (1979).
22. Bergan, P. G., "Solution Algorithms for Nonlinear Structural Problems," *Computers and Structures,* Vol. 12, pp. 497–509 (1980).
23. Bergan, P. G. and T. Soreide, "A Comparative Study of Different Numerical Solution Techniques as Applied to a Nonlinear Structural Problem," *Computer Methods in Applied Mechanics and Engineering,* Vol. 2, pp. 185–201 (1973).
24. Black, R. G., W. A. Wanger, and E. Popov, "Inelastic Buckling of Steel Struts under Cyclic Load Reversals," Report No. UCB/EERC-80/40, University of California, Berkeley, (1980).
25. Boswell, L. F., "A Small Strain Large Rotation Theory and Finite Element Formulation of Thin Curved Lattice Members," *Third International Conference on Space Structures,* Guilford, England, H. Nooshin, ed., Applied Science Publishers Ltd., London, pp. 375–380 (1984).
26. Brendel, B. and E. Ramm, "Stabilitätsuntersuchungen Weitgespannter Tragwerke mit der Methode der Finiten Elemente," *Proceedings of the International Symposium on Wide Span Surface Structures,* Universität Stuttgart, (1976).

27. Bródka, J. and A. Grudka, "Limit State Analysis and Full Scale Experiments of Double-Layer Grids," *Third International Conference on Space Structures,* Guilford, England, H. Nooshin, ed., Applied Science Publishers Ltd., London, pp. 554–560 (1984).

28. Buckley, A. G., "A Combined Conjugate-Gradient Quasi-Newton Minimization Algorithm," *Mathematical Programming,* Vol. 15, pp. 200–210 (1978).

29. Budiansky, B., "Dynamic Buckling of Elastic Structures, Criteria and Estimates," *Dynamic Stability of Structures,* G. Hermann, ed., Pergamon Press, pp. 83–106 (1967).

30. Cescotto, S., "Etude par Element Finis de Grand Displacements et Grandes Deformations," Doctor's Thesis, The University of Liège, Belgium (1977/78).

31. Chen, W. F. and T. Atsuta, "Theory of Beam-Columns," Vol. I—*In Plane Behavior and Design,* (1976); Vol. II.—*Space Behavior and Design,* McGraw-Hill, New York (1977).

32. Chen, W. F. and D. A. Ross, "Test of Fabricated Tubular Columns," *Journal of the Structural Division,* ASCE, Vol. 103, No. ST3, pp. 619–634 (1977).

33. Chen, W. F. and H. Sugimoto, "Inelastic Cyclic Behavior of Tubular Members in Offshore Structures," *Proceedings of the 8th World Conference on Earthquake Engineering,* San Francisco (1984).

34. Chu, K. H. and R. H. Rampetsreiter, "Large Deflection Buckling of Space Frames," *Journal of the Structural Division,* ASCE, Vol. 98, No. ST12, pp. 2701–2722 (1972).

35. Cichon, C., "Large Displacements in-Plane Analysis of Elastic-Plastic Frames," *Computers and Structures,* Vol. 19, No. 5/6, pp. 737–745 (1984).

36. Cichon, C., "Stability Analysis of Elastic Space Trusses," *Third International Conference on Space Structures,* Guilford, England, H. Nooshin, ed., Applied Science Publishers Ltd, London, pp. 567–570 (1984).

37. Coan, C. H. and R. H. Plaut, "Dynamic Stability of a Lattice Dome," *Earthquake Engineering and Structural Dynamics,* Vol. 11, pp. 269–274 (1983).

38. Cohn, M. Z., S. K. Ghosh, and S. R. Parimi, "Unified Approach to Theory of Plastic Structures," *Journal of the Engineering Mechanics Division,* ASCE, Vol. 98, No. EM5, pp. 1133–1158 (1972).

39. Cohn, M. Z. and T. Rafay, "Deformations of Plastic Frames Considering Axial Forces," *Journal of the Engineering Mechanics Division,* ASCE, No. EM4, pp. 725–745 (1977).

40. Collins, I. M., "An Investigation into the Collapse Behaviour of Double-Layer Grids," *Third Internation Conference on Space Structures*, Guilford, England, H. Nooshin, ed., Applied Science Publishers Ltd., London, pp. 400–405 (1984).

41. Collins, I. M. and W. J. Supple, "Experimental Post-Buckling Curves for Tubular Struts," Space Structures Research Report, Department of Civil Engineering, University of Surray (Sept. 1979).

42. Connor, J. J., D. Logcher, and S. C. Chan, "Nonlinear Analysis of Elastic Framed Structures," *Journal of the Structural Division,* ASCE, Vol. 94, No. ST6, pp. 1525–1547 (1968).

43. Connor, J. J. and N. Morin, "Perturbation Techniques in the Analysis of Geometrically Nonlinear Shells," *Proceedings of the IUTAM Colloqium on High Speed Computing of Elastic Structures,* Liège, Belgium, pp. 638–705 (Aug. 1970).

44. Creus, G. J., P. L. Torres, and A. G. Groehs, "Elastoplastic Frame Analysis with Generalized Yield Condition and Finite Displacements," *Computers and Structures,* Vol. 18, No. 5, pp. 925–929 (1984).

45. Crisfield, M. A., "A Faster Modified Newton-Raphson Iteration," *Computer Methods in Applied Mechanics and Engineering,* Vol. 20, pp. 267–278 (1979).

46. Crisfield, M. A., "Incremental/Iterative Solution Procedures for Non-Linear Structural Analysis," *Numerical Methods for Nonlinear Problems,* C. Taylor, E. Hinton, and D. R. J. Owen, eds., Pineridge Press, Swansea U.K., pp. 261–290 (1980).

47. Crisfield, M. A., "A Fast Incremental/Iterative Solution Procedure that Handles "Snap-Through," *Computers and Structures,* Vol. 13, pp. 55–62 (1981).

48. Crisfield, M. A., "An Arc-Length Method Including Line Searches and Accelerations," *International Journal for Numerical Methods in Engineering,* Vol. 19, pp. 1269–1289 (1983).

49. Crisfield, M. A., "Accelerating and Damping the Modified Newton-Raphson Method," *Computers and Structures,* Vol. 18, No. 3, pp. 395–407 (1984).

50. Dean, D. L., "Discrete Field Analysis of Structural System," CISM Courses and Lectures, No. 203, Udine, Italy, Springer-Verlag, Wien (1976).

51. Dean, D. L. and R. R. Avent, "State of the Art Discrete Field Analysis of Space Structures," *Proceedings of the 2nd International Conference on Space Structures,* Guilford, England, pp. 7–16 (1975).

52. El-Zanaty, M. H. and D. W. Murray, "Nonlinear Finite Element Analysis of Steel Frames," *Journal of Structural Engineering,* ASCE, Vol. 109, No.2, pp. 353–368 (1983).

53. Engeli, M., et al., *Refined Iterative Methods for Computation of the Solution and the Eigenvalues of Self-Adjoint Boundary Value Problems,* Birkhauser Verlag, Basel/Stuttgart (1959).

54. Flügge, W., *Stresses in Shells,* Springer-Verlag, Berlin, pp. 293–307 (1962).

55. Forman, S. E. and J. W. Hutchinson, "Buckling of Reticulated Shell Structures," *International Journal of Solids and Structures,* Vol. 6, pp. 909–932 (1970).

56. Fried, I., "Orthogonal Trajectory Accession to the Nonlinear Equilibrium Curve," *Computer Methods in Applied Mechanics and Engineering,* Vol. 47, pp. 283–297 (1984).

57. Fujimoto, et al., "Nonlinear Analysis for K-Type Braced Steel Frames," *Transactions, Architectural Institute of Japan,* No. 209, pp. 41–51, (in Japanese) (July, 1973).

58. Gadala, M. S., M. A. Dokainish and G. AE. Oravas, "Formulation Methods of Geometric and Material Nonlinearity Problems," *International Journal for Numerical Methods in Engineering,* Vol. 20, pp. 887–914 (1984).

59. Gadala, M. S. and G. AE. Oravas, "Numerical Solutions of Nonlinear Problems of Continua-I," *Computers and Structures,* Vol. 19, No. 5/6, pp. 865–877 (1984).
60. Gadala, M. S., G. AE. Oravas and M. A. Dokainish, "A Consistent Eulerian Formulation of Large Deformation Problems in Statics and Dynamics," *International Journal for Nonlinear Mechanics,* Vol. 18, pp. 21–35 (1983).
61. Gallagher, R. H., "Perturbation Procedures in Nonlinear Finite Element Structural Analysis," in *Computational Mechanics,* No. 461, Springer-Verlag, Berlin, pp. 75–96 (1975).
62. Gambolati, G., "Fast Solution to Finite Element Flow Equations by Newton Iteration and Modified Conjugate Gradient Method," *International Journal for Numerical Methods in Engineering,* Vol. 15, pp. 661–675 (1980).
63. Geradin, M., S. Idelsohn and M. Hogge, "Computational Strategies for the Solution of Large Nonlinear Problems via Quasi-Newton Methods," *Computers and Structures,* Vol. 13, pp. 73–81 (1981).
64. Gregory, W. E. and R. H. Plaut, "Dynamic Stability Boundaries for Shallow Arches," *Journal of the Engineering Mechanics Division,* ASCE, Vol. 108, No. EM6, pp. 1036–1050 (1982).
65. Grierson, D. E. and S. B. Abdel-Baset, "Plastic Analysis under Combined Stresses," *Journal of the Engineering Mechanics Division,* ASCE, Vol. 103, No. EM5, pp. 837–854 (1977).
66. Griggs, H. P., "Experimental Study of Instability in Elements of Shallow Space Frames," Research Report, Department of Civil Engineering, MIT, Cambridge, MA (1966).
67. Gugerli, H. and S. Goel, "Large Scale Tests for the Hysteresis Behavior of Inclined Bracing Members," *Proceedings of the 7th World Conference on Earthquake Engineering,* Istanbul, Vol. 7, pp. 87–94 (Sept. 1980).
68. Han, D. J. and W. F. Chen, "Buckling and Cyclic Inelastic Analysis of Steel Tubular Beam-Columns," *Engineering Structures,* Vol. 5, pp 119–132 (April, 1983).
69. Hanaor, A. and L. C. Schmidt, "Space Truss Studies with Force Limiting Devices," *Journal of the Structural Division,* ASCE, Vol. 106, No. St 11, pp. 2313–2329 (1980).
70. Hangai, Y. and S. Kawamata, "Nonlinear Analysis of Space Frames and Snap-Through Buckling of Reticulated Shell Structures," *Proceedings, 1971 IASS Pacific Symposium on Tension Structures and Space Frames,* Architectural Institute of Japan, Tokyo, pp. 803–816 (1972).
71. Hangai, Y. and N. Matsui, "Energy Criteria for Dynamic Buckling of Shallow Structures under Rectangular Loading," Bulletin of Earthquake Resistant Structure Research Center, The Institute of Industrial Science, University of Tokyo, No. 9 (1976).
72. Heki, K., "The Effect of Shear Deformation on Double Layer Lattice Plates and Shells," *Proceedings of the 2nd International Conference on Space Structures,* Guilford, England, pp. 189–198 (1975).
73. Heki, K. and Y. Fujitani, "The Stress Analysis of Grids under the Action of Bending and Shear," *Space Structures,* Blackwell Publ., Oxford, pp. 33–43 (1967).
74. Heki, K. and T. Saka, "Stress Analysis of Lattice Plates as Anisotropic Continuum Plates," *Proceedings 1971 IASS Pacific Symposium on Tension Structures and Space Frames,* Architectural Institute of Japan, Tokyo, pp. 663–674 (1972).
75. Heki, K. and T. Saka, "The Effective Strength of Double-Layer Grids in Continuum Treatment," *Proceedings of the IASS World Congress on Shell and Spatial Structures,* Madrid (1979).
76. Higginbotham, A. B. and R. D. Hanson, "Axial Hysteretic Behavior of Steel Members," *Journal of the Structural Division,* ASCE, Vol. 102, No. ST7, pp. 1365–1381 (1975).
77. Holzer, S. M., et al., "Stability of Lattice Structures under Combined Loads," *Journal of the Engineering Mechanics Division,* ASCE, Vol. 106, No. EM2, pp. 289–305 (1980).
78. Hughes, J. R., I. Levit, and J. Winget, "An Element-by-Element Solution Algorithm for Problems of Structural and Solid Mechanics," *Computer Methods in Applied Mechanics and Engineering,* Vol. 36, pp. 241–254 (1983).
79. IASS Working Group No. 8, "Analysis, Design and Realization of Space Frames," Bulletin of the International Association for Shell and Spatial Structures, Vol. XXV-1/1, n. 84/85 (1984).
80. Igarashi, S., et al., "Hysteretic Characteristic of Steel Braced Frames, Part I, The Behavior of Bracing Members under Cyclic Axial Forces," *Transactions, Architectural Institute of Japan,* No. 196 (1972).
81. Irons, B. and A. Elsawaf, "The Conjugate-Newton Algorithm for Solving Finite Element Equations," *Proceedings U.S.-German Symposium on Formulations and Algorithms in Finite Element Analysis,* K.-J. Bathe, J. T. Oden, and W. Wunderlich, eds., MIT press, pp. 656–672 (1977).
82. Issa, R. R. A. and R. R. Avent, "Superelement Stiffness Matrix for Space Trusses," *Journal of Structural Engineering,* ASCE, Vol. 110, No. 5, pp. 1163–1179 (1984).
83. Jagannathan, D., H. I. Epstein, and P. Christiano, "Nonlinear Analysis of Reticulated Space Trusses," *Journal of the Structural Division,* ASCE, Vol. 101, No. ST12, pp. 2641–2658.
84. Jain, A. K. and S. C. Goel, "Cyclic End Moments and Buckling in Steel Members," *Proceedings of the 2nd U.S. National Seminar on Earthquake Engineering,* Stanford (Aug. 1979).
85. Jain, A. K., S. C. Goel and R. D. Hanson, "Inelastic Response of Restrained Steel Tubes," *Journal of the Structural Division,* ASCE, Vol. 104, No. ST6, pp. 897–910 (1978).
86. Jain, A. K., S. C. Goel and R. D. Hanson, "Hysteretic Cycles of Axially Loaded Steel Members," *Journal of the Structural Division,* ASCE, Vol. 106, No. ST8, pp. 1777–1795 (1980).
87. Jain, A. K. and R. D. Hanson, "Hysteresis Models of Steel Members for Earthquake Response of Braced Frames," *Proceedings of the 7th World Conference on Earthquake Engineering,* Istanbul, Vol. 6, pp. 463–470 (Sept., 1980).
88. Jennings, A., "Accelerating the Convergence of Matrix

Iterative Processes," *Journal of the Institute of Mathematics and Applications,* Vol. 8, pp. 99–110 (1971).

89. Jennings, A., *Matrix Computation for Engineers and Scientists,* Wiley, London (1977).

90. Jennings, A. and G. M. Malik, "The Solution of Sparse Linear Equations by the Conjugate Gradient Method," *International Journal for Numerical Methods in Engineering,* Vol. 12, pp. 141–158 (1978).

91. Johnson, E. R. and I. K. McIvor, "The Effect of Spatial Distribution on Dynamic Snap-Through," *Journal of Applied Mechanics,* Vol. 45, pp. 612–618 (Sept., 1978).

92. Kani, I. M., R. E. McConnel and T. See, "The Analysis and Testing of a Single Layer, Shallow Braced Dome," *Third International Conference on Space Structures,* Guilford, England, H. Nooshin ed., Applied Science Publishers, London, pp. 613–618 (1984).

93. Kassimali, A., "Large Deformation Analysis of Elastic Plastic Frames," *Journal of Structural Engineering,* ASCE, Vol. 109, No. 8, pp. 1869–1886 (1983).

94. Kayal, S., "Finite Element Analysis of RC Frames," *Journal of Structural Engineering,* ASCE, Vol. 110, No. 12, pp. 2891–2908 (1984).

95. Kleiber, M., J. A. König and A. Sawczuk, "Studies on Plastic Structures: Stability, Anisotropic Hardening, Cyclic Loads," *Computer Methods in Applied Mechanics and Engineering,* Vol. 33, pp. 487–556 (1982).

96. Klimke, H. and J. Posch, "A Modified Fictitious Force Method for the Ultimate Load Analysis of Space Trusses," *Third International Conference on Space Structures,* Guilford, England, H. Nooshin, ed., Applied Science Publishers, London, pp. 589–593 (1984).

97. Koiter, W. T., "Elastic Stability and Post Buckling Behaviour," *Nonlinear Problems,* R. E. Langer, ed., University of Wisconsin Press (1963).

98. LASTRAN 80, Element Library, ISD-Report No. 281, University of Stuttgart, Germany (1980).

99. Lind, N. C., "Local Instability Analysis of Triangulated Dome Frameworks," *The Structural Engineer,* Vol. 47, No. 8, pp. 317–324 (1969).

100. Lock, M. H., S. Okubo and J. S. Whiffier, "Experiments on the Snapping of a Shallow Dome under a Step Pressure Load," *AIAA Journal,* Vol. 6, No.7, pp. 1320–1326 (1968).

101. Lock., A. C. and A. B. Sabir, "Algorithm for Large Deflection Geometrically Nonlinear Plane and Curved Structures," *Mathematics of Finite Elements and Applications,* J. R. Whiteman, ed., Academic Press, New York, pp. 483–494 (1973).

102. Madi, U. R. and D. L. Smith, "A Finite Element Model for Determining the Constitutive Relation of a Compression Member," *Third International Conference on Space Structures,* Guilford, England, H. Nooshin ed., Applied Science Publishers Ltd., London, pp. 625–629 (1984).

103. Maison, B. F. and E. Popov, "Cyclic Response Prediction for Braced Steel Frames," *Journal of the Structural Division,* ASCE, Vol. 106, No. ST7, pp. 1401–1416 (1980).

104. Makowski, Z. S., ed., *Analysis, Design and Construction of Double-Layer Grids,* Applied Science Publishers Ltd., London (1981).

105. Matthies, H. and G. Strang, "The Solution of Nonlinear Finite Element Equations," *International Journal for Numerical Methods in Engineering,* Vol. 14, pp. 1613–1626 (1979).

106. McConnel, R. E. and H. Klimke, "Geometrically Nonlinear Pin-Jointed Space Frames," *Numerical Methods for Nonlinear Problems,* Vol. 1, C. Taylor, E. Hinton, and D. R. J. Owen, eds., Pineridge Press, Swansea, pp. 333–342 (1980).

107. Meek, J. L. and H. S. Tan, "Geometrically Nonlinear Analysis of Space Frames by an Incremental Iterative Technique," *Computer Methods in Applied Mechanics and Engineering,* Vol. 47, pp. 261–282 (1984).

108. Nayak, G. C. and O. C. Zienkiewicz, "Note on the Alpha-Constant Stiffness Method for the Analysis of Nonlinear Problems," *International Journal for Numerical Methods in Engineering,* Vol. 4, pp. 579–582 (1972).

109. Nemat-Nasser, S. and H.D. Shatoff, "Numerical Analysis of Pre-and Postcritical Responce of Elastic Continua at Finite Strains," *Computers and Structures,* Vol. 3, pp. 983–999 (1973).

110. Nonaka, T., "An Elastic-Plastic Analysis of a Bar under Repeated Axial Loading," *International Journal of Solids and Structures,* Vol. 9, pp. 569–580 (1973).

111. Nonaka, T., "Approximation of Yield Condition for the Hysteretic Behavior of a Bar under Repeated Axial Loading," *International Journal of Solids and Structures,* Vol. 13, pp. 637–643 (1977).

112. Nonaka, T., "An Analysis of Large Deformation of an Elastic-Plastic Bar under Repeated Axial Loading, I and II," *International Journal of Mechanical Science,* Vol. 19, pp. 619–638 (1977).

113. Noor, A. K., "Survey of Computer Programs for Solution of Nonlinear Structural and Solid Mechanics Problems," *Computers and Structures,* Vol. 13, pp. 425–465 (1981).

114. Noor, A. K. and J. M. Peters, "Tracing Post-Limit-Point Paths with Reduced Basis Technique," *Computer Methods in Applied Mechanics and Engineering,* Vol. 28, pp. 217–240 (1981).

115. Noor, A. K. and J. M. Peters, "Instability Analysis of Space Trusses," *Computer Methods in Applied Mechanics and Engineering,* Vol. 40, pp. 199–218 (1983).

116. Nooshin, H., "Algebraic Representation and Processing of Structural Configurations," *Computers and Structures,* Vol. 5, pp. 119–130 (1975).

117. Nooshin, H., ed., *Third International Conference on Space Structures,* Guilford, England, Applied Science Publishers Ltd., London (1984).

118. Oran, C., "Tangent Stiffness in Plane Frames," *Journal of the Structural Division,* ASCE, Vol. 99, No. ST6, pp. 973–985 (1973).

119. Oran, C., "Tangent Stiffness in Space Frames," *Journal of the*

Structural Division, ASCE, Vol. 99, No. ST6, pp. 987–1001 (1973).

120. Padovan, J. and S. Tovichakchaikul, "Self-Adaptive Predictor-Corrector Algorithms for Static Nonlinear Structural Analysis," *Computers and Structures,* Vol. 15, pp. 365–377 (1982).
121. Padovan, J. and S. Tovichakchaikul, "On the Solution of Elastic-Plastic Static and Dynamic Postbuckling Collapse of General Structure," *Computers and Structures,* Vol. 16, No. 1-4, pp. 119–205 (1983).
122. Papadrakakis, M., "Post-Buckling Analysis of Spatial Structures by Vector Iteration Methods," *Computers and Structures,* Vol. 14, No. 5-6, pp. 393–402 (1981).
123. Papadrakakis, M., "A Family of Methods with Three-Term Recursion Formulae," *International Journal for Numerical Methods in Engineering,* Vol. 18, pp. 1785–1799 (1982).
124. Papadrakakis, M., "Inelastic Post-Buckling Analysis of Trusses," *Journal of Structural Engineering,* ASCE, Vol. 109, No. 9, pp. 2129–2147 (1983).
125. Papadrakakis, M. and L. Chrysos, "Inelastic Cyclic Analysis of Imperfect Columns," *Journal of Structural Engineering,* ASCE, Vol. 111, No. 6 (1985).
126. Papadrakakis, M., "Accelerating Vector Iteration Methods," *Journal of Applied Mechanics,* Vol. 53, pp. 291–297 (1986).
127. Papadrakakis, M. and P. Ghionis, "Conjugate Gradient Algorithms in Nonlinear Structural Analysis Problems," *Computer Methods in Applied Mechanics and Engineering,* Vol. 59, pp. 11–27 (1986).
128. Papadrakakis, M. and K. Loukakis, "Inelastic Cyclic Analysis of Imperfect Columns with End Restraints," *Journal of Engineering Mechanics, ASCE* (1987).
129. Paradiso, M. and G. Tempesta, "Member Buckling Effects in Nonlinear Analysis of Space Trusses," *Numerical Methods for Nonlinear Problems,* Vol. 1, C. Taylor, E. Hinton, and D.R.J. Owen, eds., Pineridge Press, Swansea, pp. 395–405 (1980).
130. Park, K. C., "A Family of Solution Algorithms for Nonlinear Structural Analysis Based on Relaxation Equations," *International Journal for Numerical Methods in Engineering,* Vol. 18, pp. 1337–1347 (1982).
131. Parke, C. A. R. and H. B. Walker, "A Limit State Design of Double-Layer Grids," *Third International Conference on Space Structures,* Guilford, England, Applied Science Publishers, Ltd., London, pp. 528–532 (1984).
132. Pian, T. H. H. and P. Tong, "Variational Formulation of Finite Displacement Analysis," *Proceedings of the IUTAM Symposium on High Speed Computing of Elastic Systems,* Liège, Belgium, pp. 43–63 (1970).
133. Popov, E. P. and R. G. Black, "Steel Struts Under Severe Cyclic Loadings," *Journal of the Structural Division,* ASCE, Vol. 107, No. ST9, pp. 1857–1881 (1981).
134. Popov, E., V. Zayas and S. Mahin, "Cyclic Inelastic Buckling of Thin Tubular Columns," *Journal of the Structural Division,* ASCE, Vol. 105, No. ST11, pp. 2261–2277 (1979).
135. Powell, G. and J. Simons, "Improved Iteration Strategy for Nonlinear Structures," *International Journal for Numerical Methods in Engineering,* Vol. 17, pp. 1455–1467 (1981).
136. Prathuangsit, D., S. C. Goel and R. D. Hanson, "Axial Hysteresis Behavior with End Restraints," *Journal of the Structural Division,* ASCE, Vol. 104, No. ST6, pp. 883–898 (1978).
137. Ramm, E., "Strategies for Tracing the Nonlinear Response Near Limit Points," *Nonlinear Finite Element Analysis in Structural Mechanics,* W. Wunderlich, E. Stein, and K.-J. Bathe, eds., Springer-Verlag, Berlin, pp. 63–89 (1981).
138. Rashed, S. M. H., et al., "Analysis of Nonlinear and Collapse Behavior of TM Space Trusses," *Third International Conference on Space Structures,* Guilford, England, H. Nooshin, ed., Applied Science Publishers Ltd, London, pp. 480–485 (1984).
139. Remseth, S. N., "Nonlinear Static and Dynamic Analysis of Framed Structures," *Computers and Structures,* Vol. 10, pp. 879–897 (1979).
140. Renton, J. D., "On the Gridword Analogy for Plates," *Journal of the Mechanics and Physics of Solids,* Vol. 13, pp. 413–420 (1965).
141. Riks, E., "The Application of Newton's Method to the Problem of Elastic Stability," *Journal of Applied Mechanics,* Vol. 39, pp. 1060–1066 (1972).
142. Riks, E., "An Incremental Approach to the Solution of Snapping and Buckling Problems," *Computers and Structures,* Vol. 15, pp. 529–551 (1979).
143. Riks, E., "Some Computational Aspects of the Stability Analysis of Nonlinear Structures," *Computer Methods in Applied Mechanics and Engineering,* Vol. 47, pp. 219–259 (1984).
144. Roeder, C. and E. Popov, "Inelastic Behavior of Eccentrically Braced Steel Frames under Cyclic Loadings," Report No. UCB/EERC-77/18, University of California, Berkeley (Aug. 1977).
145. Roorda, J., "On the Buckling of Symmetric Structural Systems with First and Second Order Imperfections," *International Journal of Solids and Structures,* Vol. 4 pp. 1137–1148 (1968).
146. Rosen A. and A. Schmit, "Design-Oriented Analysis of Imperfect Truss Structures-Part I—Accurate Analysis," *International Journal for Numerical Methods in Engineering,* Vol. 14, pp. 1309–1321 (1979).
147. Rothert, H., T. Dickel, and D. Renner, "Snap-Through Buckling of Reticulated Space Trusses," *Journal of the Structural Division,* ASCE, Vol. 107, pp. 129–143 (1981).
148. Sabir, A. B. and A. C. Lock, "The Application of Finite Elements to the Large Deflection Geometrically Nonlinear Behaviour of Cylindrical Shells," *Variational Methods in Engineering,* C.A. Brebbia and H. Tottenham, eds., Southampton University Press, pp. 7/66–7/75 (1972).
149. Saka, T. and K. Heki, "The Effect of Joint on the Strength of Space Trusses," *Third International Conference on Space Structures,* Guilford, England, H. Nooshin ed., Applied Science Publishers Ltd., London, pp. 417–422 (1984).

150. Schmidt L. C. and B. M. Gregg, "A Method for Space Truss Analysis in the Post-Buckling Range," *International Journal for Numerical Methods in Engineering,* Vol. 15, pp. 237–247 (1980).

151. Schmidt, L. C. and A. Hanaor, "Force Limiting Devices in Space Trusses," *Journal of the Structural Division,* ASCE, Vol. 105, No. ST5, 939–951 (1979).

152. Schmidt, L. C., P. R. Morgan, and J. A. Clarkson, "Space Trusses with Brittle-Type Strut Buckling," *Journal of Structural Engineering,* ASCE, Vol. 102, No. ST7, pp. 1479–1492 (1976).

153. Schmidt, L. C., P. R. Morgan, and A. Hanaor, "Ultimate Load Testing on Space Trusses," *Journal of the Structural Division,* ASCE, Vol. 108, No. ST6, 1324–1335 (1982).

154. Sherman, D. R., "Experimental Study of Post Local Buckling Behavior in Tubular Portal Type Beam-Columns," Report to Shell Oil Company, University of Wisconsin-Milwaukee (Oct., 1979).

155. Sherman, D. R., "Post Local Buckling Behavior of Tubural Strut Type Beam-Columns; An Experimental Study," Report to Shell Oil Company, University of Wisconsin-Milwaukee (June, 1980).

156. Shibata, M., "Analysis of Elastic-Plastic Behavior of a Steel Brace subjected to Repeated Axial Force," *International Journal of Solids and Structures,* Vol. 18, No. 3, pp. 217–228 (1982).

157. Shibata, M., et al., "Elastic-Plastic Behavior of Steel Braces under Repeated Axial Loading," *Proceedings of the 5th World Conference on Earthquake Engineering,* Rome (1973).

158. Singh, P. and S. C. Goel, "Hysteresis Model of Bracing Members for Earthquake Response of Braced Frames," *Proceedings of the 6th World Conference on Earthquake Engineering,* New Delhi, Vol. 11, pp. 43–48 (Jan. 1977).

159. Smith, E. A., "Space Truss Nonlinear Analysis," *Journal of Structural Engineering,* Vol. 110, No. 4, pp. 688–705 (1984).

160. Smith, E. A., "Ductility in Double Layer Grid Space Trusses," *Third International Conference on Space Structures,* Guilford, England, H. Nooshin, ed., Applied Science Publishers Ltd., London, pp. 510–515 (1984).

161. Smith, E. A. and H. I. Epstein, "Hartford Coliseum Roof Collapse: Structural Collapse Sequence and Lessons Learned," *Civil Engineering,* ASCE, pp. 59–62 (April 1980).

162. Smith, E. A. and G. D. Smith, "Collapse Analysis of Space Trusses," *Proceedings of Symposium on Long Span Roof Structures,* ASCE (Oct., 1981).

163. Stricklin, J. A. and W. E. Haisler, "Formulations and Solution Procedures for Nonlinear Structural Analysis," *Computers and Structures,* Vol. 7, pp. 125–136 (1977).

164. Stricklin, J. A., et al., "Nonlinear Dynamic Analysis of Shells of Revolution by Matrix Displacement Method," *AIAA Journal,* Vol. 9, No. 4, pp. 624–636 (April 1971).

165. Supple, W. J. and I. Collins, "Post-Critical Behaviour of Tubular Struts," *Engineering Structures,* Vol. 2, pp. 225–229 (Oct. 1980).

166. Suzuki, T., I. Kubodera, and T. Ogawa, "An Experimental Study on Load-Bearing Capacity," *Third International Conference on Space Structures,* Guilford, England, H. Nooshin, ed., Applied Science Publishers, Ltd., London, pp. 571–576 (1984).

167. Suzuki, T. and T. Ogawa, "Buckling Analysis of Double-Layer Grids, Roofs and Truss-Like Beams," *Third International Conference on Space Structures,* Guilford, England, H. Nooshin ed., Applied Science Publishers Ltd., London, pp. 474–479 (1984).

168. Tang, S. C., K. S. Yeung, and C. T. Chon, "On the Tangent Stiffness Matrix in a Convected Coordinate System," *Computers and Structures,* Vol. 12, pp. 849–856 (1980).

169. The Subcommittee on Latticed Structures of the Task Commitee on Special Structures of the Commitee on Metals of the Structural Division, "Bibliography on Latticed Structures," *Journal of the Structural Division,* ASCE, Vol. 98, No. ST7, pp. 1545–1566 (1972).

170. The Task Commitee on Latticed Structures of the Commitee on Special Structures of the Commitee on Metals of the Structural Division, "Latticed Structures: State-of-the-Art Report," *Journal of the Structural Division,* ASCE, Vol. 102, No. ST11, pp. 2197–2230 (1976).

171. The Task Committe on Latticed Structures of the Commitee on Special Structures of the Committee on Metals of the Structural Division, "Dynamic Considerations in Latticed Structures," *Journal of Structural Engineering,* ASCE, Vol. 110, No. 10, pp. 2547–2550 (1984).

172. Thompson, J. M. T., "Towards a General Statistical Theory of Imperfection Sensitivity in Elastic Post-Buckling," *Journal of the Mechanics and Physics of Solids,* Vol. 15, pp. 413–417 (1967).

173. Toma, S. and W. F. Chen, "Cyclic Analysis of Fixed-Ended Steel Beam-Columns," *Journal of Structural Division,* ASCE, Vol. 108, No. ST6, pp. 1385–1399 (1982).

174. Toma, S. and W. F. Chen, "Inelastic Cyclic Analysis of Pin-Ended Tubes," *Journal of Structural Division,* ASCE, Vol. 108, No. ST10, pp. 2279–2293 (1982).

175. Toma, S. and W. F. Chen, "Post-Buckling Behavior of Tubular Beam-Columns," *Journal of Structural Engineering,* Vol. 109, No. 8, pp. 1918–1932 (1983).

176. Truesdell, C., *A First Course in Rotational Continuum Mechanics,* Vol. 1: *General Concepts,* Academic Press, New York (1977).

177. Wagner, A. L., W. H. Mueller, and H. Erzurumlu, "Design Interaction Curve for Tubular Steel Beam-Columns," OTC Paper No. 2684, Offshore Technology Conference, Houston, pp. 755–764 (1976).

178. Wakabayashi, M., "Behavior of Braces and Braced Frames under Earthquake Loadings," *International Journal of Structures,* Vol. 2, No. 2, pp. 49–70 (April, 1982).

179. Wakabayashi, M., C. Matsui, and I. Mitani, "Cyclic Behavior of a Restrained Steel Brace under Axial Loading," *Proceedings of the 6th World Conference on Earthquake Engineering,* New Delhi, pp. 3181–3187 (1977).

180. Wakabayashi, M. and M. Shibata, "Studies on the Post-Buckling Behavior of Braces," *Part 4, Abstracts Annual Meeting Kinki Branch, Architectural Institute of Japan*, p. 201 (1976).

181. Wakabayashi, M. et al., "Hysteretic Behavior of Steel Braces Subjected to Horizontal Load due to Earthquake," *Proceedings of the 6th World Conference on Earthquake Engineering*, New Delhi, pp. 3188–3194 (1977).

182. Waszczyszyn, Z., "Numerical Problems of Nonlinear Stability Analysis of Elastic Structures," *Computers and Structures*, Vol. 17, No. 1, pp. 13–24 (1983).

183. Watson, L. T. and S. M. Holzer, "Quadratic Convergance of Crisfield's Method," *Computers and Structures*, Vol. 17, No. 1, pp. 69–72 (1983).

184. Watwood, V. B., "Mechanism Generation for Limit Analysis of Frames," *Journal of the Structural Division*, ASCE, Vol. 109, ST1, pp. 1–15 (1978).

185. Wellford, L. C. and B. Vahdani, "A Block Iteration Scheme for the Solution of Systems of Equations Resulting from Linear and Nonlinear Finite Element Models," *Computer Methods in Applied Mechanics and Engineering*, Vol. 26, pp. 33–52 (1981).

186. Wempner, G. A., "Discrete Approximations Related to Nonlinear Theories of Solids," *International Journal of Solids and Structures*, Vol. 7, pp. 1581–1599 (1971).

187. Wen, R. K. and J. Rahimzadeh, "Nonlinear Elastic Frame Analysis by Finite Element," *Journal of Structural Engineering*, ASCE, Vol. 109, No. 8, pp. 1952–1971 (1983).

188. Wolf, J. P., "Post-Buckled Strength of Large Space-Truss," *Journal of the Structural Division*, ASCE, Vol. 99, No. ST7, pp. 1708–1712 (1973).

189. Wood, R. D. and O. C. Zienkiewicz, "Geometrically Nonlinear Finite Element Analysis of Beams, Frames, Arches and Axisymmetric Shells," *Computers and Structures*, Vol. 7, pp. 725–735 (1977).

190. Wright, D. T., "Membrane Forces and Buckling in Reticulated Shells," *Journal of the Structural Division*, ASCE, Vol. 21, No. ST1, pp. 173–201 (1965).

191. Wright, E. W. and E. H. Gaylord, "Analysis of Unbraced Multi-Story Steel Rigid Frames," *Journal of the Structural Division*, ASCE, Vol. 94, pp. 1143–1163 (1968).

192. Yang, T. Y., "Matrix Displacement Solution to Elastic Problems of Beams and Frames," *International Journal of Solids and Structures*, Vol. 9, pp. 829–842 (1973).

193. Yang, T. Y. and S. Saigal, "A Simple Element for Static and Dynamic Response of Beams with Material and Geometric Nonlinearities," *International Journal for Numerical Methods in Engineering*, Vol. 20, pp. 851–867 (1984).

194. Ziegler, H., *Principles of Structural Stability*, 2nd ed., Birkhause Basel (1977).

195. Zienkiewicz, O. C., "Incremental Displacement in Non-Linear Analysis," *International Journal for Numerical Methods in Engineering*, Vol. 3, pp. 587–588 (1971).

CHAPTER 23

On the Stability of Columns Made of Time-Dependent Materials[1]

P. G. GLOCKNER* AND W. SZYSZKOWSKI**

ABSTRACT

On the basis of dynamic and static stability analyses of columns made of time-dependent materials, the behaviour of such structures is conveniently separated into two main domains: (1) a viscoelastically stable domain for which a "safe-load-limit" can be established and; (2) a viscoelastically unstable region for which the designer is interested in establishing a "safe service period."

A non-zero safe-load-limit, representing the load level below which the response of the column is bounded, can be defined for a column made of any solid type material with limited creep. A relatively simple method is presented for the determination of the safe-load-limit of columns made of such linear and nonlinear materials. The effects of imperfections and the slenderness ratio on this parameter are discussed and evaluated. For columns made of linearly viscoelastic materials, closed-form expressions are obtained for the safe-load-limit of such structures.

The safe service period is shown to be extremely sensitive to initial imperfections suggesting that estimates for the useful life span of such structures must be based on an analysis of the actual structure.

For a complete analysis of viscoelastically unstable columns an incremental approach with an associated iterative solution technique is introduced thereby providing an efficient tool for the determination of the stress-strain-deflection-time characteristics of such structures, including hereditary effects as well as various limitations on material strength and/or deflection magnitudes. Finally, a simple example is used to alert the designer to the possible danger in failing to recognize structural configurations, involving compression members, which have instantaneous unstable post-buckling load-deflection curves and consequently exhibit imperfection sensitivity, a characteristic which may affect the safe-load limit and the safe service period of the structure.

[1]The results presented here were obtained in the course of research sponsored by the Natural Sciences and Engineering Research Council of Canada, Grant No. A-2736.

*Professor and Head, Department of Mechanical Engineering, and Professor of Civil Engineering, The University of Calgary, Calgary, Alberta, Canada

**Associate Professor, Department of Mechanical Engineering, The University of Saskatchewan, Saskatoon, Saskatchewan, Canada

1. INTRODUCTION

Traditionally, creep and time-dependent behaviour of materials was a concern in the design process only for engineers using metals under high temperature environments or concrete in long span reinforced and/or prestressed structures. With the appearance of various synthetic structural materials on the market, including plastics and fibre reinforced structural materials, the introduction of improved construction techniques, the increasing number of high temperature environments, the availability of large scale computers facilitating numerical solutions, as well as the development in constitutive theory in general continuum mechanics, this "traditional" picture has changed and continues to change rapidly. Consequently, in more recent times, the trend has been to replace "time-independent" material behaviour, in which viscous effects were handled by increased safety factors, with more realistic "time-dependent" material models thereby enabling the designer to carry out a more refined analysis and predict structural response more precisely.

The intensity of time-dependent viscous/creep behaviour of materials depends, in general, primarily on two factors, namely:

1. the temperature T, of the material, and in particular, the difference $(T_m - T)$, where T_m denotes the melting temperature;
2. the magnitude of stress in the material, σ, in comparison with the yield strength, σ_y, for the given operational temperature.

When tackling design problems involving time-dependent behaviour the engineer is faced with at least two major tasks: firstly, the choice of a suitable constitutive model which, hopefully, will describe the actual material behaviour realistically, a task which is made more difficult by the scarcity of experimental data on time-dependent behaviour, particularly the response of such materials to temporally varying stress states; secondly, the choice of a solution technique which will take into account the spatial and temporal variation of all parameters, including hereditary effects, effects which may be particularly significant in problems in which parameters, such as stress and strain, vary with time even though the external load is constant, as is the case, for example, in an imperfect column subjected to a constant load. In order to side-step some of the difficulties encountered in such problems, designers have resorted to various approximations, particularly in constitutive models, approximations which in certain instances have led to quantitatively as well as qualitatively incorrect conclusions on structural behaviour.

One area of research indicating a high level of activity during the past few years deals with the stability of structures made of time-dependent materials, including the stability of linearly viscoelastic structures. Because of the time effects and the peculiar nature in which time enters as a "destabilizing" parameter, viscoelastic and creep stability problems, and in particular the post-buckling behaviour of structures made of such materials, are still very much a topic for current research. Important viscoelastic/creep stability parameters, such as the "safe load limit" and the "critical time" have been investigated extensively, and possibly satisfactorily, only for structures in which the material behaviour was modelled by using the simplest rheological spring-dashpot systems. The effect of nonlinear material behaviour on such parameters, the influence of nonlinear geometry and large deflection on the stability characteristics of columns as well as the interaction of geometrical and material nonlinearities remain, by and large, open questions. It is, therefore, not surprising that the codification of viscoelastic/creep stability problems has progressed little during the past two or three decades.

The purpose of this article is to provide a review of the stability behaviour of columns made of time-dependent materials. In view of the above comments this summary must necessarily be incomplete at this time. Nevertheless, hopefully, it will focus on the latest results of research in this field and on remaining open problems. It is thus a survey article similar, in some sense, to the one published approximately three decades ago by Hoff [27]. Due to space limitations it is impossible to review, in detail, any of the papers published in this field during the past three decades. As a compensation, a selected list of references is given at the end of the article which should help the reader find specific material of interest to him in connection with a particular design problem.

In the second section of this article we review, briefly, the constitutive models used to describe time-dependent behaviour of common engineering materials. Linear as well as nonlinear simple models are introduced and attention is focused on the so-called "limited-creep" behaviour, characteristic of the response of a number of materials when subjected to relatively low stress levels. The third section presents a dynamic stability analysis of perfect and imperfect columns made of linear time-dependent materials. It indicates possible phenomena encountered in the stability analysis of such columns and defines load ranges for which the column is dynamically unstable, viscoelastically unstable or absolutely stable. As opposed to elastic column behaviour, in which the structure is either stable or unstable, with the latter being of no particular use in practice, the viscoelastic column is shown to have three distinct states with both the absolutely stable and the viscoelastically unstable states having some relevance in design. Details of the analysis include definition of the "safe load limit," a discussion of "critical times" for linear viscoelastically unstable columns, and the importance and effects of imperfections.

In Section 4 we deal with the static stability analysis of perfect and imperfect linearly viscoelastic columns and demonstrate that this approach can be applied directly and in a straightforward manner to imperfect columns leading to results which agree with corresponding values obtained from an "exact" dynamic analysis. Application of the static method to perfect columns requires some care, particularly in the interpretation of the results. Also, the concept of "adjacent equilibrium configurations," configurations which under constant axial load are strictly speaking not possible, has to be re-examined and redefined. The analysis underlines the importance of initial imperfections which may be thought of as "permanent disturbances."

An asymptotic solution to the governing equation of the dynamic formulation is presented and discussed in Section 5 where we show that such a formulation for $t \to \infty$ leads to results indentical to those obtained from the static approach. The expression for the safe-load-limit for linearly viscoelastic hinged-hinged columns is generalized to arbitrary boundary conditions in Section 6.

Columns made of linear or nonlinear materials are treated in the seventh section. In particular, a relatively simple and convenient method for determination of the "safe-load-limit" for columns made of any material with limited creep is presented, and the results compared with those obtained in Sections 3 and 4. Section 8 deals with the time-deflection behaviour of viscoelastically unstable columns, up to failure, using an incremental formulation and an iterative solution technique. "Safe-service-periods" for such structures are defined, the effect of load level on absolutely stable and viscoelastically unstable behaviour is confirmed and the consequences of Shanley-type approximations in the constitutive law are examined and evaluated. The penultimate section deals with the effects of geometric nonlinearities on the "safe-service-period." The article is concluded by summarizing the main results, pointing to possible further directions of research and making suggestions for codification of design procedures for columns made of time-dependent materials.

2. COMMENTS ON CONSTITUTIVE RELATIONS FOR TIME-DEPENDENT MATERIALS

In deciding on a suitable constitutive model, the engineer is aided by the substantial body of experimental data available on the uniaxial stress-strain-time behaviour of various materials, particularly under constant stress and/or under various constant strain rates. Typical creep curves for such time-dependent materials are indicated in Figure 1a, for various stres levels. Note that for low stress levels, the strain tends to a maximum limiting value, ϵ_∞ as $t \rightarrow \infty$. The maximum stress level up to which such limited creep is possible is designated by σ_s, referred to as the safe stress level. It is a matter of judgement on the part of the designer to decide on how small the slope of the creep curve has to be in order to justify the "limited-creep" assumption. With increasing stress, the creep curves take the well known typical form, showing all three stages of creep, namely the primary or transient creep phase, the secondary or steady creep phase, and the tertiary or accelerating creep phase. Under very high stress levels the steady creep stage may be absent and the primary stage followed directly by the tertiary creep phase. For such stress levels, strain continues to increase as $t \rightarrow \infty$.

When dealing with stability problems, it is particularly convenient to classify time-dependent materials into two major categories:

1. materials for which there exists a stress level below which limited creep is possible, such materials being referred to as "solid type materials;"
2. materials which under any stress level will continue to deform and are referred to as "fluid type materials" [15].

In the case of structures made of the first type of material, one is interested in determining the maximum load under which strains and deflections remain limited as $t \rightarrow \infty$. This maximum load is referred to as the "safe load limit." When dealing with structures made of fluid type materials, or when stress levels in solid-type materials exceed σ_s, strains and deflections continue to increase with time and the prime concern of the designer is the determination of a time period during which the structure, under given loading and boundary conditions, may be considered to be safe.

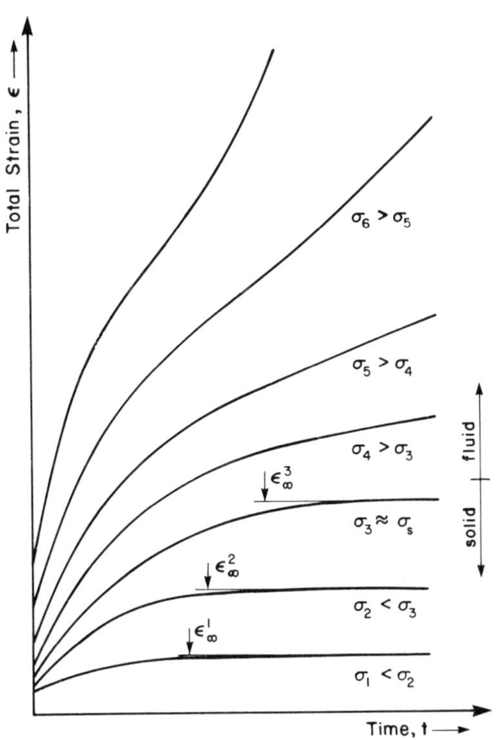

(a) Influence of Stress Level on Creep Curves

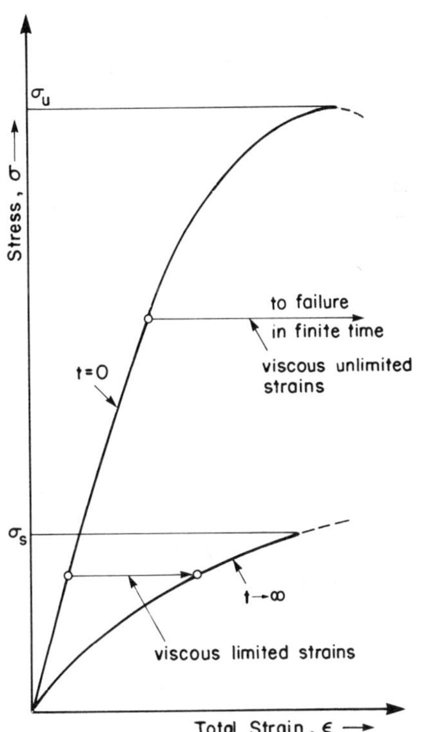

(b) Instantaneous and Limit Stress-Strain Curves

FIGURE 1. Some characteristics of time-dependent material behaviour.

This period, referred to as the "safe-service period," will depend not only on the material properties but also on such factors as initial imperfections, disturbances acting on the structure, as well as load/stress levels.

Based on the typical creep curves, shown on Figure 1a, one can draw stress-strain curves both for $t = 0$ and $t \rightarrow \infty$, as indicated on Figure 1b. The failure stress level corresponding to the instantaneous loading ($t = 0$) is designated by σ_u, while for the long-term loading ($t \rightarrow \infty$), the maximum safe stress is designated by σ_s, as stated above. In the sequel it will be asumed that for solid-type materials to be dealt with it is possible to determine stress strain curves for $t = 0$ and for $t \rightarrow \infty$.

Based on the extensive body of experimental data available on the uniaxial creep behaviour of various materials it is possible to write a fairly general uniaxial strain-stress constitutive relation, for a given constant temperature, in the form [36]

$$\epsilon(t) = \int_o^t \tilde{\mathscr{F}}[\sigma(\tau,t)]d\tau \qquad (2.1)$$

in which the lower and upper limit on the integral refers to the "virgin" state of the material and the current time, respectively, while $\tilde{\mathscr{F}}$ is a function describing the viscoelastic properties of the material, which depends on the entire stress history. This formulation of the uniaxial strain-stress relation for time-dependent materials is sufficiently general so as to allow description of all known characteristic features of such materials and, in addition, is found to be quite useful in theoretical considerations (see discussion in [57], for example). From a practical viewpoint, however, the constitutive relation must be more specific so as to become tractable and to indicate the types of experiments from which one can reasonably expect to identify the essential viscoelastic parameters (functions) of a given material. A somewhat less general but more specific form of Equation (2.1) is

$$\epsilon(t) = \frac{\sigma(t)}{E_o} + \int_o^t \bar{\mathscr{F}}[\sigma(\tau), j(t-\tau), \zeta(\tau)]\, d\tau \qquad (2.2)$$

in which the elastic deformations are separated from the viscous components and are assumed to be linear, as specified by Young's modulus, E_o. A new function, $\bar{\mathscr{F}}$, incorporates the effects of stress history, $\sigma(\tau)$, a "fading memory," defined by a shifted function $j(t-\tau)$, and the function, $\zeta(\tau)$, defining the "ageing characteristics" of the material. For most engineering materials the fading memory function, $j(t-\tau)$, depends only on crystallographic and atomic structure, while $\zeta(\tau)$ is a function only of the formation/solidification process, including any chemical reaction which may take place. Thus both of these functions may be considered to be stress independent material characteristics [9,51], as a result of which the constitutive relation, Equation (2.2), may be rewritten as

$$\epsilon(t) = \frac{\sigma(t)}{E_o} + \int_o^t \mathscr{F}[\sigma(\tau)] \cdot j(t-\tau) \cdot \zeta(\tau)\, d\tau \qquad (2.3)$$

To particularize and/or simplify this equation further we recognize that the ageing process is usually monotonic and its effects become negligible after some time [9]. Accordingly, the ageing function may be presented in the form

$$\zeta(t) = 1 + \psi(t) \qquad (2.4)$$

where $\psi(t) \rightarrow 0$ and $\zeta(t) \rightarrow 1$ for $t \rightarrow \infty$. The precise moment to be considered as the start of the material history remains an open question. If, for example, the beginning of the solidification process is chosen as the origin of the time scale ($t = 0$), then $\psi(0) \rightarrow \infty$. In the case of concrete, a very simple relation $\psi(t) = c/t$ has been suggested [9]. However, for most materials, including metals, the start of the material history is taken as the instant when its microscopic structure has been well established. In such case $\psi(t) << 1$ and the ageing process can be neglected from the outset.

The "fading memory" function, $j(t)$ may be written in the form

$$j(t) = A_1 \cdot \hat{j}(t) + A_2 \qquad (2.5)$$

where A_1 and A_2 are parameters associated with the primary hardening creep phase and the secondary steady stage of creep, respectively, while $\hat{j}(t)$, a monotonically decreasing function of time, describes the strain hardening effects during the first creep stage. Some suggested forms for $\hat{j}(t)$, based on experimental data, are also given in [5,47]. In general $\hat{j}(t) \rightarrow 0$ as $t \rightarrow \infty$. In addition $\hat{j}(t)$ is normalized so that $\hat{j}(0) = 1$. For linear materials, for example, this function is expressed as

$$\sum_{i=1}^{N} a_i e^{-\lambda_i t}$$

where a_i and λ_i are parameters to be determined experimentally [57].

The function $\mathscr{F}(\sigma)$ which defines the strain-stress nonlinearity, and which must be an odd function of stress, for a linear material is taken as $\mathscr{F}(\sigma) = \sigma$, in which case, and neglecting ageing effects, Equation (2.2) is written in the form

$$\epsilon(x,t) = \frac{1}{E_o}\left[\sigma(x,t) + \int_o^t \sigma(x,\tau) \cdot j(t-\tau)d\tau\right] \qquad (2.6)$$

which is the integral form of the hereditary uniaxial constitutive law for linear time dependent materials. Differentiating this expression, one obtains the strain rate

$$\frac{d\epsilon(x,t)}{dt} = \frac{1}{E_o}\left[\frac{d\sigma(x,t)}{dt} + j(0)\cdot\sigma(x,t) + \int_o^t \sigma(x,\tau)\cdot\frac{dj(t-\tau)}{dt}\,d\tau\right] \tag{2.7}$$

which also defines the creep rate in the form

$$\frac{d\epsilon_{cr}}{dt} = \frac{1}{E_o}\left[j(0)\cdot\sigma(x,t) + \int_o^t \sigma(x,\tau)\cdot\frac{dj(t-\tau)}{dt}\,d\tau\right] \tag{2.8}$$

Unfortunately, these expressions for creep and creep rate are very cumbersome in numerical solutions because the integrals they contain have to be evaluated for each time increment from $t = 0$ up to the current time t, indicating that the entire loading history, at each point of the structure, has to be available and therefore stored. To get around this difficulty, designers have resorted to various approximations. For example one can assume the stress in Equation (2.6) to be constant and therefore rewrite it as

$$\epsilon(x,t) = \frac{\sigma_o(x)}{E_o}\left[1 + \int_o^t j(t-\tau)d\tau\right] \tag{2.9}$$

which when differentiated w.r.t. time becomes the strain rate and simultaneously provides an approximate expression for the creep rate in the form

$$\left.\frac{d\epsilon}{dt}\right|_{\sigma=\sigma_o} = \frac{\sigma_o(x)}{E_o}j(t) \cong \frac{d\epsilon_{cr}}{dt} \tag{2.10}$$

For sufficiently smooth stress history the assumption of temporally constant stress is carried even further and this latter equation is taken to be valid for slightly variable stress, $\sigma = \sigma(x,t)$

$$\frac{d\epsilon_{cr}}{dt} \cong \frac{\sigma(x,t)}{E_o}j(t) \tag{2.11}$$

This last assumption, which results in the approximate expression for the creep rate, given by Equation (2.11), is referred to as Shanley's hypothesis which, with modifications, has been used by many authors. In stability considerations this type of approximation was used through the introduction of a "deteriorating" modulus of elasticity, $E(t)$, which from Equation (2.9) can be written as

$$E(t) = \frac{\sigma_o}{\epsilon(t)} = \frac{E_o}{1 + \mathcal{J}(t)} \tag{2.12}$$

where $\mathcal{J}(t) = \int_o^t j(t-\tau)d\tau$, an approximation which was introduced by Gerard [19] and subsequently used by others [35,37,72] in describing the stability behaviour of viscoelastic structures. Clearly, with this approximation the elastic modulus, E_o, was replaced by a time-dependent secant modulus, $E(t)$, allowing simple but dubious definitions for "critical" loads and associated "critical" times for structures made of such materials.

Unfortunately, most real materials are nonlinear. Consequently, a stress nonlinearity function, which is reasonably tractable has to be introduced. A popular form is Norton's Power Law which describes the behaviour of a fairly wide class of materials and which is written in the form

$$\mathcal{F}(\sigma) = B\sigma^n \tag{2.13}$$

where B and n are material constants. Experimental data indicates that B depends strongly on temperature while n can be assumed to be a constant, at least over a certain limited range of stress level. Values for n range from 1.0–30 [51]. In general, n decreases with decreasing stress. Therefore, in any problem in which the material is time-dependent and in which the stress levels are sufficiently low and deformations are small, a linear constitutive law may provide satisfactory results. The nonlinearity factor n also depends on the type of material. For non-metals it is usually small, approaching 1.0 for plastics, having a value of approximately 2.0 for concrete [49] and varying between 1.6–4.0 for ice [61]. For metals, n is usually much larger with a value of 7.2 being suggested in [45] for copper.

With Norton's Power Law defining the stress nonlinearity and using the above assumed form for the "fading memory" function, $j(t)$, as well as neglecting ageing effects, a relatively simple uniaxial strain-stress expression for nonlinear hereditary type materials is written as

$$\epsilon(t) = \frac{\sigma(t)}{E_o} + \frac{1}{\nu_1}\int_o^t [\sigma(\tau)]^n\cdot\hat{j}(t-\tau)d\tau + \frac{1}{\nu_2}\int_o^t [\sigma(\tau)]^n d\tau \tag{2.14}$$

in which

$$\nu_1 = 1/(A_1B) \text{ and } \nu_2 = 1/(A_2B)$$

Although nonlinear, one should recognize that this expres-

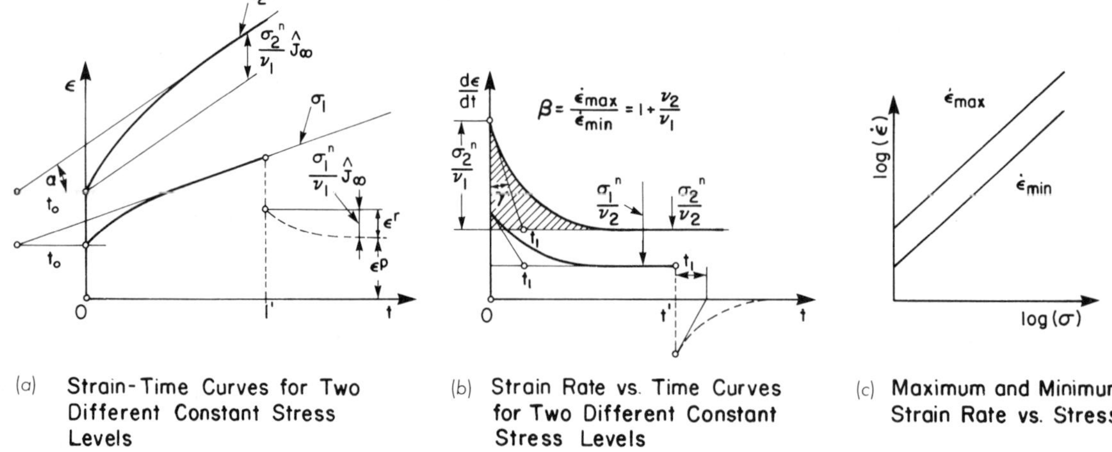

(a) Strain-Time Curves for Two Different Constant Stress Levels
(b) Strain Rate vs. Time Curves for Two Different Constant Stress Levels
(c) Maximum and Minimum Strain Rate vs. Stress

FIGURE 2. Creep test characteristics of a material obeying Norton's Power Law (from [68]).

sion is unable to simulate the experimentally observed "strain softening effects," which are usually associated with tertiary creep behaviour [61]. In many problems deformation magnitudes usually associated with tertiary creep are unacceptable and therefore, this shortcoming of the constitutive relation, Equation (2.14) is, in such cases, not too important. In fact, the designer normally considers the accelerating creep phase as an "unsafe" phase of material behaviour since it leads, very rapidly and directly, to material failure/rupture. One of the fundamental goals of any design process is, therefore, to design the structure or structural element such that its response will be kept within the first two stages of creep during its operational lifetime.

To assess the accuracy and validity of Equation (2.14) as a constitutive law for real materials, a number of tests can be designed. These tests naturally should provide data for determination of the viscous parameters n, ν_1, ν_2 as well as the function $\hat{j}(t)$. As a first check, we can use creep test data for a constant stress, σ_o, for which Equation (2.14) becomes

$$\epsilon(t) = \frac{\sigma_o}{E_o} + \sigma_o^n \left[\frac{1}{\nu_1}\hat{\mathscr{J}}(t) + \frac{1}{\nu_2}t\right] \tag{2.15a}$$

where

$$\hat{\mathscr{J}}(t) = \int_o^t \hat{j}(\tau)d\tau$$

The corresponding strain rate is written as

$$\dot{\epsilon}(t) = \sigma_o^n \left[\frac{1}{\nu_1}\hat{j}(t) + \frac{1}{\nu_2}\right] \tag{2.15b}$$

Typical strain vs. time and strain-rate vs. time curves, for various constant stress levels, can be plotted and are shown on Figure 2. These curves should exhibit certain characteristic features. For example, since $\hat{j}(t) \rightarrow 0$ as $t \rightarrow \infty$ and is normalized such that $\hat{j}(0) = 1$, the initial and final strain rates, $\dot{\epsilon}_o$ and $\dot{\epsilon}_{min} = \dot{\epsilon}(t)$, for $t \rightarrow \infty$, when plotted against stress to a log-log scale, should result in two straight lines the slopes of which define n (see Figure 2c). From the same figure ν_1 and ν_2, or alternatively, β and ν_2 can also be determined, where $\beta = (\nu_2/\nu_1) + 1$.

Some characteristic features of the function $\hat{j}(t)$ may be determined directly from creep curves, namely the time t_1, indicated on Figure 2b, which should be stress independent and given by

$$t_1 = -\frac{1}{\left.\dfrac{d\hat{j}}{dt}\right|_{t=0}} \tag{2.16a}$$

Assuming $\hat{\mathscr{J}}(t) \rightarrow \hat{\mathscr{J}}_\infty$ as $t \rightarrow \infty$, a second stress-independent parameter is established as

$$t_o = \frac{\nu_2}{\nu_1}\hat{\mathscr{J}}_\infty \tag{2.16b}$$

Clearly, t_1 and t_o are related to the hardening process during the first stage of creep, with t_1 indicating the intial rate of hardening or initial deceleration of strain rate, while t_o is proportional to the "latent" energy accumulated during the hardening process, energy which is recoverable upon unloading, producing residual strain ϵ^r (see Figure 2a).

As was discussed in [69], the function $\hat{j}(t)$ can be specified approximately by means of these two parameters, t_o and

t_1, an approximation which makes the analysis of experimental data much easier and also facilitates the numerical treatment of the Volterra type integrals in Equation (2.14) considerably.

Thus far we have discussed hereditary type constitutive laws according to which strain at a point and at a given instant of time is a function of the entire history of the material, involving Volterra type integrals. The use of such laws in actual design problems is, therefore, quite challenging in terms of the numerical analysis involved. Consequently, simpler material laws have been introduced to describe viscoelastic behaviour, laws which are written in the form

$$\tilde{\mathscr{F}}(\dot{\epsilon}_{cr}, \sigma, q_1, q_2, \ldots, q_n) = 0 \qquad (2.17)$$

an expression referred to as an "equation of state" which does not involve any integrals and in which q_i $(i = 1, \ldots, n)$ denote state parameters. Restricting the number of state parameters, for example, to a single variable, say $q_1 = \epsilon_{cr}$, one obtains the so-called strain-hardening theory. If this single parameter is time, expression (2.17) defines a time-hardening theory, a simple example of which is the Shanley hypothesis discussed above and given by Equation (2.11). Detailed discussions of such simple creep theories are given in [5,47,51]. The major disadvantage of these theories is the fact that they are unable to simulate reversible processes, while most real materials exhibit viscoelastic behaviour which is, at least partially, reversible. Consequently, such theories, when used in problems dealing with temporally variable stress, yield results which show poor correlation with corresponding experimental data. Their applicatioon to column stability problems was discussed extensively in [27]. In the present paper, attention will be focused on hereditary type constitutive theories with the strain-stress relations assumed in the form of Equation (2.14).

3. DYNAMIC ANALYSIS OF LINEAR COLUMNS

The most natural, fundamental and at the same time most general definition of stability involves notions of motion and time and is referred to as the "dynamic stability criterion," introduced by Lyapunov. In such an approach "stability" implies resistance to disturbances and includes an analysis of the response of the original undisturbed as well as the disturbed structure, both subjected to a given set of loading and boundary conditions. Generally, if the effects of the disturbance decrease in time so that they become negligible, the structure is considered to be stable. Clearly, in using such a dynamic approach, every stability investigation, including those involving only static loadings, becomes a dynamic problem requiring treatment of equations of motion which may represent a fairly sophisticated analysis problem. Columns made of simple, non-hereditary materials were analysed using this approach in [50] and were subsequently discussed in [27]. Hereditary effects were included in the dynamic stability analysis of linearly viscoelastic columns treated in [13,63], an analysis which is relatively simple and yet reveals all significant features of the stability behaviour of columns made of time-dependent materials.

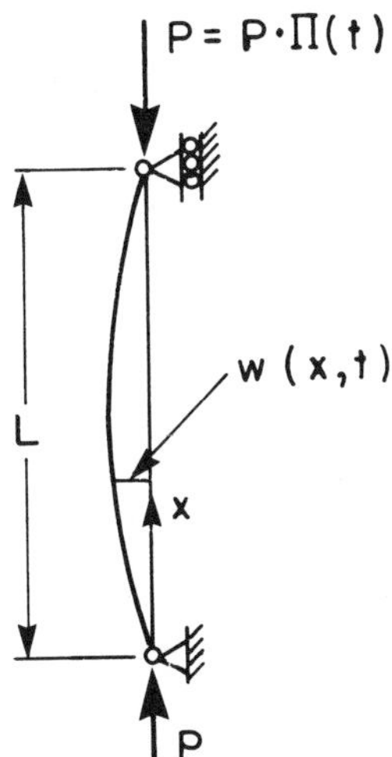

FIGURE 3. The "disturbed" simply supported viscoelastic column.

3.1 The Perfect Column

Consider a simply supported perfect column of length L and made of a linearly viscoelastic material. The column is at rest for time, $t < 0$. A constant axial load P, is applied suddenly at $t = 0$ such that $P = P \cdot \Pi(t)$ where $\Pi(t)$ is the unit step function. Simultaneously, the column is disturbed laterally, imparting to it a deflection, $w = w(x,t)$ (see Figure 3). Thus the difference between the undisturbed and the disturbed configurations is defined entirely by the function $w(x,t)$.

3.1.1 GOVERNING EQUATIONS

This dynamic column problem is defined by the following set of equations:

1. Equation of Motion

$$\frac{\partial^2 M}{\partial x^2} - P\frac{\partial^2 w}{\partial x^2} = m\frac{\partial^2 w}{\partial t^2} \qquad (3.1a)$$

where

$$M(x,t) = \int_A \sigma(x,y,t)y\,dA$$
$$P = \int_A \sigma(x,y,t)\,dA \qquad (3.1b)$$

and where m, σ and y denote mass per unit length, axial stress and a centroidal coordinate, respectively.

2. Linear Constitutive Law

$$\epsilon(x,y,t) = \frac{1}{E_o}\left[\sigma(x,y,t) + \int_o^t j(t-\tau)\sigma(x,y,\tau)d\tau\right] \tag{3.2}$$

3. Geometric Relations (Bernoulli-Euler Hypothesis)

$$\epsilon(x,y,t) = \epsilon_o(x,t) + y\varkappa(x,t) \cong \epsilon_o(x,t) - y\frac{\partial^2 w(x,t)}{\partial x^2} \tag{3.3}$$

Combining the above relations one obtains the equation of motion for the distubted structure in the form

$$E_oI\frac{\partial^4 w}{\partial x^4} + P\frac{\partial^2 w}{\partial x^2} + m\frac{\partial^2 w}{\partial t^2} + j(t) * \left(P\frac{\partial^2 w}{\partial x^2} + m\frac{\partial^2 w}{\partial t^2}\right) = 0 \tag{3.4}$$

where I denotes the centroidal moment of inertia of the cross-section and the (*) indicates convolution, i.e.

$$j(t) * f(\mathrm{x},\mathrm{t}) = \int_o^t j(t-\tau)f(x,t)d\tau \tag{3.4a}$$

Assuming the solution to Equation (3.4) in the form

$$w(x,t) = \sum_{n=1}^{\infty} F_n(t)\sin\frac{n\pi x}{L} \tag{3.5}$$

one arrives at

$$\frac{d^2F_n}{dt^2} + \omega_n^2\left(1 - \frac{P}{P_n}\right)F_n + j(t) * \left(\frac{d^2F_n}{dt^2} - \omega_n^2\frac{P}{P_n}F_n\right) = 0 \tag{3.6}$$

subject to the initial conditions

$$F_n(0) = \overset{\circ}{F}_n \tag{3.7a}$$

$$\left.\frac{dF_n}{dt}\right|_{t=0} = \overset{\circ}{\mathrm{v}}_n \tag{3.7b}$$

where

$$P_n = n^2\frac{\pi^2E_oI}{L^2} = n^2P_E$$

$$\omega_n^2 = n^4\frac{\pi^4E_oI}{L^4m} = n^4\omega_o^2$$

$$n = 1,2,\ldots\infty$$

and where P_E and ω_o denote the Euler critical load and the fundamental natural frequency of the column, respectively.

Applying Laplace transform to Equation (3.6), one obtains

$$\check{F}_n(s) = \frac{s\overset{\circ}{F}_n + \overset{\circ}{\mathrm{v}}_n}{s^2 + \omega_n^2\left[\frac{1}{1+j(s)} - p_n\right]}; \quad p_n = P/P_n \tag{3.8}$$

where $\check{F}_n(s)$ denotes the Laplace transform of the deflection parameter, which cannot be obtained without defining the creep behaviour of the material, i.e. defining the function $j(s)$.

3.1.2 SOLUTION FOR THE THREE-ELEMENT MATERIAL MODEL

The three-element model shown in Figure 4 is the simplest spring-dashpot model that can simulate the behaviour of linearly viscoelastic materials of the "solid" type with limited creep deformations when E_1 is nonzero, and of the "fluid" type with unlimited viscous deformations for $E_1 = 0$. Such a simple model for a viscoelastic material is adopted sometimes to estimate the behaviour of structures made of some plastics or organic materials and even concrete and other common structural materials in cases of low stress levels. Clearly, such a simple model requires only two viscous constants, E_1 and ν_1, to be determined exprimentally.

The kernel appearing in Equation (3.2) as well as its Laplace transform for this simple model can be written as

$$j(t) = \frac{E_o}{\nu_1}e^{-(E_1/\nu_1)t} \tag{3.9a}$$

$$j(s) = \frac{\lambda - \mu}{s + \mu} \tag{3.9b}$$

where, after [13], the coefficients λ and μ are given by

$$\lambda = (E_o + E_1)/\nu_1 \tag{3.10a}$$

$$\mu = E_1/\nu_1 \tag{3.10b}$$

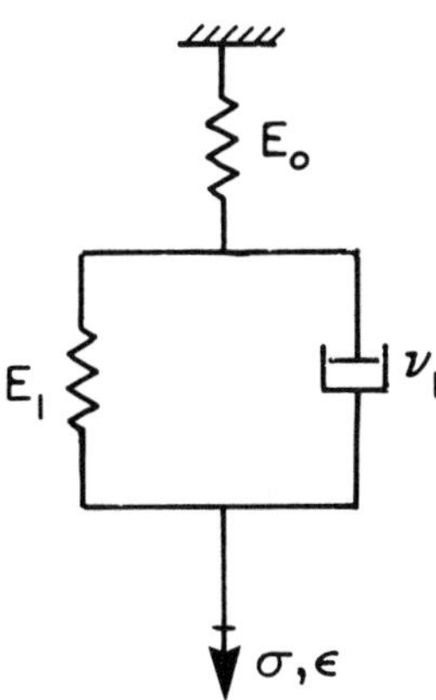

FIGURE 4. The three-element viscoelastic model.

Using Equation (3.9) in (3.8), one obtains

$$\check{F}_n(s) = \frac{(s\mathring{F}_n + \mathring{v}_n)(s + \lambda)}{s^3 + \lambda s^2 + \omega_n^2(1 - p_n)s - \omega_n^2\lambda(p_n - \mu/\lambda)} \tag{3.11}$$

For $p_n < 1$, i.e. $P < P_n$, the denominator of Equation (3.11) has one real root, S_n, and two complex roots, $-\alpha_n \pm i\beta_n$. The denominator can thus be expressed as

$$s^3 + \lambda s^2 + \omega_n^2(1 - p_n)s - \omega_n^2\lambda(p_n - \mu/\lambda) =$$

$$= (s - S_n)[s - (\alpha_n + i\beta_n)][s + (\alpha_n - i\beta_n)] \tag{3.12}$$

where

$$S_n = r_1 + r_2 - \lambda/3 \tag{3.13a}$$

$$\alpha_n = \frac{1}{2}(r_1 + r_2) + \lambda/3 \tag{3.13b}$$

$$\beta_n = (\sqrt{3}/2)(r_1 - r_2) \tag{3.13c}$$

and where

$$r_1 = [-r_3 + \sqrt{r_4}]^{1/3}$$

$$r_2 = [-r_3 - \sqrt{r_4}]^{1/3}$$

$$r_4 = (r_3)^2 + (r_5)^3$$

$$r_3 = \frac{1}{2}[(2\lambda^3)/27 - \lambda\omega_n^2(1 - p_n)/3 - \omega_n^2\lambda(p_n - \mu/\lambda)]$$

$$r_5 = [3\omega_n^2(1 - p_n) - \lambda^2]/9 = \omega_n^2(1 - p_n)/3 - \lambda^2/9$$

To draw general conclusions concerning the nature of the solution as a function of these roots is not easy. One might prefer to proceed by expanding S_n, α_n and β_n into a power series in terms of the "small" dimensionless parameter

$$\delta_n = \lambda/\omega_n = T_n/2\pi t_r \tag{3.13d}$$

where T_n is the period of elastic "free vibration," for the nth mode and t_r denotes the relaxation period for the material given by $t_r = \nu_1/(E_o + E_1)$. For some common structural materials, δ_n is very small, as for example for concrete $\delta_n \cong 1.2 \times 10^{-8}/n^2$ [13]. The power series expressions can therefore be written as

$$S_n = \lambda\frac{p_n - p_v}{(1 - p_n)}\left[1 - \frac{(p_n - p_v)(1 - p_v)}{(1 - p_n)}\delta_n^2 + \mathcal{O}(\delta_n^4)\right] \tag{3.14a}$$

$$\alpha_n = \frac{\lambda}{2}\frac{(1 - p_v)}{(1 - p_n)}\left[1 + \frac{(p_n - p_v)^2}{(1 - p_n)^3}\delta_n^2 + \mathcal{O}(\delta_n^4)\right] \tag{3.14b}$$

$$\beta_n = \sqrt{1 - p_n}\,\omega_n\left\{1 + \frac{1}{2}\frac{(1 - p_v)}{(1 - p_n)^3}\left[p_n - \frac{1}{4}(3p_v + 1)\right]\delta_n^2 + \mathcal{O}(\delta_n^4)\right\} \tag{3.14c}$$

where

$$p_v = \mu/\lambda = \frac{1}{1 + (E_o/E_1)} \tag{3.14d}$$

In view of the smallness of the parameter δ_n, values of S_n, α_n and β_n can be obtained with sufficient accuracy from the first term in Equations (3.14), neglecting all terms involving powers of this variable.

Equation (3.11) can be rewritten as

$$\check{F}_n(s) = \frac{A_n}{s - S_n} + \frac{B_n s + C_n}{(s + \alpha_n)^2 + \beta_n^2} \tag{3.15}$$

where

$$A_n = (\mathring{v}_n + S_n\mathring{F}_n)\left[\frac{2\alpha_n}{\alpha_n^2 + \beta_n^2 + S_n(-2\alpha_n + S_n)}\right]$$

$$\cong \delta_n\left[\frac{\mathring{v}_n + S_n\mathring{F}_n}{\omega_n}\right]\frac{(1 - p_v)}{(1 - p_n)^2} \tag{3.16a}$$

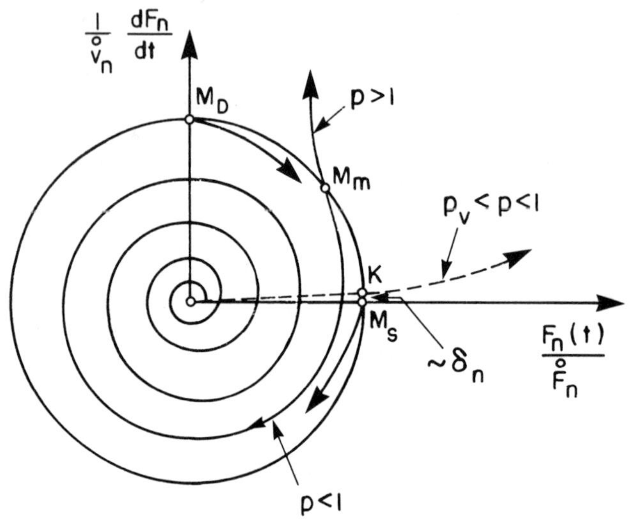

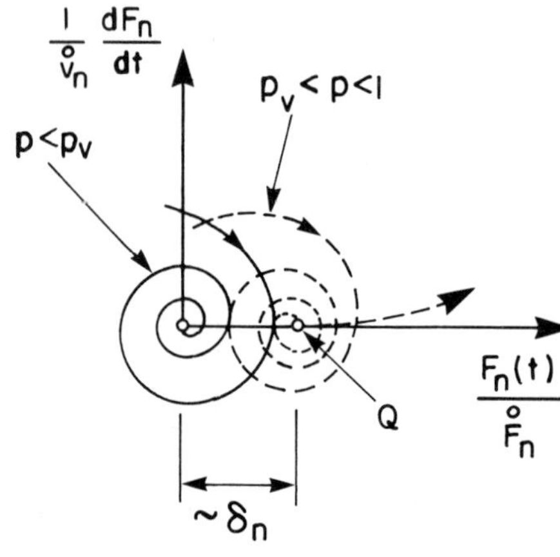

(a) **Overall Phase-Space Plot** (b) **Details of Diagram Near Origin**

FIGURE 5. Phase-space diagram for viscoelastic perfect column (from [63]).

$$B_n = \overset{\circ}{F}_n - A_n \tag{3.16b}$$

$$C_n = A_n \frac{\alpha_n^2 + \beta_n^2}{S_n} - \lambda \frac{\overset{\circ}{v}_n}{S_n} \tag{3.16c}$$

Applying the inverse Laplace transformation to Equation (3.15), one obtains

$$F_n(t) = A_n e^{\bar{s}_n \bar{t}} + e^{-\bar{\alpha}_n \bar{t}} [(\overset{\circ}{F}_n - A_n) \cos \bar{\beta}_n \bar{t} + D_n \sin \bar{\beta}_n \bar{t}] \tag{3.17}$$

where

$$D_n = \frac{\overset{\circ}{v}_n/\lambda + \overset{\circ}{F}_n \bar{\alpha}_n + A_n(\bar{s}_n - \bar{\alpha}_n)}{\bar{\beta}_n}$$

and where the following dimensionless quantities have been used

$$\bar{s}_n = S_n/\lambda = \frac{p_n - p_v}{(1 - p_n)} [1 - \ldots] \tag{3.18a}$$

$$\bar{\alpha}_n = \alpha_n/\lambda = \frac{1}{2} \frac{(1 - p_v)}{(1 - p_n)} [1 + \ldots] \tag{3.18b}$$

$$\bar{\beta}_n = \beta_n/\lambda = \frac{\sqrt{1 - p_n}}{\delta_n} [1 + \ldots] \tag{3.18c}$$

$$\bar{t} = \lambda t \tag{3.18d}$$

Note that $\bar{s}_n << \bar{\beta}_n$, $\bar{\alpha}_n << \bar{\beta}_n$, $A_n << \overset{\circ}{F}_n$ and $A_n << D_n$.

3.1.3 DISCUSSION OF STABILITY

The response of a viscoelastic column to disturbances, represented by the function $F_n(t)$, depends on the values of the parameters $\bar{s}_n$ and $\bar{\alpha}_n$. From a design point of view, the fundamental mode and the associated critical load, P_E, is clearly the most significant. Denote the dimensionless variable, p_n, for $n = 1$, by p.

For $p > 1$, i.e. $P > P_E$, all three roots of the denominator of Equation (3.11) are real and the solution consists of three exponential functions yielding a deflection function, $F_1(t)$, which is clearly monotonically increasing with time indicating instability for the column. Let us refer to this type of instability as "dynamic instability."

For $p < p_v$, i.e. $P < P_E/(1 + E_o/E_1)$, $\bar{s}_1 < 0$ and $\bar{\alpha}_1 > 0$ and therefore $F_1(t) \to 0$ as $t \to \infty$, irrespective of the type of disturbance. Thus, the column is stable and the load value, $P_v = P_E/(1 + E_o/E_1)$ will be called the "safe load limit" [72] or the "viscoelastic critical force."

If $p_v < p < 1$, i.e. $P_v < P < P_E$, $\bar{s}_1 > 0$ and $\bar{\alpha}_1 > 0$ and $F_1(t) \to \infty$ for $t \to \infty$, as in the dynamic instability case. However, since, in general, $|\bar{s}_n| < |\bar{\alpha}_n|$ and $|A_n| << |\overset{\circ}{F}_n|$ and $|A_n| << |D_n|$, the second term in Equation (3.17), representing the damped free vibration of the disturbed viscoelastic column, will decay much more rapidly than the first term in that equation will grow with time. In fact, the increase in the first term, owing to the smallness of A_n, will be very slow, at least initially. In time, this term, which contains the small but significant "memorized" portion of the initial disturbance, will grow and will ultimately be responsible for the failure of the structure. Thus, although the column is unstable in a dynamic sense, even for this load range it may fulfill a useful function during a limited time period after loading, a period which will be called the "safe service period." This type of instability will be referred to as "viscoelastic instability."

The phase plane diagram shown in Figure 5 indicates the three types of behaviour discussed above. In Figure 5a, M_s, M_D and M_m denote static, dynamic and mixed initial disturbances, respectively. Clearly, for $p > 1$, the response to any disturbance diverges dynamically. For $p < 1$, the response decays to a point close to the origin, the details of which are indicated on Figure 5b. This enlargement of the area in the vicinity of the origin indicates that for $p < p_v$, the response, indeed, decays to zero, while for $p_v < p < 1$ the response decreases in time to some small deflection of the order of magnitude δ_n, which represents the memorised portion of the initial disturbance and which slowly, but steadily, grows in time and ultimately is responsible for the instability of the structure. As is clear from Equation (3.16a), A_n and therefore the time taken for the deflection to grow beyond the magnitude of the initial disturbance are functions of the type and magnitude of the disturbance. For $p = p_v$, the deflections tend to a limit determined by the magnitude of A_n.

Note that the critical loads P_E and P_v have been determined only in terms of the "elastic" properties of the material; the Euler buckling force corresponds to the initial instantaneous elasticity ($t = 0$) of the model, when the viscous mechanism is still inactive. The force P_v, however, is related to the final elasticity ($t \rightarrow \infty$), when the dashpot does not provide any more resistance. For models of the fluid type for which $E_1 = 0$, the stiffness tends to zero as $t \rightarrow \infty$, and therefore $P_v = 0$ for columns made of such materials.

3.1.4 ANALYSIS OF VISCOELASTICALLY UNSTABLE COLUMNS

Recall that we define viscoelastic instability to be associated with the load levels $P_v < P < P_E$. Under such loads, the column may fulfill a useful role, as noted earlier. First, investigate the behaviour of such a column when subjected to a purely static disturbance, i.e. $\mathring{v}_n = 0$ and $\mathring{F}_n \neq 0$ (see point M_s in Figure 5a). The deflection function given by Equation (3.17), for paramater values used in [13] (e.g. $\lambda = 2\mu = 5.28 \times 10^{-7}$/sec, $\omega_1 = 48$ rad/sec, $\delta_1 = 1.206 \times 10^{-8}$, $p_v = 0.5$ and for p values of 0.6 and 0.8) is sketched in Figure 6.

Clearly, the vibrational part of the motion is damped out after $\bar{t} \gtrsim 5$, while the memorised disturbances become significant only after $\bar{t} \gtrsim 20$ for $p = 0.8$ and after $\bar{t} \gtrsim 150$ for $p = 0.6$. There is obviously quite a long intermediate period, when the response amplitude of the structure is extremely small, $F_1/\mathring{F}_1 = 10^{-7}$, a consequence of the very small value of the memorised portion of the disturbance that is proportional to δ_n^2 for this case.

Next, analyse the response of the structure when subjected to a dynamic disturbance (point M_D in Figure 5a), i.e. $\mathring{F}_n = 0$ and $\mathring{v}_n \neq 0$. To evaluate the response and to compare the effects of the two kinds of disturbances, let us assume that the energy of the disturbance is the same for both cases; i.e. for the dynamic disturbance, its kinetic energy is equal to the change in potential energy for the static disturbance. After some algebra [63] one arrives at

$$\mathring{v}_n = \mathring{F}_n \, \omega_n \sqrt{1 - p_n} = \mathring{F}_n \beta_n \tag{3.19}$$

which permits quantitative comparison of the column response to static and dynamic disturbances. For data analogous to that used above, the deflection function for the dynamic disturbance case is shown in Figure 7.

The overall character of the diagrams shown in Figures 6

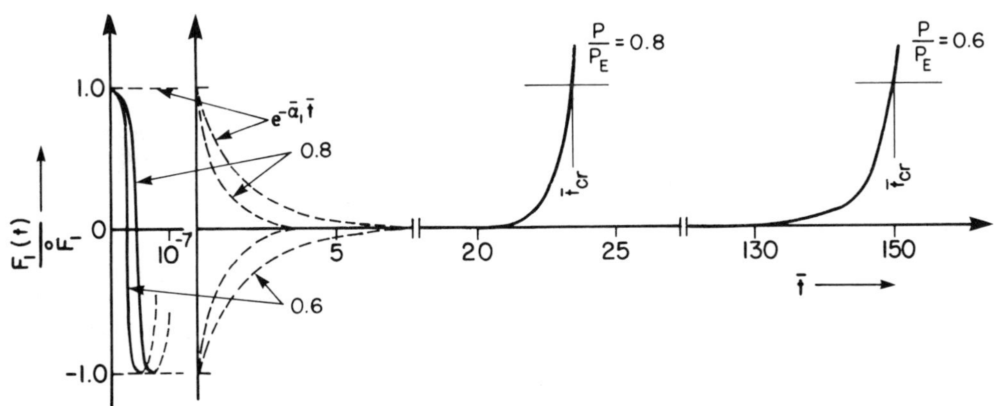

FIGURE 6. Deflection history for column subjected to "static" disturbances (from [63]).

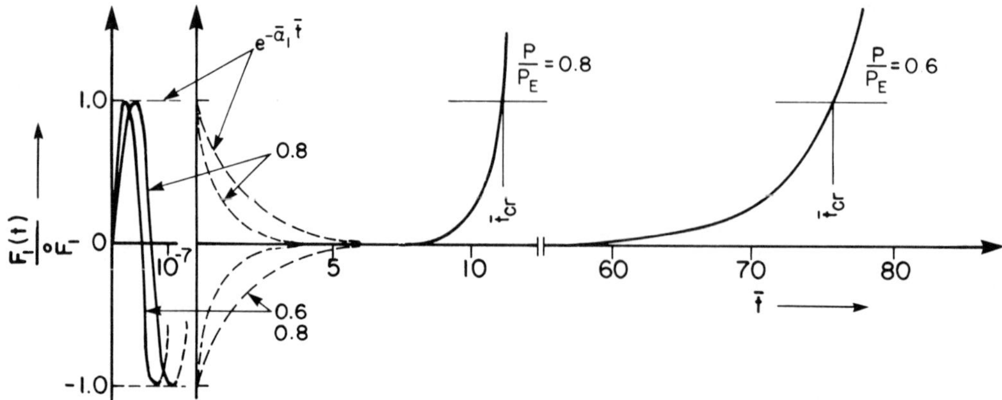

FIGURE 7. Deflection history for column subjected to "dynamic" disturbances (from [63]).

and 7 is similar. Clearly, the "intermediate period" of the dynamic disturbance case is considerably shorter than that for the static disturbance case.

On the basis of Figures 6 and 7, the response of such a structure can be divided into three main parts: (i) the damped vibration, (ii) the intermediate period, with the response amplitude $\cong 0$; and (iii) the period of rapid growth in amplitude. This type of behaviour, as noted already, is characteristic only for the structure with a load range $P_v < P < P_E$.

Having determined the main features of the behaviour of such columns, one can define a safe service period. For example, one might define such a period as the time required for the response amplitude to become equal to the magnitude of the initial disturbance (point K in Figure 5). Such a definition for the safe service period is independent of the value of the initial disturbance, whether it is static or dynamic (see Figures 6 and 7). In [63] this period was referred to as "critical time." In general, such a critical time may be found from the relation

$$\sum_n A_n e^{\bar{s}_n \bar{t}_{cr}} \cong \sum_n \sqrt{(\mathring{F}_n - A_n)^2 + D_n^2} \tag{3.20}$$

Because values of successive coefficients in this series are smaller by the factor n^{-2}, it is sufficient for practical purposes to take into account only the first term from which expressions for the critical time can be determined explicitly. Figure 8 shows the variation of critical time as a function of the axial load for both static and dynamic disturbances. From an analysis of these results (Figures 6 and 7) one concludes that the ratio of critical times for the two cases is given as

$$\frac{\bar{t}^s_{cr}}{\bar{t}^d_{cr}} \cong 2 \tag{3.21}$$

For details the reader is referred to [63].

3.2 The Imperfect Column

From a theoretical viewpoint, the stability analysis of perfect structures is interesting and important. However, in practice there are no perfect structures. Columns may have geometric imperfections or the loads may be applied imperfectly, thus resulting in a problem which is not an eigenvalue or stability problem, but rather, a load-deflection, or in the case of viscoelastic structures, a load-deflection-time problem. The question arises as to what useful information, if any, one can obtain concerning the load deflection and stability behaviour of imperfect structures from stability analyses of perfect structures. Alternatively, we are interested in what predictions we can obtain concerning the stability behaviour of perfect columns by analyzing the behaviour of real imperfect structures.

As is well known from analyses of perfect and imperfect Hookean structures, there is a close parallel between the linear stability problem of perfect columns and the corresponding load deflection behaviour of imperfect structures. One can, for example, determine the eigenvalue of a perfect column by observing the load defelction behaviour of an imperfect column and noting that the lateral deflections increase very rapidly as the axial load approaches the critical load. Also, the value of the critical load obtained from such an imperfection analysis is independent of the type of imperfection existing in the structure or in the loading.

Having established the stability behaviour of perfect viscoelastic columns using a dynamic approach, one of the purposes of this section is to compare the behaviour of such perfect structures with the behaviour of corresponding imperfect viscoelastic columns by means of a dynamic analysis. In order to keep the analysis relatively simple and tractable, the same material model, namely the three element model, is also used in this analysis.

The dynamic equilibrium equation for the column, Equation (3.4), is modified to take into account initial imperfec-

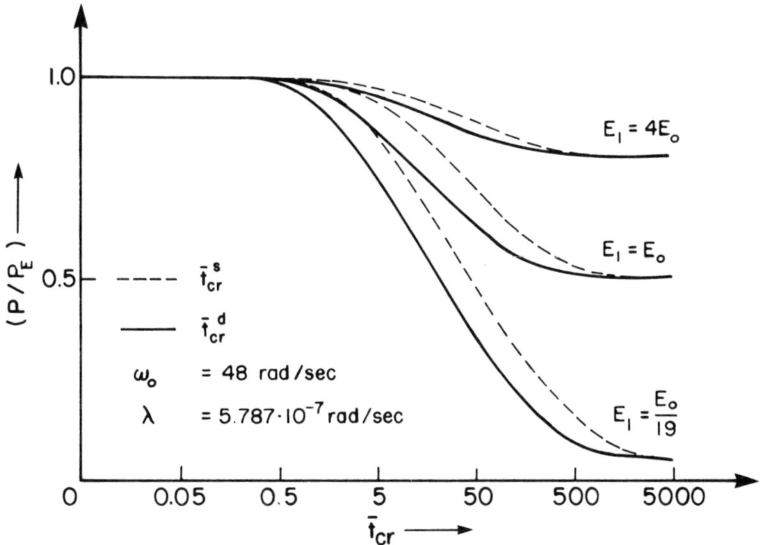

FIGURE 8. Effect of axial load on "critical time" for various model stiffnesses (from [63]).

tions, $w_o(x)$ (see Figure 9), and is consequently rewritten as

$$E_oI\frac{\partial^4 w}{\partial x^4} + P\left[\frac{\partial^2 w}{\partial x^2} + \frac{\partial^2 w_o}{\partial x^2} + j(t) * \left(\frac{\partial^2 w}{\partial x^2} + \frac{\partial^2 w_o}{\partial x^2}\right)\right] + m\left[\frac{\partial^2 w}{\partial t^2} + j(t) * \frac{\partial^2 w}{\partial t^2}\right] = 0 \qquad (3.22)$$

where $w(x,t)$ denotes the additional lateral deflection from the initial imperfect shape. The load P is again assumed to be applied suddenly at $t = 0$ such that $P = P \cdot \Pi(t)$. However, as opposed to the perfect column, no additional lateral disturbance needs to be applied to the imperfect structure at $t = 0$.

Assume the solution to Equation (3.22) to be identical in form to that used for Equation (3.4), as given by Equation (3.5). In addition, express $w_o(x)$ in a Fourier series in the form

$$w_o(x) = \sum_{n=0}^{\infty} \mathring{w}_n \sin\frac{n\pi x}{L} \qquad (3.23)$$

which when used in Equation (3.22), together with the assumed solution, leads to

$$\frac{d^2F_n}{dt^2} + F_n\omega_n^2(1 - p_n) - p_n\omega_n^2\mathring{w}_n + j(t) * \left[\frac{d^2F_n}{dt^2} - p_nF_n - p_n\omega_n^2\mathring{w}_n\right] = 0 \qquad (3.24)$$

subject to the initial conditions

$$F_n(0) = 0; \left.\frac{dF_n}{dt}\right|_{t=0} = 0 \qquad (3.25)$$

Using Laplace transformation one obtains

$$\check{F}_n(s) = \mathring{w}_n p_n \omega_n^2 \frac{1}{s}\frac{1}{\{s^2 - p_n\omega_n^2 + \omega_n^2/[1 + j(s)]\}} \qquad (3.26)$$

which can be solved after $j(t)$ has been specified. For the three-element-material model introduced above, $j(t)$ and $j(s)$ are already defined by Equations (3.9) and conse-

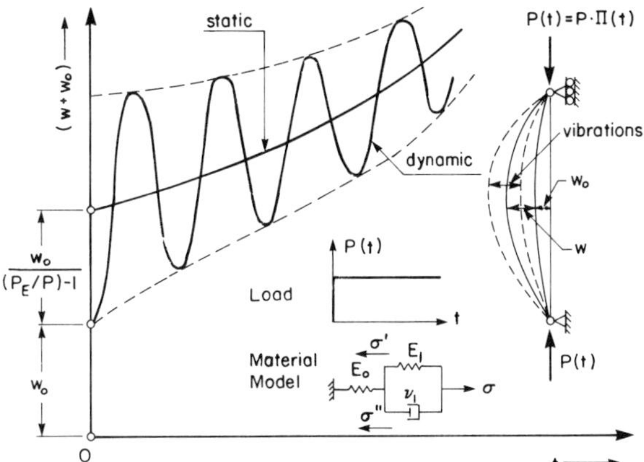

FIGURE 9. Static and dynamic response of an imperfect viscoelastic column to axial load (from [65]).

quently, Equation (3.26) for this specific material model becomes

$$\check{F}_n(s) = \overset{\circ}{w}_n p_n \omega_n^2 \frac{1}{s} \frac{s + \lambda}{[s^3 + \lambda s^2 + \omega_n^2(1 - p_n)s - \omega_n^2\lambda(p_n - p_v)]} \tag{3.27}$$

The cubic portion of the denominator within square brackets is identical to that appearing in Equation (3.11) and has been analysed in detail above. For the case of $p_n < 1$, as we saw above, the cubic portion of the denominator has only one real root, the other two being complex conjugates. Therefore, Equation (3.27) can be expressed in the form

$$\check{F}_n(s) = \overset{\circ}{w}_n p_n \omega_n^2 \frac{1}{s} \frac{s + \lambda}{(s - S_n)[(s + \alpha_n)^2 + \beta_n^2]} \tag{3.28}$$

where S_n, α_n and β_n are defined by Equations (3.14) and δ_n is given by Equation (3.13d). Applying inverse Laplace transform to Equation (3.28) leads to

$$F_n(t) = \frac{\overset{\circ}{w}_n p_n}{1-p_n} \left[\frac{(e^{\bar{s}_n \bar{t}} - 1)}{\bar{s}_n} + e^{\bar{s}_n \bar{t}} - e^{-\bar{\alpha}_n \bar{t}} \cdot \right.$$

$$\left. \cdot \left(\cos \bar{\beta}_n \bar{t} + \frac{3\bar{\alpha}_n}{\bar{\beta}_n} \sin \bar{\beta}_n \bar{t} \right) \right] \tag{3.29}$$

in which $\bar{s}_n$, $\bar{\alpha}_n$, $\bar{\beta}_n$ and $\bar{t}$ are nondimensional variables defined in Equation (3.18) and in which the last term clearly represents damped vibration about the average deflection, indicated by the first two terms as also being time-dependent. For $P < P_E$, or $p < 1$, the vibration is always damped and therefore the vibrational portion of the response for such a load range is of no further interest in this discussion on stability. The non-vibrational portion of the response depends on the value of $\bar{s}_n$, it being limited in time only for $\bar{s}_n < 0$. For $\bar{s}_n > 0$, the deflection function, $F_n(t) \to \infty$ as $t \to \infty$. For $\bar{s}_n = 0$, Equation (3.29) becomes

$$F_n(t) = \frac{\overset{\circ}{w}_n p_n}{1-p_n} \left[1 + \bar{t} - e^{-\bar{\alpha}_n \bar{t}} \cdot \left(\cos \bar{\beta}_n \bar{t} + \frac{3\bar{\alpha}_n}{\bar{\beta}_n} \sin \bar{\beta}_n \bar{t} \right) \right] \tag{3.30}$$

which, together with Equation (3.29) are plotted on Figure 10. Note that for $p_n < p_v$, the growth in initial imperfection is bounded by the value

$$\lim_{t \to \infty} F_n(t) = \frac{\overset{\circ}{w}_n}{(p_v/p_n) - 1} \tag{3.31}$$

If $p_v < p_n < 1$ the response is unbounded as $t \to \infty$, signifying instability. Thus the results from this imperfection analysis confirm the main features of column behaviour obtained from the analysis of the perfect structure. The safe-load limit, P_v, again emerges as the critical load value below which the response of the structure is bounded as $t \to \infty$,

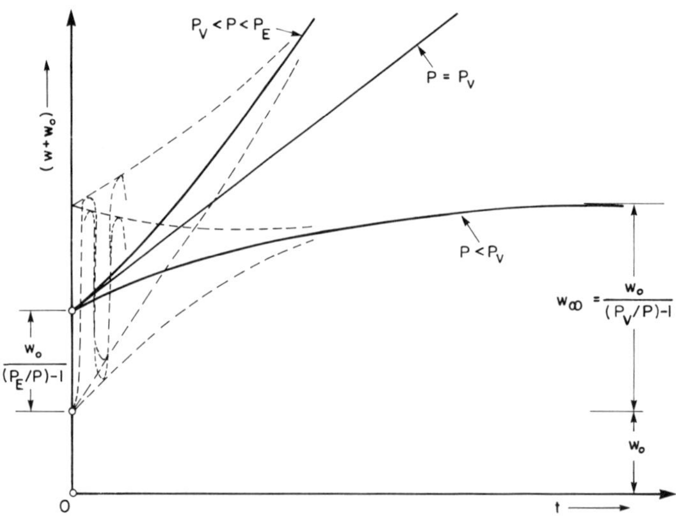

FIGURE 10. Deflection history of an imperfect viscoelastic column for various axial load ranges (from [65]).

indicating absolute stability. Also, for $P > P_v$, the column is clearly unstable in a viscoelastic sense with deflections increasing monotonically with time. A difference in behaviour between the perfect and imperfect structure is indicated for the load value $P = P_v$ for which the imperfect structure shows unbounded response as $t \to \infty$ while the perfect column was shown to be stable at this load level.

3.2.1 IMPERFECTION AS PERMANENT DISTURBANCE

The dynamic analyses for the perfect and imperfect columns gave similar results for the safe-load-limit and the overall response of the structure as a function of the load range. In this section we focus on the viscoelastically unstable domain ($P > P_v$) and compare the time periods obtained from the two analyses during which the structure may carry a load safely. In order to carry out a quantitative comparison, rewrite Equation (3.29) in the form

$$F_n(t) = \frac{\mathring{w}_n p_n}{1 - p_n} \frac{2\bar{\alpha}_n}{\bar{s}_n} e^{\bar{s}_n \bar{t}} \cdot \left(1 - \frac{e^{-\bar{s}_n \bar{t}}}{2\bar{\alpha}_n} + \Omega(t) \right) \qquad (3.32)$$

where $\Omega(t)$ denotes the damped vibrational portion of the response. The perfect column was defined to be viscoelastically unstable for $p_v < p < 1$, such instability being due to the "memorised" portion of the "disturbance." The deflection-time function for the perfect column, Equation (3.17), is rewritten as

$$F_n(t) = A_n e^{\bar{s}_n \bar{t}} + \tilde{\Omega}(t) \qquad (3.33a)$$

where $\tilde{\Omega}(t)$ represents the damped vibrational portion of the perfect column's response and where A_n depends on the type of disturbance and is given by

$$A_n = \delta_n \frac{2\bar{\alpha}_n}{\bar{s}_n} \left[\frac{\mathring{v}_n}{\omega_n} + \mathring{F}_n \delta_n \frac{p_n - p_v}{1 - p_n} \right] \cdot \frac{p_n - p_v}{(1 - p_n)^2} \qquad (3.33b)$$

and where $\mathring{v}_n$ and $\mathring{F}_n$ denote the amplitude of the velocity and displacement disturbances, respectively, imposed on the perfect structure at $t = 0$. A comparison between Equations (3.32) and (3.33) for $\bar{s}_n > 0$ leads to some interesting conclusions. First, for sufficiently large t, Equation (3.32) may be rewritten approximately as

$$F_n(t) \cong \frac{\mathring{w}_n p_n}{(1 - p_n)} \frac{2\bar{\alpha}_n}{\bar{s}_n} e^{\bar{s}_n \bar{t}} \qquad (3.34)$$

which, when compared with Equation (3.33) indicates the relation between A_n and the initial imperfection in the form

$$A_n = \frac{\mathring{w}_n p_n}{(1 - p_n)} \frac{2\bar{\alpha}_n}{\bar{s}_n}$$

from which, using results presented above, one obtains

$$\mathring{w}_n = \delta_n \left[\frac{\mathring{v}_n}{\omega_n} + \mathring{F}_n \delta_n \frac{p_n - p_v}{(1 - p_n)} \right] \frac{p_n - p_v}{p_n(1 - p_n)}$$

From Equation (3.35) one concludes that in order to get a similar response in an imperfect column as in the corresponding perfect column disturbed by dynamic "short-lasting" disturbances, the velocity disturbance of maximum amplitude $\mathring{v}_n/\omega_n$ has to be replaced by a "permanent disturbance" or imperfection, $\mathring{w}_n$, of the order of magnitude $\mathring{w}_n = \mathcal{O}(\delta_n \mathring{v}_n/\omega_n)$. In the case of a displacement-type disturbance, the corresponding imperfection, $\mathring{w}_n$, must be $\mathring{w}_n = \mathcal{O}(\delta_n^2 \mathring{F}_n)$. Since δ_n is a very small number, the magnitude of initial imperfections required or allowed for such analogous behaviour are extremely small, much below practical tolerances acceptable in ordinary manufacturing or construction processes. Therefore, the "critical times" defined and obtained above on the basis of an exact dynamic analysis of the perfect column are only of limited and primarily theoretical value. If a safe service period has to be defined for the viscoelastically unstable column and if this time period is to be of some practical value, it should be based on the results obtained for the imperfect structure.

The definition of a safe service period, introduced above for the perfect column subjected to an axial load $p_v < p < 1.0$, is, generally, not applicable to the corresponding imperfect structure, since for the latter the deflection is a monotonically increasing function of time. To indicate the sensitivity of such structures to imperfections, Figure 11 shows a few curves representing "constant amplification" for a column made of concrete with material properties taken from [13]. Thus, for example, if the axial load is $p = P/P_E = 0.7$, any "permanent disturbance" (initial imperfection) will be amplified 35 times after approximately 55 days (point A on Figure 11), and 350 times after some 124 days (point B on Figure 11). In contradistinction, if a "transitory imperfection" or disturbance is applied to an analogous perfect column, an "amplification" of unity (i.e., the deflection reaching a magnitude equal to the amplitude of the initial disturbance) is reached after 500 days in the case of a "dynamic" disturbance, and after 1100 days in the case of a "static" disturbance. A comparison of such numbers underlines the extreme sensitivity of the structure to initial imperfections, i.e., "permanent disturbances."

3.3 Discussion of Results From Dynamic Analysis

This rigorous dynamic analysis presented in the previous two subsections, permits determination of the "safe-load-limit," P_v, which separates the stable and unstable domains for viscoelastic columns. Note that identical results were obtained for P_v from the dynamic analysis of the perfect as well as the imperfect structure. However, in the case of the

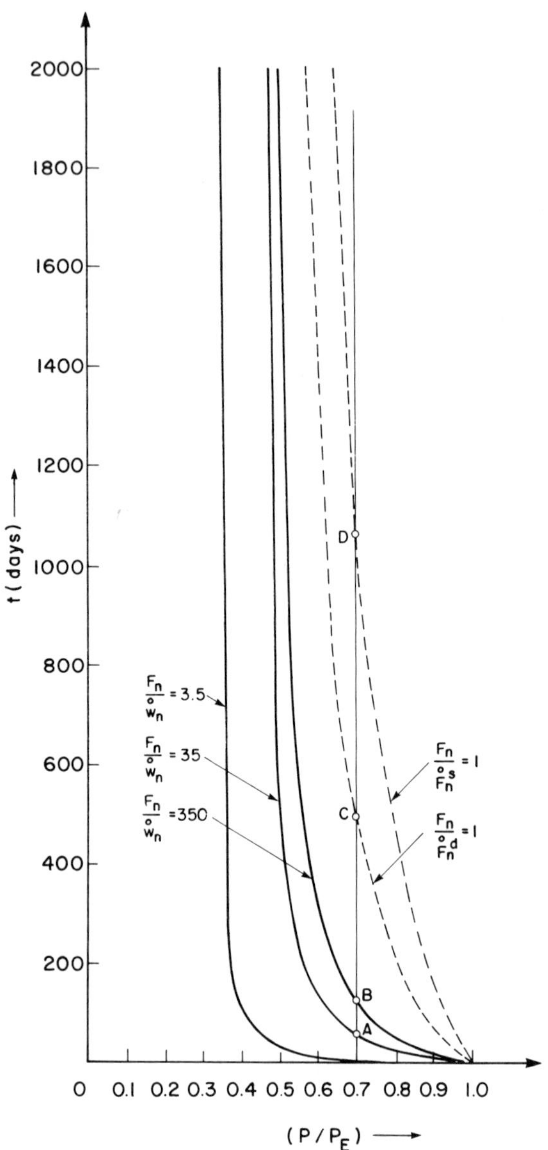

FIGURE 11. Deflection amplification in imperfect columns (from [65]).

imperfect column, the deformations occurring in the structure even under load levels less than P_v, may lead to excessive deflection/stress magnitudes which may be unacceptable.

In discussing instability of viscoelastic columns, one has to distinguish between two types of instabilities: firstly, the instantaneous or "dynamic" instability for load levels above P_E, a phenomenon which is a function only of the elastic properties of the material; and secondly, viscoelastic instability which is a phenomenon involving time during which there may be a useful "safe service period" within which the column can safely carry a given load. Considerations of such safe service periods leads to determination and definition of a "critical time," a terminology used in [63].

Results in this section indicate that for columns made of linear materials undergoing deflections and deformations sufficientlly small so as to justify a geometrically linear analysis, the instant at which the structure becomes unsafe, is not uniquely defined. The definition of "critical time" requires some additional arbitrary assumptions, as for example, an assumption with regard to the magnitude of the deflection. Unique "natural" definitions for the critical time can be introduced and are associated with geometrically and/or materially non-linear stability analysis of columns [29,66] (see Section 9) or with column behaviour/failure when plasticity and/or brittleness of the material is taken into account [70].

Results from the above dynamic analyses also indicate that in trying to estimate a safe service period, one should be careful in using the results from an analysis of an ideal structure. It was shown that the rate of deflection increase, which is monotonic and is inherently associated with viscoelastic instability, is extremely sensitive to initial imperfections. The deflection rate for any real structure with practically achievable tolerances is much higher than that for the corresponding perfect structure, and consequently the safe service period of the imperfect column is significantly shorter than that of the idealized perfect column. Since all real structures have some initial imperfections, a much more useful and from a design point of view, more reliable estimate for a safe service period can be determined from an analysis of the stability and time deflection behaviour of the imperfect structure, including its "actual" initial imperfections.

4. STATIC ANALYSIS OF LINEAR COLUMNS

Since viscous processes for most common structural materials are normally rather slow, the associated inertia forces are necessarily also small and can be neglected in the equations of dynamic equilibrium. The resulting quasi-static-analysis is, therefore, an approximate one, results from which must be evaluated carefully and cautiously and should be compared, where possible, with corresponding results from an exact dynamic analysis. However, because the omitted inertia terms are small, such an approximate analysis should yield a reasonably correct estimate of the behaviour of the viscoelastic structure. In addition, such a static analysis is normally less complex, mathematically, than the corresponding dynamic analysis and is, consequently, a preferred method of stability investigation.

There are, however, some difficulties with such an analysis which have been experienced by engineers analysing the stability of viscoelastic structures, difficulties which raised doubts about the applicability of the static aproach to this class of problems. For example, the concept of "adjacent equilibrium configuration," a necessary and fundamental notion in the static stability analysis of structures, has to be reconsidered and redefined. Clearly, even though viscoelastic processes may normally be very slow, by their very nature, they imply motion in a direction, once such motion or a tendency for motion has been initiated. Therefore, such processes justifiably raise doubts about the validity of assuming adjacent equilibrium configurations for a column subjected to a constant axial load. Also, a rigorous static formulation of the linearly viscoelastic column problem leads to a homogeneous integral equation of the Volterra type, which admits a non-trivial solution only for the critical Euler force, P_E, at time $t = 0$ [13]. Such a result, in turn, incorrectly implies that no other buckling loads, lower than P_E, exist for the structure, a result which is in contradiction to results obtained from experiments and from our dynamic analysis presented in the previous section. It was due to such contradictory results, obtained from traditional static analyses, that researchers looked for and resorted to intuitive assumptions and/or re-interpretations of the static stability approach.

In this section we show that no adjacent equilibrium configurations are possible for a linearly viscoelastic column subjected to a constant axial load, P, and for $t > 0$. We also show that adjacent deflected equilibrium configurations are possible if the axial load is decreased in time from its initial value, P_E, in a prescribed manner. Adjacent deflected shapes, which remain constant in time, are defined as adjacent equilibrium configurations. The viscoelastic critical force, P_v, emerges also from this static analysis as an important parameter in the stability behaviour of such columns.

4.1 The Imperfect Column

To indicate some peculiarities of the application of the static approach to stability problems of linearly viscoelastic columns, let us first consider the imperfect column for which an exact dynamic analysis was presented in the previous section. Omitting inertia terms in Equation (3.22), one obtains the static equilbirium equation for such imperfect columns in the form

$$E_oI\frac{\partial^4 w}{\partial x^4} + P\left[\frac{\partial^2 w}{\partial x^2} + \frac{\partial^2 w_o}{\partial x^2} + j(t) * \right.$$

$$\left. * \left(\frac{\partial^2 w}{\partial x^2} + \frac{\partial^2 w_o}{\partial x^2}\right)\right] = 0 \qquad (4.1)$$

which for a quasi-statically applied load P at $t = 0$ is subject to the initial condition

$$w(x,o) = \sum_{n=o}^{\infty} \mathring{w}_n \frac{p_n}{(1-p_n)} \sin\frac{n\pi x}{L} \qquad (4.2)$$

and where $\mathring{w}_n$ denote the Fourier coefficients of the initial deflection function. As before, assume the solution to Equation (4.1) in the form

$$w(x,t) = \sum_{n=o}^{\infty} F_n(t) \sin\frac{n\pi x}{L} \qquad (4.3)$$

which, when substituted into Equation (4.1) leads to

$$F_n(t) - p_n\,[F_n(t) + \mathring{w}_n + j(t) * (F_n(t) + \mathring{w}_n)] = 0 \qquad (4.4)$$

Applying Laplace transform to this result yields

$$\check{F}_n(s) = \mathring{w}_n \frac{p_n}{(1-p_n)} \frac{1 + j(s)}{\left[1 - j(s)\dfrac{p_n}{(1-p_n)}\right]} \cdot \frac{1}{s} \qquad (4.5)$$

which for the three-element-model material becomes

$$\check{F}_n(s) = \mathring{w}_n \frac{p_n}{(1-p_n)} \frac{s+\lambda}{s(s-s_n')} \qquad (4.6)$$

where

$$s_n' = \lambda\frac{p_n - p_v}{(1-p_n)}$$

Using the inverse transform on Equation (4.6) one obtains the Fourier coefficients of the deflection, w, in the form

$$\check{F}_n(t) = \mathring{w}_n \frac{p_n}{(1-p_n)}\left[\frac{\lambda}{s_n'}\left(e^{s_n' t} - 1\right) + e^{s_n' t}\right] \qquad (4.7)$$

a result which is very similar to Equation (3.29) except for the damped vibrational portion appearing in that expression and for the difference between S_n and s_n'. A comparison of these two parameters leads to

$$\frac{S_n}{s_n'} = 1 + \mathcal{O}(\delta_n^2) \qquad (4.8)$$

and since δ_n is a very small number, this expression shows that the static solution for the imperfect linearly viscoelastic column is, for all practical purposes, equivalent to the corresponding results obtained from a dynamic analysis. Consequently, all main features of the stability behaviour of such columns obtained by either method should be identical. In fact, the dynamic solution describes a damped vibration about the static equilibrium position defined by Equation (4.7) (see Figure 10).

4.2 The Perfect Column

The governing equation defining the static equilibrium of an "adjacent" configuration for the perfect column reads

$$E_oI\frac{\partial^4 w}{\partial x^4} + P\left[\frac{\partial^2 w}{\partial x^2} + j(t) * \frac{\partial^2 w}{\partial x^2}\right] = 0 \qquad (4.9)$$

The load P at which such adjacent equilibrium configurations are possible is referred to as the critical load. Assuming the solution to this equation in the form of Equation (4.3) one obtains

$$F_n(t)(1 - p_n) - p_n \int_o^t j(t - \tau)\, F_n(\tau)d\tau = 0 \qquad (4.10)$$

which is an integral equation of the Volterra type admitting the following possible solutions: (i) an indeterminate deflection, at $t = 0$, $F_n(0) \neq 0$ and for $p_n = 1$, which clearly represents the elastic bifurcation solution at the instant of load application; (ii) a trivial solution, $F_n(t) = 0$, for p_n = const and for $t > 0$, a solution which indicates the nonexistence of adjacent equilibrium configurations at any moment after the axial force P has been applied, irrespective of the magnitude of this load. At first sight these results would seem to suggest that the static approach cannot provide any useful information about the stability behaviour of viscoelastic columns and does not give any indication of the safe-load-limit for such structures. Upon reflection it becomes clear that such a first impression is misleading.

Let us first try to interpret properly the obtained results by examining the configurations indicated on Figure 12 for various load ranges and time domains. In Figure 12d, we show a load $P > P_E$ $(p > 1)$ applied to the column. Clearly, the response is "dynamic" instability and consequently there is no adjacent equilibrium configuration for $t \geq 0$. In other words, the bending stiffness of the column is too small for equilibrium in a deflected position. Similarly, for the configuration shown in Figure 12c, for which $P < P_E$ $(p < 1)$ an adjacent equilibrium position is ruled out for $t \geq 0$ because the bending stiffness was clearly too large at $t = 0$ and remained excessive simply because no deflected shape could persist which would have allowed viscous processes to take effect thereby reducing the bending stiffness. Figure 12a depicts a configuration at $t = 0$ and for $P = P_E$ $(p = 1)$ which clearly represents the classical elastic bifurcation solution admitting adjacent equilibrium configurations. However, deterioration in the bending stiffness due to the deflected shape will prevent this initial equilibrium configuration from persisting at any instant of time $t > 0$ (see Figure 12b). In summary, it is clear that in the static stability analysis of viscoelastic perfect columns, the time-dependent properties of the material affect only the axial stiffness, EA, of the structure while its bending rigidity, EI, remains unaltered because, in fact, no lasting bending effects exist in the column which could lead to an alteration of the original bending stiffness.

Let us therefore change our static approach to this stability analysis problem so as to admit lasting bending effects in the perfect column, an approach which, hopefully, will bring the results from this analysis into harmony with those obtained from a dynamic formulation. Instead of seeking the value of the axial load and the instant of time at which the straight column will admit deflected configurations under the action of such axial load only, let us ask what type of loading must be applied to the column to keep it in a controlled deflected configuration. Instability will then be associated with the loss of control of such a deflected shape. In this formulation we admit a bending moment and accompa-

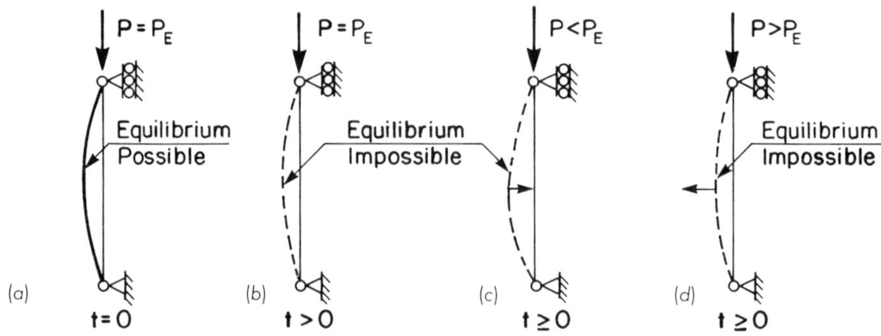

FIGURE 12. Adjacent equilibrium states for simply-supported viscoelastic column (from [64]).

nying deflections from the outset, which will allow the viscous process to affect the bending stiffness from the time of load application, i.e. from $t = 0$. There are two possible ways to achieve such controlled deflected configurations. Firstly, one can control the deflected shape of the column by varying the axial load magnitude with time, i.e., $P = P(t)$; alternatively, one can apply, in addition to a constant axial force P, some lateral loads, $q(t)$. To start with, let us examine the first possibility in detail.

The governing equation, Equation (4.9) now becomes

$$E_oI \frac{\partial^4 w}{\partial x^4} + P(t) \frac{\partial^2 w}{\partial x^2} + j(t) * \left[P(t) \cdot \frac{\partial^2 w}{\partial x^2} \right] = 0 \tag{4.11}$$

and assuming, again, the solution in the form of Equation (4.3), one obtains

$$F_n(t) - p_n(t) \cdot F_n(t) - j(t) * [p_n(t) \cdot F_n(t)] = 0 \tag{4.12}$$

in which $F_n(t)$ is assumed to be a known "control" function. Discussion of this solution for various control functions was presented in [64]. One simple solution is obtained by assuming $F_n(t) = \text{const}$, thereby keeping the deflected shape constant. For such a "controlled" configuration the axial force has to satisfy the following equation

$$1 - p_n(t) - j(t) * p_n(t) = 0 \tag{4.13}$$

from which one can specify the time variation of the axial force provided the creep function, $j(t)$ is given. One can keep the linear material specification quite general by introducing the relaxation function, $g(t)$, which appears in the constitutive law, for a given point, in the form

$$\sigma(t) = E_o \left[\epsilon(t) + \int_o^t g(t - \tau) \cdot \epsilon(\tau)\, d\tau \right] \tag{4.14}$$

and, together with $j(t)$ must satisfy the relation

$$j(t) + g(t) + j(t) * g(t) = 0 \tag{4.15}$$

Calculating $j(t)$ from this last relation and using it in Equation (4.13) one arrives at

$$p_n(t) = 1 + G(t) \tag{4.16}$$

where

$$G(t) = \int_o^t g(\tau) d\tau$$

For the three-element-model material

$$j(t) = (\lambda - \mu)e^{-\mu t} \text{ and } g(t) = -(\lambda - \mu)e^{-\lambda t}$$

which when used in Equation (4.13) or (4.16) leads to

$$p_n(t) = p_v + (1 - p_v)e^{-\lambda t} \tag{4.17}$$

This latter equation indicates that for $t = 0$ $p_n(t) = 1$ which is the familiar elastic bifurcation solution. For $t \to \infty$, $p_n(t) \to p_v$, and we recover the safe-load-limit determined earlier; i.e., for the axial force approaching the viscoelastic critical force, in the prescribed manner defined by Equation (4.17), the adjacent deflected position can persist indefinitely.

Next, let us examine the alternate possibility in which the "controlled" configuration is maintained by applying to the column, subjected to a constant axial force, P, some additional loading, say a lateral force,

$$q(x,t) = \sum_{n=o}^{\infty} q_n(t) \sin \frac{n\pi x}{L}$$

(see Figure 13). Equation (4.9) is now modified to read

$$E_oI \frac{\partial^4 w}{\partial x^4} + P \left[\frac{\partial^2 w}{\partial x^2} + j(t) * \frac{\partial^2 w}{\partial x^2} \right] = q(x,t) + j(t) * q(x,t) \tag{4.18}$$

from which, using an assumed solution in accordance with Equation (4.3), one obtains

$$F_n(t)\,(1 - p_n) - j(t) * [p_n\, F_n(t)] = q_n(t) + j(t) * q_n(t) \tag{4.19}$$

If we assume the "controlled" deflected shape again to be

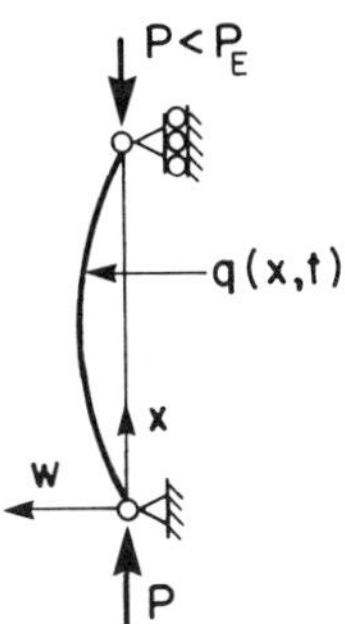

FIGURE 13. Viscoelastic column subjected to axial load and static disturbance.

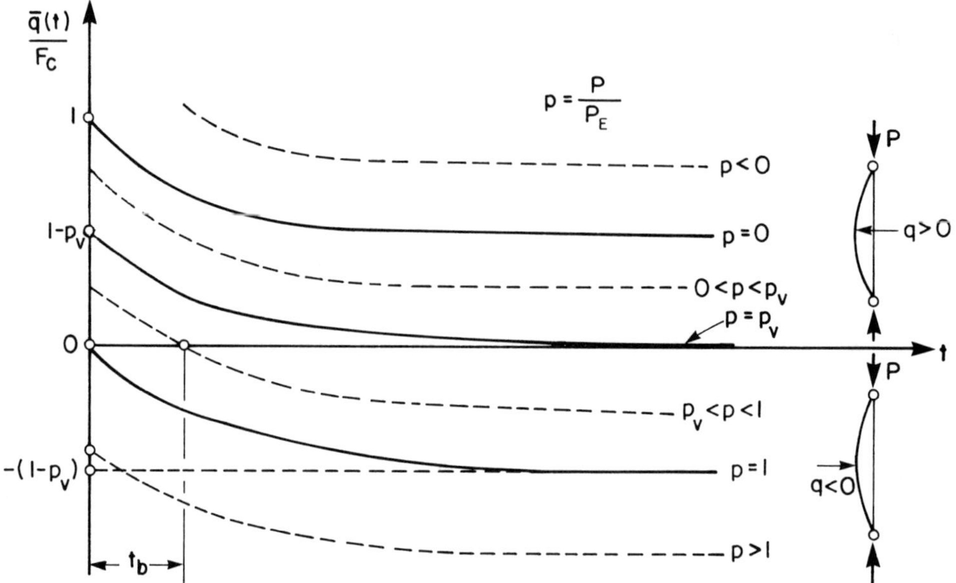

FIGURE 14. Variation of "static disturbance" history with axial load (from [64]).

constant and take $F_n(t) = \text{const} = F_c$, the equation governing the time-variation of $q_n(t)$ is given by

$$F_n\{1 - p_n[1 + \mathcal{J}(t)]\} = q_n(t) + j(t) * q_n(t) \tag{4.20}$$

which, again, for a given creep function defines $q_n(t)$. To proceed in a manner analogous to that used above, one substitutes for $j(t)$ in terms of the relaxation function, $g(t)$, to determine the time-variation of the lateral load for a linearly elastic column in a general form as

$$q_n(t) = F_c[(1 - p_n) + G(t)] \tag{4.21}$$

which for the three-element model material reduces to

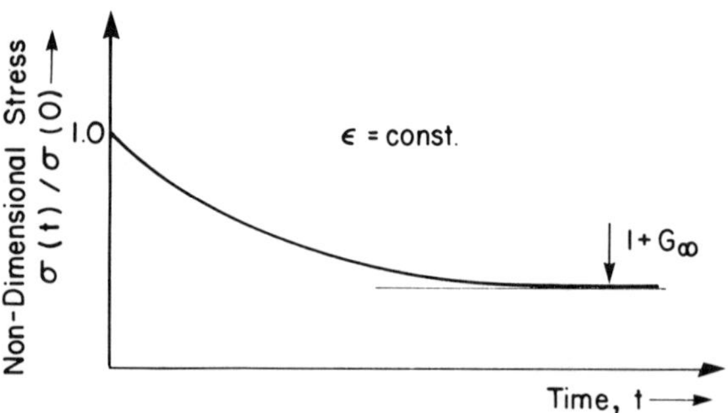

FIGURE 15. Relaxation curve.

$$q_n(t) = F_c[(p_v - p_n) + (1 - p_v)e^{-\lambda t}] \tag{4.22}$$

Depending on the magnitude of p_n, the lateral load required to maintain this constant "controlled" configuration can be positive (destabilizing) or negative (stabilizing), as shown on Figure 14. Note that for $t \to \infty$ the lateral loading $(q_n)_\infty \to F_c(p_v - p_n)$ which clearly vanishes if $p_n = p_v$. Thus for "inifinite time" and for an axial load equal to the safe-load-limit, the "controlled" deflected configuration can exist without any additional lateral loads.

Assuming $q_n(t) = 0$ in Equation (4.21), a condition which may occur at a certain instant of time, $t = t_b$ (see Figure 14) one finds the magnitude of the constant axial force, which at this instance is sufficient to keep the deflected shape in equilibrium, without any lateral forces, and is given by

$$p_n = 1 + G(t_b) \tag{4.23}$$

Interestingly, this relation is identical, in form, to Equation (4.16) which physically means that a deflected configuration which at time, t_b, is in equilibrium under the action of only the constant axial load, p_n, can also be achieved by reducing the axial load, in a prescribed manner, from its initial value of P_E at $t = 0$ to the value $p(t) = p_n$ at $t = t_b$.

Note that Equations (4.23) and (4.16) define the safe-load-limit for columns made of linearly viscoelastic materials in quite a general way in the form

$$p_v = 1 + \lim_{t \to \infty} G(t) = 1 + G_\infty \tag{4.24}$$

where G_∞ can be determined, hopefully, from a standard relaxation test (see Figure 15).

5. ASYMPTOTIC SOLUTIONS OF THE DYNAMIC STABILITY EQUATIONS

The dynamic stability analysis presented in Section 3 admitted explicit solutions for the safe-load-limit only in the case of columns for which the creep/relaxation properties of the linearly viscoelastic material had been specified. Interestingly, and perhaps surprisingly, the static approach yielded quite general expressions for the safe-load limit, Equation (4.24), for the same class of materials. The question naturally arose as to whether one could obtain expressions of equal generality by means of a dynamic formulation.

From Section 3 we know that for a perfect column made of a linearly viscoelastic material with limited creep, there exists a specific axial load value, called the safe-load-limit, under the action of which the structure tends towards a temporally constant limited deflected equilibrium configuration as $t \to \infty$, implying a steady state non-trivial solution. Thus, one can turn the problem around, assume the existence of a steady-state solution and seek the value of the axial load under which such a solution is possible with $t \to \infty$. As a result of this assumption one can write

$$\lim_{t\to\infty} w(x,t) = w_s(x); \qquad \lim_{t\to\infty} \frac{\partial^2 w}{\partial t^2} = 0 \quad (5.1)$$

in which $w_s(x)$ denotes the steady-state solution.

In order to write the equation of motion for the steady-state using Equation (3.4), one must evaluate the convolution term. For this reason we interchange the integration variables and write, for any function $w(x,t)$

$$j(t) * w(x,t) = \int_o^t j(t-\tau)\cdot w(x,\tau)d\tau$$

$$= \int_o^t j(\tau_1)\cdot w(x,t-\tau_1)d\tau_1 \quad (5.2)$$

in which the new time variable, τ_1, is indicated on Figure 16. This figure also shows that the unsteady (vibrational) state can always be eliminated by choosing a period of integration, T_o, for the steady state sufficiently large so that $j(t) \approx 0$ for $t \geq T_o$ and consequently

$$\lim_{t\to\infty} j(t) * w(x,t) = \mathcal{J}_\infty \cdot w_s(x) \quad (5.3)$$

Using Equations (5.1) and (5.3) in expression (3.4) one arrives at the equation defining the asymptotic state in the form

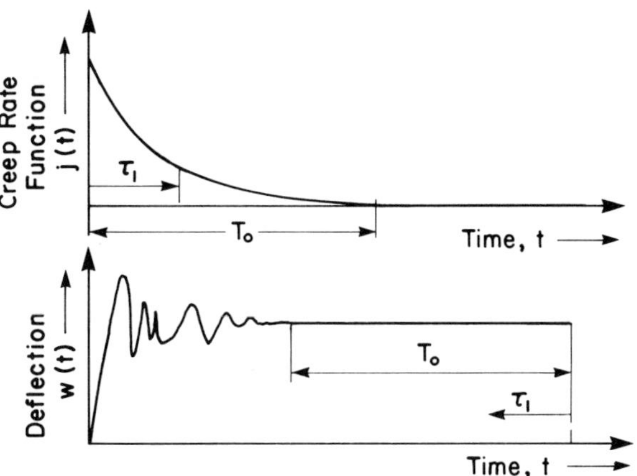

FIGURE 16. Interchange of time variable.

$$E_oI\frac{d^4w_s}{dx^4} + P(1+\mathcal{J}_\infty)\cdot\frac{d^2w_s}{dx^2} = 0 \quad (5.4)$$

a form which is identical to that of the well-known stability equation for an elastic column. The lowest eigenvalue for this equation, using simply supported boundary conditions, is obtained as

$$P(1+\mathcal{J}_\infty) = \frac{\pi^2E_oI}{L^2} = P_E \quad (5.5)$$

from which we recover, again, the nondimensional safe-load-limit for this column as

$$p_v = \frac{P}{P_E} = \frac{1}{1+\mathcal{J}_\infty} \quad (5.6)$$

We note that this result appears to be different from that given by Equation (4.24) obtained by means of the static approach. However, using Equation (4.15), one can show the following relation

$$\mathcal{J}(t) + G(t) + \int_o^t j(\tau)\cdot G(t-\tau)\,d\tau = 0 \quad (5.7)$$

from which for $t \to \infty$ one obtains

$$\mathcal{J}_\infty + G_\infty + \mathcal{J}_\infty G_\infty = 0$$

$$\therefore \frac{1}{1+\mathcal{J}_\infty} = 1 + G_\infty \quad (5.8)$$

indicating that Equations (4.24) and (5.6) are equivalent.

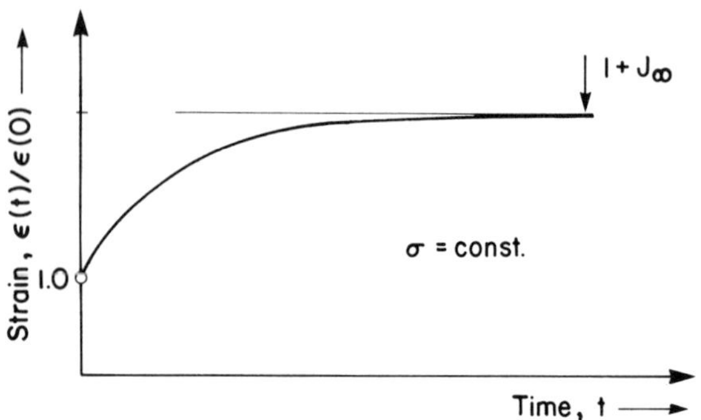

FIGURE 17. Creep curve.

Thus we have shown that with appropriate adjustment the dynamic approach provides the same results for p_v as the static approach and, consequently, the safe-load-limit for columns made of such linear materials with limited creep can equally well be determined using standard creep test data (see Figure 17).

Using the notion of "steady state" solution for $t \rightarrow \infty$ one can also study the effect of ageing on the safe-load-limit of columns made of linearly viscoelastic materials. The constitutive law for such a material, Equation (2.3) rewritten for the linear material case, is expressed as

$$\epsilon(x,y,t) = \frac{1}{E_o}\left[\sigma(x,y,t) + \int_o^t \sigma(x,y,\tau) \cdot \right.$$

$$\left. \cdot j(t - \tau) \cdot \zeta(\tau) d\tau \right] \tag{5.9}$$

in which time is counted from the instance when solidification is complete. Assuming the loading to be applied some time thereafter, say at $t = t_o$, (see Figure 18) with the perfect column being disturbed at the same time, the equation of motion can be written in the form

$$E_oI \frac{\partial^4 w}{\partial x^4} + P \frac{\partial^2 w}{\partial x^2} + m \frac{\partial^2 w}{\partial t^2} + \int_o^t \left[P \frac{\partial^2 w}{\partial x^2} + m \frac{\partial^2 w}{\partial \tau^2} \right] \cdot$$

$$\cdot j(t - \tau) \cdot \zeta(\tau + t_o) d\tau \tag{5.10}$$

where time, t, now is referenced to the instance of load application. Assuming Equations (5.1) to be valid and chang-

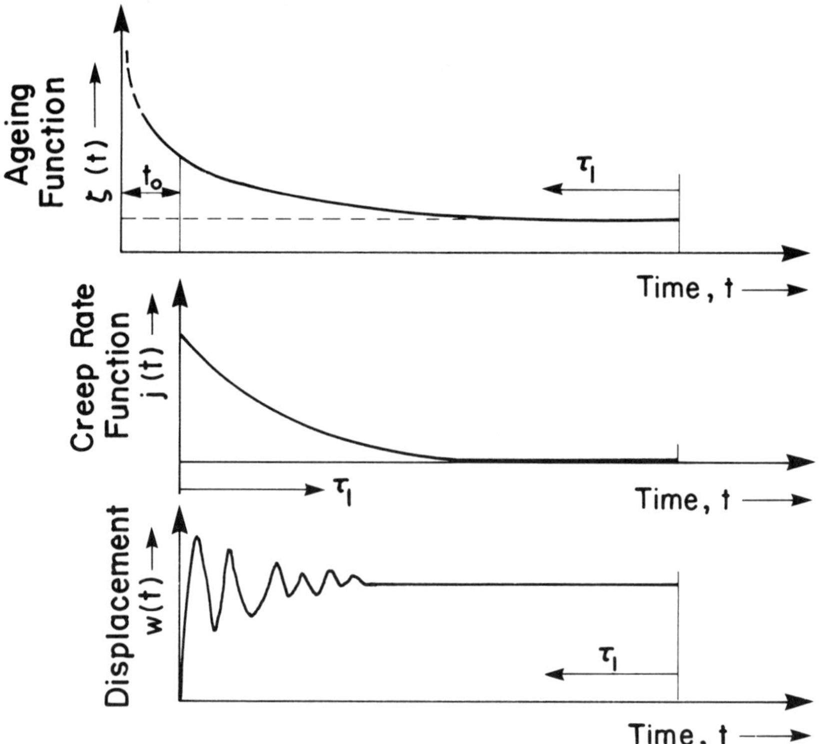

FIGURE 18. Effect of ageing on asymptotic solution.

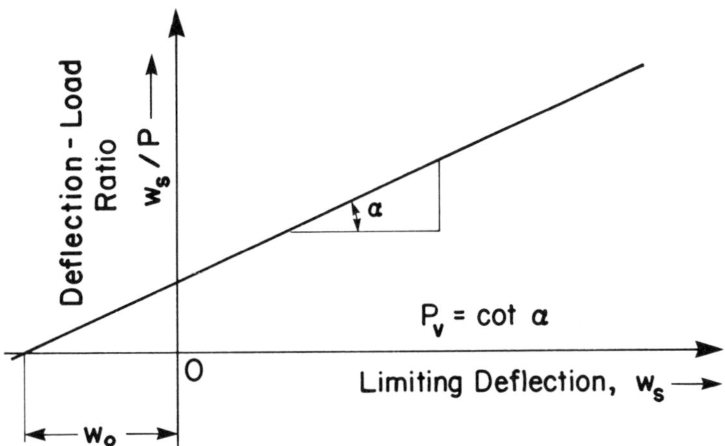

FIGURE 19. Southwell plot.

ing variables in the convolution in a manner to that used above (see Figure 18), Equation (5.10) is rewritten for the steady state ($t \to \infty$) as

$$E_oI \frac{\partial^4 w_s}{\partial x^4} + P \frac{\partial^2 w_s}{\partial x^2} + P \frac{d^2 w_s}{dx^2} \cdot \mathcal{J}_\infty \cdot [1 + \Psi(t_o)] = 0 \tag{5.11}$$

in which

$$\Psi(t_o) = \lim_{t \to \infty} \int_o^t \psi(t + t_o - \tau_1) \cdot j(\tau_1) d\tau_1$$

and from which the non-dimensional safe-load-limit is obtained in the form

$$p_v = \frac{1}{1 + \mathcal{J}_\infty [1 + \Psi(t_o)]} \tag{5.12}$$

where the effect of ageing clearly is a function of the load application, t_o. Upon reflection together with an analysis of the definition for $\Psi(t_o)$ (see also Figure 18) one concludes that for linear materials with similar creep and ageing functions the effect of ageing on the safe-load-limit will be negligible.

Finally, the asymptotic approach, can, of course, also be applied to the equation of motion for imperfect columns, Equation (3.22), which after using the procedure applied above, takes the form

$$E_oI \frac{d^4 w_s}{dx^4} + P(1 + \mathcal{J}_\infty)\left[\frac{d^2 w_s}{dx^2} + \frac{d^2 w_o}{dx^2}\right] = 0 \tag{5.13}$$

Using Equation (3.23) for $w_o(x)$ and assuming

$$w_s(x) = \sum_{n=o}^{\infty} F_n^s \sin \frac{n\pi x}{L}$$

Equation (5.13) yields

$$F_n^s = \overset{\circ}{w}_n \frac{1}{\frac{p_v}{p_n} - 1} \tag{5.14}$$

This last result shows that for imperfect columns a "steady-state" solution is possible only for $p < p_v$. If $p \to p_v$, $F_n^s \to \infty$, a result confirming the conclusions reached for this problem on the basis of the dynamic analysis.

In concluding this section we note that Equation (5.14) is identical in form to the equation defining the Southwell plot for the elastic column, with the Euler load, P_E, being replaced here by the viscoelastic critical force, P_v. Thus the safe-load-limit can be determined experimentally by plotting w_s/P vs. w_s for various axial load values $P < P_v$ (see Figure 19).

6. COLUMNS WITH VARIOUS BOUNDARY CONDITIONS

Thus far we have dealt only with simply supported columns. The purpose of this section is to see how the methods of analysis used and results obtained above are applicable and/or can be generalized to columns with arbitrary boundary conditions. Clearly, the governing equation of motion is unaffected by the boundary conditions. What is effected, however, are the mode shapes and consequently the assumed form of solution. Let us, therefore, generalize Equation (3.5) and assume the solution in the form of separated variables as

$$w(x,t) = \sum_{n=o}^{\infty} \phi_n(x) \cdot F_n(t) \tag{6.1}$$

which when substituted into Equation (3.4) leads to

$$\frac{\phi_n^{IV}}{\phi_n} + k^2 \frac{\phi_n''}{\phi_n} + k^2 \underline{\frac{\phi_n''}{\phi_n} \cdot \frac{j(t)*F_n}{F_n}} + \gamma^2 \left(\frac{\ddot{F}_n}{F_n} + \frac{j(t)*\ddot{F}_n}{F_n}\right) = 0 \tag{6.2}$$

where $k^2 = P/E_oI$ and $\gamma^2 = m/E_oI$. Unfortunately, as is clear from this expression and the underlined term, separation of variables is not possible in general. One might ask, however, under what conditions can such a solution technique be applied? Obviously, in the case of elastic columns, this troublesome term vanishes and the method is generally applicable, both for vibration and stability analysis. In the case of columns made of linearly viscoelastic materials the method is generally applicable for the static problem admitting solutions for arbitrary boundary conditions. As noted in Section 4.2, the static stability analysis of the perfect column leads to some difficulties in connection with the definition of adjacent equilibrium configurations, difficulties which are circumvented by means of the assumptions underlying the asymptotic approach discussed in Section 5. In accordance with this approach, part of the troublesome term in Equation (6.2) becomes

$$\lim_{t \to \infty} \frac{j(t)*F_n}{F_n} = \mathcal{J}_\infty \tag{6.3}$$

and consequently, the governing equation for the static case ($\ddot{F}_n = 0$) is written as

$$\phi_n^{IV} + k^2 (1 + \mathcal{J}_\infty)\phi_n'' = 0 \tag{6.4}$$

subject to given (arbitrary) boundary conditions. From this result it is clear that the safe-load-limit for any linearly viscoelastic column can be expressed in the form

$$P_v = P_E/(1 + \mathcal{J}_\infty) \tag{6.5}$$

in which P_E denotes the Euler load of the corresponding elastic column, a load which, of course, is dependent on the boundary conditions.

Finally, we note that the method of separation of variables is generally applicable to the linearly viscoelastic imperfect column analysed by means of the static approach. Since we have shown in Section 4.1 that the results for such an imperfect column from a static analysis are essentially equivalent to those obtained from a dynamic formulation, the solution technique discussed here (separation of variables) can obviously be used, quite generally, for the stability analysis of imperfect columns. To briefly demonstrate the validity of this assertion, let us represent the initial imperfection/deflection, $w_o(x)$, in the form

$$w_o(x) = \sum_{n=o}^{\infty} \mathring{w}_n\phi_n(x) \tag{6.6}$$

where $\phi_n(x)$ denotes a solution to the equation

$$\phi_n^{IV} + c_n^2\phi_n'' = 0 \tag{6.7}$$

and satisfies the prescribed boundary conditions. Substituting Equations (6.6) and (6.1) into Equation (3.22) one obtains

$$\phi_n'' \left[F_n(1 - c_n^2/k^2) + j(t)* F_n + \mathring{w}_n + j(t)* \mathring{w}_n\right] = 0 \tag{6.8}$$

Note that $c_n^2/k^2 = n^2P_E/P = n^2/p$ and P_E is the Euler load for the column with given boundary conditions. Equation (6.8) leads to

$$F_n(1 - n^2/p) + j(t) * F_n = -\mathring{w}_n \left[1 + \mathcal{J}(t)\right] \tag{6.9}$$

from which one can determine the deflection function, $F_n(t)$. Note that Equation (6.9) is identical to the corresponding expression for the simply supported column, Equation (4.4), thus indicating that this relation is valid for a linearly viscoelastic column with any boundary conditions. The solution to this equation for $t \to \infty$ confirms the safe-load-limit for columns made of this material, with arbitrary boundary conditions, to be equal to the corresponding Euler load, P_E, divided by the factor $(1 + \mathcal{J}_\infty)$.

Returning, briefly, to the general dynamic relation, Equation (6.2), instead of elimination of the underlined (troublesome) term, it can also be handled by eliminating either the spatial or the time variable from it. This, in turn, means that either $j(t)* F_n/F_n$ is set equal to or made to approach (in the limit) a constant value, as was done in the asymptotic approach [Equation (6.3)] or, alternatively, ϕ_n''/ϕ_n is set equal to a constant, say $\phi_n'' = -\tilde{c}_n^2\phi_n$. Either of these assumptions leads to complete "separation of variables" in Equation (6.2). However, the latter assumption is only valid for simply supported boundary conditions while the former, as was shown, leads to a general solution.

7. THE SAFE-LOAD-LIMIT OF NONLINEAR COLUMNS

Thus far we have treated geometrically and materially linear columns. Assumption of such complete linearity is justified only in a limited number of cases in design practice since most real materials, by their very nature, are nonlinear, except possibly for a response under very low stress levels. The designer has some control over geometrical nonlinearities in as much as he can design the structure so as to limit deflections and deformations so as to justify a geometrically linearized (approximate) analysis.

In this section we will generalize the concept of the safe-load-limit to columns and simple structures, including nonlinear geometry and nonlinear material behaviour. One could, of course, proceed to formulate the problem and derive expressions for the safe-load-limit in a manner analogous to that used in previous sections, i.e. write equations of motion, constitutive law and geometric relations, and attempt to solve such equations. Admitting nonlinearities in the geometric relations as well as in the constitutive law would make such an approach mathematically complex and/or intractable. Instead, one might proceed by considering the balance between the "active" external loads and the "passive" internal resistance of the structure. In such an approach, the exact form of the constitutive law for the material is not required; instead the material response characteristics at times $t = 0$ and $t \to \infty$ (limited creep) must be known. Thus, in analogy to the asymptotic solution technique, discussed in Section 5, this approach for determination of the safe-load-limit eliminates time as a variable from the analysis.

To introduce the concepts to be dealt with and the approach to be used let us first examine the stability behaviour of some simple rigid bar-spring-dash-pot structures, the response characteristics of which are qualitatively similar to those of columns. Such simple one-dimensional models, in which the elastic and viscoelastic elements are separated, can be very useful in gaining an understanding of some of the basic features of stability behaviour of structures made of such materials, and have, consequently, been used in research and theoretical analyses by many authors [25,29,66].

7.1 Analysis of Simple Structures

Consider the simple rigid bar-spring-dash-pot models shown in Figure 20. To model material nonlinearity, we allow the viscous element, V, to be nonlinear while the spring, S, is kept linear throughout the discussion. A linear viscous element was used in [66] in which an "exact" analysis of such simple structures was carried out in order to illustrate the effects of geometric nonlinearity on the stability and post-buckling behaviour of such structures.

Let us first treat a materially linear model for which the viscous element, V, is taken to be a linear Kelvin body defined by the material parameters, E_1 and ν_1 (Figure 3a). The load P and the rotation θ simulate external load and deflection in a column while internal stresses and strains in a real structure are modelled by the "internal" parameters, R and δ (see Figure 20).

Next, let us consider only the Kelvin body, focussing attention on the material behaviour, by plotting the relation

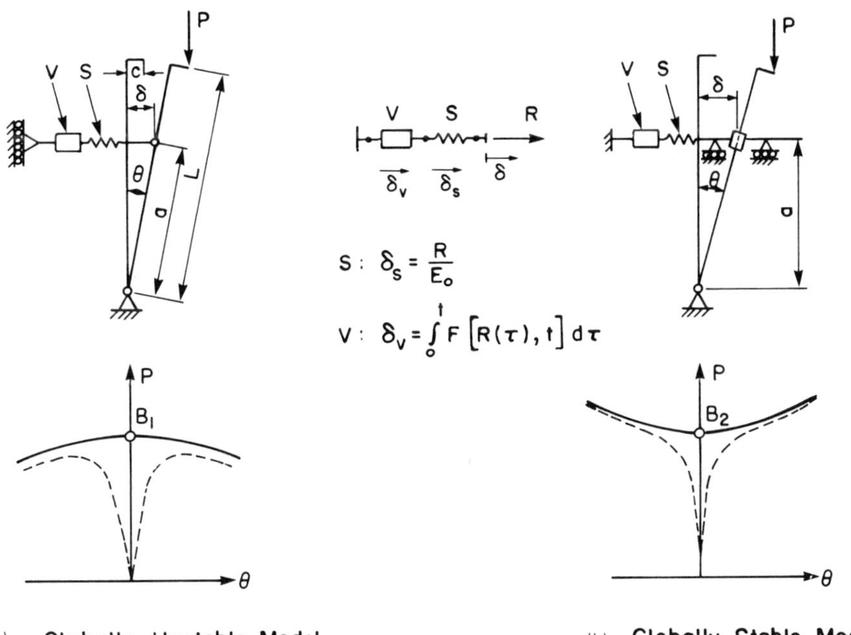

FIGURE 20. Simple rigid bar-spring/dash-pot models with symmetric bifurcation points (from [71]).

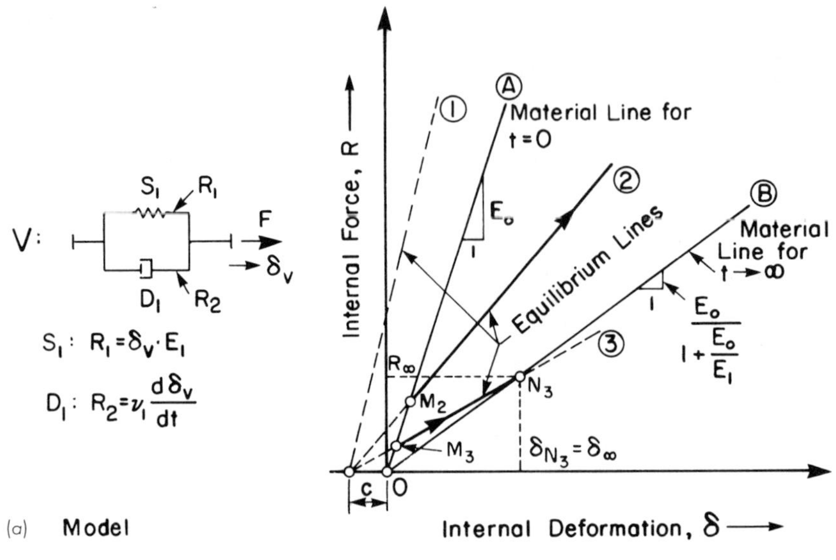

FIGURE 21. Stability analysis of completely linear viscoelastic simple structure (from [71]).

between R and δ for $t = 0$ and for $t \to \infty$, as shown by the two straight lines, OA and OB on Figure 21b. The triangularly shaped area between these two lines represents the admissible states for the internal force-deformation parameters. In addition, R and δ are constrained by the equilibrium condition for the whole structure, written for the model shown in Figure 20a as

$$R a\cos\theta = PL(\sin\theta + \bar{c}\cos\theta); \qquad \bar{c} = c/L \tag{7.1a}$$

where c denotes the load eccentricity. To eliminate θ, the geometric relation

$$\delta = a\sin\theta \tag{7.1b}$$

is used, leading to

$$R = \frac{PL}{a}\left[\frac{\bar{\delta}}{\sqrt{1 - \bar{\delta}^2}} + \bar{c}\right] \tag{7.2}$$

where $\bar{\delta} = \delta/a$. For a geometrically linearized theory (for which $\bar{\delta} << 1.0$), Equation (7.2) becomes

$$R = \frac{PL}{a}[\bar{\delta} + \bar{c}] \tag{7.3}$$

an expression which is plotted on Figure 21b for a given value of $\bar{c}$ and three different load levels, p, where $p = P/P_E$, with $P_E = E_o a^2/L$. For $p > 1$, or alternatively, $PL/a^2 > E_o$, the line defined by Equation (7.3) (line (1)) does not enter the above described "admissible" zone, which, in turn, means that static equilibrium is impossible at any time and the structure/load combination must be classified as unstable. For the load range, $1/(1 + E_o/E_1) < p < 1$, or $E_o/(1 + E_1/E_o) < PL/a^2 < E_o$, (line (2) on Figure 21b), the straight line defined by Equation (7.3) enters the "admissible" solutions zone at point M_2 and continues in this zone for all values of t. Physically this means that the viscous process continues indefinitely, with monotonically increasing deformation (δ), internal force (R) and deflection (θ), a state which in [63] was referred to as "viscoelastically unstable." Finally, for a load value $p < 1/(1 + E_o/E_1)$, or $PL/a^2 < E_o/(1 + E_o/E_1)$, the straight line defined by Equation (7.3) enters the "admissible" solutions domain at M_3 and exits from this domain at N_3 (line (3) in Figure 21b). Thus for $t \to \infty$ the deflection, $\delta \to \delta_{N3}$, indicating a limited deflection and a cessation of the viscous process. Thus, this load range, $0 < p < p_v = 1/(1 + E_o/E_1)$, is associated with stable behaviour and consequently the load, p_v, is referred to as the "safe-load-limit." Clearly, for such a geometrically and materially linear model, p_v^ℓ is independent of the imperfection parameter, $\bar{c}$, and is given as

$$p_v^\ell = \frac{1}{1 + E_o/E_1} \tag{7.4}$$

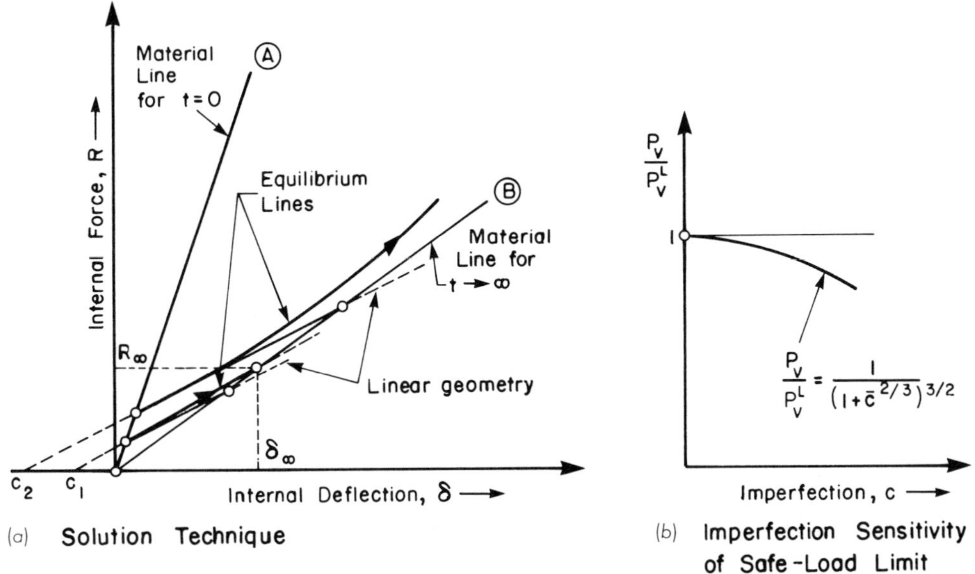

FIGURE 22. Geometrically nonlinear stability analysis of globally unstable simple viscoelastic structure (from [71]).

In general, the lines appearing on Figure 21b may be curved, indicating material or geometric nonlinearities.

Let us next discuss the effect of geometric nonlinearity, keeping the material model linear. Thus lines *OA* and *OB* on Figure 21b remain the same while the "equilibrium" lines affected by geometry, become curved. First consider the structure shown in Figure 20a exhibiting a symmetric unstable point of bifurcation and global instability. Plotting now the exact relation, Equation (7.2), on an $R - \delta$ diagram (see Figure 22a), one obtains a convex curve, with monotonically increasing slope as δ increases. From Figure 22a it is clear that the existence of a "safe-load-limit" for such a geometrically nonlinear analysis depends on the magnitude of the imperfection parameter, $\bar{c}$. For example, for $\bar{c} = \bar{c}_2$ (see Figure 22a), a geometrically linearized solution would yield a finite (limited) deflection, indicating stable behaviour, while a geometrically nonlinear analysis for the same load and imperfection values results in unlimited creep and viscoelastic instability. An analysis shows that the safe-load-limit for this case is given by

$$p_v = \frac{p_v^\ell}{(1 + \bar{c}^{2/3})^{3/2}} \tag{7.5}$$

which corresponds to a limit deflection (for $t \to \infty$)

$$\bar{\delta} = \frac{\bar{c}^{1/3}}{\sqrt{1 + \bar{c}^{2/3}}} \tag{7.6}$$

Equation (7.5) is plotted on Figure 22b indicating the effect of imperfections on the viscoelastic critical force, p_v.

Let us also consider the structure shown in Figure 20b with a symmetric stable bifurcation point and exhibiting global stability. For this structure, the relation corresponding to Equation (7.3) becomes

$$R = \frac{PL}{a} \frac{\bar{\delta} + \bar{c}}{(1 + \bar{\delta}^2)^{3/2}} \tag{7.7}$$

which, when plotted on an $R - \delta$ diagram (see Figure 23) results in concave lines with decreasing slopes for increasing δ. Upon reflection it is clear that the line *OB* (for $t \to \infty$) can be intersected by equilibrium curves corresponding to any p and $\bar{c}$ value. Naturally, the magnitude of deformation, δ, corresponding to such intersections, might be unacceptable. To keep the deflection magnitude within some reasonable range, one might wish to limit the load value to $p < p_v^\ell$ [66].

Next consider the effect of material nonlinearity on the viscoelastic critical force. To focus attention on this effect, assume linearized geometric relations in which case Equations (7.2) and (7.7) become identical and equal to Equation (7.3). Denoting the "material curve" for $t \to \infty$ by $R = \tilde{G}(\delta)$ (see Figure 24) and recognizing that on the basis of a substantial body of experimental data, the nonlinearity parameter, n, is assumed greater than unity, as a result of which the curve is concave and the safe-load-limit is found

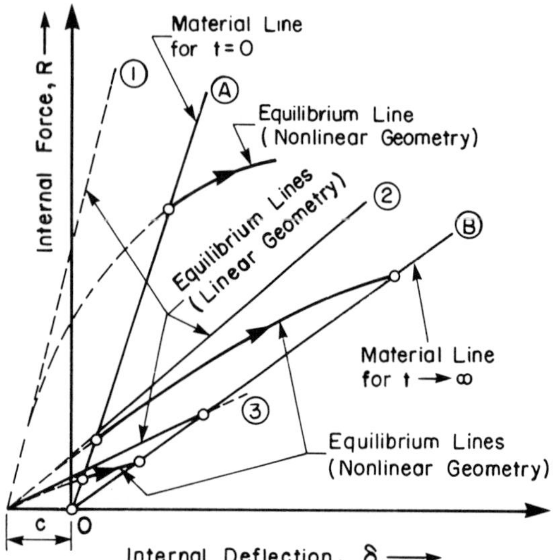

FIGURE 23. Geometrically nonlinear stability analysis of globally stable simple viscoelastic structure (from [71]).

by solving a set of two equations, namely

$$\tilde{G}(\delta) = \frac{PL}{a}\left(\frac{\delta}{a} + \frac{c}{L}\right) \tag{7.8a}$$

$$\frac{d}{d\delta}[\tilde{G}(\delta)] = \frac{PL}{a^2} \tag{7.8b}$$

from which, after elimination of δ, one arrives at the relation $p_v = p_v(c)$, indicating, again, the dependence of the viscoelastic critical force on initial imperfections. As a specific example, consider the nonlinear Kelvin body for which the strain-stress relation, shown on Figure 24b, is given as

$$\delta_v = \tilde{G}^{-1}(R) = [R + (n - 1)bR^n]/E_1 \tag{7.9}$$

in which b is a material constant, and from which, using Equations (7.8), one arrives at the safe-load-limit, p_v, in the form

$$p_v = p_v^{\ell}/[1 + p_v^{\ell}\xi(\bar{c})] \tag{7.10a}$$

where

$$\xi(\bar{c}) = D \cdot \bar{c}^{(n-1)/n} \tag{7.10b}$$

$$D = \left[\left(\frac{n}{n-1} \cdot aE_o\right)^{n-1} \cdot \frac{E_o}{E_1} bn(n-1)\right]^{1/n} \tag{7.10c}$$

It can be shown that for the linear material ($n \to 1.0$), $\xi \to 0$ and $p_v \to p_v^{\ell}$. From Equation (7.10a) one obtains

$$\frac{dp_v}{d\bar{c}} = -(p_v)^2 D \frac{(n-1)}{n \cdot (\bar{c})^{1/n}} \tag{7.11}$$

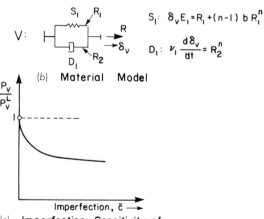

which indicates imperfection sensitivity as shown in Figure 24c. Note that the degree of sensitivity for material nonlinearity is much more severe than for a geometrically nonlinear structure (see Figure 22b).

In the case of materially and geometrically nonlinear structures, both the "material" and the "equilibrium" lines

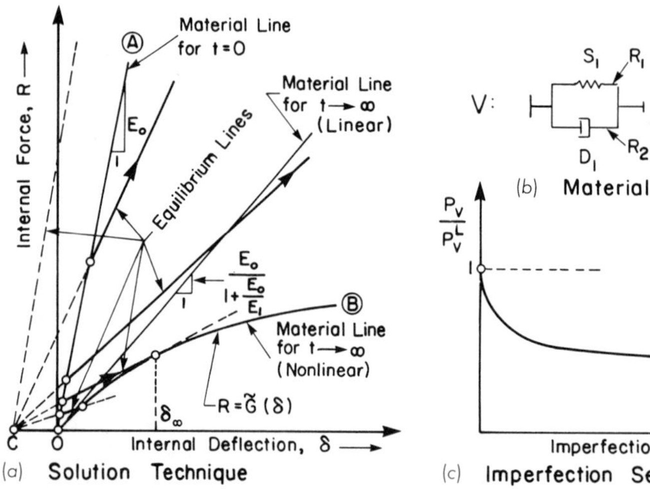

FIGURE 24. Stability analysis of materially nonlinear simple viscoelastic structure (from [71]).

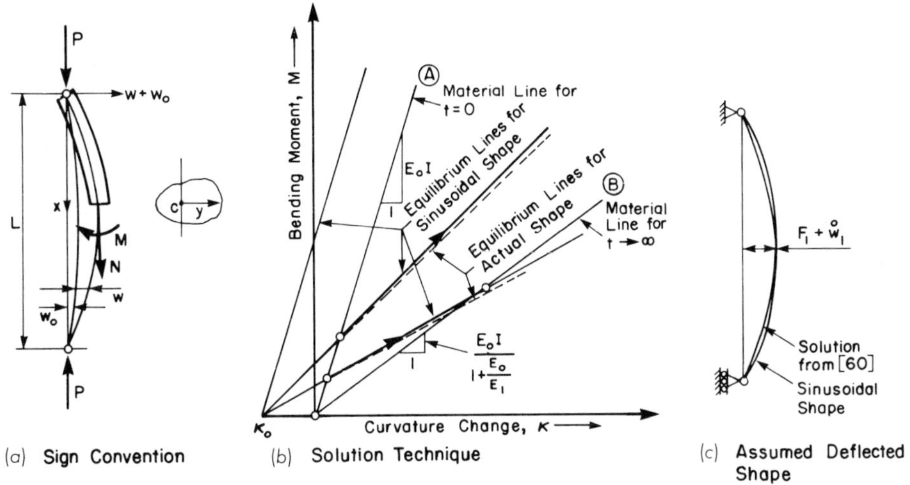

FIGURE 25. Linear stability analysis of linearly viscoelastic column (from [71]).

are curved, thereby making the algebra of the analysis more complex; the basic concepts and the method remain the same.

7.2 Safe-Load-Limit of Columns

In the case of a simply supported column made of a time-dependent material, the above outlined procedure for determination of the viscoelastic critical force is, of course, also applicable. However, as opposed to the simple, one-parameter structure treated in the previous Sub-section, such a column is a continuous body and an "exact" solution can be obtained only by satisfying the equations of equilibrium, the constitutive relation and compatibility relations at every point of the structure.

To apply the stability analysis used in Subsection 7.1 to columns, one has to, again, compare structural/material resistance with the "equilibrium effects" of the external loading, and therefore deal with curves on a bending moment, M, vs. curvature change, $\varkappa$, diagram. Taking the stress-strain relation in the form $\sigma = \tilde{G}(\epsilon)$ and using the Bernoulli-Euler hypothesis, one can define the bending moment, M, and axial force, N, as

$$M = \int_A \tilde{G}(\epsilon_o + y\varkappa)z dA \tag{7.12a}$$

$$N = \int_A \tilde{G}(\epsilon_o + y\varkappa) dA \tag{7.12b}$$

where y is the distance from the centroidal axis of the column and ϵ_o and $\varkappa$ denote axial strain along and curvature change of this axis, respectively. Since $N = -P$, it is possible to calculate ϵ_o from Equation (7.12b) in the form $\epsilon_o = \epsilon_o(\varkappa, P)$, which when used in Equation (7.12a) leads to a relation, $M = M(\varkappa, P)$.

This latter relation indicates that as opposed to the "material"/resistance curves for the simple structure treated above, which were independent of the external load P, in the case of the column, such relations/curves (the $M = M(\varkappa, P)$ curves) for $t \to \infty$ are functions of the load P except in the linear case. Thus, in the case of the column, there is a family of such "material curves," one for each different load level.

For a column made of a three-element-material model Equations (7.12) take the form

$$\left.\begin{aligned} M &= E_o I \varkappa \\ P &= -E_o A \epsilon_o \end{aligned}\right\} \quad \text{for } t = 0 \tag{7.13a}$$

$$\left.\begin{aligned} M &= \left(\frac{E_o}{1 + E_o/E_1}\right) I\varkappa \\ P &= -\left(\frac{E_o}{1 + E_o/E_1}\right) A\epsilon_o \end{aligned}\right\} \quad \text{for } t \to \infty \tag{7.13b}$$

confirming the fact that for a linear material the bending moment is independent of the axial force, P. Thus the "admissible solutions" domain on the $M - \varkappa$ diagram can be established relatively easily and is shown on Figure 25b.

The bending moment-curvature-change relations, however, are also restricted by equilibrium of the overall structure and compatibility/geometry of the deflected shape. In the case of an imperfect column with initial deflection, w_o,

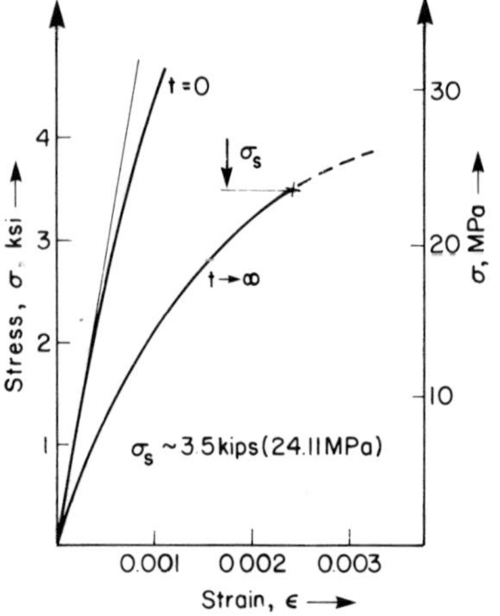

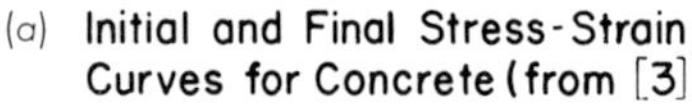
(a) Initial and Final Stress-Strain Curves for Concrete (from [3])

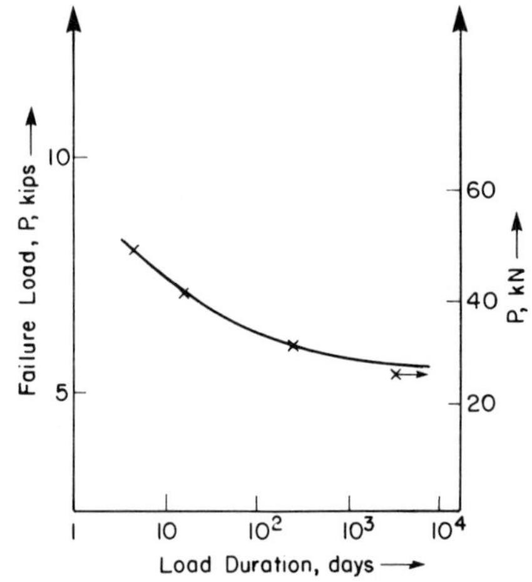

(b) Effect of Load Level on Failure Time (from [3])

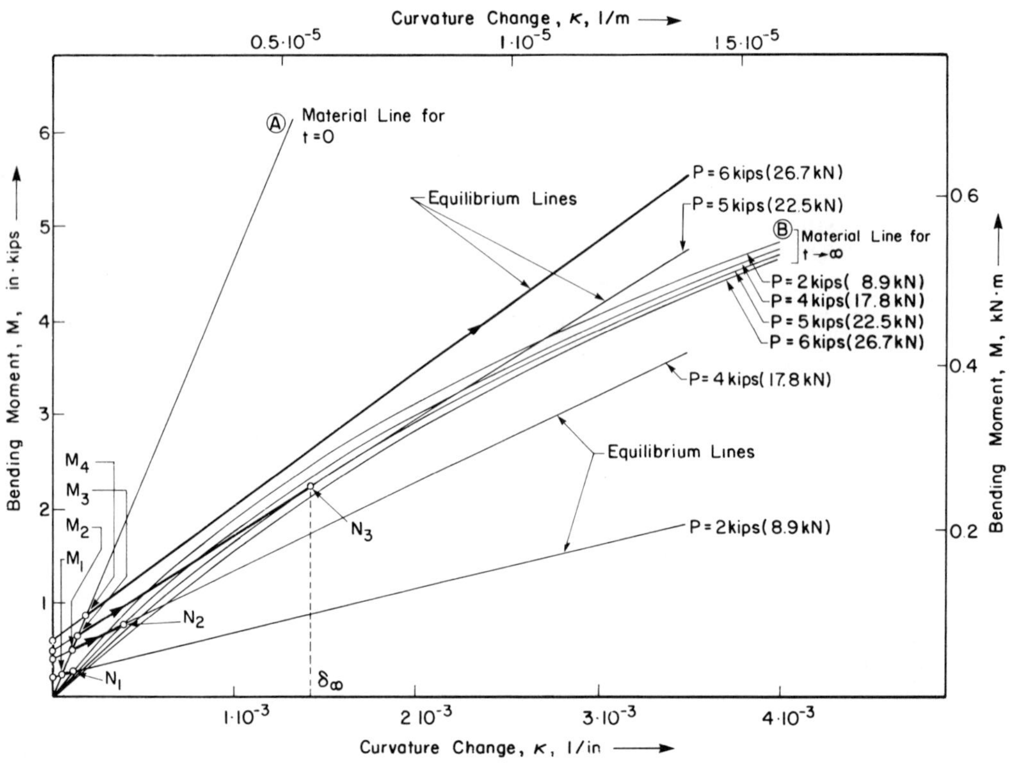

(c) Solution Technique

FIGURE 26. Stability analysis of concrete column (from [71]).

the equilibrium equation reads

$$M = P \cdot [w_o + w(\varkappa)] \tag{7.14}$$

where w denotes the additional deflection of the centroidal line. The geometric constraint, represented by the deflection-curvature change relation, $w = w(\varkappa)$, is rather complex and cumbersome in numerical treatments even for the linearized case. Consequently, we propose an approximation, justified, in part, by results presented in [60] whereby instead of the exact relation we use the expression at mid-span

$$F_1 \leq \varkappa \frac{L^2}{\pi^2} \tag{7.15}$$

a relation which recognizes the dominant effect the first harmonic has on the behaviour of imperfect viscoelastic columns as opposed to higher harmonics.

Using Equation (7.15) in relation (7.14), one obtains the expression valid at mid-span

$$M \leq M' = P \left(\overset{\circ}{w}_1 + \varkappa \frac{L^2}{\pi^2} \right) \tag{7.16}$$

which is plotted on Figure 25b. Clearly, in using the approximate bending moment, M', instead of the exact value, M, one is conservative in the assessment of the stiffness of the structure.

Comparing the slopes of the lines defined by Equations (7.13) and (7.16), it is clear that for a load $P > P_E = \pi^2 E_o I/L^2$, static equilibrium is impossible. For load values $[P_E/(1 + E_o/E_1)] < P < P_E$, the column is viscoelastically unstable with deflections increasing monotonically with time. Only for a load value, $P < P_v = P_E/(1 + E_o/E_1)$, will the deflection be limited as $t \to \infty$, confirming this value of the load as the safe-load-limit. This result is identical to that obtained in Sections 3, 4 and 5 above in which a more complex method of analysis was used.

To illustrate the use of this method for the nonlinear material case, a concrete column will be analyzed, the material properties for which are taken from [3], (see Figure 26a), in accordance with which the nonlinear "material" curve for $t \to \infty$ on a σ vs. ϵ diagram, can be approximated by the relation

$$\epsilon = G^{-1}(\sigma) \cong \frac{\sigma}{E_o} + \frac{\sigma}{|\sigma|} \frac{\sigma^2}{E_1} \tag{7.17}$$

where E_o denotes the initial tangent modulus with a value $E_o \cong 5.42 \times 10^3$ ksi (37.4 GPa), while E_1 is a material parameter associated with the material nonlinearity, the value of which, according to data in [3] is estimated as $E_1 \cong 6.02 \times 10^3$ (ksi)^2 [0.286 (GPa)^2]. The column is assumed to have a rectangular cross-section of dimensions 3.0 in by 1.5 in (76 mm × 38 mm). Its length is 48 inches (1,220 mm) with an initial imperfection amplitude at midspan of 0.1 in (2.54 mm). These dimensions are chosen so as to be able to compare the analytical results obtained here with experimental data for a corresponding reinforced concrete column of identical dimensions treated in [3].

Solving for σ from Equation (7.17), the function $\tilde{G}(\epsilon)$ appearing in Equations (7.12) is obtained in the form

$$\tilde{G}(\epsilon) = \begin{cases} -\sigma_A\left(\sqrt{1 - \epsilon/\epsilon_A} - 1\right) & \text{for } \epsilon < 0 \\ \sigma_A\left(\sqrt{1 + \epsilon/\epsilon_A} - 1\right) & \text{for } \epsilon > 0 \end{cases} \tag{7.18}$$

where $\sigma_A = E_1/2E_o$ and $\epsilon_A = E_1/4E_o^2$. Note that in this discussion of the concrete column compressive stresses/strains are considered positive.

Using Equation (7.18), expressions (7.12) become

$$\frac{M}{\frac{bh^2}{4}\sigma_A} = \begin{cases} \frac{2}{15\bar{\varkappa}^2}\left[(3\bar{\varkappa} - 2e_1)(e_1 + \bar{\varkappa})^{3/2} + (3\bar{\varkappa} + 2e_1)(e_1 - \bar{\varkappa})^{3/2}\right]; & \text{for } \bar{\varkappa} \leq \frac{\epsilon_o}{\epsilon_A} \\ \frac{2}{15\bar{\varkappa}^2}\left[(3\bar{\varkappa} - 2e_1)(e_1 + \bar{\varkappa})^{3/2} + (3\bar{\varkappa} - 2e_2)(e_2 + \bar{\varkappa})^{3/2} + 4\right] + \frac{\epsilon_o^2}{\bar{\varkappa}^2\epsilon_A^2} - 1 & \text{for } \bar{\varkappa} > \frac{\epsilon_o}{\epsilon_A} \end{cases} \tag{7.19a}$$

$$\frac{P}{bh\sigma_A} = \begin{cases} \frac{1}{3\bar{\varkappa}}\left[(e_1 + \bar{\varkappa})^{3/2} - (e_1 - \bar{\varkappa})^{3/2}\right] - 1; & \text{for } \bar{\varkappa} \leq \frac{\epsilon_o}{\epsilon_A} \\ \frac{1}{3\bar{\varkappa}}\left[(e_1 + \bar{\varkappa})^{3/2} - (e_2 + \bar{\varkappa})^{3/2}\right] - \frac{\epsilon_o}{\bar{\varkappa}\epsilon_A}; & \text{for } \bar{\varkappa} > \frac{\epsilon_o}{\epsilon_A} \end{cases} \tag{7.19b}$$

where

$$\bar{\varkappa} = \frac{h\varkappa}{2\epsilon_A}$$

$$e_1 = 1 + \epsilon_o/\epsilon_A$$

$$e_2 = 1 - \epsilon_o/\epsilon_A$$

Using Equations (7.19), the $M = M(\varkappa)$ relation is obtained which is used to calculate the structural/material "resistance" curves for given values of P as indicated on Figure 26c. The "equilibrium" lines, defined by Equation (7.16), which are straight lines for this geometrically linearized analysis, are also indicated on Figure 26c for the same values of P. These lines intersect the ordinate axis at the points $M(0) = P\mathring{w}_1$ and have a slope equal to PL^2/π^2. Note that, as opposed to the "material resistance" curves, for which the slope decreases with increasing P, these equilibrium lines become steeper as the axial load increases. By trial and error it was established that the "equilibrium" line for $P = 5.0$ kips (22.5 kN) is the only line which is tangential to the corresponding "material resistance" curve, thus indicating this value of the load as the safe-load-limit for this concrete column. Clearly, these two sets of curves intersect only for load values $P \leq 5.0$ kips; for $P > 5.0$ kips, the column is viscoelastically unstable. These results are in surprisingly good agreement with experimental data given in [3] and shown on Figure 26b.

The analysis and method of solution used here can also be applied to "perfect" columns by assuming $\bar{c} \rightarrow 0$ and recognizing that consequently $\bar{\varkappa}h << 1.0$. Expanding Equations (7.19) into a Taylor series and retaining only linear terms in $\varkappa$, one obtains

$$M \cong \frac{P_A h}{24\epsilon_A} \frac{\varkappa h}{\sqrt{1 + \epsilon_o/\epsilon_A}} \tag{7.20a}$$

$$P = P_A[\sqrt{1 + \epsilon_o/\epsilon_A} - 1] \tag{7.20b}$$

where $P_A = \sigma_A bh$. Eliminating ϵ_o one arrives at

$$M = \frac{EI\varkappa}{1 + P/P_A} \tag{7.21}$$

The safe-load-limit is calculated from the relation

$$\frac{dM}{d\varkappa} = \frac{EI}{1 + P/P_A} = P\frac{L^2}{\pi^2} \tag{7.22a}$$

$$P_v = \frac{P_A}{2}[\sqrt{1 + 4P_E/P_A} - 1] \tag{7.22b}$$

Using the data for the above concrete column example, one finds $P_A = 2.5$ kips (11.12 kN), $P_E = 19.57$ kips (87.09 kN), and $P_v = 5.87$ kips (26.12 kN), the latter of which is slightly higher than the experimentally indicated value for the imperfect column [3].

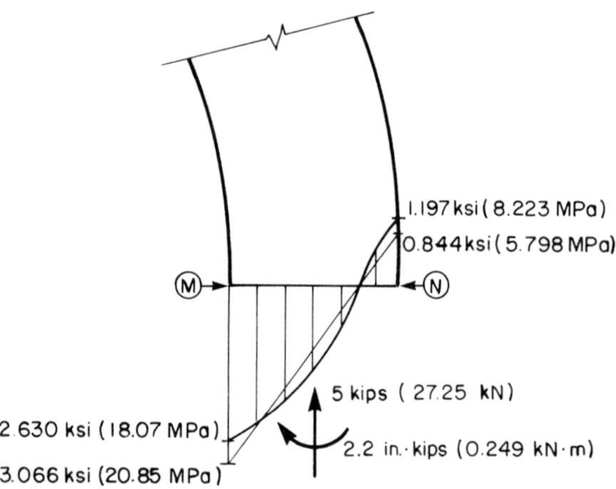

FIGURE 27. Midspan nonlinear stress distribution corresponding to point N_3 on Figure 26c and its linear approximation.

7.3 Further Approximations in the Analysis of Columns

The most complex part of the analysis presented in the previous sub-section, at least as far as the algebra is concerned, is the determination of the structural/material resistance curves, $M = M(\varkappa)$, defined, in general, by Equation (7.12). The source of such complexity can easily be traced to the nonlinear stress distribution across the column cross-section, an example of which for the concrete column treated above, is indicated in Figure 27. In view of the Bernoulli-Euler hypothesis the curvature change of the centroidal axis is given by

$$\varkappa = \frac{\epsilon_M - \epsilon_N}{h} \tag{7.23}$$

where h denotes the distance between the faces of the column. Since, in general,

$$\epsilon_K = \tilde{G}^{-1}(\sigma_k) = \tilde{G}^{-1}[\sigma_K(P,F_1,t)];\ K = M,N \tag{7.24}$$

one arrives at

$$F_1\frac{h\pi^2}{L^2} = \tilde{G}^{-1}[\sigma_M(P,F_1,t)] - \tilde{G}^{-1}[\sigma_N(P,F_1,t)] \tag{7.25}$$

an expression which allows determination of the complete midspan deflection history, for given values of P. Naturally, to use Equation (7.25) one has to know the material behaviour, i.e., the function $\tilde{G}^{-1}$ has to be specified; in addition, one also needs to know the exact stress–load relation for the column, a problem which is tractable only for linear materials.

For a column made of a three-element-model material, Equation (7.25) becomes

$$F_1(t)\frac{\pi^2 I}{L^2} = \frac{P}{E_o}[F_1(t) + \mathring{w}_1] + \frac{P}{\nu_1}\int_o^t \{[F_1(\tau) + \mathring{w}_1]e^{-E_1(t-\tau)/\nu_1}\}d\tau \tag{7.26}$$

the solution to which yields

$$F_1(\tau) = \mathring{w}_1 \frac{P}{(P_v - P)}\left(1 - \frac{P_E - P_v}{P_E - P}e^{s_1 t}\right)$$
$$s_1 = \lambda\frac{P - P_v}{P_E - P} \tag{7.27}$$

Note that an identical result was obtained in Section 4.1 [Equation (4.7)] by means of a different solution technique.

In the case of a column made of a nonlinear material, for which the stress–force relation is not known a priori, one might proceed by assuming a statically admissible stress field, $\hat{\sigma}$, which satisfies the following conditions at midspan

$$P(F_1 + \mathring{w}_1) = \int_A \hat{\sigma}\cdot y dA$$
$$P = \int_A \hat{\sigma} dA \tag{7.28a,b}$$

Naturally, this further approximation will yield solutions which are lower bounds to the exact values. A simplest stress distribution that one might assume is a linear one, as shown in Figure 27. Knowing the force and moment resultants for the "exact" nonlinear stress distribution at midspan, one can determine the extreme fibre stress levels for such a linear approximation.

Using such a straight-line approximation, one can write, instead of Equation (7.25), the following relation

$$F_1\frac{h\pi^2}{L^2} \cong \tilde{G}^{-1}\left[\frac{P}{A} + \frac{P(F_1 + \mathring{w}_1)h}{2I}\right] - \tilde{G}^{-1}\left[\frac{P}{A} - \frac{P(F_1 + \mathring{w}_1)h}{2I}\right] \tag{7.29}$$

from which, using Equation (7.17), one obtains an approximate axial force–deflection relation for the concrete column in the form

$$\bar{f} = \bar{f}_o\frac{p(1 + \alpha p)}{[1 - p(1 + \alpha p)]}$$

for

$$\bar{f} + \bar{f}_o \le \frac{1}{6} \tag{7.30a}$$

or

$$\sigma_M\cdot\sigma_N > 0$$

$$(\bar{f} + \bar{f}_o)(1 - p) - \bar{f}_o = \alpha p^2\left[3(\bar{f} + \bar{f}_o)^2 + \frac{1}{12}\right]$$

for

$$(\bar{f} + \bar{f}_o) > \frac{1}{6} \tag{7.30b}$$

or

$$\sigma_M\cdot\sigma_N < 0$$

where

$$\bar{f} = F_1/h;\ \bar{f}_o = \mathring{w}_1/h;\ p = P/P_E;\ \alpha = P_E/P_A$$

The curves defined by Equations (7.30) are shown in Figure 28 for the concrete column, for which $P_E = 19.57$ kips (87.07 kN), $P_A = 2.5$ kips (11.12 kN) and $\alpha = 7.8$. Note that the low load/deflection portion of each of the two curves on this figure, for which the extreme fibre stresses are of the same sign (i.e., compressive) is defined by Equation (7.30a). When extended beyond their domain, these curves approach a fixed value, $\mathring{p}_v$, asymptotically, irrespective of the initial imperfection deflection, an asymptotic value defined by the relation

$$1 - \mathring{p}_v(1 + \alpha\mathring{p}_v) = 0 \tag{7.31}$$

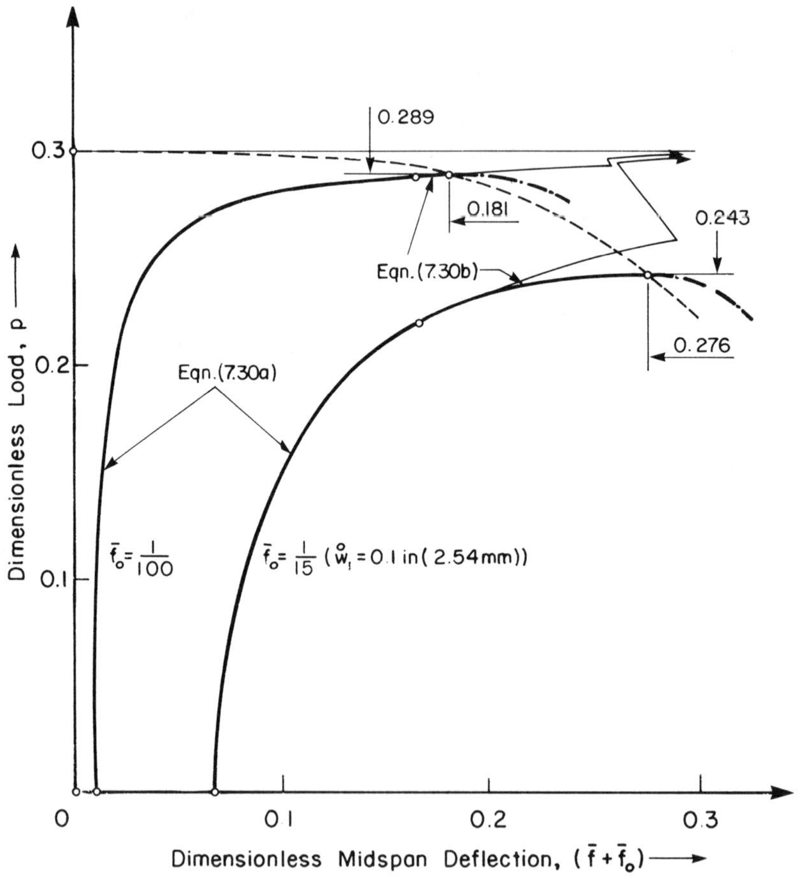

FIGURE 28. Effect of initial imperfection on load deflection behaviour of concrete column for $t \rightarrow \infty$ (from [71]).

which when solved for $\overset{o}{p}_v$ yields

$$\overset{o}{p}_v = \frac{\sqrt{1 + 4\alpha} - 1}{2\alpha} \tag{7.32}$$

a relation identical to Equation (7.22b) representing the safe-load-limit of the "perfect" column.

The second equation [Equation (7.30b)], which is valid for extreme fibre stresses of opposite sign, defines the higher load portion of each of the two curves in Figure 28. These portions exhibit an extremum (maximum), the parameters p_v and $\bar{f}_v$, which are defined by

$$(1 - p_v)^2 - \alpha^2 p_v^4 = 12\alpha \bar{f}_o p_v^2 \tag{7.33a}$$

$$\bar{f}_v = \bar{f}_o + \frac{1 - p_v}{p_v^2} \frac{1}{6\alpha} \tag{7.33b}$$

Clearly, Equation (7.33a) defines the safe-load-limit of an imperfect column with a given initial imperfection amplitude, $\bar{f}_o$. For $\bar{f}_o \rightarrow 0$, this equation reduces to relation (7.31) for the "perfect" structure.

The sensitivity of the viscoelastic critical force to initial imperfections is illustrated on Figure 29. Naturally, the curve resulting from the further, statically admissible stress field approximation, falls below the curve drawn on the basis of results obtained in Sub-section 7.2. The difference between the two curves decreases with decreasing imperfection, which is as expected, since for imperfections tending to zero, the stress distribution approaches a linear one.

It is interesting to note the difference in imperfection sensitivity obtained from a materially nonlinear stability analysis for the column (see Figure 29) as opposed to the simple structure discussed in Sub-section 7.1 (see Figure 24). Upon reflection this difference can be explained quite rationally [71].

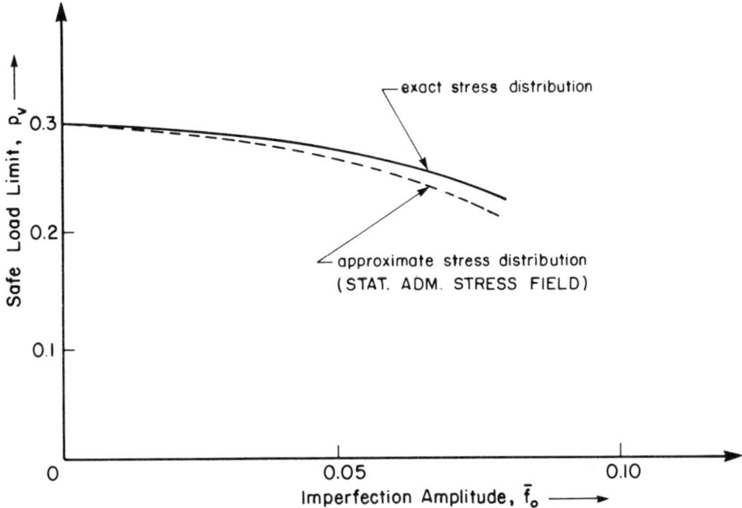

FIGURE 29. Imperfection sensitivity of safe-load limit for concrete column (from [71]).

In summary, using Equation (7.33a) and/or (7.22b), one can establish a P_v vs. slenderness ratio (S) relation which is very convenient in design and has the general form

$$P_v = P_v[P_E(S), P_A, \bar{f}_o] \tag{7.34}$$

a relation which for the concrete column treated earlier is plotted in Figure 30 for $\mathring{w}_1 = 0$ and $\mathring{w}_1 = 0.1$ in. (2.54 mm), with the curve for the perfect column ($\mathring{w}_1 = 0$) shown as a heavy line. For comparison, the corresponding experimental data, taken from [3], is also indicated and clearly exhibits the same general P_v vs. S characteristics. One must recognize, however, that the validity of Equation (7.34) is restricted to stress levels lower than σ_s (see Figure 1), for which creep is limited, or, alternatively, for slenderness ratios $\lesssim$ 60. Therefore, the curve shown in Figure 30 is replaced for small slenderness ratios by the horizontal dotted line, $P = P_s$, which may be considered to be the safe-load-limit for short columns.

The Euler load, P_E, based on the initial tangent modulus of the stress–strain curve for $t = 0$, which serves in this discussion only as a slenderness-dependent parameter, is also

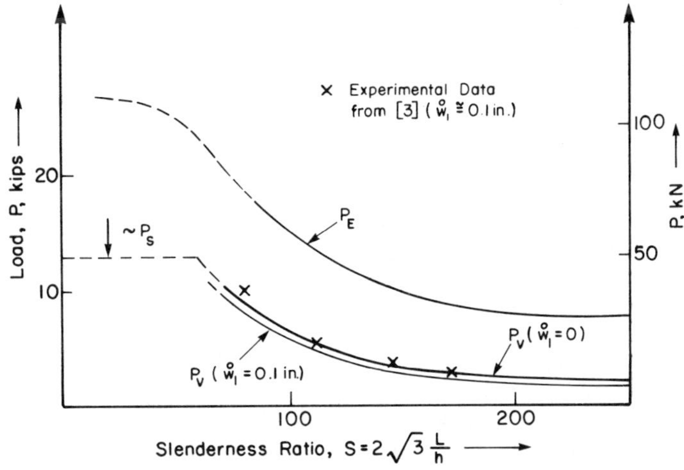

FIGURE 30. Effect of slenderness on safe-load limit (from [71]).

plotted in Figure 30 as a function of S. For columns made of materials with a nonlinear stress–strain diagram at $t = 0$, the actual instantaneous critical load will be lower since some "modified" or "tangent" modulus of elasticity will be used which is smaller than the value E_o, described above. Such stress–strain nonlinearity at $t = 0$ does not, however, affect Equation (7.34).

If the axial load value, $P_v < P < P_E$, one should, normally, expect instability/failure after some period of time, a problem to be addressed in the next section.

8. TIME-DEFLECTION ANALYSIS OF COLUMNS: AN INCREMENTAL APPROACH

When dealing with viscoelastically unstable columns the deflection-time response is a monotonically increasing curve. Consequently, one might be interested in the determination of a "safe service period" during which, despite increasing deflections, the structure can safely carry a given load. The time-deflection characteristics of such columns may differ significantly, depending on whether or not the structure is perfect or imperfect. As indicated by the analysis in previous sections, it might be advisable to treat the structure with its actual imperfections if one is to obtain a realistic estimate of the safe service period.

Results presented in Sections 3–5 indicate that in such an analysis inertial forces are negligible and, therefore, the problem may be treated as a quasi-static one. To arrive at a relatively precise value for the rate of increase in deflections, it is important that the analysis be based on a constitutive law defining the time dependent behaviour of the material fairly accurately. For various reasons, stability analyses of columns made of such materials were carried out, in the past, using relatively simple approximate constitutive models. For example, in [19,72], a time-deteriorating elastic modulus was used [see Equation (2.12)] which led to decreasing bending stiffness of the column and, consequently, allowed determination of "critical times." Results from such analyses, unfortunately, showed poor correlation with test results [3], such predictions of critical times apparently over-estimating the useful life span of columns made of time-dependent materials.

A closed-form solution for the time-deflection problem of columns, as given in Section 4.1 [see Equation (4.7)], can be obtained only for columns for which the material behaviour is described by the simplest linear models. In this section we present a more general approach, allowing the time-deflection analysis of columns made of non-linear materials as defined by Equation (2.14). The governing equations are derived in terms of rates of the corresponding variables and to facilitate the numerical solution technique, an incremental approach is introduced and adopted.

To start with, the constitutive law, Equation (2.14), is rewritten in the form

$$\dot{\epsilon}(t) = \dot{\epsilon}_e(t) + \dot{\epsilon}_v(t) \tag{8.1}$$

where

$$\dot{\epsilon}_e(t) = \dot{\sigma}(t)/E_o \tag{8.2a}$$

$$\dot{\epsilon}_v(t) = \frac{1}{\nu_1}\frac{d}{dt}\int_o^t [\sigma(\tau)]^n \hat{j}(t - \tau)d\tau + \frac{1}{\nu_2}[\sigma(t)]^n \tag{8.2b}$$

thus, formally, decomposing the strain rate into its elastic ($\dot{\epsilon}_e$) and viscous positions ($\dot{\epsilon}_v$). In this "strain-formulation," the strain rate can only be determined if the entire stress history and the stress rate at the current time are known.

The rates of bending moment and axial force (see Figure 25a) are defined by

$$\dot{M} = \int_A \dot{\sigma} y dA$$

$$\dot{N} = \int_A \dot{\sigma} dA \tag{8.3a,b}$$

while from the Bernoulli-Euler hypothesis one can write the geometric relation as

$$\dot{\epsilon} = \dot{\epsilon}_o + y\dot{\varkappa} \tag{8.4}$$

which when used with Equations (8.1)–(8.3), yields

$$\dot{\epsilon}_o = \frac{\dot{N}}{E_o A} + \frac{1}{A}\int_A \dot{\epsilon}_v dA \tag{8.5a}$$

$$\dot{\varkappa} = \frac{\dot{M}}{E_o I} + \frac{1}{I}\int_A y\dot{\epsilon}_v dA \tag{8.5b}$$

Equations (8.5) define the time–deflection behaviour of any beam-column subjected to arbitrary loading and boundary conditions.

Clearly, $\dot{M}$ and $\dot{N}$ must be determinable from those conditions. In the present case, $\dot{N} = 0$ and $\dot{M} = P\dot{w}$ so that

$$\dot{\epsilon}_o = \frac{1}{A}\int_A \dot{\epsilon}_v dA$$

$$\frac{d^2(\dot{w})}{dx^2} + \frac{P}{E_o I}\dot{w} = -\frac{1}{I}\int_A \dot{\epsilon}_v y dA \tag{8.6a,b}$$

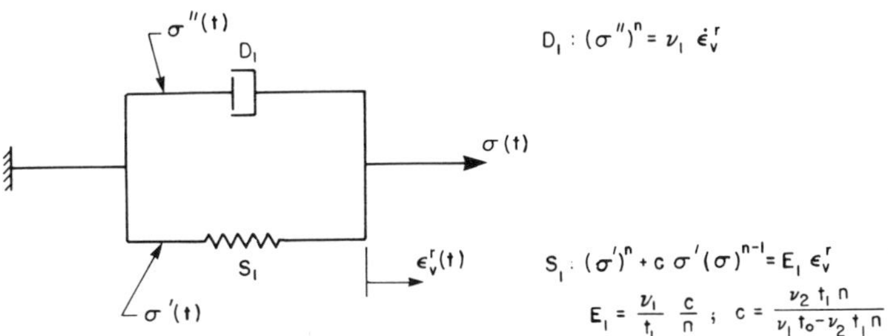

FIGURE 31. Nonlinear Kelvin body.

equations which define the rate of deformation of the column as represented by the strain and deflection rate, $\dot{\epsilon}_o(x,t)$ and $\dot{w}(x,t)$, of the centroidal axis of the structure. Note that Equations (8.6) are coupled through the integrals containing the viscous strain rate, $\dot{\epsilon}_v(x,y,t)$. Note also that these equations are linear in terms of $\dot{\epsilon}_o$ and $\dot{w}$, the material nonlinearity affecting only the creep rate expression. Consequently, their solution is tractable provided the creep rate, defined by Equation (8.2b), can be determined. In this latter expression, the second term can clearly be calculated for a given value of stress, $\sigma(x,y,t)$. Somewhat more problematic is the first term, involving an integral and representing the reversible portion of the viscous strain rate, $\dot{\epsilon}_v^r$, given by

$$\dot{\epsilon}_v^r = \frac{1}{\nu_1}\frac{d}{dt}\int_o^t [\sigma(\tau)]^n \hat{j}(t - \tau)d\tau \tag{8.7}$$

For some structures subjected to constant external load, the stress level may also be assumed to be constant temporally as for example in the case of a beam undergoing small deflections. In such cases, Equation (8.7) is rewritten as

$$\dot{\epsilon}_v^r(x,y,t) = \frac{1}{\nu_1}[\sigma(x,y)]^n \hat{j}(t) \tag{8.8}$$

For statically indeterminate structures as well as in the case of columns the stress level may vary temporally even under constant external loading and therefore, Equation (8.7) should be used. Due to the complexity encountered in evaluating this integral-differential expression, Equation (8.8) has been assumed to be valid even in cases where the stress varies temporally, i.e.

$$\dot{\epsilon}_v^r(x,y,t) \cong \frac{1}{\nu_1}[\sigma(x,y,t)]^n \hat{j}(t) \tag{8.9}$$

an approximation which was discussed in Section 2 and is referred to as Shanley's hypothesis. In a physical sense, the use of Equation (8.9) means the neglect of hereditary properties of the material. The effect of such a constitutive approximation on stability considerations was treated in detail in [67] and is to be mentioned below.

An approximate method for calculating $\dot{\epsilon}_v^r$ using Equation (8.7) was introduced in [60] and applied and discussed in [68]. The method is based on the similarity between the viscous process defined by Equation (8.7) and the characteristic features of a nonlinear Kelvin body shown in Figure 31. This model can be used to simulate the $\epsilon_v^r(t)$ vs. $\sigma(t)$ characteristics of a material defined by Equation (8.7) provided the materal parameters ν_1, ν_2, t_o and t_1, defined in Section 2, are chosen to be identical for the model and the material. The function $\hat{j}(t)$ can be defined, approximately, in terms of t_o and t_1 (see [68]).

Using the nonlinear Kelvin model, $\dot{\epsilon}_v^r$ is determined from the element D_1 (see Figure 31) as

$$\dot{\epsilon}_v^r(x,y,t) \cong \frac{1}{\nu_1}[\sigma''(x,y,t)]^n \tag{8.10}$$

in which the decomposition of $\sigma(x,y,t)$ into its constituents $\sigma''(x,y,t)$ and $\sigma'(x,y,t)$ appears as a problem. The component, σ', is determined from the relation

$$(\sigma')^n + c\sigma'\cdot\sigma^{(n-1)} = E_1\,\epsilon_v^r \tag{8.11}$$

an expression which defines the nonlinear spring, S_1. Finally, $\sigma''(x,y,t) = \sigma(x,y,t) - \sigma'(x,y,t)$. Such decomposition is clearly possible and must be done for various values of x, y, and t.

Since the problem being discussed here is nonlinear, its solution requires a numerical procedure. Consequently, the column is discretized into M by N segments in the x and y directions, respectively. The time–deflection behaviour of the structure is determined using the following procedure:

1. At a certain instant of time, $t = t_k$, the state of stress, strain (including ϵ_v^r) and deflection must be known for every node point x_i, y_j; at $t = 0$ the elastic solution represents the initial deformed state.

2. The creep rate, $\dot{\epsilon}_v(x_i, y_j, t_k)$ is calculated using the nonlinear Kelvin model discussed above.
3. Numerical integration over the column cross-section results in values for the R.H.S. of Equations (8.6) at points x_i, t_k; then, using Equations (8.6), $\dot{\epsilon}_o(x_i,t_k)$ and $\dot{w}(x_i,t_k)$ are determined.
4. Using some time interval, Δt, increments for all variables are calculated and added to the values from the previous time step to obtain up-dated magnitudes at time $t_{k+1} = t_k + \Delta t$.
5. Steps 1–4 are repeated for $t = t_{k+1}$.

This "incremental" procedure was used in [60,61,67] where the reader can also find details of the computer code and a discussion concerning convergence of the procedure. Typical results from such an analysis for the time–deflection behaviour of a viscoelastically unstable materially nonlinear ice column are shown in Figure 32. Since the model used for ice in this example depicts a material of the "fluid type," the deflection–time response is, of course, monotonically increasing, indicating viscoelastic instability for all load levels. This type of unbounded response is confirmed by the curves in Figure 32, drawn for various initial imperfections. As can be observed, the rate of deflection increases dramatically when mid-span displacements reach a value approximately equal to the column width. The time at which such displacements are attained was arbitrarily defined in the above noted publications as the "critical time." The "dashed" curves indicate solutions in which the material law was approximated by Equation (8.9), thereby neglecting hereditary effects. This approximation clearly overestimates the rigidity of the structure in as much as the "safe service period" calculated on the basis of this approximation is much longer than the actual critical time. The effect of this constitutive approximation can even lead to qualitatively erroneous conclusions concerning the stability behaviour of viscoelastic columns as was discussed in detail in [67] and is indicated in Figure 33.

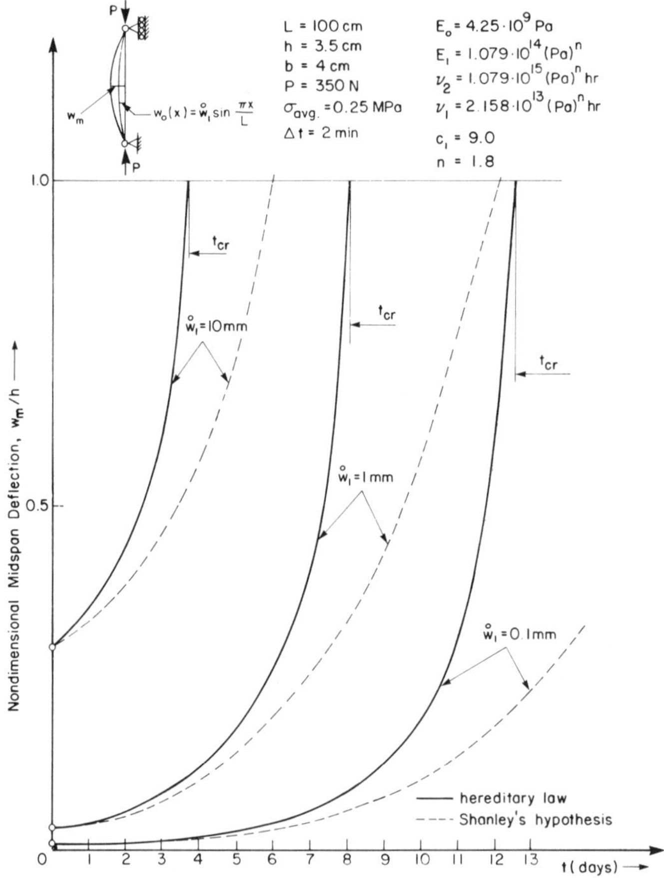

FIGURE 32. Time-deflection behaviour of ice columns with various initial imperfections (from [67]).

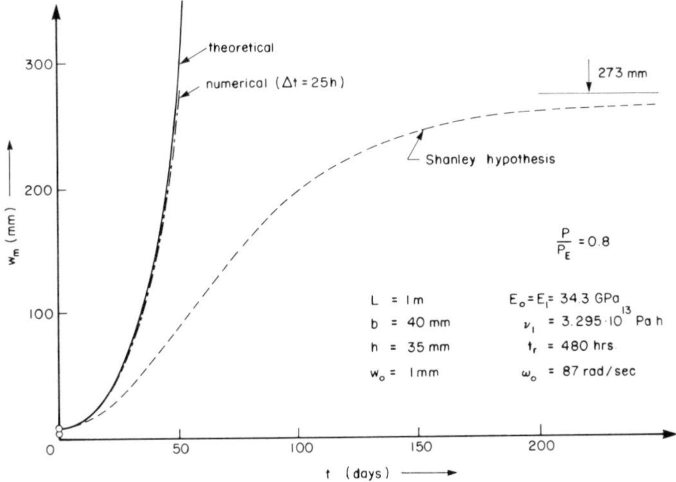

FIGURE 33. Effect of Shanley's hypothesis on deflection history of viscoelastically unstable concrete columns (from [67]).

Since the numerical "incremental" approach used here permits monitoring of the temporal variation of all variables at every point of a column up to failure, a "safe service period" can be specified in terms of material properties or other requirements. Actual failure and the associated time can therefore be based on a limiting stress, strain, or deflection magnitude. As an example, the strain–stress–deflection–time behaviour of an ice column is indicated in Figure 34. For a column made of a brittle material like ice, failure is usually triggered by initiation of cracking in the tensile zone. Assuming cracking to start at a stress level of, say, 1.6 MPa, one can calculate the failure time [61]. For materials in which failure stress is assumed to be a function of the strain rate, a damage function may be introduced to facilitate a natural definition of the instant of failure (see [69]).

The formulation and incremental solution technique used here for a homogeneous column can, of course, also be applied to unhomogeneous structures, as for example, reinforced or layered/sandwich columns. The method also admits inclusion of other effects, like shrinkage, plastic deformations, thermal influences, etc. An incremental solution technique similar to the one discussed here was used in [3] for the time–deflection analysis of a reinforced concrete column. Although heredity was neglected in that formulation, the careful simulation of the material behaviour together with the incremental solution technique resulted in analytical predictions in close agreement with their experimental data.

Finally, the incremental formulation/solution technique is a convenient and powerful tool from a practical viewpoint. It provides a solution to the quasi-static time–deflection problem of columns made of time-dependent materials.

One should note that deflection magnitudes in this discussion were limited so as to justify a geometrically linear stability analysis for the column.

9. GEOMETRICALLY NONLINEAR AND POST-BUCKLING EFFECTS

In a recent publication [66] an "exact" geometrically nonlinear analysis of simple structural models, consisting of rigid bars, springs and dashpots and chosen so as to simulate a variety of elastic post-buckling structural behaviour, was carried out. To emphasize and to focus on the geometrically nonlinear effects, the material characteristics for the models were chosen to be linear. Based on the results of that analysis, together with those presented in previous sections of this article, a number of general conclusions concerning columns and structures made of time-dependent materials can be drawn:

1. For structures made of "fluid"-type materials, subjected to a constant (non-zero) load, the deflection–time response is a monotonically increasing function of time, irrespective of the load/stress level. Specific nonlinear load deflection and/or post-buckling characteristics of the corresponding elastic structure will influence this behaviour only to the extent that it may accelerate or decelerate the rate of deformations/deflections thereby affecting the length of a so-called "safe service period" which the designer might wish to establish. However, the ultimate fate of the structure under continuing loads is unaffected by its instantaneous geometric nonlinearities or post-buckling characteristics.

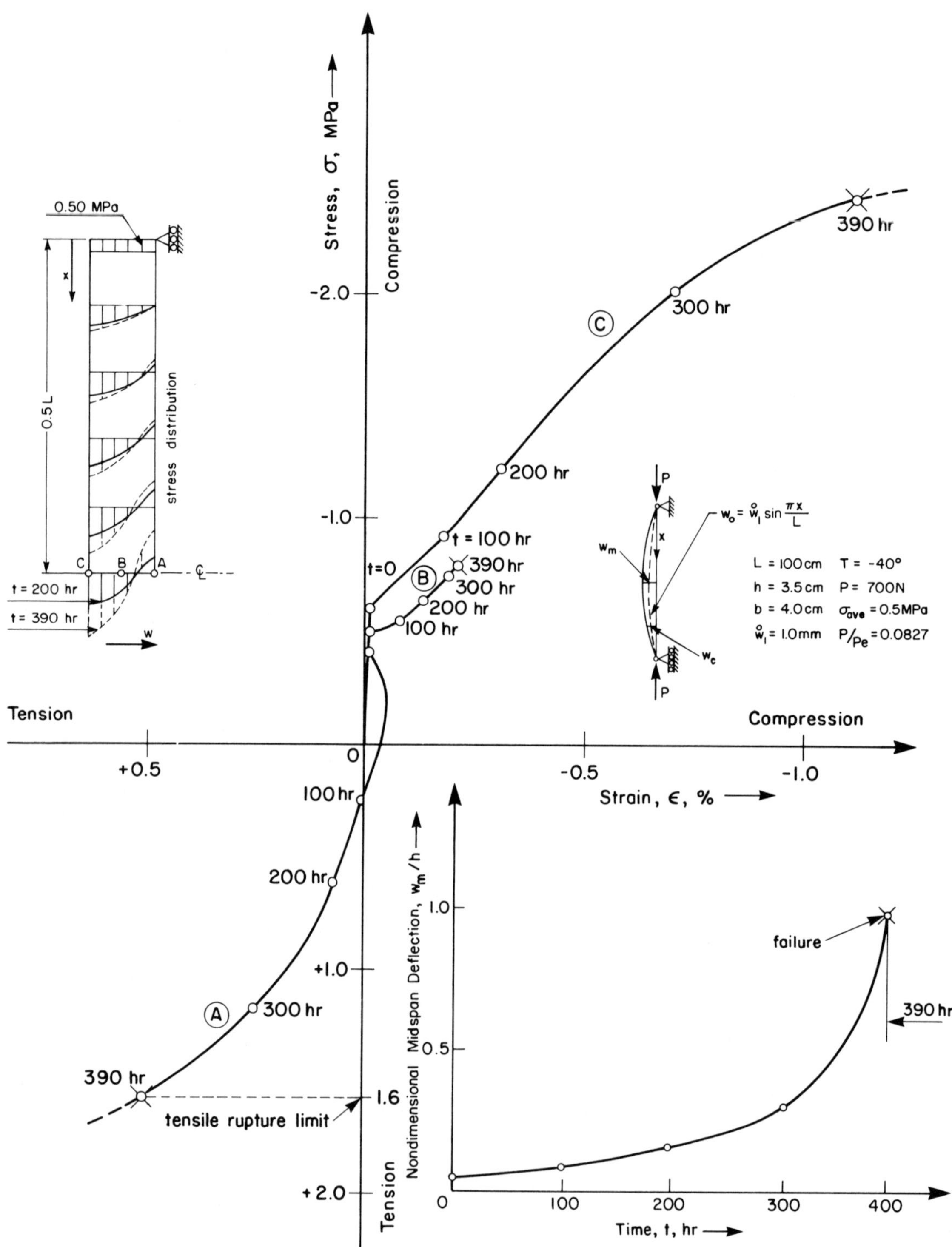

FIGURE 34. Stress-strain deflection history of ice column (from [60]).

2. A structure made of a "solid"-type material with limited creep and subjected to load levels less than the safe load limit P_v, will exhibit a deflection–time curve which shows a bounded response as $t \to \infty$, irrespective of its particular instantaneous nonlinear deflection or post-buckling characteristics. In this sense, these structures exhibit "stable" behaviour under such load levels. The magnitude of this "limited response" may be unacceptable on the basis of esthetic or serviceability criteria as a consequence of which the designer may wish to impose a "safe-service-period" even for such load levels, a period which will be based on acceptable deflection/deformation magnitudes. The value of the safe load limit, in general, is given as

$$P_v = \frac{P_M}{1 + \mathcal{J}_\infty} \tag{9.1}$$

in which P_M denotes the maximum load attained along the instantaneous load deflection curve. In the case of locally and globally stable structures, like columns, for which such maxima do not exist, P_M is taken as the instantaneous critical load for the structure, the Euler load, P_E, in the case of columns.
3. For structures made of "solid"-type materials subjected to load levels $P_v < P < P_M$ the time–deflection response is significantly influenced by the instantaneous (elastic) load deflection and post-buckling characteristics of the structure. Specifically, in the case of structures exhibiting stable equilibrium at the bifurcation point and stable post-buckling load–deflection characteristics, geometric nonlinearities appear to slow down the viscous processes so that even for such viscoelastically unstable structures the deflection–time curve shows a limiting value as $t \to \infty$, as opposed to the unbounded response obtained from a geometrically linearized analysis. On the other hand for structures exhibiting unstable equilibrium at their bifurcation point and unstable post-buckling load-deflection characteristics, geometric nonlinearities appear to accelerate the viscous process so that an unbounded response is obtained within a finite time period as opposed to infinite time predicted from a geometrically linearized analysis. Determination of such a finite time period for viscoelastically unstable structures, during which they can perform a useful function, should clearly be the prime concern of the designer and is one of the main purposes of this whole section.

The above outlined results and conclusions are illustrated in Figures 35a, b, and c, all three of which consist of two diagrams, an upper portion showing instantaneous load–deflection curves for perfect ($c = 0$) and imperfect ($c > 0$) structures, with corresponding deflection–time curves appearing on the lower diagram. Time effects for a constant load are indicated on the load–deflection plots by a horizontal line starting at point B_o on the $t = 0$ curve, a line which is the trace of the actual curve in the three-dimensional load–deflection–time domain.

Figure 35a shows typical curves for a structure exhibiting stable instantaneous post-buckling behaviour with the dashed curve on the load–deflection plot representing the trace of the load–deflection curve for a structure undergoing limited creep viscous processes and for $t \to \infty$. The horizontal line, indicating time effects, terminates at point B, a point on this dashed curve, confirming the above noted bounded response for such structures under all load levels $P < P_E$.

Figure 35b depicts analogous features for a structure exhibiting neutral equilibrium at the bifurcation point and neutral post-buckling load–deflection behaviour. In this case, as the curves also indicate, the deflections are limited only for load levels $P < P_v$, with all other loads producing unbounded response.

The third set of diagrams in this figure, Figure 35c, show instantaneous load–deflection and deflection–time curves of a structure with unstable post-buckling behaviour. In particular, the curve $A_oA_1A_2$, connecting the maxima of the various instantaneous load–deflection curves, indicates the reduction in P_M as a function of initial imperfections, thereby signifying imperfection sensitivity, a feature characteristic of this type of structure. This curve consequently separates areas of stable and unstable equilibrium in the load-deflection plane. Time effects are, once again, indicated by a horizontal line emanating from B_o and representing the trace of the actual load–deflection–time curve. The trace of the terminal point of this three dimensional curve, shown as point A_2 on the load–deflection diagram in Figure 35c, has coordinates P, δ_{cr}, and t_{cr}, where P denotes a constant load level $P > P_v$, while δ_{cr} and t_{cr} are the corresponding deflection and time, associated with this load level, beyond which equilibrium is impossible.

The deflection–time curve indicated on the lower portion of the figure shows infinite velocity at the corresponding point. Thus the terminal point of the load–deflection–time curve, the trace of which is A_2, is what was referred to in [66] as a "catastrophy point," a point along the "catastrophy curve." The time associated with this point, indicating the onset of dynamic instability, provides one natural definition for "critical time," a period which was previously defined in this article on the basis of deflection limitations and/or strength properties of the material. This critical time is calculated using the expression

$$t_{cr} = \int_{\delta_o}^{\delta_{cr}} (\dot{\delta})^{-1} \, d\delta \tag{9.2}$$

where δ_o and δ_{cr} are determined from an analysis of the in-

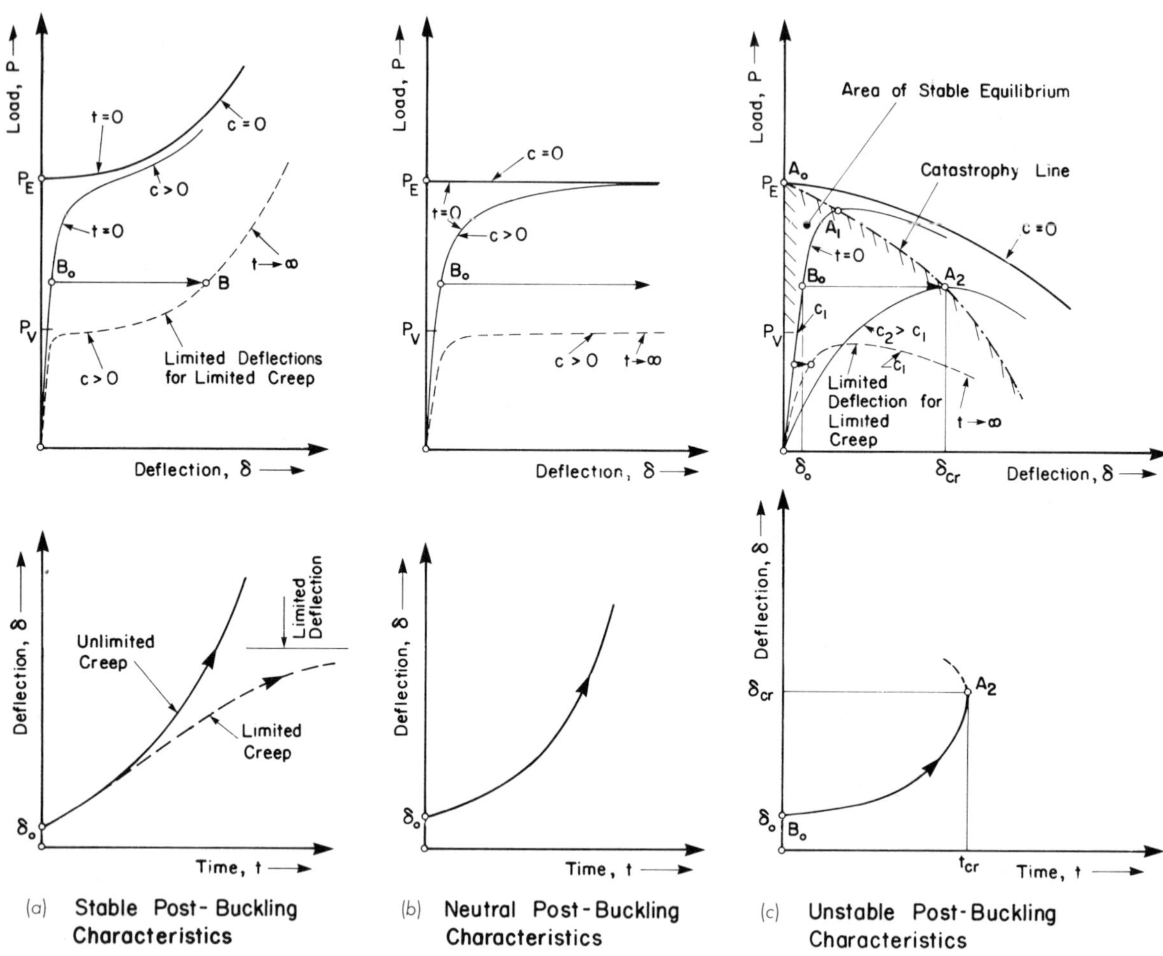

FIGURE 35. Load-deflection and deflection-time curves for structures with various post-buckling characteristics.

stantaneous response of the structure while the rate of the deflection, $\dot{\delta}$, depends on the viscous properties of the material. For viscoelastically unstable structures, $\dot{\delta} > 0$. In general, $\dot{\delta} \rightarrow 0$ only for structures made of solid-type materials subjected to load levels $P < P_v$, for which $t_{cr} \rightarrow \infty$. Critical time, as defined here, was also discussed in [25,29] for other simple structures.

Parts (a) and (b) of Figure 35 are directly relevant to columns since as is well known, their equilibrium at the bifurcation point is stable with stable post-buckling characteristics. A typical non-dimensional load-deflection curve for a column is shown in Figure 36a indicating the rather extensive relatively flat portion of the post-buckling curve in the vicinity of the bifurcation point. It is this characteristic of the post-buckling load–deflection curve which justifies its approximation by a horizontal line in the vicinity of the bifurcation point, an approximation which is used in linear stability analyses of columns and is characteristic of neutral post-buckling behaviour. An examination of the post-buckling curve of columns indicates that geometric nonlinearities are negligible for mid-span deflection magnitudes of the order of 20% of the span, $[(w_m/L_o) \lesssim 0.2]$, a deflection which corresponds to a load increase above the classical Euler load of only some 5%. Figure 36b, taken from [75], shows a non-dimensional deflection–time plot for a column subjected to a constant load $P > P_v$ the constitutive law of which is modelled by a three element solid material model. The curves on this diagram indicate significant discrepancies between results from a linear and nonlinear analysis only for deflection magnitudes $w_m/L_o \gtrsim 0.2$ with visible stiffening effects in the nonlinear analysis case, effects which increase with deflections and ultimately lead to a bounded response. These results thus confirm our conclusions concerning the effect of geometric nonlinearities on the behaviour of structures with stable post-buckling characteristics.

The applicability and relevance of Figure 35c to a discussion on stability of columns is somewhat more subtle. There are, however, structural configurations, containing columns, for which an exact geometrically nonlinear elastic analysis reveals that the load-deflection characteristics of the overall structure exhibit the type of post-buckling behaviour depicted on Figure 35c, with associated imperfection sensitivity. The degree of imperfection sensitivity of such structural configurations, together with the magnitude of imperfections normally encountered in practice, makes this post-buckling feature for structures of Hookean materials normally unimportant. However, when such structures are made of time-dependent materials, the problem may suddenly take on significance in as much as the initial imperfections may be magnified substantially by the viscous processes, thereby reducing the load carrying capacity of the structure and also decreasing its safe service period. As opposed to stable post-buckling behaviour, which the designer thinks he is dealing with, he may actually be faced with a response typical of structures with unstable post-buckling load-deflection characteristics. Thus, the second main purpose of this section is to bring to the attention of the designer such possible post-buckling behaviour in the case of structures containing columns and made of time-dependent materials.

To illustrate this type of behaviour and to facilitate tractability in the analysis, a simplest structural configuration involving a column was chosen (see Figure 37a). As is clear from this figure, this structure consists of a single inclined column, hinged at its lower (left) end and supported on rollers moving along a vertical surface at its upper (right) end. The vertical load, P, is applied at the rollers and it is assumed that the column may have an initial (sinusoidal) imperfection defined by a mid-span deflection, w_o, and that its weight is negligible. Let us first treat the elastic case and recognize that post-buckling midspan-deflections for the ideal column in the neighborhood of the bifurcation point are small; i.e. $w_m/L_o << 1.0$. Neglecting all terms higher than quadratic in (w_m/L_o), one arrives at the approximate post-buckling load-deflection relation in the form

$$\frac{P}{P_b} \cong 1 - \frac{\pi^2}{4}\left(\frac{w_m}{L_o}\right)^2\left(\cot^2\alpha_o - \frac{1}{2}\right) \tag{9.3}$$

in which $P_b = P_E[1 - \cos^2\alpha_o/(1 - \pi^2/S^2)]^{1/2}$ denotes the load value corresponding to the column's critical, Euler force, $P_E = \pi^2 E_o I/L_o^2$, and S represents the slenderness ratio. In deriving Equation (9.3) use was made of the geometric constraints imposed on the column by its supports; its shortening was calculated taking into account axial and transverse deflection effects; and its post-buckling load-deflection curve was approximated by an expression in which fourth order (and higher order) terms in (w_m/L_o) were neglected.

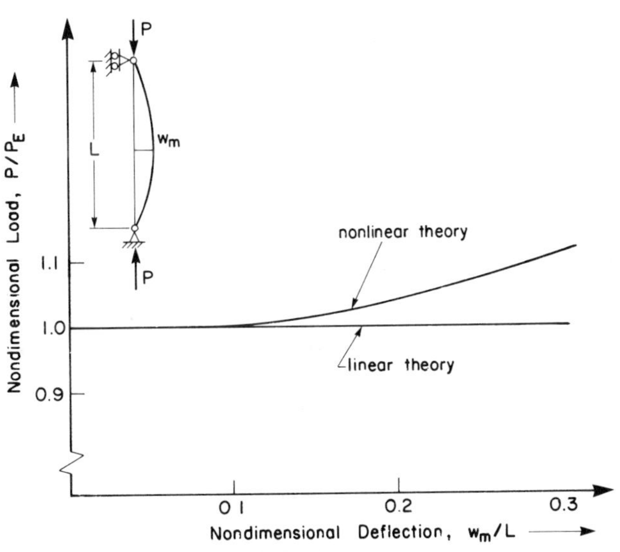

(a) Post-Buckling Curve for Elastic Column

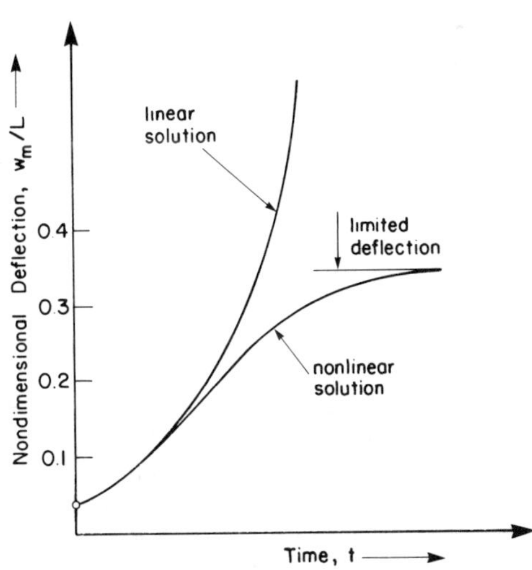

(b) Effects of Large Deflection on Deflection-Time Curve of Linearly Viscoelastic Column (from [75])

FIGURE 36. Effect of large deflections on characteristics of column behaviour.

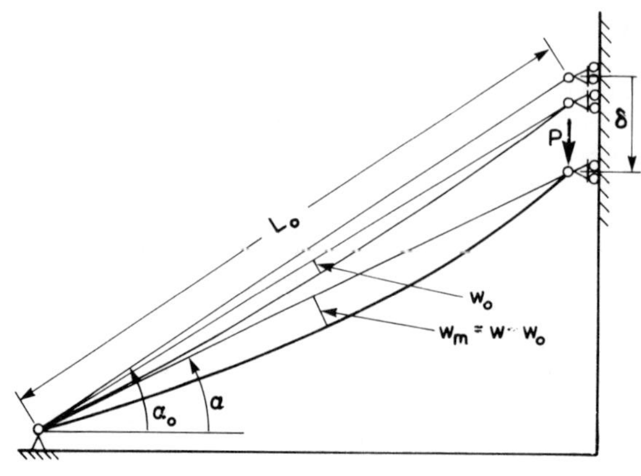

(a) Geometry

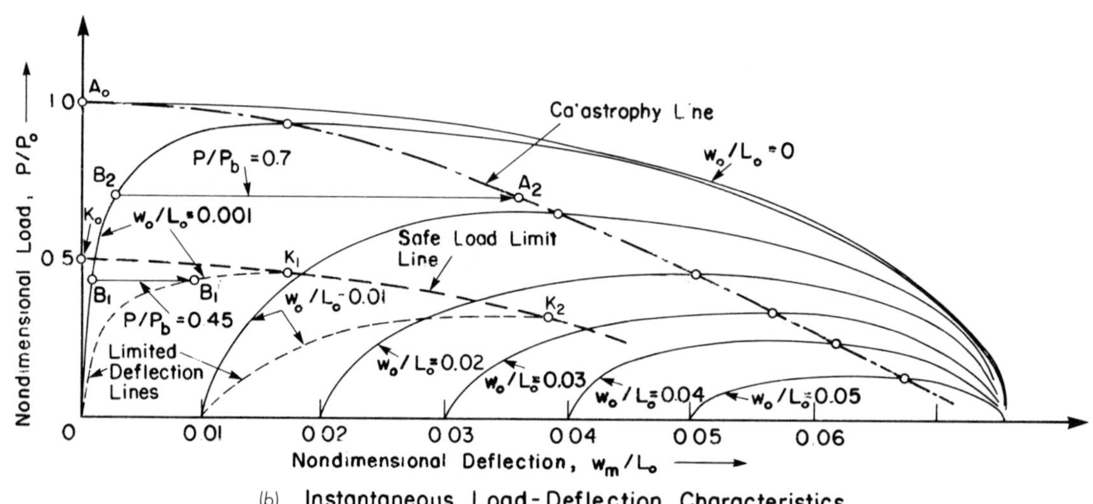

(b) Instantaneous Load-Deflection Characteristics

FIGURE 37. Column structure exhibiting unstable post-buckling behaviour.

Upon reflection, Equation (9.3) defines stable post-buckling behaviour for $\alpha_o \gtrsim 55°$ and unstable post-buckling characteristics for initial inclinations smaller than this angle. For our example structure a value $\alpha_o = 10°$ and a slenderness ratio $S = 100$ were chosen. For this structure, the elastic post-buckling load-deflection curves for various initial imperfections, are given in Figure 37b. Note that the whole deflection range on this figure is relatively small compared with the deflections indicated on Figure 36a, clearly justifying a linearized moment-curvature relation, Equation (8.5), for the analysis of the column itself. In analysing the structure as a whole, one naturally must take into account the rigid-body (chord) rotation of the column, as defined by the variation in α. Assuming the external load, P, to be constant, the axial force, N, and bending moment, M, and their rates, are given by

$$N = P/\sin\alpha; \quad \therefore \quad \dot{N} = -\frac{P\cos\alpha}{\sin^2\alpha}\dot{\alpha} \tag{9.4a,b}$$

$$M = \frac{P}{\sin\alpha}(w + w_o); \quad \therefore \quad \dot{M} = -\frac{P\cos\alpha}{\sin^2\alpha} \cdot (w + w_o)\dot{\alpha} + \frac{P}{\sin\alpha}\dot{w} \tag{9.4c,d}$$

with the use of which, Equations (8.5) become

$$\dot{\epsilon}_o = -\frac{P}{E_oA}\frac{\cos\alpha}{\sin^2\alpha}\dot{\alpha} + \frac{1}{A}\int_A \dot{\epsilon}_v\, dA \tag{9.5a}$$

$$\frac{d^2}{dx^2}(\dot{w}) + \frac{P}{E_oI}\frac{\dot{w}}{\sin\alpha} - \frac{P}{E_oI}\frac{\cos\alpha}{\sin^2\alpha}(w + w_o)\,\dot{\alpha}$$

$$+\frac{1}{I}\int_A y\dot{\epsilon}_v dA = 0 \tag{9.5b}$$

From geometry the further relation is obtained in the form

$$\dot{\alpha} = -\frac{\cos^2\alpha}{\cos\alpha_o \sin\alpha}\frac{1}{L_o}\int_o^{L_o}[\dot{\epsilon}_o + \frac{d}{dx}(w + w_o)\cdot$$

$$\cdot\frac{d}{dx}(\dot{w})]\,dx \tag{9.6}$$

Results from such an analysis of the structure shown in Figure 37a and made of a linear three-element-model material (see Figure 9) with limited creep are presented on Figure 38. The material parameters were chosen so that the safe-load-limit for the column, $P_v = 0.5\,P_b$. Consequently, for $P/P_b = 0.7$ the structure should exhibit viscoelastic instability, as is confirmed by the nondimensionalized deflection-time curve B_2A_2 which shows an infinite deflection rate at point A_2 thus defining the "critical time," t_{cr}. The trace of this constant load-deflection-time curve is also shown on Figure 37b, appearing as the horizontal line B_2A_2, the end points of which lie on the instantaneous load-deflection curve for the imperfection chosen (w_o/L_o = 0.001) and on the trace of the catastrophy line, respectively. Figure 38 also shows the deflection-time curve for this structure when the effect of chord rotation is neglected in the analysis (the dashed curve). This curve is monotonically increasing with time without any "critical" points.

For the load level $(P/P_b) = 0.45$, a load level below the safe-load-limit, one expects a bounded response as $t \rightarrow \infty$, as is indeed confirmed by the lowest of the three curves, B_1B_1', on Figure 38. The corresponding trace on Figure 37b is shown as the horizontal line B_1B_1' with the terminal point lying on the "limited deflection line" for this structure, such lines being determined on the basis of an asymptotic solution for which the ultimate $(t \rightarrow \infty)$ stiffness of the structure is reduced by the factor $1/(1 + E_o/E_1) = 0.5$. The maxima of these limited deflection lines, denoted as points K_1 and K_2, define the safe-load-limits of this structure for given imperfections, while the curve $K_0K_1K_2$, connecting these points, clearly indicates the effect of imperfections on the safe-load-limit, an effect which is analogous to the imperfection sensitivity depicted on Figure 22b for the simple rigid-bar-spring-dashpot system.

The final figure, Figure 39, is intended to compare the instantaneous (elastic) response of the structure analysed with the corresponding elastic "snap-through" problem in which buckling of the column is excluded. The nondimensional load-displacement curve from the latter analysis, shown as the dashed line, indicates the traditional nonlinear load-deflection characteristic of the snap-through "limit-point" problem, with the snap-through load, $\overline{P}_m$, an order of magnitude larger than the bifurcation load, $\overline{P}_b$. Similarly, the corresponding snap-through curve for $t \rightarrow \infty$ is indicated as a light dashed line. The results calculated above for the viscoelastic structure are also transferred to this diagram

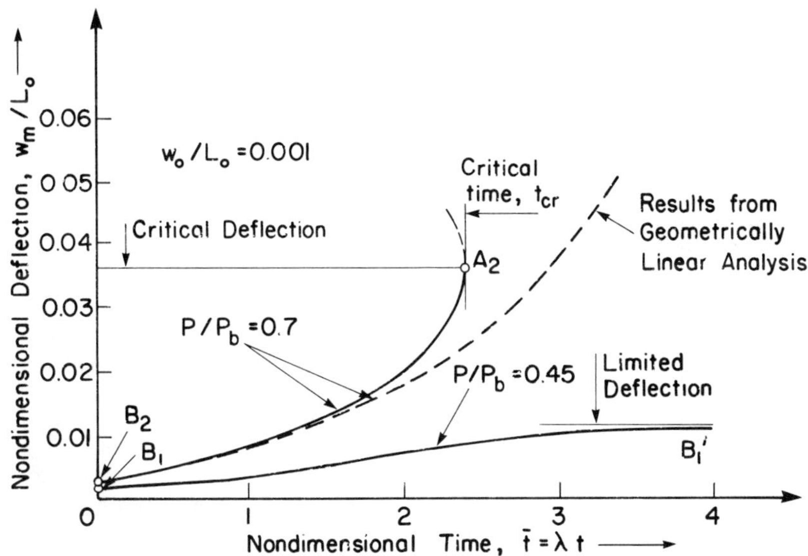

FIGURE 38. Deflection-time curves for column structure exhibiting unstable post-buckling behaviour.

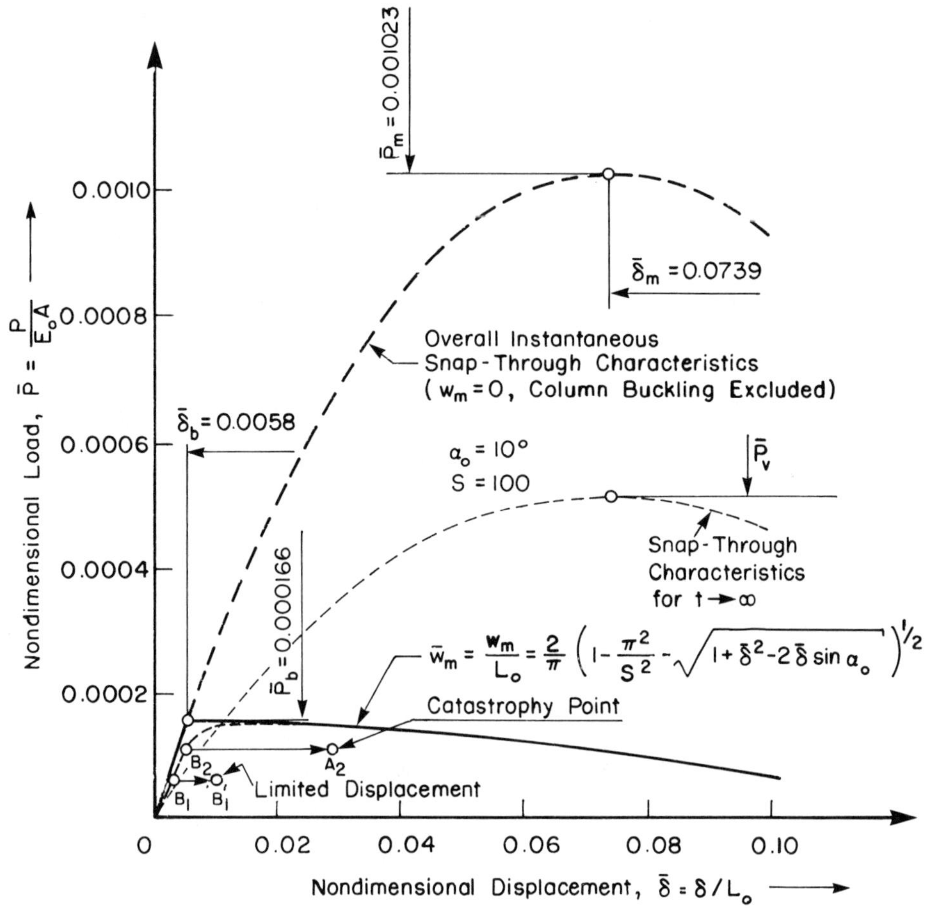

FIGURE 39. Overall load-displacement characteristics of column structure.

and are indicated by the lines B_2A_2 and B_1B_1' for the viscoelastically unstable and absolutely stable responses, respectively. Note that on Figure 39, the vertical displacement of the roller support (the point of load application) is used (nondimensionally) as the abscissa while in previous diagrams dealing with this structure, the mid-span deflection of the column w_m, was used. The snap-through problem of a viscoelastic structure was treated in [25] with analogous results.

10. CONCLUSIONS

This article presents a review and summary of the authors' present understanding and interpretation of the current state of knowledge concerning the stability behaviour of columns made of linear and nonlinear time-dependent materials. It is limited not only in extent but reflects certain gaps in the present understanding of this complex subject matter. In addition, personal preferences and/or prejudices appear in the manner of presentation, in the emphasis placed on various topics, and in deciding what was to be included within the limited space available. In view of the very nature of this article, the authors had three main goals:

1. To bring together, in one convenient publication, an up-to-date precise and concise summary of the stability behaviour of linearly and nonlinearly viscoelastic columns and associated analysis procedures, a summary to be used by designers, as well as by researchers.
2. To focus attention on specific topics or areas in which further research, either experimental or analytical, is required.
3. To help technical and professional societies and associations involved in the writing of codes, specifications and standards in making use of the latest research results and the understanding gained therefrom in defining guidelines and specifications concerning various important stability parameters of columns and structures made of time-dependent materials.

The article begins with a brief review of frequently used and/or well known uniaxial linear and nonlinear constitutive laws for time-dependent materials, exmphasizing the importance of hereditary effects in problems in which stress varies temporally even though the external load may be constant. "Fluid"- and"solid"-type materials are defined, with the former exhibiting unlimited creep for every stress level and the latter showing limited creep for stress levels below the "safe-stress" value, σ_s, as $t \rightarrow \infty$. Pitfalls associated with certain constitutive approximations, when used in stability problems, are pointed out so as to alert the designer to the possible danger associated with such approximations.

An "exact" linear stability analysis of linearly viscoelastic simply supported perfect and imperfect columns is carried out using the dynamic approach from which closed-form expressions for the safe-load-limit and an arbitrarily defined safe-service-period or "critical time" are obtained. The main results from the dynamic analysis are confirmed using a static formulation. In using the traditional static approach, a new and appropriate redefinition of "adjacent equilibrium positions" had to be introduced on the basis of which an "exact" static stability analysis was carried out and the destabilizing effect of time on such structures established. Significantly, the expression for the safe-load-limit obtained from the static approach confirmed the corresponding expression obtained earlier from a dynamic formulation. Furthermore, results from the analysis of the imperfect structure focused attention on the extreme sensitivity of viscous strain rates to imperfections suggesting that any safe-service-period or "critical time," if it is to be a practically useful and reliable design parameter, must be based on the analysis of the "real" structure, including "actual" imperfections.

The general expression obtained for the safe-load-limit of linearly viscoelastic columns from the static approach was also confirmed by means of an asymptotic solution of the dynamic stability equations, an approach which allowed generalization of the expression for the safe-load-limit of simply supported columns to columns with arbitrary boundary conditions.

Since most real materials are nonlinear, it was important to include in this article a section on the stability behaviour/analysis of columns made of nonlinear time-dependent materials. A relatively simple and convenient method for the determination of the safe-load-limit of such structures was presented and applied to the determination of the safe-load carrying capacity of a concrete column. The results from this analysis show surprisingly good agreement with corresponding experimental data. From a design point of view, it is important to note the imperfection sensitivity associated with a materially nonlinear stability analysis of columns, a feature which may significantly reduce the safe-load-limit.

The penultimate section of this report presents a completely general method for determination of the complete time-deflection behaviour (up to failure) of structures made of time-dependent materials. It is based on an incremental formulation of the governing equations and an iterative solution technique which admits inclusion of all known characteristic features of these types of materials, including hereditary effects. The technique is used to obtain the complete stress-strain-deflection history of an ice column for which the "critical time," or failure time, is based on tensile strength properties of the material. The analysis technique is also used to point out the shortcoming of a commonly used constitutive approximation, referred to as Shanley's hypothesis, which basically neglects hereditary effects and the use of which can lead even to qualitatively erroneous conclusions concerning stability behaviour.

Although columns as structural members exhibit stable equilibrium at their bifurcation point and stable post-buckling load-deflection curves, there are structural configurations, containing compression members, the overall load-deflection curves of which exhibit unstable post-buckling characteristics with associated imperfection sensitivity. The last section of this article briefly reviews this topic and alerts the designer to the possibility of encountering structural configurations in which such unstable post-buckling behaviour and associated imperfection sensitivity may become significant due to the fact that the initial imperfections may be magnified several times or orders of magnitude by the viscous processes. Thus, such post-buckling characteristics and imperfection sensitivity, which in the case of structures made of Hookean materials may not be significant from a design point of view, may have to be considered in the case of structures in which time is ever-present as a destabilizing parameter and as a catalyst for magnification of imperfections, resulting in possible significant reductions in the safe-load-carrying capacity and/or the safe-service-period of the structure.

It is in this last section, in which a geometrically nonlinear analysis is used, that an opportunity arose to introduce a "natural" definition for critical time as the time period beyond which structures exhibiting unstable post-buckling behaviour cannot be in equilibrium under the applied load. It was therefore possible to illustrate the effects of nonlinear geometry on the "critical time" of viscoelastic structures, a parameter which on the basis of a linearized analysis is predicted to be infinitely long, while for structures exhibiting unstable post-buckling behaviour, geometric nonlinearities reduce this period to a "finite" interval. Thus in the case of such structures, which are typified by monotonically softening post-buckling load deflection curves, the nonlinear geometry effects appear to accelerate the viscous process. On the other hand, in the case of structures with stable post-buckling behaviour and monotonically stiffening post-buckling load-deflection curves, such geometry effects appear to slow down the viscous process so that even for viscoelastically unstable structures made of solid-type materials the ultimate response/deflection is limited as $t \to \infty$. In the case of structures made of "fluid-type" materials, these geometric nonlinearities affect only the rate of the viscous process, accelerating it in the case of structures with unstable post-buckling behaviour and decelerating the creep rate in structures with stable post-buckling behaviour. The ultimate fate of structures made of fluid type materials, as $t \to \infty$, is, however, unaffected by such geometric nonlinearities, unbounded response being inevitable under continuing load of any magnitude.

On the basis of the material presented in this article, as well as a review of some of the references cited, it is clear that additional experimental work on the constitutive properties of time-dependent materials should be given highest priority. Researchers should determine the safe stress level, σ_s, and associated limiting strain, ϵ_∞, of solid-type materials with limited creep. Information should also be sought on stress independent parameters for this class of materials, particularly those materials which exhibit a well-defined primary and secondary creep stage under certain stress levels.

The authors hope that the present article will contribute to the efficient planning of appropriate experiments whereby such material parameters and information concerning the constitutive behaviour of time-dependent materials can be determined. We also hope that this review and summary will help the analytical researcher in planning his numerical procedures and solution techniques associated with such problems.

Finally, it is the authors' sincere wish that some of the results contained in this article trigger codification activities leading to incorporation of some important stability parameters for columns made of time-dependent materials, including safe-load-limits and safe-service-periods, in forthcoming specifications and standards.

REFERENCES

1. Ahmadi, G. and P. G. Glockner, "Dynamic Stability of a Kelvin-Viscoelastic Column," *Proc. ASCE*, 109, No. EM.4, pp. 990–999 (August 1983).
2. Bazant, Z. P. and T. Tsubaki, "Nonlinear Creep Buckling of Reinforced Concrete Columns," *Proc. ASCE*, 106, No. ST11, pp. 2235–2257 (1980).
3. Behan, J. E. and C. O'Connor, "Creep Buckling of Reinforced Concrete Columns," *Proc. ASCE*, 108, No. ST12, pp. 2799–2818 (December 1982).
4. Booker, R., B. G. Frankham and N. S. Trahair, "Stability of Viscoelastic Structural Members," *Inst. Eng. Australia, Civil Eng. Trans.*, CE16, 1, pp. 45–51 (1974).
5. Boyle, J. T. and J. Spence, "Stress Analysis for Creep," Butterworths & Co. Ltd., London (1983).
6. Carlson, R. L. and A. G. Morgan, Jr., "Column Creep Buckling with End Restraint," *AIAA Journal*, 19, No. 5, pp. 664–666 (1981).
7. Dischinger, F., "Untersuchungen Ueber Die Knicksicherheit, Die Elastische Verformung und das Kriechen des Betons bei Bogenbruecken," *Der Bauingenieur* (1937).
8. Dilger, W. and A. M. Neville, "Creep Buckling of Long Columns," *The Structural Engineer*, 49, pp. 223–226 (1971).
9. Distefano, J. N., "Creep Buckling of Slender Columns," *Proc. ASCE*, 91, No. ST3, pp. 127–150 (June 1965).
10. Distefano, J. N. and J. L. Sackman, "Stability Analysis of a Nonlinear Viscoelastic Column," *Z. Angew. Math. Mech.*, 47, H. 4, p. 349 (1967).
11. Distefano, J. N. and J. L. Sackman, "On the Stability of an Initially Imperfect Nonlinearly Viscoelastic Column," *Int. J. Solids Structures*, 4, pp. 341–354 (1968).
12. Dost, S. and P. G. Glockner, "Dynamic Stability of Viscoelastic Plates," *J. of Civil Engineering*, Ege University, Turkey, 2 (1981).
13. Dost, S. and P. G. Glockner, "On the Dynamic Stability of Viscoelastic Perfect Columns," *Int. J. Solids Structures*, 18, No. 7, pp. 587–596 (1982).
14. Dost, S. and P. G. Glockner, "On the Behaviour of Viscoelastic Columns with Imperfections," *CSME Transactions*, 7, No. 4, pp. 198–202 (1983).
15. Fluegge, W. *Viscoelasticity.* Blaisdell Pub. Comp. (1967).
16. Freudenthal, A. M., "Some Time Effects in Structural Analysis," Proc. 6th IUTAM Congress, Paris (1946).
17. Freudenthal, A. M. *The Inelastic Behaviour of Engineering Materials and Structures*. John Wiley and Sons, N.Y., p. 518 (1950).
18. Gaede, K., "Knicken von Stahlbetonstaeben unter Kurz- und Langzeit Belastung," *Deutscher Ausschuss fuer Stahlbeton*, H. 129, Berlin (1958).
19. Gerard, G., "A Creep Buckling Hypothesis," *J. Aeron. Sci.*, 23, No. 9, pp. 879–887 (1956).
20. Gerard, G. and R. Papirno, "Classical Columns and Creep," *J. Aerospace Sci.*, 29, No. 6, pp. 680–688 (1962).
21. Gioncu, V. and M. Ivan, "Fundamentals of Structural Stability Analysis," (in Roumanian), Editura Facla, Timisoara (1983).
22. Glockner, P. G., "Ice: Can It Be a Structural Material?", in *'Shelter,' Proc. 91st Annual E.I.C. Meeting*, Jasper, Alberta, pp. 70–78 (1977).
23. Glockner, P. G., "Reinforced Ice Domes: Igloos of the 21st Century?", *Proc. on 'Housing: The Impact of Economy and Technology,'* Pergamon Press, pp. 887–908 (1981).
24. Goldengershel, E. J., "On the Euler's Stability of a Visco-Elastic Rod," *PMM*, 38, No. 1, pp. 187–192 (1974).
25. Hayman, B., "Creep Buckling—A General View of the Phenomena," *3rd IUTAM Symp. on "Creep in Structures,"* Leicester, U.K., A.R.S. Ponter & R. Hayhurst (eds.), pp. 289–307 (1980).
26. Hoff, N. J., "Buckling and Stability," Forty-First Wilbur Wright Memorial Lecture, *J. Roy. Aeron. Soc.*, 58, p. 30 (1954).
27. Hoff, N. J., "A Survey of the Theories of Creep Buckling," *Proc. 3rd U.S. National Congress of Applied Mechanics*, ASME, pp. 29–49 (1958).
28. Hoff, N. J., "Creep Buckling of Plates and Shells," *Proc. 13th IUTAM Congress*, Moscow, USSR, pp. 124–140 (1972).

29. Huang, N. C., "Creep Buckling of Imperfect Columns," *J. Appl. Mech., Trans. ASME*, 1, pp. 131–136 (March 1976).
30. Hult, J.A.H., "Critical Time in Creep Buckling," *J. Appl. Mech.*, 22, p. 432 (1955).
31. Hult, J.A.H., "Creep Buckling and Instability," in *Proc. ISPRA Seminar on 'Creep of Engineering Materials'*, G. Bernasconi & G. Piatti (eds.), Applied Science Publ. Ltd., London, pp. 133–145 (1978).
32. Kempner, J., "Creep Bending and Buckling of Linearly Visco-elastic Columns," NACA Tech. Note 3136 (1954).
33. Kempner, J., "Creep Bending and Buckling of Nonlinearly Viscoelastic Columns," NACA Tech. Note 3137 (1954).
34. Kempner, J. and V. Pohl, "On the Non-Existence of a Finite Critical Time for Linear Visco-Elastic Columns," *J. Aeron. Sci.*, 20, pp. 572–573 (1953).
35. Kempner, J., "Viscoelastic Buckling," Chapter 54, *Handbook of Engineering Mechanics*, W. Fluegge (ed.), McGraw Hill, pp. 54.1–54.16 (1962).
36. Krempl, E., "On the Interaction of Rate and History Dependence in Structural Metals," *Acta Mech.*, 22, pp. 53–90 (1975).
37. Libove, C., "Creep Buckling of Columns," *J. Aerospace Sci.*, 19, pp. 459–467 (1952).
38. Lin, T. H., "Creep Stresses and Deflections of Columns," *J. Appl. Mech.*, 23, No. 2, pp. 214–218 (1956).
39. Lin, T. H., "Creep Deflections and Stresses of Beam-Columns," *J. Appl. Mech.*, 25, No. 1, pp. 75–78 (1958).
40. Lin, T. H. *Theory of Inelastic Structures*. John Wiley, New York (1968).
41. Mauch, S. and M. J. Holley, "Creep Buckling of Reinforced Concrete Columns," *Proc. ASCE*, 89, No. ST 4, pp. 451–481 (August 1963).
42. Mellor, M. and D. M. Cole, "Stress-Strain-Time Relation for Ice Under Uniaxial Compression," *Cold Regions Sci. Tech.*, 5, pp. 207–230 (1983).
43. Michel, B. *Ice Mechanics*. Les Presses de L'Universite Laval, Quebec (1978).
44. Morland, L. W., "Constitutive Laws for Ice," *Cold Regions Sci. and Technol.*, 1, pp. 101–108 (1979).
45. Mroz, Z. and H. A. Trampczynski, "On the Creep-Hardening Rule for Metals with a Memory of Maximal Prestress," *Int. J. Solids Structures*, 20, pp. 467–486 (1984).
46. Naerlovic-Verljkovic, N., "Der Einfluss des Kriechens auf die Tragfaehigkeit von Stahlbetonsaeulen," *Oesterr. Ing. Archiv.*, 14, No. 2, pp. 99–139 (June 1960).
47. Penny, R. K. and D. L. Marriott. *Design for Creep*. McGraw-Hill Book Co. (U.K.) Ltd., London (1971).
48. Pian, T. H. H. and R. I. Johnson, "On Creep Buckling of Columns and Plates," Tech. Rep. 25-24, Aeroelastic and Structures Res. Lab., MIT (1957).
49. Pomeroy, C. (editor), "Creep of Engineering Materials," *A Journal of Strain Analysis Monograph*, I. Mech. E. (1978).
50. Rabotnov, Y. N. and S. A. Shesterikov. "Creep Stability of Columns and Plates," *J. Mech. and Phys. Solids*, 6, pp. 27–34 (1957).
51. Rabotnov, Y. N. *Creep Problems in Structural Members*. North-Holland Publishing Co., Amsterdam (1969).
52. Ross, A. D. "The Effects of Creep on Instability and Indeterminacy Investigated by Plastic Models," *Structural Engineer*, 24, No. 8, p. 413, (August 1946); and 25, No. 5, p. 179 (May 1947).
53. Schapery, R. A., "A Method of Viscoelastic Stress Analysis Using Elastic Solutions," *J. of the Franklin Institute*, 279, No. 4, pp. 268–289 (1965).
54. Shanley, F. K. *Weight-Strength Analysis of Aircraft Structures*. McGraw Hill, New York, pp. 323–342, 359–385 (1952).
55. Shestopal, V. O. and P. C. Goss, "The Estimation of Column Creep Buckling Durability from the Initial Stages of Creep," *Acta Mechanica*, 52, pp. 269–275 (1984).
56. Sinha, S. C. and D. R. Pawlowski, "Stability Analysis of a Tangentially Loaded Column with a Maxwell Type Viscoelastic Foundation," *Acta Mechanica*, 52, pp. 41–50 (1984).
57. Sobotka, Z., "Rheology of Materials and Engineering Structures," Elsevier, Amsterdam (1984).
58. Stanley, R. G. and P. G. Glockner, "The Use of Reinforced Ice in Constructing Temporary Enclosures," *Marine Science Communications*, 1, No. 6, pp. 447–462 (1975).
59. Stanley, R. G. and P. G. Glockner, "Some Properties of Reinforced Ice," *Proc. Conf. on Materials Engineering in the Arctic*, M. B. Ives (ed.), ASM, Metals Park, Ohio, pp. 29–35 (1977).
60. Szyszkowski, W., S. Dost and P. G. Glockner, "A Nonlinear Constitutive Model for Ice," *Int. J. Solids and Structures*, 21, pp. 307–321 (1985).
61. Szyszkowski, W. and P. G. Glockner, "Modelling the Time-Dependent Behaviour of Ice," *Cold Regions Sci. and Technol.*, 11, pp. 3–21 (1985).
62. Szyszkowski, W. and P. G. Glockner, "An Improved Rheological Model for Ice," *Proc. IX Intl. Congress on Rheology*, Mexico, 1, pp. 509–516; 3, p. 307 (1984).
63. Szyszkowski, W. and P. G. Glockner, "The Stability of Viscoelastic Perfect Columns: A Dynamic Approach," *Int. J. Solids and Structures*, 21, pp. 545–559 (1985).
64. Szyszkowski, W. and P. G. Glockner, "On the Static Stability Analysis of Viscoelastic Perfect Columns," *CSME Transactions*, 9, pp. 156–163 (1985).
65. Szyszkowski, W. and P. G. Glockner, "The Imperfect Linearly Viscoelastic Column," *Int. J. Engng. Sci.*, 23, pp. 1113–1120 (1985).
66. Szyszkowski, W. and P. G. Glockner, "Finite Deformation Analysis of Linearly Viscoelastic Simple Structures," *Int. J. Non-Linear Mech.*, 20, pp. 177–198 (1985).
67. Szyszkowski, W. and P. G. Glockner, "An Incremental Formulation of Hereditary Constitutive Laws Applied to Uniaxial Problems," *Int. J. Mech. Science*, 20, No. 9, pp. 583–594 (1985).
68. Szyszkowski, W. and P. G. Glockner, "On a Multi-Axial Nonlinear Hereditary Constitutive Law for Non-Ageing Materials with Fading Memory," *Int. J. of Solids Structures* (in press).
69. Szyszkowski, W. and P. G. Glockner, "On a Multiaxial Consti-

tutive Law for Ice," *Mechanics of Materials*, 5, pp. 49–71 (1986).

70. Szyszkowski, W. and P. G. Glockner, "Time-Deflection Behaviour of Ice Plates." Proc. IUTAM Symp. on Inelastic Behaviour of Plates and Shells, Rio de Janeiro, 1985, L. Bevilacqua, R. Feijoo and R. Valid, (editors), Springer Verlag, 113–130 (1986).
71. Szyszkowski, W. and P. G. Glockner, "On the Safe Load Limit of Columns Made of Time-Dependent Materials," *Acta Mechanica, 65*, 181–204 (1986).
72. Vinogradov, A. M. and P. G. Glockner, "On Creep Stability of Concrete Columns," Department of Mechanical Engineering, The University of Calgary, Calgary, Alberta, Report No. 124 (1978).
73. Vinogradov, A. M. and P. G. Glockner, "Creep Stability of Concrete Plates and Shells," *Proc. I.A.S.S. Symp. on Nonlinear Behaviour of Reinforced Concrete Spatial Structures*, G. Mehlhorn, H. Ruehle and W. Zerna (editors), Werner Verlag, Duesseldorf, II, pp. 189–198 (1978).
74. Vinogradov, A. M. and P. G. Glockner, "Buckling of Spherical Viscoelastic Shells," *J. Structural Div.*, ASCE 106, No. ST1, Proc. Paper 15116, pp. 59–67 (January 1980).
75. Vinogradov, A. M., "Nonlinear Effects in Creep Buckling Analysis of Columns," *Journal of Engg. Mechanics*, ASCE 111, No. E.M.6, pp. 757–767 (1985).
76. Voitkovski, K. F., "The Mechanical Properties of Ice," Izd. Akademii Nauk SSSR, *Trans. AMS-T-R-391, Am. Met. Soc.*, Office of Tech. Serives, U.S. Dept. of Commerce, Washington, 25 (1960).
77. Volmir, A. C. *Stability of Deformable Systems.* (in Russian), Nauka, Moskow (1967).
78. Zyczkowski, M., "Geometrically Nonlinear Creep Buckling of Bars," in *Creep in Structures*, Colloquium held at Stanford University, CA, July 11–15, 1960, N. J. Hoff (ed.). Academic Press Inc. Publishers, New York, pp. 307–325 (1962).

CHAPTER 24

Wind-Buckling Approach for R/C Cooling Towers

IHSAN MUNGAN*

> We are to admit no more causes of natural things than such as are both true and sufficient to explain their appearances.
>
> Sir Isaac Newton (1643–1727)
> from "Mathematical Principles of Natural Philosophy, Book III"

INTRODUCTION

Natural draught cooling towers have had a very quick development during the last 15 years. Until 1970 cooling towers were built mostly for a plant capacity of about 350 MW. These early wet-type cooling towers had a maximum height of 120 m and a basis diameter of about 90 m. Beginning with 1975 very large cooling towers for nuclear power plant capacities up to 1300 MW have been designed and built, one of them being the wet-type cooling tower of the nuclear power plant Isar II still under construction in Bavaria, Federal Republic of Germany. This cooling tower is 165 m high and has a base diameter of about 153 m. Figure 1 shows the development of natural draught cooling towers in Germany between 1965 and 1982. In this figure, for each year the height, the base diameter and corresponding plant capacity of the largest cooling tower of that year are depicted. Similar developments are observed also in other highly developed industrial countries like the U.S.A. [1] and France.

Dedicated to Dr. E.h. Dr.-Ing. Wolfgang Zerna, Professor Emeritus, Ruhr-Universität Bochum, for the occasion of his seventieth anniversary in 1986.

*Department of Civil Engineering, Ruhr-Universität Bochum, Germany, Federal Republic; Present affiliation—Professor of Structures and Reinforced Concrete, Mimar Sinan Univeresity, Findikli, Istanbul, Turkey

For the dry-type cooling process even larger natural draught cooling towers are needed. For example, the six cooling towers of the Kendal Power Station in Republic of South Africa, each designed for 672 MW block capacity, have 163 m basis diameter and are 166 m high. This upsurge during a relatively short time could not be realized without systematic research on these largest reinforced concrete shell structures of the present day. The research was focused on six main topics:

- stress analysis
- assessment of the wind loading
- dynamics and earthquake response
- buckling behavior
- effect of reinforced concrete properties, ultimate load and safety considerations
- optimization and constructional aspects

Thanks to rapid development of computing methods there is today no problem in the stress analysis of the hyperboloidal veil of a cooling tower according to the general bending theory of shells. Preference is mostly given to Finite Element solutions, because taking layered elements and increasing the load incrementally the analysis can be extended to consider the nonlinearity and formation of cracks in reinforced concrete. The assessment of a reliable design wind loading is of vital importance for large cooling towers. Besides the axisymmetric self weight, wind is the main loading of a cooling tower shell. The stress state due to self weight is axisymmetric, but changes along the height. With the exception of a small area near the top, the hoop stresses are compression overall, whereas the meridional force is compression everywhere. The stresses induced by wind, however, change along both the height and circumference. The design of the reinforcement is based on the difference between the maximum tensile force due to wind and the compressive force due to self weight at the corresponding elevation. On the other hand, the stability of the shell is in-

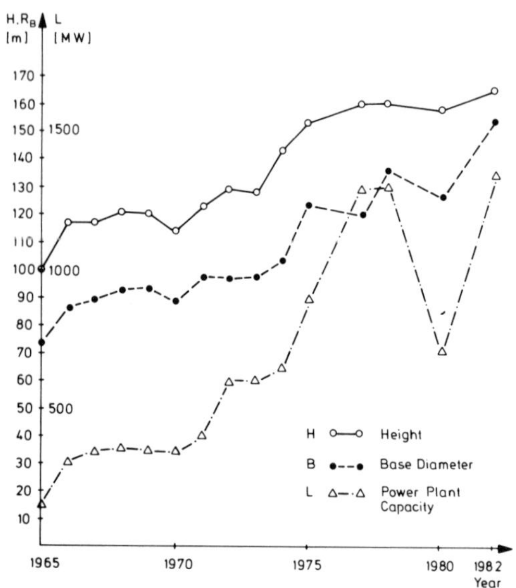

FIGURE 1. Development of cooling towers in Germany.

fluenced substantially by the compressive wind stresses added to the compression due to self weight. All these show how important it is to seize the wind effect realistically. Measurements both on aeroelastic models tested in wind tunnel and on some reinforced concrete cooling towers erected, yielded detailed information on the magnitude and distribution of the wind forces along the circumference and height of the tower, depending on the roughness of the shell surface. In Figure 2, taken from Reference 2, German Guidelines for Design of Cooling Towers, the variation of the wind pressure is depicted along the circumference. The highest lateral suction acts at an angle Θ^1 between 70° and 75° from the windward meridian. $K_{1.3}$ means the lateral suction is equal to 1.3 times of the wind pressure along the windward meridian. This high suction has to be taken, if the roughness parameter, defined as the ratio of the depth to distance of the vertical windribs, is between 0.006 and 0.010. For the surface roughness values between 0.025 and 0.100 the distribution $K_{1.0}$ is valid with a maximum lateral suction equal to the maximum of the wind pressure.

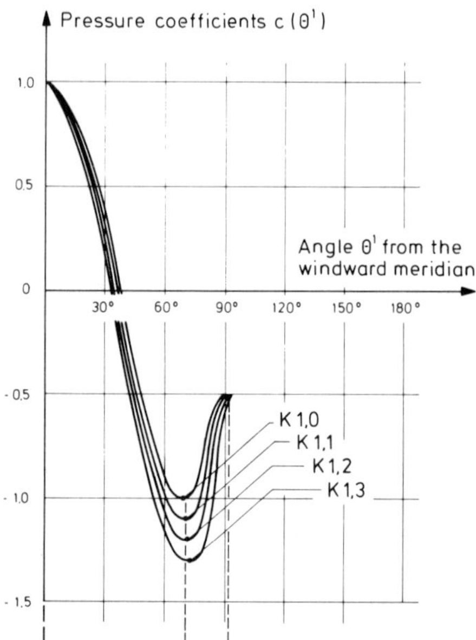

FIGURE 2. Distribution of wind load in circumferential direction.

Dynamic response of a cooling tower is of importance for two reasons. First the lowest natural frequency has to be so high that the dynamic effect of gusty wind can be taken into account through dynamic amplification factors in order to assess the so-called quasi-static wind. Natural frequencies larger than 1 Hz are in general sufficient to make use of this approximation [2]. On the other hand, so far as the seismicity is concerned, the most critical region of a cooling tower is the column part. Through earthquake large forces are induced in the connection areas between the supporting columns and the shell, in the columns themselves and in the foundation. Hence, the design of these parts is governed by the earthquake effect, if this is relevant for the site of the cooling tower, instead of by the wind [3].

The size of a cooling tower is determined according to the thermodynamical characteristics of the cooling process. Having the base and throat diameters such as the height of the cooling tower given, the structural engineer has to choose an appropriate shape for the meridian and determine the wall thickness. This first step of the structural analysis is carried out taking the prescribed lowest buckling safety factor into account. In the following, an approach will be described providing an engineering solution for the very complicated mechanical problem of the reinforced concrete cooling tower shell buckling under wind loading.

MATHEMATICAL ASPECTS OF STABILITY

According to René Descartes (1596–1650), the founder of analytical geometry, the Natural Sciences have to be Mathematics in order to be accepted as science. At that time both Physics and Mechanics were not founded as independent branches and were in nascent stage under the conduct of mathematicians. That is the reason that early investigations on structural stability came from the mathematicians.

Leonhard Euler (1707–1783), as a mathematician, was interested principally in elastic curves of slender bars. In his work "De curvis elasticis" from 1744 he studied the buckling problem of a compressed bar considering its adjacent equilibrium states [4]. He approached the problem from the point of view of variational calculus: "Since the fabric of the universe is most perfect, and is the work of a most wise Creator, nothing whatsoever takes place in the universe in which some relation of maximum and minimum does not appear." The buckling formula of columns established by Euler more than 200 years ago still has application in structural engineering.

Joseph-Louis Comte de Lagrange (1736–1813) investigated the equilibrium of a discrete conservative system and stated that the energy of a stable state had to be minimum. In his famous "Méchanique analitique" from 1788 Lagrange gives the following definition for the instability: "A state of equilibrium is unstable if, on being slightly disturbed, the system can perform oscillations which will not be small, and which will become larger and larger" [5]. Lagrange uses the energy as criterion: "If the energy of the system is minimum for a virtual displacement field compatible with the kinematic boundary conditions the equilibrium state is stable." To prove this theorem Lagrange linearizes the equations of motion of a conservative system in the neighborhood of the equilibrium state. This approach was criticized by Peter Gustav Lejeune Dirichlet (1805–1859) who did not see a logical justification of the linearization through neglecting the non-linear terms. In his paper "On the Stability of Equilibrium" from 1846 Dirichlet corrected Lagrange's proof and established the "Lagrange-Dirichlet criterion" which is known also under the name "energy test." Dirichlet points out three main aspects of the stability phenomenon. First, stability is understood as the characteristic of a state which is in equilibrium. Secondly, disturbances are expressed in terms of displacements and their velocities. Lastly, the stability is referred to certain narrow limits called neighboring states.

Jules Henri Poincaré (1854–1912) extended Dirichlet's stability concept to dynamic systems. In his 1885 paper, Poincaré introduced and defined the bifurcation concept, discussed the elements of bifurcation and the limit forms of equilibrium in detail, and finally furnished the theoretical formulation of his bifurcation concept [6]. It is certainly due to Poincaré that stability of critical solutions, dynamic stability and structural stability became three fundamental sections of a general bifurcation theory.

It was Poincaré's contemporary Alexandr Mikhaylovich Liapunov (1857–1918) who established the first connection to Poincaré's work. Liapunov published in 1892 his treatise on the stability of critical solutions [7]. His work, which is primarily a qualitative one, comprises the basis and complete idea of the stability theory of elastomechanics. Liapunov introduced two different approaches. The first method of Liapunov considers the solution of a dynamic system and checks its stability. In the second method, however, the stability investigation is carried out with help of a Liapunov functional, a functional which is chosen directly, that means without reference to the solution or solutions of the system. That is why this second method is also called "Liapunov's direct method." The main task for an application is now to find an appropriate functional. For conservative systems the Hamiltonian H which is the sum of the kinetical and potential energy of the system proves to be one of the appropriate functionals. In this case the undisturbed system is stable with respect to two positive definite metrics, namely ϱ_0, introduced to measure the disturbance, and ϱ the final state, if the following conditions are satisfied: $\dot{H}$, the derivative of H with respect of time, is negative or nil, and if for any real positive number ϵ another positive real number $\delta(\epsilon)$ can be found such that for $\varrho_0 < \delta$ the condition $\varrho < \epsilon$ is fulfilled. Or verbally, the measure of the Hamiltonian has to admit an infinitely small upper bound with respect to the metric ϱ_0. This shows the close connection between Liapunov's direct method and the classical energy criterion in the case of conservative systems.

The choice of the metrics ϱ_0 and ϱ depends on the type of the system to be investigated. If, for example, the overall or global value of the displacement u is taken to define the stability, then the average integrals within a domain Ω, i.e.,

$$\int_{\Omega} u^2 \cdot d\Omega \tag{1}$$

or

$$\int_{\Omega} |u| \cdot d\Omega \tag{1a}$$

are taken as metric or norm. On the other hand, if not the global stability but the local instability within a domain Ω has to be investigated, the norm must be the upper bound of the displacement itself:

$$\varrho = \sup_{\Omega} |u| \tag{2}$$

Sometimes a situation of conditional stability is encountered. That is the case if the system is stable only against a certain sub-family of perturbations. The conditional stability includes always the neutrally stable case called also neutral equilibrium, as will be explained considering Figure 3 in next section. Some criticism has been raised against application of the global energy criterion to two- or three-dimensional elastic continua. Due to infinite number of degrees of freedom or parameters it is in fact appropriate to make the stability test in terms of the variation of the total energy of such systems. But already in 1914 Hellinger pointed out that in the case of wave propagation in two- or three-dimensional bodies even under stable flow some local instabilities and high instantaneous velocities can occur [8]. Hamel provided an example for this case. He sup-

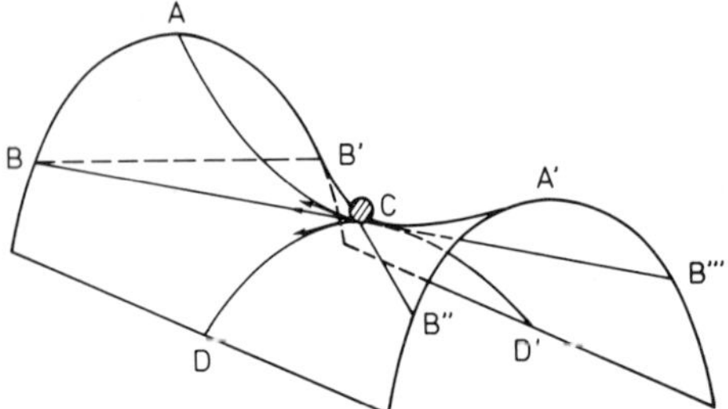

FIGURE 3. General model for stability of equilibrium.

posed a circular container in which a liquid is flowing in radial direction from the periphery to the center. Even in the case of a very slow and regular flow some drops of the liquid at the center bob up some decimeters [9]. The calculation of such an instability is demonstrated by Shield and Green on a sphere of homogeneous, isotropic and elastic material. If such a sphere is subjected to an infinitesimal initial disturbance by means of suitably selected surface forces, finite strains occur near the center [10]. All these show that the mathematical investigation of stability has to consider the physical aspects of the phenomenon.

PHYSICAL ASPECTS OF STABILITY

The main feature of the engineering research is its orientation towards application. The chain research-science-practice is in engineering a closed one. That means each link of this chain is connected with the other two. Research extends the knowledge or science which is put in practice. On the other hand, problems coming from practice give impulses for further research. The beginning of each scientific activity is observation, and the theorization of observations leads to science. The most effective tool of theorization is mathematics, first due to its consistency resulting from its linking with the logic, and secondly due to its organization and transparency. For these reasons any scientific research has two main components, namely experiment, which is physics, and theory, which is mathematics. Any scientific experiment has three steps, the first one being to think a suitable concept for the test. In the second step, the test is carried out, which means the transformation of observations in characters to be measured. The last step is the analysis and interpretation of the test results. In this last step the principle of scientific relativization is applied.

Galileo Galilei (1564–1642) can certainly be accepted as the scientist who made the first scientific experiments. Although his very simple and genius formula

$$\ell = \frac{1}{2} bt^2 \qquad (3)$$

derived from the tests with a ball rolling on an inclined plane does not include the effect of friction and other secondary factors, it demonstrates the most important mechanical law of motion under constant force. The right look into the results in order to distinguish between the main parameters and secondary effects is crucial in research.

In the case of the stability problem the phenomenon "disturbance" itself, constituting an essential part of the stability definition, is a physical aspect. Disturbances are introduced in a system through variations in the conditions of the equilibrium. Mathematically, only such disturbance functions are allowed which do not alter the regularity and smoothness conditions of the solution set and are compatible with the prescribed boundary conditions. That means, in a statical system a finite impact cannot be introduced as disturbance function, because the system will be no more static. But physically, such a disturbance is conceivable.

Physical disturbances of a system may be external or internal, or both together. The external disturbances may be put in any of the following three groups:

a) Disturbances introduced by changing the initial conditions of the system: In this case the system is subjected to slight displacements and pushed in an adjacent state. The classical demonstration object for this type of disturbance is a ball rolling under gravity force along a curvilinear path.

   Such a model is sufficient if the system has only one parameter. For multiple parameter systems like shells it is necessary to modify and extend this mechanical model, for example as the one shown in Figure 3. This time the ball is resting on a saddle surface at point C. For disturbances as infinitesimal displacements within the area CBAB' the ball exhibits a stable behavior and returns to its initial position. On the other hand, for disturbances along the characteristic lines BB''' and B'B'' the equilibrium is neutral. For any disturbance within the area CBDB'', however, the equilibrium position at point C is unstable. The sensitivity against any disturbance can be demonstrated on the same figure. The highest resistance against instability is reached if the ball is disturbed along the path ACA', whereas the lowest resistance is attained in the case of a disturbance along the tangent of the path DCD' at point C. That means the behavior of a continuous system under different types of disturbances may be completely different, the imperfection sensitivity depending on the type of disturbance.

   The ball on the saddle models the most general case of continuous systems, whereas there are some systems which can be modelled with a ball in a crater. Such systems are insensitive against any kind of disturbances. Axially compressed plates, for example, are of this type. On the other hand, the model for a spherical cap under critical lateral pressure which is highly imperfection sensitive is the ball at the apex of a hill.

b) Disturbances resulting from alterations of the forces acting on the system: The disturbing forces are completely independent from the loading of the fundamental state. Their direction can be changed in order to find out which type of disturbing forces initiate the instability. It is obvious that the disturbing forces have to fulfill the conditions prescribed in the stability definition.
c) Disturbances caused through variation of the entropy of the system: Insertion of heat energy may lead to instability.

Internal disturbances of a system may be either due to imperfections in the geometry of the system or may result from the deformation behavior of its material. Deviations between the idealized and real shape are called geometric imperfections. Such deviations which are not considered in the computations with a sufficient accuracy lead to secondary disturbances of the equilibrium as the consequence of variation of the initial conditions or of the internal forces. As numerous stability tests show, geometric imperfections are in most cases responsible for anticipated loss of stability. That is why in recent research on shell buckling, attention is focused on this aspect.

To explain the effect of imperfections let us consider the mechanical model as shown in Figure 4 consisting of two hinged links. This system is mostly investigated under the very strict assumption that bars 12 and 23 do not buckle, so that the formulas and the diagram in Figure 4 are obtained. If force *P* is applied through hanging weights, bar forces get infinitely large at the moment point 2 lies on the same line with points 1 and 3, during snap-through from position A to D of the diagram. Because the Euler-Buckling stress σ_{crEu} is in general smaller than the snap-through stress σ_{crST}, as given in Figure 4, buckling of the bars precede the snap-through of the system. This phenomenon can be demonstrated in Figure 5 which has a vertical guide allowing only vertical displacements of joint 2 in accordance with Figure 4. If bars 12 and 23 are identical, instability under weights takes place through buckling of both bars simultaneously, as Figure 6 shows. If one of the bars is weaker, as soon as the buckling stress of the weaker bar is reached this bar buckles first, but this leads to global loss of stability and then also to buckling of the stronger bar. That means the buckling stress of the weaker bar determines the stability limit of the whole system. The same is valid also if the vertical guide is removed. However, at the same load, this time only the weaker bar 23 buckles, whereas bar 12 rotates about joint 1 without buckling. Figure 7 shows this behavior. In the case that both bars have the same buckling stress they buckle simultaneously and the behavior of the system is identical with and without vertical guide, as Figures 6 and 8 demonstrate.

Compared with the system discussed the plane truss shown in Figure 9 has a higher degree of freedom. Typical for this truss is that one of its bars is weaker than all others

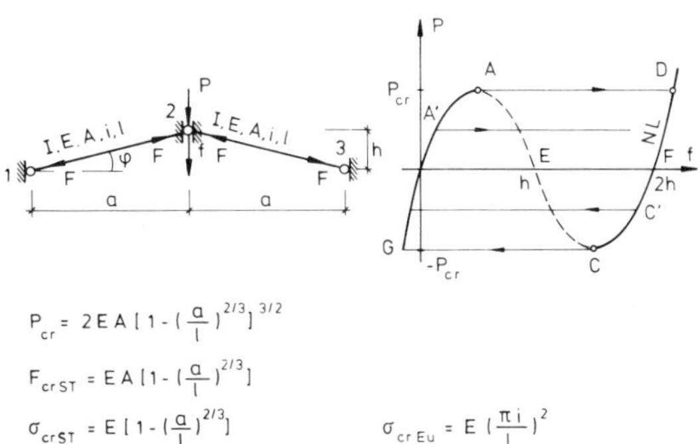

FIGURE 4. Snap-through of hinged frame with vertical guide, theory.

[11]. Under symmetrical loading the weak bar buckles, as soon as its critical force is reached. The total weight necessary for this type of local buckling is 2 × 3.850 = 7.700 kg. If the same truss is equipped with a vertical guide local buckling can no more take place even under very high asymmetrical loading, as Figure 10 demonstrates. If at each side of the truss one bar is weaker, i.e., if the truss has symmetrical imperfections, and if it is loaded symmetrically, a symmetrical buckling configuration is obtained under the same total load 7.700 kg, both with and without vertical guide (Figures 11 and 12). Due to eccentric loading, however, only the weak member under compression buckles, as soon as its buckling stress is reached. As it can be seen in Figure 13, the local buckling stress is the same, whereas the total weight is only ¼ of the cases with symmetrical loading as

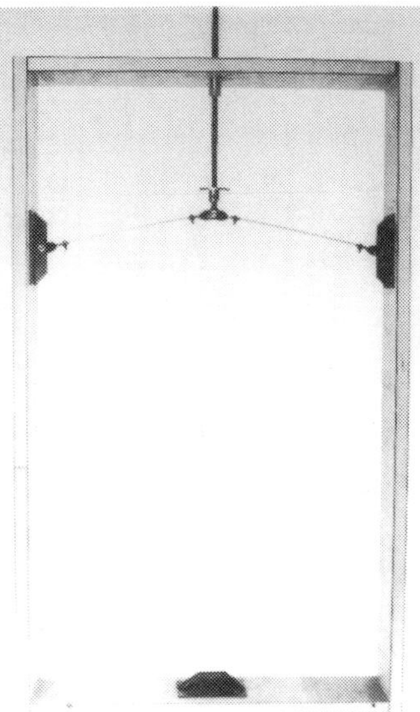

FIGURE 5. Hinged frame with vertical guide, model.

FIGURE 6. Symmetrical buckling of the hinged frame with vertical guide, model.

Figures 9, 11 and 12 demonstrate. The same behavior is still observed, if the weakness of the truss bars is due to local imperfections they have. However, as it can be seen in Figure 14, the buckling stress is this time higher.

Shells as continuous spatial structures have the highest degree of freedom and exhibit imperfections, mostly in form of deviations in the shape or in wall thickness. The

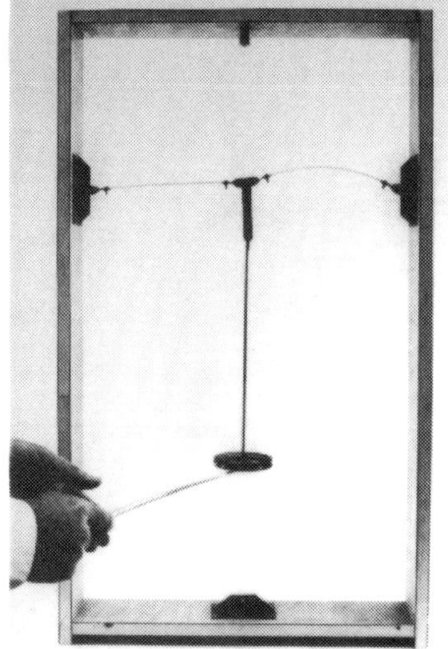

FIGURE 7. Asymmetrical buckling of the hinged frame without vertical guide due to asymmetric imperfection, model.

FIGURE 8. Symmetric buckling of the hinged frame without vertical guide, model.

relation between pre-buckling and post-buckling configurations of a shell can be explained considering the load-deformation diagram. Figure 15 shows such a diagram for an axially compressed cylinder. The straight line OT is valid for a shell without any imperfections. At point T corresponding to the theoretical bifurcation stress, the shell becomes unstable. All points of the dashed part of the post-

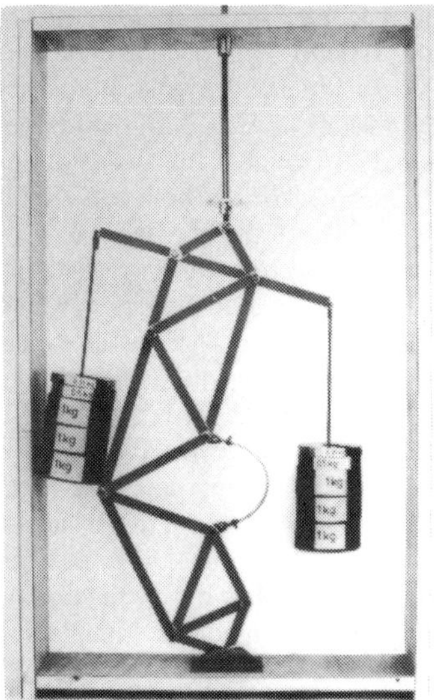

FIGURE 9. Local asymmetrical buckling under symmetric load due to asymmetric imperfection.

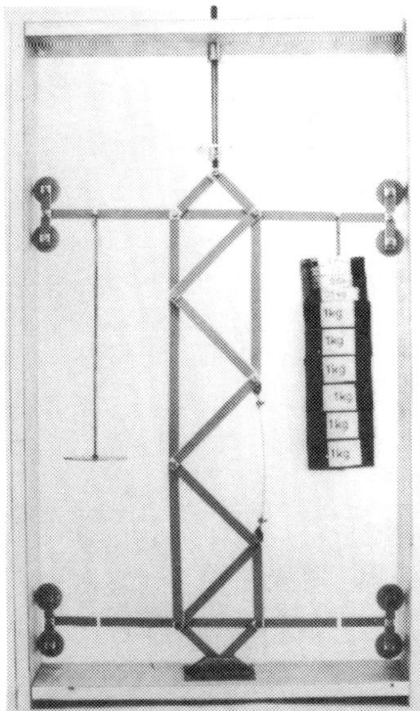

FIGURE 10. Local buckling impossible.

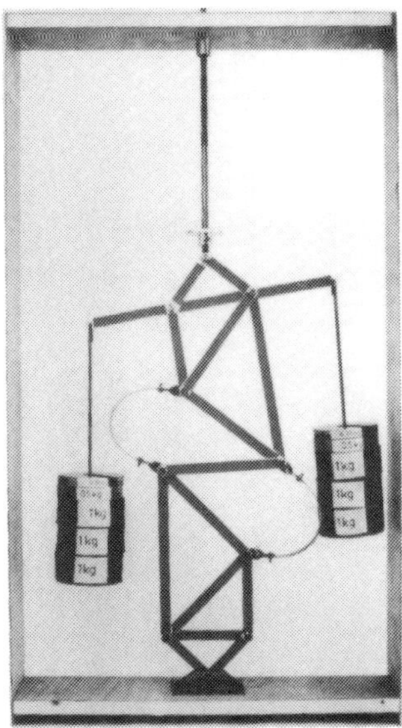

FIGURE 12. Local symmetrical buckling without deformation restriction.

buckling curve correspond to unstable equilibrium states, whereas all equilibrium states are along A_1E_1 stable. Due to imperfections present in tests the part of the curve corresponding to pre-buckling equilibrium states deviates from the theoretical straight line OT. Path OE with the number 1 is valid for a model exhibiting only slight imperfections. Reaching point E, the shell becomes unstable and goes over into one of the post-buckling equilibrium positions E_1, E_e, or A_1. Position E_1 corresponds to a testing method with weights, whereas in a very stiff hydraulic testing facility position A_1 may be reached. In most cases the post-buckling position E_e lies somewhere between A_1 and E_1, depending on the stiffness of the testing machine. The behavior is similar for a model having larger imperfections, the critical

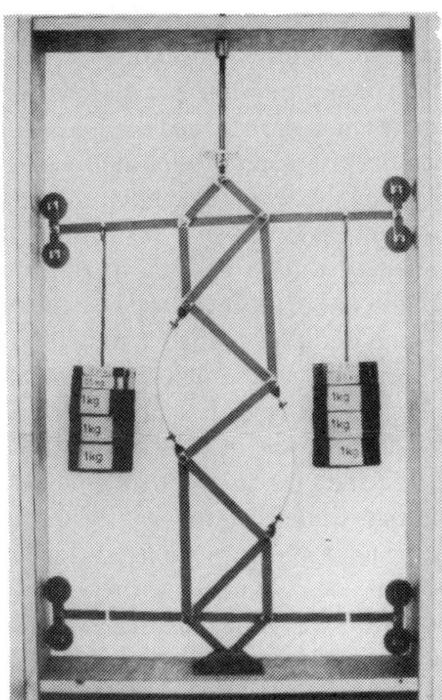

FIGURE 11. Local buckling possible due to symmetric imperfection and symmetric load.

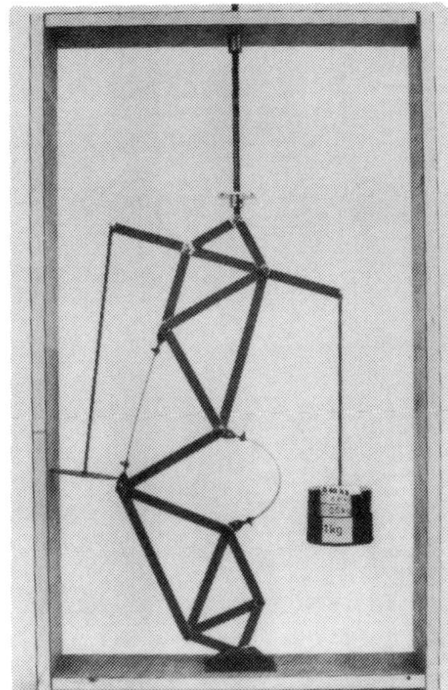

FIGURE 13. Local asymmetrical buckling due to asymmetric load.

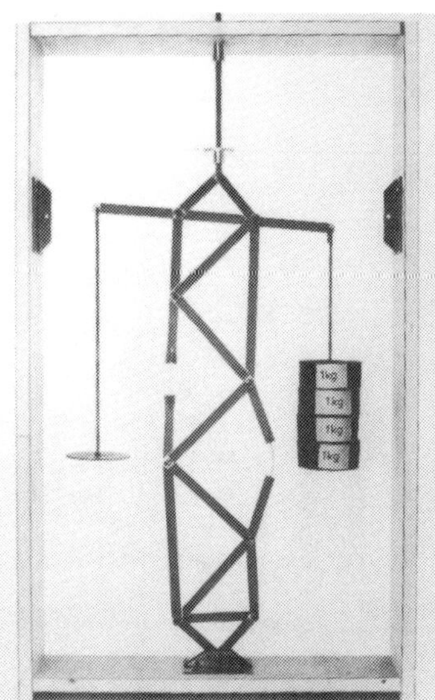

FIGURE 14. Influence of the length of the imperfection on the local buckling stress.

FIGURE 16. Wind buckling of a cylindrical shell without upper stiffening member.

stress σ_{2cr} being smaller than σ_{1cr}, as the path with number 2 in Figure 15. On the other hand, after a certain amplitude of imperfections, the transition into buckled configuration is smooth. Path 3 of Figure 15 indicates this behavior qualitatively. In this case the stiffness of the testing apparatus has no influence on the buckling configuration A_1 arrived.

Depending on the amplitude and distribution of the imperfections present, the buckling process starts locally, i.e., first a single dimple shapes. However, the succeeding development of the buckling configuration depends on the deformation capability of the shell model. Tests on complete spherical shells under volume control demonstrate this fact. Buckling starts always with a single dimple. If this dimple is prevented from getting deeper, with the help of a mandrel placed into the sphere, new dimples shape at the neighborhood. With decreasing distance between the mandrel and the shell, i.e., with decreasing depth, the size of the dimples decreases, whereas their number increases [12].

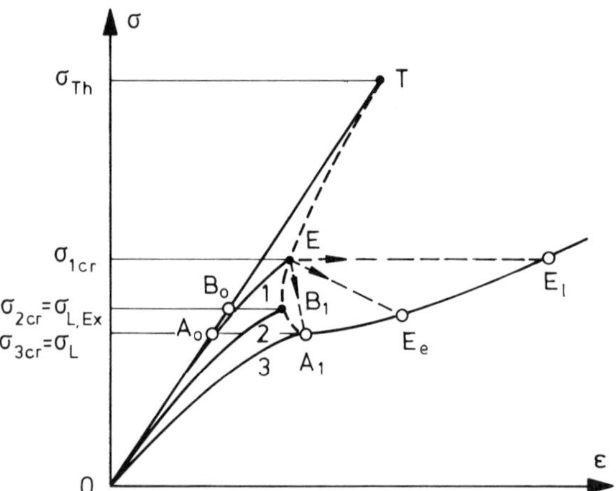

FIGURE 15. Effect of imperfections on the stability of imperfection sensitive shells.

A similar behavior is also observed in buckling tests on cylindrical shells and on hyperboloidal cooling tower models. Figure 16 shows the buckling configuration of a cylindrical shell tested in wind tunnel. The model has no stiffening edge beam at the top, and the first dimple shaping due to hoop compression at the windward meridian leads to snap through along the whole height. If the same model is stiffened at the top by means of a relatively rigid ring, the first dimple shaping at the windward meridian shifts to the middle part of the model. Its depth and extension are rather limited and some additional smaller dimples shape at the lower part of the flanks where high meridional compression is acting, as Figure 17 shows. Exactly the same behavior is observed on cooling tower models before (Figure 18) and after (Figure 19) arranging an upper stiffening ring. The formation of the first dimple occurs with and without upper stiffening ring under the same wind velocity. All these observations have one common interpretation: The magnitude of the stress initiating the buckling process is independent of the post-buckling configurations and displacements. Contrariwise post-buckling configurations of a shell depend mainly on its deformation capability and not necessarily on the level of stress initiating the buckling. In the case of shell

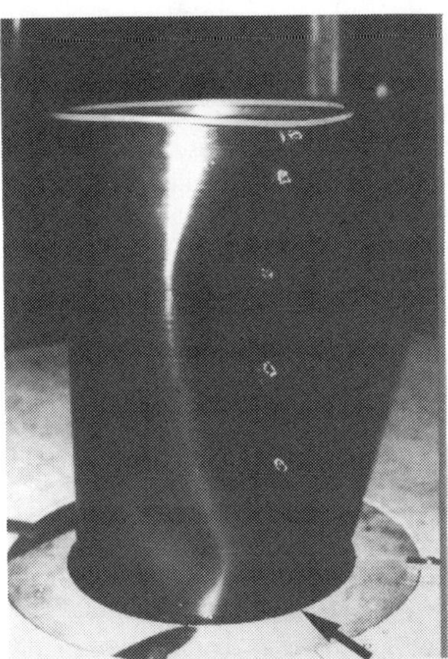

FIGURE 17. Wind buckling of a cylindrical shell with a stiff upper edge member.

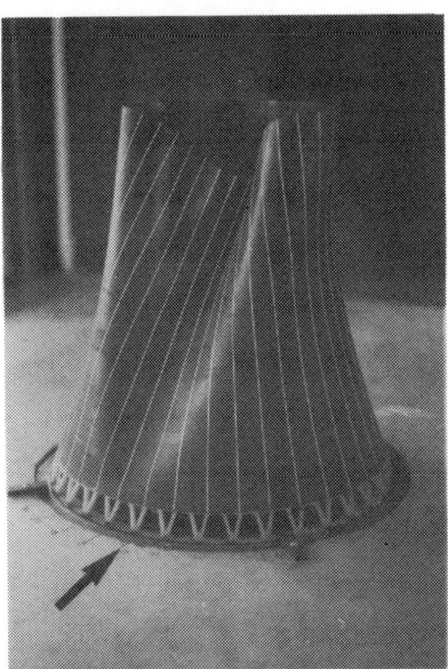

FIGURE 18. Wind buckling of a hyperboloidal cooling tower model without upper stiffening member.

structures having no or only slight restrictions with respect of their displacements the first local dimple already leads to global loss of stability. This physical phenomenon constitutes the idea behind the approach developed to design cooling tower shells against buckling: The stability limit of a shell structure without secondary supporting elements is reached as soon as local instability eventuates. In fact this criterion differs from the buckling criterion applied in design of airplanes and space vehicles where local buckling is tolerated in most cases. That is first due to the fact that local buckling of metallic shells of space vehicles occurs in elastic range. In addition, these shells constitute in most cases the skin of a space frame and local buckling remains confined within the field between the supporting stringers and rings. After unloading, the system recovers completely so that local buckling can be allowed. Such a recovery cannot be expected from reinforced concrete cooling tower shells.

BIFURCATION RESULTS FOR ELASTIC SHELLS OF REVOLUTION

Correlation of Shell Geometry and Buckling Mode

For stability investigations of shells the bifurcation theory has proved to be an efficient tool. It requires less voluminous computations and is therefore especially appropriate for parametric investigations needed to enlighten the buckling behavior of shells. The bifurcation equations are based on the nonlinear shell theory including the terms resulting from the assumption of neighboring states besides the fundamental state. Eliminating the contributions due to the fundamental state, which is for itself in equilibrium, a system of homogeneous differential equations containing the internal stresses of the buckled state supposed is obtained. At a certain load intensity called bifurcation load this internal stress state itself corresponds to an equilibrium state, provided that the adjacent state is supposed infinitesimally near the fundamental state and the boundary conditions are fulfilled. Stability equations obtained in this way constitute a system of

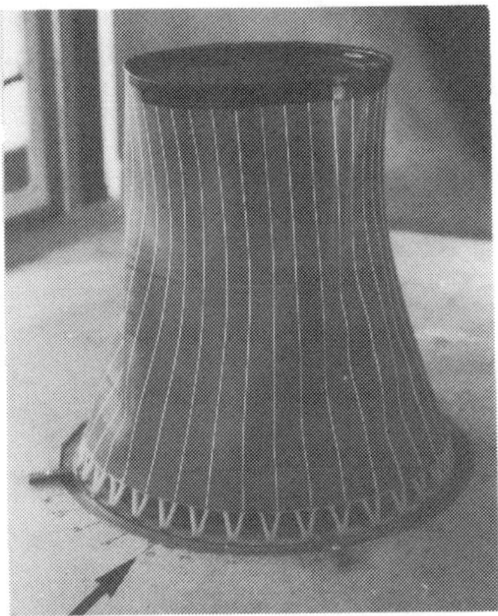

FIGURE 19. Wind buckling of a hyperboloidal cooling tower model with an upper edge member.

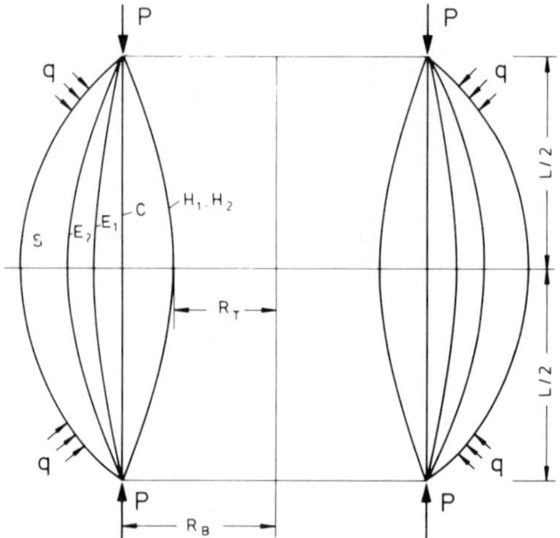

FIGURE 20. Shells of revolution investigated.

14 partial differential equations of first and second order. In the case of shells of revolution loaded axisymmetrically, trigonometric distributions can be introduced for the displacements in circumferential direction. In this way the system of partial differential equations reduces to a system of ordinary differential equations. In meridional direction the differential equations are transformed into finite difference equations. The critical load factor or eigenvalue λ is determined from the condition that the determinate of the homogeneous algebraic equation system vanishes. The eigenvalues obtained depend on the number of waves assumed in circumferential direction. The lowest eigenvalue λ_{cr} and the corresponding wave number n_{cr} define the buckling load and the buckling configuration, respectively [13].

The parametric studies carried out on six different types of shells help to enlighten some governing aspects of the buckling phenomenon such as the effects of shell geometry, stress state and boundary conditions. Figures 20 to 32 taken from a former paper of the author, Reference 14, are helpful in this respect.

Figure 20 shows the shape of the axisymmetric shells investigated. In Table 1 the types and dimensions of these shells are given, the meridian being circular. The shells are investigated once under external lateral pressure and afterwards under axial compression. In one case both boundaries are assumed to be clamped without any displacement in the buckled adjacent state. To denote the results of this case subscript *A* will be used in the diagrams. Subscript *B* is used for the case that the upper shell boundary, being still clamped, can displace in vertical direction during transition into buckled state.

In Figure 21 the lateral buckling pressure of shell H_1 is depicted for different numbers of buckling waves assumed in circumferential direction. If boundary condition *A* is considered, a buckling configuration with seven waves along the circumference yields the lowest buckling load. The number of waves reduces to five, if a vertical displacement at the upper boundary is allowed. However, in both cases a buckling configuration with six waves is possible at an only slightly higher pressure. Although buckling configurations with $n < n_{cr} - 1$ cannot develop, those with $n > n_{cr} + 1$ are very likely to shape, if the shell has imperfections corresponding to such modes. The same shell has under vertical edge loading a different behavior, as Figure 22 shows. Both for $n < n_{cr} - 1$ and $n > n_{cr} + 1$ a considerably higher buckling compression is needed.

The second shell Type H_2 of negative Gaussian curvature is studied considering only boundary condition A, however, also after stiffening by means of 5 or 9 equidistant rings of 7.5 mm width and 2.0 mm thickness, as described in a former work [15]. The results obtained are depicted in Figures 23 and 24. For lateral pressure the number of waves increases from 6, for the shell Type H_{2A} without intermediate rings, to 7 or 13, if 5 or 9 stiffening rings are arranged, respectively. The buckling pressure of the shell Type $H_{2A,9}$

TABLE 1. Characteristic Values of the Shells Investigated.

Notation	Gaussian Curvature or Shape	R_B (m)	R_T (m)	L (m)	t (mm)	E (MN/m²)	ν
H_1	negative	0.225	0.150	0.600	1.60	3630	0.35
H_2		0.300	0.200	1.200	2.00	3430	0.38
C	cylinder	0.225	0.225	0.600	1.60	3630	0.35
E_1	positive	0.225	0.268	0.600	1.60	3630	0.35
E_2		0.225	0.305	0.600	1.60	3630	0.35
S	spherical	0.225	0.375	0.600	1.60	3630	0.35

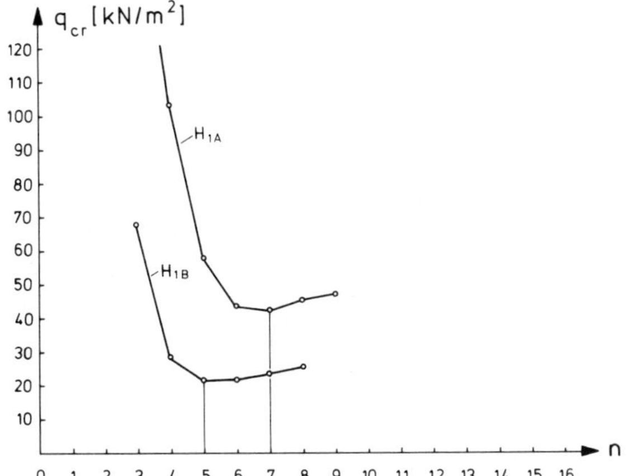

FIGURE 21. Critical lateral pressure vs. number of buckling waves for hyperboloidal shells.

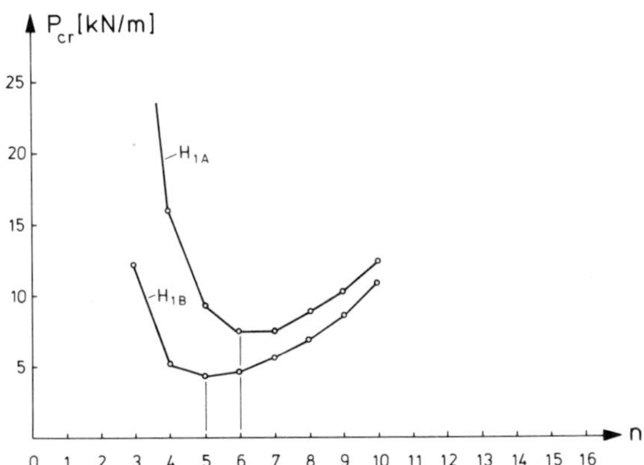

FIGURE 22. Critical edge load vs. number of buckling waves for hyperboloidal shells.

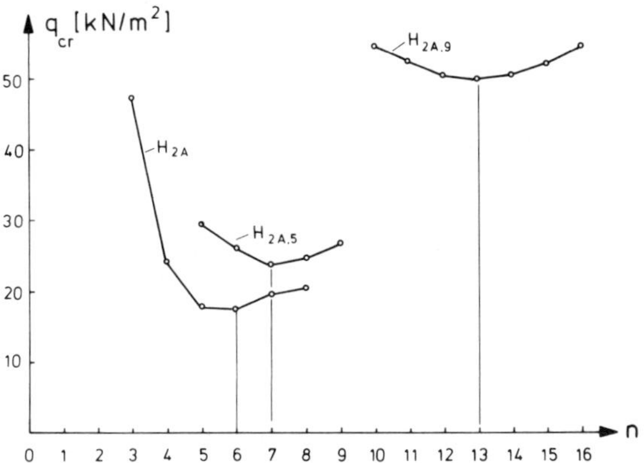

FIGURE 23. Critical lateral pressure vs. number of buckling waves for ring-stiffened hyperboloidal shells.

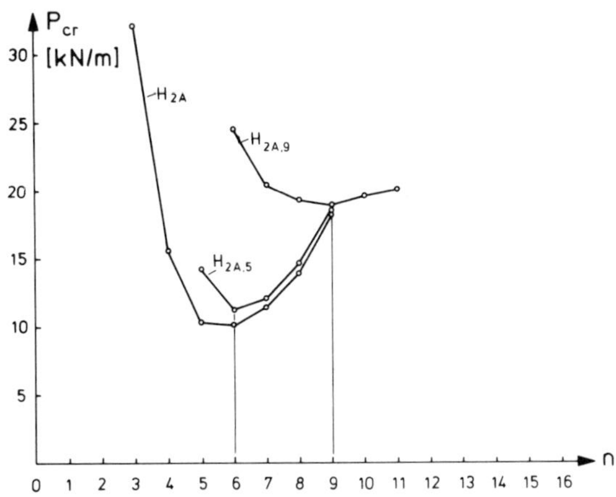

FIGURE 24. Critical edge load vs. number of buckling waves for ring-stiffened hyperboloidal shells.

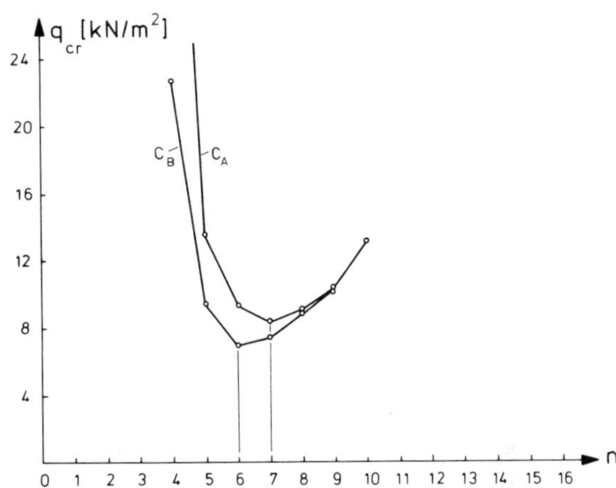

FIGURE 25. Critical lateral pressure vs. number of buckling waves for cylindrical shells.

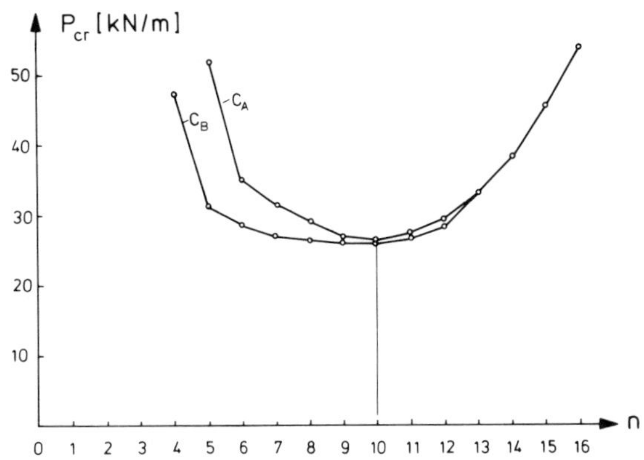

FIGURE 26. Critical edge load vs. number of buckling waves for cylindrical shells.

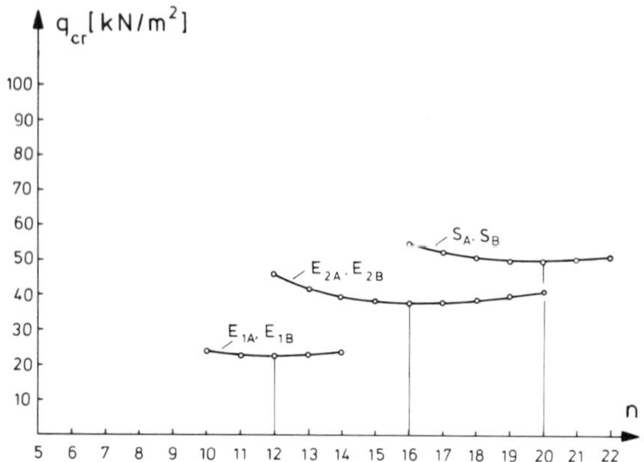

FIGURE 27. Critical lateral pressure vs. number of buckling waves for shells of positive Gaussian curvature.

having 9 stiffening rings is about 2.8-times of the buckling pressure obtained in unstiffened case; however, its variation in the neighborhood of $n_{cr} = 13$ is much less than in the case without stiffening rings. Under axial load the number of buckling waves increases from 6 to 9 and the augmentation of the buckling load is about 80% due to stiffening by means of 9 rings. Again the variation of the buckling load is less pronounced for the ring stiffened cases, as Figure 24 shows.

The diagrams in Figures 25 and 26 are for the cylindrical shell under lateral pressure and axial compression, respectively. In the case of lateral pressure the number of waves is unambiguously 7 or 6 if boundary condition A or B is considered, respectively. Under axial compression, however, especially in the case of boundary condition B, any buckling configuration with $7 \leq n \leq 11$ can shape without any remarkable variation of the critical load.

The relationship between the number of buckling waves

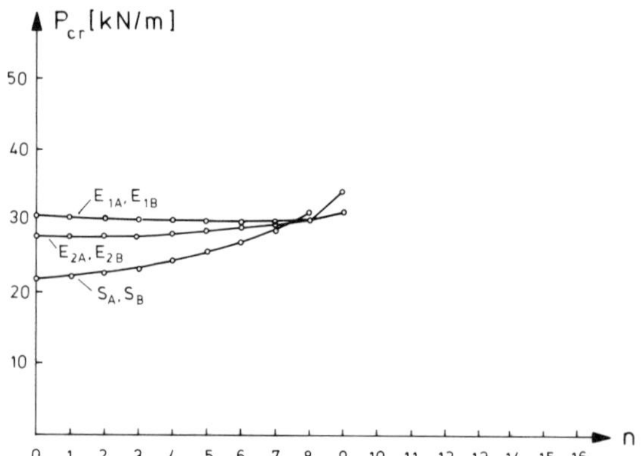

FIGURE 28. Critical edge load vs. number of buckling waves for shells of positive Gaussian curvature.

and the buckling load is in the case of shells of positive Gaussian curvature much less pronounced, both for lateral pressure q and axial compression P. Additionally, the assumption of boundary condition A or B provides the same bifurcation loads and buckling configurations in both loading cases. In Figures 27 and 28 these results are depicted. For example in the case of axially loaded shell Type E_1 any buckling configuration from $n = 0$ to 8 is expected to shape.

Correlation of Shell Geometry and Buckling Stress

To investigate the effect of the shell geometry on the buckling stress R_T/R_B-ratio is taken as parameter. Buckling loads obtained for different ranges of R_T/R_B are depicted in Figures 29 and 30. In the region of negative Gaussian curvature, i.e., $R_T/R_B < 1.00$, the increment of R_T/R_B is chosen smaller, in order to obtain the path of the curves more accurately.

According to Figure 29, the cylindrical shell has the lowest buckling resistance against lateral pressure. There fore it is justified to speak of a "cylinder-valley" in the case of hydrostatically loaded shells of revolution. With increasing absolute value of the Gaussian curvature, buckling pressure increases regularly. On the other hand, shells of negative Gaussian curvature have the lowest buckling resistance under axial compression. The reduction of R_T/R_B from 1.00 to 0.85 already leads to a buckling load which is only ¼ of the buckling load of the cylinder. For lower values of R_T/R_B, however, the buckling load decreases only slightly, as shown in Figure 30.

In Figures 31 and 32 the computational results are depicted in terms of the stresses initiating the buckling process, σ_{11cr} and σ_{22cr} being the critical stresses in hoop and meridional directions, respectively. In Figure 31 test results obtained on cylindrical shells [16] and on shells of negative Gaussian curvature [15,17] such as those obtained on shell models of positive Gaussian curvature [18] are also given. In this figure the "cylinder-valley" is still perceptible. However, in the interval $0.7 \geq R_T/R_B \geq 0.5$ the buckling stress is nearly constant.

On the other hand, the meridional buckling stresses computed exhibit a "hill" at the region of the cylindrical shell and of the shells having a slight positive Gaussian curvature, as Figure 32 illustrates. However, the theoretical results depicted need corrections in order to account for the deviations from the test results, i.e., imperfection sensitivity. In case of cylindrical shell the reduction factor 0.37, as obtained by the author [16], is realistic. Tests on shells of negative Gaussian curvature [15,17] yield for pure axial compression a reduction of about 0.90. Unfortunately there exist no buckling tests on shells of positive Gaussian curvature subjected to axial compression alone. However, with increasing Gaussian curvature under axial compression relatively high tensile stresses in circumferential direction

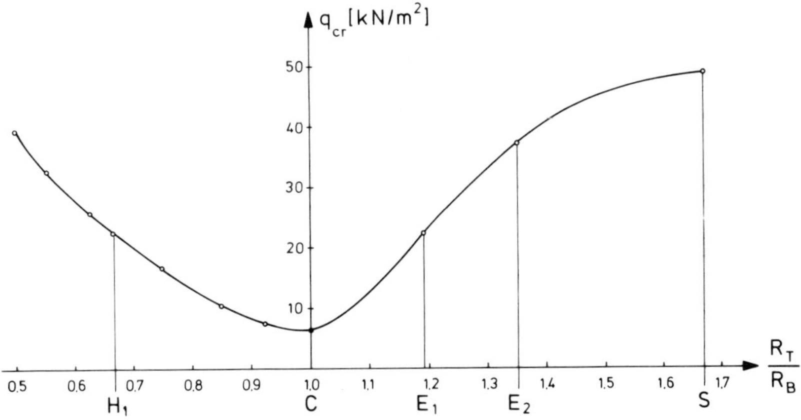

FIGURE 29. Necking (R_T/R_B) vs. lateral buckling pressure for boundary condition A.

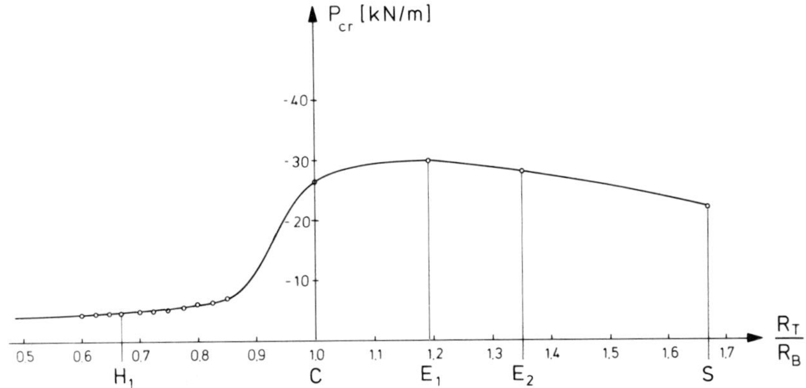

FIGURE 30. Necking vs. axial buckling load for boundary condition A.

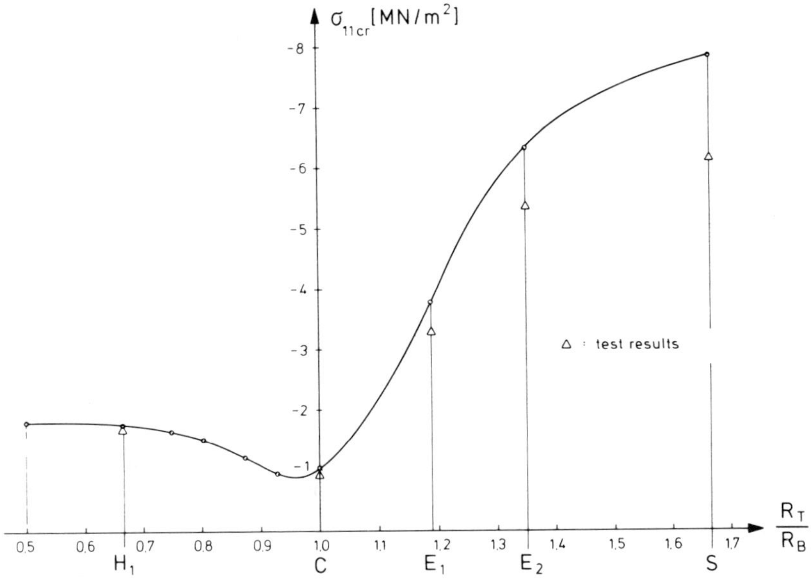

FIGURE 31. Necking vs. buckling stress in circumferential direction.

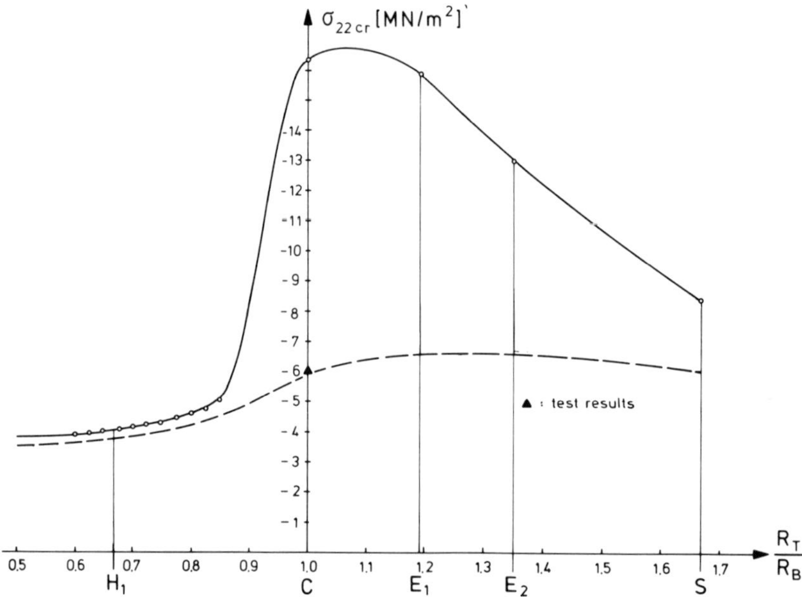

FIGURE 32. Necking vs. buckling stress in meridional direction.

are activated stabilizing the shell. Therefore, with increasing convexity of the shell or R_T/R_B-ratio a smaller correction for the load case pure axial compression is expected. With this argumentation the dashed line in Figure 32 may be accepted as realistic, at least qualitatively. That means the "cylinder-hill" coming out from the bifurcation analysis is only a "pseudo-hill" which may be eliminated easily through imperfections present in buckling tests [14].

MAIN PARAMETERS OF COOLING TOWER BUCKLING

Buckling Tests

The collapse of three of a group of seven cooling towers in Ferrybridge Power Station, England, in November 1965 gave impulse to intensive research on the buckling of hyperbolic cooling tower shells at the end of the sixties. The first extensive test series in wind tunnel on cooling tower shaped models were carried out by Der and Fidler [19]. These researchers measured the dynamic head required to buckle their electroformed copper and molded PVC models and derived the formula

$$q_{cr} = c.E\left(\frac{t}{R_T}\right)^n \quad (4)$$

for the buckling wind pressure of hyperboloids having proportions similar to their models. In Formula (4) E is modulus of elasticity, t wall thickness of the shell and R_T throat radius of the hyperboloid. Der and Fidler give an average value of 0.064 for c and propose the value 2.3 for n. The exact value of n is found later by Ewing analytically as equal to 7/3, both for axisymmetric lateral pressure and non-axisymmetric wind loading [20].

Veronda and Weingarten [21] have tested a large number of hyperboloidal shell models made of PVC sheets bonded along a longitudinal seam. Due to thermoforming the models were getting thinner with increasing distance from the throat. On the whole the models were very likely afflicted with initial imperfections in the shape due to relatively small wall thickness which was only 0.25 mm. Buckling started with a single dimple which mostly shaped near the upper boundary, where the models were thinnest [21]. Conclusion drawn from the tests and their theoretical analysis is that shells of negative Gaussian curvature exhibit a rather low grade of imperfection sensitivity. Values predicted applying the bifurcation theory are remarkably close to test results.

In both of foregoing test series no strain measurements on models were made and only the buckling load was recorded. In the buckling tests carried out in Bochum, Germany, in addition to load the strains at different locations of the models were recorded [15,17]. In this way it was possible to follow the local initiation of the first dimple. Besides, the models were relatively large and thick, t/R_T being about 1/125. In Figure 33 the geometry of these models is given. The models were cast between two molds having 5 mm spacing. After hardening, each model was machined from the external surface to the desired thickness by means of a special lathe. In this way it is possible to produce relatively perfect models without a seam. The wall thickness was measured along nine equidistant meridians at 19 equidistant elevations, i.e., at 171 points. The largest deviation from the mean value was in most of the models less than ± 10% [15]. The mean value of the thickness in the region where the first dimple was observed, i.e., critical zone, t_{cr}, is considered in the analysis of test results [15,17]. The models were tested first as symmetrical models S with 1.20 m height. The tests were repeated on the same models with reduced height 0.82 m having the cooling tower shape *C*.

The models were tested under axial load produced by a jack and hydrostatic pressure supplied pneumatically. The end of the models were bonded to the circular metal heads of the testing apparatus. The upper head was hinged allowing a vertical displacement and rotation. On internal and external surfaces of the models electrical strain gages were pasted in longitudinal and circumferential direction at five equidistant positions of the throat. The models were tested under water in order to slow down the snap-through of the model and maintain the temperature constant during each test. In each test a different ratio of axial load to hydrostatic pressure was taken and kept constant, with the load increasing in steps until the model buckled. As soon as the buckling configuration was observed, it was necessary to unload the model in order to prevent its collapse through material failure. After unloading, the buckling patterns recovered completely. This could be verified through the strain gage measurements after nearly 1 hr. A repetition of the test delivered the same buckling load and the same buckling pattern. Thus, it was possible to test the same model under various combinations of membrane stresses [15,17].

During each test the models were subjected to a dynamic disturbance, i.e., a lateral pressure of small magnitude with a frequency of 30/min. The additional circumferential stress caused in the model through the dynamic disturbance was only 2% of the critical axial stress and therefore negligible. The dynamic disturbance was applied at each loading step and the model was observed. No effect of the disturbance could be observed at the lower loading steps. However, at the critical load, periodic movements in radial direction could be observed. After 30 sec to 60 sec of pulsation the model began to buckle. At that moment the dynamic disturbance was eliminated. Nevertheless, the buckling procedure continued. In some tests the amplitude of the disturbance was doubled. In this case, the radial movements were observed at a load level which was nearly 10% lower than the buckling load. After a lengthy pulsation the model began to buckle. However, as the disturbance was eliminated, the model recovered completely and the buckling pattern disappeared. This could be verified with the strain measurements that followed. In most of the tests the disturbance was external lateral pressure. In some tests the dynamic disturbance was applied as internal pressure of the same magnitude. In both cases the same buckling load was

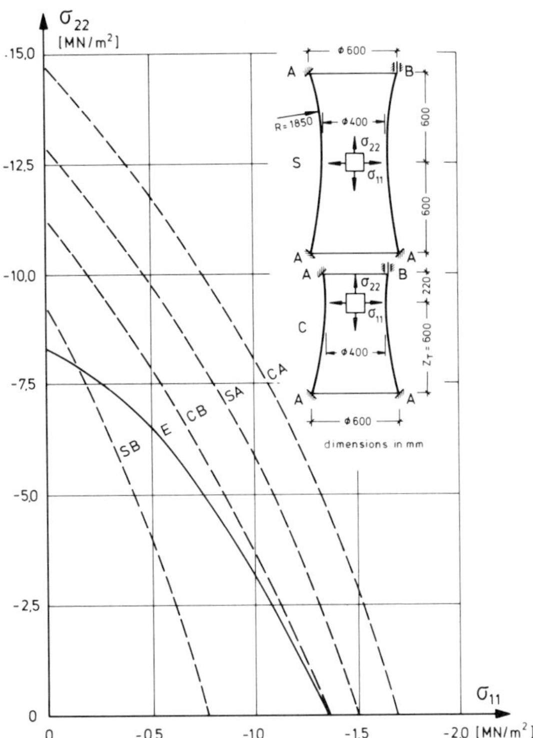

FIGURE 33. Models tested; experimental and theoretical interaction diagrams for the buckling stresses.

recorded. All these experiments show that the magnitude and direction of the periodic disturbance did not have an effect on the magnitude of the buckling load and on the buckling behavior of the models. The disturbance had affected the buckling procedure only as an initiator. In this way it was possible to reduce the deviation in test results at the highly imperfection sensitive stress states [16].

In hyperboloidal models subjected to axial compression combined with internal pressure resulting in a stress state with compressive stress in meridional direction and zero or very small tensile stress in circumferential direction buckling was initiated always through the dynamic disturbance, which points out the high imperfection sensitivity at these stress states.

The number of waves of a model observed in the tests was the same for all lateral and axial load combinations. Depending on the wall thickness five or six waves could be observed around the circumference. Only the shape, i.e., the length of the waves, changed with the ratio of the meridional and circumferential stresses. With increasing contribution of the meridional stress to buckling the waves became shorter in meridional direction. For axial compression combined with internal pressure the buckling pattern had the shape of a flat ellipse with its long axis in circumferential direction. The buckling stresses obtained in the tests had the same range, both for the symmetrical models S and for the shorter cooling tower models C. This fact can be explained through the buckling initiating effect of the unavoidable geometric imperfections which were the same in both test series. Additionally, the tests show that the increase in stiffness of the models through reduction of the height above the throat from 600 mm to 220 mm does not influence the buckling stresses and the buckling behavior [17].

Correlation of Numerical and Experimental Results

Interaction diagram E of Figure 33 given in terms of the hoop stress σ_{11} and meridional stress σ_{22} summarizes all of the experimental results obtained on eight different models. The wall thickness t_{cr} considered in the analysis of the buckling tests were 1.28 mm, 1.33 mm, 1.39 mm, 1.50 mm, 1.55 mm, 1.65 mm, 1.77 mm and 1.95 mm [15,17]. Six models were tested first in symmetrical shape S and afterwards in cooling tower shape C. One model was tested only in symmetrical shape and another directly in cooling tower shape. It turned out that, in addition to material elasticity, the main parameter influencing the buckling stress state in the tests is the value of the wall thickness at the region of the first dimple. For this reason the test results obtained on all eight models are extrapolated for a shell model having shape S or C, a wall thickness equal to 2.0 mm, as explained in the section on Wall Thickness, and made of a material having Modulus of Elasticity $E = 3{,}450$ MN/m^2 and Poisson's ratio $\nu = 0.38$.

In Figure 33 four additional interaction diagrams are given in dashed lines. These are the interaction diagrams computed applying the bifurcation theory on models having 2 mm wall thickness, $E = 3{,}450$ MN/m^2 and $\nu = 0.38$. The first capital letter indicates the shape and the second the boundary condition considered at the top of the model. Boundary condition A corresponds to built-in or fixed end in adjacent states of equilibrium. In case B the boundary is still built-in; however, displacements in vertical direction are permitted. In the tests the boundary condition at the top of the model was somewhat between boundary conditions A and B. The top of the model was attached to a rigid steel plate which could sink parallely and rotate about any axis freely, if single dimples shape only at one side. Boundary condition A does not allow both kinds of movement in the adjacent states; hence it is stiffer than the condition present in the test, whereas boundary condition B is weaker than the test condition. Because, according to boundary condition B a vertical displacement in shape of cos $n\,\Theta^1$ along the circumferential ordinate Θ^1 with any positive integer value of n is possible, whereas according to the test conditions n may be equal only to 0 or 1.

In fact symmetrical models S can be considered as cooling tower shaped models C which are supported elastically at a distance of 220 mm above the throat. This assumption is allowable, because the first dimple shapes always near the throat, i.e., away from boundaries. That means the computations yield four strongly deviating interaction diagrams SA, SB, CA and CB for the same model if the boundary condition at the upper edge is altered. That is rather surprising, because for shells of positive Gaussian curvature, calculation models A and B yield the same interaction diagram, as already shown in Figures 27 and 28.

A comparison of the theoretical interaction diagrams with the test results provides the following information: The calculations result in different interaction diagrams for the symmetrical and cooling-tower shaped models, a fact which is in contradiction to the tests. Interaction diagrams CA and SB can nearly be accepted as the upper and lower limits of buckling states, respectively. For stress states with nearly zero stress in circumferential direction, test results match with the values calculated for symmetrical models having boundary conditions B quite well. For other stress states interaction diagrams CB or SA fit better, as shown in Figure 3 of Reference 22.

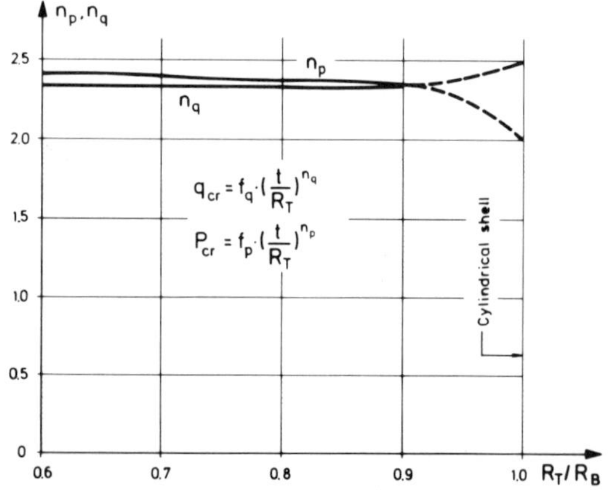

FIGURE 34. Necking vs. power of (t/R_T).

Effects of Wall Thickness, Boundary Conditions and Meridian Curve

WALL THICKNESS

In order to find out the relation between wall thickness and buckling loads or buckling stresses of a cooling tower shell, a numerical parametric study is carried out. In addition to shells with 1.6 mm wall thickness, shells with 0.8 mm and 0.4 mm thicknesses are calculated. This gives a range of shell slenderness $1/565 \leq t/R_t \leq 1/85$, because also the geometry, i.e., the necking, is varied within the range $0.60 \leq R_T/R_B \leq 1.00$, keeping $R_B = 0.225$ m constant. The results of this study are depicted in Figure 34. Within the range $0.60 \leq R_T/R_B \leq 0.90$ which is of interest for cooling tower shells, the buckling load parameters, both for the case of uniform lateral pressure and axial compression, are proportional to $(t)^{7/3}$. In order to get a nondimensional parameter the wall thickness t is divided by a length parameter, for example R_T which is the throat radius of the cooling tower shell. f_q and f_p are functions of other geometric parameters and of material constants and will be discussed later. At the range $R/R_B > 0.90$ the power n_q increases and reaches the value 2.5 for the cylindrical shell whereas n_p falls down to the value 2.0 in the case of axially compressed cylindrical shell. The buckling stresses of cooling tower themselves, i.e., at the region $0.60 \le R_T/R_B \le 0.90$, are proportional to $(t/R_T)^{4/3}$. This is the factor considered to extrapolate all test results obtained on eight models of different wall thickness in order to get interaction diagram E of Figure 33 which is valid for models S and C having 2 mm wall thickness, as mentioned in the previous section. For this end the buckling stresses measured are multiplied by the factor $(2/t_{cr})^{4/3}$ for each model [22].

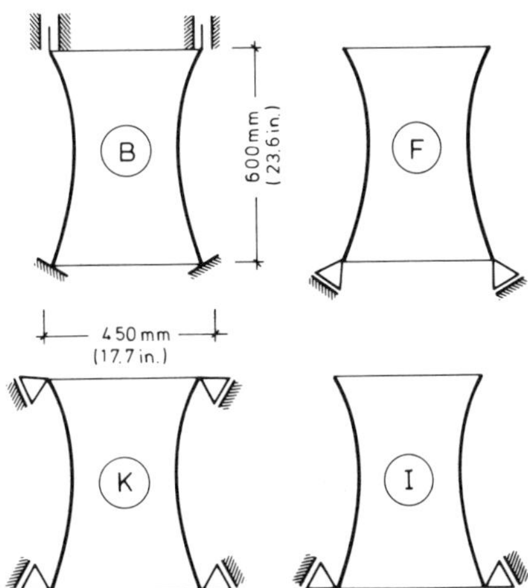

FIGURE 35. Boundary conditions investigated.

BOUNDARY CONDITIONS

The buckling loads due to lateral pressure and axial compression are obtained considering three other boundary conditions F, K and I, given in Figure 35, in addition to boundary condition B already discussed. Boundary condition F corresponds nearly to real boundary conditions of natural draught cooling tower shells, whereas the type of supporting K is the so called classical boundary condition. Boundary condition I can be considered as representative for a cooling tower standing on a soil stressed up to yield point and therefore deforming in direction of the supporting columns without any resistance, as a hypothetical lowest limit [22].

The shells calculated have a symmetrical shape with respect to the throat, the meridian curve being again circular. The height of the shell, H, and the radius at the base, R_B, are 600 mm and 225 mm, respectively. The radius at the throat, R_T, is varied in steps of $R_T/R_B = 0.025$ within the range $0.60 \le R_T/R_B \le 0.85$. The wall thickness, t, is 1.6 mm and the material constants considered are $E = 3{,}650$ MN/m² and $\nu = 0.35$.

In Figure 36 the buckling lateral pressures computed are depicted as function of R_T/R_B for all boundary conditions B, F, K and I. Also, the corresponding number of buckling waves in circumferential direction, n, is given at each part of the curves. The curves change regularly for systems SB, SF and SI, and the buckling pressure decreases with increasing R_T/R_B-ratio. In addition, the number of buckling waves changes regularly and only slightly. In contrast to these systems, SK exhibits buckling loads and wave numbers alternating strongly, even due to slight variations of R_T/R_B. This high dependence to geometry can hardly be accepted as realistic for cooling tower shells and needs, therefore, a physical interpretation.

A shell of negative Gaussian curvature having the classical boundary conditions K tends to deform quasi-inextensionally, i.e., without membrane strains. To demonstrate this, the variation of the meridional force of a hyperboloidal shell subjected to a periodical external pressure $q = \cos n\Theta^1$ is depicted in Figure 37 for different R_T/R_B-ratios. For $n = 2$ the meridional force n_{22} vanishes in the case of the shape having $R_T/R_B = 0.70$. For $n = 3$ and $n = 4$ the in-

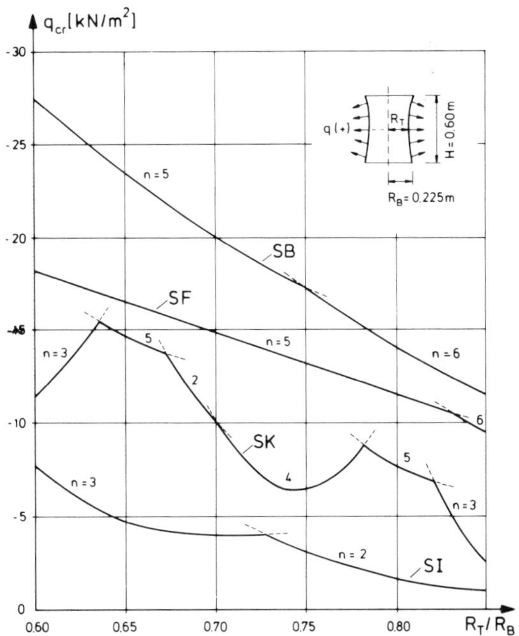

FIGURE 36. Necking vs. lateral buckling pressure for different boundary conditions.

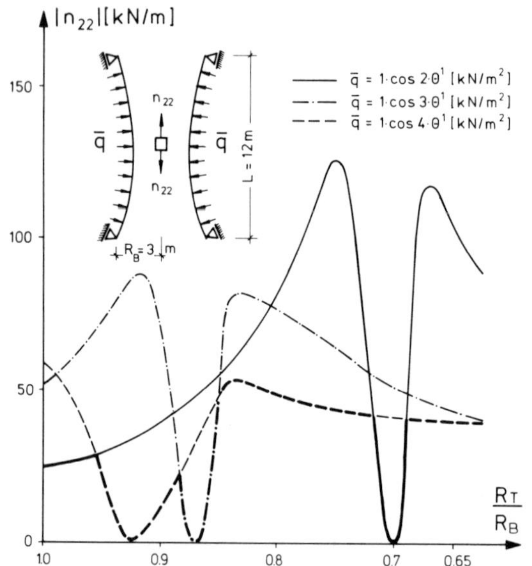

FIGURE 37. Necking vs. meridional membrane force due to periodic lateral pressure.

extensional behavior is observed at $R_T/R_B = 0.87$ and 0.93, respectively.

In the light of Figure 37 the SK-curve of Figure 38 for axially compressed hyperboloids can now be explained as follows. The low buckling loads for the cases $R_T/R_B = 0.60$

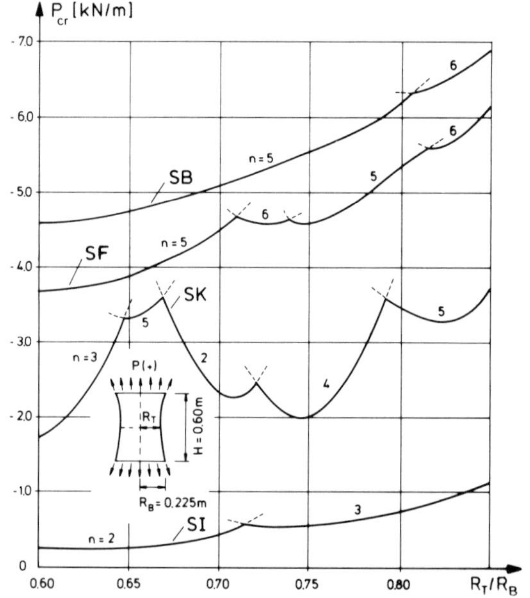

FIGURE 38. Necking vs. buckling edge load for different boundary conditions.

and 0.75 with their corresponding buckling configurations $n = 3$ and 4, respectively, are due to the dominant quasi-inextensional behaviors of these shapes. On the other hand, for $R_T/R_B = 0.65$ the deformation mode with five periodic waves in circumferential direction comprises a considerable extensional part, so that a relatively high buckling load, comparable with that of the case SF, i.e., cooling tower case, is attained.

MERIDIAN CURVE

The shape of the meridian is in natural draught cooling tower shells mostly a hyperbola. For this reason, the parametric studies are extended also on such shells having boundary condition F [22]. For comparison, the SFH-curve, which is valid for hyperbolic meridian, is depicted in Figure 39 together with SF-curve of Figure 36, which is valid for circular meridian. The fact that the critical lateral pressure in case SFH depends rather scarcely on R_T/R_B-ratio is at first sight surprising. In addition, the divergency of curves SF and SFH with decreasing R_T/R_B-ratio, i.e., increasing necking, is rather unexpected, due to only slight difference in the shape of a cooling tower with a circular or hyperbolic meridian, if the height H and R_T/R_B are the same in both cases. However, an interpretation of this at first glance contradictory behavior is possible, if the behavior at buckling tests and local buckling stresses are considered.

In tests carried out on models having circular meridian, buckling started always at the throat. This is due to constant curvature along the meridian. In the case of a hyperbola, however, the curvature decreases with increasing distance from the throat. For this reason, the location where the first dimple shapes is tending to shift to regions away from the throat. This tendency was observed also in tests of Reference 21. On the other hand, under the same lateral pressure, the circumferential stress initiating the buckling process will be nearly R/R_T-times bigger than the stress at the throat, because of the bigger hoop radius R. In Figure 39 the ratio of the buckling loads for SFH and SF is approximately equal to R_T/R_B-ratio. That means the buckling initiating stress remains nearly the same for a shell of revolution having negative Gaussian curvature, for given H, R_T and R_B, independent of the geometry of the meridian curve.

The buckling loads for the case of axial compression are given in Figure 40. This time the curves have the same tendency both for system SF and SFH, the values being larger in the case of hyperbolic meridian. Under axial loading the biggest meridional stress is acting at the throat. For this reason, it is obvious that models with circular meridian buckling always first at the throat [17] can carry only a lower axial load than the hyperbolic models buckling at regions away from the throat [21]. On the other hand, if the stresses are concerned, it turns out that again in both cases at the point where buckling starts approximately the same critical stress interaction relationship E, as given in Figure 33, holds. This fact allows the extension of the results obtained

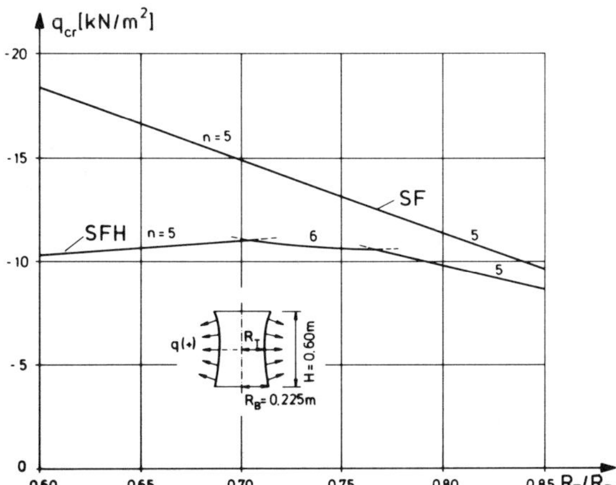

FIGURE 39. Lateral buckling pressure for circular and hyperbolic meridians.

on models with circular meridian to hyperbolic shells and further to regions other than the throat.

Comparison with Cylindrical Shells

In Figure 41 interaction diagram E of Figure 33 is given together with four interaction diagrams which are applicable to the cylindrical shells C_1 to C_4 [23]. C_1 and C_2 are cylindrical shells having a radius equal to throat radius and a length equal to the length of the cooling tower shaped models C or symmetrical models S tested, respectively, whereas C_3 and C_4 have a radius equal to the basis radius. Interaction diagrams C_1 to C_4 are based on author's results obtained on cylindrical shells by applying the same testing method [16].

It is interesting to notice that at the stress state $\sigma_{11} = 0$, uniaxial compression in meridional direction, the buckling stresses of the hyperboloidal shells and of the cylindrical shells C_1 and C_2 are equal. In shells of negative Gaussian curvature this stress state is obtained under axial compression together with internal lateral pressure. As quoted already, the first dimple forms at the throat and has the shape of a flat ellipse with its long axis in the circumferential direction. For this reason the whole dimple is restricted to an area which only slightly differs from a cylinder having the throat radius. In addition, in both cases the buckling process is initiated by means of the slight dynamic disturbance applied in the tests to check the stability in sense of Liapunov's definition.

In presence of circumferential compression, the shell of negative Gaussian curvature has a higher buckling resistance than the cylindrical shells C_1 and C_2. That means under lateral external pressure the negative curvature stiffens the shell compared to cylindrical shells. With in-

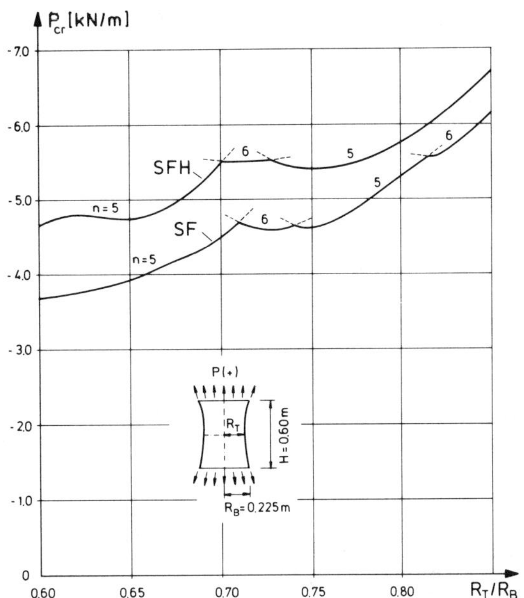

FIGURE 40. Buckling edge load for circular and hyperbolic meridians.

creasing length of the cylindrical shell the buckling resistance decreases, if circumferential compression is acting. On the other hand, the cylindrical shells C_3 and C_4 having a radius equal to base radius of the shell of negative Gaussian curvature are weaker and in no way representative for the buckling stresses of hyperboloidal shell, because of the

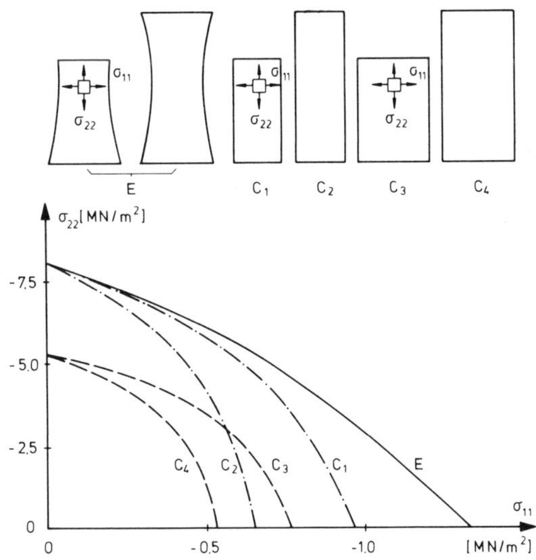

FIGURE 41. Experimental interaction diagrams for hyperboloids and cylinders.

large deviation between interaction diagram E and each of the diagrams C_3 and C_4.

BSS Approach

After the observations and the ideas explained in foregoing preparatory sections, it is now possible to develop an approach for the buckling design of cooling tower shells on local basis. That is to say the buckling design has to guarantee that local buckling is prevented everywhere within a certain safety margin. The fulfillment of this requirement leads also to the fulfillment of global or overall stability of the shell and hence constitutes a necessary and sufficient condition for stability. The fulfillment of global stability in a direct way, however, does not exclude coherently local instabilities. For that it is only a necessary condition. This is the main idea behind the Buckling Stress States-BSS-Approach explained in the following.

For generalization it is necessary to introduce nondimensional parameters in design where possible. First of all, the interaction diagram E of Figure 33 can be made nondimensional by introducing the nondimensional variables $(\sigma_{11}/\sigma_{110})$ and $(\sigma_{22}/\sigma_{220})$ indicating the level of stressing, instead of the stresses themselves, σ_{11} and σ_{22}, respectively. Herein σ_{110} and σ_{220} are the buckling stresses due to uniaxial compression in circumferential and meridional directions, respectively, whereas σ_{11} and σ_{22} are the circumferential and meridional stresses leading to buckling under biaxial stress states. The graph and the Equation (5) of the nondimensional interaction diagram E are given in Figure 42. It is interesting to notice that the expression (5) coming out from the buckling tests is symmetric with respect to the nondimensional variables $(\sigma_{11}/\sigma_{110})$ and $(\sigma_{22}/\sigma_{220})$.

Next step is the determination of the uniaxial buckling stresses σ_{110} and σ_{220}. For this end extensive numerical studies were carried out, in which the shell material, wall thickness and geometry were taken as parameters [24]. It turned out that both buckling stresses are proportional to $E/(1-\nu^2)^{3/4}$, if Hookean material with modulus of elasticity E and Poisson's ratio ν is assumed. As already discussed in a previous section and referring to Figure 34, the buckling stresses are proportional to $(t/R_T)^{4/3}$, t being the local wall thickness. The global effect of cooling tower geometry is considered by varying the necking (R_T/R_B) and slenderness (R_T/Z_T), Z_T being the height of the throat from the base. The values calculated for different shell geometries are given in Tables 2 and 3, namely k_{G11} for the uniaxial buckling stress in circumferential direction and k_{G22} in meridional direction. Finally, it is necessary to account for the deviation between theoretical values and the results obtained in buckling tests. Therefore, also reduction factors F_{11} and F_{22} are to be introduced:

$$\sigma_{110} = \frac{E}{(1-\nu^2)^{3/4}} \cdot \left(\frac{t}{R_T}\right)^{4/3} \cdot k_{G11} \cdot F_{11} \qquad (6)$$

$$\sigma_{220} = \frac{E}{(1-\nu^2)^{3/4}} \cdot \left(\frac{t}{R_T}\right)^{4/3} \cdot k_{G22} \cdot F_{22} \qquad (7)$$

In computations carried out to obtain k_{G11} the real boundary conditions of a cooling tower shell, i.e., boundary condition F of Figure 35, were considered. On the other hand k_G22 were obtained considering the boundary condition B of the same figure. This is more appropriate due to better convergency behavior of the extensive computations if boundary condition B is considered. As it can be seen in Figure 38, the transition from the values calculated for boundary condition B to those corresponding to F is quite possible. The k_{G22} values obtained are to be reduced approximately by the factor 0.85 which results from the comparison of the curves SB and SF.

The buckling tests already described deliver the values $F_{11} = 0.985$ and $F_{22} = 0.720$ [17]. If the reduction due to boundary condition F is incorporated into F_{22}, in order to use the values of Table 3, in this case F_{22} has to be taken equal to 0.612.

Having σ_{110} and σ_{220} obtained, the application of BSS Approach is straightforward. The membrane stresses σ_{11} and σ_{22} of the cooling tower shell due to dead load together with wind load are first computed at each elevation. Then the safety factor γ_B present against local buckling is obtained from the interaction relationship (5), after introducing γ_B:

$$0.80\,\gamma_b \left(\frac{\sigma_{11}}{\sigma_{110}} + \frac{\sigma_{22}}{\sigma_{220}}\right) + 0.20\,\gamma_B^2 \left[\left(\frac{\sigma_{11}}{\sigma_{110}}\right)^2 + \left(\frac{\sigma_{22}}{\sigma_{220}}\right)^2\right] = 1 \qquad (8)$$

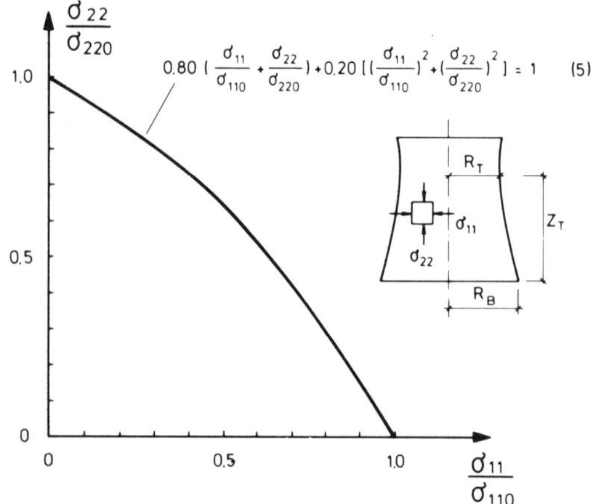

FIGURE 42. Non-dimensional interaction diagram for cooling tower models and design formulas.

TABLE 2. k_{G11}-Values of Formula 6.

R_T/Z_T \ R_T/R_B	0.571	0.600	0.628	0.667	0.715	0.800	0.833
0.250	0.105	0.102	0.098	0.092	0.081	0.063	0.056
0.333	0.162	0.157	0.150	0.138	0.124	0.096	0.085
0.416	0.222	0.216	0.210	0.198	0.185	0.162	0.151

TABLE 3. k_{G22}-Values of Formula 7.

R_T/Z_T \ R_T/R_B	0.571	0.600	0.628	0.667	0.715	0.800	0.833
0.250	1.28	1.33	1.37	1.45	1.56	1.74	1.85
0.333	1.20	1.25	1.30	1.37	1.49	1.73	1.86
0.416	1.13	1.17	1.23	1.31	1.43	1.68	1.82

STIFFENING EFFECT OF RINGS

Numerical Results

With increasing dimensions of cooling towers, the thickness of the shell wall increases over proportionally in order to maintain the buckling safety γ_B prescribed. This means much more reinforced concrete volume and workmanship are required for the shell part, but also the sizes of the supporting columns and of the foundation increase. An optimization is possible, if stiffening elements are arranged in order to improve the buckling resistance of the shell part. Stiffening elements may be either horizontal rings or meridional ribs. These two can also be combined. However, parametric studies have shown that discrete rings alone are most effective, if the augmentation of the buckling resistance is related to the additional concrete volume needed [25]. For this reason, attention will be focused now only on the stiffening effect of rings.

Again numerical and experimental parametric investigations were carried out and exclusively axisymmetric loading conditions were considered in order to reduce the volume of the computations. For the numerical analysis the shell was discretized into a series of conical frustra. The displacement field within each element was resolved into Fourier components in the circumferential direction and was expressed in terms of Hermitian polynomials in the meridional direction. At the elevation of the stiffening rings a branching of the shell in horizontal direction was supposed and a ring-beam element with its own degrees-of-freedom was introduced [26]. The parameters varied were the size, position and number of rings.

The relationship between the lateral buckling pressure and width b of the stiffening ring, as depicted in Figure 43, is based on the results obtained in Reference [27]. A 160 m high cooling tower shell is stiffened by means of two rings having 0.20 m thickness, one placed at the throat and the other in the middle of the distance between the throat and the base. q^o_{cr} and q_{cr} are the buckling pressures of the unstiffened and stiffened cooling towers, respectively. If the width b of the rings is increased proportional to local wall thickness of the shell the augmentations given in the graph of Figure 43 are obtained. The full or dashed curves are valid for the assumptions that the rings buckle in the same

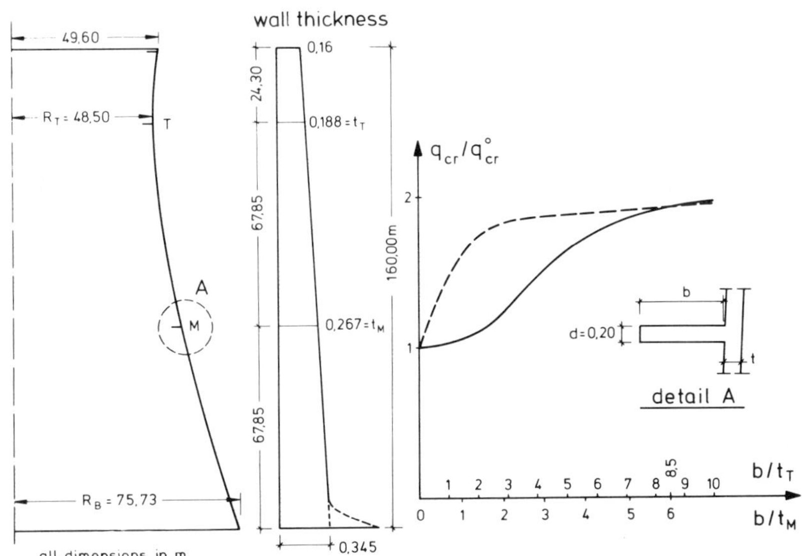

FIGURE 43. Effect of ring width on lateral buckling pressure according to Reference [27].

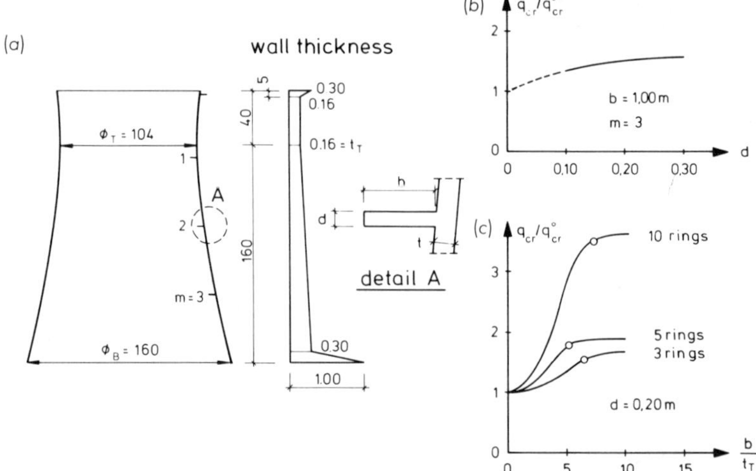

FIGURE 44. Effect of thickness, width and number of rings on lateral buckling pressure according to Reference [28].

harmonics as the shell or never at all, respectively. For a ring width corresponding to $b/t_M = 6$ or $b/t_T = 9$ both curves intersect, i.e., the ring does not buckle, even if it is allowed to.

For the cooling tower shell given in Figure 44a, curve *b* demonstrates the only slight augmentation of the buckling pressure if the ring thickness *d* is increased twofold or threefold. Effect of the ring width in case of three, five or ten rings is depicted in Figure 44c. Again a width equal to at least six times of the local shell thickness proved to be optimal [28]. In addition, the most effective position of a stiffening ring coincides with the peak of the buckling configuration of the unstiffened shell. The positions of the second and third rings can be obtained by iteration successively. For a number of rings larger than five, the equidistant positioning is efficient as well [28].

Experimental Results

In addition to numerical parametric studies, some buckling tests were carried out on models stiffened by means of 5, 9 or 19 equidistant rings [15]. The stiffened models are the same described in the section on Buckling Tests and shown in Figure 33. The width and thickness of the ring were 7.5 mm and 2.0 mm, respectively, giving a width to shell thickness ratio of about 6. The models are tested first without stiffening rings. Results obtained in this case carry the index 0. Afterwards the models are tested with the arrangement of five equidistant rings and the results carrying the index 5 are obtained. In an analogous way the number of the rings is increased to 9 and 19, on the same model. In the first test series the rings were glued to the shell along their entire circumference. In the second test series, however, the rings were glued only along 40% of their circumference, i.e., alternately 4 cm were glued to the shell wall and 6 cm were not. In both test series the same buckling loads and the same buckling patterns were observed [15].

Interaction diagrams (E) for the buckling loads obtained in the tests and those calculated applying the bifurcation theory (T), are given in Figure 45. The theoretical interaction diagram of the shell stiffened by means of 19 rings (T_{19}) is omitted in the figure, because the numerical results depend in this case substantially on the assumption of global or local loss of stability and in the second case on where the local buckling is assumed to take place. If the four experimental interaction diagrams are considered, first it is interesting to notice that E_0, E_5 and E_9 intersect in one point corresponding to axial compression P_{Lcr} together with internal lateral pressure. However, for the models stiffened by means of 19 rings the buckling load is under this load combination twice as high. Further, again under this load combination, the deviation between the experimental buckling load factors and the numerical values increases with increasing number of rings, the experimental values being only 61%, 54% or 43% of the theoretical values in the case of 5, 9 or 19 intermediate stiffening rings, respectively. This is symptomatic for increasing imperfection sensitivity of ring stiffened hyperboloidal shells and agrees with the behavior discussed in a previous section, Figure 24. On the other hand, under lateral pressure alone the experimental values are about 80% of the theoretical values for all three cases which means nearly the same but lower imperfection sensitivity.

The above analysis of test results in terms of the loads is to some extent misleading. As buckling starts locally and the position of the first dimple changes according to the number of stiffening rings, a common interpretation of the test results obtained on different models is more realistic in terms of the stresses initiating the buckling. With increasing number of stiffening rings the first dimple shifts from the throat to the upper or lower parts of the model where the radius of rotation and consequently the circumferential stress σ_{11} is greater. Figure 46 shows the position and the extension of the first buckling wave observed in tests for different number of stiffening rings, the buckling waves due to circumferential compression being in general larger than those shaping due to meridional compression. The analysis of test results in terms of the stress states initiating the buckling process yields the interaction diagrams given in Figure 46. The buckling stress (σ_{110}) under uniaxial compression in circumferential direction increases 1.40, 3.34 or even 6.30 times after stiffening the models by means of 5, 9 or 19 equidistant rings, respectively. Contrary to this, the buckling stress σ_{220} under pure meridional compression is not affected by the stiffening rings at all. This fact has the following physical interpretation: As already mentioned, buckling waves shaping mainly due to meridional compression are short in meridional direction. This is observed not only in tests under axisymmetric loading but also in our wind tunnel tests in which buckling waves shape first at the flanks due to high meridional compression. For this reason the number of stiffening rings has to be large enough, i.e., the distance between the rings sufficiently short, in order to prevent the shaping of such short dimples. On the other hand, it is partly that the slight advantage attainable arranging rings of limited number is eliminated completely through increasing imperfection sensitivity under pure meridional compression.

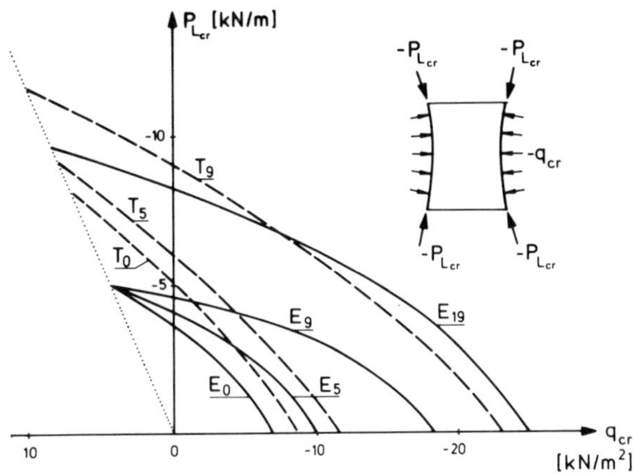

FIGURE 45. Experimental and theoretical interaction diagrams for the buckling loads of ring-stiffened hyperboloids.

The nondimensional equation of all four interaction diagrams given in Figure 46 is the same and identical with Equation (5) of unstiffened models. To get this equation the variables (σ_{11}/σ_{110}) and (σ_{22}/σ_{220}) are to be introduced, as already explained in the section on BSS Approach. In doing this the values given in Figure 46 are to be considered for σ_{110} in each case of stiffening.

BSS Approach for Ring-Stiffened Cooling Tower Shells

BSS Approach as explained in a previous section needs a modification if it is going to be applied on cooling tower

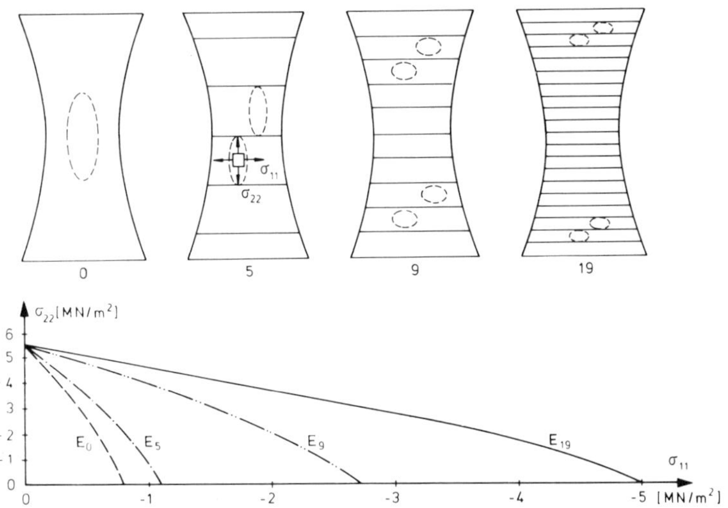

FIGURE 46. Position of the first dimple and experimental interaction diagrams for the buckling stresses of ring-stiffened hyperboloids.

TABLE 4. Amplification Factor α of Formula 6a.

	Number of Rings Symmetrical Model (Cooling Tower Model)			
	0 (0)	5 (3)	9 (6)	19 (14)
α:	1.00	1.40	3.34	6.30

shells stiffened by means of rings. The only thing to be done is to introduce an amplification factor α in Equation (6) for the uniaxial buckling stress in circumferential direction:

$$\sigma_{110} = \frac{E}{(1 - \nu^2)^{3/4}} \cdot \left(\frac{t}{R_T}\right)^{4/3} \cdot k_{G11} \cdot F_{11} \cdot \alpha \qquad (6a)$$

The rest is completely the same as in the case of cooling tower shells without stiffening rings. The values of α for different number of stiffening rings for the symmetrical models S or corresponding cooling tower shaped models (C) are given in Table 4. The final stiffening effect of rings of different number is demonstrated in Figure 47 in terms of effective augmentation of the buckling safety in case of a cooling tower shell having 165 m height, Z_T being 124 m, and 150 m basis diameter, R_T being 41.75 m. The wall thickness of the shell was first determined for a lowest local buckling safety factor $\gamma_{LB} = 5$ for the case without any intermediate stiffening rings. Arranging rings of different number in most effective positions the augmentations given in Figure 47 are obtained. It turns out that a considerable augmentation of the buckling resistance is reached first after arranging three rings. For the cooling tower shell investigated the highest augmentation rate is attainable in case of four or five rings, where the graph has a zone of inflection providing maximal inclination. Therefore, depending on the amount of augmentation needed, preference should be given to four or five intermediate rings in this case.

INFLUENCE OF R/C PROPERTIES

Buckling vs. Compression Failure

The results presented in foregoing sections are obtained on calculation and test models behaving according to Hooke's law. However, an additional important aspect to consider in buckling design of cooling tower shells is the influence of reinforced concrete's behavior. Under compression the nonlinear deformation behavior of concrete at high stress levels and its ultimate strain are decisive for the stability and strength of the shell. The effect of these parameters can be investigated after modification of the shell theory and the computation method valid for elastic shells.

The assumptions made with respect of shells in general are:

- Thickness (t) is small in comparison to the minimum curvature radius (R), i.e., $t/R << 1$.
- Displacements and their gradients are so small that their third and higher powers can be neglected.
- The magnitudes of the strain tensors $\alpha_{\alpha\beta}$ and $\omega_{\alpha\beta}$ are of the same order.

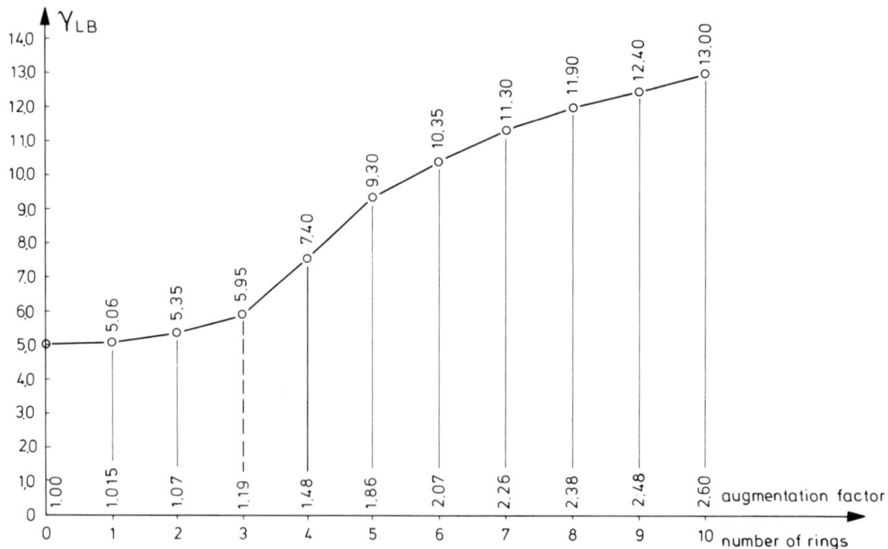

FIGURE 47. Number of stiffening rings vs. local buckling safety factor.

- Kirchhoff-Love-Hypothesis holds; i.e., the normal vectors of the undeformed middle surface remain perpendicular to the middle surface after deformation.
- Stresses normal to the middle surface are neglected; i.e., shell thickness does not change during deformation.
- Rotational inertia of the shell can be neglected in calculation of the kinetical energy.
- Shell stiffness is a continuous and smooth function of the surface coordinates within each finite element.

The modifications now are to abandon the assumptions of linear relationship between stress and strain (Hooke's Law), and the isotropy of the shell continuum [11].

Under the foregoing assumptions, Green's strain tensor takes the form

$$\gamma^{\alpha\beta} = \alpha^L_{\alpha\beta} + \alpha^N_{\alpha\beta} + \Theta^3\omega \qquad (\alpha,\beta = 1 \text{ or } 2) \qquad (9)$$

$\alpha^L_{\alpha\beta}$ and $\alpha^N_{\alpha\beta}$ being linear and nonlinear parts of the first strain tensor, respectively, and $\omega_{\alpha\beta}$ being the second strain tensor. Θ^3 is the coordinate perpendicular to the reference surface. In matrix notation the kinematic equations are:

$$\begin{bmatrix} \alpha^L_{\alpha\beta} \\ \omega_{\alpha\beta} \end{bmatrix} = [D^i_{\alpha\beta}] \cdot [\nu_i] \qquad (i = 1,2 \text{ or } 3) \qquad (10)$$

and

$$[\alpha^N_{\alpha\beta}] = [\nu_i]^T \cdot [D^{ij}] \cdot [\nu_j] \qquad (11)$$

The actions in the shell are obtained using the constitutive equations

$$\begin{bmatrix} n^{\alpha\beta} \\ m^{\alpha\beta} \end{bmatrix} = H^{\alpha\beta\rho\lambda} \begin{bmatrix} D^{\alpha\beta\rho\lambda} & K^{\alpha\beta\rho\lambda} \\ K^{\alpha\beta\rho\lambda} & B^{\alpha\beta\rho\lambda} \end{bmatrix} \begin{bmatrix} \alpha^L_{\rho\lambda} \\ \omega_{\rho\lambda} \end{bmatrix} \qquad (12)$$

in which $n^{\alpha\beta}$ and $m^{\alpha\beta}$ = the tensor of stress resultants and stress couples, respectively. $D^{\alpha\beta\rho\lambda}$, $B^{\alpha\beta\rho\lambda}$ and $K^{\alpha\beta\rho\lambda}$ = the membrane stiffness, bending stiffness, and coupling stiffness, respectively; and $H^{\alpha\beta\rho\lambda}$ = a tensor taking account for the shell geometry. In the analysis of reinforced concrete shells concrete is assumed to have a stress-induced anisotropy. Increments of the stress tensor and strain tensor are assumed to be related linearly and the proportionality coefficients are assumed to be the functions of stress tensor and strain tensor but not of their history, and not of their rates. Within the increments of the stress tensor and strain tensor the existence of an elastic potential will be assumed [11].

A material law for concrete under biaxial stress states σ_{11} and σ_{22} is given in Figure 48 in terms of the secant moduli E^1_{bs} and E^2_{bs} in the directions 1 and 2 of the principal strains. E_0 is the initial tangent modulus and β_p stands for the prism strength of concrete. The material law of Figure 47 is based

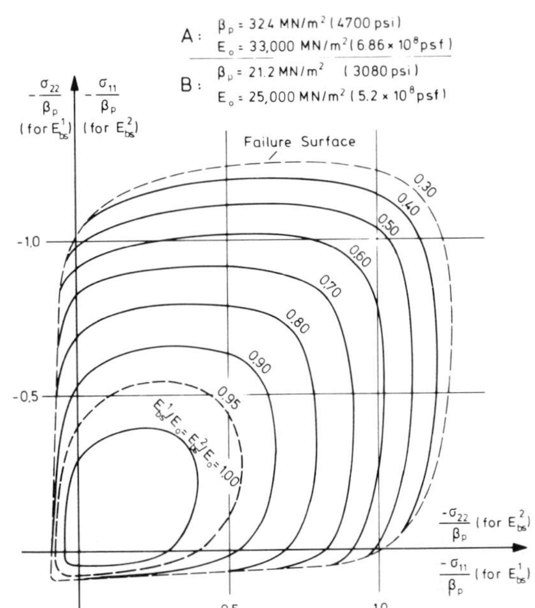

FIGURE 48. Material law for concrete under biaxial stress state.

on the tests of Kupfer [29] and has been proposed by Link [30] who introduces the condition

$$\nu^1_{bs}/E^1_{bs} = \nu^2_{bs}/E^2_{bs}; \qquad \nu^1_{bs} \text{ and } \nu^2_{bs}$$

being Poisson's ratios in the principal strain directions. The incremental elastic coefficients of the material law are now defined according to Equation (13).

$$\begin{bmatrix} \sigma_{11} \\ \sigma_{22} \\ \sigma_{12} \end{bmatrix} = \frac{E^1_{bs}}{1 - \nu^1_{bs} \cdot \nu^2_{bs}} \begin{bmatrix} 1 & \nu^2_{bs} & 0 \\ \nu^1_{bs} & \dfrac{\nu^2_{bs}}{\nu^1_{bs}} & 0 \\ 0 & 0 & \dfrac{1 - \nu^1_{bs}\,\nu^2_{bs}}{1 + 2\,\nu^1_{bs} + \dfrac{\nu^1_{bs}}{\nu^2_{bs}}} \end{bmatrix} \begin{bmatrix} \varepsilon_{11} \\ \varepsilon_{22} \\ 2\varepsilon_{12} \end{bmatrix} \qquad (13)$$

Cracking of concrete is considered by setting the modulus of elasticity in the direction perpendicular to the crack equal to zero [11]. On the other hand a certain tensile strength of concrete has to be considered, otherwise the numerical results obtained are not realistic. For reinforcing steel the

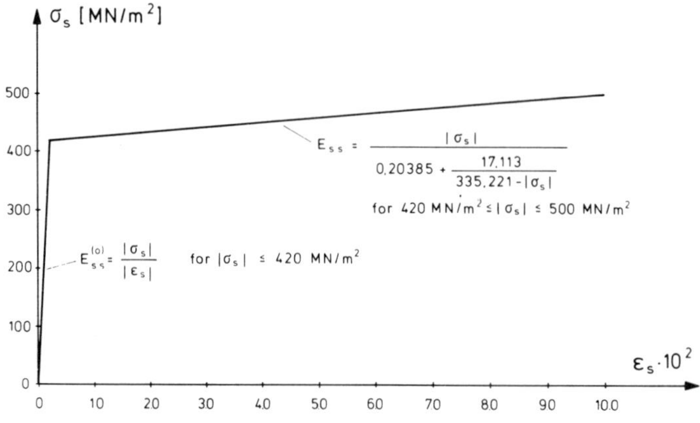

FIGURE 49. Material law for reinforcing steel.

bilinear stress-strain diagram given in Figure 49 is considered.

Because buckling starts locally and the local stress state is the governing factor, an axisymmetric loading can be substituted for the nonaxisymmetric wind load. At each elevation the most unfavorable stress state with respect to buckling has to be the same for both the actual loading, i.e., dead load plus wind load, and for the conjugate axisymmetric loading [11]. Having this done, the displacement field of the cooling tower shell can now be introduced by Fourier Series in the circumferential variable Θ_1:

$$\begin{aligned} v_1 &= v_1^n(\Theta^2) \cos n\Theta^1 \\ v_2 &= v_2^n(\Theta^2) \sin n\Theta^1 \\ v_3 &= v_3^n(\Theta^2) \cos n\Theta^1 \end{aligned} \tag{14}$$

Taking conical frustras as finite ring elements and introducing linear and cubical polynomials for the displacement components in the tangential and normal directions, respectively, the unknown functions can be determined. The finite elements are divided into 11 layers. As shown in Figure 50, the reinforcement is smeared to equivalent layers in meridional and circumferential directions [11].

To determine the variation of the buckling load factor λ_B with increasing load the stability equation

$$(\underline{K}_E + \lambda_B \underline{K}_G)\, \underline{V} = 0 \tag{15}$$

will be solved at different load levels. In Equation (15) the material stiffness matrix K_E and geometric stiffness matrix K_G depend on the level and combination of the inplane strains in different elements and layers. For this reason the service load consisting of the self weight of the reinforced concrete shell, and of the conjugate load compensating for the wind, together with the uniform internal suction, is increased incrementally. After each increment the equilibrium equation

$$\underline{K}_E \cdot \underline{V} + \underline{p} = 0 \tag{16}$$

is solved by iteration, p being the load vector.

In the iterations the secant stiffness technique was applied. The convergency of the method depends in this case on the material law considered. The convergence criterion used for stopping the iteration procedure after a load increment is based on the variation of the stiffness matrixes $\underline{K}_E^{n-1}$ and $\underline{K}_E^n$ in two successive iteration steps $(n-1)$ and (n). A variation less than 5% was considered to be sufficient and could be reached in the present case by taking sufficiently small load increments after five or six iterations. Once the secant stiffness is obtained from the equilibrium analysis, the strain dependent tangent stiffness is calculated and put into Equation (15) in order to find the lower bound of nonelastic buckling according to Shanley [31].

The method of analysis developed is applied to the cooling towers given in Figure 51. Cooling tower A is 200 m high and is designed under the wind load usually considered in Germany. The wall thickness of the shell is determined according to BSS Approach at each elevation. In German Reinforced Concrete Code DIN 1045 the same buckling safety factor is prescribed against dead load and wind load. In U.S. design codes, however, two different load factors are used, making the wind load factor higher. This deviation between the two design codes is due to the different safety concepts behind them. As Figure 51 shows the wall thickness calculated applying BSS Approach and the German recommendation to cooling tower A decreases with the height linearly from 0.331 m to 0.157 m between the lower and upper edge beams.

Cooling tower B is designed for extreme wind pressures such as those considered in some coastal areas of the U.S.

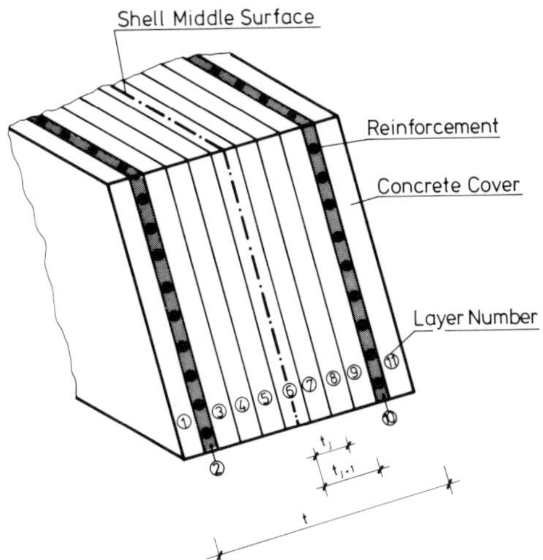

FIGURE 50. Layered R/C shell finite element.

The wall thickness of the shell is determined only under the condition of sufficiently high safety factor against global buckling. Therefore, according to U.S. practice, the dead load was kept constant and only the wind load was increased [32]. The local buckling phenomenon and BSS Approach are not taken into consideration, because at the time cooling tower B was designed BSS Approach was not developed. The variations of the wall thickness and of the wind pressure along the height are given in Figure 51 for both cooling tower shells A and B.

In order to compute the global buckling safety factors first the most unfavorable stress states with respect of local buckling due to dead load together with wind are determined in each of 60 equidistant ring elements. The nonaxisymmetric wind loading is replaced by an axisymmetric loading consisting of vertical edge loads and external or internal radial pressures acting on each ring element. In this way the most unfavorable stress state at each ring element could be generated through an axisymmetric "conjugate" load [11].

Material laws considered in calculations correspond to those given in Figures 48 and 49. In Figure 48 the compressive strength of concrete and initial tangent modulus E_0 are given for both cooling towers. The relatively low values for case B correspond to the information in Reference 32. The reinforcing steel, a minimum of 0.4% in cooling tower A and 0.35% in cooling tower B in meridional and circumferential directions, is placed half in each face.

In accordance with the German practice dead load and the conjugate load substituting the wind were incremented simultaneously until the ultimate load factor ULF 7.55 or 3.90 was reached for cooling tower A or B, respectively. At this load level the failure surface of concrete, as given in Figure 48, was reached in some layers of a ring element. This was followed by propagation of the failure to neighboring elements until most elements reached the ultimate state [11].

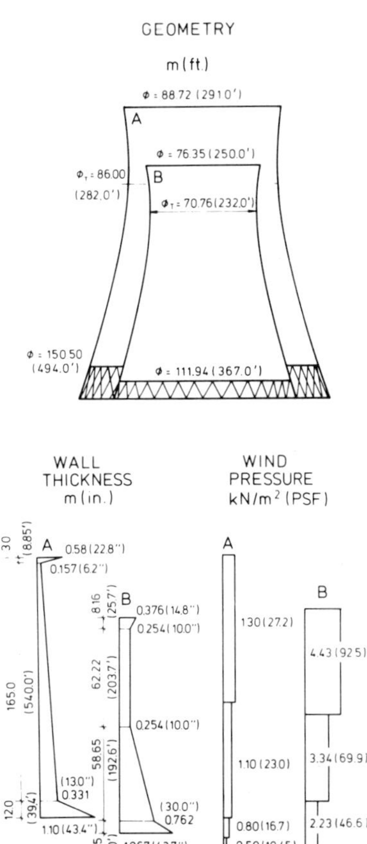

FIGURE 51. R/C cooling tower shells A and B investigated against buckling vs. compression failure.

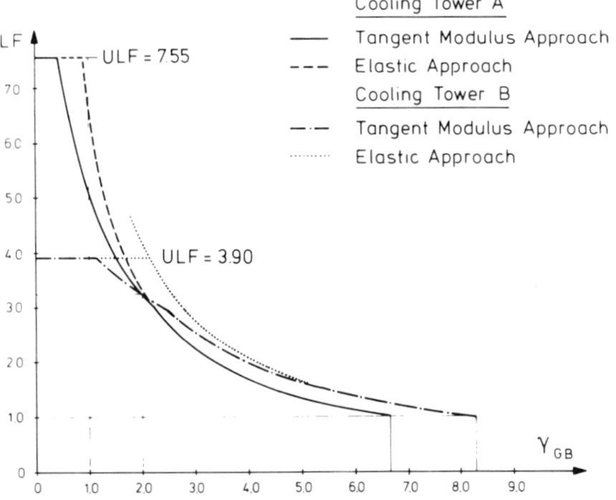

FIGURE 52. Load factor vs. global buckling safety factor (γ_{GB}) for cooling towers A and B.

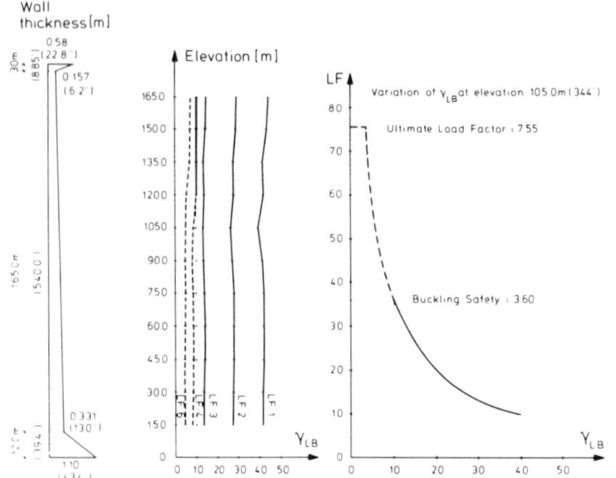

FIGURE 53. Load factor vs. local buckling safety factors (γ_{LB}) for cooling tower A.

Figure 52 shows the variation of the theoretical global buckling safety factor (γ_{GB}) calculated applying the bifurcation theory at different load factors (LF). The effect of nonlinear material behavior is found first by introducing the initial tangent modulus of elasticity (E_0) and then the strain dependent tangent modulus (E_T) at each load level. As Figure 52 shows, the buckling safety γ_{GB} drops to half of the value for elastic buckling at the near of ultimate load, if the nonlinear behaviors of concrete and of reinforcing steel are considered. Additionally, cooling tower A will buckle (γ_{GB} = 1) theoretically at LF = 5, far below the ultimate load (ULF = 7.55). Cooling tower B, however, will collapse at LF = 3.90 just before the theoretical buckling capacity at LF = 4.00 is reached [11].

If BSS Approach is applied and the tangent moduli in meridional and circumferential directions are considered the variation of the local buckling safety (γ_{LB}) at different load factors and elevations can be calculated. Figures 53 and 54 give the results of these calculations for cooling towers A and B, respectively [11].

Since the design of cooling tower A is carried out applying BSS Approach γ_{LB} varies only slightly along the height. Up to LF = 3 the local buckling safety factor is overall bigger than 1.0. At LF = 4 the whole region between the elevations 15 m and 120 m becomes unstable (γ_{LB} < 1.0). At LF = 5 the whole cooling tower is unstable. At the right part of Figure 52 the lowest local buckling safety factor is obtained as equal to 3.60 for the most critical section at El. = 105.0 m from the top of the supporting columns. On the other hand, the destruction of concrete starts at El. = 12.0 m where the thickening for the lower edge member begins.

In cooling tower B the local buckling safety factor (γ_{LB}) is altering irregularly along the height, as depicted in Figure 54. Under service load (LF = 1) γ_{LB} changes between 23 at El. = 10 m and 3.68 at El. = 67.53 m or 7.40 at El. = 97.54 m. Up to LF = 3 overall γ_{LB} > 1.0. However, at LF = 3.50 only the region in the neighborhood of El. = 67.53 m becomes unstable. At LF = 3.90, which is only a little below the ultimate load, the buckling area is still restricted to the region between El. = 56 m and El. = 73 m. On the right side of Figure 54 the lowest local buckling safety factor in the most critical section at El. = 67.53 m from the top of supporting columns is found to be minimum, γ_{LB} being 3.10. It is interesting to note also that the ultimate load is reached first at this section, as the load factor is increased to 3.90; i.e., the destruction of the shell starts in the section with the lowest local buckling safety.

The results presented are summed up in Table 5 of Reference 11. Column 2 of Table 5 gives the bifurcation results

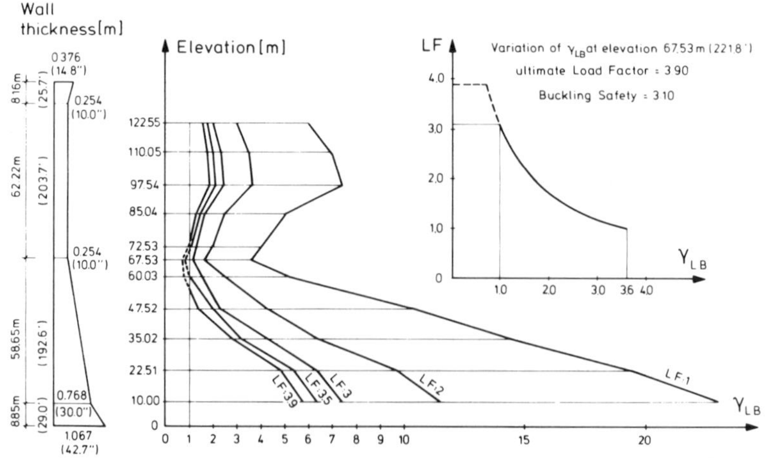

FIGURE 54. Load factor vs. local buckling safety factors (γ_{LB}) for cooling tower B.

TABLE 5. Safety Factors Against Buckling and Material Failure for Two Cooling Towers.

	Buckling Safety Factors						
	Against Global Instability γ_{GB}				Against Local Instability γ_{LB}		
	Theoretical Values		Reduced Values		Reduced Values		
Cooling tower shell (1)	Elastic approach (2)	Tangent modulus approach (3)	Elastic approach (4)	Tangent modulus approach (5)	Elastic approach (6)	Tangent modulus approach (7)	Ultimate load factor (8)
A	6.64	5.00	5.00	3.76	4.00	3.60	7.55
B	8.30	4.00	6.25	3.01	3.68	3.10	3.90

assuming linear elastic material behavior. The values of column 4, meanwhile, take account for the deviation between the bifurcation and test results. The global reduction factor 0.75 is an approximate value derived from the test results discussed in two previous sections. The approximation comes from the deviation between the geometry of the models tested and the geometries of cooling towers A and B investigated. Columns 3 and 5 give the theoretical and reduced values, respectively, of the bifurcation results, if the material nonlinearity is considered according to the strain dependent tangent modulus approach. As an approximation again the same reduction factor 0.75 is taken.

Values of columns 6 and 7 are obtained applying BSS Approach to both cooling towers in order to find out the lowest local buckling safety factors. According to Figures 53 and 54 the values 4.00 and 3.68 are obtained, if the linear elastic approach is considered for cooling towers A and B, respectively. If the nonlinear behavior is considered these values drop to 3.60 and 3.10, respectively. Finally, the ultimate load factors given in column 8 are obtained on local basis in sections where compression failure of concrete starts. Comparison of the values in columns 7 and 8 shows that both cooling towers are expected to buckle before concrete crushes [11].

Buckling vs. Tension Failure

In the cooling tower shell subjected to wind action there are some areas where the resultant action from self weight and wind loading is a tensile membrane force. In contrast to the areas compressed in both directions and which are therefore critical with respect of buckling no danger of initial buckling exists in the zones under tension. In such areas the strength against tension is the problem which is again a local phenomenon. Because buckling design yields the wall thickness required and in this way also the self weight of the cooling tower shell, there is an interaction between both problems. Increasing the buckling safety factor prescribed results in thicker shell wall, higher prestressing through self weight, lower resultant tensile force and, consequently, a higher ultimate load factor for wind.

For example, in recent German practice the design of cooling tower shells against tensile failure is carried out by keeping the self weight constant and increasing the wind load by the factor 1.75, whereas in design against buckling failure both loadings are to be increased by the same factor which has to be at least 5. The big difference between the values 1.75 and 5 makes it necessary to investigate the correlation between these two safety factors. Besides, in contrast to the usual design practice, there exists an opinion seeing no need for a proof of sufficient safety against buckling at all and accepting the cracking of concrete and yield of steel as the actual reason for failure of a reinforced-concrete cooling tower [33].

In order to get transparency in the rather complicated interaction of both failure mechanisms, cooling tower shells given in Figure 55 are investigated. These two shells have the same height, but in case of Geometry 2 the base diameter is 20% larger. The wall thickness of both cooling tower shells is obtained applying BSS Approach taking a nearly constant local buckling safety factor along the height. For each shell different values of buckling safety between 2.1 and 5 are chosen, as given at the lower right part of Figure 55 in parentheses. With exception of the upper and lower edges the amount of reinforcement is taken as equal to 0.3% in each direction, as it is generally recommended in cooling tower design practices [2,34].

Each shell designed for a certain buckling safety factor was investigated afterwards under self weight and non-axisymmetric wind load, the later increased in steps until failure. The shells were discretized by means of quadrilateral shallow shell elements, each with 32 degrees of freedom, as shown in Figure 56 and known under the name "semi-loof" [35]. In the analysis again the material laws of Figures 48 and 49 are considered for concrete and smeared reinforcing layers, respectively. Again, the shell section was

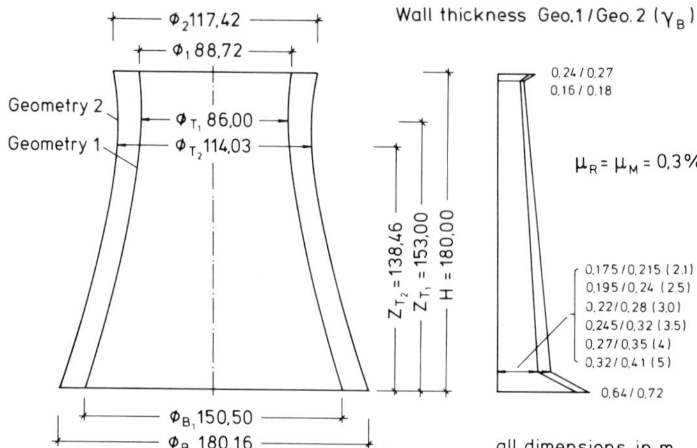

FIGURE 55. R/C cooling tower shells 1 and 2 investigated against buckling vs. tension failure.

divided into 11 layers, according to Figure 50. To limit the computational work, as a first approximation, the tensile stiffening effect was not included in the material law, i.e., concrete layers cracked remain in succeeding iterations and load increments out of consideration.

Typical variation of the radial displacement along the windward meridian and along the circumference at the elevation of the element failing first due to increasing wind load are depicted in Figure 57. These curves are valid for the cooling tower Geometry 1 designed to have the buckling safety equal to 5 [36].

Figures 58 and 59 summarize the results of all investigations for the cooling towers Geometry 1 and 2, respectively. To each design buckling safety factor a certain critical wind load factor corresponds at which the cooling tower shell fails due to tension. Failure is reached soon after concrete cracks in some layers of the most critical element. First the linear Load vs. Meridional Force and Load vs. Displacement relationships become nonlinear. Then a local redistribution of internal forces takes place after which the meridional tensile force drops abruptly, whereas the radial displacement increases rapidly [36].

The difference between the behaviors of the cooling towers Geometry 1 and 2 is that in the larger cooling tower the meridional forces under service load, i.e., g + w, are compression for all design buckling safety factors. In case of Geometry 1 ultimate load factor is always smaller than the design buckling safety factor. For example, to resist a wind load factor of magnitude 1.75 under tension a buckling design is necessary providing a safety nearly equal to 2.5. The same buckling safety factor, however, provides to cooling tower Geometry 2 a tensile strength to resist 2.44 times the wind load, as Figure 59 demonstrates.

The correlation between design buckling safety factor and

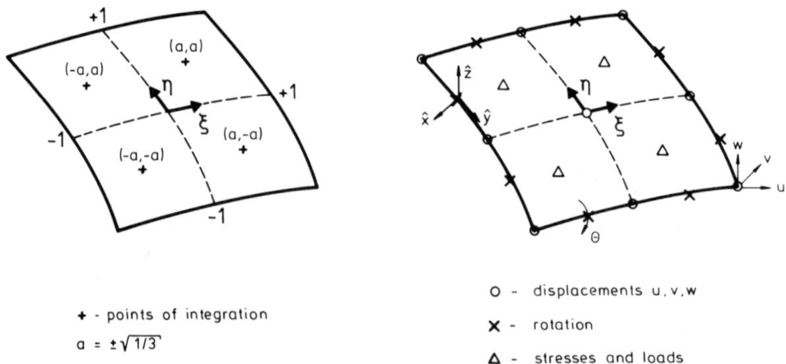

FIGURE 56. "Semi-Loof" shell element with 32 degrees of freedom according to Reference [35].

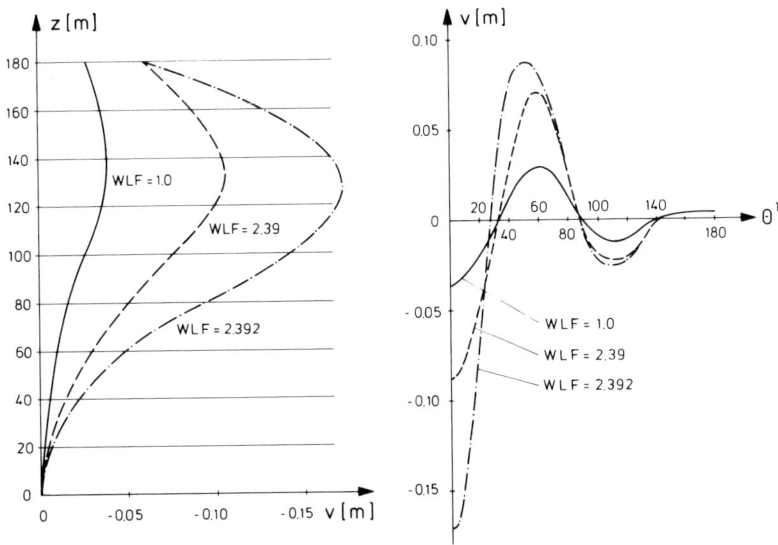

FIGURE 57. Radial displacements of cooling tower 1 at windward meridian ($\Theta' = 0°$) and along the circumference at El. = 127.5 m according to Reference [36].

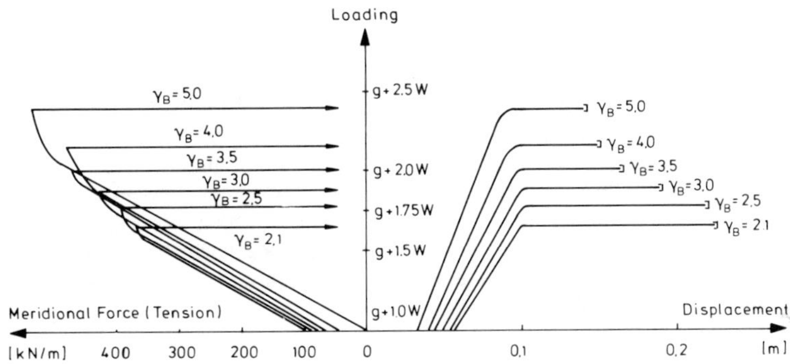

FIGURE 58. Wind load factor vs. behavior of cooling tower 1 designed for different buckling safety factors.

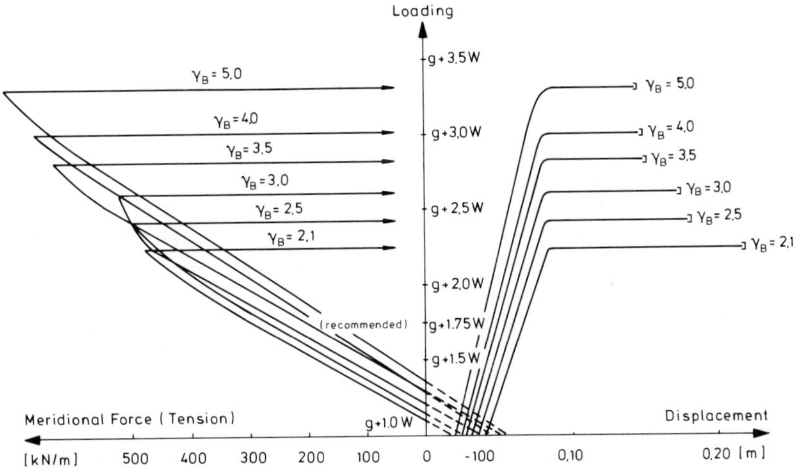

FIGURE 59. Wind load factor vs. behavior of cooling tower 2 designed for different buckling safety factors.

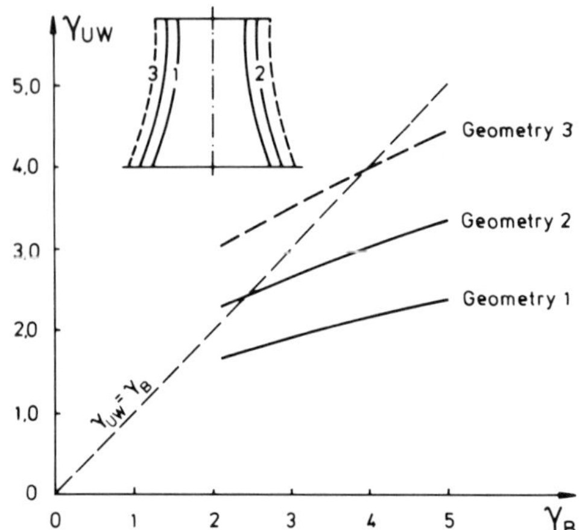

FIGURE 60. Correlation between ultimate load factor for wind and design buckling safety factor.

the ultimate wind load factor γ_{UW} is depicted in Figure 60. With increasing cooling tower diameter, for the same height, buckling is more likely to occur than tensile failure.

The tensile failure is governed by the tensile strength assumed for concrete. The effect of the reinforcement on the ultimate wind load factor is negligible, until the reinforcement is strong enough to take the whole tensile force released by concrete once cracked. This may only be the case if the amount of reinforcement $\mu > f_t'/f_s$. In the case of concrete having tensile strength $f_t' = 3.0$ MPa and reinforcing bars having yield strength $f_s = 420$ MPa, μ had to be more than 0.72%, a condition which cannot be fulfilled from point of economy.

A Design Method

In the light of the results presented in foregoing sections design of a reinforced concrete cooling tower shell can be carried out in the following steps.

1. The first step is the determination of the wall thickness of the shell. There are two criteria for this: buckling resistance and tensile strength. In the determination of the buckling safety factor due to gravity forces plus wind load, the strain dependent tangent stiffness of reinforced concrete decreasing with increasing strain should be considered. Gravity loading consists mainly of the dead weight of the shell and as such is a more or less deterministic permanent loading causing compressive membrane forces both in circumferential and meridional directions. In contrast with this the forces induced by wind are instantaneous effects whereas design wind itself has a fictitious value which is based on probabilistic considerations. For this reason it seems to be appropriate to increase both gravity forces and wind forces by the same factor in design against buckling, as this is the case in German design practice [2].

The lower limit of the safety factor against buckling failure may be assessed according to following considerations. Because buckling causes a sudden failure without warning the safety factor should be at least 2.1. This value is prescribed for the case of compression failure of concrete in German Building Code for Reinforced Concrete – DIN 1045 [37]. In German Regulation for Buckling Design of Shells an additional partial factor of safety 4/3 is introduced, to account for the deviations in buckling stresses obtained experimentally, i.e., the imperfection sensitivity [38]. Accordingly, the buckling safety factor of reinforced concrete shells in general and of cooling tower shells in particular should be at least 2.8.:

$$\gamma_{LB(g+w)} \geq 2.8 \tag{17}$$

Because of material nonlinearity the relationship between the load factor and the corresponding buckling safety factor is not proportional. If this effect cannot be considered and if there exists no test results carried out on models with a geometry near to the geometry of the cooling tower to be designed a higher buckling safety factor up to 5, as already practiced in Germany [2] and recommended in IASS-Recommendations [39], should be chosen.

The second criterion to determine the shell thickness is the load carrying capacity required against the tension failure of concrete. Because this is again a brittle failure without warning the safety factor to be applied to the loading causing tensile membrane stresses, in case of cooling towers only to the wind load, should be 2.1:

$$t \geqslant \frac{\max N_{(g+2.1w)}}{f_t'} \tag{18}$$

f_t' is the tensile strength of concrete to be considered in material law.

2. The second step is to obtain the amount of reinforcement. The safety factor against yielding of steel is taken 1.75. This factor should be applied to forces induced by wind only. Therefore, A_s, area of steel has to be:

$$A_s \geqslant \frac{\max N_{(g+1.75w)}}{f_s} \tag{19}$$

f_s is the yield stress of reinforcing bars, whereas the safety factor 1.75 is prescribed in DIN 1045 against ductile failure with warning deformations [37]. In this check which corresponds to recent German design practice the tensile strength of concrete is put equal to nil [2].

The tensile strength of reinforced concrete cooling tower shells, due either to failure of concrete or yielding of the reinforcement, can be checked against section forces obtained applying the linear bending theory of elastic shells. This is a rather conservative approach, as the computations carried out show.

The design method proposed is applicable only if the

cooling tower shell has a sufficiently perfect shape, i.e., if the amplitudes of the initial imperfections in the shape are smaller than the limits given in design recommendations for cooling towers [2,39]. Otherwise, premature failure occurs needing special consideration and additional investigations.

Differential settlement of the foundation, earthquake forces, thermal and ice forces are additional effects to be considered, if they are relevant.

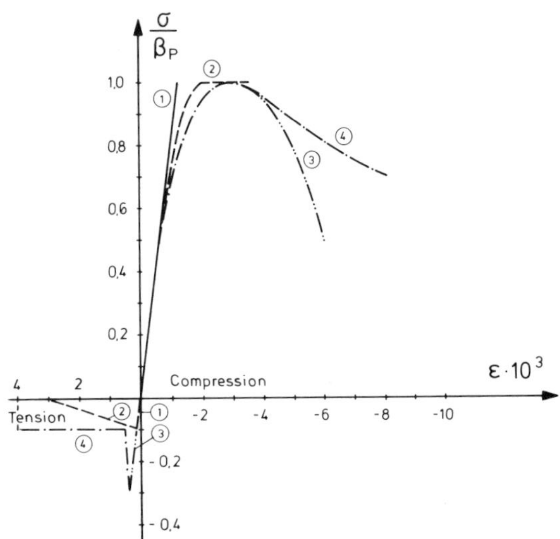

FIGURE 62. Concrete material laws considered for R/C cooling tower shells in Reference [44].

FURTHER RESEARCH AND CONCLUSION

Recent cooling tower research focuses mainly on two phenomena:

First, the correspondence between the local buckling stresses due to axisymmetric loading as developed here and under nonaxisymmetric wind loading needs an experimental verification. This is the objective of a comprehensive research project still in process. Buckling tests on aeroelastic models in wind tunnel will clear the buckling behavior of unstiffened cooling tower models and of those stiffened by means of discrete rings at different elevations. As in the case of the test series under axisymmetric loading the strains induced through wind loading are measured at different points of the models by means of electrical strain gages. In this way it is possible to determine the total stress state of the models up to the level at which buckling starts.

The results already obtained in wind tunnel tests are qualitatively in good agreement with the former test results obtained under axisymmetric loading. Local buckling is the only possible buckling behavior under wind load. In none of the tests global buckling with a periodic buckling configuration could be observed. With increasing number of rings the first dimple shifts to regions away from the throat. The effect of an upper edge beam is in case of hyperboloidal shells rather limited compared with cylindrical shells [40]. In Figure 61 the augmentation attainable in the buckling pressure due to wind is depicted for various widths of the ring beam. All these observations are indicatory to the applicability of the test results obtained under axisymmetric loading to wind loaded hyperboloidal cooling tower shells. Therefore BSS Approach constitutes a sound basis for the buckling design of reinforced concrete cooling tower shells. The BSS Approach's advantages lie in its simplicity and relatively high grade of approximation. It provides a tool cheap and easy to use in design practice. However, after the

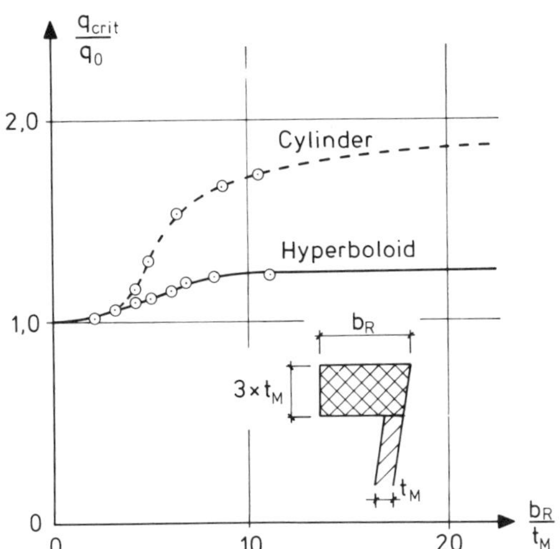

FIGURE 61. Stiffness of the upper edge member vs. wind buckling pressure according to wind tunnel tests of Reference [40].

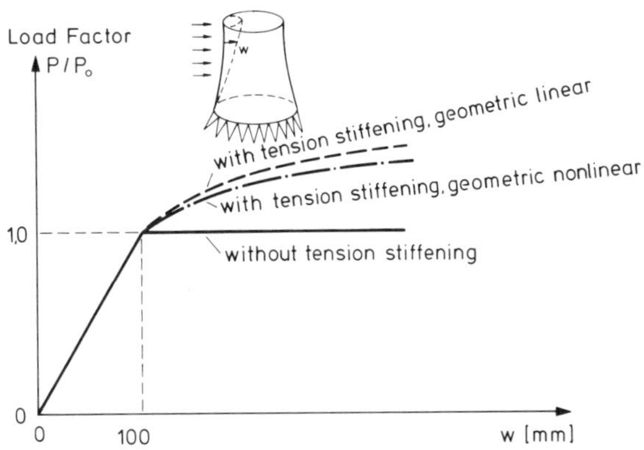

FIGURE 63. Effect of tension stiffening and geometric nonlinearity according to Reference [43].

numerical analysis of the wind tunnel tests is completed further refinement of the method may be in order.

Secondly, emphasis is laid on reinforced concrete behavior. In one of the research works the effect of the bond behavior between two cracks is considered directly [41]. In another work the effects of different material laws and tension stiffening behaviors are investigated. To estimate the upper and lower limits of such effects material laws 1 to 4 shown in Figure 62 are considered in ultimate load analysis of cooling tower shells. References 42 and 43 give results of similar investigations. In Figure 63 based on the information included in Reference 43 the ultimate load increases about 40% if a relatively high tension stiffening model is used in computations. On the other hand, the inclusion of geometric nonlinearity reduces the ultimate load only about 6%. In the computations carried out in Bochum the effect of tension stiffening coming out is less than that obtained in both References 42 and 43. Even in the case of very high tension stiffening effects corresponding to material laws number 2 or 4 of Figure 62 the augmentation attainable is only about 20%, compared with material laws 1 or 3 including no tension stiffening effect at all [44].

ACKNOWLEDGEMENTS

This chapter is based upon a part of the investigations carried out by the Research Group in Professor Zerna's "Lehrstuhl I (Chair for Reinforced and Prestressed Concrete)" at the Ruhr-Universität Bochum, Germany, between 1972 and 1985. During this time the research group was under the guidance of the author. Financial support was granted mainly by "Deutsche Forschungsgemeinschaft" and by "Ministerium für Wissenschaft und Forschung des Landes Nordrhein-Westfalen." The numerical results presented were calculated by many research assistants cited in the List of References for their Doctoral Theses. Speaking for all former students who have had the privilege and pleasure of working over the years with Professor Dr. E.h. Dr.-Ing. Wolfgang Zerna, to whom this chapter is dedicated, I express our thanks for his fascinating teaching as well as inspiring and liberal leadership in research.

REFERENCES

1. Billington, D. P., in *Natural Draught Cooling Towers*, P. L. Gould, W. B. Krätzig, I. Mungan, and U. Wittek, eds., Springer, Berlin, Heidelberg, New York, Tokyo, p. 351 (1984).
2. V.G.B. (Technical Union of Operators of Large Power Plants) *Bautechnik bei Kühltürmem: Teil 2, Bautechnische Richtlinien*, VGB-Verlag, Essen (1980).
3. Wolf, J. P., in *Natural Draught Cooling Towers*, P. L. Gould, W. B. Krätzig, I. Mungan, and U. Wittek, eds., Springer, Berlin, Heidelberg, New York, Tokyo, p. 161 (1984).
4. Euler, L. (English translation), *Isis*, 20, 1 (1933).
5. Lagrange, J.-L.C.d., *Méchanique Analitique*, Veuve Desaint, Paris (1788).
6. Poincaré, J. H., *Acta Math.*, 7, 259 (1885); see also *Oeuvres*, Gauthier-Villars, Paris, 1952, 7, p. 40.
7. Liapunov, A. M., Reprinted at Princeton, *Ann. of Math.*, 17, Princeton University Press (1949).
8. Hellinger, E., *Encykl. Math. Wiss. IV* (4), 601 (1914).
9. Hamel, G., *Theoretische Mechanik*. Springer, Berlin-Göttingen-Heidelberg (1949).
10. Shield, R. T. and A. E. Green, *Arch. Rational Mech. Anal.*, 12, 354 (1963).
11. Zerna, W., I. Mungan, and W. Steffen, "Wind-Buckling Approach for RC Cooling Towers," printed in *ASCE, Journal of Engineering Mechanics*, 109, Paper No. 18032, 836 (June 1983).
12. Berke, G. and R. L. Carlson, *Exp. Mech.*, 8, 548 (1968).
13. Almannai, A., Y. Basar, and I. Mungan, *Bauingenieur*, 54, 205 (1979).
14. Mungan, I., *Bauingenieur*, 58, 421 (1983).
15. Mungan, I., "Buckling Stresses of Stiffened Hyperboloidal Shells," printed in *ASCE, Journal of the Structural Division*, 105, Proc. Paper 14775, 1589 (August 1979).
16. Mungan, I., "Buckling Stress States of Cylindrical Shells," printed in *ASCE, Journal of the Structural Division*, 100, Proc. Paper 10965, 2289 (November 1974).
17. Mungan, I., "Buckling Stress States of Hyperboloidal Shells," printed in *ASCE, Journal of the Structural Division*, 102, Proc. Paper 12465, 2005 (October 1976).
18. Wurm, P., *Bauingenieur*, 50, 397 (1975).
19. Der, T. J. and R. Fidler, *Proc. Inst. Civ. Eng.*, 41, 105 (1968).
20. Ewing, D. J. F., Central Electricity Research Laboratories, Rep. RD–L–R 1763-23-1764 (1971).
21. Veronda, D. R. and V. I. Weingarten, *ASCE, J. Struct. Div.*, 101, 1585 (1975).
22. Almannai, A., Y. Basar, and I. Mungan, "Basic Aspects of Buckling of Cooling-Tower Shells," printed in *ASCE, Journal of the Structural Division*, 107, Proc. Paper 16121, 521 (March 1981).
23. Zerna, W. and I. Mungan, in *Buckling of Shells*, E. Ramm, ed., Springer, Berlin, Heidelberg, New York, p. 467 (1982).
24. Mungan, I., *Konstruktiver Ingenieurbau Berichte* 29–30, 75 (1977).
25. Zerna, W. and I. Mungan, "Construction and Design of Large Cooling Towers," printed in *ASCE, Journal of the Structural Division*, 106, Proc. Paper 15217, 531 (February 1980).
26. Mungan, I. and O. Lehmkämper, "Buckling of Stiffened Hyperboloidal Cooling Towers," printed in *ASCE, Journal of the Stuctural Division*, 105, Proc. Paper 14917, 1999 (October 1979).
27. Benz, H. J., Ph.D. Thesis, *Techn. Wiss. Mitt.*: 76-10, Ruhr-Universität Bochum (1976).
28. Lehmkämper, O., Ph.D. Thesis, *Techn. Wiss. Mitt.*: 78-6, Ruhr-Universität Bochum (1978).
29. Kupfer, H., Report No. 78, Lehrstuhl f. Massivbau d. TH München (1969).

30. Link, J., *Schriftenreihe DAfStb*, 270 (1976).
31. Shanley, F. R., *J. Aeron. Sci.*, 14, 261 (1947).
32. Cole, P. P., J. F. Abel, and D. P. Billington, *ASCE, J. Struct. Div.*, 101, 1205 (1975).
33. Mang, H. A., H. Floegl, F. Trappel, and H. Walter, *Eng. Struct.*, 5, 163 (1983).
34. ACI-ASCE Committee 334, J. A.C.I., 74, 22 (1977).
35. Irons, B. M., in *Finite Elements for Thin Shells and Curved Members*, John Wiley & Sons, London (1976).
36. Winter, M., "Das Grenztragverhalten von Kühlturmschalen aus Stahlbeton," Ph.D. Thesis, Ruhr-Universität Bochum (1984).
37. Deutscher Ausschuss für Stahlbeton, *DIN 1045-Beton-und Stahlbetonbau, 1978*, printed in *Beton-Kalender 1985*, Ernst & Sohn, Berlin (1985).
38. Deutscher Ausschuss für Stahlbau, *DASt-Richtlinie 013: Beulsicherheitsnachweise für Schalen*, Stahlbau Verlags GmbH, Köln (1980).
39. I.A.S.S. (International Association for Shell and Spatial Structures), *Recommendations for the Design of Hyperbolic or Other Similarly Shaped Cooling Towers*, Madrid (1979).
40. Mungan, I., J. Ruhwedel, and M. Winter, in *Natural Draught Cooling Towers*, P. L. Gould et al., eds., Springer, Berlin, Heidelberg, New York, Tokyo, p. 297 (1984).
41. Steffen, W., "Ein Beitrag zur nichtlinearen Berechnung des Tragverhaltens von Stahlbetonschalen," Ph.D. Thesis, Ruhr-Universität Bochum (1984).
42. Mang, H. A. and F. Trappel, in *Natural Draught Cooling Towers*, P. L. Gould et al., eds., Springer, Berlin, Heidelberg, New York, Tokyo, p. 279 (1984).
43. Milford, R. V. and W. C. Schnobrich, in *Natural Draught Cooling Towers*, P. L. Gould et al., eds., Springer, Berlin, Heidelberg, New York, Tokyo, p. 319 (1984).
44. Scheler, G. G., "Einfluss unterschiedlicher Stoffgesetze auf das Tragverhalten von Stahlbetonschalen," Ph.D. Thesis, Ruhr-Universität Bochum (1985).

CHAPTER 25

Theoretical and Practical Aspects of Vertical Anchor Design

EDWARD A. DICKIN* AND CHUN F. LEUNG**

INTRODUCTION

In recent years considerable attention has been given to the horizontal pull-out resistance of vertical anchors which are typically used in supporting anchored sheetpile walls. Most investigators have based their theoretical work on two-dimensional limit analyses, adopting idealisations of the complex failure mechanisms involved. Because of the assumptions made, the theories rely for their acceptability in design upon agreement between predicted and observed resistances preferably at field scale. However, much of the work reported involves conventional laboratory tests on small model plates ranging in size from 1in (25mm) to 4in (100mm) while the results from only one series of full scale tests appear to have been published.

Previous research has shown that the dimensionless anchor force coefficient $M_{\gamma q}$ reduces markedly as the size of the anchors increases. This has since been confirmed in an extensive experimental program on vertical anchors in sand using a geotechnical centrifuge. It is, therefore, inappropriate to examine the validity and limitations of the various design approaches by means of small model tests without due regard to stress level. Anchor geometry, embedment and soil condition, in addition to the scale effect, also play important roles in anchor behaviour.

In the first part of this contribution much of the theoretical design work is reviewed. The second half is concerned with existing published experimental work on vertical anchors including mechanical analogy and full scale and small scale laboratory model tests. Finally a detailed account of the writers' centrifugal model tests is presented and used to enable existing theoretical and empirical design methods to be critically assessed.

THEORETICAL INVESTIGATIONS

Typical deformation mechanisms around continuous vertical anchors at various embedments may be seen in Figure 1, obtained from two-dimensional model studies in a glass-sided tank. Failure zones around the 2in (50mm) anchors were accentuated by incorporating horizontal layers of dyed Erith sand at 0.4in (10mm) vertical intervals during test preparation. The simple failure mechanism around a shallow anchor (Figure 1a) becomes considerably more complex with increased embedment ratio (Figure 1b), although it still extends to the soil surface. However, at greater depth (Figure 1c), the failure mechanism is essentially local and rotational in nature. In consequence, theoretical limit state analyses may be categorised into those based on earth pressure theory for shallow anchors, for which failure zones extend to the soil surface, and those based on modified bearing capacity theory for deep ones.

Conventional Rankine Theory (Terzaghi [52]) has been widely adopted for the evaluation of the ultimate resistance T_u of shallow vertical continuous anchors assuming passive and active pressures, P_p and P_a respectively, to be fully developed in front of and behind the anchor (Figure 2). Thus:

$$T_u/\text{unit width} = P_p - P_a \qquad (1)$$

However, this over-simplification fails to address the vertical equilibrium conditions that arise from friction forces on the anchor. Terzaghi stated that Equation (1) is valid for embedment ratios (H/h) less than 2, and further specified that for single anchors an additional shear resistance on the side faces of the wedge pushed out by the anchor may be included. The sides of the wedge were assumed to form

*Department of Civil Engineering, University of Liverpool Liverpool, England

**Department of Civil Engineering, National University of Singapore, Singapore

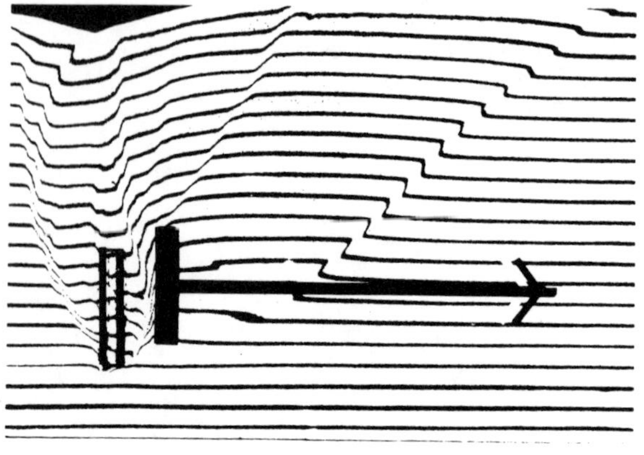

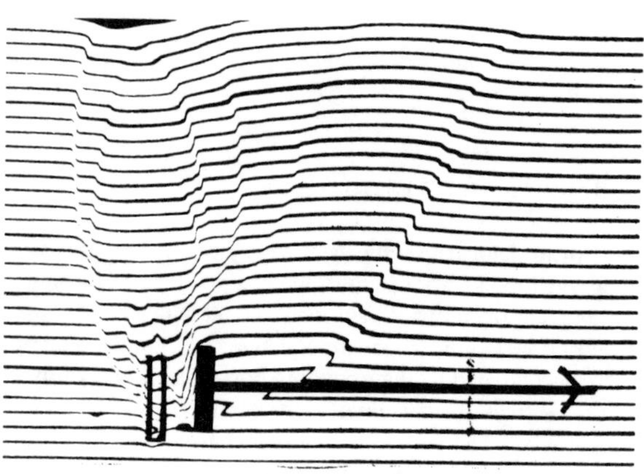

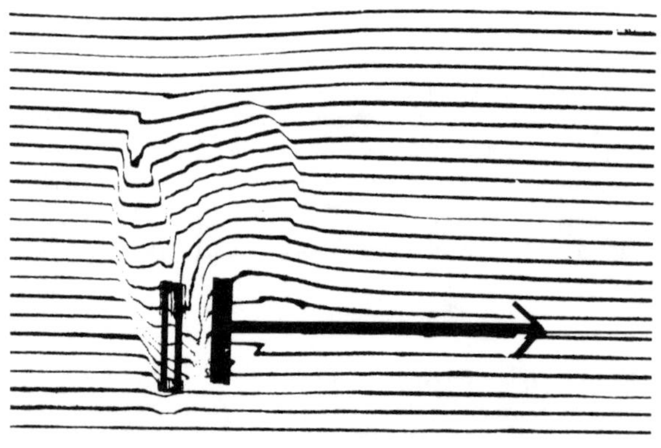

FIGURE 1. Failure mechanisms around vertical anchors in two-dimensional model tests.

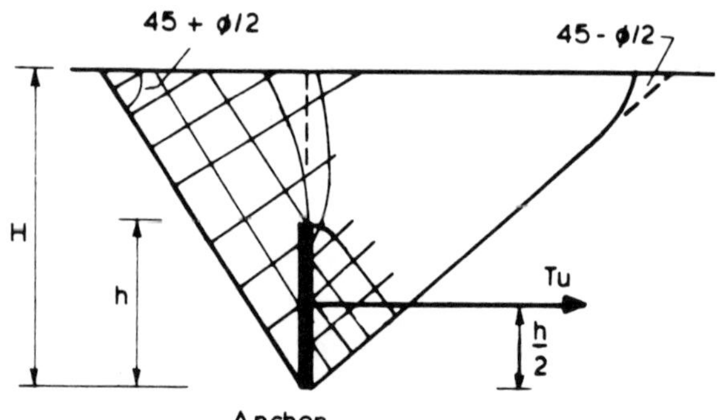

FIGURE 2. Shear pattern around a shallow anchor in sand assumed by Terzaghi (1943).

parallel to the tie rod as shown in Figure 3. Thus for a single anchor:

$$T_u = (P_p + P_s) - P_a \tag{2}$$

This approach is adopted in the British Code of Practice [16] which gives the side shear resistance for cohesionless soil as:

$$P_s = K_a \frac{\gamma H^3}{3} \tan\left(45 + \frac{\phi}{2}\right) \tan \phi \tag{3}$$

in which K_a = active earth pressure coefficient; γ = soil density; ϕ = angle of internal friction of soil; and H = depth to the base of the anchor. An expression which yields very similar values to Equation (3) is given by Teng [51] as:

$$P_s = K_o \frac{\gamma H^3}{3} (\sqrt{K_p} + \sqrt{K_a}) \tan \phi \tag{4}$$

in which K_o = coefficient of earth pressure at rest and K_p = passive earth pressure coefficient.

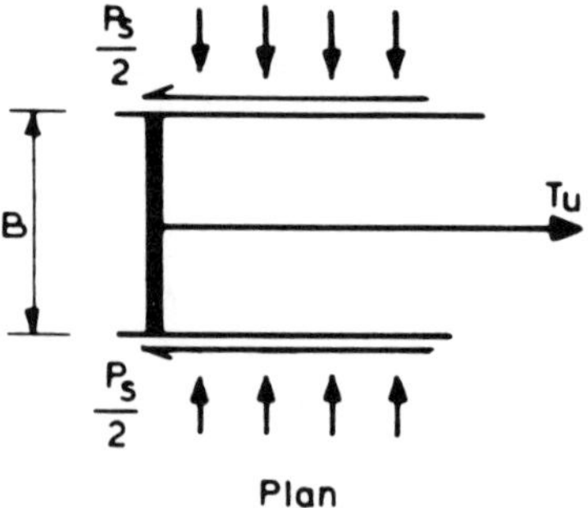

FIGURE 3. Side shear resistance P_s for a single anchor.

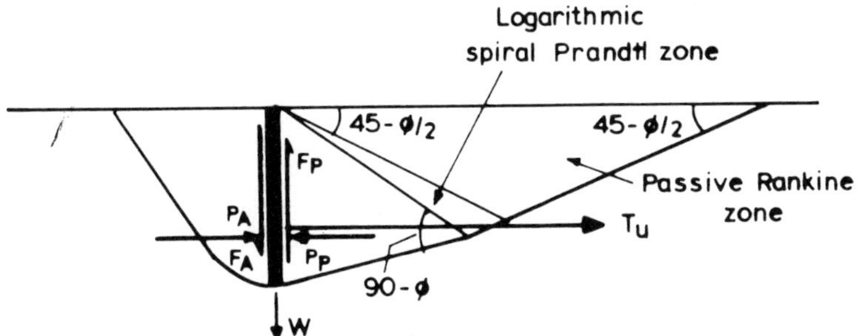

FIGURE 4. Failure mechanism for surface anchors formulated by Hansen (1953) adopted by Ovesen (1964).

If the height h of an anchor wall is small compared with the embedment depth H, Terzaghi suggested the anchor would fail by ploughing through the ground without producing a shear plane extending to the ground surface. He proposed that the force required to pull out such an anchor is approximately equal to the bearing capacity of a continuous footing with a width h whose base is located at a depth $(H - h/2)$ below the ground surface. Thus:

$$T_u = \frac{1}{2} \gamma h^2 N\gamma \tag{5}$$

in which $N_\gamma = (N_q - 1) \tan (1.4\phi)$ and $N_q = e^{\pi \tan\phi} \tan^2(45 + \phi/2)$

The complex failure mechanisms around vertical anchors were noted in early studies by Buchholz [5] using dyed sand layers. His analyses involved subdivision of the failure patterns into several zones and determination of anchor resistance from the resultant of stress vectors. Two design charts were produced giving values of earth pressure coefficients for continuous and single anchors embedded in various densities of sand. Ovesen [34] adopted a composite rupture figure proposed by Hansen [17] to derive the earth pressure in front of a continuous surface anchor, termed the "basic" case. The assumed mechanism, a straight line through the base of the anchor in combination with a rupture zone comprising Rankine and logarithmic spiral Prandtl zones is shown in Figure 4. Analyses satisfying both the horizontal and vertical equilibrium of the anchor were summarized as earth pressure coefficients given in Figure 5. Empirical reductions were applied to the "basic" result to take account of the degree of embedment and limited width of the anchor. The reduction factor for embedment, shown in Figure 6, may be expressed as:

$$R_{ov} = \frac{T_s}{T_o} = \frac{C_{ov} + 1}{C_{ov} + H/h} \tag{6}$$

in which C_{ov} = 19 and 14 for dense and loose sands respectively; T_s = pull-out resistance for the buried case; and

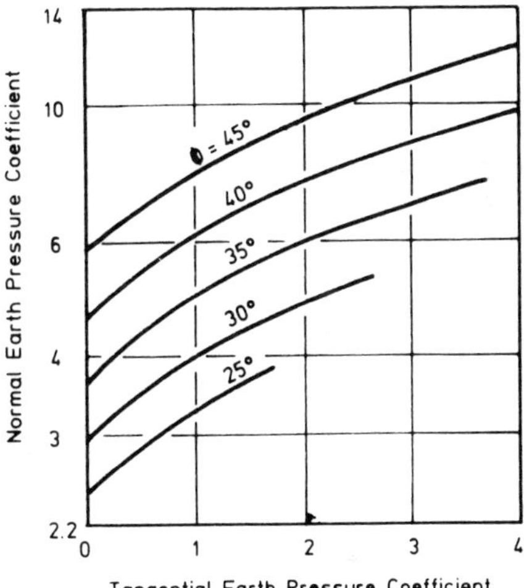

FIGURE 5. Earth pressure coefficients in front of an anchor slab (Ovesen, 1964).

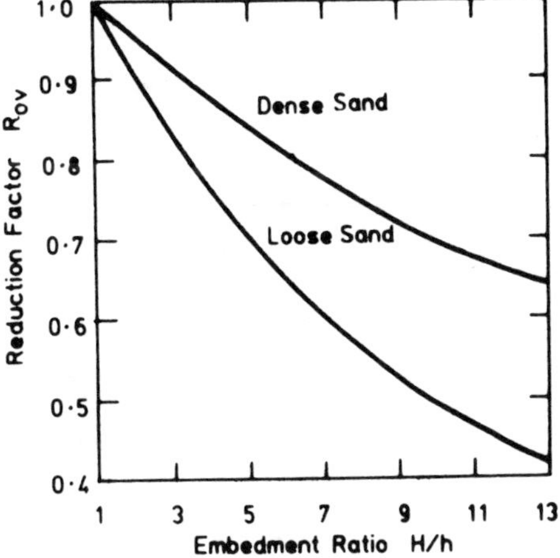

FIGURE 6. Reduction factor R_{ov} for anchor embedment (Ovesen, 1964).

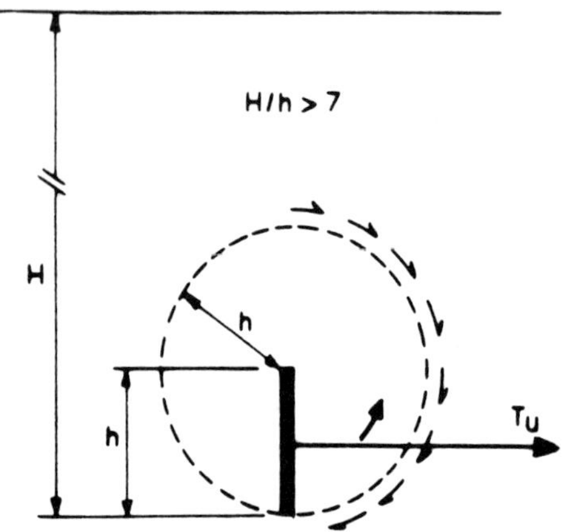

FIGURE 7. Simplified failure mechanism around deep anchors assumed by Biarez, Boucraut, and Negre (1965).

T_o = pull-out resistance for the surface case. Ovesen also derived a formula for the ultimate resistance of a deep continuous anchor, based on that of a deep strip foundation, as follows:

$$T_u/\text{unit width} = \gamma H h K_o e^{\pi \tan\phi} \tan^2(45 + \phi/2) \cdot d_c \quad (7)$$

in which $d_c = 1 + 0.35/[h/H + 0.6/(1 + 7 \tan^4 \phi)]$. In the case of very deep anchors $d_c = 1.6 + 4.1 \tan^4\phi$.

In a summary of their earlier work Biarez, Boucraut and

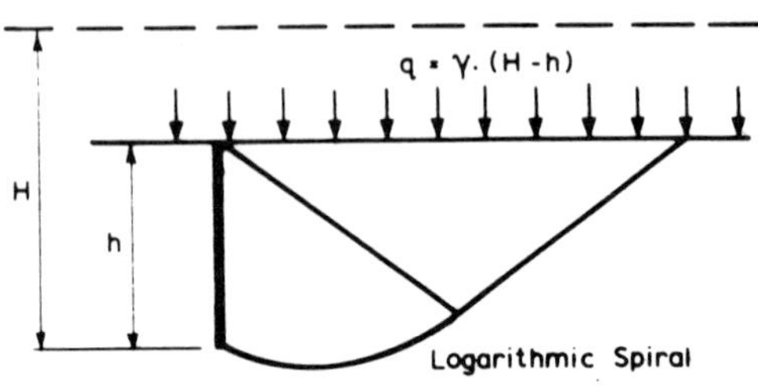

(a) Surcharge Method.

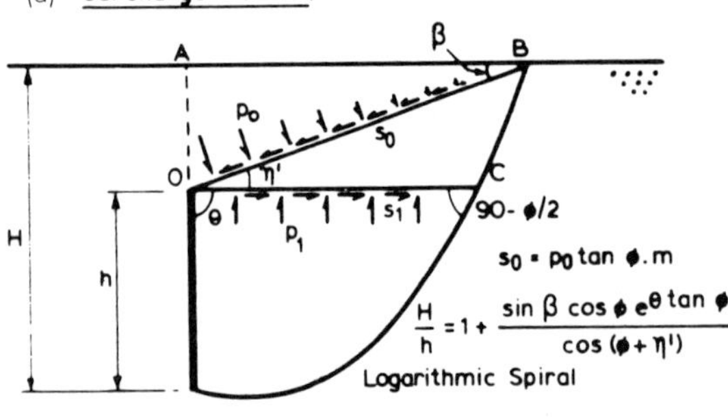

(b) Equivalent Free Surface Method.

FIGURE 8. Failure surfaces assumed by Neely, Stuart, and Graham (1973).

Negre [4] presented calculation methods for limiting equilibrium for vertical anchor piles subjected to translation or rotation. Active and passive earth pressure coefficients derived from limit analyses in which the vertical equilibrium of the anchor was satisfied were summarized on design charts. The resistance of shallow anchors ($H/h < 4$) was shown to depend upon the anchor roughness and weight. At intermediate depths ($4 < H/h < 7$) computations assuming the anchor to be smooth (angle of wall friction $\delta = 0$) gave good agreement with measured values. A simplified form of their original equation is:

$$M_{\gamma q} = (K_p - K_a)(H/h - \tfrac{1}{2}) + \frac{K_p \sin 2\phi}{2 \tan (45 + \phi/2)} (H/h - 1)^2 \quad (8)$$

in which $M_{\gamma q} = T_u/\gamma B h^2$ and B = anchor width ($B = 1$ for continuous case).

A relationship was also given for $M_{\gamma q_s}$, the resistance of a shallow single anchor, which simplifies to:

$$M_{\gamma q_s} = M_{\gamma q} + \phi\, h/B\, (\sqrt{K_p} - \sqrt{K_a})(H/h - \tfrac{2}{3}) + \tfrac{1}{2}\,(1 + \phi)\, h/B\, K_p \sin 2\phi (H/h - 1) \quad (9)$$

Biarez et al. recognised the characteristic rotational mechanism around deep continuous anchors ($H/h > 7$), as shown in Figures 7 and 1c, by considering the rotation of a cylinder of soil in front of the anchor and derived the following relation:

$$M_{\gamma q} = 4\pi\, (H/h - 1) \tan \phi \quad (10)$$

Meyerhof [29] extended his theory of uplift resistance (Meyerhof and Adams [28]) to calculate the ultimate resistance of inclined anchors. This work includes design coefficients for both shallow and deep continuous vertical anchors. Coefficients for deep square anchors were also derived using shape factors obtained from horizontally-loaded rigid piles by Hansen [18].

Neely, Stuart, and Graham [33] determined the theoretical resistance of continuous vertical anchors in sand using the theory of plasticity. Trial failure surface solutions were based on rupture zones bounded by combinations of logarithmic spirals and straight lines. They applied two methods of analysis, namely the "surcharge" and the "equivalent free surface" methods shown in Figure 8. By means of the method of characteristics, Neely et al. developed design charts for anchor resistances expressed in dimensionless form as force coefficients $M_{\gamma q}$. Neely et al. anticipated that the "surcharge" method would be relatively conservative, since no account was taken of the shear strength of the soil above the anchor. Curves in Figure 9a

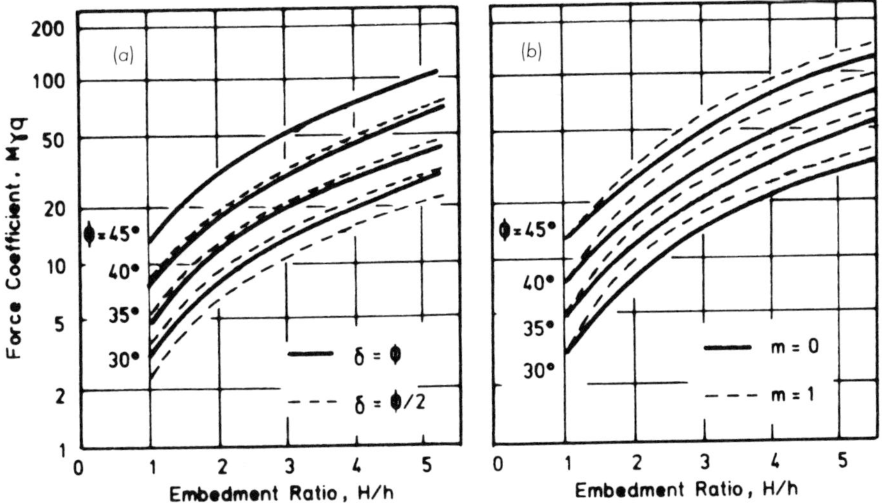

FIGURE 9. Force coefficient, $M_{\gamma q}$: (a) surcharge method; (b) equivalent free surface method. (Neely, Stuart and Graham, 1973).

are based on angles of wall friction $\delta = \phi$ and $\delta = \phi/2$. The equivalent free surface method, adapted from Meyerhof [27] who applied the concept to footing problems, requires assumptions about m, the degree of mobilization of shearing resistance along OB. Figure 9b shows a 10% increase in $M_{\gamma q}$ with $m = 1$ compared with $m = 0$.

A two-dimensional finite element solution, incorporating an elasto-plastic soil model based on the soil-structure interaction theory described by Rowe, Booker, and Balaam [47], was recently published by Rowe and Davis [48]. The resistance for a continuous vertical anchor, assumed to be thin and perfectly rigid, is given approximately by:

$$M_{\gamma q} = F_{\gamma} R_{\psi} R_R R_K \cdot h/H \qquad (11)$$

in which F_{γ} = basic anchor capacity for a smooth anchor in a non-dilatant soil with the coefficient of earth pressure at rest $K_o = 1$; and R_{ψ}, R_R and R_K = correction coefficients for the effect of soil dilatancy, anchor roughness and initial stress state, respectively.

The strong influence of H/h and critical state friction angle ϕ_{cp} on F_{γ} is seen in Figure 10. For K_o values between 0.4 and 1, the effect of initial stress state on F_{γ} is less than 10% and the assumption that $R_K = 1$ reasonable. Roughness coefficient R_R also varies with H/h in Figure 11 while Figure 12 shows the relationship between dilatancy coefficient R_{ψ} and ϕ_{cp} for an anchor in soils with associated dilatancy characteristics. Rowe and Davis found that R_{ψ} was approximately proportional to both dilatancy angle ψ and H/h. Hence dilatancy coefficients for anchors at all embedments in both associated ($\psi = \phi_{cp}$) and non-associated ($\psi < \phi_{cp}$) materials may be obtained by linear interpolation.

Appropriate values of dilatancy angle ψ may be derived for particular values of ϕ_{ps} and ϕ_{cp}, plane strain friction angles at peak and the critical state respectively, from the stress-dilatancy relation proposed by Rowe [44]:

$$\tan(45 + \psi/2) = \tan(45 + \phi_{ps}/2)/\tan(45 + \phi_{cp}/2) \qquad (12)$$

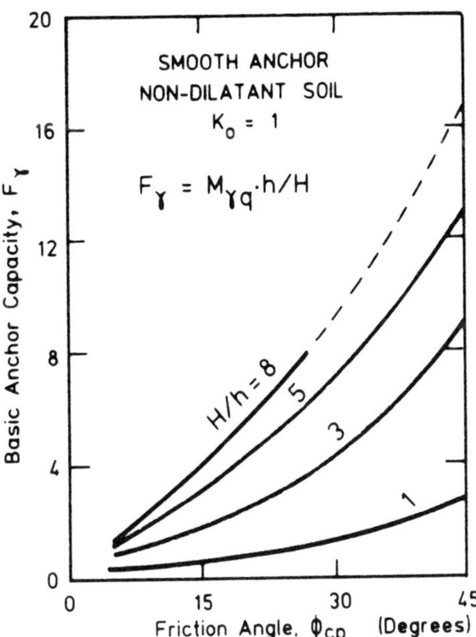

FIGURE 10. Variation of basic anchor capacity F_{γ} with friction angle Φ_{cp} (Rowe and Davis, 1982).

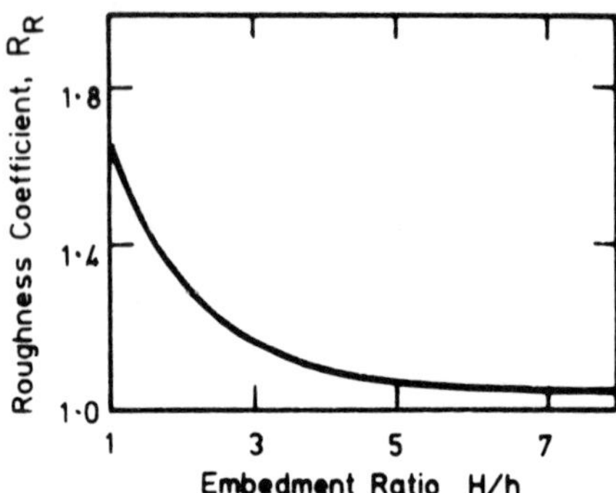

FIGURE 11. Variation of anchor roughness coefficient R_R with embedment ratio H/h. (Rowe and Davis, 1982).

An initial comparison of the different approaches for continuous anchors is given in Figure 13 which shows the variation of the dimensionless coefficient $M_{\gamma q}$ with embedment ratio H/h for a constant $\phi = 40°$. The significant disparities between alternative methods warrant a more detailed comparison with measured $M_{\gamma q}$ values.

FUNDAMENTAL EXPERIMENTAL INVESTIGATIONS

Failure Mechanisms

In addition to the two-dimension tests by Dickin and Leung [15] in Figure 1, Buchholz [5], Hueckel [19], and

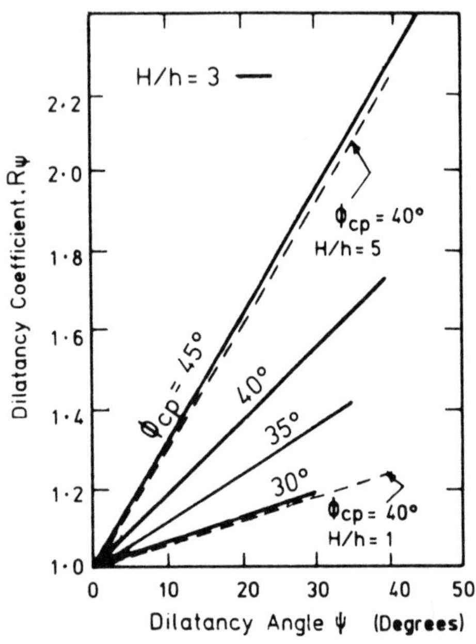

FIGURE 12. Variation of dilatancy coefficient R_ψ with dilatancy angle ψ (Rowe and Davis, 1982).

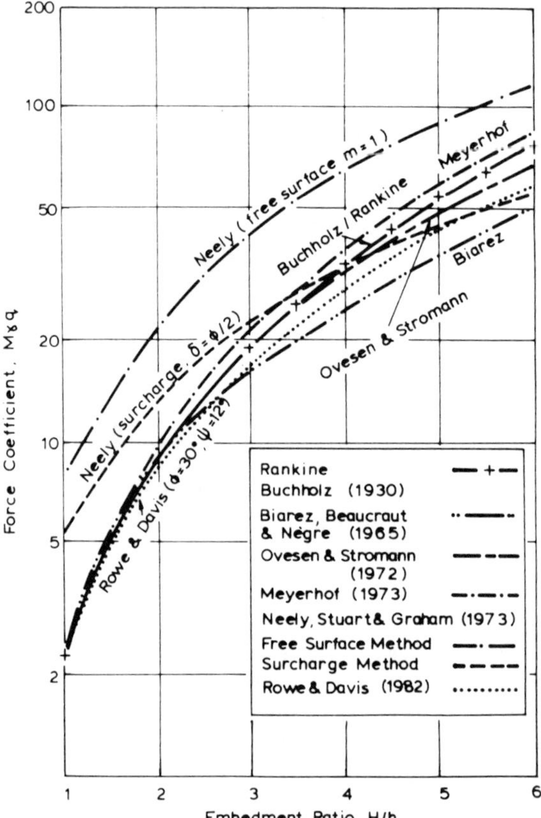

FIGURE 13. Variation of theoretical force coefficients $M_{\gamma q}$ with embedment ratio H/h for $\phi = 40°$.

Kostyukov [22] have carried out detailed studies of the rather complex failure mechanisms around vertical anchor plates.

Buchholz carried out three-dimensional model anchor tests on 6in (150mm) square plates at embedment ratios, 2, 2.7, and 4 in compacted sand incorporating 0.2in (5mm) thick horizontal layers of coloured sand at 2.4in (60mm) intervals. After failure the backfill was cut at various sections enabling the entire slip body to be carefully measured. The results, an example of which is shown in Figure 14, were used as a basis for Buchholz's theoretical computations. The observed failure planes appeared as families of curves in three dimensions with different inclinations at every point and sharp transitions from one to another. Slip planes curved upwards roughly from the top edge of the plate to the sand surface, rising steeply in every section and intersecting the surface at an angle of $\nu_o = (45 - \phi/2)$. The total mean slip-plane inclination from the lower edge of the plate to the surface of the sand was between 36–42°, the lower angle relating to smaller embedment ratios. A very similar failure pattern, resembling a calyx with a wavy outline having an

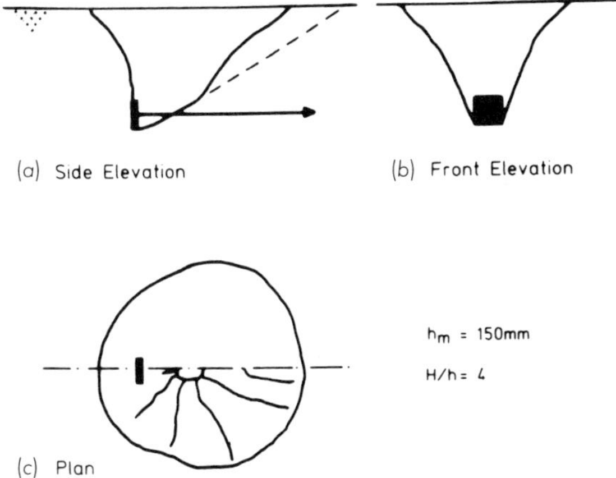

FIGURE 14. Failure pattern for square plate (Buchholz, 1930).

average inclination to the horizontal of 65° at the sides and 40° in the direction of the horizontal force, was observed by Hueckel [19]. Hueckel, Kwasniewski, and Baran [20] investigated the passive pressure distribution, in both horizontal and vertical directions, on the front surface of a 12in (300mm) rigid square vertical plate. The pressure distribution, which was essentially the same at all stages of the test, did not strictly follow the classical pattern, but was generally concave (saddle-like) in shape. This concavity was more pronounced at the lower edge of the plate.

The distribution of density in the sliding sand wedge in front of model anchor plates was examined by Kostyukov [22] using radiometry. Density determinations were carried out before loading, at an anchor force equal to half the ultimate resistance, at ultimate resistance, and after collapse of the wedge. Three zones were identified including a triangular compacted zone in front of the anchor.

Mechanical Analogy Model Studies

Ovesen [34] carried out two dimensional mechanical-analogy model tests to study the behaviour of continuous vertical anchors using steel pins to model the soil. The continuous anchor which, like the pins, spanned the box had a width of 2in (50mm) and was made of 0.5in (13mm) thick plexi-glass. The anchor was regarded as perfectly rough as the plexi-glass surface was grooved. The depth of the pin material from the base of the anchor to the surface was fixed at 4.8in (120mm). Only the height of the anchor and the tie bar level were varied. Figure 15 shows the optimum location of the tie bar level associated with the greatest resistance for various embedment ratios. Anchor resistances for the buried and basic cases, T_s and T_o respectively, are compared in Figure 16.

Typical displacement patterns found in mechanical analogy tests were reported by Akinmusuru [1]. Steel pins were

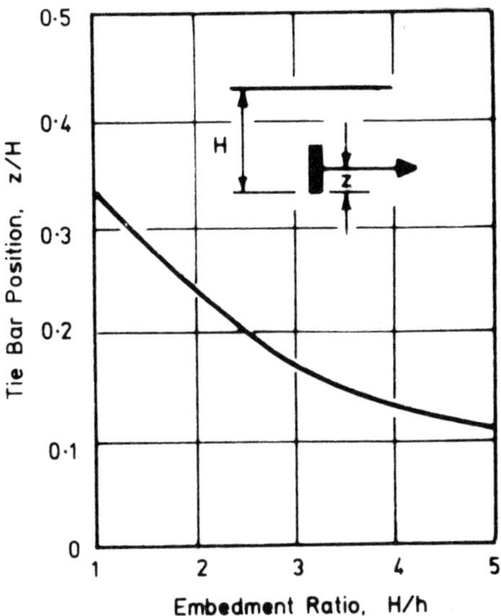

FIGURE 15. Optimum tie bar level in mechanical analogy tests. (Ovesen, 1964).

compacted to an average density of 400lb/ft^3 (62.8 kNm^{-3}) giving an angle of internal friction of 24°, and the aluminium anchor plate was 3in (76mm) wide, 2in (50mm) high and 0.5in (13mm) thick. Pin movements were photographed using a long exposure film. Displacements failed to

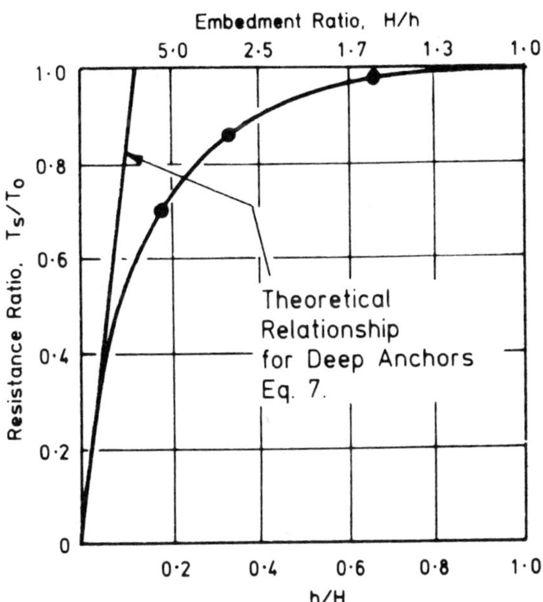

FIGURE 16. Relationship between T_s/T_o and H/h from mechanical analogy tests (Ovesen, 1964).

reach the surface for embedment ratios greater than 6.5 and were approximately circular in shape with the anchor height as radius.

FULL SCALE FIELD STUDY

A number of full scale field tests on reinforced concrete deadman anchorages at shallow depth carried out by the United States Navy were reported by Smith [49]. Horizontal loads were applied to the test slabs which ranged in size from 24in × 30in (0.61m × 0.76m) to 36in × 288 in (0.91m × 7.3m) (height/width), placed singly or in rectilinear groups of three at embedment ratios up to 3.3. Most of the anchorages were 12in (0.3m) thick with weights ranging from 0.29 to 5.94 tons (295 to 6040 kg). The largest deadman was also tested behind triangular and trapezoidal berms of different sizes. All tests were conducted in an area of sandy beach about 75ft (23m) wide by 100ft (30m) long.

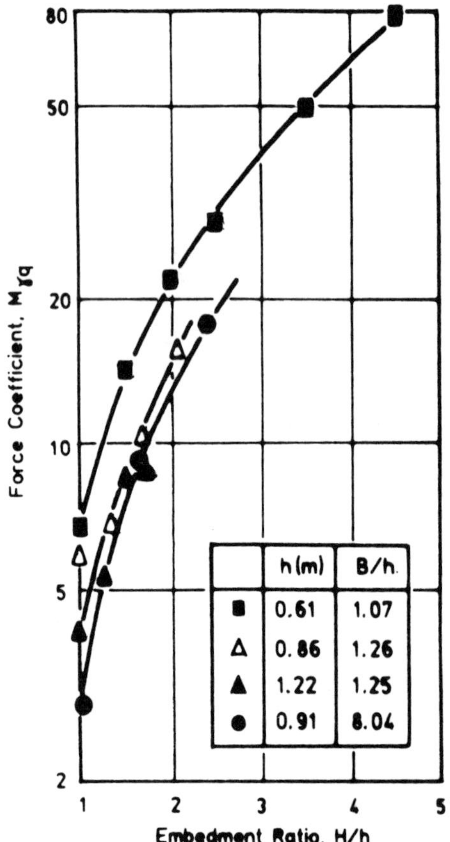

FIGURE 17. Variation of force coefficient $M_{\gamma q}$ with embedment ratio H/h for full scale tests on concrete deadmen anchors (Smith, 1962).

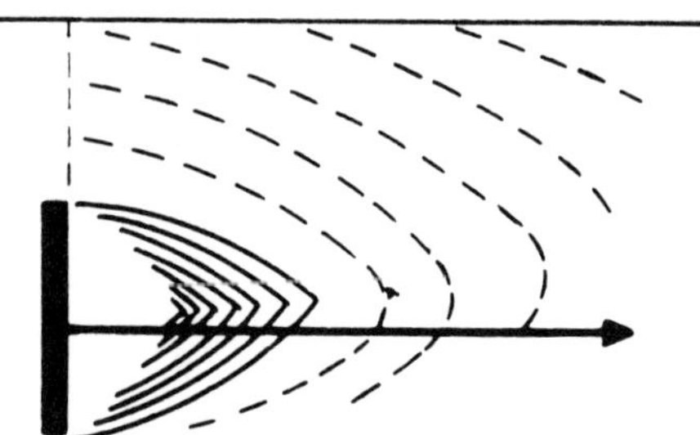

FIGURE 18. Estimated development of arrowhead-shaped wedge and activation of overburden in front of an anchor (Smith, 1962).

The angle of internal friction varied from 26° to 33° and the apparent cohesion of the sand from 0.6 to 1.3lb/in² (4.1 to 9.0 kNm⁻²), depending upon the method of preparation.

Tie rods were connected at the third height for all surface tests, but otherwise at 0.4h from the base of slabs. Applied loads were measured close to the anchors so that friction along the tie rods would not affect the results unduly. In some tests, after compacting the backfill, columns of coloured sand were placed at intervals in front of the deadman to help define the shear-failure surfaces below ground level.

Anchor ultimate resistance was defined as that load causing 4in (100mm) horizontal anchor displacement. Although his test results, shown in Figure 17, were rather scattered Smith suggested that a definite relationship existed between horizontal displacement and applied loads. He developed the following empirical formula for the ultimate resistance for $\phi = 32°$ based on a statistical analysis:

$$T_u = e^{[C_o + C_1 (H - h) + C_2 \ln (Bh) + C_3 M_1]} \tag{13}$$

where T_u is in Kips, H, h, and B are in feet, and M_1 is a conditional factor having a value of zero for a single anchor and 1.0 for anchor groups. Coefficients C_o, C_1, C_2 and C_3 are −0.83, 0.34, 0.86, and 0.30 respectively.

An equivalent metric formula for Equation (13), with T_u in kN and H, h, B in m, is:

$$T_u = 4.45\, e^{(C_o + 3.28\, C_1\, [H - h] + C_2\, [\ln (Bh) + 2.736] + C_3 M_1)} \tag{14}$$

Because of the scatter Smith further proposed the following relationship to determine the lower limit of ultimate resistance:

$$Q_L = L_1 \times T_u \tag{15}$$

where Q_L is the lower limit resistance representing a 95 percent confidence limit. From his test results, L_1 was found to be 0.72 for single anchors and 0.71 for anchor groups.

Smith observed that an arrowhead-shaped failure wedge

formed, with shear-failure surfaces occurring at the upper and lower edges of the buried slabs. Figure 18 illustrates the estimated form of the wedge and its activation of overburden in the proximity of the deadman. As loads were applied, sand immediately forward of the deadman began to consolidate. Successive failure planes formed until the entire wedge was mobilized. In surface tests the wedge was clearly defined at ground level but became less distinct at greater embedment depths.

CONVENTIONAL SMALL SCALE MODEL STUDIES

Small scale laboratory model tests have the great advantage that experimental conditions can be closely controlled, enabling behavioural trends to be established economically. Not surprisingly, these tests have been the most popular approach adopted by researchers studying vertical anchors. Buchholz [5] was the earliest investigator, while more recent work includes that of Hueckel [19,20], Ovesen [34,35], Kostyukov [22], Neely [30,31,32,33], Meyerhof [29], Das [6,7,8], Ranjan [39,40,41], Akinmusuru [1], Dickin and Leung [14], and Hoshiya [21]. The general review of experimental work on single and continuous anchors which follows includes reference to the effect of size, shape, embedment, flexibility, soil density and group interaction.

Continuous and Single Anchor Behaviour

An extensive investigation of 2in (50mm) model rigid anchors was reported by Dickin and Leung [14]. The tests were carried out in a reinforced steel bin 54in (1.37m) long, 48in (1.22m) wide and 33in (0.84m) deep containing dense Erith sand. The test equipment is shown in Figures 19 and 20 and typical load/displacement curves in Figure 21. The anchor load T_u was defined as the net resistance after applying a reduction for the tie rod resistance which was determined in separate tests under identical conditions.

The variation of ultimate resistance, expressed in terms of dimensionless force coefficient $M_{\gamma q}$ with embedment ratio H/h continuous anchors and single anchor plates with width/height ratios of 1, 2 and 5 for embedment ratios up to 13, is given in Figure 22. These results are typical and in general agreement with research reported by Neely et al. and Ranjan et al., for example. Clearly, ultimate resistance depends primarily on anchor geometry and embedment. Although somewhat scattered, failure displacements (Figure 23) increase with embedment and reduce with increased plate width also confirming other research [33,40].

The marked influence of anchor shape on the dimensionless coefficients in Figure 22 is attributable to sideshear resistance which becomes increasingly significant as the anchor width/height decreases. As in other related geotechnical problems, the effect of anchor geoemetry may be expressed quantitatively as a dimensionless shape factor S_f where:

$$S_f = \frac{M_{\gamma q}\text{ (single anchor)}}{M_{\gamma q}\text{ (continuous anchor)}} \tag{16}$$

Shape factors derived from the tests in Figure 22 and shown in Figure 24 increase with embedment but reduce with width/height ratio. Considerably lower shape factors obtained by Neely et al. in Figure 24 were for a looser sand and on the assumption that an anchor with $B/h = 5$ was effectively continuous.

FIGURE 19. Reinforced steel bin for conventional model tests.

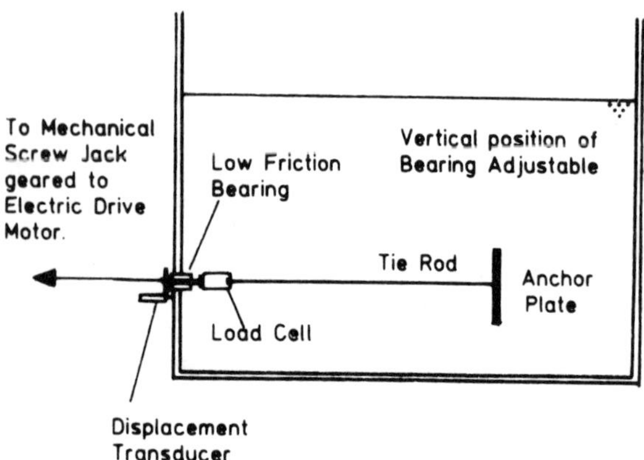

FIGURE 20. Schematic test arrangement for conventional model tests.

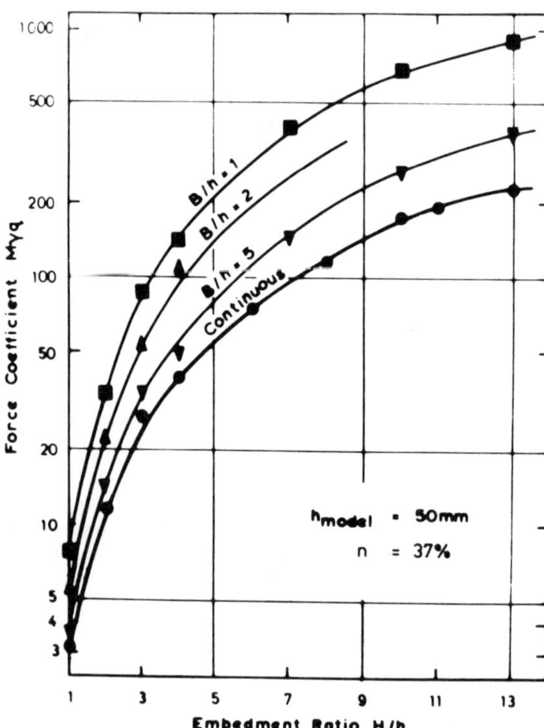

FIGURE 22. Variation of force coefficient $M_{\gamma q}$ with embedment ratio H/h for conventional tests on 50mm high anchors.

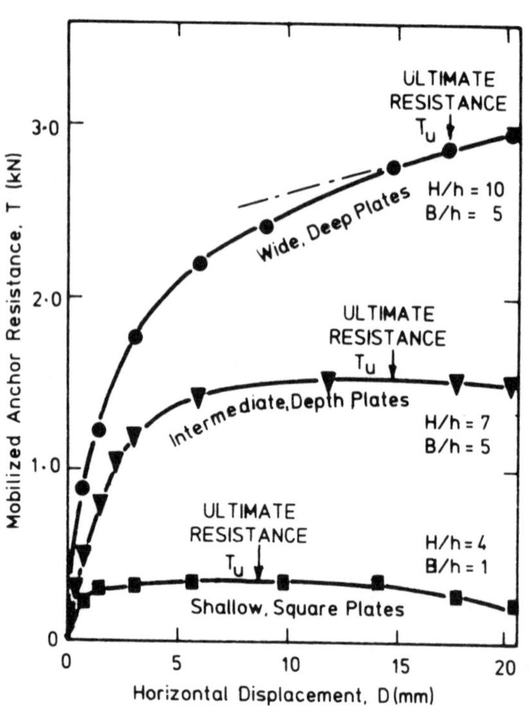

FIGURE 21. Typical resistance/displacement curves for conventional model tests on 50mm anchors.

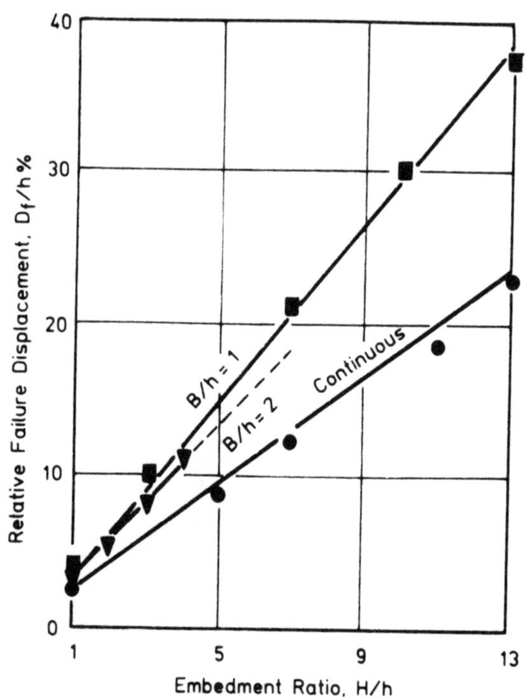

FIGURE 23. Variation of relative failure displacement D_f/h with embedment ratio H/h for 50mm high rigid plates.

Ovesen et al. [35] introduced the "effective width of anchor B^e" where:

$$T_u = T_o \cdot B^e \tag{17}$$

to take account of anchor shape. Values of B^e derived from their observations are given in Figure 25 for various anchor embedments and soil densities. Shape factors for single anchors may be obtained by taking $L = \infty$. Then:

$$S_{f(Ovesen)} = K_{ov} \frac{(H/h + 1)}{B/h} + 1 \tag{18}$$

where $K_{ov} = 0.42$ and 0.33 for dense and loose sands, respectively. These are also lower than the writers' data in Figure 24.

The ultimate resistance of a circular plate with diameter d is approximately 66% of that for a square anchor with side $h = d$, according to Das [6]. However, Akinmusuru [1] and the writers [13] concluded that circular anchors can carry the greatest load per unit area in comparison with square and rectangular plates.

Anchor size also has been shown to have a significant effect upon stress-deformation behaviour particularly in small model tests. In comparing his tests with those by Smith [49] on full scale anchors 18 times larger in almost identical well-graded sand, Neely observed a reduction of 16% in $M_{\gamma q}$. He concluded that this scale effect is due to the differing angle of internal friction ϕ at various stress levels for corresponding points in the failure zones of model and full scale structures.

Ranjan and Kaushal [39,40] reported that their experimental results were about 2 to 3 times greater than the theoretical values obtained from Rankine's theory. Meyerhof's [29] theoretical estimations in general were also lower than the test results. This difference was accounted for by the variation of ϕ with stress level and the insertion of an incorrect value in the formula used. Ranjan et al. also found that the displacements at failure load increased with an increase in the size of the plate and also with an increase in depth of embedment. This effect was attributed to the larger soil mass involved for larger anchors or anchors at greater depth.

Das, Seeley, and Das [8] found that, while the breakout factor $N_{\gamma q} = T_u/\gamma BhH$ increased with embedment ratio for shallow anchors, beyond a "critical" depth of embedment, $N_{\gamma q}$ remained constant. Their results in Figure 26 suggest that the critical embedment ratio $(H/h)_{cr}$ for square vertical anchors is dependent upon the angle of internal friction increasing from 5 for loose to about 8 for dense sands. This finding is broadly supported by the writers' two-dimensional model tests shown in Figure 1 and also by Akinmusuru who concluded that the critical embedment ratio is 6.5 in both sand and pin analogy tests. Buchholz [5] reported that $(H/h)_{cr}$ decreased as the size of the anchors increased. He also found no significant variation between anchor be-

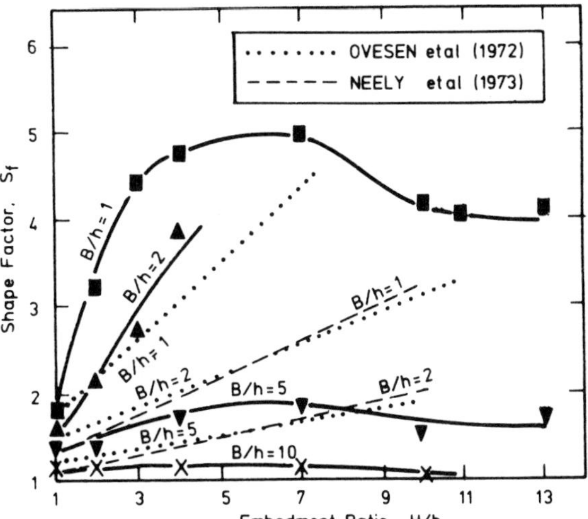

FIGURE 24. Variation of shape factor S_f with anchor geometry and embedment for 50mm anchors.

haviour in slightly damp and completely dry sand.

Considerable variations in anchor roughness are found in different researches. However, Hueckel [19] concluded that the surface roughness had no major influence on the value of the ultimate resistance; for example, a 4 × 4in (100 × 100mm height/width) steel anchor at an embedment ratio of 2 failed at 87.4lb (0.389 kN), while a concrete plate gave 77.3lb (0.344 kN). He also observed that the ultimate resistance of inclined plates having horizontal tie bars is smaller than that of vertical ones.

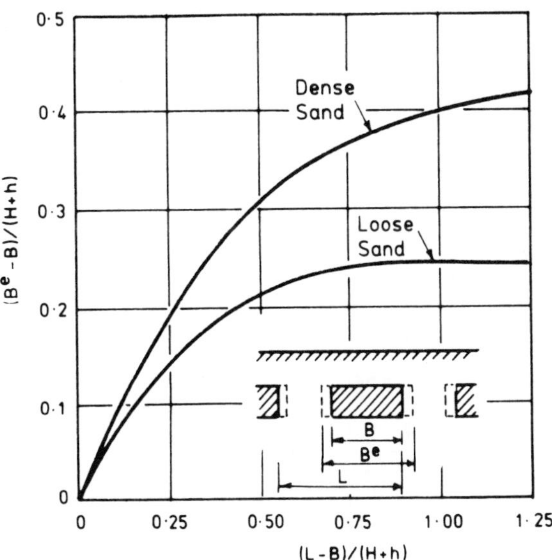

FIGURE 25. Effective length B_e of single anchors in loose and dense sand (Ovesen & Stromann, 1972).

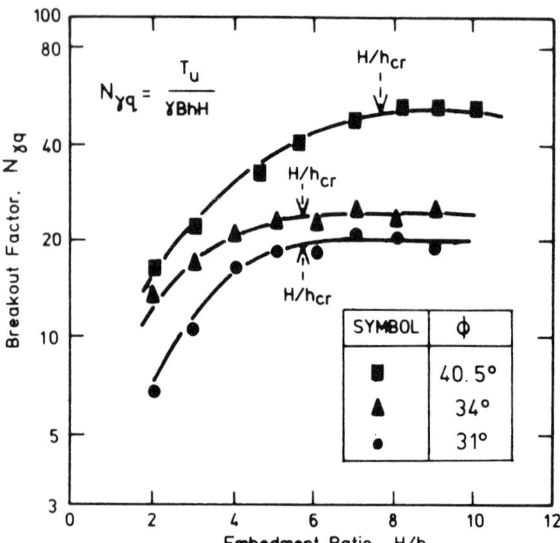

FIGURE 26. Variation of anchor breakout factor $N_{\gamma q}$ with embedment ratio H/h (Das, Seeley and Das, 1977).

Effects of Anchor Flexibility

Neely [30] examined the effect of flexibility of 10in (254mm) wide by 4in (100mm) high model anchors. Several materials and plate thicknesses were selected to give a wide range of flexibilities defined by flexibility number $\varrho = h^4/EI$ (after Rowe [43], E and I are the properties of the anchor).

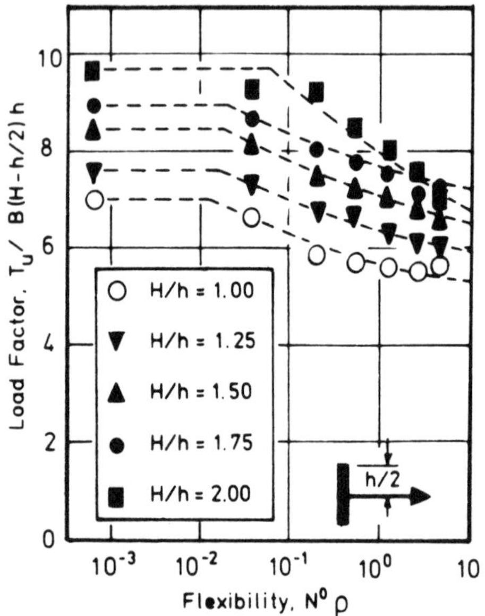

FIGURE 27. Variation of ultimate load with flexibility (Neely, 1972).

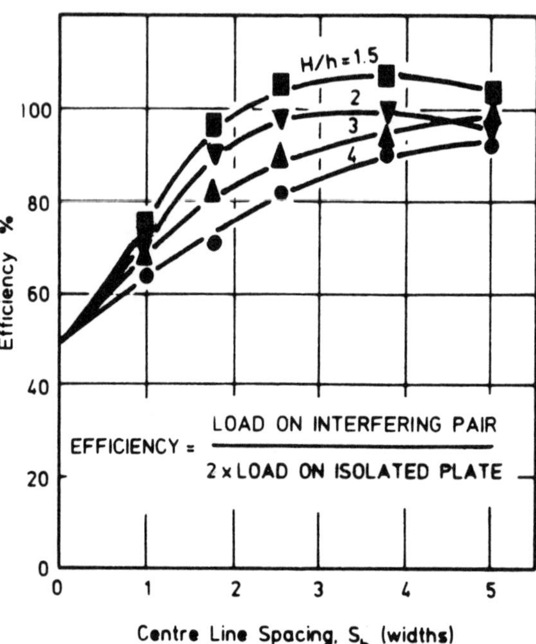

FIGURE 28. Variation of efficiency with spacing S_h for a pair of interfering square plates (Neely, 1972).

The ultimate load reduced with increased plate flexibility, as shown in Figure 27. He also concluded that the position of attachment of the rod is more important than flexibility. Flexibility effects appear to be most significant when rods are attached near the top of the anchors.

Anchor Interference

For a single anchored sheetpile wall the pull-out capacity is impaired if the anchor is located such that its associated passive wedge intersects the active wedge behind the main wall. Terzaghi [52] has specified the optimum zone for full anchor capacity in detail. Interference effects which also occur between closely spaced anchors in horizontal and vertical lines have been studied by several researchers. Hueckel defined the efficiency ξ as the ratio between the load on a group and that on a single plate of equivalent size and found that the critical horizontal spacing at which ξ falls below 100% was 2.25h.

Ovesen's empirical design charts, which incorporate the "effective width B^e" described earlier, also give the ultimate resistance of horizontal line anchor groups. Dimensionless ratios $(B^e - B)/(H + h)$ and $(L - B)/(H + h)$ in Figure 25 take account of the spacing L.

Neely [30,32] studied groups of two or four anchors at various horizontal or vertical line spacings. Figures 28 and 29 illustrate a significant reduction in efficiency for spacings

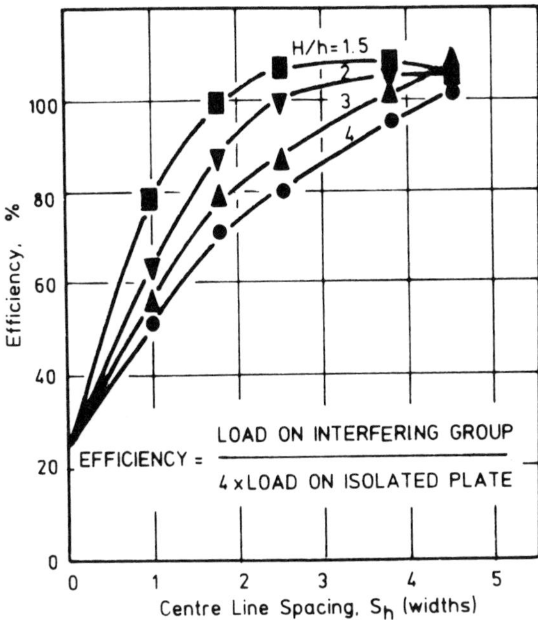

FIGURE 29. Variation of efficiency with spacing S_h for a group of 4 interfering square plates (Neely, 1972).

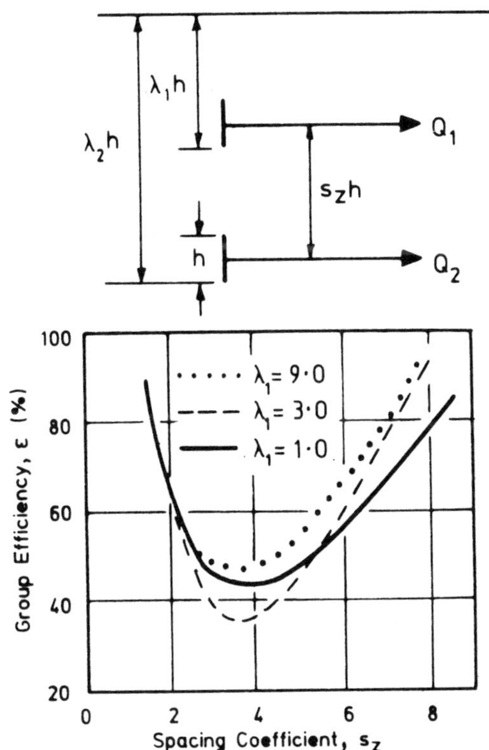

FIGURE 30. Variation of group efficiency for 2 vertically spaced plates (Akinmusuru, 1978).

less than about 2h for square anchors and an increase in critical horizontal spacing with embedment ratio. In the case of vertically spaced plates, similar variations of efficiency were observed for all embedment ratios. However, studies by Akinmusuru [2] indicate that the efficiency of a vertical line anchor group is influenced by both spacing and embedment. Efficiencies of only 40% were reported for spacings of 4h. His results, summarised in Figure 30, suggest that ξ increases not only for wider spacings but also at closer spacings due to arching effects.

CENTRIFUGAL MODEL STUDY

Theoretical considerations and differences between Smith's field data and their own laboratory results led Neely et al. [33] to conclude that the force coefficient reduces considerably with an increase in anchor size. Hence while relatively economical and valuable in establishing the relative effects of anchor geometry, embedment, etc., conventional small model tests are severely limited in predicting the performance of full scale anchors. Moreover, full scale test data are insufficient and in any case not readily obtainable. The writers employed the centrifugal modelling technique to simulate field scale anchor behaviour. This extensive investigation is considered in some detail here. The principles of centrifugal modelling and a description of the centrifuge facility at Liverpool are also included.

Principles of Centrifugal Modelling

Modelling in a centrifuge is similar to conventional physical modelling except that the model is placed under a forced gravity field set up by rotational action, enabling prototype stress levels to be achieved. An element of soil in a model rotating in a centrifuge at radius R with velocity V would, if free to move radially, accelerate away from the center of rotation at an instantaneous rate equal to V^2/R. When held at a fixed radius R each element would, therefore, exert an inertia force on the rest of the model given by the product of the mass of the element and the local acceleration. Thus the force is fully analogous to the body weight of the corresponding element in the prototype structure which is effectively under the influence of earth's gravitational field g, and the fundamental action of the centrifuge can be regarded as an increase of all bulk specific weights from $\gamma = \varrho g$ at rest, to $\varrho(V^2/R) = \varrho N g = N\gamma$ at speed. Thus, if the same soil is used in both centrifuge model and prototype and tested under the same boundary conditions, the model, with a linear scale of $1/N$ and in a gravitational field N times that experienced by the prototype, would have identical unit stresses to those at corresponding points in the prototype. Thus, the results of such a test would truly represent prototype behaviour, since all basic problems associated with scaling would be eliminated.

Consider a vertical anchor plate buried at shallow depth in sand, subscripts p and m referring to prototype and model, respectively. According to Terzaghi's theory, the total prototype resistance is given by:

$$T_{u_p} = \frac{1}{2}(K_p - K_a)\,\gamma\,H_p^2\,B_p + \frac{H_p^3}{3}K_a\,\gamma\tan\left(45 + \frac{\phi}{2}\right)\tan\phi \tag{19}$$

Subject only to earth's gravitational pull (1g), the total resistance of the model is:

$$T_{u_m} = \frac{1}{2}(K_p - K_a)\,\gamma\,H_m^2\,B_m + \frac{H_m^3}{3}K_a\,\gamma\tan\left(45 + \frac{\phi}{2}\right)\tan\phi \tag{20}$$

Assuming identical stress and boundary conditions can be achieved in model and prototype, if the centrifuge model is accelerated to Ng such that the bulk specific weight of the soil is increased to $N\gamma$, Equation (20) becomes:

$$T_{u_m} = \frac{1}{2}(K_p - K_a)\,N\gamma\,H_m^2\,B_m + \frac{H_m^3}{3}K_a\,N\gamma\tan\left(45 + \frac{\phi}{2}\right)\tan\phi$$

For a scaling factor N between the model and prototype dimensions $H_p = NH_m$ and $B_p = NB_m$. Hence,

$$T_{u_m} = N\gamma\left[\frac{1}{2}(K_p - K_a)\frac{H_p^2}{N^2}\cdot\frac{B_p}{N} + \frac{H_p^3}{3N^3}K_a\tan\left(45 + \frac{\phi}{2}\right)\tan\phi\right]$$

$$= \frac{\gamma}{N^2}\left[\frac{1}{2}(K_p - K_a)H_p^2\,B_p + \frac{H_p^3}{3}K_a\tan\left(45 + \frac{\phi}{2}\right)\tan\phi\right] \tag{21}$$

$$= \frac{T_{u_p}}{N^2}$$

An identical relationship is obtained for continuous anchors by neglecting the side shear term.

Relationships between model and prototype force coefficients and breakout factors can readily be derived. Thus:

$$M_{\gamma q_p} = \frac{T_{u_p}}{\gamma\,B_p\,h_p^2} = \frac{T_{u_m}}{\gamma\,B_m\,h_m^2\,N}$$

i.e.,

$$M_{\gamma q_p} = M_{\gamma q_m}/N \tag{22}$$

Similarly,

$$N_{\gamma q_p} = N_{\gamma q_m}/N \tag{23}$$

Applying Hooke's Law to the horizontal displacement D of a vertical anchor,

$$\sigma_p = E_p\,\epsilon_p = E_p\cdot\frac{D_p}{h_p}$$

and

$$\sigma_m = E_m\,\epsilon_m = E_m\cdot\frac{D_m}{h_m}$$

If the test results truly represent that of a prototype, $\sigma_p = \sigma_m$ and $E_p = E_m$. Hence:

$$\frac{D_p}{h_p} = \frac{D_m}{h_m}$$

and

$$D_p = ND_m \tag{24}$$

The strength of the anchor material may be defined in terms of a flexibility number $\varrho = h^4/EI$, as described earlier. For an anchor plate, the second moment of area I is proportional to the third power of its thickness t per unit width. Hence:

$$\varrho_m \alpha \frac{h_m^4}{E_m\,t_m^3}$$

and

$$\varrho_p \alpha \frac{h_p^4}{E_p\,t_p^3}$$

$$= \frac{N\,h_m^4}{E_p\,N^3 t_m^3}$$

Thus,

$$\varrho_m = \varrho_p/N$$

$$\varrho_p = Np_m \tag{25}$$

Therefore as the gravitational field is increased, the flexibility of the anchor increases. A comprehensive review of scaling laws relating to centrifugal modelling was given by Avgherinos and Schofield [3].

The Liverpool University Geotechnical Centrifuge

The machine, a model G.380.3A supplied by Triotech Inc. of California, is shown in Figure 31 and diagrammatically in Figure 32. It consists of a cylindrical steel enclosure which houses a 20 hp drive motor, drive shaft, rotor arm and test package carriages. The carriages may be either swinging buckets or fixed. The swinging buckets are 22.4in (0.57m) long, 17.7in (0.45m) wide, and 9in (0.23m) deep and facilitate the construction and testing of models in any soil. The fixed buckets are 24.8in (0.63m) long, 20.5in (0.52m) deep, and up to 18.5in (0.47m) wide, the depth available often being of advantage when testing cohesive models.

For maximum depths of soil in packages, the optimum scaling radii are 1.15m and 1.0m for the swinging and fixed buckets, respectively. These are measured to levels at a third of the depths of soil, and the corresponding maximum stress differences between model and prototype are 3.5 and 8.5%. At a maximum speed of 400rpm this gives a maximum acceleration of about 200 gravities. However the centrifugal capacity of the machine of 13.6g-tonnes limits the permissible acceleration for any particular package. The speed is measured to 0.2rpm by a magnetic pick-up which senses electrical impulses from a toothed wheel on the main drive shaft. It is varied by a rectifier/transformer which supplies a variable voltage to the armature of the motor. Very rapid acceleration is possible; for example a 330lb (150kg) package can be brought comfortably to 40g in 40 seconds. Dynamic braking enables fairly rapid deceleration down to 150rpm in about 1 minute, and the machine takes approximately 4 more minutes to come to rest.

Two models can be set up at the same time but it is usual to test only one while the other, together with additional weights, provides a counterbalance. A release mechanism allows the rotor arm and packages to be balanced statically about a horizontal pivotal shaft.

Input for mains powered equipment and output from strain gauges, etc., is obtained via 36 No. 5amp slip rings and 24 No. low signal, precision slip rings. A monochrome video camera mounted on the rotor arm close to the drive shaft allows observation of the progress of tests on a 56cm monitor. Illumination is provided by fluorescent strip lights also mounted close to the drive shaft.

Ancillary Equipment for Anchor Tests

A specially designed package arrangement was devised to enable strain-controlled anchor tests to be performed, the swinging bucket being partitioned into "anchor" and "loading" sections as shown in Figure 33. In some tests the depth of the bucket was increased using 4in (100mm) aluminum angle extension pieces. With the anchor plate placed in position, dry sand was compacted mechanically in 1in (25mm) layers using a small purpose-built hand-held vibrator. An average specific weight of 102lb/ft^3 (16 kN/m^3) was achieved using this method. Load was applied at a rate of 0.009in (0.23mm) per minute, by means of horizontal linear movement of the gearbox shaft driven by a small high-torque low-speed a-c motor. The motor was switched

FIGURE 31. University of Liverpool centrifugal test facility.

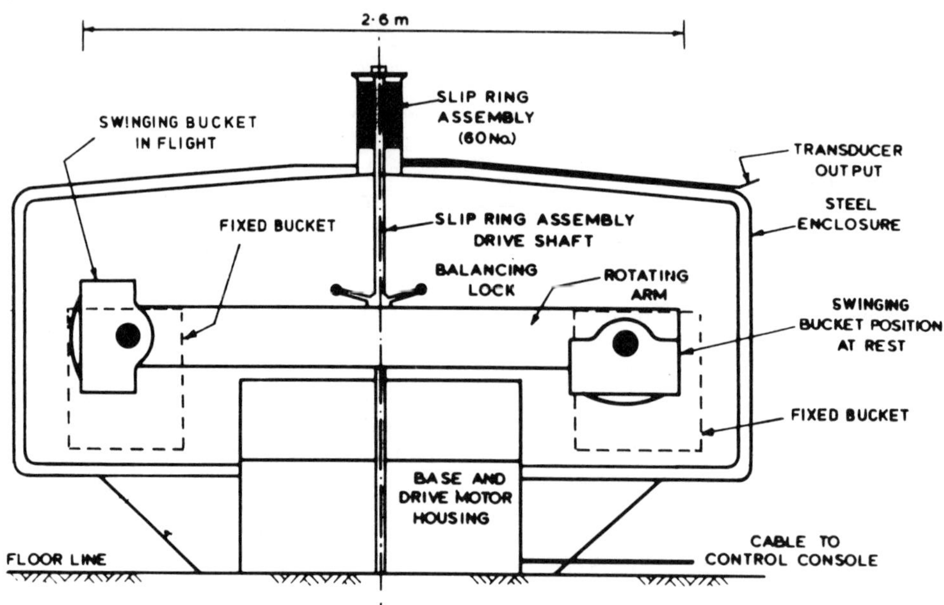

FIGURE 32. Diagrammatic representation of centrifuge.

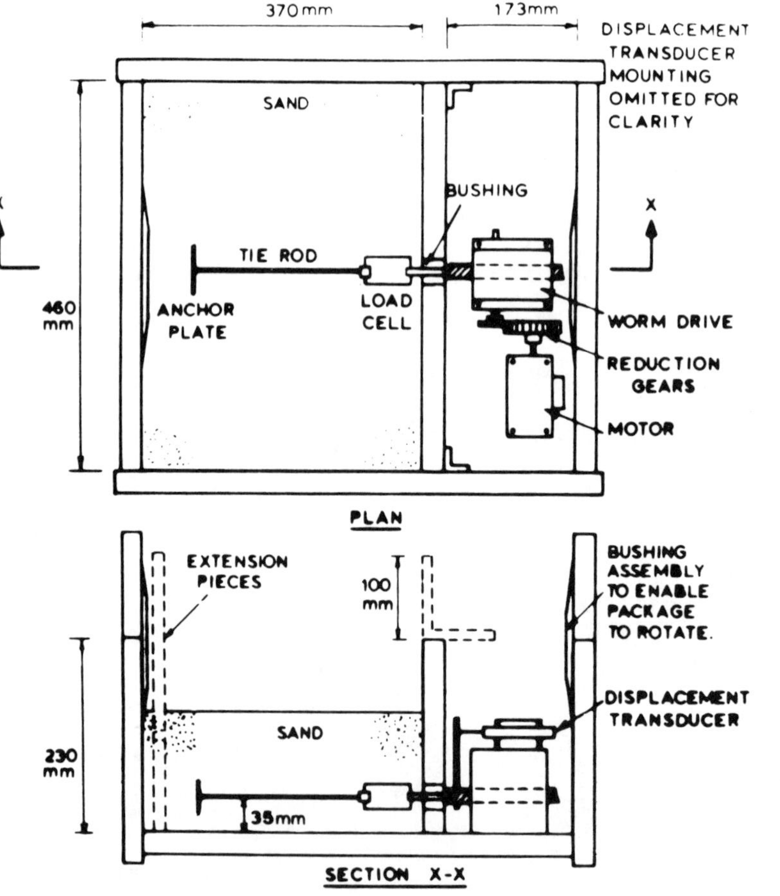

FIGURE 33. Package arrangement for testing single anchors in centrifuge.

on remotely via the slip ring system once the package had been brought to the desired rotational speed. The anchor resistance was measured by means of a Novatech tension/compression load cell of 100, 500 or 1,000lb (4.4, 22.2 or 44.5kN) range located in the "anchor" section, thereby avoiding "bushing friction" errors. The horizontal displacement of the anchor was recorded indirectly by a 1in (25mm) travel Sangamo linear displacement transducer. The 10 volt excitation voltage was applied to both transducers from a stabilized d-c power source and the output measured either on two separate remote solid-state LED voltmeters, or by a Vishay Ellis VE22 recording system through the slip ring system. Identical load cell calibration factors were obtained for static conditions and acceleration fields of 20g and 30g.

For "continuous" anchor tests, the anchor section was further subdivided using glass-faced walls 4.6in (116mm) apart, friction on the faces of which was minimized either by applying a silicone polish or fixing greased latex sheets to the glass surfaces. The anchor which spanned this reduced section was pulled out by a single tie rod as in the case of single anchors. The results of centrifugal tests with this arrangement were reported by Leung [23]. However, it was evident from the curvature of failure planes close to the glass-walls that side friction had not been entirely eradicated, nor was the effect of this interface friction upon measured resistance and displacement values known. It was also found that considerable horizontal bending of the deeper plates had occurred during these tests. A further test program in which 3 adjoining square plates were pulled out simultaneously, in a similar arrangement to that of Rowe and Peaker [46], gave substantially higher values than previously obtained due to interference effects between adjoining plates. Finally the arrangement shown in Figure 34 was adopted using a much wider test section with glass-faced walls 8in (200mm) apart and 3 equally spaced load cells monitoring the resistance of the anchor. Side friction would have a reduced effect as a result of this wider wall spacing, and was in any case evaluated by the slightly higher readings from the load cells linked to the outer tie rod. Undesirable horizontal bending encountered with a single tie rod system was also considerably reduced.

Dry Erith sand which was clean, fine and fairly uniform, having 80% of its grains between 0.005in (0.125m) and 0.01in (0.25mm), a specific gravity of 2.65 and maximum and minimum porosities of 49.5% and 34%, respectively, was used in all tests. Dickin [11] carried out an extensive program of plane strain compression tests on the sand for similar densities to that in the model tests and at cell pressures ranging from 0.4–44 lb/in² (3–300 kN/m²). The results obtained are summarised in Figure 35 and show similar trends to those found in previous work by De Beer [10] and Ponce and Bell [38] in that the angle of shearing resistance increases significantly with decreasing confining stress particularly for very low stress levels; the plane strain apparatus was described by Tang [50].

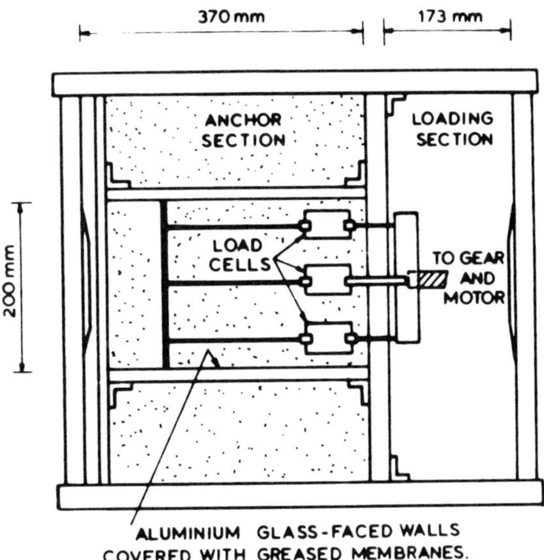

FIGURE 34. Package arrangement for testing continuous anchors in centrifuge.

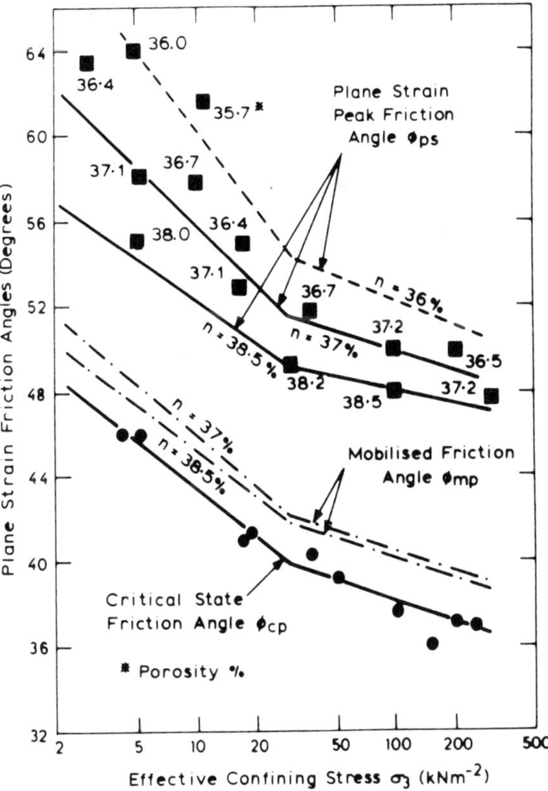

FIGURE 35. Variation of plane strain friction angles with effective confining stress for dense Erith sand.

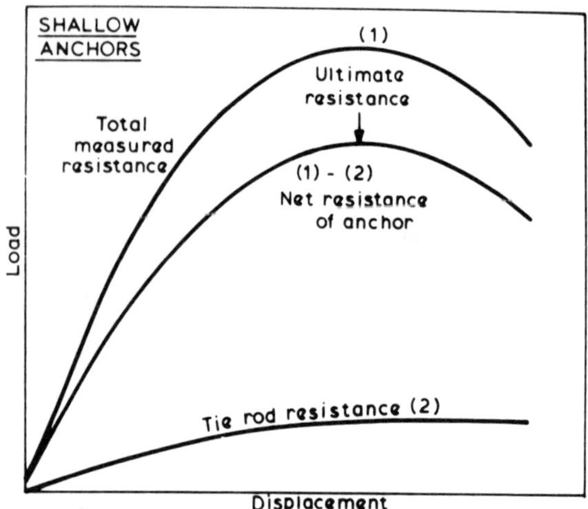

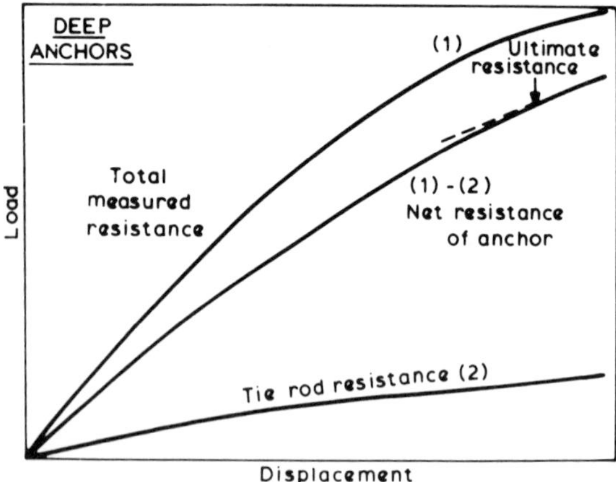

FIGURE 36. Typical load displacement curves from centrifugal model tests.

Most of the anchor plates were aluminum having a bulk specific weight of 178lb/ft³ (28 kN/m³) and modulus of elasticity of 9.8 × 10⁶lb/in² (67.4 × 10⁶ kN/m²). The faces of the plates were polished and gave a surface friction angle of 29° measured in sliding friction tests. The strength of the anchor material is best expressed by the flexibility parameter $\varrho = h^4/EI$. By suitable choice of plate thickness, all anchors tested were essentially rigid [32] in the vertical plane having flexibilities in the limited range from 2.73×10^{-5} to 6.83×10^{-6}ft³/lb (1.74×10^{-4} to 4.35×10^{-5}m³/kN). Some deeply embedded wide plates were made of stainless steel to reduce horizontal bending.

Anchor Test Program

The experimental program comprised 3 series. In Series 1, shallow 19.7in (0.5m) and 39.4in (1m) square prototype anchors were modelled by selecting appropriate combinations of acceleration and model height. In Series 2 and 3, 19.7in (0.5m) and 39.4in (1m) prototype anchors of different width/height ratios were modelled for a range of embedment ratios up to 8. Since theoretical predictions do not include the contribution to ultimate resistance of the rod, separate tests were carried out under identical conditions to determine the pull-out resistance of the 0.125in (3.2mm) diameter steel tie rods alone. The anchor load T_u was defined as the net resistance after applying a deduction for the tie rod resistance. Typical load-displacement curves shown in Figure 36 illustrate the magnitude of this tie rod contribution. Although peak resistances were well-defined in tests on shallow anchors, peaks were not attained at embedments greater than 4 and the ultimate resistance was defined as the point at which a linear resistance/displacement relation was reached.

Figures 37 shows the good agreement between alternative modellings of a 19.7in (0.5m) square surface anchor. A further comparison for all tests in Series 1 is given in Figure 38 in the manner adopted by Ovesen [32]. The good agreement exhibited by alternative modellings of the same prototype confirms the validity of the centrifugal test technique to vertical anchors. Extensive information on 19.7in (0.5m) and 39.4in (1m) prototype anchors for embedment ratios from 1 to 8 is derived from the results of tests on 1in (25mm) plates

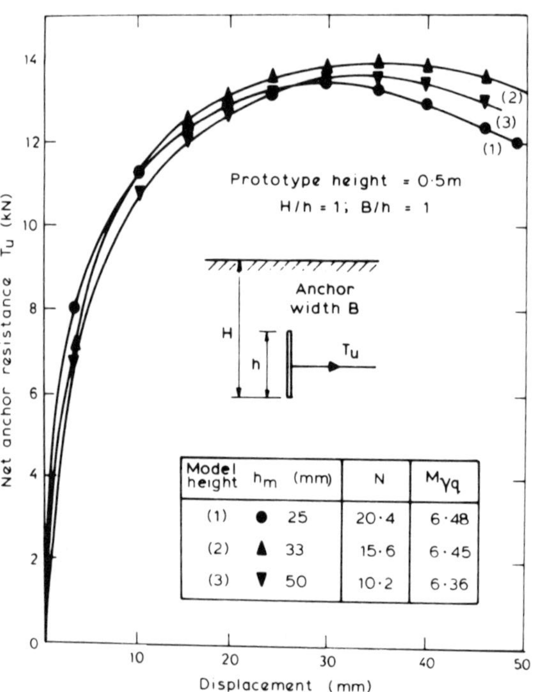

FIGURE 37. Load/displacement relationships from test series 1A.

at acceleration factors of 20g and 40g in Series 2 and 3, respectively. Square and continuous anchor tests were conducted in both programs while additional tests on anchors with width/height ratios of 2 and 5 were included in Series 3.

Influence of Shape, Size and Embedment

Figure 39 shows the marked increase in $M_{\gamma q}$ with embedment ratio plotted semi-logarithmically for 19.7in (0.5m) and 39.4in (1m) anchors. Slightly lower $M_{\gamma q}$ values were obtained for the larger prototype modelled in Series 3 for all embedment ratios and anchor geometries. The 3-dimensional nature of the pattern of failure planes around a single 2in (50mm) surface anchor with a width/height ratio of 2 after a centrifuge test is seen in Figure 40. The planes were accentuated by thin lines of dyed sand placed on the surface of the backfill before the test. A reduction in the relative magnitude of "side shear" resistance to the total anchor pull-out resistance is associated with an increase in width/height ratio and leads to a corresponding reduction in $M_{\gamma q}$ values from a maximum for square to a minimum for continuous anchors for all embedment ratios.

Figure 41 shows force coefficients from conventional tests on 1in (25mm) anchor plates with $B/h = 1,2$, and 5 and continuous anchors at embedment ratios up to 8 from tests in the centrifuge swinging bucket but with the machine static. Since the test preparation was identical to that in centrifuge tests, a direct comparison between this series and the centrifugal program is possible. The variation of breakout factor $N_{\gamma q}$ with embedment for square and continuous anchors given in Figure 42 suggests a critical embedment ratio $(H/h)_{cr}$ of approximately 10 for both these 1in (25mm) anchors and the 2in (50mm) anchors considered earlier, whereas in the centrifuge test program 19.7in (0.5mm) and 39.4in (1m) prototypes exhibited critical embedments in the region of 7. This supports earlier work of Buchholz [5] who also concluded that $(H/h)_{cr}$ reduced as the size of the anchors increased.

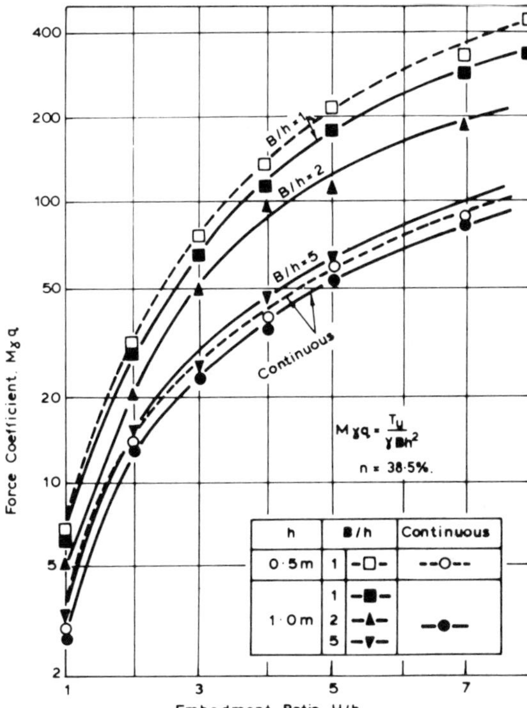

FIGURE 39. Variation of force coefficient $M_{\gamma q}$ with embedment ratio H/h for 0.5m and 1.0m prototype anchors.

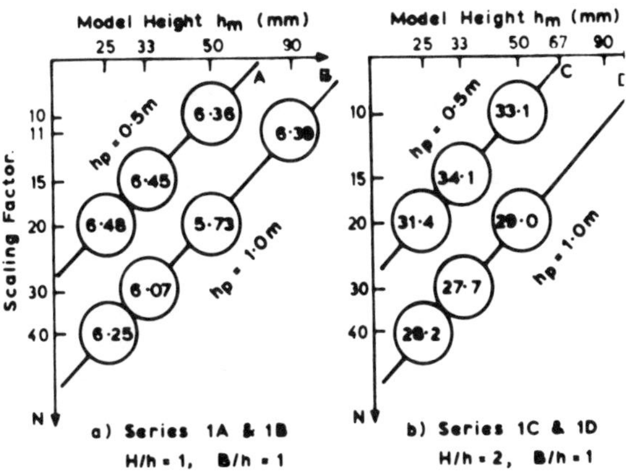

FIGURE 38. Summary of series 1 tests.

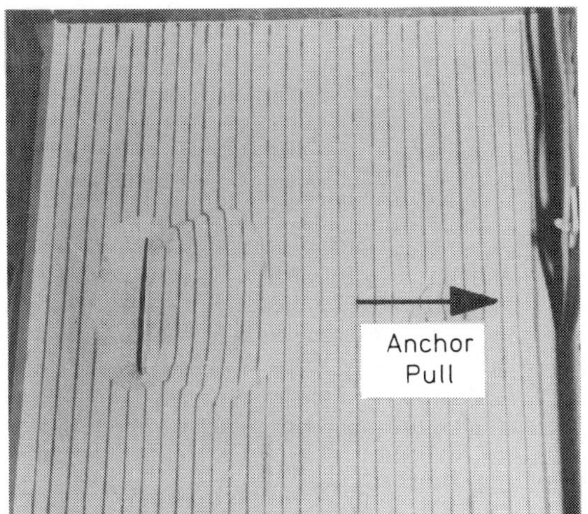

FIGURE 40. Failure planes around a 50mm high surface anchor with width/height ratio = 2 after a centrifuge test at 40g.

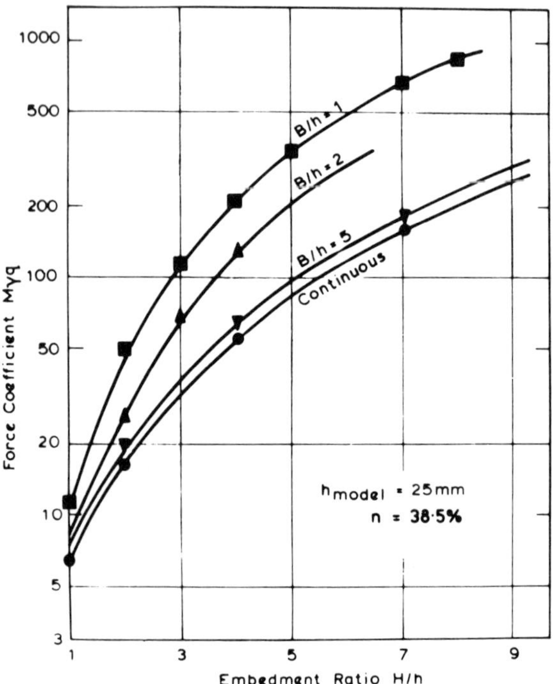

FIGURE 41. Variation of force coefficient $M_{\gamma q}$ with embedment ratio H/h for conventional tests on 25mm high anchors.

A well-defined reduction in $N_{\gamma q}$ with increased size is observed in Figure 42 for square anchors at all embedments, the 1in (25mm) plates yielding particularly high values. This clearly demonstrates the potentially dangerous overprediction of prototype anchor resistances by direct extrapolation of such small-scale data. The extreme case of 1/40th scale "modelling" of a 39.4in (1m) prototype by a conventional test on a 1in (25mm) plate would result in an approximate 80% overprediction. A 1/20th scale "modelling" of the same prototype would give 25% while a modest 1/10th scale increase from 2in (50mm) to 19.7in (0.5m) would give 15%. A similar trend is observed for continuous anchors.

Neely et al. [33] attempted to evaluate this scale error by incorporating a ϕ versus confining stress relationship for Mol Sand published by De Beer [10] into their theoretical predictions of $M_{\gamma q}$, reporting a variation of up to 31% for a 10-fold increase in plate height from 6in (150mm) to 60in (1.5m) in the case of a continuous anchor with $H/h = 4$. They attempted to further demonstrate the effect of size by comparing the results of their own tests on 2in (50mm) plates with field tests on 36in (0.9mm) concrete slabs reported by Smith [49]. On the basis of this study a 16% reduction in $M_{\gamma q}$ was suggested. However, a closer examination of Neely's comparison given in Figure 43, showing the interpolated curve for $B/h = 1.25$, suggests no significant difference in anchor force coefficients. Although the respective sands were of similar grading and relative density, the comparison was clearly limited by other differences, for example in anchor materials, thicknesses, and sand shear strengths. Using results drawn from the writers' conventional and centrifugal tests, Figure 44 demonstrates the influence of size for the range examined practically by Neely et al. [33] Reductions in $M_{\gamma q}$ of about 20% for square and 10% for rectangular B/h = 5 anchors result from a 20-fold increase in size from 2in (50mm) to 39.4in (1m).

Data from both conventional and centrifugal tests are included in Figure 45, which shows the relationship between $M_{\gamma q}$ and size for plates with an embedment ratio of 2. Similar trends exist for other embedments. The variation is most significant for anchors less than 6in (150mm) high and generally reduces as width/height ratio increases. The trend is consistent with the measured reductions in shear strength with increased confining pressure in triaxial and plane strain compression tests. Stress levels of the order of 0.5lb/in² (3.5

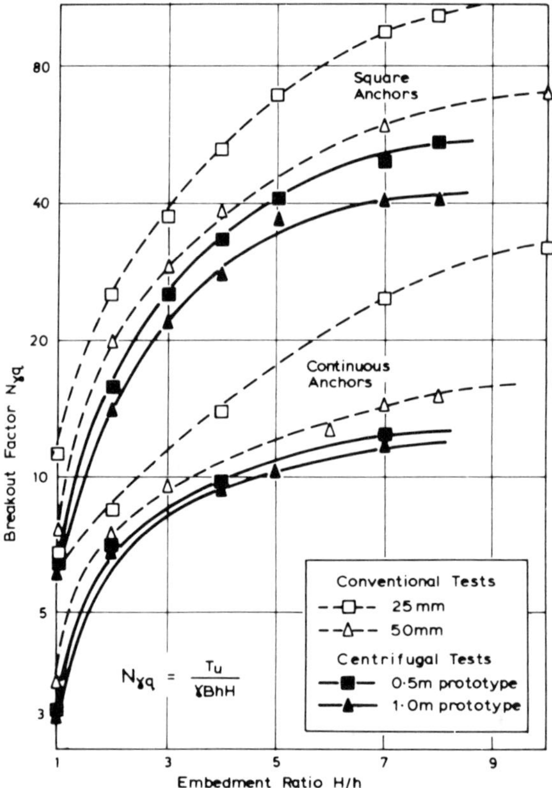

FIGURE 42. Variation of breakout factor $N_{\gamma q}$ with embedment ratio H/h for square and continuous anchors.

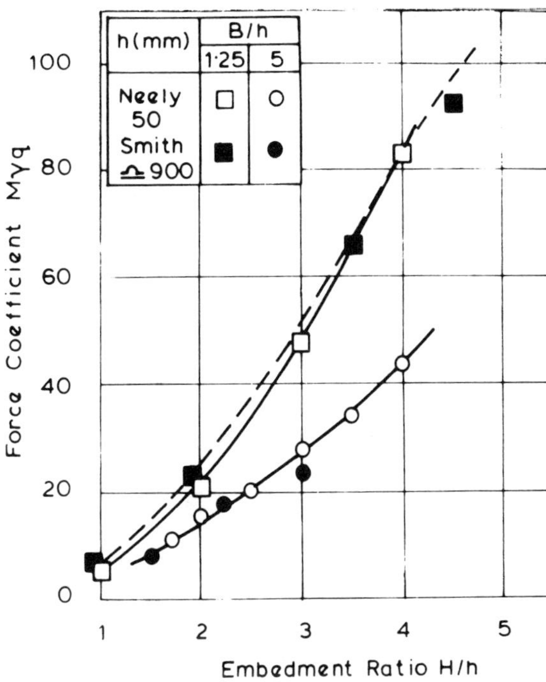

FIGURE 43. Comparison between model and field test results (Neely, Stuart and Graham, 1973).

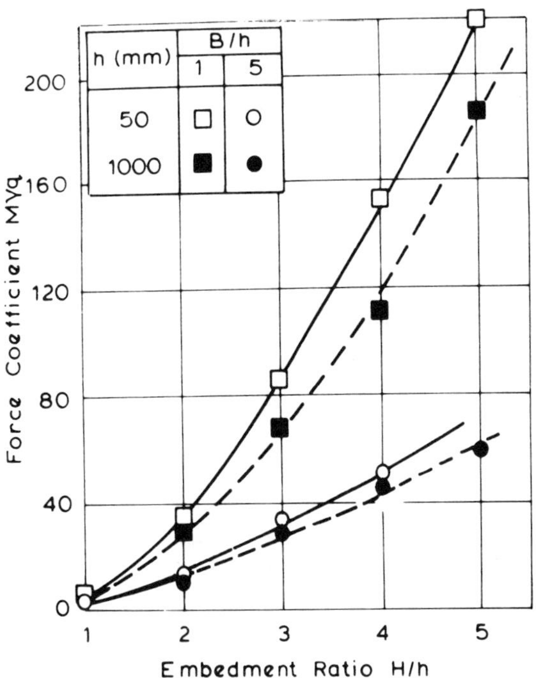

FIGURE 44. Comparison of force coefficients $M_{\gamma q}$ for 50mm and 1.0m anchors.

kN/m²) are associated with substantially higher shear strengths than encountered at field stress levels around 20lb/in² (140 kN/m²). Although at maximum resistance some regions of the backfill would contribute a reduced post-peak strength due to "progressive" failure in the dense medium, the average mobilized shear strength in the mass would still be relatively high in a small model.

Load-Displacement Relationship

The design of structures incorporating vertical anchors may often require constraints on displacement. Hence the load-displacement behaviour, rather than ultimate load, is important. Das and Seeley [7] expressed this relationship in hyperbolic form as:

$$\bar{T} = \frac{\bar{D}}{a + b\bar{D}} \tag{26}$$

where $\bar{T} = T/T_u, \bar{D} = D/D_u$, and D_u is the failure displacement. Constants a and b were 0.15 and 0.85, respectively, for conventional tests on anchors with various geometries at embedment ratios up to 4. With these values Equation (26) also fits the centrifugal data in Figure 37 reasonably well.

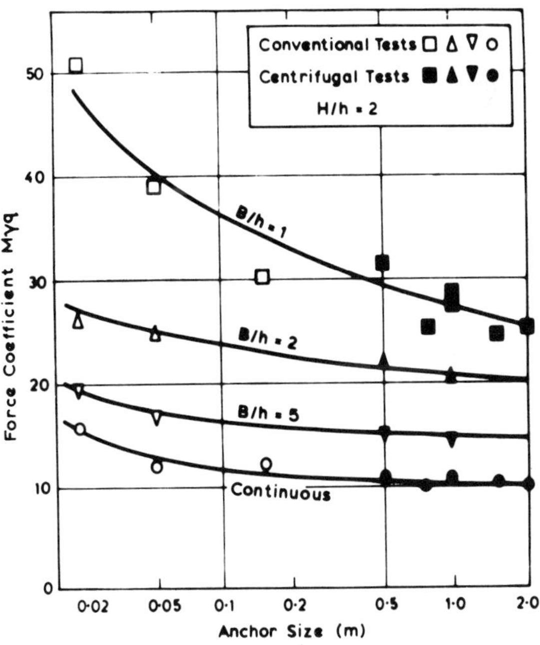

FIGURE 45. Variation of force coefficient $M_{\gamma q}$ with anchor size for all tests (H/h = 2).

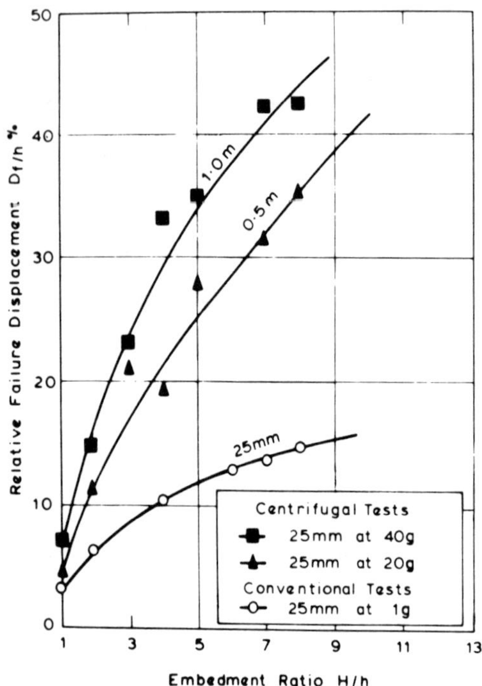

FIGURE 46. Variation of relative failure displacement D_f/h with embedment ratio H/h for square anchors.

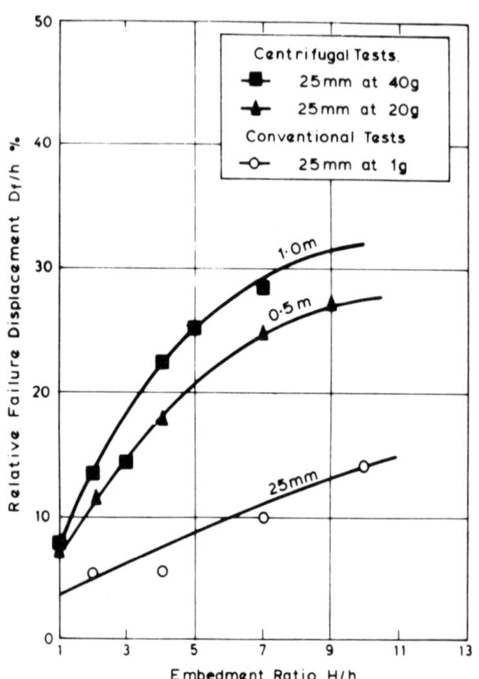

FIGURE 47. Variation of relative failure displacement D_f/h with embedment ratio H/h for continuous anchors.

However, the influence of backfill properties on a and b requires further study.

Failure displacements may also be expressed in dimensionless form in terms of *relative* failure displacement $D_f = D/h\%$. Since the centrifugal modelling laws give $D_p = ND_m$ and $h_p = Nh_m$, then $D_{fp} = D_{fm}$. Accurate evaluation of failure displacements presented difficulties in tests on shallow 1in (25mm) anchors where non-uniform densities close to the backfill surface were most significant, and in tests on deeper anchors where a peak resistance was not observed, necessitating the adoption of the failure definition shown in Figure 36. The trends given in Figures 46 and 47 for square and continuous anchors, respectively, were fairly well-defined. These figures include only results obtained in tests in the centrifuge swinging bucket either static or in flight. The 2in (50mm) anchors yielded relative displacements of similar magnitude to 1in (25mm) tests. Relative failure displacements increase with embedment ratio and with plate size. Scale errors could clearly arise in any situation where anchor displacement was the controlling criterion with small-scale test data used in the design. Relative failure displacements for continuous anchors were considerably smaller than those from single anchors with $B/h = 1$ or 2 since the backfill around the former would experience essentially plane strain conditions, whereas that around the latter would be subjected to conditions closer to triaxial. (It is well-established that, for dense media, failure deformations in plane strain are considerably smaller than in triaxial compression).

DESIGN METHODS FOR CONTINUOUS ANCHORS

A preliminary comparison between theoretical predictions for continuous anchors in a backfill with $\phi = 40°$ in Figure 13 reveals considerable variation.

Careful selection of the relevant ϕ values is a pre-requisite in any detailed comparison since the Mohr-Coulomb relationship for soil is non-linear and thus ϕ varies with stress level. Moreover, while the adoption of triaxial soil parameters in all previously reported research may be approximately valid for square anchors, plane strain friction angles would be more appropriate for the strictly two-dimensional conditions around continuous anchors. The idealized relationships between peak plane strain friction angle ϕ_{ps} and confining stress σ_3 for porosities 37% and 38.5%, shown in Figure 35, were incorporated into computations of theoretical $M_{\gamma q}$ values for 2in (50mm) and 39.4in (1m) continuous anchors respectively, with confining stress σ_3 being replaced by γH.

In general these computations yielded considerable overpredictions of the writers' experimental results as clearly demonstrated for 2in (50mm) anchors in Figure 48 and to a lesser degree for 39.4in (1m) anchors in Figure 49. This may be attributed to the effects of progressive failure because of which the maximum resistance is not mobilized simultaneously at all points in the soil mass. The phenomenon would be particularly significant for dense sand at low stress levels in the conventional tests. Allowance for progressive failure may be made by using a mobilized plane strain friction angle in the mass ϕ_{mp}, as proposed by Rowe [45], where:

$$\phi_{mp} = \phi_{ps}(1 - P_r) + P_r\,\phi_{cp} \qquad (27)$$

in which P_r = progressivity index and ϕ_{cp} = critical state friction angle.

Values of ϕ_{cp} were obtained from stress-dilatancy plots of plane strain compression test data. The idealized relationships for ϕ_{mp} shown in Figure 35 were derived from Equation (27) with $P_r = 0.8$, the value suggested by Rowe for passive resistance. Theoretical $M_{\gamma q}$ values are strongly influenced by this rather arbitrary progressivity index. Thus in the case of 39.4in (1m) anchors, taking $P_r = 0.8$ results in ϕ_{mp} up to 8° lower than ϕ_{ps}, and in consequence $M_{\gamma q}$ values are between 20% and 50% lower depending on the theory considered. A reduction in P_r from 0.8 to 0.6 typically gives between 5% and 15% higher force coefficients. In computing theoretical $M_{\gamma q}$ values in Figures 50 and 51, the writers adopted Rowe's suggested P_r value of 0.8, since this was based on a wide variety of passive pressure studies.

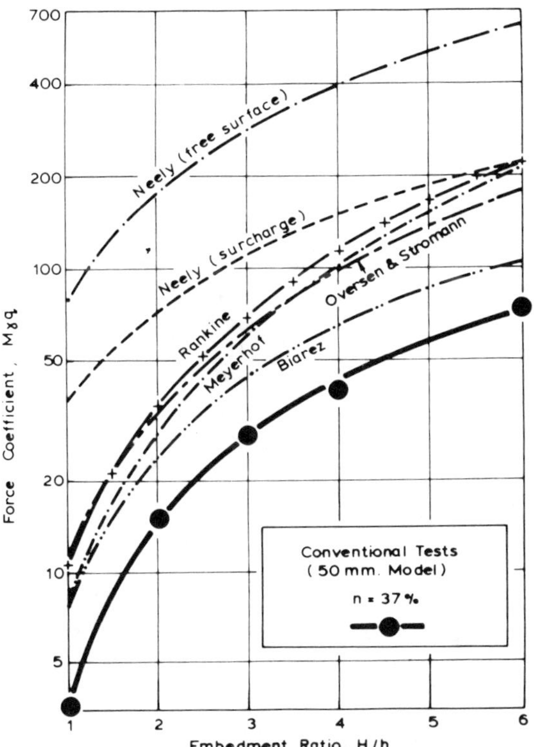

FIGURE 48. Comparison between theoretical and experimental force coefficients $M_{\gamma q}$ for 50mm high continuous anchors, (using peak plane strain friction angles ϕ_{ps}).

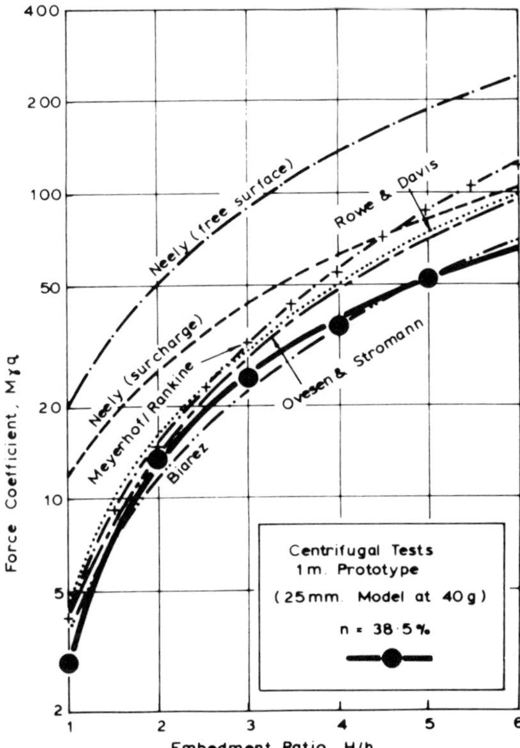

FIGURE 49. Comparison between theoretical and experimental force coefficients $M_{\gamma q}$ for 1m high continuous anchors (using peak plane strain friction angles ϕ_{ps}).

No theory gives close agreement with measured force coefficients for 2in (50mm) anchors over the full range of embedments (Figure 50). However reasonable, $M_{\gamma q}$ values are obtained for very shallow anchors from the Rankine, Meyerhof, Biarez, and Ovesen and Stromann methods. Rankine theory is only strictly applicable to surface anchors and its extension to the embedded case leads to overpredictions of up to 50%. Neely's free surface and surcharge methods predict comparatively high values for all embedments, while Biarez's formula yields conservative results.

The general disparity between theory and experiment could stem from inherent difficulties in obtaining reliable data from conventional small model tests and the relevant shear strength tests at low stress levels. Thus, conclusions drawn from comparative studies at such low stress levels should be viewed with reservation.

Predictions by Meyerhof, Rankine, and Ovesen and Stromann do match measured $M_{\gamma q}$ values for 39.4in (1m) anchors quite closely for all embedments (Figure 51). Neely's surcharge method also gives compatible values for all but

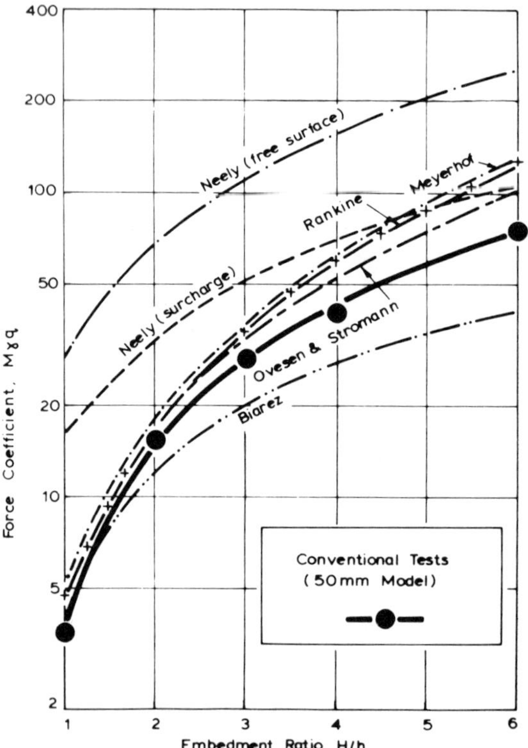

FIGURE 50. Comparison between theoretical and experimental force coefficients for 50mm high continuous anchors (using plane strain friction angles ϕ_{mp}).

the surface case. However, Neely's free surface method and Biarez's method again exhibit the greatest deviation from experimental values.

Force coefficients for 39.4in (1m) anchors based on finite element analyses by Rowe and Davis [48] are also included in Figures 49 and 51. Slightly overoptimistic $M_{\gamma q}$ values are computed when peak friction angles are used as shown in Figure 49. In this case the insertion of the appropriate ϕ_{ps} and ϕ_{cp} values in Equation (12) yields sensible ψ values in the region of 13°. Figure 53 shows an extremely good fit to the experimental data from calculations using the mobilized friction angle in the mass. However, unrealistically low values of ψ in the region of 2° are derived for $\phi_{ps} = \phi_{mp}$ in Equation (12), which casts some doubt on this refinement.

Comparison of Shape Factors

Most design methods are based on two-dimensional analyses, and their application to anchor geometries outside the continuous case requires a suitable shape factor defined by Equation (16). Figure 52 shows the variation of shape factors with embedment ratio for square anchors tested by the writers. Despite some scatter these appear essentially independent of scale.

Theoretical and empirical values of S_f are compared in Figure 53 which also includes centrifugal test data for 39.4in (1m) square anchors. Good agreement exists between the writers' tests and Ovesen and Stroman's shape factors derived from Equation (18) for embedment ratios up to 6. Understandably lower values of shape factor were obtained by Neely et al. [33] who assumed plates with $B/h = 5$ to be effectively continuous, and whose sand was not as dense as that of the writers. As previously noted [33], the shape factors for shallow anchors based on the recommendations in the British Code of Practice are extremely conservative. This is attributable to the geometry analyzed in Figure 3. Failure planes around anchors radiate outwards involving a soil mass wider than the anchor itself in the failed body. An idealization of this mechanism is shown in Figure 54.

Considering the passive resistance for the wedge abb′a′:

$$P = \int_{o}^{H} K_p\,\gamma z\,[B + 2(H - z)\tan(90 - \psi_2)\tan\psi_2]dz$$

$$= K_p\,\gamma H^2 \left[\frac{B}{2} + \frac{H}{3}\tan(90 - \psi_2)\tan\psi_2\right] \tag{28}$$

Considering normal force N on planes ab and a′b′ for ac-

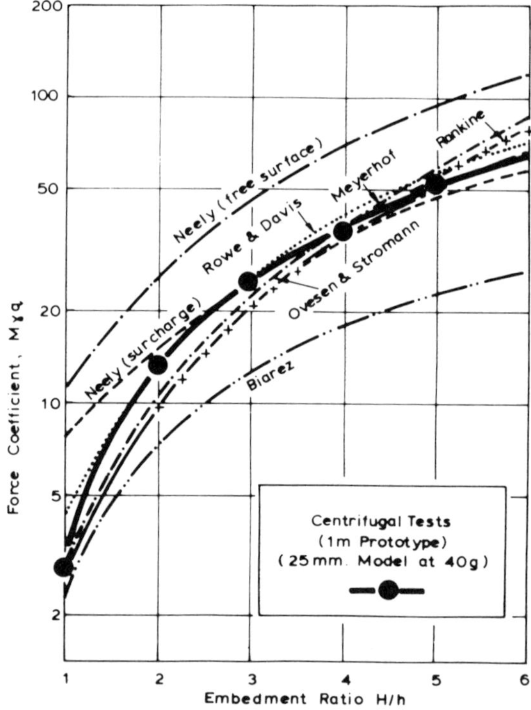

FIGURE 51. Comparison between theoretical and experimental force coefficients $M_{\gamma q}$ for 1m high continuous anchors (using plane strain friction angles ϕ_{mp}).

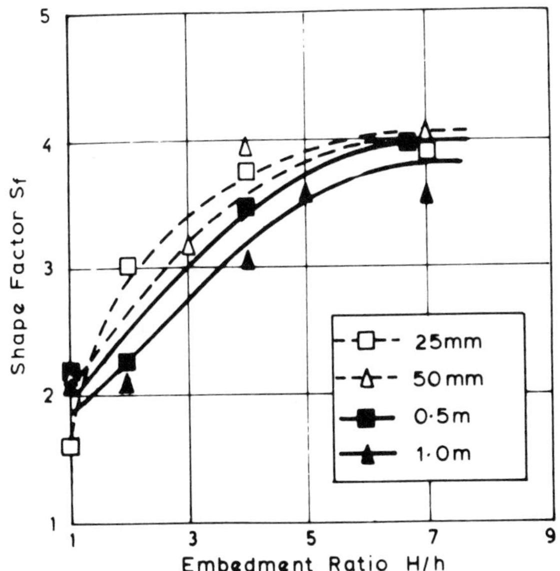

FIGURE 52. Variation of shape factor S_f with embedment ratio H/h for all tests on square anchors.

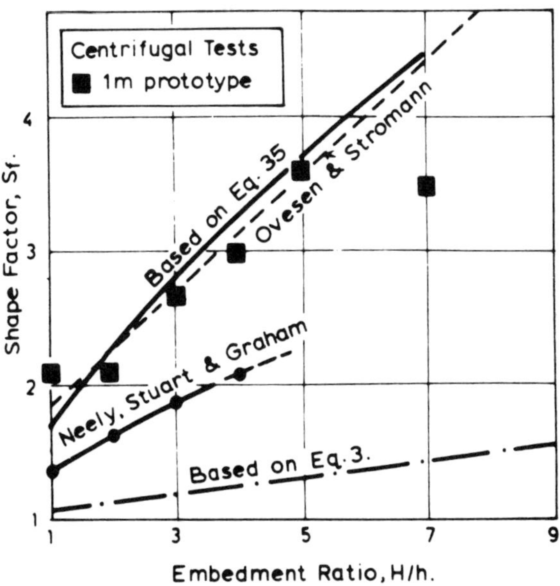

FIGURE 53. Comparison of shape factors for 1m square anchors in dense sand.

tive conditions [14]:

$$N = \int_o^H K_a \gamma z (H - z) \frac{\tan (90 - \psi_1)}{\cos \psi_2} dz$$

$$= K_a \gamma \frac{H^3}{6} \frac{\tan (90 - \psi_1)}{\cos \psi_2} \tag{29}$$

Side shear resistance for both ab and a′b′:

$$P_s = K_a \gamma \frac{H^3}{3} \frac{\tan (90 - \psi_1)}{\cos \psi_2} \tan \phi \tag{30}$$

Component of P_s parallel to tie bar:

$$P_{sT} = K_a \gamma \frac{H^3}{3} \tan (90 - \psi_1) \tan \phi \tag{31}$$

Component of N parallel to tie bar for both sides of wedge:

$$N_T = K_a \gamma \frac{H^3}{3} \tan (90 - \psi_1) \sin \psi_2 \tag{32}$$

Assuming $\psi_1 = 45 - \phi/2$ and $\psi_2 = \phi/2$ after Reese, Cox, and Koop [42]:

$$P = \frac{1}{2} K_p \gamma H^2 B + \frac{1}{3} K_p \gamma H^3 \tan (45 + \phi/2) \tan \phi/2 \tag{33}$$

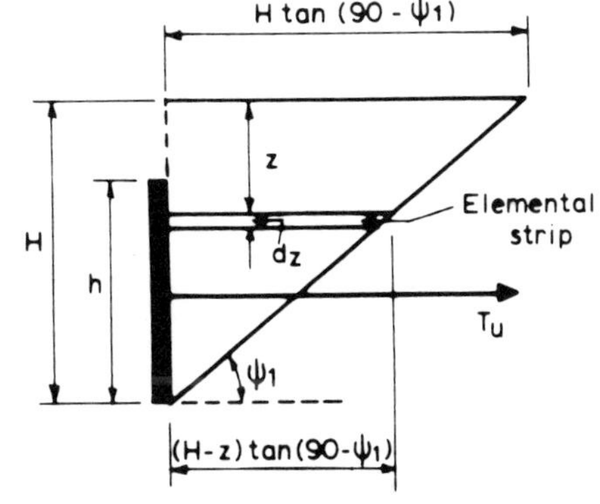

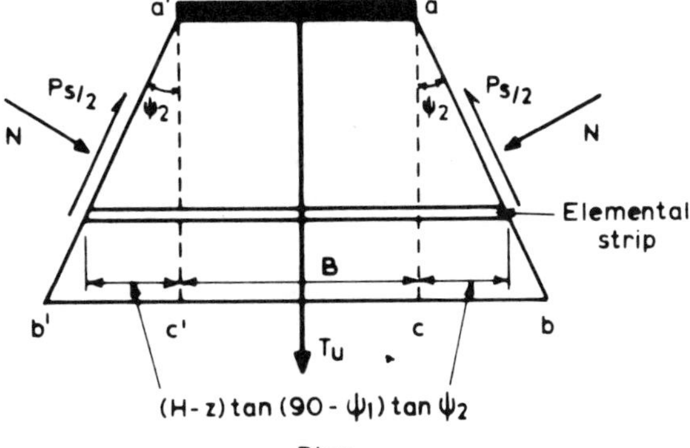

FIGURE 54. Idealised failure zone in front of a single, shallow anchor.

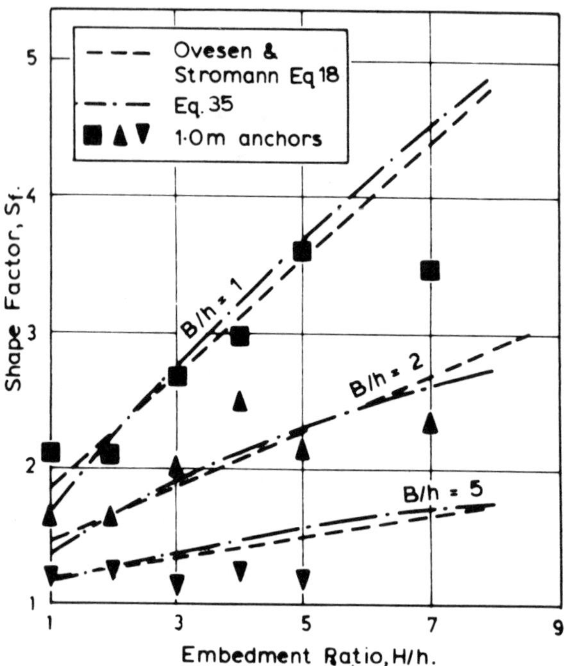

FIGURE 55. Comparison between theoretical shape factors and experimental results for 1.0m anchors.

$$P_{sT} = K_a\gamma \frac{H^3}{3} \tan (45 + \phi/2) \tan \phi \qquad (34)$$

and

$$N_T = K_a\gamma \frac{H^3}{3} \tan (45 + \phi/2) \sin \phi/2 \qquad (35)$$

Allowance for all the above components yields a shape factor:

$$S_f = \frac{P + P_{sT} - N_T}{P_p} \qquad (36)$$

in which $P_p = \frac{1}{2} K_p\gamma H^2B$.

Taking $\psi_2 = 0$, $N_T = 0$, and $P = P_p$ in Equation (36) gives shape factors in agreement with the British Code [16]. Figure 55 shows great similarity between shape factors based on Equation (36), Ovesen et al.'s empirical approach, and values derived from tests on 39.4in (lm) anchors at embedment ratios up to 5 in dense sand. Values of $\phi = \phi_{mp}$ were selected as described earlier.

DESIGN METHODS FOR SINGLE ANCHORS

Theoretical force coefficients for 2in (50mm) and 39.4in (lm) square anchors obtained by applying the appropriate shape factors to the continuous results given in Figures 50 and 51 are shown in Figures 56 and 57, respectively. Ovesen and Stromann's method, along with Rankine theory in combination with the revised shape factor [Equation (36)], and Meyerhof's curves all yield $M_{\gamma q}$ values that fit the writers' values for both anchor sizes well. Note that Rankine theory, in combination with Equation (3), leads to very conservative results. Since the shape factors of Neely et al. are significantly smaller than those derived from either Equation (18), Equation (36), or the writers' tests, their theoretical predictions are comparatively lower than for the continuous case. This leads, perhaps fortuitously, to an improved agreement between the free surface method and observation.

Rowe and Davis did not include shape factors for vertical anchors in their work. However, Figure 57 shows excellent agreement between the writers' experimental data for 39.4in (lm) anchors and values from Rowe and Davis's charts in conjunction with the revised shape factor [Equation (36)].

DESIGN METHODS FOR DEEP ANCHORS

The characteristic localised failure mechanism around a deep anchor is shown in Figure lc. This phenomenon was

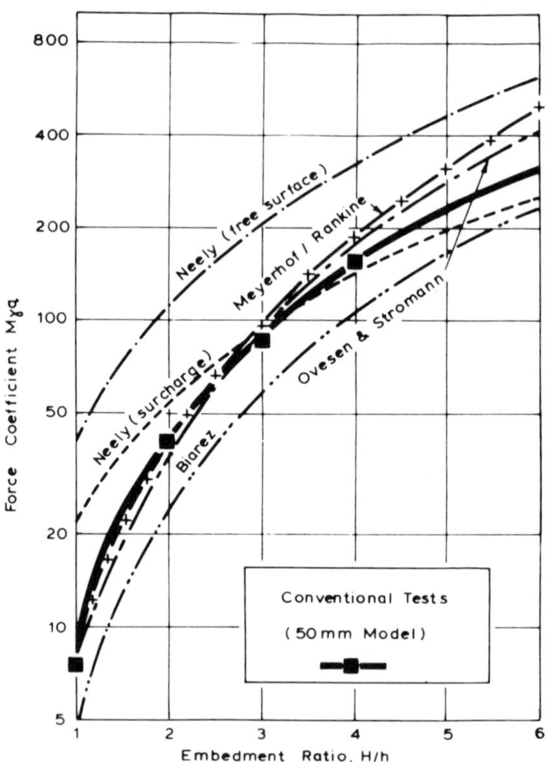

FIGURE 56. Comparison between theoretical and experimental force coefficients $M_{\gamma q}$ for 50 mm high square anchors (using ϕ_{mp}).

reported previously [1,5] and analyses based on bearing capacity theory [35,52] have been presented. Only Biarez et al. [4] recognized the essentially rotational nature of these soil displacements and determined the couple associated with such a mechanism.

Theoretical and experimental $M_{\gamma q}$ values for deep 2in (50mm) and 39.4in (1m) anchors are shown in Figures 58 and 59, respectively. Although the behavioral trends for conventional model tests on 2in (50mm) anchors are well established, an equivalent centrifugal study for 39.4in (1m) anchors was not possible due to limitations in package size. Estimated experimental curves for 39.4in (1m) anchors shown in Figure 59 are based on trends exhibited in conventional tests. Any conclusions drawn must thus be speculative. Theoretical curves were determined using a mobilized ϕ_{mp} as previously but with $P_r = 0.6$ after Rowe [45].

Most theories give extremely overoptimistic $M_{\gamma q}$ values for both sizes of square and continuous anchors considered. However, Equation (10) derived by Biarez et al. [4] yields conservative and reasonably close estimations of the experimental values for continuous anchors. Other theories previously considered for shallow anchors also give reasonable predictions of deep anchor resistance. These are

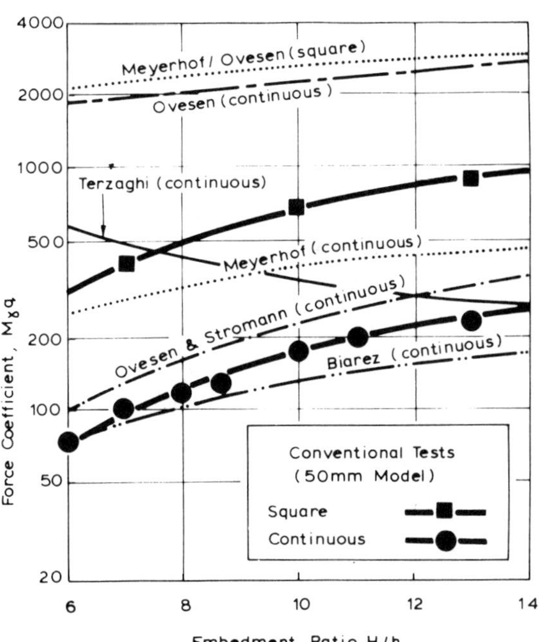

FIGURE 58. Comparison between theoretical and experimental force coefficients $M_{\gamma q}$ for deep 50mm high anchors (using ϕ_{mp}).

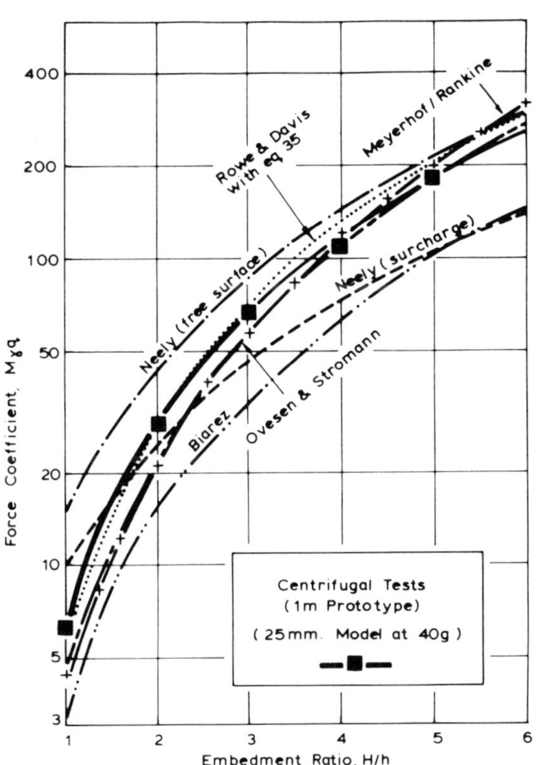

FIGURE 57. Comparison between theoretical and experimental force coefficients $M_{\gamma q}$ for 1m high square anchors (using ϕ_{mp}).

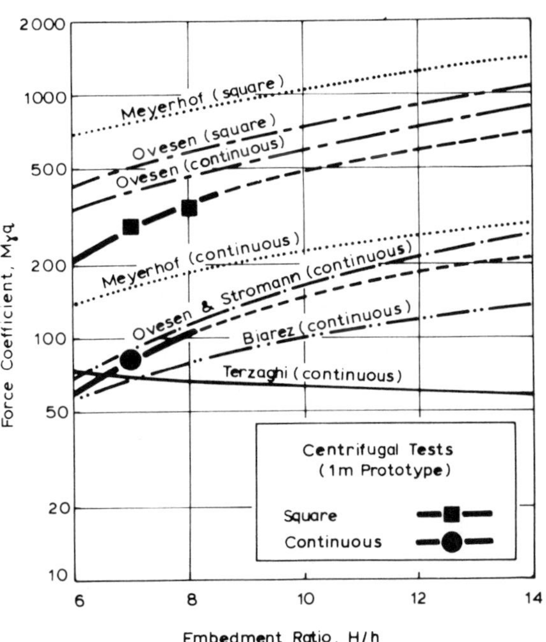

FIGURE 59. Comparison between theoretical and experimental force coefficients $M_{\gamma q}$ for deep 1m high anchors (using ϕ_{mp}).

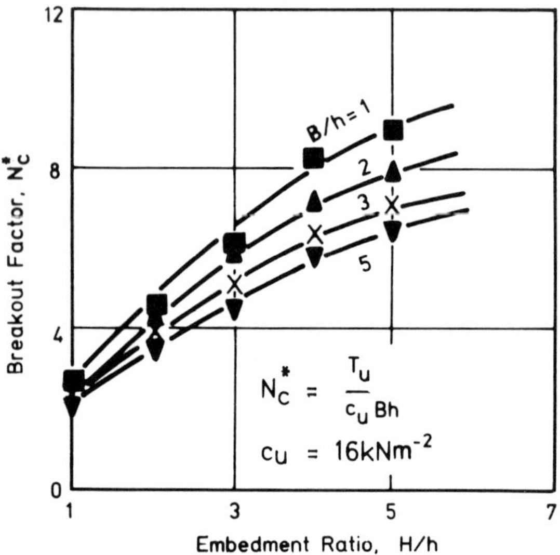

FIGURE 60. Variation of breakout factor N_c^* with embedment ratio H/h for conventional tests on 50mm anchors in clay (Das, Moreno and Dallo, 1985).

not included since they are based on analyses taken out of context.

Predictions using Ovesen and Stromann's empirical reduction factor [Equation (6)] are valid however and give better agreement with experiment than Ovesen's alternative formula for deep anchors [Equation (7)].

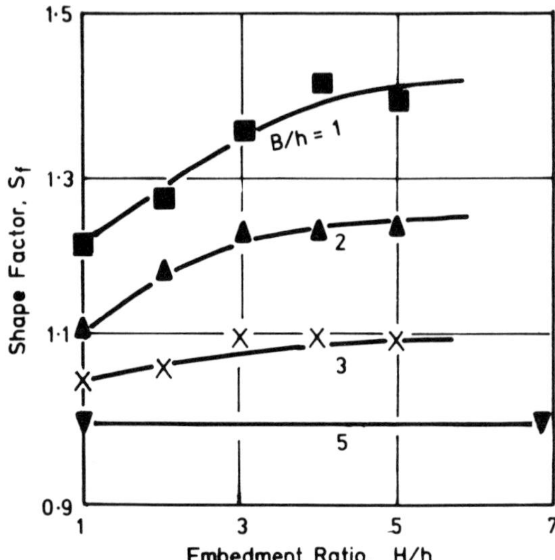

FIGURE 61. Variation of shape factor S_f with embedment ratio H/h for anchors in clay. (Das, Moreno and Dallo, 1985).

ANCHORS IN CLAY

Research on vertical anchors in clay has been very limited. However Mackenzie [26] conducted tests on continuous anchors in saturated clay which suggested that the ultimate resistance, expressed dimensionlessly as a breakout factor $N^*_c = T_u/Bhc_u$ where c_u = apparent cohesion, attains a maximum value at an embedment ratio of around 12.

More recent small scale model tests on shallow anchor plates reported by Das, Moreno, and Dallo [9] are summarised in Figure 60. Considerably lower shape factors (Figure 61) than in the case of granular soils were derived from this data. Clearly, considerably more research, both theoretical and experimental, is required before firm conclusions can be drawn for anchors in cohesive soils.

CONCLUSIONS

Comparative studies between existing design methods and recent conventional and centrifugal model tests on anchors in dense sand have shown that significant disparities exist between alternative methods. The lack of consistent agreement between theoretical and measured force coefficients for both single and continuous 2in (50mm) anchors is attributed to the inherent difficulties in obtaining reliable data from studies involving small models at low stress levels. Conclusions from such studies should be viewed with caution.

Good agreement is found for 39.4in (1m) anchors between observation and several theories, notably those of Ovesen and Stromann, Meyerhoff, and Rankine, provided a mobilized friction angle in the mass, rather than the peak value, is used in the analyses. This rather arbitrary empirical means of allowing for progressive failure appears to give satisfactory results but by its necessity demonstrates a severe limitation of limit state analyses when dense media are involved.

Design charts based on finite element computations by Rowe and Davis give slightly overoptimistic but encouraging results even when peak friction angles are used. Similar analyses incorporating a more sophisticated soil model that allows for post-peak softening in dense media should produce an improved fit.

The influence of anchor geometry is accounted for with reasonable accuracy by Ovesen and Stromann's empirical shape factor. A theoretical shape factor derived by the writers is also consistent with experimental results.

Theoretical analyses for deep anchors in dense sand give a wide disparity in values and tend to seriously overpredict measured anchor resistances. However, the analysis of Biarez et al., consistent with the observed rotational failure mechanism, predicts slightly conservative but sensible results for continuous anchors.

NOTATION

The following symbols are used in this paper:

B = width of anchor
D = horizontal displacement of anchor
D_f = relative displacement at failure
D_u = horizontal displacement of anchor at failure
d_c = depth factor
E = modulus of elasticity
F_γ = basic anchor capacity
g = earth's gravity field
H = depth of embedment
H/h = embedment ratio
$(H/h)_{cr}$ = critical embedment ratio
h = height of anchor
I = second moment of area
K_a = active earth pressure coefficient
K_o = earth pressure coefficient at rest
K_p = passive earth pressure coefficient
$M_{\gamma q}$ = anchor force coefficient
N = scaling factor
N^*_c = anchor breakout factor (in clay)
N_q = bearing capacity factor
N_γ = bearing capacity factor
$N_{\gamma q}$ = anchor breakout factor
n = porosity
n_r = relative porosity
P_a = active pressure
P_p = passive pressure
P_r = progressivity index
P_s = side shear resistance
R = radius
R_K = correction factor for initial stress state
R_{ov} = reduction factor
R_R = correction factor for anchor roughness
R_ψ = correction factor for dilatancy
S_f = shape factor
T = mobilised resistance of anchor
T_o = ultimate resistance for basic case
T_s = ultimate resistance for buried case
T_u = pull-out resistance
t = anchor thickness
γ = bulk specific weight of soil
δ = angle of wall friction
ϵ = strain
σ = direct stress
σ_3 = effective confining stress
ϱ = mass density of soil
ϕ = angle of internal friction
ϕ_{cp} = angle of internal friction (critical state)
ϕ_{mp} = angle of internal friction (mass)
ϕ_{ps} = angle of internal friction (peak-plane strain)
ψ = dilatancy angle

REFERENCES

1. Akinmusuru, J. O., "Horizontally Loaded Vertical Plate Anchors in Sand," *Journal of Geotechnical Engineering Division, ASCE,* Vol. 104, GT2, pp. 283–286 (Feb. 1978).
2. Akinmusuru, J. O., "Vertical Line Groups of Horizontal Anchors in Sand," *Journal of the Geotechnical Engineering Division, ASCE,* Vol. 104, No. GTB, Technical Note, pp. 1127–1130 (Aug. 1978).
3. Avgherinos, P. J. and A. N. Schofield, "Drawdown Failures of Centrifugal Models," *Proc. 7th Int. Conf. on Soil Mech. and Found. Eng.*, Vol. 2, Mexico, pp. 497–505 (1969).

Biarez, I., L-M Boucraut and R. Negre, "Limiting Equilibrium of Vertical Barriers Subjected to Translation and Rotation Forces," *Proc. of the 6th Int. Conf. on Soil Mech. and Foundation Engs.*, Vol. II, Montreal, Canada, pp. 368–372 (Sept. 1965).

5. Buchholz, W., "Erdwiderstand auf Ankerplatten," *Jahrbuch Hafenbautechnischen Gesellschaft,* Vol. 12, pp. 300–327 (1930).
6. Das, B. M., "Pullout Resistance of Vertical Anchors." *Journal of Geotechnical Engineering Division, Proc. ASCE,* Vol. 101, No. GT1, pp. 87–91 (Jan. 1975).
7. Das, B. M. and G. R. Seeley, "Load-Displacement Relationship for Vertical Anchor Plates," *Journal of Geotechnical Engineering Division, ASCE,* Vol. 101, No. GT7, pp. 711–715 (July 1975).
8. Das, B. M., G. R. Seeley, and S. C. Das, "Ultimate Resistance of Deep Vertical Anchors in Sand." *Soils and Foundations,* Japanese Engineering Society, Vol. 17, No. 2, pp. 52–56 (June 1977).
9. Das, B. M., R. Moreno and K. F. Dallo, "Ultimate Pullout Capacity of Shallow Vertical Anchors in Clay," *Soils and Foundations,* Japanese Engineering Society, Vol. 25, No. 2, pp. 148–152 (June 1985).

De Beer, E. E., "The Scale Effect in the Transposition of the Results of Deep Sounding Tests on the Ultimate Bearing Capacity of Piles and Caisson Foundations," *Geotechnique,* Vol. 13, No. 1, London, England, pp. 39–75 (1963).

11. Dickin, E. A., "Properties of Erith Sand at Low Stress Levels," Unpublished Internal Report, Department of Civil Engineering, University of Liverpool (1980).
12. Dickin, E. A., "The Effect of Size, Shape and Embedment on the Ultimate Resistance of Vertical Anchors in Dense Sand," *Eurochem Colloquium 134: Design Against Failure in Soils,* Technical University of Denmark, Copenhagen (Sept. 1980).
13. Dickin, E. A. and C. F. Leung, "Discussion on Horizontally Loaded Vertical Plate Anchors in Sand". *Journal of Geotechnical Engineering Division, Proc., ASCE,* Vol. 105, No. GT3, pp. 442–443 (March 1979).
14. Dickin, E. A. and C. F. Leung, "Centrifugal Model Tests on Vertical Anchor Plates," *Journal of Geotechnical Engineering Division, ASCE,* Vol. 109, No. GT12, Paper 18435, pp. 1503–25 (Dec. 1983).

15. Dickin, E. A. and C. F. Leung, "Evaluation of Design Methods for Vertical Anchor Plates," *Journal of Geotechnical Engineering, Proc., ASCE,* Vol. 111, No. 4, pp. 500–520 (April 1985).
16. "Earth Retaining Structures," *Civil Engineering Code of Practice,* No. 2, Insulation of Structural Engineers, London, England (1951).
17. Hansen, J. B., *Earth Pressure Calculations,* Danish Technical Press, Copenhagen, Denmark (1953).
18. Hansen, J. B., "A General Formula for Bearing Capacity", *Bulletin No. 11,* Danish Geotechnical Institute, Copenhagen, Denmark (1961).
19. Hueckel, S., "Model Tests on Anchoring Capacity of Vertical Inclined Plates," *Proc. 4th Int. Conf. on Soil Mech. and Found. Eng.,* Vol. 2, London, pp. 203–206 (1957).
20. Hueckel, S., J. Kwasniewski and L. Baran, "Distribution of Passive Earth Pressure on the Surface of a Square Vertical Plate Embedded in Soil," *Proc. 6th Int. Conf. on Soil Mech. and Found. Eng.,* Vol. 2, Montreal, pp. 381–385 (1965).
21. Hoshiya, M. and J. N. Mandal, "Some Studies on Anchor Plates in Sand," *Soils and Foundations, Japanese Society of Soil Mech. and Fdn., Eng.,* Vol. 24, No. 1, pp. 9–16 (Mar. 1984).
22. Kostyukov, V. D., "Distribution of the Density of Sand in the Sliding Wedge in front of Anchor Plates," *Soil Mech. and Found. Eng.,* No. 1, pp.12–13 (Jan.–Feb. 1967).
23. Leung, C. F., "The Effect of Shape, Size and Embedment on the Load-Displacement Behaviour of Vertical Anchors in Sand," Thesis presented to the University of Liverpool, at Liverpool, England, in 1981, in partial fulfillment of the requirements for the degree of Doctor of Philosophy.
24. Leung, C. F. and E. A. Dickin, "Scale Error of Conventional Model Tests," *Proc. of Int. Symposium on Geotechnical Centrifuge Model Testing,* Tokyo, pp. 133–138 (April 1984).
25. Leung, C. F. and E. A. Dickin, Discussion on "Some Studies on Anchor Plates in Sand", *Soils and Foundations, Japanese Society of Soil Mech. and Found. Eng.,* Vol. 25, No. 1, pp. 109–110 (Mar. 1985).
26. Mackenzie, T. R., "Strength of Deadman Anchors in Clay," Thesis presented to the University of Princeton, Princeton, N.J., U.S.A., in 1955, in partial fulfillment of the requirement for the degree of Master of Science.
27. Meyerhof, G. G., "The Ultimate Bearing Capacity of Foundations," *Geotechnique,* Vol. 2, No. 4, London, England, pp. 301–332 (1951).
28. Meyerhof, G. G. and J. I. Adams, "The Ultimate Uplift Capacity of Foundations," *Canadian Geotechnical Journal,* Vol. 5, No. 4, Ottawa, Canada, pp. 225–244 (1968).
29. Meyerhof, G. G., "Uplift Resistance of Inclined Anchors and Piles," *Proc. 8th Int. Conf. on Soil Mech. and Found. Eng.,* Vol. 2.1, Moscow, U.S.S.R., pp. 167–172 (1973).
30. Neely, W. J., "The Ultimate Resistance of Anchor Plates in Sand," Thesis presented to Queen's University of Belfast, in 1971, in partial fulfillment of the requirements for the Degree of Doctor of Philosophy.
31. Neely, W. J., "The Importance of Flexibility in the Design of Sheet Pile Anchors in Sand," *Ground Engineering,* Vol. 5, No. 3, London, England, pp. 14–16 (1972).
32. Neely, W. J., "Effects of Interference on the Behaviour of Groups of Anchor Plates in Sand," *Civil Eng. and Public Works Review,* Vol. 67, No. 788, London, pp. 271–273 (1972).
33. Neely, W. J., J. G. Stuart, and J. Graham, "Failure Loads of Vertical Anchor Plates in Sand," *Journal of Soil Mechanics and Foundations Division, ASCE,* Vol. 99, No. SM9, Paper 9980, pp. 669–685 (Sept. 1973).
34. Ovesen, N. K., "Anchor Slabs, Calculation Methods and Model Tests," Bulletin No. 16, Danish Geotechnical Institute, Copenhagen, Denmark (1964).
35. Ovesen, N. K. and H. Stromann, "Design Method for Vertical Anchor Slabs in Sand," *Proc. of the Speciality Conf. on Performance of Earth and Earth-Supported Structures,* Vol. 1–2, U.S.A., pp. 1418–1500 (1972).
36. Ovesen, N. K., "The Use of Physical Models in Design: the Scaling Law Relationship," *Proc. 7th European Conf. Soil Mech. Found. Eng.,* Vol. 4, Brighton, England, pp. 318–323 (1979).
37. Ovesen, N. K., "Centrifuge Tests to Determine the Uplift Capacity of Anchor Slabs in Sand," *Proceedings of 10th Int. Conf. on Soil Mech. and Found. Eng.,* Stockholm (1981).
38. Ponce, V. M. and J. M. Bell, "Shear Strength of Sand at Extremely Low Pressures," *Journal of Soil Mechanics and Foundations Division, ASCE,* Vol. 97, No. SM4, pp. 625–638 (April 1971).
39. Ranjan, G., Discussion on "Failure Loads of Vertical Anchor Plates in Sand," *Journal of Geotechnical Engineering Division, Proc. ASCE,* Vol. 100, No. GT8, pp. 954–956 (1974).
40. Ranjan, G. and Y. P. Kaushal, "Load-Deformation Characteristics of Model Anchors under Horizontal Pull in Sand," *Geotechnical Engineering,* Vol. 8, Bangkok, pp. 65–78 (1977).
41. Ranjan, G. and Y. P. Kaushal, "Behaviour of Vertical Anchors in Sand," *Proceedings of 5th Asian Regional Conference on Soil Engineering,* Vol. 1, Bangalore, India, pp. 149–152 (1975).
42. Reese, L. C., W. R. Cox, and F. D. Koop, "Analysis of Laterally Loaded Piles in Sand," Preprints, Sixth Annual Offshore Technology Conference, Vol. 2, Houston, Texas, U.S.A., OTC 2080, pp. 473–480 (1974).
43. Rowe, P. W., "Anchored Sheet Pile Walls," *Proc. the Institution of Civil Engineers,* Vol. 1, London, pp. 27–70 (1952).
44. Rowe, P. W., "The Relationship between the Shear Strength of Sands in Triaxial Compression, Plane Strain and Direct Shear," *Geotechnique,* Vol. 19, No. 1, London, England, pp. 75–86 (1969).
45. Rowe, P. W., "Progressive Failure and Strength of a Sand Mass," *Proc. of the 7th Int. Conf. on Soil Mech. and Found. Eng.,* Vol. 1, Mexico, pp. 341–349 (1969).
46. Rowe, P. W. and K. Peaker, "Passive Earth Pressure Measurements," *Geotechnique,* London, Vol. 15, No. 1, pp. 57–78 (1965).
47. Rowe, R. K., J. R. Booker, and N. P. Balaam, "Application of

the Initial Stress Method to Soil-Structure Interaction," *International Journal of Numerical Methods in Engineering,* Vol. 12, No. 5, pp. 873–880 (1978).

48. Rowe, R. K. and H. Davis, "The Behaviour of Anchor Plates in Sand," *Geotechnique,* Vol. 32, No. 1, London, England, pp. 25–41 (1982).
49. Smith, J.E., "Deadman Anchorages in Sand," Technical Report, R199, U.S. Naval Civil Engineering Laboratory (1962).
50. Tang, Y. K., "Stress-Strain Modelling of Dense Sand in Triaxial and Plane Strain Compression," Thesis presented to the University of Liverpool, at Liverpool, England, in 1979, in partial fulfillment of the requirements for the degree of Master of Engineering.
51. Teng, W. C., "Foundation Design," Prentice Hall, Englewood Cliffs, New Jersey (1962).
52. Terzaghi, K., "Theoretical Soil Mechanics," John Wiley and Sons, New York (1943).

CHAPTER 26

Adverse Wind Loads on Low Buildings

THEODORE STATHOPOULOS*

INTRODUCTION

Current standards and Codes of Practice give no guide to the designer to assess the wind loads on buildings in the presence of a nearby structure. It is well known, however, that the wind pressure distribution on a building may change drastically when a new building (or buildings) is built in its neighborhood. Obviously this is a complex problem even for a single additional building, since there is a large number of parameters, including the shape and size of the two buildings, their relative positions and the wind direction. A literature review indicates that little information is available on so critical a subject. In fact, the bulk of information used for code formulation of the wind loads on buildings is based on model tests of free-standing structures.

There are various reasons which appear to justify the adoption of this simple testing procedure for routine investigations. First, it is not possible to anticipate all the conditions of environment to which a particular building may eventually become subjected, and, secondly, according to a widely held notion, the wind loading of a building is expected to be generally less severe if it is surrounded by other structures than when it is free-standing. It is this last reason which becomes arguable in some cases, since several studies on wind loads on tall buildings and other structures [3,5,9–11,13–15] carried out either in uniform or in simulated atmospheric flow have shown quite adverse effects. These effects are likely where essentially only two structures interact and they can be encountered depending on the relative placement of the structures in the approaching wind. Introducing variations in building geometries may reduce these adverse effects significantly.

The building interaction effect has also been studied for wind loads on low buildings. Hussain and Lee [7] have found that mean drag coefficients of low buildings are lower when these are part of a large group of similar buildings. Holmes and Best [6], however, measured some adverse effects on mean pressure coefficients of grouped houses. These studies deal with the interaction effect for buildings of equal height. The buffeting effect on low buildings due to taller structures has not been considered until recently. Isyumov and Davenport [8] have reported that wind induced pressures on low buildings located at the base of the CN Tower in Toronto were significantly influenced by the presence of the tower. Peak suctions were found to be much higher than values usually associated with low buildings in an urban environment. Based on these findings a more detailed experimental study of wind loads on low buildings in the presence of a tall nearby building has been carried out in the Boundary Layer Wind Tunnel Laboratory of the University of Western Ontario as part of a major project to examine the wind loads on low buildings [4].

The purpose of this chapter is to describe the results of the experimental measurements and to report the interesting conclusions to which the analysis of the data has led in respect to the adverse effects which may be expected on the wind loading of low buildings due to buffeting induced by a taller nearby building.

EXPERIMENTAL

The Boundary Layer Wind Tunnel at the University of Western Ontario has a working section of about 80 ft (24 m) long, 8 ft (2.5 m) wide and about 7 ft (2 m) high. Most of this fetch is required for the natural production of a boundary layer which grows in a manner paralleling the atmospheric process under neutral conditions. The sur-

*Centre for Building Studies, Concordia University, Montreal, Quebec, Canada

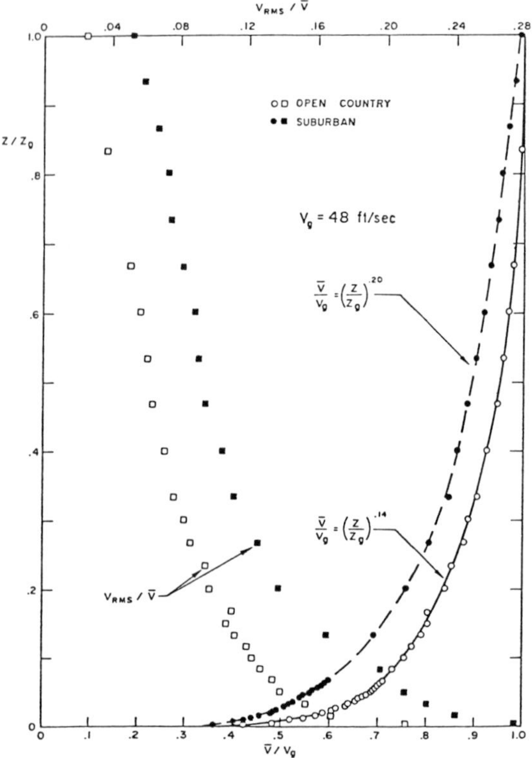

FIGURE 1. Mean wind speed and turbulence intensity profiles for the two exposures tested (Note: 1 ft = 0.305 m).

faces in the wind tunnel can be changed to represent different terrains. Boundary layer depths ranging from 2 ft–4 ft (0.6 m–1.2 m) are obtained at the test section with different surface roughness. In relation to the atmospheric boundary layer this imples that geometric scales of between 1:400 and 1:500 are most appropriate for studies of wind effects on buildings and structures; however, in the case of low buildings, models at this scale posed considerable dificulties because of their small size. Instead, most of the tests were carried out at a somewhat relaxed scale of 1:250, after an intensive set of experiments had been completed to verify that such a relaxation of scale was valid.

Two terrain models have been used. The "smooth" terrain produces a mean speed profile with height appropriate to open country conditions, whereas the "built-up" terrain corresponds to suburban conditions. Figure 1 shows typical mean speed and longitudinal turbulence intensity profiles measured at the site without the model in position for the two terrains. The mean profiles have been fitted by a convenient power-law expression. The open country exposure provides higher mean speed but less turbulence at eave height compared to the suburban exposure. These profiles are representative of exposures A and B of the National Building Code of Canada or C and B of the American National Standards Institute (ANSI) standard.

The pressure measuring system used responds to pressure fluctuations on the model of up to about 100 Hz with negligible attenuation or distortion. Several pressure measurements can be made in parallel, each of which is sampled at a rate of about 1,000 times per second. Typically, sampling is continued for a period of up to about 30 seconds in wind tunnel time, during which the computer records, for each input, the maximum and minimum values that occur, and computes the mean and root-mean-square values. The reference dynamic pressure, measured in the free stream above the boundary layer, is monitored similarly. At the end of the sampling period the measured pressures, consisting of the maximum, minimum, mean and root-mean-square pressure for each channel, are converted to pressure coeficients by dividing each by the reference dynamic pressure. Results, however, presented herein are in pressure coeficient form referenced to the mean dynamic velocity pressure at eave height. This height has been chosen because it was found that for low buildings this selection provides value not sensitive to building height. The relationship of wind speeds at eave height and gradient height can be found from the velocity profiles of Figure 1.

The maximum free stream speed, and that at which the measurements were carried out, is approx. 48 ft/sec (14.6 m/s). Similarity considerations based on the 1:500 geometric scale lead to the determination of velocity and time scales of the order of 3:10 and 1:150, respectively. Consequently, the experimental sampling rate corresponds to about seven samples per second per channel in full scale, and pressure fluctuations with frequencies up to about 0.7 Hz in full scale can be detected without distortion. Thus, peak pressures on the models correspond roughly to peaks with a 1-sec. to 2-sec. averaging time in full scale, which is probably sufficient for the purpose of cladding design. It has been shown, however [12], that if frequencies up to 2 Hz are significant, an underestimate of the measured peak values in the order of 10% to 20% may occur. Higher frequencies can hardly be significant because the associated peak pressures would be fully coherent over a very short distance. Furthermore, the use of a 30-sec. wind tunnel sample provides a statistically stable estimate of mean and root-mean-square pressures and a conservative assessment of peak values associated with a full-scale period of approximately 1 hour. This arises because the wind tunnel does not represent fluctuations in wind speed with periods of greater than a few minutes.

The basic models used for the investigation of the buffeting effect represented a 1:12 and a 12:12 roof-sloped building, 80 ft (24 m) wide and 125 ft (38 m) long. Three

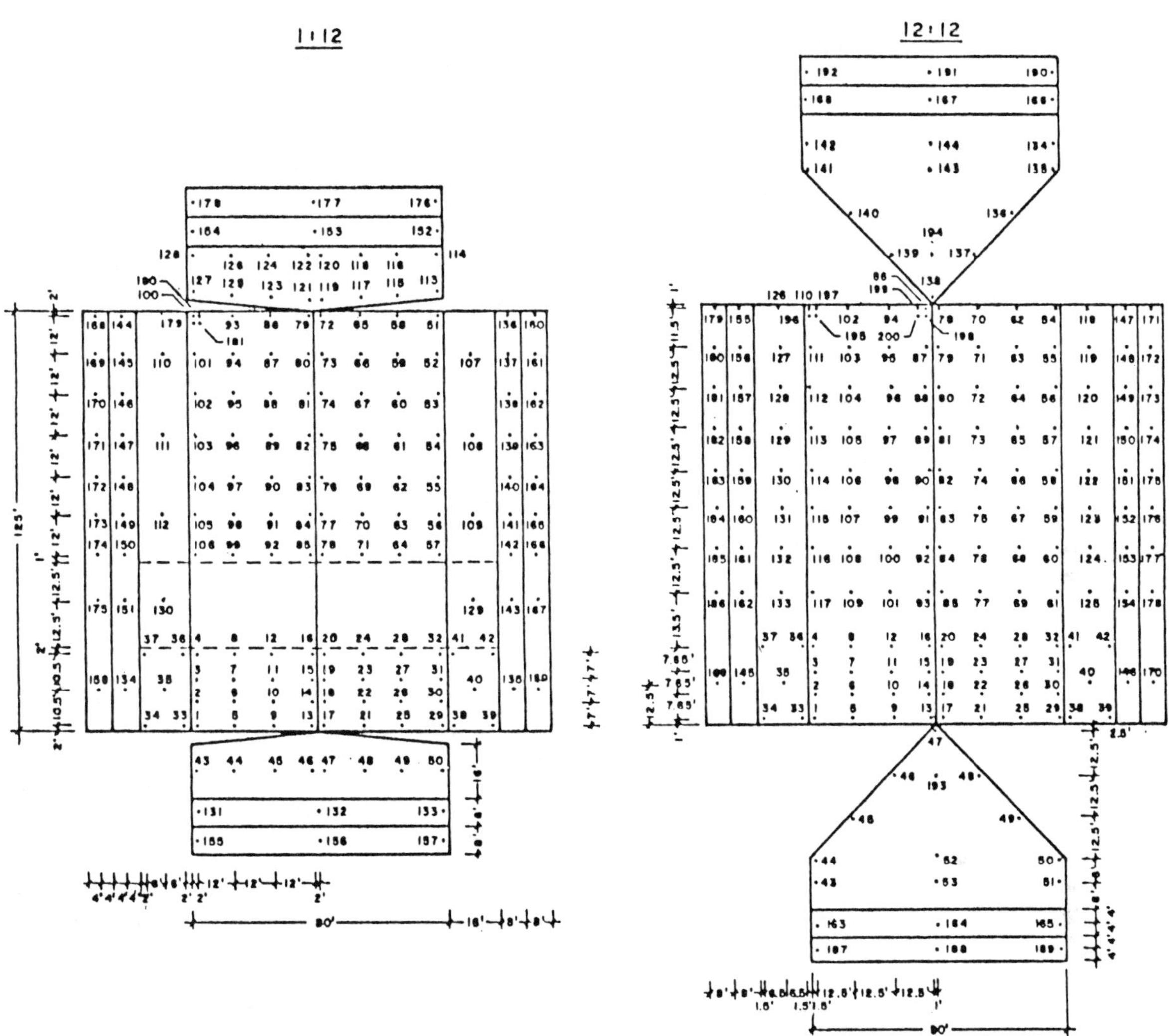

FIGURE 2. Diagrammatic view of low building models and pressure tap locations (Note: 1 ft = 0.305 m).

different eave heights were tested, representing 16 ft, 24 ft and 32 ft (5 m, 7.5 m and 10 m) in full scale. Figure 2 shows the exact model dimensions and the pressure tap locations. The dense grid of pressure taps (181 for the 1:12 model and 200 for the 12:12 model) ensures adequate collection of information regarding the pressure field influencing the low-building envelope. In the present study, local pressure coeficients measured for an isolated low building were compared with values measured under similar conditions with a single nearby structure located in various positions around the low building. As the same instrumentation system was used in the measurements with and without the nearby structure there should be no error in the ratios of peak pressures determined provided that the pressure signals have similar frequency content. If the nearby structure produces pressure fluctuations of higher frequency content the adverse effects measured for the low building may still be slightly higher in reality than those measured in the wind tunnel.

The nearby structure was chosen to be a 250 ft (76 m) high building 125 ft by 125 ft (38 m by 38 m) in plain view. Three different building configurations were examined, as Figure 3 shows diagrammatically. Most of the tests were carried out under simulated open country conditions. Only one wind direction has been considered but the relative locations of the nearby building have been selected in such a manner so that different wind effects (adverse or sheltering) on the low building are realized. Furthermore, the distance between low and interacting buildings was altered for Case B to study the variation of the adverse ef-

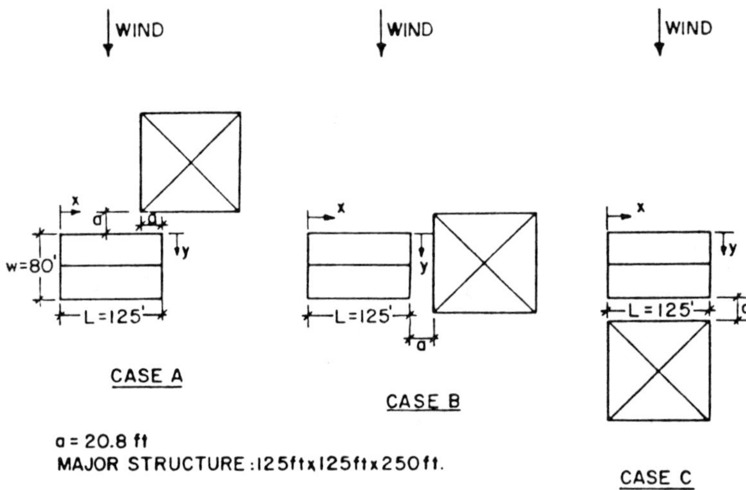

FIGURE 3. Building configurations examined (Note: 1 ft = 0.305 m).

fects for different separation distances. The latter study was carried out under simulated suburban environmental conditions only.

WIND LOADS AND DISCUSSION

The wind flow around buildings and the general modification of this flow due to nearby structures has been described in the literature [1,2,13]. For the particular building configurations considered (Figure 3) the following can be expected: In Case A the low building will be affected by the separated and accelerated flow passing around the major structure. In particular, the corner of the low building nearest to this structure will be subjected to the vortices of the wake as, at the lower level, these will squeeze to pass through the narrow gap between the buildings. Consequently, higher peak pressures are expected at the end bay roof and wall areas of the low building. In Case B the low building will be interacting with the separated and downward flow from the tall building. The size of the two buildings and their relative location will determine whether this interaction occurs for the entire envelope of the low building or part thereof. It is expected that both peak and mean pressures acting on the low building will increase but it is difficult to estimate the magnitude of this increase. Finally, in Case C it is expected that the flow around the two buildings will be governed by the geometry of the tall building. Therefore the low building will experience suctions generally lower than would develop without the nearby structure (a sheltering effect). Positive pressures, however, may be higher due to the impact of the large standing vortex on the low building envelope.

The previous assumptions are confirmed by the experimental results which also clarify further these effects. Figure 4 shows the instantaneous peak suction coefficients for the windward part of the roof of the low building. Results are presented for Cases A, B and C, as described in Figure 3, for the 32 ft (10 m) high building together with the values measured for the isolated building for comparison. The latter have been selected as the most critical (worst) values measured for the same azimuth but from all three eave height buildings to represent code or standard values. Data points are denoted by the corresponding building interaction case. Data show a significant increase, primarily for Case B and secondarily for Case A and a significant decrease (sheltering effect) for Case C. Note the high values of peak suction coefficients on the windward roof edge for Case B, particularly near the tall buildings. These values are much higher than any standard or code of practice would suggest for a low building. It should be reiterated here, however, that these pressure coefficients are referenced to the mean velocity pressure. If the fastest mile speed is to be used as reference (ANSI approach) these coefficients should be divided by a factor approx. equal to 1.64, as explained in Reference [16]. Clearly the buffeting effect deteriorates with the distance from the nearby structure; it takes approximately one tall building width for the roof suctions of Case B to reach the same or slightly higher level than that of the isolated low building.

Case A also shows an increase of roof suctions, particularly in the middle zone of the windward half of the gabled roof. The area of the roof right behind the tall building is affected more significantly. Note that interaction Cases A and B do not cause different suctions close to the ridge of the roof.

Case C decreases the level of the roof suctions significantly. This is due to the positive pressures on the low building roof induced by the large standing vortex, developed upwind of the tall building, as previously explained. Resultant peak roof suctions at the ridge area reach the zero level. This has also been observed by Bailey and Vincent [1] for mean pressures. Leeward roof suctions are very similar to those acting on the ridge areas for all building interaction cases.

Figure 5, 6 and 7 summarize the most significant data trends for the three Cases A, B and C. The low building is the same considered in Figure 4. Data for different locations are presented as ratios of the C_p values measured under conditions of buffeting over those measured for the isolated low building. Amplification ratios lower than one imply a sheltering effect for the low building, whereas negative amplification ratios indicate a change of sign on the acting wind pressure. Figure 5 shows results for instantaneous peak and mean pressures acting on the windward roof verge of the low building. For Building Interaction Case B high amplification factors of the order of 2.0 and 2.5 appear for the mean and peak suctions, respectively. In general, the amplification for peaks is higher than for means. In Interaction Case C the peak suction values are only at the 20 to 40% level of that measured for the isolated low building. The mean suctions, however, appear as pressures due to the buffeting of the tall building and the modification of the wind flow passing over the low building.

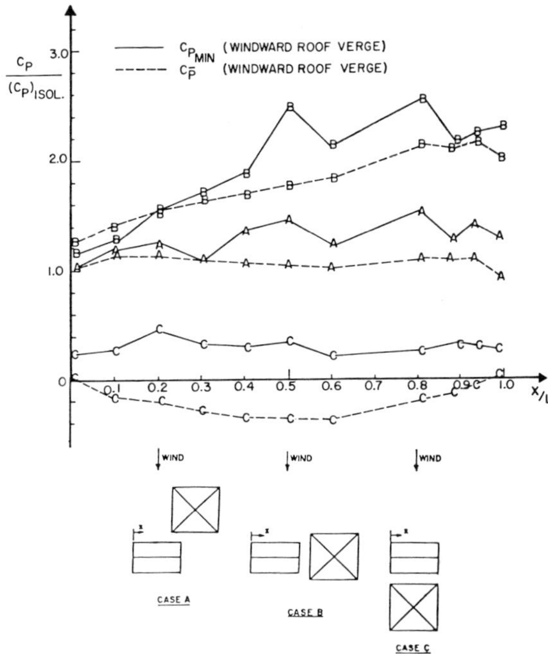

FIGURE 5. Building interaction effect on mean and peak pressure coefficients on the roof verge of a low building (open country exposure).

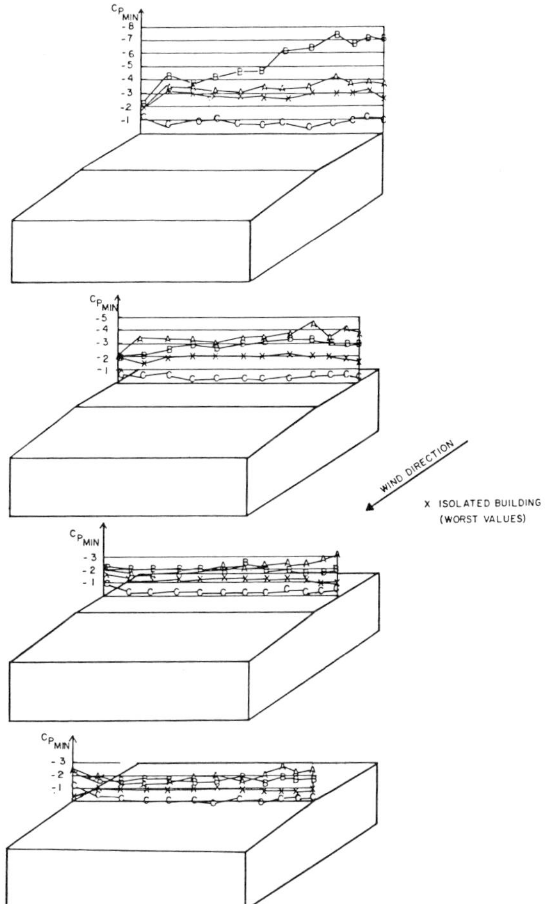

FIGURE 4. Instantaneous peak pressure coefficients for an isolated low building and under conditions of buffeting from a tall nearby building (open country exposure).

Data for peak instantaneous suctions acting on the low building gable wall and the gable roof verge nearest to the tall building are shown in Figure 6. Buffeting effects from Case A seem to be particularly critical for the leeward zones of these areas, in which suction coefficients are four times higher than those experienced by the isolated building. Case B also doubles the suction coefficients of the isolated building for these areas whereas Case C generally reduces the suctions to approximately 50% of their values in all areas except the downwind end of the gable wall which experiences amplification factors of the order of 1.5.

Figure 7 shows typical results for positive peak, negative peak and mean values of pressure coefficients acting on windward and leeward walls of the low building. Both Cases A and B show amplification factors ranging be-

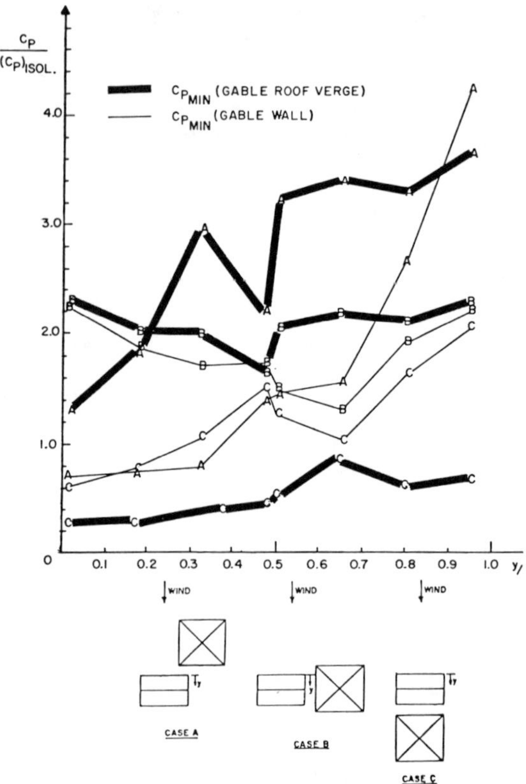

FIGURE 6. Building interaction effect on peak pressure coefficients acting on the gable wall and roof verge of a low building (open country exposure).

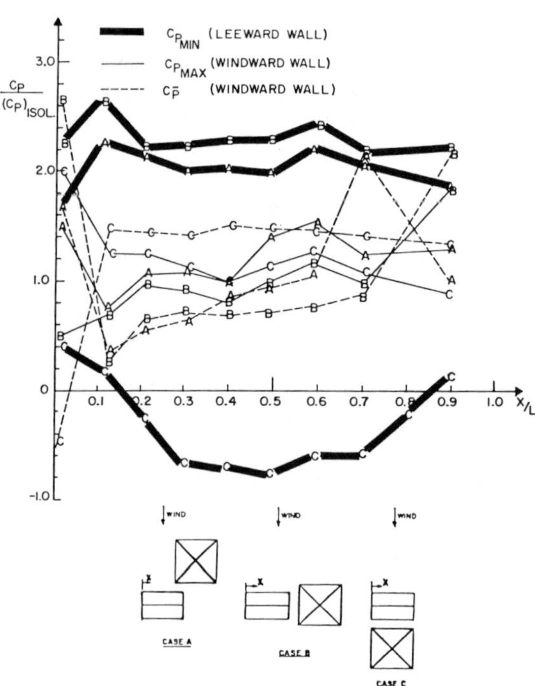

FIGURE 7. Building interaction effect on mean and peak pressure coefficients acting on the walls of a low building (open country exposure).

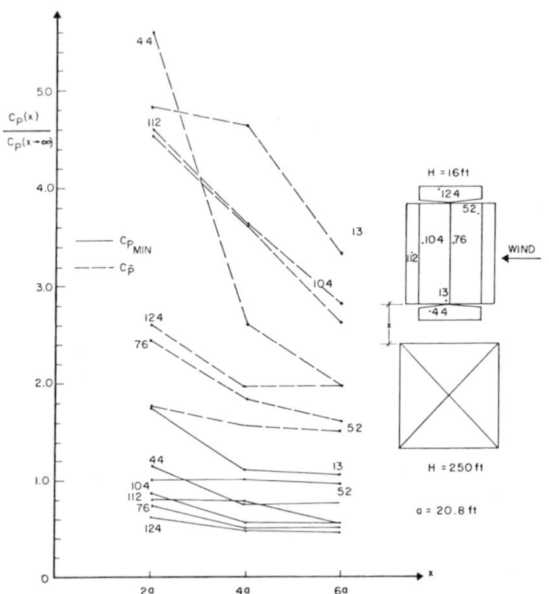

FIGURE 8. Proximity effect on mean and peak local pressure coefficients of a low building (suburban exposure).

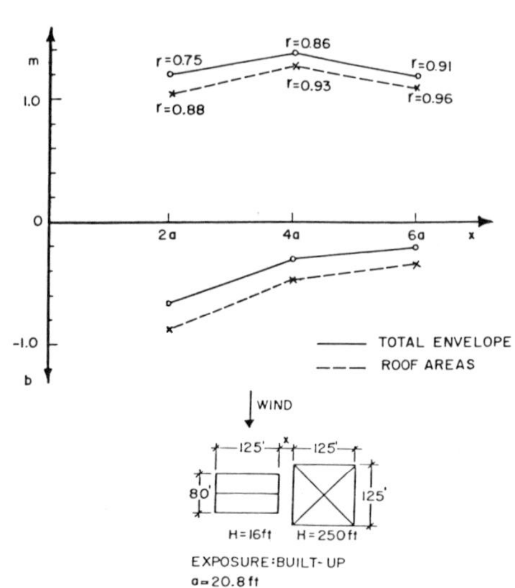

FIGURE 9. Least square fitting parameters for mean pressure coefficients acting on a low building under proximity conditions.

tween 2.0 and 2.5 for the instantaneous peak suctions of the leeward wall in comparison to the isolated building values. In contrast, results for the windward wall (positive peak and mean pressures) do not show significant adverse effects with the exception of the areas near the corners where the proximity of the tall building locally influences the flow and, consequently, the pressure distribution. For Case C the suctions of the isolated building on the leeward wall change sign and become pressures whereas the pressures of the windward wall are not affected appreciably (the mean values somewhat increase). The sudden drop of the amplification factor of the mean pressures at one end of the windward wall is not of particular concern because the pressure coefficient values for this point are very low for both the isolated building and Interaction Case C.

Data presented so far indicate the general characteristics of the pressure distributions measured for low buildings under conditions of interaction and buffeting. More data, in raw form, are available in Reference [4] including pressure coefficients measured for the low building with roof slope of 12:12. The trends of the data are generally similar to those described for the 1:12 roof-sloped building. It appears that the position of the major nearby structure is the dominant parameter for the loading pattern of the low building.

From the configurations examined in the open country exposure, Case B (i.e., the major structure beside the low building) produces the largest overall adverse effects on the low-building loading. However, it was expected that these effects would be a function of the distance between the the two buildings. Since this distance was taken as $a = 20.8$ ft (6.3 m), which may be considered rather small for real conditions, consecutive tests were run with the same configuration but varying this distance, x, in multiples of α, i.e., 2α, 4α and 6α, viz. 41.6, 83.3 and 125.0 ft (12.7, 25.4 and 38.1 m). A 16 ft (5 m) high building in a simulated suburban (built-up) exposure was considered for this series of tests. The rougher terrain was selected for the tests to examine possible exposure effects in the interaction.

Typical results are shown in Figure 8 for instantaneous peak and mean pressure coefficients measured at seven different pressure taps (for exact location see Figure 2). The data are presented for the three distances between low and tall building considered. The general reduction character of the ratio $C_p(x)/C_p(x \rightarrow \infty)$ is obvious but the reduction gradient is drastically different for different locations on the building envelope. Peak suctions do not appear affected significantly by the proximity of the tall building; in fact, lower values with than without interaction have been measured at several locations but the increased variability of the peak data inherent in the different terrain set-ups representing suburban exposures should also be considered. In contrast to the peaks, mean suctions appear affected dramatically by buffeting. At a distance of 2α, corresponding to "across the street" reality, the mean wind suctions are 2 to 5 times higher than without buffeting conditions. There is no clear explanation, at present, about the different behavior of mean and peak suctions on low building envelope under buffeting.

A more detailed comparison of pressure coefficients acting on low buildings under buffeting has been attempted by linear regression, i.e., by applying the equation

$$C_p(x) = m\, C_p(x \rightarrow \infty) + b \tag{1}$$

in which m is the slope and b is the intercept of the regression line. When m approches unity and b approaches zero the regression line shows no statistical difference for the two sets of data, i.e., the isolated and the interacted low building cases. Equation (1) is meaningful if the correlation coefficient, r, for the two sets of data is high enough—say higher than 0.75. For the particular cases studied this condition was met only for mean values—the peaks were more scattered. Results appear in Figure 9 for the entire building envelope and for the roof areas separately. The correlation coefficients are generally higher for the roof pressures than for the entire envelope. The diagram of Figure 9 provides values for the parameters m and b of Equation (1), which can be used for an approximate prediction of the expected value of mean pressure coefficients on a low building under conditions of buffeting.

In attempting the regression line fit of Equation (1) it may be assumed that the m parameter arises from the general wind speed up occurring in the vicinity of the low building and the b parameter constitutes a bias in the surrounding static pressure, again caused by the flow field of the major structure. In this perspective, the value of b may not be as significant from an overall loading point of view since it would be reflected in the low building's internal pressure. Thus, this "b" then constitutes an apparent exaggeration in the results.

Force coefficients for roof and wall areas were also calculated from the local pressure coefficients measured. Only mean values could be obtained by the equation

$$C_{\bar{F}} = \frac{1}{A} \sum_{i=1}^{n} C_{\bar{pi}} A_i \tag{2}$$

in which

$C_{\bar{pi}}$ = mean pressure coefficient at tap i
A_i = tributary area of tap i
$A = \Sigma_{i=1}^{n} A_i$ = total area of zone considered
n = total number of taps located in area A

Typical results of mean force coefficients are presented in Table 1 and Figure 10 as ratios of values with and without interaction buffeting effects. In particular, Table 1

TABLE 1. Ratios of Mean Force Coefficients With and Without Buffeting for All Interaction Cases—Roof Slope 1:12, Eave Height = 32 ft, Open Country Exposure.

Interaction Case (see Fig. 3)	$C_{\bar{F}}/(C_{\bar{F}})_{ISOL}$ Entire Roof	End Bay Roof Section
A	1.62	1.68
B	2.30	2.60
C	-1.02	-0.80

gives the ratio $C_{\bar{F}}/(C_{\bar{F}})_{ISOL}$ for the entire roof and for the end bay zone of the roof, i.e., 25 ft (7.6 m) by 80 ft (24.4 m), for Building Interaction Cases A, B and C as discussed previously. The end bay zone considered is the closest to the interacting tall building. It is interesting to see the little difference between ratios measured for the entire roof and the end bay roof section. Worst adverse effects are, again, induced by Case B whereas Case C reverses the sign of roof force from suction to downward pressure.

Figure 10 shows similar results for the proximity cases investigated in built-up (suburban) terrain. The data presented clearly indicate higher amplification factors for roof forces than for wall forces under proximity conditions. Once again, no significant difference appears between amplification factors of forces acting on the entire roof (or wall areas) and the end bay section of the low building.

IMPLICATION FOR STANDARDS AND CODES OF PRACTICE

From the results presented in the previous section it appears that a major nearby structure may cause significant adverse effects on the loading of a low building when the major structure is significantly different from its surroundings. Although not examined in the present study, it would be expected that increasing the number of nearby structures of significant size would be less serious and would, in the limit of a built-up city, lead to net shielding effects.

To obtain a feeling of the possible underestimations or overestimations which may occur to the design pressure coefficients specified for various zones of the envelope of low buildings some comparisons have been carried out by considering the zones adopted by the National Building Code of Canada (1980) and the ANSI Standard (1982). These zones are indicated in Figure 11. The use of instantaneous values of peak pressure coefficients for these comparisons may be arguable because of the high variability of such values. It is possible however, to use for each zone of the building envelope a weighted average value, $\overline{C_{PMIN}}$, defined as:

$$\overline{C_{PMIN}} = \frac{1}{A} \sum_{i=1}^{n} (C_{PMIN})_i A_i \tag{3}$$

This averaging process tends to reduce the random errors

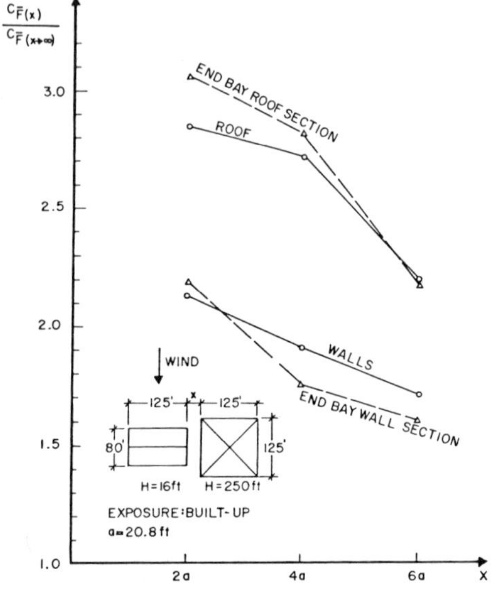

FIGURE 10. Proximity effect on mean force coefficients of a low building.

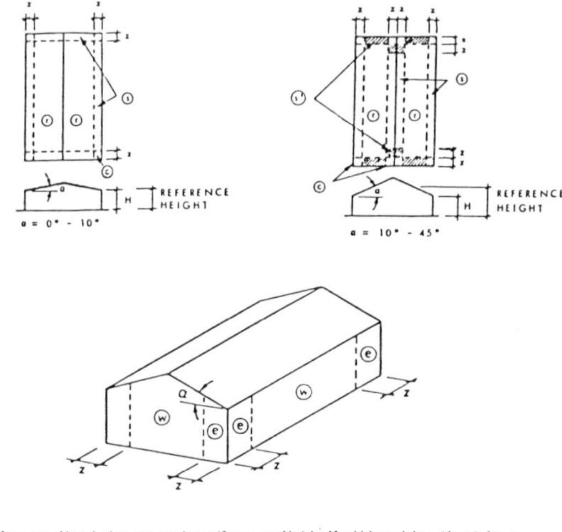

FIGURE 11. Zones of low building envelopes as determined by the National Building Code of Canada and ANSI Standard for the Evaluation of Wind Loads.

TABLE 2. Ratios of Instantaneous Peak Pressure Coefficients With and Without Buffeting for Various Building Zones (Values Averaged Over Each Zone)—Roof Slope 1:12, Eave Height = 32 ft, Open Country Exposure.

Building Zone (See Fig. 11) / Interaction Case	$\overline{C_{PMIN}} / (\overline{C_{PMIN}})_{ISOL}$ c	s	r	e	w
A	1.08	1.30	1.44	0.89	1.14
	1.08	1.58	1.78	1.50	1.75
B	1.86	1.86	1.40	1.49	1.31
	1.86	2.25	1.72	2.50	2.01
C	0.26	0.30	0.16	0.47	0.44
	0.26	0.36	0.20	0.80	0.67

Note: The first value corresponds to the pressure coefficients of the isolated building (measured for the worst azimuth) whereas the second value corresponds to pressure coefficients of the isolated building measured for the same azimuth with the building interaction case.

TABLE 3. Ratios of Instantaneous Peak Pressure Coefficients With and Without Buffeting for Various Building Zones (Values Averaged Over Each Zone)—Roof Slope 12:12, Eave Height = 32 ft, Open Country Exposure.

Building Zone (See Fig. 11) / Interaction Case	$\overline{C_{PMIN}} / (\overline{C_{PMIN}})_{ISOL}$ c	s	s′	r	e	w
A	0.63	0.71	0.66	0.62	0.77	0.75
	1.19	1.23	1.01	1.06	1.00	0.90
B	0.56	0.71	0.74	0.68	1.16	0.97
	1.06	1.22	1.14	1.16	1.50	1.16
C	0.27	0.28	0.53	0.71	0.53	0.39
	1.34	0.49	0.82	1.20	0.69	0.47

Note: The first value corresponds to the pressure coefficients of the isolated building (measured for the worst azimuth) whereas the second value corresponds to pressure coefficients of the isolated building measured for the same azimuth with the building interaction case.

associated with individual peak values and increases the reliability of these comparisons.

Table 2 presents typical comparison results for the 1:12 roof-sloped building in the open country terrain. Data are given for each building proximity configuration (see Figure 3) as ratios of the average of the peak pressure coefficients measured inside each zone over the average of instantaneous peak pressure coefficients measured in the same zone for the isolated buildings. Two values of such ratios are given. The first corresponds to the most critical pressure coefficients of the isolated building measured for the worst azimuth. This is very close to the code value. The second value corresponds to pressure coefficients of the isolated building measured for the same azimuth with the building interaction case. Therefore, the second value is always equal or higher than the first value. Consider, for example, the edge zone "s" of the roof of a low-building with a proximity tall building (Interaction Case A). The value 1.30 of Table 2 for this case indicates that the average peak suction measured in zone "s" under buffeting is 30% higher than the average peak suction measured for the isolated building for *any* azimuth (normally the code or standard value). On the other hand the value 1.58 of Table 2 for the same case implies that the average peak suction is 58% higher than the average suction measured for the isolated building for *the same* azimuth with the interaction case.

Additional comparisons for a low building of different geometry (12:12 roof slope) have been carried out and results appear in Table 3 in the same format as in Table 2. In general, lower ratio values appear in Table 3, as compared with Table 2. Note that ratios smaller than 1 in these tables imply code or standard overestimates. Finally, Table 4 indicates ratios of averaged peak pressure coefficients with and without buffeting under various proximity conditions (case B with varying x) for a 1:12 roof-sloped building, 16 ft (5 m) high in a suburban terrain. Only comparisons with values for the isolated building based on the most critical azimuth have been made. It is interesting to note, for each zone, the decrease of these ratios with the increase of the distance between the low and the interfering tall building.

The results of these comparisons can be summarized as follows:

1. Building Interaction Case B has the effect of generally increasing the mean of the weighted average of the peak values. In all proximity configurations of Case B, roof pressures are more affected than wall pressures.
2. Building Interaction Case A appears to worsen the wind loads for a 1:12 roof-sloped building and make

TABLE 4. Ratios of Instantaneous Peak Pressure Coefficients With and Without Buffeting (Values Averaged Over Each Zone)—Roof Slope 1:12, Eave Height = 16 ft, Suburban Exposure (Note: 1 ft. = 0.305 m).

Building Zone (See Fig. 11) / x (ft)	$\overline{C_{PMIN}} / (\overline{C_{PMIN}})_{ISOL}$ c	s	r	e	w
42	1.47	1.37	0.98	1.20	0.92
83	1.31	1.16	0.80	1.06	0.70
125	1.31	0.98	0.73	0.82	0.60

them lower for a 12:12 roof-sloped building (in comparison to the isolated building loads). The buffeting effect on both roof and walls is similar.

3. Building Interaction Case C generally makes wind loads lower—in some cases much lower—and obviously this is not a case of concern.

CONCLUDING REMARKS

This chapter has presented results of a wind tunnel study on the evaluation of wind loads acting on low buildings in presence of a large nearby building. The data shown have indicated significant increases of the pressure coefficients under conditions of buffeting caused by a tall building. The reader has obtained a feeling of the magnitude of overestimation or underestimation of the design pressure coefficients suggested in codes and standards for isolated low buildings.

The complexity of the problem of evaluation of wind loads on low buildings under buffeting conditions indicates that, with the present state of the art, it would be extremely difficult to treat this nearby building situation with any degree of generality. At present, for building code purposes this problem could be treated by providing a warning of possible adverse situations and recommending to seek specialist advice and possibly special wind tunnel tests.

NOTATION

A = total area of zone considered
A_i = tributary area of pressure tap i
b = intercept of regression line
C_F = force coefficient
C_p = pressure coefficient
H = height of building
L = length of low building
m = slope of regression line
n = total number of taps located in area A
r = correlation coefficient of regression data
$\overline{V}$ = mean wind velocity
V_g = gradient wind velocity
w = width of low building
x = variable distance between low and interacting buildings and abscissa in cross wind direction
y = ordinate in along wind direction
z = height above ground
z_g = gradient height
α = constant distance between low and interacting building

REFERENCES

1. Bailey, A. and N. D. G. Vincent, "Wind Pressure on Buildings Including Effects of Adjacent Buildings," *J. of the Institution of Civil Engineers*, *20*(5367), 243–275 (1943).
2. Baines, W. D., "Effects of Velocity Distribution on Wind Loads and Flow Patterns on Buildings," *Proceed. of the Symposium on Wind Effects on Buildings and Structures*, Vol. 1, National Physical Labs., Teddington, England (1963).
3. Blessmann, J. and J. D. Riera, "Interaction Effects in Neighbouring Tall Buildings," *Proc. of the 5th Intern. Conf. on Wind Engineering*, Fort Collins, Colorado, pp. 381–395 (July 1979).
4. Davenport, A. G., D. Surry, and T. Stathopoulos, "Wind Loads on Low-Rise Buildings: Final Report of Phase III—Parts 1 and 2," *BLWT Report SS4-1978*, The University of Western Ontario, London, Canada (July 1978).
5. Hamilton, G. F., "Effect of Velocity Distribution on Wind Loads on Walls and Low Buildings," *TP 6205*, University of Toronto, Toronto, Canada (Nov. 1962).
6. Holmes, J. D. and R. J. Best, "A Wind Tunnel Study of Wind Pressures on Grouped Tropical Houses," *Wind Engineering Report 5/79*, James Cook University of North Queensland, Australia (Sept. 1979).
7. Hussain, M. and B. E. Lee, "A Wind-Tunnel Study of the Mean Pressure Forces, Acting on Large Groups of Low-Rise Buildings," *J. of Wind Engineer. and Ind. Aerodynamics*, *6*, 207–225 (1980).
8. Isyumov, N. and A. G. Davenport, "A Study of Wind Induced Exterior Pressures and Suctions on Lower Accommodation Levels of the CN Tower, Toronto," *BLWT-SS2-75*, The University of Western Ontario, London, Canada (June 1975).
9. Kelnhofer, W. J., "Influence of a Neighboring Building on Flat Roof Wind Loading," *Proceed. of the Third International Conference on Wind Effects on Buildings and Structures*, Tokyo, Japan (1971).
10. Lee, B. E. and G. R. Fowler, "The Mean Forces Acting on a Pair of Square Prisms," *Building Science, 10*, 107–110 (1975).
11. Leutheusser, H. J., "Static Wind Loading of Grouped Buildings," *Proceed. of the Third International Conference on Wind Effects on Buildings and Structures*, Tokyo, Japan (1971).
12. Marshall, R. D. and T. A. Reinhold, Discussion on "Effective Wind Loads and Flat Roofs," *J. of the Structural Division, ASCE, 108* (ST2), Proc. Paper 16834, 495–498 (Feb. 1982).
13. Peterka, J. A. and J. E. Cermak, "Adverse Wind Loading Induced by Adjacent Buildings," *Journal of the Structural Division, ASCE, 102*, (ST3), Paper 11980, 533–548 (March 1976).

14. Reinhold, T. A., H. W. Tielcman, and F. J. Maher, "Interaction of Square Prisms in Two Flow Fields," *J. of Industrial Aerodynamics, 2* (3), 223–241 (Nov. 1977).
15. Saunders, J. W. and W. H. Melbourne, "Buffeting Effects of Upstream Buildings," *Proc. of the 5th Intern. Conf. on Wind Engineering*, Fort Collins, Colorado, pp. 593–606 (July 1979).
16. Simiu, E. and R. H. Scanlan, *Wind Effects on Structures: An Introduction to Wind Engineering*, 1st ed., John Wiley & Sons, Inc., New York, NY, pp. 62–63 (1978).
17. Surry, D. and W. J. Mallais, "Adverse Local Wind Loads Induced by an Adjacent Building," *J. of Structural Engineer, ASCE*, Technical Note No. 17762, Vol. 109, No. 3, 816–820 (March 1983).

SECTION FOUR

Pavement Design

CHAPTER 27

Pavement: Structural Layer Coefficient for Flexible Pavement

K. P. GEORGE*

A flexible pavement structure, or any pavement for that matter, is a layered system designed for distributing concentrated traffic loads to the subgrade soil. It is ordinarily comprised of several layers of different composition and stiffness. As the load undergoes dispersion in a layered system, it is customary to design the layers so that materials superior quality are placed at the top with the weaker materials found successively in layers below. When faced with the task of selecting two, three, or more layers, depending upon availability of materials and environmental constraints of the region, design engineers must estimate layer equivalency factors of various materials. Equivalence factor is a measure of the relative ability of the material to function as a structural component of the pavement. Structural layer coefficient is a synonym used in conjunction with the AASHO Interim Guide for design of flexible pavements. A brief survey followed by the recommended values for the two concepts is presented in the following section.

EQUIVALENCE FACTOR

The concept of equivalence factor or layer equivalency has been used in one form or another in proportioning layer thicknesses of pavement. Using this concept, it is possible to equate the load carrying capacity of one component layer of a flexible pavement to another. A variety of criteria have been employed to arrive at layer equivalency, a summary of which follows.

Making use of the plate theory, thickness equivalency may be expressed by the following equation:

$$\frac{H_1}{H_2} = \left(\frac{E_2}{E_1}\right)^{1/3} \tag{1}$$

where H_1, H_2 = thickness of any two layers;
E_1, E_2 = corresponding moduli of the two layers.

This relationship, originally conceptualized by Odemark [1], states that the thickness equivalency is proportional to cube root of the modulus of material. Brahma and Chao [2] proposed a method of finding equivalency between gravel layer of certain thickness and an asphalt concrete mat of some thickness from their respective contributions toward the shear strength. McLeod [3] adopted the surface deflection as the criterion in defining layer equivalency. Busching et al. [4] derived what they called the relative strength coefficients by comparing the stiffness modulus of each material with the average stiffness modulus of the bituminous stabilized macadam. The Asphalt Institute arrived at equivalency factors based on the results of the AASHO Road Test and the WASHO Road Test supplemented by the theoretical studies for equal vertical pressure on the subgrade. The findings tend to recommend that one inch of high-quality asphalt concrete (AC) surfacing would be equivalent to the two to three inches of good dense-graded crushed-stone base, and that one inch of asphalt concrete base would be approximately equivalent to two inches of such crushed stone. California Department of Transportation's special base study on the AASHO Road Test indicates that the cement-treated base has an equivalency of approximately 1.65 to 1. For bituminous material, with the postulate that layer equivalency is proportional to the fifth root of cohesion of mix, equivalency factors were proposed, as

*Department of Civil Engineering, University of Mississippi, University, MS

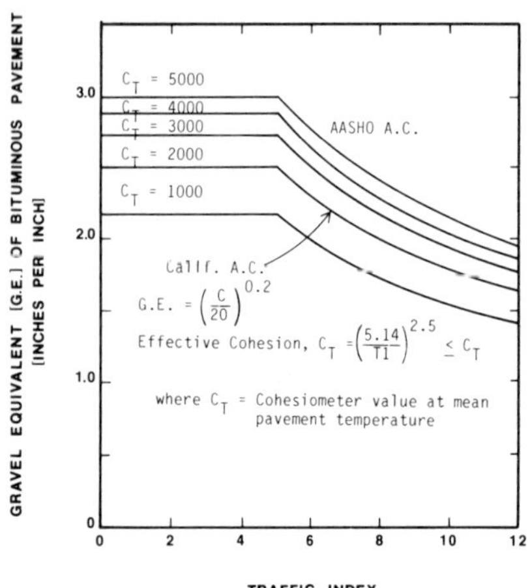

FIGURE 1. Gravel equivalent of bituminous pavement based on AASHO test road analysis—California (Source: Reference 5).

graphed in Figure 1. For comparison purposes, gravel equivalencies of AASHO AC is also graphed in the figure. In Figure 1, Traffic Index (TI) is calculated using the following equation [11].

$$TI = 9.00\left(\frac{EWL}{10^6}\right)^{0.119} \tag{2}$$

where EWL = equivalent (18,000 pounds) wheel loads.

A serious drawback of the above methods is that they neglect the fact that the pavement layers fail by repetitive application of the load. A response parameter such as elastic deflection or cohesive strength by itself cannot fully characterize the structural stability of a material. For preliminary studies, however, thickness of gravel required to equal one inch of asphalt concrete may be estimated from Figure 1.

STRUCTURAL LAYER COEFFICIENT

The concept of structural layer coefficient was first introduced in conjunction with the studies of the AASHO Road Test. In developing a relationship between the load repetitions and the pavement geometry and material characteristics, the structural number (*SN*) was conceived, which combined the effects of geometry and material properties, as expressed by the following equation:

$$SN = a_1H_1 + a_2H_2 + a_3H_3 \tag{3}$$

where

a_1, a_2, a_3 = coefficients of relative strength or structural layer coefficients
H_1 = thickness of bituminous surface course, in.
H_2 = thickness of base course, in.
H_3 = thickness of subbase, in.

Structural layer coefficients of road materials must be known when designing flexible pavements using the ASASHO Interim Guide. Interim structural layer coefficients for four different base materials were derived in the Road Test and coefficients for other base and subbase materials were estimated from the results of the special base study. Those values, listed in Table 1, are recommended by the AASHO Interim Guide [5] for flexible pavement design.

These interim values established at the AASHO Road Test are not necessarily applicable to all paving materials and environments. Accordingly, several studies have been undertaken to revise these values, and what is presented below is a brief outline of the methods and a summary of

TABLE 1. Structural Layer Coefficients Proposed by AASHO Committee on Design (Source: Reference 5).

Pavement Component	Coefficient[4]
Surface Course	
Roadmix (low stability)	0.20
Plantmix (high stability)	0.44[1]
Sand Asphalt	0.40
Base Course	
Sandy Gravel	0.07[2]
Crushed Stone	0.14[1]
Cement-Treated (no soil-cement)	
Compression strength at 7 days	
650 psi or more[2]	0.23[3]
400 psi or 650 psi	0.20
400 psi or less	0.15
Bituminous-Treated	
Coarse-graded	0.34[3]
Sand asphalt	0.30
Lime-Treated	0.15–0.30
Subbase Course	
Sandy Gravel	0.11[1]
Sandy or Sandy-Clay	0.05–0.10

[1]Established from AASHO Road Test Data.
[2]Compressive strength at 7 days.
[3]This value has been estimated from AASHO Road Test Data, but not to the accuracy of those factors marked with a number 1.
[4]It is expected that each state will study these coefficients and make such changes as experience indicates necessary.

the procedures used by a selected group of states and other agencies [5].

Arizona Highway Department, following considerable study and research, revised the AASHO interim coefficients, and, in most cases, lower values were adopted. The method used to select the coefficient for a specific material is somewhat empirical in that it primarily depends upon the gradation and/or strength properties. Using a limiting vertical subgrade pressure of 4 psi (28 kPa), researchers for Georgia Department of Highways proposed base course coefficients for use in AASHO Interim Guide. This study showed that the base coefficient value is a function of the soil support value assumed in the analysis. Illinois Department of Transportation modified the interim coefficients established by the AASHO Committee on Design to account for differences in strength of pavement materials. For instance, the coefficient for bituminous concrete was considered to vary with Marshall stability; the coefficient for granular base and subbase materials vary with CBR, and so on. According to Louisiana Department of Highways, the coefficient for surface courses is a function of Marshall stability, while the coefficients of the untreated and lime-treated base and subbase courses are a function of the Texas Triaxial Value. Coefficient for cement-stabilized base is a function of compressive strength, while that of the asphalt-stabilized base materials may vary as a function of Marshall stability. Highway Engineers of New Mexico and Wyoming vary the structural layer coefficient of surface layers as a function of Marshall stability and that of untreated granular bases and subbases as a function of R-value. The researchers at the Ohio Department of Highways concluded that there is no unique equivalence factor, and that the inclusion of a failure term is necessary for theoretical calculations of equivalency for given materials, environment and loading. Citing some earlier works, Van Til et al. [6] concluded that the layered elastic theory could be used in determining variations of structural layer coefficients under different loading, environmental, and structural layer conditions. The results of their studies, though qualitative, show that structural layer coefficient is a function of several pavement parameters, such as the elastic properties of the material, the thickness of the layer as well as those of the underlying layers. That the layer coefficient is a function of the modulus of elasticity of the material has been expressed by the following simple relationship, which is due to Takeshita [7]:

$$a = 0.00525\,(E)^{0.46} \tag{4}$$

in which E is expressed in kg/cm^2. Recently, Wang and Larson [8] from field performance data, developed limiting strain and limiting deflection criteria, and employing

TABLE 2. Structural Layer Coefficient Based on Fatigue (Source: Reference 9).

No.	Material/Layer	Recommended Layer Coefficient
1	Plant-Mix Asphalt Surface with AC-20	0.44
2	Plant-Mix Asphalt Base with AC-40	0.38
3	Plant-Mix Asphalt Binder with AC-20	0.35
4	Soil-Cement Base (7-day Compressive Strength no less than 600 psi)	0.24
5	Soil-Lime Subbase (CBR no less than 20)	0.20

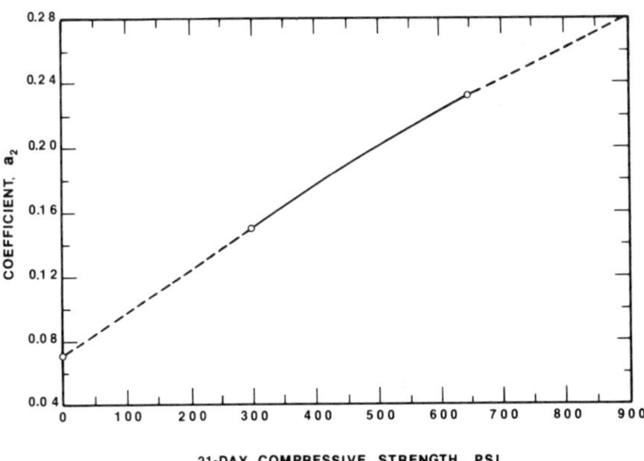

FIGURE 2. Structural layer coefficient for lime-flyash-stabilized granular material (Source: Reference 5).

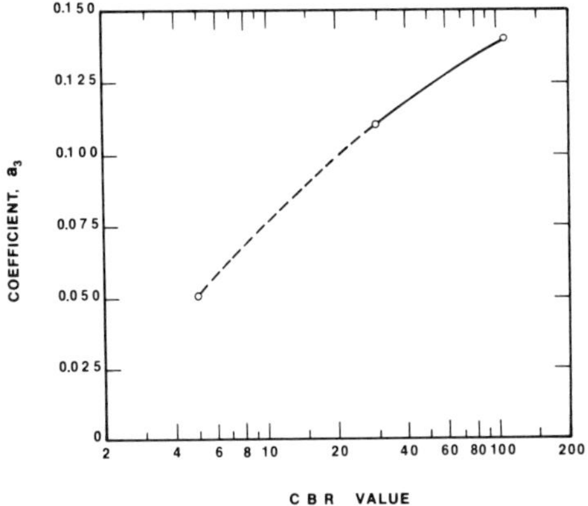

FIGURE 3. Structural layer coefficient for subbase material (Source: Reference 5).

these structural layer ceofficients of bituminous concrete base and limestone subbase were determined.

Two important observations suggest themselves from this review: (1) that the structural layer coefficient should invariably be based on some material property such as strength of stiffness or CBR, etc.; (2) that a mechanistic analysis in conjunction with an appropriate failure criterion is widely favored for layer coefficient determination. Allowing for these basic criteria and employing a probabilistic fatigue model, George [9] derived structural layer coefficients for surface, base and binder mixtures, soil-cement and soil-lime, as listed in Table 2. Note that these values are in excellent agreement with the structural layer coefficients proposed by AASHO Committee on design (Table 1).

The writer recommends that the layer coefficients proposed by AASHO Committee (Table 1) be used with AASHO Interim Guide [5] as well as revised guide [10] in designing flexible pavements. For lime-flyash-stabilized granular base materials and for subbase material, Illinois DOT coefficients, respectively graphed in Figures 2 and 3, are recommended [5]. As may be noted in Figure 2 for lime-flyash-stabilized materials, it is assumed that the coefficient varies with the 21-day compressive strength. The coefficient for subbase material in Figure 3 is correlated with CBR.

In conclusion, structural layer coefficients for flexible pavements design should be selected either from Tables 1 or 2 or from Figures 2 or 3 as appropriate.

REFERENCES

1. Odemark, N., "Investigations as to the Elastic Properties of Soils and Design of Pavements According to the Theory of Elasticity," *Staten Vaeginstitute*, Stockholm (1949).
2. Brahma, S. P. and C. S. Chao, "Equivalency Between Asphalt Concrete and Gravel by Shear Strength Criterion," *Engineering Experiment Station Report No. 12*, North Dakota State University, N.D. (1968).
3. McLeod, N. W., "The Asphalt Institute Layer Equivalency Program," *The Asphalt Institute*, Research Series No. 15 (RS-15) (1967).
4. Busching, H. W., F. L. Roberts, J. P. Rostrum, and A. E. Schwartz, "Evaluation of the Relative Strength of Flexible Pavement Components," Clemson University, Clemson, South Carolina (1971).
5. "AASHO Interim Guide for Design of Pavement Structures," American Association of State Highway and Transportation Officials, Washington, D.C. (1972).
6. VanTil, C. J., B. A. McCullough, B. A. Allerga, and R. G. Hicks, "Evaluation of AASHO Interim Guides for Design of Pavement Structures," *NCHRP Report No. 128, Highway Research Board*, Washington, D.C. (1972).
7. Takeshita, H., "Considerations on the Structural Number," *Proceedings, International Conf. Struct. Design of Asphalt Pavements*, University of Michigan, Ann Arbor (1967).
8. Wang, M. C. and T. D. Larson, "Performance Evaluation for Bituminous Concrete Pavements at Penn State Test Track," Department of Civil Engineering, The Pennsylvania State University, State College (1976).
9. George, K. P., "Structural Layer Coefficient for Flexible Pavement," *Journal of Transportation Engineering, ASCE, 110*, No. 2 (1984).
10. Finn, F. N., B. F. McCullough, W. R. Hudson, M. W. Witezak, and C. L. Monismith, "Revision of AASHO Interim Guide for Design of Pavement Structures," Prepared for NCHRP and AASHO (1984).
11. Bushey, R. W., K. L. Banmeisten, J. A. Matthews, and G. B. Sherman, "Structural Overlays for Pavement Rehabilitation," *California Research Report*, California Department of Transportation, Sacramento, California (1974).

CHAPTER 28

Elasto-Dynamic Analysis and Nondestructive Testing of Pavements

TREVOR G. DAVIES* AND MICHAEL S. MAMLOUK**

INTRODUCTION

Infrastructure repair has become, in the last several years, an increasingly important issue in the public domain. Recognition of the seriousness of the problem is also an acknowledgement of past neglect. In December 1983, the New York Academy of Sciences (and others) sponsored a three-day conference, the first of its kind, to bring together decision makers and engineers into a common forum. Of the seven technical papers presented on infrastructure reconstruction, three were devoted to pavement rehabilitation and nondestructive evaluation of infrastructure conditions [1]. The importance of infrastructure repair and the prodigious costs of carrying it out ("final price tag for New York alone will be over $35 billion,"—Governor Mario Cuomo [2]), demands that highway engineers develop accurate methods of assessing current pavement conditions and, also, that they develop reliable economical procedures for pavement repair and maintenance. This Chapter is intended to serve as a guide and critique of current procedures, particularly in regard to flexible pavements.

Static, dynamic, and seismic nondestructive tests (NDT) are reviewed and typical examples of pavement response to these tests are presented in order to demonstrate the complexity of the task. A discussion on the characterization of pavement materials is followed by a presentation of current methods of analysis which, inevitably, presume some simplification or idealization of material properties. A further article contains a critical appraisal of published claims that pavement properties can be determined accurately by backcalculation from deflection data. The concluding remarks suggest paths of inquiry which may yield definitive solutions to the problems which remain unresolved.

*Dept. of Civil Eng., University of Glasgow, Glasgow, Scotland
**Dept. of Civil Eng., Arizona State University, Tempe, AZ

NONDESTRUCTIVE TESTS

In a general paper on nondestructive evaluation of infrastructure conditions, Matzkanin et al. [3] identified 12 state-of-the-art NDT methods as candidates for assessing the degradation of solid structures and the detection of voids. These were:

1. Acoustic emission
2. Sonics
3. Vibration analysis
4. Ultrasonics
5. Acoustic holography
6. Short pulse radar
7. Frequency-modulated continuous-wave (FM-CW) radar
8. Photon backscatter
9. Photon (backscatter) tomography
10. Neutron interactions
11. Electrical resistivity
12. Infrared thermography

Practical considerations eliminated all but three of these methods, namely, FM-CW radar, electrical resistivity and electro-magnetic conductivity. While these techniques may have application to pavements, current practice is to adopt the more direct approach of measuring the load-deformation response of pavements subjected to static or dynamic loads (approximating vehicular traffic) and to use complex numerical models to infer the material properties (the inverse problem). However, some workers have advocated a technique akin to the electrical resistivity method, namely the seismic survey, to measure pavement material properties. While the seismic survey apparatus and test procedure is far more complex than that employed in the load-deformation tests, the seismic survey technique yields the material properties of the pavement structure directly (i.e.,

FIGURE 1. Benkelman beam (courtesy, The Asphalt Institute).

without iteration). These various methods are reviewed in the sequel. Some of this material is condensed from an excellent article by Smith and Lytton [4].

Static

Static loading devices have been in use for many years. However, since their productivity is low, most highway agencies prefer dynamic loading tests which can be completed in far less time and which provide high quality data under loading conditions which more closely approximate moving wheel loads. For completeness, the Benkelman Beam and its automated counterparts, the La Croix Deflectograph and the California Traveling Deflectometer are described here.

BENKELMAN BEAM

The Benkelman Beam [5] measures the deflection of pavements due to stationary truck loading (Figures 1 and 2). The Benkelman Beam is so devised that the pointer can be placed between the dual tires of a set of dual wheels. Generally, the measured values are the rebound of the pavement upon removal of the load. An 18,000 lb single axle is used on the test truck. This rebound deflection is measured by means of a dial indicator. The total length of the beam is about 12 ft. This length may not be sufficient to insure that the reference points are outside of the deflection basin. Also, since only the maximum deflection is measured with the beam, the shape and size of the deflection basin cannot be determined in one test.

In principle, the shape of the deflection basin can be ascertained by carrying out a series of tests but this is, in practice, inordinately laborious since the truck has to be moved for each measurement. The excessive labor needed to carry out this test and the limited data it provides has spurred the development of more efficient apparatus. However, the Benkelman Beam is still in widespread use as evidenced by recent publication of standardized test procedures [6,7]. The Benkelman Beam is particularly useful in developing nations where low technology and labor intensive methods are appropriate to the culture.

LA CROIX DEFLECTOGRAPH

This apparatus is a dual truck mounted set of Benkelman Beams. The beams are carried beneath the truck on a frame

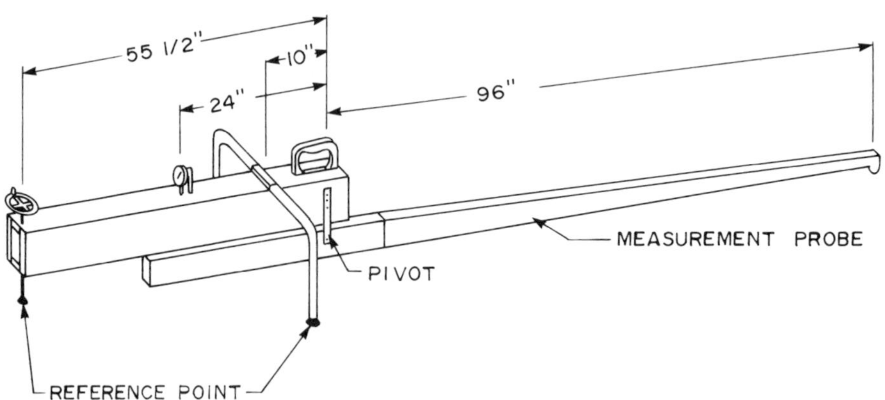

FIGURE 2. Basic components of Benkelman Beam.

which can be set down on the pavement in front of the rear wheels of the truck. As these wheels approach the apparatus, the measurement probes deflect downwards with the pavement. These deflections are measured by displacement transducers and the data are then recorded graphically on strip charts or digitized for storage on magnetic tape. The shape of the deflection basin can be determined by monitoring the deflections as the truck passes by the probes. Once the measurements are complete, the beams are lifted from the pavement surface and repositioned at the next station. The load on the rear axle of the truck can be varied between 12,000 to 26,000 lb by carrying dead weights in the truck. This automated system is obviously far superior to the simple manual Benkelman Beam apparatus described earlier. In response to this improvement, which originated in Europe, the California Department of Transportation has built its own trailer mounted version—the California Traveling Deflectometer. It has been in use for many years.

Dynamic

Dynamic loading devices can be categorized into two main divisions, namely harmonic loading devices and transient loading devices. The weight of field evidence suggests transient loading devices can better simulate moving wheel loads and do not suffer the hazards of resonance which may afflict harmonic loading devices under certain conditions. Nevertheless, harmonic loading devices are widely used and the deflection data obtained from them appears to correlate well with the data obtained from transient loading tests [8].

HARMONIC LOADING DEVICES

Harmonic loading devices apply vibratory (sinusoidal) loads to pavements by means of hydraulic actuators or a balanced pair of rotating eccentric masses. Since these devices must remain in contact with the pavement throughout the loading cycle, it is necessary to preload them with a static mass—either directly or by reaction against the test vehicle. This static preload may influence the response of the pavement in two ways: first, by increasing the stresses in the pavement, and hence changing its stiffness; and, secondly, the mass may be induced to resonate at certain loading frequencies. However, harmonic loading devices have the advantage over static loading devices in that a reference point is not needed to determine deflections. Instead, pavement accelerations under load are measured, at a number of stations, by means of velocity transducers (geophones). The corresponding deflections are computed electronically and recorded or printed as required. Two devices of this type which are widely used in practice are the Dynaflect and the Road Rater. Other vibratory testing devices are available and used by some agencies such as the 16-kip vibrator used by the U.S. Army Engineer Waterways Experiment Station for airfield pavement testing.

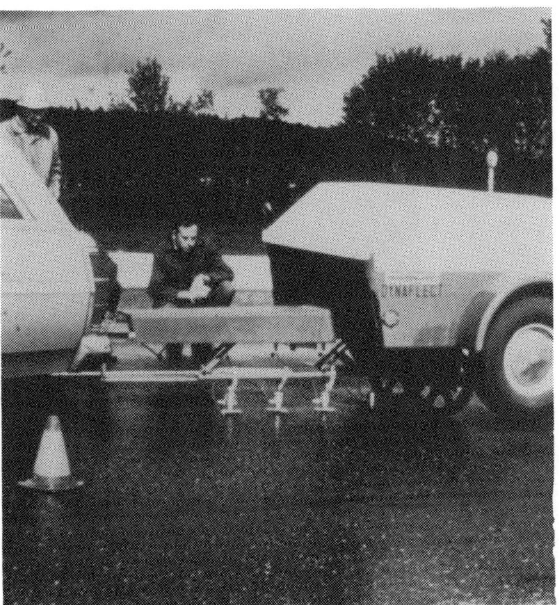

FIGURE 3. Dynaflect.

Dynaflect

The Dynaflect (Figure 3) is a trailer mounted fixed-frequency (8 Hz) device. The static weight is 2100 lb and a force generator consisting of a pair of contra-rotating unbalanced flywheels develops a sinusoidal force of 1000 lb peak-to-peak applied on two steel wheels. Deflections are measured simultaneously at five stations (Figure 4) positioned at 1 ft intervals. Data acquisition is performed by an electronic control unit which records the data on magnetic tape. The measuring devices permit accurate determination of the peak-to-peak deflection profile but the very limited load capacity and fixed frequency of this device are serious drawbacks.

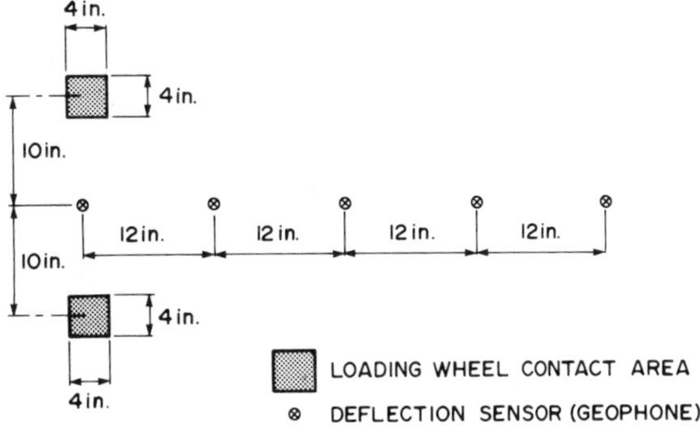

FIGURE 4. Location of loading wheels and geophones of the Dynaflect.

FIGURE 5. Road Rater Model 2008 (courtesy, Foundation Mechanics Inc.).

Road Rater

The Road Rater is available in three production models (400B, 2000, 2008) which differ primarily in their load rating (Figures 5 and 6). Unlike the Dynaflect, both the loading amplitude and frequency can be changed. The range of peak-to-peak load varies from as low as 500 lb for the smallest model to a maximum of 8000 lb for the largest. Each model is capable of exploring a wide range of load amplitudes so there is considerable overlap between the capabilities of the three devices. Loading frequencies can be varied continuously between 5–70 Hz. The preload is furnished by hydraulic actuator reactions against the trailer and loading is applied to the pavement via twin steel plates (of dimension 4 × 7 in.) for the smallest model (model 400B) and by a circular (9 in. diameter) plate for the other two models. Dynamic loading is developed by accelerating a large mass by means of a hydraulic actuator.

Deflections are measured at four stations by means of geophones located 1 ft apart (Figure 7) and may be recorded on magnetic tape or printed on paper tape as required. The device can be used to establish the shape of the deflection basin and to determine the effect of load amplitude and loading rate on pavement response. However, like the Dynaflect, the device cannot simulate the single pass

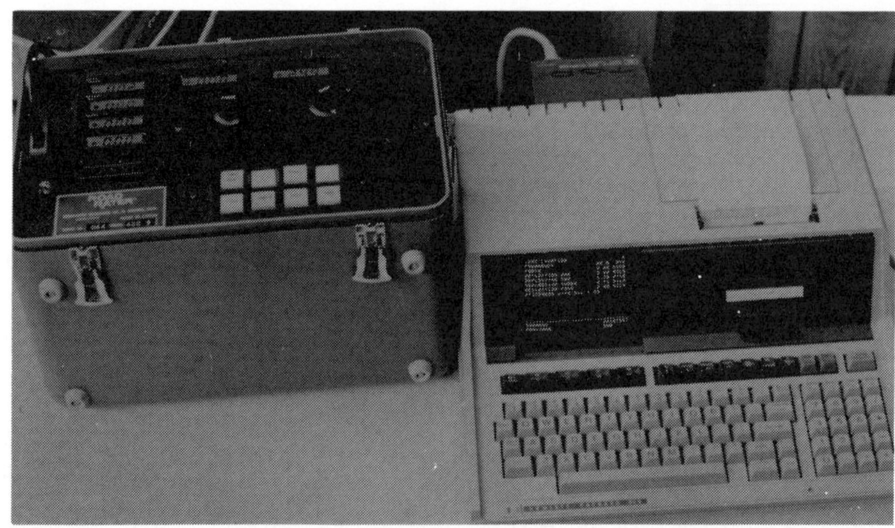

FIGURE 6. HP-85B computer for recording Road Rater data (courtesy, Foundation Mechanics Inc.).

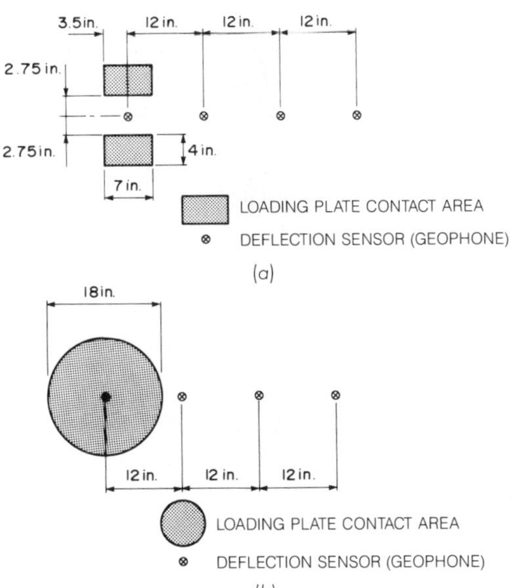

FIGURE 7. Schematic diagram of the Road Rater (a) Model 2000, (b) Model 2008.

loading of vehicular loading, and, further, the static preload may influence the test results. The device's ability to apply loads at various loading rates facilitates identification of possible resonant conditions. The operator may screen these out by recording only the test data at frequencies well away from the resonant frequency. Alternatively, knowledge of the resonant frequency may provide useful clues regarding the properties and thickness of the subgrade.

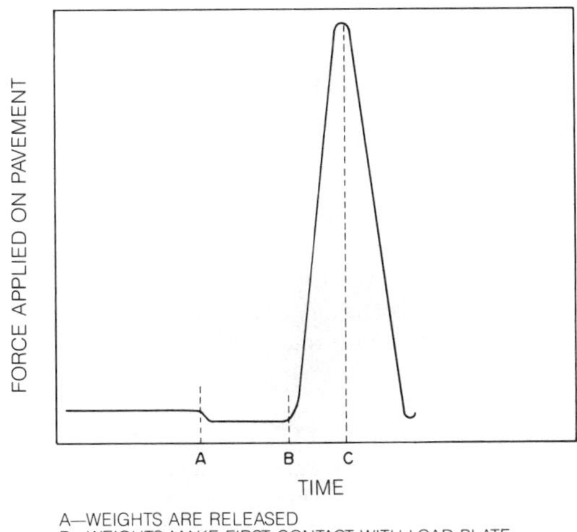

FIGURE 8. Impulse loading generated by Falling Weight Deflectometer.

FIGURE 9. Dynatest Falling Weight Deflectometer (courtesy, Dynatest Consulting, Inc.).

TRANSIENT LOADING DEVICES

Transient (impulse) loading devices develop force impulses by means of falling weights which are arrested by rubber loaded platens placed on the pavement surface. The impulse force can be varied by changing the mass of the falling weight and the drop height. The loading impulse closely resembles a half-sine wave (Figure 8) and thus models real moving wheel loadings with a fair degree of similitude, particularly since the magnitude of the loading can also be closely simulated. In principle, no preload is necessary but the self-weight of the device is typically 10% of the maximum load generated at impact. Two production models are described in the sequel.

Dynatest Falling Weight Deflectometer

This device is trailer mounted and can generate forces in the range of 1,500–24,000 lb by varying the drop heights and masses (Figure 9). The masses drop onto a rubber buffer which transmits the load impulse to the pavement via a 11.8 in. diameter loading platen. The load is measured by means of a load cell and deflections are measured at seven geophone stations located 1 ft apart (Figure 10). These data

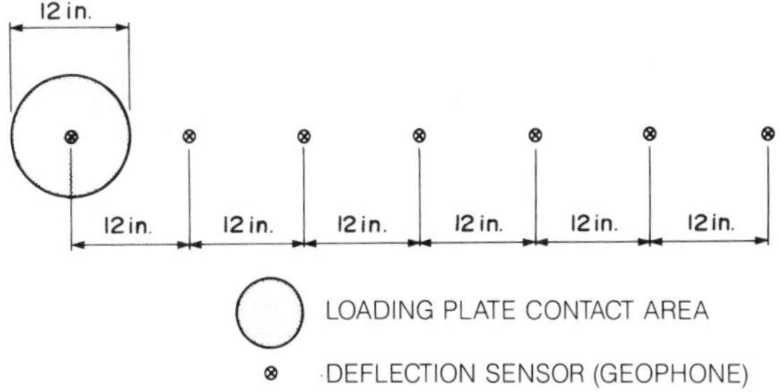

FIGURE 10. Falling Weight Deflectometer: Geophone Stations.

are recorded by a HP-85 computer on paper and magnetic tape.

KUAB Falling Weight Deflectometer

The KUAB model is also trailer mounted but it features a set of two masses which are dropped simultaneously but from different heights. The two mass falling system produces a smoother load impulse in some circumstances. The impulse load can be varied between 3,000–35,000 lb by changing the drop heights and masses. The details of the load transmission system and the data acquisition and recording devices are similar to those incorporated within the Dynatest model.

Seismic Techniques

The philosophy of NDT by means of seismic techniques is completely different from that of the load-deformation approach. The primary quantity measured during seismic tests is the velocity of propagation of wave energy away from a source of disturbance. These velocities can be correlated with the stiffness of the pavement layers. Two distinct strategies have evolved for carrying out such tests, namely, the steady-state technique and spectral analysis. Advances in this area since the late 1970's have been primarily attributable to the extensive research program carried out by Stokoe and his colleagues at the University of Texas at Austin [9–12] from which much of the following details are derived.

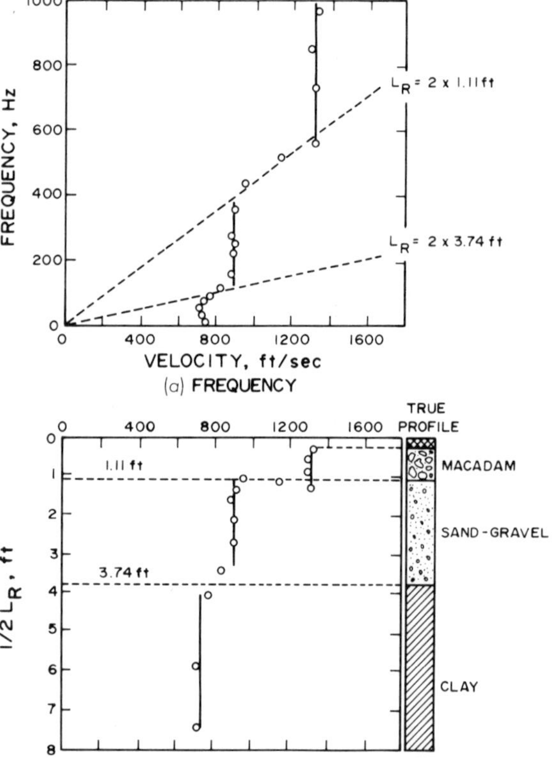

FIGURE 11. Typical velocity profile developed from steady-state Rayleigh wave testing [15].

STEADY STATE TECHNIQUES

A vertically oscillating mass is used to generate surface (Rayleigh) waves in the pavement. Their wavelengths can be determined by locating successive troughs or peaks by means of transducers located on the pavement surface. Since the frequency of vibration of the oscillating mass is predetermined, the Rayleigh wave velocity is given by the elementary equation of wave motion [13],

$$V_R = \omega L_R \tag{1}$$

where V_R is the velocity, ω is the frequency of vibration (radian/sec) and L_R is the wavelength.

At high frequencies, the Rayleigh wave penetrates only a short distance into the pavement and thus the velocity computed from Equation (1) reflects the properties of the surface layer of the pavement. With decreasing frequency, the Rayleigh wave penetrates deeper into the pavement system and the resulting velocity is thus representative of the properties of successively deeper layers. A mathematical theory [14] is required to interpret the data in this case.

Since the Rayleigh and shear wave velocities for typical pavement materials are virtually identical, the shear modulus can be determined from the Rayleigh wave velocity from the equation:

$$G = \varrho V_R^2 \tag{2}$$

where G is the shear modulus and ϱ is the mass density of the pavement layer.

Heukelom & Foster [15] and Szendrei & Freeme [16] have successfully used this technique. However, the method is prohibitively expensive in terms of labor costs and skilled personnel are required to interpret the test data. Figure 11 reproduced from reference [15] is indicative of the velocity profiles obtained by this method.

SPECTRAL ANALYSIS

The recent developments of fast Fourier transform (FFT) algorithms have opened up the way to analyze the spectrum of frequency responses which constitute the temporal response of systems to dynamic loading. In other words, it is now feasible to extract from data in the time domain the frequency content. Thus, an impulsive load, which generates a wide spectrum of Rayleigh waves, can be used to de-

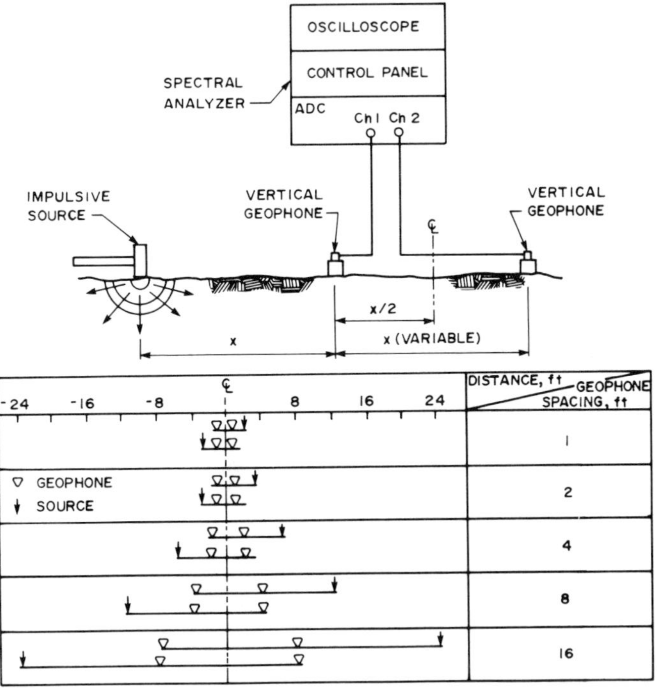

FIGURE 12. Experimental set-up for spectral analysis of surface waves [11].

termine (simultaneously) the properties of several pavement layers.

The rather complex theory has been described in detail by Heisey et al. [10] and only a brief recapitulation of the test procedure can be given here. Figure 12 shows the experimental set-up: an impulsive source (a drop hammer or Falling Weight Deflectometer (FWD)) is triggered by the spectral analyzer which records and decodes the signals received at this source and at several geophone locations. The phase information provided by the cross spectrum yields the travel times, and hence the Rayleigh wave velocity, for each frequency excited by the impulsive load. This data can be readily interpreted in terms of the layer shear moduli by means of Equation (2). To obtain data for the surface layer, very high frequencies must be excited which necessitates the use of a small hammer since relatively little energy is radiated away from a FWD blow at frequencies greater than 100 Hz. Field data must be corrected for strain effects [17,18] under such circumstances since at very small strain (load) levels, the stiffness of pavement materials are significantly higher than those measured at working load levels. The sophistication of the spectral analyzer greatly reduces the labor involved in determination of layer stiffness by seismic techniques. Yet this very sophistication has also, unfortunately, served to intimidate potential users who could, undoubtedly, profit from using this valuable technique.

PAVEMENT RESPONSE

The response of pavements to loading is a function of several interacting variables, including but not limited to, pavement layer properties and geometery, loading mode (static, harmonic or transient) and load intensity and distribution. For this reason, empirical methods of pavement evaluation must, necessarily, be of limited scope and, consequently, analyses which can accommodate these variables within a logical framework have gained increasing acceptance. Some of these factors have been discussed in detail in an excellent review by Hoffman [19] which we summarize in the sequel.

Typical Road Rater test results are shown in Figure 13 for two load levels within the frequency range of 6–30 Hz. The first and most obvious point to note is the measured deflections are frequency dependent and in several cases the variation is by no means negligible. Further, these deflections do not decrease monotonically with increased frequency. There are several cases of a distinct resonant peak which suggests that the pavement system resonates in sympathy with the forcing system. Such resonances must be attributable to the inertia of the pavement. Thus, as noted by Hoffman op. cit. the interpretation of Isihara and Kimura [20], namely, that inertial forces developed by moving wheel loads may be neglected has given way under the weight of evidence gathered from vibratory NDT devices that inertial forces

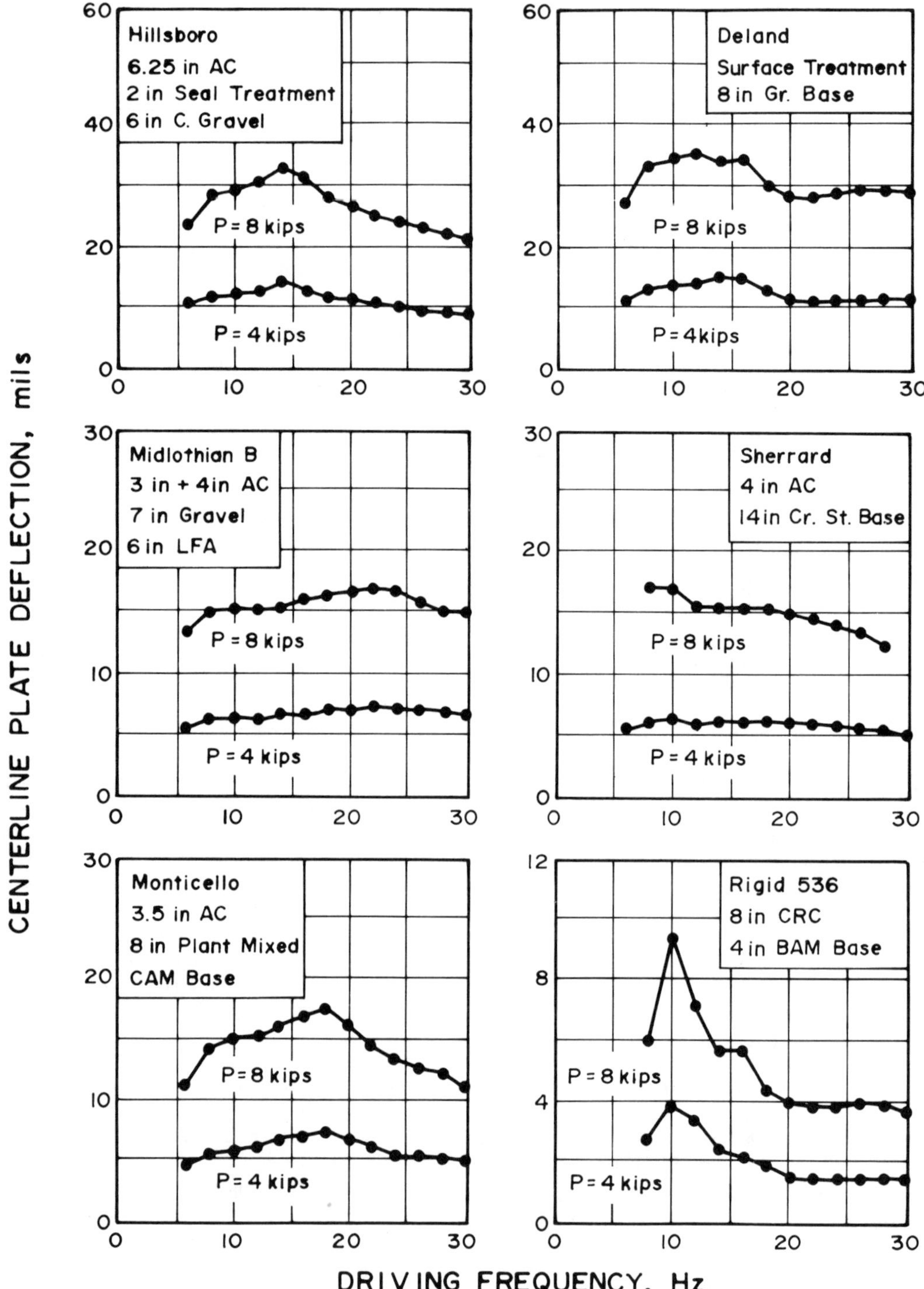

FIGURE 13. Typical Road Rater load test data [19].

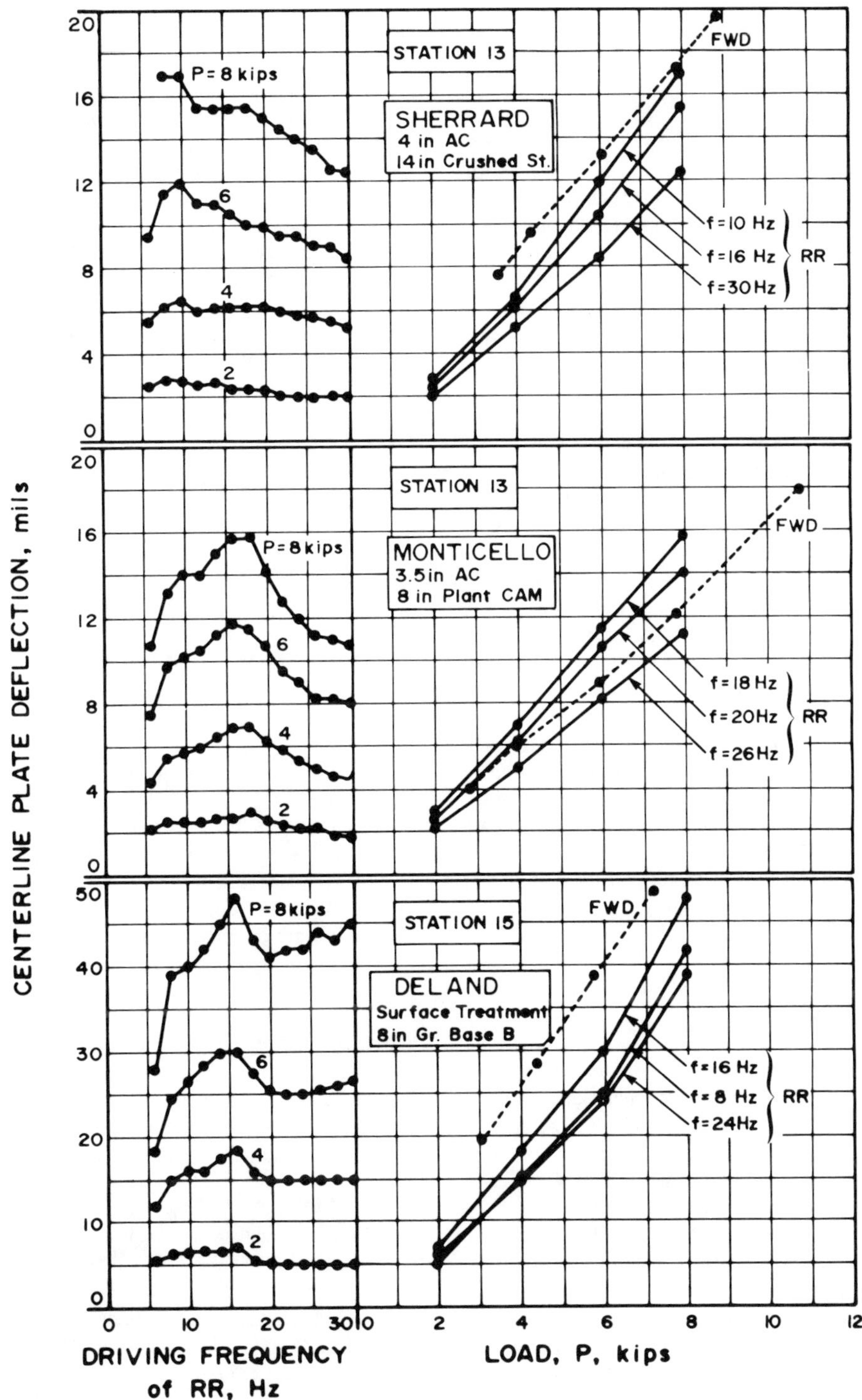

FIGURE 14. Comparison between Road Rater and Falling Weight Deflectometer test data [19].

should be considered under certain circumstances. Weiss [21] pioneered such analyses in the mid 1970s.

The second point of interest in the data shown in Figure 13 is that the load-deformation response, at a given frequency, is non-linear. This effect is, for these data, not nearly as significant as the frequency dependence discussed above; doubling the peak load from 4000 lb to 8000 lb results in a decrease of the apparent stiffness (secant modulus values) of the pavement system by typically 10–20%.

Further data are depicted in Figure 14 for the Road Rater and the Falling Weight Deflectometer. The FWD tends to give higher deflections than the Road Rater. The static preload applied during Road Rater testing, if not properly accounted for in analysis, can lead to anomalous results. In particular, since Road Rater deflections at about 15 Hz are generally magnified with respect to the deflections at very low frequencies, it may be expected that these deflections (at 15 Hz) would be greater than those obtained by means of the Benkelman Beam. Yet, the reverse is observed in practice; a phenomenon which is attributable to the stiffening of the pavement by the preload.

Hoffman op. cit. concludes that since the loading mode and load intensity are highly significant parameters in the structural evaluation of pavements, tests should be carried out to simulate real loading conditions as closely as possible. Further, material characterization back-calculated from measured pavement deflections can only be as good (or bad) as the field data. These conclusions are well supported by the evidence. We would add that the latter caution may be overly optimistic as discussed later. Further useful data are given in Reference 8 which contains a summary of extensive and carefully monitored field data obtained from a number of NDT and moving wheel load tests.

MATERIAL PROPERTIES

Highway engineers are primarily interested in the load-deformation response of pavement materials. In that respect, the only practical quantity worthy of definition is stiffness—albeit a function of temperature, time, load level and load rate. At the present time, and for the foreseeable future, analyses of pavement response will be based on either ideal elastic materials or empirical approaches. Hence, the influence of the factors mentioned above have largely been explicated by reference to field data. As a corollary of this failure of theory to predict the effects of these parameters on pavement response to loading, field tests have evolved which aim to simulate real moving wheel loads as closely as possible. The analysis of durability (fatigue) is, likewise, primitive. In this article, the basic definitions and field observations which form the basis of current analytical procedures are presented.

Stiffness

Since current methods of analysis used in conjunction with NDT employ the theories of elasticity, viscoelasticity and elastodynamics, it is implicitly recognized that pavement materials possess readily identifiable and meaningful moduli of elasticity. In practice, since the theories in current use assume isotropy, only two such constants (for each distinct layer) can be determined. These are usually: (1) the Young's modulus of elasticity, denoted by the symbol E, and (2) Poisson's ratio, denoted by the symbol ν. Other choices are possible, namely the shear modulus G, bulk modulus, K or the Lamé constants λ and μ. The interrelationships between these constants are

$$G = \frac{E}{2(1 + \nu)} \tag{3}$$

$$K = \frac{E}{3(1 - 2\nu)} \tag{4}$$

$$\lambda = \frac{E\nu}{(1 + \nu)(1 - 2\nu)} \tag{5}$$

$$\mu = G \tag{6}$$

It should be recognized that these constants have precise meanings in the linear theory of elasticity. However, in the pavement literature a plethora of moduli have been defined which, however descriptive of laboratory and field data, have no place in the classical theories of elasticity and elastodynamics. Some of the terms related to pavement materials are described in the sequel.

MODULUS OF DEFORMATION

This quantity is usually determined from a monotonic triaxial load test (Figure 15). It is the mean value of the Young's modulus of a non-linear material over the load range of interest (sometimes termed the secant value). Only if these stress-strain data are reasonably linear are linear elastic analyses of such materials justified.

RESILIENT MODULUS

Pavement materials are subjected to repetitive loading which tend to "shake-down" their response through strain hardening. Thus, after a few repetitions of loading (Figure 16), subsequent deformations become predominantly recoverable and, often, substantially linear. The resilient modulus is the ratio of axial stress to recoverable axial strain during such loading, that is, it is the Young's modulus of the material after many load repetitions have been applied. The resilient modulus is usually considerably greater than the modulus of deformation for the same load range.

COMPLEX (DYNAMIC) MODULUS TEST

The complex modulus is the complex quantity (i.e., real and imaginary number) which relates the axial stress to axial strain in a material subjected to sinusoidal loading. The imaginary part of this quantity represents the internal (material) damping property of the material while the real part characterizes its stiffness (Young's modulus). Considerable care is needed to carry out the requisite dynamic tests because of the hazards of resonance and inertial resistance in the sample and test equipment which may swamp the true response characteristics of the material, particularly at high frequencies. In classical elastodynamics [22], the stiffness of materials is assumed to be frequency invariant but frequency dependent material properties can be accommodated within such analyses by solving for each component frequency separately. Frequency dependent material damping can be treated similarly. The complex modulus is usually denoted by the symbol E^*, where

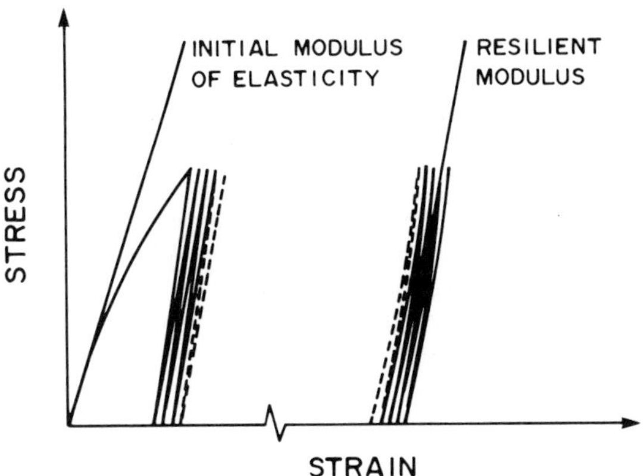

FIGURE 16. Shake-down of load deformation response [49].

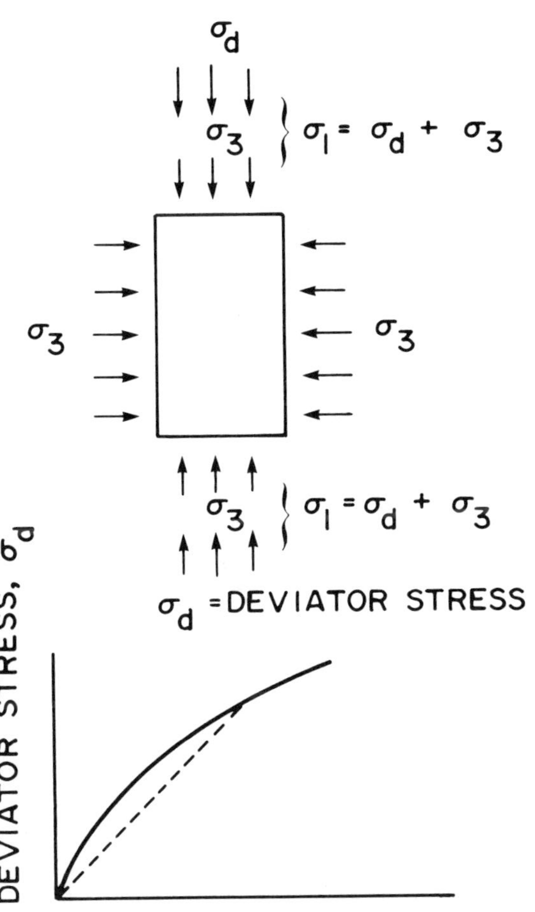

FIGURE 15. Triaxial load test.

$$E^* = \frac{\sigma \text{Sin } \omega t}{\epsilon \text{ Sin } (\omega t - \phi)} \tag{7}$$

$$= E(1 + 2i\beta) \tag{8}$$

in which ω is the angular frequency of vibration (radian/sec), ϕ is the phase difference between the stress and strain, i is the unit imaginary number and β is termed the damping ratio.

The transformation from circular functions to complex numbers is accomplished by means of the Euler transformation, whence:

$$\beta = \frac{1}{2} \text{Tan}\phi \tag{9}$$

and

$$E = \frac{\sigma}{\epsilon} \text{Cos}\phi \tag{10}$$

The absolute value (magnitude) of the complex modulus, $|E^*|$ is commonly referred to as the dynamic modulus. With respect to the parameters described above,

$$|E^*| = \frac{\sigma}{\epsilon} \tag{11}$$

Papazian [23] has measured a substantial frequency effect on the complex modulus of asphaltic materials. If this is true, and not an artifact of the test procedure as discussed above, these data should be incorporated within elastodynamic analyses of pavements.

NON-LINEAR RESPONSE

The stiffnesses of granular materials are stress and strain dependent. Obviously, at stress levels near failure the term itself is hardly meaningful, but even at design load levels a careful appraisal of the factors controlling material stiffnesses is needed. At very low strain levels ($<0.002\%$), very high moduli have been measured in granular soils by means of seismic tests (refer to Figure 17 and Reference 13). At these small strain levels, the shear modulus for round-grained sands is well represented by the empirical equation [17]

$$G = \frac{690\ (2.17 - e)^2(ap)^{1/2}}{1 + e} \tag{12}$$

where p is the mean effective stress, a is the atmospheric pressure, and e is the void ratio. The corresponding equation for angular grained sands is

$$G = \frac{320\ (2.97 - e)^2(ap)^{1/2}}{1 + e} \tag{13}$$

As a first approximation, these equations can be assumed to apply to coarser materials, too. For clays of moderate sensitivity, the result obtained from Equation (13) should be multiplied by a correction factor R^k where R is the overconsolidation ratio and the exponent k may be assumed to be one-half for clays of high plasticity (plasticity index greater than 0.7) and equal to 75% of the plasticity index for clays of lower plasticity. Figure 17 can be used to extrapolate these data to higher strain levels.

In these equations, the mean effective stress is given by

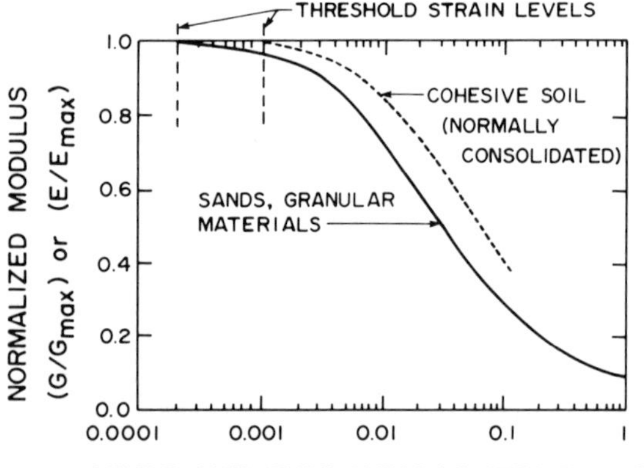

FIGURE 17. Effect of shear strain on shear modulus [13].

the expression

$$p = (\sigma_v + 2\sigma_h)/3 \tag{14}$$

where σ_v is the vertical effective stress and σ_h is the horizontal effective stress. In well-compacted or overconsolidated subgrades the horizontal stresses may be substantially greater than the vertical stresses.

The non-linearities arising from the stress and strain sensitivity of pavement materials considerably complicates the task of establishing material properties by backcalculation from deflection data. If the superimposed loading yields stress changes in the subgrade of the same order as the initial in situ stress state, the equivalent modulus of deformation over this load range may be reduced by 30% or more. Nevertheless, it may be better to conduct equivalent linear analyses in preference to incorporating unproven assumptions in a nonlinear analysis. Valuable work in this area has been done by Maree et al. [24] and Hoffman et al. [25]. The latter authors adopted a material model in which the elastic moduli were expressed as

$$E = ap^b \tag{15}$$

in which a, b = material constants. In their finite element analyses, these researchers adopted values of b in the range 0.33 – 0.52.

TEMPERATURE EFFECTS

The effect of temperature on pavement materials is a less tractable problem. Considerable data are available on the load-deformation characteristics of asphalt at different temperatures e.g., [26]. Whether its complex behavior can be usefully characterized by the simple viscoelastic model is debatable and thus it seems prudent to carry out NDT at low temperatures if it is desired to ascertain the material properties of the lower layers of the pavement structure. At the other extreme, NDT of pavements in subzero temperatures is complicated by local thaw weakening [27,28], and, in such cases, empirical methods of assessing pavement integrity based on NDT are likely to hold sway for the foreseeable future.

POISSON'S RATIO

Poisson's ratio is the ratio between lateral and axial strains in an element of a material subjected to uniaxial stress. For granular materials, Poisson's ratio may be as low as 0.3, while for saturated cohesive soils a value of 0.5, signifying incompressibility, is appropriate. This parameter has been shown to have only a minor effect on the moduli calculated by back-analysis from deflection data [29]. Poisson's ratio of asphalt concrete and asphalt treated materials is affected by temperature. Values of 0.3, 0.35 and 0.4 are commonly used at low, intermediate and high temperatures, respectively.

Material Damping

In classical theories, two forms of material damping are distinguished, namely hysteretic damping, which is frequency independent, and viscous damping which is proportional to the frequency of excitation. These forms are best understood by reference to the simple Kelvin-Voigt rheological model where the Hookean solid element is supplemented by a parallel Newtonian fluid element [30] which yields the constitutive law:

$$\sigma = G\epsilon + \eta\dot{\epsilon} \tag{16}$$

where η is the viscosity parameter and the superposed period denotes the time derivative.

For sinusoidal loading in which:

$$\epsilon = \epsilon_0 e^{i\omega t} \tag{17}$$

Equation (17) can be rewritten as

$$\sigma = G\epsilon_0 \left(1 + i\frac{\omega\eta}{G}\right) e^{i\omega t} \tag{18}$$

Clearly, the stress σ is not in phase with the strain; the phase difference, in radians, is

$$\phi = \frac{\omega\eta}{G} \tag{19}$$

and the material damping ratio is,

$$\beta = \frac{\omega\eta}{2G} \tag{20}$$

Thus, the "viscoelastic" shear modulus can be represented by the complex quantity:

$$G^* = G\,[1 + 2i\beta] \tag{21}$$

It is evident from Equation (20) that if the viscosity parameter is constant, then the material damping ratio is proportional to frequency. While this may be a fair approximation for asphaltic materials, granular materials exhibit hysteretic damping [13] characteristics (i.e., frequency invariant damping). For such materials, the "viscosity" must be inversely proportional to frequency. Experimental data [13] suggest that the material damping ratios for such materials are of the order of 5%. By far the greatest amount of damping which occurs in the field arises from radiation damping, i.e., energy is dissipated not by conversion into heat in the immediate vicinity of the loading, but by transmission outwards from the source to the far field (to be ultimately lost as heat, of course) in much the same way as waves travel outwards from a vessel passing through water. For many practical purposes, material damping in the granular layers of a pavement structure can be neglected in the analysis of dynamic pavement response to loading.

By using the principle of correspondence, Equation (21) can be used in place of the shear modulus in any analysis of elasticity in order to solve the equivalent viscoelastic problem. Further, classical elastodynamic analyses can be extended to deal with materials which exhibit internal damping by making use of the same substitution.

ANALYSIS

With the advent of more powerful and economical computers and simultaneous developments in data acquisition and analysis, rational interpretation of field data by mechanistic analyses have gained increasing popularity. Developments and applications of these techniques in practice is the main thrust of this article. The purpose of analyzing the NDT data is to evaluate the load-carrying capacity of existing pavements and/or to determine the overlay thickness that is required for the pavement to support a certain number of load applications. Some methods of analysis are based totally on empirical observations and previous experience, while other methods involve mechanistic approaches. Up to the present time no method of overlay design is available that is based completely on theoretical analyses. Empirical design relations (performance criteria) are usually needed to link pavement characteristics to the number of load applications that the pavement can support before failure, despite the fact that empirical relations are restricted to conditions similar to those under which they were originally developed.

Empirical

Several agencies have developed empirical methods for overlay design for flexible pavements based on deflection measurements. All of these methods have been in use for several years and they are usually modified when more experience with their use is gained. Two of these methods are reviewed here for completeness.

ASPHALT INSTITUTE METHOD

In this method [31], Benkelman Beam deflections are obtained at random using an 18,000 lb single axle. A nominal deflection is calculated from the equation,

$$\delta = (\bar{x} + 2s)fc \tag{22}$$

where $\bar{x}$ is the mean deflection reading over the region of interest, s is the standard deviation of the measurements, f is

TABLE 1. Factors for Estimating Standard Deviation. s = R/d or s = Rm; Where s = Standard Deviation; R = Range of Values [31].

Number of Values	Factor d	Factor m
2	1.1284	0.8862
3	1.6926	0.5908
4	2.0588	0.4857
5	2.3259	0.4299
6	2.5344	0.3946
7	2.7044	0.3698
8	2.8472	0.3512
9	2.9700	0.3369
10	2.0775	0.3249

a temperature correction factor and c is a coefficient reflecting the time of the year.

The standard deviation may be estimated from Table 1 if so desired and Figure 18 is a plot of the temperature correction factor f versus temperature. The coefficient c reflects the change in deflections during the year and should be assigned the value of unity during the critical period and lesser values at other times.

Figures 19 and 20 show the design charts developed by the Asphalt Institute. The first step is to determine the Initial Traffic Number (ITN) from the nomograph depicted in Figure 19. For low values of ITN, the result should be revised using data supplied by the Asphalt Institute. The Design Traffic Number (DTN) can then be calculated using Table 2 which accounts for the projected growth in traffic volume. Finally, Figure 20 yields the required depth of overlay.

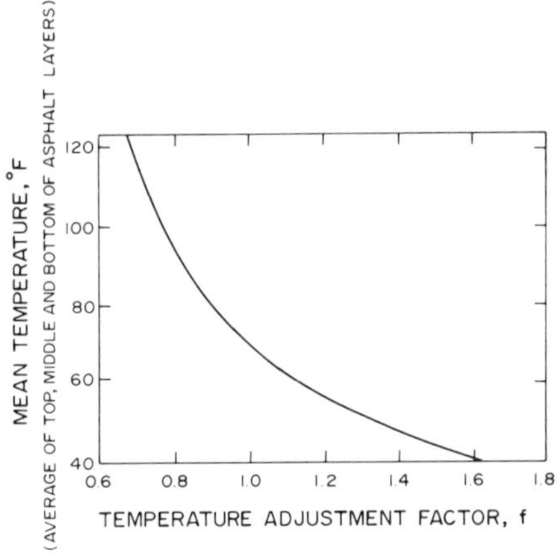

FIGURE 18. Temperature correction factor [31].

ARIZONA DEPARTMENT OF TRANSPORTATION METHOD

The Arizona Department of Transportation method [32] was developed using Dynaflect measurements from representative highway sections around the state and is referred to as the SODA (Structural Overlay Design for Arizona) method. The method is based on correlations between overlay thickness, traffic loads, road roughness, environmental factor and Dynaflect readings. The overlay thickness (T) is obtained as:

$$T = \frac{\text{Log}(L) + 0.104 \times R + 0.000578 \times Po - 0.0653 \times SIB}{0.0587 \times (2.6 + 32.0 \times D5)^{0.333}} \tag{23}$$

Where:

L = 18 kip loads in 1000's expected over the design period
R = Regional Factor
Po = Roughness (Mays-meter value), in./mile
SIB = Spreadability Index before Overlay = ((sum of the 5 Dynaflect sensor readings)/(5 × #1 sensor)) × 100
D5 = #5 Dynaflect sensor reading

The thickness should be determined at each test location and the mean value of thickness for all test locations in a design section is then used as the overlay thickness. No

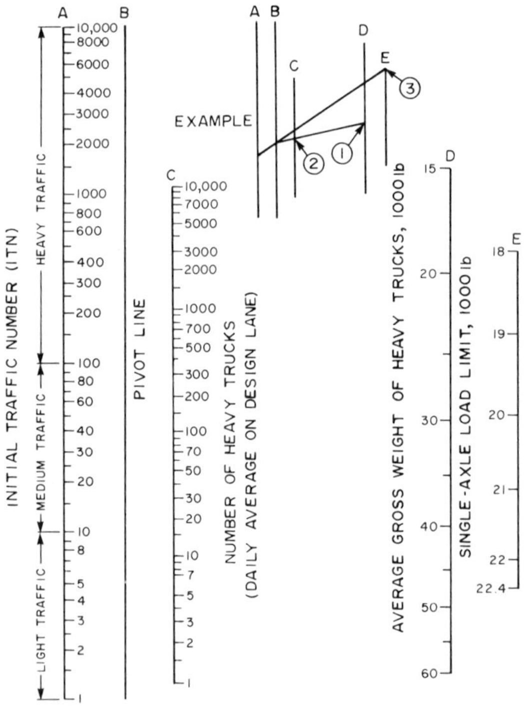

FIGURE 19. Initial traffic number nomograph [31].

statistical manipulations are needed as they were incorporated into the development of the method. Any individual test location results less than zero are assigned a value of zero and any results over six inches are assigned a value of six inches.

Static

Elastic and viscoelastic analyses of pavement deflections have been widely used for several years. They have been used extensively for the design of pavements, pavement evaluation and overlay design [33–43]. To carry out a realistic analysis of pavements, it is essential that the algorighm employed for this purpose has multi-layer capability—at least four layers for flexible pavements. Further, the algorithm should be able to evaluate stresses, strains and deflections not only directly beneath the load but also at several radial stations so that data pertaining to the shape and extent of the deflection basin may be analyzed.

All such algorithms are based on the Navier-Cauchy equations of equilibrium, which in cartesian indicial notation, take the form:

$$\mu\, u_{i,jj} + (\lambda + \mu)u_{j,ji} = 0 \tag{24}$$

where u_i is the i-th cartesian component of the displacement, the comma denotes partial differentiation with respect to the space variable, i.e.,

$$u_{i,jj} = \frac{\partial^2 u_i}{\partial x_j \partial x_j} \tag{25}$$

and μ and λ are the Lame constants (previously defined).

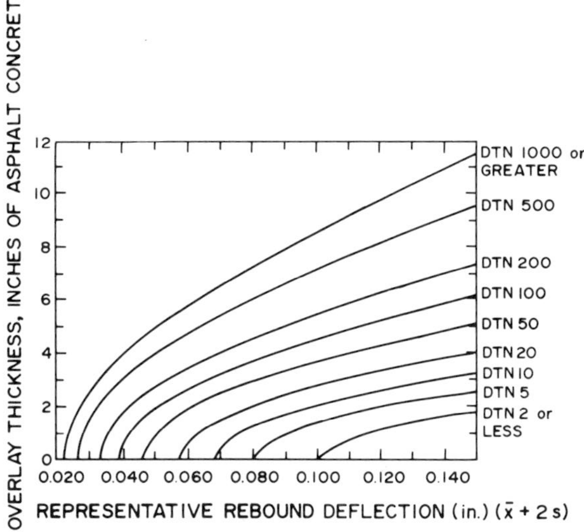

FIGURE 20. Overlay depth from design traffic number [31].

TABLE 2. Initial Traffic Number Adjustment Factors [31].

Design Period, Years (n)	Annual Growth Rate, percent (r)					
	0	2	4	6	8	10
1	0.05	0.05	0.05	0.05	0.05	0.05
2	0.10	0.10	0.10	0.10	0.10	0.10
4	0.20	0.21	0.21	0.22	0.22	0.23
6	0.30	0.32	0.33	0.35	0.37	0.39
8	0.40	0.43	0.46	0.50	0.53	0.57
10	0.50	0.55	0.60	0.66	0.72	0.80
12	0.60	0.67	0.75	0.84	0.95	1.07
14	0.70	0.80	0.92	1.05	1.21	1.40
16	0.80	0.93	1.09	1.28	1.52	1.80
18	0.90	1.07	1.28	1.55	1.87	2.28
20	1.00	1.21	1.49	1.84	2.29	2.86
25	1.25	1.60	2.08	2.74	3.66	4.92
30	1.50	2.03	2.80	3.95	5.66	8.22
35	1.75	2.50	3.68	5.57	8.62	13.55

$$\text{Factor} = \frac{(1 + r)^n - 1}{20r}$$

where r = annual growth rate
n = design period, years

In Equation (24), it is implicitly understood that the body forces are zero (or unchanged during loading) and that the load-deformation response is linear and isotropic. As noted earlier, the assumption of linearity and isotropy in pavement materials is only approximately true.

Equation (24) has to be integrated in order to determine the displacements (hence, stresses and strains) during loading. Because of the complexity of the boundary conditions, numerical methods must be used. Computer programs such as BISTRO, BISAR and Chevron, etc. have been developed to fulfill this need. The assumptions invoked in such programs should be carefully noted. If "no-slip" compatability constraints at the layer interfaces and unlimited tensile strength material properties are incorporated within such codes, their results may be suspect under certain conditions. On the other hand, it is extremely difficult to develop rigorous procedures for treating these departures from ideal elastic conditions.

Current procedures for developing NDT-based design procedures for determining asphalt-concrete overlays by means of static analyses is exemplified by the work of Thompson and Hoffman [37]. With reference to Figure 21, the principal quantities used in the design process are the parameters "AREA" and Δ. The former quantity is proportional to the cross-sectional area of the deflection basin developed by a 9000 lb moving wheel load. Thus,

$$\text{AREA} = 6[D_0 + 2D_1 + 2D_2 + D_3]/D_0 \tag{26}$$

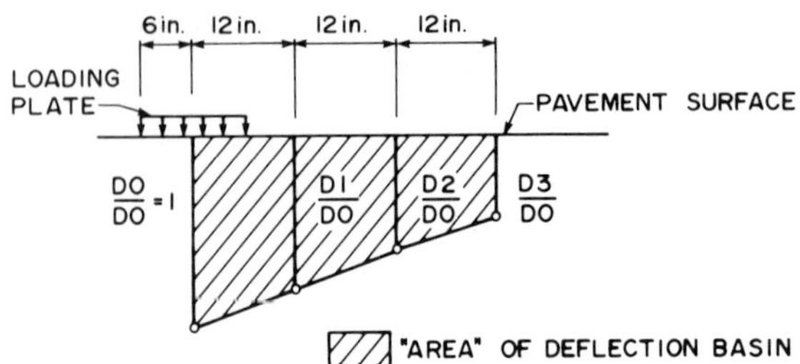

FIGURE 21. Deflection basin characterization [37].

where D_0, D_1, D_2 and D_3 are the deflections at the four geophone locations. The parameter A, which combines the data from four stations, may vary from about 11 inches to 36 inches (for a rigid pavement).

They recommend use of the Falling Weight Deflectometer for the determination of pavement response to a 9000 lb moving wheel. Alternatively, pavement deflection obtained from the Road Rater can be correlated to the moving wheel deflections. The parameter Δ is defined as the equivalent 9000 lb moving wheel load deflection (in mils).

For full-depth asphalt-concrete (AC) pavements, nomographs such as that shown in Figure 22 are used to determine the asphalt concrete modulus E_{AC} and the subgrade resilient modulus E_{ri} from the known values of Δ and A. The parameter T_{AC} is the total thickness (surface + base) of the asphalt concrete. The following equation can then be used to determine the pavement thickness required to achieve the desired surface deflection.

$$\text{Log}\,\Delta = 3.123 - 0.0273\,E_{ri} - 0.895\,\text{Log}T_{AC} - 0.359\,\text{Log}E_{AC} \tag{27}$$

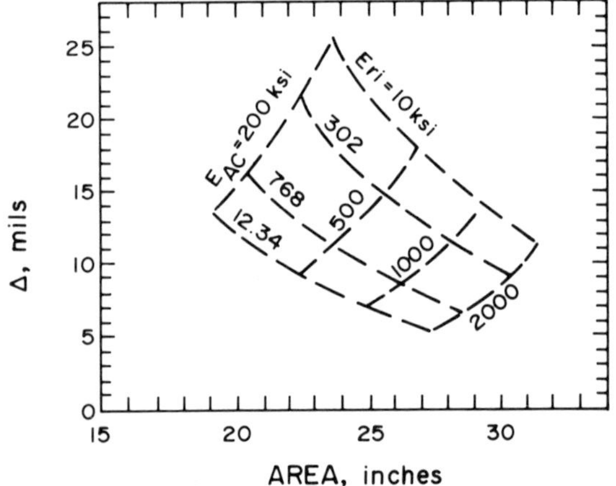

FIGURE 22. Evaluation nomograph for a nine inch thick full-depth AC pavement [37].

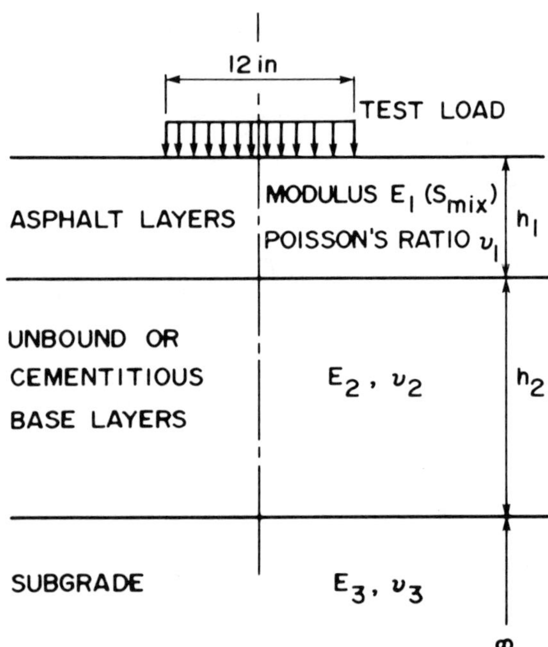

FIGURE 23. Schematic representation of pavement under a test load [39].

In the above equation, T_{AC} is measured in inches and E_{ri} and E_{AC} in ksi.

The influence of temperature and radial strains on the performance and design life of pavements is also incorporated within their design procedure, although, naturally, their effects are introduced by means of empirical factors.

Uddin et al. [12] carried out a comprehensive parametric study, using the ELYSM5 computer program of layered elastic pavement systems in order to develop procedures for extracting layer stiffnesses from deflection basin data. They concluded that the deflections were primarily a function of the stiffness of the subgrade and developed an iterative solution technique based on a scheme of successive correction of layer moduli. Their iterative algorithm was based on the equation

$$E^i_{n+1} = E^i_n(1 - C^iD^k/2) \tag{28}$$

where i refers to the i-th layer, n to the n-th iterate, C^i is a correction factor for the i-th layer and D^k is the current error in the prediction of deflection at the k-th sensor. The correction factors C^i were determined from the parametric study.

Koole [39] has described an overlay design method based on Falling Weight Deflectometer measurements in which the principles of the Shell Pavement Design Manual [44] are incorporated. The pavement structure is schematized as a three-layer model (Figure 23). The top layer represents the asphalt layer, the second layer represents the base materials (granular or cementitious) and the third, infinitely deep, layer represents the subgrade. With the aid of the BISAR program, deflection interpretation charts were derived, of which Figure 24 is an example, from which the pavement layer characteristics may be determined from a number of surface deflection measurements. However, in view of the large number of variables, this procedure cannot be readily generalized despite the fact that only three distinct layers have been assumed in the analysis.

McCullough and Taute [42] have developed a method of determining pavement material properties from Dynaflect measurements by means of an iterative algorithm based on layered elastic theory. Only three layers are assumed for simplicity. Based on an extensive parametric study, they found that the subgrade stiffness could be determined accurately from the deflection of the outermost sensor alone. Figure 25 depicts this relationship for a number of pavement structures. They also gave a nomograph for predicting basin slope for rigid pavements but concluded that this parameter was extremely sensitive to surface and subbase layer thicknesses and moduli.

Mamlouk [45] presented the results of a comprehensive parametric study of multi-layered elastic pavement systems. Several plots depicted the deflection basin profiles (for loading over a 12-inch diameter circular area) up to a radial distance of four feet from the center of loading. For given layer thicknesses and an NDT determined deflection profile, the best match with the theoretical data yields the stiffnesses of the pavement layers.

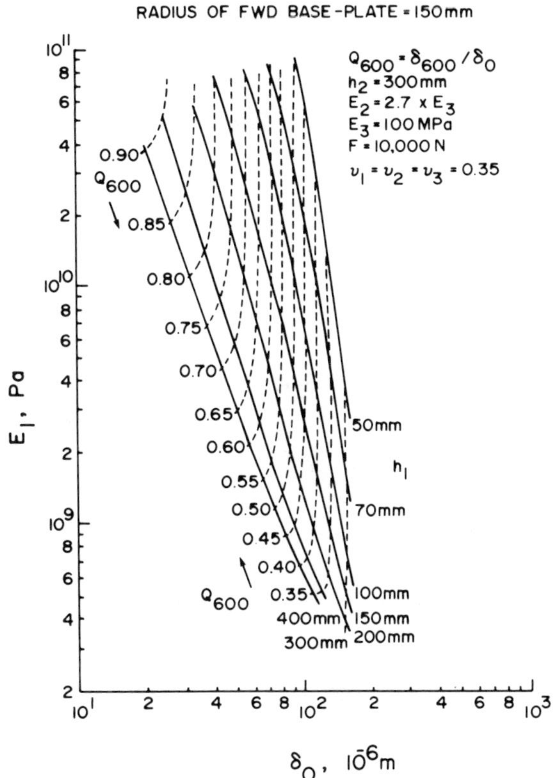

FIGURE 24. Deflection interpretation chart [39].

Dynamic

Since currently favored NDT loading devices apply dynamic loading, it is essential to establish whether dynamic (predominantly, inertial) effects are significant. Field data [25,46] have suggested that resonance may occur under certain conditions—a phenomenon which is alien to purely static analyses.

Weiss [21] has investigated the dynamic response of pavements by means of a single-degree-of-freedom (SDOF) analysis. Although continua can be interpreted in terms of such systems (e.g., the Lysmer analog [13]), a fundamental analysis by such means is impossible, particularly if the continuum is inhomogeneous since the simple constants (masses, springs, and, especially, the dashpots) of the equivalent SDOF model cannot be established a priori. Further, such models cannot provide information pertaining to deflections outside the loaded area.

By backcalculation, Weiss op. cit. demonstrated that the pavement mass mobilized by vibratory load testing devices increases with increasing pavement stiffness. However, these data are scattered even for a single loading frequency. While such data are useful in a qualitative sense, this concept of "added mass" has long since been abandoned in the

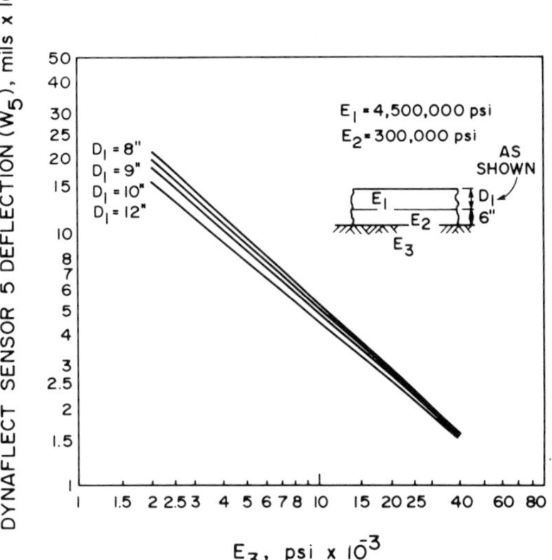

FIGURE 25. Relationship between Dynaflect sensor data and subgrade modulus [42].

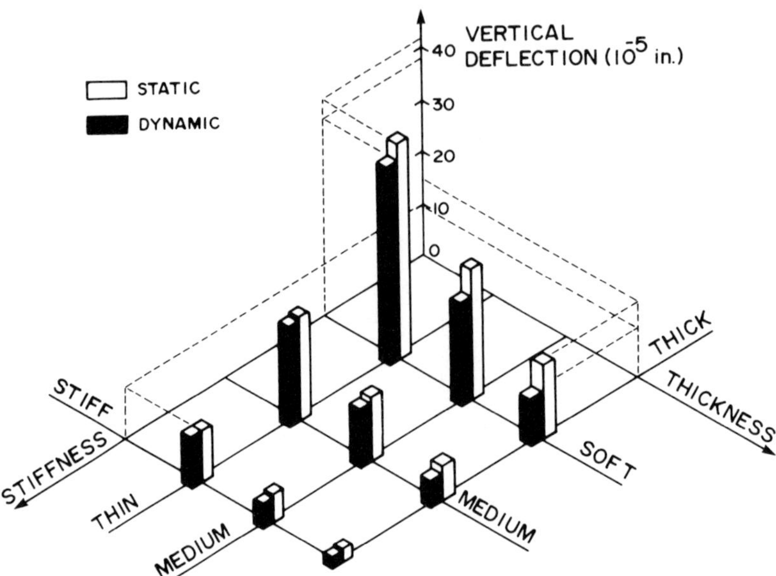

FIGURE 26. Pavement deflections under static and (25 Hz) dynamic loads adjacent to the Road Rater [48].

analysis of machine vibrations in favor of continuum analysis.

Implementation of a continuum model awaited developments in the area of seismology [47]. Recently, Davies, Mamlouk and Sebaaly [48–51] have exploited these developments to study the effects of dynamic loading on pavement response. Roesset and Shao [52] have described similar work.

The fundamental difference between statics and dynamics for ideal elastic or viscoelastic materials is the inclusion of an inertial term. The governing differential equation of harmonic (sinusoidal) motion is the Helmholtz Equation [22],

$$\mu u_{i,jj} + (\lambda + \mu)u_{j,ij} + \varrho\omega^2 u_i = 0 \tag{29}$$

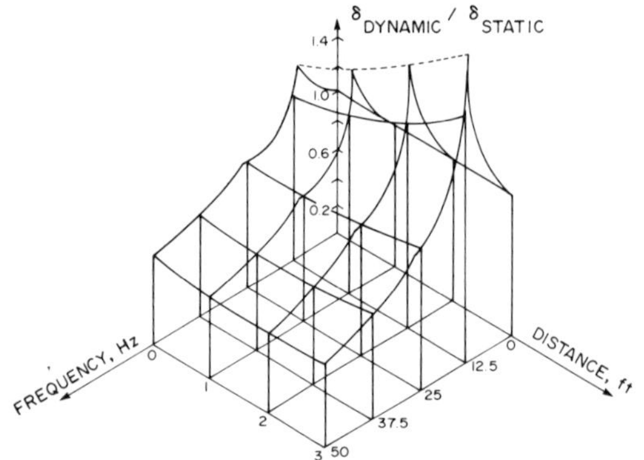

FIGURE 27. Deflection ratios at various geophone locations of the Road Rater [48].

where ϱ is the mass density. This equation is identical to the Navier-Cauchy Equation (24) except for the inclusion of the inertia term. Solution (integration) of this equation is best carried out in terms of complex numbers so that both the magnitude of the displacements and their phase with respect to some datum (typically, the loading cycle) are represented by a single quantity. Transient loading conditions (e.g., FWD impulses) can be analyzed by superimposing the spectrum of frequency responses using the methods of Fourier synthesis [50,51]. It should be clear by inspection of the Helmholtz equation that it is impossible to capture the characteristics of dynamic motion (notably resonance) by simply substituting "dynamic moduli" into the Navier-Cauchy equations.

Davies and Mamlouk [48,49,53] have carried out analyses of pavement deflections subjected to harmonic excitations, by means of the rigorous elastodynamic theory described above. They concluded that the inertial effect may be significant even at low frequencies although, in general, (in the absence of resonance) the differences between the static and dynamic deflections were more pronounced at higher frequencies. They presented their results in terms of the deflection ratio (magnification factor),

$$M = \frac{\delta_{dynamic}}{\delta_{static}} \tag{30}$$

This ratio may be significantly greater than unity of frequencies near to the resonant frequency but at higher frequencies it reduces monotonically to a value less than unity. Figure 26 shows typical values of static and dynamic deflections computed at a point located near to a Road Rater operating at 25Hz. The axis labelled "thickness" refers to the thicknesses of the individual layers of the (four-layered) pavement structure while the "stiffness" axis refers to the stiffnesses of the individual layers. The deflection ratios may be greater or smaller than those indicated here depending on the frequency of excitation. Further, the deflection ratios are not the same at all radial locations. Figure 27 shows that, for pavements of medium stiffness, the deflection ratios tend to increase with increasing distance away from the load. Yet, at some other frequencies and for other pavement structures, the reverse has been predicted. Thus, dynamic deflections resulting from Road Rater excitations are a complex function of frequency, pavement properties and geometry as well as distance from the point of application of loading.

Resonance in pavements seems to be primarily a function of subgrade thickness and stiffness, reflecting the fact that the major component of pavement deflections is deep seated. Figure 28 depicts the variation of deflection ratio with frequency for a typical pavement underlain by a subgrade of limited depth. Resonance is predicted at frequencies in the range of 7–15 Hz for this set of parameters. Thus, single frequency devices such as the Dynaflect or failure to carry out a frequency sweep with the Road Rater may give rise to misleading results. A second, and much less marked, resonance (second harmonic) occurs at frequencies approximately twice that of the fundamental.

As a working hypothesis, the numerical evidence suggests that resonance occurs when the frequency of excitation of the Road Rater satisfies the semi-empirical relation

$$f = \frac{0.4}{H}\left(\frac{G}{\varrho}\right)^{1/2} \tag{31}$$

where H is the thickness of the subgrade, G is its shear modulus and ϱ is its mass density.

Further, Davies and Mamlouk [49] suggest that increasing the operating frequency of the Road Rater by 50% above

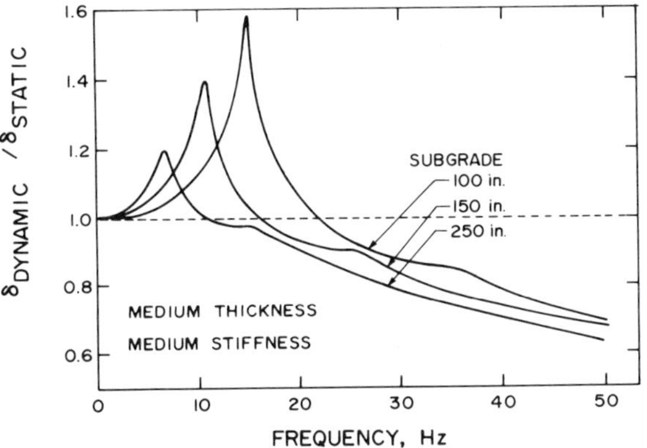

FIGURE 28. Influence of subgrade thickness on pavement resonant response [49].

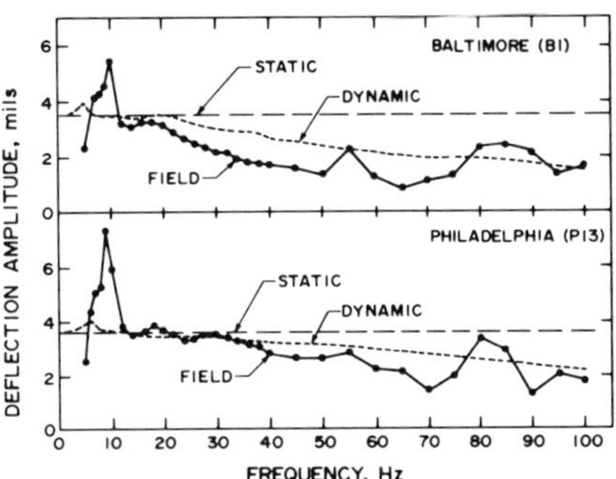

FIGURE 29. Verification of field frequency sweep using the WES 16-kip vibrator [53].

the resonant frequency (detected by field observation) should yield deflection values in close accord with the static deflections. These data may be analyzed by means of conventional elastostatic procedures.

Mamlouk [53] has examined the influence of the inertial effect on the values of layer moduli backcalculated from surface deflections. He found that significant error was incurred by using staticanalyses – ranging from 40% to 400% or more. The range of error in overlay thickness prediction varied between 40% to 110%. Field data gained from the Waterways Experiment Station (WES) 16-kip vibrator were compared with the predictions of the theoretical model. Trends in the data of decreasing deflection with increasing frequency were accurately reproduced (Figure 29). However, the (field) resonance peak at approximately 10 Hz was not captured by the analysis. It is believed that this resonance is not attributable to excitation of a fundamental mode of vibration in the pavement structure but that it is an artifact of the loading device itself. In other words, the loading system itself resonates at this frequency. Since the loading system reacts against a truck, the mass of the truck and the stiffness/damping of its shock-absorber system would have to be incorporated within the theoretical model in order to predict this phenomenon.

Davies, Mamlouk and Sebaaly [50,51] have extended their analysis of harmonic loading to deal with transient loading (such as the impulse loading of the FWD) by utilizing Fourier synthesis techniques. They showed that static analyses of pavement structures tended to overestimate deflections by, typically, 30%. Good agreement with measured truck-induced deflections and FWD measurements were obtained by this technique (Figure 30). Because resonance is less pronounced under transient loading, they concluded that FWD devices were less likely to give misleading results than vibratory loading devices.

Roesset and Shao [52] carried out similar studies to those described by Davies, Mamlouk and Sebaaly (cited above) and came to essentially identical conclusions. In particular, they showed that if the subgrade depth was less than 60 feet (Figure 31), significant resonant deflection peaks might be

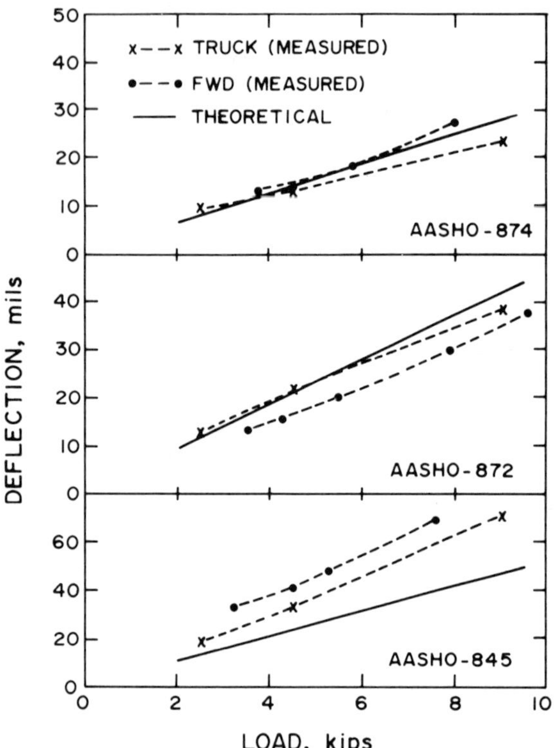

FIGURE 30. Comparison between predicted and measured deflection data for selected AASHO road test sections [50].

expected which could yield misleading results. Single frequency devices, such as the Dynaflect (which was specifically analyzed in this study), cannot provide the operator with any warning of the resonant condition.

Currently, however, no performance criteria or overlay design methods are available that incorporate dynamic analyses. Once performance criteria or overlay design methods are established, better use of the dynamic analysis can be achieved.

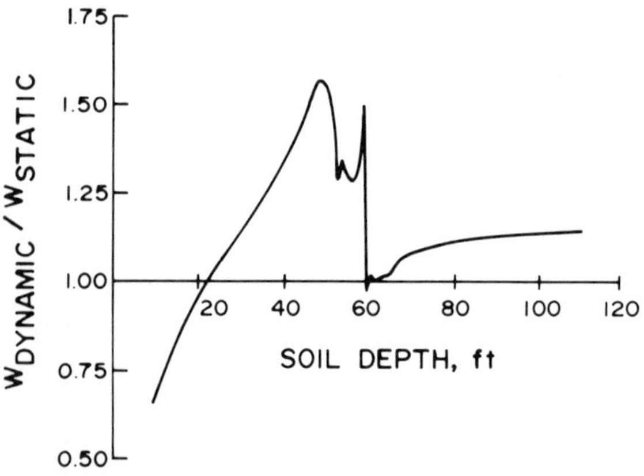

FIGURE 31. Ratio of dynamic to static deflections [52].

THE INVERSE PROBLEM

The inverse problem of determining material properties from the load-displacement response of pavements is perhaps far more difficult than current literature suggests. For a four layer flexible pavement structure, if it is assumed that the response is perfectly linearly elastic and the precise geometry of each layer is known, then (in theory at least) eight judiciously chosen and exact independent measurements of surface deflection *may* be sufficient to establish the values of the Young's moduli and Poisson's ratios of each of the four layers. Expressed in this way, the difficulties of assessing these properties come into sharp focus. So much 'noise' is introduced at each stage of the inverse process (data, modelling, solution) that the accuracy of the final results must be in considerable doubt.

Since the Poisson's ratio has little effect on surface displacement, within the practical range of values [29], this parameter is generally assumed for each layer and, in consequence, only the Young's moduli of the pavement layers are determined from (four) deflection measurements. Errors in the estimation of Poisson's ratio can easily result in errors of 5% in the calculation of surface displacements, yet this is probably the most insignificant model parameter. Thus, it might be reasonable to assume that the cumulative root-mean-square error of the several departures from reality is, at least, 25% and possibly much higher. This order of error might be acceptable if the problem were one of computing pavement deflections from the geometrical and stiffness data (the direct problem). However, the inverse problem, namely computing stiffnesses from the geometrical and deflection data is an ill-posed, ill-conditioned problem. There is no guarantee that a unique solution can be found, nor can the range of variation of solutions be determined a priori. The solutions derived in the literature from deflection data are obtained using iterative techniques which involve successive correction of initial seed values. The better iterative techniques make use of the observation that surface deflections remote from the loaded area are primarily governed by the stiffnesses of the deeper layers. This insight implies that the stiffness of the subgrade can be determined reasonably accurately. The magnitudes of the deflections in the vicinity of the loaded area are also strongly affected by the subgrade stiffness but their relative values are modified by the stiffnesses of the upper layers. In other words, the slope of the deflection basin in the vicinity of the loaded area is sensitive to the near-surface stiffness profile. If for no other reason than that these stiffnesses are functions of differences in deflection measurements, the accuracy of the solution of the inverse problem would be suspect. Yet this is only the most easily recognizable source of error in a procedure whose validity has not yet been adequately discussed, let alone verified, in the literature.

It should be noted that the suggestion that an iterative technique which yields, within a specified tolerance and from different seed values, a consistent set of layer stiffnesses is an adequate demonstration of its validity is mistaken. Since iterative techniques impose constraints on the search for solutions, additional direct computations are necessary to demonstrate that substantial and self-compensating changes in pavement stiffnesses cannot reproduce the measured profile within the bounds of observational and modelling errors. No such study has yet been reported in the literature. Until such studies have been carried out, pavement engineers would be well-advised to take a less optimistic view of the capabilities of back-analysis than that which is current in much of the literature.

CONCLUSIONS

The dynamic analysis and non-destructive testing of pavements has passed its infancy. Field observations of the operation of dynamic testing devices, notably the Road Rater and the Falling Weight Deflectometer, are capable of reproducing the response of pavements to moving wheel loading with a fair degree of accuracy. These devices can yield a wealth of data quickly and economically and in a form which can be readily marshalled for analysis. It is certain that static loading devices which do not share these advantages will pass out of favor in the near future.

The greatest difficulty that remains is data interpretation. Part of the difficulty lies in the nature of pavement material properties themselves, particularly their nonlinear response to loading. This problem can be resolved satisfactorily by only examining loading conditions which closely replicate the design loads. In effect, the pavement moduli thus determined are the secant values for the design load level. Recent developments in analysis capability have delineated the realm of applicability of purely static analyses and elucidated the nature of the dynamic response of pavements to harmonic and transient loads. More complex material models than the linear isotropic viscoelastic idealizations currently employed are unlikely to improve, by any significant extent, the predictive capacity of these analyses. Much more needs to be done, however, to verify the stability of the back-calculation procedures which are used to determine layer moduli from deflection data. Even for purely static loading conditions, this question has not been resolved. Claims that accuracies of better than 10% in the back-calculation of layer moduli can be attained should be viewed with caution. Unfortunately, until this difficulty is finally resolved in a definitive manner, the highway engineer will remain deprived of the full benefits of some potentially extremely useful devices.

Less weight has been given in this chapter to seismic nondestructive tests despite their substantial potential utility. They have not yet been adopted by any highway agencies to our knowledge. The principal reasons for this being, 1), the novelty of the method, 2), the sophistication of the equipment, and 3), the complexity of the theory underpinning the data interpretation. If the equipment becomes fully automated, these objections will become redundant. However, a further, more grave, difficulty remains; the loading conditions are completely different in magnitude and form to moving wheel loads, and therefore, considerable extrapolation is necessary to determine the layer moduli under design load conditions. It is not clear whether this problem can ever be satisfactorily resolved. Nevertheless, the seismic method can yield valuable geometric data (namely, layer thicknesses) which cannot be discerned by load tests. Perhaps, in the future, seismic and load testing devices will be combined in a single intelligent unit to yield direct readings of pavement layer stiffnesses and thicknesses – the technology for this purpose already exists.

ACKNOWLEDGEMENTS

The authors wish to thank the Asphalt Institute, Foundation Mechanics Inc. and Dynatest Consulting Inc. for providing the photographs used in this chapter. Thanks are extended to the Department of Civil Engineering and the Center for Advanced Research in Transportation at Arizona State University for making their facilities available to the authors. Dr. John P. Zaniewski is also acknowledged for reviewing the manuscript.

REFERENCES

1. Molof, A. H. and C. J. Turkstra, "Infrastructure Maintenance and Repair of Public Works," *Annals N.Y. Acad. Sci., 431* (1984).
2. Cuomo, M. in "Infrastructure Maintenance and Repair of Public Works," A. H. Molof and C. J. Turkstra (eds.), *Annals N.Y. Acad Sci., 431:* 1 (1984).
3. Matzkanin, G. A., L. S. Fountain, and O. Tranbarger, in "*Infrastructure Maintenance and Repair of Public Works,*" A. H. Molof and C. J. Turkstra (eds.), *Annals N.Y. Acad. Sci., 431:* 268 (1984).
4. Smith, R. E. and R. L. Lytton, *TRB Record 1007,* p. 1 (1985).
5. Moore, M. R., C. R. Haile, D. I. Hanson, and J. W. Hall, *TRB Circular 189* (1978).
6. AASHTO, *Standard Specifications for Transportation Research Materials and Methods of Sampling and Testing,* T256–77, Part II (1982).
7. Kennedy, C. K., P. Fevre, and C. S. Clarke, *TRRL Report 834,* United Kingdom (1978).
8. Hoffman, M. S. and M. R. Thompson, *TRB Record 852,* p. 32 (1982).
9. Hoar, R. J. and K. H. Stokoe, II, in *Dynamic Geotechnical Testing, ASTM,* STP634 (1978).

10. Heisey, J. S., K. H. Stokoe, II, and A. H. Meyer, *TRB Record 852,* p. 22 (1982).
11. Nazarian, S., K. H. Stokoe, II, and W. R. Hudson, *TRB Record 930,* p. 38 (1983).
12. Uddin, W., A. H. Meyer, W. R. Hudson, and K. H. Stokoe, II, *Proc. 3rd Int. Conf. on Concrete Pavement Design and Rehabilitation*, p. 495, Purdue Univ. (1985).
13. Richart, F. E., Jr., J. R. Hall, Jr., and R. D. Woods, *Vibrations of Soils and Foundations,* Prentice-Hall, Englewood Cliffs, N.J. (1970).
14. Jones, R., *British J. of Applied Physics, 13,* p. 21 (1962).
15. Heukelom, W. and C. R. Foster, *J. ASCE,* SM1, *86,* p. 1 (1960).
16. Szendrei, M. E. and C. R. Freeme, *J. ASCE,* SM6, *96,* p. 2099 (1970).
17. Hardin, B. O. and V. P. Drnevich, *J. ASCE,* SM7, *98,* p. 603 (1972).
18. Hardin, B. O. and V. P. Drnevich, *J. ASCE,* SM7, *98,* p. 667 (1972).
19. Hoffman, M. S., *J. ASCE,* TE5, *109,* p. 651 (1983).
20. Isihara K., and T. Kimura, *Proc. 2nd Int. Conf. on the Structural Design of Asphalt Pavements,* Ann Arbor, Mich., p. 245 (1967).
21. Weiss, R. A., *TRB Record 700,* p. 20 (1979).
22. Eringen, A. C. and E. S. Suhubi, *Elastodynamics: Linear Theory, 2,* Academic Press, N.Y. (1975).
23. Papazian, H. S., *Proc. Int. Conf. on the Structural Design of Asphalt Pavements,* Ann Arbor, Mich., p. 454 (1962).
24. Maree, J. H., N. J. W. Van Zyl and C. R. Freeme, *TRB Record 852,* p. 52 (1982).
25. Hoffman, M. S. and M. R. Thompson, *TRB Record 852,* p. 42 (1982).
26. Kingham, I. R. and B. F. Kallas, *Proc. 3rd Int. Conf. on the Structural Design of Asphalt Pavements, 1,* London, England, p. 849 (1972).
27. Stubstad, R. N. and B. Connor, *TRB Record 930,* p. 46 (1983).
28. Bibbens, R. F., C. A. Bell, and R. G. Hicks, *TRB Record 993,* p. 1 (1984).
29. Pichumani, R. *Proc. 3rd Int. Conf. on the Structural Design of Asphalt Pavements, 1,* London, England, p. 506 (1972).
30. Hardin, B. O., *J. ASCE,* SM1, *91,* p. 63 (1965).
31. Manual Series No. 17, *The Asphalt Institute* (1983).
32. *Materials Preliminary Engineering and Design Manual,* Arizona Department of Transportation (1985).
33. Burmister, D. M., *Proc. Int. Conf. on the Structural Design of Asphalt Pavements,* Ann Arbor, Mich., p. 441 (1962).
34. Kasianuk, D. A., C. L. Monismith, and W. A. Garrison, *HRB Record 291,* p. 159 (1969).
35. Dormon, G. M. and J. M. Edwards, *Proc. 2nd Int. Conf. on the Structural Design of Asphalt Pavements,* Ann Arbor, Mich., p. 99 (1967).
36. Monismith, C. L., D. A. Kasianuk, and J. A. Epps, Report TE67-4, University of California, Berkeley (1967).
37. Thompson, M. R. and M. S. Hoffman, *TRB Record 930,* p. 12 (1983).
38. Irwin, L. H., *Proc. 4th Int. Conf. on the Structural Design of Asphalt Pavements, 1,* p. 831 (1977).
39. Koole, R. C., *TRB Record 700,* p. 59 (1979).
40. McCullough, B. F. and K. J. Boedecker, *HRB Record 291,* p. 1 (1969).
41. Norman, P. I., R. A. Snowdon and J. C. Jacobs, *TRRL Report 571,* Crowthorne, United Kingdom (1973).
42. McCullough, B. F. and A. Taute, *TRB Record 852,* p. 8 (1982).
43. Sharpe, G. W., H. F. Southgate, and R. C. Deen, *J. ASCE,* TE2, *107,* p. 167 (1981).
44. *Shell Pavement Design Manual: Asphalt Pavements and Overlays for Road Traffic,* Shell Int. Petroleum Co. (1978).
45. Mamlouk, M. S., *J. of Testing and Evaluation, ASTM, 13,* p. 60 (1985).
46. Green, J. L. and J. W. Hall, U.S. Army Engineer Waterways Experiment Station, Technical Report S-75-14 (1975).
47. Kausel, E. and R. Peek, *Bull. Seismological Society of America, 72,* p. 1459 (1982).
48. Mamlouk, M. S. and T. G. Davies, *J. ASCE, 110,* p. 536 (1984).
49. Davies, T. G. and M. S. Mamlouk, *TRB Record 1022,* p. 1 (1985).
50. Sebaaly, B., T. G. Davies and M. S. Mamlouk, *J. ASCE,* TE6, *111* (1985).
51. Sebaaly, B., M. S. Mamlouk, and T. G. Davies, Presented, *65th Annual TRB Meeting* (1986).
52. Roesset, J. M. and K. Y. Shao, *TRB Record 1022* (1985).
53. Mamlouk, M. S., U.S. Army Engineer Waterways Experiment Station (1985).

SECTION FIVE

Wood Structures

CHAPTER 29

Design of Timber Structures

SHAN SOMAYAJI*

INTRODUCTION AND USES OF WOOD

Wood has been a basic construction material since the earliest days of mankind. Many classic wood structures built several years ago can still be seen around the world. Even though concrete and steel have diluted the usage of wood as the primary structural material there are probably more buildings constructed with wood than any other structural material [1]. By using the latest developments in timber design woodframe buildings of up to four storey high can be constructed.

Use of wood combines the benefits such as economy, availability, ease of workmanship, insulation, savings in time, and structural integrity. Economy comes from the fact that wood buildings can be constructed with minimal equipment combined with material cost. In addition, long-term remodeling is less expensive and relatively simple.

Wood is a renewable resource, unlike steel or concrete. United States has abundant forest resources with about 600 different tree species of which 15 are utilized in building construction [15]. Additionally about 5,000 products are being manufactured from wood [4].

Glulam (glue laminated timber) is a wood product which is used as a structural load carrying element in situations where ordinary lumber is unavailable or is too expensive. It is used in bridges, towers, marine construction, etc. It has made it possible for production of structural timbers in a wide variety of sizes and shapes. Glulams are fabricated at the plant according to job specifications and transported and erected at site, like structural steel.

Wood is used in many applications other than in residential buildings. Timber is commonly found as piles and other members in harbor construction (pressure-treated wood), in warehouse and other commercial and office buildings, in churches, etc. Recently, deterioration of old steel bridges has provoked a lot of interest in the use of timber in bridge construction [16]. Contrary to common beliefs wood can be highly weather-resistant and many species are extremely durable even when untreated. It has very low thermal conductivity and little body heat is lost by radiation [3].

Most of wood construction can be classified as a box system with horizontal framing acting as a diaphragm and vertical framing acting as a shear wall (Figure 1). It is also common to have wood horizontal system (floor or roof) supported on a masonry vertical system (shear wall).

The complete design of a box system involves the following:

1. Analysis for gravity and lateral loads.
2. Design of individual elements such as beams, columns, etc.
3. Design of horizontal and vertical subsystems (sheathing thickness, boundary elements, etc.)
4. Design of anchorage, connectors, footings, etc.

This section on timber design, due to limitations on space, describes step No. 2, namely, the design of individual elements only. Design of other components, subsystems, and systems other than box systems (arches, piles, etc.) are beyond the scope of this section. But it should be emphasized that great care should be taken in the design of connections and anchorages, and that all connections and details are given the same degree of security as the framed members [2].

PROPERTIES OF WOOD AND WOOD PRODUCTS

Structure of Wood

Wood is composed of long thin tubular cells. The cell walls are made up of cellulose and the cells are bound

*Department of Civil and Environmental Engineering, California Polytechnic State University, San Luis Obispo, CA

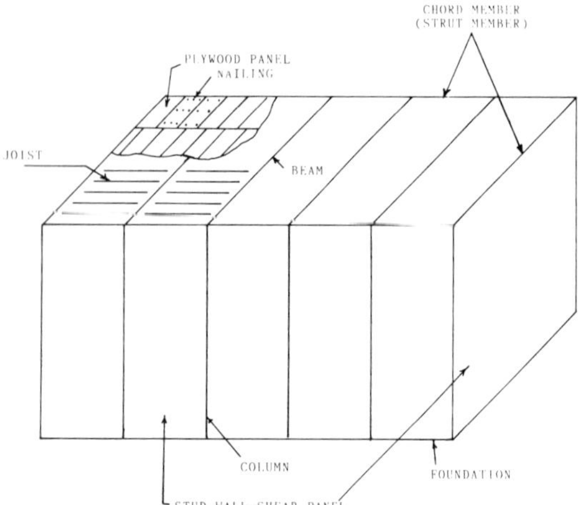

FIGURE 1. Timber box system.

together by a substance known as lignin. Most cells are oriented in the direction of the axis of truck except for cells known as rays which run radially across the trunk. Rays are present in all trees but are more pronounced in some species such as oak. In temperate countries a tree produces a new layer of wood just under the bark in the early part of every growing season. This growth ceases at the end of the growing season or during winter months. The process results in clearly visible concentric rings known as growth rings, annular rings, or annual rings (Figure 2). In tropical countries where trees grow throughout the year a tree produces wood cells that are essentially uniform.

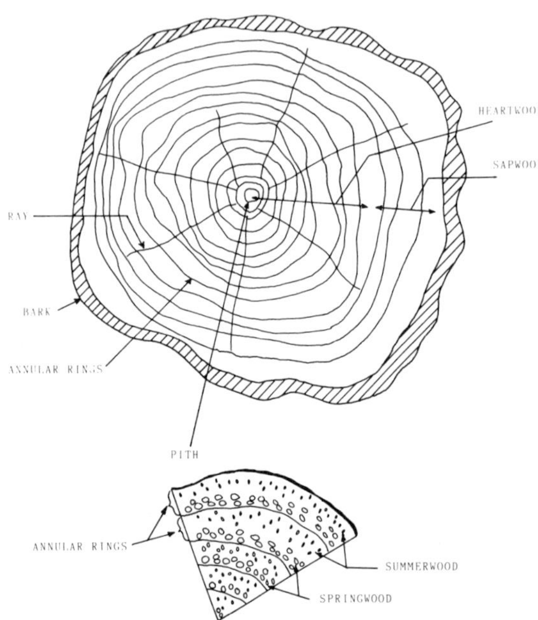

FIGURE 2. Structure of wood.

The annular band of cross section nearest to the bark is called sapwood. The central core of wood which is inside the sapwood is heartwood. The sapwood is lighter in color compared to heartwood and is 1 to 7 inches in width depending on the species. It acts as a medium of transportation of sap from roots to the leaves. On the other hand heartwood functions mainly to give mechanical support or stiffness to the trunk. But in general the moisture content, strength and weights of the two are nearly the same. Sapwood has lower natural resistance to attacks by fungi and insects and accepts preservatives more easily than heartwood.

In many trees each annular ring can be subdivided into two layers: an inner layer made up of relatively large cavities called spring wood, and an outer layer of thick walls and small cavities called summerwood (Figure 2). Since summerwood is relatively heavy, the amount of summerwood in any section is a measure of the density of wood.

Knots are a common feature of the structure of wood. It is a portion of a branch embedded by the natural growth of the tree normally originating at the center of trunk or branch.

Wood grain refers to the general direction of arrangement of fibers in wood. It is also used to describe various structural features of timber such as edge grain, flat grain, cross grain, spiral grain, and diagonal grain (Figure 3).

Hardwood and Softwood

Trees and commercial timbers are divided into two groups: hardwoods and softwoods. Hardwoods, botanically known as deciduous trees (examples: oak, maple, ash) shed their leaves in the fall. Softwoods have narrow needle-like leaves, are evergreen, and are botanically known as conifers (examples: Douglas Fir, Yellow Pine, Redwood). Most commercial hardwoods are harder than the majority of commercial softwoods. The majority of structural lumber comes from softwood category. Douglas Fir and Southern Pine are

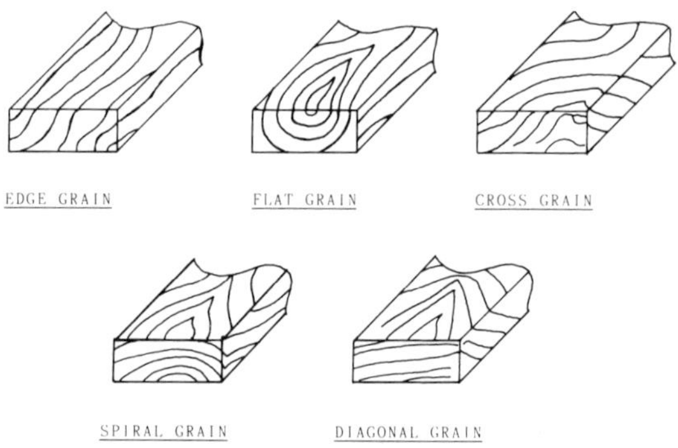

FIGURE 3. Grain patterns in wood.

two species which are widely used as structural lumber and their structural properties exceed those of many hardwoods. Domestic classification of soft and hardwoods are given in ASTM D1165 [7].

Wood Products

The following section describes in brief the most common structural and non-structural wood products.

GLULAMS

Glue-laminated timber, commonly known as glulam is a structural lumber obtained by gluing together a number of laminations of lumber ¾ in or 1.5 in thick so that the grain direction of all of them will be substantially parallel. Glulam members can be produced practically in any size, length and shape (Figure 4). Nominal 2 in. thick lumber surfaced to 1.5 in. is used to laminate straight or slightly curved members. Nominal 1 in. thick lumber surfaced to ¾ in. is commonly used to laminate curved members. Douglas Fir, Southern Pine, and Hem Fir are the most widely used species for manufacture of glulam members.

VENEERS

Veneers are thin sheets or pieces of wood obtained by rotating peeler logs in a lathe and drying to a low moisture content. Thickness of veneers range from 1/16 to 5/16 in. When used in the construction of plywood a veneer is called a ply.

PLYWOOD

Plywood is a panel made up of a number of softwood veneers. The veneers are glued together so that the grain of each layer is perpendicular to the grain of adjoining layer. In its simplest form plywood consists of 3 plies as shown in Figure 5 with grain in each ply at right angles to the grain in adjacent ply. Plywood for exterior use should be made with fully waterproof adhesive that is suitable for severe exposure conditions.

PARTICLE BOARD

A particle board is a panel made from specially prepared particles of ligno-cellulosic materials such as wood, flax, or hemp bonded with synthetic or organic binders. Due to the highly advanced mechanical techniques used in the manufacture of particle board, large flat panels can be produced with no grains or defects. It can be used for roof decking, flooring, wall-paneling, etc.

LUMBER

Lumber is the term used for the pieces of wood obtained from the trunks after sawing into lengths parallel to axis of the trunks.

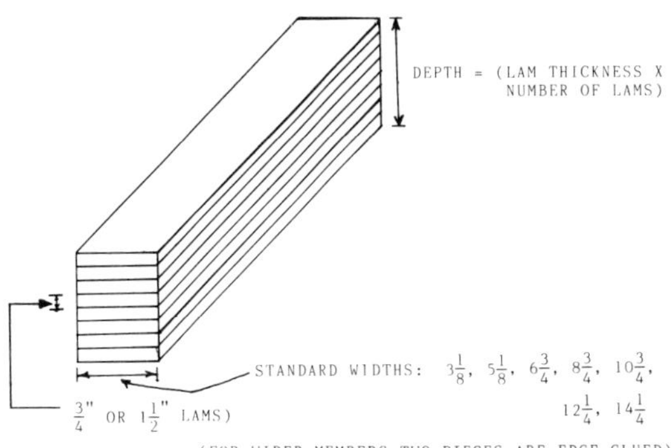

FIGURE 4. Glulam member.

TIMBER

Timber is classified as lumber with a minimum dimension of 5 in.

Moisture Content, Shrinkage, and Seasoning

A living tree may contain water as much as two times the weight of its solid material. Moisture content is the weight of water in wood expressed as a percentage of the oven dry weight. Structural lumber can take or give off moisture depending on the relative humidity of the surrounding. As the wood dries water is driven off the cell cavities. A point is reached when the cavities contain only air and cell walls are full of moisture, which is called fiber-saturation point. The moisture content at the fiber saturation point varies between 25–30%. During this drying period volume of wood remains constant as well as most mechanical properties although the density decreases. Beyond the fiber saturation point additional drying begins with the removal of water from cell walls. This shrinkage may continue linearly up to zero moisture content.

Below the fiber saturation point if the moisture is lost

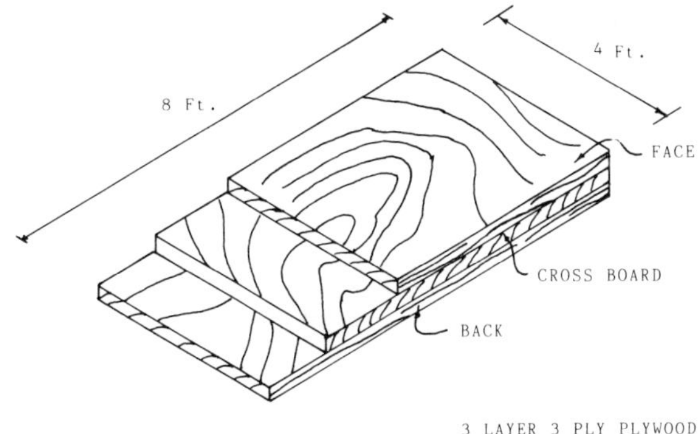

FIGURE 5. Structure of plywood.

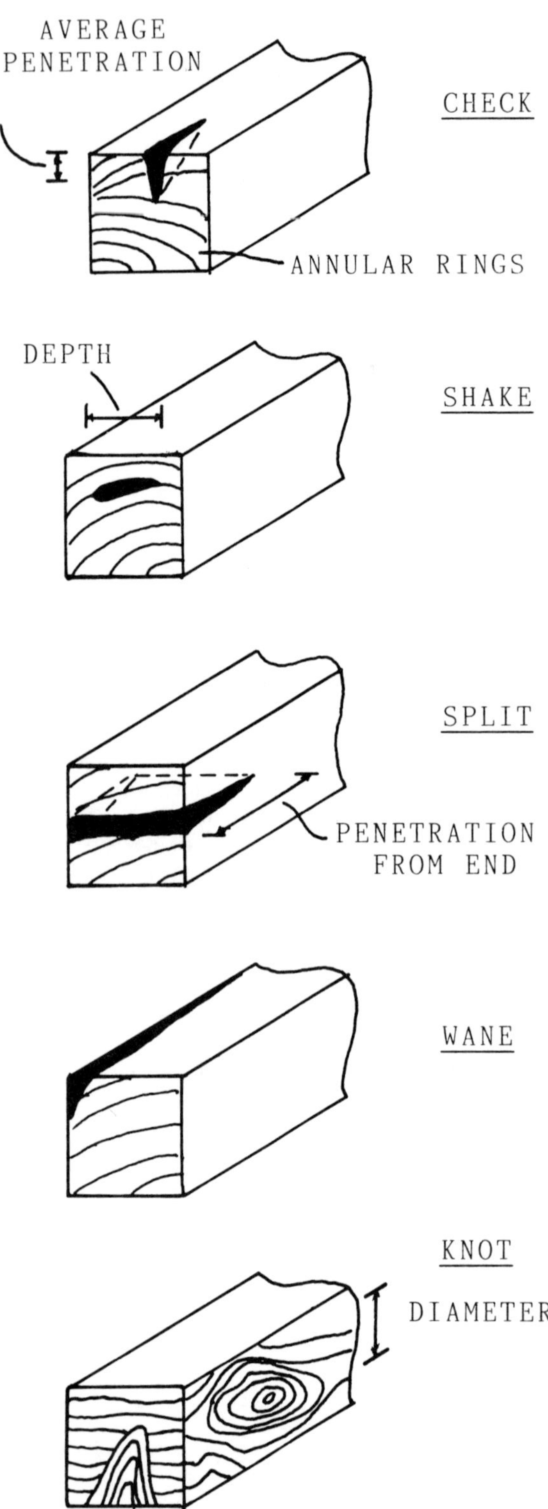

FIGURE 6. Lumber defects.

wood shrinks and if the moisture is gained it swells (primarily in the direction perpendicular to grain). Eventually wood assumes a condition of equilibrium with the final moisture content dependent on the relative humidity and ambient temperature. The moisture content at this condition, called the equilibrium moisture content of structural lumber, will range between 5 to 17% at 70°F and relative humidity between 20 to 80%.

Decrease in moisture content below fiber saturation point is often accompanied by increases in all strength characteristics of structural lumber. The process of controlled drying of lumber in order to increase structural properties is known as seasoning. It is performed by air drying or kiln drying. Cost of this artificially dried lumber is very high. The strength increase from seasoning is due to stiffening and strengthening of wood fibers. Wood is much less prone to decay if its moisture content is below 25%. On the other hand the cracks in wood resulting from drying shrinkage may affect the beneficial effects of seasoning.

Lumber with moisture content of 19% or less is called dry lumber, and lumber with moisture content of over 20% is called green lumber or unseasoned lumber. The grade stamp (as per grading rules established by various organizations such as Western Wood Products Association, National Lumber Grades Authority, etc.) on a piece of lumber indicates the moisture condition (surface dry or surface green) of the lumber at the time of fabrication.

Lumber is sawn to different cross-sectional dimensions based on the moisture condition. This means a certain size lumber fabricated from green wood will have dimensions larger in comparison with the same size lumber fabricated from surface dry wood. But at dry condition of use both pieces have nearly the same size (due to shrinkage). This means only one set of dimensions, corresponding to the dry size needs to be considered in structural calculations.

Lumber Defects and Treatment

Lumber can be extremely durable structural material; its durability and strength depends on environmental attacks and environmental conditions. The factors influencing the strength of lumber are density, moisture content, size, and presence of defects.

Environmental attacks are from various types of fungi and insects. Fungi are low forms of plant growth producing thin branching tubes which spread through the wood and use cell walls and lignin as food causing wood disintegration. The conditions required for fungal growth are proper temperature, moisture content over 19% oxygen, and food (wood fiber). The three common diseases produced by this attack are: sap stain, mold, and wood rot.

Termites and certain insects can cause wood destruction. These are normally found in warmer climates and/or environment of high humidity. Procedure for control of insects can be found in Reference [12].

Environmental conditions cause irregularity in the structure of wood, called defects. The common defects are: checks, shakes, cross breaks, compression failure, cross grain, knots, decay, splits, warp, holes, bark pockets, pitch pockets, and wane.

CHECKS

A check is a lengthwise separation of wood occurring across annular rings largely due to nonuniform shrinkage (shrinkage is greater parallel to annular rings than normal annular rings) (Figure 6).

SHAKES

A shake is a lengthwise separation of wood occurring between and parallel to annular rings and are developed in the living tree (Figure 6).

SPLITS

A split is a complete separation of wood fibers throughout the thickness of the member resulting from seasoning and handling (Figure 6).

KNOTS

A knot is a cross section or longitudinal section of branches which was cut with the lumber (Figure 6). Knots displace the clear wood and force the slope of grain to deviate, thus affecting the mechanical properties of wood. They also allow stress concentration to occur.

CROSS BREAKS

These are tension failures usually due to a local abnormal longitudinal shrinkage. Wood having these characteristics is called compression wood. It is commonly found underside of leaning trees or branches. It denotes unbalanced structure in wood and are not permitted in readily identifiable form in stress graded lumber.

WANE

Wane is bark or lack of wood on the edge and corner of a piece of lumber (Figure 6). The effect of this is to reduce the cross-sectional area of the lumber which is more pronounced on bearing area.

TREATMENT

Wood can be chemically treated for two reasons: 1) to prevent destruction due to fungi or termite attack and 2) to inhibit combustion. The treatment is carried out by injecting chemicals into wood fibers and is called pressure treatment. Such preservative treatments are expensive and may be required when wood is fully exposed to weather without any cover such as lumber in contact with ground, sill plates on ground floor slabs and footings, piles under salt water, fence posts, etc.

Glulams can be fabricated by pressure treatment of individual lams before the fabrication or after gluing.

Even without the pressure treatment wood structures can behave extremely well in a fire. Heartwood of all species has some degree of natural resistance to decay while the sapwood is susceptible to deay. Unlike other structural materials such as steel or concrete, wood does not lose its strength when the temperature is raised above 480°F. Its strength loss in a fire is mainly due to the material lost through charring of the surface. The use of fire retardant treated lumber is advisable only when building codes or the designer specifies extra measure of fire protection.

The common chemicals used for pressure treatment are: Creosote and creosote solution, penta, oil-borne chemicals, water-borne inorganic compounds, and water soluble salts (fire-retardants). Creosote solutions are used for outdoor exposure conditions whereas water-borne preservatives may be used for both exterior and interior applications.

Pressure treated lumber is used in a wide variety of structural applications. The common applications are in piles for wharves and breakwaters, utility poles, farm structures, retaining walls, decking, railroad ties, sill plates, etc.

In the case of lumber and glulam members fabricated with fire retardant treatment the specification allows lower allowable stresses. General information on pressure treated wood is found in References [13] and [14].

LUMBER SIZES AND GRADES

Nominal and Dressed Sizes

A piece of lumber is identified by its nominal size which is the commercial size designation of width and depth of standard sawn and glulam members (example: 2×4, 4×8). The dressed size (net size) which is used in all structural calculations is smaller than the nominal size. For example structural lumber of nominal size 2×8 has dressed size 1.5×7.25 ins. (Table 1). The geometrical properties such as area, moment of inertia, section modulus, etc. should be based on the net size. But actual sizes (measured sizes) of two members of the same nominal size may be different when moisture contents of the two are different. But, as mentioned in an earlier section stress grading makes it possible for the use of standard dressed size for all structural calculations, thus simplifying the design process (without a need to use measured dimensions). The standard dressed size (dry condition) or various size structural lumber given in NDS [6] are typical for lumber surfaced on all four sides (S4S).

Glulam Sizes

Size classification of glulam members is based on standard width, thickness and number of laminations. The nominal and net widths of glulam members are shown in Table 2. The net depth of a glulam member depends on the

TABLE 1. Section Properties of Sawn Lumber (From Reference [6]).

Nominal b(inches)d	Standard dressed size (S4S) b(inches)d	Area of Section A	Moment of inertia I	Section modulus S	Weight* in pounds per linear foot of piece when weight of wood per cubic foot equals:					
					25 lb.	30 lb.	35 lb.	40 lb.	45 lb.	50 lb.
1 x 3	¾ x 2½	1.875	0.977	0.781	0.326	0.391	0.456	0.521	0.586	0.651
1 x 4	¾ x 3½	2.625	2.680	1.531	0.456	0.547	0.638	0.729	0.820	0.911
1 x 6	¾ x 5½	4.125	10.398	3.781	0.716	0.859	1.003	1.146	1.289	1.432
1 x 8	¾ x 7¼	5.438	23.817	6.570	0.944	1.133	1.322	1.510	1.699	1.888
1 x 10	¾ x 9¼	6.938	49.466	10.695	1.201	1.445	1.686	1.927	2.168	2.409
1 x 12	¾ x 11¼	8.438	88.989	15.820	1.465	1.758	2.051	2.344	2.637	2.930
2 x 3	1½ x 2½	3.750	1.953	1.563	0.651	0.781	0.911	1.042	1.172	1.302
2 x 4	1½ x 3½	5.250	5.359	3.063	0.911	1.094	1.276	1.458	1.641	1.823
2 x 5	1½ x 4½	6.750	11.391	5.063	1.172	1.406	1.641	1.875	2.109	2.344
2 x 6	1½ x 5½	8.250	20.797	7.563	1.432	1.719	2.005	2.292	2.578	2.865
2 x 8	1½ x 7¼	10.875	47.635	13.141	1.888	2.266	2.643	3.021	3.398	3.776
2 x 10	1½ x 9¼	13.875	98.932	21.391	2.409	2.891	3.372	3.854	4.336	4.818
2 x 12	1½ x 11¼	16.875	177.979	31.641	2.930	3.516	4.102	4.688	5.273	5.859
2 x 14	1½ x 13¼	19.875	290.775	43.891	3.451	4.141	4.831	5.521	6.211	6.901
3 x 1	2½ x ¾	1.875	0.088	0.234	0.326	0.391	0.456	0.521	0.586	0.651
3 x 2	2½ x 1½	3.750	0.703	0.938	0.651	0.781	0.911	1.042	1.172	1.302
3 x 4	2½ x 3½	8.750	8.932	5.104	1.519	1.823	2.127	2.431	2.734	3.038
3 x 5	2½ x 4½	11.250	18.984	8.438	1.953	2.344	2.734	3.125	3.516	3.906
3 x 6	2½ x 5½	13.750	34.661	12.604	2.387	2.865	3.342	3.819	4.297	4.774
3 x 8	2½ x 7¼	18.125	79.391	21.901	3.147	3.776	4.405	5.035	5.664	6.293
3 x 10	2½ x 9¼	23.125	164.886	35.651	4.015	4.818	5.621	6.424	7.227	8.030
3 x 12	2½ x 11¼	28.125	296.631	52.734	4.883	5.859	6.836	7.813	8.789	9.766
3 x 14	2½ x 13¼	33.125	484.625	73.151	5.751	6.901	8.051	9.201	10.352	11.502
3 x 16	2½ x 15¼	38.125	738.870	96.901	6.619	7.943	9.266	10.590	11.914	13.238
4 x 1	3½ x ¾	2.625	0.123	0.328	0.456	0.547	0.638	0.729	0.820	0.911
4 x 2	3½ x 1½	5.250	0.984	1.313	0.911	1.094	1.276	1.458	1.641	1.823
4 x 3	3½ x 2½	8.750	4.557	3.646	1.519	1.823	2.127	2.431	2.734	3.038
4 x 4	3½ x 3½	12.250	12.505	7.146	2.127	2.552	2.977	3.403	3.828	4.253
4 x 5	3½ x 4½	15.750	26.578	11.813	2.734	3.281	3.828	4.375	4.922	5.469
4 x 6	3½ x 5½	19.250	48.526	17.646	3.342	4.010	4.679	5.347	6.016	6.684
4 x 8	3½ x 7¼	25.375	111.148	30.661	4.405	5.286	6.168	7.049	7.930	8.811
4 x 10	3½ x 9¼	32.375	230.840	49.911	5.621	6.745	7.869	8.933	10.117	11.241
4 x 12	3½ x 11¼	39.375	415.283	73.828	6.836	8.203	9.570	10.938	12.305	13.672
4 x 14	3½ x 13¼	46.375	678.475	102.411	8.047	9.657	11.266	12.877	14.485	16.094
4 x 16	3½ x 15¼	53.375	1034.418	135.66	9.267	11.121	12.975	14.828	16.682	18.536
5 x 2	4½ x 1½	6.750	1.266	1.688	1.172	1.406	1.641	1.875	2.109	2.344
5 x 3	4½ x 2½	11.250	5.859	4.688	1.953	2.344	2.734	3.125	3.516	3.906
5 x 4	4½ x 3½	15.750	16.078	9.188	2.734	3.281	3.828	4.375	4.922	5.469
5 x 5	4½ x 4½	20.250	34.172	15.188	3.516	4.219	4.922	5.675	6.328	7.031
6 x 1	5½ x ¾	4.125	0.193	0.516	0.716	0.859	1.003	1.146	1.289	1.432
6 x 2	5½ x 1½	8.250	1.547	2.063	1.432	1.719	2.005	2.292	2.578	2.865
6 x 3	5½ x 2½	13.750	7.161	5.729	2.387	2.865	3.342	3.819	4.297	4.774
6 x 4	5½ x 3½	19.250	19.651	11.229	3.342	4.010	4.679	5.347	6.016	6.684
6 x 6	5½ x 5½	30.250	76.255	27.729	5.252	6.302	7.352	8.403	9.453	10.503
6 x 8	5½ x 7½	41.250	193.359	51.563	7.161	8.594	10.026	11.458	12.891	14.323
6 x 10	5½ x 9½	52.250	392.963	82.729	9.071	10.885	12.700	14.514	16.328	18.142
6 x 12	5½ x 11½	63.250	697.068	121.229	10.981	13.177	15.373	17.569	19.766	21.962
6 x 14	5½ x 13½	74.250	1127.672	167.063	12.891	15.469	18.047	20.625	23.203	25.781
6 x 16	5½ x 15½	85.250	1706.776	220.229	14.800	17.760	20.720	23.681	26.641	29.601

(continued)

TABLE 1. (continued).

Nominal b(inches)d	Standard dressed size (545) b(inches)d	Area of Section A	Moment of inertia I	Section modulus S	Weight* in pounds per linear foot of piece when weight of wood per cubic foot equals: 25 lb.	30 lb.	35 lb.	40 lb.	45 lb.	50 lb.
6 x 18	5½ x 17½	96.250	2456.380	280.729	16.710	20.052	23.394	26.736	30.078	33.420
6 x 20	5½ x 19½	107.250	3398.484	348.563	18.620	22.344	26.068	29.792	33.516	37.240
6 x 22	5½ x 21½	118.250	4555.086	423.729	20.530	24.635	28.741	32.847	36.953	41.059
6 x 24	5½ x 23½	129.250	5948.191	506.229	22.439	26.927	31.415	35.903	40.391	44.878
8 x 1	7¼ x ¾	5.438	0.255	0.680	0.944	1.133	1.322	1.510	1.699	1.888
8 x 2	7¼ x 1½	10.875	2.039	2.719	1.888	2.266	2.643	3.021	3.398	3.776
8 x 3	7¼ x 2½	18.125	9.440	7.552	3.147	3.776	4.405	5.035	5.664	6.293
8 x 4	7¼ x 3½	25.375	25.904	14.803	4.405	5.286	6.168	7.049	7.930	8.811
8 x 6	7½ x 5½	41.250	103.984	37.813	7.161	8.594	10.026	11.458	12.891	14.323
8 x 8	7½ x 7½	56.250	263.672	70.313	9.766	11.719	13.672	15.625	17.578	19.531
8 x 10	7½ x 9½	71.250	535.859	112.813	12.370	14.844	17.318	19.792	22.266	24.740
8 x 12	7½ x 11½	86.250	950.547	165.313	14.974	17.969	20.964	23.958	26.953	29.948
8 x 14	7½ x 13½	101.250	1537.734	227.813	17.578	21.094	24.609	28.125	31.641	35.156
8 x 16	7½ x 15½	116.250	2327.422	300.313	20.182	24.219	28.255	32.292	36.328	40.365
8 x 18	7½ x 17½	131.250	3349.609	382.813	22.786	27.344	31.901	36.458	41.016	45.573
8 x 20	7½ x 19½	146.250	4634.297	475.313	25.391	30.469	35.547	40.625	45.703	50.781
8 x 22	7½ x 21½	161.250	6211.484	577.813	27.995	33.594	39.193	44.792	50.391	55.990
8 x 24	7½ x 23½	176.250	8111.172	690.313	30.599	36.719	42.839	48.958	55.078	61.198
10 x 1	9¼ x ¾	6.938	0.325	0.867	1.204	1.445	1.686	1.927	2.168	2.409
10 x 2	9¼ x 1½	13.875	2.602	3.469	2.409	2.891	3.372	3.854	4.336	4.818
10 x 3	9¼ x 2½	23.125	12.044	9.635	4.015	4.818	5.621	6.424	7.227	8.030
10 x 4	9¼ x 3½	32.375	33.049	18.885	5.621	6.745	7.869	8.993	10.117	11.241
10 x 6	9½ x 5½	52.250	131.714	47.896	9.071	10.885	12.700	14.514	16.328	18.142
10 x 8	9½ x 7½	71.250	333.984	89.063	12.370	14.844	17.318	19.792	22.266	24.740
10 x 10	9½ x 9½	90.250	678.755	142.896	15.668	18.802	21.936	25.069	28.203	31.337
10 x 12	9½ x 11½	109.250	1204.026	209.396	18.967	22.760	26.554	30.347	34.141	37.934
10 x 14	9½ x 13½	128.250	1947.797	288.563	22.266	26.719	31.172	35.625	40.078	44.531
10 x 16	9½ x 15½	147.250	2948.068	380.396	25.564	30.677	35.790	40.903	46.016	51.128
10 x 18	9½ x 17½	166.250	4242.836	484.896	28.863	34.635	40.408	46.181	51.953	57.726
10 x 20	9½ x 19½	185.250	5870.109	602.063	32.161	38.594	45.026	51.458	57.891	64.323
10 x 22	9½ x 21½	204.250	7867.879	731.896	35.460	42.552	49.644	56.736	63.828	70.920
10 x 24	9½ x 23½	223.250	10274.148	874.396	38.759	46.510	54.262	62.014	69.766	77.517
12 x 1	11¼ x ¾	8.438	0.396	1.055	1.465	1.758	2.051	2.344	2.637	2.930
12 x 2	11¼ x 1½	16.875	3.164	4.129	2.930	3.516	4.102	4.688	5.273	5.859
12 x 3	11¼ x 2½	28.125	14.648	11.719	4.883	5.859	6.836	7.813	8.789	9.766
12 x 4	11¼ x 3½	39.375	40.195	22.969	6.836	8.203	9.570	10.938	12.305	13.672
12 x 6	11½ x 5½	63.250	159.443	57.979	10.981	13.177	15.373	17.569	19.766	21.962
12 x 8	11½ x 7½	86.250	404.297	107.813	14.974	17.969	20.964	23.958	26.953	29.948
12 x 10	11½ x 9½	109.250	821.651	172.979	18.967	22.760	26.554	30.347	34.141	37.934
12 x 12	11½ x 11½	132.250	1457.505	253.479	22.960	27.552	32.144	36.736	41.328	45.920
12 x 14	11½ x 13½	155.250	2357.859	349.313	26.953	32.344	37.734	43.125	48.516	53.906
12 x 16	11½ x 15½	178.250	3658.713	460.479	30.946	37.135	43.325	49.514	55.703	61.892
12 x 18	11½ x 17½	201.250	5136.066	586.979	34.939	41.927	48.915	55.903	62.891	69.878
12 x 20	11½ x 19½	224.250	7105.922	728.813	38.932	46.719	54.505	62.292	70.078	77.865
12 x 22	11½ x 21½	247.250	9524.273	885.979	42.925	51.510	60.095	68.681	77.266	85.851
12 x 24	11½ x 23½	270.250	12437.129	1058.479	46.918	56.302	65.686	75.069	84.453	93.837
14 x 2	13¼ x 1½	19.875	3.727	4.969	3.451	4.141	4.831	5.521	6.211	6.901
14 x 3	13¼ x 2½	33.125	17.253	13.802	5.751	6.901	8.051	9.201	10.352	11.502
14 x 4	13¼ x 3½	46.375	47.34	27.052	8.047	9.657	11.266	12.877	14.485	16.094

(continued)

TABLE 1. (continued).

Nominal b(inches)d	Standard dressed size (545) b(inches)d	Area of Section A	Moment of inertia I	Section modulus S	Weight* in pounds per linear foot of piece when weight of wood per cubic foot equals: 25 lb.	30 lb.	35 lb.	40 lb.	45 lb.	50 lb.
14 x 6	13½ x 5½	74.250	187.172	68.063	12.891	15.469	18.047	20.625	23.203	25.781
14 x 8	13½ x 7½	101.250	474.609	176.563	17.578	21.094	24.609	28.125	31.641	35.156
14 x 10	13½ x 9½	128.250	964.547	203.063	22.266	26.719	31.172	35.625	40.078	44.531
14 x 12	13½ x 11½	155.250	1710.984	297.563	26.953	32.344	37.734	43.125	48.516	53.906
14 x 14	13½ x 13½	182.250	2767.922	410.063	31.641	37.969	44.297	50.625	56.953	63.281
14 x 16	13½ x 15½	209.250	4189.359	540.563	36.328	43.594	50.859	58.125	65.391	72.656
14 x 18	13½ x 17½	236.250	6029.297	689.063	41.016	49.219	57.422	65.625	73.828	82.031
14 x 20	13½ x 19½	263.250	8341.734	855.563	45.703	54.844	63.984	73.125	82.266	91.406
14 x 22	13½ x 21½	290.250	11180.672	1040.063	50.391	60.469	70.547	80.625	90.703	100.781
14 x 24	13½ x 23½	317.250	14600.109	1242.563	55.078	66.094	77.109	88.125	99.141	110.156
16 x 3	15¼ x 2½	38.125	19.857	15.885	6.619	7.944	9.267	10.592	11.915	13.240
16 x 4	15¼ x 3½	53.375	54.487	31.135	9.267	11.121	12.975	14.828	16.682	18.536
16 x 6	15½ x 5½	85.250	214.901	78.146	14.800	17.760	20.720	23.681	26.641	29.601
16 x 8	15½ x 7½	116.250	544.922	145.313	20.182	24.219	28.255	32.292	36.328	40.365
16 x 10	15½ x 9½	147.250	1107.443	233.146	25.564	30.677	35.790	40.903	46.016	51.128
16 x 12	15½ x 11½	178.250	1964.463	341.646	30.946	37.135	43.325	49.514	55.703	61.892
16 x 14	15½ x 13½	209.250	3177.984	470.813	36.328	43.594	50.859	58.125	65.391	72.656
16 x 16	15½ x 15½	240.250	4810.004	620.646	41.710	50.052	58.394	66.736	75.078	83.420
16 x 18	15½ x 17½	271.250	6922.523	791.146	47.092	56.510	65.929	75.347	84.766	94.184
16 x 20	15½ x 19½	302.250	9577.547	982.313	52.474	62.969	73.464	83.958	94.453	104.948
16 x 22	15½ x 21½	333.250	12837.066	1194.146	57.856	69.427	80.998	92.569	104.141	115.712
16 x 24	15½ x 23½	364.250	16763.086	1426.646	63.238	75.885	88.533	101.181	113.828	126.476
18 x 6	17½ x 5½	96.250	242.630	88.229	16.710	20.052	23.394	26.736	30.078	33.420
18 x 8	17½ x 7½	131.250	615.234	164.063	22.786	27.344	31.901	36.458	41.016	45.573
18 x 10	17½ x 9½	166.250	1250.338	263.229	28.863	34.635	40.408	46.181	51.953	57.726
18 x 12	17½ x 11½	201.250	2217.943	385.729	34.939	41.927	48.915	55.903	62.891	69.878
18 x 14	17½ x 13½	236.250	3588.047	531.563	41.016	49.219	57.422	65.625	73.828	82.031
18 x 16	17½ x 15½	271.250	5430.648	700.729	47.092	56.510	65.929	75.347	84.766	94.184
18 x 18	17½ x 17½	306.250	7815.754	893.229	53.168	63.802	74.436	85.069	95.703	106.337
18 x 20	17½ x 19½	341.250	10813.359	1109.063	59.245	71.094	82.943	94.792	106.641	118.490
18 x 22	17½ x 21½	376.250	14493.461	1348.229	65.321	78.385	91.450	104.514	117.578	130.642
18 x 24	17½ x 23½	411.250	18926.066	1610.729	71.398	85.677	99.957	114.236	128.516	142.795

*Weight in lb/Ft^3 = 62.4 × Specific gravity.

number of laminations which are of ¾ in. or 1.5 in. thick. For example, glulam beam of nominal width 6 in. and fabricated from 20 lams of 1.5 in. thick has a net size of 5.125 × 30 ins. Typically 1.5 in thick dimension lumber is used to laminate straight glulam members and ¾ in. thick lumber is used to manufacture curved members. Table 3 shows some cross-sectional properties of glulam members.

TABLE 2. Standard Widths of Glulam Members.

Nominal Width ins.	3	4	6	8	10	12	14	16
Net Width ins.	2.25	3.125	5.125	6.75	8.75	10.75	12.125	14.125

Grading

SAWN LUMBER

Lumber grading rules have been developed based on anticipated use of lumber considering its size, even though no restrictions are made for the actual use of any size lumber. This means a 6 × 6 lumber, because of its square cross section, falls under post and timber category even though it can very well be used as a beam. Table 4 illustrates the size classification of sawn lumber. Each lumber in the size category is further subdivided based on the stress grading as shown in Table 5.

TABLE 3. Section Properties of Glulam (From Reference [4]).

Use 1½″ columns in table for straight beams; use ¾″ columns in table for curved arches.

* 3⅛″ WIDTH

No. of Lams 1½″	No. of Lams ¾″	Depth d (in.)	Size Factor C_F	Area A (in.^{2})	Section Modulus S (in.^{3})	Moment of Inertia I (in.^{4})
2	4	3.00	1.00	9.4	4.7	7.0
	5	3.75	1.00	11.7	7.3	13.7
3	6	4.50	1.00	14.1	10.5	23.7
	7	5.25	1.00	16.4	14.4	37.7
4	8	6.00	1.00	18.8	18.8	56.3
	9	6.75	1.00	21.1	23.7	80.1
5	10	7.50	1.00	23.4	29.3	109.9
	11	8.25	1.00	25.8	35.4	146.2
6	12	9.00	1.00	28.1	42.2	189.8
	13	9.75	1.00	30.5	49.5	241.4
7	14	10.50	1.00	32.8	57.4	301.5
	15	11.25	1.00	35.2	65.9	370.8
8	16	12.00	1.00	37.5	75.0	450.0
	17	12.75	0.99	39.8	84.7	539.8
9	18	13.50	0.99	42.2	94.9	640.7
	19	14.25	0.98	44.5	105.8	753.6
10	20	15.00	0.98	46.9	117.2	878.9
	21	15.75	0.97	49.2	129.2	1,017.4
11	22	16.50	0.97	51.6	141.8	1,169.8
	23	17.25	0.96	53.9	155.0	1,336.7
12	24	18.00	0.96	56.3	168.8	1,518.8
	25	18.75	0.95	58.6	183.1	1,716.6
13	26	19.50	0.95	60.9	198.0	1,931.0
	27	20.25	0.94	63.3	213.6	2,162.4
14	28	21.00	0.94	65.6	229.7	2,411.7
	29	21.75	0.94	68.0	246.4	2,679.5
15	30	22.50	0.93	70.3	263.7	2,966.3
	31	23.25	0.93	72.7	281.5	3,272.9
16	32	24.00	0.93	75.0	300.0	3,600.0

* 5⅛″ WIDTH

No. of Lams 1½″	No. of Lams ¾″	Depth d (in.)	Size Factor C_F	Area A (in.^{2})	Section Modulus S (in.^{3})	Moment of Inertia I (in.^{4})
3	6	4.50	1.00	23.1	17.3	38.9
	7	5.25	1.00	26.9	23.5	61.8
4	8	6.00	1.00	30.8	30.8	92.3
	9	6.75	1.00	34.6	38.9	131.3
5	10	7.50	1.00	38.4	48.0	180.2
	11	8.25	1.00	42.3	58.1	239.8
6	12	9.00	1.00	46.1	69.2	311.3
	13	9.75	1.00	50.0	81.2	395.8
7	14	10.50	1.00	53.8	94.2	494.4
	15	11.25	1.00	57.7	108.1	608.1
8	16	12.00	1.00	61.5	123.0	738.0
	17	12.75	0.99	65.3	138.9	885.2
9	18	13.50	0.99	69.2	155.7	1,050.8
	19	14.25	0.98	73.0	173.4	1,235.8
10	20	15.00	0.98	76.9	192.2	1,441.4
	21	15.75	0.97	80.7	211.9	1,668.6
11	22	16.50	0.97	84.6	232.5	1,918.5
	23	17.25	0.96	88.4	254.2	2,192.2
12	24	18.00	0.96	92.3	276.8	2,490.8
	25	18.75	0.95	96.1	300.3	2,815.2
13	26	19.50	0.95	99.9	324.8	3,166.8
	27	20.25	0.94	103.8	350.3	3,546.4
14	28	21.00	0.94	107.6	376.7	3,955.2
	29	21.75	0.94	111.5	404.1	4,394.3
15	30	22.50	0.93	115.3	432.4	4,864.7
	31	23.25	0.93	119.2	461.7	5,367.6
16	32	24.00	0.93	123.0	492.0	5,904.0
	33	24.75	0.92	126.8	523.2	6,475.0
17	34	25.50	0.92	130.7	555.4	7,081.6
	35	26.25	0.92	134.5	588.6	7,725.0
18	36	27.00	0.91	138.4	622.7	8,406.3
	37	27.75	0.91	142.2	657.8	9,126.4
19	38	28.50	0.91	146.1	693.8	9.886.6
	39	29.25	0.91	149.9	730.8	10,687.8
20	40	30.00	0.90	153.8	768.8	11,531.3
	41	30.75	0.90	157.6	807.7	12,417.9
21	42	31.50	0.90	161.4	847.5	13,348.9
	43	32.25	0.90	165.3	888.4	14,325.2
22	44	33.00	0.89	169.1	930.2	15,348.1
	45	33.75	0.89	173.0	972.9	16,418.5
23	46	34.50	0.89	176.8	1,016.7	17,537.6
	47	35.25	0.89	180.7	1,061.4	18,706.4
24	48	36.00	0.88	184.5	1,107.0	19,926.0

6¾″ WIDTH

No. of Lams 1½″	No. of Lams ¾″	Depth d (in.)	Size Factor C_F	Area A (in.^{2})	Section Modulus S (in.^{3})	Moment of Inertia I (in.^{4})
4	8	6.00	1.00	40.5	40.5	121.5
	9	6.75	1.00	45.6	51.3	173.0
5	10	7.50	1.00	50.6	63.3	237.3
	11	8.25	1.00	55.7	76.6	315.9
6	12	9.00	1.00	60.8	91.1	410.1
	13	9.75	1.00	65.8	106.9	521.4
7	14	10.50	1.00	70.9	124.0	651.2
	15	11.25	1.00	75.9	142.4	800.9
8	16	12.00	1.00	81.0	162.0	972.0
	17	12.75	0.99	86.1	182.9	1,165.9
9	18	13.50	0.99	91.1	205.0	1,384.0
	19	14.25	0.98	96.2	228.4	1,627.7
10	20	15.00	0.98	101.3	253.1	1,898.4
	21	15.75	0.97	106.3	279.1	2,197.7
11	22	16.50	0.97	111.4	306.3	2,526.8
	23	17.25	0.96	116.4	334.8	2,887.3
12	24	18.00	0.96	121.5	364.5	3,280.5
	25	18.75	0.95	126.6	395.5	3,707.9
13	26	19.50	0.95	131.6	427.8	4,170.9
	27	20.25	0.94	136.7	461.3	4,670.9
14	28	21.00	0.94	141.8	496.1	5,209.3
	29	21.75	0.94	146.8	532.2	5,787.6
15	30	22.50	0.93	151.9	569.5	6,407.2
	31	23.25	0.93	156.9	608.1	7,069.5
16	32	24.00	0.93	162.0	648.0	7,776.0
	33	24.75	0.92	167.1	689.1	8,528.0
17	34	25.50	0.92	172.1	731.5	9,327.0
	35	26.25	0.92	177.2	775.2	10,174.4
18	36	27.00	0.91	182.3	820.1	11,071.7
	37	27.75	0.91	187.3	866.3	12,020.2
19	38	28.50	0.91	192.4	913.8	13,021.4
	39	29.25	0.91	197.4	962.5	14,076.7
20	40	30.00	0.90	202.5	1,012.5	15,187.5
	41	30.75	0.90	207.6	1,063.8	16,355.3
21	42	31.50	0.90	212.6	1,116.3	17,581.4
	43	32.25	0.90	217.7	1,170.1	18,867.4
22	44	33.00	0.89	222.8	1,225.1	20,214.6
	45	33.75	0.89	227.8	1,281.4	21,624.4
23	46	34.50	0.89	232.9	1,339.0	23,098.3
	47	35.25	0.89	237.9	1,397.9	24,637.7
24	48	36.00	0.88	243.0	1,458.0	26,244.0
	49	36.75	0.88	248.1	1,519.4	27,918.7
25	50	37.50	0.88	253.1	1,582.0	29,663.1
	51	38.25	0.88	258.2	1,645.9	31,478.7
26	52	39.00	0.88	263.3	1,711.1	33,366.9
	53	39.75	0.88	268.3	1,777.6	35,329.2
27	54	40.50	0.87	273.4	1,845.3	37,367.0
	55	41.25	0.87	278.4	1,914.3	39,481.6
28	56	42.00	0.87	283.5	1,984.5	41,674.5
	57	42.75	0.87	288.6	2,056.0	43,947.2
29	58	43.50	0.87	293.6	2,128.8	46,301.0
	59	44.25	0.87	298.7	2,202.8	48,737.4
30	60	45.00	0.86	303.8	2,278.1	51,257.8
	61	45.75	0.86	308.8	2,354.7	53,863.7
31	62	46.50	0.86	313.9	2,432.5	56,556.4
	63	47.25	0.86	318.9	2,511.6	59,337.3
32	64	48.00	0.86	324.0	2,592.0	62,208.0

8¾″ WIDTH

No. of Lams 1½″	No. of Lams ¾″	Depth d (in.)	Size Factor C_F	Area A (in.^{2})	Section Modulus S (in.^{3})	Moment of Inertia I (in.^{4})
6	12	9.00	1.00	78.8	118.1	531.6
	13	9.75	1.00	85.3	138.6	675.8
7	14	10.50	1.00	91.9	160.8	844.1
	15	11.25	1.00	98.4	184.6	1,038.2
8	16	12.00	1.00	105.0	210.0	1,260.0
	17	12.75	0.99	111.6	237.1	1,511.3
9	18	13.50	0.99	118.1	265.8	1,794.0
	19	14.25	0.98	124.7	296.1	2,109.9
10	20	15.00	0.98	131.3	328.1	2,460.9
	21	15.75	0.97	137.8	361.8	2,848.8
11	22	16.50	0.97	144.4	397.0	3,275.5
	23	17.25	0.96	150.9	433.9	3,742.8
12	24	18.00	0.96	157.5	472.5	4,252.5
	25	18.75	0.95	164.1	512.7	4.806.5
13	26	19.50	0.95	170.6	554.5	5,406.7
	27	20.25	0.94	177.2	598.0	6,054.8
14	28	21.00	0.94	183.8	643.1	6,752.8
	29	21.75	0.94	190.3	689.9	7,502.5
15	30	22.50	0.93	196.9	738.3	8,305.7
	31	23.25	0.93	203.4	788.3	9,164.2
16	32	24.00	0.93	210.0	840.0	10,080.0
	33	24.75	0.92	216.6	893.3	11,054.8
17	34	25.50	0.92	223.1	948.3	12,090.6
	35	26.25	0.92	229.7	1,004.9	13,189.1
18	36	27.00	0.91	236.3	1,063.1	14,352.2
	37	27.75	0.91	242.8	1,123.0	15,581.7
19	38	28.50	0.91	249.4	1,184.5	16,879.6
	39	29.25	0.91	255.9	1,247.7	18,247.5
20	40	30.00	0.90	262.5	1,312.5	19,687.5
	41	30.75	0.90	269.1	1,378.9	21,201.3
21	42	31.50	0.90	275.6	1,447.0	22,790.7
	43	32.25	0.90	282.2	1,516.8	24,457.7
22	44	33.00	0.89	288.8	1,588.1	26,204.1
	45	33.75	0.89	295.3	1,661.1	28,031.6
23	46	34.50	0.89	301.9	1,735.8	29,942.2
	47	35.25	0.89	308.4	1,812.1	31,937.7
24	48	36.00	0.88	315.0	1,890.0	34,020.0
	49	36.75	0.88	321.6	1,969.6	36,190.9
25	50	37.50	0.88	328.1	2,050.8	38,452.2
	51	38.25	0.88	334.7	2,133.6	40,805.7
26	52	39.00	0.88	341.3	2,218.1	43,253.4
	53	39.75	0.88	347.8	2,304.3	45,797.1
27	54	40.50	0.87	354.4	2,392.0	48,438.6
	55	41.25	0.87	360.9	2,481.4	51,179.8
28	56	42.00	0.87	367.5	2,572.5	54,022.5
	57	42.75	0.87	374.1	2,665.2	56,968.6
29	58	43.50	0.87	380.6	2,759.5	60,019.8
	59	44.25	0.87	387.2	2,855.5	63,178.1
30	60	45.00	0.86	393.8	2,953.1	66,445.3
	61	45.75	0.86	400.3	3,052.4	69.823.3
31	62	46.50	0.86	406.9	3,153.3	73,313.8
	63	47.25	0.86	413.4	3,255.8	76,918.8
32	64	48.00	0.86	420.0	3,360.0	80,640.0
	65	48.75	0.86	426.6	3,465.8	84,479.4
33	66	49.50	0.85	433.1	3,573.3	88,438.7
	67	50.25	0.85	439.7	3,682.4	92,519.9
34	68	51.00	0.85	446.3	3,793.1	96,724.7
	69	51.75	0.85	452.8	3,905.5	101,055.0
35	70	52.50	0.85	459.4	4,019.5	105,512.7
	71	53.25	0.85	465.9	4,135.2	110,099.6
36	72	54.00	0.85	472.5	4,252.5	114,817.5
	73	54.75	0.85	479.1	4,371.4	119,668.3
37	74	55.50	0.84	485.6	4,492.0	124,653.9
	75	56.25	0.84	492.2	4,614.3	129,776.0
38	76	57.00	0.84	498.8	4,738.1	135,036.6
	77	57.75	0.84	505.3	4,863.6	140,437.4
39	78	58.50	0.84	511.9	4,990.8	145,980.4
	79	59.25	0.84	518.4	5,119.6	151,667.3
40	80	60.00	0.84	525.0	5,250.0	157,500.0
	81	60.75	0.84	531.6	5,382.1	163,480.4
41	82	61.50	0.83	538.1	5,515.8	169,610.3
	83	62.25	0.83	544.7	5,651.1	175,891.5
42	84	63.00	0.83	551.3	5,788.1	182,326.0

10¾″ WIDTH

No. of Lams 1½″	No. of Lams ¾″	Depth d (in.)	Size Factor C_F	Area A (in.^{2})	Section Modulus S (in.^{3})	Moment of Inertia I (in.^{4})
7	14	10.50	1.00	112.9	197.5	1,037.0
	15	11.25	1.00	120.9	226.8	1,275.5
8	16	12.00	1.00	129.0	258.0	1,548.0
	17	12.75	0.99	137.1	291.3	1,856.8
9	18	13.50	0.99	145.1	326.5	2,204.1
	19	14.25	0.98	153.2	363.8	2,592.2
10	20	15.00	0.98	161.3	403.1	3,023.4
	21	15.75	0.97	169.3	444.4	3,500.0
11	22	16.50	0.97	177.4	487.8	4,024.2
	23	17.25	0.96	185.4	533.1	4,598.3
12	24	18.00	0.96	193.5	580.5	5,224.5
	25	18.75	0.95	201.6	629.9	5,905.2
13	26	19.50	0.95	209.6	681.3	6,642.5
	27	20.25	0.94	217.7	734.7	7,438.8
14	28	21.00	0.94	225.8	790.1	8,296.3
	29	21.75	0.94	233.8	847.6	9,217.3
15	30	22.50	0.93	241.9	907.0	10,204.1
	31	23.25	0.93	249.9	968.5	11,258.9
16	32	24.00	0.93	258.0	1,032.0	12,384.0
	33	24.75	0.92	266.1	1,097.5	13,581.7
17	34	25.50	0.92	274.1	1,165.0	14,854.1
	35	26.25	0.92	282.2	1,234.6	16,203.7
18	36	27.00	0.91	290.3	1,306.1	17,632.7
	37	27.75	0.91	298.3	1,379.7	19,143.3
19	38	28.50	0.91	306.4	1,455.3	20,737.8
	39	29.25	0.91	314.4	1,532.9	22,418.4
20	40	30.00	0.90	322.5	1,612.5	24,187.5
	41	30.75	0.90	330.6	1,694.1	26,047.3
21	42	31.50	0.90	338.6	1,777.8	28,000.1
	43	32.25	0.90	346.7	1,863.4	30,048.1
22	44	33.00	0.89	354.8	1,951.1	32,193.6
	45	33.75	0.89	362.8	2,040.8	34,438.8
23	46	34.50	0.89	370.9	2,132.5	36,786.2
	47	35.25	0.89	378.9	2,226.3	39,237.8
24	48	36.00	0.88	387.0	2,322.0	41,796.0
	49	36.75	0.88	395.1	2,419.8	44,463.1
25	50	37.50	0.88	403.1	2,519.5	47,241.2
	51	38.25	0.88	411.2	2,621.3	50,132.8
26	52	39.00	0.88	419.3	2,725.1	53,139.9
	53	39.75	0.88	427.3	2,830.9	56,265.0
27	54	40.50	0.87	435.4	2,938.8	59,510.3
	55	41.25	0.87	443.4	3,048.6	62,878.1
28	56	42.00	0.87	451.5	3,160.5	66,370.5
	57	42.75	0.87	459.6	3,274.4	69,989.9
29	58	43.50	0.87	467.6	3,390.3	73,738.6
	59	44.25	0.87	475.7	3,508.2	77,618.8
30	60	45.00	0.86	483.8	3,628.1	81,632.8
	61	45.75	0.86	491.8	3,750.1	85,782.9
31	62	46.50	0.86	499.9	3,874.0	90,071.2
	63	47.25	0.86	507.9	4,000.0	94,500.2
32	64	48.00	0.86	516.0	4,128.0	99,072.0
	65	48.75	0.86	524.1	4,258.0	103,789.0
33	66	49.50	0.85	532.1	4,390.0	108,653.3
	67	50.25	0.85	540.2	4,524.1	113,667.3
34	68	51.00	0.85	548.3	4,660.1	118,833.2
	69	51.75	0.85	556.3	4,798.2	124,153.3
35	70	52.50	0.85	564.4	4,938.3	129,629.9
	71	53.25	0.85	572.4	5,080.4	135,265.2
36	72	54.00	0.85	580.5	5,224.5	141,061.5
	73	54.75	0.85	588.6	5,370.6	147,021.1
37	74	55.50	0.84	596.6	5,518.8	153,146.2
	75	56.25	0.84	604.7	5,668.9	159,439.1
38	76	57.00	0.84	612.8	5,821.1	165,902.1
	77	57.75	0.84	620.8	5,975.3	172,537.4
39	78	58.50	0.84	628.9	6,131.5	179,347.3
	79	59.25	0.84	636.9	6,289.8	186,334.1
40	80	60.00	0.84	645.0	6,450.0	193,500.0
	81	60.75	0.84	653.1	6,612.3	200,847.4
41	82	61.50	0.83	661.1	6,776.5	208,378.4
	83	62.25	0.83	669.2	6,942.8	216,095.3
42	84	63.00	0.83	677.3	7,111.1	224,000.5
	85	63.75	0.83	685.3	7,281.4	232,096.1
43	86	64.50	0.83	693.4	7,453.8	240,384.5
	87	65.25	0.83	701.4	7,628.1	248,867.9
44	88	66.00	0.83	709.5	7,804.5	257,548.5
	89	66.75	0.83	717.6	7,982.9	266,428.8
45	90	67.50	0.83	725.6	8,163.3	275,510.8
	91	68.25	0.82	733.7	8,345.7	284,796.9
46	92	69.00	0.82	741.8	8,530.1	294,289.3
	93	69.75	0.82	749.8	8,716.6	303,990.5
47	94	70.50	0.82	757.9	8,905.0	313,902.4
	95	71.25	0.82	765.9	9,095.5	324,027.5
48	96	72.00	0.82	774.0	9,288.0	334,368.0
	97	72.75	0.82	782.1	9,482.5	344,926.3
49	98	73.50	0.82	790.1	9,679.0	355,704.5
	99	74.25	0.82	798.2	9,877.6	366,704.8
50	100	75.00	0.82	806.3	10,078.1	377,929.7

*3⅛ in. and 5⅛ in. are normal widths for Western softwood glued laminated timbers. 3 in. and 5 in. are normal widths for Southern Pine glued laminated timbers. To determine alternate section properties, multiply tabulated A, S and I values for 3⅛ in. widths by 3/3.125 and for 5⅛ in. widths by 5/5.125.

TABLE 4. Size Classification of Sawn Lumber.

Category	Sub-Category	Nominal Dimensions Ins. Thickness	Width	Example
	Light Framing	2–4	2–4	2x2, 2x4, 4x4
Dimension	Joists & Planks	2–4	≥5	2x6, 2x8, 2x10, 4x6
	Decking	2–4	≥4	2x4, 2x8, 2x10, 4x6
Beams & Stringers		≥5	>(2 + thick)	6x10, 6x12, 8x12
Posts & Timber		≥5	≤(2 + thick)	6x8, 8x8, 6x6

Note: Some of the sizes occur in more than one category.

NDS supplement [6] and UBC [5] provide tables which show allowable stresses for different species of wood in bending (F_{bt}), horizontal shear (F_{vt}), tension parallel to grain (F_{tt}), compression parallel and perpendicular to grain (F_{ct}, $F_{c\perp t}$) and modulus of elasticity (E_t). The subscript t is added to all allowable stresses to indicate that they are the tabulated values and not necessarily the allowable design values (explained in detail in a later section). It was earlier indicated that strength of wood in tension parallel to grain is larger than that in compression parallel to grain. But tabulated allowable stress in tension parallel to grain is smaller than that in compression parallel to grain because the limiting factor in tension member is usually compression or shear at supporting points. These tables suggest that the allowable stresses (tabulated values) depend on the type of species, use category and stress grade. For example 2 × 4 size No. 1 grade Douglas Fir Larch has a tabulated allowable bending stress of 1750 psi whereas 2 × 4 size standard grade has a tabulated value of 600 psi and 6 × 6 size No. 1 grade lumber has a tabulated value of 1200 psi.

GLULAM

For glulam members the designation used for bending combination are the allowable bending stresses (tabulated) in ksi followed by letter F. For example grade 22F indicates allowable bending stress of 2200 psi. Even though the allowable bending stresses are the same for all species of laminations having the same combination symbol other tabulated values, modulus of elasticity as well as allowable bending stresses for bending about minor axis depend on the type of lamination. For example 20F DF/DF (Douglas Fir lams) has allowable shear and compressive stresses of 385 and 165 psi respectively compared with 245 and 155 psi for 20F HF/HF (Hem Fir).

Stress grading of all lumber is done by seven different agencies such as Western Wood Products Association (WWP), West Coast Lumber Inspection Bureau (WCLIB), etc. Most structural lumber is stress graded by visual inspection based on ASTM provisions (Ref. 7-D245). The stamp on a piece of lumber identifies the inspection agency (Figure 7). It shows the agency mark, mill identification

TABLE 5. Typical Stress Grades for Each Size Category.

Light Framing	Joists & Planks	Decking	Beams & Stringers	Posts & Timber
Dense Select Str.	Dense Select Str.	Select	Dense Select Str.	Dense Select Str.
Select Str.	Select Str.	Commercial	Select Str.	Select Str.
Dense No. 1	Dense No. 1		Dense No. 1	Dense No. 1
No. 1	No. 1		No. 1	No. 1
Dense No. 2	Dense No. 2			
No. 2	No. 2			
No. 3	No. 3			
Appearance	Appearance			
Stud	Stud			
construction				
standard				
utility				

FIGURE 7. Typical lumber grade stamp.

(No. 12 in Figure 7), grade name (No. 2), common species (Doug Fir), and moisture condition (S-Dry).

DESIGN OF FLEXURAL MEMBERS

Assumptions

The following assumptions are made in flexural analysis and design of wood members using working stress approach.

1. Wood is solid, homogeneous and isotropic material.
2. Plane sections before bending remain plane after bending.
3. Stress-strain relationship is linear.
4. Modulus of elasticity in tension is the same as that in compression.

In reality some of the above assumptions are inaccurate. For example defects, annular rings, direction of grain etc. result in properties that are different in different directions and points. Some of the factors that influence strength properties (allowable stresses for example) are taken into consideration as adjustment factors for the tabulated allowable stresses as explained in later sections.

Criteria for Beam Design

The four criteria for beam design of sawn lumber or glulam beams are: bending moment, shear force, deflection and bearing. Bending includes a check for lateral stability of compression zone. In addition, the applied load is assumed to produce normal stresses which are parallel to the longitudinal axis of the member. Wood is very weak in tension perpendicular to the grain and thus cross-grain bending is generally not allowed. The tabulated allowable bending stresses (NDS–Reference [6]) are not applicable for bending in this direction.

The four criteria for laterally supported beam design can be summarized as below:

$$f_b \leq F_b \qquad (1)$$

$$f_\vartheta \leq F_\vartheta \qquad (2)$$

$$\Delta \leq \Delta_a \qquad (3)$$

$$f_{c\perp} \leq F_{c\perp} \qquad (4)$$

in which f_b = actual bending stress; F_b = allowable design bending stress; f_ϑ = actual shear stress; F_ϑ = allowable design shear stress; Δ = actual deflection; Δ_a = allowable deflection; $f_{c\perp}$ = bearing stress; and $F_{c\perp}$ = allowable design bearing stress. The following sections describe these terms in detail.

BENDING MOMENT

The term f_b in Equation (1) is the maximum bending stress found as below.

$$f_b = \frac{M}{S} \qquad (5)$$

in which M = maximum bending moment and S = section modulus. Tables 1 and 3 show cross-sectional dimensions and section moduli of various sawn lumber and glulam beams. F_b, the design bending stress, is the tabulated (allowable) bending stress adjusted with applicable adjustment factors. The allowable bending stresses have been tabulated in NDS for Wood Construction [6] and UBC [5] for various species of wood and size classification. They are applicable for visually graded lumber, members 12 in. or less in depth, dry condition of use (moisture content equal to or less than 19%), normal duration of load, adequate lateral support and relatively straight prismatic members. For conditions other than these the tabulated values should be modified with applicable modification factors as described in the following paragraphs.

$$F_b = F_{bt}\,(\text{CUF})\text{LDF}(C_F)C_f,\ \text{etc.} \qquad (6a)$$

or,

$$= F_{ba}(\text{LDF})C_F(C_f) \qquad (6b)$$

in which F_{bt} = tabulated allowable bending stress; CUF = condition of use factor; LDF = load duration factor; C_F = size factor; C_f = form factor; and etc. relates to any other applicable factors. The term F_{ba}, which is referred to as the allowable bending stress is the tabulated value of bending stress, adjusted with CUF.

MODIFICATION FACTORS TO TABULATED VALUES

It was indicated earlier that the tabulated allowable bending stress, F_{bt}, is for dry condition of use with moisture content equal to or less than 19%. This is true for all species except a few such as southern Pine, Virginia Pine, etc. For these species allowable bending stresses have been tabulated for 19% maximum moisture content, 14% max. m.c. and for any condition of use. As an example for Virginia Pine construction-grade lumber the tabulated bending stresses

TABLE 6. Load Duration Factors, LDF.

Type of Load	Duration	LDF
Dead	Permanent	0.9
Live	10 yrs.	1.0
Snow	2 months	1.15
Roof live	7 days	1.25
Wind and Seismic	1 day	1.33
Impact	2 secs.	2.0

are: 1100 psi for 15% max. m.c., 1000 psi for 19% max. m.c., and 825 psi for any moisture condition (single member use). For all other species the footnotes to the tables provide adjustment factors (CUF) for use other than at 19% max. m.c.

When the moisture content is equal to or less than 15% and lumber is used in conditions where moisture content does not exceed 15% the specification allows increase in allowable stress and modulus of elasticity by recommending a CUF that is larger than 1.0 (dry use factor). Similarly when the lumber is used in conditions where moisture content exceeds 19% the specification allows reduced allowable stresses and modulus of elasticity by recommending a CUF which is less than 1.0 (wet use factor). Similar wet use factors are recommended for glulam beams used in conditions where moisture content is equal to or exceeds 16%.

The term F_{ba} in Equation (6b) is the allowable bending stress after adjustment for dry or wet use conditions (CUF).

The term load duration factor (LDF) is designed to serve as an adjustment for length of time the load is applied. The normal duration of load is assumed 10 years and is attributed to live load. For this condition LDF = 1.0. For all other types of loads LDF values are given in Table 6.

The load duration factor is used to modify the tabulated stresses but not the modulus of elasticity.

When combination of loads are used in design allowable stresses should be modified using the LDF for shortest duration of load in the combination (largest factor). Thus allowable stresses in a floor member supporting a combination of roof dead load, floor dead load, and snow load is the one found using an LDF = 1.15.

The term size factor, C_F, was introduced to account for the possible reduction in moment capacity from that predicted from ordinary bending theory for beams with higher depth [8]. The tabulated values of bending stresses (bending stresses only) both for sawn lumber and glulam beams should be multiplied by size factor given below.

For depth $\leq$ 12 in., $C_F = 1.0$

$$\text{depth} > 12 \text{ in.}, \; C_F = \left(\frac{12}{d}\right)^{1/9} \tag{7}$$

Where d = depth of beam in inches. The above values of C_F are based on the following criteria:

Glulam beams: rectangular sections
uniformly loaded
simple span
span to depth ratio = 21

sawn lumber (visually graded): rectangular member
thickness (width) >4 in.

For conditions other than the above procedure for modifying the size factor is given in NDS.

Form factor is another adjustment factor that applies to bending stresses only and is used for shapes other than rectangular. Tests have shown that strength of some shapes such as I and box sections have moment capacities smaller than a rectangular section of equal depth, and few others such as round and diamond have their moment capacities larger than the rectangular section. NDS suggested shape factor for bending stresses are shown in Table 7.

Flow chart 1 describes the method of determination of various adjustment factors for square or rectangular sections.

The allowable bending stresses for sawn lumber are listed in NDS for two types of uses: 1) single member uses and 2) repetitive member uses. The repetitive members are defined as three or more parallel beams (trusses, joists, decking etc.) spaced not more than 24 in. o.c. and are joined by floor, roof or other load distributing elements adequate to support design loads. The repetitive members are awarded about 15% more allowable stresses than single member uses as they share the loads with adjacent members. However no such distinction is made in member use for glulams as they are normally spaced farther than sawn lumber beams.

For members supporting concentrated loads, Appendix D of NDS suggests a procedure by which such loads can be distributed to adjacent members.

TABLE 7. Shape Factor C_f

Shape of Beam	C_f
Rectangle	1.0
Round	1.18
Diamond	1.414

SHEAR FORCE

The shear stress along a vertical plane of cross section, f_{vy}, resulting from a vertical shear force V can be found using the following equation.

$$f_{vy} = \frac{VQ}{Ib} \tag{8}$$

in which Q = first moment of the area between the horizon-

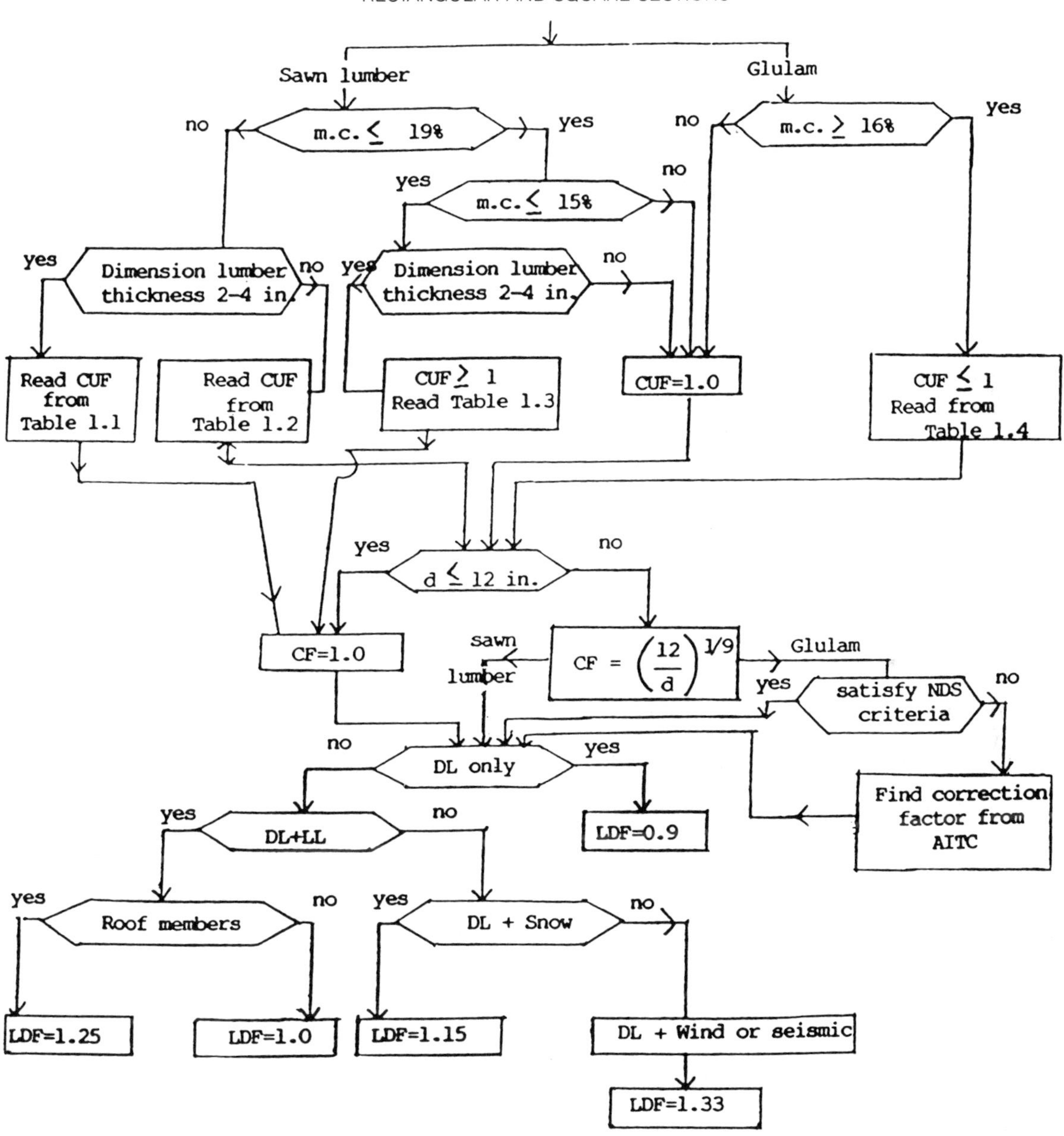

FLOW CHART 1. Modification factors for tabulated values (continued).

CUF Values

Sawn Lumber

Table	F_b	F_t	F_v	$F_{c\perp}$	F_c	E
1.1	0.86	0.84	0.97	0.67	0.70	0.97
1.2	1.00	1.00	1.00	0.67	0.91	1.00
1.3	1.08	1.08	1.05	1.00	1.17	1.05
		R E D	W O O D	O N L Y	1.15	1.04

Glulam

		Bend. about xx				Bend. about yy				Axial Load		
Table	Combination	F_b	$F_{c\perp}$	F_v	E	F_b	$F_{c\perp}$	F_v	E	F_t	F_c	E
1.4	Bending & Axial	0.8	0.53	0.875	0.833	0.8	0.53	0.875	0.833	0.8	0.73	0.833

FLOW CHART 1. Modification factors for tabulated values.

tal plane and extreme fiber about the neutral axis; I = moment of inertia of the cross section; and b = width of the horizontal plane. For rectangular sections the resulting stress distribution is that of a parabola with its maximum at the neutral axis. This maximum, f_ϑ can be found as below.

$$f_\vartheta = \frac{1.5V}{A} \tag{9}$$

in which A = area of cross section.

The criterion for shear design has been presented in Equation (2). The allowable design shear stress, F_ϑ is shown below.

$$F_\vartheta = F_{\vartheta t}\ (\text{LDF})\text{CUF, etc.} \tag{10a}$$

$$= F_{\vartheta a}\ (\text{LDF}),\ \text{etc.} \tag{10b}$$

In the above equations $F_{\vartheta t}$ is the tabulated allowable shear stress, $F_{\vartheta a}$ is the allowable shear stress after adjustment with CUF, and other terms are as defined earlier.

It should be noted that in ordinary bending horizontal shear stress occurs simultaneously with complimentary vertical shear stress. But NDS [6] lists only allowable horizontal shear stress because strength of wood in shear perpendicular to the grain is substantially higher than that along the grain.

For ordinary beams in which the load is applied on the top surface of the beam (both concentrated and uniform loads) and bottom surface rests on supports such as columns, beams or walls, NDS [6] suggests that the critical shear may be found by omitting the load over a distance d from the face of supports. But it is customary and conservative to bypass this provision and design for shear using full load on the beam.

In a beam notched at the ends (support sections) critical shear stress is found at the reduced section as shown below.

$$f_\vartheta = f_{\vartheta n}\ (K_s) \tag{11}$$

in which $f_{\vartheta n}$ is the shear stress at the notched section calculated from Equation (9) and K_s is a stress concentration factor and is equal to (d/d') where d' = depth of beam at the notch.

DEFLECTION

Generally, deflection criterion is established by the local building codes in line with local standards. The limitations recommended by UBC [5] is shown in Table 8.

The limitation on live load deflection is intended to ensure user comfort and that on total deflection is meant to avoid cracking in ceiling and other non-structural elements. The term K in the Table is a coefficient to include creep effects which depend on moisture content of wood.

The NDS Appendix F also suggests a method to find long term deflection incorporating the effects of creep. It involves using a modification factor for the modulus of elasticity (creep effect) as well as a long-term deflection factor. Appendix A of the specifications recommends that proper drainage facilities be provided to prevent ponding of roof structures and adequate camber in trusses to counteract deflection from gravity loads.

From design loads maximum deflection can be calculated using any elastic approach which can be compared with allowable deflection as per Equation (3).

BEARING

When a wood member rests on supporting member and in situations where other members such as studs frame into the beam it is required to check perpendicular to grain bearing stresses. This would insure that the supporting member or loaded edge does not get crushed resulting in settlement of the structure (Figure 8). The bearing stress thus found should be equal to or smaller than the allowable bearing stress (compression perpendicular to grain). The design criterion is as given in Equation (4) where bearing stress $f_{c\perp}$ is calculated as shown below.

$$f_{c\perp} = \frac{P}{A} \tag{12}$$

where P = reaction or load; and A = contact area. Allowable stress $F_{c\perp}$ is the tabulated stress in compression perpendicular to grain with applicable adjustment factors.

Beams Without Continuous Lateral Support

The tabulated bending stresses both for glulam and sawn lumber beams are for conditions in which flexural members are provided with continuous lateral support along the Compression side. Examples are: floor joists, roof joists, deck beams, etc., for which continuous lateral support is insured through proper connection between the flexural member and horizontal floor, roof or deck sheathing. Beams which do not have complete lateral support along the compression zone and those that are provided with lateral support at intermittent points may fail at stresses lower than tabulated bending stresses due to the tendency to buckle in transverse direction. The spacing between lateral supports is called unsupported length. Examples of these types of beams can be found over openings, floor and roof construction where top of beam is below bottom of floor or roof level, compression members of trusses which support transverse loads, etc.

There are two methods of design of laterally unsupported sawn lumber beams. One follows some rules of thumb evolved through years of experience and the other is based on reduction in allowable stresses for lateral buckling considerations. With the former approach the design allowable stresses are the tabulated values (after adjustment) and the design is identical to that of beams with full lateral support. In the second approach the beams are divided into three types based on the effective unsupported lengths—long, intermediate, and short. The design process is summarized in Flow chart 2.

The illustrated examples presented below show the step-by-step design procedure of sawn lumber and glulam beams with full lateral support and lateral support at intermediate points. Reference [9] presents a graphical method for unified design of beams without complete lateral support.

TABLE 8. Maximum Allowable Deflection for Wood Structural Members (UBC).

Type of Member	Lumber with M.C. ≤15%		Lumber with M.C. >16%	
	LL	LL + 0.5DL	LL	DL + DL
Roof members supporting plastered ceiling or floor member	$\frac{L}{360}$	$\frac{L}{240}$	$\frac{L}{360}$	$\frac{L}{240}$

Note: Long-term deflection is due to LL + K(DL).

Illustrative Examples

EXAMPLE 1

Figure 9 shows the roof framing plan of an office building. The dead loads of construction materials used are shown in the text. Design joist (A) and beam (B) for a live load of 20 psf.

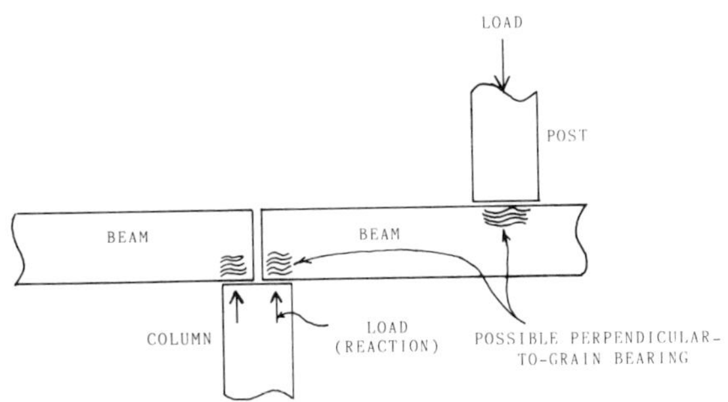

FIGURE 8. Possible bearing failures.

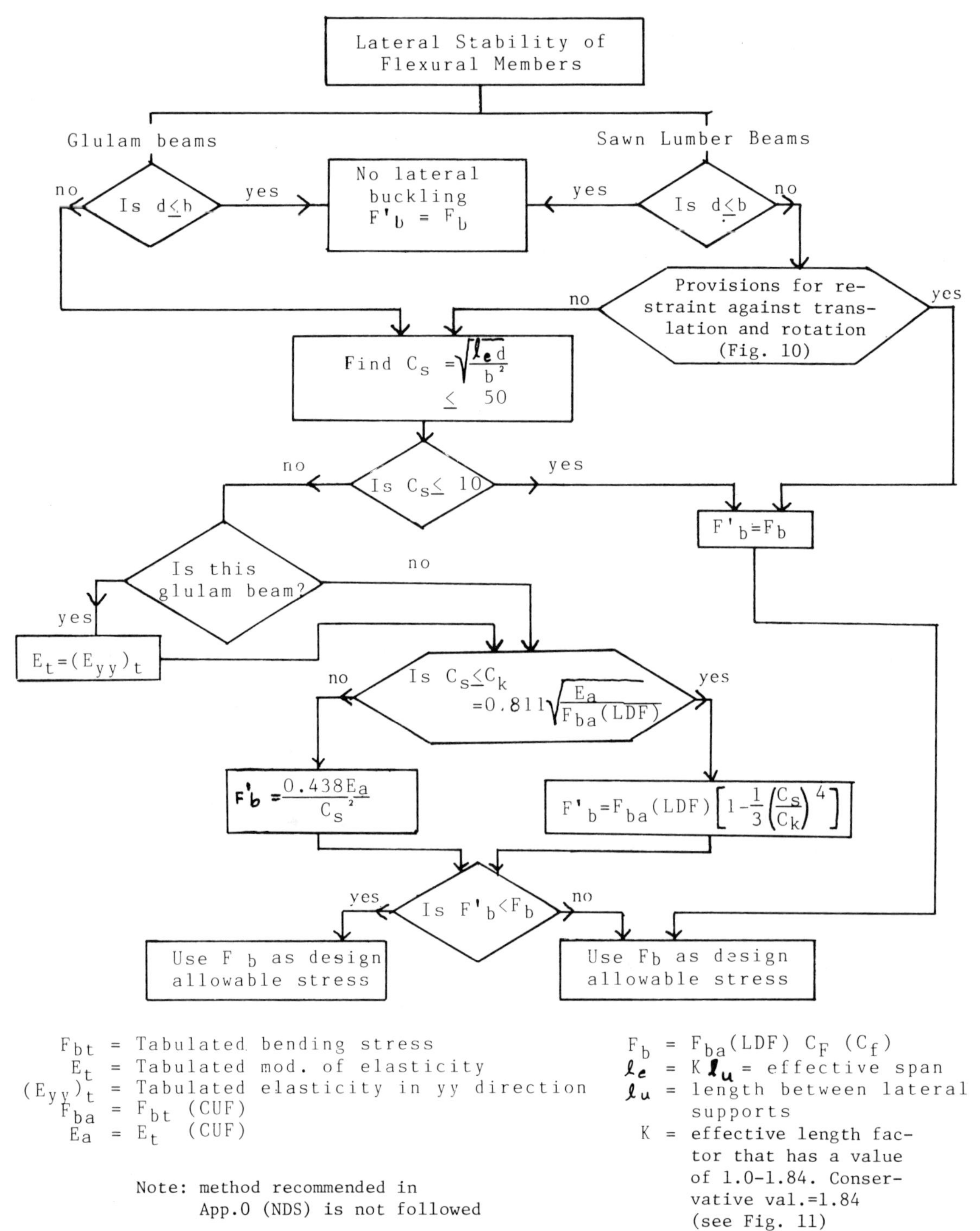
Lateral Stability of Flexural Members
Glulam beams
Sawn Lumber Beams
Is d≤b
no
yes
No lateral buckling F'b = Fb
yes
Is d≤b
no
Provisions for restraint against translation and rotation (Fig. 10)
no
yes
Find $C_s = \sqrt{\frac{l_e d}{b^2}} \leq 50$
Is $C_s \leq 10$
no
yes
$F'_b = F_b$
Is this glulam beam?
no
yes
$E_t = (E_{yy})_t$
Is $C_s \leq C_k = 0.811\sqrt{\frac{E_a}{F_{ba}(LDF)}}$
no
yes
$F'_b = \frac{0.438 E_a}{C_s^2}$
$F'_b = F_{ba}(LDF)\left[1 - \frac{1}{3}\left(\frac{C_s}{C_k}\right)^4\right]$
Is $F'_b < F_b$
yes
no
Use F b as design allowable stress
Use Fb as design allowable stress
F_{bt} = Tabulated bending stress
E_t = Tabulated mod. of elasticity
$(E_{yy})_t$ = Tabulated elasticity in yy direction
F_{ba} = F_{bt} (CUF)
Ea = E_t (CUF)
Note: method recommended in App.0 (NDS) is not followed
F_b = F_{ba}(LDF) C_F (C_f)
l_e = K l_u = effective span
l_u = length between lateral supports
K = effective length factor that has a value of 1.0-1.84. Conservative val.=1.84 (see Fig. 11)

FLOW CHART 2. *Design of beams without complete lateral support.*

Loading
Roofing material = 5.5 psf
Rigid insulation = 1.5
Plywood sheathing = 1.5
Ceiling, ½ in gypsum = 2.5

Self weight—estimated = 2.0
Total dead load = 13.0 psf
Roof live load = 20.0 psf
Total load = 33 psf

Tabulated allowable stresses:

For No. 2 Douglas-Fir Larch, 2–4 in thick and 5 in wider,

$$F_{bt} = 1250 \text{ psi—single}$$

$$= 1450 \text{ psi—repetitive}$$

$$F_{\vartheta t} = 95 \text{ psi}$$

$$(F_{c\perp})_t = 625 \text{ psi}$$

$$E_t = 1.7 \times 10^6 \text{ psi}$$

Adjustment factors

Since moisture content = 19%, CUF = 1.0
For DL + Roof LL combination, LDF = 1.25
Assuming $d < 12$ in., $C_F = 1.0$

Design of joist (A)

Obtain section:

Tributary width = 2 ft.
Span length = 12 ft.
Load = 2(33) = 66 plf

$F_b = 1.25(1.45) = 1.813$ ksi
$M = .066\ (12)^2/8 = 1.188$ ft-k.
$= 14.26$ in-k.
S-reqd. = 14.26/1.813 = 7.87 in^3

Choose 2 × 8. $b = 1.5$ in, $d = 7.25$ in.
$A = 10.875$, $S = 13.14$, weight = 2.64 plf (35 lb/ft^3)
Corrected load = 31 + 1.31 = 32.3 psf = 66 plf

Resisting moment = 1.813(13.14) = 23.88 in-k.

Actual moment = 14.26 in-k.
< 23.83 in-k.
O.K.

Check shear:

$V = 0.066\ (12)/2 = 0.40$ k
$f_\vartheta = 1.5\ (0.40)\ (1000)/(10.875) = 55.2$ psi.
$F_\vartheta = 1.25\ (95)$ psi > 55.2 psi
OK

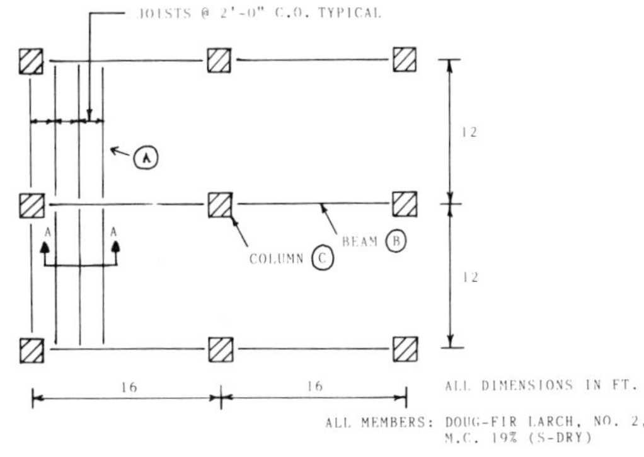

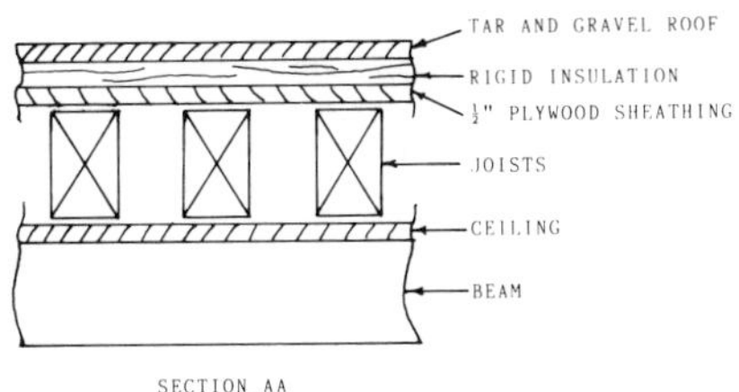

FIGURE 9. Example 1.

Check deflection:

$$\text{Midspan deflection from total load} = \frac{5\ wL^4}{384\ EI}$$

$$= \frac{5(0.066)\ 12^4\ (12)^3}{384\ (1.7)\ 10^3\ 47.64}$$

$$= 0.38 \text{ in.}$$

Allowable deflection = L/240 = 12(12)/240 = 0.60 in.
> 0.38 in.
OK

Check bearing:
Assume beam width = 3.5 in.
Bearing area = 3.5 (1.5)/2 = 2.625 in^2
Reaction = 0.404 k

$f_{c\perp} = 0.404/2.625 = 0.154$ ksi
$= 154$ psi.
$F_{c\perp} = 625\ (1) = 625$ psi
OK

Design of Beam (B)

Loading:
Load on joists = 0.066 klf
= 0.066/2 = 0.033 ksf

a. If ($\bar{d}/\bar{b}$) = 2 no lateral support required

b. For ($\bar{d}/\bar{b}$) = 3 and 4; ends are held in position by full depth blocking, bridging, hangars, nailing or bolting

c. For ($\bar{d}/\bar{b}$) = 5; one edge shall be held in line for entire length

d. For ($\bar{d}/\bar{b}$) = 6; spacing between solid blocking or bridging $\leq$ 8; or both edges held in line

e. For ($\bar{d}/\bar{b}$) = 7; both edges held in line for entire length

Note: For all of the above ends of the members should be supported to prevent lateral displacement and rotation.

$\bar{d}$ = nominal depth
$\bar{b}$ = nominal width

FIGURE 10. Provisions for restraint against translation and rotation.

Load on beam = 0.033(12)
= 0.40 klf

Self weight-estimated = 10 plf
Total load = 0.41 klf

Obtain section:

$M = 0.41\ (16)^2/8 = 13.12$ ft-k.
Fully supported $F_b = 1.25(1.25) = 1.563$ ksi

S-reqd = 13.12(12)/1.563 = 100.73 in³

Try 6 × 12

$b = 5.5$, $d = 11.5$, $A = 63.25$, $I = 697.07$, $S = 121.23$, $w = 15.4$

Corrected load = 0.40 + 0.0154
= 0.415 klf

$M = 0.415(16)^2/8 = 13.3$ ft-k.

For 6 × 2, tabulated stresses may be found for No. 1 beams and stringers.

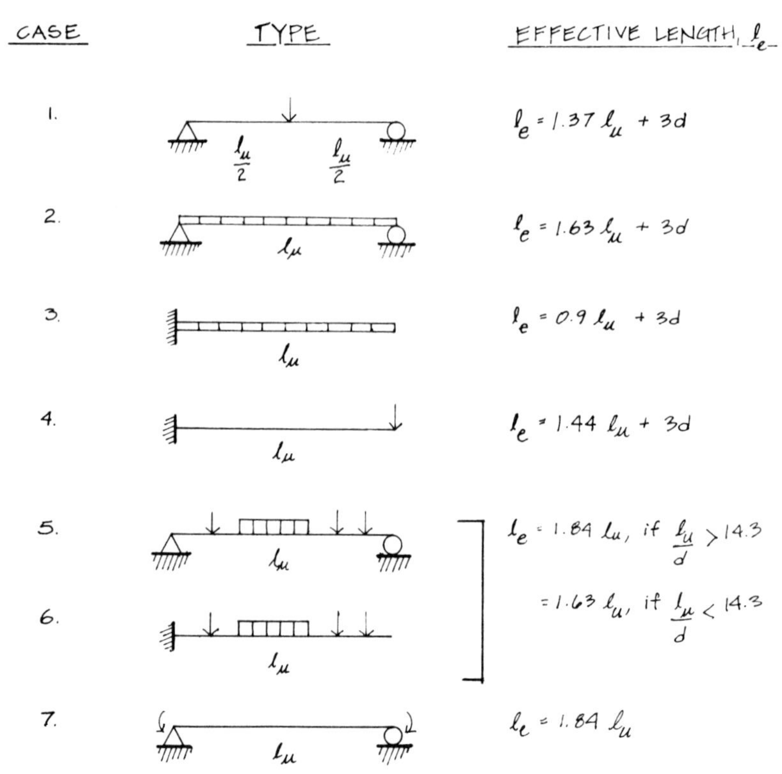

FIGURE 11. Effective length factors for beams.

$F_{bt} = 1350$, $F_{\vartheta t} = 85$, $E_t = 1.6 \times 10^6$, $F_{c\perp} = 625$ psi.
unsupported length = 2 ft = 24 in.

$L_e = 1.63(24) + 3(11.5) = 73.62$ in.
$> 1.84(24) = 44.2$ in.

$$C_S = \sqrt{\frac{44.2\ (11.5)}{5.5^2}} = 4.1$$

$$< 10$$

Note: Conservative value of $\ell_e = 1.84\ \ell_u$.

$F_b' = F_b = 1350(1.25) = 1687.5$ psi.
Resisting moment = 1687.5(121.23)/12000 = 17.05 ft-k.
> 13.3 ft-k
OK

Check Shear:

$V = 0.415(16)/2 = 3.32$ k
$f_\vartheta = 3.32(1.5)/63.25 = 0.079$ ksi
$F_\vartheta = 85(1.25) = 106.25$ psi
$> f_\vartheta$
OK

Check Deflection:

$$\text{Midspan deflection} = \frac{5(0.415)16^4(12)^3}{384(1.6)10^3\ (697.07)}$$

$$= 0.55 \text{ in}$$

Allowable deflection = 16(12)/240 = 0.80 > 0.55 in.
OK

Check bearing:

Assume that the beams are supported on 6 × 6 columns.

$$f_{c\perp} = \frac{R}{5.5\ (2.75)}$$

Reaction $R = 3.32$ k

$$f_{c\perp} = 3.32(1000)/5.5(2.75) = 219.5 \text{ psi}$$

$$F_{c\perp} = 625(1.0) = 625 \text{ psi}$$

$$> 221.48 \text{ psi}$$

OK

EXAMPLE 2

Design a 40 Ft glulam beam laterally supported at every 8 Ft. o.c. Assume dry condition of use. Beam has to support a roof dead load of 400 plf and snow load of 600 plf. Use 22F DF/DF lams.

Find load:

Total load = 400 + 600 = 1000 plf.

Find tabulated stresses and factors:

$F_{bt} = 2200$ psi, $(F_{c\perp})_t = 385$, $F_{\vartheta t} = 165$, $(E_{yy})_t = 1.6 \times 10^6$, $(E_{xx})_t = 1.7 \times 10^6$ psi.

For snow and dead load combination LDF = 1.15. CUF = 1.0

Obtain a trial size:

The unsupported length of 8 Ft is small in comparison with span length. Assume a F_b close to F_{bt}. Let $F_b = 2100$ psi.

$M = 1(40)^2/8 = 200$ ft-k
S-reqd = 200(12)1000/2100 = 1143 in³

Use 8 in nominal width beam (6.75 net)

$$\text{depth reqd.} = \sqrt{\frac{1143\ (6)}{6.75}}$$

$$= 31.88 \text{ in.}$$

Using 22-1.5 in laminations, size = 6.75 × 33 ins.

$S = 222.8$, $S = 1225.1$, $I = 20214.6$, self weight @ 35 pcf = 55 plf.

Total load = 1055 plf.

Find F_b:
$L_u = 8$ ft, $L_e = 1.84(8)12 = 176.6$ in.

$$C_S = \sqrt{\frac{176.6\ (33)}{6.75^2}}$$

$$= 11.31 < 50$$

$$> 10$$

$$C_k = 0.811\sqrt{\frac{E_a}{F_{ba}\ (\text{LDF})}}$$

$$E_a = (E_{yy})_t\ (\text{CUF}) = 1.6 \times 10^6 \text{ psi}$$

$$F_{ba} = F_{by}\ (1) = 2200 \text{ psi}$$

$$C_k = 0.811\sqrt{\frac{1.6(10)^6}{2200(1.15)}} = 20.39$$

$$C_s < C_k$$

$$F_b' = F_{ba}\,(\text{LDF})\left[1 - \frac{1}{3}\left(\frac{C_S}{C_k}\right)^4\right]$$

$$= 2200(1.15)\left[1 - \frac{1}{3}\left(\frac{11.31}{20.39}\right)^4\right]$$

$$= 2450.2 \text{ psi}$$

$F_b = F_{ba}(\text{LDF})\ C_F$

$C_f = (12/\text{d})^{1/9} = (12/33)^{1/9} = 0.894$

$F_6 = 2200(1.15)0.894$
$= 2261$ psi

Design allowable stress = 2261 psi
Resisting moment = 1225.1(2261)/12000
= 230.83 ft-k

maximum moment = $1055(40)^2/1000(8)$
= 211 ft-k
< 230.83 ft-k
OK

Check shear:

$V = 1.055(40)/2 = 21.1$ k
$f_\vartheta = 21.1(1.5)/222.8 = 0.142$ ksi
$F_\vartheta = 165(1.15) = 189.75$ psi

$> f_\vartheta$

OK

Check deflection:

$$\text{maximum deflection} = \frac{5(1.055)\ 40^4\ (12)^3}{384\ (1.7)\ 10^3\ (20214.6)}$$

= 1.768 in.

Allowable deflection = 40(12)/240 = 2.0 in

> max. deflection

OK

Check bearing:

Assuming support area of 6.75 × 9.0 ins.,

$f_{c\perp} = 21.1/6.75(9.0) = 0.347$ ksi.
$F_{c\perp} = 385(1.0) = 385$ psi
> 347 psi
OK

AXIAL FORCE MEMBERS

Timber is commonly used as axial force member or member in combined axial force and bending. Members of a truss, struts, bracing elements, arches, etc. are a few of the examples. The following section deals with design and analysis of timber elements in axial tension and compression.

Axial Tension

For both glulam and sawn lumber the criterion for design of axial tension members (struts) is as follows:

$$f_t = \frac{T}{A_n} \leq F_t \tag{13}$$

where f_t = axial tensile stress; T = axial tension; A_n = net cross-sectional area; and F_t = allowable design tensile stress which is equal to:

$$F_t = F_{tt}\ (\text{LDF})\text{CUF}(\text{TF}) \tag{14}$$

F_{tt} in the above Equation is the tabulated tensile stress (parallel to grain), TF is the tension factor that applies to dimension lumber only and the other terms are as defined before. Flow chart 3 shows the method of determination of TF.

Axial Compression

TYPES OF COLUMNS

Wood members subjected to axial compression can be classified as short, intermediate and long columns based on their slenderness ratios, and as simple, spaced, and built-up columns based on the method of manufacture. Theoretically, short columns are those which fail by compression failure rather than buckling. Long columns are those that fail by buckling at a stress smaller than that for failure of short columns, and for which Euler equation applies. The intermediate columns are the ones which fall within short and long column categories.

Simple columns are glulam or sawn lumber columns fabricated to form one piece of lumber whereas spaced columns are those built up as two or more individual members with their longitudinal axes parallel and are separated at the ends and middle points by blocking and joined (at ends) by proper timber connectors. Built-up columns are those which cannot be classified as simple or spaced columns (Figure 12).

DESIGN OF COLUMNS IN COMPRESSION

When buckling is not critical as in short columns design of a simple column can be achieved in a manner similar to that of tension members. The allowable axial load is found

Determination of TF

Sawn lumber

Glulam

TF=1.0

2-4 in. nominal thickness and nominal width ≥ 8 in.

yes

no

8" wide

yes

no

Dense select str. or select str.

yes

no

TF=0.90

stud

yes

no

TF=1.0

TF=0.80

Dense Select Str. or select str.

no

yes

stud

yes

no

TF=1.0

TF=0.60

TF=0.80

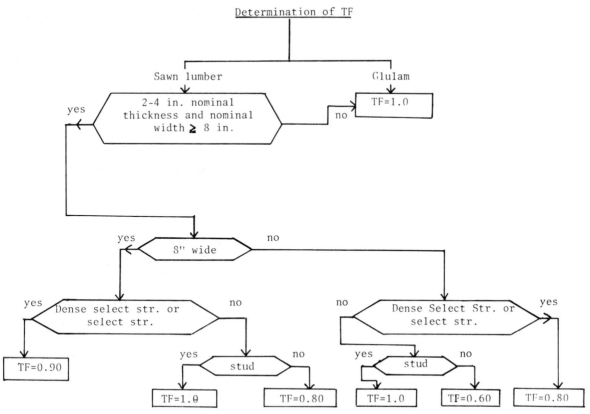

FLOW CHART 3. Method of determination of TF.

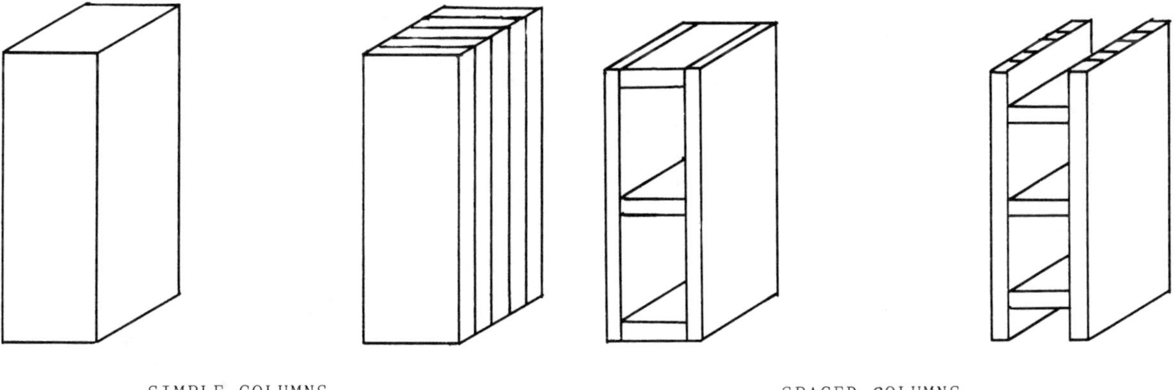

FIGURE 12. Simple and spaced columns.

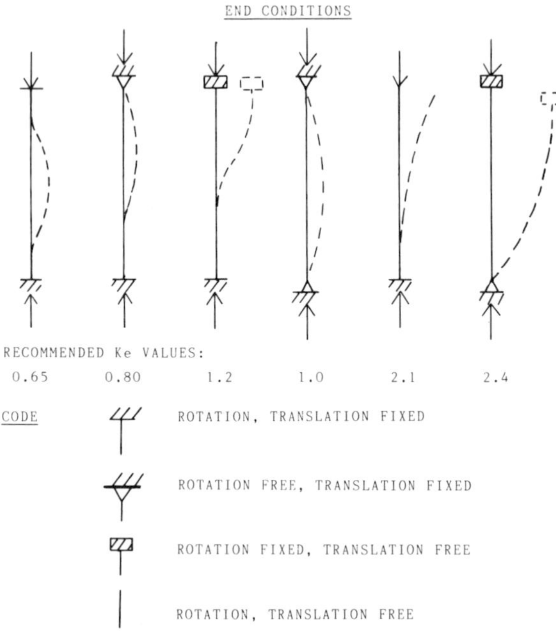

FIGURE 13. Recommended effective length factor (Ke)—NDS App. N.

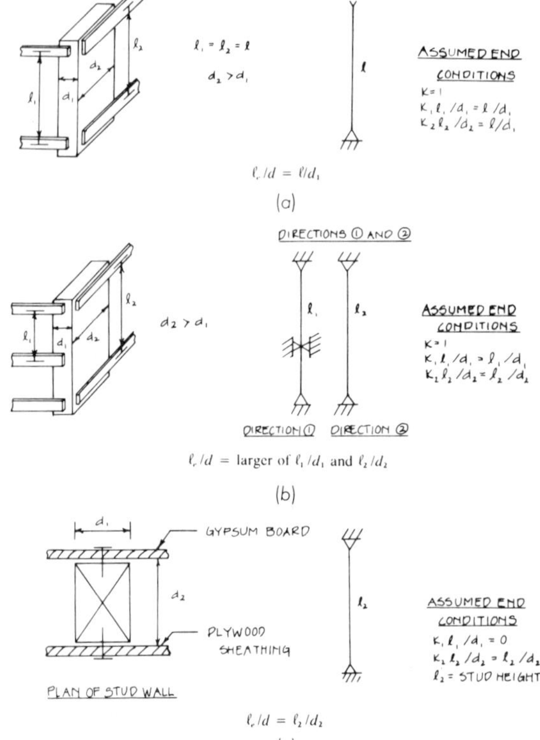

FIGURE 14. Method of finding slenderness ratio.

as net cross-sectional area (gross area minus area of any holes for connectors) times the adjusted axial stress parallel to grain, F_c. For intermediate and long columns the allowable stress may be substantially smaller than the tabulated stress (adjusted) depending on their slenderness ratios.

The slenderness ratio depends on the unsupported length, lateral dimensions, and end (support) conditions. For simple columns this value can be found as the smaller of (K_{e1} l_1/d_1) and ($K_{e2}l_2/d_2$), where l_1 and l_2 are the lengths between lateral supports in direction 1 and 2 respectively (width and thickness directions, d_1 and d_2 are the cross-sectional depths for bending in directions 1 and 2 respectively, and K_{e1} and K_{e2} are the effective length factors which depend on the end conditions. Figures 13 and 14 show the method of determination of effective length factors and slenderness ratios.

It should be pointed out that in addition to the factors mentioned above, slenderness ratio (effective length, in fact) depends on the stiffness of the frame of which the column is part of. If the frame is prevented from sidesway through bracing or other means the effective length factor is smaller than one, but is conservative and common practice to design such columns for values of $K_e = 1.0$. It is also interesting to note that the majority of compression members in typical wood frames have simple end-connections as opposed to rigid or moment-resisting connections in a reinforced concrete frame.

Flow chart 4 describes the method of design of simple columns. Here K is obtained as the minimum value of (l_e/d) at which a column can be expected to perform as long column (Euler column) combined with a safety factor for modulus of elasticity.

Illustrative examples are presented below to show the step-by-step design procedure of column struts and compression members. References [10] and [11] summarize this design process and present graphical technique of design of wood members in combined bending and compression or tension.

Illustrative Examples

EXAMPLE 1

Design column C of floor plan shown in Figure 9.

Type of wood: Douglas Fir-Larch; simple column
Height of column (from top of footing to bottom of beam) = 12 Ft.
Size of beam = 6 x 12 in.

A 6 × 6 column was chosen for the bearing check on the beam. Let us check the adequacy of this size. The column ends can be considered pin connected. $K_1 = K_2 = 1.0$.

Slenderness ratio:

$$l_e/d = K_e\, l/d = K_{e1}l_1/d_1 = K_{e2}l_2/d_2$$
$$= 1(12)12/5.5 = 26.18 < 50$$

OK

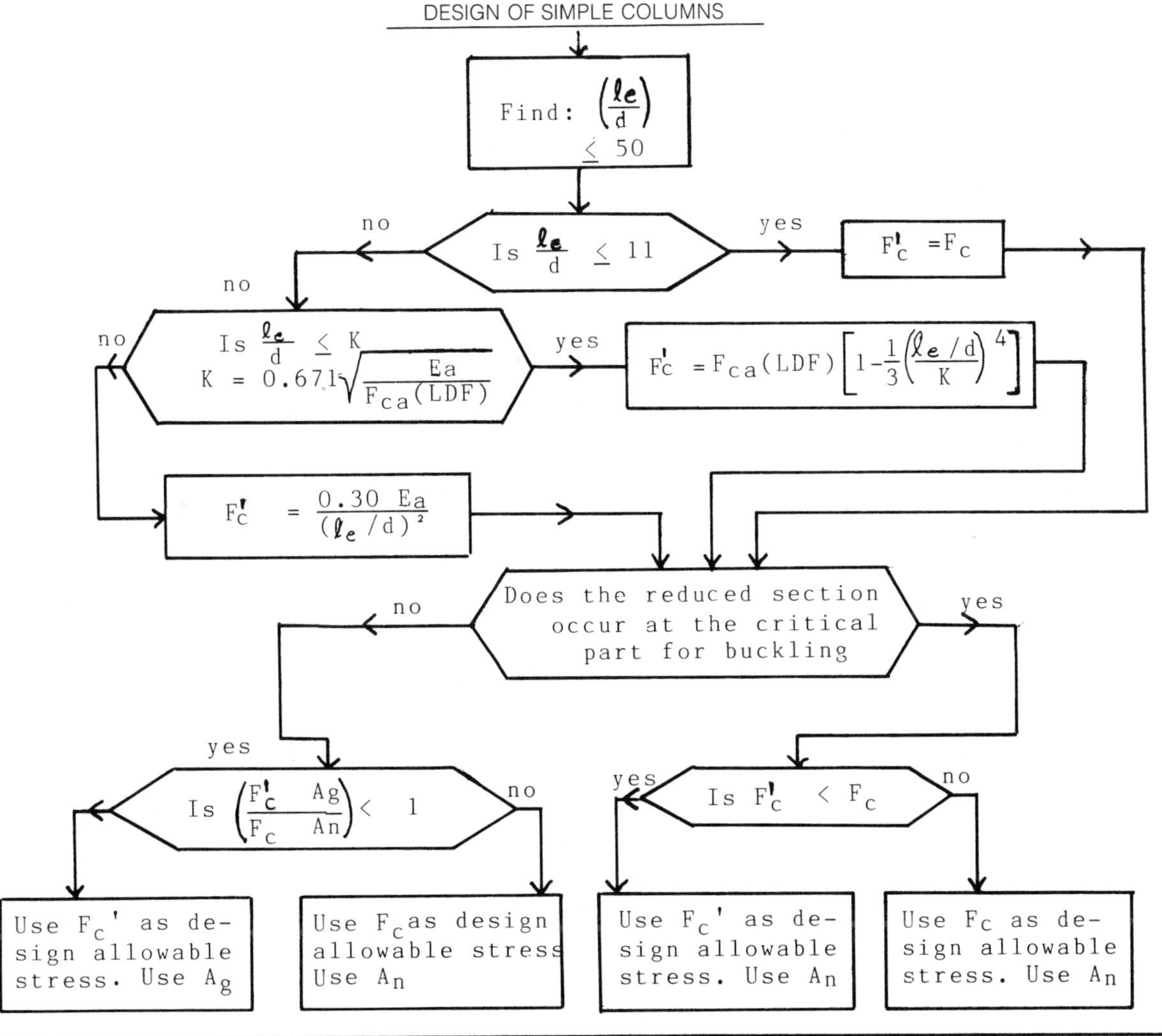

ℓ_e = $K_e\ell$ = effective column length
ℓ = length of column between lateral supports
K_e = effective length factor (=1.0 for simple end conditions)
(ℓ_e/d) = maximum slenderness ratio
F_{ca} = F_{ct} (CUF)
F_{ct} = tabulated compressive stress parallel to grain
F_c = F_{ca} (LDF)

FLOW CHART 4. *Design of compression members.*

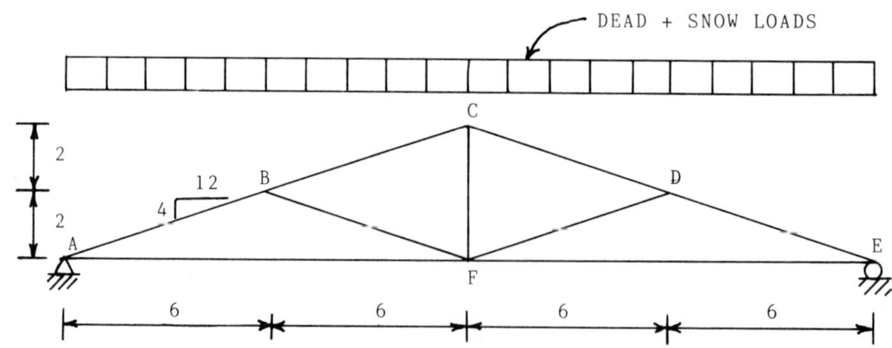

FIGURE 15. Example 5.2.

Total load:

Reaction on column from the beam = 3.35(2) = 6.7 k.
Self weight of column = 7.4 plf
= 7.4(12)/1000 = 0.089 k.
Design axial force, $P = 6.8$ k.

Find F_c':

$$K = 0.671\sqrt{\frac{E_a}{F_{ca}\,(\text{LDF})}}$$

For 6 × 6 No. 1 Douglas Fir-Larch, Post and Timber category, $F_\perp = 1000$, $E_t = 1.6 \times 10^6$ psi. LDF = 1.25 (dead load + roof live load combination), CUF = 1.0

$F_{ca} = 1000$, $E_a = 1.6 \times 10^6$ psi.

$$K = 0.671\sqrt{\frac{1.6\,(10)^6}{1000\,(1.25)}} = 24$$

$$l_e/d > K;\quad F_c' = \frac{0.30\,E_a}{(\ell_e/d)^2}$$

$$= 0.30(1.6)10^6/(26.18)^2 = 700.3 \text{ psi.}$$

Assume no bolt hole at mid height of the column and use gross area with F_c'.

$$F_c = F_{ca}\,(\text{LDF}) = 1250 \text{ psi.}$$

$$> F_c'$$

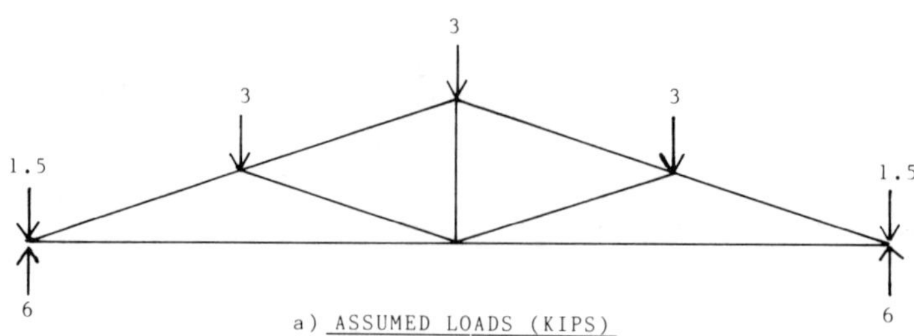

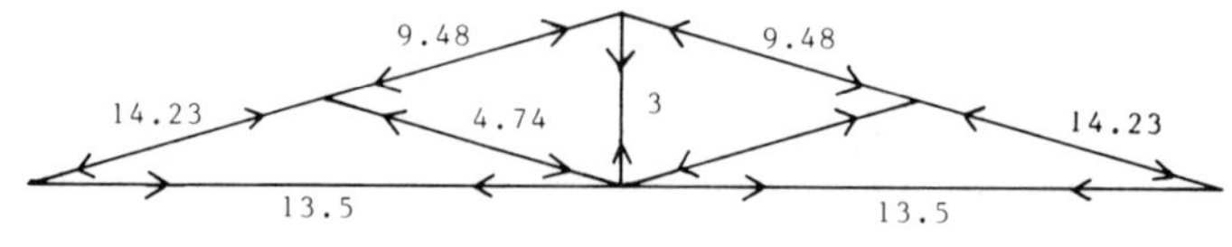

FIGURE 16. Example 5.2—Member forces.

Allowable axial load = 700.3(5.5)5.5/1000

= 21.18 k

$> P$

OK

EXAMPLE 2

Design members AB and AF of truss shown in Figure 15 using No. 1 Northern Pine. Dead load = 200 plf and snow load = 300 plf. 19% maximum moisture content.

Loading:
Dead load = 200, snow load = 300 plf.
Total load = 500 plf = 0.5 klf.

Joint loads:

Using the tributary area concept the distributed loads are converted into joint loads as shown in Figure 16(a).

Analyze the truss:

The axial forces in the members can be found using the method of sections or the method of joints. The final solution is shown in Figure 16(b). The forces in members AB and AF are 14.23 (comp) and 13.5 (ten) respectively.

Member AF:

Tension $T = 13.5$ k

Assume the ends are connected using 1 in. dia. bolts.

Tabulated values:

For 2–4 in thick and 5 in and wider No. 2 Northern Pine, $F_{tt} = 525$, $F_{ct} = 825$, $E_t = 1.3 \times 10^6$ psi.

Adjustment factors:

For dead + snow load combination, LDF = 1.15.
CUF = 1.0, TF (assumed) = 0.9.
$F_t = 525(1.15)0.90(1) = 543.4$ psi.

Find section:

Let $A_n = 0.95\ A_g$
A_g required = 13.5 (1000)/0.95(543.4) = 26.15 in²
Try 4 × 8. $A_g = 25.375$ in²
Hole dia is assumed ⅛ in larger than the bolt diameter.
Hole area = 0.625(3.5) = 2.19 in²
$A_n = 25.375 - 2.19 = 23.19$ in²
T allowable = 23.19(543.4) = 12.6 k

This force is slightly smaller than T and may be allowed. If felt unsatisfactory use the next higher size—6 × 6 (Post and Timber category).

T allowable for 6 × 6 = 20.79k

Member AB:
Compression $P = 14.23$ k
Length = 6.33 ft = 75.9 in.

Find section:
Assume a size 4 × 6.

$$l_e/d = K_e l_2/d_2 = 1(75.9)/3.5 = 21.69 > 11$$

$$K = 0.671\sqrt{\frac{1.3(10)^6}{825\ (1.15)}} = 24.84$$

$$l_e/d < K$$

Find F_c':

$$F_c' = F_{ca}\ (\text{LDF})\left(1 - \frac{1}{3}\left(\frac{\ell_e/d}{K}\right)^4\right)$$

$$= 825(1.15)\left(1 - \frac{1}{3}\left(\frac{21.69}{24.84}\right)^4\right)$$

$$= 764.9 \text{ psi.}$$

$$F_c = 825(1.15) = 948.8 \text{ psi}$$

Find allowable load:
Net section is at support
Use A_g with F_c'.
P allowable = 764.9(19.25)/1000 = 14.72 k.
> 14.23 k
OK

REFERENCES

1. Breyer, D. E. *Design of Wood Structures.* McGraw-Hill Book Co. (1980).
2. Jacoby, H. S. and R. P. Davis, *Timber Design and Construction.* John Wiley & Sons, Inc., 2nd Ed. (1930).
3. Ariba, E. W. *Wood in Building.* The Architectural Press, London (1971).
4. Glulam Systems, Amer. Inst. of Timber Constr. (1983).
5. Uniform Building Code, Int. Conf. of Building Officials, Whittier, CA (1982).
6. National Design Specifications for Wood Construction, Nat. Forest Prod. Assn., Washington, D.C. (1986).
7. American Society for Testing and Materials, Book of Standards (1983).
8. Newlin, J. A. and G. W. Trayer, "Form Factors of Beams Subjected to Transverse Loading Only," Forest Products Lab Rep. No. 1310 (1941).

9. Somayaji, S., "Design of Timber Beams: A Graphical Approach," *J. of Str. Engg., ASCE, 109*(1):271–278 (Jan. 1983).
10. Somayaji, S., "Timber Columns in Combined Bending and Tension: Design," *J. of Str. Engg., ASCE, 110*(7):1689–1694 (July 1984).
11. Somayaji, S., "Timber Columns in Combined Bending and Compression: Design," *J. of Str. Engg., ASCE, 110*(7):1694–1700 (July 1984).
12. "Insects in Wood," Western Wood Prod. Assn., Portland, Ore. (1963).
13. "Pressure Treated Wood: Design/Specs," Amer. Wood Pres. Inst., McLean, VA.
14. "Fire-Retardant Treated Wood Exterior Walls," Western Wood Prod. Assn., Portland, Or. (1982).
15. Derucher, K. N. and Heins, C. P. *Materials for Civil and Highway Engineers.* Prentice-Hall, Englewood Cliffs, NJ (1981).
16. Verna, J. R., J. F. Graham, J. M. Shannon, and P. H. Sanders, "Timber Bridges: Benefits and Costs," *J. of Str. Engg., ASCE, 110*(7):1563–1571 (July 1984).

CHAPTER 30

New Life to Timber Bridges

Frank W. Muchmore*

INTRODUCTION

Wood bridges are remarkable structures! Many of the covered wood bridges of the 18th and 19th centuries were used for more than 100 years [2]. In fact, some of them are still in service today! Wood was one of man's first building materials. However, in the 20th century, as modern technology has yielded its miracles in mass-produced steel, aluminum, concrete, and plastic construction materials, wood as a primary structural material for bridges has been overlooked by most engineers. What a shame! It is still a marvelously adaptable and competitive structural material. It:

- is simple to fabricate
- is lightweight and easy to install
- has a high strength-to-weight ratio
- has excellent sound & thermal insulation properties
- has unique aesthetic qualities
- has good shock resistance
- has good fire resistance[1]
- is immune to de-icing chemicals
- is a renewable resource
- is economical
- is long-lasting, when properly protected

[1]The unusual claim of "good fire resistance" of structural timber is well-founded. When a large structural timber is exposed to fire, there is some delay as it chars and eventually flames. As burning continues, the char layer has an insulative effect, and the burning rate slows to an average rate of about 1/40 inch (0.6 mm) per minute [or 1½ inches (38 mm) per hour], for average structural timber species. This slow rate of fire penetration means that timber structural members subjected to fire retain a high percentage of their original strength for considerable periods of time [10]. In contrast, structural steel becomes plastic when exposed to 1000°F heat and yields almost immediately.

*Northern Region, USDA Forest Service, Missoula, MT

Modern developments and techniques have advanced wood technology significantly since the days of the covered timber bridges. The widespread use of glued-laminated members, availability of effective preservative treatments, epoxies, and new prestressing techniques are but a few of the developments. Blending the old and new technologies, coupled with knowledgeable design, construction, inspection, and maintenance practices, is indeed bringing a whole "New Life to Timber Bridges."

Under proper conditions, wood will give centuries of service. If it can be protected from organisms which degrade wood, its longevity is remarkable [10]. Some techniques and processes to achieve this protection are the subject of this chapter.

WHY USE WOOD?

As a structural material, wood has the many advantages enumerated above and comparatively few disadvantages. Dr. John F. Levy of the Imperial College of Science and Technology, London, eloquently states wood's major disadvantage: "As far as a fungus is concerned, wood consists of a large number of conveniently oriented holes surrounded by food" [8]. The key to overcoming this major disadvantage of wood as a structural material is to make it inedible to decay fungi and microorganisms and to wood-boring insects and marine organisms. The use of wood preservatives basically forms a protective shell (Figure 1) which renders the wood unpalatable or toxic to these little "critters." It is important, however, to use compatible woods and treatment methods. Compare the protective shell shown in Figure 2 with that shown in Figure 1. Certain species (and subspecies) do not accept preservative treatment as well as others. Care must be exercised to specify species and treatment methods which will give proven results.

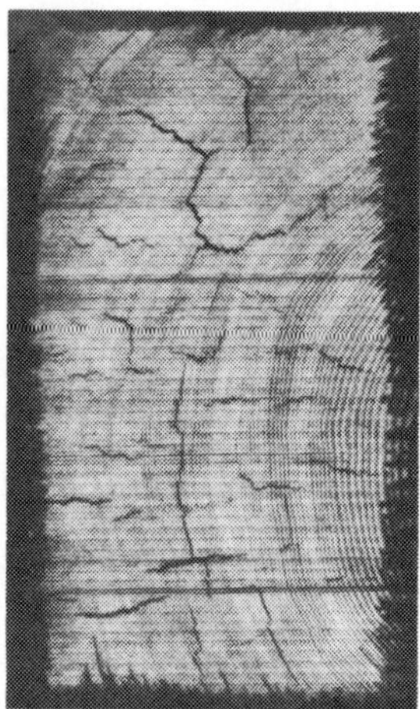

FIGURE 1. Sawed cross section of a pressure-treated coast region Douglas Fir bridge stringer. Preservation forms a protective shell.

Wood kept dry, or kept completely and continuously saturated, will not support the growth of decay fungi. The unusually long life span of the 18th and 19th century covered wood bridges can be attributed primarily to their wood being kept dry.

Treatment with the proper preservative process dramatically increases the useful life of wood exposed to the elements. For example, Richardson [8] states that in temperate climates a normal transmission pole pressure treated with creosote will have a typical life of 45–60 years, whereas an untreated identical pole will last only 6–12 years. A similarly treated railroad tie can be expected to last more than 35 years, in comparison to 8–10 years for untreated wood.

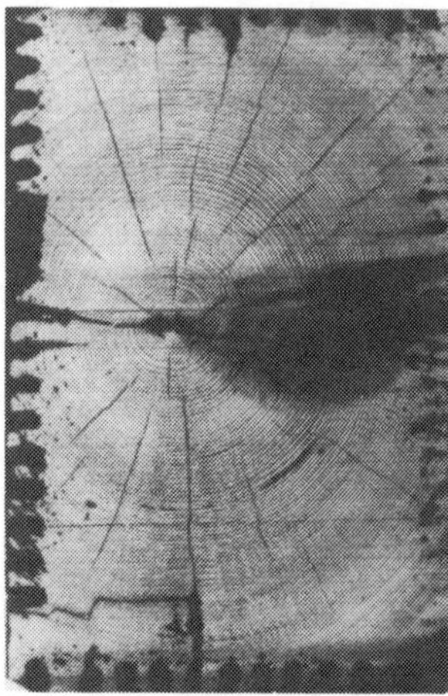

FIGURE 2. Sawed cross section of a pressure treated inland region Douglas Fir bridge stringer. Note the poor preservation penetration, typical of this sub-species, which does not form a good protective shell. Care must be taken in specifying species and treatment methods.

Richardson [8], in discussing the advantages of preserved wood, states:

> Preserved wood must be regarded as an entirely new structural material and must not be considered as just an improved form of wood, as it can be used in entirely different circumstances and certainly in more severe exposure situations. The most obvious advantage of preserved wood is that it can be used with impunity in situations where normal untreated species would inevitably decay, but it may be argued that, in many situations, this is a property that it enjoys together with many competitive materials. In fact, the use of wood has many advantages. It is extremely simple to fabricate structures from wood and, even in the most sophisticated production processes, the tooling costs are relatively low compared with those for competitive materials. Wood is ideal if it is necessary to erect an individual structure for a particular purpose but is equally suitable for small batch or mass production. When these working properties are combined with the other advantages of wood, such as its high strength-to-weight ratio, its excellent thermal insulation and fire resistance, its immunity to deicing chemicals, and the unique aesthetic properties of finished wood, it sometimes becomes difficult to understand why alternative materials have ever been considered! However, there is one feature of wood which is unique amongst all structural materials; it is a crop which can be farmed, whereas its competitors such as stone, brick, metal and plastic are all derived from exhaustible mineral sources.

With all these varied advantages, structural timber exposed to the elements, but properly selected and preserved, can give satisfactory service for 50 years and in many cases, far longer, at costs which are competitive with other structural materials, such as steel and concrete. Indeed, there is no reason that well designed and maintained timber bridges cannot survive more than a hundred years, as the covered bridges have already proven.

TIMBER BRIDGE INSPECTION

An important factor in assuring a long, useful life for timber bridges is an effective inspection and maintenance program. Of course, the effects of proper choice of materials, preservative treatments, design details, and to some degree construction practices should not be overlooked. However, once a timber structure is put in service, a systematic inspection and maintenance program will significantly increase its useful life.

There are many other factors which contribute to timber bridge deterioration, such as traffic damage from impact or heavy loads, and stream-induced damage from ice, drift, scour and flooding. However, since these factors are common to all types of bridges, they will not be discussed in detail here.

Even with the use of properly treated timber, it must be recognized that decay will eventually occur and will still be a primary cause for timber bridge deterioration. However, a preventive maintenance program including supplemental preservative treatments, coordinated with a systematic inspection program, will significantly prolong the service life of timber bridges. During the service life of the structure, the detrimental effects of decay can be minimized by alert inspectors who can locate developing decay or areas likely to decay. Armed with this information, maintenance personnel can make the necessary repairs or supplemental in-place treatments to curb the development of decay before serious structural damage occurs.

Inspecting Wood Bridges for Decay [3]

Inspectors must recognize potential problem areas where decay is likely to occur in individual bridges. They must also be able to determine the presence of and define the limits of any decay. Variations in structural designs, materials and connectors used, bridge age, and various exposure factors make this a complex assignment. Understanding the decay process and what causes it can make inspection easier and more effective.

Decay Defined [3]

Decay in wood is caused by living fungi, which are simple plants having the capability to break down and use wood cell wall material for food. These fungi are propagated by germination of fungus spores, which are functionally equivalent to the seeds of higher plants. These spores, produced in massive numbers, are microscopic. The spores are distributed so widely by wind, insects, and other means that they are commonly present on most exposed surfaces.

Most of these spores never germinate for lack of a favorable environment. All of the following conditions must prevail in order for decay fungus growth to take place:

1. Sufficient oxygen supply
2. Favorable temperaure range [32°F (0°C)–90°F (32°C)]
3. Adequate food supply (wood cells)
4. Available water (i.e., there must be "free" water in the wood cells, meaning that the moisture content must be above the fiber saturation point, which is approximately 30 percent for most structural wood species).

Once established, the fungus continues to grow at an accelerating rate as long as favorable conditions prevail. Depriving the fungus of any one or more of these required conditions will effectively curtail decay. Proper preservative treatment effectively provides a toxic barrier to the decay fungi's food supply, and thus no decay takes place.

Wood which is kept completely and continuously saturated with water will not support the growth of decay fungi, because a sufficient supply of oxygen is not available. However, care must be taken when wood is used in marine environments. Marine borers, which can quickly perforate wood immersed in sea water, become a major hazard unless the wood is properly preserved.

A common misconception is the use of the term "dry-rot," which erroneously carries the connotation that dry wood will rot, or decay. Even the growth of "dry-rot" fungi [*Merulius* (*Serpula*) *lacrymans*] requires the wood to have a moisture content above the fiber saturation point. Wood continuously maintained at moisture contents below fiber saturation can be considered safely dry from decay hazard, including "dry-rot."

Bridge Inspection Procedures [3]

Because of the many different bridge designs, ages, and site environments, detailed recommendations for inspection procedures are beyond the scope of this section. However, some basic considerations will be presented.

Visual Indicators of Decay

The inspection of timber bridges should be undertaken systematically, working from the lowest substructure members to the superstructure, or in the reverse order. One advantage of starting below is the opportunity to observe load reactions in the decking and supporting structural members from passing vehicles prior to the actual inspection of those elements. Telltale signs to watch for include abnormal deflections, looseness of joints or fasteners, and the presence of water forced from joint interfaces. These conditions indicate the presence of decay.

Visual indicators of the presence of decay may include:

1. Characteristic fungus fruiting bodies. These may be mushroom-like, or shelf- or hoof-shaped projections of flat, leathery material commonly found in partially enclosed areas (see Figure 3).
2. Abnormal surface shrinkage or "sunken" faces.
3. Abnormal discoloration.
4. Insect activity. Carpenter ants or termites can be attracted to decayed wood.

Visual indicators of conditions conducive to decay include:

1. Excessive wetting, evidenced by water marks or stains.

(a)

(b)

FIGURE 3. Examples of decay fungus fruiting bodies in growing or fresh condition.

Since a high moisture content is one of the conditions required for decay, this can be an important indicator.

2. Rust, or black iron tanate stains on wood surfaces. This also indicates possible excessive wood wetting and incipient decay.
3. Growth of vegetation on bridge members. Moss, grasses, and other plants growing in soil accumulated on any wood member not normally in soil contact also indicates potentially hazardous wetting.
4. Accumulation of soil on any wood surfaces traps water and increases decay hazard.
5. Cuts, holes, etc., made after the wood was treated, joint interfaces, mechanical fasteners, and wood adjacent to other water-trapping areas are potential sites of decay fungi growth.
6. Water-catching splits or seasoning checks in exposed wood surfaces.
7. Any point where the preservative layer has been penetrated, such as dowel or bolt holes, nails, and spikes placed after the wood was treated.

Excessive moisture accumulations are probably the most common visual indicator of conditions which promote development of decay.

FIGURE 4. Prospector's rock pick for probing and sounding timber members for decay.

Detecting Decay

It is difficult to detect and to determine the extent of decay, particularly when it is internal. External decay can usually be detected visually or by probing with a pointed tool, such as an awl, ice-pick or a prospector's rock pick (see Figure 4). The rock pick is also useful for sounding wood members. Sounding involves striking the wood member and listening to the sound produced. A dull or hollow sound may indicate internal decay. Internal moisture content may be measured with a probe-type moisture meter. Boring and coring are the most widely used techniques for detecting and defining the limits of internal decay.

Figure 5 shows an electric drill being used for boring stringers. Care must be taken so that holes resulting from boring or coring do not promote water or decay fungi spores entry into the wood. Whenever possible, holes should be oriented to let gravity aid in draining water out of the hole. Holes should always be plugged with treated wood plugs.

Figure 6 shows a hand-operated increment borer being used to cut a small diameter core, which is very useful in detecting internal decay locations.

Figure 7 shows typical internal decay locations resulting from development of seasoning checks and the use of field-installed penetrating fasteners (nails, lag bolts, drift-pins, etc.), which break the outer treated shell and establish a path for entry of decay spores into the untreated inner portions of a structural member. For maximum service life, any field fabrication (cuts, holes, daps, nails, etc.) which breaks or penetrates the "treatment envelope" should be in-place treated with a preservative compatible with the original treatment. Obviously, designs which minimize the use of penetrating fasteners will also minimize the introduction of internal decay hazard.

Strength Loss Associated with Decay [3]

There is relatively little information in the literature on strength loss due to amount or severity of decay in structural

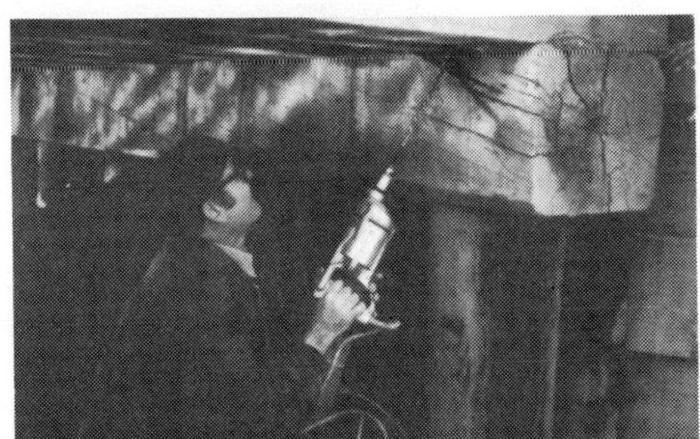

FIGURE 5. Boring timber stringer at bearing area with an electric drill.

timbers. Most approaches are ultra-conservative, assigning reduced allowable stresses to wood members with decay present. It has been shown, however, that strength loss is roughly proportional to the loss of sound wood in the section [3,6]. It is recommended that zero strength value be assigned to any decayed wood detected, and the sound wood left in the section be assigned its full strength value. However, a problem arises because it is extremely difficult to detect or determine the extent of incipient decay and thus these areas are not deducted from the sound wood.

In other words, use the section properties and normal allowable stresses of only the remaining sound wood when calculating a member's strength.

Of course, this requires a rather thorough investigation to determine the extent of the decay, by coring, boring, sounding, visual, and other means. While this may result in a conservative estimate of the residual strength of the member, it is appropriate because:

1. Incipient decay is difficult to detect.
2. The decay fungus will generally continue to grow and spread until some remedial action is taken, and such action may be delayed, resulting in a further strength loss.

BRIDGE MAINTENANCE

Timber bridge maintenance may be divided into three descriptive categories. These are:

1. Preventive Maintenance. Decay has not yet started, but conditions conducive to decay are definitely present. Preventive maintenance may include control of moisture and use of in-place preservative treatments.
2. Early Remedial Maintenance. Decay is present, but major maintenance is not required to continue the bridge in normal service. However, the bridge is likely to become more structurally damaged and load-limited if corrective measures are not applied. Measures taken may include

(a)

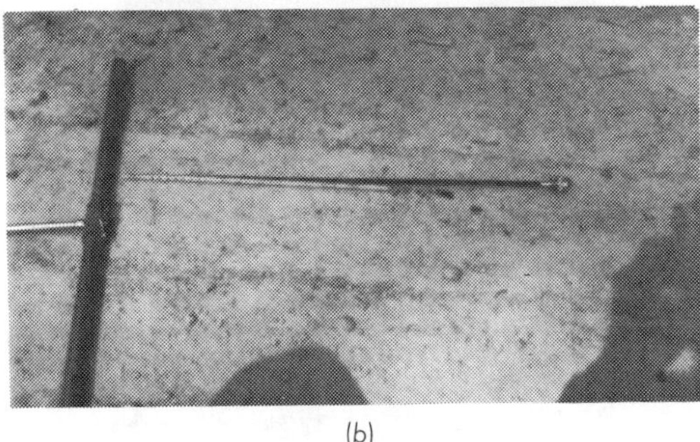

(b)

FIGURE 6. Increment borer being used to take a core sample. Cores can be saved for later reference.

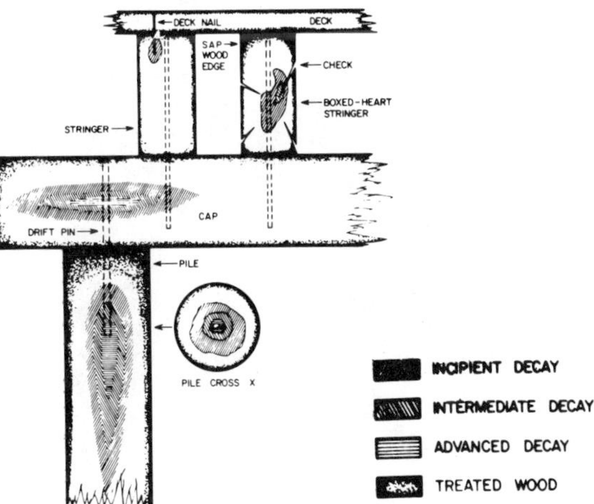

FIGURE 7. Schematic diagram of portion of pile bent, stringers, and deck showing locations where decay is most likely to occur.

(a)

(b)

(c)

FIGURE 8. Examples of structural failure where periodic inspection and preventive maintenance was not done.

the preventive maintenance measures discussed above, plus repair or replacement of decayed minor members, such as pieces of curb, guardrail running planks, or other wearing surface.

3. Major Maintenance or Replacement. Decay has progressed in major load carrying members until moderate to severe loss of structural strength makes repair or replacement mandatory and/or requires that load restrictions be placed on the bridge. The work may consist of any measures discussed above, plus replacement of portions or all of the structural deck, stringers, or timber substructures.

Unfortunately, most timber bridge maintenance is ignored until it reaches the third category, requiring major repairs or complete replacement. Figure 8 shows examples of this. Undoubtedly this neglect has contributed to the "bad press" (or "lack-of-press") that timber bridges have received as well as the general lack of perception of timber as a "permanent" bridge material.

Preventive and Early Remedial Maintenance

Control of moisture is probably the most cost-effective and practical general technique for extending the service life of new and existing timber bridges. This protective measure undoubtedly contributed significantly to the exceptionally long service life (more than 100 years in some cases!) achieved by many of the old covered wooden bridges. However, in most of the more modern timber bridge designs, preservative treatments have been relied upon to control decay, and the older principle of keeping water away from the structural members in the first place is given less (or no) emphasis.

Conversely, some attempts to control moisture have proven detrimental. Examples are:

1. Roll roofing materials as a water-diverting cover on caps, stringers, pile or post tops. As the roofing material ages, it dries out and becomes permeable to water from above. More harm than good may be done by using this type of material, because it tends to trap moisture and inhibits drying, thus promoting the very condition we are trying to avoid.
2. Sheet metal covers between pile or post tops and overlying stringer support caps are also of questionable value. The metal covers are usually installed before the timber cap is placed. When drift pins are driven down through the cap into the pile or post top, the metal cover is punctured and dimpled, creating a "funnel" for trapped water to enter the pile top and run down the drift pin past the treated top surface into the unpreserved inner wood. Again, the metal also inhibits drying, thus setting up a double jeopardy situation for a high moisture content in untreated wood, which is one of the conditions condu-

cive to the growth of decay fungi. Some results are illustrated in Figure 9.

3. Sheet metal caps on exposed pile or post tops, as shown in Figure 10, also are generally regarded as contributing to early decay. Exposed metal caps also attract vandals who cannot resist punching holes in the metal caps with sharp objects, or bullets. The net result again is the creation of "funnels" for water entrance, while inhibiting the drying of the pile or post top. An effective design criterion to promote good drainage is to slope the exposed tops of posts and other water-trapping places.

More effective moisture-proofing uses bituminous or asphaltic mastics in lieu of sheet-metal or roll roofing material as end-grain coatings, joint fillers or seals, and as check-filling compound. Care must be taken not to form water-entrapment areas under such coatings. It is also important that the end-grain or freshly cut surfaces be protected by a wood preservative prior to the application of the mastic. Also, since petroleum compounds can dry out, lose their elasticity and eventually crack, an effective maintenance program will include reapplication where feasible.

FIGURE 9. Sheet metal caps between pile tops and stringer support caps may increase decay hazard.

In-Place Treatments

Supplemental or in-place preservative treatments have been little used on timber highway bridges in the past, although the practice has been more common on timber railroad bridges. Commonly, decay is not detected until after it has become so severe that the member needs to be replaced. However, increased emphasis on periodic inspections, with attention given to early detection of decay and conditions conducive to decay, makes in-place treatments (as well as epoxy repairs) more practical as a means of extending the life of partially decayed members, rather than replacing them.

Conventional preservatives are available for in-place treating. These include various liquid- and grease-type preservative compounds. Generally, ammonium bifluoride and fluor-chrome-arsenic-phenol (FCAP) have been found to be the most effective when flooded onto the top surfaces of test units [3]. However, these waterborne preservatives can be quite toxic, so care must be taken when using them. Manufacturer's recommendations should be carefully followed. Pentachlorophenal solutions (usually petroleum oils) are not as effective for decay control, but may be required if the original treatment was "penta." Liquid preservatives can be applied to in-place timber surfaces most effectively by spraying or brushing/flooding.

Experimental work has shown that internally applied (through a series of bored holes) fumigants can successfully arrest established decay fungi for 6–10 years. Both Vapam and Chloropicrin have been registered for use as internal decay arresting fumigants in utility poles. Although these products have not been used extensively for bridge timbers, they have been successfully used to arrest decay fungi from infected Douglas fir utility poles [3] and wharf curbs [5].

Epoxy Repairs

Recent studies [1] indicate that epoxy repairs of exposed wood can be effective. However, caution is advised in initiating such repairs without careful evaluation. Epoxy repairs are difficult for a number of reasons [1]:

1. Joint deterioration may not be detectable from a visual inspection, especially around internal split ring connectors.
2. Deterioration may extend outward from a joint, where repaired strength may not be restorable to original strength by epoxy repair.
3. The epoxy repair process for seriously deteriorated joints can be extremely difficult.
4. If the source of deterioration is not removed, reoccurrence of decay problems is likely.

FIGURE 10. Sheet metal caps on exposed pile or post tops may become punctured and funnel water into pile top.

(a)

(b)

(c)

FIGURE 11. Timely inspections can detect loss of strength before major structural failures occur.

5. Without completely drying the area to be repaired, encapsulated moisture may continue to deteriorate the repaired area.

It may be more practical to replace seriously weathered and decayed timber than attempting epoxy repairs. However, using epoxies might be considered when deterioration is not extensive. The techniques discussed previously—moisture control, in-place treatments, etc.—are also applicable to epoxy-repaired areas.

Major Maintenance

When decay in treated or untreated wood is allowed to progress unchecked, ultimately complete deterioration takes place. At some stage, a structural member becomes too weak to carry the loads imposed on it. An alert bridge inspector should detect the loss of strength before these conditions advance far enough to cause structural failures, such as shown in Figure 11. When timber structural members must be replaced due to deterioration from decay, such replacement should be with properly treated members. Solid sawn or glued-laminated members such as stringers, decking, deck panels, wheel guards (curbs), rail posts and substructure members should be pressure treated with an appropriate preservative *after* fabrication (cutting, drilling, etc.) to provide economical, long-lasting service. Treating after fabrication ensures a complete protective envelope of treated wood, including the holes bored for connectors.

When major replacement of structural members is required, the designer of the repairs should try to use details which will not trap moisture, soil or other debris, which creates conditions conducive to decay. Also, field fabrication of wood members (cuts, holes, daps, etc.) should be held to an absolute minimum, since field treatment (flooding, spraying, brushing) is not nearly as effective as in-plant pressure treating after fabrication takes place. The control of moisture access to the structural members should be of primary concern. Remember the principle of the covered bridges!

Perhaps the most effective and practical method to control moisture on most wood bridge designs is to prevent water passage through the deck. In effect, make the bridge deck an impervious roof over the supporting stringers. If the timber deck is sufficiently rigid and stable, a well maintained asphalt mat (usually at least 3 inches thick), crowned to aid drainage, can be quite effective.

If the deck is not rigid or stable enough to support an asphalt mat, redecking normally will be required. There are several alternatives for producing a stable timber deck able to support an asphalt mat.

One method is to replace the deck with treated glued-laminated timber panels. This is quite effective when the laminates are oriented transverse to the bridge girders. Besides providing a more rigid support for asphalt mats than conventional nail-laminated decks, glu-lam deck panels are produced at a moisture content of about 12% and thus will be subject to very little shrinkage during service, helping to minimize crack formation in the mat [3]. Figure 12 shows glued laminated panels being installed. Experience has shown that reflective cracking of the asphalt mat usually occurs over the glued-laminated panel interfaces, particularly when the girder spacing is more than 4 feet (1.2 m). This may not be detrimental to the panels, since in the original development of deck panels, the "treatment" was considered the moisture barrier, not the asphalt. However, leakage at

FIGURE 12. Glued-laminated deck slabs being installed as replacement deck on a 12-year-old bridge.

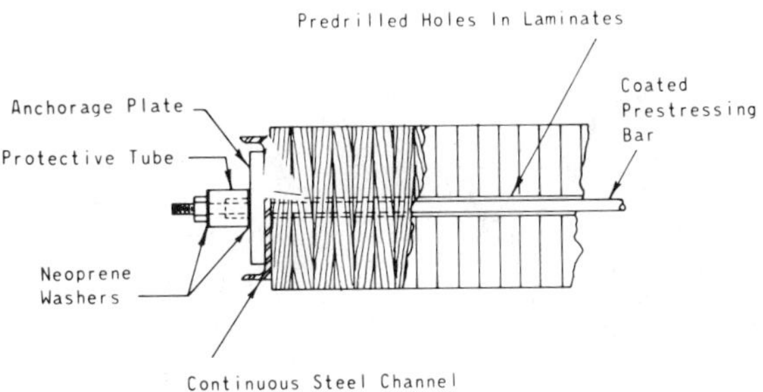

FIGURE 14. Prestressing system for new construction (courtesy of Ontario, Canada Ministry of Transportation and Communication).

the panel interfaces will create undesirable wetting of the girders and probably create unsightly stains on the exposed girder faces.

The placement of a geotextile fabric mat as an underlayment for the asphalt mat may help prevent the reflective cracking. The fabric must be "tacked down" to the deck surface with liquid asphalt; then the asphalt mat is placed by normal methods. The USDA Forest Service has placed fabric experimentally on several bridges [4], but no conclusive results have been documented to date.

Another product that may have promise as a wearing surface and "roof" is "elastomeric concrete." Developed in France for installing expansion joints in concrete decks, it is also used as a bridge deck surfacing for orthotropic steel deck bridges. It is expensive [$5.00 to $10.00 per square foot ($54.00 to $108.00 per square meter)] and difficult to install in the field, as it requires a heat "vulcanizing" process. It may be more practical if it is installed in the plant, on glulam deck panels, for instance.

A second method to reduce moisture penetration and produce a stable deck surface is to "prestress" (or "posttension") a laminated timber deck. For rehabilitation of an existing deck, the prestressing force is applied perpendicular to the laminates by a system of high-strength bars located above and below the deck. These bars are attached to heavy steel anchorage plates bearing against steel channels to form a strong, flexible clamping system (see Figure 13). Then the top bars are imbedded in the asphalt mat. This rehabilitation system is most adaptable to longitudinally laminated decks (laminates running parallel to the roadway centerline) and perhaps short transverse laminated decks (laminates perpendicular to roadway center line).

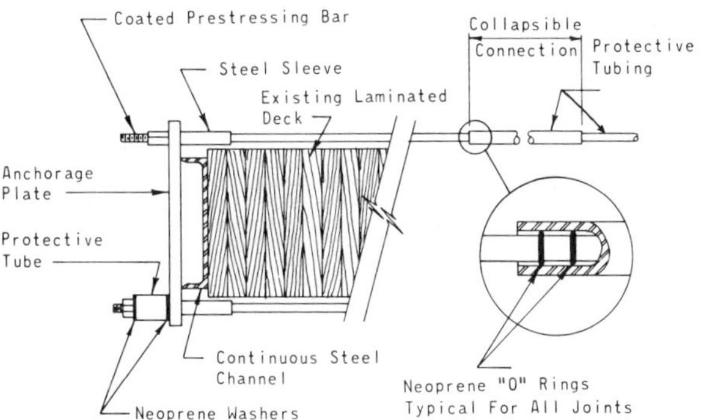

FIGURE 13. Prestressing system for rehabilitation (courtesy of Ontario, Canada Ministry of Transportation and Communication).

The prestressing system described above for rehabilitation is adapted for use in new deck installations by installing the prestressing bars through predrilled holes at mid-depth of the laminates (see Figure 14). The prestressing force is applied by anchoring the bars to steel channels and anchorages plates. This system eliminates the steel bars on top of the deck and allows prestressing (posttensioning) of either longitudinally laminated or transverse laminated decks. Transverse laminated panels can be prefabricated in the shop with steel plate bulkheads at the joining ends of the panels and can be made as large as practical for handling and transportation. When installed, the panels are prestressed together by using high-strength bar couplers and jacking against the steel plate bulkheads, in a manner very similar to segmental concrete construction. Prestressed transverse laminated decks as long as 400 feet (122 m) have been continuously tensioned by this method in Canada.

The prestressed timber deck system was developed and tested in Canada by the Research and Development Branch of the Ontario Ministry of Transportation and Communications. It has been successfully used on a number of rehabilitation projects and new bridges in Ontario, Canada. More detailed information can be found in Reference [9].

BRIDGE RENOVATION AND REPLACEMENT

The various components of composite structures frequently deteriorate at quite different rates. For example, a

FIGURE 15. Schwartz Creek Bridge across Clark Fork River near Missoula, Montana before renovation.

FIGURE 17. Schwartz Creek Bridge--completed project.

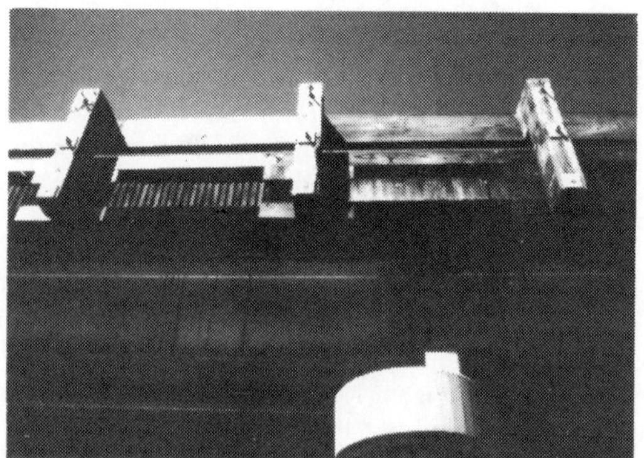

FIGURE 16a. Schwartz Creek Bridge: original nail-laminated decking at left; new glued-laminated decking at right.

FIGURE 16b. Galvanized clips holding glued-laminated deck panels to steel stringers. (Note "bleeding" of pentachlorophenol and oil as result of inappropriate treatment process—full-cell instead of empty-cell.)

bridge with steel girders or trusses and timber floor stringers and/or decking which has not been subjected to de-icing chemicals may have deteriorated timber members but still have functionally adequate steel members. When this condition exists, it may be feasible and economical just to replace the timber members. A thorough inspection and load rating analysis should be performed. If the steel members still have adequate carrying capacity and the bridge geometry still meets use requirements, it is usually economical to renovate the wood portions of the structure, providing many more years of service for a relatively modest cost.

The USDA Forest Service's nine Regions have more than 10,000 bridges on over 230,000 miles of Forest Development Roads. Of these 10,000 bridges, approximately 5,000 are constructed of wood or have a major component fabricated from wood, such as deck, stringer system, or substructure. For example, the Northern Region (Montana and northern Idaho) has about 1660 road bridges, of which about 1060 are treated timber bridges. Many of these treated timber bridges are over 50 years old and are still in service.

Many of these old bridges with major timber components have been successfully and economically renovated in recent years. Several examples follow to illustrate this trend. [For comparison, the normal range of construction costs for a new bridge in the Northern Region is about $70–$80 per square foot ($750–$860 per square meter), with an extreme range of about $50–$100 per square foot ($540–$1080 per square meter).]

Schwartz Creek Bridge Over the Clark Fork River

The Schwartz Creek bridge is a 222 ft (68 m) 4-span, single-lane 14 ft (4.3 m) width bridge across the Clark Fork River at Schwartz Creek near Missoula, Montana. It was designed for AASHTO H20-S16-44 standard highway loading when built in 1953. It consists of reinforced concrete piers and abutments supporting twin wide-flange steel longitudinal girders with a nail-laminated treated-timber deck [2 in. × 10 in. (51 mm × 254 mm) members on edge, nailed together]. The running surface consisted of steel running plates, and the bridge railing was timber posts and rails (see Figure 15).

The steel running plates were too widely spaced, and over the years traffic wear reduced the timber deck thickness, primarily on the approach spans, to the point where it was no longer adequate for legal highway loads. Decay was an insignificant factor. In order to restore the load capacity of the bridge, a contract was prepared and awarded to:

1. Replace the approach span decking with glued-laminated deck panels (Figure 16).
2. Remove the existing curb and railing and replace it with curb and railing designed to meet current Forest Service and County requirements.
3. Remove all the steel running plates and replace them with timber running planks full width (Figure 17). Treated plywood was used to make up the ½-inch (13 mm) difference in thickness between the retained main span decking and the new approach span glued-laminated panels.

The work was done for approximately $50,000 or $16.00 per square foot ($172 per square meter) in 1981. It is performing as expected and should give many more years of trouble-free service.

Simmons Creek Bridge

This bridge is located on the St. Joe River Road between Avery and Red Ives in northern Idaho. Built in 1935 for 15-ton truck loading, it was a 3-span, 86-foot (26 m) long, single-lane 12 ft (3.7 m) wide bridge. The bridge had two approach spans with treated timber stringers and a nail-

FIGURE 19. Simmons Creek Bridge: Bridge railing being installed.

laminated deck and a main span consisting of twin steel girders with a nail-laminated deck, supported on reinforced concrete piers and abutments. The bridge's inadequate load capacity (15 tons or 13.6 metric tons) and width [12 ft (3.7 m)] dictated replacement. All wood was pressure treated and decay was not a significant factor, even after 45 years!

A double-lane curved concrete replacement bridge was designed for HS30 loading (1½ times AASHTO HS20-44) in 1970. Changing management plans no longer required the heavier loading, so the double-lane concrete bridge was redesigned in 1979 for AASHTO HS20-44 loading and bids were solicited. The low bid of $190,000 ($99 per square foot or $1065 per square meter) exceeded the funds available, so bids were rejected. The Forest Service then decided to renovate the existing single-lane bridge to its current single-lane standards [HS20-44 loading and 14 ft (4.3 m) width].

New timber approach-span stringers, main span steel wide-flange girders, glued-laminated panel decking (Figure 18), timber running planks and bridge railing were installed (Figure 19) in 1980. All work was accomplished by Forest

FIGURE 18. Simmons Creek Bridge: New steel stringers and glued-laminated deck panels on existing substructure units.

FIGURE 20. Durnham Bridge over Gallatin River.

FIGURE 21. Underside of Markley Bridge before rehabilitation.

FIGURE 22. Markley Bridge: Decay in untreated deck planking. Decay at left goes completely through plank.

FIGURE 23. Markley Bridge: Cracked floor stringer. Decay present at deck-stringer interface also.

FIGURE 24. Markley Bridge: Rebuilt underside. Note glued-laminated floor stringers.

Service personnel. Total cost was $46,000 for materials, labor and equipment ($38 per square foot or $409 per square meter). This cost represents work done by Forest Service personnel; the cost to perform the same work under contract would perhaps have been slightly higher, but should not have exceeded $50 per square foot ($538 per square meter).

Markley and Durnham Bridges

These two pony-truss bridges, similar in size and design, cross the Gallatin River, several miles apart, between Bozeman and West Yellowstone, Montana. They were originally built by private owners, but through a recent land exchange have become Forest Service property. One is 75 ft (23 m) long, the other is 79 ft (24 m) long, and both are 16 ft (4.9 m) wide with steel trusses made up with welded joints from old pin-connected truss members salvaged from another unknown site (Figure 20). The floor system consisted of untreated timber floor stringers supported on steel floor beams between the trusses (Figure 21) and untreated timber blank decking, with treated timber abutments. Both bridges were posted for a 10-ton (9 metric ton) weight limit. The first Forest Service inspection in 1980 revealed the presence of serious deck decay (Figure 22) and a badly cracked stringer (Figure 23) in the Markley Bridge, with less serious deterioration noted in the untreated decking and stringers of the Durnham Bridge. The Markley Bridge weight limit was immediately reduced to 3 tons (2.7 metric tons) and a contract prepared and awarded in 1981 to:

1. Remove and dispose of all untreated timber decking and floor stringers. Approximate cost = $3,000.
2. Sandblast and repaint all structural steel. Approximate cost = $4,000.

FIGURE 25. Markley Bridge: Completed project.

FIGURE 26. Boy Scout Bridge: Advancing decay in untreated piling reduced load limit to 10 tons. In-place preservative treatment is expected to halt decay.

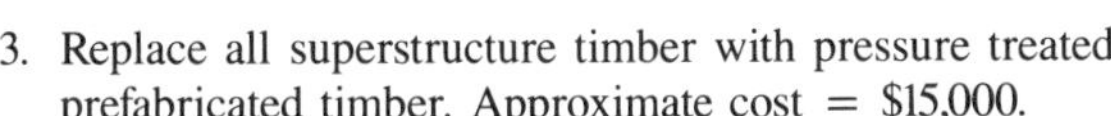

3. Replace all superstructure timber with pressure treated prefabricated timber. Approximate cost = $15,000.

Glued-laminated floor stringers, sawn floor planking, running planks and curbs were designed to increase the bridge's load capacity to an occasional 25-ton vehicle (Figures 24 and 25). The total cost of this work was $22,000 or $18 per square foot ($194 per square meter). Similar work is planned for the Durnham Bridge soon.

Clearwater River Bridge (Boy Scout Bridge)

This bridge is located at the outlet of Seeley Lake, on the Clearwater River in northwestern Montana. It was built in 1938 for H-15 truck loading, consisting of 21 simple spans at 19′0″ (5.8 m) each [total length = 400 feet (122 m)]. The superstructure consists of sawn treated timber beams with a 20′0″ (6.1 m) wide nail-laiminated deck, supported on pile bents of native untreated western larch piles with treated timber pile caps (Figure 26).

The treated timber superstructure is still in fine shape after 47 years of service, but many of the untreated larch piles are severely decayed (Figure 27). Biennial inspections have carefully monitored the advancing decay in the pilings. The deterioration of the piling became serious enough to warrant reducing the weight limit to 10 tons (9 metric tons). Further reductions in the weight limit and eventual closure to all traffic will soon be necessary if the deterioration of the piling is allowed to continue unchecked.

However, the decay can be arrested by use of in-place preservative and fumigant processes. A call for bids was let to rehabilitate this bridge rather than replace it. A new replacement bridge would cost at least $500,000. Rehabilitation of the existing structure by cutting off and replacing the severely decayed piles, in-place treating the still-serviceable

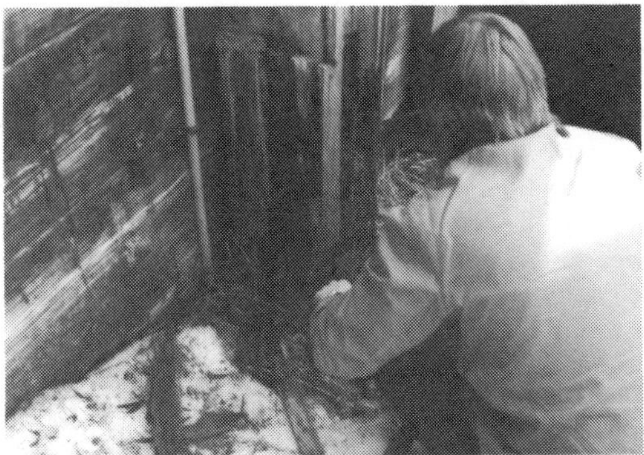

FIGURE 27. Boy Scout Bridge: Badly decayed piling will be replaced; piling still structurally sound will receive in-place preservative treatment.

FIGURE 28. Boy Scout Bridge. Completed rehabilitation. Note pilings which were cut off just above the waterline, with short pieces of new piling pinned and epoxy-bonded into place.

piling and caps with preservatives (and boring holes to internally place fumigant capsules) cost about $80,000, or $10.00 per square foot ($108 per square meter). It is expected that this work will stop the decay process in the pilings as well as upgrade the bridge's load capacity to HS20-44.

Periodic investigations and reapplication of fumigants and in-place preservative treatment in the future should extend the service life of this bridge another 40 or 50 years. It should approach the longevity of the covered bridges!

SUMMARY AND CONCLUSIONS

Wood, when properly treated to prevent early deterioration, is an economical and practical structural material for many bridges with spans in the range of 15–60 feet. Timber's inertness to de-icing chemicals, as well as some new design developments, such as glued-laminated deck panels and prestressed laminated decks, make it more attractive than ever for use in highway structures. An important factor in assuring long, useful lives for timber bridges is an effective inspection and maintenance program. Systematic and scheduled maintenance will assure a lifespan competitive with other structural materials, such as steel and concrete. Proper design considerations which minimize water-trapping details and fasteners penetrating through the treated surface into untreated wood are very important. As wood decays, loss of strength is roughly proportional to the loss of sound wood in the section. Periodic inspections and proper maintenance procedures will, in most cases, dramatically increase the useful life of timber bridges.

ACKNOWLEDGEMENTS

This chapter contains many excerpts from a paper [Reference 7] published by the American Society of Civil Engineers, and from Agriculture Handbook No. 557 [Reference 3].

REFERENCES

1. Avent, R. R., "Decay, Weathering and Epoxy Repair of Timber," *Journal of Structural Engineering, ASCE,* 111(2) (Feb. 1985).
2. Committee on History and Heritage of American Engineering, "American Wooden Bridges," ASCE Historical Publication No. 4, American Society of Civil Engineers, New York (1976).
3. Eslyn, W. E. and J. W. Clark, "Wood Bridges—Decay Inspection and Control," USDA Agriculture Handbook No. 557 (1979).
4. Faurot, R. A., "Use of Geotextiles as Bridge Paving Underlayment," *USDA Forest Service, Engineering Field Notes,* 16 (Feb.-Mar. 1984).
5. Highly, T. L. and W. E. Eslyn, "Using Fumigants to Control Interior Decay in Waterfront Timbers," *Forest Products Journal*, 32(2), 32–34 (1981).
6. Moody, R. C., et al., "Strength of Log Bridge Stringers After Several Year's Use in Southeast Alaska," Research Paper FPL 346, Forest Products Laboratory, USDA Forest Service (1979).
7. Muchmore, F. W., "Techniques to Bring New Life to Timber Bridges," *Journal of Structural Engineering, ASCE*, 110(8) (August, 1984).
8. Richardson, B. A., *Wood Preservation*. The Construction Press, Lancaster, England (1978).
9. Taylor, R. J., et al., "Prestressed Wood Bridges," Report SRR-83-01, Ontario Ministry of Transportation and Communications, Canada (1982).
10. *Wood Handbook*. USDA Agricultural Handbook No. 72 (1974).

SECTION SIX

Composites

CHAPTER 31

Composites in Construction

R. HUSSEIN* AND M. MORSI**

INTRODUCTION

There is growing interest in the development and applications of composite construction. Composite action, for example, is based on the concept of combining dissimilar materials to form a composite fulfilling specific design requirements.

A 3-layer laminated panel, as a special form of composite actions, is characterized by the use of two thin layers of strong material, denoted as faces, between which a thick layer of light-weight and comparatively weak core is sandwiched (Figure 1). In a structural panel of this type, the faces resist bending moments and in plane compressive or shear forces. The shear forces normal to the plane of the panel are resisted by the core, which also stabilizes the faces against buckling (Figure 2). This type of construction is efficient structurally due to the large stiffness achieved by spacing apart the most highly stressed elements, namely, the faces. The basic principle is much the same as that of an I-beam.

Laminated construction has found many applications in the aerospace, naval vessels, railway engineering, and in buildings.

COMPOSITE LAMINATED PANELS IN CONSTRUCTION

Answers to many problems related to construction may be found in composite laminated panels which offer several advantages. Panel virtues related to environmental requirements may include structural integrity, durable finishes, weather-tightness, dimensional stability, sound or microwave absorption. From the point of view of structural performance, 3-layer composite laminated construction is an efficient structural design due to the large stiffness achieved. Other advantages may include high strength-to-weight ratios, increased fatigue life, endurance, low moisture permeability, electrical insulation, color processability, reduced shop labour, and potentially lower costs to the house manufacture and transportation.

As any other component, there are some factors to be considered. When the core or facing materials may lead to corrosion problems, special pretreatments are required. No foam is fireproof, but many of them can be made nonflammable. Polyester 3-layer laminated panels were tested, both painted and unpainted, by the Forest Products Laboratory, and it was found that the unpainted panels deteriorated the most in three year's weathering; the edgewise compressive strength reduced by 40 percent and the flexural strength by 30 percent. Painting the other polyester panels reduced the loss in edgewise compressive strength to 22 percent, but the reduction in flexural strength was still 30 percent. Similar effects on panel strengths were observed due to the effect of moisture contents. Although several advantages can be attributed to most plastic foams, their resistance to chemical agents should be considered in the manufacturing process.

MATERIAL REQUIREMENTS

A 3-layer laminated composite panel in building systems should combine the features stated before. Constituents contribute to achieve these features as well as fulfilling the basic principle of this type of construction. A list of the required properties of each layer in a 3-layer laminated composite panel is presented in Table 1.

*University of the District of Columbia, Washington, DC
**The Arab Bureau for Design & Technical Consulting, Abbasseya, Cairo, Egypt

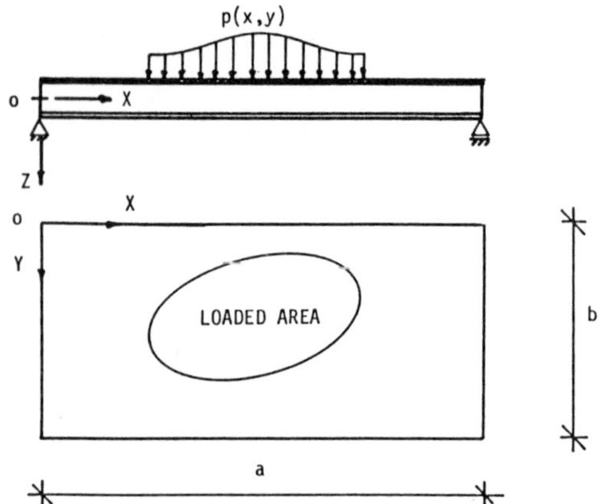

FIGURE 1. Sandwich plate.

PANEL MATERIALS

Three-layer laminated composite panel developments provide the opportunity of not favouring any one material, but rather employing most of the available materials. Therefore, a desired design requirement can be achieved by more than alternatives of material combinations.

Some of the materials that have been or might be used are listed in Table 2.

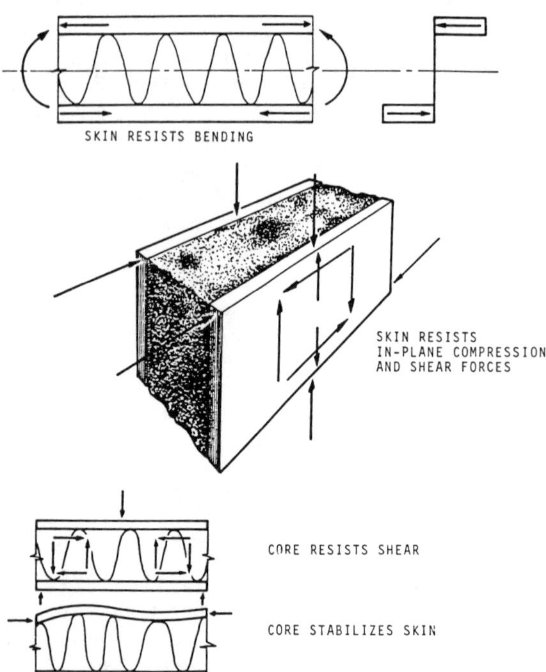

FIGURE 2. Structural action of a 3-layer laminated panel.

POLYMERS IN LAMINATED PANELS

Combining several materials into one composite unit is not a new idea. The principles of a laminated system were understood intuitively and by experience by ancient Egyptians. Spacing facings was developed by a Frenchman named Dulcan in 1820. During World War II, demands on efficient use of labor and materials imposed by the aircraft industry resulted in promoting the use of laminated panels with plywood faces and honeycomb cores. However, by today's advanced technologies, new materials, most of which are polymers, are developed and this resulted in an impact on the area of laminated construction.

Laminated panels may be made entirely of plastics or a combination of plastic with other materials. For example, the core may be made of rigid foams and skins of aluminum. A list of other plastics for the use in 3-layer laminated panels is given in Table 3.

The behavior of a laminated panel depends on its constituents, its geometry, and the applied technology in the manufacturing process. The virtues of some polymers for the use in 3-layer panels will be discussed next.

Faces

Among the synethetic materials often used for the faces of 3-layer panels are:

Poly (vinyl chloride) – PVC – and vinyl chloride copolymers
Poly (methyl methacrylate) – PMMA
Laminated wood-plastic
Fibre-reinforced plastics – FRP

POLY (VINYL CHLORIDE)

Unplasticized PVC is a hard horny material, insoluble in most solvents and not easily softened. If it is mixed while hot with certain plasticizers, it forms a rubbery material.

PVC has a tendency to liberate hydrochloric acid, particularly at temperatures approaching the upper limit of serviceability. To prevent this reaction stabilization with acid neutralizers or absorbers is necessary. PVC is nonflammable, moisture resistant, odourless and tasteless and offers exceptional resistance against a great number of corrosive media.

Advances in rigid PVC formulation and in processing equipment design have made available a new dimension in building panels to the building industry. PVC elements lend themselves admirably to use for roofing, siding, various shelter and enclosure structures and for interior panelling; in fact, they have already caused some revolutionary changes in building concepts.

Rigid PVC panels have some outstanding advantages, as follows: attractive appearance, weatherability, lightweight, corrosion and fire resistance, easy to clean and easy installa-

TABLE 1. Required Properties.

Panel	Skin	Core	Bonding	Frame	Connections
• Strength • Rigidity • Repair-ability • Durability • No loss of strength up to high temperature • Minimum weight • Fatigue • Vibration	• Strength • Resistance to local damage • Weather resistance • Nonrotting • Fire resistance • Impermeability • Durability • Reflective • Sound absorbing	• Strength • Rigidity • Impermeability • Nonrotting • Fire resistance • Sound absorption • Insulation • Durability • Minimum weight • Incombustible • Can accept adhesion	• Strength • High peel • Strength • Durability • Low creep • No loss of strength up to high temperature • Fatigue • Long life • Suitable for varied materials • Adhesion maintenance during contact with moisture or water • Chemical stability regarding both faces and core	• Strength • Rigidity • Insulation • Repair-ability • Weather resistance • Sound absorption	• Cheap • Simple • Structural • Weather-tight

tion. New rigid PVC—resistant to higher temperature—contains chlorinated PVC; these plastics may be used at temperatures up to 100°C.

Copolymerization of vinyl chloride with other vinyl monomers leads to some products with improved properties.

POLY (METHYL METHACRYLATE)

This is a clear colourless transparent plastic with a high softening point, good impact strength and weatherability. Sheets of PMMA are commonly made by extrusion.

The resistance of PMMA to outdoor exposure is outstanding and in this report this polymer is markedly superior to other thermoplastics. It has a low water absorption and the abrasion resistance is roughly comparable with that of aluminum. Like other polymers, PMMA is thermally insulating.

In sheet form, PMMA and other acrylic polymers can be softened at high temperatures and formed (usually vacuum formed) into practically any desired three-dimensional shape.

These polymers have excellent transparency (total

TABLE 2. Materials for Laminated Panels.

Faces	Core	Bonding	Frame	Connections
• Plywood • Gypsum board • High pressure laminates • painted steel sheet • Stainless steel sheet • Asbestos	• Kraft paper honeycomb • Balsa wood • Light weight concrete	• Latex • Neoprene base contact cement • Epoxy resins	• Timber • Steel • Aluminum	• Nails • Screws • Bolts • Locking devices • Cover strip

TABLE 3. Polymers for Laminated Panels.

Faces	Core	Bonding	
• Phenolic-Asbestos Laminates	• Fibre Glass	• Latex	• Phenolic-neoprene
• Epoxy-Glass Cloth Laminates	• Expanded Polystyrene	• Neoprene base contact cement	• Phenolic-neoprene
• Polyester-Glass Fiber Laminates	• Expanded Polyurethane	• Alkyd	• Phenolic-vinyl
• PVC	• Expanded PVC	• Acrylate	• Polyacrylonitrile
• PMMA	• Acrylonitrile styrene	• Caselin	• Polyamide
• Laminated wood-plastic	• Cellulose acetate	• Cellulose-nitrate	• Polyethylene
• FRP	• Epoxy	• Cellulose-vinyl	• Polyimide polyvinyl-acetate
• Polyamide (nylon)	• Methylmethacrylate styrene	• Epoxy	• Polyvinyl-butyral
• Plastic Clad Plywood	• Phenolic	• Epoxy-novalac	• Resorcinolphenol formaldehyde
	• Polyethylene	• Epoxy-phenolic	• Silicone
	• Polypropylene	• Epoxy-polyamide	• Vinyl butyral-phenolic
	• Polystyrene	• Epoxy-polysulfide	• Vinyl Copolymers
	• Polyvinyl Chloride	• Epoxy-silicone	• Urea formaldehyde
	• Silicone	• Melamine-formaldehyde	• Urethane
	• Urea Formaldehyde	• Phenolic	
	• Urethane	• Phenolic-butadience acrylonitrile	

luminous transmission up to 93%), while polystyrenes in their initial stage rarely exceed 88%–90%. Due to the latter characteristics, PMMA have been used for many lighting and glazing applications.

PLASTIC CLAD PLYWOOD

The properties of this well-known and proved product are essentially those of solid wood except that the cross-piles greatly increase strength and stability in the cross-grain direction at the expense of some reduction in the grain direction.

Plastic clad plywood touches almost all of us. This laminate is the familiar work surface of the kitchen and many restaurant tables. Again, the advantages of the plywood and light weight are enhanced for the specific use, by the hardness, decorative quality and wear resistance of the plastic outer layer.

TABLE 4. Fracture Energy.

Material	Energy to Propagate a Crack J/cm²
Glass	0.061
Epoxy resin	0.61
Metal	183.00
Glass-reinforced epoxy	
Parallel fibre	3.1
Perpendicular	430.00

COMPOSITES

Composite materials are being applied in increasing amounts for essentially the same reasons as for other applications; namely they provide properties and behaviour not attainable in single-phase materials, or they provide these features more efficiently, or at lower cost, or both.

The types of composite materials or composite structures principally found in building are: Fibrous—Fibres embedded in a continuous matrix laminar; Layers of materials blended together and possible interpenetrated by a building material particulate; Particles embedded in a continuous matrix.

An important sub-category of the laminar class of composites is the structural sandwich.

Composites have three outstanding advantages: ease of fabrication, high fracture energy, potential low cost.

The latter is particularly true for glass-reinforced resins, glass being by far the most common fibre used in resin-matrix composites. Added advantages of the resin-matrix composites often include the low density, low electrical and thermal conductivity, transluscence aesthetic, colour effects, corrosion resistance. The resin-matrix composites are highly formable, and their fracture energy is enormous. Table 4 illustrates the unusual fracture energy possibilities of resin-matrix composites.

Mechanical properties of a composite depend primarily on the type of fibres, their quantity in the laminate and their direction. The type of fibre, that is its tension strength and its modulus, determine these properties in the finished laminate and the percentage value of these fibres is directly proportional to the tensile strength and to the modulus. These are short-fibre-reinforcements such as wollastonite, asbestos, or continuous filament reinforcements such as glass fibre, basalt fibres, high modulus organic fibres, aluminum oxide and other ceramic filaments, metal filaments.

The more important fibrous materials for advanced composites are compared in Table 5.

There are several glass formulations that have been used as fibres in polymeric composites; these include A, C, D, E, M and S glass. In terms of advanced composites, E and S glass are the most important. E glass, of low alkali oxide content (composition given in Table 6), has become the most widely used formulation for both textile and industrial applications.

Between fibre and matrix (resin), there is an interfacial bond (a coupling agent) whose function seems to prevent the complete adhesion of the fibre to resin so that relative motion is possible.

The cracks, instead of penetrating through the matrix and

TABLE 5. Properties of the More Common Fibres Used in Reinforced Plastics.

Material	Tensile Strength ($N/cm^2 \times 10^3$)	Modulus ($N/cm^2 \times 10^6$)	Density (g/cm^3)
E glass	345	7.2	2.55
S glass	450	8.6	2.5
PRD-49-III	275	13	1.45
Boron	275–310	38–41	2.4
Carbon	103–310	69–62	1.4–1.9
Steel Wire	206–512	20	7.7–7.8

TABLE 6. Formulation of E Glass.

Constituent	Content (%)
Silicon dioxide	52–56
Calcium oxide	16–25
Aluminum oxide	12–16
Boron oxide	8–13
Sodium and potassium oxides	0–1
Magnesium oxide	0–6

TABLE 7. Typical Composite Efficiencies Attained in Reinforced Plastics.

Fibre Configuration	Fibre Length	Total Fibre Content (by volume) V_f	F^*_{long} ksi ($N/cm^2 \times 10^3$) F^{**}_{theor}	$F^{\dagger}_{test}$	Composite‡ Efficiency (%)
Filament-wound(un-directional)	Continuous	0.77	310(214)	180(124)	58.0
Cross laminated fibres	Continuous	0.48	197(136)	72.5(50.0)	36.0
Cloth laminated fibres	Continuous	0.48	197(136)	43.0(29.6)	21.8
Mat laminated fibres	Continuous	0.48	197(136)	57.2(39.4)	29.0
Chopped fibre systems (random)	Non-Continuous	0.13	60.7(41.8)	15.0(10.3)	24.7
Glass Flake composites	Non-Continuous	0.70	165.5(114.1)	20.0(13.8)	12.1

*F_{long} = Ultimate tensile strength in direction of greatest fibre content (longitudinal), if there is one.
**Theoretical strength based on "Rule of Mixtures": $F_{theor} = V_f S_f + (1 - V_f)S_m$
where: S_f 400.(275.8)ksi($N/cm^2 \times 10^3$)-typical boron or carbon fibre strength
S_m 10.(6.9)ksi($N/cm^2 \times 10^3$)-typical resin strength.
†F_{test} = typical experimental strength values.
‡Composite efficiency = $(F_{test}/F_{theor}) \times 100$.

TABLE 8. Characteristics and Uses of Reinforced Thermosetting Resins.

Resins	Characteristics	Uses	Limitations
Diallyl phthalate polymer	Good electrical properties; dimensional stability; chemical and heat resistance	Prepegs, ducting, randomes, aircraft, missiles	
Epoxy	Good electrical properties; chemical resistance; high strength	Printed circuit-board tooling filament winding	Require heat curing for maximum performance
Melamine formaldehyde	Good electric properties; chemical and heat resistance	Decorative, electrical (arc and track resistance), circuit breakers	
Phenolic	Low cost; chemical resistance; good electrical properties; heat resistance; nonflammable; can be used to 350°–400°F (177-204°C)	General-diverse mechanical and electrical applications	Dissolve in caustic unless specially treated
Polyester	Good all-round properties; ease of fabrication; low cost; versatile	Corrugated sheeting, seating, boats, automotive, tanks, and and piping, aircraft, tote boxes	Degraded by strong oxidizers, aromatic solvents concentrated caustic
Silicone (qv)	Heat resistance; good electrical properties	Electrical, aerospace	

directly through the fibres, are deflected along the length of the fibres.

The most common coupling agents for glass are the organosilanes and the chrome complexes; the last are cheaper and preferred in a great number of applications.

The availability of a wide variety of manufacturing techniques has undoubtedly been an important factor in the steady growth of the reinforced plastics industry, and it is very likely that it will exert considerable influence on the progress of advanced fibrough composites and laminated panels.

Table 7 shows the relative efficiencies of various forms of fibrous reinforcements: the superiority demonstrated by the continuous-oriented fibre structures is apparent.

It has already been pointed out that the properties of a fibrous composite depend upon the characteristics of fibres, matrix and coupling agent. Several polymeric matrices are used and may be divided into thermosets and thermoplastics. Sometimes, polyblends are used.

Cross-linked polyesters, phenol-formaldehyde, melamine-formaldehyde resins, epoxide and silicone polymers are among the most common in the thermosetting field. The characteristics of these polymers may be seen in Table 8.

The second group, the thermoplastics, are poor in one or more of the following engineering properties: creep resistance rigidity and tensile strength, dimensional stability, impact strength, maximum service temperatures and hardness. The use of reinforcement is one way of overcoming some of these difficulties. Table 9 shows some of the relative strength to weight ratios of various materials and illustrates well the advantages to be gained by using thermoplastics in composite form. Polyamide nylon 66 is a good example of a thermoplastic amendable to modification. A lot of properties can be improved by incorporating 20%–40% glassfibre.

In many cases, the new composites have to demonstrate their superiority over timber and its derived products, such as hardboard, chipboard and plywood. Polymer composites have the advantages that they are generally immediately ready for use and do not require finishing processes. Moreover, they usually need very little maintenance after exposure to weather.

TABLE 9. Relative Strength-to-Weight Ratios of Various Types of Materials.

Material	Strength-to-Weight Ratio Relative to Polycarbonate
Polycarbonate	1.00
Polystyrene	1.09
Nylon	1.24
Styrene-acrylonitrile copolymers	1.48
Brass (yellow cast)	1.52
Zinc alloys (cast)	1.67
Glass/polystyrene	1.71
Glass/polycarbonate	1.76
Glass/styrene-acrylonitrile copolymers	1.95
Magnesium	2.19
Aluminium	2.52
Glass/nylon	2.62

CORE

Different core materials are used: foamed or cellular plastics, honeycomb and others.

Cellular structural materials are produced by making "solid" foams from various polymers such as: phenolic resins, cellulose-acetate, polystyrene, poly (vinyl chloride), polyurethane, and others. Depending on manufacturing methods and composition, these structures may have different properties, as shown in Table 10.

TABLE 10.

1. Relative weight
2. Flexibility or rigidity
3. Preparation from thermosetting or thermoplastic polymer
4. Presence or absence of plasticizers
5. Method of cells formation
6. Nature of cells in the solid foam
7. Whether foamed in mould in place or freely-expanded

Polystyrene and polyurethane are used to make a wide variety of foamed products. Most polystyrene foams are based on foamed in-place beads, made for suspension polymerization in the presence of a foaming agent. Subsequent heating softens the resin and volatizes the foaming agent.

Polyurethane foams are expanded by a somewhat different chemical reaction. These expanded plastics may have separate interconnected or partially interconnected cells. Rigid polyurethane foams are resistant to compression and may be used to reinforce hollow structural units with a minimum of weight. In addition, they consist of closed cells and have low rates of heat transmission. They develop excellent adhesion when they are formed in voids or between sheets of material. Finally, they are resistant to oils and do not absorb appreciable amounts of water. These properties make the rigid foams valuable for prefabricated laminated structures used in the building industry for thermal insulation.

Polymers were used in an experimental study of laminated foam coextrusion using a sheet-forming die with a feedback. The polymers used in the experiment were low density polyethylene for the outer layers and ethylene–vinyl–acetate copolymer for the foamed core component. The study has shown that cell size and its distribution in the foamed core and the mechanical properties of the laminated foam product can be controlled by a judicious choice of the thickness ratio of the core to the skin components, the melextrusion temperature, and the concentration of chemical blowing agent.

The production of panels by foaming placed between two skins, takes advantage of the capability of polyurethanes to enter the moulds as a liquid or semi-expanded froth and subsequently expand to fill every section of the mould cavity. Obviously this provides a degree of freedom in panel design which is not available with lamination from slab stock. Complex shaped panels which would require an impractical amount of cutting and fitting with slab are produced rather simply by foaming in place.

Honeycomb core consists of thin foils in the form of hexagonal cells perpendicular to the faces.

Treatment of the draft paper with phenolic resin assures adequate durability if the panels are made and assembled in place to remain dry, or at least drain freely at edges and joints. Adhesives should be chosen to "wet" the honeycomb and skin, in order to form a fillet to distribute the shear stresses on the thin paper edges. Metal skins can be fully stabilized by the high strength of these hexagonal paper structures but the open cells provide little bending resistance. For many uses paper-phenolic honeycomb remains the standard for structural cores.

Bonding

These are the bonding elements between faces and the core of a laminated structure. The bond core-face allows two thin faces to be used structurally together with the core which makes the panel tough.

There are a lot of synthetic polymers which may be used to manufacture laminated panels. They may be thermosets or thermoplastics as follows:

- Urea-formaldehyde polymers—they are the cheapest but have inferior properties because of their sensitivity to moisture.
- Phenol-formaldehyde resins and their derivatives—they are cheap, but need heating or acid catalysts, which may attack faces and/or core.
- Resorcinol-formaldehyde resins—they may be used at room temperature with neutral catalysts which do not influence the panel elements.
- Unsaturated polyesters—they form the basis of glass-reinforced plastics used for constructional purposes, but are also marketed as adhesives.
- Epoxy polymers—they have a good adhesion with metals and good mechanical resistance, but are expensive. They also have the same drawbacks with the unsaturated polyesters; that is, to be made of two different components.
- Acrylic copolymers—they may be used as adhesives which cure through solvent evaporation or mixture of monomers.
- Polychloroprene—it is the most used elastomer-adhesive; it is cheap and has the advantage that there is no time restriction to be assemblage.
- Polyblends, like mixture of epoxy polymers and elastomers (polychloroprenes)—they appear good as resistant to water, but are expensive. Their use eliminates the necessity of pressing after gluing.

Adhesives for laminated panels are prepared as solutions, latices or films. With the last one a constant thickness will be needed on the glued surface to ensure that the strength is uniform. However, films are more expensive than liquid adhesives and are useful only for honeycomb cores.

Recommended curing conditions will vary depending on the facing and the core material. In many cases, modification of a given curing procedure can be brought about if advantageous to the producer. The adhesive manufacturer should be consulted for specific recommendations on his adhesive system and these recommendations should be followed exactly.

STRUCTURAL DESIGN OF 3-LAYER LAMINATED PANELS

The structural design of thin faces laminated panels will be discussed in this section. The faces are considered thin, which means that the flexural rigidity of each face about its own middle plane is neglible. The core resists only vertical shear stresses. The panel is made of two faces of equal thickness and of the same material. Materials are homogeneous, assumed isotropic and linearly elastic.

Criteria for thin faces laminated panels are:

$$\frac{h}{t_f} > 5.75 \tag{1}$$

where

t_f = face thickness
t_c = core thickness
$h = t_f + t_c$

The flexural rigidity of the core is negligible in comparison with that of faces if

$$\frac{E_f}{E_c}\frac{t_f}{t_c}\left(\frac{h}{t_c}\right)^2 > 16.70 \tag{2}$$

where

E_f = Young's modules of the faces material
E_c = Young's modules of the core material

The shear stress in the core is uniform over the depth of it if

$$\frac{E_f}{E_c}\frac{t_f}{t_c}\frac{h}{t_c} > 25 \tag{3}$$

The flexural stiffness of a panel, D_f, is given by

$$D_f = \frac{E_f t_f h^2}{2\,(1 - \nu^2)} \tag{4}$$

where

ν = Poisson's ratio of the faces material

The twisting stiffness, D_t, is given by

$$D_t = \frac{D_f}{2}(1 - \nu) \tag{5}$$

The shear stiffness of the panel, S, may be calculated from the following formula

$$S = G\frac{h^2}{t_c} \tag{6}$$

where

G = Shear modulus of the core material.

Bending of Simply Supported Laminated Panels

Consider a laminated panel simply supported along the four edges and subjected to a transverse load. The deflection at the center of this panel may be calculated from the following general formula:

$$w = \frac{P_o a^4}{D_f}K_{wb} + \frac{P_o a^2}{S}K_{ws} \tag{7}$$

a = panel length (Figure 3)
P_o = load intensity
K_{wb}, K_{ws} = deflection factors.

It can be seen from this equation that there are two terms to be added: the bending and shear components of the deflection, and each involves two factors. The first contains parameters reflecting the material properties and panel geometry and the second, the aspect ratio of the panel. To facilitate the use of Equation (7), numerical results are calculated for K_{wb} and K_{ws} and presented in the second and third columns of Tables 11–35. Five types of transverse loads are considered: uniformly distributed, hydrostatic pressure, partial load on rectangular area, concentrated load, and line load uniformly distributed across the plate width (Figure 3). In the case of concentrated and line loads,

Influence Coefficients for Simply Supported Sandwich Plates Under Uniformly Distributed Load (Tables 11–12).

$R = \frac{a}{b}$
$\nu = .30$

R	KWB	KWS	KQX	KQY	KMX	KMY	KMXY
1.0000	.0041	.0737	.3357	.3357	.0479	.0479	-.0325
1.1000	.0033	.0666	.3129	.2958	.0408	.0459	-.0267
1.2000	.0027	.0602	.2921	.2621	.0348	.0435	-.0220
1.3000	.0022	.0545	.2732	.2333	.0298	.0411	-.0182
1.4000	.0018	.0494	.2561	.2067	.0256	.0385	-.0152
1.5000	.0015	.0448	.2407	.1875	.0222	.0361	-.0127
1.6000	.0013	.0407	.2267	.1690	.0193	.0337	-.0107
1.7000	.0011	.0371	.2141	.1530	.0158	.0314	-.0091
1.8000	.0009	.0339	.2027	.1389	.0148	.0293	-.0078
1.9000	.0007	.0310	.1923	.1266	.0131	.0273	-.0067
2.0000	.0006	.0285	.1828	.1158	.0116	.0254	-.0058
3.0000	.0002	.0136	.1217	.0545	.0045	.0132	-.0018
4.0000	.0001	.0078	.0908	.0310	.0024	.0077	-.0007
5.0000	.0000	.0050	.0722	.0199	.0015	.0050	-.0004

$R = \frac{a}{b}$
$\nu = .30$

R	KWB	KWS	KQX	KQY	KMX	KMY	KMXY
1.0000	.0020	.0368	.1678	.1678	.0239	.0239	-.0162
1.1000	.0017	.0333	.1565	.1479	.0204	.0229	-.0133
1.2000	.0014	.0301	.1461	.1310	.0174	.0218	-.0110
1.3000	.0011	.0273	.1366	.1167	.0149	.0205	-.0091
1.4000	.0009	.0247	.1281	.1044	.0128	.0193	-.0076
1.5000	.0008	.0224	.1203	.0937	.0111	.0180	-.0064
1.6000	.0006	.0204	.1134	.0845	.0096	.0168	-.0054
1.7000	.0005	.0186	.1070	.0765	.0084	.0157	-.0046
1.8000	.0004	.0169	.1013	.0695	.0074	.0146	-.0039
1.9000	.0004	.0155	.0961	.0633	.0065	.0136	-.0033
2.0000	.0003	.0142	.0914	.0579	.0058	.0127	-.0029
3.0000	.0001	.0068	.0669	.0273	.0023	.0066	-.0009
4.0000	.0000	.0039	.0454	.0155	.0012	.0039	-.0004
5.0000	.0000	.0025	.0361	.0100	.0008	.0025	-.0002

Influence Coefficients for Simply Supported Sandwich Plate Under Partial Load (Tables 13–14).

$R = \frac{a}{b}$
$\xi' = .17 \quad (\xi' = \xi/a)$
$\eta' = .17 \quad (\eta' = \eta/b)$
$\nu = .30$

R	KWB	KWS	KQX	KQY	KMX	KMY	KMXY
1.0000	.0001	.0019	.0160	.0160	.0012	.0012	-.0054
1.1000	.0001	.0017	.0164	.0128	.0010	.0012	-.0044
1.2000	.0001	.0015	.0167	.0104	.0008	.0012	-.0037
1.3000	.0001	.0014	.0168	.0084	.0006	.0012	-.0031
1.4000	.0001	.0012	.0168	.0069	.0005	.0011	-.0026
1.5000	.0000	.0011	.0167	.0058	.0004	.0010	-.0022
1.6000	.0000	.0009	.0165	.0048	.0003	.0009	-.0019
1.7000	.0000	.0008	.0163	.0040	.0002	.0009	-.0017
1.8000	.0000	.0007	.0160	.0034	.0001	.0008	-.0015
1.9000	.0000	.0006	.0157	.0029	.0001	.0007	-.0013
2.0000	.0000	.0006	.0154	.0025	.0001	.0007	-.0011
3.0000	.0000	.0002	.0119	.0006	-.0000	.0003	-.0004
4.0000	.0000	.0001	.0090	.0002	-.0000	.0001	-.0002
5.0000	.0000	.0000	.0068	.0001	-.0000	.0000	-.0001

$R = \frac{a}{b}$
$\xi' = .33 \quad (\xi' = \xi/a)$
$\eta' = .17 \quad (\eta' = \eta/b)$
$\nu = .30$

R	KWB	KWS	KQX	KQY	KMX	KMY	KMXY
1.0000	.0002	.0037	.0166	.0551	.0028	.0021	-.0044
1.1000	.0002	.0035	.0158	.0465	.0024	.0021	-.0034
1.2000	.0002	.0032	.0149	.0396	.0021	.0021	-.0027
1.3000	.0001	.0029	.0139	.0340	.0018	.0020	-.0022
1.4000	.0001	.0027	.0130	.0294	.0016	.0020	-.0017
1.5000	.0001	.0025	.0120	.0256	.0013	.0019	-.0014
1.6000	.0001	.0023	.0111	.0224	.0012	.0018	-.0012
1.7000	.0001	.0021	.0102	.0197	.0010	.0017	-.0010
1.8000	.0001	.0019	.0094	.0174	.0009	.0016	-.0008
1.9000	.0000	.0018	.0086	.0155	.0008	.0015	-.0007
2.0000	.0000	.0016	.0079	.0138	.0007	.0014	-.0006
3.0000	.0000	.0008	.0032	.0052	.0002	.0008	-.0001
4.0000	.0000	.0004	.0013	.0024	.0001	.0005	-.0000
5.0000	.0000	.0002	.0006	.0013	.0000	.0003	-.0000

Influence Coefficients for Simply Supported Sandwich Plate Under Partial Load (Tables 15–18).

$R = \frac{a}{b}$
$\xi' = .50 \quad (\xi' = \xi/a)$
$\eta' = .17 \quad (\eta' = \eta/b)$
$\nu = .30$

R	KWB	KWS	KQX	KQY	KMX	KMY	KMXY
1.0000	.0003	.0046	.0114	.1064	.0037	.0023	-.0029
1.1000	.0002	.0044	.0101	.0937	.0033	.0023	-.0022
1.2000	.0002	.0041	.0089	.0832	.0030	.0023	-.0017
1.3000	.0002	.0038	.0078	.0743	.0027	.0023	-.0013
1.4000	.0001	.0036	.0068	.0668	.0024	.0022	-.0010
1.5000	.0001	.0034	.0059	.0603	.0022	.0022	-.0008
1.6000	.0001	.0032	.0051	.0548	.0020	.0021	-.0006
1.7000	.0001	.0030	.0045	.0499	.0019	.0020	-.0005
1.8000	.0001	.0028	.0039	.0457	.0017	.0019	-.0004
1.9000	.0001	.0026	.0033	.0420	.0016	.0018	-.0003
2.0000	.0001	.0025	.0029	.0387	.0015	.0018	-.0003
3.0000	.0000	.0015	.0007	.0197	.0008	.0012	-.0000
4.0000	.0000	.0010	.0002	.0118	.0005	.0008	-.0000
5.0000	.0000	.0007	.0000	.0078	.0003	.0006	-.0000

$R = \frac{a}{b}$
$\xi' = .17 \quad (\xi' = \xi/a)$
$\eta' = .33 \quad (\eta' = \eta/b)$
$\nu = .30$

R	KWB	KWS	KQX	KQY	KMX	KMY	KMXY
1.0000	.0002	.0037	.0551	.0166	.0021	.0028	-.0044
1.1000	.0002	.0033	.0536	.0142	.0016	.0026	-.0038
1.2000	.0002	.0029	.0520	.0122	.0013	.0025	-.0033
1.3000	.0001	.0025	.0503	.0104	.0010	.0023	-.0029
1.4000	.0001	.0022	.0486	.0089	.0008	.0021	-.0025
1.5000	.0001	.0019	.0469	.0076	.0006	.0019	-.0022
1.6000	.0001	.0017	.0452	.0066	.0004	.0018	-.0020
1.7000	.0001	.0015	.0435	.0057	.0003	.0016	-.0017
1.8000	.0000	.0013	.0419	.0049	.0002	.0015	-.0015
1.9000	.0000	.0011	.0404	.0042	.0002	.0013	-.0014
2.0000	.0000	.0010	.0389	.0037	.0001	.0012	-.0012
3.0000	.0000	.0003	.0267	.0010	-.0001	.0004	-.0005
4.0000	.0000	.0001	.0188	.0003	-.0000	.0002	-.0002
5.0000	.0000	.0000	.0136	.0001	-.0000	.0001	-.0001

$R = \frac{a}{b}$
$\xi' = .33 \quad (\xi' = \xi/a)$
$\eta' = .33 \quad (\eta' = \eta/b)$
$\nu = .30$

R	KWB	KWS	KQX	KQY	KMX	KMY	KMXY
1.0000	.0005	.0083	.0370	.0370	.0054	.0054	-.0051
1.1000	.0004	.0075	.0337	.0331	.0046	.0052	-.0042
1.2000	.0003	.0068	.0307	.0296	.0039	.0049	-.0035
1.3000	.0003	.0062	.0280	.0266	.0033	.0047	-.0029
1.4000	.0002	.0056	.0254	.0239	.0029	.0044	-.0024
1.5000	.0002	.0050	.0231	.0215	.0025	.0041	-.0020
1.6000	.0001	.0046	.0210	.0194	.0021	.0038	-.0017
1.7000	.0001	.0042	.0191	.0175	.0018	.0036	-.0014
1.8000	.0001	.0038	.0174	.0159	.0016	.0033	-.0012
1.9000	.0001	.0035	.0158	.0144	.0014	.0031	-.0010
2.0000	.0001	.0032	.0144	.0131	.0012	.0029	-.0009
3.0000	.0000	.0014	.0057	.0058	.0004	.0015	-.0002
4.0000	.0000	.0008	.0024	.0029	.0001	.0009	-.0001
5.0000	.0000	.0004	.0010	.0017	.0000	.0005	-.0000

$R = \frac{a}{b}$
$\xi' = .50 \quad (\xi' = \xi/a)$
$\eta' = .33 \quad (\eta' = \eta/b)$
$\nu = .30$

R	KWB	KWS	KQX	KQY	KMX	KMY	KMXY
1.0000	.0005	.0115	.0215	.0493	.0083	.0066	-.0041
1.1000	.0004	.0106	.0187	.0454	.0074	.0064	-.0032
1.2000	.0004	.0098	.0162	.0418	.0065	.0062	-.0025
1.3000	.0003	.0091	.0141	.0386	.0059	.0059	-.0020
1.4000	.0003	.0084	.0121	.0357	.0053	.0057	-.0016
1.5000	.0002	.0078	.0105	.0331	.0048	.0054	-.0013
1.6000	.0002	.0072	.0091	.0307	.0043	.0051	-.0010
1.7000	.0002	.0067	.0078	.0286	.0039	.0048	-.0008
1.8000	.0001	.0063	.0068	.0267	.0036	.0046	-.0007
1.9000	.0001	.0059	.0058	.0249	.0033	.0043	-.0005
2.0000	.0001	.0055	.0050	.0234	.0030	.0041	-.0004
3.0000	.0000	.0032	.0012	.0133	.0016	.0026	-.0001
4.0000	.0000	.0020	.0003	.0085	.0009	.0017	-.0000
5.0000	.0000	.0014	.0001	.0059	.0006	.0013	-.0000

Influence Coefficients for Simply Supported Sandwich Plate Under Partial Load (Tables 19–21).

R = a/b
ξ' = .33 (ξ' = ξ/a)
η' = .50 (η' = η/b)
ν = .30

R	KWB	KWS	KQX	KQY	KMX	KMY	KMXY
1.0000	.0005	.0115	.0493	.0215	.0066	.0083	-.0041
1.1000	.0004	.0102	.0439	.0201	.0056	.0077	-.0035
1.2000	.0004	.0091	.0392	.0187	.0047	.0071	-.0030
1.3000	.0003	.0081	.0352	.0173	.0040	.0066	-.0025
1.4000	.0002	.0073	.0316	.0159	.0034	.0061	-.0021
1.5000	.0002	.0065	.0284	.0147	.0029	.0056	-.0018
1.6000	.0002	.0059	.0256	.0135	.0025	.0052	-.0016
1.7000	.0001	.0053	.0231	.0124	.0021	.0048	-.0013
1.8000	.0001	.0048	.0209	.0114	.0019	.0044	-.0012
1.9000	.0001	.0044	.0189	.0105	.0016	.0041	-.0010
2.0000	.0001	.0040	.0172	.0097	.0014	.0038	-.0009
3.0000	.0000	.0018	.0068	.0047	.0004	.0019	-.0002
4.0000	.0000	.0009	.0029	.0025	.0001	.0010	-.0001
5.0000	.0000	.0005	.0013	.0015	.0000	.0006	-.0000

R = a/b
ξ' = .50 (ξ' = ξ/a)
η' = .50 (η' = η/b)
ν = .30

R	KWB	KWS	KQX	KQY	KMX	KMY	KMXY
1.0000	.0007	.0182	.0261	.0261	.0118	.0118	-.0037
1.1000	.0005	.0165	.0224	.0248	.0104	.0111	-.0030
1.2000	.0004	.0150	.0193	.0234	.0091	.0104	-.0024
1.3000	.0004	.0137	.0166	.0220	.0081	.0097	-.0020
1.4000	.0003	.0125	.0143	.0207	.0072	.0091	-.0016
1.5000	.0003	.0115	.0123	.0194	.0065	.0085	-.0013
1.6000	.0002	.0106	.0106	.0182	.0059	.0079	-.0011
1.7000	.0002	.0098	.0092	.0171	.0053	.0074	-.0009
1.8000	.0002	.0091	.0079	.0161	.0049	.0070	-.0007
1.9000	.0001	.0085	.0068	.0152	.0044	.0066	-.0006
2.0000	.0001	.0079	.0059	.0144	.0041	.0062	-.0005
3.0000	.0000	.0044	.0014	.0087	.0020	.0037	-.0001
4.0000	.0000	.0028	.0004	.0058	.0012	.0024	-.0000
5.0000	.0000	.0019	.0001	.0041	.0008	.0017	-.0000

R = a/b
ξ' = .17 (ξ' = ξ/a)
η' = .50 (η' = η/b)
ν = .30

R	KWB	KWS	KQX	KQY	KMX	KMY	KMXY
1.0000	.0003	.0046	.1064	.0114	.0023	.0037	-.0029
1.1000	.0002	.0040	.0994	.0103	.0018	.0034	-.0026
1.2000	.0002	.0035	.0931	.0093	.0014	.0031	-.0023
1.3000	.0002	.0030	.0874	.0083	.0011	.0028	-.0021
1.4000	.0001	.0026	.0822	.0074	.0008	.0026	-.0018
1.5000	.0001	.0023	.0774	.0066	.0006	.0023	-.0016
1.6000	.0001	.0020	.0730	.0058	.0005	.0021	-.0015
1.7000	.0001	.0017	.0690	.0051	.0003	.0019	-.0013
1.8000	.0001	.0015	.0653	.0045	.0002	.0017	-.0012
1.9000	.0000	.0013	.0619	.0040	.0002	.0016	-.0011
2.0000	.0000	.0012	.0587	.0035	.0001	.0014	-.0010
3.0000	.0000	.0003	.0366	.0010	-.0001	.0005	-.0004
4.0000	.0000	.0001	.0244	.0003	-.0001	.0002	-.0002
5.0000	.0000	.0000	.0171	.0001	-.0000	.0001	-.0001

the following equations should be used instead of Equation (7):

$$w = \frac{Pa^2R}{D_f} K_{wb} + \frac{PR}{S} K_{ws} \text{ concentrated load} \tag{8}$$

$$w = \frac{2P_oa^3}{D_f} K_{wb} + \frac{2P_oa}{S} K_{ws} \text{ line load} \tag{9}$$

where

P = the concentrated load
R = aspect ratio of the plate

Concerning the bending moments at the center of the panel and the twisting moment at its corners, for each case of the loading types, the following equations should be applied:

$M_x = P_oa^2K_{mx}$ Uniform distributed,

$M_y = P_oa^2K_{my}$ Hydrostatic pressure and partial load on rectangular area. (10)

$M_{xy} = -P_oa^2RK_{mxy}$

$M_x = PR\,K_{mx}$

$M_y = PR\,K_{my}$ Concentrated load (11)

$M_{xy} = -PR^2K_{mxy}$

$M_x = 2P_oa\,K_{mx}$

$M_y = 2P_oa\,K_{my}$ Line load (12)

$M_{xy} = -2P_oRa\,K_{mxy}$

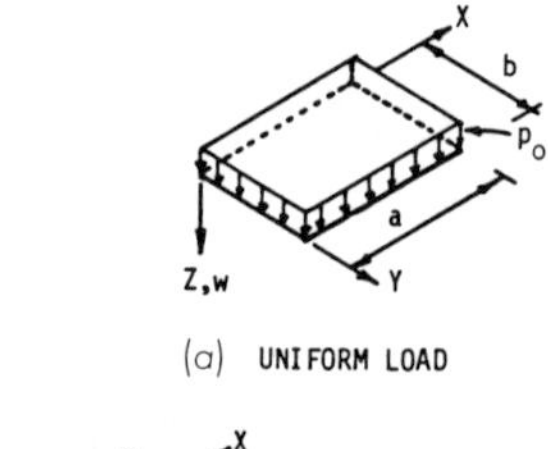

(a) UNIFORM LOAD

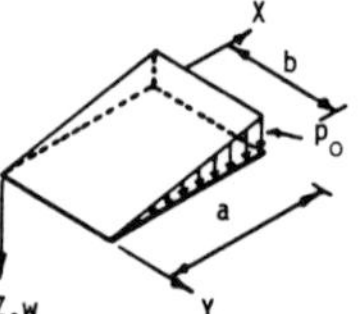

(b) HYDROSTATIC LOAD

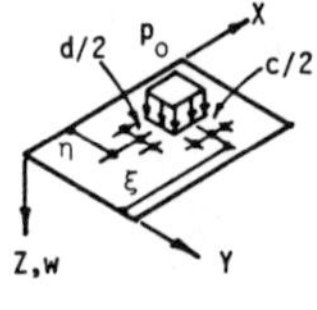

(c) PARTIAL LOAD

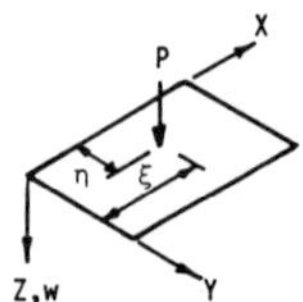

(d) CONCENTRATED LOAD

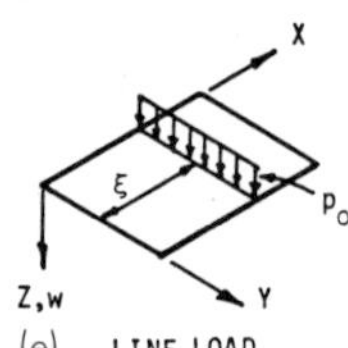

(e) LINE LOAD

FIGURE 3. Load types on simply supported sandwich plate.

Values for the factors k_{mx}, k_{my}, and k_{mxy} are obtained and presented in Tables 11–35.

Stresses are sometimes of interest in many practical problems. The inplane stresses in the faces may be obtained from the Equations (10) to (12). For example, the normal stress σ_x and σ_y are calculated by:

$$\sigma_x = \frac{M_x}{ht_f}$$

$$\sigma_y = \frac{M_y}{ht_f} \qquad (13)$$

$$\tau_{xy} = \frac{M_{xy}}{ht_f}$$

It is of interest to observe that the expressions for the stresses do not include G, nor any other term which refers to the shear stiffness of the panel. Indeed, not only are the stresses in the faces independent of the shear, but also the results in Equation (13) are identical to those determined by the classical theory of homogeneous plates. This is true due to the manner in which a simply supported 3-layer laminated panel deforms. Under the applied transverse load, the thin faces of a simply supported laminated panel undergo uniform extension or contractions as they bend about the middle plane of the whole panel. In this way, the inplane stresses are uniformly distributed across the thickness of the faces. In addition, the panel undergoes shear strains which correspond to an additional transverse deflection. The faces accommodate this extra deflection by bending about their own middle planes, as well as by displacing vertically only. This implies that, while shear deformations are taking place, the thin faces of a simple supported laminated panel do not undergo stretching or contraction, thus they retain their bending stresses unaltered.

The shear forces in a simply supported 3-layer laminated panel under each type of the loads considered can be obtained from the following formulas:

$$Q_x = P_o a\, K_{Qx} \qquad Q_y = P_o aR\, K_{Qy}$$

uniform distributed, hydrostatic pressure, and partial load on rectangular area (14)

$$Q_x = \frac{PR}{a} K_{Qx} \qquad Q_y = \frac{PR^2}{a} K_{Qy}$$

concentrated load (15)

$$Q_x = 2P_o\, K_{Qx} \qquad Q_y = 2P_o R\, K_{Qy}$$

line load (16)

Influence Coefficients for Simply Supported Sandwich Plate Under Concentrated Load (Tables 22–25).

R = a/b
ξ' = .17 (ξ' = ξ/a)
η' = .17 (η' = η/b)
ν = .30

R	KWB	KWS	KQX	KQY	KMX	KMY	KMXY
1.0000	.0023	.0306	.2556	.2556	.0199	.0199	-.1052
1.1000	.0019	.0275	.2675	.1997	.0157	.0201	-.0864
1.2000	.0015	.0245	.2762	.1561	.0122	.0197	-.0716
1.3000	.0012	.0216	.2821	.1265	.0093	.0189	-.0597
1.4000	.0010	.0190	.2855	.1022	.0070	.0178	-.0501
1.5000	.0008	.0167	.2867	.0833	.0051	.0166	-.0424
1.6000	.0007	.0146	.2859	.0684	.0036	.0154	-.0360
1.7000	.0005	.0128	.2836	.0566	.0025	.0141	-.0308
1.8000	.0004	.0112	.2799	.0471	.0016	.0129	-.0265
1.9000	.0004	.0097	.2750	.0394	.0009	.0116	-.0229
2.0000	.0003	.0085	.2691	.0331	.0004	.0107	-.0199
3.0000	.0000	.0022	.1897	.0068	-.0009	.0037	-.0059
4.0000	.0000	.0006	.1191	.0016	-.0005	.0012	-.0022
5.0000	.0000	.0002	.0720	.0005	-.0002	.0004	-.0010

R = a/b
ξ' = .33 (ξ' = ξ/a)
η' = .17 (η' = η/b)
ν = .30

R	KWB	KWS	KQX	KQY	KMX	KMY	KMXY
1.0000	.0041	.0599	.2712	.8535	.0457	.0321	-.0768
1.1000	.0034	.0557	.2585	.7061	.0393	.0331	-.0592
1.2000	.0028	.0515	.2434	.5878	.0337	.0332	-.0462
1.3000	.0023	.0473	.2269	.4924	.0288	.0327	-.0365
1.4000	.0019	.0434	.2101	.4150	.0246	.0318	-.0292
1.5000	.0016	.0397	.1934	.3518	.0210	.0306	-.0235
1.6000	.0013	.0362	.1773	.2999	.0179	.0292	-.0191
1.7000	.0011	.0331	.1619	.2570	.0153	.0278	-.0157
1.8000	.0009	.0302	.1475	.2213	.0130	.0263	-.0129
1.9000	.0008	.0276	.1341	.1915	.0111	.0248	-.0107
2.0000	.0007	.0253	.1217	.1664	.0094	.0234	-.0089
3.0000	.0002	.0107	.0442	.0495	.0015	.0124	-.0018
4.0000	.0000	.0049	.0158	.0186	-.0003	.0066	-.0005
5.0000	.0000	.0023	.0058	.0081	-.0006	.0036	-.0001

R = a/b
ξ' = .50 (ξ' = ξ/a)
η' = .17 (η' = η/b)
ν = .30

R	KWB	KWS	KQX	KQY	KMX	KMY	KMXY
1.0000	.0049	.0759	.1849	1.5505	.0628	.0358	-.0488
1.1000	.0041	.0720	.1641	1.4348	.0568	.0368	-.0365
1.2000	.0034	.0681	.1441	1.3343	.0517	.0368	-.0276
1.3000	.0028	.0642	.1257	1.2465	.0472	.0363	-.0212
1.4000	.0024	.0606	.1090	1.1692	.0433	.0354	-.0164
1.5000	.0020	.0572	.0942	1.1007	.0400	.0343	-.0128
1.6000	.0017	.0541	.0812	1.0396	.0372	.0331	-.0101
1.7000	.0014	.0512	.0699	.9848	.0347	.0318	-.0080
1.8000	.0012	.0485	.0601	.9355	.0326	.0305	-.0063
1.9000	.0011	.0461	.0516	.8908	.0307	.0292	-.0051
2.0000	.0009	.0439	.0443	.8502	.0290	.0280	-.0041
3.0000	.0003	.0294	.0096	.5837	.0190	.0191	-.0005
4.0000	.0001	.0220	.0022	.4440	.0142	.0143	-.0001
5.0000	.0001	.0176	.0006	.3571	.0114	.0114	-.0000

R = a/b
ξ' = .17 (ξ' = ξ/a)
η' = .33 (η' = η/b)
ν = .30

R	KWB	KWS	KQX	KQY	KMX	KMY	KMXY
1.0000	.0041	.0599	.8535	.2712	.0321	.0457	-.0768
1.1000	.0033	.0524	.8428	.2308	.0249	.0431	-.0674
1.2000	.0027	.0457	.8245	.1960	.0191	.0402	-.0592
1.3000	.0022	.0397	.8012	.1663	.0145	.0372	-.0521
1.4000	.0018	.0345	.7744	.1413	.0108	.0341	-.0460
1.5000	.0014	.0300	.7455	.1202	.0078	.0312	-.0407
1.6000	.0012	.0261	.7155	.1024	.0055	.0284	-.0362
1.7000	.0009	.0226	.6851	.0875	.0037	.0257	-.0322
1.8000	.0008	.0197	.6547	.0749	.0023	.0233	-.0287
1.9000	.0006	.0171	.6248	.0642	.0012	.0213	-.0256
2.0000	.0005	.0149	.5955	.0552	.0004	.0190	-.0230
3.0000	.0001	.0038	.3584	.0136	-.0015	.0065	-.0084
4.0000	.0000	.0010	.2124	.0039	-.0008	.0022	-.0035
5.0000	.0000	.0003	.1252	.0013	-.0003	.0007	-.0016

Influence Coefficients for Simply Supported Sandwich Plate Under Concentrated Load (Tables 26–29).

$R = \frac{a}{b}$
$\xi' = .33$ $(\xi' = \xi/a)$
$\eta' = .33$ $(\eta' = \eta/b)$
$\nu = .30$

R	KWB	KWS	KQX	KQY	KMX	KMY	KMXY
1.0000	.0076	.1324	.5988	.5988	.0861	.0861	-.0874
1.1000	.0062	.1196	.5450	.5365	.0721	.0834	-.0716
1.2000	.0051	.1078	.4942	.4797	.0604	.0797	-.0588
1.3000	.0042	.0970	.4471	.4285	.0508	.0756	-.0485
1.4000	.0035	.0874	.4039	.3827	.0424	.0711	-.0401
1.5000	.0029	.0787	.3645	.3420	.0356	.0667	-.0333
1.6000	.0024	.0709	.3287	.3059	.0299	.0623	-.0278
1.7000	.0020	.0640	.2964	.2738	.0251	.0581	-.0232
1.8000	.0017	.0578	.2671	.2454	.0211	.0540	-.0195
1.9000	.0014	.0523	.2407	.2202	.0178	.0502	-.0165
2.0000	.0012	.0474	.2169	.1979	.0150	.0467	-.0139
3.0000	.0003	.0191	.0768	.0736	.0020	.0228	-.0030
4.0000	.0001	.0085	.0276	.0310	-.0007	.0117	-.0008
5.0000	.0000	.0040	.0108	.0143	-.0011	.0063	-.0002

$R = \frac{a}{b}$
$\xi' = .50$ $(\xi' = \xi/a)$
$\eta' = .33$ $(\eta' = \eta/b)$
$\nu = .30$

R	KWB	KWS	KQX	KQY	KMX	KMY	KMXY
1.0000	.0092	.1894	.3499	1.2182	.1466	.0997	-.0678
1.1000	.0076	.1775	.3029	1.1021	.1328	.0979	-.0529
1.2000	.0063	.1663	.2614	1.0028	.1209	.0952	-.0415
1.3000	.0052	.1559	.2250	.9175	.1108	.0919	-.0326
1.4000	.0044	.1464	.1934	.8437	.1021	.0883	-.0258
1.5000	.0037	.1378	.1660	.7795	.0945	.0846	-.0204
1.6000	.0031	.1299	.1424	.7234	.0881	.0809	-.0163
1.7000	.0026	.1228	.1221	.6740	.0824	.0773	-.0130
1.8000	.0023	.1163	.1046	.6303	.0774	.0738	-.0104
1.9000	.0019	.1104	.0897	.5915	.0731	.0705	-.0084
2.0000	.0017	.1051	.0768	.5569	.0692	.0674	-.0068
3.0000	.0005	.0702	.0166	.3466	.0456	.0457	-.0009
4.0000	.0002	.0526	.0040	.2496	.0341	.0343	-.0001
5.0000	.0001	.0420	.0016	.1945	.0272	.0274	-.0000

$R = \frac{a}{b}$
$\xi' = .17$ $(\xi' = \xi/a)$
$\eta' = .50$ $(\eta' = \eta/b)$
$\nu = .30$

R	KWB	KWS	KQX	KQY	KMX	KMY	KMXY
1.0000	.0049	.0759	1.5505	.1849	.0358	.0628	-.0488
1.1000	.0039	.0653	1.3815	.1682	.0240	.0569	-.0440
1.2000	.0031	.0564	1.2405	.1512	.0217	.0516	-.0397
1.3000	.0025	.0488	1.1210	.1347	.0167	.0467	-.0358
1.4000	.0020	.0422	1.0164	.1192	.0126	.0423	-.0323
1.5000	.0017	.0366	.9293	.1051	.0094	.0382	-.0292
1.6000	.0013	.0318	.8512	.0924	.0069	.0345	-.0265
1.7000	.0011	.0277	.7821	.0810	.0049	.0311	-.0241
1.8000	.0009	.0241	.7206	.0709	.0033	.0281	-.0219
1.9000	.0007	.0210	.6654	.0620	.0021	.0253	-.0200
2.0000	.0006	.0184	.6155	.0543	.0011	.0226	-.0182
3.0000	.0001	.0052	.2976	.0144	-.0011	.0060	-.0077
4.0000	.0000	.0019	.1419	.0041	-.0005	.0029	-.0035
5.0000	.0000	.0009	.0574	.0013	-.0000	.0012	-.0017

$R = \frac{a}{b}$
$\xi' = .33$ $(\xi' = \xi/a)$
$\eta' = .50$ $(\eta' = \eta/b)$
$\nu = .30$

R	KWB	KWS	KQX	KQY	KMX	KMY	KMXY
1.0000	.0092	.1894	1.2182	.3499	.0997	.1466	-.0678
1.1000	.0075	.1659	1.1087	.3283	.0826	.1330	-.0581
1.2000	.0061	.1459	1.0136	.3054	.0687	.1210	-.0496
1.3000	.0050	.1289	.9303	.2825	.0573	.1103	-.0423
1.4000	.0041	.1143	.8567	.2603	.0479	.1007	-.0360
1.5000	.0034	.1017	.7914	.2393	.0402	.0920	-.0307
1.6000	.0028	.0907	.7330	.2197	.0338	.0842	-.0262
1.7000	.0023	.0812	.6807	.2015	.0285	.0771	-.0224
1.8000	.0019	.0729	.6336	.1848	.0241	.0708	-.0192
1.9000	.0016	.0657	.5912	.1695	.0204	.0650	-.0165
2.0000	.0014	.0593	.5528	.1554	.0172	.0598	-.0141
3.0000	.0003	.0238	.3149	.0674	.0030	.0279	-.0034
4.0000	.0001	.0110	.2123	.0310	-.0001	.0144	-.0009
5.0000	.0000	.0056	.1615	.0149	-.0006	.0079	-.0002

Influence Coefficients for Simply Supported Sandwich Plate Under Concentrated Load (Table 30).

$R = \frac{a}{b}$
$\xi' = .50$ $(\xi' = \xi/a)$
$\eta' = .50$ $(\eta' = \eta/b)$
$\nu = .30$

R	KWB	KWS	KQX	KQY	KMX	KMY	KMXY
1.0000	.0116	.7057	.6679	.6679	.4587	.4587	-.0610
1.1000	.0095	.6407	.5977	.6130	.4104	.4225	-.0497
1.2000	.0078	.5853	.5369	.5639	.3700	.3909	-.0403
1.3000	.0065	.5376	.4843	.5201	.3359	.3630	-.0326
1.4000	.0054	.4962	.4385	.4811	.3069	.3362	-.0264
1.5000	.0045	.4600	.3987	.4465	.2819	.3161	-.0213
1.6000	.0038	.4281	.3639	.4157	.2604	.2962	-.0173
1.7000	.0033	.3999	.3335	.3883	.2416	.2783	-.0140
1.8000	.0028	.3747	.3069	.3638	.2251	.2621	-.0113
1.9000	.0024	.3522	.2836	.3418	.2105	.2475	-.0092
2.0000	.0021	.3320	.2631	.3221	.1975	.2341	-.0075
3.0000	.0006	.2055	.1508	.2008	.1189	.1463	-.0010
4.0000	.0003	.1448	.1095	.1445	.0823	.1058	-.0002
5.0000	.0001	.1097	.0884	.1126	.0616	.0810	-.0000

Influence Coefficients for Simply Supported Sandwich Plate Subjected to Strip Load (Tables 31–33).

$R = \frac{a}{b}$
$\xi' = .10$ $(\xi' = \xi/a)$
$\nu = .30$

R	KWB	KWS	KQX	KQY	KMX	KMY	KMXY
1.0000	.0009	.0127	.4210	.0524	.0067	.0098	.0288
1.1000	.0007	.0110	.4118	.0436	.0051	.0093	.0247
1.2000	.0006	.0096	.4026	.0365	.0038	.0086	.0214
1.3000	.0005	.0083	.3934	.0306	.0028	.0080	.0187
1.4000	.0004	.0072	.3843	.0258	.0020	.0073	.0165
1.5000	.0003	.0062	.3753	.0218	.0014	.0067	.0146
1.6000	.0003	.0053	.3664	.0185	.0009	.0060	.0130
1.7000	.0002	.0046	.3576	.0157	.0005	.0055	.0116
1.8000	.0002	.0040	.3489	.0133	.0002	.0049	.0104
1.9000	.0001	.0034	.3404	.0114	.0000	.0044	.0094
2.0000	.0001	.0029	.3320	.0097	-.0001	.0039	.0085
3.0000	.0000	.0007	.2565	.0021	-.0004	.0012	.0037
4.0000	.0000	.0002	.1965	.0005	-.0002	.0004	.0019
5.0000	.0000	.0000	.1504	.0001	-.0001	.0001	.0011

$R = \frac{a}{b}$
$\xi' = .20$ $(\xi' = \xi/a)$
$\nu = .30$

R	KWB	KWS	KQX	KQY	KMX	KMY	KMXY
1.0000	.0018	.0265	.3006	.1161	.0148	.0196	.0330
1.1000	.0014	.0233	.2849	.0975	.0117	.0186	.0276
1.2000	.0012	.0204	.2696	.0823	.0091	.0174	.0233
1.3000	.0010	.0179	.2547	.0698	.0071	.0162	.0198
1.4000	.0008	.0157	.2404	.0595	.0054	.0150	.0170
1.5000	.0006	.0137	.2266	.0508	.0041	.0138	.0146
1.6000	.0005	.0120	.2135	.0436	.0030	.0126	.0126
1.7000	.0004	.0105	.2009	.0375	.0021	.0115	.0110
1.8000	.0003	.0092	.1889	.0324	.0015	.0105	.0096
1.9000	.0003	.0080	.1775	.0280	.0009	.0095	.0084
2.0000	.0002	.0070	.1667	.0243	.0005	.0086	.0074
3.0000	.0000	.0020	.0861	.0064	-.0006	.0031	.0024
4.0000	.0000	.0006	.0417	.0019	-.0004	.0011	.0009
5.0000	.0000	.0002	.0177	.0006	-.0002	.0004	.0004

$R = \frac{a}{b}$
$\xi' = .30$ $(\xi' = \xi/a)$
$\nu = .30$

R	KWB	KWS	KQX	KQY	KMX	KMY	KMXY
1.0000	.0026	.0424	.2403	.2095	.0260	.0291	.0316
1.1000	.0021	.0379	.2217	.1782	.0215	.0278	.0258
1.2000	.0017	.0339	.2041	.1526	.0178	.0263	.0213
1.3000	.0014	.0302	.1876	.1314	.0146	.0247	.0177
1.4000	.0012	.0270	.1722	.1138	.0120	.0231	.0147
1.5000	.0009	.0241	.1580	.0990	.0099	.0215	.0124
1.6000	.0008	.0216	.1449	.0864	.0081	.0199	.0104
1.7000	.0007	.0193	.1328	.0758	.0066	.0185	.0088
1.8000	.0005	.0173	.1217	.0666	.0054	.0171	.0075
1.9000	.0005	.0155	.1115	.0568	.0044	.0158	.0064
2.0000	.0004	.0139	.1022	.0520	.0035	.0146	.0055
3.0000	.0001	.0051	.0439	.0173	.0000	.0066	.0014
4.0000	.0000	.0020	.0209	.0066	-.0005	.0031	.0004
5.0000	.0000	.0009	.0120	.0028	-.0004	.0015	.0001

Influence Coefficients for Simply Supported Sandwich Plate Subjected to Strip Load (Tables 34–35).

$R = \frac{a}{b}$
$\xi' = .40 \quad (\xi' = \xi/a)$
$\nu = .30$

R	KWB	KWS	KQX	KQY	KMX	KMY	KMXY
1.0000	.0031	.0615	.1683	.3899	.0419	.0380	.0278
1.1000	.0026	.0561	.1498	.3379	.0364	.0366	.0222
1.2000	.0021	.0512	.1328	.2951	.0317	.0349	.0179
1.3000	.0018	.0468	.1175	.2595	.0277	.0332	.0145
1.4000	.0015	.0428	.1036	.2296	.0243	.0314	.0118
1.5000	.0012	.0392	.0912	.2042	.0214	.0296	.0097
1.6000	.0010	.0360	.0802	.1825	.0189	.0279	.0079
1.7000	.0009	.0331	.0703	.1639	.0168	.0263	.0065
1.8000	.0007	.0305	.0616	.1477	.0149	.0247	.0054
1.9000	.0006	.0282	.0539	.1336	.0133	.0233	.0045
2.0000	.0005	.0260	.0470	.1212	.0119	.0219	.0037
3.0000	.0001	.0130	.0103	.0528	.0043	.0125	.0007
4.0000	.0001	.0072	-.0002	.0269	.0016	.0077	.0001
5.0000	.0000	.0042	-.0032	.0151	.0004	.0050	.0000

$R = \frac{a}{b}$
$\xi' = .50 \quad (\xi' = \xi/a)$
$\nu = .30$

R	KWB	KWS	KQX	KQY	KMX	KMY	KMXY
1.0000	.0034	.0834	.1282	1.2201	.0627	.0457	.0232
1.1000	.0028	.0777	.1117	1.1016	.0568	.0443	.0182
1.2000	.0023	.0725	.0970	1.0017	.0517	.0426	.0144
1.3000	.0019	.0678	.0840	.9166	.0473	.0408	.0114
1.4000	.0016	.0635	.0727	.8432	.0436	.0390	.0090
1.5000	.0014	.0597	.0628	.7795	.0404	.0372	.0072
1.6000	.0011	.0562	.0543	.7237	.0376	.0354	.0058
1.7000	.0010	.0530	.0470	.6744	.0352	.0338	.0046
1.8000	.0008	.0502	.0407	.6307	.0330	.0322	.0037
1.9000	.0007	.0476	.0353	.5917	.0311	.0307	.0030
2.0000	.0006	.0452	.0306	.5567	.0294	.0293	.0025
3.0000	.0002	.0299	.0089	.3398	.0191	.0198	.0003
4.0000	.0001	.0222	.0044	.2365	.0141	.0148	.0001
5.0000	.0000	.0176	.0034	.1774	.0111	.0118	.0000

Numerical values for the coefficients K_{Qx} and K_{Qy} are evaluated and shown in Tables 11–35. While the transverse shear force Q_x is greatest at the middle of the sides of length b, the transverse shear force Q_x is greatest at the middle of the side of length a.

The shear stresses in the core can be obtained from:

$$\tau_{xz} = \frac{Q_x}{h} \qquad \tau_{yz} = \frac{Q_y}{h} \tag{17}$$

Continuous Laminated Panels

A rectangular 3-layer laminated panel of width a and length $(b^\ell + b^r)$, supported along the edges and also the intermediate line bb (Figure 4) forms a simply supported continuous panel over two spans. Values for the deflections, shears, and moments can be obtained by combining those for laterally loaded, simply supported panels, with those for panels under distributed moments along their edges. For this purpose, practical formulas are devised from which the deflection, moments and shear forces can be determined. The simple expressions developed contain two main factors: the first one reflects the material properties and the panel geometry, whereas the second one reflects the aspect ratio of the panels, the ratio of the right to left spans, the loading type and the shear parameter which is defined as:

$$\text{shear parameter} = \Phi^i = \frac{D}{S\, b^{i2}}$$

in which

$$i = r \text{ or } \ell$$

where r and ℓ denote the right and left panels respectively.

Numerical results, for the coefficients in the formulas, are tabulated over a certain range for each of these parameters

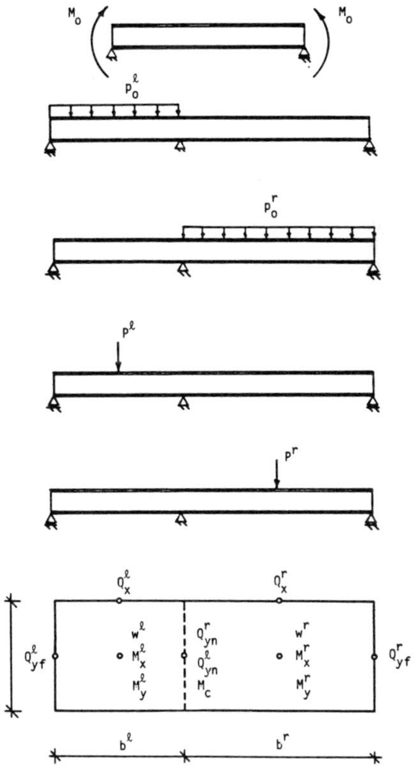

FIGURE 4. Loading types and the notations used in the practical formulas.

as follows:

(1) the ratio of the right to left spans, $k = b^r/b^\ell$, varies from 0.25 to 1.
(2) the aspect ratio of the right panel, $R^r = a/b^r$, varies from 1. to 5.
(3) the shear parameter of the right panel, Φ^r, varies from 0.0 to 0.5.
(4) Poisson's ratio of facing materials is fixed at 0.30.

Two loading types are considered: (1) a uniformly distributed load and (2) a central concentrated load. To extend further the possible application of the present formulas, the uniform and concentrated loads are considered acting on either the right or left panel. The combined effect of more than one loading type can be obtained by applying the principle of superposition. For each loading type, the numerical values of the factors in the practical formulas are tabulated in Tables 36 to 119.

In these tables, the following notations are used:

SPR = Φ^r = the shear parameter for the right panel
SPL = Φ^ℓ = the shear parameter for the left panel
= $k^2 \cdot \Phi^r$
POR, PR = p_o^r, p^r, respectively
POL, PL = p_o^ℓ, p^ℓ, respectively
where p_o^i and p_o^i are the intensity of a uniformly distributed load and the value of a central concentrated load acting on panel i, respectively
$k = b^r/b^\ell = R^\ell/R^r$
where b^i and R^i are the span and aspect ratio of panel i
POIS·R = Poisson's ratio of facing materials
N, F = Denote the intermediate support and the one parallel to it, respectively. Thus, for example, K^r_{Qyn} means that it is the factor of the shear force Q_y at the middle of the intermediate support in the right panel.
RR, RL = R^r, R^ℓ, respectively
KWR, KWL = K^r_w, K^ℓ_w, respectively
KMC = K_{mc}
KQXR, KQXL = K^r_{Qx}, K^ℓ_{Qx}, respectively
KQYRN, KQYLN = K^r_{Qyn}, K^ℓ_{Qyn}, respectively
KQYRF, KQYLF = K^r_{Qyf}, K^ℓ_{Qyf}, respectively
KMXR, KMXL = K^r_{mx}, K^ℓ_{mx}, respectively
KMYR, KMYL = K^r_{my}, K^ℓ_{my}, respectively

It should be emphasized that all the tabulated values as well as formulas presented in the following are the response of continuous panels due to only the redundant moment. The principle of superposition must be applied to determine the total effect of the applied loads.

Numerical Values for Factors in Equations (18) and (19) (Tables 36–39).

SPR......= 0.000
SPL......= 0.000
PL.......= 1.000
PR.......= 0.000
POIS.R...= .300
K........= .250

RR / RL	KWR / KWL	KMC	KQXR / KQXL	KQYRN / KQYLN	KQYRF / KQYLF	KMXR / KMXL	KMYR / KMYL
1.0000	-.0001		-.0024	-.0553	-.0011	-.0006	-.0004
.2500	-.0000	-.006	-.0001	.2210	.0000	-.0000	.0000
1.2000	-.0002		-.0060	-.0640	-.0035	-.0016	-.0014
.3000	-.0000	-.011	-.0005	.2549	.0000	-.0001	.0000
1.4000	-.0004		-.0111	-.0751	-.0081	-.0031	-.0034
.3500	-.0000	-.018	-.0017	.2970	.0000	-.0004	.0001
1.6000	-.0008		-.0169	-.0872	-.0148	-.0048	-.0064
.4000	-.0000	-.028	-.0041	.3404	.0002	-.0009	.0002
1.8000	-.0013		-.0228	-.0992	-.0232	-.0068	-.0103
.4500	-.0000	-.038	-.0078	.3806	.0005	-.0017	.0002
2.0000	-.0019		-.0284	-.1106	-.0328	-.0087	-.0148
.5000	-.0001	-.050	-.0130	.4151	.0011	-.0029	.0002
3.0000	-.0049		-.0441	-.1529	-.0822	-.0169	-.0392
.7500	-.0007	-.100	-.0496	.4966	.0124	-.0120	-.0034
4.0000	-.0070		-.0425	-.1728	-.1150	-.0212	-.0558
1.0000	-.0018	-.131	-.0778	.4860	.0340	-.0201	-.0123
5.0000	-.0080		-.0340	-.1799	-.1309	-.0231	-.0639
1.2500	-.0030	-.145	-.0864	.4539	.0547	-.0240	-.0219

SPR......= .100
SPL......= .006
PL.......= 1.000
PR.......= 0.000
POIS.R...= .300
K........= .250

RR / RL	KWR / KWL	KMC	KQXR / KQXL	KQYRN / KQYLN	KQYRF / KQYLF	KMXR / KMXL	KMYR / KMYL
1.0000	-.0000		-.0008	-.0046	-.0004	-.0000	-.0003
.2500	-.0000	-.003	-.0000	.0184	.0000	-.0000	.0000
1.2000	-.0001		-.0025	-.0106	-.0015	-.0002	-.0010
.3000	-.0000	-.006	-.0002	.0419	.0000	-.0000	.0000
1.4000	-.0002		-.0054	-.0192	-.0039	-.0006	-.0025
.3500	-.0000	-.012	-.0008	.0753	.0000	-.0001	.0000
1.6000	-.0005		-.0092	-.0297	-.0080	-.0014	-.0048
.4000	-.0000	-.019	-.0022	.1144	.0001	-.0004	.0000
1.8000	-.0008		-.0136	-.0411	-.0138	-.0023	-.0078
.4500	-.0000	-.027	-.0047	.1549	.0003	-.0009	.0000
2.0000	-.0012		-.0180	-.0528	-.0208	-.0035	-.0114
.5000	-.0001	-.036	-.0083	.1938	.0007	-.0016	-.0001
3.0000	-.0038		-.0329	-.1018	-.0626	-.0102	-.0324
.7500	-.0005	-.081	-.0378	.3196	.0095	-.0084	-.0033
4.0000	-.0057		-.0361	-.1291	-.0940	-.0149	-.0480
1.0000	-.0015	-.111	-.0642	.3474	.0280	-.0156	-.0110
5.0000	-.0068		-.0308	-.1412	-.1105	-.0172	-.0561
1.2500	-.0025	-.125	-.0747	.3380	.0469	-.0196	-.0197

SPR......= .200
SPL......= .013
PL.......= 1.000
PR.......= 0.000
POIS.R...= .300
K........= .250

RR / RL	KWR / KWL	KMC	KQXR / KQXL	KQYRN / KQYLN	KQYRF / KQYLF	KMXR / KMXL	KMYR / KMYL
1.0000	-.0000		-.0005	-.0027	-.0002	.0001	-.0003
.2500	-.0000	-.002	-.0000	.0108	.0000	-.0000	-.0000
1.2000	-.0001		-.0016	-.0066	-.0009	.0002	-.0010
.3000	-.0000	-.005	-.0001	.0262	.0000	-.0000	-.0000
1.4000	-.0001		-.0036	-.0126	-.0026	.0001	-.0022
.3500	-.0000	-.010	-.0005	.0494	.0000	-.0001	-.0000
1.6000	-.0003		-.0063	-.0203	-.0055	-.0001	-.0041
.4000	-.0000	-.016	-.0015	.0782	.0001	-.0002	-.0000
1.8000	-.0006		-.0097	-.0291	-.0098	-.0005	-.0067
.4500	-.0000	-.023	-.0033	.1097	.0002	-.0005	-.0001
2.0000	-.0009		-.0132	-.0385	-.0152	-.0011	-.0098
.5000	-.0000	-.031	-.0060	.1414	.0005	-.0010	-.0002
3.0000	-.0030		-.0275	-.0818	-.0505	-.0062	-.0282
.7500	-.0004	-.070	-.0306	.2565	.0076	-.0063	-.0032
4.0000	-.0049		-.0310	-.1085	-.0798	-.0107	-.0426
1.0000	-.0013	-.098	-.0547	.2906	.0238	-.0125	-.0102
5.0000	-.0059		-.0276	-.1213	-.0962	-.0133	-.0505
1.2500	-.0022	-.112	-.0656	.2870	.0411	-.0164	-.0180

SPR......= .300
SPL......= .019
PL.......= 1.000
PR.......= 0.000
POIS.R...= .300
K........= .250

RR / RL	KWR / KWL	KMC	KQXR / KQXL	KQYRN / KQYLN	KQYRF / KQYLF	KMXR / KMXL	KMYR / KMYL
1.0000	-.0000		-.0004	-.0019	-.0002	.0001	-.0003
.2500	-.0000	-.002	-.0000	.0077	.0000	-.0000	-.0000
1.2000	-.0000		-.0012	-.0048	-.0007	.0003	-.0009
.3000	-.0000	-.005	-.0001	.0191	.0000	-.0000	-.0000
1.4000	-.0001		-.0027	-.0094	-.0019	.0005	-.0021
.3500	-.0000	-.009	-.0004	.0368	.0000	-.0000	-.0000
1.6000	-.0002		-.0048	-.0155	-.0042	.0006	-.0038
.4000	-.0000	-.014	-.0012	.0595	.0000	-.0001	-.0001
1.8000	-.0004		-.0075	-.0226	-.0076	.0005	-.0061
.4500	-.0000	-.020	-.0026	.0850	.0002	-.0003	-.0002
2.0000	-.0007		-.0104	-.0304	-.0120	.0003	-.0089
.5000	-.0000	-.027	-.0048	.1114	.0004	-.0007	-.0003
3.0000	-.0025		-.0231	-.0685	-.0424	-.0034	-.0254
.7500	-.0004	-.063	-.0257	.2147	.0064	-.0048	-.0032
4.0000	-.0042		-.0271	-.0939	-.0693	-.0077	-.0386
1.0000	-.0011	-.089	-.0476	.2509	.0207	-.0102	-.0095
5.0000	-.0052		-.0248	-.1069	-.0853	-.0104	-.0461
1.2500	-.0020	-.102	-.0584	.2512	.0365	-.0139	-.0167

Numerical Values for Factors in Equations (18) and (19) (Tables 40–43).

SPR......= .400
SPL......= .025
PL.......= 1.000
PR.......= 0.000
POIS.R...= .300
K........= .250

RR	KWR	KMC	KQXR	KQYRN	KQYRF	KMXR	KMYR
RL	KWL		KQXL	KQYLN	KQYLF	KMXL	KMYL
1.0000	-.0000		-.0003	-.0015	-.0001	.0002	-.0003
.2500	-.0000	-.002	-.0000	.0060	.0000	.0000	-.0000
1.2000	-.0000		-.0009	-.0038	-.0005	.0004	-.0009
.3000	-.0000	-.005	-.0001	.0150	.0000	-.0000	-.0000
1.4000	-.0001		-.0021	-.0075	-.0015	.0008	-.0020
.3500	-.0000	-.008	-.0003	.0293	.0000	-.0000	-.0000
1.6000	-.0002		-.0039	-.0125	-.0034	.0010	-.0036
.4000	-.0000	-.013	-.0009	.0480	.0000	-.0001	-.0001
1.8000	-.0004		-.0061	-.0184	-.0062	.0012	-.0058
.4500	-.0000	-.019	-.0021	.0694	.0001	-.0002	-.0002
2.0000	-.0006		-.0086	-.0251	-.0099	.0012	-.0083
.5000	-.0000	-.025	-.0039	.0920	.0003	-.0004	-.0004
3.0000	-.0022		-.0199	-.0589	-.0365	-.0015	-.0234
.7500	-.0003	-.058	-.0221	.1846	.0055	-.0037	-.0031
4.0000	-.0037		-.0240	-.0829	-.0613	-.0054	-.0356
1.0000	-.0010	-.082	-.0421	.2210	.0183	-.0085	-.0090
5.0000	-.0047		-.0225	-.0957	-.0766	-.0081	-.0427
1.2500	-.0018	-.095	-.0527	.2240	.0329	-.0119	-.0156

SPR......= .500
SPL......= .031
PL.......= 1.000
PR.......= 0.000
POIS.R...= .300
K........= .250

RR	KWR	KMC	KQXR	KQYRN	KQYRF	KMXR	KMYR
RL	KWL		KQXL	KQYLN	KQYLF	KMXL	KMYL
1.0000	-.0000		-.0002	-.0012	-.0001	.0002	-.0003
.2500	-.0000	-.002	-.0000	.0049	.0000	.0000	-.0000
1.2000	-.0000		-.0008	-.0031	-.0004	.0005	-.0009
.3000	-.0000	-.004	-.0001	.0124	.0000	.0000	-.0000
1.4000	-.0001		-.0018	-.0062	-.0013	.0009	-.0019
.3500	-.0000	-.008	-.0003	.0244	.0000	-.0000	-.0000
1.6000	-.0002		-.0033	-.0105	-.0028	.0013	-.0035
.4000	-.0000	-.012	-.0008	.0402	.0000	-.0000	-.0001
1.8000	-.0003		-.0052	-.0156	-.0052	.0016	-.0055
.4500	-.0000	-.018	-.0018	.0586	.0001	-.0001	-.0002
2.0000	-.0005		-.0073	-.0213	-.0084	.0018	-.0079
.5000	-.0000	-.024	-.0034	.0783	.0003	-.0003	-.0004
3.0000	-.0019		-.0175	-.0517	-.0321	-.0000	-.0218
.7500	-.0003	-.054	-.0194	.1620	.0049	-.0029	-.0031
4.0000	-.0034		-.0216	-.0742	-.0550	-.0035	-.0332
1.0000	-.0009	-.076	-.0378	.1976	.0164	-.0071	-.0086
5.0000	-.0043		-.0206	-.0867	-.0696	-.0063	-.0398
1.2500	-.0016	-.088	-.0480	.2024	.0299	-.0103	-.0147

SPR......= 0.000
SPL......= 0.000
PL.......= 0.000
PR.......= 1.000
POIS.R...= .300
K........= .250

RR	KWR	KMC	KQXR	KQYRN	KQYRF	KMXR	KMYR
RL	KWL		KQXL	KQYLN	KQYLF	KMXL	KMYL
1.0000	-.0013		-.0574	-.3819	-.0251	-.0149	-.0091
.2500	-.0000	-.098	-.0022	1.5233	.0000	-.0004	.0002
1.2000	-.0019		-.0599	-.3507	-.0357	-.0165	-.0140
.3000	-.0000	-.102	-.0052	1.3924	.0001	-.0011	.0004
1.4000	-.0024		-.0557	-.3265	-.0428	-.0167	-.0177
.3500	-.0000	-.103	-.0089	1.2878	.0002	-.0019	.0006
1.6000	-.0026		-.0483	-.3096	-.0469	-.0163	-.0200
.4000	-.0001	-.102	-.0127	1.2124	.0005	-.0027	.0006
1.8000	-.0028		-.0397	-.2984	-.0487	-.0157	-.0212
.4500	-.0001	-.100	-.0161	1.1602	.0010	-.0035	.0005
2.0000	-.0028		-.0314	-.2910	-.0491	-.0151	-.0217
.5000	-.0001	-.099	-.0187	1.1246	.0016	-.0042	.0003
3.0000	-.0027		-.0047	-.2773	-.0457	-.0136	-.0206
.7500	-.0003	-.093	-.0215	1.0564	.0056	-.0055	-.0015
4.0000	-.0025		.0024	-.2738	-.0433	-.0133	-.0194
1.0000	-.0004	-.091	-.0161	1.0414	.0081	-.0053	-.0027
5.0000	-.0025		.0027	-.2718	-.0423	-.0132	-.0188
1.2500	-.0005	-.090	-.0100	1.0346	.0091	-.0049	-.0033

SPR......= .100
SPL......= .006
PL.......= 0.000
PR.......= 1.000
POIS.R...= .300
K........= .250

RR	KWR	KMC	KQXR	KQYRN	KQYRF	KMXR	KMYR
RL	KWL		KQXL	KQYLN	KQYLF	KMXL	KMYL
1.0000	-.0004		-.0193	-.0996	-.0084	-.0007	-.0073
.2500	-.0000	-.054	-.00[illegible]7	.3968	.0000	-.0001	.0000
1.2000	-.0008		-.0253	-.1070	-.0148	-.0021	-.0105
.3000	-.0000	-.061	-.0021	.4237	.0000	-.0003	.0001
1.4000	-.0011		-.0281	-.1075	-.0207	-.0033	-.0132
.3500	-.0000	-.064	-.0043	.4214	.0001	-.0007	.0001
1.6000	-.0014		-.0283	-.1050	-.0253	-.0041	-.0152
.4000	-.0000	-.065	-.0069	.4059	.0003	-.0012	.0001
1.8000	-.0016		-.0267	-.1017	-.0284	-.0046	-.0164
.4500	-.0000	-.065	-.0095	.3870	.0006	-.0018	.0000
2.0000	-.0017		-.0242	-.0985	-.0303	-.0049	-.0171
.5000	-.0001	-.065	-.0119	.3693	.0010	-.0023	-.0001
3.0000	-.0018		-.0095	-.0905	-.0308	-.0047	-.0172
.7500	-.0002	-.061	-.0167	.3225	.0042	-.0038	-.0015
4.0000	-.0017		-.0015	-.0886	-.0290	-.0043	-.0163
1.0000	-.0004	-.059	-.0141	.3120	.0067	-.0039	-.0026
5.0000	-.0017		.0010	-.0880	-.0281	-.0042	-.0159
1.2500	-.0004	-.059	-.0098	.3098	.0077	-.0036	-.0032

Numerical Values for Factors in Equations (18) and (19) (Tables 44–47).

SPR......= .200
SPL......= .013
PL.......= 0.000
PR.......= 1.000
POIS.R...= .300
K........= .250

RR	KWR	KMC	KQXR	KQYRN	KQYRF	KMXR	KMYR
RL	KWL		KQXL	KQYLN	KQYLF	KMXL	KMYL
1.0000	-.0003		-.0116	-.0595	-.0050	.0021	-.0069
.2500	-.0000	-.046	-.0004	.2373	.0000	-.0000	-.0000
1.2000	-.0005		-.0159	-.0669	-.0093	.0017	-.0096
.3000	-.0000	-.051	-.0014	.2650	.0000	-.0002	-.0000
1.4000	-.0008		-.0186	-.0698	-.0137	.0008	-.0117
.3500	-.0000	-.054	-.0029	.2733	.0001	-.0004	-.0000
1.6000	-.0010		-.0195	-.0701	-.0174	.0000	-.0132
.4000	-.0000	-.055	-.0048	.2705	.0002	-.0007	-.0001
1.8000	-.0011		-.0192	-.0691	-.0202	-.0007	-.0143
.4500	-.0000	-.055	-.0068	.2625	.0004	-.0011	-.0002
2.0000	-.0013		-.0180	-.0678	-.0221	-.0012	-.0149
.5000	-.0001	-.055	-.0087	.2529	.0008	-.0015	-.0003
3.0000	-.0014		-.0089	-.0628	-.0243	-.0018	-.0152
.7500	-.0002	-.053	-.0135	.2191	.0034	-.0028	-.0014
4.0000	-.0014		-.0026	-.0610	-.0232	-.0017	-.0146
1.0000	-.0003	-.052	-.0122	.2084	.0057	-.0031	-.0024
5.0000	-.0013		-.0000	-.0604	-.0225	-.0015	-.0143
1.2500	-.0004	-.051	-.0090	.2057	.0067	-.0029	-.0030

SPR......= .300
SPL......= .019
PL.......= 0.000
PR.......= 1.000
POIS.R...= .300
K........= .250

RR	KWR	KMC	KQXR	KQYRN	KQYRF	KMXR	KMYR
RL	KWL		KQXL	KQYLN	KQYLF	KMXL	KMYL
1.0000	-.0002		-.0083	-.0425	-.0036	.0033	-.0068
.2500	-.0000	-.043	-.0013	.1693	.0000	-.0000	-.0000
1.2000	-.0004		-.0117	-.0487	-.0068	.0034	-.0092
.3000	-.0000	-.047	-.0010	.1929	.0000	-.0001	-.0001
1.4000	-.0006		-.0139	-.0518	-.0102	.0029	-.0110
.3500	-.0000	-.049	-.0021	.2027	.0000	-.0002	-.0001
1.6000	-.0007		-.0149	-.0528	-.0132	.0022	-.0122
.4000	-.0000	-.050	-.0036	.2037	.0001	-.0004	-.0002
1.8000	-.0009		-.0150	-.0527	-.0156	.0015	-.0131
.4500	-.0000	-.050	-.0053	.2000	.0003	-.0007	-.0003
2.0000	-.0010		-.0143	-.0522	-.0174	.0009	-.0136
.5000	-.0000	-.050	-.0069	.1944	.0006	-.0010	-.0005
3.0000	-.0012		-.0079	-.0491	-.0202	-.0002	-.0139
.7500	-.0002	-.048	-.0114	.1697	.0029	-.0021	-.0014
4.0000	-.0012		-.0029	-.0477	-.0196	-.0002	-.0135
1.0000	-.0003	-.047	-.0107	.1601	.0049	-.0025	-.0023
5.0000	-.0011		-.0005	-.0471	-.0190	-.0001	-.0132
1.2500	-.0003	-.047	-.0082	.1572	.0060	-.0024	-.0028

SPR......= .400
SPL......= .025
PL.......= 0.000
PR.......= 1.000
POIS.R...= .300
K........= .250

RR	KWR	KMC	KQXR	KQYRN	KQYRF	KMXR	KMYR
RL	KWL		KQXL	KQYLN	KQYLF	KMXL	KMYL
1.0000	-.0001		-.0064	-.0330	-.0028	.0040	-.0067
.2500	-.0000	-.041	-.0012	.1316	.0000	.0000	-.0000
1.2000	-.0003		-.0092	-.0383	-.0054	.0044	-.0089
.3000	-.0000	-.044	-.0008	.1517	.0000	-.0000	-.0001
1.4000	-.0004		-.0111	-.0412	-.0081	.0041	-.0105
.3500	-.0000	-.046	-.0017	.1612	.0000	-.0001	-.0002
1.6000	-.0006		-.0120	-.0424	-.0107	.0035	-.0116
.4000	-.0000	-.047	-.0029	.1635	.0001	-.0002	-.0003
1.8000	-.0007		-.0122	-.0427	-.0128	.0029	-.0123
.4500	-.0000	-.047	-.0043	.1619	.0003	-.0004	-.0004
2.0000	-.0008		-.0119	-.0425	-.0144	.0023	-.0127
.5000	-.0000	-.047	-.0057	.1583	.0005	-.0006	-.0005
3.0000	-.0010		-.0070	-.0406	-.0173	.0009	-.0130
.7500	-.0001	-.045	-.0098	.1395	.0025	-.0017	-.0014
4.0000	-.0010		-.0029	-.0396	-.0171	.0008	-.0126
1.0000	-.0002	-.044	-.0096	.1312	.0044	-.0020	-.0022
5.0000	-.0010		-.0008	-.0391	-.0166	.0009	-.0124
1.2500	-.0003	-.044	-.0075	.1283	.0054	-.0020	-.0026

SPR......= .500
SPL......= .031
PL.......= 0.000
PR.......= 1.000
POIS.R...= .300
K........= .250

RR	KWR	KMC	KQXR	KQYRN	KQYRF	KMXR	KMYR
RL	KWL		KQXL	KQYLN	KQYLF	KMXL	KMYL
1.0000	-.0001		-.0053	-.0270	-.0023	.0044	-.0066
.2500	-.0000	-.039	-.0002	.1076	.0000	.0000	-.0000
1.2000	-.0002		-.0076	-.0316	-.0044	.0050	-.0088
.3000	-.0000	-.043	-.0006	.1250	.0000	.0000	-.0001
1.4000	-.0004		-.0092	-.0342	-.0068	.0049	-.0103
.3500	-.0000	-.044	-.0014	.1338	.0000	-.0000	-.0002
1.6000	-.0005		-.0101	-.0354	-.0089	.0044	-.0112
.4000	-.0000	-.044	-.0025	.1366	.0001	-.0001	-.0003
1.8000	-.0006		-.0104	-.0359	-.0108	.0038	-.0118
.4500	-.0000	-.044	-.0036	.1360	.0002	-.0002	-.0004
2.0000	-.0007		-.0101	-.0359	-.0122	.0032	-.0121
.5000	-.0000	-.044	-.0048	.1336	.0004	-.0004	-.0006
3.0000	-.0009		-.0062	-.0348	-.0151	.0017	-.0123
.7500	-.0001	-.043	-.0086	.1189	.0022	-.0013	-.0014
4.0000	-.0009		-.0028	-.0340	-.0151	.0015	-.0120
1.0000	-.0002	-.042	-.0086	.1116	.0039	-.0017	-.0021
5.0000	-.0009		-.0009	-.0335	-.0148	.0016	-.0118
1.2500	-.0003	-.042	-.0069	.1089	.0049	-.0017	-.0025

Numerical Values for Factors in Equations (18) and (19) (Tables 48–51).

SPR......= 0.000
SPL......= 0.000
PL.......= 1.000
PR.......= 0.000
POIS.R...= .300
K........= .500

RR RL	KWR KWL	KMC	KQXR KQXL	KQYRN KQYLN	KQYRF KQYLF	KMXR KMXL	KMYR KMYL
1.0000	-.0006		-.0272	-.1939	-.0118	-.0070	-.0043
.5000	-.0001	-.045	-.0118	.3868	.0010	-.0026	.0002
1.2000	-.0013		-.0403	-.2115	-.0235	-.0107	-.0093
.6000	-.0002	-.063	-.0233	.4195	.0034	-.0054	-.0004
1.4000	-.0020		-.0503	-.2201	-.0368	-.0139	-.0153
.7000	-.0004	-.079	-.0359	.4325	.0077	-.0086	-.0019
1.6000	-.0028		-.0564	-.2233	-.0497	-.0163	-.0215
.8000	-.0007	-.091	-.0477	.4329	.0137	-.0117	-.0041
1.8000	-.0035		-.0591	-.2234	-.0612	-.0180	-.0271
.9000	-.0011	-.101	-.0572	.4261	.0207	-.0144	-.0070
2.0000	-.0041		-.0591	-.2219	-.0707	-.0192	-.0320
1.0000	-.0015	-.109	-.0641	.4158	.0280	-.0166	-.0102
3.0000	-.0056		-.0410	-.2113	-.0943	-.0208	-.0445
1.5000	-.0032	-.125	-.0667	.3658	.0568	-.0206	-.0240
4.0000	-.0060		-.0218	-.2054	-.0992	-.0208	-.0473
2.0000	-.0039	-.127	-.0476	.3426	.0682	-.0202	-.0304
5.0000	-.0060		-.0103	-.2028	-.0999	-.0208	-.0477
2.5000	-.0042	-.128	-.0292	.3342	.0714	-.0197	-.0325

SPR......= .100
SPL......= .025
PL.......= 1.000
PR.......= 0.000
POIS.R...= .300
K........= .500

RR RL	KWR KWL	KMC	KQXR KQXL	KQYRN KQYLN	KQYRF KQYLF	KMXR KMXL	KMYR KMYL
1.0000	-.0002		-.0091	-.0460	-.0039	-.0003	-.0034
.5000	-.0000	-.025	-.0039	.0917	.0003	-.0004	-.0004
1.2000	-.0005		-.0167	-.0677	-.0098	-.0014	-.0069
.6000	-.0001	-.038	-.0097	.1340	.0014	-.0013	-.0011
1.4000	-.0010		-.0245	-.0859	-.0178	-.0029	-.0113
.7000	-.0002	-.051	-.0174	.1680	.0037	-.0028	-.0023
1.6000	-.0015		-.0309	-.0996	-.0270	-.0045	-.0160
.8000	-.0004	-.062	-.0259	.1918	.0074	-.0046	-.0040
1.8000	-.0021		-.0355	-.1093	-.0361	-.0061	-.0205
.9000	-.0007	-.071	-.0340	.2063	.0122	-.0064	-.0062
2.0000	-.0026		-.0382	-.1158	-.0445	-.0075	-.0245
1.0000	-.0009	-.079	-.0407	.2138	.0177	-.0082	-.0087
3.0000	-.0041		-.0341	-.1252	-.0689	-.0111	-.0361
1.5000	-.0024	-.096	-.0515	.2080	.0423	-.0130	-.0200
4.0000	-.0046		-.0214	-.1250	-.0751	-.0118	-.0391
2.0000	-.0031	-.099	-.0411	.1963	.0537	-.0135	-.0258
5.0000	-.0046		-.0116	-.1246	-.0762	-.0119	-.0396
2.5000	-.0034	-.100	-.0275	.1920	.0573	-.0133	-.0279

SPR......= .200
SPL......= .050
PL.......= 1.000
PR.......= 0.000
POIS.R...= .300
K........= .500

RR RL	KWP KWL	KMC	KQXR KQXL	KQYRN KQYLN	KQYRF KQYLF	KMXR KMXL	KMYR KMYL
1.0000	-.0001		-.0054	-.0276	-.0024	.0010	-.0033
.5000	-.0000	-.021	-.0024	.0550	.0002	-.0000	-.0005
1.2000	-.0003		-.0106	-.0427	-.0062	.0011	-.0063
.6000	-.0001	-.032	-.0061	.0844	.0009	-.0003	-.0012
1.4000	-.0006		-.0161	-.0566	-.0118	.0006	-.0100
.7000	-.0001	-.042	-.0115	.1106	.0025	-.0009	-.0024
1.6000	-.0010		-.0212	-.0681	-.0185	-.0002	-.0139
.8000	-.0003	-.052	-.0178	.1312	.0051	-.0019	-.0040
1.8000	-.0015		-.0252	-.0771	-.0257	-.0012	-.0177
.9000	-.0005	-.060	-.0241	.1456	.0087	-.0031	-.0059
2.0000	-.0019		-.0280	-.0838	-.0325	-.0023	-.0211
1.0000	-.0007	-.066	-.0298	.1547	.0129	-.0043	-.0080
3.0000	-.0033		-.0278	-.0965	-.0548	-.0063	-.0312
1.5000	-.0019	-.082	-.0414	.1592	.0338	-.0087	-.0176
4.0000	-.0037		-.0190	-.0976	-.0615	-.0075	-.0341
2.0000	-.0026	-.086	-.0351	.1507	.0445	-.0098	-.0227
5.0000	-.0038		-.0111	-.0975	-.0630	-.0078	-.0346
2.5000	-.0028	-.086	-.0247	.1467	.0482	-.0098	-.0246

SPR......= .300
SPL......= .075
PL.......= 1.000
PR.......= 0.000
POIS.R...= .300
K........= .500

RR RL	KWP KWL	KMC	KQXR KQXL	KQYRN KQYLN	KQYRF KQYLF	KMXR KMXL	KMYR KMYL
1.0000	-.0001		-.0039	-.0197	-.0017	.0016	-.0032
.5000	-.0000	-.020	-.0017	.0393	.0001	.0002	-.0005
1.2000	-.0002		-.0077	-.0312	-.0045	.0022	-.0060
.6000	-.0000	-.029	-.0044	.0616	.0007	.0002	-.0013
1.4000	-.0005		-.0120	-.0422	-.0088	.0024	-.0094
.7000	-.0001	-.038	-.0086	.0825	.0018	-.0000	-.0025
1.6000	-.0008		-.0162	-.0518	-.0141	.0021	-.0128
.8000	-.0002	-.046	-.0136	.0997	.0039	-.0005	-.0040
1.8000	-.0011		-.0196	-.0597	-.0199	.0015	-.0161
.9000	-.0004	-.053	-.0187	.1126	.0067	-.0013	-.0057
2.0000	-.0015		-.0221	-.0658	-.0256	.0006	-.0191
1.0000	-.0005	-.059	-.0234	.1215	.0102	-.0021	-.0076
3.0000	-.0027		-.0234	-.0792	-.0455	-.0032	-.0280
1.5000	-.0016	-.073	-.0346	.1302	.0282	-.0059	-.0160
4.0000	-.0032		-.0168	-.0811	-.0523	-.0047	-.0306
2.0000	-.0022	-.076	-.0304	.1241	.0381	-.0073	-.0205
5.0000	-.0033		-.0103	-.0812	-.0540	-.0051	-.0311
2.5000	-.0025	-.077	-.0221	.1206	.0417	-.0075	-.0222

Numerical Values for Factors in Equations (18) and (19) (Tables 52–55).

SPR......= .400
SPL......= .100
PL.......= 1.000
PR.......= 0.000
POIS.R...= .300
K........= .500

RR RL	KWR KWL	KMC	KQXR KQXL	KQYRN KQYLN	KQYRF KQYLF	KMXR KMXL	KMYR KMYL
1.0000	-.0001		-.0030	-.0153	-.0013	.0019	-.0031
.5000	-.0000	-.019	-.0013	.0305	.0001	.0003	-.0006
1.2000	-.0002		-.0061	-.0245	-.0035	.0028	-.0059
.6000	-.0000	-.028	-.0035	.0485	.0005	.0005	-.0013
1.4000	-.0004		-.0096	-.0336	-.0070	.0034	-.0090
.7000	-.0001	-.036	-.0068	.0658	.0015	.0005	-.0025
1.6000	-.0006		-.0130	-.0418	-.0114	.0035	-.0121
.8000	-.0002	-.043	-.0109	.0805	.0031	.0003	-.0040
1.8000	-.0009		-.0160	-.0487	-.0162	.0032	-.0151
.9000	-.0003	-.049	-.0153	.0919	.0055	-.0001	-.0056
2.0000	-.0012		-.0182	-.0542	-.0211	.0026	-.0178
1.0000	-.0005	-.054	-.0193	.1000	.0084	-.0007	-.0073
3.0000	-.0023		-.0201	-.0672	-.0390	-.0010	-.0257
1.5000	-.0013	-.067	-.0296	.1104	.0241	-.0039	-.0149
4.0000	-.0028		-.0149	-.0690	-.0455	-.0027	-.0280
2.0000	-.0019	-.070	-.0268	.1060	.0333	-.0054	-.0188
5.0000	-.0029		-.0055	-.0699	-.0473	-.0032	-.0285
2.5000	-.0022	-.070	-.0199	.1030	.0368	-.0058	-.0203

SPR......= .500
SPL......= .125
PL.......= 1.000
PR.......= 0.000
POIS.R...= .300
K........= .500

RR RL	KWR KWL	KMC	KQXR KQXL	KQYRN KQYLN	KQYRF KQYLF	KMXR KMXL	KMYR KMYL
1.0000	-.0001		-.0025	-.0125	-.0011	.0021	-.0031
.5000	-.0000	-.018	-.0011	.0250	.0001	.0004	-.0006
1.2000	-.0002		-.0050	-.0202	-.0029	.0033	-.0058
.6000	-.0000	-.026	-.0029	.0400	.0004	.0007	-.0014
1.4000	-.0003		-.0080	-.0280	-.0058	.0041	-.0087
.7000	-.0001	-.034	-.0057	.0547	.0012	.0009	-.0025
1.6000	-.0005		-.0109	-.0350	-.0096	.0044	-.0117
.8000	-.0001	-.041	-.0092	.0674	.0026	.0009	-.0039
1.8000	-.0008		-.0135	-.0411	-.0137	.0043	-.0145
.9000	-.0002	-.047	-.0129	.0776	.0046	.0007	-.0055
2.0000	-.0010		-.0155	-.0461	-.0180	.0039	-.0169
1.0000	-.0004	-.051	-.0165	.0850	.0072	.0003	-.0071
3.0000	-.0020		-.0177	-.0585	-.0341	.0007	-.0239
1.5000	-.0012	-.062	-.0260	.0959	.0211	-.0024	-.0140
4.0000	-.0025		-.0134	-.0611	-.0403	-.0012	-.0260
2.0000	-.0017	-.065	-.0239	.0927	.0295	-.0040	-.0175
5.0000	-.0026		-.0087	-.0615	-.0421	-.0018	-.0264
2.5000	-.0019	-.065	-.0180	.0902	.0329	-.0045	-.0188

SPR......= 0.000
SPL......= 0.000
PL.......= 0.000
PR.......= 1.000
POIS.R...= .300
K........= .500

RR RL	KWR KWL	KMC	KQXR KQXL	KQYRN KQYLN	KQYRF KQYLF	KMXR KMXL	KMYR KMYL
1.0000	-.0013		-.0574	-.3819	-.0251	-.0149	-.0091
.5000	-.0002	-.098	-.0250	.7617	.0022	-.0056	.0004
1.2000	-.0019		-.0599	-.3508	-.0357	-.0165	-.0140
.6000	-.0003	-.102	-.0352	.6964	.0052	-.0081	-.0006
1.4000	-.0024		-.0558	-.3267	-.0429	-.0168	-.0177
.7000	-.0005	-.103	-.0415	.6444	.0089	-.0100	-.0021
1.6000	-.0026		-.0484	-.3101	-.0470	-.0163	-.0201
.8000	-.0007	-.102	-.0441	.6074	.0128	-.0110	-.0038
1.8000	-.0028		-.0400	-.2992	-.0489	-.0158	-.0214
.9000	-.0009	-.101	-.0439	.5822	.0162	-.0115	-.0054
2.0000	-.0029		-.0319	-.2923	-.0496	-.0152	-.0220
1.0000	-.0010	-.099	-.0418	.5655	.0190	-.0116	-.0068
3.0000	-.0028		-.0060	-.2811	-.0481	-.0141	-.0217
1.5000	-.0014	-.096	-.0227	.5371	.0251	-.0105	-.0102
4.0000	-.0028		.0008	-.2790	-.0472	-.0140	-.0213
2.0000	-.0015	-.095	-.0090	.5322	.0258	-.0099	-.0109
5.0000	-.0028		.0013	-.2777	-.0470	-.0140	-.0212
2.5000	-.0015	-.095	-.0029	.5295	.0258	-.0098	-.0110

SPR......= .100
SPL......= .025
PL.......= 0.000
PR.......= 1.000
POIS.R...= .300
K........= .500

RR RL	KWR KWL	KMC	KQXR KQXL	KQYRN KQYLN	KQYRF KQYLF	KMXR KMXL	KMYR KMYL
1.0000	-.0004		-.0193	-.0996	-.0084	-.0007	-.0073
.5000	-.0001	-.054	-.0083	.1984	.0007	-.0009	-.0008
1.2000	-.0008		-.0253	-.1070	-.0148	-.0021	-.0105
.6000	-.0001	-.061	-.0146	.2119	.0021	-.0020	-.0016
1.4000	-.0011		-.0281	-.1076	-.0207	-.0033	-.0132
.7000	-.0002	-.064	-.0202	.2108	.0043	-.0032	-.0026
1.6000	-.0014		-.0283	-.1051	-.0253	-.0041	-.0152
.8000	-.0004	-.065	-.0242	.2033	.0070	-.0043	-.0038
1.8000	-.0016		-.0268	-.1020	-.0285	-.0047	-.0165
.9000	-.0005	-.065	-.0265	.1943	.0096	-.0051	-.0049
2.0000	-.0018		-.0243	-.0990	-.0305	-.0049	-.0172
1.0000	-.0006	-.065	-.0272	.1860	.0120	-.0056	-.0059
3.0000	-.0019		-.0102	-.0925	-.0321	-.0049	-.0178
1.5000	-.0010	-.063	-.0201	.1666	.0185	-.0059	-.0089
4.0000	-.0019		-.0025	-.0918	-.0314	-.0047	-.0175
2.0000	-.0011	-.062	-.0105	.1640	.0197	-.0055	-.0096
5.0000	-.0019		.0000	-.0918	-.0312	-.0047	-.0174
2.5000	-.0012	-.062	-.0044	.1638	.0198	-.0053	-.0098

Numerical Values for Factors in Equations (18) and (19) (Tables 56–59).

SPR......= .200
SPL......= .050
PL.......= 0.000
PR.......= 1.000
POIS.R...= .300
K........= .500

RR	KWR	KMC	KQXR	KQYRN	KQYRF	KMXR	KMYR
RL	KWL		KQXL	KQYLN	KQYLF	KMXL	KMYL
1.0000	-.0003		-.0116	-.0595	-.0050	.0021	-.0069
.5000	-.0000	-.046	-.0050	.1186	.0004	-.0000	-.0010
1.2000	-.0005		-.0159	-.0669	-.0093	.0017	-.0096
.6000	-.0001	-.051	-.0092	.1325	.0014	-.0004	-.0019
1.4000	-.0008		-.0166	-.0698	-.0137	.0008	-.0117
.7000	-.0002	-.054	-.0133	.1367	.0029	-.0011	-.0028
1.6000	-.0010		-.0196	-.0701	-.0174	.0000	-.0133
.8000	-.0003	-.055	-.0166	.1354	.0048	-.0018	-.0037
1.8000	-.0012		-.0192	-.0692	-.0202	-.0007	-.0143
.9000	-.0004	-.055	-.0188	.1317	.0068	-.0024	-.0046
2.0000	-.0013		-.0181	-.0680	-.0222	-.0012	-.0149
1.0000	-.0005	-.055	-.0200	.1273	.0088	-.0029	-.0055
3.0000	-.0015		-.0093	-.0640	-.0251	-.0019	-.0156
1.5000	-.0008	-.054	-.0166	.1132	.0147	-.0038	-.0079
4.0000	-.0015		-.0033	-.0631	-.0248	-.0019	-.0155
2.0000	-.0009	-.054	-.0097	.1101	.0162	-.0036	-.0086
5.0000	-.0015		-.0007	-.0630	-.0247	-.0019	-.0154
2.5000	-.0010	-.053	-.0047	.1097	.0164	-.0035	-.0088

SPR......= .300
SPL......= .075
PL.......= 0.000
PR.......= 1.000
POIS.R...= .300
K........= .500

RR	KWR	KMC	KQXR	KQYRN	KQYRF	KMXR	KMYR
RL	KWL		KQXL	KQYLN	KQYLF	KMXL	KMYL
1.0000	-.0002		-.0083	-.0425	-.0036	.0033	-.0068
.5000	-.0000	-.043	-.0036	.0846	.0003	.0004	-.0011
1.2000	-.0004		-.0117	-.0487	-.0068	.0034	-.0092
.6000	-.0001	-.047	-.0067	.0965	.0010	.0003	-.0020
1.4000	-.0006		-.0139	-.0518	-.0102	.0029	-.0110
.7000	-.0001	-.049	-.0100	.1014	.0021	-.0000	-.0029
1.6000	-.0007		-.0149	-.0528	-.0132	.0022	-.0122
.8000	-.0002	-.050	-.0127	.1020	.0036	-.0005	-.0037
1.8000	-.0009		-.0150	-.0528	-.0157	.0015	-.0131
.9000	-.0003	-.050	-.0146	.1003	.0053	-.0010	-.0045
2.0000	-.0010		-.0144	-.0523	-.0175	.0009	-.0136
1.0000	-.0004	-.050	-.0158	.0977	.0069	-.0014	-.0052
3.0000	-.0012		-.0081	-.0499	-.0207	-.0002	-.0142
1.5000	-.0007	-.049	-.0140	.0875	.0122	-.0024	-.0072
4.0000	-.0012		-.0033	-.0492	-.0207	-.0003	-.0141
2.0000	-.0008	-.049	-.0087	.0846	.0137	-.0025	-.0079
5.0000	-.0012		-.0010	-.0491	-.0206	-.0003	-.0140
2.5000	-.0008	-.049	-.0046	.0841	.0140	-.0024	-.0080

SPR......= .400
SPL......= .100
PL.......= 0.000
PR.......= 1.000
POIS.R...= .300
K........= .500

RR	KWR	KMC	KQXR	KQYRN	KQYRF	KMXR	KMYR
RL	KWL		KQXL	KQYLN	KQYLF	KMXL	KMYL
1.0000	-.0001		-.0064	-.0330	-.0028	.0040	-.0067
.5000	-.0000	-.041	-.0028	.0658	.0002	.0006	-.0012
1.2000	-.0003		-.0092	-.0383	-.0054	.0044	-.0089
.6000	-.0000	-.044	-.0053	.0759	.0008	.0007	-.0020
1.4000	-.0004		-.0111	-.0412	-.0081	.0041	-.0105
.7000	-.0001	-.046	-.0079	.0806	.0017	.0006	-.0029
1.6000	-.0006		-.0121	-.0424	-.0107	.0035	-.0116
.8000	-.0002	-.047	-.0102	.0818	.0029	.0003	-.0037
1.8000	-.0007		-.0123	-.0427	-.0128	.0029	-.0123
.9000	-.0002	-.047	-.0119	.0811	.0043	-.0000	-.0044
2.0000	-.0008		-.0119	-.0426	-.0144	.0023	-.0127
1.0000	-.0003	-.047	-.0130	.0795	.0057	-.0004	-.0050
3.0000	-.0011		-.0072	-.0412	-.0176	.0009	-.0132
1.5000	-.0006	-.046	-.0121	.0718	.0105	-.0015	-.0067
4.0000	-.0011		-.0032	-.0406	-.0179	.0007	-.0131
2.0000	-.0007	-.045	-.0078	.0693	.0120	-.0017	-.0073
5.0000	-.0011		-.0012	-.0405	-.0178	.0007	-.0131
2.5000	-.0007	-.045	-.0043	.0688	.0123	-.0016	-.0075

SPR......= .500
SPL......= .125
PL.......= 0.000
PR.......= 1.000
POIS.R...= .300
K........= .500

RR	KWR	KMC	KQXR	KQYRN	KQYRF	KMXR	KMYR
RL	KWL		KQXL	KQYLN	KQYLF	KMXL	KMYL
1.0000	-.0001		-.0053	-.0270	-.0023	.0044	-.0066
.5000	-.0000	-.039	-.0023	.0538	.0002	.0007	-.0012
1.2000	-.0002		-.0076	-.0316	-.0044	.0050	-.0088
.6000	-.0000	-.043	-.0044	.0625	.0006	.0010	-.0021
1.4000	-.0004		-.0092	-.0342	-.0068	.0049	-.0103
.7000	-.0001	-.044	-.0066	.0669	.0014	.0010	-.0029
1.6000	-.0005		-.0101	-.0354	-.0089	.0044	-.0112
.8000	-.0001	-.044	-.0086	.0684	.0025	.0009	-.0037
1.8000	-.0006		-.0104	-.0359	-.0108	.0038	-.0118
.9000	-.0002	-.044	-.0101	.0682	.0036	.0006	-.0044
2.0000	-.0007		-.0101	-.0360	-.0123	.0032	-.0121
1.0000	-.0003	-.044	-.0111	.0671	.0049	.0003	-.0049
3.0000	-.0009		-.0064	-.0351	-.0153	.0017	-.0124
1.5000	-.0005	-.043	-.0106	.0611	.0092	-.0008	-.0064
4.0000	-.0009		-.0030	-.0347	-.0157	.0015	-.0123
2.0000	-.0006	-.043	-.0071	.0589	.0106	-.0011	-.0069
5.0000	-.0009		-.0012	-.0346	-.0157	.0015	-.0123
2.5000	-.0006	-.043	-.0040	.0583	.0109	-.0011	-.0070

Numerical Values for Factors in Equations (18) and (19) (Tables 60–63).

SPR......= 0.000
SPL......= 0.000
PL.......= 1.000
PR.......= 0.000
POIS.R...= .300
K........= .750

RR	KWR	KMC	KQXR	KQYRN	KQYRF	KMXR	KMYR
RL	KWL		KQXL	KQYLN	KQYLF	KMXL	KMYL
1.0000	-.0012		-.0496	-.3157	-.0216	-.0127	-.0078
.7500	-.0006	-.082	-.0403	.4198	.0101	-.0097	-.0028
1.2000	-.0019		-.0596	-.3062	-.0350	-.0161	-.0138
.9000	-.0010	-.095	-.0537	.4055	.0194	-.0135	-.0066
1.4000	-.0026		-.0629	-.2930	-.0468	-.0179	-.0195
1.0500	-.0016	-.104	-.0614	.3856	.0292	-.0161	-.0109
1.6000	-.0031		-.0613	-.2809	-.0559	-.0187	-.0241
1.2000	-.0021	-.108	-.0639	.3669	.0380	-.0176	-.0150
1.8000	-.0036		-.0568	-.2711	-.0624	-.0190	-.0275
1.3500	-.0025	-.111	-.0627	.3515	.0452	-.0183	-.0185
2.0000	-.0039		-.0509	-.2638	-.0668	-.0190	-.0300
1.5000	-.0028	-.113	-.0590	.3398	.0506	-.0185	-.0214
3.0000	-.0044		-.0220	-.2483	-.0739	-.0184	-.0343
2.2500	-.0036	-.115	-.0319	.3141	.0617	-.0177	-.0278
4.0000	-.0044		-.0075	-.2447	-.0745	-.0182	-.0347
3.0000	-.0037	-.115	-.0134	.3084	.0632	-.0173	-.0289
5.0000	-.0044		-.0023	-.2434	-.0745	-.0182	-.0347
3.7500	-.0038	-.115	-.0050	.3066	.0634	-.0172	-.0290

SPR......= .100
SPL......= .056
PL.......= 1.000
PR.......= 0.000
POIS.R...= .300
K........= .750

RR	KWR	KMC	KQXR	KQYRN	KQYRF	KMXR	KMYR
RL	KWL		KQXL	KQYLN	KQYLF	KMXL	KMYL
1.0000	-.0004		-.0166	-.0842	-.0072	-.0006	-.0063
.7500	-.0002	-.046	-.0134	.1120	.0034	-.0010	-.0031
1.2000	-.0008		-.0248	-.1016	-.0145	-.0020	-.0103
.9000	-.0004	-.057	-.0223	.1344	.0080	-.0025	-.0058
1.4000	-.0012		-.0307	-.1110	-.0225	-.0036	-.0143
1.0500	-.0008	-.066	-.0297	.1455	.0141	-.0042	-.0088
1.6000	-.0017		-.0339	-.1150	-.0299	-.0050	-.0178
1.2000	-.0011	-.071	-.0348	.1492	.0204	-.0057	-.0117
1.8000	-.0021		-.0348	-.1159	-.0360	-.0060	-.0206
1.3500	-.0014	-.075	-.0375	.1487	.0262	-.0068	-.0144
2.0000	-.0024		-.0340	-.1153	-.0407	-.0068	-.0227
1.5000	-.0017	-.077	-.0381	.1464	.0311	-.0076	-.0166
3.0000	-.0030		-.0210	-.1109	-.0501	-.0079	-.0269
2.2500	-.0025	-.080	-.0274	.1358	.0430	-.0087	-.0222
4.0000	-.0031		-.0094	-.1099	-.0512	-.0079	-.0274
3.0000	-.0027	-.080	-.0142	.1333	.0450	-.0086	-.0233
5.0000	-.0031		-.0036	-.1097	-.0513	-.0079	-.0275
3.7500	-.0027	-.080	-.0062	.1330	.0453	-.0085	-.0235

SPR......= .200
SPL......= .113
PL.......= 1.000
PR.......= 0.000
POIS.R...= .300
K........= .750

RR	KWR	KMC	KQXR	KQYRN	KQYRF	KMXR	KMYR
RL	KWL		KQXL	KQYLN	KQYLF	KMXL	KMYL
1.0000	-.0002		-.0099	-.0505	-.0043	.0018	-.0059
.7500	-.0001	-.039	-.0081	.0671	.0020	.0007	-.0032
1.2000	-.0005		-.0156	-.0638	-.0091	.0016	-.0094
.9000	-.0003	-.048	-.0140	.0844	.0051	.0003	-.0056
1.4000	-.0008		-.0202	-.0726	-.0148	.0008	-.0126
1.0500	-.0005	-.055	-.0196	.0951	.0093	-.0004	-.0081
1.6000	-.0011		-.0232	-.0776	-.0204	-.0001	-.0154
1.2000	-.0008	-.060	-.0238	.1006	.0139	-.0014	-.0105
1.8000	-.0014		-.0247	-.0801	-.0254	-.0011	-.0176
1.3500	-.0010	-.063	-.0265	.1027	.0185	-.0023	-.0126
2.0000	-.0017		-.0248	-.0811	-.0294	-.0019	-.0193
1.5000	-.0013	-.065	-.0277	.1027	.0225	-.0031	-.0144
3.0000	-.0023		-.0173	-.0799	-.0387	-.0038	-.0229
2.2500	-.0019	-.067	-.0221	.0971	.0333	-.0049	-.0189
4.0000	-.0024		-.0088	-.0790	-.0401	-.0041	-.0234
3.0000	-.0021	-.068	-.0127	.0949	.0356	-.0051	-.0199
5.0000	-.0024		-.0038	-.0789	-.0403	-.0041	-.0235
3.7500	-.0022	-.068	-.0062	.0945	.0360	-.0051	-.0201

SPR......= .300
SPL......= .169
PL.......= 1.000
PR.......= 0.000
POIS.R...= .300
K........= .750

RR	KWR	KMC	KQXR	KQYRN	KQYRF	KMXR	KMYR
RL	KWL		KQXL	KQYLN	KQYLF	KMXL	KMYL
1.0000	-.0002		-.0071	-.0360	-.0031	.0029	-.0058
.7500	-.0001	-.036	-.0058	.0479	.0014	.0015	-.0032
1.2000	-.0004		-.0114	-.0465	-.0067	.0033	-.0089
.9000	-.0002	-.044	-.0102	.0615	.0037	.0017	-.0055
1.4000	-.0006		-.0151	-.0540	-.0110	.0030	-.0118
1.0500	-.0004	-.050	-.0146	.0707	.0069	.0014	-.0078
1.6000	-.0009		-.0177	-.0587	-.0155	.0024	-.0142
1.2000	-.0006	-.053	-.0181	.0761	.0106	.0008	-.0099
1.8000	-.0011		-.0191	-.0614	-.0196	.0016	-.0161
1.3500	-.0008	-.056	-.0205	.0787	.0143	.0001	-.0117
2.0000	-.0013		-.0195	-.0628	-.0230	.0008	-.0175
1.5000	-.0010	-.058	-.0217	.0795	.0176	-.0006	-.0132
3.0000	-.0019		-.0145	-.0632	-.0315	-.0014	-.0203
2.2500	-.0016	-.060	-.0184	.0766	.0273	-.0026	-.0169
4.0000	-.0020		-.0079	-.0627	-.0331	-.0019	-.0208
3.0000	-.0018	-.060	-.0111	.0748	.0296	-.0030	-.0177
5.0000	-.0020		-.0037	-.0625	-.0334	-.0020	-.0208
3.7500	-.0018	-.060	-.0057	.0744	.0300	-.0031	-.0178

Numerical Values for Factors in Equations (18) and (19) (Tables 64–67).

SPR......= .400
SPL......= .225
PL.......= 1.000
PR.......= 0.000
POIS.R...= .300
K........= .750

RR	KWR	KMC	KQXR	KQYRN	KQYRF	KMXR	KMYR
RL	KWL		KQXL	KQYLN	KQYLF	KMXL	KMYL
1.0000	-.0001		-.0055	-.0280	-.0024	.0034	-.0057
.7500	-.0001	-.034	-.0045	.0372	.0011	.0019	-.0033
1.2000	-.0003		-.0090	-.0366	-.0052	.0042	-.0087
.9000	-.0002	-.041	-.0081	.0484	.0029	.0024	-.0054
1.4000	-.0005		-.0120	-.0430	-.0088	.0043	-.0113
1.0500	-.0003	-.046	-.0116	.0563	.0055	.0025	-.0076
1.6000	-.0007		-.0142	-.0472	-.0125	.0039	-.0134
1.2000	-.0005	-.050	-.0146	.0612	.0085	.0022	-.0095
1.8000	-.0009		-.0156	-.0498	-.0160	.0033	-.0161
1.3500	-.0006	-.052	-.0167	.0638	.0116	.0017	-.0111
2.0000	-.0011		-.0161	-.0513	-.0189	.0026	-.0162
1.5000	-.0008	-.053	-.0179	.0650	.0145	.0011	-.0123
3.0000	-.0016		-.0124	-.0525	-.0267	.0002	-.0186
2.2500	-.0013	-.055	-.0157	.0635	.0231	-.0010	-.0154
4.0000	-.0017		-.0070	-.0522	-.0283	-.0004	-.0189
3.0000	-.0015	-.055	-.0098	.0621	.0253	-.0016	-.0161
5.0000	-.0017		-.0035	-.0521	-.0285	-.0005	-.0189
3.7500	-.0015	-.055	-.0053	.0617	.0257	-.0017	-.0162

SPR......= .500
SPL......= .281
PL.......= 1.000
PR.......= 0.000
POIS.R...= .300
K........= .750

RR	KWR	KMC	KQXR	KQYRN	KQYRF	KMXR	KMYR
RL	KWL		KQXL	KQYLN	KQYLF	KMXL	KMYL
1.0000	-.0001		-.0045	-.0229	-.0020	.0038	-.0057
.7500	-.0001	-.033	-.0037	.0305	.0009	.0021	-.0033
1.2000	-.0002		-.0074	-.0302	-.0043	.0048	-.0085
.9000	-.0001	-.040	-.0067	.0399	.0024	.0029	-.0054
1.4000	-.0004		-.0100	-.0357	-.0073	.0052	-.0110
1.0500	-.0002	-.044	-.0097	.0468	.0046	.0032	-.0074
1.6000	-.0006		-.0119	-.0395	-.0105	.0050	-.0129
1.2000	-.0004	-.047	-.0122	.0512	.0071	.0031	-.0092
1.8000	-.0008		-.0131	-.0419	-.0135	.0044	-.0144
1.3500	-.0005	-.049	-.0141	.0537	.0098	.0027	-.0106
2.0000	-.0009		-.0137	-.0434	-.0161	.0038	-.0154
1.5000	-.0007	-.050	-.0152	.0549	.0123	.0022	-.0118
3.0000	-.0014		-.0139	-.0450	-.0231	.0014	-.0173
2.2500	-.0012	-.051	-.0137	.0544	.0200	.0001	-.0144
4.0000	-.0015		-.0063	-.0448	-.0247	.0007	-.0175
3.0000	-.0013	-.051	-.0088	.0532	.0221	-.0006	-.0149
5.0000	-.0015		-.0032	-.0447	-.0250	.0005	-.0175
3.7500	-.0014	-.051	-.0048	.0529	.0226	-.0007	-.0150

SPR......= 0.000
SPL......= 0.000
PL.......= 0.000
PR.......= 1.000
POIS.R...= .300
K........= .750

RR	KWR	KMC	KQXR	KQYRN	KQYRF	KMXR	KMYR
RL	KWL		KQXL	KQYLN	KQYLF	KMXL	KMYL
1.0000	-.0013		-.0575	-.3824	-.0252	-.0149	-.0091
.7500	-.0006	-.098	-.0469	.5086	.0117	-.0114	-.0032
1.2000	-.0019		-.0602	-.3521	-.0359	-.0166	-.0141
.9000	-.0011	-.103	-.0547	.4666	.0199	-.0139	-.0067
1.4000	-.0024		-.0565	-.3292	-.0434	-.0169	-.0180
1.0500	-.0015	-.104	-.0560	.4341	.0270	-.0151	-.0100
1.6000	-.0027		-.0496	-.3138	-.0480	-.0167	-.0205
1.2000	-.0018	-.104	-.0532	.4119	.0325	-.0153	-.0127
1.8000	-.0029		-.0416	-.3040	-.0506	-.0162	-.0221
1.3500	-.0020	-.103	-.0480	.3973	.0363	-.0152	-.0147
2.0000	-.0030		-.0338	-.2979	-.0519	-.0158	-.0230
1.5000	-.0022	-.103	-.0418	.3881	.0388	-.0148	-.0162
3.0000	-.0031		-.0084	-.2885	-.0526	-.0150	-.0239
2.2500	-.0025	-.101	-.0155	.3733	.0423	-.0137	-.0186
4.0000	-.0031		-.0011	-.2866	-.0525	-.0150	-.0238
3.0000	-.0025	-.101	-.0042	.3705	.0425	-.0135	-.0189
5.0000	-.0031		.0001	-.2852	-.0525	-.0150	-.0238
3.7500	-.0025	-.101	-.0009	.3686	.0425	-.0134	-.0189

SPR......= .100
SPL......= .056
PL.......= 0.000
PR.......= 1.000
POIS.R...= .300
K........= .750

RR	KWR	KMC	KQXR	KQYRN	KQYRF	KMXR	KMYR
RL	KWL		KQXL	KQYLN	KQYLF	KMXL	KMYL
1.0000	-.0004		-.0193	-.0996	-.0084	-.0007	-.0073
.7500	-.0002	-.054	-.0156	.1324	.0039	-.0012	-.0037
1.2000	-.0008		-.0253	-.1072	-.0148	-.0021	-.0106
.9000	-.0004	-.061	-.0228	.1417	.0082	-.0026	-.0059
1.4000	-.0011		-.0282	-.1080	-.0208	-.0033	-.0133
1.0500	-.0007	-.064	-.0274	.1418	.0130	-.0039	-.0081
1.6000	-.0014		-.0286	-.1060	-.0256	-.0042	-.0153
1.2000	-.0009	-.066	-.0296	.1379	.0174	-.0048	-.0101
1.8000	-.0016		-.0273	-.1033	-.0289	-.0047	-.0167
1.3500	-.0012	-.066	-.0297	.1331	.0210	-.0055	-.0116
2.0000	-.0018		-.0250	-.1009	-.0312	-.0050	-.0176
1.5000	-.0013	-.066	-.0284	.1289	.0238	-.0058	-.0128
3.0000	-.0020		-.0114	-.0959	-.0342	-.0052	-.0189
2.2500	-.0017	-.066	-.0159	.1199	.0288	-.0059	-.0152
4.0000	-.0020		-.0035	-.0955	-.0342	-.0051	-.0189
3.0000	-.0017	-.065	-.0064	.1191	.0293	-.0057	-.0155
5.0000	-.0020		-.0007	-.0955	-.0342	-.0051	-.0189
3.7500	-.0017	-.065	-.0021	.1190	.0293	-.0057	-.0155

Numerical Values for Factors in Equations (18) and (19) (Tables 68–71).

SPR......= .200
SPL......= .113
PL.......= 0.000
PR.......= 1.000
POIS.R...= .300
K........= .750

RR	KWR	KMC	KQXR	KQYRN	KQYRF	KMXR	KMYR
RL	KWL		KQXL	KQYLN	KQYLF	KMXL	KMYL
1.0000	-.0003		-.0116	-.0595	-.0050	.0021	-.0069
.7500	-.0001	-.046	-.0094	.0791	.0023	.0008	-.0037
1.2000	-.0005		-.0160	-.0670	-.0093	.0017	-.0096
.9000	-.0003	-.051	-.0144	.0886	.0052	.0004	-.0057
1.4000	-.0008		-.0186	-.0699	-.0137	.0008	-.0118
1.0500	-.0005	-.054	-.0181	.0918	.0086	-.0004	-.0075
1.6000	-.0010		-.0197	-.0704	-.0175	.0000	-.0133
1.2000	-.0006	-.055	-.0202	.0915	.0119	-.0011	-.0090
1.8000	-.0012		-.0194	-.0698	-.0204	-.0007	-.0144
1.3500	-.0008	-.056	-.0210	.0898	.0148	-.0018	-.0102
2.0000	-.0013		-.0104	-.0688	-.0225	-.0012	-.0151
1.5000	-.0010	-.056	-.0208	.0877	.0172	-.0023	-.0111
3.0000	-.0016		-.0099	-.0657	-.0262	-.0021	-.0162
2.2500	-.0013	-.056	-.0133	.0814	.0223	-.0031	-.0131
4.0000	-.0016		-.0039	-.0652	-.0264	-.0021	-.0163
3.0000	-.0014	-.055	-.0062	.0804	.0229	-.0030	-.0134
5.0000	-.0016		-.0012	-.0652	-.0264	-.0021	-.0163
3.7500	-.0014	-.055	-.0024	.0803	.0230	-.0030	-.0134

SPR......= .300
SPL......= .169
PL.......= 0.000
PR.......= 1.000
POIS.R...= .300
K........= .750

RR	KWR	KMC	KQXR	KQYRN	KQYRF	KMXR	KMYR
RL	KWL		KQXL	KQYLN	KQYLF	KMXL	KMYL
1.0000	-.0002		-.0083	-.0425	-.0036	.0033	-.0068
.7500	-.0001	-.043	-.0067	.0565	.0017	.0017	-.0038
1.2000	-.0004		-.0117	-.0487	-.0068	.0034	-.0092
.9000	-.0002	-.047	-.0105	.0644	.0038	.0017	-.0056
1.4000	-.0006		-.0139	-.0518	-.0102	.0029	-.0110
1.0500	-.0003	-.049	-.0135	.0680	.0064	.0013	-.0072
1.6000	-.0007		-.0150	-.0529	-.0133	.0022	-.0123
1.2000	-.0005	-.050	-.0154	.0688	.0090	.0008	-.0085
1.8000	-.0009		-.0151	-.0530	-.0157	.0015	-.0131
1.3500	-.0006	-.050	-.0163	.0682	.0114	.0002	-.0095
2.0000	-.0010		-.0145	-.0527	-.0176	.0009	-.0137
1.5000	-.0007	-.050	-.0163	.0670	.0134	-.0003	-.0102
3.0000	-.0013		-.0085	-.0508	-.0213	-.0003	-.0146
2.2500	-.0011	-.050	-.0112	.0627	.0182	-.0014	-.0117
4.0000	-.0013		-.0037	-.0504	-.0217	-.0004	-.0146
3.0000	-.0011	-.050	-.0057	.0617	.0190	-.0015	-.0120
5.0000	-.0013		-.0014	-.0504	-.0217	-.0004	-.0146
3.7500	-.0011	-.050	-.0024	.0616	.0190	-.0015	-.0120

SPR......= .400
SPL......= .225
PL.......= 0.000
PR.......= 1.000
POIS.R...= .300
K........= .750

RR	KWR	KMC	KQXR	KQYRN	KQYRF	KMXR	KMYR
RL	KWL		KQXL	KQYLN	KQYLF	KMXL	KMYL
1.0000	-.0001		-.0064	-.0330	-.0028	.0040	-.0067
.7500	-.0001	-.041	-.0052	.0439	.0013	.0022	-.0038
1.2000	-.0003		-.0092	-.0383	-.0054	.0044	-.0089
.9000	-.0002	-.044	-.0083	.0507	.0030	.0025	-.0056
1.4000	-.0004		-.0111	-.0412	-.0081	.0041	-.0106
1.0500	-.0003	-.046	-.0107	.0540	.0051	.0023	-.0070
1.6000	-.0006		-.0121	-.0424	-.0107	.0035	-.0116
1.2000	-.0004	-.047	-.0124	.0551	.0073	.0019	-.0081
1.8000	-.0007		-.0123	-.0428	-.0128	.0029	-.0123
1.3500	-.0005	-.047	-.0133	.0550	.0093	.0014	-.0090
2.0000	-.0008		-.0120	-.0428	-.0145	.0023	-.0128
1.5000	-.0006	-.047	-.0134	.0544	.0111	.0010	-.0096
3.0000	-.0011		-.0073	-.0417	-.0179	.0009	-.0134
2.2500	-.0009	-.046	-.0096	.0513	.0154	-.0003	-.0108
4.0000	-.0011		-.0034	-.0414	-.0184	.0007	-.0134
3.0000	-.0010	-.046	-.0051	.0504	.0162	-.0005	-.0110
5.0000	-.0011		-.0014	-.0413	-.0185	.0007	-.0134
3.7500	-.0010	-.046	-.0023	.0503	.0163	-.0005	-.0110

SPR......= .500
SPL......= .281
PL.......= 0.000
PR.......= 1.000
POIS.R...= .300
K........= .750

RR	KWR	KMC	KQXR	KQYRN	KQYRF	KMXR	KMYR
RL	KWL		KQXL	KQYLN	KQYLF	KMXL	KMYL
1.0000	-.0001		-.0053	-.0270	-.0023	.0044	-.0066
.7500	-.0001	-.039	-.0043	.0359	.0011	.0025	-.0038
1.2000	-.0002		-.0076	-.0316	-.0044	.0050	-.0088
.9000	-.0001	-.043	-.0068	.0417	.0025	.0030	-.0055
1.4000	-.0004		-.0092	-.0342	-.0068	.0049	-.0103
1.0500	-.0002	-.044	-.0089	.0448	.0042	.0030	-.0069
1.6000	-.0005		-.0101	-.0354	-.0090	.0044	-.0112
1.2000	-.0003	-.044	-.0104	.0460	.0061	.0027	-.0079
1.8000	-.0006		-.0104	-.0359	-.0108	.0038	-.0118
1.3500	-.0004	-.045	-.0112	.0462	.0079	.0023	-.0086
2.0000	-.0007		-.0102	-.0360	-.0123	.0032	-.0121
1.5000	-.0005	-.044	-.0114	.0458	.0094	.0018	-.0091
3.0000	-.0009		-.0065	-.0354	-.0155	.0017	-.0125
2.2500	-.0008	-.044	-.0084	.0435	.0133	.0005	-.0101
4.0000	-.0010		-.0031	-.0352	-.0160	.0015	-.0125
3.0000	-.0008	-.043	-.0046	.0428	.0141	.0002	-.0102
5.0000	-.0010		-.0013	-.0351	-.0161	.0014	-.0125
3.7500	-.0009	-.043	-.0022	.0426	.0142	.0002	-.0102

Numerical Values for Factors in Equations (18) and (19) (Tables 72–75).

SPR......= 0.000
SPL......= 0.000
PL.......= 1.000
PR.......= 0.000
POIS.R...= .300
K........= 1.000

RR RL	KWR KWL	KMC	KQXR KQXL	KQYRN KQYLN	KQYRF KQYLF	KMXR KMXL	KMYR KMYL
1.0000	-.0014		-.0580	-.3848	-.0254	-.0150	-.0092
1.0000	-.0014	-.099	-.0580	.3848	.0254	-.0150	-.0092
1.2000	-.0020		-.0613	-.3566	-.0365	-.0169	-.0144
1.2000	-.0020	-.105	-.0613	.3566	.0365	-.0169	-.0144
1.4000	-.0025		-.0583	-.3354	-.0447	-.0174	-.0185
1.4000	-.0025	-.107	-.0583	.3354	.0447	-.0174	-.0185
1.6000	-.0028		-.0519	-.3212	-.0500	-.0173	-.0214
1.6000	-.0028	-.108	-.0519	.3212	.0500	-.0173	-.0214
1.8000	-.0030		-.0443	-.3121	-.0533	-.0170	-.0233
1.8000	-.0030	-.108	-.0443	.3121	.0533	-.0170	-.0233
2.0000	-.0032		-.0367	-.3064	-.0552	-.0167	-.0245
2.0000	-.0032	-.108	-.0367	.3064	.0552	-.0167	-.0245
3.0000	-.0034		-.0108	-.2969	-.0574	-.0160	-.0261
3.0000	-.0034	-.107	-.0108	.2969	.0574	-.0160	-.0261
4.0000	-.0034		-.0025	-.2948	-.0575	-.0160	-.0262
4.0000	-.0034	-.107	-.0025	.2948	.0575	-.0160	-.0262
5.0000	-.0034		-.0005	-.2934	-.0575	-.0160	-.0262
5.0000	-.0034	-.107	-.0005	.2934	.0575	-.0160	-.0262

SPR......= .100
SPL......= .100
PL.......= 1.000
PR.......= 0.000
POIS.R...= .300
K........= 1.000

RR RL	KWR KWL	KMC	KQXR KQXL	KQYRN KQYLN	KQYRF KQYLF	KMXR KMXL	KMYR KMYL
1.0000	-.0004		-.0193	-.0998	-.0084	-.0007	-.0073
1.0000	-.0004	-.054	-.0193	.0998	.0084	-.0007	-.0073
1.2000	-.0008		-.0254	-.1077	-.0149	-.0021	-.0106
1.2000	-.0008	-.061	-.0254	.1077	.0149	-.0021	-.0106
1.4000	-.0012		-.0285	-.1090	-.0210	-.0033	-.0134
1.4000	-.0012	-.065	-.0285	.1090	.0210	-.0033	-.0134
1.6000	-.0015		-.0291	-.1075	-.0260	-.0043	-.0156
1.6000	-.0015	-.067	-.0291	.1075	.0260	-.0043	-.0156
1.8000	-.0017		-.0279	-.1052	-.0296	-.0048	-.0171
1.8000	-.0017	-.068	-.0279	.1052	.0296	-.0048	-.0171
2.0000	-.0019		-.0257	-.1031	-.0321	-.0052	-.0181
2.0000	-.0019	-.068	-.0257	.1031	.0321	-.0052	-.0181
3.0000	-.0021		-.0123	-.0988	-.0359	-.0055	-.0198
3.0000	-.0021	-.068	-.0123	.0988	.0359	-.0055	-.0198
4.0000	-.0022		-.0043	-.0984	-.0362	-.0055	-.0200
4.0000	-.0022	-.068	-.0043	.0984	.0362	-.0055	-.0200
5.0000	-.0022		-.0012	-.0984	-.0362	-.0054	-.0200
5.0000	-.0022	-.068	-.0012	.0984	.0362	-.0054	-.0200

SPR......= .200
SPL......= .200
PL.......= 1.000
PR.......= 0.000
POIS.R...= .300
K........= 1.000

RR RL	KWR KWL	KMC	KQXR KQXL	KQYRN KQYLN	KQYRF KQYLF	KMXR KMXL	KMYR KMYL
1.0000	-.0003		-.0116	-.0596	-.0050	.0021	-.0069
1.0000	-.0003	-.046	-.0116	.0596	.0050	.0021	-.0069
1.2000	-.0005		-.0160	-.0671	-.0093	.0017	-.0096
1.2000	-.0005	-.051	-.0160	.0671	.0093	.0017	-.0096
1.4000	-.0008		-.0187	-.0702	-.0137	.0008	-.0118
1.4000	-.0008	-.054	-.0187	.0702	.0137	.0008	-.0118
1.6000	-.0010		-.0198	-.0708	-.0176	.0000	-.0134
1.6000	-.0010	-.056	-.0198	.0708	.0176	.0000	-.0134
1.8000	-.0012		-.0196	-.0703	-.0206	-.0007	-.0145
1.8000	-.0012	-.056	-.0196	.0703	.0206	-.0007	-.0145
2.0000	-.0013		-.0186	-.0695	-.0228	-.0012	-.0153
2.0000	-.0013	-.057	-.0186	.0695	.0228	-.0012	-.0153
3.0000	-.0016		-.0103	-.0668	-.0269	-.0021	-.0166
3.0000	-.0016	-.057	-.0103	.0668	.0269	-.0021	-.0166
4.0000	-.0016		-.0042	-.0664	-.0273	-.0022	-.0168
4.0000	-.0016	-.057	-.0042	.0664	.0273	-.0022	-.0168
5.0000	-.0016		-.0015	-.0664	-.0273	-.0022	-.0168
5.0000	-.0016	-.056	-.0015	.0664	.0273	-.0022	-.0168

SPR......= .300
SPL......= .300
PL.......= 1.000
PR.......= 0.000
POIS.R...= .300
K........= 1.000

RR RL	KWR KWL	KMC	KQXR KQXL	KQYRN KQYLN	KQYRF KQYLF	KMXR KMXL	KMYR KMYL
1.0000	-.0002		-.0083	-.0425	-.0036	.0033	-.0068
1.0000	-.0002	-.043	-.0083	.0425	.0036	.0033	-.0068
1.2000	-.0004		-.0117	-.0487	-.0068	.0034	-.0092
1.2000	-.0004	-.047	-.0117	.0487	.0068	.0034	-.0092
1.4000	-.0006		-.0139	-.0518	-.0102	.0029	-.0110
1.4000	-.0006	-.049	-.0139	.0518	.0102	.0029	-.0110
1.6000	-.0007		-.0150	-.0530	-.0133	.0022	-.0123
1.6000	-.0007	-.050	-.0150	.0530	.0133	.0022	-.0123
1.8000	-.0009		-.0151	-.0531	-.0158	.0015	-.0132
1.8000	-.0009	-.050	-.0151	.0531	.0158	.0015	-.0132
2.0000	-.0010		-.0145	-.0528	-.0177	.0009	-.0137
2.0000	-.0010	-.051	-.0145	.0528	.0177	.0009	-.0137
3.0000	-.0013		-.0086	-.0512	-.0215	-.0003	-.0147
3.0000	-.0013	-.050	-.0086	.0512	.0215	-.0003	-.0147
4.0000	-.0013		-.0038	-.0508	-.0220	-.0005	-.0148
4.0000	-.0013	-.050	-.0038	.0508	.0220	-.0005	-.0148
5.0000	-.0013		-.0015	-.0508	-.0220	-.0005	-.0148
5.0000	-.0013	-.050	-.0015	.0508	.0220	-.0005	-.0148

Numerical Values for Factors in Equations (18) and (19) (Tables 76–79).

SPR......= .400
SPL......= .400
PL.......= 1.000
PR.......= 0.000
POIS.R...= .300
K........= 1.000

RR RL	KWR KWL	KMC	KQXR KQXL	KQYRN KQYLN	KQYRF KQYLF	KMXR KMXL	KMYR KMYL
1.0000	-.0001		-.0064	-.0330	-.0028	.0040	-.0067
1.0000	-.0001	-.041	-.0064	.0330	.0028	.0040	-.0067
1.2000	-.0003		-.0092	-.0383	-.0054	.0043	-.0089
1.2000	-.0003	-.044	-.0092	.0383	.0054	.0043	-.0089
1.4000	-.0004		-.0111	-.0411	-.0081	.0041	-.0105
1.4000	-.0004	-.046	-.0111	.0411	.0081	.0041	-.0105
1.6000	-.0006		-.0120	-.0424	-.0107	.0035	-.0116
1.6000	-.0006	-.047	-.0120	.0424	.0107	.0035	-.0116
1.8000	-.0007		-.0123	-.0427	-.0128	.0029	-.0123
1.8000	-.0007	-.047	-.0123	.0427	.0128	.0029	-.0123
2.0000	-.0008		-.0119	-.0427	-.0144	.0023	-.0128
2.0000	-.0008	-.047	-.0119	.0427	.0144	.0023	-.0128
3.0000	-.0011		-.0073	-.0417	-.0179	.0009	-.0134
3.0000	-.0011	-.046	-.0073	.0417	.0179	.0009	-.0134
4.0000	-.0011		-.0034	-.0414	-.0184	.0007	-.0134
4.0000	-.0011	-.046	-.0034	.0414	.0184	.0007	-.0134
5.0000	-.0011		-.0014	-.0413	-.0185	.0007	-.0134
5.0000	-.0011	-.046	-.0014	.0413	.0185	.0007	-.0134

SPR......= .500
SPL......= .500
PL.......= 1.000
PR.......= 0.000
POIS.R...= .300
K........= 1.000

RR RL	KWR KWL	KMC	KQXR KQXL	KQYRN KQYLN	KQYRF KQYLF	KMXR KMXL	KMYR KMYL
1.0000	-.0001		-.0053	-.0270	-.0023	.0044	-.0066
1.0000	-.0001	-.039	-.0053	.0270	.0023	.0044	-.0066
1.2000	-.0002		-.0076	-.0315	-.0044	.0050	-.0087
1.2000	-.0002	-.043	-.0076	.0315	.0044	.0050	-.0087
1.4000	-.0004		-.0092	-.0341	-.0067	.0049	-.0102
1.4000	-.0004	-.044	-.0092	.0341	.0067	.0049	-.0102
1.6000	-.0005		-.0101	-.0353	-.0089	.0044	-.0112
1.6000	-.0005	-.044	-.0101	.0353	.0089	.0044	-.0112
1.8000	-.0006		-.0103	-.0358	-.0108	.0038	-.0117
1.8000	-.0006	-.044	-.0103	.0358	.0108	.0038	-.0117
2.0000	-.0007		-.0101	-.0359	-.0122	.0032	-.0121
2.0000	-.0007	-.044	-.0101	.0359	.0122	.0032	-.0121
3.0000	-.0009		-.0064	-.0352	-.0154	.0017	-.0124
3.0000	-.0009	-.043	-.0064	.0352	.0154	.0017	-.0124
4.0000	-.0010		-.0031	-.0350	-.0159	.0015	-.0125
4.0000	-.0010	-.043	-.0031	.0350	.0159	.0015	-.0125
5.0000	-.0010		-.0013	-.0349	-.0160	.0014	-.0125
5.0000	-.0010	-.043	-.0013	.0349	.0160	.0014	-.0125

SPR......= 0.000
SPL......= 0.000
FOL......= 0.000
POR......= 1.000
POIS.R...= .300
K........= .250

FR RL	KWR KWL	KMC	KQXR KQXL	KQYRN KQYLN	KQYRF KQYLF	KMXR KMXL	KMYR KMYL
1.0000	-.0006		-.0274	-.1200	-.0118	-.0069	-.0043
.2500	-.0000	-.041	-.0001	.0299	.0000	-.0000	.0000
1.2000	-.0010		-.0339	-.1133	-.0194	-.0088	-.0077
.3000	-.0000	-.048	-.0002	.0280	.0000	-.0000	.0000
1.4000	-.0015		-.0376	-.1032	-.0265	-.0098	-.0111
.3500	-.0000	-.051	-.0003	.0251	.0000	-.0001	.0000
1.6000	-.0018		-.0393	-.0922	-.0324	-.0103	-.0141
.4000	-.0000	-.053	-.0006	.0219	.0000	-.0001	.0000
1.8000	-.0021		-.0398	-.0817	-.0369	-.0103	-.0165
.4500	-.0000	-.053	-.0008	.0188	.0000	-.0002	.0000
2.0000	-.0023		-.0397	-.0724	-.0402	-.0100	-.0184
.5000	-.0000	-.053	-.0010	.0160	.0001	-.0002	.0000
3.0000	-.0026		-.0369	-.0439	-.0434	-.0076	-.0211
.7500	-.0000	-.045	-.0018	.0069	.0004	-.0004	-.0001
4.0000	-.0024		-.0354	-.0331	-.0385	-.0057	-.0192
1.0000	-.0000	-.037	-.0021	.0033	.0009	-.0005	-.0003
5.0000	-.0021		-.0347	-.0285	-.0336	-.0047	-.0169
1.2500	-.0001	-.032	-.0021	.0019	.0012	-.0005	-.0005

SPR......= .100
SPL......= .006
PUL......= 0.000
POR......= 1.000
POIS.R...= .300
K........= .250

RR PL	KWR KWL	KMC	KQXR KQXL	KQYRN KQYLN	KQYRF KQYLF	KMXR KMXL	KMYR KMYL
1.0000	-.0002		-.0091	-.0447	-.0039	-.0003	-.0034
.2500	-.0000	-.024	-.0000	.0111	.0000	-.0000	.0000
1.2000	-.0004		-.0139	-.0537	-.0081	-.0012	-.0057
.3000	-.0000	-.030	-.0001	.0133	.0000	-.0000	.0000
1.4000	-.0007		-.0178	-.0588	-.0129	-.0021	-.0081
.3500	-.0000	-.035	-.0002	.0143	.0000	-.0000	.0000
1.6000	-.0010		-.0206	-.0606	-.0177	-.0030	-.0104
.4000	-.0000	-.038	-.0003	.0145	.0000	-.0001	.0000
1.8000	-.0013		-.0223	-.0603	-.0221	-.0038	-.0124
.4500	-.0000	-.040	-.0005	.0141	.0000	-.0001	.0000
2.0000	-.0015		-.0233	-.0586	-.0258	-.0045	-.0140
.5000	-.0000	-.041	-.0006	.0133	.0001	-.0001	-.0000
3.0000	-.0021		-.0230	-.0449	-.0346	-.0057	-.0175
.7500	-.0000	-.039	-.0014	.0081	.0003	-.0003	-.0001
4.0000	-.0021		-.0217	-.0347	-.0341	-.0053	-.0170
1.0000	-.0000	-.034	-.0017	.0045	.0007	-.0004	-.0003
5.0000	-.0019		-.0211	-.0293	-.0312	-.0047	-.0155
1.2500	-.0001	-.030	-.0018	.0027	.0010	-.0004	-.0004

Numerical Values for Factors in Equations (18) and (19) (Tables 80–83).

SPR.......= .200
SPL.......= .013
PUL.......= 0.000
POR.......= 1.000
POIS.R....= .300
K.........= .250

RR	KWR	KMC	KQXR	KQYRN	KQYRF	KMXR	KMYR
RL	KWL		KQXL	KQYLN	KQYLF	KMXL	KMYL
1.0000	-.0001		-.0055	-.0269	-.0024	.0010	-.0033
.2500	-.0000	-.020	-.0000	.0067	.0000	-.0000	-.0000
1.2000	-.0003		-.0087	-.0341	-.0051	.0009	-.0052
.3000	-.0000	-.025	-.0000	.0084	.0000	-.0000	-.0000
1.4000	-.0005		-.0117	-.0391	-.0085	.0004	-.0072
.3500	-.0000	-.029	-.0001	.0096	.0000	-.0000	-.0000
1.6000	-.0007		-.0141	-.0422	-.0122	-.0002	-.0090
.4000	-.0000	-.031	-.0002	.0101	.0000	-.0000	-.0000
1.8000	-.0009		-.0158	-.0437	-.0157	-.0009	-.0106
.4500	-.0000	-.033	-.0003	.0102	.0000	-.0001	-.0000
2.0000	-.0011		-.0169	-.0441	-.0189	-.0016	-.0120
.5000	-.0000	-.034	-.0005	.0100	.0000	-.0001	-.0000
3.0000	-.0017		-.0179	-.0388	-.0282	-.0038	-.0152
.7500	-.0000	-.034	-.0011	.0072	.0003	-.0002	-.0001
4.0000	-.0018		-.0171	-.0322	-.0297	-.0044	-.0151
1.0000	-.0000	-.031	-.0014	.0045	.0006	-.0003	-.0003
5.0000	-.0016		-.0166	-.0279	-.0283	-.0043	-.0141
1.2500	-.0000	-.028	-.0015	.0029	.0009	-.0003	-.0004

SPR.......= .300
SPL.......= .019
PUL.......= 0.000
POR.......= 1.000
POIS.R....= .300
K.........= .250

RR	KWR	KMC	KQXR	KQYRN	KQYRF	KMXR	KMYR
RL	KWL		KQXL	KQYLN	KQYLF	KMXL	KMYL
1.0000	-.0001		-.0039	-.0192	-.0017	.0016	-.0032
.2500	-.0000	-.018	-.0000	.0048	.0000	-.0000	-.0000
1.2000	-.0002		-.0064	-.0250	-.0037	.0018	-.0050
.3000	-.0000	-.023	-.0000	.0062	.0000	-.0000	-.0000
1.4000	-.0003		-.0088	-.0293	-.0064	.0017	-.0067
.3500	-.0000	-.026	-.0001	.0072	.0000	-.0000	-.0000
1.6000	-.0005		-.0107	-.0323	-.0093	.0013	-.0083
.4000	-.0000	-.029	-.0002	.0078	.0000	-.0000	-.0000
1.8000	-.0007		-.0122	-.0342	-.0122	.0007	-.0097
.4500	-.0000	-.030	-.0003	.0080	.0000	-.0000	-.0000
2.0000	-.0009		-.0133	-.0351	-.0149	.0001	-.0108
.5000	-.0000	-.031	-.0004	.0080	.0000	-.0001	-.0000
3.0000	-.0014		-.0147	-.0334	-.0238	-.0025	-.0136
.7500	-.0000	-.031	-.0009	.0062	.0002	-.0002	-.0001
4.0000	-.0016		-.0143	-.0292	-.0261	-.0036	-.0136
1.0000	-.0000	-.028	-.0012	.0042	.0005	-.0003	-.0002
5.0000	-.0016		-.0139	-.0260	-.0257	-.0038	-.0129
1.2500	-.0000	-.026	-.0013	.0026	.0008	-.0003	-.0004

SPR.......= .400
SPL.......= .025
PUL.......= 0.000
POR.......= 1.000
POIS.R....= .300
K.........= .250

RR	KWR	KMC	KQXR	KQYRN	KQYRF	KMXR	KMYR
RL	KWL		KQXL	KQYLN	KQYLF	KMXL	KMYL
1.0000	-.0001		-.0030	-.0150	-.0013	.0019	-.0031
.2500	-.0000	-.017	-.0000	.0037	.0000	.0000	-.0000
1.2000	-.0002		-.0050	-.0197	-.0029	.0023	-.0048
.3000	-.0000	-.021	-.0000	.0049	.0000	-.0000	-.0000
1.4000	-.0003		-.0070	-.0234	-.0051	.0024	-.0064
.3500	-.0000	-.024	-.0001	.0057	.0000	-.0000	-.0000
1.6000	-.0004		-.0087	-.0262	-.0075	.0022	-.0079
.4000	-.0000	-.026	-.0001	.0063	.0000	-.0000	-.0000
1.8000	-.0006		-.0100	-.0280	-.0100	.0018	-.0091
.4500	-.0000	-.027	-.0002	.0066	.0000	-.0000	-.0000
2.0000	-.0007		-.0110	-.0292	-.0123	.0012	-.0101
.5000	-.0000	-.028	-.0003	.0067	.0000	-.0000	-.0000
3.0000	-.0012		-.0126	-.0293	-.0205	-.0014	-.0124
.7500	-.0000	-.028	-.0008	.0055	.0002	-.0001	-.0001
4.0000	-.0014		-.0124	-.0265	-.0233	-.0029	-.0125
1.0000	-.0000	-.025	-.0011	.0039	.0005	-.0002	-.0002
5.0000	-.0015		-.0121	-.0241	-.0234	-.0034	-.0119
1.2500	-.0000	-.024	-.0012	.0027	.0007	-.0003	-.0003

SPR.......= .500
SPL.......= .031
PUL.......= 0.000
POR.......= 1.000
POIS.R....= .300
K.........= .250

RR	KWR	KMC	KQXR	KQYRN	KQYRF	KMXR	KMYR
RL	KWL		KQXL	KQYLN	KQYLF	KMXL	KMYL
1.0000	-.0001		-.0025	-.0123	-.0011	.0021	-.0031
.2500	-.0000	-.017	-.0000	.0031	.0000	.0000	-.0000
1.2000	-.0001		-.0041	-.0162	-.0024	.0027	-.0047
.3000	-.0000	-.020	-.0000	.0040	.0000	.0000	-.0000
1.4000	-.0002		-.0058	-.0195	-.0042	.0029	-.0063
.3500	-.0000	-.023	-.0001	.0048	.0000	-.0000	-.0000
1.6000	-.0004		-.0073	-.0220	-.0063	.0028	-.0076
.4000	-.0000	-.024	-.0001	.0053	.0000	-.0000	-.0000
1.8000	-.0005		-.0084	-.0237	-.0084	.0025	-.0087
.4500	-.0000	-.025	-.0002	.0056	.0000	-.0000	-.0000
2.0000	-.0006		-.0093	-.0249	-.0105	.0020	-.0096
.5000	-.0000	-.026	-.0003	.0057	.0000	-.0000	-.0000
3.0000	-.0011		-.0110	-.0259	-.0181	-.0007	-.0115
.7500	-.0000	-.026	-.0007	.0049	.0002	-.0001	-.0001
4.0000	-.0013		-.0109	-.0242	-.0210	-.0023	-.0115
1.0000	-.0000	-.024	-.0010	.0036	.0004	-.0002	-.0002
5.0000	-.0013		-.0107	-.0224	-.0214	-.0030	-.0110
1.2500	-.0000	-.022	-.0011	.0026	.0006	-.0002	-.0003

Numerical Values for Factors in Equations (18) and (19) (Tables 84–87).

SPR.......= 0.000
SPL.......= 0.000
PUL.......= 1.000
POR.......= 0.000
POIS.R....= .300
K.........= .250

RR	KWR	KMC	KQXR	KQYRN	KQYRF	KMXR	KMYR
RL	KWL		KQXL	KQYLN	KQYLF	KMXL	KMYL
1.0000	-.0009		-.0408	-.1883	-.0177	-.0104	-.0064
.2500	-.0000	-.063	-.0001	.0469	.0000	-.0000	.0000
1.2000	-.0020		-.0630	-.2308	-.0364	-.0165	-.0144
.3000	-.0000	-.092	-.0003	.0570	.0000	-.0001	.0000
1.4000	-.0035		-.0878	-.2775	-.0631	-.0236	-.0264
.3500	-.0000	-.127	-.0008	.0677	.0000	-.0002	.0001
1.6000	-.0055		-.1144	-.3290	-.0979	-.0316	-.0425
.4000	-.0000	-.169	-.0017	.0788	.0001	-.0004	.0001
1.8000	-.0080		-.1422	-.3850	-.1404	-.0404	-.0625
.4500	-.0000	-.218	-.0030	.0901	.0002	-.0006	.0001
2.0000	-.0110		-.1704	-.4450	-.1902	-.0498	-.0864
.5000	-.0000	-.272	-.0047	.1013	.0004	-.0011	.0001
3.0000	-.0305		-.2967	-.7683	-.5072	-.1010	-.2426
.7500	-.0003	-.594	-.0195	.1472	.0049	-.0047	-.0013
4.0000	-.0510		-.3748	-1.0536	-.8343	-.1471	-.4071
1.0000	-.0009	-.908	-.0370	.1675	.0160	-.0094	-.0056
5.0000	-.0675		-.4121	-1.2595	-1.0938	-.1812	-.5390
1.2500	-.0017	-1.150	-.0505	.1683	.0308	-.0132	-.0124

SPR.......= .100
SPL.......= .006
PUL.......= 1.000
POR.......= 0.000
POIS.R....= .300
K.........= .250

RR	KWR	KMC	KQXR	KQYRN	KQYRF	KMXR	KMYR
RL	KWL		KQXL	KQYLN	KQYLF	KMXL	KMYL
1.0000	-.0003		-.0136	-.0673	-.0059	-.0005	-.0051
.2500	-.0000	-.036	-.0000	.0168	.0000	-.0000	.0000
1.2000	-.0008		-.0260	-.1023	-.0151	-.0022	-.0107
.3000	-.0000	-.057	-.0001	.0253	.0000	-.0000	.0000
1.4000	-.0017		-.0422	-.1434	-.0306	-.0050	-.0194
.3500	-.0000	-.084	-.0004	.0350	.0000	-.0001	.0000
1.6000	-.0030		-.0615	-.1899	-.0534	-.0091	-.0314
.4000	-.0000	-.118	-.0009	.0456	.0000	-.0002	.0000
1.8000	-.0048		-.0831	-.2413	-.0836	-.0143	-.0471
.4500	-.0000	-.158	-.0018	.0567	.0001	-.0003	.0000
2.0000	-.0070		-.1061	-.2970	-.1211	-.0207	-.0663
.5000	-.0000	-.204	-.0030	.0679	.0003	-.0006	-.0000
3.0000	-.0234		-.2191	-.6060	-.3892	-.0635	-.2007
.7500	-.0002	-.492	-.0148	.1173	.0037	-.0033	-.0013
4.0000	-.0424		-.2960	-.8915	-.6935	-.1096	-.3518
1.0000	-.0007	-.787	-.0305	.1437	.0132	-.0073	-.0052
5.0000	-.0584		-.3352	-1.1058	-.9483	-.1471	-.4780
1.2500	-.0014	-1.023	-.0431	.1503	.0264	-.0109	-.0111

SPR.......= .200
SPL.......= .013
PUL.......= 1.000
POR.......= 0.000
POIS.R....= .300
K.........= .250

RR	KWR	KMC	KQXR	KQYRN	KQYRF	KMXR	KMYR
RL	KWL		KQXL	KQYLN	KQYLF	KMXL	KMYL
1.0000	-.0002		-.0081	-.0405	-.0035	.0015	-.0049
.2500	-.0000	-.030	-.0000	.0101	.0000	-.0000	-.0000
1.2000	-.0005		-.0164	-.0648	-.0095	.0016	-.0097
.3000	-.0000	-.048	-.0001	.0160	.0000	-.0000	-.0000
1.4000	-.0011		-.0278	-.0950	-.0202	.0010	-.0171
.3500	-.0000	-.070	-.0003	.0232	.0000	-.0000	-.0000
1.6000	-.0021		-.0422	-.1311	-.0367	-.0005	-.0273
.4000	-.0000	-.098	-.0006	.0315	.0000	-.0001	-.0000
1.8000	-.0034		-.0590	-.1725	-.0594	-.0031	-.0406
.4500	-.0000	-.132	-.0013	.0405	.0001	-.0002	-.0000
2.0000	-.0051		-.0776	-.2188	-.0887	-.0068	-.0569
.5000	-.0000	-.172	-.0022	.0501	.0002	-.0004	-.0001
3.0000	-.0189		-.1760	-.4937	-.3149	-.0392	-.1746
.7500	-.0002	-.426	-.0120	.0958	.0030	-.0025	-.0013
4.0000	-.0361		-.2490	-.7654	-.5915	-.0814	-.3123
1.0000	-.0006	-.699	-.0259	.1240	.0112	-.0059	-.0048
5.0000	-.0514		-.2886	-.9786	-.8342	-.1193	-.4307
1.2500	-.0013	-.923	-.0377	.1340	.0232	-.0092	-.0101

SPR.......= .300
SPL.......= .019
PUL.......= 1.000
POR.......= 0.000
POIS.R....= .300
K.........= .250

RR	KWR	KMC	KQXR	KQYRN	KQYRF	KMXR	KMYR
RL	KWL		KQXL	KQYLN	KQYLF	KMXL	KMYL
1.0000	-.0001		-.0058	-.0289	-.0025	.0023	-.0047
.2500	-.0000	-.028	-.0000	.0072	.0000	-.0000	-.0000
1.2000	-.0004		-.0120	-.0473	-.0070	.0034	-.0093
.3000	-.0000	-.043	-.0001	.0117	.0000	-.0000	-.0000
1.4000	-.0008		-.0208	-.0710	-.0151	.0040	-.0160
.3500	-.0000	-.063	-.0002	.0174	.0000	-.0000	-.0000
1.6000	-.0016		-.0321	-.1000	-.0279	.0040	-.0252
.4000	-.0000	-.088	-.0005	.0240	.0000	-.0001	-.0000
1.8000	-.0026		-.0457	-.1341	-.0461	.0031	-.0370
.4500	-.0000	-.118	-.0010	.0315	.0001	-.0001	-.0001
2.0000	-.0040		-.0612	-.1730	-.0700	.0013	-.0516
.5000	-.0000	-.153	-.0017	.0396	.0002	-.0002	-.0001
3.0000	-.0159		-.1473	-.4157	-.2643	-.0226	-.1570
.7500	-.0001	-.381	-.0101	.0807	.0025	-.0019	-.0012
4.0000	-.0315		-.2155	-.6690	-.5153	-.0602	-.2829
1.0000	-.0005	-.632	-.0225	.1086	.0098	-.0048	-.0045
5.0000	-.0458		-.2544	-.8755	-.7438	-.0970	-.3936
1.2500	-.0011	-.843	-.0335	.1203	.0206	-.0078	-.0094

Numerical Values for Factors in Equations (18) and (19) (Tables 88–91).

SPR.......= .400
SPL.......= .025
POL.......= 1.000
POR.......= 0.000
POIS.R...= .300
K.........= .250

RR RL	KWR KWL	KMC	KQXR KQXL	KQYRN KQYLN	KQYRF KQYLF	KMXR KMXL	KMYR KMYL
1.0000	-.0001		-.0045	-.0225	-.0020	.0028	-.0047
.2500	-.0000	-.026	-.0000	.0056	.0000	.0000	-.0000
1.2000	-.0003		-.0094	-.0373	-.0055	.0044	-.0090
.3000	-.0000	-.041	-.0000	.0092	.0000	-.0000	-.0000
1.4000	-.0007		-.0166	-.0567	-.0120	.0058	-.0154
.3500	-.0000	-.059	-.0002	.0139	.0000	-.0000	-.0000
1.6000	-.0013		-.0259	-.0808	-.0225	.0068	-.0239
.4000	-.0000	-.081	-.0004	.0194	.0000	-.0000	-.0000
1.8000	-.0021		-.0374	-.1097	-.0377	.0071	-.0348
.4500	-.0000	-.109	-.0008	.0258	.0000	-.0001	-.0001
2.0000	-.0033		-.0505	-.1431	-.0578	.0065	-.0480
.5000	-.0000	-.140	-.0014	.0328	.0001	-.0002	-.0001
3.0000	-.0137		-.1267	-.3587	-.2277	-.0105	-.1442
.7500	-.0001	-.348	-.0087	.0697	.0022	-.0015	-.0012
4.0000	-.0279		-.1901	-.5937	-.4563	-.0437	-.2602
1.0000	-.0005	-.580	-.0199	.0965	.0086	-.0040	-.0042
5.0000	-.0413		-.2279	-.7913	-.6708	-.0789	-.3636
1.2500	-.0010	-.779	-.0302	.1090	.0186	-.0067	-.0088

SPR.......= .500
SPL.......= .031
POL.......= 1.000
POR.......= 0.000
POIS.R...= .300
K.........= .250

RR RL	KWR KWL	KMC	KQXR KQXL	KQYRN KQYLN	KQYRF KQYLF	KMXR KMXL	KMYR KMYL
1.0000	-.0001		-.0037	-.0184	-.0016	.0031	-.0046
.2500	-.0000	-.026	-.0000	.0046	.0000	.0000	-.0000
1.2000	-.0002		-.0078	-.0308	-.0045	.0050	-.0089
.3000	-.0000	-.039	-.0000	.0076	.0000	.0000	-.0000
1.4000	-.0006		-.0138	-.0472	-.0100	.0070	-.0149
.3500	-.0000	-.056	-.0001	.0115	.0000	-.0000	-.0000
1.6000	-.0011		-.0217	-.0678	-.0189	.0086	-.0230
.4000	-.0000	-.077	-.0003	.0163	.0000	-.0000	-.0000
1.8000	-.0018		-.0316	-.0928	-.0318	.0098	-.0332
.4500	-.0000	-.102	-.0007	.0218	.0000	-.0000	-.0001
2.0000	-.0028		-.0430	-.1219	-.0492	.0102	-.0456
.5000	-.0000	-.131	-.0012	.0279	.0001	-.0001	-.0001
3.0000	-.0120		-.1112	-.3154	-.2000	-.0014	-.1345
.7500	-.0001	-.324	-.0076	.0613	.0019	-.0011	-.0012
4.0000	-.0250		-.1701	-.5334	-.4095	-.0305	-.2422
1.0000	-.0004	-.539	-.0179	.0868	.0077	-.0033	-.0040
5.0000	-.0376		-.2065	-.7215	-.6108	-.0639	-.3391
1.2500	-.0009	-.726	-.0274	.0996	.0169	-.0058	-.0083

SPR.......= 0.000
SPL.......= 0.000
POL.......= 0.000
POR.......= 1.000
POIS.R...= .300
K.........= .500

RR RL	KWR KWL	KMC	KQXR KQXL	KQYRN KQYLN	KQYRF KQYLF	KMXR KMXL	KMYR KMYL
1.0000	-.0006		-.0274	-.1200	-.0118	-.0069	-.0043
.5000	-.0000	-.041	-.0029	.0597	.0003	-.0007	.0000
1.2000	-.0010		-.0339	-.1134	-.0194	-.0088	-.0077
.6000	-.0000	-.048	-.0048	.0560	.0007	-.0011	-.0001
1.4000	-.0015		-.0376	-.1033	-.0265	-.0098	-.0111
.7000	-.0001	-.051	-.0065	.0503	.0014	-.0015	-.0003
1.6000	-.0018		-.0394	-.0925	-.0325	-.0103	-.0142
.8000	-.0001	-.053	-.0079	.0440	.0023	-.0019	-.0007
1.8000	-.0021		-.0401	-.0824	-.0372	-.0103	-.0166
.9000	-.0002	-.054	-.0089	.0380	.0032	-.0022	-.0011
2.0000	-.0023		-.0401	-.0736	-.0406	-.0101	-.0186
1.0000	-.0002	-.053	-.0097	.0327	.0041	-.0024	-.0015
3.0000	-.0028		-.0386	-.0487	-.0464	-.0082	-.0225
1.5000	-.0004	-.048	-.0108	.0166	.0077	-.0025	-.0033
4.0000	-.0028		-.0380	-.0419	-.0451	-.0069	-.0224
2.0000	-.0005	-.045	-.0109	.0114	.0095	-.0022	-.0044
5.0000	-.0027		-.0380	-.0403	-.0434	-.0064	-.0217
2.5000	-.0006	-.043	-.0108	.0100	.0102	-.0019	-.0049

SPR.......= .100
SPL.......= .025
POL.......= 0.000
POR.......= 1.000
POIS.R...= .300
K.........= .500

RR RL	KWR KWL	KMC	KQXR KQXL	KQYRN KQYLN	KQYRF KQYLF	KMXR KMXL	KMYR KMYL
1.0000	-.0002		-.0091	-.0447	-.0039	-.0003	-.0034
.5000	-.0000	-.024	-.0010	.0223	.0001	-.0001	-.0001
1.2000	-.0004		-.0139	-.0538	-.0081	-.0012	-.0057
.6000	-.0000	-.030	-.0020	.0266	.0003	-.0003	-.0002
1.4000	-.0007		-.0178	-.0588	-.0129	-.0021	-.0081
.7000	-.0000	-.035	-.0032	.0287	.0007	-.0005	-.0004
1.6000	-.0010		-.0206	-.0607	-.0177	-.0030	-.0104
.8000	-.0001	-.038	-.0043	.0291	.0012	-.0008	-.0007
1.8000	-.0013		-.0224	-.0605	-.0222	-.0038	-.0124
.9000	-.0001	-.040	-.0052	.0284	.0019	-.0010	-.0010
2.0000	-.0015		-.0234	-.0591	-.0260	-.0045	-.0141
1.0000	-.0001	-.041	-.0060	.0269	.0026	-.0012	-.0013
3.0000	-.0022		-.0239	-.0475	-.0362	-.0060	-.0183
1.5000	-.0003	-.041	-.0077	.0179	.0058	-.0017	-.0027
4.0000	-.0024		-.0234	-.0402	-.0383	-.0060	-.0191
2.0000	-.0004	-.039	-.0078	.0124	.0077	-.0018	-.0037
5.0000	-.0024		-.0233	-.0373	-.0378	-.0058	-.0188
2.5000	-.0005	-.038	-.0078	.0101	.0086	-.0016	-.0041

Numerical Values for Factors in Equations (18) and (19) (Tables 92–95).

SPR.......= .200
SPL.......= .050
POL.......= 0.000
POR.......= 1.000
POIS.R...= .300
K.........= .500

RR RL	KWR KWL	KMC	KQXR KQXL	KQYRN KQYLN	KQYRF KQYLF	KMXR KMXL	KMYR KMYL
1.0000	-.0001		-.0055	-.0269	-.0024	.0010	-.0033
.5000	-.0000	-.020	-.0006	.0134	.0001	-.0000	-.0001
1.2000	-.0003		-.0087	-.0341	-.0051	.0009	-.0052
.6000	-.0000	-.025	-.0013	.0169	.0002	-.0001	-.0003
1.4000	-.0005		-.0117	-.0392	-.0085	.0004	-.0072
.7000	-.0000	-.029	-.0021	.0191	.0004	-.0002	-.0004
1.6000	-.0007		-.0141	-.0423	-.0122	-.0002	-.0090
.8000	-.0000	-.031	-.0029	.0203	.0008	-.0003	-.0007
1.8000	-.0009		-.0158	-.0438	-.0158	-.0009	-.0107
.9000	-.0001	-.033	-.0037	.0206	.0013	-.0005	-.0009
2.0000	-.0011		-.0170	-.0443	-.0190	-.0016	-.0120
1.0000	-.0001	-.035	-.0044	.0203	.0019	-.0006	-.0012
3.0000	-.0018		-.0184	-.0403	-.0292	-.0040	-.0157
1.5000	-.0003	-.036	-.0061	.0155	.0047	-.0012	-.0024
4.0000	-.0020		-.0182	-.0359	-.0325	-.0048	-.0165
2.0000	-.0004	-.034	-.0063	.0115	.0064	-.0014	-.0032
5.0000	-.0020		-.0181	-.0335	-.0330	-.0050	-.0165
2.5000	-.0004	-.033	-.0064	.0094	.0074	-.0014	-.0036

SPR.......= .300
SPL.......= .075
POL.......= 0.000
POR.......= 1.000
POIS.R...= .300
K.........= .500

RR RL	KWR KWL	KMC	KQXR KQXL	KQYRN KQYLN	KQYRF KQYLF	KMXR KMXL	KMYR KMYL
1.0000	-.0001		-.0039	-.0192	-.0017	.0016	-.0032
.5000	-.0000	-.019	-.0004	.0096	.0000	.0000	-.0001
1.2000	-.0002		-.0064	-.0250	-.0037	.0018	-.0050
.6000	-.0000	-.023	-.0009	.0123	.0001	.0000	-.0003
1.4000	-.0003		-.0088	-.0293	-.0064	.0017	-.0067
.7000	-.0000	-.026	-.0016	.0143	.0003	-.0000	-.0004
1.6000	-.0005		-.0107	-.0323	-.0093	.0013	-.0083
.8000	-.0000	-.029	-.0022	.0155	.0006	-.0001	-.0007
1.8000	-.0007		-.0123	-.0342	-.0122	.0007	-.0097
.9000	-.0001	-.030	-.0029	.0161	.0010	-.0002	-.0009
2.0000	-.0009		-.0133	-.0353	-.0150	.0001	-.0109
1.0000	-.0001	-.031	-.0035	.0161	.0015	-.0003	-.0011
3.0000	-.0015		-.0151	-.0344	-.0244	-.0025	-.0140
1.5000	-.0002	-.032	-.0050	.0133	.0039	-.0008	-.0022
4.0000	-.0017		-.0151	-.0318	-.0281	-.0038	-.0147
2.0000	-.0003	-.030	-.0054	.0104	.0055	-.0011	-.0026
5.0000	-.0018		-.0150	-.0301	-.0291	-.0043	-.0147
2.5000	-.0004	-.030	-.0054	.0086	.0064	-.0012	-.0032

SPR.......= .400
SPL.......= .100
POL.......= 0.000
POR.......= 1.000
POIS.R...= .300
K.........= .500

RR RL	KWR KWL	KMC	KQXR KQXL	KQYRN KQYLN	KQYRF KQYLF	KMXR KMXL	KMYR KMYL
1.0000	-.0001		-.0030	-.0150	-.0013	.0019	-.0031
.5000	-.0000	-.017	-.0003	.0075	.0000	.0001	-.0001
1.2000	-.0002		-.0050	-.0197	-.0029	.0023	-.0046
.6000	-.0000	-.021	-.0007	.0097	.0001	.0001	-.0003
1.4000	-.0003		-.0070	-.0234	-.0051	.0024	-.0064
.7000	-.0000	-.024	-.0012	.0114	.0003	.0001	-.0005
1.6000	-.0004		-.0087	-.0262	-.0075	.0022	-.0079
.8000	-.0000	-.026	-.0018	.0126	.0005	.0001	-.0007
1.8000	-.0006		-.0100	-.0281	-.0100	.0018	-.0091
.9000	-.0000	-.027	-.0024	.0132	.0008	-.0000	-.0009
2.0000	-.0007		-.0110	-.0292	-.0124	.0012	-.0101
1.0000	-.0001	-.028	-.0029	.0134	.0012	-.0001	-.0011
3.0000	-.0013		-.0128	-.0299	-.0210	-.0015	-.0127
1.5000	-.0002	-.028	-.0043	.0116	.0033	-.0006	-.0020
4.0000	-.0015		-.0129	-.0283	-.0247	-.0030	-.0133
2.0000	-.0003	-.028	-.0046	.0093	.0048	-.0009	-.0026
5.0000	-.0016		-.0129	-.0272	-.0259	-.0037	-.0133
2.5000	-.0003	-.027	-.0047	.0079	.0057	-.0010	-.0029

SPR.......= .500
SPL.......= .125
POL.......= 0.000
POR.......= 1.000
POIS.R...= .300
K.........= .500

RR RL	KWR KWL	KMC	KQXR KQXL	KQYRN KQYLN	KQYRF KQYLF	KMXR KMXL	KMYR KMYL
1.0000	-.0001		-.0025	-.0123	-.0011	.0021	-.0031
.5000	-.0000	-.017	-.0003	.0061	.0000	.0001	-.0001
1.2000	-.0001		-.0041	-.0162	-.0024	.0027	-.0047
.6000	-.0000	-.020	-.0006	.0080	.0001	.0001	-.0003
1.4000	-.0002		-.0058	-.0195	-.0042	.0029	-.0063
.7000	-.0000	-.023	-.0010	.0095	.0002	.0002	-.0005
1.6000	-.0004		-.0073	-.0220	-.0063	.0028	-.0076
.8000	-.0000	-.024	-.0015	.0106	.0004	.0001	-.0006
1.8000	-.0005		-.0084	-.0238	-.0084	.0025	-.0087
.9000	-.0000	-.026	-.0020	.0112	.0007	.0001	-.0008
2.0000	-.0006		-.0093	-.0250	-.0105	.0020	-.0096
1.0000	-.0001	-.026	-.0024	.0114	.0011	.0000	-.0010
3.0000	-.0011		-.0111	-.0264	-.0184	-.0007	-.0117
1.5000	-.0002	-.026	-.0037	.0103	.0029	-.0004	-.0019
4.0000	-.0013		-.0113	-.0255	-.0220	-.0024	-.0122
2.0000	-.0002	-.025	-.0041	.0085	.0043	-.0007	-.0024
5.0000	-.0014		-.0113	-.0247	-.0234	-.0032	-.0121
2.5000	-.0003	-.024	-.0042	.0072	.0051	-.0008	-.0027

Numerical Values for Factors in Equations (18) and (19) (Tables 96–99).

SPR.......= 0.000
SPL.......= 0.000
POL.......= 1.000
POR.......= 0.000
POIS.R...= .300
K.........= .500

RR RL	KWR KWL	KMC	KQXR KQXL	KQYRN KQYLN	KQYRF KQYLF	KMXR KMXL	KMYR KMYL
1.0000	-.0009		-.0399	-.1827	-.0172	-.0101	-.0063
.5000	-.0000	-.061	-.0043	.0910	.0004	-.0010	.0001
1.2000	-.0016		-.0591	-.2144	-.0341	-.0154	-.0135
.6000	-.0001	-.086	-.0084	.1060	.0012	-.0019	-.0002
1.4000	-.0031		-.0777	-.2420	-.0557	-.0208	-.0233
.7000	-.0002	-.112	-.0137	.1181	.0029	-.0032	-.0007
1.6000	-.0045		-.0944	-.2649	-.0803	-.0259	-.0349
.8000	-.0003	-.138	-.0194	.1269	.0055	-.0047	-.0017
1.8000	-.0060		-.1084	-.2835	-.1061	-.0304	-.0473
.9000	-.0005	-.163	-.0252	.1327	.0090	-.0063	-.0031
2.0000	-.0076		-.1197	-.2982	-.1318	-.0343	-.0599
1.0000	-.0007	-.186	-.0306	.1359	.0132	-.0077	-.0048
3.0000	-.0141		-.1467	-.3326	-.2341	-.0456	-.1123
1.5000	-.0021	-.268	-.0483	.1290	.0373	-.0128	-.0160
4.0000	-.0177		-.1520	-.3382	-.2891	-.0493	-.1417
2.0000	-.0033	-.307	-.0541	.1125	.0566	-.0144	-.0258
5.0000	-.0195		-.1528	-.3368	-.3149	-.0502	-.1559
2.5000	-.0041	-.323	-.0557	.0996	.0688	-.0144	-.0325

SPR.......= .100
SPL.......= .025
POL.......= 1.000
POR.......= 0.000
POIS.R...= .300
K.........= .500

RR RL	KWR KWL	KMC	KQXR KQXL	KQYRN KQYLN	KQYRF KQYLF	KMXR KMXL	KMYR KMYL
1.0000	-.0003		-.0132	-.0656	-.0057	-.0005	-.0050
.5000	-.0000	-.035	-.0014	.0327	.0001	-.0002	-.0001
1.2000	-.0008		-.0243	-.0957	-.0141	-.0020	-.0100
.6000	-.0000	-.053	-.0035	.0473	.0005	-.0005	-.0004
1.4000	-.0015		-.0372	-.1262	-.0270	-.0044	-.0171
.7000	-.0001	-.074	-.0066	.0617	.0014	-.0011	-.0009
1.6000	-.0025		-.0504	-.1549	-.0437	-.0074	-.0258
.8000	-.0002	-.097	-.0105	.0745	.0030	-.0019	-.0016
1.8000	-.0036		-.0628	-.1807	-.0630	-.0108	-.0355
.9000	-.0003	-.119	-.0149	.0850	.0053	-.0028	-.0027
2.0000	-.0048		-.0736	-.2029	-.0836	-.0143	-.0457
1.0000	-.0004	-.140	-.0192	.0931	.0083	-.0039	-.0041
3.0000	-.0107		-.1030	-.2678	-.1775	-.0290	-.0912
1.5000	-.0016	-.220	-.0355	.1059	.0280	-.0085	-.0132
4.0000	-.0144		-.1100	-.2882	-.2356	-.0372	-.1190
2.0000	-.0026	-.260	-.0417	.0986	.0455	-.0108	-.0216
5.0000	-.0164		-.1112	-.2926	-.2656	-.0410	-.1333
2.5000	-.0034	-.278	-.0434	.0891	.0572	-.0116	-.0276

SPR.......= .200
SPL.......= .050
POL.......= 1.000
POR.......= 0.000
POIS.R...= .300
K.........= .500

RR RL	KWR KWL	KMC	KQXR KQXL	KQYRN KQYLN	KQYRF KQYLF	KMXR KMXL	KMYR KMYL
1.0000	-.0002		-.0079	-.0395	-.0035	.0015	-.0047
.5000	-.0000	-.029	-.0009	.0197	.0001	-.0000	-.0002
1.2000	-.0005		-.0153	-.0606	-.0089	.0015	-.0091
.6000	-.0000	-.044	-.0022	.0300	.0003	-.0001	-.0004
1.4000	-.0010		-.0246	-.0837	-.0176	.0009	-.0151
.7000	-.0001	-.062	-.0044	.0409	.0009	-.0004	-.0009
1.6000	-.0017		-.0346	-.1070	-.0300	-.0004	-.0224
.8000	-.0001	-.080	-.0072	.0514	.0021	-.0008	-.0016
1.8000	-.0026		-.0445	-.1292	-.0448	-.0024	-.0306
.9000	-.0002	-.099	-.0105	.0608	.0038	-.0014	-.0026
2.0000	-.0035		-.0536	-.1496	-.0611	-.0047	-.0392
1.0000	-.0003	-.117	-.0141	.0687	.0061	-.0021	-.0038
3.0000	-.0085		-.0813	-.2177	-.1422	-.0180	-.0784
1.5000	-.0012	-.189	-.0283	.0865	.0224	-.0058	-.0115
4.0000	-.0121		-.0892	-.2454	-.1974	-.0277	-.1033
2.0000	-.0022	-.226	-.0344	.0848	.0379	-.0082	-.0189
5.0000	-.0141		-.0908	-.2549	-.2280	-.0333	-.1165
2.5000	-.0029	-.244	-.0362	.0787	.0488	-.0094	-.0241

SPR.......= .300
SPL.......= .075
POL.......= 1.000
POR.......= 0.000
POIS.R...= .300
K.........= .500

RR RL	KWR KWL	KMC	KQXR KQXL	KQYRN KQYLN	KQYRF KQYLF	KMXR KMXL	KMYR KMYL
1.0000	-.0001		-.0057	-.0282	-.0025	.0023	-.0046
.5000	-.0000	-.027	-.0006	.0140	.0001	.0001	-.0002
1.2000	-.0004		-.0112	-.0443	-.0065	.0032	-.0087
.6000	-.0000	-.040	-.0016	.0219	.0002	.0001	-.0005
1.4000	-.0007		-.0183	-.0625	-.0133	.0035	-.0141
.7000	-.0000	-.055	-.0033	.0306	.0007	-.0000	-.0009
1.6000	-.0013		-.0263	-.0816	-.0229	.0033	-.0206
.8000	-.0001	-.072	-.0055	.0392	.0016	-.0002	-.0016
1.8000	-.0020		-.0345	-.1005	-.0347	.0023	-.0278
.9000	-.0002	-.088	-.0082	.0473	.0029	-.0006	-.0025
2.0000	-.0028		-.0422	-.1183	-.0482	.0008	-.0354
1.0000	-.0003	-.104	-.0111	.0543	.0048	-.0010	-.0036
3.0000	-.0071		-.0674	-.1826	-.1186	-.0105	-.0699
1.5000	-.0010	-.167	-.0235	.0727	.0186	-.0040	-.0104
4.0000	-.0104		-.0755	-.2126	-.1695	-.0206	-.0919
2.0000	-.0019	-.201	-.0293	.0738	.0325	-.0063	-.0169
5.0000	-.0123		-.0773	-.2247	-.1993	-.0271	-.1039
2.5000	-.0025	-.218	-.0312	.0699	.0425	-.0077	-.0215

Numerical Values for Factors in Equations (18) and (19) (Tables 100–103).

SPR.......= .400
SPL.......= .100
POL.......= 1.000
POR.......= 0.000
POIS.R...= .300
K.........= .500

RR RL	KWR KWL	KMC	KQXR KQXL	KQYRN KQYLN	KQYRF KQYLF	KMXR KMXL	KMYR KMYL
1.0000	-.0001		-.0044	-.0219	-.0019	.0027	-.0046
.5000	-.0000	-.026	-.0005	.0109	.0000	.0001	-.0002
1.2000	-.0003		-.0088	-.0349	-.0051	.0041	-.0085
.6000	-.0000	-.038	-.0013	.0173	.0002	.0002	-.0005
1.4000	-.0006		-.0146	-.0499	-.0106	.0051	-.0135
.7000	-.0000	-.052	-.0026	.0244	.0006	.0002	-.0009
1.6000	-.0010		-.0213	-.0660	-.0185	.0055	-.0195
.8000	-.0001	-.066	-.0044	.0317	.0013	.0001	-.0016
1.8000	-.0016		-.0282	-.0821	-.0283	.0053	-.0261
.9000	-.0001	-.081	-.0067	.0387	.0024	-.0000	-.0024
2.0000	-.0023		-.0348	-.0977	-.0398	.0044	-.0330
1.0000	-.0002	-.095	-.0091	.0449	.0040	-.0003	-.0034
3.0000	-.0061		-.0576	-.1571	-.1016	-.0052	-.0638
1.5000	-.0009	-.152	-.0201	.0626	.0160	-.0027	-.0097
4.0000	-.0091		-.0655	-.1872	-.1485	-.0152	-.0834
2.0000	-.0016	-.183	-.0255	.0652	.0284	-.0049	-.0154
5.0000	-.0109		-.0675	-.2005	-.1769	-.0222	-.0941
2.5000	-.0022	-.197	-.0274	.0626	.0376	-.0064	-.0195

SPR.......= .500
SPL.......= .125
POL.......= 1.000
POR.......= 0.000
POIS.R...= .300
K.........= .500

RR RL	KWR KWL	KMC	KQXR KQXL	KQYRN KQYLN	KQYRF KQYLF	KMXR KMXL	KMYR KMYL
1.0000	-.0001		-.0036	-.0180	-.0016	.0030	-.0045
.5000	-.0000	-.025	-.0004	.0089	.0000	.0001	-.0002
1.2000	-.0002		-.0073	-.0288	-.0042	.0047	-.0083
.6000	-.0000	-.036	-.0010	.0142	.0002	.0002	-.0005
1.4000	-.0005		-.0122	-.0415	-.0088	.0061	-.0132
.7000	-.0000	-.049	-.0022	.0203	.0005	.0003	-.0010
1.6000	-.0009		-.0178	-.0553	-.0155	.0071	-.0188
.8000	-.0001	-.063	-.0037	.0266	.0011	.0004	-.0016
1.8000	-.0014		-.0238	-.0695	-.0240	.0073	-.0249
.9000	-.0001	-.076	-.0056	.0327	.0020	.0003	-.0024
2.0000	-.0020		-.0296	-.0833	-.0338	.0069	-.0312
1.0000	-.0002	-.089	-.0078	.0383	.0034	.0001	-.0034
3.0000	-.0053		-.0503	-.1378	-.0889	-.0011	-.0592
1.5000	-.0008	-.140	-.0176	.0549	.0140	-.0017	-.0091
4.0000	-.0081		-.0579	-.1670	-.1321	-.0110	-.0768
2.0000	-.0015	-.168	-.0226	.0583	.0253	-.0038	-.0143
5.0000	-.0098		-.0600	-.1808	-.1589	-.0183	-.0862
2.5000	-.0020	-.181	-.0245	.0566	.0336	-.0053	-.0180

SPR.......= 0.000
SPL.......= 0.000
POL.......= 0.000
POR.......= 1.000
POIS.R...= .300
K.........= .750

RR RL	KWR KWL	KMC	KQXR KQXL	KQYRN KQYLN	KQYRF KQYLF	KMXR KMXL	KMYR KMYL
1.0000	-.0006		-.0275	-.1202	-.0118	-.0069	-.0043
.7500	-.0002	-.042	-.0125	.0898	.0031	-.0030	-.0009
1.2000	-.0011		-.0341	-.1141	-.0195	-.0088	-.0076
.9000	-.0003	-.048	-.0170	.0847	.0061	-.0042	-.0021
1.4000	-.0015		-.0381	-.1049	-.0269	-.0099	-.0113
1.0500	-.0005	-.052	-.0204	.0770	.0095	-.0051	-.0036
1.6000	-.0019		-.0403	-.0951	-.0332	-.0105	-.0145
1.2000	-.0007	-.055	-.0226	.0687	.0128	-.0057	-.0051
1.8000	-.0022		-.0413	-.0862	-.0384	-.0107	-.0172
1.3500	-.0009	-.056	-.0239	.0611	.0159	-.0061	-.0066
2.0000	-.0025		-.0418	-.0785	-.0426	-.0107	-.0195
1.5000	-.0010	-.056	-.0247	.0544	.0186	-.0063	-.0080
3.0000	-.0032		-.0418	-.0580	-.0522	-.0094	-.0253
2.2500	-.0016	-.055	-.0257	.0357	.0266	-.0058	-.0125
4.0000	-.0033		-.0417	-.0531	-.0536	-.0085	-.0267
3.0000	-.0018	-.054	-.0257	.0306	.0293	-.0051	-.0142
5.0000	-.0034		-.0417	-.0522	-.0538	-.0081	-.0269
3.7500	-.0018	-.054	-.0257	.0293	.0300	-.0047	-.0148

SPR.......= .100
SPL.......= .056
POL.......= 0.000
POR.......= 1.000
POIS.R...= .300
K.........= .750

RR RL	KWR KWL	KMC	KQXR KQXL	KQYRN KQYLN	KQYRF KQYLF	KMXR KMXL	KMYR KMYL
1.0000	-.0002		-.0091	-.0447	-.0039	-.0003	-.0034
.7500	-.0001	-.024	-.0042	.0334	.0010	-.0003	-.0010
1.2000	-.0004		-.0139	-.0538	-.0081	-.0012	-.0057
.9000	-.0001	-.030	-.0070	.0400	.0025	-.0008	-.0018
1.4000	-.0007		-.0179	-.0591	-.0130	-.0021	-.0082
1.0500	-.0002	-.035	-.0097	.0435	.0046	-.0014	-.0028
1.6000	-.0010		-.0208	-.0613	-.0179	-.0031	-.0105
1.2000	-.0004	-.038	-.0119	.0446	.0069	-.0019	-.0039
1.8000	-.0013		-.0228	-.0616	-.0225	-.0039	-.0126
1.3500	-.0005	-.041	-.0135	.0441	.0093	-.0024	-.0050
2.0000	-.0015		-.0240	-.0607	-.0266	-.0046	-.0144
1.5000	-.0006	-.043	-.0147	.0427	.0115	-.0029	-.0061
3.0000	-.0023		-.0254	-.0518	-.0389	-.0064	-.0197
2.2500	-.0011	-.045	-.0165	.0332	.0195	-.0039	-.0097
4.0000	-.0026		-.0253	-.0463	-.0429	-.0067	-.0214
3.0000	-.0014	-.044	-.0166	.0275	.0230	-.0040	-.0113
5.0000	-.0027		-.0253	-.0443	-.0438	-.0067	-.0218
3.7500	-.0015	-.044	-.0166	.0254	.0242	-.0039	-.0120

Numerical Values for Factors in Equations (18) and (19) (Tables 104–107).

SPR......= .200
SPL......= .113
POL......= 0.000
POR......= 1.000
POIS.R...= .300
K........= .750

RR / RL	KWR / KWL	KMC	KQXR / KQXL	KQYRN / KQYLN	KQYRF / KQYLF	KMXR / KMXL	KMYR / KMYL
1.0000	-.0001		-.0055	-.0269	-.0024	.0010	-.0033
.7500	-.0000	-.020	-.0025	.0201	.0006	.0002	-.0010
1.2000	-.0003		-.0088	-.0341	-.0051	.0009	-.0052
.9000	-.0001	-.025	-.0044	.0254	.0016	.0001	-.0017
1.4000	-.0005		-.0118	-.0393	-.0085	.0004	-.0072
1.0500	-.0002	-.029	-.0064	.0289	.0030	-.0002	-.0026
1.6000	-.0007		-.0142	-.0425	-.0123	-.0002	-.0091
1.2000	-.0003	-.032	-.0081	.0309	.0047	-.0005	-.0035
1.8000	-.0009		-.0160	-.0442	-.0159	-.0009	-.0108
1.3500	-.0004	-.034	-.0095	.0317	.0065	-.0009	-.0044
2.0000	-.0011		-.0172	-.0450	-.0193	-.0016	-.0122
1.5000	-.0005	-.035	-.0106	.0317	.0083	-.0012	-.0052
3.0000	-.0016		-.0192	-.0426	-.0306	-.0041	-.0165
2.2500	-.0009	-.038	-.0126	.0275	.0153	-.0025	-.0081
4.0000	-.0022		-.0193	-.0394	-.0351	-.0052	-.0180
3.0000	-.0011	-.037	-.0128	.0237	.0187	-.0030	-.0095
5.0000	-.0023		-.0193	-.0378	-.0366	-.0055	-.0184
3.7500	-.0012	-.037	-.0128	.0218	.0201	-.0032	-.0101

SPR......= .300
SPL......= .169
POL......= 0.000
POR......= 1.000
POIS.R...= .300
K........= .750

RR / RL	KWR / KWL	KMC	KQXR / KQXL	KQYRN / KQYLN	KQYRF / KQYLF	KMXR / KMXL	KMYR / KMYL
1.0000	-.0001		-.0039	-.0192	-.0017	.0016	-.0032
.7500	-.0000	-.018	-.0018	.0144	.0004	.0004	-.0010
1.2000	-.0002		-.0064	-.0250	-.0037	.0018	-.0050
.9000	-.0001	-.023	-.0032	.0186	.0012	.0005	-.0017
1.4000	-.0004		-.0088	-.0294	-.0064	.0017	-.0067
1.0500	-.0001	-.026	-.0048	.0216	.0022	.0004	-.0025
1.6000	-.0005		-.0108	-.0324	-.0093	.0013	-.0083
1.2000	-.0002	-.028	-.0062	.0236	.0036	.0003	-.0033
1.8000	-.0007		-.0123	-.0344	-.0123	.0007	-.0098
1.3500	-.0003	-.030	-.0073	.0247	.0050	-.0000	-.0041
2.0000	-.0009		-.0135	-.0356	-.0151	.0001	-.0110
1.5000	-.0004	-.031	-.0083	.0251	.0065	-.0003	-.0047
3.0000	-.0015		-.0155	-.0357	-.0252	-.0026	-.0144
2.2500	-.0007	-.033	-.0102	.0232	.0126	-.0016	-.0071
4.0000	-.0018		-.0157	-.0339	-.0296	-.0040	-.0155
3.0000	-.0009	-.032	-.0105	.0205	.0158	-.0023	-.0082
5.0000	-.0019		-.0157	-.0327	-.0313	-.0046	-.0159
3.7500	-.0011	-.032	-.0106	.0190	.0172	-.0026	-.0087

SPR......= .400
SPL......= .225
POL......= 0.000
POR......= 1.000
POIS.R...= .300
K........= .750

RR / RL	KWR / KWL	KMC	KQXR / KQXL	KQYRN / KQYLN	KQYRF / KQYLF	KMXR / KMXL	KMYR / KMYL
1.0000	-.0001		-.0030	-.0150	-.0013	.0019	-.0031
.7500	-.0000	-.017	-.0014	.0112	.0003	.0006	-.0010
1.2000	-.0002		-.0050	-.0197	-.0029	.0023	-.0048
.9000	-.0000	-.021	-.0025	.0146	.0009	.0008	-.0017
1.4000	-.0003		-.0070	-.0234	-.0051	.0024	-.0064
1.0500	-.0001	-.024	-.0038	.0173	.0018	.0008	-.0024
1.6000	-.0004		-.0087	-.0262	-.0075	.0022	-.0079
1.2000	-.0002	-.026	-.0050	.0191	.0029	.0007	-.0032
1.8000	-.0006		-.0100	-.0281	-.0100	.0018	-.0091
1.3500	-.0002	-.027	-.0060	.0202	.0041	.0005	-.0038
2.0000	-.0007		-.0110	-.0294	-.0124	.0012	-.0102
1.5000	-.0003	-.028	-.0068	.0208	.0054	.0003	-.0044
3.0000	-.0013		-.0130	-.0306	-.0214	-.0015	-.0130
2.2500	-.0006	-.029	-.0086	.0199	.0107	-.0010	-.0065
4.0000	-.0016		-.0133	-.0296	-.0256	-.0031	-.0138
3.0000	-.0005	-.029	-.0089	.0180	.0136	-.0018	-.0073
5.0000	-.0017		-.0133	-.0288	-.0273	-.0038	-.0140
3.7500	-.0009	-.028	-.0090	.0168	.0149	-.0022	-.0077

SPR......= .500
SPL......= .281
POL......= 0.000
POR......= 1.000
POIS.R...= .300
K........= .750

RR / RL	KWR / KWL	KMC	KQXR / KQXL	KQYRN / KQYLN	KQYRF / KQYLF	KMXR / KMXL	KMYR / KMYL
1.0000	-.0001		-.0025	-.0123	-.0011	.0021	-.0031
.7500	-.0000	-.017	-.0011	.0092	.0003	.0007	-.0010
1.2000	-.0001		-.0041	-.0162	-.0024	.0027	-.0047
.9000	-.0000	-.020	-.0021	.0121	.0008	.0009	-.0017
1.4000	-.0002		-.0058	-.0195	-.0042	.0029	-.0063
1.0500	-.0001	-.023	-.0031	.0144	.0015	.0010	-.0024
1.6000	-.0004		-.0073	-.0220	-.0063	.0028	-.0076
1.2000	-.0001	-.024	-.0042	.0160	.0024	.0010	-.0031
1.8000	-.0005		-.0085	-.0238	-.0084	.0025	-.0087
1.3500	-.0002	-.026	-.0050	.0171	.0035	.0009	-.0037
2.0000	-.0006		-.0094	-.0250	-.0106	.0020	-.0096
1.5000	-.0003	-.026	-.0058	.0177	.0046	.0007	-.0042
3.0000	-.0011		-.0113	-.0268	-.0186	-.0007	-.0119
2.2500	-.0005	-.027	-.0075	.0174	.0093	-.0005	-.0059
4.0000	-.0014		-.0115	-.0262	-.0225	-.0024	-.0125
3.0000	-.0007	-.026	-.0078	.0160	.0119	-.0014	-.0066
5.0000	-.0015		-.0116	-.0257	-.0242	-.0032	-.0126
3.7500	-.0008	-.025	-.0078	.0150	.0132	-.0018	-.0069

Numerical Values for Factors in Equations (18) and (19) (Tables 108–111).

SPR......= 0.000
SPL......= 0.000
POL......= 1.000
POR......= 0.000
POIS.R...= .300
K........= .750

RR / RL	KWR / KWL	KMC	KQXR / KQXL	KQYRN / KQYLN	KQYRF / KQYLF	KMXR / KMXL	KMYR / KMYL
1.0000	-.0008		-.0348	-.1571	-.0150	-.0088	-.0055
.7500	-.0002	-.053	-.0158	.1174	.0039	-.0038	-.0011
1.2000	-.0015		-.0472	-.1666	-.0272	-.0123	-.0108
.9000	-.0005	-.068	-.0237	.1236	.0085	-.0059	-.0029
1.4000	-.0022		-.0569	-.1696	-.0405	-.0151	-.0170
1.0500	-.0008	-.080	-.0306	.1247	.0143	-.0078	-.0053
1.6000	-.0030		-.0638	-.1684	-.0537	-.0172	-.0233
1.2000	-.0011	-.091	-.0361	.1222	.0207	-.0093	-.0082
1.8000	-.0037		-.0685	-.1651	-.0658	-.0187	-.0294
1.3500	-.0015	-.099	-.0402	.1178	.0271	-.0105	-.0113
2.0000	-.0044		-.0716	-.1607	-.0765	-.0196	-.0349
1.5000	-.0018	-.106	-.0432	.1126	.0333	-.0114	-.0142
3.0000	-.0066		-.0764	-.1405	-.1097	-.0206	-.0528
2.2500	-.0032	-.122	-.0487	.0894	.0553	-.0125	-.0257
4.0000	-.0075		-.0768	-.1307	-.1217	-.0199	-.0599
3.0000	-.0039	-.126	-.0493	.0776	.0653	-.0120	-.0315
5.0000	-.0078		-.0768	-.1272	-.1255	-.0194	-.0624
3.7500	-.0043	-.127	-.0494	.0729	.0693	-.0114	-.0341

SPR......= .100
SPL......= .056
POL......= 1.000
POR......= 0.000
POIS.R...= .300
K........= .750

RR / RL	KWR / KWL	KMC	KQXR / KQXL	KQYRN / KQYLN	KQYRF / KQYLF	KMXR / KMXL	KMYR / KMYL
1.0000	-.0003		-.0115	-.0570	-.0050	-.0004	-.0043
.7500	-.0001	-.030	-.0053	.0426	.0013	-.0004	-.0012
1.2000	-.0005		-.0193	-.0756	-.0112	-.0016	-.0080
.9000	-.0002	-.042	-.0097	.0562	.0035	-.0011	-.0025
1.4000	-.0011		-.0269	-.0904	-.0195	-.0032	-.0123
1.0500	-.0004	-.053	-.0146	.0666	.0069	-.0021	-.0043
1.6000	-.0016		-.0334	-.1011	-.0289	-.0049	-.0170
1.2000	-.0006	-.063	-.0191	.0736	.0111	-.0031	-.0064
1.8000	-.0022		-.0385	-.1083	-.0385	-.0066	-.0216
1.3500	-.0009	-.071	-.0230	.0778	.0158	-.0041	-.0086
2.0000	-.0026		-.0423	-.1128	-.0476	-.0082	-.0260
1.5000	-.0011	-.078	-.0260	.0797	.0206	-.0051	-.0108
3.0000	-.0049		-.0492	-.1152	-.0806	-.0132	-.0412
2.2500	-.0024	-.097	-.0326	.0749	.0402	-.0080	-.0201
4.0000	-.0059		-.0499	-.1097	-.0955	-.0150	-.0480
3.0000	-.0031	-.102	-.0336	.0666	.0507	-.0089	-.0251
5.0000	-.0063		-.0500	-.1060	-.1011	-.0155	-.0505
3.7500	-.0034	-.103	-.0337	.0617	.0554	-.0090	-.0275

SPR......= .200
SPL......= .113
POL......= 1.000
POR......= 0.000
POIS.R...= .300
K........= .750

RR / RL	KWR / KWL	KMC	KQXR / KQXL	KQYRN / KQYLN	KQYRF / KQYLF	KMXR / KMXL	KMYR / KMYL
1.0000	-.0002		-.0069	-.0343	-.0030	.0013	-.0041
.7500	-.0000	-.026	-.0032	.0256	.0008	.0003	-.0013
1.2000	-.0004		-.0122	-.0478	-.0071	.0012	-.0072
.9000	-.0001	-.035	-.0061	.0356	.0022	.0001	-.0024
1.4000	-.0007		-.0177	-.0599	-.0129	.0006	-.0109
1.0500	-.0002	-.044	-.0096	.0441	.0045	-.0002	-.0039
1.6000	-.0011		-.0228	-.0697	-.0198	-.0003	-.0147
1.2000	-.0004	-.052	-.0131	.0508	.0076	-.0008	-.0057
1.8000	-.0015		-.0271	-.0773	-.0272	-.0015	-.0185
1.3500	-.0006	-.059	-.0162	.0555	.0112	-.0015	-.0075
2.0000	-.0020		-.0305	-.0828	-.0345	-.0027	-.0220
1.5000	-.0008	-.065	-.0188	.0586	.0149	-.0022	-.0094
3.0000	-.0036		-.0377	-.0928	-.0632	-.0082	-.0345
2.2500	-.0019	-.081	-.0251	.0605	.0315	-.0051	-.0169
4.0000	-.0048		-.0387	-.0918	-.0778	-.0112	-.0402
3.0000	-.0025	-.086	-.0263	.0561	.0412	-.0066	-.0211
5.0000	-.0052		-.0388	-.0897	-.0840	-.0125	-.0425
3.7500	-.0026	-.087	-.0264	.0526	.0459	-.0072	-.0231

SPR......= .300
SPL......= .169
POL......= 1.000
POR......= 0.000
POIS.R...= .300
K........= .750

RR / RL	KWR / KWL	KMC	KQXR / KQXL	KQYRN / KQYLN	KQYRF / KQYLF	KMXR / KMXL	KMYR / KMYL
1.0000	-.0001		-.0049	-.0245	-.0021	.0020	-.0040
.7500	-.0000	-.024	-.0023	.0183	.0006	.0006	-.0013
1.2000	-.0003		-.0089	-.0350	-.0052	.0025	-.0069
.9000	-.0001	-.032	-.0045	.0260	.0016	.0007	-.0024
1.4000	-.0005		-.0132	-.0447	-.0096	.0025	-.0102
1.0500	-.0002	-.040	-.0072	.0330	.0034	.0007	-.0038
1.6000	-.0008		-.0173	-.0531	-.0150	.0021	-.0135
1.2000	-.0003	-.046	-.0099	.0387	.0058	.0004	-.0053
1.8000	-.0012		-.0209	-.0599	-.0210	.0013	-.0168
1.3500	-.0005	-.052	-.0125	.0430	.0086	-.0000	-.0069
2.0000	-.0016		-.0238	-.0653	-.0270	.0004	-.0198
1.5000	-.0006	-.057	-.0147	.0462	.0117	-.0005	-.0085
3.0000	-.0031		-.0306	-.0771	-.0520	-.0049	-.0302
2.2500	-.0015	-.071	-.0205	.0504	.0258	-.0032	-.0149
4.0000	-.0040		-.0318	-.0783	-.0655	-.0084	-.0349
3.0000	-.0021	-.075	-.0217	.0480	.0346	-.0050	-.0183
5.0000	-.0044		-.0319	-.0774	-.0717	-.0102	-.0368
3.7500	-.0024	-.076	-.0219	.0455	.0391	-.0058	-.0200

Numerical Values for Factors in Equations (18) and (19) (Tables 112–115).

SPR.......= .400
SPL.......= .225
POL.......= 1.000
POR.......= 0.000
POIS.R...= .300
K.........= .750

RR / PL	KWR / KWL	KMC	KQXR / KQXL	KQYRN / KQYLN	KQYRF / KQYLF	KMXR / KMXL	KMYR / KMYL
1.0000	-.0001		-.0039	-.0191	-.0017	.0024	-.0040
.7500	-.0000	-.022	-.0018	.0143	.0004	.0007	-.0013
1.2000	-.0002		-.0070	-.0276	-.0041	.0032	-.0067
.9000	-.0001	-.030	-.0035	.0205	.0013	.0011	-.0024
1.4000	-.0004		-.0105	-.0357	-.0076	.0037	-.0097
1.0500	-.0001	-.037	-.0057	.0263	.0027	.0012	-.0037
1.6000	-.0007		-.0140	-.0429	-.0121	.0036	-.0126
1.2000	-.0003	-.043	-.0080	.0312	.0047	.0012	-.0051
1.8000	-.0010		-.0170	-.0489	-.0171	.0031	-.0157
1.3500	-.0004	-.048	-.0102	.0351	.0070	.0009	-.0066
2.0000	-.0013		-.0196	-.0538	-.0222	.0024	-.0183
1.5000	-.0005	-.052	-.0121	.0381	.0096	.0005	-.0080
3.0000	-.0026		-.0258	-.0658	-.0441	-.0026	-.0272
2.2500	-.0013	-.063	-.0173	.0431	.0219	-.0019	-.0135
4.0000	-.0035		-.0271	-.0681	-.0565	-.0063	-.0311
3.0000	-.0018	-.066	-.0185	.0419	.0298	-.0038	-.0163
5.0000	-.0039		-.0272	-.0679	-.0625	-.0084	-.0326
3.7500	-.0021	-.067	-.0187	.0401	.0340	-.0048	-.0177

SPR.......= .500
SPL.......= .281
POL.......= 1.000
POR.......= 0.000
POIS.R...= .300
K.........= .750

RR / PL	KWR / KWL	KMC	KQXR / KQXL	KQYRN / KQYLN	KQYRF / KQYLF	KMXR / KMXL	KMYR / KMYL
1.0000	-.0001		-.0031	-.0156	-.0014	.0026	-.0039
.7500	-.0000	-.022	-.0014	.0117	.0004	.0008	-.0013
1.2000	-.0002		-.0058	-.0227	-.0034	.0037	-.0066
.9000	-.0001	-.029	-.0029	.0169	.0010	.0013	-.0023
1.4000	-.0003		-.0087	-.0297	-.0064	.0044	-.0095
1.0500	-.0001	-.035	-.0047	.0219	.0022	.0016	-.0036
1.6000	-.0006		-.0117	-.0360	-.0101	.0046	-.0123
1.2000	-.0002	-.041	-.0067	.0262	.0039	.0017	-.0050
1.8000	-.0008		-.0144	-.0413	-.0144	.0043	-.0149
1.3500	-.0003	-.045	-.0086	.0297	.0059	.0015	-.0063
2.0000	-.0011		-.0166	-.0458	-.0189	.0038	-.0173
1.5000	-.0005	-.049	-.0103	.0324	.0082	.0013	-.0076
3.0000	-.0023		-.0223	-.0574	-.0393	-.0009	-.0250
2.2500	-.0011	-.058	-.0150	.0376	.0190	-.0009	-.0124
4.0000	-.0030		-.0236	-.0602	-.0497	-.0048	-.0282
3.0000	-.0016	-.060	-.0162	.0371	.0262	-.0028	-.0145
5.0000	-.0034		-.0238	-.0604	-.0553	-.0070	-.0293
3.7500	-.0018	-.060	-.0164	.0357	.0301	-.0040	-.0159

SPR.......= 0.000
SPL.......= 0.000
POL.......= 0.000
POR.......= 1.000
POIS.R...= .300
K.........= 1.000

RR / PL	KWR / KWL	KMC	KQXR / KQXL	KQYRN / KQYLN	KQYRF / KQYLF	KMXR / KMXL	KMYR / KMYL
1.0000	-.0006		-.0277	-.1214	-.0119	-.0070	-.0044
1.0000	-.0006	-.042	-.0277	.1214	.0119	-.0070	-.0044
1.2000	-.0011		-.0347	-.1165	-.0199	-.0090	-.0079
1.2000	-.0011	-.049	-.0347	.1165	.0199	-.0090	-.0079
1.4000	-.0015		-.0392	-.1087	-.0277	-.0103	-.0116
1.4000	-.0015	-.054	-.0392	.1087	.0277	-.0103	-.0116
1.6000	-.0019		-.0419	-.1003	-.0347	-.0110	-.0151
1.6000	-.0019	-.057	-.0419	.1003	.0347	-.0110	-.0151
1.8000	-.0023		-.0434	-.0925	-.0406	-.0113	-.0182
1.8000	-.0023	-.059	-.0434	.0925	.0406	-.0113	-.0182
2.0000	-.0026		-.0443	-.0857	-.0455	-.0114	-.0208
2.0000	-.0026	-.061	-.0443	.0857	.0455	-.0114	-.0208
3.0000	-.0035		-.0452	-.0673	-.0583	-.0106	-.0282
3.0000	-.0035	-.062	-.0452	.0673	.0583	-.0106	-.0282
4.0000	-.0038		-.0452	-.0626	-.0616	-.0098	-.0305
4.0000	-.0038	-.062	-.0452	.0626	.0616	-.0098	-.0305
5.0000	-.0039		-.0452	-.0614	-.0623	-.0095	-.0311
5.0000	-.0039	-.062	-.0452	.0614	.0623	-.0095	-.0311

SPR.......= .100
SPL.......= .100
POL.......= 0.000
POR.......= 1.000
POIS.R...= .300
K.........= 1.000

RR / PL	KWR / KWL	KMC	KQXR / KQXL	KQYRN / KQYLN	KQYRF / KQYLF	KMXR / KMXL	KMYR / KMYL
1.0000	-.0002		-.0091	-.0448	-.0040	-.0003	-.0034
1.0000	-.0002	-.024	-.0091	.0448	.0040	-.0003	-.0034
1.2000	-.0004		-.0140	-.0541	-.0081	-.0012	-.0057
1.2000	-.0004	-.030	-.0140	.0541	.0081	-.0012	-.0057
1.4000	-.0007		-.0181	-.0597	-.0131	-.0021	-.0082
1.4000	-.0007	-.035	-.0181	.0597	.0131	-.0021	-.0082
1.6000	-.0010		-.0211	-.0623	-.0182	-.0031	-.0107
1.6000	-.0010	-.039	-.0211	.0623	.0182	-.0031	-.0107
1.8000	-.0013		-.0233	-.0631	-.0230	-.0040	-.0129
1.8000	-.0013	-.042	-.0233	.0631	.0230	-.0040	-.0129
2.0000	-.0016		-.0246	-.0626	-.0274	-.0047	-.0149
2.0000	-.0016	-.044	-.0246	.0626	.0274	-.0047	-.0149
3.0000	-.0025		-.0266	-.0554	-.0412	-.0068	-.0209
3.0000	-.0025	-.048	-.0266	.0554	.0412	-.0068	-.0209
4.0000	-.0028		-.0267	-.0505	-.0462	-.0073	-.0231
4.0000	-.0028	-.048	-.0267	.0505	.0462	-.0073	-.0231
5.0000	-.0030		-.0267	-.0487	-.0476	-.0073	-.0238
5.0000	-.0030	-.048	-.0267	.0487	.0476	-.0073	-.0238

Numerical Values for Factors in Equations (18) and (19) (Tables 116–119).

SPR.......= .200
SPL.......= .200
POL.......= 0.000
POR.......= 1.000
POIS.R...= .300
K.........= 1.000

RR / PL	KWR / KWL	KMC	KQXR / KQXL	KQYRN / KQYLN	KQYRF / KQYLF	KMXR / KMXL	KMYR / KMYL
1.0000	-.0001		-.0055	-.0269	-.0024	.0010	-.0033
1.0000	-.0001	-.020	-.0055	.0269	.0024	.0010	-.0033
1.2000	-.0003		-.0088	-.0342	-.0051	.0009	-.0052
1.2000	-.0003	-.025	-.0088	.0342	.0051	.0009	-.0052
1.4000	-.0005		-.0118	-.0394	-.0086	.0004	-.0072
1.4000	-.0005	-.029	-.0118	.0394	.0086	.0004	-.0072
1.6000	-.0007		-.0143	-.0427	-.0123	-.0002	-.0091
1.6000	-.0007	-.032	-.0143	.0427	.0123	-.0002	-.0091
1.8000	-.0009		-.0161	-.0447	-.0161	-.0009	-.0109
1.8000	-.0009	-.034	-.0161	.0447	.0161	-.0009	-.0109
2.0000	-.0011		-.0174	-.0456	-.0195	-.0016	-.0124
2.0000	-.0011	-.036	-.0174	.0456	.0195	-.0016	-.0124
3.0000	-.0019		-.0197	-.0440	-.0315	-.0042	-.0170
3.0000	-.0019	-.039	-.0197	.0440	.0315	-.0042	-.0170
4.0000	-.0022		-.0199	-.0412	-.0365	-.0054	-.0187
4.0000	-.0022	-.039	-.0199	.0412	.0365	-.0054	-.0187
5.0000	-.0024		-.0199	-.0398	-.0383	-.0057	-.0193
5.0000	-.0024	-.039	-.0199	.0398	.0383	-.0057	-.0193

SPR.......= .300
SPL.......= .300
POL.......= 0.000
POR.......= 1.000
POIS.R...= .300
K.........= 1.000

RR / PL	KWR / KWL	KMC	KQXR / KQXL	KQYRN / KQYLN	KQYRF / KQYLF	KMXR / KMXL	KMYR / KMYL
1.0000	-.0001		-.0039	-.0192	-.0017	.0016	-.0032
1.0000	-.0001	-.018	-.0039	.0192	.0017	.0016	-.0032
1.2000	-.0002		-.0064	-.0250	-.0037	.0018	-.0050
1.2000	-.0002	-.023	-.0064	.0250	.0037	.0018	-.0050
1.4000	-.0004		-.0088	-.0294	-.0064	.0017	-.0067
1.4000	-.0004	-.026	-.0088	.0294	.0064	.0017	-.0067
1.6000	-.0005		-.0108	-.0325	-.0093	.0013	-.0083
1.6000	-.0005	-.028	-.0108	.0325	.0093	.0013	-.0083
1.8000	-.0007		-.0123	-.0345	-.0123	.0007	-.0098
1.8000	-.0007	-.030	-.0123	.0345	.0123	.0007	-.0098
2.0000	-.0009		-.0135	-.0357	-.0152	.0001	-.0110
2.0000	-.0009	-.031	-.0135	.0357	.0152	.0001	-.0110
3.0000	-.0015		-.0157	-.0362	-.0255	-.0026	-.0146
3.0000	-.0015	-.033	-.0157	.0362	.0255	-.0026	-.0146
4.0000	-.0019		-.0159	-.0346	-.0302	-.0041	-.0158
4.0000	-.0019	-.033	-.0159	.0346	.0302	-.0041	-.0158
5.0000	-.0020		-.0159	-.0336	-.0320	-.0046	-.0163
5.0000	-.0020	-.033	-.0159	.0336	.0320	-.0046	-.0163

SPR.......= .400
SPL.......= .400
POL.......= 0.000
POR.......= 1.000
POIS.R...= .300
K.........= 1.000

RR / PL	KWR / KWL	KMC	KQXR / KQXL	KQYRN / KQYLN	KQYRF / KQYLF	KMXR / KMXL	KMYR / KMYL
1.0000	-.0001		-.0030	-.0150	-.0013	.0019	-.0031
1.0000	-.0001	-.017	-.0030	.0150	.0013	.0019	-.0031
1.2000	-.0002		-.0050	-.0197	-.0029	.0023	-.0048
1.2000	-.0002	-.021	-.0050	.0197	.0029	.0023	-.0048
1.4000	-.0003		-.0070	-.0234	-.0051	.0024	-.0064
1.4000	-.0003	-.024	-.0070	.0234	.0051	.0024	-.0064
1.6000	-.0004		-.0087	-.0262	-.0075	.0022	-.0079
1.6000	-.0004	-.026	-.0087	.0262	.0075	.0022	-.0079
1.8000	-.0006		-.0100	-.0281	-.0100	.0018	-.0091
1.8000	-.0006	-.027	-.0100	.0281	.0100	.0018	-.0091
2.0000	-.0007		-.0110	-.0293	-.0124	.0012	-.0101
2.0000	-.0007	-.028	-.0110	.0293	.0124	.0012	-.0101
3.0000	-.0013		-.0130	-.0306	-.0214	-.0015	-.0130
3.0000	-.0013	-.029	-.0130	.0306	.0214	-.0015	-.0130
4.0000	-.0016		-.0133	-.0297	-.0257	-.0031	-.0138
4.0000	-.0016	-.029	-.0133	.0297	.0257	-.0031	-.0138
5.0000	-.0017		-.0133	-.0290	-.0274	-.0038	-.0141
5.0000	-.0017	-.029	-.0133	.0290	.0274	-.0038	-.0141

SPR.......= .500
SPL.......= .500
POL.......= 0.000
POR.......= 1.000
POIS.R...= .300
K.........= 1.000

RR / PL	KWR / KWL	KMC	KQXR / KQXL	KQYRN / KQYLN	KQYRF / KQYLF	KMXR / KMXL	KMYR / KMYL
1.0000	-.0001		-.0025	-.0122	-.0011	.0021	-.0031
1.0000	-.0001	-.017	-.0025	.0122	.0011	.0021	-.0031
1.2000	-.0001		-.0041	-.0162	-.0024	.0027	-.0047
1.2000	-.0001	-.020	-.0041	.0162	.0024	.0027	-.0047
1.4000	-.0002		-.0058	-.0194	-.0042	.0029	-.0062
1.4000	-.0002	-.023	-.0058	.0194	.0042	.0029	-.0062
1.6000	-.0004		-.0072	-.0219	-.0063	.0028	-.0076
1.6000	-.0004	-.024	-.0072	.0219	.0063	.0028	-.0076
1.8000	-.0005		-.0084	-.0237	-.0084	.0025	-.0086
1.8000	-.0005	-.025	-.0084	.0237	.0084	.0025	-.0086
2.0000	-.0006		-.0093	-.0249	-.0105	.0020	-.0095
2.0000	-.0006	-.026	-.0093	.0249	.0105	.0020	-.0095
3.0000	-.0011		-.0112	-.0265	-.0184	-.0007	-.0118
3.0000	-.0011	-.026	-.0112	.0265	.0184	-.0007	-.0118
4.0000	-.0014		-.0115	-.0260	-.0223	-.0024	-.0124
4.0000	-.0014	-.026	-.0115	.0260	.0223	-.0024	-.0124
5.0000	-.0015		-.0115	-.0255	-.0240	-.0032	-.0125
5.0000	-.0015	-.025	-.0115	.0255	.0240	-.0032	-.0125

CONTINUOUS 3-LAYER LAMINATED PANEL SUBJECTED TO CONCENTRATED LOADS AT THE CENTER OF ITS PANELS

When a central concentrated load is applied on either panel of the continuous panel, the deflection, shear forces and bending moments can be determined from the following formulas:

$$w^i = \frac{b^{i2} P^j}{D} K^{ij}_w$$

$$Q^i_x = \frac{P^j}{b^i} K^{ij}_{Qx}$$

$$Q^i_{yn} = \frac{P^j}{b^i} K^{ij}_{Qyn}$$

$$Q^i_{yf} = \frac{P^j}{b^i} K^{ij}_{Qyf} \tag{18}$$

$$M^i_x = P^j K^{ij}_{mx}$$

$$M^i_y = P^j K^{ij}_{my}$$

$$M_c = P^j K^j_{mc}$$

in which

$$j = r \text{ means } p^r = 1 \text{ and } p^\ell = 0, \text{ thus } \Phi^{j\prime}_{mc} = (\Phi'_{mc})\begin{matrix} p^r = 1 \\ p^\ell = 0 \end{matrix}$$

$$= \ell \text{ means } p^r = 0 \text{ and } p^\ell = 1, \text{ thus } \Phi^{j\prime}_{mc} = (\Phi'_{mc})\begin{matrix} p^r = 0 \\ p^\ell = 1 \end{matrix}$$

CONTINUOUS 3-LAYER LAMINATED PANEL SUBJECTED TO UNIFORMLY DISTRIBUTED LOADS

The practical formulas in this case are given by:

$$w^i = \frac{b^{i4}}{D} p^j_o K^{ij}_w$$

$$Q^i_x = b^i p^j_o K^{ij}_{Qx}$$

$$Q^i_{yn} = b^i p^j_o K^{ij}_{Qyn}$$

$$Q^i_{yf} = b^i p^j_o K^{ij}_{Qyf} \tag{19}$$

$$M^i_x = b^{i2} p^j_o K^{ij}_{mx}$$

$$M^i_y = b^{i2} p^j_o K^{ij}_{my}$$

$$M_c = b^{r2} p^j_o K^j_{mc}$$

in which

$$j = r \text{ means } p^r_o = 1 \text{ and } p^\ell_o = 0$$

$$\ell \text{ means } p^r_o = 0 \text{ and } p^\ell_o = 1$$

Laminated Panels With Bonding Having Finite Stiffness

As any other material, bonding in laminated panels have finite stiffness in resisting interlayer shear stresses. This stiffness may change with time due to creep, cyclic loads, or due to the effect of temperature changes.

It is a very common assumption in practice that bonding is rigid in shear. This assumption does not represent the real behavior of bonding materials and their effects on the responses of structural components. The deflection of a panel shows greater sensitivity to variation of the bonding stiffness when the latter is in its lower range. Beyond a certain level of stiffness, the bonding can be practically considered as rigid. An increase in the bonding stiffness is accompanied by a decrease in the normal stress of the core. The resulting loss in the resisting moment is compensated by a slight increase in the face normal stresses. As a consequence, the interlayer shear flux is practically independent of the bonding stiffness. Also, by increasing the bonding stiffness, the shear stresses in the core change from nonlinear to linear distributions with constant values.

The quantities of primary interest in structural design of 3-layer laminated panels are: the maximum normal forces or stresses in facings (which imply also moments), the maximum interlayer shear fluxes, and the maximum deflection. Practical formulas are given here to calculate these quantities for laminated panels under three types of loads: uniform distribution load, mid-span concentrated load, and axial compression.

$$N = .5\, P_o\, t_c\, F_{nu}$$

$$q = P_o F_{qu} \qquad \text{uniform distributed load} \tag{20}$$

$$W = \frac{P_o\, t_c}{2E_c b} F_{wu}$$

$$N = Q\, F_{nc}$$

$$q = \frac{Q}{a} F_{qc} \qquad \text{mid-span concentrated load} \tag{21}$$

$$w = \frac{Q}{E_{cb}} F_{wc}$$

$$P_{cr} = \frac{1}{2} E_c t_c b\, F_p \qquad \text{axial compression} \tag{22}$$

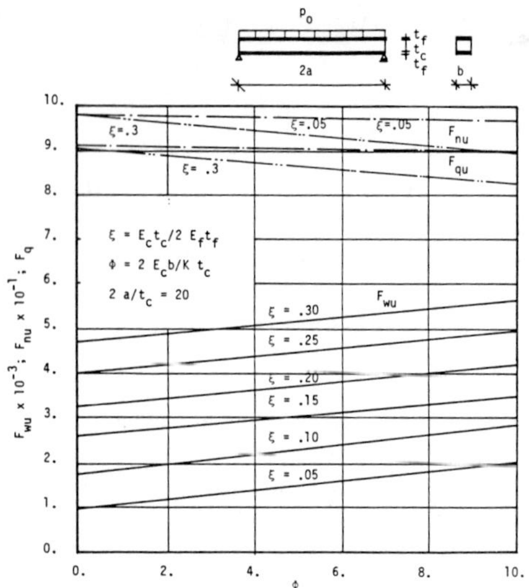

FIGURE 5. Numerical values for factors in Equations (20), (21), and (22).

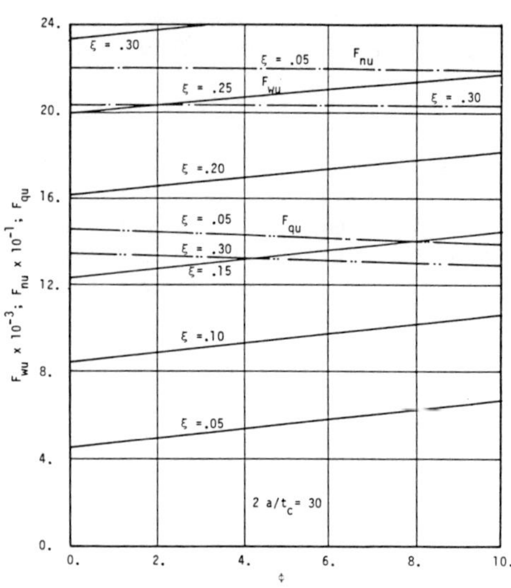

FIGURE 6. Numerical values for factors in Equations (20), (21), and (22).

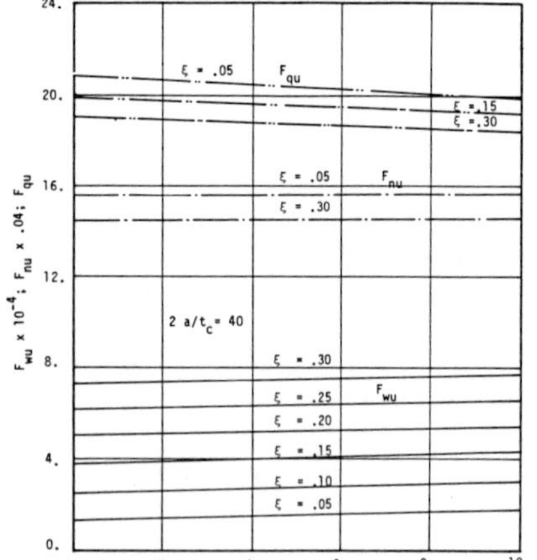

FIGURE 7. Numerical values for factors in Equations (20), (21), and (22).

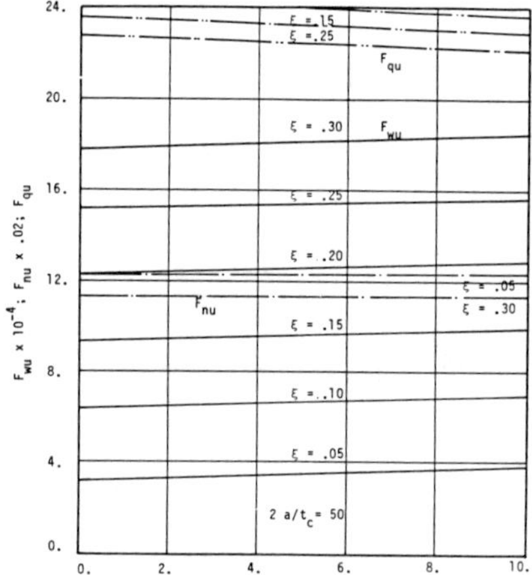

FIGURE 8. Numerical values for factors in Equations (20), (21), and (22).

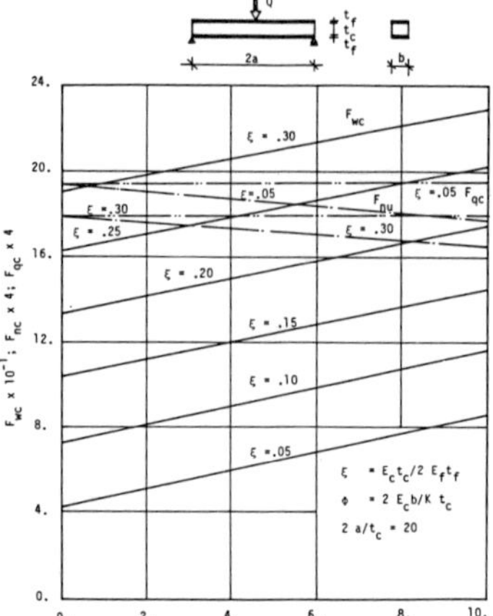

FIGURE 9. Numerical values for factors in Equations (20), (21), and (22).

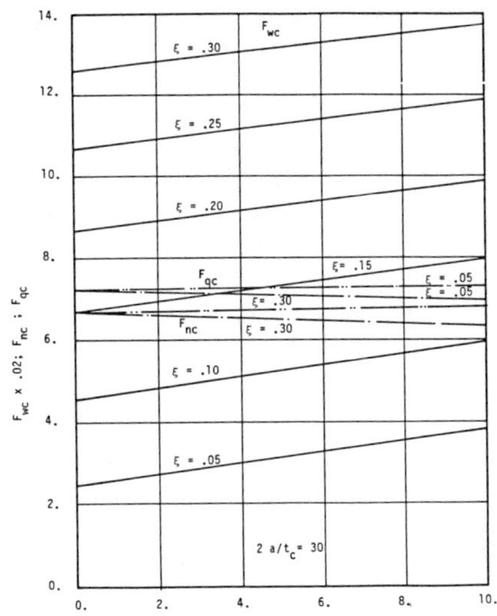

FIGURE 10. Numerical values for factors in Equations (20), (21), and (22).

where

N = the maximum normal force in facings
q = the maximum interlayer shear flux
w = the maximum deflection
P_{cr} = overall buckling load
P_o = load intensity
Q = concentrated load
b = strip width

F_{nu}, F_{qu}, F_{wu}, F_{nc}, F_{qc}, F_{wc}, and F_p are factors to be obtained from Figures 5–14.

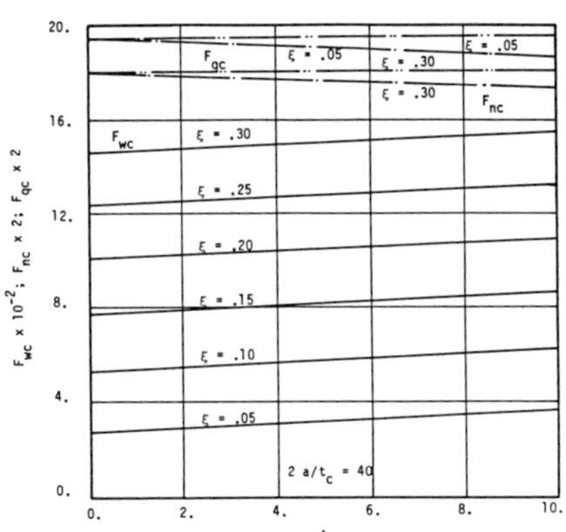

FIGURE 11. Numerical values for the factors in Equations (20), (21), and (22).

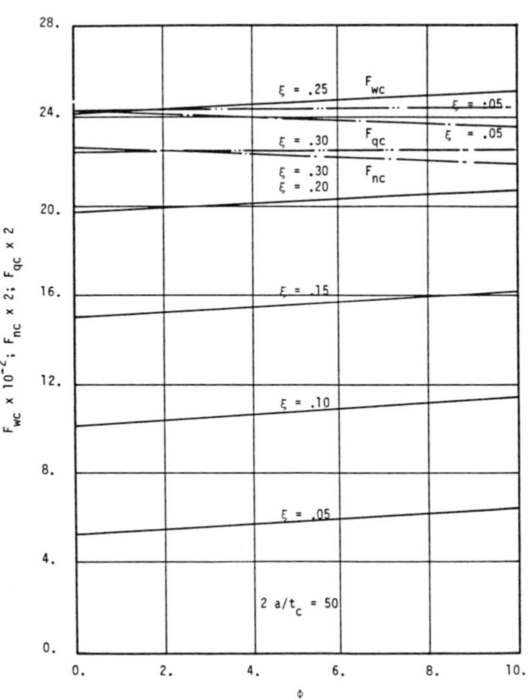

FIGURE 12. Numerical values for factors in Equations (20), (21), and (22).

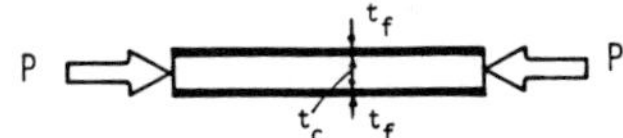

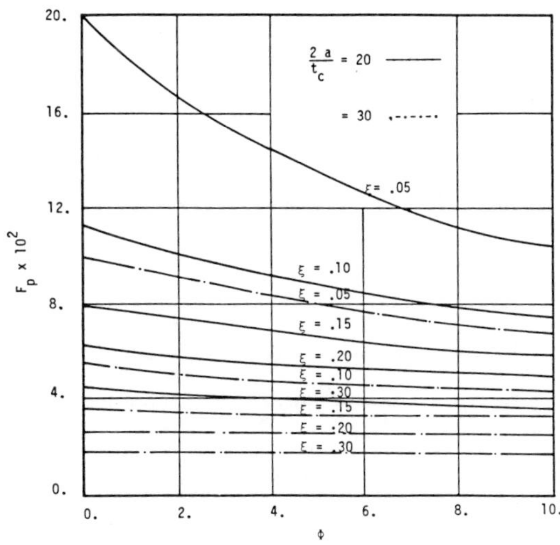

FIGURE 13. Numerical values for factors in Equation (20).

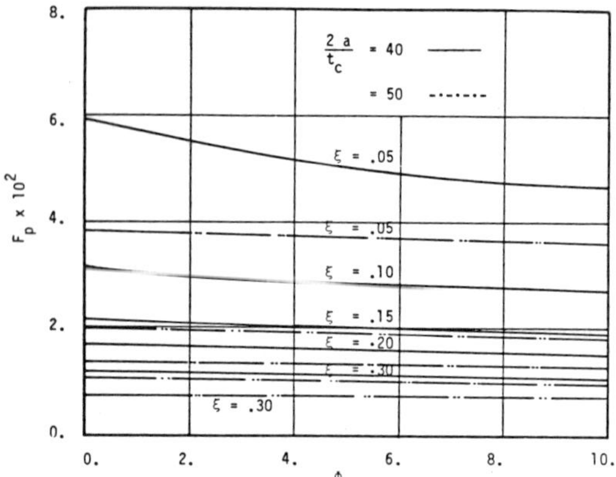

FIGURE 14. Numerical values for factor in Equation (20).

Instability of Laminated Panels

A laminated panel subjected to inplane loads may fail in different modes. The two major failure modes in which instability may manifest itself are: overall buckling and local instability.

Overall buckling occurs when a panel becomes elastically unstable under the application of inplane loads with the buckling mode characterized by long waves. Local instability of panels may be classified into three categories: dimpling, wrinkling, and crimping. Dimpling occurs in laminated construction with a honeycomb core, where in the region above the cell, the face buckles in a plate-like fashion with the cell walls acting as edge supports. Failure by wrinkling occurs in the form of short-wave lengths in the facings, and involves straining of the core material. Such failure mode could be symmetric or antisymmetric with respect to the middle plane of the panel. Finally, crimping is a shear failure mode.

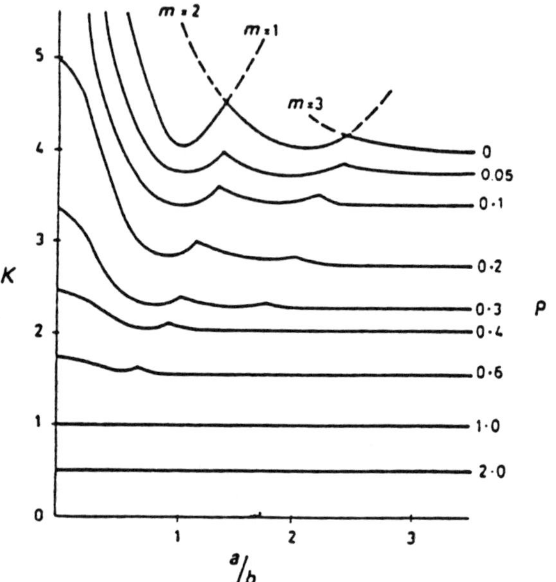

FIGURE 15. Relationship between K and a/b for known values of p.

OVERALL BUCKLING OF SIMPLY SUPPORTED 3-LAYER LAMINATED PANELS

The critical load in the x-direction of a simply supported laminated panel may be calculated from

$$P_{cr} = \frac{\pi^2 D_f}{b^2} F_{cr} \tag{23}$$

where

P_{cr} = the critical buckling edge load per unit length in the x-direction.
b = the panel dimension in the y-direction (normal to x-direction)
F_{cr} = dimensionless coefficient to be obtained from Figure 15 in terms of the ratio a/b and ϱ, where a is the panel dimension in x-direction and ϱ is defined by:
$\varrho = \pi^2/b^2$ times ratio of flexural rigidity to the shear rigidity. In this figure, m represents the number of waves in the direction of the applied load.

DIMPLING OF LAMINATED PANELS

The critical stress causing dimpling of a 3-layer laminated panel can be calculated from:

$$\sigma_{cr} = \frac{2 E_f}{(1 - \nu_f^2)} \left(\frac{t_f}{S} \right)^2 \tag{24}$$

in which

E_f = elastic modulus of the faces material
ν_f = Poisson's ratio of the faces material
t_f = the face thickness
S = the diameter of the largest circle that can be inscribed within a cell of the core.

For the case when the facings are subjected to biaxial compression, the following interaction formula may be used:

$$\sigma_x = \sigma_y = \sigma_{cr} \tag{25}$$

in which σ_x, σ_y = the applied compressive stresses in x- and y-directions, respectively.

WRINKLING STRESSES OF A PERFECTLY FLAT 3-LAYER LAMINATED PANEL

When the core is neither thin enough to cause overall buckling, nor thick enough to cause symmetric wrinkling, the possible wrinkling mode is a skew ripple. The antisymmetrical wrinkling stress can be calculated from:

$$\sigma_{cr} = 0.59 \frac{(E_f E_c t_f) \cdot 5}{t_c} + 0.387 \frac{G\, t_c}{t_f} \text{ when } t_m = \frac{t_c}{2} \quad (26)$$

$$\sigma_{cr} = 0.51\, (E_f E_c G)^{1/3} + 0.33 \frac{G\, t_c}{t_f} \text{ when } t_m < \frac{t_c}{2} \quad (27)$$

where

t_m = the width of a marginal zone
= $2.38\, t_f\, (E_f/E_c)^{1/3}$
E_c = Young's modulus for the core material

When the core is thick enough, the possible wrinkling mode is symmetrical with respect to the middle plane of the panel. The critical stress in this case may be obtained from:

$$\sigma_{cr} = 0.50\, (E_f E_c G)^{1/3} \quad (28)$$

FAILING STRESSES OF IMPERFECT 3-LAYER LAMINATED PANELS

In the case of a laminated panel with an initial imperfection in the form of a sine shape, and after applying the compressive load the deflection amplitude will be amplified by the factor:

$$\text{Amplification factor} = \left[1 - \frac{PL^2}{\pi^2 E_f I_f}\right]^{-1} \quad (29)$$

where

P = the applied axial load
L = wave length
$I_f = t_f^3/12$

As an increase in the applied compressive load will give rise to the core deformations, loading to critical situation where the ultimate strength of the core material is reached before the critical wrinkling can occur, and thus results in an internal core failure. The failing stresses may be calculated from:

$$\sigma_{fn} = \frac{\sigma_{cr}}{1 + \dfrac{E_c K w_o}{T}} \quad (30)$$

$$\sigma_{fs} = \frac{\sigma_{cr}}{1 + \dfrac{G w_o}{L_{cr} S_t}} \quad (31)$$

where

σ_{fn} = the failing stress of the core material based on its strength in compression or tension.
σ_{fs} = same meaning as σ_{fn} but for shear strength.
T = the core compressive or tensile strength.
S_t = the core shear strength.
w_o = the initial amplitude of face irregularity.

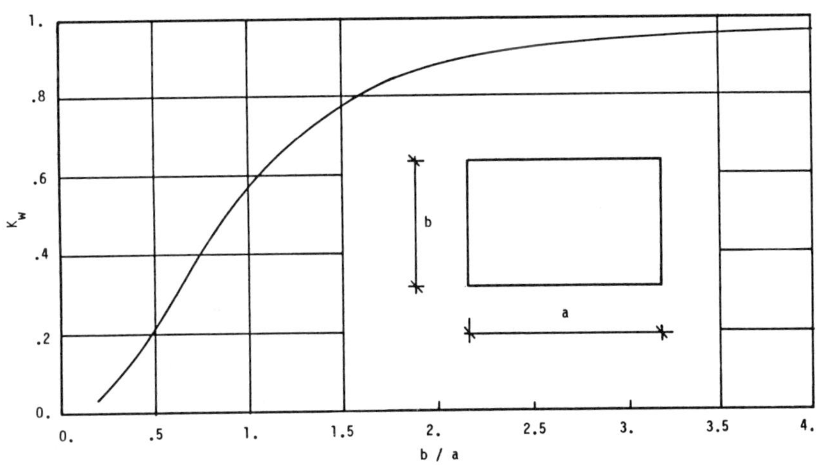

FIGURE 16. Values for K_w in Equation (34).

$$K = \frac{\pi}{L}(G/E_c)^{1/2}$$
$$L_{cr} = [2\pi^3 D/\sqrt{E_c G}]^{1/3}$$
$$D = \frac{1}{12}E_f t_f^3$$

By means of Equations (30) and (31), for any given initial imperfection and core strength, the failing stress can be determined, or conversely, for any given initial imperfection, the core strength required to sustain a specific axial load can be computed.

CRIMPING LOAD OF 3-LAYER LAMINATED PANELS

The crimping load of a laminated panel can be calculated from the following equation:

$$P_{cr} = S \tag{32}$$

where S is the shear stiffness defined before.

Thermal Stresses in Laminated Panels

Three-layer laminated panels are called upon to experience thermal stresses when subjected to a temperature change. Whether the temperature change is uniform or not, and whether the boundaries are free or not, the panel will undergo thermal stresses. This is because each layer in a laminated panel is unable to deform freely in a manner compatible with the temperature distribution through it.

Consider a laminated panel subjected to a temperature change T_c at its upper surface and T_b at its lower surface. This linear distribution through the thickness may be divided into two superimposed distributions:

$$\text{Thermal gradient } \pm T_g = \frac{T_t + T_b}{2}$$

$$\text{Uniform temperature change } T_u = \frac{T_t - T_b}{2}$$

Practical formulas are presented here to calculate the maximum normal stress in the facings and the maximum deflection in a panel under each of these two distributions.

SIMPLY SUPPORTED LAMINATED PANELS SUBJECTED TO A THERMAL GRADIENT $\pm T_g$

The maximum normal stress in the faces can be determined from:

$$\sigma_f = \frac{1}{2}\alpha_f T_g E_f \tag{33}$$

in which

α_f = thermal expansion coefficient of the faces material
T_g = temperature change at the panel surfaces

The maximum deflection in this case may be determined by the following equation:

$$w = 4\alpha_f T_g a^2 \frac{(1 + \nu_f)}{\pi^3 h} K_w \tag{34}$$

in which

a = the panel length
ν_f = Poisson's ratio of the faces materials
K_w = a factor to be obtained from Figure 16 in terms of the aspect ratio b/a.

FREE EDGES LAMINATED PANEL SUBJECTED TO A THERMAL GRADIENT $\pm T_g$

The maximum normal stress in the faces in this case may be obtained from:

$$\sigma_f = \frac{E_f \Delta_m}{1 - \nu_f + \dfrac{6 E_f t_f h}{E_c t_c^2}} \tag{35}$$

in which Δ_m = the relative free strain at the interface between facings and core due to a uniform temperature change.

FREE EDGES LAMINATED PANEL SUBJECTED TO A UNIFORM TEMPERATURE CHANGE

The maximum normal stress in the faces in this case may be obtained from:

$$\sigma_f = \frac{E_f \Delta_m}{1 - \nu_f + \dfrac{2 E_f t_f}{E_c t_c}} \tag{36}$$

SUMMARY

Demands on new developments of building systems imposed by today's advanced technologies have become so diverse and severe that they are often difficult to be fulfilled by a single material acting alone. It is sometimes necessary to combine dissimilar materials into a composite providing performance unattainable by each constituent acting alone.

Laminated panels, as a special form of composite materials, are being used in building industry. In the foregoing discussion, several features of laminated panels were discussed. The use of composites to meet with the construction requirements was elaborated through several examples. For structural design of 3-layer laminated panels, simple formulas are given. Numerical values for the coefficients in these formulas may be obtained from the tables and graphs presented.

INDEX